Energy & Fuels Preprints Presented at the 252nd ACS National Meeting & Exhibition 2016

AA004025

Division of Energy & Fuels, American Chemical Society

Energy & Fuels Preprints Volume 61 Number 2

Philadelphia, Pennsylvania, USA
21-25 August 2016

Editor:

Hong Cui

ISBN: 978-1-5108-2804-9

Printed from e-media with permission by:

Curran Associates, Inc.
57 Morehouse Lane
Red Hook, NY 12571

Some format issues inherent in the e-media version may also appear in this print version.

Printed by Curran Associates, Inc. (2016)

For permission requests, please contact American Chemical Society
at the address below.

American Chemical Society
1155 Sixteenth Street, NW
Washington, DC 20036
USA

Phone: (800) 227-5558 (US)
(202) 872-4600 (Worldwide)

help@acs.org

Additional copies of this publication are available from:

Curran Associates, Inc.
57 Morehouse Lane
Red Hook, NY 12571 USA
Phone: 845-758-0400
Fax: 845-758-2633
Email: curran@proceedings.com
Web: www.proceedings.com

TABLE OF CONTENTS

USA-CHINA SYMPOSIUM ON ENERGY

UNCONVENTIONAL ENERGY ON HEAVY OIL & SHALE GAS

PROGRESS IN COAL TO LIQUIDS & GASES

INNOVATIVE CHEMISTRY & MATERIALS FOR ELECTROENERGY PRODUCTION & STORAGE

NOVEL MATERIALS FOR GAS SEPARATION, STORAGE & UTILIZATION

WATER-ENERGY NEXUS

SOLAR FUELS: POWER TO THE PEOPLE

MESOPOROUS ZEOLITES

COMPUTATIONAL CHEMISTRY FOR ENERGY APPLICATIONS

BIOMASS

DEGRADATION OF MATERIALS FOR ENERGY & FUEL PRODUCTION

NOVEL NANOMATERIALS

HETEROGENEOUS CATALYSIS FOR SELECTIVE OXIDATION & REDUCTION TOWARD A GREEN PRODUCTION

2D MATERIALS: GRAPHENE & BEYOND & THEIR DEVICE APPLICATIONS

ADVANCES IN CHEMISTRY OF ENERGY & FUELS

ADVANCES IN ANALYTICAL METHODS IN PETROLEUM UPSTREAM APPLICATIONS

ENFL STORCH AWARD SYMPOSIUM

ENERGY & FUELS JOINT AWARD FOR EXCELLENCE IN PUBLISHING

ADVANCES IN CHEMISTRY OF ENERGY & FUELS

ENFL

ACS Division of Energy & Fuels

2016 Summer Newsletter

http://web.anl.gov/PCS/ENFL/

"Like us" on Facebook! ACS Division of Energy & Fuel

Letter from the Chair

Hello fellow ENFL members!

The Spring ACS meeting was a spectacular venue in San Diego! The ENFL program committee packed the symposia with over 427 accepted abstracts. These high quality talks were presented in 48 sessions ranging from the award symposia honoring the ENFL Distinguished Researcher in honor of Dr. Stu Soled, and the George A. Olah Symposium in honor of Dr. Mieczyslaw M. Boduszynki to symposia that deal with the core science related to ENFL's scientific interests in materials and reactions in areas such batteries, fuel cells, renewables, solar cells, computational modeling of reactions, CO_2 conversion, oil production and upgrading and methane technologies.

The symposia at the ACS 252nd in Philadelphia, will be located in the convention center and also be packed with exciting seminars, including the Storch award symposia, and Energy and Fuels Joint Award Symposia for Excellence in Publishing honoring the best publication authored by Dr. Charles Mullen, from the USDA, titled "Production of Deoxygenated Biomass Fast Pyrolysis Oils via Product Gas Recycling". Please look at the ACS website for all the programming dates and times at www.acs.org.

Congratulations to the recent winners of ENFL awards. Dr. Phillip F. Britt of Oak Ridge National lab is the recipient for the 2016 Storch award. An award symposium to honor Dr. Britt will be held at the 2016 ENFL 252nd ACS national meeting in Philadelphia. Dr. Liangbing Hu from the University of Maryland was named the ENFL 2016 Emerging Researcher. Dr. Lisa Colosi from the University of Virginia was named as the recipient of the Glenn award for the 2015 ENFL Fall meeting in Boston for best paper and presentation. Dr. Jonathan Mathews was awarded the Distinguished Service Award for his service to the ENFL division. Also, congratulations to Dr. Elise Fox, from Savannah River National Laboratory, for being recognized as a Rising Star by the Women's Chemist Committee.

We have an excellent team of volunteers in ENFL who serve in the capacity as officers and symposium organizers, and I would like to thank all of you who volunteer and have dedicated your service to our division. Without each volunteer, the division would not be able to provide quality symposia, recognize the science of our members, or contribute to local societies to enhance scientific outreach and education. It is through everyone's great contributions that allows for the successful operation of the ENFL Division.

ENFL will have a Business meeting social on Monday, August 22nd, 12pm-1pm, in the convention center, room 107A for its members. Please consider attending the Tuesday night (August 23rd, 2016) ENFL dinner at R2L in Philadelphia. Tickets can be purchased online during ACS registration.

Please take advantage of the Newsletter and the ENFL website (web address above) to post jobs, free to our members! Also, keep an eye on the website and your email for important announcements such as due dates for award nominations, ACS registration or abstracts, etc.

Thank you to Chevron, Exxon Mobil, Phillips 66 and Ted Oyama Consulting for sponsoring the ENFL dinner in San Diego!

After many years of requiring and providing preprints in addition to ENFL abstracts, it was decided that preprints will no longer be required, starting in 2017.

Finally, I wish you all the best and safe travels to Philly if you can make it!

Michelle K. Kidder

2016 Chair, ACS Energy and Fuels Division

Awards

Dr. Liangbing Hu was awarded the 2016 ENFL Emerging researcher. Dr. Hu is an associate professor at University of Maryland and has published 147 papers and with over 10,000 citations. His research interests include nanomaterials, roll-to-roll nanomanufacturing, energy storage focusing on solid-state batteries and Na ion batteries, and printed electronics. received his B.S. in physics from the University of Science and Technology of China (USTC) in 2002, where he worked with Prof Yuheng Zhang on colossal magnetoresistance (CMR) materials for three years. He did his Ph.D. in at UCLA (with George Gruner), focusing on carbon nanotube based nanoelectronics (2002-2007). In 2006, he joined Unidym Inc as a co-founding scientist.

2016 Henry H. Storch Award in Fuel Science

Phillip F. Britt
Oak Ridge National Laboratory

The Storch Award is awarded to Dr. Phil F. Britt for his elucidation of new fundamental insights into the complex mix of free-radical, ionic and concerted reaction paths of oxygen functional groups in organic energy resources, which is crucial to devising improved methods for converting fossil and biomass resources into fuels and chemicals. Since joining ORNL in 1988, Phil's research has focused on elucidating fundamental kinetic and mechanistic insights into the pyrolysis of organic molecules that serve as models for important structural features found in energy resources, such as biomass and low rank coals. Of particular note is his body of work that has focused on understanding oxygen functional group chemistry in these energy resources. This work was especially important because the role of heteroatoms in pyrolysis reactions has been less understood than traditional hydrocarbon chemistry and because heteroatoms were also thought to play an important role in the processing of the low rank coal, biomass and other oxygen-rich energy sources. He has also focused on understanding the role of oxygen functional group chemistry in retrogressive (cross-linking) reactions that are critical to conversion of low rank coals into liquid fuels and commodity chemicals. Further, he elucidated the effects of restricted mass transport on free-radical reactions. A hallmark of Phil's research has been his approach that combines meticulous experimental studies with thermochemical kinetics and computational modeling to provide valuable predictive insights into reaction rates (via estimates of Arrhenius parameters) and mechanisms. His research has resulted in over 145 peer-reviewed articles and preprints and over 200 scientific presentations, many of which were invited. He has twice won the Richard A. Glenn Award for best paper, and he has been asked to present his work at five previous Storch Award Symposia. He has also made exceptional contributions to the fuel science community, organizing over 10 symposia at national meetings, and has served as the Treasurer, Chair and Director-at-Large of the ACS Division of Fuel Chemistry, as well as Alternate Councilor for the Energy & Fuels Division. He also served as Chair of the Gordon Conference on Hydrocarbon Resources. He has received the Distinguished Service Award from the ACS Division of Fuel Chemistry, and he is a Fellow of the American Chemical Society.

Division Dinner in San Diego

The ENFL dinner was held at the beautiful Prado Inn, in Balboa Park. Between the beautiful weather, catching up with colleagues, honoring the career and scientific achievements of ENFL members. and the delicioius food, it was an absolute wonderful night.

Dr. Andrew Holmes, recipient of the 2015 Glenn Award.

Dr. Stu Soled, recipient of the Distinguished Researcher award.

Division Stats:

ENFL Membership is 2,613, which makes ENFL a class II Division (2000-3499 members). 1,844 are regular members (71%), 360 are student members (14%) 45 are retired, with 122 emeritus members / The average age is 53, 16% are female membership. We remain highly international with 24% international membership. The bulk of these members are from Canada (92), Japan (64), China (92), UK (45), and Australia (36). We have members in at least 46 nations. Our members overlap with the divisions of IEC (349), CATL (340), ENVR (183), CELL (163), and INOR (155).

Some of our ENFL dedicated Committee members

Election results for 2016

Chair-elect: Mike Reynolds (Shell)
Tresurer Kristi Fjare (Phillips 66)
Councilor: Semih Eser (Penn State Univ)
Councilor: Kathleen Carrago Gregar (Argonne National Laboratory)
Director-at-Large: Alan Chaffee (Monash University)

Tidbits:

Don't hesitate to contact an ENFL officer if you would like to become a bigger part of the governance of ENFL. It is a rewarding experience, with the biggest benefit of connecting with the scientific community at large within academics, government and industry.

ENFL is offering free job postings on its website for members, please contact Randall Winans at rwinans@anl.gov or in this newsletter kidderm@ornl.gov

ACS DIVISION OF ENERGY & FUELS
2016 OFFICERS

Chair
Michelle Kidder
Oak Ridge National Laboratory
1 Bethel Valley Rd.
Bldg. 4100, MS 6197
Oak Ridge, TN 37831
865-241-2159
kidderm@ornl.gov

Immediate Past-Chair
Yun Hang Hu
Michigan Technological University
Materials Science and Engineering
502 M&M Building
Houghton, MI 49931-1295
906-487-2261, FAX 906-487-2934
yunhangh@mtu.edu

Chair-Elect
Michael A. Reynolds
Shell Exploration and Production Co.
150 N Dairy Ashford St.
Houston, TX 77079
281.544.7286
mike.reynolds@shell.com
ENFL.Secretary@gmail.com

Secretary
Michael A. Reynolds
Shell Exploration and Production Co.
150 N Dairy Ashford St.
Houston, TX 77079
281.544.7286
mike.reynolds@shell.com
ENFL.Secretary@gmail.com

Treasurer
Kristi Fjare
Phillips 66
Ponca City Refinery
1000 S. Pine St.
Ponca City, OK 74601
kristi.a.fjare@p66.com

Directors At Large
Dady Dadyburjor (2014-2016)
West Virginia University
Dept Chemical Engineering
PO Box 6102
Morgantown, WV 26502-6102
304.293.9337
dady.dadyburjor@mail.wvu.edu

S. Ted Oyama (2015-2017)
Virginia Tech
Chemical Engineering Dept
Blacksburg, VA 24061-0001
540.231.5309
oyama@vt.edu

Alan Chaffee (2015)
Monash University
School of Chemistry
Building 23N, Clayton Campus
Clayton, 3800 Australia
+61 3 9905 4626
Alan.Chaffee@monash.edu

Councilors
Joe Allison (2015-2017)
5715 Aspen Ave.
Bartlesville, OK 74006
918.397.0290
joedallison@cableone.net

Elise Fox (2015-2017)
Savannah River National Laboratory
Materials Science & Technology
999-2W
Aiken, SC 29808
803-507-8560
elise.fox@srnl.doe.gov

Semih Eser (2016-2018)
Pennsylvania State University
101 Hosler Bldg.
University Park, PA 16802-5000
814-863-1392
seser@psu.edu

Lisa Houston (2014-2016)
PAC, LP
8824 Fallbrook Drive
Houston, TX 77064
281.940.1356
281.580.0339 (fax)
lisa.houston@paclp.com.

Alternate Councilors
Umit S. Ozkan (2015-2017)
Ohio State University
Chemical Engineering Department
151 W. Woodruff Ave.
Columbus, OH 43210-1110
614.292.6623
614.292.3769 (fax)
ozkan.1@osu.edu

Kathleen Carrado Gregar (2016-2018)
Center for Nanoscale Materials
Argonne National Laboratory
CNM 440
9700 S. Cass Ave.
Lemont, IL 60439
630.252.7968
630-252-6866 fax
kcarrado@anl.gov

Phillip Britt (2014-2016)
Chemical Sciences Division
Oak Ridge National Laboratory
Building 4100, MS-6129, P.O. Box 2008
Oak Ridge, TN 37831-6197
865-574-4986
brittpf@ornl.gov

Randall E. Winans (2016-2018)
Argonne National Laboratory
9700 S. Cass Avenue, Bldg 433
Argonne, IL 60439-4803
(630) 252-7479
rewinans@anl.gov

Awards Chair and Webmaster
Randall E. Winans
Argonne National Laboratory
9700 S. Cass Avenue, Bldg 433
Argonne, IL 60439-4803
(630) 252-7479
rewinans@anl.gov

Preprint Editor
Hong Cui
Hawaii Natural Energy Institute
School of Ocean and Earth Science and Technology, University of Hawaii at Manoa
1680 East-West Rd., POST 109
Honolulu, HI 96822
808 956-5397
hongcui@hawaii.edu

Membership
Jonathan P. Mathews
Pennsylvania State University
Energy and Mineral Eng. Dept.
126 Hosler Bldg.
University Park, PA 16802-5000
814-863-6213
jpm10@psu.edu

Newsletter Editor
Michelle Kidder
Oak Ridge National Laboratory
1 Bethel Valley Rd.
Bldg. 4100, MS 6197
Oak Ridge, TN 37831
865-241-2159
kidderm@ornl.gov

Long-Range Planning
Lisa Houston
PAC, LP
8824 Fallbrook Drive
Houston, TX 77064
281.940.1356
lisa.houston@paclp.com

Harold H. Schobert
Schobert International LLC
22401 Hunters Ridge Circle
Jordan, MN 55352
952-492-3449
schobert@ems.psu.edu

Historian
Open

Public Policy
Open

Program Secretary
TBD

Program Chair, 2014
Ranjit Koodali

EFFECT OF INTERFACIAL INTERACTION ON CATALYTIC AND ELECTROCHEMICAL REDUCTION OF CO_2

Qingfeng Ge

Department of Chemistry and Biochemistry, Southern Illinois University, Carbondale, IL 62901, USA.

Climate change resulting from anthropogenic green house gases, mainly CO_2, is considered as a major threat faced by mankind. Conversion of CO_2 into liquid fuels and other value-added chemicals is a promising alternative to controlling CO_2 emission. Developing highly selective catalysts for both catalytic and electrochemical reduction of CO_2 becomes focused research areas. In this talk, I will discuss the results from our density functional theory (DFT) computational study on mechanism of CO_2 hydrogenation to methanol on the $In_2O_3(110)$ surface and a model Pd/In_2O_3 catalyst. I will then report our DFT results on the role of SnO_x formation in the electrochemical reduction of CO_2 on Sn-based electrode. In particular, the effect of surface hydroxyls on the reaction mechanism will be discussed. Our results showed that the formation of a SnO_x monolayer on the electrode under the operating conditions promotes CO_2 reduction more effectively by forming surface hydroxyls and opens a new channel to CO formation although formic acid is still the major reduction product.

A NEW STRATEGY FOR HIGHLY EFFICIENT CO_2 REDUCTION BY WATER SPILLING WITH SOLAR/BIOMASS ENERGY-DRIVEN TWO-STEP PROCESS

*Guodong Yao, Fangming Jin**

School of Environmental Science and Engineering, Shanghai Jiao Tong University, 800 Dongchuan RD, Shanghai 200240, China
*fmjin@sjtu.edu.cn

Introduction

Reduction of CO_2 into fuels and chemicals has attracted considerable attention as a key approach to reduce greenhouse gas emissions and also to alleviate the dependence on fossil fuels. Solar energy could serve as an ideal energy for driving conversion of CO_2 to fuels. Although extensive attempts for developing solar-to-fuel conversion of CO_2 by artificial photosynthesis have been explored,[1,2] high efficient conversion with a direct use of solar energy remains a fascinating challenge. Compared to direct solar-to-fuel conversion, the integrated technology which can utilize solar energy indirectly is expected to have potential to improve the efficiency of artificial photosynthesis. The electrochemical reduction of CO_2 is considered as a typical integrated technology to increase artificial photosynthetic efficiency because electricity can generated from solar energy. More recently, an attractive technology for the dissociation of CO_2 and H_2O into chemical fuel using certain redox reactions driven by a solar two-step process has also been reported.[3] However, the product was limited to CO, and H_2 from H_2O cannot be used in the reaction.

On the other hand, the catalytic hydrogenation of CO_2 with gaseous hydrogen is regarded as the most commercially feasible approach. However, this method requires either elaborately prepared and expensive (or complex) catalysts, such as noble metal (Ir, Ru, Rh, etc.) complexes, and high-purity and high pressure hydrogen which meets hydrogen storage and transportation problem. Thus, developing a highly efficient and simple process method for CO_2 reduction is highly desired.

The abiotic synthesis of organics suggests that highly efficient dissociation of H_2O and subsequent reduction of CO_2 into organics could be achieved with metals under hydrothermal conditions. The oxidized metal can be reduced to initial metal via solar-driven thermochemical decomposition which has been well studied and is nearing practical application.[4] Thus, an integrated strategy for high-efficiency artificial photosynthesis could be developed by coupling the dissociation of H_2O and the reduction of CO_2 in the presence of metals with the solar-driven thermochemical reduction of metal oxides into metals (Figure 1, route 1).

In addition, our previous studies involving hydrothermal conversion of biomass and/or organic waste into value-added products have demonstrated that hydrogen was relatively easily generated from both biomass and organic waste.[5] These findings inspire us to study the possibility of metal oxide reduction with biomass under hydrothermal conditions for exploring another new strategy of efficient solar-to-fuel conversion (Figure 1, route 2).

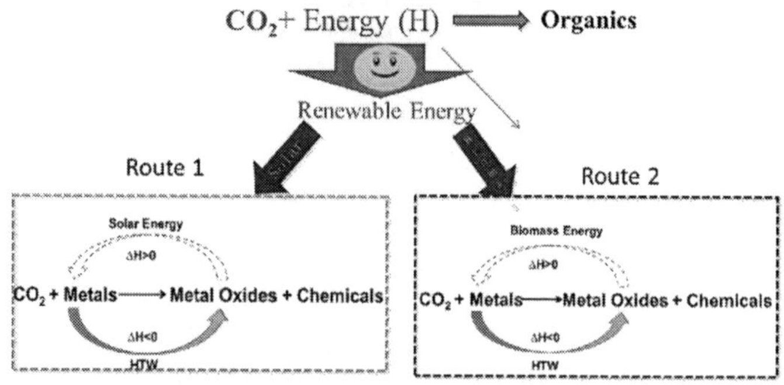

Figure 1. The proposed integrated technology for improving artificial photosynthetic efficiency (HTW = High Temperature Water).

This paper outlines some recent advances in our researches on reduction of CO_2 into value-added chemicals by water spilling with solar/biomass energy-driven two-step process.[5-8]

Experimental

Experimental procedure. $NaHCO_3$ was used as the CO_2 source in order to simplify the experiments, and glucose was employed as a model compound of biomass. Most experiments were conducted in a stainless steel SUS 316 batch reactor (3/8 in. diameter, 1 mm wall thickness, and 120 mm length) with an internal volume of 5.7 mL. The detailed description can be found elsewhere.[7] In a typical procedure, the desired amount of metal/glucose, $NaHCO_3$ with deionized water, were loaded in a batch reactor. Subsequently, the sealed reactor was put into a salt bath which had been preheated to the desired temperature. The salt bath can fast heat the batch from room temperature to 300 °C within 20 s. After the reaction, the reactor was removed from the salt bath to quench in a cold water bath. When the reactor cooling down to the room temperature, the valve was opened and the liquid sample was collected and filtered with a 0.45 mm filter membrane. Solid sample was washed with distilled water and absolute ethanol several times, and dried in air for analysis.

Product analysis. The HPLC analysis of liquid products was performed on Agilent 1260 serials equipped with UVevis detector. Two Shodex RSpak KC-811 columns were used in series. The solvent was 2 mM HClO4 with a flow rate of 1.0 mL min_1. Liquid samples also were analyzed by GCeMS on an Aglient7890A GC system equipped with a 5975C inert MSD. The samples were separated by an HP-INNOWax capillary column (Dimension: 30 m * 250 mm * 0.25 mm) using helium as a carrier gas. Gaseous samples were analyzed using HP-5890 Series II GCTCD system equipped with an HP-INNOWax packing column. Solid samples were characterized by XRD on Shimadzu XRD-6100 equipment.

Results and Discussion

CO_2 Reduction with metals. In studies of reduction of CO_2 to chemicals, various zero-valence metals, such as Fe, Mn, Zn and Al, were employed as reductants for hydrogen production by water spilling. Results showed that these metals are favor of hydrogen production by dissociation of H_2O under hydrothermal conditions, and CO_2 can be reduced efficiently by the in situ formed hydrogen. Conversion products of CO_2 mainly include formic acid, methane and methanol which are with high yields, e.g., about 92% for formic acid, about 98% for methane and about 30% for methanol, respectively (Figure 2).

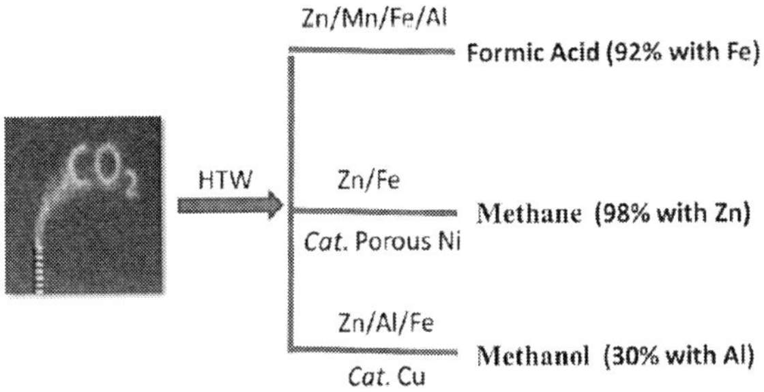

Figure 2. The products of reduction of CO_2 with metal.

Interestingly, some reactions (e.g. CO_2 to formic acid) proceed without additional catalysts. The autocatalytic mechanism was proposed based on the intermediates analysis and FT-IR observation (Figure 3).[8] Initially, in situ formed hydrogen adsorbed on the metal surface and M-H bond formed. Distinct electronegativity difference between M and H leads to strong polarization of M-H bond, resulting in negative charge enriching in H. Subsequently, $H^{\delta-}$ of M-H species attacks the $C^{\delta+}$ of HCO_3^-, followed by the hydroxyl group of HCO_3^- leaving. Finally, formate was obtained together with M hydroxyl compound, which can dehydrate H_2O to form MO.

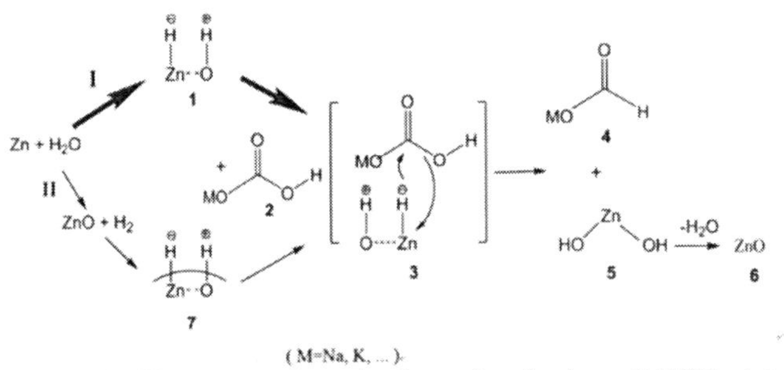

Figure 3. The proposed mechanism of reduction of HCO_3^- into formic acid with metals (Zn as example, Route I is \predominant).

An energy efficiency of 29-36% for ZnO-Zn-H_2 by solar energy has been achieved according to the literature. If we use a lower energy efficiency such as 30% for ZnO-Zn-H_2 by solar energy, then the solar-to-HCOOH energy efficiency for Zn to HCOOH could achieve a value of 5%.[8]

Metal oxides reduction with biomass. As mentioned above, our previous studies indicate biomass possess the potential for reduction of the metal oxides. To examine the possibility of metal/metal oxide cycle with biomass, expetiments of iron oxides and glycerin were conducted under hydrothermal conditions. The result showed iron oxides can be reduced into zero-valent iron metal by a biochemical carbon cycle via Fe^0/Fe_xO_y redox with glycerin (Figure 4). Beside to formic acid, methane and methanol can be also attainted. Thus, a carbon cycle can be achieved by using the oxidative potential of zero-valentmetals to reduce CO_2 to formic acid in the presence of water and the reductive potential of bio-derived chemicals such as glycerin to reduce the oxidized metal oxides to their zero-valent state for recycling.

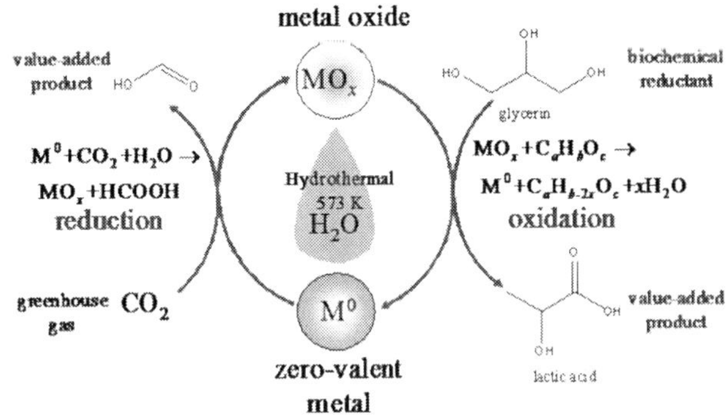

Figure 4. Carbon cycle via Fe^0/Fe_xO_y redox with glycerin.

Although challenges remain, it is promising that the hydrothermal conversion CO_2 will stimulate research on new strategies for circulating carbon while producing chemical products.

Acknowledgment. The authors thank the financial support of the National Natural Science Foundation of China (No. 21277091 & 51472159), the State Key Program of National Natural Science Foundation of China (No. 21436007) and Key Basic Research Projects of Science and Technology Commission of Shanghai (No.14JC1403100).

References

(1) Barton, E. E.; Rampulla, D. M.; Bocarsly, A. B., *J. Am. Chem. Soc.*, **2008**, 130, 6342-6344.
(2) Woolerton, T. W.; Sheard, S.; Pierce, E.; Ragsdale, S. W.; Armstrong, F. A., *Energy Environ. Sci.*, **2011**, 4, 2393-2399.
(3) Chueh, W. C.; Falter, C.; Abbott, M.; Scipio, D.; Furler, P.; Haile, S. M.; Steinfeld, A.. *Science*, **2010**, 330, 1797-1801.
(4) Huff, C. A.; Sanford, M. S., *J. Am. Chem. Soc.*, **2011**, 133, 18122-18125
(5) Jin, F. M.;Enomoto, H., *Energy Environ. Sci.*, 2011, 4,382-397.
(6) Jin, F. M.; Gao, Y.; Jin, Y. J.; Cao, J.; Wei, Z.; Smith, Jr, R. L., *Energy Environ. Sci.* 2011, 4, 881-884.
(7) Lyu, L. Y.; Zeng, X.;Yun, J.; Wei, F.; Jin, F. M., *Environ. Sci. Technol.*, **2014**, 48, 6003-6009.
(8) Jin, F. M.; Zeng, X.; Liu, J. K.; Jin, Y. J.; Wang, L. Y.; Zhong, H.; Yao, G. D.; Huo, Z. B., *Sci. Rep.*, **2014**, 4, 1-8.

INTEGRATION OF CARBON CAPTURE AND SUBSEQUENT CONVERSION

*Mei-Yan Wang[a] and Liang-Nian He[a,b]**

[a]State Key Laboratory and Institute of Elemento-Organic Chemistry, Nankai University, Tianjin, 300071, People's Republic of China.
E-mail: heln@nankai.edu.cn
[b]Collaborative Innovation Center of Chemical Science and Engineering, Nankai University, Tianjin, 300071, P. R. China.

Introduction to CO_2 Chemistry

Carbon dioxide is commonly regarded as the primary greenhouse gas, hence, CO_2 chemistry (capture and conversion) has attracted significant attention from the scientific community[1]. CO_2 capture from flue gases in industrial combustion processes is now being considered as a serious option to reduce or mitigate the so-called green-house effect. The conventional technology for the capture of CO_2 has been applied in industry through chemical adsorption by using an aqueous solution of amines. However, there are inherent drawbacks generally associated with amine-based absorbents, namely the requirement of two amine units to capture one CO_2 molecule owing to the formation of ammonium carbamate, which increases the energy consumption for regeneration. The undesired 2:1 stoichiometry and extensive energy input in desorption process would be crucial barriers to realize practical CCS (carbon capture and storage).

Introduction to CO_2 Chemistry

On the other hand, as an abundant, nontoxic, non-flammable, easily available, and renewable carbon resource, CO_2 is very attractive as an environmentally friendly feedstock for making commodity chemicals, fuels, and materials[2]. However, its inherent thermodynamic stability and kinetic inertness hinder the industrial processes that utilize CO_2 as a raw material. As a consequence, development of efficacious catalysts along with elaborately-designed strategies is crucial to successfully achieve CO_2 utilization. The challenge is to develop catalysts that are capable of activating CO_2 under low pressure (preferably at 1 bar), and thus incorporating CO_2 into organic molecules catalytically. In this regard, we for the first time in 2011 proposed the CCU (carbon capture and utilization)[2b] strategy as an alternative approach to circumvent the energy problem in CCS process. The essence of this strategy could be simultaneous activation of CO_2 upon its capture and thus in situ catalytic transformation into value-added chemicals and fuel-related under mild reaction conditions as shown in Scheme 1, avoiding additional desorption step.

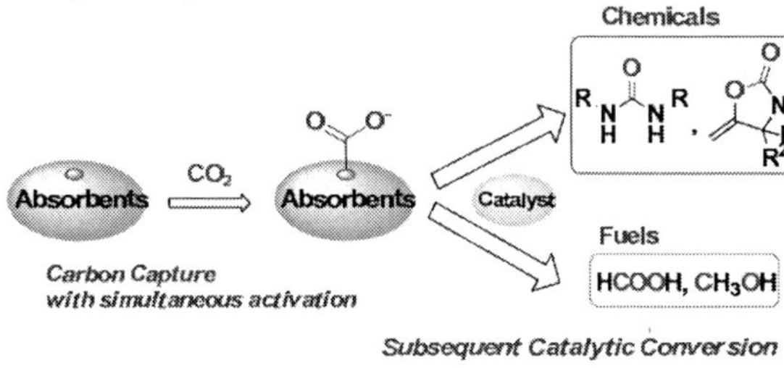

Scheme 1 CCU: Combination of Capture and *in situ* Conversion

Catalytic Conversion of the Captured CO_2 into Chemicals

As a consequence, we developed efficient absorbent to reach equimolar absorption. The capture CO_2 could simultaneously result in its activation, thus subsequent conversion into valuable compounds more easily rather than going through desorption process. Then subsequent reaction can be performed by using the captured CO_2 as a starting material and the absorbent as a catalyst to validate our strategy. For example, the liquid amidinium carbonate salt formed upon CO_2 (gas, 1 bar) absorption with the ionic liquid directly reacts with *n*-butylamine successfully to afford the target product i.e. dibutyl urea in almost quantitative yield (96%) of 1,3-dibutylurea without the use of additives[3]. It is worth mentioning that CO_2 absorption could lead to its activation, almost no additional energy input is required for desorption process with this strategy. Overall, the fixed CO_2 could be catalytically converted to value-added chemicals under extremely mild reaction condition (1 bar, 40 °C and metal-free process). In addition, sodium *N*-isopropylglycinate and potassium phthalimide were demonstrated to be excellent absorbents for rapid carbon dioxide capture with almost equimolar absorption. Based on the CO_2 capture by these absorbents, the reaction of either aziridine or propargyl amine with the captured CO_2 can proceed smoothly at room temperature to afford quantitative yield of oxazolidinones, whereas desorption process could hardly proceed at r.t.[4]. Notably, the captured CO_2 could be also sequentially hydrogenated to formate in 74% yield[5]. What is more, a simple reaction system was developed comprising ammonium carbamates as carbon dioxide surrogates, propargylic alcohols, and a silver(I) catalyst, for the effective conversion of a wide range of alcohols and secondary amines into the corresponding β-oxopropylcarbamates[6]. All of the above catalytic protocols validated the CCU strategy to supply chemical products as a part of a carbon neutral cycle.

***In Situ* Hydrogenation of the Captured CO_2 to Fuels**

On the other hand, with the CCU strategy, an elegant combination of CO_2 capture and in situ hydrogenation to fuel-related products including formic acid, methanol, and amides, as well as methylated compounds, is deemed to be an appealing strategy to supply renewable energy in a carbon neutral cycle[7]. In addition, this protocol can avoid high-pressure operations, which are commonly required for hydrogenation reactions involving CO_2. An integrated process of CO_2 capture and *in situ* hydrogenation into formate was achieved in a TON of up to 852 by the combined PEI_{600} with $RhCl_3 \cdot 3H_2O/CyPPh_2$ system[8]. Apart from an excellent absorbent which can absorb CO_2 directly from the air, PEI could also be assumed to serve as both ligand and base to promote the catalytic hydrogenation of captured CO_2, consequently acting as a "hinge base" to combine capture and hydrogenation processes. A tunable ethoxyl-functionalized amidine and Rh/bisphosphine system is also regarded as an alternative carbon capture and utilization approach to supply fuel-related products, to circumvent the energy penalty in CCS[9]. CO_2 was captured by non-volatile amidine derivatives with simultaneous activation to form zwitterionic amidinium carbonate, and subsequent hydrogenation was facilitated by Rh/bisphosphine. The adsorption capacity and hydrogenation efficiency can be optimized by tuning the ethoxyl side chain. These kinds of carbon capture and utilization pathways could be a potential energy-saving option for industrial upgrading of CO_2 from waste to fuel-related products in a carbon neutral manner.

Conclusions

Turning CO_2 into a feedstock for producing useful commodity chemicals under mild conditions would indeed circumvent most drawbacks of the energy-consuming steps of CO_2 desorption, absorbent regeneration as well as of CO_2 transportation, compression, and disposal in geological cavities, oceans or elsewhere. Consequently, it will have great significance to capture CO_2 and convert it into useful chemical products or fuels simultaneously, taking into account the atom economy and other green issues during the waste disposal process. We believe this CCU protocol will find abundant applications in CO_2 chemistry, in particular, chemical utilization of CO_2 in industry.

Acknowledgement. We are grateful to the National Natural Science Foundation of China (21472103), the Specialized Research Fund for the Doctoral Program of Higher Education (20130031110013), the MOE Innovation Team (IRT13022) of China, and the "111" Project of Ministry of Education of China (B06005).

References

(1) (a) Liu, A.-H.; Yu, B.; He, L.-N. *Greenhouse Gas Sci. Technol.* **2015**, *5*, 17; (b) Aresta, M.; Dibenedetto, A.; Angelini, A. *Chem. Rev.* **2014**, *114*, 1709 (c) Goeppert, A.; Czaun, M.; May, R. B.; Prakash, G. K. S.; Olah, G. A.; Narayanan, S. R. *J. Am. Chem. Soc.* **2011**, *133* (50), 20164; (d) He, L.- N. Carbon Dioxide Chemistry, Chinese Science Publisher (Beijing), **2013**.

(2) (a) He, L.-N.; Wang, J.-Q.; Wang, J.-L. *Pure Appl. Chem.* **2009**, *81* (11), 2069; (b) Yang, Z.-Z.; Zhao, Y.-N.; He, L.-N. *RSC Adv.* **2011**, *1* (4), 545; (c) Yang, Z.-Z.; He, L.-N.; Gao, J.; Liu, A.-H.; Yu, B. *Energy Environ. Sci.* **2012**, *5* (5), 6602.

(3) Yang, Z.-Z.; He, L.-N.; Zhao, Y.-N.; Li, B.; Yu, B. *Energy Environ. Sci.* **2011**, *4* (10), 3971.

(4) Liu, A.-H.; Ma, R.; Song, C.; Yang, Z.-Z.; Yu, A.; Cai, Y.; He, L.-N.; Zhao, Y.-N.; Yu, B.; Song, Q.-W. *Angew. Chem., Int. Ed.* **2012**, *51* (45), 11306.

(5) Zhang, S.; Li, Y.-N.; Zhang, Y.-W.; He, L.-N.; Yu, B.; Song, Q.-W.; Lang, X.-D. *ChemSusChem.* **2014**, *7* (5), 1484.

(6) Song, Q.-W.; Zhou, Z.-H.; Yin, H.; He, L.-N. *ChemSusChem.* **2015**, *8* (23), 3967.

(7) Kothandaraman, J.; Goeppert, A.; Czaun, M.; Olah, G. A.; Prakash, G. K. S. *J. Am. Chem. Soc.* **2016**, *138* (3), 778.

(8) Li, Y.-N.; He, L.-N.; Liu, A.-H.; Lang, X.-D.; Yang, Z.-Z.; Yu, B.; Luan, C.-R. *Green Chem.* **2013**, *15* (10), 2825.

(9) Li, Y.-N.; He, L.-N.; Lang, X.-D.; Liu, X.-F.; Zhang, S. *RSC Adv.* **2014**, *4* (91), 49995.

ADVANCES IN CO_2 CONVERSION

Yun Hang Hu

Department of Materials Science and Engineering, Michigan Technological University, 1400 Townsend Drive, Houghton, MI 49931-1295, USA
Email: yunhangh@mtu.edu

This presentation consists of three parts: (1) advances in CO_2 conversion will be briefly reviewed; (2) important findings for CO_2 conversion in my group will be discussed, including the thermal and the photocatalytic CO_2 conversion to fuels and the synthesis of novel solid materials from CO_2; and (3) the applications of synthesized materials for energy conversion and storage will be demonstrated.[1-4]

References

(1) Hu, Y. H., Ruckenstein, E. Catalytic conversion of methane to synthesis gas by partial oxidation and CO_2 reforming, *Adv. Catal.* **2004**, *48*, 297.
(2) B. Han, W. Wei, L. Chang, P. Cheng, Y. H. Hu, Efficient Visible Light Photocatalytic CO_2 Reforming of CH_4, *ACS Catal.* **2016**, *6*, 494.
(3) Hu, Y. H.; Huo, Y. Fast and exothermic reaction of CO2 and Li3N into C-N-containing solid materials, *J. Phys. Chem. A*, **2011**, 115, 11678.
(4) Wei, W.; Sun, K.; Hu, Y. H. Synthesis of 3D Cauliflower-fungus-like Graphene from CO_2 as Highly Efficient Counter Electrode Material for Dye-sensitized Solar Cells, *J. Mater. Chem. A* **2014**, 2, 16842.

Reduction of CO_2 into HCOOH with 2-pyrrolidinone under hydrothermal conditions

Yang Yang[1], Heng Zhong[2], Guodong Yao[1] and Fangming Jin[1,]*

[1]School of Environmental Science and Enginering, Shanghai Jiaotong University, 800 Dongchuan Road, Shanghai, China
[2]Graduate School of Environmental Studies, Tohoku University, Sendai 980-8579, Japan
*fmjin@sjtu.edu.cn

Introduction

The use of fossil fuels as major energy sources has promoted unprecedented prosperity and advancement of human development since 19th century.[1] However, the non-renewable fossil fuels will probably be run out in next century. In the meantime, combustion of fossil fuels has caused a marked increase in the atmospheric CO_2 concentration from 250 ppm to nearly 400 ppm during the past two centuries, which has led to another serious global problem of climate change.[2] Thus, conversion of CO_2 into useful chemicals and fuels has drawn lots of attention recently, since it can be one of the most potential ways to help decreasing the atmospheric CO_2 concentration and relieve the shortage of fossil fuels simultaneously.[3] Typical approaches for converting CO_2 include catalytic hydrogenation, photochemical conversion and electrochemical conversion.[4-6] Although catalytic hydrogenation has been extensively studied, the demands of pure hydrogen gas and/or noble catalysts in this method greatly lift its cost. Although photochemical or electrochemical methods can utilize renewable energies to reduce CO_2 directly or indirectly, low CO_2 and energy conversion efficiencies limits their application. Therefore, it is necessary to develop new approaches for CO_2 reduction that are highly efficient and requires neither noble catalysts nor gaseous H_2 source.

Recently, reduction of CO_2 into chemicals and fuels such as CH_4, HCOOH, CH_3OH using the *in-situ* generated hydrogen from high temperature water has been extensively studied, in which, various earth abundant metals such as Fe, Zn, Al and Mn were used as the reductants.[7-9] It has also been reported that $N_2H_4 \cdot H_2O$ can be used to reduce $NaHCO_3$ into formic acid over Ni catalyst under hydrothermal condition.[10] These researches implied that both metals and reducing chemicals can be considered as promising reducing agents to reduce CO_2 under hydrothermal conditions.

Food waste of fish and meat are commonly generated in large quantities every day, but most of them are just buried or incinerated without further utilization. The high nitrogen contents, mainly in the form of protein, in these fish and meat wastes, are potential valuable sources of nitrogen for producing chemicals, since linking nitrogen to organic substances needs high energy input. It has been reported that protein or their hydrolysis products, like amino acids, can be converted into N-containing heterocyclic compounds, such as 2-pyrrolidinone, 2-piperidinone, 2-oxoazepane, under hydrothermal conditions easily. These N-containing heterocyclic compounds may have reducing ability due to the nitrogen in these compounds are in reduced forms. Thus, the study of using 2-pyrrolidinone heterocyclic compounds to reduce CO_2 under hydrothermal condition were carried out in this research, which aims to reduce the CO_2 into value-added chemicals using valuable nitrogen resources from food wastes.

Experimental

Experimental Procedure. $NaHCO_3$ was used as the representative compound of water dissolved CO_2 in this research to simplify the reaction procedure, and 2-pyrrolidinone was selected as the representative compound of N-containing heterocyclic compounds. The experiments were conducted with a series of SUS-316 tubing reactors (9.525 mm outer diameter, 1 mm wall thickness, and 120 mm length), which were concealed with end fittings, providing an inner column of 5.7 mL. After the reactor was loaded with desired amounts of 2-pyrrolidinone, $NaHCO_3$ and deionized water, it was sealed and was put into a salt bath, which was preheated to the desired temperature (250 – 325 °C), to initiate the reaction. At the desired reaction time, the reactor was moved out from the salt bath and palced into a cold-water bath to quench the reaction.

Analytical methods. The liquid samples were collected and filtered through a 0.22-μm film filter to eliminate macromolecules before analysis. Then, the filtered liquid samples were analyzed by an Agilent 1200 HPLC analyzer, which was equipped with two KC-811 columns (SHODEX), a tunable ultraviolet/visible (UV/vis) absorbance detector adjusted to 210 nm, and a differential refractometer detector. GC-MS analysis of the filtered liquid samples was achieved with an Agilent 19091N-233 GC-MS system through an HP-INNOWax Polyethylene Glyco column (30 m x 250 μm x 0.5 μm) using helium as the carrier gas (1 mL/min). The oven temperature was initially held at 40 °C for 4 min, then increased to 150 °C at 7.5 °C/min, then increased to 220 °C at 3 °C/min, and finally held for 10 min at 220 °C. The injection sample amount was 1 μL with a split ratio as 50:1.

Definitions. The yield of the HCOOH was defined as:

Yield = (carbon in produced HCOOH, mol / carbon in original $NaHCO_3$, mol) ×100%

The water filling ratio was defined as the ratio of the volume of the solution put into the reactor to the inner volume of the reactor.

Results and Discussion

To demonstrate the reducing ability of 2-pyrrolidinone under hydrothermal condition, experiments of 1 mol/L 2-pyrrolidinone reacting with or without 1.2 mol/L $NaHCO_3$ were conducted at 325 °C for 2 h. As shown in **Figure 1**, when 2-pyrrolidinone reacted in the absence of $NaHCO_3$, no HCOOH production was observed, and the 2-pyrrolidinone concentration decreased to 0.98 mol/L after the reaction. However, when $NaHCO_3$ was added, 80 mmol/L formic acid and several other unknown products were produced, and the concentration of 2-pyrrolidinone dropped to 0.81 mol/L. These results suggest that 2-pyrrolidinone can reduce the $NaHCO_3$ into HCOOH in this reaction.

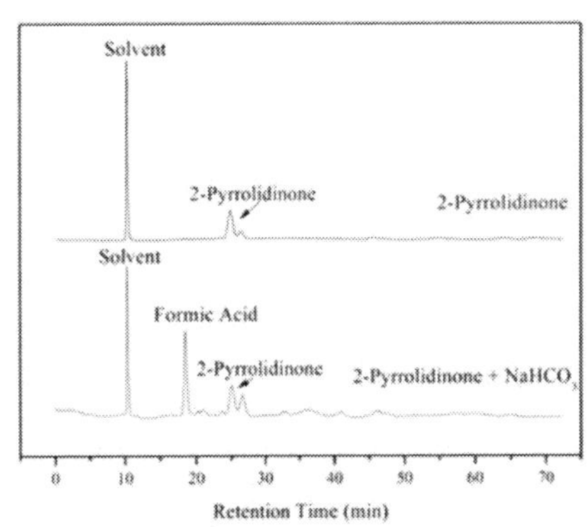

Figure 1. HPLC chromatograms of the liquid samples after the reactions of 2-pyrrolidinone with or without $NaHCO_3$. ($NaHCO_3$, 1.2 mol/L; 2-pyrrolidinone, 1 mol/L; temperature, 325 °C; time, 2 h; water filling ratio, 50%).

Subsequently, effects of different reaction conditions, such as the amount of $NaHCO_3$, reaction temperature, reaction time and water filling ratio, on the HCOOH production were investigated. **Figure 2** shows the effect of initial amount of $NaHCO_3$ on the HCOOH production. When the concentration of $NaHCO_3$ increased from 0.6 to 2.1 mol/L with a fixed concentration of 2-pyrrolidinone of 1 mol/L, the concentration of produced HCOOH increased from 28 to 103 mmol/L, which is probably due to more $NaHCO_3$ was engaged in the reaction. However, the yield of HCOOH did not increase in the same trend as the concentration of $NaHCO_3$ increasing. It dropped from 6.5% to 4.7% when the concentration of $NaHCO_3$ increased from 1.2 to 2.1 mol/L, which is probably caused by the insufficient amount of 2-pyrrolidinone.

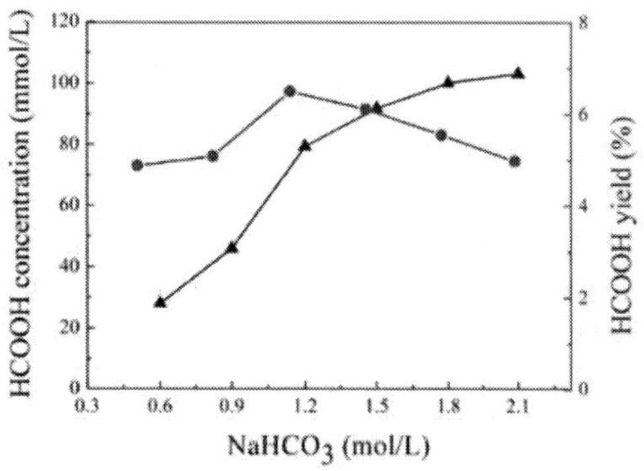

Figure 2. Effect of initial $NaHCO_3$ concentration on the yield (●) and concentration (▲) of the produced HCOOH. (2-pyrrolidinone, 1 mol/L; temperature, 325 °C; time, 2 h; water filling ratio, 50%).

Then, effects of reaction time and temperature were examined by changing the reaction temperature from 250 to 325 °C and time from 30 to 180 min. As shown in **Figure 3**, when the reaction temperature was 250 °C, the yield of HCOOH was relatively low and increased inconspicuously with time, while it increased remarkably with time when the temperature exceeded 275 °C. Hence, a relatively high temperature was favorable to the HCOOH production. The profile of HCOOH yield versus time indicated that prolonging the reaction time could increase the yield of HCOOH in the first 120 min; however, when the reaction time exceeded 120 min, the yield of HCOOH did not change significantly.

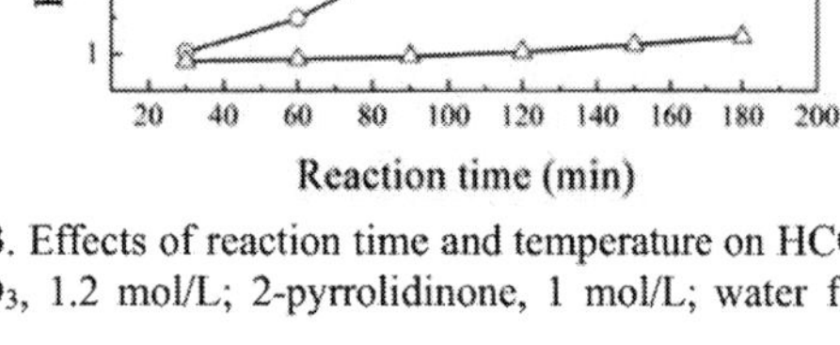

Figure 3. Effects of reaction time and temperature on HCOOH yield. ($NaHCO_3$, 1.2 mol/L; 2-pyrrolidinone, 1 mol/L; water filling ratio, 50%).

Effect of water filling ratio on the HCOOH yield was investigated by using 1.2 mol/L $NaHCO_3$ and 1 mol/L 2-pyrrolidinone with different water filing ratio varying from 30% to 55% (the absolute water amount was from 1.7 to 3.1 mL). As shown in **Figure 4**, the yield of HCOOH increased from 4.8% to 6.5% when the water filling ratio increased from 30% to 50%. When further increasing the water filling ratio to 55%, HCOOH yield did not change obviously.

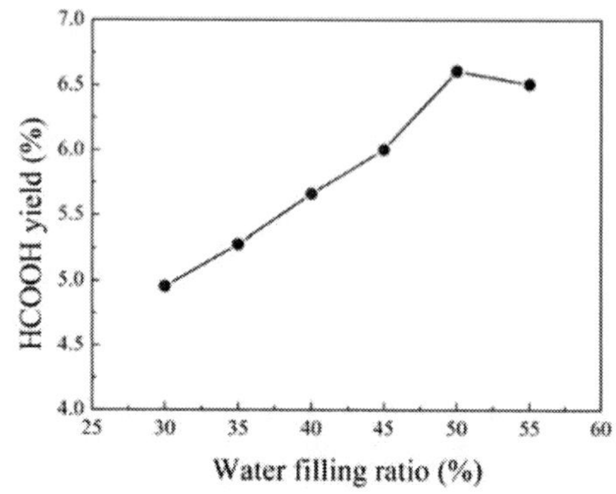

Figure 4. Effect of the water filling ratio on the HCOOH yield. ($NaHCO_3$, 1.2 mol/L; 2-pyrrolidinone, 1 mol/L; time, 2 h; temperature, 325 °C).

Above results clearly demonstrate that 2- pyrrolidinone can reduce $NaHCO_3$ into HCOOH under hydrothermal conditions, and further mechanism studies are in progress.

Acknowledgment

This work was supported by the National Natural Science Foundation of China (Grant B061104).

References

(1) G. A.Olah, A. Goeppert and G. K. S. Prakash, J. Org. Chem., 2009, 74, 487-498.
(2) X. Xiaoding and J. A. Moulijn, Energy Fuels, 1996, 10, 305-325.
(3) S. N. Riduan and Y. Zhang, Dalton transactions, 2010, 39, 3347-3357.
(4) T. Kobayashi and H. Takahashi, Energy Fuels, 2004, 18, 285-286.
(5) T. Inui, T. Yamamoto, M. Inoue, H. Hara, T. Takeguchi and J.-B. Kim, Appl. Catal., 1999, 186, 395-406.
(6) E. E. Barton, D. M. Rampulla and A. B. Bocarsly, J. Amer. Chem. Soc., 2008, 130, 6342-6344.
(7) B. Wu, Y. Gao, F. Jin, J. Cao, Y. Du and Y. Zhang, Catal. Today, 2009, 148, 405-410.
(8) L. Lyu, X. Zeng, J. Yun, F. Wei and F. Jin, Environ. Sci. Technol., 2014, 48, 6003-6009.
(9) F. Jin, X. Zeng, J. Liu, Y. Jin, L. Wang, H. Zhong, G. Yao and Z. Huo, Scientific reports, 2014, 4, 4503.
(10) F. Chen, G. Yao, Z. Huo and F. Jin, RSC Adv., 2015, 5, 11257-11260.

A MODIFIED CHOU MODEL FOR NEGATIVE TEMPERATURE DEPENDENCE IN THE HYDROGENATION OF HYDROGEN STORAGE MATERIALS

*Hongyu Long, Yuepeng Pang, Qun Luo, Qian Li**

a State Key Laboratory of Advanced Special Steels & Shanghai Key Laboratory of Advanced Ferrometallurgy & School of Materials Science and Engineering, Shanghai University, Shanghai 200072, China.

b Materials Genome Institute, Shanghai University, Shanghai 200072, China.

shuliqian@shu.edu.cn

Introduction

During the last decades, there has been an increasing interest in developing theoretical quantitative description for hydrogen storage process[1-2]. Chou model can successfully describe the absorption kinetic curves because it offers different and explicit analytic formulae which contain parameters with clear physical representation. Luo etc.[3] applied both Jander model and Chou model to investigate the hydriding kinetic behavior of Mg-Ni based alloys, and Chou model should be preferentially recommended to deal with the calculation at multi-temperatures and multi-pressures due to avoiding multistep calculation. Moreover, we can obtain the apparent activation energy of reaction through Chou model and Chou model has a function of predicting the kinetic curves of other conditions, Cui etc. testify Chou model by applying to Zr-based AB_2 hydrogen storage alloys[4].

However, the apparent activation energy of some La-Mg-Ni alloys calculated by Chou model turned out to be negative value[5-7]. Similar situation also appears in other negative temperature characteristic materials. The previous Chou model does not consider the change of chemical driving force and thermodynamic factor, which have strong effect on the reaction rate, making the calculated activation energy negative. This is inconsistent with its physical meaning and seriously hinders the in-depth understanding of the kinetic mechanism. Hydrogen pressure affects much on the chemical driven force and thus plays an important role in hydrogen absorption and desorption, and it was regard as one constant of pre-exponential factor for Arrhenius equation in the previous Chou model. Therefore, we introduced thermodynamic parameters into the new model through equilibrium pressure in order to obtain a more accurate theoretical model and reasonable activation energy for the negative temperature characteristic materials.

La-Mg-Ni ternary alloys show superior hydrogen storage properties. Ren etc. added La into Mg_2Ni-type alloys to prove that the substitution of La for Mg markedly increased the hydrogen absorption and desorption capacities and kinetics of the as-cast $Mg_{20-x}La_xNi_{10}$ (x= 0, 2, 4, 6) alloys, in their work, the sample with low content of La display the maximum hydrogen desorption capacity and most excellent kinetics[8]. However, the kinetic mechanism of the ternary alloys is still ambiguous. So applying the new Chou model to understand the kinetic behaviors of the alloys is also meaningful.

The goals of this work are: (i) based on the Chou model, combine hydrogen pressure and thermodynamic parameters to derived a new formula and calculate the activation energy, (ii) using the experiment data of $LaNi_5$ and $(LaNi_5)_{0.959}Mg_{0.041}$ to verify the new formula, (iii) to investigate what effect can adding few Mg into $LaNi_5$ do to its kinetic property.

Kinetic model

The studies[9-10] shows that diffusion or surface penetration is the rate-controlling step of the hydrogen absorption reaction mostly, then we take these two reaction steps as examples in this paper.

When diffusion of hydrogen atoms through the hydride layer is the rate-controlling step, the reaction rate ξ calculated by Chou model can be expressed as the simplest Eq. (1)

$$\xi = 1 - \left(1 - \sqrt{\frac{t}{t_{c(d)}}}\right)^3 , \qquad (1)$$

where $t_{c(d)}$ is the characteristic reaction time whose physical meaning is the required time for a complete hydriding reaction. From the equation we can know that a smaller t_c value represents a faster reaction process.

When the rate-controlling step is diffusion, Chou model define B_t as a coefficient, the reaction rate ξ with activation energy E_V (J/mol) can be described as

$$\xi = 1 - \left[1 - \sqrt{\exp(-\frac{E_{v(d)}}{RT}) \cdot \frac{t}{B_t}}\right]^3 , \qquad (2)$$

where T (K) is the reaction temperature, and R is the gas constant.

It is well known that reaction rate ξ is related to temperature and pressure, at first, we assume B_t to be a function of pressure, and we consider pressure change is too little to take into calculation so that we make B_t a constant, which is obviously not accurate now.

The reaction rate deduced by W.J.Dunning[11] from Arrhenius Formula is

$$k = k_0 \frac{\Delta G_m}{RT} exp(\frac{-\Delta E}{RT}) , \qquad (3)$$

where k_0 is equilibrium constant and ΔE is the activation energy. According to partial molar Gibbs function, the reaction rate can be expressed as

$$k = k_0 ln\left(\frac{P_0}{P_{eq}}\right) exp(\frac{-\Delta E}{RT}) , \qquad (4)$$

where P_0 is the initial effective hydrogen pressure and P_{eq} is the equilibrium hydrogen partial pressure, and it's the same form with P. D. GOODELL's formula[12].

Substituting Eq. (4) into Eq. (2) and according to Van't Hoff equation we will have

$$\xi = 1 - [1 - \sqrt{k_0\left(lnP_0 - \frac{\Delta S}{R} + \frac{\Delta H}{RT}\right) exp\left(\frac{-E_{v(d)}}{RT}\right) t}]^3 , \qquad (5)$$

where ΔS and ΔH are the entropy change and enthalpy change of the whole hydrogenation reaction respectively.

In a similar way, we can have formula for the Surface penetration controlling-step

$$\xi = 1 - \left\{1 - \left[k_0\left(lnP_0 - \frac{\Delta S}{R} + \frac{\Delta H}{RT}\right) exp\left(\frac{-E_{v(sp)}}{RT}\right) t\right]\right\}^3 . \qquad (6)$$

In the subsequent kinetic analysis, we will utilize the corresponding formula of the actual controlling-step to calculate the activation energy.

Corresponding squared correlation coefficient r^2 is used to estimate the level of agreement between theoretical data and the experimental data. The r^2 has a value range between 0-1.00, the closer to 1.00, the better agreement between experimental data and the fitting results.

Experimental

High purity raw materials are used in this work. Pure Mg (purity >99.9%), Ni (purity >99.99%), La (purity >99.9%) and pure $LaNi_5$ were purchased firstly. In order to get clean and homogeneous samples, the alloys were synthesized by high frequency levitation melting furnace (100 kHz) with a water cooled Cu crucible. Pre-evacuation protect atmosphere furnace, filling nitrogen gas until pressure is 0.1MP, melting temperature is selected in the melting sample melting point above 50 K. Annealing treatment for 50 days at 1173K temperature, and annealing for 110 days at 973K temperature, then quench the samples in an ice water mixture quickly to maintain their high temperature structure and phase structure.

In this work we labelled the $LaNi_5$ and $(LaNi_5)_{0.959}Mg_{0.041}$ as samples #1 and #2. The actual compositions were measured by 7300DV Optima type Inductively Coupled Plasma (ICP).All samples were milled into fine powders and activated at 318 K under 3 MPa for five times. Then, the hydriding temperature range was 303~333 K with hydrogen pressure 1.0~3.0 MP. The absorption kinetics was measured by isothermal constant volume method through a volumetric with high purity hydrogen (99.999%) on Suzuki Shokan PCT-2SDWIN Sievert type apparatus from Japan.

Results and Discussion

From the XRD patterns results for $LaNi_5$ and $(LaNi_5)_{0.959}Mg_{0.041}$ are presented we can confer that sample #1 is pure $LaNi_5$, sample #2 consist of $LaNi_5$ and $LaMg_2Ni_9$.

According to the Eq.(5), our new model requires the data of the entropy change and enthalpy change. While the ΔH and ΔS of the samples here have not been reported, thus we need to calculate them by the equilibrium hydrogen pressures measured by pressure-composition-temperature (PCT) absorption results.

At first, we use Chou model to fit the hydrogen absorption curves of our samples respectively to find out the rate-controlling step, and the calculated results show that the diffusion is the rate-controlling step for our samples at all condition we experimented. After that we applied Eq. (5) to calculate the activation energy. In the fitting process, ξ should divided by the maximum hydrogen absorption δ, which measured by the experiment, every curve at different temperature and pressure have different values. The fitting requires five parameters, pre-exponential factor k_0, reaction temperature T, initial hydrogen pressure P_0, entropy change ΔS and enthalpy change ΔH. Besides the activation energy E_v, the pre-exponential factor k_0 is also obtained by fitting while the other parameters are known fixed value. And the pre-exponential factor k_0, entropy change and enthalpy change are shared for a same sample while the other parameters are different (the experimental value).

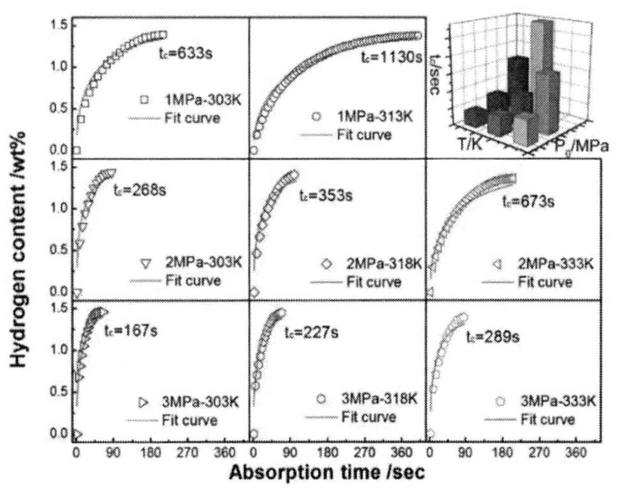

Figure 1. Global fitting calculated curves for activation energy vs. experimental data of $LaNi_5$, the upper right corner is the 3D bar graph of the t_c, and the color of the bars correspond to the hydrogen absorption curves.

Different from the previously Chou model, our formulae take both temperature and hydrogen pressure in consider, we can global fit all the data that has been obtained to get a more accurate value of activation energy. In order to get more accurate results, the value of parameters are the readout values of the experiment, i.e. as a result of different sample volume in the reaction kettle, the initial hydrogen pressure will get different values once the reaction kettle inrush of hydrogen, even the pressure is adjusted to the same value in advance.

The global fitting calculated curves for $LaNi_5$ are shown in Fig.2 respectively with their own experimental hydrogen absorption curves, and the value of characteristic reaction time t_c calculated by Eq.(1) are separately marked next to the curves.

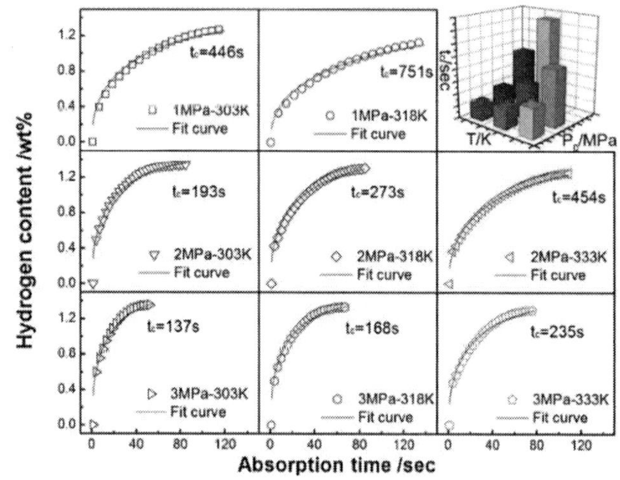

Figure 2. Global fitting calculated curves for activation energy vs. experimental data of $(LaNi_5)_{0.959}Mg_{0.041}$, the upper right corner is the 3D bar graph of the t_c. The curve and bar with the same experimental condition are drawn in the same color.

The experiment data and calculation result of $LaNi_5$ and $(LaNi_5)_{0.959}Mg_{0.041}$ are showing in Table 1. The activation energy calculated by previous Chou model are three different negative values, but one positive value for a material can be gave by our modified mod. We can confer form Fig.1 and Fig.2 that the experimental data and the theoretically calculated results have reached an excellent agreement. The value of r^2 are both above 0.96, while the all values of r^2 calculated by the previously Chou model turn out to be below 0.92. According to the characteristic reaction time t_c showing in Fig.1 and Fig.2, the reaction rate decrease with the increase of temperature at all hydrogen pressure, indicate that the $LaNi_5$ and $(LaNi_5)_{0.959}Mg_{0.041}$ are negative temperature characteristic materials typically.

Table 1. The experiment data and calculation result of $LaNi_5$ and $(LaNi_5)_{0.959}Mg_{0.041}$

	Temperature (K)	Pressure (MPa)	Hydrogen content (wt%)	E_v (kJ)	ΔH (kJ)	ΔS (J)	k	r^2
#1	305	0.75	1.47	14.12	30.44	91.4	0.61	0.96
	304	1.96	1.55					
	302	2.9	1.57					
	310	0.65	1.45					
	315	1.56	1.51					
	320	2.96	1.59					
	333	1.57	1.46					
	330	2.66	1.49					
#2	307	0.58	1.40	13.71	31.41	90.69	0.60	0.98
	307	1.5	1.42					
	307	2.9	1.45					
	320	0.6	1.35					
	321	1.2	1.42					
	322	2.3	1.41					
	337	1.2	1.44					
	337	2	1.41					

Obviously the activation energy of $LaNi_5$ is higher than $(LaNi_5)_{0.959}Mg_{0.041}$ (seeing Table 1), combine with the change of the characteristic reaction time t_c, a small addition of Mg can improve the kinetic property of $LaNi_5$ indeed.

Conclusions

(i) A simple and practical formula has been proposed to calculate the activation energy. It is an explicit function with clear physical meaning in the theoretical discussion, which taking both chemical driving force and thermodynamic factor into consideration.

(ii) The application of the modified model to $LaNi_5$ and $(LaNi_5)_{0.959}Mg_{0.041}$ shows a good agreement has been reached between experimental data and theoretical model. And we obtained a meaningful activation energy value, provide a new way to study the kinetics of negative temperature characteristics materials.

(iii) The experimental data and theoretical calculation results indicated that the small addition of magnesium can improve the kinetic property of $LaNi_5$.

Acknowledgement. Acknowledgements (if any) should be a subsection heading within "Conclusions." (9 pt)

References

(1) Chou, K. C., A kinetic model for oxidation of Si–Al–O–N materials. *Journal of the American Ceramic Society* **2006,** *89* (5), 1568-1576.

(2) Luo, W.; Gross, K. J., A kinetics model of hydrogen absorption and desorption in Ti-doped NaAlH 4. *Journal of Alloys and Compounds* **2004,** *385* (1), 224-231.

(3) Luo, Q.; An, X.-H.; Pan, Y.-B.; Zhang, X.; Zhang, J.-Y.; Li, Q., The hydriding kinetics of Mg–Ni based hydrogen storage alloys: A comparative study on Chou model and Jander model. *International Journal of Hydrogen Energy* **2010,** *35* (15), 7842-7849.

(4) Cui, X.-Y.; Li, Q.; Chou, K.-C.; Chen, S.-L.; Lin, G.-W.; Xu, K.-D., A comparative study on the hydriding kinetics of Zr-based AB 2 hydrogen storage alloys. *Intermetallics* **2008,** *16* (5), 662-667.

(5) An, X.; Pan, Y.; Luo, Q.; Zhang, X.; Zhang, J.; Li, Q., Application of a new kinetic model for the hydriding kinetics of LaNi 5− x Al x (0≤ x≤ 1.0) alloys. *Journal of Alloys and Compounds* **2010,** *506* (1), 63-69.

(6) Li, Q.; Lin, Q.; Jiang, L.; Chou, K.-C., On the characterization of La 1.5 Mg 17 Ni 0.5 composite materials prepared by hydriding combustion synthesis. *Journal of alloys and compounds* **2004,** *368* (1), 101-105.

(7) Zhao, X.; Li, Q.; Chou, K.; Liu, H.; Lin, G., Effect of Co substitution for Ni and magnetic-heat treatment on the structures and electrochemical properties of La–Mg–Ni-type hydrogen storage alloys. *Journal of Alloys and Compounds* **2009,** *473* (1), 428-432.

(8) Ren, H.-p.; Zhang, Y.-h.; Li, B.-w.; Zhao, D.-l.; Guo, S.-h.; Wang, X.-l., Influence of the substitution of La for Mg on the microstructure and hydrogen storage characteristics of Mg 20− x La x Ni 10 (x= 0–6) alloys. *International Journal of Hydrogen Energy* **2009,** *34* (3), 1429-1436.

(9) Bloch, J.; Mintz, M. H., Kinetics and mechanisms of metal hydrides formation—a review. *Journal of Alloys and Compounds* **1997,** *253*, 529-541.

(10) Li, Q.; Lin, Q.; Chou, K.-C.; Jiang, L., A study on the hydriding-dehydriding kinetics of Mg1. 9Al0. 1Ni. *Journal of materials science* **2004,** *39* (1), 61-65.

(11) Dunning, W. J., Kinetics of Reaction in Ionic system. In *Materials Science Research*, V.D.Frechette, T. J. G. a., Ed. New York, 1969.

(12) Goodell, P.; Rudman, P., Hydriding and dehydriding rates of the LaNi 5-H system. *Journal of the Less Common Metals* **1983,** *89* (1), 117-125.

A NOVEL METHOD FOR CONVERTING CARBON DIOXIDE INTO CARBON UNDER MILD CONDITIONS

*Yuqian Chen, Zhenzi Jing**

Key Laboratory of Advanced Civil Engineering Materials, Ministry of Education, Tongji University, 4800 Cao'an Road, Shanghai 201804, China

Introduction

CO_2 is considered as the main greenhouse gas blamed for raising the temperature on the earth over recent decades. The current level of atmospheric CO_2 concentration now is estimated to be more than 400 ppm which reaches a warning symbol for humans. On the other hand, CO_2 is a nontoxic, abundant, and economical substance. Therefore, developing a technology for efficiently converting CO_2 into chemicals is needed urgently, which is also very important not only to decrease the environmental impacts of greenhouse gas emissions, but also to provide new thought for carbon cycle, especially for CO_2.

At present, some methods, such as electrochemical reduction,[1] photocatalytic reduction,[2] and chemical vapor deposition[3] have been used for the CO_2 reduction. Because of the thermodynamic stability of CO_2, although these methods have been proven to be practicable for the reductions, there still existed some considerable drawbacks. The complex operation and relatively low yield cannot satisfy large scale industrial production, and the high temperature, high pressure, expensive catalyst and reductant, high energy consumption may not suit requirement of energy-saving and emission-reduction. Recently, it has been found that using metals as a reductant or catalyst in the hydrothermal reactions could reduce CO_2 to organic chemicals by hydrothermally splitting hydrogen with a high yield. For example, our previous work has showed that CO_2 could be reduced using Fe and Ni powder as reductant and catalyst respectively under hydrothermal conditions, and the increase in temperature could improve the formic acid yield under low temperature (<548K), while the formic acid favored to decompose into methane at higher temperature (at 623 K).[4] In addition, honeycomb 3D graphene-like Co_9S_8 catalyst could also reduce CO_2 into formic acid with a high yield of 3680 mg/L.[5]

In this study, instead of converting CO_2 into organic carbon materials, we tried to reduce CO_2 into inorganic carbon materials. Although some researches on the reduction CO_2 into inorganic carbon have been carried out, e.g., CO_2 was reduced to C_{60} at the temperature of 973K and pressure of 100MPa for 10h.[6] For reduction of CO_2 with a low temperature, we have reduced CO_2 into graphite at 723K using Na as the reductant. Inspired by the hydrothermal reduction of CO_2, i.e. water not only acts as solvent but also supplies pressure in the system, we have introduced a solvent to convert CO_2 so as to enhance the efficiency of the reduction.

Experimental

Materials. $NaHCO_3$ (>99.5%) powder was used as a CO_2 resource to simplify the operation and to control the amount of CO2 precisely[7]. In fact, CO_2 capture usually uses alkaline solution to dissolve CO_2, and the solution of CO_2 in alkaline aqueous (exist equilibriums among CO_2, H_2O, HCO_3^-, CO_3^{2-} and OH^-) is very similar to $NaHCO_3$ aqueous solution. All the reagents used were obtained from Sinopharm Chemical Reagent Co. Ltd, China.

Experimental procedure. Experiments were conducted as following: 0.2g metallic Na (>99.5%) and 2.1g and $NaHCO_3$ powder were loaded into an autoclave made of SUS316 stainless steel with an internal volume of 50ml. Then the autoclave was heated to 723K and kept the reaction temperature for 2h. After reaction, the black solid reaction product was dried in an oven at 353K for 24h and then analyzed by XRD, XPS, Raman and SEM.

Results and Discussion

The XRD analysis used Cu-sealed tube (Cu K X-rays of 0.154 nm) operating at 40 kV and 40 mA and 2-Theta ranged from 5° to 35° with speed of 2°/min. Result of XRD was shown in **Figure 1**, and the peak of sample appeared at 2θ=23° which means that the main production is graphite.[8,9] According to the Bragg Equation, interplanar spacing d equals to 0.386nm, the synthesized graphite therefore seemed to be partially oxidized and the formed oxygen containing functional groups should have made the interplanar spacing larger than that of graphite.

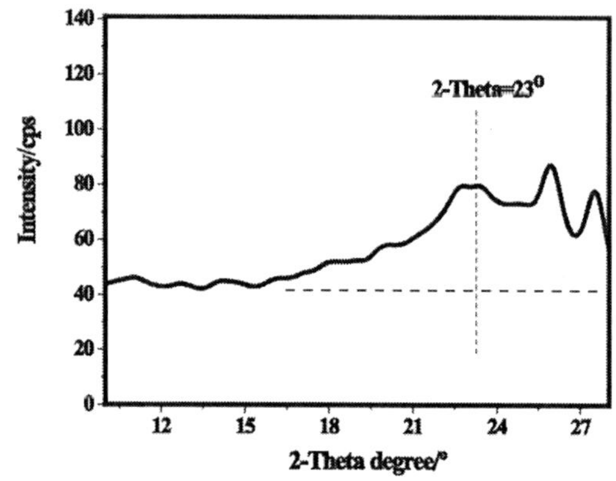

Figure 1. XRD pattern of the sample with 2.1g $NaHCO_3$ and 0.2g Na at 723K for 2h

Table 1. XPS result of C 1s spectra fitting into C chemical group for sample with 2.1g $NaHCO_3$, 0.2g Na at 723K for 2h.

	Chemical group	Binding energy(eV)	Content (%)
C 1s	sp^2	284.5	79.77
	sp^3	285.5	21.00
	C=O	288.5	8.23

The above sample was also analyzed by XPS and the result showed that the content of carbon and oxygen group atomic are 86.10% and 13.90% respectively fitting to binding energy of 284.08 eV and 532.08 eV.[10] As shown in **Table 1**, the result indicates that C 1s takes part in a large proportion in the production. The binding energy of electrons from carbon atom of different carbon hybridizations (C sp^2 and C sp^3) are within the range typical for these groups, which indicates that the sample has a high content of C sp^2 with a small amount of other groups such as C=O and C-O. C sp^2 suggests the existence of two-dimensional graphene or three-dimensional graphite, which is connected by Van der Waals force between layers[11] and the result of C-O or C=O show that part of graphite of graphene are oxidized. Therefore, above results showed that C-O bond breaking in CO_2 has resulted in C-C bond forming in graphite.

The Raman spectroscopy of the above sample is shown in **Figure 2,** and the D and G peaks at around 1330 cm^{-1} and 1580 cm^{-1} respectively are observed which belong to the main Raman features of graphite. The G peak is due to the bond stretching of all pairs of sp^2 atoms in both rings and chains, while the D peak is due to the breathing modes of sp^2 atoms in rings.[12] The peaks of the synthesized graphite seemed to shift about $10cm^{-1}$ compared with standard spectrum of graphite, which should attribute to partial oxidation of graphite.

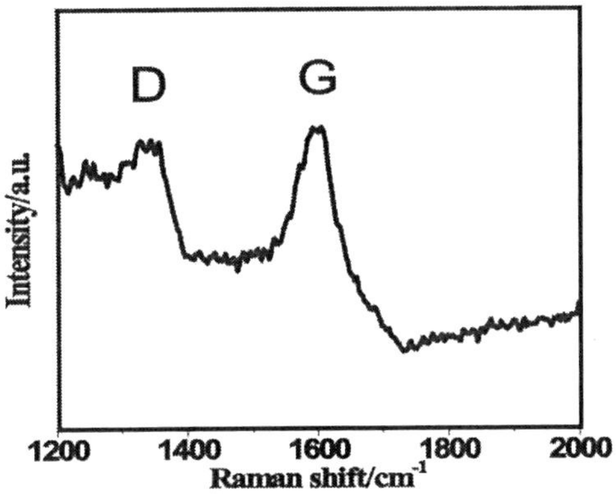

Figure 2. Raman spectroscopy of graphite :2.1g $NaHCO_3$ and 0.2g Na, at 723K for 2h

To improve the reduction and to convert more graphene, a comparative experiment was carried out with 2.1g $NaHCO_3$ and 0.2g Na but under different conditions, i.e. one without solvent toluene and one with solvent toluene (solvothermal). SEM micrographs of the synthesized graphite are shown in **Figure 3.** The different thickness graphite could be observed, reflecting different stripping degree for graphite. **Figure 3(a)** shows sheet of graphite, because of the weak Van der Waals force existing between flakes, the hydroxyl and carboxyl oxygen containing functional groups inserted into the graphite layers which made interplanar spacing larger and thus to make graphite sheet strip readily. The morphology of the graphite prepared without solvent was shown as the stacked graphite flakes, suggesting that the carbon has been reduced into graphite. **Figure 3(b)** shows an SEM image of ample with solvolthermal method, lighter color of graphite is observed which suggests that a thinner or less sheets of graphite formed corresponding to large extent exfoliation. It should be noted that the morphology of the ends of graphite appeared rolled up or wrinkled obviously. The behavior of the rolled or wrinkle morphology of graphite suggests that graphene seems to be formed readily in this process.[13]

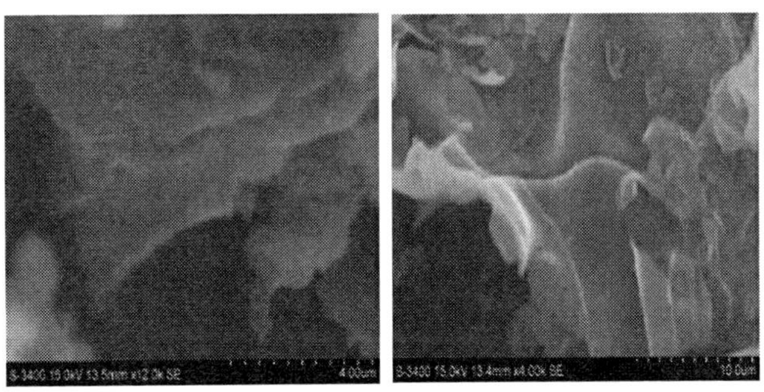

Figure 3. SEM micrographs (a):without solvent, with 2.1g $NaHCO_3$ and 0.2g Na, at 723K for 2h (b): solvothermal method, with 2.1g $NaHCO_3$, 0.2g Na and 10ml toluene solvent at 723K for 2h.

According to the above XRD, XPS, Raman and SEM results, graphite could be formed from CO_2 reduction, in which a mixture of graphite, graphite oxide and graphene has been showed to form at 723K for 2h. The graphene seemed to be formed readily from CO_2 through solvolthermal method. As such, CO_2 could be converted into graphite and further graphene with addition of metallic Na at a relatively low temperature of 723K for a short time of 2h, which is expected to provide practical information on CO_2 reduction.

Acknowledgment. The work reported here was supported by the National Natural Science Foundation of China (No. 51272180).

References

(1) D. Chery, V. Lair , M. Cassir. *Electrochimica Acta,* **2015**, *160,* 74-81.
(2) Giuseppe Marcì , Elisa I. García-López, Leonardo Palmisano. *Catalysis Communications,* **2014**, *53,* 38–41.
(3) X. J. Xu, S. M. Huang. *Materials letters,* **2007**, *61,* 4235-4237.
(4) Y. Q. Chen, Z. Z. Jing, , J. J. Miao, Y. Zhang, J. J. Fan. *International Journal of Hydrogen Energy.* Doi:10.1016/j.ijhydene.2015.11.157.
(5) B. Y. Hu, Z. Z. Jing, ,J. J. Fan, G.D. Yao, F.M. Jin. *Catalysis Today.* **2016,** *263,1,*128–135.
(6) C. L. Chen, Z. S. Lou. *Journal of Supercritical Fluids,* **2009**,*50*,42-45.
(7) B. Wu, Y.Gao, F. M. Jin, J. L. Cao, Y. X. Du, Y. L. Zhang. *Catalysis Today,* **2009**, *148,* 405-410.
(8) X. F. Mei, X. Q. Meng, F. M. Wu. *Physica E.* **2015**, *68,* 81-86.
(9) L. Stobinski, B. Lesiak, A. Malolepszy, M. Mazurkiewicz, B. Mierzwa, J. Zemek, P. Jiricek, I. Bieloshapka. *Journal of Electron Spectroscopy and Related Phenomena,* **2014**,*195,*145-154.
(10) S. H. Shim, K. T. Kim, J. U. Lee, W. H. Jo. *ACS Applied Materials & Interfaces,* **2012**, *4, 8,* 4184–4191.
(11) D. Ferrah, J. Penuelas, C. Bottela, G. Grenet, A. Ouerghi. *Surface Science,* **2013**, *615,* 47–56.
(12) Z.H. Ni, Y.Y. Wang,, T. Yu., Z.X. Shen. *Nano Research,* **2008**, *1, 4,* 273-291.
(13) E. J. Yoo, J. Kim, E. Hosono, H. Zhou, T. Kudo, I. Honma. *Nano letters,* **2008**, *ACS Publications,* 2277-2282.

SELECTIVE HYDRODEOXYGENATION CATALYSTS FOR FURFURAL AND HYDROXYMETHYLFURFURAL

Jing Luo, Cong Wang, and Raymond J. Gorte

Department of Chemical & Biomolecular Engineering,
University of Pennsylvania
Philadelphia, PA 19104
gorte@seas.upenn.edu

Abstract

Extensive studies have been carried out on the hydrodeoxygenation (HDO) of furfural to methylfuran (MF) and 5-hydroxymethylfurfural (HMF) to dimethylfuran (DMF); however, selectivity in the HDO of furfural is limited by decarbonylation, while side products in HDO of HMF tend to be open-ring and hydrogenated-ring products. We will show that differences between furfural and HMF are due to the reaction conditions that are typically employed. At the high H2 pressures used in most studies of HMF, decarbonylation is not important. Furthermore, the alloy catalysts that are selective against decarbonylation are also selective for DMF production. The reaction network will be discussed.

ORIENTED GROWTH OF $ZN_2IN_2S_4/IN(OH)_3$ HETEROJUNCTION BY A FACILE HYDROTHERMAL TRANSFORMATION AS AN EFFICIENT PHOTOCATALYST FOR H_2 PRODUCTION

Yuexiang. Li, Qinyu Fu, Yali Hou, Shaoqin Peng

Department of Chemistry
Nanchang University
Nanchang 330031
P.R. China

Introduction

The activity of a photocatalyst is strongly dependent on the transport and separation efficiencies of the photogenerated electrons and holes. It has been well-documented that a heterojunction composite which is composed of two semiconductors with matching band potentials can greatly enhance the separation efficiency of the photoexcited carrier pairs and photocatalytic activity.

Electrical conductivity of layered semiconductor $ZnIn_2S_4$ is anisotropic. The ratio of the conductivity along the layers (parallel) to that across the layers (perpendicular) is about 10^3 at 300 K.[1] When a semiconductor (semiconductor 2) couples on the (0 0 1) facet of sheet-like $ZnIn_2S_4$ (H type coupling, **Scheme 1**), the photoinduced electron transfer along z axis in $ZnIn_2S_4$ is difficult due to the high resistance, which decreases greatly the efficiency of the electron transfer from $ZnIn_2S_4$ to the semiconductor 2. Whereas, when it couples at the edges of $ZnIn_2S_4$ sheets along the layered planes (J type coupling), the transfer efficiency is very high owing to the low resistance. Thus, compared to the H type coupling, the J type heterostructure can achieve highly efficient photocatalytic activity.

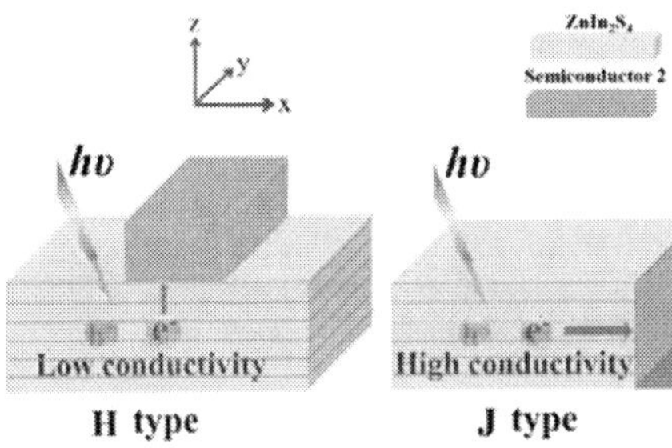

Scheme 1 Schematic of two anisotropic heterojunctions with different efficiencies of photoinduced electron transfer [along the z axis (H type) and layered planes (J type)] in layered $ZnIn_2S_4$ sheets.

Herein, $ZnIn_2S_4/In(OH)_3$ heterojunction with J type structure has been successfully fabricated by a simple hydrothermal transformation of the co-precipitation precursor obtained by using Na_2S as the sulfur source in the presence of excessive Zn^{2+} ions. $In(OH)_3$ is selectively assembled at the rims of $ZnIn_2S_4$ nanosheets. The assembling improves effectively the carrier transfer and separation. Thus, the as-prepared J type heterojunction displays highly efficient photocatalytic H_2 evolution under visible irradiation.

To the best of our knowledge, it is the first time to introduce the concept of anisotropic heterojunctions for layered photocatalysts. The concept and findings provide new insights to develop high-performance heterojunction photocatalysts for layered semiconductors.

Experimental

Synthesis of photocatalysts. 1.0 mmol of $In_2(SO_4)_3 \cdot 6H_2O$ and different amount of $ZnSO_4 \cdot 7H_2O$ (1.00, 1.08, 1.20, 1.40 and 1.60 mmol, respectively) were dissolved in 40 mL of distilled water to obtain a mixed solution. A Na_2S solution (4.0 mmol of $Na_2S \cdot 9H_2O$ in 20 mL of distilled water) was poured into the mixed solution under stirring. The mixture (the precipitate with mother solution) was transferred into a Teflon liner with 100 mL of capacity. The liner was sealed in a stainless steel autoclave, heated at 160 °C for 12 h and then slowly cooled down to room temperature. The product was centrifuged, washed with distilled water, and dried in vacuum at 80 °C for 8 h to obtain a yellow product. The product was denoted as ZIS-x, where x represents adding amount of $ZnSO_4 \cdot 7H_2O$ (mmol).

For comparison, $ZnIn_2S_4$ was synthesized by a hydrothermal method using TAA as the sulfur source. $In_2(SO_4)_3 \cdot 6H_2O$ (1.0 mmol), $ZnSO_4 \cdot 7H_2O$ (1.00 mmol) and excessive TAA (5.0 mmol) were dissolved in 60 mL of distilled water. The mixture was heated at 160 °C for 12 h. After the similar separation, washing and drying, the obtained product was named as $ZnIn_2S_4$-TAA.

Characterization. X-ray diffraction (XRD) patterns of the samples were carried out using a XD-2/3 polycrystalline X-ray diffractometer with nickel-filtered Cu K_α radiation as the X-ray source. UV–vis absorption spectra of the samples were determined on a Hitachi U-3310 spectrophotometer with $BaSO_4$ as the reference. The scan electron microscopy (SEM) images were taken on a JEOL JSM-6701F equipped with an energy dispersive spectrometer (EDS). The transmission electron microscopy (TEM) and high-resolution TEM (HRTEM) images were taken on a JEOL JEM-2100 and a JEM-2100F.

Photocatalytic reaction. The photocatalytic reaction was conducted in a closed system (Pyrex cell, 235 mL volume, side irradiation) at room temperature. 0.10 g of a photocatalyst and 100 mL aqueous solution which contained 0.59 mol/L triethanolamine (TEOA), 2.54×10^{-5} mol/L H_2PtCl_6 (the calculated Pt amount was 0.50 wt%), were added into the cell. Before the reaction, the suspension was treated in an ultrasonic bath for 5 min and bubbled with N_2 for 30 min to remove oxygen. The light source was a 350 W Xe lamp equipped with a cool water filter and band-pass filter glass to cut off the infrared ray and light of wavelength below 420 nm, respectively. The top of the cell was sealed with a silicone rubber septum. Sampling was made intermittently through the septum. During the irradiation, the photocatalyst was remained in suspension by magnetic stirring, and metal Pt was in-situ photodeposited on the photocatalyst. The photocatalytic activity was determined by measuring the amount of hydrogen evolution for 1 h irradiation on a gas chromatography (TCD detector, 13X molecular sieve column, N_2 gas carrier).

Results and Discussion

Properties of ZIS-x composites. The phases of various ZIS-x composites (after hydrothermal treatment) were determined by XRD measurements. ZIS-1.08 and ZIS-1.2 are composed of $ZnIn_2S_4$, while ZIS-1.4 and ZIS-1.6 comprise both hexagonal $ZnIn_2S_4$ and $Zn_2In_2S_5$. The peaks of cubic $In(OH)_3$ can be observed for all the ZIS-x samples (x =1.2, 1.4 and 1.6) except ZIS-1.08, and their intensity increases greatly with increasing x value. This demonstrates that the excessive Zn^{2+} ions induce the growth of $In(OH)_3$.

All of the composites can absorb visible light and have a steep absorption edge, which suggest that the absorption is relevant to the intrinsic band gap transition of the samples rather than the transition from impurity levels. The absorption edges exhibit monotonic blue shift with increase of the adding content of Zn^{2+}, which could be mainly attributed to increasing the content of $Zn_2In_2S_5$.[2]

The morphology of the as-synthesized samples was investigated via SEM. ZIS-1.2 and ZIS-1.6 consist of agglomerated sheet-like grains (**Figure 1a and c**). Interestingly, there are many small nanosheets on the surfaces (**Figure 1b and d**). For ZIS-1.2, at the most nanosheet edges (dashed circles), small and "white" particles can be observed, which could be indexed to $In(OH)_3$ (see **Figure 2b**). This indicates that $In(OH)_3$ is controllably assembled at the $ZnIn_2S_4$ nanosheet edges to form oriented heterojunctions. However, for ZIS-1.6, besides $In(OH)_3$ at the nanosheet rims, some larger white $In(OH)_3$ particles cover on the nanosheets (in the dashed circle). The results suggest that excessive Zn^{2+} ions improve formation of $Zn_mIn_2S_{3+m}$ nanosheets and assembly of $In(OH)_3$ at the edges (ZIS-1.2), but when Zn^{2+} concentration is too high (ZIS-1.6), the assembling selectivity for the heterojunction decreases significantly.

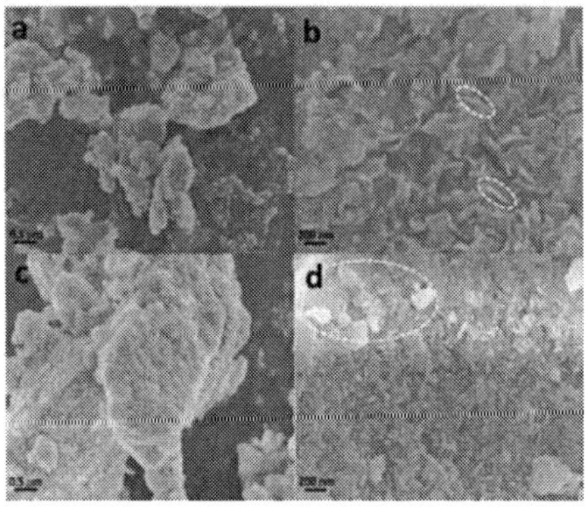

Figure 1 SEM images of ZIS-1.2 (a and b) and ZIS-1.6 (c and d).

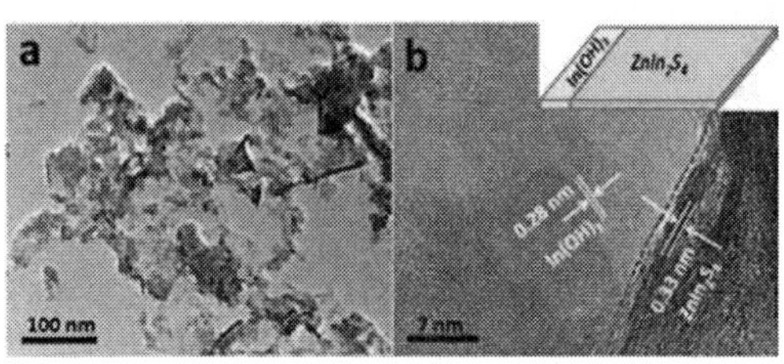

Figure 2TEM (a) and HRTEM (b) images of ZIS-1.2.

The TEM image (**Figure 2a**) indicates that it is indeed composed of nanosheets. **Figure 2b** shows that the edge part (width of ca. 15-25 nm) of the nanosheet is $In(OH)_3$, because the measured interplanar spacing is 0.28 nm which matches that of (2 2 0) plane of cubic $In(OH)_3$ Whereas, the intimate contact part with $In(OH)_3$ is hexagonal $ZnIn_2S_4$ with the exposure (0 0 1) facet, since the determined interlayer distance of 0.33 nm corresponds to the facet.[3] The results suggest that $In(OH)_3$ is selectively installed to $ZnIn_2S_4$ nanosheets along the layered planes (horizontal direction), namely J type coupling. The inset of **Figure 2b** displays schematic of the formed J type heterojunction.

Photocatalytic performance. **Figure 3** shows the effect of adding Zn^{2+} content in the preparation process on the activity of ZIS-x composites for photocatalytic hydrogen evolution. The photocatalytic activity of ZIS-1.0 is the lowest (41.2 μmol/h) among all the samples. The activity of ZIS-x increases with x value, reaches a maximum at 1.2 (147 μmol/h), and then decreases. ZIS-1.2 displays the highest activity, which can be ascribed to formation of $ZnIn_2S_4/In(OH)_3$ nanosheet heterojunction with J type superstructure. Activity of ZIS-1.6 (81.0 μmol/h) is significantly lower activity than that of ZIS-1.2, which can be attributed the two facts: (1) the visible light absorption of ZIS-1.6 is less than that of ZIS-1.2 because the former contains considerable amount of $Zn_2In_2S_5$; (2) more importantly, for ZIS-1.6, the surface active sites of $Zn_mIn_2S_{m+3}$ nanosheets are partly covered with $In(OH)_3$ to form the low efficient H type heterojunctions (Fig. 1d). The activity of ZIS-1.2 was compared with that of the sample prepared using TAA as the sulfur source. The inset shows that ZIS-1.2 exhibits much higher activity than ZIS-TAA (25.8 μmol/h), about 5.7 times as high as the latter. For ZIS-1.2, the apparent quantum yield for the photocatalytic H_2 evolution at 420 nm reaches to 38.3%.

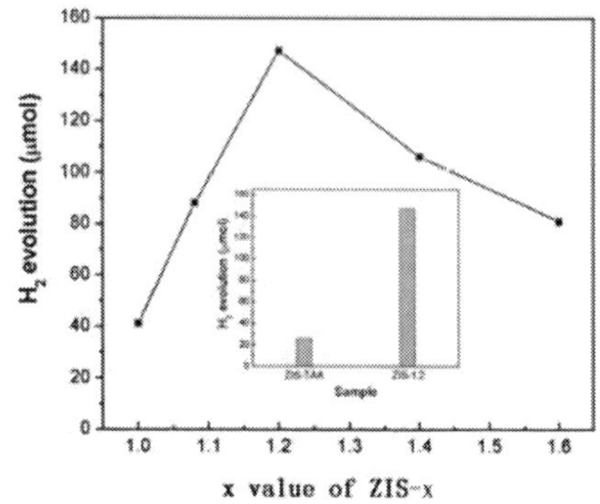

Figure 3 Effect of adding Zn^{2+} content in the preparation process on the activity of ZIS-x for photocatalytic H_2 evolution under visible light ($\lambda \geq 420$ nm) irradiation. The inset: The photoactivities of ZIS-1.2 and ZIS-TAA.

Mechanism. In^{3+} ions in an aqueous solution are hydrolyzed to mononuclear ions $In(OH)^{2+}$, $In(OH)_2^{+}$, and polynuclear ions $In[In(OH)_2]_n^{(3+n)+}$. It has been reported that even under acidic condition, In^{3+} ions can react with TAA to form $In(OH)_xS_y$. EDS analyses demonstrate that the precursors of ZIS-1.2 and ZIS-1.6 contain considerable oxygen element besides Zn, In and S, confirming that $In(OH)_xS_y$ is formed indeed.

During the hydrothermal treatment,intermediate $In(OH)_xS_y$ can also solve at the high hydrothermal temperature (160 °C) via the following reaction.

$In(OH)_xS_y = In^{3+} + x\ OH^- + yS^{2-}$ (1)

The produced In^{3+} and S^{2-} ions react with excessive Zn^{2+}, growing into $Zn_mIn_2S_{3+m}$, that is, the intermediate $In(OH)_xS_y$ transforms into stable $Zn_mIn_2S_{3+m}$. This results in spare In^{3+} ions in the solution.

Under the hydrothermal condition (160 °C), the formed In^{3+} ions can hydrolyze to produce $In(OH)_3$.

$In^{3+} + 3\ H_2O = In(OH)_3 + 3\ H^+$ (2)

Acknowledgement

This work was supported by the National Natural Science Foundation of China (No.21563019, 21163012) and the Nature Science Foundation of the Jiangxi Province (No.20151BAB203).

References

(1) Gallos, L. K.; Anagnostopoulcos, A. N.; Argyrakis A. N. *Phys. Rev. B* **1994**, *50*, 14643-14646.
(2) Shen, S.H.; Zhao, L.; Guo, L.J. *Int. J. Hydrogen Energy* **2010**, *35*, 10148-10154.
(3) Tana, C. W.; Zhua, G. Q.; Hojamberdiev, M.; Lokesh, K. S.; Luo, X. C.; Jin, L.; Zhou, J. P.; Liu, P. *J. Hazard. Mater.* **2014**, *278*, 572-583.

HYDROGEN PRODUCTION VIA STEAM REFORMING OF DIMETHYL ETHER

Zhong-Wen Liu

Key Laboratory of Applied Surface and Colloid Chemistry (MOE), Center of Applied Catalytic Research, School of Chemistry & Chemical Engineering, Shaanxi Normal University, Xi'an 710062, China

* Corresponding e-mail: zwliu@snnu.edu.cn

Among the available choices for on-board hydrogen production, dimethyl ether (DME) is recently believed to be a promising candidate due to its gas-like properties and liquid storage density, low cost, high hydrogen content (13 wt.%), low or no toxicity, etc. More importantly, comparing with other routes such as partial oxidation, steam reforming of DME (SRD) with high-quality reformate can be performed at lower temperatures of 473-673 K. Thus, since the publication of the thermodynamics feasible analysis and the pioneering work of Sobyanin and co-workers in 2001, extensive thermodynamics analyses and a few experimental studies have been done on the SRD. It is commonly accepted that the SRD consists of two series steps: the first is DME hydrolysis to methanol, and the second is steam reforming of the intermediate product of methanol. The series-reaction nature of the SRD implies that the synergetic effect between the individual catalysts may impacts greatly on the performance of the hybrid catalyst. Thus, physical mixture of solid acid and Cu-based catalyst is generally investigated, and the supported bifunctional catalysts are also reported in recent years.

Here, the synergetic effect was probed by changing the contact state of Cu/ZnO/Al_2O_3 and zeolites (ZSM-5 and USY), and the bifunctional catalysts by mixing or supporting the Cu-based catalyst with/on zeolite, alumina, or amorphous silica-alumina were optimized for the titled reaction. The reaction mechanism was released by correlating the characterization results of the catalysts with the reaction results, and a highly active and stable bifunctional catalyst for SRD was developed.

ANGSTROM AGGREGATES IN IONIC LIQUIDS FOR GREEN ENGINEERING

Suojiang Zhang

Beijing Key Laboratory of Ionic Liquids Clean Process, State Key Laboratory of Multiphase Complex System, Institute of Process Engineering, Chinese Academy of Sciences, Beijing1000190, China
Email: sjzhang@ipe.ac.cn

Ionic liquids (ILs) have been studied extensively as green solvents, catalysts, and electrolytes due to their unique properties and great potentials in green chemistry and chemical engineering. By combining experimental measurements and theoretical computational simulations, the dynamical structure and formation mechanism for ionic clusters in pure ILs and IL mixture systems have been investigated. The small ionic cluster of ILs is composed of several to tens of molecules or ions with the size of several to dozens of angstroms. Hereby, we defined such aggregate system as Ionic Angstrom Aggregates (IAA).1 IAA with the specific size at angstrom scale might possess unique structure-activity relationship, such as special arrangement in order, strong electronic effect, and excellent surface interaction, which may benefit in the area of chemical reaction, separation, and electrochemistry. Focusing on the IAA, it is found that Z-bond is one of the most important interactions for the formation of AA in ILs.2 The new force fields including Z-bond model has been developed, as well as the formation and regulation rules of AA in different concentrations of aqueous solution have been studied. The methodology of IL-based process design was proposed, which was applied to develop several novel green process technologies3. These green technologies include the coproduction of ethylene glycol and dimethyl carbonate from ethylene oxide via ethylene carbonate using supported IL as catalysts, the alkylation process by adding trace amount of ILs to replace H2SO4 partly, a green spinning process from wool keratin dissolving with ILs, and a new ILs electrolyte technology for lithium ion battery.

References

(1) Zhang, S.; Huo, F., Angstrom Science: Exploring Aggregates from a New Viewpoint. *Green Energy & Environ* **2016**, Accepted

(2) Dong, K.; Zhang, S.; Wang, J., Understanding the Hydrogen Bonds in Ionic Liquids and their Roles in Properties and Reactions. *Chem Comm* **2016,** DOI: 10.1039/C5CC10120D

(3) Zhang, S.; Sun, J.; Zhang, X.; Xin, J.; Miao, Q.; Wang, J., Ionic liquid-based green processes for energy production. *Chem Soc Rev* **2014,** *43* (22), 7838-7869.

A METAL-FREE CATHODE CATALYST FOR OXYGEN REDUCTION REACTIONS

*Zhiyi Wu, Zhenhua Yao, Zafar Iqbal, Xianqin Wang**

Department of Chemical, Biological and Pharmaceutical Engineering, New Jersey Institute of Technology, Newark, New Jersey 07102
*Corresponding author: xianqin.wang@njit.edu

Abstract

The nearly four-decades-long search for non-precious-metal electrocatalysts has so far revealed very few materials with promising oxygen reduction reaction (ORR) activity in either fundamental or in fuel cell studies, and virtually none with sufficient long-term performance stability. Recent studies found that nitrogen-containing nonmetallic materials, such as nitrogen-doped carbon nanotubes, carbon nitride (CNx), nitrogen-functionalized mesoporous carbon, and nitrogen-containing graphene are promising for replacing Pt-based oxygen reduction reaction (ORR) catalysts for low temperature fuel cells. Although Pt-based catalysts are most commonly utilized, attempts have been made to find alternative materials due to the scarcity and cost of platinum, chemical stability issues and sluggish kinetics of Pt-based catalysts. Polynitrogen (PN) species (Nn, n from 3 to 8) as highly energetic materials have attracted many theoretical calculations and predictions. Some researchers have experimentally observed N3, N4, N5 or their ions in the gas phase at high pressure and high temperature conditions. Here, a novel polynitrogen stabilized on the positively charged sidewalls of multi-walled carbon nanotubes (MWNTs) has been synthesized for the first time using cyclic voltammetry (CV) at ambient conditions and has been proved to be very active for ORR. This presentation will highlight some of our recent findings with this novel metal-free completely green cathode catalyst.

CATALYTIC HYDROTHERMAL GASIFICATION OF MICROALGAE FOR PRODUCING HYDROGEN AND METHANE-RICH GAS

Jia-Li Jiao, Pei-Gao Duan, and Feng Wang*

College of Chemistry and Chemical Engineering, Department of Applied Chemistry, Henan Polytechnic University, No. 2001, Century Avenue, Jiaozuo, Henan 454003, P.R. China
*pgduan@hpu.edu.cn

Introduction

Hydrothermal processing is one of the most promising ways for converting high-moisture microalgae into biofuel. Hydrothermal processing is typically performed under subcritical (hydrothermal liquefaction, HTL) or supercritical (hydrothermal gasification, HTG) conditions and avoids costly dehydration and drying steps, saving lots of energy. HTG, also known as supercritical water gasification (SCWG), has the potential to produce high quality and high yields of H_2 and CH_4 especially from very wet biomass feedstocks, such as microalgae. HTG comprises steam reforming, water-gas shift, and methanation reactions[1].

H_2 is primarily produced from the steam reforming and water-gas shift reactions, while CH_4 is produced from the methanation reaction[2]. Unlike HTL, HTG produces gases without nitrogen. Therefore, HTG is more suited for the conversion of high protein biomasses, such as microalgae. Non-catalyzed HTG often leads to high CO contents in the gaseous products, which is attributed to a low reaction rate in the water-gas-shift reaction. The gasification efficiency of non-catalyzed HTG is also low and typically requires high temperatures to achieve an acceptable gasification efficiency[3]. In contrast, catalyzed HTG can decrease the reaction temperature to achieve the same gasification efficiency as the non-catalyzed HTG at higher temperatures. Noble metal catalysts have typically enhanced the gasification of tar and char, thereby increasing the HTG efficiency and improving the production selectivity of H_2 and CH_4[4].

In the present study, the HTG of microalgae was examined in the presence of five noble metals supported on activated carbon (Ru/C, Pt/C, Rh, Pd/C, and Ir/C) to investigate their effects on the mole fractions and yields of H_2 and CH_4. The influences of water density and initial feedstock composition on the mole fractions and yields of H_2 and CH_4 were examined. The main objective of this work was to identify suitable catalysts, water density values, and initial feedstock compositions for the production of H_2 and CH_4.

Experimental

Materials. Four microalgae, including *Chlorella pyrenoidosa* (*CP*), *Spirulina platensis* (*SP*), *Schizochytrium limacinum* (*SL*), and *Nannochloropsis species* (*NS*) were commercially available and were received in dry powder form from Shandong Binzhou Tianjian Biotechnology Co., Ltd. (Binzhou, Shandong, North China). All microalgae were dried at 110 °C for 12 h to remove residual moisture in the cell. Five noble metals supported on activated carbon (Pt/C, Pd/C, Ru/C, Rh/C, and Ir/C) were obtained from Zhengzhou Alfachem Co., Ltd.(Zhengzhou, Henan, China) and used as HTG catalysts. All catalysts were received in powder form and had an average particle size of 25 μm. They were directly used without undergoing a pre-reduction under H_2. Freshly deionized water was used throughout the experiment.

A custom-built high-pressure stainless-steel batch reactor with a total internal volume of 37 mL was used to perform the HTG experiments. All reactors were heated using a molten-salt bath consisting of potassium nitrate and sodium nitrate at a mass ratio of 5:4.

Procedure. In a typical run, 1.0 g microalgae, 1.0 g catalyst, and 1 mL water were loaded into the reactor. These loadings resulted in a water density of 0.027 g/cm^3 at supercritical conditions. The air inside the reactor was displaced by purging the reactor headspace with helium for approximately 15 min. The purged reactor was pressurized with helium at 0.01 MPa. This added helium served as a standard in the quantification of gas yields.

After the reactor was loaded and pressurized with helium, it was placed in the molten-salt bath, which had been preheated to the 480 °C. After approximately 15 min, when the reactor reached 430 °C, the reaction time was set to zero. The isothermal molten-salt bath was maintained at 430 ± 5 °C using a temperature controller. The body of the reactor was vertically completely submerged in the molten-salt bath. The pressure inside the reactor varied between 2 and 13 MPa depending on the reaction conditions. After a reaction time of 60 min had elapsed, the reactor was removed from the molten-salt bath and immediately quenched in a cool water bath. The pressure inside the reactor varied between 22 and 55 MPa. The reactor was then removed from the cool water bath and charged with argon to an additional 5 MPa. Finally, the reactor was placed in ambient conditions for at least 4 h to allow the liquid-gas system to reach equilibrium.

Analysis. Gaseous products were analyzed with a Shanghai Techcomp Scientific Instrument Co., Ltd., model GC-7900 gas chromatograph (GC) equipped with a thermal conductivity detector (TCD). A 15-ft×1/8-in. i.d. stainless steel column, packed with 60×80 mesh Carboxen 1,000 (Supleco) was used to separate each component in the mixture. Argon (15 mL/min) served as the carrier gas for the analysis. The temperature of the column was isothermally held at 70 °C for 120 min. The injector and detector temperatures were set at 120 and 150 °C, respectively. The column pressure was 0.18 MPa, and the filament current of the detector was set at 50 mA.

Results and Discussion

Effect of catalyst. Fig. 1 shows the yields of H_2 and CH_4 formed from the microalga processing with different catalysts. Compared with the uncatalyzed reaction, the presence of a catalyst effectively improved the yields of H_2 and CH_4. The yields of H_2 and CH_4 were increased 2-3 and 3-9 times higher than the yields produced without catalyst, respectively. The Ru and Pt catalysts had the highest H_2 yields of 5.97 mmol/g and 5.75 mmol/g, respectively. An approximately 8 mmol/g H_2 yield was reported by Guan et al.[5] using a 1 g/g Ru/C loading at 410 °C, 60 min, 0.096 g/cm^3 water density, and 4.3 wt.% algae loading. This was significantly higher than the approximately 3 mmol/g yield produced from an uncatalyzed reaction at 450 °C. Despite the much higher algae loading used by the present study (4.3 wt.% vs. 50 wt.%), the lower H_2 yield was reduced because increasing the algae loading can significantly decrease the H_2 yield[5]. The CH_4 yield (2.6 mmol/g) was close to that (approximately 3 mmol/g) obtained by Guan et al.[5] because te CH_4 yield is insensitive to the algae reactor loading. The H_2 yields produced with Pd/C and Rh/C were 6.17 mmol/g and 7.19 mmol/g, respectively, which were significantly higher than the yields obtained with Ru/C (2.60 mmol/g) and Pt/C (2.44 mmol/g).

Fig. 1 presents the carbon gasification efficiency (CGE) and hydrogen gasification efficiency (HGE). The CGE and HGE were nearly the same for the uncatalyzed reaction. In contrast, the HGEs of the catalyzed reactions, particularly in the presence of Ru/C, Pt/C, and Pd/C, were higher than the CGEs, indicating the noble metals preferably to catalyze the conversion of hydrogen atoms in the algae to molecular hydrogen. Of course, some of the H atoms in the gaseous products could have originated from water molecules[5].

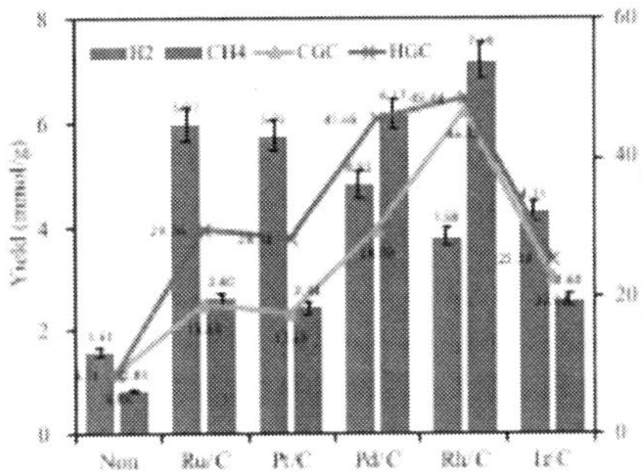

Figure 1. Influence of catalyst on gas yield (430 °C, 1 h, ρ_{H2O}= 0.027 g/cm^3, M/cat.=1:1).

Influence of water density. The reaction conditions were 430 °C, 1 h and a 1$g_{Pd/C}/g_{algae}$ catalyst load. Guan et al.[6] found that the H_2 and CH_4 yields were insensitive to the water density during the uncatalyzed hydrothermal gasification of microalgae, indicating Pd/C could significantly promote the gasification of algae. The H_2 yield increased with increasing water density and reached the highest value of 5.60 mmol/g at a water density of 0.041 g/cm^3. Further increases in water density slightly decreased the H_2 yield; this may be due to the over consumption of water during the methanation reaction. The CH_4 yield always increased with water density and reached the highest value of 38.68 mmol/g at the highest water density tested (0.054 g/cm^3). The CH_4 yield was higher than the H_2 yield as the water density increased beyond 0.027 g/cm^3, indicating methanation was significantly accelerated. Similar results were also observed by Guan et al.[5]. The H_2 yield was always much higher than the CH_4 yield due to the different biochemical compositions of the microalgae as will be discussed in a later section. Additional work with much higher water density could prove to be useful.

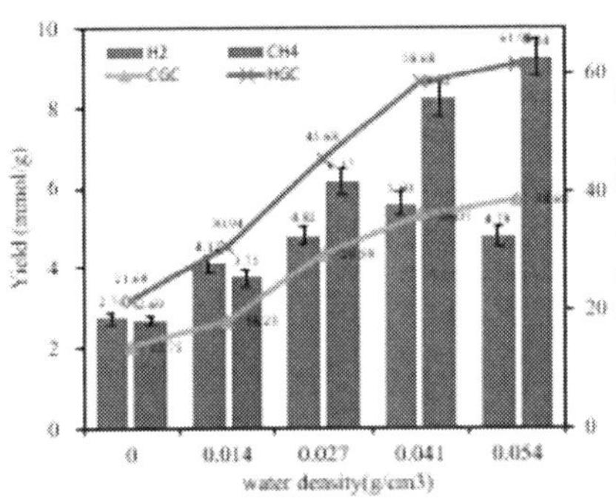

Figure 2. Influence of water density on gas yield (430 °C, 1 h, M/Pd/C=1g:1g).

The CGE and HGE were strong functions of the water density, with higher densities leading to higher H_2 and CH_4 yields. The CGE and HGE at a water density of 0.054 g/cm^3 nearly tripled their corresponding values in gasification without water. These results were consistent with the results reported for the hydrothermal gasification of algae with Ru/C[5]. A higher water density also favored the hydrolytic decomposition of algae to lower molecular-weight intermediates that could easily be gasified. The HGE was always higher than the CGE, and this difference was enlarged with increasing water density, indicating that water may be incorporated into the gaseous products. Yanik et al.[7] suggested that almost half of the H_2 production came from water rather than the biomass under supercritical water conditions.

Influence of initial feedstock. Four microalgae species, *CP*, *SP*, *SL*, and *NS*, were examined. For comparison, oil and solid residues extracted from *CP* were also used, which were defined as OEC and REC, respectively. The reaction conditions were 430 °C, 1 h and 1$g_{Pd/C}/g_{algae}$ with a water density of 0.041 g/cm^3. Fig. 3 summarizes the results.

Of those four algal biomass feedstocks, *CP* produced the highest H_2 and CH_4 yields of 5.60 and 8.22 mmol/g, respectively. Although *NS* produced the highest H_2 mole fraction, the H_2 yield was not the highest. The OEC provided higher H_2 and CH_4 yields of 9.06 and 5.99 mmol/g, whereas the REC provided lower H_2 and CH_4 yields of 4.94 and 3.85 mmol/g than those of *CP*. The CGE and HGE were also dependent on the initial feedstocks. All four microalgae except *SP* had higher HGEs than CGEs. *CP* provided the highest HGE, whereas *SP* provided the highest CGE. Compared with *SP*, the OEC and the REC provided lower HGEs and CGEs, indicating a synergistic effect existed among the biochemical components during gasification. *NS* provided the lowest HGE and CGE.

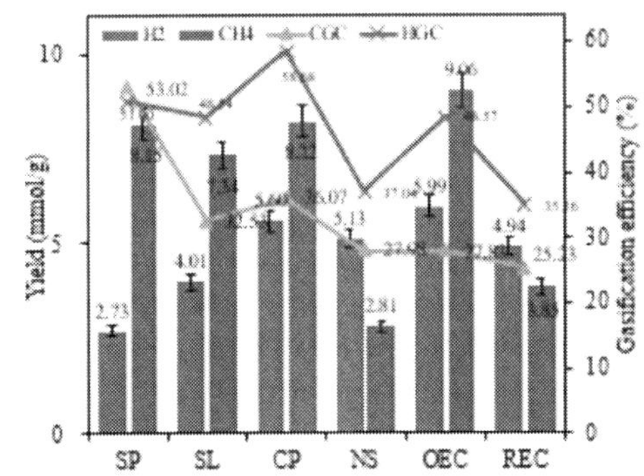

Figure 3. Influence of initial feedstock on gas yield (430 °C, 1 h, ρ_{H2O}= 0.041 g/cm^3, M/Pd/C=1:1), OEC-oil extracted from *CP*; REC-residue extracted from *CP*.

Acknowledgment. The authors gratefully acknowledge the financial support provided by the Program for New Century Excellent Talents in University (NCET-13-775) and the National Science Foundation of China (21106034).

References

(1) Chakinala, A.G., Brilman, D.W.F., van Swaaij, W.P.M., Kersten, S.R.A., *Ind. Eng. Chem. Res*, **2010**, 49, 1113-1122.

(2) Onwudili, J.A., Williams, P.T., *Appl. Catal. B, Environ.*, **2013**, 132-133, 70-79.

(3) Azadi, P., Farnood, R., Int. J. *Hydrogen Energy*, **2011**, 36, 9529-9541.

(4) Díaz-Rey, M.R., Cortés-Reyes, M., Herreraa, C., Larrubia, M.A., Amadeo, N., Laborde, M., Alemany, L.J., *Catal. Today*, **2015**, 257, 177-184.

(5) Guan, Q.Q., Wei, C.H., Savage, P.E., *Energy Fuels*, **2012**, 26, 4575-4582.

(6) Guan, Q.Q., Savagea, P.E., Wei, C.H., *J. Supercrit. Fluids*, **2012**, 61, 139-145.

(7) Yanik, J., Ebale, S., Kruse, A., Saglam, M., Yüksel, M., *Int. J. Hydrogen Energy*, **2008**, 33, 4520-4526.

PREPARATION AND ASSESSMENT OF THE ACTIVE SITES IN ANILINE-DERIVED Fe-N-C OXYGEN REDUCTION CATALYST

Nan-Hong Xie and Bo-Qing Xu

Innovative Catalysis Program, Key Lab of Organic Optoelectronics & Molecular Engineering, Department of Chemistry, Tsinghua University, Beijing 100084, CHINA
E-mail: bqxu@mail.tsinghua.edu.cn

Iron and nitrogen co-doped carbon materials (Fe-N-C) are considered as a kind of promising alternative to Pt-based electro catalysts for electrochemical oxygen reduction reaction (ORR). However, the nature of active site in these Fe-N-C catalysts has remained elusive. Diverse structure modes based on Fe-N bonding structures have been proposed to assess the possible active sites but clear observation and discussion on the most possible active site structures were always disturbed by the presence of many co-existing Fe-containing species in the earlier Fe-N-C catalysts. In this presentation, we introduce firstly a preparative approach to systematically vary the density of active structures in pyrolyzed polyaniline (PANI)-based Fe-N-C catalyst samples involving no iron sulfides and residual metallic iron residues, as well as the least possible iron carbides. Extensive characterizations of these samples with varying techniques, including X-ray diffraction, electron microscopy and high resolution HAADF-STEM, and measurements of the intrinsic ORR activity, X-ray photoelectron, near edge X-ray absorption and ^{57}Fe Mössbauer spectroscopies, are then conducted to uncover the most possible active site structures and to correlate the concentration of such structures with the kinetic (intrinsic) ORR rate. It will be shown that the intrinsic catalytic ORR rate over this type of Fe-N-C catalysts is strictly proportional to the concentration of electrochemically active Fe-species in the catalyst samples. Further comprehending of the data points to that atomically dispersed iron species in the Fe(II)-N_4 coordination structures is serving as the active sites for the ORR.

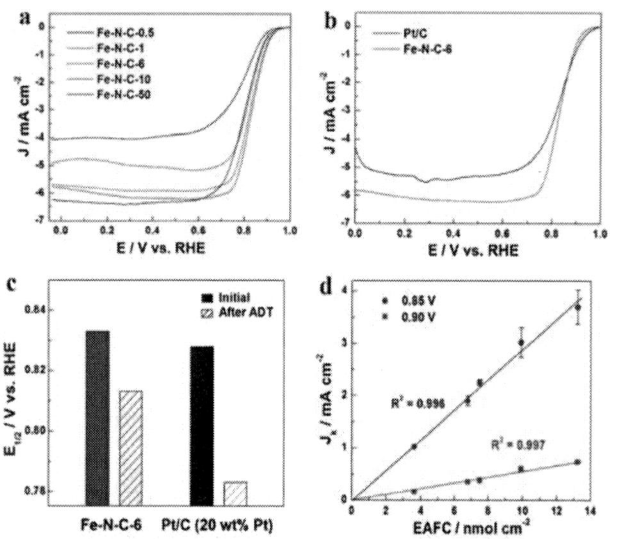

Figure 1. (a) LSV curves for the Fe-N-C-x catalysts in O2-saturated 0.1 M KOH; (b) catalytic ORR activity and (c) stability of the Fe-N-C-6 and Pt/C catalysts; (d) correlation between the kinetic ORR activity and concentration of the electrochemically active Fe-species (EAFC) in the Fe-N-C-x catalysts.

Li_2OHCL CRYSTALLINE ELECTROLYTE FOR STABLE METALLIC LITHIUM ANODE

Hui Wang,[†*] *Zachary D. Hood,*[† ‡*] *Amaresh Samuthira Pandian,*[†]
Jong Kahk Keum,[†§] *Chengdu Liang*[†]

†Center for Nanophase Materials Science, and §Chemical and Engineering Materials Division, Oak Ridge National Laboratory, Oak Ridge, Tennessee 37831, United States
‡ School of Chemistry and Biochemistry, Georgia Institute of Technology, Atlanta, Georgia 30332, United States
Corresponding Author: whui583@gmail.com

Abstract

All-solid-state batteries are promising for the next-generation battery technology due to their favorable features of stringent safety and high energy density. Solid electrolytes are critical to realize all-solid-state batteries with high performance. In a classic example of stability from instability, we show that Li_2OHCl solid electrolyte forms a stable solid electrolyte interphase (SEI) layer with a metallic lithium anode. The Li_2OHCl solid electrolyte can be readily achieved through simple mixing of LiOH and LiCl precursors at a mild processing temperature <400 °C. Additionally, we show that continuous, dense Li_2OHCl membranes can be fabricated at temperatures <400 °C, standing in great contrast to current processing temperatures of >1600 °C for most oxide-based solid electrolytes. The ionic conductivity and Arrhenius activation energy were explored for the LiOH–LiCl system of crystalline solid electrolytes, where Li_2OHCl with increased crystal defects was found to have the highest ionic conductivity and reasonable Arrhenius activation energy. The Li_2OHCl solid electrolyte displays stability against metallic lithium, even in extreme conditions past the melting point of lithium metal. To understand this excellent stability, we show that SEI formation is critical in stabilizing the interface between metallic lithium and the Li_2OHCl solid electrolyte.

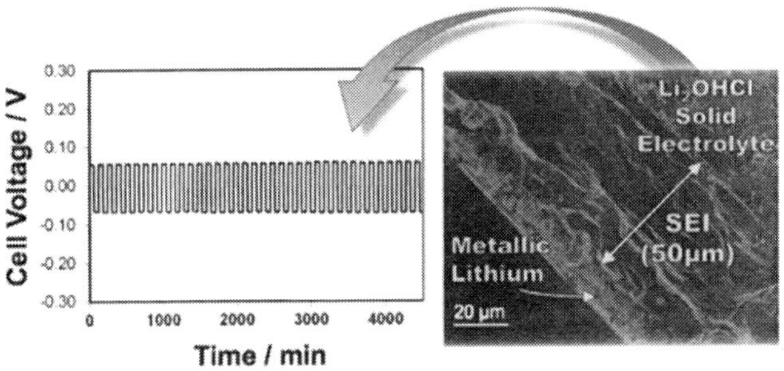

REDESIGNING THE PLATINUM ORR CATALYST FOR FUEL CELL APPLICATION

Younan Xia

Department of Biomedical Engineering, School of Chemistry and Biochemistry, Georgia Institute of Technology
younan.xia@bme.gatech.edu

Introduction

Platinum is by far the best catalyst for the oxygen reduction reaction (ORR) occurring on the cathode of a fuel cell. Its low abundance, limited supply, and ever increasing price have kept motivating researchers to maximize the use of this precious metals. One effective strategy is to increase the utilization efficiency (i.e., the fraction of Pt atoms on the surface) by reducing the particle size and thus increasing the dispersion of Pt atoms. For this reason, commercial ORR catalysts are typically based upon Pt particles of only a few nanometers in size. Despite the extensive use, it has been difficult to optimize the specific activity of such tiny particles by engineering their surface structure through facet-controlled syntheses. Such minuscule particles also tend to sinter and/or be detached from the support, resulting in major deterioration in performance during the course of operation. In this talk, I will discuss a number of strategies for increasing the mass activity and durability of Pt-based ORR catalysts, including facet-controlled synthesis, increase of dispersion by forming open structures, surface strain engineering, electronic coupling through alloying, and use of particles with optimized dimensions.

Reference

(1) Zhang, L; Roling, L. T.; Wang, X.; Vara, M.; Chi, M.; Liu, J.; Choi, S. I.; Park, J.; Herron, J. A.; Xie, Z.; Mavrikakis, M.; and Xia, Y. *Science*, **2015**, *349*, 412.

DEVELOPMENT OF BIFUNCTIONAL CATALYSTS FOR HIGHLY SELECTIVE CONVERSIONS OF SYNTHESIS GAS TO LIQUID FUELS AND LOWER OLEFINS

*Kang Cheng, Jincan Kang, Qinghong Zhang, Ye Wang**

State Key Laboratory of Physical Chemistry of Solid Surfaces, Collaborative Innovation Center of Chemistry for Energy Materials, National Engineering Laboratory of Green Chemical Production for Alcohols, Ethers and Esters, College of Chemistry and Chemical Engineering, Xiamen University, Xiamen 361005, China

*wangye@xmu.edu.cn

Introduction

Synthesis gas (syngas, CO/H_2) is a key platform for the transformation of various non-petroleum carbon resources such as coal, natural gas (also shale gas), and biomass into liquid fuels and chemicals. The direct conversion of syngas into hydrocarbons via Fischer-Tropsch (FT) synthesis has received much interest in recent years. FT synthesis typically proceeds via the dissociation of CO over Ru, Co or Fe (Fe carbides) surfaces, formation of CH_x (x = 1-3), coupling of CH_x to C_nH_m intermediates, and the hydrogenation or dehydrogenation of C_nH_m intermediates into paraffin or olefin products. The coupling of CH_x or the chain-growth process mainly determines the product distributions. However, because the C-C coupling over metal surfaces that can activate CO is uncontrollable, the hydrocarbon products generally follow a statistical distribution known as the Anderson-Schulz-Flory (ASF) distribution.[1] This distribution is non-selective for middle distillate products, which are usually the target products. For example, the maximum selectivities to gasoline-range and C_2-C_4 hydrocarbons are 45% and 58%, respectively. The development of new catalysts or new systems that can break the ASF distribution is one of the biggest challenges in syngas chemistry. Here, I will present our work to develop such catalytic systems by a strategy of reaction coupling with bifunctional catalysts, which can catalyze not only the hydrogenation of CO but also the selective C-C cleavage or the selective C-C coupling.

Bifunctional Catalysts for Conversion of Syngas to Gasoline-range Hydrocarbons

Our strategy for the selective production of liquid fuels from syngas is to integrate the hydrogenation of CO to heavier hydrocarbons and the cleavage of heavier hydrocarbons to the targeted gasoline- or diesel-range hydrocarbons on one bifunctional catalyst (Figure 1).

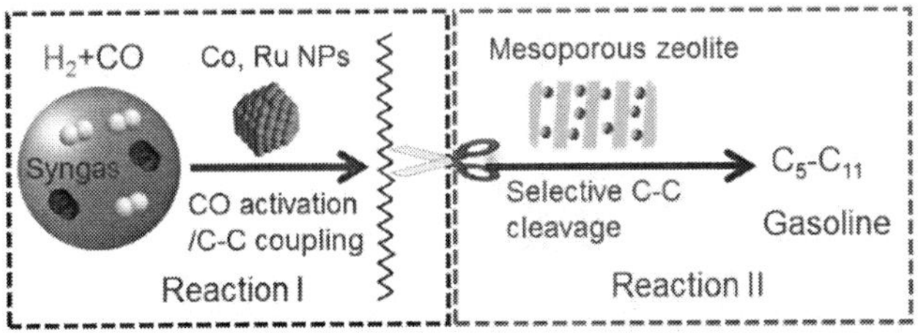

Figure 1. Reaction coupling for the direct production of gasoline from syngas by the integration of active components for CO activation/C-C coupling and selective C-C cleavage.

Zeolites with Brønsted acidity have been employed to construct bifunctional FT catalysts to produce gasoline-range hydrocarbons, but the selectivity to lower (C_1-C_4) paraffins is too high over zeolite-based catalysts.[2] We first utilized mesoporous zeolites for FT synthesis and succeeded in obtaining highly selective bifunctional catalysts for the production of gasoline-range hydrocarbons. We synthesized mesoporous ZSM-5 (meso-ZSM-5) by treating ZSM-5 in NaOH aqueous solution, and found that the size of mesopores introduced could be tuned by changing the concentration of NaOH. The increase in the NaOH concentration increased the size of mesopores. We loaded Ru nanoparticles onto meso-ZSM-5 by a simple impregnation method, and the sizes of Ru nanoparticles were ~7 nm irrespective of the size of mesopores. Our catalytic results clearly showed that the use of mesoporous ZSM-5 instead of the conventional microporous ZSM-5 as the support of Ru nanoparticles (~7 nm) significantly decreased the selectivities of CH_4 and C_2-C_4 and increased that to C_5-C_{11} hydrocarbons (Figure 2a).[3] The C_5-C_{11} selectivity could reach ~80% over the catalyst with a proper mean size of mesopores. CO conversion also increased by using mesoporous ZSM-5 (Figure 2b). The ratio of isoparaffins to n-paraffins (C_{iso}/C_n) in C_5-C_{11} was kept at ~2.7 (Figure 2b) when the size of mesopores was not too large.

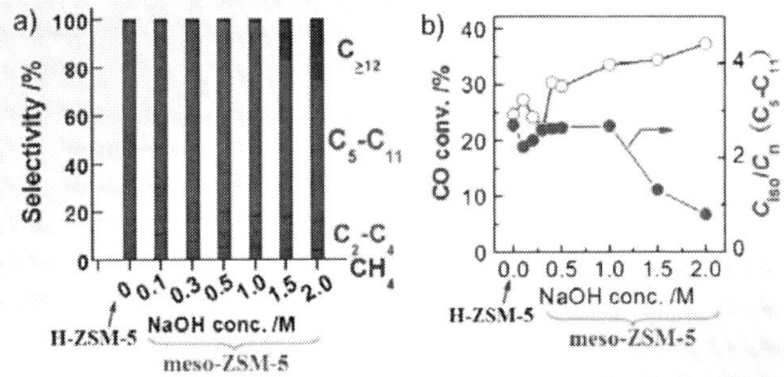

Figure 2. Conversion of syngas over Ru/meso-ZSM-5 catalysts with different mean sizes of mesopores (expressed with the concentration of NaOH used for meso-ZSM-5 synthesis).[3] a) Product selectivity. b) CO conversion and ratio of isoparaffins to n-paraffins. Reaction conditions: catalyst 0.5 g, H_2/CO=1.0, temperature 533 K, pressure 2 MPa, total flow rate 20 mLmin-1, time on stream 12 h.

Using n-$C_{16}H_{34}$ as a model molecule, we confirmed the occurrence of hydrocracking of heavier hydrocarbons under FT reaction conditions. We clarified that the Brønsted acid sites in meso-ZSM-5 were responsible for the hydrocracking of n-$C_{16}H_{34}$. Furthermore, it is of interest that the presence mesopores increases the selectivity of C_5-C_{11} during the hydrocracking of n-$C_{16}H_{34}$ by suppressing the formation of light alkanes. Thus, we can tune the product selectivity for syngas conversion by tuning the selectivity for the hydrocracking reaction.[3]

We obtained similar results over mesoporous zeolite β-supported Ru catalysts.[4] The high cost and the limited availability of Ru make the large-scale application of supported Ru catalysts less possible. The problems of zeolite-supported Co catalysts are the lower reducibility and the difficulty in controlling the Co particle size. We prepared Co nanoparticles with relatively homogeneously distributed particle sizes (mean sizes, 8-10 nm) loaded meso-ZSM-5 with different mean sizes of mesopores. We achieved a C_5-C_{11} selectivity of ~70% over Co/meso-ZSM-5 with a proper mean size of mesopores.[5] The catalyst was also found to be relatively stable in 100 h of reaction.

Bifunctional Catalysts for Conversion of Syngas to Lower Olefins

Lower (C_2-C_4) olefins are key building-blocks in the chemical industry. Many studies have been devoted to the conversion of syngas to lower olefins via FT synthesis over Fe-based catalysts, but the selectivity is limited by the ASF distribution, which shows a maximum C_2-C_4 (including both paraffin and olefin) selectivity of 58%.[6] Lower olefins may also be produced from syngas by two consecutive processes, i.e., methanol synthesis and methanol to olefins (MTO). It would be highly attractive to develop a bifunctional catalyst for the direct conversion of syngas by integrating the activation of CO to C_1 intermediates (such as CH_3OH or CH_3O) and the C-C coupling (such as MTO) (Figure 3). It can be expected that to carry out C-C coupling and CO activation on different sites may allow the coupling to proceed in a controlled manner, breaking the ASF distribution. Some studies have attempted to combine methanol synthesis and methanol to hydrocarbon catalysts in one reactor, but the major products are C_2-C_4 paraffins.

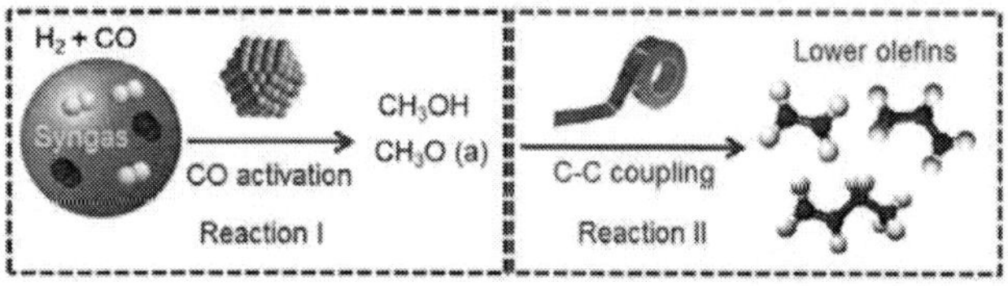

Figure 3. Reaction coupling for the direct production of lower olefins from syngas by the integration of active components for CO activation to C_1 intermediate (CH_3OH or CH_3O) and selective C-C coupling.[7]

We have demonstrated that the choice of the components for the two reactions and their integration manner are crucial to the selective of lower olefins. We chose SAPO-34, an excellent MTO catalyst, as the C-C coupling component. The combination of SAPO-34 with a conventional methanol synthesis catalyst such as Cu-Zn-Al oxide or Pd-Zn led to the formation of mainly CH_4 and C_2-C_4 paraffins. We discovered that the use of Zr-Zn binary oxide as the catalyst component for CO activation resulted in C_2-C_4 olefins with high selectivity.[7] The ability of Zr-Zn binary oxide to perform selective hydrogenation of CO in the presence of lower olefins is the key to obtaining high C_2-C_4 olefin selectivity. This has been confirmed by carrying out methanol conversion over bifunctional catalysts composed of SAPO-34 and various methanol synthesis catalyst components (including Cu-Zn-Al, Pd-Zn and Zr-Zn); only the catalysts based on Zr-Zn binary oxide showed high lower olefin selectivity. We also clarified that the acidity of SAPO-34 also affected the lower olefin selectivity. The lower density of Brønsted acid sites favored the selectivity of lower olefins because hydrogenation of olefins could be catalyzed by the Brønsted acid sites.

Furthermore, the integration manner was also a key to obtaining high catalytic performances (Figure 4).[7] The separation of the two catalyst components using the dual-bed mode only provided a much lower CO conversion because of lack of thermodynamic driving force of the MTO reaction for the methanol synthesis, which is a thermodynamically unfavorable at 673 K. On the other hand, the too close proximity of the two components would be detrimental to the olefin selectivity.

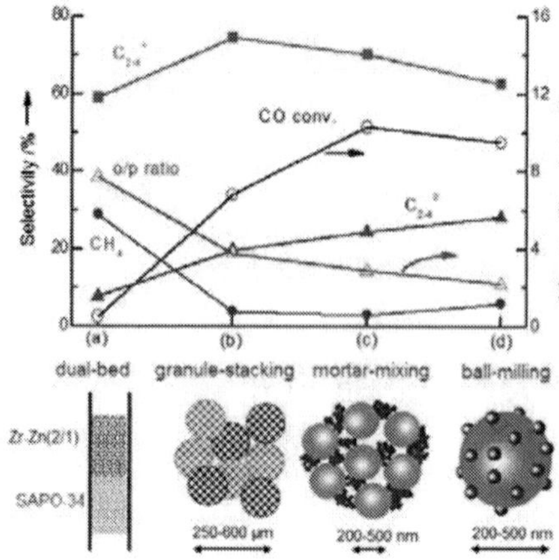

Figure 4. Effect of the integration manner on catalytic behaviors of the composite catalysts containing Zr-Zn oxide and SAPO-34.[7] (a) Dual-bed configuration. (b) Stacking of granules with sizes of 250-600 μm. (c) Simple mixing of the two components in an agate mortar. (d) Ball-milling of two components together for 24 h.

Conclusions

The development of bifunctional catalysts by reaction coupling is an effective strategy to synthesize fuels or chemicals from syngas with outstanding selectivity that breaks the ASF distribution. We have demonstrated that the combination of Ru or Co nanoparticles with an optimized mean size (6-10 nm) and mesoporous zeolites (such as meso-ZSM-5 and meso-β) with appropriate acidity and mesopores can produce gasoline-range (C_5-C_{11}) hydrocarbons with a selectivity of 70-80%. The integration of Zr-Zn binary oxide and SAPO-34 results in highly selective bifunctional catalyst for the direct conversion of syngas to C_2-C_4 olefins. The C_2-C_4 olefin selectivity can reach 74%.

Acknowledgement. This work was supported by the National Natural Science Foundation (91545203, 21433008 and 21503174), and the National Basic Research Program of China (2013CB933102).

References

(1) Zhang, Q.; Kang, J.; Wang Y. *ChemCatChem* **2010**, *2*, 1030.
(2) Bao, J.; He, J.; Zhang, Y.; Yoneyama, Y.; Tsubaki, N. *Angew. Chem. Int. Ed.* **2008**, *47*, 353.
(3) Kang, J.; Cheng, K.; Zhang, L.; Zhang, Q.; Ding, J.; Hua, W.; Lou, Y.; Zhai, Q.; Wang, Y. *Angew. Chem. Int. Ed.* **2011**, *50*, 5200.
(4) Kang, J.; Cheng, K.; Huang, S.; You, Z.; Zhang, Q.; Ding, J.; Hua, W.; Lou, Y.; Deng, W.; Wang, Y. *ACS Catal.* **2012**, *2*, 441.
(5) Kang, J.; Zhang, L.; Kang, J.; Peng, X.; Zhang, Q.; Wang, Y. *Chem. Eur. J.* **2015**, *21*, 1928.
(6) Galvis, H. M. T.; de Jong, K. P. *ACS Catal.* **2013**, *3*, 2130.
(7) Cheng, K.; Gu, B.; Liu, X.; Kang, J.; Zhang, Q.; Wang, Y. *Angew. Chem. Int. Ed.* **2016**, *55*, DOI: 10.1002/anie.201601208.

HYDROTHERMAL CARBONIZATION OF CHITOSAN FOR HIGH-PERFORMANCE SUPERCAPACITOR ELECTRODE MATERIALS

Xinhua Qi [a,*], *Linfeng Zhu* [a,b], *Feng Shen* [a]

[a] Agro-Environmental Protection Institute, Chinese Academy of Agricultural Sciences, No. 31, Fukang Road, Nankai District, Tianjin 300191, China.

[b] College of Environmental Science and Engineering, Nankai University, No. 94, Weijin Road, Nankai District, Tianjin 300071, China

[*] Corresponding author. Email: qixinhua@nankai.edu.cn (X. Qi)

Introduction

Supercapacitors, which are high-performance energy- storage and delivery components that retain electrical charge through electrochemical pseudocapacitance or by an electrostatic double-layer, have many applications in power systems and energy devices [1]. A number of materials such as activated porous carbon [2, 3] and transition metal oxides [4-6] have been used to assemble high-performance supercapacitors.

Carbon materials for use in high-performance supercapacitor electrodes need to have a high specific surface area and a large nano-porous volume that can be obtained by using methods such as hard template [7, 8], soft template and chemical activation [9]. Heteroatoms such as S and N that are embedded into the carbonaceous materials can increase the specific capacitance of the supercapacitors by improving conductivity, increasing the surface wettability and widening the accessible electrolyte ion surface area of the carbon electrode materials. Thus, preparation of nitrogen-doped porous carbons have been proposed that use in-situ (co-carbonization) or post-treatment with N-containing reagents such as NH_3, amines, melamine and cyanamide [10].

Chitosan, which is a renewable natural biopolymer, can be obtained from alkaline deacetylation of chitin that originates in crustacean seafood shells. The characteristics of chitin provide the possibility of producing porous carbon materials with high N content from chitosan, and it is possible that chitosan could be a promising nitrogen-containing precursor for many types of supercapacitor electrodes or materials. Herein, chitosan was applied to prepare N-containing carbonaceous materials by a facile two-step method in which chitosan was hydrothermally treated in the first step and then was chemically-activated with KOH in the second-step. No other post-functionalization steps were required. The prepared chitosan-derived carbonaceous materials were evaluated as electrode materials for supercapacitors and their energy storage performance was assessed.

Experimental

Preparation of chitosan-derived carbonaceous materials (CDCMs). Chitosan-derived carbonaceous materials (CDCMs) were synthesized by hydrothermal treatment of chitosan that was followed by KOH treatment of the product solids for activation. In the first step, typically 10 g chitosan and 50 mL deionized water were mixed and loaded into a 100 mL stainless steel autoclave. The vessel and its contents were heated to 250 °C and kept at that temperature for 14 h. After the given reaction time, the reactor was cooled and the black solid materials that formed were separated from the mixture and washed with ethanol and ultra-pure water repeatedly, and then the solids were dried at 80 °C in an oven overnight (denoted as HC). In the second step, the HC material was mixed with KOH (KOH/HC mass ratio 1: 1) and the mixture was transferred into an alumina crucible and heated at 600~900 °C under a nitrogen atmosphere for 2 h. The obtained solid samples were washed with 1 M hydrochloric acid aqueous solution and ultrapure water. The product carbonaceous solids were dried at 80 °C for 8 h. The resulting carbonaceous solids were labeled as CDCM-X, where X refers to the activation temperature (°C).

Results and Discussion

Characterization of the CDCM samples. Table 1 shows that the surface area, total pore volume and micropore volume of the samples increased with increasing activation temperature from 600-800 °C, and then began to decrease at 900 °C. The decrease in the micropore volume and surface area at 900 °C can be ascribed to the collapse of the formed micropores due to excess oxidation by KOH. These CDCM samples had a nitrogen content ranging from 1.1 to 6.3%.

Table 1. Physical characteristics and elemental analysis of the chitosan-derived carbonaceous materials (CDCMs)

Sample	S_{BET}[a] ($m^2\ g^{-1}$)	V_t[b] ($cm^3\ g^{-1}$)	V_m[c] ($cm^3\ g^{-1}$)	C[d] (%)	N [d] (%)	O[e] (%)
CDCM-600	815	0.45	0.24	59.6	6.3	30.5
CDCM-700	1522	0.68	0.49	63.4	2.3	29.8
CDCM-800	2169	0.99	0.67	78.3	1.3	16.0
CDCM-900	2038	1.36	0.2	87.8	1.1	6.3

Electrochemical behavior of the CDCMs samples. The electrochemical performance of the synthesized CDCM samples for supercapacitors was investigated by CV and GCD experiments in a three-electrode system that used K_2SO_4 aqueous solution (0.5 M) simulating commercial electric double layer capacitors (Fig. 1). Similar rectangular shapes for the CV spectra (Fig. 2a) showed that the electrical energy is stored in the electric double-layer that triggers the redox reaction, which is key to obtain high-performance supercapacitors. The CV profiles obtained for CDCM-800 were more rectangular in shape than those of the other samples and most likely had low internal resistances. There is an amount of Faradic movement from 0.6 to 0.8 V for all CDCMs that may be due to the polarization of the electrode that influences the redox reaction.

The GCD curves of the prepared CDCMs recorded at a current density of 0.5 A g^{-1} (Fig. 1b) showed almost isosceles triangular shapes for all samples, indicating ideal charge and discharge behavior. The specific capacitance values of the different CDCMs electrodes were calculated to range from 73 to 231 F g^{-1} in which CDCM-800 showed the highest specific capacitance (Fig. 1b). The reason that CDCM-800 had the highest specific capacitance among the samples was most likely due to its high specific surface area, favorable volume ratio of micropores to mesopores, and the presence of nitrogen and oxygenated functional groups. The mesopores have a limited role in storing charges, but they are important for promoting electrolyte ion transport and for providing effective access to active points in the micropores. Favorable proportions of micropore to mesopore volumes have a positive effect on improving electronic performance [11]. The heteroatom functional groups also play important roles in the energy storage, because these functional groups can contribute to the specific capacitance through pseudocapacitance [12]. Furthermore, they can increase the wettability of the obtained carbon materials [13]. However, although CDCM-600 has higher oxygen and nitrogen content than CDCM-800, CDCM-600 exhibited lower capacitance than CDCM-800 that is most likely due to the low surface area and pore volume of the CDCM-600 that do not provide a sufficient number of active sites for redox reactions.

When the scan rate was varied from 5 to 200 mV s^{-1}, the CV curve became less rectangular, and the corresponding specific capacitance was reduced from 237 to 139 F g^{-1} (Figs. 1c), which is due to the inability of electrolyte ions being able to contact active electrode materials at high scan rates [14].

GCD curves of CDCM-800 electrode at different current densities are illustrated in Figure 1d. It can be seen that a specific capacitance of 231 F g^{-1} was obtained at a current density of 0.5 A g^{-1}. The capacitance gradually decreased with increasing current density, but CDCM-800 could retain 154 F g^{-1} at 20 A g^{-1}. The reduction in capacitance at high current density is probably the result of electrode polarization that hinders the interactions between ions and CDCMs at high current density.

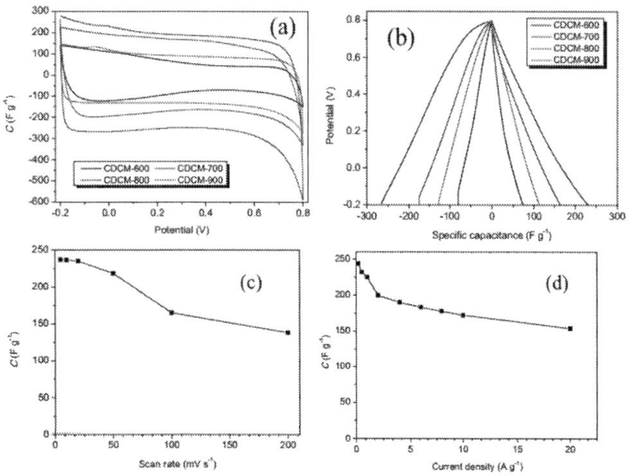

Figure 1. Electrochemical performance of chitosan-derived carbonaceous materials (CDCMs) in a three-electrode setup (0.5 M K_2SO_4): (a) cyclic voltammetry (CV) profiles of CDCMs at a scan rate of 5 mV s^{-1}; (b) galvanostatic charge–discharge (GCD) profiles of CDCMs at a current density of 0.5 A g^{-1}; (c) specific capacitance calculated from CV curves versus voltage scan rate; (d) specific capacitance calculated from GCD curves versus current density.

The CDCM-800 sample was tested in KOH aqueous solution (6 M). It can be seen that a high specific capacitance of 305 F g^{-1} could be achieved for a current density of 0.5 A g^{-1} (Fig. 2a), and a specific capacitance of about 198 F g^{-1} could be retained for a current density of 20 A g^{-1}. Compared with the K_2SO_4 electrolyte, KOH solution electrolyte exhibited higher specific capacitance because the KOH electrolyte reacts with the active substance to produce Faradaic pseudocapacitance, while K_2SO_4 only adsorbs onto the active surface. However, as a neutral electrolyte, K_2SO_4 electrolyte is preferable since it is environmentally-friendly and has a similar specific capacitance with KOH when used with the CDCM-800 synthesized in this work as electrode material.

The energy and power densities of CDCM-800 in 0.5 M K_2SO_4 and 6 M KOH aqueous solutions were calculated and are illustrated with a Ragone plot (Fig. 2b). Fig. 2b shows that the energy density of the CDCM-800 electrode materials could reach 21.4 Wh kg^{-1} with a corresponding specific power of 10 kW kg^{-1} in 0.5 M K_2SO_4 solution at a current density of 20 A g^{-1}, and a high energy density of 27.5 Wh kg^{-1} could be obtained with 10 kW kg^{-1} of specific power in electrolyte of 6 M KOH at 20 A g^{-1}. In addition, the CDCM-800 material exhibited high electrochemical stability for which capacitance remained stable at approximate 99% of its original value after 3000 cycles at a current density of 1 A g^{-1}.

Conclusions

Nitrogen-containing porous carbonaceous materials were prepared from chitosan directly with hydrothermal carbonization that was followed by KOH activation. The prepared chitosan-derived carbon materials had high surface area up to 2200 m^2 g^{-1}, large pore volume up to 1.36 cm^3 g^{-1}, and exhibited excellent energy storage performance for supercapacitor electrode materials in aqueous electrolyte of 0.5 M K_2SO_4. Maximum specific capacitances of 231 and 305 F g^{-1} could be obtained for CDCM-800 sample for a current density of 0.5 A g^{-1} in 0.5 M K_2SO_4 and 6 M KOH electrolytes, respectively, and the values were retained at 154 and 198 F g^{-1} for a current density of 20 A g^{-1}, respectively. CDCM-800 exhibited a specific energy of 21.4 and 27.5 Wh kg^{-1} at a current density of 20 A g^{-1} with a corresponding specific power of 10 kW kg^{-1} in 0.5 M K_2SO_4 and 6 M KOH electrolytes, respectively. The CDCM-800 material exhibited high electrochemical stability for which capacitance remained stable at approximate 99% of its original value after 3000 cycles at a current density of 1 A g^{-1}.

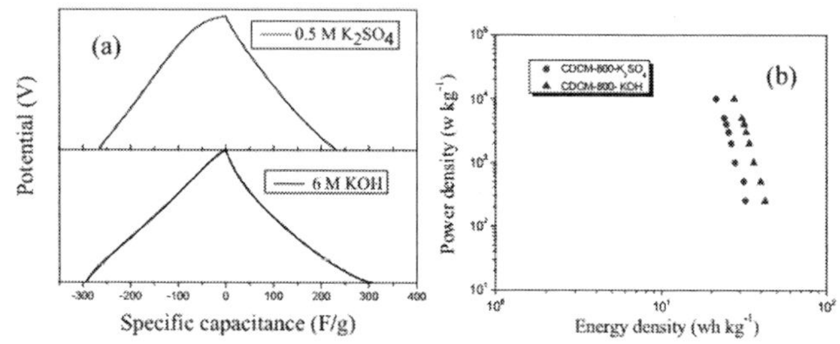

Figure 2. Electrochemical performance of chitosan-derived carbonaceous materials (CDCMs) in a three-electrode setup (in 0.5 M K_2SO_4 and 6 M KOH). (a) GCD profiles of CDCM at a current density of 0.5 A g^{-1} in different electrolytes. (b) Ragone plots of the sample in two electrolytes.

Acknowledgement. The work was supported by the National Natural Science Foundation of China (No. 21577073) and the Science and Technology Innovation Program and Elite Youth program of Chinese Academy of Agricultural Sciences.

References)

(1) S. Y. Gao, K. R. Geng, H. Y. Liu, X. J. Wei, M. Zhang, P. Wang and J. J. Wang, *Energ. Environ. Sci.*, 2015, 8, 221-229.
(2) H.-C. Lee, N. Byamba-Ochir, W.-G. Shim, M. S. Balathanigaimani and H. Moon, *J. Power Source*, 2015, 275, 668-674.
(3) T. Tooming, T. Thomberg, H. Kurig, A. Jänes and E. Lust, *J. Power Source*, 2015, 280, 667-677.
(4) S. Faraji and F. N. Ani, *J. Power Source*, 2014, 263, 338-360.
(5) S. K. Meher and G. R. Rao, *J. Power Source*, 2012, 215, 317-328.
(6) H. Yi, H. W. Wang, Y. T. Jing, T. Q. Peng and X. F. Wang, *J. Power Source*, 2015, 285, 281-290.
(7) Y. Yang, E. Shi, P. Li, D. Wu, S. Wu, Y. Shang, W. Xu, A. Cao and Q. Yuan, *Nanoscale*, 2014, 6, 3585-3592.
(8) J. W. Jeon, R. Sharma, P. Meduri, B. W. Arey, H. T. Schaef, J. L. Lutkenhaus, J. P. Lemmon, P. K. Thallapally, M. I. Nandasiri, B. P. McGrail and S. K. Nune, *ACS Appl Mater Interfaces*, 2014, 6, 7214-7222.
(9) M. Sevilla and R. Mokaya, *Energ. Environ. Sci.*, 2014, 7, 1250-1280.
(10) J. Yan, Q. Wang, T. Wei and Z. Fan, *Adv. Energy Mater.*, 2014, 4, 1300816-1300858.
(11) L. L. Zhang and X. S. Zhao, *Chem. Soc. Rev.*, 2009, 38, 2520-2531.
(12) J. W. Jeon, L. Zhang, J. L. Lutkenhaus, D. D. Laskar, J. P. Lemmon, D. Choi, M. I. Nandasiri, A. Hashmi, J. Xu, R. K. Motkuri, C. A. Fernandez, J. Liu, M. P. Tucker, P. B. McGrail, B. Yang and S. K. Nune, *ChemSusChem*, 2015, 8, 428-432.

(13) [13].G. Xu, J. Han, B. Ding, P. Nie, J. Pan, H. Dou, H. Li and X. Zhang, *Green Chem.*, 2015, 17, 1668-1674.
(14) [14].W. J. Liu, K. Tian, Y. R. He, H. Jiang and H. Q. Yu, *Environ Sci Technol*, 2014, 48, 13951-13959.

FREESTANDING NANOFIBER ELECTRODES FOR SUPERCAPACITORS AND BATTERIES

Vibha Kalra

Department of Chemical and Biological Engineering, Drexel University

Fabrication of novel three-dimensional material architectures is essential for development of energy storage devices that allow high rate operation with sufficient energy capacity. After a short description of our group's overall research, I will present our specific work on development of binder-free freestanding nanofiber-based electrodes using a simple electrospinning methodology for application in supercapacitors (EDLC/Pseudocapacitors) and lithium-sulfur batteries. I will also talk about our work on molecular dynamics (MD) simulations to understand the effect of elongational flow, a key characteristic of electrospinning, on assembly within phase separating polymer blends – a material system relevant to the fabrication of nanofiber electrodes.

We have developed porous carbon nanofibers (PCNF) that exhibit specific surface area of >1500 m^2/g and multi-levels of pore sizes in the range of micro (< 2 nm), meso (< 50 nm) and macropores (> 50 nm). We studied these materials both as electrodes in electric double layer capacitors (EDLC) (with aqueous and ionic liquid electrolytes) as a platform for polyanilene (PANi) deposition to develop hybrid supercapacitors. The unique combination of pores at different length scales allows us to achieve a battery-like energy density of up to 80 Wh/kg, while retaining high power and cycle life in EDLCs. *In-operando* infrared spectroelectrochemistry was conducted to understand the transport of ions through the carbon electrode pores and its effect on performance in ionic liquid-based EDLCs. In the hybrid supercapacitor devices, we demonstrated that the PCNF mats retain their EDLC behavior post PANi coating providing excellent electrochemical performance that integrates electric double layer and pseudocapacitive energy storage. We have also investigated the use of nanofiber materials for battery devices – Li-S, Li-O_2 and redox flow batteries. In particular, for Li-S battery devices, we have demonstrated a facile methodology to localize soluble polysulfides that results in a high initial discharge capacity and capacity retention during cycling.

3D PORE-SCALE TRANSPORT RESOLVED MODEL OF LI-AIR BATTERIES

Charles Andersen, Han Hu, and Ying Sun

Department of Mechanical Engineering and Mechanics, Drexel University, Philadelphia, PA
YSun@coe.drexel.edu

Li-air battery with its usable energy density close to 1,700Wh/kg has captured worldwide attention as a promising battery solution for electric vehicles. However, a major hurdle facing the development of Li-air battery systems is their poor round-trip efficiency owing to the formation of the electrically insulating lithium peroxide at the cathode surface. Existing models for Li-air batteries often treat the electrode/Li_2O_2 matrix as a homogenous continuum and utilize overly simplified electrode morphologies to obtain volume-averaged electrode properties and are insufficient to probe the effects of precise electrode microstructures and Li_2O_2 growth. In this work, we present a pore-scale transport resolved model of the Li-air battery that fully accounts for the electrode microstructure and peroxide growth. 3D reconstructed images of real electrode structures are used as geometric inputs to the model and two potential peroxide forming reactions are incorporated to capture different peroxide growth morphologies. A discrete element method is used to simulate the nucleation and movement of lithium peroxide. Results obtained from our pore-scale model were compared with experiments with good agreement. Through extensive 3D simulations, the effect of drawing current, ORR rate coefficients, oxygen solubility, mean pore size and distribution on Li_2O_2 growth and cell performance are systematically examined as to better understand the underlying physics of capacity fading during cycling.

MULTI-SHELLED METAL OXIDES PREPARED VIA AN ANION-ADSORPTION MECHANISM FOR LITHIUM-ION BATTERIES

Jiangyan Wang, Ranbo Yu, Huijun Zhao and Dan Wang

1. National Key Laboratory of Biochemical Engineering, CAS Center for Excellence in Nanoscience, Institute of Process Engineering, Chinese Academy of Sciences, 1 North 2nd Street, Zhongguancun, Haidian District, Beijing 100190, P. R. China
2. University of Science & Technology Beijing, 30, Xueyuan Road, Haidian District, Beijing 100083, P. R. China
3. Centre for Clean Environment and Energy, Gold Coast Campus Griffith University, Queensland 4222, Australia

Introduction

Rechargeable lithium-ion batteries (LIBs) have long been regarded as a promising candidate to meet the energy needs of a sustainable society.[1] However, the energy density of LIBs has still fallen short of expectations. The advancement has mainly been limited by the mismatch between the specific capacities of cathode and anode materials.[2]

Vanadium oxide (V_2O_5), is an attractive cathode material for next generation LIBs because of its high theoretical specific capacity and abundance. Unfortunately, the low ionic conductivity and poor structural stability upon deep discharge has greatly limited its LIB applications. A potential solution to circumvent these challenges would be to develop three-dimensional (3D) hollow microspheres with porous shells and zero-dimensional (0D) nanoparticles as basic units, to substitute the bulk structures as LIB cathode materials. Such hollow micro-/nanostructures embrace the advantages of both low-dimensional and three-dimensional nanostructures, which could enhance the LIB performance.[3,4]

Significant efforts are devoted to fabricating hollow structured V_2O_5 in recent years. However, the majority of such efforts failed. More recently, some other metal oxides of multi-shelled hollow microspheres are fabricated based on the electrostatic-attraction between negatively-charged carbonaceous microsphere (CMS) templates and metal cations.[5] Unfortunately, the approach failed when extended to synthesize multi-shelled V_2O_5 hollow microspheres because V(v) ions usually exist as anions. Although cationic V might form when decreasing pH value, the negative charges of carbon templates become too weak to be able to adsorb the V(v) ions.

Herein, an anion-adsorption by negatively-charged CMSs, is proposed and validated. As a result, Multi-shelled/cavitied V_2O_5 hollow microspheres and other MO-MS-HMSs are successfully synthesized. In addition, the triple-shelled V_2O_5 hollow microspheres (3S-V_2O_5-HMSs) show exceptionally high specific capacities and outstanding cycling stability as well as good rate capability as cathode materials for LIBs. These properties are superior compared to previously reported V_2O_5-based cathode materials.

Experimental

Synthesis of multi-shelled V_2O_5 hollow microspheres. All chemicals were analytically pure, bought from Beijing Chemical Reagent Factory, and used as recieved. CMSs were obtained by a hydrothermal method as described elsewhere[5]. Ammonium vanadate (NH_4VO_3) was adopted as V-precursor. In a representative adsorption process, freshly-prepared CMSs (0.6 g) were dispersed by ultrasonication for 10-30 minutes in 30 mL ammonium vanadate aqueous solution of 0.01-0.1 M. After that, the resulted suspension was set for 6 h in a water bath (40-70°C), filtered, washed for at least three times with deionized water, and then dried. After the adsorption process, the resultant CMS-V composite were heated up to 400°C at 1 °C min-1 in air, and stayed at 400°C for 30 minutes, then cooled naturally.

Characterization Methods. Powder XRD patterns were achieved through a Panalytical X' Pert PRO MPD [Cu Ka radiation (λ = 1.5405 Å), 40 kV, 30 mA]. SEM images were obtained through JSM-6700 microscope (5.0 kV). TEM, HRTEM and SAED images were recorded on a FEI Tecnai F20 instrument (accelerating voltage: 200 kV). Spherical aberration-corrected STEM was tested through a JEOL 2100F (JEOL, Tokyo, Japan) transmission electron microscope with a CEOS (Heidelberg, Germany) probe aberration corrector, operated at 200 keV. The nitrogen adsorption-desorption isotherms were measured on an Auantochrome Autosorb-1MP sorption analyzer.

Electrochemical measurement. The lithium-storage property of prepared V_2O_5 samples was carried out using CR2032 coin type cells. 1.0 M LiPF6 in mixture of dimethyl carbonate (DMC) and ethylene carbonate (EC) (1:1 v/v) was used as the electrolyte. Working electrodes were fabricated by painting a mixture of the active materials, conductive carbon black, and binder (sodium carboxymethyl cellulose) in a weight ratio of 7:2:1 with DI water being the solvent. The electrode capacity was measured by a galvanostatic discharge-charge method performed on a Neware Battery Testing System between 1.5-4.0 V at 50, 200, 1000, 2000 mA/g. Electrochemical impedance spectroscopy (EIS) was tested over the frequency ranging from 0.1 to 105 Hz (ac amplification voltage: 5 mV).

Results and Discussion

We demonstrate that CMSs are not only capable of adsorbing cations through electrostatic-interactions but also provide access for various anions-penetrations. A proposed adsorption mechanism is put forward in **Figure 1**. CMSs with rich –OH, –C=O and -C-O-C groups could be considered as large colloidal particles that tend to preferentially adsorb ions with similar elemental compositions. Thus OH- or other oxygen-containing groups are preferentially adsorbed by CMSs, endowing CMSs negatively-charged and following adsorption of cations through electrostatic-interactions.

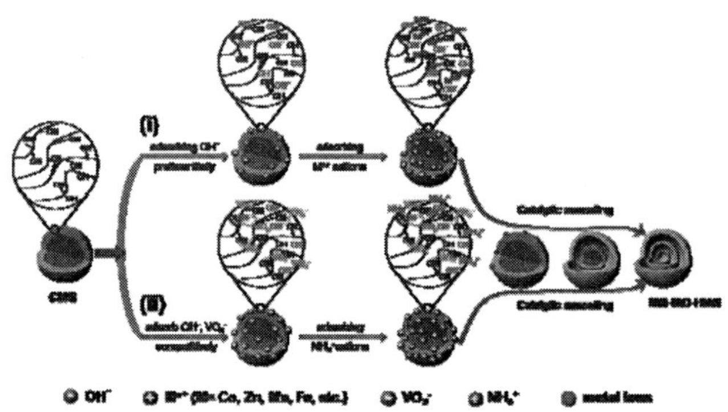

Figure 1. Schematic of two synthesis routes of multi-shelled hollow microspheres.

Modeling and calculation results show that the electrostatic repulsion energy is much smaller than the binding energy, indicating that the chemical adsorption of VO_3^- onto CMSs is thermodynamically feasible. It's also worth mentioning that, besides anions, NH_4^+ cations can also penetrate into CMSs through electrostatic-interaction or hydrogen bonding, neutralizing the negative charges and promoting the further adsorption of anions.

Lastly, CMS templates were removed to obtain MS-V_2O_5-HMSs through a Trojan catalytic combustion process: the penetrated V in CMSs act as catalysts to markedly accelerate CMSs burning process.

V_2O_5 hollow spheres with uniform morphology and structure are successfully achieved in high yield and purity. That is, thin single-, double- and thick triple-shelled hollow spheres (**Figure** 2a-c) attained by repeating adsorption cycles; smooth thick single-shelled (**Figure** 2d,g) and porous thin triple-shelled (**Figure** 2e,h) prepared through acid-treatment; and multi-cavitied hollow spheres (**Figure** 2f,i). As shown in X-ray diffraction (XRD) patterns (**Figure** 2j), all main peaks are indexed to pure orthorhombic V_2O_5 phase (JCPDS card no. 41-1426), except for widening of the peak at about 25° for the thick 1S-HMSs (curve vii in **Figure** 2j). Besides, selected area electron diffraction (SAED) and high-resolution TEM (HRTEM) characterizations prove the high crystallization of other V_2O_5 hollow microspheres, in which monocrystalline nanoparticles of orthorhombic V_2O_5 ((200) plane, d-spacing is 0.58 nm) are commonly observed.

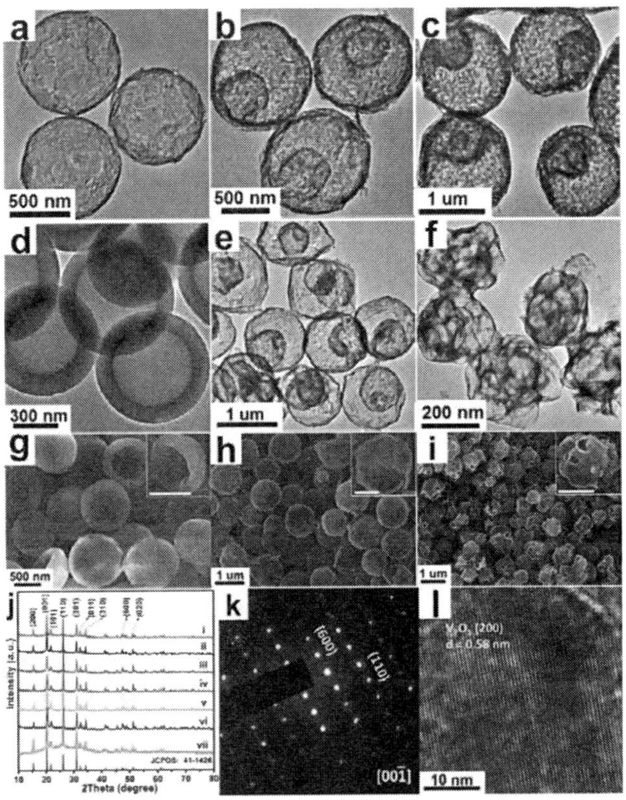

Figure 2. Morphological and structural analysis of V_2O_5 products. a,b,c,d,e,f) TEM and g,h,i) SEM images of as-prepared samples (inset is the high magnification SEM image of one single microsphere, scale bar: 500 nm). j) XRD patterns of samples. k) SAED and l) HRTEM images of 3S-V_2O_5-HMSs.

It clearly demonstrates a series of phase transformations at around 3.25, 3.06, 2.19, 1.91 V during the initial Li+-intercalation process (**Figure 3a**). Impressively, the 3S-V_2O_5-HMSs exhibit a record-high initial specific capacity of 447.9 mAh/g (theoretically 441 mAh/g for V_2O_5) at 1000 mA/g. More excitingly, the excellent cycling stability and high coulombic efficiency (CE) (almost 100%) are concurrently achieved. Besides, both thin 1S- and 2S-V_2O_5-HMSs also show a high specific capacity and good capacity retention, whereas the multi-cavitied hollow structures, thick 1S-V_2O_5-HMSs, V_2O_5 nanosheets and mixture of sphere and cube are suffered from low specific capacity and/ or poor stability.

Compared with nanosheets, 3S-V_2O_5-HMSs exhibit a dramatically enhanced rate capability. It is worth noting that an extremely high initial specific capacity is achieved by 3S-V2O5-HMSs at 50 mA/g. Obviously, this value exceeds the theoretical one, which may arise from the surface/interface lithium-storage mechanism. Spherical aberration-corrected STEM ABF (annular-bright-field) and HAADF characterizations of V_2O_5 after discharge show that there exist many defects. These defects can provide rich surfaces/interfaces and bring about abundant additional capacitive-controlled Li-storage.

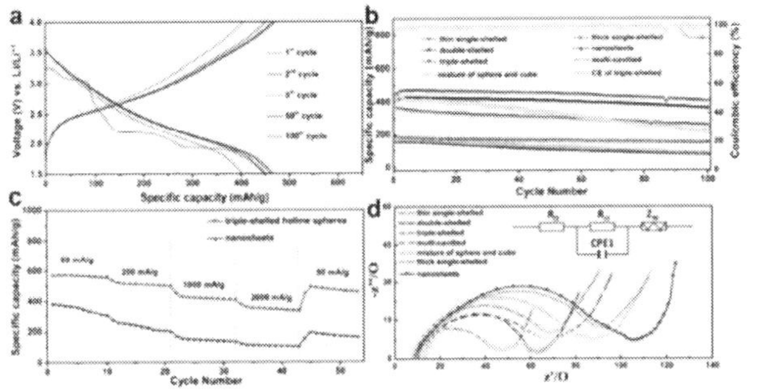

Figure 3. Electrochemical properties of as-prepared V_2O_5 products. a) Discharge-charge profiles at different cycles. b) Discharge capacity versus cycle number. c) Discharge curves at different current densities. d) EIS for V_2O_5 prodcuts (RΩ, external resistance; Rct, charge transfer resistance; CPE1, constant phase element; Zw, Warburg impedance).

Overall, the electrodes composed of 3S-V_2O_5-HMSs display significantly better LIB performance than those based on other hollow structures or V_2O_5 nanosheets. The reasons can be generalized to three parts: Firstly, the triple-shelled possesses the largest specific surface area and pore volume, which provide increased lithium-storage sites and enhanced access, thus an impressive specific capacity. Secondly, the diffusion path is greatly reduced with thin porous shells, thus shortening the diffusion time. Besides, low ion and mass transfer resistance also facilitates the electrochemical activity, guaranteeing superior rate capability (**Figure** 3d). Thirdly, the outstanding structural feature can not only take the advantage of low-dimensional nanostructures to obtain a high specific capacity, but also benefit from the better buffering of volume change and favorable distribution of the induced stress and strain to achieve excellent cycling stability.

Acknowledgment. This work was supported financially by the National Natural Science Foundation of China, National Science Fund for Distinguished Young Scholars and Australian Research Council (ARC) Discovery Project.

References

(1) Poizot, P., Laruelle, S., Grugeon, S., Dupont, L. & Tarascon, J.-M. *Nature* **2002**, *407*, 496-499.
(2) Liang, Z. et al. *ACS Nano* **2014**, 8, 5249-5256.
(3) Wang, J. Y. et al. *Angew. Chem. Int. Ed.* **2013**, *52*, 6417-6420.
(4) Xu, S. M. et al. *Energy Environ. Sci.* **2014**, *7*, 632-637.
(5) Lai, X. Y. et al. *Angew. Chem. Int. Ed.* **2011**, *50*, 2738-2741.
(6) Dong, Z. H. et al. *Adv. Mater.* **2014**, *26*, 905-909.

UNDERSTANDING AND DESIGN OF GRAPHENE-BASED MATERIALS FOR SUPERCAPACITORS

Gyeong S. Hwang, Alexander J. Pak and Eunsu Paek

Department of Chemical Engineering, University of Texas at Austin
gshwang@che.utexas.edu

Electrochemical double layer capacitors (EDLCs) are energy storage devices with high power densities and lifetimes. Nanostructured carbon materials such as graphene and ionic liquids (ILs) have garnered considerable interest as candidates for the electrode and electrolyte, respectively. The total interfacial capacitance (C_T) of graphene-like EDLCs can be determined from both electrode and electric double layer capacitances (C_D) which are in series. Furthermore, our recent studies demonstrate that the electrode capacitance (C_Q), and thereby C_T, can be significantly enhanced as a result of changes to the electronic structure through dopants, defects, and/or structural deformation. In this talk, we will present how the structural and chemical disorder in nanostructured carbon electrodes modifies the electronic structure, thereby improving the C_Q as well as the *EDL* structure and C_D at the carbon/IL interfaces. We will also discuss design strategies to further improve the performance of nanostructured carbon-based EDLCs.

STRATEGIES FOR ENGINEERING POLYMERS TO ENHANCE ENERGY CAPTURE AND STORAGE

Kenneth K. S. Lau

Department of Chemical and Biological Engineering
Drexel University

Some of the latest breakthroughs in energy capture and storage rely on electrochemical systems containing high surface area, porous, nanostructured electrodes. Such device architectures come with trade-offs that limit device performance and reduce the amount of energy available.

For example, the dye sensitized solar cell (DSSC) has a photoanode made of a sintered network of TiO_2 nanoparticles in contact with an organic liquid electrolyte. The liquid is able to wet the internal pore surface of the mesoporous photoanode. However, ions in the liquid electrolyte easily scavenge electrons traveling through the photoanode leading to charge recombination losses. There are also leakage, evaporation, and flammability issues with using a liquid. Here, we demonstrate the use of initiated chemical vapor deposition (iCVD) that enables the growth and intimate contact of polymer electrolytes within the mesoporous photoanode to replace the liquid. The polymer electrolytes lead to higher photocurrent and photovoltage compared to liquid counterparts. DSSC mathematical modeling has been valuable in understanding the impact of polymer electrolytes and guiding us towards more enhanced electrolyte materials and cell response.

In another example, the electrochemical supercapacitor has electrodes made of highly porous, conductive carbon materials. The high surface area enables significant charge storage through an electrochemical double layer of opposing charges at the electrode-electrolyte interface. The simple electrostatic interaction enables high power, rapid charge-discharge but suffers from lower energy density. Here, we demonstrate the use of oxidative chemical vapor deposition (oCVD) that enables the growth of ultrathin, redox-active, conducting polymers which conform to the porous carbon topology. In this way, the ultrathin redox polymers provide additional energy storage by means of Faradaic reactions that reversibly oxidize and reduce the polymers during discharge-charge.

We will discuss our efforts in the development of iCVD and oCVD as viable strategies in achieving significantly enhanced energy capture and storage technologies.

*Invited talk in the "USA-China Symposium on Energy", Division of Energy and Fuels

ULTRAFAST CARRIER DYNAMICS IN PHOTOVOLTAIC ABSORBER MATERIALS

Jason B. Baxter

Drexel University
Department of Chemical and Biological Engineering
ACS Fall Meeting, Philadelphia
Energy Fuel- USA-China Symposium on Energy
August 2016

Light absorption by photovoltaic absorber materials generates electron-hole pairs that must be separated and collected before they recombine. For many materials, recombination lifetimes are on the order of picoseconds to nanoseconds. Using time-resolved terahertz spectroscopy and transient reflectance spectroscopy, we can measure photoexcited carrier lifetimes and mobilities and determine dominant recombination mechanisms. Ultrafast carrier dynamics provide a link between processing, structure, and device performance that can help direct the design of more efficient solar cells. This talk will highlight our group's investigations of ultrafast carrier dynamics in several material systems, including $Cu_2ZnSn(S,Se)_4$ (CZTSSe) thin films and single crystals and PbSe nanocrystal (quantum dot) films.

Primary support for this work was provided by NSF (DMR-1507988/1508042, CBET-1333649/1335821, and ECCS-1201957).

THERMAL AND MECHANICAL INTEGRITY OF PbTe-BASED NANOSTRUCTURED THERMOELECTRIC MATERIALS

Fei Ren

Department of Mechanical Engineering, Temple University, Philadelphia, PA, USA

Abstract

Thermoelectric materials hold the promise of direct conversion from heat to electricity. However, development of commercial thermoelectric generators is limited by the relatively low energy conversion efficiency and high cost of thermoelectric materials. Furthermore, many issues related to scalable production, mechanical integrity, and thermal durability remain unsolved.

PbTe based materials have peak performance in the temperature range suitable for automotive waste heat recovery (c.a. 300 – 500°C), and thus attracted much attention in recent years. One successful strategy to enhance the efficiency is to introduce nanostructures in the PbTe matrix, which in turn can lower the thermal conductivity and increase the material figure of merit (Z). However, these materials exhibited low mechanical strength. Moreover, their thermal durability, especially the stability of the nanostructures, is unclear.

In this presentation, I will briefly introduce the synthesis strategy to form secondary phase nanostructures and discuss the techniques for scale-up production. Issues related to mechanical and thermal integrity and their mitigation methods will be presented. Furthermore, I will discuss the thermal stability of the nanoinclusions based on a series of in situ experiments including resonant ultrasound spectroscopy, in situ neutron diffraction, and small angle neutron scattering. Our results show the nanostructures do evolve with temperature and may be stabilized by certain heat treatment techniques.

POWER GENERATION AND STORAGE SYSTEMS: MODEL-GUIDED OPTIMAL DESIGN, OPERATION AND INTEGRATION

Masoud Soroush

Department of Chemical and Biological Engineering
Drexel University
Philadelphia, PA 19104
USA

The demand for energy is expected to increase for the foreseeable future, as the human population increases, the average standard of living improves, and electronics and automation find more applications in our daily life. While fossil energy resources are greatly in demand, sustainability concerns have motivated (a) energy generation from renewable and reusable energy recourses and (b) the use of more efficient energy-converting and energy-consuming systems. Consequently, for the past decade there have been major advances in harvesting energy from renewable and reusable resources and improving the efficiency of energy-consuming and energy-converting systems.

This presentation focuses on renewable power generation and storage systems, and how first-principles mathematical modeling can contribute to systematic optimal design, operation and integration of these systems. It puts in perspective how mathematical modeling has contributed to advances in solar cells, fuel cells, flow batteries and rechargeable batteries, and in the integration of these systems with power grids. As an example of a renewable power generation system, dye sensitized solar cells (DSSCs) are considered. The DSSC technology is commonly regarded as a promising next generation solar cell technology, because DSSCs have good performance and can be made relatively inexpensively without intensive purification and fabrication steps. Polymer-electrolyte DSSCs have better I-V performance than conventional liquid-electrolyte DSSCs but do not have shortcomings of liquid-electrolyte DSSCs. Currently, there is a need for a better understanding of how polymer-electrolyte chemistry affects the cell performance. It will be discussed how first-principles modeling can improve this understanding. By properly accounting for the physical and electrochemical processes occurring inside the cells, such a mathematical model can reveal competing effects within DSSC components, can identify factors that affect the overall cell performance, and can guide experimental studies. Using experimental data and the model, the effects of polymer chemistry on the overall cell performance and photochemical processes within the cell are elucidated.

MORPHOLOGY AND ACIDITY CONTROL OF SAPO-34 ZEOLITE

*Wenlong Jin, Pengfei Tuo, Baojian Shen**

State Key Laboratory of Heavy Oil Processing; The Key Laboratory of Catalysis of CNPC; College of Chemical Engineering; China University of Petroleum, Beijing 102249, P. R. China
*Corresponding author: baojian@cup.edu.cn

1. Introduction

Light olefins, especially ethylene and propylene, has an irreplaceable role in the field of petrochemical industry, organic synthesis, biological medicine and advanced materials. The disadvantage of traditional preparation technology of ethylene and propylene is overly rely on petroleum resources. Methanol-to-olefins (MTO) conversion, as a potential process for preparation light olefins, has attracted extensive attention as the crude oil becoming poorer and heavier in quality[1].

SAPO-34 molecular sieve, with 8-ring CHA structure and medium acid, exhibits excellent performance in MTO reaction, giving 100% methanol conversion and over 90% selectivity to C_2-C_4 light olefins[2]. In MTO reaction, the pore structure[3] along with acidity [4]of SAPO-34 dominate the products distribution and catalyst lifetime.

In this work, crystallization of SAPO-34 has been studied by monitoring the whole synthesis process. The regulation of morphology and chemical composition varying with crystallization conditions was investigated.

2. Experimental

SAPO-34 crystals were prepared by conventional hydrothermal method.Typically, the gel was prepared by first mixing the deionized water with phosphoric acid and together with pseudoboehmite, then the mixture was vigorously stirred to obtain a uniform gel. Then the dual-template was added dropwise to the resultant solution. After that silica sol was slowly added to the above mixture.

The reaction mixture was transferred into a Teflon-lined stainless steel autoclave and heated in an oven under different conditions. The as-synthesized products were filtered, washed with water several times, and then dried at 120 ℃ in the oven overnight, followed by calcination at 550 ℃ for 8 h to remove the templates.

Samples with different crystallization conditions named S-1 to S-8.

3. Results and Discussion

According to the SEM images (given in **Figure. 1**), samples with different crystallization conditions showed various morphology. Sample S-1 was still amorphous, pseudoboehmite bulks were clearly visible. Samples S-2, S-3 and S-4 exhibited cubic particles with rough surface and the distribution of the surface defects were obviously organized. In contrast, Sample S-5 owned hourglass-shaped hole formed from connecting the surface defects together. Sample S-6, S-7 and S-8 obtained conventional nice cubic morphology, but the particle size of sample S-8 was more regular than others, relatively.

Figure 1 SEM images of as-synthesis samples with different crystallization conditions. (a) S-1; (b) S-2; (c) S-3; (d) S-4; (e) S-5; (f) S-6; (g) S-7; (h) S-8;

From the chemical compositions and the acid amounts of samples, presented in **Table 1**, we can conclude that the silica content, mole SiO_2/Al_2O_3 and acid amounts obviously varied with crystallization conditions even though the compositions of synthesis gel were identical. It's worth pointing out, on the premise of similar silica content and mole SiO_2/Al_2O_3 in comparison to S-7, S-8 had lower acid amount.

Table 1 Chemical compositions and the acid amounts of as-synthesis samples

	Chemical compositions[a]			SiO_2/Al_2O_3 (mol)	The acid amounts[b]		
	Al_2O_3, wt%	SiO_2, wt%	P_2O_5, wt%		Weak	Medium and strong	Total
S-3	45.302	6.906	47.480	0.27	0.44	0.67	1.11
S-4	44.579	7.227	47.776	0.28	0.50	0.69	1.19
S-5	44.380	7.507	47.772	0.29	0.48	0.66	1.14
S-6	43.180	8.409	48.098	0.33	0.51	0.76	1.27
S-7	43.343	9.492	46.770	0.37	0.59	0.82	1.41
S-8	43.221	9.359	46.591	0.37	0.54	0.70	1.24

[a] Obtained by XRF . [b] Calculated from the peak area of NH_3-TPD profiles

Catalytic performance in methanol to olefins process of catalysts with different crystallization conditions were given in

Figure 2. It's obvious that sample S-5showed the longest lifetime of 1300min and sample S-4 centered (1155 min). The reason could be the exist of surface defects provides more surface for coke deposition. The lifetime of sample S-8 (1035 min) was longer than sample S-6 (855min) should be attributed to the smaller crystal size which leads to bigger external surface area, shorter diffusion path and better resistance of coke deposition.

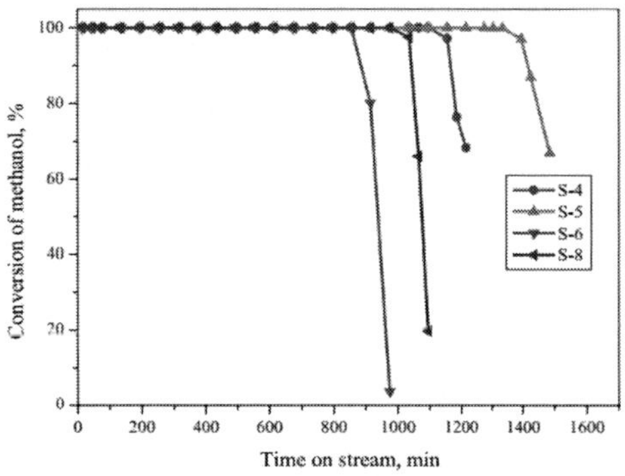

Figure 2 Methanol conversion variation with time-on-stream over SAPO catalysts with different crystallization conditions. Experimental conditions: WHSV = 1 h^{-1}, T = 460 ℃, catalyst weight = 1 g.

4. Conclusions

SAPO-34 samples with different crystallization conditions exhibited various morphology, chemical compositions and acidity. The exist of surface defects expanded the external surface area and further lengthened the crystal lifetime in methanol-to-olefins process.

Acknowledgement: The authors greatly acknowledge the funding of this project by the National Natural Science Foundation of China (U1462202) and PetroChina.

References

(1) U. Olsbye, S. Svelle, M. Bjørgen, P. Beato, T. V. W. Janssens,F. Joensen, S. Bordiga and K. P. Lillerud, Angew. Chem., Int. Ed., 2012, 51, 5810.

(2) Juan Liang, Hongyuan Li, Sugin Zhao, Wenggui Guo, Ronghui Wang, Muliang Ying. Appl. Catal., 1990, 64, 31–40.

(3) Jingrun Chen, Jinzhe Li, Yingxu Wei, Cuiyu Yuan, Bing Li, Shutao Xu, You Zhou, JinbangWang, Mozhi Zhang, Zhongmin Liu. Catal. Commun. 2014, 46 36.

(4) A.J. Marchi, G.F. Froment. Appl. Catal. 1991,71, 139.

PHOTONIC NANOSTRUCTURES FOR EFFICIENT SOLAR-TO-FUEL ENERGY CONVERSION

*Liwu Zhang**

Department of Environmental Science and Engineering, Fudan University, Shanghai 200433, China
zhanglw@fudan.edu.cn

The main reason for the low efficiency of solar fuel energy conversion is the dominant electron-hole recombination in the bulk of photoelectrode due to the short diffusion length of photoexcited minority charge carriers. To address this problem, nanostructuring has been extensively studied, reducing bulk recombination by shortening the travel length for charge carriers. However, nanostructuring can also increase surface recombination and lower the surface photovoltage. Nanophotonic structures, manipulating and confining light on the nanometer scale, provides new opportunities to improve the efficiency. Several nanophotonic structures are of particular interest for solar water splitting. An optical cavity can reduce the thickness of photoelectrodes without compromising their light absorption by trapping resonant light in ultrathin films. Photonic crystals show great potential in manipulating light based on photonic band structure concepts, in which near-bandgap resonant scattering and slow photon effects can enhance the interaction of light with a semiconductor. Plasmonic metal nanostructures with surface plasmon resonances can act as antennas to localize optical energy and control the location of charge carrier generation. The interaction of localized electric fields surrounding the plasmonic metal particles with a neighbouring semiconductor provides a pathway for the selective formation of electron-hole pairs in the near-surface region of the semiconductor.

References

(1) Zhang, L. W.; Reisner E.; Baumberg J., Al-doped ZnO Inverse Opal Networks as Efficient Electron Collectors in BiVO4 Photoanodes for Solar Water Oxidation. *Energy Environ. Sci.* 2014, 7, 1402-1408.
(2) Zhang, L. W.*, Chia□Yu Lin, Ventsislav K Valev, Erwin Reisner, Ullrich Steiner, Jeremy J Baumberg*Plasmonic Enhancement in BiVO4 Photonic Crystals for Efficient Water Splitting; *Small*, 2014, 10, 3970-8.
(3) Daniel O. Sigle, Liwu Zhang*, Sandrine Ithurria, Benoit Dubertret, and Jeremy J. Baumberg*, Ultrathin CdSe in Plasmonic Nanogaps for Enhanced Photocatalytic Water Splitting, *J. Phys. Chem. Lett.*, 2015, *6* (7), pp 1099–1103.

FUNCTIONAL ENERGY MATERIALS: FROM 1D AND 2D POLYMERS TO 3D CARBON NANOMATERIALS

Liming Dai[a,b]

[a] Case School of Energy, Case Western Reserve University, USA.
[b] College of Energy, Beijing Institute of Chemical Technology, China
lxd115@case.edu

It is estimated that the world will need to double its energy supply by 2050. With the rapid increase in the global energy consumption, there is a pressing need for clean and renewable energy alternatives. Polymers have been traditionally used as electrically insulating materials: after all, metal wires are coated in plastics to insulate them. Various conjugated macromolecules with alternating single and double bonds can now be synthesized with unusual electrical and optical properties through the □-electron delocalization along their 1D backbones. Due to the molecular rigidity of conjugated backbones, however, most unfunctionalized conjugated polymers are intractable (i.e., insoluble and/or infusible). Nevertheless, a number of synthetic methods have been devised to produce conjugated polymers with the processing advantages of plastics and the optoelectronic properties of inorganic semiconductors for optoelectronic device applications, including polymer photovoltaic cells [1].

Having conjugated all-carbon structures, carbon nanomaterials, including 1D carbon nanotubes (CNTs) and 2D graphene, also possess certain similar optoelectronic characteristics as conjugated macromolecules, apart from their unique structures and associated properties (e.g., surface/size effects) [2]. With the rapid development in nanoscience and nanotechnology, graphitic carbon nanomaterials (e.g., 1D CNTs, 2D graphene) have been playing a more and more important role in the development of efficient energy conversion and storage devices, including solar cells, fuel cells, supercapacitors and batteries [2-6]. The combination of the unique physicochemical properties of graphitic carbon nanomaterials with comparable optoelectronic properties of appropriate conjugated macromolecules has yielded some interesting synergetic effects. Therefore, considerable efforts have recently been made to utilize graphitic carbon nanomaterials, along with polymers, as energy materials, and tremendous progress has been achieved in developing high-performance energy conversion and storage devices based on graphitic carbon nanomaterials and conjugated polymers [7]. More recently, some 2D conjugated polymers and certain 3D graphitic carbon architectures (e.g., CNT-graphene pillared networks, graphene foams) have been demonstrated to show additional advantages for efficient energy conversion and storage [8,9].

In this talk, I will summarize our work on rational design and development of multi-dimensional conjugated polymers and graphitic carbon nanomaterials for efficient energy conversion and storage, including polymer solar cells containing graphitic carbon nanomaterials for improved charge transport, fuel cells and metal-air batteries with carbon nanomaterials/polymers as metal-free catalysts for oxygen reduction and evolution, and flexible supercapacitors based on CNT-/graphene-based electrodes for energy storage. A brief overview of this exciting field, along with some challenges and opportunities, will also be presented.

References

(1) Dai L. "*Intelligent Macromolecules for Smart Devices*", Springer-Verlag: Berlin, 2004.
(2) Gong K, Du F, Xia Z, Dustock M, Dai L. *Science* **2009**, *323*, 760.
(3) Yu D, Zhang Q, Dai L, Chen Y. *et al.*, *Nat. Nanotechnol.* **2014**, *9*, 555.
(4) Dai L, Xue Y, Qu L, Choi H J, Baek J B. *Chem. Rev.* **2015**, *115*, 4823.
(5) Zhang J, Zhao Z, Xia Z, Dai L. *Nat Nanotechnol.* **2015**, *10*, 444.
(6) Shui J, Wang M, Du F, Dai L. *Sci. Adv.* **2015**, *1*, e1400129.
(7) Xu J, Chen Y, Dai L. *Nat. Commun.* **2015**, *6*, 8103.
(8) Dai L. *Acc. Chem. Res.* **2013**, *46*, 31.
(9) Lu W, Baek JB, Dai L. (Eds.) "Carbon Nanotechnology for Advanced Energy Systems", Wiley, 2015.

ENERGY EFFICIENT, ENVIRONMENTALLY RESPONSIVE BUILDING SKINS

Shu Yang

Materials Science and Engineering, University of Pennsylvania, 3231 Walnut Street, Philadelphia, PA 19104
shuyang@seas.upenn.edu

Residential and commercial buildings account for nearly 40% of the total energy consumed in the US, including 72% of the nation's electricity use and 39% of carbon dioxide emission each year through their direct influence on heating, cooling, and lighting. There have been tremendous interests in design of smart roofing, skylights, and architectural windows, which can block or reflect sunlight on scorching days, and return to a transparent state at a low lighting condition to improve light harvesting and capture free heat from the sun. Although shades and blinds have been used, windows are typically static. Here, I will discuss a new form of responsive building skins from nanoparticle assemblies in a solution or in a film. We can achieve dramatic and reversible visual effects between three states, that is color display, opaqueness, and transparency. By coupling the materials-environment response at the nano- and microscale with CMOS technology, we can modulate and (re)configure building performance about sunlight, heat, and motion with feedback control as the first step toward energy efficient building skins. Further, we can improve the thermal performance of buildings by reducing water condensation and geometric design of light reflective skins that can open, close, tilt, or shift shape via strategically placed cuts.

MECHANISM OF UV DEGRADATION OF METHYLAMMONIUM LEAD IODIDE PEROVSKITES

Jin Z. Zhang

Organo-metal perovskites are highly promising materials for photovoltaic applications, yet their degradation remains a significant problem and the mechanism behind is not yet well understood. This study examines the degradation mechanism of methylammonium lead iodide ($MAPbI_3$) perovskite films subjected to light in a dry environment from a structural perspective. The effect of mesoporous titanium dioxide (mp-TiO_2) layer, commonly used in device applications, on the degradation is also investigated. The perovskite structural change was investigated using X-Ray Diffraction (XRD), UV-Vis absorption spectroscopy, Extended X-ray Absorption Fine Structure (EXAFS) spectroscopy, and Fourier Transform Infrared (FT-IR) spectroscopy. The mp-TiO_2 plays a significant role in the degradation rate of the perovskite film. We observe an unconverted PbI_2 or related complexes of about 7% in the as-prepared films with mp-TiO_2. Our results indicate that the first step in the light-induced degradation process is the deprotonation of the methylammonium cations, which results in the evaporation of HI and CH_3NH_2 molecules leaving unreacted PbI_2 behind. In addition, we have explored new strategies to stabilize organo-metal perovskites using quantum dots as a model system and unique molecules or protonated C_3N_4 as ligands. The results support the model developed for UV induced degradation.

SYNTHESIS OF 1D, 2D, AND 3D STRUCTURES OF PHOSPHORUS

*Josh Smith, Dan Hagaman, Hai-Feng JI**

Dept. Chem. Drexel University, Philadelphia, PA 19104, USA
Presenter Contact Details: hj56@drexel.edu, 01-215-895-2562

Abstract

Phosphorus exists as four different allotropes; white, red, violet, and black. White phosphorus is a four-atom molecule in a tetrahedral structure (P4). In this work, we present the growth of 1D phosphorus wires, 2D phosphorus films, and 3D phosphorus hexagons. The different morphologies were characterized with TEM, Raman spectroscopy, and electrical characterization. These techniques confirmed successful growth of ultra-long 1D fibrous phosphorus, 2D black phosphorus, and 3D red phosphorus hexagons. These materials may find a wide variety of applications, such as batteries, supercapacitors, solar cells, LED, etc.

ENGINEERING OF INORGANIC SEMICONDUCTORS FOR SOLAR FUEL GENERATION

Nianqiang Wu

Department of Mechanical & Aerospace Engineering
West Virginia University
P.O. Box 6106
Morgantown, WV 26506
E-Mail: nick.wu@mail.wvu.edu

Photocatalysts and photoelectrochemical cells are typically built on semiconductors.[1] This presentation shows how to engineer inorganic semiconductors for solar fuel generation in terms of geometry, microstructure, band structure and heterostructure. The size of nanoparticles affects the surface active sites and the surface charge traps. Tailoring the shape and orientation of nanostructures can tune the light absorption and charge transport. Engineering band structure of semiconductors can enhance the light absorption, charge separation, charge mobility and suppress the charge recombination.[2-5] Formation of a heterojunction with dissimilar materials can compensate the weakness of individual materials and bring new merits to the photocatalytic systems.

Acknowledgment. This work was supported by the National Science Foundation (CBET-1233795) and the Army Research Laboratory (W911NF-14-2-0116).

References

(1) J. Li, N. Q. Wu, *Catal. Sci. Technol.*, **2015**, 5, 1360-1384.
(2) N. Q. Wu, J. Wang, D. Tafen, H. Wang., J.-G. Zheng, J. P. Lewis, X. Liu, S. S. Leonard, A. Manivannan, *Journal of American Chemical Society*, **2010**, 132, 6679–6685
(3) J. Wang, D. N. Tafen, J. P. Lewis, Z. Hong, A. Manivannan, M. Zhi, M. Li, N. Q. Wu, *Journal of the American Chemical Society*, 2009, 131, 12290-12227
(4) B. T. Yost, S. K. Cushing, F. Meng, J. Bright, D. A. Bas, N. Q. Wu and A. D. Bristow, *Physical Chemistry Chemical Physics*, **2015**, 17, 31039-31043.
(5) F. Meng, Z. Hong, J. Arndt, M. Li, M. Zhi, F. Yang, N. Q. Wu, *Nano Research*, **2012**, 5, 213-221.
(6) J. Zhang, W. Dang, Z. Ao, S. K. Cushing, N. Q. Wu, *Physical Chemistry Chemical Physics*, **2015**, 17, 8994 – 9000.

REVEALING PYROLYSIS CHEMISTRY IN COMPLEX SYSTEMS WITH LARGE SCALE ReaxFF MOLECULAR DYNAMICS

Xiaoxia Li [a,*], *Mo Zheng* [a], *Fengguang Nie* [a], *Xiaolong Liu* [a,b], *Mingjie Gao* [a,b], *Tingting Zhang* [a,b], *Li Guo* [a]

[a] State Key Laboratory of Multiphase Complex Systems, Institute of Process Engineering, Chinese Academy of Sciences, Beijing 100190, P. R. China
[b] University of Chinese Academy of Sciences, Beijing 100049, P. R. China
xxia@ipe.ac.cn

This paper reports the efforts of methodology development of ReaxFF MD for practical pyrolysis simulation of large organic models.1 The first graphics processing unit (GPU)-enabled code (GMD-Reax) was created, which has one magnitude speed up over the CPU ReaxFF MD codes available on desktop workstations.2 The first code for comprehensive chemical reaction analysis of ReaxFF MD simulations was developed to catch the overall species and chemical reactions,3 which is critical in understanding the complex pyrolysis reaction mechanisms of polymer, biomass and coal.

The methodology has been applied to investigate the complex pyrolysis chemistry of large scale molecular systems, including HDPE4, cellulose5 and lignin, and real coal models for Hailaer brown coal, Fugu sub-bituminous coal and Liulin bituminus coal.6 The evolution tendencies of overall spectrum product with temperature are in broad agreement with various experimental observations in literatures and Py-GC/MS experiments. With this methodology, the overall product evolution and underlying reactions in the pyrolysis are readily 'observable' computationally that is hardly accessible by experiments or other computational methods.

Acknowledgements

This work was supported by NSFC (21373227, 21073195 and 91434105).

References

(1) Li, X.; Mo, Z.; Liu, J.; Guo, L., Revealing chemical reactions of coal pyrolysis with GPU-enabled ReaxFF molecular dynamics and cheminformatics analysis. *Molecular Simulation* **2015,** *41* (1-3), 13-27.

(2) Zheng, M.; Li, X.; Guo, L., Algorithms of GPU-enabled reactive force field (ReaxFF) molecular dynamics. *Journal of Molecular Graphics & Modelling* **2013,** *41*, 1-11.

(3) Liu, J.; Li, X.; Guo, L.; Zheng, M.; Han, J.; Yuan, X.; Nie, F.; Liu, X., Reaction analysis and visualization of ReaxFF molecular dynamics simulations. *Journal of Molecular Graphics and Modelling* **2014,** *53* (Sept.), 13-22.

(4) Liu, X.; Li, X.; Liu, J.; Wang, Z.; Kong, B.; Gong, X.; Yang, X.; Lin, W.; Guo, L., Study of high density polyethylene (HDPE) pyrolysis with reactive molecular dynamics. *Polymer Degradation and Stability* **2014,** *104* (6), 62-70.

(5) Zheng, M.; Wang, Z.; Li, X.; Qiao, X.; Song, W.; Guo, L., Initial reaction mechanisms of cellulose pyrolysis revealed by ReaxFF molecular dynamics. *Fuel* **2016,** *177*, 130-141.

(6) Zheng, M.; Li, X.; Liu, J.; Wang, Z.; Gong, X.; Guo, L.; Song, W., Pyrolysis of Liulin Coal Simulated by GPU-Based ReaxFF MD with Cheminformatics Analysis. *Energy & Fuels* **2014,** *28* (1), 522-534.

NANOSTRUCTURED MATERIALS: SYNTHESIS BY WET CHEMICAL AND PULSED LASER ABLATION APPROACHES AND APPLICATIONS IN CATALYSIS AND SOLAR ENERGY

Dongling Ma

Institut National de la Recherche Scientifique (INRS)
Center of Energy, Materials and Telecommunications
1650 Boulevard Lionel-Boulet, Varennes, Quebec, Canada
ma@emt.inrs.ca

Wet chemical approaches have been widely explored for the synthesis of a variety of nanomaterials of different sizes and shapes. On the other hand, nanofabrication based on advanced laser techniques is becoming a popular top-down approach for preparing metal and semiconductor nanoparticles (NPs). In this presentation, I will introduce some of our work on the preparation of metal NPs (such as Ni@Ru core@shell NPs, Au and AuPt alloy NPs) and semiconductor quantum dots (such as PbS and PbS@CdS core@shell QDs) by pulsed laser ablation (in solution or in gas phase) and/or wet chemical method, as well as their applications in catalysis, solar cells and luminescent solar concentrators [1-12]. For example, stable colloidal Au or AuPt NPs were fabricated via pulsed laser ablation in water, which led to almost naked surface that is highly beneficial for catalytic applications. They were further combined with various supports for catalytic investigations, such as nitrophenol reduction and CO oxidation reaction and exhibited remarkably higher catalytic activity in comparison to similar catalysts composed of chemically prepared Au NPs. Recently, we further applied the laser-Au NPs, showing strong surface plasmon resonance, for plasmon enhanced photocatalysis, including water splitting. Another example is about Ni@Ru core@shell NPs, which are hard to achieve with the laser approach, but can be synthesized by a wet-chemistry method. These Ni@Ru NPs showed remarkable performance (partially due to the synergistic effect) in the hydrolysis of ammonia borane and magnetic recyclability. Hollow Ce-MnOx binary nanotubes represent another example, showing strong synergistic effects. They exhibited excellent catalytic activity in CO oxidation and adsorption performance for water treatment. I will also present the use of QDs in luminescent solar concentrators and solar cells.

Acknowledgement. We thank the financial support from the Natural Sciences and Engineering Research Council of Canada, Fonds de recherche du Québec-Nature et technologies, Quebec Center for Functional Materials (CQMF), and strategic Network "Plasma-Québec".

References

(1) Xu, Z.; Quintanilla, M; Vetrone, F.; Govorov, A. O.; Chaker, M.; and Ma, D. *Adv. Funct. Mater.* **2015**, *25*, 2950 (**Front Cover**).
(2) Liang, H.; Rossouw, D.; Zhao, H.; Cushing, S. K.; Shi, H.; Korinek, A. ; Xu, H.; Rosei, F.; Wang, W.; Wu, N.; Botton, G. A.; and Ma, D. *J. Am. Chem. Soc.* **2013**, *135*, 9616.
(3) Li, S.; Zhang, J.; Kibria, M. G.; Mi, Z.; Chaker, M.; Ma, D.;* Nechache, R..;* and Rosei, F.* *Chem. Commun.* **2013**, *49*, 5856.
(4) Elbohy, H.; Kim, M.; Dubey, A.; Reza, K. M.; Ma, D.;*, Zai, J.;* Qian, X.;* and Qiao, Q.;* *J. Mater. Chem. A* **2016**, *4*, 545.
(5) Gonfa, B.; Kim, M.; Delegan, N.; Tavares, A. C.; Izquierdo, R.; Wu, N.; El Khakani, M. A.; and Ma, D. *Nanoscale* **2015**, *7*, 10039.
(6) Ka, I.; Gonfa, B.; Le Borgne, V.; Ma, D.;* El Khakani, M. A.* *Adv. Funct. Mater.* **2014**, *24*, 4042 (***Inside Front Cover***).
(7) G. Chen, F. Rosei, D. Ma, *Adv. Funct. Mater.* **2012**, *22*, 3914-3920; ***Highlighted by MaterialsViews.***
(8) J. Zhang, G. Chen, D. Guay, M. Chaker, D. Ma, *Nanoscale* **2014**, *6*, 2125-2130 (***Back Cover***).
(9) Y. Zhou *et al.*, Adv. Energy Mater. 2016, accepted.
(10) A. Benayas, *et al.*, *Adv. Funct. Mater.* **2015**, *42*, 6650.
(11) D. Wang, *et al.*, *Adv. Funct. Mater.* **2011**, *21*, 4010-4018.
(12) Liang, H.; Zhao, H.; Rossouw, D.; Wang, W.; Xu, H.; Botton, G. A.; and Ma, D. *Chem. Mater.* **2012**, *24*, 2339.

STUDY OF ANAEROBIC BIODEGRADABILITY OF RICE STRAW HYDROTHERMAL LIQUEFACTION PRODUCTS: AQUEOUS PRODUCTS AFTER EXTRACTION WITH DIFFERENT ORGANIC SOLVENTS

Huihui Chen, Gang Luo, Shicheng Zhang*, Jianmin Chen*

Shanghai Key Laboratory of Atmospheric Particle Pollution and Prevention (LAP[3]), Department of Environmental Science and Engineering, Fudan University, Shanghai 200433, China
Email: zhangsc@fudan.edu.cn

1. Introduction

Organic substrates of this lignocellulosic biomass can be converted into bio-crude through HTL meanwhile a large amount of wastewater containing various nutrients will be produced in the process [1].Nonetheless, as bio-crude oil is the most desired product from the HTL reaction and most of the previous studies were primarily focused on the characteristics and potential uses of the bio-crude oil product. Relatively, little attention has been paid to the HTL aqueous phase (HTL-ap), even though it accounts for a significant fraction of the organics [2, 3]. Typical yields of the liquid products (raw products, RP) in the direct liquefaction of biomass with near-and subcritical water are in the range of 20–50%.Efficient utilization of the aqueous co-product is a key factor affecting the overall sustainability of the process. The aqueous phase can be used in either gasification or cultivation medium in those few instances [4, 5]. It has also been proposed that an anaerobic fermentation step could convert organic carbon in the aqueous to biogas [6].

This process design integrates a thermo-chemical (hydrothermal liquefaction-HTL) rice straw conversion and a biochemical (biogas plant–BP) process. In this integrated biorefining manner, the compositions of aqueous phases from rice straw HTL were examined and their potential utility as a substrate for methane fermentation were also investigated. This study was expected to contribute to integrating biomass reuse, bio-oil production and waste water purification process. The end-products of the integrated process are a range of biofuels (biocrude, biochemicals and biogas).

2. Experimental

2.1. HTL-aps and methane production The aqueous phase was obtained from a pilot-plant for hydrothermal liquefaction. 3.0 kg of minced rice straw mixed with 39.0 kg of water were added into the reactor and then heated to 280°C at 12.0 MPa for 30 min. Five different aqueous phases were prepared before anaerobic digestion. They were HTL raw aqueous products (RP-ap), raw aqueous phase after extraction with petroleum ether (PE-ap), cyclohexane (CH-ap), dichloromethane (DM-ap) and ethyl acetate (EA-ap), respectively. The anaerobic biodegradability test for methane production was conducted in the interval flasks. The total volume was 118 mL with 15 ml seed sludge, 45 ml of the dilute aqueous phase and 58 ml headspace volume at 37 °C.

2.2. Analytical methods The organic matter concentration expressed as total organic carbon (TOC) and chemical oxygen demand (COD). To explain the differences, the organic and water phases with or without extraction before and after anaerobic digestion were analyzed by GC-MS, HPLC, FTIR and LC-MS technologies.

3. Results and Discussion

3.1 Results

3.1.1 Aqueous COD and TOC value The COD and TOC values of the five aqueous phases RP-ap, PE-ap, CH-ap, DM-ap and EA-ap were showed in Table1. The saturated solubility (in COD values) of the extractants PE, CH, DM, EA in water were 0.09, 0.22, 7.95, 146.35 g/l respectively in Table 1. The aqueous COD values were not only determined by the compositions and their content left in aqueous phase but also the solubility of the agent itself.

Table 1. The aqueous phases TOC and COD values

	Agent polarity	Saturated solvent COD(g/l)	Aqueous COD (g/l)	Extracting percent (%)	TOC (g/l)
RP-ap			20.74		5.57
PE-ap	0.01	0.09	15.99	23.33	4.33
CH-ap	0.2	0.22	17.63	16.06	4.53
DM-ap	3.1	7.95	13.28	74.3	3.53
EA-ap	4.4	146.35	148.9	87.7	40

3.1.2 Methane production Figure 1A show the methane yield trend of the five aqueous. The five aqueous methane production differ significantly with digestion time increase. EA-ap was the easiest aqueous to be anaerobic digested and 89.2% organics changed into methane after 6 days fermentation while the PE-ap, CH-ap and RP-ap fermentation variation trend were gently. Though EA had a high extracting efficiency it also had the highest solubility and most of the COD was contributed by the EA agent itself thus it can be easily converted and the maximum methane yield was 92.6%. For DM-ap, hardly any methane detected until the experiment terminated. Although DM has higher extracting efficiency and not very high solubility, the use of DM to extract some useful chemicals as pre-process doesn't work as chlorinated compounds are toxic to the microbes and even a very low dose can seriously inhibited microbial activities.

For the case of PE-ap, CH-ap aqueous, both of their methane yield were improved after extracted with the agents compared with RP-ap. At beginning 7 days, the difference were very tiny and then the superiority of aqueous extracted with PE and CH emerged. A slightly extracting efficiency change 16.07% of CH and 23.33% of PE brought a methane yield improvement from 51.4% to 54.8% and 67.4% respectively.

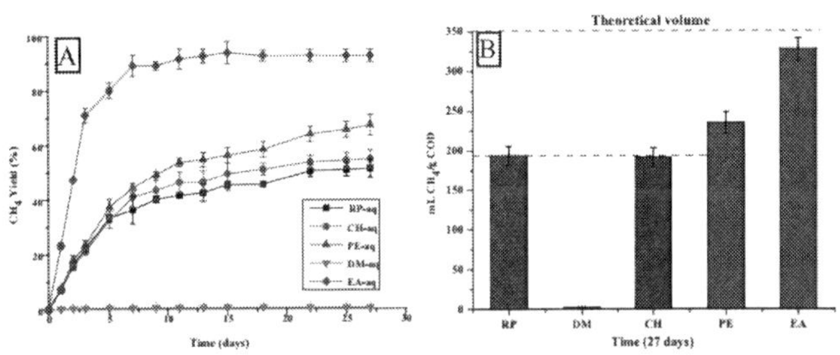

Figure1. Five aqueous methane production trend (A) and methane volume per gram COD (B) after 27 days of RP-ap, PE-ap, CH-ap, DM-ap and EA-ap in batch reactor at 37 °C with 0.75g/L COD loading.

Figure 1B shows the methane yield per gram COD of the five aqueous after 27 days fermentation. The five aqueous RP-ap, PE-ap, CH-ap, DM-ap, EA-ap had a methane yield 194, 2, 193, 235, 328 ml/g COD respectively while the theoretical yield is 350 ml/g COD in standard conditions.

3.2 Discussion

As PE and CH have lower polarity and lower solubility compared with DM and EA agent. For PE-ap, and CH-ap, the two weak polar solvents could extract most parts of the weak polar components and weaker the solvent was, much it extracted the weak polar components and smaller its COD value would be. Though DM had a higher extracting percentage, the COD value was still 13.28 g/l as DM solute during the extracting process contribute 7.95 g/l COD while PE solvent had a significant effect as much extracting agent solute in the EA-ap and the COD reached as much as 148.90 g/l. The TOC value has the same variation trend with COD value's.

To know what kinds of compounds could be easily extracted from the raw aqueous phase and what kinds of compounds were easily digested by methnogenic bacteria, eighteen kinds of three classes standard mixture solutions were prepared. Table 2 shows eighteen primary compounds in the RP-ap, PE-ap, CH-ap, DM-ap and EA-ap. Compounds left in the aqueous after extraction decreased with the increase of solvent polarity. EA has the highest polarity and the lowest compounds amount in EA-ap and followed by DM-ap. For DM-ap, about half of the acid compound still retained in the aqueous but most ketones and phenols were extracted. Most of the acid compounds like lactic acid, acetic acid and propionic acid were still in the aqueous after extraction with PE and CH but there were sightly difference of ketones and phenols amount in PE-ap and CH-ap.

Table 2. shows the HPLC characterization result of RP-ap, PE-ap, CH-ap, DM-ap and EA-ap

Name	RP-ap mg/l	PE-ap mg/l	CH-ap mg/l	DM-ap mg/l	EA-ap mg/l
Lactic acid	3722	3708	3698	1628	592
Acetic acid	1802	1792	1782	740	–
Propionic acid	680	657	657	399	–
N-butyric acid	281	289	260	59	–
Iisovaleric acid	146	142	133	–	–
Levulinic acid	291	293	285	110	–
2,5-hexanedione	242	177	208	52	20
2-hydroxy-3-methyl-2-cyclopenten-1-one	74	45	64	31	1
2-hydroxy-3-ethyl-2-cyclopenten-1-one	268	143	185	–	–
2-methyl-2-cyclopenten-1-one	366	203	260	–	–
Ethyl-2-hydroxy-2-cyclopenten-1-one	20	10	13	–	–
Acetosyringone	71	45	57	–	–
Phenol	95	60	75	–	–
2-methoxy-phenol	155	22	41	–	–
2,6-dimethoxy-phenol	68	38	39	–	–
2-methoxy-4-methyl-phenol	12	–	–	–	–
4-ethyl-phenol	21	–	–	–	–
4-ethyl-2-methoxy-phenol	25	–	–	–	–

It was perhaps important to point out that even though there was very tiny difference of compounds amount left in PE-ap and CH-ap whose COD values were 15.99 g/l and 17.63 g/l respectively, it maybe just this minor difference (1.64 g/l change in COD value) lead the methane yield improved from 55.4% to 67.3%. To identify and remove this compounds may help optimize aqueous microbial utilization to a great extent.

At the end of the fermentation, CH-ap had no methane yield improvement as PE-ap had about 40 ml/g COD enhancement and the difference of the two extract agents were their abilities to extract weak polar compounds and maybe low extracting efficiency had almost no influence on the so complex raw aqueous products. Though EA-ap can be easily converted to methane and the final yield was very close to theoretical value but the COD value was much higher than the RP-ap and it needed to be diluted severely.

We can concluded that the difference were not because of the volatile organic acid products as the acid were all still retained in the aqueous in PE-ap, CH-ap characterized by HPLC in Table 2. One most likely reason of the methane yield difference of the two aqueous were the agents' abilities to extract the phenols and ketones and other weak polar compounds. As PE indeed extracted much of phenols and ketones and maybe it was just this tiny differences promoted the anaerobic digestion and thus we may conclude that the phenols and ketones are critical components that influence fermentation to produce methane.

Conclusions

The effects of the four different organic extraction agents on aqueous methane production show significant differences. Almost 67.4% COD in the aqueous phase after extraction with PE changed into CH_4 and the CH-ap and EA-ap aqueous phases were 54.8% and 93.84% while the COD utilization percentage of HTL-ap without any extraction was 51.4%, at the same time, the DM-ap was completely inhibited by the dichloromethane dissolved during the extraction process. One most likely reason of the methane yield difference of the two aqueous were the agents' abilities to extract the phenols and ketones and other weak polar compounds as PE indeed extracted much of phenols and ketones.

Acknowledgement. The authors are thankful for the financial support from the National Key Technology Support Program (No. 2015BAD15B06), the Shanghai Talent Development Fund (No. 201414) and National Natural Science Foundation of China (51408133).

References

(1) Gai, C; Zhang, Y.H.; Chen, W.T.; Zhou, Y.; Schideman, L.; Zhang, P.; Tommaso, G.; Kuo, C.T.; Dong, Y.P. *Bioresour. Technol.*, ***2015***, *184*, 328.
(2) Tommaso, G.; Chen, W.T.; Li, P.; Schideman, L.; Zhang, Y.H. *Bioresour. Technol.*, ***2015***, *178*, 139.
(3) Nelson, M.; Zhu, L.; Thiel, A.; Wu, Y.; Guan, M.; Minty, J.; Wang, H.Y.; Lin, X.N. *Bioresour. Technol.*, ***2013***, *136:* 522.
(4) Orfield N.D., Fang A.J., Valdez P.J., Nelson M.C., Savage P.E., Lin X.N. *Sustain. Chem. Eng.* ***2014***, *2*, 867.
(5) Frank E.D.; Elgowainy A.; Han J.; Wang Z.C.; *Mitig. Adapt. Strateg. Glob. Change* **2013**, *18*, 137.
(6) Davis, R.; Aden, A.; Pienkos, P.T. *Appl. Energy* ***2011***, *88 (10)*, 3524.

LIFE CYCLE ASSESSMENT OF BIOCHEMICAL AND THERMOCHEMICAL BIO-PRODUCT PATHWAYS

Sabrina Spatari[1], Valeria Larnaudie[2], Long Nguyen[1], Yetunde Sorunmu[1], Pieter Billen[1]

[1]Civil, Architectural and Environmental Engineering, Drexel University, Philadelphia, PA, USA
[2] Depto. Bioingeniería,Facultad de Ingeniería, Universidad de la República, Montevideo, Uruguay
Ss3226@drexel.edu

Abstract

Biofuels are currently under development to meet federal and state goals for diversifying energy supply, reducing the carbon-intensity of transportation and other sectors, and stimulating economic growth in rural parts of the U.S. Policy instruments such as the federal Renewable Fuel Standard (RFS2) have been signed into law to stimulate the development of domestic renewable fuels and address climate change from those energy sectors. At present biological and thermochemical technologies are under development at laboratory and pilot scale to investigate the technological needs of scaling biofuels and value-added co-products. Ethanol, higher alcohols, and fully infrastructure compatible fuels (with 0% oxygen – physically and chemically similar to current petroleum based fuels) are being developed at different scales, including farm scale (up to 200 dry metric tons/day) and industrial scale (2000 dry metric tons/day and higher) to utilize diverse lignocellulosic sources. Life cycle assessment (LCA) is a method for evaluating the environmental performance of biofuels and biomaterials emerging through R&D, and a necessary tool for developing and judging the compliance of those biofuels under state and federal policy standards. This presentation investigates technological, environmental, and energy and resource impacts of a set of emerging lignocellulose-based biofuels from biological and thermochemical technologies at different scales using statistically based LCA methods. Uncertainties that stem from technological performance in the near and medium terms are characterized. This presentation will further highlight how analytical models for understanding the full set of environmental tradeoffs associated with emerging fuel technologies are essential inputs to guiding policy and commercial enterprise decision making for improving the overall sustainability of transportation energy.

GRAPHENE-ENABLED NANO/BIO HYBRIDS FOR CHEMICAL DETECTION AND MEDICAL DIAGNOSTICS

A.T. Charlie Johnson

Department of Physics and Astronomy, University of Pennsylvania, Philadelphia PA USA
cjohnson@physics.upenn.edu

Abstract

We have explored all-electronic chemical detectors based on bio-nano hybrids, where the biomolecule (DNA or protein) provides chemical recognition and a carbon nanotube (NT), graphene, or monolayer molybdenum disulfide (MoS_2) transistor enables electronic readout. This sensor class represents a promising approach towards sensitive and selective detection of liquid- and vapor-phase anayltes. Rapid advances in nanomaterials research and biomolecular engineering offer the prospect of optimizing the performance of each component of the hybrid to yield sensors with enhanced sensitivity and specificity for arbitrary chemical targets. Coupling chemistries have been developed that allow for the creation of a nanoelectronic and/or nanophotonic interface to a variety of different proteins including: antibodies[1,2] and antibody fragments[3], optically active proteins[4], G-protein coupled receptors (GPCRs) housed in nanoscale membrane analogues[5], and GPCR variants that are engineered for enhanced solubility and stability in aqueous solution[6]. We demonstrated the use of these bio-nano hybrids for detection of protein cancer biomarkers, antigen from various pathogens, and small molecule receptor targets at concentrations ~ 1 pg/mL. Single stranded DNA can be coupled to nanotubes and graphene through non-covalent functionalization[7]. The DNA is used for its chemical recognition for small molecule analytes rather than recognition of complementary DNA. Vapor sensors based on this approach were able to discriminate between highly similar compounds such as enantiomers[8], as well as very similar complex mixtures of vapors characteristic of humans[9,10]. More recently we have shown the promise of this system for diagnosis of disease based on volatile biomarkers (unpublished).

Acknowledgement. The work to be discussed in this talk was supported by DARPA and the US Army Research Office (W911NF1010093), the National Science Foundation (AIR ENG-1312202 and EFRI 2DARE EFMA-1542879), and Lockheed Martin.

References

(1) Lerner, M. B.; Goldsmith, B. R.; McMillon, R.; Dailey, J.; Pillai, S.; Singh, S. R.; Johnson, A. T. C., *AIP Advances* **2011,** *1* (4), 042147.
(2) Lerner, M. B.; Dailey, J.; Goldsmith, B. R.; Brisson, D.; Charlie Johnson, A. T., *Biosensors and Bioelectronics* **2013,** *45*, 163-167.
(3) Lerner, M. B.; D'Souza, J.; Pazina, T.; Dailey, J.; Goldsmith, B. R.; Robinson, M. K.; Johnson, A. T. C., *ACS Nano* **2012,** *6* (6), 5143-5149.
(4) Lu, Y.; Lerner, M. B.; John Qi, Z.; Mitala, J. J.; Hsien Lim, J.; Discher, B. M.; Charlie Johnson, A. T., *Appl. Phys. Lett.* **2012,** *100*, 033110.
(5) Goldsmith, B. R.; Mitala, J. J.; Josue, J.; Castro, A.; Lerner, M. B.; Bayburt, T. H.; Khamis, S. M.; Jones, R. A.; Brand, J. G.; Sligar, S. G.; Luetje, C. W.; Gelperin, A.; Rhodes, P. A.; Discher, B. M.; Johnson, A. T. C., *ACS Nano* **2011,** *5*, 5408-5416.
(6) Lerner, M. B.; Matsunaga, F.; Han, G. H.; Hong, S. J.; Xi, J.; Crook, A.; Perez-Aguilar, J. M.; Park, Y. W.; Saven, J. G.; Liu, R.; Johnson, A. T. C., *Nano Lett.* **2014,** *14* (5), 2709-2714.
(7) Staii, C.; Chen, M.; Gelperin, A.; Johnson, A. T., *Nano Lett.* **2005,** *5* (9), 1774-1778.
(8) Khamis, S. M.; Jones, R. A.; Johnson, A. T. C.; Preti, G.; Kwak, J.; Gelperin, A., *AIP Advances* **2012,** *2*, 022110.
(9) Kybert, N. J.; Lerner, M. B.; Yodh, J. S.; Preti, G.; Johnson, A. T., Differentiation of Complex Vapor Mixtures Using Versatile DNA-Carbon Nanotube Chemical Sensor Arrays. *ACS Nano* **2013,** *7* (3), 2800-2807.
(10) Kybert, N. J.; Egan, L.; Waldman, R. Z.; Zeng, X.-N.; Krein, M.; Preti, G.; Stuart, J. A.; Johnson, A. T. C., Analysis of sweat simulant mixtures using multiplexed arrays of DNA-carbon nanotube vapor sensors. *Journal of Forensic Science and Criminology* **2014,** *1*.

BICONTINUOUS BIPHASIC LIQUID MEDIA FOR CONTINUOUS REACTIVE SEPARATIONS

*Martin F. Haase, Nima Sharifi-Mood, Kathleen J. Stebe, Daeyeon Lee**

Department of Chemical and Biomolecular Engineering, University of Pennsylvania, Philadelphia, PA 19104
*daeyeon@seas.upenn.edu

A recent approach that has shown great promise for biomass conversion and biofuel upgrade reaction involves using biphasic emulsions stabilized with interfacially active catalytic nanoparticles. This novel methodology combines the advantages of phase transfer and heterogeneous catalysis. One drawback that limits the widespread utilization of this approach, however, is that catalyzed reaction and separation cannot be performed in a continuous mode. To enable continuous reactions and separations in a biphasic liquid mixture, we explore the formation of bicontinuous interfacially jammed emulsions (Bijels). Bijels are typically prepared by arresting spinodal decomposition of two fluid phases via interfacial attachment and jamming of colloidal particles. Their fabrication is currently limited to a few pairs of immiscible liquids, and bicontinuous morphology can be obtained within narrow composition and temperature windows, potentially limiting their use in reactions involving a specific combination of reaction/separation media, catalytic nanoparticles and reaction temperature. We explore the formation of bijels by solvent transfer-induced phase separation (STRIP) of ternary mixtures. The use of commercially available silica nanoparticles and ionic surfactants allows us to continuously form bijel fibers with controllable morphologies and domain sizes down to a few hundreds of nanometers. We study the dependence of fiber morphology on different control parameters such as particle and surfactant concentrations. We also develop a new in situ technique to characterize the mechanical properties of bijel fibers during different stages of their maturation. The potential of generating bijels that can be used in reactive separations for biofuel upgrade will be discussed.

RESEARCH ON MORPHOLOGY GENETIC MATERIALS TEMPLATED FROM NATURE SPECIES

Di Zhang

Department of State Key Laboratory of Metal Matrix Composites
Shanghai Jiao Tong University
800 Dongchuan Road. Shanghai 200240, P. R. China
E-mail: zhangdi@sjtu.edu.cn

Introduction

Morphology genetic materials are promising candidates for the utilization of natural evolution wisdom with the inheritance of the morphology of their hierarchical structures. The microstructure characteristics of the natural species have been optimized by natural selection for millions of years and can offer special functions that are much more effective than solutions built up by human beings. In addition, the final designed functional materials have more special functions that natural species cannot offer. The combination between the nature-born morphologies and the desired functional materials makes these novel materials competent for a broad range of applications. This review focuses on the use of several selected bio-templates which have been intensively researched recently, such as butterfly wings, wood tissues, plant leaves, diatoms etc.. Specific functions inherited both from the natural microstructures and coupled functional materials are highlighted in various applications, including photonic materials, solar energy applications, electromagnetic shielding, and environment-friendly materials.

Results and Discussion

There are improving understandings of the natural species since Charles Darwin published his epochal book On the Origin of Species in late 1859[1]. In his view, plants and animals on this planet have developed and have been developing complicated and precise structures, functions, and strategies, to survive themselves through a severe natural selection process. For centuries, we keep on studying and imitating the structures and behaviors of our natural-born neighbors and started a promising area known as Bionics, or Biomimetics.

However, simple imitation quickly met its bottle-neck. As researchers tried to design and fabricate similar structures or functions evolved from nature, it might be somehow awkward as compared with those natural gene-engineers. This happened especially when concerning the microcosmic scale with a rapid development in modern Nano science. For example, people have already realized that some of the butterfly wing scales are natural photonic crystals[2], an attractive term since twenty years ago, which are depicted as tiny optical devices that can control the propagation of a light beam. Nature could spontaneously and easily generate millions of these micrometer size scales for one butterfly wing controlled by their inner genes, with each composed of a complicated three dimensional (3D) nano to sub-micrometer structures. Such precise 3D structures are however unfortunately far beyond our present fabrication abilities to imitate either through a bottom-up self-organization method, or through a top-down photolithography process. This gave researchers a big push to find some new and efficient ways to effectively prepare these biomimetic microcosmic structures.

More than ten years ago, a family of novel materials known as Morphology Genetic Materials (MGM) was developed [3], which is a combination between "morphology" and "genetic", that is, the inheritance of the morphology. MGM keeps its original bio-template's structures nearly integrally, and replaces the organism components by designed functional materials. On one hand, the original species' morphologies have been optimized by natural selection for millions of years and can offer special functions that are much more effective than the existing solutions built up by human beings. On the other hand, the final designed functional components have some special functions that natural organisms cannot offer. The combination between the nature-born morphologies and the desired functional materials makes these novel materials competent for many applications.

On the technical level, the key point is how to explore such an approach. In contrast to a natural fossilization process that usually takes millions of years, an operable fossilization method should be finished in hours or days in the laboratory. Thus, several quite quick and easy methods have been developed to fabricate those functional fossils. Generally, the original organism structures are soaked into solutions containing the ions combinations that are the target composition of the final functional materials. This process is similar to the natural approach, like burying the ancient plants or animal bodies to infiltrate the mineral contents inside. To speed up the artificial fossilization process, heat or light or ultrasound are used during the synthesis. Through a careful adjustment of the fabrication parameters, the final morphologies could be controlled in a precise manner.

As to be effective, it not only means that something is easy to be fabricated, but also means an effective route from design to product. For example, the natural templates are chosen with great diversities (including several main natural components, such as chitin, cellulose and keratin), which means once a fabrication route was developed and established, bunches of various functional materials could be obtained using the same method. To avoid making the natural species endangered or extinct, all the natural templates chose are abundant and easy-to-raise. The applications of MGM will cover a wide range of fields, including photonic materials, solar energy applications and environment-friendly materials, *etc.* Figure 1 shows the collection of these various MGM types and applications. In the following part, a general view of this exciting new-born area of study will be depicted.

In the past ten years, our group have developed systematical fabrication methods of various functional materials based on three kinds of biotemplates: butterfly wings composed of chitin, plants composed of cellulose and diatoms composed of ingredient silica. The applications of these functional materials into various areas were also thoroughly investigated. In nature, many different kinds of butterflies have evolved to create highly efficient and complex photonic structures embedded into their wings. Using the MGM concept, the merits of two science (biology and materials) are combined in a synergistic approach by merging unique structural engineering in biology with state-of-the-art materials synthesis and processing. Several successful MGM samples using 3D photonic structures from butterfly wing scals as unique molds for replication into metal or inorganic compounds. [4-12] The metal materials templated from butterfly wings possess high density of plasmonic hotspots and demonstrate ulthrahigh plasmonic enhancement in SERS detection of organic molecules. In addition, the as-required functional materials with inorganic compounds have witnesssed their broad applications as structural color materials, light absorbers, responsive photonic crystals, gas sensors and so on. While the MGM methods can produce wings replicas with high fidelity, even preserving nanoscale fine-features of characteristics on the ridges, it should be noted that the inherent limitation to MGM. For example, the size of these MGM are restricted by the size of wing scales in a few hundred micrometres in size. So recently, using the Fe@Carbon MGM of wing scales as a second imprinting template, massive biomimetic butterfly wing scale structures were reproduced [13]. That is the first step in breaking out the size limitation of MGM, although the second imprinting microstructure is not as perfect as the original scale.

Bearing the enormous pressure of the environment and energy, the most feasible solution will be the recycling of these biomass to bio-energy and bio-materials. [14, 15] So the MGM concept and process will supply a potentical solution for this issue. Besides the plant leaves, wood, bamboo etc., all the lignocellulose-based biomass could be utilizated for various applications. Moreover, these MGM will be the base for the further activation and modification. In that case, the advantage of the MGM prepared together with the large supply of the natural plants, enable it a promising method for industry applications.

In addition to the two most important components, *i.e.* chitin and cellulose, the orther natural species with different components have been developed for specific application. One typical natural materials is diatom, with mainly ingredient silica), which could be applied for energy and enviroment applications. Diatoms are single-called photosynthetic microbe which can convert the solar energy into energy-rich compounds with highest efficiency.[16] Diatomite is a powerdery, non-metallic mineral, and consists of the fossilized skeletal remains of diatoms. It could be modified for the synthesis of ZnFeO4/SiO2 composite which may have applications as electromagnetic wave absorbers.[17] They are also famous for the unique biosilica cell walls (frustules) with hierarchical porous structures, which can be used to fabricate a broad range of advanced materials[18, 19]. Besides the fossilized diatomite, the living diatoms could also be used as the bio-templates for MGM fabrication. The porous metal chalcogenide MGM from one kind of the living diatom, *Coscinodiscus lineatus* was generated.[20] Similar optical properties were observed both in the living diatom and the metal chalcogenide MGM. The living diatoms are efficient energy converter, which consume 20% of global CO_2 through photosynthesis.[21] So they not only performs great optical characteristics, but also are very efficient in the greenhouse gas control. By combining with some metal-organic frameworks(MOFs), it was find that frustules, which were previously considered inert to CO_2, can be efficient CO_2 absorbents.

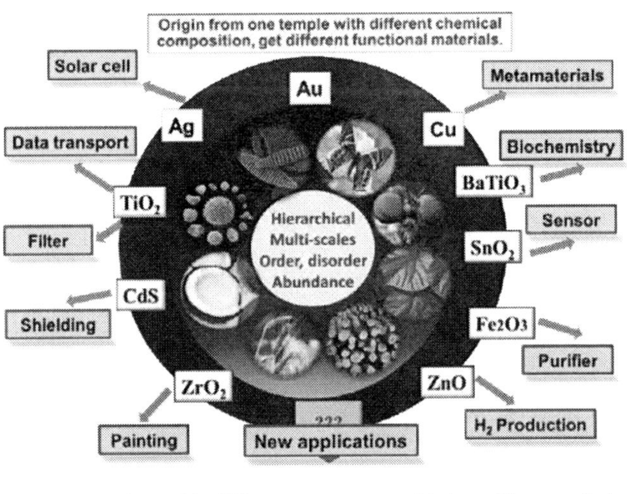

Figure 1. MGM with different components fabricated by templating from various natural species possessing hierarchical multi-scale structures, and their applications into different areas.

Conclusions

As a kind of fabrication process of sophisticated hierarchical structure of functional materials, the concept of MGM could be expanded to other kinds of natural materials (regardless the main component is cellulose or chitin or keratin). More and more efforts have been contributed into the current interdisciplinary areas, but two important concerns are still very worthy to be studied in-depth. The first one is the resolution and refining of key nano-structures of these nature materials. Assisted by the mathematic simulation and optimization, some of the key parameters of the microstructures will be proposed and highlighted for the subsequent research. The second one is to look for the most matching components for each specific microstructure for proper use. Based on the above research, these MGM will open a shortcut toward achieving the novel functional materials integrated with structural characteristics. The fabrication method may be applied to other nature substrate template and inorganic systems that could eventually lead to the production of optical, magnetic or electric devices or components as building blocks for nanoelectronic, magnetic, or photonic integrated systems. We envision that these small-scale, proof-of-principle mechanism studies can lend experimental insights into future optical, energy and environment applications, and provide the foundation for an entirely new class of advanced materials.

Acknowledgement. This work was supported by the National Natural Science Foundation of China (no. 51572169), Shanghai Rising-Star Program (16QA1402400), Shanghai Science and Technology Committee (15ZR1422400, 14JC1403300 and 14520710100), the National Basic Research Program of China (973 Program, no. 2011CB922200), Research Fund for the Doctoral Program of Higher Education of China (20120073120006 and 20120073130001).

References

(1) C. Darwin, *On the Origin of Species*, John Murray, United Kingdom **1859**.
(2) L. P. Biro, J.-P. Vigneron, *Laser & Photonics Reviews* **2011**, *5*, 27.
(3) D. Zhang, *Morphology Genetic Materials Templated from Nature Species*, Springer, **2011**.
(4) Y. Tan, J. Gu, X. Zang, W. Xu, K. Shi, L. Xu, D. Zhang, Angew. Chem.-Int. Edit. **2011**, 50, 8307.
(5) Y. Tan, J. Gu, L. Xu, X. Zang, D. Liu, W. Zhang, Q. Liu, S. Zhu, H. Su, C. Feng, G. Fan, D. Zhang, Adv. Funct. Mater. **2012**, 22, 1578.
(6) X. Zang, Y. Tan, Z. Lv, J. Gu, D. Zhang, Sens. Actuator B-Chem. **2012**, 166-167, 824.
(7) W. Zhang, D. Zhang, T. Fan, J. Ding, Q. Guo, H. Ogawa, Nanotechnology **2006**, 17, 840.
(8) Y. Tan, X. Zang, J. Gu, D. Liu, S. Zhu, H. Su, C. Feng, Q. Liu, W. M. Lau, W.-J. Moon, D. Zhang, Langmuir **2011**, 27, 11742.
(9) Y. Chen, J. Gu, D. Zhang, S. Zhu, H. Su, X. Hu, C. Feng, W. Zhang, Q. Liu, A. R. Parker, J. Mater. Chem. **2011**, 21, 15237.
(10) W. Peng, S. Zhu, W. Wang, W. Zhang, J. Gu, X. Hu, D. Zhang, Z. Chen, Adv. Funct. Mater. **2012**, 22, 2072.
(11) Q. Yang, S. Zhu, W. Peng, C. Yin, W. Wang, J. Gu, W. Zhang, J. Ma, T. Deng, C. Feng, ACS nano **2013**, 7, 4911.
(12) Y. Tan, J. Gu, W. Xu, Z. Chen, D. Liu, Q. Liu, D. Zhang, ACS Appl. Mater. Inter. **2013**, 5, 9878.
(13) Z. He, W. Zhang, W. Wang, M. Tassin, J. Gu, Q. Liu, S. Zhu, H. Su, C. Feng, D. Zhang, Journal of Materials Chemistry B **2013**, 1, 1673.
(14) H. Zhou, X. Li, T. Fan, F. E. Osterloh, J. Ding, E. M. Sabio, D. Zhang, Q. Guo, Advanced Materials **2010**, 22, 951.
(15) D. Kang, Q. Liu, J. Gu, Y. Su, W. Zhang, D. Zhang, ACS nano **2015**.
(16) X. Chen, H. Ostadi, K. Jiang, *Analytical biochemistry* **2010**, *403*, 63
(17) Z. Liu, T. Fan, H. Zhou, D. Zhang, X. Gong, Q. Guo, H. Ogawa, *Bioinspiration & biomimetics* **2007**, *2*, 30.
(18) J. Toster, K. S. Iyer, R. Burtovyy, S. S. Burgess, I. A. Luzinov, C. L. Raston, *Journal of the American Chemical Society* **2009**, *131*, 8356.
(19) M. Sumper, E. Brunner, *Advanced functional materials* **2006**, *16*, 17.
(20) H. Zhou, T. Fan, X. Li, J. Ding, D. Zhang, X. Li, Y. Gao, *Eur J Inorg Chem* **2009**, *2009*, 211.
(21) B. M. Hopkinson, C. L. Dupont, A. E. Allen, F. M. Morel, *Proceedings of the National Academy of Sciences of the United States of America* **2011**, *108*, 3830.

PRODUCING HIGH QUALITY BIO-OILS THROUGH STRUCTURED UPGRADING REACTOR

*Liaoyuan Mao, Yanxin Li, Z. Conrad Zhang**

State Key Laboratory of Catalysis, Dalian National Laboratory for Clean Energy, Dalian Institute of Chemical Physics, Chinese Academy of Sciences, 457 Zhongshan Road, Dalian 116023, Liaoning, China
*Corresponding author, zczhang@dicp.ac.cn; zczhang@yahoo.com

Bubbling sand pyrolysis followed by zeolite-coated cordierite honeycomb (ZCCH) upgrading was evaluated for lignocellulosic biomass conversion to liquid fuels. Steam with and without a low concentration of ethanol was used as carrier gas. The results showed that ZCCH catalytic upgrading effectively reduced the oxygen content and viscosity of the bio-oils. Both the yield and the higher heating value (HHV) of bio-oil were remarkably increased by the presence of a moderate amount of ethanol. Bio-oil of acceptable higher heating value (HHV around 30MJ kg^{-1}) was obtained even at low catalyst usage (WHSV = 60) and long exposure time on stream (>3 h) without regeneration. More importantly, the bio-oil was stable by maintaining low viscosity during aging at 90 °C. The HHV of bio-oil was found to linearly correlate with the oxygen concentration in the oil. Linear equations based on a series of experimental data were established and are capable of predicting bio-oil energy output from the biomass pyrolysis process. Models of bio-oil energy yield from corncob with or without ethanol addition based on unit energy yield (UEY) and yield in ethanol equivalent (YEE, in L ethanol) were developed according to the yield and oxygen content of produced bio-oils.

NANOSTRUCTURED MOLYBDENUM TRIOXIDE FOR ELECTRICAL ENERGY STORAGE

Xu Xiao[1], Jun Zhou[1], and Yury Gogotsi[2]

[1]Wuhan National Laboratory for Optoelectronics and School of Optical and Electronic Information, Huazhong University of Science and Technology, Wuhan 430074, Hubei, P. R. China
[2]Department of Materials Science and Engineering, and A.J. Drexel Nanomaterials Institute, Drexel University, Philadelphia, PA 19104, USA
gogotsi@drexel.edu

Recently, MoO_3 has attracted attention in energy storage applications due to its high theoretical capacitance (1005 C/g). However, few studies have focused on capacitive behavior of MoO_3 due to its naturally poor conductivity and rate capability, especially at high mass loading required to produce practically useful thick electrodes. This presentation provides on overview of our research focused on controlling structure, composition and particle morphology of MoO_3 with the goal of increasing it energy storage capability [1-4].

Considering the intrinsic n-type semiconductivity of MoO_3, we proposed that introduction of oxygen vacancies would increase the carrier density and the electronic conductivity of the oxygen-deficient MoO_3 (MoO_{3-x}) [1,2]. Hydrogenation was applied to reduce MoO_3 and increase its conductivity, as demonstrated by both, first principle calculations and experimental measurements on single MoO_3 nanobelts. Freestanding hydrogenated MoO_3/CNT electrodes (mass density 2.5 mg/cm^2) were fabricated and showed much improved electrochemical performance, such as a specific capacitance of ~340 F/g and a high volumetric capacitance of ~290 F/cm^3 with 64% initial capacitance retained at 10A/g.

In addition, since the energy storage mechanism of layered MoO_3 is ion intercalation into its interlayers, we hypothesized that an enlarged interlayer spacing would be favorable for charge storage, promoting the rate of ions transport. Hence, we pre-intercalated K^+ into the interlayer to increase the interlayer spacing (denoted as K_yMoO_{3-x}) by reacting MoO_3 samples with KBH_4. K_yMoO_{3-x} showed a high volumetric capacitance of 305 F/cm^3 with an excellent rate capability (57% capacitance retention from 0.5 to 100 A/g). Owing to the enlarged interlayer spacing, different cations (Na^+, K^+, and Mg^{2+}) were capable to intercalate along (010) plane and this material showed a good electrochemical performance in aqueous electrolytes, even including natural and artificial seawater [3].

To further enhance the ions transport, synthesis of two-dimensional (2D) nanosheets and then restacking them to form electrodes was used [4]. Due to the nanometer/sub-nanometer thickness of 2D oxide sheets, nearly all metal atoms may be exposed to electrolyte and potentially involved in redox reactions, resulting in pseudocapacitance approaching the theoretical values. Moreover, besides the most frequently used orthorhombic MoO_3, hexagonal MoO_3 (h-MoO_3) contains ion-accessible pores within the crystal structure--a further benefit for energy storage applications. However, due to its intrinsic non-layered structure, current strategies are not suitable to synthesize high-quality 2D h-MoO_3. Herein we report a general strategy, which uses surfaces of water-soluble salt crystals as growth templates and is applicable to various transition metal oxides, such as h-MoO_3, MoO_2, MnO and h-WO_3, not just layered compounds. Restacked 2D h-MoO_3 shows high pseudocapacitive performance in various electrolytes [4]. It reaches 300 F/cm^3 at 5 mV/s in $Al_2(SO_4)_3$ electrolyte. Such a high capacitance in an Al salt electrolyte has not been reported for any material previously. In organic electrolyte, 2D h-MoO_3 stores 1100 C/cm^3 (996 C/g) charge at 2 mV/s.

In summary, the design and control of structure, oxygen content and morphology of MoO_3 can increase its electronic and ionic conductivity, leading to high capacitance even at high charge/discharge rates.

References

(1) Xiao, X. et al. Freestanding Functionalized Carbon Nanotube-Based Electrode for Solid-State Asymmetric Supercapacitors, Nano Energy, 6, 1-9 (2014).
(2) Xiao, X. et al. Freestanding MoO3−x nanobelt/carbon nanotube films for Li-ion intercalation pseudocapacitors. Nano Energy. 9, 355-363 (2014).
(3) Xiao, X. et al. Intercalation of cations into partially reduced molybdenum oxide for high-rate pseudocapacitors. Energy Storage Mater. 1, 1-8 (2015).
(4) Xiao, X. et al. Scalable Salt-Templated Synthesis of 2D Transition Metal Oxides. Nature Comm. (2016)

MAKING CONNECTIONS: THEORY AND MODELING TO LINK MECHANICAL, ELECTRONIC, AND OPTICAL PROPERTIES OF HYBRID/HALIDE PEROVSKITES

Andrew M. Rappe

Makineni Theoretical Laboratories, Department of Chemistry,
University of Pennsylvania, Philadelphia, Pennsylvania, USA
rappe@sas.upenn.edu

Halide perovskites have attracted a high and sustained level of scientific and applied interest for solar photovoltaics because they demonstrate long carrier lifetimes, high mobility, and high power conversion efficiency. The most effective materials incorporate molecular cations on the perovskite *A* site, making them "hybrid" halide perovskites, but both hybrid and inorganic halide perovskites offer impressive optoelectronic properties and pose strong scientific challenges. Despite the promise, the perovskites face multiple challenges before industrial insertion, including hysteresis and stability issues.

In this talk, I will report on key insights gleaned from theory and modeling to understand the novel couplings between mechanical, electronic, and optical responses in this class of materials. The low energy scale for vibrations and strains necessitates their explicit incorporation in electronic and optical calculations. I will present a multi-scale approach incorporating spin-orbit Hamiltonian terms, vibrational motions, thermal disorder, first-principles modeling, tight binding, molecular dynamics, and kinetic Monte Carlo. This comprehensive scheme brings a new understanding of ionic dynamics revealing emergent anharmonicity, electron-phonon coupling variations, and carrier transport and stability.

GREEN URBANIZATION IN CHINA BASED ON ECOLOGICAL CARBON CAPTURE AND RECYCLE: BIOMASS AND MICROALGAE'S ROLE

Lingzhao Kong, Zhiyong Tang, Wei Wei, Yuhan Sun

CAS Key Laboratory of Low-Carbon Conversion Science and Engineering, Shanghai Advanced Research Institute, Chinese Academy of Sciences, Shanghai 201210, China

Introduction

In recent years, China's urbanization process has been significantly accelerated. At the end of year 2011, China's population urbanization rate is over 50%, an inflexion point, for the first time [1]. China's urbanization development must consider CO_2 emission reduction, environmental protection, and the supply of low-carbon energy and products. Strategic transformation of some conventional industry is needed for building a green modern society. We call this Green Urbanization.

Green Urbanization needs to resolve the conflict between development, resource exhausting and environmental deterioration. Apparently, there is an urgent need for excavating biomass and microalgae's potential in realizing Green Urbanization. In Green Urbanization, a low carbon system based on the ecological carbon capture and recycle (eco-CCR) should be constructed, CO_2 is fixed into biomass and microalgae via photosynthesis and then comprehensively recycled in different ways, such as distributive biomass processing, biomass/microalgae to chemicals' production, and microalgae fertilizer for soil improvement and chemical fertilizer replacement, while microalgae water purification with effective nutrient recycle could be accompanied in many circumstances.[2] Effective deployment of biomass and microalgae's potential in the process of carbon and resource recycling can help achieving Green Urbanization in the foreseeable future.

Low-carbon system based on eco-CCR

Low-carbon system based on eco-CCR is presented to meet the need of Green Urbanization in China. It can make CO_2 ecological capture and recycle come true within an effective life cycle based on carbon. The low carbon system based on ecological carbon cycle is shown in Fig. 1. The traditional carbon resources such as fossil fuels will emit significant amount of CO_2 in the process of being converted to fuels and chemicals. The utilization of fossil fuels is a carbon positive process to the whole ecosystem while the green carbon capture by biomass can achieve a net removal of CO_2 from the atmosphere. The life cycle of biomass is a carbon neutral process to the whole ecosystem in the low carbon system based on ecological carbon cycle[3]. Namely, carbon is recycled in fixation in photosynthesis and afterwards utilization of biomass, which makes the carbon recycle in the atmosphere in the form of CO_2-biomass-CO_2. This will not cause a net increase in CO_2 emission and achieve green carbon capture based on biomass.

The photosynthesis of biomass systems can fix CO_2 through the ecological cycle. Biomass systems can be a traditional agricultural or forestry system, or an artificially intensified microalgae system. The traditional agricultural or forestry biomass can be processed to fuels, chemicals and materials while microalgae can be converted to oxygenates, bio-oil, bio-materials, bio-fertilizer and so on.

In order to interpret the low carbon system based on eco-CCR, four integrated systems were proposed: the distributed biomass processing system, microalgae fuel and chemicals production system, microalgae agriculture and soil improvement system, and microalgae wastewater reclamation system.

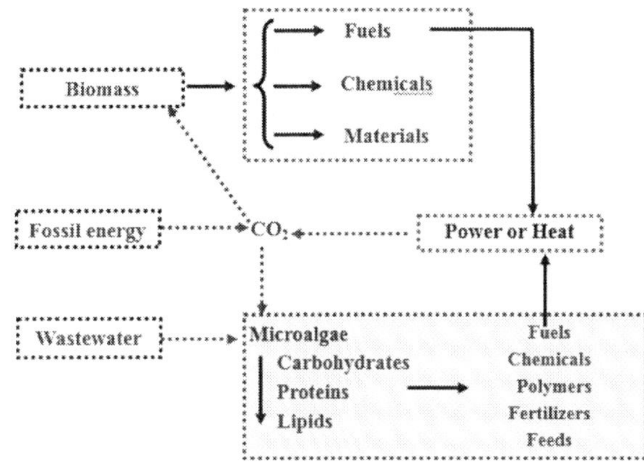

Figure 1. Low carbon system based on ecological carbon cycle.

Distributed biomass processing

For China, the total yield of agricultural residues in 2009 was close to 807 Mtons. Considering straw's feature of low energy density, high collection cost and other limiting factors, building distributed straw collection and processing system is necessary (Fig. 2). In the system, the biomass absorbs CO_2 and release O_2 through photosynthesis. High value products such as drug, paper and food can be obtained which enhance the economic efficiency for the whole system and provide the financial support for running the system. Through thermochemical conversion, the biomass waste will be transformed into biochar and biogas or will be dried and pelleting to improve the energy density of feedstock. Biochar can be back to the field to maintain soil organic matter, sequestrate carbon and improve soil fertilization. Biochar also can be used for the gasification process to obtain syngas which can provide basis for the synthetic chemicals, power and biofuel. Different from the pretreatment of biomass to biochar in local fields, the biogas and syngas should be produced in the central upgrading plant which can provide the energy supply for the distributed system. Thermochemical conversion technology is the core of distributed energy system and the technology will be focused on the enhancement of energy efficiency and economy. So, newly desiccation, carbonization and gasification technology based on small scale/mobile processing should be developing to suit the low-carbon energy supply system.

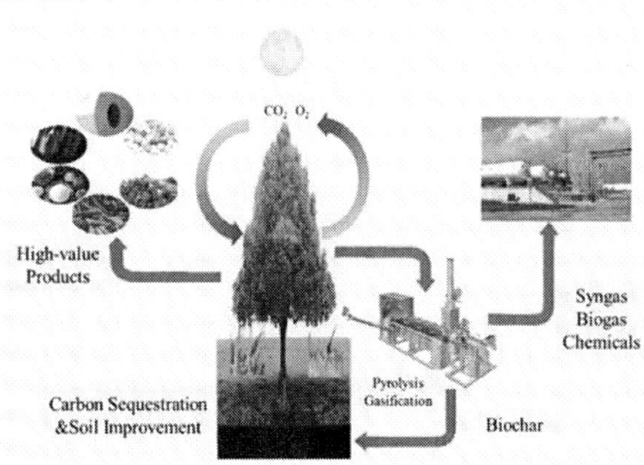

Figure 2. The schematic diagram of distributed-centralized biomass utilization model

Microalgae fuel and chemical production

CO_2 emissions from thermal power industry accounted for China's total CO_2 emissions of 35~40%, while CO_2 emissions from the steel industry has reached about 15% of the country's total emissions. Given that most thermal power plants or steel mills in China are in the suburban areas, microalgae artificial system could be a good solution to CO_2 emission reduction and related products' production. Microalgae have short growth cycle, high carbon sequestration efficiency and strong reproductive capacity, thus making them suitable for large-scale cultivation. Except of the microalgal lipids that could potentially be converted to practical biodiesel, the biomass of microalgae also contains many other valuable components, including carbohydrates, lipids, polyunsaturated fatty acid, pigments and proteins, all of which are worth developing into refined products for various applications (Fig. 3)[4].

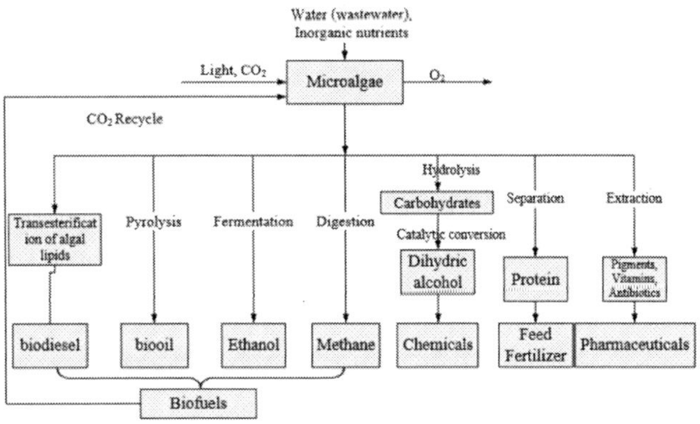

Fig. 3 Microalgae fuel and chemical production system

Microalgae fuel and chemical production system use microalgae to sequester carbon dioxide discharged by emission sources. Therefore, by coupling biotechnology, physics and chemical catalytic technology based on microalgae, CO_2 emission reduction could be achieved, while carbon could be recycled to provide low-carbon fuels, chemical, feed, bio-fertilizer and materials. Related industry chain could be developed for the good of Green Urbanization.

Microalgae agriculture and soil improvement

Ecological agriculture based on microalgae cultivation and soil reclamation system was proposed as a green solution to cope with the soil degradation and food quality problems related to the over dosage of chemical fertilizers. Cyanobacteria, one of the most primitive lives on earth, could fix nitrogen and CO_2 directly from the atmosphere and excrete amino acids, peptides, other nitrogen compounds and active substances to the surrounding environment [5]. Therefore, they can be used to improve the soil fertility, remediate alkali-saline soil, and consequently increase effective arable land in China. Meanwhile, ecological agriculture could be developed through organic vegetables' production, less pollution generated by applying microalgae fertilizers. Application of microalgae in soil remediation also contributes to tremendous CO_2 capture potential.

Microalgae could have customized formula according to different application requirements, such as organic vegetables fertilizing, soil quality improvement, saline-alkali soil remediation, livestock feed and raw materials for downstream chemical or nutritional derivatives. China has about 36.9 million hectares of saline-alkali land, which implies a huge demand for effective soil ameliorant. Microalgae could play an important role in soil improvement and modern agriculture.

Microalgae ecological agriculture and reclamation system can achieve CO_2 emission reduction, soil improvement, and value-added agricultural products' production (Fig.4). The transformation of saline-alkali and poor-fertility land to superior land resources can promote Green Urbanization, based on which more decent agricultural products, more employment, and modern agriculture industry cluster could be accomplished.

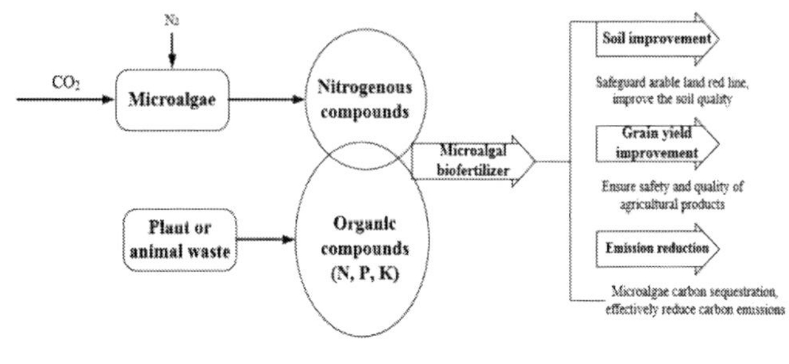

Figure 4. Ecological agriculture based on microalgae

Microalgae wastewater reclamation

More and more municipal and concentrated industrial wastewaters have been generated with the development of modern society and industrialization. The conventional wastewater treatment methods, namely, activated sludge processes, need significant energy input via aeration. Carbonaceous and nitrogenous nutrients are converted to CO_2 and N_2 through A/O processes instead of being recycled. So energy-efficient and environmental-friendly technologies are encouraged in order to be less fossil fuel dependent and cope with global warming effect. Algal wastewater treatment is a promising wastewater reclamation technology and there is substantial research on this topic in recent years [6]. For instance, concentrated animal feeding operations (CAFOs) is replacing the decentralized way during the past decades in the development process of Chinese modern agriculture, which brings about many environmental and health problems at the same time. Through microalgae treatment, which could be processed to feed, fuel or fertilizer and finish a better ecological carbon, nitrogen and phosphorus recycle, provided that relevant standards should be met.

Conclusions

This paper presents a new idea of building a low carbon system based on eco-CCR, which includes systems of distributed biomass processing, a microalgae fuel and chemical production, and a microalgae agriculture and soil improvement. A scenario layout analysis of in china is also presented. The eco-CCR system could realize coordinated development between the low carbon energy and products supply, CO_2 emission reduction and urbanization. However, there have existed technical bottlenecks and all the above-mentioned systems need to be validated through engineering demonstration at large scales, which would provide sustainable resources for Green Urbanization and realize low carbon development.

Acknowledgements. This study is supported by Shanghai Pujiang Project (No. 13PJ1407600), Science and Technology Commission of Shanghai Municipality Program (No.13DZ1205000 and No. 14DZ1203800), and the National Natural Science Fund (No.51208305 and 21406255).Acknowledgment

References

(1) Chinese academy of social sciences. social sciences academic press, 2012,1-50.
(2) Wang, L., Li, Y., Chen, P., Min, M., Chen, Y., Zhu, J., Ruan, R.R., Bioresour. Technol., 2010, 101(8), 2623-2628.
(3) Tabataa, T., Okudab, T. Japan. Energy, 2012, 45(1), 944-951.
(4) Wang, L., Li, Y., Sommerfeld, M., Hu, Q. Bioresour. Technol., 2013,129, 289-295.
(5) Kulasooriya, S. A. Ceylon Journal of Science (Bio. Sci.), 2011,40 (2), 71-88.
(6) Zhou, W.G., Min, M.Li, Y.C.Hu, B.Ma, X.C., Cheng, Y. L., Liu, Y. H.Chen, P.Ruan, R. Bioresour. Technol., 2012,110, 448-455.

CLEAN COAL TECHNOLOGIES BASED ON METHANOL PLATFORM

Yulong Zhang, Wei Di, Fan Zhang, Xiaoying Xu, Yuan Liu

National Institute of Clean and Low Carbon Energy (NICE),
Beijing, China
zhangyulong@nicenergy.com

China is the biggest coal producer and user in the world. Clean coal technologies are critical for China's economy and environment. Methanol is the smallest alcohol. It does not have a carbon-carbon bond. Methanol can be easily synthesized from synthesis gas which can be derived from natural gas, coal or any carbonaceous materials. Methanol can also be easily decomposed into synthesis gas or reformed into H2, so methanol is an excellent energy carrier.

Technologies of synthesis of methanol are relatively mature and large methanol plants (mega methanol) from coal and natural gas were built in China and other natural gas rich countries. In China, methanol production capacity is over 70 million tons annually, mainly from coal. In future, methanol could be synthesized using CO2 as carbon source and renewable energy as a way of CO2 reuse.

Methanol can be converted into gasoline through MTG (methanol to gasoline), or DME and DMMn as diesel alternatives or additives. It can be also converted into chemicals through solid acid catalysts, for example, MTO, MTP MTA etc.

Methanol platform will play a critical role in sustainable development.

APPLICATION OF AMBIENT PRESSURE X-RAY PHOTOELECTRON SPECTROSCOPY TO STUDIES OF CATALYSIS: SURFACE OF CATALYSIS IN REACTANTS

Franklin (Feng) Tao, *Luan Nguyen, and Yu Tang*

Department of Chemical and Petroleum Engineering, University of Kansas, Lawrence, KS 66045
Department of Chemistry, University of Kansas, Lawrence, KS 66045
franklin.feng.tao@ku.edu

Chemistry and structure of surface of a catalyst under a reaction condition is the crucial information for understanding catalytic mechanism since in many cases an authentic, active surface catalyzing a catalytic reaction is formed in a pretreatment or/and in a reaction between nominal catalyst and reactants. Ambient pressure X-ray photoelectron spectroscopy can be used to track surface of a catalyst under a reaction condition as the instrumentations in the last decades have made characterization of catalyst surface in a gas phase at Torr pressure or higher possible. It can characterize surface chemistry of a catalyst including surface composition, surface phase and surface oxygen vacancies and other information under a reaction condition and track their evolutions when the reaction condition is changed to another. This talk will cover instrumentation of lab-based ambient pressure XPS, notes of using lab-based AP-XPS to studies of catalysis, and the prospect of ambient pressure XPS operational at a higher pressure.

HIGH PERFORMANCE DOPED LITHIUM MANGANESE PHOSPHATE CATHODE MATERIALS FOR LITHIUM-ION BATTERIES

Pengjian Zuo[*a], Liguang Wang [a], Guangyu Cheng [a], Chunyu Du [a], Yulin Ma [a], Xinqun Cheng [a], Yunzhi Gao [a], Geping Yin[a]*

[a]*Institute of Advanced Chemical Power Source, School of Chemical Engineering and Technology, Harbin Institute of Technology, Harbin 150001, China*

* *Corresponding author: zuopj@hit.edu.cn*
Tel: 86-451-86403961; Fax: 86-451-86403961

Introduction

Lithium transition-metal phosphates of $LiMPO_4$ (M = Fe, Mn, Co, Ni) with olivine structure have attracted much attention as promising cathode materials for next generation of lithium ion batteries.[1] Compared with the traditional transition- metal oxide cathode, $LiMPO_4$ possesses the merits of lower toxicity, cost-effectiveness, environmental friendliness and high safety.[2, 3] $LiMnPO_4$ provides a moderate working voltage compatible to the present electrolyte systems and a higher energy density on account of the relatively high redox potential of ~ 4.1 V for Mn^{3+}/Mn^{2+} couple.[1] However, the pristine $LiMnPO_4$ shows a poor electrochemical performance due to the sluggish lithium diffusion kinetics within the grains caused by the low intrinsic electronic and ionic conductivity.[4, 5] Additionally, $LiMnPO_4$ also suffers from instability of the charged phase ($MnPO_4$),[6] Jahn-Teller distortion of the Mn^{3+},[7] the large volume change between $LiMnPO_4$ and $MnPO_4$ during charge-discharge process,[8] which results in the lower capacity and limited rate capability. Many efforts have been made to overcome these drawbacks so as to improve the electrochemical performance of $LiMnPO_4$ by particle-size minimization, carbon coating and cation doping.[9-14]

Here, we propose the solvothermal method to prepare the Fe/Mg-doped nanorod material, and the hierarchical hollow microsphere composites consisting of primary nanoparticles with an open 3D mesoporous structure. Furthermore, the capacity fading of the cathode materials during cycling were studied systematically.

Experimental

The doped lithium manganese phosphate nanomaterials were prepared by the solvothermal method.[15, 16] All the precursors were purchased without further pre-treatment or purification. Stoichiometric amounts of precursors were dissolved in ethanol-water solvent under stirring. Then the solution was transferred into a 100 mL Teflon-lined stainless steel autoclave, and the autoclave was put into a preset electric furnace and maintained at 200℃ for 12 h. The obtained solution was filtered and washed with ethanol and distilled water several times and finally dried at 80℃ in vacuum for 5 h. The carbon coated composite was prepared using phenolic resin or glucose as carbon resources. The mixture was heated under a Ar/H_2 ($Ar: H_2 = 95: 5$) atmosphere to obtain the final composite.

Results and Discussion

The structure and crystal orientation of the samples were investigated using TEM and electron diffraction, and the images are presented in Fig.1. The selected area electron diffraction (SAED) of the LMP nanorod (shown in Fig.1a) shows the well-defined diffraction spots as shown in Fig.1b, revealing a highly ordered single crystalline nature of the particle. Almost the same diffraction patterns can be observed in both LMFP and LMFMP nanorods. The HRTEM images show that the interplanar spacing of 2.35 Å in Fig.1c, 5.18 Å in Fig.1f and 2.34 Å in Fig.1i can be identified as the (002), (200) and (002) lattice planes, respectively, demonstrating that the nanorods grow preferentially along the *c* direction. Generally, lithium ions transport follows a zigzag one-dimensional pathway along the *b* direction in phosphate crystal with the orthorhombic olivine structure. Therefore, if a LMP nanocrystal was to grow with a short length parallel to the *b* direction, the electrochemical performance of the batteries could be significantly improved. To observe the influence of ball-milling process and the distribution of carbon pyrolyzed by glucose, the carbon coated samples were examined by TEM as well. As shown in Fig.1j, k and l, the nanorods are coated and connected by amorphous carbon derived from the glucose. Hence, the electronic conductivity as well as the electrochemical performance of the samples can be increased due to the fine carbon coating on the phosphate crystals.

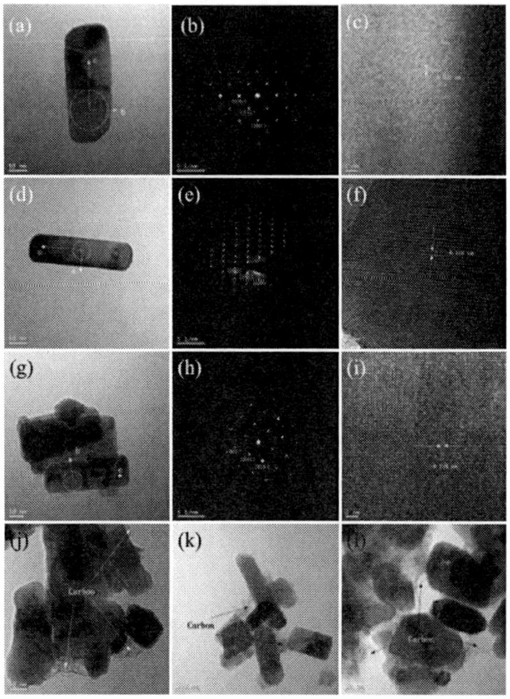

Figure 1. TEM images, SAED patterns, HRTEM images of a LMP nanorod (a–c), LMFP nanorod (d–f) and LMFMP nanorods (g–i). TEM images of (j), (k) and (l) represent carbon coated samples of LMP/C, LMFP/C and LMFMP/C, respectively.

In addition, we have prepared microspheres with an open three-dimensional nanoporous structure by a facile ion-exchange solvothermal method. The LMFP/C nano-composite exhibits a superior long-term cycling performance. After 1000 charge/discharge cycles at 5 C, the sample still remains a reversible capacity of about 105 $mAh{\cdot}g^{-1}$, corresponding to the capacity retention of 94.0% and only 0.0068 $mAh{\cdot}g^{-1}$ capacity loss per cycle compared to the maximum discharge capacity of 112 $mAh{\cdot}g^{-1}$. This slight capacity fading can be attributed to the structure changes caused by Jahn-Teller distortion of Mn^{3+} and the electrolyte decomposition.

Conclusions

The doped lithium manganese phosphates for lithium ion batteries were synthesized *via* a simple solvothermal process in water-ethanol solvent mixtures. These improved properties were ascribed to the fast reaction and ionic diffusion kinetics due to the nanometer-scale primary Fe-doped lithium manganese phosphate crystals; (ii) the good electronic contact formed by the amorphous carbon on the surface and the conductive layer; (iii) the well-crystallized primary nanocrystals and the structural stability. The antisite defects, Mn dissolution and electrolyte decomposition were the determining factors for the decreased capacity during charge-discharge process of this $LiMnPO_4$ class materials.

References

(1) Padhi, A. K.; Nanjundaswamy, K. S.; Goodenough, J. B. Phospho-olivines as Positive-Electrode Materials for Rechargeable Lithium Batteries. *Journal of The Electrochemical Society* 1997, 144, 1188-1194.

(2) Tarascon, J. M.; Armand, M. Issues and challenges facing rechargeable lithium batteries. *Nature* 2001, 414, 359-367.

(3) Whittingham, M. S.; Song, Y.; Lutta, S.; Zavalij, P. Y.; Chernova, N. A. Some transition metal (oxy)phosphates and vanadium oxides for lithium batteries. *Journal of Materials Chemistry* 2005, 15, 3362-3379.

(4) Delacourt, C.; Laffont, L.; Bouchet, R.; Wurm, C.; Leriche, J.-B.; Morcrette, M.; Tarascon, J.-M.; Masquelier, C. Toward Understanding of Electrical Limitations (Electronic, Ionic) in LiMPO4 (M = Fe , Mn) Electrode Materials. *Journal of The Electrochemical Society* 2005, 152, A913-A921.

(5) Wang, L.; Zhou, F.; Ceder, G. Ab Initio Study of the Surface Properties and Nanoscale Effects of LiMnPO4. *Electrochemical and Solid-State Letters* 2008, 11, A94-A96.

(6) Zhou, F.; Cococcioni, M.; Kang, K.; Ceder, G. The Li intercalation potential of LiMPO4 and LiMSiO4 olivines with M = Fe, Mn, Co, Ni. *Electrochemistry Communications* 2004, 6, 1144-1148.

(7) Yamada, A.; Yonemura, M.; Takei, Y.; Sonoyama, N.; Kanno, R. Fast Charging LiFePO4. *Electrochemical and Solid-State Letters* 2005, 8, A55-A58.

(8) Meethong, N.; Huang, H. Y. S.; Speakman, S. A.; Carter, W. C.; Chiang, Y. M. Strain Accommodation during Phase Transformations in Olivine-Based Cathodes as a Materials Selection Criterion for High-Power Rechargeable Batteries. *Advanced Functional Materials* 2007, 17, 1115-1123.

(9) Qin, L.; Xia, Y.; Cao, H.; Luo, L.; Zhang, Q.; Chen, L.; Liu, Z. Effects of Ti additive on the structure and electrochemical performance of LiMnPO4 cathode material. *Electrochimica Acta* 2014, 123, 240-247.

(10) Kim, M.-S.; Jegal, J.-P.; Roh, K. C.; Kim, K.-B. Synthesis of LiMn0.75Fe0.25PO4/C microspheres using a microwave-assisted process with a complexing agent for high-rate lithium ion batteries. *Journal of Materials Chemistry A* 2014, 2, 10607-10613.

(11) Oh, S.-M.; Myung, S.-T.; Park, J. B.; Scrosati, B.; Amine, K.; Sun, Y.-K. Double-Structured LiMn0.85Fe0.15PO4 Coordinated with LiFePO4 for Rechargeable Lithium Batteries. *Angewandte Chemie* 2012, 124, 1889-1892.

(12) Jo, M.; Yoo, H.; Jung, Y. S.; Cho, J. Carbon-coated nanoclustered LiMn0.71Fe0.29PO4 cathode for lithium-ion batteries. *Journal of Power Sources* 2012, 216, 162-168.

(13) Fang, H.; Yi, H.; Hu, C.; Yang, B.; Yao, Y.; Ma, W.; Dai, Y. Effect of Zn doping on the performance of LiMnPO4 cathode for lithium ion batteries. *Electrochimica Acta* 2012, 71, 266-269.

(14) Fang, H.; Dai, E.; Yang, B.; Yao, Y.; Ma, W. LiMn0.8Fe0.19$Mg_{0.01}$PO4/C as a high performance cathode material for lithium ion batteries. *Journal of Power Sources* 2012, 204, 193-196.

(15) Zuo, P.; Wang, L.; Zhang, W.; Yin, G.; Ma, Y.; Du, C.; Cheng, X.; Gao, Y. A novel nanoporous Fe-doped lithium manganese phosphate material with superior long-term cycling stability for lithium-ion batteries. *Nanoscale* 2015, 7, 11509-11514.

(16) Wang, L.; Zuo, P.; Yin, G.; Ma, Y.; Cheng, X.; Du, C.; Gao, Y. Improved electrochemical performance and capacity fading mechanism of nano-sized LiMn0.9Fe0.1PO4 cathode modified by polyacene coating. *Journal of Materials Chemistry A* 2015, 3, 1569-1579.

A NOVEL REDUCED GRAPHENE OXIDE /S COMPOSITE FOR HIGH PERFORMANCE LITHIUM/SULFUR BATTERIES

Wei Zhang [a], Pengjian Zuo[a], Junfu Hua [a], Geping Yin[a]*

[a]Institute of Advanced Chemical Power Source, School of Chemical Engineering and Technology, Harbin Institute of Technology, Harbin 150001, China

* Corresponding author: zuopj@hit.edu.cn

Tel: 86-451-86403961; Fax: 86-451-86403961

Introduction

Lithium–sulfur (Li–S) batteries have become very attractive due to the fact that elemental sulfur has a specific capacity and energy density of 1675 mAh/g and 2500 Wh/kg, respectively, which is 3 to 5 times higher than that of Li-ion batteries based on intercalation chemistry. Moreover, sulfur is abundant, low cost and environmentally friendly[1]. Nevertheless, lithium-sulfur batteries suffer from problems that prohibit their widespread application. For example, low electric conductivity of sulfur and its reduced products leads to low utilization of active material[2]. Another major hurdle is the serious capacity degradation upon cycling, which is related to solubility of the intermediate lithium polysulfides (Li_2S_x, $4 \leq x \leq 8$) in the electrolyte[3]. Moreover, the dissolved polysulfides travel between the electrodes during cycling, being oxidized and reduced, respectively, at the cathode and anode, resulting in a shuttle mechanism. The shuttle mechanism leads to low coulombic efficiency and contamination of the anode by reduced solid precipitates resulting in poor cycling stability[4].

Graphene, as a unique two-dimensional (2D) carbon, have been used to confine the sulfur compounds for Li–S batteries because of their unique electronic properties, large theoretical specific surface area, and excellent mechanical properties[5]. Herein, we are motivated to design a one-dimensional (1D) tubular structure reduced graphene oxide (RGO)/S composite (RGO/S nanoscroll composite) by shock cooling of RGO/sulfur nanoparticles dispersion with liquid nitrogen. The stable and curly 1D tubular structure of RGO can effectively inhibit the dissolution polysulfides in electrolyte. As a consequence, a high reversible capacity and excellent cyclic stability were achieved.

Experimental

Graphite oxide (GO) was synthesized by a modified Hummers method in a two oxidation steps. Then a stable RGO aqueous dispersion was obtained by reducing GO with $NaBH_4$ following by washing and ultrasonication. Then, nano-sulfur powder was simply added into above RGO dispersion to form a mixture dispersion with the help of ultrasonication. The mixture dispersion was rapidly cooled by surrounding with liquid nitrogen. The water was removed by freeze-drying machine, and the black gray product was further heated at 180°C for 2 h under argon atmosphere, removing excessive sulfur, to obtain the RGO/S nanoscroll composite.

Results and Discussion

The SEM images of RGO/S nanoscroll composite were shown in Figure1a. It is clear that the RGO nanosheets show a 1D nanoscroll structure and sulfur was encapsulated within the RGO nanoscrolls. From the element mapping (Figure 1b and c), it can be concluded that the C and S are uniformly distributed in the RGO/S nanoscroll composite. Fig. 2 shows the cycling performance and coulombic efficiency of the RGO/S nanoscroll composite at a rate of 0.2 C. A high initial discharge capacity of 1281 mAh/g was obtained and still remain a high reversible capacity of 640 mAh/g after 100 cycles. The good cycling performance can be attributed to the unique RGO nanoscrolls structure. This unique RGO nanoscrolls structure would help to mitigate the dissolution of the lithium polysulfides and accommodate the volume change of sulfur during discharge/charge process, besides, the superior electric conductivity of RGO nanocrolls can provide a fast electronic conductive network.

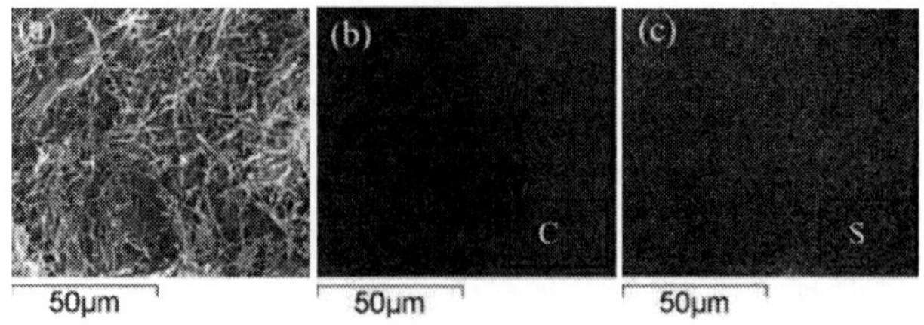

Figure 1. Morphology and chemical mapping of the RGO/S nanoscrolls composite, SEM image of RGO/S nanoscroll composite (a) and the corresponding EDS mappings of sulfur (b) and carbon (c).

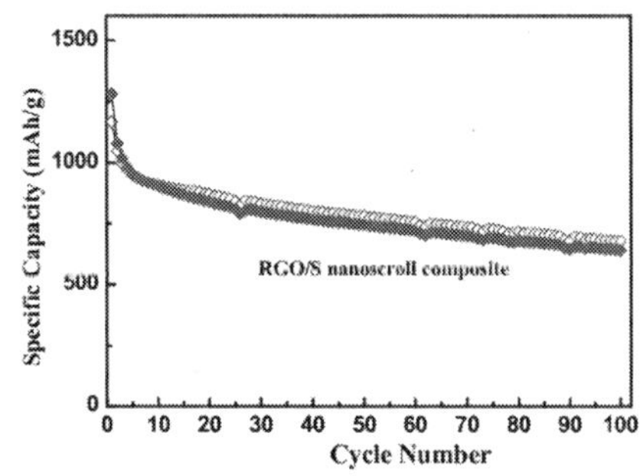

Figure 2. The cycle performance of the RGO/S nanoscroll composite at the rate of 0.2C (1 C = 1675 mA/g)

Conclusions

A unique 1D RGO/S nanoscrolls composite was successfully synthesized by shock cooling of RGO/sulfur nanoparticles dispersion with liquid nitrogen. This RGO/S composite behavior a high initial discharge capacity of 1281 mAh/g and still remain a high reversible capacity of 640 mAh/g after 100 cycles. The good cycling performance can be attributed to two aspects: i) this unique RGO nanoscrolls structure would help to mitigate the dissolution of the lithium polysulfides and accommodate the volume change of sulfur during discharge/charge process; ii) the superior electric conductivity of RGO nanocrolls can provide a fast electronic conductive network.

References

(1) [1] L. Ji, M. Rao, H. Zheng, L. Zhang, Y. Li, W. Duan, J. Guo, E.J. Cairns, Y. Zhang, Journal of the American Chemical Society, 133 (2011) 18522-18525.
(2) [2] C.X. Zu, A. Manthiram, Adv Energy Mater, 3 (2013) 1008-1012.
(3) [3] H. Xu, Y. Deng, Z. Shi, Y. Qian, Y. Meng, G. Chen, Journal of Materials Chemistry A, 1 (2013) 15142.
(4) [4] Z. Xiao, Z. Yang, L. Wang, H. Nie, M.e. Zhong, Q. Lai, X. Xu, L. Zhang, S. Huang, Advanced Materials, (2015) n/a-n/a.
(5) [5] J.-Z. Wang, L. Lu, M. Choucair, J.A. Stride, X. Xu, H.-K. Liu, Journal of Power Sources, 196 (2011) 7030-7034.

LIGHT OLEFIN SYNTHESIS FROM SHALE GAS FEEDSTOCK BY CATALYTIC ROUTES

Aditya Bhan

Dept. of Chemical Engineering and Materials Science
University of Minnesota
421 Washington Ave. SE
Minneapolis, MN 55455

This first part of this talk will detail the mechanistic pathways involved in the catalytic homologation of methanol, an intermediate platform for the conversion of shale gas feedstock, to light olefins on zeolitic catalysts. The complex, autocatalytic network of reactions describing methanol to hydrocarbons conversion on zeolites is summarized by a dual-cycle schematic where the olefins-based chemistries of methylation and β-scission are coupled with the aromatics-based chemistries of methylation and dealkylation through hydrogen transfer and cyclization. The olefins- and aromatics-based intermediates localized inside the catalyst pores, constituting the so-called 'hydrocarbon pool', act as organic co-catalysts for carbon-carbon bond and product formation. We demonstrate the practical ability to tune relative rates and product selectivities by the conscious addition of aromatics- and olefins-based hydrocarbon pool intermediates, the elucidation to the consequences of crystallite size on light olefins selectivity, and the cognizant modification of hydrocarbon pool composition via oxygenate co-feeds to improve ethylene selectivity, are all manifestations of exploits in our understanding of methanol conversion over zeotype catalysts through the lens of the dual catalytic cycles. [1-6]

The second part of this talk will discuss the non-oxidative conversion of natural gas liquids to polymer-precursors via catalytic dehydrogenation on oxidic materials, specifically, the kinetic and mechanistic details of alkane dehydrogenation on non-reducible oxides.

Acknowledgment. We acknowledge financial support The Dow Chemical Company and from the National Science Foundation (CBET 1055846)

References

(1) S. Ilias, A. Bhan, J. Catal. 290 (2012) 186-192.
(2) S. Ilias, R. Khare, A. Bhan, J. Catal. 303 (2013) 135-140
(3) R. Khare, D. Millar, A. Bhan, J. Catal. 321 (2015) 23-31
(4) R. Khare, A. Bhan, J. Catal. 329 (2015) 218-229
(5) A. Hwang, D. Prieto-Centurion, A. Bhan, J. Catal. 337 (2016) 52-56
(6) R. Khare, S. S. Arora, A. Bhan, ACS Catal. 6 (2016) 2314-2331

DEMINERALIZATION PATHWAYS FOR OIL SHALE SEMICOKE BYPRODUCT CONVERSION TO A SORBENT MATERIAL

Azat Suleimenov[1], Jillian L. Goldfarb[1,2]

1. Division of Materials Science & Engineering, Boston University, 15 St. Mary's St., Boston MA 02215; azat@bu.edu
2. Department of Mechanical Engineering, Boston University, 110 Cummington Mall, Boston MA 02215; jilliang@bu.edu

Introduction

Rapid industrial development and constantly changing oil prices, our attention has turned to alternative sources of energy, both fossil and renewable based. Oil Shale is a sedimentary rock rich in kerogen. Upon heating in an environment without oxygen, the kerogen can be separated from the rock. The resulting high molecular weight hydrocarbon is a source for crude oil and natural gas. According to US Geological Survey estimates, the US oil shale reserves account for about 2.6 trillion barrels, among which approximately 1 trillion barrels of oil equivalent can be recovered [1]. This is about three times larger than the proven reserves of Venezuela, currently thought to be the richest deposits of oil shale [2].

The future possibility of crude oil extraction from oil shales is hampered by waste production. The waste rock that is left after kerogen separation from oil shale is called semicoke, a semicarbonaceous char system. One possible way for efficient utilization of oil shales is to convert this waste to a sorbent material for flue gas treatment [2,3]. However, because the sedimentary rocks were present in the ground for a long time they are also rich in compounds such as carbonates and sulfides [2]. These substances could be removed to enable efficient flue gas capture. This paper analyzes possible conversion methods of oil shale semicoke to sorbent material for flue gas capture. The results of acid-base demineralization and activation processes are investigated as to their effect on the adsorption properties. The industrial result of this Green Technology is that a promising alternative energy source could yield its own materials for pollutant capture.

Experimental Methods

Oil shale from the Piceance Basin (Colorado, USA) was selected as a raw material and used for further transformation and characterization. A heterogeneous sample of oil shale rock was crushed with a hydraulic press into small rock samples. They were further crushed with a ball mill "VQ-N High Energy Ball Mill" (Across International) at 1200 rpm for 60 minutes. The particles were separated using a sieve shaker and standard ASTM brass sieves, particles sized 150-250 µm were chosen for this study. The obtained powder is named "Oil Shale" (OS).

Semicoke was obtained by pyrolysis of oil shale in a horizontal tube furnace (MTI Corporation GSL-1100X). The particles were first heated to 110 °C and held for 30 minutes to ensure moisture removal. Then, the temperature was increased to 600 °C and held for 1 hour to remove kerogen and produce semicoke particles. All the procedures were done at 10 °C/min heating rate and constant flow of high purity nitrogen gas at 100 mL/min to minimize the amount of oxygen in the system. Finally, the samples were cooled under N_2. The samples were named as "Semicoke" (SC).

Acid-Base Treatment. Three different activation methods were tested. According to Yan *et al.*, acid-base treatment with KOH and HCl promises valuable removal of SO_2 flue gas [4]. The first treatment method used was to immerse the particles in HCl solution (6M) for 2 hours at 80 °C. For each 1 g of a sample, 2 mL of the solution was used. After filtration the obtained samples were dried overnight at 80 °C. They are named "SC-HCl".

The second method is double HCl activation, here the HCl treatment was repeated twice. They are labeled "SC-HCl-HCl".

The third method is activation with KOH after HCl treatment. A sample of "SC-HCl" was soaked in 10M KOH solution for 2 hours at 80°C. Afterwards, to remove residual ions samples were immersed in HCl solution (1 M) for 2 hours at 80°C. After filtration and drying the obtained particles were named "SC-HCl-KOH".

Characterization. To describe the impact of acid/base treatment on the resulting oil shale semicoke, we measure the surface area (m^2/g) via the BET method. The BET surface area was determined by Nitrogen adsorption isotherms at 77.35 K on an Autosorb-1 (Quantachrome Instruments). Between 0.1 and 0.2 grams of powder were analyzed using multi-point BET, where volume of adsorbed gas was measured for 11 points over a partial pressure range of 0.05 – 0.3.

Thermogravimetric analysis (Mettler-Toledo TGA-DSC-1) was used for determination of thermal stability and volatile components of a sample. We monitor the weight change of the solid sample while increasing the temperature at a constant ramp rate. The samples were heated to 110 °C and held at this temperature for 30 minutes for moisture removal, then the temperature was increased to 900 °C and held for 1 hour under N_2 gas at 20 mL/min at a heating rate of 10°C/min. Afterwards, air was introduced into the system at 20 mL/min and samples were oxidized at 950 °C for 10 minutes. Alumina crucibles of 70 µL volume were used as sample containers, and between 5 and 15 mg of sample was used for each TGA run.

Scanning Electron Microscopy (Zeiss Supra 55VP) was used for a qualitative determination of the impact of activation conditions. The sample powders were placed on copper double-sided tape to ensure solid conductivity and grounding.

Powder X-Ray Diffraction (Bruker D8 Discover) used to study diffraction patterns of the samples to determine, semi-quantitatively, the changing mineral composition of the sample. As samples in our instrument must be aligned vertically, Kapton tape was used to stick powder to the sample holder. Then, the diffraction pattern of pure Kapton tape was subtracted from the diffraction pattern.

Results and Discussion

The change in BET surface area with corresponding sample preparation method is shown in Table 1. During pyrolysis, surface area increased about 2 times, because kerogen was separated from raw oil shale, opening the porous network. As a result, the semicoke had an initial area of 15 m^2/g, which was further altered with acid-base treatment. We note that this is already higher than many coal fly ash samples, which are commonly used as a sorbent material [7,8].

As a result of the first activation method, most of minerals within the particles were removed by hydrochloric acid [3]. The carbonates present produced carbon dioxide gas, which was seen as evolving bubbles from the solution. As a result, the BET area increased about 5.5 times. The second activation method involved KOH reaction with the sample with further removal by HCl. The result is about 1.5 times higher area than in single HCl activation. The third activation method showed the highest result. The second HCl treatment helped to remove minerals in deep from the particles which is shown by the specific surface area of 159 m^2/g.

Table 1. Impact of Activation Process on BET Surface Area

Sample	BET Surface Area, m^2/g
Raw (OS)	8.5
Pyrolyzed (SC)	15
HCl activated (SC-HCl)	89
KOH activated (SC-HCl-KOH)	123
Double HCl activated (SC-HCl-HCl)	159

Results of thermogravimetric analysis are shown in Figure 1. As suggested by Li et al. and Miao et al. [5,6], semicoke decomposes in several stages upon heating: moisture loss, release of volatile and fixed carbons, and decomposition of mineral matter. Initial weight reported excludes moisture that was present in the samples. Therefore, analysis starts at 110 °C with dry samples.

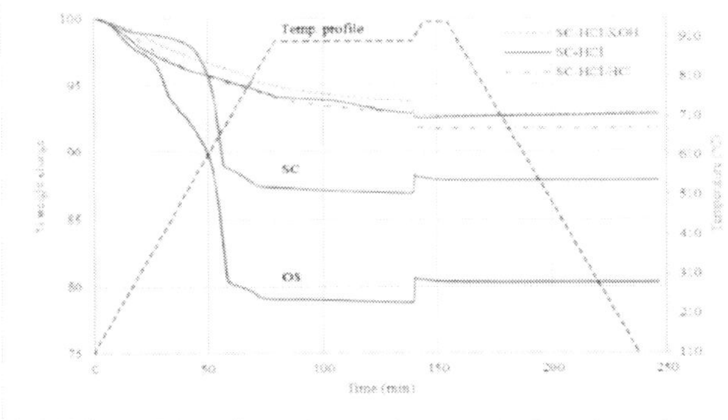

Figure 1. TGA curves for different samples

We see that the majority of mass was lost for oil shale and semicoke during the heating up to 900 °C under nitrogen atmosphere. This is due to loss of volatile carbons present in the samples. Additionally, the oil shale and semicoke samples show a slight increase in mass when temperature starts to rise from 900 °C. Here, the air is introduced in the system, and oxidation of volatile carbons happens. Table 2 shows the amount of volatile and fixed carbons present in the studied samples which are calculated from the TGA curves.

Table 2. Carbon Content Analysis

Sample	Volatile C	Fixed C	Ash
OS	21.24	0.28	78.48
SC	13.50	0.36	86.14
SC-HCl	9.37	0.30	90.33
SC-HCl-KOH	7.34	1.03	91.63
SC-HCl-HCl	9.77	1.12	89.11

From SEM images (Figure 2) we qualitatively note that much of the mineral content was removed by acid-base treatments. The porous structures of the activated samples show an increase in porosity by a corresponding decrease in amorphous carbon regions.

Figure 2. SEM images of samples SC (top left); SC-HCl (top right); SC-HCl-HCl (bottom left); SC-HCl-KOH (bottom right); scale bar indicates 1 μm

The figure below shows X-ray diffraction pattern for the samples. We see a majority of minerals such as calcite, dolomite, and pollucite present in the raw sample. Many of the minerals taken out by pyrolysis with further removal by chemical treatments, as noted by prior research in this field [9]. In addition, quartz concentration increases in the activated samples, since other components are removed by reactions in acid environment. Generally, chemically activated samples show much less mineral content than the raw sample.

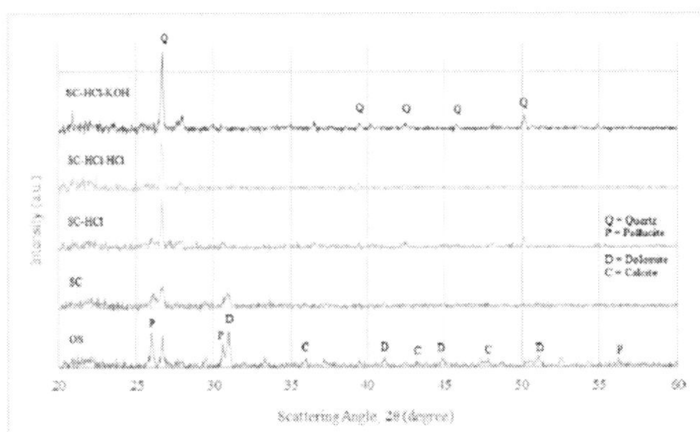

Figure 3. X-ray Diffraction pattern for raw and activated samples

Conclusions and Future Work

This work demonstrates a possible pathway of removing minerals and opening pores to efficiently use oil shale waste. The XRD pattern shows that most minerals are removed for chemically activated samples. Additionally, the SEM images show porous structures for these samples. Finally, double HCl treatment promises the highest specific surface area. This collectively can lead to efficient flue gas capture by double HCl activated oil shale waste.

Future work includes studying the amount of SO_2 absorbed by the activated semicokes to give a better perspective on the effect of the activation on chemical absorption and physical adsorption processes. This would enable us to make recommendations on the optimal activation pathway for the conversion of oil shale semicoke to a sorbent material for flue gas treatment.

Acknowledgement. The authors thank Dr. Jeffrey Bacon for assistance with XRD and Elbara Ziade for help with SEM imaging.

References

(1) Institute for Energy Research. *Oil Shale*; http://instituteforenergyresearch.org/topics/encyclopedia/oil-shale/ (accessed March 12, 2016).
(2) Brendow, K. *Oil Shale.*, **2003**, 20 (1), 81-92.
(3) Liu, Q.; Li, C.; and Li, Y. *Carbon.*, **2003**, 41 (12), 2217-2223.
(4) Yan, Z.; Liu, L.; Zhang, Y.; Liang, J.; Wang, J.; Zhang, Z.; and Wang, X. *Energy and Fuels.*, **2013**, 27 (6), 3080-3089.
(5) Sun, B.; Wang, Q.; Shen, P.; Qin, H.; and Li, S. *Oil Shale.*, **2012**, 29 (1), 63.
(6) Miao, Z.; Wu, G.; Li, P.; Zhao, N.; Wang, P.; and Meng, X. *Mining Science and Technology.*, **2009**, 19 (3), 0380-0384.
(7) Külaots, I.; Goldfarb, J. L.; Suuberg, E. M. *Fuel* **2010**, 89, 3300– 3306
(8) Trikkel, A.; Kuusik, R.; Martins, A.; Pihu, T.; Stencel, J. M. *Fuel Process. Technol.* **2008**, 89, 756–763
(9) Vyas, A.; Xue, J.; Goldfarb, J.L. *Energy & Fuels.* **2016.** DOI: 10.1021/acs.energyfuels.5b02244

CONTROLLABLE SYNTHESIS OF ZSM-5/EU-1 CO-CRYSTALLINE ZEOLITE

Liyuan Sun, Yafei Zhang, Yanjun Gong *

State Key Laboratory of Heavy Oil Processing, The Key Laboratory of Catalysis of CNPC, China University of Petroleum-Beijing, Changping, Beijing 102249, China
E-mail: sunliyuancupb@sina.com, gongyj@cup.edu.cn

Abstract

The co-crystalline zeolite and its texture property can be tailored by the co-growth process of both phases, due to adjustment of phase proportion via gel composition and structure deformation, thereby making the co-crystalline presents special catalytic performance[1,2]. ZSM-5/EU-1 co-crystalline with different phase proportions were synthesized by two-step method with two templates. In view of nucleation apparent activation energy, diquaternary ammonium was selected as the template for ZSM-5 phase due to its activation energy similar to EU-1 synthesized by hexamethonium bromide (53.9 kJ·mol $^{-1}$ vs 57.5 kJ·mol $^{-1}$), while TPABr for ZSM-5 was very lower (47.9kJ·mol $^{-1}$). The phase proportion of ZSM-5/EU-1 can be controlled by adjusting the pre-crystallization time and the gel composition of the first EU-1 forming stage, especially EU-1 template concentration. When only adding the diquaternary ammonium in the second stage, the left aluminosilica species were used to produce ZSM-5 and EU-1 phases simultaneously, showing ZSM-5 proportion was independent on the amount of ZSM-5 template existed in the second stage. XRD characteristic peaks show the change of EU-1 phase in the ZSM-5/EU-1 co-crystalline and both phases integrated closely with each other (Fig.1 A).

Compared to the mechanical mixture (MZE-39) and pure-phase zeolites, ZSM-5/EU-1 co-crystalline zeolite presented more acid sites and the largest BET surface area and highest total pore volume, especially for mesopore volume. In the reaction of n-hexane catalytic cracking, ZSM-5/EU-1 co-crystalline zeolite with EU-1 phase content of ~39% (HZE-39) exhibited the most excellent activity stability (Fig.1 B) and highest selectivity to light olefins, especially to propylene in comparison with HMZE-39 and HZSM-5 samples. At TOS of 120min, the propylene selectivity was 8.57 and 9.16 percentage points higher than the HMZE-39 and HZSM-5 control samples, respectively (Fig.1C). As a result, ZSM-5/EU-1 co-crystalline zeolite is expected to be a new kind of efficient catalytic materials.

(A)

(B)

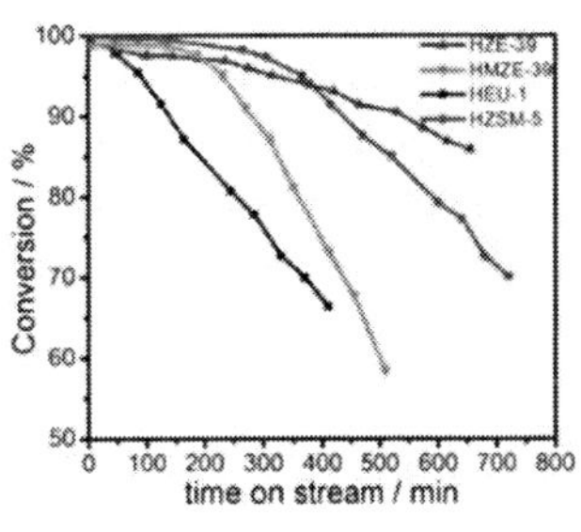

(C)

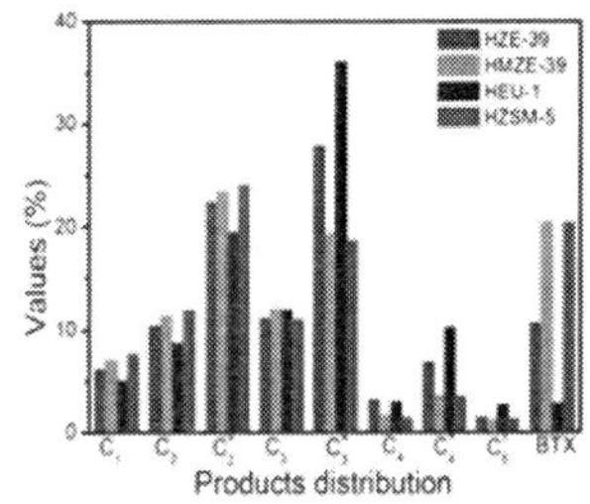

Figure 1. SEM image of ZSM-5/EU-1 zeolite (A) and n-hexane conversion (B) and products distribution (C, TOS=120min) over different zeolites (T=627°C, WHSV=3.0h^{-1}, p $_{n\text{-hexane}}$=9kPa. C_5^+ does not contain BTX.)

Thanks 973 programme (2012CB215002) and NSFC (21176255, 21276278) for financial support.

Reference

(1) Q. Meng, B. Liu, J. Piao, Q. Liu, *Journal of Catalysis* **2012**, *290*, 55-64.
(2) J. Zheng, Y. Yi, W. Wang, K. Guo, J. Ma, R. Li, *Microporous and Mesoporous Materials* **2013**, *171*, 44-52.

INFULENCE OF PORE SRUCTURE OF THE ZEOLITE USY ON CATALYTIC CRACKING OF JATROPHA CURCASL OIL

Qingqing Zheng[a], *Lin Huo*[a], *Xuliang Deng*[b], *Baojian Shen*[a]*

[a]State Key Laboratory of Heavy Oil Processing; the Key Laboratory of Catalysis of CNPC; College of Chemical Engineering; China University of Petroleum, Beijing, 102249, China
[b]Daqing Research Center of Chemical Engineering, PetroChina Research Institute of Petroleum and Petrochemicals, Heilongjiang, Daqing 163714, China
*Corresponding Author. TEL: +(8610)89733369
FAX: +(8610)89733369 EMAIL: baojian@cup.edu.cn

1 Introduction

Zeolite Y is a unique micropore zeolite and the improvement of the catalytic transformation ability toward heavy oil requires the introduction of mesopores into its framework. Additionally, its micropore system is also important for shape-selectivity giving high activity and selectivity[1]. The general preparation methods for the hierarchical zeolite production are steaming, acid leaching (dealumination), base leaching (desilication) or templating[2,3]. For Y zeolite, dealumination by steaming is still the routine method for the generation of mesopores in the crystals. Mesoporous Y zeolite with homogeneous aluminum distribution can be prepared by a sequential desilication-dealumination treatment. Catalytic cracking catalysts showed an increase in gasoline and light cycle oil selectivity when the conversion of a vacuum gas oil[4].

Vegetable oils have been suggested as important sources for the production of fuels and chemicals[5]. Hydrogen-zeolite Y was used at a temperature range of 300-500°C, Over 95 %(by mass) of the vegetable oils were converted to liquid hydrocarbons in the gasoline boiling range, light gases, and water[6]. Ultrastable Y (USY) catalyst exhibits high selectivity of gasoline and diesel oil[7].

In the work, the catalytic cracking of Jatropha Curcasl oil using USY as a heterogeneous catalyst has been studied. Firstly, USY zeolites with differert mesopore size and volume were obtained through combining of dealumination, desilication, calcination and steaming treatment. Then, catalytic cracking reactions of Jatropha Curcasl oil were investigated in fixed-fluidized bed.

2 Experimental

Feedstock. The properties of feedstock are shown in Table 1.

Table 1. Element composition and density of feedstock.

feed	Element composition/%(by mass)			d^{4}_{20}/
	H	C	O	$g\cdot cm^{-3}$
Jatropha Curcasl oil	12.43	76.40	11.17	0.9170

Sample preparation. The USY_A zeolite was prepared by desilication of parent NaY_1 zeolite in aqueous NaOH solution for 3 h, followed by ammonium exchanges, subsequent calcination once and steaming treatment once. The USY_B zeolite was prepared from parent NaY_2 zeolite by ammonium exchanges, followed by subsequent steaming treatment twice. The USY_C zeolite was prepared by desilication of parent NaY_3 zeolite in aqueous NaOH solution for 3 h, followed by ammonium exchanges, subsequent steaming treatment twice. The USY_D zeolite was prepared by desilication of parent NaY_2 zeolite in aqueous NaOH solution for 3 h, followed by ammonium exchanges, followed by subsequent steaming treatment twice.

Characterization Methods. Powder X-ray diffractometer(XRD) was performed on a Panalytical X′ pert instrument with CuKα radiation operated at 40 kV and 40 mA. The pore size distribution was determined by Nitrogen adsorption-desorption measurement on a Micromeritics TriStar 3020 instrument.

Catalysis reactions. Catalytic cracking reaction of Jatropha Curcasl oil was investigated in fixed-fluidized bed operated at atmospheric pressure, the reaction temperature of 480°C, mass ratio of catalysts/feed of 6 and residence time of 70 s.

3 Results and Discussion

3.1 The textural properties of the ultra-stable Y zeolites. The XRD patterns of the USY zeolite samples are illustrated in Fig1. All of the samples exhibit similar peaks with the faujasite zeolite diffraction patterns. The textural pore structure properties of all the zeolite samples are shown in Table 2. All of the USY zeolite samples have a similar framework silica to alumina ratio, crystallinity, and BET surface areas. However, the mesoporous volume were varied. The USY_D showed the highest mesoporous volume of 0.191 $cm^{3}\cdot g^{-1}$.

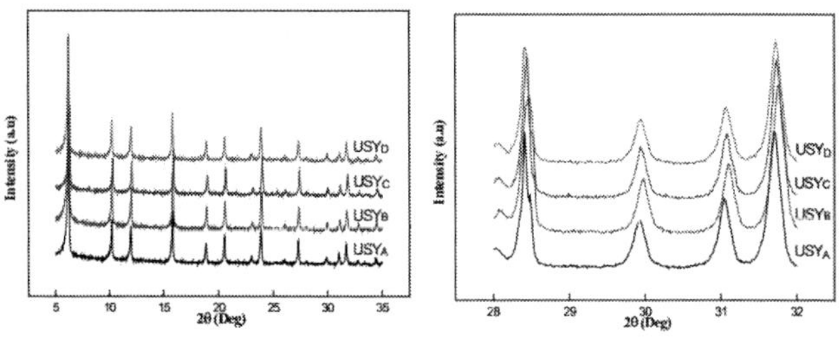

Figure 1. XRD patterns of the USY zeolite samples.

Table 2. Summary of the textural properties of USY samples.

USY	%Cryst[a]	a_0[b](Å)	SiO_2/Al_2O_3[c] (Framework)	S_{BET}[d] ($m^2\cdot g^{-1}$)	S_{exter}[e] ($m^2\cdot g^{-1}$)	V_{meso}[f] ($m^2\cdot g^{-1}$)
A	79	24.395	14.4	622	58	0.103
B	81	24.388	14.9	617	63	0.138
C	82	24.398	14.1	609	72	0.182
D	78	24.383	15.4	591	88	0.191

[a] The relative crystallinity value (%Cryst) calculated by XRD.
[b] Unit cell size (a_0) determined by XRD.
[c] The framework silica to alumina ratio obtained from a_0 via the relationship $SiO_2/Al_2O_3=(25.858-a_0)\times2/(a_0-24.191)$.
[d] BET surface area.
[e] Mesoporous surface area evaluated by the t-plot method.
[f] $V_{Meso}=V_{Total}-V_{Micro}$.

The nitrogen adsorption-desorption isotherms are presented in Fig 2. The pore size distributions from the asorption branch and desorption branch of the USY zeolite samples with different pore size are shown in Fig 3. The USY_B, USY_C, and USY_D showed a broad distribution of mesopores with two peaks.

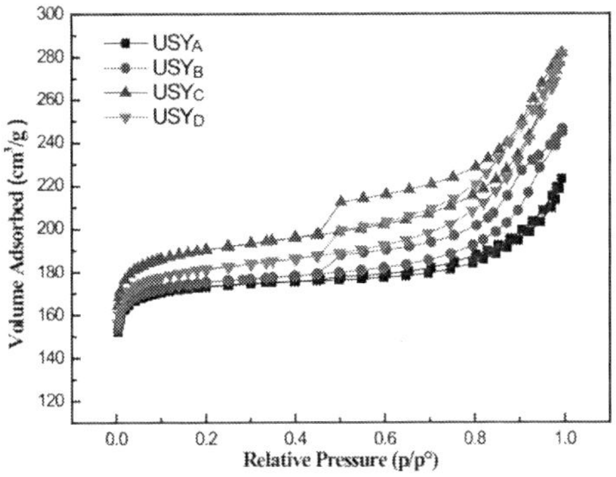

Figure 2. Nitrogen adsorption-desorption isotherms for the USY zeolite samples.

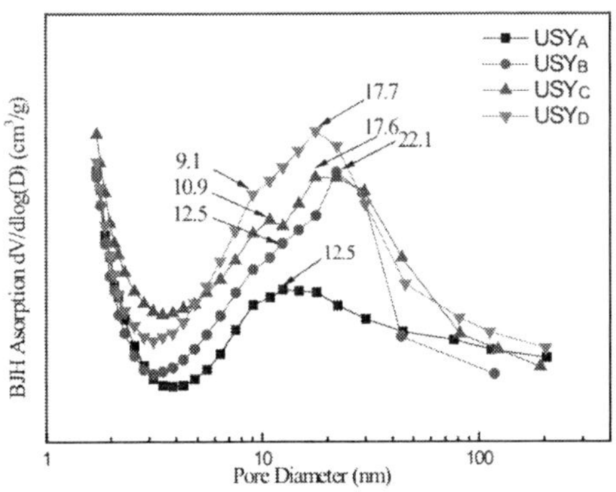

Figure 3. The pore size distributions from the asorption branch(left) and desorption branch(right) of the USY zeolite samples.

3.2 Conversion and product distribution. The activities of different catalysts for the catalytic cracking performance of Jatropha Curcasl oil on zeolites USY were shown in Table 3. Evolution of the mesopore volume of the conversion and gasoline + LCO yields are presented in Fig 4. The conversion of Jatropha Curcasl oil and gasoline+LCO yields of USY_D were 2.5 % higher than that of USY_A, as well as lower dry gas and LPG yields. Owning to the increased mesoporous volume of the USY zeolites, the catalysts showed superior catalytic performance. The catalyst with higher USY mesoporus volume showed higher conversion and higher gasoline+LCO yields. (Fig. 4).

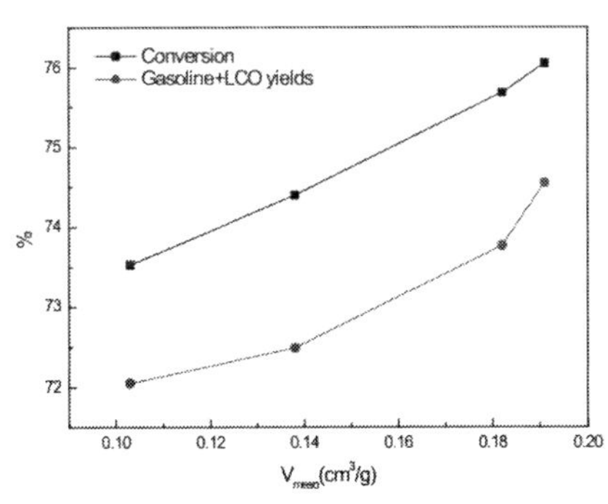

Figure 4. Evolution of the mesopore volume of the conversion and gasoline + LCO yields.

Table 3. Overall product distribution of fixed-fluidized bed (%,by mass)

Catalyst	A	B	C	D
Dry Gas	1.58	1.64	1.43	1.44
Liquid petroleum gas(LPG)	15.27	14.99	12.83	12.46
Gasoline	43.38	43.02	42.59	44.28
Light cycle oil(LCO)	28.68	29.46	31.18	30.27
Bottoms	4.34	4.69	4.46	4.11
Coke	5.14	4.18	4.56	4.88
Conversion	73.53	74.40	75.68	76.05
Gasoline + LCO	72.05	72.49	73.77	74.55
LPG + Gasoline + LCO	87.32	87.48	86.59	87.01

4 Conclusions

The USY zeolites were designed by sequential desilication, dealumination, calcination and steaming treatment. The USY with highest mesoporous volume showed the highest Jatropha Curcasl oil and gasoline+LCO yields. The higher conversion and higher gasoline+LCO yields of the catalysts were attributed to their improved pore system, which led to decreased residence times and suppressed secondary reactions. As a result, relatively larger molecule products can easily diffuse out of the zeolitic micropores via mesopores without undergoing further over cracking.

Acknowledgement. The authors greatly acknowledge the funding of this project by the National Natural Science Foundation of China(U1462202).

References

(1) Park, D. H.; Kim, S. S.; Wang, H.; Pinnavaia, T. J.; Papapetrou, M. C.; Lappas, A. A.; Triantafyllidis, K. S.; *Angew. Chem. Int. Ed.*, **2009**, *48*, 7645.

(2) De Jong, K. P.; Zečević, J.; Friedrich, H.; de Jongh, P. E.; Bulut, M.; van Donk. S.; Kenmogne, R.; Finels, A.; Hulea, W.; Fajula, F. *Angew. Chem. Int. Ed.*, **2010**, *49*, 10074.

(3) Fathi, S.; Sohrabi, M.; Falamaki, C. *Fuel*, **2014**, *116*, 529-37.

(4) Qin, Z.; Shen, B.; Gao, X.; Lin, F.; Wang, B.; Xu, C. *J. Catal.*, **2011**, *278*, 266.

(5) Dandik, L.; Aksoy, H. A.; Erdem-Senatalar, A. *Energy Fuels*, **1998,** *12*, 1148.

(6) Idem, R. O.; katikaneni, S. P. R.; Bakhshi, N. N. *Fuel Process Technol.*, **1997**, *51*, 101.

(7) Twaiq, F. A.; Zabidi, NAM; Bhatia, S.; *Energy Fuels*, 1995, *9*, 1065.

ENGINEERING THE H_2S SPLITTING CYCLE FOR OIL SANDS BITUMEN UPGRADING: CAN WE RECOVER HYDROGEN FROM H_2S TOO?

Hui Wang[1*], *Wei Zhang*[1,2], *Meisheng Liang*[1,3], *Armin Moniri*[1]

[1]Department of Chemical and Biological Engineering, University of Saskatchewan, Saskatoon, SK S7N 5A9, Canada
[2]College of Chemistry and Chemical Engineering and [3]College of Environment Science and Engineering, Taiyuan University of Technology, Taiyuan 030024, China
*Corresponding author: hui.wang@usask.ca

1. Introduction

Sulfur removal and recovery is essential for oil and gas industry, especially for upgrading of oil sands bitumen which contains 5 to 6 weight percent of sulfur. Through Claus process, elemental sulfur is recovered from H_2S but hydrogen is disposed as the form of water. However, hydrogen, not sulfur, is needing commodity in the same industry section where hydrotreaing is performed for desulfurization and denitrogenation. Inspired by the sulfur-iodine (S-I) water splitting cycle, Wang developed a chemical route of converting H_2S into H_2 via the following chemical reactions:[1]

Section 1: $H_2S + H_2SO_4 \rightarrow SO_2 + S + 2H_2O$ (1)

Section 2: $2H_2O + I_2 + SO_2 \rightarrow H_2SO_4 + 2HI$ (2)

Section 3: $2HI \rightarrow H_2 + I_2$ (3)

The process consists of three sections: H_2S oxidation, Bunsen reaction, and HI decomposition. Given all other substances H_2SO_4, SO_2, H_2O, I_2, and HI are cycled in the system, the three reactions that split H_2S into H_2 and S constitute the H_2S-splitting cycle and the net reaction is

$$H_2S \rightarrow H_2 + S \quad (4)$$

If elemental sulfur from reaction (1) can be further oxidized into SO_2,

$$O_2 + S \rightarrow SO_2 \quad (5)$$

the new cycle, denoted as H_2S–H_2O splitting or modified H_2S splitting cycle, eventually produces two moles of H_2 and one mole of sulfuric acid from one mole of H_2S, *i.e.*,

$$H_2S + O_2 + 2H_2O \rightarrow H_2SO_4 + 2H_2 \quad (6)$$

This chemistry also lays the foundation for using industrial sulfur-containing wastes for H_2 production. For example, SO_2 from stack gas can be trapped and used for producing H_2.

2. Engineering research on H_2S splitting cycle

2.1 Thermodynamics, kinetics and mass transfer studies on H_2S and sulfuric acid reaction (Section 1). The gas-liquid reactions between H_2S and concentrated sulfuric acid, denoted as section 1 in the cycle, has been studied by Wang et al.[2-6] While the oxidation of H_2S by sulfuric acid occurs, an additional reaction, the Claus reaction is also thermodynamically feasible in the reaction liquid;[2]

$$2H_2S + SO_2 \rightarrow 3S + 2H_2O \quad (7)$$

H_2S is eventually oxidized by molecular H_2SO_4 and then consecutively by produced SO_2, where the aqueous acid solution is the catalyst. Attempts were made to match the reaction rate of reactions (1) and (7) under the same conditions such that H_2S can be completely converted into elemental sulfur[3]. However, kinetic studies of these two reactions showed that in concentrated sulfuric acid solution reaction (1) overwhelms and SO_2 is mainly produced.[4,5]

Later, Wang et al.[6] studied the continuous operation of the reaction between H_2S and H_2SO_4 of various concentrations in a packed-bed or trickle-bed column reactor. The mass transfer behavior, the wetted packing surface area, and the rate-determining step between reaction and mass transfer were investigated. It has been found that the comparison between the rate constants and mass-transfer coefficient is able to determine the rate-controlling regimes in terms of operating conditions such as acid concentration, reaction temperature, and acid and gas flow rates. Tests were also carried out with gaseous compounds often found in industrial H_2S streams. The result showed that sulfuric acid does not interact with gas components such as CH_4, CO_2, COS, and CS_2. However, it consumes ethylene to some extent, with the conversion depending on acid concentration and temperature.

Moreover, various sulfur-containing compounds, which exist in different industry sectors as byproduct or waste product, were selected to produce hydrogen.[7] It has been found that H_2S-splitting cycle or H_2S–H_2O splitting cycle can be made more energy-efficient, subject to that the heat that is generated from H_2S and elemental sulfur oxidation can be recovered and used to support the energy requirements for other steps of the cycle.

To maximize SO_2 production in Section 1, different operating conditions such as temperature, stirring rate and acid concentration were investigated. The experiment results showed that higher reaction temperature and higher initial H_2SO_4 acid concentration were in favor of more SO_2 production. However, the application of stirring facilitated less SO_2 production because stirring helped the H_2S and the SO_2 gases enter bulk acid where they form elemental sulfur.

2.2 Iodine-toluene solution in Bunsen reaction (Section 2). General Atomics (GA) Company developed a Bunsen reaction stoichiometry in the 1970s.[8] It proposed that the Bunsen reaction be operated in a large excess of iodine and water in order to separate the two acid products into two immiscible liquid phases: a heavy phase of HI_x solution and a light sulfuric acid phase, shown in reaction (8).

$$9I_2 + SO_2 + 16H_2O \rightarrow (2HI + 10H_2O + 8I_2)_{heavy\ phase} + (H_2SO_4 + 4H_2O)_{light\ phase} \quad (120℃) \quad (8)$$

Although this stoichiometry allows direct separation of the two acid products by phase separation, it faces even greater challenges in further purification of the two acids for the subsequent reactions in the cycle, tremendous energy consumption in such purifications and excessive material transportations, corrosion, iodine deposition (plugging) and side reactions.

Wang et al.[9] proposed to dissolve iodine in toluene solvent to provide with fluid iodine below its melting point to the Bunsen reaction. As a result, the Bunsen reaction (Section 2) does not have to be conducted at 120℃ or higher for melting the solid iodine. Instead, it can be operated at room temperature, which is even more thermodynamically favorable. The apparent kinetics of Bunsen reaction with iodine-toluene solution was studied.[10]

Compared with the General Atomics stoichiometry and operation condition for the sulfur-iodine cycle, Iodine-toluene solution can conduct Bunsen reaction at ambient temperatures and in fluid medium which will bring about many benefits such as:

(1) Mitigating the side reactions between hydroiodic acid and sulfuric acid,
(2) Reducing the corrosion potentials,
(3) Making the product separations easier at room temperature,
(4) Avoiding iodine vapor deposition and the blockage of the set-up.
(5) Reducing the unnecessary transportation of the excessive cycling agents.

However, with introducing toluene, an organic phase to the system, the multiphase Bunsen reaction has very complicated mass transfer pathways.[10] Iodine dissolves in both toluene and HI aqueous solution. SO_2 dissolves not only in water but also in toluene.

The multi-phase mixture must lead to significant mass transfer resistance. The recently emerging Corning Advanced-Flow reactors (AFR's) can be used to facilitate the multiphase Bunsen reaction so as to improve the mixing efficiency and overcome mass transfer resistance among phases.

2.3 Corning Advanced-Flow reactors used for Bunsen reaction. Theoretically, Corning Advanced-Flow reactors can be used for the multiphase reactions in this study including reactions (1) and (2). Corning AFR's are made of ceramic or glass material, which greatly eases the corrosion concerns in handling corrosive materials such as sulfuric acid in Section 1 and the mixture of diluted hydroiodic acid and sulfuric acid in Section 2. The modulus design of the AFR's allows for easy scaling up without reducing mass transfer characteristics. The heart shape design in the micro-flow channels within the AFR's provides excellent mass transfer capabilities. We first studied the Bunsen reaction using Corning reactors.

The absorption rate of sulfur dioxide by water, toluene, and water-toluene mixture was investigated using two scales of the Corning reactors, Low Flow (LF-AFR), and G1-AFR modules. The liquid-side mass transfer coefficient (K_La) for the SO_2-water and SO_2-toluene systems was determined under such conditions that plug flow was assumed through each of the heart-shape cells. The overall mass transfer coefficient (K_La) from the two scales of the reactors resulted in the same magnitude of values. In addition, the Bunsen reaction was also studied in Corning LF-AFR, and G1-AFR modules by dissolving iodine (I_2) in toluene. The effects of the operating conditions such as gas and liquid flow rates, the water to toluene ratio, and the temperature in the ambient range (22-80^oC) on the absorption rates of SO_2 and the I_2 reaction rate were studied. Both results revealed seamless scaling-up capability of Corning reactors when the flow rates were increased twenty times.[11]

2.4 Direct electrolysis of HI-H_2SO_4 aqueous solution for hydrogen production (Section 3). Great efforts have been made for the S-I cycle in separations of H_2SO_4 and HI from their aqueous solutions after the Bunsen reaction but none is successful from the engineering sense.[12] There are also tremendous challenges when thermal decomposition or catalytic decomposition of HI vapor is used for reaction 3. Both the separation of HI and H_2SO_4 in aqueous solution and the HI decomposition are very energy-intensive.[13,14]

We proposed direct electrolysis of HI in the HI-H_2SO_4 aqueous solution in the presence of toluene – the stream from our Bunsen reaction – at room temperature. The experimental results show that controlling the decomposition voltage can render HI to be oxidized successively and H_2SO_4 to remain. The presence of toluene in the anode compartment can render the product iodine to be dissolved into organic phase, hence increasing the driving force of electrolysis and enhancing the current efficiency. After electrolysis, the I_2-toluene is circulated to the Bunsen reaction section and the H_2SO_4 aqueous solution is sent for upgrading to higher concentrations.

3. Process integration

The upper boundary thermal efficiency of the H_2S splitting cycle for hydrogen production was studied, and the values of the thermal efficiency of the two versions of the splitting cycle, $H_2S \rightarrow H_2 + S$ and $H_2S + 2H_2O + O_2 \rightarrow 2H_2 + H_2SO_4$, are found to be 0.66 and 0.70, respectively.[12]

Plant design was conducted based on this chemistry and the engineering research results. Attempts were made to see if the hydrogen recovery process can be used to replace both the Claus process for sulfur removal and recovery and the hydrogen production from natural gas stream reforming in oil sands bitumen upgrading. Figure 1 shows a simplified process flow diagram of a prototype hydrogen production plant based on the H_2S splitting cycle. The design results indicate that there are still rooms to improve the economics so that the replacement can occur.

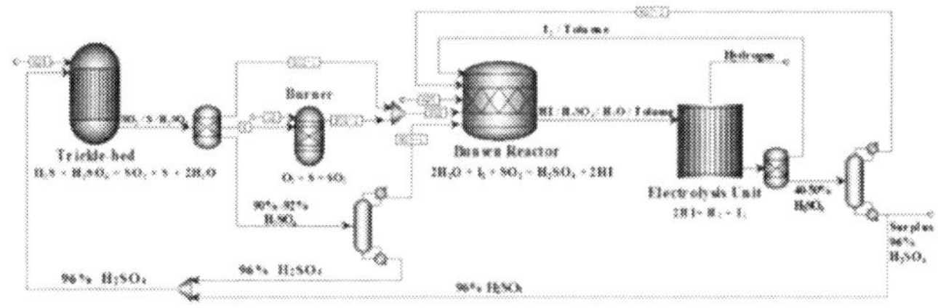

Figure 1. Overview schematic of H_2S–H_2O splitting flow sheet.

4. Conclusions

- The chemistry to recover hydrogen from H_2S through H_2S splitting cycle has been established. There are four reactions working together to generate two moles of H_2 from one mole of H_2S, which meets the hydrogen need in hydrotreating for not only desulfurization but also denitrogenation as well as aromatics removal.
- Every reaction and the associated separation processes have been found technically feasible in this research. Plant design can be made based on the experiment results.
- Economics analysis in the integration design shows that there are rooms to improve so that this hydrogen recovery process can be implemented to replace both Claus process for sulfur recovery and natural gas steam reforming for hydrogen production.

Acknowledgement: The authors gratefully acknowledge the financial support provided by the Natural Science and Engineering Research Council of Canada (NSERC) STPGP 350428 - 2007 with a Strategic Grant. We also thank Corning Reactor Technology Center (China) and Changzhou University for allowing us using their facilities.

References

(1) H. Wang, *Int. J. Hydrogen. Energ.* **2007**, *32, 3907.*
(2) H. Wang, I.G. Dalla Lana, K.T. Chuang, *Can. J. Chem. Eng.* **2003**, *81, 80.*
(3) Q. Zhang, I.G. Dalla Lana, K.T. Chuang, H. Wang, *Ind. Eng. Chem. Res.* **2000**, *39, 2505.*
(4) H. Wang, I.G. Dalla Lana, K.T. Chuang, *Ind. Eng. Chem. Res.* **2002**, *41, 6656.*
(5) H. Wang, I.G. Dalla Lana, K.T. Chuang, *Ind. Eng. Chem. Res.* **2002**, *41, 4707.*
(6) H. Wang, I.G. Dalla Lana, K.T. Chuang, *Ind. Eng. Chem. Res.* **2004**, 43, 5846.
(7) A. Moniri, P. Mertins, H. Wang, *Int. J. Hydrogen Energy.* **2012**, *37, 15003.*
(8) J.H. Norman, K.J. Mysels, R. Sharp, D. Williamson. *Int. J. Hydrogen Energy.* **1982**, *7, 545.*
(9) H. Wang, A.Le Person, X. Zhao, J. Li, P Nuncio, L.Q. Yang, A. Moniri, K.T. Chuang. *Fuel. Process. Technol.* **2013**, *108, 55.*
(10) J. Li, A. Moniri, H. Wang, *Int. J. Hydrogen Energy.* **2015**, *40, 2912.*
(11) A. Moniri. *Thesis.* University of Saskatchewan. **2015.**
(12) G. Yu, H. Wang, K.T. Chuang. *Energy & Fuels.* **2009**, *23, 2184.*
(13) M. Roth, K.F. Knoche, *Int. J. Hydrogen Energy.* **1989**, *14, 545.*
(14) D. O'Keefe, C. Allen, G. Besenbruch, L. Brown, J. Norman, R. Sharp. *Int. J. Hydrogen Energy.* **1982**, *7, 381.*

NANOSCALE DESIGN OF BIFUNCTIONAL CATALYST AND ITS IMPACT ON HYDROCONVERSION

Jovana Zečević[1], Gina Vanbutsele[2], Krijn P. de Jong[1], Johan A. Martens[2]

[1] Inorganic Chemistry and Catalysis, Utrecht University, Universiteitsweg 99, 3584CG Utrecht, The Netherlands
[2] Centre for Surface Chemistry and Catalysis, KU Leuven, Celestijnenlaan 200F, 3001 Leuven, Belgium
j.zecevic@uu.nl

Introduction

Important oil refining processes such as hydrocracking, hydrotreating and hydroisomerization rely heavily on the use of bifunctional zeolite-based catalysts [1]. Bifunctional zeolite-based catalysts typically contain metal nanoparticles (e.g. Ni/Mo or Ni/W sulfides, or Pt and Pd) which provide (de)hydrogenation function, and Brønsted acid sites located in the zeolite micropores which provide isomerization and cracking function. The balance between the metal and acid function, expressed as the ratio of number of accessible metal atoms and acid sites (n_M/n_A), is one of the factors influencing the product distribution [2]. Next to that, the intimacy criterion proposed by Weisz in 1962 [3], argues that the distance between the active sites also matters, and should be kept as short as possible. Such bifunctional catalysts are commonly prepared by depositing metal (sulfide) nanoparticles onto catalyst bodies containing zeolite component and alumina that binds zeolite crystals and insures mechanical stability of the catalyst. Therefore, in order to optimize catalyst properties, it is of great importance to understand the mechanism and establish the control over metal deposition [4-6].

By integrating the existing knowledge on interactions between Pt complexes and zeolite and other oxide supports, we devised a simple synthesis method to selectively deposit Pt on either exclusively zeolite or exclusively alumina component of zeolite-alumina (YA) extrudates [7]. This has been achieved by considering the interplay between ion exchange (IE) and electrostatic adsorption (EA) of Pt-complexes [4,5]. Advanced electron microscopy techniques were essential to unravel the catalyst structural features at nanometer length scales. Hydroconversion tests showed unexpected impact of the location of Pt (de)hydrogenation sites with respect to the acidic zeolite cracking sites on catalyst performance.

Experimental

As a support we used cylindrical extrudates containing 50/50 % weight ratio of mesoporous zeolite Y (Zeolyst CBV760, Si/Al=30 at/at) and γ-alumina. Extrudates were crushed and sieved to 0.2 - 0.4 mm fractions, here referred as YA extrudates. To deposit Pt particles exclusively in the zeolite component of YA extrudates (Pt-YA-Am), we mixed $Pt(NH_3)_4(NO_3)_2$ aqueous solution with YA fractions. After 3h of stirring at the room temperature, the suspension was filtered, washed and Pt-YA fractions were dried overnight at 120 °C and reduced at 600 °C in a flow of H_2. To deposit Pt particles exclusively in the alumina component (Pt-YA-Cl), H_2PtCl_6 solution was used as a precursor. Prior to addition of the precursor, pH of the suspension of the YA fractions was adjusted to 3. After 3h of stirring, suspension was filtered, fractions were dried overnight at 120 °C and reduced in H_2 at 600 °C. An additional heat treatment of reduced Pt-YA fractions in a N_2 with 1% O_2 flow was performed. Tecnai 20 and Talos F200X FEI microscopes were used for EDX mapping, electron tomography (ET) and high-angle annular dark field imaging (HAADF), respectively. Hydroconversion of n-nonadecane (Sigma, 98%) dissolved in heptane (Acros Organics) at 1 mol% concentration was performed at a pressure of 0.65 MPa, 14.6 molar ratio of H_2 to hydrocarbons and the space time at the reactor entrance W/F 713 kg s mol^{-1}.

Results and Discussion

70 nm thick ultramicrotomed sections of YA support were first imaged using EDX mapping. EDX maps revealed that the zeolite crystals were surrounded with at least one layer of alumina platelets and that on microscopic scale of zeolite and alumina components were well mixed, with alumina regions not exceeding 5 μm in length (Figure 1a). The intimate contact between zeolite crystals and alumina platelets was further observed using electron tomography (Figure 1b). In order to selectively deposit Pt particles on the zeolite component of the YA extrudates (Pt-YA-Am catalyst), precursor containing $Pt(NH_3)_4^{2+}$ was used. With this precursor it was assumed that IE of zeolite protons will dominate over EA on alumina providing that the pH of suspension is maintained below 9 [4,5]. To obtain Pt particles of ~3 nm, Pt-YA fractions were reduced directly in H_2, which was shown to promote the growth of Pt particles [6]. EDX map of 70 nm thick ultramicrotomed sections confirmed that Pt particles of ~2.5 nm were successfully deposited on the zeolite crystals (Figure 1c). To deposit Pt on exclusively alumina component of the YA support (Pt-YA-Cl catalyst), H_2PtCl_6 was used as a precursor. In addition, pH was lowered to ~3 with HCl, in order to ensure that the alumina surface is positively charged and EA of negatively charged $PtCl_6^{2-}$ ions could take place, while IE between zeolite protons and negatively charged $PtCl_6^{2-}$ is nil [4,5]. Direct reduction in H_2, followed by an additional heat treatment in N_2 with 1 vol% O_2, lead to Pt particles of ~ 3.5 nm. EDX mapping confirmed that Pt particles were deposited on alumina exclusively (Figure 1d). Next to comparable Pt particle size and acidity (as confirmed by NH_3 TPD), the two catalysts had similar Pt loading of 0.6 wt% and 0.7 wt%, for Pt-YA-Am and Pt-YA-Cl catalysts respectively, as shown by inductively coupled plasma (ICP) analysis.

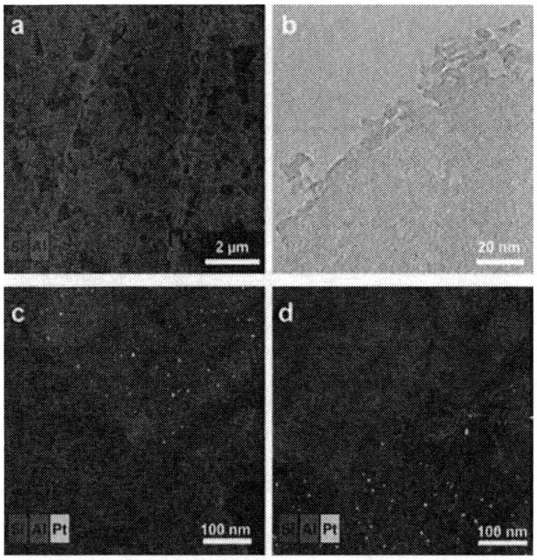

Figure 1. a) EDX elemental map of 70 nm thick ultramicrotomed section of YA extrudate, showing zeolite component in blue and alumina in red. b) Slice from electron tomography reconstruction. c) EDX elemental map of 70 nm thick ultramicrotomed section of Pt-YA-Am sample, showing zeolite component in blue, alumina in red and Pt in yellow. d) EDX elemental map of 70 nm thick ultramicrotomed section of Pt-YA-Cl sample, showing zeolite component in blue, alumina in red and Pt in yellow.

Hydroconversion tests were performed using n-nonadecane. Both catalysts showed similar activity (Figure 2a). However, remarkable difference in selectivity towards isomerized and cracked products was observed between the two catalysts (Figure 2b). This finding illustrates the importance of the relative location of hydrogenation and cracking function in the complex bifunctional catalysts. When the Pt particles were positioned inside zeolite micropores (Pt-YA-Am), in the closest proximity to acid sites of zeolite, selectivity towards cracking was promoted. The presence of Pt on alumina, and thus at nanoscale distance from zeolite acid sites, in Pt-YA-Cl led to an increase in selectivity towards isomerization products.

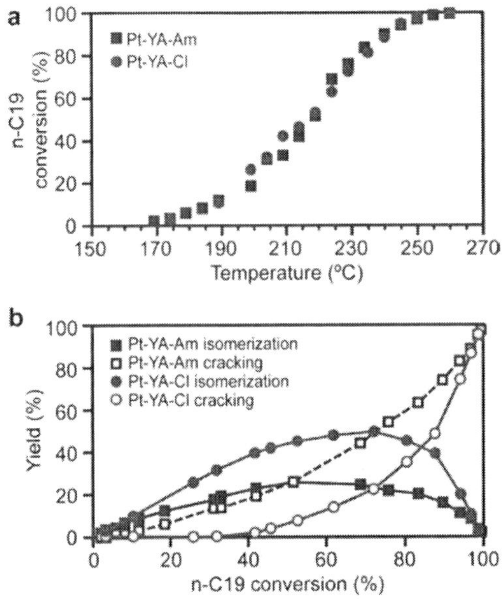

Figure 2. Hydroconversion of n-nonadecane, activity (a) and selectivity (b).

Conclusions

This study revisited the notion of intimacy criterion in bifunctional catalysts. It was shown that, opposite to what was long believed, the optimal position of metal function is not at the closest distance to the acid function (inside zeolite micropores), but rather at nanometer-scale distance, e.g. on the surrounding alumina binder of the mixed-phase support. Developed synthesis methodology allows for greater control in catalyst design and offers flexibility in tuning the catalyst properties to meet the requirements of specific processes.

Acknowledgement. This work has been supported by the NRSC-C and the European Research Council, EU FP7 ERC Advanced Grant no. 338846. (J.Z. and K.P.deJ.), and by the Flemish government via the Methusalem program (J.A.M.).

References

(1) Thybaut, J. W.; Laxmi Narasimhan, C. S.; Denayer, J. F.; Baron, G. V.; Jacobs, P. A.; Martens, J. A.; Marin, G. B. *Ind. Eng. Chem. Res.* **2005**, *44*, 5159.
(2) Alvarez, F.; Ribeiro, F. R.; Perot, G.; Thomazeau, C.; Guisnet, M. *J. Catal.* **1996**, *162*, 179.
(3) Weisz, P. B. *Adv. Catal.* **1962**, *13*, 137.
(4) Hao, X.; Spieker, W. A.; Regalbuto, J. R. *J. Colloid Interface Sci.* **2003**, *267*, 259.
(5) Spieker, W. A.; Regalbuto, J. R. *Chem. Eng. Sci.* **2001**, *56*, 3491.
(6) Zečević, J.; van der Eerden, A. M. J.; Friedrich, H.; de Jongh, P. E.; de Jong, K. P. *ACS Nano* **2013**, *7*, 3698.
(7) Zečević, J.; Vanbutsele, G.; de Jong, K. P.; Martens, J. A. *Nature*, **2015**, *528*, 245.

PREPARATION A SHAPE-SELECTIVE ZEOLITE MATERIAL AND ITS APPLICATION IN HEAVY OIL CATALYTIC CRACKING

*Penghui Zeng[1], Yun Liang[1], Xiaozhong Guo[1], Liyuan Cao[1], Qiaoxia Guo[2], Shenyong Ren[1], Xiaochun Zhu[1], Baojian Shen[1]**

[1] State Key Laboratory of Heavy Oil Processing, College of Chemical Engineering; China University of Petroleum, Beijing 102249, P. R. China

[2] College of Science, China University of Petroleum, Beijing 102249, P. R. China

*Corresponding author: baojian@cup.edu.cn

1. Introduction

The synthesis of various novel zeolites has attracted most attention since such materials are technologically important in many catalysis processes especially in the petrochemical and refining industry [1].

Recently, ITQ-13 was synthesized by Corma and co-workers [2]. It contains a unique tridirectional channels in which one of them has 9 MR (4.0 × 4.9 Å) openings and the other two channels are straight (4.7 × 5.1 Å) and sinusoidal 10 MR (4.8×5.7 Å) [2]. With these unique structure characteristics, ITQ-13 exhibits quite different catalytic activity as compared with ZSM-5 zeolite. It has been widely applied in adsorption and separation processes, hydrocarbon transformation reactions such as conversion of the aromatics and catalytic cracking of hydrocarbons for propylene boosting [3].

Incorporation of heteroatoms into the framework structure of molecular sieves is an important route to modify the nature of frameworks to impart catalytic activity. The catalytic behavior is strongly influenced by the nature, the local environment and the stability of the metal introduced [4]. Over recent decades, many efforts have been made on this aspect [5]. However, to the best of our knowledge there is no report on hydrothermal synthesis of framework substituted Ce ITQ-13 zeolite. In this study, rare earths was incorporated into the framework of Al-ITQ-13 zeolite by hydrothermal synthesis. The final zeolite was then extensively characterized by X-ray diffraction (XRD), N_2 adsorption, X-ray fluorescence, NH_3-TPD And also tested in the cracking of vacuum gasoil.

2. Experimental

The Ce-Al-ITQ-13 zeolite was synthesized under hydrothermal conditions at 180 °C according to the CN Patent 201410743244.9. An aqueous solution of hexamethonium dihydroxide ($R(OH)_2$, self-made) was used as template and the cerium nitrate hexahydrate as cerium source. The reference Al-ITQ-13 zeolite was synthesized by the same method without adding the cerium source.

Both zeolite samples were calcined in static air at 550 °C for 4 h to remove the template. The Proton forms of zeolites were obtained by double ion-exchange of the calcined zeolites with a 1 M NH_4Cl solution at 90 °C for 2 h and calcination in air at 550 °C for 4 h.

The cracking of vacuum gasoil (VGO) was performed at 530 °C and the catalyst-oil ratio was 6 in an advanced cracking evaluation (ACE). The VGO had a density of 0.93 g/mL at 20 °C, and contained only a minor amount of metals, 65% saturated hydrocarbons, 28% aromatic hydrocarbons, and 7% resin. Ce-Al-ITQ-13 and Al-ITQ-13 samples were tested as additives (10 wt%) of USY zeolite.

3. Results and Discussion

Figure 1 shows the XRD patterns of Ce-Al-ITQ-13 and Al-ITQ-13 zeolites with the ITQ-13 structure without indications of any impurity phase. The XRD pattern of Ce-Al-ITQ-13 is comparable to diffractograms of reference Al-ITQ-13 zeolite. The crystallinity, BET surface area, and micropore volume of Ce-Al-ITQ-13 sample were comparable with those of reference Al-ITQ-13 zeolite (as listed in Table 1). However, the d-spacing values of Ce-Al-ITQ-13 decreased as compared with Al-ITQ-13. These changes in d-spacing values caused by the content of heteroatoms are in accordance with earlier reports [5].

Elemental analyses of the Ce-Al-ITQ-13 zeolites show that the Ce/Si ratios in the solid products are similar to that of the corresponding gels.(see Table 1), indicating that almost all Ce cations were built into the zeolite framework.

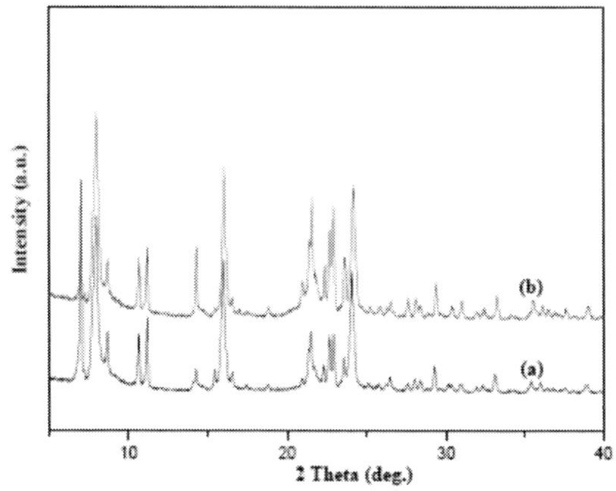

Figure 1. XRD patterns of zeolite samples: (a) Ce-Al-ITQ-13, (b) Al-ITQ-13.

Table 1. Physics and textural properties of samples.

Sample	$(Ce/Si)_{gel}$	$(Ce/Si)_{zeolite}$*	Crystallinity [%]	S_{BET} [m^2g^{-1}]	V_{total} [cm^3g^{-1}]
Ce-Al-ITQ-13	0	0	97	345	0.169
Al-ITQ-13	0.01	0.007	100	351	0.171

* determined by XRF analysis.

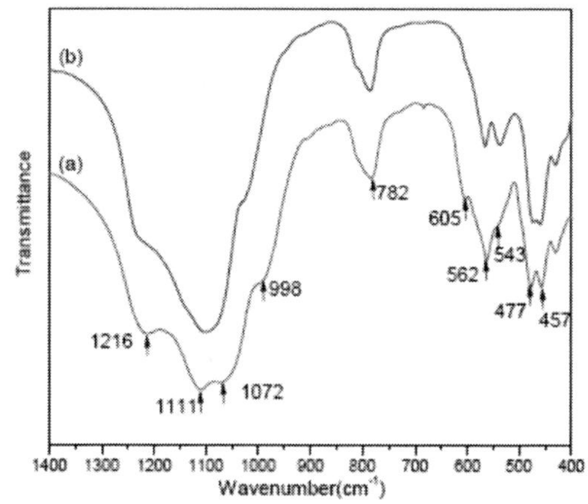

Figure 2. FT-IR spectra of samples: (a) Ce-Al-ITQ-13, (b) Al-ITQ-13.

Figure 2 shows the FT-IR spectra of Ce-Al-ITQ-13 and Al-ITQ-13.The band at ~1216 cm^{-1} is assigned to the asymmetric stretching vibrations of Si-O-Si at the external surface of zeolite sample, while the band around 1111 cm^{-1} is attributed to the internal asymmetric stretching vibrations of Si-O-Si. Moreover, the band at ~782 cm^{-1} is ascribed to the external symmetric stretching vibrations, and the bands near 543 and 562 cm^{-1} stand for the stretching vibrations of double member ring. FT-IR spectra, as supported by the XRD results, indicate the formation of ITQ-13 zeolite with a pure phase even in the case of Ce-Al-ITQ-13 sample. It should be noted that there is a distinct absorbance bands near 1072 cm^{-1} (internal T–O–T asymmetric stretch) which indicates the heteroatom substitution in the zeolite framework (Figure 2 (a)).

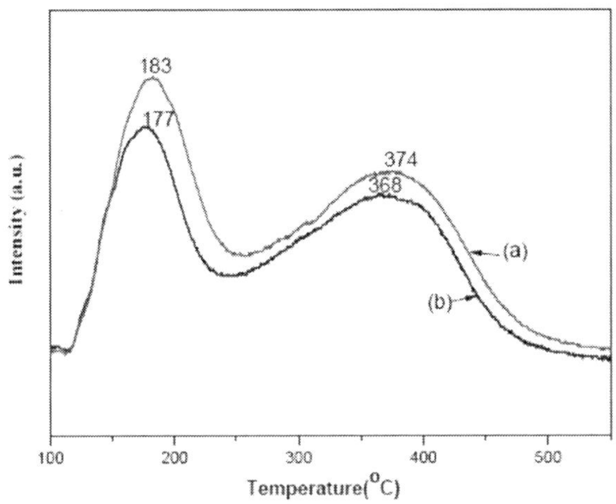

Figure 3. NH_3-TPD profiles of samples: (a) Ce-Al-ITQ-13, (b) Al-ITQ-13.

Figure 3 displays the NH_3-TPD profiles of proton form Ce-Al-ITQ-13 and Al-ITQ-13 zeolites in the temperature range between 100 °C and 600 °C. As shown in Figure 3, two peaks at around 184 and 374 °C in the H-Ce-Al-ITQ-13 curve suggest the existence of weak LAS and strong BAS, respectively. Compared with H-Ce-Al-ITQ-13, the intensity of the peak substantially decreased and shifted toward the low temperature for H-Al-ITQ-13. It indicates that there were more acid sites in H-Ce-Al-ITQ-13 than that of H-Al-ITQ-13 zeolite.

Table 2. The products distribution of VGO cracking over catalysts of USY + Ce-Al-ITQ-13(10wt.%) and USY + Al-ITQ-13(10wt.%)

CAT	USY + Ce-Al-ITQ-13	USY + Al-ITQ-13
Conversion (wt.%)	58.21	55.60
Yield (wt.%)		
Dry gas	2.07	1.95
LPG	17.03	14.46
Gasoline	33.03	33.25
LCO	27.14	28.08
Slurry	14.66	16.32
Coke	6.07	5.94
Light oil products	60.17	61.33
Liquid products	77.20	75.79
$C_2^=$	0.68	0.55
$C_3^=$	6.09	4.75
Selectivity of $C_3^=$ (wt.%)	35.76	32.84

Light oil products = gasoline + LCO.
Liquid products = LPG + gasoline + LCO.

The results of the cracking of VGO on catalyst samples containing the Ce-Al-ITQ-13 or Al-ITQ-13 additives are presented in Table 2. As expected, the conversion of the catalyst containing Ce-Al-ITQ-13 zeolite is 2.61% higher than that of the catalyst containing Al-ITQ-13. The former catalyst also show a significant decrease (1.66 wt.%) in slurry and an increase in LPG products (2.57 wt.%) compared to the later. And the former catalyst gives a increase (2.92 wt.%) in propylene selectivity.

4. Conclusions

The Ce-Al-ITQ-13 zeolite was synthesized at 180 °C under hydrothermal conditions by using an aqueous solution of hexamethonium dihydroxide as template and cerium nitrate hexahydrate as cerium source. The crystallinity, BET surface area, and micropore volume of Ce-Al-ITQ-13 sample were similar to those of Al-ITQ-13 zeolite. XRD analyses and FT-IR spectroscopy show that the cerium metal has been successfully incorporated into the framework of Al-ITQ-13 zeolite. The resulting zeolites were tested by the cracking of vacuum gasoil. Ce-Al-ITQ-13 zeolite showed higher propylene selectivity as compared with reference Al-ITQ-13 zeolite.

Acknowledgement: The authors acknowledge the funding of this project by Science Foundation of China University of Petroleum, Beijing No. 2462015YQ0309.

References

(1) Corma A. *J. Catal.* **2003**, 216, 298.
(2) Corma, A., Puche, M., Rey, F., Sankar, G., Teat, S. J. *Angew. Chem. Int. Ed.*, **2003**, 42, 1156.
(3) Castaneda R., Corma A., Fornes V., Martinez-Triguero J.,Valencia S. *J. Catal.* **2006**, 238, 79.
(4) Banani K., Anup K. Talukdar, *Mater. Chem. Phys.* **2012**, 133 713.
(5) Laha S.C., Mukherjee P., Sainkar S.R., Kumar R.J., *Catalysts*, **2002**, 207, 213.

HOLLOW ZEOLITE ENCAPSULATED NI-PT BIMETALS FOR SINTERING AND COKING RESISTANT DRY REFORMING OF METHANE

Xinwen Guo,[a] Chengyi Dai,[a] and Chunshan Song[a,b]

[a]State Key Laboratory of Fine Chemicals, PSU-DUT Joint Center for Energy Research, School of Chemical Engineering, Dalian University of Technology, Dalian 116024, P. R. China. Fax: +86-411-84986134; Tel: +86-411-84986133; E-mail: guoxw@dlut.edu.cn
[b]EMS Energy Institute, PSU-DUT Joint Center for Energy Research, Department of Energy & Mineral Engineering, Pennsylvania State University, University Park, Pennsylvania 16802, United States. Fax: 814-865-3573; Tel: 814-863-4466; E-mail: csong@psu.edu

Introduction

Dry (CO_2) reforming of methane (DRM) is a well-studied reaction that is of both scientific and industrial importance because it could transformed greenhouse gasses (CH_4 and CO_2) into the energy carrier synthesis gas (CO and H_2), which can be converted to valuable chemicals via the Fischer-Tropsch reaction.[1-2] In addition, its endothermic property makes DRM a promising method for energy storage and energy transfer. It can be used in energy transfer from solar energy to chemical energy, energy storage in the form of CO and H_2, and transporting nuclear energy.[3] For DRM, Ni-based catalysts have been regarded as the most promising because of their high activity, low cost and extensive supply.[4] According to previous studies, the sintering of Ni particles and carbon deposition under high reaction temperature (typically 800 °C) proved to be the main impediments for stable catalytic performance.[5] Therefore, preventing Ni sintering, decreasing the carbon deposition or reducing the effect of carbon deposition on the catalyst are very important for industrial progress of DRM.

In the past years, zeolites as supports have attracted interest due to their unique properties (e.g., high thermal stability, regular microporosity, and unique shape selectivity) which can be widely applied in catalytic operations.[6] Hollow zeolites have attracted considerable interest because of their superior performance in the selectivity of product, anti-leaching of active site and anti-sintering of metal.[7-9] The ability of hollow zeolite encapsulated metals to resist coke formation has not been previously reported. Herein, we report the encapsulation of Ni-Pt bimetals in hollow silicalite-1 single crystals for the first time. This material exhibits excellent metal dispersion and superior catalytic performance for sintering and coking resistant CO_2 reforming of CH_4. In addition, we report, also for the first time, a protective effect of the zeolite shell in the hollow crystals to allow continued catalytic activity in the presence of carbon.

Experimental

Preparation of silicalite-1 zeolite. Silicalite-1 (S-1) was synthesized with the clear solution method. Typically, 15.4 ml of tetraethyl orthosilicate (TEOS) was mixed with a certain amount of TPAOH solution. The molar composition of the synthesis mixture was 1 TEOS: 0.27 TPAOH: 37 H_2O. After being stirred for 3 h at 35 °C, the solution was heated at 80 °C to remove the ethanol generated during the hydrolysis of TEOS and then water was added to maintain constant volume. After crystallization at 170 °C for 3 days, the product was recovered by centrifugation and dried overnight at 100 °C. Finally, the template was removed by calcination in static air at 540 °C for 6 h.

Preparation of Ni@Hol S-1, Pt@Hol S-1 and Ni-Pt@Hol S-1. NiO/S-1 was synthesized by an incipient-wetness impregnation method. In brief, the calcined silicalite-1 was impregnated with an aqueous solution of $NiCl_2$; after drying overnight at 100 °C, the product was calcined in static air at 400 °C for 2 h. The ideal Ni loading on the NiO/S-1 were 1.5 wt%, respectively. The resulting sample was denoted as 1.5NiO/S-1.

The as-prepared NiO/S-1 was treated with 0.3 M TPAOH (20 ml of solution per gram of zeolite) at 170 °C for 72 h, then dried overnight at 100 °C and calcined in static air at 400 °C for 2 h to obtain NiO@Hol S-1. The resulting sample was denoted as 1.5NiO@Hol S-1.

Pt/S-1 was prepared by the same method as NiO/S-1, using H_2PtCl_6 as the Pt source. The ideal platinum loading on Pt/S-1 was 0.5 wt%. After treatment with 0.3 M TPAOH (20 ml of solution per gram of zeolite) at 170 °C for 72 h, drying overnight at 100 °C and calcining in static air at 400 °C for 2 h, the Pt@Hol S-1 was obtained. The resulting sample was denoted as 0.5Pt@Hol S-1.

Bimetals encapsulated in the hollow S-1 were prepared by similar methods as the single metals. $NiCl_2$ and H_2PtCl_6 were used as the Ni and Pt sources, and a co-impregnation method was used to synthesize bimetals/S-1. The ideal nickel loading on the S-1 were 1.5 wt%, while the ideal platinum loading on the S-1 was 0.5 wt%. The as-prepared NiO-Pt/S-1 was treated with 0.3 M TPAOH (20 ml of solution per gram of zeolite) at 170 °C for 72 h, then dried overnight at 100 °C and calcined in static air at 400 °C for 2 h to obtain NiO-Pt@Hollow S-1. The resulting sample was denoted as 1.5NiO-0.5Pt@Hol S-1.

After reduction under hydrogen at 800 °C for 30 min, 1.5Ni@Hol S-1, 0.5Pt@Hol S-1 and 1.5Ni-0.5Pt@Hol S-1 were obtained.

Catalytic test. The dry reforming reaction of methane was carried out in a quartz glass fixed-bed reactor at 800 °C and atmospheric pressure with a total flow rate of 240 ml/min of mixed gasses of CH_4:CO_2=1:1. A total of 200 mg of the catalyst was put inside the reactor. Before reaction, the catalyst was reduced at 500 °C for 1 h followed by introduction of CO_2 and CH_4. The composition of the product gases was analyzed with an on-line gas chromatograph equipped with a thermal conductivity detector.

Characterization. Transmission electron microscopy (TEM) images and EDX were taken on a Tecnai G2 20 S-twin instrument (FEI Company) with an acceleration voltage of 200 kV. The samples for TEM analysis were prepared by dipping the carbon-coated copper grids into the ethanol solutions of the samples and drying at ambient condition. The elemental analysis of catalysts was carried out on a Perkin Elmer OPTIMA 2000DV ICP Optical Emission Spectrometer. Thermogravimetric analysis (TGA) was performed on an SDT Q600 (TA Instruments, USA) in the temperature range of 25-800 °C under air at a heating rate of 10 °C min^{-1}.

Results and Discussion

Figure 1a shows a typical TEM image of the prepared 1.5NiO-0.5Pt/S-1. There are no observable particles on the surface of crystals. This is attributed to a high dispersion of nickel and platinum. After TPAOH treatment, several particles with average size of 3.14 nm were encapsulated in the hollow crystals (Figure 1b). These NiO-Pt particles are smaller than the encapsulated Pt particles. The amount of platinum in 1.5NiO-0.5Pt@Hol S-1 is larger than that of 0.5Pt@Hol S-1 (ICP results), confirming that the presence of nickel significantly inhibited aggregation and loss of platinum. After 1.5NiO-0.5Pt/S-1 reduction under H_2 at 800 °C, the average size of Ni-Pt bimetallic particles was 4.40 nm and there were no sintered nickel particles on the crystals (Figure 1c). The presence of platinum significantly inhibited sintering of nickel. After H_2 reaction, Ni-Pt bimetallic particles encapsulated in the hollow crystals maintained a high dispersion (4.57 nm average size, Figure 1d). According to EDX analysis, each particle contains both nickel and platinum elements, demonstrating an interaction between nickel and platinum is involved in motivation their high dispersion in the hollow crystal (Figure 1e).

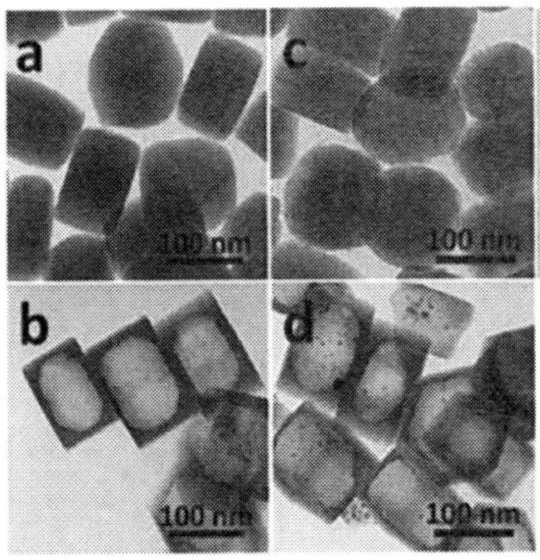

Code /#	Composition / at% Ni-K	Pt-L
1	61.82	39.18
2	74.37	25.63
3	70.73	29.27
4	69.20	30.80
5	75.22	24.78

Figure 1. TEM images of prepared (a, c) 1.5NiO-0.5Pt/S-1 and (b, d) 1.5NiO-0.5Pt@Hol S-1 after calcination in air at 400 °C for 2 h (a, b) and reduction under H_2 at 800 °C for 30 min (c, d). (e) STEM image of prepared 1.5NiO-0.5Pt@Hol S-1 after reduction under H_2 at 800 °C for 30 min, and quantified phase compositions in the Ni-Pd nanoparticles from STEM maps.

The activity and stability of catalysts during dry reforming of methane were tested under a high gas hourly space velocity (GHSV=72000 $mLg^{-1}h^{-1}$) and without any inert gas. **Figure 2A** illustrates the conversion versus time on stream for the various catalyst cycles. For 1.5Ni/S-1, the initial conversion of CO_2 and CH_4 were 76 and 73 %, respectively. However, conversion rapidly decreased to zero within one hour. 1.5Ni-0.5Pt/S-1 had similar initial activity as 1.5Ni/S-1, however, the introduction of platinum enhanced the stability of the catalyst. 1.5Ni-0.5Pt/S-1 deactivated within 6 hours. The encapsulated Ni-based catalysts showed the best catalytic performance, with conversion of CO_2 and CH_4 maintained with little deactivation past 6 hours of time on stream. The initial activities of 1.5Ni@Hol S-1 and 1.5Ni-0.5Pt@Hol S-1 were almost the same, but the 1.5Ni-0.5Pt@Hol S-1 showed better stability compared to that of 1.5Ni@Hol S-1. After 20 h reaction, both CH_4 and CO_2 conversions were higher than 60% over 1.5Ni-0.5Pt@Hol S-1, but lower than 30% over 1.5Ni@Hol S-1.

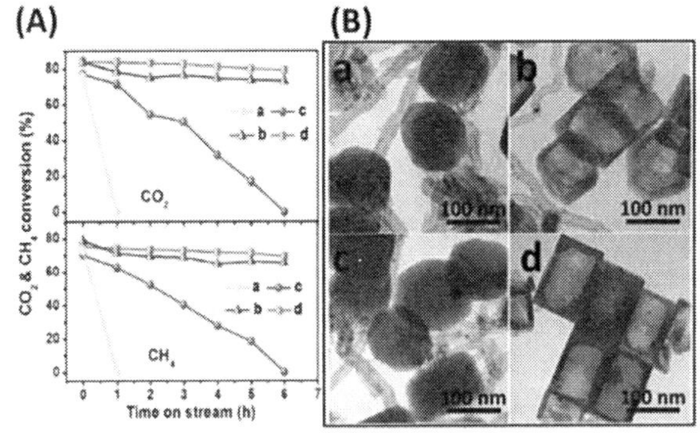

Figure 2. (A) CO_2 and CH_4 conversion as a function of time on stream over (a) 1.5Ni/S-1, (b) 1.5Ni@Hol S-1, (c) 1.5Ni-0.5Pt/S-1, and (d) 1.5Ni-0.5Pt@Hol S-1 catalysts [800 °C, atmospheric pressure, GHSV=72000 $mLg^{-1}h^{-1}$, and CH_4/CO_2=1:1(v/v)]. (B) TEM images of the spent catalysts, (a) 1.5Ni/S-1, (b) 1.5Ni@Hol S-1, (c) 1.5Ni-0.5Pt/S-1, (d) 1.5Ni-0.5Pt@Hol S-1, the reaction time was 1 h for 1.5Ni/S-1 and 6 h for the other three catalysts.

The amounts of the deposited carbon were investigated by thermal gravimetric (TG) and differential thermal gravimetric (DTG) experiments. Except for the loss of adsorbed water, the weight loss of spent 1.5Ni-0.5Pt@Hol S-1 is only 1.0 %, while the weight loss of spent 1.5Ni-0.5Pt/S-1 and 1.5Ni@Hol S-1 are 11.4 and 10.3 %, respectively. The fastest coke deposition occurs on the 1.5Ni/S-1 catalyst, with a 31.0 % weight loss measured in TG.

Although 1.5Ni-0.5Pt/S-1 and 1.5Ni@Hol S-1 have a similar amount of carbon deposition, their catalytic stabilities are very different, as shown in Figure 2A. To explain this, the deposited carbon was investigated further by TEM. Filamentous carbon formed on 1.5Ni/S-1 due to the large Ni particles on the crystal surface (**Figure 2B** a), resulting in rapid deactivation of the catalyst. For 1.5Ni@Hol S-1, filamentous carbon also formed because of the large Ni particles that were not encapsulated in the hollow crystals (Figure 2B b). However, due to the high dispersion nickel particles in the cavity, coke cannot be produced on the encapsulated particles. The shells of hollow crystals also inhibit the impact of filamentous carbon on the small nickel particles in the cavities. So, despite generating 10.3 % carbon, 1.5Ni@Hol S-1 catalyst remains highly active. Though the Ni-Pt binary metal particles are sintering resistant in 1.5Ni-0.5Pt/S-1, the small particles deactivate due to carbon build-up without the protective shell (Figure 2B c). For 1.5Ni-0.5Pt@Hol S-1, the high dispersion and protection of shell lead to almost no coke build-up after reacting for 6 hours (Figure 2B d).

The present work provides a general and robust strategy to prepare coke resistant catalysts for dry reforming of CH_4 and multi-metallic catalysts with well-dispersed nanoparticles.

Acknowledgment. This work was supported by the State Key Program of National Natural Science Foundation of China (Grant No. 21236008) and by the QianRen Type B Award to CS from Chinese government.

References

(1) Pakhare, D.; Spivey, J., *Chem. Soc. Rev.*, **2014**, *43*, 7813-7837.
(2) Song, C. S., *Catal. Today*, **2006**, *115*, 2-32.
(3) Edwards, J. H.; Maitra, A. M., *Fuel Process. Technol.*, **1995**, *42*, 269-289.
(4) Xu, L.; Song, H.; Chou, L., *ACS Catal.*, **2012**, *2*, 1331-1342.
(5) Ashok, J.; Kawi, S., *ACS Catal.*, **2013**, *4*, 289-301.

(6) Corma, A., *Chem. Rev.*, **1997**, *97*, 2373-2419.
(7) Dai, C.; Zhang, A.; Li, L.; Hou, K.; Ding, F.; Li, J.; Mu, D.; Song, C.; Liu, M.; Guo, X., *Chem. Mater.*, **2013**, *25*, 4197-4205.
(8) Dai, C.; Zhang, A.; Li, J.; Hou, K.; Liu, M.; Song, C; Guo, X., *Chem. Commun.*, **2014**, *50*, 4846-4848.
(9) Li, S.; Tuel, A.; Laprune, D.; Meunier, F.; Farrusseng, D., *Chem. Mater.*, **2015**, *27*, 276-282.

OUTLOOK OF OIL SANDS BITUMEN—CHALLENGES AND OPPORTUNITIES

Siauw H. Ng[1], Hao Ling[2], Yi Zhang[1]
Nicole E. Heshka[1], Edward Little[1], Qiang Wei[3]

[1] Natural Resources Canada, CanmetENERGY, 1 Oil Patch Drive, Devon, AB T9G 1A8, Canada; [2] State Key Laboratory of Chemical Engineering, East China University of Science and Technology, Shanghai 200237, China; [3] State Key Laboratory of Heavy Oil Processing, China University of Petroleum, Beijing 102249, China
Corresponding Author Email: Siauw.Ng@canada.ca

Introduction

Canada has the world's third-largest oil reserves, with approximately 169 billion bbl of oil sands bitumen. There has recently been increased production of nonconventional fossil fuels. Alberta refineries (comprising 22% of Canada's total refining capacity) and many US refineries receive bitumen feedstocks for processing into end products such as gasoline, diesel, and heavy fuel oil. However, oil sands bitumen-derived products are faced with a number of challenges, including increased competition from U.S. light tight oil (LTO), limited market access via pipeline infrastructure and capacity, high costs of labour, increasing environmental regulatory concerns, public scrutiny and growing challenges to social license. The recent fall of crude oil prices has created even more economic and environmental challenges for Canadian oil sands development. Therefore, research and technology development is necessary to ensure sustainable oil sands growth, improve environmental performance, address social license issues, and increase the competitiveness (in terms of price and quality) of oil sands products in world markets.

Research has taken into consideration the oil sands scenario in North America and the global market trends: 1) development of all forms of energy (e.g., fossil, renewables, LTO, and shale gas) to meet growing world energy demand; 2) heavier world crude oil supplies but higher fuel quality standards; 3) growth of oil sands crude production predominantly from increased in-situ operations; and 4) lost revenue from the oil sands industry and government arising from the current practice of delivering bitumen with diluents along pipelines. The last issue is related to the high requirements of diluent (up to 30%), which must be returned to Canada at a lower price after use and the substantial discounts (up to $15 per barrel) of bitumen-derived crudes imposed by U.S. refineries. These discounts are the results of quality deterrents such as high vacuum residue, TAN (total acid number), nitrogen and metal contents. The use of diluents occupies up to 30% of pipeline capacity and the quality deterrents further limit the volumes of bitumen-derived crudes that can be processed in typical U.S. refineries, thereby reducing the market share of oil sands crudes in the U.S. To this end, several approaches of R&D are explored at CanmetENERGY to ameliorate oil sands economics and to mitigate greenhouse gas (GHG) emissions associated with bitumen production, upgrading, and refining operation. Emphasis is placed on bitumen-derived heavy gas oil (HGO), a major portion of oil sands crude or synthetic crude oil (SCO) and a typical feed to fluid catalytic cracking unit (FCCU). The approaches are to be described as follows.

Direct Cracking of Raw Bitumen Distillate at FCCU

A 343–525°C cut heavy gas oil (HGO) from a raw bitumen was blended with various amounts of the +343°C fraction of a commercial SCO. Five SCO/bitumen HGO blends were prepared containing 100, 75, 64, 61, and 48v% of SCO HGO. The two blends with 64 and 48v% were referred to as synsynbit and synbit HGOs, respectively. Each HGO blend was catalytically cracked at 500 and 520°C using a bench-scale Advanced Cracking Evaluation (ACE) unit. Results are shown in Table 1, which demonstrates the acceptable performance of bitumen HGO when adequate amounts of SCO HGO are added. The conversion, defined as the summation of dry gas, liquefied petroleum gas, gasoline, and coke, as well as selectivities of gasoline, light cycle oil (LCO, a diesel pool component), and coke are either improved or little affected, in comparison with those of the base feed, that is 100% SCO HGO. In-house analyses supported by those reported in the literature show that the bitumen HGO actually contains less nitrogen than its counterpart (HGO from SCO) and has relatively low Ni+V and Conradson Carbon Residue (CCR). These impose no serious threat to the cracking catalysts. The major issue is the mild increase in the dry gas yield, notably due to the augmentation of H_2S, and higher sulfur levels in gasoline and diesel products. As sulfur species in bitumen HGO exist mostly as sulfides, mercaptans, and small thiophenes, they should be easily removed by mild hydrotreating or other low-cost refining process.

Table 1. Product Distributions at 72 wt% Conversion for HGO Blends

Temp. °C	v% SCO in Blend	Yield, wt%						Delta Coke[b] wt%
		Dry Gas	LPG	Gasoline	LCO	HCO[a]	Coke	
500	100	1.7	15.6	48.2	16.6	11.4	6.4	0.63
520		**2.2**	**16.7**	**47.3**	**16.0**	**12.0**	**5.9**	**0.78**
500	*75*	*2.2*	*14.7*	*48.7*	*17.2*	*10.8*	*6.3*	*0.70*
520		2.7	*15.8*	48.2	*16.5*	*11.5*	5.8	*0.82*
500	**64 (Synsyn bit)**	**2.6**	15.4	46.0	17.6	10.4	7.8	0.73-
520		**3.1**	16.4	45.3	17.2	10.8	7.3	0.80
500	*61*	*2.7*	*15.4*	*45.3*	*17.3*	*10.7*	*8.9*	*0.81-*
520		*3.3*	*16.3*	*43.1*	*17.2*	*11.1*	*8.5*	*0.80-*
500	48 (Synbit)	**2.9**	16.0	45.3	18.1	10.4	8.9	0.78-
520		**3.4**	17.1	43.4	17.5	10.7	8.8	0.81-

Direct conversion of bitumen HGO is further analyzed from practical, economical, and environmental points of view:

- Bitumen distillate (343–525°C) represents 41% of the atmospheric topped bitumen. If this fraction can be cracked directly, significant savings in operating costs and a reduction in GHG emissions will result. For example, if a refinery with 50,000 b/d FCCU capacity can accept HGO blend at an SCO/bitumen ratio of 70/30, the reduction in GHG emissions is estimated at 0.44 Mt CO_2 eq/y (compared with 46.8 Mt GHG emissions from oil sands mining, upgrading, and in-situ extraction in 2010[1]);
- The naphtha, LGO and resids (525°C+) fractions from the whole bitumen or from the SCO/bitumen blend can be sent to gasoline and diesel pools and coker or hydrocracker, respectively. If so desired, naphtha or gasoline can be desulfurized at low severity to remove sulfides and mercaptans prior to blending;
- Distillation and FCC units are standard refinery infrastructure — the proposed direct cracking process option can be adopted without major capital expenditures at refineries;
- The SCO/bitumen blend at 70/30 ratio can flow with ease to the U.S. through pipelines (where available);
- Alternatively, bitumen distillate can be sold separately at premium price due to 0% vacuum residue (VR) and low sulfur, nitrogen and aromatics compared to bitumen.

Co-processing of Biomass and Bitumen-derived/Conventional Feeds

Catalytic cracking of biomass-derived oil mixed with petroleum FCC feedstock is a potential technology to reduce carbon footprint and meet government regulatory demand on renewable transportation fuels[2]. This option can be executed using existing refining infrastructure to save capital cost. In addition, co-processing may generate synergy to improve gasoline and diesel qualities. In this comprehensive study, the bio-feeds include Canadian canola oil, and pyrolysis oil generated from Ensyn's Rapid Thermal Processing (RTP™) plant, where cellulosic biomass (wood residue and forest waste) was thermally cracked at mild temperatures in a short duration with high liquid yield. An HGO from Canadian SCO or from conventional sources were used as base oils. A series of FCC experiments were conducted to investigate the effects of incorporation of biomass feedstock into HGO on the gasoline yield and quality. The data demonstrated that the addition of biomass feed resulted in higher conversion but lower gasoline yield, at the expense of water formation. Given the characteristics of the bio-feed (lower carbon and energy content than the base HGO), gasoline yield was actually higher on an energy equivalency basis than its counterpart from HGO alone. As water formation from hydrogen and oxygen is an exothermic reaction, this leads to lower energy consumption for cracking the blend versus the pure HGO. Biomass feed addition also gave better product quality in terms of lower sulfur content in diesel and higher octane numbers in gasoline.

Co-treatment of Bitumen Feed and Light Tight Oil from Hydraulic Fracturing

As a result of novel drilling technology development, LTOs such as Bakken and Eagle Ford have become the predominant feedstocks at US refineries that are used to processing heavy crudes. Although relatively easy to process, light and paraffinic tight oil still comes with many challenges[3]. Since it is high in components boiling below 343°C, much of the naphtha from starting material will remain uncracked when processing whole crude tight oil in the FCC. Also, the expected low coke yield would result in low regenerator temperatures and possible difficulties in maintaining heat balance between the reactor and catalyst regenerator in the FCCU.

Blending LTO with Alberta oil sands derived crude such as SCO in suitable ratios is a viable option for minimizing these issues, as there is strong coherence and close affinity to co-process the two feedstocks of opposite characteristics. In addition, co-processing can use existing refining infrastructure and technologies, saving capital and operating costs. This work explores the benefits and synergies of co-processing Bakken shale oil with Canadian bitumen SCO in selected blending ratios. It appears that more LTO in the blend results in higher conversion, accompanied by increased LPG and gasoline selectivities and decreased yields of dry gas, LCO, HCO, coke, and delta coke at a given conversion. However, the octane numbers are lower. Cracking phenomena are characterized by: 1) catalyst poisoning by feed nitrogen — the effect can be reduced at higher temperatures or catalyst/oil ratios. 2) preferential skeletal isomerization of olefins to form branched compounds; and 3) oligomerization of olefins to form aromatics.

Partial Upgrading to Eliminate or Reduce Diluent Requirement

In the summer of 2015, CanmetENERGY-Devon and Alberta Innovates – Energy and Environmental Solutions, together with a number of other industrial companies and government organizations, formed a joint force to start a national partial upgrading R&D program. The overall goal of this program is to improve oil sands bitumen competitiveness and market value and to reduce associated environmental impacts and GHG emissions. Specific objectives include elimination or reduction of costly diluent use for transporting bitumen (called "dilbit" when natural gas condensate is used) along pipelines and fitting the partially upgraded bitumen (PUB) into domestic and international refineries. Prior research indicated a parabolic increase of the market value (in the sequence of dilbit, PUB, sour SCO, and sweet SCO) with upgrading cost (going through partial upgrading, coking, hydrotreating, and hydrocracking) and that partial upgrading can maximize economic returns for oil sands projects. In our National Partial Upgrading Program, the priority initiatives were identified as follows: 1) bitumen chemistry; 2) by-product value addition; 3) stability and compatibility; 4) environmental and integration systems; 5) scale-up and field demonstration. To support these initiatives and to achieve the overall goals, at CanmetENERGY under the national partial upgrading program, we are conducting R&D in the following 5 areas: 1) develop and assess partial upgrading technologies using pilot plants; 2) Expand characterization capabilities of bitumen and partially upgraded bitumen; 3) develop computational tools for molecular simulation of partially upgraded bitumen; 4) determining partial upgrading limits and predicting fouling tendency, stability and compatibility; 5) modeling, simulation, and optimization of partial upgrading processes and technologies. One of the projects in research Area 1 is to investigate the processability of partially upgraded bitumen in FCC to produce gasoline — the major and the most important end product in North American refineries.

Acknowledgement. Partial funding for this study was provided by Natural Resources Canada and government of Canada's interdepartmental Program of Energy Research and Development (PERD).

References

(1) ***"About Oil Sands, Facts and Statistics"***, Alberta Energy, January 2014.

(2) Canada Gazette, Part II: Renewable Fuels Regulations, September 1, 2010.

(3) Bryden, K.; Habib Jr., E. T.; Topete, O.; Schiller, R. Catalytic solutions for processing shale oils in the FCC. http://www.digitalrefining.com/article/1000919.

IMPACT OF PROCESS PARAMETERS ON THE DEPOSITION OF FINES IN BITUMEN-DERIVED GAS OIL ON A FIXED BED HYDROTREATER

Rachita Rana§, Sandeep Badoga§, Ajay Dalai§, and John Adjaye‡

§ University of Saskatchewan, Saskatoon SK, Canada
‡ Syncrude Research Centre, Syncrude Ltd. Edmonton, AB, Canada
akd983@mail.usask.ca

Introduction

The deposition of fine particles present in bitumen-derived gas oil, during hydrotreating, leads to problems such as catalyst fouling, poor catalyst activity and pressure drop in the reactor [1]. These problems pose a processing challenge for the oil industries and lead to economic losses as the pressure drop results in premature reactor shutdown and the fouled catalyst bed needs to be replaced without utilizing its activity completely [2]. These fine particles are clays that surface-adsorb asphaltenes. When the gas oil is fed to the hydrotreater, these fines (<20μ) cannot be trapped by the guard bed filters as their organic asphaltene coating gives them a similar surface chemistry as that of oil. Though the concentration of these fines is ~100 ppm, the successive exposure of the catalyst bed to thousands of barrels of feed with fines leads to large depositions which gradually turn the bed to a large filter for fines [3,4]. There have been studies reported for the pressure-drop due to fines deposition in the trickle-bed reactors [5]. The surface chemistry of these fines and their hydrodynamics has also been studied [6,7]. However, the impact of hydrotreating process parameters on the deposition of fine particles remains unexplored. Thus, in this work the impacts of temperature, pressure and particle loading on fines deposition have been studied using a batch system with kaolin as model fines. The trend of clay (kaolin) deposition on the catalyst is experimentally observed by maintaining the conditions in the batch system similar to the fixed bed reactor. The combined effects of these parameters on fine particle deposition and their optimization for minimum fines deposition without affecting the catalyst activity has been targeted in this research. Also, the changes in surface morphology of the catalyst due to fines deposition have been studied.

Experimental

Material. Tri-lobed gamma alumina was supplied by Sasol, Germany. Nickel nitrate, molebdenum oxide, amorphous kaolin (used as modelled fines) and butanethiol was purchased from Sigma Aldrich, Canada. Light Gas Oil feed was provided by Syncrude, Canada Ltd. 0.22μm membrane filter was purchased from Millipore.

Method. The hydrotreating experiments were designed using Central Composite Design (CCD). The feed was spiked with kaolin (model fines) in different concentrations. In each experiment, 200 ml of oil was used and the hydrotreating experiments were carried out in a batch reactor. The $NiMo/\gamma Al_2O_3$ catalyst mixed with SiC was packed in a catalyst basket installed in static mode. Temperature (355°C, 365°C, 375°C), pressure (1200 psi, 1300 psi, 1400 psi) and particle loading (1g, 1.25g, 1.5g) were the three variables used and the impact of these variables on bed deposition (BD) and sulfur conversion (SC) was studied. The interaction between the catalyst and the deposited fine particles was studied using FTIR and BET analysis. The change in the surface morphology of the catalyst due to fines deposition at different conditions was also studied using SEM. The catalyst activity in terms of nitrogen and sulfur removal was studied using NS analyzer.

Results and Discussion

The samples were collected after hydrotreating the feed for 8 h. The catalyst activity for sulfur removal was calculated for all the experimental runs. Addition of kaolin to the feed reduced the HDS activity, but the difference in fines quantity did not have any impact on the HDS activity. Figure 1 shows an illustration of the deposition of kaolin on the tri-lobed catalyst. When the spent catalyst without kaolin in the feed was seen there were no deposits observed. However, on spent catalyst for kaolin spiked feed, grey deposits were observed which were studied at higher magnification using SEM. The SEM images showed large agglomerates of fines deposited on the catalyst surface. Mass balance was conducted for the kaolin particles initially spiked in the feed to get the bed deposition (deposits on the catalyst bed). From the statistical analysis of the models for bed deposition and sulfur conversion of the feed, it was found that apart from the particle loading in the feed, the interaction between the temperature and pressure played a significant role in bed deposition. Bed deposition increased with an increase in temperature. On contrary, high pressure gave less bed deposition. FTIR spectra of the spent catalyst for kaolin spiked feed were compared with that of the fresh catalyst and also with the spent catalyst of the feed without kaolin. The spectra showed that kaolin interacted with the catalyst and hydrocarbons forming Si-O-Si, Si-O and $Si\text{-}CH_2\text{-}Si$ bonds which were absent in the fresh and spent catalyst without kaolin deposit. The surface morphology of the catalyst packed at different locations in the bed was studied to understand the change in the deposition pattern of kaolin at different locations on the bed. Also, the surface morphology of the catalyst was studied to support the mass balance results that showed larger agglomeration on the catalyst with higher bed deposition as shown in Figure 2. Figure 2 (a) shows larger agglomerate formed for a catalyst that was calculated to have maximum bed deposition and Figure 2 (b) shows small clusters being formed for catalyst that was calculated to have least bed deposition.

The models for sulfur conversion and bed deposition were developed using CCD. The optimization of the process parameters within the boundary used for the experiments was also performed and the results are shown in Table 1. As seen in the table when the particle loading in the feed was increased from 1 g to 1.5 g, the optimized results gave a reduced temperature and an increased pressure to get least bed deposition and maximum sulfur conversion.

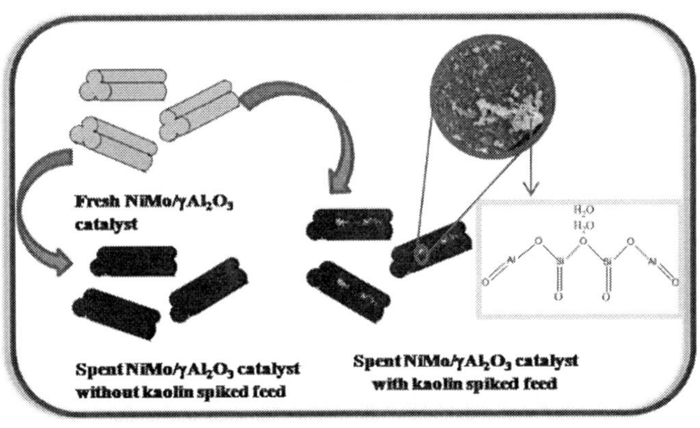

Figure 1. Schematic for fine particles deposited on the catalyst after hydrotreating

a) b)

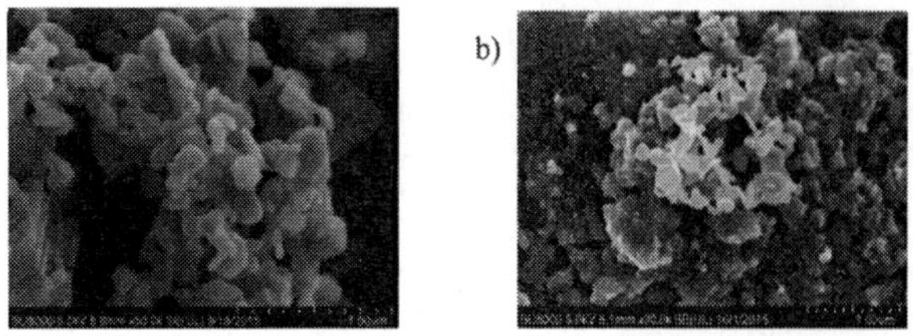

Figure 2. Scanning electron microscopy image of the catalyst for (a) maximum bed deposition and (b) minimum bed deposition

Table 1. Optimization of Results for the Developed Model

Optimization No.	Particle Loading (g)	Tempe -rature (°C)	Press -ure (Psi)	Target Bed Deposition (mg)	Target Sulfur Conversion (wt%)
1	1.5	360	1450	100	60
2	1	361	1380	100	60

Conclusions

Temperature was found to be the key parameter to affect the surface morphology of the catalyst due to fine particle deposition. Higher temperature led to higher bed deposition, though it enhanced the sulfur conversion too. The combined effects of temperature and pressure significantly influenced the bed deposition. The optimization study revealed that if particle loading of the feed was increased from 1g to 1.5g/ 200ml LGO, lower temperature (360°C) and high pressure (1450 psi) would result in less bed deposition. This optimization would not affect the HDS activity of the catalyst but reduce the probable increase in bed deposition for a high particle loading. Thus, this study addresses the pressure drop problem of the oil industries due to bed deposition of fine particles that are present in bitumen-derived gas oil feed. These findings might help to set the operating conditions of the hydrotreaters in such a way as to reduce fines deposition without deteriorating the catalyst activity.

Acknowledgement. The authors would like to acknowledge NSERC and Syncrude for funding.

References

(1) Gray, M. R.; Srinivasan, N.; Masliyah, J. H. *Can. J. Chem. Engg.* **2002**, *80* (June), 346–354.
(2) Iliuta, I.; Larachi, F.; Grandjean, B. P. a. *AIChE J.* **2003**, *49* (2), 485–495.
(3) Iliuta, I.; Ring, Z.; Larachi, F. *Chem. Eng. Sci.* **2006**, *61* (4), 1321–1333.
(4) Iliuta, I.; Larachi, F. *Chem. Eng. Sci.* **2005**, *60* (22), 6217–6225.
(5) Narayan, R.; Coury, R.; Masliyah, J. H.; Gray, M. R. *Am. Chem. Soc.* **1997**, *36* (1976), 4620–4627.
(6) Wang, S.; Chung, K. H.; Masliyah, J. H.; Gray, M. R. *Ind. Eng. Chem. Res.* **1999**, *38*, 4878–4888.
(7) Iliuta, I. *Ind.Eng. Chem. Res* **2003**, *42*, 2441–2449.

SUSTAINABLE ROUTES FOR SYNTHESIZING ZEOLITE CATALYSTS

*Qinming Wu, Na Sheng, Liang Wang, Xiangju Meng, and Feng-Shou Xiao**

Key Lab of Applied Chemistry of Zhejiang Province, Department of Chemistry, Zhejiang University, Hangzhou 310028, China
fsxiao@zju.edu.cn

Introduction

Zeolite crystals with ordered microporous structures and strong acidity as well as redox sites have been widely used as heterogeneous catalysts in the industrial processes for the production of petrochemicals and fine chemicals as well as the industrial conversion of heavy oils and shale gas currently. Notably, most zeolites are synthesized under hydrothermal conditions from silicate or aluminosilicate gels in alkaline media at temperatures between about 60 and 200 °C. However, the hydrothermal synthesis of zeolites is not green, based on the concept of sustainable chemistry, due to the use of organic templates, the presence of high pressure, low efficiency, and production of polluted waters. To overcome these disadvantages, alternative routes for synthesizing zeolites in a sustainable manner have always been sought. In this talk, it is briefly summarized our recent green routes for synthesizing zeolites from organotemplate-free and solvent-free strategies.

Results and Discussion

Organotemplate-free Synthesis of Zeolites. Modern synthetic methodologies for synthesis of zeolite catalysts normally involve in the use of organic templates, which are usually environmentally unfriendly and costly. In addition, removal of these organotemplates always produces harmful gases and energy consumption. To solve this problem, it is suggested seed-directed synthesis of zeolites in the absence of any organic templates. The first example is the organotemplate-free Beta zeolite [1], latter other important zeolites such as ZSM-5, LEV, ZSM-22, ZSM-34, and ZSM-23 are also successful in our group [2].

To understand the mechanism on the seed-directed synthesis, Beta crystallization as a typical example was investigated carefully [3]. The results suggest that the seed-directed Beta zeolites are grown from solid Beta seeds, and final Beta crystals are mainly alike core-shell structure. The core part of Beta seeds has relatively high Si/Al ratios, and the shell part grew from aluminosilicate gels has relatively low Si/Al ratios.

Solvent-free Synthesis of Zeolites. Conventionally, zeolites are necessarily synthesized in the presence of solvents such as in hydrothermal, solvothermal, and ionothermal routes. We demonstrate here a simple and generalized solvent-free route for synthesizing various types of zeolites by mixing, grinding, and heating solid raw materials [4,5]. Compared with hydrothermal route, the avoidance of solvents in the synthesis not only significantly reduces the waste production, but also greatly increases the yield of zeolite products. In addition, the use of starting solid raw materials remarkably enhances the synthesis efficiency and reduces the use of raw materials, energy, and costs. Figure 1 show S-1 zeolite (pure silica zeolite with MFI structure) crystallization process under solvent-free conditions [4]. Clearly, during the crystallization process, the samples are always in a solid phase, confirming the solidothermal synthesis. After the solidothermal treatment at 200°C for 24 h, a pure phase of S-1 zeolite with MFI structure is finally obtained. Interestingly, this solvent-free route could be extended to synthesize a series of silica-based zeolites including aluminosilicate and aluminophosphate-based zeolite catalysts [4,6]. Recently, solvent-free synthesis of tianosilicate zeolites is also successful [7].

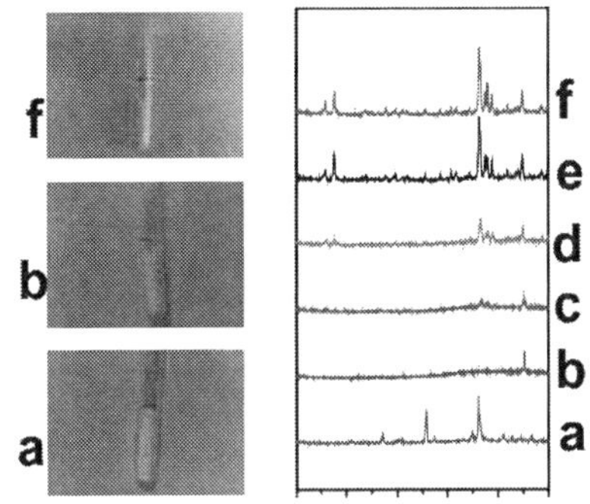

Figure 1. S-1 zeolite crystallized at crystallization time of (a) 0, (b) 2, (c) 10, (d) 12, (e) 18, and (f) 24 h at 200°C, respectively.

Combination of Organotemplate-free and Solvent-free Routes. As shown in above paragraphs, the organotemplate-free synthesis allows avoiding the use of organic templates, but water as solvent is still required; the solvent-free route on the other hand bears the potential to significantly reduce the amount of polluted waste water, but organic templates were still present. More recently, we have demonstrated a combined strategy of both organotem-plate-free and solvent-free conditions to synthesize aluminosilicate zeolites Beta and ZSM-5 (S-Beta and S-ZSM-5), two of the most important zeolites relevant for industry [8]. The results demonstrate that the S-Beta and S-ZSM-5 zeolites have almost the same textural parameters (BET surface area and pore volume) and catalytic performance in cumene cracking and m-xylene isomerization as those of conventional Beta and ZSM-5 zeolites synthesized under hydrothermal conditions. The organotemplate-free and solvent-free syntheses of S-Beta and S-ZSM-5 take place at a low-pressure regime and are free of harmful gases as well as give high product yields together with highly efficient consumption of the starting raw materials. These advantages plus the very simple procedures opened a new door to a highly sustainable zeolite synthesis protocol, compared to conventional hydrothermal methods currently.

References

(1) Xie, B.; Song, J. W.; Ren, L. M.; Ji, Y. Y.; Li, J. X.; Xiao, F.-S., ***Chem. Mater.***, **2008**, *20*, 4553.
(2) Meng, X. J.; Xiao, F.-S., ***Chem. Rev.***, **2014**, *114*, 1521.
(3) Xie, B.; Zhang, H. Y.; Yang, C. G.; Liu, S. Y.; Ren, L. M.; Zhang, L.; Meng, X. J.; Yilmaz, B.; Muller, U.; Xiao, F.-S., ***Chem. Commun.***, **2011**, *47*, 3945.
(4) Ren, L. M.; Wu, Q. M.; Yang, C. G.; Zhu, L. F.; Li, C. J.; Zhang, P. L.; Zhang, H. Y.; Meng, X. J.; Xiao, F.-S., ***J. Am. Chem. Soc.***, **2012**, *134*, 15173.
(5) Wu, Q. M.; Liu, X. L.; Zhu, L. F.; Ding, L. H.; Gao, P.; Wang, X.; Pan, S. X.; Bian, C. Q.; Meng, X. J.; Xu, J.; Deng, F.; Maurer, S.; Muller, U.; Xiao, F.-S., ***J. Am. Chem. Soc.***, **2015**, *137*, 1052.
(6) Jin, Y. Y.; Sun, Q.; Qi, C. D.; Yang, C. G.; Xu, J.; Chen, F.; Meng, X. J.; Deng, F.; Xiao, F.-S., ***Angew. Chem. Int. Ed.***, **2013**, *52*, 9172.
(7) Zhu, L. F.; Zhang, J.; Wang, L.; Wu, Q. M.; Bian, C. Q.; Pan, S. X.; Meng, X. J.; Xiao, F.-S., J. Mater. Chem. A, 2015, 2, 14093.
(8) Wu, Q. M.; Wang, X.; Qi, G. D.; Guo, Q.; Pan, S. X.; Meng, X. J.; Xu, J.; Deng, F.; Fan, F. T.; Li, C.; Maurer, S.; Muller, U.; Xiao, F.-S., ***J. Am. Chem. Soc.***, **2014**, *136*, 4019.

UTILIZATION OF HYDROGEN TRANSFER REACTION FOR CATALYTIC CRACKING OF POLYCYCLIC AROMATIC HYDROCARBONS

Iori Shimada[1], Chiaki Uno[1], Kouhei Takizawa[1], Mitsumasa Osada[1], Hiroshi Fukunaga[1], Nobuhide Takahashi[1], Toru Takatsuka[2]

[1] Faculty of Textile Science and Technology, Shinshu University 3-15-1, Tokida, Ueda, Nagano 386-8567 Japan
[2] Chiyoda Corporation, 4-6-2 Minatomirai, Nishi-ku, Yokohama, Kanagawa 220-8765 Japan
Corresponding author: iori@shinshu-u.ac.jp

Introduction

In modern petroleum refineries, the cracking of high molecular weight components is performed with the fluid catalytic cracking (FCC) process. Unfortunately, the current FCC process has been considered not to be able to directly refine polycyclic aromatic hydrocarbons (PAHs), which are one of the main components of the heavy crude oils, because of difficulties in activation of aromatic rings in an atmosphere without hydrogen.[1,2] Therefore, for the efficient utilization of the heavy crude oils, it has been considered necessary to hydrogenate the PAHs with pressurized hydrogen atmosphere so as to convert them into naphthenes before they are supplied to the FCC process.

To achieve the direct conversion of PAHs in the FCC process without hydrogen atmosphere, we focused on the hydrogen transfer reaction that proceeds in the FCC process. Hydrogen transfer reaction is a bimolecular reaction in which dehydrogenation of one molecule (hydrogen donor) and hydrogenation of the other (hydrogen acceptor) proceed simultaneously. We expected that this reaction is effective for hydrogenation of PAHs in the catalytic cracking reaction with the hydrogen donor. Therefore, we investigated the catalytic cracking reaction of PAH model compounds and hydrogen donor on the FCC catalyst with rare-earth ion exchanged USY zeolite, which is known to exhibit high hydrogen transfer activity.

Experimental

Commercial FCC catalysts consisting of rare-earth ion exchanged USY zeolite were hydrothermally deactivated under a 100% steam atmosphere at 800 °C until the unit cell size of the USY zeolites became stable. The deactivated catalysts were used in the catalytic activity tests. In addition, the equilibrium FCC catalysts were also used in some experiments.

As model compounds of 2-ring and 3-ring PAHs, 1-methylnaphthalene and 9-methylanthracene were used, respectively. These model PAHs were dissolved with 20 mol% in the hydrogen donors (*n*-hexadecane and tetralin) or the inert solvent (toluene), and the mixtures were fed into a fixed bed microactivity test (MAT) reactor. Catalytic activity tests were conducted at reaction temperatures between 450 and 516 °C and the weight hourly space velocity (WHSV) was varied between 11 and 35 h^{-1}. Reaction products were analyzed by GC-TCD, GC-FID and GC-MS.

Results and Discussion

Catalytic cracking of 2-ring and 3-ring PAHs. Reaction products of catalytic cracking of the 2-ring PAHs/*n*-hexadecane mixture, the 3-ring PAHs/*n*-hexadecane mixture and *n*-hexadecane alone were analyzed and compared.[3] The conversion of 3-ring PAHs were higher than 50%, while that of 2-ring PAHs were no more than 10% at the same reaction condition. Compared at the similar *n*-hexadecane conversion, the yield of monocyclic aromatic hydrocarbons (MAHs) from the 3-ring PAHs/*n*-hexadecane mixture was higher than that from the *n*-hexadecane alone. On the other hand, the yield of MAHs from the 2-ring PAHs/*n*-hexadecane mixture was comparable to that from the *n*-hexadecane alone. These results suggest that the 3-ring PAHs were highly reactive on the FCC catalysts and they were decomposed into MAHs.

Among the reaction products from the 3-ring PAHs/*n*-hexadecane mixture, dihydroanthracenes, dihydrophenanthrenes and biphenyls were detected and confirmed as reaction intermediates. In addition, the isomerization of anthracenes to phenanthrenes was also observed. These results indicate the reaction mechanism of the 3-ring PAH cracking as shown in Figure 1. Before the cracking of the 3-ring PAHs, the isomerization proceeds rapidly. The conversion of the 3-ring PAHs is initiated by the hydrogenation of a part of their aromatic rings via the hydrogen transfer reaction with *n*-hexadecane derivatives. Then, the saturated rings were cracked and the MAHs were produced.

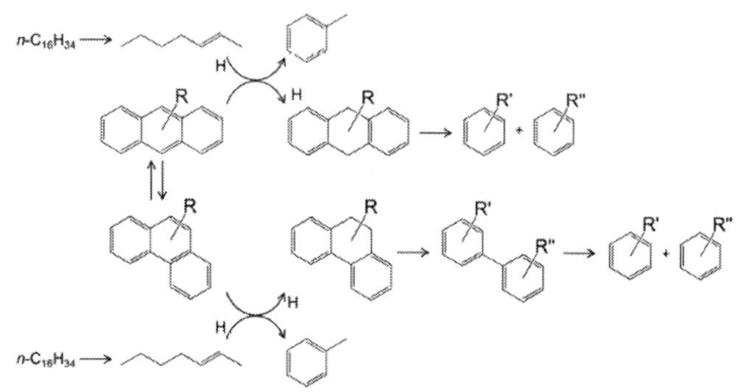

Figure 1. Schematic representation of the proposed reaction mechanism of the 3-ring PAH cracking.[3]

Effect of hydrogen donor concentration. To clarify the effect of hydrogen transfer reaction on the 3-ring PAHs conversion, the catalytic cracking of the 3-ring PAHs with different hydrogen donor concentrations were investigated. Here, the feedstocks were composed of 9-methylanthracene (3-ring PAH), tetralin (hydrogen donor) and toluene (inert solvent). Among the feedstocks, the initial concentration of the 3-ring PAHs was fixed at 20 mol%, and the initial concentrations of the tetralin and toluene were varied between 0 – 80 mol%. Figure 2 shows the conversions of the 3-ring PAHs with different initial tetralin concentrations and different contact times. The conversions of the 3-ring PAHs without hydrogen donor were much slower than those with hydrogen donor. In addition, high coke yield was observed in the reaction without hydrogen donor. These results suggest that the addition of the hydrogen donor accelerates the decomposition of the 3-ring PAHs and suppresses the coke formation.

It was surprising to us that the conversions of the 3-ring PAHs were almost the same even with different initial tetralin concentrations (20 – 80 mol%) and different contact times (0.2 – 0.8 s). We confirmed that the tetralin was still remains in the reaction products even at the longest contact time and the lowest initial tetralin concentration. The reason for the saturation of the 3-ring PAH conversion at around 60% can be attributed to the dehydrogenation reaction of the partially hydrogenated 3-ring PAHs such as dihydroanthracenes and dihidrophenanthrenes. Because the hydrogenation of the aromatic rings with the hydrogen transfer reaction and the dehydrogenation of the saturated rings to form the 3-ring PAHs proceeded faster than the cracking of the saturated rings to form the MAHs, the hydrogenation and the dehydrogenation of the 3-ring PAHs were in pseudo-equilibrium and the conversion of the 3-ring PAHs were almost constant. Consequently, to increase the conversion of the 3-ring PAHs into small fractions, not only the hydrogenation of the aromatic rings of the 3-ring PAHs with hydrogen transfer reaction but also the cracking of the saturated rings should be accelerated.

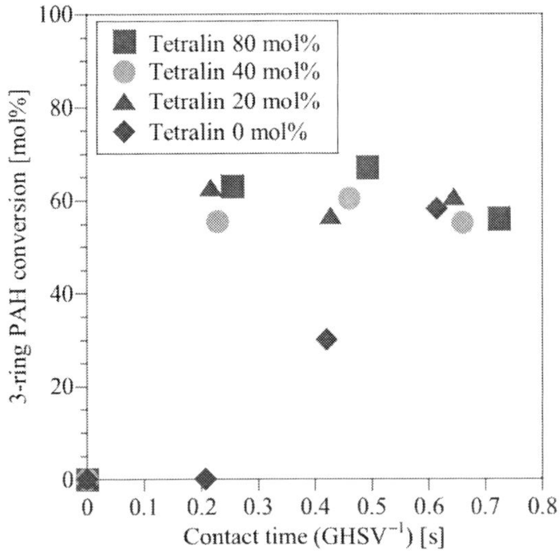

Figure 2. Conversion of 3-ring PAHs in the catalytic cracking with different initial tetralin (hydrogen donor) concentration at 450 ° C. Initial concentrations; 9-methylanthracene: 20 mol%; tetralin: 0 – 80 mol%; toluene: balance.

Conclusions

In this study, we investigated the catalytic cracking of the 2-ring and the 3-ring PAHs co-fed with hydrogen donors on the FCC catalysts with high hydrogen transfer activity. The conversion of the 2-ring PAHs proceeded only minimally, while the 3-ring PAHs were highly reactive and converted into the MAHs. This result suggests the possibility for the direct conversion of the PAHs into the MAHs in the FCC process without hydrogen atmosphere. For further increase of the PAH conversions, not only the hydrogenation of the aromatic rings with hydrogen transfer reaction but also the cracking of the saturated rings should be accelerated.

Acknowledgement. This work was financially supported by the Japan Petroleum Energy Center (JPEC) as a technological development project and the Grant for Research of The Japan Petroleum Institute, which are greatly appreciated.

References

(1) Dupain, X.; Gamas, E. D.; Madon, R.; Kelkar, C. P.; Makkee, M.; Moulijn J. A. *Fuel*, **2003**, *82*, 1559.
(2) Pujro, R.; Falco, M.; Sedran, U. *Energy Fuels*, **2015**, *29*, 1543.
(3) Shimada, I.; Takizawa, K.; Fukunaga, H.; Takahashi, N.; Takatsuka, T. *Fuel*, **2015**, *161*, 207.

MASS TRANSFER BEHAVIORS IN POROUS CATALYTIC MATERIALS: AN IMPORTANT BASIC SCIENCE FOR HEAVY OIL CATALYTIC CONVERSION

*Yucai Qin, Xiaotong Zhang, Lijuan Song**

Key Laboratory of Petrochemical Catalytic Science and Technology, Liaoning Province, Liaoning Shihua University, Fushun 113001, P.R. China

*Corresponding author: lsong56@263.net

1. Introduction

Mass transfer performance is a key factor influencing the catalytic properties of the porous petroleum processing catalysts, especially for the heavy oil catalytic conversion catalysts, such as fluid catalytic cracking and hydrocracking catalysts. Mass transfer behaviors of guest molecules in the commercial catalysts, therefore, become a widely concerned scientific problem.

These commercial catalyst particles are typical hierarchical porous materials, containing a complex porous system of micro-, meso- and macropores located in the zeolites and the matrix, in which different modes of mass transfer processes are expected, and therefore it is indistinct where is the diffusional controls for molecular transport in the whole catalyst particles. Moreover, because of the complexity of the system, the relationships between the pore structure of the porous catalysts and the transport phenomenon occurring therein are only poorly understood. Few published studies on the mass transfer processes in these catalysts can be searched due to the complexity of the systems and the lack of suitable experimental methods. Also, some consensus accepted conclusions of mass transfer processes in the commercial catalyst particles have not been achieved for such complicated system.

In view of the weakness in the traditional research contents and methods, this work will focus on the distinguishment of the complex transport behaviors of guest molecules in the hierarchical porous catalyst particles, and propose a novel philosophy for the study of mass transfer performance by a space-resolved method, which is able to spatially map dynamic phenomena in the commercial catalysts.

This work will provide the theoretical guidance and supporting data for the optimization of traditional catalysts and the development of novel heavy oil catalytic conversion catalysts with excellent mass transfer performance.

2. Experimental

A series of powdered zeolites (NaY, HY, USY, REUSY, ZSM-5 etc.) and technical preformed catalysts (FCC and hydrocracking catalysts) were utilized as the research objects. The structure and texture properties of the samples were characterized by XRD, SEM, TEM, N_2 (77.4 K) and Ar (87.3 K) adsorption, and mercury intrusion method, etc.. Mass transfer behaviors of probe molecules (benzene, cyclohexane, octane, and 1-octene, etc.) in the samples have been measured and discriminated by the Frequency Response (FR) method and an Intelligent Gravimetric Analyser (IGA). The fluorescence microscopic imaging technology as a space-resolved method is firstly explored to study the mass transfer performance. The three dimensional fluorescence microscopic images of the guest molecules within the catalyst particles can be observed by a laser confocal microscope. An ultraviolet-visible spectrophotometry can be used to detect the adsorption rate of the guest molecules into the catalysts and achieve the time-resolved data.

3. Results and discussion

The FR technique has been used to study the mass transfer behaviors of several systems of the powdered zeolites and the technical preformed catalysts (*cf.* Figure 1). The results demonstrate the discrepancy in the mass transfer processes of benzene in the contrastive samples which just possess the different texture properties and crystal structures. It can be proved that the FR technique is a very sensitive one to investigate the effects of the texture properties and crystal structure on the adsorption and diffusion behaviors of the porous catalytic materials.

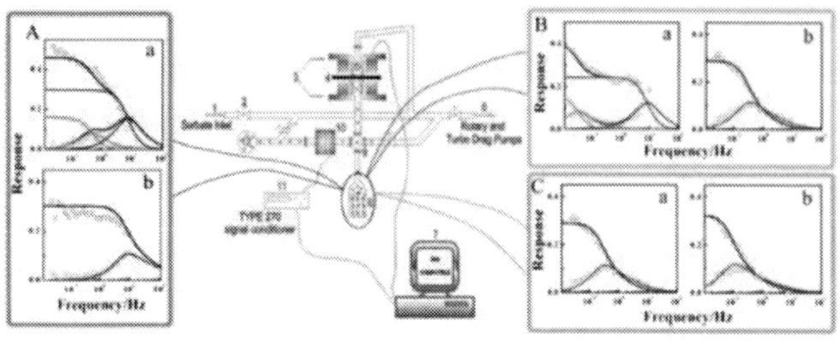

Figure 1. Schematic diagram of the FR equipment and the FR spectra of benzene in the samples used.
A: Y zeolites with micron-size (a) and nano-size (b); B: REUSY zeolite (a) and FCC catalyst particle (b); C: semi synthetic FCC catalyst (a) and in situ crystallization FCC catalyst (b).

The FR spectra demonstrate the differences in the mass transport processes of benzene in REUSY zeolite and the FCC catalyst particles under the same experimental situation (*cf.* Figure 2). The discrepancy indicates that the limiting step for the overall molecular transfer processes in FCC catalyst particles might be attributed to the mass transfer process in the meso-/macro-pore in the matrix or/and at the zeolite-matrix interface rather than the intracrystalline diffusion in the zeolite. Comparing the FR results of two kinds of FCC catalysts, it can be seen that the in situ crystallization FCC catalyst displays more excellent mass transfer behavior than the semi synthetic FCC catalyst which can be attributed to the excellent porous connectivity of the former with unique accumulation state of the highly dispersed nano Y zeolite crystals (*cf.* Figure 3).

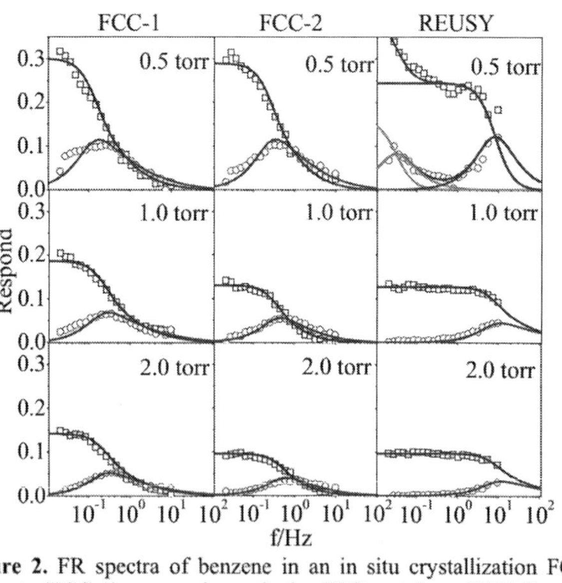

Figure 2. FR spectra of benzene in an in situ crystallization FCC catalyst (FCC-1), a semi synthetic FCC catalyst (FCC-2) and

REUSY under different pressure at 303 K. The symbols (□, ○) represent experimental in-phase and out-of-phase characteristic function data. Lines for the catalyst samples are the fits of the macroporous diffusion and adsorption model, the Yasuda sorption theoretical model, and the single microporous diffusion process model, respectively.

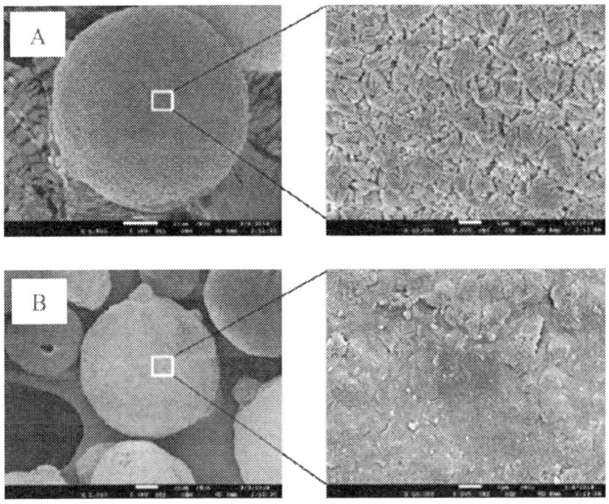

Figure 3. SEM images of an in situ crystallization FCC catalyst (A) and a semi synthetic FCC catalyst (B).

The discrepancy in the FR spectra of benzene in the novel Y-kaolin zeolite and the commercial NaY zeolite (*cf.* Figure 4) imply that the former one exhibit more excellent diffusion performance, which can be attributed to its unique texture properties and crystal structure (*cf.* Figure 5). It can be confirmed that the essence of the improvement of mass transfer performance is the reduction of the effect of the molecules rearrangement process in the pores or cages which will restrict the molecule exchange process between inside and outside of the crystalline. The mass transfer performance results have correlation with the catalytic performance of the supported NiMoS catalyst based on the novel zeolite in hydrocracking of VGO.

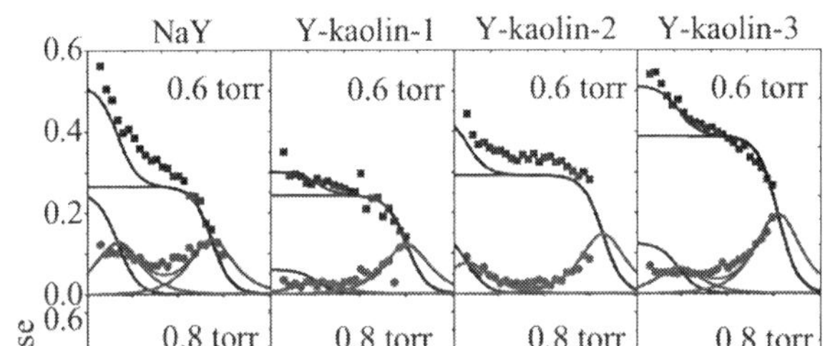

Figure 4. FR spectra of benzene in the commercial NaY zeolite (NaY), the Y-kaolin zeolite with different crystallization time (Y-kaolin-1, -2, -3) under different pressure at 303 K. The symbols (□, ○) represent experimental in-phase and out-of-phase characteristic function data. Lines are the fits of the Yasuda sorption theoretical model.

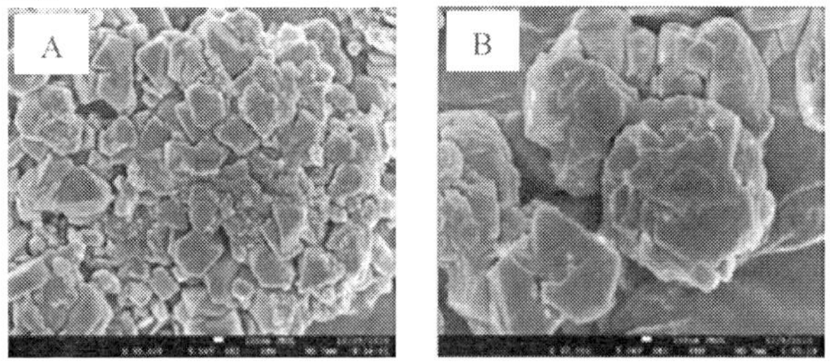

Figure 5. SEM images of a Y-kaolin zeolite (A) and a commercial NaY zeolite (B).

The fluorescence microscopic imaging technology is carried out to study the effects of the molecular properties of the guest molecules, catalyst texture properties and surface acidic properties on the molecular mass transfer behaviors of the guest molecules. The results from this novel method are of interest for it offers a unique way of both characterizing a porous solid and investigating transport phenomena within the material under identical conditions. In addition, the space-resolved data and the time-resolved data from an ultraviolet-visible spectrophotometry method can be organically integrated.

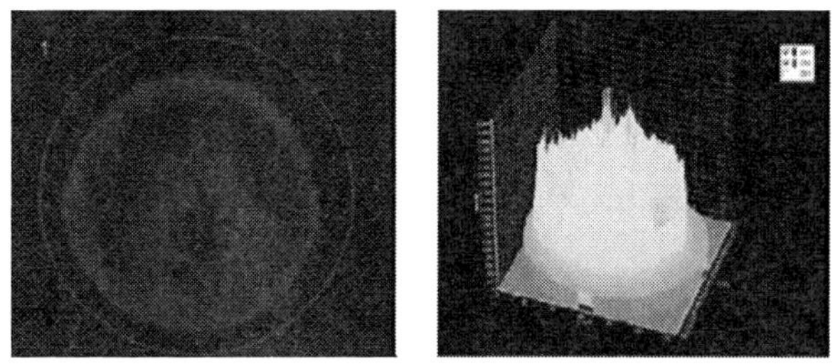

Figure 6. fluorescence microscopic image and local fluorescence intensity distribution curve of the dye molecules in a FCC catalyst.

4. Conclusions

For the commercial catalysts, the diffusional controls for molecular transport in the whole catalyst particles depend not only on the microscopic intracrystalline diffusion in zeolites but also on macroscopic intra-particle diffusion. We suggest that as well as including the enlarging of the pore structure of microporous materials, future development of the porous petroleum processing catalysts should consider the optimization of macroscopic porosity of the catalyst particles.

Acknowledgement. This work is financially supported by the National Natural Science Foundation of China (No. 21076100 and No. 21376114).

References

(1) Mitchell S.; Michels N. L.; Pérez-Ramírez J. *Chem. Soc. Rev.*, **2013**, *42(14)*, 6094.

(2) Rees, L. V. C.; Song, L. J. Recent advances in gas separation by microporous ceramic membranes, in: N.K. Kanellopoulos (Eds.), Membrane Science and Technology, Elsevier, Amsterdam, **2000**; pp.139-186.

(3) Qin, Y. C.; Gao, X. H.; Zhang,H. T.; Zhang, S. H.; Zheng, L. G.; Li, Q.; Mo, Z. S.; Duan, L. H.; Zhang,X. T.; Song, L. J. *Catal. Today*, **2015**, *245*, 147.

(4) Qin,Y. C.; Fang, X. C.; Wang, J. F.; Wang, K.; Sun, Z. L.; Song, L. J. *J. Phys. Chem. C*, Submitted.

EVALUATION OF SHALE GAS RESOURCES USING A HIGH PRESSURE WATER LABORATORY MATURATION METHOD: APPLICATION TO THE UK BOWLAND SHALE

Colin Snape[1], Clement Uguna[2], Christopher Vane[2], Vicky Moss-Hayes[2], Will Meredith[1] and Andrew Carr[3]

[1]Faculty of Engineering, University of Nottingham, Energy technologies Building, Triumph Road, Nottingham NG7 2TU, UK
[2]Center for Environmental Geochemistry, British Geological Survey, Keyworth, Nottingham, NG12 5GG, UK
[3]Advanced Geochemical Systems Ltd, 1 Towles Fields, Burton-on-the-Wolds, Leicestershire, LE12 5TD, UK
Colin.snape@nottingham.ac.uk

Introduction

The successful exploitation of shale gas plays in the United States has led to the UK starting to evaluate its shale gas reserves. The study by Andrews (2013) estimated that the shale gas resources of the entire (upper and lower unit) UK Carboniferous Bowland Shale is significant, containing 822-2281 tcf (trillion cubic feet) of gas. Andrews (2013), used the standard Rock Eval pyrolysis method that does not have the capability to measure methane or other individual hydrocarbon gases generated. Shale gas is often relatively dry (composed mainly of methane), implying that it is largely generated at high maturities, but the study by Andrews (2013), limited by the capabilities of the Rock-Eval method, did not directly address the fundamental questions of what the composition of a typical UK shale gas will be, how much methane is in the gas, and what was the source rock thermal maturity range at which gas enriched in methane is generated.

Due to the fact that actual shale gas exploration have not yet fully commenced in the UK, natural well samples are not available that can be used to address these questions. However, to properly estimate the UK shale gas resources, it is important to understand the mechanisms of shale gas formation, and the source rock thermal maturity range at which shale gas (gas enriched in methane) is generated. Also unconventional shale gas resource systems are generally slightly to highly over-pressured (Jarvie 2012), and the hydrocarbon generation and source rock maturation occur under significant water pressure conditions. In the absence of production data, it is therefore important to use laboratory pyrolysis methods that simulate conditions in geological basins to investigate shale gas generation when trying to evaluate the UK shale gas resource. To address these issues, we have used a novel high water pressure sequential pyrolysis method that simulates conditions in geological basins to investigate gas generation in shale gas reservoirs, as a function of both temperature and pressure. In this method, the source rocks were successively heated to archive source rock thermal maturity stage equivalent to the oil window, wet gas window and dry gas window. After each maturity stage, the experiment was stopped and the oil and gas generated recovered and the gas analysed. The re-heating of the same rock until VR of close to 3.0% Ro was archived, was to simulate oil and gas generation and expulsion, during successive periods of burial and uplift as experienced by the Carboniferous UK basins.

Experimental

Samples were obtained from upper Bowland shale unit within the Rempstone-1 well, (depth 665-667m, sample size 2-5 mm, initial TOC of 6.69%, hydrogen index (HI) of 410 mg/g, and initial VR of 0.58% Ro) which is Namurian (Arnsbergian) in age and located within the Pennine Basin at the southern edge of the Widmerpool Trough. The samples were pyrolysed using a 25 ml Hastalloy vessel at a water pressures in the range of 180-800 bar, at 350-420 °C for periods of 24-120 hr.

Figure 1. Pyrolysis equipment before an experiment

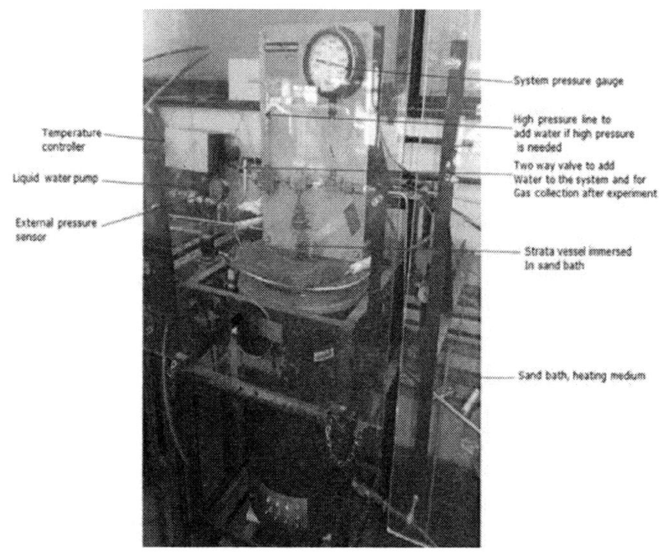

Figure 2. Pyrolysis equipment during a high pressure experiment

In this pyrolysis method, the source rocks were successively heated to achievee source rock thermal maturity stage equivalent to the oil window, wet gas and dry gas windows. After each maturity stage, the experiment was stopped and the oil and gas generated recovered and the gas analysed. The re-heating of the same rock until VR of close to 3.0% Ro was achieved, was to simulate oil and gas generation and expulsion, during successive periods of burial and uplift as experienced by the Carboniferous UK basins. The pyrolysis equipment used are shown in Figures 1 and 2.

Results and Discussion

When the shale was pyrolysed at 800 bar, the peak gas yields (methane, wet gas (C_2-C_5), and total C_1-C_5) were obtained in the late oil window (1.26% Ro), with wet (C_2-C_5) gas predominant (Figure 3). AS maturity increases (initial VR of 2.03% Ro and HI of 16mg/g), the overall gas yield is reduced markedly compared to the yield at lower maturity (1.26% Ro), which is due to the reduced remaining hydrocarbon generative potential. Interestingly methane yield starts to increase such that the gas generated (C1-C5) was 7 mg/g TOC of pyrolysed rock and contains 80% methane. This increased to 90% methane with the C1-C5 gas being 5 mg/g TOC of pyrolysed rock, when using an initial VR and HI of 2.25% Ro and 8 mg/g respectively, which is very similar to natural shale gas.

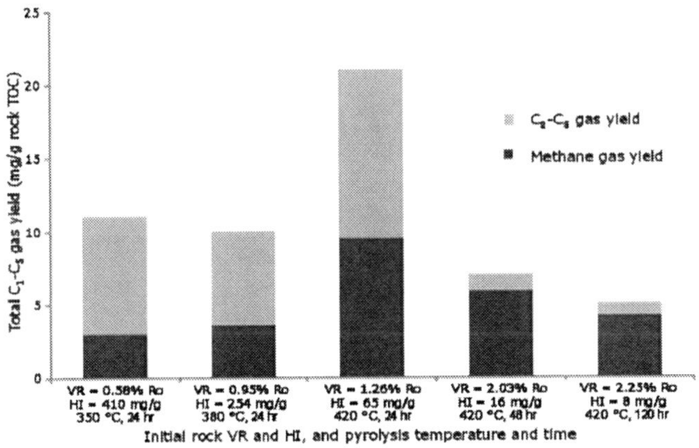

Figure 3. Gas yield and composition of Rempstone core (665-667 m depth) with increase in thermal maturity for pyrolysis at 800 bar between 24 and 120 hr and temperature range of 350-420 °C. The VR and HI values are those at the start of each experiment that generated the gases after pyrolysis.

Conclusions and implications

The upper Bowland shale unit gas in place (GIP) volume estimated from this study to that obtained by Andrews (2013) for source rock with low original hydrogen index (P90). This estimate was made using 4.2 mg (which is 2.25 X 10-4 cubic feet) of methane generated from 22.5 g of rock with an initial VR of 2.25% Ro at 800 bar pressure, and used the same shale density (2.6 g/cm3) and net shale volume (7.90 X 1011 m3) as used by Andrew (2013). Our estimate assumed that the entire upper Bowland Shale volume was at a VR of 2.25% Ro and had an initial HI of 8 mg/g TOC at the start of dry gas (over 80% methane) generation and that all the gas generated were retained in the rock. The upper Bowland Shale gas in place (GIP) volume calculated here was 21 tcf (trillion cubic feet), significantly lower than 164 tcf estimated by Andrews (2013).

This study demonstrates that shale gas (> 80% methane as found in US shale plays) will only be generated at high maturity (VR > 2.0% Ro). Although as shown here more gas will be generated at lower maturity (VR > 1.0% Ro), but the gas will certainly be wet (< 50% methane) and not dry. The fact that dry gas generation starts at high maturity with the source rock having very little generative potential remaining indicates that gas enriched in methane will be generated directly from kerogen, and not from the cracking of bitumen or oil retained in shales as suggested by Bernard et al. 2012. The reduced gas generation at high maturity found here arises from the presence of significant amounts of type III kerogen (terrestrial gas-prone) rather than dominant type II, which generates the bitumen that subsequently cracks to gas as in the Barnett Shale. The implications of this study to the Bowland Shale are that the thermal maturity range (1.1-1.9% Ro) suggested by Andrews (2013) for extensive gas generation, is actually insufficient for extensive dry gas generation. The decreasing gas yield with increasing thermal maturity suggests that the study by Andrews (2013) may have overestimated the UK Bowland Shale gas resource (based on the assumption that a typical shale gas generation starts at VR of 1.1% Ro), and a re-evaluation is required that will need to include a more reliable laboratory data set.

Acknowledgement. Dr. Clement Uguna acknowledges a fellowship funded by BGS as part of the Centre of Environmental geochemistry.

References

(1) Andrews, I.J., 2013. The Carboniferous Bowland Shale: Geology and resource estimate. British Geological Survey for Department of Energy and Climate Change, London, UK.

(2) Bernard, S., Wirth, R., Schreiber, A., Schulz, Hans-M., Horsfield, B., 2012a. Formation of nanoporous pyrobitumen residues during maturation of the Barnett shale (Fort Worth basin). International Journal of Coal Geology. 103, 3-11.

(3) Jarvie, D.M., 2012. Shale resource systems for oil and gas: Part 1–Shale-gas resource systems. In Breyer, J.A. (Ed.), shale resources– Giant resources for the 21st century: AAPG Memoir. 97, 69-87.

ZSM-5 MICROSTRUCTURE AND ITS PERFORMANCE FOR CATALYTIC CRACKING HEXANE AND NAPHTHA

Tong Ma, Bing Li, Zubao Geng, Yu Song, Guodong Li, Luoming Zhang, Yanjun Gong *

State Key Laboratory of Heavy Oil Processing, The Key Laboratory of Catalysis, China National Petroleum Corporation, China University of Petroleum-Beijing, Changping, Beijing, China 102249;

Abstract

Catalytic cracking naphtha to olefins, instead of the traditional steam cracking process, has gained much attention, due to the catalytic process producing high yield of light olefins, especially propylene and reducing energy cost and CO_2 emission[1]. Many efforts are devoted to the development of high effective catalysts which are the most important factors influencing catalytic cracking process. Zeolites synthesized by different templates exhibited different acid properties and acid distribution[2]. Herein, three kinds of ZSM-5 zeolites with similar crystal size and Si/Al ratio were synthesized using n-butylamine (NBA), tetrapropylammonium (TPA^+) and no template (NT) respectively. Applied to n-hexane cracking, ZSM-5 synthesized by NBA and TPA^+ showed a little higher initial ethylene and propylene (P+E) yields than ZSM-5 by NT. However, the lifetime of the latter was much longer than the former two (fig.1(A)). The results are attributed to the acidity properties of the ZSM-5 materials.

Except for acidity properties, coke is another important factor influencing catalytic cracking. And aromatics used to be considered as coke precursors which influence the lifetime of the catalysts as reported literature[3]. So co-feed formed by hexane and certain amount of toluene was used to investigate its inactivation, however the lifetime for cracking co-feed and pure hexane exhibited little difference over ZSM-5 with two different sizes (fig.1(B)), indicating the inactivation of cracking catalyst is not simply caused by the existence of aromatics.

The catalytic cracking performance when using two kinds of naphtha feeds were further compared with hexane. Under the same reaction condition over ZSM-5 zeolite, the P+E yields and lifetime of ZSM-5 for cracking naphtha was dependent on distribution of carbon numbers and composition of hydrocarbon, especially the naphthene and aromatics in the naphtha.

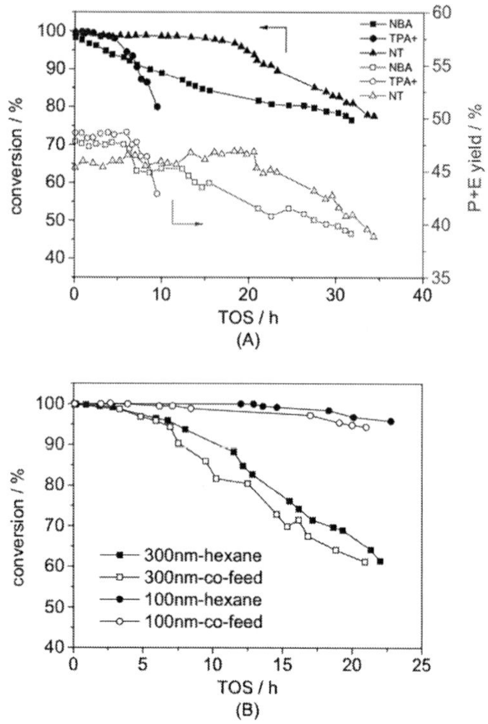

Figure 1 (A) Conversion and P+E yield of ZSM-5 zeolites synthesized by different templates; (B) Effect of hexane and toluene co-feed on the lifetime of ZSM-5 with different sizes

Acknowledgements. The State key development program for basic research (2012CB215002) and the national natural science fnoundation of China (21176255, 21276278) financially support this work.

References

(1) N. Rahimi, R. Karimzadeh, Applied Catalysis A: General. 398 (2011) 1-17.

(2) A. Hughes, K. Wilshier, B. Sexton, P. Smart, Journal of Catalysis. 80 (1983) 221-227.

(3) R. Javaid, K. Urata, S. Furukawa, T. Komatsu, Applied Catalysis A: General. 491 (2015) 100-105.

INTENSIFIED PYROLYSIS FOR ASPHALTENE UTILIZAITON IN THERMAL PLASMA

Yan Cheng, Tianyang Li, and Yi Cheng

Department of Chemical Engineering
Tsinghua University
Beijing 100084, P.R. China
Email: yicheng@tsinghua.edu.cn

Introduction

Large quantities of asphaltenes are generated during upgrading processing of heavy crude oil. However, those asphaltenes can hardly be transformed to valuable chemicals using conventional methods in petrochemical industry. Efficient utilization of asphaltenes becomes crucial to improve the profit of the entire upgrading technology.

Generally, asphaltene contains more than 60 wt% volatile matter, which makes it possible to acquire valuable products via thermochemical processes. Different from conventional technologies, pyrolysis via thermal plasma has been developed as an alternative technology to convert various kinds of carbonaceous materials, including coal [1], tar [2] and paraffins [3], as thermal plasma has the features of high enthalpy and high temperature. It is therefore anticipated that asphaltenes can be efficiently converted to gas products and carbon black based on this thermochemical method.

In the present work, thermal plasma pyrolysis experiments were carried out to prove the process feasibility. Then a particle-scale model, coupled with the heat transfer model and the asphaltene devolatilization kinetic model, was introduced to quantitatively simulate the intensified pyrolysis of asphaltenes under thermal plasma conditions. The model is validated first and then used to investigate the heat gradient inside particles and the gas-solid heat transfer behaviors at ultrahigh temperatures.

Experimental

The asphaltene particle sample was provided by Chinese University of Petroleum. A 10-kilowatt (maximum) lab-scale device was used to carry out thermal plasma pyrolysis experiments of asphaltene, which can be referred to our previous work [4]. Table 1 gives the experimental results under different working conditions. The residence time of asphaltene particle, estimated from the reactor size and the gas properties, varies from 3.4 to 3.6 ms under different operating conditions. The results show that the gas yield from asphaltene pyrolysis is 57.9% in Ar/H_2 thermal plasma, and the acetylene yield is up to 45.2%. The Ar/H_2 thermal plasma shows better pyrolysis capability than pure Ar plasma.

Mathematical model

A competition kinetic model [5] was employed in the present work. The model was further modified and the kinetics parameters were refitted according to the pyrolysis results from the asphaltene sample. A particle-scale devolatilization model, which was proposed in our previous work [6], was adopted to describe the particle heating procedures in thermal plasma. From the results in Table 1, the simulated results match the experimental data well. So the particle-scale model can be applied to simulate the pyrolysis performance under thermal plasma heating conditions.

Results and Discussion

Figure 1 shows the devolatilization time and time-mean gas-solid heat transfer coefficient under different heating conditions. The simulated results of fast pyrolysis are shown in the left part, while the right part gives the results in thermal plasma. The simulated devolatilization time in thermal plasma is ~4 ms for 1600 K argon plasma. The significant decrease of devolatilization time mainly owes to two reasons. One is the larger temperature gradient provided by higher pyrolysis temperature; the other is the larger gas-solid heat transfer coefficients, shown by the red solid line in Figure 1. These two points count for the intensification of gas-solid heat transfer effect. It can also be justified that hydrogen plasma has larger gas-solid heat transfer coefficient than argon plasma, which gives explanation to the better pyrolysis results of case-1.

Table 1. Results of thermal plasma pyrolysis

		case-1	case-2	case-3
Power	kW	1.68	1.33	1.309
Cathod Ar	L/min	8.93	10.19	10.19
Cathod H_2		0.97	0.00	0.00
Feeding	g/min	0.40	0.34	0.46
temperature	K	1625	1565	1520
C_2H_2 yield	wt%	4.5	22.7	16.9
CH_4 yield		1.5	1.4	2.0
C_2H_4 yield		1.2	3.0	3.0
Total gas yield		57.9	35.6	27.1
Simulated yield	wt%	57.8	36.6	25.1

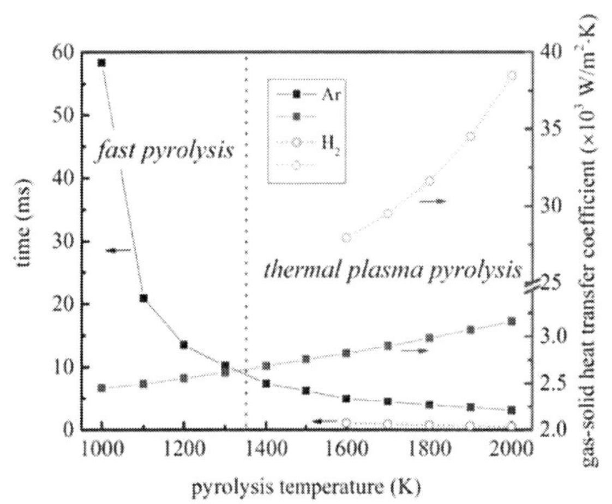

Figure 1. Effect of pyrolysis conditions on the devolatilization time and gas-solid heat transfer coefficient (particle-size: 40 μm)

Simulations were carried out under different atmosphere and particle sizes to investigate the differences caused by heating rates. The results show a strong linear correlation between the heating rate and the devolatilization time at different heating conditions, as shown in Figure 2.

The linear correlation proves that the intensified heat transfer of thermal plasma can shorten the devolatilization time for several magnitudes, and the correlation could give a quick guidance for process selection.

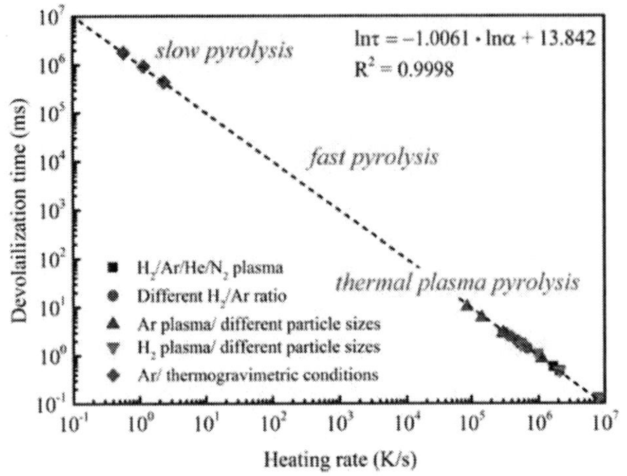

Figure 2. Correlation between the heating rate and devolatilization time (pyrolysis temperature: 2000 K; particle-size: 40 μm)

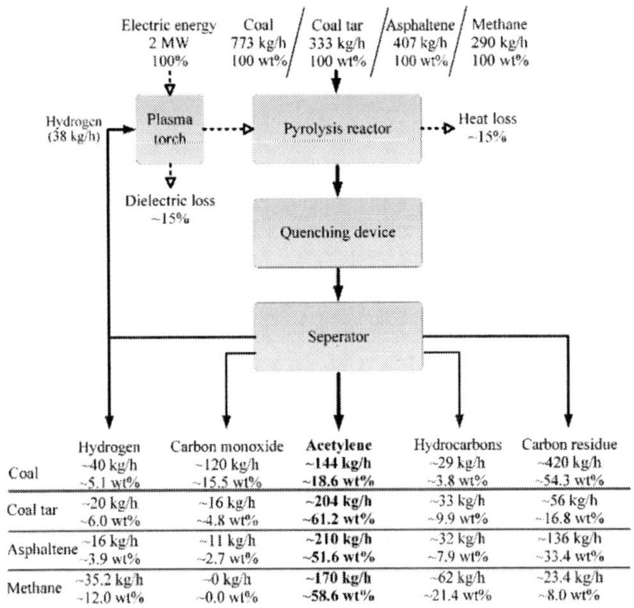

Figure 3. Comparison of thermal plasma technology to convert different feedstocks to valuable products at 2 MW pilot-plant scale

Conclusions

Thermal plasma pyrolysis has been proposed as an alternative thermochemical technology to convert asphaltenes to gas products and carbon black. In addition to the above results and discussion, we further add on more data to compare the process performances when applying the thermal plasma technology to convert different hydrocarbon feeds in Figure 3.

Acknowledgment. Financial supports from the National Basic Research Program of China (973 Program no. 2012CB720301), the National Science and Technology Key Supporting Project (2013BAF08B04), PetroChina Innovation Foundation (2013D-5006-0508) are acknowledged.

References

(1) Bond, R.L.; Ladner, W.R.; McConnell, G.I.T.; Galbraith, I.F. *Nature*, 1963, 200: 1313-1314.

(2) Cheng, Y.; Yan, B.H.; Li, T.Y.; Cheng, Y.; Li, X.; Guo, C.Y. *Plasma Chem Plasma P*, 2015, 35: 401-413.

(3) Chen, H.G.; Xie, K.C. *Petroleum Science and Technology*, 2003, 21: 709-717.

(4) Yan, B.H.; Xu, P.C.; Guo, C.Y.; Jin, Y.; Cheng, Y. *Chem Eng J*, 2012, 207: 109-116.

(5) Kobayashi, H.; Howard, J.; Sarofim, A.F. Coal devolatilization at high temperatures, in: Symposium (international) on combustion, 1977, pp. 411-425.

(6) Yan, B.H.; Cheng, Y.; Cheng, Y. *AIChE J*, 2015, 61: 913-921.

H_2 PRODUCTION VIA ENHANCED WATER-GAS SHIFT REACTION INTEGRATED WITH IN-SITU CARBON CAPTURE USING A $Mg(OH)_2$ SLURRY

*Xuejing Chen, Kyle J Fricker and Ah-Hyung Alissa Park**

Department of Earth and Environmental Engineering, Chemical Engineering
Lenfest Center for Sustainable Energy, The Earth Institute
Columbia University
500 West 120th Street, New York, NY 10027, USA
Corresponding author - E-mail address: ap2622@columbia.edu

Abstract

Syngas is a mixture of H_2 and CO can be used as feedstock to produce H_2 through water gas shift reaction (WGSR). This study investigated the integration of slurry phase $Mg(OH)_2$ carbonation with WGSR at temperatures up to 215 °C with a 0.5 wt% platinum on alumina catalyst. The removal of CO_2 by the carbonation reaction enhanced hydrogen yield of the WGSR as the equilibrium of the gas phase reaction was shifted towards products. Aqueous $Mg(OH)_2$ carbonation has been shown to occur readily, and various metastable and stable magnesium carbonates are formed depending on the reaction conditions (primarily driven by temperature). This study revealed that the H_2 yield was enhanced in the presence of $Mg(OH)_2$ slurry, and more interestingly the unexpected conversion of CO to aqueous formate ion was observed, which did limit the overall production of H_2. The presence of the basic aqueous phase within the reactor promoted a side reaction between CO and hydroxide ions. The formate can persist in solution or decompose to yield H_2 depending on the solution conditions. The solid and liquid components of the reacting slurry were carefully analyzed for carbonate and formate in order to investigate the reaction mechanisms.

Introduction

Syngas is a mixture of H_2 and CO produced via thermochemical gasification of any carboneous feedstock, including synthetic and biogenic wastes. Syngas can be upgraded into nearly any synthetic hydrocarbon fuel or chemical given specific reactor conditions and catalysts. Most downstream processes operate best at a specific H_2/CO ratio (e.g. 2:1 for the Fischer-Tropsch progress), which is typically higher than the ratio found in the coal-derived syngas. Depending on the gasification conditions and reactor arrangement, CO composition in syngas can range from 17.4 mol% (moving bed gasifier) to 60.3 mol% (entrained flow gasifier).[1] Therefore, the water gas shift reaction (WGSR) is employed to boost the H_2 levels in the syngas through the reaction of CO with water vapor (H_2O) via the following reaction.

$$CO(g) + H_2O(g) = CO_2(g) + H_2(g); \Delta H° = -41.1 \text{ kJ/mol} \quad \text{(Rxn. 1)}$$

As shown in Rxn. 1, CO_2 is the main byproduct of the WGSR and CO_2 is produced during the gasification step. By capturing CO_2 where generated, H_2 production can be enhanced according to Le Chatelier's equilibrium principle, thus improving the overall sustainability of the process. The selection of CO_2 capture media that overlaps the WGSR temperature range would be important for this scheme. The enhanced H_2 yield offers the potential to eliminate the requirement of the larger and inherently more costly, LTS reactor.

Previously, the enhanced water gas shift reaction (eWGSR) has been investigated using materials like calcium oxide, hydrotalcite, and mixed metal oxides to capture CO_2.[2] In each case however, the sorbent acted solely to separate the CO_2 because they started as carbonated materials, and therefore, the sorbent required an energy intensive sorbent regeneration step and the separated CO_2 needed to be compressed and safely disposed. The use of a sorbent derived from profuse Mg-silicate minerals, whose capacity far exceeds current and future CO_2 production,[3] provides an interesting alternative that combines carbon capture and storage steps. CO_2 reacts with the sorbent to form stable and permanent magnesium carbonate, which can be stored in confidence without long-term monitoring.

Our previous work has demonstrated the limitations of gas-solid $Mg(OH)_2$ carbonation and investigated the phenomena involved with slurry phase $Mg(OH)_2$ carbonation at moderate CO_2 pressures.[4,5] The improved kinetics and potential for near complete conversion towards carbonates were found with slurry carbonation and suggested that a $Mg(OH)_2$ slurry system would be a better fit with the eWGSR scheme. Dissolution (Rxn. 2) and hydration (Rxn. 3) of CO_2 will occur after its production by the WGSR, forming bicarbonate, carbonate and H_2 ions in solution. Acidic conditions favor the dissolution of $Mg(OH)_2$ (Rxn. 4), and subsequently carbonate and dissolved Mg^{2+} will react to precipitate anhydrous magnesium carbonate (Rxn. 5). Since the reaction temperature is high enough (> 200 °C), metastable hydrated carbonate phases should not be formed.[5]

$$CO_2(g) = CO_2(aq) \quad \text{(Rxn. 2)}$$

$$CO_2(aq)+H_2O=H_2CO_3(aq) = HCO_3^- + H^+ = CO_3^{2-} + H^+ \quad \text{(Rxn. 3)}$$

$$Mg(OH)_2(s) + 2H^+ = Mg^{2+} + 2H_2O \quad \text{(Rxn. 4)}$$

$$Mg^{2+} + CO_3^{2-} = MgCO_3(s) \quad \text{(Rxn. 5)}$$

In addition to decarbonizing the process and enhancing H_2 production, the slurry vapor pressure can provide the H_2O required by the WGSR. In a pressurized H_2O system, it has been reported that the WGSR experienced kinetic enhancement due to the pressure effect and enhanced product formation due to excess H_2O.[6] Further, $MgCO_3$(aq), a product in the proposed eWGSR system, has demonstrated moderate homogeneous catalytic activity in an aqueous WGSR of another study.[6] Thus, in this study, the effects of $Mg(OH)_2$ slurry carbonation on in-situ CO_2 capture and enhanced H_2 production were investigated while using platinum as the WGSR catalyst.

Experimental

Experimental Setup. The WGSR was performed in the headspace of a temperature controlled, high pressure batch reactor (#U761, Pressure Products Industries, Inc.) depicted in Figure 1. The lower portion of the reactor was a reservoir (300 mL) for the bulk aqueous phase for supplying the required H_2O, as well as containing the carbon capture media, $Mg(OH)_2$, in certain experiments. The reactor was equipped with temperature (type K, Omega) and pressure (Omega) data acquisition capabilities. Gas could be sampled from the headspace and liquid from the bulk solution via a manual sampling rig. During experiments, the aqueous phase was mixed well by a magnetic stirrer set to 700 rpm. In experiments which utilized a catalyst, the material was contained in a basket made of 304 stainless steel mesh 400, which was fixed in the reactor headspace. Those experiments used a 0.5 wt% platinum on alumina (Pt/Al_2O_3) catalyst (Sigma-Aldrich Co.) due to its high activity for the WGSR at relatively low temperature. A mass of 10.1 g of 3.2 mm pellets were placed in the catalyst basket.

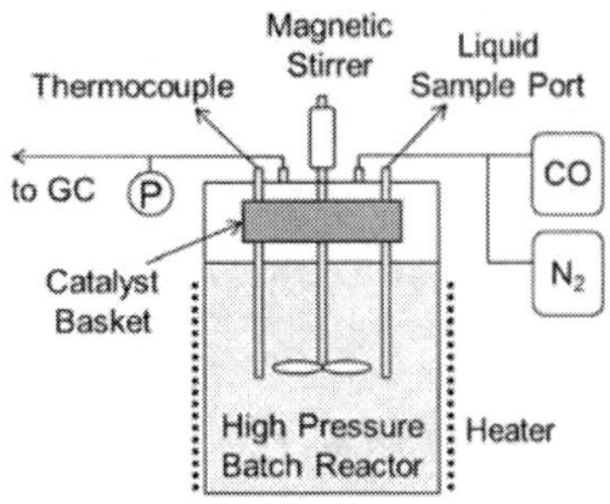

Figure 1. Schematic of high pressure batch reactor.

In eWGSR experiments, the slurry was formed by combining 10 g of dry $Mg(OH)_2$ (95% pure, Acros Organics) with 300 mL of deionized H_2O to form a 3.3 wt% slurry. After being sealed, nitrogen (N_2) flowed through the headspace for 15 min or more to purge the reactor of oxygen (O_2). Meanwhile, the temperature of the reactor was raised to ~85 °C. Upon temperature stabilization, CO was delivered into the headspace, achieving a P_{CO} of ~8.5 atm. After sealing, the temperature ramp to reaction temperature was initiated. The timer for the experiment started when the internal temperature reached 100 °C.

Four experimental cases were explored in this study. The control with neither slurry nor catalyst was referred to as "WGSR." When slurry is present without catalyst, the experiment was called "eWGSR." The catalytic WGSR and eWGSR were named "WGSR-Pt" and "eWGSR-Pt" respectively.

Analytical Methods. Gases were analyzed by a Micro GC (3000 Micro GC, Inficon, Inc.). A portion of the line consisted of a 1/4 in water trap contained in an ice water bath to prevent H_2O from entering the GC. Multiple gas samples were analyzed to ensure accurate results. The GC data provided molar concentrations of gaseous components (N_2, O_2, CO, CH_4, CO_2, H_2, C_2H_4, and C_2H_6) in the headspace other than H_2O at the end of reaction.

At the same time, liquid slurry samples (approximately 10 mL) were removed via the liquid sampling port and rig. After filtration (particle size retention: > 11 μm, GE Whatman), solids dried overnight under vacuum at 70 °C before analysis by thermogravimetric analysis (TGA) (SETSYS Evolution, Setaram Instrumentation) and X-ray diffraction (XRD) (X2, Scintag, Inc.) to determine the presence of magnesium carbonate and quantify the conversion.

The liquid fraction of the slurry was also collected and analyzed for aqueous specifies produced as a result of side reactions. Samples were diluted 50 times before ion exclusion chromatography (IC) analysis (861 Advanced Compact IC, Metrohm AG). Components were separated with a Metrosep Organic Acids 250/7.8 column using a 0.5 mM H_2SO_4 eluent and 20 mM LiCl suppressor. The IC provided the capability to quantify formate ion concentration and identify carbonate ions.

Results and Discussion

Thermodynamic Calculation. A thermodynamic model of the system shown in Figure 2 illustrates the effect of in-situ CO_2 removal (by Mg carbonation) on the equilibrium concentration of equimolar gaseous reactants ($P_{CO,i} = P_{H2O,i} = 10$ atm). WGS conversion defined as fraction of CO converted to H_2 and carbonation conversion defined as fraction of CO_2 converted to $MgCO_3$. Equilibrium constants generated with HSC Chemistry.[7] Theoretically, the integrated carbonation and WGS reactions can sustain H_2 production near 100% until the decomposition temperature of $MgCO_3$ is reached, where the decomposition temperature is dependent on the P_{CO2} within the system.

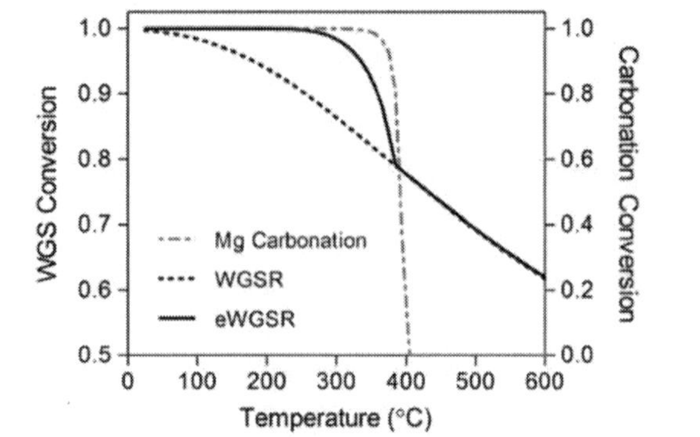

Figure 2. Thermodynamic equilibrium model of effect of carbonation on WGS reaction. $P_{CO,i} = P_{H2O,i} = 10$ atm.

Effect of Platinum Catalyst and $Mg(OH)_2$ Slurry on WGRS. Experiments were performed to investigate the fractional conversion/depletion of CO in all four reaction scenarios after 120 min. A logarithmic scale provides a better view of the instances of very low conversion, which occurred in the experiments without platinum catalyst. The WGSR experienced a CO conversion of only 0.614%, whereas the eWGSR resulted in a conversion of 15.2% of the initial CO amount. The presence of the slurry had the effect of converting CO. The experiments with the catalyst present experienced dramatically different CO conversions. WGSR-Pt and eWGSR-Pt both converted over 99.9% of the CO. The presence of the slurry did enhance the conversion of CO to H_2, by around 70%, though the H_2 production appeared limited. The disparity between CO consumption and product formation in eWGSR, WGSR-Pt and eWGSR-Pt can be explained by the production of HCO_2^-. The CO was likely converted to CHO_2^- in the basic solution and subsequently complexed to form soluble magnesium formate ($Mg(HCO_2)_2$), which was held in solution [8,9].

Conclusions

The platinum catalyst was effective and necessary to enable significant WGSR. A closer examination of the disappearance of CO and the formation of H_2 and CO_2 revealed an inconsistency, which led to the realization surrounding the creation of formate in solution. Nonetheless, the eWGS-Pt experiment produced the largest H_2 yield, likely due to the equilibrium shift cause by the aqueous carbonation reaction to form $MgCO_3$.

Acknowledgement. This work was financially supported by the National Science Foundation (CAREER Award Number 0846846).

References

(1) Stultz, S.C.; Kitto, J.B.; Steam - Its Generation and Use, 40th ed, ed. J.B. Kitto, Babcock & Wilcox, a McDermott Company, Baberton, **1992**.
(2) Harrison, D.P. *Ind. Eng. Chem. Res.* 2008, 47, 6486-6501.
(3) Lackner, K.S.; Wendt, C.H.; Butt, D.P. *Energy* **1995**, 20(11), 1153-1170.
(4) Fricker K.J.; Park, A.-H.A. *Chem. Eng. Sci.* **2013**, 100, 332-341.
(5) Fricker K.J.; Park, A.-H.A. *Ind. Eng. Chem. Res.* **2014**, 53, 18170–18179.
(6) Elliott, D.C.; Sealock,L.J *Ind. Eng. Chem. Prod. Res. Dev.* **1983**, 22, 426-431.
(7) Outotec, HSC Chemistry, in Software.
(8) Elliott, D.C.; Hallen,R.T.; Sealock, L.J. *Ind. Eng. Chem. Prod. Res. Dev.* **1983**, 22, 431-435.
(9) Shabaker, J. *J. Catal.* **2003**, 215(2), 344-352.

FISCHER-TROPSCH SYNTHESIS TARGETED TO GASOLINE-RANGE HYDROCARBONS

Chunxiang Zhu, Andrew Paluch, George M. Bollas

Department of Chemical & Biomolecular Engineering
University of Connecticut, Storrs
191 Auditorium Road, Unit 3222, Storrs, CT 06269, USA.
george.bollas@uconn.edu

Introduction

According to the U.S.DoE,[1] the high domestic gasoline demand requires processes for the production of high-octane hydrocarbons that can be a substitute for petroleum gasoline. The need to reduce U.S. dependency on foreign oil,[23] in addition to the abundance of shale and natural gas have led to an increased interest in gas/coal/biomass to liquid (XTL) processes. In this respect, Fischer-Tropsch Synthesis (FTS) is a process capable of enabling U.S. energy independence. The multiplicity of sources and types of fossil and renewable feedstocks and the fast-paced and dynamic character of the transportation fuels sector pose requirements for scalable, selective and more efficient XTL technologies, not trivial for the case of the FTS process. In this work, we explored process intensification options for FTS, selective to gasoline-range hydrocarbons.

Materials mainly of the group VIII, such as Fe, Co, Ru, and Ni, are typically used as FTS catalysts. Among them, Co stands out for its low water-gas-shift activity, moderate cost and high efficiency for industrial-scale applications. Supported Co catalysts have high activity and selectivity to linear paraffins.[4] Liquid products from Co-catalyzed FTS can be cracked and upgraded to gasoline and diesel, but this induces additional capital and operating cost, as well as increased environmental footprint of the overall process. Thus, a one-pot FTS process with high selectivity to high-octane, gasoline-range hydrocarbons is a promising alternative.

This project aims to explore the potential of structured bifunctional catalysts as candidates for intensified FTS processes, selective to high-quality gasoline production. In contrast to conventional catalysts, structured catalysts can operate at low-pressure drop, high geometric surface area, high mass-transfer coefficient, and short diffusional length. The thickness of the catalyst layer can be adjusted to achieve catalyst effectiveness close to unity.[5-7] In this work, multilayered bifunctional monolith catalysts were prepared by wash-coating monolith in Al_2O_3 slurry, cobalt nitrate solution, and ZSM-5 slurry, sequentially. These catalysts were then evaluated in terms of activity and selectivity, particularly to C5-12 hydrocarbons.

Monolith Catalyst Preparation Method

Cordierite monolith ($2MgO:2Al_2O_3:5SiO_2$ - 200 cpsi) was used as a substrate for the structured catalyst preparation. First, an alumina wash-coating solution was made by adding 5 g boehmite (Al_2O_3 20% in H2O, Alfa Aesar) and 30 g deionized water into 25 g of γ-Al2O3 (99.97%, 3 micron APS powder, 80-120 m^2/g, Sigma Aldrich). The mixture was stirred well to homogeneous slurry. Then, the monoliths, pretreated at 120 °C, were immersed in the wash-coating solution for 1 min, with the excess solution blown out gently with pressurized air. After wash-coating, the monoliths were dried at 120 °C for 4 h. Calcination was performed at 400 °C for 12 h. The weight of the monolith was then recorded. The wash-coating was repeated to control the Al_2O_3 layer thickness. After wash-coating, the active material was deposited by immersing the monolith into a $Co(NO_3)_2•6H_2O$ solution prepared by dissolving 33.3 g $Co(NO_3)_2$•6H2O (Reagent grade, 98%, Sigma Aldrich) in 25.6 ml DI water for 1 min. The excess solution was blown out gently via pressurized air. The catalyst was then dried at 120 °C for 4 h and calcined at 400 °C for 12 h. The catalyst weight was recorded again. Finally, a ZSM-5 coating was loaded by dip-coating the monolith into a NH_4-ZSM-5 slurry, made by mixing 20 g NH_4-ZSM-5 (Zeolyst international, 418 m^2/g) with 31 ml DI water. The excess solution was blown with air, dried at 120 °C for 4 h and calcined at 400 °C. For the experiments presented here, wash-coating in Al_2O3 was repeated and one Co impregnation produced a catalyst with about 5 wt% Co_3O_4 on the monolith. The notation of the catalysts reported herein represents the coating sequence (from right to left): ZSM-5-Co-Al_2O_3/M denotes a monolith, M, coated with Al_2O_3, Co and ZSM-5 sequentially.

FTS procedure

To perform the FTS reaction, a fixed bed reactor was used with 0.5" ID and a 12" long stainless steel tube. In the experiments discussed in the following, 2.90 g of the monolith catalyst with 5 wt% Co_3O_4 were loaded in the reactor. The monolith catalyst was supported with quartz wool at both ends of the reactor. The catalyst was reduced in-situ at 400 °C for 16 h, in a 50 ml/min pure H_2 flow at ambient pressure. Subsequently, the temperature was decreased to 180°C with a 5°C /min ramp rate and Ar was used to purge H_2, while increasing the reactor pressure to the temperature set-point. The reactor pressure was controlled with an Equilibar precision back pressure regulator (500 PSIG, Max, Cv: 0.07, 1/8'' NPT). When steady state was reached for the temperature (210 °C) and pressure (30 bar) in the reactor, 35 ml/min of premixed syngas (H_2:CO ratio of 2:1) was fed to the reactor. A stream of 10 ml/min of Ar was fed, as an internal standard for the Gas Chromatography analysis downstream the reactor. All of the gasses were controlled by using pre-calibrated BROOKS 5850S mass flow controllers.

Catalyst Characterization

SEM pictures of the monolith coated with Al_2O_3, Co_3O_4, and ZSM-5 are presented in Fig.1 (a), and that of the raw monolith in Fig.1 (b). Their corresponding element line mappings are shown in Figs. 1(c) and (d), respectively.

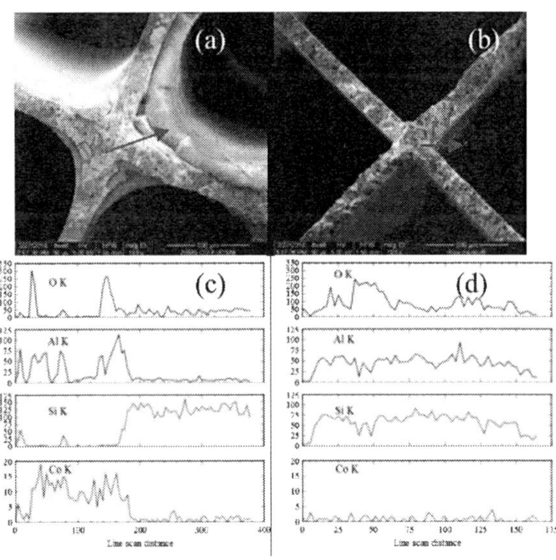

Figure 1. (a) Monolith catalyst coated with Al_2O_3, Co_3O_4, and ZSM-5. (b) Raw monolith. (c) Line mapping of the elements for (a). (d) Line mapping for (b). The line mapping direction is shown with red arrows.

It is clear in Fig.1 (a) that Al_2O_3, Co_3O_4, and ZSM-5 form different layers on the monolith. From line mapping of Fig.1 (d), it can be seen that Co_3O_4 co-exists with the Al_2O_3 phase, since it is dispersed on it. The ZSM-5 forms a distinct layer on the surface of the catalyst. Co_3O_4 was not seen to diffuse onto the ZSM-5 layer. This distinct layering of the catalyst enables accurate control of the diffusional characteristics of the gas entering and leaving the FTS active phase, which in turn allows for better control of the selectivity of the process.

FTS performance

Fig. 2 presents the selectivity of the Co-impregnated catalyst and that of the same catalyst coated with ZSM-5. The liquid product selectivity is presented in terms of carbon number, including paraffines, olefins and isomers, the latter being dominant in the ZSM-5 catalyzed experiments. From Fig. 2 it is evident that C5-C12 selectivity increases significantly with the ZSM-5-coated catalyst, while the selectivity to C20-C40 is higher for the original catalyst. This is a good indication of ZSM-5 isomerization and hydrocracking. Further analysis showed a clear contribution of the ZSM-5 catalyst on the selectivity to olefins and isomers. The impact of temperature and pressure on the conversion and selectivity with both catalysts was also studied. The hypothesis of diffusionally controlled FTS selectivity was confirmed and the selectivity to gasoline-range hydrocarbons was optimized, by tuning the operating conditions.

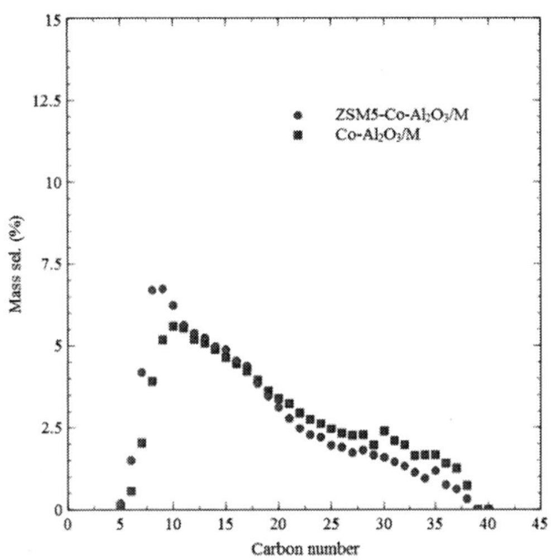

Figure 2. Liquid hydrocarbon selectivity of monolith catalysts with and without ZSM-5 coating.

Conclusions

Novel structured catalysts for *in situ* FTS product upgrading to gasoline-range hydrocarbons were successfully synthesized by coating ZSM-5 on the surface of Co/Al_2O_3/monolith structured catalysts. The bi-layered, bi-functional catalysts synthesized, were seen to improve FTS selectivity to gasoline range hydrocarbons. The simple preparation method is scalable for industrial applications. Deactivation and modification of the ZSM-5 coated monolithic catalysts, were not seen in the performance tests and are currently under investigation.

Acknowledgment. Support from the ACS PRF 53648-DNI5 is gratefully acknowledged.

References

(1) Basic Research Needs: Catalysis for Energy, Report from the U.S. Department of Energy Sciences Workshop, August 6-8, 2007.
(2) Dixon, R. K.; McGowan, E.; Onysko, G.; Scheer, R. M. *Energy Policy* **2010**, *38*, 6398–6408.
(3) Bang, G. *Energy Policy* **2010**, *38*, 1645–1653.
(4) Saib, A. M. M.; Moodley, D. J. J.; Ciobîcă, I. M.; Hauman, M. M. M.; Sigwebela, B. H. H.; Weststrate, C. J. J.; Niemantsverdriet, J. W. W.; Van De Loosdrecht, J.; Ciobîc, I. M. *Catal. Today* **2010**, *154*, 271–282.
(5) De Deugd, R. M.; Kapteijn, F.; Moulijn, J. a. *Top. Catal.* **2003**, *26*, 29–39.
(6) de Deugd, R. M.; Chougule, R. B.; Kreutzer, M. T.; Meeuse, F. M.; Grievink, J.; Kapteijn, F.; Moulijn, J. a. *Chem. Eng. Sci.* **2003**, *58*, 583–591.
(7) de Deugd, R. M.; Kapteijn, F.; Moulijn, J. a. *Catal. Today* **2003**, *79-80*, 495–501.

CO-PYROLYSIS OF COAL AND DIRECT COAL LIQUEFACTION RESIDUE

*Xiaohong LI, Daqing FU, Lili LI, Baofu LI, Wenying LI **

Key Laboratory of Coal Science and Technology (Taiyuan University of Technology), Ministry of Education and Shanxi Province, Training Base of State Key Laboratory of Coal Science and Technology Jointly Constructed by Shanxi Province and Ministry of Science and Technology, Taiyuan 030024, China
*Corresponding author. E-mail: ying@tyut.edu.cn

Abstract: An experimental study on co-pyrolysis of Shengdong coal (SD) and its direct coal liquefaction residue (DCLR) was performed in one-stage and two-stage fixed bed reactor under atmospheric pressure with nitrogen as carrier gas. The operation temperature was 600°C, and the blending ratio of DCLR and SD was 0.15:1. The results in one-stage reactor indicated that there exist synergetic effects in the co-pyrolysis of DCLR and coal. The yield of char and gas are lower than the theoretical values calculated on pyrolysis of DCLR and SD, and consequently the liquid yield such as tar and water are higher. Moreover, the Elemental analysis, the simulated distillation, FT-IR, GC-MS, NMR characterization were also performed to explore the interaction mechanisms during the co-pyrolysis process. Elemental analysis show that more hydrogen and oxygen transfer to co-pyrolysis tar and GC analysis shows there are more phenolic species in co-pyrolysis tar. The simulated distillation showed that the increased tar was mainly heavy component that boiling point higher than 300°C. The hydrogen,methane, carbon dioxide and carbon monoxide decrease in gas , which suggest that more hydrogen and oxygen may migrate to tar and water. FT-IR results show both the Har/Hal and aromaticity values decreased compared with calculated values, meaning that part of cracked fragments which would produce char in pyrolysis alone was stabilized by hydrogen to form tar in co-pyrolysis process. Two-stage fixed bed reactor was adopted to further explore interactions of SD and DCLR in co-pyrolysis process. When SD was placed in downside and DCLR in upside, the degree of interaction is maximum, suggesting the coal may the hydrogen donor of co-pyrolysis.

Keywords: Co-pyrolysis, Coal, Direct coal liquefaction residue, Synergetic effect, Tar

ALTERNATIVE WAY TO PRODUCE RENEWABLE FUEL FROM HYDRODEOXYGENATION OF PLANT OILS OVER NICKEL PHOSPHIDE CATALYSTS

Wei Li, Qingxin Guan

College of Chemistry, Key Laboratory of Advanced Energy Materials Chemistry (Ministry of Education), Collaborative Innovation Center of Chemical Science and Engineering (Tianjin), Nankai University, Tianjin, 300071, China
weili@nankai.edu.cn

Introduction

With economic development, global fossil resources are being depleted; the energy shortage has become an important factor restricting the development of the world economy. Increasing concerns have caused an intensified search for alternative sources of energy. In recent years, biomass has attracted much attention as a sustainable resource with advantages over fossil fuels such as renewability, lower expense, and a short CO_2 cycle, and protection of the environment.[1] The renewable diesel can be synthesized by hydrodeoxygenation (HDO) technology from a number of plant oils, such as jatropha oil, palm oil, castor oil, and algae oil.[2] In this paper, series of nickel phosphide catalysts were prepared and their application in the HDO reaction in a fixed-bed reactor were investigated. It was found that nickel phosphide catalysts showed high activity.

Experimental

The preparation of bulk Ni_2P is illustrated as follows. Nickel chloride ($NiCl_2$) and sodium hypophosphite (NaH_2PO_2) were dissolved in deionized water with a mole ratio of P/Ni = 1.5. After stirring for 3 h, the solution was evaporated slowly to dehydrate, and was dried at 80 °C to obtain the precursor. At the beginning of the thermal decomposition, the air in the reactor was removed by flowing Ar, after which the precursor was treated at 300 °C for 0.5 h in a static Ar atmosphere (heating rate: 5–20 °C min^{-1}). The product was cooled to ambient temperature under Ar and was washed several times with deionized water to remove ion impurities, after which the wet material was dried at 120 °C for 3 h.

The silica supported nickel phosphide catalyst was synthesized using the thermal decomposition of nickel hypophosphite precursor, the precursor being prepared using incipient wetness impregnation, as reported elsewhere. The wet precursors were evaporated slowly to dehydrate, and were dried at 80 °C for 8 h. At the beginning of the thermal decomposition, the air in the reactor was removed by flowing Ar, after which the precursor was treated at 300 °C for 0.5 h in a static Ar atmosphere. The product was cooled to ambient temperature under Ar and was washed several times with deionized water to remove ion impurities, after which the wet material was dried at 120 °C for 3 h.

Results and Discussion

In previous studies, different metal phosphides were synthesized using the thermal decomposition of metal hypophosphite precursors.[3] Among a variety of metal phosphide, nickel phosphide having the best activity and stability. To further study the influence of the molar ratio of P/Ni on the products, different nickel phosphides were prepared using the thermal decomposition of a metal hypophosphite precursor with different mole ratios of P/Ni. As shown in Figure 1, we can clearly see that Ni_2P can be synthesized in the mole ratio of P/Ni range of 1.5 to 1.75. The peaks of Ni_5P_4 were detected when the precursor mole ratio of P/Ni exceed 2.0, which indicates that P is a slight excess for the synthesis of bulk Ni_2P at this mole ratio.

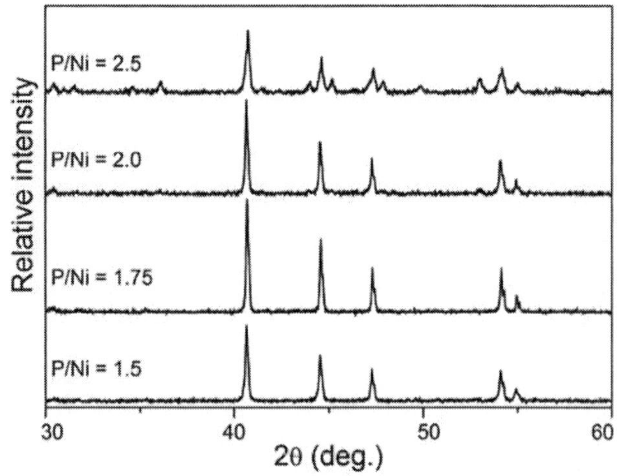

Figure 1. The XRD patterns of bulk nickel phosphides prepared using the thermal decomposition of a metal hypophosphite precursor with different mole ratios of P/Ni.

Figure 2 shows that weak diffraction peaks of Ni_2P were detected in 15 wt % Ni_2P/SiO_2 catalyst, and diffraction peaks of Ni_2P increase as the loading increases

from 15 to 25 wt %. The result shows that the increased loading will cause the sintering of nickel phosphide crystal grain.

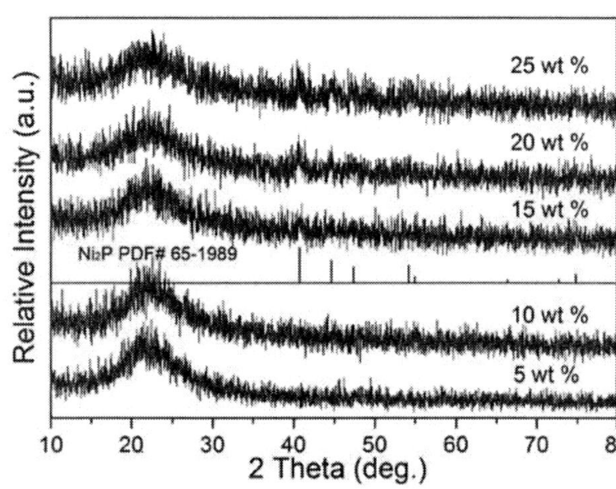

Figure 2. The XRD patterns of 10 wt % Ni_2P/SiO_2 catalysts prepared using the thermal decomposition of a metal hypophosphite precursor with different mole ratios of P/Ni.

The HDO catalytic activities were evaluated using 50 wt % methyl palmitate in decalin. To study the detailed reaction network of methyl palmitate HDO, different WHSVs, reaction temperatures, and reaction pressures were investigated in detail. Figure 3 shows the catalytic activities of 10 wt % Ni_2P/SiO_2 at different WHSVs (testing conditions: 3 MPa, 290 °C, H_2/oil = 1000). Not surprisingly, HDO conversion and selectivity of pentadecane (C15) and hexadecane (C16) were gradually decreased as the WHSV increased. Meanwhile, the amount of intermediate state products (such as pentadecene, palmitic acid, hexadecanal, and hexadecanol) gradually increased.

The previous results show that Ni_2P/SiO_2 showed significantly higher activity than Ni/SiO_2 after different pretreatment temperatures.[4] In order to investigate the influence of catalysts, different catalysts were tested using H_2-TPR. Figure 4 shows the H_2-TPR patterns of different catalysts at a heating rate of 10 °C min^{-1}. We can clearly see that the peak at 395 °C for the Ni/SiO_2 catalyst is assigned to the reduction of nickel oxide. However, at least four reduction peaks (225, 465, 550, and 600 °C) were detected for the Ni_2P/SiO_2 catalyst, which indicates that it has several different oxidation species. The reduction peak at 225 °C indicates that part of the oxidation species in Ni_2P/SiO_2 catalyst is more likely to occur redox reaction than Ni/SiO_2.

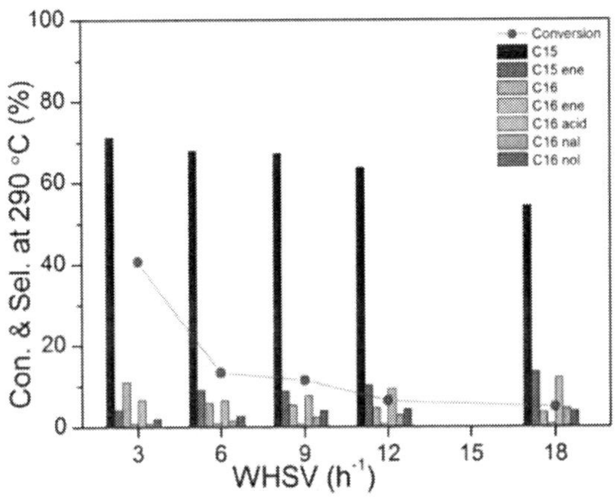

Figure 3. The catalytic activities of 10 wt % Ni_2P/SiO_2 at different WHSVs (testing conditions: 3 MPa, 290 °C, H_2/oil = 1000).

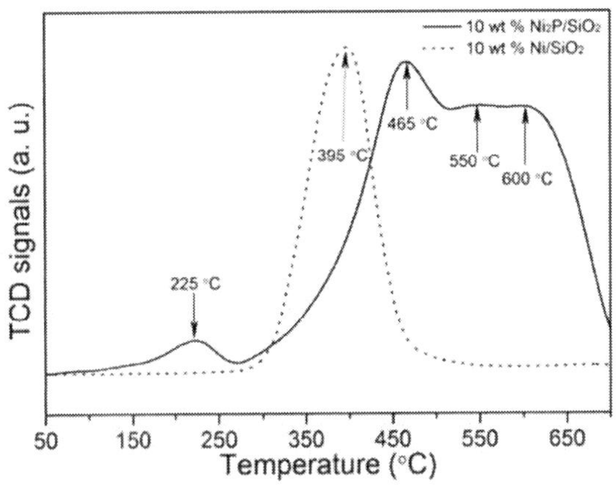

Figure 4. The H_2-TPR patterns of different catalysts at a heating rate of 10 °C min^{-1}.

Based on the analytical results of the GC–MS, we propose a possible HDO reaction mechanism of methyl palmitate on Ni_2P/SiO_2 catalysts. As shown in Figure 5, the reaction pathway of the HDO of methyl palmitate mainly includes 10 steps.

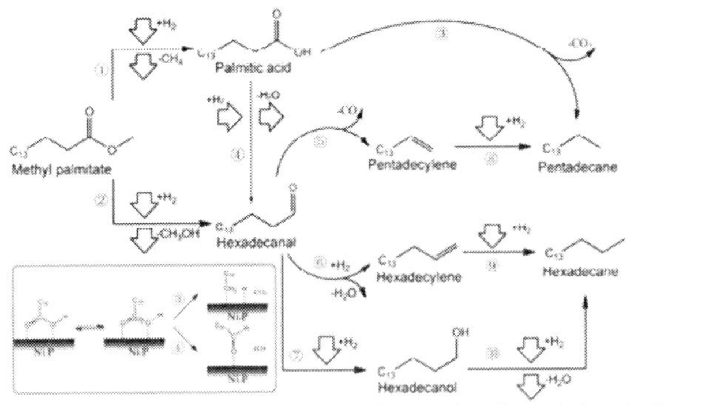

Figure 5. Reaction pathway of the HDO of methyl palmitate on Ni_2P/SiO_2 catalysts.

Conclusions

In this paper, a series of bulk and supported Ni_2P catalysts were synthesized using the thermal decomposition of metal hypophosphite precursors. The XRD results indicate that the crystal size of nickel phosphide is smaller than the detection limit of XRD and has a good dispersion on the SiO_2 support. The H_2-TPR results indicate that the Ni_2P/SiO_2 catalyst has several different oxidation species. The catalytic properties of the Ni_2P/SiO_2 catalysts for the HDO of methyl palmitate were tested in a fixed-bed reactor. The result indicates that HDO conversion and selectivity of pentadecane and hexadecane were gradually decreased as the WHSV increased. Finally, based on the analytical result of GC–MS, we propose a possible HDO reaction mechanism of methyl palmitate on Ni_2P/SiO_2 catalysts.

Acknowledgement. This work was financially supported by the NSFC (21376123, U1403293), MOE (IRT-13R30 and 113016A), and the Research Fund for 111 Project (B12015).

References

(1) Zhang, X.; Wang, T.; Ma, L.; Zhang, Q.; Huang, X.; Yu, Y., *Appl. Energy* **2013,** *112* (0), 533-538.
(2) Liu, S.; Zhu, Q.; Guan, Q.; He, L.; Li, W., *Bioresour Technol.* **2015,** *183*, 93-100.
(3) Guan, Q.; Li, W.; Zhang, M.; Tao, K., *J. Catal.* **2009,** *263* (1), 1-3.
(4) Guan, Q.; Han, F.; Li, W., *RSC Adv.* **2016**. DOI: 10.1039/C6RA02601J.

SYNGAS TO FUEL AND CHEMICALS WITH HIGH EFFICIENCY VIA FISCHER-TROPSCH SYNTHESIS

*Liangshu Zhong, Hui Wang, and Yuhan Sun**

CAS Key Laboratory of Low-Carbon Conversion Science & Engineering, Shanghai Advanced Research Institute, Chinese Academy of Sciences, Shanghai 201203, China
* sunyh@sari.ac.cn

Introduction

Fischer-Tropsch synthesis (FTS) is a heterogeneous catalytic process for sustainable production of fuel and value-added chemicals via syngas[1,2]. It is a very complicated catalytic reaction process with co-production of paraffin, olefin, alcohol and other products. The objective product is C_{5+} hydrocarbon with less methane for typical FT synthesis, C_2^+OH is required for higher alcohol synthesis (HAS), and olefins are pursued for Fischer–Tropsch to olefins (FTO) process. The product selectivity is controlled by many factors, such as catalyst systems, reactors and operation conditions. For the purpose of high selectivity to the objective product, it is very desirable to regulate the reaction network for certain reactions to take place preferentially. Obviously, one of the great challenges in FTS research is to control CO activation and then both carbon chain growth and termination for high selectivity to target product with high activity and high stability.

In the past decades, syngas to fuel and chemicals via FTS are investigated at our group from fundamental research to industrial deployment. The research directions include typical FT synthesis using Co or Fe-based catalysts, higher alcohol synthesis via bifunctional catalysis, direct production of olefins via syngas. Herein, some progress about the reaction network, structure-performance relationship and the design concept of new catalysts are discussed.

Study of Reaction Network

The reaction network for syngas conversion was investigated by the probe molecule technique over CuFe modified FT catalyst (Scheme 1)[3]. Alcohol, aldehyde, olefin as probe molecules were in-situ introduced into the reaction system. When alcohol was introduced into the reaction system, a considerable amount of esters with unique structures was formed with the formula RCO_2R', where R′ is the alkyl group from the added alcohol, and such esters followed ASF distribution with a similar chain propagation factor to that of the corresponding higher alcohols. The existence of surface acyl species indicated that the insertion of CO rather than hydroxyl was the underlying mechanism for higher alcohols synthesis over modified FT catalysts along with the support from surface simulation.

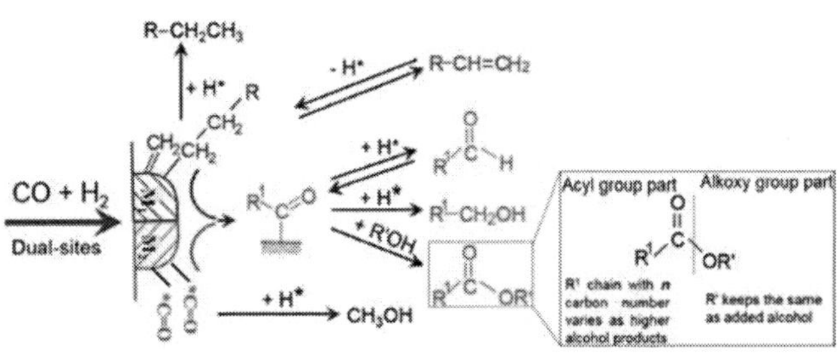

Scheme 1. Schematic depiction of the reaction network.

Study of the structure-performance relationship

CuFe bimetallic nanoparticles were prepared and used as model catalyst for deactivation study in higher alcohol synthesis[4,5]. The structural evolution of CuFe bimetallic nanoparticles during reaction was investigated. Phase separation of Cu and Fe components was observed directly by STEM-EDS and HRTEM (Figure 1). The phase separation weakened the synergism between Cu and Fe and led to decreased selectivity to alcohols with time on stream.

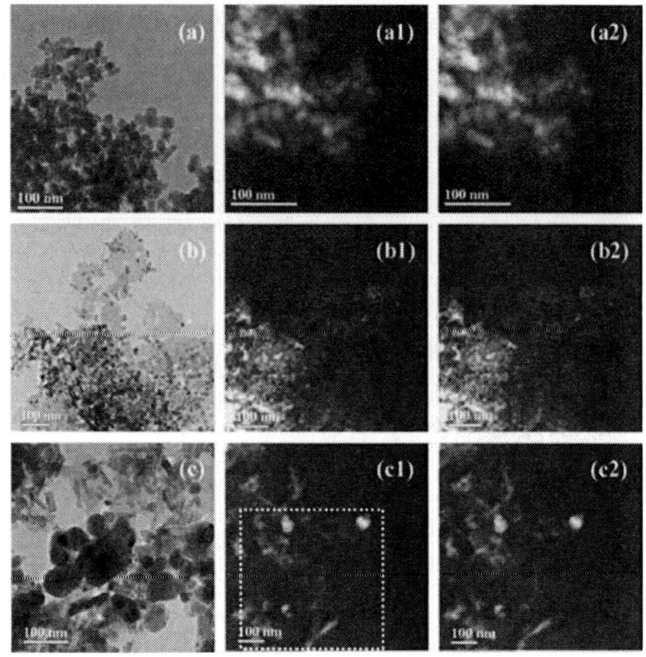

Figure 1. TEM, STEM and STEM-EDS images of fresh (a), activated (b) and spent (c) CuFe catalyst. a/b/c: TEM image, a1/b1/c1: STEM image, a2/b2/c2: combined componentsdistribution with STEM image, red color for Fe and green for Cu distribution.

Changes in catalytic performance and catalyst structural evolution during the deactivation of the CuCo model catalyst for higher alcohol synthesis via syngas were also investigated. The catalytic activity decreased continuously with time-on-stream, although there was no obvious change in total alcohol selectivity. Meanwhile, the products shifted gradually to low carbon number, and the chain growth probabilities of both alcohols and hydrocarbons decreased with reaction time (Figure 2). Using ethane hydrogenolysis as the probe reaction, the catalytic activity also declined with time-on-stream, indicating the decrease in the number of surface metallic cobalt ensembles. It was suggested particle sintering and Co_2C formation led to the catalytic deactivation for CuCo model catalyst.

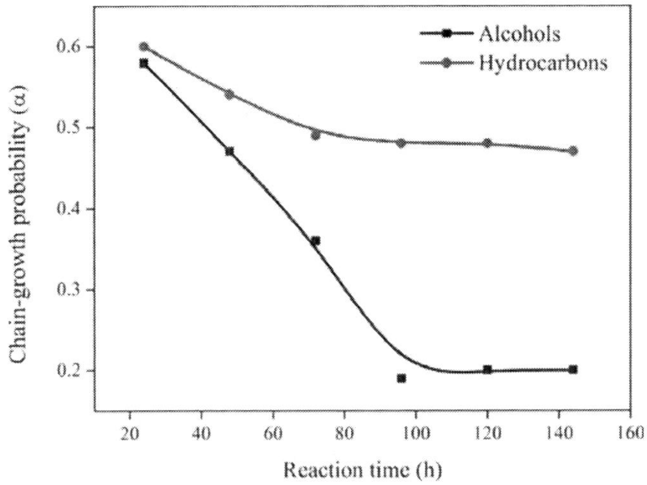

Figure 2. Changes of chain-growth probability for CuCo catalyst with reaction time.

Development of Nanocatalyst for Syngas Conversion

Based on the fundamental research concerning the nature of active site, reaction network and structure-performance relationship, several kinds of nanocatalysts are developed for typical FT synthesis, higher alcohol synthesis and olefins production with promising performance.

Low temperature Fe-based FT catalyst was developed by controlling the active site density and hindering the formation of by-product. The stability was improved by optimization of catalyst structure, heat and mass transformation. The typical catalytic results exhibited high carbon efficiency as shown as following: T = 190-210 °C, 20 Bar, H_2/CO = 1-2, CO_2 selectivity = 1.7-2.1%, CH_4 selectivity = 2-2.5 %, $C_{2-4}^{=}$ + C_{5+} selectivity > 92%. The catalyst also possessed high stability and the low selectivity to CH_4 and CO_2 was maintained even after 7000 h (Figure 3).

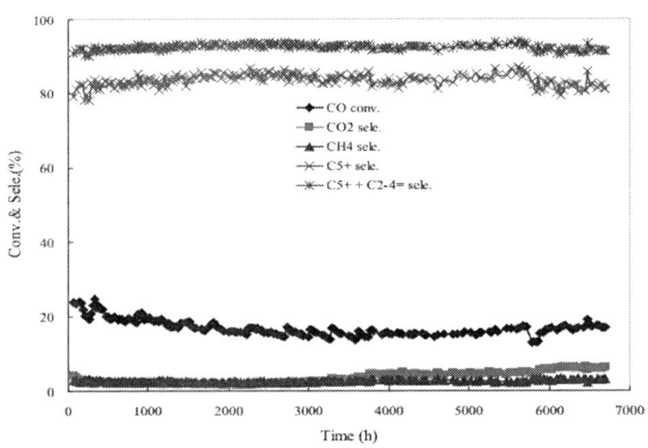

Figure 3. Catalytic performance of Fe-based FT catalyst.

Higher alcohols can be used as fuels, fuel additives and intermediates for value-added chemicals. The main by-products in higher alcohol synthesis process include CO_2 and various hydrocarbons. Obviously, the key object for HAS is to further increase the selectivity to total alcohols, especially to C_{2+}OH. In order to improve the catalytic performance for higher alcohol synthesis, it is necessary to control the reaction pathway by catalyst design and optimization of reaction process. Co-based nanocatalyst was developed for higher alcohol synthesis. The total alcohol selectivity reached 45 C% and the C_{2+}OH and C_{6+}OH selectivity in alcohol distribution achieved 90% and 60%, respectively. In the oil and wax phase, the oxygenate product accounted for more than 70%. As the product contained large amount of long-chain higher alcohol, the economical efficiency should be rather promising.

Fischer-Tropsch to olefins (FTO) process has long offered a means for direct production of lower olefins via syngas derived from coal, biomass and natural gas. However, it is impossible to obtain high selectivity toward C_{2-4} olefins and low selectivity to methane simultaneously according to the ordinary ASF model. Obviously, the main objective for FTO process is to maximize lower olefins selectivity and to reduce methane production by using new catalyst system which deviates from ASF distribution with high activity and stability. CoMn-based catalyst was developed and displayed high selectivity to lower olefins (~60 C%) in combination with low methane selectivity (~5 C%) under mild reaction condition. The O/P ratio for C_{2-4} slate was as high as 30 and the product distribution deviated greatly from the classical ASF distribution with the highest selectivity to propylene.

Outlook

Syngas conversion to fuel and chemicals via Fischer-Tropsch synthesis is one of the most challenging and attractive subjects in the field of C1 chemistry. More efforts are still needed to further reveal the reaction network and structure-performance relationship. In-situ study is important to provide deep information under working conditions. In addition, development new nanocatalysts with suitable mesostructure for syngas conversion will benefit for high activity, high selectivity and high stability.

Acknowledgement. This work has been supported by the Natural Science Foundation of China (21403278, 21573271, 91545112), Shanghai Municipal Science and Technology Commission, China (Grant No.: 15DZ1170500) along with the support from the industry Groups including China Shenhua Coal toLiquid and Chemical Company Limited, Shanxi Lu'an Coal Corporation Limited, and Shell Global Solutions International B. V.

References

(1) Mark, E. D. *Catal. Today,* **2002**, *71*, 227.
(2) Davis, B. H. *Catal. Today,* **2002**, *71*, 249
(3) Yang, Y.; Wang, L.; Xiao, K.; Zhao, T.; Wang, H.; Zhong, L.; Sun, Y. *Catal. Sci. Technol.,* **2015**, *5*, 4224.
(4) Xiao, K.; Bao, Z. H.; Qi, X. Z.; Wang, X. X.; Zhong, L. S.; Lin, M. G.; Fang, K. G.; Sun, Y. H. *J. Mol. Catal. A-Chem.*, **2013**, *378*, 319.
(5) Xiao, K.; Bao, Z. H.; Qi, X. Z.; Wang, X. X.; Zhong, L. S.; Lin, M. G.; Fang, K. G.; Sun, Y. H. *Catal. Commun.*, **2013**, *40*, 154.
(6) Yang, Y.; Qi, X.; Wang, X.; Lv, D.; Yu, F.; Zhong, L.; Wang, H.; Sun, Y. *Catal. Today,* **2015**, DOI: 10.1039/c5cy00641d.

PRODUCTION OF JP-8 JET FUEL FROM COAL

Andrew Lucero, JiaJia Meng, Brittany Koob, Kelly Mastro, and Santosh Gangwal

Southern Research, 5201 International Drive, Durham, NC, 27712 USA
lucero@southernresearch.org

Introduction

An indirect liquefaction process employing advanced process intensification approaches is being developed to reduce costs and accelerate commercialization of a coal to liquid (CTL) process for converting coal and coal-biomass mixtures to JP-8 jet fuel. The process reduces the number of parasitic syngas cleaning steps and eliminates costly upgrading of wax produced in conventional indirect liquefaction processes, resulting in improved economics and lower lifecycle greenhouse gas (GHG) emissions for CTL-based jet fuel production. These improvements directly reduce capital costs of indirect CTL processes, resulting in significantly improved economics and cost competitiveness of CTL compared to petroleum refining. This not only allows for CTL to be cost competitive at typical large scales, but also opens up opportunities for CTL to be viable at smaller scales. Research is specifically targeted towards two advanced cost-effective technologies: a compact auto-thermal reformer (ATR) from PCI to reduce syngas cleaning and water gas shift requirements, and a selective and active cobalt-zeolite hybrid Fischer-Tropsch (FT) catalyst from Chevron that predominantly produces liquid range C5-C20 hydrocarbons with high productivity (5X conventional FT catalysts) while eliminating the production and associated handling of solid C21+ wax. High fractions of the liquid product are in the desired C8 to C16 range for jet fuel. Heat management associated with the highly exothermic FT reaction is carried out using a novel fixed-bed heat exchange reactor system.

This work is focused on testing of the hybrid FT catalyst. Specific project objectives are to produce a hydrocarbon product containing at least 75% C5-C20 hydrocarbons with a selectivity of 65% jet fuel range C8-C15 hydrocarbons and no C21+ hydrocarbons, and an overall productivity greater than 0.7 $g_{C5+}/g_{cat}/h$.

Background

Fischer-Tropsch (FT) synthesis, originally discovered in Germany more than 85 years ago, is presently the preferred route for converting syngas (H2 + CO) made from coal, biomass, or natural gas into a wide variety of hydrocarbons[1-4]. For the production of liquid transportation fuels, carbon numbers in the gasoline and diesel range of C5 to C20 are preferred. Unfortunately, conventional FT is non-selective and results in a wide product slate ranging from C1 to C60+ hydrocarbons via chain growth polymerization reactions (CO + 2H2 $\rightarrow$ -CH2 - + H2O). The hydrocarbon chain growth in FT is dictated by the Anderson-Schulz-Flory (ASF) distribution parameter, α, typically ranging from 0.75 to 0.95. A catalyst with α of 0.9 or higher is preferred as it produces less gaseous hydrocarbons. Although this results in formation of significant amount of undesirable waxes (C21+), production of undesirable light gases (C1 to C4) is minimized. Also, the C5 to C10 fraction (naphtha) produced has significant paraffins, a poor gasoline blend, and needs to be upgraded elsewhere. The waxes are subjected to mild hydrocracking/hydrotreating operations to convert them to the extent possible to desirable transportation fuel-range hydrocarbons (C5 to C20). The FT and wax upgrading section can account for as much as 35 % of the capital requirement of a CTL plant. Thus, eliminating the upgrading operations can contribute significantly towards reducing the cost of CTL and CBTL processes.

Advanced Chevron FT Catalyst technology. Chevron's selective catalyst technology proposed here avoids the expensive downstream processing requirements by not producing the waxes while also producing low levels of light gases.

Chevron has developed and patented hybrid cobalt-zeolite and ruthenium-zeolite catalysts that exhibit exceptionally high selectivity to drop-in gasoline and diesel range hydrocarbons. Over 20 patents and patent applications have been issued or are pending covering various aspects of Chevron's catalyst technology [Chevron Corporation, private communication[5-8]. Chevron's patents cover fixed-bed, fluidized-bed, and slurry reactor applications.

As described in the referenced patents, the hybrid catalysts consist of FT-active metals, cobalt and ruthenium, that carryout the FT reactions. Chevron is also pursuing a ruthenium supported on zeolite hybrid catalyst[8]. These metals are deposited on specially tailored wide-pore zeolites that hydrocrack the waxes (C21+) to desirable gasoline and diesel range hydrocarbons, and reform the naphtha range hydrocarbons to high octane gasoline compounds. Development of Chevron's hybrid cobalt and ruthenium catalyst extrudates uses their extensive knowledge in preparation and activation of FT catalysts and preparation of specially tailored zeolites. In these catalysts, the FT chain growth occurs on the metals within the pores of the zeolite support. The zeolite supports are optimized with respect to acidity (silica to alumina ratio) and pore diameter to allow only up to C20 hydrocarbons to escape the pores. The C21 + hydrocarbons are too large to escape and crack to smaller desirable isoparaffins in the C5 to C20 range.

Project Phases

Full testing of the FT catalyst involves three potential phases: lab-scale experiments to screen catalysts and parameters for optimum configuration, followed by larger bench-scale testing to demonstrate characteristics of a novel fixed-bed heat exchange reactor (HXR) system, with potential scale up of the reactor system for determining modeling information required to size larger demonstration and commercial units. The information covered to date includes a lab-scale fixed-bed reactor constructed using ½ inch tube and the setup described in Figure 1. A 2" ID HXR with catalyst deposited on structured packing by Intramicron has also been constructed and initial testing completed at the National Carbon Capture Center (NCCC) located in Wilsonville Alabama. A 4" scale up version of the HXR has also been planned as discussed in future work below.

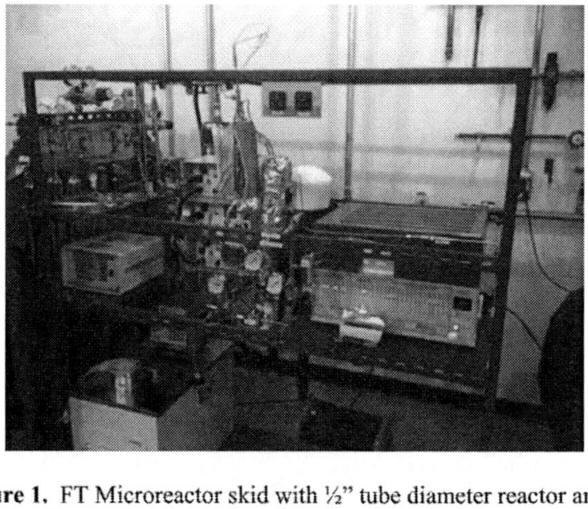

Figure 1. FT Microreactor skid with ½" tube diameter reactor and associated analytical instrumentation.

Results and Discussion

The ½" diameter lab-scale rector results were valuable in showing optimum catalyst configurations. Table 1 below displays one of the most promising catalyst configurations. Initial testing parameters were adjusted by only 10 degrees C and substantially improved the results to those shown. The corresponding data showed catalytic performance equivalent to the desired target values with nearly 74% liquids (desired was ≥75% liquids) with over 70% jet fuel selectivity (target 65%) with minimal C21+ wax formation and with liquid productivity (target >0.7 $g_{C5+}/g_{cat}/h$) at over 0.81 $g_{C5+}/g_{cat}/h$ (target >0.7 $g_{C5+}/g_{cat}/h$). This FT catalyst maintained the high reactivity for about 850 h in both fixed-bed and HXR. Several trials even exceeded 900 h without substantial catalyst deactivation at the conclusion of the tests.

Table 1. Summary of lab scale selectivity results

T °C	TOS, hr	CO Conv., %	CH_4 Sel., %	C_2-C_4 Sel., %	Liquid Sel., %	Jet Fuel Sel.,%	Liq Prod., gC_{5+}/gcat/hr
240	180-220	35.2	14.6	12.1	73.0	72.0	0.64
245	230-305	44.2	14.7	11.3	73.7	71.1	0.81
	310-430	39.2	14.2	12.9	72.6	71.5	0.72
248	760-830	45.2	16.8	11.3	71.3	71.1	0.81
250	431-525	44.3	14.6	11.7	73.4	71.0	0.81
	580-740	42.2	16.1	12.4	71.2	71.7	0.74

In addition to testing the packed bed reactor configuration in the ½" lab-scale reactor, the HXR version was also tested with catalyst deposited on a structured packing by Intramicron. The HXR addresses issues related to heat transfer, process intensification, and product selectivity encountered during FT synthesis at larger scales. HXRs provide very uniform temperature profiles inside the reactors without requiring large heat sinks in the fixed bed. They also avoid pressure drop issues encountered with fixed beds of small pellets. The reactor results below in Figure 2 show a long reliability (>800 h) with similar results to those of the fixed-bed. The large variations due to changes in space velocity are also noticeable.

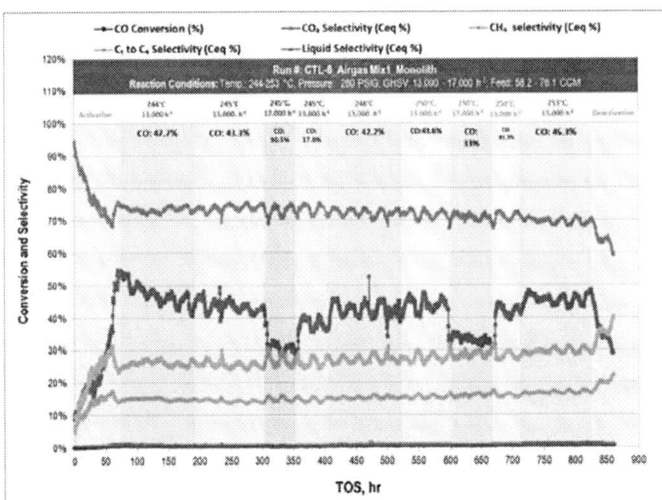

Figure 2. Summary of conversion and selectivity for microreactor test of Chevron's FT catalyst.

As shown in Figure 3 below, optimization during later trials continues to improve the parameters selected for the Chevron catalysts. Selectivity to liquid fuels is still well in excess of 70% with increased CO conversion to 45%. The jet fuel fraction has been increased to nearly 73% with the latest conditions tested below in run 15. The tests in Figures 2 and 3 also indicate the potential of the HXR to give similar performance to that of the packed bed, while improving conditions substantially at scale due to the inherent advantages previously discussed for the HXR.

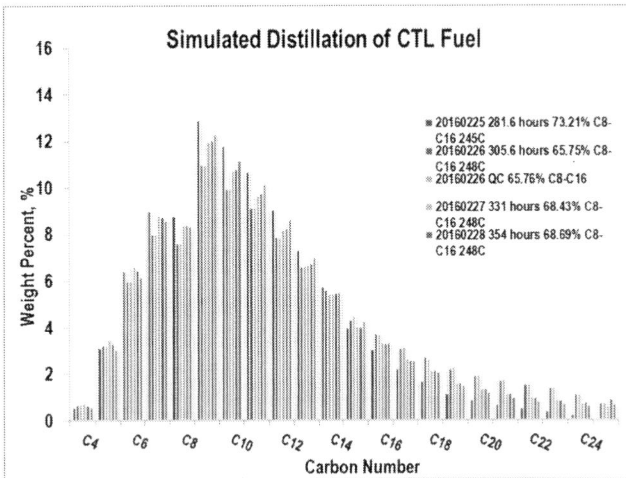

Figure 3. Current optimization of CTL fuel distillation simulation from microreactor data (run 15 conditions: Temperature: 243-248°C; Pressure: 280 PSIG; GHSV: 13,000 sccm/g/h; mix 1 = 2:1 H_2 to CO and mix 2 = 1.8:1 H_2 to CO)

Conclusions

The desired catalyst performance were exceeded in the ½ inch lab scale reactor configuration as displayed in Table 1. The FT catalyst remained active in excess of 850 h for both fixed-bed and HXR tests. The catalytic performance at temperatures between 245-253°C was similar, while space velocity (SV) increases significantly reduces the amount of straight chain hydrocarbons as well as reduced the CO conversion and liquid productivity. However, increasing the space velocity showed an increase in carbon number within the range of desired hydrocarbons desired for jet fuel production. The optimum conditions for the ½" reactor were ≥245°C, SV of 13,000 sccm/g/h and 280 psig. While these conditions only resulted in 73% liquid selectivity as opposed to the desired 75%, the selectivity to jet fuel was significantly greater than desired (>70% compared to 65% desired) and the small amounts of C_{21+} hydrocarbons (<5%) were considered small enough to consider the testing a success and continue with experiments on the 2" HXR located at the NCCC in Wilsonville, Alabama.

Future Work. After the experiments in the 2 inch reactor are completed, a limited series of experiments will be conducted with the same catalyst in a 4 inch HXR to confirm the ability of the HXR to extend to the larger diameter and supply information for process modeling. Following those experiments an extended 500+ hour test using the 4 inch HXR is planned to evaluate catalyst deactivation at scale. Reactor modeling for both the 2 inch and 4 inch reactors will be conducted to predict productivity and thermal performance. Model results will be compared to experimental results.

Acknowledgement. The authors acknowledge financial support from Department of Energy under Cooperative Agreement DE-FE0024083. Generous supply of promising hybrid catalyst samples by Chevron Energy Technology Company is gratefully acknowledged. Intramicron prepared the internals of the HXR and loaded the selected Chevron catalysts for testing.

References

(1) Adesina, A.A., Appl. Catal. A: General 138, 319 (1996)

(2) Anderson R.B., "The Fischer-Tropsch Synthesis", Academic Press, Inc., New York, 1984.

(3) Dry, M.E., in "Catalysis-Science and Technology", Anderson J.R., Boudart M. (editors) 1980, p. 160

(4) Jager, B. and Espinoza, R., Catalysis Today, 23, 17 (1995)

(5) Jothimurugesan K, Das T, Kibby C.L., and Saxton R.J., Ruthenium Hybrid Fischer-Tropsch Catalyst, and Methods for Preparation and Use Thereof, US Patent Application 20120129959 (2012)

(6) Kibby C.L., Jothimurugesan K, Das T., Saxton R.J., Burton Jr. A.W., Zeolite-supported Cobalt hybrid Fischer-Tropsch Catalyst, US Patent 7,943,674 (2011)

(7) Kibby C.L., Jothimurugesan K, Das T., Saxton R.J., Burton Jr. A.W., Zeolite-supported Cobalt hybrid Fischer-Tropsch Catalyst, US Patent Application 20110124750 (2011)

(8) Kibby C.L., Jothimurugesan K, Das T., Saxton R.J., Burton Jr. A.W., Zeolite-supported Cobalt hybrid Fischer-Tropsch Catalyst, US Patent Application 20110144219 (2011)

CLEAN SYNTHESIS OF METHYLENE DIPHENYL DIISOCYANATE WITH LOW ENVIRONMENTAL IMPACT

Fang Li, Wei Xue, Hualiang An, Dongsheng Zhang, Xinqiang Zhao and Yanji Wang

School of Chemical Engineering and Technology
Hebei University of Technology
8 Guangrongdao, Hongqiao.
Tianjin 300130, China

Introduction

Methylene diphenyl diisocyanate (MDI) is an important monomer for the preparation of polyurethane (PU), which is one of the most versatile polymers in existence and can be found in numerous forms ranging from lightweight, rigid foams to dense, solid compositions and from soft, flexible foams to tough, elastomeric moldings. The industrial production of MDI starts from benzene, and it contains the following four steps. First, the nitrification of benzene to nitrobenzene; then hydrogenation of nitrobenzene to aniline; after that, aniline and formaldehyde undergo condensation to methylene diphenyl diamine (MDA); Finally, MDA reacts with phosgene to obtain MDI. During the whole production process of MDI, aforementioned above, there are several severe problems related to environment and process safety. For example, the explosion potential of nitrobenzene and hydrogen; the corrosivity of sulfuric acid and hydrochloric acid; highly toxic phosgene, et al.

A novel production route was proposed by our group for MDI synthesis using benzene as raw material. It also consists of four steps, which include the direct synthesis of aniline from benzene, the condensation of aniline to MDA, reaction between MDA and dimethyl carbonate (DMC) to methylene diphenyl carbamate (MDC), and pyrolysis of MDC to MDI. The new process overcomes the drawbacks related in the traditional phosgene route and exhibits a prospect development.

In this paper, we will report some progresses obtained in the new synthetic route of MDI, especially the synthesis of *N*-phenyl carbamate from aniline and DMC, and the one-pot synthesis of MDC from aniline, DMC and formaldehyde.

N-Phenyl Carbamate Synthesis from Aniline and DMC

$Zn(OAc)_2$ was found to give excellent catalytic performance for MPC synthesis from aniline and DMC. However, after being used for only once it tended to lose activity because of the formation of ZnO by the reaction of $Zn(OAc)_2$ and methanol. $Zn(OAc)_2/SiO_2$ was prepared by incipient impregnation and it gave excellent catalytic performance in MPC synthesis, on which aniline conversion and MPC yield were 98.1% and 93.8%, respectively. And $Zn(OAc)_2/SiO_2$ was found to be more stable than $Zn(OAc)_2$ during the reaction. When $Zn(OAc)_2/SiO_2$ was used for the fifth time, aniline conversion and MPC yield were found to be 64.3% and 38.1%, respectively. The $Zn(OAc)_2/SiO_2$ catalyst was characterized by TG-DTA, ICP, FTIR, XRD and XPS. The characterization results suggested that the deactivation of $Zn(OAc)_2/SiO_2$ was also due to the formation of ZnO and there were two reasons for the improved stability of $Zn(OAc)_2/SiO_2$ catalyst. One was the formation of the Si–O–Zn bonds in $Zn(OAc)_2/SiO_2$ catalyst, which increased the steric hindrance of Zn and consequently retarded the reaction between $Zn(OAc)_2$ and methanol. The other cause was the dehydration between methanol and the hydroxyl group on the Sio_2 surface, which reduced the chance of a reaction between methanol and $Zn(OAc)_2$.

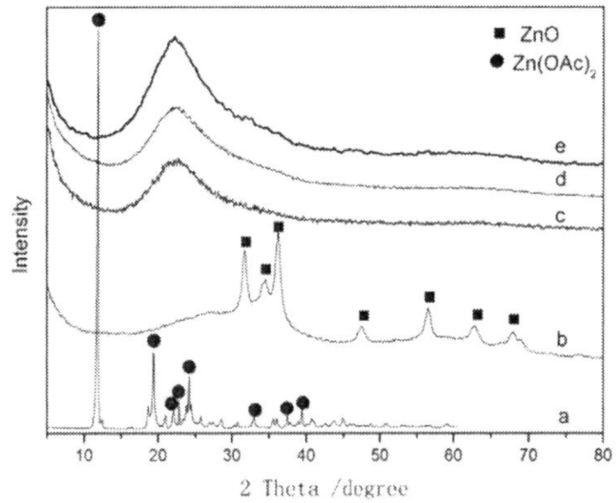

Figure 1. XRD patterns of the different catalysts. (a) $Zn(OAc)_2$, (b) $Zn(OAc)_2$ after the reaction; (c) fresh $Zn(OAc)_2/SiO_2$ catalyst; (d) $Zn(OAc)_2/SiO_2$-ME; (e) $Zn(OAc)_2/SiO_2$-DE.

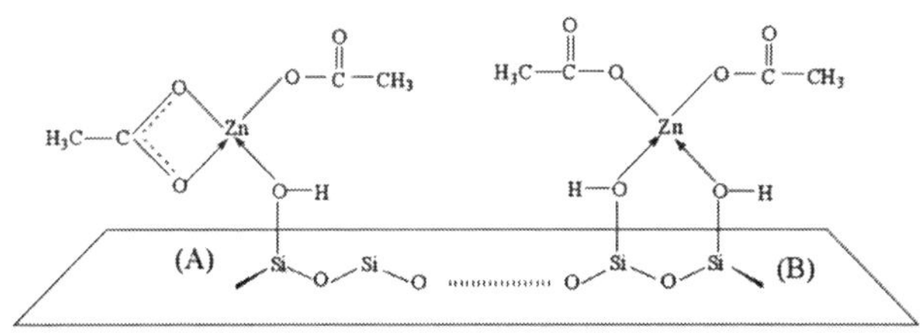

Scheme 1. Coordination mode of $Zn(OAc)_2$ on the surface of SiO_2

Condensation of MPC with Formaldehyde over Hβ Catalyst

The condensation reaction between MPC and HCHO to form MDC over different solid acid catalyst was studied. The results showed that the catalysts with surface Lewis acid sites having higher acid strength favored MDC synthesis. Among the solid acid catalysts, Hβ exhibited good catalytic activity for MPC condensation. Solvents also had an effect on the catalytic activity; protic solvents, methanol, or H_2O, were detrimental, while DMC was found to be the most appropriate reaction medium. The optimal reaction conditions consist of a reaction temperature of 160 °C, a reaction time of 3 h, n(DMC)/n(MPC)/n(HCHO)of 50:1:0.25 and 0.6 g/30 mL DMC catalyst amounts. Under these conditions, the MPC conversion is 37.2%, and the MDC yield (based on HCHO) is 76.4%. The Hβ catalyst could be easily regenerated by calcination, and no significant loss in catalytic activity was observed after five repeat uses.

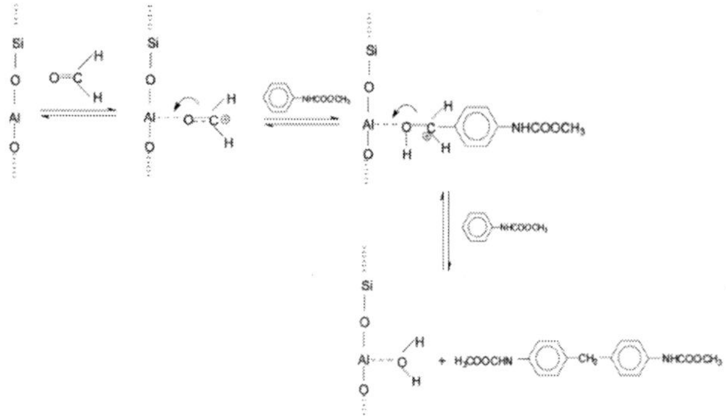

Scheme 2. Reaction Mechanism for the Condensation of MPC with HCHO over Hβ Catalyst

One-pot Synthesis of MDC from Aniline, DMC and formaldehyde

MDC is intermediate for the preparation of MDI. We conducted the one-pot synthesis of MDC using aniline, DMC, and formaldehyde as raw materials. MPC synthesis was carried out in a batch reactor using $Zn(OAc)_2$ as the catalyst, following which, without any separation, an Hβ catalyst and formaldehyde were added to continue to MPC condensation to give MDC. It was found that there were interactions between the MPC synthesis and MPC condensation, such that the presence of aniline could inhibits MPC condensation. $Zn(OAc)_2$ was also found to adsorb on the Hβ surface to form Si-O-Zn bonds during the condensation, decreasing the quantity of strong acid sites on the Hβ and reducing the catalytic performance. Under the optimal conditions for the one-pot synthesis of MDC, aniline conversion was 100% and the MDC yield was 73.9% (based on HCHO). The deactivated catalyst could be easily regenerated by treatment with acetic acid. The filter cake was dried and calcined to obtain regenerated Hβ and the filtrate was dried by vacuum rotary evaporation to obtain $Zn(OAc)_2$. No significant loss in catalytic activity was observed for the regenerated catalyst.

Table 1. Effect of reaction conditions on one-pot MDC synthesis.

Entry	Reaction conditions for MPC synthesis		Aniline conversion (%)[a]	MDC yield (%)[b]
	Zn(OAc) amount	Reaction time (h)		
1	0.03 g (0.16 mmol)	3	100	36.8
2	0.03 g	6	100	52.8
3	0.03 g	8	100	56.8
4	0.05 g (0.3 mmol)	6	100	55.7
5	0.05 g	8	100	60.4
6	0.07 g (0.4 mmol)	6	100	46.1
7	0.07 g	8	100	73.9

MPC synthesis conditions: aniline 0.7 mL (7 mmol), DMC 30 mL (350 mmol), 170 °C
MPC condensation conditions: Hβ 0.6 g, formalin 0.13 mL (1.7 mmol HCHO), 160 °C, 3 h

Acknowledgment. This work was supported by the National Natural Science Foundation of China (21236001, 21176056), the Programme for 100 Excellent Talents in Universities of Hebei Province (II) (BR2 -208), and the Natural Science Foundation of Hebei Province (B2015202228)

References

(1) Li, F., Xu, H.H., Xue, W., Wang, Y.J., Zhao, X.Q., *Chem Eng Sci*, **2015**, 135: 217-222.
(2) Li, F., Min, R., Li, J., Gao, L.Y., Xue, W., Wang, Y.J., Zhao, X.Q., *Ind Eng Chem Res*, **2014**, 53: 5406-5412
(3) Li, F., Li, W.B., Li, J., Xue, W., Wang, Y.J., Zhao, X.Q., *Appl Catal A*, **2014**, 475: 355-362
(4) Li, Z.H., Yang, Q.S., Qi, X.D., Xu, Y.Y., Zhang, D.S., Wang, Y.J., Zhao X.Q., *Chem Commun*, **2015**, 51: 1930-1932
(5) Xu, Y.Y., Li, Z.H., Gao, L.Y., Zhang, D.S., Zhao, X.Q., Wang, S.F., Wang, Y.J., *Ind Eng Chem Res*, **2015**, 54: 1068-1073
(6) Zhang, D.S., Gao, L.Y., Xue, W, Zhao, X.Q., Wang, S.F., Wang, Y.J., *Chem Lett*, **2012**, 41, 369-371

ENHANCED NEAR-ZERO-CO_2-EMISSION CHEMICALS-ORIENTED OIL PRODUCTION FROM COAL AND NATURAL GAS WITH CHEMICAL LOOPING BASED CO_2 RECYCLING

Xin Huang[a], *Xingjun Wang*[a], *Qingxi Cao*[a], *Bang Xu*[a], *Hertanto Adidharma*[a,b], *Gang Tan*[c], *Khaled A. M. Gasem*[a,b], *Ting He*[d], *and Maohong Fan*[a,b,e,*]

[a]Debarment Chemical Engineering, University of Wyoming, Laramie, WY 82071, USA
[b]Department of Petroleum Engineering University of Wyoming, Laramie, WY 82071, USA
[c]Department of Civil Engineering, University of Wyoming, Laramie, WY 82071, USA
[d]School of Energy Resources, University of Wyoming, Laramie, WY 82071, USA
[e]Idaho National Laboratory, Idaho Falls, ID 83415, USA

Introduction

People have benefited from fossil fuels for many years. However, the huge amount of CO_2 emitted from their utilizations is problematic due to the resulting climate changes. Therefore, conventional coal and natural gas utilization methods are facing serious challenges [1]. Accordingly, development of near-zero- CO_2-emission coal and natural gas utilization technologies has become critically important, and the project is designed to make progress in this area through production of oil instead of fuel. The oil can be separated into value-added and highly demanded chemicals. As shown in Fig.1, the integrated technology is realized with three steps, including CH_4/ CO_2 based syngas production, chemicals-oriented oil generation, and CO_2 recycling [2-3].

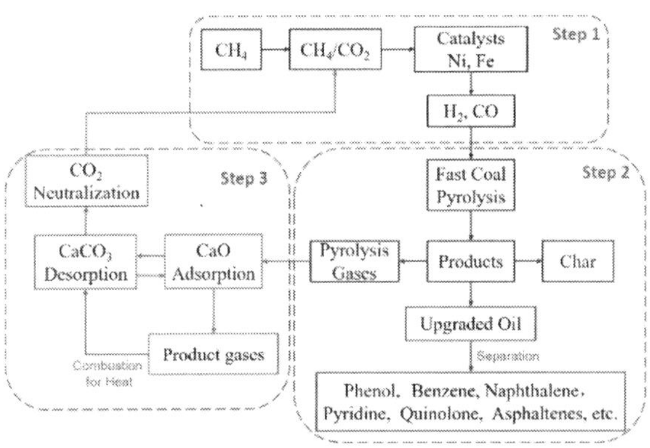

Figure 1. Flow chart of enhanced near-zero- CO_2-emission chemicals-oriented oil production from coal and natural gas with looping based CO_2 recycling

Step 1. CH_4/ CO_2 based syngas production

Ni/MgO catalyst prepared with magnesium oxides and nickel nitrate was used for CH_4/ CO_2 reforming or syngas production. Catalyst preparation process started with calcining as received MgO at 600 °C for 3 h, followed by its impregnation with nickel nitrate solution. The resulting impregnated sample in thick paste was dried at 105°C for 24h and then decomposed as well as calcined at 950 °C for 5 h. The concentration of nickel in the catalyst was 10 *wt*%.

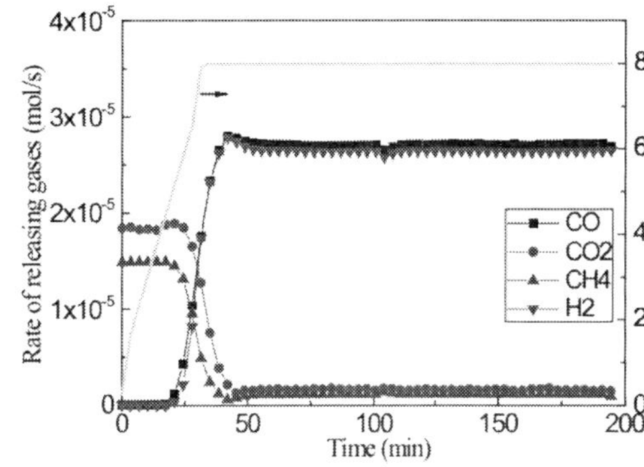

Figure 2.1. Reactivity of Ni/MgO in CH_4/ CO_2 reforming

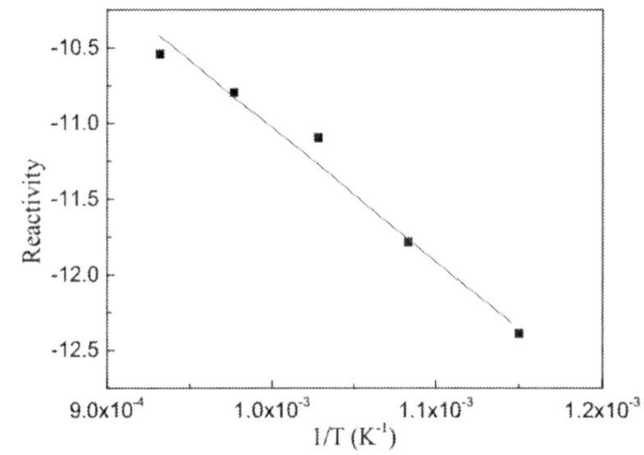

Figure 2.2. Apparent activation energy for CH_4/ CO_2 reforming with Ni/MgO

The reactivity of Ni/MgO is presented in Fig.2.1. Clearly, the catalyst has excellent performance on reforming CH_4/ CO_2 into CO/H_2 under 800°C and 1.0 atm. The concentrations of CO, H_2, CO_2, CH_4 in the obtained syngas are 48.3%, 46.9%, 2.8%, 2.0%, respectively. The conversion of methane to syngas is as high as 96.1%. In addition, the Arrhenius parameters of the CH_4/ CO_2 reforming reaction catalyzed with Ni/MgO are shown in Fig.2.2. The apparent activation energy is 73.6 $kJ \cdot mol^{-1}$.

Step 2. Chemical oriented oil production

The gas release behaviors of coal sample during fast pyrolysis process have been investigated with an in-situ Fourier transform infrared spectrometer (FTIR) [4-5]. Infrared scanning (second level) was used to analyze pyrolysis kinetics and gas species generated in pyrolysis process. Wavenumbers range from 4,000-600 cm^{-1}. The wavenumbers of CO and CO_2 are in the ranges of 2,240-2,060 cm^{-1} (C-O functional group) and 2,268-2,395 cm^{-1} (C=O functional group), respectively. Wavenumber of acetic acid, formic acid, toluene, phenol and other organic chemicals are in the range of 2,800-3,200 cm^{-1}. Three-dimensional map of the observed FTIR is shown in Fig. 3.1. Gas products CO_2, H_2O, CH_4, CO are major products of the fast pyrolysis, while other organic chemicals containing C-O-C, C=O, C-C are also present.

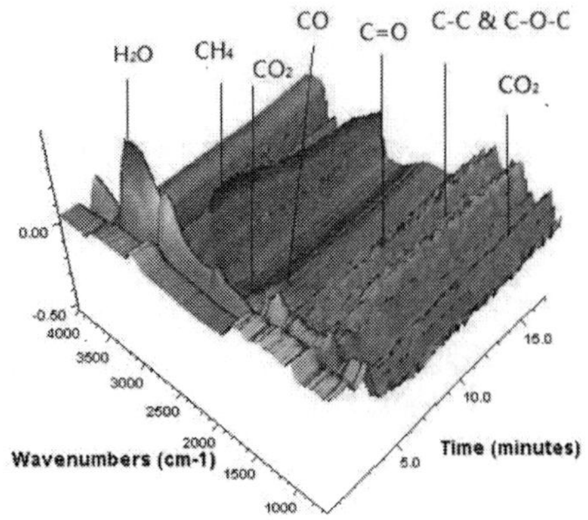

Figure 3. FTIR stack plot from the rapid pyrolysis of raw coal at 540°C (temperatures: 540°C, heating rate: 200°C/min; purging gas: helium; purging gas rate: 10 ml/min; resolution range: 4 cm^{-1}, spectra collected in every 10s).

Section 3. Chemical looping based CO_2 recycling

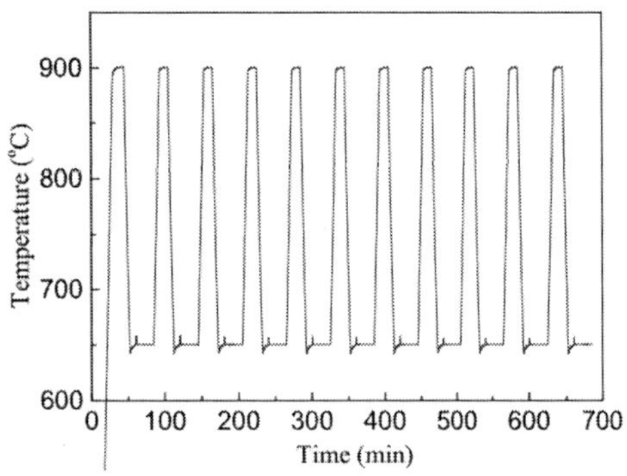

Figure 4.1. Calcination and carbonation Temperature program for CaO/MgO on TGA.

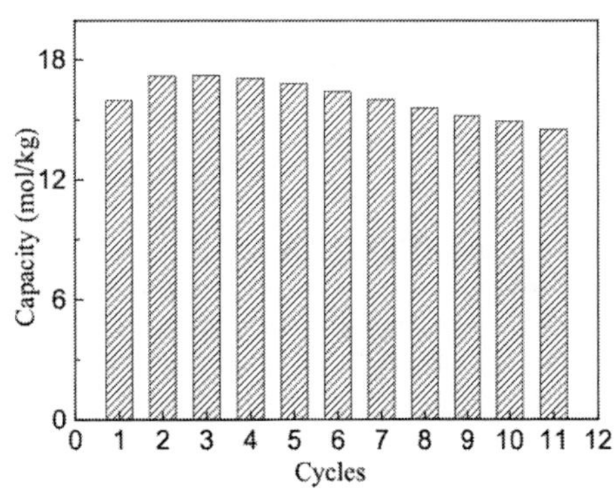

Figure 4.2. The adsorption/desorption amount of CO_2 for CaO/MgO with calcination at 900°C in N_2 and carbonation at 650°C with 20%CO_2-80%N_2.

CaO/MgO used as a sorbent for recycling the CO_2 generated from coal pyrolysis was prepared with glucose acid calcium and magnesium gluconates. To prepare the sorbent, acid calcium and magnesium gluconates were dissolved into deionized water, then the mixture was heated and stirred at 50°C for 60 min, followed by vaporizing its water at 110 °C until the water was completely removed. The solid CaO/MgO was then dried at 900°C for 6h prior to its utilization

The temperature programs of the calcination(900°C) and carbonation(650°C) processes for CaO/MgO are demonstrated in Fig.4.1. The cyclic adsorption and desorption amounts of CO_2 of the sorbent are provided in Fig.4.2. The CO_2 sorption capacity of the sorbent can be up to 14mol/kg. In addition, as shown in Figure 4.2, the CO_2 sorbent is stable because its CO_2 sorption capacity and only decreases slightly in the 11-cylces tests.

Acknowledgment. This work was supported by the State of Wyoming, and China Scholarship Council.

References

(1) Somaye Nasr, Kevin P. Plucknett. Kinetics of Iron Ore Reduction by Methane for Chemical Looping Combustion. Energy & Fuels, 2014, 28: 1387-1395

(2) P. R. Solomon, T. H. Fletcher, R. J. Pugmire. Progress in coal pyrolysis. Fuel,1993,72.587-597

(3) Jun-ichiro Hayashi, Hiroshi Takahashi, Satoshi Doi, Haruo Kumagai, Tadatoshi Chiba. Reactions in Brown Coal Pyrolysis Responsible for Heating Rate Effect on Tar Yield. Energy & Fuels, 2000, 14:400-408

(4) J.A. MacPhee , J.-P. Charland, L. Giroux. Application of TG–FTIR to the determination of organic oxygen and its speciation in the Argonne premium coal samples. Fuel Processing Technology,2006, 87:335-341

(5) Surjit Singh, Chunfei Wu, Paul T. Williams. Pyrolysis of waste materials using TGA-MS and TGA-FTIR as complementary characterization techniques. Journal of Analytical and Applied Pyrolysis,2012, 94: 99-107

SELECTIVE HYDROGENATION OF ACETYLENE OVER NaN_3-Pd/SAC CATALYSTS

Maocong Hu, Xianqin Wang

Department of Chemical, Biological and Pharmaceutical Engineering, New Jersey Institute of Technology, Newark, NJ 07102
xianqin.wang@njit.edu

Introduction

Selective hydrogenation of acetylene with Pd catalyst is commonly used for removing a trace amount of acetylene from ethylene[1]. Modification of Pd/Al_2O_3 catalysts with an additive such as silver, cobalt or zinc[2] was explored to reduce high acidity of the catalyst to minimize carbonaceous deposits on the catalyst surface and increase the catalyst lifetime. Our previous study proved the lone pairs from N_3^- are very active and could play the role as electron donor in oxygen-reduction reactions[3]. Thus, N_3^- may be employed in the Pd/AC catalyst as an additive to serve as Lewis base for adsorbing acetylene which is a weak Lewis acid[4]. Increasing acetylene coverage on the catalyst is similar to a low H_2/acetylene ratio that is used in the reactant stream, which may improve ethylene selectivity proved by experimental and simulation results[5, 6]. Therefore, it is prompt for us to examine N_3^- as an additive for selective hydrogenation of acetylene on Pd/SAC catalysts.

Experimental

In present work, we prepared spherical activated carbon supported sodium azide - palladium (NaN_3-Pd/SAC) catalysts with different N_3^- loading (0-15 wt.%) by impregnation method. The properties of the catalysts were examined by using Brunauer-Emmett-Teller (BET), temperature-programmed reduction (TPR), CO chemisorption, Fourier-transform Infrared (FTIR) spectroscopy, and temperature-programmed desorption (TPD) of hydrogen and acetylene. The selective hydrogenation of acetylene test over NaN_3-Pd/SAC catalysts was conducted in a fixed-bed reaction system.

Results and Discussion

Fig. 1 showed the TPR profiles for different catalysts. The peaks in the range of 50-200 °C correspond to the reduction of PdO_x[7]. With NaN_3 addition, this peak split into two peaks. These two peaks indicated dual functions of N_3^- species to PdOx reduction. The peak 1 can be ascribed to the reduction of well-dispersed PdO_x species which were on NaN_3 species. The peak 2 is due to the reduction of PdO_x species covered by NaN_3[8]. The 3rd peak, which appears at around 312 °C, corresponds to the decomposition of NaN_3[9]. Therefore, 200 °C was employed for the reduction procedure.

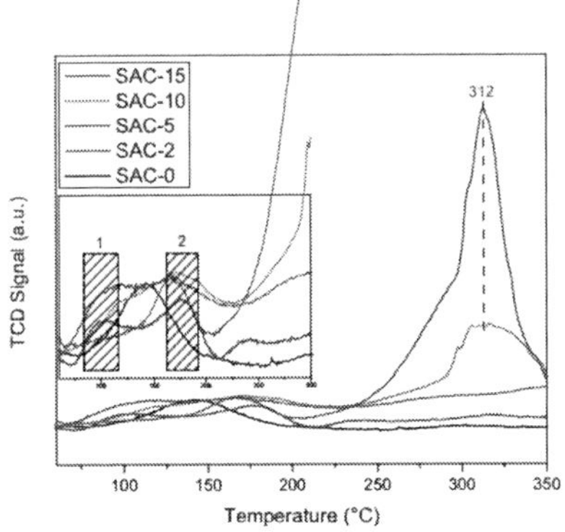

Figure 1. TPR profiles of different catalysts

The physical properties of the catalysts such as dispersion, metal surface area and crystallite size are also summarized in Table 1. The Pd dispersion is 94.5% for SAC-0, indicating that Pd atoms are highly dispersed on the support due to the large surface area of the SAC support. It is clear that both dispersion (CO/Pd) and metal area per gram of Pd decrease significantly with increase in N_3^- loadings. The results indicate Pd sites might be covered by NaN_3. In other words, it can be concluded that N_3^- addition had detrimental effect on the number of the active Pd sites. It has been found that geometric modification of Pd based catalysts for selective hydrogenation of acetylene is an effective way for the promotion of ethylene selectivity[6]. The Pd surface was decorated with the additives which blocked some neighboring Pd sites or geometrically diluted by additives and retarded polymerization step, thus improved ethylene selectivity. In present study, N_3^- species covered some Pd sites with geometric modification and therefore product distribution for the catalytic hydrogenation of acetylene would be expected to change.

Table 1. Surface and physical properties of the catalysts with different N_3^- loadings

Sample	N_3^- loading (wt.%)	BET surface area (m^2 / g)	CO uptake (μmol /g cat)	Pd Dispersion (%)	Metal area (m^2 / g(Pd))
SAC-0	0	2133	82.2	94.5	420.9
SAC-2	2	1918	67.2	83.7	373.0
SAC-5	5	1653	60.4	67.2	307.8
SAC-10	10	1352	51.0	55.9	249.2
SAC-15	15	1040	47.3	50.3	224.2

The hydrogen TPD results (not shown here) showed that the hydrogen adsorption behavior is significantly changed by N_3^- species. The amount of chemisorbed hydrogen is decreased by N_3^-, which is consistent with the decreasing Pd active sites. In view of acetylene adsorption, it has been concluded that the intermediates of acetylene adsorption that formed on the surfaces can be two types. Depending on their binding mode with Pd, one type is bound to a single Pd atom by π-bonded adsorption (type I), and the other is bound to two or more Pd atoms via σ-bonded adsorption (type II)[10]. Fig. 2 shows TPD patterns of the hydrocarbons from the catalysts after acetylene exposure at room temperature. The m/e of 15 represents C1 hydrocarbons, and the m/e of 43 represents C4 hydrocarbons. C1 hydrocarbons were from the decomposition of multiply bound C2 species on the surface (type II)[11]. With N_3^- loading increasing, both C1 and C4 species have a parabolic trend and the most desorption amounts are observed on SAC-5. Studies showed that two main C2 species that are multiply bound on the Pd surface are the di-σ-bonded type and perpendicular or parallel types that are bound to more than 3 Pd atoms respectively[12]. Of the multiply bound C2 species, the di-σ-bonded one is relatively weakly adsorbed because it is bound to two active sites, while the other species are bound to more than three active sites. As indicated above, acetylene should be adsorbed with di-σ-bond bounding to two active sites on the surface of SAC-0 while the C2 species are by perpendicular or parallel types binding to more than three active sites on NaN_3-Pd/SAC catalysts, indicating that there are more abundant adsorption sites for acetylene molecules on NaN_3-Pd/SAC catalysts except SAC-15 than those on SAC-0. It is attributed to N_3^- additives which serve as Lewis base sites because its super electron donor property helps to adsorb the Lewis acid acetylene.

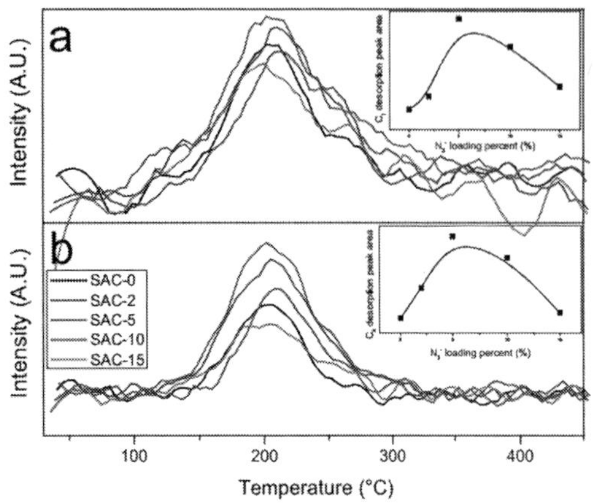

Figure. 2. TPD desorption of the (a) C1 (m/e = 15) and (b) C4 (m/e = 43) hydrocarbons obtained from different catalysts

Fig. 3 summarizes the results of selective hydrogenation of acetylene as a function of space velocity (SV). The conversions are lower for the catalysts containing N_3^- compared to SAC-0 catalyst. The conversion is decreasing with N_3^- loading increase, because the hydrogen adsorption, first step of this reaction, is decreased, which is consistent with the findings on on Si modified Pd catalyst[6]. N_3^- addition does not improve the ethylene selectivity (Fig. 3b). The negative effect by N_3^- is due to further hydrogenation of ethylene to ethane or ethylene polymerization to green oil species as shown in Fig. 3c and Fig. 3d. On the surface of SAC-0, there are two types of C2 adsorption which are single Pd atom π-bonded (type I) and multiple Pd atoms σ-bonded adsorption (type II), and the former one desorbs as ethylene without decomposition while the latter one tends to decompose to CH_3. Therefore, it's expected that ethylene from π-bonded and other products from σ-bonded adsorption were produced over SAC-0. However, all of C2 intermediates over NaN_3-Pd/SAC catalysts are mainly σ-bonded adsorption (type II) lead to other products other than ethylene. Thus, less ethylene was produced over NaN_3-Pd/SAC.

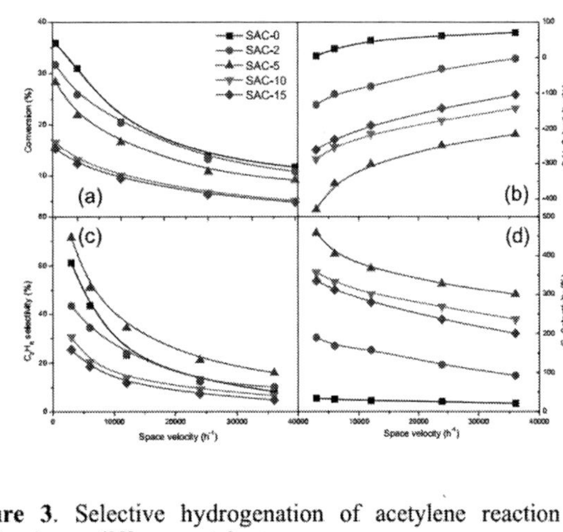

Figure 3. Selective hydrogenation of acetylene reaction results obtained over different catalysts

In summary, addition of NaN_3 showed negative effect on selective hydrogenation of acetylene although acetylene adsorption capability has been greatly promoted by tuning catalyst acidity and forming the dual function sites with N_3^- species and Pd active sites. Acetylene adsorption is changed from type I and II to dominantly type II, which affects the final product selectivity. A possible mechanism is showed in Fig. 4. Before N_3^- addition, catalyst surface is consisted of Pd and defect sites. Hydrogen could be dissociatively adsorbed by two adjacent Pd sites while acetylene is adsorbed by both type I and II. With N_3^- addition, N_3^- species cover some Pd and defect sites. Hydrogen adsorption would be suppressed since neighboring Pd sites are critical for hydrogen adsorption.

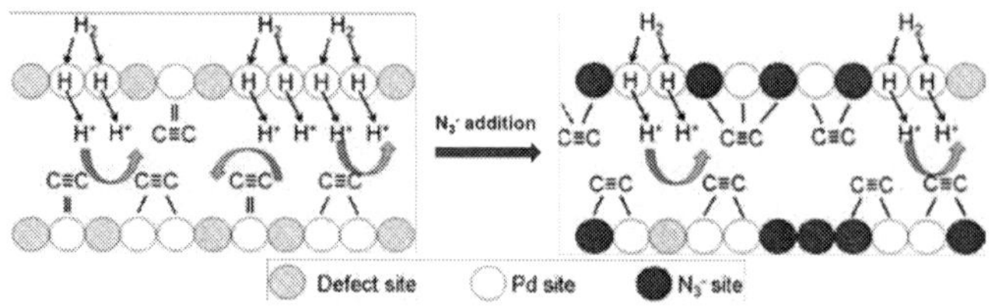

Figure 4. Proposed mechanism for acetylene hydrogenation on Pd catalyst without/with N_3^- addition[13]

Conclusion

In this study, N_3^- effect on Pd/SAC is explored for the selective hydrogenation of acetylene based on the attempt to tune catalyst acidity and further promote acetylene adsorption capacity. The N_3^- loadings on Pd/SAC have a significant influence on hydrogen adsorption capacity, acetylene adsorption capacity and types and the catalytic activity of the NaN_3-Pd/SAC samples. NaN_3 promotes acetylene adsorption but reduces hydrogen adsorption on the catalysts. The best acetylene adsorption capacity is observed over the SAC-5 sample due to the dual function sites with Pd and N_3^- species. However, acetylene adsorption types are changed from type I and II to dominantly type II. This work provided an insight in designing appropriate catalyst for selective hydrogenation of acetylene. The work confirms our hypothesis that N_3^- species promote acetylene adsorption.

Acknowledgement. This work was supported by an ACS-PRF 53582-ND10)

Reference

(1) Shao, L.; Zhang, B.; Zhang, W.; Teschner, D.; Girgsdies, F.; Schlogl, R.; Su, D. S. *Chemistry* **2012**, *18*, 14962-6.
(2) Pachulski, A.; Schödel, R.; Claus, P. *Appl. Catal., A* **2011**, *400*, 14-24.
(3) Wu, Z.; Benchafia el, M.; Iqbal, Z.; Wang, X. *Angewandte Chemie* **2014**, *53*, 12555-9.
(4) Lamberov, A. A.; Egorova, S. R.; Il'yasov, I. R.; Gil'manov, K. K.; Trifonov, S. V.; Shatilov, V. M.; Ziyatdinov, A. S. *Kinet. Catal.* **2007**, *48*, 136-142.
(5) Mei, D.; Neurock, M.; Smith, C. M. *J. Catal.* **2009**, *268*, 181-195.
(6) Moon, S. H.; Shin, E. W.; Kang, J. H.; Kim, W. J.; Park, J. D. *Appl. Catal., A* **2002**, *223*, 161-172.
(7) Fernandes, V. R.; Gustafson, J.; Farstad, M. H.; Walle, L. E.; Blomberg, S.; Lundgren, E.; Venvik, H. J.; Borg, A. *Appl. Surf. Sci.* **2014**, *313*, 794-803.
(8) Chang, J.; Li, S.; Feng, L.; Qin, X.; Shao, G. *J. Power Sources* **2014**, *266*, 481-487.
(9) Hannus, I.; Tasi, G.; Kiricsi, I.; Nagy, J. B.; Förster, H.; Fejes, P. *Thermochim. Acta* **1995**, *249*, 285-294.
(10) Kim, S. K.; Kim, C.; Lee, J. H.; Kim, J.; Lee, H.; Moon, S. H. *J. Catal.* **2013**, *306*, 146-154.
(11) Kim, S. K.; Lee, J. H.; Ahn, I. Y.; Kim, W.-J.; Moon, S. H. *Appl. Catal., A* **2011**, *401*, 12-19.
(12) Stuve, E. M.; Madix, R. J. *J. Phys. Chem.* **1985**, *89*, 105-112.
(13) Hu, M.; Wang, X. *Catal. Today* **2016**, *263*, 98-104.

DEVELOPMENT OF SULFIDE ELECTROLYTE ALL SOLID-STATE BATTERIES FOR HIGH ENERGY DENSITY

Taku Watanabe, Seitaro Ito, Takanobu Yamada, Satoshi Fujiki, Yuichi Aihara*

2-1-11 Semba Center bldg., Semba-nishi, Minoh-shi, Osaka, 560-0036, Japan
*t.watanabe@samsung.com

Introduction

All solid-state batteries (ASSBs) are a focus of intense research effort over decades because of their unique physicochemical and electrochemical properties. They include excellent cycle performance and high energy density as reported in the literature. Such high performance are often achieved by carefully controlled solid-solid interactions between the electrodes and electrolyte.

In this paper, we will report out effort on the development of ASSBs using lithium thiophosphate solid electrolytes. In particular, we will emphasize on the study of electrode-electrolyte interactions. Impact of a Li_2O-ZrO_2 (LZO) coating on the $LiNi_{0.8}Co_{0.15}Al_{0.05}O_2$ (NCA) was examined by electrochemical impedance spectroscopy (EIS) and charge-discharge cycles. Comparison of the nature of solid electrolyte (SE) with a conventional liquid electrolyte (LE) was made by a symmetric Li/Li cell. A pelletized test cell and a laminated proto-type cell were constructed to examine the electrochemical properties of these ASSBs.

Experimental

The electrochemical cells were fabricated following the procedure described in our previous publications.[2,3] A pelletized test cell was fabricated to investigate basic characteristics of the constituent materials. 15 mg of the composite cathode powder was spread across the compacted electrolyte surface in the die. The anode was either graphite or lithium depending on the measurement. A card-size laminated cell was constructed by a wet printing process. A cathode slurry was prepared by mixing LZO-NCA, Li_2S-P_2S_5, conductive carbon, and a nitrile butadiene rubber based hydrocarbon polymer binder in dehydrated xylene. The slurry was used to coat carbon coated aluminum foil. Then, the sample was dried in a vacuum oven for 8 h at 60 °C after cut into a desired size and shape. The graphite anode was prepared in the similar fashion to that in conventional lithium-ion battery using liquid electrolyte. An synthetic graphite was dispersed in PVDF dissolved NMP solution. The anode slurry was coated on a nickel fold at 10 um thick, and dried I vacuum for 18 h at 80 °C. Then the Li_2S-P_2S_5 slurry was applied on the anode layer and dried in vacuum. The composite assembly of the cathode was overlaid on the anode-electrolyte layer and compressed to form a card size (88×53 mm^2) cell.

Electrochemical impedance spectroscopy (EIS) was performed using AutoLab PGSTAT30 attached to a ESPECTH-241 temperature chamber. The measurement was done between 0.1-1MHz under a normal pressure. The cycle and rate performances were examined by TOSCAT-3000 battery tester. The charge and discharge cutoff voltages were set to 4.0 and 2.5 V, respectively.

Results and Discussion

Effect of the LZO coating on NCA Impact of the LZO coating on the cathode active material was investigated by a comparison of their electrochemical performance of the cells using bare and LZO coated NCAs. A significant reduction of the charge transfer resistance was observed in the cell using the LZO coated NCA compared to the cell adopting the bare material.[2] LZO coating of >1.0 mol% resulted in nearly 60 % reduction of the charge transfer resistance. This reduction is thought to be achieved by preventing the mutual diffusion of the constituent elements between the NCA and SE.

In accordance with the decrease in the charge transfer resistance, the specific capacity increased from ~102 to 118 mAh/g at 0.05 mA/cm^2. The rate performance showed more dramatic improvement. The specific capacity of the bare NCA cell was 52 mAh/g at 0.5 mA/cm^2, but that of the 1.0 mol% coated NCA cell reached 87 mAh/g. Similarly, 25 mAh/g at 1 mA/cm^2 of the bare NCA became 69 mA/g; nearly a factor of 2.8 increase in the capacity.[2]

The cycle performance was also improved by the application of LZO coating on NCA. (See Figure 1.) The 0.5 mol% LZO coated sample achieved 91.7 % capacity retention at 100th cycle, but the uncoated sample showed 74 %.

EIS of a symmetric Li/Li cell The nature of the SE was examined by comparison of the EIS with that of a conventional LE. Symmetric Li/Li cells were constructed and used for this study. (See Figure 2.)[3] As we all expected, the bulk resistance of the LE cell was clearly smaller than that of the SE cell. However, the trend is quite different in the charge transfer resistance. The LE cell showed approximately 7 kΩcm^2 of the charge transfer resistance, and it hardly changed before and after a charge-discharge cycle. The charge transfer resistance of the SE cell was significantly smaller than that of the LE system, and further decreased after the first cycle.

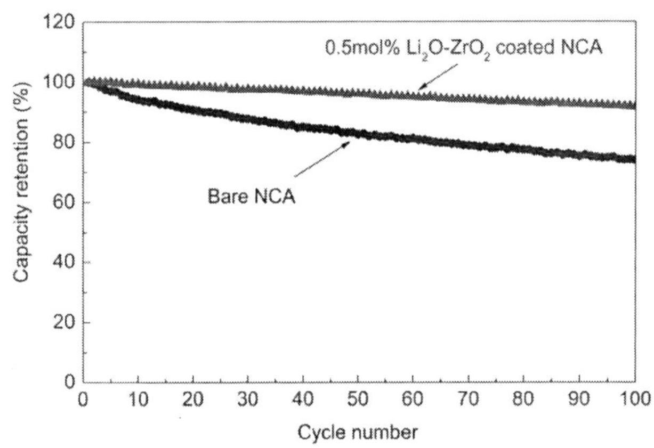

Figure 1. Cycle performance of ASSBs using bare and LZO coated NCAs at 25 °C. .

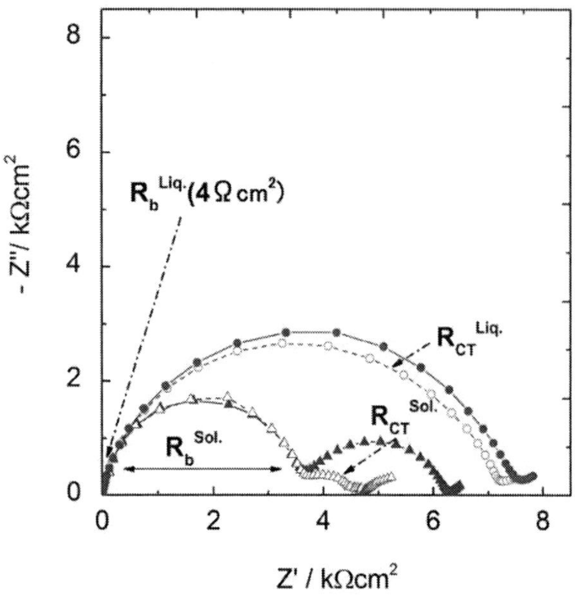

Figure 2. Nyquist plots of the liquid (red) and solid (blue) electrolyte cells using Li-Li symmetric cells. The solid and open symbols represent before and after the stripping-deposition cycles, respectively.

Conclusions

We investigated the impact of the LZO coating on the cathode active material and difference in the electrochemical characteristics of solid and liquid electrolytes against lithium-metal anodes. We observed that the cathode charge transfer resistance was significantly reduced by the LZO coating, and the total cell resistance of the cell was approximately 75 % lower than the cell using uncoated NCA. The charge transfer resistance of the SE cell was also shown to be much smaller than that of a conventional LE cell although the bulk resistance was much lower in the LE system. These findings will further facilitate the development of high performance ASSBs, which are a class of a valuable alternative for the energy storage in vehicular and portable device applications.

Acknowledgement This work was supported by the Samsung Advanced Institute of Technology.

References

(1) Machida, N.; Kashiwagi, J.; Naito, M.; Shigematsu, T., Solid State Ionics 2012, 225, 354.

(2) Ito, S.; Fujiki, S.; Yamada, T.; Aihara, Y.; Park, Y.; Kim, T. Y.; Baek, S.-W.; Lee, J.-M.; Doo, S.; Machida, N., J. Power Sources 2014, 248, 943.

(3) Yamada, T.; Ito, S.; Omoda, R.; Watanabe, T.; Aihara, Y.; Agostini, M.; Ulissi, U.; Hassoun, J.; Scrosati, B., J. Electrochem. Soc. 2015, 162, A646.

TOWARD INTIMATE IONIC CONTACTS FOR HIGH-PERFORMANCE SULFIDE-BASED ALL-SOLID-STATE LITHIUM-ION BATTERIES

*Kern Ho Park[1,2], Dae Yang Oh[1], Young Jin Nam[1], Seung M. Oh[2], Yoon Seok Jung[1]**

[1]School of Energy and Chemical Engineering, Department of Energy Engineering, Ulsan National Institute of Science and Technology (UNIST), Ulsan 689-798, South Korea
[2]School of Chemical and Biological Engineering, Seoul National University, Seoul 151-742, South Korea
ysjung@unist.ac.kr

Introduction

Application of rechargeable lithium batteries to large-scale energy storage such as electric vehicles (EVs) and energy storage systems is a main driving force for their extensive R&D efforts. Accordingly, unprecedentedly severe safety requirements are demanded. The conventional organic liquid electrolytes, however, cannot meet those requirements because of its intrinsic limitation, flammability. They also cause leakage and swelling problems. All-solid-state lithium batteries (ASLBs) fabricated by using nonflammable inorganic solid electrolytes (SEs), in this regard, can be one of the ultimate solutions.[1,2] Especially, bulk-type ASLBs based on composites of powder mixtures of SEs, active materials, and conductive additives are highly competitive in terms of scalable fabrication and high energy density.[2,3] A suitable SE candidate for realization of bulk-type ASLBs is sulfide materials because they show high ionic conductivities (10^{-4}-10^{-2} S cm^{-1}) and can be easily deformed to form two-dimensional ionic contacts with active materials by simple cold-pressing.[2,3]

Despite many research efforts including application of newly developed extremely conductive SEs (e.g., $Li_{10}GeP_2S_{12}$ (LGPS): 12 mS cm^{-1}), performances of bulk-type ASLBs are still far from that of the conventional lithium-ion batteries (LIBs) using liquid electrolytes.[2] It is believed that not only ionic conductivity of SEs but ionic contacts between SEs and active materials would be the main factor to govern overall performances of bulk-type ASLBs.[2]

Two approaches are possible to overcome the poor ionic contacts; i) including liquid electrolytes which can assist Li-ion transport at interfaces, ii) direct coating of SEs on active materials. Severe reactivity of sulfide SEs with most organic liquids is, however, a technical challenge for the former. For the latter, solution-processed preparation of SE could be a viable method. However, synthesis protocols for sulfide SEs have been so far limited to conventional methods such as solid-state synthesis and mechano-chemical methods.[2]

Experimental

Preparation of SEs. Li_3PS_4 powders were prepared by mechanical ball-milling using ZrO_2 vial and balls, followed by heat-treatment above crystallization temperature under inert atmosphere. LGPS powders were prepared by heat-treatment of stoichiometric mixtures of Li_2S, GeS_2, and P_2S_5 at 550°C under inert atmosphere. Li_4SnS_4 powders were prepared by heat-treatment of stoichiometric mixtures of Li_2S and SnS_2 at 450°C under inert atmosphere. For solution process, homogeneous solution was prepared by dissolving the as-prepared Li_4SnS_4 and LiI powders into anhydrous methanol. After the solvent was removed under vacuum, the as-obtained powders were heat-treated under vacuum. For the solution-processed SE coating on active materials was carried out by including active materials in the Li_4SnS_4- and LiI-dissolved methanol solution.

Electrochemical tests. Ionic conductivity of SEs were measured by AC impedance method using a Li-ion blocking cell. All-solid-state cells densified by cold-pressing under 370 MPa consist of composite electrode as a positive electrode, SE layers, and Li-In electrode as a counter/reference electrode.

Results and Discussion

As the first approach to solve the poor solid-solid ionic contact, addition of small amount of solvate ionic liquid, high-concentrated lithium bis(trifluoromethanesulfonyl)imide (LiTFSI) in triethylene glycol dimethyl ether (G3), in all-solid-state cells has been attempted.[4] Figure 1a shows the photo images of LGPS powders with G3 and LiTFSI-G3 solutions ($Li(G3)_4TFSI$ and LiG3TFSI) after being kept for 7 days. The yellow color of LGPS/G3 solution indicates severe reactivity of LGPS with pure G3, which is explained by nucleophilic attack of oxygen in G3 to electropositive elements in LGPS. However, as LiTFSI is added in G3, the reactivity is decreased. The LGPS/LiG3TFSI solution is transparent, demonstrating excellent compatibility of the solvate ionic liquid, LiG3TFSI, with sulfide SE. This is rationalized by significantly alleviated nucleophilicity of oxygen in G3 by strong coordination with Li ions in the solvate ionic liquid.

Figure 1b shows the second voltage profiles of $Li_4Ti_5O_{12}$/Li-In all-solid-state cells (LTO/Li-In) without and with LiG3 in the LTO electrode. Noticeable improvement in capacity by adding LiG3 is observed. The added LiG3 can wet the LTO/SE interfaces and assist facile ionic transport, enhancing the utilization of active materials.

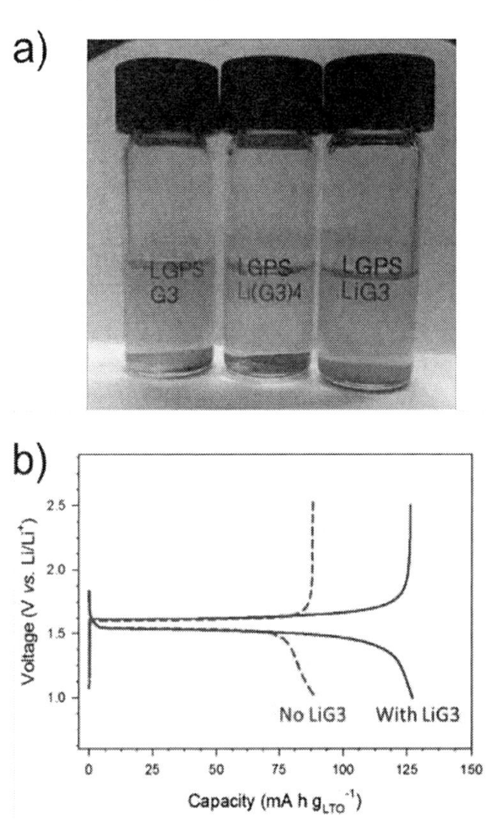

Figure 1. a) Photographic images of mixtures of LGPS powders with the liquids (G3, $Li(G3)_4$ ($Li(G3)_4TFSI$), and LiG3 (Li(G3)TFSI)). b) Voltage profiles of $Li_4Ti_5O_{12}$/Li-In (LTO/Li-In) all-solid-state cells where LiG3 is included or not in the LTO electrode.

The direct SE coating on active materials by solution process has been also attempted as the second approach to solve the poor solid-solid ionic contact.[5] Figure 2a shows the schematic diagram illustrating solution processed synthesis of LiI-Li_4SnS_4 and its coating on active materials. Li ionic conductivities of the solution-processed LiI-Li_4SnS_4 are shown in Figure 2b. The solution-processed Li_4SnS_4 exhibits higher conductivity (8.9×10^{-5} S cm^{-1}) than that prepared by conventional solid-state synthesis (1.0×10^{-5} S cm^{-1}). As the amount of added LiI in Li_4SnS_4 increases, the conductivity also increases, hitting the highest ionic conductivity of 4.1×10^{-4} S cm^{-1} for 0.4LiI-0.6Li_4SnS_4. The large ionic size and high polarizability of iodide ion is believed to be responsible for the enhancement in conductivity.

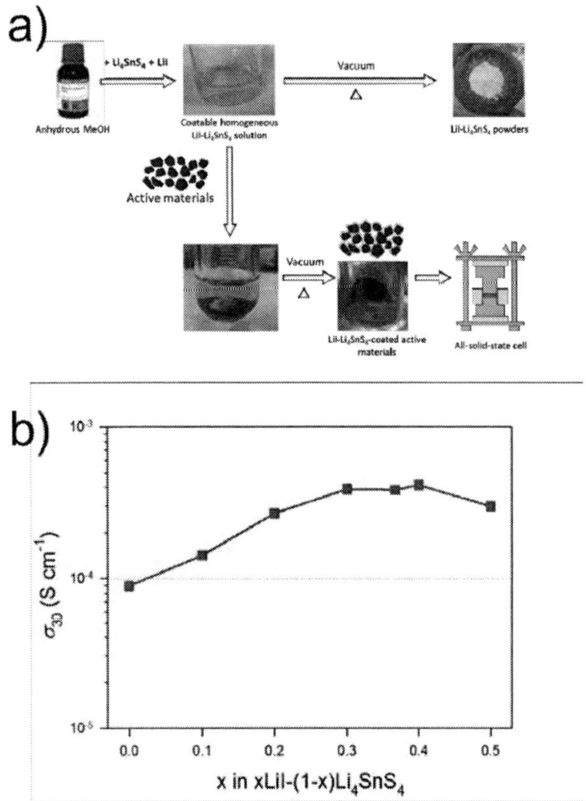

Figure 2. a) Schematic diagram illustrating solution-process of LiI-Li_4SnS_4 and its coating on active materials. b) Ionic conductivity of solution-processed xLiI-(1-x)Li_4SnS_4.

When the highly conductive 0.4LiI-0.6Li_4SnS_4 coating is applied for $LiCoO_2$, noticeable improvement in rate capability is achieved as compared with conventional mixture electrode. The SE-coated $LiCoO_2$ electrode shows the capacity retention of 83.3% at 1C as compared to the capacity at 0.1C, which is sharply contrasted by the capacity retention of 46.0% at 1C for the mixture electrode. From galvanostatic intermittent titration technique (GITT), the surface coverage of $LiCoO_2$ by solution-processed SE coating (81%) appears to be more than twice for the mixture electrode (31%), demonstrating the excellence of solution-processable SE. Moreover, the LiI-Li_4SnS_4 is dry-air stable while the conventional SE such as Li_3PS_4 degrades quickly in dry-air.

Acknowledgement. This work was supported by Basic Science Research Program through the National Research Foundation of Korea (NRF) funded by the Ministry of Education (No. NRF-2014R1A1A2058760), and by the Energy Efficiency & Resources Core Technology Program of the Korea Institute of Energy Technology Evaluation and Planning (KETEP) grant funded by the Korea government Ministry of Trade, Industry & Energy (No. 20152010103470).

References

(1) Kamaya, N.; Homma, K.; Yamakawa, Y.; Hirayama, M.; Kanno, R.; Yonemura, M.; Kamiyama, T.; Kato, Y.; Hama, S.; Kawamoto, K.; Mitsui, A. *Nat. Mater.* **2011**, *10*, 682.
(2) Jung, Y. S.; Oh, D. Y.; Nam, Y. J.; Park, K. H. *Israel J. Chem.* **2015**, *55*, 472.
(3) Nam, Y. J.; Jo, S. J.; Oh, D. Y.; Im, J. M.; Kim, S. Y.; Song, J. H.; Lee, Y. G.; Lee, S. Y.; Jung, Y. S. *Nano Lett.* **2015**, *15*, 3317.
(4) Oh, D. Y.; Nam, Y. J.; Park, K. H.; Jung, S. H.; Cho, S.-J.; Kim, Y. K.; Lee, Y.-G.; Lee, S.-Y.; Jung, Y. S. *Adv. Energy Mater.* **2015**, *5*, 1500865.
(5) Park, K. H.; Oh, D. Y.; Choi, Y. E.; Nam, Y. J.; Han, L.; Kim, J.-Y.; Xin, H.; Lin, F.; Oh, S. M.; Jung, Y. S. *Adv. Mater.* **2016**, *28*, 1874.

SUPERIONIC CONDUCTING Li_6PS_5Br SOLID ELECTROLYTES SYNTHESIZED USING ETHANOL *VIA* LIQUID-PHASE TECHNIQUE

So Yubuchi, Akitoshi Hayashi, Masahiro Tatsumisago

1-1 Gakuen-cho, Naka-ku, Sakai, Osaka, 599-8531, Japan
su108065@edu.osakafu-u.ac.jp

Introduction

All-solid-state lithium secondary batteries using nonflammable inorganic electrolytes have been expected as next-generation batteries with greater safety and reliability. In order to improve the electrochemical properties of all-solid-state batteries, studies on sulfide-based electrolytes with high ionic conductivity and favorable mechanical properties have been actively carried out. Especially, argyrodite-type Li_6PS_5X (X = Cl, Br) crystals are attracting much attention as solid electrolytes because of their high lithium-ion conductivities of more than 10^{-3} S cm^{-1} at room temperature (1). In general, sulfide-based solid electrolytes need to prepare time-consuming solid-phase processes. Moreover, a key challenge for improving the all-solid-state batteries is preparation for the solid-solid interfaces between electrode and electrolyte particles with large contact areas. In order to form favorable interfaces, a coating of active material particles with solid electrolytes is an effective process (2). A liquid-phase synthesis of solid electrolytes and preparation for the favorable interfaces *via* the homogeneous solution are thus important. We have reported that Li_2S-P_2S_5 solid electrolytes were synthesized using *N*-methylformamide (NMF) *via* homogeneous solution from starting materials (3). In addition, $LiCoO_2$ (LCO) particles were coated with Li_2S-P_2S_5 solid electrolytes using the NMF solution, and then the favorable interfaces between LCO and Li_2S-P_2S_5 solid electrolytes with large contact areas were formed. However, the prepared electrolyte showed a low ionic conductivity of 2.3×10^{-6} S cm^{-1} at room temperature. Recently, we have reported that argyrodite-type Li_6PS_5Cl electrolyte prepared by a mechanical milling technique was dissolved into ethanol, and the ionic conductivity of the reprecipitated electrolyte was more than 10^{-5} S cm^{-1}(4). In order to improve the conductivity and simplify the multi-step synthesis process, we have focused on liquid-phase synthesis of Li_6PS_5Br solid electrolyte using ethanol.

In this study, the Li_6PS_5Br electrolyte was synthesized from the mixture of starting compounds of Li_2S, LiBr and Li_3PS_4 by liquid-phase reaction in ethanol. Moreover, Li_6PS_5Br electrolyte was coated on $LiNi_{1/3}Mn_{1/3}Co_{1/3}O_2$ (Li_6PS_5Br-coated NMC) particles using the ethanol solution and then the all-solid-state batteries using Li_6PS_5Br-coated NMC were fabricated and characterized.

Experimental

Li_2S and LiBr as the starting materials were completely dissolved into dehydrated ethanol and then Li_3PS_4 glass were added into the solution. The solution was dried at 150°C under vacuum to obtain solid powders or heat-treated at 550°C in a dry argon atmosphere to enhance their crystallinity. On the other hand, NMC particles were coated with Li_6PS_5Br electrolyte using the ethanol solution. The all-solid-state batteries using the mixture of NMC and Li_6PS_5Br powders or only Li_6PS_5Br-coated NMC were fabricated and characterized. The weight ratio of NMC to Li_6PS_5Br solid electrolyte was 90/10.

Results and Discussion

The homogeneous solution was prepared by dissolution of Li_2S, LiBr and Li_3PS_4 glass into ethanol. X-ray diffraction patterns showed that the sample obtained by removing the solvent is mainly Li_6PS_5Br crystal. The Li_6PS_5Br electrolyte after drying at 150°C under vacuum or heat treatment at 550°C in a dry argon atmosphere showed 1.9×10^{-4} S cm^{-1} or 1.0×10^{-3} S cm^{-1} at room temperature.

Li_6PS_5Br-coated NMC was prepared by mixing the ethanol solution and NMC particles and subsequent removing the solvent. Fig.1 shows a scanning electron microscopy (SEM) image of the positive electrode layer using the only Li_6PS_5Br-coated NMC in all-solid-state batteries. Li_6PS_5Br coating layer was formed on NMC with the large contact areas. The all-solid-state batteries using only Li_6PS_5Br-coated NMC showed higher reversible capacity than those using the mixture of NMC and Li_6PS_5Br powders.

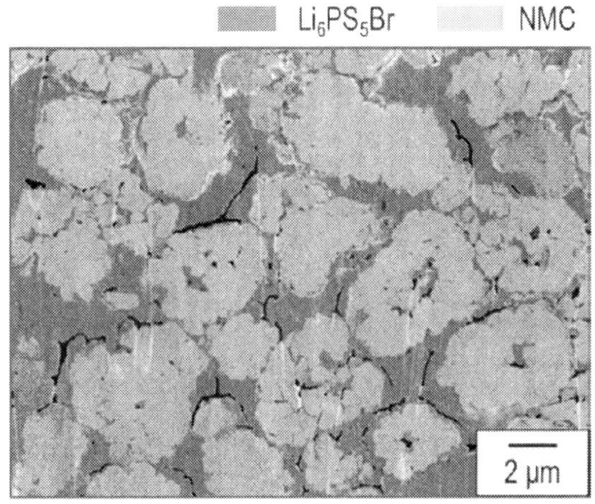

Figure 1. Cross-sectional SEM image of the positive electrolyte layer of the all-solid-state batteries using Li_6PS_5Br-coated NMC.

Conclusions

Li_6PS_5Br solid electrolyte was successfully synthesized from Li_2S, LiBr and Li_3PS_4 using ethanol and showed a high lithium-ion conductivity of more than 10^{-4} S cm^{-1} at room temperature. $LiNi_{1/3}Mn_{1/3}Co_{1/3}O_2$ particles were coated with the Li_6PS_5Br solid electrolyte using the ethanol solution. The all-solid-state batteries using only Li_6PS_5Br-coated NMC showed higher reversible capacity than those using the mixture of NMC and Li_6PS_5Br powders. This approach is a facile synthesis process of sulfide electrolytes with a high lithium-ion conductivity and it is useful for achieving favorable interfaces between electrode and electrolyte in bulk-type all-solid-state lithium batteries.

Acknowledgement. The research was financially supported by the Japan Science and Technology Agency (JST), Advanced Low Carbon Technology Research and Development Program (ALCA), Specially Promoted Research for Innovation Next Generation Batteries (SPRING) Project, and by a Grant-in-Aid for Scientific Research from the Ministry of Education, Culture, Sports, Science and Technology of Japan.

References

(1) S. Boulineau, M. Courty, J.M. Tarascon, V. Vallet, *Solid State Ionics*, **2012**, *3*, 1.
(2) A. Sakuda, A. Hayashi, S. Hama, M. Tatsumisago, *J. Am. Ceram. Soc.*, **2010**, *93*, 765.
(3) S. Teragawa, K. Aso, K. Tadanaga, A. Hayashi, M. Tatsumisago, *J. Mater. Chem. A.*, **2014**, *2*, 5095.
(4) S. Yubuchi, S. Teragawa, K. Aso, K. Tadanaga, A. Hayashi, M. Tatsumisago, *J. Power Sources*, **2015**, *293*, 941.

GARNET BASED LI-METAL BATTERIES

Liangbing Hu

Materials Science and Engineering
& Energy Research Center
University of Maryland College Park

Among different kinds of solid state electrolyte, the Li-stuffed garnets exhibit the most promising physical and chemical properties including:

- The high room temperature bulk conductivity (~10^{-3} S/cm obtained by Thangadurai with cubic $Li_7La_3Zr_2O_{12}$).
- High electrochemical stability for high voltage cathodes (up to 6 V), about 2 V higher than current organic electrolytes and about 1 V higher than the more popular LiPON.
- Excellent chemical stability in contact with elemental and molten Li anodes up to 400 °C.
- Li^+ transference number is close to the maximum of 1.00, which is critical to battery cycle efficiency, while typical polymer electrolytes are only ~0.35
- Wide operating temperature capability, electrical conductivity increases with increasing temperature reaching 0.1 Scm^{-1} at 300 °C, and maintains appreciable conductivity below 0 °C. In contrast, polymer electrolytes are flammable at high temperature
- Synthesizable as simple mixed oxide powders in air, hence easy scale up for bulk synthesis and creation of low-cost supply chain.

In this talk, I will discuss our recent research progress on Li metal batteries with garnet electrolytes, including:

(1) interface layer treatment to improve the wetting between Li metal anode and garnet electrolytes;
(2) Garnet nanofibers that allows the development of flexible, all solid state electrolyte.

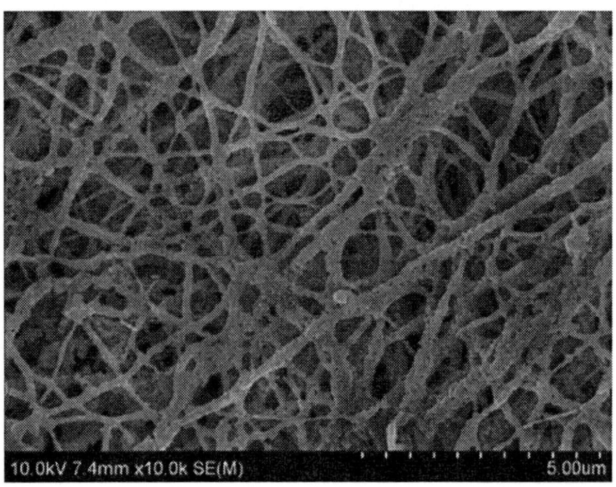

Figure 1 SEM image of garnet nanofibers.

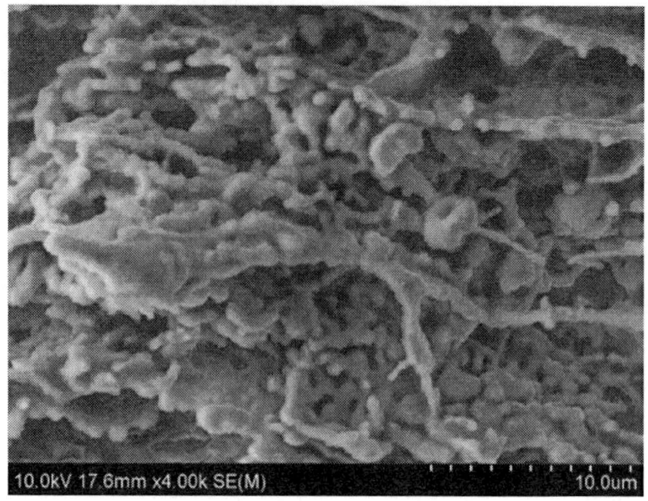

Figure 2 SEM image of garnet nanofiber-polymer hybrid electrolyte.

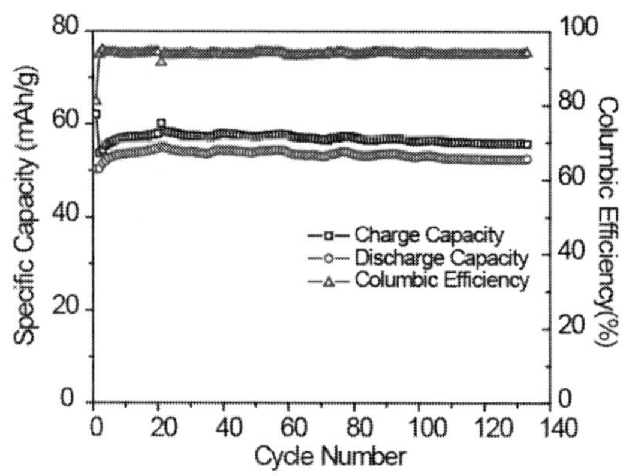

Figure 3 Full cell performance with garnet electrolyte.

References:

(1) Flexible, Solid-State Lithium Ion-conducting Membrane with 3D Garnet Nanofiber Networks, Fu, K.; Gong, Y.; Dai, J. Wachsman, E.; Hu, L. PNAS, 2016.

DEVELOPMENT OF GLASS-CERAMIC SOLID ELECTROLYTES FOR ALL-SOLID-STATE RECHARGEABLE LI OR NA BATTERIES

Akitoshi Hayashi, Masahiro Tatsumisago

Department of Applied Chemistry, Osaka Prefecture University, 1-1 Gakuencho, Naka-ku, Sakai, Osaka 599-8531, Japan
hayashi@chem.osakafu-u.ac.jp

All-solid-state rechargeable lithium batteries have attracted much attention because of their high safety, large energy density, long-cycle performance, and versatile geometry [1]. Sulfide solid electrolytes having high lithium-ion conductivity of over 10^{-2} S cm^{-1} have been developed [2,3], and the conductivity of inorganic lithium ion conductors has already reached to the same level of the conductivity of organic liquid electrolytes applied to commercially available lithium-ion batteries. We have prepared sulfide glass-ceramic electrolytes and crystalline $Li_7P_3S_{11}$ was precipitated as a metastable phase by careful heat-treatment of a mother glass [4]; the crystal was difficult to prepare by conventional solid state reaction and thus glass electrolytes are important as a precursor for precipitation of metastable phases with superionic conduction. In addition, sulfide electrolytes have a unity of lithium-transference number and a wide electrochemical window of 5 V, which are difficult to achieve in organic liquid electrolytes. Mechanical properties such as good formability and appropriate Young's modulus are also important to secure charge-discharge cycling in all-solid-state batteries. The Li_3PS_4 glass showed smaller Young's modulus of 23 GPa than corresponding oxide glasses [5]. The sulfide electrolytes have a good formability and thus a favorable electrode-electrolyte interface is formed by cold press without sintering at high temperatures.

Formation of favorable and large contact areas between electrode and electrolyte is achieved by electrolyte coating on electrode particles via gas-phase or liquid-phase techniques [6,7]. Sulfide amorphous electrolyte coating on $LiCoO_2$ particles by pulsed laser deposition is effective in increasing reversible capacity of all-solid-state batteries. Closely-attached electrolyte-$LiCoO_2$ interfaces with large contact area are achieved by electrolyte coating as shown in Fig. 1. The electrolyte coatings have a merit of reducing the amount of electrolyte in the electrode layer, which results in an increase of energy density of the batteries. Preparation of nanocomposite electrodes with solid electrolytes and nano-carbons by high-energy ball milling is useful for insulative active materials of sulfur or Li_2S with high theoretical capacity [6]. Conductivity enhancement of Li_2S by the combination of LiI contributes to the improvement of utilization of Li_2S active material in all-solid state batteries [8]. Amorphous transition metal sulfides such as amorphous TiS_3 are attractive as a mixed conductor with large capacity, and all-solid-state Li and Na batteries with high energy density are achieved [9,10].

Researches on rechargeable sodium batteries have also been boosted because they are cost-effective batteries with high energy density. We have firstly reported all-solid-state rechargeable sodium batteries with sulfide electrolytes operated at room temperature [11]. Glass-ceramic electrolytes with cubic Na_3PS_4 showed a high conductivity of over 10^{-4} S cm^{-1}, and the value was increased to 7.4 x 10^{-4} S cm^{-1} by partial substitution of Si for P in cubic Na_3PS_4. All-solid-state sodium-sulfur batteries have been developed by using sulfide glass-ceramic electrolytes.

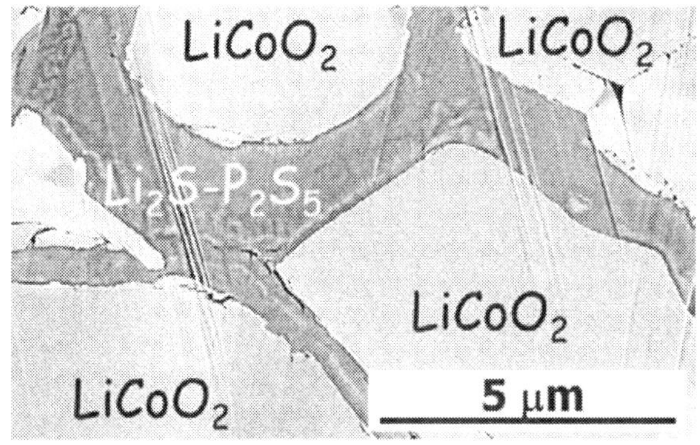

Figure 1. Cross-sectional SEM image of the positive electrolyte layer with $LiCoO_2$ particles coated with Li_2S-P_2S_5 electrolytes by the pulsed laser deposition method.

Acknowledgement: this research is supported by a Grant-in-Aid for Scientific Research from the Ministry of Education, Culture, Sports, Science and Technology (MEXT) of Japan. In particular, the research about all-solid-state Li batteries was financially supported by the Japan Science and Technology Agency (JST), Advanced Low Carbon Technology Research and Development Program (ALCA), Specially Promoted Research for Innovative Next Generation Batteries (SPRING) Project, while the research about all-solid-state Na batteries was supported by the MEXT program "Elements Strategy Initiative for Catalysts & Batteries (ESICB)".

References

(1) M. Tatsumisago and A. Hayashi, *Int. J. Appl. Glass Sci.*, **2014**, 5, 226.
(2) N. Kamaya, K. Homma, Y. Yamakawa, M. Hirayama, R. Kanno, M. Yonemura, T. Kamiyama, Y. Kato, S. Hama, K. Kawamoto and A. Matsui, *Nat. Mater.*, **2011**, *10*, 682.
(3) Y. Seino, T. Ota, K. Takada, A. Hayashi and M. Tatsumisago, *Energy Environ. Sci.*, **2014**, *7*, 627.
(4) F. Mizuno, A. Hayashi, K. Tadanaga and M. Tatsumisago, *Adv. Mater.*, **2005**, *17*, 918.
(5) A. Sakuda, A. Hayashi and M. Tatsumisago, *Sci. Rep.*, **2013**, *3* 2261.
(6) M. Tatsumisago and A. Hayashi, *J. Asian Ceram. Soc.*, **2013**, *1*, 17.
(7) S. Teragawa, K. Aso, K. Tadanaga, A. Hayashi and M. Tatsumisago, *J. Mater. Chem. A*, **2014**, *2*, 5095.
(8) T. Hakari, A. Hayashi and M. Tatsumisago, *Chem. Lett.*, **2015** *44*, 1664.
(9) T. Matsuyama, A. Hayashi, T. Ozaki, S. Mori and M. Tatsumisago, *J. Ceram. Soc. Jpn.*, **2016**, *124*, 242.
(10) N. Tanibata, A. Hayashi and M. Tatsumisago, *J. Electrochem. Soc.*, **2015**, *162*, A793.
(11) A. Hayashi, K. Noi, A. Sakuda and M. Tatsumisago, *Nat. Commun.*, **2012**, *3*, 856.

OPTIMIZING ALKALI IONIC CONDUCTIVITY IN MATERIALS

Shyue Ping Ong

Department of NanoEngineering, University of California, San Diego
S2ong@ucsd.edu

The facile conduction of alkali ions in a crystal is of crucial importance in rechargeable alkali-ion batteries, particularly in all-solid-state architectures due to the longer diffusion length scales in the alkali superionic conductor solid electrolyte. Poor alkali transport in any part of the battery – electrode, electrolyte or the interfaces between them – leads to reductions in rate capability, practical capacity and cyclability. In this talk, I will present recent insights from integrated first principles and experimental studies into doping strategies to control levers such as alkali concentration, migration barriers, and percolation probability to enhance conductivity. A key focus will be on the tight integration of first principles computations with experiments, especially how theory can provide quantifiable insights that accelerate materials development. The materials chemistry covered in this talk includes the alkali thiophosphates (e.g., $Li_7P_3S_{11}$, cubic Na_3PS_4), the NASICON family, as well as novel materials predicted by high-throughput screening.

EFFECTS OF SURFACTANT ON THE PERFORMANCE OF SAMARIUM DOPED CERIA ($CE_{0.8}SM_{0.2}O_2$) PEROVSKITES AS ELECTROLYTE FOR LOW TEMPERATURE SOFC APPLICATION

Chima B.Njoku[1], Bernard Omondi[2] and Patrick G. Ndungu[3]

[1, 2&3]School of Chemistry and Physics, Westville Campus, University of KwaZulu-Natal, Private Bag X 54001, Durban 4000, South Africa
[3]Department of Applied Chemistry, Doornfontein Campus,University of Johannesburg, P.O. Box 17011, Doornfontein 2028, Johannesburg, South Africa.
Corresponding author : kutechi02@gmail.com

Abstract

Sol-gel technique takes advantage of the structural directing properties, and templating characteristics of Pluronic F-127 surfactant to produce porous samarium doped ceria (SDC) nanoparticles. The electrolyte material was calcined at 800 °C, the crystalline nano structures and microstructures were characterized by X-ray diffraction spectroscopy and high resolution transmission electron microscopy. Textural characteristics and morphology of the particles were characterized by nitrogen sorption at 77.5 K and scanning electron microscopy. The electrochemical properties were characterised by using KITTEC SQUADRO SOFC testing equipment with air and hydrogen as gases used. Our results showed that the nature of the surfactant influenced the particle morphology, pore diameter, pore size, crystallite size, surface area and the electrochemical properties of the electrolyte.

Key words: Solid oxide fuel cell, Cerium, Samarium, Surfactant, Air, Sol gel technique

1.0 Introduction

Sol-gel technique is a unique and adaptable technique frequently used to produce ceramic oxide materials and perovskite materials with controlled structure [1, 2]. These materials are produced by two methods, hydrolysis of the metal alkoxide precursor, condensation [3] and aggregation of aqueous colloidal dispersions, containing discrete particles which lead to formation of materials with tailored microstructure [4, 5]. Addition of surfactants in synthesizing metal oxides at ambient temperatures leads to production of composite materials that are homogeneously distributed throughout the resulting matrix. This controls the interfacial tension between the non-polar and polar phases and thus controls the droplet size [6]. This causes changes in the physical characteristics such as density, pore size, nanostructure of the oxides produced and restrictions in aggregation of metallic nanoparticles into matrices by sol–gel processing [7]. The materials can also be tailored by controlling the sol–gel reaction kinetics and the relative rates of hydrolysis and condensation [8 - 12].

Samarium doped ceria (SDC) has been considered suitable as electrolytes in solid oxide fuel cell (SOFC) operation below 700 °C [13a]. At these temperatures they possess high ionic conductivity in air, have superb redox properties and contain high density of surface oxygen vacancies, resulting from the population of mixed valence states [13 - 16]. They also possess better chemical compatibility with high performance cathode materials [17 - 19]. Operating at low temperatures, the electrolyte is susceptible to reduction on the anode side, grain boundary resistance is a significant and impediment to ionic conductivity [20]. Hence surfactant was used to improve its performance at these temperatures.

Pluronic F-127 was used as surfactant, because it has been proven to provide better stabilization of the micelle walls, reduction in the inter-droplet exchange, resulting in the production of smaller particles, smaller crystals and the avoidance of the aggregation of the metallic nanoparticles into matrices [6]. Pluronic F-127 was added in different mole ratios of (0 % to 10 %) to SDC electrolyte to confirm its effect on the electrolyte material. This surfactant-assisted technique may provide an electrolyte material with appropriate pore diameter, high surface areas and a better access to the active sites [21 – 23]. The support has a fundamental role in determining the nature, number of active sites and the activity of the catalyst.

2.0 Experimental Methods

2.1 Starting materials. The Surfactant Pluronic F-127, Ceria acetylacetonate hydrate (98 % purity) and Samarium acetylacetonate hydrate (97 % purity) were purchased from sigma aldrich chemicals USA. The chemical reagents were used as acquired from their commercial sources without further purification.

2.2 Preparation of $Ce_{0.8}Sm_{0.2}O_2$ with surfactant. The synthesis was carried out by adding 0.7200 g of cerium acac and 0.3250 g of samarium acac to 50 ml of ethanol and 15 ml of deionized water as the solvent. The mixture was stirred in an ultrasonic bath (UMC 20 Ultrasonic cleaner, 50 kHz and 220 W) for 30 mins. Pluronic F127 was added to each mixture in different mole ratios of (4 % - 10 %) and the mixtures were stirred using a magnetic stirrer on a hot plate for 30 mins at a temperature of 80 °C, to form a sol. The hydrolysis, condensation and solvent evaporation was very rapid and therefore the collapse of the gel network was also rapid as a result of the surfactant. The sol was left to dry in an oven for 24 hrs and then calcined using a muffle furnace (KITTEC SQUADRO 1350) at 800 °C for 4 hrs.

2.3 Preparation of $Ce_{0.8}Sm_{0.2}O_2$ without surfactant. 0.7200 g of cerium acac and 0.3250 g of samarium acac was added to 50 ml of ethanol and 15 ml of deionized water as solvents. The mixture was sonicated for 30 mins using an ultrasonic bath (UMC 20 Ultrasonic cleaner, 50 kHz and 220 W). The mixture was stirred using a magnetic stirrer on a hot plate for 2 hrs at 80 °C, to form a sol. The hydrolysis and condensation occurred slowly and the solvent evaporation was sluggish. The gel formed was allowed to dry at atmospheric temperature for 4 hrs at 150 °C and then calcined using a muffle furnace (KITTEC SQUADRO SQ11) at 800 °C for 4 hrs. The summary of the process is stated in table 1.

Table 1: Summary of the Synthesis processes of $Ce_{0.8}Sm_{0.2}O_2$ and Pluronic F-127.

Surfactant Name	Sample ID	Molar Ratios (%)
No Surfactant	SDC 1	0 %
Pluronic F127	SDC 2	4 %
Pluronic F127	SDC 3	6 %
Pluronic F127	SDC 4	8 %
Pluronic F127	SDC 5	10 %

2.4 Characterization. X-ray diffractometer (XRD) (Phillips PW3710, 30 kv voltage and 35 mA current, Cu-Kα radiation 1.54015 wavelength) was used to analyze the crystal structure, crystal size, lattice spacing and the phase present in the powder. The XRD diffractogram was recorded from $20° \leq 2\theta \leq 90°$ at a scan speed of 2 °/ min. The mean crystallite sizes 'D' was calculated using Scherer equation:

$$D = 0.9\lambda/\beta cos\emptyset \quad (1)$$

Where: λ is the X-ray wavelength (1.5405 Å), β is the full width at half maximum of the diffraction line and Ø is the diffraction angle.

Fourier transform infrared spectroscopy (Perkin Elmer 1200 ATR- FTIR) was used to perform the infrared spectra analysis. Raman spectroscopy was analysed with DeltaNu advantage 532™ spectrometer (100 mW Nd:YAG laser with an excitation wavelength of 532 nm). The JOEL 2100 HRTEM (200 kV accelerating voltage, beam current of 110 μA and a current density of 2.4 pA/cm^2) was used to analyse the lattice fringes, low magnification and diffraction patterns in the material. The morphology and grain size of samarium doped ceria powder was investigated with ZEISS FEGSEM Ultra Plus scanning electron microscopy and the samples were coated with gold before the examination. Nitrogen adsorption and desorption (BET) was examined in a Micromeritics Tristar II 3020 V1.03 to find the surface area and porosity. For nitrogen desorption and absorption the samples were degassed using the vacuum degasser (Micromeritics VacPrep 061 sample Degas system) for 1 hr at 90 °C then increased to 200 °C to run overnight. The samples were then put in the BET instrument using Nitrogen at 77 k, to carry out the process.

2.5 Single cell fabrication. The cells were prepared by moulding the materials using a 20 Ton press and a 1.3 cm die kit. The press was made from strong steel and the die kit was made from stainless steel. For the SSFC material, 0.1435 g of the sample was placed into the die kit mould and 15 Tons of pressure was then applied using the 20 Ton press. 0.1545 g of NiO-SDC was used to make the anode and 0.2100 g of SDC- surfactant was used to prepare the electrolyte using the same method. The discs had a total active area of 1.327 cm^2. The single cell discs was made up of three layers with the electrolyte in between the anode and the cathode, then heat treated to 500 °C for 2 hrs at a steady heating rate of 5 °C/min, The electrolyte was estimated to be 100 μm thick, the cathode was 100 μm thick and the anode was 100 μm thick.

2.6 Electrochemical characterisation. The symmetrical cells were tested at 400 to 500 °C in 20 % H_2O/H_2 mixture and in air (O_2). Platinum mesh was placed on both electrodes which are used as a current and voltage collector. The gases (H_2O/H_2 and O_2) were passed through a pipe into the cell and it was controlled using flow metres. H_2 gas was passed through a gas bubbler to humidify the gas. For each cycle of the polarisation curve, the voltage was kept constant and current was measured using a Nuvant™ Powerstat 05 potentiostat and galvanostat, attached to the SOFC device.

The ionic conductivity was calculated using the following equation [24]:

$$\sigma = L/RS \qquad (2)$$

Where, L is the electrolyte thickness, S is the electrode area of the electrolyte surface and R is resistance in ohms (cell resistance).

The activation energy and the DC electrical conductivity can be calculated using the following Arrhenius relationship [25]:

$$\sigma = \sigma^{\circ}\exp[-\frac{Ea}{KT}] \qquad (3)$$

Where, Ea is the activation energy of conduction, T is the absolute temperature, σ_o is the Pre-exponential factor and K is the Boltzmann constant.

3.0 Results and discussion

3.1 Powder XRD analysis. The XRD results for samarium doped ceria (SDC) can be seen in figure 1, this confirms the structural pattern for SDC. Addition of surfactants in the sol-gel process caused changes to the structural pattern by increasing its intensity. The diffractogram showed only one main peak and smaller peaks that decreased in peak intensity with very sharp peaks from left to right as angle 2θ increased. The structure system was triclinic and the diffracted peaks were located at 2 theta values of 8.13°, 28.88°, 33.43°, 47.75°, 56.56°, 59.29°, 69.54°, 76.76°, 79.12° and 88.32° which are indexed as (111), (200), (220), (311), (222), (400), (331), (420), (JCPDS file NO. 00-034-0394). Average crystallite sizes were 3.63 nm, 5.46 nm, 13.55 nm, 15.95 nm and 17.50 nm for S1 to S5 respectively which was calculated from Scherer's equation using the most intense peak as shown in Table 2. This confirms that the sample is very crystalline and the peak intensities increased with increase in mole ratios of surfactant which also had a proportionate effect on the crystallite sizes.

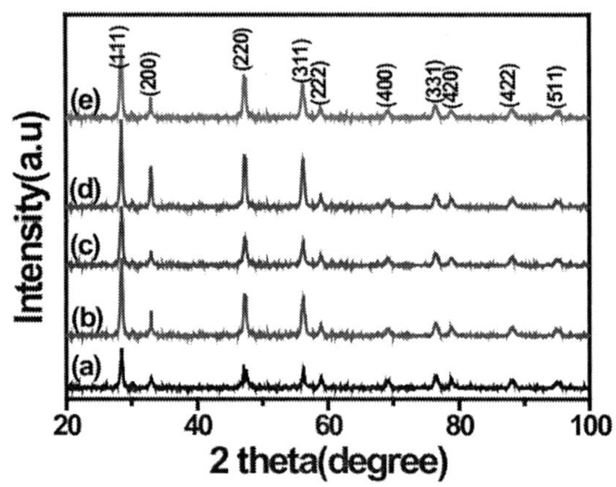

Figure 1: Shows the XRD diffractogram of (a).SDC1 (b).SDC2 (c).SDC3 (d).SDC4 and (e).SDC5 calcined at 800 °C for 4 hours.

For SDC samples after calcination, the average lattice parameter a = 5.41134 Å, which was calculated, using the following equations [26].

$$\text{Lattice parameter} = \lambda/2 \times sin\emptyset \times \sqrt{h^2} + k^2 + l^2 \qquad (4)$$

Where:

λ = wavelength of the instrument,

Ø = 2 theta value/2

The lattice parameter of the SDC material confirmed the lattice parameter of CeO_2 (JCPDS file NO. 00-034-0394) which confirms the formation of SDC compound.

The mass of the unit cell for all samples was 1.3443 x 10^{-27} mg, and this was calculated using the following equation:

$$\text{Mass of Unit Cell (M)}= \frac{(4atoms)(MolarMass)}{6.02\times10^{23}atoms/mol}[\frac{10^{-6}Mg}{g}] \qquad (5)$$

The volume density of the unit cell was 8.41mg/m^3, and the following equation was used to calculate this value:

$$\text{Volume Density} = \frac{mass(unit\ cell)}{Volume(unit\ cell)} \qquad (6)$$

Volume of unit cell was calculated to be 1.5984 x 10^{-28} m^3 (from $v = a^3$ where a = $4R/\sqrt{2}$

Table 2: Shows the difference in Crystallite Size of SDC 1, SDC2, SDC 3, SDC 4 and SDC 5.

Sample ID	FWHM (°)	Crystallite Sizes (nm)
SDC 1	2.255	3.637
SDC 2	1.501	5.461
SDC 3	0.605	13.550
SDC4	0.514	15.950
SDC5	1.092	7.509

3.2 Raman analysis. Figure 2 shows the Raman spectra of SDC solid solutions exhibiting peaks at 472 cm^{-1} and 565 cm^{-1}. The peak at 472 cm^{-1} represents F_{2g} vibrational mode Sm_2O_3 and the peak at 565 cm^{-1} is assigned to MO_8- type (M = Sm) complex [27 – 30]. It has been reported that the presence of SmO_8-type complex could be related to the different radius between Sm^{4+} and Ce^{4+} (0.097nm), since the valence state of Sm is +4 [31].

The band at 565 cm^{-1} could also be assigned to a shift resulting from the different oxidation state and ionic radius of dopants, compared to that of Ce^{4+} which leads to increase in oxygen vacancy. Therefore, it can be observed that the vibrational bands at 472 cm^{-1} increased with increase in surfactants added to the material. This is because the ordering level of the samples increased with increase in surfactant and the lattice became more perfect, leading to declining concentration of the defect sites [32].

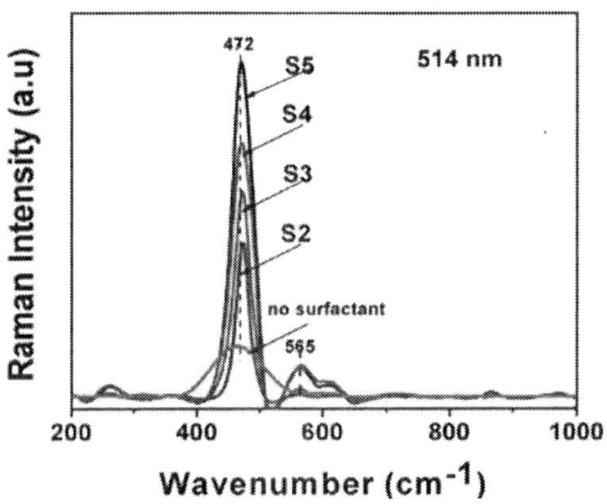

Figure 2: Shows the Raman spectra of (a).SDC1 (b).SDC2 (c).SDC3 (d).SDC4 and (e).SDC5 calcined at 800 °C for 4 hours.

3.3 HRTEM analysis. Figure 3 shows the lattice fringes and low magnification of SDC mixed with surfactants; the microstructure shows that it is crystalline with evidence of large visible particles. The representative images showed the microstructure for all samples, which showed slight differences in specific irregular shaped particles and varying crystalline shapes. It was noted that there are differences in powdered crystals, as observed in the XRD data in terms of crystal sizes.

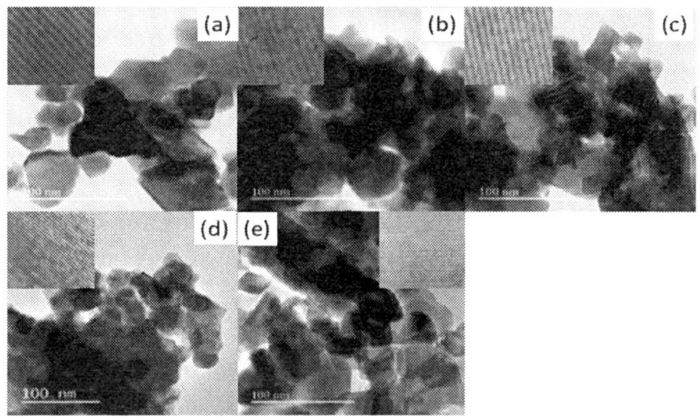

Figure 3: Shows the Low magnification and lattice fringe of the High Resolution TEM for (a.) SDC1 (b).SDC 2 (c).SDC3 (d).SDC 4 and (e).SDC 5 Calcined at 800 °C for 4 hrs

The lattice distance between the fringes for S1 was 0.183 indexed to {220}, S2 is 0.157 indexed to {222}, S3 is 0.202 indexed to {220}, S4 is 0.207 indexed to {211} and S5 is 0.190 indexed to {220}. These lattice parameters are linked to the XRD as SmO_2 structure and CeO_2 (fluorite) structures. Presence of CeO_2 fluorite structure in the HRTEM as shown in XRD revealed concentrations of O_2 vacancies on the surface. The presence of oxygen (O_2) vacancies will influence oxygen ion diffusion paths through the electrolyte thereby increasing O_2 mobility and transportation of H_2 molecule. The selected area electron diffraction patterns differed slightly in terms of the lattice and arrangements of the particles, showing slight scattering and disorder, thus displaying a decrease in crystallinity as surfactants added, to the electrolyte material decreased.

3.4 SEM analysis. Figure 4 shows the images of SDC prepared using Pluronic F-127 as surfactants, as observed in figure 4, it showed that the materials prepared with surfactant exposed a low conglomeration but a wide size distribution. This reveals an accumulation of particles bonded together to form agglomerates, these particles are held together by strong and weak bonds which exist between the metal acacs. The particles also showed separations as a result of the addition of surfactants, forming loosely coherent secondary particles. Morphology also showed the formation of holes or pore spaces on the material layer, these pore spaces increased with increase in amount of surfactants added to the material.

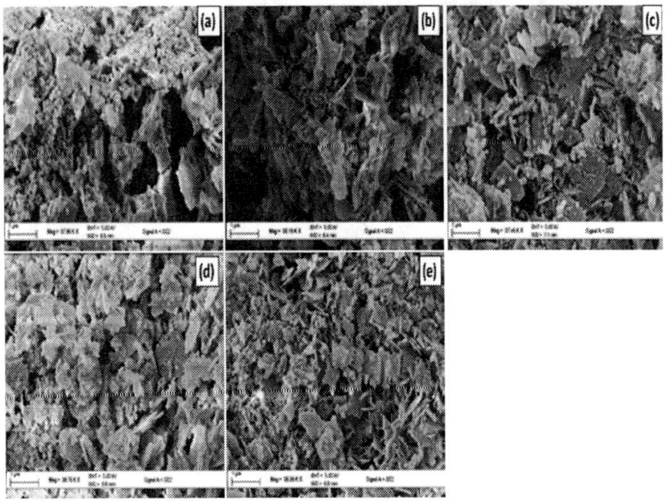

Figure 4: Shows the SEM morphology of (a).SDC1 (b).SDC2 (c).SDC3 (d).SDC4 and (e).SDC5 calcined at 800 °C for 4 hours.

The material prepared without surfactant (Figure 4(a)) showed bigger separated particles that are loosely packed with larger pore spaces. These particles bonded with other loosely packed particles and grew asymmetrically to form agglomerates. Materials formed with this method are usually prominent for its large surface to volume ratio, higher adsorption capacity and enhanced chemical reactivity [33].

3.5 Surface area analysis (BET). The nitrogen desorption of the materials were observed as follows; SDC1 exhibited the lowest surface area of 5.535 m^2/g, the highest pore size value of 39.950 nm and a low pore volume of 0.0948 cm^3/g. SDC2 isotherm plot as shown in the figure revealed that the samples are macroporous material, the surface area observed was 17.846 m^2/g, a pore size of 25.404 nm and the pore volume was 0.1325 cm^3/g. SDC 3 exhibited a higher surface area of 18.634 m^2/g, a pore size of 26.366 nm and a pore volume of 0.1421 cm^3/g. The surface area for SDC 4 was 19.044 m^2/g, the pore size was 27.131 nm and exhibited a pore volume of 0.1441 cm^3/g. The surface area for SDC 5 showed the highest surface area of 20.821 m^2/g, the pore size value was 25.759 nm and the pore volume showed 0.155 cm^3/g.

The materials observed in general, showed an increase in surface area as mole ratios of surfactant increased, the pore volume and pore size also increased. In totality the observations showed that addition of varying amounts of surfactants caused a vast increase in surface area, pore size and pore volume of the material when compared to the materials without surfactant confirming a change in the structural properties of the material with an increasing change in mole ratio of surfactants.

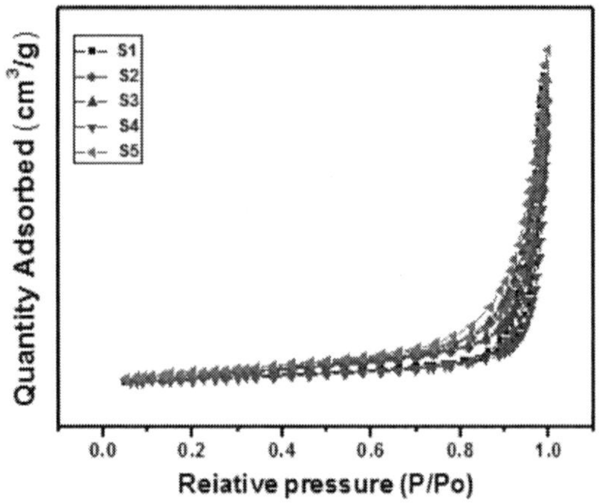

Figure 5: Shows the Nitrogen desorption Isotherm plot of (a) SDC1 (b) SDC2 (c) SDC3 (d) SDC4 and (e) SDC5 calcined at 800 °C for 4 hours.

Table 3: Shows the Surface Area, Pore diameter and Pore Volume of SDC1, SDC 2, SDC 3, SDC4 and SDC 5.

Sample ID	Pluronic F-127 Used	Surface area m^2/g	Pore diameter (nm)	Pore volume (cm^2/g)
SDC 1	No surfactant	5.535	39.950	0.0948
SDC 2	2 g	17.846	25.404	0.1325
SDC 3	3 g	18.634	26.366	0.1421
SDC 4	4 g	19.044	26.856	0.1441
SDC 5	5 g	20.821	25.759	0.1557

3.6 Electrochemical properties. The symmetrical cell was tested, using a muffle furnace made by Kittec SQUADRO and a Nuvant potentiostat and galvanostat; these instruments were used to measure current and voltage data. 50 ml of compressed air and 200 ml of humidified hydrogen gas mixed with 20 % H_2O was passed through pipes into the cell and then tested for 15 hrs.

3.6.1 Polarisation graph. Figure 6 illustrates the electrochemical properties of $Ce_{0.8}Sm_{0.2}O_2$, prepared using Pluronic F-127 as surfactants. The polarization properties were tested using air at the cathode side and humidified hydrogen (H_2+ 20 % H_2O) as fuel gas at the anode side. The cathode material used for testing was $Sm_{0.8}Sr_{0.2}Fe_{0.8}Co_{0.2}$ perovskite with a thickness of 100 μm and the anode used was NiO/SDC which was 100 μm thick and electrolyte was 150 μm thick. The symmetric cell was measured within temperature intervals of 400 – 500 °C.

The polarization graph displayed tendencies showing a linear fall in voltage with increase in current density. This resulted from the resistance to current flow, confirming that an increase in current is directly proportional to increase in voltage. The open circuit voltage (OCV) showed stability for over 15 hrs at 500 °C and showed different values for OCV. The OCV values were less than the Nernst thermodynamic values because of two reasons: hydrogen pressure reduction at the anode/electrolyte interface by the anode materials and the gas crossover flowing through the electrolyte because of the change in temperature.

The voltage showed a linear fall with corresponding increase in the current density in the polarisation graph. The linear fall was activated by these reasons: increase in Ohmic and concentration losses from the internal resistance of the cell, which was produced by the ionic conductivity and effective distance between the cell components [34]. Change in ionic conductivity can be influenced by the reduction of adsorbed molecular oxygen to O_2 ions which improves the diffusion of these ions to the electrode/electrolyte interface [35]. The increase in the overall rate of the diffusion process reduces polarisation losses and resistance; hence, low polarisation losses are observed at 400 °C.

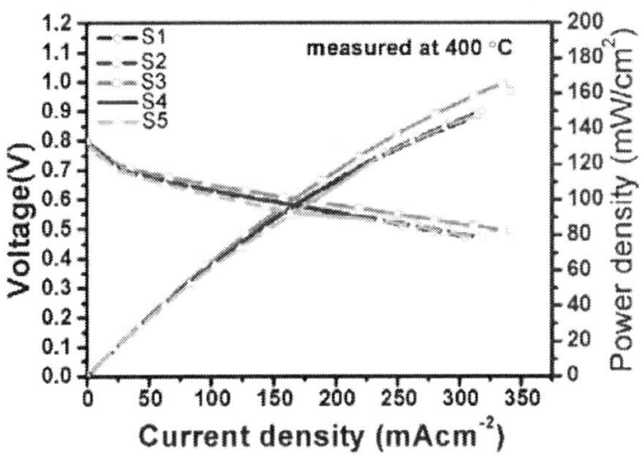

Figure 6: Shows the Polarization graph of (a).SDC1 (b).SDC2 (c).SDC3 (d).SDC4 and (e).SDC5 calcined at 800 °C for 4hours.

The maximum power density for the SDC-surfactant samples calcined at 800 °C with different wt % are shown in Figure 7. The power densities were measured at 400 °C and 500 °C and S3 presented the maximum power density of 140.61 mW/cm^2 at 500 °C and 161.24 mW/cm^2 at 400 °C. Therefore, at each test temperature, the S3 sample showed the largest power density; however the electrolyte presented a better performance at lower temperature. These results could have some implication for application, depicting that the electrolytes made, can perform better at 400 °C than 500 °C and it exhibits a moderate power output from a non-ideal single cell. Improvements can be achieved by using thin electrolyte and suitable cathode and anode materials, using tape casting or dip coating methodologies to control and optimise adhesion between the layers.

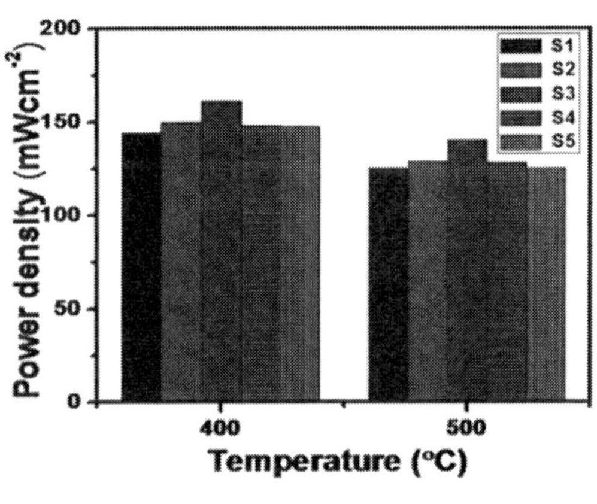

Figure 7: Shows the Power density graph of (a).SDC1 (b).SDC2 (c).SDC3 (d).SDC4 and (e).SDC5 calcined at 800 °C for 4 hours.

3.6.2 Area Specific Resistance. The area specific resistance (ASR) can be calculated from the gradient, at the centre of the linear fall representing the voltage to current graph of the IV curve where ohmic resistance is dominant. The design of the cell's electrodes, electrolyte properties and the assembly process used to make the single cell determines the ASR [36].

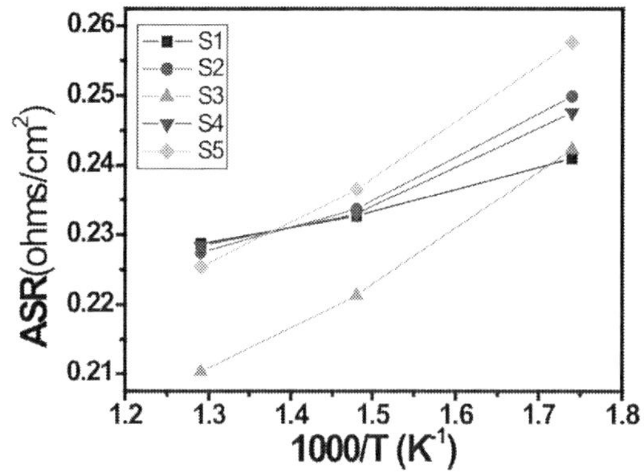

Figure 8: Shows the Arrhenius plot of the ASR of (a) S1 (b) S2 (c) S3 (d) S4 and (e) S5 calcined at 800 °C for 4 Hours measured from 300 – 500 °C.

Figure 8 shows the ASR values for electrolytes with surfactant, the graph illustrates a linear fall in resistance as the temperature increased. This can be connected to changes in material compositions and increased oxygen loss from the lattice [37, 38]. Sample S3 showed the lowest ASR values and S5 showed the highest value at 400 °C. Which could be attributed to an increase in oxygen loss from the S3 sample than S5 sample, due to increase in the mass of surfactant blocking the pores and thereby limiting oxygen ion transfer. Also reduction in the ASR values at 500 °C can be triggered by the intensification of oxygen diffusion resulting from the conversion of absorbed oxygen to O_2 ions [39]. The modification in porosity, surface area, pore volume, and morphology also influences the movement of O_2 ions at the triple phase boundary. The low ASR values by S3 sample can also be attributed to their competence to transfer O_2 ions faster than other electrolytes at the TPB [40].

Fig. 9 summarizes the resistance, Rc calculated from the resistivity measurements for the SDC electrolytes consisting of different amounts of Pluronic F-127. The resistances were measured from 300 – 600 °C using a potentiostat/galvanostat for current and voltage collection. This was collected by placing the electrolyte between platinum current collectors placed in an open flange setup situated on a muffle furnace for temperature regulation. In general, the overall resistance decreased as surfactant concentration increased to 10 % mole ratio and as the temperature increased the resistance also decreased. Further increasing the surfactant content above 10 % mole ratio might cause an increase in the resistance [41]. It is possible that the polarization resistance of the optimized electrolyte was decreased by increasing the oxygen diffusion and extending the triple phase boundary, which resulted in much lower over-potentials toward oxygen reduction.

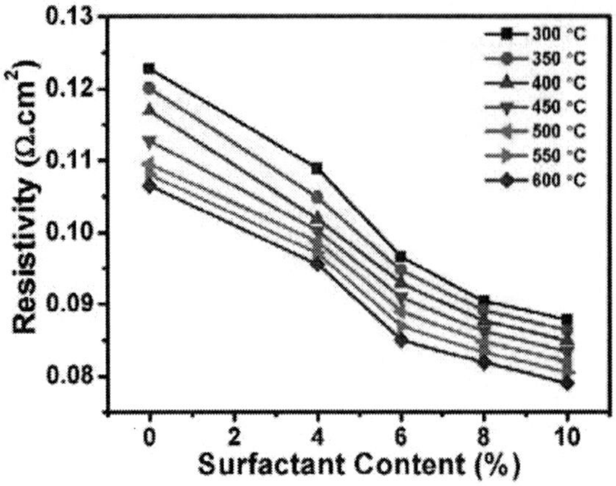

Figure 9: Shows the dependence of resistivity measurements on the surfactant content in the SDC electrolyte fired at 800 °C for 4 hours and measured at 300 – 600 °C.

Fig. 10 shows the concentration over-potential at 400 and 500 °C of these electrolytes. The electrolyte containing 6 % mole ratio of surfactant showed the lowest over-potential, depicting the best electrochemical performance at both 400 and 500 °C. The high performances observed by the electrolytes with 6 % molar ratio of surfactant can be interpreted using the effective medium percolation theory [42]. This predicts that the ambipolar transport behaviour of composite mixed ionic–electronic conductors is a function of the volume fraction of each of the randomly distributed constituent phases [43].

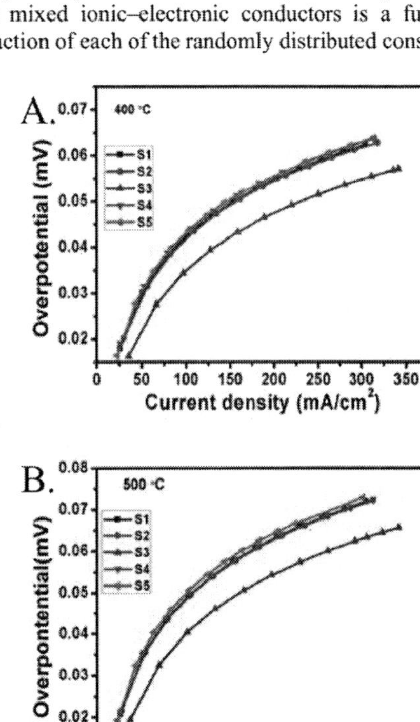

Figure 10: Overpotential curves obtained at (a.) 400 °C and (B.) 500 °C for SDC electrolyte containing 0 – 10 % molar ratio of surfactant fired at 800 °C for 4 hours.

4.0 Conclusions

Pluronic F-127 surfactant was mixed with SDC to improve its ability to perform better than conventional SDC in this experiment. The crystal structure was investigated by X-ray diffraction and the structural properties were investigated by Raman spectroscopy. The morphology, microstructure and textural characteristics were investigated by SEM, HRTEM and BET respectively. The electrochemical properties was investigated using a potentiostat and galvanostat. The XRD revealed an increase in the crystallite sizes of the materials with surfactant when compared to the material without surfactant. Nitrogen desorption analysis showed that the surface area and porosity of the material increased with increasing mass of surfactant when compared to the ordinary SDC. Raman displayed a high structural stability and high oxygen vacancy concentration. SDC combined with 6 % molar ratio of surfactant produced the highest power density of 161.24 mW/cm^2 and a current density of 325.90 mA/cm^2 at 400 °C powered for 15 hrs. When compared to contemporary SDC without surfactant, it produced a power density of 144.289 mW/cm^2 at 400 °C. This describes that the electrolyte material works best with the addition of surfactant and also works best at lower temperatures. The properties exhibited revealed that the electrolyte material mixed with surfactant can be used as a promising electrolyte for low temperature solid oxide fuel cell.

Acknowledgements. The authors thank the University of KwaZulu-Natal, for an enabling environment, ESKOM-TESP and NRF THUTUKA for their financial support.

References

(1) M. Sharma, J. Peng, Synthesis of magnesium oxide particles with stacks of plate's morphology, J.Alloys Compd. 2011 509 7881 - 7885.

(2) M. S., Mastuli., N. S. Ansaria, M. A. Nawawia, A. Maria, Effects of Cationic Surfactant in Sol-gel Synthesis of Nano-Sized Magnesium Oxide. APCBEE Procedia, 3 (2012) 93 – 98.

(3) Z. Lou, J. Peng, N. Dai, J. Qiao, Y. Yan, Z. Wang, J. Wang, K. Sun, Synthesis and characterisation of Aluminium-Doped Perovskites as Cathode Materials for Intermediate Temperature Solid Oxide Fuel Cells, Electrochem. Commun. 22 (2012) 97 – 100.

(4) N. I. Ermokhina, V.A. Nevinsky, P. A. Manorik, V. G. Ilyin, N. N. Shcherbatyuk, D. O. Klymchyuk, A. M. Puziy, Synthesis of large pore meso- porous nanocrystalline TiO_2 microspheres, Materials Letters, 75 (2012) 68 – 70.

(5) K. Tapia-Parada, G. Valverde-Aguilar, A. Mantilla, M. A. Valenzuela, E. Hernández, Synthesis and characterization of Ni/Ce-SiO_2 and Co/Ce–TiO_2 catalysts for methane decomposition, Fuel 110 (2013) 70 – 75.

(6) S. K. Das, A. K. Sinha, A. Bhaumik, Self-assembled mesoporous zirconia and sulfated zirconia nanoparticles synthesized by triblock copolymer as template, J. Phys. Chem. 113 (2009) 8918 – 8923.

(7) J. Milewski, A. Miller, Influences of the type and thickness of electrolyte on solid oxide fuel cell hybrid system performance, Journal of Fuel Cell Science and Technology 3(4) (2006) 396 – 402.

(8) S. B. Adler, Factors Governing Oxygen Reduction in Solid Oxide Fuel Cell Cathodes, Chem. Rev. 104 (2004) 4791 – 4843.

(9) A. V. Vikar, J. Chen, C. W. Tanner, J. W. Kim, The Role of Electrode Microstructure on Activation and Concentration Polarizations in Solid Oxide Fuel Cells, Solid State Ionics 131 (2000)189 –198.

(10) E. P. Murray, M. J. Sever, S. A. Barnett, Electrochemical Performance of (La, Sr) (Co, Fe) O_3– (Ce, Gd) O_3 Composite Cathodes, Solid State Ionics 148 (2002) 27 – 34.

(11) W. Guo, J. Liu, C. Jin, H. Gao, Y. Zhang, Electrochemical Evaluation of $La_{0.6}Sr_{0.4}Co_{0.8}Fe_{0.2}O_{3-\delta}$ – $La_{0.9}Sr_{0.1}Ga_{0.8}Mg_{0.2}O_{3-\delta}$ Composite Cathodes for $La_{0.9}Sr_{0.1}Ga_{0.8}Mg_{0.2}O_{3-\delta}$ Electrolyte SOFCs, J. Alloys Compd. 473 (2008) 43 – 47.

(12) F. Qiang, K. N. Sun, N. Zhang, X. D. Zhu, X. D. Lee, D. R. Zhou, Charact- erization of Electrical Properties of GDC Doped A-Site Deficient LSCF Based Composite Cathode using Impedance Spectroscopy, J. Power Sources, 168 (2007) 338 – 345.

(13) R. W. Tarnuzzer, J. Colon, S. Patil, S. Seal, Vacancy engineered ceria nanostructures for protection from radiation-induced cellular damage, Nano letters, 5 (2005) 2573 - 2577.
(14) T. Pirmohamed, J. M. Dowding, S. Singh, B. Wasserman, E. Heckert, A.S. Karakoti, J. E. S. King, W. T. Self, Nanoceria exhibit redox state-dependent catalase mimetic activity, Chemical communications, 46 (2010) 2736 - 2738.
(15) E.G. Heckert, A.S. Karakoti, S. Seal, W.T. Self, The role of cerium redox state in the SOD mimetic activity of nanoceria, Biomaterials, 29 (2008) 2705 - 2709.
(16) D. A. Andersson, S. I. Simak, B. Johansson, I. A. Abrikosov, N. V. Skorodumova, Modeling of CeO_2, Ce_2O_3 and CeO_{2-x} in the LDA+ U formalism, Physical Review B, 75 (2007) 035109.
(17) W.G. Wang, M. Mogensen, High-performance lanthanum ferrite based cathode for SOFC, Solid State Ionics, 176 (2005) 457 - 462.
(18) X. Xu, C. Xia, G. Xiao, D. Peng, Fabrication and performance of functionally graded cathodes for IT-SOFCs based on doped ceria electrolytes, Solid State Ionics, 176 (2005) 1513 - 1520.
(19) Z. Bi, M. Cheng, Y. Dong, H. Wu, Y. She, B. Yi, Electrochemical evaluation of $La_{0.6}Sr_{0.4}CoO_3$–$La_{0.45}Ce_{0.55}O_2$ composite cathodes for anode-supported $La_{0.45}Ce_{0.55}$ O_2–$La_{0.9}Sr_{0.1}Ga_{0.8}Mg$ $_{0.2}O_{2.85}$ bilayer electrolyte solid oxide fuel cells, Solid State Ionics, 176 (2005) 655 - 661.
(20) J.M. Ralph, J.A. Kilner, B.C.H. Steele, Improving Gd-doped Ceria Electrolytes for Low Temperature Solid Oxide Fuel Cells," New Materials for Batteries and Fuel Cells. Proc. Symp. San Francisco, 14 (1999) 309 - 3257.
(21) P. Blennow, J. Heljm, T. Klemens, S. Ramousse, A. Kromp, A. Leonide, A. Weber, Manufacturing and characterization of metal-supported solid oxide fuel cells, Journal of Power Sources, 196 (2011) 7117 – 7125.
(22) C. Shen, L.L. Shaw, FTIR analysis of the hydrolysis rate in the sol–gel formation of gadolinia-doped ceria with acetylacetonate precursors, J. Sol-Gel Sci Technol., 53 (2010) 571 – 577.
(23) M. C. Brasil, E. V. Benvenutti, J. R. Gregorio, A. E. Gerbase, Iron acetyl acetonate complex anchored on silica xerogel polymer. React. Funct. Polym., 63 (2005) 135 –141.
(24) C. B. Njoku, P. G. Ndungu, Effect of non-ionic, anionic and cationic surfactants on the sol gel synthesis of IrO-$Ce_{0.8}Sm_{0.2}O_2$ nanocomposite for solid oxide fuel cell application, Journal of fuel cell science and technology, 11 (2014) 101 - 109.
(25) H. Okkay, M. Bayramoglu, M. F Öksüzömer, $Ce_{0.8}Sm_{0.2}O_{1.9}$ synthesis for solid oxide fuel cell electrolyte by ultrasound assisted co-precipitation method, UltrasonicsSonochemistry, 20 (2013) 978 – 983.
(26) N. Chaubey, B. N. Wani, S. R. Bharadwaj, M. C. Chattopadhyaya, Physico chemical properties of rare earth doped ceria $Ce_{0.9}Ln_{0.1}O_{1.95}$ (Ln = Nd, Sm, Gd) as an electrolyte material for IT SOFC/ SOEC, Solid State Sciences, 20 (2013) 135 - 141.
(27) K. Kuntaiah, P. Sudarsanam, B. M. Reddy, A. Vinu, Nanocrystalline Ce_{1-x} Sm_x $O_{2-\delta}$ (x = 0.4) solid solutions: structural characterization versus CO oxidation, RSC Advances, 3 (2013) 7953 - 7962.
(28) L. Li, F. Chen, J. Q. Lu, M. F. Luo, Study of defect sites in $Ce_{1-x}M_xO_{2-\delta}$ (x = 0.2) solid solutions using Raman spectroscopy, Journal of Physical Chemistry A, 115 (2011) 7972 - 7977.
(29) A. M. Torres-Huerta, J. R.Vargas-García, M. A. Domínguez-Crespo, Preparation and characterization of IrO_2 – YSZ nanocomposite electrodes by MOCVD, Solid State Ionics, 178 (2007) 1608 - 1616.
(30) E. Barsoukov, J. R. Macdonald, Impedance Spectroscopy, Theory, Experiment and Applications 2nd Edition (2005) John Wiley & Sons,Inc.,Hoboken, New Jersey.
(31) C. B. Njoku, P. G. Ndungu, Synthesis and characterization of novel $Ce_{0.8}Sm_{0.2}Fe_{0.9}Ir_{0.03}Co_{0.07}O_{3-\delta}$ perovskite material and possible application as a cathode for low–intermediate temperature SOFCs, Materials Research Bulletin, 68 (2015) 100 -108.
(32) K. Reiner, Reducing grain boundary, dislocation line and vacancy formation energies by solute segregation: II. Experimental evidence and consequences, Acta Materialia 55 (2007) 5139-5148.
(33) Z. Linda, Y. Luo, M. Hooper, E. Hu, Removal of VOCs by photocatalysis process using adsorption enhanced TiO_2 – SiO_2 catalyst, Chemical Engineering and Processing: Process Intensification 45 (2006) 959-964.
(34) A. Patricia, C. S. Adjiman, N. P. Brandon, Anode-supported intermediate temperature direct internal reforming solid oxide fuel cell. I: model-based steady-state performance, Journal of Power Sources 138 (2004) 120-136.
(35) A. S. Arico, P. Bruce, B. Scrosati, J. M. Tarascon, W. Van Schalkwijk, Nanostructured materials for advanced energy conversion and storage devices, Nature materials, 4 (2005) 366-377.
(36) Z. Wenxia, Z. Lü, S. Li, B. Wei, J. Miao, X. Huang, K. Chen, N. Ai, W. Su, Study on $Ba_{0.5}Sr_{0.5}Co_{0.8}Fe_{0.2}O_{3-\delta}$ – $Sm_{0.5}Sr_{0.5}CoO_{3-\delta}$ composite cathode materials for IT-SOFCs, Journal of Alloys and Compounds, 465 (2008) 274 - 279.
(37) J. W. Stevenson, T. R. Armstrong, R. D. Carneim, L. R. Pederson, W. J. Weber, Electrochemical Properties of Mixed Conducting Perovskites $La_{1-x}M_x$ $Co_{1-y}Fe_yO_{3-\delta}$ (M = Sr, Ba, Ca), Journal of the Electrochemical Society, 143 (1996) 2722-2729.
(38) B. J. Wuensch, K. W. Eberman, C. Heremans, E. M. Ku, P. Onnerud, E. M. Yeo, S. M. Haile, J. K. Stalick, J. D. Jorgensen, Connection between oxygen-ion conductivity of pyrochlore fuel-cell materials and structural change with composition and temperature, Solid State Ionics 129 (2000) 111-133.
(39) F. Esch, S. Fabris, L. Zhou, T. Montini, C. Africh, P. Fornasiero, R. Rosei, Electron localization determines defect formation on ceria substrates, Science, 309 (2005) 752 - 755.
(40) Y. Y. Fang, G. M. Zeng, J. H. Huang, J. X. Liu, X. M. Xu, K. Xu, Y. H. Qu, Micellar- enhanced ultrafiltration of cadmium ions with anionic–nonionic surfactants, Journal of Membrane Science, 320 (2008) 514-519.
(41) R.Yang, J. Qu, T. Marinis, C. P. Wong, A precise numerical prediction of effective dielectric constant for polymer-ceramic composite based on effective-medium theory, Components and Packaging Technologies, IEEE Transactions on 23 (2000) 680-683.
(42) M. J. Rosen, J. T. Kunjappu, Surfactants and interfacial phenomena, John Wiley & Sons, 2012.
(43) V. Dusastre, J. A. Kilner, Optimisation of composite cathodes for intermediate temperature SOFC applications, Solid State Ionics, 126 (1999) 163-174.

EXCELLENT PERFORMANCE OF 3D CARBON NANOMATERIALS FOR ELECTRIC DOUBLE-LAYER CAPACITORS AT WIDE RANGE TEMPERATURE

Liang Chang, *Yun Hang Hu**

Department of Materials Science and Engineering, Michigan Technological University, 1400 Townsend Drive, Houghton, MI 49931-1295, USA
E-mail: yunhangh@mtu.edu

1. Introduction

Clean energy development can not only meet the industrial and life demands but also relief the greenhouse gas brought by fossil fuels [1-2]. As one type of energy storage devices, electric double-layer capacitors have advantages of high power density, long life expectation, and wide range operation conditions [3-5]. In the practical application, electric double-layer capacitors must accommodate stern climate condition and provide stable electrochemical performance. Stricter requirement is that electrode materials in the supercapacitors can be used for a wide range of operating-temperatures [6-9]. First, construction of three-dimensional structures for carbon materials can inhibit the aggregation of carbon layers to avoid the formation of graphite and provide continuous electron transfer channels. Second, suitable pore size distribution is required for the transfer of electrolyte ions inside the pores to form double layers on the electrode surface. In other words, mesopores and macropores can deserve as channels for ion transfer. Third, functional group decoration can bring redox reactions to improve the total capacity. Furthermore, electrical and chemical stabilities are indispensable for thousands of cycles and more critical operating conditions. Therefore, 3D carbon nanomaterials with suitable pore size distributions and functional groups are promising electrode materials for electrical double-layer capacitors, particularly for a wider-operating temperature range.

2. Results and discussion

The one-step synthesized carbon nanomaterials with 3D structures have large BET surface area and proper pore size distribution. Furthermore, the 3D carbon materials possess oxygen groups. Those created their excellent properties for double layer formation. The CV curves showed ideal rectangular shape at 30 mV/s scan rate and presented slight polarization even the scan rate was up to 300 mV/s. This indicates good propagation of electrolyte ions on the surface of electrodes. The symmetrically triangle charge/discharge lines further demonstrated the excellent electrochemical properties. The specific capacity was 103 F/g at 1 A/g and slight decrease of capacity can be obtained when current density increased by 10 times. Therefore, 3D carbon materials are excellent candidates for supercapacitor electrodes.

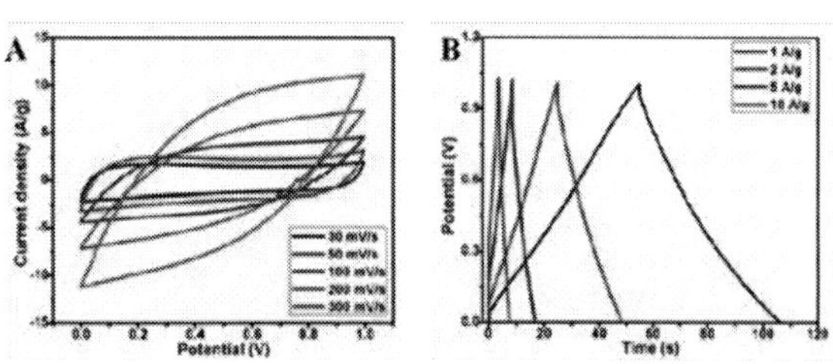

Figure 1. CV curves (A) at different scan rates and galvanostatic charge/discharge profiles (B) at different current denisties.

Reference

(1) Wang, G. P.; Zhang, L.; and Zhang, J. J.; *Chem. Soc. Rev.*, **2012**, *41*, 797-828.
(2) Larcher, D.; and Tarascon, J. M. *Nat. Chem.* **2015**; *7*, 19-29.
(3) Chmiola, J.; Largeot, C. P.; Taberna, L.; Simon, P.; and Gogotsi, Y. *Science* **2010**, *328*, 480-483.
(4) Pech, D.; Brunet, M.; Durou, H.; Huang, P. H.; Mochalin, V.; Gogotsi, Y.; Taberna, P. L.; and Simon, P. *Nat. Nanotechnol.* **2010**, *5*, 651-654
(5) Xia, J.; Chen, F.; Li, J. H.; Tao, N. J. *Nat. Nanotechnol.* **2009**, *4*, 505-509.
(6) Luo, J. Y.; Jang, H. D.; and Huang, J. X. *ACS Nano* **2013**, *7*, 1464-1471; (b) Jiang, H.; Lee, P. S.; and Li, C. Z. *Energy Environ. Sci.* **2013**, *6*, 41-53
(7) Xu, Y.; Lin, Z.; Huang, X.; Wang, Y.; Huang, Y.; and Duan, X. *Adv. Mater.* **2013**, *25*, 5779-5784
(8) Niu, Z.; Chen, J.; Hng, H. H.; Ma, J.; and Chen, X. *Adv. Mater.* **2012**, *24*, 4144-4150
(9) Yu, W. J.; Chae, S. H.; Perello, D.; Lee, S. Y.; Han, G. H.; Yun, M.; and Lee, Y. H. *ACS Nano* **2010**, *4*, 5480-5486

PREPARATION OF HIGH PURITY POLYANILINE AND POLYANILINE-CNT ELECTRODES FOR PSEDUOCAPACITORS.

Silas K. Simotwo[#], *Christopher DelRe*[#] *and Vibha Kalra**

Department of Chemical and Biological Engineering, Drexel University, 3141 Chestnut Street, Philadelphia, PA 19104, USA.
Corresponding author: Email: vk99@drexel.edu

Introduction

Polyaniline (PANI), a conducting polymer, has shown enormous promise as a pseudocapacitor, a type of supercapacitor that relies on Faradaic mechanisms for charge storage, owing to its ease of synthesis, good environmental stability, low cost, and excellent electroactivity [1]. Bulk PANI, however, is not suitable for application in electrodes of energy devices due to its low accessible surface area. Therefore, nanostructured PANI materials with large surface-area-to-volume ratios – and hence shorter ion diffusion paths – have gained interest as suitable electrode materials for supercapacitors [2]. PANI nanostructures are often synthesized through template/ substrate method. Nanomaterials such as carbon nanofibers [2-3], graphene [4], carbon nanotubes [5] and titanium oxide [6] have been employed as non-removable substrates (hard template) for polymerization of aniline to generate hybrid PANI/carbon electrodes. However, these hybrid electrodes face several shortcomings, including low or negligible capacitance of the substrate resulting in low effective specific capacitance of the complete electrode [7], a multi-step fabrication process leading to additional cost, time, and resources, and a possibility of over-estimating capacitance normalized per weight of PANI due to only a small amount of polymer deposited on the substrates. Sacrificial templates have also been employed to fabricate PANI nanostructures with the objective to eliminate electrochemically inactive (dead) weight as well as to enhance the surface area of PANI nanomaterials[8]. However the use of sacrificial templates often involves several steps before the nanostructured PANI electrodes are ready for device assembly. Moreover, many of these electrodes require the use of insulating binders prior to device assembly [9]. A straightforward, single-step technique for fabricating nanostructured PANI electrodes is therefore desirable.

Our work demonstrates a facile, single-step fabrication of freestanding, high-purity PANI and PANI/CNT nanofiber mats via electrospinning. We used high molecular weight PEO as a carrier polymer, which enabled successful electrospinning of PANI/PEO and PANI/CNT/PEO nanofibers with PEO composition of less than 10wt%. Composition of CNTs in the PANI/CNTs composite was ~12 wt%. Presence of CNTs enhanced the electrical conductivity and therefore the electrochemical performance of the nanofiber electrode. Specific capacitance over 300 F g^{-1} and 375 F g^{-1} were achieved for PANI/PEO and PANI/CNT/PEO, respectively, at a current density of 0.50 A g^{-1}. Both nanofiber mats showed high rate capability and good stability after 1000 cycles. To the best of our knowledge, no previous work has reported the preparation of PANI/CNT supercapacitor electrodes via the electrospinning technique. The bulk of the PANI/CNT research in the literature reports electrochemical or chemical deposition of polyaniline on CNTs as a template, often resulting in a high CNT fraction or dead weight in the final electrode.

Experimental Section

Materials. Polyaniline emeraldine base (PANI, M_w = 100,000 g/mole), polyethylene oxide (PEO, M_w = 8,000,000 g/mole), 10-camphorsulfonic acid (HCSA), and chloroform (>99.5%) were purchased from Sigma Aldrich. Carbon nanotubes (CNTs) (multi-walled, 8-15 nm diameter) were purchased from NanoLab Inc.

Fabrication of nanofibers based electrodes. Chloroform, PANI, and HCSA were mixed simultaneously and stirred overnight. The PANI concentration in chloroform was 10mg/ml while HCSA was added to the solution in a weight ratio of 1:1.29 PANI: HCSA. After overnight stirring, solution was sonicated for 15 mins, and PEO was added. Solution was further stirred after addition of PEO for ~6 left and then left to sit in the fume hood overnight. For the PANI/CNT solutions, CNTs were ultrasonicated in chloroform for 2 hours prior to addition of PANI and PEO. PANI concentration in chloroform and PANI/PEO weight ratio were kept unaltered in the CNT based system. Electrospinning parameters were: relative humidity <20%, tip of the needle to collector distance ~8", applied voltage 5-6 kV and flow rate of 0.6 mL hr^{-1} . 22 gauge needle from Hamilton Co. was used for electrospinning. Identical electrospinning parameters were used for the PANI/PEO and PANI/CNT/PEO solutions.

Material characterization. Scanning electron microscopy (SEM, ZEIS SUPRA 50VP) was employed to characterize the surface morphology of the electrospun fibers. The percolation of CNTs within PANI nanofibers was observed using transmission electron microscopy (TEM) of individual nanofiber strands. The specific surface area of the electrospun nanofibers was characterized using nitrogen sorption isotherms at 77 K (Autosorb-1, Quantachrome). Prior to the adsorption–desorption measurement, all samples were degassed at 60°C under vacuum for 24 h to remove impurities.

The as-made electrospun PANI/PEO and PANI/CNT/PEO nanofiber mats were punched into freestanding electrodes with a 0.50 inch diameter and tested electrochemically in a three electrode, T-type Swagelok set up. Graphite rod (diameter of 0.50 inches), Ag/AgCl and platinum mesh were used as the working, reference and counter electrodes respectively. All electrochemical testing was carried out in 1M H_2SO_4 using Gamry reference 3000 potentiostat. Electrode mass loading used for the reported data was ~2.60 mg cm^{-2}. Cyclic voltammetry (CV) experiments were conducted in the voltage range of -0.20 to 0.65 V. Cyclic charge-discharge studies were carried out in the voltage range of -0.2 to 0.60 V. Specific capacitance was calculated from CV data using the expression below.

$$C_s = \frac{1}{2}\frac{\int I\,dV}{v*m*V} \quad (1)$$

Where C_s, I, m and V represent the specific capacitance (F g^{-1}), current (A), mass of the electrode (g), and the voltage window (V), respectively. Specific capacitance from charge-discharge curves was calculated as follows:

$$C_s = \frac{\Delta t * I}{\Delta V * m} \quad (2)$$

Where C_s is the specific capacitance, I is the constant discharge current, m is the mass of the electrode, Δt is the discharge time, and ΔV is the voltage window.

Results and Discussion

Cyclic voltammetry study. Pseudocapacitive performance of the freestanding high purity electrospun PANI/PEO and PANI/CNT/PEO composite electrodes was investigated using cyclic voltammetry (CV) measurements at room temperature with a voltage window of 0.85V. Prior to CV testing, through plane EIS studies were carried out to ascertain conductivities of nanofibers mats. The electrical conductivities of PANI/PEO and PANI/CNT/PEO were established to be 0.114 and 0.154 S/cm, respectively. Both PANI/PEO and PANI/CNT/PEO exhibited a well-defined reversible pair of redox peaks in the CV curves in the voltage range of 0.10 V to 0.30 V vs. Ag/AgCl. The peaks are represented as (P_1/P_1') in figure 1 below.

These peaks are attributed to the redox transformation of polyaniline between the polaronic emaraldine state (ES) and the leucoemaraldine state (LS) [10]. Another rather diminished pair of peaks, denoted (P_2/P_2'), was observed for both systems and is believed to represent the PANI degradation by-products [11]. PANI/CNT/PEO systems showed higher redox peaks than the corresponding PANI/PEO in the CV plot. During both the forward and reverse sweeps, the conductivity of PANI/PEO system varied as the polymer transitions into different oxidation states. The polymer is at its least conductive state when in the leucoemaraldine state. This variation in conductivity of PANI may inhibit the redox reactions. Conversely, the presence of CNTs in the composite system is expected to stabilize the electrode conductivity and thereby facilitate the redox reactions, which explains the larger redox peaks observed for the PANI/CNT/PEO electrode. The PANI/CNT/PEO redox peak is marginally shifted to the right for forward scan and to the left for the reverse scan. We hypothesize this phenomenon is as a result of greater conversion of ES to LS state in the presence of CNTs. The LS state is less conductive and therefore a slight increase in resistance (electrical) is observed. The cathodic and anodic peaks show marginal shift to right and left, respectively, with increasing scan rate. The small shift suggests only a slight increase in internal resistance with increasing scan rate [12]. Both PANI/CNT/PEO and PANI/PEO at showed good specific capacitance values at 5 mVs^{-1} respectively. 67% of the specific capacitance was retained at 100 mV s^{-1} for PANI/PEO electrodes while 80% of the specific capacitance was retained for PANI/CNT/PEO indicating excellent kinetics and rate capability. Figure 2 illustrates variation in specific capacitance as a function of scan rates.

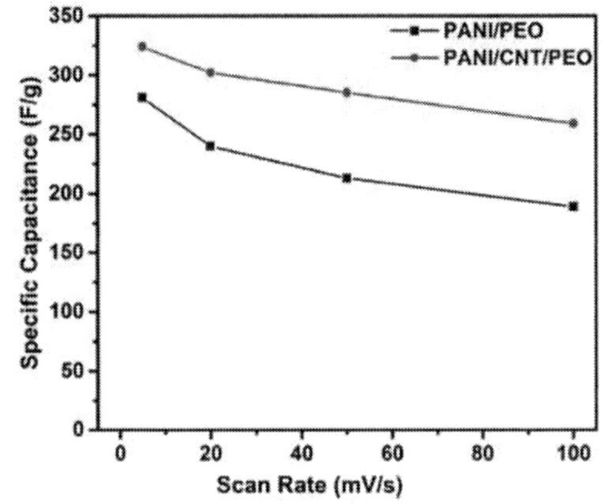

Figure 2: CV plots of PANI/PEO (a) and PANI/CNT/PEO (b) at different scan rates.

The rate capability of the fabricated systems was investigated by varying the current density from a range of 0.5 A g^{-1} to 10 A g^{-1}. Figure 4 illustrates the variation of specific capacitance as a function of current density for both PANI/PEO (black line and squares) and PANI/CNT/PEO (red line and circles). Both PANI/PEO and PANI/CNT/PEO delivered excellent highest specific capacity 0.5 A g^{-1} with good capacitance retention of 58 and 72% at 5 A g^{-1}, respectively.

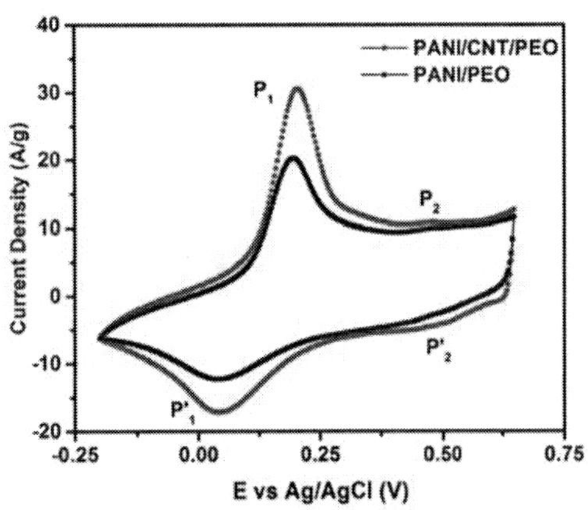

Figure 1: CV plots of PANI/PEO and PANI/CNT/PEO at 50 mV/s.

Galvanostatic charge-discharge (GCD) study of PANI/PEO and PANI/CNT/PEO nanofiber mats. GCD was studied in the potential range of -0.2V to 0.6V. These charge-discharge curves obtained were quasi-symmetrical (see figure 3) due to the redox reactions associated with transition of PANI to various oxidation states [8a].

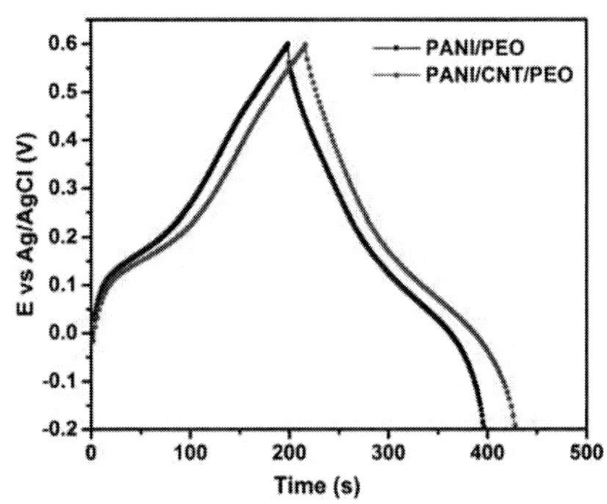

Figure 3: Charge-discharge curves for PANI/PEO and PANI/CNT/PEO at 1 A/g.

The promising electrochemical performance of our as-electrospun freestanding PANI/PEO and PANI/CNT/PEO electrodes is thought to stem from the three-dimensional non-woven fiber mat morphology, which enables quick ion transport/diffusion via inter-fiber macropores to PANI active sites. The electrode assembly process (use of PANI/binder/conductive agent slurry) in most literature works is likely to result in lack of such open, through-connected pore structure. Moreover, PANI/PEO nanofibers display some microporosity on their surface as observed in the BET data. The microporosity is possibly a result of evaporative cooling of the polymer jet due to high humidity during electrospinning. As the surface of the jet cools, water vapor from air condenses to form tiny water droplets on the fiber surface that eventually evaporate to form pores [13]. We believe that this porosity on the surface of the fibers further enhances PANI accessibility and utilization for redox reactions.

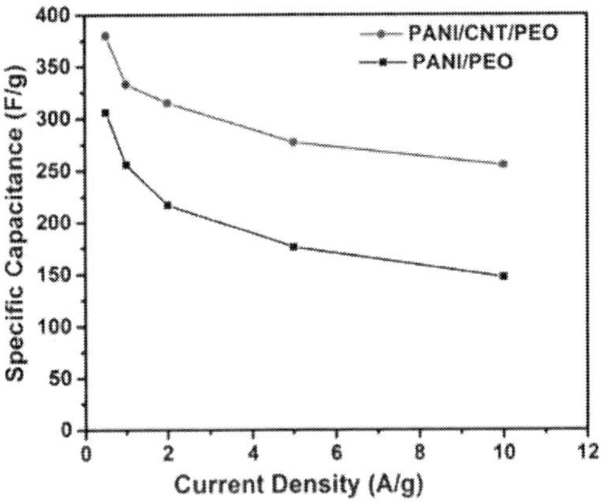

Figure 4: Cyclic charge-discharge curves of **(a)** PANI/PEO and **(b)** PANI/PEO at different current densities.

Conclusion

We have demonstrated a facile methodology to fabricate binder-free PANI and PANI/CNT nanofiber electrodes. The electrodes exhibit excellent highest specific capacitance, good rate capability and cyclability up to 1000 cycles with over 80% retention of initial specific capacitance for PANI/CNT electrode. The high electrochemical performance of the fabricated electrodes is enabled by the combination of low electrochemical impedance as well as the good inter and intra-fiber porosity, which facilitates ion diffusion to the PANI active sites. The freestanding format of the electrodes eliminates the need for electrochemically inert binders, thereby allowing us to not only minimize cell assembly procedures, but also enables us to preserve the 3D porous structure of the nanofiber mats.

Acknowledgement. We would like to thank National Science Foundation (award number: 1463170) for funding of this work. We are also grateful to Drexel University Centralized Research Facility for the use of their characterization equipment. We will also like to thank M. Boota and Prof. Y. Gogotsi at Drexel University for their help with BET surface area measurement.

Reference

This work is submitted to ACS Applied Material and Interfaces

(1) Snook, G. A.; Kao, P.; Best, A. S., Conducting-polymer-based supercapacitor devices and electrodes. *J. Power Source.*, **2011,** *196* (1), 1-12.

(2) Wei, J.; Zhang, J.; Liu, Y.; Xu, G.; Chen, Z.; Xu, Q., Controlled growth of whisker-like polyaniline on carbon nanofibers and their long cycle life for supercapacitors. *RSC Advance.*, **2013,** *3* (12), 3957.

(3) Zhou, Z.; Wu, X.-F.; Hou, H., Electrospun carbon nanofibers surface-grown with carbon nanotubes and polyaniline for use as high-performance electrode materials of supercapacitors. *RSC Advances.*, **2014,** *4* (45), 23622.

(4) Li, J.; Xie, H.; Li, Y.; Liu, J.; Li, Z., Electrochemical properties of graphene nanosheets/polyaniline nanofibers composites as electrode for supercapacitors. *J. Power Source,* **2011,** *196* (24), 10775-10781.

(5) (a) Zhou, Y.; He, B.; Zhou, W.; Huang, J.; Li, X.; Wu, B.; Li, H., Electrochemical capacitance of well-coated single-walled carbon nanotube with polyaniline composites. *Electrochem. Acta.,* **2004,** *49* (2), 257-262; (b) Gupta, V.; Miura, N., Polyaniline/single-wall carbon nanotube (PANI/SWCNT) composites for high performance supercapacitors. *Electrochimica Acta* **2006,** *52* (4), 1721-1726; (c) Sivakkumar, S. R.; Kim, W. J.; Choi, J.; MacFarlane, D. R.; Forsyth, M.; Kim, D.-W., Electrochemical performance of polyaniline nanofibres and polyaniline/multi-walled carbon nanotube composite as an electrode material for aqueous redox supercapacitors. *J. Power Sources* **2007,** *171* (2), 1062-1068.

(6) Mujawar, S. H.; Ambade, S. B.; Battumur, T.; Ambade, R. B.; Lee, S., Electropolymerization of polyaniline on titanium oxide nanotubes for supercapacitor application. *Electrochemical acta.* **2011,** *56*, 4462-4466.

(7) Wang, K.; Huang, J.; Wei, Z., Conducting Polyaniline Nanowire Arrays for High Performance Supercapacitors. *J. Phys. Chem. C.* **2010,** *114*, 8062–8067.

(8) (a) Chen, W.; Rakhi, R. B.; Alshareef, H. N., Facile synthesis of polyaniline nanotubes using reactive oxide templates for high energy density pseudocapacitors. *J. Mater. Chem. A.* **2013,** *1* (10), 3315; (b) Miao, Y. E.; Fan, W.; Chen, D.; Liu, T., High-performance supercapacitors based on hollow polyaniline nanofibers by electrospinning. *ACS. Appl. Mater. Interfaces.* **2013,** *5* (10), 4423-8.

(9) Ran, F.; Tan, Y.-t.; Liu, J.; zhao, L.; Kong, L.-b.; Luo, Y.-c.; Kang, L., Preparation of hierarchical polyaniline nanotubes based on self-assembly and its electrochemical capacitance. *Polyme. Adv. Technologies.* **2012,** *23* (9), 1297-1301.

(10) Zhang, K.; Zhang, L. L.; Zhao, X. S.; Wu, J., Graphene/Polyaniline Nanofiber Composites as Supercapacitor Electrodes. *Chem. Materials.* **2010,** *22* (4), 1392-1401.

(11) Duic, L.; Mandic, Z.; Kovac, S., Polymer-Dimer distribution in the electrochemical synthesis of polyaniline. *Electrochemical acta* **1995,** *40*, 1681-1688.

(12) Fan, W.; Zhang, C.; Tjiu, W. W.; Pramoda, K. P.; He, C.; Liu, T., Graphene-wrapped polyaniline hollow spheres as novel hybrid electrode materials for supercapacitor applications. *ACS applied materials & interfaces* **2013,** *5* (8), 3382-91.

(13) Greiner, A.; Wendorff, J. H., Electrospinning: A Fascinating Method for the Preparation of Ultrathin Fibers. *Angew. Chem. Int. Ed.* **2007,** *46*, 5670 – 5703.

ELECTROSPUN POLY(ACRYLONITRILE-*CO*-ITACONIC ACID) AS POROUS CARBON PRECURSOR FOR HIGH PERFORMANCE SUPERCAPACITOR: STUDY OF *INSITU* POROGEN ACTIVITY OF ITACONIC ACID IN COPOLYMER

Nimali C. Abeykoon, Samsuddin F. Mahmood and John P. Ferraris

Department of Chemistry and Biochemistry, The University of Texas at Dallas, 800 W. Campbell Road, Richardson, Texas -75080.
nxa124430@utdallas.edu

Introduction

Electrochemical capacitors (or supercapacitors) are energy storage devices that are becoming more prominent in applications that require high charge and discharge rates such as consumer electronics, tramways, and energy harvesting systems. Supercapacitors can deliver greater power densities than lithium-ion batteries, though they are unable to store as much energy as batteries.[1,2] Due to their applicability in a large number of energy-related technologies, improving the energy storage capabilities of supercapacitors is of great importance for technological advancement.

Utilization of *porous* carbon electrodes is one approach to increase the capacitance of EDLCs.[3] The porosity of the electrodes greatly increases their surface area, and, because capacitance is directly related to surface area, greatly increases the capacitance of the device.[4] Conventional activation methods such as chemical (KOH, $ZnCl_2$, H_3PO_4, etc.) and physical activation (CO_2, H_2O, air, etc.) have been used to produce porous materials with controlled surface area and pore size distribution from precursor materials for effective use as electrodes for supercapacitors.[5] An alternative method is the introduction of thermally labile porogens to produce porous fibers when the porogen decomposes. One of the most common carbon materials used for EDLC electrodes is derived from polyacrylonitrile (PAN). However, additional activation methods or porogens are required for the PAN fibers to produce high surface area carbon. Also, the stabilization of PAN, commonly performed at 280 °C in air, is the most crucial step when making carbon fibers. Introduction of the acidic comonomer, itaconic acid (IA), during polymerization of PAN yields the copolymer poly(acrylonitrile-co-itaconic acid) (P(AN-co-IA)). The presence of IA groups in P(AN-co-IA) causes it to initiate the cyclization of nitrile groups at a lower temperature (220 °C) than PAN, making it suitable for most porogens, which decompose at lower temperatures. Also, P(AN-co-IA) has increased spinnability as well, which favors enhancement of the carbon yield and the performance of resulting carbon fibers.[6] The use of a precursor which contains functionalities that can act as an *in-situ* porogen is useful to avoid unnecessary activation steps when fabricating carbon nanofibers (CNFs). Interestingly, the COOH group in P(AN-co-IA) can act as an *in situ* porogen to release CO_2 during carbonization to produce porous fiber.

In this study, the copolymer, P(AN-co-IA) was synthesized with different amounts of IA (2%, 5% and 8%), then electrospun into mats and finally carbonized at 1000 ^{0}C to afford porous CNFs that were characterized by SEM, surface area analysis and TGA-MS before being incorporated as electrodes in coin cell supercapacitors.

Experimental

Preparation of CNF and Activated CNFs (ACNFs). The synthesized P(AN-co-IA) with different compositions of IA was dissolved in N,N dimethylformamide (DMF) (10 wt.%) at 50 ^{0}C for 2h and then stirred for 24 hours at room temperature before electrospinning. The monomer ratio of AN/IA was selected as 98/2, 95/5 and 92/8 by weight and denoted as P(AN-co-IA)-2, P(AN-co-IA)-5 and P(AN-co-IA)-8 respectively. Also a 10 wt.% PAN solution was prepared by dissolving PAN in DMF according to the same procedure. The prepared solutions were loaded into the syringe and electrospun in to the nano fibers using a custom-built electrospinner by applying voltage of 18 kV between the needle tip attached to the syringe and a grounded drum collector wrapped with an aluminum foil rotating at speed of 300 rpm. The spinning solution was delivered a constant rate of 0.5 ml/h to the tip. The resulting free standing nanofiber mats were peeled out from the aluminum foil and stabilized at 220 ^{0}C for 3h under air with a heating rate of 2 ^{0}C/min. ACNFs were prepared by heating the stabilized mats at a rate of 5 °C/min under Helium gas flow from room temperature to 1000 °C and then kept at that temperature under CO_2 flow for 1 h. After that, the samples were cooled under Helium gas flow. CNFs were prepared according to the same procedure without purging with CO_2 at 1000 °C.

Characterization Methods. Both ^{1}H & ^{13}C NMR spectra of the synthesized copolymers were obtained on a Bruker AVANCE III™ (500 MHz) NMR spectrometer using DMSO-d6 as the solvent. The molecular weights and molecular weight distributions of the copolymers were determined via gel permeation chromatography (Viscotek GPCmax, VE2001) using two ViscoGEL columns in series (I-MBHMW 3078, Viscotek) and a Triple Array Detector (Module TDA 320, Viscotek). DMF was used as the eluent. Fourier Transform Infrared (FTIR) transmission spectra were recorded on a Nicolet 360 FTIR spectrophotometer. Thermogravimetric behavior of sample was carried out under nitrogen flow by heating from room temperature to 1000 °C (10 °C /min heating rate) on a TA Instruments SDT Q600 Analyzer. Morphological studies of samples were performed with a Zeiss-LEO (Model 1530) Scanning Electron Microscope (SEM) equipped with field emission gun accelerating at 10 keV. All the as-spun and stabilized samples were sputter coated with gold before SEM imaging whereas SEMs of carbonized samples were collected without sputter coating. The surface area of carbon fibers was measured using the Brunauer– Emmett–Teller (BET) surface area analyzer (Autosorb 1, Quantachrome Corporation) at liquid nitrogen temperature (77 K). The pore size distribution (PSD) was obtained by the Density Functional Theory (DFT). Raman spectroscopy (JobinYvon HORIBA LabRam Raman spectrometer equipped with a 633 nm He–Ne laser) was employed to characterize carbonized fiber samples.

Electrochemical tests. The electrochemical measurements were performed in a two-electrode configuration. For this, the CR2032 coin cell packaging was used. Two identical carbon electrodes were separated by Gore™ PTFE separator. Carbon coated aluminum was used as the current collectors and stainless steel spacers were placed on each side of current collector when packaging. The electrolyte (EMIFSI) was introduced to each electrode and sealed in coin cell using crimper inside the glove box. Electrochemical tests, including cyclic voltammetry (CV) and constant current charge/discharge (CDC) were carried out on Arbin Supercapacitor Testing Station (SCTS). CVs were obtained in the range of –2 to 2 V and CDC were recorded between 3.5 V and 0 V. The cells were characterized further by using electrochemical impedance spectroscopy (EIS) at 0 V DC bias in the frequency range of 100 kHz to 10 mHz using a Princeton Applied Research potentiostat/galvanostat (model 2273 A) in a two-electrode configuration.

Results and Discussion

Figure 1 shows the morphology of CNFs obtained from the PAN and P(AN-co-IA)-2 precursors. All the precursor polymers show uniform fibers without beading with diameters of 692±98 nm for PAN and 290±31 nm for the P(AN-co-IA)-2, respectively. Incorporation of the acidic comonomer causes a decrease in the diameter of the fiber due to dissociation of the carboxylic acid group in the copolymer, which results in increased solution ionic conductivity. After the thermal treatments, the carbonized fibers retained their fibrous morphology without collapsing. All the CNFs demonstrate very uniform diameters.

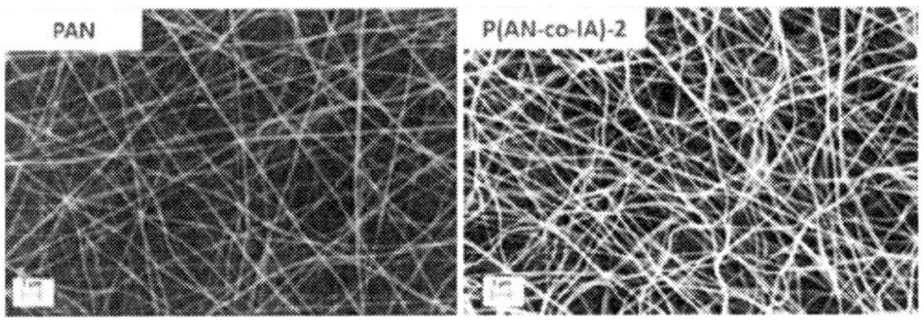

Figure 1. SEM images of CNFs of PAN and P(AN-co-IA)-2.

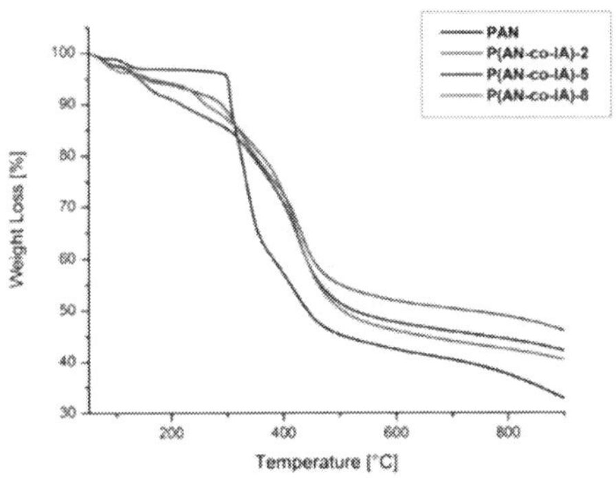

Figure 2. TGA of CNFs derived from PAN and P(AN-co-IA)-2.

Based on the TGA studies (Figure 2), IA provides a catalytic effect on the stabilization process which decreases the initiation cyclization temperature compared to the PAN homopolymer.[7,8] The stabilization temperature decreased when increasing acid content. Also, all the compositions of P(AN-co-IA) have higher carbon yields at high temperature than PAN.

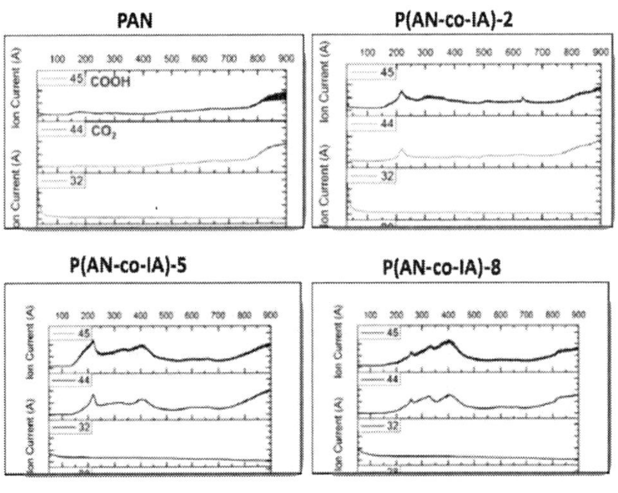

Figure 3. TGA-MS of PAN and P(AN-co-IA) precursors.

The ability of the COOH groups in P(AN-co-IA) to act as an *in situ* porogen releasing CO_2 during carbonization was confirmed by using TGA-MS (Figure 3), which clearly showed that P(AN-co-IA) starts releasing CO_2 starting around 220 ^{0}C upon stabilization whereas PAN doesn't release CO_2 in that temperature range. The evolved CO_2 causes to an *in situ* activation to increase surface area leading to high porosity (Table 1). Furthermore, the ion current for CO_2 increases with increasing IA content in the precursor.

Table 1. Surface Area and Pore Volume of PAN and P(AN-co-IA) Based Carbon Nanofibers

Precursor	Surface area (m^2/g)	Total pore Volume (cm^3/g)	$V_{micropore}$ (cm^3/g)	$V_{mesopore}$ (cm^3/g)
PAN	6	0.0129	0.0053	0.0076
P(AN-co-IA)-2	976	0.3871	0.3304	0.0567

Table 1 shows that *in situ* activation in the IA copolymer affords an increase in the surface area of the fibers. Thus CNFs prepared from P(AN-co-IA)-2 exhibit surface areas of 976 m^2/g whereas the PAN CNFs, exhibit 6 m^2/g.

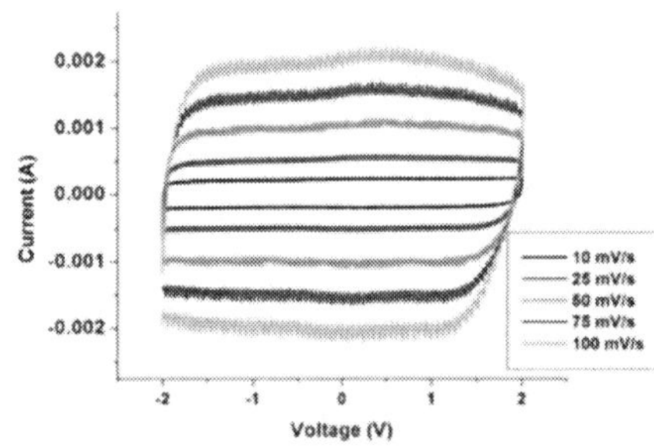

Figure 4. CV curves of P(AN-co-IA)-2 without any activation at different scan rates.

Carbonized P(AN-co-IA) without any activation shows rectangular shaped CVs (Figure 4) indicating the ideal capacitive character of the EDLC as compared to carbonized but unactivated PAN which failed to produce significant currents. The specific capacitance values obtained for the P(AN-co-IA) are higher than the PAN at all scan rates because of their higher surface area as confirmed by surface area analysis. Furthermore, the capacitance increases with increasing IA content as shown in table 2. The highest specific capacitance of P(AN-co-IA)-8 at 10 mV/s was calculated to be 92.7 F/g compared to 1.9 F/g for the PAN.

Table 2. Specific Capacitance of the CNFs Derived From the PAN and the P(AN-co-IA) at Five Different Scan Rates

Scan rate	Specific Capacitance (F/g)		
(mV/s)	PAN	P(AN-co-IA)-2	P(AN-co-IA)-8
10	1.9	64.2	92.7
25	1.8	58.9	84.1
50	1.6	53.6	78.2
75	1.3	48.5	68.5
100	0.4	46.2	66.3

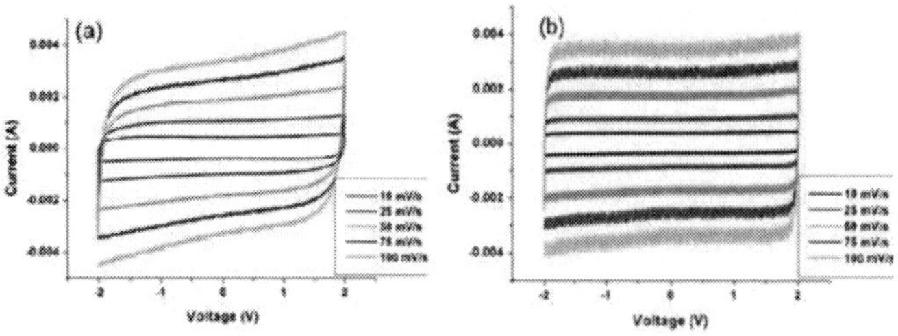

Figure 5. CV curves of (a) PAN and (b) P(AN-co-IA)-2 with CO_2 activation at different scan rate.

When these carbonized fibers were CO_2 activated P(AN-co-IA) showed very box like CV (Figure 5) even at higher current densities when compared with PAN. The rectangular shapes of CV curves were retained without dramatic distortions as the scan rate was increased from 25 mV/s to 100 mV/s, showing fast charge transfer at the electrode/electrolyte interfaces.

Table 3. Specific Capacitance of the CO_2 Activated CNFs Derived From the PAN and the P(AN-co-IA) at Five Different Scan Rates

Scan rate (mV/s)	Specific Capacitance (F/g)	
	PAN	P(AN-co-IA) -2
10	91.2	109.4
25	89.3	108.9
50	87.9	108.6
75	83.3	108.8
100	78.4	108.8

As shown in the Ragone plot (Figure 6) a very small drop in the energy density of the electrodes from P(AN-co-IA) was observed at high power density. Energy density of the electrodes from PAN drops fast with increasing current density from 1 A/g to 6 A/g . The energy and power densities of the supercapacitors assembled with CO_2 activated nanofibers at a current discharge density of 1 A/g and 10 A/g were listed in the table 4. The energy density calculated at 1 A/g was 53.1 Wh/kg for P(AN-co-IA)-2 compared to that of PAN CNFs (42.2 Wh/kg) at similar power densities.

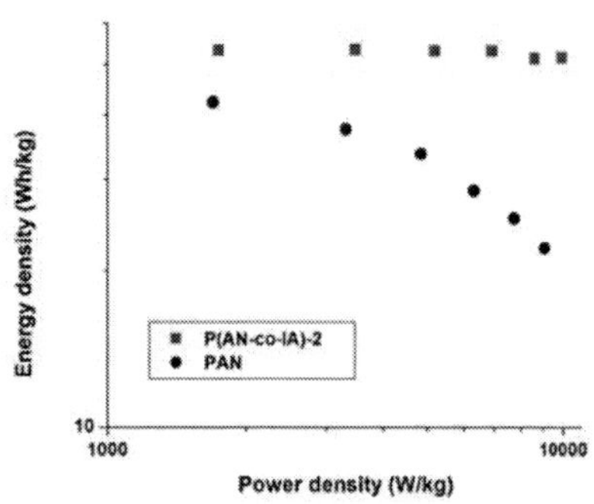

Figure 6. Ragone plots.

Table 4. Energy Densities of the CO_2 Activated CNFs Derived From the PAN and the P(AN-co-IA)

precursor	Current Density (A/g)	Energy density (Wh/ Kg)
PAN	1	42.2
	10	22.1
P(AN-co-IA)-2	1	53.1
	10	51.4

Conclusions

P(AN-co-IA) copolymers were synthesized to develop new precursor polymers for high performance carbon fibers as electrode materials for supercapacitors. Binderless flexible and free-standing carbon nanofibers were prepared by thermal treatment (and subsequent CO_2 activation) of the nanofibers mats obtained by electrospinning the copolymers. The COOH groups in P(AN-co-IA) to act as an *in situ* porogen from release of CO_2 during the carbonization process increased the surface area of the CNFs without any extra activation steps. The highest capacitance obtained for the P(AN-co-IA)-8 was 92.7 F/g at 10 mV/s. This work highlights the potential applicability of P(AN-co-IA) to produce electrode materials without any activation for high performance energy storage systems.

Acknowledgement. We acknowledge the National Science Foundation (Grant no. IIP-1127564) for financial support and Grant no. CHE-1126177 for the purchase of the Bruker AVANCE III™ 500 MHz NMR.

References

(1) Winter, M.; Brodd, R. J.; *Chem. Rev.*, **2004**, *104*, 4245–4270.
(2) Burke, A.; *J. Power Sources*, **2000**, *91*(1), 37-50.
(3) Largeot, C.; Portet, C.; Chmiola, J.; Taberna, P.-L.; Gogotsi Y.; Simon, P.; *J. Am. Chem. Soc.*, **2008**, *130*, 2730–2731.
(4) Vatamanu, J.; Bedrov, D.; *J. Phys. Chem. Lett.*, **2015**, *6*, 3594–3609.
(5) Xue, R.; Yan, J.; Liu, X.; Tian, Y.; Yi, B.; *J Appl Electrochem* , **2011**, *41*, 1357–1366.
(6) Uyen, N.; Thai, N.; Hong S. C.; *Macromolecules*, **2013**, *46*, 5882–5889.
(7) Ouyang, Q.; Cheng, L.; Wang, H.; Li K. *Polym. Degrad. Stab.*, **2008**, *93*, 1415–1421.
(8) Sen, S.; Sarac, A.S.; *Journal of Advances in Chemistry*, **2014**, *6* (2), 958-981.

2D CARBIDES AND NITRIDES OF TRANSITION METALS (MXENES): SYNTHESIS, STRUCTURE AND ENERGY STORAGE APPLICATIONS

Yury Gogotsi

Department of Materials Science and Engineering, and A. J. Drexel Nanomaterials Institute, Drexel University, Philadelphia, PA 19104, USA; nano.materials.drexel.edu

Two-dimensional (2D) solids – the thinnest materials available to us – offer unique properties and a potential path to device miniaturization. The most famous example is graphene, which is an atomically thin layer of carbon atoms bonded together in-plane with sp^2 bonds. Recently, an entirely new family of 2D solids – transition metal carbides (Ti_2C, Ti_3C_2, Nb_4C_3, etc.) and carbonitrides – was discovered by Drexel University scientists [1,2]. Selective etching of the A-group element from a MAX phase results in formation of 2D $M_{n+1}X_n$ solids, labeled "MXene". 18 different carbides and carbonitrides have been reported to date [2-5]. Structure and properties of numerous MXenes have been predicted by the density functional theory, showing that MXenes can be metallic or semiconducting, depending on their surface termination. Their elastic constants along the basal plane are predicted to be higher than that of the binary carbides. Oxygen or OH terminated MXenes are hydrophilic, but electrically conductive. Hydrazine, urea and other polar organic molecules can intercalate MXenes leading to an increase of the *c* lattice parameter of MXenes [3]. When dimethyl sulfoxide was intercalated into Ti_3C_2, followed by sonication in water, a stable colloidal solution of single- and few-layer flakes was produced. One of the many potential applications for 2D Ti_3C_2 is in electrical energy storage devices, such as batteries, Li-ion capacitors and supercapacitors [3-5]. Cations ranging from Na^+ to Mg^{2+} and Al^{3+} intercalate MXenes. Ti_3C_2 paper electrodes, produced by vacuum assisted filtration of an aqueous dispersion of delaminated Ti_3C_2, show a higher capacity than graphite anodes and also can be charged/discharged at significantly higher rates. They also demonstrate very high intercalation capacitance (up to 900 F/cm^3) in aqueous electrolytes [4].

References

(1) M. Naguib, et al, *Advanced Materials*, **23** (37), 4207-4331 (2011)
(2) M. Naguib, et al, *ACS Nano* **6** (2) 1322–1331 (2012)
(3) O. Mashtalir, et al, *Nature Communication*, **4**, 1716 (2013)
(4) M M. Ghidiu, *Nature*, **516**, 78–81 (2014)
(5) M. Naguib, Y. Gogotsi, *Accounts of Chemical Research*, **48** (1), 128-135 (2015)

HIGH-PERFORMANCE $NiCo_2S_4$@$Ni(OH)_2$ CORE-SHELL HYBRID NANOSHEET ARRAYS AS ADVANCED ELECTRODES FOR HYBRID SUPERCAPACITORS

Yefeng Yang[a,b], Ding Cheng[a], and Jie Xiong[a]*

a Department of Materials Engineering, MOE Key Laboratory of Advanced Textile Materials and Manufacturing Technology, Zhejiang Sci-Tech University, Hangzhou 310018, P. R. China
b State Key Lab of Silicon Materials, Zhejiang University, Hangzhou 310027, P. R. China
Email address: yangyf@zstu.edu.cn

Since the ever-increasing demand for energy and power in our daily life, it's been highly urgent to develop new energy storage devices. Hybrid supercapacitors (HSCs), consisting of a battery-type electrode associated with Faradaic redox reactions and an electrical double-layer capacitor (EDLC) electrode,[1] have stimulated great interests owing to their appealing advantages including fast charge-discharge capability, long life-span, high safety, much higher energy density than conventional capacitors and higher power density than batteries. Therefore, various battery-type electrode materials, including transition metal oxides,[2] hydroxides[3] and sulfides,[4] have been widely explored for advanced electrochemical electrodes.
Herein, we report a novel battery-type electrode of 3D hierarchical $NiCo_2S_4$@$Ni(OH)_2$ hybrid nanosheet arrays on Ni foam for HSCs. The interconnected $NiCo_2S_4$ nanosheets are utilized as ideal conductive scaffolds, while ultrathin $Ni(OH)_2$ shell nanosheets are controllably deposited by electrodeposition. The optimized $NiCo_2S_4$@$Ni(OH)_2$ hybrid electrode demonstrates a high areal capacity of 680 μAh/cm^2 (240.3 mAh/g) at 5 mA/cm^2, with excellent rate performance (94.9% with current density increased to 100 mA/cm^2) and cycling stability (81.4% over 2000 cycles at 40 mA/cm^2). Moreover, the HSC device assembled with $NiCo_2S_4$@$Ni(OH)_2$ and activated carbon (AC) as the positive and negative electrodes displays high energy density (53.3 Wh/kg at 290 W/kg), high power density (6420 W/kg at 32.1 Wh/kg), and excellent cycling stability (98.8% over 2000 cycles at 20 mA/cm^2), suggesting great potential of this hybrid as promising electrode materials for high-performance HSCs.

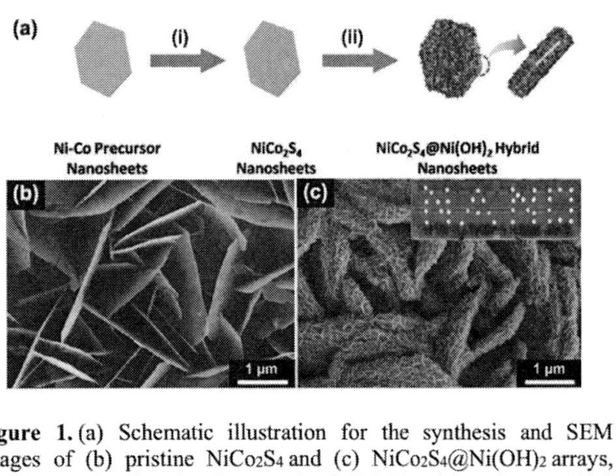

Figure 1. (a) Schematic illustration for the synthesis and SEM images of (b) pristine $NiCo_2S_4$ and (c) $NiCo_2S_4$@$Ni(OH)_2$ arrays, inset in (c) showing the photograph of 40 LEDs driven by two HSC devices connected in series.

Acknowledgment. **This work was financially supported by the National Natural Science Foundation of China (No. 51402262), Visiting Scholar Foundation of State Key Laboratory of Silicon Materials, Zhejiang University (No. SKL2014-6).**

References

(1) Ke, Q.; Zheng, M.; Liu, H.; Guan, C.; Mao, L.; Wang, J. 3D TiO_2@$Ni(OH)_2$ Core-Shell Arrays with Tunable Nanostructure for Hybrid Supercapacitor Application. *Sci. Rep.* **2015**, *5*, 13940.
(2) Cheng, D.; Yang, Y.; Xie, J.; Fang, C.; Zhang, G.; Xiong, J. Hierarchical $NiCo_2O_4$@$NiMoO_4$ Core-Shell Hybrid Nanowire/Nanosheet Arrays for High-Performance Pseudocapacitors, *J. Mater. Chem. A*, **2015**, *3*, 14348.
(3) Wang, H.; Casalongue, H.; Liang, Y.; Dai, H. $Ni(OH)_2$ Nanoplates Grown on Graphene as Advanced Electrochemical Pseudocapacitor Materials, *J. Am. Chem. Soc.*, **2010**, *132*, 7472.
(4) Chen, W.; Xia, C.; Alshareef, H. One-Step Electrodeposition Nickel Cobalt Sulfide Nanosheet Arrays for High-performance Asymmetric Supercapacitors, *ACS Nano*, **2014**, *8*, 9531.

FROM ELECTRIC DOUBLE-LAYER TO PSEUDOCAPACITANCE: A JOINT DFT APPROACH

De-en Jiang

Department of Chemistry, University of California, Riverside

Electric double-layer capacitors and pseudocapacitors are two common types of supercapacitors. Modeling a charged electrode from first principles is still challenging, especially if one also wants to incorporate the electrolyte in the system. In this talk, we apply a novel method called the joint density functional theory that treats the electrode at the electronic DFT level and the electrolyte classically. This approach allows to examine in detail the role of the electrode chemistry in influencing capacitance. With approach, we quantified the dielectric contribution to few-layer graphene electrodes and the combined EDL/redox charging of RuO_2, a prototypical pseudocapacitor.

OVERCOMING MATERIALS CHALLENGES IN ALL-SOLID-STATE LI-ION BATTERIES: INSIGHTS FROM ATOMISTIC MODELING

Yifei Mo

Department of Materials Science and Engineering, University of Maryland, College park, MD
yfmo@umd.edu

The design and discovery of new materials have been largely pursued through a trial-and-error manner based on human intuition and serendipity. This traditional materials design process is time consuming and labor intensive, significantly impeding the research and development of advanced materials critical for modern societal needs. Computational techniques based on first principles, which accurately predict materials properties with little experimental input, may revolutionize and accelerate the materials design and discovery. In this presentation, I will share our success stories of leveraging first principles computation techniques in resolving a number of key material challenges in solid-state Li-ion batteries, a new battery technology with intrinsic safety, high energy density, and enhanced cyclability. I will first present how we use first principles computation methods to design new solid electrolyte materials with enhanced properties, which are verified by multiple experimental studies. In addition, I will share our recent development in computational techniques to investigate the interfacial stability in solid-state Li-ion batteries. These techniques are applied to investigate key problems, such as interfacial degradation and high interfacial resistance, at the electrolyte-electrode interfaces. Using these first principles computation, we found that most solid electrolyte materials are not stable against Li metal or high voltages, and that the formation of solid-electrolyte-interphases are ubiquitous in all-solid-state Li-ion batteries. The properties of these solid-electrolyte-interphases would significantly affect the electrochemical performance of the all-solid-state Li-ion batteries. The mechanisms of artificial interfacial layers to improve the interfacial properties are also revealed by our computation. On the basis of unique insights from the modeling results, general design principles for the solid electrolyte materials and electrolyte-electrode interfaces will be proposed to address current challenges in all-solid-state Li-ion batteries.

SMART ELECTROLYTE FOR RECHARGEABLE BATTERIES

Ping Liu

Department of Nanoengineering, University of California, San Diego
piliu@ucsd.edu

Battery electrolytes transport ions and provide the chemical environment for reversible electrochemical reactions. In today's lithium-ion batteries, they also define the abuse tolerance of the system due to the flammable nature of the electrolytes. In general, once a battery electrolyte is selected, its properties are defined with some general trends common to most electrolytes. For example, conductivity tends to increase with temperature.

In this talk, we will analyze recent trends in battery electrolyte research, most noticeably novel aqueous electrolytes and solid electrolytes. The widening stability window of aqueous electrolytes can offer batteries with potentially competitive energy densities to lithium-ion. In the solid electrolyte area, new highly conducting phases continue to emerge. However, significant device integration challenges remain.

We will then discuss possibilities of smart electrolytes that can potentially address some of the challenges of today's electrolytes. The properties of these electrolytes can be independently manipulated inside batteries. Examples of chemical designs that can enable these smart electrolytes will be provided. Finally, potential impact of these smart electrolytes will be discussed in terms of battery energy density, life, and abuse tolerance.

CONFINED POLYMERIZATION OF PYRROLE BETWEEN 2D TITANIUM CARBIDE (MXene) FOR HIGH PERFORMANCE PSEUDOCAPACITIVE ELECTRODES

Muhammad Boota, Babak Anasori, Meng-Qiang Zhao, Michel W Barsoum, Yury Gogotsi

Department of Materials Science and Engineering and A.J. Drexel Nanomaterials Institute, Drexel University, Philadelphia, PA 19104, USA
Email: boota@drexel.edu

Redox-active organic materials are generally inexpensive, renewable and a sustainable option for small and large scale energy storage systems.[1] The key challenge which hinder their usage in commercial devices is poor cycling performance upon moderate cycling, mainly due to their low conductivity. One way to address this issue is to mix them with conductive carbon materials.[2] Recently discovered two-dimensional carbides and carbonitirides called MXenes, which exhibit metallic conductivity, may become a suitable material choice to replace carbon in pseudocapacitive electrodes to improve the electrochemical performance of organic materials.[3]

Conducting polymers are well-known organic pseudocapacitive materials.[4] They are generally synthesized by mixing monomer with excessive amounts of oxidant in acidic media. The oxidants are explosive, hazardous and also add extra cost. Here, we report a unique approach to perform confined polymerization of the conducting polymer monomer (pyrrole) between the layers of 2D titanium carbide without any oxidant at room temperature. The intercalation, chain alignment and polymerization of the pyrrole monomer was confirmed by various spectroscopic and microscopic techniques. The optimized electrodes exhibited a high gravimetric (416 F/g) and volumetric (1000 F/cm^3) capacitance at 5 mV/s. Furthermore, confined polymerization prevented polymer degradation and led to an excellent capacitance retention of 92% after 25,000 cycles at 100 mV/s.[5]

References

(1) T. Janoschka et al., *Nature* (2015)
(2) M. Boota et al., *ChemSusChem*, **8**, 3576–3581 (2015)
(3) M. Ghidiu, M. R. Lukatskaya, M.-Q. Zhao, Y. Gogotsi, and M. W. Barsoum, *Nature*, **516**, 78–81 (2014)
(4) D. Vonlanthen, P. Lazarev, K. a. See, F. Wudl, and A. J. Heeger, *Adv. Mater.*, **26**, 5095–5100 (2014).
(5) M. Boota et al., *Adv. Mater.*, **28**, 1517–22 (2015)

POTASSIUM-ION OXYGEN BATTERY BASED ON A HIGH CAPACITY ANTIMONY ANODE

*William D. McCulloch, Xiaodi Ren, Mingzhe Yu, Zhongjie Huang, and Yiying Wu**

Department of Chemistry and Biochemistry, The Ohio State University, 100 West 18th Avenue, Columbus, Ohio 43210, United States
wu@chemistry.ohio-state.edu.

Introduction

With the global energy demand expected to double by 2050, the development of an energy storage solution becomes vital.[1]. In 2013, we reported on the first K−O2 battery based on a KO2 discharge product and only a small polarization upon charging (<50 mV) was observed.[2,3] The stable superoxide allows the battery to operate through a reversible one-electron process and eliminates the need for a catalyst or redox mediators. Herein, we report the first K-ion oxygen battery utilizing a K−Sb alloy anode. From the binary phase diagram of antimony and potassium, several alloy phases may be adopted.[4] There exist four K_xSb alloys, where x = 0.5, 1, 1.25, or 3. In this work we demonstrate a nanosized antimony−carbon composite (Sb-C) can alloy with potassium forming a cubic phase K_3Sb alloy (reversible capacity of 650 mAh/g, ~ 98% of theoretical capacity). The antimony composite can also be cycled for over 50 cycles at a capacity of 250 mAh/g. Finally, we explore the application of this anode material in the form of a $K_3Sb−O_2$ cell which displays relatively high operating voltages, low over-potentials, increased safety, and interfacial stability.

Experimental

A mixture of Super P carbon black and 100 mesh antimony powder was weighed out in a 70:30 (w/w) ratio to a total of 3 g. The mixture was ball milled at 2400 rpm for 16 h. K−Sb half-cells were fabricated using coin cells in the following order: stainless steel cathode case, Sb−C electrode, glassy fiber, potassium metal, stainless steel current collector, stainless steel wave washer, and anode case. The glassy fiber separator was saturated with electrolyte consisting of 1.0 M KPF6 (Sigma- Aldrich) in propylene carbonate/ethylene carbonate, PC/EC 1:1 (w/ w; Sigma-Aldrich). The coin cells were crimped together under 800 psi. Galvanostatic discharge/charge curves were collected with constant current of 35 $mA/g_{(antimony)}$ in a potential range of 2.0 to 0.05 V vs K+/K.

The full K_3Sb-O_2 swagelok cell was constructed in the following manner. First a half-cell (coin cell) consisting of the Sb−C composite electrode was prepared according to the previous procedure. This cell was allowed to fully discharge to 0.05 V, effectively forming the K_3Sb alloy. At this time, the Sb−C electrode was removed from the coin cell and washed with DME to remove residual carbonate electrolyte. This electrode (K_3Sb alloy) was used as the anode for the oxygen cell. The full cell was assembled in the following manner: stainless steel base, K_3Sb alloy electrode, glassy fiber separator, Celguard separator, Nafion 211, glassy fiber separator carbon paper, stainless steel current collector, and two stainless steel rings for a gas diffusion layer. The separators and Nafion 211 membrane were saturated with electrolyte (0.5 M KPF_6 in 1,2-dimethoxyethane, DME)

Results and Discussion

In order to quantify the potassium storage capability of the Sb−C composite, galvanostatic discharge/charge curves for the first ten cycles of the K−Sb half-cell were collected (Figure 1). Here, the discharge/charge processes correspond to potassium alloying/dealloying reactions, respectively. During the first discharge an irreversible capacity of 230 mAh/g was recorded which is a common feature generally attributed to formation of a solid electrolyte interface (SEI) on the surface of the electrodes. A reversible capacity of 650 mAh/g was observed upon subsequent cycling indicating the formation of a K_3Sb antimonide is responsible for the high reversible capacity.

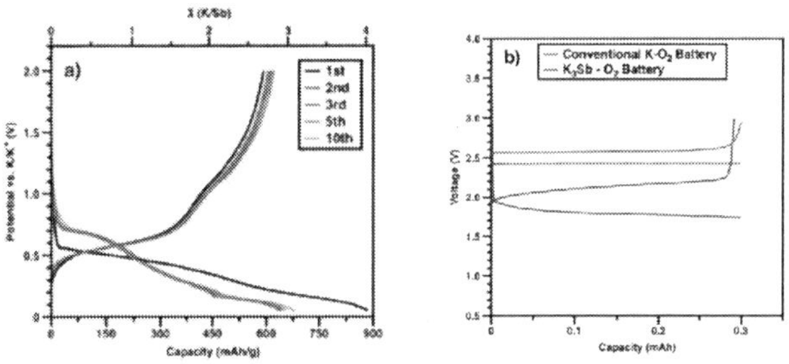

Figure 1. (a) Galvanostatic discharge/charge curves for Sb−C nanocomposite half-cell (b) Discharge/charge profiles of a conventional $K-O_2$ battery and $K_3Sb−O_2$ battery.

To demonstrate the applicability of a potassium−antimony alloy to the field of high energy density oxygen batteries, the performance of the K_3Sb alloy electrode was tested in a full oxygen cell. The fabrication of a K_3Sb−O2 cell is as previously stated. The typical discharge/ charge profile of a K_3Sb−O2 battery is compared to that of a conventional $K−O_2$ battery (K-metal anode) under the same operating conditions and is shown in Figure 2. An average discharge voltage of 1.80 V is recorded for the $K_3Sb − O_2$ battery at 20 mA/g. The overpotential demonstrated here is much lower than that of $Li-O_2$ batteries.[5,6] Here the resistance and origin of the charging overpotentials are due to the ionic conductivity of the K^+ Nafion 211 membrane used to mitigate oxygen crossover effects.[2,3]

During charge and discharge of the $K_3Sb−O_2$ battery the reactions occurring on the cathode are identical to that of a conventional $K−O_2$ battery. the cathodic reaction is confirmed using ex situ Raman measurements of a discharged and charged carbon cathode. Upon discharge, KO_2 formation shows a O−O vibrational frequency at 1143 cm^{-1}. Upon charging, the KO_2 is decomposed and no vibration at 1143 cm^{-1} is observed.

Conclusions

This study shows that potassium can reversibly alloy with antimony to a final alloying phase of cubic K3Sb with a high capacity of 650 mAh/g (~ 98% of the theoretical capacity). XRD, EDX, and galvanostatic discharge/charge curves confirm the high capacity of the Sb−C electrode can indeed be attributed to the formation of cubic K_3Sb alloy. Based on the EIS characterization of half-cells and K_xSb symmetric cells, the potassium electrolyte interface is the dominant contributor to the overall impedance in half-cell testing indicating the enhanced interfacial stability of the K−Sb electrode when compared to potassium metal. Finally, we present our results on a K3Sb−O2 cell with low overpotentials and increased safety, allowing for further development of an oxygen battery based on potassium.

Acknowledgement. This work was financially supported by the National Science Foundation (NSF Grant No. IIP-1542995)

References

(1) Dunn, B.; Kamath, H.; Tarascon, J. M. *Science* **2011**, 334, 928−935.
(2) Ren, X.; Wu, Y. *J. Am. Chem. Soc.* **2013**, 135, 2923−2926.

(3) Ren, X.; Lau, K. C.; Yu, M.; Bi, X.; Kreidler, E.; Curtiss, L. A.; Wu, Y. *ACS Appl. Mater. Interfaces* **2014**, 6, 19299−19307.
(4) Sangster, J.; Pelton, A. D. *J. Phase Equilib.* **1993**, 14, 510−514.
(5) Girishkumar, G.; McCloskey, B.; Luntz, A. C.; Swanson, S.; Wilcke, W. *J. Phys. Chem. Lett.* **2010**, 1, 2193−2203.
(6) Hartmann, P.; Bender,C.L.;Vracăr,M.;Dürr,A.K.;Garsuch, A.; Janek, J.; Adelhelm, P. *Nat. Mater.* **2013**, 12, 228−232.

IRON FLUORIDE NANOPARTICLES AS A HIGH-CAPACITY CATHODE FOR SODIUM-ION BATTERIES

*Ghulam Ali, and Kyung Yoon Chung**

[a]Center for Energy Convergence Research, Korea Institute of Science and Technology, Hwarangno 14-gil 5, Seongbuk-gu, Seoul 136-791, Republic of Korea
[b]Korea University of Science and Technology, 217 Gajeong-ro Yuseong-gu, Daejeon 305-333, Republic of Korea
kychung@kist.re.kr

Abstract

The intermittent energy sources such as solar and wind energy require effective energy storage systems to facilitate the production and steady consumption of electrical energy. In recent years, numerous efforts have been made to develop efficient rechargeable batteries to use for large scale applications such as electrical energy storage systems (ESS). Lithium-ion batteries have been the most popular and widely used batteries in portable devices like cell phones, laptops, etc. However, it has some constraints to be used for large scale applications as the cost for the raw materials for lithium is expensive. Beside other concepts, sodium-ion batteries (NIBs) can be one of the possible substitutes since it is more abundant and cheaper.

Layered oxide cathode materials are widely investigated in NIBs but their capacities are limited to use for application purpose. Iron fluoride is a potential cathode material with high theoretical capacity of 237 mAh g^{-1} with one cation insertion. Herein, we investigate iron fluoride hydrate as cathode material in NIBs which was prepared using non aqueous precipitation method. The conductivity of iron fluoride hydrate is poor which results in very low capacity. Thus, multi-walled carbon nanotubes (MWNTs) and reduced graphene oxide (rGO) were incorporated into iron fluoride to improve the conductivity of the material. The morphology studies show good adhesion of particles with rGO and MWNTs, forming nanocomposites. The nanocomposites with high surface area exhibit excellent electrochemical performances. Ex situ TEM and XAS have been employed to investigate the electrochemical reaction at different potentials during charge-discharge process. Further details about the properties of nanocomposite will be discussed at conference.

COUPLING IN-SITU TEM AND EX-SITU ANALYSIS TO UNDERSTAND HETEROGENEOUS SODIATION OF ANTIMONY

David Mitlin

Chemical & Biomolecular Engineering and Mechanical Engineering, 8 Clarkson Avenue, Potsdam NY 13699, USA
dmitlin@clarkson.edu

Introduction

Antimony is a highly promising anode material, giving a theoretical capacity of 660 mAh g^{-1} for both Li and Na. Recently, composite materials based on Sb have produced promising SIB anodes; examples include Sb/carbon nanocomposites, monodisperse Sb nanocrystals, Sb hollow Nanospheres, electrospun Sb/C Fibers, Sb_2S_3/graphene composites, bulk Sb, and Sb_2S_3. There has also been a range of novel Sb-based alloys and intermetallics in thin film and powder form, including SnSb, Cu_2Sb, AlSb, $FeSb_2$, and Mo_3Sb_7. The expansion of Sb to the fully sodiated hexagonal ($P6_3/mmc$) Na_3Sb phase ($\Delta V/V_{pristine}$) is 293%, so major changes in the stress state of the material during charging/discharging are anticipated. Work on the cycling of pure Sb has demonstrated significant study-to-study variations in the capacity decay during cycling, with an early study on micrometric Sb electrodes showing excellent cyclability, and a recent publication on thin-film electrodes showing rapid cycling-induced degradation. Although ex-situ experiments and theoretical modeling work have been carried out to study the sodiation mechanisms in the case of Sb, in-situ and operando studies on the material remain limited but will be essential for providing direct phenomenological insight into the sodiation behavior.

Experimental

For ex-situ electrochemical testing and microstructural characterizations, mirror-polished (to 50 nm), 316 stainless steel, spacer disks (MTI corporation), with a diameter and thickness of 15.8 mm and 0.5 mm, respectively, were used as the substrates for Sb thin film deposition. Prior to Sb deposition, the spacer disks were coated with a 25 nm TiN film by ALD to mimic the exact surface condition that is used for in-situ work. The spacers were weighed after each deposition step using a high accuracy microbalance (1 μg accuracy; MX5, Mettler Toledo) to calculate the Sb mass loading. For ex-situ analysis, the thickness of each Sb film was fixed at 100 nm, unless otherwise stated. The Sb electrode, along with a polyethene separator, the electrolytes, Na foil counter electrode and electrolytes (1 M $NaClO_4$ in ethylene carbonate/diethyl carbonate) were assembled in an argon-filled glovebox into coin cells (CR2032-type) for future test. Galvanostatic charge-discharge cycling was performed using a multichannel-current static system (BT2000, Arbin Instruments, TX).

Results and Discussion

We employed an in-situ electrochemical cell in the transmission electron microscope (TEM) together with ex-situ time-of-flight, secondary-ion mass spectrometry (TOF-SIMS) depth profiling, and FIB - helium ion scanning microscope (HIM) imaging to detail the structural and compositional changes associated with Na/Na^+ charging/discharging of 50 and 100 nm thin films of Sb. TOF-SIMS on a partially sodiated 100 nm Sb film gives a Na signal that progressively decreases towards the current collector, indicating that sodiation does not proceed uniformly. This heterogeneity will lead to local volumetric expansion gradients that would in turn serve as a major source of intrinsic stress in the microstructure. In-situ TEM shows time-dependent buckling and localized separation of the sodiated films from their TiN-Ge nanowire support, which is a mechanism of stress-relaxation. Localized horizontal fracture does not occur directly at the interface, but rather at a short distance away within the bulk of the Sb. HIM images of FIB cross-sections taken from sodiated half-cells, electrically disconnected and aged at room temperature, demonstrate non-uniform film swelling and the onset of analogous through-bulk separation. TOF-SIMS highlights time-dependent segregation of Na within the structure, both to the film-current collector interface and to the film surface where a solid electrolyte interphase (SEI) exists, agreeing with the electrochemical impedance results that show time-dependent increase of the films' charge transfer resistance. Since earlier theoretical work indicated that a progressive increase of the Na content in Sb results in major elastic softening, we propose that Na segregation also serves to reduce the sodiation stress.

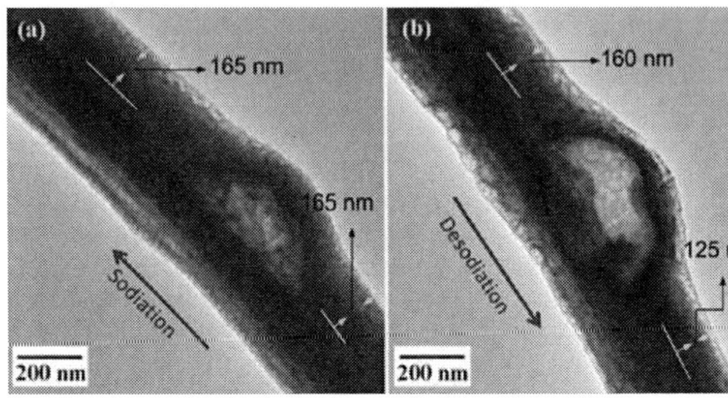

Figure 1. Sodiation – desodiation sequence as observed by TEM, highlighting the "damaged" region which does not desodiate.

Conclusions

In summary in-situ TEM and ex-situ TOF SIMS analysis have been used to provide information on time-dependent structural and compositional changes in nm-scale Sb thin films during sodiation. These changes presumably act to relax the major intrinsic and extrinsic sodiation-induced stresses in the material.

Acknowledgement. This work was performed, in part, at the Center for Integrated Nanotechnologies, an office of Science User Facility operated for the U.S. Department of Energy (DOE) Office of Science. Sandia National Laboratories is a multi-program laboratory managed and operated by Sandia Corporation, a wholly owned subsidiary of Lockheed Martin Corporation, for the U.S. Department of Energy's National Nuclear Security Administration under contract DE-AC04-94AL85000.

References

(1) Li, Zhi, Xuehai Tan, Peng Li, Peter Kalisvaart, Matthew T. Janish, William M. Mook, Erik J. Luber, Katherine L. Jungjohann, C. Barry Carter, and David Mitlin. "Coupling In situ TEM and ex situ analysis to understand heterogeneous sodiation of antimony." *Nano letters* 15, no. 10 (2015): 6339-6348.

REDOX-ACTIVE ORGANIC COMPOUNDS AS POTENTIAL ELECTRODE MATERIALS FOR RECHARGEABLE BATTERIES

*Hyun Ho Lee, Dong-Seon Shin, Sung You Hong**

School of Energy and Chemical Engineering,
Ulsan National Institute of Science and Technology (UNIST),
Ulsan 689-798, Republic of Korea
syhong@unist.ac.kr

Introduction

Rechargeable batteries provide leading storage methods to allow harnessing renewable energies. One way to furnish next-generation electrode materials from sustainable or natural resources is utilization of organic substrates having reversible redox-active properties.[1,2] Their diverse molecular scaffolds and concise synthetic methodologies permit rational design and construction of high-performance electrode materials satisfying multiple electrochemical criteria including potential range, capacity, rate capability, and cyclability.[3,4] Herein, we will present design and synthesis of redox-active organic compounds or coordination complexes as promising electrode materials for lithium- (or sodium-) ion batteries or redox flow batteries.

Experimental

Synthesis of disodium terephthalate derivatives:[1] disodium terephthalate derivatives were prepared by aqueous NaOH treatment of terephthalic acid precursors. Precipitation (ethanol/acetone) yielded the target compound as a white form.

Synthesis of coordination complexes:[3] coordination complexes were prepared by ligand-exchange reaction. Transition metal nitrate or sulfate was treated with disodium terephthalate to afford the corresponding target compound (see Figure 1). The sample was then annealed at 150 °C *in vacuo* to remove hydrates.

Electrochemical tests and characterizations: As-prepared samples were tested by gavanostatic cycling using 2032 coin cells. IR spectra were measured by a Nicolet FT-IR 200. XANES data were collected from the 7D beam line at the Pohang Light Source (Korea).

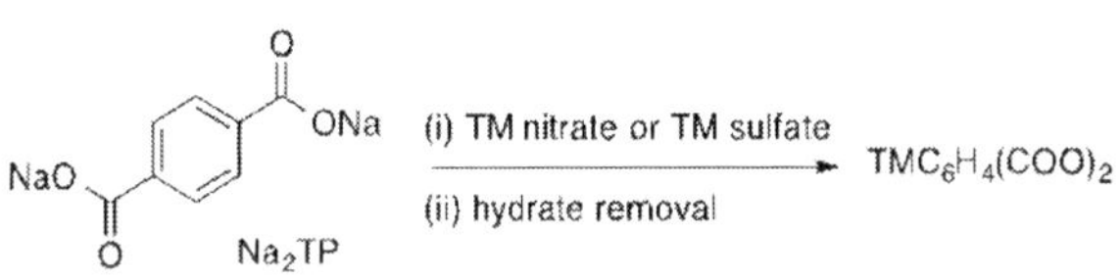

Figure 1. Synthesis of redox-active coordination complexes.[3]

Results and Discussion

Mechanistic elucidation of electrochemical (de)lithiation of redox-active organic materials is challenging due to the experimental difficulties in identifying the redox sites within organic scaffolds. Transition metal coordination complexes with terephthalate ligands reveal high capacities above ca. 1100 mAh g^{-1} with good cyclic performances. The origin of high capacities and electrochemical reaction mechanisms were investigated via synchrotron sourced X-ray absorption spectroscopy and solid-state ^{13}C NMR spectroscopy. These methods present that the electrochemical processes are composed of oxidation level changes of transition metal as well as lithium (de)insertion within the organic molecular structure (see Figure 2).

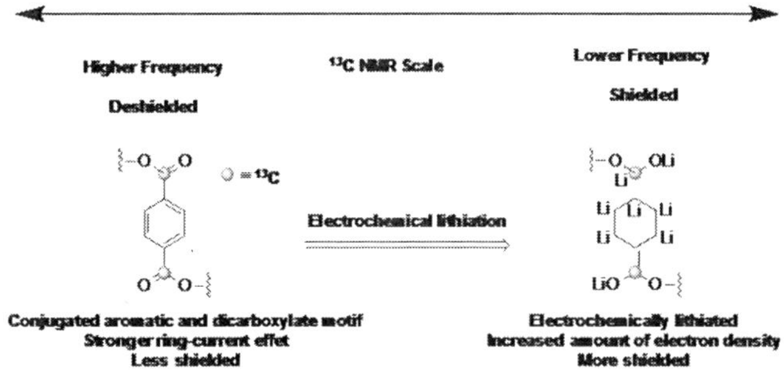

Figure 2. Lithiation of terephthalates.[3]

In addition to solid type batteries, prototype flexible and flowable rechargeable lithium batteries were built by using soluble quinone-type compounds. They showed excellent electrochemical performances even under the stressed conditions (mechanical bending and stretching). Two constitutional isomers (5,12-naphthacenequinone and 1,2-benzanthraquinone) belonging to the Wurster type redox system exhibited structure-dependent electrochemical properties.

Conclusions

We elucidated the origin of excessive capacities from non-fused aromatic compounds and coordination complexes based on NMR or XANES studies. These investigations revealed molecular structure dependent electrochemical properties. We believe the synthetic organic compounds and coordination complexes can be potential candidates for Li- (or Na-) ion batteries or redox flow systems.

Acknowledgement. This work was supported by National Research Foundation of Korea (NRF-2013R1A1A2060695).

References

(1) Park, Y.; Shin, D.-S.; Woo, S. H.; Choi, N. S.; Shin, K. H.; Oh, S. M.; Lee, K. T.; Hong, S. Y. *Adv. Mater.* **2012**, *24*, 3562.
(2) Park, M.; Shin, D.-S.; Ryu, J.; Choi, M.; Park, N.; Cho, J. *Adv. Mater.* **2015**, *27*, 5141.
(3) Lee, H. H.; Park, Y.; Kim, S. H.; Yeon, S.-H.; Kwak, S. K.; Lee, K. T.; Hong, S. Y. *Adv. Funct. Mater.* **2015**, *25*, 4859.
(4) Lee, H. H.; Park, Y.; Shin, K-H, Lee, K. T.; Hong, S. Y. *ACS Appl. Mater. Interfaces* **2014**, *6*, 19118.

MANUFACTURING OF MXENE-BASED NANOCOMPOSITE FILMS FOR ELECTROCHEMICAL ENERGY STORAGE

*Meng-Qiang Zhao, Chang E. Ren, Michael Ghidiu, Xiuqiang Xie, Michel W. Barsoum, Yury Gogotsi**

Department of Materials Science and Engineering, and A.J. Drexel Nanomaterials Institute, Drexel University, Philadelphia, PA 19104, USA
*E-mail: gogotsi@drexel.edu

Recently, we discovered a new class of two-dimensional (2D) transition metal carbides and nitrides and named them MXenes. Delamination of stacked MXenes produces single-layer nanosheets with thicknesses of ~1 nanometer and tailorable lateral sizes of the order of micrometers. The delaminated MXene flakes can be manufactured into flexible films with high electrical conductivities and superior performance as electrode materials for supercapacitors and batteries.

In this work, MXene-based nanocomposite films were fabricated in order to further improve the electrochemical performance of MXenes. Several methods, including introduction of interlayer spacers, porosity, and hybridization with metal oxides or conductive polymers, were utilized to produce the nanocomposites. Micrometers-thick flexible and highly conductive films, up to a letter-size paper sheet, were manufactured by vacuum-assisted filtration, spray coating, and direct rolling methods. These films can be directly used as electrode materials for supercapacitors, Li- and Na-ion batteries. In general, the performances of the composite films were superior to those made with only MXene flakes. This enhancement can be ascribed to the improved electrolyte accessibility and/or synergistic effects. The work provides simple, scalable, and effective strategies for the fabrication of advanced electrode materials that can be used in wearable or structural electrochemical energy storage and conversion systems.

SELF-ASSEMBLY SYNTHESIS OF ELECTRODE ARCHITECTURES

Sheng Dai[1,2]

[1]Chemical Sciences Division, Oak Ridge National Laboratory, Oak Ridge, Tennessee 37831, USA
[2]Department of Chemistry, University of Tennessee, TN 37996, USA
dais@ornl.gov

Enhancement in electrochemical energy storage resides in tailored nanotextures resulting from making electrode materials hierarchically nanoporous. The complex interplay of solvent and solute structure and dynamics at the charged interface, the transport of electrolyte ions into and out of the pores, the solvation/desolvation processes occurring in pores approaching bare-ion dimensions, and formation of interfaces via chemical reactions are all important parameters. Herein, we report several self-assembly synthesis methods for carbon and oxide composites as electrode materials for energy storage. The objective of this work is to demonstrate that mesoporous carbons and oxides derived from soft-template synthesis not only entail a high storage capacity but also, most importantly, can be made through self-assembly synthesis to have a significantly enhanced electronic conductivity and storage capacity. This enhanced electronic conductivity in 3D architectures is the key to providing high rate capability for the corresponding energy storage systems.

Acknowledgement. This work was supported by the U.S. Department of Energy (DOE), Basic Energy Sciences (BES), Materials Sciences and Engineering Division.

RECENT DEVELOPMENT OF REDOX FLOW BATTERIES.

Wei Wang wei.wang@pnnl.gov

Pacific Northwest National Laboratory

The presentation describes the materials design and chemistry development for redox flow battery technology developed at PNNL based on different redox chemistries including the soluble commodity redox active materials and transition metals, as well as the molecular structure design to improve the properties and performance.

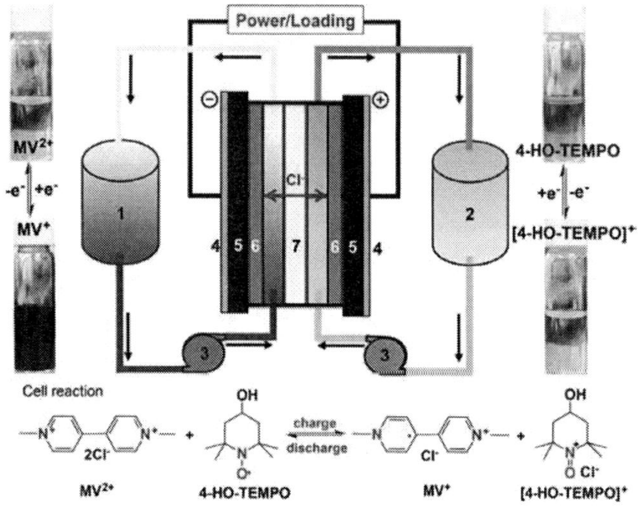

Figure 1. A schematic representation of the MV/4-HO-TEMPO organic aqueous redox flow battery (OARFB).

LITHIUM REDOX FLOW BATTERIES FOR STATIONARY ENERGY STORAGE

Yu Ding, Guihua Yu

Materials Science and Engineering, Texas Materials Institute, The University of Texas at Austin, Austin, Texas 78712
* Email: ghyu@austin.utexas.edu

As an emerging battery technology, Li-redox flow batteries inherit the advantageous features of modular design of conventional redox flow batteries and high voltage and energy efficiency of Li–ion batteries, and show promise as efficient electrical energy storage system in transportation and residential applications. The chemistry of lithium redox flow batteries with aqueous or non-aqueous electrolyte potentially provides widened electrochemical potential window thus may provide greater energy density and efficiency than conventional redox flow batteries based on proton chemistry. This talk will discuss recent advances in the field especially Li-based organic redox flow batteries, and some critical challenges/opportunities with this new promising battery system.

References

(1) Y. Zhao, Y. Ding, Y. Li, L. Peng, H.R. Byon, J.B. Goodenough, G. Yu, *Chemical Society Reviews* 2015, 44, 7968.
(2) Y. Ding, Y. Zhao, G. Yu, *Nano Letters* 2015, 15, 4108.
(3) Y. Zhao, Y. Ding, J. Song, G. Li, G. Dong, J.B. Goodenough, G. Yu, *Angewandte Chemie International Edition* 2014, 53, 11036.
(4) Y. Ding, G. Yu, *Angewandte Chemie International Edition* 2016, DOI: 10.1002/ange.201600705.

EFFECT OF pH ON THE ACTIVITY OF HYDROGEN OXIDATION REACTION/HYDROGEN EVOLUTION REACTION OVER PtRu BIMETALLIC CATALYSTS

Jared Nash[1], Jie Zheng[1], Yan Wang[2], Bingjun Xu[1], and Yushan Yan*[1]*

[1]Center for Catalytic Science and Technology, Department of Chemical and Biomolecular Engineering, University of Delaware, 150 Academy Street, Newark DE, 19716
[2]Institute of Special Chemicals, Taiyuan University of Technology, 030024, Shanxi, China
*bxu@udel.edu, *yanys@udel.edu

Introduction

With hydrogen becoming a more viable energy alternative to traditional fossil fuel, hydrogen fuel cell powered automobiles has become commercially available, as seen by the Toyota Mirai, which is implemented by proton exchange membrane fuel cells (PEMFCs). However, PEMFCs still suffer from the barrier of the cost and stability. Alternatively, hydroxide exchange membrane fuel cells (HEMFCs) are promising because they have the potential to solve the problems of catalyst cost and durability while achieving high power and energy density. Switching from an acidic to an alkaline environment, non-platinum group metals (PGMs) become stable under the operating conditions[1], opening up the possibility of employing non-PGM catalysts. However, HOR has been proven difficult with the activity decreases of two orders of magnitude for monometallic PGMs[2, 3] when switching from acid to base, which necessitates an increase in the PGM loading. The pH dependence on activity has been rationalized using two different mechanisms.

The first hypothesis is the bifunctional mechanism where adsorbed hydroxide plays a key role in alkaline conditions[4-6]. Strmcnik et al. proposed that adsorbed OH plays a role of promoting the HOR/HER activity by facilitating the removal of adsorbed H in base, with the observation that the more oxophilic monometallic Ir showed improved activity compared to monometallic Pt[4]. Recently, the improved activity of Pd/Ni and Pt/Ni bimetallics was also attributed to the bifunctional mechanism[6-8]. The bifunctional mechanism requires moderate hydrogen and hydroxide binding energies[9].

The second hypothesis is the increase in the hydrogen binding energy (HBE) moving from acid to base as can be seen by the shifting hydrogen underpotential deposition (H_{upd}) peak[2, 3, 10-12]. HBE has been used to explain the activity computationally by Norskov, as well as experimentally by Gasteiger, Yan, and Zhuang[2, 3, 10, 11, 13-15]. H_{upd} peak potential shifts positively from acid to base, indicating an HBE increase[11], which correlates with a decrease in activity for Pt, Pd, Ir, and Rh[2, 11].

In this work, the activity of bimetallic PtRu/C was studied using rotating disk electrode (RDE) method in electrolytes with different pHs. A change in the pH trend on activity is evident which indicates that pH has a significantly different effect on bimetallic catalysts.

Experimental

Catalysts and Characterization. PtRu/C (40 wt.% PtRu 1:1 atomic ratio supported on Vulcan XC-72, De Nora ETEK), Pt/C (5 wt.% Pt supported on Vulcan XC-72, Premetek Co.), and Ru/C (20 wt.% Ru supported on Vulcan XC-72, Premetek Co.) were obtained commercially and used as received. Surface composition of PtRu was determined using copper underpotential deposition (Cu_{upd}) similar to the procedure described by Green and Kucernak[16]. Scanning electron microscopy and energy dispersive X-ray spectroscopy (EDX) were performed in a JEOL JSM-7400F field emission scanning electron microscope with an EDX detector (25 keV).

Chemicals. Buffer electrolytes were prepared by mixing 4 M KOH (Sigma Aldrich pellets 99.99% purity) with DI water and phosphoric acid (85 % solution, Sigma Aldrich), boric acid powder (Alfa Aesar 99+% purity), or potassium bicarbonate powder (Fisher Scientific 99.9% purity) to the desired pH. Perchloric acid was prepared by mixing concentrated perchloric acid (70% solution, EMD Millipore) with DI water. Sulfuric acid was prepared by mixing concentrated sulfuric acid solution (Fluka, >95%) with DI water. Copper sulfate solution was prepared by mixing copper sulfate pentahydrate powder (Sigma Aldrich, >98%) with 0.1 M sulfuric acid solution. All chemicals were used without further purification.

Electrochemical Testing. Electrochemical measurements were performed in a glass cell (Pine Ins.) three-electrode setup using a Pt wire counter electrode, a saturated calomel electrode (SCE, Princeton Applied Research) reference electrode, and a glassy carbon (GC) (5 mm diameter, Pine Ins.) working electrode. GC electrodes were polished with 0.05 μm alumina polishing solution (Buehler), rinsed and sonicated in DI water. Catalyst inks were prepared by mixing the purchased catalyst with LIQUION (DuPont) in isopropyl alcohol (IPA) to produce a 0.05 wt% Nafion ink solution. The ink was sonicated (<30 °C) for 60 minutes then pipetted onto the GC electrode in 4 μL aliquots to produce 4 $\mu g/cm^2_{geo}$ or 10 $\mu g/cm^2_{geo}$ metal loading then dried in ambient conditions. All potentials reported were converted to the reversible hydrogen electrode (RHE) scale. CVs were carried out in Ar-saturated solution at a scan rate of 50 mV/s from 0.03 V to 0.9 V in order to reduce the amount of Ru dissolution. HOR/HER polarization curve measurements were carried out in a H_2-saturated solution at a rotation rate of 1600 rpm. Exchange current densities were calculated from the forward scan. The internal resistance was measured by electrochemical impedance spectroscopy at open circuit voltage to obtain the iR-free potential. Electrochemical surface areas were determined using CO stripping. The potential was held at 0.1 V for 10 minutes with CO purging, then gas was switched to Ar for 10 minutes to remove dissolved CO, and then the potential was cycled at 20 mV/s. CO stripping and H_{upd} charge densities of 420 $\mu C/cm^2_{metal}$[17] and 210 $\mu C/cm^2_{metal}$,respectively, were used. Potential windows of approximately 0.05 V to 0.35 V for PtRu and 0.05 V to 0.45 V for Pt were used. Potential ranges for HOR/HER were chosen to limit bubble formation during HER. For Ru dissolution, the potential was cycled to 1.35 V to form soluble oxides[17]. Kinetic current (i_k) is extracted from the measured current using Koutecky-Levich equation[2, 18] with the Nernstian diffusion current. i_k was fit using the Butler-Volmer equation to determine the exchange current with $\alpha_a+\alpha_c = 1$[19], where α_a, α_c are the anodic and cathodic charge transfer coefficients, respectively. Exchange current density (i_0) were calculated by normalizing the exchange current by the electrochemical surface area from CO stripping.

Results and Discussion

Exchange current density of HOR/HER on PtRu/C decreases as the pH of the electrolyte increases from 5 to 10, which is consistent with monometallic Pt-group metal (PGM) catalysts. i_0 is difficult to determine using RDE method in electrolytes with pH lower than 7 because the polarization curve overlaps with the concentration overpotential curve, which renders the extraction of i_k using the Koutecky-Levich equation impossible[20]. Above pH 10, there is an upward trend in activity (Figure 1), which has not been observed previously on monometallic catalysts. Different buffer solutions with overlapping pH values were employed and showed the same trend.

The position of the H_{upd} peak does not shift much in the pH range of 7.5-13, indicating little change in HBE. Because there is no shift in peak position with pH while the activity changes, the correlation between peak position and HOR/HER activities on monometallic catalysts does not transfer to bimetallic catalysts. The hydroxide binding energy with pH was determined using the peak position of the CO stripping[14, 21-23]. The peak potential shifts to lower potential as the pH increases, which is consistent with those on monometallic PGMs. However, the change is modest, with the total shift being only ~100 mV in 13 pH units. The hydrogen and hydroxide binding energies do not change and indicates neither hypothesis describes the activity trend.

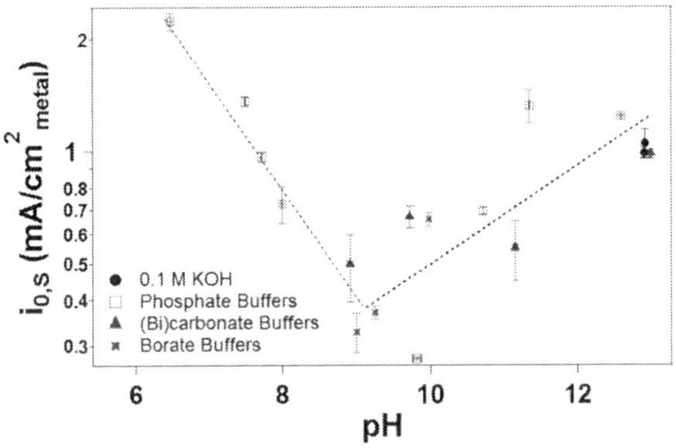

Figure 1. Exchange current density with pH. The increase in activity above pH 10 has not been seen on monometallics.

The effect of surface and bulk compositions of PtRu alloys was investigated by dissolution of Ru from the bulk catalyst. After 5 cycles the PtRu feature at 0.1 V shifts to lower potentials and diminishes in intensity while Pt-like features at 0.3 V and 0.35 V[24] appear and stabilizes after 15 cycles, resembling monometallic Pt. Bulk concentration of Ru, determined by EDX, decreases as the number of cycles increases from 45% to 15% after 20 cycles. Surface composition of Ru, determined by Cu_{upd}, indicates surface enrichment of Pt. The calculated surface coverage of Ru of the commercial PtRu/C catalyst is 24 ± 4%, and decreases below the detection limit (< 5%) after 15 cycles, which agrees with the small changes between CVs collected between 15 and 20 cycles. No further reduction in Ru content was observed after 20 cycles. This is likely due to the removal of all Ru that is accessible to the electrolyte from cycling. This result suggests that cycling removes Ru from the surface layer, and leaves behind ~15% Ru in the subsurface and bulk. Activities of the cycled catalysts were measured and correlated with the surface and bulk compositions of the catalysts. The i_0 of the cycled catalysts remains the same as the bulk Ru content reduces from 45% to 20%, below which a decrease in the i_0 is observed. Likely, the presence of subsurface Ru, which can modify the properties of the nanoparticles, increases the HOR/HER activity. α also changes with the composition of the catalyst. α approaches the value obtained on Pt(0.5) at bulk Ru content less than 15% (Figure 2). Changing α is suggestive of a change in the relative Heyrovsky-Volmer and Tafel-Volmer pathways.

By utilizing the dual pathway model by Elbert et al.[25] to fit the kinetic current measured from RDE, a Heyrovsky-Volmer(RDS) reaction mechanism was determined. The Volmer step has approximately the same energy barrier for all pH values shown. The Tafel step always has a higher energy barrier, and the Heyrovsky has the lower energy barrier making the Heyrovsky-Volmer(RDS) pathway the most favored pathway. In the lower pH values that were studied, the Heyrovsky and the Volmer steps have very comparable energy barriers. Around pH 11, there is a decrease in the energy barrier for the Heyrovsky step indicating an increase in the rate of the Heyrovsky step, which causes the Volmer to be rate determining. The increase in activity, which is an increase in the overall rate, above pH 10 is likely due to a combination of the increased Heyrovsky step with a change in the surrounding water structure to better facilitate the Volmer step.

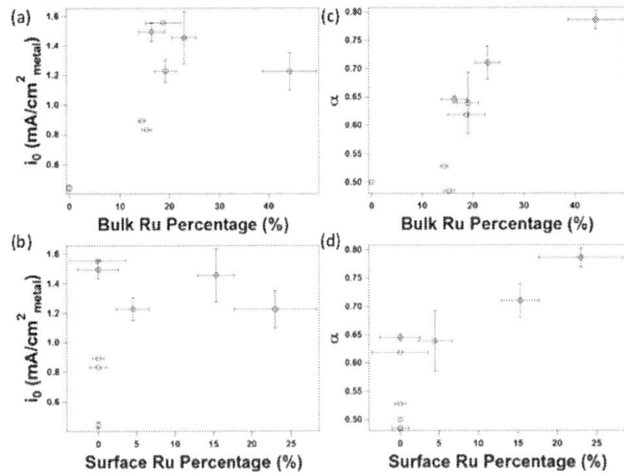

Figure 2. i_0 as a (a) bulk composition and (b) surface composition and α as a function of (c) bulk composition and (d) surface composition in 0.1 M KOH.

Conclusions

In this work, both the hydrogen and hydroxide binding energies show very little change with pH, but the activity changes significantly in the pH region studied. Below pH 10, the trend in activity with pH is similar to the monometallic catalysts that have been studied in the literature. The activity appears to remain approximately constant even when a significant amount of Ru is dissolved from the catalyst. However, α shows a constant decrease toward the 0.5 value seen on Pt. These indicate that a small amount of surface Ru can increase the activity of Pt for the HOR/HER reaction. The proposed effect of Ru affecting the mechanism of the reaction and becoming increasingly more kinetically relevant at higher pH's due to a possible lowering of the activation barrier of the Volmer step in the Heyrovsky-Volmer pathway for the HOR

Acknowledgements. This work was supported by the University of Delaware and the American Chemical Society petroleum research fund.

References

1. Gu, S., et al., *Annu Rev Chem Biomol Eng* **2014,** *5* (1), 429-454.
2. Zheng, J., et al., *Science Advances* **2016,** *2* (3).
3. Durst, J., et al., *Energy Environ. Sci.* **2014,** *7* (7), 2255-2260.
4. Strmcnik, D., et al., *Nat Chem* **2013,** *5* (4), 300-306.
5. Subbaraman, R., et al., *J PHYS CHEM C* **2012,** *116* (42), 22231-22237.
6. Subbaraman, R., et al., *Science* **2011,** *334* (6060), 1256-1260.
7. Alesker, M., et al., *J. Power Sources* **2016,** *304*, 332-339.
8. Bakos, I., et al., *Electrochimica Acta* **2015.**
9. Koper, M. T. M., *Nat Chem* **2013,** *5* (4), 255-256.
10. Zheng, J., et al., *ACS Catal* **2015,** *5* (7), 4449-4455.
11. Sheng, W., et al., *Nature communications* **2015,** *6.*
12. Sheng, W., et al., *Energy Environ. Sci.* **2013,** *6* (5), 1509-1512.
13. Nørskov, J. K., et al., *J Electrochem Soc* **2005,** *152* (3), J23.
14. Wang, Y., et al., *Energy Environ. Sci.* **2014,** *8* (1), 177-181.
15. Neyerlin, K. C., et al., *J Electrochem Soc* **2007,** *154* (7), B631.
16. Green, C. L.; Kucernak, A., *J PHYS CHEM B* **2002,** *106* (44), 11446-11456.

17. Green, C. L.; Kucernak, A., *J PHYS CHEM B* **2002,** *106* (5), 1036-1047.
18. Zhuang, Z., et al., *Nature communications* **2016,** *7*.
19. Newman, J.; Thomas-Alyea, K. E., John Wiley & Sons: 2004.
20. Zheng, J., et al., *J Electrochem Soc* **2015,** *162* (14), F1470-F1481.
21. Gómez-Marín, A. M.; Hernández-Ortiz, J. P., *J PHYS CHEM C* **2014,** *118* (5), 2475-2486.
22. Lu, C.; Masel, R. I., *J PHYS CHEM B* **2001,** *105* (40), 9793-9797.
23. Gasteiger, H. A., et al., *The Journal of Physical Chemistry* **1995,** *99* (45), 16757-16767.
24. Sheng, W., et al., *J Electrochem Soc* **2010,** *157* (11), B1529.
25. Elbert, K., et al., *ACS Catal* **2015**, 6764-6772.

IN SITU ENVIRONMENTAL TEM TECHNIQUES FOR UNDERSTANDING THE STRUCTURE OPTIMIZATIONS OF ELECTROCATALYSTS

Hong Yang

Department of Chemical and Biomolecular Engineering, University of Illinois at Urbana-Champaign, 206 Roger Adams Laboratory, 600 S. Mathews Avenue, Urbana, IL 61801, USA
hy66@illinois.edu

Introduction

High-performance metallic catalysts for fuel cell and battery electrode applications require careful design of nanostructures that form either through direct synthesis or post-synthesis processing. The conventional impregnation method typically does not have the level of control to meet various requirements, such as catalytic surface with controlled single atom position or sub-monolayer atoms. Approaches need to be developed to address preparation challenges in nanocrystal size, facet, composition and fine structures (e.g., site specific bimetallics).

Experimental

Preparation. Bimetallic Pt-Ni electrocatalysts were synthesized from $Pt(acac)_2$, $Ni(acac)_2$ in oleylamine (OAm), oleic acid (OA) and diphenyl ether, assisted by carbon monoxide (CO) gas. The detail was reported elsewhere.[1]

Characterization. Transmission electron microscopy (TEM) and high-resolution transmission electron microscopy (HR-TEM) micrographs were used to study the nanostructures of the catalysts, which were taken on a FEI TECNAI F-20 field emission microscope at an accelerating voltage of 200 kV. In situ environmental TEM (ETEM) study was conducted using a Hitachi H9500 ETEM at an accelerating voltage of 200 kV with sample holder connected to a DC power supply for resistive heating, and recorded by the Gatan Orius SC200 CCD camera, and Gatan digital micrograph software.

Results and Discussion:

Figure 1 shows the illustration of the structural changes of a sandwiched Pt-Ni electrocatalysts before and after thermal treatment under vacuum at 200 °C, after rotating disk electrode (RDE) tests for oxygen reduction reaction (ORR) performance. In suit ETEM was used to understand the structural dynamics of the PtNi ORR catalysts under heat treatment. The final structures were further studied by X-ray photoelectron microscopy (XPS). Our results suggest that there exists a clear structure-property relationship for the as-made catalyst treated under different thermal and electrochemical conditions. Less noble metal leached under RDE test, resulting a highly active ORR catalysts (**Figure 1**).

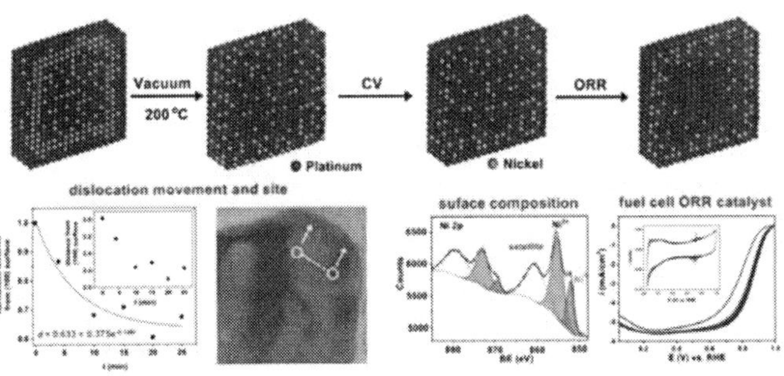

Figure 1. Illustration of the process, in situ ETEM characterizations of the dislocation movement, surface composition analysis by X-ray photoelectron microscopy, and polarization curves of Pt-Ni catalysts towards the oxygen reduction reaction (ORR).

Conclusions

This study indicates that the in situ ETEM can greatly facilitate the understanding the effect of temperature on the bimetallic surface compositions, thus the design of thermal treatment process for high performance electrocatalysts.

Acknowledgement. This research was supported in part by NSF grant (Grant No.: CHE-1213926). The ETEM was acquired through a grant support from NSF (Grant No.: MRI-1229454). We thank Prof. J-M Zuo for help.

References

(1) Pan, Y.-T., Wu, J. B., Yang, H. *AIChE J*, **2016**, 62 (2), 399-407.

INTEGRATED PHOTOELECTROCHEMICAL SOLAR ENERGY CONVERSION AND ELECTROCHEMICAL STORAGE DEVICES

Song Jin and Wenjie Li

Department of Chemistry, University of Wisconsin-Madison, 1101 University Avenue, Madison, Wisconsin 53706
*e-mail: jin@chem.wisc.edu

Renewable energy technologies generally rely on harvesting and storing energy from our most readily exploitable and only truly limitless source: the Sun. With the development for over 40 years, photovoltaic (PV) solar cells convert solar energy to electricity with increasing efficiency and decreasing cost. However, the intermittent nature of sunlight necessitates the storage of the photo-generated electricity, therefore further large-scale deployment of solar cells also depends on scalable and inexpensive grid-level energy storage solutions. Here we develop a novel type of hybrid energy conversion and storage devices by integrating liquid junction solar cells with electrochemical redox flow batteries (RFBs) using the same redox couples. Such new technology exploits a previously undeveloped connection between two major energy technologies, solar energy conversion and grid-scale energy storage, and integrates the function of solar energy harvest and conversion with the grid-scale energy storage in one simple device. By using redox couples that have facile kinetics in both liquid junction cells and scalable redox-flow batteries, not only can highly efficient solar cells be achieved, but also it enables the efficient storage of the converted solar energy into chemical energy in charging of the flow batteries that can be readily discharged to provide electricity in a simple cyclable device. We demonstrate this novel concept by first building and demonstrating integrated functional solar cell/redox flow battery devices based on aqueous organic redox couples and mature PV semiconductors to utilize the solar energy to fully charge the batteries. We further evaluate and demonstrate their excellent energy conversion efficiency, energy storage density and efficiency, and cyclability.

ABSTRACT
THE EVOLUTION OF MULTI-FUNCTIONAL NANOPOROUS METAL ELECTROCATALYSTS

Joshua Snyder

The unique physicochemical properties of nanoscale materials and the continual development of synthesis and fabrication techniques has led to a renaissance in the production of complex catalytic materials with increasing degrees of multi-functionality. Nanoporous metals are an interesting subset of nanoscale electrocatalytic materials as they possess a high surface-to-volume ratio, are intrinsically conductive, are core-shell in nature, and their morphology, composition and compositional gradient are readily tuned through a manipulation of dealloying conditions. Here we present our mechanism for the formation and evolution of nanoporous metal morphology and their application to the catalysis of several elementary electrochemical reactions including the oxygen reduction reaction (ORR) and oxygen evolution reaction (OER)[1]. Nanoporous metals are inherently metastable. In addition to the standard mechanisms of catalyst deactivation and surface area loss including Ostwald ripening and coalescence, surface energy driven coarsening is a new pathway for the loss of electrochemically active surface area in nanoporous materials. With a better understanding of the interplay between nanoporous structure coarsening and transition metal corrosion, we can propose strategies to mitigate coarsening and improve operational catalyst stability. With the insight gained from the fundamental study of nanoporous metal morphology evolution under relevant conditions, we propose and evaluate several strategies for the mitigation of this process and improvement of durability of this class of multi-functional nanostructured catalytic materials.

Reference

(1) Snyder, J.; Livi, K. & Erlebacher, J. *Adv. Funct. Mater.* **23**, 5494-5501 (2013).

METAL OXIDE-CARBON HYBRID ENGINEERED BY HETEROATOM-DOPING AS ELECTRODE MATERIALS FOR ENERGY STORAGE AND CONVERSION

*Ruizhi Yang**

College of Physics, Optoelectronics and Energy & Collaborative Innovation Center of Suzhou Nano science and Technology, Soochow University, Suzhou, Jiangsu 215006, China
E-mail: yangrz@suda.edu.cn

Metal-air batteries have attracted an increasing attention due to their high energy density, low cost and environmental benignity [1,2]. However, the performance of metal-air batteries is mainly limited by the sluggish oxygen reduction reaction (ORR) and oxygen evolution reaction (OER) at the air electrode. These two key reactions require efficient electrocatalysts to accelerate their rates. Owing to the high price and poor stability of precious electrocatalysts (such as Pt and its alloy, RuO_2 and IrO_2), the alternative low-cost, durable and highly active electrocatalysts are highly desired. Spinel based mixed valence transition metal oxide, such as Co_3O_4[3], $MnCo_2O_4$ [4,5], $Cu_xCo_{3-x}O_4$[6] and $NiCo_2O_4$ [7], have been reported to be efficient electrocatalysts for ORR and OER due to their redox stability and variable valence states. It's known that the catalytic activity of spinel oxides is limited by their low electrical conductivity.

Usually, growing or supporting the spinel oxide nanoparticles on conductive carbon based materials is an effective approach to improve the electronic conductivity of spinel oxides [8,9]. As is well known, to assure an intimate contact between metal oxide and carbon substrate, carbon is usually pretreated with harsh oxidative agents to have sufficient functional groups to coordinate the metal ions. However, excessive oxidation of carbon will inevitably destroy the sp^2 structure of carbon and decrease the electronic conductivity of carbon. Therefore, the oxidation degree of carbon needs to be carefully controlled to balance the surface functional groups and the electronic conductivity of carbon[9]. This makes the process control of metal oxide-carbon hybrid complicated. It's still an interesting and challenging issue to explore other methods to functionalize the surface of carbon for anchoring metal oxides.

Heteroatom doping is an effective way to improve the electronic conductivity of carbon. Moreover, heteroatom doping induces not only structural changes of carbon but also charge redistribution of carbon, creating charge sites of carbon[10]. It's expected that heteroatom doping in carbon could provide nucleation sites for metal oxides while improving the electronic conductivity of carbon.

Herein, we explore the doping of carbon with nitrogen (N), phosphorus (P) and nitrogen/sulfur (N/S) in engineering the coupling between metal oxide and carbon in the spinel metal oxide-carbon hybrid as electrocatalyst for metal-air batteries. It is found that engineering the chemical and electrical coupling between metal and carbon via heteroatom-doping is an effective way for enhancing the electrocatalytic activity of metal oxide. Our work provides an alternative approach for the design and preparation of electrocatalysts used in energy storage and conversion devices.

Acknowledgements. This work is supported by National Natural Science Foundation of China (Nos. 51272167, 21206101 and 51572181).

References

(1) Chen, Z.; Yu, A.; Higgins, D.; Li, H.; Wang, H.; Chen, Z. *Nano Lett.* **2012**,*12*,1946.
(2) Ge, X.M.; Liu, Y.Y.; Thomas Goh, F.W.; Andy Hor, T.S.; Zong, Y.; Xiao, P.; Zhang, Z.; Lim, S.H.; Li, B.; Wang, X. ; Liu. Z.L. *ACS Appl. Mater. Interfaces* **2014**,*6*,12684.
(3) Liang, Y.Y.; Li, Y.G.; Wang, H.L.; Zhou, J.G.; Wang, J.; Regier, T.; Dai. H.J. *Nat. Mater.* **2011**,*10*,780.
(4) Cheng, F.Y.; Shen,J.; Peng, B.; Pan, Y.D.; Chen, J. *Nat. Chem.* **2011**,*3*, 79.
(5) Oh, D.; Qi, J. F.; Han, B. H.; Zhang, D.R.; Carney, T.J.; Ohmura, J.; Zhang, Y.; Yang, S.H.; Belcher. A.M. *Nano Lett.* **2014**, 14,4837.
(6) Koninck, M.D.; Poirier, S.C.; Marsan, B. *J. Electrochem. Soc.* **2007**, 15, A381-8.
(7) Jin, C.; Lu, F. L.; Cao, X.C.; Yang, Z.R.; Yang,R.Z. *J. Mater. Chem. A* **2013**,*1*,12170.
(8) Bian,W.Y.; Yang, Z.R.; Strasser, P.; Yang, R.Z. *J. Power Sources* **2014**,*250*, 196.
(9) Liang, Y.Y.; Li, Y.G.; Wang, H.L.; Dai, H.J. *J. Am. Chem. Soc.* **2013**, *135*, 2013.
(10) Zhang, M.; Dai, L. M. *Nano Energy*, **2012**,*1*, 514.

THE ROLE OF COMPOSITION AND STRUCTURE IN MIXED METAL OXYHYDROXIDE OXYGEN EVOLUTION ELECTROCATALYSTS

Shannon W. Boettcher,[a,b] Michaela S. Burke,[a] Shihui Zou,[a] Lisa Enman,[a,b] Adam Batchellor,[a] and Matthew G. Kast,[a,b]*

[a] Department of Chemistry and Biochemistry
University of Oregon
Eugene, Oregon 97403, United States of America

[b] Center for Sustainable Materials Chemistry
University of Oregon
Eugene, Oregon 97403, United States of America

*swb@uoregon.edu

Introduction

The slow kinetics of the oxygen evolution reaction (OER, in base: $4OH^- \rightarrow 2H_2O + O_2 + 4e^-$) limit the performance of promising large-scale energy technologies based on the photo-driven or electricity-driven production of H_2 from water.[1] A major challenge in this field is understanding the surface compositions and structure, and their underlying electronic and electrochemical properties, that are responsible for catalysis. Such work is essential to guide the development of advanced catalysts.

Here we describe our progress in understanding the composition and structures of OER catalysts in basic media.

Experimental

The experimental electrocatalysis data shown here is generally collected for thin spun-cast or electrodeposited films of transition-metal oxides/hydroxides/oxyhydroxides on conductive substrates (e.g. Au/Ti on glass). Experiments are also typically performed on electrochemical quartz crystal microbalance (QCM) electrodes, so that the mass can be measured in situ. This provides not only for an estimate of the intrinsic activity on a turn-over frequency basis, but it also allows for monitoring the mass of the electrode during OER catalysis providing an experimental measure of catalyst chemical stability. The electrolytes used are typically the highest purity available (TraceSelect or Semiconductor Grade), yet even these were found to have sufficient impurities to affect the measured voltammetry. In some cases the electrolyte was first "pre-purified" using precipitated $Ni(OH)_2$ or $Co(OH)_2$ powder, which dramatically lowers the Fe impurity content.[2]

Results and Discussion

Given the high activity of Fe in Co and Ni oxyhydroxide we have previously measured,[2,3] we have recently studied in depth FeOOH, which is typically considered a poor OER catalyst.[4] We found that FeOOH is in fact a poor *electrical conductor* (Fig. 1a) and that the onsets of OER current and of electrical conductivity both occur at the same overpotential η of ~400 mV. We also found that FeOOH is chemically unstable – dissolving rapidly under anodic conditions (presumably due to FeO_4^{2-} formation), but that Fe in the CoOOH or NiOOH matrix appeared indefinitely stable. We found that FeOOH OER activity is dramatically enhanced by Au substrates, consistent with the formation of Fe-AuO_x mixed oxides under OER conditions. This work led to a new view of OER catalysis on FeOOH (and consequently on Fe oxides, which we find forms FeOOH on their surfaces). At overpotentials < 400 mV, the electrical conductivity is so poor that only the first few Fe cations that are in direct contact with the conductive support are OER active. At higher overpotentials the conductivity of FeOOH increases and a larger fraction of the film becomes OER active (Fig 1b and c).

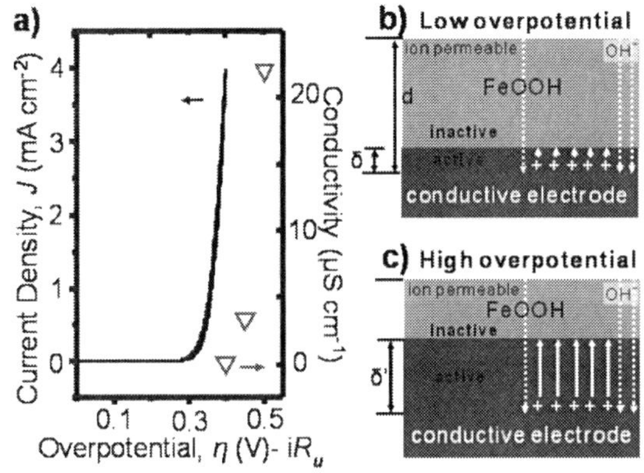

Figure 1. a) Voltammetry of FeOOH overlaid with in situ conductivity. b) Model of Fe activity at low and c) high overpotentials.

These results have thus led to a new view of OER reactivity on first-row transition metals. Fe is ubiquitous in high-activity catalysts. By rigorously removing Fe impurities, and minimizing confounding effects of conductivity, porosity, and phase instability, we established a *new intrinsic per-cation activity trend* for OER catalysts (Fig. 2):[5] $Ni(Fe)O_xH_y > Co(Fe)O_xH_y > FeO_xH_y\text{-}AuO_x > FeO_xH_y > CoO_xH_y > NiO_xH_y > MnO_xH_y$. The trend for first-row single component systems is *opposite* from other recent trends[6] (due to likely Fe contamination) and is not consistent with DFT calculations for related oxides.[7] This work provides a new measurement standard for the field and a reliable benchmark for future theory and experimental work.

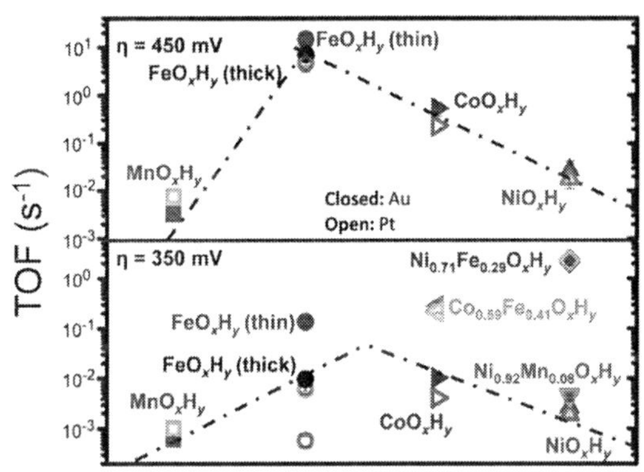

Figure 2. New activity trends for first row transition metal OER catalysts.[5] The trends are different than all other previously reported trends. Surprisingly, FeOOH is the fastest single-component OER catalyst.

Based on these discoveries we proposed a model for mixed Ni-Fe and Co-Fe oxyhydroxides, the fastest-known alkaline OER catalysts.[8] The fact that Fe is present in both (and that FeOOH is more active than previously thought) suggests Fe is needed at the active site. The NiOOH or CoOOH is essential and provides an *electrically conductive, chemically stable, and intrinsically porous/electrolyte-permeable host* for these Fe sites (Fig. 3). We know that nearly all the Ni/Fe sites are electrochemically accessible because integration of the charge in the Ni redox wave show ~ 1 e⁻ transferred for every Ni. Recent DFT calculations provide further evidence that Fe is the active site for OER,[9] although direct spectroscopic evidence of key OER intermediates on Fe are still needed.

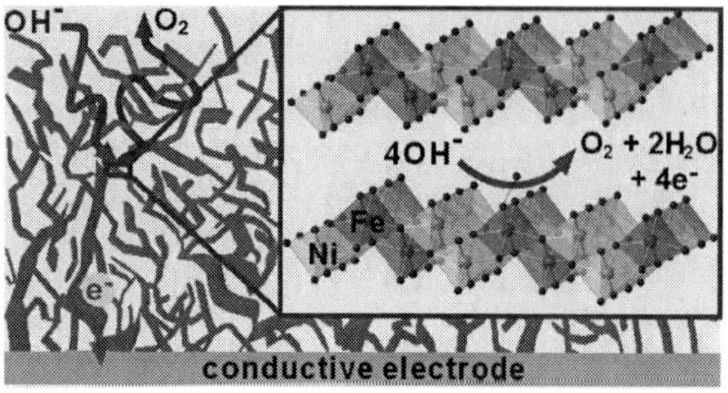

Figure 3. Schematic of active electrolyte-permeated catalyst structure. OER likely occurs at the Fe sites (orange) while the NiOOH host (grey) supports electronic transport and chemically stabilizes the Fe. Mass transport of OH^- and O_2 is facilitated by the open disordered structure and weak non-covalent interactions between the (oxy)hydroxide layers.

While the thin-film experiments provided fundamental understanding, practical electrolysis systems of interest to Proton require much higher mass loadings. At high mass loading, the intrinsic activity of each active-site can be dramatically affected by mass-transport and ionic resistances in the electrolyte/ionomer and electrical resistances of the catalyst film. We developed a pulsed electrodeposition process to deposit ~ 1 μm thick morphologies of Ni-Fe (oxy)hydroxide that retain the high intrinsic TOF of the thin-film samples used for fundamental studies (Fig. 4).[10] The pulse-electrodeposited films are denser, better electrically interconnected, and have more uniform Fe distribution than films from typical continuous electrodeposition. These films perform comparably to those with more-complicated architectures, e.g. on graphene or Ni-foam supports,[11-13] passing a current density of 120 mA cm^{-2} at 300 mV overpotential in 1 M KOH at room temperature with a loading of only 730 nmol metal cm^{-2}. We note that carbon is not a practically stable support under alkaline OER conditions, so self-supported architectures such as this are important.

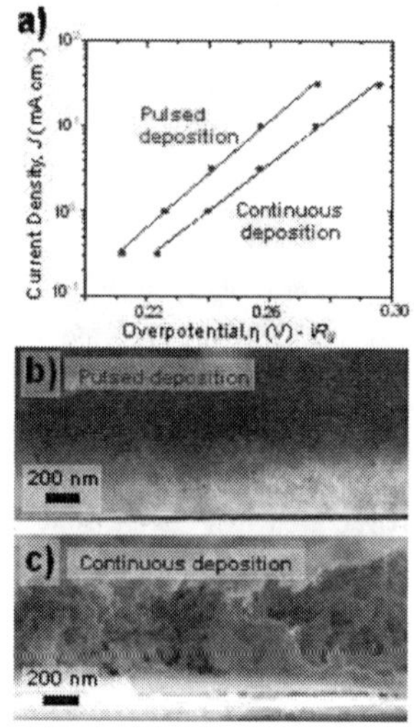

Figure 4. a) Tafel data and SEM images of b) pulsed and c) continuously deposited thick films.

Acknowledgment. This work was supported by the National Science Foundation through CHE-1301461. Initial work was also supported by the Center for Sustainable Materials Chemistry through the National Science Foundation grant CHE-1102637. The project made use of CAMCOR facilities supported by grants from the W. M. Keck Foundation, the M. J. Murdock Charitable Trust, ONAMI, the Air Force Research Laboratory (FA8650-05-1-5041), the National Science Foundation (0923577 and 0421086), and the University of Oregon. S.W.B. thanks the Research Corporation for Science Advancement for support as a Cottrell Scholar.

References

(1) Walter, M.; Warren, E.; McKone, J.; Boettcher, S. W.; Qixi, M.; Santori, L.; Lewis, N. S. *Chem. Rev.* **2010**, *110*, 6446.
(2) Trotochaud, L.; Young, S. L.; Ranney, J. K.; Boettcher, S. W. *J. Am. Chem. Soc.* **2014**, *136*, 6744.
(3) Burke, M. S.; Kast, M. G.; Trotochaud, L.; Smith, A. M.; Boettcher, S. W. *J. Am. Chem. Soc.* **2015**, *137*, 3638.
(4) Zou, S.; Burke, M. S.; Kast, M. G.; Fan, J.; Danilovic, N.; Boettcher, S. W. *Chem. Mater.* **2015**, *27*, 8011.
(5) Burke, M. S.; Zou, S.; Enman, L. J.; Kellon, J. E.; Gabor, C. A.; Pledger, E.; Boettcher, S. W. *J. Phys. Chem. Lett.* **2015**, 3737.
(6) Subbaraman, R.; Tripkovic, D.; Chang, K.-C.; Strmcnik, D.; Paulikas, A. P.; Hirunsit, P.; Chan, M.; Greeley, J.; Stamenkovic, V.; Markovic, N. M. *Nat. Mater.* **2012**, *11*, 550.
(7) Man, I. C.; Su, H.-Y.; Calle-Vallejo, F.; Hansen, H. A.; Martínez, J. I.; Inoglu, N. G.; Kitchin, J.; Jaramillo, T. F.; Nørskov, J. K.; Rossmeisl, J. *ChemCatChem* **2011**, *3*, 1159.
(8) Burke, M. S.; Enman, L. J.; Batchellor, A. S.; Zou, S.; Boettcher, S. W. *Chem. Mater.* **2015**, *27*, 7549.
(9) Friebel, D.; Louie, M. W.; Bajdich, M.; Sanwald, K. E.; Cai, Y.; Wise, A. M.; Cheng, M.-J.; Sokaras, D.; Weng, T.-C.; Alonso-Mori, R.; Davis, R. C.; Bargar, J. R.; Nørskov, J. K.; Nilsson, A.; Bell, A. T. *J. Am. Chem. Soc.* **2015**, *137*, 1305.
(10) Batchellor, A. S.; Boettcher, S. W. *ACS Catalysis* **2015**, *5*, 6680.
(11) Lu, X.; Zhao, C. *Nat. Commun.* **2015**, *6*, 6616.
(12) Gong, M.; Li, Y.; Wang, H.; Liang, Y.; Wu, J. Z.; Zhou, J.; Wang, J.; Regier, T.; Wei, F.; Dai, H. *J. Am. Chem. Soc.* **2013**, *135*, 8452.
(13) Ma, W.; Ma, R.; Wang, C.; Liang, J.; Liu, X.; Zhou, K.; Sasaki, T. *ACS Nano* **2015**, *9*, 1977.

LARGE-SIZE GRAPHENE TUBE CATALYSTS FOR SUSTAINABLE ELECTROCHEMICAL ENERGY STORAGE AND CONVERSION

Gang Wu

Department of Chemical and Biological Engineering, University at Buffalo, the State University of New York, Buffalo, New York 14260, United States

Introduction

The oxygen reduction reaction (ORR) and the oxygen evolution reaction (OER) are one pair of the most technologically important reactions for a variety of electrochemical energy storage and conversion technologies (e.g., fuel cells, metal-air batteries, and water splitting). It is important to design and develop advanced nonprecious-metal catalysts capable of facilitating these sluggish reactions at sufficient rates for practical applications. In our recent efforts,[1-7] cost-effective nitrogen-doped graphene nanocomposites have been prepared via the graphitization of nitrogen-containing polymers or compounds, combined with transition metals (Co, Ni or Fe). Among others, a new type of N-doped carbon tubes with large diameters (up to 500 nm) and relatively thin walls (less than 10 layers), which we call graphene tubes (GTs), were prepared via a scalable graphitization process of inexpensive dicyandiamide (**Figure 1**). Due to the high ratio of diameter-to-wall thickness, much increased surface area was obtained compared to conventional multi-walled carbon nanotubes (MWNTs).

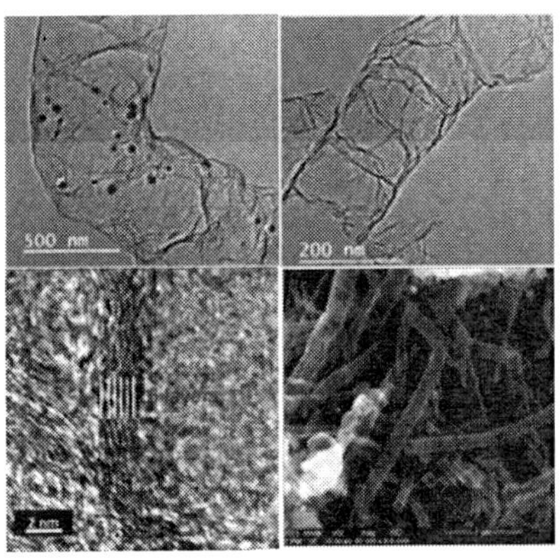

Figure 1. Typical graphene tubes with large size and thin thickness of wall (less than 10 layers)

Oxygen Reduction Reaction Electrocatalysis

An effective strategy for tuning the size of large-diameter nitrogen-doped graphene tubes (GT) from 50 to 200 nm was developed by varying the transition metals used to catalyze the graphitization. This effort demonstrates optimal manipulation of morphology and surface area, thereby providing an effective approach to further improving the performance of nonprecious metal catalysts. The catalytic activity of ORR for the graphene tube catalysts were studied in both acid and alkaline electrolytes. Fe shows the best activity in both electrolytes. Mn-based sample has the lowest activity, even the onset potential is comparable with nickel-based samples. Generally, all of catalysts exhibited much enhanced ORR activity in alkaline relative to acid. Especially, the Mn-based catalyst nearly doesn't have obvious activity in acid, but exhibiting well-defined limiting current density in alkaline. Once again, the optimal active sites for the ORR in both acid and alkaline media are likely different.

Effective Graphene Tube Supports for Pt Catalysts

Furthermore, aiming to improve the activity and stability of the state-of-the-art Pt catalysts, the ORR active GT is used as a matrix to disperse Pt nanoparticles in order to build a unique hybrid Pt cathode catalyst. This is the first demonstration of the integration of a nonprecious metal catalyst with Pt catalyst in a nanocomposite. Relative to traditional Pt/C catalyst, much improved activity and stability were achieved (**Figure 2**). This work provides a new hybrid concept to design and synthesis novel cathode catalysts for fuel cells.

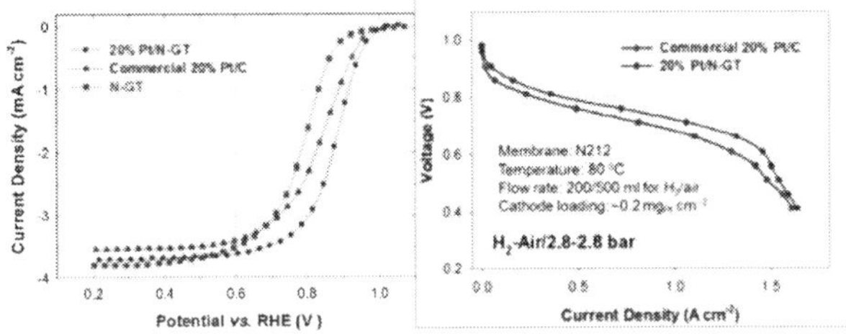

Figure 2. Enhanced catalytic activity of Pt catalysts supported on N-doped graphene tube supports, evidenced by RDE and fuel cell MEA tests.

Bifuncational Catalysts for the Oxygen Reduction and Oxygen Evolution.

Apart from dramatically enhanced ORR activity, the large-size graphene tubes demonstrated excellent OER activity comparable to state-of-the-art Ir black in alkaline media. Importantly, unlike traditional carbon blacks and 2D reduced graphene oxides, the graphene tubes can be remain stable in harsh electrochemical environments (potentials up to 1.9 V) in alkaline media (**Figure 3**). This provide a great opportunity to develop bifunctional oxygen catalysts for reversible electrochemical energy applications such as reversible alkaline media and metal-air batteries.

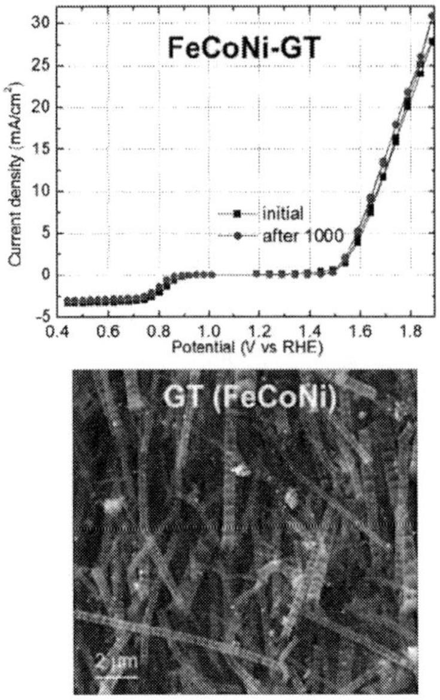

Figure 3. Good ORR/OER activity and sufficient durability observed with large-sized graphene tube bifuncational catalysts in 0.1 M NaOH solution.

Conclusions

In this work, a new nitrogen-doped graphene was prepared. The tube size can be well-controlled by using different transition metal showing different morphology and electrochemical properties Also, the graphene tube was identified as an effective nanocarbon support for deposition of Pt nanoparticle with enhanced activity and durability for fuel cell applications. In addition, a new type of nanocarbon, nitrogen-doped carbon tubes was prepared as a highly stable and ORR active nanocarbon. Future work will be focusing on the understanding the possible nonprecious metal active sites and further optimize the interaction between graphene tube supports and Pt catalysts with further enhanced activity and durability for real fuel cell applications.

Acknowledgement. This work was supported by the start-up fund from the University at Buffalo, SUNY along with U.S. Department of Energy, Fuel Cell Technologies Office (FCTO) Incubator Program (DE-EE0006960).

References

(1) G. Wu, K.L. More, C.M. Johnston, P. Zelenay, Science, 332, 443, 2011.
(2) G. Wu, P. Zelenay, Acc. Chem. Res., 46, 1878, 2013.
(3) G. Wu, N.H. Mack, W. Gao, S. Ma, J. Han, J.K. Baldwin, P. Zelenay. ACS Nano. 6, 9764, 2012.
(4) Q. Li, P. Xu, W. Gao, S. Ma, R. Cao, J. Cho, H.L. Wang, G Wu, Adv. Mater., 26, 1378,2014.
(5) Q Li, H Pan, D Higgins, R Cao, G Zhang, H Lv, K Wu, J Cho, G Wu, Small 11, 1443, 2015.
(6) Q. Li, R. Cao, J. Cho, G. Wu, Adv. Energy Mater., 4, 1301415,2014.
(7) X Wang, Q Li, H Pan, Y Lin, Y Ke, H Sheng, MT Swihart, G Wu, Nanoscale 7, 20290, 2015.

RAPID SULFUR-MELT INTEGRATION INTO ELECTROSPUN CARBON NANOFIBERS AS FREE-STANDING, BINDER-FREE CATHODES IN LITHIUM SULFUR BATTERIES

Caitlin Dillard, Vibha Kalra

Drexel University
Department of Chemical and Biological Engineering
3141 Chestnut Street, Philadelphia, PA 19104
Corresponding author: vk99@drexel.edu

Introduction

Energy storage demands are driving technology well beyond the capabilities of lithium ion (Li-ion) batteries. Out of the many emerging chemistries, lithium-sulfur (Li-S) batteries are conceivably the next step beyond Li-ion technology, packing a theoretical specific energy density ~5 times greater than Li-ion, 2,600 Wh kg^{-1} and 500 Wh kg^{-1}, respectively. Moreover, sulfur is earth-abundant, non-toxic and has a specific energy density of 1,675 mAh g^{-1}. However, the electronically insulating nature of sulfur, the dissolution and diffusion of reaction intermediates (shuttling effect), and the volume expansion during these reactions, have hindered the commercialization of sulfur cathodes.[1]

During discharge, sulfur reduces from S_8 to high-order, soluble polysulfides (S_n^{2-}, where $8 \geq n > 2$) and further reduces to low-order, insoluble polysulfides (S_n^{2-}, $n \leq 2$).[2] During this process, soluble polysulfides can diffuse and migrate to the anode side and become trapped upon charging, effectively depleting the cathode of active material upon each cycle. To this end, countless efforts have developed composites that encapsulate sulfur in a conductive matrix such as conductive polymers[3], graphene/graphene-oxides[4], and carbon scaffolds[5]. However, these methods include further powder-based processing by combining the sulfur-conductive matrix composite with binder (~10 wt%) and additional conductive carbon (~10-30%) in NMP, a toxic solvent, to cast the cathode as slurry upon a current collector. For comparison, a current collector can weigh up to 2 mg cm^{-2} in developed Li-ion batteries.[6]

Yuan *et al* [7] fabricated a free-standing, binder- and current collector-free sulfur-CNT composite via mixing, heat treatment, vacuum infiltration, and drying. The final cathode composition was ~54% sulfur by weight with high areal loading of ~6 mg cm^{-2} and resulted in an initial discharge capacity of ~995 mAh g^{-1} with an average capacity fade of 0.2% per cycle over 150 cycles. However, the tests were run at a very slow rate of C/20. Zeng *et al* produced sulfur-CNT-CNF composites via electrospinning and chemical activation of the CNF-CNT proceeded by an 8-hour sulfur infiltration. The final cathodes were approximately 40% sulfur by weight and exhibited a discharge capacity of 637 mAh g^{-1} after 100 cycles at a very low rate of 0.05 Ah g^{-1} (~C/33).

Here, we report a rapid, highly-controllable single-step sulfur-melting technique into electrospun carbon nanofibers (CNF) to produce entirely free-standing cathodes, without the use of any binder or additional current collector, therefore reducing the weight of inactive materials drastically. To this end, our novel S-CNF cathodes with typical sulfur loading of 1 mg cm^{-2} exhibited initial discharge capacities of 1,382, 1,310, and 1,140 mAh g^{-1} at C/10, C/5, and C/2 rates, respectively. After 20 cycles, the discharge capacities were 1,106, 823, and 790 mAh g^{-1}, respectively. Furthermore, highest initial discharge capacity of 1,625 mAh g^{-1} (97% of the theoretical maximum – 1,675 mAh g^{-1}) with a stable discharge capacity of 1,200 mAh g^{-1} after 50 cycles at a C/10 rate was achieved for one cathode. This performance in relation to the weight of the entire *electrode* was 330 mAh gram-electrode^{-1}, exceeding that of common Li-ion batteries (~100–200 mAh g^{-1}, which does *not* account for the additional current collector weight).[8]

The electrospun carbon nanofibers provide numerous advantages: (1) facile processing, (2) excellent electronic conductivity, (3) macropores for physical entrapment of polysulfides and diffusion of ions, and (4) electrostatic interactions to reduce polysulfide migration/diffusion. We believe the high performance from this facile cathode design will elucidate innovative cathodes to make the commercialization of lithium sulfur batteries a reality.

Experimental

Cathode Fabrication. Carbon nanofibers (CNF) were fabricated from 10 wt% polyacrylonitrile (PAN, Sigma-Aldrich) dissolved in N, N-dimethylformamide (DMF, Sigma Aldrich) were stirred at room temperature for 4–5 hours. The nanofibers were electrospun at room temperature and relative humidity below 20%. The distance between the needle tip and the grounded collector (aluminum foil) was 15 cm, a 22 gauge stainless steel needle, and an applied voltage of 13–14 kV were. The electrospun nanofibers were stabilized in air by heating to 280 °C for 5 h. The stabilized nanofibers were then pyrolyzed (carbonized) under steady nitrogen flow in a horizontal tube furnace at a heating rate of 2 °C min^{-1} to the temperature of 1000 °C and held for 1 h.[9] Elemental sulfur (S_8 Aldrich) was integrated into CNF by rapid heating of sulfur well above 150 °C combined with pressure between sulfur and CNF cathode for less than 30 seconds.

Material Characterization. Morphology of S-CNF cathodes was examined using a scanning electron microscope (SEM, Zeiss Supra 50VP) equipped with energy dispersive spectroscopy (EDS, Oxford). EDS was used to detect relative compositions of carbon, sulfur, nitrogen, and oxygen.

Electrochemical Characterization. S-CNF cathodes were used directly as cathodes in a CR-2032 coin cell. The cell assembly included the S-CNF cathode without further treatment, polypropylene separator (Celgard 3501), lithium foil anode (###), and nickel foam spacers. The electrolyte comprised of approximately 1.85 M Lithium lithium trifluoromethanesulfonate ($LiCF_3SO_3$, Sigma Aldrich) and 0.1 M lithium nitrate ($LiNO_3$, Alfa Aesar) in dimethyl ether (Sigma) and 1,3-dioxalane (DME:DOL) in a 1:1 volume ratio. Small amount of lithium nitrate was added to reduce passivation layer forming on the lithium anode.

The cycle stability and cyclic charge-discharge were tested on a battery cycler (Maccor 3000) between a voltage range of 1.6 – 2.8 V. The current was determined based on the amount of sulfur and the desired C rate, where 1C = 1,675 mAh g^{-1}.

Results and Discussion

Carbon nanofibers (CNF) were electrospun and carbonized to form a stable free-standing conductive support for Li-S cathodes. Commercial elemental sulfur was infused into the CNF mats via novel, single-step melting technique. The sulfur-melt CNF composite cathodes, denoted as S-CNF, were used directly in coin cells without further treatment. The melt procedure produced a sulfur gradient through the CNF, see Figure 1. In Figure 1, a SEM micrograph of a device-ready S-CNF cathode cross-section with EDS mapping shows highly concentrated sulfur from the point of entry. The arrows indicate the CNF side in which sulfur-melt was integrated. The carbon map in Figure 1 shows homogenous carbon content furthest from the sulfur-point-of-entry. This gradient of sulfur may be the most beneficial for long-term cyclability because the carbon-rich region may act as an incorporated interlayer. Previous work by Singhal *et al* [9] reported the incorporation of CNF interlayers between a conventional powder sulfur cathodes and the separator resulted in tremendous increases in cyclability and very high rate capability.

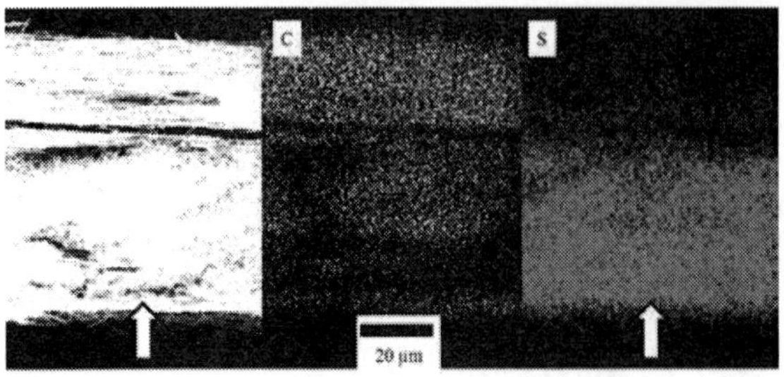

Figure 1. SEM micrograph of S-CNF (left) cathode cross-section with corresponding carbon (middle) and sulfur (right) maps. The arrow indicates the direction is which sulfur was melted into the CNF.

Electrochemical studies were performed to evaluate S-CNF cathode performance. In Figure 2, cathodes with ~30wt% (1 mg cm^{-2}) sulfur loading were tested at C/10, C/5, and C/2 rates. The initial discharge capacities were 1,382, 1,310, and 1,140 mAh g^{-1} at C/10, C/5, and C/2 rates, respectively. After 20 cycles, the discharge capacities were 1,106, 823, and 790 mAh g^{-1} (corresponding capacity retentions of 80%, 63%, and 70% respectively).

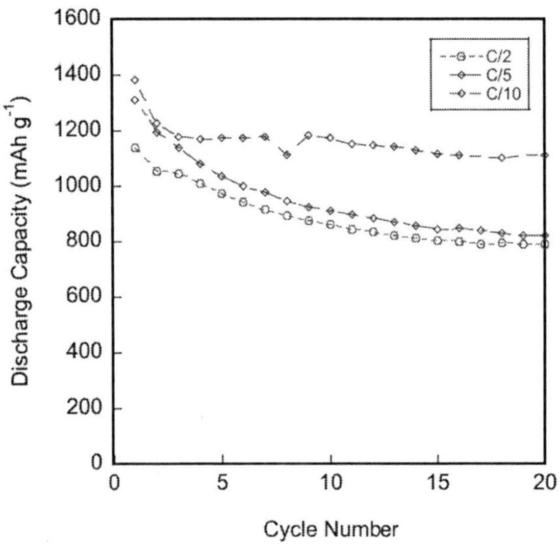

Figure 2. S-CNT cathode cyclability at C/10 (red), C/5 (blue), and C/2 (green) rates. Sulfur loading of approximately 30wt% and 1 mg cm^{-2} for each cathode. (1C = 1,675 mAh g^{-1})

The initial discharge capacity decreased as the discharge-charge rates increased from C/10 to C/2, as expected. Upon further cycling, the tests at C/5 and C/2 resulted in similarly decreasing capacities, in comparison to the C/10 test which stabilized at about 1,200 mAh g^{-1} within the first few cycles. This was also expected, as slower rates are limited by reaction kinetics, not diffusion. The slower discharge-charge rate allowed for more complete utilization of sulfur by effectively providing more time for electrolyte penetration and diffusion of reaction intermediates away from initially inaccessible sulfur. As the rate increases, diffusion, rather than kinetics, becomes the dominant limiting factor.

A cathode of similar sulfur loading tested at C/10 rate exhibited a very high initial discharge capacity of 1,625 mAh g^{-1}, meaning 97% of sulfur was utilized (theoretical maximum = 1,675 mAh g^{-1}), shown in Figure 3. The discharge capacity was stable around 1,200 mAh g^{-1} after 50 cycles.

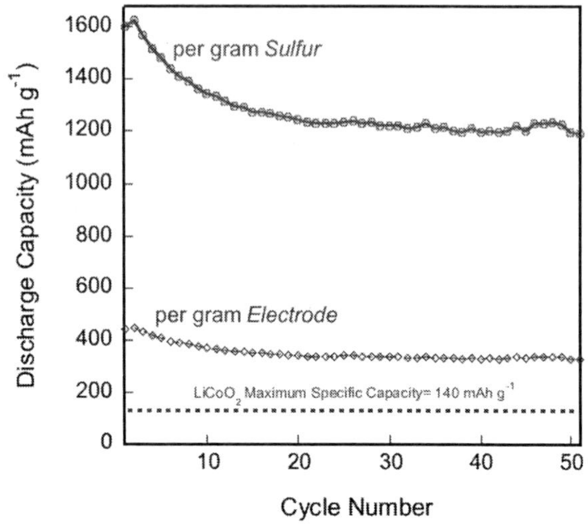

Figure 3. Discharge capacity of one S-CNF cathode at C/10 rate over 50 cycles. Sulfur loading of 28 wt% and 1.4 mg cm^{-2}. The data were analyzed on a per gram sulfur (red) and per gram electrode (blue) basis

From a device standpoint, a cathode with this performance was very promising. Relative to the weight of the *entire electrode*, a stable discharge capacity of 330 mAh gram-electrode^{-1} was achieved after 50 cycles, exceeding common Li-ion batteries (<200 mAh g^{-1}) in which the typical weight of current collector (~2 mg cm^{-2})[6] is not factored in. The discharge capacity *per gram of electrode* for this cathode and a common Li-ion (lithium cobalt oxide) capacity were also represented in Figure 3 for comparison.

Conclusions

We demonstrated a novel methodology for the integration of sulfur into an electrospun carbon nanofiber support as rational design of cathodes from lithium sulfur batteries. This design eliminates the need for current collector and binder while providing a completely free-standing, device-ready electrode. We report high discharge capacities of over 1,200 mAh g^{-1} over 20 cycles for a cathodes with ~1mg cm^{-2} sulfur loading. Furthermore, this design could produce discharge capacities of 790 mAh g^{-1} at a C/2 rate for 1.0 mg cm^{-2} loading. Furthermore, we achieved 97% of the theoretical maximum specific discharge capacity of sulfur in a cathode with 28wt% and 1.4 mg cm^{-2} with a high initial discharge of 1,625 mAh g^{-1} and a stable capacity of 1,200 mAh g^{-1} over 50 cycles. This novel design allows precise control over sulfur loading and shows promising electrochemical performance for the future of energy storage devices. Optimized cathode design will allow us to fabricate ultra-high sulfur loadings (>5 mg cm^{-2}) at varying charge-discharge rates to investigate the full potential of this cathode for lithium sulfur technology.

Acknowledgement. We would like to thank National Science Foundation for funding this research. We would also like to thank the Centralized Research Facility at Drexel University for accessibility of their characterization equipment.

References

1. (a) Li, S.; Lin, Z., Lithium-Sulfur Battery. In *Rechargeable Batteries: Materials, Technologies and New Trends*, Zhang, Z.; Zhang, S. S., Eds. Springer International Publishing: Cham, 2015; pp 587-610; (b) Etacheri, V.; Marom, R.; Elazari, R.; Salitra, G.; Aurbach, D., Challenges in the development of advanced Li-ion batteries: a review. *Energy & Environmental Science* **2011,** *4* (9), 3243-3262.
2. Zhang, S. S., Liquid electrolyte lithium/sulfur battery: Fundamental chemistry, problems, and solutions. *Journal of Power Sources* **2013,** *231*, 153-162.
3. (a) Liu, Y.; Zhang, J.; Liu, X.; Guo, J.; Pan, L.; Wang, H.; Su, Q.; Du, G., Nanosulfur/polyaniline/graphene composites for high-performance lithium–sulfur batteries: One pot in-situ synthesis. *Materials Letters* **2014,** *133*, 193-196; (b) Ma, G.; Wen, Z.; Jin, J.; Lu, Y.; Wu, X.; Wu, M.; Chen, C., Hollow polyaniline sphere@sulfur composites for prolonged cycling stability of lithium-sulfur batteries. *Journal of Materials Chemistry A* **2014,** *2* (27), 10350-10354.
4. Xu, C.; Wu, Y.; Zhao, X.; Wang, X.; Du, G.; Zhang, J.; Tu, J., Sulfur/three-dimensional graphene composite for high performance lithium–sulfur batteries. *Journal of Power Sources* **2015,** *275*, 22-25.
5. (a) Ji, X.; Lee, K. T.; Nazar, L. F., A highly ordered nanostructured carbon-sulphur cathode for lithium-sulphur batteries. *Nat Mater* **2009,** *8* (6), 500-506; (b) Zhang, J.; Xiang, J.; Dong, Z.; Liu, Y.; Wu, Y.; Xu, C.; Du, G., Biomass derived activated carbon with 3D connected architecture for rechargeable lithium – sulfur batteries. *Electrochimica Acta* **2014,** *116*, 146-151.
6. Xue, N.; Du, W.; Gupta, A.; Shyy, W.; Marie Sastry, A.; Martins, J. R. R. A., Optimization of a Single Lithium-Ion Battery Cell with a Gradient-Based Algorithm. *Journal of The Electrochemical Society* **2013,** *160* (8), A1071-A1078.
7. Yuan, Z.; Peng, H.-J.; Huang, J.-Q.; Liu, X.-Y.; Wang, D.-W.; Cheng, X.-B.; Zhang, Q., Hierarchical Free-Standing Carbon-Nanotube Paper Electrodes with Ultrahigh Sulfur-Loading for Lithium–Sulfur Batteries. *Advanced Functional Materials* **2014,** *24* (39), 6105-6112.
8. Nitta, N.; Wu, F.; Lee, J. T.; Yushin, G., Li-ion battery materials: present and future. *Materials Today* **2015,** *18* (5), 252-264.
9. Singhal, R.; Chung, S.-H.; Manthiram, A.; Kalra, V., A free-standing carbon nanofiber interlayer for high-performance lithium-sulfur batteries. *Journal of Materials Chemistry A* **2015,** *3* (8), 4530-4538.

ELECTROCATALYTIC ACTIVITY OF METAL OXIDE NANOCATALYSTS FOR LI-AIR AND NA-AIR BATTERIES

Mehmet C. Yagci, Bhushan Patil, Eda Yilmaz

Bilkent University UNAM, Bilkent, Ankara 06800, Turkey
can.yagci@bilkent.edu.tr

Introduction

Metal-air batteries. Metal-air batteries are promising candidates to replace Li-ion batteries for a more viable alternative for renewably generated energy storage and electric vehicles, due to their higher theoretical specific energies. Li-air batteries offer ten-fold increase in theoretical specific energy, however, high overpotentials, lacks cyclability and rate capability, due to electrode fouling by deposition of products and side products such as Li_2O_2, LiO_2, Li_2CO_3, LiOH etc. limit its commercialization. In addition, Na-air batteries are also advantageous in terms of cost and abundance of Na, but there are few rechargeable Na-air batteries ever realized.[1] This can be attributed to the insufficient understanding about discharge product properties and identification, whether NaO_2, Na_2O_2 or both. In order to solve aforementioned problems, rigorously designed gas diffusion cathodes (GDC) with novel bi-functional catalysts are required. Current studies focus on the synthesis and electrocatalytic activity of high surface area nanocatalysts.

Instruments and Techniques. Studies on oxygen reduction reaction (ORR) catalysts for fuel cells using Rotating Disk Electrode (RDE) setup is well accepted and became the standard technique for the field. Using RDE sweep techniques such as Cyclic Voltammetry (CV) and Linear Sweep Voltammetry (LSV) electrocatalytic activity can be monitored from measured electrode kinetics independent of mass transfer limitations.

Experimental

Electrochemical characterizations of reduction/oxidation reactions that take place during battery cycling are valuable during GDC design. Literature reveals RDE studies based on the precious metals[2] and perovskites[3] modified carbon based materials enhances electrocatalytic activity towards ORR and Oxygen Evolution Reaction (OER). In the same context present research focus on the fabrication of MOx (M=Co,Mo,Mn,Ru) nanoparticles decorated carbon nanotube (CNT) modified glassy carbon electrodes, as bifunctional electrocatalysts. Thus modified electrodes were used as working electrode and Lithium as counter electrode in the standard two electrode cells with 0.5M LiTFSI in aprotic TEGDME electrolyte. Cyclic Voltammetry (CV) was used to realize catalytic activity of selected nano metals towards ORR and OER. In addition, chronoamperometry and Linear Sweep Voltammetry (LSV) for the analysis of products formation during ORR and their oxidation during anodic scan were determined.

Results and Discussion

Among the selected catalysts CoOx shows highest electrocatalytic activity towards ORR with onset potential 2.9 V vs. Li/Li^+. LSV results demonstrate formation of Li_2CO_3 and Li_2O_2, which were oxidized at ca. >4V, 3.25-3.75V vs. Li/Li^+, respectively. In agreement with the previous studies Li_2O_2 was found to be the main product during discharge by forming two different product morphologies i.e. toroidal and amorphous with oxidation potential of 3.25V and 3.5V vs. Li/Li^+.[4]

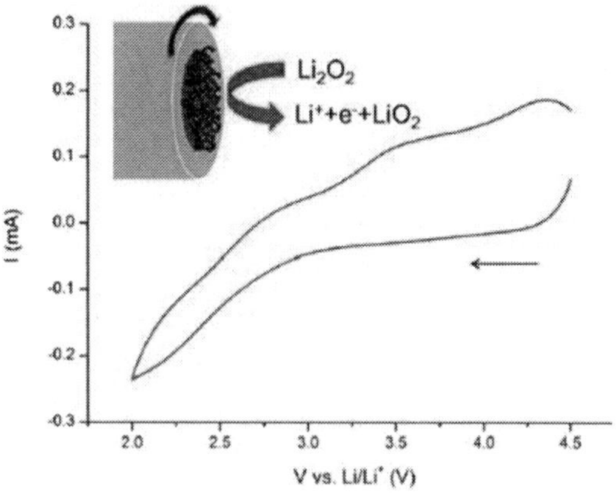

Figure 1. CV obtained at CoOx/modified CNT/GC under 0.5M LiTFSI in aprotic TEGDME electrolyte with v 10 mV s-1. Arrow shows the scan direction and inset is the schematic representation of the proposed ORR reaction at working electrode.

Conclusions

We believe electrochemical characterization of electrocatalytic activity of as-synthesized carbon-based cathodes with nano-electrocatalysts is a useful tool before proceeding into the cell preparation and testing phase of metal-air batteries. Moreover, RDE measurements can be compared with charge-discharge plateaus from the cell tests, SEM micrographs and XPS spectra for a detailed analysis of the role of the selected catalyst on the redox activity, product type, product formation pathway and morphology.

Acknowledgement. The Scientific and Technological Research Council of Turkey (TUBITAK, project 115M375) is acknowledged for funding the research.

References

(1) Sahgong, S. H.; Senthilkumar, S. T.; Kim, K.; Hwang, S. M.; Kim, Y. *Electrochem. commun.* **2015**, *61*, 53–56.

(2) Lu, Y.-C.; Gasteiger, H. A.; Crumlin, E.; McGuire, R.; Shao-Horn, Y. *J. Electrochem. Soc.* **2010**, *157 (9)*, A1016.

(3) Zhao, Y.; Xu, L.; Mai, L.; Han, C.; An, Q.; Xu, X.; Liu, X.; Zhang, Q. *Proc. Natl. Acad. Sci. U. S. A.* **2012**, *109 (48)*, 19569–19574.

(4) Ganapathy, S.; Adams, B. D.; Stenou, G.; Anastasaki, M. S.; Goubitz, K.; Miao, X.-F.; Nazar, L. F.; Wagemaker, M. *J. Am. Chem. Soc.* **2014**. 136 (46), 16335–16344

A STABLE ROOM-TEMPERATURE SODIUM-SULFUR BATTERY

*Shuya Wei[1] and Lynden A. Archer[1]**

1 School of Chemical and Biomolecular Engineering, Cornell University, Ithaca, NY 14853, United States
*laa25@cornell.edu

Abstract

High-energy and inexpensive rechargeable battery systems based on earth-abundant materials are important for both mobile and stationary energy storage technologies. Rechargeable sodium-sulfur (Na-S) batteries that are able to operate stably at room temperature are among the most sought-after of these platforms because these cells take advantage of a two-electron-redox process to yield high storage capacity from inexpensive electrode materials. Realization of practical Na-S batteries has been fraught with multiple stubborn problems ranging from unstable electrodeposition of sodium during battery recharge to rapid loss of the active cathode material by dissolution into the electrolyte. In this study, we develop a room temperature Na-S battery that uses a sodium metal anode, a carbon-sulfur composite cathode, and a liquid electrolyte containing tethered silica nanoparticles as a deposition stabilizer. We show that the Na-S cells with this configuration can cycle stably for over 100 cycles at 0.5C (1C = 1675 mAh/g) with 600 mAh/g reversible capacity and nearly 100 percent Coulombic efficiency. By means of spectroscopic and electrochemical analysis, we find that the high stability and reversibility of the cells stem from at least two sources. First, the tethered silica particles spontaneously form a Na-ion conductive film on the anode, which appears to stabilize deposition of sodium and prevent electrolyte decomposition and depletion. Second, on the cathode side, porous carbon materials play a key role that can constrain the electrochemical reaction between sodium ion and sulfur to the solid state. This combination of electrolyte and carbon substrate are shown to provide sufficiently strong association of sulfur in the cathode and at the same time stabilize the surface of the highly reactive sodium metal anode.

EXPLOITING IONOMERIC MATERIALS FOR HIGH PERFORMANCE LITHIUM-SULFUR BATTERIES

*Jinhong Lee, Jongchan Song, Hyungjun Noh, Hongkyung Lee, Yun-Jung Kim and Hee-Tak Kim**

Department of Chemical and Biomolecular Engineering, Korea Advanced Institute of Science and Technology, 291, Daehak-ro, Yuseong-gu, Daejeon 305-701, Republic of Korea
yeetak.kim@kaist.ac.kr

Introduction

Recently, lithium–sulfur (Li–S) battery is considered a promising candidate to succeed LIBs because of its high theoretical energy density of 2500 Wh·kg^{-1}, non-toxic nature, low cost, and natural abundance.[1,2] The conversion chemistry between solid sulfur and solid Li_2S can proceed in reversible manner, in spite of their electronically insulating nature, via the formation of soluble polysulfides (PSs) which renders a liquid-solid mechanism working.[3] However, ironically, the soluble PSs cause challenges including low sulfur utilization, short cycle life, low cycling efficiency, and high self-discharge rate. These are mainly attributed to a process known as the polysulfide (PS) shuttle[4,5]; the soluble PSs diffuse to the Li anode where they directly react with the Li metal to produce lower order PS species, which diffuse back to the sulfur cathode to regenerate higher PS forms.[4,5] The PS shuttle leads to incomplete charging of the sulfur electrode and corrosion of the Li electrode, thus resulting in poor battery performance. Prevention of the PS shuttle is therefore extremely important for the practical use of Li–S battery. Previous efforts to mitigate the PS shuttle were mainly focused on the following[2]: 1) suppressing the diffusion of the dissolved PS out from the cathode by the geometrical trapping of PS in porous carbon or PS adsorption on oxides, conducting polymers, and by ion-selective polymer and 2) protecting the Li anode from reaction with PS by forming a passivation layer on the Li metal surface or a hybrid anode structure.

Against the backdrop, we report a Li-S battery design with exploiting ionomer materials to suppress PS shuttle while maintaining a fast redox kinetics via liquid-solid mechanism. This approach features the use of hybrid single ion conductor (HSIC) as an electrolyte medium or a protective layer for Li-S battery. The HSIC, which consists of a polymeric single ion conductor swollen with lithium salt-free organic solvent, or lithium salt-containing provides a Li^+ hopping through a nanoscale conducting channel (Fig. 1a) and suppresses PS dissolution according to Donnan exclusion principle (Fig. 1b). In this work, lithiated perfluorinated sulfonic acid (Li-PFSA) polymer and lithiated α-lipoic acid were strategically employed as a proof-of-concept. The mixed organic solvent dissociates Li-PFSA and forms Li^+ conducting phase in PFSA matrix (Fig. 1a). The dissociated fixed negative charged sulfonate anions create a Donnan potential, which lowers PS solubility of the negatively charge PSs.

Experimental

Lithiation of ionomer. A PFSA membrane was lithiated by immersing the membrane in a solution of 2.0 M LiOH in H_2O: EtOH (2:1 by weight) mixture at 80°C for 6 h under vigorous stirring. Then the resulting lithiated membrane was rinsed in deionized water to remove residual salt. The Li^+ exchanged membrane is used as a separator for HSIC system.

Fabrication of Cathodes. A Ketjen black carbon/sulfur composite was fabricated as the active material by melting method. A slurry which consists of 20wt% KB/S composite, 10wt% vapor grown carbon fiber (VGCF), and 70wt% Li-PFSA was casted on the etched Al foil. In the other configuration, a Li-PFSA layer was formed on the surface of a sulfur cathode. The casted slurry was dried in 60°C oven for 24h.

Results and Discussion

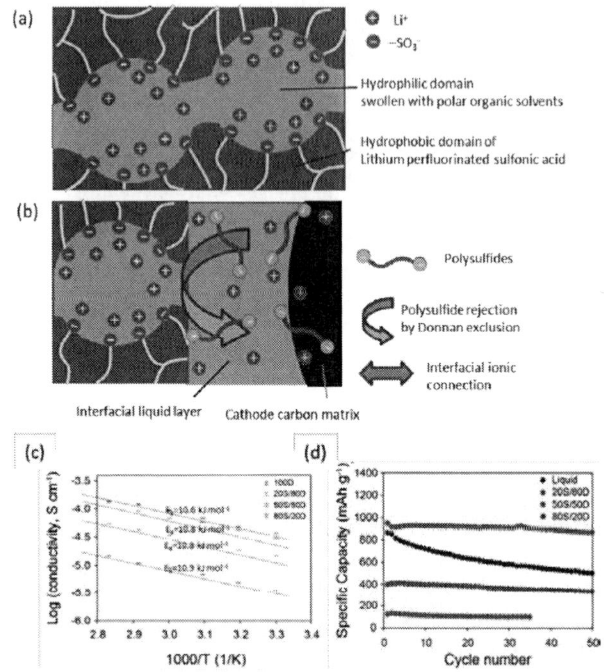

Figure 1. (a) Model for a nanophase separation in hybrid single ion conductor, (b) polysulfide rejection by Donnan exclusion at the interface between HSIC and liquid electrolyte, (c) temperature dependency of HSICs and (d) capacity retentions for the Li-S cells with HSICs.

The HSIC presented in this work was inspired by hydrated PFSA polymers which is featured by heterogeneous membrane structure of non-conducting hydrophobic polymer phase and nano-scale solvent-filled phase surrounded by the fixed sulfonate anion groups within which ionic transfer takes place. In order to design organic solvent-based HSIC which can dictate the morphological feature and the benefits of the hydrated PFSA polymers, the Li-PFSA polymer was equilibrated in various organic solvents frequently used for lithium batteries, and their lithium ion conductivities and solvent uptakes were measured. Among the solvents, the mixture of EC/PC showed the highest ionic conductivity (1.5 X 10^{-4} S cm^{-1}) and the highest solvent up-take amount (91%) at 25°C, however, carbonate solvents were excluded for the use in Li-S batteries because of their irreversible reaction with high order PS. DME and DOL, generally employed for Li-S battery, are not compatible with the Li PFSA polymer; they have low swelling ratios and consequent low Li ion conductivities. Interestingly, sulfolane showed a high swelling of 71% and reasonably high Li^+ conductivity above its melting point of 27°C. For providing room temperature Li^+ conductivity, diglyme, which is compatible with sulfur and lithium metal electrode, was employed as a co-solvent. The HSICs fully swollen with sulfolane/diglyme mixtures of various mixing ratios (0/100, 20/80, 50/50, and 80/20 in volume ratio) exhibited an increased solvent uptake from 14 to 69 % and increased Li^+ conductivity from 0.595 to 3.88×10^{-5} S cm^{-1} with increasing sulfolane content from 0 to 80%. The temperature dependency of the ionic conductivities follows

Arrhenius equation, which is typical of hopping mechanism (Fig. 1c). Interestingly, the activation energies were found to be in the narrow range of 10.6~10.8 KJ mol^{-1} in spite of the differences in sulfolane/digylme ratio. It indicates that the interaction between solvated Li^+ and fixed SO_3^- and the hopping distance between neighboring SO_3^- group, which mainly influence the activation energy for Li ion hopping, are not much different among these HSICs.

The liquid electrolyte cells showed overcharging behavior during charging and shortening of the upper plateau and lower plateau with cycling regardless of sulfolane/diglyeme ratio. These behaviors indicate a significant PS loss due to PS dissolution from the sulfur cathode. For the 50S/50D liquid electrolyte cell, the upper and lower voltage plateaus in discharge profile became shorten upon extended cycling, resulting in the fast capacity fade in the early stage of cycling (Fig. 1d), By contrast, the HSIC-based Li-S batteries showed profoundly improved capacity retention than the liquid electrolyte-based references. In particular, HSIC-50S/50D cell delivered a high discharge capacity of 863 mAh g^{-1} after 50 cycles. The HSIC-based Li-S batteries (20S/80D, 50S/50D, 80S/20D) revealed coulombic efficiencies higher than 92% coulombic efficiency for whole cycle ranges, which are far higher than those of the liquid electrolyte based batteries (~70%). These results demonstrate that the HSIC approach can pave a new way to address the current problems of Li-S batteries.

Acknowledgement. This work was supported by Korea Institute of Science and Technology (2E26291) and Basic Science Research Program through the National Research Foundation of Korea (NRF) funded by the Ministry of Education (NRF-2014R1A1A2056199).

References

(1) Bruce P.G.; Freunberger S.A.; Hardwick, L.J.; Tarascon, J.M., *Nat. Mater.* **2012**, *11*, 19.
(2) Manthiram, A; Fu, Y.; Chung, S.H.; Zu, C.; Su, Y.S., *Chem. Rev.* **2014**, *114*, 11751.
(3) Yamin, H.; Gorenshtein, A.; Penciner, J.; Sternberg, Y.; Peled, E. *J. Electrochem. Soc.* **1988**, *135*, 1045.
(4) Cheon, S.E.; Ko, K.S.; Cho, J.H.; Kim, S.W.; Chin, E.Y.; Kim, H.T., *J. Electrochem. Soc.* **2003**, *150*, A800.
(5) Cheon, S.E.; Ko, K.S.; Cho, J.H.; Kim, S.W.; Chin, E.Y.; Kim, H.T., *J. Electrochem. Soc.* **2003**, *150*, A796.

FUNCTIONAL ORGANOSULFIDE ELECTROLYTE FOR HIGH PERFORMANCE LITHIUM–SULFUR BATTERIES

Shuru Chen, Yue Gao, Donghai Wang

Department of Mechanical and Nuclear Engineering
The Pennsylvania State University
University Park, PA 16802

Lithium-sulfur (Li-S) batteries have attracted great attention because of the very high theoretical specific energy of sulfur cathodes, ~2,600 Wh kg^{-1}. Dissolution, diffusion and side-reaction of lithium polysulfide intermediates and irreversible deposition of Li_2S are known to be the main technical barriers to achieving high-energy-density, long-cycling Li-S batteries. Most research has thus focused on improving cell cycle life by hindering the diffusion and side-reaction of dissolved lithium polysulfide intermediates. Today, despite some progress, these issues have not been intrinsically solved. This sharp focus has also largely ignored another persistent issue: the huge excess of electrolyte required for adequate Li-S cell performance and resultant drawdown in the energy density.

In this work, we report a new electrochemical pathway for Li-S batteries using dimethyl disulfide (DMDS) as a functional electrolyte co-solvent.[1] DMDS promotes the discharge of sulfur by formation and subsequent reduction of soluble dimethyl polysulfide species to new intermediates/products including lithium organosulfides. This largely avoids formation of soluble lithium polysulfides and has two major advantages. First, DMDS boosts cell capacity while avoiding the extreme increase in electrolyte viscosity and cell resistance associated with lithium polysulfide dissolution, enabling high performance with much less electrolyte. Second, the new intermediates/products trigger an automatic shutoff mechanism that stops discharge before cell-killing agglomerates of lithium sulfide can form as final reduction products, thus enhancing cell cycle life. Combined, these advantages allow cells to reach unparalleled performance: a stable capacity of around 1,000 mAh g-1 sulfur (700 mAh g^{-1} based on the carbon/sulfur cathode composite) at a very low electrolyte/sulfur ratio of 5 mL g^{-1} using practical, high-sulfur-loading (4 mg cm^{-2}) cathodes, almost double the capacity obtained with conventional electrolytes under the same harsh conditions.

The work enclosed here makes major inroads on the technical issues for Li-S batteries, presenting a new electrochemical pathway for greatly enhancing their energy density and cycle life.

Acknowledgment. This work was supported by the Assistant Secretary for Energy Efficiency and Renewable Energy, Office of Vehicle Technologies of the U.S. Department of Energy under Contract No. DE- EE0005475.

Reference

(1) Chen, S. R., Dai, F., Gordin, M. L., Yu, Z. X., Gao, Y., Song, J. X., Wang, D. H. *Angewandte Chemie International Edition* **2016**, 55(13), 4231–4235.

SUSTAINABLE POWER SOURCES BASED ON HIGH EFFICIENCY THERMOPOWER WAVE DEVICES

Sayalee G. Mahajan[1,ⱡ], Albert Tianxiang Liu[1,ⱡ], Anton L. Cottrill[1], Yuichiro Kunai[1], David Bender[1], Javier Castillo Jr[1], Stephen L. Gibbs[1], and Michael S. Strano[1]*

[1] 77 Massachusetts Ave, Department of Chemical Engineering, Massachusetts Institute of Technology, Cambridge, MA, USA

ⱡ These authors contributed equally to this work.

* Corresponding author's email address: strano@MIT.edu

Abstract

There is a pressing need to find alternatives to portable power sources such as Li-ion batteries, which contain toxic metals, present recycling difficulties due to harmful inorganic components, and rely on elements in finite global supply. Thermopower wave (TPW) devices, which convert chemical to electrical energy by means of self-propagating reaction waves guided along nanostructured thermal conduits, have the potential to address this demand. Herein, we demonstrate orders of magnitude higher chemical-to-electrical conversion efficiency of thermopower wave devices, in excess of 1%, with sustainable fuels such as sucrose and NaN_3 for the first time, that produce energy densities on par with Li-ion batteries operating at 80% efficiency (0.2 MJ/L versus 0.8 MJ/L). We show that efficiency can be increased significantly by selecting fuels such as sodium azide or sucrose with potassium nitrate to offset the inherent penalty in chemical potential imposed by strongly p-doping fuels, a validation of the predictions of Excess Thermopower theory. Such TPW devices can be scaled to lengths greater than 10 cm and durations longer than 10s, an over 5-fold improvement over the highest reported values, and they are capable of powering a commercial LED device. Lastly, a mathematical model of wave propagation, coupling thermal and electron transport with energy losses, is presented to describe the dynamics of power generation, explaining why both unipolar and bipolar waveforms can be observed. These results represent a significant advancement toward realizing TPW devices as new portable, high power density energy sources that are metal-free.

Introduction

Current battery technology is typically comprised of toxic metals, presenting problems for recycle, and environmental exposure.[1] There is a pressing need for portable power sources that are sustainable, preferably using materials that are not in finite supply or mined from the earth, such as lithium.[2] Over the past decade, researchers have explored many potential alternatives, namely secondary batteries made with other metals,[2-5] pure organic materials,[6] graphitic carbon nitride,[7] and even exotic bio-organic materials like DNA for energy conversion and storage.[8] Numerous reports have given high merits for nanotechnologies (i.e. functionalized graphene[9] and strongly coupled inorganic-nano-carbon hybrid materials[10]) in this field. Such approaches benefit from their use of metal free, organic materials, but have not yet demonstrated power or energy densities commensurate with power sources such as the Li-ion battery, at any size or time scale. The scaling with electrode area fundamentally limits specific power and energy output. Thermopower waves are a potential solution because the approach uses a self-propagating reaction wave to control the duration of energy release, accessing high power densities from simple, chemical fuels. First reported in 2010, thermopower waves convert chemical energy in the form of an adsorbed fuel directly to electrical current by means of self-propagating reaction waves guided along nanostructured thermal conduits. Such TPW devices have been shown to exhibit high power density comparable to that of Li-ion batteries (80,000 μW/mm³ versus about 10,000 μW/mm³ demonstrated for Li microbattery).[11] The high power and energy densities of such devices provide attractive scaling for Micro-Electro-Mechanical Systems (MEMS),[12] smart dust applications,[13] or wireless sensor network nodes.[14] Conduit types have included multi-walled carbon nanotubes (MWNTs);[11] single-walled carbon nanotubes (SWNT);[15] SWNT yarns; ZnO;[16,17] Bi_2Te_3 coated on alumina and terracotta;[18,19] Sb_2Te_3 coated on alumina, terracotta and MWNTs;[19,20] and MnO_2 coated alumina[21]. Fuels employed for TPW studies, on the other hand, have included 2, 4, 6-Trinitroaniline (TNA), picramide and nitrocellulose (with sodium azide),[15,22] and nanothermites consisting of aluminum (Al) iron oxide (Fe_2O_3).[23] However, the efficiency of energy conversion is an issue that has not been significantly addressed to date, and reported data have ranged from 10^{-6} to 10^{-3} %. In this work, we show that such efficiency limitations have been dominated by radiative losses, convective losses and fuel selections that necessarily set up a partially-cancelling thermopower wave, limiting power generation. Improved design and fuel complexation gives rise to devices that exceed efficiencies of 1 % for the first time, commensurate with experimental power source technologies.[24] These results are inspired by a mathematical wave model that allows one to describe the transient voltage profile of the traveling thermal wave, capturing many important aspects of the wave dynamics. In many ways, this conceptual advance allows us to push the boundaries of the future portable power sources comprised of carbon nanomaterials, which challenges the current state of the art from both an energy and a sustainability point of view.

Main Text

Electrical power generation with the propagation of the concomitant reaction wave is described by the theory of excess thermopower, which describes the induced voltage as the sum of two distinct sources, namely thermoelectricity and the transient chemical potential gradient. The former is driven by the temperature difference existing across the thermal conduit, while the latter arises because of the chemical potential difference existing across the conduit.[25]

$$\Delta V_{TPW} = S\left(T_{back} - T_{front}\right) - \frac{1}{e}\left(\mu_{back} - \mu_{front}\right) \qquad (1)$$

Here, S is the Seebeck coefficient of the device; T is the temperature; μ is the chemical potential of the dominant charge carriers in the system, holes or electrons, and the subscript *front* and *back* denotes the spatial frontend and backend of the SWNT conduit, respectively. One can then define a chemical-to-electrical conversion efficiency (η) as the ratio of the output electrical energy to the input chemical energy into the system:

$$\eta = \frac{\text{Output Electrical Energy}}{\text{Input Chemical Energy}} = \frac{\int_{t_{start}}^{t_{end}} \frac{\Delta V_{out}^{\;2}}{R_{ext}} dt}{m_{fuel}\left(-\Delta H_{rxn}\right)}$$

Here, t is the time coordinate, with the subscript indicating the start and end of the experiment; m_{fuel} is the mass of the chemical fuel used and ΔH_{rxn} represents its specific heat of reaction. The value of η does not include the initiation energy as it is generally orders of magnitude smaller than the denominator in equation (2). Historically η has not exceeded approximately 10^{-3}% in the literature, primarily due to limitations in fuel coupling[25] (as shown below), radiative and convective losses. Figure 1a shows a circuit diagram of a TPW device as a source producing an open circuit voltage ΔV_{OC} with an internal resistance R_{int} which produces an output voltage ΔV_{out} over an external resistance R_{ext}.[26] Figure 1b shows a typical (bipolar) voltage output obtained from a nitrocellulose-SWNT yarn TPW device. One consequence of the excess thermopower theory represented in equation (1) is that the additive chemical potential contribution, which is a fuel property, resulting significant variance in the $\Delta\mu$ contribution. Different fuels also change the adiabatic reaction temperature, affecting the first term in equation (1) as well.[27] A screening of potential fuels produced two of notable interest, namely sodium azide (NaN_3) and a carbohydrate fuel in the form of sucrose with a premixed oxidizer, potassium nitrate (KNO_3). We observed that these fuels led to a substantially improved efficiency as compared to a standard nitrocellulose fuel (Figure 1c). Up to 100 times improvement was observed for NaN_3 despite the fact that at 320 J gm^{-1} it has a heat of combustion 1/10th of nitrocellulose (4200 J gm^{-1}).[28] Sucrose with KNO_3 dissolved in water at 35:65 ratio also consistently demonstrated more than a factor of 10 higher efficiency, despite having a similar heat of combustion to nitrocellulose.

With a decreased heat of combustion (therefore a reduced contribution from the thermoelectric effect), the amplified output electrical energy is accounted for by the enhanced excess thermopower ($\Delta\mu$ contribution) led by the favorable chemical doping.[25] To further investigate the effect of doping on the SWNT by the different chemical fuels tested, Raman spectroscopy analyses were performed (Figure 1d) where the G peak Raman shifts were converted to Fermi energies using the data reported by Farhat *et al.* and a method outlined in Abrahamson *et al.* (modified below).[25,29,30] With the presence of nitro groups, the electron withdrawing nitrocellulose is known to p-dope graphene and SWNT.[25] With this insight, we calculate a Fermi energy of -0.21 eV for the bare SWNT (1577 cm^{-1}) (known to be p-doped by O_2), and -0.28 eV for the nitrocellulose/NaN_3 doped SWNT (1584 cm^{-1}). As was found from model regression of experimental data (below). This p-doping by chemical fuel actually generates an opposing electronic wave, reducing efficiency. It is noteworthy that with a Raman shift around 1579 cm^{-1}, NaN_3 and sucrose/KNO_3 either lower the p-doping losses (-0.23eV) or supportively n-dope (+0.40 eV) the SWNT as compared with nitrocellulose/NaN_3. In either case, the Fermi level difference (and hence the chemical potential gradient) of at least 0.05 eV between the doped states of SWNT justifies the higher electrical energy output observed for using solely NaN_3 or sucrose/KNO_3 as chemical fuels. This link between fuel properties and efficiency, supported by the excess thermopower theory, promises further improvements in fuel design for TPW devices.

It is also noted that the temperatures of propagating TPW wavefronts can approach the adiabatic limit (2427 K for nitrocellulose) leading to substantial heat losses from the device. We find that suspending TPW devices within an airgap flanked by two parallel Bi_2Te_3 thermoelectric (TE) harvesters is effective for capture of radiative and convective energy losses (Figure 1e). Combined with fuels such as sucrose-KNO_3 and NaN_3, efficiencies in excess of 1 % can be achieved in this configuration (Figure 1f). This scheme necessarily generates more electrical energy than conventional thermophotovoltaic or combustion schemes by adding the high power density TPW.[31,32] There does not appear to be a correlation between adiabatic reaction temperature (760 K for NaN_3 and 1043K for sucrose/KNO_3) and efficiency, an indication that TPW devices are generally not limited by the conventional thermoelectric term, and operate distinctly from thermophotovoltaics.[24]

Reaction waves such as TPWs are theoretically solitons with highly nonlinear properties arising from the exponential thermal source term associated with reaction. This exponential term partially cancels any dissipation that would be observed from thermal diffusion, predicting infinite propagation along an infinite conduit. To test this empirically, we generated TPW devices of increasing length. Sample voltage outputs from NaN_3-based and sucrose/KNO_3-based TPW devices are shown in Figure 2a, displaying opposite polarity voltage pulses from opposing propagation directions, as expected. As seen from Figure 2b, the duration of energy generation scales linearly with length, and its wave propagation is captured in Figure 2c, with notable cooling behind the advancing wavefront. The discovery of the linear scaling of voltage duration versus device length suggests a potential route to the scaling-up of TPW devices. Indeed, with this knowledge in hand, we demonstrate the longest SWNT-based TPW device reported to date, both in terms of length (over 10 cm) and duration of energy generation (7 s). The wave propagation of said device is presented in Figure 2c, and this linear scaling of voltage duration to device length clearly goes beyond the limited length window investigated in the past (Figure 2d).

The results demonstrated in this work are a clear advance over the state of the art, and systems examined in the literature that have focused on conduit type. Figure 2e charts the increase in efficiency from approximately 10^{-6} % for SWNT-based devices in 2010 and subsequent increase as conventional TE materials were examined in the years following by other researchers.[15-23,33-35] While higher Seebeck coefficient conduit types do maximize the conventional thermoelectric term in equation (1),[18] this current work highlights the importance of the excess thermopower through constructive chemical doping toward increasing efficiency. Initially discovered by us, the use of nitrocellulose as a solid state fuel due to its high combustion enthalpy has been adopted by many researchers.[16,18,19,21,35] This work further reveals a criteria for selecting better performing fuels for TPW propagation, namely NaN_3 and sucrose/KNO_3, by taking advantage of their favorable electronic interactions with SWNTs. The observed increase in energy output by reducing the nitrocellulose level (as indicated by the decreased nitrocellulose/NaN_3 ratio in Figure 2f) can be attributed to both the enhanced maximum peak voltage, as well as the prolonged energy output duration. To clarify, our current work (Figure 2e, red) demonstrates the highest TPW device efficiencies to date with or without the use of external energy harvesting. These advances pave the way for future chemical fuel optimization through rational design for SWNT-based TPW or even broader applications. As a demonstration of a practical use of the current high efficiency device, we use three NaN_3-fueled SWNTs connected in series with thermoelectric residual heat harvesters to charge a voltage step-up converter (Figure 2g). For the first time, we are able to reach 1.5 V, power and sustain a commercial light-emitting diode (LED) for over 20 s (Figure 2h), a practical milestone that promises additional gains in energy and power density.

Lastly, these observations are coupled to a detailed mechanistic model to provide insights on how conduit type, fuel and thermal losses contribute to energy and power density, as well as efficiency. Using the excess thermopower concept outlined in equation (1) as a starting point, this model is capable of describing the transient voltage output of TPW by simultaneously solving the heat and mass transport equations that govern the TPW propagation along the SWNT fiber. The transient temperature and mass profiles are then used to predict both terms in equation (1), including the time-dependent variation of doping levels as fuel reacts along the wavefront.[25] As shown previously, a 1D self-propagating TPW can be analyzed by studying the corresponding heat and mass transfer for a one-dimensional reaction system of continuous fuel layers.[36]

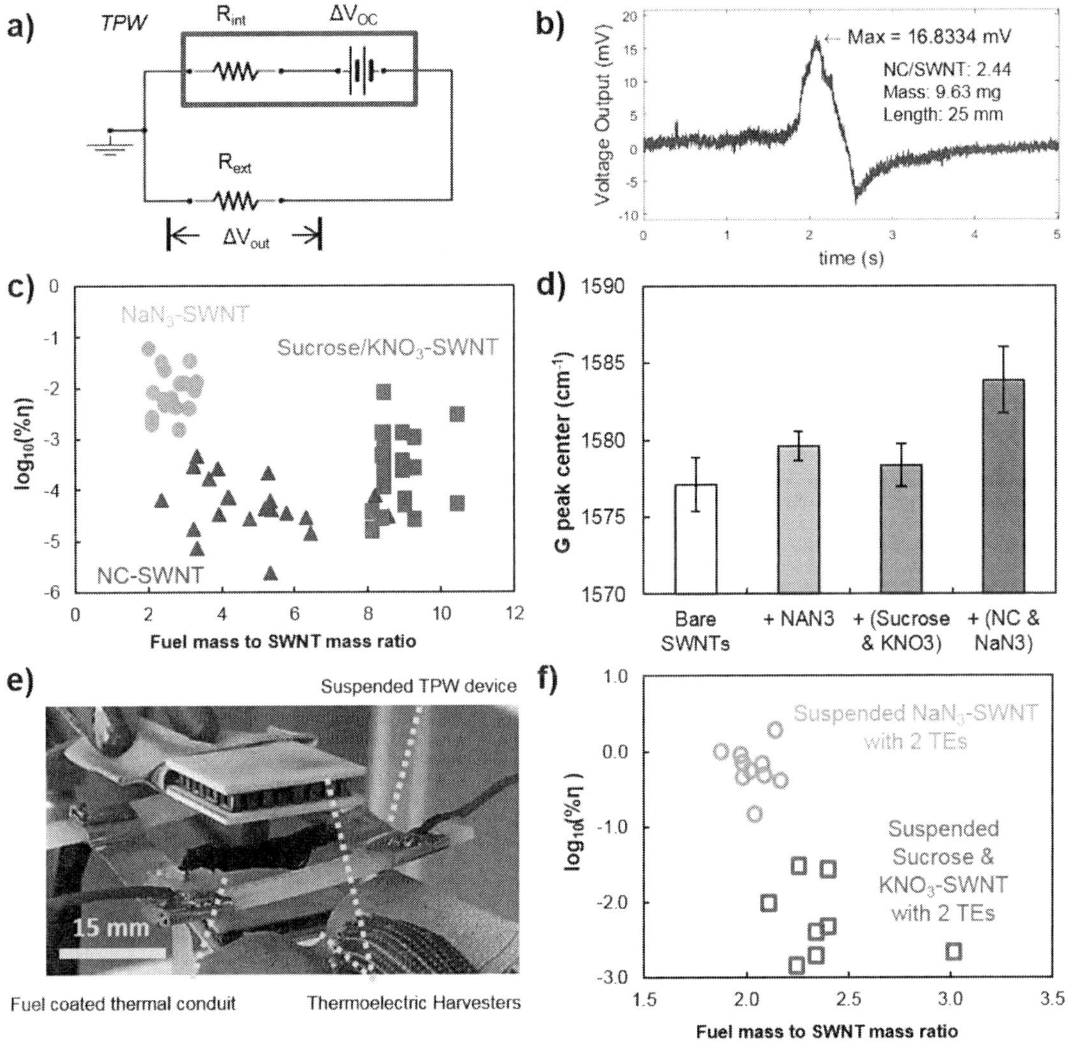

Figure 1. a) Circuit diagram for the electrical analysis of thermopower wave (TPW) devices where we connect an external load R_{ext} to measure the voltage drop ΔV_{out} across it. b) Sample of an actual thermopower wave voltage output. The voltage output for this 25 mm long reaction wave lasts for about 1.2 s. c) Chemical to electrical conversion efficiency of TPW devices using different fuels: Nitrocellulose (NC), Sodium azide and Sucrose with KNO_3. d) Doping analysis using Raman spectroscopy for different fueled and unfueled SWNT samples. The error bars represent 95% confidence intervals. e) A picture of an experimental setup involving TE harvesters (seen as white squares) suspended above and below a TPW device. f) Chemical to electrical conversion efficiencies of the updated TPW wave setups (i.e., including the electrical output from the reaction wave as well the TE harvesters).

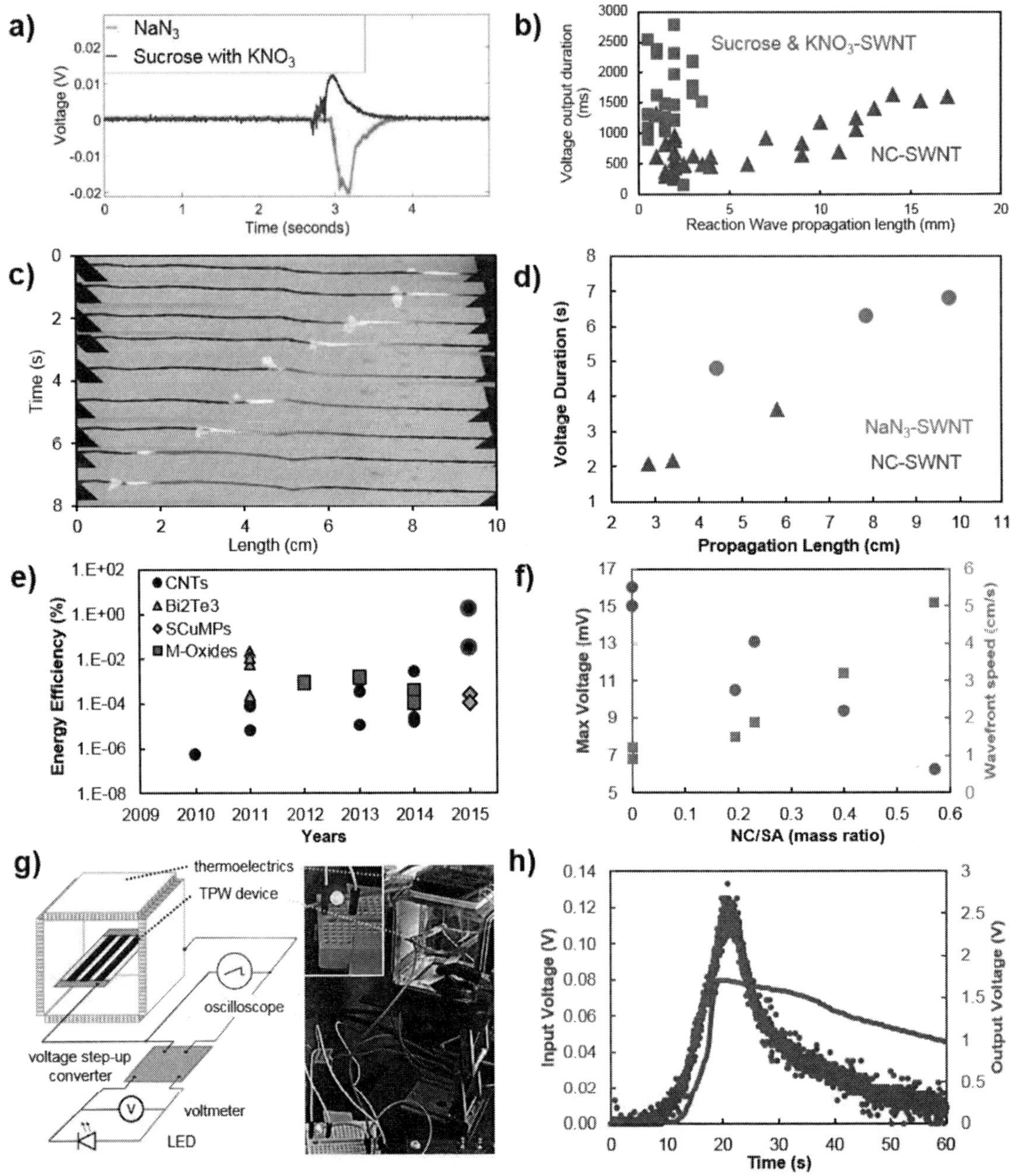

Figure 2. a) Sample voltage output from SWNT thermopower reaction waves launched using sodium azide and sucrose/KNO_3. b) Voltage output duration plotted against device length for nitrocellulose (NC) coated SWNTs and sucrose/KNO_3 coated SWNTs in the sub-centimeter regime. c) Time elapsed pictures displaying the reaction wave propagation on a 10 cm long NaN_3–SWNT TPW device. d) Decimeter-scale NaN_3 and nitrocellulose (NC) coated SWNTs showing linear scaling between voltage durations and wave propagation length. e) Master plot for chemical to electrical conversion efficiencies of all thermopower wave device setups published so far. The highest efficiency highlighted points in 2015 correspond to this work. f) Maximum voltage and speed of the reaction wavefront plotted against increasing nitrocellulose (NC) to sodium azide (SA) ratio. g) Schematic of the TPW-thermoelectric harvester setup coupled with a voltage step-up converter that illuminated a red LED (shown on the right) for 20 s. h) The input and output voltage measured across the step-up converter (a DC-to-DC power converter with an output voltage greater than its input voltage) plotted as a function of time. The LED stays on while the output voltage is above 1.5 V.

The system is initially at ambient temperature T_{amb}. The heat of reaction ($-\Delta H$) is assumed to be temperature invariant, with the reaction modeled as first order with an Arrhenius activation energy E_a and attempt frequency prefactor k_0. This formulation assumes a fast thermal equilibrium between the thermal conduit and the fuel layer.[36] The non-dimensional forms of the heat and mass balance equations are shown in equations (3) and (4), which can be solved using method of lines, with the space domain discretized as shown in Figure 3a.

$$\frac{\partial u}{\partial \tau} = \frac{\partial^2 u}{\partial \xi^2} + y e^{-\frac{1}{u}} - w_{rad}\left(u^4 - u_{amb}^4\right) - w_{conv}\left(u - u_{an}\right) \quad (3)$$

$$\frac{dy}{d\tau} = -\beta y e^{-\frac{1}{u}} \quad (4)$$

where u and y represent the non-dimensional temperature and fuel concentration, both of which are a function of the dimensionless space (ξ) and time (τ); w_{rad} and w_{conv} correspond to the dimensionless net radiative and convective heat transfer coefficients, respectively. Non-dimensionalization of variables also gives rise to an important factor □, the inverse dimensionless adiabatic temperature rise:[37]

$$\beta = \frac{C_{p,CNT} E_a}{(-\Delta H) R} \quad (5)$$

where $C_{p,CNT}$ is the heat capacity of the SWNT conduit and R is the gas constant.

TPW devices can be modeled as voltage sources giving rise to a transient open circuit voltage difference (ΔV_{TPW}) across its ends, given in equation (1). When connected to an external electrical load or electrical resistance, we obtain useable output voltage (ΔV_{out}) and current flow through the load. If we can estimate the magnitude of the thermal gradient and the chemical potential gradient, for a system with a known Seebeck coefficient S, we can calculate the expected open circuit voltage (ΔV_{TPW}) for a given pair of fuel-thermal conduit TPW device. Using the values of the internal resistance of the TPW device (R_{int}) and the resistance of the load attached (R_{ext}), we can compute the expected output voltage (ΔV_{out}) as shown in Equation (6). In Figure 3b, we present a schematic of a circuit diagram for a TPW device with the expected voltage output.

$$\Delta V_{out} = \frac{S\left(T_{back} - T_{front}\right) - \frac{1}{e}\left(\mu_{back} - \mu_{front}\right)}{1 + \frac{R_{int}}{R_{ext}}} \quad (6)$$

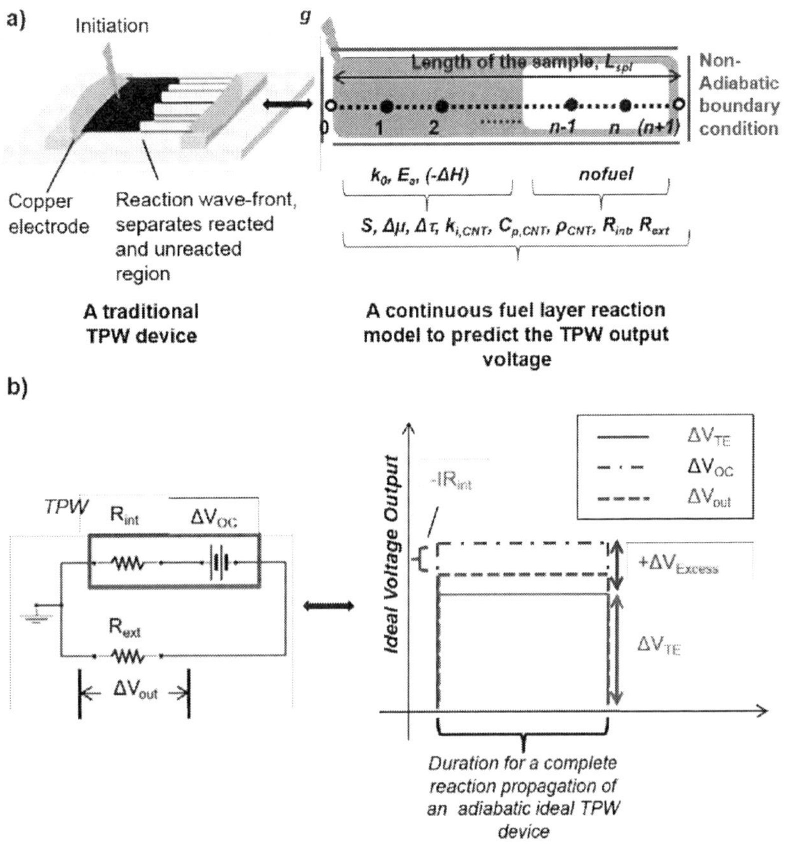

Figure 3. a) The left shows a schematic of a traditional thermopower wave device with a reaction wave propagation along a fuel coated thermal conduit connected to copper tape electrodes and supported on a glass slide. The right is a visualization of a model of a continuous layer of a fuel undergoing reaction to predict or fit the voltage output from a thermopower wave device. All of the properties (except the device's external heat loss properties) needed to completely define a thermopower wave device using such a scheme have been noted. b) The left shows a schematic of the electrical circuit diagram for analyzing the output voltage ΔV_{out} across an external resistor R_{ext} powered by a thermopower wave device with internal resistance R_{int}. The right shows an ideal voltage output profile for an adiabatic thermopower wave device undergoing complete reaction propagation. The open circuit voltage ΔV_{OC} from a thermopower wave device is contributed to by the temperature gradient and the chemical potential gradient.

This model allows us to gain intuitive understandings about the different trends observed in voltage output profiles.[38] From selected sample fits (SWNTs fueled with nitrocellulose), shown in Figure 4a-4f, it is clear that this model predicts the experimental results with fairly high accuracy and consistency. When the reaction wave undergoes complete propagation, we expect a double polarity thermoelectric voltage output contribution, where the voltage output peak after complete wave propagation is smaller in magnitude, such as what is seen in Figure 2b. The effect of excess thermopower will either boost or oppose the first of the double polarity peaks, depending on the sign of the Seebeck coefficient of the SWNT cluster and the type of doping induced by the fuel. When the thermoelectric and the excess thermopower voltage contributions oppose each other, the first peak which is ideally higher in magnitude (as compared to the reverse polarity second peak) is repressed, as shown in Figure 4c. In the case of partial propagation, we expect a single polarity thermoelectric voltage output (e.g. Figure 4f), since the reaction wave stops in the middle of the conduit and never reaches the other end to create the opposite temperature gradient. Again, depending on whether the effect of excess thermopower boosts or opposes the thermoelectric voltage output, one expects to see 'kinks' or multiple peaks in the single polarity peak voltage output. Examples of these situations can be found in Figure 4a, 4b, 4d, and 4e. The most interesting takeaway from these fits is that for the 20 nitrocellulose-SWNT TPW samples that were fitted, 17 samples show the effect of excess thermopower opposing the voltage output due to the thermoelectric effect. This finding agrees with the Raman results discussed earlier in Figure 1d, where we concluded that the p-doping nitrocellulose generates an excess thermopower voltage opposite to its thermoelectric counterpart, reducing the overall energy output of the nitrocellulose TPW devices. This qualitative agreement between our model and experimental Raman measurements strengthens our confidence in the current theoretical construct and its prospects in aiding further TPW device optimization.

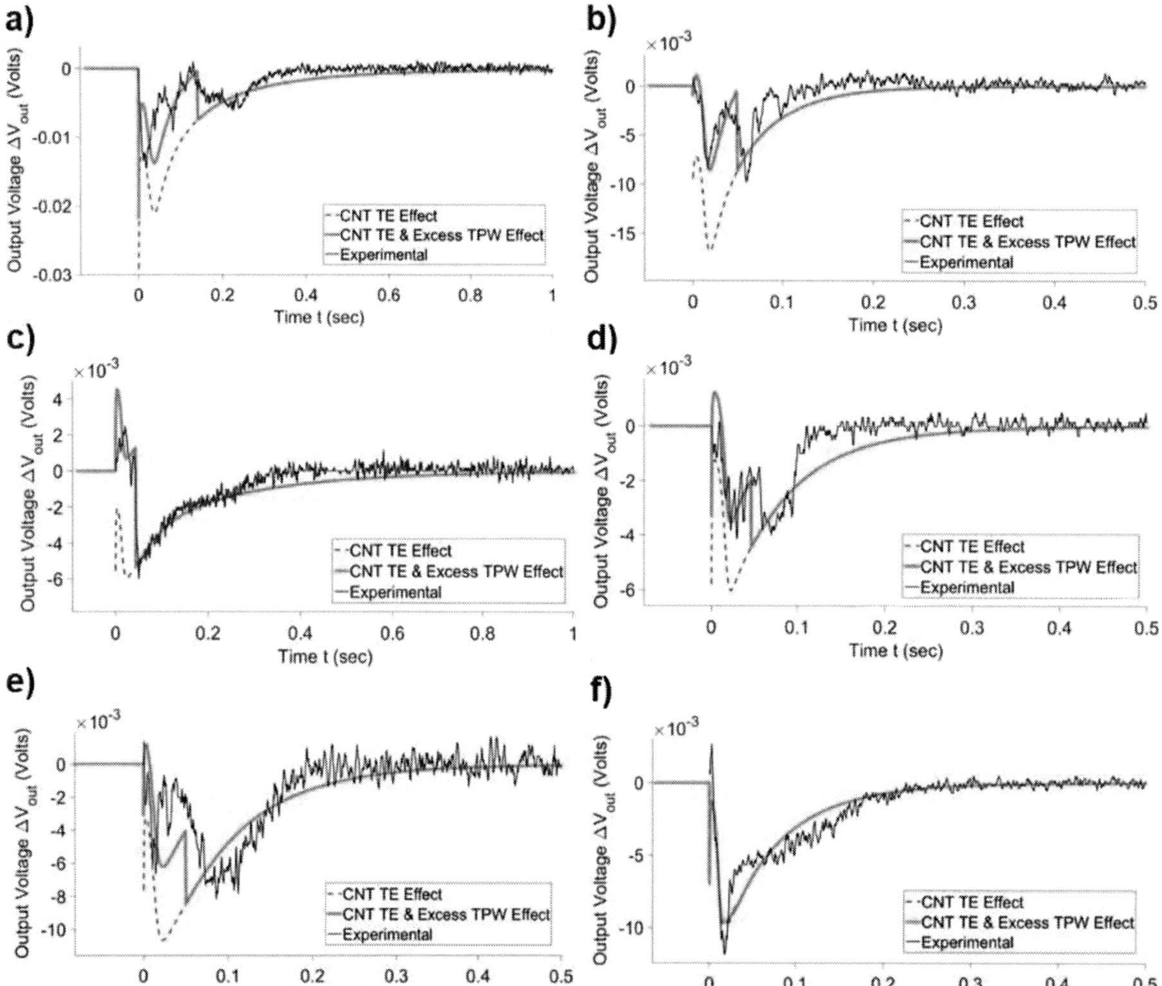

Figure 4 a-f) Thermopower wave voltage output prediction for experimental samples. The experimental output is shown in black, the model-predicted thermoelectric output is shown in red dotted lines, and the final model-predicted output is shown in green.

In summary, on the basis of the theory of excess thermopower, the chemical to electrical conversion efficiency of thermopower wave devices was improved from 10^{-3} % to 10^{-2} % by using novel fuels such as NaN_3 and sucrose/KNO_3. Combining this improved performance of thermopower waves with external thermoelectric harvesting, we demonstrate a net energy conversion efficiency of up to 1 % and successfully illuminate an LED for 20 s. We propose a detailed 1D heat and mass transfer model to be used in combination with the theory of excess thermopower to explain the experimental voltage outputs from thermopower wave devices. The theoretical analyses for the voltage outputs from nitrocellulose-SWNTs, in combination with Raman spectroscopy and a variety of controlled experiments, suggest that choosing n-doping fuels might further boost the efficiency of these SWNT fiber-based thermopower wave devices.

Acknowledgement. We thank the Air Force Office of Scientific Research and the Office of Naval Research Neptune Program for sponsoring this work.

References

(1) Melot, B. C., Tarascon, J. M. Design and preparation of materials for advanced electrochemical storage. *Accounts of Chemical Research* **46**, 1226–1238 (2013).

(2) Orikasa, Y. *et al.* High energy density rechargeable magnesium battery using earth-abundant and non-toxic elements. *Scientific Reports* **4**, 5622 (2014).

(3) McKerracher, R. D., Ponce de Leon, C., Wills, R. G. A., Shah, A. A., Walsh, F. C. A review of the Iron–Air secondary battery for energy storage. *ChemPlusChem* **80**, 323–335 (2015).

(4) Zhao, X., Ma, L. Recent progress in hydrogen storage alloys for nickel/metal hydride secondary batteries. *International Journal of Hydrogen Energy* **34**, 4788–4796 (2009).

(5) Kim, H. *et al.* Metallic anodes for next generation secondary batteries. *Chemical Society Reviews* **42**, 9011–9034 (2013).

(6) Häupler, B., Wild, A., Schubert, U. S. Carbonyls: powerful organic materials for secondary batteries. *Advanced Energy Materials* **5**, (2015).

(7) Gong, Y., Li, M., Wang, Y. Carbon nitride in energy conversion and storage: recent advances and future prospects. *ChemSusChem* **8**, 931–946 (2015).

(8) Irimia-Vladu, M., Sariciftci, N. S., Bauer, S. Exotic materials for bio-organic electronics. *Journal of Materials Chemistry* **21**, 1350–1361 (2011).

(9) Dai, L. Functionalization of graphene for efficient energy conversion and storage. *Accounts of Chemical Research* **46**, 31–42 (2013).

(10) Wang, H., Dai, H. Strongly coupled inorganic-nano-carbon hybrid materials for energy storage. *Chemical Society Reviews* **42**, 3088–3113 (2013).

(11) Choi, W., Abrahamson, J. T., Strano, J. M, Strano, M. S. Carbon nanotube-guided thermopower waves. *Materials Today* **13**, 22–33 (2010).

(12) Walther, D. C., Ahn, J. Advances and challenges in the development of power-generation systems at small scales. *Progress in Energy and Combustion Science* **37**, 583–610 (2011).

(13) Sailor, M. J., Link, J. R. "Smart dust": nanostructured devices in a grain of sand. *Chemical Communications*, 1375–1383 (2005).

(14) Roundy, S., Steingart, D., Frechette, L., Wright, P., Rabaey, J. in *Wireless Sensor Networks* Vol. 2920 *Lecture Notes in Computer Science* (eds Holger Karl, Adam Wolisz, & Andreas Willig) Ch. 1, 1–17 (Springer Berlin Heidelberg, 2004).

(15) Abrahamson, J. T. *et al.* Synthesis and energy release of nitrobenzene-functionalized single-walled carbon nanotubes. *Chemistry of Materials* **23**, 4557–4562 (2011).

(16) Walia, S. *et al.* ZnO based thermopower wave sources. *Chemical Communications* **48**, 7462–7464 (2012).

(17) Lee, K. Y., Hwang, H., Choi, W. Advanced thermopower wave in novel ZnO nanostructures/fuel composite. *ACS Applied Materials & Interfaces* **6**, 15575–15582 (2014).

(18) Walia, S. *et al.* Oscillatory thermopower waves based on Bi_2Te_3 Films. *Advanced Functional Materials* **21**, 2072–2079 (2011).

(19) Walia, S. *et al.* Sb_2Te_3 and Bi_2Te_3 based thermopower wave sources. *Energy & Environmental Science* **4**, 3558–3564 (2011).

(20) Hong, S. *et al.* Enhanced electrical potential of thermoelectric power waves by Sb_2Te_3-coated multiwalled carbon nanotube arrays. *The Journal of Physical Chemistry C* **117**, 913–917 (2013).

(21) Walia, S. *et al.* MnO_2-Based thermopower wave sources with exceptionally large output voltages. *The Journal of Physical Chemistry C* **117**, 9137–9142 (2013).

(22) Yeo, T. *et al.* Effects of chemical fuel composition on energy generation from thermopower waves. *Nanotechnology* **25**, 445403–445412 (2014).

(23) Hong, S. *et al.* Surface thermochemical reaction control utilizing planar anisotropic thermal conduit. *Energy & Environmental Science* **4**, 2045–2049 (2011).

(24) Lenert, A. *et al.* A nanophotonic solar thermophotovoltaic device. *Nat Nano* **9**, 126–130 (2014).

(25) Abrahamson, J. T. *et al.* Excess thermopower and the theory of thermopower waves. *ACS Nano* **7**, 6533–6544 (2013).

(26) Johnson, D. H. Origins of the equivalent circuit concept: the current-source equivalent. *Proceedings of the IEEE* **91**, 817–821 (2003).

(27) Abrahamson, J. T., Strano, M. S. Analytical solution to coupled chemical reaction and thermally diffusing systems: applicability to self-propagating thermopower waves. *The Journal of Physical Chemistry Letters* **1**, 3514–3519 (2010).

(28) Kim, C., Thomas, S. W., Whitesides, G. M. Long-duration transmission of information with infofuses. *Angewandte Chemie International Edition* **49**, 4571–4575 (2010).

(29) Farhat, H. *et al.* Phonon softening in individual metallic carbon nanotubes due to the Kohn anomaly. *Physical Review Letters* **99**, 145506–145509 (2007).

(30) The (μ_{end} - μ_{start}) term in equation (1) corresponding to the chemical potential driving force of the excess thermopower is given by $\Delta\mu = \Delta E_F = p_m \times E_{F,m} + p_s \times E_{F,s}$, where ΔE_F is the Fermi level shift between the start and the end of the SWNT conduit, and p_m and p_s are coefficients representing the fractions of metallic and semi-conductive SWNTs in unsorted fibers. Theses fractions are determined by the rolling vectors of the SWNTs, about 1/3 and 2/3, respectively, since the manufacturer uses a generic chemical vapor deposition process. Raman spectroscopy and the G peak frequency can be used to estimate the average Fermi level of electrons in the SWNT fiber quantitatively.

(31) Datas, A., Algora, C. Development and experimental evaluation of a complete solar thermophotovoltaic system. *Progress in Photovoltaics: Research and Applications* **21**, 1025–1039 (2013).

(32) Rephaeli, E., Fan, S. Absorber and emitter for solar thermo-photovoltaic systems to achieve efficiency exceeding the Shockley-Queisser limit. *Opt. Express* **17**, 15145–15159 (2009).

(33) Choi, W. *et al.* Chemically driven carbon-nanotube-guided thermopower waves. *Nat Mater* **9**, 423–429 (2010).

(34) Lee, K. Y., Hwang, H., Shin, D., Choi, W. Enhanced thermopower wave via nanowire bonding and grain boundary fusion in combustion of fuel/CuO-Cu_2O-Cu hybrid composites. *Journal of Materials Chemistry A* **3**, 5457–5466 (2015).

(35) Hwang, H. *et al.* Investigation of the effect of the structure of large-area carbon nanotube/fuel composites on energy generation from thermopower waves. *Nanoscale Research Letters* **9**, 536–544 (2014).

(36) Abrahamson, J. T. *et al.* Wavefront velocity oscillations of carbon-nanotube-guided thermopower waves: nanoscale alternating current sources. *ACS Nano* **5**, 367–375 (2010).

(37) Mahajan, S. G. *et al.* Superadiabaticity in reaction waves as a mechanism for energy concentration. *Energy & Environmental Science* **7**, 3391–3402 (2014).

(38) Ideally, the chemical potential gradient should also be just dependent on the fuel-thermal conduit pair and should remain constant as long as there is uniform reaction wave propagation. However factors such as non-uniform coating of fuel layer can lead to variable chemical potential gradient across samples. Also, this chemical potential gradient stops existing as soon as the reaction has completely propagated or has stopped after partial propagation. For partial propagation, we hypothesize that the charge carriers internally diffuse towards equilibration within the thermal conduit layer when the reaction wave

has halted. Depending on the internal heat transfer properties of the system and the rate at which it loses heat to the surroundings, there is a possibility of reversal of the temperature gradient leading to opposite polarity voltage output contribution because of the thermoelectric effect. Interplay of factors such as the time dependence of the temperature gradient and the chemical potential gradient and practical conditions for device operation (e.g., non-adiabatic system and device-to-device differences) can explain the variety of types of voltage outputs.

NOVEL MALLEABLE THERMOSETS FOR SOLID-STATE BATTERY APPLICATION

Wei Zhang, Philip Taynton, Justin Whiteley, Sehee Lee

Department of Chemistry and Biochemistry, Department of Mechanical Engineering, University of Colorado, Boulder, CO 80309, USA
Wei.Zhang@Colorado.EDU

Introduction

Conventional linear polymers typically have melting points and can be reprocessed and recycled, but are not durable enough for many applications. In contrast, cross-linked polymers, commonly called thermosets, are mechanically tough, but are hard to reprocess and not able to be recycled. Malleable polymers[1] are a new class of materials, wherein reversible covalent bonds[2] are employed. These materials typically include catalysts and can maintain mechanical properties consistent with thermosets under ambient conditions, yet at elevated temperatures they can be reprocessed and recycled through a cross-link exchange and rearrangement process. This talk will be focused on the development of a catalyst-free malleable network polymer, which can be completely recycled through simple heat treatment (~ 100 °C).[3] This material takes advantage of reversible metathesis reaction to achieve its unique process-ability while retaining tough mechanical properties. Application of such malleable thermosets toward composite fabrication,[4,5] including solid-state electrolyte membrane for battery applications, will also be discussed.

Acknowledgement. We thank National Science Foundation MAST Center for funding support.

References

(1) Denissen, W.; Winne, J. M.; Du Prez, F. E. *Chem. Sci.* **2016**, *7*, 30-38.
(2) Jin, Y.; Yu, C.; Denman, R. J.; Zhang, W. "Recent Advances in Dynamic Covalent Chemistry" *Chem. Soc. Rev.* **2013**, *42*, 6634-6654.
(3) Taynton, P.; Yu, K.; Shoemaker, R.; Jin, Y.; Qi, H. J.; Zhang, W. *Adv. Mater.* **2014,** *26*, 3938–3942.
(4) Whiteley, J. M.; Taynton, P.; Zhang W.; Lee, S.-H. *Adv. Mater.* **2015**, *27*, 6922-6927.
(5) Taynton, P.; Ni, H.; Zhu, C.; Yu, K.; Loob, S.; Jin, Y.; Qi, J.; Zhang, W. *Adv. Mater.* **2016**, 28, in press (DOI: 10.1002/adma.201505245).

UNUSUAL ONE-STEP FORMATION OF Fe_2O_3/MnO_2 CORE-SHELL HOLLOW NANORODS AS A HIGH PERFORMANCE ANODE MATERIAL FOR LITHIUM ION BATTERIES

Shiji Hao[a,b], Yizhong Huang[b]*

[a]Energy Research Institute @ NTU, Nanyang Technological University, Singapore 639798, Singapore
[b]School of Materials Science & Engineering, Nanyang Technological University, Singapore 639798, Singapore
YZHUANG@ntu.edu.sg (Yizhong Huang)

Introduction

Recently, the core-shell heterostructured nanomaterials have attracted much attention for their fascinating synergetic properties and multifunctionalities through reinforcement or modification of components[1,2]. To a great extent, intrinsic properties of heterostructured materials also depends on their interior structure[3-5]. However, for the central construction of in-situ tuning, there is still lack of an effective and facile approach. Herein, the Fe_2O_3/MnO_2 core-shell nanorods with different interior structure was built with the inward movement of Mn element under different anealing temperature.

Experimental

First of all, the β-FeOOH nanorods was prepared by the water bath reaction with aqueous solution of $FeCl_3 \cdot 6H_2O$ and PVP (5,5000 MW). Then, the $FeOOH/MnO_2$ core-shell nanorods was obtained utilizing the chemical deposition reaction between $Mn(NO_3)_2 \cdot 4H_2O$ and $KMnO_4$. Finally, the Fe_2O_3/MnO_2 core-shell hollow and $Fe_2O_3/(Fe_{0.42}Mn_{0.58})_2O_3/MnO_2$ core-shell nanorods were prepared after the precursor was annealed at 500 and 600 °C.

Electrochemical Measurement. The galvanostatic charge-discharge cycling test was conducted with the Neware battery test system (Neware, BTS-5V10mA, China). Cyclic voltammetry (CV) test was performed by using an electrochemical workstation (autolab, PGSTAT302N).

Results and Discussion

In the present paper, the abnormal interior structural change was observed, corresponding to the inward flow of Mn molar under annealing treatment in the air. As shown in Fig. 1a, firstly, the hierarchical MnO_2 shell uniformly grows on the surface of FeOOH nanorods under the assisting of surfactant PVP with chemical deposition reaction. Then, the Fe_2O_3/MnO_2 core-shell porous or hollow nanorods is formed under the annealing condition of low temperature (400 or 500 °C) in the air, which is assigned to the decomposition of the internal FeOOH with the removal of H_2O and organic species. When the calcination temperature increases to 600 °C, the sample undergoes phase transformation: the phase of $(Fe_{0.42}Mn_{0.58})_2O_3$ is produced in the middle layer, ascribing to the partially solid state reaction between Fe_2O_3 and MnO_2 by the inward movement of Mn element from the shell component (MnO_2). At the same time, the internal cavity turns into the solid structure. Due to their unique intrinsic morphology, the Fe_2O_3/MnO_2 nanorods show exceptional electrochemical performances As anode materials for ithium ion batteries under a high current density (5 C). In the measurement, the capacity retention is measured to be as high as 706 mAh g^{-1} after 2000 cycles (Fig. 1b). On the one hand, the interior cavity could not only increase the stability of the material's structure but also decrease the diffusion length for Li ion during cycling. On the other hand, the core-shell structure brings out synergistic effects such as lowering the energy barrier for reaction and thus preserves the integrity upon the consecutive Li ion insertion/extraction cycling.

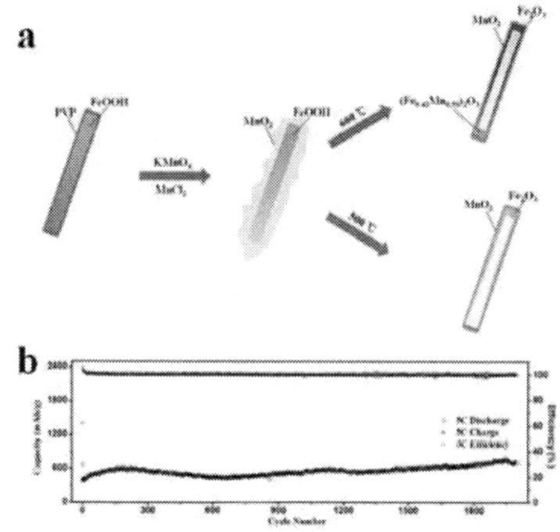

Figure 1. (a) Schematic of the fabrication process of the Fe_2O_3/MnO_2 hollow and $Fe_2O_3/(Fe_{0.42}Mn_{0.58})_2O_3/MnO_2$ solid core-shell nanorods and (b) cycling performance of the Fe_2O_3/MnO_2 nanorods at a current density of 5C

Conclusions

In summary, the Fe_2O_3/MnO_2 nanorods with different interior structure are constructed. The formation of the interior hollow, porous and solid triple layer structures are results of the decomposition of FeOOH nanorods and the inward flow of the Mn element. It is demonstrated that the core-shell nanorods display superior performance as anode materials for LIBs , attributed by the unique core-shell with complex interior structures.

Acknowledgement. This research was support by SUG (Start-up funding in NTU), Tier 1 (AcRF grant MOE Singapore M401992), Tier 2 (AcRF grant MOE Singapore M4020159) and the Chinese Natural Science Foundation (Grant 60906053, 62174118 and 51308050309).

References

(1) Tian, Q. W.; Hu, J. Q.; Zhu, Y. H.; Zou, R. J.; Chen, Z. G.; Yang, S. P.; Li, R. W.; Su, Q. Q.; Han, Y.; Liu, X. G. *J Am Chem Soc* **2013**, *135*, 8571.
(2) Zhang, Y. X.; Ding, H. L.; Liu, Y. Y.; Pan, S. S.; Luo, Y. Y.; Li, G. H. *J Mater Chem* **2012**, *22*, 10779.
(3) Kim, S.; Fisher, B.; Eisler, H. Y.; Bawendi, M. G. *Abstr Pap Am Chem S* **2002**, *224*, U443.
(4) Jiang, H.; Hu, Y. J.; Guo, S. J.; Yan, C. Y.; Lee, P. S.; Li, C. Z. *Acs Nano* **2014**, *8*, 6038.
(5) Yang, Y.; Kim, D. S.; Knez, M.; Scholz, R.; Berger, A.; Pippel, E.; Hesse, D.; Gosele, U.; Zacharias, M. *J Phys Chem C* **2008**, *112*, 4068.

LI-AIR FLOW BATTERY FOR ACCELERATED SEPARATION OF O_2 REDUCTION REACTION AND DISCHARGE PRODUCT STORAGE

Kensuke Takechi[1], Byoungsu Kim[2], Sichao Ma[3], Sumit Verma[2], Amit Desai[2], Fuminori Mizuno[1] and Paul JA Kenis[2]*

[1] Materials Research Department,
Toyota Research Institute of North America
[2] Department of Chemical & Biomolecular Engineering,
University of Illinois at Urbana-Champaign,
[3] Department of Chemistry,
University of Illinois at Urbana-Champaign,
kensuke.takechi@toyota.com

Introduction

Despite having an ultra-high theoretical energy density, non-aqueous Li-air battery suffers from drawbacks such as significant capacity fade and low current density. These issues can be attributed to the deposition of the discharge product on the cathode surface. The deposition results in the cathode pores getting clogged and thus prevents a continuous and steady discharge. Moreover, the discharge product is difficult to decompose electrochemically on the cathode after charge due to chemical and/or physical reasons.[1] In other words, a fundamental difficulty that resides in the Li-air battery architecture can be described as: *'electrochemical reaction and discharge product storage take place at the same location, on the cathode'.* Based on the discussion, we came to the conclusion that we need to separate the *'electrochemical reaction'* and *'discharge product storage'.* In prior work, we had proposed and demonstrated a new Li-air battery concept where the discharge product (Li_2O_2) can be stored in the electrolyte instead of on the cathode and the new battery architecture could be realized by using Li-ion free ionic liquid.[2] In this presentation, we further propose a new electrolyte flow based device architecture to boost the performance of the battery.

Experimental

Materials and Setup. Cell: custom made flow cell, Cathode: Carbon paper, Anode: Li metal, Electrolyte for cathode (catholyte): *N,N*-diethyl-*N*-(2-methoxyethyl)-*N*- methylammonium bis(trifluoromethylsulphonyl) imide (DEME-TFSI), Electrolyte for anode(anolyte): 1 M LiTFSI/PC, Solid state electrolyte: Li-ion conducting glass ceramics(LIC-GC, OHARA).

Evaluation. Catholyte flow rate: 0.1 ml/min, Catholyte supply: Push-pull type syringe pump, Supplied gas: dried air (1.4 RH%), Gas flow rate: 10 sccm, Discharge current density: 0.013 mA/cm^2.

Results and Discussion

We previously demonstrated that Li-ion free ionic liquid (DEME-TFSI) as an electrolyte facilitates the formation of the discharge product (Li_2O_2) in the electrolyte by avoiding direct reaction between the reduced O_2 and the Li-ion on the cathode. Although the discharge behavior became pseudo-continuous due to the steady ORR on cathode, the discharge capacity was still limited because of the slow diffusion of the reduced O_2, superoxide radical. Here, we design and fabricate an electrolyte flow architecture, (previously shown to have improved the performance of fuel cells)[3] by applying a solid state Li-ion conducting separator to accommodate non-aqueous electrolytes (Fig. 1). The discharge behavior of the new design is shown in Fig. 2. The new architecture dramatically improved the capacity as well. The flow architecture was quite effective to promote the mass transport of all the reactive and reacted species around the cathode and successfully allowed the discharge product precipitation in the electrolyte. The separately stored Li_2O_2 from the cathode can be indirectly charged (decomposed) by appropriate redox mediators.[1, 4] We also investigated the capability to operate this battery with dried air supply, not only with pure O_2, and confirmed its effective and flexible operation.

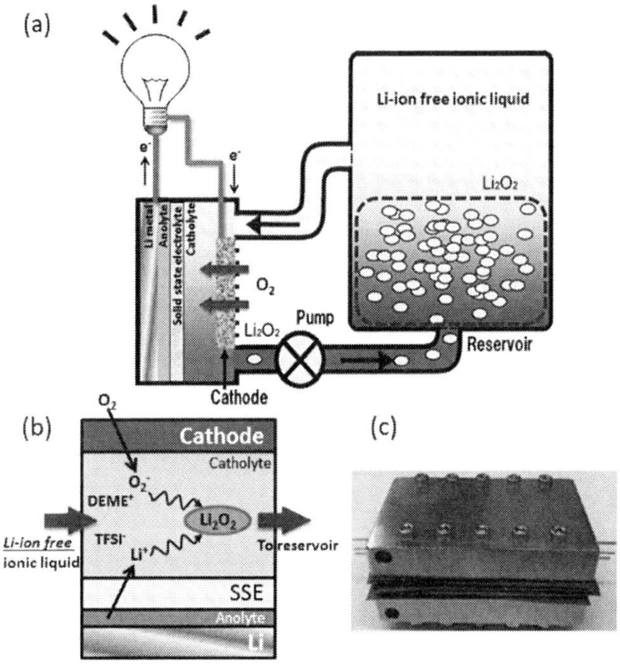

Figure 1. (a) Basic concept of the Li-air flow battery with a Li-ion free ionic liquid, (b) mechanistic illustration of the cathodic reaction and (c) actual appearance of the flow cell.

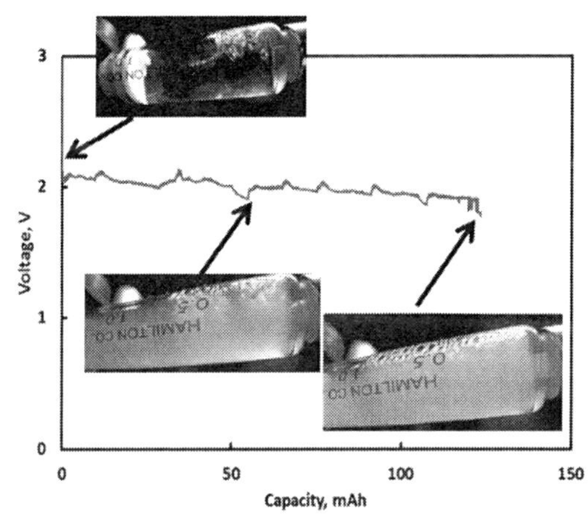

Figure 2. Voltage behavior of Li-air flow battery and appearance of the electrolyte for cathode(catholyte) during discharge.

References

(1) Hase, Y.; Ito, E.; Shiga, T.; Mizuno, F.; Nishikoori, H.; Iba, H.; Takechi, K. *Chem. Commun.*, **2013**, 49, 8389.
(2) Takechi, K.; Mizuno, F. *Abstract of 229th Electrochemical Society Meeting*, **2016**, A02-0338.
(3) (a) Jayashree R.S.; Mitchell, M.; Natarajan, D.; Markoski L.J.; Kenis P.J.A. *Langmuir*, **2007**, 23(13), 6871, (b) Jhong, H.R.; Brushett, F.R.; Yin, L.; Stevenson, D.M.; Kenis, P.J.A. *J. Electrochem. Soc.*, **2012**, 159(3), B292.

(4) Hase, Y; Seki, J.; Shiga, T.; Mizuno, F.; Nishikoori, H.; Iba, H.; Takechi, K., **2016**, *submitted.*

POLY(ARYLENE ETHER) BASED SINGLE-ION CONDUCTORS FOR LITHIUM ION BATTERIES

Guang Yang, Hyukkeun Oh, Kui Xu, Chalathorn Chanthad, Ismail Alperen Ayhan, Qing Wang

Department of Materials Science and Engineering, The Pennsylvania State University, University Park, Pennsylvania 16802, United States
wang@matse.psu.edu

Introduction

Lithium-ion batteries have been the clean energy source for portable electronics, stationary grids and hybrid/electric vehicles.[1] Almost all of liquid and polymer electrolytes that currently prevail in both prototype and commercial lithium-ion batteries have been binary ion conductors where both lithium ions and their counteranions migrate between electrodes during charging and discharging process.[2] Since the conductivity of these electrolytes are dominated by the anions' motion. The lithium ion transference number (T_{Li^+}) is usually lower than 0.3. The buildup of the anions at the electrode/electrolyte interface causes concentration polarization leading to loss of power from the battery. To suppress the free movement of the anions, single ion polymer conductors that can immobilize anions were proposed and studied. The employment of single-ion conductors minimizes the concentration gradient and polarization effect.[3] Since the ionic conductivity of polymer electrolyte is highly influenced by segmental motion of polymer chains,[4] most single-ion polymer conductors contain low glass transition portions such as poly(ethylene oxide). A group of single-ion conductors comprising poly(ethylenimine)-based polymer blends has been reported to show a state-of-the-art conductivity of 4×10^{-4} S cm^{-1} at ambient temperature.

In this work, we present the synthesis and characterization of single-ion conductors based on high-T_g aromatic poly(arylene ether)s with pendant lithium perfluoroethyl sulfonates. The prepared single-ion conductors soaked with organic carbonates exhibit excellent conductivities over a wide temperature range, that is, 3.1 and 0.74×10^{-3} S cm^{-1} at 25 and −20 °C, respectively, and excellent electrochemical properties while maintaining outstanding mechanical and thermal stabilities to serve as both separator and conductor between electrodes in the battery cells. The single-ion conducting feature of the membrane has been investigated. Remarkable battery performance has been demonstrated in the cells assembled from the prepared single-ion conductors at room temperature.

Experimental

Polymerization. In a three-neck flask equipped with a Dean−Stark trap and gas inlet were added hydroquinone (7 mmol), 4,4′-biphenol (3 mmol), K_2CO_3 (12 mmol), anhydrous DMAc (25 mL) and toluene (10 mL) under argon atmosphere. The mixture was heated at reflux for 2 h before the removal of toluene/water azeotrope from the Dean-Stark trap. Monomer TFP (10 mmol) was then added to the flask, and the reaction was stirred at 140 °C for 16 h. After cooling to room temperature, the reaction mixture was precipitated in water, and the precipitate was collected by filtration and washed thoroughly with water. The obtained polymer was then converted from sodium salt form into lithium salt form by immersing in 5 M lithium triflate aqueous solution at 40 °C for 24 h. After filtration and being washed thoroughly with water, the resulting PAE-LiPFS ionomer was dried under vacuum at 80 °C for overnight. The chemical structure and composition of PAE-LiPFS were analyzed by NMR and elemental analysis. The molar ratio of hydroquinone unit to 4,4′-biphenol unit in the PAE-LiPFS ionomer obtained was found to be 7:3, which was estimated from the integral data of ^{1}H NMR spectrum. In addition, the elemental analysis results also agree well with calculated values on the basis of the polymer structure with a hydroquinone/4,4′-biphenol unit ratio of 7:3. Anal. Calcd: C, 46.0; H, 1.7; F, 24.0; O, 20.2; S, 6.7. Found: C, 46.1; H, 1.9; F, 23.6; O, 20.5; S, 6.6. Intrinsic viscosity (in DMF/0.05 M LiBr at 25 °C): 1.6 dL/g.

Membrane Fabrication. A solution of PAE-LiPFS (0.7 g) and poly(ethylene glycol) dimethyl ether (PEG, Mw = 1000, 0.3 g, Sigma-Aldrich) in DMF (10 wt %) was prepared and casted onto a glass plate in an oven. The solvent was evaporated by heating the glass plate to 55 °C for 2 h and 75 °C for 4 h. The solid film obtained was then immersed in deionized water at 50 °C for 12 h to extract PEG. After being washed with deionized water thoroughly, the resulting porous PAE-LiPFS film was dried in vacuum oven for 24 h. The thickness of the film obtained is around 45 μm. The porosity (P) is calculated from the densities of the PAE-LiPFS porous film (ρf) and pure polymer (ρp) according to P = 1 − (ρf/ρp).

Cell Fabrication. The composite positive electrodes were fabricated by casting a slurry of $LiFePO_4$ (LFP, provided by Süd Chemie), Super P (a carbon additive to enhance electrode conductivity), and PAE-LiPFS ionomer (as a binder) mixture dispersed in NMP on an etched Al foil. The electrode plate with this composition was dried overnight in air at room temperature, cut into circular disks (12 mm in diameter), and then dried overnight in a vacuum oven at 120 °C. Before the cell was assembled, the membrane disks were immersed in a solution of EC + PC + DEC (1:1:1, v/v/v) for 6 h in an argon-filled glovebox. CR2032-type coin cells were assembled in the glovebox using lithium counter electrodes, carbonate-soaked polymer membranes, and the LFP/PAE-LiPFS composite electrodes.

Results and Discussion

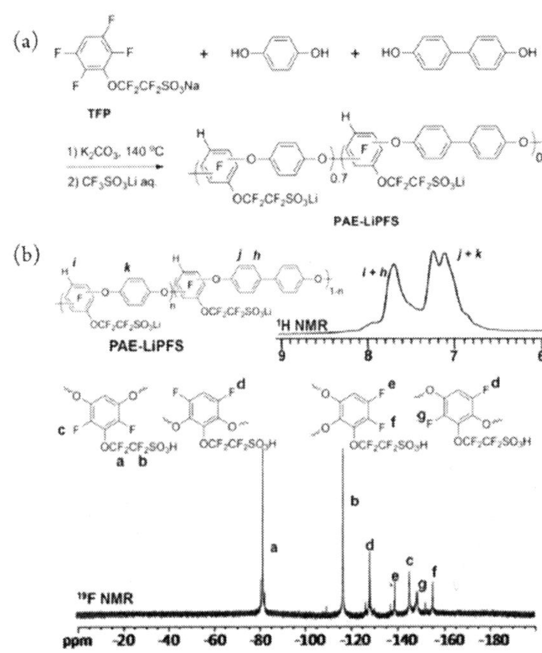

Figure 1. (a) Synthetic scheme of PAE-LiPFS. (b) ^{1}H and ^{19}F NMR spectra of PAE-LiPFS.

Polymer Synthesis. The synthesis of PAE-LiPFS, an aromatic high-T_g poly(arylene ether) based lithium single-ion electrolyte, is outlined in Figure 1a. The degree of nucleophilic substitution of fluoride groups in TFP by phenolates can be readily controlled by adjusting the reaction temperature. It has been found that only the first two phenyl fluorides of TFP are reactive when temperature is below 145 °C, yielding a linear polymer that is soluble in polar aprotic solvents such as dimethylformamide (DMF) and N-methylpyrrolidone (NMP). The remaining fluorines show much lower reactivity and are only substituted above 160 °C, which is probably attributed to the deactivation effect from the electron-donating ether oxygens. The fluorines left in polymer backbone are anticipated to afford chemical and thermal stability to the polymer normally associated with fluoropolymers. The synthesis of PAE-LiPFS includes the polycondensation of sodium 2-(2′,3′,5′,6′-tetrafluorophenoxy)-perfluoroethanesulfonate (TFP) with hydroquinone and 4,4′-biphenol followed by ion-exchange with lithium salt. The chemical structure and composition of the polymer have been confirmed by ^{1}H and ^{19}F NMR (Figure 1b).

Membrane Preparation and Characterization. Thin films with a thickness of around 45 μm were cast from a mixture of PAE-LiPFS and poly(ethylene glycol) dimethyl ether (PEG) (30 wt %) in DMF. Subsequent extraction of PEG from the films by water gave rise to a microporous film. It has been documented that a pore size of less than 1 μm is most beneficial for Li-ion batteries to avoid both lithium dendrite formation and solvent leakage.

Ionic Conductivity. Whereas the dry PAE-LiPFS films display very low room-temperature Li^+ conductivity (i.e., < 10^{-7} S cm^{-1}), as shown in Figure 2, high Li^+ conductivities have been obtained in the PAE-LiPFS film soaked with DEC + EC + PC (1:1:1 by volume, 92 wt %) in a temperature range of −20−60 °C. For example, the hybrid film displays a room- temperature conductivity of 3.1×10^{-3} S cm^{-1}, which is among the highest conductivity values achieved for single-ion conductors. This value is greater than those of traditional liquid electrolytes incorporated with polyolefin separators, that is, $0.1^{-1} \times 10^{-3}$ S cm^{-1}, depending on the structures of salts and solvents.

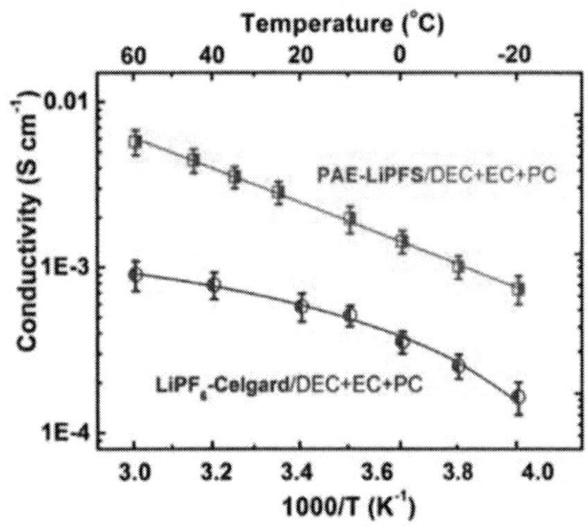

Figure 2. Arrhenius plot of the ionic conductivity of the PAE LiPFS/DEC + EC + PC film (1:1:1 by volume, 92 wt %) and 1M $LiPF_6$–Celgard/ DEC + EC + PC (1:1:1 by volume).

As shown in Figure 2a, 1 M $LiPF_6$ in DEC + EC + PC (1:1:1 by volume) with Celgard 2325 exhibits a room temperature conductivity of ~0.62×10^{-3} S cm^{-1}, which is consistent with literature results. For example, 1 M LiPF6 in DMC + EC using Celgard 2500 as separator was reported to have an ionic conductivity of 0.85×10^{-3} S cm^{-1} and a t_{Li+} value of 0.32 at room temperature. More notably, the organic carbonate swollen PAE-LiPFS film retains sufficiently high conductivities at low temperatures, for example, 7.4×10^{-4} S cm^{-1} at −20 °C, implying its potential for wide temperature range operation.

Battery Performance. From Figure 3a, typical flat-shaped curves around 3.39−3.48 V are observed in the first galvanostatic charge−discharge voltage profile of the cell measured at room temperature, which are consistent with the reported two phase coexistence reaction for $LiFePO_4$ cathode. The cell delivers a discharge capacity of 153 mA h g^{-1}, which amounts to 90% of the theoretical value (170 mA h g^{-1}) and exhibits a small overpotential value. More remarkably, as depicted in the inset of Figure 4a, a Coulombic efficiency of about 100% is maintained after 40 cycles, suggesting its excellent cyclability.

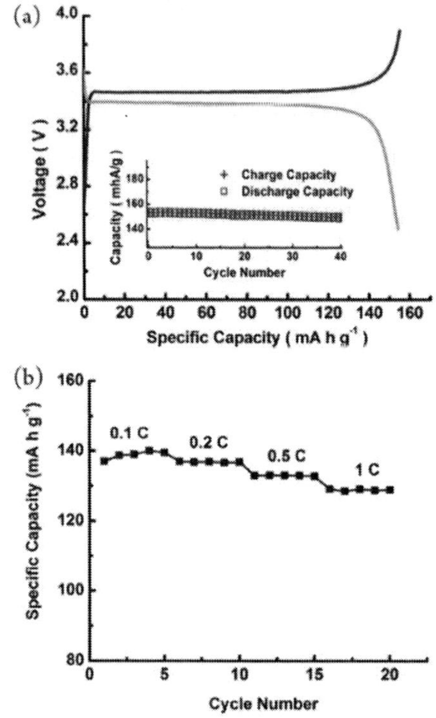

Figure 3. (a) (a) First galvanostatic charge−discharge voltage profile of the $LiFePO_4$/electrolyte film/Li cell at room temperature. The inset shows the Coulombic efficiency and cyclability of the cell. (b) Discharge capacities of the $LiFePO_4$/electrolyte film/Li cell at different discharge rates.

Acknowledgment

This work was supported by the U.S. National Science Foundation under grant number CBET#1235761.

References

1. Scrosati, B.; Garche, J. *Journal of Power Sources* **2010,** 195, (9), 2419-2430.
2. Meyer, W. H. *Advanced materials* **1998,** 10, (6), 439-448.
3. Thomas, K. E.; Sloop, S. E.; Kerr, J. B.; Newman, J. *Journal of Power Sources* **2000,** 89, (2), 132-138.
4. Gorecki, W.; Jeannin, M.; Belorizky, E.; Roux, C.; Armand, M. *Journal of Physics: Condensed Matter* **1995,** 7, (34), 6823.

NANOSCALE ORGANIZATION OF CATHODE MATERIALS FOR Li-ion BATTERY VIA BIOTEMPLATING

Evgenia Barannikova, Mark Allen

University of Maryland Baltimore County
1000 Hilltop Cr, Baltimore, MD, 21250
Evgenia1@umbc.edu

Continuous research in the field of energy storage systems has enabled rapid progress in the use of Li-ion batteries in a wide variety of applications, ranging from small electronic devices to electric and hybrid vehicles and space crafts. Recent improvements in the performance of Li-ion batteries include the use of nanostructured materials, which offers several advantages like shorter pathway for Li-ion and electron diffusion and larger surface area for potential electrode/electrolyte interactions. However, morphological changes during the cell cycling result in a loss of conductivity and lower battery performance. Thus, there is a need to develop new methodologies that will allow precise control over the assembly of nanostructured electrodes in order to improve their electrochemical stability and cyclability for the system. In this study, we present a peptide-templated approach that incorporates heterofunctional nanostructured materials during cathode assembly via specific interactions between the material and organic functional groups of the peptide. In particular, we isolated solid binding peptides (SBP) that have an affinity for one of the promising high-voltage cathode materials $LiNi_{0.5}Mn_{1.5}O_4$ (LMNO). We also designed multifunctional peptides that add to the LMNO peptide a sequence of amino acids that have a high binding affinity to multiwalled carbon nanotubes, which are known to improve conductivity of Li-ion battery cathodes. Our preliminary data indicates that the cathode materials assembled with dual-affinity polypeptide and tested with Li-foil as negative electrode exhibited a decrease in charge transfer resistance, as well as an increase in higher specific capacity during the initial cycles. Ultimately, fabrication of nanostructured electrodes through biotemplating with multifunctional peptides will impart specificity for the target materials, and potentially offer a stable system with higher capacity and improved cyclability at high discharge rates.

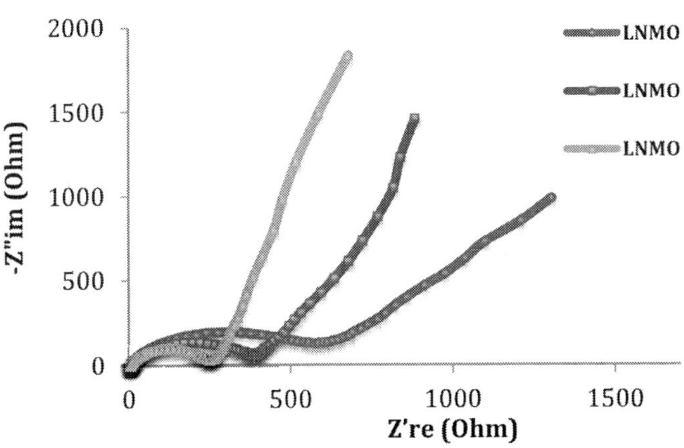

Figure 1. Electrochemical impedance spectroscopy of $LiNi_{0.5}Mn_{1.5}O_4$ cathode shows smaller charge transfer resistance (R_{ct}) after addition of specific bifunctional peptide to the slurry composition.

NATURALLY SYNTHETIC: A BIOLOGICAL TOOLBOX FOR NON-BIOLOGICAL CHALLENGES

Evgenia Barannikova, Scott Riley, Alexander Winton, ***Mark A. Allen***

Department of Chemistry, University of Maryland Baltimore County, Baltimore, MD 21250
allenm@umbc.edu

Abstract

Throughout the past it seems that nature has been confronted with countless challenges that eventually were surmounted through the application of natural selection. These challenges have involved thriving in toxic environments and developing ways of finding water in the driest of places; biology has always applied a diverse toolkit to handle every challenge. The tools in the toolbox include nucleic acids, lipids, polysaccharides, and proteins, each of which have been used for direct applications far different from their defined purposes to make sensors or other electronic devices. Proteins in particular represent nature's most diverse polymer with a range of functionality determined by 20 naturally encoded amino acids. In this presentation I will discuss the application of proteins or polypeptides for the purpose of addressing challenges that biological systems have never previously had to hurdle with an emphasis on bio-templated lithium ion batteries.

While natural selection ends at modifications to the environment that allow generations of organisms to continue, our lab uses a form of artificial selection called phage display in order to address technological problems that are not commonly thought of as relevant to nature. The toolbox remains the same however the application of the tools is to identify solid binding polypeptides that have very strong but specific interactions with electroactive materials and to identify and exploit these interactions in order to improve devices with an emphasis on the improvement of lithium ion batteries.

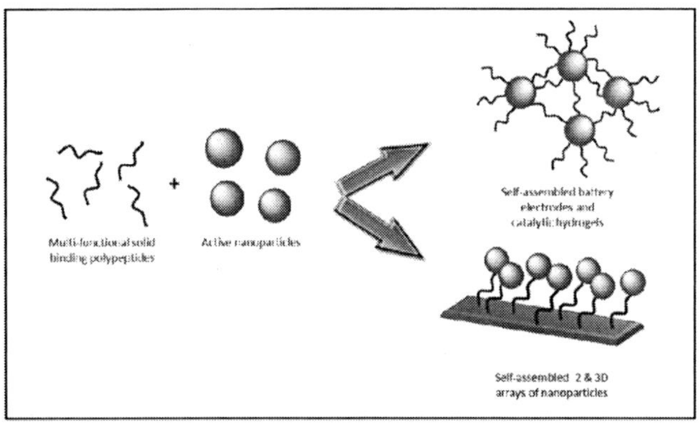

Figure 1. In this schematic multifunctional polypeptides serve as a tethering agent that allow for the self-organization of nanomaterials onto a surface, or the self-assembly of functional materials into an electrode

ANTHRAQUINONE FUNCTIONALIZED REDUCED GRAPHENE OXIDE AS ELECTRODE MATERIAL FOR RECHARGEABLE BATTERIES

Burak Esat[a], *Sumeyye Bahceci Sertkol*[a]; *Anton Momchilov*[b]

[a]Fatih University, Faculty of Science and Art, Department of Chemistry, 34500 Buyukcekmece / Istanbul-Turkey
[b]Bulgarian Academy Of Sciences Acad. Evgeni Budevski Institute Of Electrochemistry And Energy Systems (IEES), Sofia -Bulgaria
besat@fatih.edu.tr

Introduction

The ever-increasing demand for high-performing, light weight, economical, and safe power storage for high–tech portable devices and electric vehicles leads to augmented research efforts in the field of organic or organic/hybrid materials to be used as electrodes in energy storage devices [1-14]. An anode material based on anthraquinone-functionalized reduced graphene oxide (AQ-RGO) is synthesized and its electrochemical properties have been investigated. The initial studies against Li metal revealed that a quite reversible capacity of 200 mAh/g was obtained based on the weight of electro-active anthraquinone moieties when cycled between 1.8and 3.2 V at 0.3C rate. The energy density was found to be around 450 mWh/g. The AQ-RGO has been used also as an anode material against in an aqueous battery against NiOOH (AQ-rGO | 30% NaOH | NiOOH). This battery provided a capacity of 140 mA/h at 0.83V.

Experimental

An azide-functionalized AQ derivative (AQ ester azide) was prepared as follows. 0.271 g (1mmol) anthraquinone-2-carbonyl chloride, 0.27 g (1.2 mmol) 2-azido ethanol, 0.303 g (3mmol) kuru Et_3N and 14.66 mg (0.12 mmol) DMAP were stirred in CH_2Cl_2 under Ar(g) overnight. Reaction was quenched with 1M HCl and organic phase was washed again with 1M HCl , 10% $NaHCO_3$, H_2O and brine. The compund was obtained after evaporation of solvent with 65% yield.

AQ functionalization of RGO was achieved by the following method.Typically, RGO (45mg) was dispersed in 8.5ml dried DMAc in 25 ml reaction flask in an ultrasonic bath for 1h under Ar(g). Then, varied amounts of azide-functionalized AQ derivative were added. The mixture was refluxed under Ar(g) with constant stirring for 24 hour. After cooling to room temperature, the reaction mixture was filtered through Teflon filter (0.22µm) and washed throughly with acetone, CH_2Cl_2 and ethanol. Added amount of azide substituent AQ derivative varied as 50, 200 and 500% by weight relative to the mass of RGO used in the reaction, and the samples are referred to as RGO-AQ 50%, RGO-AQ 200% and RGO-AQ 500%, respectively.

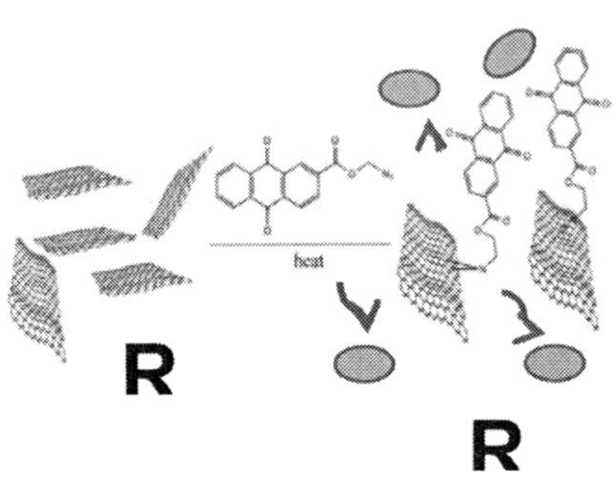

Scheme 1. Reaction pathway of reduced graphene oxide (RGO) with AQ precursor via nitrene chemistry.

Results and Discussion

RGO and AQ-functionalized RGO (RGO-AQ) samples were characterized by FT-IR, Raman, XPS, TGA, SEM, Four-probe conductivity, elemental and X-Ray analysis.

Table1. Atomic Composition of Graphite, GO, RGO and RGO AQ Samples Through XPS Analysis

Atomic Composition %						
Sample	**C**	**O**	**N**	**S**	**C/O**	**C/N**
Graphite	88.23	8.69	-	-	10.15	-
GO	73.52	25.97	-	0.51	2.83	-
RGO	87.42	8.49	4.09	-	10.3	21.4
RGO AQ 200%	84.75	10.06	5.19	-	8.4	16.3
RGO AQ 500%	85.28	9.37	5.36	-	9.1	15.9

Table 2. Fitted Results of C1s Core Level of XPS Spectra of Graphite, GO, RGO and Functionalized Samples.

Sample	sp^2-C	sp2 C-N ring	Ester
RGO	30.9	7.7	
RGO AQ 200%	48.0	9.6	6.0
RGO AQ 500%	41.0	12.4	7.74

Table 3. Elemental Analysis of GO, RGO and RGO AQ Samples.

Atomic Composition %			
Sample	**C**	**N**	**H**
GO	42.49	-	2.35
RGO	78.74	4.27	2.07
RGO AQ 50%	76.56	4.41	2.12
RGO AQ 200%	75.85	5.35	2.27
RGO AQ 500%	72.74	5.56	2.23

As can be seen from data presented in Tables 1-3, the AQ functionalization was successful and its level increases, although at diminishing ratios, with increasing level of loading of the AQ-functionalized azide during RGO-AQ synthesis. This was also supported by other experimental data.

The investigations of electrochemical properties of AQ-functionalized RGO composites (prepared by addition of 15% teflonized acetylene black) were performed in a two electrode cell system containing 1M $LiClO_4$ in PC and 1M $LiPF_6$ in EC: DMC (1:1) against Li metal as reference and counter electrode. Loading level of composites on Al current collector changes 0.7-1.3 mg/cm^2. The reversible single electron redox peak of AQ group on the RGO surface in CV analysis (Figure 1) verifies presence of AQ group on RGO surface. (this peak is absent in CV of RGO itself as shown in inset in Fig.1).

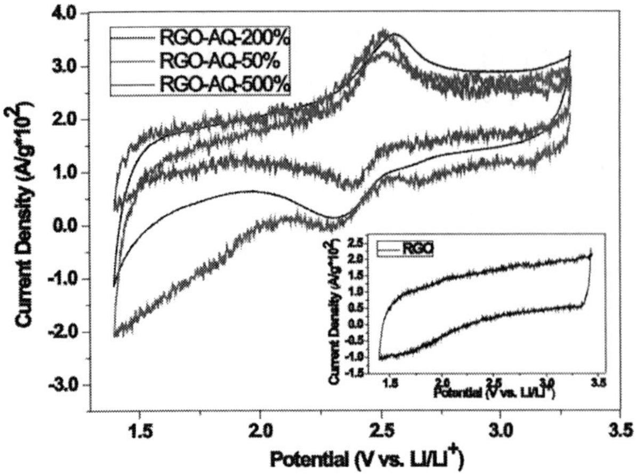

Fig. 1. Cyclic voltammetry results of AQ functionalized RGO (RGO-AQ) electrodes vs. Li metal in 1M LiClO4/PC with a scan rate of 0.07mV/s between 1.4-3.2V. Inlay: Cyclic voltammetry of RGO in the same conditions.

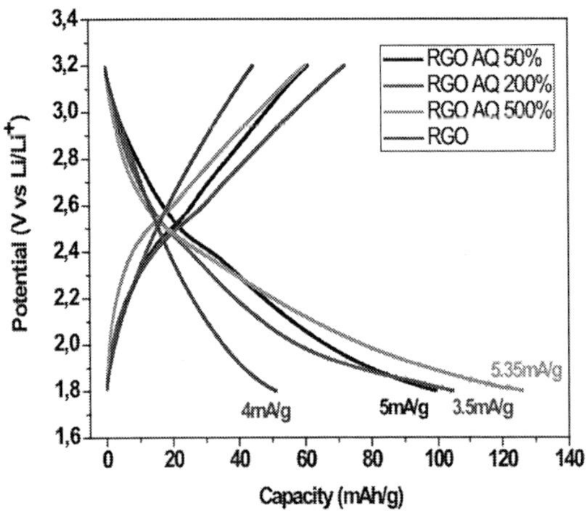

Fig 2. Charge-discharge/ profiles of RGO and RGO AQ samples during the 2. cycle of each at nearly same current densities between 1.8-3.2 V .in 1M $LiPF_6$ in EC: DMC (1:1)

Figure 2 presents charge-discharge profiles of RGO and RGO-AQ samples between 1.8-3.2V vs. Li/Li^+ at nearly same current densities. AQ attached RGO samples displayed distinguishable plateaus at about 2.5 V compared to RGO with increasing capacity from 51 mAh/g to 99, 105 and 126 mAh/g for RGO, RGO-AQ 50, 200 and 500% composites, respectively. There is possibly some capacity contribution to from RGO surface groups. Actually, it is not so accurate to say that this contribution content is exactly 51 mAh/g, as obtained for RGO itself, because surface reactivity of RGO might change during functionalization by nitrene chemistry which involved heating. The cell with RGO-AQ 500% showed quite good rate capabilities as shown in Fig. 3.

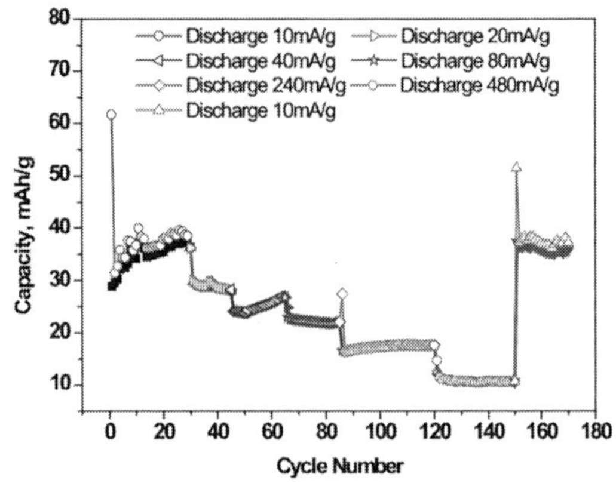

Figure 3. Specific charge-discharge capacities of RGO AQ 500% in 1M $LiPF_6$/EC: DMC(1:1)

The cell constructed with RGO-AQ 500% and 1M $LiClO_4$/PC electrolyte also showed a quite high capacitiy when cycled between 1.3 and 3.6V as can be seen in Fig. 4..

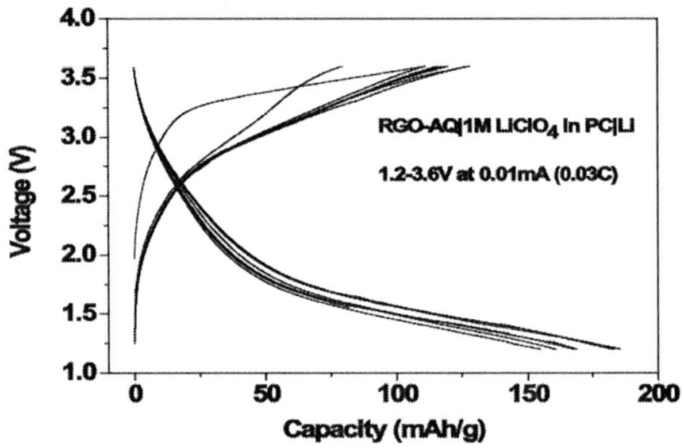

Figure 4. Specific charge/discharge capacities of RGO AQ 500% in in 1M $LiClO_4$/PC between 1.3-3.6V at current density of 6 mA/g .

RGO-AQ was also utilized as an anode material against NiOOH cathode in an aqueous battery containing 30% aq. NaOH as the electrolyte (Fig. 5). This cell provided a capacity of 140 mA/h at 0.2C rate and a working voltage of 0.83V.

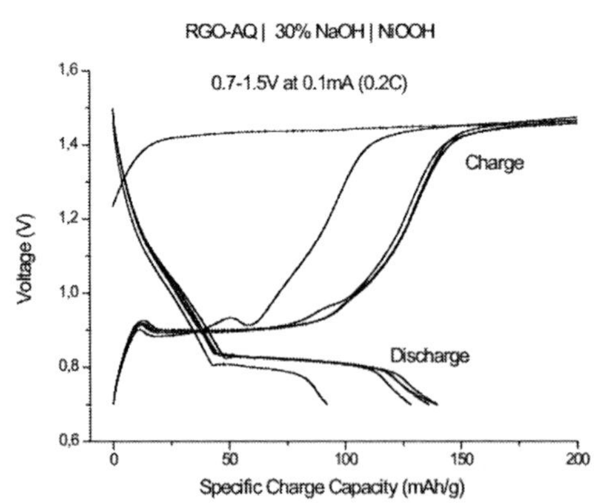

Conclusions

RGO was successfully functionalized with AQ electro-active groups via nitrene chemistry. AQ-functionalized RGO was demonstrated to be good candidate for electrode materials in both Li-batteries and aqueous batteries.

Acknowledgement. This work was financially supported by TUBITAK (Project no. 114Z295).

References

(1) Liang ,Y. ; Tao, Z.; Chen, *J. Adv. Energy Mater.* **2012**, 2, 742–769.

(2) Novák, P.; Müller, K. ; Santhanam , K. S. V. ; Haas , O. *Chem. Rev.* **1997** , 97 , 207.

(3) Cheng , F. Y. ; Tang , W. ; Li , C. S. ; Chen , J. ; Liu , H. K. ; Shen , P. W. ; Dou , S. X. *Chem. Eur. J.* **2006**, 12 , 3082.

(4) Neves , S. ; Polo Fonseca , C. *J. Power Sources* **2002** , 107 , 13 .

(5) Janoschka, T. ; Hager, M. D. ; Schubert, U. S. *Adv. Mater.* **2012**, 24, 6397–6409

(6) Sukegawa, T. ; Kai, A. ; Oyaizu, K. ; Nishide, H. *Macromolecules* **2013**, 46, 1361−1367

(7) Choi, W. ; Harada, D. ; Oyaizu, K. ; Nishide, H. *J. Am. Chem. Soc.* **2011**, 133, 19839–19843

(8) Suga T., Ohshiro H., Sugita S., Oyaizu K., Nishide H., *Adv. Mater.* **2009** , 21 , 1627

(9) Tang Z. Y., Xu G. X., *Acta Phys. -Chim. Sin.* **2003** 19 , 307 .

(10) Yao M., Senoh H., Sakai T., Kiyobayashi T., , *Int. J. Electrochem. Sci* **2011**. 6 , 2905 .

(11) Yao M., Senoh H., Sakai T., Kiyobayashi T., *J. Power Sources* **2012**,202, 364.

(12) Song Z., Zhan H., Zhou Y., , *Chem. Commun.* **2009**, 448 .

(13) Song Z., Zhan H., Zhou Y., *Angew. Chem. Int. Ed.* **2010** , 49 , 8444 .

(14) Himadri Chakraborti, Suman Kalyan Pal, *Chem. Phys. Lett.* **2014**,600 , 118–122

MOLECULAR TRIANGLES FOR ORGANIC RECHARGEABLE BATTERIES

Dong Jun Kim, Keith Hermann, Aleksandrs Prokofjevs, Michael T. Otley, Cristian Pezzato, J. Fraser Stoddart

Department of Chemistry, Northwestern University,
2145 Sheridan Road, Evanston, Illinois 60208, United States
Stoddart@northwestern.edu

Introduction

Organic rechargeable batteries (ORBs), using redox-active small molecules or polymers, have been receiving a significant amount of attention because of their versatility in design, sustainability, flexibility, environmental benignity and availability from abundant natural sources. While recent investigation of ORBs have shown improved electrochemical performances, there still exists a dilemma in the design of organic electrodes.

Results and Discussion

Many small redox-active organic molecules suffer from dissolution or side reactions in the presence of electrolyte solutions or as a result of repeated redox cycles. This situation triggers contact loss between organic molecules and the current collector as well as irreversible reactions, which can lead to severe capacity decay during cell cycling.[1] In order to overcome those issues, a number of investigations have been directed toward either (i) incorporating conjugated redox-active molecules into polymers or by (ii) employing them as an insoluble polymer backbone. Although polymeric materials have better electrochemical cyclability compared with small molecules, it is difficult to achieve a constant potential because of electron-donating effects by reduced functional groups during the discharge process, resulting in a less well-defined voltage profile.[2] So far, it has been a major challenge to achieve both a single well-defined voltage profile and a cycle life in redox-active organic molecules.

By introducing pyromellitic diimide (PMDI) into a molecular triangle[3], cyclic through-space electron delocalization occurs. This property results in increased cycle life time and, more importantly, a single well-defined output voltage. The PMDI triangle exhibits a one voltage at 2.2 V, compared with two peaks at (i) 2.2 and 1.7 V for reduction and (ii) 2.0 and 2.7 V for oxidation of PMDI monomeric analogues. Furthermore, the PMDI triangle delivers over 100 mAh g^{-1} capacity after 100 cycles.

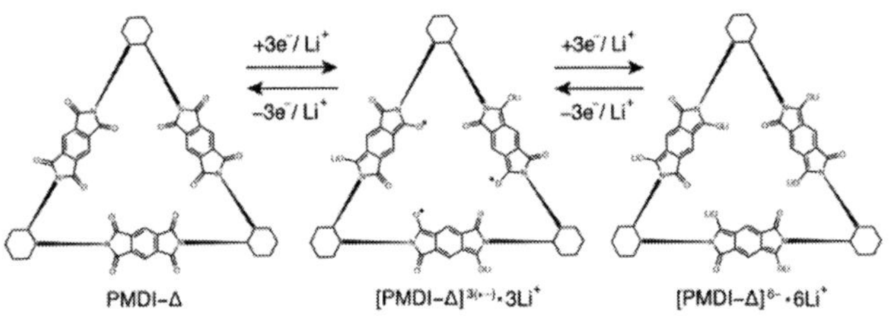

Figure 1. Structural formula and redox process for pyromellitic diimide triangle.

Conclusions

Our results suggest that an efficient delocalization of electrons within the triangle is important for obtaining good organic battery performance. It is not unlikely that this property can be harnessed in other redox-active macrocyclic organic compounds.

References

(1) Häupler. B.; Wild. A,; Schubert, U. S. Carbonyls: Powerful Organic Materials for Secondary Batteries. *Adv. Energy Mater.* **2015**, *5*, DOI: 10.1002/aenm.201300036.

(2) Song, Z.; Zhou, H. Towards Sustainable and Versatile Energy Storage Devices: An Overview of Organic Electrode Materials. *Energy Environ. Sci.* **2013**, *6*, 2280–2301.

(3) Chen, D.; Avestro, A. J.; Sun. J.; Wang. S.; Xiao. M.; Erno. Z.; Algaradah. M. M.; Nassar. M. S.; Amine. K.; Meng. Y.; Stoddart. J. F. A Rigid Naphthalenediimide Triangle for Organic Rechargeable Lithium-Ion Batteries. *Adv. Mater.* **2015**, *27*, 2907–2912.

POROUS 2D TRANSITION METAL CARBIDES (MXenes) FOR HIGH-PERFORMANCE LITHIUM-ION STORAGE

*Chang E. Ren, Meng-Qiang Zhao, Taron Makaryan, Joseph Halim, Muhammad Boota, Michel W. Barsoum, and Yury Gogotsi**

Department of Materials Science and Engineering and A.J. Drexel Nanomaterials Institute, Drexel University, Philadelphia, PA 19104, USA

*Corresponding author, email: gogotsi@drexel.edu

Abstract

Introduction of well-designed micro-, meso- or macro-porous structures into functional building blocks, generally leads to significantly improved or unexpected physicochemical properties and performance. MXenes are a family of two-dimensional (2D) transition metal carbides, which have shown great promise as electrodes in lithium-ion batteries and supercapacitors. However, the accessibility of electrolyte ions into MXenes is limited due to the compact stacking of the 2D flakes, hindering the full utilization of their energy storage performance. Herein, we successfully introduce pores into MXene flakes and demonstrate the significant improvement of their Li-ion storage capability.

The pores were introduced by an easy and controllable chemical etching method using different transition metal salts (e.g. $CuSO_4$, $CoSO_4$ or $FeSO_4$), which yielded intrinsically porous, free-standing and flexible Ti_3C_2 films (p-Ti_3C_2). The presence of the pores were confirmed by gas sorption and microscopic analyses. The electrochemical performance of p-Ti_3C_2 was evaluated using p-Ti_3C_2/CNT composite film as an anode material in Li-ion batteries. At 0.5 C, a capacity of ≈750 mAh g^{-1} was achieved, along with excellent cycling stability. This value was significantly higher than the ≈220 mAhg^{-1} obtained with non-porous Ti_3C_2/CNT films. In addition to the high capacity, the p-Ti_3C_2/CNT electrodes showed impressive rate performance. After a pre-cycling step, the capacity was ≈1250 mAh g^{-1} at 0.1 C, and ≈330 mAh g^{-1} at 10 C. This chemical etching method, applicable to other MXene chemistries such as V_2C and Nb_2C, provides a simple yet effective method to improve the physicochemical properties of MXenes, and possibly other kinds of 2D materials.

DYNAMIC SURFACE STRESS RESPONSE DURING REVERSIBLE Mg ELECTRODEPOSITION AND STRIPPING

Yeyoung Ha,[†] *Zhenhua Zeng,*[‡] *Christopher J. Barile,*[†] *Jinho Chang,*[†] *Ralph G. Nuzzo,*[†] *Jeffrey Greeley,*[*,‡] *and Andrew A. Gewirth*[*,†]

[†] Department of Chemistry, University of Illinois, Urbana, IL 61801, U.S.A.
[‡] School of Chemical Engineering, Purdue University, West Lafayette, IN 47907, U.S.A.
* jgreeley@purdue.edu; agewirth@illinois.edu

There is much interest in developing Mg batteries due to their high theoretical volumetric capacity originating from the divalent nature of Mg. Moreover, Mg is abundant and does not suffer from harmful dendrite formation during the deposition process. However, developing a suitable electrolyte for reversible Mg deposition and dissolution is challenging due to the formation of passivation films on the electrode surface.[1,2]

Recently, several electrolyte systems have shown promise, but the interfacial chemistries are complicated.[3-8] Monitoring the evolution of surface stress during deposition and dissolution of Mg provides information at the interfacial region, which could help in developing understanding of the mechanisms and leading to better electrolytes.

In this work, the dynamic surface stress response during reversible deposition and stripping of Mg from four different electrolytes are examined: Grignard (PhMgCl) and Lewis-acid ($AlCl_3$) based APC[3]; all-magnesium and Lewis acid-free (DTBP)MgCl–$MgCl_2$[6]; inorganic salt ($MgCl_2$ and $AlCl_3$) solution MACC[4,5]; and halide-free $Mg(BH_4)_2$+$LiBH_4$[7,8]. Moreover, periodic DFT calculations are employed to further interrogate the origin of the experimentally measured surface stress changes.

Together with the experimental and computational analyses of the surface stress development, we propose a general Mg deposition and stripping scheme.

References

(1) Muldoon, J.; Bucur, C. B.; Oliver, A. G.; Sugimoto, T.; Matsui, M.; Kim, H. S.; Allred, G. D.; Zajicek, J.; Kotani, Y. *Energy Environ. Sci.* **2012**, *5*, 5941.
(2) Yoo, H. D.; Shterenberg, I.; Gofer, Y.; Gershinsky, G.; Pour, N.; Aurbach, D. *Energy Environ. Sci.* **2013**, *6*, 2265.
(3) Mizrahi, O.; Amir, N.; Pollak, E.; Chusid, O.; Marks, V.; Gottlieb, H.; Larush, L.; Zinigrad, E.; Aurbach, D. *J. Electrochem. Soc.* **2008**, *155*, A103.
(4) Doe, R. E.; Han, R.; Hwang, J.; Gmitter, A. J.; Shterenberg, I.; Yoo, H. D.; Pour, N.; Aurbach, D. *Chem. Commun.* **2014**, *50*, 243.
(5) Liu, T.; Shao, Y.; Li, G.; Gu, M.; Hu, J.; Xu, S.; Nie, Z.; Chen, X.; Wang, C.; Liu, J. *J. Mater. Chem. A* **2014**, *2*, 3430.
(6) Pan, B.; Zhang, J.; Huang, J.; Vaughey, J. T.; Zhang, L.; Han, S.-D.; Burrell, A. K.; Zhang, Z.; Liao, C. *Chem. Commun.* **2015**, *51*, 6214.
(7) Mohtadi, R.; Matsui, M.; Arthur, T. S.; Hwang, S.-J. *Angew. Chem., Int. Ed.* **2012**, *51*, 9780.
(8) Shao, Y.; Liu, T.; Li, G.; Gu, M.; Nie, Z.; Engelhard, M.; Xiao, J.; Lv, D.; Wang, C.; Zhang, J.-G.; Liu, J. *Sci. Rep.* **2013**, *3*, 3130.

NON-PRECIOUS METAL OXYGEN REDUCTION CATALYSIS IN FUEL CELL: RATIONAL DESIGN FROM CATALYST TO ELECTRODE

Di-Jia Liu

Chemical Sciences & Engineering Division
Argonne National Laboratory
9700 S. Cass Ave.
Argonne, IL 60439
djliu@anl.gov

Introduction

Finding inexpensive replacements for the platinum group metals (PGMs) has been the ultimate goal for proton exchange membrane fuel cell (PEMFC) catalyst research. Since the electrode/catalyst materials contribute to nearly half of a fuel cell stack cost, there is an urgent need to reduce or replace PGM usage. Among all the non-PGM candidates explored so far, transition metal doped nitrogen-carbon (TM-N-C) composites appear to be the most promising ones. Generally, these materials are prepared by forming TM-N_x molecular complex over amorphous carbon support, followed by thermal activation. Non-PGM catalysts are known to have lower turn-over frequency *per catalytic site* compared to Pt. Their active site densities must be therefore substantially higher to deliver a comparable performance. Using the catalytically inert carbon support dilutes the active site density therefore lowering the overall activity. To maximize the utilization of non-PGM catalytic centers also requires an electrode structure that ensures efficient mass transport of reactant and product to and from the active site. Furthermore, the catalyst must be durable under accelerated aging conditions.

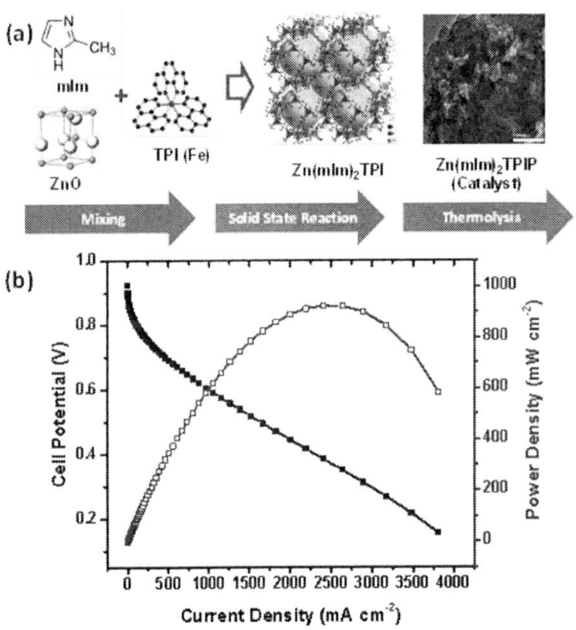

Figure 1. (a) One-pot synthesis of ZIF-derived non-PGM catalyst; (b) Representative fuel cell performance with MOF-derived non-PGM catalyst at cathode (one-bar O_2).

MOF and POP Derived ORR Catalysts

Since the initial discovery of N-chelated transition metal as oxygen reduction catalyst over a half century ago, [1] the precursors used for non-PGM catalyst have been predominantly limited to molecular systems with the square-planar configurations. Such molecular precursors require high surface area conductive materials, such as amorphous carbon, to support and disperse, in order to reach optimal exposure to the gas phase reactant. Using catalytically inert support reduces the volumetric and gravimetric catalytic site densities and limits the potential to maximize the non-PGM function overall. At Argonne National Laboratory, we pioneered the researches of using MOFs and POPs as the precursors to prepare "support-free" non-PGM catalyst for cathodic oxygen reduction reaction (ORR) .[2,3,4] The ORR catalysts prepared from these precursors have been proven highly active in both rotating disk electrode (RDE) and under actual fuel cell tests. This presentation provides a brief review on our effort in design and synthesis of MOF and POP based non-PGM catalysts.

MOF/ZIF based non-PGM Catalyst Approaches. Meta-organic frameworks, or MOFs, have clearly-defined lattice structures through the regioselective metal-ligand coordination chemistry. The TM-N_4 entities can be grafted into MOFs with very high ligation site densities. During thermal activation, the organic linkers are converted to electro-conductive carbon while maintaining porous framework, leading to catalysts with high surface areas and uniformly distributed catalytic sites. Our team at Argonne was the first to demonstrate that zeolitic imidazolate framework (ZIF), subclass of MOF, can used to prepare low-cost, highly active non-PGM catalysts.[2] More recently, we developed a "one-pot" solid-state synthesis method which produces non-PGM catalyst using low-cost material without the need of solvent and separation (**Figure 1a**). Such method will substantially reduce the cost of non-PGM catalyst. Furthermore, the cathode catalysts prepared by MOF or POP have produced some major catalytic activity breakthroughs, generating the highest current and power densities among those reported in the literature. (**Figure 1b**)

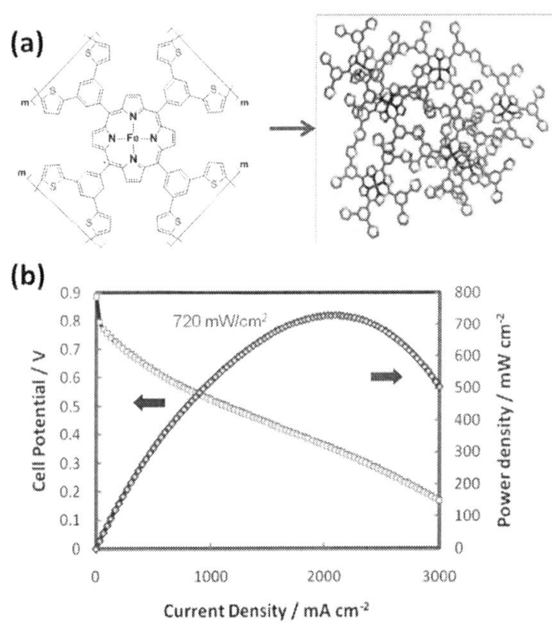

Figure 2. (a) An example of Fe-containing polyporphyrin POP prepared from cross-linking reaction; (b) the fuel cell performance in oxygen with the POP-derived non-PGM catalyst at cathode.

POP based non-PGM Catalyst Approaches. Parallel to MOFs, we also developed another class of non-PGM catalysts using the porous organic polymers (POPs).[4] POPs are prepared by cross-linking the monomers containing strong TM-N binding site and stereo contorted core through polymerization chemistry. Similar to MOFs, POPs have high surface areas and uniformly distributed active sites inside the pours framework. Different from MOFs, however, POPs add some additional advantages. Since the metals are not required as the building blocks, the metalation of POPs can be controlled at monomer level before synthesis or through post ligation after polymerization. Therefore the pyrolysis temperature can be significantly reduced and acid wash step can be avoided in respect to excess metal removal. As the result, the catalyst yield is often higher. **Figure 2a** shows a representative polyporphyrin POP prepared in our laboratory and corresponding fuel cell performance when tested under oxygen. (**Figure 2b**)

Electrode with Nanofibrous Architecture

To improve mass and charge transfers for non-PGM catalysts, we also investigated a new method to produce non-PGM catalyst with nano-network electrode architecture.[6] The MOF-based catalysts are incorporated into individual nano-fibers connected by a graphitic network. High micro-pore volume and surface area are maintained whereas the meso-pores in the conventional powder catalysts were no longer necessary and eliminated. Mass transport is improved through macro-pores inside the nano-network while the charge transfer is accomplished through the network of graphitic fibers. (**Figure 3**) The new nano-network non-PGM catalyst (Fe/N/CF) achieved excellent fuel cell performances in both activity and durability. Catalyst durability under fuel cell operating condition represents another critical challenge facing non-PGM and low-PGM catalyst development. We have recently developed a new approach to stabilize the catalyst performance. The new catalyst demonstrated less than 12% loss of mass activity at 0.9 V after 30,000 voltage cycles in a fuel cell test following DOE test protocol.

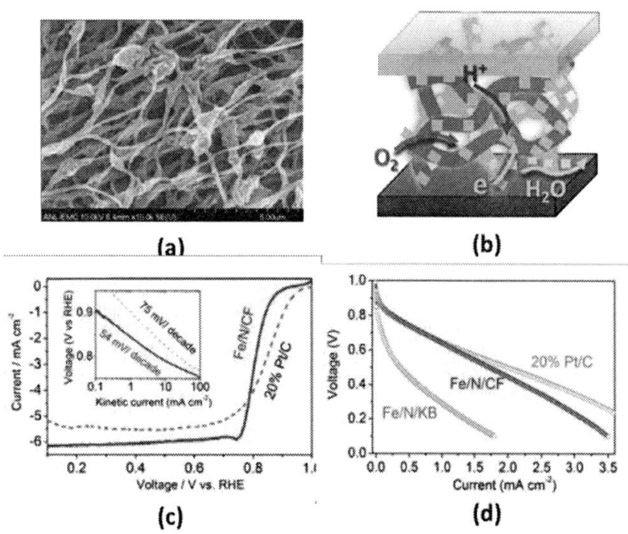

Figure 3. (a) The "string-knot" morphology of the nanofibrous catalyst produced from electrospin followed by pyrolysis; (b) the schematic electrode structure with nanofibrous architecture; (c) comparison of catalytic activities of Fe/N/CF with 20% Pt/C tested by the rotating disk electrode and (d) in fuel cell cathode under oxygen.

Acknowledgement. The financial supports from the U. S. Department of Energy, Office of Energy Efficiency and Renewable Energy, Fuel Cell Technologies Office and Office of Science are gratefully acknowledged. Argonne National Laboratory is a U.S. Department of Energy Office of Science Laboratory operated under Contract No. DE-AC02-06CH11357 by UChicago Argonne, LLC.

References

(1) Jasinski, R., *Nature*, **1964**, *201*, 1212
(2) Ma, S.; Goenaga, G.; Call, A; and Liu, D.-J. *Chemistry: A European Journal* **2011**, *17*, 2063. B. M.
(3) Zhao, D.; Shui, J.-L.; Chen, C.; Chen, X.; Reprogle, B. Wang, D.; and Liu, D.-J. *Chem. Sci.*, **2012** *3* (11), 3200.
(4) Yuan, S.; Shui, J.-L.; Grabstanowicz, L.; Chen, C.; Commet, S. M.; Reprogle, B.; Xu, T.; Yu, L.; and Liu, D.-J. *Angew. Chem. Int. Ed.*, **2013** *52*(32), 8349.
(5) Zhao, D.; Shui, J.-L.; Grabstanowicz, L. R.; Chen, C.; Commet, S. M.; Xu, T.; Lu, L.; and Liu, D.-J. *Advanced Materials*, **2014** *26* 1093.
(6) Shui, J.; Chen, C.; Grabstanowicz, L. R.; Zhao, D. and Liu, D.-J. *Proceedings of National Academy of Sciences*, **2015**, *112*, (34), 10629

ELECTROCHEMICAL SYNTHESIS OF AMMONIA UNDER AMBIENT CONDITIONS USING NOVEL SOLVENTS WITH HIGH NITROGEN SOLUBILITY AND WIDE ELECTROCHEMICAL WINDOW

Kwiyong Kim[a], Hyung Chul Yoon[b], Jong-In Han[a]

[a] Department of Civil and Environmental Engineering, Korea Advanced Institute of Science and Technology, 291 Daehak-ro, Yuseong-gu, Daejeon 34141, Republic of Korea
[b] Petroleum and Gas Laboratory, Korea Institute of Energy Research, 152 Gajeong-ro, Yuseong-gu, Daejeon 34129, Republic of Korea
hanj2@kaist.ac.kr

Introduction

Electrochemical synthesis of NH_3, a widely-used fertilizer and also potential energy carrier, is a promising way that can replace energy-intensive Haber-Bosch process. The success of this approach at ambient conditions lies chiefly in the proper selection of an electrolyte medium. In this study, some organic solvents with high N_2 solubility and wide electrochemical window were selected and their effectiveness in the NH_3 electrosynthesis were evaluated.

Experimental

Materials. The organic solvents, LiCl, and AgCl were purchased from Sigma Aldrich.

Cell components. For preliminary tests, a one-chamber cell was constructed. A porous Ni plate, Pt plate, and Ag^+/Ag were used as a cathode, anode, and reference electrode, respectively. An advanced two-chambered system with ion exchange membrane was also used for NH_3 synthesis.

Electrochemical measurements. Cyclic voltammetry was carried out in the solvents to identify their electrochemical window.

Electrolysis. Galvanostatic electrolysis was carried out using DC power supply (H3005D, Finepower).

Analytical method. The concentration of NH_3 before and after electrolysis was analyzed using the indophenol methods [1].

Results and Discussion

Organic solvent 1 is an alcohol-based electrolyte. From CV measurement it was found that this solvent possesses wider electrochemical window compared to water, implying suppressed H_2 evolution. The solvent exhibited considerably enhanced NH_3 synthesis compared to water due to increased N_2 solubility and suppressed H_2 evolution (Fig.1) [2].

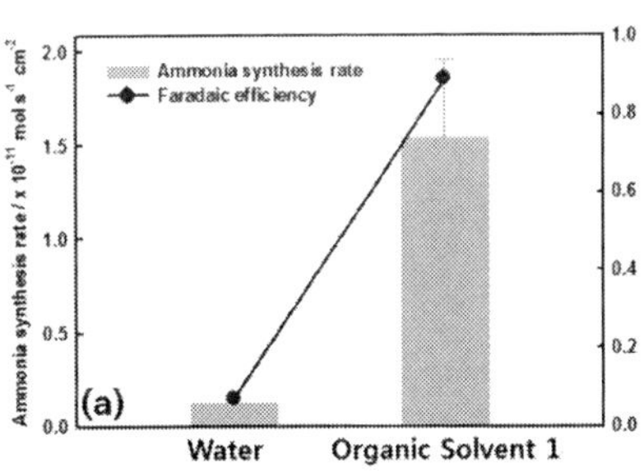

Figure 1. Effect of electrolyte medium (water vs. organic solvent 1) on NH_3 synthesis and Faradaic efficiency.

Organic solvent 2, whose electrochemical window is wide enough to allow cathodic Li deposition (Fig. 2), was found and NH_3 synthesis in this solvent are being actively pursued at present, which will be covered in the presentation.

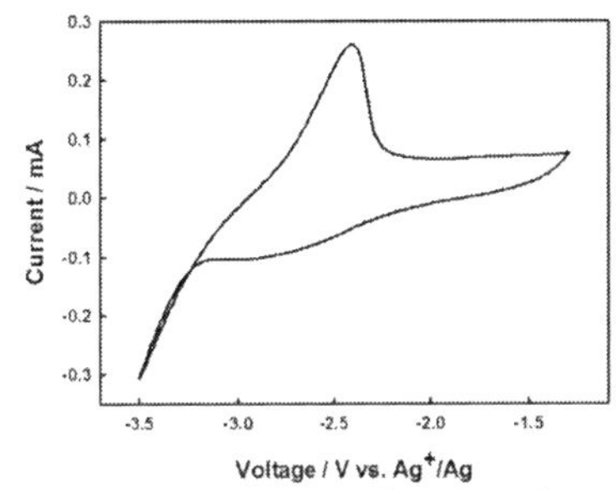

Figure 2. Cyclic voltammogram in organic solvent 2.

Conclusions

We found several organic solvents with high N_2 solubility and wide electrochemical window which enable efficient electron transfer toward N_2 on cathode, and they showed more than 10 times higher NH_3 synthesis compared to aqueous solvents.

Acknowledgement. This work was conducted under the framework of the Research and Development Program of the Korea Institute of Energy Research (KIER) (B6-2442).

References

(1) Banait, J. S.; Singh, B.; Kanur, H. *Indian. J. Chem. Sect A.* **2007**, *46*, 266.
(2) Köleli, F.; Röpke, T. *Appl. Catal. B Environ.* **2006**, *62*, 306.

ELECTROSPUN CARBON NANOFIBER AEROGEL MEETS POLYIMIDE-DERIVED CARBON SHEETS: A RATIONAL STRUCTURAL DESIGN FOR ENVIRONMENTAL PROTECTION AND ENERGY STORAGE

Feili Lai,[a] Yue-E Miao,[a] Lizeng Zuo,[a] and Tianxi Liu[a,]*

[a]State Key Laboratory of Molecular Engineering of Polymers, Department of Macromolecular Science, Fudan University, Shanghai 200433, P. R. China.
E-mail address: txliu@fudan.edu.cn

1. Introduction

Carbon aerogels, a family of three-dimensionally (3D) porous materials, characterized by low density, large internal surface area, small pore size, and large pore volume, have been a rising star for both scientific researches and industrial applications.[1] Recently, various nanofibers (such as cotton, bacterial cellulose) show their greatly potential values for 3D carbon aerogel fabrication.[2] As a mature technique, electrospinning is widely considered to fabricate two-dimensional membrane consisting of continuously fibrous materials with controllable diameter, high specific surface aera, robust mechanical strength and excellent flexibility instead of 3D ones.[3] In virtue of the dual-processes of electrospinning and freeze-shaping techniques proposed by Ding et al., three-dimensionally crosslinked nanofibrous aerogels were reconstructed from bicomponents of polyacrylonitrile/bifunctional benzoxazine nanofibers and silicon dioxide nanofibers, showing low densities, rapid recovery from deformation, as well as efficient energy absorption for a wide range of applications in terms of thermal insulation, sound absorption, emulsion separation and elasticity-responsive electric conduction.[4] However, the absence of strong framework support in the inner structure of fibrous carbon aerogels leads to their poor stress tolerance under harsh conditions. Therefore, how to introduce efficient chemical/physical bonds between electrospun fibers is still the unavoidable problem before its commercial production.

Herein, we present a novel carbon aerogel with rational structure consisting of one-dimensional carbon nanofibers and two-dimensional carbon sheets derived from polyacrylonitrile nanofibers and poly(amic acid) respectively, by combining electrospinning technique, freeze-shaping technique and carbonization process. Thus obtained carbon aerogels (denoted as oP) can not only be directly used as versatile adsorbents for oil/organic pollutants with a maximal sorption capacity (62 times of its original weight), but also act as support for fabrication of high-performance electrode materials such as activated carbon materials (denoted as A-oP) and oP/MnO_2 hybrid carbon aerogel. Consequently, the assembled oP/MnO_2//A-oP asymmetric supercapacitor exhibits high energy density of 30.3 Wh kg^{-1} at a power density of 501.3 W kg^{-1}, which is more excellent than most reported MnO_2 based asymmetric supercapacitors. So, the strategy in our work provides an innovative method to achieve novel carbon aerogel material for both versatile adsorbents and supports of high-performance supercapacitor electrode materials, exhibiting potential values for further practical applications.

2. Experimental Section

The preparation procedure of oP/MnO_2 hybrid carbon aerogel is schematically shown in Figure 1a. Firstly, electrospun polyacrylonitrile (PAN) nanofibers[5] and water-soluble poly(amic acid) (PAA)[6] are mixed in water with the assistance of triethylamine to form PAN/PAA mixed dispersion. After freeze drying, PAN/PAA mixed dispersion is dried into PAN/PAA composite aerogel. Then, oP carbon aerogel is obtained via imidization and carbonization processes of PAN/PAA composite aerogel. At last, a certain amount of acid-treated oP carbon aerogel is immersed into $KMnO_4$ solution, and keeps at 80 °C for 2 h to achieve the successful in-situ oxidation of MnO_4^- and carbon. The obtained products are denoted as oP/MnO_2 hybrid carbon aerogel.

3. Results and Discussion

The preparation process of oP/MnO_2 hybrid carbon aerogel is displayed in Figure 1a. Attributing to the strong hydrogen bonds between -COOH of poly(amic acid) (PAA) and -OH/-NH_2 of oxidized polyacrylonitrile nanofibers, a novel three-dimensionally cellular structure of oP carbon aerogel is presented in Figure 1b with a cellular pore size distribution of 5-20 □m, one-dimensional (1D) carbon nanofibers and two-dimensioanl (2D) carbon sheets. The detailed linkages between carbon nanofibers and sheets are further revealed through TEM image in Figure 1c.

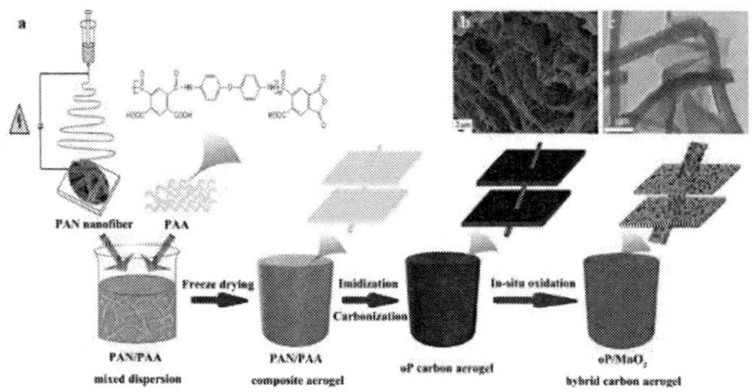

Figure 1. (a) The schematic for the preparation of PAN/PAA composite aerogel, oP carbon aerogel, and oP/MnO_2 hybrid carbon aerogel; FESEM and TEM images (b and c) of oP/MnO_2 hybrid carbon aerogels.

Benefiting from the well-organized and interconnected porous structures of oP carbon aerogels, they can facilely absorb a series of common pollutants in our daily life as well as from industries, including commercial oils (e.g., pump oil), fats (e.g., colza oil), water-miscible solvents (e.g., acetone, ethanol), aromatic compounds and their derivatives (e.g., toluene, benzyalcohol). The results show that oP carbon aerogel exhibits a high adsorption capacity for all of these organic liquids ranging from 20 to 62 times of its original weight (Figure 2a), which is attributed to the enhanced capillary effect with the existence of thousands of electrospun carbon nanofibers among carbon sheets (Figure 2b). Furthermore, oP carbon aerogels possess outstanding recyclability with high retention of adsorbing capacity after five cycles (Figure 2c and 2d), which indicates its great possibility for wide applications in the cleanup fields.

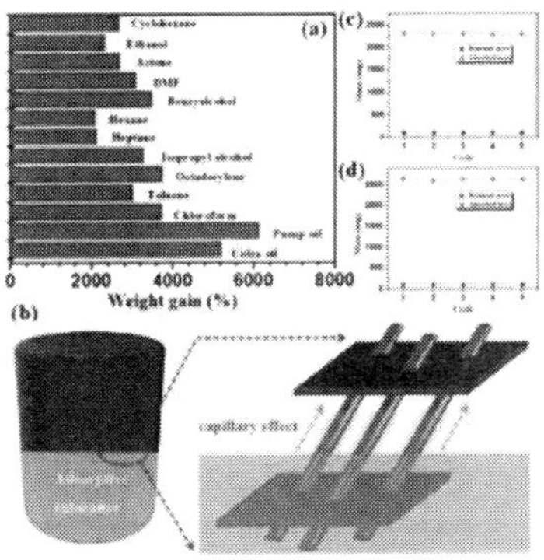

Figure 2. (a) Adsorption efficiency of oP carbon aerogel on various organic liquids; (b) The illustrated schematic for the probable mechanism of enhanced capillary effect for oP carbon aerogels; (c and d) Recycling performance of oP carbon aerogel.

Meanwhile, oP carbon aerogel is also an outstanding template for further loaing of nanoparticles. Hence, MnO_2 nanosheets are grown on surface of oP carbon aerogel by oxidizing its exterior carbon, based on the redox reaction:

$$4KMnO_4 + 3C + H_2O = 4\ MnO_2 + 2KHCO_3 + K_2CO_3$$

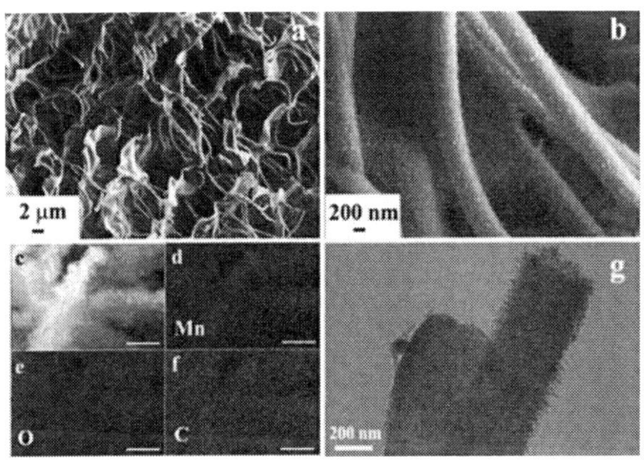

Figure 3. Low- and high-magnification FESEM images of (a and b) oP/MnO_2 hybrid carbon aerogels; The corresponding EDS elemental mappings of (c) oP/MnO_2 hybrid carbon aerogels: (d) Mn, (e) O, (f) C (the scale bar is 2.5 □m); TEM images of (g) oP/MnO_2 hybrid carbon aerogels.

As displayed in Figure 3a, the 3D interconnected structure of oP carbon aerogel is still well maintained after the coverage of MnO_2 nanosheets for oP/MnO_2 hybrid carbon aerogel. The detialed surface information is displayed in Figure 3b, and MnO_2 nanosheets are homogeneously deposited on the surface of both carbon nanofibers and carbon sheets without obvious self-agglomeration, demonstrating the stability of carbon template structures during deposition processes. Through the EDS elemental mappings (Figure 3c-f), it once again confirms the successful growth of MnO_2 nanosheets on the surface of carbon template uniformly and becomingly.

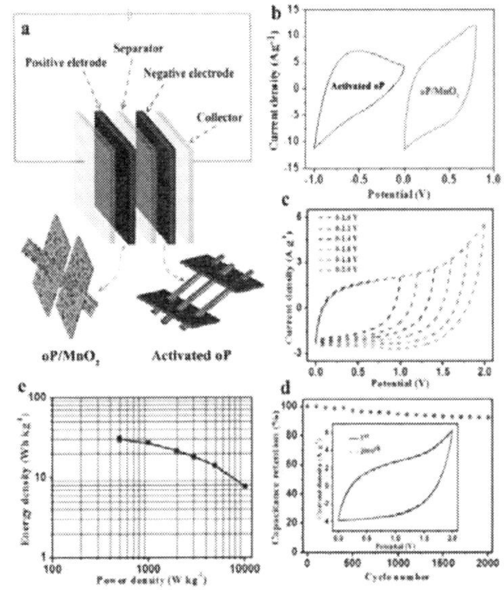

Figure 4. (a) The schematic of the assembled asymmetric supercapacitor (ASC); (b) Comparative CV curves of oP/MnO_2 hybrid carbon aerogel and A-oP carbon aerogel electrodes in a three-electrode set-up at a scan rate of 50 mV s^{-1}; (c) CV curves of oP/MnO_2//A-oP ASC within different operation voltages at a scan rate of 50 mV s^{-1}; (d) Cycling performance of oP/MnO_2//A-oP ASC (the inset showing the CV curves at the 1st and 2000th cycles); (e) Ragone plots of oP/MnO_2//A-oP ASC.

Finally, asymmetric supercapacitors (ASC) are assembled using A-oP as negative electrode and oP/MnO_2 hybrid carbon aerogel as positive electrode respectively, and denote as oP/MnO_2//A-oP ASC device (Figure 4a). As shown in Figure 4b and 4c, the operating cell voltage is greatly extended for oP/MnO_2//A-oP ASC device, which exhibits good capactive behaviors. Additionally, our assembled ASC device also exhibits excellent cycle performance. As shown in Figure 4d and its inset, 92% of its initial capacitance is retained after 2000 cycles at a scan rate of 50 mV s^{-1}, indicating its good cycle stability as asymmetric system. The energy densities and power densities for oP/MnO_2//A-oP ASC device at various current densities are shown in the Ragone plots (Figure 4e), delivering maximum energy density of 30.3 Wh kg^{-1} at a power density of 501.3 W kg^{-1}.

4. Conclusions

In summary, oP carbon aerogels with unique structure are innovatively fabricated by incorporating electrospinning technique, freeze-shaping technique and carbonization process, which is consisted of one-dimensional carbon nanofibers and two-dimensional carbon sheets derived from oxidized polyacrylonitrile nanofibers and poly(amic acid) respectively. Benefiting from the multi-dimensional structure and enhanced capillary effect, thus obtained oP carbon aerogels can be applied as versatile adsorbents for various pollutants in our daily life as well as from industries, showing maximal sorption capacity of 62 times of its original weight and excellent recyclability. Interestingly, oP carbon aerogels can also act as support for asymmetric supercapacitor (ASC) electrode materials, such as activated oP (A-oP) carbon aerogels (negative electrode) and oP/MnO_2 hybrid carbon aerogel (positive electrode). The assembled oP/MnO_2//A-oP ASC device exhibits a high energy density of 30.3 Wh kg^{-1} at power density of 501.3 W kg^{-1}, which is much higher than most reported MnO_2 based asymmetric supercapacitors. Hence, our work fabricated a novel carbon aerogel material with multi-functions (such as adsorbents, supercapacitor electrode, and even nanoparticle supports), which owns immense potential for both scientific researches and practical applications.

Acknowledgement. The authors are grateful for the financial support from the National Natural Science Foundation of China (51125011, 51373037, 51433001).

References

(1) Worsley, M. A.; Pauzauskie, P. J.; Olson, T. Y.; Biener, J.; Satcher, J. H.; Baumann, T. F. *J. Am. Chem. Soc.* **2010**, *132*, 14067.

(2) Bi, H. C.; Yin, Z. Y.; Gao, X. H.; Xie, X.; Tan, C. L.; Huang, X.; Chen, B.; Yang, Q. L.; Bu, X. Y.; Lu, X. H.; Sun, L. T.; Zhang, H. *Adv. Mater.* **2013**, *25*, 5916.

(3) Si, Y.; Yu, J. Y.; Tang, X. M.; Ge, J. L.; Ding, B. *Nat. Commun.* **2014**, *5*, 5802.

(4) Liu, Y.; Zhou, J. Y.; Chen, L. L.; Zhang, P.; Fu, W. B.; Zhao, H.; Ma, Y. F.; Pan, X. J.; Zhang, Z. X.; Han, W. H.; Xie, E. Q. *ACS Appl. Mater. Interfaces* **2015**, *7*, 23515.

(5) Lai, F. L.; Miao, Y. E; Huang, Y. P.; Chung, T. S.; Liu, T. X. *J. Phys. Chem. C* **2015**, *119*, 13442.

(6) Zhang, Y. F.; Zuo, L. Z.; Huang, Y. P.; Zhang, L. S.; Lai, F. L.; Fan, W.; Liu, T. X. *ACS Sustainable Chem. Eng.* **2015**, *3*, 3140.

ELECTROCATALYSIS ON TRANSITION METAL-OXIDE FILMS EPITAXIALLY GROWN ON SINGLE CRYSTAL SUBSTRATES

Jin Suntivich

Materials Science and Engineering,
Cornell University, Ithaca NY USA
jsuntivich@cornell.edu

The electrocatalysis of the oxygen evolution reaction (OER) is a critical process in the water-splitting technologies. Intensive research efforts have spent on identifying strategies for controlling the surface oxygen interactions by using, as an example, cationic substitutions[1], structural engineering[2], and support interactions[3]. While these efforts have led to successful developments of new electrocatalysts for the OER, most studies have thus far focused on polydisperse materials. The reliance on the polydisperse materials complicates the process of connecting the surface structure with the electrochemical mechanism as one cannot isolate whether the activity trend is due to the presence of different terminations, structural defects, or even other presenting (secondary) phases.

We present the synthesis and the characterizations of transition-metal oxide films grown on single-crystal oxide substrates as a model system for studying the OER electrocatalysis. These transition metal oxides are single crystals and thus eliminate surface heterogeneities that are present in polydisperse oxides. In the past decade, advances in deposition technologies and substrate availabilities have enabled the growth of single-crystal transition-metal oxide films with high structural perfection. We use these advances, specifically Molecular-Beam Epitaxy (MBE), to prepare epitaxial ruthenate and iridate films on single-crystal oxide substrates. Iridates and ruthenates are among the most studied OER electrocatalysts for water splitting. The use of epitaxial, single-crystal films allow us to link the features from our electrochemical characterizations to the physical process, which we will compare to first-principle calculations. We will further discuss strategies for controlling these surface adsorption processes. Finally, we use this information to reveal insights into how to tune the surface physicochemical properties on these oxide materials and discuss the implications on the OER and other electrocatalysis.

References

(1) Suntivich, J.; May, K. J.; Goodenough, J.B.; Gasteiger, H.A., and Shao-Horn, Y. *Science*, **2011**, *334*, pp. 1383-1385.

(2) Robinson, D. M.; Go, Y. B.; Mui, M.; Gardner, G.; Zhang, Z. J.; Mastrogiovanni, D.; Garfunkel, E.; Li, J.; Greenblatt, M.; and Dismukes, G. C. *J. Am. Chem. Soc.* **2013**, *135* (9), pp. 3494-3501.

(3) Gorlin, Y.; Chung, C. J.; Benck, J.D.; Nordlund, D.; Seitz, L.; Weng, T. C.; Sokaras, D.; Clemens, B. M..; and Jaramillo, T. F. *J. Am. Chem. Soc.* **2014**, *136* (13), pp. 4920-4926.

HIGHLY DENSE Cu NANOWIRES FOR LOW-OVERPOTENTIAL CO_2 REDUCTION

David Raciti, Kenneth Livi, Chao Wang

3400 N Charles Street, Baltimore, MD 21218
chaowang@jhu.edu

Electrochemical reduction of CO_2, an artificial way of carbon recycling, represents one promising solution for energy and environmental sustainability. It is, however, challenged by the lack of active and selective catalysts. Here we report a two-step synthesis towards highly dense Cu nanowires as advanced electrocatalysts for CO_2 reduction. CuO nanowires were first grown by oxidation of Cu mesh in air and then reduced by either annealing in the presence of hydrogen or applying a cathodic electrochemical potential to produce Cu nanowires. The two reduction methods generated Cu nanowires with similar dimensions but distinct surface structures, which have provided an ideal platform for comparative studies of the effect of surface structure on the electrocatalytic properties. In particular, the Cu nanowires generated by electrochemical reduction were highly active and selective for CO_2 reduction, requiring an overpotential of only 0.3 V to reach 1 mA/cm^2 electrode current density and achieving Faradaic efficiency toward CO as high as ~60%. Our work has advanced the understanding of the structure-property relationship of Cu-based nanocatalysts, which could be valuable for the further development of advanced electrocatalytic materials for CO_2 reduction.

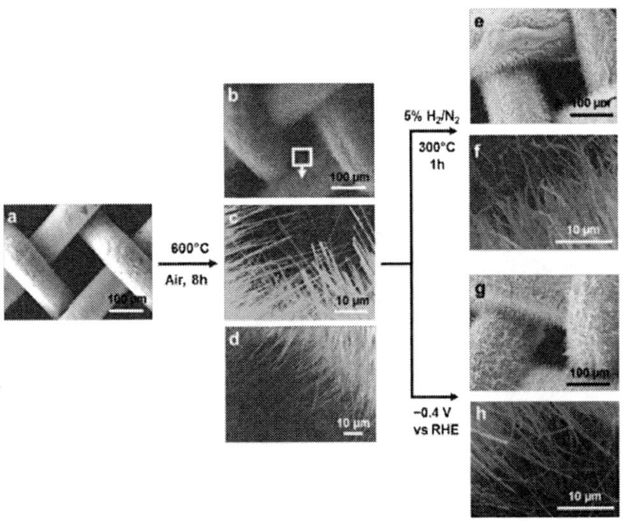

Figure 1. Reaction schematic showing the process for producing Cu nanowires: CuO nanowires were first grown by oxidation of Cu mesh in air at 600 °C, which were then reduced to Cu by either annealing in the presence of hydrogen at 300 °C or electrochemical reduction at −0.4 V vs. RHE.

FUNCTIONALIZATION OF GRAPHENE FOR EFFICIENT ENERGY CONVERSION AND STORAGE

Liming Dai

Center of Advanced Science and Engineering for Carbon (Case4Carbon)
Departments of Macromolecular Science and Engineering, Case Western Reserve University, Cleveland, Ohio, 44106, USA

ABSTRACT

As a building block for carbon materials of all other dimensionalities (such as 0D buckyball, 1D nanotube, 3D graphite), the two-dimensional (2D) single atomic carbon sheet of graphene has emerged as an attractive candidate for energy applications due to its unique structure and properties. Like other materials, however, a graphene-based material that possesses desirable bulk properties rarely features the surface characteristics required for certain specific applications. Therefore, surface functionalization is essential, and researchers have devised various covalent and non-covalent chemistries for making graphene materials with the bulk and surface properties needed for efficient energy conversion and storage. Judicious application of these *site-selective* reactions to graphene sheets has opened up a rich field of graphene-based energy materials with enhanced performance in energy conversion and storage.

In this talk, I will provide some rational concepts for the controlled functionalization of graphene for the development of efficient energy conversion and storage devices (*e.g.*, solar cells, fuel cells, supercapacitors, and batteries), along with an overview on recent progresses in this emerging research area.

A COMPARATIVE STUDY OF THE INFLUENCE OF *N,N'*-DIALKYL *VS N,N'*-DIARYL ELECTRON DONORS ANCILLARY LIGANDS ON THE PHOTOCURRENT AND PHOTOVOLTAGE IN DYE-SENSITIZED SOLAR CELLS (DSSCS)

Saba Ashraf[a,c], *Dalia El-Sherbiny*[c], *Humaira M. Siddiqi*[a], *Javeed Akhtar*[b], *Ahmed El-Shafei*[c]*

[a] Department of Chemistry, Quaid-I-Azam University, Islamabad 45320, Pakistan

[b] Department of Physics, Polymers & Materials Synthesis Lab (PMS), COMSATS, Institute of Information Technology, Chak Shahzad Islamabad, Pakistan

[c] Polymer and Color Chemistry Program, North Carolina State University, Raleigh, NC, 27695, USA

(*) Corresponding author: Ahmed El-Shafei@ncsu.edu

Abstract

In this work, we report synthesis of a novel heteroleptic Ru (II)-sensitizer *namely* Ru(2,2'-bipyridine-4,4'-dicarboxylic acid)-4,4'-*bis* (4-piperidin-1-yl)phenyl ethenyl)-2,2'-bipyridine) $(NCS)_2$ denoted as SD-I in comparison to K77-7[1](*N,N'*-diaryl Ru-sensitizer namely Ru(2,2'-bipyridine-4,4'-dicarboxylic-acid)-4,4'-*bis*(2-(4-*N,N'*-diphenylaminophenyl)ethenyl)-2,2' bipyridine) $(NCS)_2$. The molecular structure of SD-I was confirmed using FTIR, ^{1}H-NMR and mass spectrometry. The ground state oxidation potentials of SD-I & K77-7 were measured by cyclic voltammetry and were found to be thermodynamically favorable for dye regeneration and electron injection into the conduction band edge of TiO_2. The photovoltaics performance of SD-I outperformed that of N-719 and K77-7 with overall efficiency of 8.5 % compared to 7.9% of N719 and 8.0% K77-7 under the same experimental device conditions. These results clearly demonstrated that the enhanced light-harvesting ability of SD-I can be attributed to the sp^3-nitrogen, which is tethered to sp^3-carbons (dialkyl) not diaryls (K77-7), demonstrating stronger electron donating power of the former.

Keywords. DSSC, photocurrent, cyclic-voltammetry, HOMO, quantum efficiency

METAL ORGANIC FRAMEWORK MEMBRANES FOR Kr/Xe SEPARATION

Moises A. Carreon [1], Praveen K. Thallapally [2]

1 Chemical and Biological Engineering Department, Colorado School of Mines, Golden, CO 80401 United States
2 Physical and Computational Science Directorate, Pacific Northwest National Laboratory, Richland, WA 99352 United StatesOverwrite
email: mcarreon@mines.edu

Introduction

The separation of krypton (Kr) from xenon (Xe) is a relevant problem for nuclear industries. Separating Kr from Xe is a critical step in removing radioactive 85Kr during treatment of spent nuclear fuel.[1]The conventional method to separate these two gases is fractional distillation at cryogenic temperatures, which is an energy intensive process. In addition, even after cryogenic distillation, trace levels of radioactive Kr in the Xe-rich phase are too high to permit further use. In this respect, membrane technology could play a key role in making this separation less energy intensive and therefore economically feasible. Membrane separation processes have several advantages over conventional fractional distillation; for instance, it is a viable energy-saving method, since it does not involve any phase transformation, furthermore, the required membrane process equipment is simple, easy to operate, control and scale-up. Metal organic frameworks (MOFs) [2] consisting of metal cations or metal-based-clusters linked by organic molecules forming a crystalline porous network, have emerged as a novel type of crystalline porous materials which combine highly desirable properties, such as uniform micropores, high surface areas, and exceptional thermal and chemical stability, making them ideal candidates for molecular gas separation applications. Herein, we present the development of MOF membranes which show great promise for Kr/Xe separation.

Acknowledgement. We gratefully acknowledge the financial support by the Department of Energy (DOE) Nuclear Energy University Program (NEUP) under Grant No. DE-NE0008429.

References

(1) Izumi, J. In Handbook of Zeolite Science and Technology; Auerbach, S. M., Carrado, K. A., Dutta, P. K., Eds.; CRC Press: New York, 2003.
(2) (a) Li, H.; Eddaoudi, M.; O'Keeffe, M.; Yaghi, O. M. Nature 1999, 402, 276. (b) Eddaoudi, M.; Kim, J.; Rosi, N. L.; Vodak, D. T.; Wachter, J.; O'Keeffe, M.; Yaghi, O. M. Science 2002, 295, 469. (c) Uemura., T; Yanai., N.; Kitagawa, S. Chem. Soc. Rev. 2009, 38, 1228.

NOVEL CUSTOMIZED AMORPHOUS FLUOROPOLYMER (CAF) MEMBRANES FOR OLEFIN-PARAFFIN SEPARATION

H. Murnen, S. Majumdar, Y. Koizumi, K. Loprete, K. Pennisi, S. Nemser, A. Feiring, N. Shangguan

Compact Membrane Systems, Newport, DE 19804
smajumdar @compactmembrane.com

Introduction

Olefins are the most important building blocks of the petrochemical industry. Ethylene and propylene are major chemical industry raw materials in petrochemical and polymer manufacturing such as polyethylene, polypropylene and styrene. Olefin-paraffin (O-P) separations are some of the costliest, most energy intensive, and most technically difficult separations in the industry. Distillation is the technology of choice for O-P separation and is estimated to consume 250 trillion BTU/yr of energy.

A hybrid distillation-membrane process that can efficiently separate and recover the olefin from the paraffin would provide substantial economic benefit. Membranes have been evaluated previously for separating olefins from paraffins. In many cases, silver salt facilitating agents have been incorporated into membranes to selectively transport olefin. Historically, while good separations have been demonstrated in the laboratory, stability problems have prevented commercial implementation of membrane technology.

Experimental Results

A thorough long-term investigation of CMS custom amorphous fluoropolymers (CAF) membrane for separating propylene from propane has been carried out. The performance results for over 270 days for propylene-propane system without any impurities is shown (Figure 1). The modest initial drop (~ 0 to 10 days) in performance is attributed to physical aging of thin film high free-volume glassy polymer [1, 2]. We have also demonstrated membrane stability in presence of poisons such as hydrogen sulfide, acetylene and hydrogen.

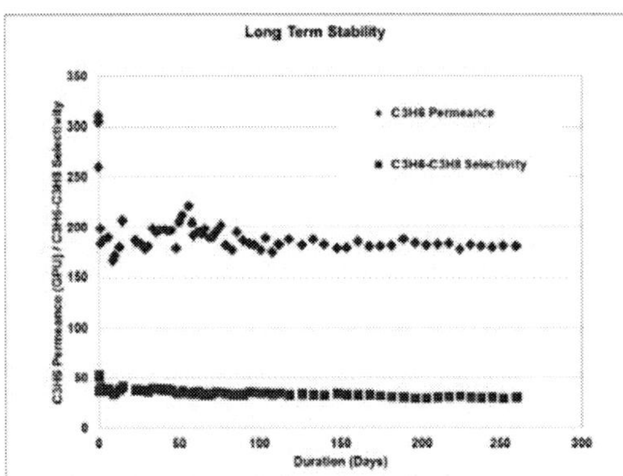

Figure 1. Long term membrane performance: propylene-propane mixture separation

Figure 2 provides an industrial plot mapping out flux/selectivity (does not include stability) of historical O-P separation membranes and shows CMS CAF O-P membranes are dramatically superior. It has been stated that a stable membrane with propylene/propane selectivity of 6–10 and a propylene permeance of 20–40 GPU would be very useful for a number of industrial processes [3]. Note that CMS values are based on mixed gas experiments and also include data containing H_2S while most of the data that are used to generate the Robeson line are based on single gas evaluation at zero time with no foulant exposure.

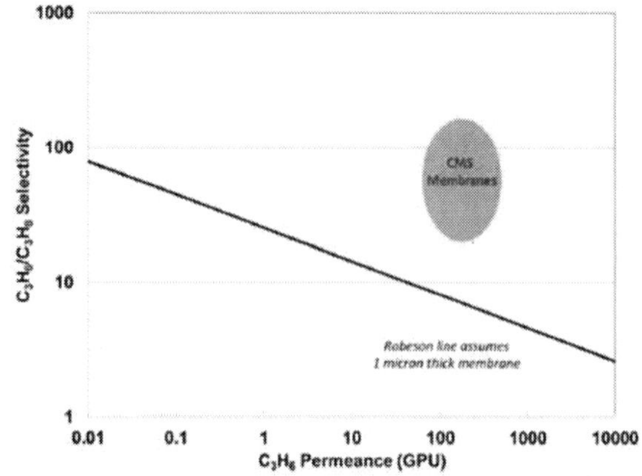

Figure 2. Flux/selectivity map for propylene-propane system (Robeson Plot)

Conclusions

New customized amorphous fluoropolymer (CAF) membranes have been developed by Compact Membrane Systems for stable olefin-paraffin separation with very good propylene flux and propylene/propane selectivity. The permeance and selectivity combination of the membranes are above the industry Robeson curve. In addition results are stable when exposed to high levels of conventional poisons such as hydrogen, acetylene or sulfur containing gas. Separation of ethylene/ethane has also been demonstrated recently. These results along with economic modeling show very low payback times and high IRRs for a variety of streams encountered in petrochemical processes.

References

(1) Huang, Y., Paul, D.R., *Ind. Eng. Chem. Res*. **2007**, 46 (8), 2342.

(2) Tiwari, R.R., Smith, Z.P., Lin, H., Freeman, B.D., Paul, D.R., *Polymer, 2014*, 55, 5788.

(3) Baker, R.W. and Low, B.T., *Macromolecules*, **2014**, 47, 6999.

MULTISCALE SIMULATION AND CONTROLABLE SYNTHESIS OF POROUS ADSORPTION MATERIALS

Dapeng Cao and Xiaofei Zeng*

State Key Lab of Organic-Inorganic Composites, Beijing University of Chemical Technology, Beijing 100029, China, Email: caodp@mail.buct.edu.cn

Abstract

It has become increasingly clear that hydrogen as a viable energy carrier may play an important role in future energy plans. However, development of reasonably safe and reversible hydrogen storage materials for transportation is still a great challenge. In this work, we present the scheme of designing novel materials and address the multiscale simulation method, in which the first-principles calculation is performed to calculate the interaction energy between adsorbate and adsorbent, and then the interaction is input to grand canonical Monte Carlo (GCMC) simulation for evaluation of hydrogen storage capacities.[1-2] By using the multiscale simulation method proposed, a series of novel materials are designed by simulating synthesis. These materials include new carbon allotrope - diamondyne, [3] porous organic polymers, [4] Li-doped covalent organic frameworks (COFs),[1] porous aromatic frameworks (PAFs).[5] Furthermore, hydrogen storage capacities of these materials are also investigated. Although all these materials show excellent hydrogen storage properties, modification of these materials can further improve their hydrogen storage capacities. Toward this end, we also used the multiscale simulation method to screen the metal dopants to modify these materials, and found that the lithium atom is the best modifier of COF materials for enhancing gas storage. More importantly, we also used experiments to synthesize the designed materials to confirm the multiscale simulation results.[6] Besides hydrogen storage materials, we also investigated the nitrogen-doped graphene analogue materials for oxygen reduce reaction.[7-9]

References

(1) D.P. Cao, J. Lan, W. Wang, B. Smit, ***Angew. Chem. Int. Ed.***, 2009, 48, 4730.
(2) J. Lan, D. P. Cao, W. Wang, *et al.*, ***ACS Nano***, 2010, 4, 4225.
(3) L. Huang, Z. Xiang, D.P. Cao, ***J. Mater. Chem. A***, 2013, 1, 3851.
(4) Z. Xiang, R. Mercado, J. Huck, H. Wang, Z. Guo, W. Wang, D. P. Cao*, et al., ***J. Am. Chem. Soc.***, **2015**, 137, 13301
(5) T. Ben, H. Ren, S. Ma, D.P. Cao, et al., ***Angew. Chem. Int. Ed.***, 2009, 48, 9457
(6) Z. Xiang, Z. Hu, D. P. Cao, *et al.*, ***Angew. Chem. Int. Ed.***, 2011, 50, 491
(7) P. Zhang, F. Sun, ... D. P. Cao, ***Energy Environ. Sci.***, 2014, 7, 442.
(8) Z. Xiang, D. P. Cao, *et al*, ***Adv Mater.*** 2014, 26, 3315.
(9) Z. Xiang, Y. Xue, D. P. Cao, L. Huang, J. Chen, L. Dai, ***Angew. Chem. Int. Ed.***, 2014, 53, 2433

NANOPOROUS MATERIALS FOR GAS STORAGE AND SEPARATION: THEORETICAL ASPECTS

Rodion V. Belosludov[1] and Y. Kawazoe[2]

(1) Institute for Materials Research, Tohoku Univesity; JAPAN;
(2) New Industry Hatchery Center, Tohoku University, JAPAN
rodion@imr.tohoku.ac.jp

Introduction

In order to realize materials with desirable characteristics it is important to have a good understanding of the atomic-scale chemical and physical properties of materials. Thus, the computer-aided materials design is one of the important factors in the fabrication of novel materials to be applied in urgent technological needs for achieving a sustainable society. The recent advent of metal–organic framework materials (MOFs), as new functional adsorbents has attracted the attention due to scientific interest in the creation of unprecedented regular nano-sized spaces and in the finding of novel phenomena, as well as commercial interest in their application for storage and separation.[1,2] For MOFs the structural versatility of molecular chemistry has allowed the rational design and assembly of materials having novel topologies and exceptional host-guest properties that are important for urgent sustainable applications. Due to the regularity of MOF structures, in a computer simulation we are easily able to build structural models of MOFs that are very helpful to find new materials with desired characteristics. As an example, the high sorption ability for acetylene on specific MOF material was determined, using both different experimental measurements and first-principles calculations which ascribe to the double hydrogen bond support between the acidic acetylene proton and its acceptor basic site on the channel surface.[3] It was shown that the concept using a designable regular MOF material could be applicable to a highly stable, selective adsorption system.

The aim of this study is detailed theoretical analysis the adsorption of targeted molecules into selected nanoporous materials in order to accelerate the experimental realization of novel gas storage materials.

Computational Methods

First-principles calculations have been performed within the framework of density functional theory (DFT), as implemented in the Vienna Ab initio Simulation Package (VASP).[4] Electron exchange-correlation was treated by the generalized gradient approximation (GGA) with the Perdew, Burke, and Ernzerhof (PBE) parameterization,[5] and interactions between the ion cores and valence electrons were modeled by the all-electron projector augmented wave (PAW) method.[6] The plain-wave cutoff energy was 400 eV and Brillouin zone integrations were performed using the Monkhorst-Pack k-point mesh. The XCrySDen code was used for visualizing the obtained results.[7]

Results and Discussion

We have performed the accurate systematic simulation of MOF properties including host-guest interaction, guest dynamics, and thermodynamic stability of empty host. Our theoretical analysis of MOF structures for gas storage and separation have included: (a) the modification of organic linkers by including different functional groups; (b) the effect of open or unsaturated metal sites on sorption properties of selected MOF structures; (c) the catenation (interpenetration) effect on thermodynamic stability of MOF structures with high pore size in order to realize the optimal ratio between pore volume and surface area. Here, our recent studies have been reported.

The adsorption of C_2H_2, CO_2 and CO on urotropine $Zn_4(dmf)(ndc)_4(ur)_2$ has been investigated. This material is a rare example of a microporous compound with open basic sites decorating the inner surface of the MOF host. It has been found that the inclusion of the urotropine ligand into the MOF framework enhances the adsorption of acetylene and carbon dioxide. The adsorption energy of a CO_2 molecule increases in value from -10.4 kJ/mol to -18.3 kJ/mol. The same tendency has been observed for acetylene, with an increase of 4.8 kJ/mol in adsorption energy. It is interesting to see that the adsorption energy of carbon monoxide is almost the same on the adsorption site of the N atom of urotropine and on the surface of naphtathalene dicarboxylic acid. The adsorptions of these molecules are associated with the charge transfer as shown in Figure 1b and 1c, respectively. In the case of the C_2H_2 molecule, the most favorable attraction between the guest molecule and the host structure can be explained by the formation of a weak C-H···N hydrogen bond between the H atom of acetylene and the N atom of urotropine. In the case of the CO molecule (see Figure 1d), the charge density isosurface shows only a charge redistribution in the adsorbed molecule indicating weak interaction between the MOF framework and the adsorbed non-coordinated gas molecules. The use of urotropine as a co-bridging ligand in metal–carboxylate MOFs may therefore result in a variety of other porous materials with notable basic functionality derived from non-coordinated N-centers.[8]

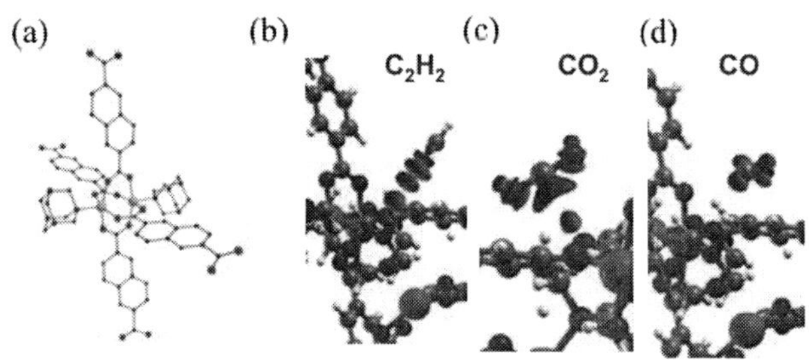

Figure 1. (a) The fragment of $Zn_4(dmf)(ndc)_4(ur)_2$ and charge density isosurface for interactions of (b) C_2H_2, (c) CO_2 and (d) CO with urotropine moiety of $Zn_4(dmf)(ndc)_4(ur)_2$. Red represents an accumulation of electrons and blue a depletion of electrons.

The MOF structures, such as HKUST-1,[9] have been taken a lot of scientific attentions due to specific interactions of the open or unsaturated metals sites. This feature may play an important role in gas separation, sensors and storage.[10] We have studied the CO/N_2 separation using the nanoporous materials with accessible Cu^{2+} metal sites. In the case HKUST-1 structure, the CO molecule interacts with the Cu^{2+} site by energy of 24.5 kJ/mol. The analysis of electron density for the Cu^{2+}-CO interaction indicates the formation of the weak chemical bond between Cu^{2+} and CO as shown in Figure 2a. The N_2 adsorption on Cu^{2+} site was found to be significantly weaker (-7.2 kJ/mol) as compared with the CO adsorption. However, the next step of adsorption process is completely different. The CO adsorption inside HKUST-1 channel was found to be significantly weaker that also confirmed by analysis of electron density. The value of -0.56 kJ/mol was calculated for one CO adsorbed inside pore of CO-coordinated HKUST-1. There is no significant difference in electron charge density for interactions of CO and N_2 inside CO-coordinated HKUST-1 structure and the calculated adsorption energy of N_2 molecule was found to be -0.28 kJ/mol.

Recently, a new soft porous coordination polymer Cu^{2+}(5-azidoisophthalate) with adaptable pores that selectively adsorbs CO from CO/N_2 gas mixture has been discovered.[11]

(a) (b)

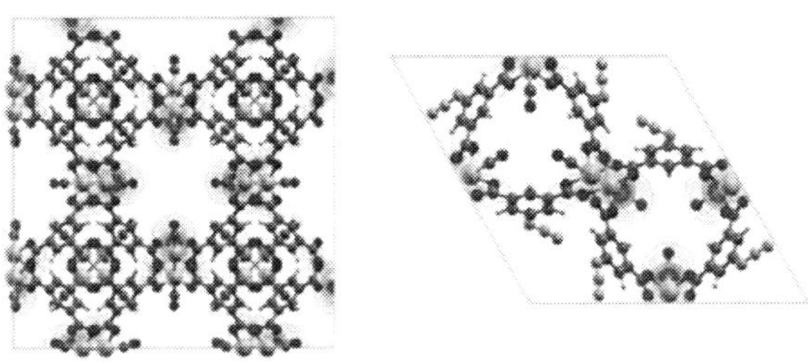

Figure 2. The electron charge density contours for Cu^{2+}-CO interaction in (a) HKUST-1 and (b) PCP. The planes that cut through Cu-C bonds were chosen.

The key feature of this structure is the presence of two channels, small (*S*) and large (*L*), having different pore size and the size of small one increased due to interaction of CO molecules with the Cu^{2+} sites. It has been found in this case the interaction of CO molecule with the Cu^{2+} site is smaller in energy (-21.1 kJ/mol) as compared with interaction of CO with copper ion in HKUST-1. The analysis of electron density for Cu^{2+}-CO interaction indicates almost the same features as in the case HKUST-1 structure as shown in Figure 2b. Moreover, such framework topology significantly reduces the interaction of nitrogen molecules with the Cu^{2+} sites and this is improved the separation process. Moreover, the increasing size of small channel also allows one to control the next step of separation process and the calculation results showed a remarkable distinction in adsorption properties of the CO-coordinated PCP towards CO and N_2 separation. It was found that the adsorption behavior of guest molecules differed between *S* and *L* channels. Thus, CO molecules preferably adsorbed in channel *S* by 7.3 kJ/mol as compared with N_2 adsorption in the same channel.

For the adsorption of the CO molecule in the small channel it was found that the excess charge localized on adsorbed molecules, and electron depletion observed on the CO-coordinated PCP framework indicated the favorable attraction between the guest molecule and the host structure (see Figure 3a).

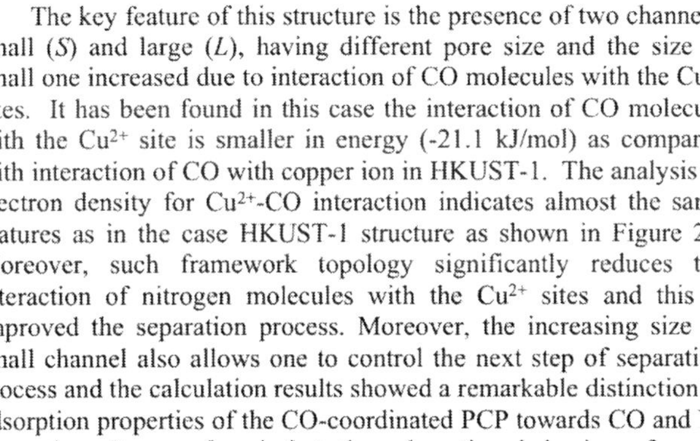

Figure 3. Charge density distribution for CO-coordinated PCP with (a) CO and (b) N_2 in small channels. Red represents an excess and blue a depletion of electrons.

As shown in Figure 3b, the electron depletion localized on both N_2 and CO-coordinated PCP. This shows the repulsion between nitrogen molecules and the PCP framework.

Thus, the high selectivity has been achieved by the synergetic effect of the local interaction between CO and accessible Cu^{2+} metal sites and a global transformation of the framework. This transformable crystalline material realized the separation of CO from a mixture with nitrogen, a gas that is the most competitive to CO. [11]

Due to the van der Waals interaction between MOF framework and adsorbed gas molecules the realization of optimal ratio between pore volume and surface area is an important aspect toward gas storage and separation at temperatures higher than cryogenic one. One of the possible solutions controlling this ratio is the formation of MOF structures having two identical frameworks with high pore size by interpenetration.[12] In order to study the interpenetration effect, we have selected the well-known binuclear carboxylate complexes having an octahedral geometry ($Zn_2(R)_4L_2$). The biphenyl connected to the Zn metals through pyridine molecules and terphenyldicarboxylate (TPDC) are selected as *L* and *R* ligands, respectively. For increasing the pore size the replacement of biphenyl on porphyrin in *L* ligand has been considered in this MOF structure. Thermodynamic stability of selected MOF structures was investigated using molecular dynamics method. It was shown that the both frameworks are stable at temperatures required for practical gas storage. The calculated interaction energies between two frameworks have shown that the interpenetrating structures are energetically more favorable due to attractive interaction between the MOF ligands. In the case of *R* ligand containing biphenyl moiety, the interaction is larger because the closer distances between TPDC of different frameworks which can leads to interweaving between two frameworks and hence a number of adsorption sites is decreased due to formation of thicker walls. The substitution of biphenyl by porphyrin moieties can prevent such adhesion due to the increasing distance between TPDC and results in weaker interaction between two interpenetrating frameworks.

Acknowledgement. The work was supported by E-IMR center of the Institute for Materials Research, Tohoku University. The authors would like to express their sincere thanks to the staff of the Center for Computational Materials Science of the Institute for Materials Research, Tohoku University for their continuous support of the SR16000-M1 supercomputing facilities.

References

(1) Kitagawa, S.; Kitaura, R.; Noro, S. *Angew. Chem. Int. Ed.* **2004**, *43*, 2334.
(2) Yaghi, O. M.; O'Keeffe, M.; Ockwig, N. W.; Chae, H. K.; Eddaoudi, M.; Kim, J. *Nature* **2003**, *423*, 705.
(3) Matsuda, R.; Kitaura, R.; Kitagawa, S.; Kubota, Y.; Belosludov, R.V.; Kobayashi, T. C.; Sakamoto, H.; Chiba, T.; Takata, M.; Kawazoe, Y.; Mita, Y. *Nature* **2005**, *436*, 238.
(4) Kresse G.; Furthmüller, J. *Comput. Mater. Sci.*, **1996**, *6*, 15.
(5) Perdew, J. P.; Burke, K.; Ernzerhof, M. *Phys. Rev. Lett.*, **1996**, *77*, 3865.
(6) Kresse, G.; Joubert, D. *Phys. Rev. B*, **1999**, *59*, 1758.
(7) Kokalj, A. *Comput. Mater. Sci.*, **2003**, *28*, 155.
(8) Sapchenko, S. A.; Dybtsev, D. N.; Samsonenko, D. G.; Belosludov, R. V.; Belosludov, V. R.; Kawazoe, Y.; Schroder, M.; Fedin, V. M. *Chem. Comm.* **2015**, *51*, 13918.
(9) Chui, S. S.-Y.; Lo, S. M.-F.; Charmant, J. P. H.; Orpen, A. G.; Williams, I.D. *Science*, **1999**, *283*, 1148.
(10) Bordiga, S.; Regli, L.; Bonino, F.; Groppo, E.; Lamberti, C.; Xiao, B.; Wheatley, P. S.; Morris, R. E.; Zecchina, A. *Phys. Chem. Chem. Phys.*, **2007**, *9*, 2676.
(11) Sato, H. ; Kosaka, W.; Matsuda, R. ; Hori, A.; Hijikata, Y. ; Belosludov, R. V. ; Sasaki, S.; Takata, M.; Kitagawa, S. *Science*, **2014**, *343*, 167.
(12) Rowsell J. L. C.; Yaghi O. M. *Angew. Chem. Int. Ed.* **2005**, *44*, 4670.

IMPROVEMENT OF GAS SEPARATION PROPERTIES OF POLYBENZIMIDAZOLE MEMBRANES FOR GAS SEPARATIONS AT HIGH PRESSURE AND HIGH TEMPERATURE THROUGH THERMAL TREATMENT

*Edson V. Perez, Kenneth J. Balkus, Jr., John P. Ferraris, and Inga H. Musselman**

Department of Chemistry and Biochemistry
The University of Texas at Dallas, BE26
Richardson, TX 75080
imusselm@utdallas.edu

Introduction

The majority of the world's energy supply currently comes from fossil fuels, including oil, coal, and natural gas.[1] It is estimated that the demand for energy will continue to increase by as much as 56% by 2040 with fossil fuels continuing to supply much of the energy until new sources of energy can replace them.[2] The utilization of fossil fuels ensures a low cost supply of energy either as a fuel for the generation of electricity or as a feedstock in the gasification of coal and the water-gas shift reaction (WGSR, Equation 1). The WGSR is an important reaction not only for the production of fuels from syngas, but also for the Haber-Bosch process. The production of ammonia from hydrogen and nitrogen[3,4] (Equation 2) uses the hydrogen produced in the WGSR.

$$CO + H_2O \rightarrow CO_2 + H_2 \qquad = -41.2 \text{ kJ mol}^{-1} \qquad (1)$$

$$N_2 + 3H_2 \rightarrow 2NH_3 \qquad = -91.8 \text{ kJ mol}^{-1} \qquad (2)$$

The use of these natural resources has been implicated in climate change due to the excessive emission of greenhouse gases (i.e., CO_2) as a result of the lack of an efficient technology to separate and capture these gases. Research to address this problem has focused on the application of new polymer-based membrane materials for the purification of H_2 from CO_2 in the WGSR.[5,6] A membrane capable of efficiently separating H_2 from CO_2 *in situ* at low cost is more advantageous than other separation methods (e.g., amine-based CO_2 separation), since it requires less energy to operate. The current demands for *in situ* H_2 separation require a temperature >250 °C, a P_{H2} > 7 bar, and a feed pressure of up to 27 bar.[7] Therefore, testing the performance and mechanical stability of membranes under high pressures and high temperatures is a requirement for screening membranes for real world applications.

Polybenzimidazole (PBI) membranes are mechanically and chemically resilient making them suitable for gas separations at high pressure and high temperature. Hydrogen/carbon dioxide selectivities from 3 to 20 have been reported[8] for PBI membranes measured at temperatures up to 250 °C, which illustrates significant temperature-dependent permselectivity properties. We investigated the effect of drying temperature and testing conditions on the gas permeation properties of PBI membranes. Four sets of PBI membranes were cast and then subjected to a final drying temperature of 250, 300, 350, or 400 °C. Permeability experiments with pure and mixed gases were conducted at temperatures that ranged from 35 to 300 °C with pressures from 5 to 45 bar to reproduce real world conditions.

Experimental

Materials. Molecular sieves 4A were purchased from Aldrich and activated at 400 °C for 1 d before use. *N,N*-dimethylacetamide (DMAc) was purchased from Fisher Scientific and dried over activated molecular sieves 4A for 1 d before use. Polybenzimidazole (PBI) 26 wt% dope in DMAc with 1.5 wt% LiCl was obtained from PBI Performance Products, Inc. Mylar® sheets (25 µm) were obtained from Active Industries. Grafoil® graphite sheets 0.5 mm thick were obtained from Lamons and cut to fit the permeation cell and the diameter of the membrane exposed to gas. For gas permeation experiments, H_2, CO_2, NH_3, and N_2 gases (>99.999%) and the 50/50 H_2/CO_2 mixture were purchased from Airgas.

Instrumentation. Single gas and gas mixture permeation measurements and permeate composition analyses were performed using a custom-built high pressure, high temperature permeameter[9] equipped with a calibrated SRI-3610 gas chromatograph with a thermal conductivity detector. GC calibration curves were generated for H_2, CO_2, and NH_3 using N_2 as carrier gas for H_2 and He as carrier gas for CO_2 and NH_3 due to limitations in thermal conductivities. Analyzing the composition of the permeate for low flux membranes was achieved by increasing the sampling loop volume of the GC to 2 mL while maintaining vacuum to allow sampling at pressures lower than the ambient pressure. The sampling loop was connected to an evacuated manifold equipped with a pressure transducer and metering valves in order to control and reproduce the pressure of the sample between analyses. With this set up, permeate samples at pressures between 50 and 760 Torr were introduced from the permeate side of the permeameter to the GC's sampling loop without the use of a carrier gas, increasing the response of the detector.

To mount the membrane, a flat membrane permeation cell was designed to accommodate the membrane between 3 graphite gaskets compressed between the flat surfaces of two flanges.[9] This configuration left a distance of 0.5 mm between the membrane and the feed side flange that increases the flow of the feed across the surface of the membrane and minimizes concentration polarization during gas mixture experiments. The retentate was connected to a pressure regulator and mass flow controllers to adjust the stage cut of the membrane for optimal separation of gas mixtures.

Membrane Preparation. Flat PBI membranes were cast at room temperature onto Mylar® substrates from 12 wt% PBI solutions in DMAc using a calibrated blade attached to an automatic applicator (Sheen 1133N). The membranes on the Mylar® support were then transferred to a preheated drying table purged with N_2 to dry for 2 to 3 h at 50 °C. The dried membranes were then removed from the substrate and dried in a programmable vacuum oven between glass plates. The temperature program started at 80 °C for 12 h then increased to 150 °C for 1 d, 200 °C for 12 h, and to a final temperature of either 250, 300, 350, or 400 °C for 1 d. The thicknesses of the membranes ranged from 15 to 21 µm.

Gas Permeation. The permeability of each gas was measured at least 4 times, and the average permeability calculated using the last 3 measurements. In a typical experiment, a 5.0 cm^2 membrane was mounted in a stainless steel cell that exposed one side of the membrane to a pressurized feed of 5, 15, 30, or 45 bar and the other side to an evacuated line. The entire system was evacuated for at least 6 h at the experimental temperatures of 35, 100, 200, or 300 °C before a 1 h leak rate test was performed. Permeability experiments with single gases at 35 °C lasted from 2 to 5 h for fast diffusing gases and up to 1 d for slow diffusing gases. Degassing times between H_2 runs lasted for 5 h while CO_2 and NH_3 degassing times increased to 10 h. This testing protocol was repeated for each pressure increment. Permeability experiments with a 50/50 H_2/CO_2 mixture with an initial stage cut of 1 at 30 bar and at 35 °C were performed for 30 d reaching a downstream pressure of 1.3 bar. During the first 7 d, the downstream was evacuated to ensure the permeate composition corresponded to the steady state.

Results and Discussion

PBI membranes dried up to 250 °C and tested at 35 °C exhibited changes in H_2/CO_2 ideal selectivities that ranged from 18 to 33 as the feed pressure increased from 5 to 30 bar (Figure 1). This suggests that high pressures compacted the membrane and limited the mobility of larger gas molecules resulting in an increase in selectivities. The same membranes tested at 300 °C exhibited relatively constant H_2/CO_2 selectivities from 28 to 30 with increasing pressures from 5 to 30 bar (Figure 1) that could be attributed to a better packing of the polymer. Permeabilities for H_2 and CO_2 (Figure 2) also increased with increasing temperature, but at temperatures close to 300 °C, CO_2 permeabilities showed smaller increments than H_2 resulting in a higher H_2/CO_2 selectivity.

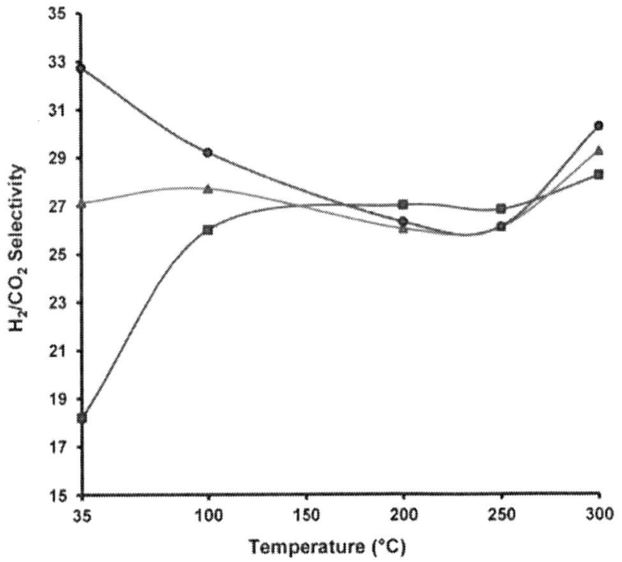

Figure 1. H_2/CO_2 ideal selectivities of PBI membranes dried at 250 °C and tested at temperatures from 35 to 300 °C and at 5 (■), 15 (▲), and 30 (●) bar.

Membranes dried at 300 °C exhibited improved separation properties when tested at 35 °C, showing a H_2/CO_2 ideal selectivity of 49 at 5 bar, 53 at 30 bar, and 47 at 45 bar, a 96% increase in selectivity over the membranes dried at 250 °C. Using the 50/50 H_2/CO_2 gas mixture, the H_2/CO_2 selectivity of this membrane at 35 °C and 30 bar was even higher at 72, surpassing reported selectivity values of PBI membranes dried at 150 °C (H_2/CO_2 = 1 at 25 °C and 20 at 250 °C).[8] Our results suggest that PBI membranes need to be dried at high temperatures to fully maximize their separation properties. Higher temperatures may remove remaining casting solvent molecules more effectively resulting in a more compact chain conformation. According to molecular dynamics simulations,[10] a compact PBI chain conformation can be associated with higher selectivities than for a more relaxed conformation. This may be the case for the membranes dried at 250 and 300 °C, where the membrane dried at higher temperature could exhibit a more compact chain conformation than the membrane dried at lower temperature.

Testing the PBI membranes dried at 300 °C with NH_3 at 35 °C and 4 bar resulted in a NH_3/H_2 selectivity of 13 and NH_3/N_2 selectivity of 13,000. These high selectivities indicate PBI membranes are chemically stable to NH_3.

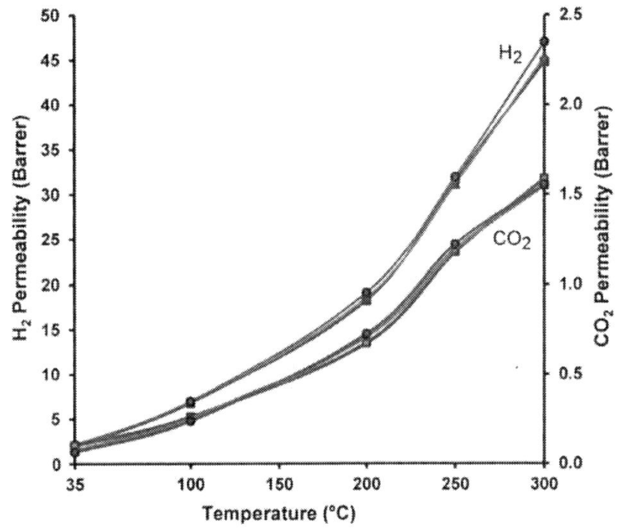

Figure 2. H_2 and CO_2 permeabilities of PBI membranes dried at 250 °C and tested at temperatures from 35 to 300 °C and at 5 (■), 15 (▲), and 30 (●) bar.

Conclusions

Permeability measurements show that the gas transport properties of PBI membranes are sensitive to thermal treatment. Our results demonstrate that it is possible to enhance the separation properties of PBI membranes by drying them in vacuum at temperatures >250 °C. Although the exact mechanisms responsible for the increased selectivity of the membranes due to the thermal treatment described in this work are not yet clearly understood, the permeation results indicate the membranes have the potential to be used in H_2 and NH_3 separations at high pressure and high temperature. Furthermore, the chemical stability of PBI to NH_3 makes this polymer a good candidate for NH_3 separations in the Haber-Bosch process.

Acknowledgement. This material is based upon work supported by the National Science Foundation under Grant no. CBET-0933563.

References

(1) Annual energy outlook 2014 early release overview. EIA, Ed. U.S. Energy Information Administration: 2014.
(2) Conti, J.; Holtberg, P.; Beamon, J. A.; Napolitano, S.; Schaal, A. M.; Turnure, J. T.; Westfall, L., International energy outlook 2013. EIA, Ed. U.S. Energy Information Administration: 2013.
(3) Hellman, A.; Honkala, K.; Dahl, S.; Christensen, C. H.; Nørskov, J. K., Ammonia synthesis: State of the bellwether reaction. In *Comprehensive Inorganic Chemistry II*, Second ed.; Poeppelmeier, J. R., Ed. Elsevier: Amsterdam, 2013; pp 459-474.
(4) Habgood, D. C. C.; Hoadley, A. F. A.; Zhang, L., *Chem. Eng. Res. Des.* **2015,** *102*, 57-68.
(5) Shoko, E.; McLellan, B.; Dicks, A. L.; da Costa, J. C. D., *Int. J. Coal Geol.* **2006,** *65* (3–4), 213-222.
(6) Gallucci, F.; Fernandez, E.; Corengia, P.; van Sint Annaland, M., *Chem. Eng. Sci.* **2013,** *92*, 40-66.
(7) Driscoll, D.; Morreale, B.; Headley, L., NETL test protocol: Testing of hydrogen separation membranes. U.S. Department of Energy - National Energy Technology Laboratory: 2008.
(8) Pesiri, D. R.; Jorgensen, B.; Dye, R. C., *J. Membr. Sci.* **2003,** *218* (1–2), 11-18.
(9) Perez, E. V.; Balkus, Jr., K. J.; Ferraris, J. P.; Musselman, I. H., *Rev. Sci. Instrum.* **2013,** *84* (6), 065107-7.
(10) Wang, Y.; Rashkeev, S. N.; Klaehn, J. R.; Orme, C. J.; Peterson, E. S., *J. Membr. Sci.* **2011,** *384* (1–2), 176-183.

HYDROGEN SULFIDE REMOVAL FROM CARBON DIOXIDE-CONTAINING GAS MIXTURE ON "MOLECULAR BASKET" SORBENTS

*Wenying Quan [1,2], Xiaoxing Wang [1], Chunshan Song [1]**

[1] Clean Fuels & Catalysis Program, EMS Energy Institute, The Pennsylvania State University, University Park, PA 16802, USA
[2] John and Willie Leone Family Department of Energy and Mineral Engineering, The Pennsylvania State University, 209 Academic Projects Building, University Park, PA 16802, USA
Email: csong@psu.edu

Introduction

Hydrogen sulfide (H_2S), a malodorous gas, exists in natural gas and biogas. Its presence causes sulfur-poisoning to catalysts in downstream and fuel cell applications. Moreover, during the fuel combustion, sulfur oxides produced from H_2S is not only toxic and environmentally harmful but also corrosive to the equipment and pipelines. Therefore, H_2S removal is the priority requirement for the utilization of fuel gases.

"Molecular basket" sorbents (MBS) is a novel sorbent for acid gas separation which achieves H_2S sorption/desorption at ambient conditions. [1-3] However, the coexisting CO_2 in bio-gas and natural gas strongly inhibits H_2S sorption on MBS due to its stronger acidity.[3] As is known, primary, secondary, and tertiary amines have different chemical interactions with CO_2 or H_2S. CO_2 cannot interact with the tertiary amine, while H_2S can interact with tertiary amine through N-H bond. Therefore, loading tertiary amine on SBA-15 may improve the sorption selectivity of H_2S from the gas mixture containing CO_2, and thus selective H_2S sorption on MBS could be achieved.

In order to study H_2S removal from the gas mixture containing high concentration of CO_2 over tertiray-amine-based MBS, polyallylamine (PA), polyethylenimine (PEI), and tetramthyl hexanediamine (TMHDA), representing primary, secondary, and tertiary amine, respectively, were loaded on SBA-15, and the H_2S sorption performance in the presence of CO_2 was tested and compared. To investigate the H_2S sorption performance on TMHDA loaded SBA-15, a series of experiments were conducted in a fixed-bed flow system. The effects of TMHDA loading level, gas flow rate, temperature, and inlet gas component concentration (H_2S and CO_2) were examined as well.

Experimental

Sorbent Preparation. SBA-15 was synthesized following the procedure published in the literature. [4,5] Triblock copolymer Pluronic P123 ($EO_{20}PO_{70}EO_{20}$, MW=5800, Aldrich) and tetraethyl orthosilicate (TEOS) were mixed homogeneously in hydrochloric acid under stirring at 40 °C for 20 hours, and then aged at 100 °C for 24 hours. The solid resultant was filtered, washed, and dried at 100 °C overnight, and finally calcined at 550 °C for 6 hours.

For the sorbents preparation, a wet impregnation method in methanol was used. [1-3] To compare the H_2S sorption performance on different amine structures, 30 wt% of polyallylamine (10 wt% in water with Mw of 15,000 from Polyscience Inc.), polyethyleimine (linear PEI with Mw of 1,800, from Polyscience Inc.), and N, N, N′, N′-Tetramethyl-1, 6-hexanediamine (99%, from Aldrich) was loaded on SBA-15, respectively. Other evaluation experiments used a series of TMHDA/SBA-15 sorbents. Typically, a prescribed amount of TMHDA was dissolved in 20 g methanol under stirring for 30 min, then 2 g SBA-15 was added into the mixture, stirring for 8 hours. After the liquid evaporated, the solid resultant was dried in a vacuum oven at 40 °C overnight.

Characterization. The series of TMHDA/SBA-15 sorbents were characterized by N_2 physisorption and FTIR. N_2 physisorption was conducted on a Micromeritics TriStar II analyzer to examine the porous properties of each sample. The IR for each sample and the bare support was carried out by a Nicolet NEXUS 470 FT-IR spectrometer (Thermo Electron Corp.) with the resolution of 4 cm^{-1} using KBr as background.

H_2S Sorption Measurement. The H_2S sorption from CO_2-containing gas mixture was conducted using a fixed-bed flow system under atmospheric pressure. Typically, 0.3 g of a sorbent was packed into the reactor for the sorption/desorption tests. The residual space was filled with glass wool. The sample was pretreated using nitrogen gas flow (25 ml/min) at 110 °C and held overnight to remove moisture prior to the experiment. After the pretreatment, a model gas, containing 500 ppmv H_2S and 10 vol% CO_2 balanced with N_2, was introduced for sorption/desorption experiment.

Results and Discussion

The comparison of H_2S sorption performance over PA/SBA-15, PEI/SBA-15, and TMHDA/SBA-15 in the absence and presence of CO_2 is shown in **Figure 1**. The ratio of H_2S sorption capacity in the presence of CO_2 over that in the absence CO_2 represents the H_2S sorption selectivity on the sorbents from CO_2-containing gas mixture. On primary and secondary amine loaded SBA-15, the ratio started to drop at 16 min and decreased to less than 0.5 at 34 min, while the ratio remained stable at 0.98 on tertiary amine loaded SBA-15. Moreover, as listed in **Table 1**, the capacity drop with CO_2 on primary and secondary amine loaded SBA-15 was 97 %, but this value was only 4 % of that on tertiary amine loaded SBA-15. The result indicates that tertiary amine loaded SBA-15 can dramatically improve the H_2S selectivity from CO_2-containing gas mixture.

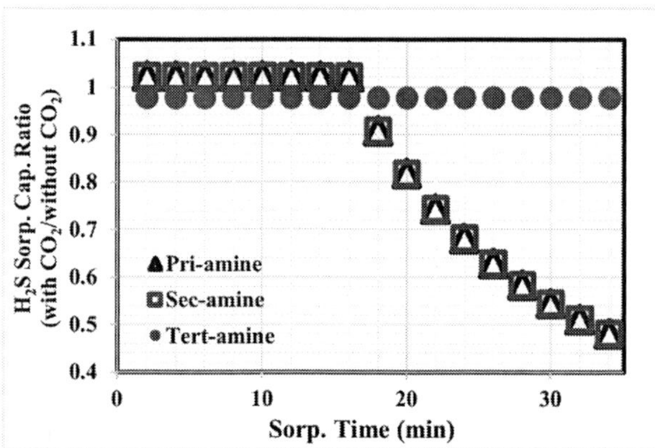

Figure 1. The H_2S selectivity on PA/SBA-15, PEI/SBA-15 and TMHDA/SBA-15 with and without CO_2. Conditions: TMHDA loading, 30 wt%; wt. of sorbents, ≈ 0.33g; *T*, 25 °C; feed-gas flow rate, 15 ml/min, 500ppmvH_2S/10 vol%CO_2/N_2.

Table 1. H_2S Breakthrough Sorption Capacity in the Absence and Presence on PA/SBA-15, PEI/SBA-15, and TMHDA/SBA-15

Sample	BT Cap. (mg-H_2S/g-sorb)		Cap. Drop
	w/o CO_2	w/ CO_2	(%)
PA/SBA-15	3.16	0.11	97
PEI/SBA-15	4.94	0.15	97
TMHDA/SBA-15	0.25	0.24	4

Conclusions

The tertiary amine loaded MBS was able to dramatically improve the selectivity of H_2S sorption from CO_2-containing gas mixture. 15 wt% was the optimal TMHDA loading level to obtain the maximum breakthrough and saturation capacities. The H_2S sorption on TMHDA/SBA-15 appears to be chemical sorption which was fitted to Langmuir adsorption isotherm.

References

(1) Wang, X.; Ma, X.; Song, C. *Green Chem.* **2007**, *9*, 695-702.

(2) Wang, X.; Ma, X.; Xu, X.; Sun, L.; Song, C. *Top. Catal.* **2008**, *49*, 108-117.

(3) Ma, X.; Wang, X.; Song, C. *J. Am. Chem. Soc.* **2009**, *131*, 5777-5783.

(4) Zhao, D.; Feng, J.; Huo, Q.; Melosh, N.; Fredrickson, G. H.; Chemlka, B. F.; Stucky, G. D. *Science*, **1998**, *279*, 548-552.

(5) Wang, X.; Zhang, Q.; Yang, S.; Wang, Y.; *J. Phys. Chem. B*, **2005**, *109*, 23500-23508.

DEVELOPMENT OF FLEXIBLE FAUJASITIC ZEOLITE MEMBRANES FOR CO_2 CAPTURE

*Bo Wang, Subhrakanti Chakraborty, Prabir K. Dutta**

Department of Chemistry and Biochemistry, the Ohio State University, 100 West 18th Avenue, Columbus, Ohio, 43210
*Corresponding Author's Email Address: dutta.1@osu.edu

Introduction

Energy from fossil fuels essentially sustains our modern standard of living. However, the product of burning fossil fuel, carbon dioxide (CO_2), has been accumulating in atmosphere and recently crossed a level of 400 ppm, and is responsible for global climate change. To negate the effects of high atmospheric CO_2 concentrations, both long term and short term technologies need to be developed. Solar and nuclear energy result in negligible CO_2 emissions, but development of this infrastructure is going to take years. Thus, completely replacing fossil fuels in the short term in not feasible. Another option is to decelerate atmospheric CO_2 concentration increase by implementation of CO_2 capture methods, primarily from stationary sources. Currently, CO_2 capture technologies are amine scrubbing, adsorption, absorption, chemical looping and membranes. Because CO_2 capture is in scale of tens of gigatons per year, cost of capturing CO_2 is the key criteria to evaluate these technologies. Membranes can be cost effective on an industrial scale, and does not involve energy expensive temperature or pressure cycling. Polymer membranes are manufactured on large scale with roller assemblies and have been demonstrated for pilot scale CO_2 capture[1]. Cost will be reduced if the gas permeance of polymer membranes and the selectivity can be increased, however, polymers typically have an inverse relationship between permeance and selectivity, represented by the Robeson upper bound limit[2].

Zeolites are a family of aluminosilicate materials with well-defined microporous structure. Uniform and interconnected microporous structure of zeolites leads to predictions of extraordinary gas separation performance, and not limited by Robeson upper bound[3]. However, cost to produce conventional zeolite membranes are high (\$5000-\$13000 per m^2, 2 orders of magnitude higher than polymers)[4]. Zeolite synthesis typically takes hours to days, and the membranes usually grown on inflexible alumina supports are done in a batch fashion, and are also defective[4]. Our group has reported a series of papers of zeolite/polymer membranes. Crystallization process of zeolite Y has been decreased to 1 hour, which is compatible with roller assemblies[5]. Expensive conventional alumina support was replaced with polymer support, which is expected to significantly decrease the cost of flat-sheet zeolite membranes[6]. A bendable zeolite membrane platform was developed with high reproducibility and bendability[7].

This preprint demonstrates a continuous process for zeolite membrane fabrication with roller assemblies, proving the potential for large scale membrane fabrication and application for CO_2 capture. Membranes fabricated with the roller assembly were characterized with X-Ray Diffraction (XRD) and Scanning Electron Microscopy (SEM). Their gas separation performance was evaluated with transport measurements. The mechanical properties of the membranes was evaluated with atomic force microscopy (AFM), and a basic understanding of their flexibility based on the elastic modulus was developed.

Experimental

Materials. Aluminum isopropoxide ($Al(O\text{-}CH(CH_3)_2)_3$, 98%), tetramethylammonium bromide ($(CH_3)_4NBr$, 98%), Ludox HS-30 colloidal silica (SiO_2, 30%) and Ludox SM-30 colloidal silica (SiO_2, 30%) were purchased from Aldrich (Milwaukee, WI, USA). Aluminum hydroxide ($Al(OH)_3$, 76.5%) was purchased from Alfa Aesar. Tetramethylammonium hydroxide ((TMAOH), 25% aqueous solution) was purchased from SACHEM Inc. Sodium hydroxide (NaOH, 99.0%) was purchased from Fisher Scientific. Polyethersulfone (PES) 300kDa membrane was purchased from MILLIPORE Biomax. Helium, carbon dioxide and nitrogen were purchased from Praxair. Dehesive 944 Polydimethylsiloxane (PDMS) was provided by Wacker Silicones, Inc. H_2O used in this study was purified by a Millipore ultrapure water system. All chemicals were used as received without further purification.

Synthesis. *Zeolite Y Nanoparticle Synthesis* Nanozeolite Y seeds were synthesized according to literature with composition of 0.048 Na_2O: 2.40 $(TMA)_2O(2OH)$: 1.2 $(TMA)_2O(2Br)$: 4.35 SiO_2: 1.0 Al_2O_3: 249 H_2O, where TMA^+ is tetramethylammonium cations. In solution A, 26.2 g Ludox HS-30 and 10.46 g TMAOH were mixed in a sealed bottle and stirred at room temperature for 30 min. In solution B, 12.5 g aluminum isopropoxide was dissolved in mixture of 76.5 g H_2O and 52.3 g TMAOH solution, and heated in a water bath at 70 °C until complete dissolution. After cooling to room temperature, 13.1 g TMABr was added to alumina source solution followed by mixing with the silicon source. The clear solution was aged at room temperature with stirring for 3 days and then in an oil bath at 100°C for 4 days. After synthesis, nanozeolite particles were separated by ultracentrifugation, and washed until pH of supernatant was 7. *Deposition of Zeolite Y Nanoparticle on PES support* Nanozeolite seed particles were deposited on PES supports by vacuum assisted dip-coating. A ~25 psi vacuum was applied on the back of the PES support to pull the seeds onto the support. After coating, the support was dried at room temperature overnight and stored in plastic sample bags.

Roll-to-Roll Synthesis of Zeolite Membranes A roller assembly setup was designed[8]. A nonwoven fabric band was twisted around the fix column and taped onto the 2 identical rollers with diameter of 5.1 cm. In rolling process, the band was driven by the rollers to move at constant speed. Nanozeolite seed coated PES support was stapled onto the band. Reactor temperature was controlled by the thermocouple, temperature controlling box and four heaters inserted in the holes at the bottom. Zeolite membrane growth gel composition is: 8.3 Na_2O: 1 Al_2O_3: 6.4 SiO_2: 483.9 H_2O. 4.416 g of $Al(OH)_3$ and 14.58 g NaOH were dissolved in 170.48 g H_2O. After complete dissolving, 27.7 g Ludox SM-30 was added to the gel. Mixed gel was aged at room temperature for 4 hours and then transferred to a dehydration/rehydration hydrothermal setup for removal of half the water in 1 hour. This partially dehydrated hot gel was then immediately transferred to the reactor shown in figure 1b and heated to 100 °C. In 1 hour of zeolite growth process, water was added to the gel, diluting the gel from 120 mL to 200 mL. After zeolite growth, membrane sample was washed with flowing water and rubbed with a fur brush, soaked in water to remove residual surface species and dried for further study.

Polydimethylsiloxane (PDMS) coating Zeolite membranes were coated with PDMS (Dehesive 944 kit) before gas transport measurement. Commercial PDMS is in 3 separate bottles of PDMS monomer, catalyst and cross-linker (the exact nature of these chemicals is proprietary, and the procedure followed was supplied by the vendor). PDMS monomer solution was firstly diluted to 2wt% with heptane. After complete dispersion, cross linker and catalyst were added with the ratio of 100:1:0.5 (PDMS: Cross linker: Catalyst) to prepare PDMS precursor solution. Spinning includes 2,000 rpm for 5 seconds and 4,000 rpm for 1 min. After coating, PDMS was polymerized at room temperature overnight.

Characterization. XRD pattern of membranes were collected with Bruker D8 X-Ray Diffractometer with CuKα ($\lambda = 1.5405$ Å). FEI Helios Nanolab 600 Dual Beam Focused Ion Beam/Scanning Electron

Microscope (FIB/SEM) was employed to characterize surface morphology of zeolite membrane samples. Gas separation performance of membranes were characterized with transport measurements with feed gas composition of 20% CO_2 and 80% N_2 and pure He as sweep gas. Pressure at both feed side and permeate side are 1 atm. SRI 310C gas chromatograph equipped with a Hysep D column and TCD detector was employed to measure the compositions of gas samples. AFM measurements were performed using a Nanoscope V Icon Dimension atomic force microscope (Bruker AXS Corp., Santa Barbara, CA). Images were acquired peak force tapping mode (PeakForce-Quantitative Nano-Mechanics PF-QNM).

Results and Discussion

Zeolite Membrane Fabrication. Before zeolite membrane synthesis, commercially available PES support was coated with nanozeolite seed particles with 10 μg/ml nanozeolite seed dispersion. Zeolite synthesis composition was 8.5 Na_2O: 1 Al_2O_3: 10.9 SiO_2: 974 H_2O as reported in previous paper.[5,7] After aging at room temperature for 4 hours, gel was moved to the dehydration/rehydration setup. In 1 hour of heating, 80 mL water was evaporated from the gel and condensed on the funnel. The concentrated gel was then moved to roller assembly for membrane fabrications. After 1 hour of heating and rolling, zeolite membrane synthesis was complete.

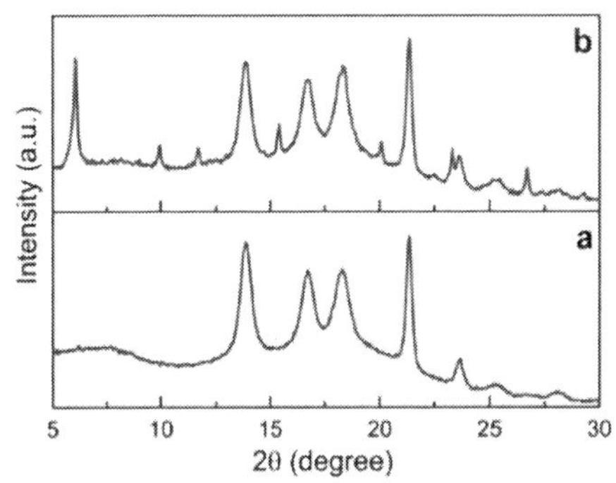

Figure 1. XRD patterns of (a) bare PES support and (b) zeolite membrane

Zeolite Membrane Characterization. To confirm the zeolite synthesis process in the reactor of roller assembly, samples after 1 hour synthesis was collected and characterized with XRD (figure 1). All characteristic zeolite Y peaks were observed on zeolite membrane (PES peaks are evident in Figure 1a). Zeolite membrane surface was crack-free in entire surface range. High magnification SEM image indicates that only individual zeolite crystals were grown on top of PES support, evidenced by the visibility of PES support. To study the morphology of zeolite layer, PES support of zeolite membrane was dissolved with N-methylpyrrolidone (NMP). After PES dissolving, zeolite sheet in centimeter scale was observed, which proved the interconnection of zeolite particles within PES support.

Gas Separation Performance of Zeolite Membrane As-synthesized zeolite membrane was spin coated with polydimethylsiloxane (PDMS) and cross-linked at room temperature overnight to seal micron sized defects before gas transport measurements. Five samples prepared with identical operations have CO_2 permeance of 1841 ± 181 GPU and CO_2/N_2 selectivity of 35 ± 4. As control experiments, PDMS coated PES support only has CO_2/N_2 selectivity of 14. Zeolite layer is the CO_2/N_2 selective layer in this composite membrane.

Mechanical Properties of Zeolite Membrane. AFM besides performing atomic scale scan of the sample surface to produce high resolution topographic image can also by force-volume (F-V) imaging provide information about the mechanical property of sample surfaces.[9] Especially, Peak Force Tapping (PFT) mode of scanning, in which the tip is brought intermittently in contact (tapping) with the sample surface maintaining an amplitude of 100-300 nm. The intermittent contact is modulated with a sinusoidal wave at a frequency of 2 kHz, with amplitude defined by the peak force. In Peak force quantitative nanomechanical measurement (PFQNM) technique, the individual force curves obtained during PFT scan data acquisition can be analyzed for mechanical properties, including adhesion, deformation, elastic modulus. From the elastic modulus map, the average elastic modulus of the zeolite/polymer membrane is 10 GPa, with a standard deviation of 3.7 GPa. The elastic modulus is lower than the reported value for zeolite Y of 38-41 GPa, and it is because of this lower modulus, the zeolite membrane has flexibility.

Conclusions

Demonstration of flexible ductile zeolite membrane under roll-to-roll synthesis conditions opens up the potential of large scale manufacture of zeolite membranes with decreased costs.

Acknowledgement. This work is funded by The U. S. Department of Energy under Award Number DE-FE0007632. However, any opinions, findings, conclusions, or recommendations expressed herein are those of the author(s) and do not necessarily reflect the views of the DOE.

References

(1) Merkel, T. C.; Lin, H.; Wei, X.; Baker, R. Power Plant Post-Combustion Carbon Dioxide Capture: An Opportunity for Membranes. *J. Membr. Sci.* **2010**, *359* (1–2), 126–139.

(2) Robeson, L. M. The Upper Bound Revisited. *J. Membr. Sci.* **2008**, *320* (1–2), 390–400.

(3) Krishna, R.; van Baten, J. M. In Silico Screening of Zeolite Membranes for CO2 Capture. *J. Membr. Sci.* **2010**, *360* (1–2), 323–333.

(4) Gascon, J.; Kapteijn, F.; Zornoza, B.; Sebastián, V.; Casado, C.; Coronas, J. Practical Approach to Zeolitic Membranes and Coatings: State of the Art, Opportunities, Barriers, and Future Perspectives. *Chem. Mater.* **2012**, *24* (15), 2829–2844.

(5) Severance, M.; Wang, B.; Ramasubramanian, K.; Zhao, L.; Ho, W. S. W.; Dutta, P. K. Rapid Crystallization of Faujasitic Zeolites: Mechanism and Application to Zeolite Membrane Growth on Polymer Supports. *Langmuir* **2014**, *30* (23), 6929–6937.

(6) Wang, B.; Sun, C.; Li, Y.; Zhao, L.; Ho, W. S. W.; Dutta, P. K. Rapid Synthesis of Faujasite/polyethersulfone Composite Membrane and Application for CO2/N2 Separation. *Microporous Mesoporous Mater.* **2015**, *208*, 72–82.

(7) Wang, B.; Ho, W. S. W.; Figueroa, J. D.; Dutta, P. K. Bendable Zeolite Membranes: Synthesis and Improved Gas Separation Performance. *Langmuir* **2015**, *31* (24), 6894–6901.

(8) Wang, B.; Jackson, E. A.; Hoff, J. W.; Dutta, P. K. Fabrication of Zeolite/polymer Composite Membranes in a Roller Assembly. *Microporous Mesoporous Mater.* **2016**, *223*, 247–253.

(9 Young, T.; Monclus, M.; Burnett, T.; Broughton, W.; Ogin, S.; Smith, P., The use of the PeakForceTM quantitative nanomechanical mapping AFM-based method for high-resolution Young's modulus measurement of polymers. *Measurement Science and Technology* **2011**, *22* (12), 125703.

NOVEL POLYETHERS FOR MEMBRANE CO_2/N_2 SEPARATION

Junyi Liu, Chong Cheng, Haiqing Lin

Department of Chemical and Biological Engineering, University at Buffalo, The State University of New York, Buffalo, NY, USA
junyiliu@buffalo.edu haiqingl@buffalo.edu

Carbon capture and sequestration (CCS) could be an important approach to mitigate CO_2 emission to the air. Membrane technology has been widely explored for CO_2 separation and capture from coal power plant derived flue gas, due to its low cost and high energy-efficiency. The key to the success of membrane technology is membranes with high CO_2 permeability and high CO_2/N_2 selectivity. The current leading materials for membrane CO_2/N_2 separation are poly(ether oxide) (PEO) containing polymers, because ether oxygens can interact favorably with CO_2, leading to high CO_2 sorption and permeability. Given the importance of this application, any improvement in CO_2/N_2 separation properties can significantly decrease the cost of CO_2 capture. The goal of this work is to design and synthesize a series of new polymers containing higher content of ether oxygens than PEO to further improve CO_2 permeability and CO_2/N_2 selectivity. More specifically, we have synthesized poly (1, 3 dioxolane) (PDXL) with different molecular weights of 510 g/mol and 1005 g/mol with a ratio of oxygen to carbon of 0.67, which is higher than that in PEO (0.5). These PDXL oligomers were also functionalized with acrylate groups and were thoroughly characterized using NMR, MS and FTIR. These oligomers were then polymerized by photo-polymerization and the resulting polymers were characterized for gas sorption and permeation. These new polyethers demonstrate their promise for membrane CO_2/N_2 separation. For example, PDXL (n=5) derived polymer exhibits CO_2/N_2 selectivity of 82 with CO_2 permeability of 100 Barrers at 35°C. The CO_2/N_2 selectivity is much higher than that of crosslinked PEO analogues (with CO_2/N_2 selectivity of 52). This presentation will also compare these novel polyethers with the state-of-art membrane polymers for CO_2/N_2 separation.

NOVEL POROUS ORGANIC LIGAND POLYEMRS (POLS) AS HIGHLY EFFICIENT HETEROGENEOUS CATALYSTS

*Zhifeng Dai, Qi Sun, Liang Wang, Xiangju Meng, and Feng-Shou Xiao**

Key Lab of Applied Chemistry of Zhejiang Province, Department of Chemistry, Zhejiang University, Hangzhou 310028, China
fsxiao@zju.edu.cn

Introduction

Because of the fully accessible and highly tunable active species, the homogeneous catalysts have important advantages in catalytic activity and selectivity over the heterogeneous catalysts. Nevertheless, the heterogeneous catalysts are more desirable for industrial applications owing to the convenience of recovering and recycling of the catalysts. To solve this problem, the strategies such as supporting homogenenous catalysts have been developed, but their catalytic properties cannot compare with those of homogeneous catalysts.

Organic ligands such as triphenylphosphine (PPh_3) and diphosphine [bis(diphenylphosphino)ethane, dppe] are of critical importance for synthesis of homogeneous catalysts used for organic transformation. However, their recycles are complicated. Inspired by hydrothermal synthesis of microporous zeolite from inorganic silica-based monomers, we show a generalized methodology for synthesizing a series of porous organic ligand polymers (POLs) by solvothermal polymerization from vinyl-functionalized organic ligand monomers. Interestingly, these POLs have high surface area, large pore volume, hierarchical porosity, superior thermal stability, adjustable concentration of organic ligands, strong coordination ability, and quantitative yields. As a typical example, porous organic ligand diphosphine polymer (POL-dppe) was synthesized from solvothermal polymerization of vinyl-functionalized dppe ligand monomer in an autoclave. Interestingly, the POL-dppe supported Rh as a heterogeneous catalyst (M/POL-dppe) exhibit unprecedentedly high activities and excellent recyclabilities in catalytic hydroformylation. These diphosphine ligand constructed porous polymers may provide new platforms for the hydroformylation of olefins in the future.

Results and Discussion

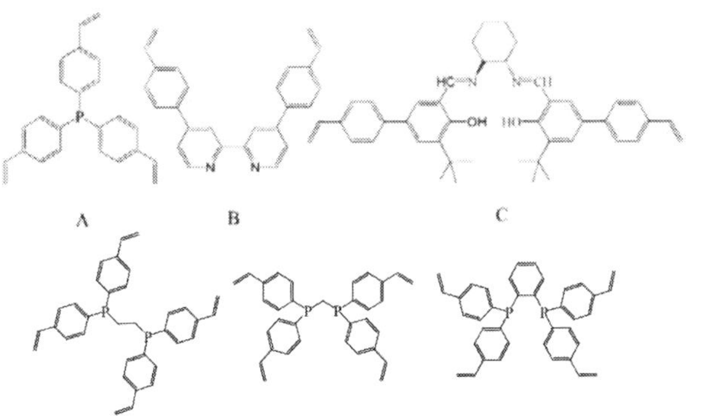

Scheme 1. Structures of vinyl-functionalized organic ligand monomers of (A) triphenylphosphine (PPh_3), (B) 2,2'-bipyridyl, (C) salen, (D) 1,2-bis(diphenylphosphino)ethane (dppe), (E) bis(diphenylphosphino)methane (dppm), and (F) 1,2-bis(diphenylphosphino)benzene (dppb).

As typical examples, six organic ligand monomers have been synthesized (Scheme 1). After solvothermal polymerization, it was obtained porous organic ligand polymers (POLs) with high surface areas, large pore volume, and hierarchical porosity. As typical example in Figure 1, it is shown N_2 sorption isotherms and SEM image of the POL-dppe. These results confirm the presence of hierarchical porosity in the sample.

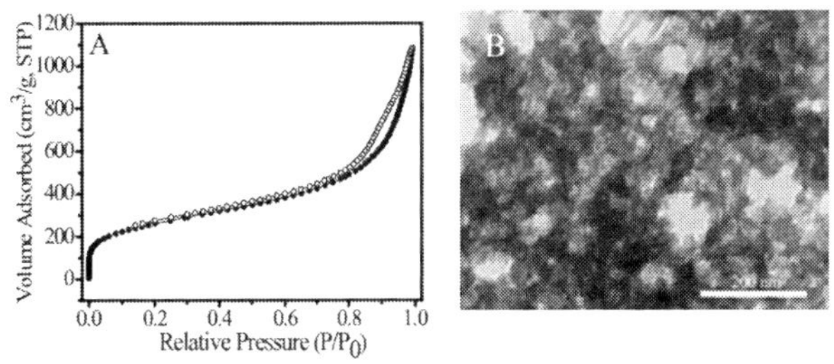

Figure 1. (A) N_2 sorption isotherms and (B) SEM image of POL-dppe polymer.

After loading various metals, POLs still show very high surface areas. For example, POL-dppe supported $Rh(CO)_2(acac)$, denoted as Rh/POL-dppe, shows the BET surface area and pore volume at 856 m^2/g and 1.41 cm^3/g. HAADF-STEM image and EDX mappings show that the dispersion of Rh species in the sample are very high and uniform.

As typical example for evalution of catalytic properties, Table 1 presents catalytic performance in the hydroformylation of long-chain 1-octene, the heterogeneous catalysts of Rh/POL-dppe not only can be easily separated from the reaction system by filtration, but also can exhibit much higher selectivity for the corresponding aldehydes than those of homogeneous catalysts of $Rh(CO)_2(acac)$ and mixtures of $Rh(CO)_2(acac)$ with dppe ligand under similar conversions. Even if the catalyst recycles for 5 times, the selectivity of aldehyde can still reach 98.9%.

Table 1. Catalytic data in 1-octene hydroformation over various catalysts.[a]

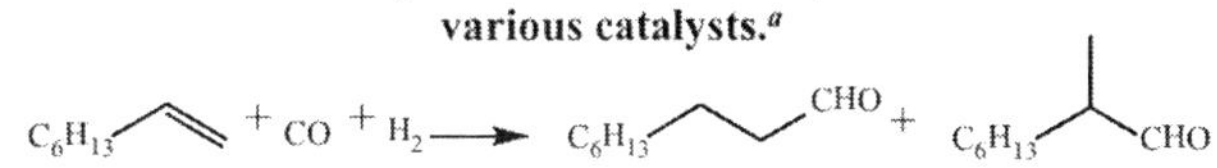

Run	Catalyst	Conv. (%)	Aldehydes (%)
1	$Rh(CO)_2(acac)$, homo	98.8	50.1
2	Rh/POL-dppe, hetero	96.9	99.3
3[b]	Rh/POL-dppe, hetero	96.7	98.9
4	Rh/dppe, homo	99.5	52.3
5	Rh/10dppe, homo	54.7	97.9

[a]90 °C, CO/H_2=1:1 (2.0 MPa), S/C=2000, 2.5 h for *n*-octene and 3.0 h for *n*-dodecene, Rh loading at 2.0 wt.%; [b]Recycles for 5 times.

Conclusions

We report a facile and universal route for synthesizing various porous organic ligand polymers. After supported with various metal species, the obtained heterogeneous catalysts exhibit not only high activities, but also excellent recyclabilities in their classic organic transformation. The use of POLs might open a new door for preparing a series of highly efficient heterogeneous catalysts in the future.

ADVANCED MEMBRANES FOR GAS SEPARATIONS: ENTROPY ENGINEERING AT THE SUB-NANOMETER SCALE

William J. Koros

School of Chemical & Biomolecular Engineering
Georgia Institute of Technology
Atlanta, GA 30327
wjk@chbe.gatech.edu

Gas separations are highly energy-intensive processes when based upon traditional phase-change paradigms such as distillation and sorption processes. Membranes avoid this paradigm by integrating molecularly selective partitioning and diffusive discrimination between components into a single device. This integrative ability is now changing the separations landscape by offering a full order of magnitude lower energy-intensity compared to phase-change separation approaches. The selective partitioning feature is understandable in terms of well-known thermodynamic factors. Engineering the diffusion selectivity feature usually relies upon tuning relative diffusion activation energies of permeating species within the membrane. Unfortunately, higher relative diffusion activation energies *between* similar penetrants are correlated with higher absolute diffusion activation energies for *both* components. This correlation creates a natural tradeoff in which dramatically lower permeabilities are typically seen for highly selective membranes.

A new approach, tuning the *diffusion activation entropy* associated with size-similar penetrants, can avoid major productivity losses if done properly. Examples will be discussed of advanced materials that can be economically transformed into large scale membranes, with tunable diffusion activation entropies, for high productivity and energy-efficient gas separations.

TITLE: LOW-VISCOUS AND HIGHLY EFFICIENT CO_2 CAPTURE BY AQUEOUS ABSORBENT OF MULTI AMINE FUNCTIONALIZED IONIC LIQUID

Dr. Bi-Hong Lv, Zuo-Ming Zhou and Guo-Hua Jing*

Director College of Chemical Engineering,
Huaqiao University China

Abstract

The concept of carbon dioxide (CO_2) capture and sequestration (CCS) has attracted considerable interest over the past two decades, as a result of the emerging problem of climate change. To date, ionic liquids (ILs) have received an advantage in various applications including in CO_2 capture. More interestingly, introducing some special groups to the anion or the cation of ILs to obtain functionalized ILs can achieve excellent performance in CO_2 capture. Thus design and synthesis of new ionic liquids have been attracting remarkable interests. In this work, a novel multi-amine functionalized IL ([APmim][Lys]) had been developed for CO_2 capture in the present work, in which the amine and lysine functionalized groups were both introduced to the imidazole ionic liquid (IL). The absorption capacity of [APmim][Lys] was found to be 1.80 mol CO_2 / mol ILs (6.71 mol CO_2 / kg IL, 26.7 wt. % CO_2) , which was much higher than other functionalized ILs solutions. The viscosity of [APmim][Lys] (0.5 mol/L) was found to be 0.8 mPa/s at 323.15 K, which was just 1.25 times of that of MEA solution, and was much lower than that of pure ILs. The optimal regeneration temperature of the solvent was 120 oC, and the regeneration efficiency after 5th thermal desorption even kept at 99.1%. The structures of [APmim][Lys] with or without water were estimated by quantum chemical calculations, and the difference in the structures was reflected by the viscosities of the ILs. The results of the characterization by FT-IR and 13C NMR indicated that CO_2 capture into [APmim][Lys] with or without water was different processes. In pure ILs, CO_2 reacted directly with [APmim][Lys] to form carbonate salt. In aqueous solution, it started with an exothermic reaction to form carbamate, followed with a hydrolysis of carbamate, and then finished with an endothermic reaction by hydrolysis.

Biography

Dr. Bi-Hong Lv, Environmental Engineering Doctor (Ph.D.-Engineering), now is an instructor of College of Chemical Engineering from Huaqiao University. She got her Environmental Engineering Doctor's degree (Ph.D.) at Zhejiang University in 2014. Currently Dr. Bi-Hong Lv, researches focus at air pollution control and treatment, and carbon dioxide capture.

PERMANENT CO_2 STORAGE VIA ENHANCED EX-SITU MINERALIZATION USING HEAT TREATED SERPENTINE

Chengchuan Zhou and Ah-Hyung Alissa Park

Department of Earth and Environmental Engineering,
Department of Chemical Engineering,
Lenfest Center for Sustainable Energy,
Columbia University, New York, NY 10027
ap2622@columbia.edu

Abstract

Carbon mineralization is one of the unique CO_2 storage schemes that can be done both in-situ and ex-situ. Since gaseous CO_2 is fixed into solid matrix, there is negligible risk associated with accidental leakage and costly long-term monitoring. Unfortunately the cost of carbon mineralization is relatively high compared to other carbon storage methods. Thus, efforts have been carried out by many researchers to reduce such costs by enhancing the mineral dissolution, which has been considered as the rate limiting step. Heat treatment of hydrous magnesium silicate mineral (serpentine) is one of the approaches being considered. The extents of carbonation of heat-treated serpentine were limited to 50% even at high CO_2 pressures and high reaction temperatures when the reaction was performed in a single-step mode. The highest carbonation conversion achieved for heat-treated serpentine I, heat-treated serpentine II and heat-treated antigorite (Cedar Hill) were 42%, 40% and 48%, respectively. It was speculated that heat-treated serpentine is significantly more reactive (soluble) compared to untreated serpentine. Thus, the initial mineral dissolution rate of heat-treated serpentine would have been very fast, and this would have led to a high Mg concentration near the mineral particle surface. Therefore, the growth of Mg-carbonate layer on serpentine particle surface would have been fast as well, which eventually limited further dissolution of serpentine. This phenomenon is similar to some of our previous mineral dissolution experiments performed in a batch reactor. It was concluded that a two-step carbon mineralization process would be a better option for the carbonation of heat-treated serpentine instead of a single-step process. The subsequent experiments for this project were performed using a two-step reactive system.

Introduction

The ever increasing atmospheric CO_2 concentration since the industrial revolution has become a significant challenge for the future development of human society. Considering that fossil fuels will still be our major energy source in the near future, we need to employ carbon capture, storage and utilization (CCUS) technology to reduce the CO_2 emissions before a carbon-free energy source becomes our major energy supply. Mineral carbonation is one option of the CCUS technology to permanently stabilize CO_2, with which CO_2 is reacted with Mg- or Ca-bearing silicate to produce thermally stable and environmentally benign magnesium or calcium carbonate products (Lackner *et al.*, 1995). Serpentine ($(Mg, Fe)_3(OH)_4Si_3O_5$) is one of the most abundant magnesium silicate minerals which has a considerable capacity to store CO_2 (Alexander *et al.*, 2007). However, there is also challenge to apply serpentine in mineral carbonation process, because the reactivity of raw serpentine minerals is quite limited and thus they cannot react with CO_2 efficiently. In the engineered mineral carbonation process, people have taken different measures to accelerate the reaction kinetics, especially the kinetics of the extraction of Mg ions from the Serpentine minerals. Heat treatment is one of these activation measures, with which the dehydration and dehydroxylation process can break the crystal structure of the mineral, releasing Mg ions from the lattice, and create porous structures inside the minerals (Dlugogorski & Balucan, 2014).

In this study, heat-treated Serpentine is characterized and the key reaction parameters are investigated in terms of the extent of Mg extraction. The experimental findings were compared to those obtained using raw Serpentine and thermodynamic calculations.

Experimental design

In order to study the mineral dissolution kinetics and select the best chelating agent for the subsequent two-step mineral carbonation experiments, the differential bed reactor shown in Figure 1 was used to conduct the mineral dissolution experiments.

In each experiment, 20 mg of heat-treated serpentine received from MCi was spread onto a filter paper (pore size = 11 μm) and the loaded filter paper was placed in a sample holder installed at the tip of the syringe acting as a solvent reservoir. A syringe pump was used to provide a constant solution flow of 10 ml/min through the sample holder. The leaching solution exiting the filter paper was then re-filtered using a membrane with the pore size of 0.2 μm in order to remove any residual mineral particles. The collected liquid samples were analyzed using Inductively Coupled Plasma Optical Emission Spectroscopy (ICP-OES: ACTIVA-M) to determine the elemental concentrations of Mg and Si in each sample.

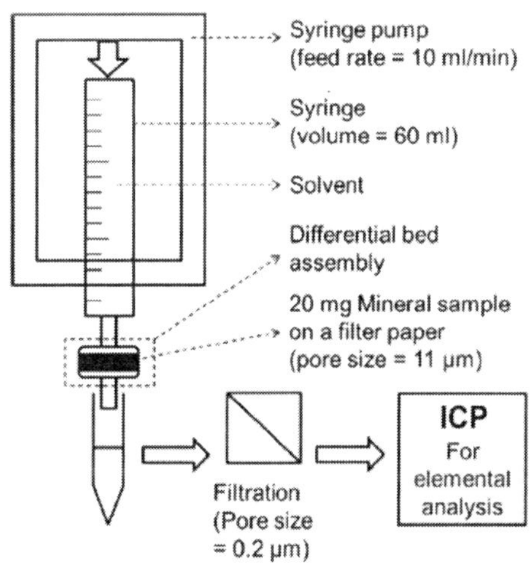

Figure 1. Schematic of the custom-made differential bed reactor

Results and discussions

Based on our previous studies on the dissolution of olivine and serpentine (raw, not heat-treated), four candidate Mg-targeting ligands including oxalate, acetate, citrate and glutamate were selected for this study. While the dissolution trend of heat-treated serpentine was speculated to be similar to that of untreated serpentine, it was also suggested that the changes in surface functional groups of heat-treated serpentine may affect its dissolution behavior.

In all cases, 0.01 M solution of each chelating agent was used, while fixing the initial solution pH to 3 by adding HNO_3 in order to isolate the ligand effect from the pH effect. The mineral dissolution is a strong function of pH and 0.01 M solutions of various chelating agents have different pH levels. All the experiments were conducted under ambient pressure and temperature unless noted. The extents of Mg and Si extraction were calculated based on the initial MgO and SiO_2 contents in heat-treated serpentine, which were 39.29% and 44.80%, respectively.

One of the first differences in dissolving untreated and heat-treated serpentine was the extent of Mg and Si extraction during the initial fast kinetic regime. Generally, the disordered surface of mineral particles created during grinding results in a significantly faster dissolution rate than the actual surface reaction kinetics. Under low pH conditions, the dissolution of serpentine plateaus relatively quickly and reaches the mass transfer regime due to the buildup of the unreactive silica layer. Initially the dissolution experiments were performed only for 6 minutes each, but since the dissolution of heat-treated serpentine was still significant after 6 minutes, and thus, the reaction time was extended to 60 minutes. As expected, serpentine dissolution was incongruent and the extent of Si dissolution was lower than that of Mg dissolution in the presence of Mg-targeting chelating agent.

In order to further investigate the effect of heat treatment on mineral dissolution, XRD analyses were performed as shown in Figure 2. It can be clearly seen that the crystal structures of the Antigorite mineral gradually disappeared with the increase of the heat treatment temperature.

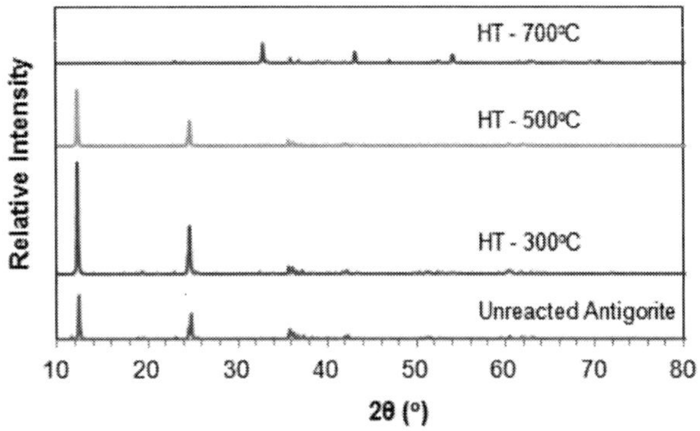

Figure 2. Changes in XRD patterns due to heat-treatment of antigorite. The heat treatments were performed with 100ml/min N_2 gas flow through a reactor in a moderate temperature furnace. The temperature was first increased to the target value (300, 500, 600, and 700 °C) in 15 min, and then kept constant for 6 hr.

Finally, a series of experiments were performed to evaluate the overall carbonation potential of heat treated serpentine. As shown in Figure 3, the heat treatment can lead to a significant increase in the extent of serpentine carbonation. The carbonation extent of the raw Antigorite is only 12.5%. Under the same reaction conditions, with the increase in the degree of the heat treatment, the extent of carbonation also increased significantly, indicating that the increase in the activities of the heat-treated Serpentine minerals.

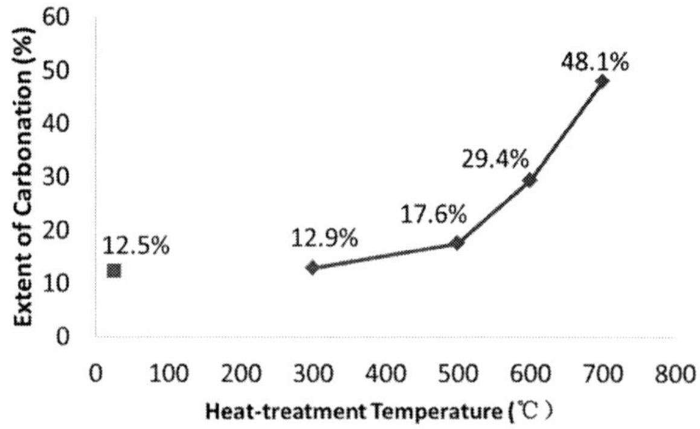

Figure 3. Effect of heat-treatment temperature on the extent of antigorite carbonation. These experiments were performed at 185 °C, P_{CO2} = 139 atm for a reaction time for 3 hours in 1.0 $NaHCO_3$ solutions.

Conclusions

In this study, the efforts were focused on the evaluation of heat-treated serpentine in terms of its carbons storage potential. The custom-made differential bed reactor was used to collect liquid samples with a fast sample rate while eliminating the need for buffering. Relevant thermodynamic analyses were also performed since the activities of ligands are strongly influenced by the solution pH. It was found that citrate was the best chelating agent in terms of both Mg and Si extraction among the Mg-targeting ligands under study, including oxalate, acetate, glutamate and citrate. These ligands were used in a form of Na-salts, and it was also found that the ionic strength also affects the dissolution of serpentine. Overall, the heat treatment of serpentine resulted in a great improvement in mineral dissolution and carbonation kinetics.

Acknowledgement. This work is financially supported by the Mineral Carbonation International (MCi) Ltd in Australia.

References

(1) Lackner, K. S.; Wendt, C. H.; Butt, D. P.; Joyce, E. L.; Sharp, D. H. *Energy*, **1995**, *20*(11), 1153-1170.

(2) Alexander, G.; Mercedes Maroto-Valer, M.; Gafarova-Aksoy, P. *Fuel*, **2007**, *86*, 273–281.

(3) Dlugogorski, B. Z.; Balucan, R. D. *Renew. Sustain. Energy Rev.* **2014**, *31*, 353–367.

SYNTHESIS OF NITROGEN-RICH AND ULTRA-MICROPOROUS ORGANIC POLYMERS FOR SEPARATION OF CARBON DIOXIDE FROM GAS MIXTURES

*Jacob J Taylor, Allison L Byrd, Nicholas L Loes, Barry G Graffagna, Katherine A VonArx, Mohammad G Rabbani**

Department of Chemistry, University of Wisconsin-Platteville,
1 University Plaza, Platteville WI 53818
*Email: rabbanim@uwplatt.edu

Introduction

Energy production and use account for two-thirds of the world's greenhouse-gas (GHG) emissions. At present, the vast majority of energy consumed by the human population is derived from burning fossil fuels because of their natural abundance and remarkably high energy density. This burning of carbon based fossil fuels not only gives off useful energy but also results in a chemical byproduct, CO_2, which is a GHG and is considered a primary cause for global warming.[1] In 2013, CO_2 accounted for about 82% of all U.S. greenhouse gas emissions from human activities.[2] At the Paris climate conference (COP21) in December 2015, 195 countries adopted the first-ever universal, legally binding global climate deal to control the emissions of CO_2.[3] Therefore, separation of CO_2 from gas mixtures is an essential and important task for protecting our environment and generating economic benefits. Recently, porous solid materials have emerged as fascinating materials in gas storage and gas separation applications because of their high surface areas, low density, and particularly easy regeneration process.[4] Among them, porous organic polymers pay significant attention because of their chemical and thermal stability as well as diverse synthetic routes to control the atomic heterogeneity though the reactions of varieties of functional groups. The presence of heteroatoms, particularly, the presence of nitrogen offers an important advantage by providing polar nature to the surface which is considered to affect the CO_2 adsorption significantly. Tuning of pore size is another important tool to control the selective gas adsorption in an effort to separate the gas mixtures.[5] It has been reported that smaller pores are more important than surface area in gas separation. However, controlling the pore sizes in subnanometer scale which is compatible with the sizes of small gas molecules is a big challenge. We report here the synthesis of a highly nitrogen-rich porous organic polymers by condensation reaction between *ortho*-diketone and *ortho*-diamine through phenazine-ring formation. Synthesized polymer possesses ultra-small pores in the range of 0.25-0.6 nm within the frame apertures, which significantly enhanced the CO_2 sorption selectivity over methane (CH_4) and nitrogen (N_2).

Experimental

All chemicals were purchased from commercial suppliers (Sigma-Aldrich, Acros Organics, and Frontier Scientific) and used without further purification, unless otherwise noted. Air-sensitive samples and reactions were handled under an inert atmosphere of nitrogen using either glovebox or Schlenk line techniques. FT-IR spectra were obtained using Attenuated Total Reflectance sampling on PerkinElmer FT-IR spectrometer. Sorption experiments were collected using a NOVA-1000 series analyzer using adsorbates of UHP grade. In a typical experiment, a sample was loaded into a 9 mm large bulb cell of known weight and then hooked up to NOVA series analyzer and degassed at 120°C for 12h. The degassed sample was refilled with nitrogen, weighed precisely and then transferred back to the analyzer. The temperatures for adsorption measurements were controlled by using refrigerated bath of liquid nitrogen (77 K), or temperature controlled water bath (273K, 288K, 298K and 313K). CO_2 and CH_4 isotherms were collected at 273, 288, 298 and 313K. N_2 isotherms were collected at 77, 273, 298 and 313K. Pore size distribution (PSD) was calculated using nonlocal density functional theory (NLDFT) model. PSD in mesoporous and microporous regions were calculated from N_2 isotherm collected at 77 K, while the PSD in ultra-micropore (0.35 - 1.5 nm) region was estimated from CO_2 isotherm collected at 273K.

Synthesis of phenazine-linked-polymer (PLP-1): A 100 mL Schlenk flask was charged with stoichiometric amount of hexaketocyclohexane and 3,3'-diaminobenzidine and refluxed in acetic acid for 3 days to form PLP-1 according to the reaction scheme in FIG. 1. Solid product was filtered, washed with water and then ethanol, and finally dried under vacuum at 150°C overnight. A light brown colored solid product was obtained in high yield (85%).

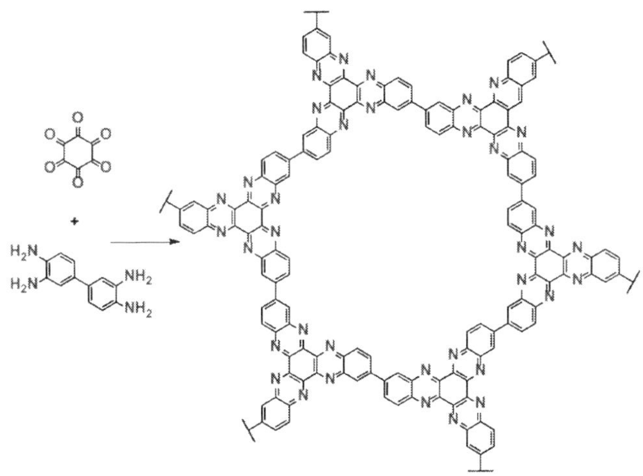

Figure 1. Reaction scheme for synthesis of PLP-1.

Results and Discussion

PLP-1 was synthesized by the polycondensation reaction as shown in FIGS. 1. The synthesized polymers are stable in 2M HCl, 2M NaOH and in all common organic solvents which indicates the chemical stability of the respective polymers. The chemical connectivity within the polymer was investigated by FT-IR spectroscopic technique. The formation of phenazine ring in PLP-1 was supported by the appearance of characteristic C=N and C-N stretching bands at 1615 cm-1 and 1358 cm^{-1}, respectively. The N-H stretching bands at around 3200-3400 cm^{-1} in amine monomer were disappeared upon polymerization which confirmed the consumption of amine in phenazine ring formation. The weak shoulder at 1660 cm^{-1} is presumably due to the residual C=O groups in PLP-1.

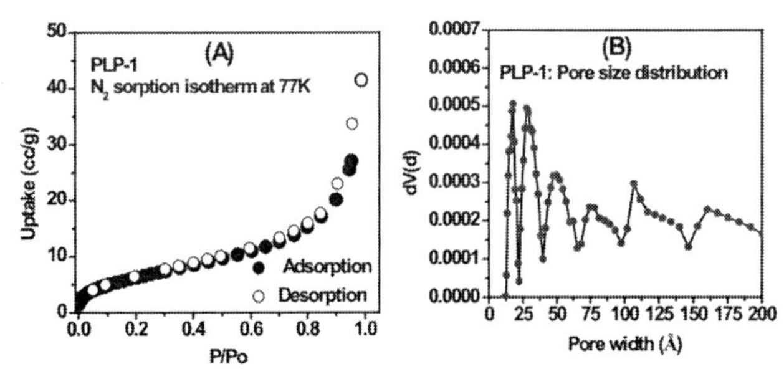

Figure 2. N_2 sorption isotherm at 77 K (A), and pore size distribution (PSD) from N_2 (77K) sorption isotherm (B).

The porous nature of PLP-1 was studied by nitrogen sorption-desorption isotherms collected at 77 K as shown in FIG. 2A. This is a type II isotherm which suggests the strong adsorbate-adsorbent interactions. Initial sharp uptake suggests the microporous nature of PLP-1. The pore size distribution (PSD) was calculated from N_2 isotherm which reflected very small percentage of mesoporosity (higher than 2 nm) and higher percentage of microposity (less than 2 nm) as shown in FIG. 2B. The Brunauer–Emmett–Teller (BET) surface area calculation resulted in value of 24 m^2g^{-1}. This significantly low surface area suggested one of the two following possibilities: i) the polymer is just nonporous; ii) the polymer is porous with a pore diameter significantly small to prevent penetration of nitrogen (N_2) gas molecules. In order to confirm the second possibility, adsorption isotherms were collected using CO_2 as an adsorbate gas molecule which has smaller kinetic diameter (0.33 nm) compared to N_2 (0.36 nm). Surprisingly, PLP-1 adsorbs significantly high amount of CO_2 at different temperatures and the isotherms are as shown in FIG. 3A. High CO_2 adsorption indicates the presence of ultra-small pores in the frames which are large enough for entering smaller CO_2 gas molecules but not slightly larger N_2 gas molecules. The CO_2 isotherm at 273K was fitted by the NLDFT model to calculate the PSD in ultra-micropore region, as shown in FIG. 3B. The dominant PSD was found to be centered on around 0.35 nm and 0.5 nm. Although the kinetic diameter of N_2 gas molecules apparently falls within the average PSD of PLP-1, the narrower pores would probably block the N_2 gas molecules from entering freely through the pores. In addition to the ultra-small pores, the presence of basic-nitrogen centers in the pore apertures as shown in the structure of PLP-1 (FIG. 1) is expected to facilitate the adsorption of the acidic and relatively smaller CO_2 gas molecules onto the polymer PLP-1.

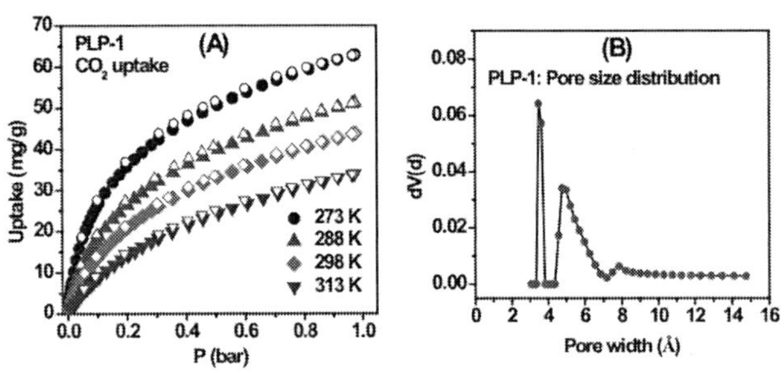

Figure 3. CO_2 adsorption-desorption isotherms at 273, 288, 298 and 313K (A) and the pore size distributions (PSD) calculated from CO_2 isotherm at 273K using NLDFT model (B).

In order to understand the high CO_2 affinity for the surface of PLP-1, isosteric heats of adsorption (Q_{st}) was calculated using the virial method from the isotherms measured at 273 and 298K. The Q_{st} value of 36 kJ/mol for PLP-1 is among the highest for any reported porous organic polymers. This high Q_{st} value suggests the significant surface affinity for CO_2 gas molecules due to the presence of ultra-micropores and surface heterogeneity.

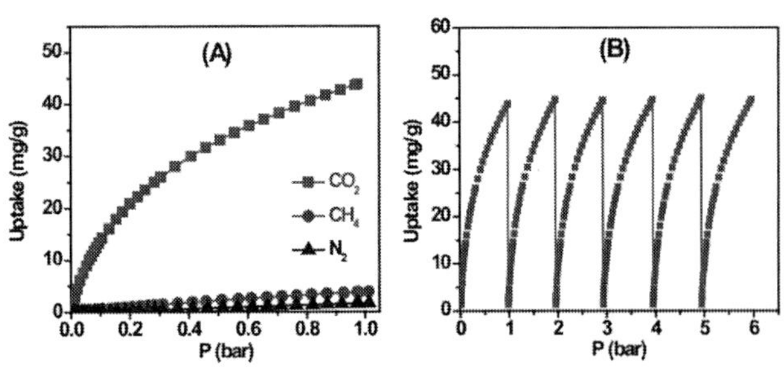

Figure 4. Selective adsorption of CO_2, CH_4 and N_2 at 298K (A) and the successive CO_2 adsorption-desorption cycles at 298K (B) for PLP-1.

Given the high CO_2 uptake per unit surface area, and desirable binding affinity, we studied the selective CO_2 capture over CH_4 and N_2 in order to understand the practical use of PLP-1 in CO_2 separation from an input gas mixtures, such as a landfill gas or a flue gas, among others. PLP-1 showed significantly low uptakes for both CH_4 and N_2 compared CO_2 uptake (FIG 4A). The adsorption selectivity (mg/mg) was estimated for CO_2 over CH_4, and CO_2 over N_2 from Henry's law on initial slope calculations using the single component gas adsorption isotherms at 298 K. PLP-1 has 140 times greater selectivity for CO_2 over N_2 (CO_2/N_2 = 140) and 35 times greater selectivity for CO_2 over CH_4 (CO_2/CH_4 = 35) at 298 K. These observed selectivity values for PLP-1 at 298 K surpass most of the reported porous organic polymers.[6]

In order to be industrially applicable for CO_2 separation from an input gas, such as a flue gas or landfill gas, the regeneration property of PLP-1 were evaluated for CO_2 at 298 K. The results shown in FIG 4B show successive six adsorption-desorption cycles which were measured successively without any extra efforts/steps employed for degassing the sample. The measurements were repeated once the prior measurement was completed. Sample was automatically degassed for few minutes once each cycle was completed. No extra effort was given either for a longer evacuation time or any heat treatment. Further, the last cycle was measured one week later after the first five consecutive cycles. As shown in FIG. 2B, there is no loss of CO_2 uptake within any of the six adsorption/desorption cycles which indicate the stability and the energy saving regeneration of the synthesized polymers.

Conclusions

An ultra-microporous organic polymer, PLP-1 was synthesized using phenazine linker which provides high percentage of nitrogen heterogeneity in the frame. The ultra-micropores effectively controlled the selective CO_2 gas sorption over N_2 or CH_4. Adsorption selectivity and its easy regeneration make PLP-1 a promising candidate for using in membranes-based gas separation application.

Acknowledgement. This work was supported by the WiSys Technology Foundation, Inc., Applied Research-WiSys Technology Advancement Grant (AR-WiTAG), under Award number 106-Y-06-8000-4. M.G.R is also grateful to the UW System Regent Scholar Grant, under Award number IUJAMH0067.

References

(1) (a) Overview of Greenhouse Gases: http://www3.epa.gov/climatechange/ghgemissions/gases/co2.html; (b) Causes of Climate Change: http://www3.epa.gov/climatechange/science/causes.html.

(2) Global Greenhouse Gas Emissions Data: *United States Environmental Protection Agency (EPA)*, http://www3.epa.gov/climatechange/ghgemissions/global.html.

(3) (a) Paris Agreement: ec.europa.eu/clima/policies/international/negotiations/future/index_en.htm; (b) The EU Emissions Trading System: ec.europa.eu/clima/policies/ets/index_en.htm.

(4) (a) Sreenivasulu, B.; Sreedhar, I.; Suresh, P.; Raghavan, K. V., Development Trends in Porous Adsorbents for Carbon Capture. *Environmental Science & Technology* **2015,** *49* (21), 12641-12661; (b) Rabbani, M. G.; El-Kaderi, H. M., Synthesis and Characterization of Porous Benzimidazole-Linked Polymers and Their Performance in Small Gas Storage and Selective Uptake. *Chemistry of Materials* **2012,** *24* (8), 1511-1517; (c) Li, B.; Wen, H.-M.; Zhou, W.; Chen, B., Porous Metal–Organic Frameworks for Gas Storage and Separation: What, How, and Why? *The Journal of Physical Chemistry Letters* **2014,** *5* (20), 3468-3479.

(5) Dawson, R.; Cooper, A. I.; Adams, D. J., Nanoporous organic polymer networks. *Progress in Polymer Science* **2012,** *37* (4), 530-563.

(6) Sekizkardes, A. K.; Culp, J. T.; Islamoglu, T.; Marti, A.; Hopkinson, D.; Myers, C.; El-Kaderi, H. M.; Nulwala, H. B., An ultra-microporous

organic polymer for high performance carbon dioxide capture and separation. *Chemical Communications* **2015,** *51* (69), 13393-13396.

NON-AQUEOUS AMINE-BASED SOLVENTS FOR CARBON DIOXIDE CAPTURE WITH LOW REGENERATION ENERGY

Phillip K. Koech, Deepika Malhotra, David J. Heldebrant, Abhijeet Karkamkar, Feng Zheng, Mark E. Bearden*

Pacific Northwest National Laboratory, Energy Processes & Materials Division, 902 Battelle Blvd, MS K2-44, Richland, WA, 99352, USA
Phillip.Koech@pnnl.gov

Climate change is partially attributed to anthropogenic release of greenhouse gases such as carbon dioxide (CO_2) to the atmosphere from both stationary and mobile sources. Carbon dioxide capture and storage (CCS) of CO_2 from point sources like power plants is a potentially an effective way of mitigating adverse environmental effects from CO_2 emission.1 To this end numerous CO_2 capture processes have been developed which include membrane separations, 2 absorption liquids both physical and chemical, adsorption solids,3-4 cryogenics5 and microbial algal systems. 6 The most advanced of these technologies utilizes aqueous solutions of primary and secondary alkanolamines such as MEA or diethanolamine (DEA) for chemical absorption of CO_2. These amines have rapid absorption kinetics at low partial pressures due to the facile formation of stable carbamates salts.7 However, aqueous alkanolamines have several limitations that include high regeneration temperature (>120 °C), high steel corrosion, and high solvent evaporative losses.8 One way to decrease the regeneration temperature below 100 °C is by eliminating water a major contributor of the high regeneration temperature due to its high heat capacity. To this end, several water-lean transformational solvents systems have been developed which include nano-material organic hybrids (NOHMs),[9] task-specific,[10] protic, and conventional ionic liquids,[11] phase-change solvents[12] and CO_2-binding organic liquids (CO_2BOLs).[13] These water-lean solvent systems can be energetically viable technologies with potentially improved net power output, but capital costs are astronomically high due to increased viscosity and low thermal conductivity.

Non-aqueous alkanolamines solvents like 2-[(1,1-dimethylethyl)amino]ethanol (TBAE), 1-[(1,1-dimethylethyl)amino]-2-propanol (TBAP) consisting of a sterically hindered amine moiety have been developed.14 These alkanolamines reversibly bind CO_2 in a 1:1 molar ratio exclusively via the hydroxyl group, forming a solid ammonium alkylcarbonate zwitterion dissolved in organic solvents such as ethylene glycol or methanol. Non-aqueous sterically hindered alkanolamine solvent system consisting of 2-amino-2-methyl-1-propanol (AMP)/piperazine (PZ) in ethyl glycol mono methylether (EGME) have been reported to form less stable carbamate or carbamic acid that can be regenerated at lower temperature (80 °C).15 These materials bind CO_2 through the conventional carbamate mechanism; however, significant carbonic acid products are formed resulting in loading capacity greater than 50 mol% and the viscosity of the CO_2-rich material is low (<30 cP). Most of these amine based solvents require a co-solvent to dissolve the carbamate salt, thus a non-viscous CO_2-rich solvent is required for improved CO_2 capacity.

As part of an ongoing program in our group we have developed a new water lean non-aqueous amine-based solvent system which has the potential to be energy an efficient CO_2 capture material. These solvents bind CO_2 at standard temperature and pressure (STP) to form liquid carbamates salts. The CO_2 gravimetric capacity for these solvent is high 17-20 wt% at 25 °C and the regeneration temperature is exceptionally low (70-100 °C) as determine using an automated gas burette. This is one of the few available examples of non-aqueous amine-based solvents that react with CO_2 to form liquid carbamates and release CO_2 at low temperature below 100 °C. The solvent viscosity increase with CO_2 loading but the solutions remains liquid even upon storage for months. The recyclability of these solvents over multiple CO_2 absorption and regeneration cycles was evaluated through CO_2 uptake experiments at 25 °C and regeneration at 100 °C over 5 cycles. We demonstrated that these amine-based solvents can be recycled over multiple runs without loss of activity as confirmed ^{1}H and ^{13} C NMR spectra of the recovered material which was identical to the pristine solvent.

Acknowledgement. The authors would like acknowledge the U.S. Department of Energy (DOE), Office of Fossil Energy, for funding this project through NETL FWP Award # DE-FE0007466 and the Pacific Northwest National Laboratory (PNNL) for facilitics. PNNL is proudly operated by Battelle for the U. S. Department of Energy.

References

(1) (a) Rao, A. B.; Rubin, E. S.; *Environ. Sci. Technol.* **2002**, *36*, 4467. (b) MacDowell, N.; Florin, N.; Buchard, A,; Hallett, J.; Galindo, A.; Jackson, G.; Adjiman, C. S.; Williams, C. K.; Shah, N.; Fennell, P. *Energy Environ. Sci.* **2010**, *3*, 1645

(2) Zhang, Y.; Sunarso, J.; Liu, S.; Wang, R.; *Int. J. Greenhouse Gas control* **2013**, 12, 84

(3) D' Alessandro, D. M.; Smit, B.; Long, J. R. *Angew. Chem. Int. Ed.* **2010,** *49,* 6058

(4 Wang, Q,; Luo, J.; Borgna, A. *Energy Environ. Sci.* **2011**, 4, 42.

(5) Tuinier, M. J.; Hamers, H. P.; Van Sint Annaland, M. *Int. J. Greenhouse Gas control* **2011**, *5*, 1559.

(6) Sayre, R. *Bioscience* **2010**, *60*, 722.

(7 (a) Xie, H.-B.; Zhou, Y.; Zhang, Y.; Johnson, J. K. *J. Phys. Chem.* **2010**, 114, 11844. (b) Barzagli, F.; Mani, F.; Peruzzini, M. *Energy Environ. Sci.* **2010**, 3, 772

(8) (a) Sigh, P.; Niederer, J. P. M.; Versteeg, G. F. *Chem. Eng. Res. Des.* **2009**, *87*, 135 (b) Peeters, A. N. M.; AFaaij, A. P. C.; Turkenburg, W. C.; *Int. J. Greenhouse Gas Control*, **2007**, *1*, 396.

(9) K. Lin and A. Park, *Environ. Sci. Technol.*, **2011**, *45*, 6633.

(10 (a) C. Wang, S. M. Mahurin, H. Luo, G. A. Baker, H. Li and S. Dai, *Green Chem.*, **2010**, *12*, 870; (b) E. D. Bates, R. D. Mayton, I. Ntai and J. H. J. Davis, *J. Am. Chem. Soc.*, **2002**, *124*, 926.

(11) M. Ramdin, T. de Loos and T. Vlugt, *Ind. Eng. Chem. Res.*, **2012**, *51*, 8149.

(12) R. Perry, B. Wood, S. Genovese, M. O'Brien, T. Westendorf, M. Meketa, R. Farnum, J. McDermott, I. Sultanova, T. Perry, R. Vipperla, L. Wichmann, R. Enick, L. Hong and D. Tapriyal, *Energy Fuels*, 2012, 26, 2528.

(13) (a) Heldebrant D. J.; Koech, P. K.; Yonker, C. R. *Energy Environ. Sci.* **2010**, *3,* 111. (b) Heldebrant D. J.; Koech, P. K.; Ang, M. T. C.; Liang, C.; Rainbolt, J. E.; Yonker, C. R.; Jessop, P. G. *Green Chem.* **2010**, *12*, 713.

(14) (a) Im, J.; Hong, S. Y.; Cheon, Y.; Lee, J.; Lee, J. S.; Kim, H. S.; Cheong, M.; Park, H. *Energy Environ. Sci.* **2011**, *4*, 4284. (b) Cheon, Y.; Jung, Y. M.; Lee, J.; Kim, H.; Im, J.; Cheong, M.; Kim, H. S.; Park, H. S. *ChemPhysChem* **2012**, *13*, 3365. (c) Hong, S. Y.; Lee, J. S.; Cheong, M.; Kim, H. S. *Energy Procedia* **2014**, *63*, 2190. (d) Choi, Y.-S.; Im, J.; Jeong, J. K.; Hong, S. Y.; Jang, H. G.; Cheong, M.; Lee, J. S.; Kim, H. S. *Environ. Sci. Technol.* **2014**, *48*, 4163.

(15) Barzagli, F.; Di Vaira, M.; Mani, F.; Peruzzini, M. *ChemSusChem* 2012, *5*, 1724.

CO_2 SPECIATION AND SOLVENT STRUCTURE OF CO_2 ANHYDROUS CAPTURE FLUIDS

David J. Heldebrant, Deepika Malhotra, Martin Owen Jones, Tom Headen, David Cantu, Vassiliki-Alexandra Glezakou, Roger Rousseau.

Water-free CO_2 capture solvents have been gaining notoriety for CO_2 capture, with the promise of increased efficiency due to the reduced energy loss associated with the boiling and condensing of water. CO_2-Binding Organic Liquids (CO_2BOLs) are a single-component variant of the water-lean solvents. CO_2BOLs are a class of switchable ionic liquids specifically designed for CO_2 separations. These solvents were discovered by the Jessop group in 2005 as a solvent that "switches" its polarity (*i.e.*, dielectric constant) between non-polar and polar forms.[1] As the chemical trigger for the switch is CO_2, these solvents were perfectly suited as CO_2 separating solvents.

CO_2BOLs, like all water-lean solvent systems, retain the chemiselective binding of CO_2 as conventional water-based amine CO_2 capture solvents, utilizing a non-nucleophilic (amidine or guanidine) base and an alcohol used instead of nucleophilic primary and secondary amines and water. The neutral alkanolguanidine (alcohol and base) shown in Figure 1a, reacts with CO_2, forming a zwitterionic alkylcarbonate ionic liquid (Figure 1b), with CO_2 being released thermally, up to 150 °C.[2-3] The molecular structure and all associated physical and thermodynamic properties, change as the solvent continuously changes between ionic and non-ionic forms of the fluid. These liquids are never entirely ionic or neutral, unlike the "switchable" descriptor. These liquids are always undergoing gradient changes in CO_2 loading, changing the size and shape of the CO_2-rich ionic domains, in turn altering the physical/thermodynamic properties of the liquid. These gradient changes in structure are vastly different from conventional molecular solvents that remain neutral, or established ionic liquids that remain fully ionic whether CO_2 is chemically bound or not.

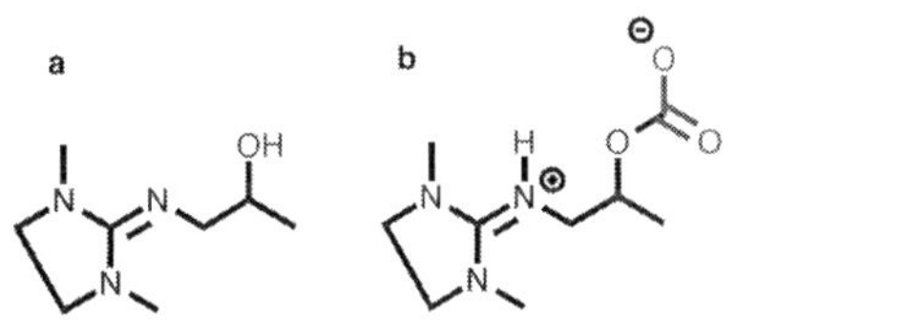

Figure 1. IPADM-2-BOL: (a) alkanol- guanidine (without CO_2) and (b) guanidinium alkylcarbonate (with CO_2).

These solvents have been hypothesized to have a unique and ever-changing 'heterogeneous' molecular structure, differing prominently from other classes of water-based CO_2-capture solvents (Figure 2).[4,5] Liauw, *et al.*, were the first to propose such a structure, using ultraviolet (UV)-visible and infrared (IR) spectroscopy to see heterogeneous domains ebb and flow within the liquid structure.[4,5] Such a structure would contain discrete microenvironments of unequal polarity, though the solvent's bulk dielectric would indicate otherwise. A 'heterogeneous' molecular structure would be in direct contrast to other CO_2-capture solvents, where the CO_2-rich ions are solvated and uniformly distributed in solution by co-solvents. This 'heterogeneous' molecular structure also differs substantially from conventional ionic liquids where the solvent remains ionic regardless of the presence of CO_2.[6-8, 9]

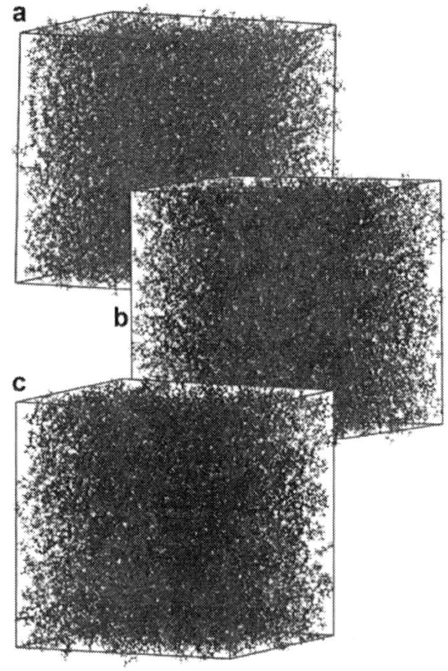

Figure 2. MD simulations of the liquid phase structure of IPADM-2-BOL, as a function of CO_2 loading (mol%): (a) pure solvent, (b) 25% CO_2, and (c) 50% CO_2. The purple color represents CO_2-bearing species.[10]

Many studies of CO_2BOLs have been performed, measuring many unique phenomena with respect to viscosity and mass transport, which we attribute to this heterogeneous solvent structure. CO_2BOLs have been shown to exhibit a distinctive non-linear increase in viscosity as a function of CO_2 loading, with the non-linearity occurring near 75 percent of the maximum molar uptake of CO_2. Drastic increases in viscosity limit any practical utilization for CO_2-capture, as the viscosity increases to thousands of centipoise at full CO_2 loading. Another unique phenomenon is that the rate of CO_2 diffusion into solution was observed to *increase* with a *decrease* in temperature. This has been attributed to the presence of a passivated diffusion barrier, made up of CO_2-rich ions at the gas-liquid interface[4,5] where physical dissolution of CO_2 (known to vary inversely with temperature) would be required to get CO_2 into the un-reacted solvent prior to chemical fixation. We hypothesize that these unique observed phenomenon are directly linked to the molecular solution structure, where molecular-level interactions between the 'heterogeneous' domains change as a function of loading and temperature. Our group has set out to further shed light on the heterogeneous solvent structure and how it might impact the physical and thermodynamic properties of the solvent, CO_2 diffusion in and out of the liquid and the mechanism of CO_2 capture and release.

Our group has performed experimental and computational measurements to observe the proposed 'heterogeneous' solvent structure. Time-of-flight secondary ion mass spectrometry (ToF-SIMS) measurements were conducted, directly observing non-uniform concentration and speciation of CO_2 in a representative CO_2BOL solvent at different sampling areas within the fluid environment.[10] Our team has also performed molecular dynamics (MD) simulations, showing a heterogeneous solvent structure for the first time.[9,11] These simulations predicted that the CO_2-rich ionic domains are ever changing in size and shape as a function of CO_2 loading (Fig. 2) and temperature. To corroborate this phenomenon, we performed total neutron scattering experiments, directly measuring the shape and orientation of the CO_2-rich domains in solution changing as a function of temperature and CO_2 loading.

Ultimately, the heterogeneous molecular structure of CO_2BOL solvents has been experimentally observed, and we link this 'heterogeneous' molecular-level structure to previously measured physical and thermodynamic phenomena. We hypothesize that other water-lean solvent systems for CO_2 capture would exhibit similar molecular structure, as we would predict aggregation of CO_2-rich ions in an organic solution. Ultimately, this knowledge helps develop more advanced and better performing solvents[9] for CO_2 capture.

References.

(1) Jessop et al. *Nature* **2005**, *436*, 1102-1102.
(2) Heldebrant et al. *Energy Environ. Sci.* **2008**, *1*, 487-493.
(3) Koech at al. *RSC Adv.* **2012**, *3*, 566-572.
(4) Hardy et al. *Analyst* **2013**, *138*, 819-824.
(5) Hardy et al. *Chem. Eng. J.* **2013**, *233*, 292-296.
(6) Bates et al. *J. Am. Chem. Soc.* **2002**, *124*, 926-927
(7) Gurkan et al. *J. Phys. Chem. Lett.*, **2010**, *1*, 3494-3499.
(8) Wang et al. *Angew. Chem. Int. Ed.* **2010**, *34*, 5978-5981.
(9) Cantu et al. *J. Chem. Phys Lett.*, **2016**, **DOI:** 10.1021/acs.jpclett.6b00395
(10) Yu et al. *Surface Science Spectra,* **2016**, *in press.*
(11) Heldebrant et al. *Energ. Proc.* GHGT-12, **2014**, *63, 8144-8152.*

METHANE CONVERSION TO SYNGAS USING DOPANT MODIFIED METAL OXIDE COMPOSITES IN CHEMICAL LOOPING REFORMING

Lang Qin[1], Zhuo Cheng[1], Mengqing Guo[1], Mingyuan Xu[1], Jonathan A. Fan[2], and L.-S. Fan[1]

[1]Department of Chemical and Biomolecular Engineering, The Ohio State University
[2]Department of Electrical Engineering, Stanford University

Abstract

Metal oxide composites in bulk form are used as enabling materials for operation in many energy conversion systems including shale gas and coal gas reforming using chemical looping systems. When metal oxide composites are involved in redox reactions, ionic diffusion takes place which is accompanied by the creation and diffusion of defects that decide such their morphology and chemical properties. Rare earth element dopants can affect the redox reactions by providing alternative reaction sites, thus change the ionic diffusivity and reaction reactivity of metal oxides. In this study, we examine at nanoscale to delineate the characteristics of undoped and doped iron oxide that undergo methane reforming reaction. 0-2% lanthanum doped iron oxides are fabricated by modified wet chemical synthesis. The reactions are examined at 500-1000°C. Experimental results show that lanthanum dopant dramatically enhances the reactivity of metal oxide during redox reaction. XRD and SEM/EDS characterization implies that the dopant changes lattice parameters of metal oxides whose surface energy is changed. The atomistic thermodynamics methods and ab initio calculations are carried out to investigate the interaction between methane molecules and undoped/doped iron oxide. It is found that dissociation of H atom through breaking of C-H bonds in the CH-La species is the key step for CH_4 partial oxidation reaction, and low-fold O atoms neighboring dopants in metal oxide composites could more readily dissociate from external surface to create oxygen vacancies. Moreover, the presence of rare earth element dopants could decrease activation energy and reaction energy of CH_4/La-Fe_2O_3 system, hence the reaction between CH_4 and iron oxide is promoted. Furthermore, La dopant is found to exhibit excellent compatibility with the iron oxide structure at the concentration range of 0-2%. The fundamental insight into the mechanism of cyclic redox reactions of La dopant modified metal oxide composites at nanoscale revealed in this study can be applied to other rare earth element dopant modified metal oxide systems and to guide future efforts to improve oxygen carriers for methane conversion.

CHARACTERIZATION AND TESTING OF HIERARCHAL CARBON-BASED PHYSICAL SORBENTS FOR CO_2 CAPTURE

Jiajun He and Jennifer Wilcox

Department of Energy Resources Engineering
367 Panama Street, Stanford, CA 94305
Email: jen.wilcox@stanford.edu

Abstract

Carbon capture and sequestration from point sources is an important component in the CO_2 emission mitigation portfolio. In particular, sorbents with both high capacity and selectivity are required for reducing the cost of CO_2 capture. Although physisorbents have the advantage of low energy consumption for regeneration, it remains a challenge to obtain simultaneous high capacity and sufficient CO_2/N_2 selectivity. The controlled synthesis of a novel N-doped hierarchical carbon that exhibits very high Henry's law CO_2/N_2 selectivity among physisorptive carbons while having a high CO_2 adsorption capacity will be discussed. The low-temperature carbonization and activation processes allow for the development of ultra-small pores (d <0.5 nm) and preservation of nitrogen functionalities, which are essential for enhanced CO_2 affinity. Furthermore, our described work provides a strategy to initiate the dadvancement of rationally-designed porous conjugated polymer structures and carbon-based materials for various potential applications.

THERMODYNAMIC COMPLEXITY OF CO_2 CAPTURE IN METAL-ORGANIC FRAMEWORK SORBENTS

Di Wu[1,2], Jeremiah J Gassensmith[3], Thomas M. McDonald[4], Xiaofeng Guo[2,5], Zewei Quan[6], Sergey V. Ushakov[2], Peng Zhang[7], Jeffrey R. Long[4] and Alexandra Navrotsky[2,]*

[1]Gene and Linda Voiland School of Chemical Engineering and Bioengineering, Washington State University, Pullman, WA 99164, United States
[2]Peter A. Rock Thermochemistry Laboratory and NEAT ORU, University of California, Davis, One Shields Avenue, Davis, CA 95616, United States
[3]Department of Chemistry, University of Texas, Dallas, Richardson, TX 75080, United States
[4]Department of Chemistry, University of California, Berkeley, CA 94720, United States
[5]Earth and Environmental Sciences Division, Los Alamos National Laboratory, Los Alamos, NM 87545, United States
[6]Department of Chemistry, South University of Science and Technology of China, Shenzhen, Guangdong, 518055, China
[7]School of Materials Science and Engineering, State Key Laboratory for Metallic Matrix Composite Materials, Shanghai Jiao Tong University, Shanghai, 200240, China
anavrotsky@ucdavis.edu

Abstract

Understanding the thermodynamics of CO_2 adsorption on metal-organic frameworks (MOFs) is crucial for development of future carbon capture materials. We have performed calorimetric studies on CO_2 adsorption in two model MOFs, including a carbohydrate MOF, CD-MOF-2, and an alkylamine-appended MOF, mmen-Mg_2(dobpdc), employing gas adsorption calorimetry. For both materials, the enthalpy of adsorption as a function of CO_2 loading suggests three different adsorption regions: the strongest exothermic initial chemisorption at near zero coverage, the majority moderate chemisorption at intermediate loading and the weakest physisorption at highest coverage. We interpreted the CO_2 adsorption mechanism and discussed the reaction evolution and change of partial molar properties as a function of gas loading and of temperature. The potential impact and competitive role of H_2O vapor in the CO_2 capture process were analyzed. Our studies reveal that CO_2 adsorption on functionalized MOFs is a complex process involving multiple thermodynamic factors reflecting changes in surface phase and structure, chemical bonding and degree of disorder as temperature and gas loading vary. These fundamental thermodynamic insights may help optimize the design, synthesis and application of MOF-based CO_2 sorbents.

RATIONAL DESIGN AND DEVELOPMENT OF ENERGY EFFICIENT CO_2-BINDING ORGANIC LIQUIDS FOR FLUE GAS CLEAN UP

Deepika Malhotra, Phillip K. Koech, David J. Heldebrant, David C. Cantu, Vassiliki-Alexandra Glezakou, Roger Rousseau, Feng (Richard) Zheng

Pacific Northwest National Laboratory, Energy Processes & Materials Division, 902 Battelle Blvd, MS K2-44, Richland, WA, 99352, USA

1. Introduction

Carbon dioxide is the most important anthropogenic greenhouse gas and its concentration in the atmosphere has significantly increased from 280 ppm in mid-18th century to 402.5 ppm today. It's critical to capture and sequester CO_2 in order to mitigate the potential impacts on environment as well as climate change. In this context, liquid systems lie at the heart of CO_2 capture, and aqueous amines, such as monoethanolamine (MEA), are amongst the forerunners for CO_2 capture from flue gas. Although, they have low CO_2-rich viscosity, fast kinetics and favorable thermodynamics, their energy requirements are prohibitively high.1 The high heat capacity of water in these solvents leads to tremendous increase in costs of thermal solvent regeneration. Water-lean transformational solvent platforms have been identified as an alternative to address these critical challenges. These systems include nano-material organic hybrids (NOHMs),2 task-specific,3 protic and reversible ionic liquids,4 and phase change solvents.5 They use current knowledge and infrastructure, and therefore present attractive substitute for MEA technology. In the past few years our group has developed, water-lean transformational solvents known as CO_2 binding organic liquids (CO_2BOLs) that have the potential to be energetically-viable CO_2 capture solvents. The CO_2BOL solvent platforms operate on the principles of switchable ionic liquid technology, which enables the distinctive Polarity Swing Assisted Regeneration (PSAR) to facilitate the regeneration of CO_2BOL solvents (Figure 1) by decreasing reboiler temperatures as much as 73 °C.[8] These zwitterionic liquids are formed when non-ionic alkanolguanidines react with CO_2 to form guanidinium alkyl carbonate as ionic liquids with high polarity. These solvent systems have shown excellent CO_2 uptake, acceptable mass transfer, and favorable thermodynamics, albeit at higher viscosities.6

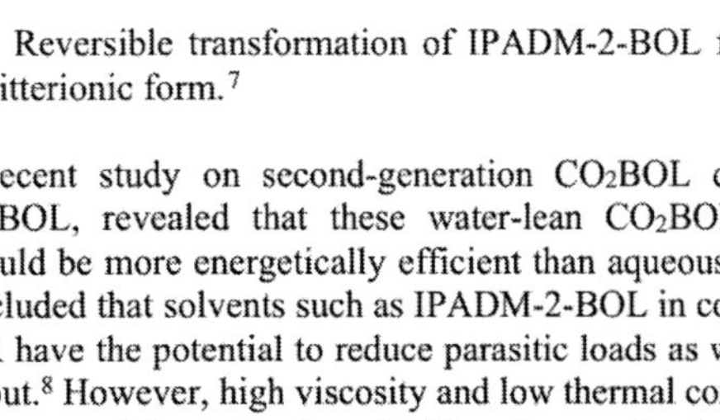

Figure 1: Reversible transformation of IPADM-2-BOL from non-ionic to zwitterionic form.[7]

Our recent study on second-generation CO_2BOL derivative, IPADM-2-BOL, revealed that these water-lean CO_2BOL solvent systems could be more energetically efficient than aqueous solvents. It was concluded that solvents such as IPADM-2-BOL in conjunction with PSAR have the potential to reduce parasitic loads as well as net power output.[8] However, high viscosity and low thermal conductivity lead to huge capital costs, thus limiting the application of this formulation in the commercial scale.[8b] Currently, our group is focused on identifying a less-viscous solvent system to advance this technology to commercialization. Our synergistic synthesis and computational efforts are dedicated towards the development of fundamental understanding of molecular level interactions that control viscosity in these solvents with ultimate aim of designing higher-performance solvents. Moreover, the knowledge gained from our study will be translated to other CO_2-capture solvent classes. In this contribution, we present the rational design, synthesis, and CO_2-capture properties of CO_2BOL with the focus on reducing the viscosity of CO_2-rich solvent.

2. Experimental

General: All reactions were conducted under an atmosphere of nitrogen, unless otherwise indicated. Solvents were transferred via gas-tight syringe. Flasks were oven-dried and cooled under vacuum. 2-Chloro-1,3-dimethylimidazolinium chloride, and dichloromethane (DCM), were purchased from Sigma-Aldrich Chemical Company and used without purification. Selected target molecules are synthesized via our already-established, one-step protocol involving the reaction of Vilsmeier salts with alkanolamines in the presence of triethylamine utilizing dichloromethane as a solvent to furnish alkanolguanidine analogues.[6] Alkanolamines were either purchased from commercial sources or prepared using literature procedures. The CO_2-free and -rich solvents were characterized using standard spectroscopic techniques such as ^1H and $^{13}C\{^1H\}$ NMR, and FTIR. Gravimetric CO_2 uptake and viscosity measurements were obtained using standard procedures. For gravimetric CO_2 uptake studies, 500 mg samples of each CO_2BOL were added into an oven-dried pre-weighed GC vials. Each preweighed vial was charged with CO_2 by bubbling CO_2 through the sample for 1 hour and then each sample was re-weighed for CO_2 uptake. The amount of CO_2 is calculated based on the difference in mass.

3. Molecular design principles

Our strategy involves building a modular approach to develop a fundamental understanding of how molecular structure impacts the viscosity in water-lean solvent systems. The two basic elements of molecular design include retaining the strongly basic guanidine core to ensure a high enough heat of solution and cyclic guanidine ring to prevent hydrolysis. In order to evaluate the impacts of viscosity, libraries of molecules were designed based on molecular design principles also known as *viscosity modifying factors*. In the CO_2-rich phase, hydrogen bonding and cation-anion electrostatic interactions play a key role. In order to attain a less-viscous solvent system such interactions need to be minimized. By incorporating the individual molecular design elements, libraries of molecules based on a cyclic guanidine core has been designed to evaluate the influence of hydrogen bonding, cation-anion interactions and ion solvation effect towards reducing the viscosity (Figure 2).

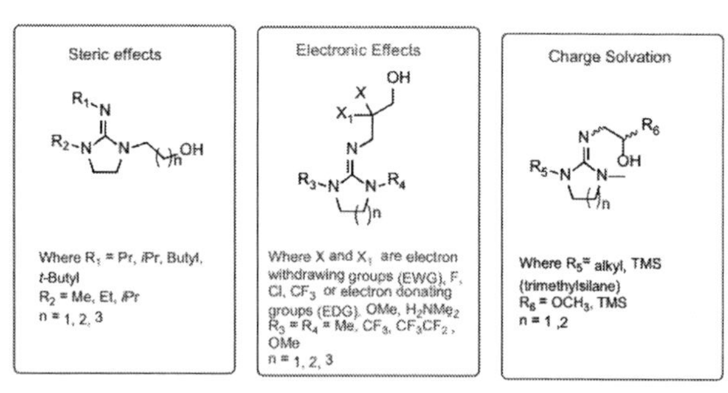

Figure 2: Design strategies for viscosity reduction in CO_2BOLs.

Since, it is not practical to synthesize hundreds of molecules in reasonable timeframe, computational tools were utilized to narrow down our synthesis targets. Representative molecules from each library were selected in order to systematically study the impact of each design element on viscosity reduction.

4. Results and Discussion

Each molecular library is based on different design elements/functionalities such as inter vs. intramolecular hydrogen bonding, electronic tuning, sterics, broken symmetry and heteroatoms like ethers. As a next step, we are synthesizing representative target molecules from each respective category to evaluate their physical and thermodynamic properties (Figure 3). Our current CO_2BOL formulations are based on the fundamental insight gained from the ongoing study on molecular design coupled with molecular dynamic simulations towards identifying energetically-efficient, less-viscous CO_2-rich solvent.

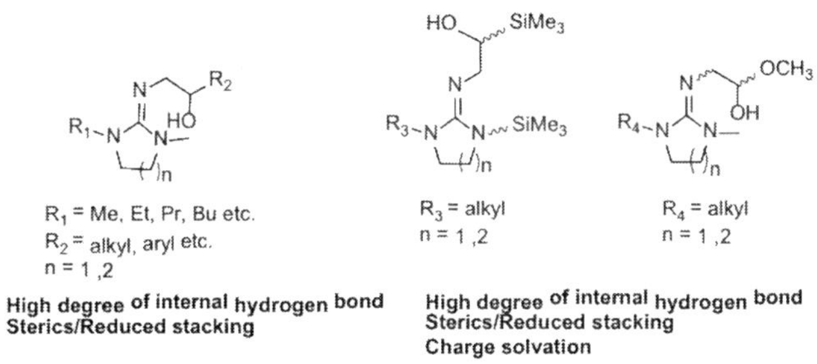

Figure 3: CO_2BOL formulations based on molecular design principles.

One of our initial design elements is focused onthe introduction of steric bulk on the anion to fine-tune the electronics of the anion. By making a weaker coordinating anion, the cation-anion interactions are reduced. To test our hypothesis, we synthesized different analogues and identified the role of alkyl and aryl groups on the ring as well as on the side chain containing alcohol. The insight gained from studying these molecular targets is helping us investigate the impact of the steric environment of both cation and anion centers. The results of this study are being utilized to guide our efforts towards designing an optimal synthetic model for the identification of viable CO_2BOLs with lower viscosities.

In our second molecular design strategy, we explore the impact of solvation of charges on viscosity, by attaching a silane moiety to the CO_2BOL molecules. The logic behind designing silane-based compounds stems from the fact that silicon centers are oxo-philic and thus are expected to form a dative bond with the carbonate anion. This dative bond formation weakens the cation-anion interactions and presumably contributes towards reduced viscosity of the molecular system. Our first silane-based CO_2BOL, 1-TMSIPADM-2-BOL (CO_2 capture capacity of 5.5 wt%), has displayed promising results in terms of viscosity reduction; however, we are doing molecular refinement to identify optimal performance candidate. In parallel, we also explored the effect of cation charge solvation on viscosity by incorporating ether functionalities, which are known to lower the viscosity of liquids due to the solvation of charges by the coordinating ether moieties. One of our derivatives has shown significant reduction in viscosity (CO_2 capture capacity of 7.2 wt%). The percentage drop in viscosity of CO_2-rich form for the ether CO_2BOL variant is approximately 60% with respect to our base compound IPADM-2-BOL. This is highly encouraging result and provides an initial proof of concept for our rational molecular design approach.

Currently, our synthetic efforts are focused on combining the important learnings of sterics, stacking and solvation of charges towards designing a less-viscous, optimal CO_2BOL guided by simulation studies. Another design element we are exploring is tuning the acid/base properties by incorporating fluorine motif. Fluorine-based solvents have been shown to have lower viscosity, therefore, CO_2BOL derivatives based on fluorine motifs are presently being synthesized to reduce the electrostatic cation-anion interactions in CO_2-rich solvent. Moreover, molecular simulations carried out by our collaborators predict nearly 30% viscosity reduction for fluorine derivatives of IPADM-2-BOL. Combining the insight gained from studying these molecular targets will direct in developing a basic understanding towards designing an optimal synthetic model to identify viable CO_2BOLs with lower viscosities. In addition, scale-up of some of the promising CO_2BOL candidates is underway to provide enough material for comprehensive measurements of properties such as viscosity, density, vapor pressure, heat capacity, and kinetic and thermodynamic bench scale testing

5. Conclusions

Our systematic design strategy involves exploring the impact of viscosity-modifying factors towards viscosity reduction using our synthetic methodology as well as high throughput computational simulations. We are gaining fundamental knowledge of the solvent structure and its interactions, by studying different molecular templates based on cation-anion electrostatic interactions, electronics, steric effects and charge solvation. In order to identify an optimum carbon capture solvent system with a CO_2-loaded viscosity below 50 cP, fine tuning and structural modification of the current molecules is desired to develop cost-effective CO_2 capture technology. Synthesis, characterization, along with viscosity profile and thermodynamic data of analogues of CO_2BOL will be discussed in detail.

Acknowledgements. The authors would like acknowledge the U.S. Department of Energy (DOE), Office of Fossil Energy, for funding this project through NETL FWP Award # DE-FE0007466 and the Pacific Northwest National Laboratory (PNNL) for facilities. PNNL is proudly operated by Battelle for the U. S. Department of Energy.

References

(1) Peeters, A. N. M.; Faaij, A. P, C.; Turkenburg, W. C. *Int. J. Greenhouse Gas Control*, **2007**, *1*, 396.

(2) Lin, K.; Park, A. *Environ. Sci. Technol.*, **2011**, *45*, 6633.

(3) (a) Wang, C.; Mahurin, S. M.; Luo, H.; Baker, G. A.; Li H.; Dai, S. *Green Chem.*, **2010**, *12*, 870; (b) Bates, E. D.; Mayton, R. D.; Ntai, I.; Davis, J. H. J. *J. Am. Chem. Soc.*, **2002**, *124*, 926.

(4) Ramdin, M.; de Loos, T. W.; Vlugt, T. J. H. *Ind. Eng. Chem. Res.*, **2012**, *51*, 8149.

(5) Perry, R.; Wood, B.; Genovese, S.; O'Brien, M.; Westendorf, T.; Meketa, M.; Farnum, R.; McDermott, J.; Sultanova, I.; Perry, T.; Vipperla, R.; Wichmann, L.; Enick, R.; Hong, L.; Tapriyal, D. *Energy Fuels*, **2012**, *26*, 2528.

(6) Koech, P. K.; Zhang, J.; Kutnyakov, I. V.; Cosimbescu, L.; Lee, S.-J.; Bowden, M. E.; Smurthwaite, T. D.; Heldebrant, D. J.; *RSC Adv.* **2013**, *3*, 566.

(7) (a) Heldebrant D. J.; Koech, P. K.; Ang, M. T. C.; Liang, C.; Rainbolt, J. E.; Yonker, C. R.; Jessop, P. G. *Green Chem.* **2010**, DOI: 10.1039/B924790D. (b) Heldebrant D. J.; Koech, P. K.; Yonker, C. R. *Energy Environ. Sci.* **2010**, *3*, 111.

(8) (a) Mathias, P. M.; Afshar, K.; Zheng, F.; Bearden, M. D.; Freeman, C. J.; Andrea, T.; Koech, P. K.; Kutnyakov, I.; Zwoster, A.; Smith, A. R. *Energy Environ. Sci.*, **2013**, *6*, 2233. (b) Zheng, F.; Heldebrant D. J.; Mathias, P. M.; Koech, P. K.; Bhakta, M.; Freeman, C. J.; Zwoster, A. *Energy Fuels*, **2016**, *30*, 1192.

NOVEL LIQUID-LIKE NANOPARTICLE ORGANIC HYBRID MATERIALS (NOHMS) BASED ELECTROLYTE FOR COMBINED CO_2 CAPTURE AND CONVERSION

Ming Gao and Ah-Hyung Alissa Park

Department of Earth and Environmental Engineering,
Department of Chemical Engineering
Lenfest Center for Sustainable Energy
Columbia University in the City of New York
ap2622@columbia.edu

Abstract

A new advancement in the field of carbon capture and storage involves utilization of the captured CO_2 as a chemical feedstock. While a lot of research have demonstrated excellent carbon capture materials and carbon conversion methods, respectively, a medium capable of combined carbon capture and conversion is still in demand. Thus, a novel Nanoparticle Organic Hybrid Materials (NOHMs) based electrolyte is developed. NOHMs consist of inorganic cores with a grafted polymeric canopy, with negligible vapor pressure, great thermal stability, and interesting chemical and physical tunability. NOHMs have exhibited good carbon capture capacity thanks to the entropic and enthalpic interactions between NOHMs and CO_2, as well as good regeneration ability. In this study, it has been found that NOHMs not only have interesting CO_2 capture properties but can also serve as an electrolyte for subsequent reduction of CO_2. Thus, NOHMs' electrolytic properties in the presence of various secondary fluids and salts are evaluated to design them as a dual-purpose reactive media for combined CO_2 capture and conversion. Preliminary data have shown interesting electrochemical properties of NOHMs based electrolyte, and Small-angle X-ray Scattering (SAXS) has also revealed unique fractal structure within NOHMs based electrolyte, which is crucial for understanding the ion transport and CO_2 diffusion pathway. An integrated process of carbon capture and conversion is developed based on NOHMs electrolyte under the carbon capture and utilization scheme. This novel approach has a great potential to minimize the parasitic energy consumption and offers a new perspective for carbon management.

Introduction

Carbon dioxide (CO_2) is a dominant anthropogenic greenhouse gas and there is a focus on limiting and further reducing its atmospheric concentration to prevent climate change. Many proposals have a goal of limiting the concentration to 500±50 ppm (less than double the preindustrial concentration of 280 ppm)[1], and meanwhile the atmospheric concentration is increasing at an annual rate of about 2 ppm[2]. Carbon management and the development of efficient CO_2 capture and utilization technologies are solutions that can effectively lower the atmospheric CO_2 concentrations given the existing carbon-based energy infrastructure. Amine scrubbing is a common technology employed for CO_2 capture. It utilizes amine-based solvents such as monoethanolamine (MEA) to capture post-combustion CO_2[3–5]. MEA has a high CO_2 capture capacity and fast reaction kinetics, but it also has a high vapor pressure that can result in corrosion and material loss. Furthermore, the concentration of MEA is limited to 15-30% in water and it is energy intensive to regenerate CO_2[5]. Both inorganic and organic materials as alternative materials to MEA for CO_2 capture have also been studied including ionic liquids[6], organic solvents[7], amine functionalized solid mesoporous sorbents[8–10], ion exchange resin[11] and alkaline solutions[12].

Enlightened by study on ionic liquid[13–22], a class of liquid-like organic-inorganic hybrids called nanoscale ionic materials (NIMs) was developed by Archer et al[23]. These organic-inorganic hybrid nanoparticles were comprised of solid, inorganic nanoparticle cores functionalized with charged oligomers to form coronas that oppositely charged polymers were tethered to in order to form canopies. Such NIMs was further developed and classified by Park et al. as referred to Nanoparticle Organic Hybrid Materials (NOHMs). Such liquid-like materials have shown near zero vapor pressure, high thermal stability, chemical tunability and good capture capacity[2,24]. The CO_2 capture mechanism via NOHMs were also extensively studied by Park et al. in that both enthalpy and entropy contributed to carbon capture and the induced structural change was revealed by different techniques, such as NMR, Raman and FTIR[24,25].

As utilization was incorporated into the Carbon Capture, Utilization and Storage (CCUS) scheme, electrochemical conversion of CO_2 into valuable products became more attractive to researchers. Electrochemical reduction of CO_2 has been studied extensively with respect to types of electrodes, catalysts, electrolytes, operating voltages and so forth[26–30]. However, these studies used CO_2 gas stream as the feedstock, and the regeneration of CO_2 from the capture medium would bring about additional energy penalty if combined with carbon capture scheme. Therefore, a call for a combined carbon capture and conversion process was established in that the captured CO_2 could be converted into valuable products within the same medium via electrochemical methods, to avoid the energy cost associated with CO_2 regeneration.

NOHMs have been considered a candidate for such combined carbon capture and conversion. Archer et al. mixed various kinds of NOHMs with salt and reported promising electrochemical properties in terms of conductivity, electrochemical stability and other aspects[3132]. A Small-angle X-ray Scattering (SAXS) was also conducted by Archer et al. to have an insightful view on how the structure affected the NOHMs poly-dispersity[33]. However, the system by Archer et al. was pure NOHMs, which were too viscous for application purpose. Thus, in this work, some secondary fluids were mixed with NOHMs and salt, reducing the viscosity of the system while partially retaining the CO_2 capture ability. Electrochemical properties were tested for such NOHMs electrolyte and supplemented by SAXS characterization.

Experimental

Design and synthesis of NOHMs. Nanoparticle Organic Hybrid Materials (NOHMs) were designed and synthesized via a previously reported method by Park et al[2]. Two types of NOHMs were selected for investigation --- NOHM-I-PEI and NOHM-I-HPE. For synthesis of NOHM-I-PEI, silica nanoparticle suspension (Sigma-Aldrich) was protonated after running through an ion exchange resin column. Afterwards, the protonated silica nanoparticle suspension was fully reacted with polyethylenimine (PEI) (Polysciences, Inc.). The neat NOHM-I-PEI sample was obtained after drying under vacuum overnight. The synthesis of NOHM-I-HPE involved one more step of linker grafting. Silica nanoparticle suspension was thoroughly reacted with 3-(trihydroxysilyl)-1-propane-sulfonic acid (Gelest, Inc.) or the linker, after pH adjustment. Dialysis was performed to completely remove the unreacted linker. The resulted solution was also protonated via ion exchange resin and further titrated by Jeffamine M-2070 (HPE) (Huntsman) for complete tethering. The pure NOHM-I-HPE was good to use after vacuum drying overnight.

Design and synthesis of NOHMs electrolyte. In addressing the combined carbon capture and conversion demand, a NOHMs based electrolyte was designed and synthesized. NOHMs and a secondary fluid, such as water and polyethylene glycol methyl ether (mPEG)

(M_n=550, Sigma-Aldrich), were mixed at a certain weight ratio. A 1M bis(trifluoromethane)sulfonimide lithium salt (LITFSL) (Sigma-Aldrich), considering its high conductivity, was also added to ensure a certain conductivity of the electrolyte. The resulting electrolyte was stored under inert environment and fully mixed with stirring for 3h before use.

Conductivity. A Princeton Applied Research PARSTAT MC potentiostat was used. Electrochemical Impedance Spectroscopy (EIS) was employed with frequency range from 100 kHz to 0.1 Hz. Working and counter electrodes were platinum disk and platinum wire, respectively. The reference electrode was Ag|AgCl. Conductivity was calculated from impedance data[31,34].

Small-angle X-ray Scattering (SAXS). Small-angle X-ray Scattering were performed at station 12-BM in Advanced Photon Source (APS) in Argonne National Laboratory. NOHMs based electrolyte were placed in a customized cell with glass windows holding the sample. The samples were measured at multiple spots to ensure uniformity.

Results and Discussion

The conductivity of NOHMs based electrolyte was higher than that of the corresponding polymer based electrolyte. And the conductivities for the two kinds of NOHMs based electrolyte varied. These were hypothesized to be due to the structural configurations of NOHMs based electrolyte. The different chemistry between NOHMs polymer and secondary fluids would induce different ion transport pathways, and different CO_2 diffusion routes for subsequent carbon conversion. This was indicated by the SAXS plot in figure 1. Different scattering behaviors of NOHM-I-PEI and NOHM-I-HPE based electrolytes implied different structural configurations within the electrolyte and supported the conductivity difference. The understanding of the dependence of electrochemical properties on the electrolyte structures would help better design the electrolyte for better electrochemical performances and potential carbon conversion within the same medium.

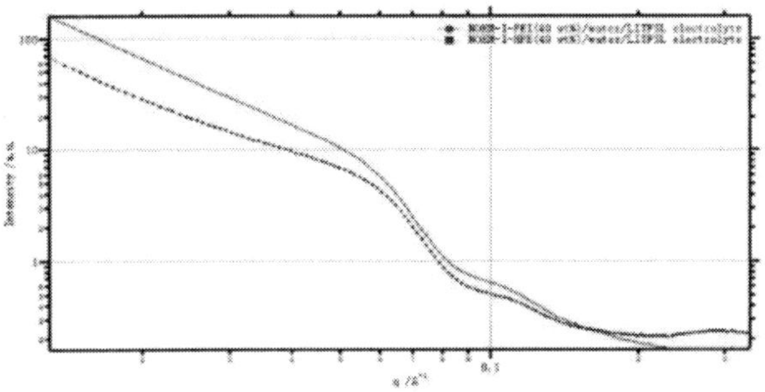

Figure 1. Small-angle X-ray Scattering (SAXS) of NOHM-I-PEI (40 wt%) and NOHM-I-HPE(40 wt%) in the presence of mPEG and salt

Conclusions

Nanoparticle Organic Hybrid Materials (NOHMs) have exhibited excellent carbon capture ability[2], and after transformed into an electrolyte, it demonstrated interesting electrochemical properties. The unique structures of NOHMs based electrolyte contributed to an improved conductivity compared to the polymer based electrolyte. In addition, the different electrochemical performances of NOHM-I-PEI and NOHM-I-HPE based electrolyte indicated different chemistry of tethered polymers with secondary fluids. This was preliminarily investigated via Small-angle X-ray Scattering (SAXS). Future work is needed to understand the structure and chemistry of NOHMs based electrolyte so that the tuning of ion transport and CO_2 diffusion pathway could better facilitate the combined carbon capture and conversion application via NOHMs based electrolyte.

References

(1) Pacala, S.; Socolow, R. *Science* **2004**, *305*, 968–972.
(2) Lin, K.; Park, A. *Environ. Sci. Technol.* **2011**, *45*, 6633–6639.
(3) Rochelle, G. T. *Science* **2009**, *325* (5948), 1652–1654.
(4) Rao, A. B.; Rubin, E. S. *Environ. Sci. Technol.* **2002**, *36*, 4467–4475.
(5) da Silva, E. F.; Svendsen, H. F. *Ind. Eng. Chem. Res.* **2004**, *43*, 3413–3418.
(6) Anthony, J. L.; Anderson, J. L.; Maginn, E. J.; Brennecke, J. F. *J. Phys. Chem. B* **2005**, *109* (13), 6366–6374.
(7) Heldebrant, D. J.; Yonker, C. R.; Jessop, P. G.; Phan, L. *Energy & Environmental Science*. 2008, pp 487–493.
(8) Choi, S.; Gray, M. L.; Jones, C. W. *ChemSusChem* **2011**, *4* (5), 628–635.
(9) Bollini, P.; Brunelli, N. A.; Didas, S. A.; Jones, C. W. *Ind. Eng. Chem. Res.* **2012**, *51*, 15153–15162.
(10) Harlick, P. J. E.; Sayari, A. **2007**, 446–458.
(11) Wang, T.; Lackner, K. S.; Wright, A. *Environ. Sci. Technol.* **2011**, *45* (15), 6670–6675.
(12) Stolaroff, J. K.; Keith, D. W.; Lowry, G. V. *Environ. Sci. Technol.* **2008**, *42* (8), 2728–2735.
(13) Zhang, X.; Zhang, X.; Dong, H.; Zhao, Z.; Zhang, S.; Huang, Y. *Energy Environ. Sci.* **2012**, *5* (5), 6668.
(14) Karadas, F.; Atilhan, M.; Aparicio, S. *Energy & Fuels* **2010**, *24* (11), 5817–5828.
(15) Bates, E. D.; Mayton, R. D.; Ntai, I.; Davis, J. H. *J. Am. Chem. Soc.* **2002**, *124* (6), 926–927.
(16) Zhang, J.; Zhang, S.; Dong, K.; Zhang, Y.; Shen, Y.; Lv, X. *Chemistry* **2006**, *12* (15), 4021–4026.
(17) Feng, Z.; Cheng-Gang, F.; You-Ting, W.; Yuan-Tao, W.; Ai-Min, L.; Zhi-Bing, Z. *Chem. Eng. J.* **2010**, *160* (2), 691–697.
(18) Scovazzo, P.; Kieft, J.; Finan, D.; Koval, C.; Dubois, D.; Noble, R. *J. Memb. Sci.* **2004**, *238* (1-2), 57–63.
(19) Cadena, C.; Anthony, J. L.; Shah, J. K.; Morrow, T. I.; Brennecke, J. F.; Maginn, E. J. *J. Am. Chem. Soc.* **2004**, *126* (16), 5300–5308.
(20) Gutowski, K. E.; Maginn, E. J. *J. Am. Chem. Soc.* **2008**, *130* (44), 14690–14704.
(21) Yu, G.; Zhang, S. *Fluid Phase Equilib.* **2007**, *255* (1), 86–92.
(22) Gurkan, B.; Goodrich, B. F.; Mindrup, E. M.; Ficke, L. E.; Massel, M.; Seo, S.; Senftle, T. P.; Wu, H.; Glaser, M. F.; Shah, J. K.; Maginn, E. J.; Brennecke, J. F.; Schneider, W. F. *J. Phys. Chem. Lett.* **2010**, *1* (24), 3494–3499.
(23) Rodriguez, R.; Herrer, R.; Bourlinos, A. B.; Li, R.; Amassian, A.; Archer, L. A.; Giannelis, E. P. In *Applied Organometallic Chemistry*; 2010; Vol. 24, pp 581–589.
(24) Park, Y.; Decatur, J.; Lin, K.-Y. A.; Park, A.-H. A. *Phys. Chem. Chem. Phys.* **2011**, *13* (40), 18115–18122.
(25) Petit, C.; Park, Y.; Lin, K. A.; Park, A. A. **2011**.
(26) Hori, Y.; Murata, A.; Takahashi, R. *J. Chem. Soc. Faraday Trans. 1 Phys. Chem. Condens. Phases* **1989**, *85* (8), 2309.
(27) Hori, Y.; Wakebe, H.; Tsukamoto, T.; Koga, O. *Electrochim. Acta* **1994**, *39* (11-12), 1833–1839.
(28) Lu, Q.; Rosen, J.; Zhou, Y.; Hutchings, G. S.; Kimmel, Y. C.; Chen, J. G.; Jiao, F. *Nat. Commun.* **2014**, *5*, 3242.
(29) Morris, A. J.; McGibbon, R. T.; Bocarsly, A. B. *ChemSusChem* **2011**, *4* (2), 191–196.
(30) Whipple, D. T.; Finke, E. C.; Kenis, P. J. A. *Electrochem. Solid-State Lett.* **2010**, *13* (9), B109.
(31) Nugent, J. L.; Moganty, S. S.; Archer, L. a. *Adv. Mater.* **2010**, *22* (33), 3677–3680.
(32) Lynden A. Archer; Laura Lynne Olenick; Jennifer Lyn Schaefer; Alexandra Elena Corona. Nanoparticle Organic Hybrid Materials (NOHMS), February 16, 2012.
(33) Yu, H.-Y.; Srivastava, S.; Archer, L. a; Koch, D. L. *Soft Matter* **2014**, *10* (45), 9120–9135.
(34) Jacobs, P. W. M.; Lorimer, J. W.; Russer, a.; Wasiucionek, M. *J. Power Sources* **1989**, *26* (3-4), 503–510.

NOVEL CATALYTIC MATERIALS FOR THE ACTIVATION OF CO_2: TUNNING ACTIVITY AND SELECTIVITY AT THE METAL-OXIDE AND METAL-CARBIDE INTERFACES

José A. Rodriguez,

Chemistry Department
Brookhaven National Laboratory
Upton, NY 11973, USA

Introduction

This talk will discuss new approaches for the preparation of highly active catalysts for the activation of CO_2 and its conversion to CO or methanol. Methanol is a key commodity used to produce acetic acid, formaldehyde, and a number of key chemical intermediates.[1] It is synthesized industrially from mixtures of H_2, CO_2, and CO at elevated pressures (50 to 100 atm) and temperatures (450 to 600K) with a $Cu/ZnO/Al_2O_3$ catalyst.[2,3] Of particular interest is the synthesis of methanol from CO_2[2,4] not only as a way to mitigate this greenhouse gas but also because of the potential use of CO_2 as an alternative and economical feedstock.[2,4,5]

The hydrogenation of CO_2 to methanol and water

$$CO_2 + 3H_2 \rightarrow CH_3OH + H_2O \qquad (1)$$

is an exothermic reaction (ΔH= - 11.8 kcal/mol at 25 °C), but it is a challenging process due to the high stability and low reactivity of the CO_2 molecule. The splitting of a C-O bond in CO_2 is characterized by a large energy barrier. Thus, the effective activation of CO_2 is a requirement for improving the reaction kinetics for methanol synthesis.

Metal-oxide and metal-carbide interfaces can be manipulated yield highly active catalysts for the conversion of CO_2. In recent studies, we have found that the deposition of Cu on MoC(001) and $CeO_x/TiO_2(110)$ surfaces generates catalysts for the $CO_2 \rightarrow CH_3OH$ conversion which are highly active and much more efficient than model $Cu/ZnO(000\bar{1})$ catalysts [5,6,7] This is a consequence of the formation of Cu-carbide and Cu-oxide interfaces which bind well the CO_2 molecule.[5,6,7]

This talk will focus on three areas: (i) *in-situ* studies to determine catalyst structure, oxidation state and chemistry under reaction conditions; (ii) studies of relevant model systems, primarily based on nanoparticles supported on single crystal substrates; and (iii) computational modeling. By closely linking these three research thrusts, our goal is to provide strongly validated mechanistic and structural conclusions for the future design and optimization of nanostructured catalysts for achieving the conversion of CO_2 to CO and/or CH_3OH conversion.

Experimental

The photoemission studies for the model catalysts were done at beamlines U7A and U12 of the National Synchrotron Light Source (NSLS). The microscopy studies were carried out in an Omicron variable temperature STM system. Additional characterization studies were carried out in a system which combines a batch reactor and a ultrahigh vacuum (UHV) chamber.[6,7] The UHV chamber (base pressure ~ 1 x 10^{-10} Torr) was equipped with instrumentation for X-ray photoelectron spectroscopy (XPS), low-energy electron diffraction (LEED), ion-scattering spectroscopy (ISS), and thermal-desorption mass spectroscopy (TDS). The sample could be transferred between the UHV chamber and reactor without exposure to air. Typically, it was transferred to the batch reactor at room temperature, the reactant gases were introduced (0.5 atm of CO_2 and 4.5 atm of H_2), and then the sample was rapidly heated to the reaction temperature (500, 525, 550, 575 or 600 K). Product yields were analyzed by a gas chromatograph.[6,7] The amount of molecules produced was normalized by the active area exposed by the sample.

Results and Discussion

A. A novel oxide support : Properties of $CeO_x/TiO_2(110)$ surfaces. Mixed-metal oxides play a very important role in many areas of chemistry, physics, materials science and geochemistry. At BNL, STM, photoemission, and density-functional calculations were used to study the behavior of ceria nanoparticles deposited on a $TiO_2(110)$ surface.[8] The titania substrate imposed non-typical coordination modes on the ceria nano-particles. The results of scanning tunneling microscopy show the formation of Ce_2O_3 dimers which form small linear arrays on top of the $TiO_2(110)$ substrate (see Figure 1). An increase in the stability of the Ce^{3+} oxidation state leads to an enhancement in the chemical and catalytic activity of the ceria nanoparticles.[8,9] As, we will see below $CeOx/TiO_2(110)$ is an excellent support for the dispersion of copper and the generation of a highly active catalyst for the conversion of CO_2 to methanol.

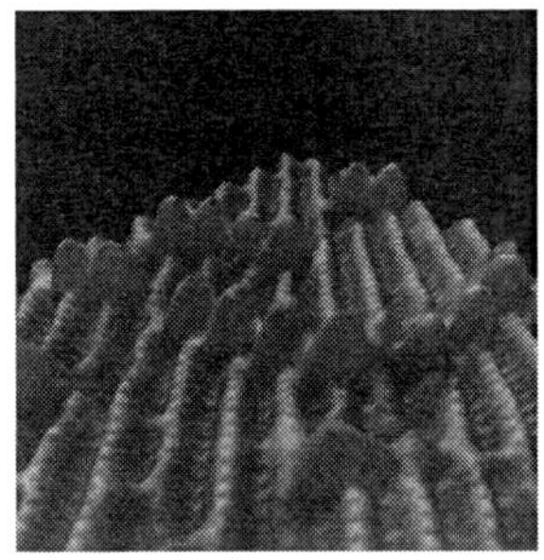

Figure 1. STM image for the deposition of CeO_x on $TiO_2(110)$[6]

B. Conversion of CO_2 on metal/oxide catalysts. The activity for methanol synthesis of a series of catalysts which contain copper promoted con ZnO and CeOx is compared in Figure 2.[7,9] Extended surfaces of pure copper, like Cu(111), bind the CO_2 molecular poorly and exhibit a very low activity for the hydrogenation of CO_2 to methanol. A drastic increase in the catalytic activity is observed in Figure 2 after depositing nanoparticles of ceria on Cu(111)[9] or after

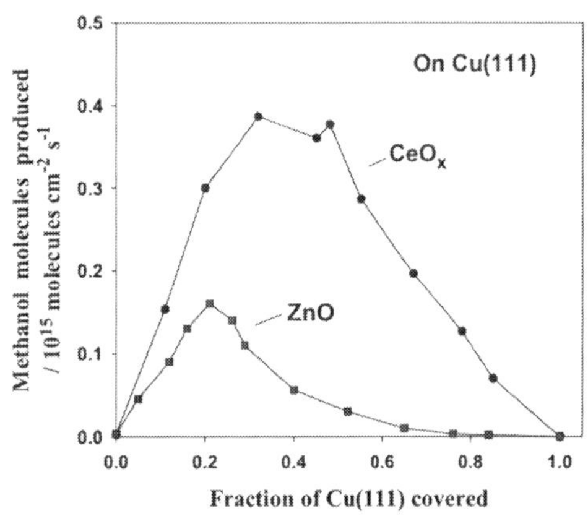

Figure 2. Activity of Cu(111) for the synthesis of methanol from CO_2 after the deposition of ZnO_x and CeO_x.[9]

co-depositing nanoparticles of Cu and ceria on $TiO_2(110)$.[7] The combination of metal and oxide sites in the copper-ceria interface affords complementary chemical properties that lead to special reaction pathways for the $CO_2 \rightarrow CH_3OH$ conversion.[7] Thus, the rate of methanol production on $CeO_x/Cu(111)$ is ~ 200 times faster than on plain Cu(111). Furthermore, the rate of methanol production on $Cu/CeO_x/TiO_2(110)$, where Cu and ceria nanoparticles are present

(Figure 3), is ~ 1280 times faster than on Cu(111). CeO_x/Cu(111) and Cu/CeO_x/TiO_2(110) are also much better catalysts than Cu/ZnO(000$\bar{1}$).[7,9] The results of ambient-pressure X-ray photoelectron and infrared spectroscopies, plus density-function calculations, point to a reaction mechanism that involves the initial formation of a OCOH species which can yield CO or HCO (formyl), H_2CO (formaldehyde), H_3CO (methoxy) and methanol upon hydrogenation.[7,9] The calculated energy barriers for these processes were all lower than 14 kcal/mol and the reaction intermediates were simultaneously bonded to Cu and ceria sites at the metal-oxide interface.[7]

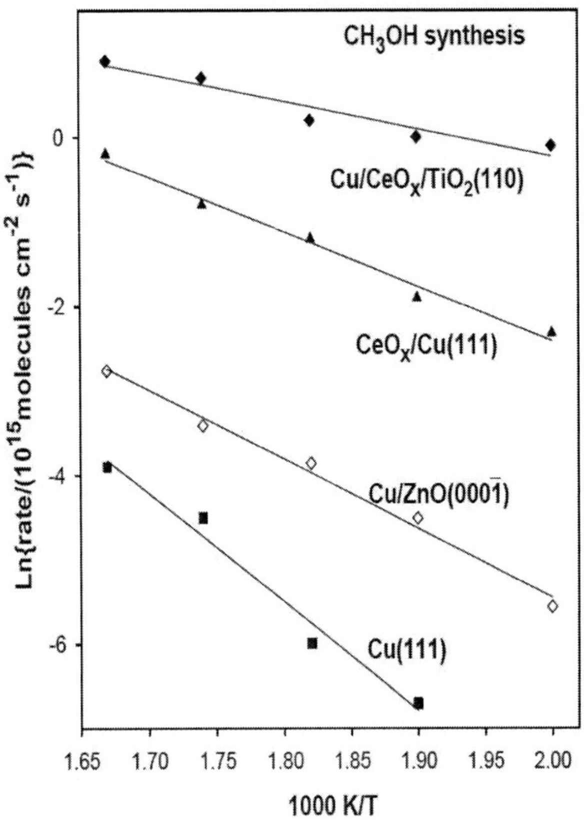

Figure 3 Arrhenius plots for the rates of methanol production from CO_2 hydrogenation on copper-oxide catalysts. P_{CO2}= 0.5 atm, P_{H2}=4.5 atm.[7]

C. A metal carbide support : Properties of Cu/TiC(001) and Cu/MoC(001) surfaces. Two-dimensional Cu clusters in contact with TiC(001) or MoC(001) undergo a significant polarization of electrons that enhances their reactivity towards CO_2.[5,10]

D. Conversion of CO_2 on metal/carbide catalysts. Excellent performance for the $CO_2 \rightarrow CH_3OH$ conversion was also observed after depositing Cu on TiC(001) or a MoC surface, Figure 4.[5,6,10] Cu adatoms bind the CO_2 molecule very well and a OCOH species is again a key intermediate in the production of methanol.[5,6]

Neither Cu/TiC(001) nor Cu/MoC exhibited problems of stability.[5,6,10] On these catalysts there was no full decomposition of CO_2 and deposition of oxygen on the carbide surface. They were highly selective towards the formation of CO and methanol.[5,6,10] On the other hand the deposition of Cu on Mo_2C(001) produced a catalysts that fully decomposed CO_2 and the deposited C atoms were hydrogenated yielding methane.[10]

Acknowledgement. This work was financed by the US Department of Energy (US-DOE), Chemical Sciences Division (DE-SC0012704). The National Synchrotron Light Source is supported by the Divisions of Materials and Chemical Sciences of US-DOE.

References

(1) Olah, G.A., Goeppert, A., and Surya Prakash, G.K. *Beyond Oil and Gas: The Methanol Economy* . Wiley-VCH, Weinheim, Germany, 2011.
(2) Behrens M. *et al, Science*, **2012**, *336*, 893 .
(3) Spenser, M.S. *Top. Catal.* **1999**, *8*, 259 .
(4) Aresta, M. (editor), *Carbon Dioxide as Chemical Feedstock*. Wiley, New York, 2010.
(5) Vidal, A.B. *et al, J. Phys. Chem. Lett.* **2012**, *3*, 2275.
(6) Rodriguez, J.A. *et al, J. Catal.* **2013**, *307*, 162.
(7) Graciani, J. *et al, Science*, **2014**, *345*, 546.
(8) Park, J.B., et al., *Proceedings of the National Academy of Sciences (PNAS)*, **2009**. *106*, 4975-4980.
(9) Senanayake et al, J. Phys. Chem. C, **2016**, *120*, 1778-1784.
(10) Posada et al, *Catal. Sci. & Technol.* **2016**, in press.

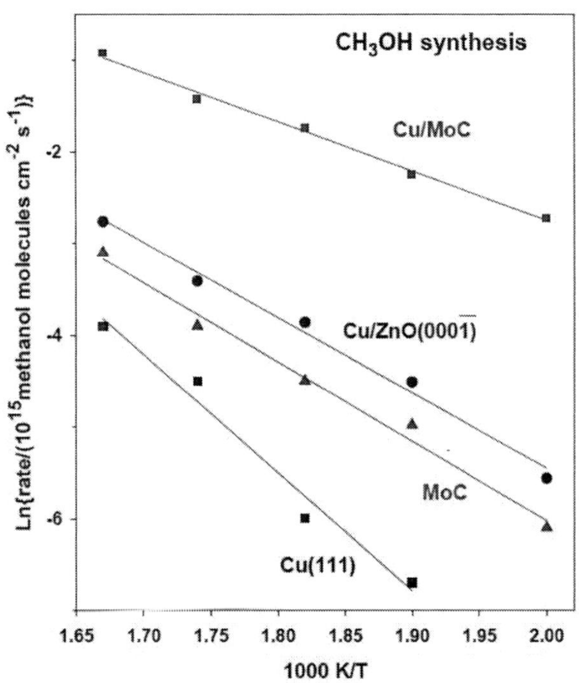

Figure 4 Arrhenius plots for the rates of methanol production from CO_2 hydrogenation on Cu(111), copper-carbide and copper-oxide catalysts. P_{CO2}= 0.5 atm, P_{H2}=4.5 atm.[10]

EFFECT OF OPERATING PARAMETERS IN A HIGH RATE ACTIVATED SLUDGE A-STAGE SYSTEM ON CARBON CAPTURE FOR METHANE PRODUCTION

Maureen N. Kinyua[1], Warner Thomas[2], Matt Elliott[2], Bernhard Wett[3], Sudhir Murthy[4], Kartik Chandran[1], Charles B. Bott[5]*

[1]Earth and Environmental Engineering, Columbia University, New York, NY, [2]Environmental Engineering, Old Dominion University, Norfolk, VA, [3]ARA Consult GmbH, Unterbergerstraße 1, Innsbruck, Austria, [4]DC Water Authority, Washington DC, [5]Hampton Roads Sanitation District, Virginia Beach, VA
*Corresponding Author: mk3855@columbia.edu

INTRODUCTION

The adsorption high rate activated sludge (HRAS) A-stage process, is operated at a hydraulic retention time (HRT) of 30 minutes, solids retention time (SRT) between 0.2 and 1 day and dissolved oxygen (DO) concentration below 1 mg/L. This aggressive operation results in minimal particulate (pCOD) and colloidal (cCOD) carbon oxidation, and that material can be captured/bio-flocculated and directed to an energy recovery system such as anaerobic digestion.

The organic portion of the waste activated sludge (WAS) proteins, carbohydrates and lipids, which can be classified as intra- and extracellular carbon fractions. The intracellular carbon fractions include polyhydroxyalkanoates (PHA), phospholipids, glycogen, cytoplasmic proteins and polysaccharides. Operating parameters (DO, HRT and SRT) have been shown to impact the production of intra- and extracellular carbon fractions which affects carbon capture and subsequent energy recovery (1, 2).

Previous researchers have investigated the effect of operating parameters on PHA, glycogen and lipid accumulation in fully-aerobic A-stage processes (aerobic storage not associated with anaerobic/aerobic poly-P accumulating and glycogen accumulating organisms) (1,3). However, to the best of our knowledge, there has been no previous study that has examined the effect of the aggressive operating parameters in a HRAS A-stage process on intracellular carbon fractions. Therefore, the objective of this research is to quantify the effect of DO, SRT and HRT on PHA, glycogen, lipid, and protein concentration and subsequent methane production potential. The rationale is that quantifying these simultaneous effects may enable us to identify a favorable operating condition for A-stage at which the WAS produced can achieve maximum methane yield, while the A-stage COD removal is maximized or alternatively, the A-stage effluent provides a desired carbon to nitrogen ratio for the following biological nitrogen removal system.

MATERIALS AND METHODS

Pilot A-stage: The HRAS A-stage pilot operated at 20°C consisted of three bioreactors in series followed by a separate clarifier. The bioreactors had a total volume of 511 L with a side water depth of 3.4 m. The bioreactors are currently being operated at a HRT of 30 and 60 minutes and DO concentration of 0.5, 1.0 and 1.5 mg/L. The SRT is maintained between 0.1 - 0.5 days.

Analytical methods: Influent and effluent total suspended solids (TSS), volatile suspended solids (VSS), and COD were monitored daily according to Standard Methods (4).Cells were lysed using Thermo Scientific B-PER Bacterial Protein Extraction Reagent (Grand Island, NY) and analyzed for total proteins (5). Polysaccharides were extracted by boiling a sludge pellet in 30% KOH for 30 minutes and analyzed for total polysaccharides using a modified anthrone method. The intracellular proteins and polysaccharides were then calculated by subtracting the LB-EPS and TB-EPS components from the total fraction. It should be noted that the definition of intracellular proteins and polysaccharides in this study is based solely on the extraction and lysis methods applied to the sludge. Poly-beta-hydroxybutyrate (PHB) and poly-beta-hydroxyvalerate (PHV) were analyzed according to Braunegg (6). Glycogen was determined according to Roe and Dailey (7). Detailed PHA and glycogen methods are provided in the methods document.

RESULTS AND DISCUSSION

Overall A-stage Performance. Average influent and effluent concentrations comparing A-stage HRAS operated at 0.3 $\pm$ 0.1 day SRT (HRT = 30 minutes) and a SRT of 0.5 $\pm$ 0.2 day (HRT = 60 minutes) receiving the same RWI at a DO set point of 0.5 mg/L over a 3 week period are shown in Table 1 and the average removals are reported in Figure 1. From these results, two conclusions can be made in regards to how the SRT impacts the performance of the A-stage HRAS process.

The removal of TSS and all COD fractions increase with increasing SRT (Figure 1) suggesting that the bacteria have more time to consume sCOD for cell growth as well as increase the extent of bioflocculation resulting in more pCOD and cCOD removal. These values are similar to those reported by Jimenez et al. (2) who reported an increase in percent removals from approximately 63% to 75%, 15% to 22% and 15% to 35% for sCOD, cCOD and pCOD operated at 0.3 and 0.5 day SRT, respectively. The sCOD removals in this study were similar to those of Jimenez et al. (2015) whereas pCOD and cCOD removals were much higher than their study. This could be attributed to the fact that their study maintained a DO concentration of 1.0 mg/L. Only TSS, tCOD and pCOD removal were found to be significantly different ($p<0.05$) between the two operating conditions.

The second conclusion was that SRT had an influence on the settleability of the activated sludge. The SVI values increased from 75.6 $\pm$ 9.4 mL/gTSS to 147.1 $\pm$ 45.1 mL/gTSS when operated at a SRT of 0.3 $\pm$ 0.1 days and 0.5 $\pm$ 0.2 days, respectively. This suggests that increasing the SRT (increased bioflocculation) resulted in larger flocs entering the intermediate clarifier capturing more solids settling with a sweep floc action. This led to decreased ESS value at a higher SRT. Previous literature has suggested that systems with shorter retention times are bioflocculation limited due to less extracellular polymeric substances (EPS) production (8).

Table 1: A-stage HRAS Average Influent and Effluent Concentrations Over 3 Weeks (n=22) Operated at SRT'S of 0.3 ± 0.1 and 0.5 ± 0.2 Days (HRT = 30 and 60 minutes, respectively).

		Influent	Effluent	
			SRT = 0.3 ± 0.1 d	SRT = 0.5 ± 0.2 d
Sample	Units	Average	Average	Average
C/N		17.8 ± 1.9	8.7 ± 1.3	7.0 ± 1.1
tCOD	mg/L	575.1 ± 67.0	228.6 ± 36.1	175.1 ± 24.0
sCOD	mg/L	149.3 ± 17.9	48.7 ± 4.6	43.1 ± 11.6
pCOD	mg/L	345.3 ± 43.0	122.7 ± 28.5	88.7 ± 20.9
cCOD	mg/L	85.0 ± 19.6	61.3 ± 16.8	50.5 ± 7.8
TSS	mg/L	210.8 ± 26.4	85.6 ± 15.2	65.8 ± 14.4
VSS	mg/L	196.9 ± 29.0	78.9 ± 9.6	61.1 ± 13.2
ESS	mg/L	-	69.3 ± 27.2	51.8 ± 12.4
Turbidity	NTU	-	64.1 ± 23.8	44.0 ± 11.8
SVI	mL/g	-	75.6 ± 9.4	147.1 ± 45.1

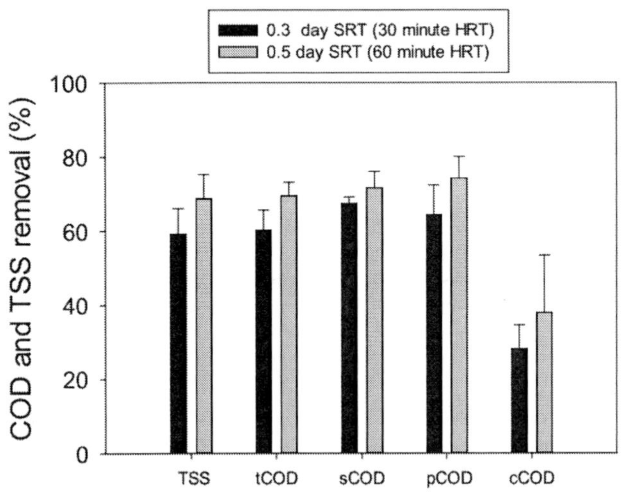

Figure 1. Average removal performance of A-stage HRAS operated at SRT's of 0.3 ± 0.1 and 0.5 ± 0.2 days over 3 weeks (n=22)

Storage Compounds Production and Methane Yield Potential. A summary of intracellular storage products at DO = 0.5 mg/L and SRT of 0.3 day (30 minutes HRT) and 0.5 day (60 minute HRT) in the pilot A-stage are shown on Figure 2. From these results, three conclusions can be drawn. First, the A-stage process was able to achieve PHA and glycogen accumulation similar to anaerobic-aerobic systems. In the raw wastewater, both acetate and propionate were present at concentrations of 54.5 ± 5.7 mg COD/L and 14.1 ± 1.6 mg COD/L, respectively. The lower PHV content compared to PHB could be due to the lower concentration of propionate in the influent compared to acetate. However, the mechanisms resulting in microbial selection to uptake acetate or propionate are still unclear. Jiang et al. (9) hypothesized that specific microbial transport pathways may play a role in the selection of substrate to store. Although glycogen was stored, the glycogen concentrations were lower than those reported by Carta et al. (10). The low glycogen concentration could be due to two reasons: one is that the influent wastewater had low polysaccharides concentrations (19.4 mg/L) and two; microbes may store more glycogen when exposed to anaerobic/aerobic conditions.

The second main conclusion was that the 0.30 day SRT (30 minute HRT) resulted in better production of intracellular storage products compared to the 0.50 day SRT (60 minute HRT). The 0.50 day SRT resulted in increased sCOD removal indicating that by increasing the SRT, the microbes oxidized more of the readily biodegradable COD and used it for growth instead of storing it in the cell. To further evaluate the differences in the metabolism/respiration rate of the activated sludge from the two SRT reactors, a respirometric oxygen uptake rate (OUR) test was carried out for four hours using raw influent as the carbon source. These results are shown on Figure 3. The OUR results indicate that the differences in the respiration rate between the sludge from the two SRTs was not statistically significant (p = 0.29). Although the OUR tests did not show differences in the respiration rate, this approach may be relevant in understanding storage yields and substrate consumption.

Lastly, the methane production potential of the WAS is affected by the substrate characteristics (COD fractions) and the hydrolysis rate. A comparison of 0.3 day SRT (30 minute HRT) and 0.5 day SRT (60 minute HRT) ML and WAS intracellular protein and polysaccharide concentrations is shown on Figure 4. Based on stoichiometry, 1g of polysaccharides, proteins, lipids and PHAs can produce 0.44 L, 0.55 L, 0.75 L and 0.65 L of CH_4, respectively (11). Therefore, the higher the concentrations of these COD fractions in the WAS, the higher the methane production potential. An estimated cumulative methane yield of 322.7 and 181.6 L CH_4/kg VS was calculated from the COD fractions for the 0.30 and 0.50 day SRT, respectively from Figure 4. These results are within range to those reported by Wang et al. (12) who reported a potential methane yield of 192 L CH_4/kg VS from anaerobic-aerobic WAS. The shorter SRT resulted in a higher potential yield due to an increase the carbon that is captured and can be redirected for energy production in an anaerobic digester. However, the hydrolysis rate of each of these COD fractions has to be considered to provide more accurate predictions of methane yield. In addition, from Figure 4, it is clear that there is microbial activity in the clarifier due to the release of intracellular carbon fractions. The clarifier receiving sludge from the 0.3 day SRT had greater release of these fractions compared to the 0.5 day SRT. More research needs to be done to evaluate the mechanisms leading to this.

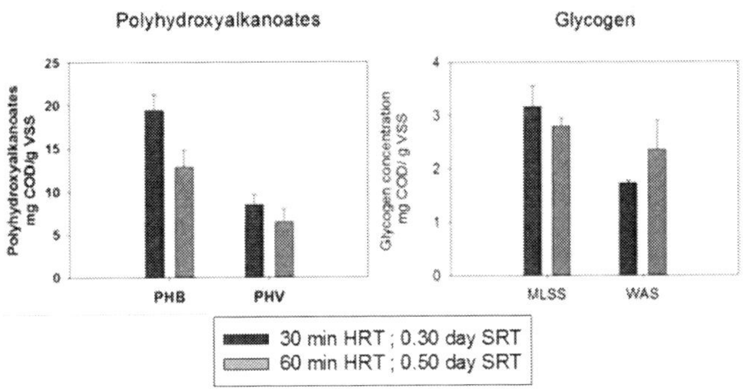

Figure 2. Average PHA and glycogen concentrations comparing 0.3 day SRT (30 minute HRT) and 0.5 day SRT (60 minute HRT) over a 3 week period (n=3). Only ML results are shown for PHA.

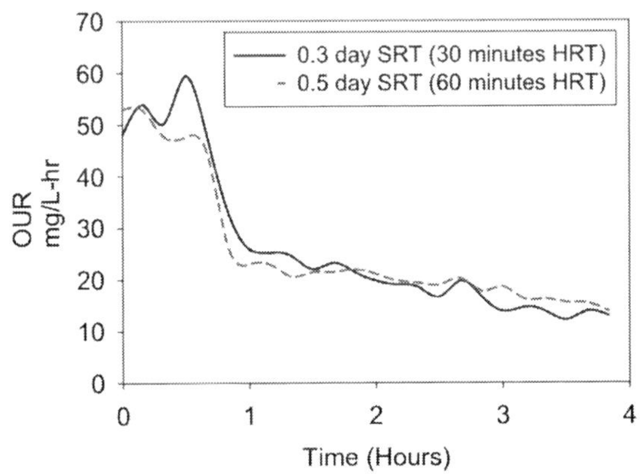

Figure 3. Comparison of 0.3 day SRT (30 minute HRT) and 0.5 day SRT (60 minute HRT) OUR using raw influent as a carbon source.

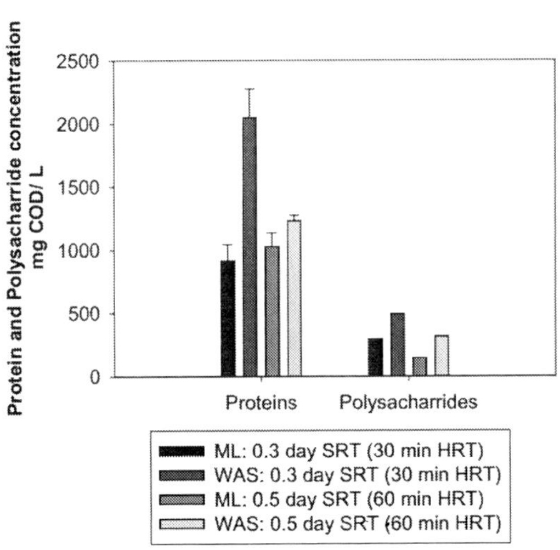

Figure 4. Comparison of 0.3 day SRT (30 minute HRT) and 0.5 day SRT (60 minute HRT) ML and WAS intracellular protein and polysaccharide concentrations (n=3).

CONCLUSIONS

The following is a summary of the major findings of the storage compounds production objective so far:

- Increasing the SRT results in better TSS and COD removal indicating more complete bioflocculation.
- The A-stage process was able to achieve PHA and glycogen accumulation similar to anaerobic-aerobic systems.
- The 0.30 day SRT resulted in better production of intracellular storage products compared to the 0.50 day SRT but the microbial respiration rate did not differ between the two SRTs.
- The WAS from the 0.3 day SRT at a DO concentration of 0.5mg/L resulted in a higher potential methane yield.
- Acknowledgements: This material is based upon work supported by the Water Environment Research Foundation (WERF) under Grant No. U3R14 and Hampton Roads Sanitation District (HRSD). Any opinions, findings, or conclusions expressed in this paper are those of the authors and do not necessarily reflect the views of the WERF and HRSD.

REFERENCES

(1) Reyes, P., Urtubia, A., Schiappacasse, M., Chamy, R., Montalvo, S., Borja, R. Effect of dissolved oxygen and temperature on macromolecular composition and PHB storage of activated sludge, *Journal of Environmental Science and Health, Part A: Toxic/Hazardous Substances and Environmental Engineering*, **2014**, 49 (7), 857-862, DOI: 10.1080/10934529.2014.882682

(2) Jimenez, J., Miller, M., Bott, C., Murthy, S., De Clippeleir, H., Wett, B. High-rate activated sludge system for carbon management – Evaluation of crucial process mechanisms and design parameters. *Water Research, 2015, 87;476-482* http://dx.doi.org/10.1016/j.watres.2015.07.032.

(3) Carucci, A., Dionisi, D., Majone, M., Rolle, E., Smurra, P. Aerobic Storage by activated sludge on real wastewater. Water Research, **2001**, 35 (16), 3833-3844

(4) APHA. Standard Methods for the Examination of Water and Wastewater, APHAAWWA-WEF, Washington, D.C., **2012**.

(5) Frølund, B., Griebe, T. and Nielsen, P.H. Enzymatic activity in the activated-sludge floc matrix. *Appl Microbiol Biotechnol,* ***1995****,* 43(4), 755-761.

(6) Braunegg, G., Sonnleitner, B., Lafferty, R.M. A rapid gas chromatographic method for the determination of poly-beta-hydroxybutyric acid in microbial biomass. *Applied microbiology and biotechnology*, **1978**, 6, 29-37.

(7) Roe, J.H., Dailey, R.E. Determination of glycogen with the anthrone reagent. *Analytical Biochemistry*, **1966**, 15, 245-250

(8) Miller, M. Optimizing high-rate activated sludge: Organic substrate for biological nutrient removal and energy recovery. Virginia Polytechnic Institute of Technology, Blacksburg, Virginia, **2015**.

(9) Jiang, Y., Hebly, M., Kleerebezem, R., Muyzer, G., van Loosdrecht, M. Metabolic modeling of mixed substrate uptake for polyhydroxyalkanoate (PHA) production. *Water Research*, **2011**, 45, 1309-1321.

(10) Carta, F., Beun, J., van Loosdrecht, M., Heijnen, J. Simultaneous storage and degradation of phb and glycogen in activated sludge cultures. *Water Research*, **2011**, 35 (11), 2693-2701.

(11) Huete, E., de Gracia, M., Ayesa, E., Garcia-Heras, J.L. ADM1-based methodology for the characterisation of influent sludge in anaerobic reactors. *Water Science and Technology*. **2006**, 54 (4), 157-166

(12) Wang et al. Polyhydroxyalkanoates in waste activated sludge enhances anaerobic methane production through improving biochemical methane potential instead of hydrolysis rate. *Scientific Reports*, **2016**. DOI: 10.1038/srep19713

HIGHLY ACTIVE PT-NI AND NI CATALYST FOR CATALYZING THE DECOMPOSITION OF HYDROGEN IODIDE, AS PART OF THE SULFUR IODINE CYCLE FOR HYDROGEN PRODUCTION FROM WATER

Amit Singhania, Venkatesan V. Krishnan* @, Ashok N. Bhaskarwar* $, Bharat Bhargav#, D. Parvatalu#, Satinath Bannerjee#*

**, Indian Institute of Technology, Delhi, New Delhi, INDIA*
#, ONGC Energy Centre, New Delhi, INDIA
@, Currently at Teesside University, Middlesbrough, UK
$, Ashok N. Bhaskarwar, Corresponding Author
Email: anbhaskarwar@gmail.com

Introduction

HI decomposition is an integral part of the 3-step Sulfur Iodine cycle used for the thermochemical splitting of water that has been touted as an industrially viable process for producing large quantities of hydrogen, utilizing high grade nuclear heat. This is an important step in creating a supply chain for Hydrogen for the Hydrogen economy of the future. The Sulfur Iodine cycle comprises 3 reactions, as follows:

$SO_2 + I_2 + 2H_2O \rightarrow 2HI + H_2SO_4$ (T= 20-120°C; ΔH= -75±15kJ mol^{-1}) - Bunsen reaction
$H_2SO_4 \rightarrow H_2O + SO_2 + \frac{1}{2} O_2$ (T= 800-900°C; ΔH= 186±3kJ mol^{-1}) - H_2SO_4 decomposition
$2HI \rightarrow H_2 + I_2$ (T= 300-500°C; ΔH~12kJ mol^{-1}) - HI decomposition

HI decomposition which is a highly endothermic and equilibrium limited reaction provides hydrogen product, and iodine. Both HI and I_2 produced are highly corrosive, due to which, the choice of temperatures is limited by materials of construction. It is imperative therefore also to identify a catalyst, with a suitable support, that is stable at temperatures of 450-550 deg. C (where Hastelloy compositions can be used as MoC), to sintering and to corrosion (1-4).

Results and Discussion

In this work, the catalytic activity of highly active metallic nanoparticles synthesized by modifying existing synthesis methods such as impregnation and solution combustion, are measured. Some of the catalysts include - Bimetallic Ni-Pt nanoparticles supported on four different supports of different BET surface areas (Activated carbon (AC))> γ-alumina > zirconia > ceria) and non-Noble metal Ni-ZrO_2 supported catalyst (made by solution combustion). The catalysts were also characterized by powder XRD and TEM.

In case of impregnation, the synthesis process was modified as follows, wherein in-situ precipitation and chemical reduction were achieved (Figure 1), generating nano-particles of Pt, Pt-Ni, as per the composition of the catalyst.

High metal dispersions (Ni-Pt) were obtained with particle sizes of the order of 10 nm or less, with the HI decomposition broadly following the dispersion of the bimetallic phase on the high surface area support, i.e., Ni-Pt/AC > Ni-Pt/γ-alumina > Ni-Pt/Zirconia > Ni-Pt/Ceria. Bimetallic Ni-Pt/AC also showed excellent stability for 100 h in HI decomposition reaction. Ni-Pt /AC catalysts reached about 24% conversion at space velocities of about 12.9 hr $^{-1}$ at temperatures of 550 deg. C, which is very close to the equilibrium values.

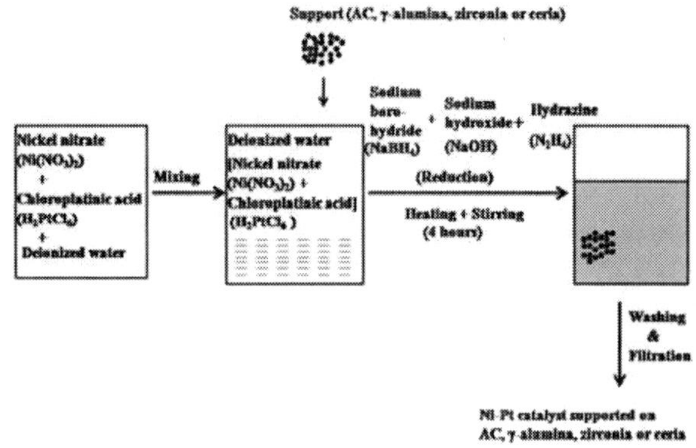

Figure 1: Modified impregnation method with low temperature reduction

Likewise, NiO-ZrO_2 catalysts, prepared by the Solution Combustion method (Figure 2) also showed adequate catalytic activity (near equilibrium values at 550 deg. C) at high Ni loadings, demonstrating their effectiveness for the reaction as well under the 500-550 deg. C temperature range (Figure 3). These catalysts were found to be stable for at least 50 hours Figure 4). The NiO initially forms a solid solution with ZrO_2, and found to precipitate in the form of nanoparticles of NiO after calcination, which upon in-situ reduction, formed a highly active and well dispersed catalyst for HI decomposition.

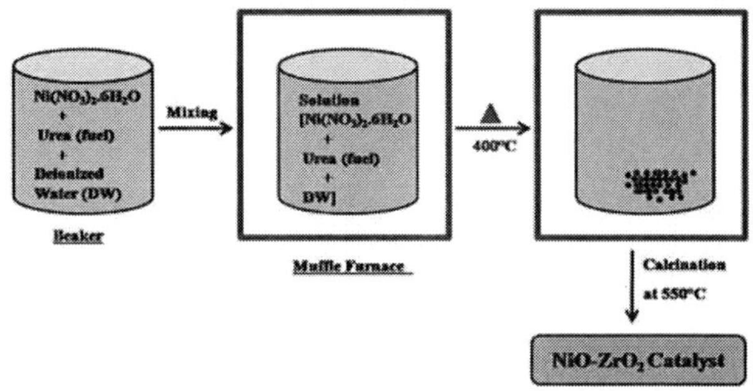

Figure 2: Solution Combustion method for NiO-ZrO_2 non PGM catalyst

NiO-ZrO_2 catalysts are also a viable alternative to the high performing Pt-Ni bimetallic systems, due to their considerably lower costs, and comparable catalytic activity, arising out of the ex-solution of Ni on calcination and reduction.

The catalytic activity is correlated to the particle morphology as seen in TEM and the verifiable solid solution observed by XRD and Raman Spectroscopy. In this publication, the authors hope to discuss the modification in catalyst synthesis methods, to generate stable nano-particles of Pt-Ni and Ni, that show maximum catalytic activity in a highly corrosive environment.

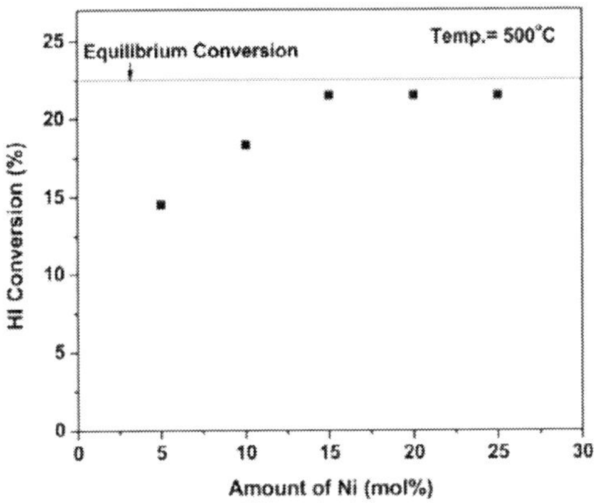

Figure 3: Optimal NiO content in ZrO_2 support, to achieve maximum conversion (near equilibrium)

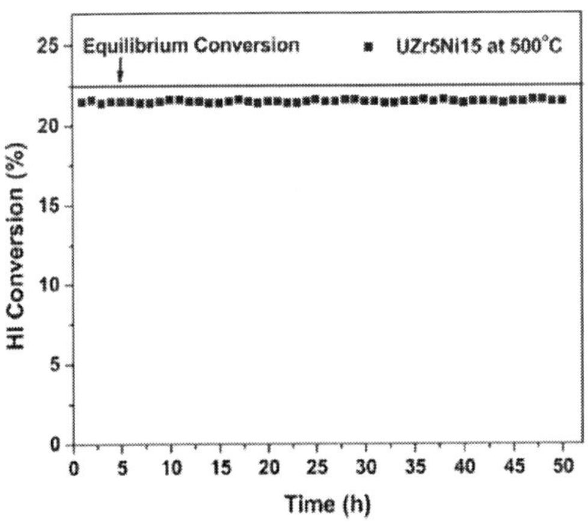

Figure 4: Initial Stability testing over time with 15 mol % NiO-ZrO_2

Acknowledgement. The authors acknowledge the Oil and Natural Gas Commission, India, for financial support for this project.

References

(1) Winter, C. J.; *Int. J. Hydrogen Energy*, **2009**, *34*, S1.
(2) Moriarty, P.; and Honnery D.; *Int. J. Hydrogen Energy*, **2009**, *34*, 31.
(3) Maeda, K.; and Domen, K.; *J. Phys. Chem. C.*, **2007**, *111*, 7851.
(4) Goldstein, S.; Borgard, J. M.; and Vitart, X.; *Int. J. Hydrogen Energy* **2005**, *30*, 619.

FABRICATION OF Cu/Ti NANOELECTRODE FOR ELECTROCHEMICAL DENITRIFICATION OF GROUNDWATER

Fang Liu, Miao Li, Lele Wang, Guoyu Ding, Xiang Liu

School of Environment, Tsinghua University, Beijing 100084;

Abstract

Nitrate contamination of groundwater is a widespread environmental problem. Nitrate-contaminated water above the permissible drinking water limit is toxic to human health, specifically to children. In this study, a novel Cu/Ti nanoelectrode was fabricated and its nitrate reduction efficiency was investigated. With the Cu/Ti bilayer nanoelectrode, nitrate removal efficiency was 71.8% after 90 min electrolysis at a current density of 30 mA/cm^2. Under the same conditions using a Ti nanoelectrode, the removal efficiency was only 28.2%. Moreover, the Cu/Ti nanoelectrode had good performance for electrochemical denitrification under different conditions. The nitrate reduction rate increased with increasing current density and pH in the range of 5–11 had little effect on nitrate reduction.

PROBING NANOSCALE CHARACTERISTICS OF HYDROPHILIZED POLYETHERSULFONE MEMBRANES

Wanyi Fu[1], *Somenath Mitra*[2], *Wen Zhang*[1*]

[1]John A. Reif, Jr. Department of Civil & Environmental Engineering, New Jersey Institute of Technology, Newark, NJ 07102
[2]Department of Chemistry and Environmental Science, New Jersey Institute of Technology, Newark, NJ 07102
[*]Corresponding author: e-mail: wzhang81@njit.edu; Phone: 1-973-596-5520; Fax: 1-973-596-5790.

During the last few decades, membrane filtration has been extensively used in water and wastewater treatment, desalination, dairy making, biomass/water separation, and recovery of rare metals. However, membrane fouling is one of the major challenges in the industrial applications of various membrane processes. Hydrophilic membranes are known to suffer less from membrane fouling in water treatment in general. Thus, hydrophilized polymer membranes are broadly manufactured for industrial applications. After the hydrophilization process, membrane properties could significantly differ from the unmodified materials besides the hydrophilicity. This study investigated the effects of modifying additives or chemical blends (i.e., polyvinylpyrrolidone (PVP) and polyethylene glycol (PEG)) on membrane characteristics (e.g., surface morphology, structures, functional groups, zeta potential and hydrophobicity) of polyethersulfone (PES) membranes developed by industrial manufacturers. Visualization of membrane surface and cross section morphology was achieved by atomic force microscope (AFM) and scanning electron microscopy (SEM). The results show that by blending PES membranes with macromolecular additives (PVP or PEG), the above-assessed membrane properties were significantly altered, which offers new insights into novel membrane development, surface modification, membrane defouling and other related applications.

PHOTOELECTROCHEMICAL HYDROGEN GENERATION USING MULTI-BAND III-NITRIDE NANOWIRE ARRAYS

Hieu P. T. Nguyen, hieu.p.nguyen@njit.edu, Moab R. Philip, Dipayan D. Choudhary, and Mehrdad Djavid,

Electrical and Computer Engineering Department, New Jersey Institute of Technology, Newark, NJ 07102, United States

Recently, group III-nitride semiconductor has been intensively studied as a new class of materials for solar water splitting and hydrogen generation. This material system exhibits several significant advantages, compared to the conventional oxide-based photoelectrodes which have been limited by the bandgap energy and fundamental properties of such materials. III-nitride semiconductors exhibit high carrier mobility, relatively high absorption co-efficient, and extreme chemical stability. Moreover, group III-nitride semiconductor, whose energy bandgap, or absorption spectrum can be varied across the entire solar spectrum. It is the only known material whose energy band edges can straddle the redox potentials deep in the visible and infrared wavelength. Significantly enhanced photocatalytic activities and energy conversion efficiency are further expected from nanowire structures, due to the reduced defect densities and enhanced charge separation, compared to their thin-film counterparts.

In this study, multi-band InGaN/GaN core-shell nanowire photoanodes have been investigated for photoelectrochemical hydrogen generation under ultraviolet and visible light illumination. The nanowire heterostructures were grown by plasma assisted molecular beam epitaxy on Si substrates. The unique multi-band core-shell nanowire structures were efficiently used to protect the InGaN segments from corrosion and oxidization, leading to relatively high incident-photon-to-current-conversion (~ 27%). Moreover, the InGaN/GaN core-shell nanowire photoanode shows virtually no degradation after an intensive reaction in HBr. Our studies also suggest that the III-nitride nanowire semiconductor is a highly promising candidate for future low cost, high performance solar-fuel systems.

EXCITON DISSOCIATION AND PLASMON INDUCED HOT ELECTRON TRANSFER IN SEMICONDUCTOR/METAL QUANTUM ROD HETEROSTRUCTURES

Tianquan Lian

Department of Chemistry, Emory University
tlian@emory.edu

Long Distance Charge Separation and Efficient H2 Generation in Nanorod/Pt Heterostructures

Quantum confined semiconductor nanocrystals (0D quantum dots, 1D quantum rods and 2D quantum platlets) have been intensively investigated as light harvesting and charge separation materials for photovoltaic and photocatalytic applications. The efficiency of these semiconductor nanocrystal-based devices depends on many fundamental processes, including light harvesting, carrier relaxation, exciton localization and transport, charge separation and charge recombination. The competition between these processes determines the overall solar energy conversion (solar to electricity or fuel) efficiency. Semiconductor nano-heterostructures, combining two or more material components, offer unique opportunities to control their charge separation properties by tailoring their compositions, dimensions and spatial arrangement. Further integration of catalysts (heterogeneous or homogeneous) to these materials form multifunctional nano-heterostructures. Using 0D, 1D and 2D CdSe/CdS/Pt heterostructures as model systems, we directly probe the above-mentioned fundamental exciton and carrier processes by transient absorption and time-resolved fluorescence spectroscopy. We are examining how to control these fundamental processes through the design of heterostructures to achieve long-lived charge separation and efficient H_2 generation. In this talk, we will discuss mechanism of 1D exciton transport and dissociation in nanorods, and key factors limiting H_2 generation efficiency in CdSe/CdS/Pt nanorod heterostructures.[1]

Efficient plasmon induced hot electron transfer

Surface plasmon resonance in metal nanostructures has been widely used to enhance the efficiency of semiconductors and/or molecular chromophore based solar energy conversion devices by increasing the absorption or energy transfer rate through the enhanced local field strength. In more recent years, it has been shown that excitation of plasmons in metal nanostructures can lead to the injection of hot electrons into semiconductors and enhanced photochemistry. This novel mechanism suggests that plasmonic nanostructures can potentially function as a new class of widely tunable and robust light harvesting materials for solar energy conversion. However, plasmon-induced hot electron injections from metal to semiconductor or molecules are still inefficient because of the competing ultrafast hot electron relaxation processes within the metallic domain.

In this talk we discuss a recent study on the plasmon-exciton interaction mechanisms in colloidal quantum-confined semiconductor-gold nanorod heterostructures.[2,3] In CdSe NRs with Au tips, the distinct plasmon band of the Au nanoparticles was completely damped due to strong interaction with the CdSe domain. Using transient absorption spectroscopy, we show that optical excitation of plasmons in the Au tip leads to efficient hot electron injection into the semiconductor nanorod. In the presence of sacrificial electron donors, this plasmon induced hot electron transfer process can be utilized to drive photoreduction reactions under continuous illumination. We propose that the strong metal/semiconductor coupling in CdSe/Au hetersostructures leads to a new pathway for this surprising efficient hot electron transfer. In this plasmon induced interfacial charge transfer transition (PICTT) the a plasmon decay by direct excitation of an electron from the metal to semiconductor, bypassing the competition with hot electron transfer in metal. Ongoing studies are examining the generality of this mechanism and exploring possible approaches for improving its efficiency through controlling the size and shape of the plasmonic and excitonic domains.

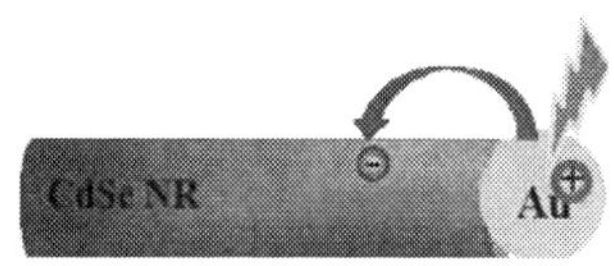

Figure 1. Efficient plasmon induced hot electron transfer from Au tip to CdSe nanorod.

Acknowledgement. This material is based upon work supported by the U.S. Department of Energy, Office of Science, Office of Basic Energy Sciences, Solar Photochemistry Program under Award Number (DE-FG02-12ER16347)

References

(1) Kaifeng Wu, Haiming Zhu, Tianquan Lian, "Ultrafast Exciton Dynamics and Light-driven H2 Evolution in Colloidal Semiconductor Nanorods and Pt-tipped Nanorods", ***Acc. Chem. Res. (2015)***, 48, 851-859.
(2) Kaifeng Wu, Jinquan Chen, James R. McBride, Tianquan Lian, "Efficient hot-electron transfer by a plasmon-induced interfacial charge-transfer transition", ***Science*** (2015), 349 (6248): 632.
(3) Kaifeng Wu, William E. Rodríguez-Córdoba, Ye Yang, and Tianquan Lian, "Plasmon-Induced Hot Electron Transfer from the Au Tip to CdS Rod in CdS-Au Nanoheterostructures", **Nano Lett. (2013)**, 13(11), 5255-5263.

SOLAR HYDROGEN EVOLUTION FROM RU-BIPYRIDYL DYE SENSITIZED AND TITANIUM DIOXIDE NANOCLUSTERS DISPERSED ON PERIODIC CUBIC MCM-48 SILICA MESOPOROUS MATERIAL

Shivatharsiny Rasalingam, Rui Peng, Chia-Ming Wu, Kadarkaraisamy Mariappan, and Ranjit T. Koodali

Department of Chemistry, University of South Dakota, Vermillion, South Dakota 57069
E-mail address: Ranjit.Koodali@usd.edu

Abstract

A series of titanium dioxide containing cubic mesoporous MCM-48 materials were synthesized at room temperature. The powder X-day diffractometry studies indicate that the titania species are present as nanoclusters and well dispersed within the MCM-48 matrix. A Ru-based dye, $[Ru(bpy)_2(dcbpy)][PF_6]_2$ (Ru-4,4' dye) was adsorbed to the TiO_2-MCM-48 mesoporous materials. The ability of the titania supported mesoporous materials to generate hydrogen under visible light illumination in the presence of a sacrificial agent was explored. The photocatalytic studies indicate that the Ru-4,4' dye is anchored strongly on the titania containing MCM-48 support and the photocatalysts show excellent recycling activity.

PLASMONIC METAL-SEMICONDUCTOR HETEROJUNCTIONS FOR SOLAR ENERGY CONVERSION

Nianqiang Wu[1], *Scott Cushing*[1], *Deryn Chu*[2], *Jiangtian Li*[1]

[1]Department of Mechanical & Aerospace Engineering
West Virginia University
P.O. Box 6106
Morgantown, WV 26506, USA
E-Mail: nick.wu@mail.wvu.edu

[2]Sensors and Electron Devices Directorate
US Army Research Laboratory
Adelphi, Maryland 20783-1197, USA.

Introduction

Solar energy can be converted by three means: solar thermal power generation devices, photovoltaics and photocatalysts (or photoelectrochemical cells (PECs)). Both photovoltaics and photocatalysts are built on the semiconductor. The conversion of solar radiation by semicondcutors is fundamentally limited by the heating losses in charge carriers with excess energy above the band gap of the semiconductor.[1] The balance of increased light absorption with increased thermalization losses makes a band gap of ~1.3 eV most efficient for photovoltaics according to the Schockley-Queisser limit[1], and ~1.9 eV for water splitting according to the energy necessary to drive water photolysis.[2,3]. Hence single semiconductor-based photovoltaics and photocatalysts cannot harvest a large portion of solar radiation in the visible-light and near-infrared light regions. Formation of heterojunctions is an effective route to mitigate this problem. Plasmonic metal nanostructures can be incorporated with the semiconductor to form a heterojunction, in which the plasmonic metal can serve as the photosensitizer.

Results and Discussion

In the plasmonic metal-semiconductor heterojunctions, plasmon can extend the light absorption spectral range; transfer the energy from the metal to the semiconductor via three mechanisms: light scattering (or trap), plasmon-induced resonance energy transfer (PIRET) [4,5], and hot electron injection to the semiconductor[6,7].

The overall efficiency of solar energy conversion for the plasmonic metal-semiconductor heterojunctions depends on the coupling between the metal and the semiconductor, and on the utilization of three mechanisms[1,8,9]. Simple mixing of plasmonic nanoparticles with semiconductor may result in poor performance of the metal-semiconductor heterojunction. It is necessary to conduct careful design to make large enhancement of the plasmonic energy transfer and the resulting solar energy conversion efficiency[10,11,12].

Conclusions

Formation of plasmonic metal-semiconductor heterojunctions provides an alternative route to enhance the solar energy conversion. However, careful design of heterojunctions based on the three mechanisms is needed to improve the efficiency of solar energy conversion.

Acknowledgment. The resource and facilities used in this work were supported by the US Army Research Laboratory (W911NF-14-2-0116) and National Science Foundation (CBET-1233795), and the West Virginia University Research Corporation and the West Virginia EPSCoR Office. The use of WVU shared facilities is greatly appreciated.

References

(1) Shockley, W. & Read, W. T., *Phys. Rev.*, **1952**, 87, 835.
(2) J. Li, N. Q. Wu, *Catal. Sci. Technol.*, **2015**, 5, 1360-1384.
(3) S. K. Cushing, N. Q. Wu, *J. Phys. Chem. Lett.*, **2016**, 7, 666-675.
(4) Cushing, S. K.; Li, J.; Meng, F.; Senty, T. R.; Suri, S.; Zhi, M.; Li, M.; Bristow, A. D.; Wu, N. Q. *J. Am. Chem. Soc.*, **2012**, 134, 15033–15041.
(5) J. Li, S. K. Cushing, F. Meng, T. R. Senty, A. D. Bristow, N. Q. Wu, *Nature Photonics*, **2015**, 9, 601-607
(6) Tian, Y.; Tatsuma, T. *Chem. Commun.* **2004**, 1810-1811.
(7) Y. Tian and T. Tatsuma, *J. Am. Chem. Soc.*, **2005**, 127, 7632- 7637.
(8) S. K. Cushing, A. D. Bristow, N. Q. Wu, *Phys. Chem. Chem. Phys.*, **2015**, 17, 30013-30022.
(9) Cushing S. K.; and Wu, N. Q. *Interface*, **2013**, 22(2), 63-67.
(10) Li, J.; Cushing, S. K.; Zheng, P.; Meng, F.; Chu, D.; Wu, N. Q. *Nature Communications*, **2013**, 4, 2651.
(11) J. Li, S. K. Cushing, P. Zheng, T. Senty, F. Meng, A. D. Bristow, A. Manivannan, N. Q. Wu, *J. Am. Chem. Soc.*, **2014**, 136, 8438-8449.
(12) Li, J.; Cushing, S. K.; Bright, J.; Meng, F.; Senty, T. R.; Zheng, P.; Bristow, A. D.; Wu, N. Q. *ACS Catalysis*, **2013**, 3, 47-51.

ON ESTABLISHING A CONSORTIUM TO PRINT POLYMER PHOTOVOLTAIC CELLS

Andrew B. Holmes

School of Chemistry, Bio21 Institute
University of Melbourne
Melbourne, Vic 3010,
Australia

Introduction

Organic electronic materials are showing real promise in a range of applications from light emitting devices through thin film transistors and photovoltaics to biological devices.[1] My own interest in this field developed as a result of the serendipitous discovery in collaboration with physics colleagues at the University of Cambridge that a thin film of a highly fluorescent semiconducting polymer (poly-*p*-phenylene vinylene) could be made to fluoresce through the process of electroluminescence when sandwiched between charged electrodes, one of which was transparent.[2] This was a fortunate opportunity and for the first year the group was able to focus on developing the technology and protecting the work through patent filing. As a result it was possible to develop red,[3] green[2] and blue light emitting devices.[4] Thus equipped and with intellectual property protected Cambridge Display Technology was founded and is still trading today, albeit under the ownership of Sumitomo Chemical.[5,6,7]

The experience with the development of thin film organic light emitting devices (OLEDs) led to the hypothesis that the process could be reversed. Electroluminescence is the generation of fluorescence through double charge injection, subsequent recombination, and electronic excitation to a fluorescent state. The question was asked whether the process could be run in reverse - light excitation, charge separation and production of an electric current when the photoelectrodes were connected by a wire. This is a process that Nature also finds difficult, hence the phenomenon of photosynthesis. Photoexcitation of the green chlorophyll pigment in plants results in electron-hole charge separation through a complex cascade of redox couples, the electron going on to reduce carbon dioxide to carbohydrates and the positive charge oxidizing water to oxygen in the presence of protons. Ultimately in collaboration with Cambridge physics it became possible to generate a photocurrent using a blend of cyano-PPV with an electron donor polymer.[8] However the really big breakthrough emerged when Wudl and Heeger and colleagues made use of a solution-processible electron-accepting fullerene derivative in a blend with MEH-PPV to generate a bulk heterojunction, leading to a modest organic photovoltaic device.[9]

On my return to Australia in 2004 I sought to identify collaborations that differentiated themselves from the polymer OLED work in Cambridge. In partnership with various universities and CSIRO (then Molecular Sciences) we agreed to form a consortium to concentrate the effort on thin film organic photovoltaics. We were fortunate in forming a strong international consortium (International Consortium on Organic Solar Cells - ICOS) that comprised an Australian partnership with the University of Melbourne, Monash University, University of Wollongong, University of Sydney, University of Newcastle, and University of Queensland working in partnership with Imperial College, Merck (Southampton), Cornell University, Georgia Tech, A*-STAR Singapore and the University of Otago, with various industrial partnerships. This consortium helped place the Australian OPV effort on the world stage.

Following the successful initial consortium the Australian partners were forced to fragment owing to the scale of funding required to support such a large consortium. In Victoria we formed a strong group (Victorian Organic Solar Cell Consortium - VICOSC consisting of The University of Melbourne, Monash University, CSIRO Materials Science and Engineering (now CSIRO Manufacturing Flagship), in partnership with the Australian polymer banknote company (Securency, now Innovia Security) and Bosch SEA Pte. This consortium set itself the task of reaching an in-principle proof of solution-deposited organic bulk heterojunction solar cells as well as dye-sensitised solar cells. The bulk of the financial support (AUD $5m) came from the Victorian State Government. The project successfully reached its target.

As a result of the success of the initial phase of the VICOSC consortium, further funding (*ca.* AUD $10m) became available form the Victorian State Government and the Australian Federal Government, as well as the previous industrial partners and a new industrial partner that had a large investment in steel roofing materials (BlueScope Steel). The target of the three-year project was to deliver laboratory-sized (1 x 1 cm^2) organic bulk heterojunction solar cells and dye-sensitised solar cells as well as hybrid solar cells demonstrating, ideally, 10% energy conversion efficiency. In addition the consortium set out to deliver roll-to-roll fully printed polymer bulk heterojunction cells of the area of an A4 (20 x 30 cm^2 sheet). The remainder of this presentation will summarise the high points of the delivery of these objectives.

Efficient Laboratory Sized Solar Cells

Polymer Devices. The polymer CP3 **1** was synthesized at scale. Material of high molar mass could be obtained by controlled Suzuki polycondensation.

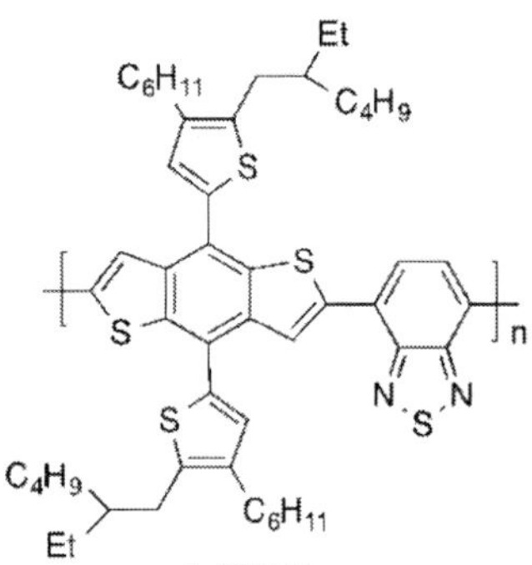

1 (CP3)

Hero cells showing an energy conversion efficiency of 9.4% were fabricated using a blend of CP3 and $PC_{70}BM$ (1:2) in an inverse device configuration.[11]

Molecular Materials. The molecular material BTR **2** was synthesized at scale using cross-coupling reactions. It demonstrated a nematic liquid crystalline phase. This is a beautiful solution-processible material that, when blended with $PC_{70}BM$ in a bulk heterojunction cell, delivered energy conversion efficiencies in excess of 9.3%.[12]

Printed Solar Cells

Perovskite Devices Colleagues at CSIRO have developed methods for printing perovskite solar cells from solution.[13]

They deposited zinc oxide nanoparticles in EtOH solution by slot die-coating on ITO-glass that was air dried, then annealed at 120° C. PbI_2 was coated from DMF and dried under nitrogen, followed by

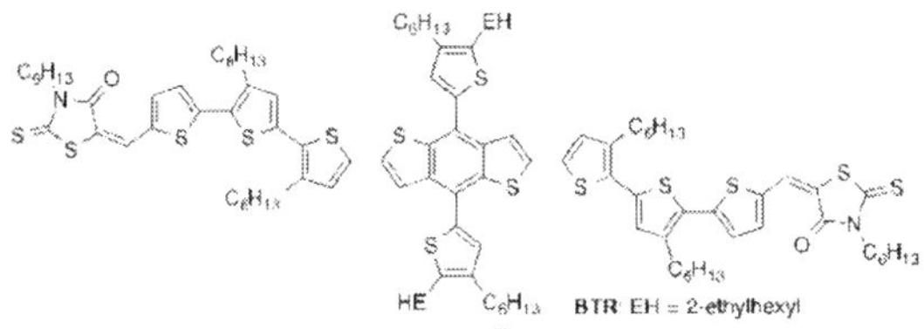

2

vapour annealing in a chamber. Methyl ammonium iodide in 2-propanol was then coated onto the PbI_2 film at 70° C. P3HT in dichlorobenzene with $LiNTf_2$ in MeCN and 4-^{t}Bu-pyridine was slot die-coated on the perovskite. Finally, a silver electrode was deposited by metal evaporation. The resulting device exhibited an energy conversion efficiency of 10.1% with the best device exhibiting almost 12%.

Large Area Roll-to-Roll Printed Modules The Consortium has succeeded in efficiently depositing fully printed 10 x 10 cm^2 modules of a blend of poly-3-hexylthiophene with PCBM (1:1) by a combination of reverse gravure and screen printing on patterned ITO on polyethylene terephthalate (PET). Efficiencies are not yet as high as would be predicted, but it is expected that sophisticated deposition techniques will lead to more efficient devices.

Durabilty Preliminary studies on lifetimes under all weather exposure conditions have been initiated, and it has been shown that fully encapsulated modules (against air and water) exhibit some promise in extended operating lifetimes under ambient conditions.

Acknowledgment. This work was made possible by funding to the Victorian Organic Solar Cell Consortium (VICOSC) from the Victorian State Government (DSDBI) and the Australian Renewable Energy Agency (ARENA, Project 2-A018). Current funding from ARENA is administered through the Australian Centre for Advanced Photovoltaics (ACAP). The work of contributing researchers is acknowledged in their co-authorship of the references cited below.

References

(1) Liao, C,; Zhang, M.; Yao, M. Y.; Hua, T.; Li, L.; Yan, F., *Adv. Mater.*, **2014**, *27*, 7493-7527.
(2) Burroughes, J. H.; Bradley, D. D. C.; Brown, A. R.; Marks, R. N.; Mackay,K.; Friend, R. H.; Burn, P. L.; Holmes, A. B. *Nature*, **1990**, *347*, 539-541.
(3) Greenham, N. C.; Moratti, S. C.; Bradley, D. D. C.; Friend, R. H.; Holmes, A. B., *Nature*, **1993**, *365*, 628-630.
(4) I. D. Rees, K. L. Robinson, A. B. Holmes, C. R. Towns and R. O'Dell, *MRS. Bull.*, 2002, **27**, 451-455.
(5) Grimsdale, A. C.; Chan, K. L.; Martin, R. E.; Jokisz, P. G.; Holmes, A. B., *Chem. Rev.*, **2009**, *109*, 897-1091.
(6) From the Synthesis of Acetylenic Natural Products to Seeing the Light with Polymers, Holmes, A. B.; Burn, P. L.; Kraft, A.; White, J. M.; Wong, W. W. H. in Turning Points in Solid-State, Materials and Surface Science, Harris, K. D. M. and Edwards, P. P., Eds. RSC Publishing, Cambridge, Chapter 20, **2008**, pp. 334-345.
(7) See www.sumitomo chem.co.jp/english/newsreleases/docs/20070731_3.pdf, 31 Jul, 2007.
(8) Halls, J. J. M.; Walsh, C. A.; Greenham, N. C.; Marseglia, E. A.; Friend, R. H.; Moratti, S. C.; Holmes, A. B., *Nature*, **1995**, *376*, 498-500.
(9) Jannsen, R. A. J.; Hummelen, J. C.; Lee, K.; Pakbaz, K.; Sariciftci, N. S.; Heeger, A. J.; Wudl, F., *J. Chem. Phys.*, **1995**, *103*, 788-795.
(10) See www.vicosc.unmelb.edu.au.
(11) Subbiah, J.; Purushothaman, B.; Chen, M.; Qin, T.; Gao, M.; Vak, D.; Scholes, F.; Chen, X.; Watkins, S. E.; Wilson, G. J.; Holmes, A. B.; Wong, W. W. H.; Jones, D. J., *Adv. Mater.*, **2015**, *27*, 702-705.
(12) Sun, K.; Xiao, X.; Lu, S.; Zajaczkowski, W.; Pisula, W.; Hanssen, E.; White, J. M.; Williamson, R. M.; Subbiah, J.; Ouyang, J.; Holmes, A. B.; Wong, W. W. H.; Jones, D. J., *Nat. Commun.*, **2014**, *5*, 6013.
(13) Hwang, K.; Jung, Y.-S.; Heo, Y.-J.; Scholes, F. H.; Watkins, S. E.; Jones, D. J.; Kim, D.-Y.; Vak, D., *Adv. Mater.*, **2015**, *27*, 1241-1247.

POLY(*PARA*-PHENYLENE) TYPE POLYMERS FOR PHOTOCATALYTIC WATER SPLITTING

Reiner Sebastian Sprick, Duncan J. Woods, Baltasar Bonillo, Pierre Guiglion, Benjamin J. Slater, Martijn A. Zwijnenburg, Dave J. Adams, and Andrew I. Cooper

Department of Chemistry and Centre for Materials Discovery, University of Liverpool, Crown Street, Liverpool, L69 7ZD (UK). Department of Chemistry, University College London, 20 Gordon Street, London, WC1H 0AJ (UK).

aicooper@liverpool.ac.uk

Introduction

Due to the worldwide increased energy consumption and the resulting increased usage of fossil fuels, cleaner and renewable energy sources are needed to lower emissions in an effort to ease environmental problems. There is much interest in using visible light to generate solar fuels, e.g. photocatalytic water splitting. Mostly inorganic catalysts have been explored.[1] Recently, organic materials and particularly graphitic carbon nitride have gained significant attention[2,3] and we have developed conjugated microporous polymers for photocatalytic hydrogen evolution from water.[4,5] Despite the wealth of knowledge in the field of linear conjugated polymers there have been far fewer examples of these in phototcatalysis. The earliest example an all organic photocatalyst for hydrogen evolution, poly(*p*-phenylene) (PPP), was reported in 1985,[6] but PPP is only modestly active for hydrogen production in the UV in the presence of a sacrificial electron donor. Our approach to improve photocatalytic performance was to apply the concept of chromophore planarization, as used previously to enhance charge-carrier mobilites and the separation of excitons.[7] Also we have started to explore how we can make soluble and processable analogues of these polymers and how we can lower the HOMO and LUMU levels even further by substitution of benzenes with nitrogen containing analogues.

Experimental

Synthesis. All materials were synthesized using previously published reaction conditions.[7] The 2-ethyl-hexyl-substituted carbazole-co-phenyl polymer was synthesized using standard Suzuki polycondensation conditions[8] and fractionated via Soxhlet extraction with methanol, ethyl acetate and chloroform.

Measurements. For hydrogen evolution experiments a flask was charged with the polymer powder (25 mg), water (7.5 mL), triethylamine (7.5 mL), methanol (7.5 mL), and sealed with a septum. The resulting suspension was ultrasonicated until the photocatalyst was dispersed before degassing by N_2 bubbling for 30 minutes. The reaction mixture was illuminated with a 300 W Newport Xe light-source. Gas samples were taken with a gas-tight syringe, and run on a Bruker 450-GC gas chromatograph equipped with a Molecular Sieve 13X 60-80 mesh 1.5 m × ⅛" × 2 mm *ss* column at 50 °C with an argon flow of 40 mL min^{-1}. Hydrogen was detected with a thermal conductivity detector referencing against standard gas with a known concentration of hydrogen. Time-correlated single photon counting was performed on a Horiba system. The apparatus was used with a 404 nm diode (instrument response 200 ps fwhm), with a 435 nm high pass filter for emission detection. Polymer suspensions were prepared by ultrasonicating 2 mg of the polymer in 5 mL of THF for 10 min and diluting them until the stop count rate was below 2% (typically around 1% was used). The prompt was measured for THF at the excitation wavelength without filter. Decay times were fitted in the DAS 6 software using suggested lifetime estimates. The UV-visible absorption spectra of the polymers were recorded on a Shimadzu UV-2550 UV-Vis spectrometer as powders. The fluorescence spectra of the polymer powders were measured with a Shimadzu RF-5301PC fluorescence spectrometer.

Results and Discussion

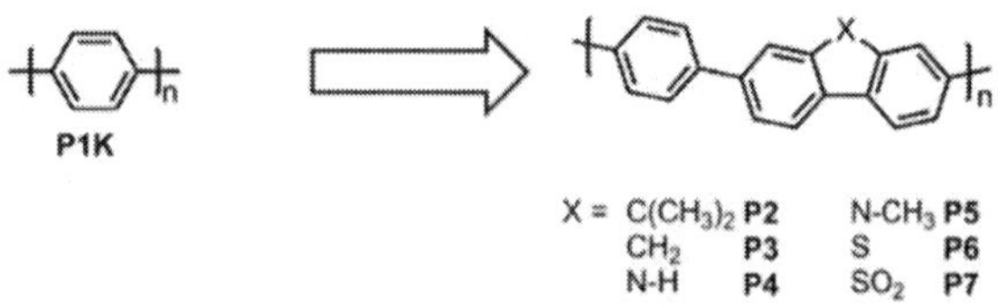

Figure 1. Structures of poly(*p*-phenylene)s (P1K, left) and their planarized fluorene-type analogs (P2–P7; right).

The introduction of planar fluorene, carbazole, dibenzothiophene, or dibenzothiophene sulfone units is a known strategy to enhance charge-carrier mobilites and the separation of excitons. This strategy was found greatly enhance the H_2 evolution rates of PPP-derived copolymers from water in the presence of triethylamine as a sacrificial electron donor.[7] This is not only the case for the hydrogen evolution rate under UV light, but also the performance under visible light (> 420 nm) was significantly enhanced. Polymer P7 was found to be 46 times more active under visible light than PPP. This is also reflected in an increase of the lifetime of the excited state as measured by time-correlated single photon counting (TCSPC, Table 1).

Table 1. TCSPC results of polymer suspensions in THF and photocatalytic performance of the polymers.[7]

Polymer	T1 Lifetime / ns	T2 Lifetime / ns	HER > 420 nm / µmol h^{-1}	HER > 295 nm / µmol h^{-1}
P1K	0.31	1.25	2.0	10.3
P2	0.38	1.54	8.3	43.6
P4	-	-	7.8	35.1
P6	0.52	2.09	26.6	102.4
P7	1.22	4.89	92.0	145.0

As processability of 2D-materials and conjugated microporous polymers remains a significant challenge making soluble analogues of these linear polymers is a very attractive research goal. Linear conjugated polymers, such as polycarbazoles, can be made processable by the introduction of alkyl chains on the bridge-head. Polymerization of a 2-ethyl-hexyl substituted carbazole-co-phenyl polymer using standard Suzuki polycondensation condition gives a photocatalyticlly material. When fractioned by Soxhlet extraction a chloroform soluble material is obtained (SP4-$CHCl_3$) as well as an all organic insoluble material. The performance of the soluble material SP-$CHCl_3$ is very similar to its unsubstituted analogue P4 (see Fig 2.). Interestingly, the insoluble fraction of the alkylated polymer (SP4-insoluble) significantly outperforms the unsubstituted polymer P4. This could be an indication that the molecular weights archived are higher in the case of the alkylated polymer as the growing chain will precipitate out of solution at a later state of the polycondensation reaction compared to P4.

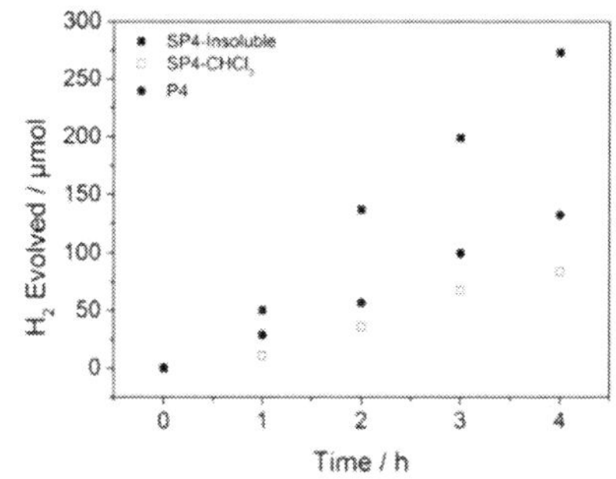

Figure 2. Photocatalytic performance of the two SP4 fractions and P4 under cominbed UV and visible light (> 295 nm).

Besides the substitution of the bridge-head atom in fluorene type materials the substitution of carbons for heteroatoms in the benzene ring is another strategy to tune the photophysical properties of conjugated polymers. We have synthesized a series of pyridine, pyrazine and pyrimidine co-polymers (Fig. 3).

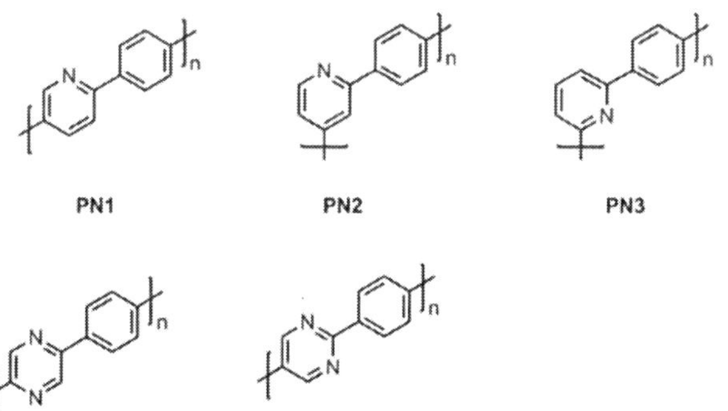

Figure 3. Structures of the pyridine, pyrazine and pyrimidine co-polymers.

Table 2. Optical properties and photocatalytic performance of the pyridine, pyrazine and pyrimidine co-polymers.

Polymer	Optical gap / eV	λ_{em} / nm	HER > 420 nm / µmol h^{-1}	HER > 295 nm / µmol h^{-1}
P1K	2.79	456, 483	2.0	10.3
PN1	2.72	456, 482	34.9	68.5
PN2	3.35	382	< 0.1	9.7
PN3	3.24	432	-	1.7
PN4	2.23	462	53.8	75.3
PN5	2.78	477	4.3	9.2

The substitution of benzene in P1K with pyridine in PN1 results in a red-shift of the absorption on-set and a significantly higher hyrdrogen evolution rate both under UV and visible light. The introduction of *meta*-linkages was found not to be beneficial as the effective conjugation length is shorted: When the linkage in the pyridine is varied from a 2,5 to a 3,5 or 2,6-linkage the performance of the material drops. A total loss of activity under visible light is observed as the absorption on-set is blue shifted and a lower activity under UV light compared to PN1 was found. The pyrazine co-polymer also shows enhancement of the performance compared to P1K under both UV and visible light, while the pyrimidine co-polymer shows a similar performance under UV and slightly enhance visible activity.

The introduction of heteroatoms into linear poly(*p*-phenylene) type polymers significantly enhances the photocatalytic performance of these materials. Two strategies have been explored and we will focus in further experiments on mechanistic understanding of the photocatalytic process using transient absorption spectroscopy and other methods.

References

(1) Kudo, A.; Miseki, Y. *Chem. Soc. Rev.* **2009**, *38*, 253.
(2) Chen, X.; Shen, S.; Guo, L.; Mao, S. S. *Chem. Rev.* **2010**, *110*, 6503.
(3) Wang, X.; Maeda, K.; Thomas, A.; Takanabe, K.; Xin, G.; Carlsson, J. M.; Domen, K.; Antonietti, M. *Nat. Mater.* **2009**, *8*, 76.
(4) Sprick, R. S.; Jiang, J.-X.; Bonillo, B.; Ren, S.; Ratvijitvech, T.; Guiglion, P.; Zwijnenburg, M.; Adams, D. J.; Cooper, A. I. *J. Am. Chem. Soc.* **2015**, *137*, 3265.
(5) Vyas, V. S.; Lotsch, B. V.; *Nature* **2015**, *521*, 41.
(6) Yanagida, S.; Kabumoto, A.; Mizumoto, K.; Pac, C.; Yoshino, K. *J. Chem. Soc., Chem. Commun.* **1985**, 474.
(7) Sprick, R. S.; Bonillo, B.; Clowes, R., Guiglion, P.; Brownbill, N. K.; Slater, B. J.; Blanc, F.; Zwijnenburg, M.; Adams, D. J.; Cooper, A. I. *Angew. Chem. Int. Ed.* **2016**, *55*, 1792.
(8) Behrendt, J. M.; Wang, Y.; Willcock, H.; Wall, L.; McCairn, M. C.; O'Reilly, R. K.; Turner, M. L. *Polym. Chem.* **2013**, *4*, 1333.

MOLECULAR ANALOGS OF MoS_2 EDGES FOR HYDROGEN-EVOLUTION ELECTROCATALYSIS AND THEIR APPLICATIONS IN DYE-SENSITIZED SOLAR FUELS

Yiying Wu

Department of Chemistry and Biochemistry, The Ohio State University, 100 W 18th Avenue, Columbus, Ohio 43210, USA
wu@chemistry.ohio-state.edu

Proton reduction is one of the most fundamental and important reactions in nature. MoS_2 edges have been identified as the active sites for hydrogen evolution reaction (HER) electrocatalysis. Designing molecular mimics of MoS_2 edge sites is an attractive strategy to understand the underlying catalytic mechanism of different edge sites and improve their activities. Herein I will discuss our recent progress in this aspect and their applications in dye-sensitized photocatalytic water reduction. The first example is a dimeric molecular analog $[Mo_2S_{12}]^{2-}$, as the smallest unit possessing both the terminal and bridging disulfide ligands.[1] Our electrochemical tests show that $[Mo_2S_{12}]^{2-}$ is a superior heterogeneous HER catalyst under acidic conditions. Computations suggest that the bridging disulfide ligand of $[Mo_2S_{12}]^{2-}$ exhibits a hydrogen adsorption free energy near zero (-0.05 eV). The second example is a group of compounds with the general formular of $MoO(S_2)_2bpy$.[2] By varying substitution on the bpy ligand, we demonstrate the induction effect on the S—S ligand, thereby tuning the rate constants for HER and overpotentials. We have studied homogenous electrocatalytic hydrogen production performance metrics of three catalysts with different bipyridine substitutions. By varying the electron-donating abilities, we present the first demonstration of using the ligand to tune catalytic properties of the S—S bond in molecular MoS_2 edge site mimics. These works can shed light on the relationship between structure and electrocatalytic activity of molecular MoS_2 catalysts and thus is of broad importance from catalytic hydrogen production to biological enzyme functions. At the end, I will present the use of MoS2 molecular analogs coupled with our design of an organic donor–acceptor dye for dye sensitized photocatalytic water reduction.[3] The dye prevents both dye desorption and semiconductor degradation by mimicking the hydrophobic/hydrophilic properties of lipid bilayer membranes. The dual-functional photosensitizer allows for efficient light harvesting while also protecting the semiconductor surface from protons and water via its hydrophobic π linker. The protection afforded by this membrane-mimicking dye gives this system excellent stability in extremely acidic (pH 0) conditions. The result is of interest to tandem dye-sensitized photoelectrochemical cells (DSPECs) for total water splitting.

References

(1) Huang, Z. J., Luo, W. J., Ma, L., Yu, M. Z., Ren, X. D., He, M. F., Polen, S., Click, K., Garrett, B., Lu, J., Amine, K., Hadad, C., Chen, W. L., Asthagiri, A. & Wu, Y. Y. Dimeric [Mo2S12](2-) Cluster: A Molecular Analogue of MoS2 Edges for Superior Hydrogen-Evolution Electrocatalysis. *Angew Chem Int Edit* **54**, 15181-15185 (2015).

(2) Garrett, B. R., Polen, S. M., Click, K. A., He, M., Huang, Z., Hadad, C. M. & Wu, Y. Tunable Molecular MoS2 Edge-Site Mimics for Catalytic Hydrogen Production. *Inorg Chem* **in press** (2016).

(3) Click, K. A., Beauchamp, D. R., Huang, Z. J., Chen, W. L. & Wu, Y. Y. Membrane-Inspired Acidically Stable Dye-Sensitized Photocathode for Solar Fuel Production. *J Am Chem Soc* **138**, 1174-1179 (2016).

CO/TIO_2 AS CATALYST FOR CO_2 REFORMING OF METHANE

*Wei Wei and Yun Hang Hu**

Department of Materials Science and Engineering, Michigan Technological University, 1400 Townsend Drive, Houghton, MI 49931-1295, USA
yunhangh@mtu.edu

Abstract

CO_2 reforming of methane is a unique approach to convert those two green-house gases into syngas (mixture of CO and H_2). However, the large energy requirement of the reforming process has obstructed its application for a century. Solar energy is a solution for the issue. This presentation will show the excellent performance of inexpensive Co/TiO_2 catalysts for visible light photocatalytic CO_2 reforming of methane at relative low temperatures.

Introduction

Carbon dioxide (CO_2) is the typical greenhouse gas. The increase of CO_2 content in the atmosphere has been identified as a primary cause for the global warming over the past century. Methane (CH_4), which can absorb heat ~200 times more efficiently than CO_2, is the second-most prevalent greenhouse gas from human activities. Furthermore, CH_4 is the main component of natural gas. The CO_2 reforming of methane (CRM) (eq. 1), which was investigated as early as 1888(1), would be the most effective way of utilizing these two greenhouse gases.

$$CH_4 + CO_2 = 2CO + 2H_2 \quad (1)$$

CRM produces a synthesis gas (a mixture of H_2/CO), which can be used as a feedstock for the synthesis of liquid fuels and chemicals(2,3). Those economic and environmental incentives promoted intensive research and development (R&D) efforts to develop effective catalysts for the heterogeneous catalytic CRM over the past century(2−6). Unfortunately, no industrial technology has been established for CRM, because of its large energy cost caused by the ultrahigh stability of CO_2 and CH_4. Such a thermodynamically unfavorable feature requires a high temperature (over 1000 K) for the reaction to proceed. A photocatalytic process for CRM would be a promising technology, which can not only lower the reaction temperature, but also exploit solar energy. The research demonstrated that Co/TiO_2 catalyts exhibited excellent performance for visible light photocatalytic CO_2 reforming of methane.

Experimental

Silicon dioxide and quartz wool were mixed with water at weight ratio of 30:30:1 and then heated to 1100 °C for 2 h to form a piece of white solid. The solid piece was shaped as a rectangle disk substrate (1.4 cm × 2.9 cm). The earth abundant element coated TiO_2 catalyst (15 mg) was synthesized with impregnation method and then was dispersed on the light-diffuse-reflection-surface of a 4 cm^2 SiO_2 substrate. Then, the catalyst/substrate was located in a quartz tube reactor and reduced in situ in pure H_2 flow at 500 °C for 2 h. At this temperature reduction of metal oxide would occur, and TiO_2 phase transformation would not happen (7,8). The color of all TiO_2 based catalyst after hydrogen treatment turned into black, which is consistent with Li group's result(9). After cooling down to room temperature, CH_4/CO_2 mixture gas (1:1 ratio and GHSV=40000 ml/gcat/h) flow was introduced into the reactor. Then the catalysts were heated to a selected temperature by an electrical furnace. All products were analyzed by Gas Chromatograph (GC) and Mass Spectrometer (MS).

Results and Discussion

Reaction without light illumination hardly occurred at 500 °C or below. This is a typical thermocatalytic CRM, which requires a high temperature. In contrast, when the catalyst was irradiated by AM 1.5 global sunlight, the reaction started at 400 °C, which is 100 °C lower than that without light illumination. The case is similar when the catalyst was irradiated by visible light, in which the UV light was completely filtered from AM 1.5G global light. Since an in situ thermocouple (contacted to the catalyst surface) for accurate temperature control was employed to eliminate the effect of light illumination on the temperature of the catalyst, the large enhancement of H_2 and CO yields by light illumination is due to the photocatalytic reaction. Mass loading effect of the catalyst were also investigated. It was found that low loading can lessen the light blocking effect of the co-catalyst, thus improving the light utilization by black TiO_2. The quantum efficiency increased with increasing Co metal mass loaded and reaches a maximum when the amount of Co loaded is 20 wt%. A further increase in the amount of Co metal results in a decrease in the activity.

Conclusions

In summary, the combination of Co/black TiO_2 catalyst with the light-diffuse-reflection surface of a SiO_2 substrate created an efficient visible light photocatalytic process for CO_2 reforming of methane. This provides a novel photocatalytic approach for efficient conversions of the two greenhouse gases using solar energy.

Acknowledgement. This work was supported by the U.S. National Science Foundation (NSF-CBET-0931587) and the ACS Petroleum Research Fund (PRF-51799-ND10). Hu also thanks Charles and Carroll McArthur for their great support.

References

(1) Lang, J. Z. *Phys. Chem.*, **1888**, *2*, 161−183.
(2) Ashcroft, A. T.; Cheetham, A. K.; Green, M. L. H.; and Vernon, P. D. F. *Nature*, **1991**, *352*, 225−226.
(3) Hu, Y. H.; and Ruckenstein, E. *Adv. Catal.*, **2004**, *48*, 297−345.
(4) Pakhare, D.; and Spivey, *J. Chem. Soc. Rev.*, **2014**, *43*, 7813−7837.
(5) Bradford, M. C. J.; and Vannice, M. A. *Catal. Rev.: Sci. Eng.*, **1999**, *41*, 1−42.
(6) Hu, Y. H.; and Ruckenstein, E. *Catal. Rev.: Sci. Eng.*, **2002**, *44*, 423− 453.
(7) Yuliati, L.; and Yoshida, H. *Chem. Soc. Rev.*, **2008**, *37*, 1592−1602.
(8) Teramura, K.; Tanaka, T.; Ishikawa, H.; Kohno, Y.; and Funabiki, T. *J. Phys. Chem. B*, **2004**, *108*, 346−354.
(9) Morris, A. J.; Meyer, G. J.; and Fujita, E. *Acc. Chem. Res.*, **2009**, *42*, 1983−1994.

FLOW-ENABLED SELF-ASSEMBLY OF FORMAMIDINIUM LEAD IODIDE LARGE GRAINS FOR HIGH-PERFORMANCE PEROVSKITE SOLAR CELLS

Zhiqun Lin, Ming He, Bo Li, Beibei Jiang, Yanjie He*

School of Materials Science and Engineering, Georgia Institute of Technology, Atlanta, GA 30332
Email: zhiqun.lin@mse.gatech.edu

Introduction

Organolead halide perovskite solar cells have been recognized as promising candidates for high-efficiency solar cells with low costs, the power conversion efficiencies (PCEs) of which have been exceeding 20 %.[1-4] Compared to conventional methylammonium lead iodide ($MAPbI_3$) perovskites, formamidinium lead iodide ($FAPbI_3$) perovskites with low bandgaps, good stability, and negligible hysteresis have recently been widely explored.[3, 5] The major energy loss in organolead halide perovskite solar cells is correlated with the non-radiative charge carrier recombination due to the formation of charge trap states at grain boundaries of perovskite crystals.[6-7] Therefore, fabricating high-quality perovskite thin films with large crystal sizes is beneficial for decreasing grain boundaries, reducing structural defects, suppressing non-radiative recombination loss, and eventually enhancing the photovoltaic performances.[8] For solution-processable perovskite thin films, the key to large-sized perovskite grains is controlling the nuclear generation and crystal growth during the solvent evaporation. Controlled evaporation induced self-assembly (CESA) has been widely utilized to regulate solvent evaporation process, thereby entailing highly-ordered deposition structures of various nonvolatile nanomaterials (i.e., DNA,[9-10] micelles,[11] colloidal nanoparticles,[12] and block copolymers[13]). As a movable CESA, flow-enabled self-assembly offers promising route to controlling the formation of large-area thin films. Herein, we report a facile strategy of fabricating large $FAPbI_3$ crystal grains by FESA, leading to high-efficiency perovskite solar cells. The crystal growth kinetics of the $FAPbI_3$ large grains in the FESA process were also elucidated. This strategy can stand out as an effective strategy for fabricating large-sized, high-quality, and good-stability organolead halide perovskite thin films.

Experimental

For the fabrication of photovoltaic devices, FTO glasses were selectively patterned with Zn powder and HCl, followed by washing with acetone, methanol, and isopropanol. Then FTO glasses were further cleaned by O_3-plasma treatment for 15 min. PEDOT:PSS solution was spin coated on FTO glasses at 5000 rpm for 30s, followed by drying at 100 °C for 30 mins. 0.17 g formamidinium lead, 0.46 g lead iodide, 0.05 g methylammonium bromide, and 0.18 g lead bromide were dissolved in dimethyl sulfoxide by stirring at 70 °C for 2 h to prepare $FAPbI_3/MAPbBr_3$ perovskite solution. The perovskite solution was then subjected to evaporate in a two-nearly parallel-plate geometry constructed by placing a fixed upper plate slightly angled with respective to the lower movable FTO glass substrate that can be precisely controlled by a motorized linear translational stage (Parker Hannifin Corp, mode: MX80LVixBL2b), leading to the formation of $FAPbI_3/MAPbBr_3$ thin films with large grain sizes. 50 mg/ml PCBM in 1,2-Dichlorobenzene was spin coated on the perovskite layer as the electron-transport layer. The Ag electrode was coated by thermal deposition.

Results and Discussion

The evaporation rate of dimethyl sulfoxide was related with the height distance between two-nearly parallel-plate geometry constructed by placing a fixed upper plate slightly angled with respective to the lower movable FTO glass substrate. A radial flow was generated and transported the non-volatile solute (i.e., $FAPbI_3/MAPbBr_3$) towards to the drying front to compensate the volume loss at the contact line where the evaporation rate maximized, leading to the formation of crystal nucleus at the front contact area and promoting the crystal growth with more dimethyl sulfoxide was evaporated. The evaporation of solvent can be controlled by adjusting the moving speed of the substrate, the temperature of the substrate, and the height distance between the top and the bottom plates. Through tuning the evaporate rate of dimethyl sulfoxide, we can control the density of nucleus sites, as well as the growth rates these nucleuses, making the formation of large crystals from these nucleus sites. The $FAPbI_3/MAPbBr_3$ grains could grow over 50 μm, and the PCEs exceeded 16 %. The kinetics analysis revealed that the nucleus site density was mainly determined by the height distance of the top and the bottom plates of FESA and the crystal growth rate of a single grain was determined by the numbers of other grains surrounded it.

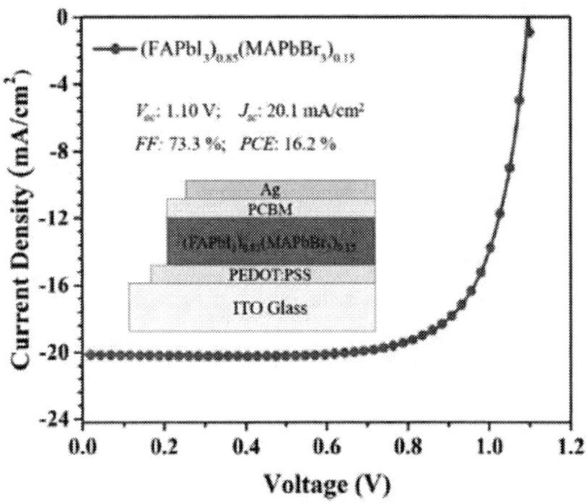

Figure 1. (1) Schematic representation of fabricating formamidinium lead iodide perovskite solar cells by FESA; (2) The *current density-voltage (J-V)* curve of formamidinium lead iodide perovskite solar cells.

References

(1) Kojima, A.; Teshima, K.; Shirai, Y.; Miyasaka, T., *J. Am. Chem. Soc.* **2009,** *131*, 6050.

(2) Liu, M. Z.; Johnston, M. B.; Snaith, H. J., *Nature* **2013,** *501*, 395.

(3) Yang, W. S.; Noh, J. H.; Jeon, N. J.; Kim, Y. C.; Ryu, S.; Seo, J.; Seok, S. I., *Science* **2015,** *348*, 1234.

(4) He, M.; Pang, X.; Liu, X.; Jiang, B.; He, Y.; Snaith, H.; Lin, Z., *Angew. Chem. Int. Ed.* **2016,** *55*, 4280.

(5) Eperon, G. E.; Stranks, S. D.; Menelaou, C.; Johnston, M. B.; Herz, L. M.; Snaith, H. J., *Energy Environ. Sci.* **2014,** *7*, 982.

(6) Yamada, Y.; Nakamura, T.; Endo, M.; Wakamiya, A.; Kanemitsu, Y., *J. Am. Chem. Soc.* **2014,** *136*, 11610.

(7) Ponseca, C. S.; Hutter, E. M.; Piatkowski, P.; Cohen, B.; Pascher, T.; Douhal, A.; Yartsev, A.; Sundstrom, V.; Savenije, T. J., *J. Am. Chem. Soc.* **2015,** *137*, 16043.

(8) Yang, B.; Dyck, O.; Poplawsky, J.; Keum, J.; Puretzky, A.; Das, S.; Ivanov, I.; Rouleau, C.; Duscher, G.; Geohegan, D.; Xiao, K., *J. Am. Chem. Soc.* **2015,** *137*, 9210.

(9) Li, B.; Han, W.; Byun, M.; Zhu, L.; Zou, Q. Z.; Lin, Z. Q., *ACS Nano* **2013,** *7*, 4326.

(10) Li, B.; Zhang, C. C.; Jiang, B. B.; Han, W.; Lin, Z. Q., *Angew. Chem. Int. Ed.* **2015,** *54*, 4250.
(11) Han, W.; Byun, M.; Li, B.; Pang, X. C.; Lin, Z. Q., *Angew. Chem. Int. Ed.* **2012,** *51*, 12588.
(12) Han, W.; Li, B.; Lin, Z. Q., *ACS Nano* **2013,** *7*, 6079.
(13) Han, W.; He, M.; Byun, M.; Li, B.; Lin, Z. Q., *Angew. Chem. Int. Ed.* **2013,** *52*, 2564.

ENGINEERING BISMUTH VANADATE NANOSTRUCTURES FOR PHOTOCATALYTIC AND PHOTOELECTROCHEMICAL WATER SPLITTING

Yun Hau Ng

Particles and Catalysis Research Group (PARTCAT), School of Chemical Engineering, The University of New South Wales, Sydney 2052, Australia
yh.ng@unsw.edu.au

Introduction

Bismuth vanadate ($BiVO_4$) has been an ideal visible light-driven semiconductor with narrow band gap energy of 2.4 eV ($\lambda < 520$ nm).[1] It intrinsically has sufficient absorption within the solar spectrum and stability against photocorrosion. Furthermore, it is inexpensive, environmentally benign, and can be synthesized using numerous facile methods. Sayama et al. prepared $BiVO_4$ thin film on fluorinated tin oxide (FTO) transparent electrode with an improved contact between $BiVO_4$ and the FTO using metal–organic decomposition method.[2] They also reported that the silver ion treatment of $BiVO_4$ enhanced the visible light photocurrent generation by ~3 times. Kisch and coworkers modified $BiVO_4$ with p-type cobalt oxide to form a p–n junction within the composite that improved the charge separation efficiency.[3] Grimes recently demonstrated that the photoelectrochemical properties of $BiVO_4$ could be adjusted by changing the structure of $BiVO_4$ from nanowire to nanopyramid.[4] In general, pure $BiVO_4$ falls short on one main principal barrier, namely, the high charge-carrier's recombination rate. Transport of photogenerated electrons across the particle network randomly exposes them to many grain boundaries and recombination centers prior to their collection at the electrode surface. This work aims to introduce a few considerations when $BiVO_4$ is engineered to achieve better water splitting efficiency.

Experimental

$BiVO_4$ was synthesised using solid-liquid state reaction by mixing equimolar amount (2.5 mmol) of bismuth (III) oxide (Bi_2O_3) and vanadium (V) oxide (V_2O_5) in 25 mL of nitric acid (HNO_3) with two different concentrations: 0.5 M and 1.0 M. The mixture was stirred for four days at room temperature. The vivid yellow $BiVO_4$ product was collected by suction filtration, washed with distilled water and dried at 110 °C.

Photocatalytic O_2 evolution from aqueous $AgNO_3$ solution was executed in an enclosed top-irradiation cell with a Pyrex window via visible light irradiation of 100 mg of $BiVO_4$ suspended in 100 mL of 0.05 M $AgNO_3$ solution. Prior to illumination, the suspension was purged with argon gas for 30 minutes. The amount of O_2 generated was quantified by gas chromatography (TCD, Shimadzu GC-8A, Ar carrier gas). For quantification of the amount of Ag deposited on $BiVO_4$ during water oxidation, a separate but similar reaction was carried out with continuous deaeration of the suspension using argon gas. The suspension was sampled intermittently and the photoreduced product (Ag-loaded $BiVO_4$) was retrieved using centrifugation and dried at 60 °C.

Photoelectrochemical water splitting was performed using a H-cell reactor consist of a $BiVO_4$ working electrode and a Pt wire counter electrode. Hydrogen and oxygen evolved from each compartment reactor are quantified by a Chromatography.

Results and Discussion

Different photoreactivities of different crystal facets on a photocatalyst have commonly been associated with surface reactivity and selective reactant adsorption. Nevertheless, recent observations of facet-selective photooxidation and photoreduction processes on a photocatalyst suggest that photogenerated electrons and holes are spatially driven to different facets, which function distinctively as active reduction or oxidation site. Such phenomenon was first demonstrated by Ohno et al. with titanium dioxide (TiO_2), a benchmark photocatalyst.[5] Interestingly, Li et al. also proved the presence of similar redox functional facets on the visible-light-responsive m-$BiVO_4$, in which {010} facets provided the reduction sites and {110} facets functioned as the oxidation sites.[6] m-$BiVO_4$ with a relatively small band gap of 2.4-2.5 eV has been widely shown to be active for visible-light-triggered organic degradations and oxygen evolution from water, attributing to its deep valence band energy level. Understanding the intrinsic charge properties of m-$BiVO_4$ in relation to its crystal facets is therefore of great interest to allow advanced design of high-performing m-$BiVO_4$.

Despite {110} being proven as the oxidation functional facets, photooxidation performance of m-$BiVO_4$ has been suggested to correlate with its {010} facets. For instance, m-$BiVO_4$ materials with exposed {010} facets exhibited enhanced photooxidation of water and organic molecules. One might argue that the two-dimensional m-$BiVO_4$ nanoplates and nanosheets do not justify the advantage of {010} for photooxidation compared to other facets. It is unclear why photooxidation performs better on {010} dominant m-$BiVO_4$ although {010} facet is the reduction functional surface. Are the beneficial roles of {010} facets in water photooxidation mainly attributed to its higher charge mobility, preferential water adsorption and lower energy barrier than that of {110} facets or could the underlying reasons be due to manipulation of surface charge population and charge transfer efficiency from varying the relative exposure extent of {010} and {110} facets on m-$BiVO_4$?

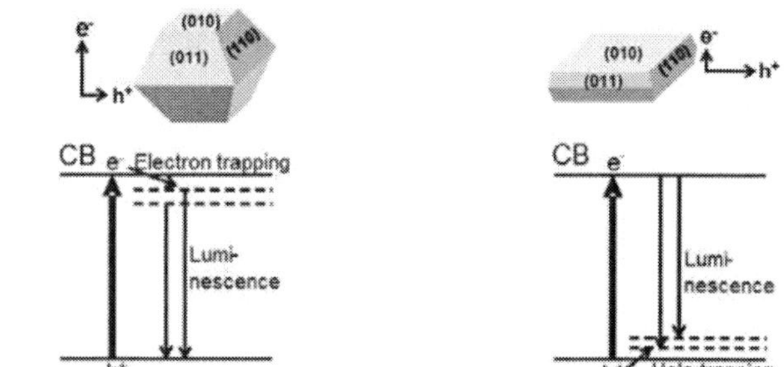

Figure 1. Charge dynamic in BiVO4 with different ratio of exposed planes

In our work, the {010} and {110} crystal facets of monoclinic bismuth vanadate (m-$BiVO_4$) has been demonstrated to be the active reduction and oxidation sites, respectively. Here, we show using dual-faceted m-$BiVO_4$ with distinctly different dominant exposed facet, one which is {010} dominant and the other {110} dominant, contrary to prediction, the former m-$BiVO_4$ exhibits superior photooxidation activities. The population of photogenerated electrons and holes on the surface are revealed to be proportional to the respective surface area of {010} and {110} exposed on m-$BiVO_4$, as evidenced by steady-state photoluminescence (PL) measurements in the presence of charge scavenger. The better photoactivity of {010} dominant m-$BiVO_4$ is attributed to prompt electron transfer facilitated by the presence of more photogenerated electrons on the larger {010} surface. Additionally, the greater extent of electron trapping in {110} dominant m-$BiVO_4$ also deteriorates its photoactivity by inducing electron-hole pair recombination.Each table should have a number and heading, formatted as follows above the table:

Conclusions

In conclusion, although platelike $BiVO_4$ has considerably larger reduction functional {010} facets, it is shown to have superior photooxidation activities than truncated bipyramid $BiVO_4$ with dominating oxidation functional {110} facets. The difference in photocatalytic activity is ascribed to different electron transfer efficiency, which relies on the number of surface available photogenerated electrons that is proportional to the absolute area of $BiVO_4$ {010} facets. The small {010} surface and greater degree of electron trapping of truncated bipyramid $BiVO_4$ hampers efficient electron extraction that results in its poor photoactivity due to augmented holes loss through electron-hole pair recombination. This work demonstrates that crystal facets greatly affect the charge population on the surface of a photocatalyst and suppression of electron trapping through enlargement of {010} is indispensable to producing an effective photocatalyst.

Acknowledgement. We thank the Australian Research Council for financial support. We acknowledge the UNSW Mark Wainwright Analytical Centre for providing access to all the analytical instruments.

References

(1) Ng, Y. H.; Iwase, A.; Kudo, A.; Amal, R. *J. Phys. Chem. Lett.* **2010**, *1*, 2607.
(2) Sayama, K.; Nomura, A.; Zou, Z.; Abe, R.; Abe, Y.; Arakawa, H. *Chem. Commun.* **2003**, 2908.
(3) Long, M.; Cai, W.; Kisch, H. *J. Phys. Chem. C* **2008**, *112*, 548.
(4) Su, J.; Guo, L.; Yoriya, S.; Grimes, C. A. *Cryst. Growth Des.* **2010**, *10*, 856.
(5) Ohno, T.; Sarukawa, K.; Matsumura, M. *New J. Chem.* **2002**, *26*, 1167.
(6) Li, R.; Zhang, F.; Wang, D.; Yang, J.; Li, M.; Zhu, J.; Zhou, X.; Han, H.; Li, C. *Nature Commun.* **2013**, *4*, 1432.

THE IMPACT OF SURFACE DEFECTS ON THE PHOTOACTIVITY OF PHOTOCATALYSTS

Chuanyi Wang, Guohui Dong, Yunqing Zhu, and Jie Zhao*

Laboratory of Environmental Sciences and Technology, Xinjiang Technical Institute of Physics & Chemistry, Chinese Academy of Sciences, Urumqi 830011, China
Email address: cywang@ms.xjb.ac.cn.

Introduction

Semiconductor photocatalysis is gaining increased worldwide attention in the fields of water or air purification, water splitting for hydrogen production, and CO_2 photoreduction etc. Over the past decades, various photocatalysts, have been developed to utilize solar energy for environmental purification and energy conversion. Among these materials, TiO_2 is the most extensively studied one due to its excellent physicochemical properties as well as its earth abundance, nontoxicity and stability. However, the large band gap limits its utilization of sunlight and practical applications. Besides TiO_2, graphitic carbon nitride (g-C_3N_4) is another promising photocatalyst because of its high thermal and chemical stability, semiconductivity, and desirable band gap of 2.7 eV. Unfortunately, the visible light photocatalytic activity of g-C_3N_4 is still low because of the fast recombination of the photogenerated charge carriers. Therefore, it is essential to design and develop new strategies which could efficiently improve the photocatalytic activity of TiO_2 and g-C_3N_4.

Surface defects are very important in the process of photocatalysis. They play an important role in modifying the electronic structure and surface structure and consequently the photocatalytic activity of semiconductor photocatalysts. In this study, the effects of different surface defects on the photocatalytic activity over TiO_2 and g-C_3N_4 were systematically explored. Our work demonstrates that the photocatalytic activity of TiO_2 and g-C_3N_4 could be well turned by the introduction of surface defects.

Experimental

Reduced TiO_2 synthesis. Different amounts of L-ascorbic acid were added to distilled water and stirred for 10 min. Then, 3.1 mL of $TiCl_3$ was added and a purple solution was formed. Subsequently, NaOH solution (1 mol/L) was added to raise the pH to 4. After stirring for 30 min, above solution was transferred to a 100 mL Teflonlined stainless steel autoclave and heated at 180 °C for 12 h.

g-C_3N_4 synthesis. g-C_3N_4 was synthesized by directly heating melamine based on previous reports. Typically, melamine was placed in a covered crucible and heated to 500 °C for 2 h with a heating rate of 20 °C/min, followed by further heat treatment at 520 °C for 2 h.

Defective g-C_3N_4 synthesis. Nitrogen vacancy contained g-C_3N_4 (Nv-g-C_3N_4) was synthesized by calcining the as-prepared g-C_3N_4 in a tube furnace at 520 °C under high purity N_2 flow for 2 h. Carbon vacancy contained g-C_3N_4 (Cv-g-C_3N_4) was synthesized by calcining the as-prepared g-C_3N_4 in a tube furnace at 520 °C under high purity argon gas flow for 2 h.

Characterization Methods. X-ray diffraction (XRD) patterns of the samples were collected on Bruker D8 Advance powder diffractometer using Cu Kα radiation. The electron paramagnetic resonance (EPR) spectra were obtained with Bruker E500 Spectrophotometer. X-ray photoelectron spectra (XPS) of the samples were measured using a Kratos Analytical AMICUS XPS instrument.

Results and Discussion

By controlling the amount of L-ascorbic acid in the precursor of TiO_2, we introduced oxygen vacancies (OVs) in the framework of TiO_2. The defective TiO_{2-x} showed high photocatalytic efficiency in pollutants degradation and in hydrogen evolution under visible light (Figure 1).[1] The cycling test of TiO_{2-x} for hydrogen evolution indicating the good stability of defective TiO_{2-x} under the visible light irradiation.

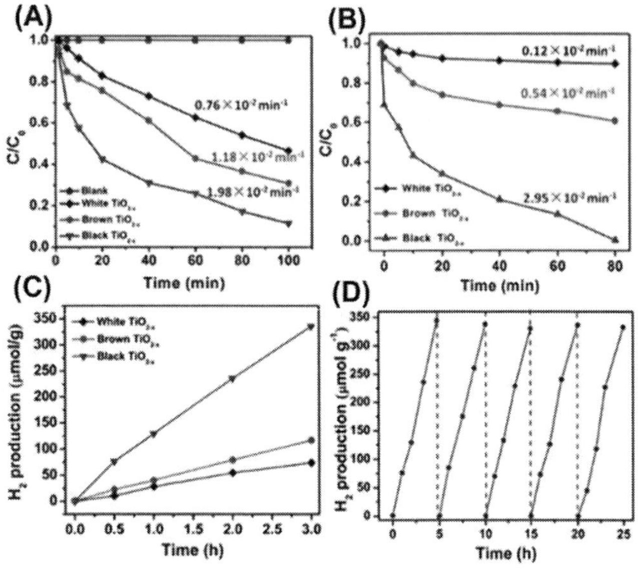

Figure 1. Degradation of MB (A) and phenol (B), hydrogen evolution (C) and cycle test (D) of the defective TiO_{2-x} nanocrystals under visible light.

Based on the results of above research, nitrogen vacancies (NVs) and carbon vacancies (CVs) were respectively introduced on the surface of a metal-free polymer semiconductor, graphitic carbon nitride (g-C_3N_4). It was very interesting to notice that NVs could endow g-C_3N_4 with the photocatalytic N_2 fixation ability (Figure 2).[2] The Photocatalytic N_2 fixation induced by NVs is free from the interference of other gases. Additionally, CVs greatly improved the activity of g-C_3N_4 for H_2O_2 production in the absence of organic scavenger (Figure 3).[3] Moreover, CVs could change the H_2O_2 generation pathway from a two-step single-electron indirect reduction to a one-step two-electron direct reduction.

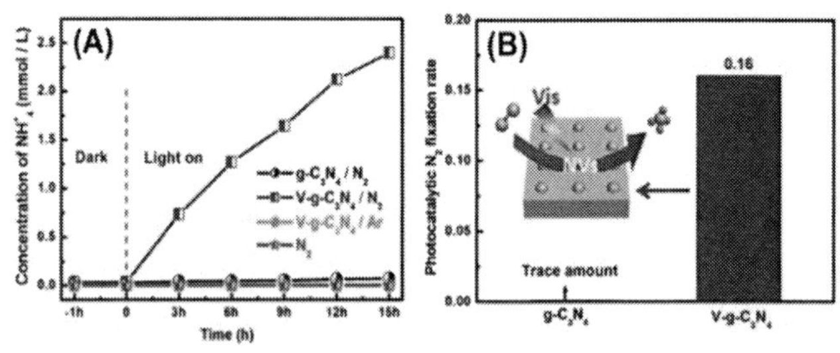

Figure 2. The concentration of generated NH_4^+ in different systems (A); photocatalytic N_2 fixation of g-C_3N_4 and V-g-C_3N_4 (B).

A series of experiments were designed to identify the roles of surface defects in the process of photocatalysis. The presence of surface defects could reduce the symmetry of photocatalysts and narrow down its band gap, thus extending the visible light absorption and increasing the excitable electrons. Moreover, surface defects have been found to be able to trap photogenerated electrons, thus inhibiting the recombination of photogenerated electrons and holes. Meanwhile, they can also enhance the adsorption and activation of gaseous molecules because of their abundant localized electrons.

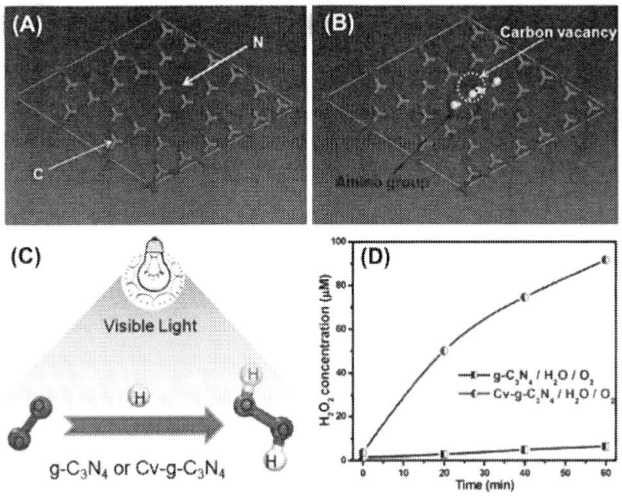

Figure 3. (A) Schematic illustration of g-C_3N_4; (B) carbon vacancy in g-C_3N_4; (C) H_2O_2 production is accomplished by photocatalysis; (D) the concentration of H_2O_2 generated in g-C_3N_4 and Cv-g-C_3N_4 systems.

Acknowledgment. This work was supported by the National Science Foundation of China (Grant No. 21473248), "Western Light" Program of Chinese Academy of Sciences (2015-XBQN-B-06), and the CAS/SAFEA International Partnership Program for Creative Research Teams.

References

(1) Shah, M. W., Zhu, Y. Q., Fan, X. Y., Zhao, J., Li, Y. X., Asim, S., Wang, C. Y., *Sci. Rep.* **2015**, *5*, 15804-15811.

(2) Dong, G. H., Ho, W. K., Wang, C. Y., *J. Mater. Chem. A*, **2015**, *3*, 23435-23441.

(3) Li, S. N., Dong, G. H., Hailili, R., Yang, L. P., Wang, F., Li, Y. X., Zeng, Y. B., Wang, C. Y., *Appl. Catal., B Environ.* **2016**, *190*, 26-35.

COLD THOUGHTS ON PEROVSKITE FEVER

Tao Xu, Jue Gong

Department of Chemistry and Biochemistry
Northern Illinois University
DeKalb, IL60115, United States

Introduction

Benchtop solar cells based on hybrid organic-inorganic lead triiodides perovskite materials have recently been reported with over 20% efficiency, which emerges as the most promising candidate for the next generation photovoltaics.[1] However, despite the ongoing perovskite fever, some cold thoughts are definitely desired to address the dark side of this rising star and to construct the mechanistic foundations for better photovoltaic mateials-by-design. Current grand challenges include the stability issue of $CH_3NH_3PbI_3$,[2] the use of environment-hazardous lead,[3] the cost of complex organics as hole-blocking layer[4], and the use of expensive noble metals as back cathode.[5]

Experimental

$CH_3NH_3PbI_3$ precursory solution: 0.289g of PbI_2 (from Aldrich) and 0.1g CH_3NH_3I were dissolved in 0.52 ml γ-butyrolactone t 60 °C under stirring for 3 hours. $(CH_3NH_3)_2Pb(SCN)_2I_2$ precursory solution: 0.203g $Pb(SCN)_2$ and 0.15g CH_3NH_3I were dissolved in 0.6ml dimethylformamide (DMF) under stirring at 60 °C for 3 hours. The hole transport material (HTM) layer was prepared by dissolving 92 mg 2,2',7,7'-tetrakis(N,N-di-p-methoxyphenylamine)-9,9'-spiro-bifluorene (spiro-MeOTAD), 7.2 mg Li bis-trifluoromethane sulfonimide, and 12 mg 4-tert-butylpyridine in 1 ml chlorobenzene.

Fabrication of the corresponding perovskite solar cells: the perovskite precursory solution was spin-coated at 2000 rmp for 1 min in dry air onto the photoanodes, fabricated as we described previously,[5-7] followed by annealing at 105°C in air for 30 minutes. The HTM layer was spin-coated at 2500 rpm. A layer of 80 nm thick Au film was thermally coated atop the HTM layer using an Edward 306A thermal evaporator. Ni film was coated by sputtering.

Characterization: Potentiostat (Gamry Reference 600) and solar simulator (Photo Emission Inc. CA, model SS50B) were used to acquire the current density-voltage (*J-V*) curves. Scan electron microscope (SEM) images were taken on a Tescan SEM (model Vega II SBH). Diffusive reflectance of the devices was taken on a Filmetrics F20-EXR thin film analyser with an integrating sphere for measuring the reflectance. XRD spectra were collected on a Miniflex X-ray diffractometer.

Results and Discussion

In order to maximize the attainable open circuit voltage (V_{oc}), low chemical potential, namely, high work-function noble metals, such as gold, and silver,[1] are generally used as back cathode. Thermal evaporation of gold is a very costly and wasteful process because only a tiny portion of gold is eventually deposited onto the devices. Therefore, replacing gold with earth-abundant elements as the cathode in perovskite solar cells, while still retaining their high V_{oc} and energy conversion efficiency is a pivotally critical step towards the cost-effective production of perovskite solar cells. In quest for high work-function metals, noble metals such as gold, platinum, or silver often come up as the routine choice. Herein, we break down this stereotype selection rule by perovskite-type solar cells using nickel as the back cathode.[5] Nickel has a work-function of ~5.04 eV, very close to that of gold (~5.1 eV), but with a unit price less than 0.03% of that of gold. As configured in Figure 1, the cell contains a ~900 nm-high rutile TiO_2 nanowires (NWs) array on FTO glass as photoanode, which is filled with a layer of spinning-coated $CH_3NH_3PbI_3$, a 220 nm-thick spinning-coated spiro-MeOTAD (2,2′7,7′-tetrakis-(N,N-di-p-methoxyphenylamine)-9,9′-spirobifluorene) as hole transport material (HTM), capped by a sputtering-coated nickel film as the cathode.

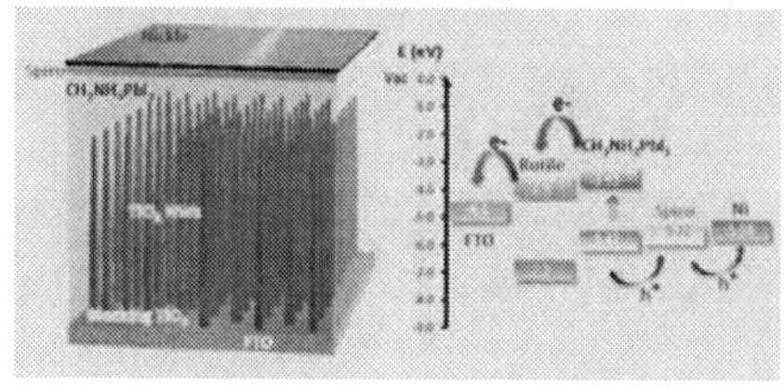

Figure 1. A schematic diagram of a perovskite solar cell consisting of a FTO anode covered with a blocking TiO_2 thin film, rutile TiO_2 nanowires, spin-coated $CH_3NH_3PbI_3$ layer and sprio-MeTAD layer, followed by a sputtered nickel cathode. The left panel shows the energy diagram of the device. [5]

Figure 2 shows the photocurrent density vs photovoltage (J-V) curves of cells using 900 nm TiO_2 nanowires as photoaondes. It can been seen that the cell using 300 nm nickel film as cathode exhibits comparable energy conversion efficiency (η=10.4%) to that of the cell using 80 nm gold film as cathode (η=11.6%). However, the cells using 150 nm nickel film as the cathodes exhibits notable drop in efficiency (η=7.7%). To troubleshoot the source of energy loss in 150 nm nickel-cathoded cells, short circuit current density (J_{sc}) for all cells are first compared. However, both cells exhibit almost the same J_{sc}, about 20 mA/cm^{-2}. Furthermore, V_{oc} of the cell using 150 nm nickel as cathode is 0.752V, only slightly lower than that of the cells using 80nm gold or 300 nm Ni film as cathode. Then, compared to the fill factor (FF) of 68% for the cell using gold as cathode, it is the relatively low FF (50%) found in the cell using nickel as cathode that is responsible for its low conversion efficiency. Since there is no significant difference in V_{oc} and J_{sc} in comparison to the other two cells, we attribute the low FF in cell with 150 nm nickel cathode mainly to its high series resistance. Indeed, shunt resistance (R_{sh}) at J_{sc} for 150 nm nickel-backed cell is 11794Ω, which is about 80% of that for gold-backed cell; while the series resistance (Rs) at V_{oc} for the nickel-backed cell is 264Ω, over 3 times higher than that of the gold-backed cell.

As we further increased the thickness of the nickel film to 300 nm thick, the FF of the cell increases to 0.61 and the conversion efficiency reaches 10.4%. The two-probe measurement on the edge-to-edge resistance of the 300 nm thick nickel pads is measured to be 3Ω and the four-probe sheet resistance of the 300 nm-thick nickel cathodes is measured to be 3~5Ω/square.

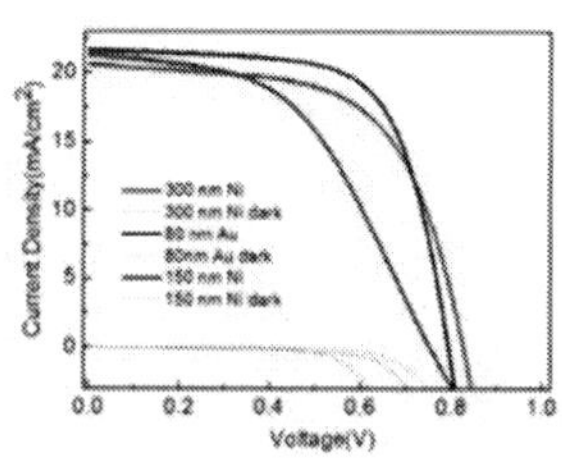

Figure 2. Comparison of J-V characteristic under one Sun, 1.5 global air mass) and in dark for gold-backed and nickel-back cells.[5]

The stability of the photovoltaic performance of the cells using nickel cathodes are also studied. Figure 3a and 3b respectively show the J-V curves of the cells using 150nm Ni and 300 nm Ni film as the cathode. Explicitly, for the 150 nm Ni-cathoded cell, the Jsc decreased from 22.4 mA/cm^{-2} on day 1 to 20.4 mA/cm^{-2} on day 5, and FF increased from 43% on day 1 to 50% on day 5. For the 300 nm Ni-cathoded cell, Voc changed from the 0.70 V on day 1 to 0.75 V on day 5, J_{sc} changed slightly from 20.1 mA/cm^{-2} on day 1 to 20.6 mA/cm^{-2} on day 5, and FF increased from 50% on day 1 to 61% on day 5. We think that the decrease in J_{sc} for the 150 nm Ni-cathoded cell is likely due to the absorption of moisture in the $CH_3NH_3PbI_3$. The overall efficiencies increase from 6.7% on day 1 to 7.7% on day 5 for the 150 nm Ni-cathoded cell, and 7.8% on day 1 to 10.4% on day 5 for the 300 nm Ni-cathoded cell.

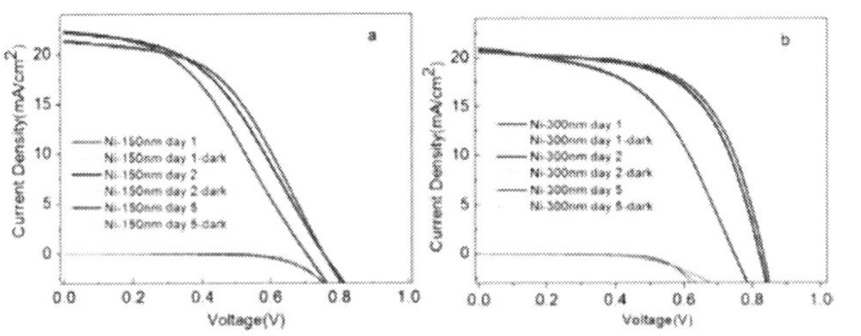

Figure 3. J-V curves for nickel-cathoded cells, (a) 150nm thick Ni and (b) 300 thick nickel, measured on 1 day 1, day 2 and day 5, respectively. Between each measurement, the devices are stored in a desiccator.[5]

Furthermore, perovskite materials decomposes in the moisture air, which is the main barrier for its future commercialization. Pseudohalides have similar chemical behaviors and properties to true halide.[16] In this study, we developed a new perovskite material (CH3NH3)2Pb(SCN)2I2 for perovskite solar cells by replacing some iodides in $CH_3NH_3PbI_3$ with o thiocyanate ions. Figure 4 shows the XRD pattern of $(CH_3NH_3)_2Pb(SCN)_2I_2$ along with $CH_3NH_3PbI_3$ and CH_3NH_3I for comparison.

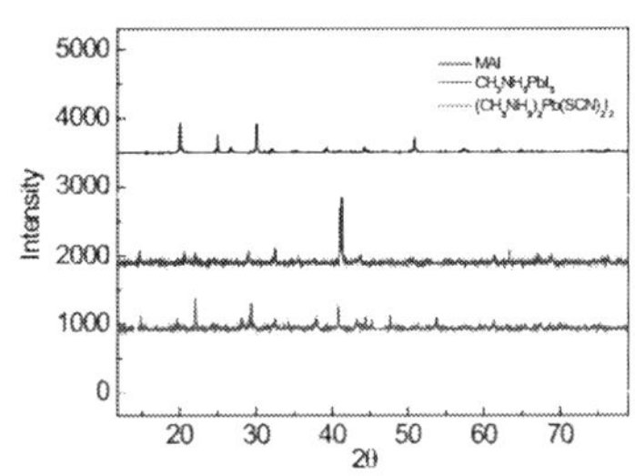

Figure 4. XRD patterns of $(CH_3NH_3)_2Pb(SCN)_2I_2$ along with $CH_3NH_3PbI_3$ and CH_3NH_3I for comparison.[7]

According to the crystal structure of $CH_3NH_3PbX_3$ perovskite material, lead ion and halides form the frame of the perovskite structure.[17] The first step of degradation in moisture involves a formation of a hydrated intermediate containing isolated PbX_6^{4-} octahedra.[13b] The formation constant of lead-halide complex is essentially its equilibrium constant, which reflects the binding tightness between the halides and the central Pb^{2+}. The formation constant K_4 (cumulative formation constant $\beta_n = K_1 \cdot K_2 \cdots K_n$) is calculated to be only 3.5 for PbI_4^{2-},[18] which is in the range of weak interaction between I^- and Pb^{2+}. In the case of $(CH_3NH_3)_2Pb(SCN)_2I_2$, the interaction between Pb^{2+} and SCN^- is much stronger and the formation constant K_4 is up to 7 for $Pb(SCN)_4^{2-}$.[19] Comparing the spheric shape I^- with the linear shape SCN^- as indicated by their Lewis structures, the lone pair of electrons from S and N in SCN- can form stronger ligand with Pb, which in turn stabilizes the frame structure of $(CH_3NH_3)_2Pb(SCN)_2I_2$. We show that direct improvement of moisture tolerance for $(CH_3NH_3)_2Pb(SCN)_2I_2$ has been observed in accelerated stability test compared with traditional $CH_3NH_3PbI_3$ perovskite material.

The accelerated moisture-tolerance tests for perovskite materials were performed by monitoring the reflection of the corresponding perovskite films in 95% relative humidity air at room temperature. Increase of reflection in Figure 3a for $CH_3NH_3PbI_3$ is due to the decomposition of perovskite structure in moisture.[8b] As can be seen, perovskite type $CH_3NH_3PbI_3$ started to decompose immediately after being exposed to moisture. After 1.5 hours, most of the black film has been decomposed to the yellowish color, originated from PbI_2, and the corresponding reflection increases from about 10% to about 20%. In the case of $(CH_3NH_3)_2Pb(SCN)_2I_2$ as shown in Figure 3b, no obvious reflection increase was observed even after 4 hours exposure to moisture. The corresponding reflection increases only 2% after 4 hours of exposure in moisture, from about 6% to 8%. Some tiny white patches emerge at the corners, which are likely due to $Pb(SCN)_2$. In addition, the black color of $(CH_3NH_3)_2Pb(SCN)_2I_2$ film can last for months in air with 40% relative humidity or below. The lower inserted images in figure 3a and 3b compare the $CH_3NH_3PbI_3$ film and $CH_3NH_3Pb(SCN)_2I$ film after 30 days in air at 20~40% relative humidity.

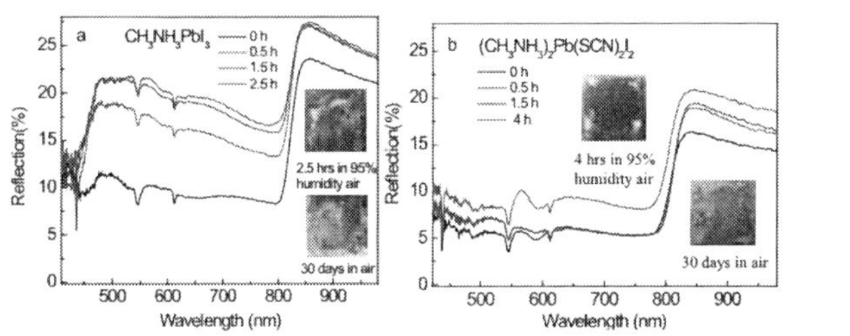

Figure 5. Accelerated stability test for perovskite films (a) $CH_3NH_3PbI_3$ and (b) $(CH_3NH_3)_2Pb(SCN)_2I_2$ in air with 95% relative humidity.[7]

Acknowledgment. This work was supported by the U. S. National Science Foundation (CBET-1150617).

References

(1) Liu, X.; Zhao, W.; Cui, H.; Xie, Y.; Wang, Y.; Xu, T.; Huang, F. *Inorganic Chemistry Frontiers*, **2015**, *2*, 315-335.
(2) Leijtens, T.; Eperon, G. E.; Pathak, S.; Abate, A.; Lee, M. M.; Snaith, H. J., *Nat. Commun.* **2013**, *4*, 2885.
(3) Hao, F.; Stoumpos, C. C.; Cao, D. H.; Chang, R. P. H.; Kanatzidis, M. G., *Nature Photonics* **2014**, *8*, 489-494.
(4) Christians, J. A.; Fung, R. C.; Kamat, P. V., *J. Am. Chem. Soc.* **2014**, *136*, 758-764.
(5) Jiang, Q.; Xia, S.; Shi, B.; Feng, X.; Xu, T. *J. Phys. Chem. C.*, **2014**, *118*, 25878–25883.
(6) Jiang, Q.; Sheng X.; Li, Y.; Feng, X.; Xu, T. **2014**, *50*, 14720-14723.
(7) Jiang, Q.; Rebollar, D.; Gong, J.; Piacentino, E. L.; Zheng, C.; Xu, T. *Agnew. Chem. Int. Ed.*, **2015**, *54*, 7617-7620.

ELEMENTAL LOSS IN THIN-FILM PV ORIGINATED FROM NANOPARTICLES PRECURSORS

Daniela R. Radu[1,2], *Dominik Berg*[1], *Mimi Liu*[1], *Kevin Dobson*[3], *Po-Yu Hwang*[1] *and Cheng-Yu Lai*[1]

[1]Department of Chemistry, Delaware State University, Dover, DE
[2]Department of Materials Science and Engineering, University of Delaware , Newark, DE
[3]Institute of Energy Conversion, University of Delaware,Newark, DE
dradur@desu.edu

Introduction

Nanoparticle (NP) precursors have been found useful in generating the absorber layer in Gen 2 thin-film solar photovoltaics (PV) due to their melting point depression as a result of the nano-confinement. With melting occurring at several hundred degrees lower than the melting point in bulk, the thermal treatment of NP precursors enables fabrication of uniform, crystalline thin-films on inexpensive substrates that only sustain moderate annealing temperatures (such as soda lime glass, SLG).

The operating hypothesis is that the final crystalline film will mirror the nanoprecursors composition prior the thermal treatment, and therefore, tremendous efforts have been reported toward rigorous control of nanoparticles composition. Unfortunately, such efforts often proved themselves futile, as the reactivity owed to nano-confinement also leads to elemental loss during annealing. Our focus is on ternary p-type semiconductors, in the class of Cu and Fe chalcogenides, toward enabling Earth-Abundant PV materials. The presented work, while addressing this general concern regarding the nanoprecursors approach in PV, it will specifically address two materials: iron germanium sulfide (Fe_2GeS_4) and copper tin sulfide (Cu_2SnS_3). Cu_2SnS_3 has been investigated[1] and solar cells have been reported with a power conversion efficiency of 3.66 % in the nanoparticles approach, it seems that the material is far behind its quaternary relative Cu_2ZnSnS_4 (CZTS).

The consideration of Fe_2GeS_4 (FGS) and the isostructural Fe_2SiS_4 stemmed from the large effort of rehabilitating pyrite as a solar material. Recently, a large team of scientists from NREL and Oregon State University[2] has investigated the phenomena related to lack of performance in FeS_2 and pointed to an intrinsic therma instability of the material. The team highlighted that to provide a ligand-field splitting of sufficient magnitude for effective solar absorption the Fe^{2+} ion must be bound by at least six S atoms thus assuring a sufficiently large band gap, which generally requires Fe^{2+} in an octahedral site. The hypothesis that a third element with an electronegativity that favors strong covalent bonding with sulfur can stabilize such a site, lead to investigating Fe_2SiS_4 and Fe_2GeS_4. The analytical evaluations (summary in Table 1, compiled from ref. [2]) drove to the conclusion that the two materials have excellent performance in PV.

Table 1. Fe_2SiS_4 **and** Fe_2GeS_4 **analytical evaluation results [ref.2].**

Cpd.	Calc. Direct Bg(eV)	Meas. Direct Bg (eV)	Enthalpy of Decomp. in Binary Sulfides (eV)	TGA Mass Loss Starting Point (°C)	Calc. Abs. Coeff. (cm^{-1})
Fe_2SiS_4	1.55	1.54	0.59	1000	$>10^5$
Fe_2GeS_4	1.4	1.36	0.64	725	$>10^5$

Experimental

Synthesis of CTS and FGS have followed modified reported procedures, with variations in temperatures, reactants and reaction time, toward obtaining high purity precursors.[3] The nanoprecursors have been characterized for crystal structure and composition by X-Ray diffraction and X-ray fluorescence, while the particles size and morphology were analyzed by HR-TEM. Figure 1 illustrates the typical morphology of FGS nanoparticles. Given that the size of the particles exceeds 100 nm in one dimension, the annealing has proven difficult and the synthesis has been further optimized toward generating particle size in the range of 1-10 nm.

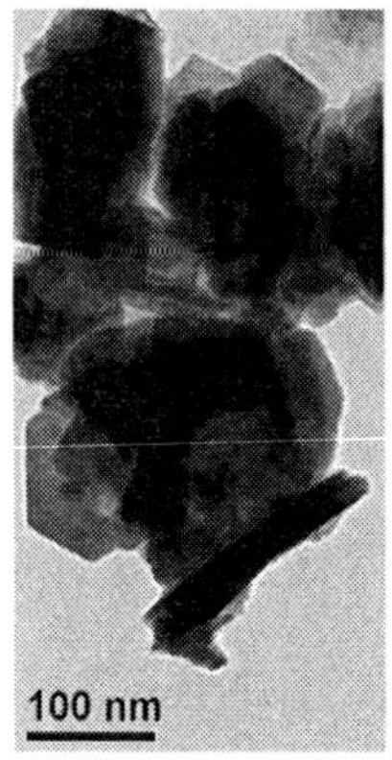

Figure 1. TEM Micrograph of FGS Nanoparticles

The XRD of the nanoparticles, upon annealing in powder, indicated that secondary phases, typically pyrrhotite, a form of magnetite (Figure 2).

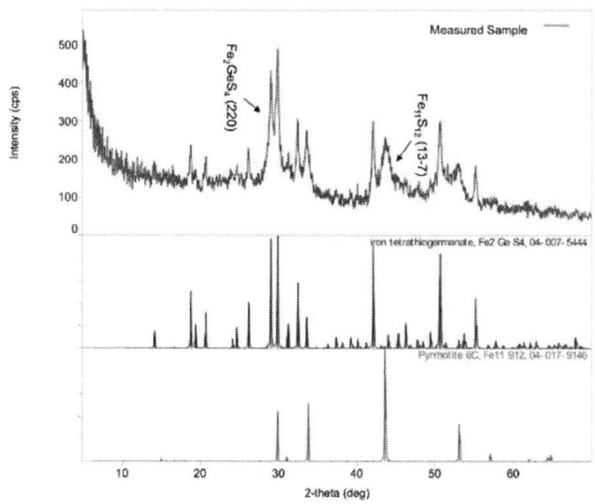

Figure 2. XRD of FGS nanoparticles upon thermal treatment in bulk

Results and Discussion

To our surprise, the annealing of nanoparticles precursors in powders and films, at temperatures as low as 450 °C revealed large Ge loss, albeit the reported high stability of the material in bulk. (Table 1)

Prior reports on chalcogenide absorber fabrication in materials like CZTS(e) pointed to Sn losses through desorption of SnS(e) from CZTS(e) at elevated temperatures (above 400 °C) which impede thorough control of film composition and film homogeneity. The comparison with CTS was originated from similar observation on annealing CTS nanoparticles, where Sn loss has been seen during films annealing.

Mitigation strategies include annealing in the presence of Ge for FGS and SnS and Sn respectively. The strategy have been fruitful for CTS which, when annealed in the presence of SnS, results in films with high purity (Figure 3).

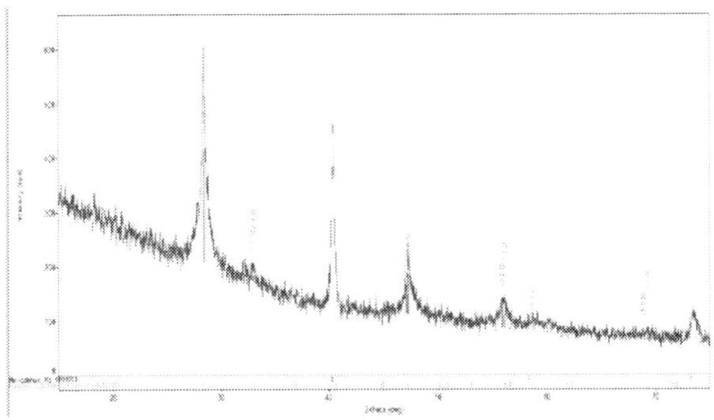

Figure 3. CTS nanoparticles upon annealing on a Mo coated substrate (Mo – unassigned peak)

Conclusions

We have interrogated a variety of reaction conditions that enabled high purity of FGS and CTS and concluded that high temperature reactions, above 280 °C result in increased particle purity and decreased particle size. Unfortunately, the small particle size showed difficulty in XRD pattern identification, due to peak broadening. To understand the precursor crystallinity, such samples have been further annealed in powder and the XRD compared with the original. The composition before and after annealing has been determined by XRF. This approach lead us to discover aforementioned elemental loss, which we have been able to mitigate for FGS and completely eliminate for CTS.

Acknowledgement. The authors acknowledge funding from DOE SunShot, Award # DE-EE0006322 and the generous help from Dr. Robert Birkmire of Institute of Energy Conversion, University of Delaware.

References

(1) Lokhande, A. C.; Gurav, K. V.; Jo, E.; Lokhande, C. D.; Kim, J. H., Chemical synthesis of Cu2SnS3 (CTS) nanoparticles: A status review. *J. Alloys Compd.* **2016,** *656*, 295-310.

(2) Yu, L.; Lany, S.; Kykyneshi, R.; Jieratum, V.; Ravichandran, R.; Pelatt, B.; Altschul, E.; Platt, H. A. S.; Wager, J. F.; Keszler, D. A.; Zunger, A., Iron Chalcogenide Photovoltaic Absorbers. *Advanced Energy Materials* **2011,** *1* (5), 748-753.

(3) (a) Radu, D. R.; Caspar, J. V.; Johnson, L. K.; Rosenfeld, H. D.; Malajovich, I.; Lu, M. Copper zinc tin chalcogenide nanoparticles. WO2010135622A1, 2010; (b) Radu, D. R.; Caspar, J. V.; Lu, M.; Johnson, L. K.; Cao, Y.; Ionkin, A. S.; Malajovich, I.; Rosenfeld, H. D.; Sun, F.; Tassi, N. G., CZTS absorber layer in thin film p-n junction solar devices from quaternary nanoparticle precursors. *Prepr. Symp. - Am. Chem. Soc., Div. Fuel Chem.* **2011,** *56* (Copyright (C) 2013 American Chemical Society (ACS). All Rights Reserved.), 179; (c) Cao, Y.; Denny, M. S.; Caspar, J. V.; Farneth, W. E.; Guo, Q.; Ionkin, A. S.; Johnson, L. K.; Lu, M.; Malajovich, I.; Radu, D.; Rosenfeld, H. D.; Choudhury, K. R.; Wu, W., High-Efficiency Solution-Processed Cu2ZnSn(S,Se)4 Thin-Film Solar Cells Prepared from Binary and Ternary Nanoparticles. *Journal of the American Chemical Society* **2012,** *134* (38), 15644-15647.

LOW OVER-POTENTIAL Ni(111) PARTICLES IN METAL-ORGANIC FRAMEWORKS FOR VISIBLE PHOTOCATALYTIC H2 GENERATION

Gongxuan Lu

Lanzhou Institute of Chemical Physics, Chinese Academy of Sciences, Lanzhou, 730000, China. gxlu@lzb.ac.cn

Here, we reported that high efficient small-sized nickel particles embedded in the frameworks of MOF-5 as a co-catalyst with low over-potential for visible photocatalytic hydrogen evolution. The electrochemical measurements showed a low over-potential of -0.37V, which have the similar over-potential as Pt@MOF-5. Such low over-potential is attributed to their high dispersion (41.8%), small-sized nickel particles (~9 nm), and high specific surface area of MOF-5 (2973 m^2/g). As evidenced by electrochemical impedance spectra (EIS) measurements, Ni nanoparticles with exposed (111) facet were more benefited the electrons transfer from MOF-5 than that of (200) facet one. In addition, Ni@MOF-5 exhibited the larger transient photocurrent and longer fluorescence lifetime. These results led to the high photocatalytic activities and stability for hydrogen evolution sensitized by Eosin Y (EY) over Ni@MOF-5. The rate of hydrogen evolution reached 30.22 $mmol{\cdot}h^{-1}{\cdot}g^{-1}{}_{[Ni]}$ (at pH 11) over Ni@MOF-5 irradiated under visible light irradiation ($\lambda \geq 420$nm) in 2h. The apparent quantum efficiency (AQE) of 16.7% over EY-Ni@MOF-5 was achieved at 430 nm. MOF-5 might promote the photogenerated electrons transfer from excited EY to the hydrogen evolution active sites (Ni), and consequently enhance photocatalytic hydrogen evolution efficiency.

BIOINSPIRED ENGINEERING OF PHOTOTHERMAL MATERIALS FROM BUTTERFLY WINGS

Wang Zhang, Junlong Tian, and Di Zhang *

Department of State Key Laboratory of Metal Matrix Composites
Shanghai Jiao Tong University
800 Dongchuan Road. Shanghai 200240, P. R. China
zhangdi@sjtu.edu.cn (D. Zhang)

Abstract

A macro centimeter-scale Au-CuS combination nanoparticle system integrated with a sub-micrometer antireflection quasi-photonic structure (SAPS) has been synthesized for the first time by utilizing *T.helena* forewing (T_FW) as a biomimetic template (Au-CuS_T_FW). As-fabricated Au-CuS_T_FW exhibits enhanced broadband infrared absorption and lowered reflectance performance. Such an improvement is attributed to the plasmon-to-exciton/plasmon coupling effect among the Au and CuS nanoparticles as well as the coherent coupling between adjacent resonant systems integrated with the SAPS of the T_FW. Moreover, the Au-CuS_T_FW not only possesses a predominant infrared photothermal conversion performance (30.56%), but also is effective for solar photothermal conversion, for the low temperature applications (T< 60 °C). Furthermore, the plasmon coupling effect integrated with the SAPS to enhance broadband infrared absorption and the coherent coupling between adjacent resonant systems enhance hot power yield, which were demonstrated by numerical simulation. Our results offer a new research direction for absorbing and utilizing infrared light, and will open a novel pathway to improve the absorber of the solar thermal collectors for sunlight photothermal and photoelectric conversion, especially, over the infrared range.

Introduction

Along with the development of society and an increasingly serious energy crisis, more and more attention has been attracted towards the development and application of new energy. Solar energy as a clean and renewable energy source provides an effective way to relieve the energy crisis. Over the past decades, numerous attempts have been made to maximize the utilization of solar energy, such as solar cells,[1-6] thermal solar collectors[7] and hybrid photovoltaic-thermal solar collectors[8]. However, most of the current utilization of solar energy only focuses in the range of visible light. In fact, about 99% of solar energy focuses in the wavelength range of 250-4000 nm. And, infrared (IR) lights account for 46.40% of the solar energy, almost the same amount as the visible region (46.41%). Furthermore, much more than 99% of IR irradiance energy focuses in the near IR (NIR) and mid IR (MIR) wavelength ranges, which can directly produce photothermal effect.[9] Thus, it is significant to develop solar energy infrared photothermal conversion owing to its outstanding potential utilization of solar energy. Although solar thermal collectors have been commercialized, further improvement of the solar collectors for power generation is currently limited by their low sunlight absorption rate and conversion efficiency. In addition, to our knowledge, few attentions focused on solar energy infrared photothermal conversion. Evidently, it is necessary to overcome this limitation to improve the absorber and conversion film, which is the heart of a solar collector, with intensive sunlight absorption and effective IR photothermal conversion.

In order to enhance the absorption of sunlight over a broadband wavelength range and promote the IR photothermal conversion, a material with intensive absorption and effective IR photothermal conversion must be obtained. In the past few years, numerous studies have been conducted to obtain effective photothermal agents with more intensive absorption and relatively high photothermal conversion efficiency, such as Au nanoshells (13%), Au nanorods (21%) and copper selenide (Cu2-xSe) nanocrystals (22%) irradiated with 800 nm light,[10] hydrophilic Cu9S5 nanocrystals (25.7%) and Au nanorods (23.7%) irradiated with 980 nm laser[11]. Nevertheless, these photothermal agents only consist of a unitary component, which limited their powerful multifunctionality for various applications. Presently, a considerable number of multicomponent nanoparticle (MNP) systems, including metal-metal,[12, 13] metal-semiconductor[14-17] and semiconductor-semiconductor[18-20] combinations, have been synthesized and attracted a great deal of attention due to their powerful multifunctionality for applications in optics, biomedicine, catalysis, solar energy conversion, electronics, magnetism[14, 21] and photothermal conversion[17, 22]. In addition, a structure with high absorption and low reflectance capability is essential. The self-assembly of these MNPs into thin films and bulk materials, particularly with a related functional sub-micrometer structure, can give these MNPs film tremendous design freedom, geometric complexity and multifunctionality.[21] However, it is difficult to fabricate that with such functional sub-micrometer structure up to macro-scale as a result of the limitations of traditional technologies, such as spontaneous epitaxial nucleation and growth of a second and third component onto seed nanoparticles (NPs).[12, 21]

Biomimetic template synthesis is an efficient method to surmount the limitations of the traditional synthesis technologies, since nature has created a considerable amount of extraordinary structures with given functions.[23, 24] Butterflies are one of the most delicate species, and some of them possess subtle scales that bear complex periodic architecture at the sub-micrometer level. These scales are typical photonic structures which cause an interaction between the light and the physical structures of the scales.[25, 26] Metal[27] or semiconductor NPs[28] assembled into photonic structures using the butterfly wing as a template can be used for surface-enhanced Raman scattering detection,[29] controlling of light propagation[30] and enhancement of light harvesting efficiency,[3] due to the coupling effect of the NPs and the functional structure of the template. Nonetheless, these materials just apply single component NPs coupled with relational functional structure at a macro centimeter-scale. To address these challenges, in this study, a combined metal-semiconductor (Au-CuS) NP system integrated with the sub-micron antireflection quasi-photonic structure (SAPS) was synthesized for the first time at a macro centimeter-scale by using the Troides helena (Linnaeus) (T.helena) forewing (T_FW) as the biomimetic template (Au-CuS_T_FW). Because the black T_FW is extremely absorbing in the visible range of the solar spectrum and effectively convert the absorbed energy into heat energy to warm up the muscles before flying.[31] This synthesis approach does not require expensive equipment and energy intensive processes. As-fabricated Au-CuS_T_FW possesses enhancement broadband light absorption, specifically, over the red and IR ranges, and achieves efficient photothermal conversion performance. Furthermore, we have affixed the Au-CuS_T_FW to the surface of the Cu substrate which is the same as the Cu substrate of BlueTec eta plus, to manufacture the solar absorber (Au-CuS_T_FW_APCF). The solar absorptance of this material is up to 98% over the whole wavelength of 300-2500 nm. In addition, the Au-CuS_T_FW is effective for solar photothermal conversion, for the low temperature applications (T< 60 °C). This work provides an effective approach for the synthesis of a combined metal-semiconductor multicomponent NP film with SAPS on a macro centimeter-scale, which is promising and significant for utilizing IR. These materials can be used as efficient absorbers of solar collectors for photothermal and photoelectric conversion, especially, over the IR range.

Results and Discussion

The black forewing of the T.helena exhibits extremely absorbing in visible light, through the coupling effect of melanin/chitin composite and SAPS of its scales, which has attracted extensive interests.[31, 32] The wing is composed of overlapped tiny elongated rectangular black scales, with two or more layers, forming a roof-tile-like arrangement (Figure 1a). Quasi-periodic triangular roof-type ridges run the length of the scale, as shown in Figure 1b-c. The triangular roof-type ridges can focus light into the scale interior via multiple antireflections, and form the antireflection quasi-photonic structure that facilitates the capture of the light. Declining microribs run down the sides of the ridge that induces internal light scattering assisted in trapping light. A double-row staggered quasi-periodic lattice works are present between every two ridges formed a series of windows. The windows elongate the effective light path and the energy density distribution interspace, implying a potentially important linkage between this feature and the light-harvesting capacity. The ridges, microribs, and windows construct the SAPS. These elements are favorable for light-harvesting that will be explained in this paper via FDTD simulation. The dimensions (d1-d8 and θ, as shown in Figure 1b-c) for the scale of the T_FW were obtained by statistical analysis of several images obtained from the SEM and TEM, as listed in Table S1.[33]

Further details of the results and discussion will be supplied in future work.

Conclusions

In this work, we synthesized a macro centimeter-scale metal-semiconductor (Au-CuS) combination NP system coupling with the SAPS via a simple and promising method that combines chemosynthesis with biomimetic techniques. This synthesis does not require expensive equipment or raw materials and energy intensive processes. Au-CuS_T_FW achieved more intensive enhancement broadband sunlight absorption and lower reflectance, specifically over the red and NIR ranges. Furthermore, the enhancement absorption was also obtained over the MIR range. The reason of this result is that the SPR of the Au NPs generates additional oscillator strength to enhance the excitonic transition of CuS NPs and the free carries SPR of CuS NPs, meanwhile the plasmon-to-exciton/plasmon coupling[35, 36] and the coherent coupling between adjacent resonant systems synergistically combine with the SAPS. Due to the coupling effect of the efficient light-absorbing SAPS (physical factor) and the excellent IR light absorption and photothermal conversion material (chemical factor), the Au-CuS_T_FW not only possesses a predominant infrared photothermal conversion performance (30.56%), but also is effective for photothermal conversion of solar energy, for the low temperature applications (T< 60 °C). By numerical simulation based on FDTD method and the Joule effect, we find that the more intensive heat source density distribute on the adjacent region between two plasmonic structures and the coherent coupling between adjacent resonant systems enhance hot power yield. In addition, the most hot power yields of the Au-CuS_T_W arise from the photothermal materials which covered on the surface of the ridges of the T_FW. We believe that this work provides a guideline for synthesizing metal-semiconductor combinations multicomponent NP systems integrated with the SAPS on a macro centimeter-scale, which is promising and significant for utilizing IR. For the high solar absorptance (98%), the Au-CuS_T_FW_APCF provides a new perspective for manufacturing of an efficient solar absorber, especially over infrared range, with performance-excellence and cost-effectiveness which can be applied in high-end and small devices. The super properties of broadband sunlight absorption and efficient photothermal conversion also can be utilized to the photovoltaic systems, thermophotovoltaic systems, IR imaging and more.

Acknowledgement. This work was supported by the National Natural Science Foundation of China (no. 51572169), Shanghai Rising-Star Program (16QA1402400), Shanghai Science and Technology Committee (15ZR1422400, 14JC1403300 and 14520710100), the National Basic Research Program of China (973 Program, no. 2011CB922200), Research Fund for the Doctoral Program of Higher Education of China (20120073120006 and 20120073130001).

References

(1) X. Li, W.C. Choy, L. Huo, F. Xie, W.E. Sha, B. Ding, X. Guo, Y. Li, J. Hou, J. You. *Advanced Materials* **2012**, 24, 046-3052.

(2) W. Wang, S. Wu, K. Reinhardt, Y. Lu, S. Chen. *Nano letters*, **2010**, 10 ,2012-2018.

(3) W. Zhang, D. Zhang, T. Fan, J. Gu, J. Ding, H. Wang, Q. Guo, H. Ogawa., *Chem. Mater.* **2008**, 21, 33-40.

(4) M. Helgesen, R. Søndergaard, F.C. Krebs., *Journal of Materials Chemistry*, **2010**, 20, 36-60.

(5) T. Ameri, P. Khoram, J. Min, C.J. Brabec., *Advanced Materials* 25 (2013) 4245–4266.

(6) H.A. Atwater, A. Polman., *Nature materials* 9 (2010) 205-213.

(7) F. Motte, G. Notton, C. Cristofari, J.-L. Canaletti., *Applied Energy* 102 (2013) 631-639.

(8) N. Amrizal, D. Chemisana, J.I. Rosell., *Applied Energy* 101 (2013) 797-807.

(9) M. Iqbal. An introduction to solar radiation, *Academic Press*,Orlando, FL, 1983.

(10) C.M. Hessel, V. P. Pattani, M. Rasch, M.G. Panthani, B. Koo, J.W. Tunnell, B.A. Korgel., *Nano Lett.* 11 (2011) 2560-2566.

(11) Q. Tian, F. Jiang, R. Zou, Q. Liu, Z. Chen, M. Zhu, S. Yang, J. Wang, J. Wang, J. Hu., *ACS Nano* 5 (2011) 9761-9771.

(12) X. Huang, S. Tang, B. Liu, B. Ren, N. Zheng., *Adv. Mater.* 23 (2011) 3420-3425.

(13) X. Teng, W. Han, Q. Wang, L. Li, A.I. Frenkel, J.C. Yang. *J. Phys. Chem.C* 112 (2008) 14696-14701.

(14) R. Costi, A.E. Saunders, U. Banin., *Angew. Chem. Int. Ed.* 49 (2010) 4878-4897.

(15) Y.-T. Yen, C.-W. Chen, M. Fang, Y.-Z. Chen, C.-C. Lai, C.-H. Hsu, Y.-C. Wang, H. Lin, C.-H. Shen, J.-M. Shieh., *Nano Energy* (2015).

(16) X. Ding, C.H. Liow, M. Zhang, R. Huang, C. Li, H. Shen, M. Liu, Y. Zou, N. Gao, Z. Zhang., *Journal of the American Chemical Society* 136 (2014) 15684-15693.

(17) S.B. Lakshmanan, X. Zou, M. Hossu, L. Ma, C. Yang, W. Chen., *Journal of Biomedical Nanotechnology* 8 (2012) 883-890.

(18) S. Kim, B. Fisher, H.-J. Eisler, M. Bawendi, *J. Am. Chem. Soc.* 125 (2003) 11466-11467.

(19) Z. Chen, J. Moore, G. Radtke, H. Sirringhaus, S. O'Brien., *J. Am. Chem. Soc.* 129 (2007) 15702-15709.

(20) L. Carbone, P.D. Cozzoli. Colloidal heterostructured nanocrystals: Synthesis and growth mechanisms, *Nano Today* 5 (2010) 449-493.

(21) W. Shi, H. Zeng, Y. Sahoo, T.Y. Ohulchanskyy, Y. Ding, Z.L. Wang, M. Swihart, P.N. Prasad., *Nano Lett.* 6 (2006) 875-881.

(22) C. Yang, L. Ma, X. Zou, G. Xiang, W. Chen., *Cancer Nanotechnology* 4 (2013) 81-89.

(23) A.R. Parker., *Philos Transact A Math Phys Eng Sci* 367 (2009) 1759-1782.

(24) H.-B. Yao, H.-Y. Fang, X.-H. Wang, S.-H. Yu, *Chem. Soc. Rev. 40 (2011) 3764-3785.*

SOLAR ENERGY FOR FUEL PRODUCTION

Yun Hang Hu

Department of Materials Science and Engineering, Michigan Technological University, 1400 Townsend Drive, Houghton, MI 49931-1295, USA
Email: yunhangh@mtu.edu

In the recent years, great efforts were made to develop novel materials and processes for solar energy conversion by my group. For production of fuels by solar energy, highly efficient temperature-induced visible light photocatalytic processes for H_2 production and CO_2 conversion were invented. Large apparent quantum efficiencies above 60% was obtained for entire visible light range.[1-3]

References

(1) B. Han, W. Wei, L. Chang, P. Cheng, Y. H. Hu, Efficient Visible Light Photocatalytic CO_2 Reforming of CH_4, *ACS Catal.* **2016**, *6*, 494.
(2) B. Han, Y. H. Hu, Highly Efficient Temperature-Induced Visible Light Photocatalytic Hydrogen Production from Water, *J. Phys. Chem. C*, **2015**, *119*, 18927.
(3) Hu, Y. H. Highly Efficient Photocatalyst - Hydrogenated Black TiO_2 for Solar Splitting of Water to Hydrogen, *Angew. Chem. Int. Ed.* **2012**, *51*, 12410.

PHOTOELECTROCHEMICAL SOLAR ENERGY CONVERSION USING NEW EARTH-ABUNDANT ELECTROCATALYSTS AND SEMICONDUCTORS

Song Jin

Department of Chemistry, University of Wisconsin-Madison, 1101 University Avenue, Madison, Wisconsin 53706
*e-mail: jin@chem.wisc.edu

The scale of renewable energy challenges not only calls for highly efficient photovoltaic (PV) or photoelectrochemical (PEC) technologies but also abundant, inexpensive, and robust materials. However, significant challenges in material performance, bulk and interface defects, and controlled doping and physical properties must be overcome before the potentials of earth-abundant materials can be fulfilled. New semiconductors for efficient light harvesting and charge separation, and highly active electrocatalysts must be discovered to enable the most efficient and sustainable production of energy using solar cells and PEC water splitting cells. Towards these goals, we have comprehensively elucidated the intrinsic defects in earth-abundant semiconductor iron pyrite (FeS_2) and conclusively explained the origin of its poor solar performance, which will enable us to rationally improve its solar performance. We also report several new earth-abundant electrocatalysts for the hydrogen evolution reaction (HER) and significantly enhanced their catalytic performance. We further established ternary pyrite-type cobalt phosphosulfide (CoPS) as *the best earth-abundant HER catalyst to date* that does not contain expensive noble metals. Nanostructured CoPS can achieve catalytic activity for HER very close to that of platinum with outstanding long-term operation stability. These earth-abundant HER catalysts have been integrated with semiconductors to enable the most efficient solar-driven hydrogen generation devices using earth-abundant materials.

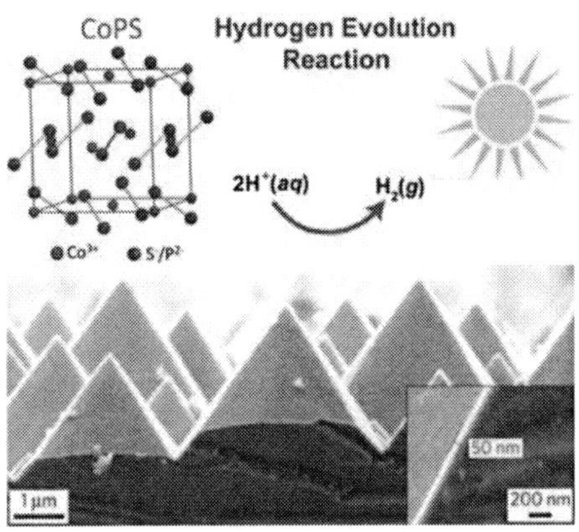

Figure 1. Chemical structure of CoPS as the catalyst for hydrogen evolution reaction (HER) in electrochemical and photoelectrochemical hydrogen evolution driven by solar energy.

References

1. Cabán-Acevedo, M.; Stone, M. L.; Schmidt, J. R.; Thomas, J. G.; Ding, Q.; Chang, H.-C.; Tsai, M.-L.; He, J.-H.; Jin, S.; Efficient Hydrogen Evolution Catalysis Using Ternary Pyrite-Type Cobalt Phosphosulphide *Nature Materials* **2015**, *14*, 1245-1251., DOI:10.1038/nmat4410.
2. Ding, Q.; Zhai, J.; Cabán-Acevedo, M.; Shearer, M. J.; Li, L.; Chang, H.-C.; Tsai, M.-L.; Ma, D.; Zhang, X.; Hamers, R. J.; He, J.-H.; Jin, S.; "Designing Efficient Solar-Driven Hydrogen Evolution Photocathodes Using Semi-Transparent MoQ_xCl_y (Q=S, Se) Catalysts on Si Micropyramids" *Adv. Mater.* **2015**, 27, 6511–6518. DOI: 10.1002/adma.201501884.
3. Zhou, J.; Cabán-Acevedo, M.; Liang, H.; Samad, L.; Ding, Q.; Fu, Y.; Li, M.; Jin, S.; "High-Performance Electrocatalysis for Hydrogen Evolution Reaction Using Se-Doped Pyrite-Phase Nickel Diphosphide Nanostructure" *ACS Catal.* **2015**, *5*, 6355–6361.
4. Faber, M. S.; Dziedzic, R.; Lukowski, M. A.; Kaiser, N. S.; Ding, Q.; Jin, S.; High-Performance Electrocatalysis Using Metallic Cobalt Pyrite (CoS_2) Micro- and Nanostructures *J. Am. Chem. Soc.*, **2014**, *136*, 10053–10061. DOI: 10.1021/ja504099w.
5. Faber, M. S.; Lukowski, M. A.; Ding, Q.; Kaiser, N. S.; Jin, S.; "Earth-Abundant Metal Pyrites (FeS_2, CoS_2, NiS_2, and Their Alloys) for Highly Efficient Hydrogen Evolution and Polysulfide Reduction Electrocatalysis" *J. Phys. Chem. C* **2014**, *118*, 21347–21356. **DOI:** 10.1021/jp506288w.
6. Faber, M. S.; Jin, S.; Earth-Abundant Inorganic Electrocatalysts and Their Nanostructures for Energy Conversion Applications *Energy Environ. Sci.*, **2014**, *7*, 3519-3542. **DOI:** 10.1039/C4EE01760A.
7. Cabán-Acevedo, M.; Kaiser, N. S.; English, C. R.; Liang, D.; Thompson, B. J.; Chen, H.-E.; Czech, K. J.; Wright, J.C.; Hamers, R. J.; Jin, S.; Ionization of High-Density Deep Donor Defect States Explains the Low Photovoltage of Iron Pyrite Single Crystals *J. Am. Chem. Soc.* **2014**, *136*, 17163.

TOWARDS HARVESTING MORE PHOTONS BY DEVELOPING NANOSTRUCTURED MATERIALS AND ASSERMBLIES

Fuqiang Ren, Jianming Zhang, Belete Gonfa, Zhenhe Xu, and Dongling Ma

Institut National de la Recherche Scientifique (INRS)
Center of Energy, Materials and Telecommunications
1650 Boulevard Lionel-Boulet, Varennes, Quebec, Canada
ma@emt.inrs.ca

Introduction

Efficiently harvesting visible and near infrared (NIR) photons represents an attractive approach to improve the efficiency of solar-to-electricity conversion, solar-to-fuel conversion and photocatalysis. Plasmonic nanostructures with unique surface plasmon resonance have recently been explored for enhancing solar energy harvesting in the visible and NIR regimes. On the other hand, NIR quantum dots (QDs) with size tunable bandgaps and high potential for multiple exciton generation represent a class of promising materials for new generations of solar cells. In this talk, I will present our recent work on the synthesis of plasmonic nanostructures, NIR QDs, and related assemblies as well as their applications in solar cells, solar fuel and photocatalysis.

Experimental

PbS QDs were synthesized by the hot injection method.[1] Briefly, in this type of synthesis, sulfur precursor (such as bis(trimethylsilyl) sulfide) will be swiftly injected into lead precursor (such as lead acetate) in the presence of oleic acid, trioctylphosphine and octadecene at an elevated temperature, followed by quenching. The QDs were then purified by repeated centrifugation and re-dispersion. PbS@CdS core@shell QDs were synthesized by a cation exchange approach.[2,3] Briefly, the solution of CdO in the mixture of oleic acid and octadecene was injected into PbS QDs/toluene solution and allowed to react at different temperatures for certain time to obtain the core@shell QDs of different shell thickness. Depleted bulk heterojunction solar cells were fabricated by spin coating QD solution layer by layer on the film of TiO_2 nanorod arrays grown via a hydrothermal method.[4]

Au nanoparticles (NPs) were fabricated by laser ablation with a KrF excimer laser (wavelength: 248 nm; repetition rate 20 Hz).[5] The beam was focused by an objective lens with a focal length of 7.5 cm onto a gold plate target. Such prepared Au NPs were then mixed with $NaYF_4$:Yb^{3+}, Er^{3+}, Tm^{3+} (NYF)-core@porous-TiO_2-shell core@shell microspheres and loaded into the porous TiO_2 shell to realize plasmon enhanced photocatalysis.[6] Both monodisperse NYF microspheres and porous TiO_2 shells were synthesized by hydrothermal approaches, while in separate steps.

Au nanostars were synthesized in two steps.[7] Au NP seeds of 20 nm in diameter were prepared by standard citrate reduction. The Au NP seeds were then coated with poly(vinylpyrrolidone) (PVP). For the growth of nanostars, the solution of PVP in N,N-dimethylformamide was mixed with $HAuCl_4$ aqueous solution, followed by the addition of PVP-coated Au seed solution. The reaction was allowed to proceed for certain time at room temperature under constant stirring. The Au nanostars were finally centrifuged, collected and re-dispersed in water. These Au nanostars were spin coated and integrated into QD-based solar cells for plasmon-enhanced solar cell applications.

Solar cell devices were characterized by conducting current-voltage (I-V) measurements using an Agilent B2901A instrument in ambient atmosphere in dark and under calibrated AM1.5 simulated sunlight from an Oriel Sol3A Class AAA Solar Simulator (Oriel Instruments, Newport Corporation). The external quantum efficiency of the solar cell devices was measured using Oriel® IQE-200™ quantum efficiency measurement system. The photocatalytic activity of the NYF@TiO_2-Au core@shell microspheres was investigated by monitoring the photodegradation of methyl orange using an absorption spectrometer under illumination of a commercial photoreactor (LUZ-4V, Luzchem), a 300 W xenon lamp (with an optical filter, $\lambda > 420$ nm), a 2 W 980 nm diode laser or a 150 W solar simulator.

Results and Discussion

As the first demonstration, PbS@CdS QDs were deposited in ambient atmosphere on vertically grown TiO_2 nanorod arrays to make bulk heterojunction solar cells.[4,8] The devices showed about 40% higher efficiency than similar devices involving PbS QDs processed in inert atmosphere, clearly showing the advantages of the core@shell QDs. For comparison, solar cells based on PbS QDs fabricated via laser deposition will also be briefly presented.[9] The integration of Au nanostars further increased the power conversion efficiency by over 10%, due to enhancement of absorption via both the near field and scattering effects.

With the combination of the NIR-absorbing upconverting core, UV-absorbing porous TiO_2 shell further loaded with visible-absorbing plasmonic Au NPs, the novel Au NP decorated NYF@TiO_2 core@shell microspheres can harvest solar photons over a wide spectral range from UV to NIR and exhibited significantly better photocatalytic performance than the benchmark Degussa P25.[6] The enhanced activity is attributed to synergistic effects from nanocomponents arranged into the nanostructured architecture in such a way that favours the efficient charge/energy transfer among nanocomponents and largely reduced charge recombination.

Conclusions

Hybrid nanostructures based on quantum dots, plasmonic and catalytic nanostructures were developed for solar cell and photocatalysis applications and showed enhanced properties.

Acknowledgement.

We thank the financial support from the Natural Sciences and Engineering Research Council of Canada, Fonds de recherche du Québec-Nature et technologies, Quebec Center for Functional Materials (CQMF), and strategic Network "Plasma-Québec".

References

(1) Hines, M.A.; and Scholes, G.D. *Adv. Mater.* **2013**, *15*, 1844.
(2) Zhao, H.; Chaker, M.; Wu, N.; and Ma, D. *J. Mater. Chem.* **2011**, *21*, 8898.
(3) Zhao, H.; Chaker, M.; and Ma, D. *J. Mater. Chem.* **2011**, *21*, 17483.
(4) Gonfa, B.; Zhao, H.; Li, J.; Qiu, J.; Saidani, M.; Zhang, S.; Izquierdo, R.; Wu, N.; El Khakani, M. A.; Ma, D.; *Sol. Energ. Mat. Sol. Cells* **2014**, *124*, 67.
(5) Li, S.; Zhang, J.; Kibria, M. G.; Mi, Z.; Chaker, M.; Ma, D.;* Nechache, R..;* and Rosei, F.* *Chem. Commun.* **2013**, *49*, 5856.
(6) Xu, Z.; Quintanilla, M; Vetrone, F.; Govorov, A. O.; Chaker, M.; and Ma, D. *Adv. Funct. Mater.* **2015**, *25*, 2950 (**Front Cover**).
(7) Elbohy, H.; Kim, M.; Dubey, A.; Reza, K. M.; Ma, D.;*, Zai, J.;* Qian, X.;* and Qiao, Q.;* *J. Mater. Chem. A* **2016**, *4*, 545.
(8) Gonfa, B.; Kim, M.; Delegan, N.; Tavares, A. C.; Izquierdo, R.; Wu, N.; El Khakani, M. A.; and Ma, D. *Nanoscale* **2015**, *7*, 10039.
(9) Ka, I.; Gonfa, B.; Le Borgne, V.; Ma, D.;* El Khakani, M. A.* *Adv. Funct. Mater.* **2014**, *24*, 4042 (***Inside Front Cover***).

ON THE DESIGN OF NEW LOW COST PHOTOCATALYSTS FOR EFFICIENT SOLAR WATER OXIDATION

Lionel Vayssieres

International Research Center for Renewable Energy (IRCRE), Xi'an Jiaotong University, 710049 Xi'an, P.R. China
lionelv@xjtu.edu.cn

Introduction

An extensive rightfully rehabilitated interest in the field of artificial photosynthesis and solar water splitting has occurred worldwide within the last decade due to global climate change seriously compromising our environment as well as the exponential increase in energy demand from emerging countries[1]. Transition metal oxides offer a wide range of applications for renewable energy systems and solar energy conversion. Most oxides are stable, non toxic, earth-abundant, easy to prepare on a large scale and at relatively low cost in various forms ranging from dispersions, powders, thin films, dense ceramics and ordered arrays at various dimensional scales, i.e. from nano- to micro- to macro-scales. They also exhibit the widest range of electrical, optical, and electronic properties and are highly suitable for low cost solar water splitting implementation in our societies[2].

Design strategy

Given that conventional technologies which attempt to improve the performance of existing materials and devices for photocatalysis and photovoltaics by further development along the same incremental approach are reaching their limits, it is crucial to develop novel multifunctional materials where bulk limitations are overcome by changing the fundamental underlying physics and chemistry, by e.g. quantum confinement and interfacial effects. Our approach is to develop novel design strategies for the purpose-built fabrication of metal oxide semiconductor heteronanostructures consisting of oriented arrays of quantum dots and quantum rods of high purity entirely produced from aqueous chemical growth at low temperature without surfactant nor template[3] and with precise controlled size and water-interfacial chemistry[4,5] with intermediate bands for high visible light energy conversion, highly quantized band structure for bandgap and optimized band edge for stability against photocorrosion and operation conditions in aqueous solutions at neutral pH without sacrificial agent[6]. Such unique characteristics, combined with the in-depth investigation of their controlled dimensionality and size-dependent[7] and interfacial[8] electronic structure[9-13], optical, photoelectrochemical[14-16] as well as structural and conductivity[17] properties do provide a substantial advance in fundamental understanding as well as structure-efficiency relationships for a cost effective and sustainable generation of hydrogen from the two most abundant and geographically-balanced free resources, i.e. the sun and seawater.

The use of co-catalysts for hydrogen and oxygen evolution reactions, both in molecular and solid-state form, has become frequent to increase the efficiency of the devices[18]. To combine molecular co-catalysts and inorganic semiconductors in highly ordered nanoarchitectures, we have developed a new design strategy based on the spontaneous photoelectric-field enhancement inherent in pyramid-like morphologies[19]. Indeed, according to simulations, strong electrical fields are spontaneously generated at the very tip/apex of nanopyramidal/nanocone structures. Consequently, a localized enhanced charge separation could be achieved and thus, highly specific site-selective depositions could be reached at the tip of nanopyramids. These hierarchical vertically oriented arrayed structures would exhibit an active molecular co-catalyst selectively deposited at the apex of a visible light-active large bandgap semiconductor nanopyramid-arrays. With this unique architecture, enhancing spatially the photogenerated charge carriers separation, efficient photoelectrochemical water oxidation activity and stability could be achieved in neutral aqueous medium and without sacrificial agents. A very successful example of this new design strategy has been achieved by the low cost fabrication of very large arrays of highly oriented $BiVO_4$ nanopyramids where only their apex is coated with photodeposited spherical nanoparticles of amorphous Co-Pi, a well known molecular co-catalyst for oxygen evolution. Results show good efficiency and stability for water oxidation at pH 7[19].

Conclusions

A new materials fabrication strategy has been demonstrated utilizing the spontaneous photo-induced electric field enhancement effect inherent in pyramidal-like morphologies. Novel visible light-active heteronanostructures consisting of oriented nanopyramid-arrays (of a large bandgap semiconductor) with (molecular co-catalyst) nanoparticles grown exclusively on the apex were successfully fabricated. These new hierarchical highly ordered and homogeneous structures combining stable and earth abundant elements and simple fabrication techniques are setting a new standard in smart (photocatalysts) design. This novel strategy was demonstrated here for the large scale and low cost development and operation for heterojunction-based photoelectrodes which offers great expectations for environmental remediation and water purifications as well as solar water splitting in neutral solutions. Furthermore, given the rather universal capability of this approach, innovative multifunctional heteronanostructures could easily be generated for various classes of inorganic materials combining redox molecular chemistry, photochemistry and hydrothermal synthesis as well as for the design of model systems for better fundamental understanding of structure-property relationships.

Acknowledgement. This work was supported by the Thousand Talent Plan, National Natural Science Foundation of China (No. 51202186).

References

(1) Y. Tachibana, L.Vayssieres, J. R. Durrant, *Nat. Photonics* **2012**, *6*, 511
(2) C. X. Kronawitter *et al*, *Energy Environ. Sci.* **2011**, *4*, 3889
(3) L. Vayssieres, *Appl. Phys. A* **2007**, *89*, 1; *Int. J. Nanotechnol.* **2007**, *4*, 750; *Int. J. Nanotechnol.* **2004**, *1-2*, 1; *Angew. Chem. Int. Ed.* **2004**, *43*, 3666; *Adv. Mater.* **2003**, *15*, 464; *J. Phys. Chem. B* **2003**, *107*, 2623; *Nano Lett.* **2002**, *2*, 1393; *Chem. Mater.* **2001**, *13*, 4395; *Chem. Mater.* **2001**, *13*, 233; *J. Phys. Chem. B* **2001**, *105*, 3350; *J. Nanosci. Nanotechnol.* **2001**, *1*, 385; *Pure Appl. Chem.* **2000**, *72*, 47
(4) L. Vayssieres, *Int. J. Nanotechnol.* **2005**, *2*, 411; *J. Colloid Interface Sci.* **1998**, *205*, 205
(5) L. Vayssieres, *J. Phys. Chem. C* **2009**, *113*, 4733
(6) *On Solar Hydrogen & Nanotechnology*, L. Vayssieres ed. (Wiley, 2010), Chapter 17, pp. 523-558
(7) L. Vayssieres *et al*, *Appl. Phys. Lett.* **2011**, *99*, 183101; *Adv. Mater.* **2005**, 17, 2320
(8) C. X. Kronawitter *et al*, *Nano Lett.* **2011**, *11*, 3855; *Phys. Rev. B* **2012**, *85*, 125109; *J. Phys. Chem. C* **2012**, *116*, 22780
(9) C. X. Kronawitter *et al*, *Energy Environ. Sci.* **2014**, *7*, 3100
(10) F. J. Himpsel *et al*, *J. Electron Spectrosc. Relat. Phenom.* **2013**, *190*, 2
(11) C.X. Kronawitter *et al*, *PhysChemChemPhys* **2013**, *15*, 13483
(12) C. L. Dong *et al*, *Phys. Rev. B* **2004**, *70*, 195325
(13) J.-H. Guo *et al*, *J. Phys.: Condens. Matter* **2002**, *14*, 6969
(14) N. Beermann *et al*, *J. Electrochem. Soc.* **2000**, *147*, 2456
(15) K. Keis et al, *J. Electrochem. Soc.* **2001**, *148*, A149
(16) T.Lindgren *et al*, *Sol. Energy Mater. Sol. Cells* **2002**, *71*, 231
(17) J. Engel *et al*, *Adv. Funct. Mater.* **2014**, *24*, 4952
(18) J.Z. Su and L.Vayssieres, *ACS Energy Lett.* **2016** *Invited Review*

(19) Y.K. Wei *et al*, *Nano Res.* **2016** DOI 10.1007/s12274-016-1050-9

TUNING THE CHEMISTRY OF MIXED LEAD HALIDE PEROVSKITES

Prashant V. Kamat and Seog Joon Yoon

Department of Chemistry & Biochemistry and Radiation Laboratory,
University of Notre Dame, Notre Dame, Indiana 46556, USA
Email:pkamat@nd.edu, **website**: http://kamatlab.com
twitter:kamatlabND

Understanding chemistry of organic-inorganic lead halide perovskites is important for achieving high photovoltaic performance. By tuning precursor stoichiometry crystallization kinetics can be tailored to yield improved thin film homogeneity. An interesting property of mixed halide lead perovskites (e.g.,$CH_3NH_3PbI_{3-x}Br_x$) is phase segregation to create Iodine-rich and Bromide- rich regions when subjected to visible irradiation. This intriguing aspect of halide ion movement in these mixed halide films can be tracked from the changes in the photoluminescence and absorption spectra. In order to probe the reversibility of this effect we employed emission and transient absorption spectroscopy tools and succeeded in elucidating the time scale with which such separation occur under laser (405 nm) irradiation followed by dark recovery. Implication of such phase segregation in mixed halide perovskite solar cell will be discussed.

HIGH EFFICIENCY AT LOW APPLIED VOLTAGE FROM Sb-DOPED $SnO_2/BiVO_4$ CORE/SHELL NANOROD-ARRAY PHOTOANODES

*Lite Zhou[1,2], Lyubov V. Titova[3] and Pratap M. Rao[1,2]**

[1]Department of Mechanical Engineering, Worcester Polytechnic Institute, Worcester, MA 01609, USA
[2]Materials Science and Engineering Graduate Program, Worcester Polytechnic Institute, Worcester, MA 01609, USA
[3]Department of Physics, Worcester Polytechnic Institute, Worcester, MA 01609, USA
*pmrao@wpi.edu

Introduction

$BiVO_4$ has become the top-performing semiconductor among photoanodes for photoelectrochemical water oxidation. However, $BiVO_4$ photoanodes are still limited to a fraction of the theoretically-possible photocurrent at low applied voltages because of modest electron transport properties and a trade-off between light absorption and charge separation efficiencies. One effective strategy to improve $\eta_{abs} \times \eta_{sep}$ is to coat a thin layer of $BiVO_4$ onto high-surface-area nanowire-arrays composed of much more conductive oxides, such as WO_3.[1-3] In the resulting core/shell nanowires, η_{abs} is enhanced due to long light absorption paths in $BiVO_4$ along the nanowire axial direction and η_{sep} is enhanced both by the small thickness of $BiVO_4$, which allows a short transport distance for electrons in $BiVO_4$ and facile axial transport of electrons to the current collector through the more conductive nanowires. In this study, we investigate photoanodes composed of thin layers of $BiVO_4$ coated onto Sb-doped SnO_2 (Sb:SnO_2) nanorod-arrays (Sb:$SnO_2/BiVO_4$ NRAs - Figure 1a), and demonstrate that Sb:SnO_2 nanorods are more favorable as a core material than the previously-investigated WO_3 nanowires, leading to the highest $\eta_{abs} \times \eta_{sep}$ product reported to date for any $BiVO_4$-based photoanode.

Experimental

Synthesis. The Sb:$SnO_2/BiVO_4$ NRA photoanodes were synthesized by a combination of hydrothermal synthesis of Sb:SnO_2 NRAs and drop-casting of $BiVO_4$ onto the NRAs. 0.474 g of $SnCl_2$ and 0.25 g of PEG 6000 were dissolved into 10 mL of 2-methoxyethanol to prepare the SnO_2 seed layer solution. Then, the substrates were seeded by spin coating the SnO_2 seed layer solution and annealed at 550 °C for 2 hours. The recipe for the hydrothermal growth of Sb:SnO_2 NRAs was adapted from a previous report.[4] Two pieces of seeded substrates of dimensions 2 cm x 2.5 cm were placed face-down against the wall of a Teflon-lined autoclave (50 mL) containing a mixture of 15 mL of DI water, 15 mL of ethanol, 1.2 mL of HCl (38%), 135 mg of $SnCl_4.5H_2O$ and 0.043 mL Sb-dopant solution for hydrothermal synthesis of Sb:SnO_2 NRAs. The dopant solution was prepared by adding 0.0228 g of $SbCl_3$ in 10 mL ethanol. The hydrothermal reaction was performed 10 times at 200 °C for 12 h each to obtain Sb:SnO_2 NRAs of the desired length. Finally, the nanorod films were annealed at 550 °C for 30 minutes in air. Each 2 cm x 2.5 cm piece was then cut into two pieces to yield four pieces measuring 2 cm x 1.25 cm each. The solution for drop-casting $BiVO_4$ was composed of 0.1225g $Bi(NO_3)_3.5H_2O$ and 0.0663g $OV(C_5H_7O_2)_2$, dissolved in 5mL acetic acid and 0.25mL acetylacetone by 30 minutes sonication. 2 drops of 5 μL solution were then dropped onto the Sb:SnO_2 NRAs, which was then dried on a hotplate in air at 500°C for 10mins to make one $BiVO_4$ drop-casting layer. After 3 layers of drop-casting, the sample was annealed in a furnace in air at 550°C for 2 hours to crystallize the $BiVO_4$ and form the core/shell Sb:$SnO_2/BiVO_4$ NRA photoanode.

Instrumentation and Measurements. The Sb:$SnO_2/BiVO_4$ NRA photoanodes were characterized by x-ray diffraction (XRD), scanning electron microscopy (SEM), and optical absorption measurements using an integrating sphere. Photoelectrochemical behavior of the Sb:$SnO_2/BiVO_4$ NRAs was evaluated by measuring the current density-voltage (J-V) curve at a scan rate of 10mV/s and wavelength-dependent incident photo-to-current conversion efficiency (IPCE) under back-side (glass-side) illumination in a three-electrode configuration using a potentiostat (Model SP-200, BioLogic), with the photoanode as working electrode, a saturated calomel electrode (SCE) as the reference electrode and a Pt wire as the counter electrode. The entire exposed area of the photoanode (1.44cm^2) was used in all calculations of photocurrent density. Since our goal is to measure the $\eta_{abs} \times \eta_{sep}$ product rather than the photocurrent for water oxidation (J_{H2O}), both the J-V curve and the IPCE were measured in 0.5M potassium phosphate aqueous solution buffered to pH 7 with 1M sodium sulfite (Na_2SO_3) added as a hole scavenger. Because of the fast oxidation kinetics of sulfite, the surface charge transfer efficiency of the photoanode (η_{trans}) is nearly equal to 100%, allowing the $\eta_{abs} \times \eta_{sep}$ product to be calculated from $\eta_{abs} \times \eta_{sep} \approx J_{sulfite}/J_{max}$, where $J_{sulfite}$ is the photocurrent density for sulfite oxidation. The simulated solar illumination was provided by a Xe lamp (Model 6258, Oriel) equipped with an AM1.5G filter (Model 81094, Oriel), and the spectral irradiance was measured by a spectrometer (Model USB2000+Rad, Ocean Optics). The electrical conductivity of the Sb:SnO_2 and undoped SnO_2 NRA films was measured using terahertz time-domain spectroscopy (THz-TDS).[5] THz pulses were generated by optical rectification of 800 nm, 100 fs pulses from a 1 kHz amplified Ti:sapphire laser source in a [110] ZnTe crystal, and coherently detected by free-space electro-optic sampling in a second [110] ZnTe crystal.

Results and Discussion

Monoclinic $BiVO_4$ and tetragonal SnO_2 are the only phases detected by XRD for the Sb:$SnO_2/BiVO_4$ NRAs grown on FTO substrates. According to SEM images (Figure 1b), the SnO_2 nanorods are quasi-aligned perpendicularly to the FTO substrate and have diameters ranging from 50 nm to 250 nm with an average of about 100 nm, lengths up to 1.3 μm with an average of about 1.0 μm, and an average inter-rod (axis-to-axis) spacing of about 170 nm. After 3 layers of $BiVO_4$ are drop-casted and annealed (Figure 1c), the Sb:$SnO_2/BiVO_4$ NRAs have a similar length as the uncoated Sb:SnO_2 NRAs, which indicates that most of the $BiVO_4$ was coated on the sidewalls of the Sb:SnO_2 NRs rather than on top. The $BiVO_4$ coating thickness ranges from 15 nm to 100 nm.

As shown in the J-V curve (Figure 2a), $J_{sulfite}$ under back-side illumination with dark current subtracted is 4.08 mA/cm^2 and 5.3 mA/cm^2 at 0.6 V_{RHE} and 1.23 V_{RHE}, respectively, which correspond to $\eta_{abs} \times \eta_{sep}$ products (Figure 2b) of 51.2% and 66.5% based on the J_{max} value of 7.97 mA/cm^2 obtained by the integration of the measured spectral irradiance of the Xe lamp solar simulator. For comparison, $J_{sulfite}$, $\eta_{abs} \times \eta_{sep}$ and η_{abs} are also plotted for the same mass of $BiVO_4$ (3 layers) drop-casted directly onto a planar FTO substrate without any NRAs for both front- and back-side illumination (Figure 2a-c). This photoanode consists of a porous $BiVO_4$ film with 400 nm total thickness and 250 nm average particle size. At 0.6 V_{RHE}, the photocurrent of this porous film without any NRAs is ~14 times lower than that of the Sb:$SnO_2/BiVO_4$ core/shell NRA film under front-side illumination.

The IPCE of the Sb:$SnO_2/BiVO_4$ NRAs for sulfite oxidation at 0.6 V_{RHE} (Figure 2d) was calculated from IPCE = [$J_{sulfite}$(mA/cm^2) × 1240(volt×nm)] / [P_{mono} (mW/cm^2) × λ (nm)], where P_{mono} is the intensity and λ the wavelength of the incident monochromatic light. The IPCE for sulfite oxidation is itself a measure of the $\eta_{abs} \times \eta_{sep}$

product at each wavelength, and peaks at 450nm, at a value of 76%. The IPCE at wavelengths below 350 nm is very low because the light is absorbed in the FTO layer of the substrate. This matches other studies of $BiVO_4$ in which light is incident through the FTO layer.[6] The photocurrent at 0.6 V_{RHE} obtained by integrating the product of the measured IPCE and the spectral irradiance of the Xe lamp is ~4.1 mA/cm^2, which matches closely with the photocurrent of 4.08 mA/cm^2 obtained from the J-V curve under the simulated illumination. Moreover, the photocurrent at 0.6 V $_{RHE}$ obtained by integrating the product of the measured IPCE and the spectral irradiance of the standard AM 1.5G spectrum is ~3.9 mA/cm^2, which is also very similar.

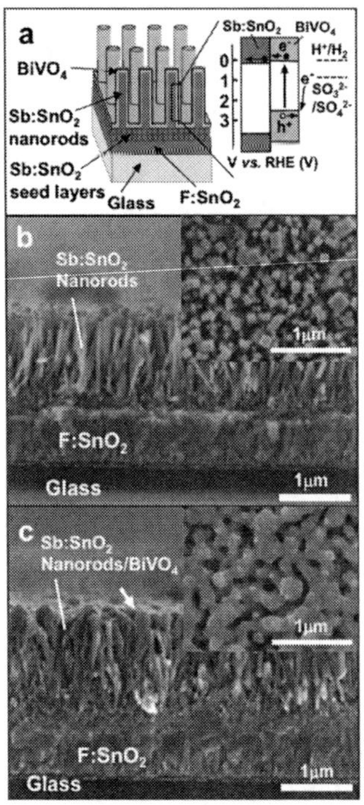

Figure 1. (a) Structure schematic and energy band diagram of the Sb-doped SnO_2 (Sb:SnO_2)/$BiVO_4$ core/shell nanorod arrays (NRAs) on FTO substrate. Sulfite is used as a hole scavenger for the measurement of charge separation efficiency. Scanning electron microscopy cross-section images (and inset top-view images) of (b) Sb:SnO_2 NRAs, and (c) 3 layers of $BiVO_4$ drop-casted onto the Sb:SnO_2 NRAs to form Sb:SnO_2/$BiVO_4$ core/shell NRAs.

WO_3 is commonly used as the core nano- wire or rod material for core/shell $BiVO_4$ photoanodes. Therefore, another appropriate comparison is to WO_3/$BiVO_4$ core/shell nanowire array photoanodes, for which the highest reported $J_{sulfite}$ at 0.6 V_{RHE} and 1.23 V_{RHE} under 100 mW/cm^2 simulated solar illumination is ~3.6 mA/cm^2 and ~5.35 mA/cm^2, respectively, for helix-shaped WO_3 nanowires.[2] The photocurrent of the Sb:SnO_2/$BiVO_4$ NRAs studied here is higher at 0.6 V_{RHE} and similar at 1.23 V_{RHE}. We have measured the electrical conductivity of the Sb:SnO_2 NRAs to be ~33 S/cm, which is significantly higher than the value of ~1 S/cm measured for axial electron transport across an individual single-crystal WO_3 nanowire in a previous study,[1] which can partially account for improved η_{sep} in the Sb:SnO_2/$BiVO_4$ core/shell NRAs compared to WO_3/ $BiVO_4$ nanowires photoanodes due to smaller voltage required to drive the electron drift current in Sb:SnO_2 compared to WO_3.

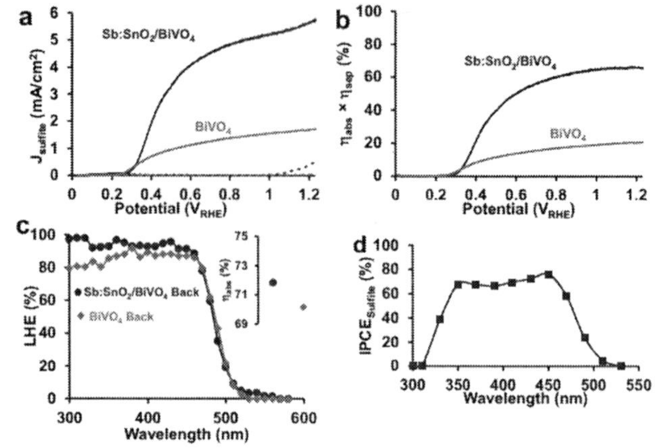

Figure 2. Optical and photoelectrochemical performance of Sb:SnO_2 NRAs coated with 3 layers of $BiVO_4$ (Sb:SnO_2/$BiVO_4$), and same-mass $BiVO_4$ film deposited directly onto FTO substrate without any NRAs ($BiVO_4$). a) Photocurrent for sulfite oxidation ($J_{sulfite}$). Dark currents for Sb:SnO_2/$BiVO_4$ and $BiVO_4$ are is shown as blue and green dashed lines, respectively. b) Product of light absorption and charge separation efficiency ($\eta_{abs} \times \eta_{sep}$) versus potential, with dark currents subtracted. c) Light harvesting efficiency (LHE) and inset showing integrated light absorption efficiency (η_{abs}). d) Charge separation efficiency (η_{sep}) with dark current subtracted. e) Incident photon-to-current efficiency (IPCE) of Sb:SnO_2/$BiVO_4$ core/shell NRAs at 0.6 V_{RHE}.

Conclusions

In conclusion, Sb:SnO_2 nanorods are a promising conductive scaffold for $BiVO_4$-based photoanodes, and provide higher charge separation efficiency compared to the previously investigated WO_3 nanostructures at low applied voltages, which is important for the future integration of $BiVO_4$ photoanodes with photocathodes to achieve overall water-splitting. To the best of our knowledge, the AM 1.5G-integrated photocurrent of ~3.9 mA/cm^2 and the corresponding $\eta_{abs} \times \eta_{sep}$ product of ~52% generated here by the Sb:SnO_2/$BiVO_4$ NRAs for sulfite oxidation at 0.6 V_{RHE} are the highest values achieved for any oxidation reaction by any reported $BiVO_4$ photoanode at this voltage. Moving forward, the Sb:SnO_2 nanorod scaffold can be combined with other promising strategies such as doping and the addition of oxygen evolution catalysts to further increase the performance of $BiVO_4$ photoanodes for PEC water oxidation, and to enable the integration of $BiVO_4$ photoanodes with photocathodes for efficient overall PEC water-splitting.

References

1. Rao, P. M.; Cai, L.; Liu, C.; Cho, I. S.; Lee, C. H.; Weisse, J. M.; Yang, P.; Zheng, X. *Nano Lett.* **2014**, *14*, 1099-1105.
2. Shi, X.; Choi, I. Y.; Zhang, K.; Kwon, J.; Kim, D. Y.; Lee, J. K.; Oh, S. H.; Kim, J. K.; Park, J. H. *Nat. Commun.* **2014**, *5*, 4775.
3. Pihosh, Y.; Turkevych, I.; Mawatari, K.; Uemura, J.; Kazoe, Y.; Kosar, S.; Makita, K.; Sugaya, T.; Matsui, T.; Fujita, D.; Tosa, M.; Kondo, M.; Kitamori, T. *Sci. Rep.* **2015**, *5*, 11141.
4. Sun, Y.; Chemelewski, W. D.; Berglund, S. P.; Li, C.; He, H.; Shi, G.; Mullins, C. B. *ACS Appl. Mater. Interfaces* **2014**, *6*, 5494-5499.
5. Titova, L. V.; Cocker, T. L.; Cooke, D. G.; Wang, X.; Meldrum, A.; Hegmann, F. A. *Phys. Rev. B* **2011**, *83*, 085403.
6. Liang, Y.; Tsubota, T.; Mooij, L. P. A.; van de Krol, R. *J. Phys. Chem. C* **2011**, *115*, 17594-17598.

CHARGE TRANSPORT THROUGH ORGANIC MOLECULAR WIRES EMBEDDED IN ULTRATHIN INSULATING INORGANIC LAYER

Eran Edri and Heinz Frei

Physical Biosciences Division, Lawrence Berkeley National Laboratory, University of California, Berkeley, CA 94720

Introduction

Closing the artificial photosynthesis cycle on a nanometer scale can be beneficial for broadening the spectrum of materials and designs that can be used in future solar fuels devices. On the other hand, at this length scale spatial separation that includes a membrane between the reaction sites for fuel generation and water oxidation is needed if high yields are sought. Currently, a nanometer thick membrane that can separate the reaction sites, block intermediates and products, transmit protons and conduct electrons is unavailable.[1] An organic-inorganic membrane made by combining the permeability properties of ~2 nm thick silica with the electronic properties of ~2 nm long molecular wires can fulfill these requirements. The molecular wires are anchored to the surface of a water oxidation catalyst with designed end-groups, and are embedded in silica using low temperature atomic layer deposition. The hetero-structure is studied using electron microscopy, spectroscopic methods and cyclic voltammetry. We find that such hybrid materials combination allows electronic charge transport (hopping dominated) only through the wires, while blocking molecular transport. This is an important step towards closing the artificial photosynthesis cycle on a nanometer scale.

Experimental

PV3 attachment to CoO_x.[2] PV3_tripod and HCTU were added to a 20 mL vial. 4 mL anhydrous acetonitrile and 1 mL 1,2-difluorobenzene were added. The solution was sonicated for 10 min and the substrate (facing down) and 0.2 mL N,N-diisopropylethylamine were added. After 24 h stirring at ambient temperature, the substrate was transferred to a fresh acetonitrile for 10 min sonication followed by drying in a stream of dry Ar.

Atomic Layer deposition (ALD) of cobalt oxide was carried out using Plasma Enhanced Atomic Layer Deposition at 100°C using $CoCp_2$ as precursor. $CoCp_2$ dose was 5 sec long, followed by 5 sec of purging. Oxygen plasma half-cycle was 1 sec pre-plasma, 5 sec plasma (300 W) and 15 sec purging. 3.5 nm CoO_x layer were usually deposited.

ALD of silica was conducted at 40°C using hollow cathode plasma source in a chamber adapted with a grounding grid to reduce charged particle bombardment. Under 5 sccm flow of ultrapure Ar (chamber pressure of ~80 mTorr) a 0.05 sec pulse of tris-dimethylaminosilane was used as Si precursor alongside closing the chamber exhaust valve for 40 sec. The chamber was purged with 100 sccm Ar for 15 sec. Before plasma was started, the chamber was purged with 5 sccm of oxygen flow (alongside 3 sccm Ar flow, chamber pressure ~ 200 mTorr) for 20 sec and then O_2 plasma (250 W) was introduced to the chamber for 30 sec followed by 35 sec of Ar purging (15 sec at 5 sccm and 20 sec at 100 sccm). These steps constitute a single ALD cycle. A growth rate at 40°C was 1.3 Å/cycle, respectively. Typically 18 cycles were applied.

Cyclic voltammetry (C-V) measurements were carried out using Ag/AgCl as reference electrode, and Pt counter electrode. Acetonitrile with 0.01M ferrocene and 0.1M tert-butylammonium perchlorate solvent.

Dye sensitization. Similar samples as made for C-V were used. A 0.5 mM $Ru(bpy)_3Cl_2$ (bpy = 2,2'-bipyridine) and 0.1 M sodium persulfate aqueous solution was purged with ultra pure Ar for 10 min prior to start of the experiment. Alternatively, 0.5 mM Sn porphyrin (Sn(P)) and 0.1 M tri-ethylamine were used. The current through the electrodes was measured with a picoammeter. For $[Ru(bpy)_3]^{+2}$ excitation, a 476 nm Ar laser beam (370 mW) expanded to ~1 cm diameter beam was directed at the stirred solution. For Sn porphyrin excitation 514 nm Ar laser beam (400 mW) was used in similar way. In both cases the photo current was read directly from the picoammeter.

Results and Discussion

Structural aspects of PV3 molecular wire attachment to cobalt oxide, embodiment in silica, and quantification of wire surface density were studied using STEM-EDX. The deposition of cobalt oxide, the PV3 attachment to the cobalt oxide surface, and encapsulation in silica were carried out directly on a TEM grid. In **Figure 1** (HAADF image) the structure and the diffraction patterns of the deposited layers are shown (cobalt oxide, 3.5 nm; silica with embedded wires, 2.3 nm). When deposited on a silicon wafer, a similar morphology is seen (not shown). The identifiable diffraction rings seen in the inset correspond to 220, 311, 331 of Co_3O_4 spinel crystal structure (PDF no. 01-071-0816).

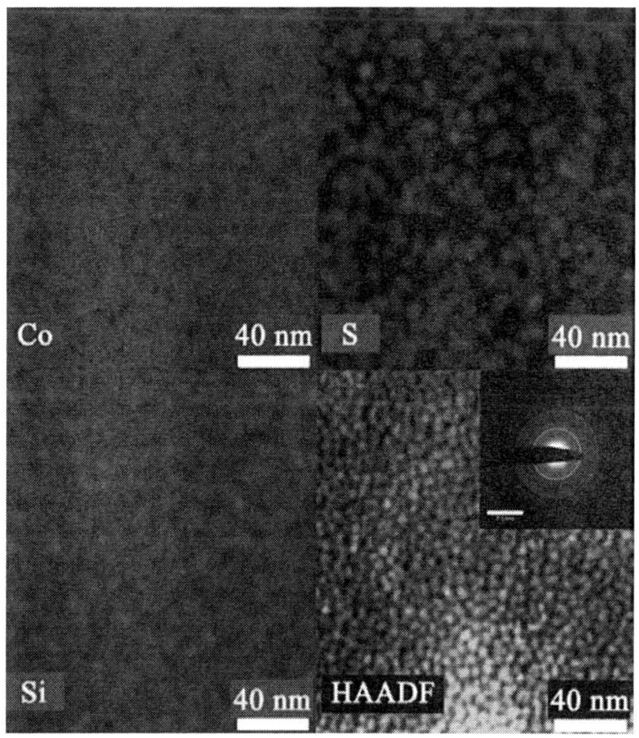

Figure 1. STEM-EDX maps of Co (blue, top left) S (red, top right) Si (green bottom left) of CoO_x PV3 wires embedded in silica on TEM grid. A dark-field image is added for comparison (bottom right) alongside selected area diffraction pattern.

Using EDX (**Figure 1**) to map the distribution of Co, Si and S (the latter originating from the PV3 sulfonate group), we find that Co coats the surface uniformly, and is to the most part accompanied by Si signal. We find no areas lacking Si or Co, which allows us to conclude that the amorphous silica coat is uniformly and continuously covering the CoO_x layer. Further support for this conclusion through electrochemical measurements is described below. Unlike Co and Si, the S signal is less dense and has less uniform distribution with gaps of a few to ten nm, separating small (3-4 nm) S-containing regions. We did not find large areas (> 10 nm^2) of specifically higher or lower S content. It is clear from the results that the coverage is substantially below a monolayer and consists of small patches of molecular wires attached to the CoO_x. Because the Si pattern is uniform and uncorrelated to the S pattern, we conclude that the wires are embedded in the silica. In particular, it does not seem as if the silica is depleted from the environment of the wires. Electrochemical measurements presented below confirm that the silica layer provides completely gap-free coverage of the entire CoO_x deposit.

Grazing angle attenuated total reflection FT-IR (GAATR-FT-IR) was used to further study the molecular wires attachment to CoO_x and encapsulation in silica. In **Figure 2** the difference spectra in the 1220-900 cm^{-1} region of CoO_x with PV3 (trace 2) and CoO_x with PV3 embedded in silica (trace 1) are shown together with the spectrum of free PV3 in KBr (trace 3). PV3 bands of the fingerprint region at 961, 971, 1008, 1036, 1106, 1122, 1181, 1194 cm^{-1} agree with previous spectra of this wire molecule in KBr, and the bands of the GAATR-FT-IR spectrum of CoO_x with attached PV3 coincide with those of the spectrum of powders of Co_3O_4 nanoparticles with attached PV3 reported previously.[3]

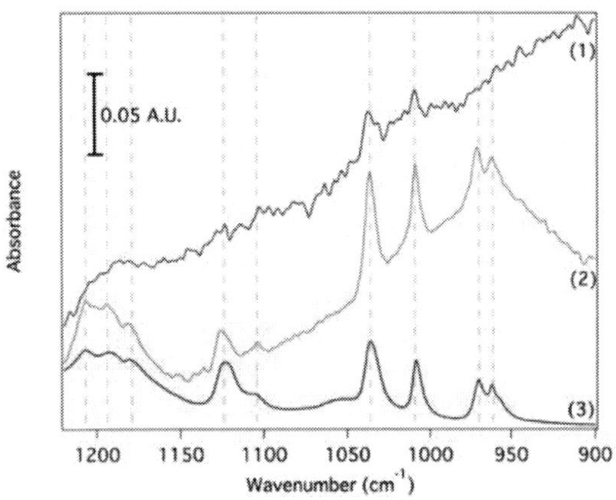

Figure 2. Grazing Angle ATR-FTIR spectra. PV3 molecular wires in KBr pellet (red trace 3), signal reduced by a factor of 25 for ease of comparison. Difference of PV3 molecular wires attached to CoO_x from bare CoO_x (orange trace 2). Difference of silica embedded PV3 wires on CoO_x from bare CoO_x (blue trace 1, multiplied by 5 for clarity). All ATR FTIR measurements were done at an angle of 62.5° and 4 cm^{-1} resolution. Baseline was measured as ATR crystal/air. Red dashed vertical traces indicate positions of main PV3 bands.

Figure 3 shows XPS spectra of CoO_x with PV3, and CoO_x with PV3 embedded in silica (both samples were deposited on Si). The doublet observed before silica casting at 167.9 and 169.0 eV are sulfonate S 2p peak. The N 1s region (not shown) for the sample before silica deposition shows a peak at 399.7 eV and one centered at 402.1 according to spectral deconvolution shown in the center panel of **Figure 5**. The peak at 399.7 is attributed to amide bond nitrogen, *i.e.* to the tripod-anchoring group of the PV3 molecules. The smaller N 1s peak at higher binding energy has the proper position for alkyl ammonium, probably from HCTU coupling agent. The XPS results further support our STEM-EDX and FT-IR findings of the presence of PV3 wires before and after silica deposition. Taken together, all spectroscopic and microscopic measurements allow us to conclude that PV3 molecular wires are attached to ALD-deposited CoO_x layers using HCTU coupling method, and that they remain intact after deposition of silica by ALD.

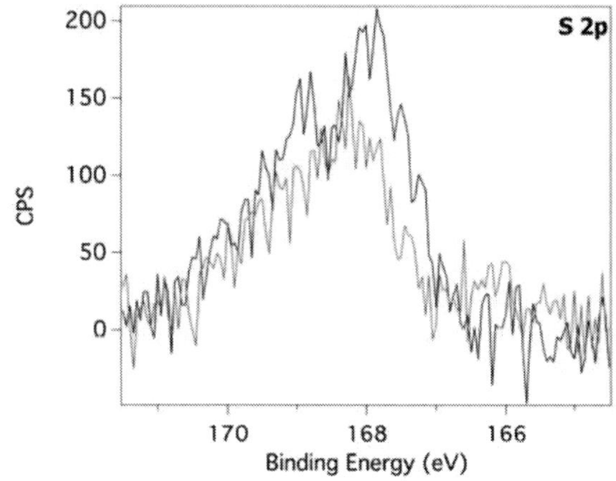

Figure 3. X-ray photoelectron spectra of CoO_x with PV3 before (blue) and after deposition of 2.3 nm silica (green). Top: Sulfur 2p spectrum (background subtracted). Middle: Nitrogen 1s showing spectral deconvolution. Inset compares XPS before (blue) and after deposition of silica (green). Bottom: Cobalt 2p spectra. XPS spectral analysis was done with software CasaXPS V2.3. CPS stands for counts per second.

To test for the presence of pinholes in the silica layer we use cyclic voltammetry (CV) with a redox molecule added to the solution (ferrocene). Since the silica is insulating, a redox wave caused by ferrocene reacting on CoO_x/Pt would be an indication of pinholes in the silica. A characteristic redox wave of ferrocene on bare CoO_x/Pt in acetonitrile is shown in **Figure 4** (red trace, offset for clarity) along with CV scans of CoO_x with silica with and without wires (blue and green traces, respectively). Note the 3 orders of magnitude difference in the current density scale for bare and silica coated CoO_x. It can be seen that silica coated CoO_x, both with and without wires shows no ferrocene redox wave. The result indicates that the silica layer is tight enough to prevent ferrocene from reaching and reacting on the CoO_x.

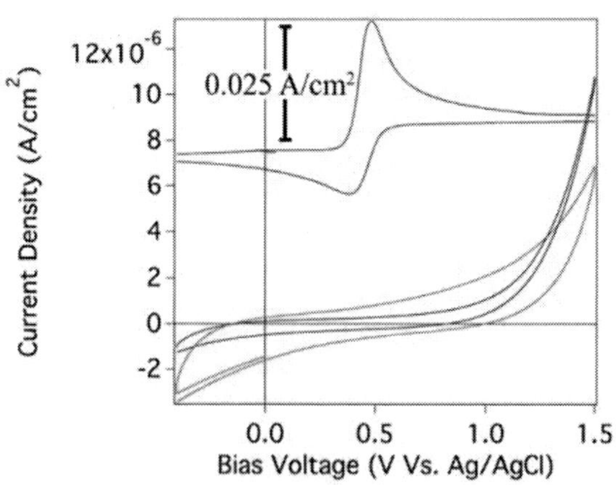

Figure 4. Cyclic voltammetry. Acetonitrile containing 0.01 M ferrocene and 0.1 M tert-butylammonium perchlorate. Bare CoO_x/Pt, red trace (offset for clarity; note the different current density scale).

CoO_x with 2 nm thick SiO_2 layer (no wires), green trace. CoO_x with SiO_2 layer containing PV3 wires, blue trace. In all scans 5 cycles were performed and only the last, stable scan is shown. Scan rate is 0.1 V/sec.

To measure charge transfer through the wires we measured short circuit photocurrent originating from charge transfer from photo-excited dye molecules in solution to the electrode (**Figure 5**).

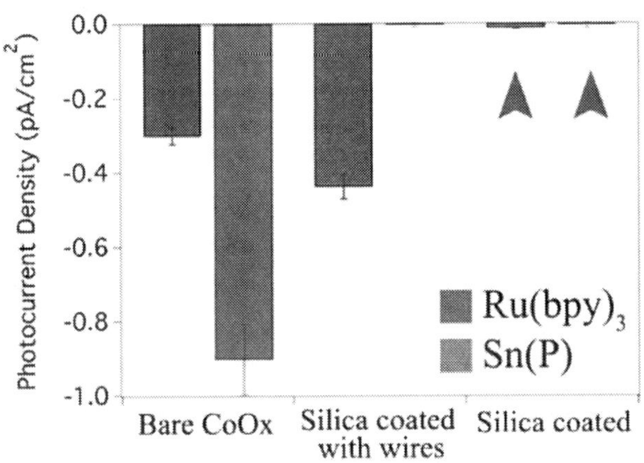

Figure 5. Photocurrent from dye sensitized electrodes featuring bare CoO_x, silica-coated CoO_x, and CoO_x with silica embedded PV3 molecular wires.

Charge transport of photo-generated holes through the wires embedded in silica is quantified by directly measuring the light-induced short circuit current. $[Ru(bpy)_3]^{2+}$ is used as visible light sensitizer in aqueous solution containing persulfate ($Na_2S_2O_8$) as electron acceptor. Upon excitation of $[Ru(bpy)_3]^{2+}$ at 476 nm, electron transfer from the excited metal-to-ligand charge-transfer state to the electron acceptor generates a $[Ru(bpy)_3]^{3+}$. This species has sufficient positive potential to inject a hole into the molecular wires, or directly into CoO_x. Whether directly or through a wire, when a hole is transferred to the CoO_x, it is collected at the Pt electrode and generates a photocurrent. As is shown in **Figure 5**, samples with bare CoO_x produce photocurrent while samples of CoO_x coated with silica do not. The latter observation shows that the 2.3 nm silica layer blocks hole transfer from $[Ru(bpy)_3]^{3+}$ to CoO_x. When CoO_x is coated with silica with embedded PV3 wires, a photocurrent is produced as well. Photosensitization experiments were conducted to test whether selecting dye molecules with enough potential in the excited state to inject a charge into the wire is required to observe photocurrent. When excited in aqueous solution containing an electron donor (tri-ethylamine), Sn porphyrin (Sn(P)) has sufficient potential to inject an electron to CoO_x but not to the LUMO of the PV3 wire molecule. As shown in **Figure 5**, a photocurrent is generated using bare CoO_x as electrode but not when using silica-coated CoO_x even in the presence of embedded wires. We conclude that enough driving force for charge injection is necessary to observe photocurrent; only in the case with $[Ru(bpy)_3]^{2+}$ does the charge-injecting species, here $[Ru(bpy)_3]^{3+}$, have sufficient potential to transfer a charge to the molecular wire, in this case a hole to the HOMO of PV3.

Acknowledgment. This work was supported by the Director, Office of Science, Office of Basic Energy Sciences, Division of Chemical, Geological and Biosciences of the U.S. Department of Energy under Contract No. DE-AC02-05CH11231. Portions of this work (plasma enhanced atomic layer deposition, e-beam evaporation, ellipsometry), were performed as a User Project at The Molecular Foundry, Lawrence Berkeley National Laboratory, which is supported by the Office of Science, Office of Basic Energy Sciences. Electron microscopy part of this work was performed as a User Project at The National Center for Electron Microscopy, Lawrence Berkeley National Laboratory, which is supported by the Office of Science, Office of Basic Energy Sciences. Portion of the work (XPS measurements, GAATR-FT-IR) was performed at the Joint Center for Artificial Photosynthesis, a DOE Energy Innovation Hub, supported through the Office of Science of the U.S. Department of Energy under Award No. DE-SC0004993.

References

(1) Edri, E.; Frei H. Charge Transport through Organic Molecular Wires Embedded in Ultrathin Insulating Inorganic Layer. *J. Phys. Chem. C.* **2015**, *119*, 28326–28334.

(2) Soo, H. S.; Agiral, A.; Bachmeier, A.; Frei, H. Visible Light-Induced Hole Injection into Rectifying Molecular Wires Anchored on Co_3O_4 and SiO_2 Nanoparticles. *J. Am. Chem. Soc.* **2012**, *134*, 17104–17116.

(3) Agiral, A.; Soo, H. S.; Frei, H. Visible Light Induced Hole Transport from Sensitizer to Co_3O_4 Water Oxidation Catalyst across Nanoscale Silica Barrier with Embedded Molecular Wires. *Chem. Mater.* **2013**, *25*, 2264–2273.

TOP-DOWN AND BOTTOM-UP APPROACHES TO NANOSIZED AND HIERARCHICAL ZEOLITES

Tatsuya Okubo

Department of Chemical System Engineering
The University of Tokyo
7-3-1 Hongo, Bunkyo-ku, Tokyo 113-8656, Japan
okubo@chemsys.t.u-tokyo.ac.jp

Abstract

Zeolites are microporous crystalline materials possessing pores and channels at molecular dimensions. Because of the unique, structural features, zeolites have been widely applied in industrial processes. However, the presence of only micropores in zeolite frameworks often limits the molecular diffusion, resulting in the restricted transport of bulky molecules. We have developed two approaches to solve the problems. One is the preparation of nanosized zeolite by top-down approach, and the other is that of hierarchical zeolite by bottom-up approach. In this presentation, our recent progresses will be introduced.

Top-down approach to nanosized zeolite [1)]

A top-down methodology was developed to prepare nanosized zeolites with tunable size by combining post-synthesis milling with fast recrystallization. By introducing the rapid heating process, the preparation was completed in several minutes (10 min for SSZ-13 and 5 min for AlPO-5). A continuous-flow recrystallization process was demonstrated to further enhance the overall productivity.

Bottom-up approach to hierarchical zeolite[2)]

A new approach for the construction of hierarchically organized zeolites by sequential intergrowth was developed. A complex and unusual morphology was observed for MFI zeolites. These zeolites possessing micro-, meso-, and macroporosity have been prepared by a simple, single-step, hydrothermal synthesis using a small OSDA without meso- or macroporogens. The hierarchical structure was achieved via sequential intergrowth of MFI zeolite.

Conclusions

In addition to traditional bottom-up synthetic approach, top-down approach was introduced for the production of functional zeolites. There still remain several routes for the preparation of functional zeolites.

References

(1) Liu, Z.; Nomura, N.; Nishioka, D.; Hotta, Y.; Matsuo, T.; Oshima, K.; Yanaba, Y.; Yoshikawa, T.; Ohara, K.; Kohara, S.; Takewaki, T.; Okubo, T.; and Wakihara, T. *Chemical Communications,* **2015**, *51*, 12567.

(2) Chaikittisilp, W.; Suzuki, Y.; Mukti, R.R.; Suzuki, T.; Sugita, K.; Itabashi, K.; Shimojima, A.; and Okubo, T. *Angew.Chem. Int. Ed.*, **2013**, *52*, 3335.

ORGANOTEMPLATE-FREE SYNTHESIS OF MESOPOROUS ZSM-5 ZEOLITE

Liang Wang[1], *Haiyan Zhang*[2], *and Feng-Shou Xiao*[1,*]

[1] Key Lab of Applied Chemistry of Zhejiang Province, Department of Chemistry, Zhejiang University, Hangzhou 310028, China
fsxiao@zju.edu.cn
[2] College of Electronic Science and Engineering, Jilin University, Changchun 130013, China

Introduction

Mesoporous zeolites have been attracted much attention recently due to combining the advantages of both zeolites (shape-selectivity) and mesoporous materials (high mass transfer and conversion of bulky molecules) since the pioneering works from soft-templating routes. In these cases, a mixed template of mesoscale molecules and small organic cations with the aim of simultaneously directing the formation of mesoporous and microporous structure in the zeolites has been used. However, the competition between the two different templating systems in the synthesis easily results in the formation of amorphous mesoporous materials by the phase separation. To overcome this problem, it is developed a new class of the templates featuring multi-ammonium groups with which they synthesized mesostructured and mesoporous zeolites without the involvement of small templates. In these examples, the organic templates are necessary.

Despite much encouraging progress in recent years, the practical applications of mesoporous zeolites are still lower than those of conventional zeolites. The drawback is probably the use of high-cost mesoscale template and relatively complex procedures for synthesizing mesoporous zeolites. Considering recent successful examples for organotemplate-free synthesis of zeolites, it is strongly desirable to synthesize mesoporous zeolites in the absence of any organic templates, which would not only avoid the use of expensive templates but also save the energy for removing the templates by calcination at high temperature. We show here an organotemplate-free and seed-directed synthesis of aluminosilicate MFI zeolite from aluminosilicate gels with relatively low Si/Al ratios. By rationally adjusting synthetic parameters, it is successfully obtained aluminosilicate MFI crystals with both low Si/Al ratio and mesoporous structure, designated as MFI-SDS. As expected, the catalytic cracking of cumene shows that the MFI-SDS is more active than conventional ZSM-5 zeolites synthesized in the absence and presence of organic template (TPA^+) designed as ZSM-5-OTF and ZSM-5-TPA, which is due to the presence of mesoporosity and abundant 4-coordinated Al species in the MFI-SDS.

Results and Discussion

MFI-SDS was prepared by the seeds-directed and organotemplate-free (SDS) method. H-form of ZSM-5 zeolites was obtained by ion-exchange with 1 M NH_4NO_3 solution at 80 °C for 3 h, followed by calcination at 550 °C for 3 h. This process was repeated one time. Figure 1 shows XRD, SEM image, N_2 sorption isotherms, and ^{27}Al MAS NMR spectrum of the MFI-SDS. The XRD pattern confirm the structure of MFI, the SEM image show the presence of mesoporosity formed by aggregative nanorods, the N_2 sorption isotherms suggest the presence of both microporosity and mesoporosity, and the ^{27}Al MAS NMR spectrum demonstrates that only tetrahedrally coordinated aluminum species exist in the zeolite framework. Furthermore, the sample Si/Al ratio analyzed by the inductively coupled plasma optical emission spectroscopy (ICP-OES) is 9.5, which is much lower than that of the ZSM-5 zeolite reported previously.

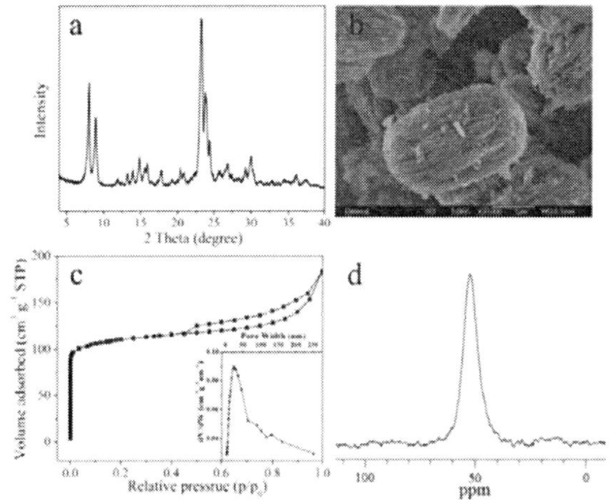

Figure 1. (a) XRD pattern, (b) SEM image, (c) N2 sorption isotherms, and (d) 27Al MAS NMR spectrum of the MFI-SDS. Inset in Figure 1c: BJH pore size distribution.

As a model reaction catalyzed by acidic sites, catalytic cracking of cumene was chosen. Table 1 presents catalytic data in cumene cracking over MFI-SDS, ZSM-5-OTF, and ZSM-5-TPA catalysts. Notably, for the first 5 injections, MFI-SDS gives the average conversion at 93.2%, which is obviously higher than those of ZSM-5-OTF (87.2 %) and ZSM-5-TPA (77.8 %). These results are reasonably related to their difference in acidic concentration. MFI-SDS with rich tetrahedral Al species has higher acidic concentration than ZSM-5-OTF and ZSM-5-TPA. Furthermore, it is observed that the MFI-SDS exhibits slower deactivation than the catalysts of ZSM-5-OTF and ZSM-5-TPA. For example, after reactions for 20 injections, MFI-SDS still exhibits the conversion of cumene at 84.4%, while ZSM-5-OTF and ZSM-5-TPA give the conversions at 71.8% and 66.6%, respectively. This phenomenon is no doubt contributed to that the MFI-SDS has the mesoporosity, which is very favorable for mass transfer.

Table 1. Catalytic performances of ZSM-5-TPA, ZSM-5-OTF, and MFI-5-SDS[a]

[a] Si/Al ratios in the final products. [b] The catalytic tests were performed by pulse injections, and the data presented in this table are

	Sample	Si/Al	Conv. Cumene, %[b]
1	ZSM-5-TPA	25	77.8[c]/66.6[d]
2	ZSM-5-OTF	14	87.2[c]/71.8[d]
3	MFI-SDS	9.5	93.2[c]/84.4[d]

the average values of 5 injections. In each run, 35 mg of catalyst was used, pulse injection of the reactant was 0.4 μL, and the reaction flow rate was 53.7 mL/min. [c] The data are the average values of first 5 injections. [d] The data are the average values of 20 to 25 injections.

Conclusion

In summary, MFI-SDS zeolite with mesoporosity, rich tetradedral Al species, and high crystallinity, has been successfully synthesized from an organotemplate-free and seed-directed approach. These features are of critical importance for designing and developing novel industrial zeolite catalysts in the future.

ADVANCED MICROSCOPY FOR MESOPOROUS ZEOLITE CHARACTERIZATION: INTER- AND INTRAPARTICLE HETEROGENEITY

Maarten B.J. Roeffaers

Center for Surface Chemistry and Catalysis, KU Leuven
Celestijnenlaan 200F
B-3001 Heverlee, Belgium
maarten.roeffaers@biw.kuleuven.be

Introduction

The performance of zeolites as solid acid catalysts is largely determined by an intricate interplay between acid site strength, distribution and accessibility. As such, structural variations within and between particles potentially have an important impact on the performance, since synthetic zeolites commonly grow as complex aggregates of small nanocrystallites. Furthermore, post-synthetic modifications, routinely applied to enhance catalyst performance, add to this heterogeneity: besides acid leaching or hydrothermal treatments, which increase active site accessibility, ion exchange procedures can be applied to introduce rare-earths into the microporous structure to increase acidity and stability.

Traditional ensemble-averaging tools for characterizing heterogeneous catalysts are unable to directly relate the structural properties to the catalytic performance at the relevant nanometer length scales, since either the spatial resolution or the molecular sensitivity is lacking.[1,2] The effect of these heterogeneities on catalyst performance is therefore often neglected and catalyst improvements are based on trial and error.

Optical microscopy is a recent addition to the toolbox for the *in situ* study of solid catalytic materials. It combines ease of use with non-invasiveness, high sensitivity and good temporal resolution.[3-5] However, due to the diffraction of light, an optical microscope has a limited spatial resolution, which is insufficient to distinguish the nanosized domains in zeolites. By recording and localizing single catalytic turnovers using fluorescence microscopy we can now generate diffraction unlimited images of working zeolites catalysts.[6,7] This approach, which is experimentally simple, not only yields super-resolution images but also delivers a nanoscale reactivity map.

In order to rationalize the performance of nanostructured catalysts it also is important to link these detailed catalytic performance maps to the compositional and structural properties of the catalyst, if possible obtained at the same small length scales. While correlating electron to optical micrographs can give direct insight into the structural aspects of the catalyst in its influence on the performance, detailed understanding of catalytic site distribution and accessibility is also critical. Rather than looking at the inorganic compositions, as in electron and X-ray microscopy,[8] vibrational microspectroscopy with organic probes can be used to assess the distribution, strength and accessibility of catalytic sites.[7]

In this contribution results on H-Mordenite (H-MOR) zeolites will be discussed as example. Similar results have also been obtained on other zeolites such as H-ZSM-5, H-ZSM-11 and USY.

Experimental

Fluorescence Microscopy. Synthetic zeolites (H-MOR, H-ZSM-5, USY, ...) deposited and thermally activated (450°C) on cover glass were exposed to a 0.7 wt% aqueous furfuryl alcohol solution and measured on an inverted epi-fluorescence microscope using a 100X oil-immersion objective (1.4 NA), with 532 nm excitation and an EM-CCD camera. Fluorescence imaging (545 nm long pass filter) and further single molecule identification, localization and activity map generation was performed with IgorPro v.6.34A software.[7]

Stimulated Raman Scattering (SRS) Microscopy. Acid sites distibution and mass transport in calcined mordenites were probed using stimulated Raman microscope with deuterated acetonitrile (d-ACN) or benzonitrile (BZN) vapors. The setup used a 25x (1.05 NA) objective and two synchronously pulsed lasers, which coherently excite molecular vibrations in the nitrile region. Via this approach molecular distributions were captured with submicron resolution in three dimensions inside individual zeolite crystals.[7] Dynamic imaging with 3 frames per second also allowed to determine variations in molecular diffusion.

Integrated Fluorescence Electron Microscopy. The integrated microscope consists of a FEI Quanta 250 FEG environmental SEM, equipped with a modified chamber door in which a wide-field fluorescence microscope is integrated (SECOM, Delmic BV). The stage of this SECOM platform holds a high NA oil immersion objective lens (Nikon 100x 1.4NA). Additionally, this commercial setup has been adapted in-house to enable sensitive single molecule fluorescence detection with a EM-CCD camera (Hamamatsu C9100-13) for single molecule detection and a fiber coupled multiline laser system (LightHUB, Omicron) for the proper excitation. The acid catalysed conversion of the non-fluorescent furfuryl alcohol into strongly fluorescent oligomer products inside the zeolite catalyst is used to generate nanoscale reactivity maps based on single molecule localization.

Results and Discussion

To investigate the effect of the dealumination process on the catalytic performance at the individual H-MOR crystal level, we recorded the location of individual catalytic turnovers using fluorescence microscopy and furfuryl alcohol (FFA) as a fluorogenic reactant. These NASCA microscopy[6] experiments revealed that for the parent, non-dealuminated small-port H-mordenite (SP-MOR; Si/Al = 6.2) the catalytic activity is limited to the outer surface of the crystals (Figure 1A).

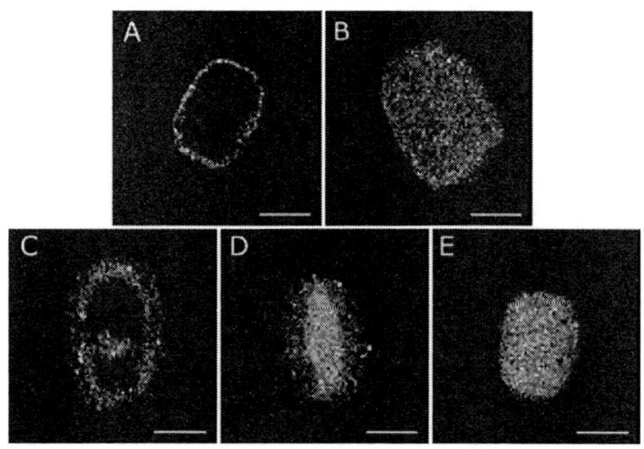

Figure 1. Super-resolution fluorescence microscopy based image single H-MOR particle reactivity. (Scale bars are 5 μm). A) Non-dealuminated SP-MOR (Si/Al=6,2); B-D) mildly dealuminated mordenite (Si/Al=11); E) severly dealuminated mordenite (Si/Al=65).

Severely dealuminated particles (SD-MOR, Si/Al = 65) show a homogeneous activity throughout the whole crystal volume (Figure 1E). In contrast, particles from the mildly dealuminated batch (MD-MOR, Si/Al = 11) shows a very broad distribution in activity when comparing different particles (Figure 1B-D). Also within individual MD-MOR crystals a large variability in activity was recorded. Some crystals and crystal zones show no catalytic activity whereas other crystals show a highly active crystal center with reduced activity near the crystals edges.

To rationalize these results we resorted to SRS microscopy. With d-ACN we can probe the distribution of the acid sites as well as the degree of dealumination, whereas BZN gives direct insight into the pore accessibility enabling to discriminate between small and large port behavior. In SP-MOR, d-ACN and hence the active sites are homogeneously distributed while the absence of BZN signals confirms the SP behavior. In the SD-MOR BZN can access the whole crystal volume, however most acid sites are removed, resulting in low d-ACN signals. In the mildly dealuminated samples a large interparticle heterogeneity was observed both for d-ACN and BZN. About one third of the crystals still shows the SP behavior; another third more closely resembled the SD-MOR however with higher acid site density; the last group contains crystals with highly accessible crystal rims with very small amounts of acid sites and a highly acidic crystal core with intermediate accessibility. Quantification of the d-ACN signals allows to determine the local acid site density. By fast monitoring of the nitrile uptake we can also determine the dynamic acid site accessibility (Figure 2).

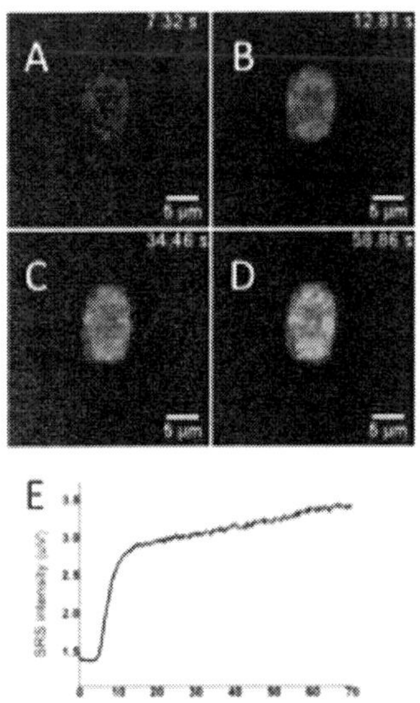

Figure 2. Dynamic monitoring of BZN uptake in a single SD-MOR crystal (Si/Al=65). A-D) SRS intensity (grayscale) images after 7.32, 12.81, 34.46 and 58.86 seconds; E) Time dependent intensity averaged over the whole H-MOR crystal.

The majority of zeolite crystals and crystal features are however substantially below one micrometer. Here complementary information to the nanoscale reactivity maps, as recorded from fluorescence microscopy can come from the corresponding electron micrographs. In order to investigate the complex relationship between catalyst structure and activity, we have applied correlative light- and electron microscopy to a commercial small-port H-MOR catalyst. It is demonstrated that the catalytic activity is limited to the outer 300 nm of the 12MR channels, clearly revealing the presence of mass transport limitations (Figure 3). Strikingly, interfaces between intergrown crystallites show a similar catalytic activity, which is unexpected since these crystallite regions are believed to be inaccessible. Experiments with linear polarized excitation light have additionally shown that the formed fluorescent product molecules are still mostly confined within the one dimensional porous structure, indicating that shape selectivity is largely maintained in these disordered crystal parts.

Acid dealumination of the small-port H-MOR sample using oxalic acid, led to a significant activity increase in catalytic turnovers at the intergrown regions compared to the other parts of the mordenite crystallites. By relating these single particle insights to the performance of the acid treated sample in the bulk scale naphthalene isopropylation reaction, their bulk scale relevance is validated, as these structural features are the main cause of the increased bulk scale catalyst activity.

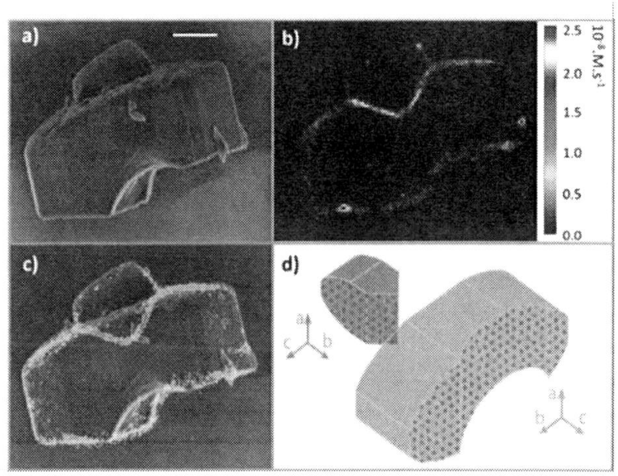

Figure 3. Scanning electron micrograph of two intergrown H-MOR crystallites (a), corresponding quantitative activity map recorded via NASCA microscopy (b), catalytic turnovers overlaid onto the scanning electron micrograph (c) and schematic representation of the catalyst indicating the crystallographic orientation with the 12MR channels (d). Scale bar: 1 micrometer

Conclusions

In this work we show how single molecule fluorescence microscopy in combination with electron microscopy and stimulated Raman enable rationalization of zeolite catalytic performance. For the first time we can directly assign observed variations in catalytic performance to the properties of the catalyst. Following this approach we uncovered a remarkable heterogeneity at the intercrystal and subcrystal level. This heterogeneity in performance caused by variations in acid site distribution and accessibility seems to be universally present in zeolite based catalyst powders, besides mordenites also other zeolites like ZSM-5,[9] ZSM-22 and USY have been studied in detail.

Acknowledgement. Financial support from European Research Council for financial support (ERC Starting Grant 307523) is gratefully acknowledged.

References

(1) Janssen, K.P.F. et al. *Chem. Soc. Rev.* **2014**, *43*, 990.
(2) Buurmans, I.; Weckhuysen, B. *Nature Chem.* **2012**, *4*, 873.
(3) Roeffaers, M.B.J. et al. *Nature* **2006**, *439*, 572.
(4) Roeffaers, M.B.J. et al. *Proc. Natl. Acad. Sci. U.S.A.* **2007**, *104*, 12603.
(5) Weckhuysen, B. *Angew. Chem. Int. Ed.* **2009**, *48*, 4910.
(6) Roeffaers, M.B.J. et al. *Angew. Chem. Int. Ed.* **2009**, *48*, 9285.
(7) Liu, K.L.; Kubarev A. et al. *ACS Nano* **2014**, *8*, 12650.
(8) de Smit, E. et al. *Nature* **2008**, *456*, 222.
(9) Ristanovic, Z. et al. *J. Am. Chem. Soc.* **2015**, *137*, 6559.

POST-SYNTHESIZED SMALL CRYSTAL Y ZEOLITE WITH ULTRAHIGH MESOPORES AND CONTROLLABLE PORE SIZE DISTRIBUTIONS

Cheng Li, Delin Yuan, Penghui Zeng, Baojian Shen*

State Key Laboratory of Heavy Oil Processing; the Key Laboratory of Catalysis of CNPC; College of Chemical Engineering; China University of Petroleum, Beijing 102249, P. R. China

*Corresponding author: baojian@cup.edu.cn

Introduction

Diverse strategies can be applied to prepare mesoporous materials, that is, through structure breaking or so-called "top-down" modification pathways, as well as by structure building or so-called "bottom-up" synthesis pathways [1, 2]. Top-down strategies (including dealumination and/or desilication) possess flexible production process, low cost, less environmental problems and many other advantages compared with bottom-up methods, as a result, it is more promising in industrial application [1]. Through post-synthesis techniques to create mesopores in the structure of Y zeolite is of great importance in oil refining industry, especially for heavy oil conversion.

On the other hand, small crystal Y zeolite possesses high external surface area, short pore channel and adjustable porosity [3]. So it is beneficial to cut down mass and heat transfer resistances, improve the catalytic selectivity and reduce coke formation [4], which in turn increases light oil (gasoline and diesel) yield and quality in FCC reaction. Especially to deserve to be mentioned, small crystal zeolite can also be used to construct mesoporous materials for several applications, such as catalysts, membranes and others [5].

In this work, considering the unique properties of small crystal Y zeolite, coupled with the advantages of desilicication and dealumination treatments, we offered a proposal to obtain small crystal Y zeolites with numerous mesopores through proper post-treatments.

Experimental

Sample Preparation. Here, we adopted conventional crystallization method with controlling the parameters of synthesis process to obtain small crystal NaY zeolites. Then, specific post-treatment was carried out to obtain samples with numerous mesopores and different mesopore size distributions. The as-synthesized samples were named MSY-1, MSY-2, MSY-3, and MSY-4, respectively.

The contrast sample (denoted as USY) was prepared from parent small crystal NaY by ammonium exchanges, followed by subsequent steaming treatment twice.

Characterization. X-ray diffraction patterns of the zeolites were obtained using a PANalytical X'Pert Powder X-ray diffractometer with monochromatized Cu Ka radiation (40 kV, 40mA). Nitrogen physisorption measurements were performed at liquid-nitrogen temperature with a Micromeritics Tristar II 3020 apparatus. The morphology of the parent and treated zeolites were visualized by SEM (FEI Quanta 200F). The existence and the morphology of the mesopores were evaluated using a JEOL JEM-2100 transmission electron microscope (TEM) operated at 200 kV.

Catalytic testing. Zeolite activity was determined with 1, 3, 5-triisopropylbenzene as the probe molecule, using a flow-type apparatus equipped with a fixed-bed reactor.

Results and Discussion

An interesting phenomenon found from table 1 is that the relative crystallinities of the modified samples increase with the rising of treating solution concentrations. Meanwhile, XRD determined SiO_2/Al_2O_3 ratios of the samples gradually decrease. It is because that non-framework species and/or amorphous species are removed from the blocked pores of the samples. All the MSY samples possess lower SiO_2/Al_2O_3 ratios than that of USY, which demonstrate their higher acid amount.

Table 1. Relative crystallinity and chemical composition of the zeolites

Samples	Relative crystallinity, %	SiO_2/Al_2O_3 by XRD
MSY-1	55	8.61
MSY-2	62	8.57
MSY-3	74	8.41
MSY-4	93	8.13
USY	70	13.69

Nitrogen adsorption-desorption isotherms and the corresponding BJH pore size distributions from the adsorption branch and desorption branch of the MSY samples are shown in Figure 1. Hysteresis loops which represented the existence of mesopores can be observed in all nitrogen adsorption-desorption isotherms of the samples. The shape of hysteresis loops and isotherms alter with the varying of treating solution concentrations, which prove the change of pore structures of the modified MSY samples. On the other hand, we delightedly discover that pore size distributions of MSY samples can be tuned precisely through controlling the treating solution concentrations.

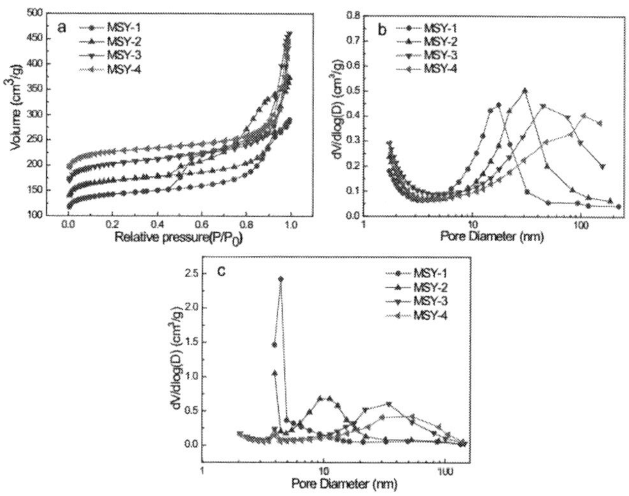

Figure 1. Nitrogen adsorption-desorption isotherms (a) and the corresponding BJH pore size distributions from the adsorption branch (b) and desorption branch (c) of the zeolites.

Table 2. Nitrogen adsorption-desorption data of the zeolites

Samples	Surface area (m^2/g)			Pore volume (cm^3/g)		
	BET[a]	Micropore[b]	External	Total[c]	Micropore[b]	Mesopore
MSY-1	464.3	342.8	121.5	0.448	0.167	0.281
MSY-2	551.8	413.7	138.1	0.578	0.202	0.376
MSY-3	659.5	497.9	161.6	0.713	0.243	0.470
MSY-4	731.9	594.2	137.7	0.688	0.290	0.398
USY	598.8	515.6	83.2	0.418	0.253	0.165

[a] BET method, [b] t-plot method, [c] Volume adsorbed at p/p_0=0.99

Nitrogen adsorption-desorption data of the zeolites are listed in table 2. With the increasing of treating solution concentrations, BET surface area, micropore surface area, total pore volume and micropore volume all raise to different extent. It is consistent with the changing trends of relative crystallinity in table 1. The highest external surface area and mosopore volume are 161.6 m^2/g and 0.470 cm^3/g, respectively, which suggest that there must be an optimal reaction condition. By precisely adjusting a series of factors, mesopore volume could reach 0.662 cm^3/g, while the total pore volume and micropore volume were 0.828 cm^3/g and 0.166 cm^3/g, respectively, which meant that its crystal structural was well preserved. Another satisfactory result is that the mesopores proportion of MSY-3 (66%) is much higher than that of USY counterpart (39%).

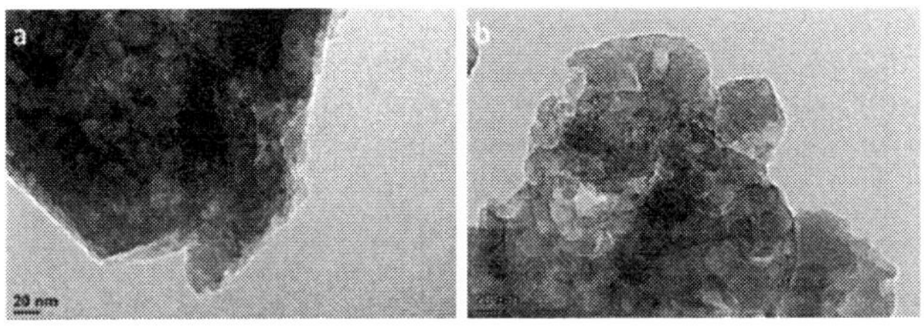

Figure 2. TEM images of MSY-3 (a) and MSY-4 (b) samples.

The TEM images (Figure 2) provide us with a direct observation of the morphology of MSY-3 and MSY-4 samples. The mesopores are spread all over the crystal particles of the samples under the electron beams. These TEM images offer us an evident clue that MSY samples possess numerous mesopores, which coincides with the nitrogen adsorption-desorption results exhibited in Table 2. The pore size distributions from desorption branch reveals that the introduced mesopores are centered at around 30 nm for MSY-3 and 40 nm for MSY-4, which are in line with the TEM observations.

Cracking of 1, 3, 5-triisopropylbenzene was investigated to compare the catalytic performance of the MSY zeolites with that of the USY zeolite and to evaluate the application potential of the MSY zeolites. The MSY zeolites showed higher initial activity and lower deactivation tendency in the case of 1, 3, 5-triisopropylbenzene cracking owing to their numerous mesopores and higher acid amount than USY sample.

Conclusions

Small crystal Y zeolites with ultrahigh mesopores and controllable pore size distributions were successfully prepared by a top-down strategy. The lower SiO_2/Al_2O_3 ratios compared with USY implied its higher acidity and this material might possess great application prospects.

Acknowledgement. The authors greatly acknowledge the funding of this project by the National Natural Science Foundation of China (U1462202) and PetroChina.

References

(1) Robin Chal, Corine Gérardin, Metin Bulut, Sander van Donk. *ChemCatChem*, **2011**, *3*, 67
(2) Danny Verboekend, Javier Pérez-Ramírez. *Catal. Sci. Technol.*, **2011**, *1*, 879
(3) Sarah C. Larsen. *J. Phys. Chem. C*, **2007**, *111*, 18464
(4) Shiguo Bao, Guozhu Liu, Xiangwen Zhang, Li Wang, Zhentao Mi. *Ind. Eng. Chem. Res.*, **2010**, *49*, 3972
(5) Limin Huang, Zhengbao Wang, Jinyu Sun, Lei Miao, Quanzhi Li, Yushan Yan, Dongyuan Zhao. *J. Am. Chem. Soc.*, **2000**, *122*, 3530

HIERARCHICAL ZEOLITES WITH UNIFORM MESOPORES PREPARED BY A MIXED APPROACH

José. M. Escola[1], David P. Serrano[1,2], Raúl Sanz[1], Rafael A. García[1], Ángel Peral[1], Inés Moreno[1], María Linares[1]

[1]Department of Chemical, Energy, Environmental and Mechanical Technology, ESCET, Rey Juan Carlos University, Móstoles (Spain),
[2]IMDEA Energy Institute, Móstoles (Spain)
josemaria.escola.saez@urjc.es

Very recently, our group has demonstrated the viability of synthesizing zeolites with uniform mesoporosity by a mixed strategy [1] combining protozeolitic unit silanization followed with a surfactant rearrangement method [2]. In this work, we show that the later approach can be applied successfully to three different zeolites: ZSM-5, Beta and TS-1. Moreover, the attained uniform mesoporosity leads toward enhanced catalytic properties in a diverse set of reactions.

Hierarchical ZSM-5, Beta and TS-1 zeolites were prepared by crystallization of silanized protozeolitic units (h-ZSM-5 / h-Beta / h-TS-1). Cumulative pore volume and pore size distribution of these materials are depicted in Figure 1. In addition to the intrinsic zeolitic microporosity, secondary micro- and mesoporosity (15-70 Å) is clearly appreciated in all the three zeolites. These zeolites were subsequently treated with a CTABr containing alkaline solution, adjusting the pH to values lower than 11.5 in order to avoid desilication. After this mesopore narrowing treatment (MNT), uniform mesoporosity results in the three zeolites (h-ZSM-5 (MNT) / h-Beta (MNT) / h-TS-1 (MNT)) (Figure 1), with a clear maximum around 40-50 Å. It is noteworthy that this second treatment is rather less effective with non-hierarchical zeolites. Thus, the presence of a previous non-uniform mesoporosity is required in order to attain uniform mesopores.

These zeolites having uniform mesopore size distribution exhibit catalytic advantages in terms of activity and selectivity in different reactions. Hence, in the catalytic cracking of low density polyethylene h-ZSM-5 (MNT) shows a 22% higher conversion than h-ZSM-5, while a conversion 25% higher was attained over the h-Beta (MNT) compared to h-Beta in the acylation of veratrole with acetyl chloride

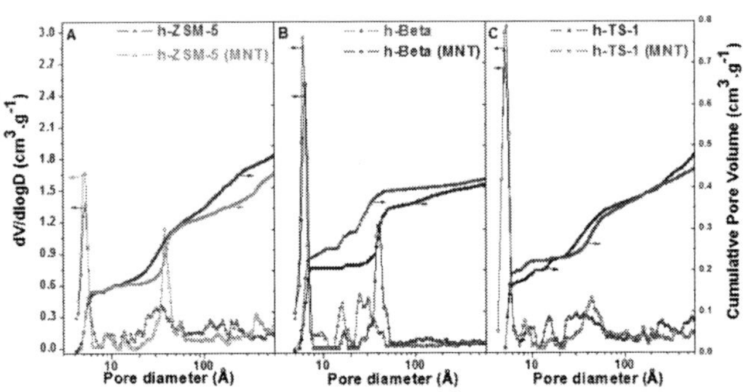

Figure 1. Cumulative pore volumes and pore size distribution of the different zeolites (NLDFT, Ar sorption).

References

(1) Serrano, D. P.; J. M. Escola, Sanz, R.; Garcia, R. A.; Peral, A. Moreno, I.; Linares, M. *New J. Chem.*. **2016**, *Advance Article*, doi: 10.1039/C5NJ02856F.

(2) Ying, J. Y.; García-Martínez, J. *US Patent* **2009**, *7,589,041*.

TEMPLATE-FREE SYNTHESIS OF HIERARCHICAL ZEOLITE BY SOLVENT-FREE CRYSTALLIZATION

Yuxin Wang, Jianjun Song, Nathan C. Baxter and Shengnian Wang

Chemical Engineering, Louisiana Tech University
911 Hergot Avenue, Ruston, LA, 71272, USA
swang@latech.edu

Introduction

Zeolites are widely used as catalyst in petrochemical processes because of their strong acidity and crystal-defined micropores for shape-based selectivity. However, when bulky molecules are involved, this is often accompanied with serious diffusion problem inside zeolite crystalline, leading to low utilization of active sites and consequent pore blocking issues. Introducing mesopores in zeolites is considered as a promising solution to such molecule transport limitations in zeolites catalyzed reactions.[1-3] Here, we report a new route to effectively synthesize hierarchical zeolites with MFI structure that requires no help on template material or solvent-involved crystallization environment. The synthesis procedure is illustrated schematically in Figure 1 and detailed in Experimental section. In this approach, aluminosilicate nanogels are firstly produced using only precursor compounds. The dry nanogels are then assembled by solid-solid interactions and transformed into crystals slowly using only water stored in individual nanogels. The stacking of these nanocrystals with jointed edges creates many inter-lattice mesoscale space inside the monolithic zeolites.

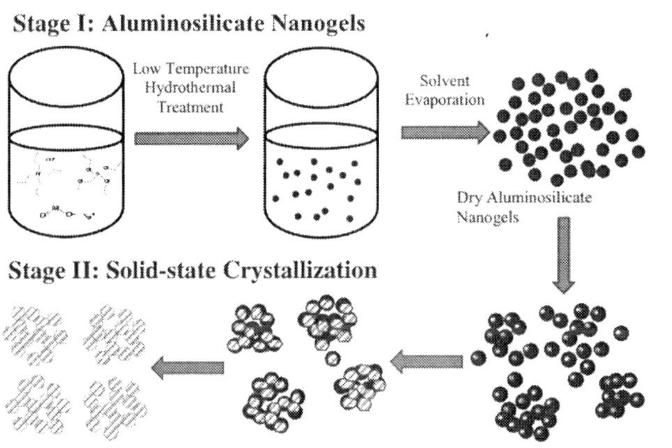

Figure 1. Scheme Caption. Schematics of the template-free, solvent-free synthesis route towards hierarchically zeolites, Meso-ZSM-5.

Experimental

Materials. Tetrapropyl ammonium hydroxide (TPAOH), sodium aluminate ($NaAlO_2$), tetraethylorthosilicate (TEOS), benzene, and benzyl alcohol were purchased from Sigma-Aldrich and used as received without further purification.

Synthesis. Meso-ZSM-5 was synthesized by mixing TPAOH, $NaAlO_2$, TEOS, and H_2O according to a molar ratio of 0.25TPAOH: $0.03Al_2O_3$: SiO_2: $80H_2O$. The mixture was kept at 80 °C until a clear solution was obtained. The received solution was then left overnight for solvent evaporation. The as-prepared dry nanogels (white powders) were further transferred to a Teflon-lined autoclave to carry out the crystallization without introducing additional water or other solvents to obtain Meso-ZSM-5 samples.

Characterization. XRD, TEM, and SEM were used to characterize the crystal pattern and morphology of zeolites. Nitrogen adsorption isotherms were done to measure the surface area and porosity and temperature-programmed desorption of ammonia (NH_3-TPD) was tested on acidity. The hydrothermal stability was tested at 150 °C under saturated vapor pressure. Zeolites were converted to H-form and tested for their catalytical performance with the alkylation of benzene with benzyl alcohol (BA) at 80 °C, with 0.15 g catalyst, 34 mL benzene, and 0.5 mL BA in a three-necked round-bottom flask equipped with a reflux condenser. The sample was analyzed by GC.

Results and Discussion

It takes around 12 hours to transform the amorphous aluminosilicate nanogels with an average size of ~20-30 nm into uniform spherical particles of ~400 nm with very rough surface (Figure 2a). The high magnification SEM image (inset of Figure 2a) reveals that they are highly mesoporous, attributed to the assembly of many nanoparticles. The high-resolution TEM (HRTEM) image further tells that these are nanosized crystals (Figures 2b). Considering their similar size to the original aluminosilicates nanogels, we believe each monolithic zeolite nucleates simultaneously from many individual nanogels. The formed small nanocrystals exhibit lattice fringes of the same orientation (Figures 2b), suggesting that these nanocrystals have been grown and merged into a single large, monolithic crystal rather than randomly packed nanocrystal aggregates. The electron diffraction spot pattern of a selected-area in HRTEM image (inset of Figure 2b) evidences this fact. With the XRD pattern of these zeolites matching that from a conventional microporous ZSM-5 sample and in good agreement with the database MFI structure (Figure 2c), we confirm the successful synthesis of mesoporous ZSM-5 with our approach (denoted as Meso-ZSM-5 in later discussions). The jointed nanocrystals create many inter-lattice mesopores in zeolite crystalline frameworks, showing features of both type I and type IV isotherms with two steep steps, one at $P/P_0 < 0.02$ and the other at $0.45 < P/P_0 < 0.90$ locations, responding for micropore filling and mesopore capillary condensation, respectively (Figure 2d). It also has a much higher BET surface area and much larger total pore volume when compared to microporous ZSM-5 with a similar Si/Al ratio.

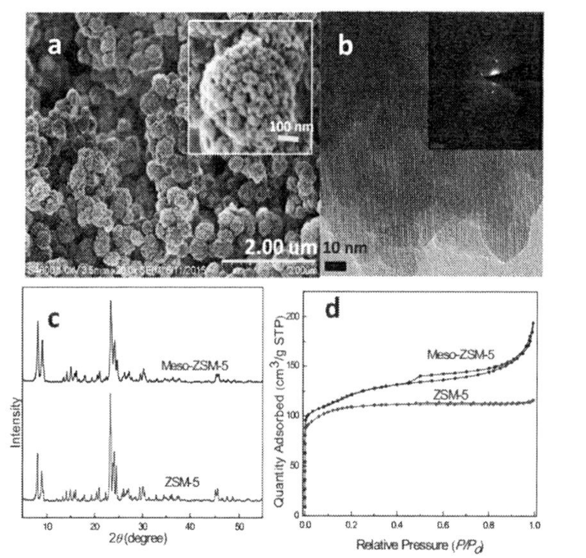

Figure 2. Figure 2. (a) SEM and (b) HRTEM images of Meso-ZSM-5; (c) XRD patterns, and (d) N_2 sorption isotherms of Meso-ZSM-5 and conventional ZSM-5. The inset in (a) is a high magnitude

SEM image of Meso-ZSM-5. The inset in (b) is the corresponding electron diffraction pattern of a selected area.

The catalytic performance of Meso-ZSM-5 for benzene alkylation was also evaluated. As shown in Figure 3, Meso-ZSM-5 shows much higher catalytic conversion of benzyl alcohol than conventional ZSM-5 after 6 h reaction at 80 °C. Considering their similar Si/Al ratio and acidity, the superior catalytic activity of Meso-ZSM-5 is mainly attributed to its mesoporous structure, which not only provides more accessible active sites, but also reduces the diffusion barrier for large reactants/products. Unlike mesopores in physically aggregated nanosized crystals, the hierarchical structure from fused nanocrystals of Meso-ZSM-5 shows excellent stability after being further hydrothermally treated at 150 °C for 72 h: the same diffraction patterns and peak intensity are preserved, so as their size and morphology and the catalytic activity in Friedel-Crafts alkylation. These results indicate that the monolithic Meso-ZSM-5 synthesized by this template-free and solvent-free approach has good hydrothermal stability. It also proves that the solid-state crystallization process is completed after the 12-hour calcinations and the produced Meso-ZSM-5 zeolites indeed reach their stable crystalline structure, not intermediates taken from the middle of a slow crystallization process.

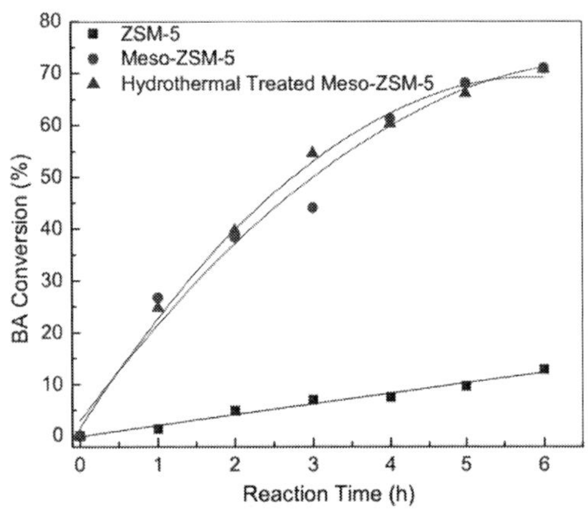

Figure 3. Benzyl alcohol conversion over Meso-ZSM-5 and conventional ZSM-5.

Conclusions

In summary, hierarchical zeolites ZSM-5 is successfully synthesized in a new two-stage solid crystallization process without templates. The resulting mesoporous zeolites show excellent hydrothermal stability and superior catalytic performance in reactions like Friedel-Crafts alkylations. Its unique template-free and solvent-free synthesis conditions largely simplify the synthesis routes of zeolites, reduce the production cost, and avoid common pollution issues in zeolite industry. Its superior catalytic performance and economically favorable synthesis route may stimulate further exploration interest for its large-scale production and wide effectiveness to a variety of applications requiring hierarchical zeolites.

Acknowledgement. We acknowledge financial support from National Science Foundation under Grant No. 1130468 and Louisiana Harrelson Family Professorship.

References

(1) Y. Wang, J. Wu, S. Wang, *RSC Adv.* **2013**, 3, 12635.

(2) Y. Wang, T. He, K. Liu, J. Wu, Y. Fang, *Bioresour. Technol.* **2012**, 108, 280.

(3) Y. Wang, Y. Fang, T. He, H. Hu, J. Wu, *Catal. Commun.* **2011**, 12, 1201.

MESOPOROSITY IN ZEOLITES: A TOOL TO MITIGATE THEIR DEACTIVATION BY COKING AND OPTIMIZE THEIR REGENERATION

J-P. Gilson, L. Lakiss, F. Ngoye, C. Canaff, S. Laforge, Y. Pouilloux, Z. Qin, M. Tarighi, K. Thomas, V. Valtchev, A. Vicente, L. Pinard, C. Fernandez

Laboratory Catalysis & Spectrochemistry, ENSICAEN, Université de Caen, CNRS, 6 Bd Maréchal Juin, F-14050 Caen
gilson@ensicaen.fr

Abstract

Many zeolite catalysts currently used commercially are already hierarchical: FCC, Hydroisomerization and Hydrocracking being important examples in oil refining. In these cases, hierarchization is brought by steam and/or acid treatments where Al is selectively removed from the framework. Recently, other methods have been rediscovered (caustic treatment where Si is removed to a greater extent than Al), studied in-depth by modern techniques and new methodologies created (fluoride treatment where Si and Al are removed at and equal rate). Moreover, the production of nanosized zeolite crystals is equivalent to hierarchization as a sizable (mesoporous) surface is developed on their external surface.

I will focus on the effect of hierarchization on coking and regeneration of ZSM-5 zeolite catalysts in the ethanol to hydrocarbons transformation. Two parents ZSM-5, one made of micron-sized crystals and the other containing nano-sized crystals will be studied as well as offsprings produced by caustic and fluoride treatments. The time on stream behavior of the various catalysts shows the advantages of hierarchization and particle size. The deactivated catalysts are extensively characterized by their residual acidity and pore volumes, nature and quantity of the carbonaceous residues and ease of decoking.

One important feature is that nanosized and hierarchical zeolite crystals produce less of the very toxic coke molecules (located at channels intersections) because most coke precursors diffuse efficiently to the external surface where their deleterious effect is minimal. Nanosized ZSM-5 are systematically better than any hierarchical zeolites produced by post-synthesis treatments due, among other factors, to the much lower diffusion pathways molecules have to follow in the microporous channels. Another important feature of hierarchical and nanosized zeolites is a much easier removal of coke by combustion: temperatures required to completely remove coke from hierarchical or nanosized zeolite catalysts are up to 150 K lower than on their parent.)

BIFUNCTIONAL ZEOLITES FOR THE ADSORPTIVE DESULFURIZATION OF MODEL FUELS

Kevin X. Lee and Julia A. Valla

Department of Chemical & Biomolecular Engineering, University of Connecticut, Storrs
191 Auditorium Road, Unit 3222, Storrs, CT 06269, USA.
ioulia.valla@uconn.edu

Introduction

The Environmental Protection Agency (EPA) mandates the level of sulfur in transportation fuels to be less than 10 ppmw in gasoline and less than 15 ppmw in diesel.[1]. Furthermore fuel cells, which utilize H_2 from fossil-based fuels can only operate below 0.1-5 ppmw to avoid poisoning of expensive catalyst and electrodes.[2] To meet the above goals significant scientific research has been devoted on the development of effective, energy efficient and cost competitive technologies for the deep desulfurization of transportation fuels. One of the most attractive technologies is adsorption using zeolites. Zeolites are attractive materials for adsorptive desulfurization due to their high surface area and unique pore structure. However, the microporous nature of zeolites may cause diffusion and selectivity limitations, especially for refractory sulfur compounds. To overcome these limitations mesoporosity can be introduced to the zeolites to give pore access to the active sites, while metals can be ion-exchanged to induce selectivity for sulfur compounds via either π-complexations or direct S-M σ bond between zeolite and sulfur. The incorporation of these two entities form bifunctional zeolites. The objective of this study is to understand the role of mesoporosity and active sites in Y zeolites on the adsorption of thiophene, benzothiophene and dibenzothiophene from liquid fuels.

Experimental

Preparation and Characterization of Y Zeolite. Fresh NH_4Y zeolite was purchased from Zeolyst International. Mesoporous Y has been prepared using two top-down approaches: a) desilication (DS) and b) the surfactant-assisted method (SA), as described in literature.[3,4] Ion-exchanged Y and mesoporous Y were prepared by ion-exchange with 5 wt% of Cu or Ce. All materials were subjected to calcination or reduction at elevated temperature prior to testing. Crystallinity and microporosity of the materials have been characterized using X-ray diffraction (XRD). Pore structure and surface area were characterized by N_2 adsorption (BET method). Brönsted (BAS) and Lewis acid sites (LAS) and location of metals were characterized using pyridine and CO adsorption and monitored by Diffuse Reflectance Infrared Spectroscopy (DRIFTS-FTIR), respectively. The metal loading was determined using an inductively coupled plasma attached to mass spectrometer (ICP-MS) while the presence of metal ions were confirmed using a UV-Vis spectroscopy.

Fixed-bed adsorption experiment. All dynamic adsorption experiments were conducted at ambient conditions in a custom-made quartz fixed-bed adsorption column equipped with built-in frit. The apparatus consisted of a HPLC pump, a feed tank and a collection vial. 0.25 g of zeolite was weighed and packed inside the column which, was connected to the adsorption setup. The sorbent was first activated in-situ and then allowed to cool to room temperature. A model fuel containing either 100 ppmw of thiophene (T), benzothiophene (BT), DBT or 4,6-DMDBT was fed into the column at a flow rate of 0.05 mL/min and the effluent was collected at regular intervals until saturation was reached. The collected samples were quantified for their sulfur content using a gas-chromatograph (Agilent 7890 GC)-sulfur chemiluminescent detector (Agilent 355 SCD). Breakthrough adsorption curves were generated by plotting the concentration of sulfur against the cumulative volume per unit weight of adsorbent.

Heat of adsorption calculation. Adsorption isotherms were collected by a batch method. 5 mL of solutions containing varying amounts of sulfur and 50 mg of sorbent were prepared inside a flask. The sorbent and sorbate were allowed to mix for 5 hours at two different adsorption temperatures. The supernatant liquid was collected and sulfur concentration was determined by GC-SCD.

Results and Discussion

Material Characterization. Table 1 shows the surface areas as well as total, micropore and mesopore volume for all the zeolites prepared. All the mesoporous zeolites showed a significant increase in mesopore volume and a decrease in micropore volume.

Table 1. Surface areas and micropore and mesopore volume for each material

Material	S_{tot} (m^2/g)	S_{micro} (m^2/g)	S_{meso} (m^2/g)	V_{tot} (cm^3/g)	V_{micro} (cm^3/g)	V_{meso} (cm^3/g)
NH_4Y	574	530	43.9	0.282	0.246	0.036
SAY	693	373	319	0.373	0.172	0.201
DSY	646	446	201	0.368	0.206	0.162
CeY	550	500	50.1	0.273	0.232	0.041
CuY	628	394	57.2	0.281	0.229	0.052
CeSAY	648	413	235	0.342	0.190	0.152
CuSAY	637	411	226	0.336	0.190	0.146

Table 2. Brönsted and Lewis acid sites, and metal content in each material.

Material	BAS[1] (μmol / g cat)	LAS[1] (μmol / g cat)	Ratio	Ce[2] (wt%)	Cu[2] (wt%)
NH_4Y	113.9	30.2	3.77	-	-
SAY	98.2	35.0	2.81	-	-
CeY	82.1	20.5	4.00	5.5	-
CuY	43.7	69.3	0.63	-	4.9
CeSAY	-	-	-	5.6	-
CuSAY	-	-	-	-	4.6

1. Pyridine adsorption
2. ICP-MS

Pyridine adsorption was used to quantify BAS and LAS which are reported in Table 2. Mesoporous SAY exhibits slightly higher LAS compared to the parent Y. Incorporation of Ce decreases the LAS, while CuY shows otherwise. Similar results have been reported in the past. In addition, CO adsorption was conducted to locate the metal ions in the zeolite. The FTIR spectra revealed that Ce prefers the internal sites I and I' whereas Cu occupies the more accessible sites II and II'. The presence of pre-reduced metals were confirmed by UV-Vis spectroscopy prior to testing.

Fixed-bed adsorption results. Figure 1 shows the breakthrough curves of different thiophenic compounds on various materials. Figure 1(a) suggests metal-exchanged zeolites increase capacity for thiophene (T).

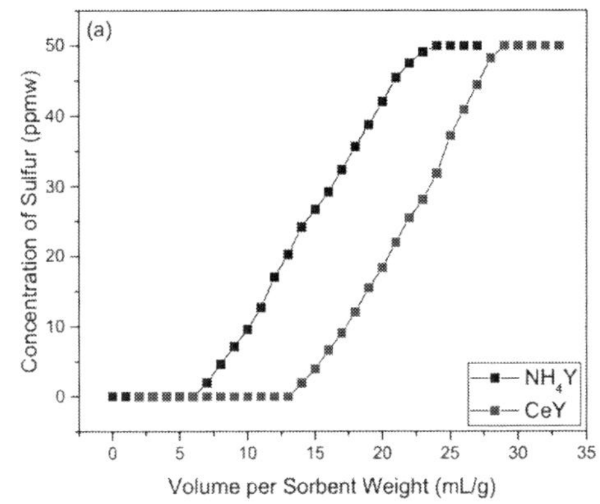

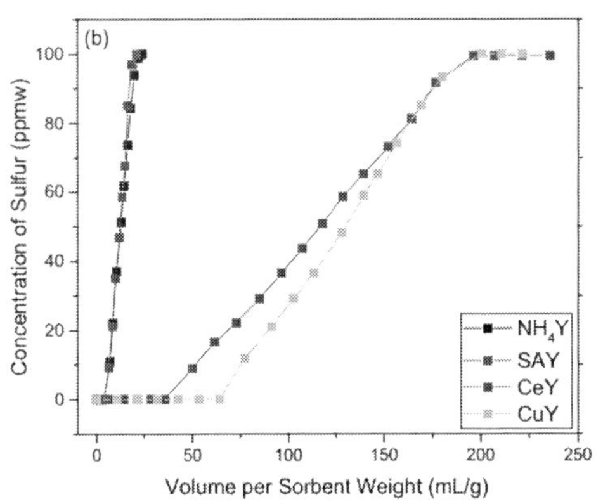

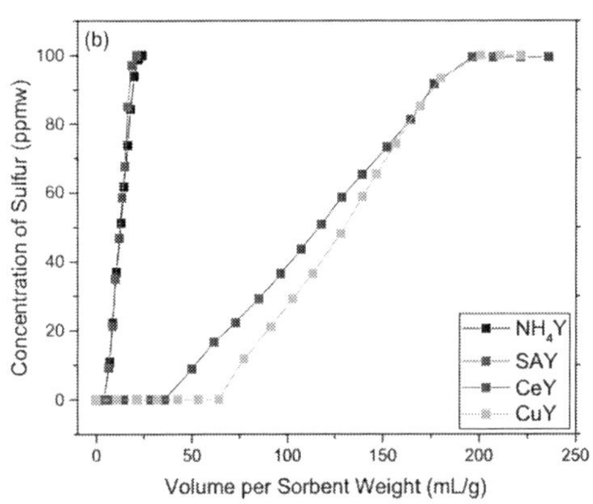

Figure 4. Breakthrough curves of (a) T, (b) BT, and (c) DBT on different materials.

Figure 1(b) shows the breakthrough curves of benzothiophene (BT). CeY and CuY were again the most effective sorbents due to enhanced selectivity and capacity for sulfur. SAY did not improve the adsorption of BT, which suggests that diffusion limitation does not exist. Figure 4(c) demonstrates that the best sorbents for Dibenzothiophene (DBT) removal are the bifunctional zeolites CuSAY and CeSAY. Both materials contain active sites that can form relatively strong π-complexations or direct S-M σ bonds with DBT. CuSAY, for instance, can produce up to 125 mL/g of sulfur-free fuel. This result also suggests that the adsorption of DBT is accompanied by diffusion limitations, which can be overcome by the mesoporous zeolites. The selectivity towards sulfur can be further improved using metals.

Heat of adsorption calculation. Isosteric heats of adsorption were calculated and compared for all materials. It was found that CuY exhibits the highest ΔH value. CuSAY reported a lower value which suggests that the heat of adsorption is only a descriptor of the strength of interaction. The presence of mesoporosity in bifunctional zeolites could decrease the ΔH values due to the loss of framework alumina.

Conclusions

Breakthrough measurements of adsorptive desulfurization have been investigated on parent Y, mesoporous Y, metal-exchanged Y and bifunctional Y zeolites. CuSAY exhibits the highest DBT adsorption capacity of 125 mL/g followed by CeSAY with a capacity of about 100 mL/g. Bifunctional zeolites enhance both accessibility to the active sites and selectivity for thiophenic compounds with high kinetic diameter. This enhancement is driven by the presence of metals which create more active sites to bind with the sulfur either via π-complexation or the direct S-M σ bond. The incorporation of mesoporosity using SA method allows for uniform distribution of mesopores responsible for solving diffusion problems. The breakthrough performance was also verified with theoretical calculations of heats of adsorptions. Finding the optimum balance between the pore structure and metals in a bifunctional zeolite is essential for adsorptive desulfurization of hydrocarbons fuels.

Acknowledgement. This work was funded by ACS PRF# 55900-DN15.

References

(1) Song, C.; Ma, X. New design approcahes to ultra-clean diesel fuels by deep desulfurization and deep dearomatization. *Appl. Catal. B Environ.* **2003**, *41*, 207–238.
(2) Bhandari, V. M.; Ko, C. H.; Park, J. G.; Han, S. S.; Cho, S. H.; Kim, J. N. Desulfurization of diesel using ion-exchanged zeolites. *Chem. Eng. Sci.* **2006**, 61, 2599–2608.
(3) Li, K.; Valla, J.; Garcia-Martinez, G. Realizing the commercial potential of hierarchical zeolites: new opportunities in catalytic cracking. *Chem. Cat. Chem.* **2014**, 6, 46–66.
(4) Garcia-Martinez, J.; Johnson, M.; Valla, J.; Li, K.; Ying, J. Y. Mesostructured zeolite Y - high hydrothermal stability and superior FCC catalytic performance. Catal. Sci. Technol. **2012**, 2, 987–994.

CATALYTIC FAST PYROLYSIS OF BIOMASS DERIVED FURANS OVER HIERARCHICAL ZSM-5 CATALYSTS

Jinsheng Gou, Xiaoduo Qi, Vivek Vattipalli and Wei Fan

Chemical Engineering, University of Massachusetts Amherst, 686 North Pleasant St.
Amherst, MA 01003

Renewable fuels and chemicals are of growing importance in satisfying environmental and economic concerns over fossil fuel usage. Liquid bio-oil produced by the fast pyrolysis of lignocellulose has the potentials to provide transportation fuels and commodity chemicals. In particular, the production of renewable aromatics (e.g. benzene, toluene, xylene) and olefins by the catalytic fast pyrolysis (CFP) of lignocellulose over ZSM-5 catalysts is highly desired due to the importance of these commodity chemicals in the manufacture of several polymer materials.

There are, however, still several challenges on the selective conversion of lignocellulose into aromatics using ZSM-5 catalysts, such as fast catalyst deactivation and coke formation. The fast deactivation of zeolite catalysts in the CFP is partially due to the slow diffusion of bulky molecules and limited deoxygenation activity. The diffusion limitation leads to the fact that more consecutive reactions can occur between the products, so that poly aromatic hydrocarbons (PAHs) and coke can finally form in the micropores as well as on the external surface of the catalysts, resulting in catalyst deactivation. The objective of this work is to fundamentally understand how mesoporous structures of hierarchical ZSM-5 affects the CFP of lignocellulose biomass. In this study, a simple biomass derived oxygenated compound, furan, was used in the CFP process over ZSM-5 catalysts with varying particle size and mesoporosity in order to simplify the pyrolysis chemistry and provide insights into the catalyst design.

Our study suggests that the use of hierarchical ZSM-5 and ZSM-5 nanocrystals not only increases the reaction rate, but also leads to higher selectivity to aromatics and olefins as well as lower coke formation compared to conventional ZSM-5 catalyst. This high selectivities to high-value chemicals are mainly due to the enhanced mass transport. It was found that the coke were mainly formed in the micropores of the ZSM-5 catalysts. However, the mass transport properties of the zeolite catalyst were not affected by the formation of coke inside the micropores even when 30% of micropore was occupied by the coke. The counter-intuitive phenomena might be because of the three dimensionally interconnected microporous structure of ZSM-5 which provides alternative diffusion pathways for molecules.

3D NANOSCALE IMAGING OF ZEOLITE-BASED CATALYSTS

Jovana Zečević,[1] Cedric Gommes,[2] Heiner Friedrich,[1] Petra E. de Jongh[1] and Krijn P. de Jong[1]

[1] Inorganic Chemistry and Catalysis, Utrecht University, Universiteitsweg 99, 3584CG Utrecht, The Netherlands
[2] Department of Chemical Engineering, University of Liège B6A, Allée du 6 août 3, B-4000 Liège, Belgium
[3] Laboratory of Materials and Interface Chemistry, Department of Chemical Engineering and Chemistry, Eindhoven University of Technology, Den Dolech 2, 5612 AZ Eindhoven, The Netherlands
j.zecevic@uu.nl

Introduction

A wide range of important industrial processes rely heavily on zeolites and zeolite supported metal catalysts. For example, bifunctional zeolite Y supported Pt catalysts are used in oil refining processes such as hydroisomerisation and hydrocracking. The catalytic performance of bifunctional catalyst depends greatly on the accessibility of its active sites [1]. Acid sites, which are located in the micropores of zeolite Y (~1 nm diameter), provide cracking function, while Pt particles whose location and dispersion might vary with preparation methods provide (de)hydrogenation function. Prior to addition of Pt particles, zeolite Y is typically treated by steaming and acid leaching in order to introduce mesopores (2-50 nm diameter), which facilitate molecular transport to and from the acid sites. It is not solely the presence of mesopores, but also their connectivity, shape and overall architecture that play an important role in catalysis [2]. Next to the porosity, revealing the metal particle size distribution and their location within the zeolite support is crucial for understanding the reaction mechanisms as well as for optimizing preparation procedures [3-4]. Bulk characterization techniques, such as EXAFS, NMR, H_2-chemisorption, are commonly used to assess the structural parameters of such bifunctional zeolite-based catalysts. Although these techniques provide structural information averaged over milligrams of catalyst sample, local nanoscale properties and/or heterogeneities thereof, which are crucial for catalyst performance, are often overlooked. Transmission electron microscopy (TEM) is the most direct technique which enables visualization and measurement of local nanoscale structural features, including mesopore size and shape, and metal particle size distribution [5]. However, due to overlapping features in 2D projections of TEM micrographs, determination of exact shape and connectivity of mesopores, or metal particle location is hampered. The need for 3D structural information has led to the development of electron tomography (ET), or 3D-TEM, which when coupled with image analysis, proved to be an excellent tool for a detailed quantitative analysis of the complex nanoscale structures, such as zeolite supported metal catalysts [6].

Experimental

For our study of zeolite mesopore network we used commercially available mesoporous zeolite Y powder in H-form and with Si/Al atomic ratio of 30 (Zeolyst, CBV760) [2]. To study bifunctional Pt loaded zeolite catalyst, we impregnated mesoporous zeolite Y powder (Zeolyst, CBV760) by incipient wetness impregnation with an appropriate amount of aqueous solution of $Pt(NH_3)_4(NO_3)_2$ to obtain 1 wt% Pt loading [3]. Impregnated powder was dried and submitted to different heat treatment conditions, either calcination under air flow and reduction in H_2, or direct reduction in H_2. In an additional study [4], to assess the impact of using Pt precursor that restricts the interaction with acidic sites (ion exchange with zeolite protons) we used H_2PtCl_6 solution. Such prepared catalysts were characterized using different bulk characterization techniques, including N_2 physisorption and EXAFS. To assess the nanoscale structural features, tilt series for electron tomography study were acquired on a Tecnai 20 transmission electron microscope (FEI). Alignment and reconstruction of the acquired tilt series has been performed in IMOD. Image segmentation and analysis were done using DipLib toolbox (TU Delft) in Matlab, while visualization of the segmented volumes was conducted in either Matlab or Amira.

Results and Discussion

Electron tomography (ET) study coupled with image analysis of mesoporous zeolite Y, provided an unprecedented insight into the structure of mesopore network created by steaming and acid leaching of zeolite Y [2]. This set of tools allowed us to discriminate open from closed mesopores, whereby open mesopores are the ones that can be accessed from the surface of the zeolite crystal through the mesopore network, while the closed ones are the ones accessible only through the micropores (Figure 1). The mesopore network volumes of two randomly selected zeolite Y crystals derived from quantitative analysis of tomograms, were strikingly in line with the results obtained from the bulk N_2 physisorption measurements. Furthermore, the tortuosity of 1.3 of the mesopore network has been determined, a property important for mass transfer efficiency. Finally, ET and image analysis allowed for the first time quantitative analysis of the size of the microporous domains within zeolite crystal after the introduction of mesopores. Having an access to such feature is beneficial for understanding molecular diffusion through mesoporous zeolites and for judging the effectiveness of the introduced mesopore network.

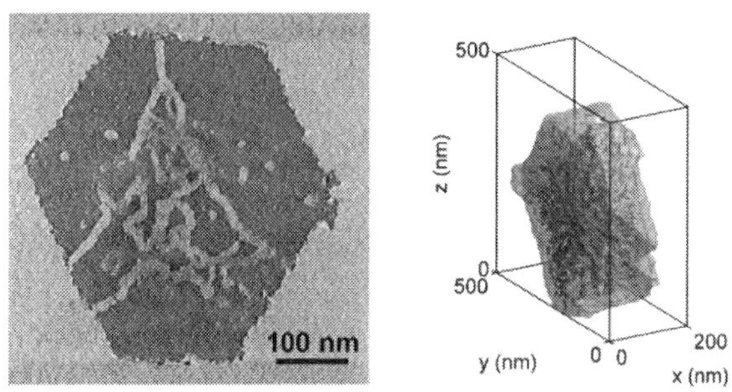

Figure 1. Slice through the tomographic reconstruction of one zeolite crystal (on the left) shows the presence of complex mesopore network. Volume rendered representation (on the right) of open mesopores (green) and closed mesopores (red) within the zeolite crystal (grey) shown on the left.

In the study of the Pt loaded zeolite Y, where $Pt(NH_3)_4(NO_3)_2$ was used as a precursor, ET and image analysis provided a unique opportunity to visualize thousands of Pt particles located within the zeolite crystals and to unravel morphological differences between different Pt-zeolite Y crystals [3]. It has been shown that there are large heterogeneities with respect to Pt loading between different zeolite Y crystals of the same batch of sample with nominal loading of about 1 wt% (as confirmed by EXAFS). The Pt loading varied from 0.2 to 7.1 wt%, while the Pt particle size distribution remained similar and with a mean particle size of ~ 1.2 nm (corresponding to the size of zeolite Y micropore) when impregnated zeolite powder was calcined and reduced (Figure 2). We speculate that such large heterogeneity in zeolite crystal loading could stem either from uneven distribution of Pt precursor upon impregnation, or differences in Si/Al ration between zeolite crystals. Upon direct reduction, however, average Pt particle size was 3 nm, and the additional experiment where non-mesoporous zeolite Y was impregnated with Pt showed that, although exceeding the size of the micropores, these Pt particles can still be found within the microporous region.

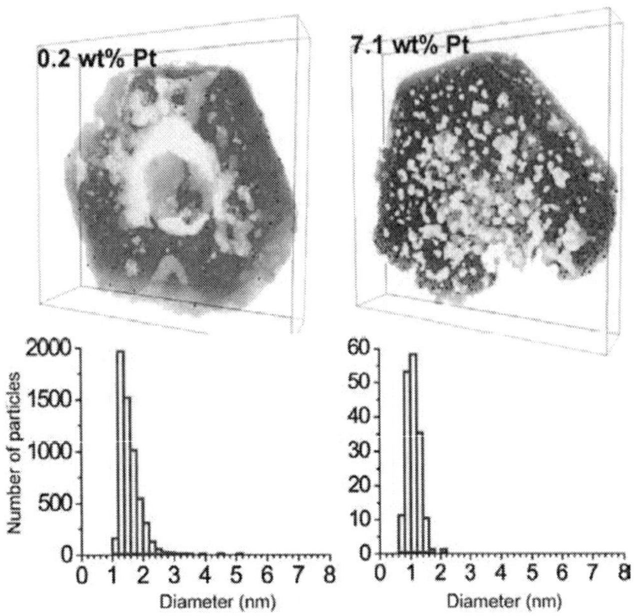

Figure 2. Volume rendered representations (top) of the two Pt-zeolite crystals studied with electron tomography, with microporous region in orange, mesopores in white and Pt particles in blue. Pt particle size distributions are shown at the bottom of the corresponding Pt-zeolite crystals.

Separate study where H_2PtCl_6 was used as a precursor showed that ET is indispensable tool to unravel the true Pt particle size distribution [4]. While EXAFS measurements pointed to Pt size of 1.9 nm, ET and image analysis revealed that the actual size of Pt particles varies, with the size distribution being bimodal. With such prepared catalyst, two distinct sizes of Pt particles were obtained – small Pt particles of ~ 1.2 nm and large Pt particles of ~ 3.5 nm (Figure 3). Clearly ET was crucial to study local properties that can help understanding the precursor-zeolite interaction and help improving the preparation procedures.

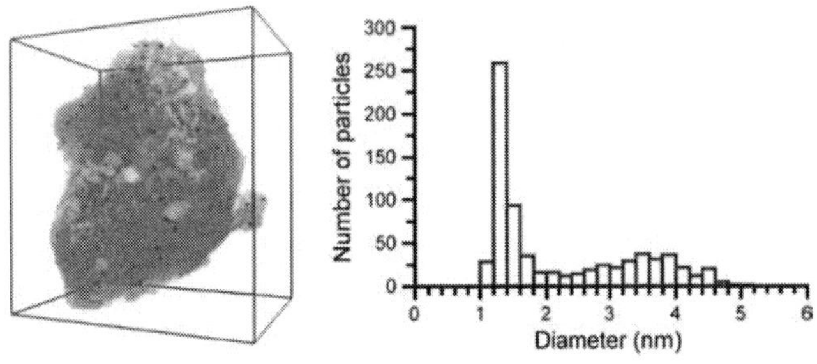

Figure 3. Volume rendered representation of Pt-zeolite crystal (on the left) and Pt size distribution (on the right).

Conclusions

With increased complexity of the nanoscale structures of modern catalysts, advanced characterization tools are becoming ever more important [1-4]. Electron tomography coupled with image analysis has been shown to provide unprecedented insights into the nanoscale structures of industrially relevant zeolite-based catalysts. Obtained results can be further used to modify and adjust zeolite catalyst preparation procedures and enhance their performance.

Acknowledgement. J.Z., H.F. and K.P.deJ. acknowledge support from the Dutch National Research School Combination Catalysis (NRSCC) and from NWO-CW; C.J.G. is supported by the Fonds de la Recherche Scientifique (F.R.S.-FNRS, Belgium).

References

(1) Zečević, J.; Vanbutsele, G.; de Jong, K. P.; Martens, J. A. *Nature*, **2015**, *528*, 245.
(2) Zečević, J.; Gommes, C.; Friedrich, H.; de Jongh, P. E.; de Jong, K. P. *Angew. Chem. Int. Ed.* **2012**, *51*, 4213.
(3) Zečević, J.; van der Eerden, A. M. J.; Friedrich, H.; de Jongh, P. E.; de Jong, K. P. *ACS Nano* **2013**, *7*, 3698.
(4) Zečević, J.; van der Eerden, A. M. J.; Friedrich, H.; de Jongh, P. E.; de Jong, K. P. *Micro. Meso. Mater.* **2012**, *164*, 99.
(5) Pandya, K. I.; Heald, S. M.; Hriljac, J. A.; Petrakis, L.; Fraissard, J.; *J. Phys. Chem.* **1996**, *100*, 5070.
(6) Wei, Y.; Parmentier, T.; de Jong, K. P.; Zečević, J. *Chem. Soc. Rev.* **2015**, *44*, 7234.

SYNTHESIS AND CHARACTERIZATION OF NANOGIBBSITE WITH OR WITHOUT MCM-41 IN AN AMBIENT ENVIRONMENT

Viktor Dubovoy, Michael Stranick, Long Pan

Colgate-Palmolive Company, 909 River Road, Piscataway, NJ
longrutgers@yahoo.com

Introduction

In this article, through extensive characterization, we identified that the guanidinium-containing amino acid L-arginine prevents metal hydroxide aggregation to stabilize $Al(OH)_3$ nanoparticles with average hydrodynamic diameter 10-30nm. It is proposed that the amphoteric and zwitterionic nature of arginine effectively shielded the surface charge of $Al(OH)_3$ nanoparticles during the mild hydrolysis, yielding an environment disfavoring extensive aggregation. Although arginine itself could not further reduce particle size in this "open" system, taking advantage of cage-effect confinement in a mesoporous material MCM-41, we detect ultrafine metal hydroxide nanoparticles formed within mesopores as well as framework alumination of MCM-41.

Results and Discussion

Nanogibbsite Synthesis. Size-exclusion chromatography (SEC) can resolve the wide array of antiperspirant clusters into five domains, arbitrarily designated peaks 1-5.[1] To our best knowledge, no report exists pertaining to synthesis or characterization of the enormous Al species that elute under SEC peak 1 with the smallest elution time. In this study, we report that $Al(OH)_3$ nanoparticles ranging 10-30 nm in diameter are a component of possibly several discrete molecules that elute under SEC peak 1. Structural (PXRD) and spectroscopic (^{27}Al NMR and FTIR) analysis unambiguously identified a $Al(OH)_3$ structure with minor cationic impurities exhibiting $[Al(OH)_xCl_{3-x}]$ stoichiometry (where x is the molar hydrolysis ratio ranging 0-3). SEC data demonstrates that the nanoparticles can be synthesized at a conversion rate of 82%.

Characterization. Vibrational spectroscopy (FTIR-ATR) is an invaluable analytical tool, which was utilized herein to identify the aluminum hydrolysis product as the gibbsite polymorph based on the high frequency OH stretch at 3620 cm^{-1} (gibbsite), which varies substantially with 3650 cm^{-1} (bayerite). The IR bands observed in the OH stretch (3617, 3523, 3453 cm^{-1}) and AlOH bending region (1023, 970, 918 cm^{-1}) are characteristic of gibbsite.[2] Arginine infrared bands were not observed suggesting effective purification from GPC and a lack of coordination between arginine and aluminum.

At Arg/Al ratio of 2.75, distribution of the gibbsite particles was 10-30 nm average diameter as determined via static light scattering method. Crystalline size (*ca.* 8nm) was determined by applying the Scherrer equation to the PXRD pattern. TEM images show particles with diameters *ca.* 5-15 nm.

^{27}Al (MAS) NMR analysis was performed to elucidate the Al environment and coordination modes in the prepared material. The aluminum nucleus is largely limited to octahedral (O_h) and tetrahedral (T_d) coordinations. The dispersed nano-$Al(OH)_3$ exhibits an ^{27}Al NMR signal in aqueous environment suggesting this phase can be directly probed using common aqueous analysis techniques due to substantial interaction at the solid-liquid interface. In Figure 1, NMR spectra are plotted versus Arg/Al molar ratio. Aluminum chloride monomer (Al^{3+}) exhibits a characteristic sharp NMR peak at 0 ppm. ^{27}Al NMR data demonstrates some tetrahedral aluminum Keggin clusters, specifically $Al_{13\text{-mer}}$ and $Al_{30\text{-mer}}$, were formed in the sample with Arg/Al 2.25 and exhibited characteristic ^{27}Al NMR tetrahedral Al signals at 63 and 70 ppm, respectively *(see Fig. 1b)*. As arginine content increased from Arg/Al 0 to 2.25, the concentration of Al_{13}-type compounds increased until reaching a maximum at 2.25. Maximum turbidity was between Arg/Al ratio 2.0-2.5 suggesting electrostatic instability. At Arg/Al ratio above 2.25, suspended $Al(OH)_3$ nanoparticles were produced with the concentration reaching a maximum at 2.75. At the climax of nanoparticle concentration, liquid NMR spectrum exhibits a single Gaussian-type resonance at 8ppm. The relatively sharp nature of the NMR peak suggests presence of only a single O_h specie.

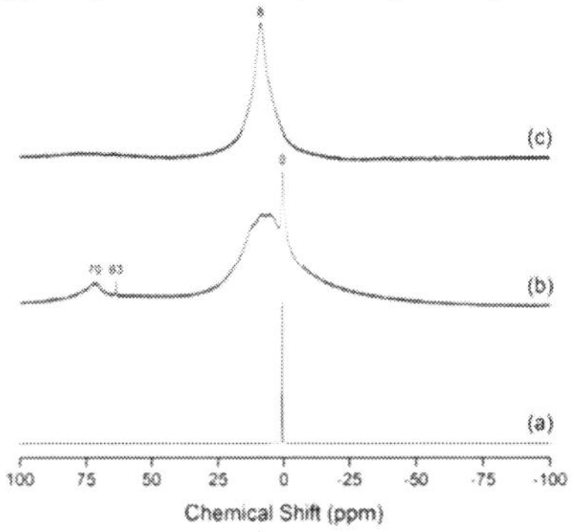

Figure 1. ^{27}Al NMR of Arg/Al 0 (a), 2.25 (b), and 2.75 (c)

GPC technique was implemented to isolate nano-$Al(OH)_3$ from the synthetic byproducts (i.e. arginine and soluble cationic Al compounds). Elution from the polyacrylamide packing produced two distinct peaks with adequate resolution for separation. The $Al(OH)_3$ suspension is collected with 99% purity, exhibits pH of 6.7, and undetectable amounts of arginine (via ICP-OES). Zeta potential, often conducted parallel to potentiometric or conductivity studies, is a common technique imperative for investigating solid-liquid interface in heterogeneous systems. The purified nanoparticle suspension exhibits a zeta potential of +8.9mV with electrophoretic mobility of 0.7 μmcm/Vs and conductivity of 0.7 mS/cm. The positive zeta potential can be explained by the presence of cationic aluminum species indicated by a small concentration of Cl^- from XPS analysis. Chloride is present at an Al:Cl ratio of 35:1, suggesting presence of minor aluminum (hydroxide) chloride impurities. Based on XPS data, Al binding energy was 74.3 eV for both purified material and reagent grade $Al(OH)_3$. Stoichiometric ratios of aluminum to oxygen (1:3.3) were in agreement with $Al(OH)_3$. Furthermore, zeta potential and electrophoretic mobility show only a slight deviation from deionized water. The large difference observed in conductivity is attributed to the cationic aluminum species. Initial hypothesis of aluminum hydroxide formation was formulated upon freeze drying the GPC-purified solution containing Arg/Al 2.75. It is noteworthy that the dehydrated $Al(OH)_3$ powder was insoluble in water suggesting complete neutralization of Al^{3+}.

Al-MCM-41 Synthesis and Characterization. Purely siliceous MCM-41 with 2.7 nm pore size was selected to investigate the interaction of MCM-41 with $Al(OH)_3$. MCM-41 was activated in a vacuum oven at 120°C to remove atmospheric contaminants prior to aqueous aluminum chloride addition. Adequate (1h) magnetic stirring of heterogeneous system ensured maximum Al^{3+} concentration within MCM-41 void space. Local flocculation was observed during the slow addition of arginine, which was allowed to dissipate prior to further arginine addition. Cage-effect confinement of $Al(H_2O)^{3+}$ hydrolysis within mesoporous framework is able to produce ultrafine $Al(OH)_3$ nanoparticles smaller than 2.7nm. Product formation in the bulk solution was carefully monitored via SEC and ^{27}Al NMR. Results indicate the Al^{3+} reactant has been effectively converted to predominantly $Al(OH)_3$ phase. We predicted the cage effect of the rigid mesoporous framework would constrain the $Al(OH)_3$ particle growth. ^{27}Al MAS NMR of the guest Al molecules demonstrates presence of both octahedral (*ca.* 2 ppm) and tetrahedral (*ca.* 57 ppm) Al environments. Mokaya *et al.* reported incorporation of tetrahedral aluminum into the MCM-41 framework by grafting with aluminum chlorohydrate (ACH).[3] In addition to $Al(OH)_3$ incorporated into the MCM-41 channels, consistent with their results, our NMR data suggests a portion of the Al was incorporated into the SiO_2 framework. However, our synthesis led to a significantly higher content of octahedral Al (O_h/T_d = 1.4) compared to Mokaya and coworkers ($O_h/T_d < 1$).

Al loading into MCM-41 void space was additionally confirmed via BET N_2 adsorption method. Figure 2 shows the amount of adsorbed nitrogen at 77 K versus the relative pressure. The adsorption isotherm is type IV with a well-defined inflection at approximately 0.30-0.40 and 0.25-0.35 P/P° for MCM-41 and Al-MCM-41, respectively. Incorporation of $Al(OH)_3$ nanoparticles within the pores significantly reduced nitrogen physisorption capacity within the mesoporous SiO_2 matrix. The structure of unmodified MCM-41 exhibited BET surface area of 997 m^2/g with a 0.932 cm^3/g pore volume and 2.7 nm pore size. After the Al grafting, reduction of BET surface area (20.4%), pore volume (30.4%), and pore width (22.2%) is observed. The drastic change in measured internal morphology suggests the channels within MCM-41 contained aluminum particles. ICP analysis of HNO_3-digested Al-MCM-41 suggests Al is present at *ca.* 10% by weight. EDS analysis confirmed Al was present at 11 wt. % with Si/Al 2.90.

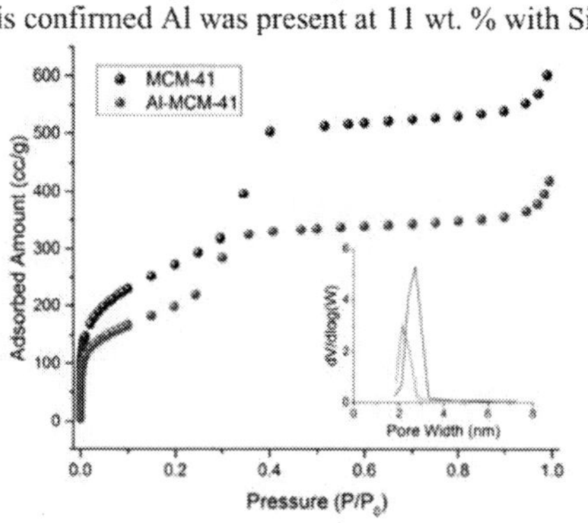

Figure 2. N_2 sorption isotherms of MCM-41 and Al-MCM-41. Inset is the corresponding pore size distribution

Small angle x-ray diffraction (SAXRD) pattern of MCM-41 before and after Al insertion was obtained and indexed based on apparent hexagonal symmetry *(see Fig. 3)*. The presence of 100 (2.2°), 110 (3.9°), 200 (4.4°), and 210 (5.8°) reflections in both samples indicates that long range structural ordering is maintained in the Al-MCM-41 sample despite having a relatively high Al content (Si/Al ratio 2.9).

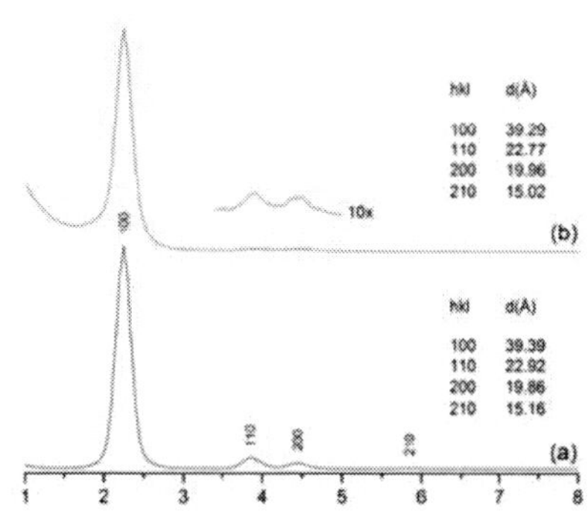

Figure 3. SAXRD pattern of MCM-41 (a) and Al-MCM-41 (b)

Incorporation of Al caused a downfield shift (*ca.* 1 ppm) for the predominant 3.1 ppm peak observed in the 1H MAS NMR spectrum of MCM-41. Furthermore, a sharp isolated signal emerges at 0.9 ppm in Al-MCM-41 sample. Since this proton peak is not observed in the starting material and exhibits stronger shielding, we assign the 0.9 ppm peak to hydroxyl groups coordinated to aluminum atoms. Previous ^1H MAS NMR studies of acidic zeolites have tentatively assigned the *ca.* 0.9 and 2.8 ppm signals to terminal Al-OH groups with octahedral or tetrahedral coordination, respectively.[4]

High resolution transmission electron microscope (HRTEM) experiments suggest MCM-41 maintained its integrity of highly ordered hexagonal channels during internal $Al(OH)_3$ nanoparticle formation. Even though the bulk of the functional surface area is contained within the pores, it should be noted that Si-OH groups are also present on the external surface which may have immobilized some $Al(OH)_3$. Therefore, we predicted to observe aluminum particles anchored on the outside Si-OH groups. Large $Al(OH)_3$ particles were not observed near the external surface of MCM-41, which also provides further evidence that the aluminum is contained predominantly within the channels. Attempts to directly observe the host particles inside the channels were ineffective via TEM method, due to the weak contrast between the silica framework and aluminum hydroxide. Although, consistent with similar work, some color contrast, brightness variation, and pore grid distortions were visible upon closer examination of TEM images. This was attributed to nonhomogeneous distribution of guest molecules.

EDS analysis was conducted on an area of the prepared solid and detected significant amounts of Al, Si, and O. Elemental composition was 8.02% Al, 23.26% Si, and 68.70% O. Surface elemental composition, measured via XPS, consists of 6.13% Al, 21.75% Si, and 66.36% O. Aluminum was not detected in the original MCM-41 material. Consistent with $Al(OH)_3$ stoichiometry, the bulk O/Al atomic ratio, corrected for SiO_2, is 2.77. However, the corrected surface O/Al ratio (3.83) suggests a high concentration of $Al(O)_4$ tetrahedrons as also evidenced by T_d NMR resonance at 57 ppm. Si/Al ratio was 2.9 and 3.6 as determined by EDS and XPS, respectively. A higher ratio of Si/Al measured using XPS versus EDS indicates more Al penetrated into the pores versus building up on the surface. Chloride was not detected in stoichiometric concentrations using either method.

Conclusion

In this research, we prepared aluminum hydroxide nanoparticles with 10-30 nm average diameter via arginine-stabilized technique from Al^{3+} precursor. A transparent suspension of $Al(OH)_3$ exhibiting neutral pH was isolated by GPC column. Using this mild synthesis methodology, $Al(OH)_3$ was loaded into the 2.7 nm mesoporous cavities of MCM-41 resulting in significant reduction of surface area and pore size, as determined by N_2 adsorption experiments. After the post-synthesis alumination of MCM-41, Al was detected at 11.00 wt. % with Si/Al 2.90.

References

(1) Laden, K. Antiperspirants and Deodorants, 2nd ed.; Marcel Dekker, Inc.: New York, **1999**
(2) Kumara, C.K.; Ng, W.J.; Bandara, A.; Weerasooriya, R. *J. Colloid Interface Sci.*, **2010**, 352, 252
(3) Mokaya, R.; Jones, W. *J. Mater. Chem.* **1999**, 9, 555
(4) Zeng, Q.; Nekvasil, H.; Grey, C.P. *J. Phys. Chem. B.* **1999**, 103, 7406

Photo courtesy of the ACS

ELECTRON DYNAMICS OF LARGE SYSTEMS FROM REAL-TIME TDDFTB

Bryan M. Wong, M. Belén Oviedo, and Niranjan V. Ilawe

Department of Chemical & Environmental Engineering
Materials Science & Engineering Program
University of California-Riverside, Riverside, CA
e-mail: bryan.wong@ucr.edu

Introduction

Electronic excitations in the condensed phase play a central role in several fields of modern chemical research, ranging from light-harvesting chromophores in biochemical systems[1-4] to ultrafast excitations in all facets of time-domain spectroscopy. In each of these areas, the chemical system exhibits a multitude of interactions (both classical and quantum) between different chemical interfaces that give rise to the experimentally-observed spectrum. These many-body interactions are particularly crucial in the liquid phase, as these effects are significantly more complex than in either gas-phase or solid-state systems.

In this talk, a new microscopic picture of solute-solvent electron dynamics and solvatochromic effects is presented, which is uncovered using a new real-time quantum dynamics approach for large solvated systems.[5] In particular, we show that a complex interplay of quantum interactions underlies our observations of solute-solvent effects. While this presentation focuses on a prototypical C_{60}:dimethylaniline donor-acceptor system (C_{60}:dimethylaniline is also utilized in new photovoltaic materials[6,7]), our main qualitative results are expected to apply to a broad range of other solvated donor-acceptor systems. Our ability to probe extremely large systems allows us to highlight the unusual many-body mechanisms that mediate solvatochromic shifts in both nonpolar and polar solvent – *without recourse to empirically-derived continuum models*. Moreover, our time-domain studies of these solvated systems provide a natural approach to probe in microscopic detail the time-dependent electron dynamics and couplings between the solute and solvent in emergent order.

Real-Time Quantum Dynamics of Large Systems

To address the massive sizes of our solvated complexes (up to 205 explicit solvent molecules and 3,594 orbitals) we carried out our electronic structure calculations using the self-consistent density functional tight-binding (SCC-DFTB) formalism[8]. Our use of the DFTB formalism is crucial to this study since the immense sizes of our systems *cannot* be routinely probed with conventional DFT/TD-DFT or continuum methods.

For our real-time electron dynamics calculations, we implemented and utilized a locally-modified real-time, time-dependent DFTB (RT-TDDFTB) code for probing the optical properties in all of our solvated systems. This method is based on the real-time quantum-mechanical propagation of the one electron density matrix in the presence of external time-varying electric fields to obtain the optical response of the system. We carry out our real-time electron dynamics calculations by applying a time-dependent electric field to the initial ground state density matrix. Upon application of an arbitrary time-dependent field, the density matrix, ρ, evolves according to the Liouville-Von Neumann equation of motion. For example, to calculate the photo-induced electron dynamics in the time domain, a sinusoidal electric field can be applied to the Hamiltonian with a frequency ω tuned to a specific absorption peak for the system of interest. The distinct advantage of our approach is that we directly calculate the electron-hole dynamics in the presence of these non-perturbative, time-dependent fields which *cannot* be achieved by a simple time-evolution propagation of an excited state. This capability allows us to probe the intricate real-time electron dynamics of these complex systems at a level of detail that is beyond the reach of traditional linear-response TDDFT approaches.

Many-Body Interactions in Explicit Solvent

To generate the representative solute-solvent geometries (cf. Fig. 1) for our electron dynamics studies, we obtained relaxed configurations via a series of molecular dynamics (MD) calculations. The solvated C_{60}:DMA complex was placed in a periodic box of side length 55 Å, where the solvent was composed of either 5,590 explicit water or 829 toluene molecules. A short energy minimization of 10,000 (steepest descent) steps was first applied, subsequently followed by a cooling step via a constrained simulated annealing from 400 to 293 K for 1 ns.

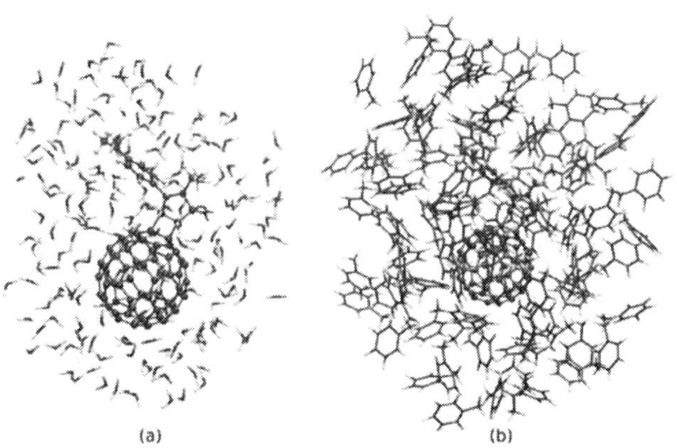

Figure 1. Representative configurations of the C_{60}:DMA complex surrounded by (a) 205 water and (b) 91 toluene explicit solvent molecules.

With these relaxed configurations, a series of several RT-TDDFTB calculations were carried out to obtain the absorption spectra of C_{60}:DMA in water and toluene. These series of calculations are summarized in Fig. 2a and naturally result in the well-known inhomogeneous broadening of the absorption spectra that can be directly compared to experiment[9]

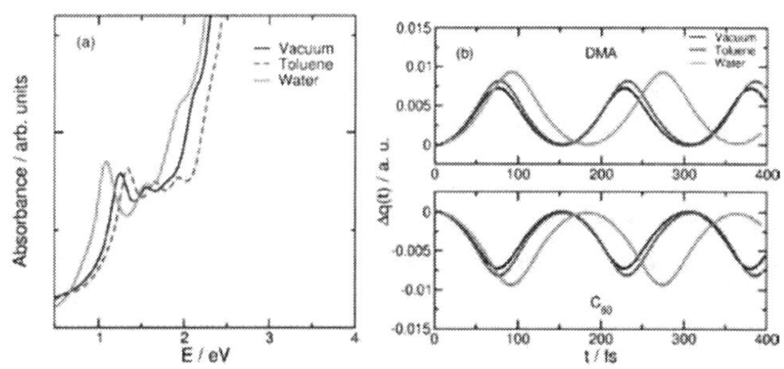

Figure 2. (a) Frequency-domain absorption spectra of C_{60}:DMA in explicit aqueous and toluene solvent. (b) Changes in Mulliken charges as a function of time for both DMA and fullerene in aqueous and toluene solvent.

Indeed, our RT-TDDFTB simulations are in excellent agreement with condensed-phase spectroscopic experiments and show a red-shift of the excitation energies for the polar H_2O solvent relative to vacuum. Conversely, for the non-polar toluene solvent, we observe a solvatochromic blue shift of the absorption band. It is important to point out that one *cannot* obtain these results with conventional Casida-type TDDFT methods due to the immense size of these systems (up to 3,594 orbitals), whereas the entire spectrum can be directly obtained with RT-TDDFTB dynamics. Most importantly, we naturally obtain the solvatochromic shifts and inhomogeneous broadening effects by carrying out our real-time dynamics with *both* the solvent and solute being treated on the same footing at the quantum-mechanical level.

Most importantly, we can obtain additional mechanistic insight by probing the intramolecular excited-state solvent dynamics in the time domain. To probe these effects, we apply a continuous laser perturbation to the entire solute-solvent system with ω tuned to the first charge-transfer excitation. Our resulting time-domain calculations in Fig. 2b clearly show that the Mulliken charge dynamics are strongly dependent on the polarity of the solvent. Mechanistically, as the solvated system interacts with the applied electric field, an electron-hole pair is generated within the C_{60}:DMA complex. This electron-hole pair interacts dynamically with the electrostatic environment of the solvent, resulting in an oscillation period/lifetime that directly increases with solvent polarity (cf. Fig. 2b). On a deep mechanistic level, the RT-TDDFTB calculations show that this intramolecular charge-transfer excitation is stabilized by the polar character of the solvent, and recombination of the photo-induced electron-hole pair in aqueous C_{60}:DMA is diminished relative to the nonpolar toluene.

In addition to intramolecular charge-transfer excitations *within* the C_{60}:DMA complex, we can apply the same real-time quantum techniques to probe intermolecular interactions *between* C_{60}:DMA and the surrounding solvent. These intermolecular dynamics are significantly more complicated as they inherently involve complex many-body interactions throughout the entire system. In other words, the spectrum of a solvated system is the result of *collective* electronic excitations acting in concert between the solute and each of the surrounding solvent molecules. To probe these delicate effects, we applied a continuous laser field (with the same charge-transfer excitation frequency used previously) to the entire C_{60}:DMA-solvent system. Figs. 3a and 3b show the time-dependent charge fluctuations between C_{60}:DMA and the surrounding water and toluene molecules, respectively. The resulting charge-population analyses clearly show that the dynamics for water and toluene solvent are qualitatively different: in Fig. 3a, electron injection from water into C_{60}:DMA occurs, while toluene plays the *opposite* role of the electron acceptor in Fig. 3b. Moreover, the amount of charge transferred between toluene and C_{60}:DMA is an order of magnitude smaller compared to the electron dynamics in water.

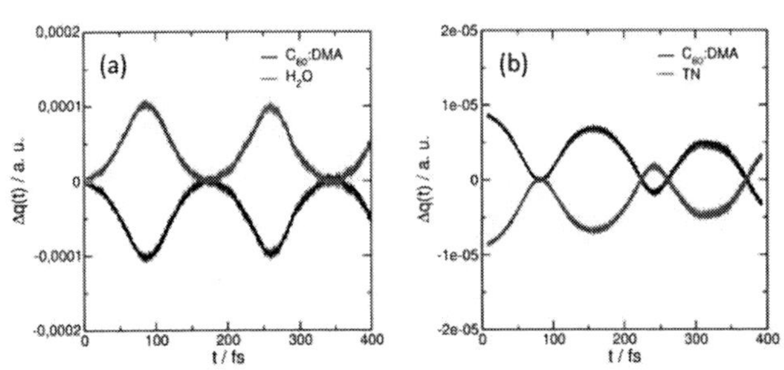

Figure 3. Time-dependent Mulliken charge-transfer between C_{60}:DMA and explicit (a) water and (b) toluene molecules. The amount of charge-transfer in toluene is an order of magnitude smaller than the electron dynamics in water.

It is important to mention that while all of these charge-transfer processes are a natural outcome of our real-time dynamics approach, these intricate mechanisms *cannot* be obtained from either conventional PCM or simple-point-charge solvent models. Specifically, capturing these subtle electron-hole interactions is impossible in either the PCM or simple-point-charge models since the solvent in these methods is treated at a lower level of fidelity, and much of the detailed quantum dynamics between the solvent and solute become irreversibly lost. It is only by treating both the solvent and the solute at the same footing within the RT-TDDFTB approach that we are able to recover the intricate, full many-bodied dynamics of these complex, solvated systems.

Conclusions

In summary, we have demonstrated a new framework for probing the intrinsic many-bodied nature of solute-solvent electron dynamics in complex solvated environments. By treating both the solvent and the solute *on the same footing* at a quantum-mechanical level, we demonstrate that the electron dynamics in these solvated systems are surprisingly rich.

Moreover, it is also important to mention that while all of these charge-transfer processes are a natural outcome of our real-time dynamics approach, these intricate dynamical mechanisms *cannot* be obtained from either continuum or simple-point-charge solvent treatments. By directly probing the real-time electron dynamics in explicit toluene and water solvent (up to 3,594 orbitals), we give crucial insight into the vastly different mechanisms that mediate solvatochromic shifts in both nonpolar and polar solvent – without recourse to empirically-derived continuum models. Most notably, we show that the presence of explicit solvent molecules has a non-trivial effect on controlling non-equilibrium charge-transfer processes, as demonstrated by their vastly different time-dependent charge dynamics. Moreover, we find that electron-hole interactions in these complex environments gives fundamental insight into controlling the electron dynamics of these solvated systems.

Acknowledgement. Support from the National Science Foundation - Institute for Complex Adaptive Matter (NSF-ICAM) is gratefully acknowledged.

References

(1) Scholes, G. D.; Fleming, G. R.; Olaya-Castro, A.; van Grondelle, R. *Nat Chem* **2011**, *3*, 763.
(2) Fleming, G. R.; Scholes, G. D. *Nature* **2004**, *431*, 256.
(3) Francesca, F.; Alexandra, O.-C. *New Journal of Physics* **2010**, *12*, 085006.
(4) Rebentrost, P.; Mohseni, M.; Aspuru-Guzik, A. *The Journal of Physical Chemistry B* **2009**, *113*, 9942.
(5) Oviedo, M. B; Wong, B. M. *Journal of Chemical Theory and Computation* **2016**, in press, http://dx.doi.org/10.1021/acs.jctc.5b01019.
(6) Liu, S.-G.; Shu, L.; Rivera, J.; Liu, H.; Raimundo, J.-M.; Roncali, J.; Gorgues, A.; Echegoyen, L. *The Journal of Organic Chemistry* **1999**, *64*, 4884.
(7) Hasobe, T.; Kamat, P. V.; Absalom, M. A.; Kashiwagi, Y.; Sly, J.; Crossley, M. J.; Hosomizu, K.; Imahori, H.; Fukuzumi, S. *The Journal of Physical Chemistry B* **2004**, *108*, 12865.
(8) Elstner, M.; Porezag, D.; Jungnickel, G.; Elsner, J.; Haugk, M.; Frauenheim, T.; Suhai, S.; Seifert, G. *Phys Rev B* **1998**, *58*, 7260.
(9) Seabra, G. d. M.; Walker, R. C.; Elstner, M.; Case, D. A.; Roitberg, A. E. *The Journal of Physical Chemistry A* **2007**, *111*, 5655.

THEORETICAL DESIGN OF HYDROGEN-EVOLVING MOLECULAR ELECTROCATALYSTS

Sharon Hammes-Schiffer

Department of Chemistry, 600 South Mathews Avenue, University of Illinois at Urbana-Champaign, Urbana, Illinois 61801
shs3@illinois.edu

Introduction

Molecular electrocatalysts play an essential role in a wide range of energy conversion processes. The objective of electrocatalyst design is to maximize the turnover frequency and minimize the overpotential for the overall catalytic cycle. Typically the catalytic cycle for hydrogen evolution is dominated by key proton-coupled electron transfer (PCET) processes comprised of sequential or concerted electron and proton transfer steps. Theoretical studies of the mechanisms, thermodynamics, and kinetics for molecular electrocatalysts that produce hydrogen will be presented.

Results and Discussion

Calculations on cobalt complexes with diglyoxime or dithiolene ligands have revealed linear correlations that can be used to predict the impact of modifying the ligands, substituents, or metal centers. Studies of nickel complexes with pendant amines have illustrated the significance of the proton relay between the pendant amine and the metal center and the importance of molecular motions that decrease this proton transfer distance. Investigations of nickel-iron hydrogenase models have indicated that ligand isomerization at the nickel center is essential to produce a thermodynamically accessible and significantly more basic two-electron mixed valence species. Calculations on cobalt and nickel hangman porphyrins have provided insights into the roles of pendant proton relays as well as ligand non-innocence and dearomatization. Specifically, the calculations predicted a mechanism that proceeds via a phlorin intermediate, which was subsequently observed experimentally.

Conclusion

All of these theoretical studies are assisting in the interpretation of experimental data and the design of more effective molecular electrocatalysts.

References

(1) B. H. Solis and S. Hammes-Schiffer, Theoretical analysis of mechanistic pathways for hydrogen evolution catalyzed by cobaloximes, *Inorg. Chem.* **50**, 11252-11262 (2011).

(2) B. H. Solis and S. Hammes-Schiffer, Substituent effects on cobalt diglyoxime catalysts for hydrogen evolution, *J. Am. Chem. Soc.* **133**, 19036-19039 (2011).

(3) L. E. Fernandez, S. Horvath, and S. Hammes-Schiffer, Theoretical analysis of the sequential proton-coupled electron transfer mechanisms for H_2 oxidation and production pathways catalyzed by nickel molecular electrocatalysts, *J. Phys. Chem. C* **116**, 3171-3180 (2012).

(4) S. Horvath, L. E. Fernandez, A. V. Soudackov, and S. Hammes-Schiffer, Insights into proton-coupled electron transfer mechanisms of electrocatalytic H_2 oxidation and production, *Proc. Nat. Acad. Sci. USA* **109**, 15663-15668 (2012).

(5) B. H. Solis and S. Hammes-Schiffer, Computational study of anomalous reduction potentials for hydrogen evolution catalyzed by cobalt dithiolene complexes, *J. Am. Chem. Soc.* **134**, 15253-15256 (2012).

(6) L. E. Fernandez, S. Horvath, and S. Hammes-Schiffer, Theoretical design of molecular electrocatalysts with flexible pendant amines for hydrogen production and oxidation, *J. Phys. Chem. Lett.* **4**, 542-546 (2013).

(7) S. Horvath, L. E. Fernandez, A. M. Appel, and S. Hammes-Schiffer, pH-dependent reduction potentials and proton-coupled electron transfer mechanisms in hydrogen-producing nickel molecular electrocatalysts, *Inorg. Chem.* **52**, 3643-3652 (2013).

(8) B. H. Solis, Y. Yu, and S. Hammes-Schiffer, Effects of ligand modification and protonation on metal oxime hydrogen evolution electrocatalysts, *Inorg. Chem.* **52**, 6994-6999 (2013).

(9) S. Ghosh, S. Horvath, A. V. Soudackov, and S. Hammes-Schiffer, Electrochemical solvent reorganization energies in the framework of the polarizable continuum model, *J. Chem. Theory Comput.* **10**, 2091-2102 (2014).

(10) B. H. Solis and S. Hammes-Schiffer, Proton-coupled electron transfer in molecular electrocatalysis: Theoretical methods and design principles, *Inorg. Chem.* **53**, 6427-6443 (2014).

(11) M. T. Huynh, D. Schilter, S. Hammes-Schiffer, and T. B. Rauchfuss, Protonation of nickel-iron hydrogenase models proceeds after isomerization at nickel, *J. Am. Chem. Soc.* **136**, 12385-12395 (2014).

(12) M. T. Huynh, W. Wang, T. B. Rauchfuss, and S. Hammes-Schiffer, Computational investigation of [FeFe]-hydrogenase models: Characterization of singly and doubly protonated intermediates and mechanistic insights, *Inorg. Chem.* **53**, 10301-10311 (2014).

(13) D. K. Bediako, B. H. Solis, D. K. Dogutan, M. M. Roubelakis, A. G. Maher, C. H. Lee, M. B. Chambers, S. Hammes-Schiffer, and D. G. Nocera, Role of pendant proton relays and proton-coupled electron transfer on the hydrogen evolution reaction by nickel hangman porphyrins, *Proc. Nat. Acad. Sci. USA* **111**, 15001-15006 (2014).

(14) B. H. Solis, A. G. Maher, T. Honda, D. C. Powers, D. G. Nocera, and S. Hammes-Schiffer, Theoretical analysis of cobalt hangman porphyrins: Ligand dearomatization and mechanistic implications for hydrogen evolution, *ACS Catal.* **4**, 4516-4526 (2014).

(15) S. Ghosh and S. Hammes-Schiffer, Calculation of electrochemical reorganization energies for redox molecules at self-assembled monolayer modified electrodes, *J. Phys. Chem. Lett.* **6**, 1-5 (2015).

(16) S. Hammes-Schiffer, Proton-coupled electron transfer: Moving together and charging forward, *J. Am. Chem. Soc.* **137**, 8860-8871 (2015).

(17) G. M. Chambers, M. T. Huynh, Y. Li, S. Hammes-Schiffer, T. B. Rauchfuss, E. Reijerse, and W. Lubitz, Models of the Ni-L and Ni-SI_a states of the [NiFe]-hydrogenase active site, *Inorg. Chem.* **55**, 419-431 (2016).

(18) B. H. Solis, A. G. Maher, D. K. Dogutan, D. G. Nocera, and S. Hammes-Schiffer, Nickel phlorin intermediate formed by proton-coupled electron transfer in hydrogen evolution mechanism, *Proc. Nat. Acad. Sci. USA* **113**, 485-492 (2016).

(19) S. Raugei, M. L. Helm, S. Hammes-Schiffer, A. M. Appel, M. O'Hagan, E. S. Wiedner, and R. M. Bullock, Experimental and computational mechanistic studies guiding therational design of molecular electrocatalysts for production and oxidation of hydrogen, *Inorg. Chem.* **55**, 445-460 (2016).

(20) A. K. Harshan, B. H. Solis, J. R. Winkler, H. B. Gray, and S. Hammes-Schiffer, Computational study of fluorinated diglyoxime-iron complexes: Tuning the electrocatalytic pathways for hydrogen evolution, *Inorg. Chem.* DOI: 10.1021/acs.inorgchem.5b02857.

COMPUTATIONAL STUDIES OF DECOMPOSITION MECHANISMS OF CATECHOL COMPOUNDS FOR AQUEOUS REDOX FLOW BATTERY APPLICATION

Lei Cheng[a], Kenneth D. Smith[b], Paul Redfern[a] and Larry A. Curtiss[a]

[a]Materials Science Division, Argonne National Laboratory, 9700 S. Cass Ave. Lemont IL 60439
[b]United Technologies Research Center, 411 Silver Ln, East Hartford, CT 06118

Introduction

One of the main challenges in developing new batteries beyond Li-ion is the selection and discovery of suitable electrolyte materials. High-throughput computational screening techniques have been used as an effective alternative to expensive experimentation for discovery of new materials to select electrolyte materials based on multiple property evaluations.(1) By using high-throughput quantum chemical predictions of properties, the hieratical computing scheme was shown to effectively eliminate many unpromising candidates and uncover structure-property relationships for developing design principles. However, experimental validation work for the selected molecule candidates shows that most of the computationally selected compounds have irreversible redox activity, suggesting the instability of these molecules in either neutral or charged states is the key limitation of their application in electrolyte system. Unfortunately, stability is too complicated to be addressed using high-throughput computation as there are many degradation mechanisms possible and there is no simple descriptor available for prediction. Therefore, the most rational approach to address stability problem is to study stability of different classes of molecules under the assumption that similar degradation mechanism applies to the same class of compounds. Structure-reactivity correlation can then be established to further predict stability of more compounds of the same class.

In this presentation, as an example of correlating structure to redox reversibility, we study decomposition mechanisms of a class of catechol-based compounds. Catechol compounds have been studied as potential active catholyte materials in aqueous redox-flow battery. A series of catechol derivatives with different functional groups have been shown experimentally to have very different redox reversibilities, making the class of compounds an idea example for stability study.

Results and Discussion

Catechol compounds with different functional groups have been studied experimentally and computationally as redox-active compounds in aqueous flow battery. We found that functional groups have profound effect on oxidation potential as well as reversibility of the compound. Several decomposition mechanisms reported in literature for catechol-based compounds were investigated computationally as reactions likely to occur in flow battery system. H_2O addition reaction as proposed as activation mechanism for tironin in Figure 1, was found to be the most probable reaction path. The first 1.4-addition step was found to be the rate-limiting step in this reaction path. We then compared the reaction free energies and barriers this reaction from several catechol compounds with different functional groups to establish structure-reactivity relationship. This correlation also has indication on predicting stability/reversibility of other classes of compounds with similar structural differences for their applications in battery systems.

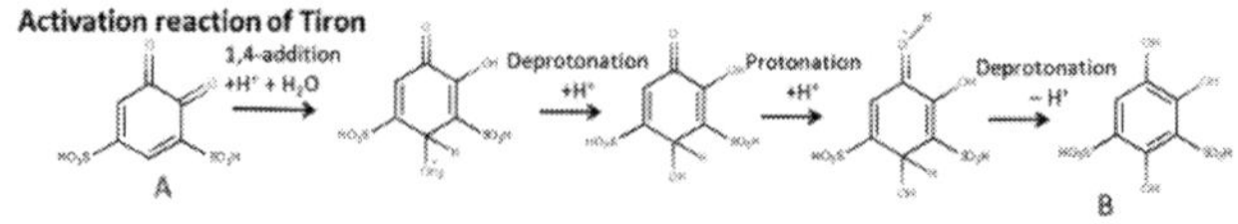

Figure 1. Water addition pathway as illustrated on tiron activation.

Conclusions

Decomposition mechanisms of a class of catechol compounds have been investigated and water addition was found to be the most probably pathway. Comparison of reaction energy profiles of different compounds were used to establish structure-activity relationship for the stability of these compounds to be used as redox-active materials in flow battery.

Acknowledgement. This work was supported as part of the Joint Center for Energy Storage Research, an Energy Innovation Hub funded by the US Department of Energy, Offi ce of Science, Basic Energy Sciences. The submitted manuscript was created by UChicago Argonne, LLC, Operator of Argonne National Laboratory ("Argonne"). Argonne, a US Department of Energy Offi ce of Science laboratory, is operated under Contract DE-AC02-06CH11357.

References

(1) L. Cheng, R. S. Assary, X. Qu, A. Jain, S. P. Ong, N. N. Rajput, K. Person and L. A. Curtiss *J. Phys. Chem. Lett.*, **2015**, *6*, 283–291.

EXCITED STATE DYNAMICS IN NANOSCALE MATERIALS FOR SOLAR ENERGY HARVESTING

Oleg Prezhdo

University of Southern California, Los Angeles, USA

Photo-induced processes play key roles in photovoltaic and photo-catalytic applications of numerous novel nanoscale materials. They require understanding of the material's dynamical response to the photo-excitation on atomic and nanometer scales. Our non-adiabatic molecular dynamics techniques, implemented within time-dependent density functional theory, allow us to model such non-equilibrium response in real time. The talk will focus on photo-initiated charge and energy transfer in several classes of nanoscale materials. Examples include T_iO_2 sensitized with organic molecules, water, semiconductor quantum dots, graphene and perovskites, a GaN/water interface, carbon nanotube bundles, mixtures of C60 with inorganic particles, etc. Photo-induced charge and energy transfer, energy losses and charge recombination create many challenges due to large differences between molecular and periodic, and organic and inorganic matter. Our simulations provide a unifying description of quantum dynamics on the nanoscale, characterize the rates and branching ratios of competing processes, resolve debated issues, and generate theoretical guidelines for development of novel systems for solar energy utilization.

THEORETICAL INSIGHTS INTO MULTISCALE ELECTRONIC PROCESSES IN ORGANIC PHOTOVOLTAICS

Sergei Tretiak

Theoretical Division, Center for Nonlinear Studies and Center for Integrated Nanotechnologies,
Los Alamos National Laboratory, Los Alamos NM, 87545
E-mail: serg@lanl.gov

Present day electronic devices are enabled by design and implementation of precise interfaces that control the flow of charge carriers. This requires robust and predictive multiscale approaches for theoretical description of underlining complex phenomena. Combined with thorough experimental studies such approaches provide a reliable estimate of physical properties of nanostructured materials and enable a rational design of devices. From this perspective I will discuss first principle modeling of small-molecule bulk-heterojunction organic solar cells and push-pull chromophores for tunable-color organic light emitters. The emphasis is on electronic processes involving intra- and intermolecular energy or charge transfer driven by strong electron-phonon coupling inherent to pi-conjugated systems. Finally I will describe how precise manipulation and control of organic-organic interfaces in a photovoltaic device can increase its power conversion efficiency by 2-5 times in a model bilayer system. Applications of these design principles to practical architectures like bulk heterojunction devices lead to an enhancement in power conversion efficiency from 4.0% to 7.0%. These interface manipulation strategies are universally applicable to any donor-acceptor interface, making them both fundamentally interesting and technologically important for achieving high efficiency organic electronic devices.

References

(1) J. E. Coughlin, R. C. Bakus II, T. S. van der Poll, G. C. Welch, A. Zhugayevych, S. J. Teat, G. C. Bazan, and S. Tretiak, J. Phys. Chem. C, 118, 15610 - 15623 (2014).
(2) A. Zhugayevych and S. Tretiak, Annu. Rev. Phys. Chem., 66, 305-330 (2015).
(3) W. Nie, G. Gupta, B.K. Crone, F. Liu, D.L. Smith, P. Ruden, C. Kuo, H. Tsai, H.-L. Wang, H. Li, S. Tretiak, and A.D. Mohite, Adv. Sci., 2, 1500024 (2015).
(4) C.-Y. Kuo, Y. Liu, D. Yarotski, H. Li, S. Tretiak, and H.-L. Wang, (2016, submitted).

ELECTRON TRANSFER REACTIONS IN CONDENSED PHASE

Valerie Vaissier, Michael Mavros and Troy VanVoorhis

Department of Chemsitry
Massachusetts Institute of Technology
77 Massachusetts Avenue, Cambridge, MA 02139
tvan@mit.edu

Introduction

Electron transfer is one of the most fundamental reactions in biology, chemistry and physics. Therefore, it is key to many research fields, and in particular to the design of efficient systems for energy conversion and/or storage. In concert with the growth of our computational ressources comes the incorporation of the environment in our simulation of systems subject to electron transfer. Ultimately, a powerful predictive tool should account for the variations in the thermodynamic and kinetic quantities of electron transfer reactions associated with changes in the surrounding medium. Only then could we derive design rules for efficient electronic systems.

In this work, we develop and test efficient computational methods that compute the driving force and rate of electron transfer reactions in systems in interaction with their environment. First, we propose a method that overcomes the prohibitive cost of QM/MM calculations of redox potential of aqueous ions. Second, we develop a theoretical framework to calculate non adiabatic reaction rates in cases where the condon approximation breaks. We apply it to the calculation of non radiative decay rate through conical intersection in organic molecules.

Computational methods

Redox Potential calculations. We make use of the linear approximation to calculate redox potential as the average of the vertical energy gap between the oxidised and reduced species. The average is over the NVT ensemble that we sample with QM/MM as also described in (1). To adequately account for solvation effects, we build a 40×40×40 A^3 water box at the center of which sits the solute. The QM region consists of the metal ion and its first solvation shell. All QM/MM calculations were performed with the CP2K software package. Non polarizable TIP3P force field was used for the MM region made of 2141 water molecules. The B3LYP functional in combination with the SZV basis set was used for the production of the QM/MM trajectories while DZVP was used for QM/MM energy calculations. We use two calculation schemes as shown in Figure 1. In the regular QM/MM calculation scheme (full solvation), the electrostatic potential due to the MM waters is included in the Hamiltonian from which the electron density of the solute is computed self consistently at each step. In our adiabatic approximation case however, we keep the electron density of the solvent fixed, allowing only for the relaxation of the MM solvent molecules.

Non adiabatic dynamics. Electron transfer rates were computed with a stanndard algorithm for solving Volterra integrodifferential equations of the second kind.(2) As also described in (3), we use spectral densities of the energy gap and electronic coupling between two states to calculate kernels that enters the master equation governing the population dynamics. The sampling was performed on the NVE ensemble with molecular dynamic simulations.

Conical intersections. The electronic coupling between the ground and excited state of cytosine and its derivative was calculated using the generalised Hush-Mulliken formula for which we require transition dipole moments (TDDFT) and energy gaps (ROKS and DFT for the excited state and ground state respectively). Throughout this study we used B3LYP functional in combination with 6-311G(d,p) (Qchem).

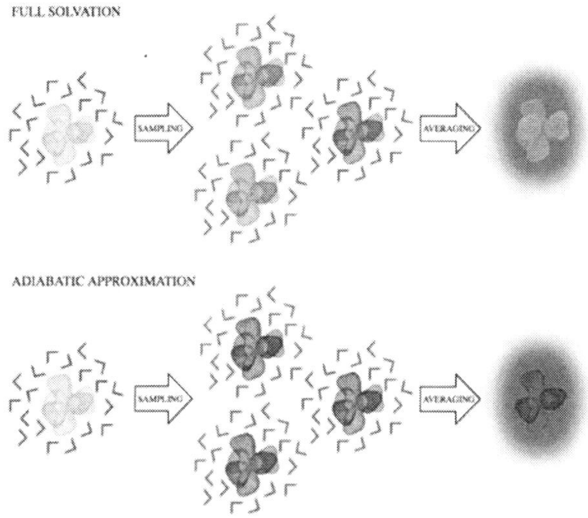

Figure 1. Comparison of the two QM/MM calculation schemes: the full solvation (regular QM/MM) and the adiabatic approximation. The QM region (the solute) is pictured by a wire-frame while the MM water molecules are thick lines when considering a specific sample and a solid color background after averaging.

Results and discussion

We propose a QM/MM calcualtion scheme (described in the computational method section) that makes use of the adiabatic approxination to speed up the sampling process. We assume that the solute and solvent degrees of freedom can be separated by such extent that a single solute electron density is representative of all configurations adopted by the solvent molecules. We test this approximation by applying it to the calculation of redox potential of transition metal ions in water, relevant for catalysis. Figure 2 shows redox potentials calculated with this and the traditional QM/MM method, compared with experimental data.

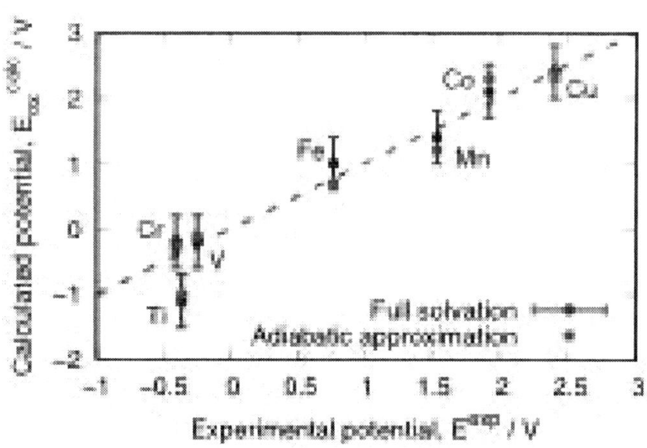

Figure 2. Comparison of the calculated and experimental redox potential of transition metal aqueous ions. The dotted line (y=x)

serves as a guide for the eye and the error bars for the computed values are one standard deviation of the full solvation case (in black).

We report a theory-to-theory root mean square deviation (RMSD) of 0.17 V and a theory-to-experiments RMSD of 0.31V. Moreover our adiabatic approximation QM/MM calculations are up to a factor 10 times faster than the conventional scheme. This means that we can calculate redox potentials at reduced cost without loosing chemical accuracy.

In terms of kinetics, we would like to be able to compute the rate of electron transfer reactions accounting for environmental effects. In particular, we are interested in photoinduced processes that involve electronic transition, ubiquitous in systems for energy applications. In these cases we cannot use Marcus equation because it does not allow for variations in electronic coupling, driving the reaction. One important example is the non-radiative decay to the ground state through conical intersections in organic molecules. To study this process, we look at cytosine and its derivative for which the litterature provides us with experimental data. Figure 3 shows our calculated potential energy curve for the ground (black) and excited state (green) near the conical intersection. At first approximation we estimate the rate of non-radiative decay with the barrier height between the minimum of the excited state and the conical intersection. The ordering of our cytosine derivative is in agreement eith experimental data, demonstrating that we have the tools to adequately compute the electronic structure of these systems.

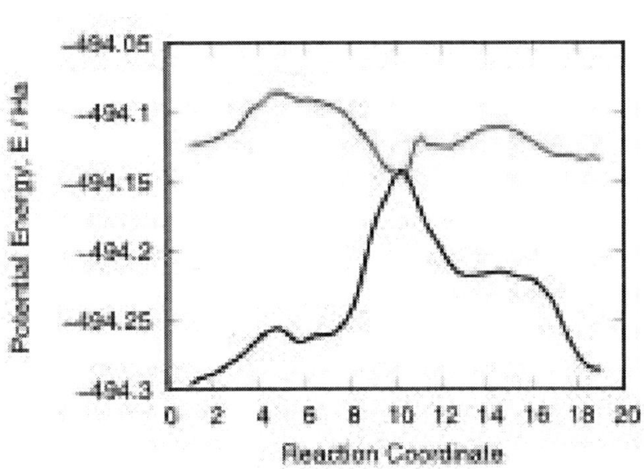

Figure 3. Energy potential curves of the ground state (black) and excited state (green) of the DNA base cytosine near a conical intersection. The reaction coordinates span the nuclear configuration space between the optimised ground state geometry (rc=1), the CI (rc=10) and the optimised excited state geometry (rc=19). The paths were computed using the Freezing String Method (FSM).

To quantitatively compute the rate of these reactions, we developped a theoretical framework based on a two level model of a system coupled to a bath through two degrees of freedom: the electronic coupling and the energy gap. This is similar to a spin-boson model where the (non-constant) electronic coupling would be determined from quantum chemical calcualtions. We generate spectral densities for these two degrees of freedom to compute the population dynamics. We tested this approach with a simple self-exchange reaction between iron ions in water. We observed that the variations in electronic coupling significantly affect the short term dynamics of the reaction when the average coupling and the amplitude of the fluctuations are of the same magnitude.

Figure 4 shows the electronic coupling variations for the DNA base cytosine at the conical intersection geometry during a short ab-initio molecular dynamic trajectory. Here, the average coupling is 0.18 eV while the amplitude of the fluctuations are approximately 0.20 eV revealing that non-condon effects (expected in electronic transitions through conical intersections) are well captured by our method.

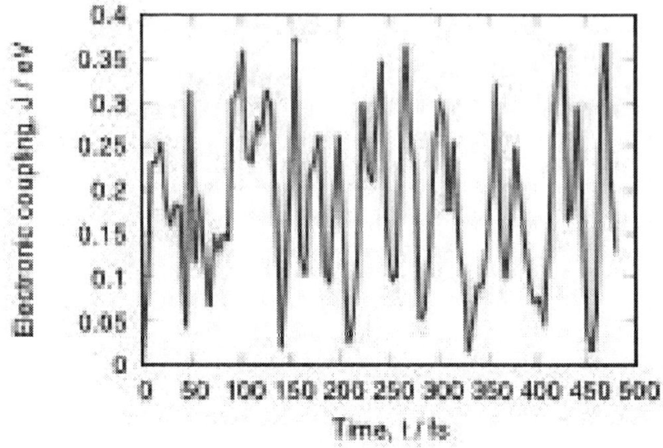

Figure 4. Short time generalized Hush-Mulliken energy coupling between the ground and excited state of the DNA base cytosine. The dynamics span the NVE ensemble.

Conclusions

In this paper, we exploited the separation of the degrees of freedom between a solute (QM region) and a solvent (MM region) to produce QM/MM NVT trajectories at lower computational cost. We tested this method by calculating the redox potential of transition metal ions in water and comparing it with conventional QM/MM calculations. The two procedures give identical results, meaning that the adiabatic approximation works well for aqueous ions.

Furthermore, we have shown that for DNA bases the electronic coupling between the ground and first excited state fluctuates significantly on short timescales. These fluctuations are necessary in order to predict photochemical dynamics through conical intersections accurately. Even in simple electron transfer systems, the Condon approximation will often result in short-time dynamics that are not quantitatively accurate.

Acknowledgement. The authors would like to thank NSF, Solar Frontiers Research Program and MRSEC progran for funding.

References

(1) Wang, L. P.; Van Voorhis, T. *J. Chem. Theory Comput.*,**2012**, *8*, 610-617.
(2) Mavros, M. G.; Hait, D. and Van Voorhis, T. *In preparation*
(3) Mavros, M. G.; Van Voorhis, T. *J. Chem. Phys.*,**2014**, *141*, 054112.

ACHIEVING ACCURATE REDUCTION POTENTIAL PREDICTIONS FOR ANTHRAQUINONES IN WATER AND APROTIC SOLVENTS: EXPLICIT SOLVENT AND ION PAIR

*Hyungjun Kim, Theodore Goodson, III, and Paul M. Zimmerman**

Department of Chemistry, University of Michigan, Ann Arbor, Michigan, 48109, United States
paulzim@umich.edu

Introduction

There is growing interest in utilizing redox flow batteries for storing energy from intermittent renewable sources. In particular, organic redox active materials show important advantages over metal complexes, for instance high solubility and low cost.[1] One of the promising organic materials are quinones, which can accept and store two electrons.[2] In this work, the factors affecting accurate computation of solution phase reduction potentials for anthraquinone (AQ) derivatives are investigated.

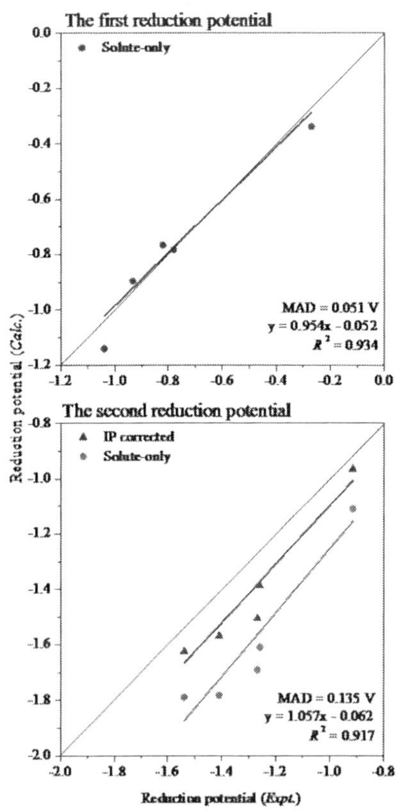

Figure 1. The 1st and 2nd reduction potential of the selected 5 derivatives in polar aprotic solvent.

Results and Discussion

The reduction of 19 AQ derivatives to hydroanthraquinones was investigated using density function theory and implicit solvation models. While most of these redox potentials are successfully predicted to within 0.1 V MAE, the remainder had substantially larger errors. These errors were traced to specific H-bonding interactions, which cannot be captured by implicit solvent models, which do not include chemically specific intermolecular interactions. QM/MM computations were therefore used to introduce explicit solvent, remedying this effect and producing high accuracy in predicting redox behavior of the entire set of AQs.

Having successfully predicted aqueous redox potentials, we constructed a library of 225 AQs for examination in aprotic solvent. While the 1st reduction potential could be predicted with high accuracy, the 2nd reduction potential was found to be more challenging. We show that the latter errors are caused by ion pair interactions between the dianion and the electrolyte, which must be explicitly accounted for. (Figure 1)

Conclusion

In summary, this presentation will discuss the process of accurately predicting AQ redox behavior, and suggest chemical principles to guide design and synthesis of new organic materials with tunable redox properties.

Acknowledgement. The authors thank the University of Michigan Energy Institute for a postdoctoral fellowship that supports this work.

Reference

(1) Wang, W.; Sprenkle, V.; *Nature Chemistry*, **2016**, *8* (3), 204
(2) Huskinson, B.; Marshak, M.; Suh, C.; Er, S.; Michael, R. G.; Cooper, J. G.; Xudong, C.; Alán, A.-G.; Roy, G. G.; Michael, J. A.; *Nature*, **2014**, *505* (7482), 195

PHOTON-ABSORBING CHARGE-BRIDING STATES IN OGRANIC PHOTOVOLTAIC DEVICES OFDIKETOPYRROLOPYRROLE-BASED DONOR AND PCBM

Mikiya Fujii, Shohei Koda, and Koichi Yamashita

Department of Chemical System Engineering, The University of Tokyo, Tokyo 113-8656, Japan
yamasita@chemsys.t.u-tokyo.ac.jp

Introduction

Guiding principles to design highly optimized organic photovoltaic devices that achieve high power conversion efficiencies are strongly desired. To obtain a useful guideline, we have investigated the photo- and electrochemical properties of five diketopyrrolopyrrole (DPP) derivatives and four DPP derivatives containing double DPP blocks by *ab initio* calculations (1,2). To investigate the factors that influence the IQE, investigations on charge recombination, hole transfer, and charge transfer induced by photoabsorption were conducted for the complexes of each DPP derivative and PCBM. We may suggest that the change in the dipole moment in the complex of donor and acceptor molecules is a promising guideline for designing new donor molecules to improve the short-circuit current and power conversion efficiencies.

Comutational Method

Theoretical calculations were conducted based on the Hartree-Fock (HF), density functional theory (B3LYP), and time-dependent density functional theory (TDDFT) with the basis function 6-31G*, and the semiempirical quantum chemical method PM3. For supermolecules such as pairs of donor and acceptor molecules and dimers of donor molecules, the exchange - correlation functional B3LYP-D2 were adopted to include van der Waals interactions. For the analysis of charge recombination and hole-transfer processes, we investigated electronic couplings between the electronic states. Finally, to explore electronically excited states arising upon photoexcitation, we mainly adopted configurational interaction singles (CIS) based on the semiempirical quantum method PM3.

Results and Discussion

Figure 1 plots the experimental short-circuit current J_{SC} of the organic photovoltaic devices fabricated by each DPP derivative (see Fig.2) and $PC_{71}BM$ and changes in the dipole moments $\Delta\mu_{D/A}$ between the ground and the prominent excited states of the complexes of each DPP derivative and $PC_{61}BM$. In the case of the complex, unlike just the DPP derivative, we can confirm a linear correlation between J_{SC} and $\Delta\mu_{D/A}$. Namely, complexes that exhibit a larger change in the dipole moment $\Delta\mu_{D/A}$ show more suitable electronic structures to promote charge separation. In particular, BDT-DPP/$PC_{61}BM$ exhibits the largest change in the dipole moment $\Delta\mu_{D/A}$. This is because the photoexcitation of BDT-DPP/$PC_{61}BM$ is the partial charge transfer from HOMO on the BDT-DPP to the charge-bridging orbital, which is delocalized over both BDT-DPP and $PC_{61}BM$. We, therefore, can conclude that changes in the dipole moments of complexes of donor and acceptor molecules, $\Delta\mu_{D/A}$, are a promising guideline for designing new donor molecules to improve the PCE in organic solar cells. We validated exchange-correlation functional dependencies of the correlation by adopting LC-BLYP to estimate the changes in dipole moments. The charge-bridging orbitals from which the electron and hole can separate easily enhance the short-circuit current.

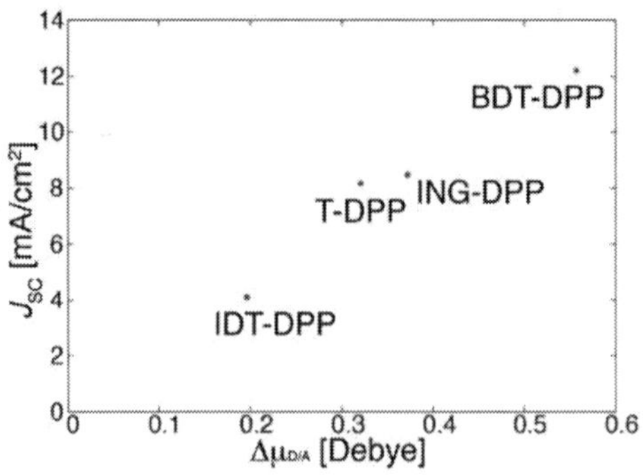

Figure 1. Experimental short-circuit currents of the organic photovoltaic devices fabricated by each DPP derivative and $PC_{71}BM$ vs. changes in the dipole moment between the ground and excited states of complexes of each DPP derivative and $PC_{61}BM$.

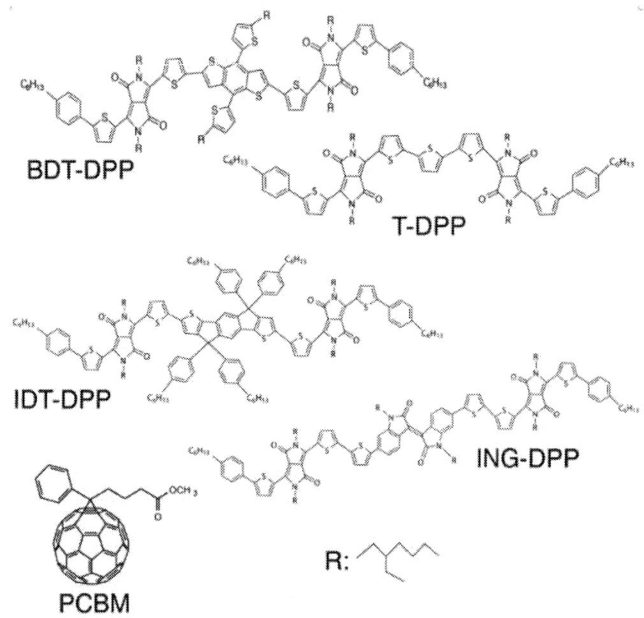

Figure 2. Chemical structures of donor (diketopyrrolopyrrole derivatives) and acceptor (PCBM) molecules studied.

Conclusions

We have found that the short-circuit current correlates with changes in dipole moments of complexes of donor and acceptor molecules rather than with the changes in dipole moments of donor molecules only. This is because of photoinduced partial charge transfer from the HOMO on the donor to the *charge-bridging orbital* that is delocalized over both donor and acceptor molecules.

Acknowledgement. This work was supported by JSPS KAKENHI Grant No. 24750012, CREST JST. Theoretical calculations in this study were performed using facilities at Supercomputer Center in Institute for Solid State Physics (ISSP) of University of Tokyo and at Research Center for Computational Science in Institute for Molecular Science (IMS) in Okazaki, Japan.

References

(1) Fujii, M.; Shin, W.; Yasuda, T.; Yamashita, K. *Phys. Chem. Chem. Phys.*, **2016**, *accepted 7th March.*

(2) Koda, S.; Fujii, M.; Hatamiya, S.; Yamashita, K. *Theo. Chem. Acc.*, **2016**, *accepted 22nd March.*

MODELING SOLID-SOLID INTERFACES RELEVANT TO LITHIUM ION BATTERIES

Kevin Leung

Sandia National Laboratories, Albuquerque, New Mexico 87185
kleung@sandia.gov

Introduction

Solid-solid interfaces are ubiquitous at lithium ion battery (LIB) interfaces. They are present not just in all-solid batteries with solid electrolytes, but also at the interfaces between electrodes and solid-electrolyte-interphase ("SEI") films invariably covering the electrodes in organic-solvent based LIB. Little is known about their structures and properties at atomic lengthscale due to the limited experimental resolution associated with imaging buried interfaces. Electronic structure (i.e., DFT) modeling of plausible interfacial structures of multiple materials presents the opportunity to correlate structural motifs responsible for low-impedance, chemically stable, non-dendrite forming LIB solid-solid interfaces.

A key paradigm to be investigated is the role of the electric double layer (EDL, or equivalently the space-charge layer). A thick EDL can lead to Li depletion over a wide interfacial region and lower Li-conductivity, but a very sharp EDL implies a strong electric field that can lead to electrochemical reactions that degrade the interface. Spatial inhomogeneity at the interface is also expected to be associated with dendrite formation on anodes. This latter phenomenon needs to be probed at lengthscale larger than traditional DFT methods can handle.

Method

We apply electronic structure DFT, dispersion forces-corrected DFT, and DFT+U techniques to obtain the atomic lengthscale structures of interfaces. The Li-occupancy at each interfacial site is examined by comparing the site energy with Li-chemical potential appropriate to different electrode systems. Lithium metal/Li(2)O, Li(s)/Li(2)CO(3), and Li(x)CoO(2)/LIPON interfaces (both pristine and "SEI"-covered) will be examined.

The DFT predictions are fed into coarse-grained lattice-gas models which represent Li^+ vacancies or Li^+ interstitials interacting with each other with coulomb repulsions. Disorder and spatial inhomogeneities will also be incorporated into such models.

Results and Discussions

We predict that Li(2)CO(3), a major organic electrolyte-generated SEI component, decomposes on Li metal surfaces within a 1-hour time frame to form Li(2)O (Fig. 1a). The reaction is thermodynamically exothermic. Hence Li(2)O should be the dominant component of passivating film on Li metal surfaces (Fig. 1)[1].

A DFT-fitted Ising model is used to describe the space charge layer thickness between Li(2)O and Li-metal as a function of applied voltage. The effect of spatial inhomogeneity will also be examined.

Interfaces between Li(x)CoO(2) (104) and model LIPON[2] is examined. The possibility of solid-electrolyte SEI formation arising from LIPON decomposition products[3,4], and how this modify electronic and Li-diffusive properties compared to the interface without solid-electrolyte SEI products will be considered.

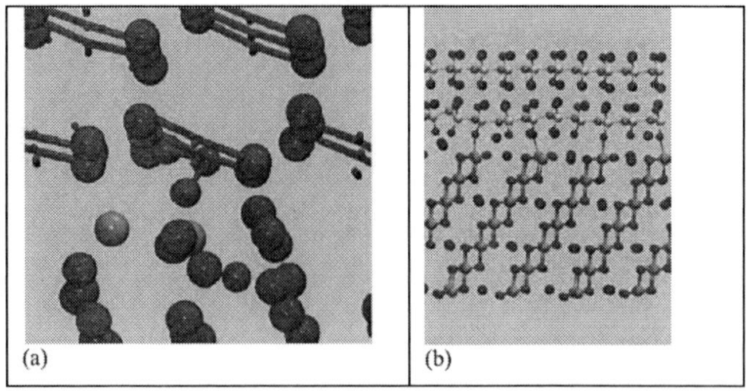

Figure 1. (a) Li(2)CO(3) decomposes on Li(001) surface. (b) As-yet undecomposed model LIPON on LiCoO(2) (104) surface. Blue, red, grey, orange, green, and yellow are Li, O, C, Co, N, and P atoms.

Conclusions

Our DFT –predictions suggest that battery electrode interfaces often contain interlayers, even for seemingly stable SEI components. The solid-state electric double layer or space charge thickness is found to depend on applied voltages, dielectric mismatch, and many other factors.

Acknowledgement. Sandia National Laboratories is a multiprogram laboratory managed and operated by Sandia Corporation, a wholly owned subsidiary of Lockheed Martin Corporation, for the U.S. Department of Energy's National Nuclear Security Administration under contract DE-AC04-94AL85000. The anode work was supported by the Assistant Secretary for Energy Efficiency and Renewable Energy, Office of Vehicle Technologies of the U. S. Department of Energy under Contract No. DE-AC02-05CH11231, Subcontract No. 7060634 under the Advanced Batteries Materials Research (BMR) Program. The cathode work was supported by by Nanostructures for Electrical Energy Storage (NEES), an Energy Frontier Research Center funded by the U.S. Department of Energy, Office of Science, Office of Basic Energy Sciences under Award Number DESC0001160.

References

(1) Leung, K.; et al. *J. Phys. Chem. C* (to appear).
(2) Du, Y.A..; Holzwarth, N.A.W. *Phys. Rev. B.* **2010**, *81*, 184106.
(3) Zhu, Y.; He, X.; Mo, Y. ACS Appl. Mater. Interfaces, 2015, 7, 23685.
(4) Jacke, S.; Song, J.; Cherkashinin, G.; Dimesso, L.; Jaegermann, W. Ionics **2010**, *16*, 769.

ELECTRON DELOCALIZATION IN AROMATIC RING ON HYDRODEOXGENATION OF PHONEL ON PT(111)

Dan Liu,[†] *Gaofeng Li,*[†] *Hua Wang,*[†] *Jinyu Han,*[†] *Xinli Zhu*[*,†] *and Qingfeng Ge*[*†‡]

[†]Collaborative Innovation Center of Chemical Science and Engineering, School of Chemical Engineering and Technology, Tianjin University, Tianjin 300072, China
[‡]Department of Chemistry and Biochemistry, Southern Illinois University, Carbondale, Illinois 62901, United States

Catalytic conversion of lignin-based biomass to fuels and chemicals is of importance for the sustainable development of our society. Phenol has the simplest representative structure of phenolic compounds and has been widely used as a model compound for lignin-based biomass in the study of hydrogenation and hydrodeoxygenation of lignin-derived compounds to produce fuels and chemicals. Among these researches the oxygen removal was considered to be the most difficult and rate-limiting step. To better understand the hydrodeoxygenation mechanism, we examined the elementary surface hydrogenation/dehydrogenation steps and deoxygenation steps. Based on energetics of the elementary steps from DFT calculations and microkinetic analysis, we mapped out the most favorable route in deoxygenation from six distinct reaction routes. In particular, we used the Electron Localization Function of the hydrogenation intermediates to evaluate the extent of the electron location in the aromatic ring on the subsequent hydrogenation and deoxygenation reactions. Our results showed that hydrogenating partially the benzene ring reduces electron delocalization and enhances the deoxgenation activity. The kinetic parameters from the DFT calculations were used in the microkinetic modeling. Our results supported a hydrogenation-deoxygenation-dehydrogenation as the major route of benzene formation.

NON-ADIABATIC EFFECTS AND ELECTRONIC EXCITATIONS DURING DISSOCIATION ON CATALYTIC SURFACES

Matthew M. Montemore, Robert A. Hoyt, Efthimios Kaxiras

Harvard University
Cambridge, Massachusetts
kaxiras@physics.harvard.edu

Introduction

The Born-Oppenheimer approximation, which assumes that the electrons are always in their ground state, is widely used in density functional theory (DFT) studies of catalytic processes. These calculations have been quite successful in explaining a large number of catalytic phenomena, but it is unclear when the Born-Oppenheimer approximation may break down. Metal surfaces form an important class of catalysts, and electronic excitations are easily induced in these surfaces, due to the lack of a band gap. Here, we explore the possibility that processes on catalytic surfaces may induce electronic excitations, which are not captured by simulations using the Born-Oppenheimer approximation.

It has previously been shown that surface processes can induce electronic excitations in metals, at least in certain cases. For example, adsorption on metal surfaces can produce a measurable current, presumably due to electronic excitations.[1] Surface chemical processes can require the dissipation of a large amount of energy on short timescales, raising the question of how much energy is dissipated into electronic degrees of freedom, and how much is dissipated into phonons.

Since the methodology for studying electronic excitations in metal surfaces has not been widely implemented or utilized, many questions remain about electronic excitations in catalysis, including their frequency, nature, and effect on surface chemistry. It is not clear under what circumstances excitations are induced, how large the excitations are, and their form. We focus on two factors: (1) determining how energetic the electronic excitations are in particular cases, and (2) determining the timescale for electron-hole pair excitation, relaxation into "hot electrons" (where the electrons are characterized by a Fermi-Dirac distribution with a characteristic temperature different than that of the ionic motion), and relaxation into the ground state. Previous work has shown that hot electrons and electronic excitations can affect surface chemistry.[2]

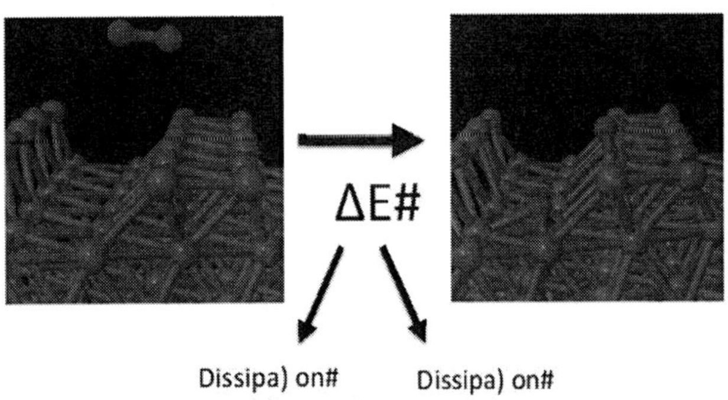

Figure 1. When N_2 dissociates on a Ru surface, the potential energy released can be dissipated into phonons or into electrons.

Methods

Non-adiabatic dynamics were performed using real-time, time-dependent DFT, in which the Kohn-Sham wavefunctions are explicitly propogated in time. Ehrenfest dynamics and the ΔSCF method have been implemented in the SIESTA code, which efficiently performs periodic DFT using localized numerical orbitals and pseudopotentials. As a test system of considerable scientific interest, we study N_2 dissociation on Ru nanoparticles.

Results and Discussion

We focus on a simulation of an energetic N_2 molecule, with approximately 1 eV of initial kinetic energy directed toward a Ru_{13} nanoparticle. The simulation was run for 140 fs, with a timestep of approximately 25 as. Every 100 timesteps, the coordinates from the non-adiabatic simulation were extracted, and a ground state calculation was performed using these coordinates. The potential energy of both the non-adiabatic simulation and the ground state calculations are shown in Figure 2. The difference between these energies is the amount of energy dissipated into electronic degrees of freedom during the non-adiabatic calculation. Initially, while the N_2 is approaching the nanoparticle, no electronic excitation occurs. However, during and after dissociation, more than 1 eV of energy is dissipated into electronic excitations.

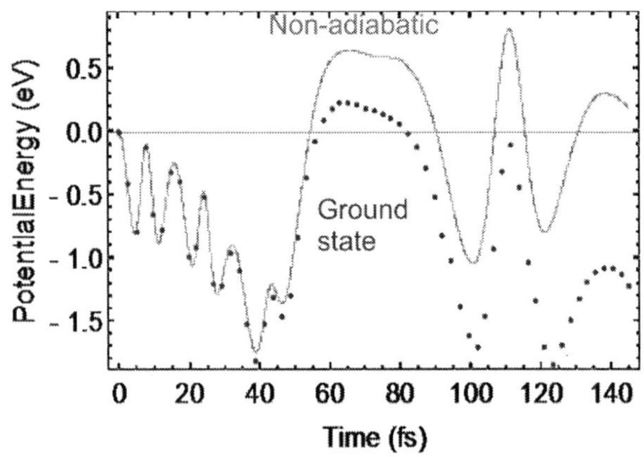

Figure 2. Comparison of the non-adiabatic potential energy to the ground state potential energy at the same atomic coordinates. At the end of the simulation, approximately 1.3 eV of energy is dissipated into the electronic degrees of freedom.

The excitations take the form of electron-hole pairs, but after tens of fs they relax into hot electrons. Figure 3 shows representative plots of the occupation of electronic states as a function of their energy. Part (a) shows a case where there is an electron-hole pair excitation, while part (b) shows a case where the electrons can be characterized by a Fermi-Dirac distribution, with a characteristic temperature of >900 K.

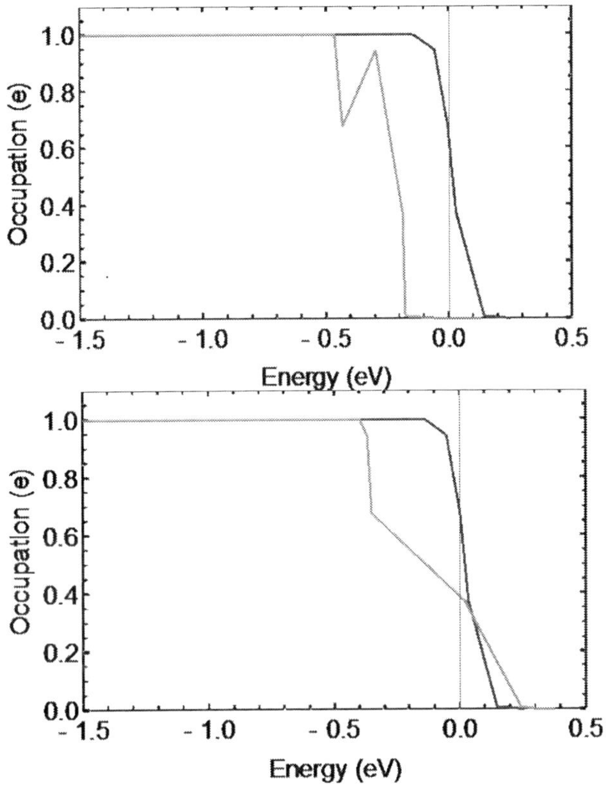

Figure 3. Occupation of electronic states as a function of energy, for (a) t=87 fs and (b) t=119 fs. Orange corresponds to the non-adiabatic simulation, while blue is a ground state calculation at the same geometry.

Conclusions

Non-adiabatic calculations based on Ehrenfest dynamics were performed for N_2 dissociation on a Ru nanoparticle. These calculations show that a significant amount of energy can be dissipated into electronic degrees of freedom, comparable to the amount of energy dissociated into ionic vibrations. This suggests that the electronic structure of catalytic surfaces may, transiently, be quite different from the ground state.

Acknowledgement. This work was supported as part of the Integrated Mesoscale Architectures for Sustainable Catalysis, an Energy Frontier Research Center funded by the U.S. Department of Energy, Office of Science, Basic Energy Sciences under Award # DE-SC0012573. The computations in this work were run on the Odyssey cluster supported by the FAS Division of Science, Research Computing Group at Harvard University.

References

(1) Gergen, B.; Nienhaus, H.; Weinberg, W. H.; McFarland, E. W. Chemically Induced Electronic Excitations at Metal Surfaces. *Science* **2001**, *294* (5551), 2521–2523.

(2) Bonn, M.; Funk, S.; Hess, C.; Denzler, D. N.; Stampfl, C.; Scheffler, M.; Wolf, M.; Ertl, G. Phonon- Versus Electron-Mediated Desorption and Oxidation of CO on Ru(0001). *Science*. **1999**, *285* (5430), 1042–1045.

(3) Lindenblatt, M.; Pehlke, E. Ab Initio Simulation of the Spin Transition during Chemisorption: H/Al(111). *Phys. Rev. Lett.* **2006**, *97* (21), 1–4.

TESTS OF SCALING RELATIONS AT BOUNDARY SITES OF OXIDE SUPPORTED METAL CATALYSTS

Prateek Mehta, William F. Schneider

Department of Chemical and Biomolecular Engineering, University of Notre Dame, Notre Dame, IN, 46556
wschneider@nd.edu

Introduction

Metal nanoparticles dispersed on metal oxide supports are ubiquitous in heterogeneous catalysis. For a large number of energy relevant chemical transformations, including CO oxidation, water gas shift, etc, the active sites are believed to be at the boundary between the metal and the oxide. Such interfacial sites, where reactants can interact with both the metal and the oxide, have often been linked to unique reactivity for a compared to the metal or the oxide in isolation. This dual-nature of the metal-oxide interface presents a challenge to the classical surface science approach that has been very successful in understand and predict catalyst performance for transition metals [1, 2]. Since the chemistry of the underlying support in generally ignored within such an approach, the exact nature of the metal-oxide perimeter sites, and their role in the catalysis remain poorly understood. In this paper, we use density functional theory (DFT) calculations to examine how the inclusion of a support phase can lead to can lead to deviations in reactivity compared to a metal-only model.

Computational Methods

The DFT calculations were performed using the Vienna ab-initio Simulation Package (VASP) using the PBE functional within the PAW core model. Several late transition metals were studied (M=Rh, Ir, Pd, Pt, Ag, Au), with MgO(100) as the support. The metals were represented (see Fig. 1) as quasi one dimensional nanowires similar to the model proposed by *Molina et al* [3]. The adsorbates were placed at the metal-oxide boundary, on the bridge between two metal atoms and on top of an Mg atom. An oxygen atom adsorbed at this site has been shown in Fig. 1.

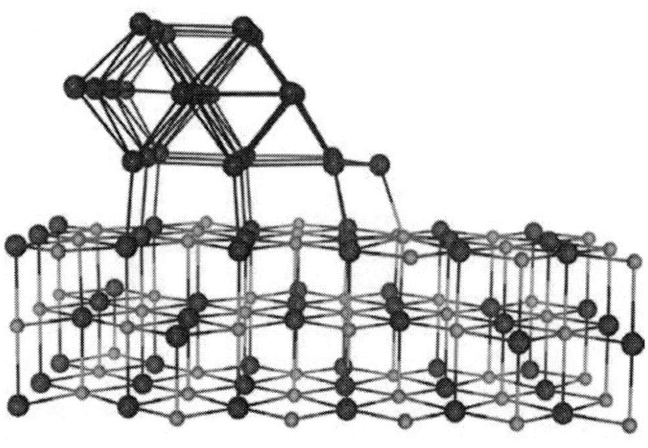

Figure 1. Structural model of a Pd nanowire supported on MgO(100). An oxygen atom adsorbed at the metal-oxide boundary is also shown.

Results and Discussion

Our results show that the role of the oxide is highly variable and strongly depends on the choice of metal and adsorbate. For some metal-adsorbate combinations, the presence of the support results in enhanced adsorption at the perimeter sites compared to the unsupported metal nanowire. In other cases, adsorption is destabilized when the metal is supported by the MgO. We have traced the origins of these stabilization/destabilization effects to be a combination for several structural and electronic perturbations, including steric interactions, a shift in the metal d-band center, and redox energy gained by charge transfer facilitated by the oxide. Moreover, we show that in many cases, the effect that the oxide has on the adsorption energy of an atomic adsorbate A, does not correlate with the adsorption energy of its hydrogenated form AH_x. In fact, while a C atom is repelled away from the metal-oxide boundary, CH_2 is stabilized at interfacial sites. We use these results to discuss possible strategies to circumvent metal-only scaling relations at the metal-oxide interface.

Conclusions

We identify which metals are most affected by the support, and propose simple rules of thumb for site preferences of different adsorbates. Implications for scaling relationships are also discussed.

Acknowledgement. We acknowledge partial support from the U.S. Department of Energy under contract DE-FG02-03ER15466. PM acknowledges a graduate student fellowship from the Center for Sustainable Energy at the University of Notre Dame.

References

(1) Norskov, J. K.; Abild-Pedersen, F.; Studt, F.; & Bligaard, T. (2011) *PNAS*, (2011) 108(3), 937–943.
(2) Abild-Pedersen, F.; Greeley, J.; Studt, F.; Rossmeisl, J.; Munter, T.; Moses, P.; Skulason E.; Bligaard, T.; Norskov, J. K. *Phys. Rev. Lett.*, (2007) 99(1), 016105.
(3) Molina, L.; & Hammer, B. *Phys. Rev. Lett.*, (2003) 90(20), 206102.

MECHANISTIC INSIGHTS INTO AQUEOUS PHASE ACETIC ACID KETONIZATION OVER MONOCLINIC ZIRCONIA

Qiuxia Cai[1,2], *Juan A. Lopez-Ruiz*[1], *Alan R. Cooper*[1], *Karl O. Albrecht*[1,*], *Jian-guo Wang*[2,*], *Donghai Mei*[1,*]

[1] Institute for Integrated Catalysis, Pacific Northwest National Laboratory, Richland, WA 99352, USA
[2] College of Chemical Engineering, Zhejiang University of Technology, Hangzhou, Zhejiang 310014, China
donghai.mei@pnnl.gov; karl.albrecht@pnnl.gov; jgw@zjut.edu.cn

Introduction

The bio-oil generated during biomass liquefaction process generally contains a large amount of oxygenates such as sugars, phenolic compounds and carboxylic acids.[1] In order to improve the physiochemical properties and stability of bio-oil, upgrading of these molecules is needed to reduce the oxygen content. Among various catalytic upgrading routes, ketonization of carboxylic acids provides an efficient method for facilitating carbon-carbon bond formation and oxygen removal as carbon dioxide and water.[2-3] Numerous experimental studies have explored the vapor-phase ketonization of carboxylic acids on a series of metal oxide catalysts.[4-5] Among various metal oxides, monoclinic ZrO_2 has been regarded as one of the most efficient catalysts for vapor phase carboxylic acid ketonization.[6] In the present work, DFT calculations and AIMD simulations were performed to study the effects of an aqueous phase on the ketonization of acetic acid over a monoclinic zirconia (m-ZrO_2) catalyst. The aqueous phase was modeled by explicit water molecules. In particular, we focused on gaining fundamental understanding of the role of surrounding water molecules in the surface ketonization reaction over metal oxide catalysts using acetic acid ketonization on monoclinic zirconia as a demonstration case.

Computational

All calculations were performed using the spin-polarized, gradient-corrected functional of Perdew, Burke, and Ernzerhof (PBE) implemented in the CP2K package.[7-8] The wave functions were expanded in an optimized double-ζ Gaussian basis set. An auxiliary plane wave basis set of 360 Ry energy cutoff was used for the electrostatic energy calculation. The van der Waals correction was included in all calculations. The m-ZrO_2 catalyst surface was modeled by a periodic (2 x 2) ZrO_2(-111) surface slab with four layers. A total of 111 water molecules were filled in this region to represent the aqueous phase with a density of 0.93 g/cm^3. AIMD simulations of the aqueous phase/ZrO_2(-111) system at 550 K for a time period of 10 ps were performed to obtain the equilibrium configuration of the aqueous phase/ZrO_2(-111) interfacial structure. To account for important entropic contribution and zero-point energy corrections, in addition to the reaction enthalpy, Gibbs free energy (ΔG) along reaction pathways using standard thermodynamic method were calculated.[9]

Results and Discussion

The existence of the aqueous phase may not only affect the stabilities of reactants, products and reaction intermediates but also change the stabilities of transition states of elementary reaction steps, thus affecting the surface reaction rate and selectivity. Acetic acid ketonization reaction pathway involving the formation of β-keto acid intermediate shown in Figure 1 in the aqueous phase was investigated. We found that the α-H abstraction of acetate is spontaneous in aqueous phase via proton transfer (blue in Figure 2). Most importantly, the new C-C coupling channel with molecular acetic acid in the aqueous phase directly coupled with CH_2COO was identified (red route in Figure 2).

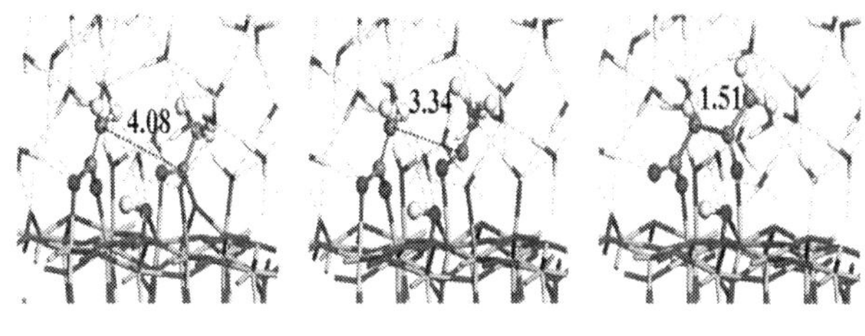

Figure 1. The β–keto acid formation over the ZrO_2(-111) surface in aqueous phase.

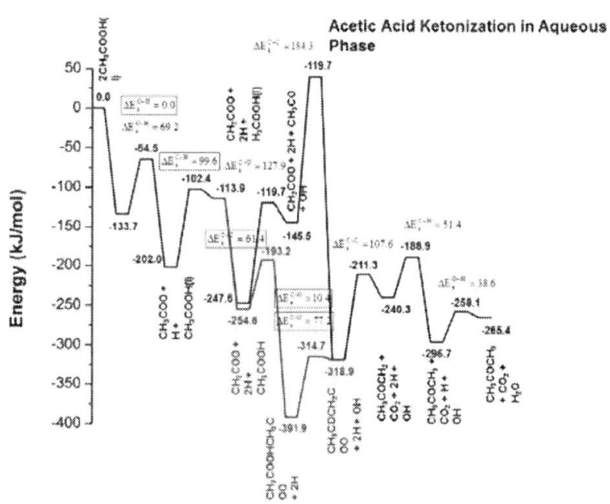

Figure 2. Acetic acid ketonization reaction pathways over the ZrO_2(-111) surface in aqueous phase.

Conclusions

The effects of the aqueous phase on the acetic acid ketonization reaction over a monoclinic zirconia catalyst were investigated using first principles-based DFT calculations. On one hand, the presence of aqueous phase in the ketonization of acetic acid on the ZrO_2(-111) surface decreases the number of active sites due to the formation of hydroxylated surface. On the other hand, several key reaction steps including α-H abstraction and C-C coupling are dramatically facilitated via proton transfer mechanism. New reaction channels assisted with the enhanced stability of acetic acid in the aqueous phase is also identified. Specifically, the adsorption of reactant and desorption of products may become the rate-limiting steps in aqueous phase ketonization of acetic acid over zirconia catalysts.

Acknowledgement. This work was financially supported by the United States Department of Energy (DOE)'s Bioenergy Technologies Office (BETO) and performed at the Pacific Northwest National Laboratory (PNNL). PNNL is a multi-program national laboratory operated for DOE by Battelle Memorial Institute. Computing time and advanced catalyst characterization use was granted by a user proposal at the William R. Wiley Environmental Molecular Sciences Laboratory (EMSL) and by the National Energy Research Scientific Computing Center (NERSC). EMSL is a national scientific user facility sponsored by the Department of Energy's Office of Biological and Environmental Research and located at PNNL.

References

(1) Huber, G. W.; Iborra, S.; Corma, A. *Chem. Rev.* **2006,** *106*, 4044-4098.

(2) Pacchioni, G. *ACS Catal.* **2014,** *4*, 2874-2888.

(3) Pham, T. N.; Sooknoi, T.; Crossley, S. P.; Resasco, D. E. *ACS Catal.* **2013,** *3*, 2456-2473.

(4) Glinski, M.; Kijenski, J.; Jakubowski, A. *Appl. Catal. A* **1995,** *128*, 209-217.

(5) Gaertner, C. A.; Serrano-Ruiz, J. C.; Braden, D. J.; Dumesic, J. A. *Ind. Eng. Chem. Res.* **2010,** *49*, 6027-6033.

(6) Pulido, A.; Oliver-Tomas, B.; Renz, M.; Boronat, M.; Corma, A. *ChemSusChem* **2013,** *6*, 141-151.

(7) Perdew, J. P.; Burke, K.; Ernzerhof, M. *Phys. Rev. Lett.* **1996,** *77*, 3865-3868.

(8) VandeVondele, J.; Krack, M.; Mohamed, F.; Parrinello, M.; Chassaing, T.; Hutter, J. *Comp. Phys. Commun.* **2005,** *167*, 103-128.

(9) Psofogiannakis, G.; St-Amant, A.; Ternan, M. *J. Phys. Chem. B* **2006,** *110*, 24593-24605.

HYDROGENOLYSIS MECHANISM OF FURFURAL ON Ru/RuO$_x$ CATALYST

Alexander V. Mironenko, Dionisios G. Vlachos

Catalysis Center for Energy Innovation
Department of Chemical and Biomolecular Engineering
221 Academy Street,
Newark, DE 19716

Introduction

Facile C-O bond scission is essential in various industrially important processes. For example, in Fischer-Tropsch synthesis, which experiences revival due to the revolution of shale gas, C-O bond activation can potentially govern the hydrocarbons-to-alcohols product ratio. Selective C-O cleavage is also crucial in production of biofuels, as efficient oxygen removal is necessary for integration of biomass into the downstream oil refining infrastructure.

In comparison with reforming and dehydrogenation, which involve C-C and C-H bond breaking, a significant challenge exists for the design of a selective C-O scission catalyst. Activity and selectivity maps for small oxygenates on transition metals reveal the best deoxygenation metal catalysts to be orders of magnitude slower than good reforming catalysts at the same reaction conditions [1]. Similarly, if metal oxides are used as catalysts, high temperatures (>320 °C) are required to obtain appreciable C-O scission activity [2]. It is clear that monofunctional catalysis is ineffective for C-O bond activation.

Combination of transition metals and metal oxides results in a surprisingly high activity for selective C-O bond cleavage at moderate temperatures, <200°C. Selected examples include glycerol hydrogenolysis to either 1,2-propanediol on Ru/ReO$_x$ [3], and deoxygenation and ring-opening reactions in diols, triols, furans, and pyrans on Rh-ReO$_x$ and Rh-MoO$_x$ [4]. The mechanism explaining the synergy between metallic and metal oxide functionalities remains poorly understood and occasionally controversial.

In this contribution, we employ multiscale chemical kinetics approaches to reveal the mechanism and compare to experimental data.

Methods and Results

Here, we elucidate the metal/oxide synergy for C-O bond activation using a model reaction system: catalytic transfer hydrogenolysis of furfural on a Ru/RuO$_2$ catalyst using 2-propanol as a hydrogen donor. Yields of 2-methyl furan (sustainable platform chemical) up to 76% have been reported [5]. By combining density functional theory (DFT), microkinetic modeling, and electronic structure analysis, we gain insights into why neither the Ru nor the RuO$_2$ phase alone is capable of catalyzing C-O bond hydrogenolysis.

DFT energy profiles on Ru(0001) and RuO$_2$(110) surfaces (Figure 1) reveal that metallic sites are active in breaking the C-O bond, but become poisoned by oxygen, due to high Ru oxophilicity. On the other hand, metal oxide Lewis acid sites are effective for oxygen removal as water, but cannot break the C-O bond. These findings are consistent with the experimental data.

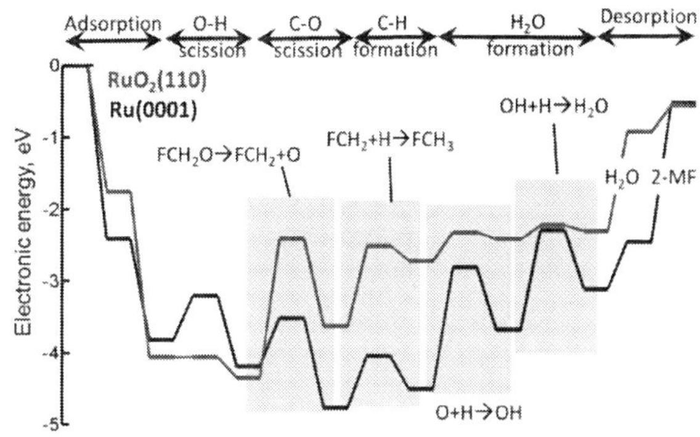

Figure 1. Minimum energy DFT reaction pathways for furfuryl alcohol conversion to 2-methyl furan on Ru(0001) and RuO$_2$(110) surfaces.

DFT and microkinetic modeling (not shown) indicate a trifunctional catalytic mechanism involving both Ru and RuO$_2$ to be responsible for low temperature catalyst activity. For example, hydrogenation of the carbonyl group of furfural proceeds via an intermolecular hydride transfer in a concerted step between furfural and the hydrogen donor, 2-propanol. Electronic structure analysis indicates that the mechanism is mediated by radical intermediates and is in quantitative agreement with H/D tracing and recent kinetic isotope effect measurements[6].

Our results provide for the first time strong support for a trifunctional reaction mechanism operative at temperatures < 200 °C during furfural conversion to 2-methyl furan on the Ru/RuO$_2$ catalyst. This finding provides opportunities for reaction pathway and multifunctional catalyst design.

Acknowledgment. The work was supported by the Catalysis Center for Energy Innovation (CCEI), an Energy Frontier Research Center funded by the U.S. Department of Energy, Office of Science, Office of Basic Energy Sciences under Award # DE-SC0001004.

References

(1) Salciccioli, M. and D.G. Vlachos, ACS Catalysis, 2011. **1**(10): p. 1246-1256.
(2) Prasomsri, T., et al., Energy & Environmental Science, 2014. **7**(8): p. 2660-2669.
(3) Ma, L. and D. He, Catalysis Today, 2010. **149**(1): p. 148-156.
(4) Chia, M., et al., Journal of the American Chemical Society, 2011. **133**(32): p. 12675-12689.
(5) Panagiotopoulou, P., N. Martin, and D.G. Vlachos, Journal of Molecular Catalysis A: Chemical, 2014. **392**: p. 223-228.
(6) Gilkey, M.J., et al., ACS Catalysis, 2015.

THEORETICAL INVESTIGATION OF THE NON-AQUEOUS OXYGEN REDUCTION REACTION IN Li-O2 BATTERIES

*William C. McKee, Saurin Rawal, and Ye Xu**

Department of Chemical Engineering, Louisiana State University, Baton Rouge, LA 70803, USA
* yexu@lsu.edu

Introduction

The non-aqueous Li-air (O_2) battery has attracted a great deal of research interest because it can potentially far exceed the gravimetric energy storage density than the current rechargeable battery technologies. The oxygen reduction reaction by Li (Li-ORR) is a key electrochemical process at a Li-O_2 cathode. Presently Li-O_2 batteries suffer from poor round-trip efficiency, low current density, and capacity fading,[1-3] much of which is likely due to parasitic reactions involving impurities that exist in almost all of the non-aqueous electrolytes, including residual protons and proton-containing compounds that can react with reduced oxygen species. To alleviate the parasitic reactions to an acceptable extent and improve the cyclability of Li-O_2 batteries, it is crucial to have a fundamental understanding of the reactive O_2 species involved in the ORR on different electrodes.

Methods

Spin-polarized periodic density functional theory (DFT) calculations have been performed using the Vienna Ab initio Simulation Package. The DFT results are combined with the electrochemical modeling approach of Nørskov et al.[4] to account for the dependence of reaction energetics on the electrode potential.[5,6]

Results and Discussion

Recent studies have shown that the inclusion of metals in the cathode of Li-O_2 batteries can alter the cathode efficiency and possibly alter the nature of the discharge product.[1,2,7-11] We have performed a series of theoretical calculations of Li-ORR on metals and other materials. Our results show that the discharge potential is controlled by the formation of intermediates on metal surfaces. For instance, the initial O_2 reduction on Au occurs associatively with the formation of molecular lithium superoxide (LiO_2) at an equilibrium potential of ca. 2.0 V (Fig. 1).[12] The *intrinsic* activity of Li-ORR on different metals (including Au, Ag, Pt, Pd, Ir, and Ru) forms a volcano-like trend with respect to the adsorption energy of oxygen (Fig. 1).[12] Similar to the hydrogen variant of the ORR, Pt and Pd are found to be the most active metals and are located at the top of the volcano, which has been borne out experimentally.[13] This trend can be traced to the competition between the electrode on one hand and Li on the other hand for bonding with the oxygen. As a selection criterion, the adsorption strength of atomic O on a metal should be close to that on Pt and Pd in order to for it to have promotional effects for the Li-ORR. Metal particles with more edges and defects are predicted to be more active as measured by the onset of ORR activity.

The aprotic nature of typical Li-O_2 cells, however, makes the one-electron reduction of O_2 to the superoxide anion (O_2^-) possible (i.e., without concurrent Li^+ transfer), and allows O_2^- to have appreciable lifetime, particularly on a relatively inert metal such as Au. O_2^- formation occurs at appreciably higher potential than molecular LiO_2 on many of the metals (Fig. 1). DFT calculations, in conjunction with spectro-electrochemical studies,[14,15] therefore suggest that O_2^- formation needs to be taken into account when predicting the observed activity of ORR in Li-O_2 cells. On Au, for instance, O_2^- formation is the key process that corresponds to the discharge of a Li-O_2 cell at low overpotentials (Fig. 2) and determines how and where the solid discharge products are formed.[15] Our findings contribute to an improved understanding of the ORR in non-aqueous cells and to the further development of Li-O_2 batteries.

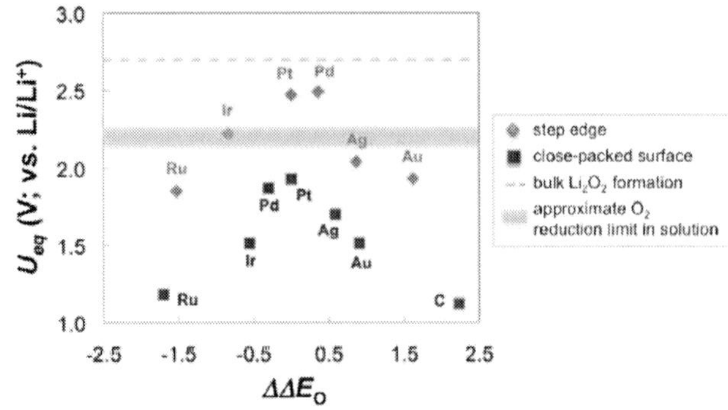

Figure 1. Equilibrium formation potential (U_{eq}) of the first e^- transfer in Li-ORR on different metals plotted against the adsorption energy of atomic oxygen relative to that on Pt ($\Delta\Delta E_O$). Adapted from Ref. 12.

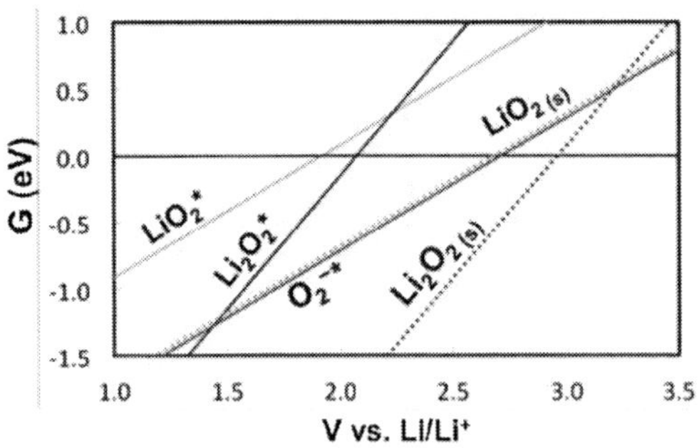

Figure 2. Free energy vs. electrode potential phase diagram for Li-ORR on Au(111) in DMSO. Zero on the y-axis represents gas-phase O_2 at room temperature and 0.1 bar. Adapted from Ref. 15.

References

(1) Abraham, K. M.; Jiang, Z. *J. Electrochem. Soc.* **1996**, *143*, 1.
(2) Débart, A.; Bao, J.; Armstrong, G.; Bruce, P. G. *J. Power Sources* **2007**, *174*, 1177.
(3) Girishkumar, G.; McCloskey, B.; Luntz, A. C.; Swanson, S.; Wilcke, W. *J. Phys. Chem. Lett.* **2010**, *1*, 2193.
(4) Nørskov, J. K.; Rossmeisl, J.; Logadottir, A.; Lindqvist, L.; Kitchin, J. R.; Bligaard, T.; Jonsson, H. *J. Phys. Chem. B* **2004**, *108*, 17886.
(5) Hummelshøj, J. S.; Blomqvist, J.; Datta, S.; Vegge, T.; Rossmeisl, J.; Thygesen, K. S.; Luntz, A. C.; Jacobsen, K. W.; Nørskov, J. K. *J. Chem. Phys.* **2010**, *132*, 071101.
(6) Xu, Y.; Shelton, W. A. *J. Chem. Phys.* **2010**, *133*, 024703.
(7) Allen, C. J.; Mukerjee, S.; Plichta, E. J.; Hendrickson, M. A.; Abraham, K. M. *J. Phys. Chem. Lett.* **2011**, *2*, 2420.
(8) Lu, Y.-C.; Gasteiger, H. A.; Shao-Horn, Y. *J. Am. Chem. Soc.* **2011**, *133*, 19048.
(9) Peng, Z.; Freunberger, S. A.; Chen, Y.; Bruce, P. G. *Science* **2012**, *337*, 563.
(10) Lu, J.; Lei, Y.; Lau, K. C.; Luo, X. Y.; Du, P.; Wen, J. G.; Assary, R. S.; Das, U.; Miller, D. J.; Elam, J. W.*et al. Nat. Commun.* **2013**, *4*, 2383.
(11) Lu, J.; Jung Lee, Y.; Luo, X.; Chun Lau, K.; Asadi, M.; Wang, H.-H.; Brombosz, S.; Wen, J.; Zhai, D.; Chen, Z.*et al. Nature* **2016**, *529*, 377.
(12) Dathar, G. K. P.; Shelton, W. A.; Xu, Y. *J. Phys. Chem. Lett.* **2012**, *3*, 891.

(13) Lu, Y.-C.; Gallant, B. M.; Kwabi, D. G.; Harding, J. R.; Mitchell, R. R.; Whittingham, M. S.; Shao-Horn, Y. *Energ. Environ. Sci.* **2013**, *6*, 750.
(14) Peng, Z.; Chen, Y.; Bruce, P. G.; Xu, Y. *Angew. Chem. Int. Edit.* **2015**, *54*, 8165.
(15) Zhang, Y.; Zhang, X.; Wang, J.; McKee, W. C.; Xu, Y.; Peng, Z. *J. Phys. Chem. C* **2016**, *120*, 3690.

FIRST-PRINCIPLES GUIDED CATALYST OPTIMIZATION AND KINETIC EVALUATION FOR HYDROGEN FUEL PRODUCTION

Bin Liu and Mingxia Zhou

Department of Chemical Engineering
Kansas State University
Manhattan, KS 66503
mingxia@ksu.edu

Catalytic hydrogen production still remains as a challenging catalyst design and optimization problem, as the working catalysts need to be sufficiently active for the dehydrogenation of hydrocarbon or oxygenated hydrocarbon sources. In the mean time, optimal catalysts also need to be selective by effectively prohibiting side reactions, such as hydrogenolysis involving C-O bond scission. Furthermore, water-gas shift reaction (WGSR) can boost hydrogen production by removing CO from reforming in the concomitant catalytic process. The trial-and-error approach to explore the vast catalyst design parameter space has been a limiting factor to meet the demands of clean energy needs. The theoretical alternative, based on density functional theory, can significantly accelerate the design and optimization process via highly efficient screen strategies. Aided by the elementary reaction mechanisms derived from periodic density functional theory methods for WGSR, methanation, and reforming, the thermochemistry and kinetics for catalyst design parameters relevant to structure sensitivity and compositions that maximize hydrogen production rates have been calculated. Microkinetic modeling was then performed under various temperature and pressure conditions to further evaluate the rate and selectivities of identified catalysts.

UNDERSTANDING HOW pH AND ALKALI CATIONS AFFECT CYCLIC VOLTAMMOGRAMS AND THE HYDROGEN OXIDATION REACTION ON TRANSITION METAL SURFACES

Ian T. McCrum, Michael J. Janik

Department of Chemical Engineering, Pennsylvania State University, University Park, PA
itm107@psu.edu

Introduction

The structure and composition of the complex electrochemical interface between an electrode surface and electrolyte have a direct effect on the rates and mechanisms of electrocatalytic reactions. For example, the rate of the hydrogen oxidation reaction on a platinum electrode is strongly dependent on electrolyte pH, and this effect had tentatively been correlated to a shift in hydrogen binding strength with pH (beyond the expected Nernstian shift) as measured by cyclic voltammetry[1].We have previously examined the adsorption of hydrogen, hydroxide, water, and alkali metal cations on the low index facets of platinum with density functional theory (DFT), to show that the low potential adsorption processes measured by cyclic voltammetry correspond to the co- or competitive adsorption of hydrogen and hydroxide on Pt(110) and Pt(100) facets or steps[2]. We have shown that this competitive adsorption process shifts to higher potentials as the pH increased, due to the specific adsorption of alkali metal cations, and their weakening of hydroxide adsorption[2].

In this talk, I will overview our work using density functional theory to examine hydrogen, hydroxide, and alkali metal cation adsorption on the low index facets and higher order stepped surfaces of platinum. Recent work extending this analysis to other transition metals surfaces will be discussed. Specific topics that will be addressed include the effects of cation adsorption and pH on hydrogen and hydroxide adsorption across transition metals, the effect on measured cyclic voltammograms and electrochemically active surface area measurements, and their possible effects on catalysis, in particular, the hydrogen oxidation reaction.

Computational Methods

Density functional theory calculations were performed using the Vienna ab initio Simulation Package,[3-5] a plane-wave basis set DFT code. Adsorption potentials are calculated using the energy calculated with DFT and corrected to Gibbs free energies by including contributions from entropy using statistical mechanics. The effects of surface solvation on adsorption are evaluated using co-adsorbed explicit water molecules.

Results and Discussion

We have examined the adsorption of hydrogen, hydroxide, water, and the alkali metal cations on a variety of transition metal electrode surfaces. On platinum, we have already shown, in support of a growing body of experimental evidence, that the low potential peaks measured by cyclic voltammetry on high order single crystal electrodes and polycrystalline platinum correspond to the competitive adsorption of hydrogen and hydroxide on Pt(110) and Pt(100) and that these peaks shift to higher potentials with increasing pH, due to the weaken of hydroxide adsorption by the specific adsorption of alkali metal cations[2].

Figure 1 illustrates that the weakening of hydroxide adsorption in the presence of specifically adsorbed cations is not unique to platinum. In the presence of specifically adsorbed potassium, the OH* formation potential is driven to more positive potentials, indicating weaker adsorption. We also consider the adsorption of hydrogen on these same surfaces, and the effect of the alkali metal cation on hydrogen adsorption, which we find to be minimal. We also examine the adsorption of hydrogen and hydroxide on the other low index facets of these transition metal surfaces, and can use this information to evaluate experimentally measured cyclic voltammograms. We find that the competitive adsorption of hydrogen and hydroxide can occur on many surfaces that bind both adsorbates (H* and OH*) more strongly than platinum.

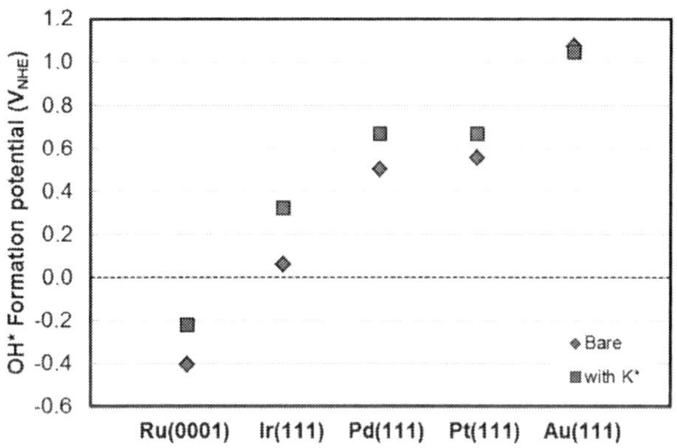

Figure 1. The formation potential of adsorbed hydroxide (OH*) from adsorbed water on Ru(0001), Ir(111), Pd(111), Pt(111), and Au(111) in the absence (blue diamond) and the presence (red square) of specifically adsorbed potassium (K*).

These results are particularly important in understanding the rate and mechanism of many electrocatalytic reactions which involve adsorbed hydrogen and adsorbed hydroxide as intermediates, such as hydrogen oxidation and oxygen reduction. We consider our results in light of recent experimental work which has shown that the rate of the hydrogen oxidation reaction is strongly dependent on pH, not only on platinum electrodes[1], but across other transition metal electrode surfaces as well,[6-7] where it has been suggested that adsorbed hydroxide may play an important role in dictating the rate[7].

Conclusions

DFT can be used to examine the thermodynamics of hydrogen, hydroxide, and alkali cation specific adsorption to explain experimentally measured cyclic voltammograms and better understand the rates and mechanisms of electrocatalytic reactions.

References.

(1) Sheng, W.; Zhuang, Z.; Gao, M.; Zheng, J.,; Chen, J. G.; Yan, Y., Nat. Commun. **2015**, 6, 5848.
(2) McCrum, I.T.; Janik, M.J., J. Phys. Chem. C. **2016**, 120 (1), 457-471.
(3) Kresse, G.; Furthmuller, J., Comput. Mater. Sci. **1996**, 6 (1), 15-50.
(4) Kresse, G.; Furthmuller, J., Phys. Rev. B **1996**, 54 (16), 11169-11186.
(5) Kresse, G.; Joubert, D., Phys. Rev. B **1999**, 59 (3), 1758-1775.
(6) Strmcnik, D.; Uchimura, M.; Wang, C.; Subbaraman, R.; Danilovic, N.; van der Vliet D.; Paulikas, A.P.; Stamenkovic, V.R.; Markovic, N.M., Nat. Chem. **2013**, 5, 300-306.
(7) Durst, J.; Siebel, A.; Simon, C.; Hasché, F.; Herranz, J.; Gasteiger, H.A., Energy Environ. Sci. **2014**, 7, 2255-2260.

HYDROTHERMAL AND MECHANICAL STABILITY OF METAL-ORGANIC FRAMEWORKS: INFLUENCE OF LINKER FUNCTIONALIZATION AND PRESENCE OF DEFECTS

Fabien Trousselet, François-Xavier Coudert

Chimie ParisTech, PSL Research University, CNRS,
Institut de Recherche de Chimie Paris, 75005 Paris, France
fx.coudert@chimie-paristech.fr

Metal–organic frameworks are proposed for a vast number of technological applications, with two specific areas being the focus of much recent research. First, flexible MOFs — also known as soft porous crystals — that respond by large changes in structure and properties to small external stimuli, such as adsorption of guest molecules and changes in temperature or pressure. Secondly, defect-engineered MOFs show great promise for increased catalytic activity and adsorption properties.

While the introduction of flexibility and defects in MOFs is highly desirable in applications such as sensing and actuation, their extreme flexibility can also be synonymous with decreased chemical, mechanical and thermal stability — which can drastically limit their performance in industrial environments, e.g. by a lack of stability at elevated temperatures and in the presence of water.

Here, we show how the combination of computational and experimental techniques can help better understand and improve the stability of MOFs. In particular, we demonstrate how DFT-based molecular dynamics simulations helps investigate the hydrothermal stability of MIL-53(Ga)[1] and the mechanical stability of UiO materials[2]. We also use quantum mechanical calculations to investigate the process of defect formation in UiO-66 and shed light into the trade-off between performance and stability[3,4] that is key to engineering MOFs that are "defective by design".

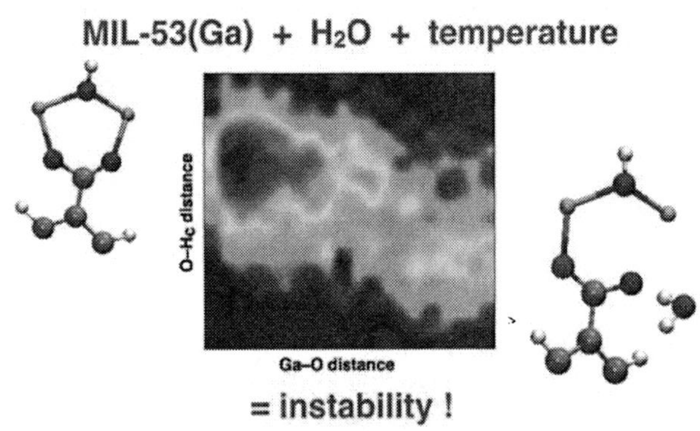

Figure 1. Free energy map, obtained from ab initio molecular dynamics, of water displacing the organic linker in MIL-53(Ga), showing the root cause of hydrothermal instability.

References

(1) Haigis, V.; Coudert, F.-X.; Vuilleumier, R.; Boutin, A.; Fuchs, A. H. *J. Phys. Chem. Lett.* **2015**, *6*, 4365–4370.
(2) Hobday, C. L.; Marshall, R. J.; Murphie, C. F.; Sotelo, J.; Richards, T.; Allan, D. R.; Düren, T.; Coudert, F.-X.; Forgan, R. S.; Morrison, C. A.; Moggach, S. A.; Bennett, T. D. *Angew. Chem. Int. Ed.* **2016**, *55*, 2401–2405.
(3) Thornton, A. W.; Babarao, R.; Jain, A.; Trousselet, F.; Coudert, F.-X. *Dalton Trans.* **2016**, *45*, 4352–4359.
(4) Cliffe, M. J.; Hill, J. A.; Murray, C. A.; Coudert, F.-X.; Goodwin, A. L. *Phys. Chem. Chem. Phys.* **2015**, *17*, 11586–11592.

AUTOMATED TOOLKIT FOR DISCOVERY IN INORGANIC CHEMISTRY

Efthymios I. Ioannidis[1], *Terry Z. H. Gani*[1], *and Heather J. Kulik*[1,*]

[1]Department of Chemical Engineering, Massachusetts Institute of Technology, Cambridge, MA 02139
*hjkulik@mit.edu

Transition metals are central to nearly all of catalysis, be it biological, molecular, or heterogeneous in nature. Transition metals also impart unique properties to materials systems and enlarge the palette for materials discovery. In heterogeneous catalysis and in materials science, computational chemistry plays a key role in the prediction of mechanism and function and in the design of properties. Density functional theory (DFT) remains a workhorse for these discovery efforts due to its first-principles description of electronic structure along with affordable computational cost. We present an automated approach for the first-principles screening and discovery of new inorganic molecules and intermolecular complexes. Inorganic complexes exhibit a large variability in coordination number and oxidation state, which creates challenges for automated screening approaches. Our molSimplify tool overcomes well-known challenges in the automatic generation of candidate inorganic molecule structures through a divide and conquer tactic that combines force-field preoptimization of organic fragments with alignment to first-principles-trained and rules-based metal-ligand geometries. Exploration of chemical space is enabled through random generation of ligands and intermolecular complexes from large databases. Finally, seamless generation of input files, completion of electronic structure calculations, and postprocessing for each generated catalyst structure elucidates underlying chemical and energetic trends. We demonstrate the applicability of our approach with diverse applications in catalysis and computational chemistry ranging from tuning the reactivity of precursors employed in quantum dot synthesis to designing redox active ferrocenium complexes capable of selective anion recognition and binding. In all cases, our screening approach uncovers new design principles that are explained through the interplay of electronic and geometric features of the complexes.

Overview of the code

The framework is comprised of over 13,000 lines of code and includes three main modules described in greater detail in the rest of this article:

1. The main structure generation module builds and modifies inorganic complexes and supramolecular complexes. It also generates the appropriate input files and job scripts to run first-principles quantum chemistry calculations on the generated structures.
2. The database module interacts with external chemical databases for ligand search and retrieval to be fed back into the structure generation module.
3. The post-processing module contains routines that monitor progress of quantum chemistry calculations, analyze their outcomes, and post-process wavefunction and charge density data to elucidate structure-property correlations. This module also interacts with external codes for some electronic structure analysis.

Conclusions

While computational chemistry has been a vital tool in the accurate prediction of catalytic activity and materials properties, several outstanding challenges remain. One such challenge is the ability to rapidly build, screen, and discover new complexes. We have presented an open source toolkit and approach that accelerates discovery. Ongoing extensions include to periodic systems and to reaction network discovery.

Acknowledgement. This work was supported by the National Science Foundation under grant number ECCS-1449291 and an MIT Energy Initiative seed grant. H.J.K. holds a Career Award at the Scientific Interface from the Burroughs Wellcome Fund. This work was carried out in part using computational resources from the Extreme Science and Engineering Discovery Environment (XSEDE), which is supported by National Science Foundation grant number ACI-1053575. This research also partially used the computational resources of the Center for Nanoscale Materials (Carbon cluster), an Office of Science user facility, was supported by the U. S. Department of Energy, Office of Science, Office of Basic Energy Sciences, under Contract No. DE-AC02-06CH11357.

DESIGN OF FUNCTIONALIZED METAL ORGANIC FRAMEWORKS FOR CO2 HYDROGENATION: THE EFFECT OF MOF TOPOLOGY AND FUNCTIONAL GROUP

Jingyun Ye[1], *J. Karl Johnson*[1,2]

[1]Department of Chemical & Petroleum Engineering, University of Pittsburgh, Pittsburgh, Pennsylvania, 15261, USA
[2]Pittsburgh Quantum Institute, Pittsburgh, Pennsylvania 15261, United States

Abstract:

CO_2 capture coupled with chemical recycling of CO_2 to renewable fuels and valuable chemicals is an effective approach to reduce carbon emissions and ultimately close the natural carbon loop. Metal organic frameworks (MOFs), a class of porous crystalline materials, have great potential for CO_2 capture and storage. Frustrated Lewis pairs (FLPs), which are organic molecules possessing both Lewis acid and base sites as catalytic active sites can bind CO_2 and heterolytically dissociate H_2. Our strategy is to combine the advantages of MOFs and FLPs into a single material. In our previous work, we design a novel CO_2 hydrogenation catalyst through functionalization of UiO-66 (Zr based MOF) with Lewis pair (P-BF_2) to create heterogeneous catalysts, UiO-66-P-BF_2. This new catalyst retains the catalytic activity of FLPs and also avoids the limitations typical for homogeneous catalysts, such as difficulty with catalyst recycling and product separation.[1,2] The energy favorable pathway for CO_2 hydrogenation to produce formic acid (FA) over UiO-66-P-BF_2 has two steps as revealed by density functional theory (DFT) calculation: (i) dissociation of H_2 to hydridic and protic hydrogen and (ii) CO_2 reacts with hydridic and protic hydrogens to produce FA.[1,2]

However, one drawback of UiO-66-P-BF_2 is that CO_2 binds much stronger with Lewis acid-base sites than H_2, which could poison the catalytic active sites for H_2 dissociation. Additionally, LPs might quench each other inside the pore of MOFs which thus deactivates the catalyst. In order to overcome this difficulty, we propose to use other MOFs in combination with new LPs. In this work, we design a UiO-67-NBF_2 catalyst which has high selectivity for H_2 dissociation and activity for FA production based on DFT study. In addition, our DFT results suggest UiO-67-$(NBF_2)_4$ has a high activity for methanol production from CO_2 hydrogenation. Going further, we have examined the effect of MOF topology on CO_2 hydrogenation by incorporation of NBF_2 on other MOFs. In parallel, we have also screened hundreds of LPs for promising candidates for CO_2 hydrogenation based on their ability in H_2 dissociation and CO_2 adsorption.

References:

(1) Ye, J.; Johnson, J. K. Design of Lewis Pair-Functionalized Metal Organic Frameworks for CO2 Hydrogenation. *ACS Catal.* **2015**, 5(5), 2921-2928.

(2) Ye, J.; Johnson, J. K. Screening Lewis Pair Moieties for Catalytic Hydrogenation of CO_2 in Functionalized UiO-66. *ACS Catal.* **2015,** *5* (10), 6219-6229.

CHARGE SCREENING EFFECT IN ZEOLITE-CATALYZED C-C COUPLING REACTIONS

Bin Wang

Center for Interfacial Reaction Engineering (CIRE), School of Chemical, Biological and Materials Engineering, The University of Oklahoma, Norman, OK 73019
Wang_cbme@ou.edu

Introduction

Zeolites are remarkable materials with adjustable confinement at the nano scale and Brønsted acid strength that catalyze reactions of vital importance.[1] For example, Brønsted zeolites catalyze a variety of C-C coupling reactions, including alkylation, acylation, Diels-Alder cycloaddition, aldol condensation, and ketonization, all of which are important reactions for production of fuels and chemicals.[2]

Many aspects of the zeolite structure, such as Si/Al ratio, pore size and shape, and surface area, play a role in determining the catalytic performance. An issue that has attracted the attention of researchers for a long time is the role that the zeolite characteristics, such as intrinsic acid strength, structural confinement, as well as van der Waals and Coulomb interactions between the molecules and the zeolite framework (Figure 1), may have on the catalytic phenomena[3]. In many cases, the stability of reaction intermediates and transition states that leads to differences in catalytic activity and selectivity is a combination of these factors.

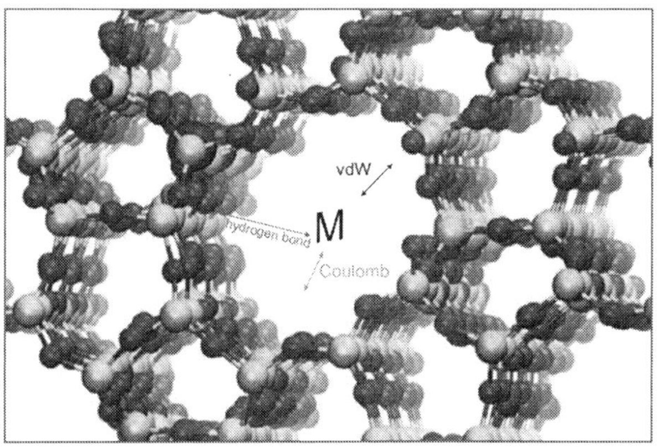

Figure 1. Schematic figure to show the active sites in HZSM-5. The molecule (M) may interact with the zeolite via the hydrogen bond with the acid site, van der Waals (vdW) interaction, and Coulomb interaction when the molecule is charged.

In this study, we focus on the important influence of the charge screening effect in zeolite catalysis using acylation of biomass-derived furanic compounds as an example. Acylation is one of the most widely studied reaction involving C-C coupling reactions. As illustrated in Figure 2, the significance of this reaction in organic chemistry is the production of aromatic ketones that are used in the manufacture of several important products ranging from pharmaceuticals to flavors and fragrances.[4] The use of zeolites in acylation reactions has received increasing attention recently as has been discussed in several review articles.[5]

The acylation reaction over Brønsted zeolites is believed to occur via the formation of acylium ions in equilibrium with surface acyl species, which can be formed by decomposition of an anhydride, an acyl chloride, carbonylation of a methoxy species, or dehydration of a carboxylic acid. Figure 2 shows an example of a typical acylation reaction, as well as the common chemical intermediates present on the zeolite surface.

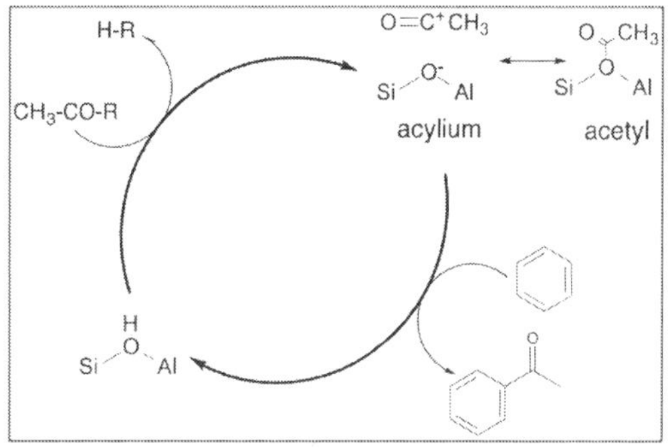

Figure 2. Acylation reaction over Brønsted zeolites via formation of an acylium ion in equilibrium with a surface acyl species. The R group in the reaction cycle represents Cl in acyl chlorides that release HCl, an O-(C=O)-C in the case of acetic anhydride that releases acetic acid, or OH in the case of acetic acid that releases H_2O.

In this study, we focus on the C-C coupling step between the surface acyl species and biomass-derived furanic compounds. In particular we show the important influence of the aromatic species on the stabilization of the charge of the acylium ion at the transition state, which facilitates C-C bond formation. This charge screening effect may be generally true for other zeolite-catalyzed C-C coupling reactions.

Computational methods

Density functional theory (DFT) calculations were carried out using the VASP package[6]. The PBE-GGA exchange-correlation potential [7] was used, and the electron-core interactions were treated in the projector augmented wave (PAW) method[8]. The van der Waals interaction has been taken into account through the so-called DFT-D3 semi-empirical methods via a pair-wise force field [9]. Structures have been optimized using a single Γ point of the Brillouin zone with a kinetic cut off energy of 400 eV. The nudged elastic band (NEB)[10] method has been used to find the transition states, which have been further verified by vibrational calculations. Hybrid functional calculations have also been performed[11].

Results and Discussion

Figure 3. DFT calculation of the surface acyl species.

We find that the whole acylation reaction occurs on two oxygen atoms bound to the same Al atom. First, acetic acid adsorbs on HZSM-5 by forming H-bonds with the Brønsted acid site, and subsequently, acyl species and water form with an apparent barrier of 16 kJ/mol due to very exothermic adsorption of acetic acid resulting from both the H-bond and the structural confinement. We obtain an enhanced apparent barrier of 40 kJ/mol using the hybrid functional. The calculated structure of the surface acyl species is shown in Figure 3.

Direct desorption of the acyl species to become an acylium ion is very endothermic with an energy cost of 116 kJ/mol. When 2-methyl furan (MF) is introduced, MF reacts with the surface acyl group to form an intermediate species, which is deprotonated subsequently to recover both the molecular aromaticity and the Brønsted acid site leading to a large energy gain. The true activation barrier of formation of 5-acetyl 2-methy furan from MF is 76 kJ/mol involving an acylium-ion-like transition state. The results imply that the presence of MF stabilizes the charged acylium ion species.

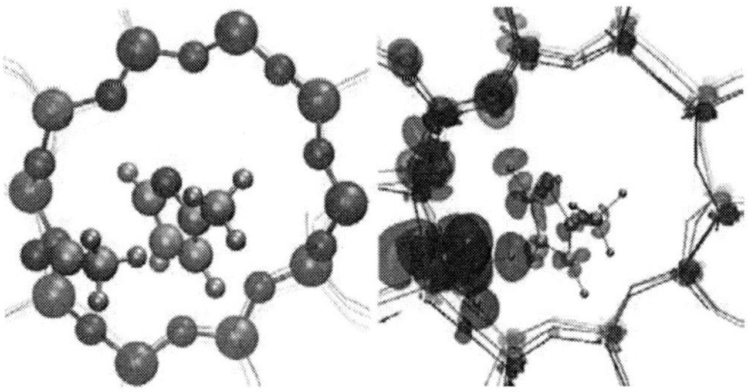

Figure 4. DFT calculated transition state for the C-C coupling step for acylation of MF over HZSM-5. (left) Calculated transition state structure, and (right) charge redistribution between the acylium ion and zeolite framework as well as the furanic compound. Red and Blue indicates charge accumulation (negative charge) and charge depletion (positive charge). The isosurface corresponds to a charge density of ±0.015 eÅ^{-3}.

Further insight is obtained through the charge redistribution calculations. Figure 4 shows the charge transfer between the complex (the desorbed acyl and MF) and the zeolite framework in the transition state of the C-C coupling step. To compensate the charge in the zeolite framework, a certain number of electrons flow from the complex to zeolite, evidenced by the charge depletion in the desorbed acyl and the MF molecule and charge accumulation at the oxygen atoms bonded to the Al (Figure 4). Integration of the charges via Bader analysis gives a total number of electron transfer of 1.2 electrons to the zeolite framework including 0.6 electrons from the desorbed acyl (acylium ion) and 0.6 electrons from the MF. In the absence of MF, the integrated charge of 1.2 electrons is all located at the acylium ion. This charge delocalization and electronic hybridization caused by MF in the proximity stabilizes the acylium ion and lowers the activation barrier in the C-C coupling step. This charge analysis implies that the ability of the methylfuran molecule to compensate the charge of the acylium ion in the transition state may significantly reduce the barrier for C-C coupling. Similar results have been obtained for other aromatic molecules to interact with the surface acyl species in different zeolites.

Conclusions

DFT calculations have been performed for C-C coupling reactions of biomass-derived model compounds over zeolite. Insight into the transition state obtained from calculations demonstrates the importance of the ability of methylfuran to stabilize the charge of the acylium ion, which may be generally true for other C-C coupling reactions involving a charged species in the reaction pathway.

Acknowledgement. This work is supported by the U.S. Department of Energy, DOE/EPSCOR (Grant DESC0004600). The computation was performed at DOE National Energy Research Scientific Computing Center (NERSC) and the OU Supercomputing Center for Education & Research (OSCER) at the University of Oklahoma.

References

(1) Kubička, D.; Kubičková, I.; Čejka, J., Application of molecular sieves in transformations of biomass and biomass-derived feedstocks. *Catalysis Reviews* **2013,** *55* (1), 1-78.

(2) Resasco, D. E.; Wang, B.; Crossley, S., Zeolite-catalysed C–C bond forming reactions for biomass conversion to fuels and chemicals. *Catal. Sci. Technol.* **2016**, 10.1039/C5CY02271A.

(3) (a) Jones, A. J.; Iglesia, E., The Strength of Bronsted Acid Sites in Microporous Aluminosilicates. *Acs Catalysis* **2015,** *5* (10), 5741-5755; (b) Jones, A. J.; Zones, S. I.; Iglesia, E., Implications of Transition State Confinement within Small Voids for Acid Catalysis. *Journal of Physical Chemistry C* **2014,** *118* (31), 17787-17800; (c) Jones, A. J.; Carr, R. T.; Zones, S. I.; Iglesia, E., Acid strength and solvation in catalysis by MFI zeolites and effects of the identity, concentration and location of framework heteroatoms. *Journal of Catalysis* **2014,** *312*, 58-68.

(4) (a) Franck, H.-G.; Stadelhofer, J. W., *Industrial Aromatic Chemistry: Raw Materials· Processes· Products*. Springer Science & Business Media: 2012; (b) Horsley, J. A., Producing bulk and fine chemicals using solid acids. *Chemtech* **1997,** *27* (10), 45-49; (c) BauerK, G., SurbergH, Common Fragrance and Flavor Materials. WHC Vertagsgesellschaft, Weinheim: 1990.

(5) (a) Venuto, P. B., Organic catalysis over zeolites: a perspective on reaction paths within micropores. *Microporous Materials* **1994,** *2* (5), 297-411; (b) Tanabe, K.; Hölderich, W. F., Industrial application of solid acid–base catalysts. *Applied Catalysis A: General* **1999,** *181* (2), 399-434; (c) Corma, A., Inorganic solid acids and their use in acid-catalyzed hydrocarbon reactions. *Chemical Reviews* **1995,** *95* (3), 559-614; (d) Sartori, G.; Maggi, R., Update 1 of: Use of Solid Catalysts in Friedel–Crafts Acylation Reactions. *Chemical reviews* **2011,** *111* (5), PR181-PR214; (e) Bejblova, M.; Prochazkova, D.; Čejka, J., Acylation reactions over zeolites and mesoporous catalysts. *ChemSusChem* **2009,** *2* (6), 486-499; (f) Martín-Aranda, R. M.; Čejka, J., Recent advances in catalysis over mesoporous molecular sieves. *Top Catal* **2010,** *53* (3-4), 141-153.

(6) Kresse, G.; Furthmuller, J., Efficient iterative schemes for ab initio total-energy calculations using a plane-wave basis set. *Phys Rev B* **1996,** *54* (16), 11169-11186.

(7) Perdew, J. P.; Burke, K.; Ernzerhof, M., Generalized gradient approximation made simple. *Phys Rev Lett* **1996,** *77* (18), 3865-3868.

(8) (a) Blochl, P. E., Projector Augmented-Wave Method. *Phys Rev B* **1994,** *50* (24), 17953-17979; (b) Kresse, G.; Joubert, D., From ultrasoft pseudopotentials to the projector augmented-wave method. *Phys Rev B* **1999,** *59* (3), 1758-1775.

(9) (a) Grimme, S.; Antony, J.; Ehrlich, S.; Krieg, H., A consistent and accurate ab initio parametrization of density functional dispersion correction (DFT-D) for the 94 elements H-Pu. *J Chem Phys* **2010,** *132* (15), 154104; (b) Grimme, S.; Ehrlich, S.; Goerigk, L., Effect of the Damping Function in Dispersion Corrected Density Functional Theory. *J Comput Chem* **2011,** *32* (7), 1456-1465.

(10) Henkelman, G.; Uberuaga, B. P.; Jonsson, H., A climbing image nudged elastic band method for finding saddle points and minimum energy paths. *J Chem Phys* **2000,** *113* (22), 9901-9904.

(11) (a) Heyd, J.; Scuseria, G. E.; Ernzerhof, M., Hybrid functionals based on a screened Coulomb potential (vol 118, pg 8207, 2003). *J Chem Phys* **2006,** *124* (21); (b) Heyd, J.; Scuseria, G. E.; Ernzerhof, M., Hybrid

functionals based on a screened Coulomb potential. *J Chem Phys* **2003**, *118* (18), 8207-8215.

REDICTING LIGHT HYDROCARBON DIFFUSION THROUGH ZEOLITIC IMIDAZOLATE FRAMEWORKS FOR SEPARATION APPLICATIONS

Ross J. Verploegh, Sankar Nair, David S. Sholl

School of Chemical and Biomolecular Engineering,
Georgia Institute of Technology
Atlanta, GA 30332-0100
david.sholl@gatech.chbe.edu

The use of nanoporous materials for light paraffin and olefin separations has the potential to disrupt the current industry standard, energy intensive cryogenic distillation. The ability to predict the diffusion characteristics of these nanoporous materials would further the design of either thermodynamic or kinetic separations. We focus on a class of nanoporous materials called zeolitic imidazolate frameworks (ZIFs), a subfamily of metal-organic frameworks (MOFs), for which accurate and efficient predictions of hydrocarbon diffusivities are challenging. The small pore sizes of ZIFs (e.g. ZIF-8), coupled with adsorbates having kinetic diameters equal to or larger than the pore size, produce a wide range of diffusion time scales that cannot be measured directly with equilibrium molecular dynamics simulations. We have computationally measured the hopping rates of 15 different molecules in ZIF-8 via dynamically corrected transition state theory (dcTST).[1] Umbrella sampling, an enhanced molecular dynamics sampling method, along with the one-dimensional weighted histogram analysis method (WHAM) was used to calculate the diffusion free energy barriers. Comparison of the computed diffusivities to extant experimental data[2,3] shows remarkable agreement within an order of magnitude.

Both the umbrella sampling and dynamical correction calculations included ZIF-8 flexibility, which was found to be critical in accurately describing molecular diffusion for larger adsorbates capable of deforming the ZIF-8 window.[1] The dcTST method was also applied to study the effect of hydrocarbon loadings on the transport diffusion coefficients. We will present self and transport diffusion coefficients for methane, ethane, ethylene, propane, propylene, n-butane, and 1-butene at room temperature and loadings from infinite dilution to liquid-like loadings. Our simulated transport diffusion coefficients along with adsorption isotherms from Grand Canonical Monte Carlo (GCMC) simulations were used to quantitatively predict single-component performance of thin-film ZIF-8 membranes. Lastly, we will comment on the application of our methodology to study hydrocarbon diffusion in other ZIFs.

Acknowledgement. This work was supported by the Phillips 66 Company.

References

(1) Verploegh, R. J.; Nair, S.; Sholl, D. S. *J. Am. Chem. Soc.* **2015**, *137*, 15760-15771.
(2) Zhang, C.; Lively, R. P.; Zhang, K.; Johnson, J. R.; Karvan, O.; Koros, W. J. *J. Phys. Chem. Lett.* **2012**, *3*, 2130-2134.
(3) Eum, K.; Jayachandrababu, K. C.; Rashidi, F.; Zhang, K.; Leisen, J.; Graham, S.; Lively, R. P.; Chance, R. R.; Sholl, D. S.; Jones, C. W.; Nair, S. *J. Am. Chem. Soc.* **2015**, *137*, 4191.

INVESTIGATION OF $A_xA'_{1-x}Mn_yB_{1-y}O_3$ PEROVSKITES FOR CHEMICAL LOOPING WITH OXYGEN UNCOUPLING (CLOU) THROUGH FIRST PRINCIPLES CALCULATIONS

Amit Mishra, Nathan Galinsky, Fanxing Li, Erik E. Santiso**

amishra4@ncsu.edu
fli5@ncsu.edu, eesantis@ncsu.edu

Introduction

Given the current energy demands and concerns over global climate change, there is an urgent need for efficient energy production from carbonaceous fuel conversion with facile CO_2 capture. Chemical looping combustion (CLC) offers a unique solution to this problem through a redox material capable of fossil fuel conversion with its lattice oxygen. In this two-step process, a metal oxide based redox material is first reduced with a fuel and subsequently re-oxidized to its original state with air. This process offers two main advantages 1) a pure oxygen source through a metal oxide, acting as an "oxygen carrier" and 2) exergy recuperation as result of chemical looping reactions. The first advantage is particularly promising as it results in CO_2 capture with minimal energy penalty.[1] From a practical standpoint; however, this process can be challenging for solid fuel conversions due to kinetic limitations for solid-solid reactions. A potential solution to this problem is through a material that possesses CLOU properties, i.e. spontaneous release of gaseous oxygen for solid fuel combustion. While there have been over 700 oxygen carriers tested for chemical looping[2], there has been lack of promising oxygen carriers for CLOU, spurring the needs for an efficient method for materials screening. In this work, the $A_xA'_{1-x}Mn_yB_{1-y}O_3$ class of materials has been investigated through first principle calculations. In particular, vacancy formation in the dilute limit was determined for $CaMnO_{3-\delta}$, $BaMnO_{3-\delta}$, $Ca_{0.75}Sr_{0.25}MnO_{3-\delta}$, $CaMn_{0.75}Fe_{0.25}O_{3-\delta}$. The effects of defect properties, Hubbard U parameters, and oxide magnetism are also investigated. It is determined that the vacancy formation energy can be used as an efficient screening parameter in terms of correlating with CLOU properties of mixed metal oxides.

Computational methodology and experimental

Computational Methodology: DFT Calculations. Geometry optimizations were performed using spin polarized generalized gradient approximations (GGA) and the projector augmented wave (PAW) approach as implemented in VASP. Various magnetic configurations were tested for stability (with spin polarization), and calculations were converged in k-point density. A cutoff energy of 425 eV was used for all systems. The energy to create a vacancy was found using the difference between the energy of a supercell with a single vacancy plus half a diatomic oxygen and a pristine supercell. The supercell size was increased until the energy to form an oxygen vacancy became constant, corresponding to the dilute limit. The system sizes used were: 80, 160, and 320 atoms for $BaMnO_3$, and 40, 80, and 160 atoms for $CaMnO_3$, $CaMn_{0.75}Fe_{0.25}O_3$, and $Ca_{0.75}Sr_{0.25}MnO_3$. Doped structures were made by replacement of either Ca or Mn in a 20 atom supercell of $CaMnO_3$.

Experimental: CLOU Properties Characterizations. Isothermal redox experiments and temperature programmed desorption experiments were conducted using thermogravimetric analysis to determine the oxygen vacancy concentrations at various temperatures and the temperature of oxygen release. Temperature programmed desorption experiments were conducted using inert helium (5.0 grade) with a ramping rate of 20oC/min, up to 1000oC. Isothermal redox experiments were conducted at various temperatures between 650 oC and 1000 oC, cycling between inert gas and 10% O2 (Extra Dry).

Results and Discussion

The vacancy formation energy in the dilute limit was determined for all materials, showing two main findings: 1) Ca is preferred over Ba for A site of the perovskite in terms of oxygen vacancy formation and 2) doping of Sr in the A site and Fe in the B site of $CaMnO_3$, lowers the energy of vacancy formation even further. These results are summarized the Figure 1. These results correspond well to experimental findings, with both dopants lowering the initial temperature of oxygen uncoupling (shown in Figure 2). In addition, the results provide insight into the nature of the vacancy sites within a doped structure. In $Ca_{0.75}Sr_{0.25}MnO_3$, the vacancy preferentially forms when the closest cation neighbors are consisted of either Ca or Mn. With respect to $CaMn_{0.75}Fe_{0.25}O_3$, the vacancy preferentially forms when an Fe cation is seated close to the vacancy.

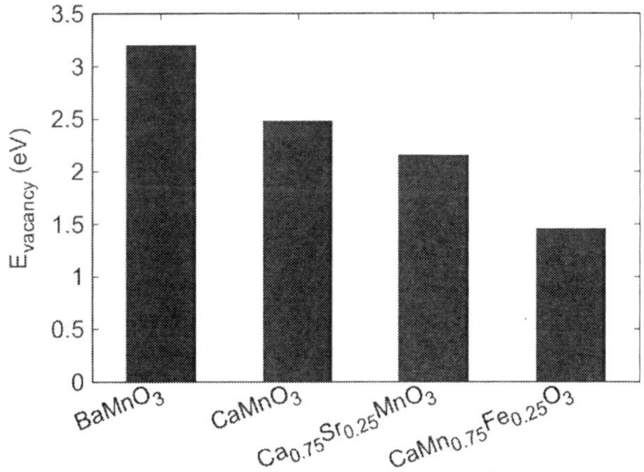

Figure 1: Energy of vacancy formation for $CaMnO_3$, $BaMnO_3$, $Ca_{0.75}Sr_{0.25}MnO_3$, and $CaMn_{0.75}Fe_{0.25}O_3$ in the dilute limit.

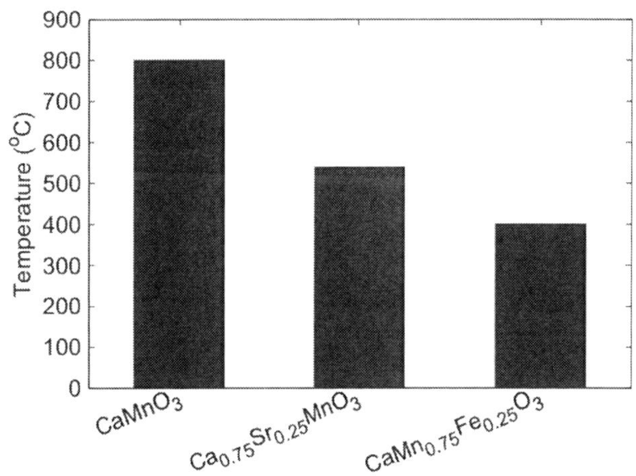

Figure 2: Oxygen uncoupling temperature of $CaMnO_3$, $Ca_{0.75}Sr_{0.25}MnO_3$, and $CaMn_{0.75}Fe_{0.25}O_3$ ($BaMnO_3$ shows negligible oxygen uncoupling properties up to 1,000°C).

Conclusions

The current study indicates that vacancy formation energy can be used as a screening parameter in determining materials suitable for CLOU. Specifically, the results of this study show that the energy of vacancy formation in four materials: $CaMnO_3$, $BaMnO_3$, $Ca_{0.75}Sr_{0.25}MnO_3$, and $CaMn_{0.75}Fe_{0.25}O_3$ correspond to their CLOU properties, i.e. lower vacancy formation energy generally leads to lower oxygen uncoupling temperature. Further investigation of doping effect reveals that the doped structures preferentially form vacancies in contrasting fashion; oxygen vacancy formation is favored in locations away from a Sr cation and close to a Fe cation.

Acknowledgement. This work was supported by the US Department of Energy (Award No. FE0011247) and Kenan Institute. The authors acknowledge North Carolina State University for the use of their Analytical Instrumentations Facility (Supported by the State of North Carolina and NSF)

References

(1) Figueroa, J. D.; Fout, T.; Plasynski, S.; McIlvried, H.; Srivastava, R. D. *International Journal of Greenhouse Gas Control* **2008**, *2* (1), 9–20.

(1) Adanez, J.; Abad, A.; Garcia-Labiano, F.; Gayan, P.; de Diego, L. F. *Progress in Energy and Combustion Science* **2012**, *38* (2), 215–282.

SIMULATIONS OF WATER-SOLID INTERFACES

*Henrik H. Kristoffersen, Ru-Fen Liu, Joan Shea, and Horia Metiu**

Department of Chemistry & Biochemistry
University of California, Santa Barbara
*metiu@chem.ucsb.edu

We use ab initio, constant temperature, molecular dynamics simulations to examine some aspects of the liquid/solid interface and the catalytic properties of molten metal solutions. In one study we examine whether calculations of the binding energy of a molecule to a surface in vacuum have any relationship to the binding energy of the same molecule and the same surface covered by a water film. The molecules we examined were catechol and HCl and the surface was rutile $TiO_2(110)$. We calculated the binding energy of the catechol, water and HCl to the oxide surface in vacuum. Then we calculated, by ab initio molecular dynamics, the mean energy of the catechol or HCl adsorbed at the oxide/water film interface. By subtracting the energy of the water film on the oxide and the energy of gas phase catechol, we obtain the binding energy of catechol at the water/oxide interface.

Next we inquired whether there might be a simple way of estimating the binding energy of the catechol at the water/oxide interface by performing only calculations of the binding energy in vacuum. We found that the binding energy of catechol at the water/oxide interface is well approximated by the binding energy of catechol to the oxide in vacuum minus the binding energy of two water molecules to the oxide in vacuum, plus the solvation energy of two water molecules. The factor of two comes in because catechol is a large molecule and the sites to which it binds can adsorb two water molecules. Essentially the same simple formula gives a good estimate of the binding energy of HCl at the water/oxide interface except that HCl, being small, replaces only one water molecule.

It is tempting to assume that this simple rule gives a reasonable approximation of the binding energy of other molecules to other oxides. One requirement is that the surface should be such that the water and the molecule bind to the same surface sites. We examine the chemical properties of Li dissolved in molten LiCl, which results in the formation of a "solvated electron" that is chemically active. H_2 and N_2 dissolve in the melt forming negative ions. We also investigate the properties of Pt atoms dissolved in molten Sn. We hope to create single-atom Pt catalysts.

INSIGHTS INTO URANIUM SPECIATION AND EXTRACTION FROM THE SEA

De-en Jiang

Department of Chemistry, University of California, Riverside

The sustainability of nuclear energy depends on a long-term, continuous supply of uranium for centuries to come. Seawater provides the ultimate resource and reassurance due to the vast amount of dissolved uranium (~4.5 billion tons), despite its low concentration (~3 ppb). In this talk, I will discuss our modeling efforts in understanding the fundamentals of uranium speciation and ligand binding in seawater. Quantum mechanical (QM), molecular mechanical (MM), and the hybrid QM/MM approach all will be used to address the problem. We will show that the common ions in the sea such as Ca and Na ions play an important role in the aqueous uranium chemistry.

DFTMD SIMULATIONS OF PYRIDINIUM CATALYZED CO_2 REDUCTION

Xueting Fan[1], *and Jun Cheng*[1,2,*]

[1]College of Chemistry and Chemical Engineering, Xiamen University, Xiamen 361005, China
[2]Department of Chemistry, University of Aberdeen, Aberdeen AB24 3UE, UK
*chengjun@xmu.edu.cn; jcheng@abdn.ac.uk

Introduction

Pyridinyl radical has been believed to play an important role in pyridinium catalyzed reduction of CO_2[1]. However, according to the redox potential of pyridinium calculated theoretically[2-4], the step of pyridinium reduction in solution is not thermodynamically feasible.

Previous theoretical work used semi-empirical implicit solvtion model. Here, we report the thermochemistry of proton coupled electron transfer (PCET) of pyridinium reduction by using a density functional theory based molecular dynamics (DFTMD) method[5-7] that treats the solute and solvent at the same level of electronic structure theory.

It was shown that p-GaP is a good photocatalyst for CO_2 reduction with a high selectivity towards methanol. [8,9]. We've also studied the band alignment of GaP at the aqueous interface that is crucial to understand the photocatalytic activity of the material.

Method

The method we use combine DFTMD simulations and free energy perturbation theory (FEP)[5-7]. The free energy of reversible insertion of a proton or electron in a periodic DFTMD model system is computed by integrating vertical energy gaps of the reactant and product states. The pK_a and redox potential (U^o) vs. SHE can be obtained by using the following equations[7].

$$2.3k_BTpK_a = \Delta_{dp}A_{XH} - \Delta_{dp}A_{H_3O^+} + k_BT\ln\left[c^0\Lambda^3_{H^+}\right] \quad (1)$$

$$e_0U^\circ = \Delta_{ox}A_{X^-} + \Delta_{dp}A_{H_3O^+} - \mu^{g,o}_{H^+} - \Delta E_{zp} \quad (2)$$

$$\Delta_{dh}G^\circ = \Delta_{dh}A_{XH} - \mu^{g,o}_{H^+} - \Delta E_{zp} + k_BT\ln[c^o\Lambda^3_{H^+}] \quad (3)$$

where $\Delta_{dp}A_{XH}$ is the deprotonation integral of the XH, $\Delta_{dp}A_{H_3O^+}$ is the deprotonation integral of H_3O^+, $\Delta_{ox}A_{X^-}$ denotes the oxidation integral of X^-, $\Delta_{dh}A_{XH}$ denotes the dehydrogenation integral of XH, $\mu^{g,o}_{H^+}$ is the standard chemical potential of H^+, ΔE_{zp} is the zero point energy of O(X)-H bond, c^o is the standard concentration of 1 mol/L, Λ_{H^+} is the thermal wavelength of proton. For $\mu^{g,o}_{H^+}$ we use the experimental value of 15.81eV. The two correction terms, ΔE_{zp} and $k_BT\ln\left[c^o\Lambda^3_{H^+}\right]$, are 0.35 and -0.19eV respectively.

The DFTMD calculations were run using the freely available program package CP2K/QUICKSTEP[10-11]. Analytic Goedecker-Teter-Hutter pseudopotentials were used. The density cutoff in the reciprocal space is set as 400 Ry. TZV2P basis sets were used for homogeneous solution, while for the interface model DZVP basis sets were employed. The NVT condition was controlled by a Nosé-Hoover thermostat at 330 K.

For comparison, the redox potentials of pyridinium and CO_2 were also computed using implicit solvation model based on the CPCM approach. molecular cavities defined by Hartree-Fock (UAHF) radii, and DFT-B3LYP/aug-cc-pVDZ electron densities were employed.

Results and Discussion

Acidity Constant and Redox Potential of Pyridinium in Solution. The thermodynamic integrals of Eqs.1, 2 and 3 are calculated by time accumulative averages of vertical energy gap for a minimum number of values of the coupling parameter η. Each η denotes a mixing state of reactant and product state. As can be seen from Figure1, 4 sampling states are used for estimating the oxidation energy. In the case of deprotonaton and dehydrogenation energy, more mixing states are used. The number of values of η associated with the extent of deviation from the liner response of solvent. The vertical energy gaps are converged within 5-10 ps.

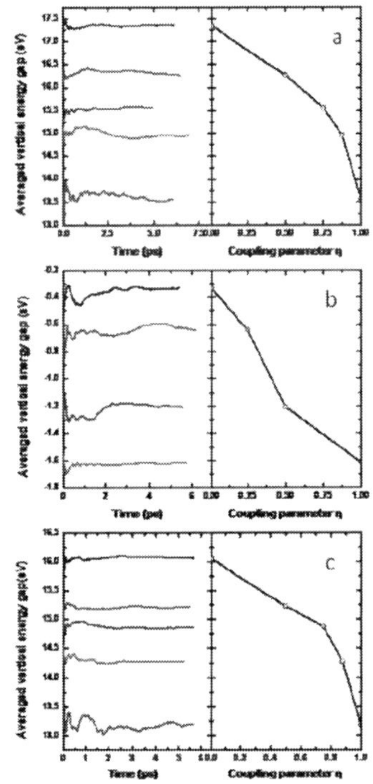

Figure 1. Time accumulative averages of vertical energy gaps against coupling parameter η (<ΔE>η) for a) deprotonation of the pyridinium ion, b) oxidation of a pyridinyl radical, c) dehydrogenation of a pyridinyl radical.

By integrating the vertical energy gaps, the values of $\Delta_{dp}A_{PyH^+}$, $\Delta_{ox}A_{PyH^\bullet}$ and $\Delta_{dh}A_{PyH^\bullet}$ are obtained and then substituted into the equations 1,2 and 3 calculate the pK_a, redox potential and dehydrogenation free energy. The results are shown in figure 2.

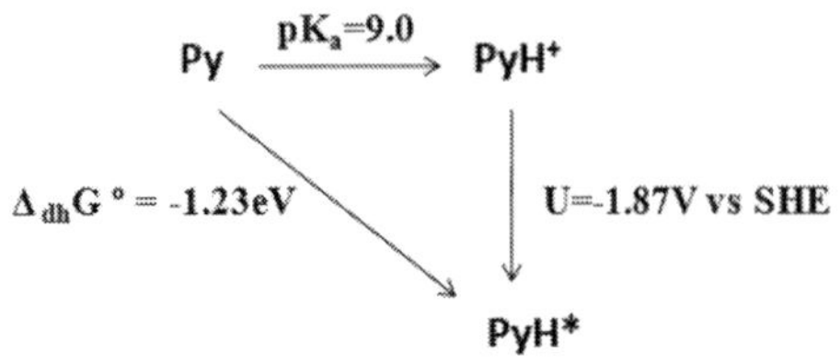

Figure 2. Thermochemistry of proton coupled electron transfer of pyridinium hydrogenation.

According to Hess's law, the dehydrogenation energy of pyridinyl radical equals to the sum of deprotonation energy of pyridinium and oxidation energy of pyridinyl radical. Our calculation shows that the Hess's law is satisfied, and the statistical uncertainty is 0.11 eV. This gives a strong test for the statistical accuracy in the method we used.

Redox potential we obtained is nearly 1.5 V more negative than the experimentally observed onset potential (-0.34 V vs. SHE). This is consistent with the previous theoretical results using implicit solvation model, and our calculation support the previous argument that the pyridinyl radical cannot be the active species that reduces CO_2 in the homogeneous environment.

On the other hand, our computed potential is about 0.7V more negative than the previous values. Possible reasons are discussed as follows: (i) The implicit solvation models don't account for the band structure of water, which was found recently to be important for the calculation of redox potentials due to the hybridization between the solute level and the electronic band of water. We has also calculated the redox potential of pyridinium by using the CPCM solvent model with a different number of explicit water molecules added (see Figure 3a). There is a tendency of the computed potentials with increased number of explicit water approaching to our results, as shown in Figure 3. (ii) Our results are calculated using GGA-BLYP functional. Although the delocalization error in GGA functionals may potentially cause an error in our calculation, we found it has minimal effect on CO_2 redox potential due to the accurate positioning of water CBM by this functional, and would expect it should apply to pyridinium reduction. To be certain, we are currently calculating the redox potential using a hybrid functional.

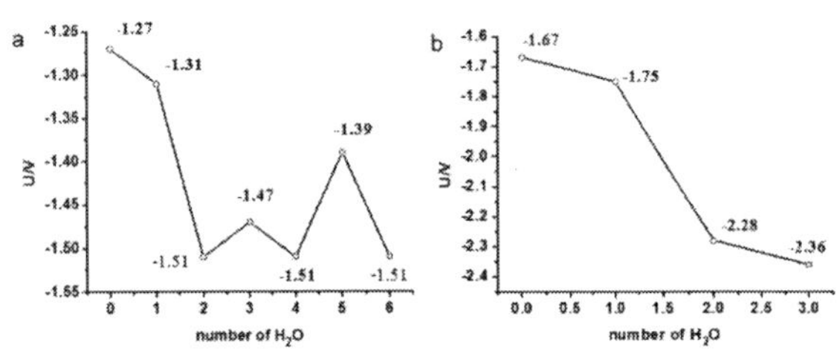

Figure 3. The computed redox potentials of a) pyridinium and b) CO_2 using CPCM solvent model with a different number of explicit water included.

Role of GaP Electrode in the Pyridinium Catalyzed Reduction of CO_2. The band positions of GaP(110) was calculated. To see the effect of solvent on electronic structure of GaP(110) surface, we calculated the band position of monolayer water molecular adsorbed GaP as well. As can be seen from Figure 4, the great changes of CBM due to adsorption of water indicate that solvent will affect the band positions of GaP(110) surface. The calculation of band positions of a full GaP(110)/H_2O interface is underway.

Simply comparing the CBM of GaP with redox potential of pyridinium in solution, it may be concluded that pyridinium reduction could not occur by transfer of a photoexcited electron from the GaP photocathode.

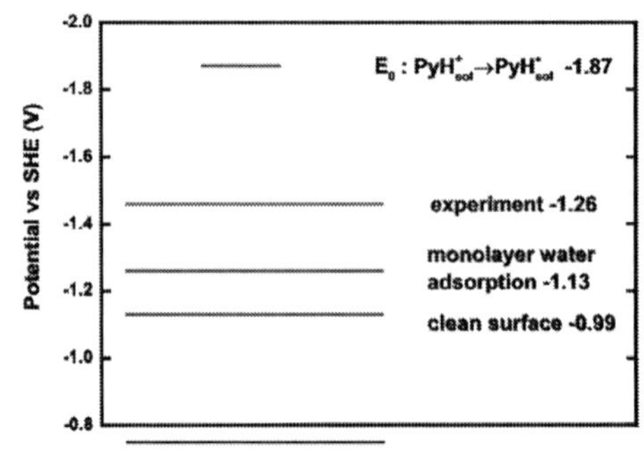

Figure 4. Comparison of computed energy levels relative to SHE.

Conclusions

The redox potential of pyridinium we calculated (-1.87V vs SHE) using a DFTMD method gives support to the claim from the previous theoretical work that pyridinyl radical is unlikely to be the active species in reducing CO_2 in homogeneous solution. We are running the calculation using a hybrid functional to understand the observed discrepancy between our value and the literature numbers obtained by implicit solvation model.

The role of GaP photocathodes in pyridinium-catalyzed CO_2 reduction is also studied. The calculation shows that the interaction with water affect the band positions of the GaP(110) surface significantly. We are currently calculating the energy levels of GaP(110)/H_2O interface. Based on the preliminary results of 1 ML water adsorption model, we find that reduction of pyridinium could not occur by transfer of photoexcited electrons from the GaP photocathode.

Acknowledgement. This work was supported by Natural Science Foundation of China.

References

(1) Barton Cole, E.; Lakkaraju, P. S.; Rampulla, D. M.; Morris, A. J.; Abelev, E.; Bocarsly, A. B. *J. Am. Chem. Soc.* **2010**, *132*, 11539-11551.
(2) Tossell, J. A. *Comp. Theor. Chem.* **2011**, *977*, 123-127.McCullen, J. B.; Higgins, J. B.; Schlenker, J. L. *J. Am. Chem. Soc.* **1992**, *114*, 10834.
(3) Lim, C. H.; Holder, A. M.; Musgrave, C. B. *J. Am. Chem. Soc.* **2013**, *135*, 142-154.
(4) Keith, J. A.; Carter, E. A. *J. Am. Chem. Soc.* **2012**, *134*, 7580-7583.
(5) Cheng, J.; Sulpizi, M.; Sprik, M. *J. Chem. Phys.* **2009**, *131*, 154504/154501-154504/154520.
(6) Cheng, J.; Sprik, M. *Phys. Chem. Chem. Phys.* **2012**, *14*, 11245-11267
(7) Cheng, J.; Liu, X.; VandeVondele, J.; Sulpizi, M.; Sprik, M. *Acc Chem Res.* **2014**, *47*, 3522-3529.
(8) Barton, E. E.; Rampulla, D. M.; Bocarsly, A. B. *J. Am. Chem. Soc.* **2008**, *130*, 6342-6344.
(9) Keith, J. A.; Muñoz-García, A. B.; Lessio, M.; Carter, E. A. *Top. Catal.* **2015**, *58*, 46-56.
(10) VandeVondele, J.; Krack, M.; Mohamed, F.; Parrinello, M.; Chassaing, T.; Hutter, *J. Phys. Commun.* **2005**, 167, 103−128.
(11) The CP2K developers group, http://www.cp2k.org, **2008**.

CFD SIMULATION OF COMPRESSIBLE FLOW IN TRANSFER LINES

*Ya Qin, Xingang Li**

1. Research Institute of Petroleum Processing, SINOPEC, Beijing 100083, China;
2. School of Chemical Engineering and Technology, Tianjin University, Tianjin 300072, China
qinya.ripp@sinopec.com

Introduction

The internal compressible flow in pipelines is commonly encountered in chemical processes[1-3], such as the fluid in transfer lines. Transfer lines are used for petroleum transportations in petrochemical industry, which refers to a pipeline connection between the furnace and the vacuum column in the crude oil distillation unit. A well-designed transfer line can increase the evaporation rate of the feed in the vacuum distillation tower, improve oil product quality and save energy for vacuum distillation unit. To better design new pipelines or optimize the existing ones, it is necessary to describe the hydrodynamic behaviors of the multi-component fluids flow in the pipelines. However, due to the complexity of the system, the model for analysis of the flow inside the transfer line has been paid less attention. In this study, we will focus on developing a mathematical model to get a better understanding of the complex internal compressible gas flow inside the transfer line using CFD models.

Modeling method

A commercial computational fluid dynamics (CFD) package was used to develop a three-dimensional, fully turbulent flow model for the simulations of the multicomponent mixture across a pipeline. In this work, the internal compressible flow fields of the transfer line from a refinery in SINOPEC were simulated by the proposed CFD model.

Results and discussion

The 3D flow fields of pressure, velocity and Ma number distributions in the transfer line are obtained by the model. Figure 2 illustrates the simulated pressure profiles for the compressible flow across the transfer line, at two characteristic sections which are the horizontal and vertical center planes, when gas feed superficial velocity is 69 $m \cdot s^{-1}$. It is found that the pressure drop mainly occurred in the bends and the junction site, while straight pipe with large diameter does not cause noticeable losses. The acceleration pressure drop in bends is due to the increase of velocity, which is induced by significant gas density variations in the pipes. At the junction site, sudden expansion losses and turning losses may be the main reasons for pressure drop [4].

It is worth mentioning that the maximum velocity can reach critical velocity. The critical condition means a singular discontinuity where the finite pressure and/or velocity change corresponds to an infinitesimal change in space coordinate [5], which should be avoid in practice because that it may lead to large pressure drop and other operation problems.

Conclusions

A CFD model for compressible flow inside pipes carrying hydrocarbon mixtures is presented. As an example, a transfer line in petrochemical field is presented. The 3D flow fields of pressure, velocity and Ma number distributions in the transfer line are obtained by the model. It is found that the pressure drop mainly occurred in the transition section and the junction site, while low-speed section leads to relatively lower pressure drop.

The results indicate that CFD simulations can be used to understand the nature of the compressible flow field and to aid the design of pipelines in petrochemical industry.

References

(1) Diego, A. A.; Timothy, A. Shedd. *J. Fluids Eng.* **2007**,129(9) ,1193.
(2) Perez-Garcia J.; Sanmiguel-Rojas E.; Hernandez-Grau J.; Viedma A. *Expe.Therm. Fluid Sci.* **2006**,31(2):61.
(3) Souza B.A.; Matos E.M.; Guirardello R.; Nunhez J.R. *J Petrol. Sci. Eng.* **2006**, 51,138.
(4) Lightstone L.; Osamusali S.I.; Chang J. *AIChE J.* **1991**, 37, 111-122.
(5) Konoutek J., Zachoval J., Odstrcil M. and Stehlik P. *Heat Transfer. Eng.* **2001**,22 , 32-40.

CFD SIMULATIONS OF IMPREGNATING AND DRYING EQUIPMENT FOR S-ZORB CATALYSTS

Zhenxing Zhu, Lufeng Lv, Zhihong Tian, Shuandi Hou

No.18, Xueyuan Road, Haidian District, Beijing, 100083, China
Zhenxing Zhu: Zhuzx.ripp@sinopec.com

Introduction

S-Zorb technology is a highly effective approach for desulphurization of FCC gasoline. It was developed by ConocoPhillips, bought out and improved by SINOPEC. Through S-Zorb, the concentration of sulfur of gasoline had been decreased within 10mg•kg^{-1} in many refining industries of SINOPEC. Impregnating and drying process are the key steps in preparation of S-Zorb catalysts, which is account for the final strength and sphericity of catalysts. Furthermore the flowing behavior in impregnating and drying equipment is quite important for the quality of S-Zorb catalysts.

CFD simulations of impregnating and drying equipment

An equipment coupled flash drying with continuous impregnating process was introduced in this study. The diameter of the equipment is 800mm, the height is 3.5m. There is a impeller with three layers of blades in the bottom of the equipment, and the rotating speed is 150rpm. The hot gas was pumped in the equipment from the bottom and the impregnated catalysts were transferred into the equipment in the middle of the equipment. Then the flowing behavior of both the gas and the catalysts were analysis by CFD simulations with Euler-Euler multiphase model.

Optimization of the equipment. The depth of catalyst inlet into the equipment was an important parameter, which had a profound effect on the flowing uniformity of the catalysts inside the equipment.

Results and Discussion

The velocity and volume fraction distributions of both catalysts and gas were obtained in the equipment with deferent depth of catalyst inlet, 20mm, 80mm and 140mm. With the increase of the depth of catalyst inlet, the uniformity of distribution of catalysts became better and better, which is good for the final strength and sphericity of catalysts.

Conclusions

With the increase of the depth of catalyst inlet, the uniformity of distribution of catalysts became better and better, which is good for the final strength and sphericity of catalysts.

References

References should use ACS style. References may be inserted and numbered manually, or automatically by using the endnote/footnote function of the word processor. For manual insertion, format the references as follows:

(1) Bohn M S, Siebarth J E. Dynamic Settler. WO Pat, WO 02/097007 A2. 2002.
(2) BP Exploration Operating Company Limited, Davy Process Technology Limited. Filtration Process for Separating Liquid Hydrocarbons from a Particulate Fischer-Tropsch Catalyst. WO Pat Appl, WO 2002-GB2337. 2002.
(3) Concophillips Company. Solid/liquid Separation System for Multiphase Converters. WO Pat Appl, WO 2003089102 A2. 2003.
(4) University of Kentucky Research Foundation. Bubble Column Apparatus for Separating Wax from Catalyst Slurry. US Pat, US 6762209 B1. 2004..

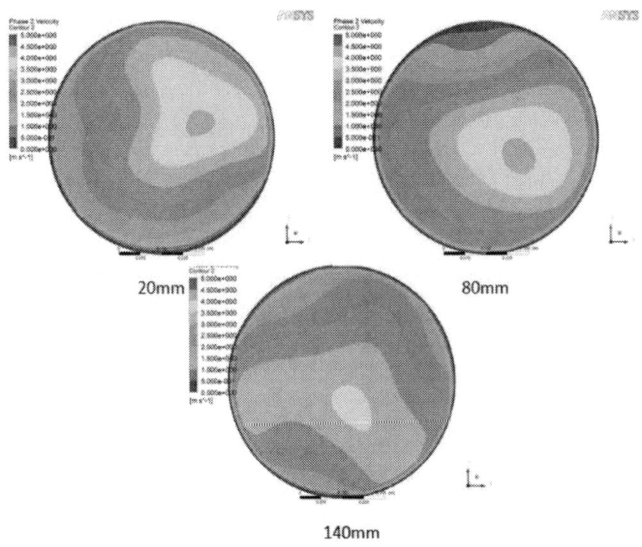

Figure 1. The radial velocity distributions of catalysts at the cross section at 2.6m

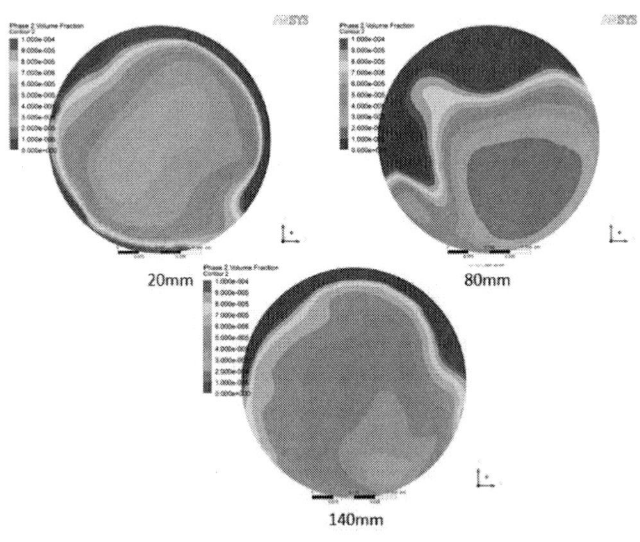

Figure 2. The radial volume fraction distributions of catalysts at the cross section at 2.6m

(5) Sasol Technology Proprietary Limited. Process for Separating a Catalyst from a Liquid. WO Pat Appl, WO2004-IB2243. 2004.
(6) Espinoza R L. Catalyst-liquid Separation in a Slurry Reactor. EP Pat Appl, 03255751.4. 2003.
(7) Gidaspow D. Multiphase Flow and Fluidization. Boston: Academic Press, 1994: 56~60.

MECHANISTIC INVESTIGATION ON THE La_2O_3-CATALYZED OXIDATIVE COUPLING OF METHANE REACTION

Shenggang Li, Changqing Chu, Shibin Wang, Linna Cong, Yonghui Zhao, Yuhan Sun*

CAS Key Laboratory of Low-Carbon Science and Engineering, Shanghai Advanced Research Institute, Chinese Academy of Sciences, 100 Haike Road, Shanghai, China 201210
*Email: lisg@sari.ac.cn

Introduction

Oxidative coupling of methane (OCM) is a promising route for the efficient conversion of methane into ethylene, and La_2O_3-based catalysts have been shown to have good catalytic activity, selectivity and stability. Despite the numerous experimental and computational studies on this reaction over the past three decades, there remain debates on the nature of the active sites and the detailed catalytic mechanism for this reaction. We have used both cluster and surface models to study the activation and conversion of methane on the La_2O_3 catalysts. In this contribution, we examined the reactivity of the different oxygen species with methane,[1,2,3] and built a comprehensive reaction network for the OCM reaction. In addition, we also investigated the effects of the acid-base properties on the catalytic activity. Reactivities of the different La_2O_3 surfaces were also studied.

Computational

With the cluster model, we used the stoichiometric La_2O_3 clusters as the catalyst models, from the small La_4O_6 cluster to the large $La_{32}O_{48}$ cluster resembling the (001) surface. With the surface model, we used the periodic slabs built from the bulk structure of La_2O_3.

Cluster model calculations. Geometries were optimized with the PBE and B3LYP density functional theory (DFT) methods with double-zeta quality basis sets, and energies were calculated with up to the CCSD(T) level with triple-zeta quality basis sets. Reaction energy as well as energy barrier were calculated for each elementary steps to build the reaction network. Acid-base properties of the La-O pair site were calculated as the fluoride and proton affinities as well as CO_2 chemisorption energies.

Surface model calculations. Geometries were optimized with the PBE method with PAW pseudopotential plane wave basis sets. Physisorption and chemisorption energies as well as energy barriers were calculated for methane activation on different La_2O_3 surfaces.

Results and Discussion

Using the La_4O_6 cluster as the catalyst model, we examined the reactivity of the different oxygen species with methane. As shown in Figure 1, the oxygen radical site ($O^{\bullet-}$) was predicted to be the most reactive with energy barrier of <1 kcal/mol. The lattice oxygen site (O^{2-}) in the $La^{3+}–O^{2-}$ acid-base pair has significant reactivity with energy barrier of ~24 kcal/mol. The superoxo radical site ($O_2^{\bullet-}$) is also fairly reactive with energy barriers from 28 to 36 kcal/mol depending on its structure. The peroxide site (O_2^{2-}) is the least reactive with energy barrier of ~46 kcal/mol leading to the one-step formation of CH_3OH, presumably a by-product of the OCM reaction.

The reaction network in Figure 1 constructed from the complete catalytic cycles predicted from the first principles calculations suggests the most favorable pathway for the formation of methyl radical ($CH_3^{\bullet}$) is not that via methane activation at the highly reactive $O^{\bullet-}$ site. Instead, the $CH_3^{\bullet}$ radical can be formed by first methane activation at the $La^{3+}–O^{2-}$ acid-base pair site, followed by O_2 activation of the $La^{3+}–CH_3^-$ intermediate. The second $CH_3^{\bullet}$ radical can be formed by methane activation at the $O_2^{\bullet-}$ site. At high temperatures, the different oxygen species are all likely to be involved in this catalytic reaction to some extent.

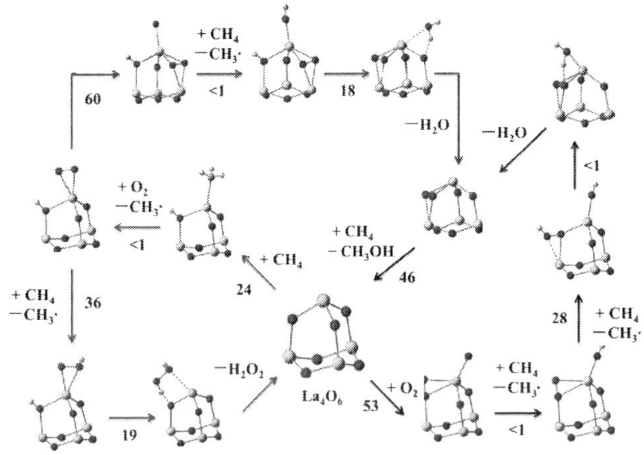

Figure 1. Reaction network of the OCM reaction predicted with the B3LYP method using the La_4O_6 cluster as the catalyst model. Energy barriers were shown in kcal/mol. The pathway shown in red is kinetically the most favorable for the formation of the methyl radical.

In addition to the above study on the OCM reaction network, we also studied the effects of the acid-base properties of the $La^{3+}–O^{2-}$ acid-base pair site on its energy barrier for methane activation. Linear relationships were predicted to exist between the acid-base properties and the energy barrier, and stronger CO_2 chemisorption leads to lower energy barrier for methane activation.

We also studied the reactivities of the different La_2O_3 surfaces with methane. From the calculated energy barriers, we were able to show that the most stable (001) surface is not surprisingly also the least reactive surface, whereas the (011), (100), and (110) surfaces have similar reactivity.

Conclusions

We have used cluster and surface models to investigate the mechanism of the La_2O_3-catalyzed OCM reaction, and built a comprehensive reaction network based on first principles calculations. The acid-base properties of the La-O acid-base pair site were found to directly influence its reactivity with methane. The reactivities of the different La_2O_3 surfaces with methane were also predicted.

Acknowledgement. This work is supported by the Hundred Talents program of the Chinese Academy of Sciences (Y224591401), the National Science Foundation of China (21473233), the Frontier Science program of Shell Global Solutions International B.V. (PT32281).

References

(1) Lei, Y.; Chu, C.; Li, S.; Sun, Y. *J. Phys. Chem. C* **2014**, *118*, 7932.
(2) Chu, C.; Zhao, Y.; Li, S.; Sun, Y. *J. Phys. Chem. C* **2014**, *118*, 27954.
(3) Chu, C.; Zhao, Y.; Li, S.; Sun, Y. *J. Phys. Chem. C* **2016**, *120*, 2737.

BIO-OIL SEPARATION AND STABILIZATION BY NEAR-CRITICAL PROPANE FRACTIONATION

Daniel M. Ginosar, Lucia M. Petkovic, Foster A. Agblevor

Idaho National Laboaratory
Idaho Falls, ID 83415
daniel.ginosar@inl.gov

Introduction

Bio-oils produced by thermal process, such as fast pyrolysis, catalytic fast pyrolysis (CFP), and hydrothermolysis are promising sources of sustainable, low greenhouse gas alternative fuels. These thermal processes are also well suited to decentralized energy production due to low capital and operating costs.

Algae feedstocks for bio-oil production are of particular interest, due in part to their high-energy growth yields. Further, algae can be grown in non-arable areas in fresh, brackish, salt water, or even waste water. These feedstocks have faster growth rates and productivities than any agricultural feedstock and require much less land than terrestrial crops [1,2].

Unfortunately, bio-oils produced by thermal processes present significant stability challenges [3,4]. These oils have complex chemical compositions, are viscous, reactive, and thermally unstable [1]. Further, the components within the oils are difficult to separate by fractional distillation. By far, the most effective separation and stabilization method has been solvent extraction. However, liquid phase extraction processes pose two main obstacles to commercialization; they require a significant amount of energy to remove and recover the solvent from the product, and they have a propensity for the solvent to become contaminated with minerals from the char and ash present in the original bio-oil.

Separation and fractionation of thermally produced bio-oils using supercritical fluids (SCF) offers the advantages of liquid solvent extraction while drastically reducing energy demands and the predisposition to carry over solids into the extracted phase. SCFs are dense fluids with liquid-like solvent properties and gas-like transport properties. Further, SCF density and solvent strength can be tuned with minor adjustments in pressure, co-solvent addition, or gas anti-solvent addition.

Experimental

Catalytic pyrolysis oils were produced from *Scenedesmus dimorphus* algae using a fluid catalytic cracking catalyst. The catalytic pyrolysis studies were conducted in the 4 kg/h pilot scale fluidized bed pyrolysis reactor. The catalyst used was proprietary material supplied by BASF Inc. About 1,000 g of catalyst was used per run and the dried *S. dimorphus* algae biomass ground to pass 2-mm mesh was used for the studies. Catalytic pyrolysis was carried out at 400 °C. FCC catalyst was used as both heat transfer medium and catalyst.

Bio-oil produced from catalytic fast pyrolysis (CFP) was separated using critical fluids. The system consisted of two high-pressure syringe pumps, a temperature controlled stainless steel 32 mL high-pressure extraction vessel, an automated micro-metering valve, a high-pressure sight glass collection vessel, and a back pressure regulator. The system was protected from over-pressurization with two spring loaded relief valves and back flow was prevented with one-way check valves. Near-critical propane extraction was performed at 65 °C at a fluid reduced pressure of 2.0 (85 bar) using an eight to one solvent to feed ratio by weight.

Results and Discussion

About 83 % of the bio-oil was extracted using propane, leaving about 17 % as raffinate. Propane extracts were different from the parent raw oil. For example, elemental analyses showed that the content of nitrogen in the fraction extracted with propane was about half of the original nitrogen present in the parent raw oil. The propane extract had a lower viscosity than the parent.

Infrared analyses (Figure 1) showed that aliphatic chains (peaks within the 2600-3100 cm^{-1}) and low-polarity carbonyl-containing species, such as ketones and/or esters (1700 cm^{-1}) were preferentially extracted with propane while hydrogen bonded OH species (broad band around 3000 cm^{-1}) were preferentially left in the raffinate. In addition, higher-polarity carbonyl-containing species such as acids, amides, and alcohols, were also left in the raffinate.

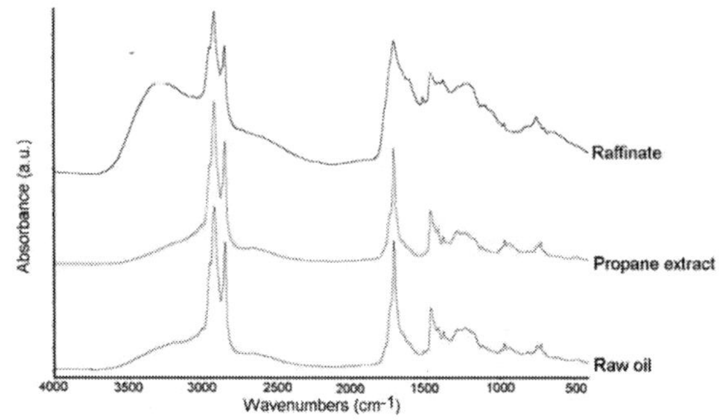

Figure 1. Extraction using propane at 65 °C.

Infrared findings were confirmed by NMR analyses. ^{13}C NMR analyses are shown in Figure 2. Carbonyl and C–O contributions are higher in the raffinate. C–C–C contribution is higher in the extract. ^{1}H NMR analyses are shown in Figure 3. Exchangeable OHs are reduced in the propane extracted fraction. These exchangeable OH include water, carboxylic acids, and many alcohols.

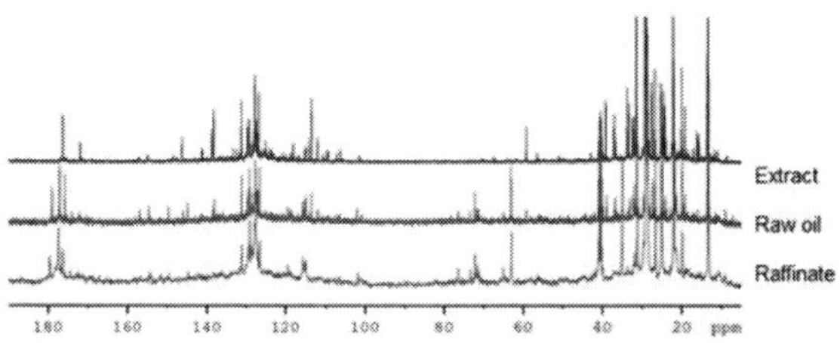

Figure 2. ^{13}C NMR analyses. Extraction using propane at 65 °C.

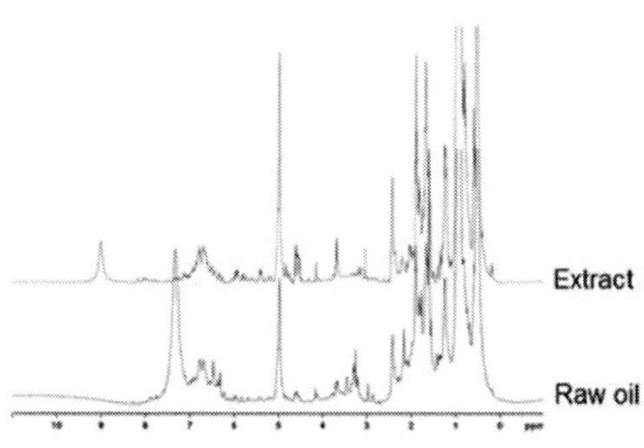

Figure 3. ^{1}H NMR analyses. Extraction using propane at 65 °C.

MALDI-TOF analyses showed that the propane extract had a lower average molecular weight than the parent oil. The raffinate had a higher average molecular weight. The propane extract was also more volatile as measured by temperature programmed volatilization under flowing helium. About 80 % of the raw oil and propane extracts volatilized below 250 °C. The last 20 % of the original oil mass required higher temperatures for the parent oil than for the extract. The raffinate seemed to be much more active when heated and sudden bursts of volatiles did not allow for a steady profile even at the low heating rate of 1 K/min that was applied. The raffinate required temperatures above 500 °C for the last 20 % to volatilize.

Aging of parent oil, propane extract, and propane raffinate. Samples of the parent oil, propane extract, and propane raffinate were submitted to accelerated aging which consisted in keeping bio-oil aliquots in closed vials for 24 hours and 2 weeks at 80 °C. Next, the aged samples were analyzed to determine the effects of aging on sample physicochemical properties.

The results of viscosity measurements at different temperatures showed that the propane extract was more stable than the raw oil. For example, the viscosity at 30 °C of the parent oil, which was 8.07 cP when fresh, became 68.1 cP after 2 weeks. In contrast, the propane extract, which had a viscosity of 6.76 cP when fresh, showed a viscosity of 29.6 cP after 2 weeks. In other words, the parent oil viscosity increased about 8 times with aging and the propane extract viscosity increased only 4 times. The amount of raffinate collected was insufficient for viscosity analyses.

To understand the processes that may be affecting the stability of these oils, additional analyses were performed.

Infrared analyses performed on fresh and aged samples did not show significant differences. This is an indication that the species that produced such important changes in viscosity may be present in low quantities. The molecular mass distribution shifted to higher molecular masses with aging in all samples. In general, volatility also decreased with aging.

Parent oil and propane extract were submitted to an in-situ NMR aging study for 2 weeks at 80 °C in the NMR instrument. The results are shown in Figures 4-6. The shift of the peaks was more significant in the parent oil than in the propane extract. For example, the peak at 176-177 ppm shifted -0.45 ppm in the parent oil and -0.15 in the propane extract. This shows that the propane extract was chemically more stable than the parent oil. In addition, the parent oil started with more exchangeable protons which declined more rapidly with aging than the propane fraction. The exchangeable protons in the propane fraction displayed a downfield shift compared to the parent oil suggesting a larger fraction of acid in the exchangeable protons. A general shift of the spectrum upfield with aging suggested the consumption of acid over time in both samples. This suggests that aging was likely due to polymerization of esters with alcohols in presence of acid and base that are present within the oil.

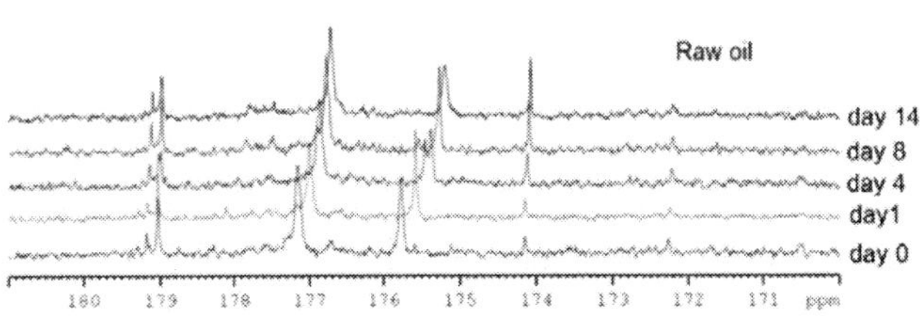

Figure 4. In situ ^{13}C NMR aging analyses of parent oil.

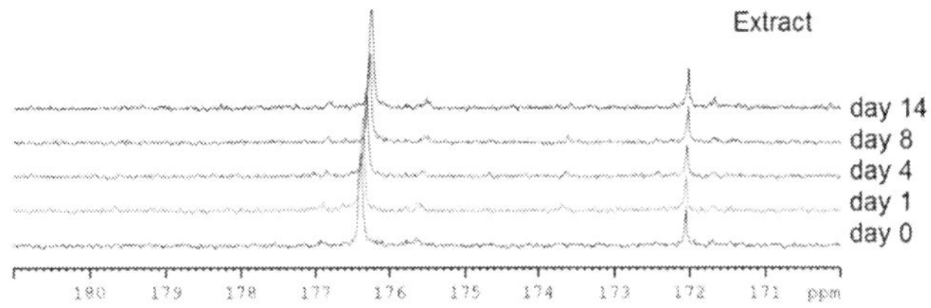

Figure 5. In situ ^{13}C NMR aging analyses of propane extract.

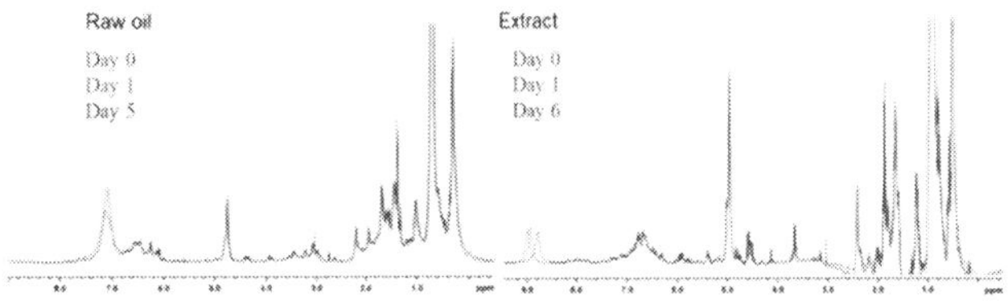

Figure 6. In situ ^{1}H NMR aging analyses of parent oil (left) and propane extract (right).

Conclusions

Extraction of catalytic fast pyrolysis oil with near critical propane produced an oil extract that was physically and chemically different from and more stable than the original oil. The propane extract displayed lower viscosity and lower average molecular weight. The species present in the propane extract were likely the less polar that would be expected from using a non-polar solvent (propane). Carbonyl containing species in the extract were likely ketones and esters. The raffinate contained a higher amount of OH bonded species along with the more polar more polar acids, amides, and alcohols. The higher concentration of nitrogen in the raffinate may confirm the presence of amides.

Viscosity of the propane extract increased only half as much as that of the CFP bio-oil. Further, In situ NMR aging studies showed that the propane extract was more stable than the raw oil. In conclusion, propane extraction is a promising method to decrease the nitrogen content of bio-oils and to improve the stability of bio-oils obtained by the catalytic pyrolysis of algae based biomass.

Acknowledgement. This work was funded by U.S. Department of Energy, Energy Efficiency and Renewable Energy, Bioenergy Technology Office. The work was performed through Battelle Energy Alliance under contract number DE-AC07-05ID14517 with U.S. Department of Energy.

References

(1) A.F. Clarens, E.P. Resurreccion, M.A. White, L.M. Colosi, Environmental Life Cycle Comparison of Algae to Other Bioenergy Feedstocks. Environmental Science & Technology 44(5) (2010) 1813-1819.

(2) A. Campanella, R. Muncrief, M.P. Harold, D.C. Griffith, N.M. Whitton, R.S. Weber, Thermolysis of microalgae and duckweed in a CO2-swept fixed-bed reactor: Bio-oil yield and compositional effects. Bioresource Technology 109 (2012) 154-162.

(3) Q. Lu, W.-Z. Li, X.-F. Zhu, Overview of fuel properties of biomass fast pyrolysis oils. Energy Conversion and Management 50(5) (2009) 1376-1383.

(4) M. Balat, An Overview of the Properties and Applications of Biomass Pyrolysis Oils. Energy Sources Part a-Recovery Utilization and Environmental Effects 33(7) (2011) 674-689.

MOLYBDENUM CARBIDE CATALYZES THE CONVERSION OF BIOMASS PYROLYSIS VAPORS TO PARAFFINIC AND AROMATIC COMPOUNDS

Erick A. White, Calvin Mukarakate, Michael Griffin, Connor Nash, Matthew M. Yung, Mark R. Nimlos, Daniel A. Ruddy, Joshua A. Schaidle

National Bioenergy Center, National Renewable Energy Laboratory, Golden, Colorado 80401, United States
Erick.White@nrel.gov

Introduction

The transportation sector consumes more than a quarter of the energy produced worldwide. While several renewable alternatives, such as wind and solar generators, are displacing conventional fossil fuel-based generators for stationary energy applications, there are currently few commercially viable alternatives for transportation applications that require high gravimetric energy density (e.g., aviation). Petroleum-derived fuels are especially well suited for transportation applications because they are abundant, energy-dense liquids that are safely stored and transported at ambient conditions. While several technologies with the potential to displace liquid petroleum-based fuels are on the horizon (e.g., hydrogen fuel cells and batteries), they require large investments in infrastructure to become ubiquitous and significant technological advances to displace liquid fuels for high energy density applications. In the near term, the efficient conversion of inedible, lignocellulosic biomass feedstocks to energy dense liquid fuels that are compatible with the current infrastructure is an important pathway to increasing the renewable energy portfolio for the transportation sector.

The pyrolysis of lignocellulosic biomass is a potential pathway for converting biomass into transportation fuels which involves heating processed biomass to ca. 500 °C for about 2 seconds in the absence of reactive gasses. Biomass pyrolysis produces a carbonaceous solid phase (i.e., char) and a vapor phase comprising monomers and other decomposition products of cellulose, hemicellulose and lignin. The vapor phase, referred to here as pyrolysis vapors, is a complex mixture of 300-400 fuel precursors that retains the bulk chemical composition of the original biomass ($CH_{1.9}O_{0.7}$), and, that when condensed, is acidic, reactive and immiscible with refinery feedstocks. The overall goal of this work is to investigate Mo_2C as an economical multifunctional catalyst to deoxygenate, hydrogenate and increase the carbon number of pyrolysis vapors, prior to condensation, in order to displace petroleum-derived refinery feedstocks for transportation fuels[1].

Experimental

Catalyst preparation and characterization. Mo_2C was synthesized and characterized as described in Schaidle et al.[2] by heating ammonium paramolybdate ($(NH_4)_6Mo_7O_{24}\cdot 4H_2O$, Alpha-Aesar) to 590°C in flowing gas comprising 15% CH_4 in H_2 for 2 h then cooling to room temperature. The resulting Mo_2C catalyst was passivated in flowing 1% O_2 in He at room temperature for 12 h. The Mo_2C catalysts were reduced in H_2 at 400°C *in situ* prior to experiments with pine pyrolysis vapors and model compounds. Catalysts were characterized by XRD, N_2 physisorption, H_2 chemisorption, NH_3 temperature programmed desorption and X-ray photoelectron spectroscopy to determine crystallite size, BET surface area, metal-like site density, acidic site density and surface composition, respectively[2].

Deoxygenation of whole pyrolysis vapors. The ability of Mo_2C to deoxygenate whole pyrolysis vapors was demonstrated at NREL on two systems that each had a two-stage reactor in series with a detector. In both systems, pine chips were periodically fed in pulses to the first stage for pyrolysis at 500°C and ambient pressure. A continuously flowing He/H_2 gas mixture swept the pyrolysis vapors from the first stage to the second stage containing the Mo_2C upgrading catalyst at either 350 or 400°C. In one system, the resulting vapors were swept to a molecular beam mass spectrometer (MBMS) and in the other system to a gas chromatograph/mass spectrometer (GCMS). Due to the requirements of the detectors, pine chips were fed to the MBMS-equipped reactor every two minutes while pine chips were fed to the GCMS-equipped reactor every 30 minutes.

Deoxygenation of model compounds. Model compound experiments using acetic acid and guaiacol were conducted in continuously fed, fixed bed reactors at 350 and 400 °C. The acetic acid experiments were performed at 0.14 MPa with a flowing reactive gas mixture comprising 0.4% acetic acid and 2.4% H_2 in He (i.e., H_2/acetic acid = 6). The guaiacol experiments were performed at 0.5 MPa with an H_2/guaiacol ratio of 12. For both reactions the molar ratio of H_2 to reactant was twice the stoichiometric amount required for complete deoxygenation and hydrogenation of the reactant.

Results and Discussion

The initial reaction products of whole pine pyrolysis vapors over Mo_2C, measured with the GCMS system, at 350 and 400 °C are different than those produced over the conventional deoxygenation catalyst, HZSM-5, at the typical operating temperature of 500 °C. This is illustrated in Figure 1 which shows that the average of the first three pine pulses over Mo_2C produced predominantly two-carbon through seven-carbon paraffinic compounds (abbreviated in Figure 1 as C2 and C7, respectively) while HZSM-5 initially produces aromatic compounds and C2-C3 olefins (note that carbon-carbon bond types are not explicitly shown in Figure 1). Also, the total carbon yields for whole pine pyrolysis vapors are greater over Mo_2C at 350 and 400 °C (approximately 30% and 22%, respectively) than the total carbon yield over HZSM-5 (approximately 18%, though the observed carbon yield may be lower than actual due to the limited ability to detect and quantify heavier aromatics on the GCMS system).

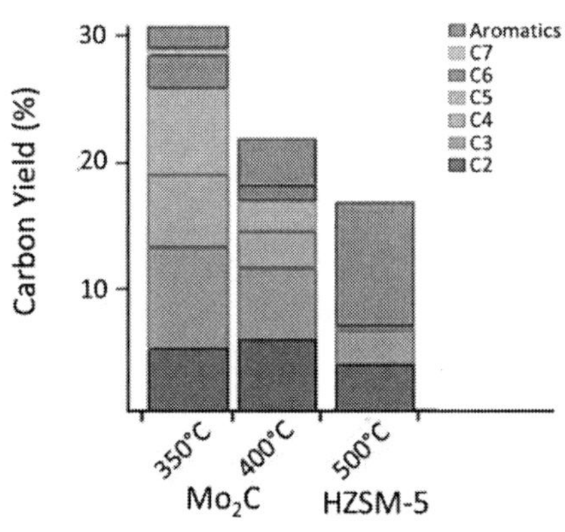

Figure 1. The initial carbon yields of pine pyrolysis vapors upgraded over Mo_2C in the GCMS system were higher and more selective towards two-carbon through seven-carbon (i.e., C2 and C7, respectively) paraffinic compounds at 350 and 400 °C than over HZSM-5 at 500 °C

Additional information about the function of Mo_2C can be deduced from the model compound studies. For example, the product distribution for reaction of acetic acid with H_2 over Mo_2C at 350 and

400 °C, illustrated in Figure 2 by the percent selectivity after 270 min. time-on-stream, shows that deoxygenation of acetic acid to acetaldehyde is favored at 350 °C while C-C bond cleavage to form methane, carbon monoxide and carbon dioxide is favored at 400 °C. Additionally, limited hydrogenation was observed at either temperature which was illustrated by the higher selectivity to ethylene than ethane.

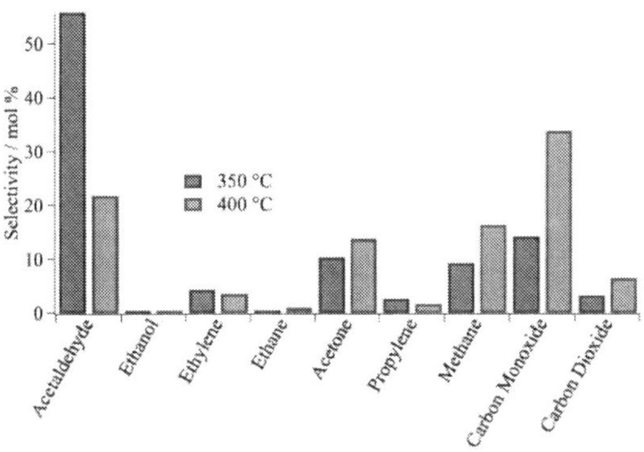

Figure 2. Product selectivity for acetic acid over Mo_2C at 350 and 400 °C indicates that C-O cleavage is favored at 350 °C while C-C cleavage is favored at 400 °C. Minimal hydrogenation was observed at both temperatures.

The product distribution for the reaction of guaiacol and H_2 over Mo_2C at 350 and 400 °C is illustrated by the plot of the organic phase selectivity shown in Figure 3. Consistent with the experiments using pine pyrolysis vapors, guaiacol was completely deoxygenated to form benzene at the lower WHSV and 350 °C. The higher WHSV yielded primarily phenol, a partially deoxygenated product, at both 350 and 400 °C. Like the results of the acetic acid experiments, limited hydrogenation was observed at all conditions. Previously published surface characterization studies and numeric modeling suggest that surface bound hydroxyl groups are responsible for the observed 8:1 ratio of acidic sites to metal-like sites[1], which explains the observed propensity towards deoxygenation reactions with minimal hydrogenation for the model compound studies.

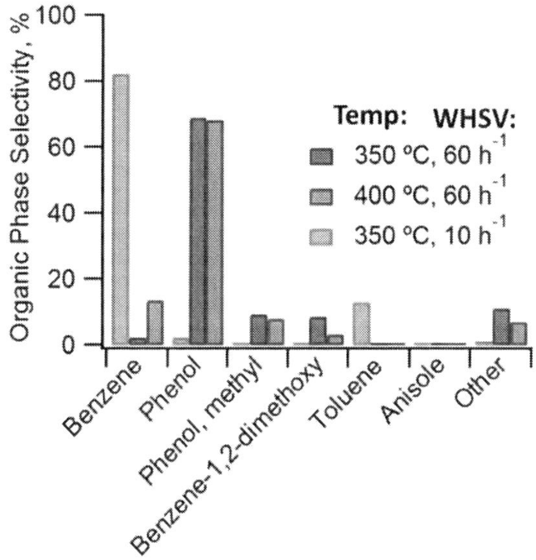

Figure 3. The organic phase selectivity for guaiacol over Mo_2C indicates that complete deoxygenation to benzene is achieved at low weight-hourly space-velocity (WHSV) and 350 °C while partial deoxygenation to phenol is favored at high WHSV for both 350 and 400 °C.

Evidence towards the mode of Mo_2C catalyst deactivation is obtained by plotting the MBMS ion signal compared to the GCMS ion signal as a function of the biomass-to-catalyst ratio in Figure 4. The aromatic signal decreased with biomass-to-catalyst ratio in the MBMS system, for which pine chips were fed every 2 minutes, while the aromatic signal is relatively constant for the GCMS system, for which pine chips were fed every 30 minutes. Because H_2 was continuously flowing between biomass pulses in both experiments, it appears that Mo_2C is at least partially re-generable in H_2.

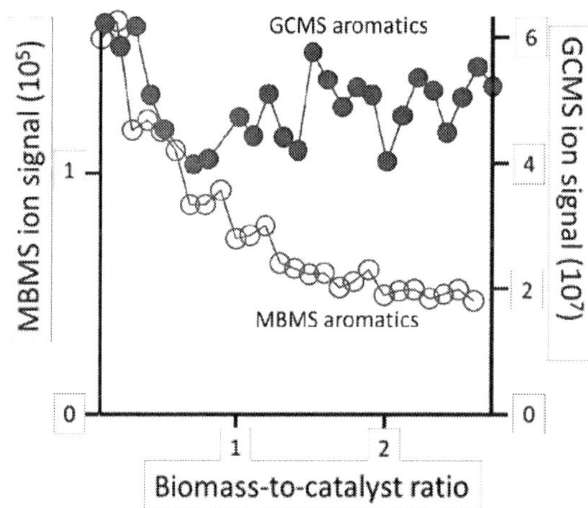

Figure 4. The aromatic signal as a function of the biomass-to-catalyst ratio for pine pyrolysis vapors over Mo_2C in the MBMS and GCMS systems.

Conclusions

This work demonstrates that Mo_2C is a promising catalyst for completely deoxygenating whole pyrolysis vapors. These results are corroborated by model compound studies using acetic acid and guaiacol as representative monomers of cellulose and lignin, respectively. Model compound studies show that Mo_2C facilitates primarily deoxygenation with minimal hydrogenation under the studied conditions. Finally, experimental results suggest that Mo_2C deactivates in the presence of whole pyrolysis vapors and is potentially regenerated by flowing H_2.

Acknowledgement. This work was supported by the Department of Energy Bioenergy Technologies Office under Contract no. DE-AC36-08-GO28308.

References

(1) Ruddy, D. A.; Schaidle, J. A.; Ferrell Iii, J. R.; Wang, J.; Moens, L.; Hensley, J. E., Recent advances in heterogeneous catalysts for bio-oil upgrading via "ex situ catalytic fast pyrolysis": catalyst development through the study of model compounds. *Green Chemistry* **2014,** *16* (2), 454-490.

(2) Schaidle, J. A.; Blackburn, J.; Farberow, C. A.; Nash, C.; Steirer, K. X.; Clark, J.; Robichaud, D. J.; Ruddy, D. A., Experimental and Computational Investigation of Acetic Acid Deoxygenation over Oxophilic Molybdenum Carbide: Surface Chemistry and Active Site Identity. *ACS Catalysis* **2016,** *6* (2), 1181-1197.

FRACTIONATION OF BIO-OILS USING SUPERCRITICAL FLUIDS

Lucia M. Petkovic, Daniel M. Ginosar, Catherine M. Hrbac, Soe Lwin

Idaho National Laboratory
P.O Box 1625, Idaho Falls, ID 83415-3632
Lucia.Petkovic@inl.gov

Introduction

Several methods can be applied to separate bio-oils into more stable fractions. These methods include adding water or aqueous salt solutions to form aqueous and non-aqueous phases, vapor phase filtration, fractional condensation, vacuum distillation, and solvent extraction. Although effective, solvent extraction poses two main obstacles to commercialization, i.e., the significant amounts of energy required to remove the solvent from the product and the propensity of the solvent to be contaminated with minerals from the char or ash.(1)

Separation and fractionation of thermally produced bio-oils using supercritical fluids (SCF) offers the advantages of liquid solvent extraction while drastically reducing energy demands and the predisposition to carry over solids into the extracted phase. SCFs are dense fluids with liquid-like solvent properties and gas-like transport properties. Further, SCF density and solvent strength can be tuned with minor adjustments in pressure, co-solvent addition, or gas anti-solvent addition.(2)

This contribution presents the results of SCF fractionation using carbon dioxide of bio-oil simulant mixtures representative of bio-oils produced by multi-stage thermal fractionation of woody biomass at 270 and 360 °C and residence times of 20 min and 5 min, respectively.

Experimental

Mixtures representatives of bio-oils obtained by multi-stage thermal fractionation were prepared using pure components in the amounts shown in Table 1.

Table 1. Composition of Bio-Oil Simulant Mixtures

Component	Stage 1 (wt%)	Stage 2 (wt%)
water	70.35	38.85
acetic acid	28.15	3.50
furfural	0.70	5.90
guaiacol	0.40	4.70
m-cresol	0.40	2.70
acetol		8.30
levoglucosan		33.25
furan		2.80

Figure 1 shows a sketch of the experimental system used. It includes a high pressure syringe pump (SP), an extractor vessel (E) electrically heated, and a sight glass (SG, Jerguson tube) for collection of extract. Ten mL of bio-oil simulant was loaded in the extractor vessel. After the desired temperature was stable, the vessel was filled with carbon dioxide, and allowed to rest for 10 min. Next a flow of carbon dioxide was started while the extract was dropped out of solution in a collection tube in the sight glass. Conditions applied ranged from near-critical to supercritical. Composition of feeds, extracts, and raffinates were determined by gas chromatography.

Simulation of the process was built on Aspen Plus V8.6 in conjunction with data from the NIST Chemistry Database. Operating conditions were set to 2100 psi 45°C for the first flash tank and atmospheric conditions for the second tank. These conditions mirror conditions used in experiments to analyze the oil's separation from the aqueous phase. The property method used was SR Polar, due to the super critical state of the oil in the first tank (~2Pc) and the mixture of polar and nonpolar components. The binary parameters of the components of the bio-oil were estimated by linking Aspen Plus to the NIST Databank. The simulation produced stream separation results and model energy usage results that were compared to experimental results.

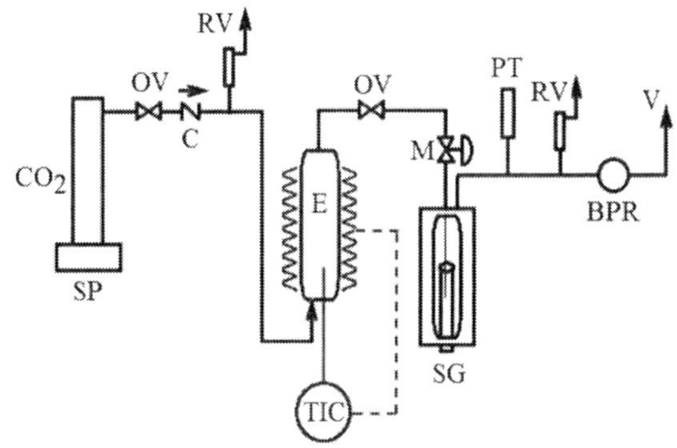

Figure 1. Sketch of experimental setup. Syringe pump (SP), on-off valve (OV), check valve (C), extractor (E) micrometering valve (M), pressure transducer (PT), relief valve (RV), back pressure regulator (BPR), sight glass (SG), temperature measurement and control (TIC), vent (V).

Results and Discussion

The effects of temperature and pressure on the amount of material extracted from Stage 1 simulant are shown in Figures 2 and 3.

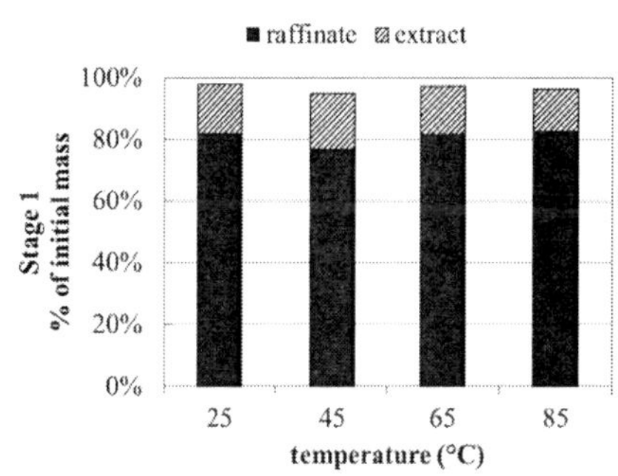

Figure 2. Effect of temperature on total amount extracted at 2128 psig.

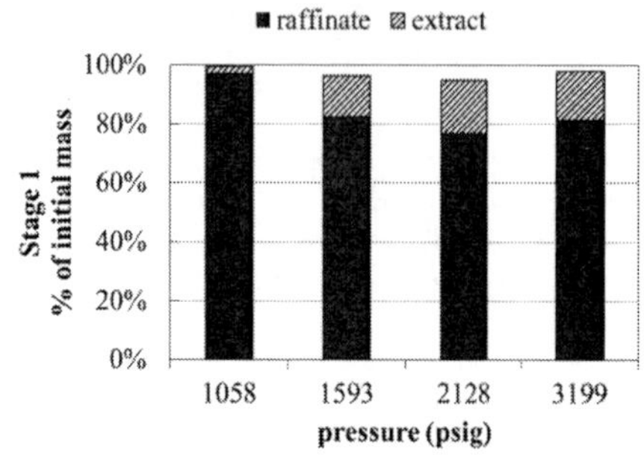

Figure 3. Effect of pressure on total amount extracted at 45 °C.

Figure 2 shows that about 15-20 wt% of material was extracted by carbon dioxide at twice its critical pressure, i.e., 2128 psig. Figure 3 shows that the amount extracted at 1058 psig was much lower than at higher pressures. This behavior correlates directly with the sharp increase in carbon dioxide density when pressure is increased above its critical point. In consequence the enhanced solvent properties facilitated a higher extraction. Thus, 45 °C and 2128 psig were selected for further experimentation.

It should be noted that the most important parameters to determine the feasibility of fractionation using supercritical carbon dioxide are the distribution of each components into the two fractions: raffinate and extract.

Figures 4 and 5 show the mass percentages of each component that formed each of the two fractions. For example, Figure 4 shows that most of the water remained in the raffinate fraction and most of the furfural, guaiacol, and m-cresol formed the extract fraction. Acetic acid distributed between the two fractions. This trend was confirmed when testing the Stage 2 simulant. In this case most of the polar compounds such as water and levoglucosan remained in the raffinate and the less polar formed the extract. However, the separation was not complete and further optimization of the conditions must be achieved.

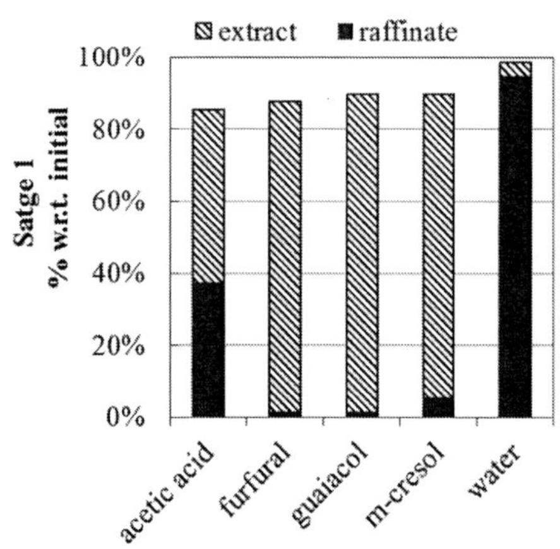

Figure 4. Distribution of Stage 1 components between extract and raffinate at 45 °C and 2128 psig.

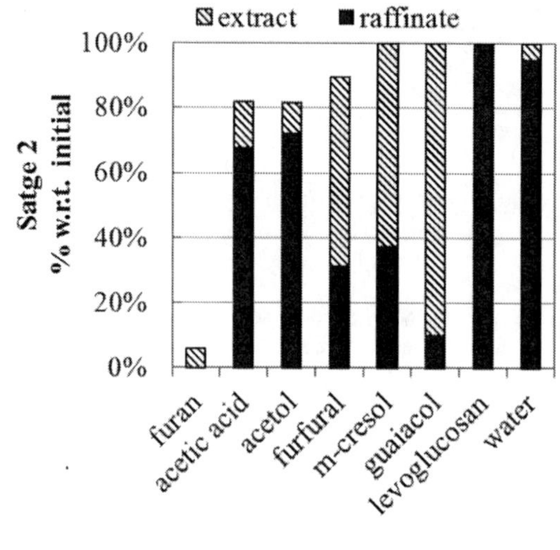

Figure 5. Distribution of Stage 2 components between extract and raffinate at 45 °C and 2128 psig.

The results of simulation showed similar separation to the experimental results (see Figure 6). Success for the model was determined to be 20% of the heating energy of the oil. Each stage composition met these energy and separation standards according to the model.

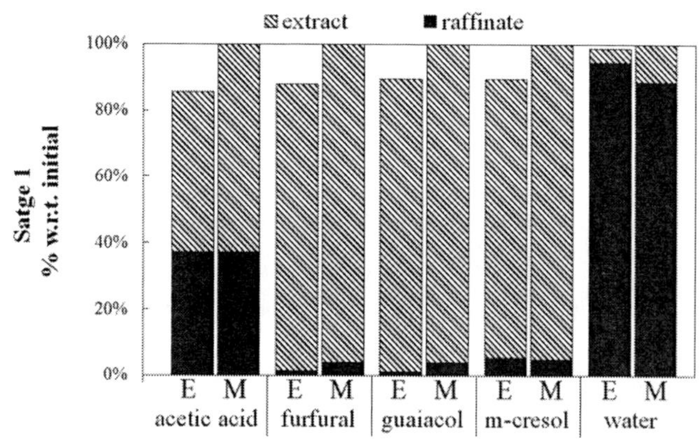

Figure 6. Comparison between experimental (E) and model (M) results for fractionation of Stage 1 simulant at 45 °C and 2128 psig.

Conclusions

Water, acetol, and levoglucosan were preferentially separated as raffinates. Furan, furfural, and guaiacol were preferentially separated as extracts. Further preferential separation of compounds would be possible by adjusting temperature and pressure. The results of ASPEN simulation showed that the energy required to effect the separation was less than 20% of the calorific value of the oil.

Acknowledgement. This work was funded by U.S. Department of Energy, Energy Efficiency and Renewable Energy, Bioenergy Technology Office. This manuscript has been authored by Battelle Energy Alliance, LLC under Contract No. DE-AC07-05ID14517 with the U.S. Department of Energy. The United States Government retains and the publisher, by accepting the article for publication, acknowledges that the United States Government retains a nonexclusive, paid-up, irrevocable, world-wide license to publish or reproduce the published form of this manuscript, or allow others to do so, for United States Government purposes.

References

(1) Mohan, D.; Pittman C.U.; Steele, P.H. *Energy & Fuels*, **2006,** *20*(3) 848.
(2) Savage, P.E.; Gopalan, S.; Mizan, T.I.; Martino, C.J.; Brock, E.E. *AICHE Journal*, **1995**, 41(7) 1723.

SUPPORT AND PROMOTER EFFECTS ON BIOFUEL UPGRADING

*Zhong He, Maocong Hu, Xianqin Wang**

Department of Chemical, Biological and Pharmaceutical Engineering, New Jersey Institute of Technology, Newark, NJ, 07102, USA.
Fax: +1-973-596-8436; Tel: +1-973-596-5707; E-mail: xianqin@njit.edu
* to whom all correspondence should be addressed.

Abstract

This work will discuss the support and promotor effects on the biofuel upgrading. Upgrading pyrolysis oils via hydrodeoxygenation (HDO) process is carried out in this work over a series of Pt/transition metal oxides and non-sulfided conventional catalysts. The reaction mechanisms of two representative compounds of pyrolysis oils, acetic acid (HAc) and guaiacol (GUA), are investigated in detail. The required catalytic properties for HDO process are explored. This work provides the insights to develop a highly selective and active catalytic system for upgrading raw bio-oils. The observations from HAc and GUA work indicate that the surface acidity of the catalysts is one of the critical points in achieving highly selective productions of hydrocarbons.

A NEW PROCESS USING LOW COST WASTE WOODY MATERIALS TO PRODUCE BIOFUELS

Dr. Anne M. Gaffney

Idaho National Laboratory
P.O. Box 1625, MS 2203
Idaho Falls, ID 83415
Anne.Gaffney@inl.gov

INL researchers have developed a process for using low cost waste woody materials from the logging industry to produce biofuels. These materials currently are not utilized because they contain contaminants that interfere with conversion to biofuels. The INL process utilizes air classification and leaching to remove these contaminants. An economic analysis of this process demonstrated that this material could be delivered to a biorefinery for 19% lower cost than the preferred pulpwood feedstock and still meet all of the specifications required of pulpwood. In addition, there is enough of this waste material available to increase woody biomass delivery to biorefineries by 200%.

OXYGEN REMOVAL FOR BIOMASS ON TRANSITION METAL PHOSPHIDE CATALYSTS

Tao Li [1]*; Randall E. Winans* [1]*; Brandon O'Neill* [2]*; James Dumesic* [2]

[1]. X-Ray Science Division, Argonne National Laboratory, Lemont, Illinois, United States

[2]. Chemical and Biological Engineering, University of Wisconsin, Madison, Wisconsin, United States

Taoli@aps.anl.gov; rewinans@anl.gov.

Abstract

Biomass derived molecules are being converted catalytically to useful products. For example, hydrogenation of furfural to furfuryl alcohol using nanocatalyst. However, many of these reactions are conducted under harsh conditions where the catalyst will rapidly sinter and the reactivity will drop quickly. Recently, it has been shown that over coating with alumina on metal nanoparticles using atomic layer deposition (ALD) will reduce the sintering. In addition, the lifetime and reactivity of catalyst increase dramatically compared to the catalyst without ALD over coating. It was believed that the nanopore was formed during the heating.

Therefore, understanding the mechanism of the pore formation and further controlling the pore size are necessary for the development of the ALD overcoated catalyst. This work demonstrates that SAXS is very powerful technique to characterize these ALD overcoated catalysts. It was found that ALD filled the pores of the particles and the thermal treatment resulted in the formation of ~ 3.5 nm pores, which agrees with the BET analysis. It is possible to control the size of the pores in the overcoat by following in situ pore formation by SAXS as a function of heating rates, final temperature and gas used. The porosity can be tuned to a desired size by using these three variables during heating.

FUNDAMENTAL UNDERSTANDING OF ACID-BASE CATALYSIS FOR THE UPGRADING OF BIOMASS-DERIVED FEEDSTOCKS

Yong Wang

The Gene & Linda Voiland School of Chemical Engineering and Bioengineering, Washington State University, Pullman, WA 99164, USA
Institute for Integrated Catalysis, Pacific Northwest National Laboratory, Richland, WA 99354, USA
Wang42@wsu.edu

Biomass-derived feedstocks can be upgraded to fuels and chemicals by C-C bond formation and deoxygenation. In this talk, we report that a balanced Lewis acid-base pair on ZnxZryOz can achieve cascade (cross) coupling and self-deoxygenation reactions. Specifically, groups of small oxygenates including alcohols, aldehydes, ketones, and carboxylic acids can be converted to olefins with up to 80 mol% carbon selectivity to C3=-C6=. This study provides significant advantages over many of the current catalysts by its unique way to permit selective C-C bond formation and oxygen removal suitable for upgrading a wide range of biomass derived feedstock and intermediates.

BIOMASS DEPLOYMERIZATION VIA ACIDIFIED MOLTEN SALT HYDRATES

Sunitha Sadula, Basudeb Saha and Dionisios G. Vlachos**

Catalysis Center for Energy Innovation, University of Delaware, Newark, Delaware – 19716, USA.
bsaha@udel.edu; vlachos@udel.edu

Introduction

Lignocellulosic biomass, includes agricultural residues such as corn stover and wheat stalks, energy crops such as poplar trees and switchgrass, and municipal solid waste is a promising alternative resource for sustainable energy and renewable chemical production.[1-2] According to the U.S. Billion-Ton update: 'Biomass supply for Bioenergy and Bioproducts industry'[3] U.S. could produce over 1 billion dry tons of biomass per year which offers a unique opportunity for a sustainable society with unprecedented impact on U.S. economy, energy security and independence. While biomass conversion to fuels and chemicals involves multiple steps, depolymerization of carbohydrates is the crucial step. Primary methods of depolymerization are based on acid treatment[4], enzymatic[5-6] and ionic liquids[7] which are expensive and hazardous to the environment. Thus, efficient depolymerization of biomass is still challenging and remains a barrier to cellulosic biorefining processes economically viable. To address this challenge, we have developed an integrated process for efficient hydrolysis of crystalline cellulose and hemicellulose in Acidified Molten Salt Hydrates (AMSH) media with >95% yields of water soluble sugar in 30 min[8], and a one pot process for subsequent dehydration of sugars to furfurals. This process has also achieved high yield of glucose (90%) directly from woody biomass.

Experimental

Initial screening hydrolysis experiments of crystalline cellulose, beechwood hemicellulose and poplar wood biomass using were conducted in high pressure glass tubes sealed with aluminum crimp caps in oil bath or Al-heating blocks at temperatures up to 85 °C. Kinetic experiments were carried out in three neck R.B. flasks with refluxing. After hydrolysis step, an organic solvent and a Lewis acid catalyst were added to the C_6 and C_5 sugar reaction mixture and heated at 140 °C. Furfurals were extracted in the extracting solvent and the recyclability of the MSH phase was investigated.

Results and Discussion

Initially, depolymerization of crystalline cellulose and hemicellulose (xylan) was carried out using 0.1-0.5 wt% sulfuric acid and 40-69 wt% LiBr in the temperature range of 70-105 °C. Under optimum reaction conditions cellulose gave >90% yield of water soluble sugars including 82% glucose and hemicellulose yielded >95% xylose (Figure 1). The results indicate that the enhanced hydrolysis rates are due to the synergistic effect of LiBr and dilute acid which makes the medium more acidic and breaking the crystalline structure through swelling. Analysis of the kinetic data indicates that hydrolysis rates are first order in cellulose and in H_2SO_4 concentration. Both cellulose and hemicellulose hydrolysis reactions progressed through sequential depolymerization via oligomers of Dp 2-6. Xylan hydrolysis kinetics were very fast compared to cellulose and ~80% xylose yield were achieved at lower acid (0.01 M) and lower LiBr (40%) for 2 h. With a good understanding of cellulose and hemicellulose hydrolysis in AMSH, extended the process to woody biomass and achieved higher glucose yields (90%).

For further conversion of sugars to furans, the hydrolysis mixture containing soluble sugars and AMSH was heated in a biphasic solvent in the presence of a Lewis acid at 140 °C for 2 h. C_6 sugar mixture gave 55% HMF yield and C_5 mixture gave 70% furfural yield in 2 h and the carbon mass loss is presumably accounts for humins by-products,[9] which agrees with the appearance of darker color in the aqueous phase. Glucose and xylose dehydration was progressed through fructose and xylulose isomers formation. Upon separation of humins by simple filtration, the recovered solution was reused for hydrolysis of cellulose and hemicellulose.

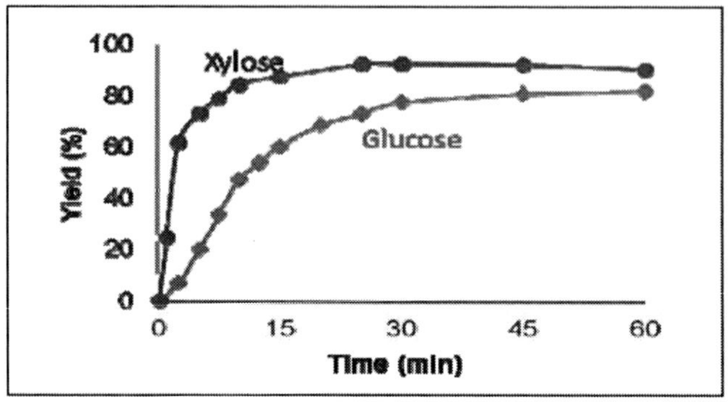

Figure 1. A comparison of glucose and xylose yields from crystalline cellulose and xylan in AMSH.

Conclusions

We developed an integrated one-pot process for producing sugars and furans from biomass without involving complex and expensive separation steps. The results show that > 90% C_6 and C_5 sugars can be produced from crystalline cellulose and hemicellulose, which can then be converted to furans in the same pot with high yields. This process has also achieved high yield of glucose (90%) directly from woody biomass.

Acknowledgement. The work was supported "by the U.S. Department of Energy, Office of Science, and Office of Basic Energy Sciences as part of an Energy Frontier Research Center".

References

(1) Bohre, A.; Dutta, S.; Saha, B.; Abu-Omar, M.M. *ACS Sus. Chem. Eng.* **2015**, 3, 1263
(2) Saha, B.; Abu-Omar, M.M. Chem. Sus. Chem. 20158, 1133
(3) U.S. Billion-Ton update: *Biomass supply for Bioenergy and Bioproducts Industry*, Department of Energy, August 2011
(4) Xiang, Q.; Kim, J. S.; Lee, Y. Y. *Applied Biochemistry and Biotechnolog.* **2003,** *105*, 337
(5) Shi, J.; Gladden, J. M.; Sathitsuksanoh, N.; Kambam, P.; Sandoval, L.; Mitra, D.; Zhang, S.; George, A., Singer, W. S.; Simmons, B. A.; Singh, S. *Green Chem.* **2013**, 2579
(6) Ju, X.; Bowden, M.; Engelhard, M.; Zhang, X. *Appl Microbiol Biotechnol.* **2014**, 89, 4409.
(7) Raj, T.; Kapoor, M.; Semwal, S.; Sadula, S.; Pandey, V.; Gupta, R. P.; Kumar, R.;Tuli, D. K.; Das, B. P. *Journal of Cleaner Production.* **2016**,1005
(8) Deng, W.; Kennedy, J. R.; Tsilomelekis, G.; Zheng, W.; Nikolakis, V. *Ind. Eng. Chem. Res.* 2015,54, 5226
(9) Tsilomelekis, G.; Orella, M. J.; Lin, Z.; Cheng, Z.; Zheng, W.; Nikolakis, V.; Vlachos, D. G.; *Green Chem.* DOI 10.1039/C5GC01938A.

MOLECULAR DESIGN OF COOPERATIVE INTERACTIONS FOR HETEROGENEOUS CATALYTIC MATERIALS TO TUNE CATALYTIC RATES AND SELECTIVITIES

Nitish Deshpande, Aamena Parulkar, Mariah Whitaker, Nicholas A. Brunelli

The Ohio State University, 151 W. Woodruff Ave, Columbus, OH 43210
Brunelli.2@osu.edu

Introduction

Cooperative catalysis is achieved through judiciously combining two catalytic species in close proximity.[1] While nature routinely employs cooperative interactions, these are challenging to achieve in heterogeneous catalytic materials because synthetic techniques tend to produce a distribution of catalytic sites. We take a molecular approach to create well-defined catalytic sites in silica materials.[2,3] We use these materials to gain greater mechanistic insights on cooperative interactions for important catalytic reactions.

Our work is relevant to a wide variety of biomass-related reactions, specifically the dehydration of fructose to 5-hydroxymethyl furfural. The production of HMF is a critical step in valorization of biomass,[4] but typically suffers from low reaction selectivity.[5] Researchers have employed numerous strategies to increase the catalytic selectivity for this reaction, including creating biphasic reaction systems and mixing miscible organic solvents.[6,7] The addition of a second solvent in addition to water adds a considerable cost to the process. We have instead sought to create a bifunctional material containing both acid sites and a second molecule that allows us to influence catalytic activity and selectivity. In this talk, we will describe our efforts to synthesize, characterize, and test in catalytic reactions these bifunctional catalytic materials.

Experimental

We use a combination of organic and inorganic synthetic techniques to produce a catalytic material inspired by enzymes. We synthesize molecular precursors through known methods. The precursors are further incorporated into material synthesis through either grafting or co-condensation to produce new types of catalytic sites that are active for a broad array of chemical reactions.

Material Characterization. The catalyst was characterized using a battery of standard techniques, including XRD, SS DNP NMR, nitrogen physisorption, titration for acid content, elemental analysis, thermogravimetric analysis (TGA), diffuse reflectance UV-vis spectroscopy (DR-UV-vis), diffuse reflectance Infrared Fourier Transform Specotroscopy (DRIFTS), and scanning electron microscopy (SEM).

Catalytic Testing. The materials were tested in catalytic reactions. We use a 10 wt%% solution of fructose in water with a catalyst mass corresponding to 2 mol% sulfonic acid sites. The reaction mixtures are heated in sealed pressure tubes for a fixed period of time before removing from the heat source. The reaction mixtures were filtered and characterized using HPLC to determine conversion and selectivity. We perform a number of control experiments to understand the impact of our bifunctional catalytic material on reaction selectivity. In addition, we compare catalysts in different solvent mixtures to determine how the bifunctional catalyst synthesis can be further refined to increase catalytic selectivity for this reaction.

Results and Discussion

We have successfully synthesized a series of catalytic materials containing acid sites and a second functional group through a mild sequential grafting method. Specifically, we have created three materials of interest to this study: (1) sulfonic acid sites (0.25 mmol/g; SBA-SA0.25); (2) sulfonic acid (0.25 mmol/g) and second site (0.25 mmol/g; SBA-SA0.25-S0.25); and (3) sulfonic acid (0.25 mmol/g) and second site (1.0 mmol/g; SBA-SA0.25-S1.0). This is achieved through first grafting 3-mercaptopropyl trimethoxysilane on the surface of SBA-15. The thiol is oxidized using hydrogen peroxide. We then graft the second functional group on the surface using a similar procedure at two loadings to enable systematic investigation of the effect of the second functional group on catalyst activity and selectivity.

The materials are characterized using a variety of standard techniques, including titration to determine acid site content, FTIR, and solid state NMR. Most importantly, we find that the titration results demonstrate for a theoretical loading of 0.25 mmol H^+/g that we obtain 0.22 mmol H^+/g, consistent with previous literature. As we add additional organic content from the second organosilane, the amount of acid decreases proportionally. This observation is corroborated using TGA measurements.

We further characterize these materials in catalytic reactions in pure water as the solvent. Using SBA-SA0.25, we find at low conversion of 20% a selectivity to HMF of 30%, similar to previous literature. Using our bifunctional catalytic materials, we find the selectivity at a similar conversion of 26% increases to a selectivity for HMF of 85%. Current work is seeking to characterize the stability of this catalyst.

Conclusions

We present a novel series of bifunctional catalytic materials capable of increasing the selectivity for the dehydration of fructose to HMF. The increase in catalytic selectivity provides insight on a process that could be commercially viable method to produce HMF from fructose.

Acknowledgement. We wish to acknowledge financial support from the Ohio State Institute of Materials Research (IMR), the ACS-PRF, and the National Science Foundation.

References

(1) Margelefsky, E. L.; Zeidan, R. K.; Davis, M. E. Cooperative catalysis by bifunctionalized mesoporous silica. *Chem. Soc. Rev.* **2008**, *37*, 1118–1126.
(2) Brunelli, N. A.; Didas, S. A.; Venkatasubbaiah, K.; Jones, C. W. Tuning Cooperativity by Controlling the Linker Length of Silica-Supported Amines in Catalysis and CO2 Capture. *J. Am. Chem. Soc.* **2012**, *134*, 13950–13953.
(3) Brunelli, N. A.; Jones, C. W. Tuning Acid-Base Cooperativity to Create Next Generation Silica-Supported Organocatalysts. *J. Catal.* **2013**, *308*, 60–72.
(4) Gallezot, P. Conversion of biomass to selected chemical products. *Chem. Soc. Rev.* **2012**, *41*, 1538–1558.
(5) van Putten, R.-J.; van der Waal, J. C.; de Jong, E.; Rasrendra, C. B.; Heeres, H. J.; De Vries, J. G. Hydroxymethylfurfural, a versatile platform chemical made from renewable resources. *Chem. Rev.* **2013**, *113*, 1499–1597.
(6) Saha, B.; Abu-Omar, M. M. Advances in 5-hydroxymethylfurfural production from biomass in biphasic solvents. *Green Chemistry* **2014**, *16*, 24.
(7) Román-Leshkov, Y.; Chheda, J. N.; Dumesic, J. A. Phase modifiers promote efficient production of hydroxymethylfurfural from fructose. *Science* **2006**, *312*, 1933–1937.

PATHWAYS TO SELECTIVELY FORM INDIVIDUAL AROMATIC PRODUCTS FROM ETHANOL

David W. Flaherty

University of Illinois at Urbana-Champaign, Urbana, IL, 61801
dwflhrty@illinois.edu

Introduction

Ethanol derived from the fermentation of biomass can be catalytically converted into higher molecular weight hydrocarbons and oxygenates to produce advanced biofuels and lubricants. The Guerbet reaction, which involves dehydrogenation, aldol condensation, and hydrogenation, is one pathway for this process. Cascades of subsequent self- and cross-condensation reactions between ethanol and $C_{\geq 4}$ alcohol derived intermediates provide a wide-range of possible higher carbon number (C_n) products. Successful production of these products requires high selectivities for dehydrogenation and condensation reactions to avoid the loss of carbon by forming undesirable products (e.g., C_1-, C_2-hydrocarbons) but also requires the ability to form specific molecules which can serve as fuels and platform chemicals. Here, we show how bifunctional catalysts, comprised of amphoteric oxides (MgO, TiO_2, hydroxyapatites, and hydrotalcites) and Cu clusters promote these networks of reactions by measuring formation rates of more than 30 distinct products as functions of reactant concentrations, temperature (523 - 623 K), and the ratios of the numbers active sites for hydrogen transfer and those for condensation reactions.

Experimental

Oxide catalyst (TiO_2, MgO, hydroxyapatite, modified faujasite) were synthesized by simple precipitation and ion-exchange methods. The crystallographic structure and surface areas of the catalysts were determined by XRD and N_2 physisorption, respectively. Steady-state catalytic rates were measured in plug-flow reactors and the products were characterized by GC and GC-MS, whereas, in FTIR and isotope labeling experiments were used to determine the prevalent surface intermediates and the mechanisms for hydrogen transfer reactions.

Results and Discussion

The selectivity with which ethanol is incorporated into growing chains is limited by the selectivity of both catalytic functions, while the isomeric structure of the products is determined by the concentrations of the unsaturated oxygenates present on the oxide surfaces and differences between the deprotonation energies of C-atoms α to the carbonyl groups of these species. Reactions among product mixtures that predominantly contain alcohols and aldehyde reactants form broad distributions of fatty alcohols and whose product selectivities follow predictions from step-growth polymerization models. These diverse mixtures, however, are only suitable as fuels.

Aldol condensations primarily form unsaturated aldehydes, which exist at much higher concentrations than their saturated analogs in the absence of effective hydrogen transfer catalysts. Large differences in the deprotonation energies between the α C-atoms for aldehydes and unsaturated aldehydes facilitate the sequential condensation reactions of acetaldehyde with enals, which grow with increasing residence time and C_2 conversion until forming 2,4,6-octratrienal (Figure 1). This linear trienal, and minority branched trienals, undergo rapid dehydrocyclization to form a very small set of C_8 aromatic species (*o*- and *p*- methyl benzyl aldehydes and alcohols), which are potential renewable precursors to phthalic acids and anhydrides.

The steric protection about the carbonyl of the aromatic product formed effectively halts further condensation reactions and provides a unique opportunity to make a desirable platform chemical within in this complex network of reactions. The isomeric structures of the C_8 aromatics are sensitive to the presence or absence of steric hindrance about the acid-base site pairs that catalyze the reaction. Coupling hydroxyapatite, which lacks microporosity produces predominantly ortho-C_8 products with minor (< 10%) amounts of the para-isomers. These selectivities can be inverted, however, when the reactions proceed within the microporous environments of dealuminated faujasite framework containing lewis acidic metal oxide clusters.

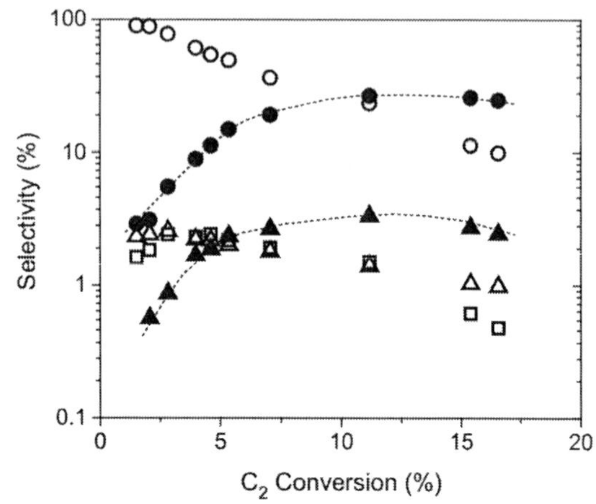

Figure 1. Formation of C_4–C_8 enals (○, 2-butenal, C_4; Δ, 2,4-hexadieneal, C_6; □, 2,4,6-octatrienal, C_8) and aromatic C_8 products (●, *o*-methylbenzyl aldehyde and *o*-methylbenzyl alcohol; ▲, *p*-methylbenzyl aldehyde and *p*-methylbenzyl alcohol) as a function of C_2 reactant conversion over Ca-HAP (1 kPa C_2H_5OH, 0.35 kPa C_2H_4O, H_2 balance, 548 K, 1.67 Ca/P). Dashed lines in the figure are included to guide the eye.

Conclusions

These findings provide evidence that the Guerbet reaction proceeds by dehydrogenation and aldol condensation steps and elucidates and the complex sets of surface reactions that dictate the formation rates of certain categories of products. This knowledge will help guide further research into reaction schemes to convert alcohols into larger oxygenates and hydrocarbons.

Acknowledgements. This work was funded by the Energy Biosciences Institute at the University of Illinois, Urbana-Champaign.

ADSORPTION-ENHANCED HYDROLYSIS OF GLUCAN OLIGOMERS INTO GLUCOSE OVER THREE-DIMENSIONALLY ORDERED MESOPOROUS CARBON CATALYSTS

Paul Dornath[a], *Paul Dauenhauer*[b] *and Wei Fan*[a]

a. Chemical Engineering, University of Massachusetts Amherst, MA 01003

b. Chemical Engineering and Material Science, University of Minnesota, Minneapolis, MN 55455.

Introduction

The production of glucose from crystalline cellulose, an abundant biopolymer consisting of anhydroglucose units (AGUs) connected via β (1→4) glycosidic linkages, is paramount to the efficient utilization of lignocellulosic biomass for production of fuels and value added chemicals. Amorphous carbon materials with functional surface groups have recently received a great deal of interest as catalysts and adsorbents for aqueous phase hydrolysis of cellulose to glucose due to their low cost, hydrothermal stability and unique adsorption properties. Amorphous carbons with high carbon content (>80%) are composed of polycyclic aromatic domain in irregular arrangements. The surface of amorphous carbon may also contain a large variety of oxygen containing groups which can tune the carbon surface properties from hydrophobic to hydrophilic. The adsorption can be enthalpically and/or entropically driven, which increases with increasing the chain length of β (1→4) glucan oligomer due to an increase in the interaction between the glucan and carbon surface and the number of water molecules being released per AGU unit. While it has been shown that carbon materials used as adsorbents and catalysts are highly attractive for the hydrolysis of cellulose, the effects of surface property and polycyclic aromatic domain size on their adsorption and catalytic properties are not well understood.

The objective of this study is to develop a structure-property relationship for the three dimensionally ordered mesoporus (3DOm) carbon catalyst so that the catalysts can be rationally tailored for the selective hydrolysis of glucan oligomers. 3DOm carbon catalysts with well-controlled mesoporous structures and surface functional groups (SO_3H) were synthesized. The adsorption of glucose and glucan oligomers with different chain lengths on the 3DOm carbon with and without sulfonic acid groups were measured and correlated to their surface structures including the domain size and surface functional groups.

Experimental

3DOm carbon with a cage of 35 nm was prepared according to literature [1]. 24 Surface functionalization of the 3DOm carbon with sulfonic acid groups was achieved by reacting with benzenesulfonic acid radicals derived from diazonium salts under mild conditions [2]. Aqueous phase adsorption of glucose, cellobiose and long glucan oligomers on the carbon samples was performed at 0, 20, 50 and 70 °C with concentrations from 1 to 50 mg mL^{-1}. All adsorption solutions were analyzed using high performance liquid chromatography (HPLC) (LC-20AT, Shimadzu) equipped with refractive index (RID-10A) and UV-Vis (SPD-2AV) detectors at oven temperature of 85 °C. An Agilent Hi-Plex-Na HPLC column (7.7 mm (ID), 300 mm (Length), 10 μm (particle size)) with a guard column was used. HPLC grade water (Fisher) was used as a mobile phase with a flow rate of 0.3 mL min^{-1}.

Hydrolysis reactions were performed using 1 wt.% cellobiose or glucan solution. 30 mg of reactants were mixed with 20 mg of SO_3H-3DOm carbon catalyst and 3 mL of deionized water, and sealed in a 5.0 mL borosilicate microwave heating vial (Chemglass). The reaction was performed on a Dynabloc aluminum stir plate with 22 (ID) × 35.6 (depth) mm holes (Ace Glass) at 120 °C for up to 2.0 h.

Results and Discussion

The morphologies and surface properties of the 3DOm carbon catalysts before and after the addition of sulfonic acid groups were characterized. The sulfonated 3DOm carbon after surface functionalization contains 0.5 mmol SO_3H g^{-1}. The textural properties of the two carbon samples were characterized using a nitrogen physisorption at -196 °C and SAXS as shown in **Figure 1**. The N_2 physisorption isotherms as well as the SAXS patterns of SO3H-3DOm carbon are nearly identical to parent-3DOm carbon, suggesting that the ordered porous structures of 3DOm carbon remain intact after the functionalization by diazomium salts.

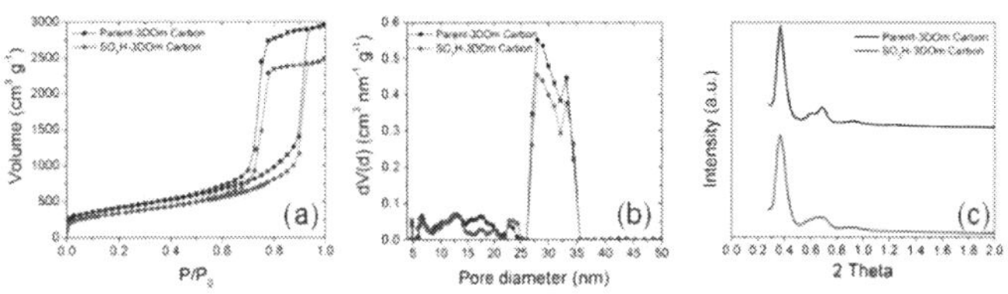

Figure 1. Textural properties of parent-3DOm and SO3H-3DOm carbons. (A) Nitrogen adsorption/desorption isotherms. (B) Pore size distribution calculated from the adsorption branch using QSDFT model. (C) SAXS patterns.

SO_3H-3DOm carbon catalysts with different quantities of sulfonic acid groups were prepared in order to understand the effects of acid sites on the adsorption properties of the carbon catalysts. It was observed that the number of sulfonic acid sites on the carbon catalysts can be easily controlled by varying the amount of diazonium sulfonate generated from the sulfanilic acid added in the synthesis solution. The results suggest that the number of sulfonic acid sites on the carbon catalyst can be varied from 0.14 to 1.31 mmol SO3H g^{-1} as shown in **Figure 2A**. The adsorption sites were estimated by the maximum adsorption capacity of cellobiose at 0 °C. **Figure 2B** shows the total number of adsorption and acid sites for the 3DOm carbon samples with different quantities of sulfonic acid groups. It was observed that the total number of sulfonic acid and adsorption sites is approximately constant for each sample, with each acid site annihilating one adsorption site. The linear reduction in adsorption sites with the addition of acid sites is likely due to the reaction between the carbon surface and diazonium sulfonate, which annihilates the basal-plane π-bonds responsible for the adsorption of saccharides. Additionally, the sulfonic acid group may reduce electron density from the surrounding carbon surface, an effect which is proportional to the number of acid sites. This electron-poor region may not be able to adsorb saccharides strongly. The SO_3H-3DOm carbon catalysts were optimized by testing the TOF for hydrolysis of cellobiose at low conversion (< 20 %) as shown in **Figure 2C**. The TOF for the hydrolysis of cellobiose over SO_3H-3DOm carbon is highest for the sample containing 0.50 mmol SO_3H g^{-1}, indicating that an optimized ratio between adsorption and reaction sites is critical for the carbon catalysts. Based on this result the SO_3H-3DOm carbon sample with 0.50 mmol SO_3H g^{-1} was used for all of the remaining experiments.

To compare the adsorption strength and capacity of glucan oligomers on the 3DOm carbon samples, adsorption isotherms of glucose, cellobiose and mixed glucan oligomers were measured at room temperature. The composition of the mixed glucan oligomers produced from the ball-milling of cellulose co-impregnated with glucose using our previous method [3] is as follows: 3.0 wt.% levoglucosan, 6.3 wt.% glucose, 7.0 wt.% dimers, 4.3 wt.% trimers, 4.3 wt.% tetramers, 3.8 wt.% pentamers, 4.1 wt.% hexamers, 3.9 wt.% heptamers, and 63.2 wt.% of the large glucan oligomers with chain length greater than 7 AGUs. The structure of the large glucan oligomers was obtained by 1H-NMR analysis. The relative peak areas of α and β reducing ends as well as the α(1→6) and β(1→4) linkage peaks from the large glucan oligomers were used to calculate the chain length and degree of branching. It was found that the large glucan oligomers contain an average of 14.50 AGUs with 1.65 branches per chain. The large glucan oligomer is referred to as the large oligomer fraction (L-OF) while the remaining fraction contains G1-G7 species is referred to as the small oligomer fraction (S-OF). Adsorption isotherms of glucose, cellobiose and mixed glucan oligomers on parent-3DOm and SO_3H-3DOm carbon are shown in **Figure 3A** and **Figure 3B**, respectively. The adsorption isotherms were fitted using the Langmuir adsorption model. The adsorption capacities, Q_m, and Langmuir adsorption constants, K_L, for cellobiose and glucose are shown in Table 1. The result clearly demonstrates that large glucan oligomers preferentially adsorb on the carbon compared to the small glucan oligomers. At high concentration (50 mg mL^{-1}) the adsorption on the carbon sample is dominated by large glucan oligomers, with 90.2 wt.% L-OF and 9.8 wt.% S-OF adsorbed on the surface. In the solution with 10 mg mL^{-1} of the mixed glucan oligomers 79 wt.% of the L-OF and 18 wt.% of the S-OF are adsorbed on the carbon sample. When the concentration of the mixed glucan oligomers decreases to 1.0 mg mL^{-1}, there are 93 wt.% of the L-OF and 78 wt.% of the S-OF adsorbed on the carbon sample. The portion of the S-OF adsorbed on the carbon sample increases at the low concentration of the mixed glucan oligomers. The result is easy to understand since the surface coverage on the carbon catalysts by the L-OF is not sufficiently high under the low concentration of the mixed glucan oligomers so that the S-OF can be still adsorbed on the surface. However, under the high concentration of the mixed glucan oligomers, the surface of the carbon catalyst is mainly covered by the L-OF due to the preferred adsorption. Thus the Q_m (mmol/g) for the mixed glucan oligomer adsorbed on the carbon sample (Table 1) is estimated using molecular weight of L-OF.

The adsorption enthalpy and entropy of glucans with different AGUs on the carbon adsorbents were also measured. The correlation with the surface structure of the carbon catalysts will be discussed on during the presentation.

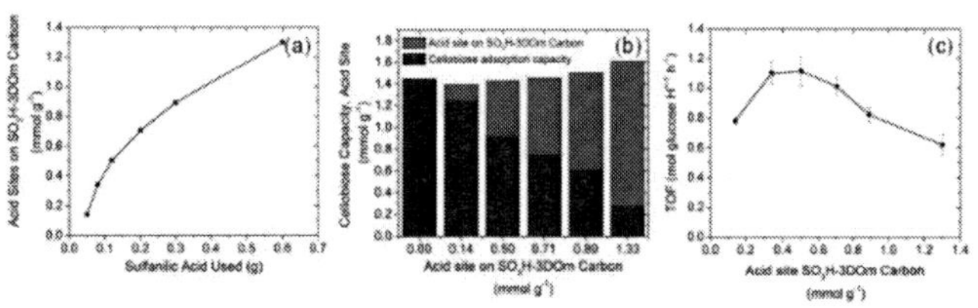

Figure 2. Optimization of acid sites and adsorption properties of SO3H-3DOm catalyst. (a) Control of the number of acid sites on 3DOm carbon by tailoring the amount of sulfanilic acid in the synthesis solution. (b) Correlation between cellobiose adsorption capacity and sulfonic acid groups on various SO3H-3DOm carbons. The number of sulfonic groups plus the number of cellobiose adsorption sites is approximately constant. (c) TOF obtained from the hydrolysis of cellobiose over SO3H-3DOm carbons with different amounts of sulfonic acid sites.

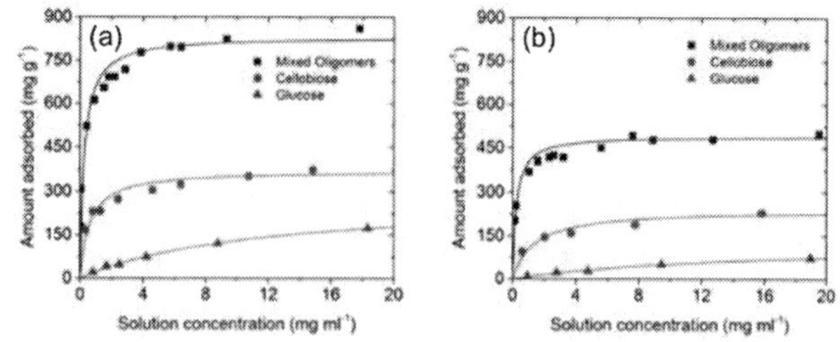

Figure 3. Adsorption isotherms of glucose, cellobiose and mixed oligomers on (a) parent-3DOm carbon and (b) SO_3H-3DOm carbon at room temperature. The points are from experimental measurement. The lines are from Langmuir adsorption fitting.

Table 1. Langmuir adsorption constants for the adsorption of glucose and glucan oligomers with different chain lengths on carbon samples.

Adsorbate	Parent-3DOm carbon				SO_3H-3DOm carbon			
	Q_m (mg g^{-1})	Q_m (mmol g^{-1})	K_L (L mol^{-1})	R^2	Q_m (mg g^{-1})	Q_m (mmol g^{-1})	K_L (L mol^{-1})	R^2
Mixed oligomers	834	0.35	N/A	0.95	492	0.21	N/A	0.94
Cellobiose	357	1.04	658	0.90	225	0.66	323	0.97
Glucose	285	1.58	14.8	0.99	120	0.67	14.0	0.99

Conclusions

It was found that the adsorption capacity of the 3DOm carbon increased with increasing oligomer chain length. The adsorption capacity and domain size measurements indicated that two cellobiose molecules or three glucose molecules could fit on each domain of the parent 3DOm carbon. The increased adsorption mass was attributed to more efficient packing of large glucan oligomers on the carbon surface as well as to the likelihood that large glucan oligomers could adsorb across multiple domains, thus allowing them more area to adsorb compared to glucose and other small oligomers that may not be able to adsorb on areas of domain edges. The adsorption enthalpy increased with increasing chain length but did not change with the surface properties of the carbon samples. Adsorption entropy also increased with increasing chain length, but the adsorption entropy was higher for cellobiose on parent-3DOm carbon than that of cellobiose on SO_3H-3DOm carbon. The number of acid sites on the 3DOm carbon catalyst was optimized for the hydrolysis of cellobiose and larger glucan oligomers. It was found that 3DOm carbon catalysts containing 0.50 mmol g^{-1} sulfonic groups exhibit the best catalytic performance, resulting from balancing adsorption strength, number of adsorption site and acid site.

Acknowledgment. This work is financially supported as a part of the Catalysis Center for Energy Innovation, an Energy Frontier Research Center funded by the U.S. Department of Energy, Office of Science, Office of Basic Energy Science under Award Number DE-SC0001004.

References

(1) Fan, W., Synder, M. A., Sandeep, K., Lee, S. P., Yoo, C. W., McCormick, A. V., Penn, L. R., Stein, A., and Tsapatsis, M. Nat. Mater., **2008**, 7, 984-991.
(2) Wang, X.; Liu, R.; Waje, M. M.; Chen, Z.; Yan, Y.; Bozhilov, K. N.; Feng, P. Chem. Mater. 2007, 19, 2395.
(3) Dornath, P.; Cho, H. J.; Paulsen, A.; Dauenhauer, P.; Fan, W. Green Chem. 2015, 17, 769

SOLVENT EFFECT IN CATALYTIC UPGRADING OF BIOMASS-DERIVED FURANICS

Bin Wang, Reda Bababrik, Zheng Zhao, Daniel Resasco

Center for Interfacial Reaction Engineering (CIRE), School of Chemical, Biological and Materials Engineering, The University of Oklahoma, Norman, OK 73019
Wang_cbme@ou.edu

Introduction

In liquid-phase heterogeneous catalysis, solvent provides an additional degree of freedom for tuning the catalytic activity and selectivity. Solvent may have multiple effects, which may work simultaneously, on the catalytic reactions, and the dominant effect varies among different reactions. For example, solvent may affect the kinetics of the reactions when the rate-determining step involves a solvated transition state[1]. When reactions are conducted in aqueous phase, water can affect the molecular adsorption by redistributing the electrons at the solid/water interface[2], and directly participate the reaction[3]. Moreover, the expanding utilization of water/oil biphasic system for reactions has depicted simultaneously separation of the water- and oil-soluble products while the reaction keeps occurring at the water/oil interface, which minimizes the catalyst deactivation[4].

Aqueous-phase upgrading of small oxygenates is particularly desirable for biomass conversion[5] since pyrolysis of biomass produces large amounts of water and water-soluble small oxygenates[6]. Furfural, one of these small oxygenates, is an important platform chemicals produced from pyrolysis of biomass at moderate temperature by dehydration of xylose[7]. Hydrogenation of furfural produces furfuryl alcohol (FOL), tetrahydrofurfural (THFAL) and tetrahydrofurfuryl alcohol (THFOL) as shown in Scheme 1. Further hydrodeoxygenation of FOL produces methyfuran over metal catalysts[8] while hydrolysis of FOL under aqueous acids results in levulinic acid[9]. Both methyfuran and levulinic acid are desirable and important chemicals in biorefineries[10].

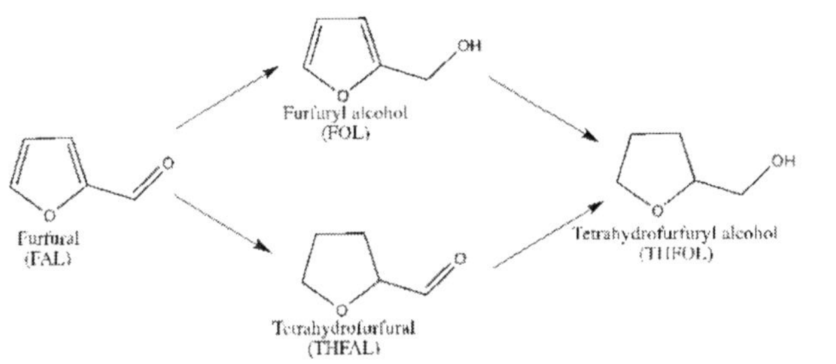

Scheme 1. Reaction pathway of furfural hydrogenation.

Fundamental understanding of the solvent effect will be valuable for further promoting the liquid-phase heterogeneous catalysis. Here we will focus on furfural (FAL) and discuss the solvent effect on its catalytic reactions through a synergistic combination of experiment, theory and computation. We find rate enhancement of furfural conversion and improved selectivity towards hydrogenation of the carbonyl group in an aqueous phase as compared to reactions in organic solutions. We show that the water solvent has multiple, distinct effects: (1) the aqueous phase stabilizes the intermediate and final products by forming hydrogen bonds and changes the rate-determining step. (2) hydrogen can be transferred by a water molecule adsorbed locally in the proximity of the furfural molecules, which reduces the activation barrier, and (3) hydrogen (protons) can be shuttled via the water network and hydrogenates the oxygen atom in the carbonyl group.

Computational Methods

Density functional theory (DFT) calculations were carried out using the VASP package[11]. The PBE-GGA exchange-correlation potential [12] was used, and the electron-core interactions were treated in the projector augmented wave (PAW) method [13]. The van der Waals interaction has been taken into account through the so-called DFT-D3 semi-empirical method[14]. All the calculations have been performed using a four-layer metal slab with a vacuum space (or filled with a liquid phase) of 15 Å between adjacent supercells. Structures have been optimized using a single Γ point of the Brillouin zone with a kinetic cut off energy of 400 eV. The uppermost metal layer and the molecules were fully relaxed. The nudged elastic band (NEB)[15] method has been used to find the transition states, which have been further verified by vibrational calculations.

Results and Discussion

Furfural hydrogenation reactions were carried out in non-polar solvents, decalin and cyclohexane, and water on Pd/α-Al_2O_3. The FAL conversion and product yields are compared in Figure 1. The disappearance of FAL in decalin should be related to the poor solubility of FAL and catalyst deactivation. The product yields of FAL hydrogenation in cyclohexane illustrates that the furanyl ring hydrogenation is preferred and the major product is THFAL, while in water, the carbonyl group hydrogenation is preferred leading to high yield of FOL. In addition to the varied selectivity, the conversion was increased significantly in water as compared to cyclohexane.

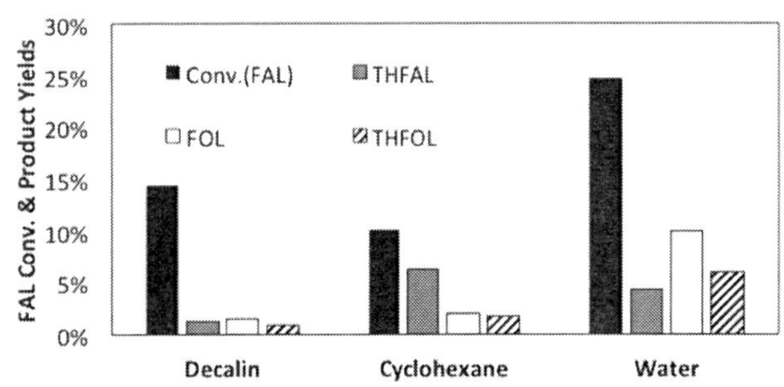

Figure 1. Conversion of FAL hydrogenation and products yields in different solvents. Reaction conditions: initial concentration of FAL is 0.1M, T=313 K, P=5.52 MPa, catalyst amount = 250mg 3% Pd/α-Al_2O_3, reaction time=0.8h.

We explored the role of solvents by performing first-principles DFT calculations on the adsorption and hydrogenation of FAL at the liquid/Pd interface. On a clean Pd (111) surface, the overall activation barrier is 92 kJ/mol (Fig. 2). In the presence of liquid water, the overall activation decreases to 68 kJ/mol with the oxygen hydrogenations step as the rate-determining step. The overall reaction becomes slightly exothermic in water, as compared to endothermic on a clean Pd(111) surface, very largely because the water phase stabilizes the intermediate states (the product after first-step oxygen hydrogenation) and FOL.

Water can also directly participate through proton shuttling in the reaction as shown in other studies[2]. We find one water molecule, which adsorbs in the proximity of the FAL and forms an hydrogen bond with the FAL carbonyl group, in the liquid phase can take the hydrogen atom directly from the surface, as shown in Fig. 3. In this case, this water molecule acts as a bridge between the carbonyl group and the surface hydrogen atom as compared to direct hydrogenation from the surface, resulting in an activation barrier of 41 kJ/mol for the first step (oxygen hydrogenation).

In a different scenario, one water molecule, which doesn't form hydrogen bonds with the carbonyl group, can also pick up a surface hydrogen atom to form H_3O^+ leaving an electron in the metal electron reservoir. This scenario further reduces the activation barrier for the first step (oxygen hydrogenation) to 32 kJ/mol.

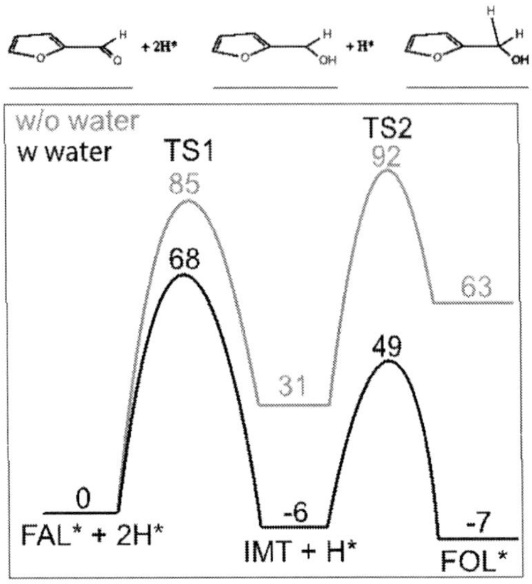

Figure 2. DFT calculations of hydrogenation of the carbonyl group of furfural with or without liquid water. In both cases, the hydrogen atom comes from the Pd surface.

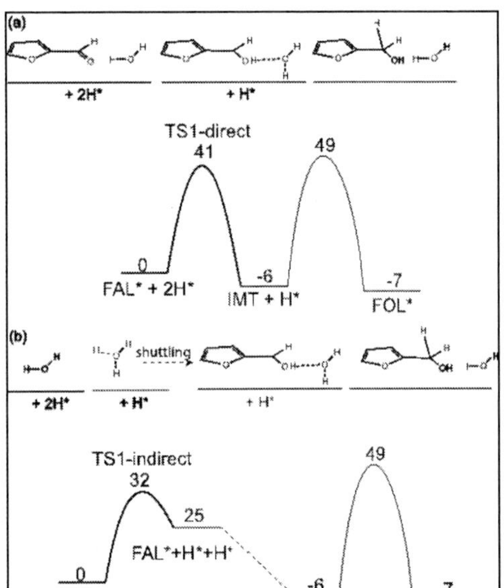

Figure 3. DFT calculations of hydrogenation of the carbonyl group of FAL in the presence of water. (a) Reaction path via a local water molecule in the proximity of the furfural molecule, (b) hydrogen is picked up by water and shuttled via the water network.

Conclusions

DFT calculations have been performed for both the adsorption and reaction of biomass-derived furfural, and the results have been compared to experimental results in a liquid phase. We find that the solvent can either perturb the molecular adsorption or participate directly in the reaction through proton transfer. This synergistic combination of experiment, theory and computation thus provides atomic-scale understanding of the reaction mechanism of bio-oil upgrading in a complicated liquid environment.

Acknowledgement. We acknowledge Dr. Sanwu Wang. Dr. Tawan Sooknoi, and Dr. Steven Crossley for fruitful discussion. This work is supported by the U.S. Department of Energy, DOE/EPSCOR (Grant DESC0004600). The computation was performed at the Extreme Science and Engineering Discovery Environment (XSEDE) and the OU Supercomputing Center for Education & Research (OSCER) at the University of Oklahoma.

References

(1) (a) Madon, R. J.; Iglesia, E., Catalytic reaction rates in thermodynamically non-ideal systems. *J Mol Catal a-Chem* **2000,** *163* (1-2), 189-204; (b) Mellmer, M. A.; Sener, C.; Gallo, J. M. R.; Luterbacher, J. S.; Alonso, D. M.; Dumesic, J. A., Solvent Effects in Acid-Catalyzed Biomass Conversion Reactions. *Angew Chem Int Edit* **2014,** *53* (44), 11872-11875.

(2) Yoon, Y.; Rousseau, R.; Weber, R. S.; Mei, D. H.; Lercher, J. A., First-Principles Study of Phenol Hydrogenation on Pt and Ni Catalysts in Aqueous Phase. *J Am Chem Soc* **2014,** *136* (29), 10287-10298.

(3) (a) Hibbitts, D. D.; Loveless, B. T.; Neurock, M.; Iglesia, E., Mechanistic Role of Water on the Rate and Selectivity of Fischer-Tropsch Synthesis on Ruthenium Catalysts. *Angew Chem Int Edit* **2013,** *52* (47), 12273-12278; (b) Zope, B. N.; Hibbitts, D. D.; Neurock, M.; Davis, R. J., Reactivity of the Gold/Water Interface During Selective Oxidation Catalysis. *Science* **2010,** *330* (6000), 74-78.

(4) (a) Crossley, S.; Faria, J.; Shen, M.; Resasco, D. E., Solid Nanoparticles that Catalyze Biofuel Upgrade Reactions at the Water/Oil Interface. *Science* **2010,** *327* (5961), 68-72; (b) Shi, D. C.; Faria, J.; Pham, T. N.; Resasco, D. E., Enhanced Activity and Selectivity of Fischer-Tropsch Synthesis Catalysts in Water/Oil Emulsions. *Acs Catal* **2014,** *4* (6), 1944-1952.

(5) Chheda, J. N.; Huber, G. W.; Dumesic, J. A., Liquid-phase catalytic processing of biomass-derived oxygenated hydrocarbons to fuels and chemicals. *Angew Chem Int Edit* **2007,** *46* (38), 7164-7183.

(6) Pham, T. N.; Shi, D. C.; Resasco, D. E., Evaluating strategies for catalytic upgrading of pyrolysis oil in liquid phase. *Appl Catal B-Environ* **2014,** *145*, 10-23.

(7) (a) Resasco, D. E.; Sitthisa, S.; Faria, J.; Prasomsri, T.; Ruiz, M. P., 5. Furfurals as chemical platform for biofuels production. In *Solid Waste as a Renewable Resource: Methodologies*, Albanese, A. J. F.; Ruiz, M. P., Eds. 2011; p 103; (b) Lange, J. P.; van der Heide, E.; van Buijtenen, J.; Price, R., Furfural—a promising platform for lignocellulosic biofuels. *ChemSusChem* **2012,** *5* (1), 150-166.

(8) Sitthisa, S.; Resasco, D. E., Hydrodeoxygenation of Furfural Over Supported Metal Catalysts: A Comparative Study of Cu, Pd and Ni. *Catal Lett* **2011,** *141* (6), 784-791.

(9) Maldonado, G. M. G.; Assary, R. S.; Dumesic, J.; Curtiss, L. A., Experimental and theoretical studies of the acid-catalyzed conversion of furfuryl alcohol to levulinic acid in aqueous solution. *Energ Environ Sci* **2012,** *5* (5), 6981-6989.

(10) Serrano-Ruiz, J. C.; Luque, R.; Sepulveda-Escribano, A., Transformations of biomass-derived platform molecules: from high added-value chemicals to fuels via aqueous-phase processing. *Chem Soc Rev* **2011,** *40* (11), 5266-5281.

(11) Kresse, G.; Furthmuller, J., Efficient iterative schemes for ab initio total-energy calculations using a plane-wave basis set. *Phys Rev B* **1996,** *54* (16), 11169-11186.

(12) Perdew, J. P.; Burke, K.; Ernzerhof, M., Generalized gradient approximation made simple. *Phys Rev Lett* **1996,** *77* (18), 3865-3868.

(13) (a) Blochl, P. E., Projector Augmented-Wave Method. *Phys Rev B* **1994,** *50* (24), 17953-17979; (b) Kresse, G.; Joubert, D., From ultrasoft pseudopotentials to the projector augmented-wave method. *Phys Rev B* **1999,** *59* (3), 1758-1775.

(14) (a) Grimme, S.; Antony, J.; Ehrlich, S.; Krieg, H., A consistent and accurate ab initio parametrization of density functional dispersion correction (DFT-D) for the 94 elements H-Pu. *J Chem Phys* **2010,** *132* (15), 154104; (b) Grimme, S.; Ehrlich, S.; Goerigk, L., Effect of the Damping Function in Dispersion Corrected Density Functional Theory. *J Comput Chem* **2011,** *32* (7), 1456-1465.
(15) Henkelman, G.; Uberuaga, B. P.; Jonsson, H., A climbing image nudged elastic band method for finding saddle points and minimum energy paths. *J Chem Phys* **2000,** *113* (22), 9901-9904.

COUPLING BIOLOGICAL AND CHEMICAL CATALYSIS FOR THE PRODUCTION OF BIO-BASED POLYAMIDES

John E. Matthiesen, Miguel Suastegui, Jack M. Carraher, Zengyi Shao, Jean-Philippe Tessonnier

Chemical and Biological Engineering, Iowa State University, Ames, IA 50010 (USA)
NSF Engineering Research Center for Biorenewable Chemicals (CBiRC), Ames, IA 50010 (USA)
tesso@iastate.edu

Introduction

Recent progress in metabolic engineering has provided new microorganisms able to convert carbohydrates and waste lignin into platform chemicals that can be further diversified using chemical catalysts.(1-3) Among the various platform chemicals investigated, muconic acid (MA) holds high promise as an intermediate for the production of commercially important commodity chemicals including adipic and terephthalic acids (Fig. 1).(4,5) However, to achieve economically sustainable conversion, it is crucial to streamline the fermentation and downstream catalytic processing steps.(6) This generates new challenges as biogenic impurities inherent to the fermentation broth readily deactivate Pt group metal catalysts.(1) As a result, substantial purification and catalyst engineering are required to mitigate catalyst deactivation. An alternative approach would be to utilize novel conversion schemes using robust base metals that are stable in the presence of biogenic impurities. We show that electrocatalytic hydrogenation (ECH) represents a promising approach for the direct conversion of unpurified broths.(6)

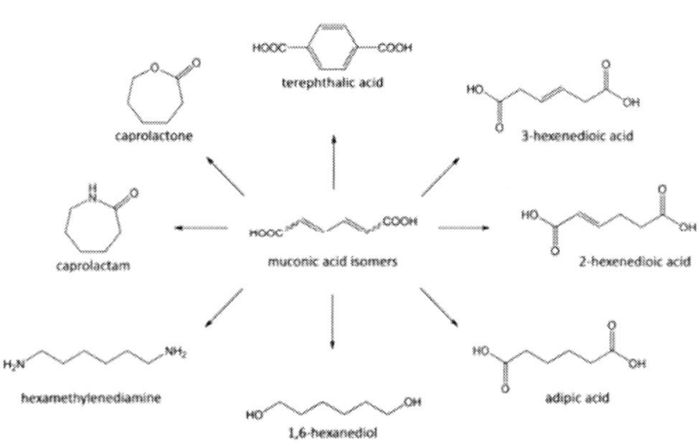

Figure 1. Value-added products obtained from the catalytic diversification of bio-based muconic acid.

Experimental

To establish the *Saccharomyces cerevisiae* production platform, the three-gene heterologous pathway was introduced into the host to convert dehydroshikimic acid (DHS) to muconic acid (7). Further manipulations to increase the precursor availability included the overexpression of the transketolase gene (*tkl1*) and a feedback insensitive 3-dehydroxy-D-arabino-heptulose-7-phosphate (DAHP) synthase ($aro4_{K229L}$) to remove the repression caused by tyrosine in the medium. To increase the carbon flux into the shikimic acid pathway while reducing the conversion of DHS into the downstream metabolites, a mutant version of the pentafunctional *AROM* enzyme was engineered. This was achieved by site-directed mutagenesis to substitute the aspartic acid residue 1409 to alanine ($aro1_{D1409A}$). Implementation of oxygen-controlled fermentations at a medium level (10-20% dissolved oxygen) were necessary to mitigate the bottleneck step caused by the enzyme protocatechuate decarboxylase (*aroY*). These manipulations together led to production of MA at a titer of 559.5 mg L^{-1} from glucose, improving the production 4-fold compared to the previously reported strains.(7)

The electrochemical studies were conducted in a three-electrode electrochemical cell using a BioLogic VSP-300 potentiostat. The Ag/AgCl reference electrode and platinum counter electrode were purchased from BioLogic Science Instruments. Catalyst screenings were conducted with 50 ml solutions of 3.52 mM *ct*MA and 0.26 M formic acid (electrolyte). Formic acid was preferred over sulfuric acid to more closely resemble the hydrogenation of MA in a yeast fermentation broth and for ease in preparation for product analysis by ultra-performance liquid chromatography coupled with photodiode array (PDA) and QDa mass detectors (UPLC-PDA-MS). The working electrodes were purchased as follows: Sigma (Pd, Mo), Flinn Scientific (Cu, Fe, Pb, Zn), and Science Company (Ni). During chronoamperometry experiments, 0.5 ml aliquots of the reaction medium were collected at 5, 15, 30, and 60 min for analysis.

ECH of the fermentation medium was performed as follows. Directly after the fermentation, 11 mL of broth was put in a small volume cell. At the start of the ECH, the surface of a lead electrode was polished with a kimwipe and used as the electrocatalyst. In-between the successive batch reactions, the lead electrode was gently dipped in deionized water, dried, and put in another 11 mL solution of the fermentation medium. In addition, the counter and reference electrodes were washed with deionized water in-between each run. To achieve a 94% yield of 3-hexenedioic acid (HDA) the pH of the fermentation broth was decreased to 2.0 by dropwise adding 0.5 M H_2SO_4. Additionally, the concentration of MA was increased to 2.4 mM.

Samples were analyzed either by UPLC-PDA-MS or ^{1}H NMR. For NMR analysis, the samples were dried at room temperature, reconstituted in deuterium oxide, and analyzed with a Bruker 600 MHz NMR spectrometer (AVIII600). UPLC analysis was performed with a Waters H-Class Acquity chromatograph equipped with a HSS C18 Column (1.8 μm, 2.1 x 100 mm).

Results and Discussion

The initial screening of a variety of early, late, and post transition metal catalysts for the ECH of muconic acid revealed trends consistent with the free energy of hydrogen binding ΔG_H for these metals. Specifically, weak hydrogen binding on high overpotential metals strongly favored ECH over the competing hydrogen evolution reaction (HER). The highest yields were observed with lead (Pb) and, therefore, this metal was subsequently tested for the electrocatalytic processing of real fermentation broths.

The water, salts, and acidic impurities already present in the spent fermentation broth provided a suitable electrolyte and Pb performed well under these conditions despite the presence of a myriad of biogenic impurities and other fermentation byproducts. Protons already present in the solution were electro-reduced to supply *in situ* hydrogen for the desired hydrogenation reaction. Muconic acid was converted to 3-hexenedioic acid, an important monomer for advanced polyamides and polyesters, with a 94% yield and 100% faradaic efficiency after optimizing the reaction conditions. Repeated runs showed no deactivation, thus confirming that lead is unaffected by the N- and S-containing biogenic impurities present in the broth. The synthesized 3-hexenedioic acid was subsequently isolated, purified, and tested in the synthesis of bio-based unsaturated polyamide 6,6, a bio-advantaged polymer analogue to Nylon 6,6. The obtained polymer consisted of a transparent, partly crystalized material with comparable physical and chemical properties to petrochemical nylon. Polymers based on blends of hexenedioic and adipic acids were also achieved to enable different levels of tenability, for example by grafting sites to custom-tailor the existing nylon grades with desired crosslinkability, paintability, flame retardance, etc.

Conclusions

This work demonstrates a new strategy to bridge the gap between biological and chemical catalysis in biorefineries.(6) Electrocatalytic processing of fermentation broths allows the efficient conversion of platform intermediates obtained by fermentation to the desired commodity chemicals in a single step and under ambient conditions while, at the same time, minimizing water and chemical utilization. Here, muconic acid was converted to 3-hexenedioic acid, a monomer for unsaturated polyamides.

Acknowledgement. This material is based upon work supported in part by the National Science Foundation Grant Numbers EEC-0813570, EPSC-1101284, and CBET-1512126. Research at the Ames Laboratory was supported by the U.S. Department of Energy-Laboratory Royalty Revenue through Contract No.DE-AC02-07CH11358.

References

(1) Schwartz, T. J.; Johnson, R. L.; Cardenas, J.; Okerlund, A.; Da Silva, N. A.; Schmidt-Rohr, K.; Dumesic, J. A. *Angew. Chem. Int. Ed.* **2014**, *53 (47)*, 12718.

(2) Schwartz, T. J.; O'Neill, B. J.; Shanks, B. H.; Dumesic, J. A. *ACS Catal.* **2014**, *4 (6)*, 2060.

(3) Schwartz, T. J.; Shanks, B. H.; Dumesic, J. A. *Curr. Opin. Biotechnol.* **2016**, *38*, 54.

(4) Vardon, D. R.; Franden, M. A.; Johnson, C. W.; Karp, E. M.; Guarnieri, M. T.; Linger, J. G.; Salm, M. J.; Strathmann, T. J.; Beckham, G. T. *Energy Environ. Sci.* **2015**, *8 (2)*, 617.

(5) Vardon, D. R.; Rorrer, N. A.; Salvachua, D.; Settle, A. E.; Johnson, C. W.; Menart, M. J.; Cleveland, N. S.; Ciesielski, P. N.; Steirer, K. X.; Dorgan, J.; Beckham, G. T. *Green Chem.* **2016**, in press.

(6) Suastegui, M.; Matthiesen, J. E.; Carraher, J. M.; Hernandez, N.; Rodriguez Quiroz, N.; Okerlund, A.; Cochran, E. W.; Shao, Z.; Tessonnier, J.-P. *Angew. Chem. Int. Ed.* **2016**, *55 (7)*, 2368.

(7) Curran, K. A.; Leavitt J. M.; Karim, A. S.; Alper, H. S. *Metabolic Engineering* **2012**, *15*, 55.

KINETICS OF C-C COUPLING OF CARBOXYLIC ACIDS VIA KETONIZATION AND ACYLATION OVER ZEOLITES

*Abhishek Gumidyala, Manasa Godavarthy, Bin Wang and Steven Crossley**

School of Chemical, Biological, and Materials Engineering
University of Oklahoma

Introduction

The efficient conversion of molecules present in streams derived from lignocellulosic biomass to useful fuel and chemicals is a daunting task. Thermochemical routes such as pyrolysis transform the polymers in lignocellulosic biomass to monomeric species yielding a complex mixture of furanics, phenolics, and small oxygenates such as carboxylic acids[1] that when stored together recombine to form polymers at room temperature.[2]

The acidity inherent to these streams caused by acetic acid facilitates the polymerization of furanic species at room temperature, creating obvious storage and transportation challenges. This thermal instability also leads to challenges in separating the various species, as traditional separation methods such as vacuum distillation result in significant amounts of polymerization in the process.

Stabilization through severe hydrodeoxygenation decomposes small molecules, such as acetic acid, to low-value C_1 and C_2 containing species. While this approach yields a stable product, the excessive requirement of hydrogen and limited amount of products that are liquid at room temperature hinders the economics of the overall process.

This complexity can be mitigated to some degree by upfront separation strategies. Staged thermal approaches, such as torrefaction, can selectively decompose hemicellulose and cellulose.[3] While this approach minimizes the amount of phenolics in the stream, the resulting product is a blend consisting primarily of furanic and carboxylic acid species that are difficult to separate. The acetic acid present in these streams is largely responsible for the storage and transportation challenges, especially when furanic and phenolic components are also present. Acetic acid is the most abundant single compound in the bio-oil, and its acidity is known to present corrosion problems and promote polymerization of other species present in pyrolysis oil. The resulting solids are not readily upgraded and accelerate catalyst deactivation, which is a major concern for upgrading pyrolysis oil. A strategy to selectively couple this acetic acid to more stable molecules would be highly advantageous. If the resulting chemistry converts the carbon contained in acetic acid to larger transportation fuel precursors, this is even more desirable. There is a practical requirement of conducting this chemistry in the presence of water and furanic species that will inherently be present in these streams.

Figure 1: Decarboxylative ketonization.

The ketonization reaction (Figure 1) is a well-known and effective method for carboxylic acid upgrading as it converts two acids to a ketone while removing oxygen in the form of water and CO_2. This reaction is very appealing as the stable ketones formed can be further converted to longer-chain hydrocarbons via aldol condensation followed by hydrogenation. This reaction is known to occur over basic and reducible oxides. The mechanism for this reaction over oxides is commonly believed to occur via a dehydration between two acids to form a β-ketoacid intermediate as the rate-determining step (RDS) as highlighted by a recent review from our group.[4]

As will be discussed in more detail in section 2, the same ketonization reaction can be catalyzed by Brønsted acid sites. The primary difference is in the reactive intermediates involved. In this contribution, we explore the critical role of surface acyl species on the overall ketonization rate. We will use a combination of survace Temperature Programmed techniques and spectroscopy coupled with reaction kinetics and DFT calculations to explore the surface kinetics of this reaction.

The role of stable acyl intermediates in this reaction enables a second C-C coupling pathway via acylation. Several of the oxygenated species inherently present in bio oils can serve as acyl acceptors. An example of the acylation of a biomass derived methylfuran molecule with acetic acid is shown in **Figure 2** below.

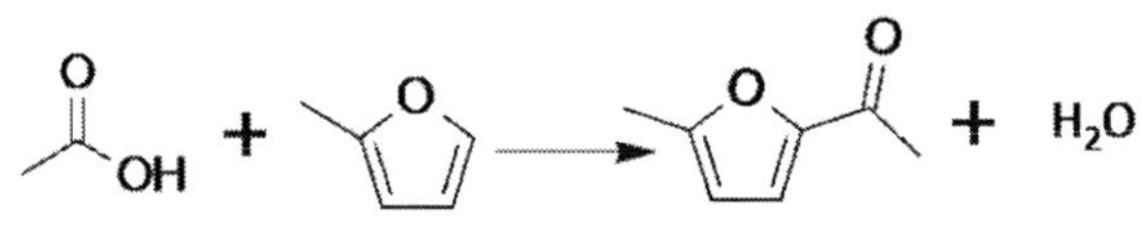

Figure 2: Acylation reaction pathway with methylfuran.

An inherent challenge associated with the conversion of these molecules is catalyst deactivation. If acyl species are indeed important surface intermediates, they will be in equilibrium with acylium ions as well as gas-phase ketene species. Ketenes are known coke precursors, so conditions that favor high abundance of ketenes must be avoided. Biomass derived acyl acceptor molecules such as methylfuran depicted above can also deactivate catalysts via polymerization over Brønsted zeolites. Therefore, detailed understanding of the intrinsic rate of each competing reaction as well as the coverage of each of the surface species is necessary to achieve a catalyst that is selective and stable for this reaction.

Finally, any reaction that involves the conversion of biomass derived oxygenates should consider the influence that water will have on the reaction. Water could play several roles on these reactions by competing for surface sites, accelerating reactions involved in proton transfer, as well as interacting with surface intermediates. The influence of water on each of these reactions, both in terms of intrinsic rates and catalyst stability are discussed.

Experimental

A typical procedure for gas-phase flow reactions is as follows. The catalyst was packed between quartz wool in a 1/4 inch quartz reactor and pretreated at 300 °C for one hour in helium flow. A thermocouple was attached on outside wall of the reactor near the catalyst bed to maintain the reactor temperature. After pretreatment, the temperature was adjusted to the required reaction temperature and reactants were fed through a syringe pump. Inlet of the reactor is heated to create vaporization zone for the reactants and out of the reactor up to six port valve is heated to 250 °C to prevent condensation of products. Downstream of the six port is connected to HPGC equipped with flame ionization detector and innowax column for analysis of products. Temperature programmed desorption studies are performed in a heated reactor system equipped with a syringe port and size port valve for injection of reactants, as well as an online MKS MS for off gas analysis.

Results and Discussion

Ketonization kinetics. An initial TPD experiment shows two distinct regimes. Upon heating the catalyst sample that has been exposed to acetic acid, water is evolved at temperatures below 220°C. Further heating reveals the simultaneous evolution of CO_2 and acetone. The higher heat of adsorption of water than CO_2 over Brønsted sites eliminates any contribution of readsorption and other bed effects on the conclusion that carboxylic acids first dehydrate and subsequently couple, with higher temperatures required for the latter.

A kinetic rate expression can be developed by assuming that an adsorbed acid and an adsorbed acyl interact during the rate-determining step to form acetone.

$$RCOOH / RCO^{+} \xrightarrow{k} RCOR \bullet H^{+} + CO_2$$

An expression highlighting the second order dependence of this rate-determining step results in the following equation:

$$rate = \frac{k K_{Cou} K_{equilibrium} K_{Acid} P^{2}_{Acid}}{P_{H_2O}\left(1 + K_{Acid}P_{Acid} + \frac{K_{equilibrium}K_{Acid}P_{Acid}}{P_{H_2O}} + \frac{K_{Cou}K_{equilibrium}K_{Acid}P^{2}_{Acid}}{P_{H_2O}} + K_{Ketone}P_{Ketone} + K_{H_2O}P_{H_2O} + K_{CO_2}P_{CO_2}\right)}$$

Several Flow reactions have been conducted across a wide range of temperatures, and the equation above provides an adequate fit of the data as shown by the lines fit to the data in **Figure 3** below. It should be noted that initial estimates for heats of adsorption for several of the species, such as acetic acid, water, CO_2, and acetone, were taken either from published data in the literature over MFI catalysts or microcalorimetery data obtained in our lab, with final fitted values not deviating more than 10% from the initial estimates.

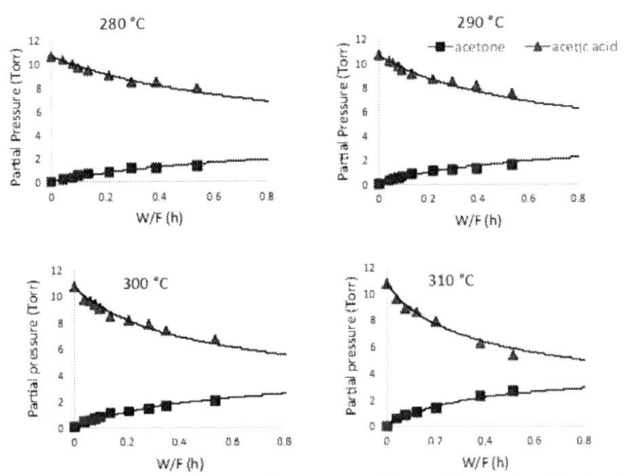

Figure 3: Preliminary kinetic fitting of self-coupling of acids in the presence of water by assuming the second adsorbed acid interacting with an acyl as the RDS.

These results coupled with TPD and other isotopic labeling results (not shown) demonstrate the importance of acyl intermediates in this reaction, but a 2nd order coupling RDS. The coupling RDS leads one to question the role of water in this reaction, as water will undoubtedly be present biomass derived streams. Reactions co-fed with water illustrate a -0.46 of the ketonization rate with respect to water. This order was taken at initial time on stream values. At longer time on stream, the apparent order with respect to water significantly diminishes, with an apparent -0.25 order with respect to water observed after only 80 minutes time on stream. This is due to the significant enhancement in catalyst stability brought about by the introduction of water to the system. While water can compete for sites, as well as accelerate the equilibrium concentration of surface acyl species, the negative consequences are balanced by an improvement in stability, likely a result of the reaction of water with gas phase ketene coke precursors.

Acylation. The coupling of acetic acid with other biomass derived species such as phenolic and furanic compounds should occur under conditions identified above where the dehydration of carboxylic acids to form surface acyl species is kinetically favored while C-C coupling with other carboxylic acids is not. A simple test to investigate this kinetic regime is shown in **Figure 4**. The acetic acid saturated surface was subsequently heated to 220°C to evolve water and expose surface acyl species. These species are then subjected to isothermal pulses of methylfuran at 220°C. Results indicate that acetylmethylfuran is produced exclusively, further confirming the presence of acyl surface species under these conditions. The production of acetylmethylfuran diminishes with subsequent pulses as the surface acyl species are consumed.

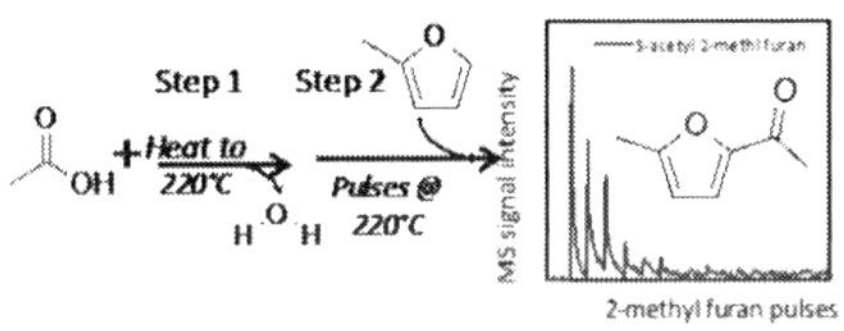

Figure 4 Acylation of methylfuran with acetic acid via a pulse experiment following the generation of surface acyl species

Flow reactions confirm the high selectivity to this reaction under conditions with appropriate surface coverages of both species. The reaction exhibits a -0.7 order with respect to water, but a similar enhancement in catalyst stability is observed. The apparent reaction energy barrier of 24 kJ/mol agrees well with DFT calculations, implying that the formation of acyl species is the rate-determining step in this reaction, both for the acylation of methylfuran as well as several phenolic species. This is very promising for biomass conversion, as the light acids can be coupled with other species to grow chains of uninterrupted C-C bonds without sacrificing carbon as CO_2 in the process.

Conclusions

Acetic acid conversion over Brønsted zeolites reveals that the reaction proceeds through surface acyl species. These species can couple with other acids to produce acetone and CO_2, which can be fit well to a Langmuir-Hinshelwood rate expression assuming a second order coupling RDS. Water inhibits the reaction with an order of -0.46, but water also has a significant positive influence on catalyst stability by reacting with ketene species.

Surface acyl species can also readily be coupled with acyl acceptors, exhibiting a lower barrier than self coupling via ketonization. Acylation with methylfuran exhibits a reduced apparent barrier of 24 kJ/mol, which implies an initial dehydration RDS. The selectivity for each route after acyl formation can be tuned by manipulating the temperature and partial pressure of the acyl acceptors. These reactions offer promising routes for the coupling of abundant carboxylic acids to form fuel precursors from biomass.

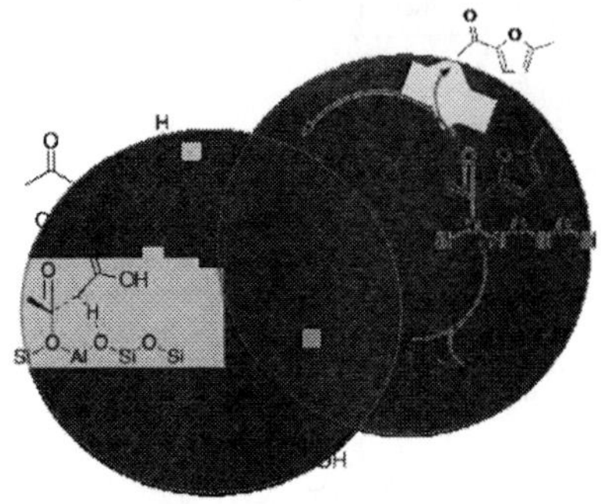

Acknowledgements. Financial support from the American Chemical Society (grant PRF 54107-DNI5) and the Department of Energy (Grant DE-EE0006287) is gratefully acknowledged.

References

(1) Huber, G. W.; Corma, A., *Angewandte Chemie-International Edition* **2007,** *46* (38), 7184-7201.
(2) (a) Bridgwater, A. V.; Peacocke, G. V. C., *Renew Sust Energ Rev* **2000,** *4* (1), 1-73; (b) Wan, S.; Waters, C.; Stevens, A.; Gumidyala, A.; Jentoft, R.; Lobban, L.; Resasco, D.; Mallinson, R.; Crossley, S., *ChemSusChem* **2015,** *8* (3), 552-559.
(3) Resasco, D. E.; Crossley, S. P., *Catalysis Today* **2014**.
(4) Pham, T. N.; Sooknoi, T.; Crossley, S. P.; Resasco, D. E., *Acs Catalysis* **2013,** *3* (11), 2456-2473.

EXPLAINING OBSERVED SOLVENT EFFECTS ON RU SURFACES: A MICROKINETIC ANALYSIS

Jesse Q. Bond and Omar A. Abdelrahman

Biomedical and Chemical Engineering
Syracuse University
jqbond@syr.edu

Levulinic acid (LA) hydrogenation over supported Ru catalysts is frequently reported to exhibit strong solvent effects, where water provides the greatest promotional effect amongst solvents tested. Interpretation of solvent effects are commonly undertaken without any comparison to a solvent-free condition, kinetic measurements are also complicated by the presence of a solvent. Kinetic studies of LA hydrogenation would therefore be clearer in the absence of a solvent, gas phase hydrogenation, unfortunately the low volatility and reactive nature of LA prohibits such a study. Here we present a comparative kinetic study of 2-pentanone hydrogenation in the gas phase and in various solvent environments, where 2-pentanone serves as the monofunctional ketone analog of LA that can be readily vaporized. Kinetics of the gas phase hydrogenation of C_3-C_5 ketones over Ru/SiO_2 catalysts was examined over a range of partial pressures and temperatures ranging from 303-476 K. A single thermodynamically consistent microkinetic model is developed to fit the observed experimental trends over all conditions and chain length, where a step-wise addition of hydrogen Horuiti-Polanyi mechanism is adopted. The reaction is found to proceed through a reactive alkoxide intermediate on the surface, where the addition of hydrogen to the alkoxide forming the alcohol on the surface is found to be the rate determining step. The developed model is then expanded to include the presence of multiple solvent functionalities on the surface including water, alcohols, alkanes and ethers.

CATALYTIC CONSEQUENCES OF MICROPORE TOPOLOGY, MESOPOROSITY, AND ACIDITY ON THE HYDROLYSIS OF BIOMASS OVER ZEOLITE CATALYSTS

Yao He, Thien Nguyendo, Thomas Hoff, Yiqing Wu and Dongxia Liu

Department of Chemical and Biomolecular Engineering, University of Maryland, College Park, MD, 20742
Dongxia Liu: liud@umd.edu

Introduction

Biomass attracts considerable attention as a renewable feedstock due to its huge reservoir of renewable carbon and ''CO_2 moderate" impact on the environment. The hydrolysis of biomass such as sucrose and inulin allows its conversion into glucose and fructose sugars, representing the starting point for the synthesis of chemicals or fuels by acid-catalyzed reactions from polysaccharides. Meso-/microporous zeolites contain both meso- and micropores, which allow the facile pore access and fast transport of bulky molecules and thereby enable beneficial effects on the activity, selectivity, and/or stability in a wide range of catalyzed reactions.[1] A systematic study on the effects of zeolite microenvironment (micropore topology, mesoporosity, and acidity) of the conventional microporous zeolites and the emerging meso-/microporous zeolites on the sucrose and inulin hydrolysis reaction has not been reported.

In the present study, we assess the catalytic consequences of micropore topology, mesoporosity, and acidity of a range of microporous and meso-/microporous zeolite catalysts on the hydrolysis of sucrose and inulin. The commercially available zeolite catalysts, ferrierite (FER), MOR, BEA, MFI, and FAU with a range of Si/Al ratios, were employed to examine the effects of zeolite micropore topology and acidity on the sucrose hydrolysis reactions. The effect of mesoporosity on the sucrose hydrolysis reactions was investigated by using the meso-/microporous pillared MWW (PMWW) and pillared MFI (PMFI) zeolite catalysts. The present study leads to a systematic understanding of the effects of zeolite microenvironment (topology, mesoporosity, and acidity) on conversion of sucrose, inulin and other related biomass molecules under environmentally friendly conditions.

Experimental

Preparation of zeolite catalysts. The conventional microporous FER, MFI, MOR, BEA, and FAU with different acidity (Si/Al ratio) were purchased from Zeolyst. MWW and PMWW were prepared by using the method described by Maheshwari et al.[2] Pillaring of multilamellar MFI was done as reported by Na et al.[3] to produce PMFI. All zeolite samples in their NH_4^+-form were treated in dry air (100 mL min^{-1}, ultrapure, Airgas) by increasing the temperature from ambient temperature to 823 K at 1.45 K min^{-1} and holding for 4 h to thermally decompose NH_4^+ to NH_3 and H^+. To differentiate the same type of zeolite with different Si/Al ratios, each catalyst is named by its structure type and Si/Al ratio in the remainder of this paper.

Catalytic hydrolysis of sucrose. The liquid phase catalytic hydrolysis of sucrose (Sigma-Aldrich, 99.5% purity) in DI water was carried out in a three-neck round-bottom flask (100 mL) equipped with a reflux condenser and heated in a temperature controlled oil bath under atmospheric pressure and magnetic stirring (1" stirring bar, 500 rpm stirring speed) conditions. In a typical experiment, 20 mL of DI water was added to the desired amount (typically 0.2 g) of zeolite catalyst. The as-obtained mixture was maintained for 0.5 h at the required reaction temperature and stirring conditions and then 1.0 g of sucrose was added. Liquid samples were withdrawn at regular intervals and analyzed by a high performance liquid chromatography (Agilent 1100 HPLC) equipped with a Bio-Rad Aminex HPX-87H column connected to an autosampler and a refractive index detector to calibrate and separate the reactants and products.

Results and Discussion

Topological properties of zeolite catalysts. Small, medium, and large pore zeolites are classified according to their pore openings that are 8-, 10-, and 12-membered ring (MR), respectively. Table 1 summarizes the types of zeolites and graphical representation of the pore systems of each zeolite catalyst. FER zeolite contains one-dimensional chann2els of 10-MR (0.42 x 0.54 nm) and one-dimensional channels of 8-MR (0.35 x 0.48 nm) that are perpendicularly intersected. MFI zeolite consists of two interconnected 10-MR channel systems: one is straight running along the *b*-axis direction (0.53 x 0.56 nm) and the other is zigzag running parallel to the *a*-axis direction (0.51 x 0.55 nm). The PMFI contains mesopores created by the inorganic pillar species sitting between

Table 1. Topology and graphical representation of pore systems of the medium-pore, large-pore, and mesoporous zeolite catalysts used in the biomass hydrolysis reactions

Zeolite		Si/Al ratio	Pore structure		Graphical representation of pore topology
			Pore shape	Pore size (nm)	
Medium-pore	FER	28	8 MR 10 MR	0.35 x 0.48 0.42 x 0.54	
	MFI	12 40 70	10 MR 10 MR	0.51 x 0.55 0.53 x 0.56	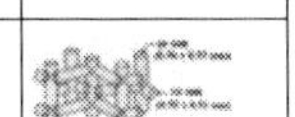
Large-pore	MOR	45	8 MR 12 MR	0.34 x 0.48 0.65 x 0.7	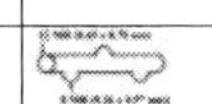
	MWW	20 30	10 MR 10 MR supercage side pocket	0.41 x 0.55 0.41 x 0.51 0.71 x 1.82 0.71 d x 0.9 h	
	BEA	12 19 100	12 MR 12 MR	0.56 x 0.56 0.66 x 0.67	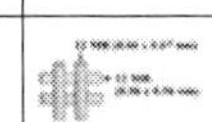
	FAU	15 40	12 MR 12 MR	0.74 x 0.74 0.74 x 0.74	
Mesopore	PMFI	70 150 200	10 MR 10 MR mesopore'	0.51 x 0.55 0.53 x 0.56 2.8	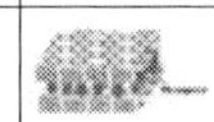
	PMWW	30	10 MR supercage side pocket mesopore'	0.41 x 0.51 0.71 x 1.82 0.71 x 0.9 1.8	

MFI layers, parallel to the zigzag channels and perpendicular to the straight channels within the layers. MWW structure contains two independent pore systems. One system is defined by sinusoidal 10-MR channels with dimensions of 0.41 x 0.51 nm and the other system consists of supercages delimited by 12-MR channels with dimensions of 0.71 x 0.71 x 1.81 nm. The consecutive supercages are connected through slightly distorted elliptical 10-MR windows (0.40 x 0.55 nm). The PMWW was created by pillaring the MWW layers by SiO_2, containing the 10-MR sinusoidal channels and hourglass shaped pores (half of the supercages in MWW) within the intact layers and mesopores between the layers. Table 1 also shows that MOR zeolite has the 8-MR pocket (0.34 × 0.48 nm) and 12-MR (0.7 × 0.65 nm) channel systems, with the 8-MR pockets sitting in the walls of 12-MR channels. BEA zeolite structure consists of 12-MR straight channels of a free aperture of 0.66 x 0.67 nm viewed along *a*-axis and 12-MR zigzag channels of 0.56 x 0.56 nm viewed along *c*-axis. FAU zeolite contains 12-MR pore mouths (0.74 x 0.74 nm) that are 3-dimensionally connected with internal supercages (~ 1.3 nm). The topological analysis above indicates that FER and MFI are medium-pore zeolites, MOR, MWW, BEA, and FAU are large-pore zeolites, and PMWW and PMFI are mesoporous zeolites, respectively.

Sucrose hydrolysis reaction. In water solvent, sucrose (A) was consumed simultaneously via hydrolysis (A+ H_2O → B + C) reactions to form fructose (B) and glucose (C). Because an excess amount of water was used, the hydrolysis reaction can be approximated as pseudo-first order in the sucrose. The profiles of sucrose conversion as a function of reaction time and $\ln(C_A^0/C_A)$ as a function of reaction time in the hydrolysis of sucrose for different temperatures over the investigated catalysts with variable micropore topology, mesoporosity, and acidity. The effect of zeolite micropore topology, acidity, and mesoporosity on the hydrolysis of sucrose reactions and the corresponding kinetic parameters (measured activation energies and entropies) are discussed in sequence.

Figure 1 shows the rate constants (k_{hydro}) of sucrose hydrolysis at 352 K over zeolites with similar Si/Al ratio but different micropore topology. The sucrose hydrolysis rate on FER-28 and MWW-30 (with similar Si/Al ratio of ~30) increases with increasing zeolite micropore sizes from 8-MR x 10-MR channels in FER to 10-MR channels x 12-MR supercages in MWW. The increase in reaction rate with increasing micropore sizes of zeolites is also observed for the hydrolysis reaction over zeolite MFI, MOR, and FAU with Si/Al ratio of ~40, zeolite MFI and BEA with Si/Al ratio of ~12, and BEA, MWW, and FAU with Si/Al ratio of ~20, respectively. These results clearly indicate that the hydrolysis of sucrose preferentially occurs in large-pore zeolites. In contrast, spatial constraints imposed by small pores of zeolites may mainly allow surface catalysis to occur.

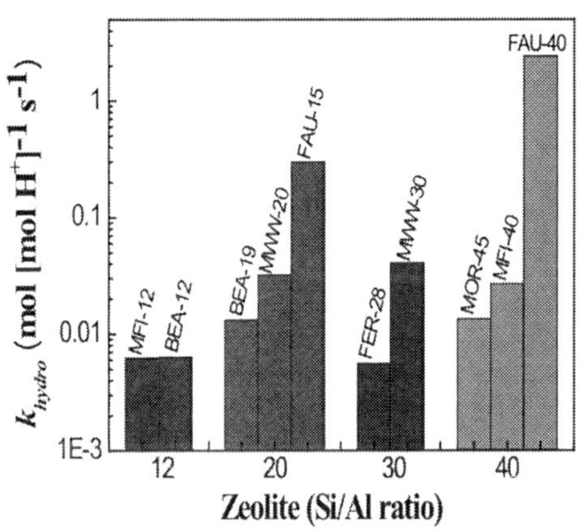

Figure 1. Rate constant of sucrose hydrolysis over zeolite catalysts with the similar Si/Al ratio and different micropore topology (T = 352 K).

The comparison of sucrose hydrolysis reaction rates of the meso-/microporous zeolites (PMFI and PMWW) and the large-pore zeolites (BEA and FAU) shows that the meso-/microporous zeolites have comparable activity to those of BEA-type zeolites, but lower activity (~1 order of magnitude) than that of FAU-type zeolites. This result indicates that the hydrolysis of sucrose is dominated by the zeolite microporous topological properties. PMFI and PMWW contain mesopores between the zeolitic layers and micropores within the zeolite layers. The pillaring causes the increase in mesoporosity, but not on zeolite microporosity.

Figure 2 compares the activation energies (ΔE_{meas}) and entropies (ΔS_{meas}) for the sucrose hydrolysis over all the investigated zeolite catalysts. Activation energies are comparable for the medium-pore, large-pore, and mesoporous zeolites, in the range of 102-138 kJ mol^{-1}. The activation energies are consistent with that of sucrose hydrolysis in FAU zeolites reported by Buttersack[4] and more general hydrolysis of other disaccharides in acid solutions. The activation entropies increases from 15 to 135 J mol^{-1} K^{-1} as the zeolite category changes from the medium-pore to large-pore zeolites. The increases in Si/Al ratios in the same type of zeolite resulted in the increase in activation entropies by a factor ~2. The presence of mesopores in the meso-/microporous PMWW and PMFI zeolite leads to the increase in activation entropies by a factor ~ 1.5 compared to their microporous analogues. The changes in the reaction entropies are consistent with the variation on the zeolite microenvironment (micropore topology, mesoporosity, and acidity), indicating that the hydrolysis of sucrose over zeolite catalysts is primarily driven by the activation entropies, in accordance with reports by BeMiller et al.[5] Adsorption and diffusion should contribute to the observed reaction kinetics. The absence of adsorption equilibrium constants and diffusion coefficients of sucrose in each zeolite catalyst limits the further analysis on the intrinsic kinetic parameters of sucrose hydrolysis in the microenvironments of the zeolite catalysts.

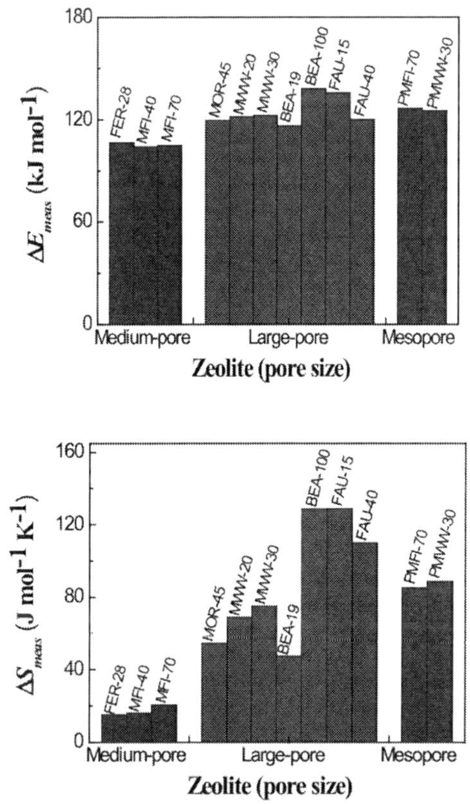

Figure 2. The measured activation energy (ΔE_{meas}) (a) and measured entropy (ΔS_{meas}) at 352 K (b) of sucrose hydrolysis reactions over zeolite catalysts with different micropore topology, mesoporosity, and acidity (Si/Al ratio).

Conclusions

The catalytic consequences of micropore topology, mesoporosity, and acidity on the hydrolysis of sucrose over zeolite catalysts were examined systematically. The results confirmed that the large-pore zeolites have higher activity than medium-pore zeolites. Meso-/microporous zeolites have higher activity compared to their microporous analogues, possibly due to the mitigated diffusion constraints and enhanced acid site accessibility, but still lower than that of large-pore FAU zeolites. The activity per acid site of the zeolite catalysts was found to be increased with decreasing the amount of acid sites in each zeolite. Similar activation energies are observed for all the investigated catalysts, while the entropies vary with each zeolite features. This indicates that activation of sucrose over the zeolite catalysts is driven primarily by the reaction entropies, which is influenced by the zeolite topological properties and acidity. The authors gratefully acknowledge support from the ACS-Petroleum Research Fund (ACS-PRF). We acknowledge the support of the Harry K. Wells Graduate Fellowship for this research.

References

(1) Cejka, J.; Centi, G.; Perez-Pariente, J.; Roth, W. J. *Catal. Today*, **2012**, *179*, 2.
(2) Maheshwari, S.; Jordan, E.; Kumar, S.; Bates, F. S.; Penn, R. L.; Shantz, D. F.; Tsapatsis, M. *J. Am. Chem. Soc.*, **2008**, *130*, 1507.
(3) Na, K.; Choi, M.; Park, W.; Sakamoto, Y.; Terasaki, O.; Ryoo, R. *J. Am. Chem. Soc.*, **2010**, *132*, 4169.
(4) Buttersack, C.; Laketic, D.; *J. Mol. Catal.*, **1994**, *94*, L283.
(5) BeMiller, J. N. *Adv. Carbohydr. Chem. Biochem.*, **1967**, *22*, 25.

STERICALLY PROTECTED AND ELECTRONICALLY ACTIVATED AZAMACROCYCLE CATALYSTS FOR LIGNIN DEPOLYMERIZATION: A NEW APPROACH TO BIOMASS VALORIZATION

Mukund S. Chorghade

CSO, THINQ Pharma
14 Carlson Circle, Natick, MA 01760

We have developed innovative, cost-effective, proprietary technology solutions for pre-treatment of biomass with synthetic catalysts for effective depolymerization of lignin. We have synthesized, efficient, sterically protected and electronically activated organic biomimetic catalysts that depolymerize lignin, at ambient temperatures and pressures, to a mixture of phenolics, hemicellulose and cellulose in excellent yield.

Salient features of our technology are:

- Structural scaffolds incorporate the aza macrocycle into the primary structure.
- Depolymerization of lignin is complete in primarily aqueous systems; limited amounts of organic solvents are used. The co-oxidants that are used are effective in neutral to mildly alkaline conditions.
- Reduction/elimination of enzyme in saccharification through in-situ enzyme production with recombinant microbes further reduces the overall production cost
- Use of plant waste that is geographically abundant and available locally as feedstock
- Improved enzyme efficiency across different environments (pH and temperature)
- Genetically modified efficient microbes to gain higher yield in the fermentation process
- Adopted Sustainable and Eco-Friendly Alternatives
- Used a wide variety of plant waste

EFFECT OF BIOCHAR SURFACE CHEMISTRY ON PYROLYTIC PHENOLIC COMPOUND PRODUCTION FROM LIGNIN

Kyung A Jung and Jong Moon Park*

Bioenergy Research Center, POSTECH, Pohang 37673, South Korea
*E-mail: kajung@postech.ac.kr

Introduction

Lignin, the main component of biomass, is a potential resource of simply producing phenolic compounds through pyrolysis reactions because of its phenylpropane units such as hydroxyphenyl (H), guaiacyl (G), and syringyl (S) units. However, bio-oil from lignin pyrolysis is composed of various phenolic compounds and its composition could change according to various factors such as biomass type, operation conditions, and catalysts (1). Unlike raw biomass, biochar, a carbonized byproduct of biomass pyrolysis, has acidic or basic surface chemistry according to reaction conditions and it is more stable during the reaction. Thus, this study characterized lignin biochar at different temperatures and estimated the potential of biochar addition to the raw lignin pyrolysis for selective phenolic compound production.

Experiments

Lignin pyrolysis. Lignin (alkali, Sigma-Aldrich Co.) pyrolysis was carried out from room temperature to different temperature (400, 600, and 800 °C) at 10 °C/min in a fixed-bed pyrolysis system. N_2 gas was flowed at 100 ml/min. The pyrolytic product was separately collected in the form of bio-oil, biochar, and gas. The lignin biochar obtained at the different temperature was added to the pyrolysis of a raw lignin feedstock to explore effect of lignin biochar surface chemistry on formation of lignin-derived phenolic compounds. Phenolic compounds in the lignin bio-oil were analyzed using a GC/MS instrument (Varian CP-3800GC/320MS). The spectra of each sample were identified by retention time and NIST mass spectral libraries through an AMDIS32 computer software.

Characterization of biochar. The raw lignin and lignin biochar obtained at different temperatures were analyzed using a FTIR spectrometer (Nicolet 6700, Thermo Scientific.), a Mastersizer Hydro 2000S particle size analyzer (Malvern Inst.), and a SEM instrument (JSM-6510, JEOL ltd.). Total acidity of biochar was determined by pH and the Boehm titration method (2).

Results and Discussion

Yields and properties of pyrolytic lignin products significantly depended on final temperature because primary pyrolysis and secondary reaction of lignin occurred according to temperature. As the final temperature increased, the bio-oil yield increased, but the biochar yield decreased. The main phenolic compounds in the lignin bio-oil were 2-methoxyphenol, creosol, 4-ethyl-2-methoxyphenol, methyl dehydroabietate, homovanillic acid, and catechol; however, the relative content of the main compounds changed according to the different final temperature.

The raw lignin feedstock had strong intensities in region of the FTIR spectra, however, lignin biochar lost the typical band of lignin corresponding to C=O stretching conjugated to the aromatic rings, and CH-H bending of methyl and methylene groups, as temperature increased. The lignin biochar particle was converted larger than the raw lignin due to agglomeration effects; however, over 400 °C, the lignin biochar particles at 600 and 800 °C became smaller than that due to more harsh thermal degradation and loss of tensile strength (Fig.1a). As temperature increased, lignin biochar formed smaller holes and carbonized structure (Fig. 1b). At 800 °C, fine pores under 10 μm-diameter newly appeared due to further secondary reaction.

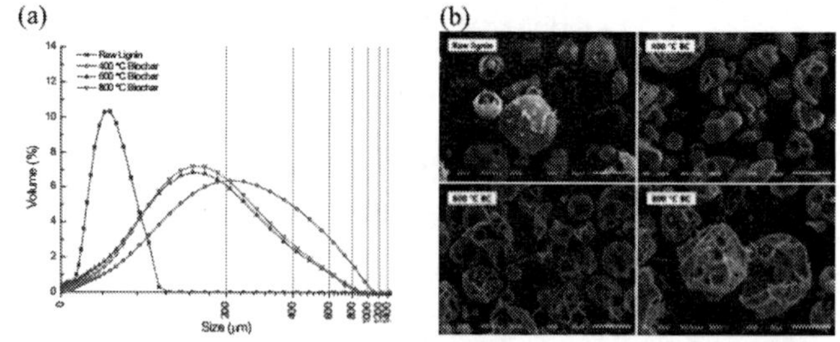

Figure 1. Characterization of lignin biochar produced at different temperatures: particle size distribution (a), and SEM images (b).

The pH of lignin biochar also changed according to the final temperature even though the raw lignin had neutral pH 7.10. At 400 °C, the pH of biochar slightly decreased (6.74); however, the pH of lignin biochar continuously increased up to pH 10.39 due to separation of alkali salts from lignin structure during thermal degradation. As temperature increased, the acidic surface groups on lignin biochar gradually decreased, but the basic surface groups suddenly increased at 800 °C (Fig. 2a). When the lignin biochar was added to the raw lignin pyrolysis, the relative content of phenolic compounds in the bio-oil changed according to the biochar production temperature (Fig. 2b). In the presence of surface basic lignin biochar, the radical-induced reactions of lignin pyrolysis might be inhibited by the weak basic sites on the lignin biochar (3).

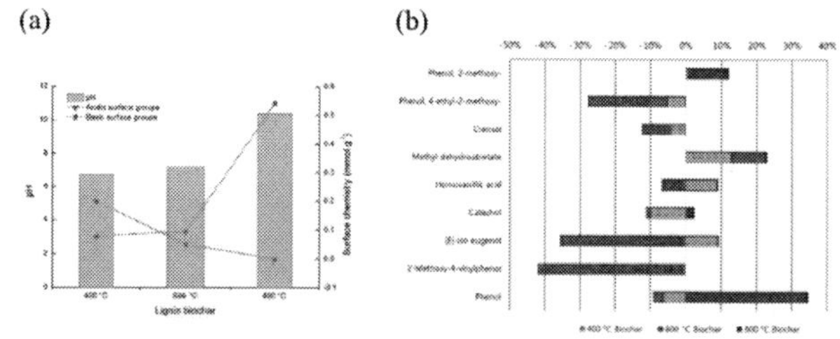

Figure 2. Surface chemistry of lignin biochar (a) and the effect of the biochar addition on the relative content of phenolic compounds in the bio-oil (b).

Conclusions

Lignin biochar produced at high temperature changed the relative content of phenolic compounds in the bio-oil of raw lignin pyrolysis due to different surface chemistry of the biochar. Thus, the lignin biochar could be utilized as an internal additive for production of lower-molecular-weight phenolic compounds from lignin pyrolysis.

References

(1) Jung K.A.; Woo S.H.; Lim S.R.; Park J.M. *Chem. Eng. J.*, **2015**, 259, 107.

(2) Novak J.M.; Lima I.; Xing B.; Gaskin J.W.; Steiner C. *Ann. Environ. Sci.*, **2009**, 3, 1939.

(3) Montes-Morán M.A.; Suárez D.; Menéndez J.A.; Fuente E. *Carbon*, **2004**, 42, 1219.

MODELING CATALYTIC VAPOR PHASE UPGRADING USING FIRST PRINCIPLES

Rajeev S. Assary, Cong Liu, Lei Cheng, Larry A. Curtiss*

Materials Science Division, Argonne National Laboratory, 9700 S. Cass Avenue, Argonne, IL, USA, 60439
assary@anl.gov

Introduction

The aim of the contribution is to provide guidance to develop efficient carbon-carbon bond forming catalysts for vapor phase upgrading of bio-oil vapors. This task involves a detailed investigation of the reaction mechanisms and factors controlling the catalytic gas/liquid phase transformation of primary and secondary products obtained from catalytic fast pyrolysis to distillate chemicals. This project is an integral part of the Computational Pyrolysis Consortium (CPC), funded by BETO (Bioenergy Technology Office). This subtask as part of the consortium would enable us to discover new and improved catalysts essential for the bio-oil upgrading and hence the BETO target of $3/gallon for biofuel by 2022.

Computational Details

All calculations presented in this work are performed using density functional theory (DFT). The cluster calculations are performed used Gaussian 09 Software and periodic DFT calculations are performed using VASP software.

Results and Discussion

In this contribution, results from two computational projects are described. First, from 'deep dive' computations that enabling us to understand mechanisms of a representative carbon-carbon coupling reaction. Second, quantum chemical screening to understand the properties of catalytic centers upon modifications of metal centers.

Pyrolysis produces aromatics (for eg.furans) and small molecular weight carbohydrate molecules (C1 to C3 molecules). Combining these molecules to form value added distillate molecules are a challenging problem in catalysis and essential to reduce the cost of bio-oil. Catalytic C-C coupling[1] and deoxygenation reactions are essential for upgrading of biomass-derived oxygenates to fuel-range hydrocarbons. Detailed understanding of mechanistic and energetic aspects of these reactions is crucial to enable and improve the catalytic upgrading of small oxygenates to useful chemicals and fuels. Using a combined computational (ANL) and experimental (NREL) approach, we have investigated the interaction between furan and acetaldehyde in a HZSM-5 zeolite catalyst, a representative system associated with the catalytic upgrading of pyrolysis vapors. Comprehensive energy profiles[2] (See Figure 1 for major reaction pathways and their barriers) were computed using periodic density functional theory for self-reactions (i.e., acetaldehyde coupling and furan coupling) and cross-reactions (i.e., acetaldehyde + furan) of this representative mixture. The results show that furan interacts with acetaldehyde in HZSM-5 via an alkylation mechanism, which is more favorable than the self-reactions, indicating that mixing furans with aldehydes could be a promising approach to maximize effective C-C coupling and dehydration while reducing the catalyst deactivation (e.g., coke formation) from aldehyde condensation.

Second task, is to generate a library of information towards catalytic properties of zeolitic centers. For example, relative Bronsted acidity, relative Lewis acidity, Stability of the catalysts, interaction with oxygenates, etc. In Figure 2, trivalent metal ions are introduced into the zeolitic-framework. Based on the computed deprotonation energy (lower deprotonation energy means higher Bronsted acidity.

Conclusions

Here, simulations based on first principle are used to understand the reaction mechanisms of carbon-carbon coupling reaction of acetaldehyde and furan. The HSZM-5 catalyst used for the reaction is modeled and modified to improve the catalytic efficacy using first principles.

Acknowledgement. This work was supported as part of the Computational Pyrolysis Consortium, funded by Eenergy Efficiency and Renewable Energy (EERE). We gratefully acknowledge the computing resources provided on "Fusion," a 320-node computing cluster operated by the Laboratory Computing Resource Center at Argonne National Laboratory. Use of the Center for Nanoscale Materials was supported by the U. S. Department of Energy, Office of Science, Office of Basic Energy Sciences, under Contract No. DE-AC02-06CH11357

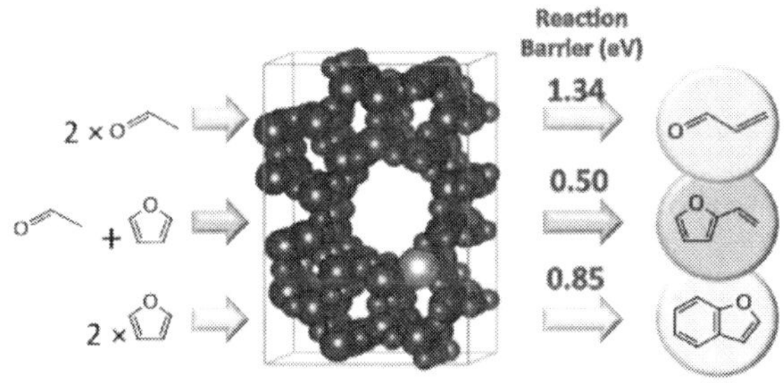

Figure 1. . Representative reactions simulated in HZSM-5 and the DFT computed reaction barriers in the gas phase

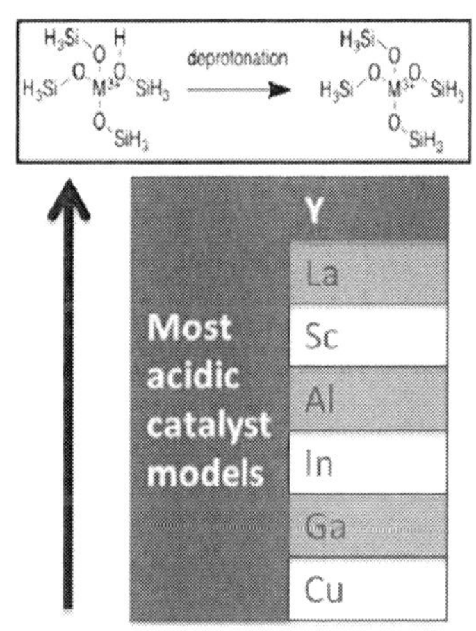

Figure 2. Based on the doping by Trivalent metal ions and computation deprotonation energy (gas phase) more acidic sites can be proposed.

References

(1) Liu, C.; Assary, R. S.; Curtiss, L. A., Computational studies of C-C coupling to increase the carbon content of furans with zeolite catalysts. *Abstr Pap Am Chem S* **2014**, *247*.

(2) Liu, C.; Evans, T. J.; Cheng, L.; Ninnlos, M. R.; Mukarakate, C.; Robichaud, D. J.; Assary, R. S.; Curtiss, L. A., Catalytic Upgrading of Biomass-Derived Compounds via C-C Coupling Reactions: Computational and Experimental Studies of Acetaldehyde and Furan Reactions in HZSM-5. *Journal of Physical Chemistry C* **2015**, *119*, 24025-24035.

MULTI-DIMENSIONAL PHASE DIAGRAMS FOR THE COADSORPTION OF AROMATIC OXYGENATES AND HYDROGEN ON METALLIC SURFACES

Vassili Vorotnikov, Glen Ferguson, Kenny Gruchalla, Nicholas Wunder, Jared Clark, Timothy Bartholomew, David Robichaud, Gregg Beckham

15523 Denver West Parkway, Golden, CO 80401, United States
Gregg Beckham

Introduction

Catalytic deoxygenation is required for valorization of pyrolysis vapors. One promising technology involves the use of metallic catalysts under various reaction conditions, including hydrotreating (HT: T ~ 500K, P_{H2} ~ 80 bar) and ex-situ catalytic fast pyrolysis (CFP: T ~ 650K, P_{H2} ~ 5 bar). However, the catalyst and operating conditions choice is currently driven by high-throughput experiments and is not well understood at the molecular level. Computational work can enhance this understanding by defining factors influencing activity, selectivity and stability of the catalysts. Recent density functional theory (DFT) studies have shown the effect that surface coverage has on reaction thermodynamics and kinetics.[1-2] For instance, hydrogenation of chemisorbed m-cresol on Pt(111) is found to be prohibitively difficult (effective E_A ~1.9 eV).[3] In contrast, hydrogenation of m-cresol physisorbed on a hydrogen-covered surface is found to be kinetically favorable (E_A ~ 0.7 eV), consistent with experiments performed at higher H_2 pressures.[2]

Coadsorption of Guaiacol and Hydrogen on Pt(111)

In this work, we performed DFT calculations to investigate the coadsorption of hydrogen and guaiacol, an oxygenate present in pyrolysis vapors, on Pt(111). Using surface free energy formulation for multiple adsorbates, we compute ab initio phase diagrams as a function of temperature, hydrogen and guaiacol pressures to investigate equilibrium surface coverage. We find that operating conditions have a significant effect on coadsorption. Under CFP conditions, guaiacol chemisorbs on the surface with sub-monolayer hydrogen coverage. On the other hand, guaiacol physisorbs on Pt(111) with at least full-monolayer hydrogen coverage under HT conditions. We demonstrate that conditions-driven surface characteristics may be the key factor in understanding the mechanism in pyrolysis vapors' upgrading.

Freely Available Computational Tool

Recognizing the need to visualize surface composition in multi-component systems, we developed a web-based tool, called Surface Phase Explorer (SPE). SPE is open-source, compilation-free and is capable of producing interactive phase diagrams from standard DFT output. We emphasize that this tool can be used to explore not only Pt(111), but the coadsorption of any molecular fragments on crystalline material surfaces.

Conclusions

In this work, we performed DFT calculations to investigate the coadsorption of hydrogen and guaiacol on Pt(111). Using these calculations, we computed ab initio phase diagrams and identified stable configurations under hydrotreating and ex-situ catalytic fast pyrolysis conditions. We find that chemisorbed guaiacol is preferred under CFP conditions while physisorbed guaiacol is preferred under HT conditions. Finally, we introduce an open-source web-based tool called Surface Phase Explorer capable of producing interactive phase diagrams using standard DFT output.

Acknowledgement. This work was supported by the Department of Energy Bioenergy Technologies Office under Contract No. DE-AC3608GO28308. Computer time was provided by Engineering Discovery Environment Grant MCB-090159 and by the National Renewable Energy Laboratory Computational Sciences Center and the Texas Advanced Computing Center under the National Science Foundation Extreme Science.

References

(1) Wang, S.; Vorotnikov, V.; Vlachos, D. G., *ACS Catalysis* **2015**, *5*, 104-112.
(2) Griffin, M. B.; Ferguson, G. A.; Ruddy, D. A.; Biddy, M. J.; Beckham, G. T.; Schaidle, J. A., *ACS Catalysis* **2016**, 2715-2727.
(3) Tan, Q.; Wang, G.; Nie, L.; Dinse, A.; Buda, C.; Shabaker, J.; Resasco, D. E., *ACS Catalysis* **2015**, *5*, 6271-6283.

CHEMICAL VIABILITY OF BIO-OILS PYROLYZED FROM DIFFERENT FEEDSTOCKS

Rebecca L. Ware[1], *Steven M. Rowland*[2], *Ryan P. Rodgers*[1,2,3], *Alan G. Marshall*[1,3]

[1] Department of Chemistry and Biochemistry, 95 Chieftain Way, Florida State University, Tallahassee, FL 32306 USA
[2]Future Fuels Institute, 1800 East Paul Dirac Drive, Tallahassee, FL 32310 USA
[3] Ion Cyclotron Resonance Program, National High Magnetic Field Laboratory, Florida State University, 1800 East Paul Dirac Drive, Tallahassee, FL 32310 USA
rbeasley@magnet.fsu.edu

Introduction

Decreased availability of desirable petroleum crude oils and rapid climate change has increased the need for renewable energy sources. Industrialization and growing global populations not only require more energy sources, but also need environmentally friendly waste management. Pyrolysis of waste materials addresses both issues by providing a crude bio-oil that can be used to supplement or replace petroleum and allow for waste disposal that does not negatively affect the environment.

As for petroleum crude oils that are compositionally different depending on their source, bio-oils are unique based on their feedstock. Although many analysis techniques have been used for characterization of a range of different bio-oils,[1–4] a direct comparison of bio-oils pyrolyzed by comparable pyrolysis methods but from different feedstocks has not been done.

Here, we present a comparison by Fourier transform ion cyclotron resonance (FT-ICR) mass spectrometry (MS), elemental analysis, and Fourier transform infrared spectroscopy (FTIR) of three fast pyrolysis bio-oils produced from municipal solid waste (MSW), recycled plastics waste (Plastics), and pine forestry residue (Pine). We also report the analysis of a silica gel fractionation for the separation of hydrocarbon species from polar species.

Experimental

Sample preparation. Bio-oil samples were obtained from PER North America (MSW, Duluth, GA), GenAgain Technologies, L.L.C. (Plastics, Lithia Springs, GA), and The University of Georgia Driftmier Engineering Center (Pine, Athens, GA). Prior to all analyses, bio-oil samples were dried over sodium sulfate (anhydrous) to remove water. Samples were also desolvated under N_2 to remove volatile components.

Elemental Analysis. Bulk elemental analysis of each of the bio-oils was performed with a Thermo Finnigan elemental analyzer (Flash EA 1112) to determine the weight percent of C, H, N, and O. C, H, and N were determined with a sample mass between 1 mg and 2 mg. S content for these bio-oils is below the limit of detection. The weighed sample was placed in a tin cup which was crushed and placed in the autosampler. O was determined in a second run with a sample mass between 1 mg and 2 mg. The weighed sample was placed in a silver cup which was crushed and placed into the autosampler. All samples were run in quintuplicate to calculate the average elemental percentage and standard deviation. Calibration of the instrument was provided by analysis of 2,5-bis(5-tert-butyl-2-benzoxazolyl)thiophene (BBOT) standard (CE Elentech, Lakewood, NJ). Each sample run also included BBOT to ensure accuracy.

FTIR. Infrared analysis was performed with a Perkin Elmer Sectrum 100 FTIR spectrometer. Sufficient sample was used to cover the zinc selenide crystal and scanned in the frequency range 4000 cm^{-1} to 500 cm^{-1}.

Sample Preparation for FT-ICR MS. MSW and Plastics samples were dissolved to 1 mg/mL in toluene and the Pine sample was dissolved to 1 mg/mL in toluene:methanol (50:50, v/v). All samples were further diluted to 100 μg/mL in toluene for positive (+) atmospheric pressure photoionization (APPI).

Instrumentation. Samples were analyzed with a custom built 9.4 T FT-ICR mass spectrometer equipped with a modular software package for data acquisition (Predator).[5,6] Multiple (50) time-domain transients were co-added, Hanning-appodized, zero-filled, and fast Fourier transformed to yield a mass spectrum.[7]

Mass Calibration and Data Analysis. Calibration and imaging is done with custom software, PetroOrg©.[8] Peak lists were generated from peaks with magnitude greater than 6 standard deviations of rms baseline noise. All spectra were internally calibrated from homologous series, whose composition differs by multiples of CH_2. Ion mass was then converted into Kendrick mass and grouped by degree of alkylation and assigned an elemental composition. Data was visualized with heteroatom class distribution histograms and isoabundance-contoured plots of double bond equivalents (DBE = number of rings and double bonds to carbon) versus carbon number for individual classes.

Silica gel fractionation. Hydrocarbons were separated from polar species on activated silica gel. Silica gel was dried overnight at 100 °C prior to column packing. After preconditioning, about 20 mg of sample was loaded onto the column and allowed to equilibrate. Saturated hydrocarbons were eluted first with cyclohexane. Next, a mix of cyclohexane and dichloromethane (DCM) (90:10, v/v) eluted the aromatic hydrocarbons. The polars were eluted with DCM, DCM and methanol (50:50, v/v), and MeOH. All fractions were dried under N_2.

Results and Discussion

Elemental analysis shows that the major differences between bio-oils produced from cellulosic waste (Pine) and those pyrolyzed from municipal wastes (MSW and Plastics) are the oxygen and carbon content (**Table 1**). A desirable bio-oils contains a high carbon content with low oxygen. MSW and Plastics contain 72.0 wt% and 82.0 wt% carbon, whereas Pine is 41.1 wt% carbon. The oxygen content of Pine is 40.4 wt% which is considerably greater than MSW and Plastics at 4.60 wt% and 1.35 wt%.

Table 1. Elemental Analysis of MSW, Plastics, and Pine Bio-Oils

	Wt% Carbon	Wt% Hydrogen	Wt% Nitrogen	Wt% Oxygen
MSW	**72.0 ± 1.60**	**9.44 ± 0.21**	**0.48 ± 0.04**	**4.60 ± 0.10**
Plastics	**82.0 ± 0.50**	**12.7 ± 0.40**	**0.30 ± 0.02**	**1.35 ± 0.06**
Pine	**41.1 ± 1.40**	**6.01 ± 0.34**	**0.19 ± 0.07**	**40.4 ± 0.40**

FTIR spectra in **Figure 1** confirm the differences in carbon and oxygen content between Pine and MSW/Plastics and reveals some functional information. Oxygen peaks, seen at 3400 cm^{-1} and 1050 cm^{-1}, are much more prominent for Pine than MSW or Plastics. The peak at 3400 cm^{-1} is from the O-H stretch; however because all water has been removed, the oxygen indicated in this region is from organic molecules. The peak at 1050 cm^{-1} (C-O) is broad and large due to the contributions from multiple oxygen functionalities. The carbonyl peak at 1700 cm^{-1} has a negligible abundance for Plastics, in accord with the imperceptible acidity of this bio-oil. The hydrocarbon content for Pine is more aromatic than MSW or Plastics, illustrated by the aromatic peak at 1600 cm^{-1} and weak alkyl peaks at 2900 cm^{-1} and 1450 cm^{-1}. Multiple peaks at 2900 cm^{-1} are due to both straight chain and branched hydrocarbons.

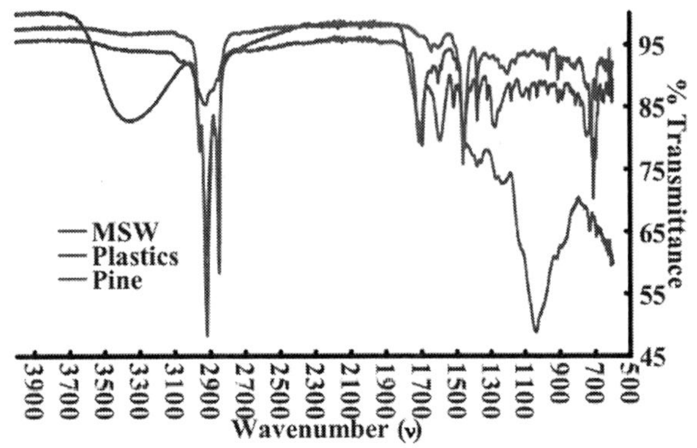

Figure 1. FTIR of MSW, Plastics, and Pine bio-oils.

FT-ICR MS accounts for the difference in bulk carbon and oxygen content and provides information regarding molecular structures. **Figure 2** shows that most of the oxygen in Pine is found in molecules that contain multiple oxygens (O_2-O_{10}). In contrast, O_1 in MSW and Plastics has the highest relative abundance of the oxygen classes, with negligible relative abundance for heteroatom classes above O_5. MSW and Plastics exhibit considerable hydrocarbons relative abundance relative to Pine. The low relative abundance of the hydrocarbon class in Pine indicates that most of the carbon is contained in molecules with oxygen.

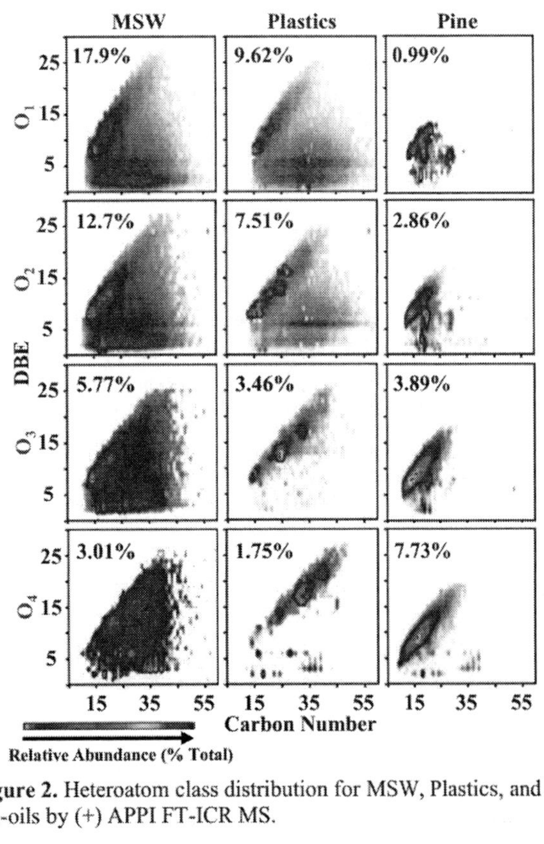

Figure 2. Heteroatom class distribution for MSW, Plastics, and Pine bio-oils by (+) APPI FT-ICR MS.

Compositional coverage displayed in DBE versus carbon number plots (**Figure 3**) shows that MSW contains higher relative abundance of aromatic hydrocarbons, whereas Plastics contains more alkylated hydrocarbons. The high relative abundance of aromatic components, presumably due to higher ionization efficiency, is not expected for MSW because the weight percent of aromatic hydrocarbons is less than that for saturates (**Table 2**). **Figure 4** shows broad compositional coverage for oxygen classes in MSW, comprised of both aromatic and alkylated regions. Pine and Plastics contain oxygen functionalities within more aromatic structures, illustrated by the area covered. Oxygen functionalities in aromatic structures are more difficult to remove and are therefore undesirable.

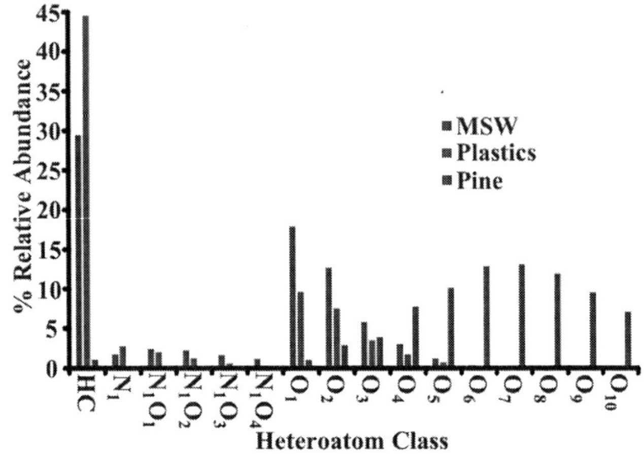

Figure 3. Isoabundance-contored plots of DBE versus carbon number for the hydrocarbon heteroatom class from MSW, Plastics, and Pine bio-oils based on (+) APPI FT-ICR MS.

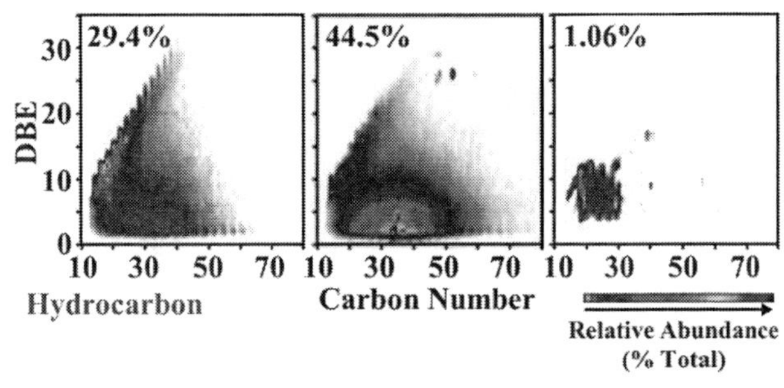

Figure 4. Isoabundance-contoured plots of DBE versus carbon number for selected oxygen-containing heteroatom classes from MSW, Plastics, and Pine bio-oils based on (+) APPI FT-ICR MS.

Fractionation on activated silica gel separates the hydrocarbon species from the polar species in each bio-oil. This fractionation also separates the saturated hydrocarbons from the aromatic hydrocarbons. Mass recovery (**Table 2**) shows that MSW and Plastics are predominantly saturated hydrocarbons, the most desirable components for fuel. In contrast, most of the mass recovery for Pine is in the polar fractions (74.5%), as expected for the high oxygen content in Pine. Errors causing >100% mass recovery include column bleed and inseparable solvent.

Table 2. Mass Recoveries for Silica Gel Fractionation of MSW, Plastics and Pine Bio-Oils

	MSW	Plastics	Pine
Saturated Hydrocarbons	49.0%	92.2%	9.7%
Aromatic Hydrocarbons	20.4%	10.8%	15.2%
Polar Species	37.8%	8.8%	74.5%
Total Recovery	107.2%	111.8%	99.4%

Conclusions

Interest in bio-oil has grown as the need for renewable fuel sources expands. Comparison of three fast pyrolysis bio-oils produced from different feedstocks shows some significant differences. First, the hydrocarbon content in MSW and Plastics is in much greater abundance and is mostly saturates. Pine is aromatic and contains mainly polar species that contain a large number of oxygen functionalities. Further work is needed to determine the functionalities present in oxygen heteroatom classes and structures for the hydrocarbon species.

Acknowledgments. Work supported by NSF Division of Materials Research (DMR-1157490) and the State of Florida. The authors thank Nathan K. Kaiser, Greg T. Blakney, and John P. Quinn for continued assistance in instrument maintenance and data analysis. The authors also thank Yuri E. Corilo for data processing and imaging software.

References

(1) Jarvis, J. M.; Page-dumroese, D. S.; Anderson, N. M.; Corilo, Y.; Rodgers, R. P. *Energy & Fuels* **2014**, *28*, 6438–6446.
(2) Demirbas, A. *J. Anal. Appl. Pyrolysis* **2004**, *72* (1), 97–102.
(3) Singh, S.; Wu, C.; Williams, P. T. *J. Anal. Appl. Pyrolysis* **2012**, *94*, 99–107.
(4) Pütün, A. E.; Özean, A.; Pütün, E. *J. Anal. Appl. Pyrolysis* **1999**, *52*, 33–49.
(5) Kaiser, N. K.; Quinn, J. P.; Blakney, G. T.; Hendrickson, C. L.; Marshall, A. G. *J. Am. Soc. Mass Spectrom.* **2011**, *22* (8), 1343–1351.
(6) Blakney, G. T.; Hendrickson, C. L.; Marshall, A. G. *Int. J. Mass Spectrom.* **2011**, *306* (2-3), 246–252.
(7) Xian, F.; Hendrickson, C. L.; Blakney, G. T.; Beu, S. C.; Marshall, A. G. *Anal. Chem.* **2010**, *82* (21), 8807–8812.
(8) © PetroOrg Software, The Florida State University, All rights reserved, http://software.petroorg.com

ELUCIDATION OF ETHANOL TO 1,3-BUTADIENE REACTION MECHANISM: COMBINED EXPERIMENTAL AND DFT STUDY

William E. Taifan and Jonas Baltrusaitis

Department of Chemical and Biomolecular Engineering
Lehigh University
111 Research Drive Bethlehem, PA 18015

Introduction

Since the rapid expansion of coal industry in the 18th century, world has relied on non-renewable sources to generate organic chemicals.[1] Biomass offers a renewable solution as an alternative carbon source. Since directly producing high value chemicals from biomass is not yet economically feasible ethanol catalytic upgrading has been proposed as a viable route[2] of indirect biomass valorization. Furthermore, 1,3-butadiene, the most important monomer for synthetic rubber, had been produced via catalytic processing of ethanol during World War II by USSR and USA using Lebedev and Ostromislensky processes, respectively.[3]

Recently, several catalysts for a one-step, i.e. Lebedev process, are reported to have achieved ~50-60% yield,[4] while the two-step process, i.e. Ostromislensky, could attain over 60% yield with purity of about 98-99% at lower temperature of 300-350°C. Recent abundance of shale gas resulted in a different catalytic cracker product distribution dominated by ethylene.[4] This caused a worldwide shortage of C4 hydrocarbons, such as 1,3-butadiene. Since ethanol can be produced using variety of biomass processing routes including fermentation and gasification, it recently reemerged as the green route to catalytically form 1,3-butadiene[2,4].

Rational design of catalysts can only be achieved via kinetic control and understanding of reaction mechanism. The generally accepted mechanism involves complex steps of intermolecular proton transfer and C-C carbon coupling via aldol condensation.[5] In addition, Ostromislensky and Fripiat proposed two other possible reaction pathways.[4] Current development, however, could not unambiguously confirm the true reaction mechanism of ethanol to 1,3-butadiene.[6,7]

In this study, we identified the reaction mechanisms by combining both theoretical and experimental study by means of DFT, *in-situ* FTIR, and TPSR.

Experimental

Computational details. Periodic DFT calculations have been performed using the VASP code.[8] The Kohn–Sham equations have been solved variationally in a plane-wave basis set using the projector-augmented-wave (PAW) method of Blochl,[9] as adapted by Kresse and Joubert. The exchange-correlation energy was described by the PBE generalized gradient approximation.[10] Brillouin-zone was sampled using 2x2x1 k-point mesh. Transition states have been identified using the DIMER method,[11] as recently improved by Heyden *et al.*[12] and using a quasi-Newton method adapted for the saddle-point relaxation, implemented in program GADGET.[13] Atomic positions were considered to be relaxed if all forces acting on the atoms were smaller than 0.005 eV/Å. The intrinsic reaction coordinates[14,15] (IRCs) for the forward and backward reaction steps were identified using the damped velocity Verlet algorithm.[16]

***In-situ* FTIR study**. IR spectra were recorded using Nicolet iS50 FT-IR spectrometers, equipped with Harrick Praying Mantis accessory with Harrick gas Cell HVC-DRM, using ZnSe windows. Spectra were recorded using a mercury cadmium telluride (MCT) detector in 96 scans with 4 cm^{-1} resolution. About 30-40 mg of MgO catalyst was used. Catalyst was initially pretreated under air flow at 750°C at 10°C/min rate for two hours. Ethanol was then continuously bubbled to the gas cell at room temperature using nitrogen at 30 cc/min, with the temperature increased up to 450 °C at 10 °C/min while spectra was taken every minute.

Temperature-programmed Surface Reaction (TPSR). TPSR was carried out using an Altamira AMI-200 spectroscope equipped with a Dycor Dymaxion DME200MS online quadrupole mass spectrometer. About 80 mg of catalyst was loaded into a U-type quartz tube and initially treated in flowing 10% O_2/Ar to 500 °C (30 mL/min; heating rate of 10 °C/min) for 60 min to remove any organic impurities and dehydrate the sample. The catalyst was then cooled to 100°C under 10%O_2/Ar. He was then flown through ethanol bubbler for 1 hour to ensure surface saturation followed by temperature programmed reaction up to 500°C at 10 °C/min, while mass spectra were taken under continuous ethanol flow condition as a function of catalyst temperature.

Results and Discussion

Surface studies on MgO, both TPSR and *in-situ* FTIR, were initially carried out without participation of gas-phase ethanol molecules. However, the experimental data showed no evidence of aldehyde nor olefin formation (not shown). This finding indicated participation of gas-phase molecules is necessary for the reaction to take place. Subsequently experiments were carried out under continuous reactant flow and the spectra taken shown in Figure 1.

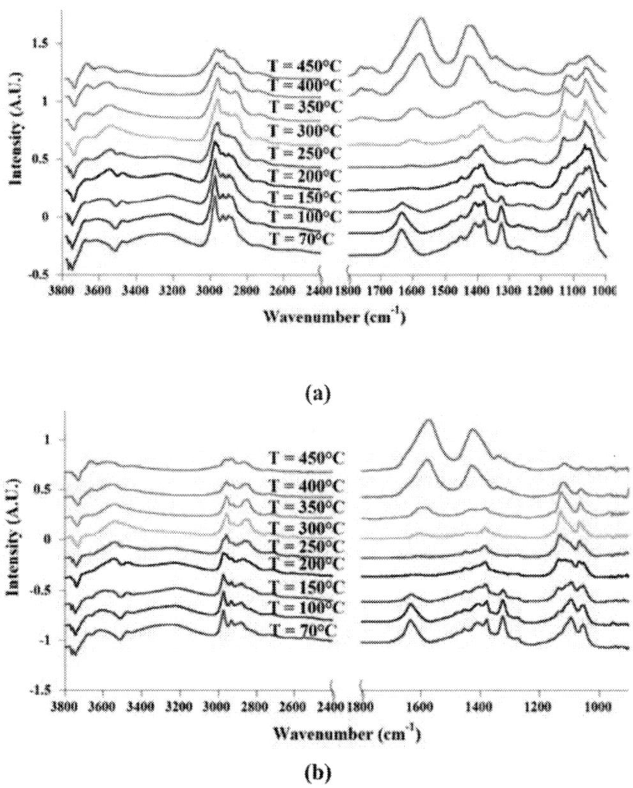

Figure 1. (a) *In-situ* FTIR study of ethanol on MgO under continuous reactant flow. (b) with *in-situ* IR spectra of ethanol on KBr subtracted.

Figure 1a shows that at higher temperature, i.e. ≥ 350°C, peaks associated with gas-phase aldehyde, ca. 1750 cm^{-1}, appear. All previously proposed mechanisms involved the formation of both acetaldehyde and crotonaldehyde as reactive intermediates.[4] Triplet peaks associated with ethanol ca. 1060, 1400, and 2970 cm^{-1} also noticeably decreased in intensity. On an important note, one peak at 2948 cm^{-1} increased simultaneously as ethanol peaks decreased. This peak is associated with a C-H stretch, indicating a presence of another hydrocarbon. To exclude the participation from gas-phase contribution, another experiment was carried out over KBr, which is unreactive. The subtraction spectra can be seen in **Figure 1b.**

Figure 1b shows that the gas phase aldehyde peaks are now negligible. Unfortunately, this also means that aldehydes were also made in the gas-phase without the presence of catalyst. Another interesting note is that a peak at 1128 cm^{-1} appeared at higher temperature starting at 200°C at the expense of the peak at 1134 cm^{-1}. Peak in this region is associated with C-O stretches and can originate from either alkoxide species or the C=O bond opening. With the proposed mechanisms this peak can either belong to acetaldehyde, crotonaldehyde or crotyl alcohol For the former two species, another way to confirm it is to find a unique O=C-H stretch, which generally resides at ~3000 cm^{-1}. One weak peak at this region, ca. 3024 cm^{-1} appeared starting at higher temperature of ≥ 350°C.

TPSR experiments can further prove intermediates and products being made during the process. The spectra for the TPSR experiment is shown in the **Figure 2**.

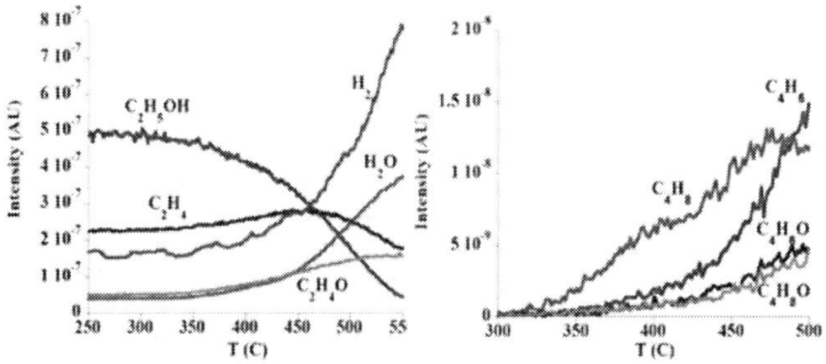

Figure 2. TPSR spectra of ethanol reaction over MgO: (left) C2 intermediates; (right): C4 intermediates and products.

Figure 2 shows that H_2 production steadily increased starting from ~300°C and strongly above 450°C. The first half of its production aligns well with the produced acetaldehyde. However, the latter is produced significantly slower than the former, even at higher temperature. Butenes are also produced starting at 300°C with two peaks, one at 400°C, and the other at 470°C. At higher m/z number, the C4 oxygenates, i.e. crotonaldehyde and crotyl alcohol, did not precede the 1,3-butadiene. However, when they both started increase more significantly, the 1,3-butadiene production also increased. This indicates two different mechanisms possibly taking place, one via C_4H_8, and the other via aldol condensation, not excluding the gas-phase decomposition at high temperature.[17]

The enhanced formation of 1,3-butadiene at 450°C also coincides with H_2 accelerated production at the same temperature. Starting at ~300°C, water and ethylene were being produced together with butene, before 1,3-butadiene formation took place. That alternative possible mechanism here is the formation of butene from ethylene via Cosse-Arlman mechanism where the low coordinated metal atom open the ethylene bond and reacts with a gaseous ethylene molecule. 1-butene subsequently loses hydrogen to give 1,3-butadiene, presumably in gas phase.

Mechanistic study was also carried out using DFT study, with VASP code. Shown in the **Figure 3** is the comparison between dehydration and dehydrogenation reactions of ethanol.

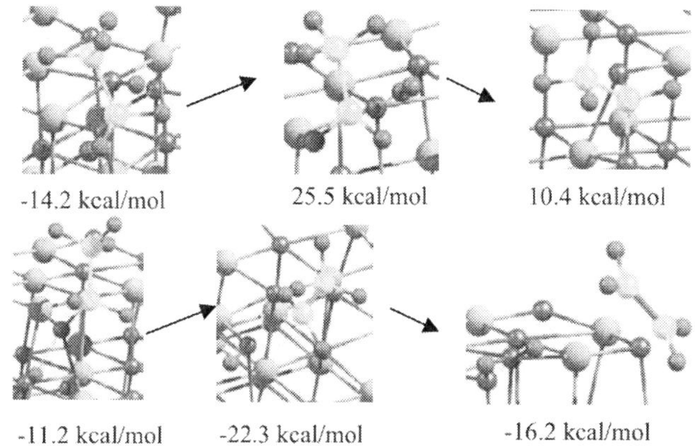

Figure 3. DFT calculation of ethanol dehydrogenation (top) and dehydration (bottom)

The reactions investigated by DFT shows that activation energies of both mechanisms are comparable and, surprisingly, the dehydration reaction is an exergonic compared to the endergonic nature of acetaldehyde formation from ethanol. This can be the reason why the route to 1,3-butadiene could take place via butene intermediate.

Acknowledgments. We thank Minghui Zhu and Prof. Israel Wachs from Lehigh University for providing the equipment for TPSR experiments and Dr. Tomas Bucko from Comenius University for his insights on running DFT calculation. Lehigh University is acknowledged for startup funding.

References

(1) Matar, S. & Hatch, L. F. Chemistry of Petrochemical Processes. Chemistry of Petrochemical Processes (Elsevier, 2001).
(2) Sun, J. & Wang, Y., *ACS Catal.*, 2014. 4: p. 1078–1090.
(3) Lebedev, S. V., *Zh Obs. Khim.*, 1922: p. 698–717.
(4) Angelici, C., Weckhuysen, B. M. & Bruijnincx, P. C., *ChemSusChem.*, 2013. 6: p. 1595–1614.
(5) Quattlebaum, W. M., Toussaint, W. J. & Dunn, J. T., *J. Am. Chem. Soc.*, 1947. 69: p. 593–599.
(6) Chieregato, A. et al., *ChemSusChem*, 2014. 8: p. 377–388.
(7) Zhang, M., Gao, M., Chen, J. & Yu, Y., *RSC Adv.*, 2015. 5: p. 25959–25966.
(8) Kresse, G. & Furthmüller, J., *Phys. Rev. B*, 1996. 54: p. 11169–11186.
(9) Blöchl, P. E., *Phys. Rev. B*, 1994. 50: p. 17953–17979.
(10) Perdew, J. P., Burke, K. & Ernzerhof, M., *Phys. Rev. Lett.*, 1996. 77: p. 3865–3868.
(11) Henkelman, G. & Jónsson, H., *J. Chem. Phys.*, 1999. 111: p. 7010–7022.
(12) Heyden, A., Bell, A. T. & Keil, F. J., *J. Chem. Phys.*, 2005. 123: p. 224101/1–224101/14.
(13) Bučko, T., Hafner, J. & Ángyán, J. G., *J. Chem. Phys.*, 2005. 122: p. 124508/1–124508/10.
(14) Fukui, K., *J. Phys. Chem.*, 1970. 74: p. 4161–4163.
(15) Fukui, K., *Acc. Chem. Res.*, 1981. 14: p. 363–368.
(16) Hratchian, H. P. & Schlegel, H. B., *J. Phys. Chem. A*, 2001. 106: p. 165–169.
(17) Morris, H. E., *Chem. Rev.*, 1932. 10 (3): p. 465–506

EXPLORING THE MECHANISMS OF FAST PYROLYSIS OF HEMICELLULOSES VIA TANDEM MASS SPECTROMETRY AND QUANTUM CHEMICAL CALCULATIONS: A SYNTHETIC MODEL COMPOUND STUDY

*Priya Murria[1]; John C. Degenstein[2]; Mckay Easton[2]; Hanyu Zhu[1]; John J. Nash[1]; Rakesh Agrawal[2]; W. Nicholas Delgass[2]; Fabio H. Ribeiro[2]; Hilkka I. Kenttämaa[1]**

[1]Department of Chemistry, Purdue University, West Lafayette, IN
[2]Department of Chemical Engineering, Purdue University, West Lafayette, IN
Corresponding author e-mail: hilkka@purdue.edu (Hilkka I. Kenttämaa)

Introduction

Fast pyrolysis is a promising method for biomass depolymerization with the ultimate goal of obtaining fuels and value-added chemicals.[1] Rational design of fast pyrolysis reactors requires mechanistic understanding of the pyrolytic reactions. Since product mixtures from fast pyrolysis of biomass are extremely complex, delineation of the pyrolytic reaction mechanisms requires studies on model compounds.[2] Although numerous studies have been carried out on fast pyrolysis of cellulose and lignin by using model compounds,[3–7] no such mechanistic studies have been published on fast pyrolysis of hemicelluloses. This can be attributed to the structural complexity and diversity of hemicelluloses.[8,9] Hemicelluloses are the second most abundant component of lignocellulosic biomass. They are defined as alkali-soluble polysaccharides in plant cell walls that have β-(1→4)-linked pyranose backbones with an equatorial configuration.[9] Xylans (glucuronoxylan and glucuronoarabinoxylan) are the most abundant type of hemicelluloses in the secondary cell walls of plants. In this study, we employed a pyroprobe-mass spectrometry set-up[10,5] to examine the initial products of fast pyrolysis of unsubstituted xylan model compounds - xylobiose and xylotriose. Quantum chemical calculations were performed to explore reaction mechanisms of fast pyrolysis.

Materials and Methods

Chemicals. Xylobiose (≥95%) and xylotriose (≥95%) were purchased from Carbosynth. The chemical structures for these compounds are shown in Figure 1. Methanol (LC/MS grade; ≥99.9%) and water (LC/MS grade) were purchased from ProteoChem. Ammonium hydroxide (28−30% as NH_3) was purchased from Mallinckrodt Chemicals and a compressed nitrogen cylinder (≥99.9%) was purchased from Indiana Oxygen. All chemicals were used without further purification.

Pyroprobe. All pyrolysis experiments were performed using a CDS Analytical (Oxford, PA) Pyroprobe 5200. The pyroprobe utilizes a resistively heated platinum ribbon with dimensions 2.1 mm × 35 mm × 0.1 mm on which the model compounds to be pyrolyzed were loaded. The ribbon was heated at a rate of 1000 °C s^{-1} to a final temperature of 600 °C and maintained at that temperature for 1 s.

Mass Spectrometry. Products of fast pyrolysis were identified using a Thermo Scientific (Waltham, MA) Orbitrap XL hybrid ion trap-orbitrap mass spectrometer that was coupled to a Finnigan Surveyor Liquid Chromatograph (LC) MS Pump Plus. Mass spectrometer was operated in the positive ion mode. An atmospheric pressure chemical ionization (APCI) source was used for ionization of the pyrolysis products. Ammonium hydroxide: water solution (75:25 v/v) was pumped into the APCI source by using a syringe pump at a rate of 10 µL min^{-1} with a solution of methanol/water (50: 50 v/v) t-infused from the MS pump at a rate of 200 µL min^{-1}. For collisionally activated dissociation experiments, an isolation window of 2 m/z was used and ions were fragmented using a q value of 0.25 and activation time of 30 ms. Data processing was performed using Xcalibur 2.1 software.

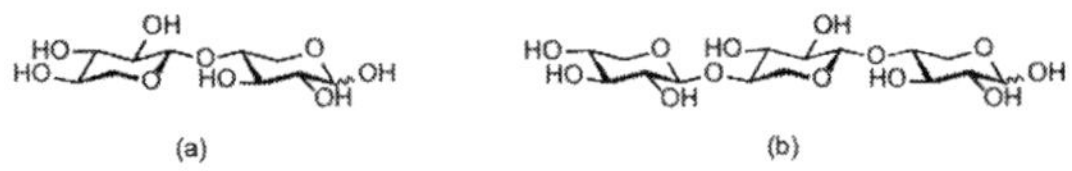

Figure 1. Structures of model compounds used in this study: (a) β-1,4-xylobiose and (b) β-1,4-xylotriose.

Quantum Chemical Calculations. Density functional theory calculations were performed with the Gaussian09 program using the M06-2X functional and 6-311++G(d,p) basis set. All geometry optimizations were followed by a full Hessian calculation to guarantee that the minima had all positive eigenvalues and the transition states had exactly one negative eigenvalue. Intrinsic reaction coordinate calculations were performed to confirm that appropriate reactants and products were associated with each transition state.

Results and Discussion

A CDS Analytical pyroprobe was coupled with a Thermo Scientific LTQ linear quadrupole ion trap mass spectrometer (combined residence and analysis time of ~125 ms).[11] Model compounds were loaded on the platinum ribbon of the pyroprobe and the initial products of fast pyrolysis that evaporated from the surface of the ribbon were quenched in nitrogen atmosphere at 100 °C at atmospheric pressure. Atmospheric pressure chemical ionization with ammonium hydroxide as a dopant was employed to ionize the pyrolysis products. This method ionizes carbohydrate analytes by attaching an ammonium ion to each compound without fragmentation.

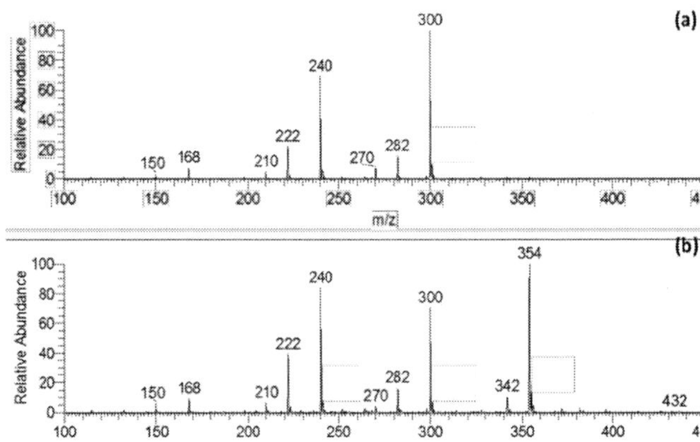

Figure 2. Pyrolysis mass spectra of (a) β-1,4-xylobiose and (b) β-1,4-xylotriose. All peaks shown in the mass spectra correspond to ammonium adducts of products generated from pyrolysis.

Pyrolysis mass spectra of β-1,4-xylobiose and β-1,4-xylotriose are shown in Figure 2. These mass spectra were obtained by averaging individual mass spectra obtained during the pyrolysis event. Xylobiose and xylotriose generated many identical products upon fast pyrolysis. This was confirmed by MS^2 experiments (isolating ions and subjecting them to collisionally activated dissociation). Ions of m/z 300 shown in Figure 2(a) are ammonium adducts of xylobiose. The major product of fast pyrolysis of xylobiose (ions of m/z 240) is proposed to be β-D-xylopyranosyl-glyceraldehyde, likely formed by ring opening of the reducing end followed by loss of an ethenediol (or isomeric) molecule. Other products observed in significant abundance correspond to ions of m/z 222 (likely formed by loss of a water molecule and an ethenediol or isomeric molecule from xylobiose) and ions of m/z 282 (formed upon loss of a water molecule from xylobiose). Minor products from xylobiose fast pyrolysis include those that correspond to ions of m/z 270 (formed upon loss of a formaldehyde molecule from xylobiose), ions of m/z 210 (likely formed upon loss of a formaldehyde molecule and an ethenediol or isomeric molecule from xylobiose), ions of m/z 168 (xylose) and ions of m/z 150 (anhydroxylose) formed upon cleavage of a glycosidic linkage.

Table 1. Elemental compositions, proposed structures as well as m/z values of ammoniated compounds for selected products of fast pyrolysis xylobiose and xylotriose.

m/z of product ion	Elemental composition	Proposed structure
150	$[C_5H_8O_4+NH_4^+]$	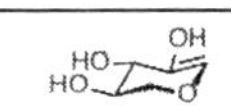
168	$[C_5H_{10}O_5+NH_4^+]$	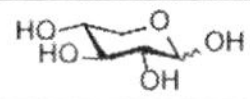
210	$[C_7H_{12}O_6+NH_4^+]$	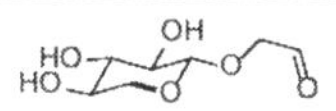
222	$[C_8H_{12}O_6+NH_4^+]$	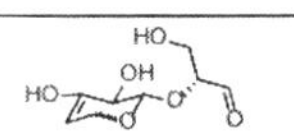
240	$[C_8H_{14}O_7+NH_4^+]$	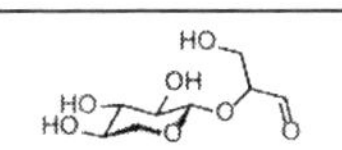
270	$[C_9H_{16}O_8+NH_4^+]$	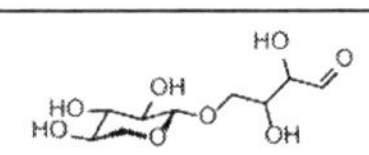
282	$[C_{10}H_{16}O_8+NH_4^+]$	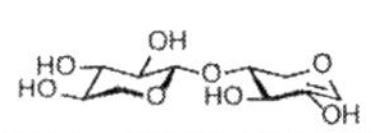
300	$[C_{10}H_{18}O_9+NH_4^+]$	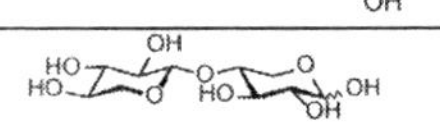

Ions of m/z 432 in the pyrolysis mass spectrum of xylotriose (Fig. 2(b)) correspond to the ammonium adducts of xylotriose – the reactant. Major products from fast pyrolysis of xylotriose were found to yield ions of m/z 354 (likely formed upon loss of an ethenediol or isomeric molecule and a water molecule from xylotriose), ions of m/z 300 (xylobiose) and ions of m/z 240 (β-D-xylopyranosyl glyceraldehyde – identical product observed during fast pyrolysis of xylobiose). Other products observed yielded ions of m/z 282, ions of m/z 222, ions of m/z 210, ions of m/z 168, and ions of m/z 150 – which were also observed for fast pyrolysis of xylobiose. Products yielding ions of m/z 342 are likely formed by loss of an ethenediol or isomeric molecule and a formaldehyde molecule from xylotriose.

Quantum chemical calculations at 600 °C at the M06-2X/6-311++G(d,p)//M06-2X/6-311++G(d,p) level of theory revealed a low-energy pathway for the concerted elimination of an ethenediol molecule from xylobiose, producing β-D-xylopyranosyl-glyceraldehyde (XGRA). The free energy barrier at 600 °C for ring opening of the reducing end of xylobiose to produce open chain xylobiose (XBO) was found to be 49.8 kcal mol^{-1}. The free energy barrier for concerted elimination of ethenediol from open chain xylobiose (XBO) was found to be 42.3 kcal mol^{-1}. This reaction was found to be exoergic by 12.5 kcal mol^{-1}.

Further fragmentation of β-D-xylopyranosyl glyceraldehyde via loss of formaldehyde to produce ions of m/z 210 is not a major pathway, which is in agreement with a relatively high calculated free energy barrier of 76.6 kcal mol^{-1}.

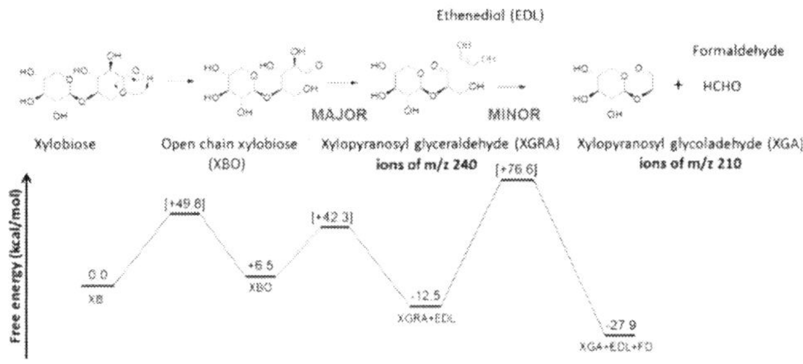

Figure 3. Calculated free energies (kcal mol^{-1}) of intermediates and transition states (square brackets) for the formation of xylopyranosyl glyceraldehyde from xylobiose via concerted elimination of an ethenediol molecule at 600 °C (M06-2X/6-311++G(d,p)//M06-2X/6-311++G(d,p) level of theory).

Free energy barriers for 1,2-dehydration and cleavage of a glycosidic linkage (via Maccoll elimination) were also calculated. Cleavages of glycosidic bonds and aglyconic bonds were found to be equally likely based on the energy barriers, suggesting that xylose and anhydroxylose are equally likely to be produced from both the reducing and the non-reducing ends. Glycosidic bond cleavages were not as dominant for xylobiose and xylotriose as for cellobiose,[5] likely because of the absence of a hydroxymethyl group on the 6-position.

Conclusions

Pyrolysis mass spectra of model compounds of xylans – xylobiose and xylotriose – were found to generate many identical products. Ring opening at the reducing end and subsequent elimination of ethenediol molecules from the reducing end via concerted mechanisms were found to be the major pyrolysis pathway. Other pathways include 1,2-dehydration and cleavage of glycosidic linkages via concerted Maccoll elimination mechanism. Quantum chemical calculations revealed a feasible low-energy pathway for some of the products observed.

Acknowledgement. The authors gratefully acknowledge financial support by the Center for Direct Catalytic Conversion of Biomass to Biofuels (C3Bio), an Energy Frontier Research Center funded by the U.S. Department of Energy, Office of Science, and Office of Basic Energy Sciences.

References

(1) Bridgwater, A. V. *Biomass Bioenergy* **2012**, *38*, 68-94.
(2) Mettler, M. S.; Vlachos, D. G.; Dauenhauer, P. J. *Energy Environ. Sci.* **2012**, *5*, 7797-7809.
(3) Mettler, M. S.; Paulsen, A. D.; Vlachos, D. G.; Dauenhauer, P. J. *Green Chem.* **2012**, *14*, 1284-1288.
(4) Vinu, R.; Broadbelt, L. J. *Energy Environ. Sci.* **2012**, *5*, 9808-9826.
(5) Degenstein, J. C.; Murria, P.; Easton, M.; Sheng, H.; Hurt, M.; Dow, A. R.; Gao, J.; Nash, J. J.; Agrawal, R.; Delgass, W. N.; Ribeiro, F. H.; Kenttämaa, H. I. *J. Org. Chem.* **2015**, *80*, 1909-1914.
(6) Britt, P. F.; Buchanan, A. C.; Cooney, M. J.; Martineau, D. R. *J. Org. Chem.* **2000**, *65*, 1376-1389.
(7) Jarvis, M. W.; Daily, J. W. Carstensen, H-H; Dean, A. M.; Sharma, S.; Dayton, D. C.; Robichaud, D. J.; Nimlos, M. R. J. Phys. Chem. 2011, 115, 428-438.
(8) Caffall, K. H.; Mohnen, D. *Carbohydr. Res.* **2009**, *344*, 1879-1900.
(9) Scheller, H. V.; Ulvskov, P. Hemicelluloses. *Annu. Rev. Plant Biol.* **2010**, *61*, 263-289.
(10) Hurt, M. R.; Degenstein, J. C.; Gawecki, P.; Borton II, D. J.; Vinueza, N. R.; Yang, L.; Agrawal, R.; Delgass, W. N.; Ribeiro, F. H.; Kenttämaa, H. I. *Anal. Chem.* **2013**, *85* (22), 10927-10934.
(11) Degenstein, J. C.; Hurt, M.; Murria, P.; Easton, M.; Choudhari, H.; Yang, L.; Riedeman, J.; Carlsen, M. S.; Nash, J. J.; Agrawal, R.; W. N.; Ribeiro, F. H.; Kenttämaa, H. I. *Eur. J. Mass Spectrom.* **2015**, *21*, 321-326.

BIOFUEL SYNTHESIS IN SUPERCRITICAL FLUIDS

Jaehoon Kim[1], Young Haeng Lee[2]*

[1]School of Mechanical Engineering,
SKKU Advanced Institute of Nano Technology,
Sungkyunkwan University
2066, Seobu-Ro, Jangan-Gu, Suwon, Gyeong Gi-Do, 440-746,
Republic of Korea
[2]Center for Water Resource Cycle Research,
Korea Institute of Science and Technology,
Hwarangno 14-gil 5, Seongbuk-gu, Seoul 136-791,
Republic of Korea
*Jaehoonkim@skku.edu

Introduction

Supercritical fluids can offer environmentally benign and facile reaction conditions for the production of biofuels owing to their unique physical properties, including low viscosity, fast diffusion, zero surface tension, tunable physical properties, and high reactivity. This paper will cover biomass liquefaction in supercritical fluids. An effective utilization of the unique hydrogen-donating ability associated with scEtOH involving both donation of an α-hydrogen in the hydride form and proton transfer through the Meerwein–Ponndorf–Verley reduction mechanism, can be very effective in the depolymerization of various types of biomass including lignocellulosic biomass, lignin, macroalgae. This hydrogen-donating ability would also play a role in deoxygenation to improve the calorific value of upgraded oil, and in retarding repolymerization and tar formation. In addition to the hydrogen generation, scEtOH is very effective in the non-catalytic esterification of carboxylic acids (e.g., free-fatty acids to produce biodiesel).

Results and Discussion

Lignocellulosic biomass conversion in supercritical ethanol. Figure 1 shows the effect of reaction temperature on pinewood sawdust conversion, yields of biocrude, solid residue, and water + gas + light hydrocarbons, gas composition, and C/H/O content in biocrude and solid residue. Supercritical ethanol (scEtOH) was used as the liquefaction solvent. With temperature increases from 280 to 400 °C, a significant increase in conversion from 34.0% to 98.1% and biocrude yield from 15.8 to 59.9 wt%, as well as a significant decrease in solid residue from 66.0 to 1.9 wt% was observed. Close to 100% conversion of solid biomass to liquid and gas was observed at high reaction temperature. Char/tar formation at a high reaction temperature, often caused by condensation, cyclization, and/or polymerization in scH_2O-based liquefaction, did not ensue in scEtOH-based liquefaction[1]

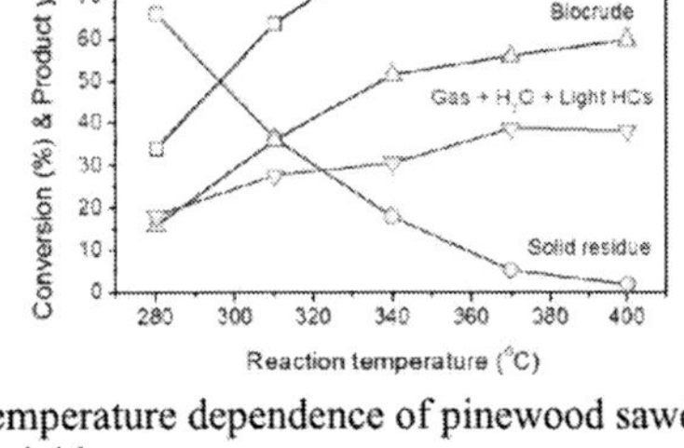

Figure 1. Temperature dependence of pinewood sawdust conversion and product yield

Heating Rate Effect. When $subH_2O$ was used, no significant effect of heating rate at the final temperatures of 250-280 °C was observed; the conversions obtained from both heating rates are close to each other at 45%-47% and 67%-72%. When the final temperature increased above 315 °C, a significant impact of heating rate was observed; higher conversions of 86%-88% were obtained with the high heating rate of 20 °C/min, whereas the conversion obtained at the low heating rate of 2 °C /min stagnated at 72%. In sharp contrast to $subH_2O$-based liquefaction, only a marginal effect of heating rate on conversion and biocrude yield was observed when scEtOH was used as the liquefaction solvent (Figure 2). Somewhat similar values of the conversion and biocrude yield with increasing temperature were observed at both the heating rates of 2 and 20 °C/min. This suggests that the decomposition of the solid matrix into smaller oligomers and monomers was strongly dependent on the final temperature, but not on the heating rate.[2]

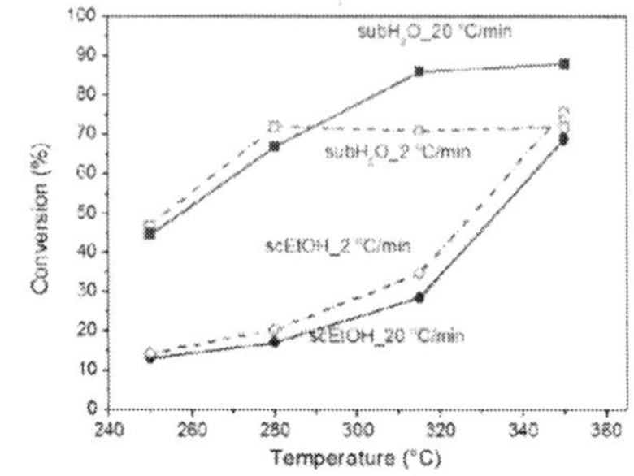

Figure 2. Influence of heating rate on biomass conversion

Lignin Liquefaction in Supercritical Ethanol. The presence of formic acid significantly increases the depolymerization of strong acid hydrolysis lignin, as shown in Figure 3. The conversion and yield of bio-oil steadily increase reaching maximum values of 92% and 85 wt%, respectively, at a FL mass ratio of 1.5. The solid residue yield decreases rapidly from 40 to 14 wt% as the FL mass ratio increases from 0 to 1.5. The overall role of formic acid is for the production of active hydrogen, which acts as a radical scavenger to retard the recombination reactions of intermediate species. Greater formic acid decomposition under the reaction conditions may be the major reason for the increase in gas yield up to ~10 wt% at the high FL mass ratio of 1.5.[3]

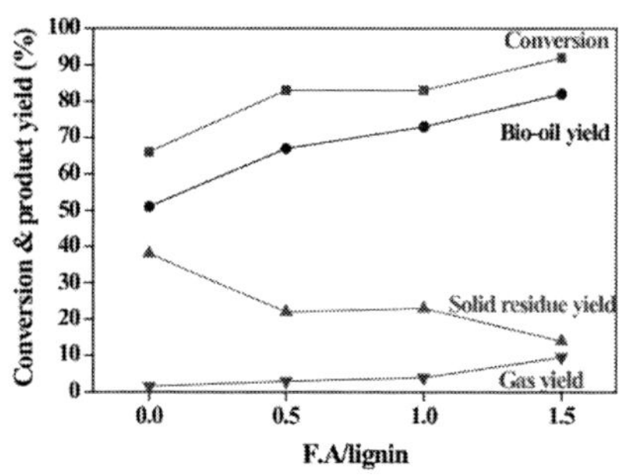

Figure 3. Effect of formic acid in the depolymerization of lignin

Fast-Pyrolysis Bio-Oil Upgrading in Supercritical Ethanol. The relationship between the H/C and O/C molar ratios with increasing reaction temperature was plotted as a van Krevelen diagram, as shown in Figure 4. At 300 °C, the H/C ratio increased by approximately four times (from 0.18 to 0.92) while the O/C ratio decreased up to 42% (from 0.354 to 0.204). The H/C molar ratio was further increased to 1.10 and the O/C ratio was significantly decreased to 0.12 at a higher temperature of 400 °C. This indicates that the hydrodeoxygenation reaction associated with scEtOH was enhanced at the high reaction temperature to reduce the oxygen content and to increase the hydrogen content of the liquid product. Overall, the upgraded oils exhibited high calorific values and coal-like C/H/O content at the high reaction temperature used.[4]

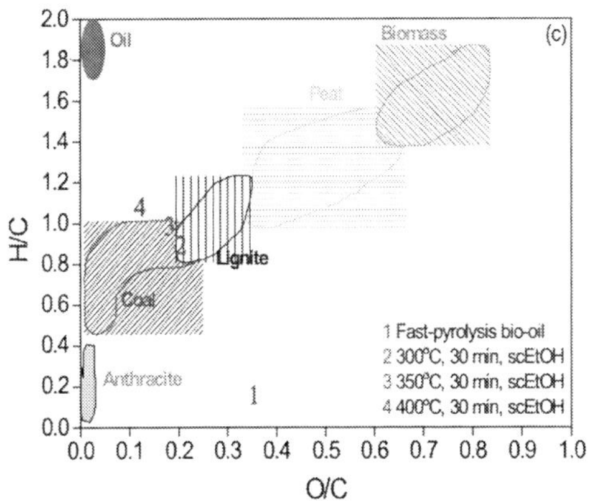

Figure 4. Van Krevelen diagram of upgraded oil

Acknowledgement. This work was supported by the New & Renewable Energy Core Technology Program (No. 20143030090940 and No. 20143030100980) of the Korea Institute of Energy Technology Evaluation and Planning (KETEP) financed by the Ministry of Trade, Industry, and Energy, Republic of Korea

References

(1) Brand, S.; Susanti, R. F.; Kim, S. K.; Lee, H.-S. and Kim, J., *Energy* **2013**, *59* 173.
(2) Brand, S.; Hardi, F.; Suh, D. J.; and Kim, J., *Energy* **2014**, *68* 420.
(3) Riaz, A.; Kim, C. S.; Kim, Y.; and Kim, J., *Fuel* **2016**, *98* 213.
(4) Prajitno, H.; Insyani, R.; Park, J.; Ryu. C.; and Kim, J., *Applied Energy* **2016**, *accepted.*

HYDRODEPOLYMERIZED CELLULOSIC DIESEL FUEL (HDCD): CHARACTERIZATION, DEVELOPMENT OF A SURROGATE FUEL MIXTURE, AND ENGINE COMBUSTION

Dianne Luning Prak[1], Bridget Lee,[1] Paul Trulove,[1] and Jim Cowart[2]

(1) Chemistry Department
U.S. Naval Academy
572 M Holloway Road
Annapolis, MD, 21402
(2) Mechanical Engineering Department
U.S. Naval Academy
590 Holloway Rd
Annapolis, MD, 21402
prak@usna.edu

Introduction

Alternative fuels can be derived from the fermentation of sugars and starches to alcohols, the esterification of vegetable oils or animal fats with postprocessing procedures such as hydrotreating, or from the conversion of lignoellulose materials to fuels via methods such as pyrolysis, gasification, and enzymatic hydrolysis. Recently, a new fuel derived from cellulose was tested for use in military engines. This fuel, Hydrodepolymerized Cellulosic Diesel (HDCD) Fuel, had a low Derived Cetane Number (DCN = 27) and could not be run neat in a diesel engine.[1] Blends of HDCD with Navy petroleum-based F-76 diesel fuel containing 70% or more of F-76, however, were able to meet the Navy requirement that the engine must reach rated speed within 10 seconds of starting.[1] To better understand the combustion of this fuel, the goal of this work was to characterize HDCD, develop simple mixtures that matched the physical properties of HDCD fuel, and compare the combustion of the surrogates with HDCD.

Experimental

Materials. Hydrodepolymerized Cellulosic Diesel (HDCD) Fuel, and petroleum-based Navy diesel fuel (F-76) were provided by Naval Fuels and Lubricants Cross Function Team at Patuxent River, Maryland (PAX River). Surrogate mixtures containing two components were formulated using bicyclohexyl (TCI, 99.5 pure), 1,2,3,4-tetrahydronaphthalene (tetralin, Aldrich, 98.95 pure) and trans-decahydronaphthalene (trans-decalin, TCI, 99.5% pure).

Composition and Physical Property Measurements. To determine the composition of the HDCD, GC/MS analysis was conducted. In the analysis, an Agilent 5975 inert mass selective detector for a quadrapole mass separator was used in conjunction with an Agilent 6890N GC system. An electron impact ionization (EI) method was used for the mass spectrometer with an m/z scan from 30 to 600. The GC/MS system was equipped with software that matches the mass spectral pattern found at a specified retention time with patterns in the NIST/EPA/NIH Mass Spectral Library (Version 2.0 g) to determine the most likely compound for that retention time. Viscosity was measured using an SVM 3000 Stabinger Viscometer (Anton Paar), and density and speed of sound were measured using a DSA 5000 density and sound analyzer (Anton Paar).

Engine Tests. The start-up behavior of the HDCD and the surrogate mixture was tested using a laboratory installed Yanmar L100V single-cylinder diesel engine coupled to a 6 kW electrical generator. The engine configuration is described by Bermudez et al.[1]

Results and Discussion

Composition of HDCD. The HDCD had components that span a smaller boiling point range than those found in the petroleum-based F-76 as suggested by the GC/MS scan in Figure 1. The majority of components elude before 13 min. Three compounds found in the NIST/EPA/NIH Mass Spectral Library had mass fragmentation patterns similar to components found in HDCD: bicyclohexyl, tetralin, and trans-decalin, suggesting that these compounds are present in the HDCD.

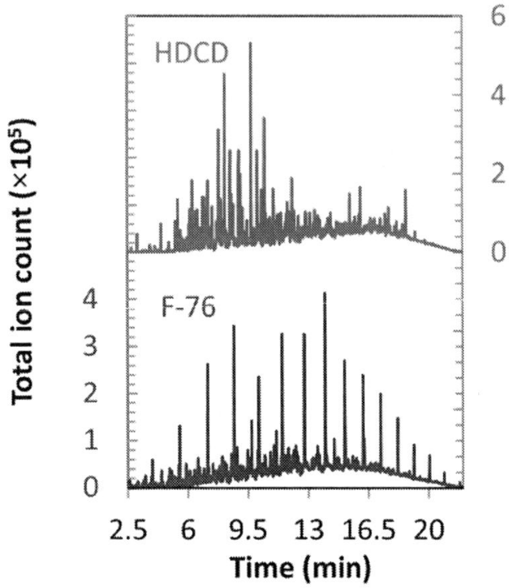

Figure 1. GC/MS scan of F-76 and HDCD

Physical Properties of ATJ and Development of Surrogate Mixture to Match Physical Properties. Two-component surrogate mixtures were prepared containing bicyclohexyl, tetralin, and trans-decalin. To find the best 2-component surrogate mixture for HDCD, the density, bulk modulus, and viscosity of the surrogate mixtures were compared with those of the HDCD. The bulk modulus was calculated from the density (ρ) and speed of sound (c) by

$$E_v/\text{Pa} = (c^2 /\text{m}^2\cdot\text{s}^{-2})(\rho/\text{kg}\cdot\text{m}^{-3}) \qquad (1)$$

These parameters were used for comparison because viscosity and density are required measurements for military fuels, and density and bulk modulus have been found to impact the start of injection and start of ignition in diesel engines.

No surrogate mixture was able to match all three properties. The viscosity of HDCD at 293 K (green line) was matched by the mixtures containing bicyclohexyl as shown in Figure 2. The y-axis is the mass fraction of the first component in the second component (e.g. mass fraction of trans-decalin in bicyclohexyl for blue triangles). The viscosities of the trans-decalin/tetralin mixtures were too low. The density of HDCD at 293 K was best matched by mixtures containing tetralin as shown in Figure 3. The trans-decalin/bicyclohexyl mixtures had density values that were too low. The speed of sound of HDCD at 293 K was matched by mixtures containing trans-decalin. The speed of sound of tetralin/bicyclohexyl mixtures was too high.

Since no 2-component mixture had properties that matched all three of the properties of HDCD, the transdecalin/tetralin surrogates were selected for further testing because they matched the density and bulk modulus of HDCD, and these two properties have been found to have the greatest impact on the startup behavior in the engines used in this study.

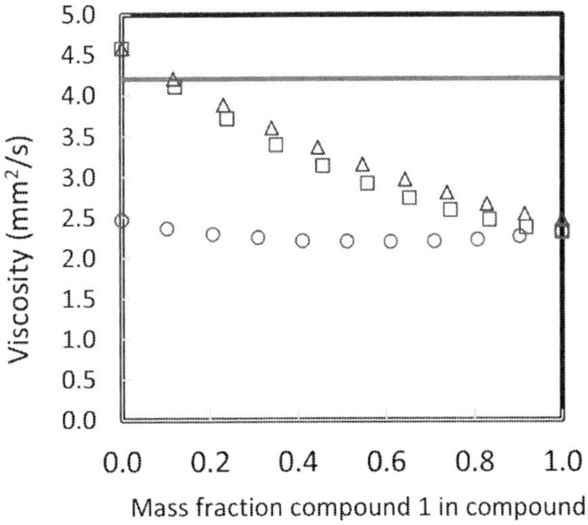

Figure 2. Viscosity of Δ trans-decalin in bicyclohexyl[1]; □ tetralin in bicyclohexyl[1]; and ○ tetralin in trans-decalin.[3] Green line is HDCD.

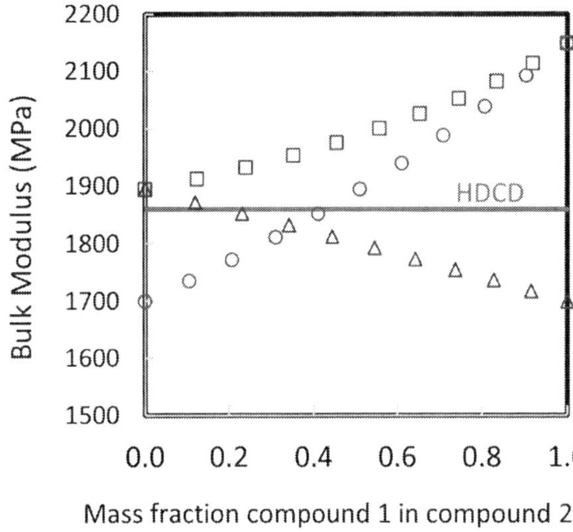

Figure 3. Bulk Modulus of Δ trans-decalin in bicyclohexyl[1]; □ tetralin in bicyclohexyl[1]; and ○ tetralin in trans-decalin.[3] Green line is HDCD.

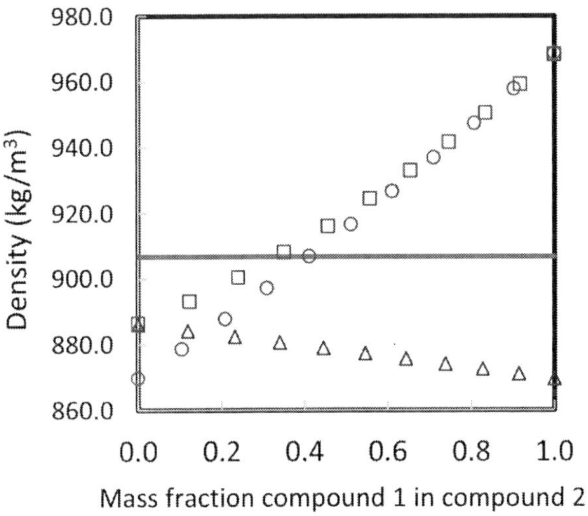

Figure 3. Density of Δ trans-decalin in bicyclohexyl[1]; □ tetralin in bicyclohexyl[1]; and ○ tetralin in trans-decalin.[3] Green line is HDCD.

Preliminary Diesel Engine Results for F-76 mixed with HDCD Surrogate Mixtures. Since engines have failed to start when neat HDCD was used as a fuel,[1] it was likely that the surrogate would also fail to combust. For combustion testing, F-76 was combined with tetralin/trans-decalin mixtures at compositions of 80% F-76 and 60%-F-76. Preliminary results suggest that the mixture containing the higher concentration of the tetralin surrogate starts later than the mixture containing less of the surrogate. This start-up behavior is similar to the behavior reported for the mixtures of F-76 with HDCD. Additional analysis is needed to confirm that this surrogate can match the combustion behavior of HDCD.

Conclusions

Based on GC/MS analysis, bicyclohexyl, tetralin, and decalin are likely components of HDCD. Analysis of the viscosity, density, and bulk modulus of 2-component mixtures of these compounds showed that each set of mixtures matched only two of the three physical properties. Since bulk modulus and density have been found to be more important than viscosity in the startup behavior in the some diesel engines, the mixture system that matched both of these properties was selected for engine testing. This mixture system was tetralin/trans-decalin. Preliminary engine tests of F-76 mixtures with 20% of the tetralin/trans-decalin surrogate and 40% of tetralin/trans-decalin surrogate indicate the mixture with the larger percentage of the surrogate takes longer to reach rated speed. This result is similar to that found for HDCD, suggesting that a mixture of tetralin/trans-decalin may be a good surrogate mixture for HDCD. Further analysis is needed to confirm this prediction.

Acknowledgement. This work was funded by the Office of Naval Research (NEPTUNE program) and Naval Air Systems Command.

References

(1) Bermudez, E., Luning Prak, D., Hamilton, L., Cowart, J., McDaniel, A., Dickerson, T., S., Start-up and Steady-State Performance of a New Renewable Hydro-processed Depolymerized Cellulosic Diesel (HDCD) in Multiple Diesel Engines, *Journal of Engineering for Gas Turbines and Power* **2016**, DOI: 10.1115/1.4032992.

(2) Luning Prak, D. J., Cowart, J.S., Trulove, P.C. Density and Viscosity from293.15 to 373.15 K, Speed of Sound and Bulk Modulus from 293.15 to 343.15 K, Surface Tension, and Flash Point of Binary Mixtures of Bicyclohexyl and 1,2,3,4-Tetrahydronaphthalene or Trans-decahydronaphthalene at 0.1 MPa, *J. Chem. Eng. Data* **61**, 650–661.

(3) Luning Prak, D. J., Lee, B. Density, Viscosity, Speed of Sound, Bulk Modulus, Surface Tension, and Flash Point of Binary Mixtures of 1,2,3,4- Tetrahydronaphthalene and Trans-decahydronaphthalene, submitted to *J. Chem. Eng. Data* Dec 21, 2015, in review.

MODIFYING LIQUID-SOLID INTERFACES FOR THE PROCESSING OF RENEWABLE FUELS

David C. Cantu, Yang-Gang Wang, Yeohoon Yoon, Vassiliki-Alexandra Glezakou, Roger Rousseau, Robert S. Weber[1]

Pacific Northwest National Laboratory, Institute for Integrated Catalysis, PO Box 999, MS-IN K2-12, Richland, WA, 99352, United States, 509-372-4748, 509-371-6498, robert.weber@pnnl.gov

Introduction

The transformation of biomass into renewable fuels necessarily involves liquid phases both because liquids are produced directly during pyrolysis (or hydrothermal liquefaction) and because upgrading the gaseous products without condensation produces mainly light hydrocarbons that will not satisfy upcoming demands for transportation fuels. The condensed phases contain a broad mixture of lipophilic, hydrophilic and ambiphilic species whose interactions with the surfaces of contact catalysts generate local compositions governed by both liquid-liquid interactions and liquid-solid interactions.

In a previous report we described the role of entropy and enthalpy in determining speciation at the solid/liquid interface, with the goal of assisting the design of catalyst supports that could, in principle, enhance activity in hydrotreating of bio-oil intermediates and products. Here, we report our use of molecular dynamics to probe the near-surface composition of the fluid phase in the cybotactic region adjacent to surfaces whose solvophilicity has been varied between hydrophilic to lipophilic by adjusting the density of surface hydroxyl groups.

Methods

As previously we employ surface geometries optimized with *ab initio* molecular dynamics simulations using the CP2K package. These simulations can provide accurate estimates of surface adsorption and reactivity, but are limited by system size and simulation time and thus cannot be used for large-scale simulations (ns-□s). To overcome this limitation, data from these simulations (atomic charges, binding energies) will be used to construct and tune classical potentials for large-scale liquid simulations. Classical molecular dynamics (MD) will be used to simulate the surface-solvent mixture systems, using Gromacs software. The OPLS all atom force field will be used for organic atoms (CHO) except for water molecules modeled with TIP3P. The positions of all surface atoms will be frozen during the NVT simulations performed to observe how solvent mixtures behave near the surfaces.

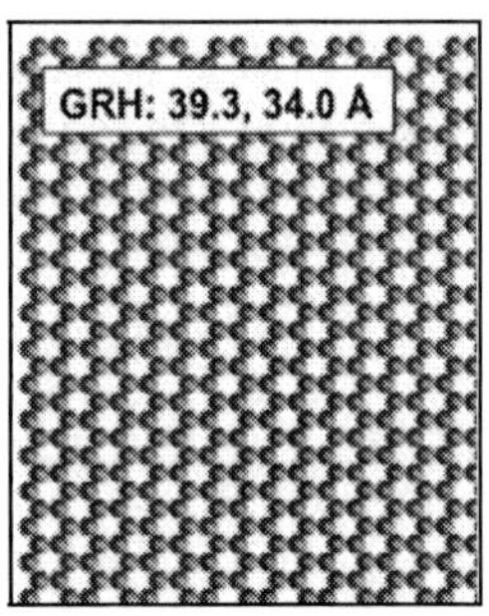

Figure 1: Base surface model employed in this study.

Our analysis focuses on a solvent mixture containing 2 mol% phenol (PHE) in water and its adsorption at 50-250 °C, which is relevant to low temperature and high temperate stages in the hydrotreating process. The surface is graphite, decorated with patches of isolated and comingled –OH and –O-CH_3 groups.

Results and Discussion

We will present maps of the composition of the cybotactic region as a function of the density the hydroxyl groups and as a function of temperature to explore whether the surface can be tuned to exhibit lateral structuring of the cybotactic region, complementing the z-structuring (perpendicular to the surface) that we reported previously. Implications for the design and characterization of catalysts will be discussed.

Acknowledgement. This work is supported by the US Department of Energy, Office of Energy Efficiency and Renewable Energy, Bioenergy Technologies Office. Pacific Northwest National Laboratory (PNNL) is a multiprogram national laboratory operated for DOE by Battelle.

GREEN BIOREFINERY CONCEPT AS A FUTURE OF SUSTAINABLE BIOMASS UTILIZATION

Ana R. C. Morais[1,2], *André M. da Costa Lopes*[1,2], *R. Bogel-Lukasik*[1,2*]

[1]Laboratório Nacional de Energia e Geologia, I. P., Unidade de Bioenergia, 1649-038 Lisbon, Portugal
[2]LAQV/REQUIMTE, Departamento de Química, Faculdade de Ciências e Tecnologia, Universidade Nova de Lisboa, 2829-516 Caparica, Portugal
*rafal.lukasik@lneg.pt

Introduction

The current challenges and demands for lignocellulosic feedstock processing in the frame of the biorefinery concept lead to seek novel and innovative technologies that fulfill green chemistry concept. One of the main principles of Green Chemistry[1-2] is the use of green solvents (e.g. ionic liquids and carbon dioxide) that can contribute to act as more sustainable solvents, catalysts and reagents.[3-8]

Ionic liquids are a unique class of salts, liquid at or close to room temperature. The negligible vapour pressure of ILs[9] allows recycling and their reuse in diverse processes. In addition, ILs offer solution for solvent emission and catalyst recycling problems occurring during the classical chemical reactions. The recent works debunk the myth about the price of ILs[10] and toxicities[11] making them even more interesting green solvents for industrial applications including biomass processing.

Imidazole is an alternative to ionic liquids as it is broadly used as precursor of imidazolium-based ionic liquids. Up to now imidazole is also used as solvent for starch dissolution.[12] Imidazole is characterised by low toxicity and high boiling point and as such due to the negligible vapour pressure is easy to be handled and to be recycled, being an interesting cheap alternative to other more expensive novel solvents used in biomass processing. As such, imidazole can contribute to greener biomass valorisation by reducing the need of hazardous chemicals, integrated biomass fractionation, delignification and depolymerisation and the use of safer solvents and auxiliaries.

Carbon dioxide is another green solvent broadly used in industry. CO_2 is one of the supercritical fluids. Small changes in temperature of pressure close to critical point can result in up to 100-fold changes in solubility simplifying separation. Due to this remarkable solvent power adjustment, scF are considered and applied in several industries as solvent for extraction processes or co-solvent in the gas-liquid hydrogenation reactions.

The use of green solvents helps to accomplish the integrated biomass valorization to produce diverse pivot compounds such as oligosaccharides, monosaccharides, furans and polyphenolics.

Results and Discussion

One of the major challenges in the biomass valorisation is the pre-treatment step that aims to produce fractionated biomass which is more susceptive for further chemical or biological conversion.

The biomass fractionation with ILs is recently one of the most developing technologies.[3, 10] It is mostly due to a great potential and the solvent designer properties of ILs because the *ideal* IL(s) can be developed and later used in the biomass processing. One of the approaches is the fractionation of biomass into cellulose, hemicellulose and lignin (Figure 1).

Alternative can be the biomass fractionation into cellulose and lignin fraction and simultaneous selective hemicellulose conversion towards xylose or furfural.[8, 13] Inexpensive acidic imidazolium-based ILs[10] permits to achieve the selective fractionation and conversion to two different products. As an example 1-butyl-3-methylimidazolium hydrogen sulphate, due to the presence of anion arising from strong inorganic acids (H_2SO_4) and, at the same time, cheap reagent, permits to produce selectively xylose or furfural. Other important factors affecting the selective recovery of solid constituted by mixture of cellulose and lignin and simultaneous production of a liquor rich in hemicellulose hydrolysed products namely xylose or furfural are temperature, time and water content.[13]

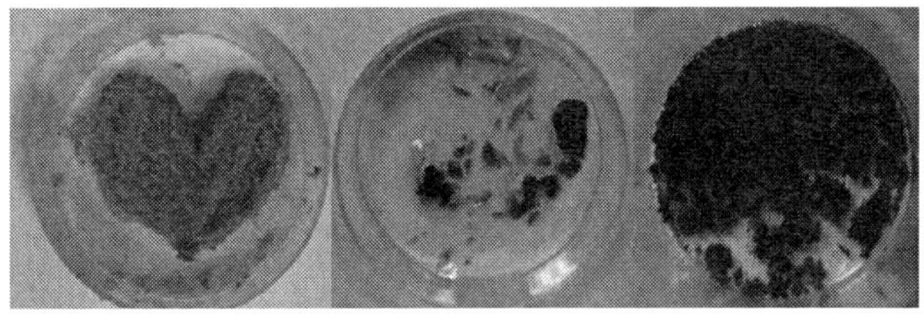

Figure 1. Cellulose, hemicellulose and lignin obtained after the biomass fractionation with 1-ethyl-3-methylimidazolium acetate.[3-4]

Among the mentioned factors water content is very important because of the equilibrium between furfural and xylose. Furfural is the product of dehydration of xylose thus water governs the equilibrium between these two products. Other product of the IL technology is a solid rich in cellulose and lignin. The processing of biomass with ILs showed that cellulose is more susceptible for enzymatic hydrolysis due to the crystallinity reduction.[3] Acidic ILs promotes the recovery of lignin in amount which is surprisingly high; much higher than initial lignin content in the raw feedstock.[13] This extra lignin is the effect of humins formation occurring at temperatures above the lignin liquefaction.[14]

Imidazole is a novel solvent for biomass processing. Similarly to ionic liquids, e.g. 1-ethyl-3-methylimidazolium acetate,[3-4] the successful fractionation of biomass into cellulose- and hemicellulose-rich fractions can be achieved. Additionally, imidazole as an organic base provokes the delignification of the recovered materials, enhancing the purity of these fractions. Furthermore, the biomass fractionation with imidazole causes structural and morphological changes, promoting reduction of crystallinity and advanced lignin removal, enhancing the performance of enzymatic digestibility of the produced cellulose-rich samples. The overall concept of imidazole-based biomass pre-treatment is depicted in Figure 3.

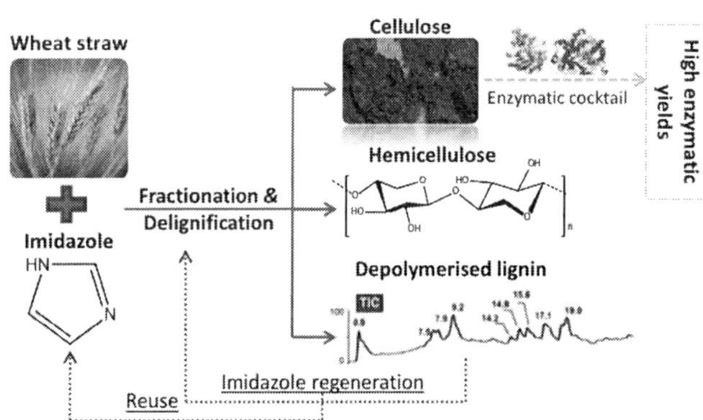

Figure 3. The concept of biomass fractionation with imidazole.[15]

Moreover, the presence of value-added phenolic compounds from lignin depolymerisation (e.g. vanillin, rosmarinic acid, vanillic acid) makes the employment of imidazole an inexpensive, economically favourable, efficient and environmentally benign process of biomass processing, within the green biorefinery concept.

Supercritical CO_2 is a fluid with easily achievable supercritical conditions. Additionally, CO_2 in contact with water makes advantageous conditions, since it permits the *in-situ* formation of carbonic acid which dissociates liberating hydronium ions stimulating the advanced acidic hydrolysis of foremost hemicellulose.[5] Additionally unlike in the acid-hydrolysis, the acidity of CO_2 produced medium does not represent an environmental problem since after the depressurisation, CO_2 is no longer present in the reaction environment.[16] At the same time, it favours the production of xylo-oligosaccharides with concentration superior to than autohydrolysis.[6] At more severe conditions (especially at higher temperature), the resulted liquor contains sugars in the monomeric form – xylose and solid which is rich is cellulose together with lignin, either native or humins. One of the major advantages of the proposed CO_2 technology is that all the products can be obtained simultaneously without additional volatile solvents, chemicals and catalysts.[17]

Another application of CO_2 is a conversion of biomass into furfural. As mentioned above, the C_5-sugars extraction from biomass occurs generally in aqueous stream hence, the ideal process of furfural production would preferably take place in water in the presence of an environmentally benign catalyst. The proposed concept involves a series of reactions to produce furfural from polysaccharides containing pentose monomer units. These reactions start with the hydrolysis of hemicellulose (pentosans) to obtain pentoses as either oligomers or monomers as discussed before. The produced pentose-rich hydrolysates in the presence of high-pressure CO_2 can undergo further dehydration to produce furfural as shown in Figure 4.[18]

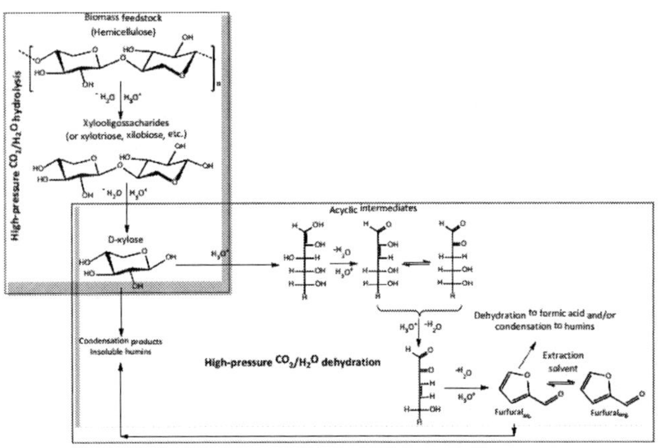

Figure 4. The reaction pathways involving high-pressure CO_2-H_2O extraction of hemicelluloses from biomass and CO_2-catalysed dehydration of pentoses to furfural in the presence of an extracting solvent.[19-20]

Conclusions

The constant aim towards more green and sustainable processes and products motivates the numerous researches in the field of green biorefinery. This increases of global awareness about the importance of this field. To promote the further progress in development of this research future integrated biorefineries are need to involve those processes in order to give solutions assisting our quest towards these aspirations. In general, neoteric solvents or solvent-catalyst systems will most likely be a part of the solution but only if the principles and the spirit of the 'Green Chemistry' are respected. Therefore, appropriate choice of concepts and chemistries, efficient recycling and separation concepts, re-designed benign chemical derivatives as building blocks for our commodities and daily consumables are the trends of today.

Acknowledgement: This work was supported by the Fundação para a Ciência e Tecnologia (FCT/MEC, Portugal) through SFRH/BD/90282/2012, SFRH/BD/94297/2013 and IF/00424/2013 grants and by the Associated Laboratory for Sustainable Chemistry - Clean Processes and Technologies - LAQV which is financed by national funds from FCT/MEC (UID/QUI/50006/2013) and co-financed by the ERDF under the PT2020 Partnership Agreement (POCI-01-0145-FEDER - 007265). Special acknowledgment is addressed to FCT and CAPES for financing Bilateral Cooperation Portugal (FCT)/Brazil (CAPES) project (FCT/1909/27/2/2014/S) CAPES for PVE 155/2012 project.

References

(1) Anastas, P. T.; Zimmerman, J. B. *Environ. Sci. Technol.* **2003,** *37* (5), 94a.
(2) Anastas, P. T.; Warner, J. C. Oxford University Press: New York, 1998.
(3) da Costa Lopes, A. M.; Joao, K. G.; Rubik, D. F.; Bogel-Lukasik, E.; Duarte, L. C.; Andreaus, J.; Bogel-Lukasik, R. *Bioresource Technol.* **2013,** *142*, 198.
(4) Magalhães da Silva, S. P.; da Costa Lopes, A. M.; Roseiro, L. B.; Bogel-Lukasik, R. *RSC Adv.* **2013,** *3* (36), 16040.
(5) Magalhães da Silva, S. P.; Morais, A. R. C.; Bogel-Lukasik, R. *Green Chem.* **2014,** *16* (1), 238.
(6) Morais, A. R. C.; Mata, A. C.; Bogel-Lukasik, R. *Green Chem.* **2014,** *16* (9), 4312.
(7) Morais, A. R. C.; da Costa Lopes, A. M.; Bogel-Lukasik, R. *Chem. Rev.* **2015,** *115* (1), 3.
(8) da Costa Lopes, A. M.; Bogel-Lukasik, R. *ChemSusChem* **2015,** *8* (6), 947.
(9) Earle, M. J.; Esperanca, J. M. S. S.; Gilea, M. A.; Lopes, J. N. C.; Rebelo, L. P. N.; Magee, J. W.; Seddon, K. R.; Widegren, J. A. *Nature* **2006,** *439* (7078), 831.
(10) Chen, L.; Sharifzadeh, M.; Mac Dowell, N.; Welton, T.; Shah, N.; Hallett, J. P. *Green Chem.* **2014,** *16* (6), 3098.
(11) Coleman, D.; Gathergood, N. *Chem. Soc. Rev.* **2010,** *39* (2), 600.
(12) Jordan, T.; Schmidt, S.; Liebert, T.; Heinze, T. *Green Chem.* **2014,** *16* (4), 1967.
(13) Carvalho, A. V.; da Costa Lopes, A. M.; Bogel-Lukasik, R. *RSC Adv.* **2015,** *5*, 47153.
(14) Rasmussen, H.; Sorensen, H. R.; Meyer, A. S. *Carbohydr. Res.* **2014,** *385*, 45.
(15) Morais, A. R. C.; Vaz Pinto, J.; Nunes, D.; Roseiro, L. B.; Oliveira, M. C.; Fortunato, E.; Bogel-Lukasik, R. *ACS Sustain. Chem. Eng.* **2016,** *4*, 1643.
(16) Carvalheiro, F.; Silva-Fernandes, T.; Duarte, L. C.; Gírio, F. M. *Appl. Biochem. Biotechnol.* **2009,** *153* (1), 84.
(17) Hoang, T. M. C.; van Eck, E. R. H.; Bula, W. P.; Gardeniers, J. G. E.; Lefferts, L.; Seshan, K. *Green Chem.* **2015,** *17*, 959.
(18) Feather, M. S.; Harris, D. W.; Nichols, S. B. *J. Org. Chem.* **1972,** *37* (10), 1606.
(19) Morais, A. R. C.; Bogel-Lukasik, R. *Green Chem.* **2016,** 10.1039/C5GC02863A
(20) Morais, A. R. C.; Matuchaki, M. D. J.; Andreaus, J.; Bogel-Lukasik, R. *Green Chem.* **2016,** 10.1039/C6GC00043F.

QUANTITATIVE MEASUREMENT OF BIOMASS PYROLYSIS EFFICIENCY USING GCxGC/MS

Kristyn M. Johnson, Salene M. Rowe, Christopher A. Zarzana, Gary S. Groenewold, Rachel Emerson, Tyler Westover, Amber Hoover

Idaho National Laboratory, Idaho Falls, ID
kristyn.johnson@inl.gov

Abstract

Accurate characterization of pyrolysis products can provide valuable guidance for thermal processing and product up-blending but is challenging to achieve because the large number of products can make measurements ambiguous. Ambiguity in compound identification was reduced by developing a pyrolysis/two-dimensional gas chromatography/mass spectrometry (PY/GCxGC/MS) analysis, which eliminates most of the chromatographic co-elution and enables compound identification using mass spectral library searching with a higher level of confidence. Quantitative measurement was performed for 33 pyrolysis products using an internal standard method. Biomass sample analyses were performed by adding a known quantity of the internal standard, which together with the pyrolysate compounds was volatilized and transferred to the GCxGC/MS when pyrolysis was initiated. This enabled direct measurement of the pyrolysis product concentrations. The concentrations of additional compounds were measured in a semi-quantitative fashion using response factors of structurally related calibrated compounds corrected for the relative intensities of the quant ions. The PY/GCxGC/MS method was used to evaluate parameters affecting pyrolysis efficiency for miscanthus samples. Increased sample mass produced increased quantities of pyrolysate compounds, but the production efficiency actually decreased. Standard sample mass loading of 0.6 g represented the best compromise between compound detection, pyrolysis efficiency and instrument operation; the latter deteriorated when the total quantity of pyrolysate (particularly acetic acid) was too great. The effect of pyrolysis temperature on pyrolysis efficiency was found to vary by compound. Lignin-derived compounds like vinyl guaiacol increased in an exponential fashion from 350 to 500°C, however carbohydrate-derived compounds such as furfural were comparatively unaffected. The results indicate that compound-explicit efficiency-versus-temperature profiles can be produced using the pyrolysis/GCxGC/MS method that may offer more informed guidance for thermal processing, upblending of lower-value feedstocks, and thermal value of the fuel eventually produced.

LIGNOFORCE™ LIGNIN CHARACTERIZATION AND POTENTIAL HIGH-VALUE APPLICATIONS

Michael Paleologou, Lamfeddal Kouisni, Kirsten Maki, Yaolin Zhang and Martin Feng*

FPInnovations
570 Boulevard Saint-Jean, Pointe Claire, QC, H9R-3J9
*michael.paleologou@fpinnovations.ca

Introduction

Lignin represents the largest reservoir of natural aromatic compounds available on earth. Softwoods (SW) and hardwoods (HW) are composed of about 25-35wt% and 15-25wt% lignin, respectively. During alkaline pulping processes, the lignin is chemically broken down and partially separated from the carbohydrate portion of the furnish. Following pulping, lignin and other residual chemicals are removed from the fibers through washing with water or evaporator condensates. The washing filtrate, commonly referred to as weak black liquor (WBL), is usually concentrated from about 20wt% to about 70wt% solids through evaporation and then fired into the recovery boiler to produce steam, electricity and pulping chemicals for internal mill use. As many chemical pulp mills have been increasing pulp production over the last 30 years, the recovery boiler has, in most cases, become the production bottleneck. A cost-effective way of offloading the recovery boiler with respect to calorific load is to remove a portion of the lignin from the black liquor (BL). For every tonne of lignin that is removed, a typical chemical pulp mill can produce an additional tonne of pulp. Furthermore, such mills can generate additional revenue by selling the lignin for use as a cost-effective, renewable substitute for petroleum-based chemicals and materials in several low-, medium- and high-value applications.

Several processes exist for the recovery of lignin from BL. These include: the Westvaco process developed over 60 years ago, the LignoBoost™ process, developed by STFI (now called Innventia) and licensed to Valmet and the LignoForce system™ jointly developed by FPInnovations and NORAM. A simplified schematic of the latter process is shown in Figure 1.

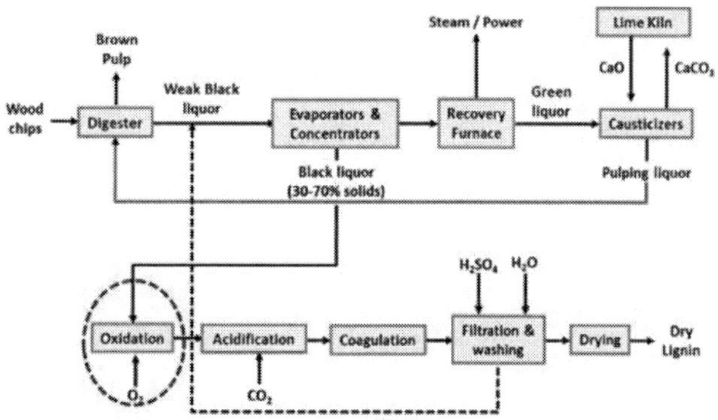

Figure 1. The LignoForce System™ for lignin recovery from black liquor (*US Patent No. 8,771,464*)

A unique feature of the LignoForce™ process is that, prior to the addition of CO_2, the BL is oxidized under controlled conditions, with respect to O_2 charge, temperature and time [1,2]. Under these conditions, the chemical requirements are reduced, lignin filterability is improved and pure lignin at high solids is obtained. Another particular advantage of the LignoForce System™ is that the emission of malodorous sulphur compounds from both process and product are significantly reduced [3]. For the purpose of evaluating lignin recovery from a wide range of black liquors (e.g. SW, HW and Eucalyptus (EU)), in 2012, FPInnovations and NORAM built a lignin demo plant at a kraft pulp mill in Thunder Bay, Ontario, Canada for the production of 100 kg of lignin per working day. Based on the design data generated at the Thunder Bay demo plant, in 2015, a commercial lignin plant, using the LignoForce™ technology, was built by West Fraser at its Hinton pulp mill in Alberta, Canada, for the production of 30 tonnes of SW lignin per day [4]. This system is currently being started up. The main objective of this paper is to discuss the main features of LignoForce™ lignin from three different black liquor sources with respect to chemical composition, main functional groups, molecular weight distribution and thermal properties. Based on these results, the technical feasibility of using these lignins in several low, medium and high-value applications will also be discussed.

Experimental

Black liquor characterization. For the purpose of evaluating the recovery of lignin from a wide range of BLs, the LignoForce™ pilot plant in Pointe-Claire, Quebec, Canada and the LignoForce™ demo plant in Thunder Bay, Ontario, were used in this study. Black liquor samples were collected in different campaigns from several SW, HW and Eucalyptus kraft pulp mills. These samples were analyzed for total solids, residual effective alkali (REA), sulphide, ash, sodium, sulphur, organics, lignin, sugars, calorific value, soap, and elemental Carbon, Hydrogen, Nitrogen and Oxygen content.

Lignin recovery. In the first step of the LignoForce System™, a sufficient amount of O_2 was added at a high temperature (80 to 100°C), to oxidize the BL to about 0–0.5 g/L residual sulfide. In a second step, a sufficient amount of CO_2 was added at a temperature of 65 to 75°C to reduce the pH of the oxidized liquor to about 9.5, at which point, lignin precipitated out of solution in the form of colloidal particles. Following the acidification, the lignin slurry was subjected to a coagulation/aging step under gentle mixing, at a temperature of 60-65°C, to improve its filterability. In a fourth step, the lignin slurry was filtered using a filter press. A lignin cake was thus obtained which was washed with sulphuric acid and water on the filter itself. Finally, the lignin cake was squeezed and air-blown to obtain a high-solids, purified lignin product.

Lignin characterization. The recovered lignin samples were analyzed for dry solids, UV-lignin, calorific value, carbohydrates, sodium, sulphur, and ash content. In addition, these lignins were characterized with respect to weight average molecular weight (Mw) and number average molecular weight (Mn), as described in [5], and main functional groups, as described in [6]. With respect to thermal properties, differential scanning calorimetry (DSC) was used to measure the glass transition temperature of the various lignins produced while thermogravimetric analysis (TGA) was used to measure the initial decomposition temperature.

Results and Discussion

The lignin characterization results are shown in Table 1. Particularly notable is the low ash and sulphur content of all lignins as well as their relatively high calorific value. These data suggest that these lignins are of sufficient purity and calorific value to be used as a fuel in the lime kiln, the power boiler or most other combustion processes. It is also worth noting here that, in most cases, the solids content of the lignin cake recovered from the filter press was quite high. This suggests that the cost of drying the lignin to over 95% solids (if needed) will be relatively low as compared to the lignin cakes recovered from a belt filter (e.g. conventional lignin recovery processes) which are usually much lower in solids (32-35wt%) [7].

Table 1. Lignin characterization

Lignin type	SW Lignin	HW lignin	Eucalyptus lignin
Wet lignin solids, wt%	45-65	45-68	45-66
Total solids (air-dried), wt%	94-99	95-98	87-96
pH @ 15wt% solids	1.7-5.9	1.9-3.8	1.8-3.9
Ash, wt%	0.10-1.5	0.04-1.2	0.3-0.8
Na, wt%	0.05-0.60	0.0-0.30	0.03-0.1
S, wt%	1.3-2.1	1.7-2.6	1.7-2.1
UV-lignin, wt%	92-98	96-97	95-97
Carbohydrates, wt%	1.2-2.4	0.26-1.8	0.25-1.1
Carboxylic hydroxyl, mmol/g	0.20-0.50	0.11-0.40	0.28-0.30
Aliphatic OH, mmol/g	1.4-2.1	1.3-1.6	1.1-1.2
Phenolic OH, mmol/g	1.3-1.8	2.2-3.2	3.1-3.3
Condensed unit OH, mmol/g	1.1-1.6	0.31-0.50	0.33-0.38
Total OH groups, mmol/g	4.3-5.7	4.1-5.6	4.7-5.2
Glass transition temp., °C	162-185	133-152	118-128
Initial decomposition temp., °C	212-358	220-260	255-266
Molecular weight (Mw) using the UV detector	6000-12500	2635-6249	2100-2700
Polydispersity index	3.0-4.6	2.4-4.1	1.9-2.9
Calorific Value, BTU/lb	11200-11800	10200-11200	10800-11200

The SEC analysis showed some differences in the Mw, Mn and the polydispersity index (Mw/Mn) of the three types of lignins. The Mw of lignins recovered from the HW and EU BLs is about half of the Mw of lignins recovered from SW BLs while the polydispersity index appears to be the lowest for the EU lignins. From this analysis, we also observe that the total hydroxyl groups in HW and EU lignins is about the same as that obtained in lignins recovered from SW BLs. However, the distribution of hydroxyl groups (e.g. carboxylic, phenolic, aliphatic and condensed unit hydroxyl groups) was different between the three lignins. The HW and EU lignins had, generally, a lower content of aliphatic and, especially, condensed unit hydroxyl groups but a significantly higher content of phenolic hydroxyl groups as compared to the SW lignins. Of the latter groups, however, in the SW lignin case, over 95% were of the guaiacyl type while in the case of the HW and EU lignins only about 17-26% were of this type. Based on these results, one would expect to achieve higher phenol substitution rates in phenol-formaldehyde (PF) resins using SW lignins as compared to HW or EU lignins. In contrast, because of the lower Mw of HW and EU lignins as compared to SW lignins one would expect to achieve higher substitution rates when these are used as a replacement of petroleum-based polyols in the preparation of polyurethane (PU) materials (e.g. foams) since this would lead to higher solubility in and lower viscosity of the resulting petroleum polyol-lignin mixtures. Based on prior work in our lab, it appears technically possible to replace as much as 50wt% of phenol with SW kraft lignins in PF resins [7] and as much as 25wt% of polyols with HW kraft lignins in PU foams.

The thermal properties of lignin are very important in relation to the processability and thermal stability of lignin-polymer blends. Based on the data shown in Table 1, it appears that the glass transition temperature of all lignins is in the same range as the processing temperatures of most major polymeric materials (e.g. polypropylene (PP)) while the first decomposition temperature is sufficiently high to allow, for example, for the conversion of these lignins to carbon fiber at high product yields. In particular, as shown in Table 1, the lignins recovered from SW BLs have higher glass transition and initial decomposition temperatures than those recovered from HW and EU BLs. Based on prior work in our lab, it appears technically possible to incorporate as much as 20wt% kraft SW or HW lignin into PP without any major impact on physical properties as long as suitable coupling agents and plasticizers are used.

Conclusions

In this paper, we described the main features of the LignoForce System™ and the lignins produced from three different feedstocks: SW, HW and EU BLs. Based on the chemical and physical properties of these three lignins, it appears to be technically possible to use these lignins in several high-value applications including: as a replacement of phenol in PF resins and polyols in PU foams as well as a component in thermoplastic materials such as PP.

References

(1) Kouisni, L.; Paleologou, M., U.S. Pat. No. 8,771,464, **2014**.

(2) Kouisni, P.; Holt-Hindle, P.; Maki, K.; Paleologou, M., *Journal of Science & Technology for Forest Products and Processes*, **2012**, *2 (4)*, 6-10.

(3) Kouisni, L.; Holt-Hindle, P.; Maki, K.; Chan, C.; Paleologou, M., 6th NWBC Conference Proceedings, Helsinki, Finland, October 20-22, 2015, p. 193-199.

(4) Wells, K.; Pors, D.; Foan, J.; Maki, K.; Kouisni, L.; Paleologou, M., PACWEST Conference Proceedings, Whistler, BC, 2015.

(5) Gagné, A.; Kouisni, L.; Paleologou, M., 5th NWBC Conference Proceedings, Stockholm, Sweden, March 25-27, 2014.

(6) Argyropoulos, D., *Journal of Wood Chemistry and Technology*, **1994**, *14 (1)*, 65-82.

(7) Kouisni, L.; Fang, Y.; Paleologou, M.; Ahvazi, B.; Hawari, J.; Zhang, Y.; Wang X.-M., *Cellulose Chemistry and Technology*, **2011**, *45 (7-8)*, 515-520.

ENHANCEMENT OF BIODIESEL PRODUCTION FROM MICROBIAL COMMUNITY OF WASTE ACTIVATED SLUDGE BY nZVI and $FeCl_3$

*Jia LIU, Yi SHEN, Lijun LU, Kaiming PENG, Xiangfeng HUANG**

College of Environmental Science and Engineering, State Key Laboratory of Pollution Control and Resource Reuse, Tongji University, Shanghai 200092, China.
Corresponding author: hxf@tongji.edu.cn

Introduction

Biodiesel is produced by the transesterification of vegetable oils, animal fats or microbial oils with an alcohol such as methanol to yield the corresponding esters of the fatty acids comprising the lipid feedstock. Biodiesel production has attracted much more attention in recent years. However, its production costs are currently higher than petroleum biodiesel due to the expensive raw material as feedstocks[1]. The large amount of waste activated sludge(WAS) contains lipids absorbed from wastewater as well as microbial phospholipids, which can be used as cheap and widely availaraw raw material for biodiesel production with the advantages of decreasing the biodiesel cost. One popular method is to directly extract lipids in microbial cells and lipids adsorbed on the flocs in the WAS. However, the major compounds in the lipid extracts were PHAs, wax esters, sterylesters, triacylglycerides, free fatty acids, free sterols and phospholipids[2], and only a fraction of them which is saponifiable can be converted into fatty acid methyl ester, which results in low yield of biodiesel ranging only 2–5% of the sludge dry weight[3]. Results showed that limited nutrient relative to an excess carbon (sugar) could elicit oleaginous behavior among the constituents of this natural microbial consortium. Thereafter, it is necessary to increase the content of saponifiable lipid, which will be helpful to strengthen competitiveness of WAS as a feedstock for biodiesel production.

Experimental

Activated sludge samples were obtained from a full-scale sewage treatment plant in Shanghai, China. Sludge samples were cultivated in 500-mL flasks with working volumes of 100 mL for the batch cultivation experiments on a gyratory shaker at 200 rpm at 25 ℃ (room temperature). The cultivation medium contained the following components (L^{-1} of deionized water): 1.95 g NaH_2PO_4. $2H_2O$,1.0 g K_2HPO_4, 5 mL trace mineral solution. Glucose and ammonium sulfate were added as carbon and nitrogen sources respectively, in amounts corresponding to initial carbon-to-nitrogen ratio (C/N) based on mass and the glucose loading was fixed at 60 g/L. The pH of the culture was treated with 8 mol/L NaOH and 6 mol/L HCl to maintain the pH approximately throughout the duration of the experiment.

Two different forms of iron, nanoscale zero valent iron (nZVI) and ferric trichloride ($FeCl_3$) were added to the medium respectively. Five levels of iron were added: 0, 0.1, 0.25, 0.5, 1.0 and 2.5 mmol/L. All initial sludge concentrations in mixed liquor suspended solids (MLVSS) were 2.65 g/L. The total incubation time for the batch culture was 7 days and culture broth samples were collected at regular time intervals for analysis.

Biomass accumulation was determined by gravity weighing. Lipid was extracted by modified acid-heating extraction. The biodiesel yield and fatty acid profile of the lipid extracts were estimated as FAMEs via transesterification, which was catalyzed using a methanolic BF_3 solution. FAMEs were detected by gas chromatography using a gas chromatograph (GC) (6890N, Agilent Technologies, Santa Clara, CA, USA) equipped with a flame ionization detector (FID). Microbial community was analyzed using high throughput pyrosequencing and sequence analysis.

Results and Discussion

It was efficient to increase the content of saponifiable lipid by controlling the pH during fermentation process and adjusting the initial C/N of the media. The biodiesel content was increased to 13.06% under the optimal condition of pH=4 and C/N=100. Under that condition, nZVI and $FeCl_3$ was separately added in the media to enhance the lipid accumulation. As Figure 1 showed, 0.25 mmol/L nZVI improved the biodiesel content to 3.1 g/L and 215.3 mg/g dry sludge (21.53%). Compared with that, 2.5 mmol L^{-1} $FeCl_3$ further promoted the highest biodiesel production at 4.76 g/L and 274.9 mg/g dry sludge (27.49%), indicating that Fe(III) was more efficient in elicit oleaginous ability of the WAS.

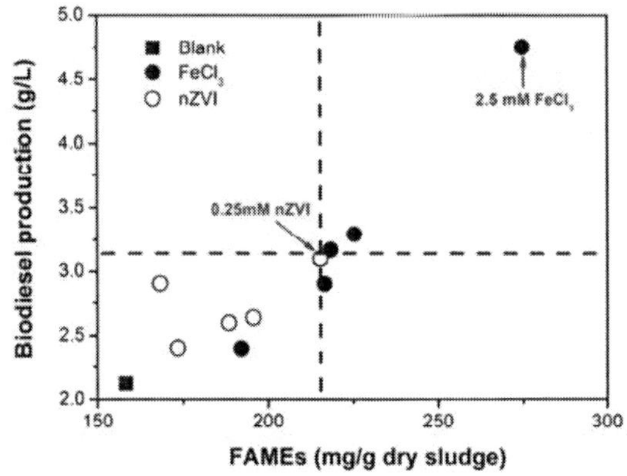

Figure 1. The effect of nZVI and $FeCl_3$ on the biodiesel production from waste activated sludge.

In order to investigate the specific influence of different forms of iron on the characteristics of microbial communities, bacteria in sludge was identified by high-throughput sequencing technology to further reveal the influence of different forms of iron on the composition of microbial communities. Results showed that the predominant organism group of the control group was mainly *Gluconobacter* at the end of fermentation. The predominant organism groups of group $FeCl_3$-2.5 mmol/L Fe were mainly *Gluconacetobacter* and *Gluconobacter*, meanwhile, the abundance of *Gluconacetobacter* was far higher than that of *Gluconobacter*. While, the predominant organism group of group Fe-0.25 mmol/L Fe was mainly *Gluconobacter*. Microbial species of all test groups tended to be single dominant and community composition tended to be simplistic. The genus *Gluconobacter*, as well as *Gluconacetobacter* belong to acetic acid bacteria, which is usually isolated from vinegar, with the ability to tolerate high acid environment. Previous studies showed these two genuses contained the straight-chain unsaturated acid of C18:1 as a major constituent[4, 5], which accouts for approximately 60~70% of the total fatty acid. An increasing relative abundance of the genuses is responsible for the the dramastic increase in the proportion of C18:1 at the end of cultivation, which was also observed in other glucose-fed activated sludge with different culture conditions[1]. In addition, they may contribute to the high proportion of saponifiable lipid at the end of cultivation. It can be speculated that biodiesel yield of WAS was enhanced because of that the microbial community was largely changed by adding $FeCl_3$.

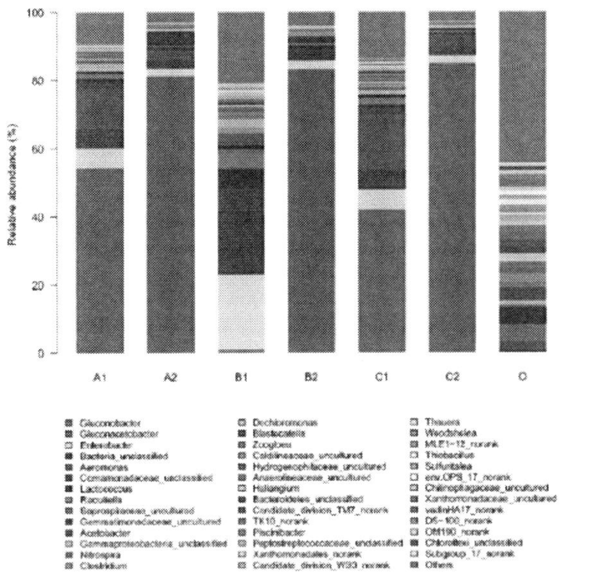

Figure 2. The effect of nZVI and $FeCl_3$ on the changes of microbial community of waste activated sludge. A:blank, B:2.5 mM $FeCl_3$, C:0.25 mM nZVI, O: raw sludge, 1: fermented for 3d , 2: fermented for 7d)

The composition of fatty acid was analyzed shown in Table1. For raw activated sludge(RAS), the fatty acids are mainly composed of saturated fatty acids palmitate (16%) and stearate(22%) followed by the unsaturated fatty acids palmitate(9%) and stearate(9%), with 36% of unclassified fatty acids. After cultivation for 7 days, the proportion of unsaturated fatty acids (including C16:1, C18:1, C18:2, C18:3) were nearly threefold those from RAS (24% vs. around 68%) while the percentage of unclassified fatty acids was reduced to around 6% regardless of the treatments. A higher percentage of unsaturated fatty acids is responsible for better cold flow properties which is an essential characteristic of biodiesel. The higher percentage of saturated fatty acids could promote oxidative stability and cetane number of the biodiesel[6]. These properties can be well balanced by C18:1, the high level of which can provide the greater stability for biodiesel[7]. The addition of $FeCl_3$ was in favour of the improvement of the content of C18: 1, C18: 2 and C18: 3, while the addition of nano zero-valent iron was good for the improvement of the content of C16: 1, C18: 1 and C18 : 2. In addition, biodiesel indicators including cetane number, iodine value, viscosity, cold filter plugging point were able to meet the standards calculated based on the fatty acid composition.

Table 1. Fatty acids composition (% w/w total saponifiable lipids) of activated sludge cultured for 7d

Distribution of fatty acids (%)	RAS	nZVI	$FeCl_3$
Palmitic acid,C16:0	16.4	11.4	11.1
Palmitoleic acid,C16:1	9.68	5.94	5.44
Stearic acid,C18:0	22.8	13.3	14.3
Oleic acid,C18:1	9.48	51.7	52.3
Linoleic acid,C18:2	4.61	8.95	10.2
Linolenic acid,C18:3	0.89	1.26	1.52
Unknowns	36.2	7.45	5.20

Conclusions

$FeCl_3$ was more efficient in the enhancement of biodiesel yield compared with nZVI. Their stimulating effects were complex, one of the reasons maybe changing the microbial community. The lipids produced contained nealy 60% of unsaturated fatty acids, including palmitic acid, heptadecanoic acid, ginkgolic acid, stearic acid, oleic acid, and linoleic acid, showing the adequacy of biodiesel production.

Acknowledgement. This work was supported by National Science and Technology Pillar Program (No. 2012BAC11B04) and the National Natural Science Foundation of China (Grant No. 51108333).

References

(1) Sun Q, A Li, M Li, et al. Effect of pH on biodiesel production and the microbial structure of glucose-fed activated sludge. International Biodeterioration & Biodegradation, 2015, **104**:224-230.

(2) Revellame E D, R Hernandez, W French, et al. Lipid storage compounds in raw activated sludge microorganisms for biofuels and oleochemicals production. Rsc Advances, 2012, **2**(5):2015-2031.

(3) Gaeta-Bernardi A and V Parente. Organic municipal solid waste (MSW) as feedstock for biodiesel production: A financial feasibility analysis. Renewable Energy, 2016, **86**:1422-1432.

(4) Heefner D and G Claus. Lipid and fatty acid composition of Gluconobacter oxydans before and after intracytoplasmic membrane formation. Journal of bacteriology, 1978, **134**(1):38-47.

(5) YAMADA Y, M NUNODA, T ISHIKAWA, et al. The cellular fatty acid composition in acetic acid bacteria. The Journal of General and Applied Microbiology, 1981, **27**(5):405-417.

(6) Singh B, A Guldhe, I Rawat, et al. Towards a sustainable approach for development of biodiesel from plant and microalgae. Renewable and sustainable Energy reviews, 2014, **29**:216-245.

(7) Bringe N A, B J Calabotta, and D A Morgenstern, *Soybean oil composition for biodiesel*. 2005: AOCS Press: Champaign, IL.

KINETICS AND MATHEMATICAL MODELING OF AGAVE BAGASSE HEMICELLULOSE USING HYDROTHERMAL PRETREATMENT UNDER BIOREFINERY CONCEPT

Elisa Zanuso[1], Jingqian Chen[2], Rosa Ma. Rodriguez-Jasso[1], Arturo Sanchez [3], Heather Trajano[2] and Héctor A. Ruiz[1]

[1]Biorefinery Group, Food Research Department, School of Chemistry, Autonomous University of Coahuila, 25280 Saltillo, Coahuila Mexico
[2]Department of Chemical and Biological Engineering, University of British Columbia, 2360 East 6 Mall, Vancouver, BC, Canada, V6T 1Z3.
[3]Unidad Guadalajara de Ingeniería Avanzada, Centro de Investigación y Estudios Avanzados (CINVESTAV), Zapopan, Jalisco, México.

Introduction

Agave bagasse (AGB) is a residual fiber left behind after the production of alcoholic beverages such as tequila, mezcal or sotol in Mexico. The annual agave consumption for tequila production in 2015 was around 8.09 x 10^5 tons. AGB represents usually 40% of the processed agave on a dry weight basis[1]. This agroindustrial residue is mainly by cellulose 30-50%, hemicellulose 20-40%, lignin 15-25% and ashes 3-10%. Thus, this residue represents an important source for the biorefinery processing for the production of high-added value compounds and sugar rich substrates for bioethanol production[2]. Hydrothermal pretreatment is a chemical-free process that only involves water and lignocellulosic biomass. This pretreatment is a fast, economical and eco-friendly operation. In this pretreatment, the hemicellulose gets depolymerized in oligomers, as xilo-oligosaccharides (XOS), and monomers. XOS can be extracted from the liquid phase and the solid phase can be used as substrate for enzymatic hydrolysis, after this step, bioethanol can be produced[2]. Batch reactors are used as a conventional heating system to perform the hydrothermal pretreatment and conventional heating is generally used for biorefineries of second generation. According to Vallegos et al.[3], the mathematical modelling of the changes in the composition of the lignocellulosic materials during the hydrothermal processing is applied to design and optimize the processes in the production of high-added value compounds as the oligomers from hemicellulose. Kinetic models and severity parameters are different mathematical methods to describe the depolymerization of hemicelluloses. This work deals with the kinetic modelling of the hydrothermal pretreatment on AGB. The effects of pretreatments on hemicellulose (xylan) and on xylan-reaction products have been established, and a kinetic model (based on consecutive, pseudo-homogeneous, first-order reactions) has been applied to explain the variation of hemicellulosic carbohydrates released during the pretreatment.

Experimental

Agave bagasse was donated by a Tequilera in Guadalajara, Jalisco, Mexico. AGB were chopped using a cutter mill to a particle size of 0.3 – 1 mm. Samples were milled to pass a 0.5 mm screen for the physicochemical characterization where ashes, moisture and extractives were determinate. The chemical composition (cellulose, hemicelulose and lignin content) of the raw material was determined according to the methodology reported by Ruiz et al.[4].

Hydrothermal pretreatment. The hydrothermal pretreatment was performed in a stainless steel 316 pressurized reactor with a volume work of 750 mL adapted with PID control and speed agitation, heat resistance, manometer for internal pressure (Figure 1). For the pretreatments a ratio of 1:10 and 1:15 w/v (biomass:water) was used with different residence time (0, 9, 27, 40 and 100 min) and temperature (160, 180, 200°C).

Characterization of solid and liquid phase after hydrothermal pretreatment. At the end of pretreatment, the liquid and solid phases were separated by filtration and the solid residues were washed with distilled water. The liquid was analyzed by HPLC with a MetaCarb 87H (300×7.8 mm, Varian, USA) column. The samples were analyzed for glucose, xylose arabinose, 5-hydroxymethylfurfural (HMF), furfural and acetic acid. A second aliquot of liquors (20 ml) was subjected to quantitative post-hydrolysis (with 4% H_2SO_4 at 121°C during 60 min) before HPLC analysis for the determination of oligomers from hemicellulose. The pretreated solids were analyzed as described above.

Fitting of data. The kinetic of the reactions generating products in the liquor, from the hemicellulose (xylan) in the solid material, was determined by adapting the model proposed by Mehlberg and Tsao [5] and cited by Garrote et al.[6]. The experimental data were fitted to the proposed kinetic models by minimization of the sum of squares using commercial software with a built-in optimization routine dealing with the Newton's method using the Polymath software.

Results and Discussion

The results of raw material characterization were: glucan, xylan and lignin with 31 % and 16 % and 18.4 % respectively. This helps to predict the amount expected of oligosaccharides

Figure 1. Hydrothermal pretreatment reactor.

This chemical composition is in good agreement with other values found in the literature for this material[1].

The maximum resident reactions were selected to allow a complete observation of the behavior of XOS concentration, and the temperature range covers the usual values used in hydrothermal processes. The mechanism of hemicelulose (xylan) depolymerization during the second step (acid pre-hydrolysis) proposed by Mehlberg and Tsao[5] included the assumption that two xylan fraction exist in the feedstock (with different susceptibility towards hydrolytic depolymerization). Hemicellulose (xylan) is converted into xylooligosaccharides, which give xylose and this pentose is dehydrated to furfural[6]. The more detailed results about the kinetics and mathematical modeling of agave bagasse hemicellulose will be presented and discussed.

Acknowledgment. We gratefully acknowledge support for this research by the CONACYT-SENER- SUSTENTABILIDAD ENERGÉTICA CEMIE-BIO in the bioalcohols cluster: Biocombustibles Lignocelulósicos para el Sector Autotransporte (CONACYT-SENER 249564).

References

(1) Perez-Pimienta, J.A., Flores-Gómez, C.A., Ruiz, H.R., Sathitsuksanoh, N., Balan, V., Da Costa Sousa, L., Dale, B.E., Singh, S., Simmons, B.A., *Bioresource Technology*. 2016. 211: p. 216-23.
(2) Ruiz., H.A., Rodríguez-Jasso, R.M., Fernandes, B.D., Vicente A.A., Teixeira J.A., *Renewable and Sustainable Energy Reviews*, 2013. 21: p. 35-51.
(3) Vallegos, M.E., Felissia, F.E., Kruyeniski, J., *Industrial Crops and Products*, 2015. 67: p. 1-6.
(4) Ruiz, H.A., Ruzene, D.S., Silva, D.P., Macieira da Silva, F.M., Vicente, A.A., Teixeira, J.A., *Applied Biochemistry and Biotechnology*, 164: P. 629-41.
(5) Mehlberg, R., Tsao, G.T., *178th ACS National Meeting*.1979. Washington.
(6) Garrote, G., Domínguez, H., Parajó, J.C., *Process Biochemistry*. 36, 571-78.

LIGNIN S/G RATIO FACILITATES *POPULUS* SOLUBILIZATION DURING CONSOLIDATED BIOPROCESSING

*Dumitrache A, **Akinosho H**, Rodriguez M Jr, Meng X, Yoo CG, Natzke J, Engle NL, Sykes RW, Tschaplinski TJ, Muchero W, Ragauskas AJ, Davison BH, Brown SD.*

1 Bethel Valley Rd Building 1520 Office 225
Oak Ridge, TN 37830
akinoshoho@ornl.gov

Introduction

Consolidated bioprocessing (CBP) is an integrated processing scheme that combines the traditionally separate cellulase production, enzymatic hydrolysis, fermentation stages. Several microorganisms are used in CBP applications including the *Clostridium thermocellum*. This anaerobic, thermophilic bacterium possesses impressively high enzymatic hydrolysis rates in Avicel; however, enzymatic hydrolysis rates are much higher in Avicel in comparison to biomass [1].

In fungal enzyme cocktails, biomass recalcitrance has been identified as a hindrance to bioethanol production. S/G ratio has been found to be correlated with recalcitrance, where high S/G ratio biomass demonstrates more favorable deconstruction properties[2]. This study similarly investigates lignin S/G ratio instead as a facilitator or obstacle to CBP with *C. thermocellum*.

Experimental

The detailed experimental procedure for this project is described in (Dumitrache et. al, 2016) [3]. Each subsection represents a highlighted but truncated version of the method.

Lignin molecular weight. Lignin was isolated from air-dried, ball-milled *Populus trichocarpa* via enzymatic hydrolysis. The isolated lignin was purified with 96% w/w *p*-dioxane and derivatized with 1:1 w/w pyridine:acetic anhydride. The derivatized lignin was recovered and separated from the liquid mixture and as dissolved in THF. The dissolved samples were filtered and analyzed by GPC to determine molecular weight.

Cellulose molecular weight. Cellulose was isolated from *Populus trichocarpa* holocellulose. Cellulose was derivatized with a mixture of 8:1 w/w phenyl isocyanate and pyridine. The derivatized cellulose was precipitated and washed prior to dissolution in THF. The dissolved samples were filtered and analyzed by GPC to determine molecular weight.

Cellulose crystallinity. Cellulose for crystallinity analysis was prepared from acid-hydrolyzed holocellulose. The recovered cellulose was washed and filtered. Crystallinity was determined by FTIR and NMR.

Simons' stain. The method for the Simons' stain was taken from (Chandra et. al, 2008) [4]. No modifications were made to the procedure.

Results and Discussion

Table 1 describes the natural variants used in this study in conjunction with solubilization and fermentation data. The table indicates that there are two S/G ratio variant extremes that exhibit distinct solubilization and fermentation properties. To further explore this observation, we assessed select plant features that have been determined to influence recalcitrance to account for their involvement in the observed trends.

The relationships between structure and S/G ratio varied with the feature tested. For example, cellulose crystallinity and cellulose molecular weight appeared to share little, if any, relationship with S/G ratio. Biomass accessibility and lignin molecular weight appeared to be linked to S/G ratio. AJ21 was associated with higher molecular weight lignin and more accessible biomass as well as higher solubilization and fermentation than A377. The results of this study are the first known investigation into differences in the CBP of low and high S/G ratio. This research is also the first known quantitative investigation into the relative MW differences between low and high S/G ratio biomass. Prior to this study, low S/G ratio lignin was believed to have higher molecular weights than high S/G ratio biomass[5].

Table 1. Lignin S/G ratio and its Associated Solubilization and Fermentation Properties

Sample	Lignin S/G Ratio	Solubilization (%)	Ethanol Yield (mg/g glucan)
A377	1.2	12	7
AJ21	2.1	20	22

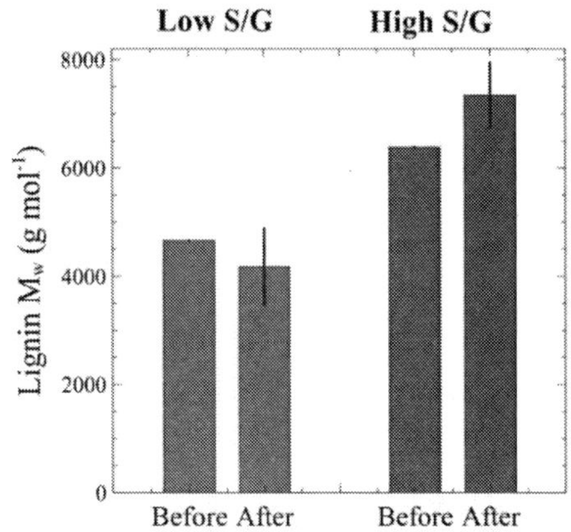

Figure 1. Lignin molecular weights before and after consolidated bioprocessing in low and high S/G ratio lignin

Conclusion

Lignin S/G ratio has been confirmed as an influential characteristic of enzymatic hydrolysis. This study has revealed that this trait extends to consolidated bioprocessing when *C. thermocellum* is employed. The lower molecular weight and the higher accessibility of high S/G ratio lignin may partially contribute to the performance characteristics observed.

Acknowledgement. We graciously acknowledge the US DOE for their support in this research through the BioEnergy Science Center. Furthermore, we would like to thank the Renewable Bioproducts Institute at Georgia Institute of Technology for their financial support (graduate assistantship) for H. Akinosho.

References

(1) (a) Shao, X.; Jin, M.; Guseva, A.; Liu, C.; Balan, V.; Hogsett, D.; Dale, B. E.; Lynd, L., Conversion for Avicel and AFEX pretreated corn stover by Clostridium thermocellum and simultaneous saccharification and fermentation: insights into microbial conversion of pretreated cellulosic biomass. *Bioresour Technol* **2011,** *102*; (b) Bernardez, T. D.; Lyford, K. A.; Lynd, L. R., Kinetics of the extracellular cellulases of Clostridium thermocellum acting on pretreated mixed hardwood and Avicel. *Applied Microbiology and Biotechnology 41* (5), 620-625.

(2) Davison, B. H.; Drescher, S. R.; Tuskan, G. A.; Davis, M. F.; Nghiem, N. P., Variation of S/G ratio and lignin content in a Populus family influences the release of xylose by dilute acid hydrolysis. *Appl Biochem Biotechnol* **2006,** *130.*

(3) Dumitrache, A.; Akinosho, H.; Rodriguez, M.; Meng, X.; Yoo, C. G.; Natzke, J.; Engle, N. L.; Sykes, R. W.; Tschaplinski, T. J.; Muchero, W.; Ragauskas, A. J.; Davison, B. H.; Brown, S. D., Consolidated bioprocessing of Populus using Clostridium (Ruminiclostridium) thermocellum: a case study on the impact of lignin composition and structure. *Biotechnology for Biofuels* **2016,** *9* (1), 1-14.
(4) Chandra, R.; Ewanick, S.; Hsieh, C.; Saddler, J. N., The characterization of pretreated lignocellulosic substrates prior to enzymatic hydrolysis, part 1: a modified Simons' staining technique. *Biotechnol Prog* **2008,** *24*.
(5) Studer, M. H.; DeMartini, J. D.; Davis, M. F.; Sykes, R. W.; Davison, B.; Keller, M.; Tuskan, G. A.; Wyman, C. E., Lignin content in natural Populus variants affects sugar release. *Proc Natl Acad Sci USA* **2011,** *108*.

ETHANOL PRODUCTION FROM CANE BAGASSE USING DIFFERENT PROCESS CONFIGURATIONS

Douglas H. Fockink[1], *Luiz P. Ramos*[1,2*]

[1] CEPESQ, Department of Chemistry, UFPR – Brazil
[2] INCT in Energy & Environment, UFPR – Curitiba, Brazil
* luiz.ramos@ufpr.br

Introduction

Steam explosion is one of the most widely used method for biomass pretreatment because it is tunable to different types of biomass, it produces substrates with low hemicellulose content at high glucan yields, it can be operated in high substrate total solids (TS) and it produces lignified substrates that are susceptible to hydrolysis and fermentation(1). Steam explosion can be carried out without (auto-hydrolysis) or with the addition of an acid catalyst. In auto-hydrolysis, biomass solubilization is mostly due to the acetic acid that is produced by hemicellulose deacetylation but these reactions are considerably accelerated when the biomass is pre-impregnated with an exogenous acid catalyst such as dilute H_2SO_4 or H_3PO_4. To facilitate the comparative analysis of such pretreatment configurations, Chum et al.(2) developed the combined severity factor (CSF) in which the influence of temperature, time and pH are used to describe response variables such as pretreatment yields and cellulose accessibility. Also, successful enzymatic hydrolysis at high TS is one of major barriers existing nowadays for the full commercialization of cellulosic ethanol, with Cellic CTec3 being one of the most advanced cellulase preparations developed for this purpose. In this work, the properties of steam-exploded substrates that were derived from cane bagasse at equivalent CSF were compared, as well as their susceptibility to enzymatic hydrolysis at high TS and the easiness with which cellulosic ethanol could be produced by both separate hydrolysis and fermentation (SHF) and pre-hydrolysis for 12 h followed by simultaneous saccharification and fermentation (pSSF).

Experimental

Sugarane bagasse (SCB) was obtained from the Cane Technology Center (CTC, Piracicaba, SP, Brazil). Cellic CTec3 was donated by Novozymes Latin America (Araucária, PR, Brazil). The industrial strain of *Saccharomyces cerevisiae* (PE-2) was obtained from LNS Latin America (Bento Gonçalves, RS, Brazil).
Steam explosion was carried out using different pretreatment configurations at equivalent CSF of around 0.77 (Table 1). Steam-exploded bagasse was named SEB while sulfuric acid- and phosphoric acid-catalyzed steam-exploded bagasse were named SEB/SA and SEB/PA, respectively. Water washing was applied to remove both hydrolysis and fermentation inhibitors from steam-exploded substrates. The composition of untreated and treated materials was determined according to the NREL methods. The carbohydrates were analyzed by high performance liquid chromatography (HPLC) after dilute sulfuric acid hydrolysis.

Enzymatic hydrolysis of steam-treated materials was carried out in shake flasks using 15 wt% TS and 13.2 FPU of Cellic CTec3 g^{-1} TS. Reactions conditions were carried out for 72 h at 50 °C and 150 rpm in acetate buffer 50 mol L^{-1} pH 5.2. Hydrolysis experiments were also carried out using a 3.6-L Labfors 5HT Bioethanol bioreactor (Infors, Switzerland). Apparent viscosities were obtained in flow experiments before and during enzymatic hydrolysis using a controlled-stress rheometer (TA instruments, New Castle, DE, USA) with vane-in-cup geometries. Ethanol was produced by SHF and pSSF using 1 g L^{-1} of the *S. cerevisiae* strain PE-2 at 37 °C and 150 rpm. Quantitative analysis was always carried out by HPLC.

Results and Discussion

The different pretreatment conditions used in this study presented FSC values ranging from 0.74 to 0.79. The total mass recovery (water soluble plus water insoluble fractions) were the highest for SEB/SA and SEB/PA, which corresponded to 85.2% and 83.4%, respectively. Steam explosion of bagasse (autohydrolysis) resulted in total mass recoveries of 76.6% for SEB-1, 76.3% for SEB-2 and 79.7% for SEB-3, probably due to the use of higher temperatures and reaction times (Table 1). Glucan recovery after pretreatment was above 90% in all steam-exploded materials. Acetyl groups and anidroarabinose were below the detection limit in the compositional analysis of all pretreated materials. It is important to observe that, by applying pretreatment conditions with equivalent FSC values, substrates with similar glucan and xylan contents were produced (Table 1).

Table 1. Experimental Conditions for Pretreatment and Chemical Composition of Sugarcane Bagasse.

Sample	Conditions			Glucan	Xylan[2]
	T (°C)	t (min)	H^+ (mg g^{-1})[1]	content	content
SCB	-	-	-	34.6	16.8
SEB-1	195	15	-	50.8	2.9
SEB-2	205	7.5	-	49.7	2.9
SEB-3	200	10.5	-	51.5	3.6
SEB/SA	195	7.5	2.5	49.4	3.3
SEB/PA	195	7.5	9.5	49.7	2.3

[1]Concentration given in mg of acid per gram of dry substrate; [2]Summation of anhydroxylose, anhydroarabinose and acetyl group contents.

Enzymatic hydrolysis of steam-exploded substrates was carried out with Cellic CTec3. All materials had a similar hydrolysis profile, indicating that equivalent FSC produced materials with the same enzyme accessibility. Nearly 75 g L^{-1} glucose was produced in 72 h (around 90% of glucan conversion) at relatively high TS and low enzyme loadings. Also, 50% glucan conversion was already achieved in 12 h (or ~40 g L^{-1} glucose), which is good for pSSF experiments. Hydrolysis in the 3.6-L Labfors 5HT bioreactor with proper impellers under fed-batch feeding produced even higher glucose yields.
The apparent viscosity (at 10 s^{-1}) at the reaction beginning was lower for SEB and SEB/PA (around 21 Pa s) while SEB/SA showed values considerably higher at 30.5 Pa s. Cellic CTec3 led to a subtle decrease in observed apparent viscosity, which was 70% lower in average after only 3 h of hydrolysis. All hydrolysates were easily fermented by SHF, giving ethanol yields of 44% and productivities of 0.25 g L^{-1} h^{-1} by considering 72 h of hydrolysis plus 24 h of fermentation. In pSSF using pre-hydrolysis for 12 h, the resulting ethanol yields were around 39% for ethanol productivities of 0.62 g L^{-1} h^{-1} (12 h of pre-hydrolysis plus 36 h of SSF).

Conclusions

The substrates produced at equivalent CSF presented high glucan yields, equivalent susceptibilities to enzymatic hydrolysis and good fermentation efficiencies in both SHF and pSSF trials. Cellic CTec3 reduced the substrate apparent viscosity very quickly, reaching 1 Pa s in 24 h for all steam-exploded materials. pSSF was the best ethanol productivity because nearly the same ethanol concentrations were achieved in shorter reaction times.

Acknowledgements. The authors are grateful to the financial help of CNPq and to CAPES for providing a D.Sc. scholarship to D.H.F.

References

(1) Ramos, L. P. *Quim. Nova*, **2003**, 26, 863-871.
(2) Chum, H. L.; Johnson, D. K.; Black, S. K.; Overend, R. P. *Appl Biochem. Biotechnol.* **1990**, 24, 1-14.

THE LONG WAY TO INDUSTRIAL ENZYMES IN COMMERCIAL BIOMASS CONVERSION INTO CELLULOSIC ETHANOL (10 PT)

Armindo Ribeiro Gaspar[*,1)], *Viviane Serpa Müller*[2)], *Eduardo Borges da Silva*[2)], *Prashant Iyer*[1)], *Jim Liu*[1)]

[1)]Novozymes North America Inc., 77 Perrys Chapel Church Road, Franklinton NC 27525 United States; [2)]Novozymes Latin America Ltda., Rua Professor Francisco Ribeiro 683, 83707-660 Araucária - PR Brazil
*agsp@novozymes.com

Biomass conversion into ethanol is a commercial reality today. The bio-route process to transform biomass into ethanol includes de-assembling of polysaccharides using enzymatic hydrolysis to produce sugars, which are fermented into ethanol. The enzymes development has taken a long way to the whole commercialization and has been crucial for the success of the new Biomass Conversion industry today. This Biomass bio-conversion is just the beginning of the Sugars as the new petroleum era, contributing with specifics and alternatives to advance even further into the sustainable bio-economy.

"FISSION-FUSION" SYNTHESIS OF NANOALLOYS FROM INDIVIDUAL BULK METALS ASSISTED BY MOLTEN LITHIUM

Heather M. Barkholtz[a, b], *James R. Gallagher*[b], *Tao Li*[c, *], *Yuzi Liu*[d], *Randall E. Winans*[c], *Jeffrey T. Miller*[b], *Di-Jia Liu*[b], *Tao Xu*[a, *]

[a] Department of Chemistry and Biochemistry, Northern Illinois University, DeKalb Illinois 60115, USA
[b] Chemical Science and Engineering Division, Argonne National Laboratory, Argonne, Illinois 60439, USA
[c] Advanced Photon Source, Argonne National Laboratory, Argonne, Illinois 60439, USA
[d] Center for Nanoscale Materials, Argonne National Laboratory, Argonne, Illinois 60439, USA
Overwrite Corresponding taoli@aps.anl.gov;txu@niu.edu

Introduction

We develop an alternative strategy for the synthesis of ligand-free metallic nanoalloys that can be used for the key catalytic reactions in biofuel, fuel cell, battery and other energy-related applications. The synthesis of nanoalloys is directly from monometallic bulk metals by dissolving them in molten lithium, as shown in **Figure 1**.[1]

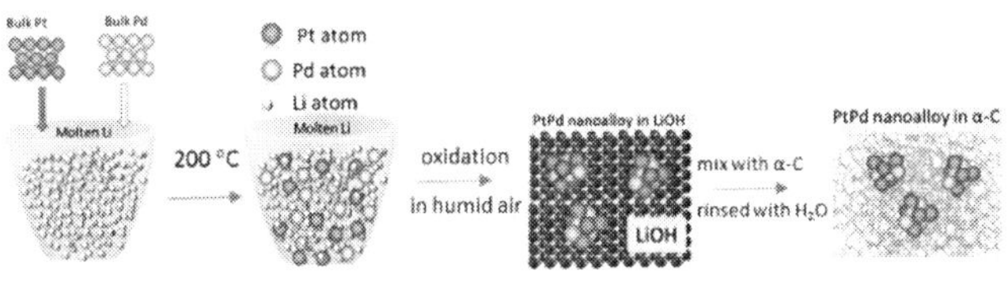

Figure 1. Schematic diagram of the fission-fusion synthesis of PtPd nanoalloys in molten lithium and their transfer from LiOH to a non-water-soluble support, e.g. amorphous carbon (α-C).

When two pure bulk metals such as Pd and Pt are placed in molten lithium (~200 °C), they undergo a dissolution process. The molten lithium ruptures the metal-metal bonds of the bulk metals, resulting in the existence of individual Pd and Pt atoms, confirmed from EXAFS data. These atoms further undergo an alloying process, to aggregate into nanoalloys upon converting metallic lithium to LiOH in humid air. Furthermore, the metal nanoparticles in LiOH powder can be mixed with any water-insoluble support. The LiOH in the mixture can be selectively leached off with copious amounts of water under ambient condition, allowing the formed nanoalloys on the support. This lithium assisted "dissolution-alloying" method bypasses many complications intrinsic to conventional ion reduction-based nanoalloys synthesis including the necessity of ligated metal ions, the use of proper reducing agents and dispersing surfactants, and the presence of segregated phases due to different reduction potentials of the constituent metal ions. In addition, this method is cost-effective since the leached lithium is recyclable in the format of $LiOH/Li_2O$, which is exactly the raw material for lithium production from electrolysis of LiCl.

Experimental

Sample Preparation. Synthesis of PdPt nanoalloys was carried out in an Ar filled glove box (oxygen level < 0.3 ppm) similarly to previously report liquid lithium assisted synthetic methods. [2] Using a nickel crucible, 0.29 mol of lithium was heated to 200 °C, to which 0.45 mmol of palladium foil and 0.45 mmol of platinum foil or 0.45 mmol of zinc metal were simultaneously added. A tip ultrasonic homogenizer was used to ensure a homogenous dispersion of metal particles while the lithium melt was maintained at 200 °C for four hours. Formation of a solid solution was achieved by rapidly pouring the molt onto a clean 316 stainless steel plate to quench the melt and avoid segregation of metal components. The Li-Pd-Pt melt (0.15 mol% PdPt) solid solutions were removed from the glove box and, in the case of unsupported nanoalloys, immediately placed in double distilled water. The black precipitates were collected and washed three times with double distilled water and dried in a desiccator overnight. In the case of amorphous carbon supported nanoalloys, once Li-Pd-Pt solid solutions were removed from the glove box, they were cut into small pieces and slowly converted from Li to LiOH using humidified air at 100 °C. The ensuing PdPt-NP/LiOH materials were combined with the desired amount of carbon black (XC-72R) and mix with an agate mortar and pestle until homogenous. The LiOH can be leached out with copious amounts of double distilled water leaving the nanoalloys imbedded in the amorphous carbon support.

Characterization. TEM samples were prepared by dispersing 4 mg of as-prepared PdPt/LiOH in a 1 to 1 (by volume) water and ethanol solution which was sonicated for one hour. The carbon support film backed copper TEM grid was dipped in the solution and dried on filter paper overnight in air. Powder X-ray diffraction (PXRD) measurements were carried out on a Rigaku Miniflex diffractometer with Cu Kα radiation (λ = 1.5406 Å). Samples were prepared by placing them on a silicon zero diffraction plate with amorphous carbon grease. Small angle x-ray scattering (SAXS) experiments were performed at the APS 12ID-B station. PdPt-NP/C catalysts were loaded into 1.5 mm diameter quartz capillary tubes for the SAXS measurements. X-ray absorption measurements (XAS) on PdPt-NP and PdPt-NP/C were acquired at the 10-BM beamline of the Materials Research Collaborative Access Team (MRCAT) at the Advanced Photon Source, Argonne National Laboratory.

Results and Discussion.

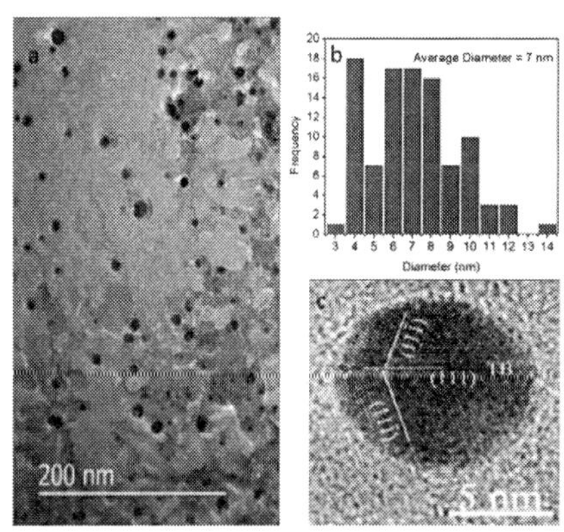

Figure 2. (a) TEM images (bar = 200 nm) of PdPt-NP on an amorphous carbon supporting film. (b) particle size histogram (total 100 particles). Averaged particle diameter is 7 nm. (c) High-resolution TEM of an individual PdPt nanoalloy.

TEM images of PdPt nanoalloys are shown in Figure 2a. The particle size distribution shown in Figure 2b indicates an average particle diameter of 7 nm. The high resolution TEM micrograph shown in Figure 2c indicates the nanoparticle is well crystalized but with (111) twin boundary. The PdPt particle sizes were further evaluated using SAXS measurements on PdPt-NP/C. Unlike TEM, which only provides particle sizes of a small amount of samples locally, SAXS provides an average particle size on a large scale of the sample. The average particle size extracted from the SAXS measurements for the PdPt nanoparticles is 5 nm.

Figure 3 is the PXRD analysis of PdPt nanoparticles compared to bulk Pd and Pt. As can be seen, the XRD peaks of Pd and Pt are located very close to those of PdPt nanoparticles due to the very similar nature of Pd and Pt's crystal structure. Upon closer inspection, one can see that the (111) peak position of Pt is 40.7°, Pd is 40.5°, and the PdPt alloy peak falls in between the positions Pd and Pt at 40.6°, indicating alloy formation. The atomic radius of Pd is 137 pm and Pt is 139 pm, both forming face-centered cubic crystalline structured giving rise to very similar PXRD peaks. Pd and Pt also form substitutional alloys possessing the same FCC structure across the full compositional range; therefore there is potentially significant overlap between the peaks of the pure metals and the binary alloy. As such, it is difficult to conclusively determine whether a phase-pure alloy has been formed or whether there is a mixture of metallic phases. Furthermore, the peaks of PdPt nanoparticles are noticeably broadened due to the nano-scale diameter of prepared particles.

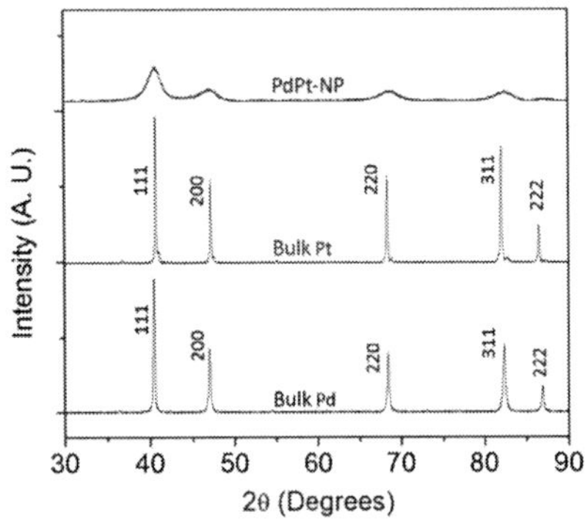

Figure 3. Powder XRD patterns of PdPt-NP. Pt and Pd bulk metals were filed from their corresponding bulk foils. Peak broadening of PdPt-NP is due to their nano-scale size.

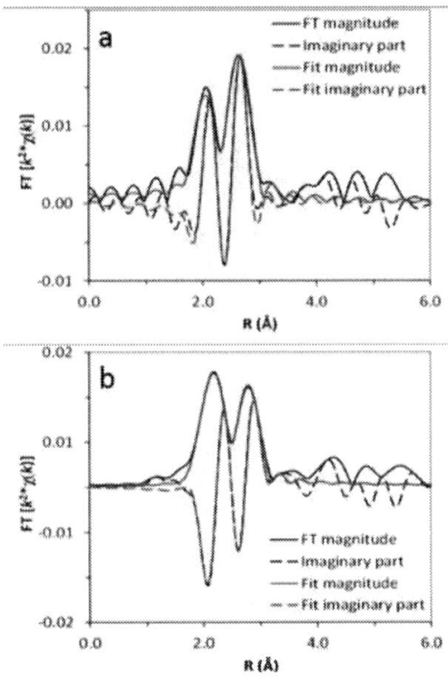

Figure 4. (a) Fitting results of the Pd K-edge R-space EXAFS spectrum of PdPt-NP/C. Fitting range 2.9 Å^{-1} < k < 11.0^{-1} Å, 1.7 Å < R < 3.0 Å. (b) Fitting results of Pt L_3-edge R-space EXAFS spectrum of PdPt-NP/C. Fitting range 2.9 Å^{-1} < k < 11.1 Å^{-1}, 1.8 Å < R < 3.2 Å.

To verify whether particles were truly alloyed, extended X-ray absorption fine structures (EXAFS) studies were performed on both unsupported PdPt nanoalloys and PdPt nanoalloys supported on amorphous carbon, denoted as PdPt-NP and PdPt-NP/C, respectively. Figure 4a exhibits the Pd K-edge R-space EXAFS spectrum of PdPt-NP/C, which was reduced prior to measurement in 3.5% H_2/He at 250 °C. Figure 4b shows the Pt L_3-edge R-space EXAFS spectrum for the same sample. The shape of the EXAFS at both edges is indicative of alloy formation and this is further confirmed by the fitted parameters. A poor fit was obtained when only homonuclear scattering was used (i.e. assuming only Pt-Pt scattering or Pd-Pd scattering), while a good fit to measured data was obtained when both homonuclear and heternuclear scattering were used (i.e. assuming Pt and Pd are atomically intermixed giving Pd-Pd, Pt-Pt, and Pd-Pt scattering). The Pd-Pt and Pt-Pd coordination numbers are within error (5.2 and 4.4, respectively) as are the bond distances, demonstrating that the fitting at both edges are consistent. The heteronuclear coordination numbers (i.e. Pd-Pt = 5.2) are close to the homonuclear coordination numbers (i.e. Pd-Pd = 6.6) and therefore there is little segregation.

Conclusions

In summary, we have demonstrated an unprecedented "fission-fusion" synthesis of nanoalloys directly from pure individual bulk metals by dispersing these pure metals molten lithium and removing the lithium content with water at room temperature. The PdPt nanoalloys have been exemplified using this method. TEM studies demonstrate that the average particle diameter is 5~7 nm. X-ray diffraction and X-ray absorption studies imply that the resulting Pd-Pt nanoalloys are predominantly metallic and of an alloy structure. Due to the expansion of this method to include alloy systems of dissimilar bulk individual metal crystalline structure and large difference in reduction potentials, this method offers a simple and robust alternative to ionic reduction based methods for synthesizing metallic nanoalloys and mitigates the metal segregation within the nanoalloys. This method could potentially reduce the overall cost and environmental impact of preparing nanoalloys by circumventing the need to oxidize bulk metals into metal salts which are then reduced back into metal particles with the use of stabilizing agents or surfactants.

Acknowledgement. TX acknowledges the support from National Science Foundation (CBET-1150617). X-ray absorption spectroscopy studies by JTM and JRG were supported by the Institute for Atom-efficient Chemical Transformations (IACT), an Energy Frontier Research Center funded by the U.S. Department of Energy, Office of Science, Office of Basic Energy Sciences. Use of the Advanced Photon Source (APS) and Center for Nanoscale Materials (CNM) is supported by the U.S. Department of Energy, Office of Science, Office of Basic Energy Sciences, under contract DE-AC02-06CH11357. MRCAT operations are supported by the Department of Energy and the MRCAT member institutions. We also acknowledge the technique support and insightful discussion from Dr. Dale L. Brewe at APS.

References

(1) Barkholtz, H. M.; Gallagher, J. R.; Li, T.; Liu, Y.; Winans, R. E.; Miller, J. T.; Liu, D.-J.; Xu, T. Lithium Assisted "Dissolution - Alloying" Synthesis of Nanoalloys from Individual Bulk Metals. *Chem. Mater.* **2016**.

(2) Xu, T.; Lin, C.; Wang, C.; Brewe, D. L.; Ito, Y.; Lu, J. Synthesis of Supported Platinum Nanoparticles from Li–Pt Solid Solution. *J. Am. Chem. Soc.* **2010**, *132*, 2151-2153.

OPERANDO IR INVESTIGATIONS ON CATALYSTS FOR FUEL SYNTHESIS AND CONVERSION: REACTION MECHANISMS AND DEACTIVATION

P. Bazin[1], O. Marie[1], A.-C. Roger[2], S. Thomas[2] and M. Daturi[1,]*

[1] Normandie Univ, ENSICAEN, UNICAEN, CNRS, Laboratoire Catalyse et Spectrochimie, 14000 Caen, France
[2] Laboratoire des Matériaux, Surfaces et Procédés pour la Catalyse, Groupe Energie et Carburants pour un Environnement Durable, UMR CNRS 7515, Université de Strasbourg, Strasbourg, France
* marco.daturi@ensicaen.fr

Introduction

Increasing concerns about environmental problems and fossil fuels increasing scarcity, together with the industrial development of large countries on Earth urge us to find sustainable solutions for energy transformation. The outbreak of mixed solutions for energy harvesting obliges us to change the energy distribution modes and its use as well. For example, intermittent power production devices, such as wind farms or solar plants imply the development of efficient methods for energy storage, out of the scarcely sustainable batteries. Typically, hydrogen production is a way to store energy into a medium which can be used efficiently in converters such as fuel cells. In the same way, new generation fuels produced by biomass can be mixed to fossil fuels or directly converted to produce hydrogen. The second option allows avoiding the problems related with the combustion of oxygen-rich molecules, badly adapted to the current engines and generating new classes of pollutants. Additionally, the availability of large amounts of hydrogen not coming from fossil sources favors the hydrogenation processes, which can lead to the transformation of different organic molecules to obtain valuable chemical platforms, and to the reuse of CO_2, thus helping to reduce the carbon footprint of numerous processes. The use of CO_2 as a reactant, in fact, could contribute to the decrease of greenhouse gas emissions through chemical recycling. Its use as chemical reactant (in parallel to its sequestration and storage) is limited at the moment to a few processes [1] although several studies on its catalytic conversion into liquid hydrocarbons have been performed [2]. CO_2 methanation remains the most advantageous reaction with respect to thermodynamics when compared to the production of other hydrocarbons or alcohols. The methane production can be inserted into a global scheme where decarbonized electricity, from solar or wind power, would be used in delocalized flexible units to perform water electrolysis to produce hydrogen, which would react with CO_2 capture from emission sources. Hence, the methane produced can be used directly as a fuel, indirectly as an energy source to generate electricity in a power plant or be injected into the existing natural gas network.

Therefore, fuels transformation and hydrogen production are key points in the development of eco-friendly and sustainable solutions for the future energy and chemical sectors.

In the present communication we will address different aspects of the aforementioned problems. In all the cases, hydrogen production and use will go through catalytic steps. The design of the most efficient catalysts and processes requires a deep comprehension of the reaction mechanism, as well as the understanding of deactivation phenomena. *In situ* and *operando* spectroscopies are techniques particularly adapted for this kind of investigations, this is why we have developed devices able to simulate relevant reaction conditions, in which it is possible to monitor heterogeneous catalysts at work. Here we will show the knowledge gained through this approach in three cases: i) mechanistic investigation on ethanol steam reforming for hydrogen production; ii) H_2 on-board production from gasoline for increasing the car engine tolerance to EGR; iii) mechanism of CO_2 methanation under stoichiometric conditions.

Experimental

Samples synthesis. Three catalysts are considered here: $Ce_2Zr_{1.1}Co_{0.9}O_{8-\delta}$ for ethanol steam reforming; $1\%Rh(Co)/Ce_{0.65}Pr_{0.08}Zr_{0.27}/Al_2O_3$ for gasoline reforming; $5\%Ni/Ce_{0.5}Zr_{0.5}O_2$ for CO_2 hydrogenation. They were synthesized using a pseudo sol–gel method, based on the thermal decomposition of propionate precursors. Metal were deposed on the supports by wet impregnation. All the samples were dried, and then submitted to calcinations at 800°C for 10 hours.

Samples characterization. The sample were characterized by the current structural and morphological techniques, prior to submitting them to surface analyses. Samples were pressed into self-supported disks and placed in a quartz cell equipped with KBr windows. A movable quartz sample holder permits one to adjust the pellet in the infrared beam for spectra recording and to displace it into a furnace at the top of the cell for thermal treatment. The cell was connected to a vacuum line for evacuation and for the introduction of gases into the infrared cell. Spectra can be recorded at room temperature or at ~100 K by cooling the sample holder with liquid nitrogen after quenching the sample from the thermal treatment. The addition of well known doses of gases in the cell was possible via a calibrated volume connected to a pressure gauge for the control of the gas pressure. A Nicolet Nexus spectrometer equipped with a Mercury Cadmium Telluride (MCT) cryodetector and an extended KBr beam splitter was used for the acquisition of spectra recorded in the 600–5500 cm^{-1} range.

Operando IR measurements were performed in a DRIFT Spectratech commercial cell or in a homemade transmission IR reactor-cell [3]. The IR *operando* system is composed of four main parts: the infrared spectrometer, the IR reactor-cell, the gas flow set-up and the exhaust gas analyzers [4]. The cell is connected to the *operando* gas-system including mass flow controllers for the introduction of gases and vapors into the lines. The system allows investigating the composition at the outlet of the cell by a Quadrupole Mass Spectrometer (Pfeiffer Omnistar GSD 301). IR spectra (64 scans per spectrum, optical resolution of 4 cm^{-1}) were collected at the wished temperature via a Thermo Scientific Nicolet 6700 spectrometer, equipped with a MCT detector. Spectra were subtracted from the reference sample at the corresponding temperature. Samples were pressed as self supported wafer of around 20 mg and activated *in situ* before introducing the reaction mixture in a stream of Ar as carrier gas. The hourly space velocity could vary between around 120000 and 30000 h^{-1}.

In parallel, experiments in a classical PFR reactors were performed for comparison, using the same gas compositions and contact times.

Results and Discussion

Ethanol steam reforming over a $Ce_2Zr_{1.1}Co_{0.9}O_{8-\delta}$ catalyst. Combining the IR monitoring of the catalyst with that of the gas phase (also analyzed by MS), cyclically substituted by molecules isotopically labeled on the carbon or hydrogen sites, it was possible to determine the reaction pathway. It was demonstrated that only a minor part of hydrogen was coming from ethanol dehydrogenation, whereas the majority was produced by water dissociation on the reduced sample, kept in its reduced state by the hydrocarbon itself. Coherently, a lacunar mechanism for hydrogen production was proposed on this Ce-based catalyst. Different byproducts and intermediates (such acetaldehyde and acetates, for example) were identified as well. In particular, ethylene (generated by ethanol dehydration) was found as responsible for catalyst deactivation upon reaction on the cobalt site, with the subsequent production of a carbon deposit on the mixed oxide surface. The catalyst was found to be reactivated by both reducing and oxidizing treatments, yielding to methane and CO_2 (then carbonates) respectively [5]. The high selectivity provided by the catalyst was also linked to its capability to perform WGS reaction, able to decrease the amount of CO among the products.

Steam reforming of isooctane over a 1%Rh(Co)/$Ce_{0.65}Pr_{0.08}Zr_{0.27}/Al_2O_3$ catalyst. Different samples based on cerium-zirconium-praseodymium oxide, supporting or not Rh and Co, were tested for isooctane (to emulate gasoline) steam reforming. These materials were synthesized on the basis of a long experience on reforming reactions, looking for the most active metal supported on an oxide able to provide the right basicity to promote the hydrocarbon conversion. On another hand, basicity has the tendency to fix CO_2 as carbonates, which may partially inhibit the reaction. Therefore, enhanced oxygen mobility was searched as well, in order to destabilize the carbonates. The best compromise was found in the Ce- Zr-Pr solid solution: a basic oxide with a great oxygen mobility. Cobalt was then added to increase further the reducibility of the catalyst, as well as for its qualities in reforming.

Little activity was observed over the bare support, showing the primary role of rhodium in this reaction. The most active sample tested was 1%Rh/$Ce_{0.65}Pr_{0.08}Zr_{0.27}/Al_2O_3$, producing slightly more hydrogen than the cobalt containing catalyst. The differences between the formulations were discussed also in terms of side reactions, such as methanation and Water Gas Shift. Finally it was evidenced that, if the presence of cobalt effectively enhances the redox properties of the material, it also changes the chemical-physics properties of the surface. In fact, during the reaction, cobalt aluminate is formed, as shown by X-ray diffraction. This new compound is responsible for the partial coverage of both acid and basic sites at the material surface and for a decrease in the strength of basic sites, as observed from the *in situ* IR characterizations. The cobalt-containing sample also showed a lack of acid-base pair sites involving the electronegative cations of the support, so changing strikingly the chemistry of the surface versus the coordination of the reactive molecules.

CO_2 methanation under stoichiometric conditions. A mechanistic investigation on the methanation reaction from CO_2/H_2 under stoichiometric conditions was performed over a Ni-ceria-zirconia catalyst (in comparison with conventional Ni/silica systems), discussing in particular the role of CO in the global reaction scheme. Despite its much lower specific surface, the 5%Ni/$Ce_{0.5}Zr_{0.5}O_2$ sample was found much active than the 5%Ni/SiO_2 catalyst. To understand this behavior, *operando* IR studies were performed in a transmission cell under the conditions of CO_2 methanation. As carbon monoxide is the main by-product and is discussed in literature as a reaction intermediate in the CO_2 methanation, the reaction of CO methanation was also studied. Combining observations belonging to steady state and transient conditions of reaction, a mechanism for CO_2 methanation was proposed. The main reaction pathway does not involve CO as an intermediate. Carbon dioxide adsorbs on sites of mild basicity to form covalent carbonates, hydrogen carbonates and then bidentate carbonates. These species are further reduced and hydrogenated by H atoms formed by dissociation of hydrogen at the surface of Ni^0 particles to form formates and then probably formaldehyde-type and methoxy species (not observed) and to finally release methane. Conversely, CO is not an intermediate: a parallel pathway exists where CO is formed by a redox cycle on reduced ceria, oxidized by CO_2. This pathway may decrease the catalyst activity by CO-H_2 competitive adsorption on metal particles and enhance the deactivation by Ni sintering through the formation of di- or tri-carbonyls and by the formation of stable polycarbonates, which block the CO_2 adsorption sites.

The 5%Ni/$Ce_{0.5}Zr_{0.5}O_2$ catalyst behaves differently from 5%Ni/SiO_2 catalyst because of the particular interaction between Ni^{2+} cations and the ceria-zirconia support, which prevents the Ni^0 particles from sintering. In fact, if nickel is introduced on the sample by impregnation (instead of by direct sol-gel synthesis), the catalyst deactivates quickly. This strong deactivation can be explained by the secondary pathway of CO_2 reaction, which releases CO and become predominant when the metal-support interface is decreased.

Thanks to the evidences provided by IR spectroscopy it appeared that on the Ni-ceria-zirconia catalysts, the main mechanism for CO_2 methanation does not require CO as reaction intermediate (contrarily to what proposed in the literature). Hence, H_2 would be dissociated on Ni^0 sites, while CO_2 would be activated on the ceria-zirconia support to form carbonates, which would be hydrogenated into formates and further into methoxy species. This mechanism involves weak basic sites of the support for the adsorption of CO_2 and implies a stable metal-support interface [6]. It explains the much better activity of these catalysts compared to Ni/silica, on which both CO_2 and H_2 are activated on Ni^0 particles.

Conclusions

The use of *in situ* and *operando* IR spectroscopy has allowed unraveling several reaction mechanisms. Active sites and intermediates, as well as spectator species have been identified and quantified, leading to the description of a global reaction pathway and, in some cases, even to a kinetic model. Spectroscopic evidences also clarified catalyst deactivation phenomena, thus providing elements for the formulation of alternative materials for a more efficient heterogeneous catalysis.

References

(1) Centi, G.; Perathoner, S. *Catal. Today* **2009**, *148*, 191
(2) Benson, E.E.; Kubiak, C.P.; Sathrum, A.J.; Smieja, J.M. *Chem. Soc. Rev.* **2009**, *38*, 89
(3) Wuttke, S.; Bazin, P.; Vimont, A.; Serre, C.; Seo, Y.K.; Hwang, Y.K.; Chang, J.S.; Férey, G.; Daturi, M. *Chem. Eur. J.* **2012**, *18*, 11959.
(4) Lesage, T.; Verrier C., Bazin, P.; Saussey, J.; Daturi, M. *Physical Chemistry Chemical Physics* **2003**, 5, 4435.
(5) Verrier, C. ; Bazin, P. ; Marie, O. ; Daturi, M. In *Science and Technology in Catalysis 2006*; Eguchi, K.; Machida, M.; Yamanaka, I., Eds., Studies in Surface Science and Catalysis 172, Kodansha, Tokyo, 2007; pp. 297-300.
(6) Ussa Aldana, P.A.; Ocampo, F.; Kobl, K.; Louis, B.; Thibault-Starzyk, F.; Daturi, M.; Bazin, P.; Thomas, S.; Roger, A.C. *Catal. Today* **2013**, *215*, 201

FORMATION OF PLATINUM OXIDE FILMS ON THE SURFACE OF PLATINUM NANOPARTICLES DURING PROPANE OXIDATION

Casey P. O'Brien[a], *Glen R. Jenness*[b], *Dionisios G. Vlachos*[b], *Ivan C. Lee*[a*]

[a]U.S. Army Research Laboratory, Sensors and Electron Devices Directorate, 2800 Powder Mill Road, Adelphi, MD 20783, USA
[b]Department of Chemical and Biomolecular Engineering, Catalysis Center for Energy Innovation and Center for Catalytic Science and Technology, University of Delaware, DE 19716, USA
*ivan.c.lee2.civ@mail.mil

Introduction

It has been known for decades that platinum oxide formation has a significant impact on the catalytic activity of platinum for many different reactions, including oxidation of CO,[1] NO,[2] and propane.[3] However, there are still many aspects of platinum oxide formation—including their structure, the kinetics and mechanisms of their formation, and their catalytic activity—that are not well understood. In this work, the formation of platinum oxides on the surface of platinum nanoparticles during propane oxidation is investigated by diffuse reflectance infrared Fourier-transform spectroscopy (DRIFTS) of CO adsorbed on platinum nanoparticles. We show that metallic (111) facets on the surface of platinum nanoparticles are slowly transformed into a specific platinum oxide phase (PtO) during propane oxidation, which causes a slow deactivation of the catalyst over hundreds of minutes. The extent of PtO formation decreases with increasing temperature and, as a result, catalyst deactivation is not observed at higher temperatures. We also believe that different oxide phases are formed at different O_2/C_3H_8 ratios, which interact differently with CO.

Experimental

The experimental procedure has been described in detail previously.[4] The Pt/Al_2O_3 catalyst was prepared by incipient wetness impregnation of γ-Al_2O_3 (Davicat AL 2100, Davison Catalysts) with Pt(IV) nitrate solution (Pt 15% w/w, Alfa Aesar) followed by calcination in air at 700 °C for 16 h. All experiments were performed in a Harrick Praying Mantis™ High Temperature Reaction Chamber. The reaction chamber was placed inside of a Praying Mantis™ Diffuse Reflectance Accessory (Harrick Scientific Products, Inc.) in the sample compartment of a Bruker Vertex 70 FTIR spectrometer. Propane oxidation experiments were performed by flowing Ar (62.4 mL/min, Airgas, 99.999%), O_2 (Airgas, 99.999%) and propane to the reaction chamber held at constant temperature at atmospheric pressure. Propane was fed to the reactor by flowing 15.0 mL/min of a (4.0±0.1)% propane in Ar balance gas mixture (Airgas), giving a propane flow rate of 0.6 mL/min. Two different O_2/C_3H_8 ratios were used; the O_2 flow rate was 3.0 mL/min with O_2/C_3H_8 equal to 5 and 6.0 mL/min with O_2/C_3H_8 equal to 20. The product gas exiting the reaction chamber was analyzed by an Agilent 3000A Micro gas chromatograph (GC). DRIFTS of adsorbed CO was performed by saturating the platinum nanoparticles with CO by flowing 50 mL/min of a 10% CO in Ar gas mixture (Airgas) at 25 °C. After 10 minutes of CO exposure, the reaction chamber was purged with 50 mL/min of Ar and DRIFTS spectra were collected until no more gas-phase CO was detected in the spectra.

Results and Discussion

Propane oxidation over Pt/Al_2O_3 was performed over several hours in the 200-300 °C temperature range to demonstrate the influence of reaction time and temperature on the deactivation of the catalyst. Figure 1(a) shows the CO_2 production rate versus time during propane oxidation over Pt/Al_2O_3 at 200 °C. The CO_2 production rate is ~1.0 site^{-1}s^{-1} at the beginning of the reaction and then decays to ~0.5 site^{-1}s^{-1} after ~100 minutes of the reaction.

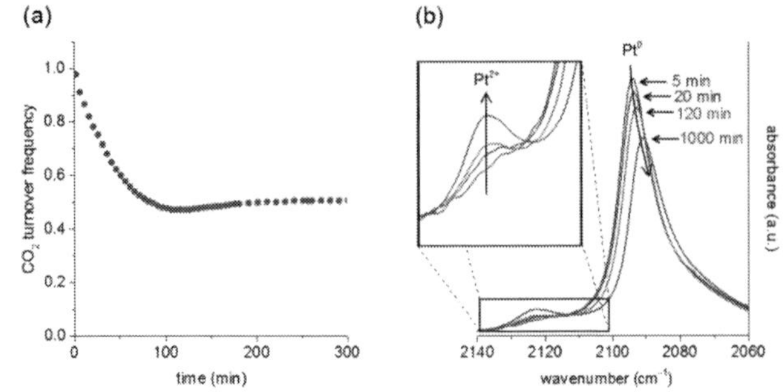

Figure 1. (a) CO_2 production rate during propane oxidation over Pt/Al_2O_3 at 200 °C. (b) DRIFTS of CO adsorbed on Pt/Al_2O_3 after 5, 20, 120, and 1000 minutes of propane oxidation at 200 °C.

To determine whether the deactivation shown in Figure 1(a) results from reconstruction of the platinum nanoparticles, DRIFTS of adsorbed CO was performed to track the evolution of the platinum nanoparticle structure over the course of the reaction. Figure 1(b) shows CO-DRIFTS spectra collected after 5, 20, 120, and 1000 minutes of propane oxidation at 200 °C. The spectrum of CO adsorbed on the clean platinum nanoparticles before the reaction (not shown) exhibits one main band centered at ~2096 cm^{-1}, which is characteristic of CO adsorbed on large nanoparticles that are composed mainly of (111) facets.[5] After 5 minutes of propane oxidation, the main band in the CO-DRIFTS spectrum is shifted to 2094 cm^{-1}. With further increasing propane oxidation time, the main band decreases in intensity and shifts to lower wavenumber. The shift in the CO absorption band to lower wavenumber is not consistent with sintering of platinum nanoparticles, which would result in a shift to higher wavenumbers.[6] Starting from about 20 minutes of propane oxidation, a band at 2123 cm^{-1} appears which increases in intensity with increasing propane oxidation time. This band has been observed in previous studies[7] of CO adsorption on platinum, and has been assigned to CO adsorbed on, or near, oxidized platinum sites. These results show that the platinum nanoparticles are significantly reconstructed during the propane oxidation reaction by the formation of platinum oxides, and the formation of platinum oxides may be responsible for the deactivation displayed in Figure 1(a).

There are many different platinum oxide phases (e.g. PtO, Pt_3O_4, PtO_2) and it is not clear from the literature which of these phases is associated with the CO-DRIFTS band at 2123 cm^{-1}. To assign the band at 2123 cm^{-1} to a specific platinum oxide phase, we carried out density functional theory calculations of the vibrational frequencies and binding energies of CO adsorbed on platinum and the lowest energy surfaces of several platinum oxides (PtO, Pt_3O_4, α-PtO_2, β-PtO_2). The results are summarized in Table 1. Taking the experimentally measured vibrational frequency of CO adsorbed on Pt(111) of 2096 cm^{-1} as our reference, a scaling factor of 1.015 is applied to the calculated vibrational frequencies to match the experimentally measured vibrational frequency. The calculated vibrational frequency of CO adsorbed on PtO(101) and Pt_3O_4(100) at 2122.3 and 2121.1 cm^{-1}, respectively, are both similar to the experimentally measured frequency of 2123 cm^{-1}. However, the binding energy of CO on Pt_3O_4(100) is very low (-0.31 eV) and CO would not be stable on this oxide at room temperature. The binding energy of CO on PtO(101) (-2.18 eV) is much higher than that on Pt_3O_4(100) and even Pt(111) (-1.65 eV) indicating that CO would be stable on PtO(101) at room temperature. Therefore, we can confidently assign the band at 2123 cm^{-1} to CO adsorbed on PtO.

Table 1. Calculated CO vibrational frequencies (ν_{CO}) and binding energies (E_{BE}) on platinum and platinum oxide surfaces.

System	ν_{CO} (cm^{-1})[a]	E_{BE} (eV)
Pt(111)	2065.2 (2096.0)	−1.65
PtO(101)	2091.1 (2122.3)	−2.18
α-PtO_2(0001)	2127.3 (2159.0)	−0.01
β-PtO_2(110)	1847.9 (1875.5)	−1.78
Pt_3O_4(100)	2089.9 (2121.1)	−0.31

[a]Results in parentheses are scaled by a factor of 1.015 in order to reproduce the experimental CO-Pt(111) vibrational frequency.

The reaction temperature has a significant influence on the formation of platinum oxides and the deactivation of the catalyst. Figure 2(a) shows the CO_2 production rates during propane oxidation over Pt/Al_2O_3 at 200 °C, 250 °C, and 300 °C. In contrast to the strong deactivation observed at 200 °C, the CO_2 production rates at 250 °C and 300 °C are nearly constant over the entire 300 minutes of the reaction at ~1.5 $site^{-1}s^{-1}$ and ~1.8 $site^{-1}s^{-1}$, respectively. The CO-DRIFTS spectra collected after 1000 minutes of propane oxidation at 200 °C, 250 °C, and 300 °C are displayed in Figure 2(b). As the reaction temperature increases, the main CO absorption band increases in intensity and shifts to higher wavenumber. The band associated with CO adsorbed on PtO at 2123 cm^{-1} decreases in intensity with increasing temperature. These results indicate that the extent of PtO formation decreases with increasing temperature in the 200-300 °C range, and as a result, catalyst deactivation is less severe at higher temperatures.

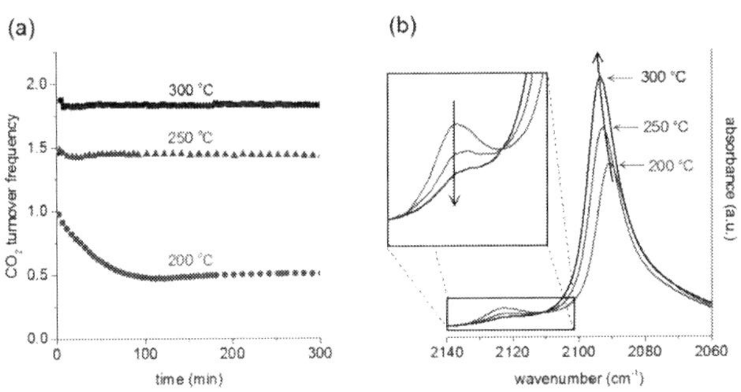

Figure 2. (a) CO_2 production rate during propane oxidation over Pt/Al_2O_3 at 200 °C (red), 250 °C (blue), and 300 °C (black). (b) DRIFTS of CO adsorbed on Pt/Al_2O_3 after 1000 minutes of propane oxidation at 200 °C (red), 250 °C (blue), and 300 °C (black).

In addition to the strong influence of the reaction temperature on the formation of PtO, the O_2/C_3H_8 ratio also influences the oxidation behavior of platinum. Figure 3(a) shows the CO_2 production rate during propane oxidation over Pt/Al_2O_3 at 200 °C with O_2/C_3H_8 equal to 5 and 20. Increasing the O_2/C_3H_8 ratio from 5 to 20 resulted in a more significant deactivation of the catalyst in a shorter time; about 80% of the catalyst activity was lost within the first 30 minutes of the reaction. Figure 3(b) shows the CO-DRIFTS spectra collected after 1000 minutes of propane oxidation over Pt/Al_2O_3 at 200 °C with O_2/C_3H_8 equal to 5 and 20. Unexpectedly, the intensity of the band associated with CO adsorbed on PtO is less intense with O_2/C_3H_8 equal to 20 than that with O_2/C_3H_8 equal to 5. This result suggests that the oxidation behavior of platinum is significantly different for O_2/C_3H_8 equal to 20 and 5. We believe that a different oxide phase is formed with O_2/C_3H_8 equal to 20 (e.g. Pt_3O_4, PtO_2), that is not detectable by CO-DRIFTS either because CO does not stick to these oxides at 25 °C, or these oxides are reduced to metallic platinum during CO exposure.

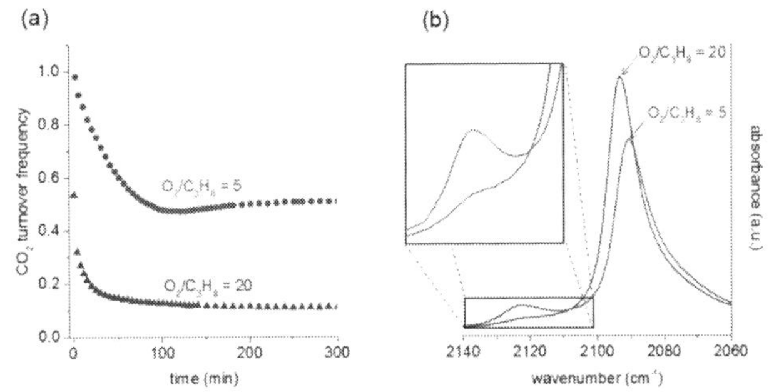

Figure 3. (a) CO_2 production rate during propane oxidation over Pt/Al_2O_3 at 200 °C with O_2/C_3H_8 equal to 5 (red) and 20 (blue). (b) DRIFTS of CO adsorbed on Pt/Al_2O_3 after 1000 minutes of propane oxidation at 200 °C with O_2/C_3H_8 equal to 5 (red) and 20 (blue).

Conclusions

This work shows that platinum oxide films grow on the surface of platinum nanoparticles during propane oxidation over several hours, causing slow deactivation of the catalyst. Using density functional theory calculations, the platinum oxide phase has been positively identified as PtO. The extent of PtO formation increases with decreasing temperature, and as a result, catalyst deactivation is more severe at lower temperatures. At higher O_2/C_3H_8 ratios, PtO gives its place to a different oxide phase that is not detectable by CO-DRIFTS at 25 °C.

Acknowledgement. Research was sponsored by the U.S. Army Research Laboratory and was accomplished under Cooperative Agreement Numbers W911NF-12-2-0019 and W911NF-10-0047. The views and conclusions contained in this document are those of the authors and should not be interpreted as representing the official policies, either expressed or implied, of the Army Research Laboratory or the U.S. Government. The U.S. Government is authorized to reproduce and distribute reprints for Government purposes notwithstanding any copyright notation herein. The DFT calculations were conducted at UD and were supported as part of the Catalysis Center for Energy Innovation, an Energy Frontier Research Center funded by the US Dept. of Energy, Office of Science, Office of Basic Energy Sciences under award number DE-SC0001004.

References

(1) Carlsson, P. A.; Osterlund, L.; Thormahlen, P.; Palmqvist, A.; Fridell, E.; Jansson, J.; Skoglundh, M., *Journal of Catalysis* **2004,** *226* (2), 422-434.
(2) Hauff, K.; Tuttlies, U.; Eigenberger, G.; Nieken, U., *Applied Catalysis B-Environmental* **2012,** *123*, 107-116.
(3) Yoshida, H.; Yazawa, Y.; Hattori, T., *Catalysis Today* **2003,** *87* (1-4), 19-28.
(4) O'Brien , C. P.; Jenness, G.R.; Dong, H.; Vlachos, D.G.; Lee, I.C., *Journal of Catalysis* **2016,** 337, 122-132.
(5) Allian, A. D.; Takanabe, K.; Fujdala, K. L.; Hao, X.; Truex, T. J.; Cai, J.; Buda, C.; Neurock, M.; Iglesia, E., *J. Am. Chem. Soc.* **2011,** *133* (12), 4498-4517.

(6) Rivallan, M.; Seguin, E.; Thomas, S.; Lepage, M.; Takagi, N.; Hirata, H.; Thibault-Starzyk, F., *Angewandte Chemie-International Edition* **2010,** *49* (4), 785-789.
(7) Heyne, H.; Tompkins, F. C., *Transactions of the Faraday Society* **1967,** *63* (533P), 1274-1285.

CASCADE ALDOLIZATION AND SELF-DEOXYGENATION OVER $Zn_xZr_yO_z$ MIXED OXIDES: THE EFFECT OF SURFACE ACIDITY ON THE CATALYST DEACTIVATION

Junming Sun,[1] Rebecca A. Baylon,[1] Donghai Mei,[2] Kevin J. Martin,[3] Padmesh Venkitasubramanian,[3] Yong Wang[1,2]

[1] The Gene and Linda Voiland School of Chemical Engineering and Bioengineering, Washington State University, Pullman, WA 99164, United States
[2] Institute for Integrated Catalysis, Pacific Northwest National Laboratory, Richland, WA 99352, United States
[3] Archer Daniels Midland Company, 1001 N Brush College Road, Decatur, IL 62521 (USA)
Email: Junming.sun@wsu.edu; yong.wang@pnnl.gov

Introduction

The depleting petroleum reserves combined with increased energy demand has led to extensive research interests on the sustainable alternatives. Particularly, biomass conversion has been recognized as the most readily viable technology to produce biofuel concerning our reliance on liquid fuels for transportation.[1] The interest in biofuel has also resulted in a greater abundance and reduced costs of biomass derived small oxygenates such as alcohol, ketone, and carboxylic acid, which is subject to further upgrading to produce valuable fuels or alternative chemicals.[1-3] Current catalysts/strategies developed are still suffering either low activity or stability,[2] it is still highly demanded to develop new catalysts and processes for efficient biomass conversion. We have developed a $Zn_xZr_yO_z$ catalyst with balanced acid-base site pairs, on which small oxygenates such as ethanol, ketone, and carboxylic acid, can be converted into olefins via a cascade aldolization and self-deoxygenation pathway.[4-6] In this presentation, we focus on the effect of the type of acidity on the catalyst performances. Specifically, the $Zn_xZr_yO_z$ catalyst with and without co-existing Brønsted acid sites have been compared in the rate determining acetone-to-isobutene reactions in the ethanol-to-isobutene conversion. The possible roles of the Brønsted acid site and Lewis acid-base pair as well as its effect on the reaction pathway and deactivation of catalyst will be discussed.

Experimental

Two different methods were used to synthesize the $Zn_xZr_yO_z$ catalysts, namely the hard template method[5] ($Zn_xZr_yO_z$-H) and incipient wetness impregnation without using hard template($Zn_xZr_yO_z$-I).[4] Catalysts were characterized using X-ray diffraction, N2-sorption, DRIFTS analysis of adsorbed pyridine, CO_2, D_2O, and acetone. The catalysts were evaluated in a home-built test unit described elsewhere.[4]

Results and Discussion

DRIFTS analysis of adsorbed pyridine (Figure 1) and CO_2 (not shown) that the $Zn_xZr_yO_z$-H catalysts prepared by hard template exhibit co-existing Brønsted acid site, however the $Zn_xZr_yO_z$-I show only Lewis acid-base pairs. It is found that acetone act as the key intermediate in the direct ethanol-to-isobutene conversion, and the reaction is limited by the secondary acetone aldolization and self-deoxygenation reactions.[4] Although Brønsted acid sites on the $Zn_xZr_yO_z$-H catalysts are active for acetone-to-isobutene reaction, it also catalyzes IB isomerization reaction to form 1&2-butenes, as well as other side reactions like polymerization and coke deposition on the $Zn_xZr_yO_z$-H catalyst. This leads to the continuous coking and thus deactivation of active sites for the secondary acetone-to-isobutene reaction (Figure 2, hollow mark). On the other hand, Lewis acid-base pairs on ZrO_2 catalyst can catalyze the acetone aldolization reaction to form mesityl oxide as well as parallel acetone decomposition. The formed mesityl oxide, however, strongly adsorbs and blocks the Lewis acid-base active site, due to its stronger Lewis acidity. The strongly adsorbed mesityl oxide also inhibits the forward reaction toward isobutene, while leading to the dominant acetone decomposition (Scheme 1, red highlighted pathway), as well as other possible side reactions like polymerization. Most importantly, ZnO addition significantly modifies the properties of surface Lewis acid-base pairs. As a result, the cascade acetone aldolization and self-deoxygenation reactions are significantly accelerated (Scheme 1), leading to the highly active and stable $Zn_xZr_yO_z$-I catalyst for both acetone-to-isobutene and ethanol-to-isobutene (Figure 2, solid mark).

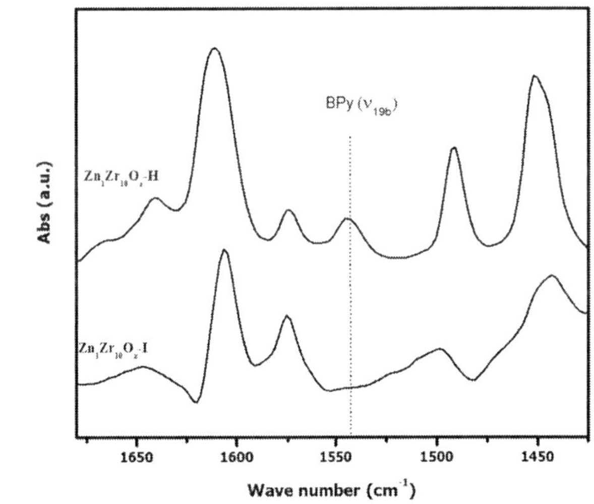

Figure 1. DRIFTS analysis of $Zn_xZr_yO_z$ catalysts

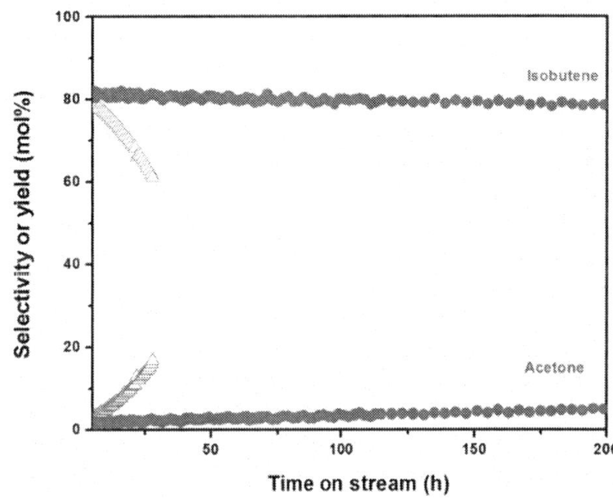

Figure 2. Stability test for ethanol-to-isobutene over $Zn_1Zr_8O_z$-I (Solid mark) and $Zn_1Zr_8O_z$-H (hollow mark). Dark yellow marked isobutene yield and magenta marked acetone selectivity. $P_{ethanol}$ =83 KPa S/C=2.5, WHSV=0.3 $g_{ethanol}.g_{catal}^{-1}.h^{-1}$. T=450 °C.

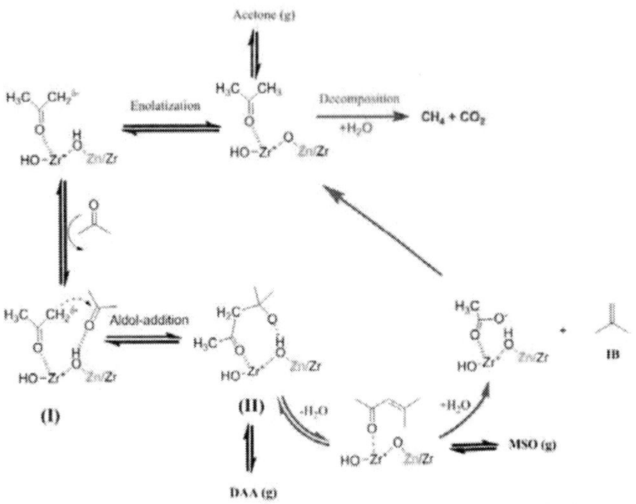

Scheme 1. Proposed reaction pathway for acetone-to-isobutene reaction on $Zn_xZr_yO_z$-I.

Conclusions

A cascade acetone aldolization and self-deoxygenation has been identified over $Zn_xZr_yO_z$-I catalysts with balanced Lewis acid-base pairs, which leads to the high selectivity and durability of the catalyst in the small oxygenate to olefin conversion.

Acknowledgement. We acknowledge the financial support from the US Department of Energy (DOE), Office of Basic Energy Sciences, Division of Chemical Sciences, Geosciences, and Biosciences (DE-AC05-RL01830, FWP-47319) and Archer Daniels Midland Company (WSU002206).

References

(1) Huber, G. W.; Chheda, J. N.; Barrett, C. J.; Dumesic, J. A. *Science* **2005**, *308*, 1446.
(2) Gayubo, A. G.; Aguayo, A. T.; Atutxa, A.; Aguado, R.; Olazar, M.; Bilbao, J. *Ind. Eng. Chem. Res.* **2004**, *43*, 2619.
(3) Sun, J.; Wang, Y. *Acs Catal.* **2014**, *4*, 1078.
(4) Sun, J.; Baylon, R. A.; Liu, C.; Mei, D.; Martin, K. J.; Venkitasubramanian, P.; Wang, Y. *J. Am. Chem. Soc.* **2016**, *138*, 507.
(5) Sun, J. M.; Zhu, K. K.; Gao, F.; Wang, C. M.; Liu, J.; Peden, C. H. F.; Wang, Y. *J. Am. Chem. Soc.* **2011**, *133*, 11096.
(6) Baylon, R. A. L.; Sun, J.; Martin, K. J.; Venkitasubramanian, P.; Wang, Y. *Chem. Commun.* **2016**, DOI:10.1039/C5CC10528E.

CONTROLLED SYNTHESIS OF SINTERING-RESISTANT CATALYSTS THROUGH NANOSTRUCTURED MATERIALS

Sheng Dai

Chemical Sciences Division, Oak Ridge National Laboratory, Oak Ridge, TN 37831-6201
Department of Chemistry, University of Tennessee, Knoxville, TN 37966

Nanostructured materials are critically important to energy storage and production and to meeting the environmental quality mission of promoting the development and utilization of clean, efficient, and reliable energy resources. It is essential to understand the relationships between the atomic and nanoscale structure of metal nanoparticles and catalyst supports and the crucial role these play in promoting or altering chemical and catalytic pathways. The key focus of this talk lies in the controlled synthesis of porous oxides and metallic catalysts with unique metal-support interactions and nanostructured nonmetallic catalysts for energy-related applications. The scientific questions we want to investigate are: a) Where are active sites located on metal cores or oxide interfaces? and b) If located on the oxide interfaces, what are the scaling properties of catalytic activities with the oxide interfacial structures? Critical issues and emerging science and technology in heterogeneous processes related to energy conversion and catalysis will be discussed in context of controlled synthesis.

Acknowledgment. This work is sponsored by U.S. Department of Energy (DOE), Office of Science, Basic Energy Sciences (BES), Chemical Sciences, Geosciences and Biosciences Division.

STRATEGIES FOR PROTECTING THE ACTIVE SITE IN CHEMICAL CONVERSION ON Ni-CeOx CATALYSTS: INSIGHTS FROM IN-SITU STUDIES OVER MODELS AND POWDERS

Sanjaya D. Senanayake,[1] *Zongyuan Liu,*[1] *David C. Grinter,*[1] *Dimitry Vovchok,*[1] *Curtis Guild,*[2] *Steven Suib*[2] *and Jose A. Rodriguez*[1]

[1] Chemistry Department Brookhaven National Laboratory
[2] Dept of Chemistry, University of Connecticut
ssenanay@bnl.gov

Deactivation of Ni-CeOx catalysts have been observed in numerous reactions that we have studied including the Water-gas Shift (WGS), Alcohol Steam Reforming, and very recently in CO_2 Dry Reforming. The ubiquitous bulk properties of Ni for CO coupling and subsequent coking (Figure 1) is a big challenge to overcome and is often a major pre-requisite to achieve good stability and sustained catalytic activity for all aforementioned reactions. Metal-support interactions can be one way to mitigate this problem, by carefully loading Ni as sma ll particles onto a CeOx support and tuning electronic, chemical and support effects that result. We have observed through in situ investigations (AP-XPS, XRD, XANES, TEM/STEM) that decoupling of this metal-support interaction can lead to sintering of Ni particles and often result in the loss of selectivity and cokeing through steps that originate at NixC type of surface species. We propose several strategies to mitigate this problem and offer a way to protect the active site: 1. through preparation of solid solutions of NiCeOx, where Ni is protected inside the CeOx lattice; 2. Addition of interfacial agents that sustain metal-support interactions, such as the addition of W; and finally 3. Robust high surface area mesoporous CeOx supports that can offer sites that bind and disperse Ni very well and protect the catalyst from the corrosive environment of the reaction. Our studies support the phenomenological benefits of sustained metal-support interactions for reaction stability and improved catalytic production of fuels.

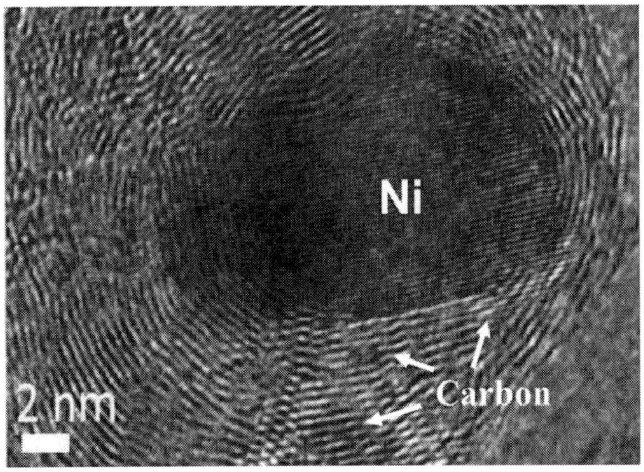

Figure 1. TEM of carbon encapsulated and sintered Ni particle taken post reaction Ni-CeOx catalyst.

EFECT OF SURFACE STRUCTURE ON CO_2 ADSORPTION ON TIO_2 NANOPARTICLES: EXPERIMENTAL AND THEORETICAL INVESTIGATIONS

Uma Tumuluri[a], *Joshua Howe*[b], *Willam Mounfield*[b], *Meijun Li*[c], *Krista S. Walton*[b], *David Sholl*[b], *Sheng Dai*[a c], *and Zili Wu*[a d*]

[a] Chemical Sciences Division, Oak Ridge National Laboratory, Oak Ridge, Tennessee, 37831.
[b] School of Chemical and Biomolecular Engineering, Georgia Institute of Technology, Atlanta, Georgia 30332.
[c] Department of Chemistry, University of Tennessee, Knoxville, TN 37996
[d] Center for Nanophase Material Science, Oak Ridge National Laboratory, Oak Ridge, Tennessee, 37831.
Email: wuz1@ornl.gov

The effect of surface structure of TiO_2 nanoparticles on the structure and strength of adsorbed CO_2 was investigated using in situ IR spectroscopy and mass spectrometric techniques along with first-principle calculations based on density functional theory (DFT). TiO_2 rods (010)+(101)+(001), disks (001) and truncated rhombi (101)+(001)+(010) were used to represent different TiO_2 structures. Carboxylates, and carbonates (bridged, monodentate) are formed on TiO_2 rods and disks, whereas only bidentate and monodentate carbonates are formed on TiO_2 truncated rhombi during CO_2 adsorption. In general, the order of thermal stability of the adsorbed CO_2 species is: carboxylates ≈ monodentate carbonates > bridged carbonates > bidentate carbonates ≈ bicarbonates. TiO_2 rods and disks which have large number of O-vacancies adsorb CO_2 more strongly than TiO_2 truncated rhombi. The effect of SO_2 was also investigated by in situ IR on the nature, amount and strength of CO_2 species on the different TiO_2 nanoshapes.

Acknowledgements: This work is supported by the Center for Understanding and Control of Acid Gas-Induced Evolution of Materials for Energy (UNCAGE-ME), an Energy Frontier Research Center funded by U.S. Department of Energy, Office of Science, Basic Energy Sciences. Part of the work including the IR and Raman was conducted at the Center for Nanophase Materials Sciences, which is a DOE Office of Science User Facility. This research used computational resources of the National Energy Research Scientific Computing Center, which is supported by the U.S. DOE Office of Science under Contract No. DE-AC02-05CH11231.

ASSESSING DEACTIVATION PROCESSES OF SUPPORTED VANADIUM OXIDE CATALYSTS AT A MOLECULAR-LEVEL; AN *OPERANDO* RAMAN METHODOLOGY APPROACH.

M. V. Martínez-Huerta,[1] *M. O. Guerrero-Perez,*[2] *and Miguel A. Bañares,*[1]

[1] Insitute for Catalysis, ICP-CSIC, Madrid, SPAIN
[2] Escuela de Ingenierías, Universidad de Málaga, Málaga, SPAIN

Supported vanadium oxide species molecularly disperse on oxide supports. These, serve as vanadia carriers, facilitating vanadia exposure to reactants. While its interaction with the support modulates their structure and reactivity, solid-state reaction with support may create active phases or transform active species into inert phases. In this work, we summarize the use of *operando* Raman methodology to assess the molecular basis of catalyst deactivation due to solid-state reaction among catalyst components for alkane oxidative dehydrogenation (ODH) and ammoxidation reactions. Cases will be commented for alumina, titania and ceria supports during ethane ODH, and for mixed V-Sb-Nb-O for propane ammoxidation.

COMPUTATIONAL CHARACTERIZATION OF DEFECTS IN ZEOLITIC IMIDAZOLATE FRAMEWORKS

Chenyang Zhang[†], *Chu Han*[‡], *David S. Sholl*[§], *and J. R. Schmidt*[*†]

[†] Theoretical Chemistry Institute and Department of Chemistry, University of Wisconsin-Madison, 1101 University Avenue, Madison, Wisconsin 53706, United States
[‡] School of Chemistry & Biochemistry, Georgia Institute of Technology, Atlanta, Georgia 30332, United States
[§] School of Chemical & Biomolecular Engineering, Georgia Institute of Technology, 311 Ferst Drive NW, Atlanta, Georgia 30332-0100, United States
schmidt@chem.wisc.edu

Overview

Metal-organic frameworks (MOFs) have attracted considerable attention for diverse applications ranging from gas adsorption to catalysis. Nonetheless, the long-term stability of MOF materials has received significantly less scrutiny, particularly under "working" conditions that may involve high temperatures and exposure to water and/or acid gases. I will present computational results focusing on point defect structures (including vacancies, substitutions, and "dangling" linkers) within zeolitic imidazolate frameworks (ZIFs), which may serve as "nucleation" sites for subsequent ZIF degradation. Considering both postsynthetic (gas-phase) and synthetic (solution-phase) conditions, we find that several of the defect structures lie low in energy relative to the defect-free parent crystal, with barriers to defect formation that are large but surmountable under relevant temperatures. I will also discuss results for the interaction of acid gases with these defective ZIFs and the spectroscopic signatures of these interactions.

Acknowledgement. This work was financially supported by the Center for "Understanding and Control of Acid Gas-induced Evolution of Materials for Energy", an Energy Frontier Research Center funded by DOE, Office of Science, BES under Award #DESC0012577. Computational resources were provided in part by National Science Foundation Grant CHE-0840494 and using the compute resources and assistance of the UW-Madison Center for High Throughput Computing (CHTC) in the Department of Computer Sciences. The CHTC is supported by UW-Madison, the Advanced Computing Initiative, the Wisconsin Alumni Research Foundation, the Wisconsin Institutes for Discovery, and the National Science Foundation, and is an active member of the Open Science Grid, which is supported by the National Science Foundation and the U.S. Department of Energy's Office of Science. J.R.S is a Camille Dreyfus Teacher-Scholar.

References

(1) C. Zhang, C. Han, D. S. Sholl and J. R. Schmidt, *J. Phys. Chem. Lett.*, 2016, 7, 459

AB INITIO STUDY OF SMALL MOLECULE INTERACTIONS IN MIXED-METAL METAL-ORGANIC FRAMEWORKS

Joshua D. Howe[1], Cody R. Morelock[1], Yang Jiao[1], Krista S. Walton[1], David S. Sholl[1]*

[1]School of Chemical and Biomolecular Engineering, Georgia Institute of Technology, Atlanta, Georgia, 303332-0100, USA
*david.sholl@chbe.gatech.edu

Introduction

Metal-organic frameworks (MOFs) are a novel class of inorganic-organic hybrid materials comprised of metal cations coordinated by organic linkers that are of interest in energy applications for their tunable chemistries and topologies as well as their high gravimetric surface areas. It has been shown that the existence of undercoordinated "open-metal" sites in these materials can generate strong adsorption sites for energy-relevant molecules. MOF-74 has an especially high volumetric and gravimetric density of open-metal sites exposed within its pores. Mg-MOF-74, in particular, has been shown to have exceptional CO_2 uptake.[1] Like many MOFs with open-metal sites, however, Mg-MOF-74 experiences a sharp reduction in adsorption surface area when exposed to water. In contrast, Ni-MOF-74 shows much lower surface area reduction when exposed to water. We have shown that generating a MOF-74 with a mixture of metals, so-called mixed-metal-, or MM-MOF-74s, can improve stability to water exposure; for example, incorporation of as little as 17 molar percent of Ni in Mg-MOF-74 results in surface area reduction upon exposure to 90% relative humidity that is similar to that of Ni-MOF-74.[2] In this talk, we investigate this phenomenon computationally, employing density functional theory to understand the nature of water adsorption at open-metal sites in Mg-MOF-74, Ni-MOF-74, and MgNi-MOF-74s.

Acknowledgement. This work was supported by the Center for Understanding and Control of Acid Gas-Induced Evolution of Materials for Energy (UNCAGE-ME), an Energy Frontier Research Center funded by the U.S. Department of Energy, Office of Science, Basic Energy Sciences and by the BES Catalysis Center Program.

References

(1) Yazaydın, A. O.; Snurr, R. Q.; Park, T.-H.; Koh, K.; Liu, J.; LeVan, M. D.; Benin, A. I.; Jakubczak, P.; Lanuza, M.; Galloway, D. B.; Low, J. J.; Willis, R. R. *J. Amer. Chem. Soc.*, **2009**, *131*, 18198.
(2) Jiao, Y.; Morelock, C. R.; Burtch, N. C.; Mounfield III, W. P.; Hungerford, J. T.; Walton, K. S. *Ind. Eng. Chem. Res.*, **2015**, *54*, 12408.

TOWARDS THE THERMALLY STABLE SITE-ISOLATED PD CATALYSTS

Mar Piernavieja-Hermida,[1] Zheng Lu,[1] Zili Wu,[2] Yu Lei[1]

[1]Department of Chemical and Materials Engineering, University of Alabama in Huntsville, 301 Sparkman Dr., Huntsville, AL 35899, USA
[2]Center for Nanophase Materials Sciences and Chemical Science Division, Oak Ridge National Laboratory, 1 Bethel Valley Rd., Oak Ridge, TN 37831, USA
yu.lei@uah.edu

Abstract

When the dimensions of supported precious metals approach the nanometer and sub-nanometer scale, unique catalytic properties emerge related to under-coordinated surface atoms, electronic structure, and diffusion barriers of surface species.[1-4] The sub-nanometer and especially single atom catalysts have shown unexpectedly high activity and selectivity to valuable products under realistic catalytic reaction conditions. Despite several successful examples, the thermal stability of precious metal single site catalysts under reaction conditions remains a significant challenge. The synthesis of highly active single-atom catalyst with improved thermal stability attracts both fundamental and industrial interests. Here we demonstrate an approach to improve the thermal stability of uniformly dispersed Pd_1 single atom catalysts using ultrathin metal oxide protective coatings.

The Pd catalysts were prepared by means of atomic layer deposition (ALD). Spherical alumina was used as the substrates without further treatment. The TiO_2 coating used alternating exposures to $TiCl_4$ and deionized water at 150 °C. For the TiO_2 overcoat, The Pd ALD used exposures to Pd (II) hexafluoroacetylacetonate at 100 °C.

DRIFTS measurements on CO adsorption were performed using a Thermo Nicolet Nexus 670 spectrometer, with a MCT/A detector with a spectral resolution of 4 cm^{-1}, in diffuse reflectance mode while the exiting stream was analyzed by QMS. All reported IR spectra are difference spectra referenced to a background spectrum collected after pretreatment but prior to CO adsorption.

Pd K-edge (23.564 keV) XAS was performed at the 10-BM beamline at the Advanced Photon Source (APS) at Argonne National Laboratory. Pd L_3-edge (3.173 keV) XAS was carried out at the 9-BM beamline at the APS. The sample was measured in the fluorescence mode.

By combining Pd and templated-metal oxide ALD, the as-prepared site-isolated Pd catalysts showed improved stability under both oxidation and reduction conditions. Agglomeration was observed when carbon monoxide was introduced to the site-isolated Pd. The site-isolated catalysts were evaluated for methanol decomposition reactions. Single atom Pd_1 showed poor reactivity towards methanol decomposition due to the strong adsorption of CO on palladium causing self-poisoning effect as illustrated by the *in situ* Pd L-edge XAS measurements.

Acknowledgement. We gratefully acknowledge support from National Science Foundation (CBET-1511820). Part of the work including the methanol decomposition reaction and *in situ* DRIFTS study was conducted at the Center for Nanophase Materials Sciences, which is a DOE Office of Science User Facility. Use of the Advanced Photon Source was supported by the U. S. Department of Energy, Office of Science, Office of Basic Energy Sciences, under Contract No. DE-AC02-06CH11357. MRCAT operations are supported by the Department of Energy and the MRCAT member institutions.

References

(1) Haruta, M., Size- and Support-Dependency in the Catalysis of Gold. *Catal. Today* **1997**, *36*, 153-166.
(2) Guzman, J.; Carrettin, S.; Fierro-Gonzalez, J. C.; Hao, Y. L.; Gates, B. C.; Corma, A., CO Oxidation Catalyzed by Supported Gold: Cooperation between Gold and Nanocrystalline Rare-Earth Supports Forms Reactive Surface Superoxide and Peroxide Species. *Angew. Chem., Int. Ed.* **2005**, *44*, 4778-4781.
(3) Tsung, C. K.; Kuhn, J. N.; Huang, W. Y.; Aliaga, C.; Hung, L. I.; Somorjai, G. A.; Yang, P. D., Sub-10 nm Platinum Nanocrystals with Size and Shape Control: Catalytic Study for Ethylene and Pyrrole Hydrogenation. *J. Am. Chem. Soc.* **2009**, *131*, 5816-5822.
(4) Yang, M.; Li, S.; Wang, Y.; Herron, J. A.; Xu, Y.; Allard, L. F.; Lee, S.; Huang, J.; Mavrikakis, M.; Flytzani-Stephanopoulos, M., Catalytically Active Au-O(OH)$_x$-Species Stabilized by Alkali Ions on Zeolites and Mesoporous Oxides. *Science* **2014**, *346*, 1498-1501.

ENGINEERING NANOPOROUS MATERIALS WITH INCREASED ACID GAS RESISTANCE

*Guanghui Zhu, Simon Pang, Yang Liu, Uma Tumuluri, Zili Wu, David S. Sholl, Sankar Nair, Christopher W. Jones, and Ryan P. Lively**

Guanghui Zhu, Dr. Simon H. Pang, Yang Liu, Dr. David S. Sholl, Dr. Sankar Nair, Dr. Christopher W. Jones, and Dr. Ryan P. Lively
School of Chemical and Biomolecular Engineering
Georgia Institute of Technology
311 Ferst Drive NW, Atlanta, GA 30332
Dr. Uma Tumuluri and Dr. Zili Wu
Chemical Sciences Division and Center for Nanophase Materials Sciences
Oak Ridge National Laboratory
1 Bethel Valley Road, Oak Ridge, TN 37831
Corresponding Author E-mail: ryan.lively@chbe.gatech.edu

Introduction

Microporous materials such as metal-organic frameworks (MOFs), zeolitic imidazolate frameworks (ZIFs) and porous organic cages (POCs) have significant potential to enhance and revolutionize molecular separation processes. The stability of the material in the face of aggressive contaminants such as H_2S, SO_X, NO_X, and water, is an important challenge not often discussed in the literature. Here, we assess the resistance of ZIF-8 and the POC CC3 to SO_2 exposure. The chemical and physical degradation mechanism of the two materials is discussed, and promising routes to engineering improved acid gas resistance are developed.

Experimental

Synthesis of CC3 samples: In a typical synthesis[1], 243 mg triformylbenzene was added to a conical flask. A solution of 255 mg (1R,2R)-1,2-diaminocyclohexane in 10 mL anhydrous DCM was added slowly to the flask so that the triformylbenzene powder was not disturbed. 10 μL trifluoroacetic acid was added to the mixture as the iminization catalyst. The flask was capped and kept still at room temperature for 3 days. The yellow reaction mixture was filtered and washed to white using a 95% ethanol / 5% DCM mixture. The product was dried in a conventional oven at 80 °C and then in a vacuum oven at 80 °C overnight. In the CC3-S synthesis, (1S,2S)-1,2-diaminocyclohexane was used instead of (1R,2R)-1,2-diaminocyclohexane. In the CC3-trans synthesis, a racemic mixture of (1R,2R)-1,2-diaminocyclohexane and (1S,2S)-1,2-diaminocyclohexane was used. In the CC3-racemic synthesis, a mixture of cis-diaminocyclohexane, (1R,2R)-1,2-diaminocyclohexane and (1S,2S)-1,2-diaminocyclohexane was used.

Synthesis of ZIF-8 samples: 300 μm ZIF-8 was synthesized solvothermally in methanol using sodium formate as a modulating ligand.[2] 1 μm rhombic dodecahedra and truncated rhombic dodecahedra were synthesized in room temperature water for varying amounts of time to control the final particle morphology. 100 nm ZIF-8 truncated cubes were synthesized hydrothermally using CTAB as a surface-specific capping ligand.[3,4]

In situ IR Studies. CC3-R and ZIF-8 were pretreated at 110°C for 1 h with 25 cm3/min He in the DRIFT cell (diffuse reflectance Fourier transform infrared spectroscopy, Pike Technologies HC-900). The temperature was cooled down to 25°C, after temperature equilibrium was reached, 210 ppm SO_2 in He was supplied to the cell at 25 cm3/min for 60 min. The residual gas was purged with He flow at 25 cm3/min for 10 min. A temperature programmed desorption was followed after the purge. The CC3-R and ZIF-8 samples was heated to 120°C with a ramping rate of 10°C/min; the temperature of the sample was held at 120°C for 15 min, and was then cooled down to 25°C at a ramping rate of 10°C/min. Single beam IR spectra were collected continuously during the SO_2 adsorption step using a Thermo Nicolet Nexus 670 spectrometer in diffuse reflectance mode. The outlet gases from the DRIFTS reactor were analyzed using a quadrupole mass spectrometer (Omnistar GSD-301 O_2, Pfeiffer Vacuum).

Acid stability experiments: The acid gas stability was examined by exposing CC3 and ZIF-8 particles to aqueous solutions of the acid gases at varying concentrations. The concentration of exposure is reported as the vapor phase concentration in equilibrium with the aqueous solution, according to Henry's law. A solution at the desired concentration was introduced into a 60 mL glass vial with approximately 200 mg of ZIF-8 or CC3 and sealed. Smaller particles were sonicated in the acid gas solutions to ensure good dispersion. After the desired length of time, particles were filtered or centrifuged, and washed with DI water and methanol and finally dried under vacuum at 60 °C.

DFT calculations. All DFT calculations, including molecule geometry optimization and analysis of vibrational modes, were performed by using the B3LYP functional with the 6-31G(d,p) basis set in the Gaussian09 package. Each calculation examined a single CC3-R molecule or its derivative structure. GaussView was used for visualizing molecule structure and its vibrational modes. In order to better compare calculated vibrational frequencies to experimentally measured IR spectra all the calculated vibrational modes were scaled by a factor of 0.9613, which has previously been established as appropriate for the B3LYP functional.

Results and Discussion

CC3. Although synthesized under the same conditions, CC3-trans and CC3-racemic crystals have smaller sizes and smoother crystal surfaces than CC3-R and CC3-S crystals. Racemic cage pair interactions have been noted to be more favorable than homochiral cage pair interactions.[5] In the CC3-trans and CC3-racemic cases, although each cage contains different chiral linkers, the packing is still more favorable than for enantiopure cages. The lack of strong interactions in CC3-R and CC3-S molecular packing results in more mesoscale defects such as grain boundaries in the crystal, which were visualized by fluorescent confocal microscopy. The images taken at different depths in the CC3 crystals clearly show grain boundaries in CC3-R and CC3-S crystals, which are filled with the dye solution (green color in the **Figure 1**). However, no grain boundaries are observed throughout the CC3-trans and CC3-racemic crystals. This difference, we believe, plays a critical role in the distinct interaction behavior of humid SO_2 with the various CC3 crystals.

All samples were challenged with humid and aqueous SO_2 conditions. CC3-trans and CC3-racemic exhibited only minor surface layer exfoliation compared to crack formation in CC3-R and CC3-S. Although CC3-racemic exhibits some surface morphology changes, the bulk chemical structure has not changed. The gas sorption properties of CC3-racemic were also retained after the treatment.

A potential chemical degradation mechanism was developed utilizing in situ IR experiments and both the vibrational modes and corresponding IR frequencies of various species that are formed during degradation were calculated using DFT methods. First, the imine bond in pristine CC3-R is attacked by HSO_3^- and H^+. This intermediate structure becomes a broken cage with a dangling amine group and aldehyde group. Finally, SO_2 can easily interact with the newly-generated amine groups to form $-NH_2$--SO_2 complexes. In general, good agreement between the calculated and experimental results was observed, although some minor deviations are still present.

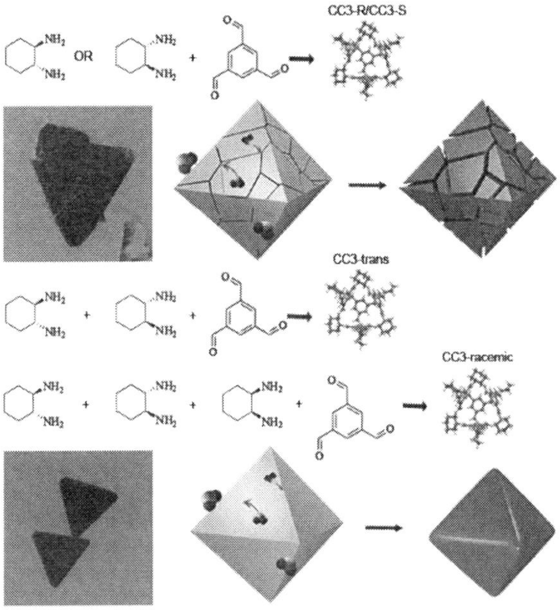

Figure 1. CC3 synthesized from different diamine linkers shows distinct physical degradation

The decomposition of CC3 species in the presence of SO_2 and H_2O vapor or liquid happens at the molecular level (i.e., individual cages). The imine bond connecting the two linkers within an individual cage dissociates when challenged with high concentrations of humid acid gas or aqueous acid solutions. Although the total water uptake of CC3-trans and CC3-racemic is comparable, the minor difference in the extent of surface erosion is also related to the significant difference in water adsorption at the same relative humidity. Indeed, "liquid-like" water does not occur in CC3-racemic and CC3-trans until high relative humidity. The combination of elimination of grain boundaries as well as decrease in water uptake both contribute to the increase in chemical resistance of CC3-racemic relative to CC3-R or CC3-S.

ZIF-8

Microporous materials such as metal-organic frameworks (MOFs) and zeolitic imidazolate frameworks (ZIFs) are typically described by their bulk properties, such as crystal structure, internal surface area, and metal-ligand connectivity. However, little consideration is given to the external surfaces of these materials, which are necessarily undercoordinated compared to the bulk structure and which become increasingly significant as the crystal sizes approach the nanoscale. It is currently unclear how these surfaces are terminated and how the specific external termination affects the stability of a given crystal facet and bulk structure as a whole. It is expected that the external surface will play a large role in the crystal stability under exposure of these materials to acid gases such as SO_2.

After exposure to aqueous SO_2 we observe selective degradation in the surface region of ZIF-8 without affecting the bulk properties of the microporous material. Bulk properties of the crystals were probed via x-ray diffraction, cryogenic nitrogen physisorption, and pulsed-field gradient nuclear magnetic resonance experiments to probe the intracrystalline diffusion properties; these properties are essentially unchanged before and after aqueous SO_2 exposure. The external surface of the ZIF-8 crystals was probed via scanning electron microscopy, x-ray photoelectron spectroscopy, and macroscopic diffusion measurements. These results, paired with vibrational

Figure 2: Facet-dependent degradation of ZIF-8 crystals in the presence of aqueous SO_2

spectroscopy studies, suggest selective removal of 2-methylimidazolate linkers from the surface region of the crystals, and replacement by hydroxyl species. Furthermore, these investigations provide evidence that the (110) facet of ZIF-8 is selectively degraded, whereas the (100) facet appears to be more resistant to aqueous SO_2 attack (**Fig. 2**). Samples synthesized with only (100) terminations qualitatively exhibit enhanced acid resistance relative to (110) terminations.

Conclusions

In summary, the degradation of organic and organic-inorganic nanoporous materials under SO_2 exposure conditions was studied. The synthesis of CC3 variants with mixed linkers at the molecular level is reported. The linkers with different chirality increased the stability of the resulting cages packed into the solid state. The morphological differences in the two sample types and their acid-induced degradation was probed by fluorescent confocal microscopy; mesoporous grain boundaries in CC3-R and CC3-S (but not CC3-racemic and CC3-trans) were observed. These mesoporous grain boundaries in CC3-R and CC3-S accelerated the corrosion due to reaction with SO2 and water, as the acid can access surfaces small crystal grains throughout the CC3-R crystals.

Acknowledgement. This work was supported as part of UNCAGE-ME, an Energy Frontier Research Center funded by the U.S. Department of Energy, Office of Science, Basic Energy Sciences under Award # DE-SC0012577. The in situ IR work was conducted at the Center for Nanophase Materials Sciences, which is a DOE Office of Science User Facility.

References

(1) D. Holden, K. E. Jelfs, A. Trewin, D. J. Willock, M. Haranczyk, A. I. Cooper, *J. Phys. Chem. C* **2014**, *118*, 12734-12743
(2) C. Zhang, J.A. Gee, D.S. Sholl, R.P. Lively, *J. Phys. Chem. C* **2014**, *118* (35), 20727–20733.
(3) Y. Pan, D. Heryadi, F. Zhou, L. Zhao, G. Lestari, H. Su, Z. Lai, *CrystEngComm* **2011**, *13* (23), 6937.
(4) Z. Li, C.H. Zeng, *Chem. Mater.* **2013**, *25* (9), 1761–1768.
(5) T. Hasell, S. Y. Chong, K. E. Jelfs, D. J. Adams, A. I. Cooper, *J. Am. Chem. Soc.* **2012**, *134*, 588-598.

INTERACTIONS OF ACID GASES WITH METAL OXIDE NANOCLUSTERS AND MODELS OF MOF-2

Luis Flores,[1] Julia Murphy,[1] Zongtang Fang,[1] Michael Outlaw,[1] Mingyang Chen,[2] Tjerk Straatsma,[2] Joshua D. Howe,[3] David S. Sholl,[3] and David A. Dixon[1]

Department of Chemistry, The University of Alabama, Tiscaloosa, AL 35487-0336; National Center for Computational Sciences, Oak Ridge National Laboratory, Oak Ridge, Tennessee, 37831; School of Chemical and Biomolecular Engineering, Georgia Institute of Technology, Atlanta GA, 30332-0100
dadixon@ua.edu

Introduction

Acid gases play an important role in a range of large-scale energy applications. CO_2 is the most important greenhouse gas affecting climate change, and also represents a potentially important carbon feedstock. In addition, SO_x, NO_x, and H_2S are important components of energy-related gas mixtures. For example, there are significant amounts of SO_x and NO_x in flue gas from power plants and modern flue gas management involves sorbents and catalysts that convert these species to less harmful products, both of which can be degraded by these gases. H_2S is a common contaminant in natural gas deposits and is also present in landfill gases. In addition, H_2O is always present for any of these applications. Acid gas conversion can make use of reactive sorbents, and these sorbents as well as catalysts must not substantially degrade under operating conditions. For example, the dispersion and activity of the sites depend on complex interactions with the support, which can be easily disrupted by corrosion. Such degradation effects are often decisive factors in the practical use of materials such as sorbents for carbon capture, acid gas conversion, and natural gas purification. In addition, the reactions of H_2O, both as an acid gas and as a solvent, are important. There is also interest in the reactions of CO_2 with heavy element oxides in terms of the role of formation of carbonates. The interactions of acid gases with complex materials are poorly understood. In order to improve our understanding of such materials, we have been using electronic structure methods to better understand these interactions in terms of 'physisorption' and 'chemisorption' type processes.

Computational Approaches

A variety of computational electronic structure methods were used. The predominant method was density functional theory (DFT) with a range of exchange-correlation functionals including generalized gradient approximations and hybrids, and basis sets. The DFT results were benchmarked by using correlated molecular orbital theory at the coupled cluster CCSD(T) level with the augmented correlation consistent basis sets including the use of effective core potentials for the heavier elements to handle the scalar relativity effects. Geometries were optimized and second derivatives were calculated to show that the structures were minima as well as to provide data for comparison to experiment, for example, ligand vibrational frequencies.

Results and Discussion

Metal Oxide Nanocluster Reactions The interactions of Group IV and VI transition metal oxide nanoclusters with Lewis acid gases have been calculated. This work builds on our work on the reactions of H_2O with such nanoclusters.[1,2,3] The reactions of H_2O with TiO_2 nanoclusters shows that one can readily convert the M=O bonds to OH groups exothermically.[4] Dissociative adsorption of H_2O on Ti with coordination number (CN) for Ti < 4 is much more exothermic than molecular adsorption of H_2O onto the same Ti center, whereas molecular adsorption of H_2O on Ti with CN(Ti) = 4 to form Lewis acid-base Ti-OH_2 bond is ~ 5 kcal/mol more exothermic than the dissociative adsorption reaction. The number of H_2O molecules required to eliminate all of the under-coordinated Ti centers and the 4-coordinated Ti centers with distorted geometry is the key for the hydrolysis limit, i.e. the maximum equivalents of H_2O that can be completely dissociatively adsorbed in favor of being molecularly adsorbed. The hydrolysis limits are predicted to be 2, 2, 3, 2, and 3 for $(TiO_2)_n$, n = 1, 2, 3, 4 and 8, respectively at 0 K.

Bulk RuO_2 is a metal and it is of interest to understand the reactions of RuO_2 nanoclusters with H_2O as the ground state of the nanoclusters changes from singlet to triplet. For $(RuO_2)_n$, the ground state is a singlet with ΔE(S-T) = 9.5 kcal/mol, for n = 2, the ground state for n =1 is a singlet with ΔE(S-T) = 2.2 kcal/mol, and for n =3, the ground state is a triplet with ΔE(S-T) = -22.9 kcal/mol The potential energy surfaces of the hydrolysis reactions on $(RuO_2)_n$ (n = 1 to 3) nanoclusters have been studied with the density functional theory (DFT) with the B3LYP functional and correlated CCSD(T) MO theory. The reactions of H_2O with the metal site having a Ru=O bond and/or Ru–O bonds as well as H transfer to both terminal =O atoms and bridge –O atoms were studied. The hydrolysis reactions of the nanoclusters with water are exothermic. The chemisorption energies on the singlet clusters are generally more exothermic than on the triplet clusters. For the monomer and dimer, the physisorption energies for the first addition of H_2O to the clusters are predicted to be -10 to -20 kcal/mol and the physisorption energies on the singlet clusters are more exothermic than for the triplet. The physisorption energies on the trimer are less exothermic, generally <10 kcal/mol. The reaction barriers for the proton transfer to a bridge –O atom are smaller than to a terminal =O atom for the dimer and trimer. The predicted barriers have a broad range and are dependent on the reaction and the cluster. In comparison with the hydrolysis of Group IV ZrO_2 nanoclusters, the physisorption and chemisorption energies are less exothermic and the barriers are larger for the hydrolysis reactions of RuO_2 nanoclusters.

The adsorption of CO_2, SO_2, and NO_2 to Group IV $(MO_2)_n$ and Group VI $(MO_3)_n$ metal oxide nanoclusters have been studied in addition ot our prior work on the hydrolyis of these clusters. Group VI $(MO_3)_n$ nanoclusters are predicted to have a low affinity for CO_2, preferentially physisorbing the gas to the metal through the Lewis basis oxygen site. The Lewis acid-base adduct energies are significnat for n =1 (~20 kcal/mol) but are barely bound for larger clusters. Chemisorption leads to the formation of a carbonate (reaction of the CO_2 with an O^{2-} either terminal or bridging) and this is not energetically favored. In contrast, the Group IV nanoclusters are predicted to have a higher affinity for CO_2, with both chemisorption (carbonate formation) and physisorption being energetically favored. Hf and Zr metal oxides preferentially chemisorb CO_2, while Ti metal oxides are expected to preferentially physisorb CO_2. The chemisorption energies can be as high as 30 kcal/mol for HfO_2 nanoclusters.

The reaction of SO_2 is of interest because the SO_2 can physisorb through the O to a metal center and react with an O^{2-} ,either terminal or bridging, to form an SO_3^{2-} just as CO_2 forms a carbonate, and undergo a redox reaction to generate SO_4^{2-}. Mo and W are predicted to physisorb SO_2, again through the O, but Cr is predcited to chemisorb SO_2 by sulfate formation via a redox reaction. This is most likely due to the higher reducibility of Cr. The Group IV metal oxides have a higher affinity for SO_2 than do the Group VI metal oxides, with chemisorption to form sulfite preferred with no redox. Chemisorption energies can reach as high as 60 kcal/mol of the Hf oxides.

NO_2 can react by physisorption or by formation of a nitrate. The NO_2 is an open shell species and the radical electron can remain on the ligand or transfer to the metal. Chemisorption is preferred for the Cr clusters, but physisorption is preferred for the W clusters. For the Group IV nanoclusters, physisorption is preferred.

MOF-2 Reactions The interactions of the acid gases shown in Table 1 were studied with two models of the core MOF as shown in Figure 1. Ligand binding energies and vibrational frequencies with IR and Raman intensities were calculated for naked MOF-2 cluster models and with one or two acid gases bound to them. Two clusters were used to model Lewis acid gas interactions with MOF-2. The first, larger model is composed of two M^{2+} (M = Zn, Cu, and Co), atoms connected by 4 bridging p-benzenedicarboxylate (BDC) ligands, as a model for the metal sites in MOF-2. The second, smaller model composed of two M^{2+} and 4 bridging formate ligands was used to employ higher levels of theory. The two models give similar results. For all three metals, the ligand dissociation energies for both models were in good agreement with each other at the density functional theory level with the B3LYP and ωB97X functionals within 2 kcal/mol of each other, showing that the smaller model is a useful model. Calculations on the small model with a range of methods (B3LYP/DZVP2, ωB97X/DZVP2, ωB97XD/DZVP2, PW91/DZVP2, BP86/DZVP2, and MP2/aug-cc-pVDZ) are in good agreement with each other. SO_2, and NO_2 only bind through oxygen to the metal. The binding strength of a single ligand to the metal center for the Zn and Cu clusters follow the trend $H_2O > H_2S > SO_2 > CO > CO_2$. The Co clusters show a slightly different trend, $H_2O > CO > H_2S > SO_2 > CO_2$, indicating that Co has a higher affinity for CO than the other two metals. The cluster geometry of the inner coordination sphere for both Zn centers is closer to a flattened tetrahedron, whereas the Zn centers in the crystal structure adopt a geometry which is closer to square planar. Further calculations were performed on both clusters by constraining the cluster to D_{4h} symmetry which resulted in a higher energy cluster with a square planar metal center. The constrained cluster showed the same binding trend to the gas molecules as the stable clusters, and all gases had a stronger binding energy to the constrained cluster. This suggests that the metal centers in the bulk are constrained in a higher energy structure leading to a higher binding energy to the acid gases.

Table 1. Ligand Dissociation Energies (ΔH_{298}) in kcal/mol at the ωB97X/DZVP2 level for the $(H_2O)(L_2)$–M_2BDC_4 MOF-2 Adsorption Models

L_2	bidentate		monodentate		hydrolyzed	
	L_2	H_2O	L_2	H_2O	L_2	H_2O
Cu^{2+}						
H_2O	13.7	13.7	11.4	11.4	-0.1	-0.1
SO_2	11.7	14.5	8.9	11.8	-2.0	0.9
H_2S	7.1	10.0	7.3	10.2	-1.0	2.0
CO	a	a	7.0	10.9	-4.0	-0.1
CO_2	8.2	15.4	6.2	12.7	-6.3	0.2
NO_2	8.1	14.6	6.4	12.5	-5.9	0.3
Zn^{2+}						
H_2O	a	a	17.7	17.7	18.2	18.2
SO_2	a	a	13.3	18.2	8.3	13.3
H_2S	a	a	12.8	18.1	12.6	18.0
CO	a	a	10.7	17.9	6.1	13.3
CO_2	4.9	14.5	7.6	17.3	3.5	13.2
NO_2	a	a	8.2	17.4	4.0	13.2

a Optimizes to monodentate

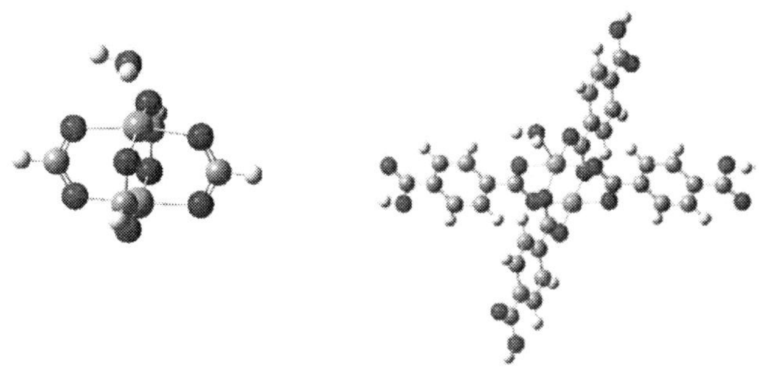

Figure 1. MOF-2 models. Right: H_2O-$Cu_2(OOCH)_4$ and Left: H_2O-$Cu_2(BDC)_4$

The addition of water as a second ligand can lead to the BDC linker forming a monodentate interaction with the other binding site on the metal occupied by a ligand. In addition, hydrolysis of the BDC linker can occur although this is not energetically favored for Cu^{2+} as the metal. In contrast, the monodentate and hydrolyzed structures are almost isoenergetic for Zn^{2+} as the metal.

Conclusions

Electronic structure calculations are beginning to provide a detailed look at the interactions of acid gases with models of sorbents and catalysts. The interactions of H_2O with either metal oxide nanoclusters or with models of MOFs show that the clusters can be degraded. The interactions of CO_2 with metal oxides lead to weak interactions with Group VI metal oxide nanoclusters and to the formation of carbonates for Group IV metal oxide nanoclusters. Similarly, the interactions of SO_2 and NO_2 with the Group VI nanoclusters except for Cr favor physisorption. However, for Cr, the reducibility of the metal plays a role leading to sulfate and nitrate formation, respectively. Sulfites are preferred for the Group IV nanoclusters and physisorption is preferred for the Group IV interactions with NO_2. The presence of two ligands on MOF-2 can lead to monodentate binding of a BDC ligand to the metal center for Cu^{2+}. Bidentate binding only occurs of the interaction of CO_2 and H_2O for the Zn^{2+} MOF-2 model. For all other ligand combinations with H_2O, only mondentate binding was predicted and hydrolysis of the BDC ligand was predicted to be essentially isoenergetic with monodentate binding.

Acknowledgement. This work was supported by the Center for Understanding and Control of Acid Gas-Induced Evolution of Materials for Energy (UNCAGE-ME), an Energy Frontier Research Center funded by the U.S. Department of Energy, Office of Science, Basic Energy Sciences and by the BES Catalysis Center Program.

References

(1) Fang, Z.; Outlaw, M.; Smith, K.; Gist, N.; Li, S.; Dixon, D. A.; Gole, J. L. *J. Phys. Chem. C*, **2012**, *116*, 8475-8492.
(2) Li, S.; Guenther, C. L.; Kelley, M. S.; Dixon, D. A. *J. Phys. Chem. C*, **2011**, *115*, 8072–8103.
(3) Wang, T.-H.; Fang, Z.; Gist, N. W.; Li, S.; Dixon, D. A.; Gole, J. L. *J. Phys. Chem. C*, **2011**, *115*, 9344-9360.
(4) Chen, M.; Straatsma, T. P.; Dixon, D. A. *J. Phys. Chem. A*, **2015**, *119*, 11406−11421.

HYDROTHERMAL STABILITY OF ZSM-5 ZEOLITE

Michael T. Timko and Alex R. Maag

Worcester Polytechnic Institute
mttimko@wpi.edu

Introduction

Zeolites are microporous aluminosilicate structures commonly used for catalysis in environmental, fuel upgrading and biomass industries. These heterogeneous catalysts are easily recoverable and more size selective relative to liquid acid catalysts. Zeolites can have high acid site densities due to their large surface area and improved product selectivity due to the steric nature of their pore sizes. Their strong Brønsted acidity can be tuned by the amount Al incorporated in the structure due to catalytic protons balancing the tetrahedrally coordinated internal AlO_4 sites. To date, most zeolite work has been performed under vapor phase conditions. Recent advances to improve process intensity and energy efficiency motivate interest in zeolite catalysis in the presence of liquid water or water-rich liquid reaction mixtures. However, many zeolites are prone to framework degradation under aqueous phase conditions, even at temperatures less than about 200 °C. Therefore, further work is required to understand zeolite stability and degradation mechanisms.

We focus our attention on ZSM-5, a commonly used zeolite for hydrothermal catalytic cracking of fatty acids and model fuels that has proven to be more robust than other zeolites such as HY and Beta. In particular, we were interested in understanding ZSM-5 decrystallization at temperatures greater than 250 °C and for exposure times longer than 6 hours (the most aggressive conditions reported in the literature). Many chemical reactions of interest require temperatures >250 °C and industrially relevant catalyst materials must exhibit framework stability for thousands of hours to be economically viable. Literature studies to date had found ZSM-5 to be stable for at least 6 hours at 250 °C; however, we hypothesized that ZSM-5 would degrade under more aggressive conditions, allowing us to investigate the degradation mechanism and determine methods for stabilizing ZSM-5. In particular, we were interested to understand ZSM-5 stability at temperatures greater than water's critical point as the familiar properties of liquid water (i.e., high dielectric and auto-ionization) shift dramatically near the critical point. Therefore, we further hypothesized that studying ZSM-5 hydrothermal stability both above and below water's critical temperature might provide deep insight into the role of water in ZSM-5 degradation.

Experimental

ZSM-5 with Si/Al=38 was obtained from ACS Materials. The catalyst was initially heated to 100 °C for at least one hour followed by calcination at 550 °C for at least 16 hours. Zeolite degradation process was done in a continuous flow study, where 0.25g of ZSM-5 was packed into a ¼" tube with a 0.2 um frit containing it inside. Preheated water was fed into the reactor at 4 ml/min. A back pressure regulator kept the system at 3600 psi.

Zeolites were characterized using a variety of complementary techniques. SEM analysis was performed on a JSM-7000F SEM instrument using a LaB_6 ion source. Samples were gold sputtered prior to imaging to improve their conductivity and mounted to the analysis stage using conductive carbon tape. A Thermo Nicolet Magna 560 with a SpectraTech DRIFTS cell was used to obtain IR spectra of ZSM-5 samples. Samples were analyzed over the range from 4000 to 400 cm^{-1}, at a resolution of 2 cm^{-1}, and an accumulation of 96 scans. The DRIFT cell was loaded and flattened with ZSM-5 before purging with N_2 for 10 minutes. 20 °C increments at 10-15 min intervals were ran until 100 °C, where it was held for 30 minutes before raising 50 °C increments to 550 °C. Argon Sorption on a ASIQ IQ Quantachrome Instrument was used to determine BET area, micropore volume and area. X-ray diffraction was performed over the range 5-80° with a 0.5 step size and 1s dwell time. Degree of crystallinity loss were determined by comparison of integrated peak areas measured for treated samples compared to a calcined sample used as standards.

Results and Discussion

The key to zeolite catalytic activity is its high surface area, a result of its highly ordered, crystalline, and microporous structure. Therefore, crystallinity is a key metric of zeolite integrity. The crystallinity of hydrothermally treated ZSM-5 was characterized using XRDS. Figure 1 provides representative XRDS spectra and Figure 2 provides relative crystallinity index data obtained by integration of the original XRDS spectra. Figure 1 makes qualitatively clear that ZSM-5 loses crystallinity after 3 hours of exposure to liquid water at temperature > 300 °C.

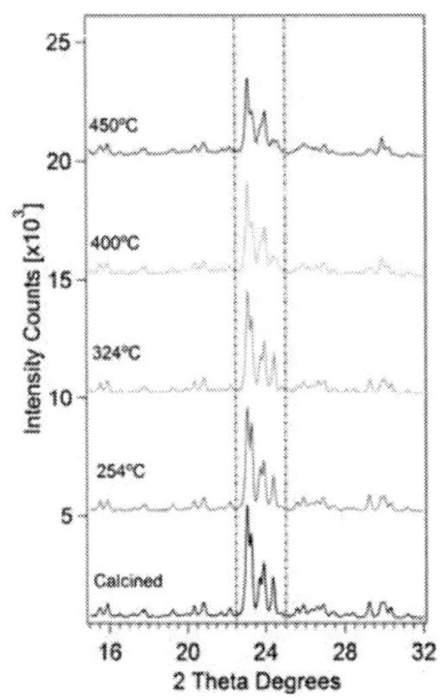

Figure 1. Representative XRDS specta obtained for ZSM-5 after calcination and after 3 hours of hydrothermal treatment at 250 to 450 °C.

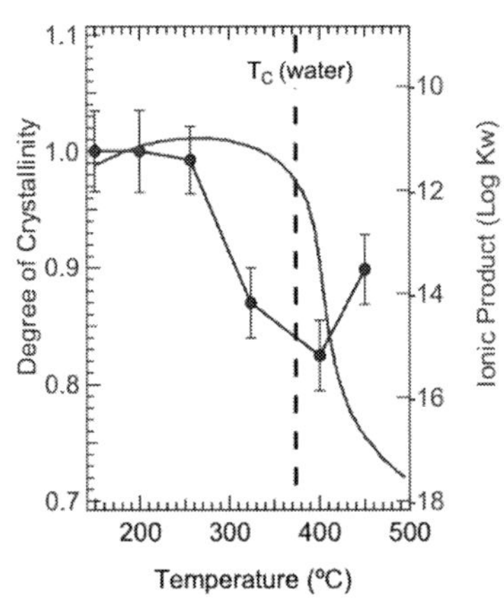

Figure 2. Degree of crystallinity measured from XRDS data for hydrothermally treated ZSM-5 (blue) superimposed on water's dielectric constant (red).

Figure 2 quantifies this trend, showing an abrupt decrease in crystallinity for ZSM-5 treated at 325 °C. Interestingly, we found that ZSM-5 retains crystallinity when exposed to "dense" water above its critical point. Figure 2 superimposes water's dielectric constant over the crystallinity data, showing an apparent correlation between the two measurements. Previous reports on liquid-phase stability of zeolites suggest that degradation occurs due to ionic reactions, particularly those that are acid-base catalyzed. As liquid water above its critical point becomes incapable of supporting ions, including H^+ and OH^-, we hypothesize becomes a less aggressive environment for ZSM-5 degradation.

To understand the degradation mechanism more fully, we performed additional characterization tests. In particular, we attempted to understand if the degradation mechanism was primarily de-silication (as had been reported previously as the degradation method for zeolites exposed to liquid water) or de-alumination (which has been reported for zeolites under steaming conditions, but not previously under liquid-phase conditions). We performed IR, gas sorption, TEM, SEM, magic-angle spinning NMR, Raman, and XPS characterization methods. Data from all of these are consistent with one another. Here, we focus on IR and TEM data.

Figure 3 provides representative IR spectra obtained using the DRIFTS technique. The calcined ZSM-5 is shown as a reference along with spectra for samples treated at 250 to 450 °C. The calcined sample exhibits clear Brønsted acid sites, along with surface silanol groups. With increasing exposure temperatures, we find loss of Brønsted acid sites in parallel with the appearance of extra framework alumina (EFAL) and disappearance of silanol groups. From Figure 3, we surmise that framework degradation involves formation of EFAL. The appearance of EFAL is suggestive of a de-alumination mechanism; however, de-silication can also lead to formation of EFAL on zeolite surfaces and we performed additional tests to refine our preliminary conclusion.

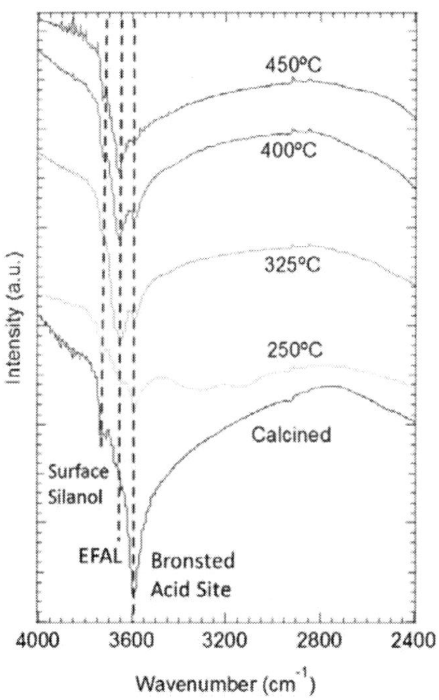

Figure 3. IR data for calcined ZSM-5 and for ZSM-5 treated at increasing temperatures in liquid phase water.

Next, we performed electron microscopy analysis of the zeolites using SEM and TEM. SEM images (not shown) indicate the evolution of needle-like microparticles on the ZSM-5 particle surfaces. We surmise that the needle-like microparticles are primarily Al_2O_3, a conclusion supported by an increased abundance of Al on the zeolite surface indicated by XPS measurements. Next, we obtained TEM images of the hydrothermally treated zeolite. Figure 4 shows retention of clear zeolite crystal planes at regular lattice spacing. Figure 4 shows no evidence of mesopore formation, as would be indicated by internal de-silication. TEM does show increased surface roughness; along with the apparent loss in crystallinity in XRD, surface roughness indicates a surface desilication mechanism between 250-350 °C. Based on Figure 4, we surmise that ZSM-5 internal degradation is dominated by de-alumination in high-temperature liquid water. This conclusion suggests future work to increase ZSM-5 stability.

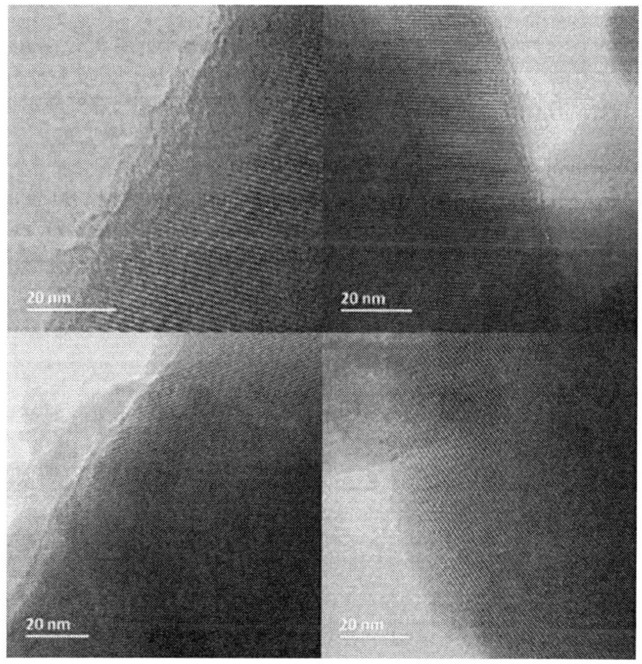

Figure 4. TEM data for hydrothermally treated ZSM-5 showing intact crystal planes and no evidence of internal mesoporosity.

Conclusions

We have investigated the stability of ZSM-5 in high-temperature liquid water at temperatures above and below its critical point. Degradation is dominated internally by de-alumination with surface desilication at higher subcritical temperatures. At temperatures greater than the critical point, we find that ZSM-5 exhibits improved crystallinity stability, which we attribute to reduced de-silication rates due to the weaker ability of water above its critical point to promote ionic reactions.

Acknowledgements. Prof. Gisele Azimi (University of Toronto) provided TEM images. Prof. Luis Smith (Clark University) provided MAS-NMR data. Saudi Aramco and ACS Petroleum Research Fund each provided partial support for this work.

CATALYTIC ETHANOLYSIS OF LIGNIN OVER ZEOLITES OF VARIOUS POROSITY AND ACIDITY

*Yuxin Wang, Nathan C. Baxter, Heath A. Barnett, and Shengnian Wang**

Chemical Engineering, Louisiana Tech University
911 Hergot Avenue, Ruston, LA, 71272, USA
swang@latech.edu

Introduction

Lignin has long been a high-volume byproduct of the paper and pulp industry. The challenge comes from the difficulty on effective cleavage of the many C-O-C and C-C linkages in lignin to release the rich aromatics in its chemical structure. With the increasing interests in the use of biomass, depolymerization of lignin becomes one of a few major focus processes recently. To produce phenol-type, high-valuable feedstocks, hydrolysis, oxidation, and reduction of lignin with the help of catalysts has been investigated and many catalysts including NaOH, nickel-based catalysts, MoO3/SiO_2, and $CuMgAlO_x$ have been tested. Considering the superior performance of zeolites in heavy hydrocarbon cracking and the enhanced mass transport with hierarchical porous structure, we report here the use of various mesoporous zeolites in lignin depolymerization. Within both supercritical ethanol and water-alcohol mixture conditions, we investigate the important of the acidity and mesoporosity in the depolymerization of lignin. Different zeolite types (e.g., zeolite Y, Beta, and ZSM-5) are examined in a broad range of Si/Al ratios. Special focus is made on the conversion of lignin, yield, and selectivity on the high-value chemicals aromatics. Our results show that zeolites could help increase the conversion of lignin and deeply break it into desired high-valued aromatics, especially phenols such as guaiacol and alkane substituted guaiacols (ethylguaiacol and methylguaiacol).

Experimental

Materials. Tetrapropyl ammonium hydroxide (TPAOH, 1 M in H_2O), sodium aluminate ($NaAlO_2$, ~8% H_2O, >99%), tetraethylorthosilicate (TEOS, >99%) were purchased from Sigma-Aldrich and used as received without further purification. Zeolite Y (Si/Al=30), Beta(Si/Al=38), and ZSM-5(Si/Al=30) were purchased from Zeolyst International.

Synthesis. ZSM-5 with different Si/Al ratios are synthesized by mixing TPAOH, $NaAlO_2$, TEOS, and H_2O according to a molar ratio of 0.2TPAOH: 1/xAl_2O_3: 1.0 SiO_2: 64H_2O, x is Si/Al ratio and result zeolite named as ZSM-5-X. Meso-ZSM-5 is synthesized according to a molar ratio of 0.25TPAOH: 0.03Al_2O_3: SiO_2: 80H_2O. The mixture is fist kept at 80 °C to get a clear solution. After solvent evaporation, the received dry nanogels are further transferred to a Teflon-lined autoclave for crystallization without introducing additional water or other solvents to obtain Meso-ZSM-5 samples. All H-type zeolites are obtained through ion-exchanged with a 0.1 M NH_4NO_3 solution at 90 °C for 90 min three times. In order to investigate the important of the acidity, Y zeolite with ion-exchanged one time is also prepared and name as HNaY.

Characterization. X-ray diffraction (XRD) analysis was done on a Bruker D8 Advance X-ray diffractometer, using Cu-Kα radiation. Scanning electron microscopy (SEM) images were taken on a Hitachi S-4800. Transmission electron microscopy (TEM) and selected-area electron diffraction (ED) measurements were carried out on a JEOL 2010F TEM. Nitrogen adsorption isotherms were obtained at -196 °C on a Micromeritics ASAP 2020 Surface Area and Porosity Analyzer. Samples were degassed at 300 °C for 6 h before measurement. Temperature-programmed desorption of ammonia (NH_3-TPD) was tested on a Micromeritics AutochemII Chemisorption Analyzer.

Catalytic reaction tests. The catalytic conversion of Kraft lignin (Sigma–Aldrich) was carried out at 280 °C for 6 h in a high-pressure batch reactor (Col-Int Tech, 200 mL). For each run, lignin (1.0 g), catalyst (0.5 g), and ethanol (100 mL) were first loaded in the reactor. The sealed reactor was then purged with high-purity nitrogen gas for three times at room temperature. The reactor temperature was then risen to 280 °C with the reactor pressure reaching ~10 MPa. After the reaction was carried out for 6 h, the solution was filtered to remove solid residues and the liquid sample was taken for product identification and analysis on a GC–MS instrument (Agilent 7890A-5975C). The lignin conversion was calculated by measuring the weight of the unconverted lignin.

Results and Discussion

As shown in Figure1, among three different zeolites (H-Beta, H-ZSM-5, and H-Y), H-Y zeolites has highest aromatics selectivity (up to 40%) in lignin depolymerization. The major difference of the three types zeolites lies on the largest pore size and the strongest acidity of zeolites Y than ZSM-5 and Beta. It suggests the importance of pore size and acidity in lignin depolymerization.

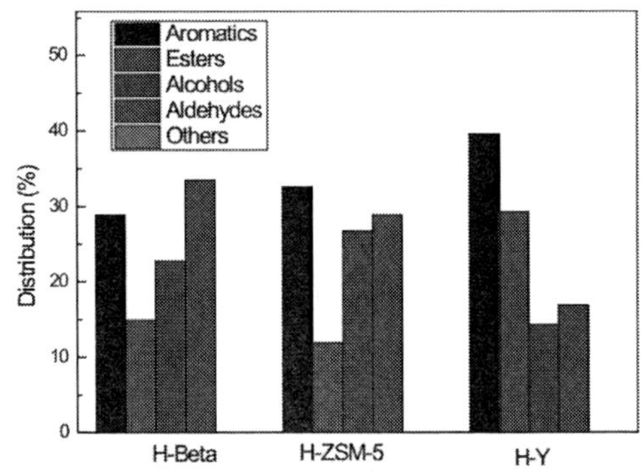

Figure 1. Products distribution of lignin depolymerization over different zeolite types

Table 1. Lignin conversion and the product distribution over Y zeolite with different acidity levels

Catalyst	Lignin Conversion (%)	Product distribution (peak area %)			
		aromatics	esters	alcohols	aldehyde
NaY	60.6	16.11	17.35	40.11	26.43
HNaY	66.1	29.76	26.9	29.03	14.31
HY	70.7	39.57	29.25	14.32	16.86

In order to further separate the effect of zeolite's acidity and porosity, zeolites with different levels of acidity are further prepared by adjusting the proton exchange level of original NaY zeolites. With Na-type Y (NaY) as the parent zeolites, the sample after three times proton exchange process is named as H-Y and the sample with once proton exchange process is named as HNaY. As the acid form of zeolites is largely attributed to the replacement of Na ions with ammonium followed by further decomposition, the acidity varies with the exchange level (i.e., increases as proton exchange level increases). As shown in Table 1, as the acidity increases, the lignin conversion increases, and so does the aromatics selectivity. As the acidity increases, it is more effective to cleave the C-O-C and C-C linkages in lignin to release more aromatic derivatives. Beside different proton exchange levels, zeolites with different Si/Al ratios are also tested to further confirm the importance of acidity.

Table 2. Textural Properties of Meso-ZSM-5 and the Conventional ZSM-5

Sample	S_{BET} (m^2/g)	V_{total} (cm^3/g)	V_{Meso} (cm^3/g)	Pore sized (nm)
Meso-ZSM-5	418.4	0.27	0.18	4.8
ZSM-5	371.8	0.18	0.06	2.9

To further investigate the porosity effect, mesoporous zeolites are synthesized. Table 2 shows the textural properties of Meso-ZSM-5 and the conventional ZSM-5. Compared with the conventional ZSM-5, the Meso-ZSM-5 shows much larger surface area, pore volume, and pore size. Its catalytic activity performance of lignin depolymerization is also much higher than the conventional ZSM-5 (72.5% over Meso-ZSM-5 versus 62.6% over ZSM-5). Meso-ZSM-5 also shows much higher aromatic selectivity, especially for the phenols, as shown in Figure 2. Considering their similar Si/Al ratio for Meso-ZSM-5 and conventional ZSM-5, the better catalytic activity must be attributed to the mesoporous structure which facilitates the diffusion of bulky fragments from lignin depolymerization and accessible to more activity sites.

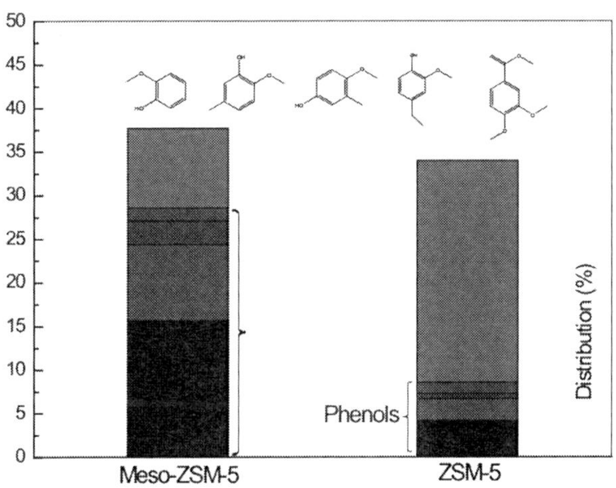

Figure 2. Products distribution of lignin depolymerization over different zeolite types

Conclusions

Lignin depolymerization over a series of zeolites is investigated here. Our results show that zeolites could help increase the conversion of lignin and deeply crack it into desired high-valued aromatics, especially phenols such as guaiacol and alkane substituted guaiacols (ethylguaiacol and methylguaiacol). Zeolites with high acidity and hierarchical pore structure are preferred for more effective depolymerization of lignin.

Acknowledgement. We acknowledge financial support from National Science Foundation under Grant No. 1130468 and Louisiana Harrelson Family Professorship.

References

(1) Y. Wang, J. Wu, S. Wang, *RSC Adv.* **2013**, *3*, 12635.
(2) Y. Wang, T. He, K. Liu, J. Wu, Y. Fang, *Bioresour. Technol.* **2012**, *108*, 280.
(3) R. Ma, W. Hao, X. Ma, Y. Tian, Y. Li, *Angew. Chem. Int. Ed.* **2014**, *53*, 7310.

BIMETAL-ORGANIC FRAMEWORK SELF-ADJUSTED SYNTHESIS OF SUPPORT-FREE NONPRECIOUS ELECTROCATALYSTS FOR EFFICIENT OXYGEN REDUCTION

*Bo You and Yujie Sun**

Department of Chemistry and Biochemistry, Utah State University, 0300 Old Main Hill, Logan, Utah 84322 (USA)
yujie.sun@usu.edu

Introduction

Electrochemical oxygen reduction (ORR) is a crucial and limiting process for several emerging renewable energy technologies.[1] Great efforts have been devoted to developing nonprecious ORR catalysts employing earth-abundant metal alloys, transition metal oxides, and metal-free heteroatom-doped carbons, among which, transition metal-nitrogen-carbon (M-N-C) composites have attracted increasing interest.[2] However, most reported M-N-C electrocatalysts still require higher ORR overpotential than that of Pt in alkaline media, partially due to the non-uniform and insufficient exposure of catalytic active sites resulting from the lack of accurate control of the catalyst composition and/or structure. In addition, the usually tedious post-treatment procedures, such as longstanding hot acid leaching and/or iterative high temperature annealing, significantly increase their cost of scalable production.[3]

Resulting from the unique features of high specific surface area, controllable pore texture, and diverse composition, metal-organic frameworks (MOFs) offer an opportunity to obtain competent ORR electrocatalysts after carbonization.[4] Herein, we report a facile and novel Zn and Co bi-MOF self-adjusted route to produce self-supported Co-N-C nanopolyhedrons as superior ORR electrocatalysts by direct pyrolysis of $Zn_xCo_{1-x}(MeIM)_2$ frameworks (MeIM = 2-methylimidazole and "x" indicates the molar ratio of Zn in the initial bi-MOF). The presence of Zn species not only form a spatial isolation for Co to suppress its sintering, but also increase the specific surface area of the resulting catalysts due to the evaporation of Zn during pyrolysis. Fine tuning of the original Zn content allowed us to optimize its specific surface area, porosity, Co nanoparticle size, and hence ORR activity.[5]

Experimental

Synthesis of Co-N-C catalysts. 12 mmol $Co(NO_3)_2{\cdot}6H_2O$ or $Zn(NO_3)_2{\cdot}6H_2O$ or their mixture were dissolved in 120 mL methanol, and the mixture was poured into 40 mL methanol containing 48 mmol 2-methylimidazole (MeIM). After incubation, the as-obtained precipitates were centrifuged and washed with ethanol, resulting in $Zn_xCo_{1-x}(MeIM)_2$ bi-MOF. The powder of $Zn_xCo_{1-x}(MeIM)_2$ was later annealed at desired temperature (800, 900 and 1000 °C) for 3 h under Ar to obtain the self-supported Co-N-C-x-y, wherein "y" represents the carbonization temperature.

Characterizations. Scanning electron microscopy (SEM) and scanning transmission electron microscopy (STEM) measurements were collected on a FEI QUANTA FEG 650 (FEI, USA). High-angle annular dark-field scanning transmission electron microscopy (HAADF-STEM) and element mapping were collected on a JEM-2800 scanning transmission electron microscope (JEOL, Japan). The X-ray absorption spectra were collected at the Advanced Light Source at Lawrence Berkeley National Laboratory on beamline 10.3.2.

Electrocatalytic measurements. A certain amount of catalyst was carefully loaded on the polished glassy carbon rotating disk electrode (RDE) or rotating ring disk electrode (RRDE), leading to a desirable catalyst loading. In the basic electrolyte (0.1 M KOH), the nonprecious catalyst loading was 0.283 mg cm^{-2}. Electrochemical measurements were performed by a computer-controlled Gamry Interface 1000 electrochemical workstation and a RRDE-3A apparatus with a three-electrode cell system.

Results and Discussion

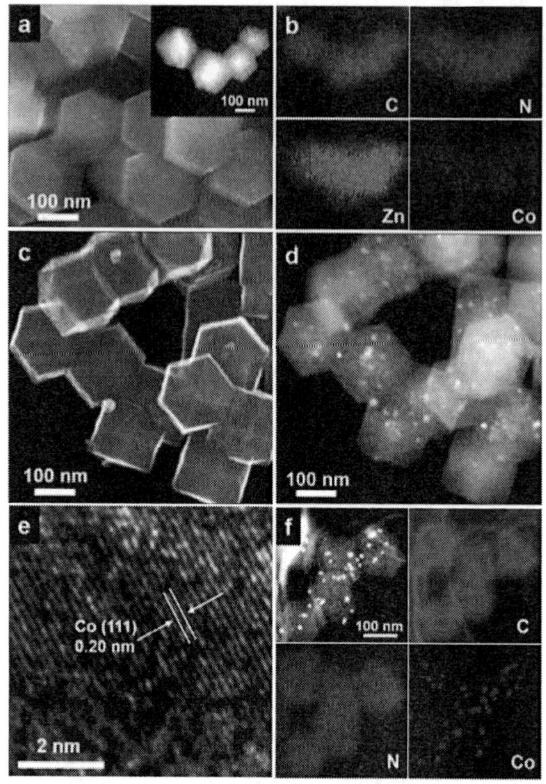

Figure 1. (a) SEM and STEM (inset) images of $Zn_{0.8}Co_{0.2}(MeIM)_2$. (b) The corresponding element mapping images of (a) inset. (c) SEM and (d) STEM images of Co-N-C derived from $Zn_{0.8}Co_{0.2}(MeIM)_2$. (e) HR-TEM image of Co-N-C. (f) STEM and the element mapping images of Co-N-C.

As shown in Fig. 1a-b, uniform polyhedron-like bi-MOF could be obtained. Element mapping analysis suggested the well-proportioned distribution of Zn, Co, C, and N. It is clear that the uniformly dispersed Zn species spatially isolated the Co species and diluted the Co concentration. Energy-dispersive X-ray spectroscopic analysis corroborated the molar ratio of Zn and Co close to 4. Upon carbonization, the resulting Co-N-C inherited the overall polyhedron-like morphology of the parent $Zn_{0.8}Co_{0.2}(MeIM)_2$ (Fig. 1c). Remarkably, the corresponding STEM images revealed that Co-N-C consisted of fine Co nanoparticles with a mean diameter of ~9.5 nm sparsely and uniformly embedded in the porous carbon matrices. In sharp contrast, the control samples with lower Zn contents resulted in severe Co aggregation. These results clearly demonstrate that the presence of Zn species at high content could effectively suppress Co sintering during carbonization due to a spatial isolation of Co by Zn species. High resolution TEM (HR-TEM) images revealed a lattice fringe of 0.20 nm for an embedded Co nanoparticle (Fig. 1e), corresponding to the (111) plane of face-centered cubic metallic Co (PDF: 15-0806). High-angle annular dark-field STEM and the corresponded element mapping (Fig. 1f) confirmed the uniform distribution of C, N, and highly dispersed Co nanoparticles. EDX analysis exhibited the presence of Co, N, and C, and the absence of Zn. It's anticipated that the well-dispersed Co nanoparticles with small size in the porous carbon skeleton and high N content with large percentages of pyridinic (47%) and pyrrolic (43%) N can reduce the energy barrier of O_2 adsorption and accelerate the rate-limiting first electron transfer step of ORR; simultaneously the large surface area of Co-N-C will promote the accessibility of active sites, thus synergistically resulting in superior electrocatalytic ORR activity.

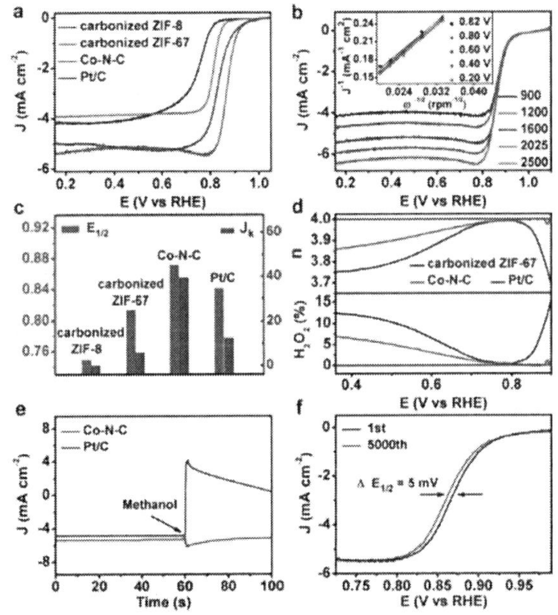

Figure 2. (a) RDE polarization curves of carbonized ZIF-8, carbonized ZIF-67, Co-N-C, and Pt/C at 10 mV s^{-1} and 1600 rpm in 0.1 M KOH. (b) RDE polarization curves of Co-N-C at different rotating speeds. Inset: K-L plots at different potentials. (c) Comparison of $E_{1/2}$ and J_k for different catalysts. (d) Electron transfer number (n) (top) and H_2O_2 yield (bottom) versus potential. (e) Chronoamperometric response of Co-N-C and Pt/C upon the addition of 2.0 M methanol at 0.70 V vs RHE. (f) RDE polarization curves of Co-N-C (near mixed-diffusion kinetic region) before and after 5000 potential cycles ranging from 1.15 to 0.15 V vs RHE.

The electrocatalytic ORR performance of Co-N-C was first studied by steady-state linear sweep voltammetry on RDE in O_2-saturated 0.1 M KOH. The best ORR activity was achieved at Zn molar content of 0.8 and carbonization temperature of 900 °C (denoted as Co-N-C for brevity), as judged by the onset (E_{onset}) and half-wave ($E_{1/2}$) potentials of the RDE polarization curves. Co-N-C exhibited significantly higher activity than those of carbonized ZIF-8 and carbonized ZIF-67 (Fig. 2a and c). More importantly, the E_{onset} of Co-N-C was almost the same as that of 20 wt % Pt/C (0.982 V vs RHE); while the $E_{1/2}$ of Co-N-C (0.871 V vs RHE) was more positive than those of Pt/C (0.841 V), indicating much higher electrocatalytic ORR activity of our Co-N-C. The Koutecky-Levich (K-L) plots from RDE polarization curves at different rotating speeds for Co-N-C (Fig. 2b) between 0.82 and 0.20 V resulted in nearly parallel fitting lines and the calculated electron transfer number (n) was from 3.84 to 4.00, suggesting a four-electron ORR pathway. Furthermore, the kinetic current density (J_k) derived from the intercept of K-L plots (Fig. 3c) was 39.3 mA cm^{-2} at 0.80 V, substantially higher than those of carbonized ZIF-8 (~ 0 mA cm^{-2}), carbonized ZIF-67 (5.8 mA cm^{-2}), and Pt/C (12.4 mA cm^{-2}). The RRDE tests also showed a high electron transfer number above 3.8 and a low H_2O_2 yield below 7.0 % for Co-N-C, both of which were superior to those of carbonized ZIF-67 and close to those of Pt/C (Fig. 2d). In addition, Co-N-C exhibited excellent methanol tolerance (Fig. 3e) and great stability (Fig. 2f). In sharp contrast, an immediate response in chronoamperometric curve (Fig. 2e) and a 25 mV loss of $E_{1/2}$ for Pt/C under the same conditions were observed.

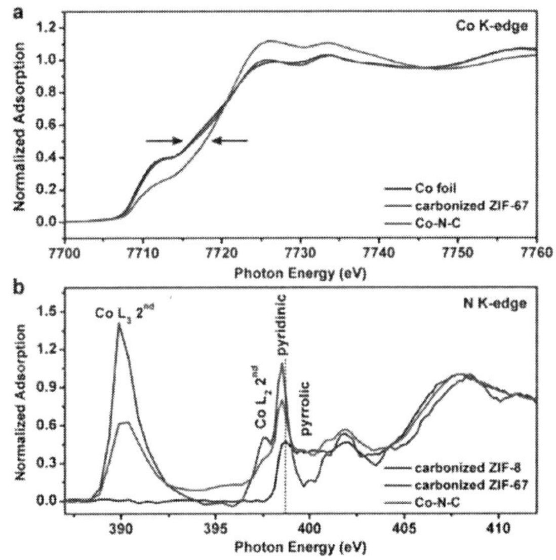

Figure 3. (a) Co K-edge XANES spectra of Co foil, carbonized ZIF-67 and Co-N-C. (b) N K-edge XANES spectra of carbonized ZIF-8, carbonized ZIF-67 and Co-N-C.

To gain deeper insights about the superior ORR activity of Co-N-C, the X-ray absorption near-edge structure (XANES) spectra of related samples were collected (Fig. 3). The drastic difference in the Co K-edge of the Co-N-C compared to cobalt oxides (CoO and Co_3O_4) unambiguously excludes the major involvement of the latter in Co-N-C (Fig. 6a). In fact, the Co K-edge of carbonized ZIF-67 is very similar to that of Co foil, indicating the metallic Co nature of those large Co particles. However, the Co K-edge of Co-N-C was shifted to higher energy relative to that of carbonized ZIF-67, implying a much stronger interaction between Co and coordinated N, which would cause more electron transfer from Co to N. Such a Co-N interaction would also lead to a negative shift of the pyridinic π^* transitions in the N K-edge, as verified by the results in Fig. 3b. It was also found that the total content of ORR active pyridinic and pyrrolic N (7.65 wt%)[6] in Co-N-C was much higher than those of carbonized ZIF-8 (6.73 wt%) and ZIF-67 (<1.1 wt%), which further promoted ORR activity. Collectively, the small Co nanoparticle size, intense interaction between Co and N, high content of ORR active pyridinic and pyrrolic N as well as large specific surface area synergistically render Co-N-C outstanding ORR activity.

Acknowledgement. We thank the support of Utah State University.

References

(1) Debe, M. K. *Nature* **2013**, *486*, 43-51.
(2) Guo, S.; Zhang, S.; Sun, S. *Angew. Chem. Int. Ed.* **2013**, *52*, 8526-8544.
(3) Wu, G.; More, K. L.; Johnston, C. M.; Zelenay, P. *Science* **2011**, *332*, 443-447.
(4) Xia, W.; Mahmood, A.; Zou, R.; Xu, Q. *Energy Environ. Sci.* **2015**, *8*, 1837-1866.
(5) You, B.; Jiang, N.; Sheng, M.; Drisdell, W. S.; Yano, J.; Sun, Y. *ACS Catal.* **2015**, *5*, 7068-7076.
(6) Leinweber, P.; Kruse, J.; Walley, F. L.; Gillespie, A.; Eckhardt, K. U.; Blyth, R.; Regier, T. J. Synchrotron Rad. 2007, 14, 500-511.

UNDERSTANDING THE ACTIVE SITE IN ELECTROCATALYSIS

Aravind Asthagiri[1]

[1]William G. Lowrie Department of Chemical & Biomolecular Engineering, The Ohio State University
asthagiri.1@osu.edu

Introduction

Electrocatalysts will play an important role in the transition from fossil fuels to a renewable and sustainable energy economy. Important applications, such as fuel cells, CO_2 utilization, and batteries, depend on improvements in electrocatalysts. Because of the complex environment (electrode-electrolyte interactions, applied potential, electron and/or proton transfer steps, and complex electrode materials) it is difficult to resolve the elementary steps and the active site responsible for the observed activity or selectivity. First-principles modeling, in particular density functional theory (DFT), has played an important role in our efforts to understand the nature of the active sites in electrocatalysts. In this talk, I will discuss three efforts in our group as examples of the role of first-principles modeling: (1) CO_2 electroreduction on Cu surfaces, (2) hydrogen evolution reaction (HER) on Mo_xS_y cubane clusters, and (3) oxygen reduction reaction (ORR) on carbon-based catalysts. While the chemistry is different in these three applications, the DFT-based electrochemical methods used to study these materials are the same and we will discuss the application of both potential dependent free energies and barriers in attempting to resolve fundamental questions on the electrocatalyst performance.

Computational Methods

To examine the electrochemical reduction steps, potential-dependent reaction free energies were determined by the linear free energy method [1] as shown in Equation (1). All the elementary steps that we examined involve a coupled proton-electron transfer step and the potential-dependent kinetic barriers were calculated using the method reported in our previous papers [2,3] as described in Equation (2).

$$\Delta G = G(\mathrm{AH^*}) - G(\mathrm{A^*}) - [\tfrac{1}{2}G(\mathrm{H_2}) - eU] \quad (1)$$

$$E_{act}(U) = E_{act}{}^{0}(U^{0}) + \beta'(U - U^{0}) \quad (2)$$

The critical feature of Eqn. 2 is that the calculated barrier is referenced to the equilibrium potential (U^o) associated with the initial state (consisting of the reactant and a H atom) of the barrier calculation. This approach ensures that the choice of high energy initial states does not lead to unrealistic small barriers, since such states will require a large equilibrium potential. While solvation can effect the stability of the intermediates and the transferring H atom, for most calculations we have used only one water molecule and for important steps this has been extended to 2 water molecules.

DFT calculations were performed with Vienna *Ab initio* Simulation Package (VASP) with the Perdew, Burke, and Ernzerhof (PBE) functional. Activation barriers for all elementary steps were calculated using the climbing image nudged elastic band (CI-NEB) approach. The minimum energy path is examined using 5-10 images, including the initial and final state, during the transition state search. Each transition state was confirmed to have a single imaginary vibrational frequency along the reaction coordinate.

Results and Discussion

CO_2 electroreduction on Cu(111) and (100). Electrochemical conversion of CO_2 to fuels is an attractive approach to store renewable but intermittent sources of energy, such as solar or wind energy, in a chemical form that can be used in existing transportation and energy infrastructure. While CO_2 electroreduction on various metal and semiconductor electrodes has been examined experimentally, only Cu has shown an ability to reduce CO_2 to hydrocarbon products at reasonable current densities and Faradaic efficiencies [4]. The primary products are methane and ethylene with no methanol observed. Electrokinetic experiments on Cu(100) and Cu(111) single crystal electrodes have shown strong facet dependence with Cu(111) producing methane and ethylene through a common pathway around 0.9 V-RHE, while Cu(100) shows, in addition to a similar high potential pathway as observed on Cu(111), a low potential pathway (~0.4 V-RHE) that produces only ethylene [5,6]. To better understand the differences in the reduction pathways on Cu(100) and Cu(111) we have used density functional theory (DFT) calculations to evaluate the potential-dependent barriers for several possible reaction paths on these two surfaces.

Figure 1 shows the pathways we have examined for Cu(100) [7].

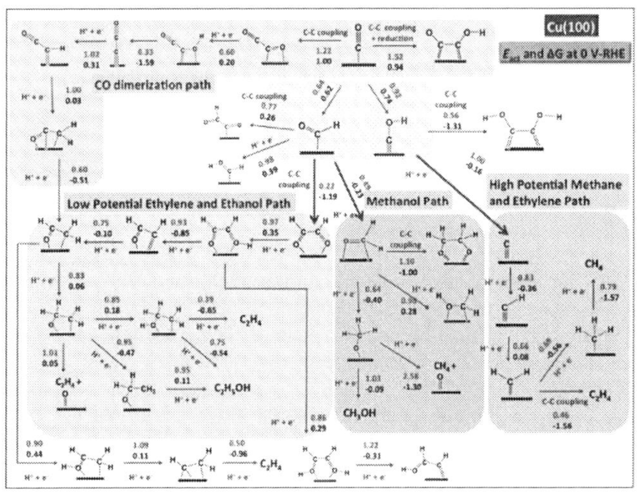

Figure 1. The reaction networks on Cu(100) examined starting with CO* with barriers (E_{act}) at 0 V-RHE given in red numbers and reaction energy (ΔG) at 0 V-RHE given in black numbers.

We find a low potential pathway that goes through CHO* coupling that occurs after initial hydrogenation of CO*. Interestingly, our earlier study on Cu(111) showed that on that facet CO* hydrogenation leads to COH* surface intermediate. We associate the pathway through COH* with the high potential pathway that produces both methane and ethylene on Cu(100) and Cu(111). We propose that this switch in pathways due to potential occurs through changes in CO* coverage. Using a simple competitive reaction model, we have found that at ~ -0.6 V-RHE that the Cu(100) surface would have a 0.5 ML coverage of CO. In the presence of that high of a CO coverage, we find that the hydrogenation of CO* proceeds via COH* instead of CHO*. This coverage effect would explain the observed potential-dependent change in CO2 reduction pathways.

Returning to the pathway through CHO*, from Figure 1 we can find pathways to C2 surface intermediates that would show a much lower barrier (and hence a lower potential) than that observed on Cu(111). Unfortunately, the reduction of C2 surface intermediates such as CHO-CHO require the breaking of C-O bonds which leads to relatively large barriers of ~ 1 eV. Despite attempts at several different pathways and the presence of additional water molecule we could not identify a low potential pathway to C2 products. Our study shows that the large barriers to any low-potential pathways will come later in the reduction path instead of the initial C-C coupling, which is very feasible for hydrogenated CO species on Cu(100). While we cannot provide any conclusive resolution of this issue, we hypothesize that experimentally reported Cu(100) reconstruction at the potentials for CO_2 reduction may play a role in providing low coordinated structures that reduce the barrier to C-O bond breaking.

HER on Mo_xS_y clusters. Hydrogen evolution electrocatalysts are critical for water splitting applications through the use of renewable electricity. MoS_2 edges have been identified as an attractive HER catalyst with performance very close to Pt. Designing molecular mimics of MoS_2 edges provides an opportunity to optimize the active site while gaining fundamental understanding of the link of the molecular structure to the HER performance.

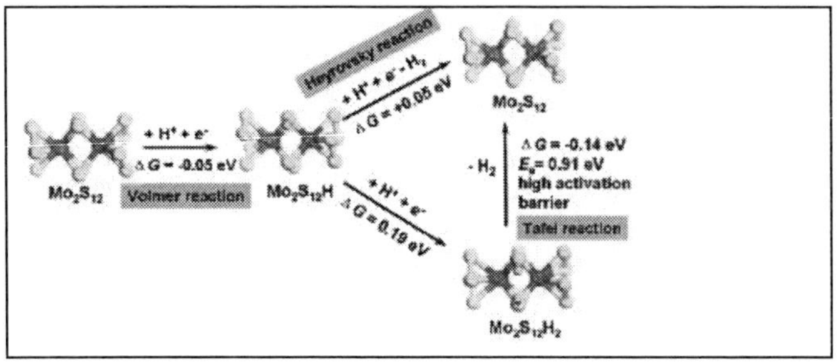

Figure 2. Reaction pathways and energies of the HER on Mo_2S_{12} with two different reaction mechanisms.

Our experimental colleagues in Prof. Yiying Wu's group at Ohio State have synthesized Mo_2S_{12} clusters and shown superior performance even to Pt and MoS_2 catalysts [8]. We have used DFT calculations to evaluate the free energy change (ΔG) of the elementary steps as shown in Fig. 2. Based on our DFT calculations we predict a limiting potential of -0.05 V-RHE and that the HER proceeds via a Volmer-Heyrovsky mechanism with the Heyrovsky step as the rate limiting step, both predictions which match experimental results well.

ORR on Carbon-based catalysts. Two classes of carbon-based catalysts, FeCN and CN_x, have been shown to have similar activity to Pt for ORR, but there is still debates about the exact structure of the active site in these materials. In collaboration with Prof. Ozkan's group at Ohio State University, we have examined CO poisoning of these two catalysts. Experimentally, they find that CO has no effect on the CNx catalysts while they observe a partial poisoning of the FeCN catalysts. We have applied DFT to examine the adsorption of CO versus O2 on a total of 15 proposed active site models. For all the CNx models we find weak adsorption of CO, which matches the experimental results. For FeCN we find variation in the CO versus O_2 adsorption strength with a few active sites showing stronger CO adsorption. In particular, the FeN4C pyridinic and FeN4C pyrrolic show stronger CO binding and these have been proposed to be strong candidates for ORR active sites. Our DFT calculations suggest that FeCN likely has a range of active sites with some of these sites being poisoned by CO. Future work on evaluating the ORR elementary steps on these sites will assist us in relating the electronic structure to the ORR activity. Furthermore, other poisons are being examined currently to see if ones that are particularly active in poisoning these two sites can be identified. Our preliminary work has shown that poisoning studies might play a key role in unraveling the range of active sites available in these materials.

Conclusions

We have applied DFT to examine the electrochemistry of three important process, CO2 reduction, HER, and ORR, with the goal of resolving the elementary mechanism and the nature of the active site. For CO2 electroreduction on Cu surfaces we have identified a high potential pathway that goes through COH* to form methane and ethylene. On Cu(100) this pathway is triggered by high coverage of CO* observed at higher potentials, but at lower potentials an alternate pathway through CHO* is found to be favored. This low potential pathway allows for C2 intermediates but we have not been able to resolve a complete pathway to the experimentally observed ethylene.

For HER on molecular Mo_xS_y clusters, we have used DFT to identify the rate limiting step on Mo_2S_{12} clusters that matches well with experimental evidence. These clusters are the smallest MoS cubanes that can produce hydrogen at near zero overpotential. Finally we have performed DFT studies of the adsorption of CO versus O_2 on a range of proposed active site models on FeCN and CNx catalysts. In comparison to experiment we confirm that CNx should show no poisoning effect due to CO, but predict that some active site models for FeCN show poisoning while others do not. We hypothesize that the experimental FeCN catalysts have a range of different active sites and this hypothesis would explain the experimental observation of some but not complete poisoning of these catalysts with CO. Future work will examine other poisons and explore the ORR mechanisms on the most promising active site models to resolve the long standing debate on the nature of the active site in these complex materials.

Acknowledgement. We acknowledge the Ohio Supercomputing Center for providing the computational resources. We also acknowledge the US Department of Energy, Office of Basic Energy Sciences, under Contract No. DE-FG02-07ER15896 for partial support for the work presented in this talk.

References

(1) Norskov, J. K.; Rossmeisl, J.; Logadottir, A.; Lindqvist, L.; Kitchin, J. R.; Bligaard, T.; Jonsson, H. *J. Phys. Chem. B* **2004**, *108*, 17886.
(2) Nie, X.; Esopi, M. R.; Janik, M. J.; Asthagiri, A. *Angew. Chem. Int. Ed.* **2013**, *52*, 2459.
(3) Nie, X.; Luo, W.; Janik, M. J.; Asthagiri, A. *J. Catal.* **2014**, *312*, 108.
(4) Hori, Y. In *Modern Aspects of Electrochemistry*; Vayenas, C. G., White, Ralph E., Gamboa-Aldeco, M. E., Eds. 2008; Vol. 42, p 89.
(5) Schouten, K. J. P.; Qin, Z. S.; Gallent, E. P.; Koper, M. T. M. *J. Am. Chem. Soc.* **2012**, *134*, 9864.
(6) Schouten, K. J. P.; Kwon, Y.; van der Ham, C. J. M.; Qin, Z.; Koper, M. T. M. *Chem. Sci.* **2011**, *2*, 1902.
(7) Luo, W.; Nie, X.; Janik, M. J.; Asthagiri, A. *ACS Catal.* **2016**, *6*, 219.
(8) Huang, Z.; Luo, W.; Ma, L.; Yu, M.; Ren X.; He, M.; Polen, S.; Click, K.; Garrett, G.; Lu, J.; Amine, K.; Hadad, C.; Chen, W.; Asthagiri, A.; Wu, Y. Angew. Chem. Int. Ed. **2015**, *54*, 15181.

METAL-MODIFIED CARBIDES AS LOW-COST AND IMPURITY-TOLERANT ELECTROCATALYSTS

Brian M. Tackett and Jingguang G. Chen

Department of Chemical Engineering
Columbia University
New York, NY 10027
jgchen@columbia.edu

Introduction

Low-cost and impurity-tolerant electrocatalysts for the hydrogen evolution reaction (HER) and hydrogen oxidation reaction (HOR) are essential for achieving widespread use of water electrolyzers for renewable energy storage and hydrogen fuel cells. In general, platinum group metals are the most active for HER/HOR, but the high cost and scarcity of such metals make them impractical for very large-scale applications. To overcome this obstacle, alternative electrocatalysts that decrease or eliminate the use of Pt are desired.

Transition metal carbides (TMCs) have gained attention recently as cost efficient electrocatalysts materials, due to their stability under a wide range of reaction conditions and Pt-like bulk electronic structure.[1,2] TMCs alone, however, are less active than Pt for the HER/HOR. This could be attributed to their stronger hydrogen binding energies (HBEs) compared to Pt. HBE has been shown to correlate with HER/HOR activity via a well-known volcano relationship, with the Pt HBE at the activity peak.[3] Density functional theory (DFT) calculations for TMCs modified with an atom-thick monolayer of Pt revealed similar HBEs to that of Pt.[4] Experimentally, TMC thin films modified with a Pt monolaryer (ML) have shown improved, Pt-like HER/HOR activity.[3] Thus, metal modified TMCs are promising electrocatalysts for HER/HOR. Now, it is necessary to apply these concepts to two industrially relevant research areas: (1) impurity tolerant catalysts, and (2) high surface area powder catalysts.

The effect of water impurities on HER electrocatalyst performance needs to be understood to enable wide-spread electrolyzer use. Deionized water is currently required for commercial membrane electrolyzers, but this is an inconvenience and an added cost.[5] Electrocatalysts should at least be resistant to common impurities found in tap water to make electrolyzers less expensive and practical for residential use. We studied the HER performance of ML Pt on niobium carbide (NbC) thin films in the presence of tap water impurity ions in both acidic and alkaline electrolytes. NbC is an attractive substrate for HER because it can be easily synthesized as a high surface area powder. Also, Pt/NbC has a HBE close to that of Pt.

Beyond impurity tolerance, the metal modified TMC concept must be extended to powder electrocatlysts. We examined the alkaline HER activity of low loadings of Pt nanoparticles dispersed on commercially available carbides, including niobium (NbC), tantalum (TaC), vanadium (VC), zirconium (ZrC), titanium (TiC), and tungsten (WC). The effect of different carbides on metal dispersion and the minimum loading of Pt on carbide supports were also explored.

Experimental

Monolayer Pt/NbC Thin Film Synthesis. NbC thin films were synthesized from Nb foil via a temperature programmed reaction technique in a horizontal quartz tube furnace under a flowing CH_4/H_2 (20% CH_4 by volume) environment with a maximum temperature of 1273K. Pt monolayers were added to the thin films via physical vapor deposition in an ultra-high vacuum chamber also equipped with X-ray photoelectron spectroscopy (XPS) equipment that was used to quantify Pt overlayer thickness.

Synthetic Tap Water Composition and Testing. Synthetic tap water was prepared by adding ions commonly found in drinking water to deionized water. The composition is described in Table 1.

Table 1. Synthetic Tap Water Composition

Ion	Concentration (mM)
Na^+	~3.5
Ca^{2+}	1.3
Mg^{2+}	0.3
Cl^-	2.9
HCO_3^-	2.5
SO_4^{2-}	0.5
NO_3^-	0.01

Acidic (0.5M H_2SO_4) and basic (0.1M KOH) electrolytes were prepared from the synthetic tap water and used for electrochemical testing in a standard 3-electrode, single cell setup. Impurity tolerance of Pt/NbC was measured by calculating the electrochemical activity in the presence of tap water ions and comparing it to the activity in pure electrolytes. Electrochemical activity was described by exchange current density, which was calculated from a Tafel analysis of electrochemical linear scan voltammetry (LSV).

Pt on TMC Powder Synthesis and Testing. Pt on commercial TMC powders with a weight loading of 5% and 1% were synthesized via a wet impregnation method. Synthesized powders were dispersed in liquid solution, and specific quantity was dropped onto a glassy carbon rotating disk electrode. Samples were then tested electrochemically in a standard 3-electrode, single cell with pure alkaline (0.1M KOH) electrolyte. Activity was characterized by the exchange current density calculated either from Tafel analysis or from fitting LSV curves to the Butler-Volmer equation. Roughness factor calculations for each electrode were obtained by an electrochemical copper stripping technique.

Results and Discussion

Figure 1a shows the electrochemical activity (exchange current density, i_o) plotted for various coverage of Pt on NbC thin films in pure acidic electrolyte and in acidic electrolyte with tap water impurity ions.[6] The HER activity increased with increasing coverage of Pt, until a plateau was reached around 2.5 ML. The $\log_{10}(i_o)$ value of -3.2 was similar to bulk Pt and matched well with previous measurements. The activity of 1ML Pt/NbC in the presence of tap water ions (red circle) showed similar activity to its pure water counterpart in the acidic electrolyte.

Figure 2b shows the same analysis in alkaline electrolyte. Again, the catalyst activity increased as Pt coverage increased, reaching a plateau near the activity of bulk Pt around 2 ML. The 1 ML Pt/NbC sample also appeared to be unaffected by tap water impurities in the alkaline environment.

These results showed that Pt/NbC was capable of reducing the amount of Pt required for the HER in acid and alkaline electrolysis without sacrificing activity. This also confirmed that DFT calculated hydrogen binding energy, which predicted high activity for ML Pt/NbC, is a useful predictor of HER activity for metal modified TMCs. XPS measurements before and after electrochemical testing also confirmed that the Pt overlayer was stable for HER at both high and low pH. Pt/NbC also appeared to be impurity tolerant to ions commonly present in tap water. Resistance to impurity ions is likely due to a low ion binding energy on the Pt/NbC surface. This would allow hydrogen to preferentially adsorb to the electrocatalyst surface at low potentials, permitting the HER to proceed uninhibited. Thus, future exploration of impurity tolerant HER catalysts may benefit from DFT screening of impurity binding energies.

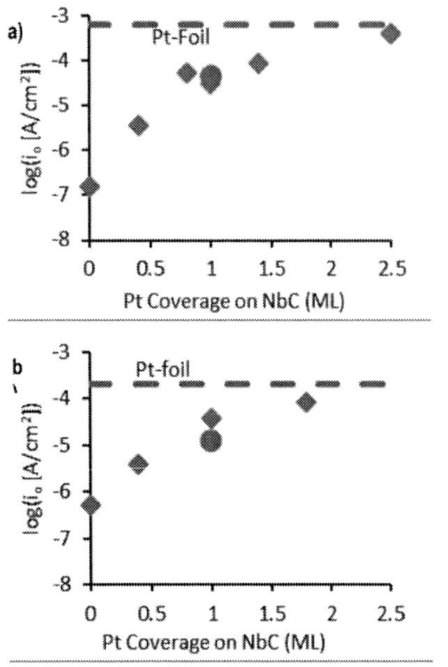

Figure 1. Activity ($\log_{10}(i_0)$) vs. Pt overlayer coverage plots for Pt/NbC in pure water (blue diamond) and synthetic tap water (red circle) in 0.5M H_2SO_4 (a) and 0.1M KOH (b). From Reference [6].

Next, HER activity on industrially relevant Pt on TMC powders were explored. Figure 2 shows the alkaline HER/HOR LSV curves for 5% Pt on each of the six carbide powder catalysts, as well as commercial 46% Pt/C for reference.[7] The HOR branches of the curves were corrected for H_2 mass transfer by using the Koutecky-Levich equation. 5% Pt on VC, TiC, and WC all showed Pt-like activity, while the other 3 samples were much less active. The roughness factor calculations, which describe the electrochemical surface area of each catalyst, revealed that the three samples with low activity (5% Pt on NbC, ZrC and TaC) had much lower roughness factors than the active samples. The normalized exchange current densities for each sample were, in fact, all similar to that of Pt.

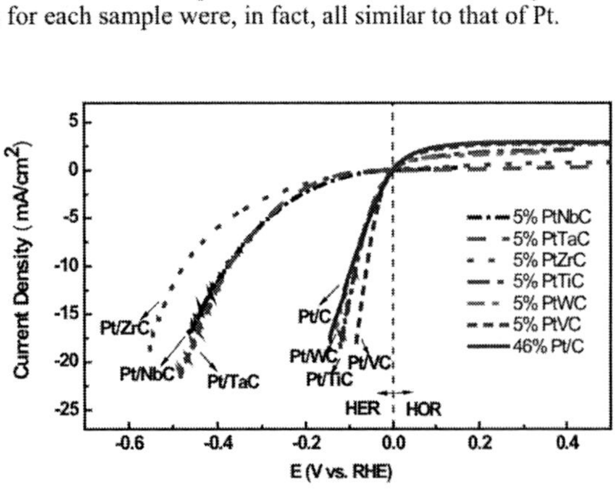

Figure 2. Current densities of each 5% Pt/TMC catalyst compared to 46% Pt/C (blue full line) in 0.5M KOH. From Reference [7].

This showed that HER kinetic activity of Pt nanoparticles on TMC powder did not change with respect to the TMC substrate. However, the TMC drastically impacted the dispersion of metal nanoparticles, which impacted the overall electrocatalyst mass activity. VC was found to promote the best Pt nanoparticle dispersion. This was confirmed by roughness factor calculations and TEM imaging. It was also found that 1% by mass was the minimum loading of Pt on VC that maintained Pt-like alkaline HER/HOR activity.

Conclusions

In order to develop inexpensive and impurity resistant electrocatalysts for water electrolysis and hydrogen fuel cells, metal modified TMC thin films and powders were synthesized and tested electrochemically. ML Pt on NbC thin films show Pt-like activity in both acid and alkaline environments and are tolerant of common impurity ions present in tap water. This makes Pt/NbC a promising candidate to reduce the amount of Pt required for electrolysis cells. Low loadings of Pt nanoparticles dispersed on TMC powders via wet impregnation have Pt-like activity in alkaline conditions. While Pt on all tested TMC substrates show similar exchange current densities, the choice of TMC drastically effects the Pt dispersion and consequent mass activity of the electrocatalyst. VC in particular shows good ability to disperse Pt nanoparticles, and Pt/VC maintains Pt-like HER/HOR activity with as little as 1% Pt by mass. The impurity tolerance and high activity of the inexpensive metal modified TMC electrocatalysts presented here make them attractive options to replace bulk Pt in hydrogen fuel cells and water electrolyzers. Future work should focus on analyzing the impurity tolerance of metal modified TMC powders and optimizing metal nanoparticle dispersion on TMCs.

Acknowledgement. This work is supported by the US Department of Energy (DE-FG02-13ER16381).

References

(1) Levy, R. B.; Boudart, M. *Science* **1973**, *181* (4099), 547.
(2) Kelly, T. G; Chen, J. G. *Chem. Soc. Rev.* **2012**, *41*, 8021.
(3) Nørskov, J. K.; Bligaard, T.; Logadottir, A.; Kitchin, J. R.; Chen, J. G.; Pandelov, S.; Stimming, U. *J. Electrochem. Soc.* **2005**, *152* (3), J23.
(4) Esposito, D. V.; Chen, J. G. *Energy Environ. Sci.* **2011**, *4* (10), 3900.
(5) Lehner, M.; Tichler, R.; Steinmüller, H.; Koppe, M. *Power-to-Gas: Technology and Business Models*; SpringerBriefs in Energy; Springer International Publishing: Cham, 2014.
(6) Tackett, B. M.; Kimmel, Y. C.; Chen, J. G. *Int. J. Hydrogen Energy*, **2016**, DOI: 10.1016/j.ijhydene.2016.01.167.
(7) Wang, L.; Mahoney, E. G.; Zhao, S.; Yang, B.; Chen, J.G. *Chem. Comm.* **2016**, *52*, 3697.

DURABILITY IMPROVEMENT THROUGH GRADIENT CATHODE CATLYST LAYER

Radenka Maric[1,4]
Haoran Yu[1], *Andrea Baricci*[2], *Laure Guetaz*[3], *Andrea Casalegno*[2],

[1] Department of Chemical and Biomolecular Engineering, University of Connecticut, 191 Auditorium Road, Unit 3222, Storrs, CT 06269
[2] Department of Energy, Politecnico di Milano,
[3]Grenoble Electron Microscopy @ Minatec, CEA-Grenoble
[4]Department of Material Science and Engineering, University of Connecticut
maric@engr.uconn.edu

Introduction

A goal critical to achieving current automotive commercialization targets for proton exchange membrane fuel cells, PEMFC, is improved durability of the catalyst supports. Typical carbon blacks corrode easily under start-stop conditions when the cathode experiences high transient potentials approaching 1.4 V [1]. The structural change of the catalyst layer can cause changes in the PEM fuel cell performance due to carbon corrosion and loss of catalyst utilization, electronic/protonic conductivity, and changes in the mass transport of the gases.

In this study, we used Reactive Spray Deposition Technology, RSDT, to deposit catalyst-coated membranes (CCMs) with Pt supported on Ketjen Black (KB)) to elucidate the effects of cathode catalyst particle size, dispersion, support physicochemical properties, and cell operating conditions on the catalyst performance. RSDT has proven to be a *cost-effective method for producing high performance PEMFC membrane electrode* assemblies (MEAs) for both low temperature (<100 °C) [2,3] and high temperature (>160 °C) operations [4,5]. RSDT is a flame-based method that combines catalyst synthesis and processing in a single step and has the flexibility in optimizing the catalyst composition, making gradient and other complex structures possible. This has allowed our team to reduce the cathode Pt loading below 0.1 mg/cm^2 with enhanced performance.

RSDT allows the stable anchoring of the separate catalyst particles due to the high velocity of the particles coming from the flame and high crystallinity of the particles. The particles' size, shape, and ionomer content can be optimized on the fly to have the highest possible adhesion on the support. In this study we controlled the particle sizes and deposited a gradient layer with a 3 µm thick layer of 5 nm particles toward the Nafion® membrane and 7 µm thick layer of 2 nm particles toward GDL.

Electrochemical measurements were made in both the rotating disk electrode (RDE) and membrane electrode assembly (MEA) environments to evaluate the activity for oxygen reduction reaction (ORR) and PEMFC performance and durability.

Experimental

RSDT. The Pt precursor (Pt acetylacetonate, Colonial Metals. Inc) was dissolved in a combination of 75 wt. % xylene (Sigma Aldrich, ACS reagent, ≥ 98.5 %) and 25 wt. % acetone (Sigma Aldrich, HPLC ≥ 99.9 %) at 10 mM Pt concentration. Liquid propane was added to the Pt precursor solutions prior to deposition until it comprised 15% of the solvent mass (xylene+acetone+propane) in order to aid atomization and increase the combustion heat during deposition. The support, Ketjen Black (KB) EC-600JD (AzkoNobel), was dispersed in methanol at a concentration of 2.5 mg mL^{-1}, and Nafion® (5 wt.% 1100 EW, Ion Power, Inc.) was used as the binder. The I/C ratio used in this study was optimized based on previous test results. For Pt/KB, the ionomer/carbon weight ratio (I/C ratio) was kept at 0.5 for anode and 1.0 for cathode. The Pt/C electrodes were deposited onto N212 membrane (50 µm thickness, Ion power, Inc.,) with 5x5 cm^2 active area. The air quench, at flow rate of 70-75 L min-1, was used to rapidly cool the combustion zone and to maintain the deposition temperature between 100 and 110 °C.

Characterization Methods. The surface morphology and catalyst layer thickness were characterized using a FEI Quanta FEG 250 scanning electron microscopy (SEM) with a field emission source and Everhart-Thornley (secondary electron) detector. The size and distribution of Pt nanoparticles on the freshly prepared supports was revealed by TEM images. Figure 1a shows the HADDF image of The XEDS mapping of the Pt/KB agglomerate (Figure 1a) shows the spatial relationship between Pt and uniform distribution of fluorine on carbon(Figure 1b) .

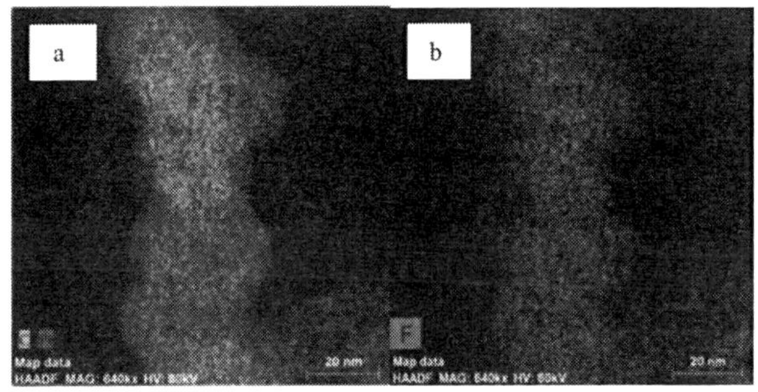

Figure 1. HAADF images: 80 kV, FEI Metrios, STEM TEM BF: FEI Tecnai T12, 120 kV, a) Pt supported on Ketjen Black b) Fluorine mapping.

Rotating disk electrodes. Glassy carbon electrodes were directly deposited through the RSDT process onto glassy carbon for 30 min and the estimated Pt loading was 20 µg cm^{-2}. Cyclic voltammograms (CV) were collected in a nitrogen-purged cell between 0.05 V and 1.2 V (vs. RHE) at a scan rate of 20 mV s-1 for 30 cycles. The polarization curves for ORR were recorded at a scan rate of 10 mV s^{-1} from 0.05 V to 1.05 V (vs. RHE) under 1600 RPM rotation.

MEA testing The CCMs were assembled with SGL-25BC gas diffusion layers and a pair of graphite flow fields with triple serpentine channels on the cathode and single serpentine channels on the anode. The active area was 25 cm2. The cell was conditioned overnight at 0.6 V in 100% humidified H2/Air (0.2 L min-1/0.2 L min-1) at a stoic of 2. The 30,000 potential cycles (H_2/N_2 50/50 mL min^{-1}, 0.6-1.0 V, 50 mV/s, 80°C, single cell 25 cm^2) which includes various potential ranges between 0.6 to 1.0 V vs. RHE.

Results and Discussion

In this study, we used the RSDT process to deposit catalyst on a membrane to make a catalyst –coated membrane (CCM) with Pt supported on Ketjen Black (KB). Pt/C electrocatalyst with mean particle sizes ranging from 2 nm to 5 nm and Pt loading of 0.1 mg/cm^2 have been successfully prepared and evaluated in situ as cathodic electrocatalyst in PEMFC. SEM cross-section (cryofracture) images of the MEA after 30,000 cycles is shown in Figure. 2. It is evident that there is no delamination of either anode or cathode layer, and a good integrity after cycling, however, a Pt band was observed in the membrane near the cathode side. This indicates that Pt underwent dissolution and re-deposition during cycling, see Figure 3. Pt on the anode remained small with an average particle size 1.5-2.5 nm.

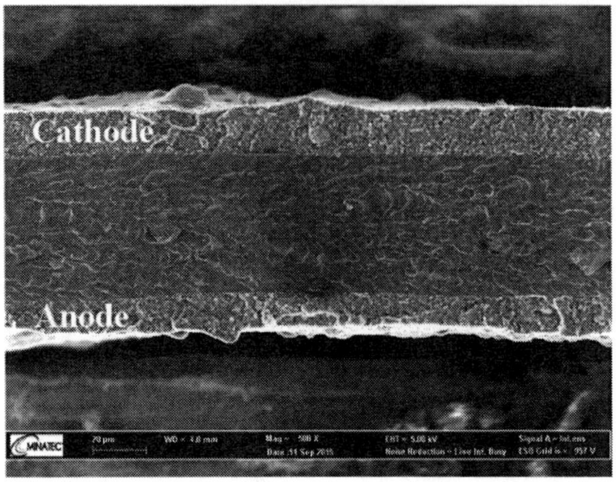

Figure 2. SEM image of Catalyst: PtKB; Loading: Anode/cathode, 0.05/0.1 mg cm^{-2}

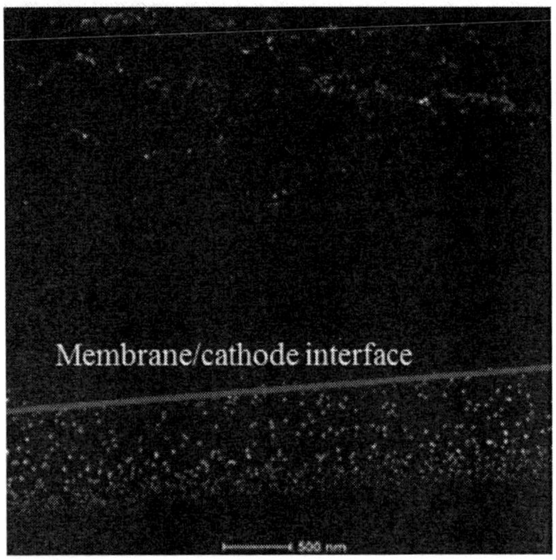

Figure 3. HAADF/STEM image of an ultramicrotome sample.

As shown in Figure 4 the Pt nanoparticles showed a clear particle size effect on both the oxygen reduction activity and the electrochemical stability for the electrocatalyst. At the beginning of testing (BOL), Figure 4a, the smaller particle size of 2 nm has much higher electrochemically active surface area of 75 m^2/g_{Pt} up to 100 cycles, however at the end of life (EOL), after 30,000 cycles, all samples tend to a similar final value of electrochemically active surface area ECSA of 15 m^2/g_{Pt}. In additional testing, when we deposit a gradient catalyst layer with larger particles closer to interface between membrane and the catalyst layer and smaller particles toward the gas diffusion layer, while still keeping the catalyst loading constant, a significant improvement in stability was obtained. For this configuration, the final value of ECSA was 50% after 30,000 cycles. After 30k cycles, 5 nm and 2+5 nm double layer increase significantly, 3 nm double layer slightly increases, while 2 nm double layer remains quite the same.

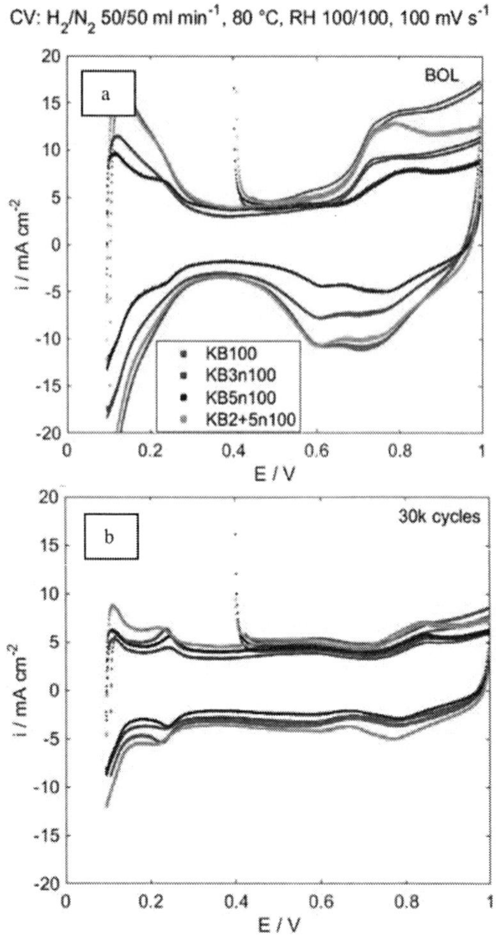

Figure 4 Cyclic voltametry (a), ORR polarization curve at the beginning of life, b) after 30 k cycles.

Conclusions

The degradation of the studied MEA is mostly due to cathode degradation. For nanoparticles averaged 2, 3, and 5 nm in the cathode layer, 80% loss of ECSA is observed after 30k cycles (0.6-1.0V). By applying a gradient layer with larger particles, 5nm, toward membrane and smaller particles 2 nm toward GDL we observed less dissolution of Pt in membrane and higher stability.

Acknowledgment. This work was supported by the National Science Foundation (award number CMMI-1265893).

References

(1) Reiser, C.A; Bregoli, L.; Patterson, T.W. Yi, J.S. Yang, J.D.; Perry, M.L.;Jarvi, T.D *Electrochem. Solid-State Lett.* **2005**, *8*, A273

(2) Roller, J.M; Arellano-Jimenez, M.J. ;. Yu, H ; Jain, R ; Carter, C.B. ; Maric, R. Electrochim. Acta **2013**, 107, 632.

(3) Yu, H. ;. Roller, J.M; Mustain, W.E. ; Maric, R. J Power Sources **2015**, 283 84.

(4) Yu, H. ;. Roller, J.M; Kim, S. ; Wang, Y.;. Kwak, D.; Maric, R. J. Electrochem. Soc. **2014**, 161 F622.

(5) Kim, S. ;Myles, T.D.; Kunz, H.R. ; Kwak, D. ; Wang, Y. ; Maric, R. Electrochim. Acta ,2015, 177,190.

NOVEL ELECTRODE MATERIALS FOR ENERGY CONVERSION AND STORAGE DEVICES

Yun Hang Hu

Department of Materials Science and Engineering, Michigan Technological University, 1400 Townsend Drive, Houghton, MI 49931-1295, USA
Email: yunhangh@mtu.edu

Electrode materials play a critical role in energy conversion and storage devices. In this talk, the design of novel electrode nanomaterials will be discussed, and their successful synthesis will be demonstrated. Furthermore, their excellent performances will be presented for various devices, such as solar cells, supercapacitors, fuel cells, and capacitive deionization cells.[1-7]

References

(1) Wang, H.; Hu, Y. H. Graphene as a counter electrode material for dye-sensitized solar cells, *Energy Environ. Sci.*, **2012**, *5*, 8182.
(2) Wang, H.; Sun, K.; Tao, F.; Stacchiola, D.; Hu, Y. H. 3D Honeycomb-Like Structured Graphene and Its High Efficiency as Counter Electrode Catalysts for Dye-sensitized Solar Cells, *Angew. Chem. Int. Ed.* **2013**, *52*, 9210.
(3) Wei, W.; Sun, K.; Hu, Y. H. Synthesis of 3D Cauliflower-fungus-like Graphene from CO_2 as Highly Efficient Counter Electrode Material for Dye-sensitized Solar Cells, *J. Mater. Chem. A* **2014**, 2, 16842.
(4) Chang, L.; Wei, W.; Sun, K.; Hu, Y. H. 3D flower-structure graphene from CO_2 for Supercapacitors with ultrahigh capacitance at high current density, *J. Mater. Chem. A*, **2015**, 3, 10183.
(5) Qi, J.; Benipal, N.; Wang, H.; Chadderdon, D.; Jiang, Y.; Wei, W.; Hu, Y. H., and Li, W. Metal-Free Anion Exchange Membrane Fuel Cells with 3D Graphene as Anode Catalyst, *ChemSusChem* **2015**, 8, 1147.
(6) Wei, W.; Wang, H.; Hu, Y. H. Unusual particle-size-induced promoter-to-poison transition of ZrN in counter electrodes for dye-sensitized solar cells, *J. Mater. Chem. A*, **2013**, 1, 14350.
(7) Wang, H.; Wei, W.; Hu, Y. H. Efficient ZnO-based counter electrodes for dye-sensitized solar cells, *J. Mater. Chem. A*, **2013**, *1*, 6622.

WELL-DEFINED NANOSTRUCTURED METAL OXIDE ELECTROCATALYSTS FOR PHOTOELECTROCATALYTIC WATER SPLITTING

Zhu Chen and Bruce E. Koel

Department of Chemical and Biological Engineering,
Princeton University, Princeton, NJ 08544
bkoel@princeton.edu

Introduction

Progress toward a fundamental understanding of relationships between the materials properties of photoelectrocatalysts and their performance has been inhibited by the inherent inhomogeneity of practical heterogeneous photoelectrocatalysts that causes difficulty in general but in particular for identifying the most active and selective sites for reactions at the surface. Adopting a rational approach to aid the search for high performance photoelectrocatalysts, our emphasis is on synthesizing surface-science based model photoelectrocatalysts with well-controlled surface composition and surface structure, and on the synthesis of well-defined metal oxide nanomaterials. Such systems enable important comparisons and validation and extensive exploration of a variety of key factors that impact catalytic properties. For example, photoelectrochemical water oxidation on metal oxides has been extensively studied, yet the relationship between the various facets exposed, heteroatom doping, and associated electrocatalytic activity has not been adequately explored in many cases. Our research is currently investigating pure and modified transition metal oxide and oxyhydroxide photoelectrocatalysts of interest for photoelectrocatalytic water splitting and establishing clear structure-activity relationships in oxygen evolution reaction (OER) electrocatalysis.

Experimental Methods

This research utilized carefully tailored hydrothermal synthesis methods to obtain well-defined nanostructured catalysts and evaluated their catalytic performance in an electrochemical half cell setup. The nanostructured catalysts reported herein were characterized by scanning and transmission electron microscopy (SEM and TEM), X-ray diffraction (XRD), BET surface area analysis, Raman spectroscopy, and X-ray photoelectron spectroscopy (XPS). A number of electrochemical evaluations, including cyclic voltammetry (CV), rotating ring disc electrode (RRDE) voltammetry, and electrochemical impedance spectroscopy (EIS), were used to investigate catalyst performance in alkaline conditions for the O_2 evolution reaction.

Investigations reported herein concern three well-defined nanostructured catalysts: (i) pure and Ni-doped α-Fe_2O_3 nanoplates with different proportions of basal plane exposure; (ii) faceted Co_3O_4 nanocrystals; and (iii) pure and M-modified CoOOH nanowires (M=Ni or Mn). These model electrocatalysts were synthesized, characterized, and evaluated for their OER catalytic activity.

Pure and Ni-doped α-Fe_2O_3 nanoplates as electrocatalysts for OER.[1] Hematite (α-Fe_2O_3) nanocrystals were synthesized with continuous tuning of the aspect-ratio and fine control of the surface area ratio of the (0001) facet with respect to other surfaces. Nickel was used to dope these samples by diffusion of Ni clusters into the subsurface region, which forms a uniformly doped $Ni_xFe_{2-x}O_3$ surface overlayer that was confirmed using the combined results of HRTEM, SEM, XRD, Raman, BET, and XPS measurements. Electrochemical testing of water oxidation catalysis was carried out on the doped and shape-controlled hematite nanocrystals. Nickel doping improved the electrocatalytic activity of hematite for water oxidation, with the enhancement increased as the surface area ratio of the (0001) facet of hematite nanocrystals increased, showing a preferential activation of the (0001) surfaces for water oxidation by Ni doping.

Well-defined Co_3O_4 nanocrystals as electrocatalysts for OER.[2] Our studies found that the OER activity of Co_3O_4 nanocrystals increased with increasing proportion of the exposed (111) surface. In particular, for Co_3O_4 nanooctahedra where (111) surfaces are dominant, 45% greater current density was obtained compared to that for nanocubes where the (100) surface is the predominant facet. CV data show more Co^{3+} ions on the nanooctahedra compared to the nanocubes, which could be one of the reasons for greater OER activity. Although not probed in this work, the local coordination geometry of Co ions also greatly influences OER activity. Slight catalyst degradation was observed for nanooctahedra and nanocubes despite no change in the catalyst morphology or chemical composition, which might indicate the formation of surface amorphous layers. Results from this study illustrate the importance of surface crystal facets in optimization of OER activity for Co_3O_4 catalysts in alkaline conditions.

Pure and M-modified CoOOH (M=Ni or Mn) nanowires as electrocatalysts for OER.[3] Cobalt oxyhydroxide (CoOOH) has been shown to be the equilibrium phase during OER for cobalt-based catalysts, and have demonstrated high OER activity. In this study, we synthesized higher surface area CoOOH nanowires directly on a conductive substrate and explored their surface modification with Ni and Mn. Ni modification significantly improves the OER activity of CoOOH, resulting in excellent OER activity comparable to that of iridium oxide-based catalysts. In contrast, Mn modification did not alter the OER performance of CoOOH. We propose that Ni modification changes the rate determining step of OER, and it also stabilizes the surface species more so compared to pure and Mn modified CoOOH.

Conclusions

In summary, we have utilized carefully tailored hydrothermal synthesis methods to obtain well-defined nanostructured metal oxide and oxyhydroxide materials and evaluated their electrocatalytic performance in alkaline conditions for the O_2 evolution reaction (OER). In studies of pure and Ni-doped α-Fe_2O_3 nanoplates with different proportions of basal plane exposure, we found Ni doping improved the electrocatalytic activity of hematite for OER, with a preferential activation of the (0001) surfaces. For faceted Co_3O_4 nanocrystals we found that Co_3O_4 nanooctahedra where (111) surfaces are dominant vastly outperform nanocubes where (100) surfaces are predominant for OER. In investigations of pure and M-modified CoOOH nanowires (M=Ni or Mn), we found Ni modification significantly improved the OER activity of CoOOH, in contrast to Mn modification that did not alter the activity.

Acknowledgement. This material is based upon work supported by the National Science Foundation under Grant No. CHE-1465082.

References

(1) Zhao, P., Wu, F., Kronawitter, C. X., Chen, Z., Yao, N., and Koel, B. E., *Phys. Chem. Chem. Phys.*, **17**, 26797–26803 (2015).
(2) Chen, Z., Kronawitter, C. X., and Koel, B. E., *Phys. Chem. Chem. Phys.*, **17**, 29387-29393 (2015).
(3) Chen, Z., Kronawitter, C. X., Yeh, Y.-W., Yang, X., Zhao, P., Yao, N., and Koel, B. E., *ACS Catalysis*, submitted.

NANOSTRUCTURED THIN-FILM ELECTROCATALYSTS FOR WATER SPLITTING

Yang Yang

Nanoscience Technology Center, Department of Materials Science and Engineering, University of Central Florida, Orlando, FL, US

Abstract

Porous materials with sufficiently exposed surface associated with abundant active sites for electrochemical reactions have been identified as efficient catalysts for water splitting in which hydrogen evolution reaction (HER) and oxygen evolution reaction (OER) are involved. Conventional porous catalyst powders may deactivate during long-term utilization when carbonaceous additives are added to prepare electrodes for water electrolysis. Here, we developed a facile and simple route to prepare self-organized metal compound porous films without any additives as electrocatalysts for water splitting. The exposed catalytically active sites and interconnected porous structure makes these porous films deliver excellent HER and OER activities with long-term durability. This nanoscale engineering fabrication approach to tailor the catalytic activity of thin-films opens a new direction in designing efficient catalysts.

CRYSTAL PHASE-CONTROLLED SYNTHESIS OF NOVEL NOBLE METAL NANOMATERIALS

Hua Zhang

School of Materials Science and Engineering, Nanyang Technological University, 50 Nanyang Avenue, Singapore 639798, Singapore
HZhang@ntu.edu.sg

In this talk, I will summarize the recent research on the crystal phase-controlled synthesis of novel noble metal nanomaterials in my group. It includes the first-time synthesis of hexagonal-close packed (*hcp*) Au nanosheets (AuSSs) on graphene oxide,[1] surface-induced phase transformation of AuSSs from *hcp* to face-centered cubic (*fcc*) structures,[2] alternating *hcp*/*fcc* Au square-like plates from AuSSs,[3] ultrathin Au nanowires containing *hcp* phase,[4] synthesis of ultrathin *fcc* Au@Pt and Au@Pd rhombic nanoplates through the epitaxial growth of Pt and Pd on the *hcp* AuSSs, respectively,[5] the first-time synthesis of 4H hexagonal phase Au nanoribbons (NRBs) and their phase transformation to *fcc* Au RNBs, the epitaxial growth of Ag, Pt, Pd, PtAg, PdAg, PtPdAg on 4H Au NRBs to form the 4H/*fcc* Au@metal core–shell NRBs,[6] and the synthesis of 4H/*fcc*-Au@metal sulfide core-shell NRB heterostructures.[7] In addition, the concept of crystal-phase noble metal heterostructures is proposed.[8]

References

(1) X. Huang, et al., *Nat. Commun.* **2011**, *2*, 292.
(2) Z. X. Fan, et al., *Nat. Commun.*, **2015**, *6*, 6571.
(3) X. Huang, et al., *Angew. Chem. Int. Ed.* **2011**, *50*, 12245.
(4) X. Huang, et al., *Adv. Mater.* **2012**, *24*, 979.
(5) Z. X. Fan, et al., *Angew. Chem. Int. Ed.*, **2015**, *54*, 5672.
(6) (a) Z. X. Fan, et al., *Nat. Commun.*, **2015**, *6*, 7684. (b) Z. X. Fan, et al., *J. Am. Chem. Soc.*, **2016**, *138*, 1414.
(7) Z. X. Fan, et al., *J. Am. Chem. Soc.*, **2015**, *137*, 10910.
(8) H. Zhang, *ACS Nano*, **2015**, *9*, 9451.

NOVEL NANOSTRUCTURED MATERIALS FOR ENERGY APPLICATIONS

Feng Jiao

Center for Catalytic Science and Technology, Department of Chemical and Biomolecular Engineering, University of Delaware, Newark, DE 19716, USA
jiao@udel.edu

Introduction

Nanoporous metal-based solids are of particular interests, because they combine very large internal surface areas, interconnected porous networks, nano-sized crystalline walls, and open d-shells, leading to unique physical and chemical properties for a variety of applications. Among all the synthetic methods, nanocasting is the most widely method to synthesize nanoporous metal oxides with ordered porous structures. A silica material with a three-dimensional porous structure is often employed as a hard template, in which empty pores are filled with an inorganic precursor through impregnation, followed by a thermal conversion to form the desired phase and a template removal process using a strong base or a hydrofluoric acid solution. While nanocasting is very effective for the synthesis of nanoporous metal oxides, it is difficult to prepare metal-based materials beyond simple metal oxides.

In this presentation, our recent work on nanoporous metals, hierarchical porous metal oxides, nanoporous metal sulfides and carbides will be used as examples to facilitate the discussion of new synthetic strategies beyond traditional nanocasting approach. Leaching of aluminum from Al-rich metal alloys is able to produce both monometallic and bimetallic materials with highly porous structures. Combining nanoporosity and highly curved internal surface, the as-made nanoporous metals exhibited exceptional electrocatalytic properties in a variety of reactions, including electrochemical carbon dioxide reduction and hydrogen evolution from water. The selective leaching concept was also successfully applied to the fabrication of hierarchical nanoporous cobalt oxide. A nanoporous Mg-doped Co_3O_4 was first prepared using a traditional nanocasting approach, followed by a leaching process of Mg in an acidic environment. The selective dissolution of Mg resulted in a hierarchical porous structure with an ultrahigh surface area (~250 $m^2 g^{-1}$). In addition to the selective leaching strategy, a general synthetic method for nanoporous metal sulfides was developed by combining nanocasting and "oxide-to-sulfide" transformation to address the large volume change issue associated with the synthesis of nanoporous metal sulfide using a nanocasting route. Three nanoporous metal sulfides, i.e., FeS_2, CoS_2, and NiS_2, have been prepared, demonstrating the feasibility and generality of this new method. Further attempts to synthesize nanoporous metal carbides using the post-nanocasting transformation idea were unsuccessful, mainly due to the slow diffusion of carbon during the carbonization process. To bypass the technical challenge, an alternative strategy was introduced based on an *in situ* carbonization treatment using methane as both carbon source and reductant, leading to the successful preparations of two nanoporous metal carbides with ordered porous structures.

Results and Discussion

Nanoporous Ag. Reduction in greenhouse CO_2 emissions from fossil fuel utilization is critical for human society. Ideally, one would like to convert CO_2 produced in power plants, refineries and petrochemical plants to fuels or other chemicals through renewable energy utilization. This desired solution imposes major technological challenges because CO_2 is a fully oxidized and thermodynamically stable molecule. A suitable catalyst for CO_2 reduction is essential to achieve a cost effective process with high efficiency and selectivity. In the past two decades, electrocatalytic CO_2 reduction has attracted much attention because the required electricity may be obtained at a low cost from renewable energy sources, such as wind, solar, and wave. While researchers have identified several potential catalysts that are able to reduce CO_2 electrochemically in aqueous electrolytes, Searching for abundant catalysts with high selectivity is crucial for commercializing CO_2 reduction processes by reducing costs associated with catalyst fabrication and product separation.

The selective conversion of CO_2 to CO is a promising route for clean energy. The CO product can be used as feedstock in the Fischer-Tropsch process, a well-known and well-characterized process that has been used in industry to produce chemicals and synthetic fuels from syngas (CO + H_2) for many decades. By coupling the catalytic reduction of CO_2 to CO with the Fischer-Tropsch process to produce synthetic fuels and industrial chemicals, the estimated maximum reduction of atmospheric CO_2 emissions is 40%.

Silver is an attractive CO_2 reduction electrocatalyst, because it is able to reduce CO_2 to CO with a good selectivity (~81%) and it also costs much less than other precious metal catalysts. Recently, a nanoporous silver catalyst was discovered (Figure 1), which is able to reduce CO_2 electrochemically to CO in a highly efficient and selective way.[1] Not only the porous structure creates an extremely large surface area for catalytic reaction, but also the curved internal surface generates a large number of highly active step sites for CO_2 conversion, resulting in an exceptional activity that is over three orders of magnitude higher than that of the polycrystalline counterpart at a moderate overpotential of < 500 mV. More importantly, such a remarkable activity for CO_2 electroreduction has been achieved with a CO Faradaic efficiency of 92%.

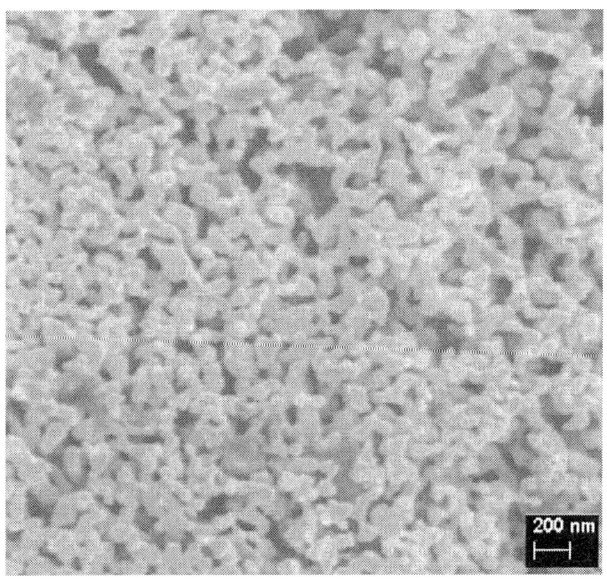

Figure 1. A typical scanning electron microscopy image of nanoporous Ag.

Nanostructured Zn. While nanoporous Ag exhibited an excellent performance in CO_2 reduction, the rarity and cost of Ag is a significant factor that must be taken into considered when designing catalysts for commercial purposes. Ideally one would also like to develop catalysts using less noble materials. Therefore we have also recently focused our efforts on developing new catalysts using less noble transition metals such as Zn.

Figure 2. A typical scanning electron microscopy image of Zn dendrites.

Electrochemical deposition has shown promise as a method of synthesizing highly active nanostructured CO_2 reduction catalysts. In the case of Zn, the rate of deposition has been shown to have a dramatic effect on catalyst morphology yielding various types of nanostructured materials. Taking advantage of this technique, we were able to develop an electrodeposited Zn catalyst with dendrite morphology able to selectively reduce CO_2 at rates around an order of magnitude higher than bulk Zn with great long term stability (Figure 2).[2] It is likely higher surface area and defect sites formed in the electrochemical deposition of Zn show improved CO_2 reduction activity and tolerance to the reaction conditions. Using in-situ x-ray absorption spectroscopy we are able to characterize the behavior of the Zn catalyst during CO_2* electrolysis giving further insight into the oxidation state and coordination of the dendritic Zn catalyst.

Nanoporous CuTi. Bridging the gap between noble and non-noble metal catalysts for hydrogen-evolution reaction (HER) is a critical step for achieving hydrogen economy for clean and sustainable energy future. Recently we found that an appropriate arrangement of copper-titanium (Cu-Ti) atoms produces highly active surface with a high density of reaction sites exhibiting an optimum hydrogen binding energy similar to that of platinum (Pt).[3] A non-noble element Cu-Ti bimetal catalyst with such surface was designed with monolithic and bimodal pore structure to achieve over doubled improvement in total activity of HER comparing to state-of-the-art carbon supported Pt (Pt/C) catalyst. The morphological control of material pore sets simultaneously enhanced the exposed catalytic surface area and transportation of electrolytic ions and gas bubbles; the intact metallic structure provided high-conductivity and eliminated the necessity of using supporting substrate, both of which enabled the realization of much higher mass loading for industrial application than traditional particle-based electrocatalysts.

A robust and efficient non-precious metal catalyst for hydrogen evolution reaction (HER) is one of the key components for CO_2-free hydrogen production. Here, a hierarchical nanoporous copper-titanium (np-CuTi) bimetallic electrocatalyst is able to produce hydrogen from water under a mild overpotential at a rate more than two times higher than that of the current state-of-the-art carbon-supported Pt catalyst. Although both Cu and Ti are known to be poor HER catalysts, the combination of these two elements creates unique Cu-Cu-Ti hollow sites, which have a hydrogen binding energy very similar to that of Pt, resulting in an exceptional HER activity. Additionally, the hierarchical porosity of the np-CuTi catalyst also contributed to its high HER activity, because it not only provides a large surface area for electrocatalytic HER, but also improves the mass transport properties. Moreover, the np-CuTi catalyst is self-supported, eliminating the overpotential associated with the catalyst/support interface.

Conclusions

In short, our recent work on nanostructured metallic materials demonstrated the superior properties of these materials as potential electrocatalysts for energy-related applications.

References

(1) Lu, Q.; Rosen, J.; Zhou, Y.; Hutchings, G. S.; Kimmel, Y. C.; Chen, J. G.; Jiao, F. *Nature Commun.* **2014**, 5, 3242.
(2) Rosen, J.; Hutchings, G. S.; Lu, Q.; Forest, R. V.; Moore, A.; Jiao, F. *ACS Catal.* **2015**, 5, 4586-4591.
(3) Lu, Q.; Hutchings, G. S.; Yu, W.; Zhou, Y.; Forest, R. V.; Tao, R.; Rosen, J.; Yonemoto, B. T.; Cao, Z.; Zheng, H.; Xiao, J. Q.; **Jiao, F.**; Chen, J. G. *Nature Commun.* **2015**, 6, 6567.

PEROVSKITE QUANTUM DOTS (PQDS): A MODEL SYSTEM FOR UNDERSTANDING THE ORIGIN OF INSTABILITY OF ORGANO-METAL PEROVSKITES

Jin Z. Zhang

UCSC

Organo-metal perovskites have attracted significant attention due to their outstanding performance in photovoltaic solar cells. However, device and material degradations caused by factors such as light, oxygen, water, and oxide layers used in devices present a major challenge to practical applications. We have synthesized various organo-metal perovskite quantum dots (PQDs) and used them as a good model system to understand the origin of the materials instability. We have found the photogenerated electron and hole can cause material degradation through redox reactions that damage the ammonium ions acting as localized trap sites, which is likely the origin of instability from light. We also found that species that are likely to react with or extract the protons from the ammonium ions will cause degradation, such as water or alcoholic solvents. Therefore, a more acidic environment is desired for better stability. The new insight gained from our studies is important for developing new perovskite materials for solar cell and other applications such as LEDs and sensors.

NANO-STRUCTURED Cu/SiO_2 CATALYSTS FOR ESTER HYDROGENATION SYSTEMS WITH BALANCED Cu^0 AND Cu^+ ACTIVE SPECIES

Xinbin Ma, Yue Wang, Shouying Huang, Yujun Zhao, Shengping Wang*

Key Laboratory for Green Chemical Technology of Ministry of Education, Collaborative Innovation Center of Chemical Science and Engineering, School of Chemical Engineering and Technology, Tianjin University; Tianjin 300072, China
*Email: xbma@tju.edu.cn

Introduction

Alcohols (e.g., ethanol, ethylene glycol) are attractive products because of the versatile applications in the synthesis of chemicals, fuel alternatives or additives, polymers, etc.[1,2] For alcohol synthesis, ester hydrogenation has received considerable attentions as an important potential substitute approach due to its environmentally benign and atom-economic advantages. Cu species are considered as the excellent catalytic active sites accounting for the selective hydrogenolysis of C-O bonds, which have been widely used in vapor-phase hydrogenation of esters to alcohols.[3]

Despite extensive studies on ester hydrogenation reactions, there is still a debate regarding the nature of active sites of Cu-based catalysts. Only Cu^0 species were believed to be the active sites in early reasearch.[12] However, our previous work and the literatures indicated a synergetic effect between Cu^0 and the oxide components were the vital for high activity and stability in hydrogenation of methyl acetate (MA)[13], dimethyl oxalate (DMO)[14-15], furfural[10], etc. Thus it is essential to fabricate catalysts in a controlled manner with stable structures and coexisting Cu^0 and Cu^+ active sites for a better understanding of the structure-activity relationship. Besides, even Cu-based catalytsts showed excellent initial catalytic properties, the tendency of metallic copper nanoparticles to grow into larger crystallites through migration and coalescence of particles at high temperature is still an impediment to stable performance. And the strongly reducing H_2 make the Cu^+ species unstable in the reaction.

Therefore, we developed the ammonia evaporation hydrothermal (AEH) method here to produce nano-lamella structured catalysts, which makes the copper species (particularly for Cu^+ species) very stable when the hydrogenation reaction temperature is blow 673 K.[4] Adjusting the synthesis parameters, a series of Cu/SiO_2 catalysts with both highly similar textures and substantially different distributions of Cu species were obtained, which enable us to elucidate the precise roles of the active sites.[5]

Furthermore, we constructed core-sheath nanotube[6] and hollow sphere nanoreactors with copper phyllosilicate phases, which exhibited remarkable stability and efficiency in DMO hydrogenation. And the special structures allowed us to control the depth of hydrogenation and selectively obtain different products.

Experimental

Ammonia evaporation hydrothermal method was applied to produce Cu/SiO_2 catalysts with phyllosilicate phases as follows. Silica sol was added to the copper ammonia complex solution and the initial pH of the suspension was 11~12. After stirred and aged at room temperature for several hours, the suspension was heated to allow for the evaporation of ammonia until its pH was decreased to 6~7. Then, the mixture was sealed in autoclave and heated at 463~483 K for 12 h. The obtained precipitate was filtrated, washed and dried followed by calcination at 673 K.

To obtain the catalysts with remarkable different copper species distribution, the copper content (5~40 wt%) and aging time (4~40 h) were adjusted in certain ranges respectively. Another adjustment was using the ammonia treated silica sol solely as the aging procedure at 303 K or 353 K before the copper nitrate solution was added. The following steps of preparation were the same as described previously.

The copper-phyllosilicate core-sheath nano-catalysts were synthesized as follows. First, the copper-phyllosilicate nanotubes (CuPSNT) were prepared by hydrothermal treatment of a mixture of copper nitrate, ammonia, sodium silicate, and ethanol. Then, the Cu@CuPSNTs were prepared by filling the CuPSNTs with copper metals on the basis of the capillary forces. By varying the different impregnation solutions, Cu nanoparticles could be placed in or out the CuPSNTs.

A type of copper silicate hollow spheres (CuSHS) assembled by copper silicate nanotubes with different copper loadings were synthesized by a hydrothermal method, which contained mixing of three substances (copper nitrate, ammonia and silica spheres), aging for 3 h, heating in autoclave at 413 K and calcination at 723 K for 4 h. The silica spheres were fabricated according to Stöber method.

The catalytic performance of the catalysts was carried out in a continuous-flow stainless steel fix-bed reactor inside a horizontal furnace with a temperature controller. Before measurement, the catalysts were pretreated under H_2 flow under 573 K.

H_2-TPD, N_2O titration, in situ FTIR, TEM, XRD, XPS and so on were carried out to characterize the valence, dispersion, and crystal phase as well as the morphology of the nano-catalysts.

Results and Discussion

As shown in Figure 1, the formation of copper phyllosilicate has been confirmed by the lamellar structure in TEM image of calcined Cu/SiO_2 catalyst prepared by AEH method. The structure is formed via reaction of the Cu^{2+} complex with the silanol groups of the silica surface via hydrolytic adsorption to form $\equiv SiOCu^{II}$ monomer and subsequent further polymerization. Moreover, no particles shown in the TEM image indicates that the adsorbed CuO species are highly dispersed. During reduction, the well dispersed CuO turns to Cu^0 and the copper phyllosilicate turns to Cu^+. Because of the strong interaction between the metal and SiO_2, this structure could supply stable Cu species on the surface and remain intact after calcination, reduction and reaction.

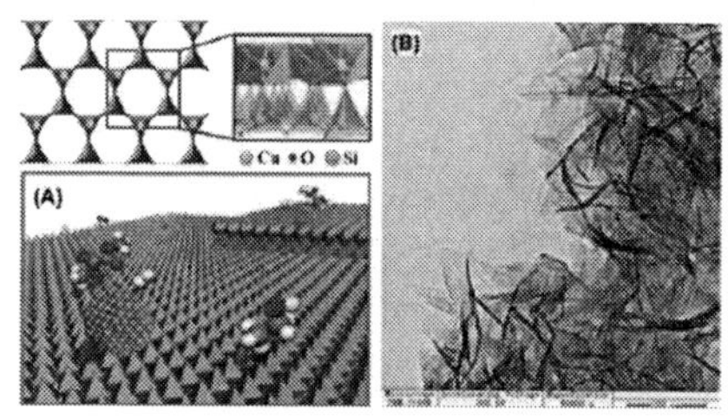

Figure 1. (A) The schematic diagram and (B) TEM image of the Cu/SiO_2 catalyst prepared by AEH method.

Based on this structure, the Cu/SiO_2 catalyst performs remarkable stability and efficiency (shown in Figure 2). The selectivity for ethanol and ethylene glycol (EG) can be tuned simply by regulating the reaction temperature. Additionally, the catalyst is quite stable at 553 K, with a constant conversion of DMO (100%) and selectivity of ethanol (~83%). This indirect synthetic methodology using inexpensive Cu-based catalysts under facile conditions is a promising process that has the potential to enable a sustainable ethanol and EG synthesis using syngas from practical point of view.

However, the respective contributions of Cu^0 and Cu^+ to the catalytic activity remain in dispute and several researchers even have contradictory opinions about the proper surface distribution that most benefits the catalytic performance.[7,8] Thus, we adjusted the synthesis parameters and obtained a series of Cu/SiO_2 catalysts with the lamellar structures. After multiple and mutually corroborated characterizations, these catalysts are demonstrated with high comparability due to their full-range distribution of Cu species and similar general morphologies. According to the results in Figure 3, the balancing effect of these two active Cu sites on enhancing the catalytic performance is revealed: when the accessible metallic Cu surface area is below a certain value, the catalytic activity of hydrogenation linearly increases with increasing Cu^0 surface area, whereas it is primarily affected by the Cu^+ surface area. Furthermore, combined the in situ FTIR experiments and DFT calculations, the precise contributions of Cu^0 and Cu^+ active sites are illuminated: the Cu^+ sites adsorb the methoxy and acyl species, while the Cu^0 facilitates the H_2 decomposition. This insight into the precise roles of active species can lead to new possibilities in the rational design of catalysts for hydrogenation of esters.

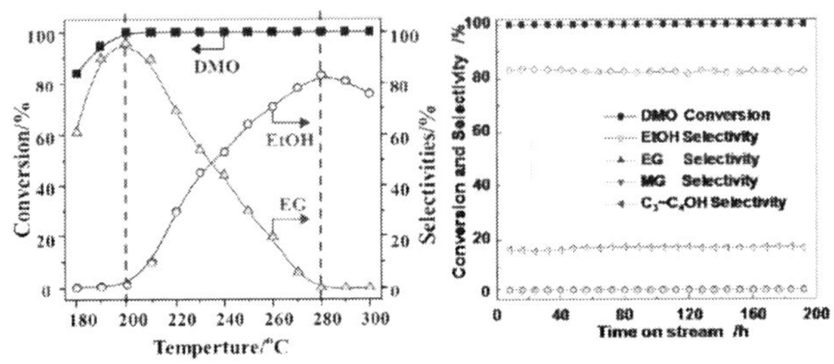

Figure 2. The performance of the catalytic hydrogenation of DMO over Cu/SiO_2 catalyst prepared by AEH method. Reaction conditions: 2.5MPa, LHSV=2.0h^{-1}, H_2/DMO=200.

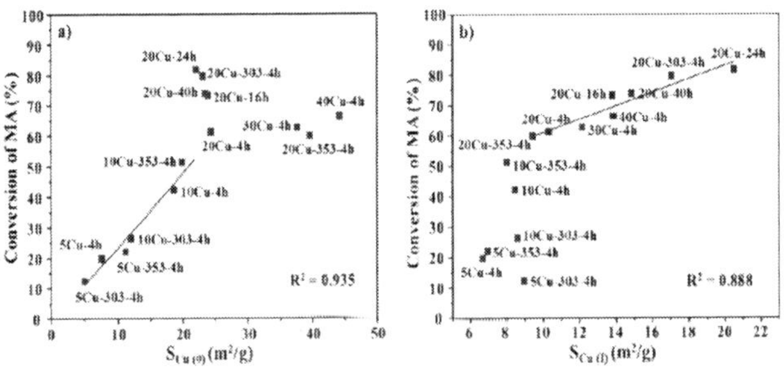

Figure 3. The balancing effect of Cu^0 and Cu^+ active species on the catalytic performance of methyl acetate hydrogenation. Reaction conditions: 488 K, 2.5MPa, LHSV=2.0h^{-1}, H_2/MA=80.

To make the catalysts more stable and manipulate the balance of Cu^0 and Cu^+ active sites more easily, we built a core (copper) – sheath (copper phyllosilicate) nanoreactor which presented superior high efficiency (an ethanol yield of 91%) and steady performance (> 300 h at 553 K). As shown in Figure 4, there was no obvious change in the yield of ethanol for the three catalysts prior to the thermal treatment. However, the ethanol yield of Cu@CuPSNT-out catalyst decreased from 70 to 47%, whereas that of Cu@CuPSNT-in catalyst decreased from 92 to 85%. The Cu@CuPSNT-in catalyst also exhibited a higher stability than our previous Cu/SiO_2 catalyst. This superiority is due to the more stability of Cu^0 and Cu^+ species during calcination, reduction and reaction.

Moreover, the inner diameter and length of CuPSNTs are tunable. Confinement of reactants and reaction intermediates within such nanochannels could prolong the contact time of these species with the active sites inside the CuPSNTs, which favors the deep hydrogenation reaction and leads to a higher selectivity to EG or ethanol.

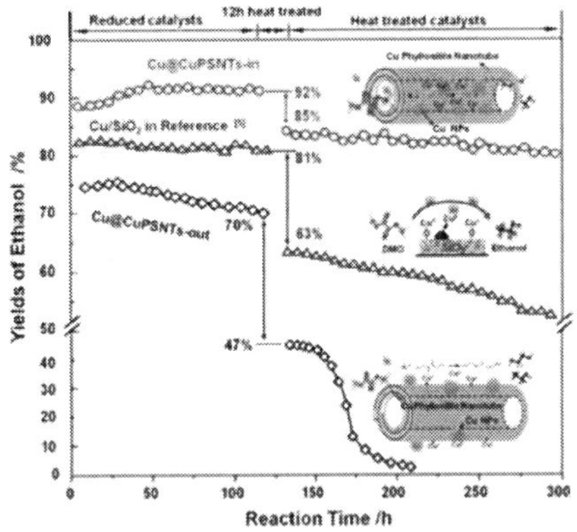

Figure 4. Comparison of catalytic performance and stabilities. Reaction conditions: 553 K, 2.5MPa, LHSV=2.0h^{-1}, H_2/DMO=200.

To control the depth of hydrogenation and selectively synthesize different products more effectively, a copper silicate hollow sphere nanoreactor was constructed. What's more, the well-defined nano-sized channels provide spatial restriction on the reactants, which can increase the H_2 concentration along with the DMO converting. As a result, the H_2/ester ratio in the hydrogenation of DMO to EG can be reduced from 80 to 30 compared to our previous Cu/SiO_2 catalyst, which is meaningful in industry practice for significantly decreasing the circulation of H_2.

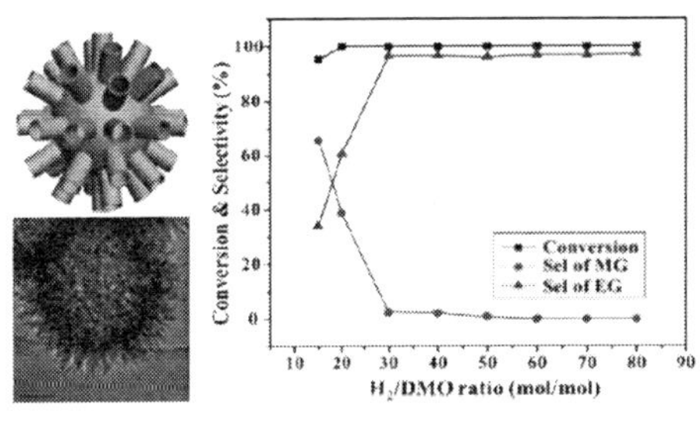

Figure 5. Reduced H_2/ester ratio in hydrogenation of DMO over CuSHS nanoreactor.

Conclusions

In summary, we developed several nano-structured catalysts with copper phyllosilicate phases which can provide balanced and stable Cu^0 and Cu^+ species in high-temperature hydrogenation reactions. These unique structures performed superior reactivity and the morphology with spatial restriction allowed us to selectively synthesize the hydrogenation products and significantly reduce the circulation of H_2, which provided a new guidance for catalyst design.

Acknowledgement. This research was supported by the National Natural Science Foundation of China (91434127, 21325626, 21276186, U1510203).

References

(1) Goldemberg, J. *Science* **2007**, *315*, 808.
(2) Yue, H.; Zhao, Y.; Ma, X.; Gong, J. *Chem. Soc. Rev.* **2012**, *41*, 4218.
(3) Yue, H. R.; Ma, X. B.; Gong, J. L. *Acc. Chem. Res.* **2014**, *47*, 1483.
(4) Gong, J.; Yue, H.; Zhao, Y.; Zhao, S.; Zhao, L.; Lv, J.; Wang, S.; Ma, X. *J. Am. Chem. Soc.* **2012**, *134*, 13922.
(5) Wang, Y.; Shen, Y.; Zhao, Y.; Lv, J.; Wang, S.; Ma, X. *ACS Catal.* **2015**, *5*, 6200.
(6) Yue, H. R.; Zhao, Y. J.; Zhao, S.; Wang, B.; Ma, X. B.; Gong, J. L. *Nature Commun.* **2013**, *4*.
(7) Yin, A.; Guo, X.; Dai, W.-L.; Fan, K. *J. Phys. Chem. C* **2009**, *113*, 11003.
(8) Zheng, X.; Lin, H.; Zheng, J.; Duan, X.; Yuan, Y. *ACS Catal.* **2013**, *3*, 2738.

A COMPUTATIONAL STUDY ON THE SURFACE STABILITY OF SPINEL $MgAl_2O_4$ MATERIALS

Qiuxia Cai[1,2], *Jian-guo Wang*[2,*], *Yong Wang*[1,3,*], *Donghai Mei*[1,*]

[1] Institute for Integrated Catalysis, Pacific Northwest National Laboratory, Richland, WA 99352, USA
[2] College of Chemical Engineering, Zhejiang University of Technology, Hangzhou, Zhejiang 310014, China
[3] Voiland School of Chemical Engineering and Bioengineering, Washington State University, Pullman, WA 99164, USA
donghai.mei@onnl.gov; yong.wang@pnnl.gov; jgw@zjut.edu.cn

Introduction

Spinel $MgAl_2O_4$ has shown the superior ability as a refractory support material to stabilize small metal nanoparticle catalysts in harsh operating conditions due to its remarkable physicochemical properties such as high melting temperature, thermal stability and hydrothermal stability.[1-2] It has been found that the surface structure, acidic–base strengths as well as compositions of $MgAl_2O_4$ play important roles in the sizes and morphologies of metal nanoparticles. In particular, the exposed surface structure or termination of the spinel $MgAl_2O_4$ support is highly related to the stability of metal nanoparticle.[3-4] Therefore, fundamental understanding of the stability of surface structure of $MgAl_2O_4$ under various reaction conditions is essential to develop highly stable and selective metal catalysts for industrially catalytic applications. In the present work, functional theory (DFT) calculations and thermodynamic analysis were combined to calculate surface Gibbs free energies of various surface terminations of three low-index structures (100, 110, and 111) for the spinel $MgAl_2O_4$ material. On the basis of the calculated surface Gibbs free energy of the most stable termination for each low-index structure, the shapes and surface structures of the spinel $MgAl_2O_4$ nanoparticles under different oxygen partial pressures and variation of aluminum chemical potential were predicted using Wulff construction.[5]

Computational

All DFT calculations were performed using the Vienna ab-initio simulation package (VASP).[6-9] The projector augmented wave (PAW) method combined with a plane wave basis set at a cutoff energy of 500 eV was used to describe core and valence electrons. The Perdew-Burke-Ernzerhof (PBE) functional was used in all the calculations. The ground-state atomic geometries of bulk and surface were obtained by minimizing the forces on each atom to below 0.05 eV/Å. A (4×4×1) Monkhorst-Pack k-point mesh were used in the surface structure terminations. Three low-index structures (100, 110, and 111) with a total of 12 terminations of spinel $MgAl_2O_4$ bulk were taken into account.

Results and Discussion

The most stable surface termination has the lowest surface Gibbs free energy. The stability of each surface termination of spinel $MgAl_2O_4$ material is investigated using ab initio thermodynamics approach.[10] Each line in the right side of Figure 1 represents a specific oxygen partial pressure. For example, the dash crossing line from the right side to the left side indicates only the 100_AlO2 termination (shown in green) is the most stable surface structure at T=1100 K and ultra-high vacuum regardless of Al-poor or Al-rich environment. With increasing oxygen pressure, the 111_O2(Al) termination (shown in blue) then becomes the most stable surface in the relatively Al-rich environment. This explains that why the (100) surface structure was theoretically predicted as the most stable surface in the previous studies[11] while the (111) termination was observed as the dominant surface structure for the natural $MgAl_2O_4$ material.[12] Clearly, the oxygen environment in equilibrium with the formed $MgAl_2O_4$ material has to be taken into account when the stability of surface structure is determined.

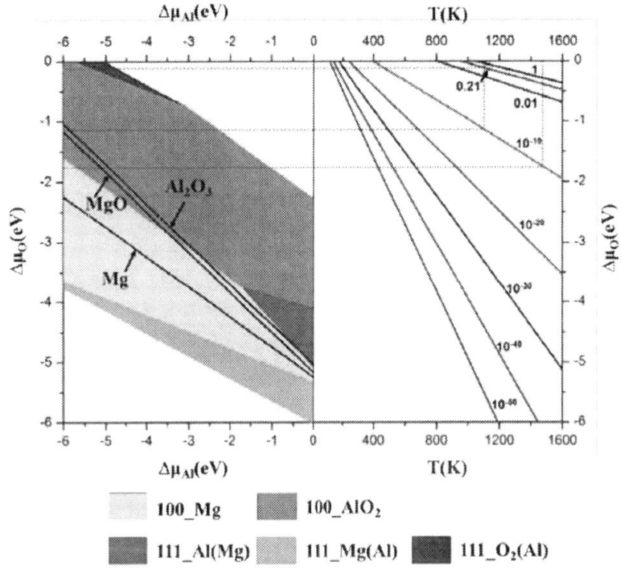

Figure 1. Phase diagrams: the regions of stability of $MgAl_2O_4$ surfaces with different terminations as a function of aluminum and oxygen atoms chemical potential variations. The arrows point to the lines where the oxides begin to precipitate. The right side of figure is a family of $\Delta\mu_O$ as a function of temperature at various oxygen gas pressures. The label m on the lines in the right subfigure specifies the oxygen partial pressure P_{O2} = m (atm).

Using Wulff construction,[5] we can predict the shape and relatively exposed surface areas of each termination over the $MgAl_2O_4$ with the most stable termination for each low-index surface structure. As shown in Figure 2, the 100_AlO2 is the dominant surface termination over the $MgAl_2O_4$ particle under ultra-high vacuum condition at 1100 K, especially at the Al-rich environment. With decreasing Al content, the exposed area of the 110_AlO2 termination rapidly grows. The exposed areas of the AlO2 termination of (100) and (110) structures are relatively close (56.7% and 46.2%) at $\Delta\mu_{Al}$=−6 eV. The (111) surface structure is negligible under ultra-high vacuum condition. However, under atmospheric condition (P_{O2}=0.21 atm), the (110) surface structure is negligible. The 100_AlO2 is predominant (81.6%) over the 111_O2(Al) termination (18.4%). While the opposite is found in the Al-rich environment. The 111_O2(Al) termination becomes the dominant surface structure over the $MgAl_2O_4$ particle with the surface area ratio of 97.9%. The surface occupancy of the 100_AlO2 termination is minimal (2.1%). Out prediction is in agreement with previous experimental and theoretical studies.[3-4]

Conclusions

The $MgAl_2O_4$ spinel is widely used as the supporting materials for transition metal catalysts due to its unique capability against sintering of metal nanoparticles. The effects of oxygen partial pressure and the Al environment on the surface stability are studied. It is found that the 100_AlO2 termination is the most stable surface structure under the ultra-high vacuum condition at T=1100 K, while the 111_O2(Al) termination becomes the most stable surface in the

Al-rich environment as the oxygen pressure increases. The oxygen vacancy formation is thermodynamically favorable over five surface terminations of (100) and (111) orientations. Using Wulff construction with calculated surface Gibbs free energies for both perfect and defective surface terminations, the 100_AlO_2 and 111_O_2(Al) are found to be the most dominant surface structures under atmospheric condition. The theoretical predictions of this work are consistent with our previously experimental observation.

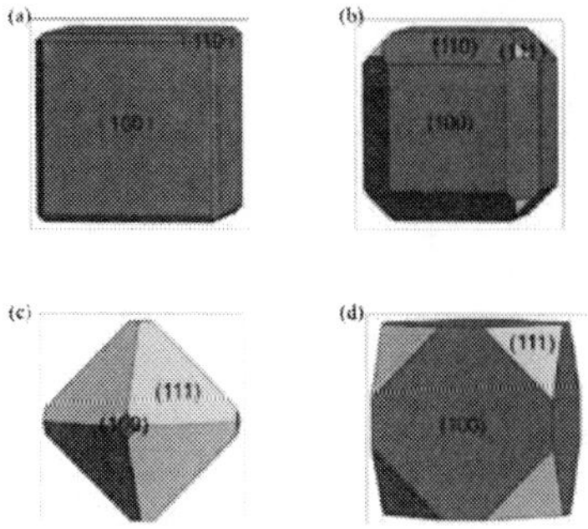

Figure 2. Predicted the exposed areas of perfect surface terminations and the shape of the $MgAl_2O_4$ particle at 1100 K via Wulff construction. (a) $P_{O_2}=10^{-10}$ atm, $\Delta\mu_{Al}=-5$ eV (Al-rich); (b) $P_{O_2}=10^{-10}$ atm, $\Delta\mu_{Al}=-6$ eV (Al-poor); (c) $P_{O_2}=0.21$ atm, $\Delta\mu_{Al}=-5$ eV (Al-rich); (d) $P_{O_2}=0.21$ atm, $\Delta\mu_{Al}=-6$ eV (Al-poor).

Acknowledgement. This work was supported by a Laboratory Directed Research and Development (LDRD) project of the Pacific Northwest National Laboratory (PNNL). PNNL is operated by Battelle for the US Department of Energy (DOE). The computing time was granted by the National Energy Research Scientific Computing Center (NERSC). Part of computing time was also granted by a scientific theme user proposal in the William R. Wiley Environmental Molecular Sciences Laboratory (EMSL), which is a U.S. Department of Energy national scientific user facility located at PNNL in Richland, Washington.

References

(1) Villa, A.; Gaiassi, A.; Rossetti, I.; Bianchi, C. L.; van Benthem, K.; Veith, G. M.; Prati, L., *J. Catal.* **2010,** *275*, 108-116.

(2) Liu, P.; Hensen, E. J. M., *J. Am. Chem. Soc.* **2013,** *135*, 14032-14035.

(3) Li, W.-Z.; Kovarik, L.; Mei, D.; Liu, J.; Wang, Y.; Peden, C. H. F., *Nature Comm.* **2013,** *4*, 2481.

(4) Li, W.-Z.; Kovarik, L.; Mei, D.; Engelhard, M. H.; Gao, F.; Liu, J.; Wang, Y.; Peden, C. H. F., *Chem. Mater.* **2014,** *26*, 5475-5481.

(5) Center for Theoretical and Computational Materials Science, N. *CTCMS Wulffman Program*, 1.2.4; 2002.

(6) Kresse, G.; Furthmuller, J., *Comput. Mater. Sci.* **1996,** *6*, 15-50.

(7) Kresse, G.; Hafner, J., *Phys. Rev. B* **1993,** *47*, 558-561.

(8) Kresse, G.; Hafner, J., *Phys. Rev. B* **1994,** *49*, 14251-14269.

(9) Kresse, G.; Furthmuller, J., *Phys. Rev. B* **1996,** *54*, 11169-11186.

(10) Shishkin, M.; Ziegler, T., *Surf. Sci.* **2012,** *606*, 1078-1087.

(11) Rasmussen, M. K.; Foster, A. S.; Hinnemann, B.; Canova, F. F.; Helveg, S.; Meinander, K.; Martin, N. M.; Knudsen, J.; Vlad, A.; Lundgren, E.; Stierle, A.; Besenbacher, F.; Lauritsen, J. V., *Phys. Rev. Lett.* **2011,** *107*.

(12) Goniakowski, J.; Finocchi, F.; Noguera, C., *Rep. Prog. Phys.* **2008,** *71*, 016501.

CHEMICAL CONTROL OVER STABLE SUPPORTED HIGH-COVERAGE NANOPARTICLE LAYERS

Mackenzie G. Williams and Andrew V. Teplyakov

Department of Chemistry and Biochemistry
University of Delaware
Newark, DE 19716
andrewt@udel.edu

Introduction

Design of nanoparticulate monolayers for applications in catalysis, energy conversion, solar energy harvesting, sensing, and spintronics has undergone tremendous growth over the last decade. An important consideration for this development is achieving high-coverage stable monolayers of predesigned nanoparticles. Previously developed self-assembly approaches yield excellent coverages of nanoparticles but the produced interfaces lack stability.[1] On the other hand, covalent binding based on efficient reactions and novel approaches to surface modification has indeed been demonstrated to produce very stable systems;[2] however, reaching a high coverage limit with this approach proved very difficult.

This paper will present a modified approach to reach both goals and to produce very high coverages of surface-modified nanoparticles on flat substrates. The produced monolayers are stable in ambient and with respect to mild thermal annealing. The surface modification strategy follows the applicability of copper-catalyzed "click chemistry" between azide and alkyne functional groups that leads to the formation of a stable triazole ring,[3] as confirmed by infrared spectroscopy and X-ray photoelectron spectroscopy (XPS). The formation of high-coverage monolayers for model silica and Fe_3O_4 nanoparticles is demonstrated with the help of scanning electron microscopy (SEM) and atomic force microscopy (AFM). Density functional theory (DFT) provides the explanation of the high efficiency of the proposed modification scheme and explores the possible reaction intermediates responsible for such efficiency.[4,5]

This approach has the potential to be expanded to include other materials and substrates, to produce conformally-filled monolayers that can in the future be used in microfluidic devices, to build controlled 3D materials starting with these well-defined monolayers, and to have applications well beyond currently envisioned fields.

Experimental

Monolayer Formation. Gold wafers cleaned in piranha solution were immersed in 11-azidoundecanethiol in ethanol for 24 hours, followed by washing in a variety of common solvents. Silica nanoparticles (80 nm) were stirred with tetramethoxy orthosilicate for 24 hours; 4-pentyn-1-ol was added and the solution was stirred for a further 24 hours at 80 °C, followed by washing in methanol. Iron oxide nanoparticles (20 nm) were sonicated with 10-undecynyl-phosphonic acid in toluene for 30 minutes, followed by washing in hexanes.

To create the silica monolayer, the azide-terminated gold substrate was immersed in a solution of methanol, water, alkyne-modified silica nanoparticles, and catalytic amounts of copper sulfate pentahydrate and sodium ascorbate. The mixture was sonicated for 24 hours and the substrate was subsequently rinsed in a variety of common solvents and sonicated in ethanol to remove physisorbed particles. To create the magnetic monolayer, the azide-terminated gold substrate was stirred in a solution of dimethyl sulfoxide, water, alkyne-modified iron oxide nanoparticles, and catalytic amounts of copper sulfate pentahydrate and sodium ascorbate for 48 hours, then rinsed in a variety of common solvents and sonicated in ethanol.

Characterization. Completion of the layers was confirmed by a Zeiss Auriga 60 SEM, operating at an accelerating voltage of 3 keV (silica) or 5 keV (iron oxide) and a working distance of 5.0 mm. Monolayer height was determined with a gallium focused ion beam with an energy of 30 kV and current of 120 pA or by AFM using a Veeco Multimode SPM with either a Nanoscope Dimension 3100 controller or a Nanoscope V controller. Single point attenuated total reflectance Fourier-transform infrared (ATR-FTIR) spectroscopy was used to monitor the "click" reaction with a Bruker Optics Vertex 70 FT-IR with a Bruker Hyperion 2000 Microscope attachment. Formation of a triazole ring was confirmed with either a PHI 5600 XPS or a Thermo Scientific K-Alpha+ XPS, both equipped with an Al kα source. In both cases, the spectra were calibrated to the Au $4f_{7/2}$ peak. DFT calculations were performed using Gaussian 09 with the 6-311G+ (d,p) basis set and B3LYP functional.

Results and Discussion

The proposed approach to build high-coverage monolayers of nanoparticles is based on the copper-catalyzed "click reaction" between the azide and alkyne functionalities summarized in Figure 1. The choice of appropriate anchor group and conditions for modification of silica and iron oxide (Fe_3O_4) nanoparticles is indicated in the figure caption.

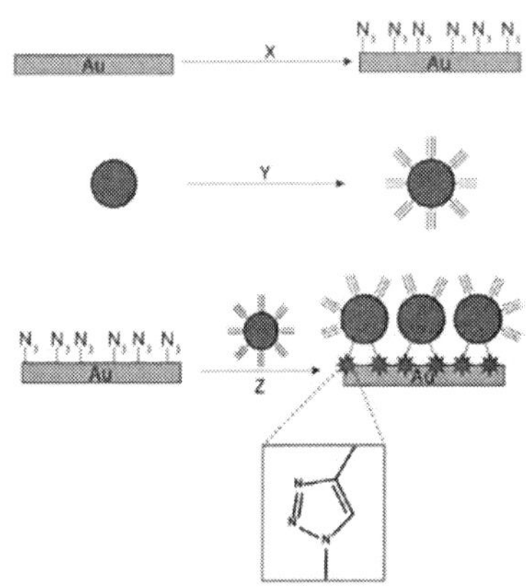

Figure 1. Functionalization scheme of substrate and nanoparticles, and monolayer formation. X = 11-azidoundecanethiol in ethanol. For silica nanoparticles, Y = tetramethoxy orthosilicate and 4-pentyn-1-ol in toluene and Z = $CuSO_4{\bullet}5H_2O$, sodium ascorbate, methanol, water. For iron oxide nanoparticles, Y = 10-undecynyl phosphonic acid in toluene and Z = $CuSO_4{\bullet}5H_2O$, sodium ascorbate, dimethyl sulfoxide, water.

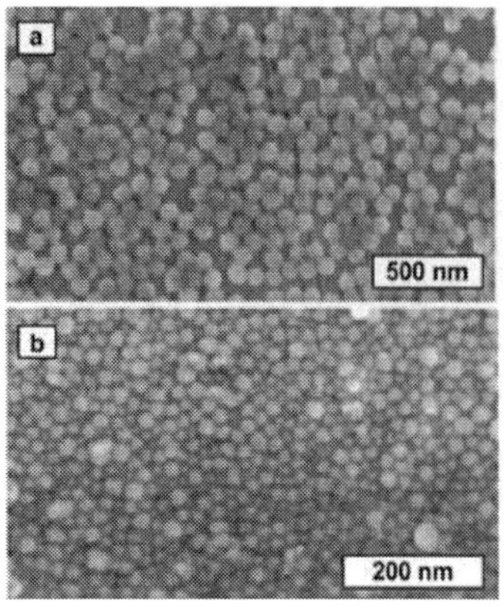

Figure 2. SEM image of silica (a) and iron oxide (b) nanoparticle monolayers.

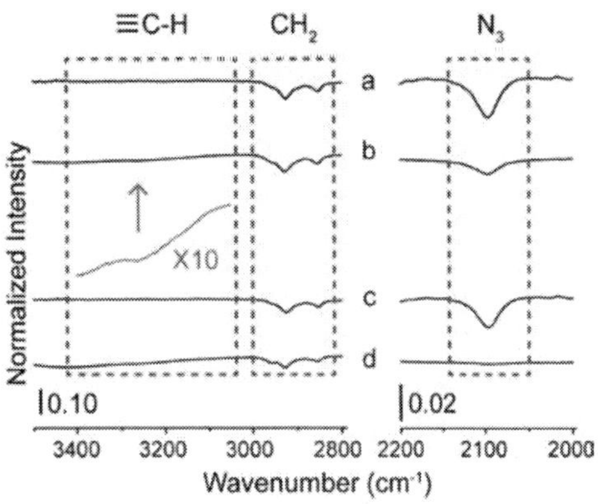

Figure 3. Informative IR spectral regions of the azide-modified substrate before (a) and after (b) addition of the alkyne-modified silica nanoparticles, and the same surface before (c) and after (d) addition of the alkyne-modified iron oxide nanoparticles. The grey line shows the alkyne C-H region of spectrum (b) magnified ten times.

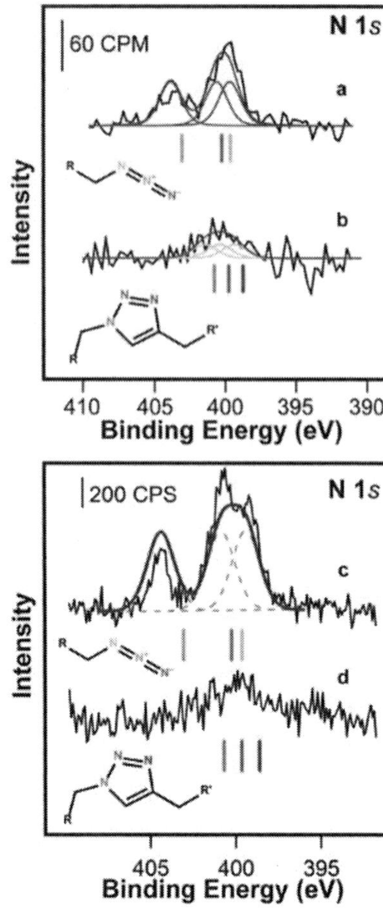

Figure 4. XPS N 1*s* region of the azide-modified substrate before (a) and after (b) addition of the alkyne-modified silica nanoparticles, and the same surface before (c) and after (d) addition of the alkyne-modified iron oxide nanoparticles. The colored lines below the spectra indicate computationally-predicted binding energies of the corresponding color atoms in the models.

Figure 2 summarizes the SEM studies of the resulting systems in both cases and confirms that very high coverage is reached for this chemical modification approach. These resulting monolayers are stable in air and under common solvents; however, spectroscopic methods are needed to confirm that they are covalently bound to the substrate surface.

This is achieved by examining infrared spectra summarized in Figure 3 for both types of nanoparticles, where the formation of the triazole ring and the removal of azide group are followed as a function of the monolayer formation. This result is also confirmed by the XPS studies summarized in Figure 4 for the same systems. Here the formation of the triazole ring and the changes in the N 1s spectral region are also compared with the results of DFT calculations for models shown in the figure. Similar observations are recorded for the silica nanoparticles.

Conclusions

A simple approach for creating monolayers of nanoparticles on solid support materials based on "click chemistry" was modified to ensure the yield of very high-coverage layers that are covalently bound to the surface of the substrate and stable in ambient conditions. Spectroscopic investigations with infrared spectroscopy and XPS confirmed the efficiency of the "click" reaction and the formation of strong covalent bonds between the nanoparticles and the underlying substrate. The coverage was estimated based on the electron microscopy and atomic force microscopy investigations. The microscopy studies also confirmed the formation of a single monolayer of nanoparticles as opposed to multiple layers or ad layers. It was shown that the specifics of nanoparticle and substrate chemical functionalization determine the nanoparticle coverage. The explanation of this observation is offered with a help of DFT investigations.

Acknowledgement. Funding of this work was provided by the National Science Foundation (CHE 1057374) and the Petroleum Research Fund, administered by the American Chemical Society. The authors would like to thank Professor T. P. Beebe, Jr., Ms. R. Pupillo, and Mr. Z. Voras (Surface Analysis Facility, University of Delaware) for XPS support, as well as the NSF (CHE 1428149) and the NIH NIGMS COBRE program (P30-GM110758) for partial support of the activities in the University of Delaware Surface Analysis Facility. They also would like to thank Professor C. Ni and Dr. F. Deng (Keck Electron Microscopy Facility, University of Delaware) and Mr. Y. Duan (Department of Chemistry, University of Delaware) for SEM and AFM support. Acknowledgement is also given to Professor K. Booksh and Dr. D. Boyne (Department of Chemistry, University of Delaware) for their support of the single point ATR instrument, as well as to Professor J. Koh and Mr. E. Zhang (Department of Chemistry, University of Delaware) for help with synthesis of the 11-azido-undecanethiol.

References

(1) Updahyay, A. P.; Behara, D. K.; Sharma, G. P.; Bajpai, A.; Sharac, N.; Ragan, R.; Pala, R. G. S.; Sivakumar, S. *ACS Appl. Mater. Interfaces*, **2013**, *5*, 9554.

(2) Toulemon, D.; Pichon, B. P.; Cattoën, X.; Wong Chi Man, M.; Begin-Colin, S. *Chem. Commun.* **2011**, *47*, 11954.

(3) Kolb, H. C.; Finn, M. G.; Sharpless, K. B. *Angew. Chem. Int. Ed.* **2001**, *40*, 2004.

(4) Liu, Y.; Williams, M. G.; Miller, T. J.; Teplyakov, A. V. *Thin Solid Films*, **2016**, *598*, 16.

(5) Williams, M. G.; Teplyakov, A. V. *Appl. Surf. Sci.*, [Online early access.] DOI: 10.1016/j.apsusc.2015.11.212. Published Online: Nov. 25, 2015.

DISCOVERY AND OPTIMIZATION OF MULTICOMPONENT METAL OXIDE CATALYSTS USING DENSITY FUNCTIONAL THEORY

Michael J. Janik

Department of Chemical Engineering, Pennsylvania State University, University Park, PA
mjanik@psu.edu

Introduction

Multicomponent metal oxides can offer tunable redox catalysis properties by controlling the identity and concentration of two or more metal atoms. A host oxide may accommodate a second, "dopant" metal atom into its crystal structure that alters the oxide reducibility, oxygen mobility, surface energy, and catalytic activity. If the "dopant" metal atom surface segregates within a less expensive host, ideal dispersion may be reached where each "dopant" is available on the surface to promote reaction chemistry.

In this talk, I will overview our group's density functional theory studies of multicomponent oxides. Specific topics that will be addressed include dopant and oxygen vacancy distribution in the bulk and surface of doped oxides, the impact of a second metal atom on bulk and surface reducibility, the stability of surface mixed-oxide phases, and guiding the selection of mixed oxide for hydrocarbon selective oxidation and combustion. Throughout the talk, challenges in accurate representation of energetics with DFT methods will be highlighted.

Computational Methods

Density functional theory calculations were run using the Vienna ab initio Simulation Package,[1-3] a plane-wave basis set DFT code. GGA functionals typically do not properly alleviate self-interaction errors such that highly localized d and f states on metal atoms are often incorrectly delocalized. We, therefore, apply the DFT+U (GGA+U) approach to representing the electronic structure.

Results and Discussion

Our group has examined a series of metal-doped CeO_2 surfaces, demonstrating that the reactivity for activation of C-H bonds correlates with the reducibility of the doped surface.[4] Figure 1 demonstrates that this correlation holds reasonably well across a wider series oxides and doped oxides, including TbO_x, TiO_2, ZnO, M-CeO_2, and M-MgO surfaces.[5] This correlation, though with noticeable noise, holds across different oxides, surface facets, and with varying U value.

The correlation in Figure 1 suggests a trade-off in the design of combustion catalysts. Catalysts to the bottom left have a low activation barrier for C-H activation of methane, suggesting a more rapid rate-limiting step for methane oxidation. These catalysts, however, also have a more favorable oxygen vacancy formation energy, suggesting they are easy to reduce and will be slow to re-oxidize. Optimal combustion catalysts will therefore balance these two factors, and the optimal material will depend on the oxygen chemical potential, as well as the character (methyl, primary, secondary, tertiary) of the C-H bond to be activated.

The correlation in Figure 1 also holds for the adsorption energy of methyl radicals, with more reducible surfaces binding methyl radicals more strongly. This leads to a trade-off in design of catalysts for oxidative coupling of methane. Catalysts to the bottom left have higher activity with lower barriers for methane activation. These catalysts, however, will also be more likely to adsorb the resulting methyl radical, leading to non-selective further oxidation.

The sources of deviations from the trend shown in Figure 1, as well as the accuracy with which DFT+U can capture the position of a specific material along this trend, will be discussed.

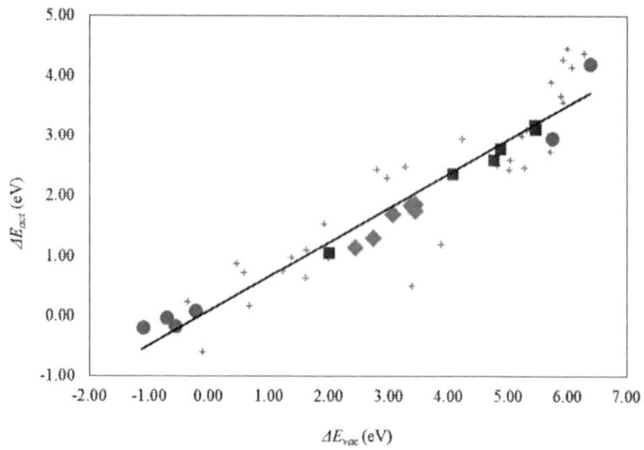

Figure 1. The C-H activation reaction energy plotted against the oxygen vacancy formation energy, the correlation for which holds when varying the U-value for Ce in CeO_2 (◆), the surface facets for TiO_2 (■) and the oxidation state for TbO_x (●). Other data (+) includes ZnO as well as doped CeO_2 and MgO surfaces.

Conclusions

DFT+U can be used to help guide the design of mixed oxide catalysts. In addition to the discussion of optimizing doped systems for hydrocarbon oxidation, this talk will discuss how DFT+U can be used to evaluate the stability of mixed oxide structures and aid in discovery of potentially stable structures.

References.

(1) Kresse, G.; Furthmuller, J., Comput. Mater. Sci. **1996**, 6 (1), 15-50.
(2) Kresse, G.; Furthmuller, J., Phys. Rev. B **1996**, 54 (16), 11169-11186.
(3) Kresse, G.; Joubert, D., Phys. Rev. B **1999**, 59 (3), 1758-1775.
(4) Krcha, M. D.; Mayernick, A. D.; Janik, M. J., J. Catal. **2012**, 293, 103-115.
(5) Kumar,G.; Lau, S. L. J.; Krcha,M. D.; Janik, M. J. ACS Catal. 2016, 6, 1812-1821.

ANHARMONIC EFFECTS ON MATERIAL PROPERTIES AT HIGH TEMPERATURES FROM AB INITIO MOLECULAR DYNAMICS SIMULATIONS

Mal-Soon Lee, Roger Rousseau, and Vassiliki-Alexandra Glezakou*

Fundamental and Computational Sciences Directorate,
Pacific Northwest National Laboratory, Richland, WA 99352
*Mal-Soon.Lee@pnnl.gov

Introduction

The *ab initio* molecular dynamics (AIMD) simulation technique is a promising tool to study various properties of vibrational spectra, entropy, free energy, etc. even at high temperatures. With the harmonic phonon model, all interatomic forces are treated as purely harmonic so that the equilibrium distance between atoms is independent of temperature. For the low temperature regime, this model, commonly used in most quantum chemistry codes, can approximately model vibrational properties and free energetics. At higher temperatures, however, the harmonic model can break down since it cannot explain thermal expansion, where the role of anharmonicity becomes very important in the description and interpretation of the system. In these cases, the quasi-harmonic approximation can be used to better describe properties of the systems at higher temperatures. In this work, we report a few examples showing how the quasi-harmonic approximation can be used to rigorously evaluate the vibrational density of states (VDOS) at elevated temperatures. From the obtained VDOS, the quantitative evaluation of thermodynamic properties such as entropy, enthalpy, free energy, and equilibrium constants can then be inferred. A variety of systems ranging from water/scCO_2/mineral interfaces (carbon sequestration), ethanol-adsorption in zeolites (catalysis), and Tc-incorporation in magnetite (nuclear remediation) are used to illustrate the power of this approach.

Computational Details

Three different case studies are presented here: (i) H_2O/scCO_2/anorthite interfaces (carbon sequestration), (ii) ethanol-adsorbed zeolite (catalysis), and (iii) Tc-incorporated magnetite (nuclear remediation). We performed *ab initio* molecular dynamics simulations with periodic boundary conditions (3D PBC) as implemented in the CP2K package. The Perdew, Burke and Ernzerhof (PBE) generalized gradient approximation was used for the exchange-correlation functional. In case of the Tc-incorporated magnetite systems, spin-polarized DFT calculations with GGA-U scheme were used. To study free energetics of water film formation and carbonation on the anorthite surface, we also performed Blue-Moon ensemble simulations of the reaction free energy profile-namely the potential of mean force PMF. Detailed method of these calculations can be found in Refs. 1,2.

Results and Discussion

H_2O/scCO_2/anorthite interfaces. The boundary layer at solid-liquid interfaces is a unique reaction environment that poses significant scientific challenges to characterize and understand by experimentation alone. Using *ab initio* molecular dynamics (AIMD) methods, we studied the structure and dynamics of boundary layer formation, cation mobilization and carbonation under geologic carbon sequestration scenarios (T=323 K and P=90 bar) on a prototypical anorthite (001) surface. We first focus on water transport to, from and across scCO_2/anorthite interfaces, which may be relevant to steel corrosion in water-lean scCO_2 or preferential sulfite formation in carbonate minerals. We computed the potential of the mean force (PMF) from constrained AIMD simulations by moving a single H_2O from the middle of the liquid slab through the box, such that its immediate environment changes from being solvated in the scCO_2 to being bound on the Ca-rich surface. This process was repeated by adding water molecule one by one up to 14 H_2O, completion of 1 water layer on the surface. These simulations showed that water film growth begins with a single water molecule as a nucleation site followed by the addition of subsequent molecules, to form an island. Our simulations also indicate that even very low water concentrations (water solubility in scCO_2 is ~10^{-4} %/mol) can result in a water film formation on the surface by showing lowest free energies at the surface (Fig. 1a). The presence of a water layer, adjacent to the surface, induces large disorder in the Ca positions. Free energy calculations of a single Ca atom extraction from the surface into the water layer, involves a two-step equilibrium between an inner and an outer complex as shown in Fig. 1b By examining the free energetics of the Ca dissolution process, we find that the free-energy barrier for dissolution is 30 kJ/mol indicating that at 323 K this process will be relatively fast (Fig. 1b). The change of Helmholtz free energy, ΔA, between the two minima is 14 kJ/mol demonstrating that at this temperature there will be an equilibrium between inner and outer sphere Ca^{2+} complexes with a $K_{eq} = \exp(-\Delta A/k_BT) \sim 10^{-3}$. Note that this complex shows a positive increase in free energy as it moves out into the scCO_2 layer, as long as the first coordination sphere of Ca^{2+} consists only of water. However, complexes with mixed coordination (H_2O and CO_2) may further enhance Ca transport away from the surface. Ca^{2+} dissolution results in Ca-vacancies where carbonate formation is expected to be spontaneous as shown in Fig. 1c, in a process similar to sulfite formation.

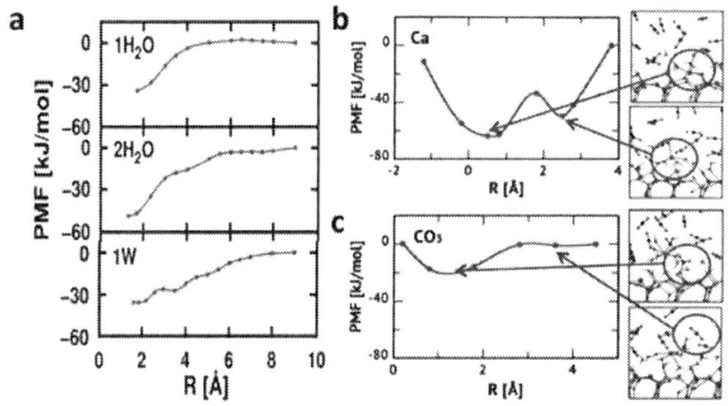

Figure 1. PMF for water layer formation (a), for Ca dissolution (b), for subsequent carbonation at the newly formed Ca vacancy (c).

Ethanol-adsorbed zeolite. To account for thermal and entropic effects caused by the dynamics of the motion of the reaction intermediates, ethanol adsorption on the Brønsted acid site of the H-ZSM-5 catalyst has been studied at different temperatures ($100 \leq T \leq 700$ K) and ethanol loadings using *ab initio* molecular dynamics (AIMD) simulations, infrared (IR) spectroscopy and calorimetric measurements. At low temperatures ($T \leq 400$ K) and ethanol loading, a single ethanol molecule adsorbed in H-ZSM-5 forms a Zundel-like structure where the proton is equally shared between the oxygen of the zeolite and the oxygen of the alcohol. At higher ethanol loading, a second ethanol molecule helps stabilize the protonated ethanol at all temperatures by acting as a solvating agent. To calculate free energetics using quasi-harmonic approximation (QHA), unlike harmonic approximation where vibrational modes are obtained from the lowest energy structure, vibrational density of states (VDOS) are calculated using velocity autocorrelation function with obtained velocities from AIMD simulations. QHA enables us to include the effect of configurational change at a given temperature. With obtained VDOS vibrational enthalpies and entropies are calculated. The calculated VDOS using QHA are in excellent agreement with measured IR spectra for C_2H_5OH, C_2H_5OD and C_2D_5OH isotopomers where the results support the existence of both monomers and dimers. Employing a QHA analysis, applied to the VDOS, provides estimates of adsorption free energy within chemical accuracy (~10 kJ/mol) of the experimentally determined quantities, whereas the traditional approach, employing harmonic frequencies from a single ground state minimum, strongly overestimates the adsorption free energy by as much as 16~43 kJ/mol. This discrepancy is traced back to the inability of the harmonic approximation to represent the contributions to the vibrational motions of the ethanol molecule upon confinement in the zeolite.

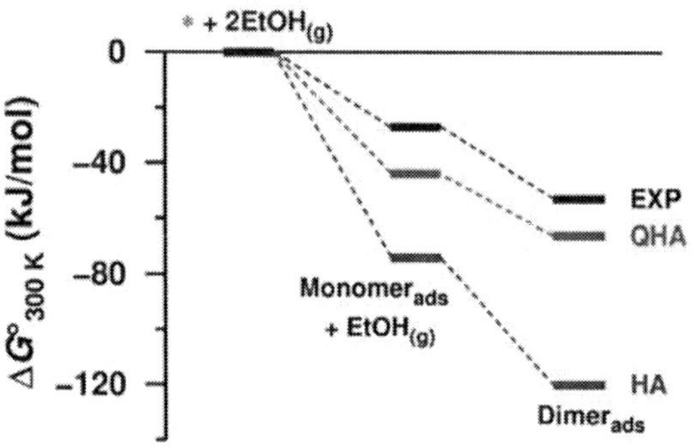

Figure 2. Gibbs free energy profile at 300 K for ethanol adsorption on the Brønsted acid site (*) of H-ZSM-5 (HA: harmonic approximation, QHA: quasi-harmonic approximation, EXP: experiment).

Tc-incorporated magnetite. Technetium (^{99}Tc) is an abundant, long-lived radioactive fission product whose mobility in the subsurface is largely governed by its oxidation state. Immobilization of Tc is crucial for radioactive waste management and environmental remediation. Tc(IV) incorporation in spinels has been proposed as a novel method to increase Tc retention in glass waste forms during vitrification where Fe(II) is acting as an effective reducing agent for Tc(IV) incorporation. However, experiments with Tc-magnetite, under high temperature and oxic conditions, showed reoxidation of Tc(IV) to volatile pertechnetate, Tc(VII). To understand AIMD simulations show that Tc stays at the surface with the prevalent oxidation state of Tc(IV) at 25 °C, while at 600 °C Tc oxidation is in process, commensurate with the tetrahedrally coordinated Tc by moving out above the surface as it transitions to TcO_4^-. Motivated by experimental observations showing stabilization of Fe(II) in magnetite with 1 wt % of 1st row transition metal,[3] we proposed systems with Co/Ni/Zn dopants to stabilize Tc(IV) in magnetite. Our simulations indeed showed that at elevated temperatures, doping can significantly enhance Tc retention in magnetite in the order Co > Zn > Ni. Subsequent experiments with doped spinels at T = 700 °C provided quantitative confirmation of increased Tc retention in the same order predicted by theory. To best connect with the experimental observations, estimates of the Gibbs free energy are obtained from the relation $\Delta G = -RT \ln K_{eq}$, where K_{eq} = $[Tc_{in}]/[Tc_{out}]$ were calculated using populations of the two different oxidation states of Tc(IV) ($[Tc_{in}]$) and Tc(VII) ($[Tc_{out}]$). These results show a remarkable agreement between the theoretical prediction and experimental validations, not only in terms of relative order, but also in magnitude (Table 1). The underlying reason is based on the increase in the reducing capacity of the Tc-containing spinels upon doping. This can be quantified by the difference in energy between the Fermi level, E_f, and the highest occupied Tc d states, ΔE_{gap}, where Zn and Co show a much higher stabilization that ultimately hinders Tc re-oxidation (Table 1).

Table 1. Equilibrium constants, free energy estimates compared with experiments considering doping effects.

Doping	K_{eq}	ΔG (kJ/mol)	Exp. Retention (%wt)*	ΔE_{gap} (eV)
Tc	0.15	14.2	--	0.15
Tc/Ni	0.56	4.3	4	0.35
Tc/Zn	2.79	-7.7	12	1.03
Tc/Co	21.80	-23.1	29	1.12

* Experimental values correspond to the amount of Tc remaining in the doped magnetite after exposure to 700 °C.

Conclusions

Current AIMD simulations are able to address complex system models (ca 1000 atoms) for time scales on the order of 10-100 ps at elevated temperatures. These methods allow us to approach problems of high chemical complexity. The works summarized here highlight the power of modern, state-of-the-art simulations to provide essential insights that can be used to guide the development of real materials for topical applications in catalysis and environmental technology and generate theory-inspired design criteria of complex materials.

Acknowledgement. This works were supported by the US department of Energy, Office of Basic Energy Science, Division of Chemical Sciences, Geosciences & Biosciences, and Office of River Protection, Waste Treatment and Immobilization Plant. Computational resources were provided at W. R. Wiley Environmental Molecular Science Laboratory (EMSL), a national scientific user facility sponsored by the Department of Energy's Office of Biological and Environmental Research located at PNNL, the National Energy Research Scientific Computing Center (NERSC) at Lawrence Berkeley National Laboratory.

References

(1) Lee, Mal-Soon; McGrail, B. P.; Rousseau, R.; Glezakou, V.-G. *Sci. Rep.* **2015**, 5, 14857.

(2) Alexopoulos, K.; Lee, M.-S.; Liu, Y.; Zhi, Y.; Liu, Y.; Reyniers, M.-F.; Marin, G. B.; Glezakou, V.-A.; Rousseau, R.; Lercher, J. A. *J. Phys. Chem. C* DOI: 10.1021/acs.jpcc.6b00923.

(3) Sidhu, P. S., Gilkes, R. J., Posner, A. M. Mechanism of the low temperature oxidation of synthetic magnetites. *J. Inorg. Nucl. Chem.* **39**, 1953-1958 (1977).

FABRICATION OF PALLADIUM CONCAVE NANOCRYSTALS VIA AUTO-CATALYTIC TIP OVERGROWTH

Na Su, Xueying Chen, * *Bin Yue and Heyong He*

Department of Chemistry, Shanghai Key Laboratory of Molecular Catalysis and Innovative Materials, Collaborative Innovation Center of Chemistry for Energy Materials, Fudan University, Shanghai 200433, P. R. China
E-mail: xueyingchen@fudan.edu.cn

Introduction

Shape-controlled synthesis of noble-metal nanocrystals has attracted tremendous attentions for decades owing to their fantastic properties and promising applications.[1,2] As a major limitation, these nanocrystals commonly investigated are enclosed by convex surfaces. However, noble-metal nanocrystals with concave surfaces have been demonstrated with enhanced catalytic properties as compared to their convex counterparts because of the higher density of atomic steps and kinks than the convex surfaces.[3,4]

The great challenge in the synthesis of noble metal concave nanocrystals can be largely attributed to the presence of concave surfaces which are usually not favored by thermodynamics owing to the involvement of the high-index facets with higher surface energy. One promising method to synthesize scalable concave nanocrystals is the colloid method, where the use of a facet-selective capping agent, the control of the reduction kinetics, the concentration gradient and the surface diffusion of atoms have been proven to be effective factors for forming concave nanocrystals. However, the interplay of these important factors involved in the formation of concave nanocrystals is still not clear. Understanding of the interplay of these important parameters involved in the synthesis can offer deep insight into the growth mechanism of concave nanocrystals, which is crucial for the rational design of nanostructures with novel properties.

In this work, we systematically studied the detailed growth mechanism of concave palladium arrow-headed tripods and revealed how the formation of the concave Pd nanocrystals was controlled by the interplay of reduction kinetics, concentration gradient of Pd precursors, and surface diffusion of atoms.

Experimental

Sample synthesis. 160 mg of PVP and 50 mg of $Pd(acac)_2$ were dissolved in 10 mL of benzyl alcohol. Certain amount of benzaldehyde was added in the reaction mixture. The obtained yellowish solution was transferred to oil bath preset at a specific temperature and heated under nitrogen protection without stirring. The resulting products were collected by centrifugation and washed with ethanol and acetone to remove the excessive PVP. The final products were dispersed in ethanol for further characterization. In a typical preparation, 40 μL of benzalehyde was used and the synthesis system was maintained at 70 °C for 6 h.

Characterization Methods. High-resolution scanning electron microscopy (HRSEM) images were taken on a Zeiss Ultra 55 field-emission scanning electron microscope. The scanning transmission electron microscopy (STEM) images were acquired on a FEI Tecnai G^2 F20 S-Twin field-emission transmission electron microscope. The transmission electron microscopy (TEM) images and selected-area electron diffraction (SAED) patterns of the samples were acquired on a JEOL 2011 transmission electron microscope. The analysis of the solution composition was conducted by GC9560 gas chromatography (Shanghai Hua-Ai Chromatography Analysis Co. Ltd.) equipped with a flame ionization detector and a HP-5 capillary column (0.32 mm × 30 m × 0.25 μm, Agilent, USA). The products were confirmed by the retention time of the standard samples.

Results and Discussion

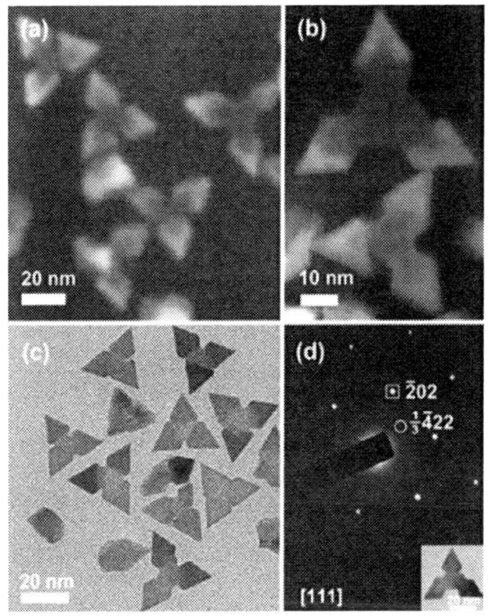

Figure 1. (a) HRSEM, (b) STEM, (c) TEM and (d) SAED images of Pd arrow-headed tripods obtained using typical preparation procedure.

Figure 1 presents the typical HRSEM, STEM and TEM images of the Pd arrow-headed tripods with an average tip-to-tip distance of ca. 40 nm synthesized using the typical preparation procedure. The three arms of the planar tripods are 120° apart and the tips of the tripods are interestingly in the shape of truncated or concave tetrahedron (**Figure 1a-c**). The SAED pattern of an individual Pd tripod lying flat on the TEM grid (**Figure 1d**) shows a 6-fold rotational symmetry, indicating that the tripod is single-crystalline and the basal facets of the tripod are {111} planes. The observation of 1/3{422} reflection in the SAED pattern indicates the presence of stacking faults in the {111} plane perpendicular to the electron beam.

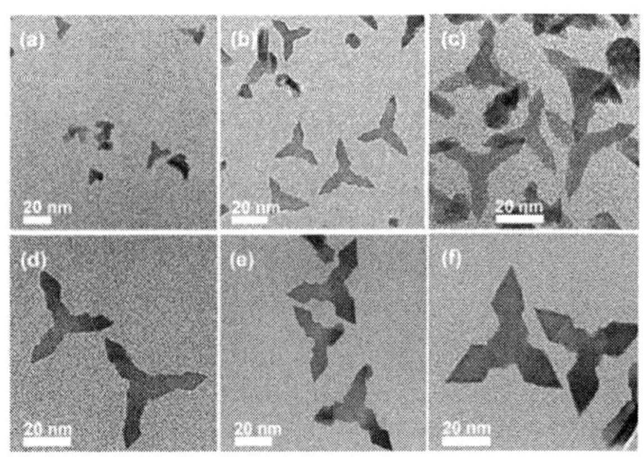

Figure 2. TEM images of representative Pd nanocrystals obtained using the typical preparation procedure except for the use of 20 μL of benzaldehyde at different reaction times: (a) 0.5 h, (b) 1 h, (c) 2 h, (d) 3 h, (e) 9 h, and (f) 24 h

TEM images of the specimens collected at different reaction stages are presented in **Figure 2**. After a reaction time of 0.5 h, the products were mainly triangular nanoplates (**Figure 2a**). As the reaction time proceeded to 1 h, the nanoplates grew along three <211> directions and three short arms appeared to form tripod (**Figure 2b**). When the reaction time was extended to 2 h, the length of the arms increased gradually and many troughs appeared on the arms (**Figure 2c**). When the reaction time was increased to 3 h (**Figure 2d**), some tips of the tripods became slightly darker than the other part of the tripods, indicating that the thickness at the tips is larger than that at the other part. When the reaction time was prolonged to 9-24 h, the troughs on the arms gradually disappeared and the tips of the tripods gradually became arrow-headed shapes (**Figure 2e-f**).

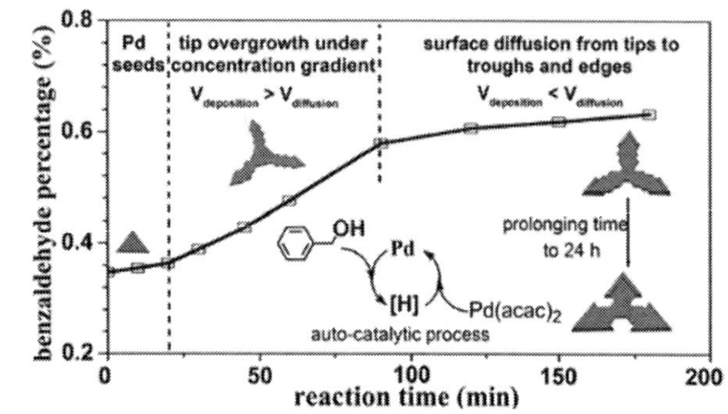

Figure 3. The evolution of benzaldehyde and the Pd nanocrystals in the reaction mixture as a function of reaction time using the typical preparation procedure except for the use of 20 μL of benzaldehyde.

The solution composition of the reaction mixture at different reaction stages was also recorded through GC analysis. In our N_2 protected synthesis system, only benzyl alcohol, benzaldehyde and very small amount of toluene and benzoic acid were detected in the solution. **Figure 3** shows the evolution of benzaldehyde as a function of reaction time. The evolution plot can be divided into three stages based on the slope of the plot. At the initial stage of reaction ($t < 20$ min), the percentages of benzaldehyde only changed very slightly, suggesting that $Pd(acac)_2$ was reduced at a very slow rate to form triangular nanoplate seeds. At the second stage of the reaction (30 min $< t <$ 90 min), the Pd seeds could auto-catalyze the dehydrogenation of benzyl alcohol to generate hydrogen atom [H] and benzaldehyde, both of which can reduce $Pd(acac)_2$ and accelerate its reduction rate to form Pd in return. Thus, sharply increased percentage of benzaldehyde was observed. The fast consumption of $Pd(acac)_2$ facilitated the introduction of concentration gradient of Pd precursors in our non-stirring reaction system and the kinetically-controlled tip overgrowth to form tripod. Under the condition where the rate of Pd atom deposition (denoted as $V_{deposition}$) was much larger than the rate of surface diffusion (denoted as $V_{diffusion}$), the continuous overgrowth of the tips led to the formation of the tripod with many troughs on the three arms. At the late stage ($t > 90$ min), the auto-catalytic process was depressed with the increase of the tripod size and the change of the percentages of benzyl alcohol became much smaller. Accordingly, the concentration gradient of Pd precursors was less apparent and the rate of Pd atom deposition was decreased dramatically. Under the condition where the rate of Pd atom deposition was smaller than the rate of surface diffusion ($V_{deposition} < V_{diffusion}$), the deposited atoms would be more likely to undergo surface diffusion from tips to troughs on the arms or to the edges of the main {111} planar surface of the tripods. As a result, the majority of the troughs gradually disappeared and the arrow-headed tips gradually formed.

By tuning the relative rate of atom deposition and surface diffusion of atoms, various Pd nanostructures could be successfully synthesized. As shown in **Figure 4a-b**, if the reduction rate was accelerated by replacing $Pd(acac)_2$ with Na_2PdCl_4 or by replacing benzaldehyde with KBH_4, faster reduction led to the formation of large amount of Pd grains which tended to aggregate together to form nanodendrites or nanochains. When the benzyl alcohol was changed to the mixture of benzene and ethanol which is impossible to undergo dehydrogenation reaction under our present reaction conditions, only quasi-spherical nanoparticles with the average diameters of ~7 nm were obtained (**Figure 4c**). When the benzyl alcohol was replaced by benzaldehyde, the Pd nanocrystals obtained were truncated tetrahedra associated with some quasi-spherical nanoparticles (**Figure 4d**), instead of tripod structure. When the reaction temperature was increased to 80 °C to enhance the surface diffusion or stirring was introduced to depress the concentration gradient of Pd precursors, the dominant products were concave or truncated tetrahedral (**Figure 4e-f**). Our results demonstrate the importance of the auto-catalytic process, and the interplay of the reduction kinetics, the concentration gradient of Pd precursors and the surface diffusion in the shape-controlled synthesis of nanocrystals with concave surface. Owing to the unique morphology and the presence of the high-index facets, these concave Pd nanocrystals are expected to have wide applications in catalysis, nanoelectronics and other related fields.

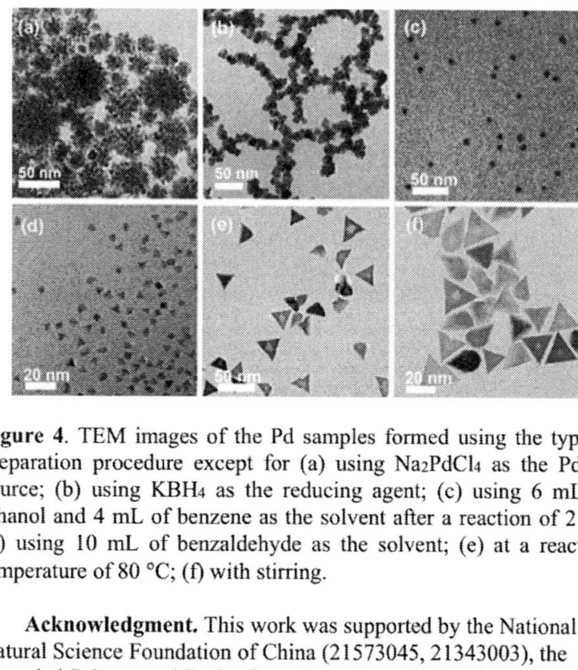

Figure 4. TEM images of the Pd samples formed using the typical preparation procedure except for (a) using Na_2PdCl_4 as the Pd(II) source; (b) using KBH_4 as the reducing agent; (c) using 6 mL of ethanol and 4 mL of benzene as the solvent after a reaction of 21 h; (d) using 10 mL of benzaldehyde as the solvent; (e) at a reaction temperature of 80 °C; (f) with stirring.

Acknowledgment. This work was supported by the National Natural Science Foundation of China (21573045, 21343003), the Shanghai Science and Technology Committee (12ZR1401400) and the Ministry of Education (the Scientific Research Foundation for the Returned Overseas Chinese Scholars).

References

1. Burda, C.; Chen, X. B.; Narayanan, R.; El-Sayed, M. A., *Chem. Rev.* **2005,** *105*, 1025-1102.
2. Chen, H. J.; Shao, L.; Li, Q.; Wang, J. F., *Chem. Soc. Rev.* **2013,** *42*, 2679-2724.
3. Dai, Y.; Mu, X. L.; Tan, Y. M.; Lin, K. Q.; Yang, Z. L.; Zheng, N. F.; Fu, G., *J. Am. Chem. Soc.* **2012,** *134*, 7073-7080.

4. Zhou, Z. Y.; Tian, N.; Li, J. T.; Broadwell, I.; Sun, S. G., *Chem. Soc. Rev.* **2011,** *40*, 4167-4185.

METAL-ORGANIC FRAMEWORKS: SYNTHESIS AND OXYGEN ELECTROCATALYSIS

*Hao Wang, Fengxiang Yin, and Biaohua Chen**

College of Chemical Engineering
Beijing University of Chemical Technology
Beijing 100029, P.R. China
chenbh@mail.buct.edu.cn

Introduction

The increasing environmental pollution and the emergent energy crisis prompt the development of highly efficient energy storage and conversion systems, such as regenerative fuel cells and rechargeable metal-air batteries. In such systems, the oxygen reduction reaction (ORR) and oxygen evolution reaction (OER) are the core components in determining the practical performance. Unfortunately, both ORR and OER involve multiple elementary reaction steps, resulting in the sluggish kinetics. The sluggish ORR/OER will cause a low performance of storage and conversion systems, which cannot meet the requirements of practical application. It is very essential to accelerate the ORR and OER by effective electrocatalysts and further reduce overpotentials, finally improving the performance of energy storage and conversion technologies.

Metal-organic frameworks (MOFs) are porous crystalline materials self-assembled with metal nodes and organic linkers. MOFs possess various compositions, large specific surface areas and tunable pore structures, and they are easily functionalized. Therefore, many researches on MOFs have been focused on the applications in energy conversion and storage, especially as electrocatalysts for ORR/OER.[1] In this work, we systemically investigated MOFs as effective ORR/OER electrocatalysts in the following ways: (1) MOFs in pure phase, (2) hybrids of MOFs with sustainable carbon, (3) MOFs decorated with transition-metal ions/oxides, and (4) materials derived from MOFs.

Experimental

Sample synthesis. (1) A MOF(Fe/Co) with the mixed metal nodes was synthesized through a hydrothermal method.[2] (2) As for the ZIF-67@NPC-2 hybrids, N-doped pomelo-peel-derived carbon (NPC) was first synthesized from pomelo peels as carbon source and melamine as nitrogen source. Then ZIF-67 was then integrated with NPC through a hydrothermal process. (3) A MIL-101(Cr) with various surface Co^{III} and Co^{II} contents was prepared via an impregnation process, followed by an oxidation or a reduction treatment.[3] (4) An ε-MnO_2/MOF(Fe) composite was synthesized by thermal decomposition of manganese nitrate on MOF(Fe) matrix.[4]

Electrochemical Measurements. All the electrochemical measurements were conducted at room temperature in 0.1 M KOH using a carbon paper test cell, rotating disk electrode (RDE), or a rotating ring-disk electrode (RRDE) apparatus. The apparatus was connected to a computer-controlled CHI 760E electrochemical workstation (CH Instruments, China). The electrochemical measurements of the working electrodes were performed in a conventional three-electrode system. A KCl-saturated Ag/AgCl electrode and a Pt wire were used as the reference and counter electrodes, respectively. The KCl-saturated Ag/AgCl electrode was calibrated with respect to a reversible hydrogen electrode (RHE) according to our previous work. [4]

Characterization Methods. The N_2 adsorption–desorption isotherms of the samples were collected using an Autosorb-iQ2-MP automatic analyser (Quantachrome, USA). The specific surface areas (SSAs) were measured using the Brunauer–Emmett–Teller (BET) model. Field-emission scanning electron microscopy (FESEM) images of the samples were obtained using a SUPRA55 microscope (CARL ZEISS, Germany). The samples' transmission electron microscopy (TEM) images, high-resolution TEM (HRTEM) images and high-angle annular dark field scanning TEM (HAADF-STEM) images were observed using a Tecnai G^2 F20 electron microscope (FEI, USA).

Results and Discussion

The unique structures and design flexibility highlight the potential of MOFs as ORR/OER electrocatalysts directly. We first synthesized a mixed metal ions MOF, MOF(Fe/Co).[2] MOF(Fe/Co) showed excellent crystalline structures, high specific surface area, and well-defined micropores. We tested the bifunctional catalytic activities of MOF(Fe/Co) by coating the catalyst on carbon papers as working electrodes in a carbon paper test cell. Compare with Super P (SP), MOF(Fe/Co)+SP afforded excellent catalytic performance toward ORR/OER (Figure 1C-D). The preliminary analysis indicates that the numerous unsaturated Fe or Co nodes and open structures are favorable for ORR/OER. Additionally, high specific surface area and numerous micropores also play key roles in bifunctional catalytic activities.

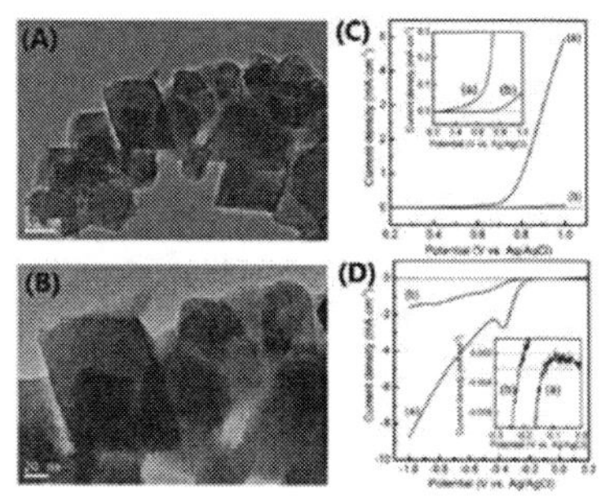

Figure 1. (A-B) TEM images of the MOF(Fe/Co) samples. (C) OER and (D) ORR catalytic activities of the samples: (a) MOF(Fe/Co)+SP and (b) SP.[2]

MOFs directly as electrocatalysts are still limited by their intrinsic poor electronic conductivity and need to be improved. An effective strategy to improve the electrical conductivity of MOFs is to integrate MOFs with carbons. The synthesis of carbon materials is traditionally from fossil fuels-based precursors, and requires harsh synthetic conditions and high energy consumption. Therefore introducing sustainable concepts into the preparation to obtain sustainable carbon materials has aroused great interests.

In our following work, we prepared N-doped pomelo-peel-derived carbon (NPC) from widely available food waste—pomelo peels—as carbon source and melamine as nitrogen source. The synthetic NPC exhibited well-defined porosities and a high doped-N content and thus afforded excellent ORR catalytic activities in alkaline electrolytes. NPC was integrated with ZIF-67 to form ZIF-67@NPC hybrids through hydrothermal reactions. The hybrid catalysts showed substantially enhanced ORR catalytic activities comparable with the activity of commercial 20 wt% Pt/C and also exhibited excellent OER catalytic activities. Among all of the prepared ZIF-67@NPC hybrids, the optimal composition denoted as ZIF-67@NPC-2(2:1) exhibited the best ORR and OER catalytic performance and thus the best bifunctional catalytic performance with the smallest ΔE (E_{OER}@10 mA·cm^{-2}-E_{ORR}@-1 mA·cm^{-2}) value, desirable 4-electron transfer pathways and superior catalytic stabilities. The Co-N_4 in ZIF-67, electrochemical active surface area, and the strong interactions between ZIF-67 and NPC were found to be the main contributors to the bifunctional catalytic activities of the hybrid.

Another prevailing strategy is to load active species on MOFs. Decorating MOFs with transition metal active species not only takes the advantages of high specific surface areas and high porosities of MOFs, but also significantly limits the agglomeration of transition metal species, thus improving the dispersion of them on MOFs and the ORR/OER activities.

The ORR catalytic activities of Co-based electrocatalysts are considered to be related to the Co oxidation states. We decorated MIL-101(Cr) with Co species with various oxidation states via using KBH_4 reducing agent or H_2O_2 oxidant (Figure 2).[3] The results showed that the OER activities increased with the surface Co^{III} contents, whereas the ORR activities were positively affected by Co^{II} contents, and higher Co^{III}/Co^{II} ratio in the electrocatalyst resulted in higher overall bifunctional activity. The porous and open structures of MIL-101(Cr) support favored the contact between oxygen species and the active Co sites. Additionally, the interactions between Co species and MOF support also enhanced the catalytic durability of sample. This work is helpful for preparing efficient electrocatalysts based on MOFs via facilely incorporation of highly active species.

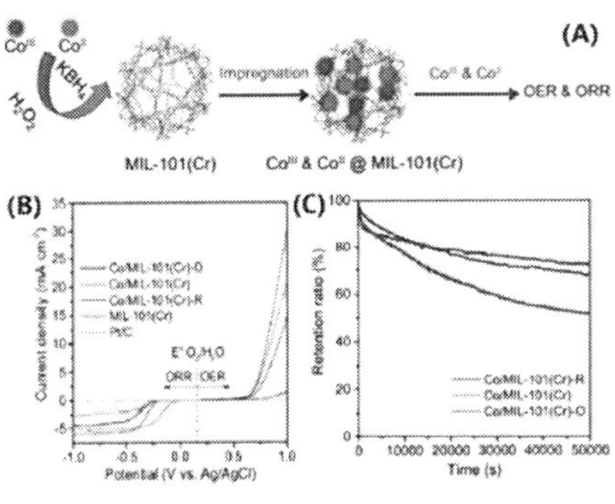

Figure 2. (A) The illustration of design and preparation of the samples. (B) Overall bifunctional LSV curves and (C) chronoamperometric curves of the samples.[3]

Manganese oxides, especially MnO_2, have been demonstrated to be active toward ORR, even comparable to the commercial Pt/C. However, MnO_2 nanoparticles tend to agglomerate during preparation and electrochemical reaction, likely leading to larger MnO_2 particles with lower specific surface areas and lowering the catalytic activity and stability. It is expected to find effective supporting materials to enhance the MnO_2 performance.

We prepared a MOF(Fe) via hydrothermal reaction to support ε-MnO_2.[4] In the ε-MnO_2/MOF(Fe) composite, ε-MnO_2 nanorods were firmly anchored onto the MOF(Fe) matrix (Figure 3A) and showed better ORR catalytic activity than bare MOF(Fe), ε-MnO_2 and the physical mixture of ε-MnO_2+MOF(Fe). The ε-MnO_2/MOF(Fe) composite favored an apparent 4-electron transfer pathway in which oxygen was first reduced to hydroperoxide, which was further chemically decomposed into primarily OH^- and O_2 (Figure 3C). The ε-MnO_2/MOF(Fe) composite showed excellent methanol crossover stability and ORR catalytic durability (Figure 3D).

Apart from MOFs with or without added materials, MOFs can be transformed into porous carbon or metal@carbon (the metal can be in the metallic, oxide or carbide phase) with novel structures due to the organic-inorganic hybrid structures at the atomic scale of MOFs. Our further study focused on a metal@C hybrid derived from a mixture of ZIF and graphite oxide (GO) through a pyrolysis process. The initially well-dispersed metal centers in ZIF agglomerated into small metal particles which were then uniformly anchored on the GO. The metal@C hybrid afforded much better bifunctional catalytic activity towards ORR/OER than the pristine MOF, and one of the reasons is the enhanced electronic conductivity of the MOF-derived material.

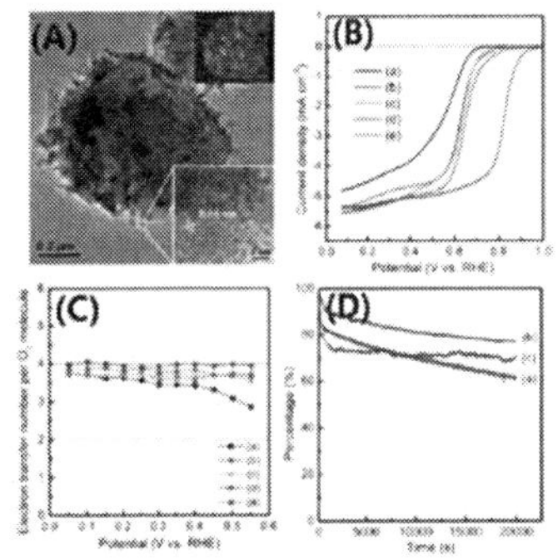

Figure 3. (A) TEM image of the ε-MnO_2/MOF(Fe) sample (inset: HRTEM image performed on a ε-MnO_2 nanorod and element mapping performed on the particle). (B) ORR LSV curves and (C) electron-transfer number of the samples: (a) MOF(Fe), (b) ε-MnO_2, (c) ε-MnO_2+MOF(Fe), (d) ε-MnO_2/MOF(Fe) and (e) 20wt% Pt/C. (D) Chronoamperometric curves of the samples: (a) ε-MnO_2, (b) ε-MnO_2/MOF(Fe) and (c) 20wt% Pt/C.[4]

Our primary results suggest that MOFs show some catalytic activity towards ORR/OER, but suffers from poor electronic conductivity. Despite of having good electronic conductivity, MOFs-derived materials are at the expense of the MOFs' properties. For instance, the specific surface areas of the MOFs are reduced and their ordered porosities are destroyed; both of these effects are disadvantageous for ORR/OER catalysis. A fast ORR/OER response requires strong interactions in between catalytic site, electrolyte and reactants, thus further requires suitable surface area and pore structure of catalysts for reactant and electrolyte diffusion. Therefore, further design, construction and optimization of MOFs-based materials should be focused on the following aspects: (1) effective construction of active sites within MOFs, (2) electronic conductivity improvement, (3) high specific surface area and suitable pore structures to contain enough electrolytes and provide accessible mass-transfer pathways and (4) better understanding of ORR/OER mechanism on such structures.

Acknowledgment. We gratefully acknowledge the Natural Science Foundation of China (Project No. 21276018) for its financial support.

References

(1) Morozan, A.; Jaouen, F., *Energy & Environ. Sci.* **2012**, *5* (11), 9269-9290.
(2) Wang, H.; Yin, F.; Li, G.; Chen, B.; Wang, Z., *Inter. J. Hydrogen Energy* **2014**, *39* (28), 16179-16186.
(3) He, X.; Yin, F.; Li, G., *Inter. J. Hydrogen Energy* **2015**, *40* (31), 9713-9722.
(4) Wang, H.; Yin, F.; Chen, B.; Li, G., *J. Mater. Chem. A* **2015**, 3, 16168-16176.

RATIONAL DESIGN OF REDOX CATALYSTS FOR HYDROCARBON OXIDATION, WATER-SPLITTING, AND CO_2 UTILIZATION

Fanxing Li

Department of Chemical and Bimolecular Engineering
North Carolina State University
Raleigh, NC 27695
Email: fli5@ncsu.edu

Introduction

Unlike conventional fuel conversion processes which oxidize carbonaceous molecules with gaseous oxidants, the chemical looping strategy seeks to intensify fossil fuel conversion processes using lattice oxygen of transition metal oxides as the oxidant.[1] In a typical chemical looping scheme, carbonaceous molecules are oxidized by active lattice oxygen (O^{2-}) in the transition metal oxide, which functions as an oxygen carrier, a.k.a. redox catalyst. In a subsequent step, the O^{2-} deprived oxygen carrier is replenished by a gaseous oxidant. In addition to fossil fuel combustions, tailored redox catalysts can be applied for a variety of applications including selective oxidation, water-splitting, and CO_2 reduction.[2-5]

The current study exemplifies the use of iron and manganese containing mixed oxides for methane partial oxidation, oxidative dehydrogenation (ODH) of ethane, water-splitting, and CO_2 reduction. A number of redox catalysts composed of iron and/or manganese oxides are synthesized and characterized. Rational strategies to improve the redox activity, product selectivity, and/or water-splitting and CO_2 reduction conversions are investigated through the assistance of *ab-initio* calculation tools and mechanistic investigations. Compared to traditional heterogeneous catalysts, the redox catalysts in the current study show the potential to be more selective. They also possess excellent redox stability. Close to 100% water and CO_2 reduction efficiencies are shown to be achievable possible through rational selection of mixed metal oxides.

Experimental

Redox catalysts were synthesized using solid state reaction (SSR), sol-gel, and impregnation methods. In terms of methane oxidation and water-splitting, iron oxide is used as the primary oxygen carrier with a mixed conductive support. Iron-containing mixed oxides were also investigated for water-splitting and CO_2 reduction. To investigate oxidative dehydrogenation of ethane, manganese oxide was used as the primary oxide with both inert and mixed-conductive supports. The effects of promoters on activity and selectivity of these redox catalysts were also investigated.

Various characterization techniques were used to analyze the structural, surface, and morphological properties of the catalyst samples. XRD was used to analyze the crystalline phases present in each sample before and after redox tests. XRD spectra were obtained using a Rigaku SmartLab X-ray diffractometer with Cu-Kα (λ=0.1542) radiation source operating at 40 kV and 44 mA. X-ray Photoelectron Spectroscopy (XPS, SPECS-XPS) with an Mg-Kα excitation (1254 eV) and a PHOIBIS 150 hemispherical analyzer at 10^{-10} mbar were used to analyze the elemental compositions on the particle surface. The specific surface areas of the samples were measured using a BET surface area analyzer (Quantachrome, Autosorb-1). Reducibility of the samples was tested using a Quantachrome ChemBET Pulsar temperature programmed reduction/desorption (TPR/TPD) instrument. 5% hydrogen mixed with argon was used as the reducing gas. Temperature ramping during the TPR experiment is 5°C/min. Sample morphologies were characterized using Scanning Electron Microscopy (SEM, Hitachi S3200). The same instrument was used to analyze the elemental composition of the samples through Energy Dispersive X-ray Spectroscopy (EDS). Transmission Electron Microscopy (TEM, JEOL JEM 2010F) and Energy Filtered imaging (EFTEM) with 200 keV accelerating voltage were also used to obtain elemental mapping.

Redox experiments were conducted in both differential and integral bed reactors. The reactor system was composed of a computer-controlled panel for gas mixing and delivery, a TGA (STARAM SETSYS Evolution) and a U-tube reactor for redox reactions, and a quadrupole mass spectrometer (QMS, MKS Cirrus 2) for gaseous product analyses. Cyclic studies were also conducted to evaluate the stability of the redox catalysts. Carbon formation during the reduction of the redox catalyst was characterized by combustion analysis.

Density functional theory (DFT) calculations were performed using spin polarized generalized gradient approximations (GGA) and the projector augmented wave (PAW) approach as implemented in VASP. Various magnetic configurations were tested for stability (with spin polarization), and calculations were converged in k-point density. A cutoff energy of 425 eV was used for all systems. The energy to create a vacancy was found using the difference between the energy of a supercell with a single vacancy plus half a diatomic oxygen and a pristine supercell. The supercell size was increased until the energy to form an oxygen vacancy became constant, corresponding to the dilute limit. A CI-NEB method was used to calculate the transition state of vacancy migration.

Results and Discussion

In terms of methane partial oxidation and CO_2 reduction, cyclic redox operations are carried out in the fixed bed reactor. F-T ready syngas is produced from the methane oxidation cycle whereas CO is produced from the CO_2 reduction cycle. To illustrate the stability of the redox catalyst over multiple redox cycles, syngas productivity of the redox catalyst (in the methane partial oxidation step) is shown in Figure 1. As can be seen, the redox catalyst is highly recyclable, showing minimal deactivation from cycle 10 onwards. Moreover, an exceptionally high syngas selectivity was achieved, as evidenced by the negligible amount of complete oxidation products, e.g. CO_2, in the product stream. Figure 2 illustrates the gas phase product distributions during the CO_2 reduction half-cycle. As can be seen, the reduced redox catalyst is capable of converting nearly 100% CO_2 into CO. Such a high conversion is achieved by tailoring the thermodynamic and kinetic properties of the redox catalyst.

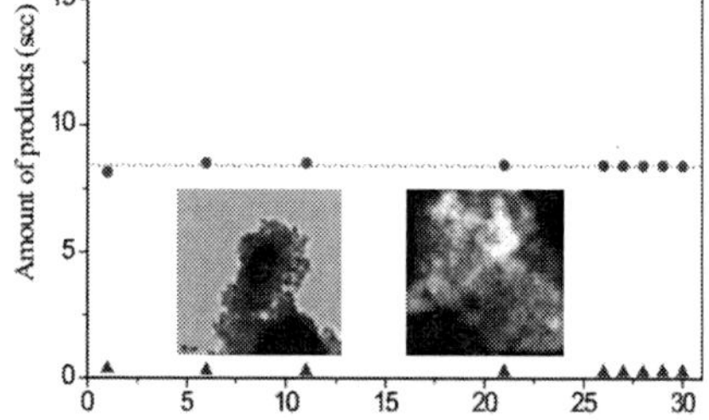

Figure 1. Overall product gas yields from methane partial oxidation reaction over 30 redox cycles.

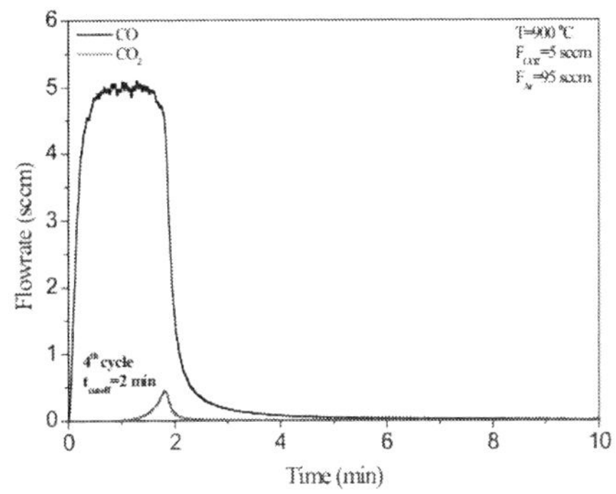

Figure 2. Product gas profile from a typical CO_2 reduction half cycle

Figure 3 summarizes the ODH performance of several promoted, Mn-based redox catalysts for ethane oxidative dehydrogenation. As can be seen, phase of the oxygen carrying metal oxide and its promoters play a critical role for activity and selectivity of Mn-based redox catalysts. High ethylene yield is achieved with tailored metal oxides and proper selection of support. Metal oxide selection rationale and optimization strategy are illustrated through fundamental understanding of the redox and surface reaction mechanism.

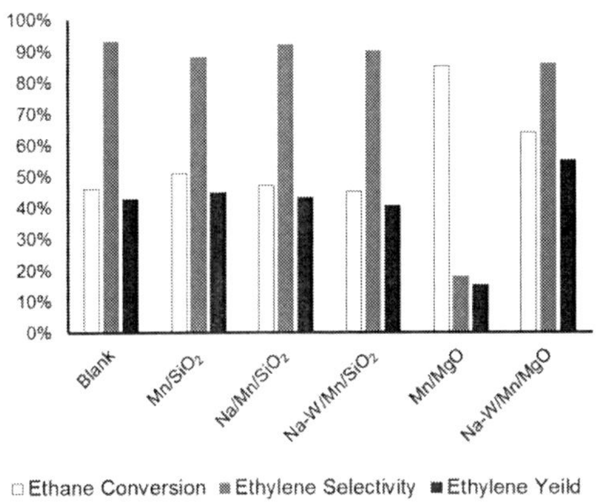

Figure 3. Conversion, selectivity and yield for ODH of redox catalyst

Conclusion

The present study investigates first-row transition metal oxide based redox catalysts for partial oxidation of light paraffins including methane and ethane as well as their applications for water-splitting and CO_2 reduction. Novel, mixed-oxide redox catalysts are found to be highly effective for these reactions of interest. Both support and promoter can play crucial roles for the redox activity and product selectivity of these redox catalysts.

References

(1) Li, F. and L.-S. Fan, *Clean Coal Conversion Processes – Progress and Challenges*. Energy & Environmental Science, 2008. **1**(2).

(2) He, F. and F.X. Li, *Perovskite Promoted Iron Oxide for Hybrid Water-Splitting and Syngas Generation with Exceptional Conversion*. Energy & Environmental Science, 2015. **8**(2): p. 535-539.

(3) He, F., et al., *A Hybrid Solar-Redox Scheme for Liquid Fuel and Hydrogen Coproduction*. Energy & Environmental Science, 2014. **7**(6): p. 2033-2042.

(4) Shafiefarhood, A., et al., *Fe_2O_3@$La_xSr_{1-x}FeO_3$ Core- Shell Redox Catalyst for Methane Partial Oxidation*. ChemCatChem, 2014. **6**(3): p. 790-799.

(5) Neal, L., Yusuf, S., Sofranko, J.A., Li, and F. *Oxidative Dehydrogenation of Ethane: A Chemical Looping Approach*. Energy Technology, 2016: DOI: 10.1002/ente.201600074

HIGHLY EFFICIENT SYNTHESIS OF HYDROGEN STORAGE MATERIAL OF FORMIC ACID FROM CO_2 AND WATER

Heng Zhong[1], *Guodong Yao*[2] *and Fangming Jin*[2,*]

[1]Graduate School of Environmental Studies, Tohoku University, Sendai 980-8579, Japan
[2]School of Environmental Science and Engineering, Shanghai Jiao Tong University, 800 Dongchuan RD, Shanghai 200240, China
*Email: fmjin@sjtu.edu.cn

Introduction

The depletion of fossil fuels, which is also known as the energy crisis, is one of most serious problems faced by human beings in the 21st century. Developing renewable energies, such as solar, wind and biomass energies, to replace the fossil fuels to achieve a sustainable development of the human societies is under the spotlight of recent scientific researches.[1] However, due to the problems of instability and unequal distribution of the renewable energies, methods for conversion, storage and transportation of the renewable energies are indispensible. Since most of the renewable energies can be used to generate electricity, H_2 produced from the dissociation of water using renewable energies derived electricity is regarded as one of most promising materials for storing the renewable energies.[2] H_2 is a non-toxic and environmental benign chemical fuel, which has the potential to replace fossil fuel in the future. However, since H_2 is hard to liquefy and has relatively low energy density, H_2 is not a suitable material for energy storage and transportation.

Recently, the conversion of H_2 into HCOOH, which is regarded as a H_2 carrier, for H_2 transportation and utilization has been extensively studied.[3] This is because HCOOH is a stable and low toxic liquid at ambient conditions which is easy to handle, and most importantly, the only byproduct of the H_2 regeneration from HCOOH is the CO_2, which can be recovered and hydrogenated to HCOOH again to achieve the CO_2-HCOOH cycle. On the other hand, CO_2 is a well known greenhouse gas that causes the problem of global warming due to the combustion of large amount of fossil fuels, which is another serious environmental problem face by the human beings in the 21st century. Thus, conversion of the CO_2 into HCOOH can be one promising method to help decreasing the atmospheric CO_2 concentration and alleviating the problem of global warming.

In the traditional methods of hydrogenation of CO_2 into HCOOH, gaseous H_2 source and/or noble metal catalysts are required, which greatly lifts its cost.[4] Photochemical and electrochemical ways for CO_2 reduction can utilize the renewable energies directly or indirectly, and can use the hydrogen from water directly. However, the low conversion efficiency still remains the problem. On the other hand, the hydrothermal method has shown a great potential in the reduction of CO_2 into HCOOH with earth abundant metals as the reductant and water as the hydrogen source recently.[5] With the *in-situ* formed hydrogen from the reaction of metal reductants with hot water, the hydrothermal method can achieve very high efficient in the CO_2 reduction into HCOOH. Since the metals can be reduced from its oxidized state by solar energy, renewable energies derived electricity or biomass, the metal reductants can be recycled for the hydrothermal reduction of CO_2 and can be regarded as an input of renewable energies. Thus, a sustainable method for conversion of CO_2 into HCOOH with the earth abundant metals as energy input and H_2O as hydrogen source can be achieved. This paper gives an overview of some recent strategy of hydrothermal conversion of CO_2 into formic acid by using earth abundant metals as the reductants as well as hot water as the hydrogen source.[5-9]

Experimental

Both $NaHCO_3$ and gaseous CO_2 were used as the source of CO_2. Commercial available Fe, Al, Zn and Mn powders were used as the reductants directly without any further treatment.

Experiments were conducted in a series of SUS-316 tubular reactors with an inner volume of 5.7 mL. The reaction temperature ranged from 250 to 325 °C and the reaction time varied from 5 to 150 min. The water filling ratios were 20 – 60%.

Liquid samples were analyzed by HPLC (Agilent 1260), GC-MS (Agilent 7890A + 5975C MSD), GC-FID (Agilent 7890A) and TOC (Shimadzu TOC-V CPH). Gaseous samples were analyzed by GC-TCD and the solid samples were characterized by XRD (Bruker D8 Advance, Shimadzu XRD-6100), XPS (EXCALAB 250) and dispersive Raman microscope (RAM, Senterra R200-L).

Results and Discussion

Reduction of CO_2 into HCOOH with Fe, Al, Zn and Mn. Firstly, the yields of HCOOH from CO_2 with different metal reductants at temperatures of 250 – 325 °C and reaction times of 5 – 150 min were investigated, and the results are shown in Fig. 1. When the temperature was lower than 300 °C, increasing the reaction temperature can effectively promote the yield of HCOOH, which is defined as the percent of obtained formic acid and the initial CO2 based on carbon base, in all the cases. However, when the temperature exceeded 300 °C, decrease in the yield of HCOOH was observed in the cases of Al or Zn using as the reductant, which was probably caused by the decomposition of HCOOH at high temperature. For the effect of reaction time, as can be seen in Fig. 1b, the yields of HCOOH increased continuously with the reaction time in all the cases. However, when Zn or Fe was used as the reductant, the increase in the yield of HCOOH with time slow down after 60 min, which means the reactions were close to the equilibrium, suggesting that Zn and Fe have better abilities to reduce the CO_2 into HCOOH than Al and Mn. It can also be seen that the yields of HCOOH obtained with Fe and Zn were higher than those obtained with Al and Mn generally.

Other reaction conditions, such amount of reactants, water filling ratio, also have great effects on the yield of HCOOH from CO_2. However, the details will not be discussed here. After investigating the effects of above reaction conditions on the yield of HCOOH, the highest yields of HCOOH obtained from the CO_2 reductions using different metal reductants are summarized in Table 1. The results showed that the earth abundant metals of Fe, Zn, Al and Mn exhibited great abilities to convert the CO_2 into HCOOH with very high yields. The highest yield of HCOOH was 92%, which was obtained when Fe was used.

Proposed reaction mechanism. After analyzing the gaseous samples with GC-TCD, large amount of H_2 gas were detected. Therefore, H_2 gas produced from reaction of metal with hot water was first assumed as the reductant of CO_2 reduction into HCOOH. However, substitution of the metal reductants with gaseous H_2 led to a great decline in the yield of HCOOH as shown in Table 2. XRD analysis of the precipitates after the reactions showed that the metal reductants of Fe, Zn, Al and Mn were oxidized into Fe_3O_4, ZnO, AlO(OH) and MnO+Mn_3O_4, respectively. Considering that these metal oxides might catalyze the CO_2 reduction into HCOOH during the reaction, effects of gaseous H_2 plus metal oxides on the yield of HCOOH were examined. However, the yields of HCOOH were still much lower than those obtained using metal reductants directly (Table 2). Thus, it is probably the *in-situ* formed H from water, rather than H_2 gas, played the key role in CO_2 reduction. A possible reaction path way of the reduction of HCO_3^- into HCOOH with *in-situ* formed H from the reaction of metal reductants with hot water is proposed in Fig. 2.

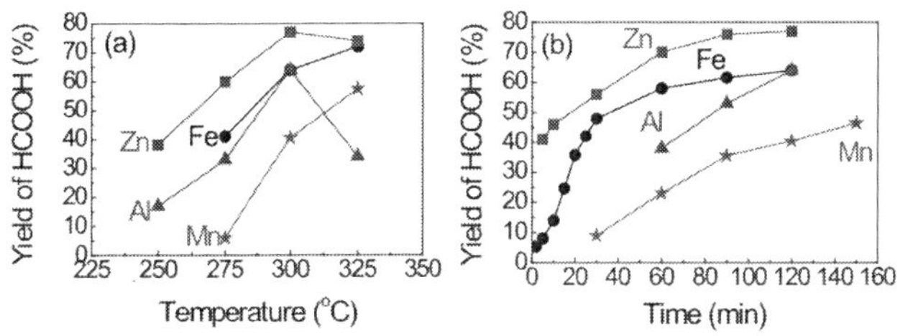

Figure 1. Effects of (a) temperature (at 120 min) and (b) time (at 300 °C) on the yields of HCOOH obtained with different metal reductants.

Table 1. Highest yields of HCOOH obtained with different metal reductants and the related reaction conditions.

Reductant		$NaHCO_3$ (mmol)	Temp. (°C)	Time (min)	Water filling (%)	Formic acid yield (%)
Name	Amount (mmol)					
Fe	24	2	300	120	53	92
Al	6	1	300	120	35	64
Zn	10	1	300	120	35	78
Mn	8	1	325	60	55	76

Table 2. Comparison of the effects of metal reductants and gaseous H_2 on the CO_2 reduction into HCOOH.

Entry	Metal additives		H_2 (mmol)	$NaHCO_3$ (mmol)	Yield of HCOOH
	Name	Amount (mmol)			
1	Zn	6	-	1	57.5
2	-	-	6	1	13
3	ZnO	6	-	1	16
4	Mn	8	-	1	76
5	-	-	6	1	2
6	Mn_3O_4	6	6	1	2
7	MnO	6	6	1	9

Reaction conditions: Entries 1 – 3 were conducted at 300 °C for 2 h; Entries 4 – 7 were conducted at 325 °C for 1 h.

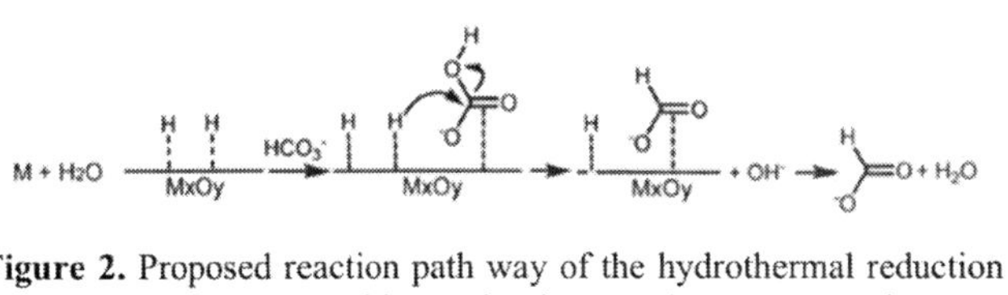

Figure 2. Proposed reaction path way of the hydrothermal reduction of HCO_3^- into HCOOH with metal reductants (M represents the zero-valent metal reductants; M_xO_y represents the related metal oxides).

Standard enthalpy of the reaction. The standard enthalpies of the reduction of HCO_3^- into HCOOH with different metal reductants were examined and the results are summarized in Table 3. It can be seen that no matter what metal reductant was used, the standard enthalpies of all the reactions were negative, suggesting that these reactions are exothermic and can be potentially self-supported. This means, as discussed in the Introduction, the metal reductants can be regarded as an input of the renewable energies to some extent.

Table 3. Standard enthalpies of the reduction of HCO_3^- into HCOOH with different metal reductants.

	Reaction	ΔH_r^Θ (298.15 K) (kJ/mol)
1	$Zn (cr) + HCO_3^- (aq) = ZnO (cr) + HCOO^- (aq)$	-86.1
2	$3Fe (cr) + 4HCO_3^- (aq) = Fe_3O_4 (cr) + 4HCOO^- (aq)$	-61.1
3	$2Al (cr) + 2HCO_3^- (aq) + 2H_2O (l) = 2AlO(OH) (cr) + 2HCOO^- (aq) + H_2 (g)$	-905.1
4	$Mn (cr) + HCO_3^- (aq) = MnO (cr) + HCOO^- (aq)$	-120.9

ΔH_r^Θ: standard enthalpy of the reaction.

Conclusions

The commercially available earth abundant metals showed excellent abilities to convert the CO_2 into HCOOH, a widely accepted H_2 storage material, with very high conversion efficiencies. Since the metal reductants can be easily obtained from the reduction of their natural oxides by solar energies, renewable energy derived electricity or biomass, the proposed method provided a new potential sustainable way to achieve a quick reduction of CO_2 into chemical fuels to solve problems of both energy crisis and global warming. This method is simple, efficient, and environmentally friendly as evident from the following points:

(i) it requires neither elaborately prepared noble catalysts nor highly purified reagents;
(ii) it requires no gaseous H_2 and can generate highly activated hydrogen *in-situ* without using any catalyst
(iii) it can achieve both CO_2 conversion and HCOOH yield over 90% within 2 h;
(iv) it only generates the by-product of metal oxides, which can be reduced to zero-valent again and recycled by using renewable energies

Acknowledgment. The authors thank the financial support of the National Natural Science Foundation of China (No. 21277091 & 51472159), the State Key Program of National Natural Science Foundation of China (No. 21436007) and Key Basic Research Projects of Science and Technology Commission of Shanghai (No.14JC1403100).

References

(1) Dincer, I. *Renew Sust Energ Rev* **2000**, *4*, 157-175.
(2) Ahmad, H.; Kamarudin, S. K.; Minggu, L. J.; Kassim, M. *Renew Sust Energ Rev* **2015**, *43*, 599-610.
(3) Enthaler, S.; von Langermann, J.; Schmidt, T. *Energy Environ Sci* **2010**, *3*, 1207-1217.
(4) Nguyen, L. T. M.; Park, H.; Banu, M.; Kim, J. Y.; Youn, D. H.; Magesh, G.; Kim, W. Y.; Lee, J. S. *Rsc Adv* **2015**, *5*, 105560-105566.
(5) Jin, F.; Gao, Y.; Jin, Y.; Zhang, Y.; Cao, J.; Wei, Z.; Smith Jr, R. L. *Energy Environ Sci* **2011**, *4*, 881-884.
(6) Duo, J.; Jin, F.; Wang, Y.; Zhong, H.; Lyu, L.; Yao, G.; Huo, Z. *Chem Commun* **2016**, *52*, 3316-3319.
(7) Jin, F.; Zeng, X.; Liu, J.; Jin, Y.; Wang, L.; Zhong, H.; Yao, G.; Huo, Z. *Sci Rep* **2014**, *4*, 4503.
(8) Lyu, L.; Zeng, X.; Yun, J.; Wei, F.; Jin, F. *Environ Sci Technol* **2014**, *48*, 6003-6009.
(9) Yao, G.; Zeng, X.; Jin, Y.; Zhong, H.; Duo, J.; Jin, F. *Int J Hydrogen Energy* **2015**, *40*, 14284-14289.

COBALT NANOCRYSTAL ASSEMBLED HOLLOW NANOPARTICLES FOR ELECTROCATALYTIC HYDROGEN GENERATION FROM NEUTRAL-PH WATER

Bingrui Liu and Mingming Ma

Department of Chemistry
University of Science and Technology of China
96 Jinzhai Rd. Hefei
Anhui P. R. China 230026

Introduction

Electrolysis of water driven by electricity that is derived from renewable energy sources could provide hydrogen as an ideal energy carrier for clean and sustainable energy technologies.[1] To achieve a high energy conversion efficiency, active catalysts for electrochemical hydrogen evolution reaction (HER) are required.[2] Nature's hydrogenase enzymes drive HER in neutral-pH solution with high efficiency and activity, but suffer from a low density of active sites and poor stability, which limit their practical use.[3] A few non-noble metal compounds have been investigated as catalysts with moderate activity for HER in neutral-pH solution. Therefore, highly active and stable electrocatalysts for HER in neutral-pH solution are highly desired, but remain very difficult to achieve.[4]

Cobalt metal nanocrystals (Co-NC) that are protected by surface oxides and borates can be facilely prepared by borohydride reduction methods,[5] which have been widely utilized as hydride transfer catalysts for organic synthesis and hydrogen generation by chemical reactions, but have not been investigated as electrocatalysts for HER. Herein we report a facile synthetic approach to cobalt nanocrystal assembled hollow nanoparticles (Co-HNP), serving as a highly active and stable electrocatalysts for HER from neutral-pH water. Tested in neutral-pH phosphate buffer, the electrode comprised of Co-HNP on a carbon cloth (1.0 mg/cm^2 mass loading) produced cathodic current densities of 10 and 100 mA/cm^2 at overpotentials of -85 mV and -237 mV, respectively. The electrode activity was essentially unchanged after either 1000 cyclic voltammetric sweeps, or 20 h hydrogen generation at a high current density of 150 mA/cm^2.

Experimental

Synthesis of Co-HNP. 0.119 g $CoCl_2 \cdot 6H_2O$, 0.099 g $Na_2WO_4 \cdot 2H_2O$ and 0.136 g $NaH_2PO_2 \cdot H_2O$ was dissolved in ultrapure water separately and quickly mixed together under ultrasonic radiation to give a pink turbid solution containing amorphous $CoWO_4$ NPs (20 mL). Cooled in an ice water bath, 0.073 g solid $NaBH_4$ as a reducing agent was slowly added into the solution under ultrasonic radiation. Ultrasonic radiation was continued until no more bubble formation in the solution. The black product was collected by centrifugation (8000 rpm for 5 min), and washed with ultrapure water for three times.

Synthesis of Co-SNP. 0.119 g $CoCl_2 \cdot 6H_2O$ and 0.136 g $NaH_2PO_2 \cdot H_2O$ were dissolved in ultrapure water to give the solution A (15 mL). 0.073 g $NaBH_4$ and 0.099 g $Na_2WO_4 \cdot 2H_2O$ was dissolved in water to give the solution B (5 mL). The solution B was slowly added into the solution A under ultrasonic radiation in an ice water bath. Post-processing was all the same as that for Co-HNP.

Electrochemical tests. All of the electrochemical measurements were performed in a three-electrode system on an electrochemical workstation (CHI 660E) at room temperature. Catalysts were coated on a hydrophilic carbon cloth with 1.0 mg/cm^2 loading as the working electrode. A stainless steel plate and SCE (saturated calomel electrode) was used as counter and reference electrode, respectively. The electrolyte was 1 M phosphate buffer solution at pH = 7. Tafel and polarization curves were obtained at a scanning rate of 1 mV·s^{-1}. The electrochemical impedance spectroscopy (EIS) measurement was performed over a frequency range from 100 kHz to 0.05 Hz at a potential of -1000 mV vs. SCE, with a sinusoidal voltage of 5 mV. The cyclic voltammetry cycling test was conducted in the non-Faradaic potential range of -0.6 to -0.7 V vs. SCE at a sweep rate of 50 mV·s^{-1} for 1000 cycles. The long-term hydrogen generation test was performed at a high current density of 150 mA/cm^2 for 20 h.

Characterization Methods. X-ray powder diffraction (XRD) was carried out on a Rigaku D X-ray diffractometer with Cu Kα radiation (λ=1.54178 Å). Field emission scanning electron microscope (FE-SEM, JEOL JSM-6700F) was used to carry out the SEM images. TEM/HRTEM images, SAED patterns were taken with a JEOL JEM-2010 transmission electron microscope with an acceleration voltage of 200kV. The nitrogen sorption isotherms were measured by using automatic volumetric adsorption equipment (Micromeritics ASAP 2020).

Results and Discussion

Co-HNP was prepared using a "sacrificial template" method (**Figure 1a**). Amorphous $CoWO_4$ nanoparticles (NPs) were prepared by a rapid precipitation reaction between $CoCl_2$ and Na_2WO_4 in water solution (~ 80 nm in diameter). Under an ultrasonication-assisted reduction reaction by sodium borohydride, CoB alloy deposited on the surface of $CoWO_4$ NPs and converted to protect Co-NC. The extreme but transient local conditions caused by acoustic cavitation could facilitate the uniform deposition of CoB alloy onto the surface of $CoWO_4$ NPs, the conversion from CoB alloy to Co-NCs, and the fast dissolution of amorphous $CoWO_4$ NPs. Tungsten and phosphorous as doping elements were introduced to help to protect Co-NC and also improve the catalytic activity.

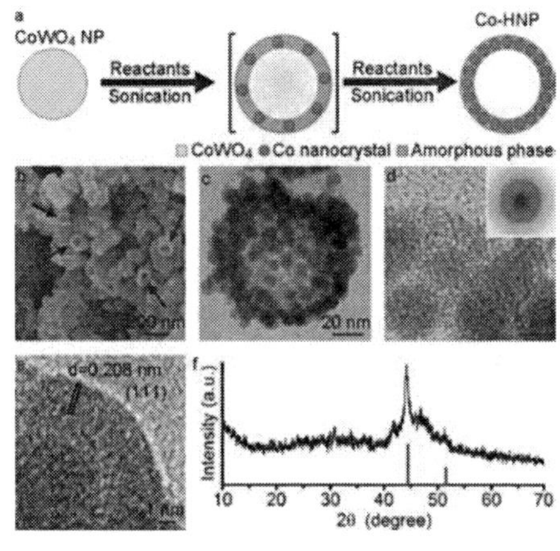

Figure 1. a) Schematic representation of the synthetic route Co-HNP. b) SEM and c) TEM images of Co-HNP. d) HRTEM image showing the shell of Co-HNP is composed by nanocrystals and amorphous materials. Insert: SAED pattern. e) HRTEM image of one nanocrystal showing Co fcc phase. f) XRD patterns of Co-HNP

Examined by scanning electron microscopy (SEM), the optimized Co-HNP appeared as nanoparticles with a diameter around 100 nm (**Figure 1b**), with some broken ones showing the hollow structure. Transmission electron microscopy (TEM) image showed that the diameter of the inside cavity was around 80 nm (**Figure 1c**), consistent with the size of $CoWO_4$ NPs that served as sacrificial templates. The thin shell of each Co-HNP was composed by multiple 5-10 nm nanoparticles (the dark dots in **Figure 1c**). High-resolution TEM (HRTEM) image revealed the crystalline structure of these small nanoparticles (the dark particles in **Figure 1d**), which were surrounded by amorphous materials (the lighter region in **Figure 1d**). Selected area electron diffraction (SAED) (**Figure 1d** insert) and HRTEM image (**Figure 1e**) identified these NCs with an inter-plane spacing of 0.208 nm, corresponding to the (111) plane of cobalt face-centered cubic (fcc) phase. The two diffraction rings in SAED pattern recorded from these randomly oriented Co-NCs can be assigned as the (111) and (110) planes of Co fcc phase, which was consistent with the powder X-ray diffraction (XRD) data (Figure 1f). By using Scherrer equation, the mean diameter of Co-NCs was estimated based on the peak at $2\theta = 44°$ to be 5.5 nm, consistent with the observation by HRTEM (**Figure 1d**).

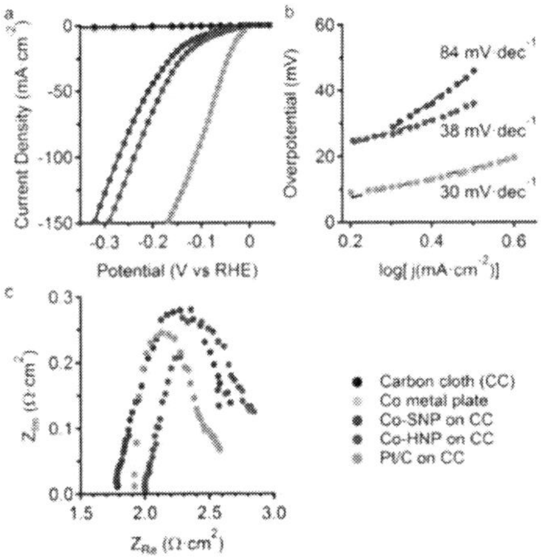

Figure 2. a) Polarization curves, b) Tafel plots and c) Nyquist plots of cobalt metal plate, Pt/C, Co-SNP and Co-HNP.

Co-HNP was loaded on carbon cloth (CC) with a 1.0 mg/cm^2 mass loading as the working electrode, through a facile drop-and-dry method without using any binder. Its catalytic activity for HER was tested in 1 M pH = 7 phosphate buffer at room temperature. Blank CC, cobalt metal plate and commercial Pt/C catalyst (20 wt.% Pt on Vulcan XC-72) were also examined. As shown in **Figure 2a**, both blank CC and cobalt metal plate show negligible HER activity, while Pt/C shows superior HER activity with near-zero overpotential (η). Despite the low HER activity of bulk cobalt metal, Co-HNP exhibits an excellent activity with a low onset η of 23 mV and a small Tafel slope value of 38 mV/dec (**Figure 2b**), which is superior to previously reported non-noble HER catalysts working in neutral-pH solution. The Tafel slope of 38 mV/dec suggests that the catalytic process follows a Volmer–Heyrovsky mechanism. The Co-HNP catalyst needs small η of 180 and 237 mV to drive current densities of 50 and 100 mA/cm^2, respectively. The comparison with previously reported results demonstrates the superior activity of Co-HNP for HER in neutral-pH water.

To probe the durability of the Co-HNP as electrocatalyst, continuous cyclic voltammetry (CV) was performed on Co-HNP electrode at a scanning rate of 50 mV/s. The polarization curve of Co-HNP/CC electrode after 1000 CV cycles overlays almost exactly with the initial one (**Figure 3a**). In contrast, the polarization curve of Pt/C electrode after 1000 CV cycles shifts to the left, indicating a reduced activity. The practical operation of Co-HNP catalyst was examined by a long-term hydrogen generation experiment at a high current density of 150 mA/cm^2, and compared with the performance of Pt/C catalyst at the same condition (**Figure 3b**). For the Co-HNP electrode, a potential of -300 mV vs RHE was needed to drive the hydrogen generation. The potential gradually shifted to -330 mV in 5 h and remained there up to 20 h. For the Pt/C electrode, a smaller potential of -190 mV was needed at the beginning, while the potential quickly shifted to -550 mV in 3 h and remained around -530 mV for the rest of time. The comparison clearly demonstrates the superior stability of Co-HNP over Pt/C as electrocatalyst at a heavy-loading operation condition in neutral-pH water.

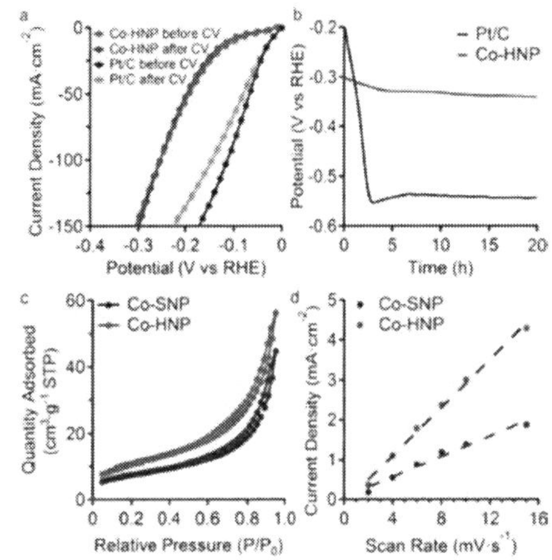

Figure 3. a) Polarization curves of Co-HNP or Pt/C electrodes before and after 1000 CV sweeps. b) The overpotentials applied on Co-HNP or Pt/C electrodes during hydrogen generation at a static current density of 150 mA/cm^2. c) Nitrogen adsorption /desorption isotherms and d) electrochemically active surface area measurements

Since the Co-HNP and Co-SNP showed similar R_{ct} values, indicating similar intrinsic catalytic activity, the superior HER performance of the Co-HNP was attributed to the beneficial effect of hollow nanostructure over solid one. The nitrogen adsorption/desorption isotherm (**Figure 3c**) of Co-HNP gave a Brunauer–Emmett–Teller (BET) surface area of 38 m^2/g, while the BET surface area of Co-SNP was only 20 m^2/g. we measured the double-layer capacitance of electrodes by using cyclic voltammetry method. The areal specific capacitances of Co-HNP and Co-SNP on CC electrode were calculated as 290 and 130 mF/cm^2, respectively (**Figure 3d**). The comparison demonstrates that the hollow nanostructure provides a larger catalytically active surface area, due to sufficiently accessible active sites at both interior and exterior surfaces of the nanoparticles, which accounts for the higher activity of Co-HNP over Co-SNP.

Acknowledgment. This work was supported by the National Natural Science Foundation of China (21474094) and the Recruitment Program of Global Experts.

References

(1) J. Luo, J.-H. Im, M. T. Mayer, M. Schreier, M. K. Nazeeruddin, N.-G. Park, S. D. Tilley, H. J. Fan and M. Grätzel, *Science*, **2014**, 345, 1593-96.
(2) M. G. Walter, E. L. Warren, J. R. McKone, S. W. Boettcher, Q. Mi, E. A. Santori and N. S. Lewis, *Chem. Rev.*, **2010**, 110, 6446-73.
(3) K. A. Vincent, A. Parkin and F. A. Armstrong, *Chem. Rev.*, **2007**, 107, 4366-413.

(4) Y. Sun, C. Liu, D. C. Grauer, J. Yano, J. R. Long, P. Yang and C. J. Chang, *J. Am. Chem. Soc.*, **2013**, 135, 17699-702.
(5) G. Arzac, T. C. Rojas and A. Fernández, *ChemCatChem*, **2011**, 3, 1305-13.

TUNABLE METALLIC FOAMS AS PLATFORMS FOR SYNTHESIZING HIGHLY ACTIVE OXYGEN REDUCTION AND ALCOHOL OXIDATION CATALYSTS

Eric Coleman, Heewon Choi, Joshua Billy and Anne Co

Department of Chemistry and Biochemistry
The Ohio State University, Columbus, OH 43210
co@chemistry.ohio-state.edu

Introduction

Metallic foams are materials with unique properties, including high surface area, good electrical and thermal conductivity, and enhanced catalytic activity; making them good candidates as fuel cell catalysts and support for battery electrodes. In this work, copper and bimetallic Cu foams are prepared by chemically etching aluminum rich metal alloys in sodium hydroxide. The metal foams are characterized using scanning electron microscopy (SEM), Brunner-Emmet-Teller (BET) surface area analysis, and electrochemical methods to evaluate the performance of the Cu foams and their alloys for the catalysis of oxygen reduction and alcohol oxidation reactions.

Polymer electrolyte fuel cells (PEMFCs) are energy conversion devices that catalyzes the conversion of hydrogen or hydrocarbon fuel at the anode, and oxygen at the cathode, to water or carbon dioxide, respectively, at high conversion efficiencies compared to conventional technologies based on heat engines. The performance of PEMFCs is essentially limited by the catalytic reduction of oxygen (ORR) to water or hydrogen peroxide. Here we synthesized a highly active PtCu ORR catalyst that has an average specific activity of 720 μA per cm^2 of active Pt at 900 mV at room temperature. Electron microscopy and BET surface analysis indicate our catalyst has extremely high porosity (> $10 m^2/g$), which acts to maximize mass transport, increase active surface area.

Experimental

A copper/aluminum alloy was made in-house by mixing high purity metals (>99.95%), with composition ranging between 20 to 50 wt% Cu, and melted in an induction furnace. The molten solution was poured at various temperature ranging between 600 to 900 °C, into rod-shape molds. The cooling rate is approximately 10 °C/min. Once cooled, the rods were cut into discs between 0.1 to 1 mm thick, depending on the experiment.

High surface area Cu foams were created by removing the aluminum through etching in 1 to 6 M KOH, or electrochemically by controlling the current and temperature at which Al is removed. Upon etching, the nanoporous support (Figure 1) provides a high surface area of 10-40 m^2/g.

Cu and bimetallic copper was prepared by galvanically displacing a second metal to the Cu structure. For example, for PtCu catalyst, Pt was deposited onto the support via a galvanic displacement reaction at a constant rotation rate to ensure mass transfer control for 0.5 to 60 min, resulting in a Pt:Cu ratio between 0.5 to 2.

ORR, methanol, ethanol and formic acid oxidation activity was evaluated using a rotating ring disc electrode (PINE Instruments) as working electrodes, and Pt and a reversible hydrogen electrode as the counter and reference electrode, respectively. All experiments are performed at room temperature in 0.1 M $HClO_4$, 0.1 M H_2SO_4 and 0.1 M KOH solutions.

Results and Discussion

A nanoporous Cu support was formed by etching Al from an in-house made CuAl alloy. Upon etching, the nanoporous support (Figure 1) provides a high surface area of 10-40 m^2/g.

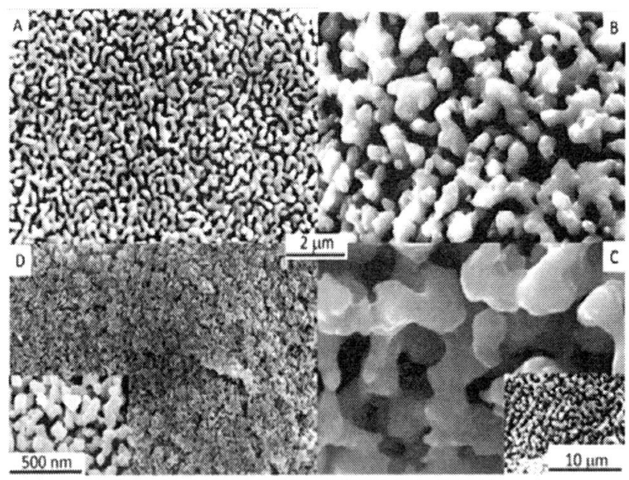

Figure 1. SEM images of nanoporous, open cell copper forms prepared using four different etching regimes: panel A (top left) 30 at% CuAl alloy etched in 6 M NaOH, at 80°C and 10 mA/g; panel B (top right) 30 at% CuAl alloy etched in 6 M NaOH, at 80°C and 0 mA/g; panel C (bottom right) 30 at% CuAl alloy etched in 6 M NaOH, at 80°C and 100 mA/g; and panel D (bottom left) 17 at% CuAl alloy etched in 6 M NaOH, at 80°C and 0 mA/g.

The open circuit potential (OCP) during the galvanic displacement process shows an appreciable change within the first 5 min of deposition (Figure 2), reaching a plateau at around 0.9 V vs. the standard hydrogen electrode (SHE).

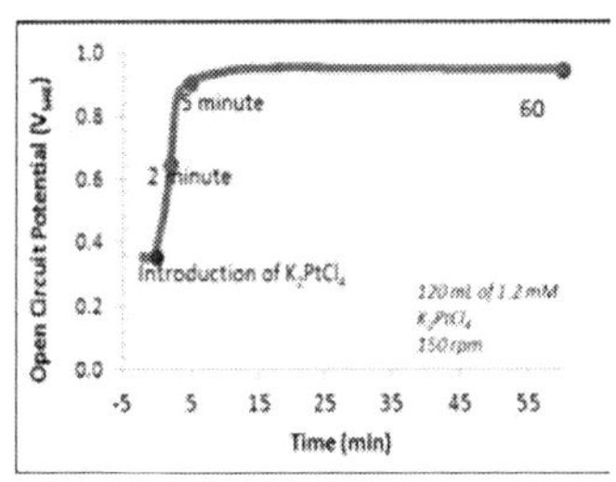

Figure 2. Open circuit potential as a function of galvanic displacement reaction time.

After normalizing activities to the electrochemical active area from the H_{upd} peaks of Pt, the resulting PtCu specific activities at 0.9, 0.95 and 1.0 V are shown in Figure 2. The 2 min PtCu catalyst has a mean activity 18% higher than Pt black at 0.9V and 10% higher than Pt black at 0.95 V. On average, the specific activity of our 2 minute catalyst exceeds the 2015 DOE target for platinum group metal specific activity (720 μA cm^{-2} @ 900 mV in 80 °C) by as much as 31% in a few cases. As a general trend, the specific activity seems to decrease as deposition time increases. This trend may be explained by considering that shorter Pt deposition times should normally translate to a thinner Pt overlayer. Observed ORR enhancement may be due to lattice strain effects [1,2].

Mass activities @ 0.9 V for the catalysts are summarized in Table 1. Mass data was obtained from a Perkin-Elmer Elan 6000 inductively coupled plasma mass spectrometer (ICP-MS) after electrochemical measurements were taken and electrodes stripped of catalysts with aqua regia. The current mass activity of our 2 min deposition is around 0.2 A/mg Pt, which compares well with commercially available Pt black standard.

Table 1. Summary of ORR activity at 0.9 V for Pt catalysts

Catalyst	Pt loading (μg)	Mass Activity (A mg^{-1})	Specific Activity (μA cm^{-2})
2 min PtCu	3.6	0.21	680 – 990
5 min PtCu	4.7	0.10	680
Pt black	30	0.18	880
2015 DOE target (80 °C)	----	≥0.44	720

Figure 2 illustrates the stability of our 60 min deposition by cycling the potential from 0-1.2 V in 0.1 M H_2SO_4 under 500 rpm rotation control for 24 hours. The largest loss in activity occurred during the first 12 hours, after which the activity seems to approach a steady value. We are confident that we will be able to address the stability and mass activity through fine tuning the nanostructures. Moreover, our mass activity and catalytic activity were obtained at room temperature. Typical fuel cell activities are compared at the operating temperature of 80 °C. Better performances are usually observed at higher temperatures due to improved kinetics. Our next set of experiments will focus on obtaining mass activities at higher temperature conditions.

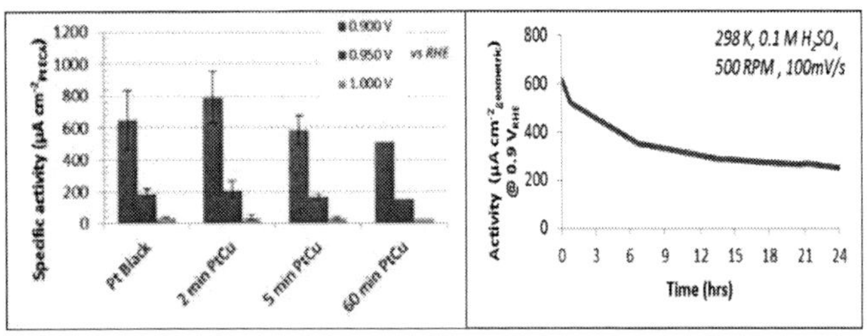

Figure 3. (left) ORR activity on PtCu at 0.9 V, 0.950 V, and 1.0 V vs. RHE measured with RDE in 0.1 M $HClO_4$ at 298 K. (right) stability at 0.9 V at 500 rpm between 0 and +1.2 V at 100 mV/s over 24 hrs

Acknowledgement. Authors would like to thank Sarum Boonmee from the Department of Materials Engineering for discussions and preparation of various Cu-based alloys.

References

(1) Cui, C.-H.; Li, H.-H.; Liu, X.-J.; Gao, M.-R.; Yu, S.-H. *Acs Catalysis* **2012**, *2*, 916.

(2) Stephens, I. E. L.; Bondarenko, A. S.; Perez-Alonso, F. J.; Calle-Vallejo, F.; Bech, L.; Johansson, T. P.; Jepsen, A. K.; Frydendal, R.; Knudsen, B. P.; Rossmeisl, J.; Chorkendorff, I. *Journal of the American Chemical Society* **2011**, *133*, 5485.

A STABLE SINGLE CO ATOMS/N-DOPED POROUS CARBON SUPERIOR ORR CATALYST

Yadong Li

Department of Chemistry, Tsinghua University, Beijing 100084 (China)
ydli@mail.tsinghua.edu.cn

Developing new synthetic methods to single atoms and cluster is a significant challenge in material science.[1] Here we report a new strategy to achieve stable Co single atoms/nitrogen-doped porous carbon with high metal concentration over 4%. This strategy is based on a pyrolysis process of designed bimetallic Zn/Co metal-organic frameworks, which Co can be reduced by carbonization of the organic linker and Zn will be selectively evaporated away at high temperature above 800 °C. The spherical aberration correction electron microscopy and extended x-ray absorption fine structure measurements both confirm the atomic dispersion of Co atoms stabilized by as-generated N-doped porous carbon. This synthetic strategy can be readily applied to other metal single atoms such as Ni, Pt and Ru etc. Surprisingly, the obtained Co-N_x single sites exhibit superior ORR performance with half-wave potential (0.877 V), which is more positive than commercial Pt/C (0.812 V) and most reported non-precious metal catalysts. Durability test revealed the Co single atoms exhibit outstanding chemical stability during the electrocatalysis and thermal stability at 900 °C. Our findings open up a new routine for general and practical synthesis of a variety of single atoms material, which shed light to new discovery in condensed materials at atomic scale.

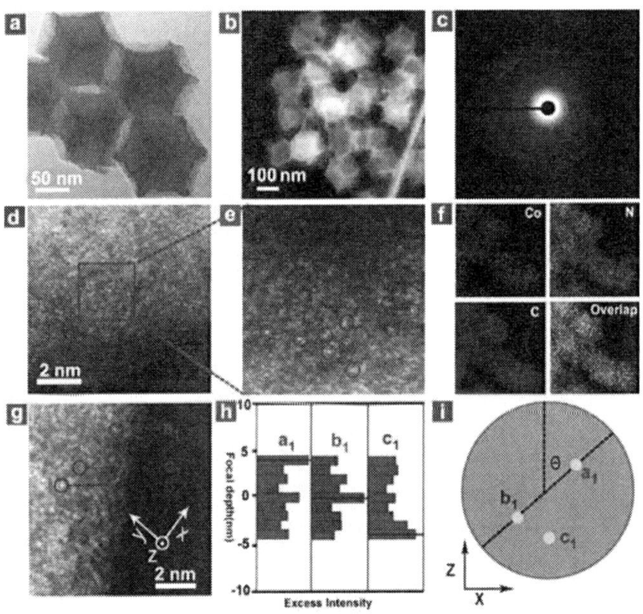

Figure 1. Characterizations of Single Co Atoms/N-doped Porous Carbon.

In summary, a general strategy enabling the practical and controlled synthesis of metal SAs catalyst with atomically dispersion and high atom utilization has been developed and applied to a series of metal SAs/N-C. The stabilization of SAs is strictly related to the N-doped porous carbon support and the mixing of Zn as protection. This finding may open up a new family of single metal or single molecule materials which show great potential to their fantastic properties in condensed science.

Acknowledgement. This work was supported by the Fundamental Research Funds for the Central Universities (WK2060190043), National Natural Science Foundation of China (21521091, 21131004, 21390393, U1463202). This work made use of the resources of National Synchrotron Radiation Laboratory in Beijing and Shanghai.

References

(1) Judai, K.; Abbet, S.; Wörz, A. S.; Heiz, U.; Henry, C. R., Low-temperature cluster catalysis. *J. Am. Chem. Soc.* **2004,** *126* (9), 2732-2737;
(2) Lei, Y.; Mehmood, F.; Lee, S.; Greeley, J.; Lee, B.; Seifert, S.; Winans, R. E.; Elam, J. W.; Meyer, R. J.; Redfern, P. C., Increased silver activity for direct propylene epoxidation via subnanometer size effects. *Science* **2010,** *328* (5975), 224-228;
(3) Qiao, B.; Wang, A.; Yang, X.; Allard, L. F.; Jiang, Z.; Cui, Y.; Liu, J.; Li, J.; Zhang, T., Single-atom catalysis of CO oxidation using Pt1/FeOx. *Nat. Chem.* **2011,** *3* (8), 634-641;
(4) Kyriakou, G.; Boucher, M. B.; Jewell, A. D.; Lewis, E. A.; Lawton, T. J.; Baber, A. E.; Tierney, H. L.; Flytzani-Stephanopoulos, M.; Sykes, E. H., Isolated metal atom geometries as a strategy for selective heterogeneous hydrogenations. *Science(Washington)* **2012,** *335* (6073), 1209-1212;
(5) Guo, X.; Fang, G.; Li, G.; Ma, H.; Fan, H.; Yu, L.; Ma, C.; Wu, X.; Deng, D.; Wei, M., Direct, nonoxidative conversion of methane to ethylene, aromatics, and hydrogen. *Science* **2014,** *344* (6184), 616-619.

UNCONVENTIONAL CATALYSTS FOR ELECTROCHEMICAL HYDROGEN EVOLUTION AND CARBON DIOXIDE REDUCTION

Nikolay Kornienko and Peidong Yang

University of California, Berkeley
Nick.kornienko.berkeley@gmail.com

Introduction

Electrochemical and photoelectrochemical generation of hydrogen and hydrocarbon based fuels from water and carbon dioxide is a promising approach towards solving the global energy crisis. Despite significant progress in this direction, continual development of light absorbers and catalysts remains a vital direction. In this talk, I will highlight recent developments in hydrogen evolution and carbon dioxide reduction electrocatalysts from the Peidong Yang group.

Amorphous Hydrogen Evolution Catalysts

Amorphous cobalt sulfide (CoSx) hydrogen evolution catalysts were found to be superior to their crystalline analogues. Subsequent in situ spectroscopic investigations have revealed that such catalysts undergo spontaneous transformations into molecular CoS2 like clusters under operating conditions. The abundance of under-coordinated sulfide peripheral atoms is proposed to function as high density active sites that give rise to the high performance of this material.

Porous Carbon Dioxide Reduction Catalysts

Ordered, porous metal organic and covalent frameworks (MOFs and COFs) are next utilized as carbon dioxide electrocatalysts. By functionalizing molecular linkers that are bound in a 3D network, MOF and COF catalysts combine favorable attributes of both heterogeneous and homogeneous catalysts. In addition, every active site is exposed to reactants in the electrolyte yet electrically connected to the electrode. These developments open up an entirely new class of catalyst which can be subsequently modified via a modular approach to give rise to improved materials and generate structure-function relationships.

Acknowledgement. Acknowledgements (if any) should be a subsection heading within "Conclusions." (9 pt)

References

(1) "Operando Spectroscopic Analysis of an Amorphous Cobalt Sulfide Electrocatalyst" N. Kornienko, J. Resasco, N. Becknell, C. Jiang, Y. Liu, K. Nie, X. Sun, J. Guo, S. Leone, P. Yang, J. Am. Chem. Soc., 137, 7448 (2015)

(2) "Mesoscopic Constructs of Ordered and Oriented Metal-Organic Frameworks on Plasmonic Silver Nanocrystals" Y. Zhao*, N. Kornienko*, Z. Liu,C. Zhu, S. Asahina, T. Kuo, W. Bao, C. Xie, O. Terasaki, P. Yang, O. Yaghi, J. Am. Chem. Soc., 137, 2199, 2015

(3) "Metal-Organic Frameworks for Electrocatalytic Reduction of Carbon Dioxide" N. Kornienko*, Y. Zhao*, C. Kley, C. Zhu, D. Kim, S. Lin, C. Chang, O. Yaghi, P. Yang, J. Am. Chem. Soc., 137, 14129 (2015)

(4) "Covalent Organic Frameworks Comprising Cobalt Porphyrins for Catalytic CO2 Reduction in Water" S. Lin*, C. Dierks*, Y. Zhang*, N. Kornienko, E. Nichols, Y. Zhao, A. Paris, D. Kim, P. Yang, O. Yaghi, C. Chang, Science, 349, 1208 (2015)

NOVEL APPROACHES AND NANOMATERIALS FOR BIOFUEL CELLS

Zafar Iqbal

New Jersey Institute of Technology,
University Heights, Newark, New Jersey 07102
iqbal@njit.edu

Abstract

Two novel approaches, namely mechanical compression of enzymes on the surface of Nafion-coated carbon nanopaper sheets supported on commercial Toray carbon paper, and the fabrication of flexible biofuel cell patches on enzyme functionalized carbon nanopaper sheets sandwiching an electrolytic gel with the patches enclosed in a biocompatible polymer with electrical access points, have been studied. Both approaches have provided power densities in excess of 100μW/cm^2, and the former approach has also increased biofuel cell lifetimes. Another approach to enhance biofuel cell lifetime is to replace the oxygen reduction cathodic enzyme lacasse with a polynitrogen catalyst supported on carbon nanotubes. Early results on this new catalyst in a biofuel cell are discussed.

Introduction

A biological or biofuel cell employs oxidation and reduction reaction enzymes to catalyze the conversion of a biofuel, such as glucose, related sugars or alcohol, to produce power for implantable or patchable devices. Biofuel cells were first fabricated in 1964 [1] to provide power for an implantable cardiac pacemaker. The big issue for commercialization was due to the fact that the power generated was small. In1980s, the biological fertilizer was used to provide fuel for battery, and then biological enzyme was recognized not only catalyst for human body, but also for biofuel cell [2]. With the aim of enhancing the biofuel cell power, enzymes such as glucose oxidase, laccase and bilirubin oxidase were evaluated in biofuel cell [3-6]. In this paper, glucose oxidase (GOx) was employed to oxidize glucose at anode and laccase was employed to reduce oxygen at the cathode.

Carbon nanotubes have been considered as ideal conducting nanowires in biochemistry [7] due to their excellent electrical conductivity and high chemical stability. Thus, carbon nanotubes were employed to be part of a biofuel cell to establish efficient direct electron transfer between active sites of an enzyme and electrode surface [8-10]. Wang et al obtained 1μW at external load of 400kΩwith a potential of 357mV whereas Katz et al reached 4μW/cm^2 at 600 mV using mediator-supported gold electrodes [8]. Topcagic and Minteer employed alcohol dehydrogenase and bilirubin as enzymatic catalysts with a redox mediator to construct a biofuel cell that was stable for about 30 days with power densities of up to 0.46mW/ cm^2 [11].

Castorena-Gonzalez et al assembled enzyme modified electrodes in a biofuel cell which were first characterized in human serum solution before placing on exposed rat tissue. A steady-state open circuit voltage of 140±30 mV and short circuit current of 10±3πA were achieved in the *in vivo* operating biofuel cell [12]. Zebda et al incorporated glucose oxidase and laccase in carbon nanotube disks followed by mechanical compression and obtained 1mW/ cm^2 power density under physiological conditions [13]. To date, this value is the best performance in a membrane-free, mediatorless biofuel cell, but the bioelectrode fabrication process is likely to be expensive and difficult to scale-up.

Here, a new approach was used to immobilize the enzyme in a "sandwich" structure comprised of carbon nanotube paper. The enzymes were positioned between conventional carbon and carbon nanotube paper as a sandwich structure. The "sandwich" was then placed under mechanical compression. The carbon paper was able to protect the CNT paper and enzymes during the mechanical compression. The advantage of compression lies in increased attachment between enzymes and nanotubes resulting in immobilization of the enzymes. Biofuel cells also require porous anode and cathode structures that support fuel transport to the catalyst reaction sites[14]. The bioanodes and biocathodes described is special in that both glucose oxidase and laccase are immobilized within a carbon nanotube paper or on nanopaper modified Nafion membrane layer that entraps and stabilizes the enzymes. In addition, preliminary results on the use of a novel polynitrogen oxygen reduction reaction catalyst stabilized on carbon nanotube paper [15] replacing lacasse, are also presented.

Experimental

Carbon paper (E-TEK,Inc) was used as-received. Multiwall carbon nanotubes (NanoLab Inc) were also used as-received. Nafion(5 wt.%), ethanol and 0.05M phosphate buffer solution (PBS, pH4) were purchased from Fisher Scientific Inc. Glucose oxidase (Gox, 200 U/mg, Aspergillus niger, Sigma), Laccase (Lac, 11.08U/mg, Pleurotus ostreatus, Sigma) and Glucose (Sigma Aldrich) were used as-received.

Carbon nanotube paper (CNT paper) was fabricated as follows: Carbon nanotube powder was mixed with sodium dodecyl sulfate (SDS) surfactant followed by horn sonication and a vacuum filtration process to produce free-standing carbon nanotube paper sheets. To prepare a bioelectrode the CNT paper sheets covered with Nafion were cut to an area of 1 sq.cm before use. Gox was evenly distributed on the CNT/Nafion structure to prepare the anode as shown in Fig. 1 followed by carbon paper on the top and on the bottom of CNT paper to form a sandwich structure before mechanical compression was applied. The top and the bottom layers are carbon paper and the middle layer is nafion. The cathode structure was similarly prepared using Lac instead of Gox. In preliminary experiments polynitrogen was deposited instead of Lac according to ref. 15 on CNT paper to form the oxygen reduction electrode.

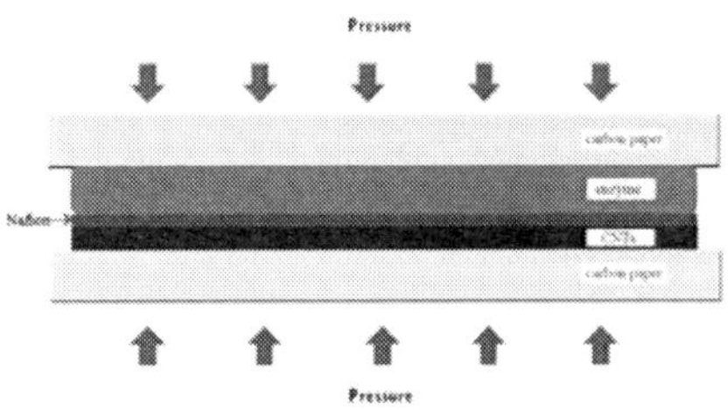

Figure 1: Schematic showing the geometry used to fabricate a bioelectrode..

Compressed electrodes without a separate membrane or mediator formed the glucose biofuel cells. The test of biofuel cells were performed at room temperature (25°C). The bioanode and biocathode sandwiches with glucose oxidase and laccase were used to form a biofuel cell. The current density – voltage outputs of the biofuel cell were obtained at various loads with glucose in buffer solution with cónstant stirring and bubbling of air into the cell. The biofuel cells consisted of anodes and cathodes positioned 10 mm apart in the cell.

Results and Discussion

Table 1 provides details about the biofuel cells (BFCs) prepared and tested. The highest power densities changing with time of BFC-1 and BFC-2 are 6.736μW/cm^2 and 9.443μW/cm^2, respectively, indicating that the enzymes are trapped and therefore immobilized in the nanotube paper structure resulting in a substantial increase of the power density just by using hand pressing in BFC 2 compared to no compression in BFC-2.

Table 1. Biofuel Cells (BFCs) Tested

BFC Name	Pressure (metric tons)	Amount of enzymes (mg)	Glucose concentration (mM)
BFC-1	0	50	5
BFC-2	Hand-compressed	50	5
BFC-3	10	50	10
BFC-4	11	80	10
BFC-5	11	50	5

The lifetimes of the BFCs also improved with compression because it provided more channels for electron transfer. The power density of BFC-1 went down substantially to 1.672μW/cm^2. In contrast, the lifetime of BFC-2 had a lifetime of 30 hours. At the beginning, BFC-1 and BFC-2 both increased in power. It took more time for BFC-2 to get stable compared with BFC-1, and it lasted several hours, due to the enzymes dissolving into the electrolyte, the power density went down continuously. To improve the ability to trap the enzymes, a Nafion membrane was used as a porous medium for enhanced immobilization of the enzymes. BFC-3 was fabricated using a Nafion membrane between the CNT paper and enzymes (see Fig.1). In addition, machine compression was used instead of hand compression. The performance of BFC-3 showed excellent improvement in power density and lifetime. Before setting up BFC-3, the performance of membrane was examined, using polarization. The highest point of the polarization curve is at a power density of 17.8 μW/cm² for a 2000 Ω load.

During the test of BFC-3, the power density stayed stable above 100μW/cm² from the third to fifth hour and the highest point was observed at the fourth hour with a power density of 167.12μW/cm² at 407.6μA/cm² and 0.41 V. After 20 hours, BFC-3 became unstable and the power density went down to less than 10μW/cm². In order to ensure that the biofuel cell unit had enough glucose, 10 ml glucose was added at concentrations of 5 mg/ml every 24 hours. The power density of BFC-3 went up over 20% after adding glucose. BFC-3 worked for 90 hours down to a power density of over 1μW/cm². The stability of BFC-3 was better than BFC-2, which was fabricated using hand compression. It can therefore be concluded that in a certain range, with applied load increasing, the contact between the CNT paper and enzymes is improved.

In order to improve biofuel cell performance, BFC-4 was fabricated using 11 metric tons – a 10% greater pressure than that used for fabricating BFC-3. From previous research, enzyme concentration also had a large impact on power density [16]. In order to investigate the influence of enzyme content, the enzyme entrapped into the CNT paper was increased to 80 mg for BFC-4. BFC-4 took 25 hours to get to the highest point in its current density which was 143.22μW/cm² with a corresponding current density and voltage of 267.5μA/cm² and 0.53V respectively. After 48 hours there was a sharp decrease of power density, but the value was still higher than that of BFC-3. It should be noted that the trend to decrease in lifetime decreased. The lifetime of BFC-4 showed an excellent performance than the other three groups of BFCs prepared shown in Table I. It is due to the increasing amount of enzymes and higher pressure applied on the "sandwich" structure shown in Fig.1.

The ultimate purpose of BFCs is to implant into human bodies, so the last group (BFC-5) of biofuel cells were fabricated at a glucose concentration of 5mM close to that of the human blood glucose level [16]. Although the concentration of glucose in BFC-5 was reduced it still got an excellent performance. The highest power density of BFC-5 was 143.97μW/cm² with corresponding current density and voltage of 268.4μA/cm² and 0.54V respectively, but with a lower lifetime than BFC-4. BFCs were also produced with Lac replaced by an electrochemically prepared polynitrogen as oxygen reduction reaction catalyst. These preliminary BFCs produced power levels in the 5μW/cm², but with lifetimes of a few hours. Lifetime and power improvement should provide BFCs not dependent on intrinsically unstable enzymes on the cathode side.

Another novel approach developed in this work was to produce flexible BFCs as patches. Using a similar sandwich structure consisting of anode and cathode comprising of gold/silver coated flexible CNT sheets and a gel electrolyte containing glucose or related sugar and packaged in a polymer with openings for electrical connections, a flexible biofuel cell was fabricated with power density above 100μW/cm².

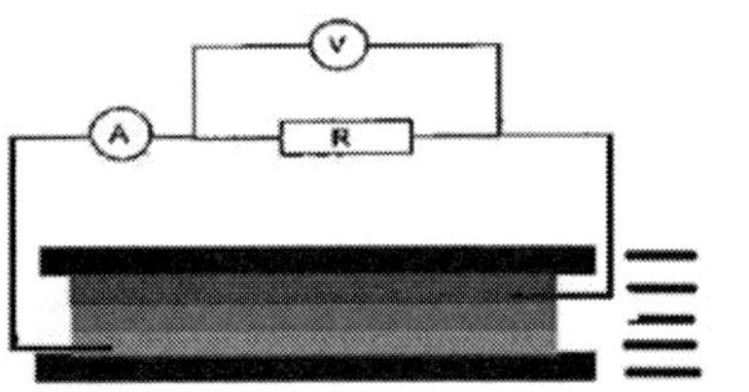

Figure 2: Schematic showing architecture of a flexible patch biofuel cell.

References

(1) Yahiro, A.; Lee, S.; D. Kimble, D. *Biochimica Biophysica Acta,* **1964**, *88,* 375.

(2) Moehlenbrock, M.J; Minteer, S.D. *Chemical Society Reviews,* **2008**, *37,*1188.

(3) Mano, N.; Mao, F.; Heller, A. *Jour. Amer. Chem. Soc.,* **2002**, *124,*12962.

(4) Barrière, F.; Kavanagh, P; Leech, D. *Electrochimica Acta,* **2006**, *51,*5187.

(5) Tsujimura,S.; Murata,K. *Electrochimica Acta,* **2015**,*180,*555.

(6) Barton, S.C.; Gallaway, J.; Atanassov, P., *Chemical Reviews,* **2004**, *104,*4867.

(7) Guiseppi-Elie, A.; Lei, C.; Baughman, R.H. *Nanotechnology* **2002**, *13,* 559.

(8) Wang, S.C.; Yang, F.; Silva, M.; Zarow, A.; Wang, Y.; Iqbal, Z., *Electrochemistry Communications*, **2009**, *11*, 34..

(9) Stolarczyk, K.; Kizling, M.; Majdecka, D.; Żelechowska, K.; Biernat, J.F.; Rogalski, J.; Bilewicz, R. *Journal of Power Sources*, **2014**,*249*, 263.

(10) Castorena-Gonzalez, J.A.; Foote, C.; MacVittie, K.; Halámek, J.; Halámková, L.; Martinez-Lemus, L.A.; Katz, E. *Electroanalysis*, **2013**,*25*, 1579.

(11) Topcagic, S.;Minteer, S.D. *Electrochimica Acta,* **2006**, *51* 2168.

(12) Castorena-Gonzalez, J.A. ; Foote, C.; MacVittie, K.; Halámek, J.; Halámková, L.; Martinez-Lemus, L.A.; Katz, E. *Electroanalysis,* **2013**, *25,* 1579.

(13) Zebda, A.; Gondran, C.; Le Goff, A.; Holzinger, M.; Cinquin, P.; Cosnier, S. *Nat Commun.*, **2011**, *2*, 370.

(14) Minteer, S.D.; Liaw, B.Y.; Cooney, M.J. *Current Opinion in Biotechnology,* **2007**,*18,* 228.

(15) Wu, Z.; Benchafia, E.M.; Iqbal, Z.;Wang, X. *Angew. Chem. Int. Ed.* **2014,** *53,* 12555.

(16) Bergmeyer, H.-U *Methods of enzymatic analysis,* Elsevier:New York, 2012.

SYNTHESIS AND SENSITIVITY OF PLASMONIC NANOPARTICLES

Guoliang (Greg) Liu, Assad Khan

Department of Chemistry, Virginia Tech
900 W. Campus Dr., Blacksburg, VA 24061

Although there are many existing methods of using plasmonics to detect chemical, biomolecules, and viruses, the major drawback is insufficient sensitivity. In search of highly sensitive materials, theoretical studies have suggested that the sensitivity (S) of metal nanorods depends largely on aspect ratio (R), but it is not experimentally proven. Moreover, the parameter that governs sensitivity of other shaped nanoparticles is still unveiled. In this talk, we will comprehensively survey the sensitivities of nanoparticles of various shapes, sizes, and compositions. By measuring sensitivities of >20 types of nanoparticles and summarizing >25 reported values in the literature, we find that nanoparticle sensitivity is mainly determined by aspect ratio and follows a linear relationship. Other parameters including metal compositions, shapes, bulk plasmon resonance frequency, and cross-sectional area have little effect on nanoparticle sensitivity. This work provides a foundation for integrating plasmonic nanoparticles with polymers in sensing applications.

BIO-TEMPLATED PT CATALYST FOR METHANOL ELECTROCHEMICAL OXIDATION

1st Wei Li,[a,*] 2nd Shengdong Dai,[a] 3rd Jinli Zhang [a]

[a]School of Chemical Engineering and Technology, Tianjin University,Tianjin,300350,China
*Corresponding author: liwei@tju.edu.cn

1. Introduction

Pt nanoparticles with both excellent stabilities and ultrasmall sizes have been successfully synthesized by using nucleation templates such as dendrimers, proteins and nucleic acids. For example, an i-motif DNA-stabilized Pt nanozyme with a size of 2.9 nm can react faster than native horseradish peroxidase[1]. Direct methanol fuel cells (DMFCs) have been considered as the promising power source for future energy security over the past few decades, where nano-Pt works as the catalyst in plenty of literature. However, it is still a challenge to explore high efficient electrocatalysts for DMFCs with higher activity, low-cost and tolerance to CO-like intermediate species[2].

2. Experimental

Reduced GSH was purchased from J&K Scientific. Pt nanoparticles were synthesized through the reduction of K2PtCl4 by Dimethyl-amine borane (DMAB). First, different concentrations of GSH were dissolved in 10 mM PBS buffer (pH 5.0) and incubated with 300 mM K2PtCl4 for 2 h at 25 °C. Then freshly prepared DMAB aqueous solution was added to the above mixture to a final [DMAB]/[K2PtCl4] ratio of 5. The as-prepared GSH-Pt can be used after 12 h reduction at 25 °C. For the methanol electrochemical oxidation, GSH-Pt/G composite catalyst was prepared with the Pt loading content of 20%. The electrochemical performance of GSH-Pt/G catalyst was assessed in a 0.5 M nitrogen-saturated H_2SO_4 solution.

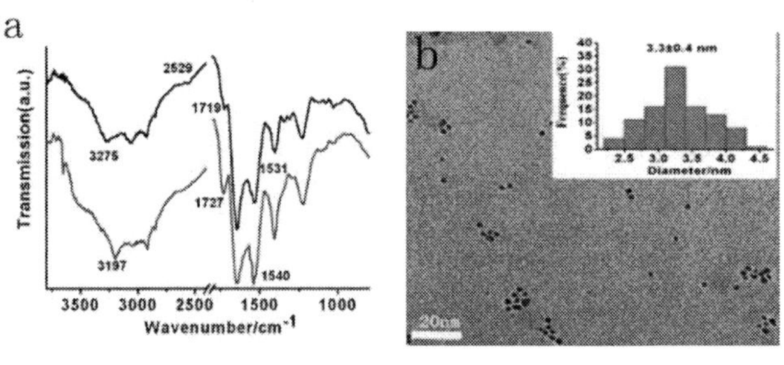

Figure 1. (a) FTIR spectra of GSH (red) and GSH-Pt^{2+}(black); (b) TEM images of GSH-Pt ($[K_2PtCl_4]/[GSH]=6$)

As shown in Fig.2a, the electrochemical surface area values of GSH-Pt/G catalysts equal 220.15, 181.23, 238.95m^2/g_{Pt} for the $[K_2PtCl_4]/[GSH]$ ratio of 24:4, 24:3, 24:2 respectively, which are much higher than that of Pt/C (55.41 m^2/g_{Pt}) catalyst, according to the area of the hydrogen desorption peaks. Fig.2b shows a typical CV curves of methanol oxidation, the GSH-Pt/G catalyst shows a higher oxidation activity with the $[K_2PtCl_4]/[GSH]$ ratio of 24:4, 24:3, 24:2, the forward peak current I_f is respectively 3.4, 2.6 and 3.6 times higher than that of the Pt/C (246.9Ag^{-1}). It is suggested that the GSH-Pt/G composite can function as an efficent catalyst for the methanol electrochemical oxidation, probably owing to the uniform small particle size, and ligand protection by GSH.

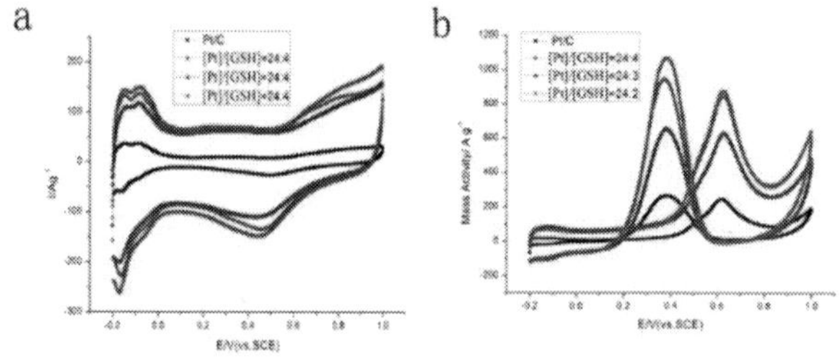

Figure 2. CV curves for GSH-Pt/G catalyst in (a) 0.5M H_2SO_4 (b) 0.5M H_2SO_4 +1 M methanol at a scan rate of 50 mVs^{-1}.

4. Conclusions

Glutathione (GSH) can function as an efficient bio-template to adjust the particle size and surface charge state of Pt nanoparticles. GSH-Pt3.3 consisting of 67% metallic Pt and 33% Pt (II) possesses the highest peroxidase-like activity, providing the linear response range of 5 ~ 50 nM Hg (II) with the LOD of 0.53 ± 0.003 nM (20 °C) in the colorimetric assays. In the case of GSH-Pt/Graphene composite, it is indicated that the electrochemical surface area value of the optimal catalyst GSH-Pt/G is at least 4 times higher than that of the commercial Pt/C catalyst, calculated by the area of the hydrogen desorption peaks.

Acknowledgement. The work is partially supported by the Major State Basic Research Development Program (No.2012CB720300).

References

(1) Fu Y., Zhao X. Y., Zhang J. L. and Li W., *J. Phys. Chem. C*, 2014, 118, 18116–18125
(2) Zhang G. L., Yang Z. Z., Huang C. D., Zhang W., Wang Y. X. *Nanoscale* 2015, 7 (22), 10170-10177
(3) 3.Fu Y., Zhang H. X., Dai S. D., Zhi X.; Zhang J. L., Li W. *Analyst* 2015, 140 (19), 6676-6683.

MICROWAVE ENABLED RAPID FABRICATION OF PHOSPHORS AND PHOSPHORS CO-DOPED GRAPHITIC CARBON WITH HIERARCHY POROUS STRUCTURES

Mehulkumar Patel[1], Keerthi Savaram[1], Feixiang Luo[2], M.Reza Khoshi[1], Carol Flach[1], Richard Mendelsohn[1], Eric Garfunkel[2], Michal Szostak[1], and Huixin He[1]*

[1]Department of Chemistry, Rutgers University, Newark, NJ 07102, USA; [2]Department of Chemistry and Chemical Biology, Rutgers University, Piscataway, NJ 08854, USA

Correspondence Author: huixinhe@ rutgers.edu

Introduction

The pivotal role of catalytic materials in various industries is unambiguously demonstrated by the fact that up to 90% of commercially available chemical products involve the use of catalysts at some production stage.[1] However, most catalytic materials were developed based on toxic and/or precious metals, which are unsustainable and possess environmental risks. In the drive towards green and sustainable chemistry, there is an increasing interest in developing new carbon-based materials that are benign, abundant, readily available, and metal free to act as catalysts for chemical synthesis. Compared to traditional metal based catalysts, carbon based materials provide additional advantages because of the existence of giant π structures promotes strong interactions with various reactants. More importantly, its physicochemical and electronic properties, which in principle determine the catalytic properties of a material, can be tailored and fine-tuned by molecular engineering and/or heteroatomic doping.[1, 2]

A plethora of reports have demonstrated that doping of heteroatoms into graphene and other carbon based materials gives rise to enhanced performance in electrocatalytic oxygen reduction reaction (ORR), when compared to their undoped analogues.[2] Compared to ORR, studies that use doped and/or co-doped carbon materials as catalysts for selective organic synthesis are in their early stages of development, although a great potential has already been demonstrated. Majority of the research has been focused on graphene oxide (GO) and Nitrogen doped carbon materials.[3, 4] All these limited research areas are largely due to lack of generic access, in large scale to other heteroatom (such as P and S)-doped carbon materials, which can be fine-tuned in terms of their molecular, electronic and geometric structures. Tremendous efforts have been made for the fabrication of carbon materials and heteroatom doped carbon materials, including graphene and carbon nanotubes. However, most of the approaches are time consuming and lack energy and cost effectiveness to quickly generate sufficient quantities of materials with tailored structures for synthetic studies.

In this report, an extremely simple and rapid (seconds) approach is developed to directly synthesize gram quantities of P-doped graphitic porous carbon materials with controlled P bond configuration.[3, 6] For the first time, it is demonstrated that the P-doped carbon materials can be used as a selective metal free catalyst for aerobic oxidation reactions. The work function of P-doped carbon materials, its connectivity to the P bond configuration, and the correlation with its catalytic efficiency are studied and established. A unique catalytic mechanism is demonstrated, which differs from both GO and N-doped graphene obtained by high temperature nitrification. The unique and unexpected catalytic pathway endows the P-doped materials with not only good catalytic efficiency but also recyclability. Furthermore, this approach is easily extended for the synthesis of other heteroatom-doped, dual doped, and even triple-doped porous carbon materials. The excellent catalytic performance combined with a rapid, energy saving approach that permits fabrication on a large scale, suggests that the P-doped, and P-codoped porous materials are promising materials for "green catalysis" due to their higher theoretical surface area, sustainability, environmental friendliness and low cost.

Experimental

The P doped carbon material (PGc) was synthesized by heating the phytic acid (Sigma Aldrich, 50 w/w% in water, 1ml) in domestic microwave oven (1100W, Sanyo-EM-S9515W, 2.45GHz) for 40seconds. P doped carbon with different P type was synthesized by heating the PGc powder into domestic microwave for different time with intermit mode, such as 30 (PGc-30) and 180s (PGc-180). P-codoped carbon catalyst can be synthesize by heating the mixture of a phytic acid and pre-weighed amount of other heteroatom source (such as amorphous sulfur or ammonium hydroxide or boric acid as S, N and B source, respectively) in microwave. We can also synthesize P, B and N tri-doped carbon material (PBN-Gc) and P, N and S tri-doped carbon material (PNS-Gc) by mixing the phytic acid with multiple suitable heteroatom precursors. To synthesize individual heteroatom doped carbon, 200.0 mg of inositol is mixed with concentrated phosphoric acid/sulfuric acid and heteroatom source and heated under microwave. For the catalytic alcohol oxidation reaction the catalyst is mixed with alcohol and sealed in desired atmosphere and heated in oil bath at predetermined temperature and time. After completion of the reaction, the reaction mixture was filtered via 0.02μm syringe filter and analyzed by 1H NMR spectroscopy.

Material Characterization. The morphology of all samples were studied using the scanning electron microscopy (SEM, Hitachi S-4800). The heteroatom doping and atomic % of all elements in the porous carbon was analyzed by Energy Dispersive X-ray Spectroscopy (EDS) and X-ray photoelectron spectroscopy (XPS).

Results and Discussion

In this approach, a novel sustainable biomass molecule, phytic acid, a well-known anti-nutrient molecule in food was chosen as our starting material. The phytic acid strongly absorbs microwave energy. Therefore the as-purchased phytic acid solution can be directly used for the fabrication of P doped graphitic carbon product (PGc) with microwave energy in very short time (40s). Using microwave heating instead of traditional heating ensures that the approach is both sustainable and low energy cost. Furthermore, the fabrication can be performed in air, under ambient conditions without the requirement of inert environment, which makes this approach even more cost effective and convenient. Figure 1 shows a scanning electron microscope (SEM) image of the as-fabricated PGc which is obtained by subjecting the as-purchased phytic acid solution (50 wt% in water) to short (40 seconds) microwave irradiation. PGc has a very unique structure in which a porous carbon monolith sandwiched by two highly wrinkled graphene-like sheets (Figure 1). The wrinkled structure is possibly the result of P doping and the larger diameter of P atoms compared to C atoms which induce local geometrical distortion in the carbon network. Indeed, Energy-dispersive X-ray spectroscopy (EDS), X-ray photoelectron spectroscopy (XPS) measurement (Figure 1B, C and D) shows that PGc incorporates 4.9 atomic % P, demonstrating that P is doped in the carbon matrix.

We first studied the catalytic efficiency of PGc for selective oxidation of benzyl alcohol to benzaldehyde without any solvent in which neat benzyl alcohol was heated to 80 °C for 48 hours in the presence of different catalyst, under 1 atm of oxygen (Table 1). If the reaction is performed under the same conditions, the catalytic efficiency of PGc is better compared to GO and N doped carbon[4]. Moreover, these PGc catalyst can be recycled for at least 8 times without losing its catalytic efficiency and selectivity towards the aldehyde which is very different from the GO[7] and N doped carbon catalysts[4]. We have also compared the catalytic activity of P -coped catalyst in Table -1 and found that S and P co-doped carbon shows the best synergistic effect.

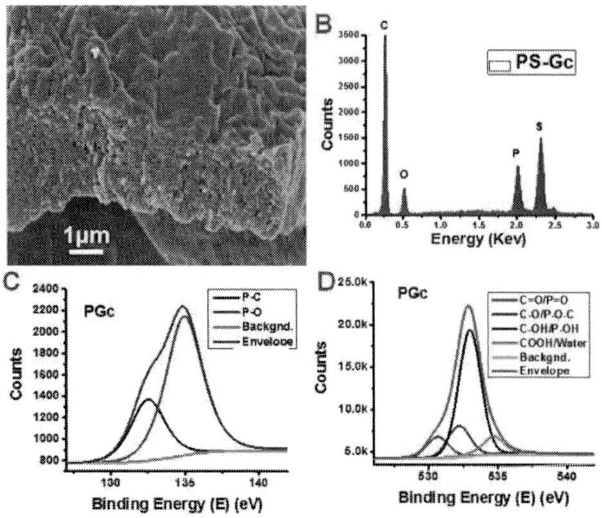

Figure 1. SEM (A), EDS (B) of P and S codoped carbon (PS-Gc). High resolution XPS P2p (C) and O1s (D) peak of PGc material.

Table 1. The catalytic activity of PGc in the oxidation of different alcohols.

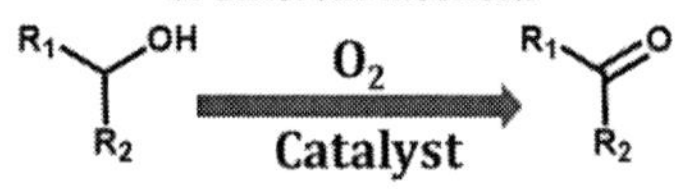

Entry	Catalyst	R1	R2	% Conversion	% Selectivity
1[a]	No catalyst	-H	$-C_6H_5$	0.06	100
2[a]	PGc	-H	$-C_6H_5$	21.67	98
3[a]	PGc-30s	-H	$-C_6H_5$	14.95	97
4[a]	PGc-180s	-H	$-C_6H_5$	8.35	94
5[a]	GO	-H	$-C_6H_5$	12.97	95
6[b]	PGc	-H	$-C_6H_4-OCH_3$	99.0	99.9
7[b]	PGc	-H	$-C_6H_4-CH_3$	52.45	94.2
8[b]	PGC	-H	$-C_6H_4-NO_2$	2.2	64.1
9[b]	PGc	$-CH_3$	$-C_6H_5$	23.3	95
10[b]	PS-Gc	-H	$-C_6H_5$	53.84	87.67
11[b]	PN-Gc	-H	$-C_6H_5$	35.43	97.17
12[b]	PB-Gc	-H	$-C_6H_5$	3.78	94.66

Reaction conditions: [a] 25mg PGc catalyst, 0.5mmol alcohol, 1atm O_2, 80 °C, 48 hours. [b] reaction was performed at 100 °C for 24 hours. % conversion to the alcohol and % selectivity with respect to benzaldehyde calculated using ^1H NMR.

We have performed several control experiments and concluded following important points regarding the catalytic mechanism of alcohol oxidation by PGc. (1) Molecular oxygen is not involved in the first step of the reaction, which is also different from N-doped graphene. However the presence of oxygen is critical for regenerating the active sites on the PGc catalysts. (2) The co-existence of P=O, P-OH functionalities along with the graphitic regions on PGc play a pivotal role in the catalytic alcohol oxidation reaction. (3) A radical inhibitor, BHT, does not inhibit the oxidation reaction. (4) There are no detectable H_2O_2 as byproduct generated during the reaction. (5) In addition, the reaction is associated with loss of water when reaction was performed under oxygen as demonstrated by the presence of water peak in ^1H NMR spectroscopy of the final product mixture proposed the following mechanism (Scheme 1). We propose the following mechanism as shown in Scheme 1 where primary benzyl alcohol is used as an example. In the first step of catalysis, condensation between the alcohol and P=O moieties on PGc takes place, and an alcoholate intermediate is formed. The condensation is likely facilitated by the interaction of the alcohol with the PGc surface by π-π interactions with the graphitic domains and hydrogen bonding with the polar groups (such as P-OH). In the second step of the reaction, a rate determining H transfer takes place, possibly through a cyclic transition state. The aldehyde product and a water molecule are released simultaneously. Next, the generated P (III) groups on PGc react with molecular oxygen to regenerate the P (V) centers for further reactions, thus completing the catalytic cycle (Scheme 1).

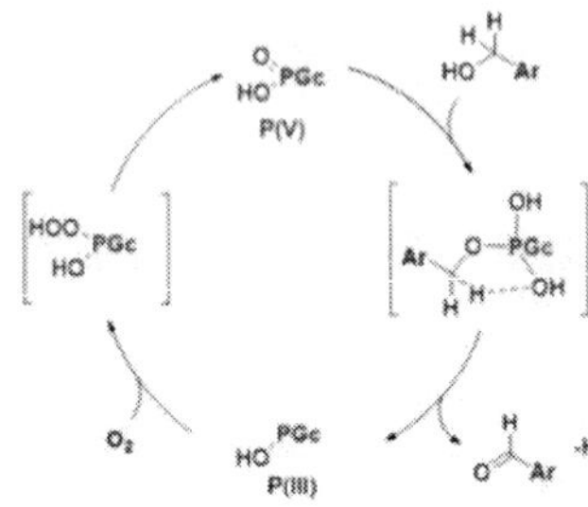

Scheme 1. Proposed mechanism of benzyl alcohol oxidation catalyzed by PGc in presence of oxygen as an oxidant.

Conclusions

In summary, we have reported an extremely simple, energy effective, and scalable approach to rapidly fabricate P doped and co-doped carbon materials with controlled P bond configuration. For the first time, we demonstrated that the P doped carbon materials can be used as a selective metal free catalyst for aerobic oxidation of benzylic alcohols with >98% selectivity to benzaldehydes. Based on our extensive experimental studies, a unique catalytic mechanism seem to be operating when PGc catalysts are used, which is different from both GO and N-doped graphene obtained by high temperature nitrification.

Acknowledgment. Financial support was provided the National Science Foundation (CBET 1438493 and DMR 1507812). The Bruker 500 MHz spectrometer used in this study was supported by the NSF-MRI grant (CHE-1229030). Work function studies with Kelvin Probe Microscope were supported by NSF MRI 1429062.

References

(1) Hu, H.; Xin, J. H.; Hu, H.; Wang, X.; Kong, Y. *Appl. Catal., A* **2015**, *492*, 1.

(2) Wang, X.; Sun, G.; Routh, P.; Kim, D.-H.; Huang, W.; Chen, P. *Chem. Soc. Rev.* **2014**, *43*, 7067.

(3) Hu, F.; Patel, M.; Luo, F.; Flach, C.; Mendelsohn, R.; Garfunkel, E.; He, H.; Szostak, M. *J. Am. Chem. Soc.* **2015**, *137*, 14473.

(4) Long, J.; Xie, X.; Xu, J.; Gu, Q.; Chen, L.; Wang, X. *ACS Catal.* **2012**, *2*, 622.

(5) Dhakshinamoorthy, A.; Primo, A.; Concepcion, P.; Alvaro, M.; Garcia, H. *Chem. Eur. J.* **2013**, *19*, 7547.

(6) Patel, M. A.; Luo, F.; Khoshi, M. R.; Rabie, E.; Zhang, Q.; Flach, C. R.; Mendelsohn, R.; Garfunkel, E.; Szostak, M.; He, H. *ACS Nano* **2016**, *10*, 2305.

(7) Dreyer, D. R.; Jia, H.-P.; Bielawski, C. W. *Angew. Chem. Int. Ed.* **2010**, *49*, 6813.

DEFECT-TOLERANT MATERIALS DESIGN FOR EFFICIENT AND STABLE SOLAR-FUEL PRODUCTION

Shu Hu, Maoqing Yao, Ke Sun, P. Daniel Dapkus, Nathan S. Lewis

[1] Department of Chemical and Environmental Engineering, School of Engineering and Applied Sciences, Yale University, New Haven, CT
[2] Division of Chemistry and Chemical Engineering, California Institute of Technology, Pasadena, CA
[3] Department of Electrical Engineering, University of Southern California, Los Angeles, CA

Introduction

Efficient photoelectrochemical (PEC) water splitting with solar-to-hydrogen (STH) conversion efficiencies > 20% requires a tandem light-absorber design that incorporates two materials of 0.7 – 1.2 eV and 1.5 – 1.9 eV band gaps.[1] To date, tandem light absorbers stablized by a protective coating strategy have shown multiple intences of >10% STH efficiencies: thus far, all these are planar lattice matched tandems e.g. 1.84 eV $In_{0.49}Ga_{0.51}P$/1.42 eV GaAs[2]. While lattice-matched planar structure limit the ability to achieve optimal band gap combinations, advances in nanoscale materials engineering allows for integration engineering allows for integration of dissimilar materials via bottom-up growth of nanowires, e.g. 1.7 eV GaAsP on Si [3]. Therefore, STH efficiency could be further improved by nanowire-on-planar design, for beyond 20%.

All these targets have been enabled via a thin-film protective coating that stabilize these efficient but otherwise unstable III-V light absorbers against photocorrosion.[4,5] But thus far, most of planar light absorbers (other than Si) have shown limited stability to only 100 hours, still less than the thousand-hour stability targeted by US Department of Energy. Essentially, no corrosion protection strategy is inherently robust unless the underlying light absorbers is either immune to corrosion (like TiO_2) or self-passivate (like Si) thus preventing further corrosion and dissolution. As illustrated in **Figure 1**, geometric isolation of such defective regions by nanowire-on-planar design would mitigate this failure mechanism.

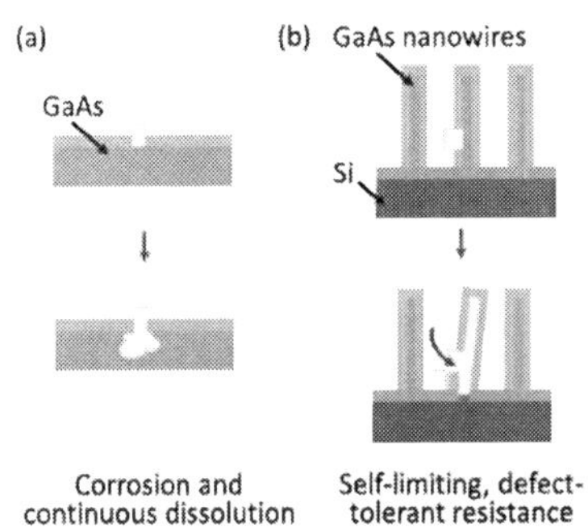

Figure 1. Schematics of (a) continuous corrosion and dissolution for failure of TiO_2-protected planar GaAs photoanodes, and (b) self-limiting, defect-tolerant behavior for TiO_2-protected nanowires on Si.

Here, we report the design and fabrication of a tandem III-V nanowire/Si planar device architecture to investigate whether such systems can provide defect tolerance for III-V materials in solar-driven water-oxidation processes. We further confirm the voltage addition behavior that has been demonstrated for GaAs nanowire on Si planar solar cell [6], which is crucial for PEC water splitting.

Experimental

Tandem GaAs Nanowire/Si Planar Photoelectrodes. Arrays of n-type or np^+-radial junction GaAs nanowires were grown by metal-organic chemical-vapor deposition (MOCVD) on either GaAs (111)B or on np^+-Si(111) substrates in which the emitter was formed by degenerate doping with boron. Arrays of GaAs nanowires were then coated and protected with ALD-grown TiO_2, followed by conformal deposition of a film of nickel oxide (NiO_x) that served as the oxygen-evolution catalyst. (**Figure 2**)

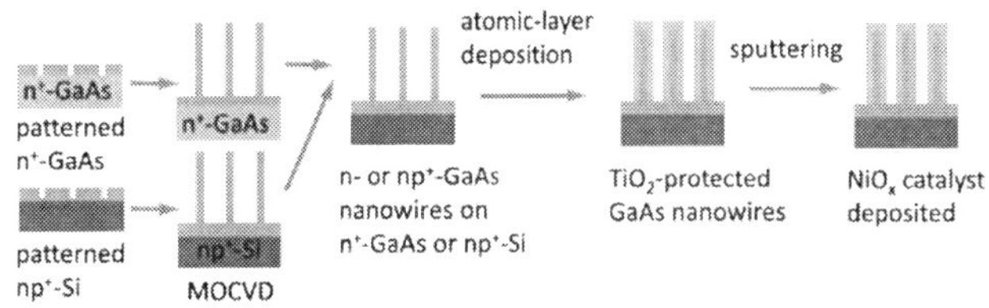

Figure 2. Schematic fabrication procedure of GaAs NW on Si planar.

TiO_2-Protected, Catalyst-Coated Photoelectrodes. Savannah S200 atomic-layer deposition (ALD) system was used to deposit ~100 nm TiO_2 conformally on as-grown GaAs NW-arrays. Each ALD cycle consisted of a 0.015 s pulse of N_2-purged H_2O, followed by a 0.10 s pulse of tetrakis-dimethylamido titanium (TDMAT, Sigma-Aldrich, 99.999%, used as received). A 15 s purge under a constant flow of high-purity $N_2(g)$ was performed between each precursor pulse. Then, NiO_x was deposited on TiO_2-protected GaAs nanowire arrays by reactive RF magnetron sputtering (AJA International) with a metallic Ni target sputtered at 150 W. The deposition was performed at room temperature and at a pressure of 5 mTorr, with a continuous flow of 20 sccm $N_2(g)$ and 1 sccm of $O_2(g)$. The deposition time was 15 min, which resulted in a thickness of ~ 30 nm planar equivalent.

PEC Characterization. Simulated Air Mass (AM) 1.5 illumination was used by setting up a Xe lamp with an AM 1.5 filter. The illumination intensity was calibrated by a calibrated Si photodiode. For intensity-dependent measurements, the light intensity was attenuated with a series of neutral-density filters. All PEC measurements were performed using a Biologic SP-200 potentiostat. An aqueous PEC cell with a three-electrode configuration was used to evaluate the current density vs potential behavior, spectral response, stability and Faradaic efficiency of the photoelectrodes. A Pt wire (0.5 mm diameter, 99.99% trace metals basis, Alfa-Aesar) was used as a reference electrode and a Pt gauze was used as the counter electrode.

Results and Discussion

Figure 3a shows a plan-view scanning-electron microcopy (SEM) micrographs of as-grown square arrays of GaAs nanowires heteroepitaxially grown on (111)-oriented Si. All SEM micrographs were taken by the FEI Nova NanoSEM. GaAs nanowire arrays were high fidelity on both Si and GaAs epitaxial substrates with >99% wire vertical by fine tuning the MOCVD growth conditions. The measured diameter of the GaAs NWs on Si or GaAs substrates was ~ 180 nm in diameter and ~ 1.5 μm long for a 30 min growth, and the pitch between neighboring wires was 600 nm as designed by lithography. The nanowire arrays were then conformally coated with ALD-grown TiO_2. **Figures 3b and 3c** show the SEM micrographs of the TiO_2-protected GaAs nanowire arrays before and after deposition of NiO_x electrocatalysts (15 min sputtering, thickness of ~ 30 nm planar equivalent), respectively. No significant non-uniformity for

NiO_x deposition was found by comparing tops vs bottoms of TiO_2-protected nanowires.

The intrinsic energy-conversion properties of the GaAs nanowire photoanodes were characterized using non-aqueous PEC experiments. **Figure 4a** shows the light-dependent *J-E* behavior of np^+-GaAs NW-array photoanodes grown on n^+-GaAs (111)B substrates (upper route in **Fig. 2**) in contact with the CH_3CN-$Fe(Cp)_2^{+/0}$ redox system (0.5 mM $Fe(Cp)_2^+$– 90 mM $Fe(Cp)_2$), under 0.25 – 1 Sun illumination intensities as provided by tungsten-halogen lamps and neutral-density filters. The as-grown np^+-GaAs nanowire arrays produced J_{ph} = 12.2 mA cm^{-2} and exhibited an open-circuit potential of −0.64 V *vs.* the Nernstian potential of the non-aqueous electrolyte ($E(Fe(Cp)_2^{+/0})$). At 1 Sun of intensity, the ideal regenerative cell energy-conversion efficiency, η_{IRC}, of np^+-GaAs NW-array photoanodes on n^+-GaAs (111)B substrates was 1.8%, with a fill factor (*ff*) of 0.34.

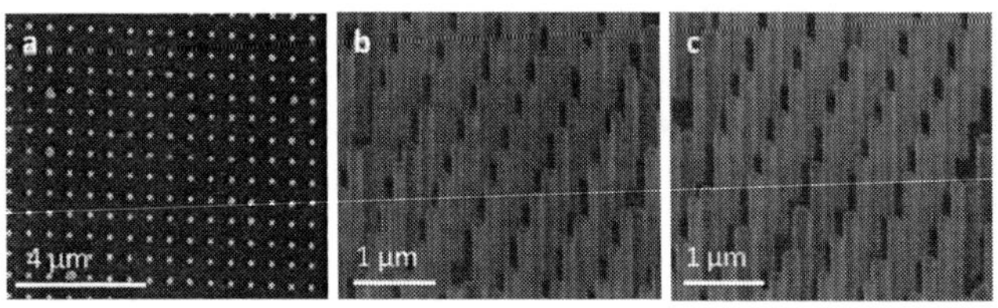

Figure 3. SEM micrographs of (a) plan-view, as-grown GaAs nanowires on np^+-Si substrates, (b) 45° from plan view of TiO_2-protected GaAs nanowires in (a), and (c) 45° from plan view of TiO_2-coated GaAs with sputter deposited NiO_x OER electrocatalysts.

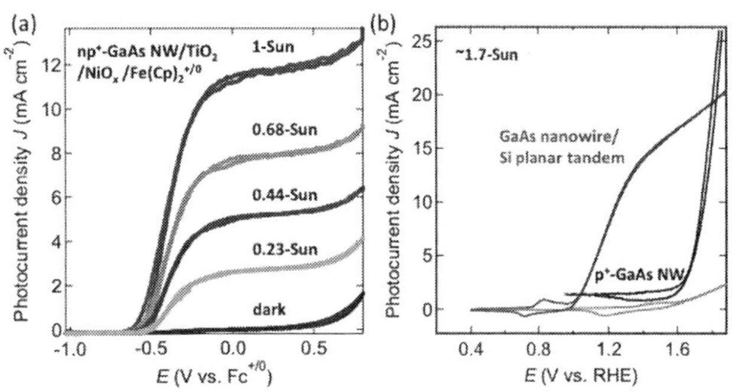

Figure 4. Photoelectrochemical (a) light-dependent *J-E* behavior of np^+-GaAs nanowires with a non-aqueous $Fe(Cp)_2^{+/0}$ redox system, and (b) *J-E* behaviors of np^+-GaAs NW-array/np^+-Si planar/TiO_2 /NiO_x photoanodes under ~1.7-Sun intensity of illumination (red curve) and in complete dark (black curve), respectively.

With this property understood, we investigated the water-oxidation performance of TiO_2-protected, catalyst-deposited GaAs NW-array/Si planar tandem photoanodes (lower route in **Fig. 2**). **Figure 4b** shows the *J-E* behavior of np^+-GaAs nanowire-array (NW-array)/np^+-Si planar/TiO_2-coating/NiO_x catalyst photoanodes under ~1.7-Sun intensity of illumination (red curve) and in the dark (green curve), respectively. Under 1.7-Sun in 1.0 M KOH(aq), the tandem GaAs NW-array/Si planar photoelectrode produced J_{ph} = 15.4 mA cm^{-2} and E_{oc} = 0.45 V *vs.* RHE, corresponding to a photovoltage of 0.78 V relative to $E^{0\prime}(O_2/H_2O)$. The onset potentials for water oxidation in the dark using the p^+-GaAs NW-array/TiO_2/NiO_x anodes on p^+-GaAs (111)B (black curve) or using the tandem GaAs NW/Si planar photoanodes under leaky reverse bias conditions (green curve) were ~1.65 V vs. RHE. The leakage operation is consistent with the performance of the np^+-junction configuration shown in **Figure 4a**.

The measured photovoltage of 0.78 V is in accord with the negative shift of the onset potentials, 0.98 V *vs.* RHE under illumination, with respect to 1.65 V *vs.* RHE in dark. However, both p^+-GaAs NW/TiO_2/NiO_x anodes and n-GaAs NW/TiO_2/NiO_x photoanodes grown on GaAs corroded in 1.0 M KOH(aq). The non-zero anodic current of p^+-GaAs NW/TiO_2/NiO_x anodes suggests nanowires on GaAs substrates corrode almost inevitably. For all the nanowire photoanodes on GaAs described above, an open-circuit corrosion potential of 0.35 ± 0.16 V *vs.* RHE was observed in 1.0 M KOH (aq).

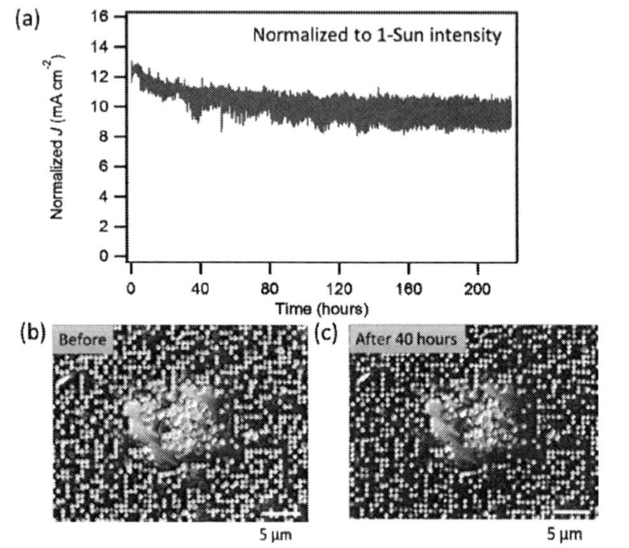

Figure 5. >200 hours stability of GaAs NW/Si planar tandem photoelectrodes for water oxidation in 1 M KOH (aq).

In contrast with nanowire on GaAs photoanodes, TiO_2-protected GaAs NW-array/np^+-Si planar photoanodes exhibited defect-tolerate water-oxidation stability. The expected self-limiting termination mechanism that is specific to substrates like Si is at work. **Figure 5a** shows the stable water oxidation of the tandem np^+-GaAs nanowire – np^+-Si/TiO_2/NiO_x photoelectrodes at 1.50 V *vs.* RHE (0.60 V *vs.* Hg/HgO) under simulated AM 1.5 1-Sun illumination for >200 hours. **Figures 5b and 5c** compare the NW photoelectrode morphology of the same location before and after water oxidation for 40 hour: the morphology of defect-affected regions remains the same. In a control study, corrosion of unprotected GaAs NW-arrays was immediately observed after 3 cyclic-voltammetry scans in $Fe(CN)_6^{3-/4-}$(aq) electrolyte at neutral pH. This suggest that the nanowire arrays on Si are defect tolerate, due to the discrete geometry of the wire arrays and the self-limiting and self-passivating nature of underlying Si substrates.

Acknowledgment. This work was supported through the Office of Science of the U.S. Department of Energy under Award No. DE-SC0004993 to the Joint Center for Artificial Photosynthesis, a DOE Energy Innovation Hub, and under Award Number DE-SC0001013 to the Center for Energy Nanoscience, an Energy Frontier Research Center (EFRC).

References

(1) Hu, S.; Xiang, C.; Haussener, S.; Berger, A. D.; Lewis, N. S. *Energy Environ. Sci.* **2013**, *6* (10), 2984–2993.

(2) Verlage, E.; Hu, S.; Liu, R.; Jones, R. J. R.; Sun, K.; Xiang, C.; Lewis, N. S.; Atwater, H. A. *Energy Environ. Sci.* **2015**, *8*, 3166–3172.

(3) Wu, J.; Li, Y.; Kubota, J.; Domen, K.; Aagesen, M.; Ward, T.; Sanchez,

A.; Beanland, R.; Zhang, Y.; Tang, M.; Hatch, S.; Seeds, A.; Liu, H. *Nano Lett.* **2014**, *14* (4), 2013–2018.
(4) Hu, S.; Shaner, M. R.; Beardslee, J. A.; Lichterman, M.; Brunschwig, B. S.; Lewis, N. S. *Science* **2014**, *344* (6187), 1005–1009.
(5) Hu, S.; Lewis, N. S.; Ager, J. W.; Yang, J.; McKone, J. R.; Strandwitz, N. C. *J. Phys. Chem. C* **2015**, *119* (43), 24201–24228.
(6) Yao, M.; Cong, S.; Arab, S.; Huang, N.; Povinelli, M. L.; Cronin, S. B.; Dapkus, P. D.; Zhou, C. *Nano Lett.* **2015**, *15* (11), 7217.

DEALLOYED NANOPOROUS MATERIALS FOR ENERGY STORAGE /CONVERSION APPLICATIONS: CHALLENGES AND OPPORTUNITIES

*Xizheng Liu, Yi Ding**

Tianjin Key Laboratory of Advanced Functional Porous Materials, School of Materials Science and Engineering, Tianjin University of Technology, Tianjin 300384, China.
E-mail: yding@tjut.edu.cn, Tel: +86-22-60216546

Introduction

With the global warming and the depletion of fossil fuel resources, it is highly urgent to develop new energy storage/conversion technologies and materials to cope with the ever increasing energy demand. Among various materials solutions, dealloyed nanoporous materials that possess excellent conductivity and 3D open nanoporosity, have shown great promise in energy storage/conversion devices [1-3]. In this talk, we will highlight the great progress for these new materials for applications as high-performance electrodes for fuel cells, lithium/sodium ion batteries, lithium air batteries and supercapacitors.

Results and Discussion

Nanoporous materials for fuel cells: Proton exchange membrane fuel cells (PEMFCs) are known for their high energy utilization efficiency and low emissions. However, the high dependence on Pt electro-catalysts hampers their practical applications. Dealloyed nanoporous gold (NPG) can be made into very thin free-standing membranes and is catalytically active for a series of electrochemical reactions related to fuel cell technologies. The electrochemical performance can also been enhanced by surface modification and optimizing the porous structure. Tremendous efforts have been devoted on the development of various NPG/Pt(Ru,Pd) leaves with controllable precious mass loading toward electro-oxidation of fuel molecules [4] and direct liquid fuel cells [5]. For instance, with nearly perfect Pt utilization and desired surface ensemble structure, NPG/Pt electrode can demonstrate orders of magnitude improvement in electrochemical performance. Further alloying with other elements is also an effective way to further improve the electrode performance. For example, NPG/Pt/Bi thin films can be fabricated by further decorating the NPG/Pt core/shell nanostructure with a sub-monolayer of Bi. The obtained hybrid film allows a dramatic decrease in Pt usage down to 3 μg/cm^2 while showing superior electrode activity, power performance and durability in actual direct formic acid fuel cells [6]. A Pt/Ni alloy core with a nearly pure Pt surface have been prepared by dealloying Pt/Ni/Al ternary alloy and investigated as ORR catalyst. In the absence of any catalyst support, nanoporous alloys show great improved durability and electrocatalytic activity for ORR. The remarkable specific activity of 1.23 mA/cm^2 has been achieved at a high potential of 0.9 V (vs. RHE) [7].

Nanoporous materials for rechargeable batteries: The unique structural properties of mechanical rigidity, superiority electrical conductivity endorse nanoporous materials great promise in rechargeable batteries such as lithium/sodium ion batteries and lithium air batteries. We will introduce our recent progress in the design and fabrication of series of nanoporous anode materials. Nanoporous CuSi and CuSn alloys have been achieved by leaching out Al from ternary alloys of CuSiAl or CuSnAl. Both of them show improved rate and cycle performance compared with the bulk Si or Sn materials. The bicontinuous structural features facilitate the charge transfer and endow the high tolerance towards the volume expansion. Nanoporous Sb has also been reported as electrode in sodium ion battery which exhibits high capacity retention of 573.8 mAh/g even after 200 cycles [8]. The nano-flowers of MnO_x series have also been carefully investigated by dealloying MnAl alloy and followed a smartly controlled temperature sequences and gas conditions. Hierarchically porous MnO_x, pure MnO and MnO_2 have been obtained and used for anodes in LIBs [9-10]. The MnO_x microspheres display a high specific capacity of 1081 mAh/g (at 100 mA/g) and nearly no capacity fading after 100 cycles. NPG based electrode has also been investigated in lithium air battery and showed high durability towards the reversible formation/decomposition of Li_2O_2 [11].

Nanoporous materials for supercapacitors: By introducing nanoporous structures into electrode materials, the 3D bicontinuous porous structure and the high conductivity of the nanoporous metal-based framework benefit the electron-proton transport and electrolyte permeation, giving rise to ultrahigh specific capacitance. Conductive polymers are promising electrodes for pseudo-capacitors due to the high theoretical capacitance values. Here, we introduce an ultrathin flexible all-solid-state supercapacitor based on NPG/PPy by decorating PPy on conductive nanoporous metal framework [12]. For the entire supercapacitor including both the electrolyte and electrodes, the volumetric power and energy of the flexible supercapacitor device are estimated to be 56.7 W cm^{-3} and 2.8 mWh cm^{-3}, respectively. The fast capacitive response of the solid state supercapacitor further indicates the efficient ion and electron transport within the porous nanocomposite.

Conclusions

This talk summaries the recent progress in design and fabrication of advanced energy materials by using dealloyed nanoporous materials. The features of controllable porous structures, large surface area and good conductivity endorse their great potentials for energy conversion/storage devices including fuel cells, lithium/sodium ion batteries, lithium air batteries and supercapacitors. The challenges and opportunities will be descripted in the presentation.

Acknowledgement. This work was supported by the 973 Program Project of China and the National Science Foundation of China.

References:

(1) Ding, Y.; Chen, M. W. *MRS Bull.* **2009**, *34*, 569.
(2) Ding, Y., Chen, M. W., Erlebacher, J.; *J. Am. Chem. Soc.* **2004**, *126*, 6876.
(3) Xu, C. X.; Su, J. X.; Xu, X. H.; Liu, P. P.; Zhao, H. J.; Tian, F. ; Ding, Y. *J. Am. Chem. Soc.* **2007**, *129*, 42.
(4) Wang, R. Y.; Wang, C.; Cai, W. B.; Ding, Y. *Adv. Mater.* **2010**, *22*, 1845.
(5) Wang, R.; Liu, J.; Liu, P.; Bi, X.; Yan, X.; Wang, W.; Ge, X.; Chen, M.; Ding, Y. *Chem. Sci.* **2014**, *5*, 403.
(6) Wang, R.; Liu, J.; Liu, P.; Bi, X.; Yan, X.; Wang, W.; Meng, Y.; Ge, X.; Chen, M.; Ding, Y. *Nano Res.* **2014**, *7(11)*, 1569.
(7) Wang, R. Y.; Xu, C. X.; Bi, X. X.; Ding, Y. *Energy Environ. Sci.* **2012**, *5*, 5281.
(8) Liu, S.; Feng, J.; Bian, X.; Liu, J.; Xu, H. *Energy Environ. Sci.* DOI:10.1039/C5EE03699B.
(9) Jiang, X.; Wang, Y.; Yang, L.; Li, D.; Xu, H.; Ding, Y. *J. Power Sour.* **2015**, *274*, 862.
(10) Jiang, X.; Yu, W.; Wang, H.; Xu, H.; Liu, X.; Ding, Y. *J. Mater. Chem. A*, **2016**, *4*, 920.
(11) Peng, Z.; Freunberger, S. A.; Chen, Y.; Bruce, P. G. *Science* **2012**, *337*, 563.
(12) Meng, F., Ding, Y. Adv. Mater. 2011, 23, 4098.

SEMICONDUCTOR NANOMATERIALS FOR SOLAR WATER SPLITTING AND CELL INTERFACES

Gengfeng Zheng

Laboratory of Advanced Materials, Department of Chemistry, Fudan University, Shanghai, 200433, China
Email: gfzheng@fudan.edu.cn

The solar energy-driven photoelectrochemical (PEC) water splitting and artificial photosynthesis have been continuing to drive substantial developments in new materials and methods for optimizing the energy conversion and utilization efficiency. Meanwhile, these studies have been inspiring research efforts in utilizing the PEC conversion platform to investigate the chemical/biological targets in life sciences. Here, we would like to present our recent works in designing and fabrication of semiconductor nano-heterostructures that are capable of being tailored during synthesis for enhanced solar energy-driven PEC energy conversion and utilization, including efficient photoabsorption, charge separation, and charge transfer. In addition, attributed to their suitable band alignment and excellent photoactivity, these nanowire-based structures are capable of post surface functionalization for molecular recognition and PEC catalysis, leading to solar energy-driven, real-time, sensitive detection of a variety of biologically important molecules for cellular signaling and enzymatic functions. Moreover, the chemical stability and biological benignity of these nanowire arrays further enable direct interfacing with primary neuron networks and in-situ photoelectrochemical neuron stimulation for artificial retinas. These hybrid nanowire photoelectrodes suggest new structures with enhanced photoactivity and sensitivity, which may allow for new exciting studies of interfacing of nanomaterials and biomaterials.

BIO-INSPIRED NANOSTRUCTURES FOR SOLAR ENERGY CONVERSION STUDY

Nurlan Bakhranov[a], Joseph Johnson[b], Sarkyt Kudaibergenov[c], and Nurxat Nuraje [b]

a) Kazakh-British Technical University, Almaty, Kazakhstan
b) Department of Chemical Engineering, Texas Tech University Lubbock, TX, USA
c) Kazakh National Technical University, Almaty, Kazakhstan

Introduction

Fossil fuel energy has a huge impact on the environment which cause global warming, and is also considered as a depleting energy source. Alternative energy sources are actively searched to replace the above fossil fuels [1, 2]. These alternative sources of energy have to meet certain requirements including renewable, environmental concerns, and, at the same time, be affordable. Therefore, solar energy is chosen as one of the choices of renewable energy for meeting the world's demand because of its enormous magnitude – approximately 10^5 terawatts, and no impact on the environment. However, the solar energy techniques including photovoltaic and solar fuels used presently still have the main concerns of harnessing incident photons, lowering production costs, enhancing efficiency, and storing energy.

Specifically, in the solar energy conversion including photovoltaic and storage of solar energy (water splitting), the critical steps include solar energy harnessing, charge separations, and catalytic reaction which only exists in water splitting. For better understanding the processes, Natural photosynthetic system always inspire us to mimic and design the artificial photosynthetic systems. One of the great examples is Purple photosynthetic bacteria [3] (**Figure 1**). It possesses unique structure of a photosynthetic complex with a new type of light-harvesting protein and the transport channels that allow to efficiently transfer energy and transport electron-transporting. To study the fundamentals, nanotechnology provides us a way of building these nanostructures existing in natural photosynthetic systems.

This study focuses on developing artificial photosynthetic systems for understanding the fundamentals of energy transfer, and electron transport in both water splitting, and photovoltaic devices with uniquely designed nanostructures. In the artificial design systems, dye-sensitized solar cell systems were a base for investigating the above fundamental problems. For the efficiency of solar energy harnessing and better electron transport systems, uniquely assembled nanostructures of ZnO, TiO_2 , and an acceptor/donor energy system were fabricated using electrochemical and chemical vapor deposition (CVD) approaches. The nanostructures including nanorods, nanosheets and nanowries were investigated for electron transfer using characterization methods (such as: SEM, TEM, XRD), spectroscopic methods (such as: UV-vis spectrometer, time-resolved photoluminscence spectroscopy), and electrochemical potentiostat methods. This fundamental study will lead us to better design artificial photosystems.

Experimental

Materials. The chemicals used in the synthesis of nano-wires are 99.998% Zn, 99.998% $Zn(NO_3)_2$.6 H_2O, ITO glass, and KCl from Sigma-Aldrich chemical company. TiO_2 nanoparticles (Degussa-25). Chitosan and Sodium dodecyl sulfate(SDS).

Nano-Wires (NWs) by Chemical Vapor Deposition (CVD). NWs where made using a indium-doped tin oxide (ITO) substrate half coated with a chitosan-SDS complex. The vapor deposition was preformed using a 99.998% pure Zn source, and oxygen and argon as carrier gases at a ratio of 1:1. The CVD was heated to 850°C and left to run for 5 min at a pressure of 0.04 MPa with a substrate temperature of 150°C. After the CVD cooled, the substrate with NWs was rinsed with ultrapure water, and dried under a gentle N_2 stream.

Nano-Structures by Electrochemical deposition(ECD). One-dimensional nano-rods (NRs) were grown via electrochemical deposition for 1 h at 70°C in a 0.001M $Zn(NO_3)_2$.6 H_2O aqueous solution with 0.1M KCl at -1.1V. The substrate was then rinsed with ultra pure water, and gently blown dry with N_2.

Two-dimensional nano-sheets (NSs) were grown via electrochemical deposition for 30 min at 50°C in a 0.05M $Zn(NO_3)_2$.6 H_2O aqueous solution with 0.1M KCl at -1.1V. The substrate was then rinsed with ultra pure water, and gently blown dry with N_2.

Characterization Methods. Scanning electron microscopy (SEM) and X-ray diffractometer (XRD) were utilized to record the morphology and confirm the crystallinity of synthesized nanostructures respectively. Gas chromatography was used to measure the evolved hydrogen and oxygen gases from the artificial system. The luminescence and the absorption spectrum of assembled nanostructures were measured on a highly sensitive fiber-optic spectrophotometer (AvaSpec-2048) at room temperature. The photocurrent-voltage characteristics were measured using a source meter instrument (Keithley 2400) by irradiating with simulated solar light that is 100 mW/cm^2 (PET PHOTO Emission TECH., INC).

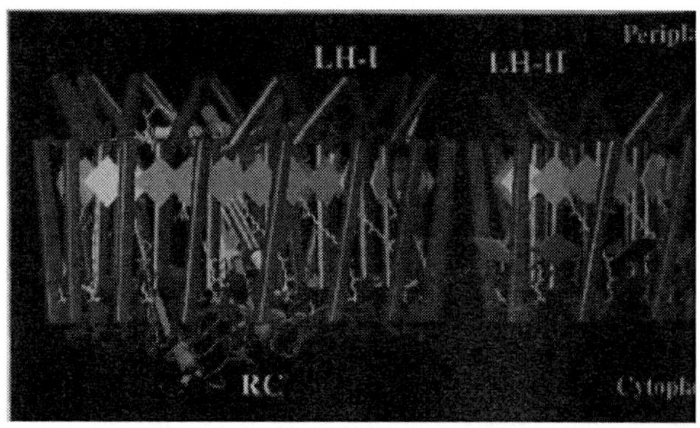

Figure 1. Modeled photosynthetic units including Förster energy transfer and electron transport made of nanosheets, nanorods, and nanowires shown in a purple bacteria [1]. Reproduce with Permission.

Results and Discussion

To study the energy transfer and electron transport in the natural photosynthetic system of purple bacteria, two types of artificial systems were built and studied. A donor–acceptor system consisting of a xanthene dye (donor of energy) and a polymethine dye (acceptor of energy) on the surface of TiO_2 films in a dye-sensitized solar cell[4] was applied to mimic Förster resonance energy transfer system in purple bacteria. The semiconductor film sensitized by the donor–acceptor compounds leads to a doubling of energy conversion efficiency in comparison to the cell sensitized only by the polymethine dye. The enlargement of the total energy brought in the increased generation of charge carriers at the interface of TiO_2 and the dyes.

In the design of electron transport pathways for mimicking photosynthetic purple bacteria, various ZnO nanostructures with rods, wires, and sheets were fabricated. CVD and ECD techniques were applied to fabricate the nanowires, nanorods, and nanosheets of ZnO.

The NWs, NSs, and NRs were examined in an SEM (**Figure 2**). The NWs were found to be ~50 nm in diameter, and seemed only to grow on the chitosan-SDS complex coated side of the ITO substrate.

The NSs are seen to be about ~3x3 μm, and ~100 nm thick. The NRs were found to be ~150nm in diameter.

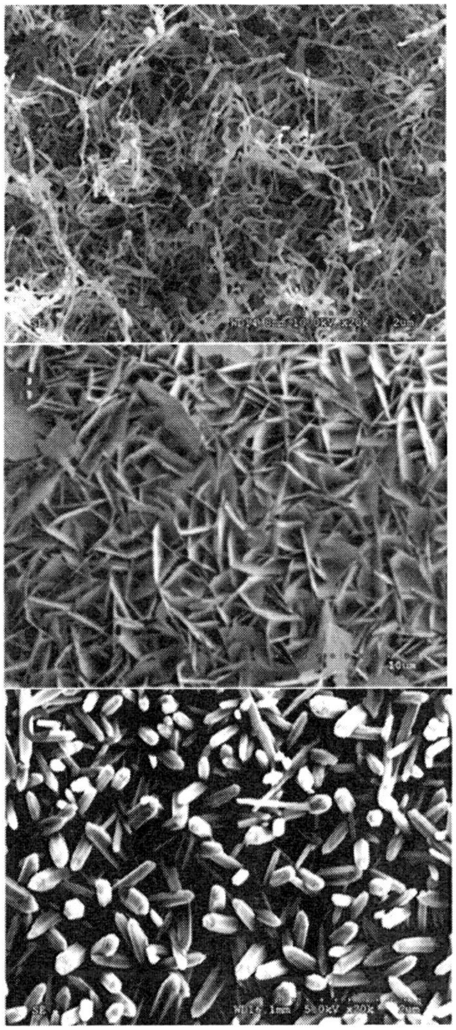

Figure 2. Comparison of various morphologies made by CVD, and ECD. (A)- Nano-Wires made by CVD. (B)- Nano-Sheets made by ECD. (C)- Nano-Rods made by ECD.

Photovoltaic properties of the DSSC cells with nanorod, and nanosheet arrays were measured [5]. This study shows that DSSC made of nanosheets have higher current compared to the DSSC assembled from nanorods (**Table 1**). The further study was carried out by electrical impedance (EI) method, and the above results were explained by diffusion lengths obtained by EI.

Table 1. Photovoltaic performances of DSSCs consisting of the nanorod, and nanosheet arrays of ZnO.

	J_{sc} mA/cm^2	U_{oc}	*FF*
DSSC based on the nanorod array	0.68	0.66	0.52
DSSC based on the nanorod array	3.72	0.37	0.36

The ZnO nanostructures showed in **Figure 2** were assembled for water splitting studies. The electron transport mechanism of these systems decorated with visible active photocatalysts such as copper oxide and iron oxide are under investigation in combination with hydrogen evolution experiments using photoelectrochemistry approach. At the same time, photoelectrochemical properties incorporated with metallic plasmonic metal nanoparticles are being studied.

In conclusion, uniquely assembled nanostructures developed in our lab are useful to create bio-mimicked photosynthetic structures for further investigation of the fundamental issues in solar energy conversion. Synthesis of various nanostructures by CVD and ECD, and their assembled nanostructures help us to design natural photosynthetic system. Time-resolved spectroscopy and electrochemical techniques in combination with other characterization methods were applied to study the energy transfer, and electron transport properties. This fundamental study will help us to design better artificial photosynthetic systems.

Acknowledgment. Authors greatly acknowledge financial support from the Texas Tech University.

References

(1) Nuraje, N.; Asmatulu, R.; Kudaibergenov, S., Current Inorganic Chemistry 2 (2), 124-146

(2) Moniruddin, MD.; Kudaibergenov, S.; Nuraje, N., Title of the book: "*Green Photo-active Nanomaterials: Sustainable Energy and Environmental Remediation*" edited by Nuraje, N., Asmatulu, R., Mul, G., RSC Publisher 2015, pp-

(3) Hu, X.; Schulten, K., *Physics Today*, 1997, 50, 28-34.

(4) Ibrayev, N., Seliverstova, E., Nuraje, N.; Ishchenko, A., *Material Science in Semiconducting Process,* **2015**, 31, 358-362

(5) Ilyassov, B.; Ibrayev, N.; Nuraje, N., *Material Science in Semiconducting Process, **2015**, 40, 885–889*

CONFINED SYNTHESIS OF THREE-DIMENSIONALLY ORDERED MESOPOROUS-IMPRINTED ZEOLITES AND THEIR APPLICATIONS IN CATALYSIS AND SEPARATION

Zhuopeng Wang, Chun-Chih Chang and Wei Fan

Chemical Engineering
University of Massachusetts Amherst
686 North Pleasant St.
Amherst, MA 01003

Introduction

Due to the intrinsic ordered micropore structures and strong acidity, zeolites are extensively used as heterogeneous acid catalysts in crude oil refineries and petrochemical processes. However, the micropore structures and high intrinsic activities frequently lead these materials to be subject to diffusion limitations that restrict reactant accessibility to the active sites on the interior surfaces of zeolites and inhibit the full utilization of zeolite catalysts.

Nanofabrication of hierarchical zeolite catalysts with mesoporosity is a proven strategy for integrating shape selectivity provided by the intrinsic micropore structures and efficient mass transport facilitated by the mesopore structures [1]. Considerable efforts have been devoted to the synthesis of zeolite catalysts with mesoporosity. Among the various approaches, the synthesis of hierarchical zeolites by the confined synthesis in hard templates holds exciting implications in term of creating ordered mesopore structures and controlling mesopore sizes. In principle, any desired mesopore structure can be achieved by manipulating the structure of the hard template. These advantages provide the unique opportunities to design the hierarchical pore structure and quantitatively investigate the effects of mesopore structure on the catalytic performance of zeolite catalysts.

Herein, we reported a general synthesis route for the nanofabrication of hierarchical mesoporous zeolites within 3-dimensionally ordered mesoporous (3DOm) by combining hydrothermal synthesis method with confined synthesis method. The approach exhibits versatile abilities for controlling the topologies, mesoporosity and composition of hierarchical zeolites.

Experimental

As a hard template in the synthesis process, 3DOm carbon with a pore size from around 10 to 40 nm was obtained by replication from colloidal crystals composed of size tunable silica nanoparticles. For the confined growth of zeolites in the 3DOm carbon, it is essential to allow the nutrients needed for zeolite formation to access the mesopores. For this reason, clear synthesis solutions were used in the syntheses. For the case of 3DOm-i zeolite beta (3DOm-i BEA), 3DOm carbon was immersed into a clear solution with a composition of $0.35Na_2O:4.5(TEA)_2O:0.25Al_2O_3:25SiO_2:330H_2O$, followed by hydrothermal synthesis at 100°C for 4 days. After the first synthesis cycle, zeolite precursors or seeds were formed within the 3DOm carbon. The 3DOm carbon containing zeolite seeds or precursors was added into a fresh synthesis solution for subsequent crystal growth. After repeating the synthesis process 4 to 6 times, 3DOm-i zeolite beta was successfully grown within the 3DOm carbon without forming on the external surface of the template. By tuning the initial composition, a series of 3DOm-i zeolites with controllable microporous structures, mesoporosities and acidities were synthesized. The synthesis scheme is shown in **Figure 1**.

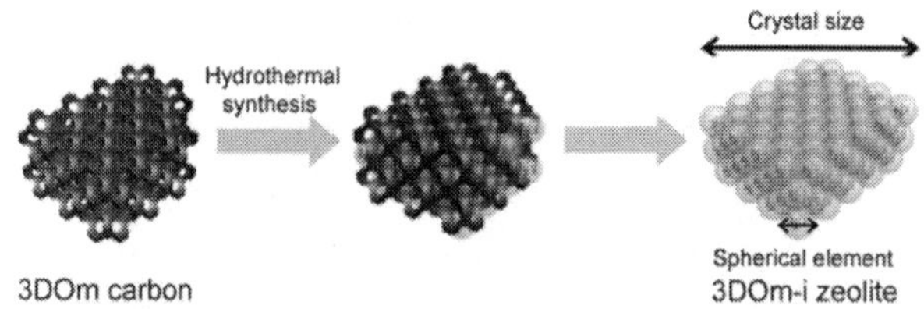

Figure 1. Schematic of formation of 3DOm-i zeolites confined in the pore space of 3DOm carbon.

Results and Discussion

For the case of 3DOm-i zeolite beta (3DOm-i BEA), XRD at wide angles shows weak diffraction peaks from the sample after the first synthesis cycle. The peaks become more intense after the second cycle, and continuously increase in intensity with additional synthesis cycles. The XRD patterns of **Figure 2a** confirm that 3DOm-i BEA can be prepared after 4 HT cycles in 3DOm carbons with mesopores ranging from 10 nm to 40 nm. It is found that the diffraction peak at 22.7°, associated with the (302) reflection of zeolite beta, is slightly broadened with the decrease in the pore size of the 3DOm carbon. The broad peak at around 7.5° is composed of the (100) and (101) reflections of zeolite beta, indicating a highly faulted intergrown structure of A and B polymorphs. The crystal structure is well retained after removal of 3DOm carbon template at 550 °C, although a decrease in the intensity of the peak at 22.7° is observed.

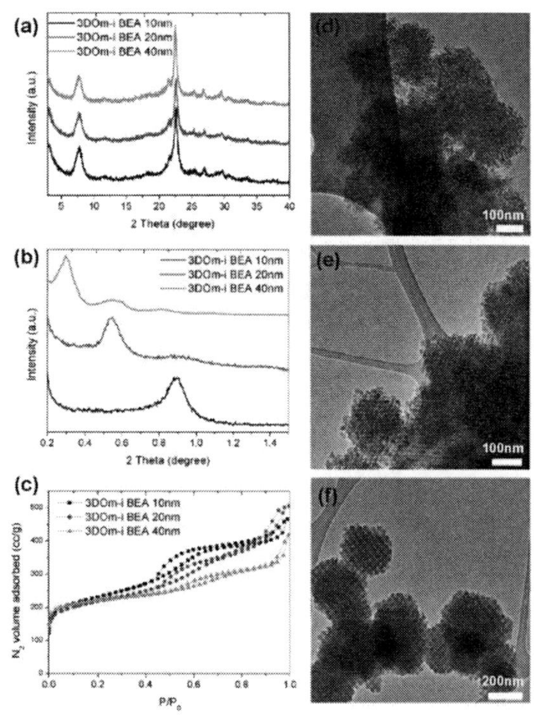

Figure 2. Characterization of the 3DOm-i BEA formed within the 3DOm carbon templates. Wide angle XRD of the 3DOm-i BEA samples (a) before the removal of carbon template, low angle XRD (b), and nitrogen adsorption-desorption isotherms (c), after the carbon was removed by calcination. TEM images of the 3DOm-i BEA formed within 10 nm (d), 20 nm (e), and 40 nm (f) 3DOm carbon, respectively.

The XRD at low angles after the carbon removal (**Figure 2b**) reveals the 3DOm-i BEA samples retain high degree of mesostructural ordering imprinted from the 3DOm carbon. The low angle XRD patterns of the 3DOm-i BEA samples formed after each HT cycle also reveal the gradual development of an ordered mesostructure during the synthesis. After the diffraction peaks at low angle were indexed according to a face-centered-cubic (FCC) structure, the center-to-center distances between the spherical zeolite elements were calculated (from Bragg's law using the lowest angle peak as (111)) to be 12, 20 and 37 nm, in agreement with the corresponding pore size of the 3DOm carbon template used and the expected nominal spherical element size of 3DOm-i BEA confirmed by transmission electron microscopy (TEM) (**Figure 2d, e and f**).

To further explore the feasibility of this synthesis approach, attempts have been devoted to the synthesis of 3DOm-i zeolite LTA, FAU and LTL in the 3DOm carbon template by changing the initial synthesis solutions and synthesis parameters. The mesostructure is clearly observed in the representative TEM images of 3DOm-i LTA FAU and LTL (**Figure 3**). XRD patterns at wide and low angle reveal the signature reflections of highly crystalline LTA and of an ordered mesostructure, respectively. The selected area electron diffraction (SAED, **Figure 3a** inset) of the 3DOm-i LTA sample clearly shows a single crystal diffraction pattern. As discussed before, the formation of these single zeolite crystals probably starts from the nucleation in one or more adjacent cages of the 3DOm carbon, and it is followed by growth propagating from cage to cage through the connecting windows to near complete filling of the neighboring cages. A similar result was observed for the case of 3DOm-i FAU and 3DOm-i BEA as shown in the insets of **Figure 3b and c**.

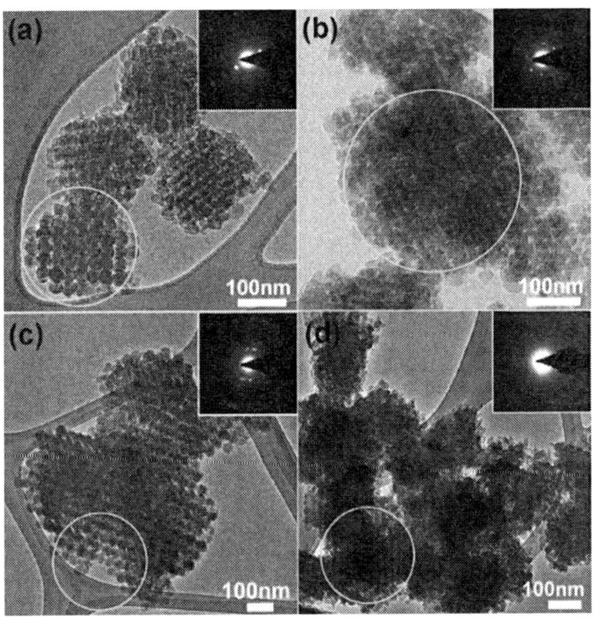

Figure 3. 3DOm-i zeolite crystals. TEM images of 3DOm-i LTA (a), FAU (b), BEA (c), and LTL (d) grown within a 40nm 3DOm carbon template (inset: electron diffraction patterns from the marked areas in the corresponding TEM images).

It is well known that the acidity of zeolites varies with the amount of heteroatoms incorporated in the framework. For example, the Si/Al ratio of ZSM-5 can be tailored from infinite to about 12 by changing hydrothermal synthesis conditions. With using appropriate synthesis solutions, we demonstrate that 3DOm-i ZSM-5 catalysts with well-controlled Si/Al ratio from 32 to 80 can be prepared within the 3DOm carbon using the approach proposed in this study. The framework Al in the catalysts was demonstrated by NH_3-TPD measurement and ICP. The adsorbed NH_3 is higher for the 3DOm-i ZSM-5 with lower Si/Al, indicating different amount of framework Al was incorporated in the samples. The acid site concentration measured by the ammonia TPD is shown in Table 1.

Table 1. Textural properties and acid concentration of 3DOm-i ZSM-5 samples with different Si/Al ratios

	S_{BET} (m^2/g)	S_{ext}[a] (m^2/g)	V_{mic}[b] (cm^3/g)	V_{meso}[b] (cm^3/g)	V_{total}[c] (cm^3/g)	Acid concentration[d] (mmol/g)
3DOm-i ZSM-5 (SAR=32)	445	259	0.13	0.15	0.37	0.491
3DOm-i ZSM-5 (SAR=53)	485	267	0.14	0.15	0.44	0.411
3DOm-i ZSM-5 (SAR=80)	497	258	0.14	0.15	0.35	0.291

[a] Calculated using *t*-plot method. [b] Calculated using α_s-plot method. [c] Calculated at P/P_0=0.95. [d] Measured by NH_3-TPD

The catalytic performance of the 3DOm-i ZSM-5 with different Si/Al ratios were tested with the self-etherification of benzyl alcohol, a reaction severely limited by the diffusivity of reactants and products in the micropores of zeolite catalysts. It was revealed that the 3DOm-i structure can significantly improve the reaction rate by alleviating diffusion limitation. The applications of the hierarchical materials in catalysis and separation will be discussed during the presentation.

Acknowledgment. This work is financially supported as a part of the Catalysis Center for Energy Innovation, an Energy Frontier Research Center funded by the U.S. Department of Energy, Office of Science, Office of Basic Energy Science under Award Number DE-SC0001004.

References

(1) Hartmann, M. *Angew. Chem., Int. Ed.* **2004,** *43,* 5880.
(2) Chen, H.Y., Wydra, J., Zhang, X.Y., Lee, P.S., Wang, Z.P., Fan, W., and Tsapatsis, M. *J.Am. Chem. Soc.*, **2011**, 133, 12390-12393
(3) Fan, W., Synder, M. A., Sandeep, K., Lee, S. P., Yoo, C. W., McCormick, A. V., Penn, L. R., Stein, A., and Tsapatsis, M. *Nat. Mater.*, **2008,** 7, 984-991, 2008

HOLLOW NANO-STRUCTURED ZEOLITE CRYSTALS AS MULTI-FUNCTIONAL CATALYSTS AND MATERIALS FOR ENERGY AND ENVIRONMENTAL APPLICATIONS

Chengyi Dai,[a] Anfeng Zhang,[a] Xinwen Guo[a] and Chunshan Song[a,b]**

[a] PSU-DUT Joint Center for Energy Research, State Key Laboratory of Fine Chemicals, School of Chemical Engineering, Dalian University of Technology, Dalian 116024, P. R. China.

[b] EMS Energy Institute, PSU-DUT Joint Center for Energy Research, Department of Energy & Mineral Engineering, and Department of Chemical Engineering, The Pennsylvania State University, 209 Academic Projects Building, University Park, PA 16802, United States.

*E-mail: csong@psu.edu

Hollow nano-structured zeolite single crystals have great potential as multi-functional catalysts and nano-confined materials for energy and environmental applications. In this invited lecture, we will present results from our recent collaborative research on the synthesis and catalytic applications of hollow nanomaterials.[1-5] Hollow Silicalite-1 and hollow ZSM-5 single crystals with silicon-rich exterior surface were prepared by a "dissolution-recrystallization" strategy in tetrapropylammonium hydroxide solution.[1,5] Selective dissolution and exterior recrystallization causes the silicon components to migrate from the inside to outside, resulting in a regular void in the interior of the crystal, increased Brönsted acid sites and a silicon-rich external surface of ZSM-5.

More interestingly, hollow ZSM-5 single crystals with double shells were successfully prepared by layer-by-layer technique followed with dissolution-recrystallization strategy. Furthermore, hollow ZSM-5 encapsulating iron and carbon nanotubes were successfully synthesized, as illustrated in Figure 1. Hollow Silicalite-1 or ZSM-5 nano-sized crystals with the interior functionalized as bimetallic nanoparticles (such as Ni-Pt and Cu-Pd) were also successfully synthesized.[2]

Some catalytic applications will be presented that demonstrate the unique features and versatility of these novel materials. The as-prepared hollow ZSM-5 exhibits excellent acid catalysis with enhanced shape selectivity, as shown in biphenyl methylation for synthesis of 4,4'-dimethyl bipheneyl inside the nano-confined hollow space, which is attributed to the silicon-rich external surface and thus the inhibition of isomerization on external surface. The catalytic reactions where hollow Silicalite-1 or ZSM-5 zeolite crystals were used as support include in-situ synthesized heteropoly acid catalyst for Suzuki-Miyaura coupling reaction inside the hollow space, and highly stable catalysis for CO_2 conversion over monometallic and bimetallic nanoparticles encapsulated inside the nano-confined hollow space.

Figure 1. Scheme representing the synthesis of hollow ZSM-5 with silicon-rich external surface, double shells, and functionalized interior with metallic nanoparticles and carbon nanotubes

Acknowledgement. Support from the Chinese Government with the QianRen B Jihua award to CS is gratefully acknowledged.

References

(1) Dai, C.Y.; Zhang, A.F.; Liu, M.; Guo, X.W.; Song, C.S. Hollow ZSM-5 with Silicon-Rich Surface, Double Shells, and Functionalized Interior with Metallic Nanoparticles and Carbon Nanotubes. *Advanced Functional Materials*, **2015**, 15, 7479–7487.

(2) Dai, C.Y.; Zhang, S.H.; Zhang, A.F.; Song, C.S.; Shi, C.; Guo, X.W. Hollow Zeolite Encapsulated Ni-Pt Bimetals for Sintering and Coking Resistant Dry Reforming of Methane. *Journal of Materials Chemistry A*, **2015**, 3, 16461-16468.

(3) Dai, C.Y.; Li, X.M.; Zhang, A.F.; Liu, C.; Song, C.S.; Guo, X.W. Pd and Pd-Cu Nanoparticles in Hollow Silicalite-1 Single Crystals for Enhancing Selectivity and Activity for the Suzuki-Miyaura Reaction. *RSC Advances*, **2015**, 5, 40297-40302.

(4) Dai, C.Y.; Zhang, A.F.; Li, J.J.; Hou, K.K.; Liu, M.; Song, C.S.; Guo, X.W. Synthesis of Yolk-Shell HPW@Hollow Silicalite-1 for Esterification Reaction. *Chemical Communications*, **2014**, 50, 4846-4848.

(5) Dai, C.Y.; Zhang, A.F.; Hou, K.K.; Liu, M.; Song, C.S.; Guo, X.W. Synthesis of Hollow Nanocubes and Macroporous Monoliths of Silicalite-1 by Alkaline Treatment. *Chemistry of Materials*, **2013**, 25 (21), 4197-4205.

DEVELOPMENT OF OPEN FRAMEWORK CHALCOGENIDE MATERIALS FOR ENERGY RELATED APPLICATIONS

Xitong Chen[a], Qipu Lin[a], Xianhui Bu[b] and Pingyun Feng[a]

[a] Department of Chemistry
University of California, Riverside, Riverside, CA 92521
[b] Department of Chemistry and Biochemistry,
California State University, Long Beach, CA 90840

Introduction

Great efforts have been devoted to developing crystalline porous materials (CPMs) in the past two decades. Among them, open framework chalcogenides represent a unique family of materials that integrate crystalline semiconducting properties with high surface area and uniform porosity. Open framework chalcogenides covered in this talk are based on metal chalcogenide molecular clusters with well defined chemical composition and uniform size. These materials bridge semiconducting nanoclusters and open framework materials and are well characterized by single crystal analysis. These could play a vital role in correlating the materials structural features with physical properties, which could lead to practical applications.

Superlattices, A series of novel open framework chalcogenides involving monovalent, divalent metal cations (such as Cu^+, Li^+, Zn^{2+}), trivalent, or tetravalent main group metal cations (Ga^{3+}, In^{3+}, Ge^{4+}, Sn^{4+}) and chalcogenide anions (S^{2-}, Se^{2-}, Te^{2-}) have been synthesized and structurally characterized. The molecular semiconducting clusters can be organized into one, two or three-dimensional covalent superlattices with different topologies[2] or can exist as superlattices through weak interactions. The resulting superlattices are usually based on supertetrahedral clusters (denoted as T_n, where n is the number of metal cation layers), which are actually tetrahedrally shaped fragments of cubic ZnS-type lattices. The size of the cluster can be rationally tuned by applying various combinations of metal cations with different charges in the self-assembly process. Because of the local charge balance requirement, the use of high-valent metal cations such as trivalent and tetravalent metal ions will result in small clusters and the addition of mono- or divalent metal cations to trivalent metal ions reaction system will induce larger clusters such as T4 and T5[3,4]. Up to now, the largest known supertetrahedral cluster in either isolated clusters or polymeric covalent superlattices, is T5. The tetrahedral clusters can be assembled into open frameworks by sharing the corner sulfur sites. Thus, the supertetrahedral cluster can be treated as pseudo-tetrahedral coordination sites, forming 3D zeolite-like structures with different topologies.

Supertetrahedral clusters assembling into extended frameworks are usually realized by sharing corner bi-coordinated sulfur anions. In addition, three T4 clusters can be connected by sharing one corner sulfur site, forming UCR-8 with cubic-C_3N_4 structure type. Furthermore, four supertetrahedral clusters can be linked together to form super-supertetrahedral clusters and their superlattices[6]. It is also possible for different sizes of clusters to co-crystallize into hybrid superlattices. For example, T3 and T4 clusters alternatively connect together, forming UCR-19 with double-diamond type structure[7]. Addition of tetravalent metal cations can induce smaller cluster such as T2. In OCF-42, alternating T2 and T4 clusters are found[8]. CIS-52 has an unusual structure, involving co-assembly of the largest and the smallest metal chalcogenide supertetrahedral clusters - T5 and T2[9].

Besides sharing the anionic sulfur sites at four corners of each cluster, the molecular clusters can be assembled into covalent superlattices by utilization of di- or polytopic organic linkers. Pyridine- and imidazolate-type ligands have been used for construction of open frameworks[10].

Highly Stable Superlattices. So far, at least 8 topological types have been realized in the materials based on supertetrahedral clusters. Among these, four different topological types were discovered in M^{4+}/M^{3+} reaction system, one of which possess the sodalite-type structure and was denoted as UCR-20[11]. In UCR-20GaGeS-TAEA, ion exchange with Cs^+ can help remove almost all of the amines. The Cs^+-exchanged product exhibits the type I isotherm characteristic of a microporous solid and has a high BET surface area of 807 m^2/g. However, the ratio of M^{4+}/M^{3+} in these materials ranging from 1.3 to 0.21 can be considered as low silica type zeolite materials with limited hydrothermal and photostability. Recently by introducing divalent metal cations, a family of high silica like zeolite materials such as CPM-120-ZnGeS-AEM was successfully synthesized. The materials possess the same structural topology as UCR-20 but with a much higher M^{4+}/M^{2+} ratio[14]. As shown in **Figure 1**, it consists of T2 clusters which are assembled into 3D superlattice by sharing corner sulfur sites. This framework with high concentration of tetravalent germanium represents one of the high-silica-zeolite-like materials and possesses much high thermal and photostability. It can be stable up to 610 °C in nitrogen flow and 310 °C for an hour in the air. After ion exchange with Cs^+, nitrogen sorption experiment shows type I isotherm with a BET surface area of 599 m^2/g.

Properties and Potential Applications. Due to the integration of porosity with semiconducting property, the synthesized crystalline porous semiconductors can be ideal photocatalysts for photocatalytic reactions. There are several advantages for porous chalcogenides for photocatalysis. Compared with conventional dense phase photocatalysts in which the available active sites are only on the surface, the open framework chalcogenides possess large numbers of reachable active sites in the internal channel due to high surface area. As the reactants can diffuse into the channels of the porous materials and be adjacent to the internal active sites, the generated electron-hole pairs may not need to travel a long distance to the external surface in order for the reaction to occur, thus reducing the possibility for the electron-hole pair recombination. In addition, the band gap of porous chalcogenides can be tuned by varying the chemical compositions and topologies. The robust framework, high porosity and semiconducting property make CPM-120 an excellent candidate for photocatalytic application. It is firstly tested for photocatalytic hydrogen production. Under UV-vis light irradiation, it exhibits a hydrogen release amount of 200 μmol h^{-1}/ (0.1g) in 0.1M Na_2SO_3 and 0.25M Na_2S solution (**Figure 2**) with the turnover number larger than 50. Even after illumination of 200 hours, the hydrogen evolution amount is almost the same without obvious decrease. It is well known that the photocorrosion is one of the drawbacks in metal sulfide materials for photocatalysis. This robust framework shows the excellent resistance to photocorrosion and circumvents this problem.

Besides photocatalytic hydrogen evolution, the CPM-120 materials also show good performance for CO_2 reduction. By integrating porosity with semiconducting property, various photocatalytic reactions with charged organic species can be evaluated in such a tunable platform. In another study, we have successfully proved the negative framework of CPM-120 can be a platform for studying porous photocatalyst with specific selectivity for different organics.

Different from common amine-templated porous chalcogenides, a series of hydrated open framework chalcogenides with hydrated inorganic cations as charge balancing species were synthesized in aqueous solution[12]. These frameworks show high ionic conductivity due to the richness of inorganic cations in the channel. For example, the specific conductivity of ICF-5 CuInS-Na is 1.2×10^{-2} ohm^{-1} cm^{-1}

at 18 °C under 100% relative humidity. In the lithium ion balanced framework, ICF-21 InS-Li shows a high conductivity up to 1.1×10^{-2} ohm^{-1} cm^{-1} at 24 °C under 31.7% relative humidity.

With the tunable pore size and structural features, it is expected a great potential can be realized in the open framework chalcogenides that combine porosities and semiconducting properties.

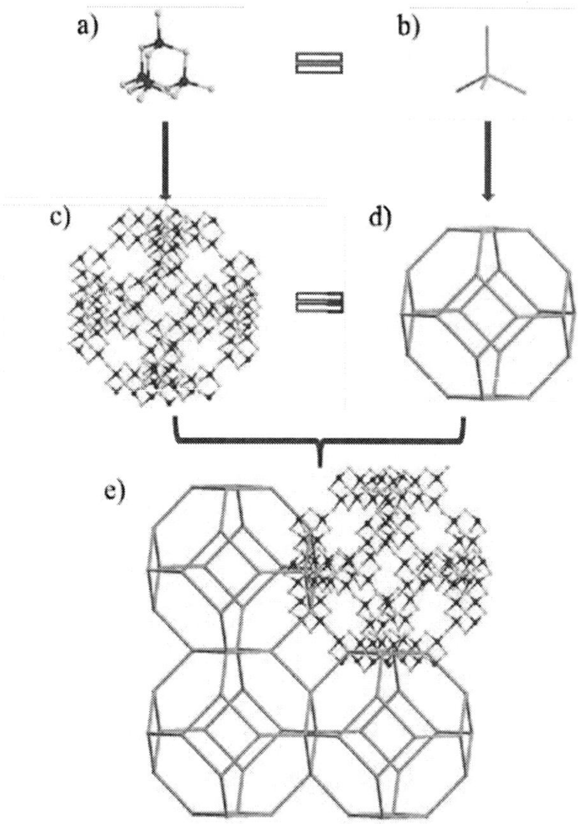

Figure 1. Structure of CPM-120. a) and b) T2 cluster. c) and d) supersodalite cage in CPM-120.

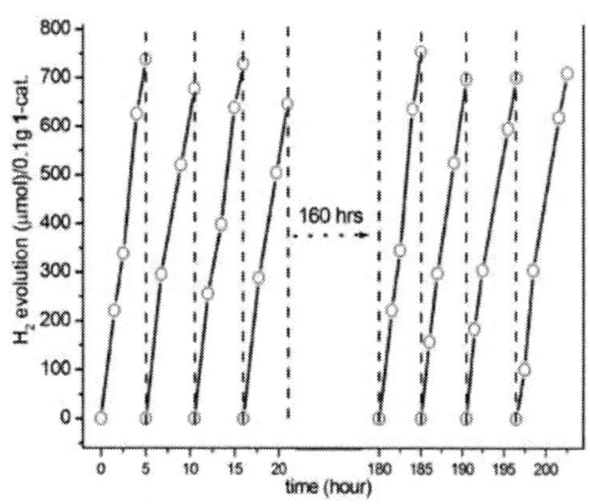

Figure 2. Photocatalytic H_2 production over 0.10 g catalyst of CPM-120 in 0.1 M Na_2SO_3 and 0.25 M Na_2S without loading cocatalysts.

Acknowledgment. This research was supported by National Science Foundation (NSF) (P.F., DMR-1506661).

References

(1) Feng, P. Y.; Bu, X. H.; Zheng, N. F. *Acc. Chem. Res.* **2005**, *38*, 293.
(2) Bu, X. H.; Zheng, N. F.; Feng, P. Y. *Chem.- Eur. J.* **2004**, *10*, 3356.
(3) Wu, T.; Bu, X.; Liao, P.; Wang, L.; Zheng, S.-T.; Ma, R.; Feng, P. *J. Am. Chem. Soc.* **2012**, *134*, 3619.
(4) Wu, T.; Bu, X.; Zhao, X.; Khazhakyan, R.; Feng, P. *J. Am. Chem. Soc.* **2011**, *133*, 9616.
(5) Wang, C.; Li, Y. Q.; Bu, X. H.; Zheng, N. F.; Zivkovic, O.; Yang, C. S.; Feng, P. Y. *J. Am. Chem. Soc.* **2001**, *123*, 11506.
(6) Bu, X. H.; Zheng, N. F.; Li, Y. Q.; Feng, P. Y. *J. Am. Chem. Soc.* **2003**, *125*, 6024.
(7) Zheng, N. F.; Bu, X. H.; Feng, P. Y. *J. Am. Chem. Soc.* **2003**, *125*, 1138.
(8) Wu, T.; Wang, X.; Bu, X.; Zhao, X.; Wang, L.; Feng, P. *Angew. Chem., Int. Ed.* **2009**, *48*, 7204.
(9) Wang, L.; Wu, T.; Bu, X.; Zhao, X.; Zuo, F.; Feng, P. *Inorg. Chem.* **2013**, *52*, 2259.
(10) Wu, T.; Khazhakyan, R.; Wang, L.; Bu, X.; Zheng, S.-T.; Chau, V.; Feng, P. *Angew. Chem., Int. Ed.* **2011**, *50*, 2536.
(11) Zheng, N. F.; Bu, X. G.; Wang, B.; Feng, P. Y. *Science* **2002**, *298*, 2366.
(12) Zheng, N. F.; Bu, X. H.; Feng, P. Y. *Nature* **2003**, *426*, 428.
(13) Zheng, N.; Bu, X. H.; Vu, H.; Feng, P. Y. *Angew. Chem., Int. Ed.* **2005**, *44*, 5299.
(14) Lin, Q.; Bu, X.; Mao, C.; Zhao, X.; Sasan, K.; Feng, P. *J. Am. Chem. Soc.* **2015**, *137*, 6184.

NOVEL MATERIALS FOR SAVING PRECIOUS METALS IN (SOLID) HETEROGENEOUS CATALYSTS

Bo-Qing Xu

Innovative Catalysis Program, Key Lab of Organic Optoelectronics & Molecular Engineering, Department of Chemistry, Tsinghua University, Beijing 100084, CHINA
E-mail: bqxu@mail.tsinghua.edu.cn

Precious metal elements constitute an important class of active components in many existing and emerging solid (heterogeneous) catalysts. Irreducible high surface area supporting materials are usually required in this class of catalysts to disperse and stabilize the precious metal entities for improving their utilization efficiency and mass activity of the precious metal elements for catalytic reactions. These precious metal containing catalysts are widely applied in energy conversion, chemical and petrochemical industries, waste abatement and environmental protection, as well as fine chemical and pharmaceutical syntheses. This presentation intends to highlight the approaches practiced in our laboratory for saving precious metals in solid thermal and electro catalysts by making the most possible use of the more precious elements or seeking nonprecious metal alternative catalysts in some catalytic reactions. The main discussion will be made on materials categorized as below:

1) Precious metal elements dispersed as isolated single atoms (cations) at the surface of the supporting materials.[1]
2) Precious metal-on-another metal (like Pt-on-Au or Pt-on-Ag) nanostructures.[2-8]
3) Core-shell nanostructures with the more precious metal enriched to the outmost layers.[9,10]
4) Transition metal-containing heteroatom doped carbon materials as the alternatives of precious metal catalysts.[11,12]

Acknowledgements. This work was financially supported by the National Basic Research Program of China (grant: 2013CB933103), the National Natural Science Foundation of China (grant: 21221062), and Tsinghua University (grant: 20131089311).

References

(1) Zhang, X.; Shi, H.; Xu, B.-Q. *Angew. Chem. Int. Ed.* **2005**, *44*, 7132.
(2) Sun, K.-Q.; Hong, Y.-C.; Zhang, G.-R.; Xu, B.-Q. *ACS Catal.* **2011**, *1*, 1336.
(3) Zhao, D.; Xu, B.-Q. *Angew. Chem. Int. Ed.* **2006**, *45*, 4955.
(4) Zhao, D.; Xu, B.-Q. *Phys. Chem. Chem. Phys.* **2006**, *8*, 5106.
(5) Zhao, D.; Wang, Y.-H.; Xu, B.-Q. *J. Phys. Chem. C* **2009**, *113*, 20903.
(6) Zhao, D.; Wang, Y.-H.; Yan, B.; Xu, B.-Q. *J. Phys. Chem. C* **2009**, *113*, 1242.
(7) Feng, Y.-Y.; Zhang, G.-R.; Xu, B.-Q. *RSC Adv.* **2013**, *3*, 1748.
(8) Zhang, G.-R.; Zhao, D.; Feng, Y.-Y.; Zhang, B.-S.; Su, D.-S.; Liu, G,; Xu, B.-Q. *ACS Nano* **2012**, *6*, 2226.
(9) Zhang, G.-R.; Wu, J.; Xu, B.-Q. *J. Phys. Chem. C* **2012**, *116*, 20839.
(10) Shen, L.-L.; Zhang, G.-R.; Miao, S.; Liu, J.; Xu, B.-Q. *ACS Catal.* **2016**, *6*, 1680.
(11) Yan, X.-H.; Zhang, G.-R.; Xu, B.-Q. *Chin. J. Catal.* **2013**, *34*, 1992.
(12) Yan, X.-H.; Xu, B.-Q. *J. Mater. Chem. A* **2014**, *2*, 8617.

NANO/MICROSCALE METAL-ORGANIC FRAMEWORK MATERIALS WITH SELECTIVE ADSORPTION AND SENSING PROPERTY

Wei-Yin Sun, Zhao-Peng Qi, Ji-MinYang, and Qing Liu

Coordination Chemistry Institute, State Key Laboratory of Coordination Chemistry, School of Chemistry and Chemical Engineering, Nanjing University, Nanjing 210023 (China)
E-mail: sunwy@nju.edu.cn

Introduction

In the past years, porous metal-organic frameworks (MOFs) have attracted remarkable research interests not only because of their intriguing framework structures, but also due to their interesting properties and potential applications such as adsorption, storage and separation, heterogeneous catalysis, chemical sensing and drug delivery.[1,2] In addition, not only the chemical compositions and structures, but also the specific size and morphology dependent properties make porous MOF nano/micromaterials more attractive for the applications in some specific fields compared with the bulk MOFs.[3,4] Furthermore, although the chemical composition and structure of MOF nano/micromaterials will dominate their properties, it has been found that the properties and applications of MOF nano/micromaterials also depend on their crystal morphology and size. Therefore, the controlled synthesis of porous MOF nano/micromaterials with specific morphology and size is attractive, however, remains still a great challenge.

We focus our attention on controlled syntheses and properties of nano/microscale MOF materials.[4] In this presentation, we will report porous MOF nano/micromaterials with selective adsorption and sensing properties. For example, a typical porous MOF - DUT-23(Cu) [$Cu_6(BTB)_4(BPY)_3$] was obtained by solvothermal reaction of 1,3,5-benzenetribenzoic acid (H_3BTB), 4,4'-bipyridine (BPY) with $Cu(NO_3)_2·3H_2O$ using different volume ratio of DMF and H_2O as solvent.[5] The results show that its nano/microcrystals obtained with volume ratio of DMF and H_2O of 18 : 2 exhibit excellent adsorption capacity for methylene blue (MB) with high selectivity as compared with other studied dyes of methyl orange, rhodamine B, and acid chrome blue K. While in the case of MIL-103(Eu) [Eu(BTB)],[6] which can be facilely and environmental friendly fabricated at room temperature, it is found that the addition of sodium acetate and the concentration of the reactants have important impact on determining the morphology and size of resulted MIL-103(Eu) crystals. Gas adsorption measurements reveal that the MIL-103(Eu) nanorods can selectively adsorb CO_2 over N_2 at ambient conditions, and display sensitive and selective sensing for Cu(II) ions and acetone molecules in solution.

Experimental

Materials and Measurements. All commercially available chemicals and solvents are of reagent grade and were used as received without further purification. Powder X-ray diffraction (PXRD) data were collected on a Bruker D8 Advance X-ray diffractometer with Cu-Kα (λ = 1.5418 Å) radiation at room temperature. Scanning electron microscopy (SEM) measurements were performed on a Hitachi S-4800 field emission scanning electron microscope at an accelerating voltage of 5 kV. Thermogravimetric analyses (TGA) were performed on a Mettler-Toledo (TGA/DSC1) thermal analyzer under the nitrogen atmosphere with a heating rate of 10 K min^{-1}. Adsorption experiments were carried out on a Belsorp-max volumetric gas sorption instrument. Surface areas were determined by the Brunauer-Emmett-Teller (BET) method. UV-Vis spectra were measured on a UV-Vis spectrophotometer (Shimadzu, UV-3600). Photoluminescence spectra were measured on an Aminco Bowman Series 2 spectrofluorometer with a xenon arc lamp as the light source.

Synthesis of DUT-23(Cu). In a typical experiment, H_3BTB (0.02 mmol), BPY (0.01 mmol) and $Cu(NO_3)_2•3H_2O$ (0.05 mmol) were mixed in 12 mL DMF and 8 mL H_2O (the volume ratio of DMF and H_2O is 12:8), and sealed into a Teflon-lined stainless steel container and heated at 85 °C for 4 h. After cooling to room temperature, blue-green precipitates were collected by centrifugation, washed with DMF and ethanol 3 times, and then dried under vacuum at 85 °C for 4 h. The activation of DUT-23(Cu) samples was carried out according to the published procedure using a conventional vacuum-drying process. The samples were soaked in methylene chloride for 2 d, the solvent was refreshed three times during the soaking process. The wet sample was transferred to a BET sample cell, and evacuated at room temperature under vacuum for 1 d.

Synthesis of MIL-103(Eu) [Eu(BTB)]. In a typical experiment, H_3BTB (0.05 mmol) was dissolved in 16 mL of ethanol and 4 mL of deionized water under ultrasonic wave conditions and then 0 - 4 equiv of sodium acetate with respect to H_3BTB was added with stirring to give clear solution. Then $Eu(NO_3)_3·3H_2O$ (0.05 mmol) was added into the solution with stirring at ambient temperature. The mixed solution was stirred for 30 min at room temperature and white precipitates were subsequently centrifuged, washed with distilled water and absolute ethanol several times, and dried in a vacuum at 60 °C for 5 h.

Results and Discussion

The PXRD data and SEM images of the samples obtained by using different volume ratios of mixed solvent of DMF and H_2O confirm the formation of the DUT-23(Cu) with different morphology and size. Ill-defined particles were obtained without addition of water, namely the volume ratio of DMF and H_2O is 20:0, however, when the volume ratio of DMF and H_2O is 18:2, the formation of homogeneous particles with size of *ca.* 100 nm was observed. With increasing the amount of water, morphology changes from cube with truncated edges of about 500 nm in size (DMF : H_2O = 15 : 5) to uniform cube crystals with particle size of *ca.* 2 μm (DMF : H_2O = 12 : 8) was achieved. It is noteworthy that blue precipitate appeared with irregular sheet shape when the content of DMF decreases to or lower than half of the mixed solvent, which shows weak and inconsistent PXRD patterns compared to that of bulk crystals, implying the degradation of the framework. The results indicate that DUT-23(Cu) nano/micromaterials with different morphology and size were obtained by tuning the DMF and H_2O volume ratio from 20:0 to 12:8.

Table 1. Adsorption Capacities for MB by Different Porous Materials

Material	Adsorption Capacity (mg g^{-1})	Ref.
DUT-23(Cu) nanoparticle	814	This work
ZJU-24	902	8
Amino-MIL-101(Al)	762	9
UMCM-150 (flower-like nanostructure)	560	10
$PW_{11}V$@MIL-101	371	11
MIL-101	195	9
MOF-235	187	12
HKUST-1	15.3	13

Considering that nano/micro-sized MOFs with different shape and size may exhibit unique and/or enhanced properties,[7] DUT-23(Cu) nano/micromaterials were tested for removal of different dye molecules in the aqueous solution. It is interesting to find that the DUT-23(Cu) nano/micromaterials show excellent adsorption capacity for methylene blue (MB), particularly, the materials obtained with the DMF and H_2O volume ratio of 18:2 exhibited the highest adsorption speed and uptake for the MB (814 mg g^{-1}). This is among the highest value of the porous MOF materials for MB removal reported to date (Table 1).[8-13] The high adsorption capacity of the materials is ascribed to the high surface area and porosity as confirmed from the nitrogen sorption experiments. The BET surface areas of the DUT-23(Cu) nano/micromaterials obtained with the DMF and H_2O volume ratio of 18:2, 15:5 and 12:8 are 1940.8, 1835.0 and 1630.8 m^2 g^{-1}, respectively. Further investigation on selective adsorption and separation of dyes was carried out for not only unitary dye molecule, but also binary dyes. The results show that in unitary dye solution, the materials obtained with the DMF and H_2O volume ratio of 18:2 has moderate adsorption ability for the positively charged rhodamine B due to its large size, however, no adsorption for the anionic dyes of methyl orange and acid chrome blue K due to the electrostatic repulsion. Furthermore, the DUT-23(Cu) nano/micromaterials exhibit selective adsorption of MB in binary dye solutions such as MB/methyl orange, MB/rhodamine B, and MB/acid chrome blue K. The results imply that the DUT-23(Cu) nano/micromaterials have superior removal efficiency for MB compared to the other dyes.

In the preparation of MIL-103(Eu) [Eu(BTB)] nano/microscale crystals, sodium acetate (NaOAc) was added in the reaction as alkaline additive to control the crystal nucleation and growth since the NaOAc can facilitate the deprotonation of the H_3BTB. The results of PXRD and SEM images show that the MIL-103(Eu) nano/microcrystals can be obtained with addition of 0 - 3 equiv of NaOAc (given as equivalent with respect to H_3BTB). Furthermore, the MIL-103(Eu) particle size decreases gradually and the morphology changes from pillar-like microrods to nanorods, and finally to spherical nanoparticles with increasing the amount of added sodium acetate. In addition, it was found that the concentration of the reactants also has influence on the particle morphology and size of MIL-103(Eu) crystals since higher concentration of the reactant can accelerate nucleation rates so as to give smaller particles. MIL-103(Eu) nanorods with relative small size and the highest BET surface area was used to study the selective adsorption and sensing. The results of N_2 and CO_2 adsorption isotherms at 273 K and 298 K, reveal that MIL-103(Eu) nanorods show gradual uptake of CO_2 to 1 atm but almost no adsorption of N_2. The calculated CO_2/N_2 selectivity is 39 at 273 K and 26 at 298 K implying that MIL-103(Eu) nanorods may have potential application in the separation of CO_2/N_2 mixtures at room temperature. Further studies show that that MIL-103(Eu) nano/microscale crystals exhibit red emission corresponding to the $^5D_0 \rightarrow ^7F_2$ transition of the Eu(III) ion, and the MIL-103(Eu) nanorods can sensitively and selectively sense Cu(II) ions and acetone molecules in solution.

Conclusions

Our study shows that nano/microscale MOF materials with different morphology and size can be facilely achieved by tuning the reaction conditions. In the particular cases of DUT-23(Cu) nano/microcrystals obtained with volume ratio of DMF and H_2O of 18 : 2 exhibiting selectively adsorption of methylene blue (MB) and MIL-103(Eu) nanorods displaying selectively adsorption of CO_2 over N_2 at ambient conditions, and sensitively and selectively sensing for Cu(II) ions and acetone molecules in solution.

Acknowledgement. This work was financially supported by the National Natural Science Foundation of China (grant no. 21331002 and 21573106). This work was also supported by a Project Funded by the Priority Academic Program Development of Jiangsu Higher Education Institutions.

References

(1) Lu, W.-G.; Wei, Z.-W.; Gu, Z.-Y.; Liu, T.-F.; Park, J.; Tian, J.; Zhang, M.-W.; Zhang, Q.; Gentle, T.; Bosch, M. and Zhou, H.-C. *Chem. Soc. Rev.*, **2014**, *43*, 5561.

(2) He, C.-B.; Liu, D.-M. and Lin, W.-B. *Chem. Rev.*, **2015**, *115*, 11079.

(3) Sindoro, M.; Yanai, N.; Jee, A. and Granick, S. *Acc. Chem. Res.*, **2014**, *47*, 459.

(4) Jin, L. N.; Liu, Q. and Sun, W. Y. *CrystEngComm*, **2014**, *16*, 3816.

(5) Klein, N.; Senkovska, I.; Baburin, I. A.; Grünker, R.; Stoeck, U.; Schlichtenmayer, M.; Streppel, B.; Mueller, U.; Leoni, S.; Hirscher, M. and Kaskel, S. *Chem. Eur. J.*, **2011**, *17*, 13007.

(6) Devic, T.; Serre, C.; Audebrand, N.; Marrot, J. and Férey, G. *J. Am. Chem. Soc.*, **2005**, *127*, 12788.

(7) Stock, N. and Biswas, S. *Chem. Rev.*, **2012**, *112*, 933.

(8) Zhang, Q.; Yu, J.-C.; Cai, J.-F.; Song, R.-J.; Cui, Y.-J.; Yang, Y.; Chen, B.-L.; and Qian, G.-D. *Chem. Commun.*, **2014**, *50*, 14455.

(9) Haque, E.; Lo, V.; Minett, A. I.; Harris, A. T.; and Church, T. L. *J. Mater. Chem. A*, **2014**, *2*, 193.

(10) Yang, J.-M.; Liu, Q.; Kang, Y.-S.; and Sun, W.-Y. *CrystEngComm*, **2015**, *17*, 4825.

(11) Yan, A.-X.; Yao, S.; Li, Y.-G.; Zhang, Z.-M.; Lu, Y.; Chen, W.-L. and Wang, E.-B. *Chem. Eur. J.*, **2014**, *20*, 6927.

(12) Haque, E.; Jun, J. W. and Jhung, S. H. *J. Hazard. Mater.*, **2011**, *185*, 507.

(13) Lin, S.; Song, Z.-L.; Che, G.-B.; Ren, A.; Li, P.; Liu, C.-B. and Zhang, J.-S. *Microporous Mesoporous Mater.*, **2014**, *193*, 27.

INVESTIGATION OF POROSITY AND CONNECTIVITY IN MESOPOROUS SILICON SPONGE MATERIALS USING HYPER-POLARIZED ^{129}XE NMR SPECTROSCOPY

Yougang Mao,[†] *Dokyoung Kim,*[‡] *Russell Hopson,*[†] *Jinmyoung Joo,*[‡] *Michael J. Sailor,*[‡] *and Li-Qiong Wang*[*,†]

[†] Department of Chemistry, Brown University, Providence, RI 02912,
[‡] Department of Chemistry and Biochemistry, University of California, San Diego, CA 92093-0358
*Corresponding Author: li_qiong_wang@brown.edu

Introduction

Porous silicon, or silicon sponge, is considered as a promising anode material for lithium ion batteries because of its preformed meso/macro porous structures that can accommodate large volume expansion during the lithiation process and its superior electrochemical performance.[1-3] To increase the conductivity, the MSS materials are carbon coated for the preparation of Si anodes in lithium ion batteries. It is important to examine the porosity and connectivity of the pores in these MSS materials since the porosity and connectivity influence the electrochemcial properties of the electrodes.

The pore geometry in most porous materials, even in ordered mesoporous silicon, is complex and characterized by interconnected cages, channels and micropores. As a result of these complex topographies, characterizing the inter-connectivity of pores in nano- or meso-porous materials is often challenging and requires application of several methods. Techniques such as small angle x-ray or neutron scattering, nano x-ray tomography, transmission electron microscopy and N_2 absorption, however, do not provide direct information on how the channels and cages are connected.

In this study, we have domonsrated that temperature dependent hyper-polarized (HP) ^{129}Xe NMR is a powerful tool for probing structures, uniformity and interconnectivity of the pores, that is complementary to other characterization techniques. Temperature dependent (173–353 K) hyper-polarized (HP) ^{129}Xe NMR is applied to investigate the structure and porosity of bare mesoporous silicon sponge (MSS) materials with varying pore and particle sizes, and carbon coated MSS materials.

Experimental

MSS material preparation. Mesoporous silicon sponge materials, 20n50µ (20 nm pore size, >50 µm particle size), 20n5µ (20 nm pore size, 5 µm particle size), 50n50µ (50 nm pore size, >50 µm particle size), and 50n5µ (50 nm pore size, 5 µm particle size), were prepared by electrochemical etching under different conditions as described previously.[1]

Carbon coating on MSS particles was conducted using a vacuum chemical vapor deposition system.[1] The MSS sample was heated at 600 °C at a rate of 10 °C min^{-1}, and at 600 °C, the precursor gas (argon: acetylene 1⁄4 9:1) was introduced. The furnace temperature was increased to 690 °C and was maintained at that temperature for 30 min. After carbon deposition, the furnace was cooled to room temperature in pure argon.

Hyperpolarized (HP) ^{129}Xe NMR. HP ^{129}Xe NMR experiments were carried out on a Bruker Advance DSX 300 spectrometer operating at 82.98 MHz (magnetic field 7.05 T) using a variable temperature static probe with a continuous flow (CF) of HP xenon. The HP ^{129}Xe is produced by the homemade ^{129}Xe NMR polarizer.[4]

For HP ^{129}Xe NMR experiments, a single-pulse (SP) Bloch-decay method was used. The HP ^{129}Xe NMR experiments were performed without spinning. SP spectra were collected with a 1 µs ^{129}Xe pulse, and a repetition delay of 0.5 sec.

Results and Discussion

SEM (Scanning electron microscope) images and BET pore size distribution plots are given in Figure 1. From the SEM images, pore sizes for the 20 nm pore size materials (20n50µ, 20n5µ) appear to be more uniform than those of the 50 nm pore size materials (50n50µ, 50n5µ). In addition, BET data dispay a narrow pore size distribution for the 20 nm pores but a very wide distribution for the 50 nm pores regardless of particle sizes.

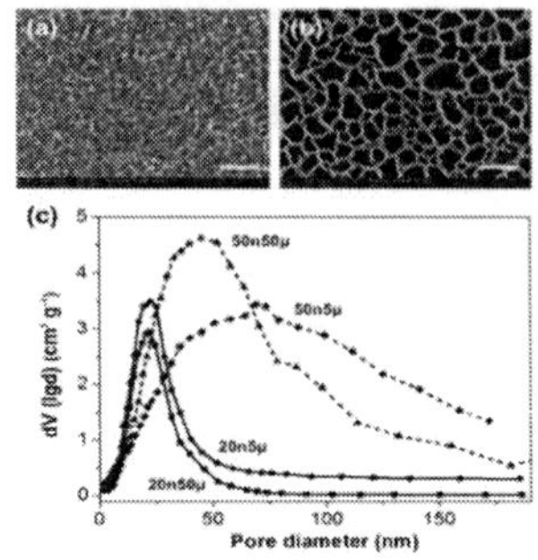

Figure 1. Scanning Electron Microscopy (SEM) and nitrogen adsorption analysis. (a, b) Plan view of 20 nm (20n5µ) and 50 nm (50n5µ) pore size materials, respectively. Scale bar: 200 nm. (c) Pore size distribution of materials using BJH adsorption model: dV/dLog(D) vs. pore diameter values.

HP ^{129}Xe NMR spectra are given in Figs. 2 and 3 for four MSS samples taken at 300 K, and at variable temperatures, respectively. The observed ^{129}Xe NMR signal for the adsorbed Xe is a weighted average among various pore environments sampled by a Xe atom within its characteristic diffusion distance. The narrow 0 ppm peak is associated with the free Xe gas and the broader peaks at higher chemical shifts are associated with adsorbed Xe on the MSS materials. In general, for a given particle size, the higher chemical shift corresponds to the smaller pore size. Interestingly for a given pore size, smaller size particles are associated with the lower chemical shift, which is due to the faster exchange of adsorbed Xe inside the smaller MSS particles with the gas phase Xe. The chemical shifts follow the trend: 73.3 ppm (20n50µ) > 62.6 ppm (20n5µ) > 60.8 ppm (50n50µ) > 42.5 ppm (50n5µ), indicating that the chemical shift decreases with increasing pore size, and for a given pore with decreasing particle size. Overall temperature dependent spectra shown in Fig. 3 for all MSS samples look similar with some variations in line widths and peak positions, in contrast with the N_2 adsorption data showing a large difference between MSS samples with 20 and 50 nm pores.

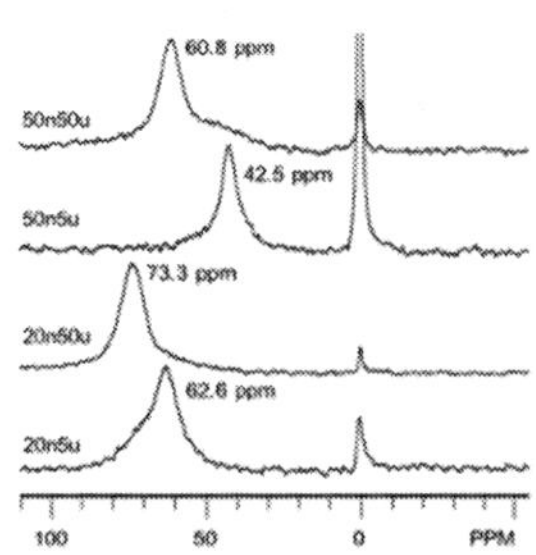

Figure 2. HP ^{129}Xe NMR spectra taken at 300 K for four MSS materials

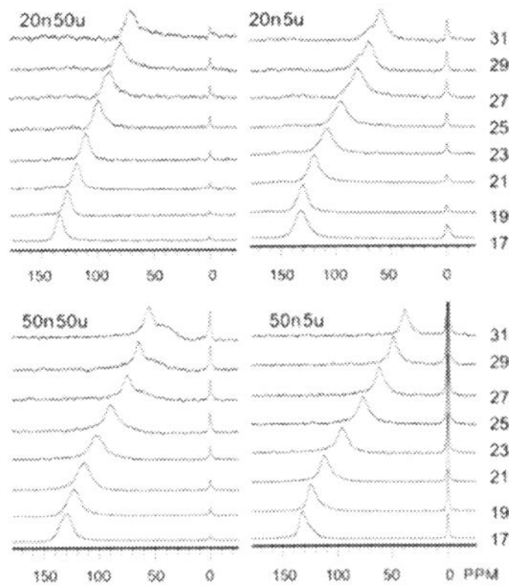

Figure 3. Variable temperature HP ^{129}Xe NMR spectra for four MSS materials.

Table I lists physical parameters obtained from fitting the temperature dependent chemical shift plots based on the Henry's Law, as described previously.[5] The values of δ_s among all four MSS materials range from 131 ppm to 148 ppm. The H_{ads} values observed for all the materials are 12–16 kJ/mol, which is consistent with what is generally seen for physical adsorption of Xe on solid surfaces. Pore diameters derived from the ^{129}Xe NMR are much smaller than those from SEM and BET measurements, indicating the presence of microspores in addition to the mesoporous channels in the framework of the MSS materials (shown in Fig. 4)

Table 1. Parameters derived from temperature-dependent chemical shifts of MSS materials.[a]

Material	ΔH (kJ/mol)	δ_s (ppm)	D_χ (nm)
20n50μ	12.2	139	8
20n5μ	12.8	148	12
50n50μ	15.4	131	10
50n5μ	15.2	138	20

[a] Term δ_s is characteristic chemical shift derived from fit to eq. 1. D_χ is pore size derived from geometric fit as described in the text.

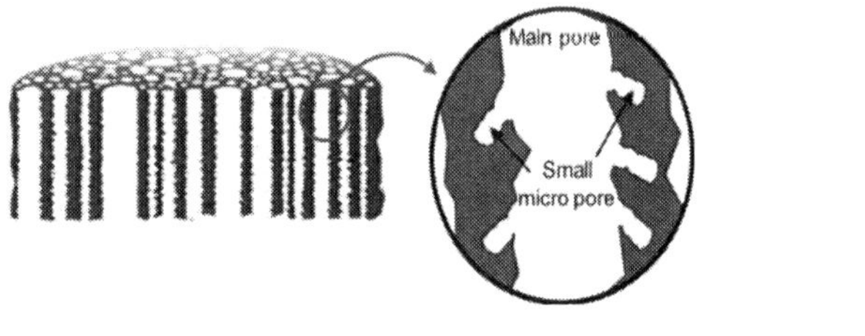

Figure 4. Illustration of proposed pore morphology in MSS structures, depicting the presence of small micro-pores embedded in larger mesoporous channels.

The temperature dependent line-shape and line-width of absorbed Xe peaks can be used to assess the comparative uniformity and interconnectivity of the pores among the four MSS sample types. Since at low temperatures Xe principally occupies smaller micro-pores and the broadness of the peaks results most likely from pore heterogeneity, the order in line- widths suggests that the order of the homogeneity in the micropores in these samples is: 50n50μ < 20n5μ < 50n5μ < 20n50μ. As the temperature increases, Xe enters the larger mesopores and undergoes fast exchange with the different pore environments. Thus the line- widths at higher temperature are related to the uniformity of larger mesopore channels and the interconnectivity among the different pores. The broader line-widths and shoulder peaks found in the 20n5μ and 50n50μ materials at higher temperatures indicate there are at least two different mesopore environments in these materials. Narrower peaks observed at higher temperatures for 20n50μ and 50n5μ than those for 20n5μ and 50n50μ suggest greater uniformity and interconnectivity among the different pore environments in the 20n50μ and 50n5μ samples than in the 20n5μ and 50n50μ materials.

Since the electrodes are made of carbon coated MSS materials, it is important to examine the porosity for MSS materials before and after carbon coating. The HP ^{129}Xe NMR spectra taken at 300 K for MSS materials before and after carbon coating are shown in Figure 5 where the chemical shift for the carbon coated pores is smaller than that for the bare MSS materials. In principal, when pores are coated with the carbon, the pore size should be getting smaller, which normally results in a higher chemical shift. However in this case, since the MSS materials has a unique pore structure shown in Fig. 4, the smaller chemical shift suggests that coated carbon may block the smaller microspores present in the mesoporous channels, resulting in a larger averaged pore size. The fact of removal of or blocking the smaller microspores is further supported by the temperature dependent ^{129}Xe NMR data given in Fig. 6, where the narrower line-widths observed for bare MSS samples at both high and low temperature regions indicate that that more uniform pores for carbon coated MSS materials.

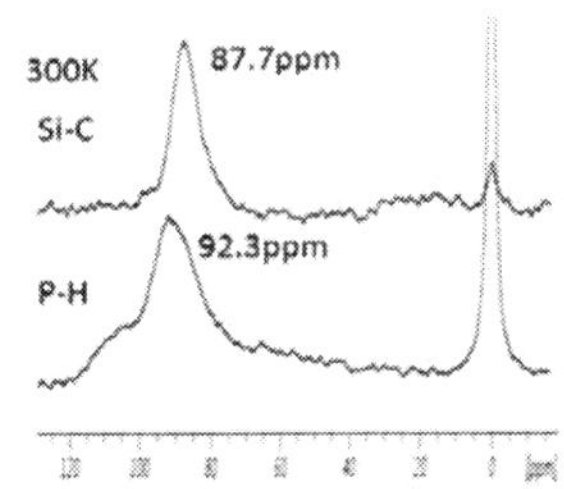

Figure 5. ^{129}Xe NMR spectra for bare and carbon coated MSS materials at 300 K

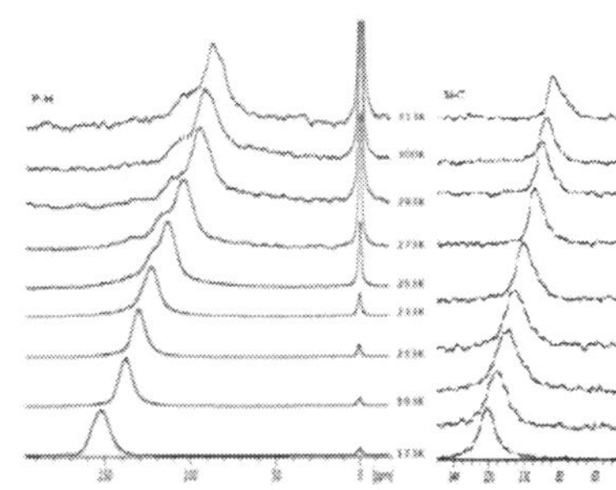

Figure 6. Variable temperature HP ^{129}Xe NMR spectra for the MSS materials before and after carbon coating.

In summary, temperature dependent HP ^{129}Xe NMR data from this study provide a fingerprint of the structure and interconnectivity of the pores for each material. It was found that the carbon coating removes the surface roughness or micropores inside the main mesochannels. Such information is useful to gain a better understanding of the relationship between porous structures and their corresponding electrochemical properties in the design of novel electrodes.

Acknowledgment. The HP ^{129}Xe NMR work was supported by the EPSCoR Implementation grant DE-SC0007074, Office of Basic Energy Sciences, U.S.DOE. MMS material preparation and analysis work was supported by the NSF under Grant No. DMR-1210417 and by DARPA under Cooperative Agreement HR0011-13-2-0017. The authors thank the advance grant #HRD-0548311 for the purchase of new equipment used in this study.

References

(1) Li. X.; Gu, M.; Hu, S.; Kennard, R.; Yan, P.; *et al.* Mesoporous Silicon Sponge as an Anti-pulverization Structure for High-performance Lithium-ion Battery Anodes. *Nat Commun.* **2014**, *5*, 4105.

(2) Armand, M.; Tarascon, J.-M. Building Better Batteries. *Nature* **2008**, *451*, 652-657.

(3) Whittingham, M. S. Materials Challenges Facing Electrical Energy Storage. *MRS Bull.* **2008**, *33*, 411-419.

(4) Y. Mao, M. Song, R. Hopson, N.K. Karan, P. R. Guduru, and L-Q. Wang, " Hyperpolarized ^{129}Xe Nuclear Magnetic Resonance Studies of Si Nanocomposite Electrode Materials," Energy & Fuels, 30 (2), 1470 (2016).

(5) Terskikh, V. V.; Moudrakovski, I. L.; Mastikhin, V. M. 129Xe Nuclear Magnetic Resonance Studies of the Porous Structure of Silica Gels. *J. Chem. Soc. Faraday Trans.* **1993**, *89*, 4239-4243.

IRON BASED BIMENTALLIC SULFIDE AND ITS HYDRODESULFURIZATION OF DBTS: A STUDY ON STRONG SYNERGY EFFECT

*Hao Li, Jiancong Li, Lei Li, Yandan Wang, Baojian Shen**

State Key Laboratory of Heavy Oil Processing; the Key Laboratory of Catalysis of CNPC; College of Chemical Engineering, China University of Petroleum,
Beijing 102249, P. R. China
*Corresponding author: baojian@cup.edu.cn

Introduction

Cobalt (nickel) prompt molybdenum (tungsten) sulfides catalysts play a key role in fuel hydrotreating process in past decades [1]. However, the insufficiency of supply and high cost of these metals is still a challenge to the growing needs for hydrogenation catalyst. As a cheaper and widely available transition metal, iron gets some attention. In 1980s, Pecoraro and coworkers [2] investigated the hydrodesulfurization (HDS) activity of most unsupported transition metal sulfides (TMS), which showed iron sulfide has a poor HDS activity. Similar conclusion was obtained from the research of alumina supported TMS [3].

Meanwhile, several hypotheses [4-7] studied the promoter effect of Fe on the HDS activity of MoS_2, most of these researches proposed that a FeMoS phase was responsible for the weak synergy effect of FeMo catalyst.

For the present work, alumina supported iron based bimetallic sulfides were prepared by impregnation-sulfidation method. The HDS of dibenzothiophene (DBT) and 4,6-dimethydibenzothiophene (DMDBT) were used as model reactions. This iron based bimetallic sulfides catalyst showed an obvious synergy effect compared with iron and specific metal (M) mono-sulfide.

Experimental

Catalyst preparation. The oxide samples were prepared by the incipient wetness impregnation method using an aqueous solution of iron nitrate and titanium, copper, vanadium, zinc, lanthanum, manganese salts etc., respectively. After impregnation, a dry and calcination procedure was performed. A reference Mo based catalyst was prepared via same procedure by using ammonium molybdate tetrahydrate as precursor solution.

Catalyst characterization. X-ray photoelectron spectroscopy (XPS) analysis of the samples was conducted on a PHI Quantera SXM photoelectron spectrometer with a monochromated A1 anode. The charge effect was minimized by using a charge neutralizer. The C 1s peak at 284.6 ev was taken as the reference. Since the sulfide samples were easy to re-oxidized upon exposure to air, the samples were all prepared and kept in a glove box before transfer to the XPS measurement.

Catalytic activity test. The HDS activity tests were carried out in a continuous flow fixed-bed micro-reactor under the conditions: 4.0 MPa, 360 ℃, respectively , the catalyst loading of 1.0 g, 0.58 wt.% DBT and 0.67 wt.% 4,6-DMDBT in decane as the model feed, respectively , the feed flow rate of 0.20 cm^3/min, the H_2 flow rate of 60 cm^3/min. For each run, the weighed catalyst was diluted with quartz powder, and the mixture was placed in a stainless steel reactor (Ø8 mm). The catalysts were presulfided in situ with the sulfiding feed of 5 wt.% CS_2 in decane, 4.0 MPa and 360 ℃ K for 4 h. After steady-state conditions were reached, the liquid effluents were periodically collected. And the conditions of catalytic activity evaluation and the method of products analysis have been described in detail elsewhere.

Results and Discussion

X-ray photoelectron spectroscopy (XPS) of Fe 2p is presented in Fig. 1. The binding energy (BE) values of 710.1 and 707.0 eV for Fe 2p3/2 in mono-Fe catalyst (refered as F) can be assigned to FeS and FeS_2. Interestingly, a new peak which has a BE equal to 708.6 eV is observed when a specific metal was introduced (refered as FMx), characteristics a new Fe species. There is little information about this species except for a Fe species in a FeMoS phase which has a similar BE value of 708.9-709.1 eV [7]. Combined with the prompt metal 2p2/3 XPS and HRTEM-EDS mapping (not show), we propose a bimetallic sulfide is formulating instead of a simple mixture of separate phases of iron and M sulfides.

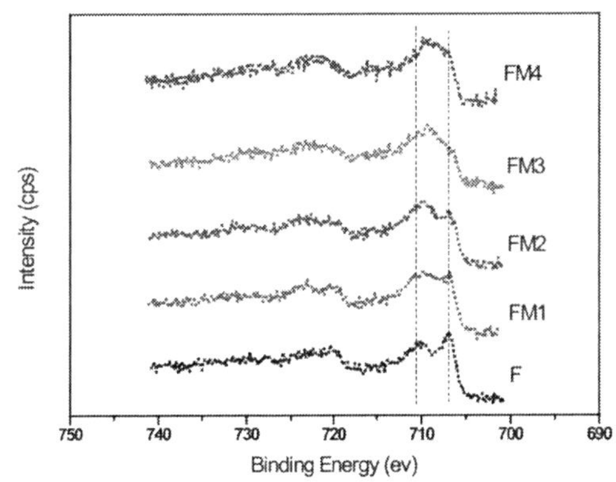

Figure 1. The Fe 2p XPS of mono Fe and series Fe-M bimetallic sulfide catalysts..

The reaction results of DBT HDS over F, M (mono-M catalyst), bimetallic sulfide and reference Mo catalysts are shown in Figure 2. While F had a moderate DBT HDS activity, the M showed poor catalytic activity. However, the iron-M sulfides showed an obvious synergy effect, in an optimal M/Fe ratio, the activity was approach MoS_2 based catalyst. In addition, compared with F and M, F-M samples showed a low BIP selectivity, as we know, BIP is the main product of direct desulfurization route (DDS).

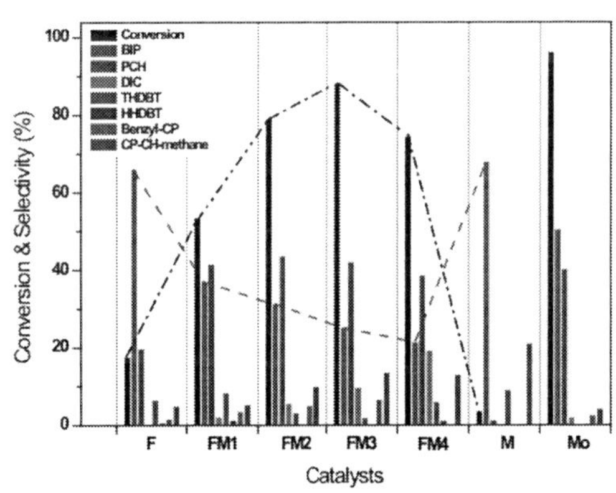

Figure 2. Conversion and product distribution in HDS of DBT over different catalysts.

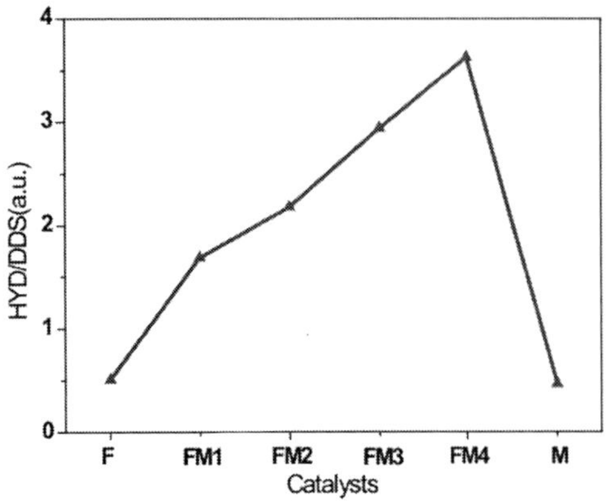

Figure 3. A HYD/DDS ratio relation over different catalysts.

The DBT HDS reaction can take place via either the DDS or the HYD reaction pathway. The reaction pathway selectivity indexed as HYD/DDS ratio could be determined from product selectivity as HYD/DDS ratio =Selectivity of (1-BIP) / selectivity of BIP, and used to identify the preferential reaction pathway. The HYD / DDS ratios of F, M and Mo (not show) were about 0.5 (Figure 3), indicating a DDS preferably. In other hand, as increasing M ratio, a high HYD / DDS ratio could be observed, thus, M can promote the HYD pathway by means of the presence of bimetallic sulfide.

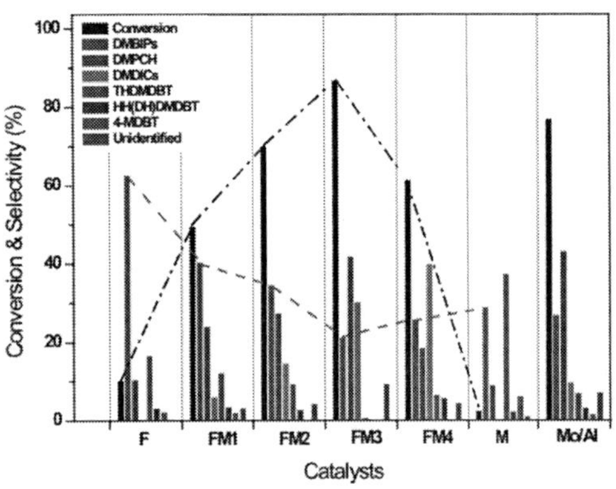

Figure 4. Conversion and product distribution in HDS of 4,6-DMDBT over different catalysts..

To confirming above proposal, 4,6-DMDBT HDS reaction was performed. Since it's widely accepted that sterically hindered molcules are preferably desulfurized through the HYD pathway. As show in figure 4, a more pronounced synergy effect was observed. The activities even exceed traditional MoS_2 based catalyst. M greatly enhanced the HDS reaction of 4,6-DMDBT though a HYD pathway according to a low BIP selectivity.

Conclusion

Specific metal was used to modify iron sulfide catalyst. It was found that a bimetallic sulfide is formulating instead of a simple mixture of separate phases of iron and prompt metal sulfides. A strong synergy effect was observed, when it used in HDS of DBT especially 4,6-DMDBT, which was due to the high hydrogenation activity.

Acknowledgment. The authors greatly acknowledge the funding of this project by the PetroChina.

References

(1) S. Harris, R. R. Chianelli. *J. Catal.* **1986**, 98, 17.
(2) T. A. Pecoraro, R. R. Chianelli.. *J. Catal.* **1981**, 67, 430.
(3) J. Quartararo, S. Mignard, S. Kasztelan. *J. Catal.* **2000**, 192, 307.
(4) W. L. T. M. Ramselaar, M. W. H. Craje, E. Gerkema, V. H. J.De Beer, van der Kraan. *Bull. Soc. Chim. Belg.* **1987**, 96, 931.
(5) H. Topsøe, R. Candia, N. Y.Topsøe, B. S. Clausen. *Bull. Soc. Chim. Belg.* **1984**, 93, 783.
(6) B. S. Clausen, S. Morup, H. Topsøe, R. Candia. *J. Phys. Colloq.* **1976**, C6-249, 146.
(7) M. Karroua, J. Ladriere, H. Matralis, P. Grange, B. Delmon. *J. Catal.* **1992**, 138, 640.

PLASMON-ENHANCED SPECTROSCOPY WITH SHELL-ISOLATED MODE

Jian-Feng Li

College of Chemistry and Chemical Engineering, Xiamen University, Xiamen 361005, China
Li@xmu.edu.cn

When light of a resonant wavelength is incident upon metal nanostructure surface, it will initiate collective oscillations in electron density that gives narrow regions of enhanced electromagnetic field strength at the surface or nanogaps, as surface plasmon resonance (SPR). This extremely enhanced electromagnetic field can enhance the Raman signals of nearby molecules. The effect was called surface-enhanced Raman scattering (SERS),[1,2] which can be used for in situ investigation of trace chemical species and identification with single-molecule sensitivity. However, the SERS enhancement is limited to a few noble metals such as Ag, Au, or Cu with roughened or nanostructured surfaces, which had limited the breadth of practical applications of SERS in a variety of fields.

Shell-isolated nanoparticle-enhanced Raman spectroscopy (SHINERS)[3] was therefore invented to break the long-standing limitation of SERS. In SHINERS, Au/Ag@SiO_2 core-shell nanoparticles were rationally designed (Figure 1). The Au/Ag core as plasmonic antenna provides high electromagnetic field to enhance the Raman signals of probed molecules, while the ultra-thin, uniform and pinhole-free silica shell separates the Au/Ag cores from the system under study and ensures no interference from processes involving the SERS-active cores, thus, providing the original information from the target systems. SHINERS has already been applied to many challenging systems.[4,5]

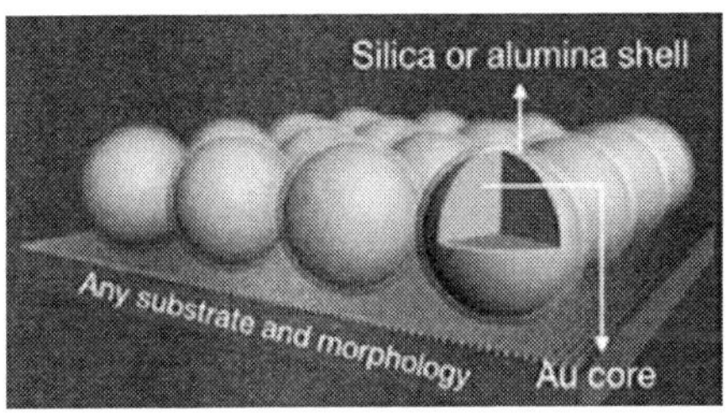

Figure 1. Schematic of working principles of SHINERS.

Different SHIN types are used to meet different analysis requirements. Gold cores with a 55 nm diameter (Figure 2a) are usually employed because these SHINs are easy to make and provide enough Raman signal enhancement for most applications. Gold cores with a 120 nm diameter (Figure 2b) are prepared if more enhancement is required. 3D-FDTD simulations have shown that a 120 nm gold core diameter is optimal for use with a 633 nm excitation wavelength. Gold nanocube and nanorod SHINs (Figures 2c and 2d) were developed because their SPR absorption maxima can be tuned from the visible to the near infrared by adjusting nanocube size and nanorod aspect ratio. This quality is essential for biosensor applications. SHINs with a mostly silver core (Figure 2e) were synthesized because silver is a more efficient plasmonic substrate than gold, and because the SPR of silver nanostructures can be forced to any wavelength in the visible region of the spectrum. On the other side, the shell material should be chemically and electrically inert to prevent chemical interaction and electrical contact. For example, the shell material of SHINs can be TiO_2 (Figure 2G), Al_2O_3 (Figure 2H), MnO_2 (Figure 2I), and Ag_2S (Figure 2J), which meet the demands of various chemical environments (e.g., stability in different pH solutions).

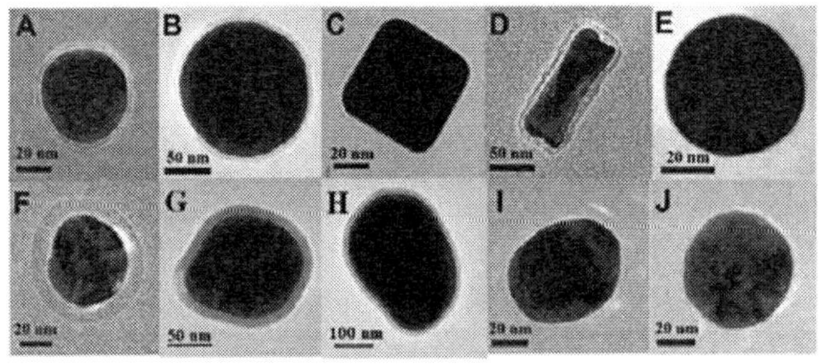

Figure 2. HR-TEM images of various SHINs. Au@SiO_2 NPs with a 55 nm spherical core (A), a 120 nm spherical core (B), a nanocube core (C), a nanorod core (D), a shell-isolated nanoparticle with a Ag core (E), 55 nm Au@SiO_2 NP with 20 nm silica shell (F), Ag@TiO_2 (G), Ag@Al_2O_3 (H), MnO_2 (I), and Ag_2S (J).

Shell-isolated nanoparticles (SHINs) can be spread as "smart dusts" over surfaces with diverse compositions and morphologies, and it has already been applied to a number of challenging systems, such as hydrogen and CO on Pt(hkl) and Rh(hkl), which can't be realized by traditional SERS. Combining with electrochemical methods, we have in-situ monitored the surface electro-oxidation at Au(hkl) electrodes,[6] and oxygen reduction reaction at Pt(hkl) surfaces. Hydroxyl, peroxide and superoxide were directly observed as intermediates which proved the long-standing speculation in electrochemistry. These pioneering studies demonstrate convincingly the ability of SHINERS in exploring correlations between structure and reactivity at the well-defined interfaces. SHINERS method has also been used to probe structure and processes on materials of different composition and morphology, from metal single-crystals to semiconductors and from food samples to living cells.

The concept of shell-isolated nanoparticle-enhancement is being applied to other spectroscopies such as infrared absorption and sum frequency generation. For instance, the shell-isolated mode was applied to fluorescence spectroscopy with a thousand time enhancement on fluorescent signals.[7] The concept was also extended to second-harmonic generation (SHG) to build up a gap-mode nanoruler.[8] New scanning probe microscopy tip structures inspired by SHINERS, such as shell-isolated tip-enhanced spectroscopy (SITERS) with nanoscale spatial resolution, are now under development. The shell-isolated mode is extremely flexible and of use in many areas of science and technology.

References

(1) Moskovits, M. *Rev. Mod. Phys.* **1985,** *57*, 783-826.
(2) Willets, K. A.; Van Duyne, R. P. *Annu. Rev. Phys. Chem.* **2007,** *58*, 267-297.
(3) Li, J. F.; Huang, Y. F.; Ding, Y.; Yang, Z. L.; Li, S. B.; Zhou, X. S.; Fan, F. R.; Zhang, W.; Zhou, Z. Y.; Wu, D. Y.; Ren, B.; Wang, Z. L.; Tian, Z. Q. *Nature* **2010,** *464*, 392-395.
(4) Li, J. F.; Tian, X. D.; Li, S. B.; Anema, J. R.; Yang, Z. L.; Ding, Y.; Wu, Y. F.; Zeng, Y. M.; Chen, Q. Z.; Ren, B.; Wang, Z. L.; Tian, Z. Q. *Nat. Protoc.* **2013,** *8*, 52-65.
(5) Li, J. F.; Anema, J. R.; Wandlowski, T.; Tian, Z. Q. *Chem. Soc. Rev.* **2015,** *44*, 8399-8409.
(6) Li, C. Y.; Dong, J. C.; Jin, X.; Chen, S.; Panneerselvam, R.; Rudnev, A. V.; Yang, Z. L.; Li, J. F.; Wandlowski, T.; Tian, Z. Q. *J. Am. Chem. Soc.* **2015,** *137*, 7648-7651.
(7) Li, C. Y.; Meng, M.; Huang, S. C.; Li, L.; Huang, S. R.; Chen, S.; Meng, L. Y.; Panneerselvam, R.; Zhang, S. J.; Ren, B.; Yang, Z. L.; Li, J. F.; Tian, Z. Q., *J. Am. Chem. Soc.* **2015,** *137*, 13784-13787.

(8) Shen, S. X.; Meng, L. Y.; Zhang, Y. J.; Han, J. B.; Ma, Z. W.; Hu, S.; He, Y. H.; Li, J. F.; Ren, B.; Shih, T. M.; Wang, Z. H.; Yang, Z. L.; Tian, Z. Q. *Nano Lett.* **2015**, *15*, 6716-6721.

IDENTIFICATION AND COMPARISON OF THE HYDROGEN-EVOLUTION ACTIVITY OF BRIDGING AND TERMINAL DISULFIDE IN MoS_x CATALYSTS UNDER ACIDIC ENVIRONMENT

Chengyi Hu, Zhening Chen, Gang Fu, Nanfeng Zheng

Collaborative Innovation Center of Chemistry for Energy Materials, State Key Laboratory for Physical Chemistry of Solid Surfaces, College of Chemistry and Chemical Engineering, Xiamen University, Xiamen 361005, China.
nfzheng@xmu.edu.cn

Introduction

Water electrolysis is considered to be a clean method to produce hydrogen. Efficient catalysts are needed for hydrogen evolution reaction (HER) and oxygen evolution reaction (OER) in electrocatalytic or photoelectrocatalytic water splitting. MoS_2 has been emerging as one of the most efficient HER electrocatalysts since the first discovery in 2005 by Nørskov and coworkers.[1] In 2007, Chorkendorff *et al.* experimentally demonstrated that only the edges of MoS_2 are catalytically active, while the basal planes are inert.[2] Afterwards numerous efforts have been made to prepare MoS_2 nanostructures with more exposed edge sites to improve the activity. However, it is always ignored that the crystalline MoS_2 have two kinds of edge sites: the (-1 0 1 0) S-edge and the (1 0 -1 0) Mo edge. It is difficult to distinguish the activity between the different edge sites which are always coexist in the MoS_2.[3] Great efforts have been made to understand the active sites in amorphous MoS_x catalysts. But it remains elusive whether terminal or bridging disulfide is catalytically active due to the amorphous structures.[4,5] The directly design of MoS molecules with specific edge site has been proved to be a promising method to investigate the structure-catalysis correlation. $[(PY5Me_2)MoS_2]^{2+}$ can serve as a molecular mimic of the sulfided Mo-edge site and can catalytically generate hydrogen from acidic organic solvent and from acidic water.[6] Recently, some progress has been achieved to identify the active sites using MoS_x clusters to mimic the MoS_2 edge sites. However, it is still difficult to compare the activity of the two edge sites due to the coexistence of bridging ($(S\text{-}S)_b$) and terminal disulfide ($(S\text{-}S)_t$) in the $[Mo_3S_{13}]^{2-}$ and $[Mo_2S_{12}]^{2-}$ clusters.[7,8] Herein we prepared the $[Mo_3S_7Cl_6]^{2-}$ cluster with only $(S\text{-}S)_b$ in the unit. This simple and unique structure provided us an ideal model to directly investigate the HER activity of $(S\text{-}S)_b$ and $(S\text{-}S)_t$ under acidic environment, nicely simulating the corresponding S-edge and sulfided Mo-edge of crystalline MoS_2 (**Figure 1**).

Experimental

Synthesis of $(PPh_4)_2[Mo_3S_7Cl_6]$ cluster. $(NH_4)_2[Mo_3S_{13}]$ nH_2O and $(NH_4)_2Mo_2S_{12}$ nH_2O clusters were synthesized according to the reported procedures.[7,8] The $(PPh_4)_2[Mo_3S_7Cl_6]$ cluster was prepared by adding 0.5 g $(NH_4)_2[Mo_3S_{13}]$ nH_2O into 50mL concentrated HCl solution and heated at 60 °C for 30 minutes. Adding PPh_4Cl to the filtrate and left undisturbed for one day. The crude products were recrystallized in CH_2Cl_2/hexane mixture for single crystal characterization.[9]

Electrochemical characterization. The electrochemical measurements were carried out by CHI 760e electrochemical workstation in a three-electrode cell, consisting of an SCE as the reference electrode and a graphite foil as the counter electrode. The catalysts ink was made by mixing clusters, 10 mg Vulcan XC-72 carbon and 30 mL 5 wt% Nafion solution in 1 mL acetonitrile or methanol and sonicated for 30 min to form a homogeneous ink. The working electrode was prepared by drop-casting the catalyst ink on GC rotating disk electrode with a loading amount of 100 nmol cm^{-2} clusters (confirmed by ICP-MS). The HER performance was measured in 0.5 M H_2SO_4 electrolyte under continues purging with N_2 gas with a scan rate of 5 mV/s and a rotating rate of 1600 r.p.m.

Results and Discussion

$(PPh_4)_2[Mo_3S_7Cl_6]$ was synthesized by reacting $(NH_4)_2[Mo_3S_{13}]$ with concentrated HCl solution. After the precipitation with PPh_4Cl and recrystallization, the single crystal of $(PPh_4)_2[Mo_3S_7Cl_6]$ was obtained. As revealed by the single crystal X-ray diffraction, the Mo_3 core with one apical S and three bridging disulfides were well reserved, while the three terminal disulfides were replaced by six chloride under this strongly acidic condition resulting in the formation of $[Mo_3S_7Cl_6]^{2-}$ clusters. Moreover the $(PPh_4)_2[Mo_3S_7Br_6]$ cluster with the similar structure was also obtained by reacting $(NH_4)_2[Mo_3S_{13}]$ with concentrated HBr solution.

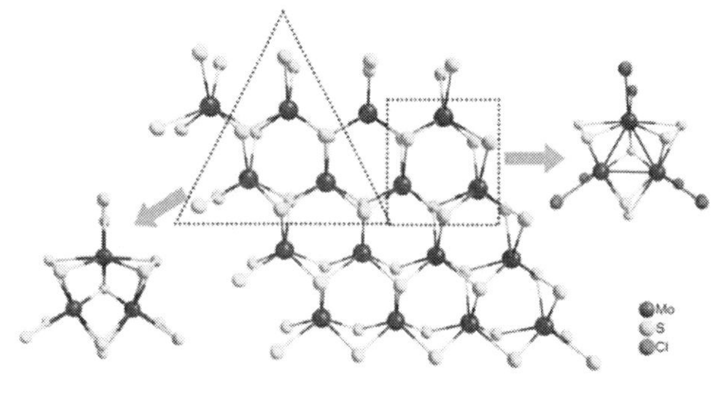

Figure 1. Crystal structure of MoS_2 and the corresponding $[Mo_3S_{13}]^{2-}$ and $[Mo_3S_7Cl_6]^{2-}$ clusters.

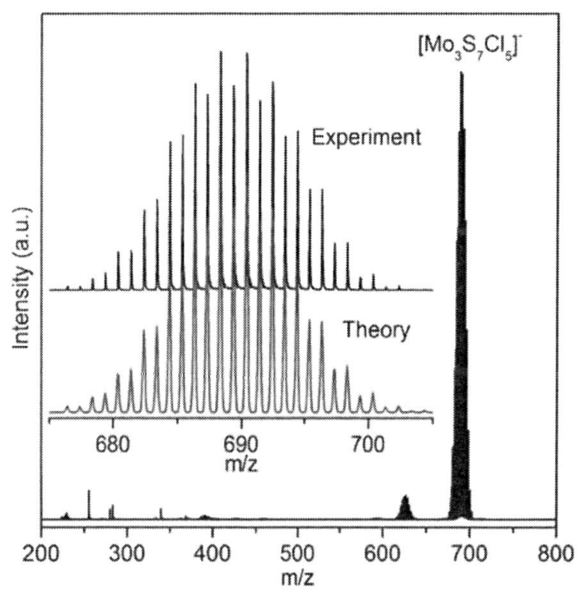

Figure 2. ESI-MS spectra of the $[Mo_3S_7Cl_6]^{2-}$.

In order to identify the composition, the as-prepared cluster was characterized by electrospray ionization time-of-flight mass spectrometry (ESI-TOF-MS). As shown in **Figure 2**, the as-prepared cluster displayed one main peak at m/z = 689.35. Detailed analysis on the high-resolution mass data suggested that the strongest peak at m/z = 689.35 belonged to $[Mo_3S_7Cl_5]^-$ with the isotopic pattern matched the simulation perfectly. One chloride might get lost during the ionization process. The ESI-MS result indicated the high-purity of the products. **Figure 3** shows the Raman spectrum of different MoS_x clusters. In $[Mo_3S_{13}]^{2-}$ cluster, vibration of bridging disulfide ($\nu(S\text{-}S)_b$) and terminal disulfide ($\nu(S\text{-}S)_t$) were found at 548 and 512 cm^{-1}, respectively. Vibrations of molybdenum sulfide bonds (ν(Mo-S)) were found at 385-286 cm^{-1}. $[Mo_2S_{12}]^{2-}$ displayed similar vibrations with $\nu(S\text{-}S)_b$ at 530 cm^{-1} and $\nu(S\text{-}S)_t$ at 544 cm^{-1}. In the $[Mo_3S_7Cl_6]^{2-}$, only one vibrational peak of $\nu(S\text{-}S)_b$ at 565 cm^{-1} was observed, indicating the successfully remove all of the $(S\text{-}S)_t$ in the $[Mo_3S_7Cl_6]^{2-}$, while the $(S\text{-}S)_b$ were nicely reserved.

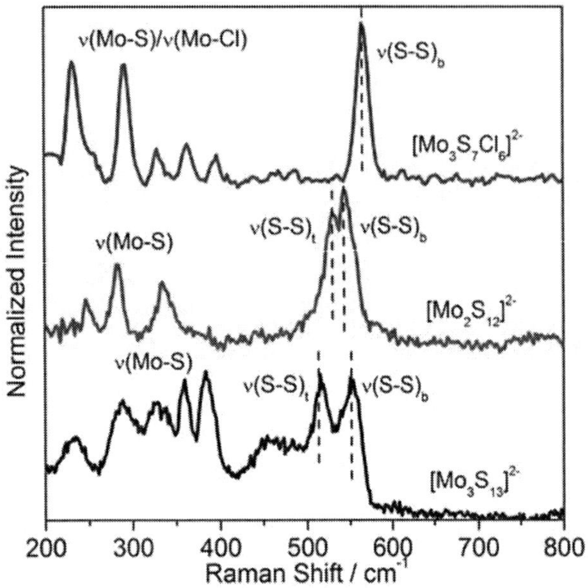

Figure 3. Raman spectra of the $[Mo_3S_7Cl_6]^{2-}$, $[Mo_3S_{13}]^{2-}$ and $[Mo_2S_{12}]^{2-}$ clusters.

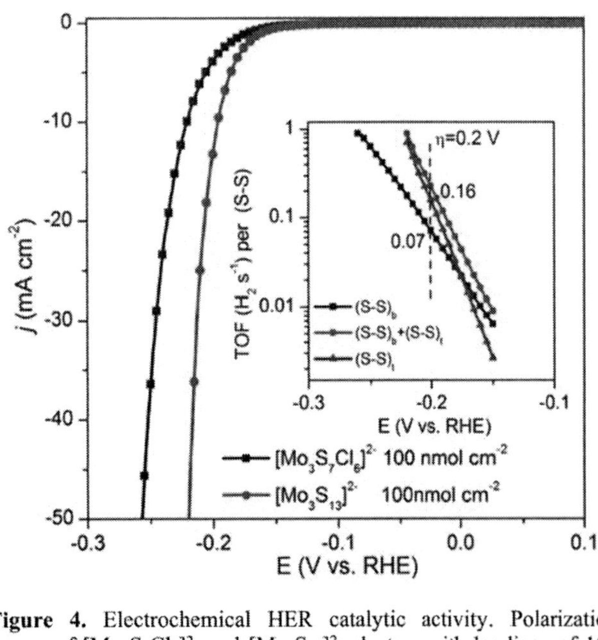

Figure 4. Electrochemical HER catalytic activity. Polarization curves of $[Mo_3S_7Cl_6]^{2-}$ and $[Mo_3S_{13}]^{2-}$ clusters with loadings of 100 nmol cm^{-2}. (Inset: TOFs per bridging disulfide $(S\text{-}S)_b$ and terminal disulfide $(S\text{-}S)_t$.)

The electrocatalytic HER activities of MoS_x clusters were investigated using a three-electrode electrochemical setup in 0.5 M H_2SO_4 under N_2. **Figure 4** shows the linear sweep voltammetry of $[Mo_3S_7Cl_6]^{2-}$ and compared with that of the state-of-the-art catalyst $[Mo_3S_{13}]^{2-}$ at a scan rate of 5 mV/s with iR compensation. The polarization curves shows that the activity of $[Mo_3S_7Cl_6]^{2-}$ was much lower compared with that of $[Mo_3S_{13}]^{2-}$ with equivalent catalysts loading of 100 nmol cm^{-2}. $[Mo_3S_{13}]^{2-}$ required an overpotential of 196 mV to reach a current density of 10 mA cm^{-2}, while $[Mo_3S_7Cl_6]^{2-}$ required 220mV. In the previous reports, the TOF often normalized per Mo atom because it was difficult to distinguish different sulfur edge sites in the catalysts. However, since there are only $(S\text{-}S)_b$ in the $[Mo_3S_7Cl_6]^{2-}$, the TOF of $(S\text{-}S)_b$ can be obtained exactly. Since the TOF per Mo atom of $[Mo_3S_{13}]^{2-}$ is equal to the TOF per $(S\text{-}S)_b$+$(S\text{-}S)_t$, the TOF of $(S\text{-}S)_t$ was obtained by subtracting the TOF of $(S\text{-}S)_b$ from $(S\text{-}S)_b$+$(S\text{-}S)_t$. Finally, the TOFs of $(S\text{-}S)_b$ and $(S\text{-}S)_t$ was calculated as 0.07 and 0.16 s^{-1}, respectively. Consequently, the activity of $(S\text{-}S)_t$ is higher than $(S\text{-}S)_b$ at high overpotential.

Conclusions

In conclusion, $(PPh_4)_2[Mo_3S_7Cl_6]$ cluster was successfully synthesized and used to investigate the active sites of HER. The as-prepared $[Mo_3S_7Cl_6]^{2-}$ was characterized by single crystal X-ray diffraction, ESI-MS and Raman, confirmed the presence of $(S\text{-}S)_b$ without any other kinds of sulfur in the cluster. The HER activity of $[Mo_3S_7Cl_6]^{2-}$ under acidic environment was lower than $[Mo_3S_{13}]^{2-}$ with both $(S\text{-}S)_b$ and $(S\text{-}S)_t$, and showed a higher Tofel slope. Owing to the single kind of sulfur site in $[Mo_3S_7Cl_6]^{2-}$, the TOFs per $(S\text{-}S)_b$ and $(S\text{-}S)_t$ were obtained for the first time. The results showed that the activity of $(S\text{-}S)_t$ is twice higher than $(S\text{-}S)_b$ at the overpotential of 200 mV. The use of $[Mo_3S_7Cl_6]^{2-}$ with single kind of sulfur site helped us to compare the activity of $(S\text{-}S)_b$ and $(S\text{-}S)_t$, and also shed light on the rational design of MoS_x catalysts for HER.

Acknowledgement. This work was supported by MOST of China (2015CB932303) and the NSFC of China (221420102001, 21131005, 21390390, 21333008).

References

(1) Hinnemann, B.; Moses, P. G.; Bonde, J.; Jorgensen, K. P.; Nielsen, J. H.; Horch, S.; Chorkendorff, I.; Norskov, J. K. *J. Am. Chem. Soc.* **2005**, *127*, 5308.
(2) Jaramillo, T. F.; Jorgensen, K. P.; Bonde, J.; Nielsen, J. H.; Horch, S.; Chorkendorff, I. *Science* **2007**, *317*, 100.
(3) Lauritsen, J. V.; Kibsgaard, J.; Helveg, S.; Topsoe, H.; Clausen, B. S.; Laegsgaard, E.; Besenbacher, F. *Nat. Nanotechnol.* **2007**, *2*, 53.
(4) Lassalle-Kaiser, B.; Merki, D.; Vrubel, H.; Gul, S.; Yachandra, V. K.; Hu, X.; Yano, J. *J. Am. Chem. Soc.* **2015**, *137*, 314.
(5) Ting, L. R. L.; Deng, Y.; Ma, L.; Zhang, Y.-J.; Peterson, A. A.; Yeo, B. S. *ACS Catal.* **2016**, 861.
(6) Karunadasa, H. I.; Montalvo, E.; Sun, Y.; Majda, M.; Long, J. R.; Chang, C. J. *Science* **2012**, *335*, 698.
(7) Kibsgaard, J.; Jaramillo, T. F.; Besenbacher, F. *Nat. Chem.* **2014**, *6*, 248.
(8) Huang, Z.; Luo, W.; Ma, L.; Yu, M.; Ren, X.; He, M.; Polen, S.; Click, K.; Garrett, B.; Lu, J.; Amine, K.; Hadad, C.; Chen, W.; Asthagiri, A.; Wu, Y. *Angew. Chem. Int. Ed.* **2015**.
(9) Fedin, V. P.; Sokolov, M. N.; Mironov, Y. V.; Kolesov, B. A.; Tkachev, S. V.; Fedorov, V. Y. *Inorg. Chim. Acta* **1990**, *167*, 39.

MECHANISTIC INSIGHTS INTO METAL LEWIS ACID-MEDIATED CATALYTIC TRANSFER HYDROGENATION REACTIONS

Matthew Gilkey, Alexander Mironenko, Dionisios Vlachos and Bingjun Xu

Catalysis Center for Energy Innovation
Department of Chemical and Biomolecular Engineering
University of Delaware, 150 Academy St., Newark, DE

Introduction

The multifunctional nature of biomass-derived feedstocks demands intricate interaction between the reactant molecule and the catalyst surfaces to remove the excess oxygen and partially hydrogenate specific unsaturated groups for many relevant chemicals and fuels, a process known as hydrodeoxygenation (HDO). In doing so, one needs to avoid over-hydrogenation and unselective cracking the C-C bonds of the molecule. This tandem of reactions typically employs a Brønsted acid catalyst, such as sulfuric acid, to remove oxygen as water, and a metal catalyst, such as Pt, to carry out the hydrogenation using high-pressure hydrogen.[1,2] Such dual metal/Brønsted acid catalysts can be environmentally harsh, and strong Brønsted acid catalysts can lead to side reactions and reduce selectivity. Development of more selective processes and catalysts can have a profound impact on the viability of biomass upgrade.

A selective route to remove excess oxygen-containing functional groups in biomass-derived feedstocks is via catalytic transfer hydrogenation (CTH), which employs renewable organic compounds such as alcohols and organic acids as the hydrogen donor. Since biomass is typically available in remote areas that lack a transportation system, CTH is an attractive alternative to alleviate the need for high-pressure hydrogen that requires large-scale infrastructure, available only in integrated chemical plants. CTH has been demonstrated in the upgrade of various biomass molecules, including cellulose, sugars, and furanics often employing a supported metal or metal oxides as a hydrogenation catalyst.

Herein, we present a combined experimental and computational study on the reaction pathways in the CTH of furfural over a mildly oxidized Ru/C catalyst, referred to as Ru/RuO$_x$/C, by detailed mass fragmentation analysis with isotopically labeled chemicals and rigorous kinetic studies. Furfural is an ideal biomass derived "platform" chemical and its hydrogenolysis product, MF, has a high octane number and can be employed for the production of renewable toluene. We show that the intermolecular hydride transfer is the dominant pathway for the hydrogenation of the carbonyl group of furfural to furfuryl alcohol (FA). Further, hydrogenolysis via ring activation is a major, if not the dominant, pathway in the C-OH bond cleavage in the conversion of FA to MF. Our results provide molecular level understanding of the role of the cooperativity of metal and Lewis acid sites in mediating the CTH of biomass derivatives, which can guide the rational design of efficient bifunctional catalysts by suitable choice and proper atomic level arrangement of Lewis acid and metal sites. Further, we demonstrate that detailed mass fragmentation analysis is a powerful technique in mechanistic investigations of complex reactions involving biomass-derived molecules.

Experimental

Materials and catalyst preparation. All chemicals were obtained from Sigma Aldrich and used without further purification. The catalyst used was 5 wt.% Ru/C (Sigma-Aldrich), which, prior to each experiment, was treated for 3 h at 300 °C in H_2 flow of 40 cc/min followed by mild oxidation for 3 h at 130 °C in 5%O_2/He flow (40 cc/min).

Catalyst evaluation

Catalytic transfer hydrogenation of furfural was carried out in a 100 mL Parr batch reactor. In a typical experiment, the reactor was charged with 24 mL of a 10 vol% 2-propanol (unlabeled (-d0) or perdeuterated(-d8))/toluene solution of furfural (1 wt%) and 0.1 g of Ru/RuO$_x$/C catalyst, pressurized to 300 psi (2.04 MPa) with N_2 and heated to a predetermined temperature. It typically takes approximately 30 min for the reactants to reach the desired temperature, which is not included in the reported reaction time. Reaction was quenched by soaking the reactor in an ice bath upon reaching the desired reaction time. The suspension in the reactor was collected after it reached room temperature, filtered, stored in sealed vials and analysed.

Computational methods

Density functional calculations have been performed using VASP software. Core electrons were represented using PAW formalism, and valence electrons were modeled with the exchange-correlation functional of Perdew, Burke, and Ernzerhof (PBE). Ground state energies were calculated self-consistently using a plane wave basis set with a kinetic energy cutoff of 400 eV, an electronic energy threshold of 10^{-6} eV, and a force tolerance of 0.05 eV/Å.

Results and Discussion

Isotopic labeling studies of catalytic transfer hydrogenation of furfural to MF. *Hydrogenation of furfural to FA.* The mass fragmentation pattern of FA produced with 2-propanol-d_0 as the hydrogen donor at 140 °C in toluene matches well that in the NIST database[34]. The parent ion of FA (98 amu) is the most intense peak (red bars in Figure 1a), and the small signal in 99 amu is attributed to the natural isotope abundance of ^{13}C. A clear mass shift by 1 amu was observed when 2-propanol-d_8 was used as the hydrogen donor under otherwise identical reaction conditions (blue bars in Figure 1a), indicating that the produced FA molecule contains one D. In both hydrogenation mechanisms (Scheme S1), the FA molecule contains two Ds using 2-propanol-d_8. The discrepancy between the predicted and observed mass shifts (2 amu and 1 amu, respectively) can be rationalized by the fact that the deuterium in the OD group of FA can exchange the OH groups in the capillary gas chromatograph (GC) column, as evidenced by the injection of neat 2-propanol-d_8 (Figure S2). Therefore, we conclude the 1 amu mass shift in the FA formed with 2-propanol-d_8 as the hydrogen donor originates from the D bonded to the hydroxymethyl carbon (Figure 1d).

Since FA formed via either mechanism has an identical isotopic structure, replacing 2-propanol-d_0 with 2-propanol-d_8 cannot differentiate the two mechanisms. The key difference between the two reaction pathways (Figure 1d and e) is whether the H being transferred adsorbs on the catalyst surface. The MVP mechanism proceeds via the formation of a six-membered ring configuration, through which the β-H of 2-propanol is transferred to the carbonyl C atom of furfural. The H transfer occurs in a concerted step, in which the H being transferred never adsorbs on the surface of the catalyst. Therefore, when 2-propanol-d_8 is used as the hydrogen donor, *the β-D will be directly transferred to the carbonyl C atom in furfural.* In contrast, two consecutive reaction steps are involved in the metal-mediated hydrogenation: first dehydrogenation of 2-propanol occurs on metallic sites to form adsorbed H atoms and acetone, followed by the addition of the adsorbed hydrogen atoms to the carbonyl bond. In

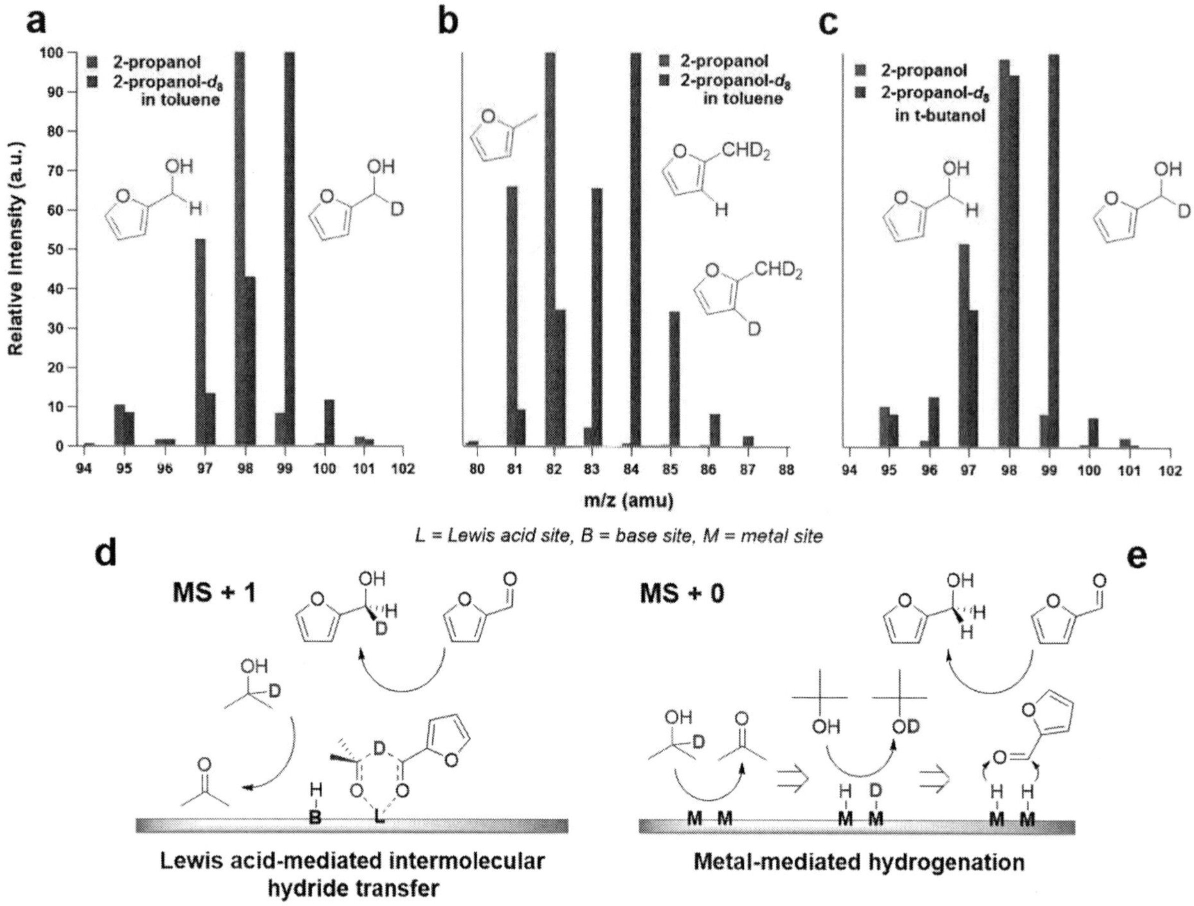

Figure 1. Mass fragmentation analysis of the products of CTH of furfural. Mass spectra (all intensities scaled to 100%) of (a) FA, (b) MF obtained after furfural hydrogenolysis in toluene, and (c) FA obtained from furfural hydrogenolysis in t-butanol. Schematic of the reaction mechanism of FA formation via (d) Lewis acid mediated intermolecular hydride transfer and (e) metal mediated hydrogenation.

this pathway, *all hydrogen atoms transferred are adsorbed on the metallic sites prior to being added to the carbonyl group of furfural.* In order to differentiate these two possible pathways, we replaced toluene with t-butanol as solvent in the reaction because: 1) as a tertiary alcohol, the absence of β-H prevents it from acting as a hydrogen donor; and 2) the H in the OH group of t-butanol is able to exchange with active H atoms on the catalyst, e.g., O-H and Ru-H; while leaving H in C-H bonds untouched. Since t-butanol is in great excess (~90 mol%) with respect to 2-propanol, most of the OD in 2-propanol-d_8 will be exchanged into OH due to the fast OH/OD exchange between alcohols. Thus, 2-propanol-d_7 ($CD_3CD(OH)CD_3$) is the majority of the D(or H) donor when 2-propanol-d_8 is used. In the case of an MVP pathway, the only D transferred to furfural is the β-D of 2-propanol-d_7, because it is unexchanged and is transferred via a concerted pathway without adsorbing on a metallic site (Figure 1d). Thus, a 1 amu MS shift of the FA formed is expected. In contrast, FA formed following the metal-mediated hydrogenation is expected to be free of D (no mass shift), because the fast H/D exchange between t-butanol and adsorbed D will replace most of the adsorbed D with H (Figure 1e). FA formed in the hydrogenation of furfural with 2-propanol-d_0 and -d_8 in t-butanol clearly shows a mass shift of the parent ion by 1 amu (Figure 1c), indicating that the concerted intermolecular hydride transfer mechanism is the dominant pathway in the hydrogenation of the carbonyl group of furfural.[3,4]

DFT calculations confirm that the reaction barrier for the hydrogenation of furfural to FA is lower for the MPV mechanism, while oxygen vacancies on the RuO_x are likely to play an important role in catalyzing the selective C-O bond cleavage in the conversion of FA to MF. A complete catalytic cycle is proposed.

Acknowledgment. We acknowledge support from the Catalysis Center for Energy Innovation, an Energy Frontier Research Center funded by the U.S. Department of Energy, Office of Science, Office of Basic Energy Sciences under Award number DE-SC0001004.

References

(1) Sutton, A. D.; Waldie, F. D.; Wu, R.; Schlaf, M.; Louis, A., III. *Nature Chemistry*. 2013, pp428-432.
(2) Zhu, X.; Lobban, L. L.; Mallinson, R. G.; Resasco, D. E. *Journal of Catalysis*. 2011, pp 21–29.
(3) Gilkey, M. J.; Panagiotopoulou, P.; Mironenko, A. V.; Jenness, G. R.; Vlachos, D. G.; Xu, B. *ACS Catalysis*. American Chemical Society June 3, 2015, pp 3988–3994.
(4) Mironenko, A. V.; Gilkey, M. J.; Panagiotopoulou, P.; Facas, G.; Vlachos, D. G.; Xu, B. *Journal of Physical Chemistry C*. 2015, pp 6075–6085.

FUNCTIONALIZED CARBON NANOTUBE-SUPPORTED METAL NANOPARTICLES AS EFFICIENT CATALYSTS FOR GREEN CHEMICAL REACTIONS

*Weiping Deng, Qinghong Zhang, Ye Wang**

State Key Laboratory of Physical Chemistry of Solid Surfaces, Collaborative Innovation Center of Chemistry for Energy Materials, National Engineering Laboratory of Green Chemical Production for Alcohols, Ethers and Esters, College of Chemistry and Chemical Engineering, Xiamen University, Xiamen 361005, China
*wangye@xmu.edu.cn

Introduction

Carbon nanotubes (CNTs) possess many advantages as catalysts or catalyst supports due to their uniform 1D sp^2-bonded carbon structure, tunable surface properties by functionalization, controllable interactions with supported metals and high external surface areas suitable for transformation of large molecules.[1] Here, we present our studies using functionalized CNT-supported metal NPs for green catalytic reactions involving biomass conversions and aerobic oxidation reactions.

Results and Discussion

We found that metal NPs loaded on CNTs with acid functional groups (e.g., –COOH) could work as efficient bifunctional catalysts for the conversion of cellulose or cellobiose, a model of cellulose.[2,3] For examples, Ru and Au NPs loaded on CNTs with acid functional groups were highly efficient for the hydrogenation and oxidation of cellobiose or cellulose into sorbitol and gluconic acid, respectively (Figure 1).

The aerobic oxidation of HMF into FDCA is an attractive reaction for establishing biomass-based sustainable chemical processes. CNT-supported Au-Pd alloy NPs were highly efficient and recyclable catalysts for HMF to FDCA in water under base-free conditions.[4] Besides the alloy effect, we demonstrated that the organic groups (in particular ketone/quinone) on CNTs could facilitate the adsorption of HMF and intermediates from the reactant solution, accelerating FDCA formation. Pt NPs loaded on properly functionalized CNTs also exhibited excellent performances for this reaction.[5]

The Au-Pd/CNT was found to be a highly efficient catalyst for aerobic oxidation of amines, providing the highest TOFs for oxidative homocoupling of benzylamine or oxidative dehydrogenation of dibenzylamine. Our studies revealed that the ketone/quinone group on CNTs functioned as a redox mediator in cooperation with the Au-Pd alloy NPs to accelerate the C-H activation and hydride transfer.

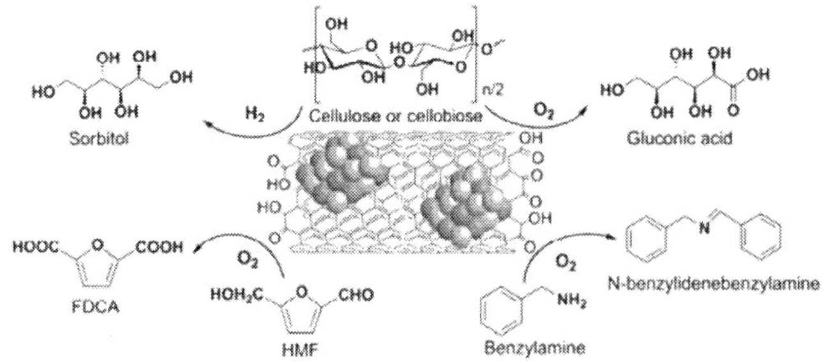

Figure 1. Functionalized CNT-supported metal nanoparticles for some green chemical reactions.

Conclusions

CNTs are highly attractive materials for green catalysis. The combination of CNTs and metal nanoparticles represents a powerful system for many bifunctional catalytic reactions in liquid media. We have demonstrated that the CNT can not only work for the dispersion of metal nanoparticles but also can provide functional groups for acid or redox catalysis. The peculiar surface of CNT can further enhance the adsorption of organic reactants or intermediates from the reaction media such as aqueous solution.

References

(1) Su, D. S.; Perathoner, S.; Centi, G. *Chem. Rev.* **2013**, *113*, 5782.
(2) a) Deng, W.; Tan, X.; Fang, W.; Zhang, Q.; Wang, Y. *Catal. Lett.* **2009**, *133*, 167. b) Deng, W.; Liu, M.; Tan, X.; Zhang, Q.; Wang, Y. *J. Catal.* **2010**, *271*, 22.
(3) Tan, X.; Deng, W.; Liu, M.; Zhang, Q.; Wang, Y. *Chem. Commun.* **2009**, 7179.
(4) Wan, X.; Zhou, C.; Chen, J.; Deng, W.; Zhang, Q.; Yang, Y.; Wang, Y. *ACS Catal.* **2014**, *4*, 2175.
(5) Zhou, C.; Deng, W.; Wan, X.; Zhang, Q.; Yang, Y.; Wang, Y. *ChemCatChem* **2015**, *7*, 2853.

FABRICATING UNIQUE INTERFACES FOR BASE-FREE HYDROGEN PRODUCTION FROM FORMALDEHYDE SOLUTION AT MILD CONDITIONS

*Shihui Zou, Renhong Li, Juanjuan Liu, Xiaoqing Yan, Hisayoshi Kobayashi, Liping Xiao, Jie Fan**

Department of Chemistry, Zhejiang University, Hangzhou, China, 310027

jfan@zju.edu.cn

Introduction

Hydrogen is expected to play a significant role in the future energy system due to its high energy density (142 MJ kg^{-1}), renewable source and environment-friendliness.[1] However, the large-scale replacement of fossil fuels by hydrogen is still restricted owing to the technical barriers in controllable storage and release of hydrogen. To overcome this problem, great efforts have been devoted to develop advanced catalysts for selective hydrogen production from H_2-rich chemicals, such as methanol, ammonia borane, formic acid, formaldehyde and water. Among them, formaldehyde aqueous solution, generally in the form of methanediol, is potentially interesting for mobile applications because of its long-term stability/recyclability, non-flammability and considerable theoretic H_2 weight efficiency (8.4 wt%).[2] Nevertheless, the traditional catalytic systems for dehydrogenation of formaldehyde solution are always limited by the low H_2 production efficiency, unclear catalytic mechanism as well as the addition of high concentration soluble bases (e.g. NaOH).[3] From the "H_2-economy" perspective, it is of great interest to develop efficient catalysts for base-free hydrogen evolution from formaldehyde solution at mild conditions. In addition, the corresponding catalytic mechanism also merits further study.

Herein, we proposed that the interfaces of catalysts play a vital role in hydrogen evolution reaction (HER). A core-shell structured $ZnO@Bi(NO_3)_3$ composite was successfully fabricated via surface Lewis-acid combination. The interface of ZnO and $Bi(NO_3)_3$ was shown to have a unique electronic structures where electrons trend to transfer from Zn to Bi. The unprecedented charge transfer benefits the adsorption and dehydrogenation of $HCHO/H_2O$, which results in a superior HER performance over the bare ZnO and $Bi(NO_3)_3$. On the other hand, the AgNPs/MgO catalyst activates the molecular O_2 to form a stabilized MgO/Ag–•OOH interface, which exhibits a surprisingly high TOF for H_2 production from formaldehyde solution. Combining the reaction kinetics, spectroscopic evidence and DFT calculations, we for the first time revealed the origin of molecular O_2 as a promoter for HER.

Experimental

BN-ZnO-X samples. 1 mmol ZnO and X mmol $Bi(NO_3)_3 \cdot 5H_2O$ were mixed in 10 mL water. After 1 h stirring at 25 °C, the precipitates were collected by centrifuging and drying, denoted as BN-ZnO-X (X is the molar ratio of Bi/Zn).

AgNPs/MgO synthesis. MgO support was synthesized by the control hydrolysis of magnesium methoxide at 80 °C for 6 h, followed by 5 h calcinations at 450 °C. AgNPs was prepared according to the literature.[4] To fabricate 0.8 wt% AgNPs/MgO catalysts, desired amount of AgNPs and MgO were mixed in 25 mL chloroform and stirred for 30 min. The solid product was centrifuged, dried, and then calcined in 5% (v/v) H_2/Ar mixed gas at 450 °C for 5 h to obtain AgNPs/MgO nanocatalysts. HAADF-STEM analysis revealed that AgNPs were well dispersed on the surface of MgO with an average size of ca. 5 nm.

Catalytic test. Catalytic H_2 production from $HCHO/H_2O$ solution was carried out with 20 mg of catalysts suspended in 5 mL of aqueous HCHO solution (1 M) in a 55 mL Pyrex text tube under stirring (400 ± 10 rpm). The oxygen content in the reaction tube was adjusted by bubbling with pure O_2 gas for calculated time, and the tubes were finally sealed with silica gel stoppers. GC-TCD was employed for evaluating the gas evolution amount, including H_2, O_2, CO_2 and CO. The TOF reported here is an apparent TOF value based on the surface number of Ag atoms in catalyst, which is calculated by the following equation: TOF = $(n_{H2}/n_{Ag} \times t) \times (d/1.3)$, where n_{H2} and n_{Ag} represent the molar amounts of evolved H_2 within t h and Ag catalyst, respectively, and d is the average diameter of AgNPs (~20.4 nm); $1.3/d$ is the empirical surface atom dispersion degree for nanoscale particles.

Results and Discussion

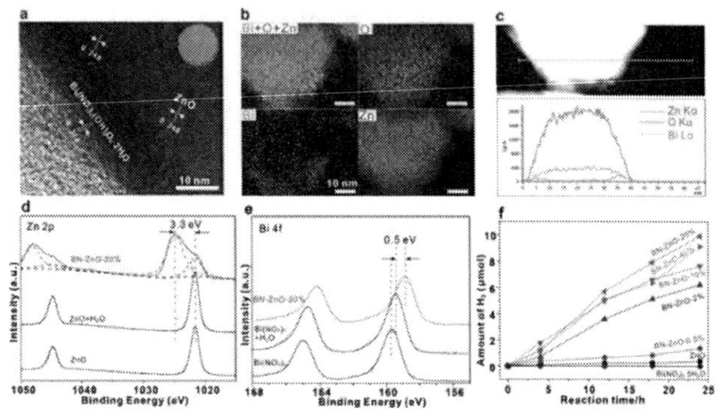

Figure 1. a) HRTEM, b) EDX elemental mapping and c) Line-scanning profile of BN-ZnO-2%. d) Zn2p and e) Bi4f XPS spectra for different samples. f) H_2 evolution activity on the samples with different $ZnO:Bi(NO_3)_3 \cdot 5H_2O$ molar ratio at 50 °C in air.

HER catalyzed by unique interface of BN-ZnO. Figure 1a-1c shows the microstructure of as-prepared BN-ZnO-2%. The lattice fringes (0.314 nm) of $Bi_6(NO_3)_4(OH)_2O_6 \cdot 2H_2O$ appear in the edge of the particle while that (0.248 nm) of ZnO dominates the core of the particle, indicating a possible $ZnO@Bi(NO_3)_3$ (for the convenience of writing, the real formula is $Bi_6(NO_3)_4(OH)_2O_6 \cdot 2H_2O$) core-shell structure. The presence of $Bi_6(NO_3)_4(OH)_2O_6 \cdot 2H_2O$ rather than $Bi(NO_3)_3 \cdot 5H_2O$ is likely due to the partially hydrolysis of $Bi(NO_3)_3 \cdot 5H_2O$ on ZnO surface, which is also confirmed by X-ray diffraction. The core-shell structure is further confirmed by EDX-mapping and line-scanning profile. As shown in Figure 1b, Bi signals only enrich in the edge region while O signals and Zn signals well distribute in the center region, which is in line with the typical core-shell structure. Wide-ranging EDX-mapping accomplished by SEM indicate that most $Bi(NO_3)_3$ are well attached to the surface of ZnO. No visible isolate $Bi(NO_3)_3$ particle is detected.

The well dispersion of $Bi(NO_3)_3$ on ZnO and the formation of $ZnO/Bi(NO_3)_3$ core/shell structure is likely due to the surface Lewis acid-base combination between ZnO and $Bi(NO_3)_3 \cdot 5H_2O$. The ability of $Bi(NO_3)_3$ to accept electrons motivate it to interact with electron-rich species, ZnO, and thus achieve a good dispersion. The interaction between ZnO and $Bi(NO_3)_3$ is further verified by X-ray photoelectron spectroscopy. As shown in Figure 1D, the binding energies of $Zn2p_{3/2}$ for pure ZnO and ZnO dispersed in H_2O solution (denoted as $ZnO+H_2O$) are both 1021.6 eV. However, after interacting with $Bi(NO_3)_3 \cdot 5H_2O$, a significant upshift (3.3 eV) of the major (75%) peak to the higher binding energy side is observed (1024.9 eV). Such a significant upshift is unprecedented and is thus expected to exhibit unique properties.

It is interesting to find that the HER activity of ZnO or $Bi(NO_3)_3 \cdot 5H_2O$ is negligible while BN-ZnO-X exhibit steady H_2 evolution. The optimized ratio of Bi to Zn is 20%, which exhibit >10 times higher catalytic performance than that of ZnO and $Bi(NO_3)_3 \cdot 5H_2O$. Interestingly, once the Lewis acid-base interaction between ZnO and $Bi(NO_3)_3 \cdot 5H_2O$ is destroyed by annealing, the steady catalytic performance expire. In addition, the catalytic performance could be further enhanced by visible light irradiation or increasing the contact area (i.e. the surface area of ZnO) between ZnO and $Bi(NO_3)_3 \cdot 5H_2O$. These results lead us to the conclusion that the unique interface structure plays a significant role in HER. To further understand the reaction mechanism, DFT calculations were conducted to investigate the electronic structure of BN-ZnO as well as the reaction pathway and energy profile. The unique electronic structure of BN-ZnO is suggested to benefit the adsorption and dehydrogenation of $HCHO/H_2O$ and thus acts as the real active site of HER.

O_2 dependent HER on AgNPs/MgO. As for the AgNPs/MgO catalyzed HER from $HCHO/H_2O$ solution, the most significant finding is that the reaction rate depends on p_{O2}, which is in stark contrary to common HER processes where O_2 should be pre-eliminated by inert gas (e.g., N_2 and Ar) to avoid the reverse H_2 oxidation process. Figure 2a shows that no H_2 was produced in the absence of O_2 (0 atm), while the HER rate was significantly promoted as the p_{O2} increased from 0 to 1 atm, corresponding to a TOF variation from 0 to 1969 h^{-1}. Besides, the HER rate could be further enhanced in pressurized O_2 conditions (Figure 1b). Notably, when p_{O2} was set at 5 atm, which is the highest pressure can be obtained using the home-made quartz reactor, the TOF reached 6641 h^{-1} and the average rate of H_2 production was determined to be *ca.* 1378 L H_2 h^{-1} g_{Ag}^{-1}. To the best of our knowledge, this reaction constant is the highest value ever reported for HER from $HCHO/H_2O$ solution using a heterogeneous catalyst even with additives or at elevated temperatures. Interestingly, MgO support alone or replacing the support by silica do not show any HER activity, indicating that there might be synergic effect between MgO and AgNPs.

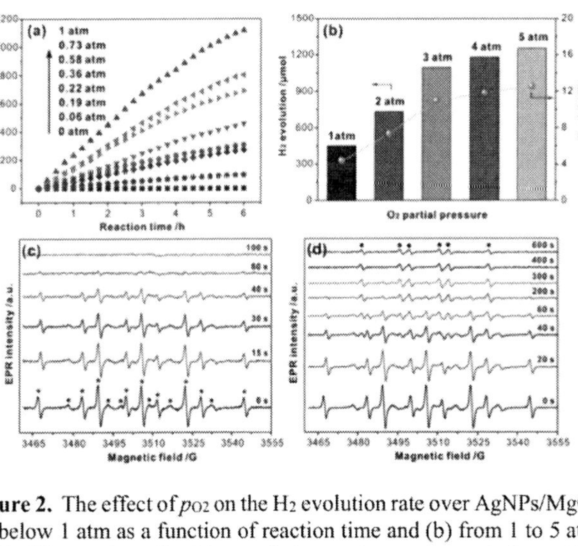

Figure 2. The effect of p_{O2} on the H_2 evolution rate over AgNPs/MgO: (a) below 1 atm as a function of reaction time and (b) from 1 to 5 atm within 2 h reaction; all of the catalytic reactions were carried out at 25 °C. Operando EPR spectra of DMPO adducts recorded in the ternary systems containing HCHO, H_2O and AgNPs/MgO catalyst (c) in air and (d) in an O_2 flow.

The surprising finding that O_2 is not consumed during the reaction leads us to the hypothesis that the oxygen might undergo a reversible conversion between molecular oxygen and active oxygen species. To make it clear, we identified the surface intermediates on AgNPs/MgO using operando liquid phase EPR spin-trapping experiments. This examination was initially carried out with a suspension containing AgNPs/MgO catalyst, $HCHO/H_2O$ mixture, and DMPO in air. As shown in Figure 2c, a pronounced ninefold signal labeled by the stars characteristic of the DMPO–•H radical was produced as soon as the EPR recorded. In addition, a group of signals labeled by squares were also emerged, which can be assigned to the DMPO–$O_2^{\cdot-}$ adduct (α_N = 19.4 G; α_H = 16.7 G). The intensity of the radical signal decreased as a function of operation time, and completely disappeared after 100 s, implying that •H radicals have been consumed to form H_2. It is noted when the suspension was purged with N_2 in advance, no prominent DMPO–•H signal was found under otherwise identical reaction conditions, demonstrating the vital role of O_2 to generate •H radical. Importantly, when the same suspension was placed under a constant O_2 flow, in addition to the DMPO–•H radical, a spin adduct with coupling constants of a_N = 14.8 G, a_β^H = 14.7 G, and a_γ^H = 1.6 G was observed from the mid-term to the end of 10 min reaction, which is consistent with an oxygen-centered species, most likely •OOH radicals, over the catalyst surface (Figure 1d). Notably, the intensity of the •OOH radicals was enhanced thereafter the disappearance of •H radicals, and kept almost constant with prolonged reaction time. Thus, the resulting •OOH radicals are supposed to be the main active species for the dehydrogenation of $HCHO/H_2O$ mixture. Importantly, no active radical species were produced in the absence of either AgNPs or MgO, in line with the catalytic results, confirming the synergic effect between MgO and AgNPs. DFT calculations further verified that the •OOH was generated in the interface of AgNPs/MgO, which are subsequently involved in the dehydrogenation reaction via an anticipated MgO/Ag–•OOH complex.

Conclusions

We successfully fabricated $ZnO@Bi(NO_3)_3$ core-shell structure by utilizing the surface Lewis-acid combination between ZnO and $Bi(NO_3)_3$. Interestingly, the strong interaction between ZnO and $Bi(NO_3)_3$ results in a unique interface electronic structure where electrons trend to transfer from Zn to Bi. A superior HER performance was thus observed on BN-ZnO over bare ZnO and $Bi(NO_3)_3$. On the other hand, the AgNPs/MgO catalyst activates molecular O_2 to form a stabilized MgO/Ag–•OOH interface, which then catalyzes HER in a radical pathway at mild conditions. The HER rates of AgNPs/MgO catalysts exhibit strong dependence on the p_{O2}. A TOF as high as 6641 h^{-1} is obtained when p_{O2} was set to 5 atm. The present study provides a valuable insight into the catalyst-interface which benefits the rational design of catalysts in the future.

Acknowledgement. This work was supported by Shell Global Solutions International B.V. (PT37712), National Natural Science Foundation of China (21271153, 21373181, 21222307,U1402233), Major Research Plan Of National Natural Science Foundation of China (91545113) , Fok Ying Tung Education Foundation (131015) and the Fundamental Research Funds for the Central Universities (2014XZZX003-02).

References

(1) Schlapbach, L.; Zuttel, A. *Nature* **2001**, *414*, 353.
(2) Heim, L. E.; Schlörer, N. E.; Choi, J.-H.; Prechtl, M. H. G. *Nat. Commun.* **2014**, *5*, 3621.
(3) Ashby, E. C.; Doctorovich, F.; Liotta, C. L.; Neumann, H. M.; Barefield, E. K.; Konda, A.; Zhang, K.; Hurley, J.; Siemer, D. D. *J. Am. Chem. Soc.* **1993**, *115*, 1171.
(4) Zheng, N.; Fan, J.; Stucky, G. D. *J. Am. Chem. Soc.* **2006**, *128*, 6550.

LOW TEMPERATURE "TITRATION" INVESTIGATION OF THE ROLE OF FORMATE IN METHANOL SYNTHESIS - PARTICIPANTS OR SPECTATORS?

Y. Yang,[a,b,d,*] *D. Mei,*[a] *C.H.F. Peden,*[a] *C.T. Campbell*[b], *C.A. Mims,*[c]

[a]Institute for Integrated Catalysis, Pacific Northwest National Laboratory, Richland, WA 99354 USA
[b]University of Washington, Department of Chemistry, Seattle, WA 98195 USA
[c]Departments of Chemical Engineering & Chemistry, University of Toronto, Toronto, ON, Canada, M5S 3E5
[d]ShanghaiTech University, Shanghai, 201210 China
*Corresponding author: yangyong@shanghaitech.edu.cn

1. Introduction

It has long been considered that surface formate is a key intermediate in the mechanism of copper catalyzed methanol synthesis from H_2/CO_2 inputs [1, 2]. In-situ infrared spectroscopy has shown formate to be the dominant surface species on the steady state catalytic surface during low temperature methanol synthesis in gas compositions above. Combining online simultaneous MS with IR, we have previously shown that no methanol is produced by exposure of formate containing adlayers (both from formic acid and by H_2:CO_2 catalysis) to 6 bar dry hydrogen [3]. These experiments ruled out the possibility that a simple hydrogenation of formate to form methanol. On the other hand, *operando* IR-MS also revealed that the reactivity of formate could be altered substantially by the presence of co-adsorbed byproduct species under the catalytic conditions, such as water [4]. Small amount of water also catalyzes CO:hydrogen (syngas), (CO:H_2:H_2O = 1:3:0.001), makes methanol at steady state rates similar to those in CO_2:hydrogen at the temperatures of this study. To test the effects of co-adsorbates on formate reactivity we have performed titration kinetics transient experiments similar to the earlier formate hydrogenation with an array of different reactive gases which include all of the species present during catalysis mentioned above.

2. Experimental

The experimental details are available in our recent publications [3,5]. Both supported (10% Cu-SiO2) and unsupported copper catalysts were used. Reactant gas streams were prepared in a flow controlled manifolds. . The reactivity of surface adsorbed species present on copper catalysts during methanol synthesis at low temperatures was studied by simultaneous infrared spectroscopy (IR) and mass spectroscopy (MS) measurements during "titration" (transient surface reaction) experiments with, if necessary, isotopic tracing. The methanol yields in surface reaction (or "titration") experiments were measured by mass spectrometry while the coverage by surface formate (and its isotopic composition) was simultaneously measured by IR. The titration experiments involved (1) establishment of a surface adlayer, followed by (2) a surface reaction period in chosen gas mixtures, all at 6 bar total pressure. Titration was often preceded by a brief reactor purge in Ar. Reaction temperatures ranged from 413K and 463 K - lower than commercial conditions. Both H and D isotopomers (hereafter noted simply by H) were used. Carbon tracing with 13C was used in cases where the titration contains the same carbon species as the adlayer

3. Results and discussion

Figure 1 is both MS and formate IR titration traces catalysis adlayer on Cu/Silica after 6 bar H_2/CO_2 saturated exposure. A broad variety of titration gases combinations were applied to test the activity of the formates dominant surface. We also performed a series of similar process starting with formates related adlayer from different reactant input combinations. More details are in our recent publication[5]. The results show that formate cannot be converted to methanol by any titration recipe (table1).

Table 1. Summary of titration experiment results:

Adlayer method / Titration gas	CO_2:H_2 catalytic adlayer (formate visible)	Formic Acid chemisorption adlayer (formate visible)	CO:H_2:H_2O catalytic adlayer (no visible formate)
Dry H_2	No	No	Yes
Moist H_2	Yes	No	Yes
Moist Ar	No (trace)	No	No
Dry H_2 post N_2O	Yes	No	-
Dry H_2 : CO	No	No	-
*CO_2:H_2:H_2O	Yes (SSITKA)	No	-

Cell contents indicate whether appreciable methanol is formed during titration with the gas composition given by the row descriptor from adlayers prepared by the method in the column descriptor.

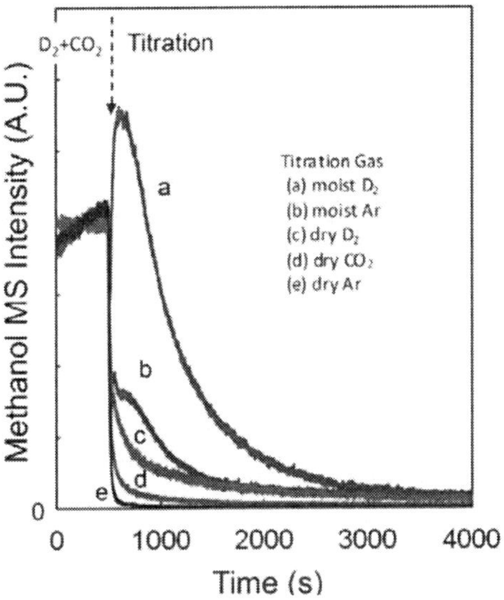

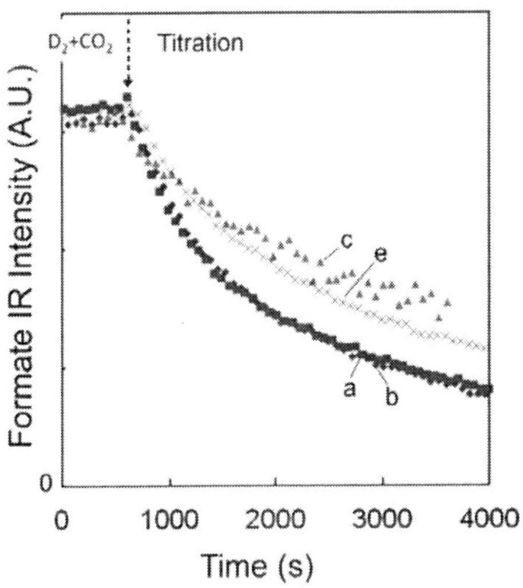

Figure 1. *Left panel:* Transient methanol production (m/e = 34 signal) from a Cu/SiO_2 catalyst upon switching from steady state catalytic conditions (D_2:CO_2 = 0.75:0.25, 6.0 bar total pressure, T = 413 K) to: (a) 6.0 bar D_2:Ar:D_2O = 0.75:0.25:0.0010; (b) 6.0 bar Ar:D_2O = 1:0.0040; (c) 6.0 bar D_2:Ar = 0.75:0.25; (d) 6.0 bar Ar:CO_2 = 0.75:0.25; (e) 6 bar Ar. *Right panel:* Simultaneously obtained IR intensities of the DCOO stretch (1330 cm^{-1}) of Cu-bound bidentate formate during experiments in the upper panel. Dashed arrow indicates change of gas composition.

4. Conclusions

Formates behavior in MeOH synthesis process on metallic and silica supported Cu catalysis is thoroughly studied with both reactants combination of H_2/CO_2 and H_2/CO. Water plays a critical role in both cases, which is a major feature in our results. Non-formate intermediates exist during low temperature methanol synthesis on copper which can be converted to methanol gas by titration with pure H_2 plus water in sufficient quantities for that intermediate to be observable by IR. However, formate itself is only a "spectator" in this reaction and gives no observable methanol upon any titration we performed with H_2 or H_2 plus water.

References

(1) R. Farrauto, S. Hwang, L. Shore, W. Ruettinger, T. Giroux, Y. Liu, O. Ilinich, Annual review of Material Research, 33 (2003), 1.

(2) L. C. Grabow, M. Mavrikakis, ACS Catal.,1, (2011) 365.

(3) Y. Yang, C.A. Mims, R.S. Disselkamp, Ja-Hun Kwak, C.H.F. Peden, C.T. Campbell, J. Phys Chem. C, 114 (2010) 17205.

(4) Y.Yang, C.A. Mims, D.H. Mei, C.H.F. Peden, C.T. Campbell, J.Catal. 298, 10 (2013).

(5) **Y. Yang,** C. A. Mims, D. Mei, C. H. F. Peden, C. T. Campbell, *ACS Catalysis*, DOI 10.1021 (2015).

NOVEL FUNCTIONAL CARBON CATALYSTS FOR ADVANCED BIOFUEL PRECURSORS

*Ashish Bohre, Saikat Dutta and Basudeb Saha**

Catalysis Center for Energy Innovation
University of Delaware
Newark, Delaware 19716, USA

Introduction

The rapid global consumption of petrochemical products and fossil fuels and high price volatility caused by demand-supply imbalance has necessitated the development of renewable resources for energy and chemicals. Among several renewable alternatives, sustainable and inexpensive non-food biomass such as, wood chips, waste biomass and municipal wastes, is a promising carbon neutral source. The United States Department of Energy (DOE) has targeted to produce about 36 billion gallons of 2nd generation biofuels by 2022 of which 16 billion gallons is to be derived from cellulosic feedstock.[1] As a result, the production of high octane advanced biofuels such as 2,5-dimethylfuran (DMF), 5-ethoxymethylfurtutal (EMF), γ-valerolactone (GVL) and diesel and jet fuel ranged long chain hydrocarbons from lignocellulosic biomass has received momentum.[2-3] The transformation of low carbon containing biomass derived platform molecules into higher carbon chain precursors for diesel and jet fuels involves the formation of new carbon–carbon bonds, which can be done via aldol condensation or hydroxyalkylation/alkylation. The catalytic hydroxyalkylation /alkylation (HAA) of sylvan (2-methyfuran, 2-MF) with biomass or non-biomass based aldehydes to produce precursors for long chain hydrolcarbon fuels has been reported using homogeneous catalysts.[4] However, homogeneous acid catalysts are corrosive, promote cross and self-condensation reactions and poses concerns in waste disposal and treatment. Herein, we report the use of an improved graphene oxide (IGO) of enriched oxygen functionalities as a solid carbonaceous acid catalyst for the production of fuel precursors via C-C bond formation between 2-MF and carbonyls (Figure 1). We found IGO containing oxophilic surface, high surface acidity and hydrophilicity and enhanced interlayer spacing of 0.967 nm is an effective oxidized carbon framework for HAA reactions under solvent-free conditions.

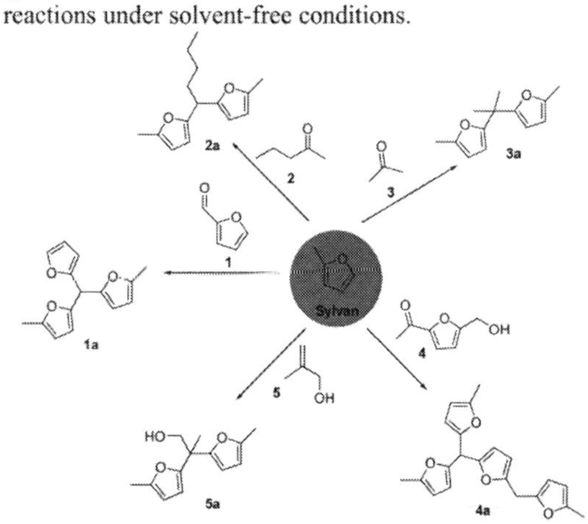

Figure 1. IGO catalyzed HAA of 2-MF (Sylvan) with biomass derived carbonyl compounds.

Experimental

Catalyst synthesis. IGO was prepared from commercial graphite powder by using a modified Hummers method. Briefly, graphite powder (3.0 g) was suspended in mixture of concentrated H_2SO_4/H_3PO_4 (360:40 mL) under magnetic stirring for 0.5 h. Then the mixture was quenched in an ice bath before $KMnO_4$ (18 g) was slowly added, during which temperature was kept below 10 °C. Subsequently, the mixture was transferred in a 500 mL round bottom flask and heated at 50 °C under constant stirring for 12 h. After cooling down to room temperature, the solution was poured into chilled water (~450 mL) containing 3 mL of 30 % H_2O_2. The mixture was centrifuged to obtain solid which was washed several times with water, 30% HCl, and ethanol.

Catalyst characterization. Powder X-ray diffraction (XRD) pattern of IGO, recycled IGO and commercial GO were collected at room temperature on a Bruker D8 Advance X- ray diffractometer system equipped with a monochromatized Cu Kα radiation (λ = 1.54056 Å) source operating at 45 V and 40 mA. Thermal gravimetric analysis (TGA) analysis was run using a TA Discovery TGA instrument. TEM images were collected on a JEM-2010F (Keck CAMM) aberration corrected transmission electron microscope operating at an accelerating voltage of 120 kV. Samples for TEM observation were prepared by applying one drop of dilute suspension of as-synthesized IGO nanocrystal dispersion in ethanol onto the carbon coated Cu grid and allowing the solvent to slowly evaporate at room temperature. Raman analysis was carried out using a Renishaw Raman RM 1000 spectrometer (Renishawplc, Gloucestershire, UK), equipped with a Leica research microscope. A 20X magnifying lens was used for analysis. All measurements were performed at excited wavelength of 514.5 nm using an argon ion laser source (maximal output 20 mW). FT-IR spectra (KBr disk, 4000–400 cm^{-1}) were recorded on a Perkin-Elmer FTIR 2000 spectrophotometer. Ammonia desorption analysis was performed on ALTAMIRA Q-800 instrument equipped with thermal conductivity detector and analyzer. HAA product was analyzed by JEOL JNM ECX-400 P 400 MHz nuclear magnetic resonance spectrometer and GC.

Catalysis. For a typical experiment, calculated amount of furfural, 2-methylfuran and IGO catalyst were charged in a 4 mL screw-cap vial equipped with a magnetic spin bar. The reaction mixture was sealed and stirred for desired time at a temperature in the range of 25 -60 °C. After completion of reaction, the reaction mixture was cooled down to room temperature, catalyst was separated by centrifugation and resultant liquid HAA product was analyzed by NMR, GC and GC-MS.

Results and Discussion

Characterization data. The powder XRD pattern of IGO shows a strong and characteristic peak at 2θ = 9.20° for 002 plane of graphene oxide. The characteristic peak of graphite at 26.5° was not observed in the XRD pattern of IGO, suggesting the presence of oxygen functionalities in IGO upon oxidation. The Raman spectrum of IGO shows the characteristic D and G bands. The D band at about 1352 cm^{-1} is associated with breathing mode of sp^2 carbon and G band (~1600 cm^{-1}) is generally assigned to the E_{2g} phonon of the sp^2 bond of carbon atoms.[5] I_D/I_G intensity ratio of IGO was 0.97 which indicates an extensive disorder in graphitic structure, due to the introduction of oxygen functionalities. The FT-IR spectrum of IGO shows strong vibration band at 3400 cm^{-1} corresponding to O-H starching vibrations. Bands at 1718 cm^{-1} and 1620 cm^{-1} are assigned to -COOH and C=O vibration. The bands at 1071 cm^{-1} and 820 cm^{-1} for C-O starching of epoxy groups and at 1470 cm^{-1}, 1015 cm^{-1} for C–O stretching of hydroxyl groups were also onserved. X-ray photoelectron spectroscopic (XPS) analysis (Figure 2) support an order of overall oxidation (degree of oxidation) which may also be correlated with XRD interlayer spacing ($d_{(200)}$) which is high (9.67 Å) for IGO.

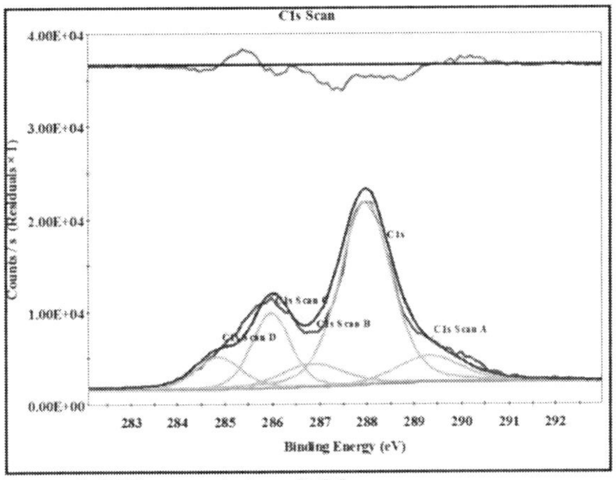

Figure 2. XPS spectral plot of IGO.

HRTEM images confirm the extended sheet like structure of IGO which is multilayer thick (7 nm with 15-20 layers) (Figure 3). The sample also exhibits wrinkle and folded regions, which could be attributed to sp^3 carbons attached with oxygen functionalities in basal plane.[6] High-resolution images (Figure 3(c) and 3(d)) shows the multilayer information at the edges of the extended IGO sheets.

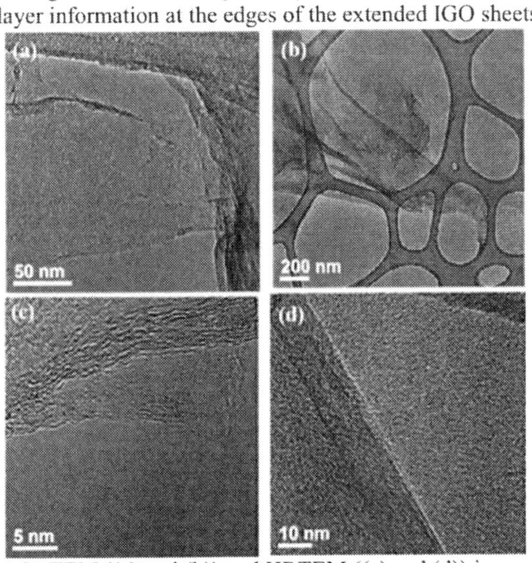

Figure 3. TEM ((a) and (b)) and HRTEM ((c) and (d)) images of multilayered sheets of IGO.

We examined the catalytic activity of IGO for HAA condensation of 2-MF (1.45 g) with furfural (0.77 g) at varying reaction conditions of catalyst loading (10-60 mg), reaction time (1-12 h), and reaction temperatures (25-60 °C) without using any solvent. The results show that the catalyst is effective, giving 95% yield of 5,5'-bis(2-methylfuranyl)furan-2-ylmethane (BMFFM) at quantitative conversion of 2-MF. Interestingly, the reaction at room temperature also achieved high yield of BMFFM (71%) for 6 h. This yield is expected to increase further for a longer reaction. The activity of IGO was also compared with a set of most commonly used and structurally diverse porous catalysts ((HY(80), HY(30), H-BEA(300)) and ZIF-8s) containing framework acidic functions. Among the Faujasite and Beta zeolites, higher Si/Al ratio containing zeolite (HY-80) exhibit improved HAA performance under solventless as compared to other HY and H-beta zeolites. Depending on the origin of Brönsted acidity of zeolitic imidazolate framework (ZIFs) which is –NH- fragment in the framework, exhibit no HAA activity in spite of containing high surface area and microporous

Table 1. Performance of IGO catalyst for the synthesis of BMFFM from furfural (0.770 g) and 2-methylfuran (1.450 g).

Entry	Catalysts	Loading (mg)	Time (h)	Temperature (^{0}C)	Yield (%)
1	IGO	60	12	60	95
2	IGO	60	6	60	95
3	IGO	60	3	60	91
4	IGO	60	1	60	83
5	IGO	10	6	60	63
6	IGO	30	6	60	71
7	IGO	60	6	RT	71
8	IGO	60	6	40	71
9	IGO	60	6	80	95
10	HY80	60	12	60	48
11	HY30	60	12	60	42
12	BEA-300	60	12	60	35
13	HY80	60	12	200	63

framework (Table 1). From the results presented in Table 1 on using the framework-acid catalyst for HAA of 2-MF with furfural, it is clear that the framework-type acid catalyst exhibits weaker activity as compared to IGO under optimized conditions. It is also evident that the IGO is active even at room temperature and exhibit much improved activity in terms of higher product yields upon increase of HAA process time (h) (Table 1, entry 2, 3, 4) and loading of IGO (mg) (Table 1, entry 4, 5, 6). Similar results were obtained for the HAA of 2-MF and butyraldehyde under solventless using a series of zeolites and IGO as catalysts.

Acknowledgment

This work was supported as part of the Catalysis Center for Energy Innovation, an Energy Frontier Research Center funded by the U.S. Department of Energy, Office of Science, Office of Basic Energy Sciences under Award Number DE-SC0001004.

References

1. W. Coyle, *In Next-Generation Biofuels: Near-Term Challenges and Implications for Agriculture*, U.S. Department of Agriculture, Economic Research Service, June **2010**, *8*, 20-17.
2. Bohre, A.; Saha, B.; Abu-Omar, M. M. *ChemSusChem.*, 2015, **8**: p.4022-9.
3. Bohre, A.; Dutta, S.; Saha, B.; Abu-Omar, M. M. *ACS Sus. Chem. Eng.*, 2015, **3**: p. 1263-77.
4. Corma, A.; de la Torre, O.; Renz, M. *Energy Environ. Sci.*, 2012, **5**: p. 6328–44.
5. Ambrosi, A.; Chua, A.K.; Bonanni, A.; Pumera, M. Chemistry of Materials, 2012, 24: p. 2292-8
6. Wei, Z.; Yang, Y.; Hou, Y.; Liu, Y.; He, X.; Demg, S. ChemCatChem, 2014, 6: p.2354-63.

RHODIUM- AND IRIDIUM-MEDIATED FUNCTIONALIZATION OF THE CARBORANE $[CB_{11}H_{12}]^-$ ANION FOR THE SYNTHESIS OF NOVEL BORON CLUSTER-BASED MATERIALS

*Yunjun Shen, Yani Pan, Jiyong Liu, and Simon Duttwyler**

Department of Chemistry
Zhejiang University
38 Zheda Road, 310027 Hangzhou, P. R. China

Introduction

The carborane anion $[CB_{11}H_{12}]^-$ and the neutral carboranes $C_2B_{10}H_{12}$, of which three isomers exist, are boron-based cage compounds in which one or more boron-hydrogen moieties are replaced by a carbon-hydrogen vertex (Figure 1).[1–4] Their B–H and C–H units are connected in an icosahedral arrangement, which provides them with unique geometrical properties. For example, $[CB_{11}H_{12}]^-$ and appropriately substituted congeners exhibit five-fold rotational symmetry when viewed through their C(1)–B(12) axis.

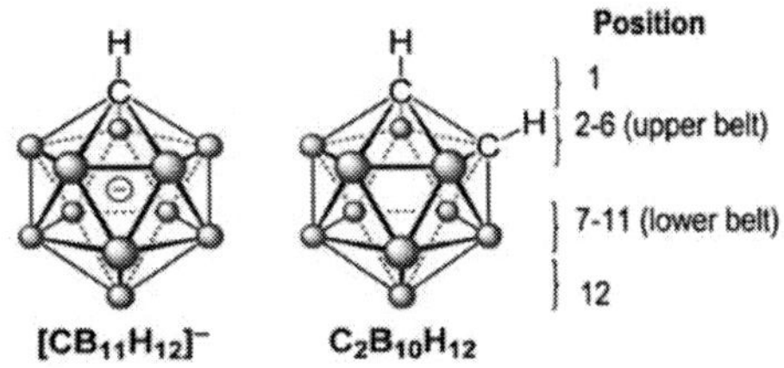

Figure 1. Carborane cages $[CB_{11}H_{12}]^-$ and $C_2B_{10}H_{12}$ (only 1,2-isomer shown).

With regard to sterics, carboranes possess C–B and B–B bond lengths in the order of 1.7–1.8 Å, as compared to 1.54 and 1.40 Å for typical C–C distances in saturated and aromatic organic molecules, respectively (Figure 2). This leads to a Van der Waals volume of approximately 150 Å^3 for $[CB_{11}H_{12}]^-$ when hydrogen atoms are included, about 10% bigger than adamantane.[5]

In order to take advantage of the potential of boron cluster compounds for applications as novel materials, new methodologies for their selective derivatization are required. To that end, we report on a B–H activation strategy to synthesize functionalized carboranes.

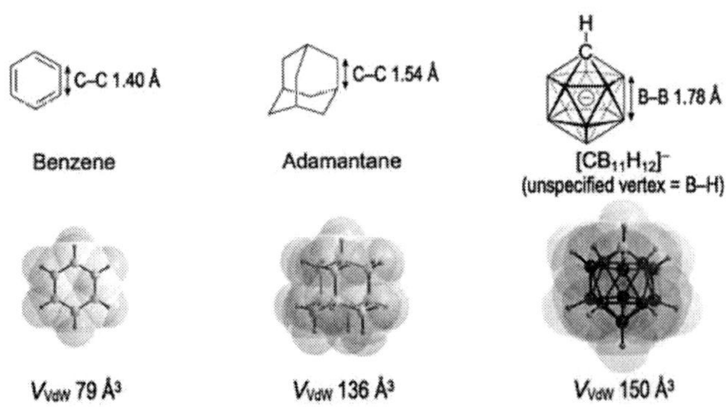

Figure 2. Comparison of benzene, adamantane and the $[CB_{11}H_{12}]^-$ anion; Lewis structures with typical bond lengths and approximate Van der Waals volumes including hydrogen atoms.

Results and Discussion

C–H derivatization of carboranes can be achieved by deprotonation and reaction of the resulting metalate with an electrophile or under cross coupling conditions. Because of the hydridic nature of the B–H bond, substitution at boron vertices requires different conditions. Direct electrophilic substitution or halogenation combined with cross coupling are typical methods; however, limited control over regioselectivity and degree of substitution and moderate overall yields are drawbacks of this approach. As an alternative, metal-catalyzed B–H activation represents an attractive and step-economical strategy. The successful use of amide directing groups is well-precedented in the area of transition metal-mediated C–H bond functionalization.[6,7]

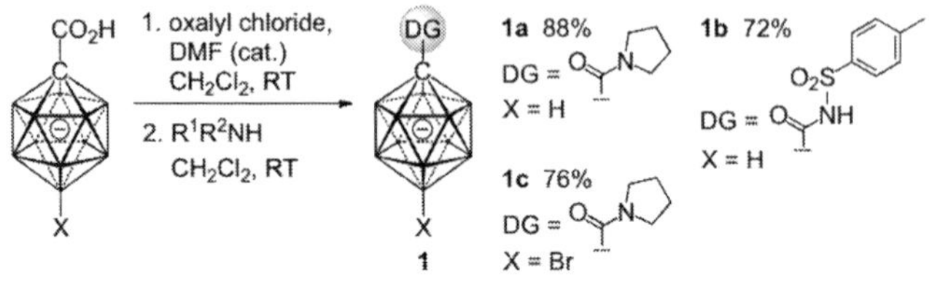

Scheme 1. Synthesis of carborane amides **1**.

In order to achieve regioselective B–H activation, we first synthesized carborane amides **1** starting from the corresponding carboxylic acids (Scheme 1). The products bearing pyrrolidine and tosyl amide units were obtained in good overall yields and characterized by NMR ($^1H\{^{11}B\}$, ^{11}B, $^{11}B\{1H\}$, $^{13}C\{1H\}$) spectroscopy and mass spectrometry. Analysis of **1a** and **1c** revealed slow rotation about the C(cage)–C(O) and C(O)–N bonds on the NMR time scale at room temperature, resulting in broad resonances. However, spectra with much sharper signals could be obtained at 80 °C. For **1a** and **1c**, the molecular structures were in addition elucidated by X-ray crystallography (Figure 3).

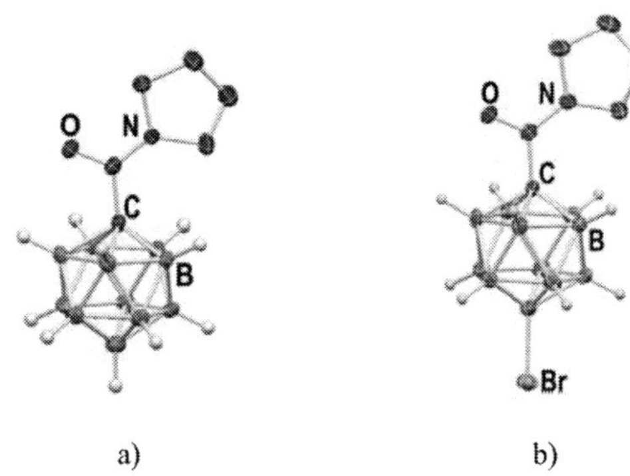

Figure 3. X-ray crystal structures of **1a** (a) and **1b** (b); 25% displacement ellipsoids, cations and hydrogen atoms except for B–H omitted for clarity.

In order to establish conditions for efficient intermolecular coupling, we studied the model reaction between a carborane amide **1** and an excess of electrophile **2** to give functionalized product **3** with potential multiple substitution (Table 1). More specifically, we chose the reaction between **1a** and ethyl acrylate for screening. The experiments were conducted under air in sealed vials. Based on these screening results, conditions of entry 5 and 13 were chosen for the synthesis of a variety of products on a preparative scale.

Table 1. Optimization of reaction conditions for the reaction of carborane pyrrolidine amide 1a wih ethyl acrylate.

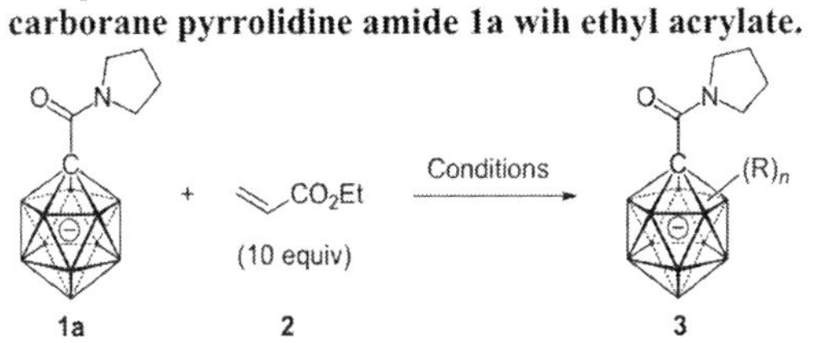

Entry	Catalyst[a]	Additive[b]	Solv.[c]	T [°C]	Result/substitution degree[d]
1	$[RhCp^*Cl_2]_2$	none	DMA	100	no reaction
2	$[RhCp^*Cl_2]_2$	Ag(I)	DMA	100	tri / tetra
3	$[RhCp^*Cl_2]_2$	Cu(II)	DMA	100	tri / tetra
4	$[RhCp^*Cl_2]_2$	Ag(I)/Cu(II)	DMA	60	tri / tetra
5	$[RhCp^*Cl_2]_2$	Ag(I)/Cu(II)	DMA	100	tetra
6	$[RhCp^*Cl_2]_2$	Ag(I)/Cu(II)	DMA	120	tetra
7	$[RhCp^*Cl_2]_2$	Ag(I)/Cu(II)[e]	DMA	100	tetra
8	$[RhCp^*Cl_2]_2$	Ag(I)/KOAc[f]	DMA	100	di / tri / tetra
9	$[RhCp^*Cl_2]_2$[g]	Ag(I)/Cu(II)	DMA	100	tetra
10	$[RhCp^*Cl_2]_2$	Ag(I)/Cu(II)	MeCN	100	di / tri / tetra
11	$[RhCp^*Cl_2]_2$	Ag(I)/Cu(II)	NMP	100	tetra
12	$[IrCp^*Cl_2]_2$	none	DMA	100	no reaction
13	$[IrCp^*Cl_2]_2$	Ag(I)/Cu(II)	DMA	100	di
14	$[Ru(cy)Cl_2]_2$	Ag(I)/Cu(II)	DMA	100	no reaction

[a] 10 mol% of [TM] dimer was used, and the reaction was run in a sealed vial under air; [b] Ag(I) = 40 mol% $AgSbF_6$, Cu(II) = 1 equiv $Cu(OAc)_2{\cdot}H_2O$; [c] DMA = dimethylacetamide, MeCN = acetonitrile, NMP = *N*-methyl-pyrrolidinone; [d] Assessed by ESI-MS and ^{11}B NMR spectroscopy, di / tri / tetra = di-/tri-/tetra-substitution in positions B-(2–6); [e] Using 8 equiv $Cu(OAc)_2{\cdot}H_2O$; [f] 2 equiv KOAc; [g] Reaction was set up in a glove box and run under N_2.

Amides **1** could be substituted with a range of organic groups (Table 1). The experiments were set up on the bench and did not require air-free conditions. After purification by silica gel chromatography, characterization comprised NMR and MS analysis. For a number of compounds, we also obtained X-ray crystal structures; for a representative example, see Figure 3.

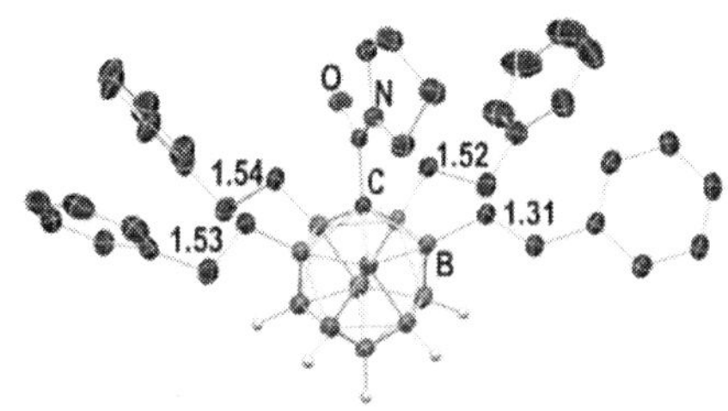

Figure 3. X-ray crystal structures of **3b**; 25% displacement ellipsoids, cations and hydrogen atoms except for B–H omitted for clarity. Numbers are C–C distances in Å.

Table 2. Synthesis of substituted carboranes 3.

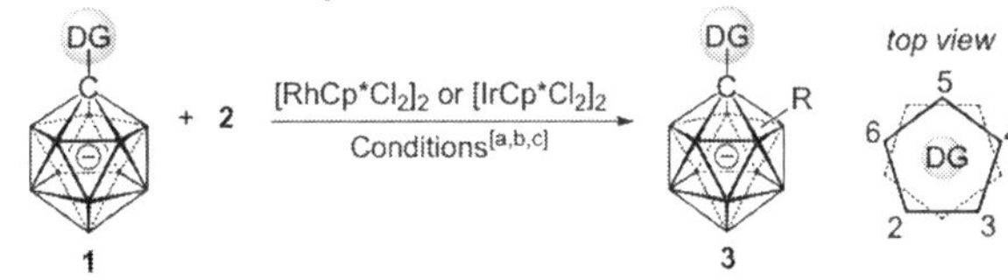

Entry	1	2	3[d]
1[a]	1a	CO_2Et (acrylate)	3a 74% R = 2,3,4-(CH₂CH₂CO₂Et)₃ 5-(CH=CHCO₂Et)
2[a]	1a	Ph (styrene)	3b 61% R = 2,3,4-(CH₂CH₂Ph)₃ 5-(CH=CHPh)
3[a]	1a	F (4-fluorostyrene)	3c 57% R = 2,3,4-(CH₂CH₂C₆H₄-4-F)₃ 5-(CH=CHC₆H₄-4-F)
4[b]	1a	F (4-fluorostyrene)	3d 22% R = 2-(CH₂CH₂C₆H₄-4-F) 4-(CH=CHC₆H₄-4-F)
5[a]	1a	Ph—≡—Ph	3e 50% R = 2,3,4,5-(C(Ph)=CHPh)₄
6[c]	1a	Ph—≡—Ph	3f 55% R = 2-(C(Ph)=CHPh)
7[c]	1a	$N_3SO_2C_6H_4CH_3$ = N_3-Ts	3g 60% R = 2,4-(NHTs)₂
8[c]	1b	$N_3SO_2C_6H_4CH_3$	3h 43% R = 2,4-(NHTs)₂

[a] **1** (0.1 mmol), **2** (0.6 mmol), $[RhCp^*Cl_2]_2$ (0.005 mmol), $AgSbF_6$ (0.02 mmol), $Cu(OAc)_2{\cdot}H_2O$ (0.1 mmol), in dimethylacetamide (1.5 mL) at 100 or 130 ¡C; for further details, see the SI; [b] In MeCN (2 mL) at 100 ¡C; [c] Using $[IrCp^*Cl_2]_2$ without Cu(II) additive, in 1,2-dichloroethane (2 mL) at 80 or 100 ¡C; [d] Isolated yields after purification by silica gel chromatography.

In conclusion, our methodology allows for the regioselective derivatization of the $[CB_{11}H_{12}]^-$ anion; research on functionalized carboranes into supramolecular structures and novel materials is currently carried out in our laboratory.

Acknowledgment. This work was supported by the Natural Science Foundation of China (grant 107305-N11412), the Chinese "1000 Young Talents Plan", and the National Basic Research Program of China (973 Project, grant 2015CB856500).

References

(1) Grimes, R. N., Carboranes (2nd ed.), Elsevier: Amsterdam, 2011.
(2) Hosmane, N. S., Boron Science: New Technologies and Applications, Taylor & Francis Books: Boca Raton, FL, 2011.
(3) Grimes, R. N. *Dalton Trans.* **2015**, *44*, 5939–5956.
(4) Douvris, C; Michl, J. *Chem. Rev.* **2013**, *113*, PR179–233.
(5) Scholz, M.; Hey-Hawkins, E. *Chem. Rev.* **2011**, *111*, 7035–7062.
(6) Chen, Z.; Wang, B.; Zhang, J.; Yu, W.; Liu, Z.; Zhang, Y. *Org. Chem. Front.* **2015**, *2*, 1107–1295.
(7) Song, G.; Li, X *Acc. Chem. Res.* **2015**, *48*, 1007–1020

X-RAY ABSORPTION SPECTROSCOPIC STUDY OF N-DOPING EFFECT IN CARBON CATHODES FOR LITHIUM-SULFUR BATTERIES

Yongsheng Chen[1], Donghai Wang[2], Jiangxuan Song[2]

[1]Department of Mechanical and Automation Engineering
The Chinese University of Hong Kong
Shatin, N.T., Hong Kong SAR, China
[2]Department of Mechanical and Nuclear Engineering
The Pennsylvania State University
University Park, PA 16802, USA
Corresponding Author Email: yschen@mae.cuhk.edu.hk

Abstract

Carbon is a versatile material and finds various applications in energy storage and conversion. Oxygen-containing surface functional groups on carbon materials have great impacts on their properties. X-ray absorption near edge structure (XANES) spectroscopy was shown to be a reliable technique to characterize these surface functional groups.(1) Doping of carbon materials is a common way to fine tune their properties for different applications, for example, N-doped carbon cathodes had improved sulfur retention in lithium-sulfur batteries.(2) XANES study has shown that N dopants did not provide extra binding sites for sulfur but they rather made the oxygen functional groups more reactive towards sulfur, thus, sulfur was attracted more strongly to carbon cathode through oxygen.(3)

References

(1) Kim, K; Zhu, P; Li, N; Ma, X; Chen, Y. *Carbon*, 49 (5) 2011, 1745-1751.
(2) Zhu, P; Song, J; Lv, D; Wang, D; Jaye, C; Fischer, D. A; Wu, T.; Chen, Y. *J. Phys. Chem. C*, 118 (15) 2014, 7765–7771.
(3) Song, J; Xu, T; Gordin, M.L; Zhu, P; Lv, D; Jiang, Y.-B; Chen, Y; Duan, Y; Wang, D. *Adv. Funct. Mater.*, 24 (9) 2014, 1243–1250.

TRANSITION-METAL-DOPED, NANOSTRUCTURED LITHIUM ZIRCONATE AS A HIGH-SPECIFIC-CAPACITY LITHIUM-ION BATTERY CATHODE

Donald G. Truhlar, Shuping Huang, Yuan Fang, Bo Wang, Benjamin E. Wilson, Nam Tran, William H. Smyrl,[a] and Andreas Stein

Department of Chemistry, Chemical Theory Center, and Supercomputing Insitute, University of Minnesota, Minneapolis, Minnesota 55455-0431
[a]Department of Chemical Engineering & Materials Science, University of Minnesota, Minneapolis, Minnesota 55455-0132
Corresponding author email: truhlar@umn.edu

Introduction

Lithium-ion batteries (LIBs) are promising devices for high capacity, rechargeable electrical energy storage; however, LIBs are currently limited by the low specific capacity of the cathode compared to the anode. Doped Li8ZrO6 (LZO) is a pseudo-layered material under consideration for lithium-ion battery cathodes and solid electrolyte coatings (1–3).

Results

Specific capacities of the doped LZO depend on the transition metal and on the Li:Zr ratio used in the synthesis, but they are generally higher than in similarly prepared undoped LZO. The effects of doping LZO with Ce, Ti, Mg, Nb, and Y on structure, band gaps, conductivity, and activation energy for ion migration have been investigated both experimentally and by quantum mechanical calculations. Fe(III)-doped LZO/C composites showed the highest specific capacities, with an initial discharge capacity specific capacity of 175 mAh/g maintained after 140 cycles. The effects of doping were also investigated by density functional calculations of dopant locations, band gaps, and delithiation energies. While all dopants slightly reduce the electronic conductivity at room temperature, doping with Mg or Nb increases ionic conductivity by an order of magnitude. Introducing a high loading of Nb into LZO decreases the activation energy for Li-ion diffusion. Calculations on lithium-ion diffusion in LZO show that it occurs by a polaron–vacancy complex mechanism. The energy barrier is lowest for the lithium hopping in a zigzag fashion between tetrahedral voids within adjacent layers (see Figure 1). The diffusion barrier is reduced as the number of Li vacancies increases during battery charging. We find that the most stable surface is the (001) surface with the tetrahedral Li layer being exposed, and the delithiation energy on this) surface is slightly higher than that for the bulk. The diffusion barrier from the surface tetrahedral layer to the subsurface octahedral layer on the (001) surface is lower than for the most favorable path in the bulk material.

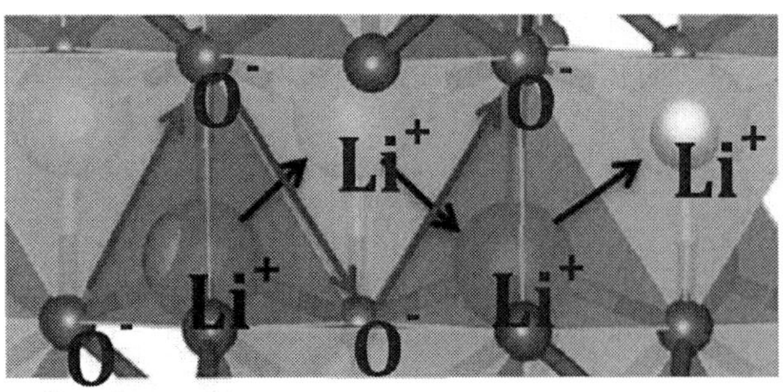

Figure 1. Zigzag path of lithium ions diffusing between tetrahedral sites.

Concluding Remark

These characterizations of surface and doping effects will assist future materials design.

Acknowledgment. This work was supported in part by the U. S. Department of Energy, Office of Basic Energy Sciences, under grant no. DE-SC0008662.

References

(1) Huang, S.; Wilson, B. E.; Wang, B.; Fang, Y.; Buffington, K.; Stein, A.; Truhlar, D. G. Y-doped Li_8ZrO_6: A Li-Ion Battery Cathode Material with High Capacity. *J. Am. Chem. Soc.* **2015**, *137*, 10992-11003.

(2) Huang, S.; Wilson, B. E.; Smyrl, W. H.; Truhlar, D. G.; Stein, A. Transition-Metal-Doped M-Li_8ZrO_6 (M = Mn, Fe, Co, Ni, Cu, Ce) as High-Specific-Capacity Li-Ion Battery Cathode Materials: Synthesis, Electrochemistry, and Quantum Mechanical Characterization. *Chem. Mater.* **2016**, *28*, 746-755.

(3) Huang, S.; Fang, Y.; Wang, B.; Wilson, B. E.; Tran, N.; Truhlar, D. G.; Stein, A. Conduction and Surface Effects in Cathode Materials: Li_8ZrO_6 and Doped Li_8ZrO_6, to be published.

RATIONAL DESIGN OF INORGANIC/CARBON COMPOSITE NANOSTRUCTURES FOR ENERGY STORAGE AND CONVERSION

Zhong Jin[1,*], *Jie Liu*[1,2]

[1] Key Laboratory of Mesoscopic Chemistry of MOE and Collaborative Innovation Center of Chemistry for Life Sciences, School of Chemistry and Chemical Engineering, Nanjing University, Nanjing, 210093, China
[2] Department of Chemistry, Duke University, Durham, NC 27708, USA
zhongjin@nju.edu.cn; j.liu@duke.edu

Introduction

Carbon nanomaterials, such as carbon nanotubes, graphene and carbon aerogels, have attracted intensive interests owing to their extraordinary electric, mechanical and chemical properties for the applications of clean energy. Nanostructures composed of carbon and inorganic materials (transition metal oxides, chalcogenides, etc.) can exhibit very high conductivity, surface area, flexibility, electrochemical and catalytic performances. However, it is still a great challenge to rationally design and controllably prepare these ordered nanostructures for high-efficiency energy storage and conversion, such as lithium-ion batteries, photocatalysis and solar cells. Recently, we developed several feasible approaches to prepare 3D hierarchical carbon and/or inorganic based architectures with precisely-designed components and structures for energy applications.[1-4]

1. CuO encapsulated mesoporous carbon multi-yolk-shell octahedra as high-stability anodes for lithium ion batteries

Although transition metal oxides have attracted considerable attention for their high energy density as anode materials of lithium-ion batteries, they suffer from large volume expansion during lithiation process, which usually causes fast capacity degradation.

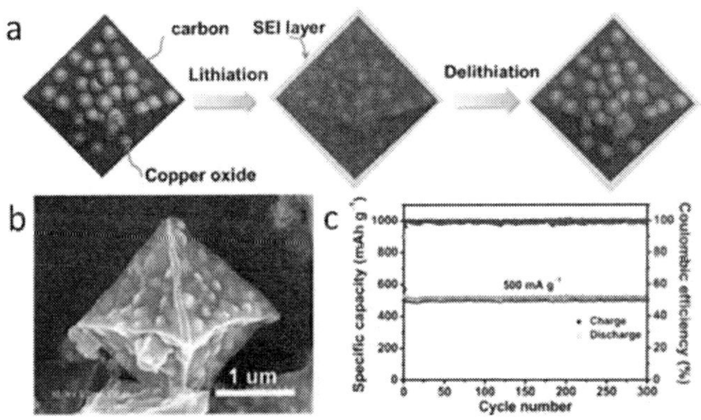

Figure 1. (a) Schematic comparison of the morphological and volumetric changes of CuO@C multi-yolk-shell octahedra during lithiation/delithiation processes. (b) SEM image of CuO@C multi-yolk-shell octahedra. (c) Cycling performance of CuO@C multi-yolk-shell octahedra electrode at 500 mA·g^{-1} for 300 cycles.

Herein, we report a rational design and facile preparation strategy of copper oxide encapsulated mesoporous carbon multi-yolk-shell octahedra, in which multiple CuO nanoparticles are well-confined in the compartments of micro-scale octahedral carbon scaffolds. The advantages of the novel multi-yolk-shell design are that the 3D carbon scaffolds can buffer the volume change and prevent aggregation of CuO nanoparticles during the charge/discharge cycles, provide pathways for electron transport and Li^+ diffusion, and restrict the thin solid-electrolyte interphase layer to the outer surface of carbon shells. The results demonstrate how the electrochemical properties of anodes can be significantly improved by the multi-yolk-shell nanostructures with greatly enhanced structural stability and electrochemical actuation. Moreover, the micrometer-size CuO@C octahedra reduce the relative quality of SEI, resulting in high Coulombic efficiency and long cycling stability. In Li-ion cells, the CuO@C multi-yolk-shell octahedra anodes deliver a highly-reversible capacity of 598 mAh·g^{-1} at 250 mA·g^{-1}, excellent rate capacity of 365 mAh·g^{-1} at 3000 mA·g^{-1} and exhibit long-term cyclability with a capacity of 512 mAh·g^{-1} after 300 cycles at 500 mA·g^{-1}.

2. Flexible and binder-free $Li_3V_2(PO_4)_3$ encapsulated carbon nanofabric cathodes for fast charging lithium-ion batteries

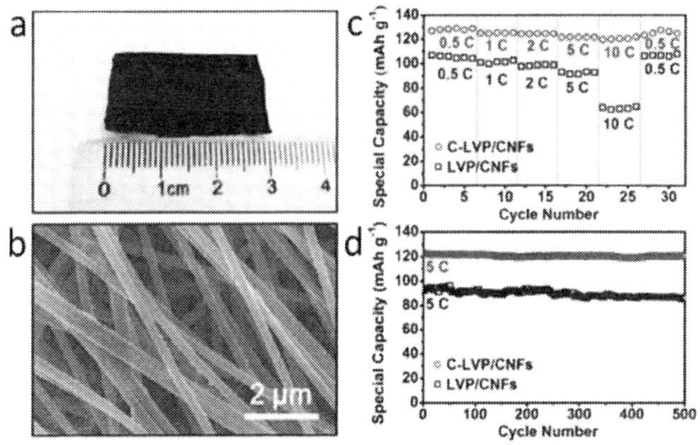

Figure 2. (a) Optical image and (b) SEM image of carbon coated $Li_3V_2(PO_4)_3$/Carbon nanofabric cathode (C-LVP/CNFs). (c) Rate capability of C-LVP/CNFs and LVP-CNFs. (d) Long-term cycling performance of C-LVP/CNFs and LVP-CNFs at 5 C.

Lithiated transition metal phosphates with large theoretical capacities have emerged as promising cathode materials for rechargeable lithium-ion batteries. However, the poor kinetic properties caused by their low intrinsic electronic and ionic conductivity greatly hint their practical applications. In this work, we demonstrate a novel strategy to prepare monoclinic lithium vanadium phosphate nanoparticles implanted in carbon nanofibers as the cathodes of Li-ion cells with high capacity, flexibility, long cycle stability and significantly improved high-rate performance. The composite nanofibers were obtained by electrospinning using polyacrylonitrile and $Li_3V_2(PO_4)_3$ nanoparticles, following by annealing and coating with a thin layer of carbon by plasma enhanced chemical vapor deposition. The Li3V2(PO4)3 nanocrystals with monoclinic phase were uniformly distributed in the composite nanofibers. Electrochemical performances of as-prepared binder-free fibrous cathodes were characterized by potentiostatic and galvanostatic tests. At a rate of 0.5 C in the range of 3.0–4.3 V, the composite displayed an initial discharge capacity of 127.9 mAh·g^{-1} (96.2% of the theoretical capacity). A discharge capacity of 120.3 mAh·g^{-1} was still observed even at a high rate of 10 C, and a capacity retention of 98.9% was maintained after 500 cycles at 5C, indicating excellent high-rate capability and capacity retention. Compared to the control samples without carbon outer-layer, the composite nanofibers with carbon coating demonstrated much better electrochemical performances. It indicates carbon coating can greatly improve electronic conductivity profiles without hindering the Li-ion mobility and also can prevent undesired side reactions with electrolyte.

3. Surface plasmon resonance enhanced p-n junction nano-shish-kebabs for overall water splitting and hydrogen generation

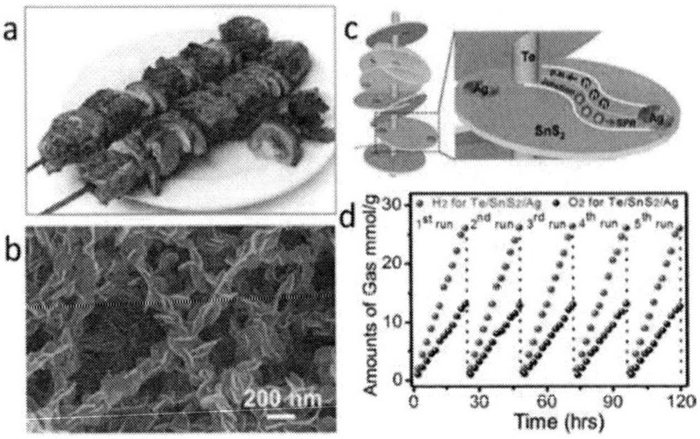

Figure 3. (a) Real shish-kebabs in the dish. (b) SEM image of Te/SnS_2/Ag inorganic NSKHs. (c) Schematic of Te/SnS_2/Ag NSKH with high way of "SnS_2-Junction Interface-Te NW-H_2O" for the rapid charge transport. (d) Gas evolution under visible light irradiation in pure water, the hydrogen production rate was found to be 1.10 mmolg^{-1}h^{-1} as compared of plantinized P25 about 0.106 mmolg^{-1}h^{-1} (2% methanol).

Visible-light sensible and stable photocatalysts for the generation of clean H_2 fuel to dealing with serious energy crises is desperately demanded nowadays. Here we present a novel nano-shish-kebab heterostructure (NSKH) to utilize visible light using a facile hydrothermal method. Both p-n junction and silver (Ag) nanoparticle promoted surface plasmon resonance (SPR) are exploited to synergistically achieve overall water splitting. The highly conductive Te nanowires facilitated the charge transferring, consequently leading to much enhanced water splitting performance. Ag nanoparticle as a localized SPR source has also been included to greatly maximize the quantum yield. Eventually, the high way of "SnS_2-Junction Interface-Te NW-H_2O" for charge separation and transfer builds up. The Te/SnS_2/Ag NSKH achieves a much higher hydrogen production rate of 1.10 mmolg^{-1}h^{-1} without any sacrificial reagent as compared with control experiments and commercialized P25 (0.106 mmolg^{-1}h^{-1}). It maintains the original unique structure after 5 cycles measurements verifying its good stability.

4. One-step fabrication of ultrathin MoS_2 nanofilms with high catalytic activity for dye-sensitized solar cells

Ultrathin MoS_2 nanofilms are grown directly on the FTO glass by a one-step and low-cost hydrothermal method. Scanning electron microscope and transmission electron microscope images exhibit these MoS_2 nanofilms just have few layers and spread on the FTO glass uniformly. Electrochemical Impedance Spectroscopy and cyclic voltammetry reveal that the MoS_2-based counter electrode displays an excellent catalytic activity toward triiodide reduction. When the MoS_2-based counter electrode is used in the dye-sensitized solar cell (DSSC), it shows an impressive photovoltaic performance of 8.3%, which is equivalent to or even higher than the DSSC based on Pt-based counter electrode. Therefore, MoS_2-based counter electrode is believed to be a desirable alternative for the traditional DSSCs due to its simplified preparation process, low-cost, and superior photovoltaic performance.

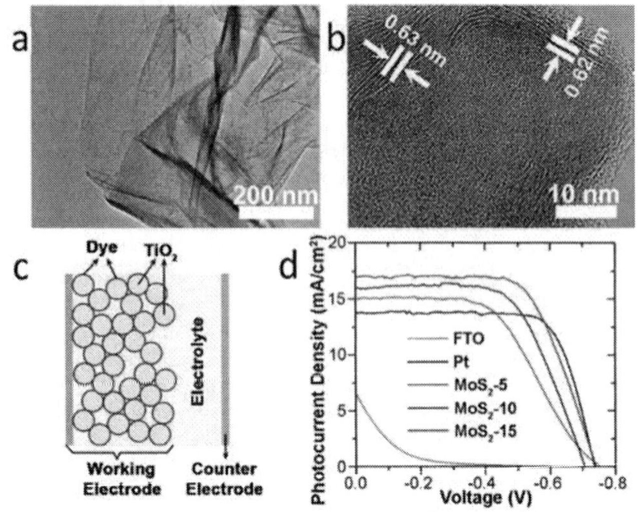

Figure 4. (a) TEM and (b) HRTEM images of the ultrathin MoS_2 nanofilms. (c) A schematic diagram of a DSSC. (d) *J-V* plots of the DSSCs using the pristine, platinized, MoS_2-5-based, MoS_2-10-based, and MoS_2-15-based FTO glasses as the counter electrodes.

Conclusions

In summary, we recently have developed several novel carbon and/or inorganic based composite nanostructures with high performances for energy storage and conversion applications, such as new cathode and anode materials for fast-charging and long-life lithium-ion batteries, SPR-enhanced p-n junctions for water splitting and hydrogen generation, and ultrathin 2D MoS_2 nanofilms for high-efficiency DSSCs. New principles to design and prepare novel nanostructures for clean energy was established, promising applications of these new nanocomposites are expected.

Acknowledgement. This work is supported by the National Thousand Young Talents Program of China, the Young Scientists Project of National Basic Research Program of China (973 Program No. 2015CB659300) and the National Natural Science Foundation of China (NSFC Grants No. 21403105 and No. 21573108). This work was also supported by the Fundamental Research Funds for the Central Universities, the Natural Science Foundation for Young Scholars of Jiangsu Province, China (Grants No. BK20150571 and No. BK20150583) and a project funded by the Priority Academic Program Development of Jiangsu Higher Education Institutions (PAPD).

References

(1) Chen T.; Hu Y.; Cheng B.R.; Chen R.P.; Lv H.L.; Ma L.B.; Zhu G.Y.; Wang Y.R.; Yan C.Z.; Tie Z.X.; Jin Z.; Liu J. *Nano Energy*, 2016, 20, 305-315.
(2) Sun P.P.; Zhao X.Y.; Chen R.P.; Chen T.; Fan Q.; Lv H.L.; Hu Y.; Jin Z.; Xu Q.Y.; Liu J. *Nanoscale*, 2016, in press.
(3) Yan C.Z.; Zhang W.J.; Xue X.L.; Li X.J.; Jin Z; Liu J. in preparation.
(4) Liang J.; Li J.; Zhu H.F.; Han Y.X.; Zhang G.M.; Wang Y.R.; Wang C.X.; Jin Z.; Liu J. in preparation.

TAILORING ZNO NANOCRYSTAL THIN FILMS AS ELECTRON-TRANSPORTING LAYERS FOR SOLUTION-PROCESSED LIGHT-EMITTING DIODES AND SOLAR CELLS

Yizheng Jin

Center for Chemistry of High-Performance & Novel Materials, State Key Laboratory of Silicon Materials, Department of Chemistry, Zhejiang University, Hangzhou 310027, People's Republic of China
Email: yizhengjin@zju.edu.cn

Introduction

Solution-processed light-emitting diodes (LEDs) and solar cells are promising as next-generation technologies to fabricate cost-effective, large-area and flexible devices. ZnO thin films have been widely adopted in solution-processed LEDs and solar cells as electron-transporting layers (ETLs) due to their appealing properties, such as excellent visible transparency, high electron mobility, and environment-friendly nature. Employment of colloidal ZnO nanocrystals, which decouples the oxide-synthesis step from the film-deposition step, is an attractive strategy. Owing to the excellent solution dispersibility, colloidal ZnO nanocrystals can be readily integrated into a variety of devices such as polymer LEDs, organic solar cells, perovskite LEDs and quantum-dot LEDs.

In this talk, I shall provide four examples[1-4] that tailoring the ZnO nanocrystal ETLs via rational design of material synthesis and post-synthesis modifications leads to controlled electronic and interfacial properties and thereby improved device performance.

Results and Discussion

ZnO nanocrystal ETLs for polymer LEDs: tunable work function by chemical doping. Work function is one of the most important parameters for charge-transporting layers (CTLs). The work function of oxide CTLs and the electronic structures of the contacting layers govern the interfacial energy-level alignment and hence the charge injection/extraction barriers. The work function of oxide CTLs influences the built-in field and thus affects important physical processes, such as charge transport and charge recombination.

We adopt the strategy of controlled doping to alter the electronic properties of ZnO nanocrystals. Colloidal indium-doped ZnO (IZO) nanocrystals were used as a model system to study the doping effects on the work function of the oxide ETLs (Figure 1).[1] Mechanism studies reveal that the combination of employing the reverse-injection approach and using zinc stearate and indium 2-ethylhexanoate as the cationic precursors ensures both controlled reaction pathways and balanced relative dopant-host precursor reactivity and hence high-quality IZO nanocrystals. The successful incorporation of indium dopants into host oxide nanocrystals accompanied by the generation of high density of free electrons leads to oxide thin films with lower work function. Polymer light-emitting diodes with ETLs based on IZO nanocrystals exhibit enhanced device characteristics, i.e. lower turn-on voltage, higher maximum luminance, and higher efficiency compared with those of the LEDs with ETLs based on ZnO nanocrystals. The better device performances were attributed to the improved electron-injection characteristics induced by the lower work function of the IZO nanocrystal ETLs, as supported by scanning Kelvin probe microscopy (SKPM) investigations on the bilayers of emissive polymer/oxide nanocrystal ETLs.

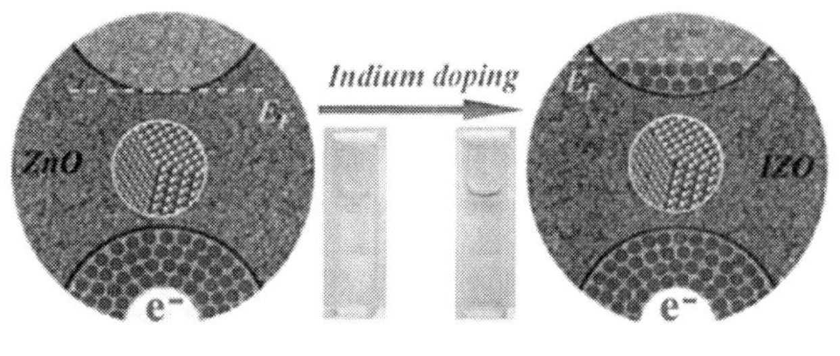

Figure 1. Indium doping results in ZnO nanocrystals with high density of free electrons and ETLs with lower work function.

ZnO nanocrystal ETLs for perovskite LEDs: surface modification to control work function and interfacial properties. The work function of oxide ETLs can also be modulated by introducing interfacial dipole layer. The use of interfacial modifier may change the surface properties.

We report a new perovskite-LED structure which incorporates a multifunctional polyethyleneimine (PEI) interlayer between the ZnO nanocrystal ETL and the perovskite emissive layer (Figure 2).[2] This interfacial engineering approach provides a high-energy hydrophilic surface that allows formation of high-quality perovskite thin films from solution and affords a low work-function cathode for ohmic electron injection into the high-lying conduction band of the hybrid perovskites. These two merits substantially improve device performance, leading to $CH_3NH_3PbI_{3-x}Cl_x$ based near-infrared (NIR) perovskite LEDs with an internal quantum efficiency of ≈15.7% at a high radiance of 28 W sr^{-1} m^{-2} and a low voltage of 2.2 V. These results represent one of the most efficient NIR devices produced using solution-processed emitters. Using the same interfacial engineering approach, green-emitting $CH_3NH_3PbBr_3$ based perovskite LEDs with a luminous efficacy of ≈4.0 lm W^{-1} at a luminance of ≈20 000 cd m^{-2} and a low voltage of 2.8 V are also demonstrated.

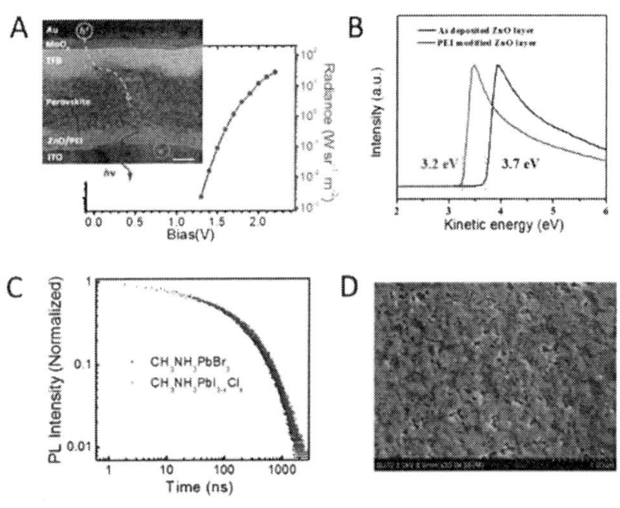

Figure 2. Interfacial control of ZnO nanocrystal ETLs towards efficient perovskite LEDs. A) Device structure and voltage-radiance characteristics. B) UPS spectra indicating that PEI treatment reduced the work function of the ZnO ETLs. C) Time-resolved PL decay transient and D) SEM imaging indicating low defect densities and good surface coverage of the perovskite films formed on the PEI-modified ZnO nanocrystal ETLs, respectively.

ZnO nanocrystal ETLs for quantum-dot LEDs: inserting an insulating layer to achieve balanced charge injection and to preserve emissive properties of quantum dots. ZnO nanocrystal thin films were employed as ETLs for quantum-dot LEDs because of the benefit of efficient electron injection into the quantum dot layers. However, the mainstream hole-transporting layers for quantum-dot LEDs are organic materials, which generally offer less efficient hole injection. Achieving efficient and balanced charge injection became the critical challenge for quantum-dot LEDs.

We developed a conceptually new device structure by the insertion of an insulating layer between the quantum dot layer and ZnO nanocrystal ETL (Figure 3).[3] Adjusting the thickness of the inserted insulating layers modulates the electron injection from the ZnO ETLs to the quantum dot layers and eliminate excess electron currents in the QLEDs. Furthermore the insulating layers help to maintain charge neutrality of quantum dot emitters and preserve their superior emissive properties. These merits led to quantum-dot LEDs with superior performance and reproducibility. It exhibits colour-saturated deep-redemission, sub-bandgap turn-on at 1.7 V, high external quantum efficiencies of up to 20.5 %, low efficiency roll-off (up to 15.1 % of the external quantum efficiency at 100 mA cm^{-2}), and a long operational lifetime of more than 100,000 hours at 100 cd m^{-2}, making this device the best-performing solution-processed red LED so far, comparable to state-of-the-art vacuum-deposited organic LEDs.

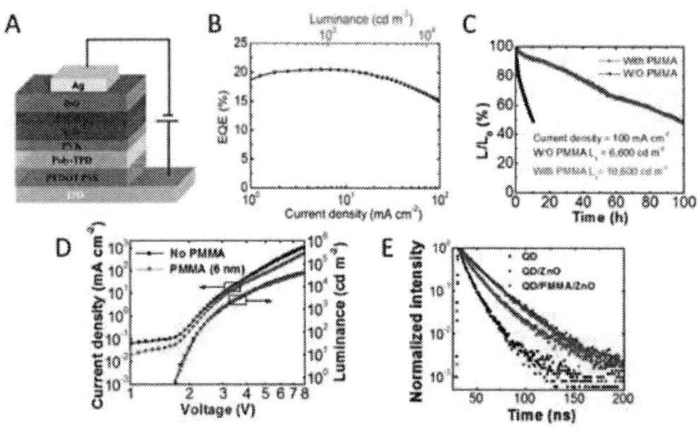

Figure 3. Quantum-dot LEDs with an insulating PMMA layer between the quantum dot layer and the ZnO ETL. A) Device structure. B) EQE versus current density and luminance for the device with the best efficiency. C) Stability data for the quantum-dot LED with and without the PMMA layer. D) Current density and luminance versus voltage characteristics for quantum-dot LEDs without and with the 6 nm PMMA layer. E) Time-resolved photoluminance decay for the quantum dot films contacting different layers.

ZnO nanocrystal ETLs for organic solar cells: surface passivation to modulate intragap states. A major challenge of applying ZnO nanocrystal ETLs in solution-processed optoelectronic devices lies in how to control the intragap states induced by surface defects. When the ZnO nanocrystal films are used as ETLs in inverted organic solar cells, the intragap states act as interfacial recombination centers for photogenerated charges and thereby degrade the device performance.

We demonstrated a simple passivation method based on ethanedithiol (EDT) treatment, which effectively removes the surface defects of the ZnO nanocrystal films by forming zinc ethanedithiolates (Figure 4).[4] The passivation by EDT treatment modulates the intragap states of the ZnO films and introduces a new intragap band. When the EDT-treated ZnO nanocrystal films are used as ETLs in inverted organic solar cells, both the power conversion efficiency and stability of the devices are improved. The control studies show that the solar cells with EDT-treated ZnO films exhibit reduced charge recombination rates and enhanced charge extraction properties. These features are consistent with the fact that the modulation of the intragap states results in reduction of interfacial recombination as well as the improved charge selectivity and electron transport properties of the ETLs. It is further demonstrated that the EDT treatment-based passivation method can be extended to ZnO films deposited from sol–gel precursors.

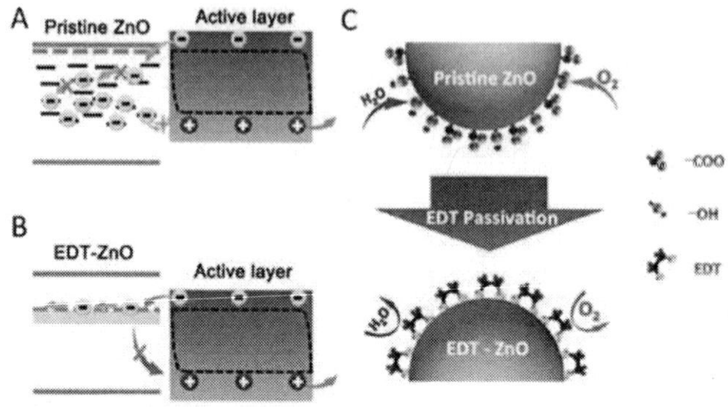

Figure 4. The impact of EDT treatment on the ZnO films and devices with films of ZnO nanocrystals as ETLs. A) The pristine ZnO films have various intragap states, which act as recombination centers for photogenerated charges. B) After EDT treatment, the various intragap states due to surface defects are modified to a new intragap band. This new intragap band facilitates electron transport in the ETLs, thereby suppressing the interfacial bimolecular recombination and enhancing the charge extraction properties of the devices. C) Schematic representations showing that the various surface groups are removed and EDT molecules are covalently bound onto ZnO. The excellent passivation makes the ZnO films less susceptible to oxygen and water molecules.

Conclusions

We show that developing new chemistry for colloidal ZnO oxide nanocrystals, together with the design of new device structure shall greatly benefit to the realization of high-performance solution-processed LEDs and PVs. We expect that these knowledge shall be transferred to tailor the properties of inorganic oxide hole-transporting layers.

Acknowledgement. This work was supported by National Natural Science Foundation of China (Grant Nos. 91433204, 51172203), and the Fundamental Research Funds for the Central Universities (2015FZA3005).

References

(1) Liang, X.; Ren, Y.; Bai, S.; Zhang, N.; Dai, X.; Wang, X.; He, H.; Jin, C.; Ye, Z.; Chen, Q.; Chen, L.; Wang, J.; Jin, Y. *Chem. Mater.* **2014**, *26*, 5169-5178.

(2) Wang, J.; Wang, N.; Jin, Y.; Si, J.; Tan, Z. K.; Du, H.; Cheng, L.; Dai, X.; Bai, S.; He, H.; Ye, Z.; Lai, M. L.; Friend, R. H.; Huang, W. *Adv Mater* **2015**, *27*, 2311-6.

(3) Dai, X.; Zhang, Z.; Jin, Y.; Niu, Y.; Cao, H.; Liang, X.; Chen, L.; Wang, J.; Peng, X. *Nature* **2014**, *515*, 96-9.

(4) Bai, S.; Jin, Y.; Liang, X.; Ye, Z.; Wu, Z.; Sun, B.; Ma, Z.; Tang, Z.; Wang, J.; Würfel, U.; Gao, F.; Zhang, F. *Advanced Energy Materials* **2015**, *5*, 1401606.

NANOSTRUCTURED ELECTRODE MATERIALS FOR Li/Na ION STORAGE

Xianhong Rui, Dan Yang, Yu Zhang, Qingyu Yan

School of Materials Science and Engineering
Nanyang Technological University
alexyan@ntu.edu.sg

Among various electrochemical energy storage technologies, lithium ions batteries were considered as most practical ones due to their high energy densities and high stability. Recently, Na-ion batteries (NIBs) have attracted great attention because of the abundance of sodium and low cost in comparison to lithium ion batteries. However, NIBs still have many technical issues, such as larger volume change of the electrode materials, lower specific capacities, poorer rate capability, as well as shorter cycling life due to larger ionic radius and molar mass of Na^+ ions than that of Li^+ ions.

We will present our recent activities on synthesis of nanostructured electrdoe materials for Na/Li ion storage applications. We carried out works on phosphate[1], V_2O_5, $Na_3V_2(PO_4)_3$ (NVP)[2], black phosphorus (BP), and sulfides by various preparation process. From materials design point of view, we aim to increase the specific capacity, volumetric capacity, 1st-cycle Coulumbic efficiency, cycling life and rate performance. Many of these materials show high cycling stability and good rate capability, which is suitable for high power and fast charging applications.

References

(1) D. Yang, Z. Lu, X. Rui, X. Huang, H. Li, J. Zhu, W. Zhang, Y. M. Lam, H. H. Hng, H. Zhang, Q Yan, "Synthesis of two-dimensional transition metal phosphates with highly ordered mesoporous structures for lithium-ion battery application", *Angew. Chem.* **2014**, 53 (35), 9352

(2) X. Rui, W. Sun, C. Wu, Y. Yu, Q. Yan, "High-Performance and Highly Recyclable $Na_3V_2(PO_4)_3$ Cathodes", *Adv. Mater,* **2015,** 27 (42), 6670

IONIC CONDUCTORS SOR SOLID-STATE DYE-SENSITIZED SOLAR CELLS

Hong Wang, Juan Li, Yan Li, Feng Gong, and Zhong-Sheng Wang

Department of Chemistry, Lab of Advanced Materials, Collaborative Innovation Center of Chemistry for Energy Materials
Fudan University
2205 Songhu Road
Shanghai 200438, P. R. China

Introduction

The dye-sensitized solar cell (DSSC) is composed of dye-loaded nanocrystalline TiO_2 film as the working electrode, a Pt-coated conductive glass as the counter electrode, and a redox couple I^-/I_3^- based electrolyte.[1] Although the highest power conversion efficiency has been achieved using a volatile organic liquid electrolyte,[2,3] DSSCs containing liquid electrolyte are limited for outdoor applications in view of the need for robust encapsulation of the volatile organic liquid. For this reason, considerable efforts have been made to replace the liquid electrolyte with solid-state electrolytes.

Imidazolium iodides have been proven to be the promising material as ionic liquid electrolytes for high-performance DSSCs due to their wide electrochemical window, high electrical conductivity, negligible vapor pressure, and nontoxicity. However, they are still a type of liquid materials and the leakage during long-term operation is unavoidable. Therefore, solid-state DSSCs (ssDSSCs) with solid-state imidazolium iodides having high conductivity are much desired for practical applications.

By introducing functional groups to the imidazolium ring, we have designed and synthesized a series of ionic conductors for use as solid-state electrolytes of ssDSSCs, and the ionic conductivity is enhanced significantly as compared to the alkyl-substituted imidazolium iodides.[4-6] When applied as the solid-state electrolytes of ssDSSCs, the solid-state ionic conductors achieve high efficiency and good long-term stability.

Experimental

Synthesis of Ionic Conductors. A functional group was first linked to one nitrogen atom of imidazole, and then ionic conductors were synthesized by quaternization with RI (R = alkyl or functional group).

Characterization Methods. Ionic conductors were characterized by ^{1}H-NMR (Varian 400 MHz NMR spectrometer) and elemental analysis (Vario EL3, Germany). Mass spectra were obtained on an HP-5988A spectrometer by direct inlet at 70 eV. The X-ray diffraction data of the single crystal were collected on a Bruker SMART APEX (II)-CCD area detector diffractometer using graphite monochromatic Mo *Kα*. The performance of ssDSSC was tested by recording the current density–voltage (J - V) curves with a Keithley 2420 Source Meter (Oriel) under illumination of simulated AM1.5G solar light coming from an AAA solar simulator (Newport-94043A). Action spectra of the incident monochromatic photon-to-electron conversion efficiency (IPCE) for the solar cells were recorded using an SM-250 system (Bunkoh-Keiki, Japan).

Results and Discussion

The structures of representative ionic conductors (PMIm and PMPi) are shown in Figure 1a. The as-prepared propargyl-functionalized ionic conductors are solid state (Figure 1b) below 80 °C and have good solubility in common organic solvents such as methanol and ethanol, which is advantageous to solution processing for solid state electrolyte filling in ssDSSCs.

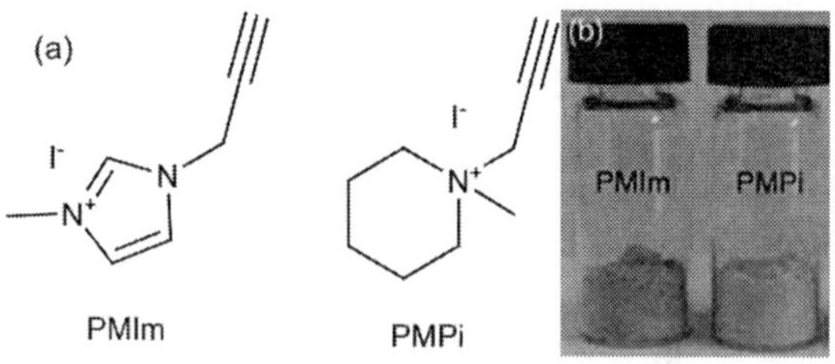

Figure 1. The structures of the propargyl-functionalized ionic conductors (a) and their photograph (b).

XRD measurements provide insight into molecular packing arrangement of the ionic conductors. Figure 2 shows the XRD patterns for these ionic conductors, where Bragg peaks are clearly observed at 11.7 and 9.6° for PMIm and PMPi, respectively. The inter-layer distances are 7.6 Å and 9.2 Å for PMIm and PMPi, respectively. PMPi showed larger inter-layer distance due to the twisting of the piperidinium ring while PMIm showed smaller inter-layer distance due to the planarity of the imidazolium ring. The layer structured feature of these propargyl-functionalized ionic conductors is beneficial to fast charge transport along the polyiodide chain. As compared to PMPi, PMIm exhibited a stronger Bragg peak, suggesting that the latter has better crystallinity than the former.

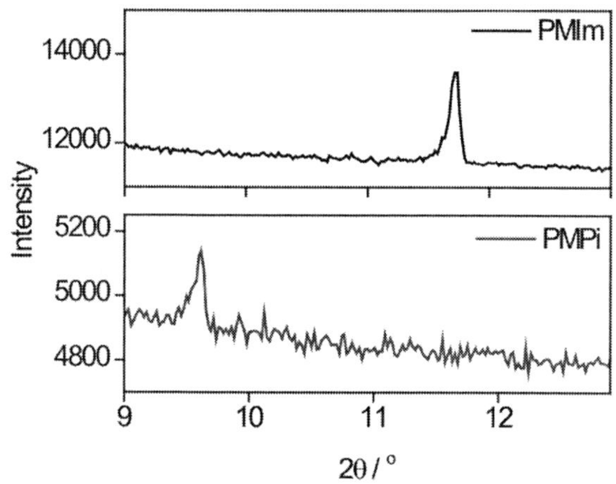

Figure 2. XRD patterns of the propargyl-functionalized ionic conductors.

The temperature dependence of conductivity for the as-prepared ionic conductors is shown in Figure 3. The linear dependence of lnσ on $1/T$ indicates that the conductivity of these propargyl-functionalized ionic conductors is attributed to the ionic conduction, which originates from the charge transfer along the polyiodide chain according to the relay-type Grotthus mechanism. The conductivity at RT for PMIm was higher than that for PMPi. This can be explained by the smaller inter-layer distance and better crystallinity for PMIm as compared to PMPi, revealed by the XRD peak position and intensity. Impressively, the conductivity of PMIm was as high as 40 mS cm^{-1} at RT, which was even higher than that (25 mS cm^{-1}) for the solution of 0.4 M LiI + 0.04 M I_2 in acetonitrile. To emphasize the effect of propargyl group on the conductivity, we compared PMIm with 1,2-dimethyl-3-propylimidazolium iodide (DMPImI, solid state at RT). The introduction of propargyl to the imidazolium ring enhanced the conductivity by 40000-fold. This indicates that our molecular design is successful to achieve high conductivity.

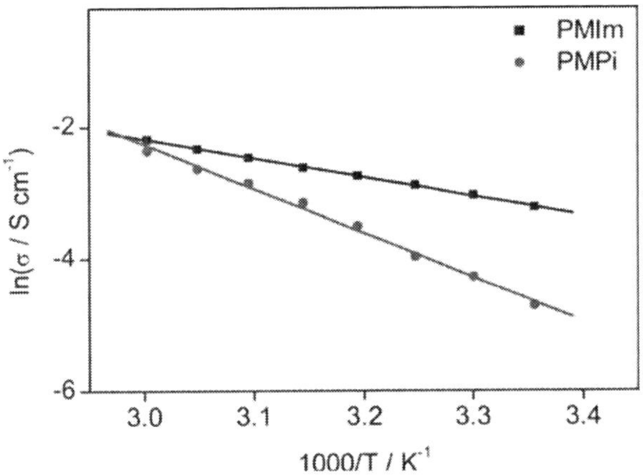

Figure 3. The temperature dependence of ionic conductivity for the two propargyl-functionalized ionic conductors.

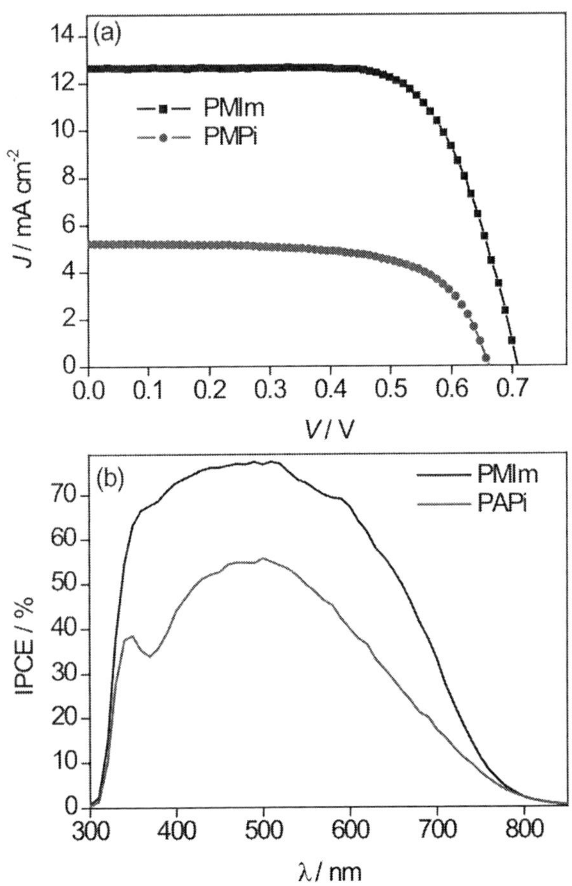

Figure 4. Photocurrent density–voltage characteristics (a) and IPCE as a function of incident wavelength (b) for ssDSSCs based on PMIm and PAPi, respectively.

ssDSSCs with these propargyl-functionalized ionic conductors as single-component solid electrolytes were evaluated under illumination of AM1.5G simulated solar light (100 mW cm^{-2}). Figure 7a shows the $J - V$ curves with the corresponding photovoltaic data summarized in Table 1. PMIm produced η of 6.3% (J_{sc} = 12.65 mA cm^{-2}, V_{oc} = 0.71 V, FF = 0.70), while PMPi produced η of 2.2% (J_{sc} = 5.21 mA cm^{-2}, V_{oc} = 0.66 V, FF = 0.65). PMIm generated much higher J_{sc} than PMPi since the former has higher conductivity than the latter. As DMPImI that has a very low conductivity, it cannot generate photocurrent and efficiency. The obtained V_{oc} value is high considering that no 4-tertbutylpyridine (TBP), which is an often-used additive in the electrolyte for V_{oc} improvement, is present in the device. As iodine is not included in this solid-state electrolyte, only trace amount of triiodide exists in the device due to oxidation of iodide by oxygen in air, which is responsible for the high V_{oc}. In a DSSC system, charge recombination rate decreases with the decrease in concentration of triiodide. A slower charge recombination contributes to higher V_{oc}. Even in the absence of TBP, therefore, high V_{oc} values can be achieved using the single-component solid-state electrolyte, in which no iodine is added. The value of FF is also impressive for ssDSSCs. The high FF value is the result of high conductivity for the solid-state electrolytes; higher conductivity of the solid state electrolyte results in higher FF when other materials and conditions are the same.

It is impressive for an ssDSSC with a single-component solid electrolyte to achieve such high power conversion efficiency. We emphasize herein that the ssDSSCs based on these propargyl-functionalized ionic conductors work well without the addition of any additives such as iodine, lithium salt, or TBP to the solid-state electrolyte. In a DSSC system, a redox couple is usually needed to reduce dye cations and accept electrons at the counter electrode. For example, the iodide alone such as LiI or DMPImI cannot operate a DSSC in the absence of iodine because the formed triiodide resulted from the oxidation of iodide cannot transport fast enough to the counter electrode to accept the electron there due to the low conductivity, though trace amount of iodine or triiodide is already present in the bulk electrolyte owing to the oxidation of iodide. Therefore, appropriate amount of iodine should be included in the typical electrolyte in order to complete the electric circuit. However, high photocurrent can be generated with PMIm alone as the solid-state electrolyte. This is attributed to its high conductivity; the iodide can efficiently regenerate the dye and the triiodide can move quickly, due to the efficient charge transport, to the counter electrode to accept electrons for circuit completion. The hole transfer from the dye to the ionic conductor corresponds to the oxidation of iodide to I_3^-, while the electron transfer from the counter electrode to the ionic conductor corresponds to the reduction of I_3^- to iodide. The transfer processes of hole and electron just look like movements of iodide and I_3^- in opposite directions. For this reason, high efficiency can be achieved using this single-component solid electrolyte with high conductivity.

Acknowledgment. This work was supported by STCSM (12JC1401500).

References

(1) O'Regan, B. C.; Grätzel, M., *Nature* **1991**, 353, 737-40.
(2) Gong, F.; Wang, H.; Xu, X.; Zhou, G.; Wang, Z.-S., *J. Am. Chem. Soc.* **2012**, *134*, 10953-58.
(3) Li, Y.; Wang, H.; Feng, Q.; Zhou, G.; Wang, Z.-S., *Energy Environ. Sci.*, **2013**, *6*, 2156–65.
(4) Li, J.; Wang, H.; Zhou, G.; Wang, Z.-S., *Chem. Commun.* **2013**, *49*, 9446-8.
(5) Wang, H.; Zhang, X.; Gong, F.; Zhou, G.; Wang, Z.-S., *Adv. Mater.* **2012**, *24*, 121-4.
(6) Wang, H.; Li, J.; Gong, F.; Zhou, G.; Wang, Z.-S., *J. Am. Chem. Soc.* **2013**, *135*, 12627-33.

INORGANIC NANOGELS FOR ULTRAFAST CHARGING LITHIUM ION BATTERIES

Xiaodong Chen

School of Materials Science and Engineering, Nanyang Technological University, 50 Nanyang Avenue, 639798, Singapore
Email: chenxd@ntu.edu.sg

Rechargeable lithium-ion batteries (LIBs) are important electrochemical energy storage devices for consumer electronics and emerging electrical/hybrid vehicles. However, one of the formidable challenges is to develop ultrafast charging LIBs with the rate capability at least one order of magnitude higher than that of the currently commercialized LIBs. In this talk, I will present the state-of-the-art developments of ultrafast charging LIBs based on the rational design of inorganic nanogels. First of all, fundamental electrochemistry and related ionic/electronic conduction theories identify that the rate capability of LIBs is kinetically limited by sluggish solid-state diffusion process in electrode materials. Then, several aspects on the intrinsic materials, materials engineering and processing, and electrode materials architecture design towards maximizing both ionic and electronic conductivity in electrode with short diffusion length are deliberated. Finally, the future trends and perspectives for the ultrafast rechargeable LIBs are discussed. Continued rapid progress in this area is essential and urgent to endow LIBs with ultrafast charging capability to meet huge demands in the near future.

Reference

(1) Tang, Y.; Zhang, Y.; Rui., X.; Qi, D.; Luo, Y.; Leow, W. R.; Chen, S.; Guo, J.; Wei, J.; Li, W.; Deng, J.; Lai, Y.; Ma, B.; Chen, X.* "Conductive Inks Based on Lithium Titanate Nanotube Gel for High Rate Lithium-ion Batteries with Customized Configuration" *Adv. Mater.* **2016**, *28*, 1567.

(2) Tang, Y.; Zhang, Y.; Li, W.; Ma, B.; Chen, X.* "Rational Material Design for Ultrafast Rechargeable Lithium-ion Batteries" *Chem. Soc. Rev.* **2015**, *44*, 5926.

(3) Tang, Y.; Zhang, Y.; Deng, J.; Qi, D.; Leow, W. R.; Wei, J.; Yin, S.; Dong, Z.; Chen, Z.;* Chen, X.* "Unravelling the Correlation between the Aspect Ratio of Nanotubular Structures and Their Electrochemical Performance To Achieve High-Rate and Long-Life Lithium-Ion Batteries" *Angew. Chem. Int. Ed.* **2014**, *53*, 13488.

(4) Tang, Y.; Zhang, Y.; Deng, J.; Wei, J.; Tam, H. L.; Chandran, B. K.; Dong, Z.; Chen, Z.; Chen, X.* "Mechanical Force-driven Growth of Elongated Bending TiO_2-based Nanotubular Materials for Ultrafast Rechargeable Lithium-ion Batteries" *Adv. Mater.* **2014**, *26*, 6111.

IN-SITU TEM&X RAY ABSORPTION STUDY OF LITHIUM SULFUR BATTERY

Hong-Gang Liao

Chemistry and Chemical engineering Department, Xiamen University, Xiamen, Fujian, China
hgliao@xmu.edu.cn

Introduction

A huge effort has been made to shift the world's energy landscape towards clean and renewable energy. Energy storage devices play a vital role in this movement to address the issue with climate change, energy sustainability, electric transportation and energy security. Lithium sulfur battery is one of the promising energy storage devices due to its high theoretical gravimetric capacity (1600 mAh/g) and high energy density (2600 Wh/Kg) with cheap non-toxic sulfur as cathode. [1-3] However, two primary problems of Li-S battery are the short life time and shuttle effect. To understand the chemical species producing at the failure point of the battery, which is important to understand the short life time problem, is still missing. Some In situ X ray diffraction study shows crystalline Li2S never forms at the end of discharging of lithium sulfur battery. No convincing evidence shows the existence of Li2S2 from ex situ X ray photoemission spectroscopy or ex situ X ray diffraction. Thus, it is important to identify the chemical species at the failure point of the battery, which could be Li2S or Li2S2. Additionally, long chain polysulfide migrates to the anode and gets reduced to insoluble species or short chain polysulfide. Short chain polysulfide diffuses back to the cathode and gets re-oxidized to the long chain polysulfide. This is so called shuttle effect. The shuttle effect of the polysulfide causes the low utilization of active materials, poor columbic efficiency and degradation of lithium anode. Manthiram et. al proposed to cycle the lithium sulfur battery during the second plateau of discharging curve to limit the shuttle effect and confine the sulfur near the cathode area. However, the chemical species dynamics during the second plateau is still unknown. Here we report the phase transformation between Li2S2 and Li2S during discharging and recharging on the second plateau of the lithium sulfur battery by in situ transmission electron microscopy (TEM). The chemical species at the failure point of the lithium sulfur battery has been confirmed to be Li2S by in situ near edge X ray absorption (NEXAS). These insights will give a guideline to solve the short life time and shuttle effect problems of lithium sulfur battery for the future energy storage devices.

Experimental

Figure 1 shows the schematics of in situ NEXAS and TEM method. Figure 1 (a) shows the electrochemical liquid cell for in situ NEXAS measurement. Two platinum wires go through the bottom of the cell as the counter and reference electrode. The silicon chip with Si3N4 window was coated with aluminum as the current collector and window for in situ X ray absorption measurement. The cathode material is coated onto the Al current collector as the working electrode, which was connected with the Cu ring after the assembly. The Cu ring is connected to the Cu base via the Cu wire. Thus, the Cu base is connected to the cathode materials as the working electrode. Figure 1(b) shows a schematic for in situ X ray absorption of lithium sulfur battery. In a typical sample preparation process, sulfur particles are mixed with super-P carbon black and PVDF binder to form slurry in NMP. The slurry was dropped cast onto the aluminum as the working electrode. During the assembly of the liquid cell, lithium foil is first put at the bottom of the liquid reservoir and connected to the Pt wires. After filling the electrolyte into the liquid reservoir, a separator was used to separate sulfur cathode and electrolyte. Then the working electrode was put on the top of the separator to seal the cell. During the measurement, x ray fluorescence mode (photon in and photon out) was operated to detect the evolution of chemical species during the operando of lithium sulfur battery via an external potential station. During the analysis of the of lithium sulfur battery, LiClO4/LiNO3/TEGDME was used as the electrolyte, which is shown in figure 1(b), and cycled the battery from elemental sulfur cathode to the failure point of the battery. The reason why LiClO4 was chosen for as the salt for the electrolyte is because the sulfur signal from the electrolyte should be eliminated. During the investigation on the second plateau of the lithium sulfur battery, shown in figure 1(c), Li2S4 in TEMGDME is synthesized as the electrolyte is because Li2S4 is the initial point of the second plateau of discharging curve. No elemental sulfur was used in this experiment. Figure 1(d) shows the schematic of the in situ TEM as our previous report. (The fabrications of TEM electrochemical liquid cell and sample preparation methods for in situ TEM are not shown here.)

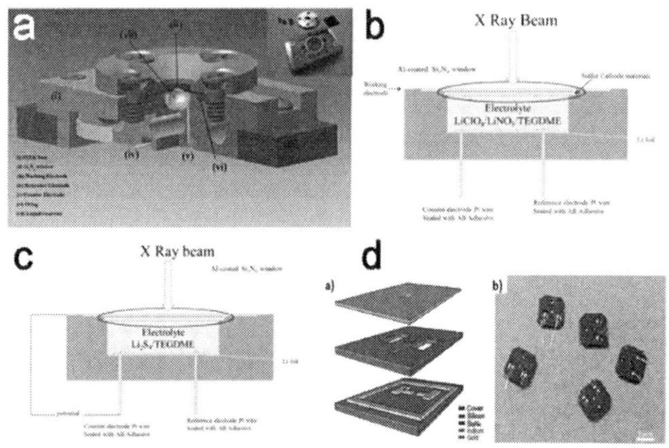

Figure 1. Methods: (a) liquid cell for in situ X ray absorption (XAS) measurement. (b) The schematic setup for in situ XAS of Li-S battery with full discharging curve. (c) The schematic setup for in situ XAS of Li-S battery within the second plateau of the discharging curve. (d) The schematic setup for in situ TEM measurement.

Results and Discussion

In order to identify the chemical species produced in the lithium sulfur battery at different discharge state. In situ chronopotentiometry was applied to dischrage lithium sulfur battery, shown in figure 2(a)-(g). The initial voltage of the battery was 2.52 V. Two plateaus were observed during discharging. The first plateau was between 2.52 V to 2.4 V, shown in figure (a)-(c), and the second plateau was between 2.1 V to 2.0 V, shown in figure (e)-(g). The third plateau between 1.7 V to 1.6 V, shown in figure 2(g), is due to the LiNO3 as the additive. During the in situ X ray absorption measurement, the chronopotentiometry was stopped at the end point of each graph in figure 2(a)-(g) and measure the XAS there. The in situ NEXAS spectra with a function of time are shown in figure 2(h). At initial point (0 s) before discharging the cell, shown in the black line in figure 2(h), the elemental sulfur was shown a sulfur main peak at 2473.14 eV and no pre-peak was observed. This indicates that only elemental sulfur exists on the cathode side before discharging. After discharging the battery for 10860 s with a corresponding voltage at 2.46 V, corresponding to figure 2(a), no distinct pre-peak was observed besides sulfur main peak in the red line figure 2(h). This is because the sulfur atoms in Li2S8 have the similar chemical bonding environment as the elemental sulfur. Thus, the pre-peak would not be clearly observed for Li2S8. The light blue line figure 2(h) shows the XAS spectrum after discharging the battery for 14320 s with a corresponding voltage of 2.4 V, corresponding to figure 2(b). Besides the sulfur main peak at 2473.14 eV, a pre-peak appeared at 2470.94 eV. This indicates that partial of the elemental sulfur has been converted to long chain polysulfide L2S6 at 2.4 V as the dominant species on the first plateau. After discharging the battery for 25890 s, shown in the dark green line in figure 2(h), besides the sulfur main peak, the pre-peak was shifted from 2470.94 eV to 2471.14 eV, which indicates the formation of Li2S4 at the end of the first plateau at 2.4 V, corresponding to figure 2(c). The dark green, pink, yellow, and dark blue line all have the same wavelength peak as the light green line at the top, which is the synthesized Li2S4 in TEGDME. This indicates that Li2S4 is the dominant species between 2.4 V to 2.0 V, which is including the slopping region between the first and second plateau, corresponding to figure 2(d), and the second plateau, corresponding to figure 2(e)-(f). After discharging the cell for 162330 s (=45 hours), the lithium sulfur battery reached the failure point at 1.5 V, corresponding to figure 2(g). The corresponding NEXAS spectrum is shown as the brown line in figure 2(h). The comparison between failure point species and standard Li2S in TEGDME is shown in figure 2(i), which both has the same first sulfur K edge peak at 2473.94 eV, which is about 0.8 eV shift from the sulfur main peak at 2473.14 eV. This shift is consistent with the previous measurement. Thus, it indicates the formation of Li2S at the failure point of the lithium sulfur battery. This is consistent with the theory prediction. The evolution from Li2S8 to Li2S6, Li2S4 will lead the pre-peak will shift to the high energy level while the sulfur main peak will remain the same. The complete convert from soluble polysulfide to insoluble polysulfide (Li2S or Li2S2) would lead the disappearance of the sulfur main peak. The elemental sulfur peak disappeared completely in the brown line in figure 2(h). This indicates the depletion of sulfur cathode. Based on this in situ XAS study, a chemical dynamics during the discharging curve was plotted in figure 2(j). We can confirm that the lithium sulfur battery fully discharged as the formation of Li2S as an insulation layer. However, we don't know the morphology and crystallinity of Li2S.

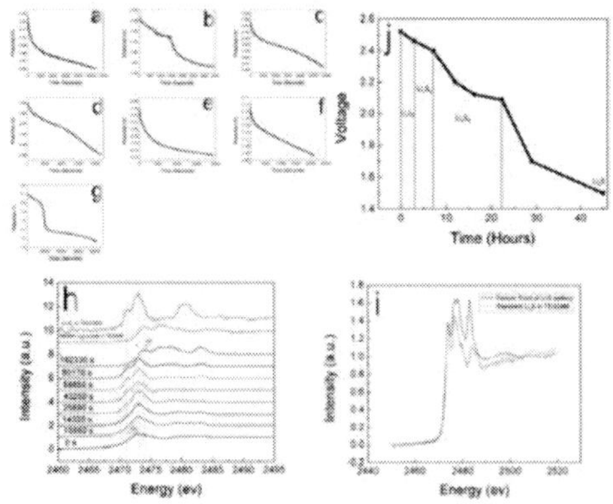

Figure 2. In situ Chronopotentiometry and XAS for analyzing the failure point species: (a)-(g) in situ chronopotentiometry of Li-S battery. (h) In situ XAS of Li-S battery with full discharging curve corresponding to (a)-(g). The black line is corresponding to the initial point before discharging. The red line is corresponding to the point of 2.46 V at the end of (a). The light blue line is corresponding to the point of 2.4 V at the end of (b). The dark green line is corresponding to the point of 2.4 V at the end of (c). The pink line is corresponding to the point of 2.2 V at the end of (d). The yellow line is corresponding to the point of 2.12 V at the end of (e). The dark blue line is corresponding to the 2.0 V at the end of (f). The brown line is corresponding to the failure point at 1.5 V at the end of (g). The dark red line at the top is the standard Li2S powder in TEGDEM. The light green line at the top is the synthesized Li2S4 in TEGDME. (i) The failure point species analysis. (j) The chemical dynamics during the discharging curve.

In order to answer whether crystalline Li_2Sx/Li_2S has formed during the discharging curve, we first performed in situ TEM. Based on our in situ X ray absorption measurement, Li_2S will appear at the end of the second plateau. Thus, we focus to examine the chemical dynamics on the second plateau. From the in situ XANES study, we confirm that Li2S4 would be the starting point of the second plateau. Hence, an electrolyte with Li2S4 in TEGDME was synthesized for the in situ TEM study. An ex situ cyclic voltammetry and chronopotentiometry with electrochemical liquid TEM cell were performed using this electrolyte with lithium as the anode. It shows the discharging curve with a plateau from 2.1 V to 1.9 V, which is not shown here. This design allows us to image the dynamics directly on the second plateau from the very beginning. The in-situ TEM movie showing the discharging and recharging process can be found in the supporting information. The sequential images of the discharging process can be found in figure 3(a)-(h), The dendritic growth at the electrode-electrolyte interface is clearly seen in the discharging process on the second plateau. In contrast, the dendrites were dissolved during the recharging process on the second plateau, leaving an amorphous layer on the electrode. This amorphous layer could be polysulfide species. Some round shape particles can also been seen during the recharging process.

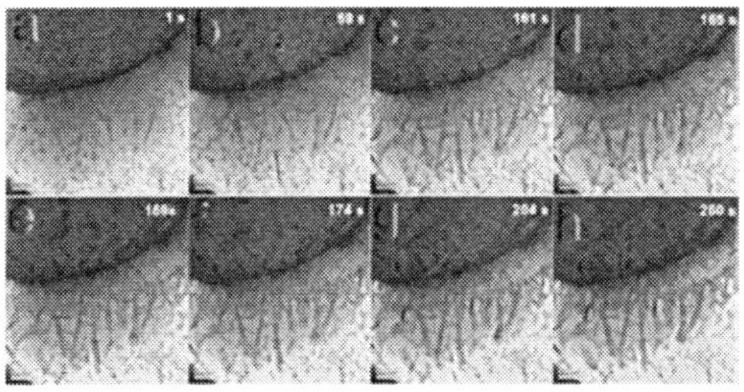

Figure 3. In situ TEM study of Li-S battery.

Conclusions

In conclusion, we performed the in situ TEM study of lithium sulfur battery. A phase transformation between Li2S with Li2SX intermediate was observed both during discharging and recharging process on the second plateau. Combining with in situ XANES study, it is confirmed that the formation of solid Li2Sx intermediate phase formed and transfer to crystalline dendrite shape Li2S at the end of the discharge with the voltage of 1.5 V. These insights will give a guideline to solve the short life time and shuttle effect problems of lithium sulfur battery for the future energy storage devices.

Acknowledgement. This work was supported by State Key Laboratory of Physical Chemistry of Solid Surfaces, China and the Office of Basic Energy Sciences, U.S. Department of Energy.

References

(1) Choi, N.-S.; Chen, Z.; Freunberger, S. A.; Ji, X.; Sun, Y.-K.; Amine, K.; Yushin, G.; Nazar, L. F.; Cho, J.; Bruce, P. G. Angew. Chem., Int. Ed. **2012**, 51, 9994

(2) Bruce, P. G.; Freunberger, S. A.; Hardwick, L. J.; Tarascon, J.-M. Nat. Mater. **2012**, 11, 19

(3) Evers, S.; Nazar, L. F. Acc. Chem. Res. **2012**, 46, 1135

OPTOELECTRONIC APPLICATION OF NANOGUMBOS

Noureen Siraj, Tony Karam, Louis Haber, and Isiah M. Warner

Department of Chemistry, Louisiana State University, Baton Rouge LA-70803

Ionic liquids (ILs) are fused salts with a melting point below 100 °C. ILs have received considerable attention due to their outstanding properties such as tunability and low vapor pressure. We have recently introduced a new class of solid phase organic salts termed as GUMBOS (***g***roup of ***u***niform ***m***aterials ***b***ased on ***o***rganic ***s***alts) in our laboratory. GUMBOS encompass the category of frozen ILs and organic salts with melting point ranging from 25 °C up to 250 °C. Nanoparticles prepared from GUMBOS are known as nanoGUMBOS. NanoGUMBOS have been found to have beneficial characteristics for optoelectronic applications. This talk highlights various aspects for employing nanoGUMBOS for optoelectronic applications. As a result of their unique tailoring and tunable properties, some properties can be enhanced simply by changing nanoparticle size. Transient absorption spectroscopy studies revealed that we are able to tune the excited state lifetime by changing nanoparticles size. Long-lived acoustic phonon oscillations are observed with the large size crystalline nanoparticles that are very beneficial for effective charge transport into a semiconductor electrode. This study provides new insights into the photodynamics of these novel nanoGUMBOS for potential advances in optoelectronic applications.

TiN_2: PREDICTION OF A NOVEL STOICHIOMETRY OF SEMICONDUCTING TITANIUM NITRIDE 2D SHEETS WITH LOW METAL CONTENT

Veronica Barone

Department of Physics, Central Michigan University, Mt. Pleasant, MI 48859
v.barone@cmich.edu

Introduction

Two-dimensional (2D) materials have been the focus of extensive studies due to their enormous potential for technological applications. Recently, 2D layers of early transition metal elements (M) and carbon or nitrogen (X) (MXenes) with hexagonal symmetry have been synthesized from the layered bulk form dubbed MAX phase.[1,2,3] These novel graphene-like MXenes follow the general formula $M_{n+1}X_n$ and have been studied for a variety of applications.

In this work, we present a screened exchange hybrid density functional theory (DFT) study of a different type of 2D structure with hexagonal symmetry (MoS_2 structure) and stoichiometry TiX_2 with X=C,N.

Results

DFT calculations are carried out using the Gaussian 09 program.[4] All structures have been fully relaxed performed using the PBE functional and a modified 6-31G* basis set for which Gaussian exponents smaller than 0.1 are removed.[5] We first establish the relative stability of these materials by computing the formation energies of the structures with respect to the pure constituents according to the formula,

$$E_f = E_T(MX_2) - E(M) - 2E(X)$$

where E_T represents the total energy per atom of the phase under consideration, $E(M)$ is the total energy per atom of the pure metal M, and $E(X)$ is the total energy per atom of the N_2 molecule for X=N and graphene for X=C. For comparison purposes, we also calculate the formation energy of known stable bulk forms of MX as well as the cohesive energy of each structure (atomization energy). Schemes of the 2D and 3D structures under consideration are shown (for the nitrides) in Figure 1 and formation and cohesive energies are presented in Table 1.

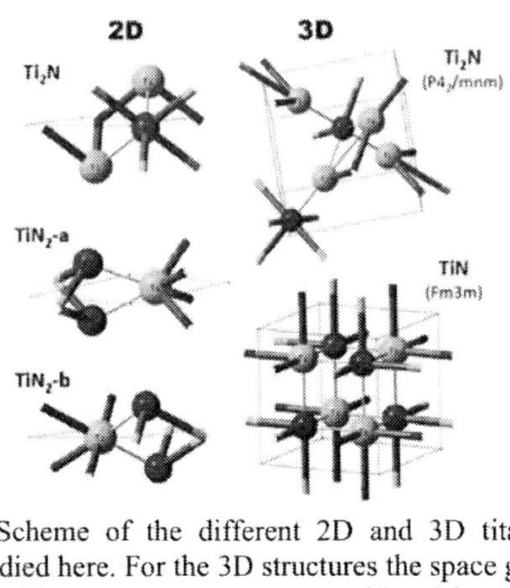

Figure 1. Scheme of the different 2D and 3D titanium nitride structures studied here. For the 3D structures the space group is noted in parenthesis. The same structures are considered for carbides.

As expected, all the bulk phases are highly stable while the 2D sheets are somewhat less stable. 2D carbides are generally less stable than nitrides, especially for the case of TiC_2 for which the calculated formation energy is highly positive (metastable). However, for nitrides, the formation energy is negative (stable) and the composition TiN_2 is slightly more stable than Ti_2N, suggesting that these sheets can be realized experimentally.

Table 1. Calculated formation and cohesive energies for the different 2D and 3D carbides and nitrides using the PBE functional.

2D	**E_f(eV/atom)**	**E_c(eV/atom)**
Ti_2C	0.05	-6.38
TiC_2	0.93	-6.26
Ti_2N	-0.56	-6.06
TiN_2-a	-0.58	-5.92
TiN_2-b	0.07	-5.26
3D	**E_f (eV/atom)**	**E_c (eV/atom)**
TiC (F m3m)	-0.94	-7.75
Ti_2N (P42/mnm)	-1.43	-6.93
TiN (F m3m)	-1.88	-7.3
Ti_2C (P42/mnm)	-0.69	-7.12

Due to the significantly greater stability of the nitrides over the carbides, we will limit our discussion of the electronic properties of the 2D sheets to nitrides only and compare their band structure with the band structure of the Ti_2N MXene. It has been shown that the HSE functional provides a much more reliable picture of the electronic properties of materials.[5,6] Therefore, in Figure 2, we present PBE (black bold lines) as well as HSE (dashed red lines) bands for Ti_2N and TiN_2. As seen in Figure 2, while Ti_2N is metallic, TiN_2 is a direct semiconductor with a bandgap of 1.5 eV (obtained with HSE).

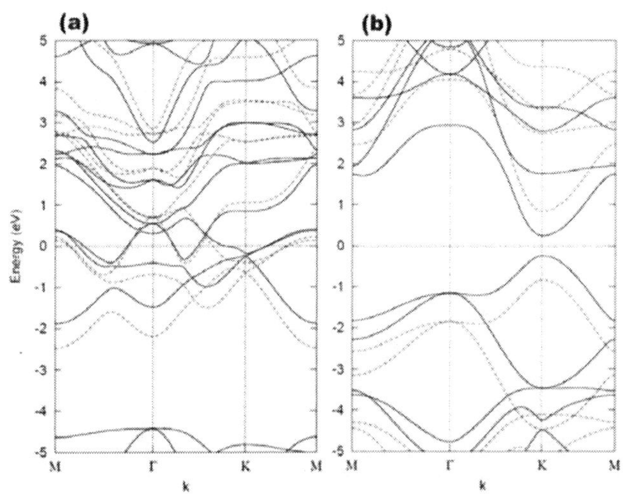

Figure 2. Calculated band structure with PBE (solid black lines) and HSE (dashed red lines) for (a) Ti_2N and (b) TiN_2.

Conclusions

We have studied the relative stability of different titanium nitride and carbide bulk materials and compared them with 2D sheets of different compositions. We have found that while 2D TiC_2 structures are metastable, TiN_2 sheets are as stable as the recently synthesized Ti_2N MXenes, suggesting that these sheets can be realized experimentally. While the latter exhibit a metallic behavior, Ti_2N sheets are direct semiconductors with a bandgap of approximately 1.5 eV.

Acknowledgement. This work was supported by NSF CBET-1335944.

References

(1) Naguib, M.; Mochalin, V. N.; Barsoum, M. W.; Gogotsi, Y., 25th Anniversary Article: MXenes: A New Family of Two-Dimensional Materials. *Advanced Materials* **2014,** *26* (7), 992-1005.

(2) Naguib, M.; Mashtalir, O.; Carle, J.; Presser, V.; Lu, J.; Hultman, L.; Gogotsi, Y.; Barsoum, M. W., Two-Dimensional Transition Metal Carbides. *ACS Nano* **2012,** *6* (2), 1322-1331.

(3) Zhang, X.; Ma, Z.; Zhao, X.; Tang, Q.; Zhou, Z.; Zhang, X.; Ma, Z.; Zhao, X.; Tang, Q.; Zhou, Z., Computational studies on structural and electronic properties of functionalized MXene monolayers and nanotubes. **2015**.

(4) Frisch, M., et al. Gaussian Inc. 2008.

(5) Heyd, J.; Scuseria, G. E., Efficient hybrid density functional calculations in solids: Assessment of the Heyd-Scuseria-Ernzerhof screened Coulomb hybrid functional. *Journal of Chemical Physics* **2004,** *121* (3), 1187-1192.

(6) Barone, V.; Hod, O.; Peralta, J.; Scuseria, G., Accurate Prediction of the Electronic Properties of Low-Dimensional Graphene Derivatives Using a Screened Hybrid Density Functional. *Accounts of Chemical Research* **2011,** *44* (4), 269-279.

ENHANCED ETHANOL PRODUCTIVITY THROUGH THE USE ELECTROSPUN ALGINATE THREADS

Akira Nordmeier[1] *and Dev Chidambaram*[1]*

[1]Chemical and Materials Engineering, University of Nevada, Reno, Reno, NV 89557-0388.
*dcc@unr.edu

Introduction

Current production methods use batch-based reactors or fermenters, which allow for large collection volumes and yields.[1] However, batch-based processes have long processing times because the reactor must be cleaned before a new batch is started, and then the microorganism of choice must be allowed to grow before ethanol production can occur. Continuous flow reactor (CFR) could offer significant advantages compared to traditional batch reactor. However, current CFRs have lower yields than those of the traditional batch reactors.[2] CFRs with enhanced rates of productivity are needed. Alginate threads have been shown to have higher diffusion rates over that of the typically used beads.[3] Electrospun threads could further reduce diffusion rates as well as interactions between the polymer and the microorganism allowing for increased productivity. This is possible as electrospinning can form threads on the nanometer scale which immobilize the microorganisms within the threads.

Experimental

Zymomonas mobilis strain ZM4 (ATCC 31821) immobilized in Ca-alginate was used for fermentation in a CFR. Ethanol production of *Z. mobilis* immobilized in electrospun Ca-alginate threads was studied. The production of ethanol and the consumption of glucose were monitored using a high performance liquid chromatograph.

Results

Millimeter sized beads and threads were able to obtain similar ethanol concentrations. As shown in Figure 1, at the given flow rate, the threads in the CFR consumed all the glucose that was provided, while the beads were unable to fully ferment all the glucose present. Further reduction of the size of the threads through the use of electrospinning could further increase the productivity of the CFR.

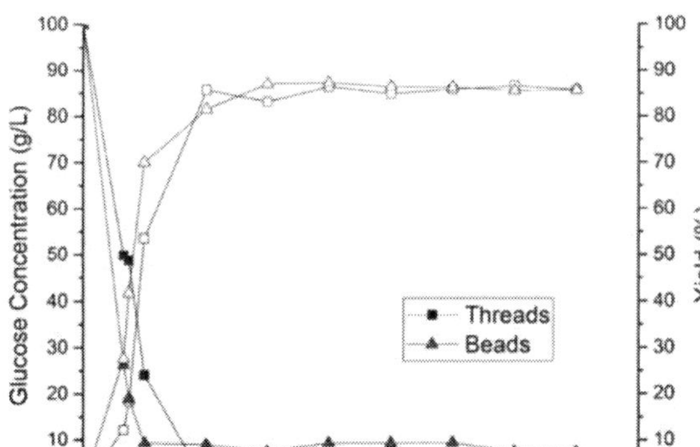

Figure 1. The comparison of the concentration of glucose (solid symbols) to the yield of ethanol (open symbols) for CFRs containing Threads and Beads of approximately 2mm in size.

Acknowledgments

This material is based upon work supported by the National Science Foundation under Grant No. 1159772.

(1) Remer, D. S. C., L.H., Design cost factors for scaling-up engineering equipment. *Chemical Engineering Progress* **1990,** *86* (8).
(2) Li, T.; Chen, X. b.; Chen, J. c.; Wu, Q.; Chen, G. Q., Open and continuous fermentation: Products, conditions and bioprocess economy. *Biotechnology journal* **2014,** *9* (12), 1503-1511.
(3) Que, F., Using a thread type of alginate gel particles as cell-immobilised support and some concept of packed bed fermenter design. *Biotechnol Tech* **1993,** *7* (10), 755-760.

One-Pot, One-Step Direct Fabrication of Spinel/Layered Li excess cathode materials for High Performance Rechargeable Lithium Ion Batteries

*Keerthi Savaram**, *Jianqing Zhao*[ʃ], *Ruiming Huang**, *Feng Wang*[ʃ], *Huixin He***

[ʃ]Brookhaven National Lab, 2nd Center street, Upton, NY-11973

To a greater extent, development of new-generation lithium ion batteries is limited by the low energy density, low operating voltage and poor rate capability of cathode materials.[1,2,3] Recently, the emerging Li-excess layered oxides have attracted a great deal of research efforts due to their high capacities. These cathode materials can be cycled over a broad voltage range between 2.0 and 4.8 V *vs.* Li/Li^+ and deliver specific capacities higher than 250 $mAhg^{-1}$. Compared to the widely used cathode materials such as $LiCoO_2$ (140 mA h g^{-1}), these cathode materials also offer many other advantages such as low cost, environmental benignity, and safety.[4-9] Li[Li0.2Mn0.54Ni0.13Co0.13]O_2 (marked as LMNCO) is one of the most promising Li excess layered cathode materials, which has an impressive theoretical capacity of 321 mAh/g with operating voltage up to 4.8 V *vs.* Li/Li^+.[10] However, the unavoidable layered-to-spinel phase transformation of LMNCO during in-situ electrochemical cycling causes drastic capacity degradation and short cycle life, which limits its practical applications. Recently, we developed an ex-situ activation approach for complete phase conversion of Li-excess layered Li[Li0.2Mn0.54Ni0.13Co0.13]O_2 to a $Li_4Mn_5O_{12}$-type spinel phase.[11,12] More impressively, the newly-formed spinel material delivers an initial discharge capacity of 313.6 mAh/g at 0.1C (1C = 250 mA/g), much higher than 211.3 mAh/g from pristine Li-excess layered cathode. However, this process is very time consuming and energy intensive. The requirement for long time, high reaction temperature is against the original idea of energy saving and sustainability. In this work, we report our efforts to directly and rapidly fabricate lithium excess cathode materials with a mixture of layered and spinel structure without requirement of high temperature annealing. Using microwave irradiation instead of traditional convection heating, the approach not only dramatically shortens the fabrication time from hours to minutes, but also tremendously improves the quality. It exhibits excellent performance with a discharge capacity of 444.8mAh/g at 1C, almost double the discharge capacity of the spinel structures transformed from LMNCO. Further, adding a small amount carbon source into the metal salt mixtures before microwave irradiation improve the recycle stability of the materials even though with a slight decrease of the initial discharge capacity of 340.9mAh/g at 1C. The as obtained product does not need further annealing, making the reaction time for the synthesis of the materials very short. This facilitates the use of the as prepared material for the next generation lithium ion batteries.

Experimental procedure

Synthesis: The Li excess cathode material is synthesized by dissolving the LiOH: Mn: Ni: Co in 1:0.54:0.13:0.13. The as obtained solution is microwaved for 180sec and is washed with water and ethanol and dried in the oven overnight. This sample is referred to as MWLMNCO (Microwaved Lithium excess Manganese Nickel Cobalt oxide cathode materials).As for the sample with Carbon source, 0.24ml of the carbon source is added and then microwaved according to the procedure mentioned above. This sample is referred to as MWLMNCO+C (Microwaved Lithium excess Manganese Nickel Cobalt oxide cathode materials with carbon source).

Characterization: The electrochemical measurements were performed by assembling a coin cell composed of 80wt % of the MWLMNCO or MWLMNCO+C, 10 wt% conductive carbon and 10wt% polyvinylidene fluoride as the binder. Lithium foil is used as an anode with $LiPF_6$ as the electrolyte. Galvanostatic charge and discharge were performed in a voltage range of 2.0 to 4.8V *vs.* Li/Li^+ using an 8-channel battery analyzer. Theoretical capacities are all set to 250mAh/g with the current density at 1C corresponding to 1mAh/g. Cyclic voltammetry curves of the cathodes were recorded at a scanning rate of 0.1mVs^{-1} between 2.0 and 4.8V *vs.* Li/Li^+ using an electrochemical analyzer.

Results

X-ray powder Diffraction (XRD) technique was used to study the crystal structure of the MWLMNCO particles and follow the evolution of the structural changes as a function of microwave irradiation time for 300sec Fig 1. The XRD analysis clearly indicates the evolution of the layered/spinel phase over microwave time. The layered structure was evolved at around 45sec microwave time. The highest degree of layered with the spinel structure evolution was observed at 180sec and further increase in the microwave time did not enhance the crystallinity of the particles. We found that with increasing the microwave irradiation time, the amount of spinel structure increased gradually. However, there is still small amount of metal oxide residues, which could not be eliminated even after further microwaving the sample upto 300sec.

Scanning electron microscope was used to characterize the morphology of the materials obtained with microwave irradiation of 180 seconds. The Fig 2(a,b) indicates the SEM image of the as synthesized MWLMNCO at 180sec. The size of the nanoparticles are in the range of <500nm with some thin layer like structure embedded in between. Using this material as cathode and Li foil as the anode, a coin cell battery was assembled. The Galvanostatic charge and discharge studies were conducted in the voltage range of 2.0 to 4.8V. The results are depicted in the Fig 2(c). These studies were conducted at 1C (1C= 250$mAhg^{-1}$). The initial discharge capacity was around 444 $mAhg^{-1}$ with a slow decrease rate of 0.01$mAhg^{-1}$. The capacity decreased to 275$mAhg^{-1}$ after 100cycles, which is still higher than the discharge capacity obtained from the layered/spinel phase by the sol gel method i.e. 204$mAhg^{-1}$ as shown in Fig 2(c). The Cyclic voltammetry was conducted to study the initial charge and discharge curves. The results were interpreted in Fig 2(d) which indicates the initial activation of the nanomaterial during the first cycle and repeated cycleability in the second and third cycle. This data is in close agreement with the layered/spinel structure which was conducted in our group[13].

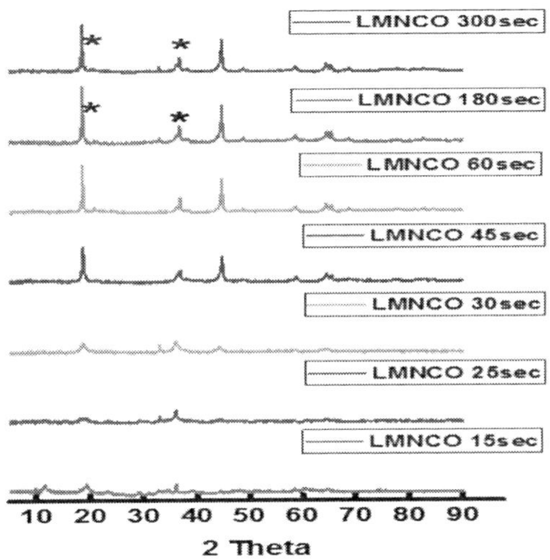

Figure 1: The XRD study of the MWLMNCO studied at various microwave reaction time, where the emerging of the spinel/layered phase can be depicted from 45sec microwave irradiation time.

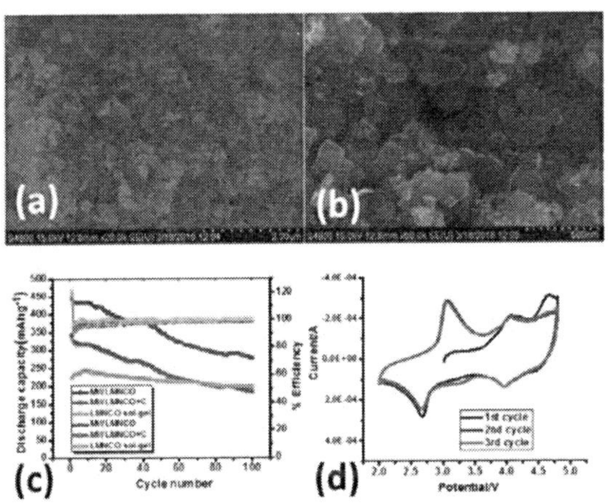

Figure 2: SEM images of Li $[Li_{0.2}Mn_{0.54}Ni_{0.13}Co_{0.13}]$ O_2 (MWLMNCO) particles fabricated via microwave irradiation of mixture of metal salts (a) and the Zoom in image of Fig a (b). The Discharge capacity and the efficiency of the MWLMNCO and MWLMNCO+C coin cell when studied the Galvanostatic charge and discharge in a voltage range of 2.0 to 4.8 *vs.* Li/Li^+ and its comparison with the LMNCO for 100cycles (c). The Cyclic voltammetry curves of the cathode recorded at a scanning rate of $0.1mVs^{-1}$ between 2.0 and 4.8V *vs.* Li/Li^+ for 3 successive cycles with the first cycle contributing to the initial activation of the material(d).

Conclusion

A simple, rapid and energy efficient approach is developed to fabricate Li excess cathode materials. A mixture of layered and spinel structured was directly obtained in one pot, one step reaction without requiring further annealing at high temperatures. The as synthesized compound i.e. MWLMNCO exhibits a high discharge capacity of $444.8mAhg^{-1}$ over 100cycles. The MWLMNCO+C exhibited a discharge capacity of $340.9mAhg^{-1}$ at 1C. These materials need to be further analyzed using high resolution Transmission electron Microscopy (TEM) and XRD for appropriate analysis.

ACKNOWLEDGMENT. Financial support was provided the National Science Foundation (CBET 1438493 and DMR 1507812).

References:

(1) YangKook Sun; Seung-Taek Myung; Byung-Chun Park; Jai Prakash; Ilias Belharouak; Amine, K. *Nature materials* **2009**, *8*, 320-324.
(2) Manthiram, A.; Chemelewski, K.; Lee, E.-S. *Energy & Environmental Science* **2014**, *7*, 1339-1350.
(3) Lung-Hao Hu, B.; Wu, F. Y.; Lin, C. T.; Khlobystov, A. N.; Li, L. J. *Nature communications* **2013**, *4*, 1687.
(4) Jiang, M.; Key, B.; Meng, Y. S.; Grey, C. P. *Chemistry of Materials* **2009**, *21*, 2733-2745.
(5) Yabuuchi, N.; Yoshii, K.; Myung, S. T.; Nakai, I.; Komaba, S. *Journal of the American Chemical Society* **2011**, *133*, 4404-4419.
(6) Hy, S.; Felix, F.; Rick, J.; Su, W. N.; Hwang, B. J. *Journal of the American Chemical Society* **2014**, *136*, 999-1007.
(7) Mohanty, D.; Kalnaus, S.; Meisner, R. A.; Rhodes, K. J.; Li, J.; Payzant, E. A.; Wood, D. L.; Daniel, C. *Journal of Power Sources* **2013**, *229*, 239-248.
(8) Lee, E.-S.; Huq, A.; Chang, H.-Y.; Manthiram, A. *Chemistry of Materials* **2012**, *24*, 600-612.
(9) Johnson, C. S.; Li, N.; Lefief, C.; Vaughey, J. T.; Thackeray, M. M. *Chemistry of Materials* **2008**, *20*, 6095–6106.
(10) Wu, Y.; Manthiram, A. *Electrochemical and Solid-State Letters* **2006**, *9*, A221-A224.
(11) Zhao, J. Q.; Huang, R. M.; Gao, W. P.; Zuo, J. M.; Zhang, X. F.; Misture, S. T.; Chen, Y.; Lockard, J. V.; Zhang , B. L.; Guo , S. M.; Khoshi , M. R.; Dooley, K.; He, H. X.; Wang, Y. *Adv. Energy Mater.* **2015**, 1401937.
(12) Zhao, J. Q.; Huangb, R. M.; He, H. X.; Wang, Y. *ECS Transactions* **2015**, *64*, 77-86.
(13) Zhao, J.; Huang, R.; Gao, W.; Zuo, J. M.; Zhang, X. F.; Misture, S. T.; Chen, Y.; Lockard, J. V.; Zhang, B.; Guo, S. ; Wiley Online Library, 2015; Vol. 5.

NANO-STRUCTURES AND THIN FILMS BY PULSED LASER DEPOSITION FOR ENGINEERING AND TECHNOLOGY APPLICATIONS

D.Ravinder

Department of Physics, Osmania Universitry, Hyderabab-500007
E:mail: ravindergupta28@rediffmail.com

Abstract

Ferrite nano thin films were deposited from a target of Cu-Zn ferrite onto a sapphire substrate using XeCl excimer laser operating 308 nm with an energy of 225 mJ and a frequency of 30 Hz. Films were deposited from the target onto sapphire (001) substrates heated to 650^0 C in an oxygen atmosphere of 100 m Torr. The laser beam was incident On the target face at an angle of 45^0. Studies on crystal structure were done by X-ray diffactometry (XRD).The surface texture, cross-section morphology and grain size was observed by JEOL-JSM-6400 scanning electron microscopy, atomic force microscopy(AFM) and magnetic force microscopy (MFM). The observed results are useful in engineering, technological and potential applications. Potential applications such as (1) high density information storage in computers, (2) ferrofluid technology, magnetocaloric refrigeration, magnetic resonance imaging (MRI) enhancement, magnetic guided drug delivery, microwave devices and magnetic recording media and magnetic sensors.

COPPER DOPED MESOPOROUS CeO_2 CATALYSTS FOR THE WATER GAS SHIFT REACTION

Dimitriy Vovchok[1,2], Curtis Guild[3], Jordi Llorca[4], Wenqian Xu[5], Sanjaya D. Senanayake[1,], Jose A. Rodriguez[1,2], and Steven L. Suib[3].*

1. Chemistry Department, Brookhaven National Laboratory, Upton, NY 11973 (USA)
2. Chemistry Department, State University of New York at Stony Brook, Stony Brook, NY 11794
3. Department of Chemistry, University of Connecticut, 55 N. Eagleville Rd., Storrs, CT 06269-3060
4. Institute of Energy Technologies. Technical University of Catalonia. Barcelona 08028, Spain
5. Advanced Photon Source, Argonne National Laboratory, Argonne, IL, 60439

* Corresponding Author: ssenanay@bnl.gov

Introduction

Novel mesoporous[1] copper-ceria[2] materials are presented as excellent catalysts for the water shift reaction ($CO + H_2O \dashrightarrow CO_2 + H_2$), prompting the potential for the use of this reaction as a hydrogen production pathway. A variety of powerful in situ and operando techniques are used to resolve the structural, electronic, and chemical processes underlying the high catalytic activity of these materials.

Experimental

Operando synchrotron X-ray diffraction studies elucidate the behavior of the cerium oxide and copper lattices, while in situ X-ray absorption studies allow for determination of electronic properties. Operando DRIFTS (Diffuse reflectance infrared Fourier transform spectroscopy) is used to probe the surface chemical environment[3] of the catalysts in an effort to gain insights on the catalytic mechanism.

Results and Discussion

The nature of the catalytic activity of mesoporous Cu/CeO_2 catalysts for the WGS reaction was examined through the use of in-operando synchrotron XRD and XANES. At the conditions and temperatures necessary for the catalytic activity of the mesoporous Cu/CeO_2 catalysts, we observed the presence of copper as metallic Cu0. Additionally, the reduction of a portion of Ce^{4+} in the CeO_2 lattice to Ce^{3+} was recorded, implying the formation of oxygen vacancy[4] sites in the lattice. Both of these changes occurred when the catalyst material was heated to 150°C under WGS reaction conditions. These findings are in accordance with previous determinations of metallic copper and CeO_2 lattice oxygen vacancies as active sites in non-porous Cu/CeO_2 catalysts for the WGS reaction[3].

References

(1) Poyraz, A. S.; Kuo, C.-H.; Biswas, S.; King'ondu, C. K.; Suib, S. L., A general approach to crystalline and monomodal pore size mesoporous materials. *Nat Commun* **2013,** *4*.
(2) Mullins, D. R., The surface chemistry of cerium oxide. *Surface Science Reports* **2015,** *70* (1), 42-85.
(3) Yao, S. Y.; Xu, W. Q.; Johnston-Peck, A. C.; Zhao, F. Z.; Liu, Z. Y.; Luo, S.; Senanayake, S. D.; Martinez-Arias, A.; Liu, W. J.; Rodriguez, J. A., Morphological effects of the nanostructured ceria support on the activity and stability of CuO/CeO2 catalysts for the water-gas shift reaction. *Phys Chem Chem Phys* **2014,** *16* (32), 17183-95.
(4) Wang, X.; Rodriguez, J. A.; Hanson, J. C.; Gamarra, D.; Martinez-Arias, A.; Fernandez-Garcia, M., In situ studies of the active sites for the water gas shift reaction over Cu-CeO2 catalysts: complex interaction between metallic copper and oxygen vacancies of ceria. *J Phys Chem B* **2006,** *110* (1), 428-34.

STACKED CORE-SHELL NANOROD ARRAY ELECTROCATALYST BY A HIPS-GLAD METHOD FOR ENHANCED ORR IN PEMFCS

Fatma M. Yurtsever[1], Mahbuba Begum[1], Mesut Yurukcu[2], Tansel Karabacak[2]*

[1] Department of Chemistry, University of Arkansas at Little Rock, Little Rock AR, 72204, USA
[2]Department of Physics and Astronomy, University of Arkansas at Little Rock, Little Rock AR, 72204, USA
*fmyurtsever@ualr.edu

Introduction

Low Temperature Polymer Electrolyte Fuel Cells (LT-PEMFC) are one of the most promising renewable energy devices in twenty-first century. It has many advantages in terms of zero carbon emission, quite operation mode, high power density, higher efficiency than Carnot engines, and in addition it requires less warm up & fast start-up time. However in order to make PEMFCs more practical, further development in catalyst is needed in order to decrease the cost, enhance stability, and improve sluggish oxygen reduction reaction (ORR) kinetics.

Experimental

In this study, we developed new design of stacked core-shell nanorod array electrocatalyst to improve ORR kinetics and catalyst stability. For this purpose, we fabricated a two-layer nanorod structure, where each layer is composed of a tungsten carbide (WC) core and a platinum-nickel (Pt-Ni) alloy shell as illustrated in Figure 1. WC nanorods were fabricated by a glancing angle deposition (GLAD). A conformal Pt-Ni alloy was coated around the WC nanorods by a new high pressure sputtering (HIPS) method. This process was repeated two times in order to obtain longer stacked core-shell nanorods. We investigated three different Pt:Ni ratios. Cyclic voltammetry (CV) and rotating disk electrode (RDE) methods were used for electrochemical characterizations. 0.1 M $HClO_4$ was chosen as the electrolyte solution.

Morphological and crystallographic analysis were performed using scanning electron microscopy (SEM) and X-ray diffraction (XRD) respectively. Pt mass loading values were measured using quartz crystal microbalance (QCM).

Results and Conclusions

We report that electrochemically active surface area (ECSA) values for different Pt:Ni compositions change in the order of 1:3>1:1>3:1. Specific activity (SA) and mass activity (MA) were observed to be higher for Ni-rich compositions, which are further enhanced after accelerated stability tests.

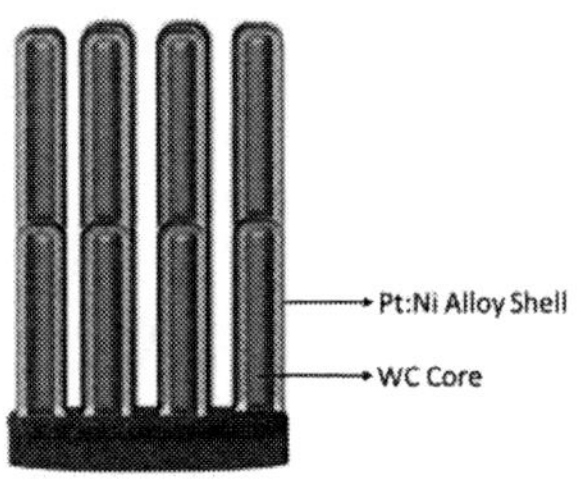

Figure 1. Illustration of Stacked Pt:Ni on WC Core-Shell Nanorod Arrays

Acknowledgement. The work at the University of Arkansas at Little Rock (UALR) was financially supported by the National Science Foundation under grant numbers EPS-1003970 and 1159830. The authors thank UALR Center for Integrative Nanotechnology Sciences for helping with SEM images.

References

(1) Debe, M. K. Electrocatalyst Approaches and Challenges for Automotive Fuel Cells. **2012**.
(2) Gasteiger, H. A.; Kocha, S. S.; Sompalli, B.; Wagner, F. T. Activity Benchmarks and Requirements for Pt, Pt-Alloy, and Non-Pt Oxygen Reduction Catalysts for PEMFCs. **2005**, *56*, 9–35.
(3) Kariuki, N. N.; Khudhayer, W. J.; Karabacak, T.; Myers, D. J. GLAD Pt – Ni Alloy Nanorods for Oxygen Reduction Reaction. **2013**.
(4) Stephens, Ifan E. L.; Bondarenko, A. S.; Grønbjerg, U.; Rossmeisl, J.; Chorkendorff, I. Environmental Science Understanding the Electrocatalysis of Oxygen Reduction on Platinum and Its. **2012**, 6744–6762.
(5) Vliet, D. F. Van Der; Wang, C.; Tripkovic, D.; Strmcnik, D.; Zhang, X. F.; Debe, M. K.; Atanasoski, R. T.; Markovic, N. M.; Stamenkovic, V. R. Mesostructured Thin Films as Electrocatalysts with Tunable Composition and Surface Morphology. *Nat. Mater.* **2012**, *11* (12), 1051–1058.

CRYSTALLINE AND AMORPHOUS TiO2 LOW ENERGY BANDGAP NANOMATERIALS AND THEIR ENERGY AND ENVIRONMENTAL APPLICATIONS

*Youngmin Kim, Luyang Wang and Hyoyoung Lee**

Center for Integrated Nanostructure Physics, Institute for Basic Science, Department of Chemistry and Department of Energy Science, Sungkyunkwan University, Suwon 440-746, Korea
hyoyoung@skku.edu

Until now, the most widely applied photocatalyst oxide is titanium dioxide (TiO2) due to its low cost, high photocorrosion stability and excellent activity under UV irradiation. Even though black TiO2 which can absorb visible and infrared solar light has been known for a long time, there is no report on the selective crystalline and amorphous TiO2 materials.

Here, we demonstrate a selective conversion of white TiO2 into black TiO2 from simple room temperature solution processing. The reducing agent, lithium in ethylenediamine (Li-EDA), can reduce only white rutile phase of Degussa P-25 samples mixed with anatase and rutile phases to produce a disordered rutile black TiO2 that has a low bandgap energy. Electrons and holes generated from the reduced black TiO2 under solar light can be efficiently separated through the white crystalline anatase, exhibiting the world best photocatalytic hydrogen evolving reaction.[1]

In addition, electrons generated from the reduced black TiO2 can effectively generate ROS under solar light and successfully remove a bloom of algae. Amazingly, only reduced TiO2 materials can generate ROS under solar light, which was confirmed by electron spin resonance. Among the three different types of Li-EDA treated TiO2 (anatase, rutile and both phased TiO2), the both phased rTiO2 showed the best performance to produce ROS. The generated ROS effectively removed the common green algae Chlamydomonas.[2] This is the first report on algae degradation under solar light, proving the feasibility of commercially available products for disinfection.

Acknowledgement This work was supported by the Center for Integrated Nanostructure Physics (CINAP), Institute for Basic Science (IBS) (IBS-R011-D1).

References

(1) Zhang, K.; Wang, Luyang; Kim, J. K.; Ma, M.; Veerappan, G.; Lee, C.-L.; Kong, K.-j.; Lee, H.* and Park, J.H.*, *Energy & Environmental Science*, **2016**, *9*, 499.
(2) Kim, Y.; Hwang, H. M.; Wang, L.; Kim, I; Yoon, Y.; Lee, H.*, *Scientific Reports*, Revision, **2016**.

SYNTHESIS OF MAGNETIC CELLULOSE NANOCRYSTALS FOR HIGH PERFORMANCE APPLICATIONS

*Prodyut Dhar, Surendra Singh Gaur, Amit Kumar and Vimal Katiyar**

Department of Chemical Engineering,
Indian Institute of Technology Guwahati, Guwahati, Assam, India – 781039
*Corresponding Author: vkatiyar@iitg.ernet.in

Introduction

Multiple applications of iron oxides nanoparticles have been reported in literature such as electrode in lithium ion batteries, in biotechnology as enzyme immobilization, catalyst and electrocatalyst in multiple reactions[1,2,3,4]. Most of these applications require excellent electrical conductivity, high specific surface area, and easy accessibility to the active sites, so having iron oxide particles in nano dimension would surely improve the performance. In this study we report a green one-step approach of fabricating magnetic cellulose nanocrystals (CNCs) through the chemical reduction of the iron nanoparticles onto its surface. The prepared nanocomposites were characterized through both chemical and structural analysis to confirm the presence of the iron nanoparticles. The fabricated nanocomposites were found to be magneto-responsive with electrochemical/ bio-catalytic activity towards the pollutant remediation.

Experiments

Fabrication of the Cellulose Nanocrystals. Cellulose Nanocrystals (CNCs) were fabricated from the cellulose extracted from the bamboo pulp as precursor. The bamboo pulp were pretreated with the sodium hydroxide followed by peroxide treatment and bleaching with sodium hypochlorite to extract the purified cellulose as reported in our previous studies[1]. The extracted cellulose was acid hydrolyzed using 64 % sulphuric acid at room temperature under stringent magnetic stirring conditions. The acid was neutralized with excess acid followed by centrifugation and dialysis to remove the excess acid. The neutral solution of CNCs precipitates were freeze dried to obtain the dried CNC powders.

Fabrication of the Iron-oxide/Cellulose Nanocrystals. CNC supported iron nanoparticles (Fe-CNC) were prepared from anhydrous $FeCl_3$ and CNCs, by using $NaBH_4$ reducing agent. $FeCl_3$ and CNCs were taken in 2:1 weight ratio in 100ml water and kept under stirring at 100°C under N_2 atmosphere for 3 hours. After three hours, 500 mg solution of $NaBH_4$ solution in water was added in the system and the reaction was continued for further 3 hours under inert conditions. The pale yellow iron-CNC solution turned to the dark black color in the presence of the reducing agent which confirms the formation of the iron oxide nanoparticles on the surface of the CNCs (Figure 1 (a)&(b)). The resulted nanoparticles were separated using the electromagnets and washed with ethanol for several times and were dried under vacuum at 50°C.

Characterizations Techniques. Morphology of the fabricated CNCs and Fe-CNCs were characterized using the Field Emission Electron Microscopy (FESEM). The dried powders of the samples were placed on a carbon tape followed by gold coating inside a sputtering unit. X-ray diffraction (XRD)(D8 Advance diffractometer, Bruker, Germany) studies were carried out to understand the type of crystal structure prevalent in the CNCs and Fe-CNCs. The chemical structure analysis of the fabricated nanocomposites were done through Fourier transform infrared spectroscopy (FTIR) analysis (Shimadzhu IR) in the spectral range of wavenumber 400-4000 cm^{-1}. Raman spectroscopy for the powdered samples were measured using Horiba Jobin Vyon (LabRam HR, Japan) equipped with a 1W, 1064nm Nd: YAG diode-pumped laser at an excitation wavelength of 514 nm. Vibrating sample magnetometer (VSM) (Lakeshore, Model: 7410 series) was used to measure the magnetic properties of the fabricated Fe-CNCs.

Results and Discussion

CNCs fabricated using sulphuric acid hydrolysis using the cellulose extracted from bamboo pulp shows rod-shaped morphology with a dimensions of length 550±85 nm and diameter of 25±12 nm, an aspect ratio of ~22 (Figure 1 (c)). Reduction of the iron salt/CNC suspension with the dropwise addition of the sodium borohydride solution led to the growth of the iron nanoparticles on the surface of the CNC rods. As per our previously reported studies, chemical structure analysis of the iron salt-CNC solution shows the presence of enhanced interaction between the hydroxyl groups of CNCs and the metal salt solution. Presence of such interaction is probably due to the presence of the electronegative oxygen moieties present in the hydroxyl groups of CNCs and with the positively charged iron salt. FESEM analysis of the Fe-CNCs shows the presence of the ribbon-like structure with its surface modified by the high density of the spherical iron nanoparticles. The grafted spherical nanoparticles were of dimensions ~30-45 nm and the morphology of the CNC backbone was significantly modified from cylindrical rods to ribbon like structure (Figure 1 (c)).

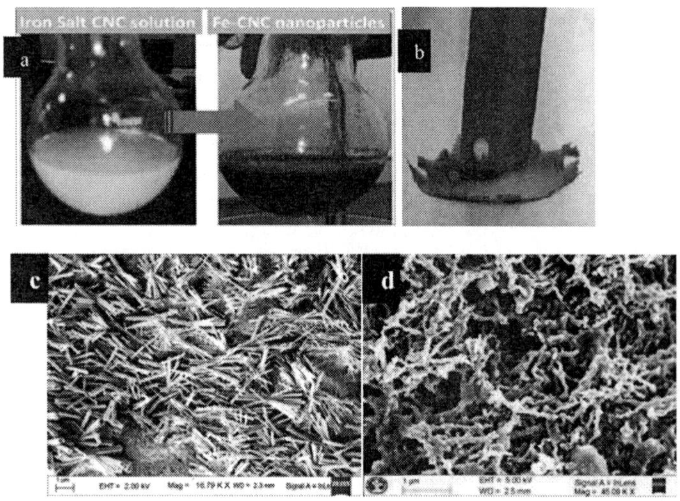

Figure 1. (a) Conversion of the iron-CNC salt solution into Fe^{o} nanoparticles using reducing agents, (b) free standing films of Fe-CNCs attracted towards magnets, FESEM micrographs of (b) CNCs from bamboo pulp and (c) Fe-CNC with grafted iron nanoparticles.

The crystal structure analysis of the prepared CNCs shows the presence of the cellulose I crystal structure with representative peaks at 14.4, 15.2 and 22.5° respectively. The % crystallinity index (C.I) of the CNCs calculated along the (200) plane was found to be~85%. XRD analysis of the Fe-CNCs clearly shows the presence of the zero iron peaks which corresponds to the sharp intense peak at ~44.9° (Figure 2). However, the diffractograms shows the presence of the other oxide peaks, with representative peaks at 31.8 and 55.2 ° which corresponds to formation of Fe_2O_3, the since drying of the Fe-CNCs were not carried out in inert conditions intentionally. The formation of Fe_2O_3 layer on the surface of the Fe^0 nanoparticles is probably due to unstable nature of Fe^0 is in normal atmosphere and hence converts to Fe_2O_3. However, some fraction of the iron still remains in zero oxidation state, which is visible from Fe^0 peak in XRD graph. The presence of the low intense crystalline peaks of CNCs suggest the probable conversion of the crystalline CNCs into amorphous segments during the heat treatment in presence of the iron salts and reducing agents. The crystallite size of the Fe^0 nanoparticles grafted on the CNCs are calculated using Scherrer equation are found to be ~32nm, which lie within the range of dimensions of Fe-CNCs calculated from FESEM studies. From our previously reported studies, the presence of the CNC backbone have shown to increase the oxidation stability of the zerovalent irons nanoparticles that are attached with the CNC surface[2]. These significantly helps in increasing the storage capacity thereby the shelf life of the catalyst, thereby, it could retain its catalytic activity for prolonged period.

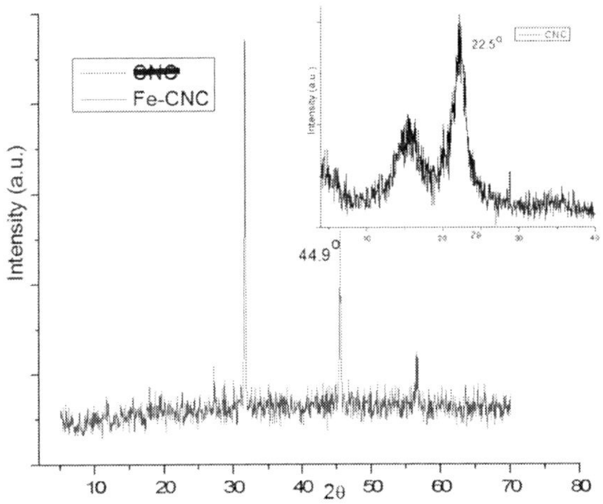

Figure 2. XRD graph of the CNC supported Fe_2O_3 nanoparticles, containing small amount of Fe^0 nanoparticles.

The chemical structure analysis of the Fe-CNCs are carried out using FTIR and Raman spectroscopic studies to confirm the presence of iron nanoparticles. Figure 3 shows the FTIR analysis of CNC supported Fe_2O_3 nanoparticles. The presence of the peak around ~619cm^{-1} corresponds to the Fe–O bonding, which confirms the presence of iron oxide nanoparticles. The –OH peak which usually occurs ~ 3270 cm^{-1} in CNC has shifted to around 3410 cm^{-1}, which is probably due to the grafting of the iron nanoparticles on the hydroxyl surface of the CNCs (Figure 4). Raman spectroscopic studies also confirmed the presence of the Fe^0 nanoparticles with the presence of the peaks at ~225 and 282 cm^{-1} respectively. The Fe^0 nanoparticles showed new representative peaks at 627 and 1389 cm^{-1}, whose intensity increased with time, representing the formation of oxide layer on ZVIs. Interestingly, the presence of peaks at 380 cm^{-1} suggests that the formation of the Fe^0 nanoparticles have only taken place on the surface of the CNCs (Figure 5). The fabricated Fe-CNCs are found to have high magnetic moment of ~44 emu/g with ferro-magnetic behavior of the iron oxide nanoparticles grafted on its surface. The mass fraction of the iron nanoparticles grafted onto the surface of the CNCs are found from the electron dispersive X-ray (EDX) analysis and thermogravimetric (TGA) thermograms.

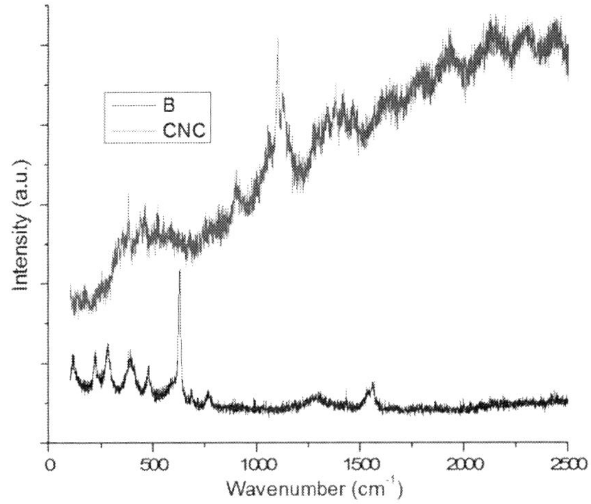

Figure 4. Comparison of the FTIR spectra of the CNCs and CNC supported Fe_2O_3 nanoparticles (Fe-CNCs)

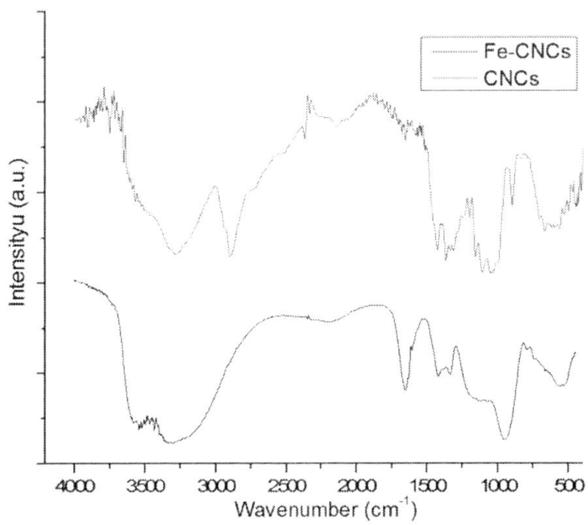

Figure 5. Comparison of the Raman spectra of the CNCs and CNC supported Fe_2O_3 nanoparticles (Fe-CNCs).

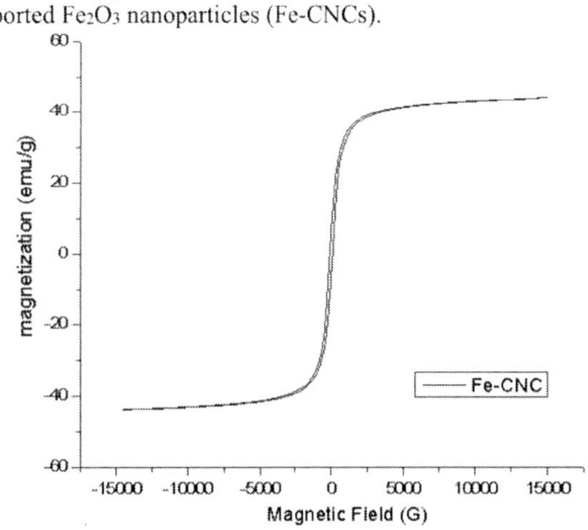

Figure 6. Magnetic properties of the fabricated Fe-CNCs.

The fabricated Fe-CNCs have the capability to form free standing stable magnetic films which leads to its potential application in the field of electronics, sensors, paper based actuators and also potent targeted drug delivery vehicles. In this we demonstrate two different applications of the Fe-CNCs: (i) the Fe^0 nanoparticles embedded on the surface of the CNCs could effectively degrade the Methylene Blue (MB) dye completely within few minutes and (b) as an efficient biocatalyst with high electrochemical activity with potential charge storage capability. For the former case, redox reactions occurring at the surface of the Fe^0 nanoparticles leads to generation of radicals which acts as an active agent for degradation of the dye. The degraded dye products were also analyzed with GCMS. Interestingly, it is found that the degraded MB dye products have the capability to get absorb on the CNC surface through ionic interactions thereby leading of complete removal of the dye as well as the degraded harmful products. In the latter case for the applications of Fe-CNCs as biocatalyst with high electrochemical activity, the Fe-CNCs were calcined under inert conditions such that an iron based graphitic carbon complex is formed. Such Fe-C system was found to be porous with very high surface area and also showed high charge storage capacity.

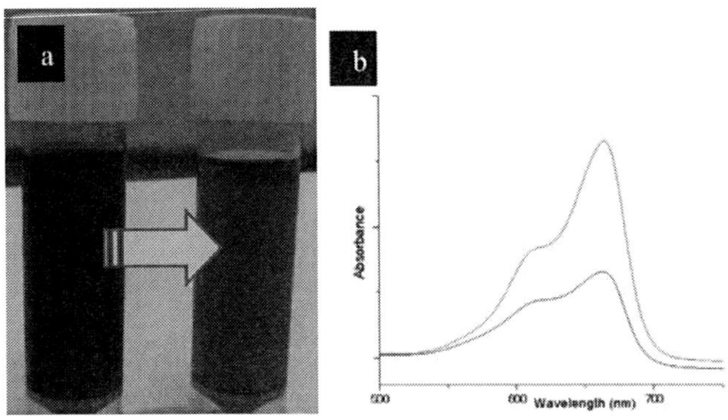

Figure 5. (a) Degradation of the Methylene Blue using the Fe-CNC catalyst and (b) UV analysis of the degraded MB dye after 60s of reaction with Fe-CNCs.

Conclusions

The iron nanoparticles are successfully grafted onto the hydroxyl functional groups of the CNC surface through the application of the reducing agents. The Fe^0 nanoparticles of spherical morphology were found to be uniformly distributed onto the surface of the CNCs with almost uniform size distribution. The chemical structure and the crystal structure of the Fe-CNC was confirmed from the XRD and FTIR studies. The fabricated Fe-CNC nanocomposites films were magneto responsive with a very high magnetic moment of ~42 emu/g. The ability of the Fe-CNC to form films finds it as potential application as a membrane for waste water remediation. It showed very high redox capacity for degradation of the dyes as well as with high electrochemical activity. Therefore, this study provide us a novel approach of fabricating iron nanoparticle embedded CNCs with high catalytic activity and magneto responsive behavior for potential high performance applications.

Acknowledgement: We sincerely acknowledge the Centre of Excellence for Sustainable polymers (CoE-SusPol) at IIT Guwahati, for providing the research facilities. We also thank to the Central Instruments Facility (CIF), IIT Guwahati for providing characterization facilities.

References

(1) Chen, Z.; Xu, W.; Jin, L.; Zha, J.; Tao, T.; Lin, Y.; Wang, Z. *J. Mater. Chem. A*, **2014**, 2, 18339.

(2) Jiang, X.; Yang, X.; Zhu, Y.; Yao, Y.; Zhao, P.; Li, C. *J. Mater. Chem. A*, **2015**, 3, 2361.

(3) Xu, D.; Xie, Y.; Song, Y. J.; Deng, W. Q. *J. Mater. Chem. A*, **2015**, 3, 16072.

(4) Wang, Y.; Zhao, H.; Zhao, G. *Electroanalysis*, **2016**, 28, 169.

THE INFLUENCES OF ALLOYING AND SURFACE MODIFICATION ON THE DIRECT SYNTHESIS OF H_2O_2 ON METAL CLUSTERS

David W. Flaherty

University of Illinois at Urbana-Champaign, Urbana, IL, 61801
dwflhrty@illinois.edu

Introduction

H_2O_2 is an environmentally-friendly and selective oxidant useful for epoxidations, bleaching, and disinfection, yet, its use is limited because H_2O_2 production by the anthraquinone oxidation process is economically viable only at very large scales. Direct synthesis of H_2O_2 ($H_2 + O_2 \rightarrow H_2O_2$) could enable H_2O_2 to be produced cost-effectively on-site, and even *in situ*, which would curtail the use of chlorine as a selective oxidant in many applications. H_2O_2 formation rates and selectivities increase when halides[1] or Au[2] are added to Pd clusters. Theoretical investigations have suggested that these additions change energy barriers for elementary steps by modifying the densities of states on surface Pd atoms.[3] Such proposals have not been tested, moreover, the mechanism and identity of the reactive intermediates are a topic of investigation.[4,5,6] Here, we combine rate measurements and in situ time-resolved infrared spectroscopy on Pd and Pt clusters and also Pd- and Pt-based bimetallic clusters to determine the mechanism of this reaction and to understand the reasons why changes in the size and composition of metal clusters increase H_2O_2 selectivities.

Experimental

A series of Pd and Pd-alloy clusters (1-10 nm mean diameter) were synthesized by combinations of strong electrostatic adsorption, electroless deposition, and colloidal methods. Clusters were subsequently dispersed on SiO_2 via incipient wetness impregnation. Turnover rates and H_2O_2 selectivities (at differential conversions) were measured at 268-323 K, 5-400 kPa H_2 and 10-400 kPa O_2, with 400-4,000 kPa N_2 as a diluent within aqueous methanol mixtures in a fixed bed, up flow reactor. H_2O_2 concentrations were determined by colorimetric titration using neocuproine-$CuSO_4$ solutions and UV-vis absorption measurements while a gas chromatograph with a thermal conductivity detector was used to measure concentrations of H_2 and O_2 in the effluent. The Madon-Boudart criterion was used to show that rate measurements were not affected by mass transfer. *In situ* time-resolved infrared spectra were collected using a modulation excitation approach by flowing H_2 and O_2 saturated solutions through an attenuated total reflectance infrared cell, collecting time-resolved spectra with rapid scan spectroscopy (Bruker Vertex 70), and extracting the spectral features of *only* the active surface intermediates using phase-sensitive detection methods.

Results and Discussion

The change in H_2O_2 and H_2O formation rates with H_2 and O_2 pressures (Fig. 1) together are not consistent with a Langmuir-Hinshelwood mechanism, rather, they suggest a reaction between a liquid-phase intermediate and pools of O_2-derived surface species. H_2O_2 formation is sensitive to the nature of the reaction solvent; most notably, H_2O_2 rates in protic solvents are 10^3 larger than those measured in aprotic liquids. Figure 2 shows that increases in the H_2O_2 concentration over time within a semi-batch reactor are much greater in protic solvents (e.g., methanol, water) than in aprotic solvents (e.g., acetonitrile, dimethyl sulfoxide, and propylene carbonate). These initial turnover rates calculated using this data (Fig. 2) show that the H_2O_2 turnover rates in protic solvents are at least 10^3 times greater than those in aprotic solvents (whose H_2O_2 turnover rates fall below the detection limit of the experiment, $< 1.7 \cdot 10^{-6}$ (mol H_2O_2)(mol Pd_s $s)^{-1}$). These turnover rates depend on the protic-aprotic nature of the solvent but do not depend strongly on the dielectric constants of the solvents.

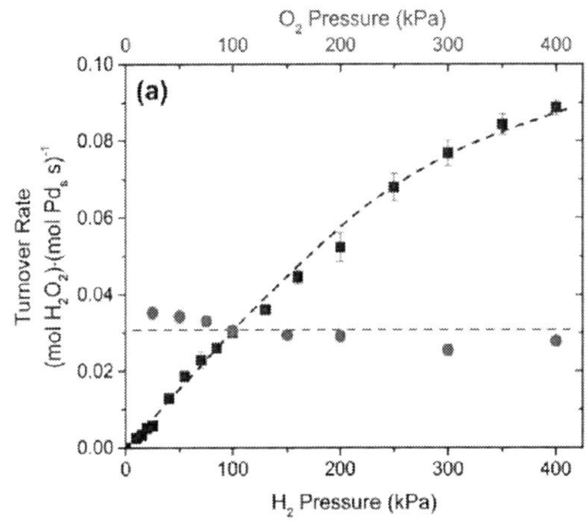

Figure 1. H_2O_2 and turnover rates as functions of H_2 pressure at 60 kPa O_2 (black, ■) and O_2 pressure at 60 kPa H_2 (red, ●) on 0.7 nm Pd clusters (277 K, 10 cm^3 min^{-1} CH_3OH). Lines are intended to guide the eye.

Collectively, these observations strongly suggest that proton-electron transfer processes, reminiscent of the two electron oxygen reduction reaction, form H_2O_2 on silica supported Pd and Pd-alloy clusters. Charge conservation requires that these metal clusters must also catalyze both heterolytic hydrogen oxidation ($H_2 \rightarrow 2H^+ + e^-$) and oxygen reduction ($O_2 + 2e^- + 2H^+ \rightarrow H_2O_2$).

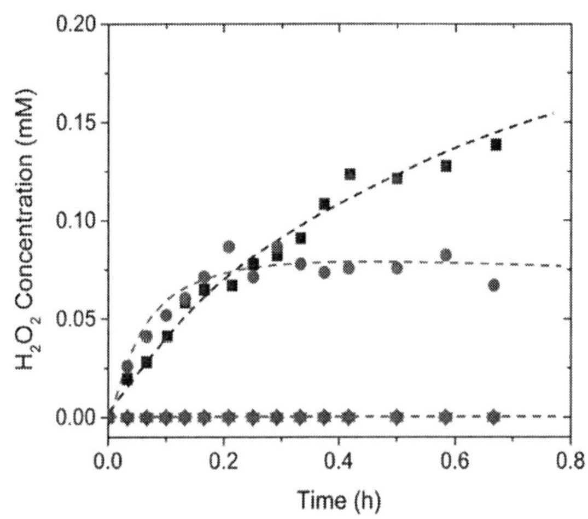

Figure 2. H_2O_2 concentrations as functions of time during direct synthesis in a well-mixed semi-batch reactor using protic (methanol (black ■) and water (red ●)) or aprotic (dimethyl sulfoxide (green ▲), acetonitrile (blue ▼), and propylene carbonate (magenta ♦)) solvents (4.2 kPa H_2, 4.2 kPa O_2, 40 cm^3 solvent, Pd-SiO_2, 295 K).

These reactions occur on surfaces saturated with O_2*- and OOH*-intermediates, confirmed by the appearance of absorption features between 1000-1200 cm^{-1} during *in situ* ATR-IR using modulation excitation methods. These spectra directly show that the active sites (i.e., those that turnover during the catalytic cycle) are saturated by superoxo (O_2^-) species, although much of the surface may in fact be covered by –OH* intermediates and be inactive. The O-O bonds within O_2* and OOH* chemisorbed intermediates cleave homolytically (with rates comparable to those for proton-electron transfer), which ultimately forms H_2O. Consequently, the H_2O_2 selectivities are determined by competition between heterolytic and homolytic reactions involving O_2 at liquid-solid interface.

These mechanisms persist also on monometallic and bimetallic clusters comprised of Pd, Au, Pt, Ag and other late transition metals. Incorporating Au, other metals, or halides into Pd clusters does not change the mechanism for H_2O_2 or H_2O formation. These modifications do, however, change barriers for the formation of H_2O (significantly) with a minimal effect on barriers for steps that lead to H_2O_2. We anticipate that ATR-IR spectra will show that the extent of charge transfer from the metal to O_2* species is strongly affected by changes in the cluster composition. The concurrent changes in catalytic selectivities and activation barriers indicate that H_2O_2 formation does not reflect these changes, whereas, O-O bond rupture is highly sensitive to the electronic structure of the cluster. Overall, our work presents a clear mechanism for H_2O_2 formation on metal clusters and explains the roles of cluster size and composition and proton donors in determining H_2O_2 selectivities.

Conclusions

This study presents a clear mechanistic view of how the direct synthesis of H_2O_2 proceeds at steady-state and under relevant conditions, and, provides guidance as how the selectivity to H_2O_2 may be improved by individually controlling the rates of competing heterolytic and homolytic reaction pathways.

Acknowledgements. We gratefully acknowledge support from the University of Illinois and a CAREER Award from the National Science Foundation.

References

(1) Lunsford, J. H. et al. *Appl. Catal., A.* (2006) *314*, 94; *J. Catal.* (2003) *216*, 455.
(2) Hutchings, G. J. et al. *Science* (2009) *323*, 1037; *ACS Catal.* (2013) *3*, 487.
(3) Hwang et al. *J. Phys. Chem. Lett.* (2009) *113*, 12943; *Catal. Today* (2011) *165*, 138.
(4) Nijhuis, T. A. et al. *Appl. Catal. , A.* (2015) *505*, 249
(5) Voloshin, Y. et al. *Chem. Eng. Sci.* (2010) *65*, 1028
(6) Flaherty, D. W. et al. *J. Am. Chem. Soc.* (2016) *138*, 574

ENCAPSULATED INTERMETALLIC NANOPARTICLE FOR SELECTIVE FURFURAL HYDROGENATION

Wenyu Huang[*, †, ‡] *Raghu V. Maligal-Ganesh,*[†] *Linlin Wang,*[‡] *Duane D. Johnson,*[‡,§] *Shiran Zhang,*[#] *and Franklin (Feng) Tao,*[#] *and Chaoxian Xiao,*[†]

[†]Department of Chemistry, Iowa State University, Ames, Iowa 50011, United States

[‡]Ames Laboratory, U.S. Department of Energy, Ames, Iowa 50011, United States

[§]Departments of Materials Science & Engineering and Chemical & Biological Engineering, Iowa State University, Ames, IA 50011, United States

[#]Department of Chemical & Petroleum Engineering, Department of Chemistry, University of Kansas, Lawrence, KS 66045, United States

Abstract

Intermetallic compounds have attracted significant scientific interests as efficient catalysts with improved catalytic properties compared to their monometallic counterparts.[1-4] The modified surface geometric and electronic structures of intermetallic compounds make them unique catalytic materials because they will alter the configuration and binding energies of surface adsorbates and thus change catalytic properties.[5,6,7] However, the synthesis of these intermetallic compounds usually involves high temperatures and an inability to exert control over size or morphology, resulting in lower surface areas for catalysis. To address the aggregation of intermetallic nanoparticles during high temperature treatment, we used a ship-in-a-bottle approach to synthesize monodisperse nanoscale intermetallic compounds confined in a thermally stable inorganic shell. Using the Pt nanoparticles encapsulated in a mesoporous silica shell ($mSiO_2$) as the seeds and the metal precursor of the additional metal, we synthesized intermetallic compound nanoparticles confined within the $mSiO_2$ shell. This approach addresses the biggest synthesis dilemma of intermetallic nanoparticles. Since the formation of intermetallic phase usually requires high temperature, the traditional methods using organic capping agents could not avoid the aggregation of nanoparticles during the high temperature annealing. With the protection of the inorganic $mSiO_2$ shell, we could routinely anneal the bimetallic core at high temperature, which will lead to the formation of intermetallic nanoparticle while prevent their aggregation at the same time. Intermetallic NPs synthesized by this method were well confined within the mesoporous silica environment. Moreover, the mesoporous silica encapsulation did not pose any hindrance during catalytic reactions. Among the intermetallic NPs synthesized by this approach, $PtSn@mSiO_2$ exhibits the best performance in furfural hydrogenation, requiring only one-tenth of the quantity of Pt used in $Pt@mSiO_2$ for similar activity and near 100% selectivity to furfuryl alcohol. We measured the Pt dispersion using chemisorption on the $mSiO_2$-encapsulated PtSn and Pt NPs, and found the specific activity per surface Pt site on $PtSn@mSiO_2$ is 40 times of that on $Pt@mSiO_2$.

References

(1) Studt, F.; Abild-Pedersen, F.; Bligaard, T.; Sørensen, R. Z.; Christensen, C. H.; Nørskov, J. K. *Science* **2008**, *320*, 1320-1322.

(2) Xiao, C.; Wang, L.-L.; Maligal-Ganesh, R. V.; Smetana, V.; Walen, H.; Thiel, P. A.; Miller, G. J.; Johnson, D. D.; Huang, W. *J. Am. Chem. Soc.* **2013**, *135*, 9592-9595.

(3) Studt, F.; Sharafutdinov, I.; Abild-Pedersen, F.; Elkjaer, C. F.; Hummelshoj, J. S.; Dahl, S.; Chorkendorff, I.; Norskov, J. K. *Nat. Chem.* **2014**, *6*, 320-324.

(4) Armbrüster, M.; Kovnir, K.; Friedrich, M.; Teschner, D.; Wowsnick, G.; Hahne, M.; Gille, P.; Szentmiklósi, L.; Feuerbacher, M.; Heggen, M.; Girgsdies, F.; Rosenthal, D.; Schlögl, R.; Grin, Y. *Nat. Mater.* **2012**, *11*, 690-693.

(5) Armbrüster, M.; Schlögl, R.; Grin, Y. *Sci. Tech. Adv. Mater.* **2014**, *15*, 1-17.

(6) Armbrüster, M.; Kovnir, K.; Behrens, M.; Teschner, D.; Grin, Y.; Schlögl, R. *J. Am. Chem. Soc.* **2010**, *132*, 14745-14747.

(7) Armbrüster, M.; Wowsnick, G.; Friedrich, M.; Heggen, M.; Cardoso-Gil, R. *J. Am. Chem. Soc.* **2011**, *133*, 9112-9118.

UNDERSTANDING THE IMPORTANT ROLE OF REACTANT-ADSORPTION ON THE SURFACE OF SUPPORTED AU CATALYSTS FOR SELECTIVE OXIDATION OF BENZYL ALCOHOL

*Shihui Zou, Yuheng Zhou, Wuzhong Yi, Liping Xiao, Jie Fan**

Department of Chemistry, Zhejiang University, Hangzhou, China, 310027
jfan@zju.edu.cn

Introduction

Selective oxidation of alcohols is one of the most challenging reactions in the emerging field of green chemistry.[1] Generally, the alcohol oxidations were often performed using toxic chromate or permanganate as oxidant, which technically can hardly satisfy the increasing requirements of environmental protection and economic efficiency. Aerobic oxidation of alcohols has attracted considerable attention due to its green nature by using air or oxygen as oxidant and high atom-efficiency by utilizing recyclable catalysts.[2,3] The development of stable, selective, and efficient heterogeneous catalysts, which can aerobically oxidize a variety of alcohols to their corresponding aldehydes and ketones, is of both economic and environmental significance. Among catalysts been explored for the selective oxidation of alcohols, supported gold catalyst often exhibiting high selectivity and reactivity, seems to be a promising one. In practice, Au could catalyze the selective oxidation of alcohols in both liquid-phase and gas-phase, depending mainly on the thermal stability and volatility of the reagents and products. The convenience of catalyst separation and solvent-free conditions make a gas-phase process more attractive for industrial applications. Although great efforts have been devoted to design advanced Au-based catalysts for gas-phase selective oxidation of alcohols, a comprehensive understanding on the reaction mechanism still remains challenge.

Herein, we systematically investigated the important role of reactant-adsorption on the surface of supported Au catalysts for gas-phase selective oxidation of benzyl alcohol. By tuning the surface environments of Au nanoparticles, we successfully achieved a reversible multiply-twinned nanoparticles (MTPs) to single crystals (SCs) structural transformation. Interestingly, Au nanoparticles (AuNPs) of twin crystal structure exhibited four-times higher activity than that of single crystal ones. After extensive characterizations, we confirmed that the significant difference results from the easy adsorption of alcohol molecules on corners and edges of AuNPs. On the other hand, we developed a facile seed-mediated solid-state alloying strategy to introduce nickel gold, which further improve the catalytic performance by enhancing the adsorption and activation of molecular oxygen.

Experimental

Au/SiO_2. AuNPs and SiO_2 support were synthesized according to literatures.[4] For the synthesis of Au/SiO_2, 5.0 mg of AuNPs were dissolved in 25 mL chloroform or toluene, followed by the addition of 1.0 g of silica microspheres. After 30 min stirring, the solid product was centrifuged and dried in air. The powder was then calcined at 350 °C for 5 h to remove the surface capping agents before the use in catalytic reactions.

Post-treatment of Au-SiO_2. 50 mg Au/SiO_2 samples were washed by using 2 mL water (methanol, ethanol, chloroform and toluene) for 3 times and centrifuged at 10000 rmp for 5 min, followed by drying and annealing at 350 °C for 5 h, denoted as Au/SiO_2-W(or M, E, C, T).

AuNi/FDU-12. AuNPs and EP-FDU-12 support were synthesized according to literatures.[4,5] The synthesis of AuNi/FDU-12 includes two steps. First, certain amount of nickel nitrate was loaded onto FDU-12 by ethanol solvent evaporation method, dried in 40 °C vacuum oven overnight. Second, the as obtained powder was added to AuNPs-chloroform solution and stirred for 30 min. After centrifuging and drying, AuNi/FDU-12 were obtained by annealing at different temperature under H_2/Ar (5% *v/v*) for 4 h, denoted as Au_xNi_y-Z (x,y: the initial molar ratio of Au and Ni; Z: the calcination temperature). The loading weight of Au was 1 wt% for all samples.

Catalytic test. The oxidation of benzyl alcohol was carried out in a fixed bed vertical glass reactor (h = 250 mm, d = 12 mm), fitted with a glass frit carrying the catalyst (20 mg for 0.5 wt% Au-SiO_2 microsphere, 5 mg for AuNi/FDU-12, mixed with certain amount of quartz sand) and provided with an electronically controlled furnace. The oxygen stream (30 mL min^{-1}) was controlled by a mass flow instrument and the liquid reagent (1.8 mL h^{-1}) was supplied through a syringe pump.

Results and Discussion

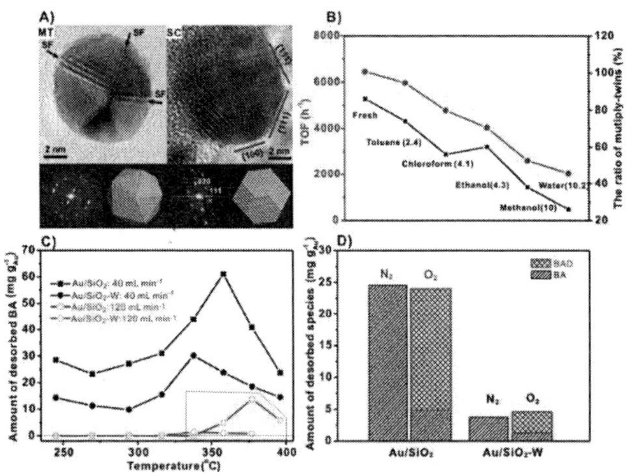

Figure 1. A) HRTEM of multiply-twinned(left) gold nanoparticles and gold single crystals (right), B) TOFs and the ratio of mutiply-twins for Au/SiO_2 treated by different solvents. C) TPD of BA on Au/SiO_2 and Au/SiO_2-W at different carrier gas flow rate in N_2. D) TPD of BA on Au/SiO_2 and Au/SiO_2-W under different atmosphere at a gas flow rate of 120 mL min^{-1}.

Reversible Multiply-Twinned Nanoparticle–to-Single Crystal Transformation. Figure 1A show the typical microstructures of Au/SiO_2 and Au/SiO_2-W. The Au NPs in Au/SiO_2 are clearly identified as five-fold MTPs from the FFT pattern and the diffraction contrast arising from the twin boundaries (TBs). Meanwhile, HRTEM image and FFT diffractogram on Au/SiO_2-W confirmed the single-crystal nature of Au NPs, which had a truncated octahedra morphology and expose the {111} and {100} facets. By carefully counting the frequency of MTPs and SCs in Au/SiO_2 and Au/SiO_2-W, we found that the MTP concentration is dramatically decreased from ca. 90% to ca. 25%, indicating that the bonding of water to the fresh supported AuNPs and followed heating process leads to a dramatic structural change from MTPs to SCs. The water-driven structural evolution of gold icosahedra was further confirmed by in-situ HRTEM at 350 °C.

It is interesting to note that post-treatment with different solvent results in different concentration of MTPs. The higher polarity, the less MTPs.

Different microstructures lead to different catalytic performance. As shown in Figure 1B, samples with different MTPs concentrations exhibited different turnover frequencies (TOFs), with the fresh sample achieving the best performance and Au/SiO_2-W the worst. More interestingly, a good correlation between the concentration of MTPs and TOFs was obtained. These results lead to the conclusion that MTPs have better catalytic performance in selective oxidation of BA than SCs. Kinetic studies, however, showed that apparent activation energies on those samples were almost the same, which support the hypothesis that the difference in catalytic performance is likely due to the different amount of active sites. Notably, Au/SiO_2 and Au/SiO_2-W share the same apparent reaction orders of O_2. Whereas the order of BA on Au/SiO_2-W is more than twice of that on Au/SiO_2. According to the classic catalytic theory, the lower reaction order indicates the stronger adsorption on the surface of the catalyst. We thus hypothesize that the different catalytic performances of Au/SiO_2 and Au/SiO_2-W result from the different ability to adsorb BA molecules . To confirm this idea, we conducted the temperature programmed desorption of BA at various atmosphere. As can be seen from Figure 1C, the amount of desorbed BA molecules on Au/SiO_2 is always more than that on Au/SiO_2-W. The significant difference is observed especially when the gas flow rate is 120 mL min^{-1}. We ascribed these BA molecules to chemisorbed BA and verified this idea by performing a TPD experiment in O_2. As shown in Figure 1D, when N_2 is replaced by O_2, the adsorbed BA molecules converted to BAD. It is obvious that Au/SiO_2 are more favorable for BA-chemisorptions than Au/SiO_2-W, which perfectly confirmed our hypothesis. Besides, the important role of BA-adsorption was also verified by solid-NMR and DFT calculations. It is also important to note that treating Au/SiO_2-W by thiol could partially convert the SCs to MTPs and thus recover the catalytic performance.

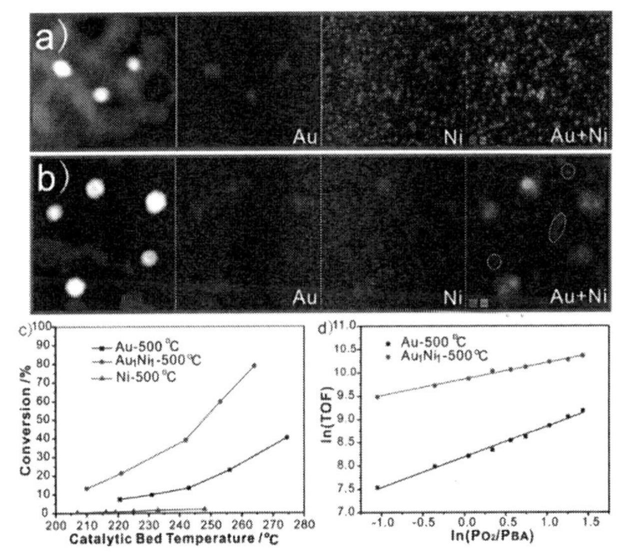

Figure 2. a) HAADF-STEM EDS Mappings of a) Au_1Ni_1/FDU-RT sample and b) Au_1Ni_1/FDU-500 °C sample; c) catalytic performance of different samples and d) oxygen pressure dependence at 240 °C over different samples.

AuNi bimetallic catalysts prepared by seed-mediated solid-state alloying strategy. From the above discussion, we observed that BA-adsorption could greatly influence the catalytic performance. And it is generally accepted that Au is not a good catalyst for O_2 activation. To further improve the selective oxidation of BA on Au-based catalyst, it is thus necessary to introduce an effective component (such as nickel) to activating oxygen, which lead us to develop efficient strategies for rational design of AuNi bimetallic catalysts.

As shown in Figure 2a, pre-loaded nickel nitrate and AuNPs (5.1 ± 0.2 nm) with initial molar ratio of 1:1 were uniformly dispersed throughout EP-FDU-12 before calcination (denote as Au_1Ni_1-RT). The distribution of Au and Ni evolved after calcining in H_2/Ar at 500 °C (Au_1Ni_1-550 °C). It was surprising to find that most nickel signal in EDS mapping overlapped with Au signal (Fig. 1b). And only a small amount of Ni aggregated into monometallic Ni clusters through self-nucleation and growth (yellow circle in Fig. 2b). These results led us to the hypothesis that during the high-temperature reduction process, AuNPs played a role of seeds for most of Ni reduction-growth and alloyed with it. The alloy structure was confirmed by XRD patterns where Au peaks shifted from 38.2° to 38.5° after high-temperature calcinations. The alloyed Au_1Ni_1/FDU-12 exhibited a much better catalytic performance than Au/SiO_2, just as expected (Figure 2c). Kinetic studies verified that the introduction of Ni enhanced the adsorption and activation of oxygen (Figure 2d). Besides, the composition of AuNi bimetallic catalysts could be conveniently tuned by changing the precursors concentration. The TOF of the optimized Au_3Ni_1/FDU-12 catalyst was identified to be 53000 h^{-1}. To the best of our knowledge, this TOF is the highest value ever reported for both liquid-phase and gas-phase selective oxidation of BA.

Conclusions

We systematically investigated the important role of reactant-adsorption on the surface of supported Au-based catalysts for gas-phase selective oxidation of benzyl alcohol. The multiply-twinned gold nanoparticles facilitate the adsorption of BA and thus achieve better catalytic performance for selective oxidation of BA than gold single crystals. On the other hand, introducing nickel into supported Au catalyst could further improve the catalytic performance by promoting O_2 adsorption and activation. The present study identifies the L-H model of BA selective oxidation, which benefit the rational design of advanced catalysts for this reaction.

Acknowledgement This work was supported by Shell Global Solutions International B.V. (PT37712), National Natural Science Foundation of China (21271153, 21373181, 21222307,U1402233), Major Research Plan Of National Natural Science Foundation of China (91545113) , Fok Ying Tung Education Foundation (131015) and the Fundamental Research Funds for the Central Universities (2014XZZX003-02).

References

(1) Poliakoff, M.; Licence, P. *Nature*, **2007**, 450, 810-812
(2) Enache, D. I.; Edwards, J. K.; Landon, P.; Solsona-Espriu, B.; A. Carley, F.; Herzing, A. A.; Watanabe, M.; Kiely, C. J.; Knight, D. W.; Hutchings, G. J. *Science*, **2006**, 311, 362-365.
(3) Liu, Y.; Tsunoyama, H.; Akita, T.; Xie, S.; Tsukuda, T. *ACS Catal.*, **2011**, 1, 2-6.
(4) Zheng, N.; Fan, J.; Stucky, G.D. *J. Am. Chem. Soc.* **2006**, 128, 6550.
(5) Ma, G.; Yan, X.; Li, Y.; Xiao, L.; Huang, Z.; Lu, Y.; Fan, J. *J. Am. Chem. Soc.* **2010**, 132, 9596.

THE ROLES OF RE AND CS AS PROMOTERS FOR AG/A-AL$_2$O$_3$ IN HIGH SELECTIVITY ETHYLENE EPOXIDATION CATALYSTS

John R. Monnier, Weijian Diao, Christopher D. DiGiulio[1], Melanie T. Schaal, and Shuguo Ma

Department of Chemical Engineering, University of South Carolina, Columbia, SC 29208 USA

[1]Present address: UOP Honeywell Co., 25 E. Algonquin Rd, Des Plaines, IL 60016.

Abstract

The global production of ethylene oxide (EO) currently exceeds 60B lbs/yr and generates 50% of all economic value for catalytic, hydrocarbon oxidation reactions. Since 1937, when Union Carbide opened the first EO plant, the selective, partial oxidation of ethylene has been carried out using molecular oxygen and supported Ag catalysts. Until the mid-1980's the preferred catalyst was Cs-promoted Ag with 75 - 80% selectivity. At that time a new generation of catalysts with selectivities as high as 85-86% that also contained high-valent oxyanions of ReO_4 and other high-valent co-promoters, such as MoO_4, WO_4, and even SO_4, were developed [1]. With continued refinement, selectivities as high as 90-92% are currently claimed. While Cs-only, promoted catalysts have been continuously studied for the past 70 years, the newer generation Re-containing catalysts have been relatively ignored in the open literature. Until the recent publication from our group [2], only three other manuscripts had been published, even though essentially all commercial catalysts employ Re as a key component; this is very surprising given the commercial importance of this reaction. One theme that has emerged from the study of Cs-promoted, Ag catalysts is the importance of the electronic properties of O adsorbed on the Ag surface during reaction conditions. Relatively electron deficient, or electrophilic, O is required for selective addition to the e-rich C=C moiety to form EO. Alternatively, electron-rich Ag-O species preferentially abstract H from adsorbed ethylene in a non-selective, reaction pathway to form CO_2 and H_2O. Of course, the acid-catalyzed isomerization of gas phase EO to form acetaldehyde, which undergoes facile combustion to CO_2 and H_2O represents the well-known consecutive pathway for non-selectivity [3].

In this presentation we describe characterization and performance of several series of Cs-Ag, Re-Ag, Cs-Re-Ag and Cs-Re-Mo-Ag catalysts, all supported on low surface area α-Al_2O_3, for ethylene epoxidation to determine the mechanism of selectivity enhancement for the current generation of high selectivity ethylene oxide (EO) catalysts. Optimal amounts of Re (as NH_4ReO_4) and Mo [as $(NH_4)_2MoO_4$] on Cs-promoted, Ag/α-Ai_2O_3 increase EO selectivity from approximately 78% to 83.5%; addition of $(NH_4)_2SO_4$ boosts selectivity to 85.4%. Figure 1 shows clearly the selectivity enhancements for a similar series of catalysts prepared by a method slightly different from the methodology used in this presentation. Analyses by XPS and SEM suggest that both Re and Cs promotions are electronic in nature. SEM shows that neither Re or Mo changes the morphology or particle sizes of the Ag particles. XPS analyses, however, show that Re shifts Ag 3d BE to higher values which enhances electrophilic attack by oxygen adsorbed on Ag (Ag-O) at the electron-rich C=C bond of C_2H_4. In the subsequent step, Cs lowers the Ag 3d BE to facilitate desorption of the EO precursor to form gas phase EO. Thus, Re and Cs promote different steps of the mechanism for EO formation. This is supported by the observation that Ag catalysts promoted only by Re (no Cs) are much less active and selective than even un-promoted Ag catalysts. Thus, Re is necessary, but not sufficient for high selectivity. The Cs effect for Re-modified EO catalysts appears to be similar to that observed for the selective epoxidation of butadiene to form 3,4-epoxybutene (EpB), whereby the highly polarizable cations of Cs^+, Rb^+, or Tl^+ lowered the desorption energy of an adsorbed EpB precursor to give both higher activity and selectivity to EpB [4,5].

Finally, a modified reaction scheme detailing the mechanism of EO formation over Re-modified, Cs-promoted Ag catalysts is also presented within the framework of the traditional, parallel-consecutive mechanism for EO formation.

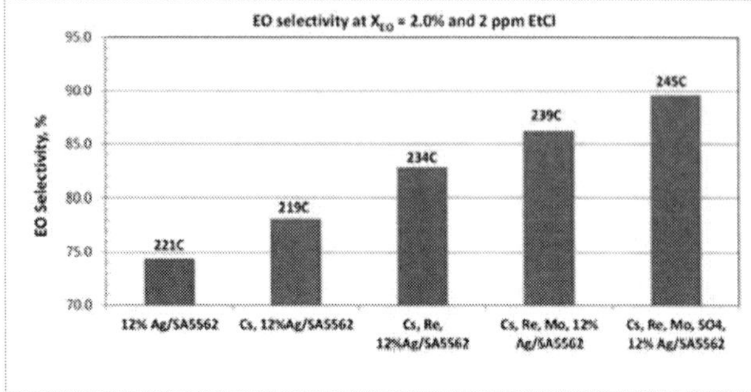

Figure 1. Cumulative effect of high-valent oxyanions. Feed: 8% O_2, 25% C_2H_4, 1-3 ppm EtCl, balance CH_4. GHSV = 4500hr^{-1} and P = 250 psig. Reaction temperatures needed to attain EO concentrations of 2.0% are shown above each datum bar.

References

(1) A.M. Lauritzen, US Patent 4,766,105, "Catalysts and process for the production of ethylene oxide," (1984), to Shell Oil Company.
(2) W. Diao, C.D. DiGiulio, M.T. Schaal, S. Ma, and J.R. Monnier, J. Catal., 322 (2015) 14.
(3) C.H. Bartholomew, R.J. Farrauto, Fundamentals of Industrial Catalytic Processes, John Wiley and Sons, New York, 2006, p. 597.
(4) J.R. Monnier, Stud. Surf. Sci. Catal. 110 (1997) 135.
(5) J.R. Monnier, J.W. Medlin, M.A. Barteau, J. Catal. 203 (2001) 362.

NB-INCORPORATED MESOPOROUS SILICATES FOR SELECTIVE ETHYLENE EPOXIDATION

Swarup K. Maiti,[a] *Wenjuan Yan,*[a] *Anand Ramanathan,*[a] *Pansy D.Patel,*[b] *Ward Thompson,*[b] *Bala Subramaniam*[a]

[a]Center for Environmentally Beneficial Catalysis, University of Kansas, Lawrence, KS 66047, United States
[b]Department of Chemistry, University of Kansas, Lawrence, KS 66045, United States

Abstract

We recently reported that tungsten and niobium incorporated KIT-6 are selective catalysts for liquid phase epoxidation of ethylene with no CO_2 formation.[1] However, complete leaching of W and 30 - 75% leaching of Nb was observed during 5 h batch runs. Further, the oxidant, H_2O_2 decomposed due to Brønsted acidity significantly reducing its utilization towards epoxide formation. In contrast, Nb-TUD-1 exhibits higher productivity towards ethylene oxide (EO) formation.[2] The measured EO productivity over Nb-TUD-1 materials (342–2539 g EO h^{-1} kg^{-1} Nb) spans a greater range than those observed with Nb-KIT-6 (234–794 g EO h^{-1} kg^{-1} Nb), Nb-KIT-5 (273–867 g EO h^{-1} kg^{-1} Nb) and Nb-MCM-48 (71–219 g EO h^{-1} kg^{-1} Nb) materials at similar operating conditions.[2]

Computational studies implicated the presence of Brønsted acid sites as a causative factor for both H_2O_2 decomposition and metal leaching.[2] Indeed, highly isolated tetrahedral Nb sites (achieved at relatively low Nb loadings) showed a total acidity of 0.13 mmol NH_3/g with much reduced Brønsted acidity. With such a catalyst, a dramatic increase in EO productivity of 4304 g EO h^{-1} kg^{-1} Nb was observed. Nevertheless, slow leaching of Nb was still observed. Various passivation techniques that almost eliminate the Brønsted acidity while minimizing H_2O_2 decomposition and metal leaching will be presented.

References

(1) Yan, W.; Ramanathan, A.; Ghanta, M.; Subramaniam, B., *Catal. Sci. & Technol.* **2014,** 4, (12), 4433-4439.
(2) Yan, W.; Ramanathan, A.; Patel, P. D.; Maiti, S. K.; Laird, B. B.; Thompson, W. H.; Subramaniam, B., *J. Catal.* **2016,** 336, 75-84.

COMPUTATIONAL STUDIES OF ETHYLENE EPOXIDATION ON A MESOPOROUS METAL-SUBSTITUTED SILICA HETEROGENOUS CATALYST

Pansy D. Patel, Krista G. Steenbergen, Brian B. Laird, and Ward H. Thompson

Department of Chemistry, University of Kansas, Lawrence, KS 66045 and
Center for Environmentally Beneficial Catalysis, University of Kansas, Lawrence, KS 66047
wthompson@ku.edu

Abstract

Selective oxidation of ethylene using metal-substituted silica mesopores has recently been shown to have promise as a cheaper and environmentally friendly catalyst to produce ethylene oxide. In particular, this system, with hydrogen peroxide as the oxidant, has a dramatically reduced carbon footprint compared to the conventional industrial scale process. To gain insight into the reaction mechanisms and aid in rational design of the catalyst, this system has been examined by computational simulation from two points of view. First, density functional theory calculations have been used to examine the reaction pathways of ethylene epoxidation with hydrogen peroxide, including undesired side reactions observed in experimental measurements. Second, classical Monte Carlo simulations are used to understand the phase behavior of the gas-expanded solvent used in the reaction. The results of these studies are discussed in the context of design principles for this catalytic system.

SELECTIVE HYDRODEOXYGENATION OF ANISOLE ON A COBALT OXIDE-BASED BI-FUNCTIONAL CATALYST

Shiran Zhang, Yu Tang, Franklin (Feng) Tao

Department of Chemical and Petroleum Engineering and Department of Chemistry, University of Kansas, Lawrence, KS, 66045, USA
franklin.feng.tao@ku.edu

Abstract

Selective hydrodeoxygenation (HDO) of anisole was chosen as a model reaction for the developing catalyst active for upgrading bio-oil molecules to biofuels. A bi-functional catalyst consisting of doped Co_3O_4 nanoparticles with anchored Pd atoms was prepared. It exhibits high conversion of anisole with a selectivity > 80% for production of benzene at 150°C. The ready cleavage of C-O bond of anisole results from binding of oxygen atom of anisole to a surface oxygen vacancy. The hydrogen atoms formed through dissociation of molecular H_2 on Pd atoms couples with the intermediate, phenyl formed through C-O cleavage on a surface oxygen vacancy. The high selectivity and activity for production of benzene on this catalyst at a temperature as low as 150°C result from a synergistic effect of the two catalytic sites.

References

(1) Alonso, D. M.; Wettstein, S. G.; Dumesic, J. A., *Chem. Soc. Rev.* **2012**, *41 (24)*, 8075.
(2) Lee, W.-S.; Wang, Z.; Wu, R. J.; Bhan, A., *J. Catal.* **2014**, *319*, 44-53.

SYNGAS CONVERSION TO HIGHER ALCOHOLS: APPLICATION OF NOVEL KCORHMO CATALYSTS SUPPORTED OVER CARBON NANOHORNS (CNHs) AND ITS BY-PRODUCTS (OCP & OCP$_f$)

Philip E. Boahene [a], Ajay K. Dalai [a], Ramaswami Sammynaiken [b]*

[a] Catalysis and Chemical Reaction Engineering Laboratories, University of Saskatchewan, Saskatoon, SK, S7N 5A9, Canada
[b] Saskatchewan Structural Sciences Centre, University of Saskatchewan, Saskatoon, SK, S7N 5A9, Canada
**Corresponding author: ajay.dalai@usask.ca (Ajay K. Dalai)*

Introduction

Gasoline blends with up to 10% ethanol (E10) and higher alcohols are commercially available at gasoline pump stations in the USA, Canada, and some parts of Europe [1]. The production of these alcohols (C_2 and C_{2+}) promises immense potential to replace other additives utilized as octane boosters in automotive fuels. Unlike gasoline and diesel fuels, higher alcohols contain oxygen components that allow gasoline-blended fuels to combust more completely; thereby, increasing the combustion efficiency and reducing air pollution, and can also be renewable resource depending on the origin of syngas feedstock (via biomass gasification). To improve selectivity and activity of these alcohol products, the catalyst metals and supports used for this reaction play a vital role [2].

In the catalysis of hydrogenation reactions on heterogeneous surfaces, whereby metal precursors dispersed on a porous support act as catalyst, the support plays a crucial role due to the probable hydrogen spillover effect [3]. That notwithstanding, in the CO hydrogenation reaction catalysis, interaction of the support and active metal components can significantly alter selectivity to ethanol and higher alcohols. For instance, Surisetty et. al [4] investigated the influence of different textural properties of activated carbon-supported KCoRhMoS$_2$ catalysts for higher alcohols synthesis and compared selectivity to MWCNT-supported catalyst. A space time yield and selectivity of total alcohols of 0.296 g/($g_{cat.}$/h) and 35.6%, respectively, was reported for the KCoRhMo/MWCNT catalyst at similar metal loadings. They inferred that support textural properties (pore size and volume) directly influenced the synthesis of mixed alcohols from synthesis gas. Another porous carbon material that has of recent times attracted significant importance in the field of heterogeneous catalysis is carbon nanohorns (CNH). Though, to the best of our knowledge, no work has yet been reported on CO hydrogenation reaction over these supports for higher alcohols synthesis, they proved successful in the hydrogenation of petroleum feedstock in hydrotreating reactions [5]. Thus, the motivation of the present work was to explore the potential catalytic applications of CNH and its by-products (OCP$_f$ and OCP) supports for higher alcohols synthesis via syngas conversion in a laboratory fixed bed reactor under industrial conditions and evaluate its performance with the known MWCNT counterpart at similar metals loading.

Experimental

The pristine supports (CNH, OCP$_f$ and OCP) were synthesized by the submerged arc discharge in liquid nitrogen technique described elsewhere [6], at a fixed current of 90A and with other modifications. Surface modification of as-made CNH and its by-products (OCP$_f$ and OCP) were achieved using 7 M HNO_3 at 110°C under reflux condition for 30 min, filtered, thoroughly rinsed with de-ionized water until a neutral pH, and then dried overnight prior to particles separation and supports alkalinisation with the required amount of aqueous solution of K_2CO_3. This was followed by supports stabilization at 300°C in argon flow of 50 mL/min at a heating rate of 10°C/min for 4 h, and consequently, the co-impregnation of respective metal (Co, Mo, Rh) species using ammonium heptamolybdate tetra hydrate (Alfa-Aesar), potassium carbonate (Sigma-Aldrich), cobalt acetate tetra hydrate (Alfa-Aesar), and rhodium chloride hydrate (Sigma-Aldrich) as precursors for Mo, K, Co, and Rh, respectively. The oxidic form of the dried powder K-doped CoRhMo samples were obtained after 4 h of oven drying at 120°C, followed by stabilization at 450°C under argon flow of 50 mL/min at a heating rate of 10°C/min for 12 h. The final catalyst composition (wt.%) of 9% K, 15% Mo, 4.5% Co and 1.5% Rh, respectively, is to be attained. Henceforth, the fine powder form of the CNH, OCP$_f$ and OCP-supported KCoRhMo catalyst samples are designated as Cat-CNH, Cat-OCP$_f$ and Cat-OCP, respectively. The feedstock used in this study was the commercially available syngas mixture procured from Praxair Canada.

Results and discussion

Table 1 is a summary of the textural properties of the pristine and HNO3 treated supports (CNH, OCP$_f$, and OCP) as well as their corresponding KCoRhMo-supported catalysts. 30wt.% HNO_3 pre-treatment of the pristine CNH, OCP$_f$, and OCP supports resulted in respective increases of the specific surface areas. This observation is consistent with a previous work conducted by Aryee et al., 2014 [5]. It is also worth mentioning that acid pre-treatment of a similar carbon support (pure MWCNTs) has been noted to open the closed ends and caps of the pristine material and also significantly improves its textural properties. In this study, acid pre-treatment resulted in the opening of the tips of the carbon nanohorns; thus, creating nano-windows as a result, connecting the individual horn-shaped carbon materials. With the incorporation of metal precursors onto the support matrix via the conventional co-impregnation method, a resultant decrease in both the surface areas and pore volumes of the catalysts was observed to follow the order: Cat-CNH>Cat-OCP$_f$>Cat-OCP; however, changes in the pore diameter of the supports was minimal. At this metals loading for the KCoRhMo/CNH, KCoRhMo/OCP$_f$ and KCoRhMo/OCP catalysts formulation, one can conclude that the incorporation of metals precursors on the supports slightly affected the porosity and the overall textural properties of the Cat-CNH and significantly on Cats-OCP$_f$ & OCP counterparts. This observation is quite evident from the NS_{BET} values, which gives an indication of minimal pore blockage in the CNH-supported catalyst as compared to the OCP$_f$ and OCP-supported catalyst at similar metals loading; thus, affecting the catalytic performance for the HAS reaction. Though, all catalysts investigated showed maximum activity toward ethanol, at the same conditions, superior alcohol productivity was observed for the CNH-supported catalyst as compared to its OCP$_f$ and OCP counterparts. Total alcohol productivity for all catalysts at the optimum temperature (330°C) followed the trend: Cat-CNH (92.4%)>Cat-OCP$_f$ (70.1%)>Cat-OCP (54.6%). Nonetheless, for all catalysts evaluated at the various temperatures, the main by-products observed for the CO hydrogenation reactions included carbon dioxide and hydrocarbons (mainly methane and ethane). These by-products tend to be more favorable at higher temperatures for these KCoRhMo catalysts studied.

Sample ID	BET Analysis SSA (m^2/g)	PV (cm^3/g)	BJH$_{des}$ (nm)	NS_{BET}
SUPPORTS				
Pris-CNH	89	0.34	18.5	-
Pris-OCP$_f$	34	0.10	14.6	-
Pris-OCP	10	0.03	11.9	-
Fun-CNH	499	0.49	9.5	-
Fun-OCP$_f$	82	0.11	8.7	-
Fun-OCP	19	0.05	9.2	-
CATALYSTS				
Cat-CNH	322	0.27	8.8	0.92
Cat-OCP$_f$	42	0.06	9.3	0.74
Cat-OCP	9	0.03	8.9	0.70

Conclusions

In summary, CNH support and its by-products (OCP_f and OCP) were synthesized by the submerged arc discharge in liquid nitrogen technique, functionalized with 30wt.% nitric acid under reflux conditions, and employed in the formulation of K-doped $CoRhMoS_2$ catalysts for HAS applications. All catalysts prepared using these novel supports showed activity towards higher alcohols via the CO hydrogenation route. The results of this study provide insight to elucidate the crucial role different carbon materials play as catalyst support in their application for chemical reactions.

References

(1) Geidl, J.; Walton, H.; Mayes, F.; Fennell, D.; Disbrow, J.; Gordon, J.; Lowe, R., DOE/EIA-0585/O, **1994**

(2) Surisetty, V.R., Ph.D Thesis, University of Saskatchewan, **2010**, pp. 30-35.

(3) Filikov, A.V.; Myasoedov, N.F. J. Phys. Chem. 90, **1986**, 4915-4916.

(4) Surisetty, V.R.; Tavasoli, A.; Dalai, A.K. Appl. Catal. A, 393, **2011**, 50-58.

(5) Aryee, E.; Dalai, A.K.; Adjaye, J. Top Catal., 57, **2014**, 796–805.

(6) Wang, H.; Chhowalla, M.; Sano, N.; Jia, S.; Amaratunga, G. Nanotechnology, 15, 2004, 546–550.

METHANE-TO-METHANOL CONVERSION IN THE ZEOLITE SSZ-13

Florian Göltl[1], Philippe Sautet[2], Ive Hermans[1]

[1]University of Wisconsin, Madison – Department of Chemistry
[2]Ecole Normale Superieure de Lyon, Laboratoire de Chimie
fgoeltl@chem.wisc.edu

Transitionmetal-oxo sites in zeolites show high activity for the conversion of methane-to-methanol. Despite remaining challenges encouraging results have been obtained for these systems in recent years. However, up until today it is still not fully understood how different zeolite framework types influence the respective activity of the material. In this contribution we investigate two different aspects of this problem in SSZ-13, namely the nature of Cu_xO_y clusters exchanged SSZ-13 and the impact of confinement effects on the reaction enthalpies in the methane-to-methanol conversion over a Fe-oxo site in SSZ-13.

In recent literature Cu_xO_y clusters in zeolites have drawn significant attention in the methane-to methanol conversion. While initially larger pore zeolites such as ZSM-5 and Mordenite were studied, more recently a high activity of SSZ-13 was observed. Typically a combination of Raman- and UV-vis spectroscopy is used to characterize the active sites. Here we model a set of over 20 possible active sites, calculate the respective spectra and compare them to literature. We find that a novel Cu_2O_2 cluster shows excellent agreement with the experimental spectra and furthermore perform a thermodynamic analysis to prove that it is stable under realistic conditions.

Besides the different active sites, also confinement effects, caused by van der Waals (vdW) interactions between reactant and framework influence the reaction enthalpies. However, up until today a detailed understanding has not fully been achieved. To gain more detailed insights we model the methane-to-methanol conversion over a Fe-oxo center in SSZ-13 using different levels of theory. Comparing them to cluster calculations reveals that vdW-interactions change with the coordination of the carbon group. This allows to establish a simple, yet intuitive rule how confinement, not only in methane-to-methanol conversion, but more generally, affects reaction enthalpies.

HIGHLY ACTIVE AND SELECTIVE THALLIUM DOPED CERIA SUPPORTED GOLD CATALYST FOR BENZYL ALCOHOL AEROBIC OXIDATION

Chiranjit Santra,[1] Juan J. Bravo-Suárez,[2] Franklin (Feng) Tao,[2,3] Bojja Sreedhar,[4] Jai K. Pandey,[5] Biswajit Chowdhury[1]

[1]Department of Applied Chemistry, Indian School of Mines, Dhanbad, India
[2]Department of Chemical & Petroleum Engineering, The University of Kansas, Lawrence, KS 66045, USA
[3]Department of Chemistry, The University of Kansas, Lawrence, KS 66045, USA
[4]Inorganic and Physical Chemistry Division, Indian Institute of Chemical Technology, Hyderabad, India
[5]Respiratory Protection Laboratory, Central Institute of Mining and Fuels Research, Dhanbad 826015, India

Abstract

Gold based catalysts have been explored extensively for many reactions in gas and liquid phase including acetylene hydrochlorination, glycerol oxidation, and CO oxidation. The selective oxidation of organic compounds with molecular oxygen is an industrially relevant reaction due to economic and green potential. This work will present novel Tl-CeO_2 supported gold catalysts for the selective oxidation of benzyl alcohol with molecular oxygen at high alcohol conversion rates, mild temperature (343-403 K), and ambient pressure conditions. The Au/Tl-CeO_2 catalysts were characterized by having small Au nanoparticle (NP) size (~4.7 nm), free ceria, presence of oxide vacancies, and mixed valence of Ce^{3+}/Ce^{4+} species which appear to be required for high alcohol conversion and selectivity. These supported Au NPs were extremely efficient in the conversion (>95%) of benzyl alcohol to the corresponding aldehyde with selectivities above 99% and TOF values as high as 0.305 molecules converted (Au surface atoms)$^{-1}$ s^{-1}. Kinetic measurements at constant oxygen supply indicated that the reaction rate order dependency on benzyl alcohol was nearly zero at intermediate benzyl alcohol concentrations (0.9~1.3 M) and negative at high benzyl alcohol concentrations (>1.3 M) with an apparent activation energy of 41.1 kJ mol-1. These Au/Tl-CeO_2 catalysts also demonstrated to be quite reactive towards the oxidation of other aliphatic and aromatic alcohols such as 2-octanol, geraniol, cinnamyl alcohol, and 4-methylbenzyl alcohol at similar reaction conditions. This study will highlight the synergistic role of gold nanoparticles, Tl-CeO_2 support, and oxygen vacancies required for highly active and selective oxidation catalysts with green oxidants such as molecular oxygen.

ENHANCED TWO-DIMENSIONAL DISPERSION OF GROUP V METAL OXIDES ON SILICA

J. T. Grant[1], *C. A. Carrero*[1], *A. M. Love*[1], *R. Verel*[3], *P. Mueller*[1], *I. Hermans*[1,2]

[1] University of Wisconsin-Madison, Department of Chemistry, 1101 University Avenue, Madison, Wisconsin 53706, United States
[2] University of Wisconsin-Madison, Department of Chemical and Biological Engineering, 1415 Engineering Drive, Madison, Wisconsin 53706, United States
[3] ETH Zurich, Department of Chemistry and Applied Biosciences, Vladimir Prelog Weg 2, 8093 Zurich, Switzerland
hermans@chem.wisc.edu

Abstract

Catalytic performance of supported metal oxides is often dependent on the existence of two- or three-dimensional surface structures. Silica, one of the most utilized oxides for supported metal oxide catalysts, has historically allowed the lowest quantity of supported 2D surface species to exist before triggering the formation of 3D nanoparticles, despite its high surface area. We show that by adding low amounts of sodium as a promoter to the SiO2 surface the allowable surface density of 2D supported Group V metal oxides increases to equivalent levels of these metal oxides dispersed on all other common oxide supports.1,2 This enhanced 2D dispersion on SiO2 is confirmed with Raman and DRUV-vis spectroscopy, 51V MAS NMR, and the oxidative dehydrogenation of propane utilized as a probe reaction. Supported 2D vanadia catalysts show notably higher selectivity to propylene than their 3D counterparts, and propane consumption rate increases proportionally to vanadium surface density. It is hypothesized that the role of sodium is to catalytically deprotonate surface silanol groups to create more-reactive anchoring sites for the incoming Group V metal oxide precursor. Metal oxide anchoring mechanisms and reaction kinetics, in particular, will be discussed.

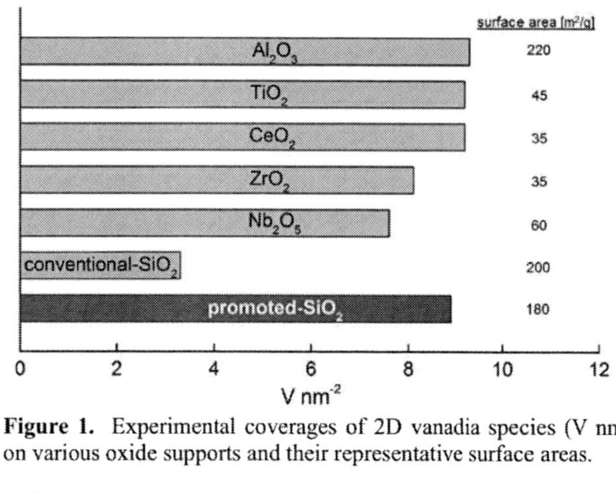

Figure 1. Experimental coverages of 2D vanadia species (V nm^{-2}) on various oxide supports and their representative surface areas.

References

(1) Grant, J.; Carrero, C.; Love, A.; Verel, R.; Hermans, I.; *ACS Catal.*, **2015**, *5*, 5787
(2) Grant, J.; Carrero, C.; Hermans, I.; US Patent 62/112,689, February 5, **2016**

C-H BOND OXIDATION BY MN(IV)-OXO SPECIES AND CATALYTIC APPLICATIONS

Timothy A. Jackson

Department of Chemistry, University of Kansas, Lawrence, KS, 66045, USA
taj@ku.edu

Abstract

Substrate oxidation, including C-H bond oxidation, by manganese-based catalysts has seen an increased focus in recent years. The transition metal manganese displays several promising criteria for a metal catalyst. Manganese is earth-abundant; is relatively non-toxic; displays a broad range of accessible oxidation states; and can catalyze a number of important chemical transformations, including oxidation of stain molecules (textile and fabric bleaching) and chemo- and stereo-selective substrate oxidations. We have reported a Mn(IV)-oxo complex supported by the pentadentate N4py ligand (N4py = N,N-bis(2-pyridylmethyl)-N-bis(2-pyridyl)methylamine) that is capable of oxidizing relatively strong C-H bonds.[1] To determine electronic and geometric contributions to reactivity, we are currently employing N4py derivatives with systematically varied properties to support a series of Mn(IV)-oxo complexes. This presentation will describe the reactivity of these complexes in both the stoichiometric and catalytic oxidation of C-H bonds, correlating structure and reactivity. Catalytic reactivity is focused on the oxygenation of cyclohexane and other hydrocarbons.

References

(1) Leto et al., *Chem. Commun.*, **2013**, *49*, 5378

CONTROLLING ENVIRONMENTS FOR CATALYTIC HYDROGEN AND OXYGEN TRANSFER ON SOLID SUPPORTS

Andrew Palermo,[a,b] Alexander Okrut,[a] Nicolas Grosso-Giordano,[a] Richard Ouyang,[a,c] Andrew Solovyov,[a] Daniel Ertler,[a] Stacey I. Zones,[a,c] Bruce C. Gates,[b] Alexander Katz[a]

[a]Department of Chemical and Biomolecular Engineering, University of California, Berkeley, California 94720, United States
[b]Department of Chemical Engineering and Materials Science, University of California at Davis, One Shields Avenue, Davis, California 95616, United States
[c]Chevron Energy Technology Company, Richmond, California 94804, United States
askatz@berkeley.edu

Abstract

In the first portion of the talk, control of ligand environment at a supported tetrairidium cluster site on silica is used, in order to understand attributes of the active site that are most conducive to hydrogenation catalysis. This is enabled by a comparative study of a systematic series of P-ligands on the tetrairidium cluster core, which affects olefin hydrogenation rates by affecting strength of the metal-H bond. In the second part of the talk, progress on the synthesis of dispersed grafted Ti sites on a variety of silaceous supports will be described. Emphasis will be placed on elucidating the role of the structure of the support by comparing amorphous and crystalline supports. Catalytic characterization will consist of epoxidation of terminal olefins such as 1-octene using organic hydroperoxides as oxidant.

SELECTIVE TANDEM CATALYSIS BY METAL NANOCLUSTERS ENCAPSULATED IN METAL-ORGANIC FRAMEWORKS

Xinle Li, [†] *Chaoxian Xiao,*[†] *Tian Wei Goh,*[†] *and Wenyu Huang*[*, †, ‡]

[†]Department of Chemistry, Iowa State University, Ames, Iowa 50011, United States

[‡]Ames Laboratory, U.S. Department of Energy, Ames, Iowa 50011, United States

Abstract

Tandem catalysis, in which multi-step chemical transformations are catalyzed with a multifunctional catalyst, has attracted increasing research attention. In tandem catalysis, selective multifunctional catalysts will make a multi-step reaction go to the desired products in one reactor without the need for separation, purification, and transfer of intermediates produced in each step.[1-8] The development of multi-step catalytic processes using highly selective multifunctional catalysts can also minimize the amount of by-products/waste produced during multi-step chemical reactions and achieve atomic efficiency. Conversely, traditional multi-step catalytic processes involve a series of conversions in different reactors with intermediate purification steps to achieve a desired product. Therefore, traditional multi-step catalytic processes usually require a high-energy input for the separation of intermediates and cause a low yield of the final products due to the losses of intermediates within each individual step of the processes.

Metal–organic frameworks (MOFs) could serve as a perfect platform for designing tandem catalysts. MOFs are constructed by coordinating metal ions/clusters with organic linkers, which could contain acidic/basic sites for catalysis. The usage of MOFs as novel hosts for metal nanoparticles has also been extensively investigated.[9-13] By combining the redox active sites of metal particles encapsulated in MOFs and the acidic/basic sites of MOFs, we could construct multifunctional catalysts for tandem catalysis. We have developed a heterogeneous tandem catalyst for the oxidation of benzyl alcohol and the subsequent acetalization with ethylene glycol catalyzed by UiO-66-NH_2 supported Pd nanoclusters. This bifunctional Pd@UiO-66-NH_2 catalyst exhibits high selectivity to give acetal products in this tandem oxidation-acetalization reaction and can be reused 5 times without any loss in activity and selectivity. On the contrary, Pd@UiO-66 showed deactivation at the third recycle run and more severe aggregation of Pd NCs was observed. We also found that using the nonpolar solvent, cyclohexane, is essential for the high selectivity to acetal, which could promote the acetalization reaction and prevent the further oxidation of aldehydes to the corresponding esters. This work demonstrates the potential of MOF-based catalysts in tandem heterogeneous catalysis. Under appropriate conditions, both high selectivity and stability of MOF-based catalysts can be achieved. Further work on exploring new tandem catalytic reactions with these catalytic systems and optimizing the performance of the multifunctional MOFs via preparing bimetallic nanoclusters is currently underway.

References

(1) Shiju, N. R.; Alberts, A. H.; Khalid, S.; Brown, D. R.; Rothenberg, G. *Angew. Chem.-Int. Edit.* **2011**, *50*, 9615.
(2) Grondal, C.; Jeanty, M.; Enders, D. *Nat. Chem.* **2010**, *2*, 167.
(3) Shin, J. Y.; Jung, D. J.; Lee, S.-g. *ACS Catal.* **2013**, *3*, 525.
(4) Guan, Y.; Zhang, D.; Wang, Y. *Catal. Lett.* **2012**, *142*, 1225.
(5) Kim, J. H.; Lee, S.-g. *Org. Lett.* **2011**, *13*, 1350.
(6) Shylesh, S.; Wagener, A.; Seifert, A.; Ernst, S.; Thiel, W. R. *Angew. Chem.-Int. Edit.* **2010**, *49*, 184.
(7) Nicolaou, K. C.; Edmonds, D. J.; Bulger, P. G. *Angew. Chem.-Int. Edit.* **2006**, *45*, 7134.
(8) Yang, G.; Kawata, H.; Lin, Q.; Wang, J.; Jin, Y.; Zeng, C.; Yoneyama, Y.; Tsubaki, N. *Chem. Sci.* **2013**, *4*, 3958.
(9) Aijaz, A.; Karkamkar, A.; Choi, Y. J.; Tsumori, N.; Rönnebro, E.; Autrey, T.; Shioyama, H.; Xu, Q. *J. Am. Chem. Soc.* **2012**, *134*, 13926.
(10) Hu, P.; Morabito, J. V.; Tsung, C.-K. *ACS Catal.* **2014**, *4*, 4409.
(11) Na, K.; Choi, K. M.; Yaghi, O. M.; Somorjai, G. A. *Nano Lett.* **2014**, *14*, 5979.
(12) Lu, G.; Li, S.; Guo, Z.; Farha, O. K.; Hauser, B. G.; Qi, X.; Wang, Y.; Wang, X.; Han, S.; Liu, X. *Nat. Chem.* **2012**, *4*, 310.
(13) Chen, Y.-Z.; Xu, Q.; Yu, S.-H.; Jiang, H.-L. *Small* **2015**, *11*, 70.

CONTROL CATALYTIC PROPERTIES OF NANOCLUSTERS USING ISORETICULAR METAL-ORGANIC FRAMEWORKS

Xinle Li,[†] *Tian Wei Goh,*[†] *Chaoxian Xiao,*[†] *and Wenyu Huang*[*, †, ‡]

[†]Department of Chemistry, Iowa State University, Ames, Iowa 50011, United States

[‡]Ames Laboratory, U.S. Department of Energy, Ames, Iowa 50011, United States

Abstract

Metal–organic frameworks (MOFs) are constructed by coordinating metal ions/clusters with organic linkers. The usage of MOFs as novel hosts for metal nanoparticles has been extensively investigated.[1-5] Unlike other porous materials, such as zeolites and mesoporous silica, the properties of MOFs are rationally 'designable' to study the effect of the chemical environment on the encapsulated metal nanoparticles (NPs) or nanoclusters (NCs) because the metal ions and the organic linkers of MOFs can be fine-tuned by either utilizing different building units or by the post-synthetic modification of pre-synthesized MOFs.[6-8]

Most of MOFs contain sub-nanometer windows that have been used to regulate and separate molecules of different sizes. The introduction of different functional groups to these sub-nanometer windows can dramatically affect the gas sorption capacity and uptake selectivity,[9,10] luminescence,[11] water stability,[12] and mechanical properties of the MOFs.[13] However, studies on the effects of introducing different functional groups to these sub-nanometer windows on catalysis have been rarely reported.[14-16] In pioneering work carried out by De Vos' and Jhung's group, it was found that Lewis acid catalysis, can be affected by changing the organic linkers in Zirconium-based UiO-66 MOFs.[17,18] Nonetheless, varying product selectivity controlled through the well-defined chemical environment of isoreticular MOFs have scarcely been reported in heterogeneous catalysis to date.[19,20]

We have developed Pd NCs encapsulated in isoreticular MOFs of different chemical environments (Pd@UiO-66-X, X = H, NH_2, and OMe), which were used to catalyze the reaction between benzaldehyde and ethylene glycol. Notably, the linker modification in MOFs can atomically tune the product distribution in this reaction. We found that Pd@UiO-66-NH_2 favors the formation of acetal, while Pd@UiO-66 and Pd@UiO-66-OMe have a higher selectivity to ester. DRIFTS studies revealed that –NH_2 groups coordinate to the Pd surface and donate electrons to the Pd NCs in Pd@UiO-66-NH_2. These NH_2-Pd interactions decrease (but do not completely eliminate) the oxidation capability of encapsulated Pd NCs and lead to the high acetal selectivity by Pd@UiO-66-NH_2 in the reaction. DFT calculations show the Pd-NH_2 NCs system processes higher chemical potential and thus weaker oxidation capability than the Pd-OMe NCs, which agrees with the experimental results. This strategy can be extended to gas-phase furfural hydrogenation and linker modification in MOFs is decisive to the product distribution as well. The present work demonstrates the control of heterogeneous catalytic sites through their surrounding chemical environment at an atomic level, thus opening up fascinating opportunities for the development of tunable metal-MOF catalysts for selective chemical transformations.

References

(1) Aijaz, A.; Karkamkar, A.; Choi, Y. J.; Tsumori, N.; Rönnebro, E.; Autrey, T.; Shioyama, H.; Xu, Q. *J. Am. Chem. Soc.* **2012**, *134*, 13926.
(2) Hu, P.; Morabito, J. V.; Tsung, C.-K. *ACS Catal.* **2014**, *4*, 4409.
(3) Na, K.; Choi, K. M.; Yaghi, O. M.; Somorjai, G. A. *Nano Lett.* **2014**, *14*, 5979.
(4) Lu, G.; Li, S.; Guo, Z.; Farha, O. K.; Hauser, B. G.; Qi, X.; Wang, Y.; Wang, X.; Han, S.; Liu, X. *Nat. Chem.* **2012**, *4*, 310.
(5) Chen, Y.-Z.; Xu, Q.; Yu, S.-H.; Jiang, H.-L. *Small* **2015**, *11*, 70.
(6) Corma, A.; Garcia, H.; Llabres i Xamena, F. X. *Chem. Rev.* **2010**, *110*, 4606.
(7) Wang, Z.; Cohen, S. M. *Chem. Soc. Rev.* **2009**, *38*, 1315.
(8) Zhang, Z.; Zhang, L.; Wojtas, L.; Nugent, P.; Eddaoudi, M.; Zaworotko, M. J. *J. Am. Chem. Soc.* **2011**, *134*, 924.
(9) Yang, Q.; Guillerm, V.; Ragon, F.; Wiersum, A. D.; Llewellyn, P. L.; Zhong, C.; Devic, T.; Serre, C.; Maurin, G. *Chem. Commun.* **2012**, *48*, 9831.
(10) Biswas, S.; Zhang, J.; Li, Z.; Liu, Y.-Y.; Grzywa, M.; Sun, L.; Volkmer, D.; Van Der Voort, P. *Dalton Trans.* **2013**, *42*, 4730.
(11) Bauer, C. A.; Timofeeva, T. V.; Settersten, T. B.; Patterson, B. D.; Liu, V. H.; Simmons, B. A.; Allendorf, M. D. *J. Am. Chem. Soc.* **2007**, *129*, 7136.
(12) Burtch, N. C.; Jasuja, H.; Walton, K. S. *Chem. Rev.* **2014**, *114*, 10575.
(13) Tan, J. C.; Cheetham, A. K. *Chem. Soc. Rev.* **2011**, *40*, 1059.
(14) Goh, T. W.; Xiao, C.; Maligal-Ganesh, R. V.; Li, X.; Huang, W. *Chem. Eng. Sci.* **2015**, *124*, 45.
(15) Shen, L.; Liang, R.; Luo, M.; Jing, F.; Wu, L. *Phys Chem Chem Phys* **2015**, *17*, 117.
(16) Esken, D.; Turner, S.; Lebedev, O. I.; Van Tendeloo, G.; Fischer, R. A. *Chem. Mater.* **2010**, *22*, 6393.
(17) Vermoortele, F.; Vandichel, M.; Van de Voorde, B.; Ameloot, R.; Waroquier, M.; Van Speybroeck, V.; De Vos, D. E. *Angew. Chem.-Int. Edit.* **2012**, *51*, 4887.
(18) Timofeeva, M. N.; Panchenko, V. N.; Jun, J. W.; Hasan, Z.; Matrosova, M. M.; Jhung, S. H. *Appl. Catal., A* **2014**, *471*, 91.
(19) Choi, K. M.; Na, K.; Somorjai, G. A.; Yaghi, O. M. *J. Am. Chem. Soc.* **2015**, *137*, 7810.
(20) Zhang, D.; Guan, Y.; Hensen, E. J. M.; Xue, T.; Wang, Y. *Catal. Sci. Tech.* **2014**, *4*, 795.

ROLE OF STRAIN AND LIGAND EFFECT IN DECOMPOSITION OF FORMIC ACID ON THE BIMETALLIC PD/M CATALYSTS

*Jinwon Cho[1], Sangheon Lee[2], Jonghee Han[1], Sung Pil Yoon[1], Suk Woo Nam[1], Kwan-Young Lee[1], Hyung Chul Ham[1]**

[1]Fuel Cell Research Center, Korea Institute of Science and Technology (KIST), Seoul, Republic of Korea
[2]Chemical Engineering and Materials Science, Ewha Womans University, Seoul, Republic of Korea
[3]Department of Chemical and Biological Engineering, Korea University, Seoul, Republic of Korea
*hchahm@kist.re.kr

Introduction

Formic acid (HCOOH) is considered as one of the most promising compounds for chemical hydrogen (H_2) storage particularly for low-temperature fuel cell applications.[1] Formic acid is an excellent in situ source of hydrogen for fuel cells, for it offers high energy density and remains as a liquid at standard pressure and temperature. HCOOH readily decomposes into H_2 and CO_2 in the presence of a suitable catalyst.[2] The decomposition of HCOOH also includes a side reaction, resulting H_2O and CO.[1] The gas stream from the decomposition of HCOOH has to be free from CO gas (< 10 ppm), or the catalytic performance of the fuel cells will be degraded significantly due to the contamination of electrodes by CO.[2] Therefore, catalysts that are highly selective (suppressing the formation of CO) and active at low temperatures for the decomposition of HCOOH into H_2 and CO_2 are indispensable to HCOOH-based low-temperature fuel cells.

Pd based catalysis are widely used in formic acid decomposition because Pd it is abundant, inexpensive, and tolerant to CO.[1] Currently, it was found that among the noble metal catalysts, the activity for the H_2 production via HCOOH decomposition is enhanced on Pd catalysts, however, the selectivity to H_2 production is still to be improved. To solve such issue, the bimetallic alloy Metal-Pd core-shell (Pd/M) catalyst has been proposed to improve both the catalytic activity and selectivity of Pd toward H_2 production from HCOOH decomposition. In the bimetallic core-shell alloy catalyst, the alloying effect can be attributed to the strain caused by the mismatch between the two different elements, so called strain effect, and electronic state changes due to metal-metal interaction, which is called ligand effect.[8] Understanding these alloying effects is important in order to further develop effective catalyst; however, they occur simultaneously so it is nearly impossible to separate one from the other in experiment. Computational approaches, particularly the prediction using density functional theory (DFT) can be a powerful and flexible source for unraveling the role of each of the alloying effects in the bimetallic alloy catalysts. In this study, the reactions of HCOOH were examined on the Pd/M catalysts owing to the following fundamental questions; a role of Pd shell thickness, the role of strain and ligand effect in determining the dehydrogenation and dehydration of HCOOH.

Calculation Details

The calculations reported herein were performed on the basis of spin polarized DFT within the PW91 functional, as implemented in the Vienna Ab-initio Simulation Package (VASP).[3] The projector augmented wave (PAW) method with a planewave basis set was employed to describe the interaction between ion cores and valence electrons. An energy cutoff of 350eV was applied for expansion of the electronic eigenfunctions.[4] For the Brillouin zone integration, we used a (5×5×1) Monkhorst-Pack mesh of k points to determine the optimal geometries and total energies of systems.[5] For the electronic structure and d-band center calculation, we increase the k point mesh to (12×12×1). Reaction pathways and barriers were determined using the climbing-image nudged elastic band method (c-NEBM) with six intermediate images for each elementary step. For model surfaces, we used a supercell slab that consists of a rectangular 2x2 surface unit cell with five atomic layers of which contains 4 atoms.[6] The bottom 4 layers are late transition FCC metals, such as Cu, Rh, Pd, Ir, Pt, Au, and Ag (listed in increasing order of lattice constant) and the top layer is consisted of Pd atoms. The slab is separated from its periodic images in the vertical direction by a vacuum space corresponding to seven atomic layers. While the bottom two layers of the five-layered slab are fixed at corresponding bulk positions, the upper three layers are fully relaxed using the conjugate gradient method until residual forces on all the constituent atoms become smaller than 5×10^{-2} eV/A. The lattice constant for bulk Cu, Rh, Pd, Ir, Pt, Au, and Ag is predicted to be 3.60eV, 3.84eV, 3.86eV, 3.89eV, 3.92eV, 4.08eV and 4.09eV respectively, which are virtually identical to the previous calculations. Stability of Pd/M models are expected to be stable at the operating condition as there is an energy barrier required for surface segregation as a minimum of 0.29eV to 1.89eV at maximum.

Results and Discussion

We initially modeled the gas-phase reaction on the (111) surfaces of Pd/Ag structure as a starting point to understand catalytic decomposition of HCOOH and thickness effect of Pd shell as well. For dehydrogenation pathway, the first dehydrogenation from HCOOH via the O-H cleavage [D1], then proceeds from HCOO via the C-H cleavage [D2].[8] For dehydration pathway the decomposition of HCOOH via the C-O cleavage, yielding HCO and OH [I1], and the formation of CO and H_2O [I2] by the combination reaction of I1 (see Figure 1).[8]

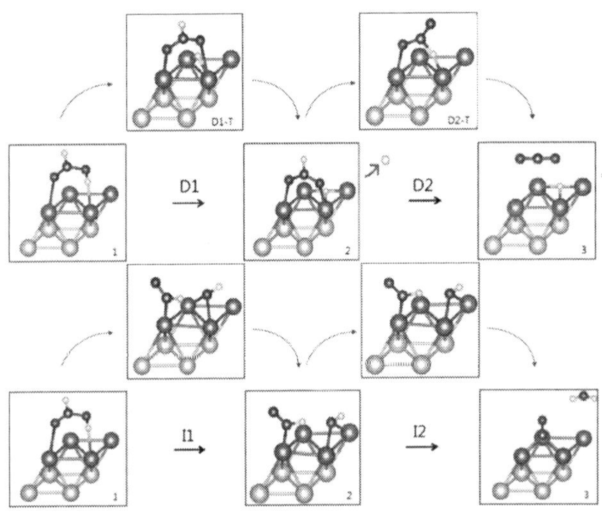

Figure 1. Two possible decomposition pathways of HCOOH oxidation on the Pd surface. One is dehydrogenation, presented in the top half, proceeds to HCOO + H (D1) and then to CO_2 to H (D2). The other, presented in the bottom half, dehydration proceeds to HCO + OH (I1) and followed by CO and H_2O (I2).

As in Table 1, our calculations predict that in dehydrogenation reaction [(D2)], the Pd monolayer on Ag substrate [Pd_{Mono}/Ag(111)] turns out to be energetically most favorable (ΔE= –0.68eV [Pd(111)], –0.84eV [Pd_{Mono}/Ag(111)], –0.56eV [Pd_{Bi}/Ag(111)], –0.63eV

[Pd_{Tri}/Ag(111)], and –0.65eV [Pd(111)$_{Str}$]) and in turn have the lowest barrier compared to the pure Pd(111), Pd_{Bi}/Ag(111), and Pd_{Tri}/Ag(111)cases (Note the order of barriers for step D2: 0.85eV [Pd(111)] > 0.83eV [Pd_{Tri}/Ag(111)] ≈ 0.84eV [Pd_{Bi}/Ag(111)] ≈ 0.84eV [Pd(111)$_{Str}$] > 0.76eV [Pd_{Mono}/Ag(111)]). Comparing the barriers in a rate-determining step (which is the step D2 on all Pd surfaces we considered), we find that the Pd_{Mono}/Ag(111) surface can greatly enhance the reaction rate in dehydrogenation pathway compared to the Pd (111) surface case, while for the Pd_{Bi}/Ag(111), Pd_{Tri}/Ag(111) and Pd(111)$_{Str}$ surfaces, the improvement is insignificant.[8] Expecting that the barrier variation of a rate-determining step between dehydrogenation and dehydration reaction is related to the selectivity in formic acid oxidation, we also compared the activation barrier difference (ΔE_a): ΔE_a = 0.27 eV, [Pd(111)$_{Str}$)], 0.25 eV [Pd_{Tri}/Ag(111)], 0.14 eV [Pd_{Bi}/Ag(111)], 0.58 eV [Pd_{Mono}/Ag(111)], 0.31eV [Pd(111)], indicating that the Pd_{Mono}/Ag(111) case greatly enhances the selectivity to dehydrogenation route, while for the Pd_{Bi}/Ag(111), Pd_{Tri}/Ag(111) and Pd(111)$_{Str}$ cases the opposite is true.[8]

Table 1. The Calculated Total Energy Changes (ΔE) and Activation Barriers (E_a in parenthesis) for Dehydrogenation and Dehydration Steps on the Pure Pd and Pd/Ag Catalysts.

ΔE /(E_a) (eV)	Pd_{Mono}/Ag(111)	Pd_{Bi}/Ag(111)	Pd_{Tri}/Ag(111)	Pd_{Str}/(111)	Pd(111)
(D1) HCOOH to HCOO + H	0.13/(0.75)	-0.43/(0.45)	-0.25/(0.58)	-0.24/(0.58)	0.01/(0.69)
(D2) HCOO to CO_2 + H	-0.84/(0.76)	-0.56/(0.84)	-0.63/(0.83)	-0.65/(0.84)	-0.68/(0.85)
(I1) HCOOH to HCO + OH	1.19/(1.34)	0.57/(0.98)	0.68/(1.08)	0.78/(1.11)	0.97/(1.16)
(I2) HCO + OH to CO and H_2O	-2.02/(0.01)	-1.71/(0.25)	-1.85/(0.03)	-1.75/(0.17)	-1.79/(0.07)

This would be consistent with a recent experimental study where the highest activity of the Ag-Pd core shell catalysts toward the HCOOH decomposition is achieved when the Pd shell thickness is the thinnest[9].

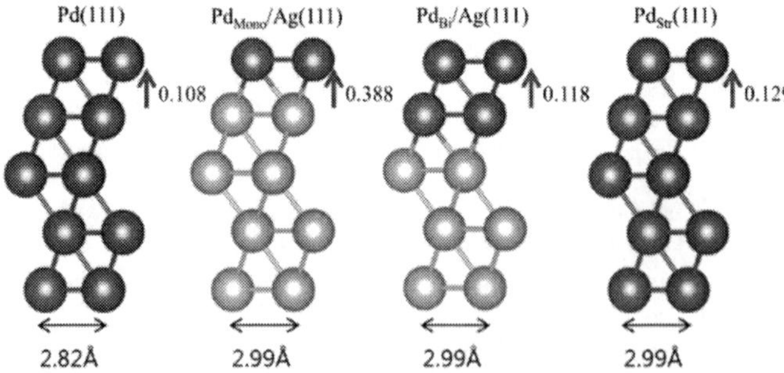

HCOO	Pd(111)	Pd_{Mono}/Ag(111)	Pd_{Bi}/Ag(111)	Pd_{Str} (111)
Binding Energy	-2.82eV	-2.68eV	-3.11eV	-2.97eV
Strain Effect	-	-0.15	-0.15	-
Ligand Effect	-	0.29	-0.14	-
Charge Transferred				
Surface	0.108	0.388	0.118	0.129
Subsurface	-0.090	-0.348	0.142	-0.109
3rd Layer	-0.049	0.035	-0.260	-0.08

Figure 2. The amount of strain effect and ligand effect on Pd_{mono}/Ag and Pd_{bi}/Ag is displayed as well as the charge polarization on Pd(111) and Pd/Ag structures.

As in Figure 2, This demonstrates that the number of Pd atomic layer on Ag substrate plays an important role in determining the dehydrogenation reaction since the strong Ag-Pd ligand effect (which significantly reduces the density of state (especially, $d_{z^2-r^2}$ orbital) near the Fermi level) resulting from the charge transfer to the Pd surface from the subsurface Ag in the Pd_{Mono}/Ag(111) surface substantially reduces the surface binding strength of specific intermediates such as HCOO and HCO [note the insignificant change of adsorption energy for the other key intermediates (CO,OH and H)], which is responsible for the enhanced reaction rate toward H_2 production from HCOOH decomposition.[8]

Based on the abovementioned effect of Pd shell thickness, we performed the formic acid decomposition over the Pd monolayer based Pd/M catalysts to find the most economical use of rather expensive noble metal and to further understand a role of alloying effect such as strain and ligand effect towards hydrogen production rate and selectivity simultaneously (Table 2) .

Table 2. Calculated Energy Changes in a Reaction and Activation Energies (E_a) for Dehydrogenation and Dehydration Pathways on Pd/M. All Values are Given in eV.

	Metals used for M substrates						
ΔE/(E_a)	Rh	Pt	Ir	Cu	Pd	Au	Ag
DH-I HCOOH → HCOO +H	0.06 /(0.73)	-0.33 /(0.51)	-0.07 /(0.71)	0.21 /(0.78)	-0.14 /(0.68)	-0.18 /(0.69)	-0.04 /(0.72)
DH-II HCOO → CO_2+H	-0.30 /(0.87)	-0.25 /(0.94)	-0.18 /(0.92)	-0.94 /(0.76)	-0.41 /(0.90)	-0.44 /(0.85)	-0.58 /(0.81)
DCO-I HCOOH → HCO+OH	1.15 /(1.12)	0.93 /(0.99)	1.07 /(1.09)	2.16 /(2.02)	0.90 /(1.08)	1.01 /(1.06)	0.98 /(1.19)
DCO-II HCO+OH → CO+H_2O	-1.60 /(0.00)	-1.44 /(0.00)	-1.67 /(0.00)	-2.56 /(0.00)	-1.79 /(0.00)	-1.67 /(0.00)	-2.02 /(0.00)

The overall activation energy of dehydrogenation pathway for Pd/Cu is the lowest (E_a = 0.78eV) whereas for Pd/Pt is the highest (E_a =

0.94eV) while the energy barrier of dehydration pathway is the highest for Cu-Pd (E_a = 2.02eV) and the lowest for Pd/Pt (E_a = 0.99eV). Considering CO contamination is detrimental to the Pd surface, the ideal condition for catalyst is to have low dehydrogenation activation energy barrier and high dehydration activation energy barrier as possible. Accordingly, we also calculated the barrier difference (ΔE_a) between dehydrogenation pathway and dehydration pathway which is related to the selectivity in formic acid decomposition, and the selectivity of Pd/Cu core-shell structure (ΔE_a = 1.26eV) is revealed to be enhancing the selectivity to H_2 production, while for Pd/Pt (ΔE_a = 0.05eV) the opposite is true. At this point, we asked ourselves for what reasons Cu and Ag shows better H_2 productivity whereas Pt has worse selectivity than Pure Pd.

The answer to the question is the strong understanding of structure and underlying mechanism in the bimetallic core-shell alloy catalyst is essential for determining the catalytic performance. Particularly, the interatomic mixing between the surface and subsurface layer and the mismatch of lattice parameter between the two atoms is one of the most important factors that can significantly affect the surface reactivity of alloy catalysts. It has been difficult to separate the strain and ligand effects, even in surface science studies, because they usually occur simultaneously, however, by using DFT calculations, we can observe how each effect influences on the decomposition of formic acid. Experimentally, for thin layers of one metal on top of another it can be difficult to isolate the strain effects because they are folded with the effects due to the interaction of the overlayer with the substrate, however with DFT computational simulation, we were able to examine the effect of changing the lattice parameter to the surface of the Pd(111) slab to focus directly on the strain effects. As a result, there is a considerable variation in the activation energy of dehydration increases as the lattice parameter decreases. This tendency also hints that the Pd surface strain should be compressive as possible in order to achieve the maximum H_2 selectivity from HCOOH decomposition. On the other hand, the amount of calculated ligand contribution is directly proportional to the amount of charges on the surface, which is transferred from the subsurface M, more the charge transfers, and more the ligand contribution on the surface. This relationship also suggests that the ligand contribution on the surface is determined by the transferred charges to the surface since the gained charges mixes with the charges of Pd on the surface. There is also tendency to a decrease in energy barrier of dehydrogenation as there are more charges transferred to the surface. In order to reduce the activation energy of dehydrogenation barrier, Pd shell is in need of as much as charge gains possible, since the charge transfer to the shell ultimately mixes up with the original Pd shell charges.

Conclusions

In this study, the DFT calculations have been performed to understand the role of Pd shell thickness, strain effect and ligand effect in the bimetallic Pd/M catalysts in the decomposition of HCOOH. For such purposes, we separated strain and ligand effect. Here, the interatomic mixing between the surface and subsurface layer, the ligand effect, is responsible for dehydrogenation since excessive charge on the Pd surface reduces energy barriers, such as Pd/Cu, Pd/Ag, and Pd/Rh, and the mismatch of lattice parameter between the two different atoms, the strain effect, is key to dehydration since it ultimately raises CO production energy barriers. Our theoretical calculation offers the fundamental mechanism of HCOOH decomposition on Pd/M catalysts and also provides physical and chemical intuition for the next generation bimetallic catalyst.

Acknowledgement. The current work was financially supported by the New and Renewable Energy Core Technology Program of the Korea Institute of Energy Technology Evaluation and Planning (KETEP) granted financial resource from the Ministry of Trade, Industry & Energy, Republic of Korea (No. 20133030011320), and also supported by KIST institutional program for Korea Institute of Science and Technology. One of the authors (S.L.) wishes to acknowledge the financial support from the Basic Science Research Program through the National Research Foundation (NRF) of Korea funded by the Ministry of Education (NRF-2013R1A6A3A04059268) and the supercomputing resources from the Supercomputing Center/Korea Institute of Science and Technology Information (KSC-2013- C3-047).

References

(1) (a) Ojeda, M.; Iglesia. Angew. Chem. 2009, 48 (26), 4800−3. (b) Haan, J. L.; Masel, R. I. Electrochim. Acta 2009, 54 (16), 4073−4078.
(2) Tedsree, K.; Li, T.; Jones, S.; Chan, C. W.; Yu, K. M.; Bagot, P. A.; Marquis, E. A.; Smith, G. D.; Tsang, S. C. Nat. Nanotechnol. 2011, 6 (5), 302−7.
(3) Kresse, G.; Furthmuller, J. Vienna University of Technology: Vienna, Austria, 2001.
(4) Blochl, P. E. Phys. Rev. B 1994, 50 (24), 17953−17979
(5) Blochl, P. E.; Jepsen, O.; Andersen, O. K. Phys. Rev. B 1994, 49 (23), 16223−16233
(6) Henkelman, G.; Uberuaga, B. P.; Jonsson, H. J. Chem. Phys. 2000, 113, 9901.
(7) Hammer, B.; Hansen, L. B.; Norskov, J. K. Phys. Rev. B 1999, 59 (11), 7413−7421.
(8) Cho, J.; Lee, S.; Han, J.; Yoon, S. P.; Nam, S. W.; Choi, S. H.; Lee, K.-Y.; Ham, H. C. *J. Phys. Chem. C* **2014**, *118* (39), 22553-22560.
(9) Cho, J.; Lee, S.; Han, J.; Yoon, S. P.; Nam, S. W.; Choi, S. H.; Ahn, S. A.; Lee, K. -Y.; Ham, H. C. *J. Nanosci. Nanotechnol.* **2015,** *15* (10), 8233-8237.
(10) Lee, S.-Y.; Jung, N.; Cho, J.; Park, H. -Y.; Ryu, J.; Jang, I.; Kim, H.-J.; Cho, E.; Park, Y.-H.; Ham, H. C.; Jang, J. H.; Yoo, S. J. *ACS Catal.* **2014,** *4* (8), 2402-2408.
(11) Mavrikakis, M.; Hammer, B.; Nørskov, J. Phys. Rev. Lett. 1998, 81.
(12) Maark, T. A.; Peterson, A. A. J. Phys. Chem. C 2014, 118 (8), 4275−4281.
(13) Ham, H. C.; Manogaran, D.; Lee, K. H.; Kwon, K.; Jin, S. A.; You, D. J.; Pak, C.; Hwang, G. S. J. Chem. Phys. 2013, 139 (20).

COMPUTATIONAL DISCOVERY AND PREDICTION OF SYNTHESIS CONDITIONS OF NOVEL 2D MATERIALS FOR ENERGY TECHNOLOGIES AND SPINTRONIC DEVICES

Richard G. Hennig

Department of Materials Science and Engineering,
University of Florida, Gainesville, Florida 32611

rhennig@ufl.edu

Introduction

The rapid rise of novel two-dimensional (2D) materials, presents the exciting opportunity for materials science to explore an entirely new class of materials. This comes at the time when mature computational methods provide the predictive capability to enable the computational discovery, characterization, and design of single-layer materials and provide the needed input and guidance to experimental studies.

This work reviews our our data-mining and high-throughput approach, *MPInterfaces*,[1,2] and our evolutionary algorithm package, *GASP*,[3] to identify novel 2D materials with low formation energies and show how unexpected structures emerge when a material is reduced to sub-nanometers in thickness. These open-source software packages, including manuals, examples, and tutorials, are freely distributed on Github at https://github.com/henniggroup/ under open-source licenses.

Computational Methods

The computational discovery, characterization, and design of 2D materials takes advantage of accurate and efficient density-functional theory (DFT) methods. The results discussed in this work are all obtained using the Vienna *ab-initio* simulation package (VASP).[4,5] To predict and characterize 2D materials we developed a computational framework, MPInterface, and a genetic algorithm package, GASP.

MPInterfaces. MPInterface, is an open-source python framework for the discovery and study of 2D materials and materials interfaces.[2] MPInterfaces extends the Materials Project software package[6] to automate the high-throughput computational screening and study of interfacial systems. The framework encompasses the creation and manipulation of interface structures for solid/solid heterostructures, solid/implicit solvents systems,[7] nanoparticle/ligands systems, and the creation of simple system-agnostic workflows for in-depth computational analysis using density-functional theory or empirical energy models. The package leverages existing open-source high-throughput tools and extends their capabilities towards building knowledge of interfacial systems.

GASP. GASP is a genetic algorithm package for structure and phase prediction of bulk and 2D materials.[3,8-11] It couples a grand-canonical genetic algorithm to ab-initio or (semi-) empirical methods to accurately predict how atoms arrange into crystal or molecular structures and to determine the composition of the ground state phases of materials without any prior information about the system. Our approach is unique in that it not only predicts the structure of materials but also the composition. We recently extended our GASP algorithm to enable unbiased searching for the compositions and structures of novel 2D materials with low formation energies and finite layer thicknesses.[3]

Results and Discussion

Figure 1 illustrates the computational workflow to identify and characterize 2D materials. We apply the computational tools to (i) screen 2D materials candidates by their stability, (ii) predict the

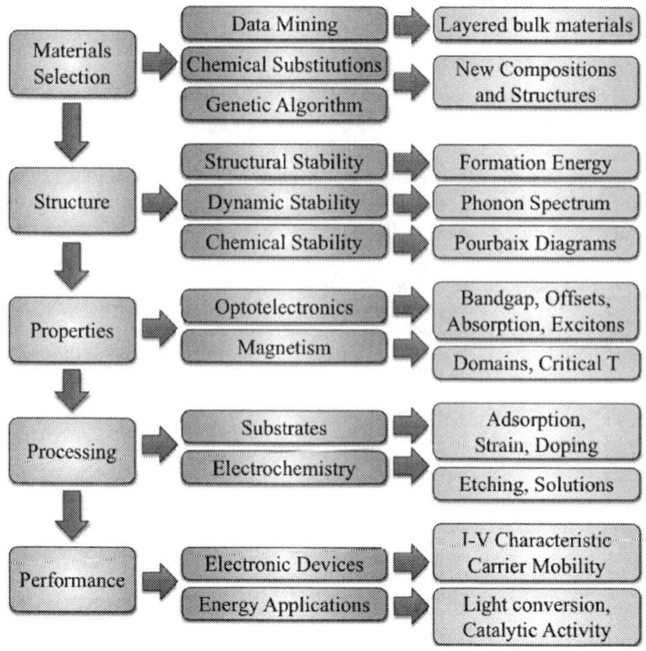

Figure 1. Flowchart for the computational discovery, characterization, and design of novel 2D materials.

properties of the 2D materials, (iii) identify possible synthesis conditions. The Materials Project and MPInterfaces framework provide the software tools to automate the generation of various 2D candidate structures, identify the competing phases, and prepare the input files for the DFT calculations of the energetic stability and properties of the materials, including the potential stabilization on substrates and electrochemical synthesis conditions.

2D materials selection. Prediction of promising 2D materials candidates occur by one of four general approaches: (i) data-mining of bulk crystal structure databases for layered structures that are either held together by van der Waals interactions or that can be electrochemically etched into 2D layers, (ii) chemical substitutions of known 2D materials, and (iii) direct computational searches of the configuration and composition space of 2D materials.

Stability of 2D materials structures. The stability of 2D materials is controlled by their formation energy. All 2D materials are thermodynamically metastable as the energy of any 2D material can always by lowered through attractive van der Waals interactions leading to layered bulk compounds. In fact, searching for layered bulk compounds that exhibit a van der Waals gap is one of the most promising approaches to identify potential 2D materials. In our work, we consider the formation energy, ΔE_f, of a 2D material relative to the energies of competing bulk material structure(s). Empirically, all 2D materials that have experimentally been synthesized as free-standing films exhibit formation energies below 200 meV/atom compared to the energy of the competing bulk phase(s). In our studies, we employ the empirical bound of 200 meV/atom as a criterion to screen candidate 2D materials.

Density-functional calculations can assess the energy of candidate 2D materials and their competing phases. For materials systems with layered bulk structures, it is important to include van der Waals interactions when considering the formation energies. In addition to screening a 2D material by its formation energy, it is helpful to also determine its dynamic stability by calculating the phonon spectrum and identifying possible unstable modes, which could lead to lower-energy structures.

Furthermore, many applications require the 2D material to be exposed to air and humidity or other aqueous environments. In those cases, it is important to determine the thermodynamics stability of the 2D material as a function of pH and applied potential. This can be achieved by estimating the Pourbaix diagram of the 2D material.

Property characterization. DFT and many-body methods can accurately predict the electronic, optical, and magnetic properties of materials as well as the coupling coefficient important for properties such as piezoelectricity, magnetostriction, *etc*. Calculations of electronic bandgaps and band edge positions require methods beyond local and semilocal exchange-correlation functionals. Hybrid functionals and *GW* methods can provide these electronic properties with sufficient accuracy for device applications.[12] Optical properties such as adsorption spectra and the exciton binding energy needed for calculations of the optical bandgap require the use of the Bethe-Salpeter equation.[13]

Predicted 2D materials structures, properties, and processing. We discovered many 2D materials in the families, of oxides,[14] chalcogenides,[15-17] pnictides,[18] and carbides[19,20] with properties suitable for electronic devices and solar energy technologies, such as photocatalysis.[15,21-23] For many of these materials we determine their electronic properties and piezoelectric coefficients[24] using the hybrid functional HSE06 or the G_0W_0 many-body method.

For several chalcogenide compounds we find that ferromagnetic order emerges at temperatures accessible to experiments. This opens the opportunity to investigate the interplay of magnetic order and reduced dimensionality and may provide materials for spintronics applications. In the 2D VS_2 compound,[25] our calculations predict that the magnetic moments are localized on the V atoms and couple ferromagnetically due to the super-exchange interactions mediated by the S atoms. The magnetic anisotropy shows an easy plane for the magnetic moment and the critical temperature, T_c, for the Berezinsky-Kosterlitz-Thouless transition of 2*H* single-layer VS_2 is predicted to be about 90 K. Another 2D material, the Fe_3GeTe_2 compound,[26] exhibits a magnetic moment that originates from a Stoner instability. Interestingly, the single-layer Fe_3GeTe_2 exhibits a significant uniaxial magnetocrystalline anisotropy energy of 920 μeV per Fe atom originating from spin-orbit coupling. This shows that 2D Fe_3GeTe_2 should exhibit a ferromagnetic phase transition. The anisotropy is further enhanced by biaxial tensile strains, resulting in strong magnetostriction with a sizable magneostrictive coefficient.

Using our python-based high-throughput framework, we identify electrochemical synthesis conditions and suitable substrates that can stabilize several of these as-yet hypothetical materials.[23,27-29]

Conclusions

The search for 2D materials with unique and useful properties is an ongoing effort in our and other groups. The rapid computational prediction of novel functional 2D materials can provide guidance for experimental synthesis efforts and future searches of materials suitable for applications in energy technologies and electronic devices.

Acknowledgement. This work was supported by the National Science Foundation under the CAREER Award No. DMR-1056587 and under Award No. ACI-1440547. This research used computational resources provided by the University of Florida Research Computing, the Texas Advanced Computing Center under Contract No. TG-DMR050028N, and the Extreme Science and Engineering Discovery Environment (XSEDE), which is supported by National Science Foundation Grant No. ACI-1053575.

References

(1) Zhuang, H. L.; Hennig, R. G. *JOM* **2014**, *66* (3), 366–374.
(2) Singh, A. K.; Mathew, K.; Davydov, A. V.; Hennig, R. G.; Tavazza, F. *Proc. of SPIE* **2015**, *9553*, 955316.
(3) Revard, B. C.; Tipton, W. W.; Yesypenko, A.; Hennig, R. G. *Phys. Rev. B* **2016**, *93* (5), 054117.
(4) Kresse, G.; Hafner, J. *Phys. Rev. B* **1993**, *47* (1), 558–561.
(5) Kresse, G.; Furthmüller, J. *Phys. Rev. B* **1996**, *54* (16), 11169–11186.
(6) Ong, S. P.; Jain, A.; Hautier, G.; Kocher, M.; Cholia, S.; Gunter, D.; Bailey, D.; Skinner, D.; Persson, K. A.; Ceder, G. *The Materials Project*; 2011.
(7) Mathew, K.; Sundararaman, R.; Letchworth-Weaver, K.; Arias, T. A.; Hennig, R. G. *J. Chem. Phys.* **2014**, *140* (8), 084106.
(8) Tipton, W. W.; Bealing, C. R.; Mathew, K.; Hennig, R. G. *Phys. Rev. B* **2013**, *87* (18), 184114.
(9) Tipton, W. W.; Hennig, R. G. *J. Phys.: Condens. Matter* **2013**, *25* (49), 495401.
(10) Tipton, W. W.; Matulis, C. A.; Hennig, R. G. *Comput. Mater. Sci.* **2014**, *93*, 133–136.
(11) Revard, B. C.; Tipton, W. W.; Hennig, R. G. In *Prediction and Calculation of Crystal Structures*; Atahan-Evrenk, S., Aspuru-Guzik, A., Eds.; Topics in Current Chemistry; Springer International Publishing: Cham, 2014; Vol. 345, pp 181–222.
(12) Heyd, J. J.; Peralta, J. E.; Scuseria, G. E.; Martin, R. L. *J. Chem. Phys.* **2005**, *123* (17), 174101.
(13) Salpeter, E.; Bethe, H. *Phys. Rev.* **1951**, *84* (6), 1232–1242.
(14) Zhuang, H. L.; Hennig, R. G. *Appl. Phys. Lett.* **2013**, *103* (21), 212102.
(15) Zhuang, H. L.; Hennig, R. G. *J. Phys. Chem. C* **2013**, *117* (40), 20440–20445.
(16) Singh, A. K.; Hennig, R. G. *Appl. Phys. Lett.* **2014**, *105* (4), 042103.
(17) Zhuang, H. L.; Johannes, M. D.; Blonsky, M. N.; Hennig, R. G. *Appl. Phys. Lett.* **2014**, *104* (2), 022116.
(18) Zhuang, H. L.; Hennig, R. G. *Appl. Phys. Lett.* **2012**, *101* (15), 153109.
(19) Ashton, M.; Mathew, K.; Hennig, R. G.; Sinnott, S. B. *J. Phys. Chem. C* **2016**, *120* (6), 3550–3556.
(20) Ashton, M.; Hennig, R. G.; Sinnott, S. B. *Appl. Phys. Lett.* **2016**, *108* (2), 023901.
(21) Zhuang, H. L.; Hennig, R. G. *Chem. Mater.* **2013**, *25* (15), 3232–3238.
(22) Zhuang, H. L.; Hennig, R. G. *Phys. Rev. B* **2013**, *88* (11), 115314.
(23) Singh, A. K.; Mathew, K.; Zhuang, H. L.; Hennig, R. G. *J. Phys. Chem. Lett.* **2015**, *6* (6), 1087–1098.
(24) Blonsky, M. N.; Zhuang, H. L.; Singh, A. K.; Hennig, R. G. *ACS Nano* **2015**, 150911095522002.
(25) Zhuang, H. L.; Hennig, R. G. *Phys. Rev. B* **2016**, *93* (5), 054429.
(26) Zhuang, H. L.; Kent, P. R. C.; Hennig, R. G. *Phys. Rev. B in print*.
(27) Singh, A. K.; Hennig, R. G. *Appl. Phys. Lett.* **2014**, *105* (5), 051604.
(28) Singh, A. K.; Zhuang, H. L.; Hennig, R. G. **2014**, *89* (24), 245431.
(29) Singh, A. K.; Hennig, R. G.; Davydov, A. V.; Tavazza, F. *Appl. Phys. Lett.* **2015**, *107* (5), 053106.

MXENE-TiO_2 COMPOSITES AS EFFECTIVE PHOTOCATALYSTS FOR HYDROGEN PRODUCTION UNDER VISIBLE LIGHT IRRADIATION

Hui Wang[†*a], *Rui Peng*[†a], *Zachary D. Hood*[a][b], *Michael Naguib*[c], *Shiba P. Adhikari*[d], *and Zili Wu*[*a]

[a] Center for Nanophase Materials Sciences, Oak Ridge National, Laboratory, Oak Ridge, TN 37830

[b] School of Chemistry and Biochemistry, Georgia Institute of Technology Atlanta, Georgia, 30332, USA

[c] Materials Science Technology Division, Oak Ridge National Laboratory, Oak Ridge, TN, 37831, USA

[d] Department of Chemistry, Wake Forest University Winston-Salem, NC, 27109, USA

E-mail: wanghl@ornl.gov, wuzl@ornl.gov

Introduction

Hydrogen production triggered from the splitting of water by harnessing unlimited solar energy is a fascinating scenario and ultimate solution for the energy crisis and its accompanied environmental issues. Since the ground-breaking work reported by Fujishima and Honda in 1972, in which hydrogen was photoelectrochemically generated by utilizing n-type TiO_2 and Pt as electrodes, research has expanded extensively in this field.[1] A myriad of semiconductors, such as TiO_2, ZnO, CdS etc., have been developed for the application of solar hydrogen production.[2] However, pure TiO_2 photocatalyst exhibits only moderate activity towards solar hydrogen generation reaction. This can be attributed to the fast charge carrier recombination in bulk TiO_2. To address this pitfall, several strategies have been proposed to effectively enhance the charge carrier separation in TiO_2. Coupling TiO_2 with another semiconductor with suitable band gap and band edge position has been widely adopted to extend the light absorption range to visible light and minimize the electron hole recombination for enhanced photcatalytic activity. Another strategy is to load a noble metal as a co-catalyst onto TiO_2, which enhances the overall photocatalytic activity. The most commonly employed noble metal co-catalyst contains Au, Pt, Ag, Pd etc.[3] However, these noble metals are scarce and expensive. Thus, the wide application and large-scale manufacturing of the noble metal containing materials is forbidden. Therefore, it is indispensable to explore competent alternatives for the precious metal co-catalyst. Recently, a number of metal analogous materials, including graphene and MoS_2, have been developed and employed as co-catalyst for photocatalytic hydrogen evolution reaction.

MXenes, a new family of two-dimensional (2D) transition metal carbides and carbonitrides, has recently attracted intense attention.[4] MXenes are produced by etching out the A layers from MAX phases, where M stands for an early transition metal (Ti, V, Nb, Ta), A represents an A-group (mostly groups 13 and 14) element, and X corrsponds to carbon or nitrogen. Since the etching is carried out in aqueous system containing fluorides, the as synthesized MXenes have a mixture of surface terminations (O, OH, and F), hence after will be referred as T_x. Since the first MXene was reported in 2011 ($Ti_3C_2T_x$ from Ti_3AlC_2 by etching Al), the MXene family encountered a fast expansion to more than 15 compounds, including Ti_2CT_x, Nb_2CT_x, V_2CT_x, Ti_3CNT_x, $Ta_4C_3T_x$, and so on. Beside the discovery of new MXenes, both simulation modeling and experimental characterizations were employed to understand the structure and properties of MXenes.[5] This work expands the application of MXenes as co-catalysts for photocatalytic hydrogen evolution from water splitting by promoting the transportation of photogenerated electrons from TiO_2 to metallic analogous metal carbides and facilitating the electron-hole separation in photocatalysis processes. Herein, we mainly focus on $Ti_3C_2T_x$ metal carbide, the first reported and the most studied MXene, as a potential co-catalyst to assist rutile TiO_2 for visible-light-driven hydrogen evolution from the photocatalytic splitting of water. $TiO_2/Ti_3C_2T_x$ composite photocatalyst was prepared by a rapid and facile hydrothermal synthetic route. Moreover, the composite photocatalyst exhibits significantly enhanced activity in terms of the solar hydrogen production compared to the pure TiO_2 under the identical experimental conditions. We also explored two other MXenes (Ti_2CT_x and Nb_2CT_x) to broaden the scope and to provide more insight as to their merits in the application of photocatalytic hydrogen generation reaction. To the best of our knowledge, this work is the first report that employs the emerging transition metal carbide MXenes as co-catalysts to assist TiO_2 for *visible-light-driven hydrogen evolution from the photocatalytic splitting of water.*

Experimental

Synthesis. A hydrothermal method was employed to synthesize $TiO_2/Ti_3C_2T_x$ composites by using titanium tetrachloride ($TiCl_4$, 98%, Sigma) as the precursor. A desired amount of $TiCl_4$ was dissolved in delaminated $Ti_3C_2T_x$ suspension (50 mL in a 150 mL container) in an ice-water bath. The mixture was stirred at a high speed while $TiCl_4$ was added dropwise. The mixture was heated to 95 °C and held at this temperature for 4 h, then cooled down to room temperature in air. The suspension was kept stirring vigorously during the entire process. After centrifugation at 8000 rpm for 10 min, the sediment was collected and dried at 80 °C.

Characterizations

In this work, powder X-ray diffraction (PXRD), Raman spectroscopy, UV-vis spectroscopy, scanning electron microscopy (SEM), and transmission electron microscopy (TEM) X-ray photoelectron spectroscopy (XPS), and photoluminescence techniques are adopted to examine the structure of the limited layered MoS_2.

Conclusions

In summary, we successfully demonstrated that MXenes, a new family of 2D transition metal carbides, serve as novel co-catalysts with TiO_2 for visible-light-induced hydrogen production. We also broaden the interest and provide more insights on the photocatalytic mechanism of the composite by exploring Nb_2CT_x and Ti_2CT_x MXenes, which demonstrates that MXenes in general can be employed as efficient co-catalysts to assist TiO_2 for solar hydrogen production.

Acknowledgement. This research was conducted at the Center for Nanophase Materials Sciences, which is a DOE Office of Science User Facility.

References

(1) Fujishima, A.; Honda, K., Electrochemical Photolysis of Water at a Semiconductor Electrode. *Nature* **1972,** *238* (5358), 37-38.

(2) Chen, X.; Shen, S.; Guo, L.; Mao, S. S., Semiconductor-based Photocatalytic Hydrogen Generation. *Chemical Reviews* **2010,** *110* (11), 6503-6570.

(3) Peng, R.; Baltrusaitis, J.; Wu, C.-M.; Koodali, R. T., Pd–Ti-MCM-48 cubic mesoporous materials for solar simulated hydrogen evolution. *International Journal of Hydrogen Energy* **2015,** *40* (2), 905-918.

(4) Naguib, M.; Kurtoglu, M.; Presser, V.; Lu, J.; Niu, J. J.; Heon, M.; Hultman, L.; Gogotsi, Y.; Barsoum, M. W., Two-Dimensional Nanocrystals Produced by Exfoliation of Ti3AlC2. *Advanced Materials* **2011,** *23* (37), 4248-4253.

(5) Khazaei, M.; Arai, M.; Sasaki, T.; Chung, C. Y.; Venkataramanan, N. S.; Estili, M.; Sakka, Y.; Kawazoe, Y., Novel Electronic and Magnetic Properties of Two-Dimensional Transition Metal Carbides and Nitrides. *Advanced Functional Materials* **2013,** *23* (17), 2185-2192.

CONTROLLING CHEMICAL AND ELECTROCHEMICAL PROPERTIES OF LOW-DIMENSIONAL NANOSTRUCTURES USING MECHANICAL STRAIN

Cary L. Pint, Nitin Muralidharan, Landon Oakes, and Rachel Carter*

Department of Mechanical Engineering, Vanderbilt University, Nashville TN 37235
*cary.l.pint@vanderbilt.edu

Traditional foundations of chemistry and electrochemistry of materials are forged in the absence of an understanding for how mechanical stresses are incorporated into chemical or electrochemical properties. Therefore, conventional routes to modulate chemical or electrochemical response of materials default to the modification of the atomic composition of a material or nanostructure. Here, I will demonstrate that strain engineering is a versatile tool to modify and control chemical processes in low-dimensional (2-D) materials. This is presented in the context of two different electrochemical systems: (1) 2-D MoS_2 nanosheets, where applied strain leads to modifications of the energetics for chemical conversion upon lithium insertion that activates the capability of this material to be a practical battery cathode. (2) ultrathin 2-D V_2O_5 materials produced by atomic layer deposition, where the thermodynamics and kinetics of Li^+ insertion and diffusion are significantly modified in strained materials. Our experimental observations open a window into a poorly explored intersection between mechanics and chemistry/electrochemistry of materials that has promise to provide an exciting future landscape to engineer materials, especially 2-D materials, for chemical and electrochemical applications.

IN-DEPTH STUDY OF Sn-BASED COMPOSITE MATERIALS AS ANODE FOR HIGH ENERGY Na-ION BATTERIES

Chuze Ma[a], Baihua Qu[b], Ying Ching Lu[c], Jim Yang Lee[b], Shigeto Okada[d], Ying Shirley Meng[a]*

a. Department of NanoEngineering, University of California San Diego, La Jolla, CA 92037, USA
b. Department of Chemical and Biomolecular Engineering, National University of Singapore, 10 Kent Ridge Crescent, 119260, Singapore
c. Department of Chemical Engineering, Interdisciplinary Graduate School of Engineering Sciences, Kyushu University, Fukuoka 819-0395, Japan
d. Institute for Materials Chemistry and Engineering, Kyushu University, Fukuoka 816- 8580, Japan.

Introduction

Following their commercialization in 1991, Li-ion batteries (LIBs) have been widely adopted for portable electronic devices and electric vehicles. Quickly this technology has become ubiquitous in societies' daily life. Although initially LIBs have been used in portable electronic devices, their use has extended to large-scale applications such as battery-assisted bicycles and electric vehicles. The advantages of LIBs over nickel-hydride and lead-acid batteries are high cell voltage and high energy density. In addition, due to their high charge/discharge efficiency and long cycle life, LIBs have recently been considered as a temporary energy storage system for unstable renewable energy sources. However, the limited lithium resources and high cost are drawbacks for large-scale applications. For this reason, Na-ion batteries (SIBs) have recently gained attention as a potential alternative to LIBs because of their lower cost and similar electrochemical behavior to LIBs. Nevertheless, potential electrode candidates for SIBs are limited, particularly for the negative electrodes. Some carbonaceous anodes have been investigated for SIBs using a similar analogy to LIBs.[1,2] However, thus far, the maximum rechargeable capacity of a hard carbon anode for sodium-ion batteries has been less than ca. 320 mA h g^{-1}.[3] On the other hand, several Na alloy anodes, such as tin (Sn),[4] antimony (Sb),[5] and phosphorus (P),[6] have been proposed to improve the rechargeable capacity. Among these, Sn is a very promising anode material for SIB anodes due to its high theoretical capacity ($Na_{15}Sn_4$: 854 mA h g^{-1}), good chemical stability, and low toxicity. However, the operating voltage is extremely close to the Na plating/stripping potential. Moreover, the large volume change and particle agglomeration during the charge and discharge cause capacity degradation of the metallic Sn. To reduce the volume change and prevent the particle agglomeration of Sn, several compounds and nanostructures have been proposed. Our team looked into various compounds such as SnO, SnO_2, SnS and SnS_2, and then combine them with various carbon-based matrix to further improve the properties of these active materials.[7-9] The studies turn out to be very positive results, indicating good promise of these composite materials in future energy storage applications. For instance, the 2D SnS_2 - reduced graphene oxide (rGO) composite demonstrates high reversible capacity (630 mAh g^{-1} at 0.2 A g^{-1}), good rate performance (544 mAh g^{-1} at 2 A g^{-1}) and long cycle-life (500 mAh g^{-1} at 1 A g^{-1} for 400 cycles). A comprehensive study on this material by in situ XAS, synchrotron XRD, TEM and Raman spectroscopy reveals the detailed reaction mechanism, which offer a well-balanced approach for sodium storage to create high capacity and cycle-stable anode materials.[10]

Experimental

Material Synthesis and characterization. The graphite powder from Alfa Aesar, tin (IV) chloride pentahydrate ($SnCl_4$•$5H_2O$), thioacetamide (TAA) and Na-alginate from Sigma-Aldrich, were used as received without further purification. GO was prepared from the graphite powder by a modified Hummers' method.[11] A typical synthesis of SnS_2-RGO composite is given as follows: 0.7015 g $SnCl_4$•$5H_2O$ (2 mmol) and 0.6014 g thioacetamide (TAA) (8 mmol) were added to 40 ml of 1.5 mg ml^{-1} GO suspension under stirring. The mixture was sonicated for 30 min before it was transferred to a 50ml Teflon-lined stainless steel autoclave, and heated at 160 °C for 12 h. After cooling to room temperature, the solid product was recovered by centrifugation, washed with deionized water and absolute alcohol, and dried in vacuum at 60 °C overnight. The dried product was then heated in Ar to 400 °C and kept at this temperature for 4 h to get the final product.

Synchrotron X-ray Diffraction (XRD). High-quality XRD patterns were collected in transmission mode at the 11-BMB beamline of the Advanced Photon Sourse (APS) using multiple single-crystal analyzer detectors. Electrode samples were prepared for *ex situ* measurement at a wavelength of 0.459 Å. XRD patterns were collected between 0.3° and 50.0° in 2θ angles.

X-ray Absorption Spectroscopy (XAS). The XAS work was performed at 20-BM-B at Argonne's Advanced Photon Source (APS). Transmission spectra were collected along with a simultaneous spectrum on a reference foil of metallic Sn to ensure consistent energy calibration. Energy calibration was carried out using the first derivative of the spectra of the Sn metal foil. The data were analyzed and refined using the Ifeffit package.

Electrochemical tests. Electrodes were prepared from slurries containing 80 wt% of active material, 10 wt% of Na-alginate as a binder, and 10 wt% Super P carbon black as a carbon additive in deionized water. The slurries were then casted on aluminum foils and dried in vacuum at 80 °C. Typical loading of the active material was around 2 mg cm^{-2}. A glass fiber GF/F (Whatman) filter was used as the separator and 1 M $NaPF_6$ in a 1:1 (v/v) mixture of ethylene carbonate (EC) and diethylene carbonate (DEC) solution was used as the electrolyte. Battery assembly was carried out in an MBraun glovebox (H_2O < 0.1ppm). Galvanostatic discharge and charge at various current densities were performed on an Arbin BT2000 battery cycler.

Results and Discussion

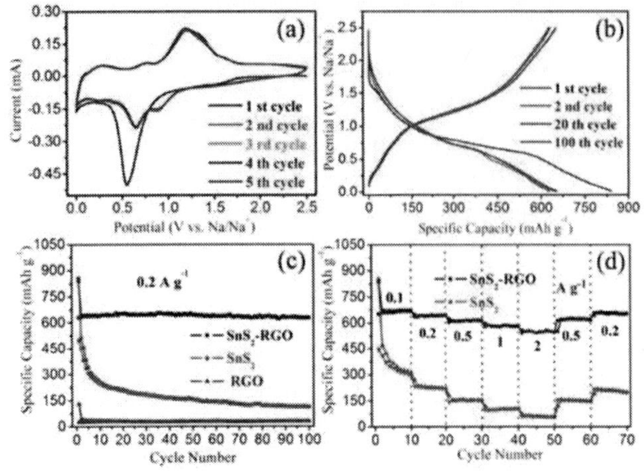

Figure 1. (a) The first five cycle voltammograms of the SnS_2-rGO electrode at 0.1 mV s^{-1}. (b) Voltage profiles of a SnS_2- rGO electrode in the 1st, 2nd, 20th and 100th cycles. (c) Cycling performance of SnS_2-rGO, SnS_2 and graphene electrodes at 0.2 A g^{-1} for 100 cycles, respectively. (d) Rate performance of SnS_2-rGO and SnS_2 electrodes.

Cyclic voltammetry (CV) was performed in the 0 to 2.5 V potential region at 0.1 mV s^{-1} (Figure 1a). In the first cycle, a reduction peak around 1.7 V can be assigned to the intercalation of Na-ions into SnS_2 (0 0 1) plane. The major reduction peaks around 0.6 V should correspond to the synergetic conversion and alloying reactions. The peaks around 0.3 V and 0.01 V could be ascribed to the formation of Na_2S_2 and $Na_{15}Sn_4$ alloy phase. During the charging, the oxidation peaks around 0.3 V and 0.8 V could originate from the desodiation of $Na_{15}Sn_4$ and Na_2S_2. While the broaden peak from 1.0 V to 1.5 V should be attributed to the desodiation of Na-Sn alloy phases as well as the reformation of SnS_2. The following electrochemical tests indicate excellent electrochemical properties of the composite material - high reversible capacity (630 mAh g^{-1} at 0.2 A g^{-1}), good rate performance (544 mAh g^{-1} at 2 A g^{-1}) and long cycle-life (500 mAh g^{-1} at 1 A g^{-1} for 400 cycles).

To obtain direct evidence of the phase transformations and reaction pathway during electrochemical cycling, we carried out ex situ SXRD and in situ XAS analysis. As shown in Figure 2 (a), tin metal first precipitates from the conversion reaction of Na and SnS_2 during the discharge. The XRD patterns of discharged sample at 0.3 V well matches the diffraction patterns of tin metal. The sample after fully discharge to 0.01 V shows diffraction peaks that can be associated with Na_2S_2 and $Na_{15}Sn_4$ phases, although the diffraction peaks from $Na_{15}Sn_4$ are weak. During the charging process, $Na_{15}Sn_4$ desodiates and transforms to tin metal, while Na_2S_2 also disappear gradually. At 1.5 V, tin metal reforms according to the diffraction result. After fully charged to 2.5 V, the material turns into amorphous state, which does not go back to the crystalline SnS_2 phase.

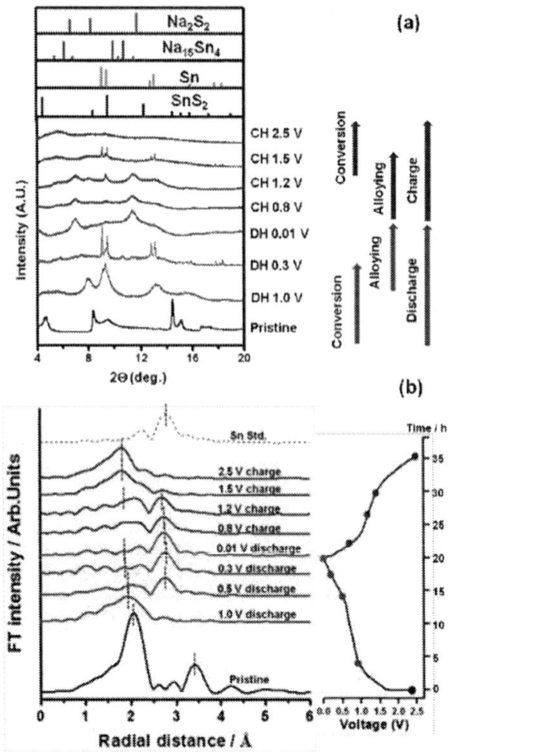

Figure 2. (a) Synchrotron X-ray diffraction results of SnS_2-rGO electrodes at different discharge/charge states. (b) Sn K-edge EXAFS spectra of SnS_2-rGO electrodes at different voltages in the first cycle.

In Figure 2 (b), it shows the EXAFS results of the SnS_2-rGO electrodes at different discharge/charge states. The shells around 1.9 Å and 3.4 Å in the pristine sample are related with the Sn-S and Sn-Sn bonds respectively. During the first discharge, the intensity of both peaks decrease. At the same time, a shell around 2.8 A appears, which belongs to the Sn-Sn bond in tin metal. With the formation of Na-Sn alloy phase, the Sn-Sn shell slight shifts from 2.8 A to 2.7 A at the fully discharged state. During the first charge, the Sn-S shell begins to appear again, accompanied by the disappearing of Sn-Sn shell. However, it is obvious that the Sn-S shell has less radial distance and intensity comparing with the pristine SnS_2-rGO sample. The Sn-Sn shell of the pristine sample does not show up at the fully charged state. These results indicate the formation of nano-clustered SnS_2, instead of the crystalline SnS_2 phase as it is in the pristine state. By linking the XRD and EXAFS results, we proposed a synergistic mechanism that takes advantages of both conversion and alloying reactions, ensuring the high specific capacity, long cycle life and fast kinetics. The details are discussed in our publication.[10]

Conclusions

In our study, various carbon materials were used to improve the electrochemical performance of Sn based anode compounds. Among them, the 2D reduced graphene oxide exhibits the best performance as a conductive and buffer matrix that greatly improves the performance of many materials. As a result, the SnX-rGO composite materials demonstrate much better properties comparing with the Sn metal. These novel Na anode materials have also been comprehensively studied by using various material characterization techniques such as SXRD, XAS, TEM, Raman, etc. New reaction mechanism has been proposed and will shed light on future material design and development.

Acknowledgement. We acknowledge the funding support from the National Science Foundation under Award Number DMR1057170.

References

(1) M. M. Doeff, Y. P. Ma, S. J. Visco and L. C. Dejonghe, *J. Electrochem. Soc.* **1993**, *140*, L169
(2) R. Alcantara, J. M. Jimenez-Mateos, P. Lavela and J. L. Tirado, *Electrochem. Commun.* **2001**, *3*, 639.
(3) D. A. Stevens and J. R. Dahn, *J. Electrochem. Soc.* **2000**, *147*, 1271.
(4) S. Komaba, Y. Matsuura, T. Ishikawa, N. Yabuuchi, W. Murata and S. Kuze, *Electrochem. Commun.* **2012**, *21*, 65.
(5) A. Darwiche, C. Marino, M. T. Sougrati, B. Fraisse, L. Stievano and L. Monconduit, *J. Am. Chem. Soc.* **2012**, *134(51)*, 20805.
(6) J. Qian, X. Wu, Y. Cao, X. Ai and H. Yang, *Angew. Chem., Int. Ed.* **2013**, *52*, 4633.
(7) Y. C. Lu, C. Ma, J. Alvarado, N. Dimov, Y. S. Meng, and S. Okada, *J. Mater. Chem. A* **2015**, *3*, 16971–16977
(8) Y. C. Lu, C. Ma, J. Alvarado, T. Kidera, N. Dimov, Y. S. Meng, and S. Okada, *J. Power Sources* **2015**, *284*, 287–295
(9) B. H. Qu, C. Ma, G. Ji, C.H. Xu, J. Xu, Y.S. Meng, T. H. Wang, J. Y. Lee, *Adv. Mater.* **2014,** *26* (23), 3854-3859.
(10) C. Ma, J. Xu, J. Alvarado, B. H. Qu, J. Somerville, J. Y. Lee, Y. S. Meng, *Chem. Mater.* **2015**, *27*, 5633–5640
(11) W. S. Hummers, R. E. Offeman, *J. Am. Chem. Soc.* **1958,** *80* (6), 1339-1339.

BUCKLED ATOMIC SHEETS: ADVENTURES WITH PHOSPHORENE AND SILICENE FOR LOW-ENERGY SYSTEMS (INVITED)

Deji Akinwande

Microelectronics Research Center, The University of Texas – Austin, USA.
*deji@ece.utexas.edu

Abstract

Layered atomic sheets beyond graphene have garnered great interest over the past 5years, primarily motivated by their unique properties beyond what graphene can offer [1, 2]. In particular, black phosphorus (BP) and silicene, the phosphorus and silicon analogues of graphene, respectively, have been a central focus. BP offers both high mobility and sizeable bandgap and can enable optoelectronics and sensors that afford both higher speeds and lower operation energies compared to graphene [3]. Silicene is very promising for quantum phenomena [4], including the quantum spin Hall effect, popularly known as the 2D topological insulator state that promises a reinvention of computation with unprecedented low-energy operation. We will present our work on experimental BP and silicene materials and transistors, progress, breakthroughs, concepts and contemporary challenges.

Acknowledgement. This work is supported in part by the Army Research Office (ARO) and the Office of Navy Research (ONR).

References

(1) Bhimanapati, G. R.; Lin, Z.; Meunier, V.; Jung, Y.; Cha, J. J.; Das, S.; Xiao, D.; Son, Y.; Strano, M. S.; Cooper, V. R.; Liang, L.; Louie, S. G.; Ringe, E.; Zhou, W.; Sumpter, B. G.; Terrones, H.; Xia, F.; Wang, Y.; Zhu, J.; Akinwande, D.; Alem, N.; Schuller, J. A.; Schaak, R. E.; Terrones, M.; Robinson, J. A. ACS Nano 2015.

(2) Akinwande, D.; Petrone, N.; Hone, J. Nature Communications 2014, 5.

(3) Zhu, W.; Yogeesh, M. N.; Yang, S.; Aldave, S. H.; Kim, J.; Sonde, S. S.; Tao, L.; Lu, N.; Akinwande, D. Nano Letters 2015, 15, 1883–1890.

(4) Tao, L.; Cinquanta, E.; Chiappe, D.; Grazianetti, C.; Fanciulli, M.; Dubey, M.; Molle, A.; Akinwande, D. Nature Nanotechnology 2015, 10, 227–231.

APPLICATION OF TWO DIMENTIONAL STRUCTURES FOR PEROVSKITE SOLAR CELLS

Jinsong Huang

Department of Mechanical and Materials Engineering and Nebraska Center for Materials and Nanoscience, University of Nebraska-Lincoln, Lincoln, Nebraska 68588-0656, USA
Email: jhuang2@unl.edu

The instability of hybrid perovskite materials to water and moisture arises as one major challenge to be addressed before any practical application after the demonstration of very high efficiency perovskite solar cells. Here I will report our recent progress in application of two dimensional structure perovskites for stability improvement of hybrid perovskite solar cells. The challenges on the charge transport and optical properties for two dimensional perovskite will also be briefed, and their prospectus on future applications will be critically commented.

OPTOELECTRONIC PROPERTIES OF MONOLAYER MoS_2 AND WSe_2-MoS_2 LATERAL HETEROSTRUCTURES PROBED USING PHOTOCURRENT SPECTRAL ATOMIC FORCE MICROSCOPIC IMAGING

Michael S. Strano

Carbon P. Dubbs Professor
MIT, 66-566 Department of Chemical Engineering
77 Massachusetts Avenue
Cambridge, MA 02139-4307
Email: strano (at) MIT.EDU
phone: (617) 324-4323 fax: (617) 258-8224
http://web.mit.edu/stranogroup/

In the pursuit of two-dimensional (2D) materials beyond graphene, enormous advances have been made in exploring the exciting and useful properties of transition metal dichalcogenides (TMDCs)—such as a permanent band gap in the visible range and the transition from indirect to direct band gap due to 2D quantum confinement—and their potential for a wide range of device applications. In particular, recent success in the synthesis of seamless monolayer lateral heterostructures of different TMDCs via chemical vapor deposition (CVD) methods has provided an effective solution to producing an in-plane p-n junction, which is a critical component in electronic and optoelectronic device applications. However, spatial variation of the electronic and optoelectonic properties of the synthesized heterojunction crystals—throughout the homogeneous as well as the lateral junction region—and the charge carrier transport behavior at their nanoscale junctions with metals remain unaddressed. In this work, we use photocurrent spectral atomic force microscopy (PCS-AFM) to image the current (in the dark) and photocurrent (under illumination) generated between a biased PtIr tip and a monolayer WSe2-MoS2 lateral heterostructure. Current measurements in the dark in both forward and reverse bias reveal an opposite characteristic diode behavior for WSe2 and MoS2, owing to the formation of a Schottky barrier of dissimilar properties. Notably, by changing the polarity and magnitude of the tip voltage applied, pixels that show the photoresponse of the heterostructure are observed to be selectively switched on and off, allowing for the realization of a hyper-resolution array of the switchable photodiode pixels. This experimental approach has significant implications toward the development of novel optoelectronic technologies for regioselective photodetection and imaging at nanoscale resolutions. Comparative 2D Fourier analysis of physical height and current images shows high spatial frequency variations in substrate/MoS2 (or WSe2) contact that exceed the frequencies imposed by the underlying substrates. These results should provide important insights in the design and understanding of electronic and optoelectronic devices based on quantum confined atomically thin 2D lateral heterostructures.

GRAPHENE OXIDE BASED CATHODE FOR LITHIUM-SULFUR BATTERIES

Juchen Guo

University of California, Riverside, CA 92521
jguo@engr.ucr.edu

Introduction

Lithium-sulfur (Li-S) batteries have attracted great attention because of their high theoretical energy density (2600 Wh kg^{-1}), in addition to the usage of nontoxic and low cost sulfur cathode[1]. However, due to its low conductivity and the shuttling of lithium polysulfide, the performance of Li-S batteries are hindered by the low utilization and rapid capacity fading[2]. In order to solve these problems, varies carbon materials (such as carbon nanotube[3], porous carbon[2], carbon nanofibers[4]) have been used as carbon matrices to form carbon-sulfur hybrids. These carbon matrices have good electrical conductivity, high porosity and large specific surface area, which can store sulfur and restrict the shuttling of the polysulfide. However, aimed to reach both high energy and power densities, the pores in these carbon materials need to be well designed.

Graphene oxide, as a single-atom-thick carbon material, has been used as a promising carbon matrix for sulfur cathode. This 2D structure was reported to serve both as a good electron conductor, also as may confine the dissolution of polysulfide. In recent report, a chemical reaction-deposition strategy was used to immobilize sulfur and lithium polysulfide, this strong interaction between graphene oxide and sulfur or polysulfide enabled the lithium/sulfur cells with a high reversible capacity[1]. Despite this progress, it is still a challenge to obtain a high loading of sulfur and long cyclic stability simultaneously.

In this work, porous carbon with sandwich structure was carefully designed for lithium-sulfur batteries by using graphene oxide as a shape-directing template and conductive filler. These graphene sheets were uniformly coated with carbon on both sides, which provided enough space for sulfur and polysulfide, also shortened the ion transfer distance, ensuring a good performance of the carbon-sulfur hybrid. In general these sandwich carbons show high specific surface areas and high pore volume, demonstrating an excellent rate performance and long cycling performance.

Experimental

Preparation of the sandwich carbon materials. An aqueous dispersion of GO was purchased from Graphenea without further treatment. The GO/amorphous carbon composites were prepared by in situ chemical coating of melamine-resorcinol-formaldehyde resin onto GO sheets, followed by carbonization and activation. Typically, 2.86 g resorcinol and 4.22 g formaldehyde (37 wt % solution) were dissolved in 30 mL deionized water (solution **A**), 3.28 g melamine and 6.32 g formaldehyde were dissolved in 30 mL deionized water (solution **B**) at 80 °C. Then certain amount of the aqueous dispersion of GO was added into the mixed solution of A and B under ultrasound treatment for 10 min. The mixture was then stirred at 80 °C for 24 h. The obtained polymer-coated GO sheets were filtered and dried overnight. The GO/MRF nanocomposites were carbonized at 1000 °C in Ar flow for 3 h at the rate of 3 °C/min to obtain the GO/carbon composites.

The CO_2 activation was performed by placing a ceramic boat with 0.1 g of GO/carbon composites in the tube furnace under flowing argon with a heating rate of 5 °C/min up to 900 °C. After this temperature was reached, the activating gas was introduced for different time from 2 hour to 6 hours.

Characterization. The morphologies of the nanocomposites were observed under scanning electron microscopy (FEI NNS450). The microstructures were observed under transmission electron microscopy (FEI Tecnai 12) operated at 120 kV. Nitrogen adsorption / desorption was performed using an ASAP 2020 instrument at 77. The specific surface area and the pore size distribution were calculated using the Brunauer-Emmett-Teller (BET) and density functional theory (DFT) methods, respectively. Thermal gravimetric analysis (TGA) was conducted on a TG instrument (Q500) under nitrogen protection at a heating rate of 10 °C/min from room temperature to 600 °C.

Electrochemical Test. The electrochemical tests were carried out with 2032 coin cell by mixing 80 wt% active material, 10 wt% carbon black (super C65) and 10 wt% polymer binder dissolved in deionized water. The slurry was coated onto a carbon-coated aluminum foil with a doctor blade and dried at 50 oC overnight. The electrolyte was 1M bis(trifluoromethane) sulfonamide lithium salt (LiTFSI) dissolved in a mixture of $PYR_{14}TFSI$, 1, 2-dioxolane (DOL) and dimethoxymethane (DME) (2:1:1 by volume) with 1.5 wt% of $LiNO_3$. A Celgard membrane 2400 was selected as the separator. The GCD experiments of Lithium-Sulfur batteries were performed on an Arbin battery test station, and the CV analysis was conducted on a Gamry Interface 1000 with a scan rate of 0.1 mV s^{-1}.

Results and Discussion

The key to the fabrication of the uniform coating of melamine-resorcinol-formaldehyde polymer onto the surface of GO sheets lies in the electrostatic interaction. GO with a large number of oxygenated groups on the surface serves as nucleus, which absorbed uniform deposition of the MRF polymer. After the phase separation process, the obtained GO/polymer sheets were chemically activated with CO_2 for certain hours. As shown in Fig. 1, the GO acts as a shape-directing template and facilitates the uniform coating of MRF polymer. After carbonization and activation, the GO/carbon maintains the sheet-like structure. The TEM image also indicates that the GO is fully coated by the carbon layer.

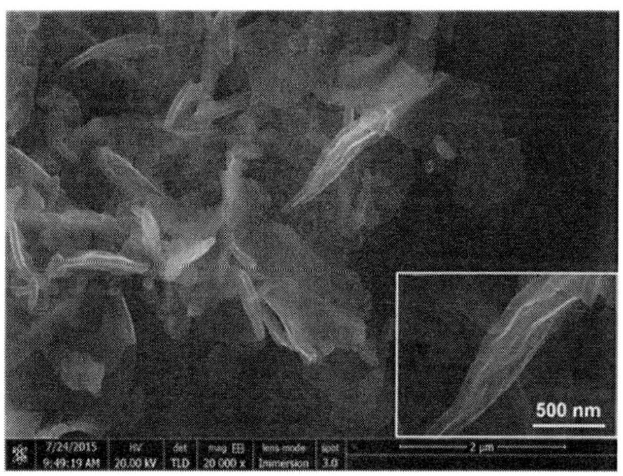

Figure 1. SEM image of GO/carbon sheets after carbonization.

To further investigate the pore structure of GO/carbon sheets, N_2 adsorption measurements were performed and the results are shown in Fig. 2. The direct carbonization of GO/polymer sheets created considerable micropores, while CO_2 activation process produced higher surface area and pore volume. The mild oxidant CO_2 can eliminate carbon atoms via the reaction of C and CO_2, creating a carbon sheets with high porosity. All five samples exhibit type I characteristic, suggesting the existence of abundant micropores. The pore size distribution (PSD) calculated by the density functional theory (DFT) shows that the pores in the GO/carbon sheets are smaller than 4 nm, increased with increasing activation time.

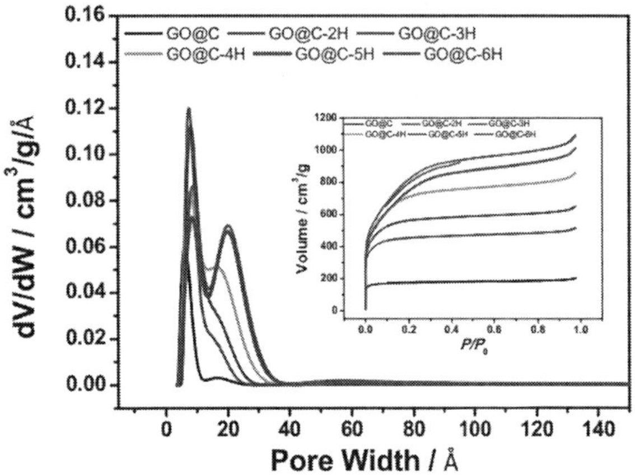

Figure 2. N_2 adsorption-desorption isotherms (inset) and pore size distribution of GO@C sheets.

The S-GO@C composites were prepared through the heat melt-diffusion method. The electrochemical performance of the GO/carbon-S hybrid was evaluated using coin cells (2032). As shown in Fig. 3, different GO@C host demonstrated different overall capacity (based on overall S-GO@C mass) and capacity retention: The GO@C-5h (activation for 5 hours) demonstrated the best overall capacity and cycle stability.

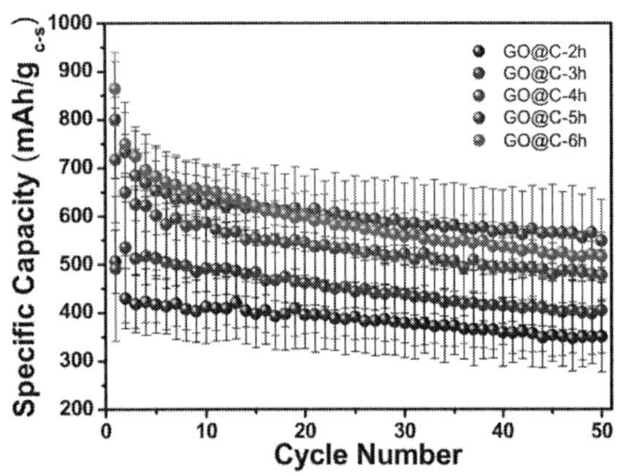

Figure 3. Cycle stabilities of the S-GO@C composites.

In addition to the effect of carbon host structure, we further investigated the effect of binders. Polymer binder should play an active role in addition to the function of binding. The discharge process of sulfur cathode is to electrochemically reduce polysulfide anions on the negatively charged surface, therefore the electrostatic repulsion of the polysulfide anions from the surface could be a significant problem that has been largely overlooked by the community. Based on this concept, we proposed to use positively charged polymers (ionomer) as the binders to mitigate the electrostatic repulsion. Scheme 1 shows the three polymers used in this study: polyvinylpyrrolidone (PVP), polyquaternium (D11) and Poly(diallyldimethyl ammonium chloride) (PDADMAC). The majority of D11 backbone is as same as PVP, containing a positively small segment (x:y = 26:1). PDADMAC contains more positive charge than D11, and the original Cl- was exchanged to triflate anions avoid potential corrosion. The cycle stability results of S-GO@C-5H cathode using these three binders in Fig. 4 clearly demonstrated the superior performance of positively charged ionomers. PDADMAC, which contains the highest charge density, shows the best performance.

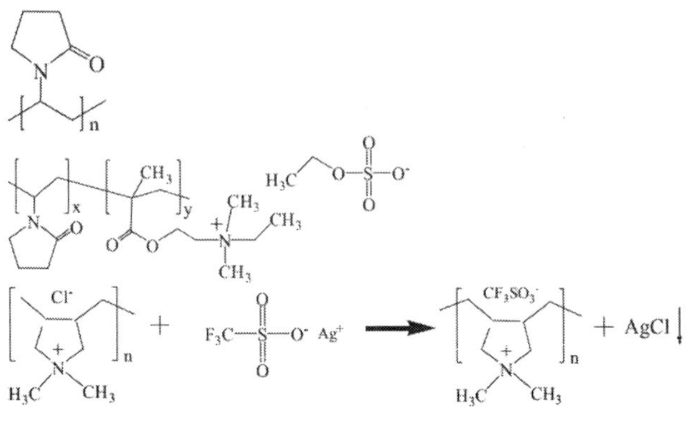

Scheme 1. Three polymer binders used in this sutdy: PVP, D11 and PDADMAC.

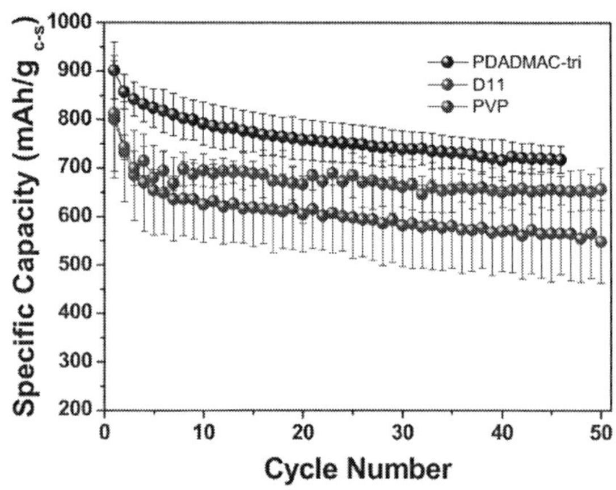

Figure 4. Cycle stability of S-GO@C-5H cathode with different binders including PVP, D11 and PDADMAC.

In conclusion, the GO@C sheets showed great advantages as the sulfur hosts for lithium-sulfur batteries. Positively charged ionomer binders also demonstrated improved performance.

References

(1) Yang, X.; Zhang, L.; Zhang, F.; Huang, Y.; Chen, Y., Sulfur-infiltrated graphene-based layered porous carbon cathodes for high-performance lithium–sulfur batteries. *ACS nano* **2014,** *8* (5), 5208-5215.

(2) Zhou, W.; Wang, C.; Zhang, Q.; Abruña, H. D.; He, Y.; Wang, J.; Mao, S. X.; Xiao, X., Tailoring Pore Size of Nitrogen-Doped Hollow Carbon Nanospheres for Confining Sulfur in Lithium–Sulfur Batteries. *Advanced Energy Materials* **2015.**

(3) Zhao, Y.; Wu, W.; Li, J.; Xu, Z.; Guan, L., Encapsulating MWNTs into Hollow Porous Carbon Nanotubes: A Tube-in-Tube Carbon Nanostructure for High-Performance Lithium-Sulfur Batteries. *Advanced Materials* **2014,** *26* (30), 5113-5118.

(4) Miao, L.; Wang, W.; Yuan, K.; Yang, Y.; Wang, A., A lithium–sulfur cathode with high sulfur loading and high capacity per area: a binder-free carbon fiber cloth–sulfur material. *Chemical Communications* **2014,** *50* (87), 13231-13234.

INTERLAYER EXCITON BETWEEN GRAPHENE AND MONOLAYER MoS_2

*ZHANG, Qicheng; NAYLOR, Carl; GAO, Zhaoli; WU, Ruizhe; LUO, Zhengtang**

Department of Chemical and Biomolecular Engineering, Hong Kong University of Science and Technology, Clear Water Bay, Kowloon, Hong Kong
keztluo@ust.hk

Introduction

Novel physics in reduced dimensions renders the van der Waals interfaces of graphene-like two-dimensional (2D) semiconductor crystals interesting. For example, poorly screened Coulomb potential resulted from the low dielectric constants or the 2D geometry of semiconducting transition-metal dichalcogenide (TMDC) interfaces, leads to a tightly bound interlayer exciton[1]. Due to the reduced dimension, interfacial exciton is thought to highly localized in space and thus uncertain in momentum space, rising up to unexpected independence of the orientation of the heterostructure[1]. Besides, type-II semiconductor junction band alignment gives to tunable gap[2], which can also be achieve from strong electric field even for two identical TMDC layers[3].

In this work, it was found that interfacial exciton exists even between MoS_2 and long thought "semi-metallic" graphene, which is frequently used as photoluminescence quencher, and in this case was found to "quench" MoS_2 but not the interfacial exciton. The electrostatics of the heterostructure explains the linear energy variation of the interfacial exciton with gate voltage.

Experimental Section

Sandwich of MoS_2/graphene or MoS_2/h-BN/graphene was made from CVD grown monolayers. Electrochemical gating was used and in-situ photoluminescence was conducted.

Results and Discussion

The energy of the observed photoluminescence red-shift with increasing of the Fermi level by applying electrochemical gate voltage. Photoluminescence corresponding to exciton and trion of MoS_2 was found, whose behavior towards upheaval of Fermi level is compliant to previous report[4]. However, another photoluminescence was observed and the different gate dependence from exciton or trion of MoS_2 indicates that it's not from photon-replica. We attribute this photoluminescence to interlayer exciton and the electrostatic analysis of the heterostructure explains the linear energy shift towards gate voltage (**Figure 1**). The failing of quenching of graphene to MoS_2 may arise from the bandgap opening at Dirac point due to periodical potential from MoS_2[5] (**Figure 1 left**).

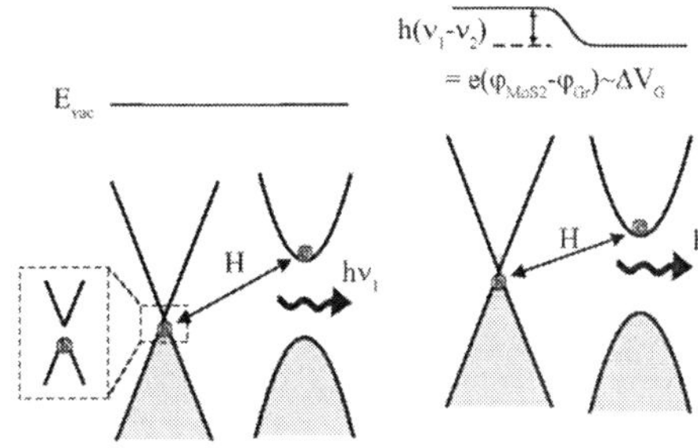

Figure 1. Scheme of the interfacial exciton between graphene and MoS_2. Left) periodic potential of MoS_2 opens a very small gap at K-point of graphene, making graphene more a semiconductor than a metal. Right) applying gate voltage results in potential difference between graphene and MoS_2, and thus the variation of the interfacial exciton energy, which varies proximately linear to the applied gate voltage.

Conclusions

Unexpected interlayer exciton between graphene and MoS_2 was found and the electrostatics of the heterostructure renders a tunable gap.

Acknowledgement. This project was supported by the Research Grant Council of Hong Kong SAR. Technical assistance from the Materials Characterization and Preparation Facilities is greatly appreciated.

References

1. Zhu, X. Y.; Monahan, N. R.; Gong, Z. Z.; Zhu, H. M.; Williams, K. W.; Nelson, C. A., Charge Transfer Excitons at van der Waals Interfaces. Journal of the American Chemical Society 2015, 137 (26), 8313-8320.
2. Rivera, P.; Schaibley, J. R.; Jones, A. M.; Ross, J. S.; Wu, S. F.; Aivazian, G.; Klement, P.; Seyler, K.; Clark, G.; Ghimire, N. J.; Yan, J. Q.; Mandrus, D. G.; Yao, W.; Xu, X. D., Observation of long-lived interlayer excitons in monolayer MoSe2-WSe2 heterostructures. Nature Communications 2015, 6.
3. Chu, T.; Ilatikhameneh, H.; Klimeck, G.; Rahman, R.; Chen, Z. H., Electrically Tunable Bandgaps in Bilayer MoS2. Nano Letters 2015, 15 (12), 8000-8007.
4. Mak, K. F.; He, K. L.; Lee, C.; Lee, G. H.; Hone, J.; Heinz, T. F.; Shan, J., Tightly bound trions in monolayer MoS2. Nature Materials 2013, 12 (3), 207-211.
5. Ma, Y. D.; Dai, Y.; Guo, M.; Niu, C. W.; Huang, B. B., Graphene adhesion on MoS2 monolayer: An ab initio study. Nanoscale 2011, 3 (9), 3883-3887.

SCALABLE HOLEY GRAPHENE ELECTRODES FOR ENERGY STORAGE

Yi Lin,[1] *Jae-Woo Kim,*[1] *John W. Connell*[2]

[1]National Institute of Aerospace, 100 Exploration Way, Hampton, Virginia 23666, [2]Advanced Materials and Processing Branch, NASA Langley Research Center, Hampton, Virginia 23681-2199
Email: yi.lin-1@nasa.gov

Practical energy storage applications require sufficient capacity in each individual device. At the materials research level, most reports have only discussed the energy storage performance of a small amount of active materials. When scaled up to device level, the true performance would significantly be reduced due to the overwhelming deadweight of other device components.[1] Therefore, whether a material is eventually useful in most energy storage applications depends upon their scalability in the synthesis as well as the performance in thick electrode formats.

Recently, a few methodologies have been developed in our laboratories to prepare holey graphene materials at a large scale with or without the use of catalysts.[2-4] With randomly distributed through-thickness holes on the large surface area of atomically thin conductive graphene nanosheets, holey graphene is an exciting derivative of graphene with such unique morphology that allows for effective transport of molecular species such as electrolyte ions through the two-dimensional graphene plane. In addition, a large number of active functional groups or doping can be achieved at or near the hole edges, making holey graphene an attractive platform for various energy storage applications.[3-5]

In this presentation, the scalability in both holey graphene synthesis and effective processing of this material into high capacity electrodes for supercapacitors and lithium-air batteries will be discussed. While the materials were first studied electrochemically in small quantities, holey graphene electrode platforms with thicknesses of tens of micrometers toward practical applications have recently been demonstrated which present an entirely different set of challenges that will be the main focus of the discussion.

References:

(1) Simon, P.; Gogotsi, Y. *Acc. Chem. Res.* **2013**, *46*, 1094-1103.
(2) Lin, Y.; Kim, J.-W.; Baggett, D. W.; Working, D. W.; Connell, J. W. *Nanoscale* **2013**, *5*, 7814-7824.
(3) Han, X.; Funk, M. R.; Shen, F.; Chen, Y.-C.; Li, Y.; Campbell, C. J.; Dai, J.; Kim, J.-W.; Liao, Y.; Connell, J. W.; Barone, V.; Chen, Z.; Lin, Y.; Hu, L. *ACS Nano* **2014**, *8*, 8255-8265.
(4) Lin, Y.; Han, X.; Campbell, C. J.; Kim, J.-W.; Zhao, B.; Luo, W.; Dai, J.; Hu, L.; Connell, J. W. *Adv. Func. Mater.* **2015**, *25*, 2920-2927.
(5) Xu, J.; Lin, Y.; Connell, J. W.; Dai, L. *Small* **2015**, *11*, 6179-6185.

REGULATION OF ELECTRIC BEHAVIOR IN TWO-DIMENSIONAL INORGANIC SOLIDS FOR ENERGY APPLICATIONS.

Changzheng Wu[1,*]

[1] Hefei National Laboratory for Physical Sciences at the Microscale & Department of Chemistry, University of Science and Technology of China, No.96, Jinzhai Road, Hefei, 230026
*Email: czwu@ustc.edu.cn

Two-dimensional (2D) inorganic solids have emerged as an important class of inorganic functional-material system owing to their advantages in precise crystal structure, easy-to-scale-up applications and versatile chemical regulation ways.

Summarized herein are our group's recent reports on regulations of electric behavior of 2D inorganic solids, especifically focusing on tuning the intrinsic physical properties such as electric transport properties and spin behavior, in order to provide insights into the investigation of how intrinsic physical properties influence the performance of energy storage and energy conversion[1]. In order to sustain a continuous "stream" of electrons for optimizing the electric behavior, we developed novel high electric conductivity in energy solids by combining the concepts of "trenching" and "irrigation". On one hand, we constructed new electric transport channels through atomic correlation and electronic spin (trenching) by developing metallic 2D VS_2 nanosheets[2] and 2D TiTeI nanosheets with large negative magnetoresistance[3]. On the other hand, we obtained new metallic inorganic solids and optimized electric transport behavior in conventional inorganic solids by surface modification to inject extra electrons (irrigation), for example, surface chemical modification of hydrogen atoms on TiS_2 could form H_yTiS_2 ultrathin nanosheets, reaching best conductivity among the solution assembled thin films of 2D inorganic nanomaterials[4].

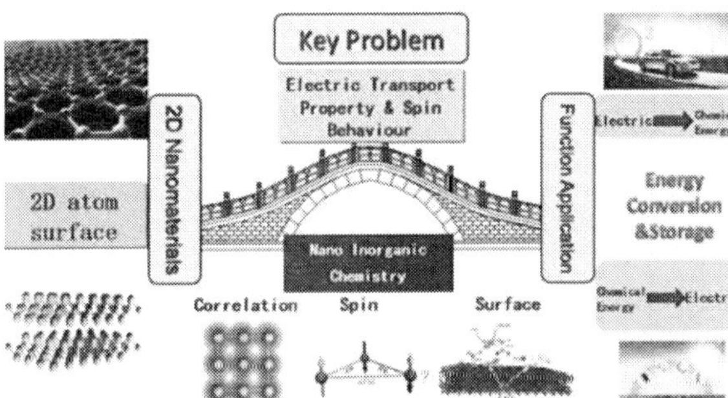

Figure 1. Regulation of electric behavior in 2D inorganic solids for energy applications.

Inspired by concept of electrical behavior regulation, we further developed the 2D conducting inorganic solids for various energy applications. For example, highly efficient electrochemical supercapacitors and electrocatalytical systems, via synergizing their intrinsic electric transport and electrochemical activities:

Electrochemical supercapacitors. Inspired by metallic nature of layered VS_2 derived from in-plane 2D S-V-S layer, we constructed in-plane supercapacitors by assembling a highly c-oriented VS_2 thin film by utilizing synergic advantages of high conductivity and high specific area; the devices demonstrated a remarkable improved specific capacitance for supercapacitor as energy storage devices. Furthermore, we discussed electrochemical supercapacitors assembled by 2D VN and TiN mesocrystal nanosheets with controllable facets exposed, highlighting the synergistic advantages of superior conductivity and highly open nanostructures for pursuing better electrochemical performance[5].

Electrocatalyst systems. The performance of electrocatalysts is largely dependent on their structures and morphologies. Moreover, the electrical conductivity is another critical factor to influence the performance of electrocatalysts[6, 7]. Taking 2H-MoS_2 as an example, we developed a novel intralayer vanadium-doping strategy to produce semimetallic vanadium-doped MoS_2 ultrathin nanosheets with tunable conductivity and higher carrier concentration within the pristine MoS_2 structure, exhibiting enhanced HER catalytical performance compared with that of a pure MoS_2 system, providing a new perspective for broadening the designing space of 2D materials[8]. Furthermore, we developed 2D metallic Ni-based material with high electrical conductivity as an efficient OER catalyst. Because of the enhanced metallic electrical conductivity behavior and disorder structure, Ni_3N nanosheets exhibit enhanced OER catalytic activity compared with bulk Ni_3N and NiO nanosheets, with earlier onset potential and a smaller Tafel slope, paving a new pathway for designing high efficient OER electrocatalysts based on conducting materials, as well as more inspirations on scientific interests in 2D metallic materials catering for the field of energy storage and conversion[9].

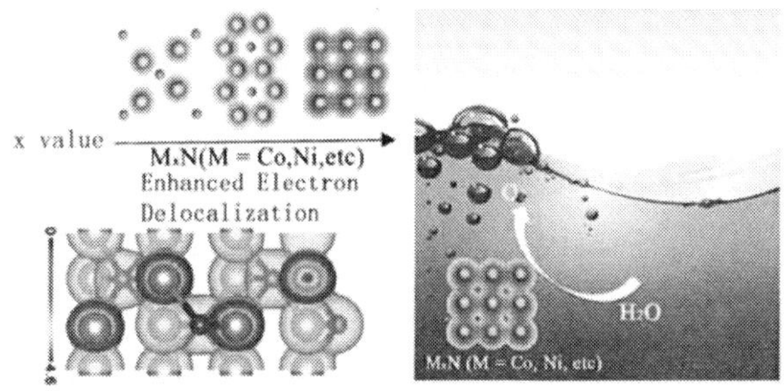

Figure 2. Metallic transition metal nitride Ni_3N with enhanced intrinsic conductivity and disordered Ni sites for OER.

In summary, optimized electric-transport properties have provided a solid material foundation towards applications of 2D inorganic solids in superior conducting electrodes, planar spintronics as well as designs of highly efficient energy storage materials and conversion catalysts.

Acknowledgement. This work was financially supported by the National Basic Research Program of China (Grant No. 2015CB932302), Natural Science Foundation of China (Grant Nos. 21222101).

References

(1) Guo, Y.; Xu, K.; Wu, C.; Zhao, J.; Xie, Y. *Chemical Society Reviews* **2015**, *44* (3), 637-646.
(2) Feng, J.; Sun, X.; Wu, C.; Peng, L.; Lin, C.; Hu, S.; Yang, J.; Xie, Y. *J. Am. Chem. Soc.* **2011**, *133* (44), 17832-17838.
(3) Guo, Y.; Dai, J.; Zhao, J.; Wu, C.; Li, D.; Zhang, L.; Ning, W.; Tian, M.; Zeng, X. C.; Xie, Y. *Phys. Rev. Lett.* **2014**, *113* (15), 157202.
(4) Lin, C.; Zhu, X.; Feng, J.; Wu, C.; Hu, S.; Peng, J.; Guo, Y.; Peng, L.; Zhao, J.; Huang, J. *J. Am. Chem. Soc.* **2013**, *135* (13), 5144-5151.
(5) Bi, W.; Hu, Z.; Li, X.; Wu, C.; Wu, J.; Wu, Y.; Xie, Y. *Nano Research* **2015**, *8* (1), 193-200.
(6) Guo, Y.; Tong, Y.; Chen, P.; Xu, K.; Zhao, J.; Lin, Y.; Chu, W.; Peng, Z.; Wu, C.; Xie, Y. *Advanced Materials* **2015**, *27* (39), 5989-5994.

(7) Chen, P.; Xu, K.; Zhou, T.; Tong, Y.; Wu, J.; Cheng, H.; Lu, X.; Ding, H.; Wu, C.; Xie, Y. *Angew. Chem.* **2016**. *128*, 2534 –2538
(8) Sun, X.; Dai, J.; Guo, Y.; Wu, C.; Hu, F.; Zhao, J.; Zeng, X.; Xie, Y. *Nanoscale* **2014**, *6* (14), 8359-8367.
(9) Xu, K.; Chen, P.; Li, X.; Tong, Y.; Ding, H.; Wu, X.; Chu, W.; Peng, Z.; Wu, C.; Xie, Y. *J. Am. Chem. Soc.* **2015**, *137* (12), 4119-4125.

CHEMISTRY OF MXenes: A NEW LARGE FAMILY OF 2D SOLIDS WITH EXCEPTIONAL PROMISE

Michel W. Barsoum

Department of Materials Science and Engineering
Drexel University, Philadelphia, PA 19104
barsoumw@drexel.edu

By now it is well-established that the layered, hexagonal carbides and nitrides with the general formula, $M_{n+1}AX_n$, (MAX) where n = 1 to 3, M is an early transition metal, A is an A-group (mostly 13 and 14) element and X is either C and/or N combine some of the best attributes of metals and ceramics.[1] More recently, we have shown that by simply immersing MAX phase powders, at room temperature, in HF or HCl and LiF solutions, the A-layers – mostly Al - are selectively etched to produce 2D materials that we labeled MXenes to emphasize the loss of the A-group element and their similarities to graphene (Fig. 1).[2-4]

At some level, our discovery is paradigm shifting since it shows that 2D solids can be synthesized from layered solids in which the layers are *not* weakly bonded together. For example, the Ti-Al bonds in Ti_3AlC_2 expand at roughly the same rate as the Ti-C bonds in the same compound.[1] This MAX phase is typically synthesized at temperatures > 1400 °C.[1]

Unlike hydrophobic graphene, MXenes are hydrophilic and behave as "conductive clays", a hitherto unknown combination.[3] Along the same lines, because the density of states at the Fermi level is substantial, and the electric conductivity, for the most part, is metal-like, MXenes can be considered as 2D metals that are surface terminated by OH and/or similar groups. Said otherwise, they possess a metallic core with a hydroxide/F shell.

Quantifying the surface terminations is non-trivial. However, using careful XPS analysis it is possible to do so.[5] The surface terminations are a combination of O, OH, and/or F (Fig. 2). For the most part we showed that the sum of the negative charges corresponding to the surface terminations depended on the MXene chemistry. The ramifications of these insights on ion exchange capabilities, delamination, and interlayer spacings are important and will be discussed.

The potential of using MXenes – already the largest family of 2D materials known with more than 15 members (Table 1), and with new ones being discovered almost monthly - in energy storage,[6] as transparent conductive electrodes,[7] among many other applications will be discussed.

Table 1. Table 1: List of MXenes Compositions Known to Date. First Column lists the values of n in $M_{n+1}X_nT_x$.

1	Ti_2C	V_2C	Nb_2C	TiNbC	Mo_2C	$(TiV)_2C$	
2	Ti_3C_2	Ti_3CN	Zr_3C_2	$(V_{0.5}Cr_{0.5})_3C_2$	Mo_2TiC_2	$(TiV)_3C_2$	Cr_2TiC_2
3	Ta_4C_3	Nb_4C_3	$(Nb,Ti)_4C_3$	$(Nb_{0.8}Zr_{0.2})_4C_3$	$Mo_2Ti_2C_2$		

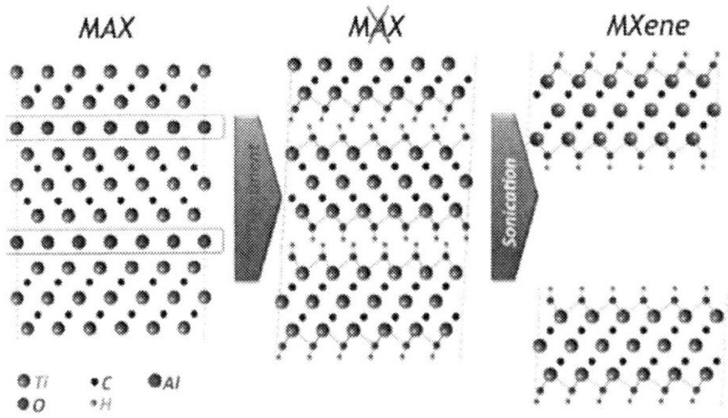

Figure 1: a) Schematic of MAX to MXene transformation.

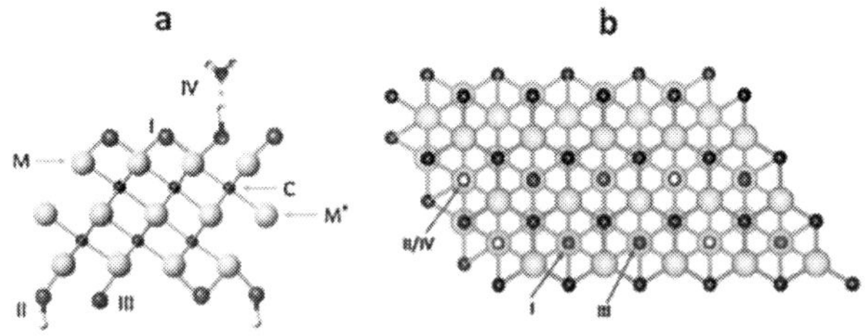

Figure 2: Schematic of terminations observed on MXenes, a) side view, and b) top view. I is a bridging oxygen, II is OH, III is F and IV is an OH termination strongly attached to a water molecule.

Acknowledgement. This work was funded by the Ceramics program of the Division of Materials Research, NSF. DMR-1310245

References:

(1) M.W. Barsoum, MAX Phases: Properties of Machinable Carbides and Nitrides, Wiley VCH GmbH & Co., Weinheim, 2013.
(2) M. Naguib, M. Kurtoglu, V. Presser, J. Lu, J. Niu, M. Heon, L. Hultman, Y. Gogotsi, M.W. Barsoum, Advan. Mater., 23 (2011) 4248-4253.
(3) M. Ghidiu, M.R. Lukatskaya, M.Q. Zhao, Y. Gogotsi, M.W. Barsoum, Nature, 516 (2014) 78-81.
(4) M. Naguib, O. Mashtalir, J. Carle, J. Lu, L. Hultman, Y. Gogotsi, M.W. Barsoum, ACS Nano, 6 (2012) 1322-1331.
(5) J. Halim, K.M. Cook, M. Naguib, P. Eklund, Y. Gogotsi, J. Rosen, M.W. Barsoum, Appl. Surf. Sci., 362 (2016) 406-417.
(6) M. Lukatskaya, O. Mashtalir, C.E. Ren, Y. Dall'Agnese, P. Rozier, P.L. Taberna, M. Naguib, P. Simon, M.W. Barsoum, Y. Gogotsi, Science, 341 (2013) 1502.
(7) A.D. Dillon, M. Ghidiu, A. Krick, J. Griggs, S.J. May, Y. Gogotsi, M.W. Barsoum, A.T. Fafarman, Adv. Funct. Mater., In print (2016).

2D LAYERED GRAPHENE FLAKES, TRANSITION METAL DICHALCOGENS AND THEIR ENERGY APPLICATIONS

Hyoyoung Lee*

Center for Integrated Nanostructure Physics, Institute for Basic Science, Department of Chemistry and Department of Energy Science, Sungkyunkwan University, Suwon 440-746, Korea
hyoyoung@skku.edu

Recently, 2 dimensional (D)-based graphene (G), Graphene oxide (GO), Graphene flake (GF) and transition metal dichalcogen (TMD) material like MoS2 exhibit great potential for various applications in energy storage including supercapacitor and energy generation such as hydrogen evolving reaction due to their unique structure and outstanding intrinsic physical properties including extraordinarily high electrical conductivity, large surface area, and/or various chemical functionalities. For the energy storage such as supercapacitors and batteries, both high surface area and high pore volume at the same time becomes a crucial issue. Here, we like to introduce new idea on how to increase the surface area of GO/rGO and the pore volume of the non-stacked GO/rGO 3D structure since the energy charge storage is a surface process.[1,3] And we also like to report new concept on how to fabricate highly dense and vertically aligned reduced graphene oxide (VArGO) electrodes involving simple hand-rolling and cutting processes. Due to their vertically aligned and opened-edge graphene structure, VArGO electrodes displayed high packing density and highly efficient volumetric and areal electrochemical characteristics, very fast electrolyte ion diffusion with rectangular CV curves even at a high scan rate, and the highest volumetric capacitance among known rGO electrodes.[2,4] In addition, an introduction of MoS2 to carbonaceous architecture materials showed superior electrochemical performance for anode materials of a lithium ion battery.[5,6]

Acknowledgement This work was supported by the Center for Integrated Nanostructure Physics (CINAP), Institute for Basic Science (IBS) (IBS-R011-D1).

References

(1) Yoon, Y.; Lee, K.; Baik, C.; Yoo, H.; Min, M.; Park, Y. Lee, S. M.; Lee, H.*, *Advanced Materials*, **2013**, *25*, 4437.

(2) Yoon, Y.; Lee, K.; Kwon, S.; Seo, S.; Yoo, H.; Kim, S.; Shin, Y.; Park, Y.; Kim, D.; Choi, J-Y.; Lee, H.*, *ACS Nano*, **2014**, *8(5)*, 4580.

(3) Lee, K.; Kim, D.; Yoon, Y.; Yang, J.; Yun, H.-G.; You, I.-K.; Lee, H.*, *RSC Advances*, **2015**, *5*, 60914.

(4) Yoon, Y.; Lee, K.; Lee, H.*, *Nanotechnology*, *In-print*, **2016**.

(5) Tiwari, A. P.; Yoo, H.; Lee, J.; Kim, D.; Park, J.; Lee, H.*, *Nanoscale*, **2015**, *7*, 11928.

(6) Yoo, H.; Tiwari, A. P.; Lee, J.; Kim, D.; Park, J.; Lee, H.*, *Nanoscale*, **2015**, *7*, 3404.

ION-EXCHANGE AND CATION SOLVATION REACTIONS IN Ti_3C_2 MXene

Michael Ghidiu,[†] *Joseph Halim,*[†,§] *Sankalp Kota,*[†] *David Bish,*[‡] *Yury Gogotsi,*[†,∥] *and Michel W. Barsoum*[†,*]

[†] Department of Materials Science and Engineering, Drexel University, Philadelphia, PA 19104
[‡] Department of Geological Sciences, Indiana University, Bloomington, IN 47405
§ Thin Film Physics Division, Department of Physics, Chemistry and Biology (IFM), Linköping University, SE-58183 Linköping (Sweden)
[∥] A. J. Drexel Nanomaterials Institute, Drexel University, Philadelphia, PA 19104
*barsoumw@drexel.edu

Introduction

MXenes – two-dimensional transition metal carbonitrides of the $M_{n+1}X_nT_x$ composition, where M is an early transition metal, X is carbon and/or nitrogen, n = 1,2,3, and T_x represents a variable surface termination such as O, OH, F – are a rapidly-growing class of materials featuring hydrophilicity, high electrical conductivity, and tuneable chemistry and structure.[1,2] They have been explored in applications from energy storage to water purification to transparent conductive electrodes.[3,4,5] Furthermore, spontaneous intercalation of organic molecules (urea, hydrazine, DMF, etc.), as well as spontaneous and electrochemical intercalation of various cations (Li^+, Na^+, K^+, NH_4^+, Mg^{2+}, Al^{3+}) were previously reported.[6,7,8] Understanding the intercalation chemistry of 2D materials is a prerequisite to many applications. Based on the ability to intercalate cations and the reversible increase in the *c* parameter (c-LP) with H_2O uptake, MXenes have been compared with expanding clay minerals.[9] Although the reversible expansion of clay minerals in H_2O and the effects of cation exchange have been studied extensively, a fundamental study of the structural effects resulting from cation substitution in MXenes in aqueous, pH-neutral environments has not been undertaken. This is especially pertinent now that a number of MXene applications have been explored involving interactions with salt solutions; for example, when MXenes are used in supercapacitors with aqueous electrolytes,[7,10] as sorbents for Pb or Cr ions,[11,12] or membranes for ion sieving, the nature of the cation can play a large role in affecting the sorption efficiency or structure and permeability of the membranes.[4] It has been shown that spontaneous cation intercalation described in reference 7 may lead to changes in MXene surface chemistry,[13] but no systematic studies have been reported. If intercalated cations are structurally inherent in MXenes, controlling their concentrations and understanding their effects on interlayer separation can be crucial for tuning electrochemical energy storage, optical, and ion-exchange properties. The drastic changes noticed when Li^+ ions were presumably intercalated into the MXene structure during synthesis[9] led us to examine this more carefully in a model system, using $Ti_3C_2T_x$, to date the most widely studied and best understood MXene member.

Experimental

$Ti_3C_2T_x$ HF10 was Made according to the literature.[14] $Ti_3C_2T_x$ intercalated with ions: Similar to the $Ti_3C_2T_x$ HF10 procedure, Ti_3AlC_2 powder was added to an etching mixture in the same ratio. In this case, however, the etchant was a mixture of 10% HF and LiCl, in a molar ratio 5 LiCl : 1 Ti_3AlC_2. The mixture was stirred for 24 h at 25 °C followed by washing as in Ref. 14, with the exception of having three washes of 6 M HCl (Fisher TraceMetal) before the water washes to remove traces of LiF impurity. Before the samples were allowed to dry, salt solutions (1 N LiCl, NaCl, KCl, $MgCl_2$, or $CaCl_2$ in distilled water) were added in a ratio of roughly 0.5 g MXene to 40 mL solution. After shaking for 2 min, the mixtures were allowed to sit for 1 h. The samples were then centrifuged to settle the powders, and the supernatants were decanted and replaced with fresh solutions. The samples were again shaken and allowed to sit for 24 h. Then they were centrifuged, the supernatants were discarded, and water was added, followed by agitation and centrifugation. After decanting, the sediment was collected via filtration, and washed with distilled water (2 x 5 mL) followed by drying in air (roughly 50% relative humidity) to yield desired powders K-$Ti_3C_2T_x$, Li-$Ti_3C_2T_x$, Na-$Ti_3C_2T_x$, Mg-$Ti_3C_2T_x$, and Ca-$Ti_3C_2T_x$.

***In-situ* XRD measurements** were carried out on a Bruker D8 Advance X-Ray powder diffractometer under controlled humidity on a thin layer slurry mount.

Thermogravimetric Analysis was performed on a TA Instruments SDT 2960 Simultaneous DSC-TGA in alumina crucibles under dry nitrogen (Zero Grade 99.998%, passed through a molecular sieve drying column at 100 cc min^{-1}), with a standard of alumina powder.

Results and Discussion

The relative humidity (RH) was varied between ~ 0 and 95%, with 40 minutes given for equilibration before each measurement. The results are shown in Fig. 1. 3D plots in Fig. 1a show how the 2θ of the 002 reflection for Ti_3C_2 exchanged with various cations shifted as the RH dropped from 95% to ~ 0% and then increased again to the initial humidity. The same data are presented superimposed as more typical intensity *vs.* 2θ XRD patterns in Fig. 1b so that the shape of the peaks can be seen. The positions of the reflections, at maximum intensity, are summarized as a function of RH in Fig. 1c. It is immediately clear that: (1) the reflections do not shift continuously; there are only two major reflections, with no peaks in between and, (2) different cations result in very different responses. The *c* of all samples at ~ 0% RH is roughly 25 Å, which contrasts with that of MXene under similar conditions produced using HF alone at ~ 20 Å, a difference corresponding roughly to the diameter of an H_2O molecule. We believe this 25 Å-phase is intercalated with a single layer of H_2O molecules. Almost all samples, with the exception of K-$Ti_3C_2T_x$, experienced a discontinuous shift in *c* at high humidities of the order of +3 Å per interlayer, most probably due to the presence of a bilayer H_2O structure, as has been observed in other materials.[15]

These results imply that the RH at which expansion from the monolayer structure occurs is related to the hydration enthalpy of the cation, with the K^+ samples showing almost no expansion and the Mg^{2+} samples showing almost no single-H_2O-layer form until very low humidities. The simplest physical explanation is that there is an energy requirement to separate the $Ti_3C_2T_x$ sheets, which is overcome by the hydration of the intercalated cation; the divalent cations have a sufficiently high driving force (related to hydration enthalpy) to form a bilayer H_2O structure in the MXene interlayer at moderate humidity, whereas Li^+ and Na^+ require higher RH due to their lower hydration energies, and K^+ is unable to stabilize a bilayer. This behavior is parallel with the behavior noted in numerous studies of smectites, showing a direct link to hydration energy of interlayer cations, with a secondary effect from the charge on the individual silicate layers.[16] In addition, XRD data, for all but the K-exchanged sample, reveal significant hysteresis between the hydration and dehydration branches, which is also consistent with this suggested hydration behavior. This can be seen, for example, in the marked difference in the RH value required for Ca-$Ti_3C_2T_x$ to change between the monolayer and bilayer structure on hydration (~ 50% RH) and dehydration (~ 20% RH).

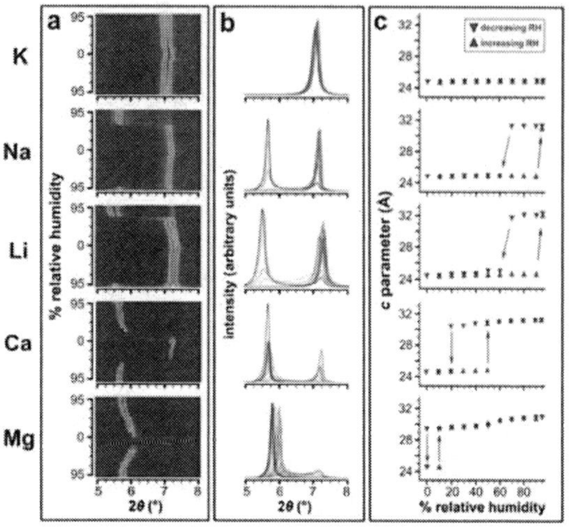

Figure 1. a) 3D plots of the 002 region of *A*-$Ti_3C_2T_x$ (cation *A* denoted at left), as RH was varied from 95% to 0 and back to 95%. The color scheme represents diffraction intensity, with red being highest and blue lowest. b) the same data, with the peaks superimposed and presented as intensity versus 2θ plots. Solid blue traces denote the initial condition at 95% RH, red traces denote the material at 0% RH, and dashed blue traces denote the material after return to 95% RH. c) *c* (2 x d_{002}) extracted from the peaks shown in a and b, as a function of RH.

We performed TGA on Li-$Ti_3C_2T_x$, Na-$Ti_3C_2T_x$, K-$Ti_3C_2T_x$, and Ca-$Ti_3C_2T_x$ (Fig. 2). All samples were first equilibrated at ~ 50% RH at 25 °C, and TGA was subsequently performed in various temperature steps (Fig. 2a), ramping stepwise up to 120 °C. In order to compare results, we made the assumption that most of the H_2O associated with cations (H_2O of hydration) was removed after heating using the temperature program shown in Fig. 2a. Normalizing the weight loss results by the weight 250 minutes after the start of the experiments gave the mass-loss curves shown in Fig. 2a. Plotting the weight loss at 27 °C against cation hydration enthalpy (Fig. 2b) gives a clear picture of the relation between the cation properties and water content.

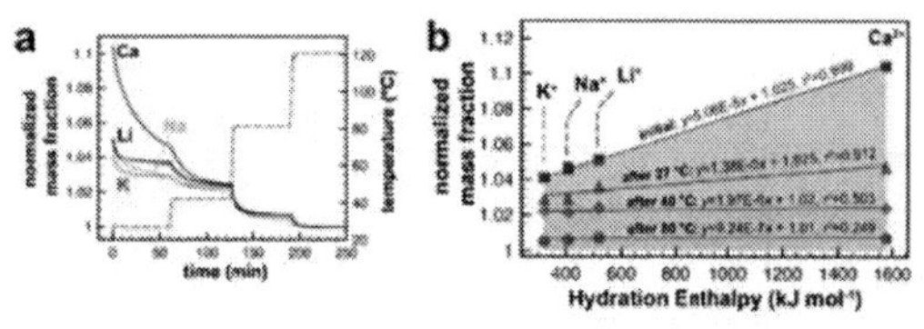

Figure 2. a) Temperature during TGA runs (right axis). Mass fraction for selected samples is shown on left axis. b) The same mass fraction data from (a) initially, and after equilibration steps, plotted against cation hydration enthalpy. Linear regression fits are shown above each data set.

Conclusions

We have demonstrated that etching of Ti_3AlC_2 to produce $Ti_3C_2T_x$ with the combination of HF and LiCl can yield MXenes with different characteristics than those produced with HF alone, and we provide evidence that this material behaves like swelling clay minerals and metal dichalcogenides. The intercalated cations give rise to quite different structural changes in response to RH, as evidenced by XRD, involving a discontinuous structural expansion in the direction normal to the basal plane as H_2O molecules intercalate. The findings from this study potentially impact the use of MXenes in many applications, from electrochemical energy storage to water purification or desalination.

Acknowledgement. This work was supported by the US National Science Foundation under Grant No. DMR-1310245 and by the National Science Foundation Graduate Research Fellowship under Grant 283036-3304. M.W.B. and J.H. were supported through Swedish Research Council (Project Grant No. 621-2012-4430), the Swedish Foundation for Strategic Research through the Synergy Grant FUNCASE Functional Carbides for Advanced Surface Engineering. We thank the Drexel Core Facility and staff for assistance with characterisation involving XRD. Collaboration between Drexel University and Indiana University was supported by the Fluid Interface Reactions, Structures and Transport (FIRST) Center, an Energy Frontier Research Center funded by the U.S. Department of Energy, Office of Science, and Office of Basic Energy Sciences. The authors are grateful to Dr. David J. Wesolowski from the Oak Ridge National Laboratory, who initiated this collaboration.

References

(1) Naguib, M.; Mashtalir, O.; Carle, J.; Presser, V.; Lu, J.; Hultman, L.; Gogotsi, Y.; Barsoum, M. W. *ACS Nano* **2012**, *6* (2), 1322–1331.
(2) Naguib, M.; Mochalin, V. N.; Barsoum, M. W.; Gogotsi, Y. *Adv. Mater.* **2014**, *26*, 992–1005.
(3) Xie, Y.; Dall'Agnese, Y.; Naguib, M.; Gogotsi, Y.; Barsoum, M. W.; Zhuang, H. L.; Kent, P. R. C. *ACS Nano* **2014**, *8* (9), 9606–9615.
(4) Ren, C. E.; Hatzell, K. B.; Alhabeb, M.; Ling, Z.; Mahmoud, K. A.; Gogotsi, Y. *J. Phys. Chem. Lett.* **2015**, *6* (20), 4026–4031.
(5) Halim, J.; Lukatskaya, M. R.; Cook, K. M.; Lu, J.; Smith, C. R.; Näslund, L.-Å.; May, S. J.; Hultman, L.; Gogotsi, Y.; Eklund, P.; Barsoum, M. W. *Chem. Mater.* **2014**, *26* (7), 2374–2381.
(6) Mashtalir, O.; Naguib, M.; Mochalin, V. N.; Dall'Agnese, Y.; Heon, M.; Barsoum, M. W.; Gogotsi, Y. *Nat. Commun.* **2013**, *4*, 1716.
(7) Lukatskaya, M. R.; Mashtalir, O.; Ren, C. E.; Dall'Agnese, Y.; Rozier, P.; Taberna, P. L.; Naguib, M.; Simon, P.; Barsoum, M. W.; Gogotsi, Y. *Science* **2013**, *341* (6153), 1502–1505.
(8) Levi, M. D.; Lukatskaya, M. R.; Sigalov, S.; Beidaghi, M.; Shpigel, N.; Daikhin, L.; Aurbach, D.; Barsoum, M. W.; Gogotsi, Y. *Adv. Energy Mater.* **2014**, *5* (1), 1400815.
(9) Ghidiu, M.; Lukatskaya, M. R.; Zhao, M.-Q.; Gogotsi, Y.; Barsoum, M. W. *Nature* **2014**, *516*, 78–81.
(10) Wang, X.; Kajiyama, S.; Iinuma, H.; Hosono, E.; Oro, S.; Moriguchi, I.; Okubo, M.; Yamada, A. *Nat. Commun.* **2015**, *6*, 6544.
(11) Ying, Y.; Liu, Y.; Wang, X.; Mao, Y.; Cao, W.; Hu, P.; Peng, X. *ACS Appl. Mater. Interfaces* **2015**, *7* (3), 1795–1803.
(12) Peng, Q.; Guo, J.; Zhang, Q.; Xiang, J.; Liu, B.; Zhou, A.; Liu, R.; Tian, Y. *J. Am. Chem. Soc.* **2014**, *136* (11), 4113–4116.
(13) Dall'Agnese, Y.; Lukatskaya, M. R.; Cook, K. M.; Taberna, P.-L.; Gogotsi, Y.; Simon, P. *Electrochem. Commun.* **2014**, *48*, 118–122.
(14) Wang, H.-W.; Naguib, M.; Page, K.; Wesolowski, D. J.; Gogotsi, Y. *Chem. Mater.* **2016**, *28* (1), 349–359.
(15) A. Lerf; R. Schöllhorn. *Inorg. Chem.* **1977**, *16* (11), 2950–2956.
(16) Bish, D. L. In *Handbook of Clay Science*; Elsevier, 2013; pp 243–345.

GROUP IV GRAPHANE MATERIALS AS A CANVAS; PAINTING PROPERTIES VIA COVALENT CHEMISTRY

Joshua Goldberger

Department of Chemistry and Biochemistry
The Ohio State University
100 W. 18th Ave
Columbus, OH 43210
goldberger@chemistry.ohio-state.edu

Introduction

2D materials have generated considerable research interest over the past decade, in part because of the new physical phenomena that emerge in these atomically thin framework, as well as their potential for impacting diverse sets of technologies including electronics, optoelectronics, sensing, and catalysis. One of the major lessons of these studies is that the surrounding environment, the substrate flatness, and the local dielectric constant can significantly influence the properties and reactivity of these single/few-atom thick materials. Establishing 2D crystals in which the framework and immediate chemical environment can be modified via covalent functionalization, gives a powerful handle towards designing novel properties. Here, we will describe our recent developments in the creation of hydrogen and organic-terminated group IV graphane analogues, from the topochemical deintercalation of precursor Zintl phases, such as $CaGe_2$.[1-8] We will discuss established the fundamental guidelines behind systematically controlling the electronic structure by substituting either the surface ligand or via alloying with other Group IV elements. Finally, we will describe our recent efforts on the creation and properties of Sn-containing graphane alloys in order to access novel topological insulator phenomena predicted to occur in these materials.

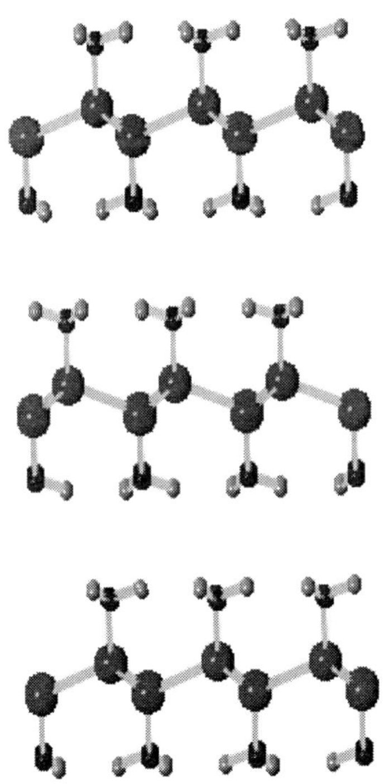

Figure 1. Crystal structure of $GeCH_3$ (Ge=blue, C=Black, H=grey)

Acknowledgement. J. G. would like to acknowledge NSF DMR-1420451, and the Camille and Henry Dreyfus Foundation.

References

(1) Amamou, W.; Odenthal, P. M.; Bushong, E. J.; O'Hara, D. J.; Luo, Y. K.; Baren, J. v.; Pinchuk, I.; Wu, Y.; Ahmed, A. S.; Katoch, J.; Bockrath, M. W.; Tom, H. W. K.; Goldberger, J. E.; Kawakami, R. K. *2D Materials* **2015**, *2*, 035012.
(2) Arguilla, M. Q.; Jiang, S.; Chitara, B.; Goldberger, J. E. *Chemistry of Materials* **2014**, *26*, 6941.
(3) Bianco, E.; Butler, S.; Jiang, S.; Restrepo, O. D.; Windl, W.; Goldberger, J. E. *ACS Nano* **2013**, *7*, 4414.
(4) Butler, S. Z.; Hollen, S. M.; Cao, L.; Cui, Y.; Gupta, J. A.; Gutierrez, H. R.; Heinz, T. F.; Hong, S. S.; Huang, J.; Ismach, A. F.; Johnston-Halperin, E.; Kuno, M.; Plashnitsa, V. V.; Robinson, R. D.; Ruoff, R. S.; Salahuddin, S.; Shan, J.; Shi, L.; Spencer, M. G.; Terrones, M.; Windl, W.; Goldberger, J. E. *ACS Nano* **2013**, *7*, 2898.
(5) Jiang, S.; Arguilla, M. Q.; Cultrara, N. D.; Goldberger, J. E. *Accounts of Chemical Research* **2015**, *48*, 144.
(6) Jiang, S.; Butler, S.; Bianco, E.; Restrepo, O. D.; Windl, W.; Goldberger, J. E. *Nature Communications* **2014**, *5*.
(7) Pinchuk, I. V.; Odenthal, P. M.; Ahmed, A. S.; Amamou, W.; Goldberger, J. E.; Kawakami, R. K. *Journal of Materials Research* **2014**, *29*, 410.
(8) Restrepo, O. D.; Mishra, R.; Goldberger, J. E.; Windl, W. *J. Appl. Phys.* **2014**, *115*, In Press.

POLYCRYSTALLINE 2D MATERIALS: A THEORIST'S VIEW

Oleg V. Yazyev

Institute of Physics, Ecole Polytechnique Fédérale de Lausanne (EPFL), CH-1015 Lausanne, Switzerland
E-mail: oleg.yazyev@epfl.ch

Grain boundaries and dislocations are intrinsic topological defects of polycrystalline materials, which inevitably affect their physical properties. In my talk, I will discuss the structure of topological defects in two-dimensional (2D) materials such as graphene and monolayer transition metal dichalcogenides.[1] Special attention is devoted to the effect of topological defects on the electronic transport properties of these 2D materials.

I will first introduce a general approach for constructing dislocations in graphene characterized by arbitrary Burgers vectors and grain boundaries covering the complete range of possible misorientation angles (see **Figure 1** for an example of a grain boundary structure). By means of first-principles calculations we address the thermodynamic properties of grain boundaries revealing energetically favorable large-angle configurations as well as dramatic stabilization of small-angle configurations via the out-of-plane deformation, a remarkable feature of graphene as a two-dimensional material.[2] Both the presence of stable large-angle grain-boundary motifs and the out-of-plane deformation of small-angle configurations have recently been observed by scanning tunneling microscopy.[3]

In the rest of my talk, I will focus on the electronic transport properties of polycrystalline 2D materials. Ballistic charge-carrier transmission across periodic grain boundaries is governed primarily by momentum conservation. Two distinct transport behaviors of such grain boundaries in graphene are predicted – either perfect reflection or high transparency with respect to low-energy charge carriers depending on the grain boundary periodicity.[4] It is also shown that certain periodic line defect structures can be engineered and offer opportunities for generating valley polarized charge carriers.[5] Beyond the momentum conservation picture we find that the transmission of low-energy charge carriers can be dramatically suppressed in the small-angle limit.[6] Unlike graphene, monolayer transition metal dichalcogenides (MoS_2 and alike) combine a two-valley electronic band structure with strong spin-orbit effects. The latter can be employed for creating spin-polarized currents and adds yet another conservation law in the electronic transport across regular defects such as the frequently observed inversion domain boundaries.[7,8]

These results demonstrate that dislocations and grain boundaries are intrinsic topological defects that dramatically affect the transport properties of 2D materials and can also be used for engineering novel functional devices.

Figure 1. Atomistic model of a representative grain boundary defect in graphene.

Acknowledgement. This work has been supported by the Swiss NSF, ERC and the Graphene Flagship.

References

(1) Yazyev, O. V.; Chen, Y. P.: Polycrystalline graphene and other two-dimensional materials. *Nat. Nanotechnol.* **2014**, *9*, 755-767.
(2) Yazyev, O. V.; Louie, S. G.: Topological defects in graphene: Dislocations and grain boundaries. *Phys. Rev. B* **2010**, *81*, 195420.
(3) Tison, Y.; Lagoute, J.; Repain, V.; Chacon, C.; Girard, Y.; Joucken, F.; Sporken, R.; Gargiulo, F.; Yazyev, O. V.; Rousset, S.: Grain Boundaries in Graphene on SiC(000$\bar{1}$) Substrate. *Nano Lett.* **2014**, *14*, 6382-6386.
(4) Yazyev, O. V.; Louie, S. G.: Electronic transport in polycrystalline graphene. *Nat. Mater.* **2010**, *9*, 806-809.
(5) Chen, J. H.; Autès, G.; Alem, N.; Gargiulo, F.; Gautam, A.; Linck, M.; Kisielowski, C.; Yazyev, O. V.; Louie, S. G.; Zettl, A.: Controlled growth of a line defect in graphene and implications for gate-tunable valley filtering. *Phys. Rev. B* **2014**, *89*, 121407.
(6) Gargiulo, F.; Yazyev, O. V.: Topological Aspects of Charge-Carrier Transmission across Grain Boundaries in Graphene. *Nano Lett.* **2013**, *14*, 250-254.
(7) Pulkin, A.; Yazyev, O. V.: Spin- and valley-polarized transport across line defects in monolayer MoS_2. *Phys. Rev. B* **2016**, *93*, 041419.
(8) Lehtinen, O.; Komsa, H.-P.; Pulkin, A.; Whitwick, M. B.; Chen, M.-W.; Lehnert, T.; Mohn, M. J.; Yazyev, O. V.; Kis, A.; Kaiser, U.; Krasheninnikov, A. V.: Atomic Scale Microstructure and Properties of Se-Deficient Two-Dimensional $MoSe_2$. *ACS Nano* **2015**, *9*, 3274-3283.

FUNCTIONAL FLEXIBLE/WEARABLE SUPERCAPACITORS

Chunyi Zhi

Department of Physics and Materials Science, City University of Hong Kong
cy.zhi@cityu.edu.hk

Wearable electronic textiles that store capacitive energy are a next frontier in personalized electronics [1-6]. However, the lack of industrially weavable and knittable conductive yarns in conjunction with high capacitance, limits the wide-scale application of such textiles. Here pristine soft conductive yarns are continuously produced by a scalable method with the use of twist-bundle-drawing technique, and are mechanically robust enough to be knitted to a cloth by a commercial cloth knitting machine. Subsequently, we demonstrate a combination of textile industry available conductive yarn and conducting polymers can form a great basement for wearable energy storage devices. For example, a combination of metal oxide and conductive polymer can great enhanced tolerance of stretch-induced performance degradation of stretchable supercapacitors [2-3]. In case of self-healable PU sheel applied, a yarn supercapacitor can be self-healable [5]. In addition, we demonstrate a new electrolyte comprising polyacrylic acid dual cross-linked by hydrogen bonding and vinyl hybrid silica nanoparticles (VSNPs-PAA) that addresses all the superior functions and provide an ultimate solution to the intrinsic self-healability and high stretchability of a supercapacitor. Supercapacitors with VSNPs-PAA as the electrolyte are self-healed, achieving an excellent healing efficiency of ~100% even after 20 cycles of breaking/healing. By a designed facile electrode fabrication procedure, they are stretched up to 600% strain with performance enhanced. Our research represents a solid progress in portable and wearable multifunctional devices with extreme self-healability and stretchability [1].

References

(1) Huang Y.; Zhong M.; Huang Y.; Zhu M.; Pei Z.; Wang Z.; Xue Q.; Xie X.; Zhi C.; *Nature Communications* **2015**, *6*, 10310.
(2) Huang Y.; Hu H.; Huang Y.; Zhu M.; Meng W.; Liu C.; Hao C.; Wang Z.; Zhi C.; *ACS Nano* **2015**, *9*, 4766.
(3) Huang Y.; Tao J. Y.; Meng W. J.; Zhu M.; Huang Y.; Fu Y. Q.; Gao Y. H.; Zhi C. Y.; *Nano Energy* **2015**, *11*, 518.
(4) Zhu M.; Meng W.; Huang Y.; Huang Y.; Zhi C. Y.; *Acs Applied Materials & Interfaces* **2014**, **6**, 18901.
(5) Huang Y.; Huang Y.; Zhu M.; Meng W.; Pei Z.; Liu C.; Hu H.; Zhi C. Y.; *ACS Nano* **2015**, *9*, 6242.
(6) Meng W. J.; Chen W.; Zhao L.; Huang Y.; Zhu M.; Huang Y.; Fu Y. Q.; Geng F. X.; Yu J.; Chen X. F.; Zhi C. Y.; *Nano Energy* **2014**, *8*, 133. 2014.

METAL INTERCALATION IN GRAPHITES: A COMPUTATIONAL PERSPECTIVE

Veronica Barone

Department of Physics, Central Michigan University, Mt. Pleasant, MI 48859
v.barone@cmich.edu

Introduction

Two dimensional materials are those layered structures in which the interaction between layers is sufficiently weak as to permit the easy exfoliation of the individual layers. Graphene is perhaps the most widely known member of this family, as it is commonly used in a variety of technologically important applications.[1] The properties of graphene can be engineered for specific applications by creating defects, adding functional groups, reducing its dimensionality, and by substitutional doping or a combination of the above modifications.[2] The dimensionality of graphene can be reduced to a quasi-one-dimension giving raise to edges, namely, armchair and zigzag. The presence of these edges produces a variety of electronic behaviors and affects not only the reactivity of the carbon material toward the adsorption of adatoms, such as Li or Na, but also their diffusion characteristics.[3] Much work has been devoted to understanding the intercalation of Li in graphite and its derivatives. Here we discuss computational insights into the relatively less understood processes of Na and K intercalation.[4]

Results

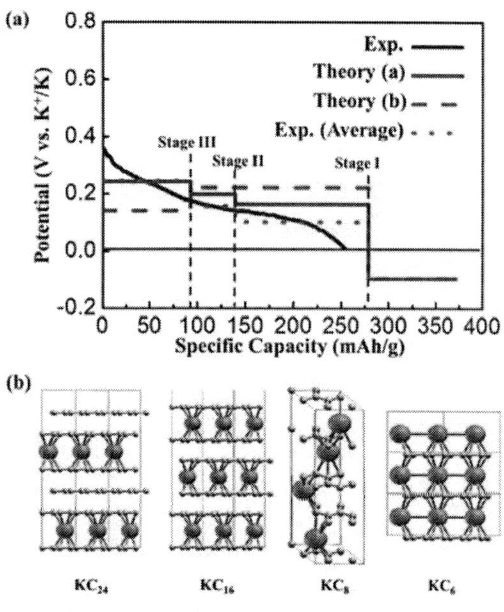

Figure 1. (a) Calculated potential profile for K intercalation into graphite for different staging scenarios. Blue line (Theory (a)) corresponds to an intercalation staging: KC_{24}(Stage III) → KC_{16} (Stage II) → KC_8 (Stage I). Green dotted line (Theory (b)) corresponds to calculated values for the previously reported staging: KC_{24} (Stage II) → KC_8 (Stage I). Red dotted line corresponds to the averaged experimental data shifted by 26 mAh/g to correct the capacity contribution from SEI formation. (b) Scheme of the different stages of K-intercalated graphite, K shown in blue and C in yellow.

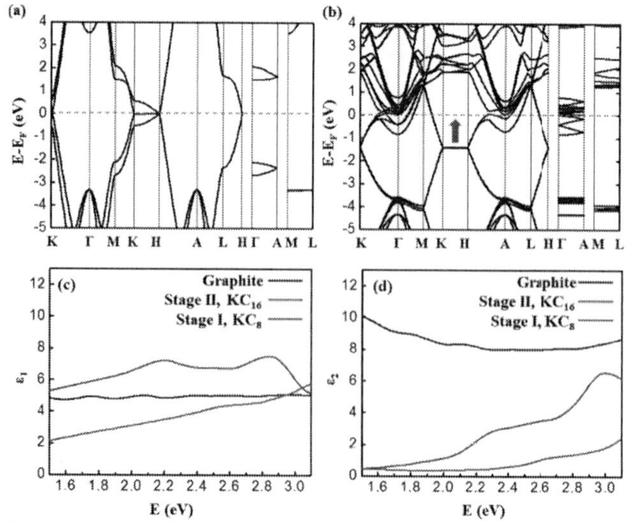

Figure 2. Calculated band structure of (a) graphite and (b) KC_8 (Stage-I), where the arrow between K-H indicates the shift in the Fermi level with respect to graphite; (c) ε_1 and (d) ε_2 calculated within the Random Phase Approximation in the visible range for pristine graphite as well as K intercalated graphite.

Conclusions

We show that the electrochemical intercalation process displays the formation of stages in K-GICs. These electrochemical intercalation stages are different than the ones reported in gas phase experiments and are supported by our first-principles calculations as well as experiments.

Acknowledgements. We acknowledge the experimental work done by Prof. Hu's group.[4]
We gratefully acknowledge the support from NSF-CBET Grant Nos. 1335944 and 1335979.

References

(1) Geim, A. K.; Novoselov, K. S., The rise of graphene. *Nature Materials* **2007,** *6* (3), 183-191.

(2) Barone, V.; Hod, O.; Peralta, J.; Scuseria, G., Accurate Prediction of the Electronic Properties of Low-Dimensional Graphene Derivatives Using a Screened Hybrid Density Functional. *Accounts of Chemical Research* **2011,** *44* (4), 269-279.

(3) Bhardwaj, T.; Antic, A.; Pavan, B.; Barone, V.; Fahlman, B. D., Enhanced Electrochemical Lithium Storage by Graphene Nanoribbons. *Journal of the American Chemical Society* **2010,** *132* (36), 12556-12558; Uthaisar, C.; Barone, V., Edge Effects on the Characteristics of Li Diffusion in Graphene. *Nano Letters* **2010,** *10* (8), 2838-2842.

(4) Luo, W.; Wan, J.; Ozdemir, B.; Bao, W.; Chen, Y.; Dai, J.; Lin, H.; Xu, Y.; Gu, F.; Barone, V.; Hu, L., Potassium Ion Batteries with Graphitic Materials. *Nano letters* **2015,** *15* (11), 7671-7.

INVESTIGATION OF ANHYDROUS PROTON TRANSPORT ON FUNCTIONALIZED GRAPHENE

Abhishek Bagusetty1, J. Karl Johnson, Pabitra Choudhury[2], Bridget S. Derksen[1], Elizabeth Gatto[1]

[1]University of Pittsburgh, Department of Chemical & Petroleum Engineering, Pittsburgh, PA 15261, USA
New Mexico Tech, Department of Chemical Engineering, Socorro, NM 87801, USA
karlj@pitt.edu

Introduction

Proton transport at interfaces and surfaces is of tremendous importance in diverse areas such as proton exchange membranes in fuel cells, biological membranes and proton pump channels. We are specifically interested in the transport of protons at interfaces for a better design of new materials for proton exchange membrane fuel cells. The practical engineering motivation for developing new materials is to extend the temperature and humidity operating range of fuel cells through designing materials that can conduct protons under anhydrous conditions. We have used density functional theory to design and evaluate various configurations of functionalized graphene for proton transport membrane materials.

Methods

We used periodic density functional theory to compute energetic, structural and dynamic properties of functionalized graphene for proton transport. Ab initio molecular dynamics has been used to compute the dynamics of proton transport on functionalized graphene. We have used the nudged elastic band method to compute transition states and barrier heights for the ground state proton hopping processes.

Results and Discussion

We show that hydroxylated graphene having a contiguous hydrogen bonded network is capable of facile transport of protons in the complete absence of water. The network of hydrogen bonded OH groups facilitate proton transport through a "Grotthuss-like" mechanism. We have computed activation energy barriers and investigated mechanisms for proton transport reactions on the surfaces of functionalized graphene with excess protons.

Acknowledgement. Computations were carried out at the University of Pittsburgh Center for Simulation and Modeling and at the Extreme Science and Engineering Discovery Environment (XSEDE) under project TG-DMR110091.

FLEXIBLE MICRO-SUPERCAPACITORS BASED ON GRAPHENE

Xiaodong Chen

School of Materials Science and Engineering, Nanyang Technological University, 50 Nanyang Avenue, 639798, Singapore
Email: chenxd@ntu.edu.sg

Micro-supercapacitors with unique two-dimensional (2D) structures are gaining attention due to their small size, high energy density and potential applications in on-chip and portable electronics. Compared to the sandwich structure of conventional supercapacitors, the 2D structure of micro-supercapacitors enables a reduction in the ionic diffusing pathway, and more efficient utilization of the surface area of electrode materials. Meanwhile, emerging wearable electronics require the property of stretchability in addition to flexibility for application on the soft and curved human body that is covered with highly extensible skins. Micro-supercapacitors, as a candidate for essential integrated energy conversion and storage units on wearable electronics, ought to be capable of accommodating large strain while retaining their performance. In this talk, I will present our recent development of highly stretchable micro-supercapacitors with stable electrochemical performance. The excellent stretchable and electrochemical performance relies on the out-of-plane wavy structures of graphene micro-ribbons. It decreases the stain concentration on the electrode fingers, so that the detaching and cracking of the electrode materials could be prevented. In addition, it ensured the electrode fingers keeping relative constant distance, so the stability of the micro-supercapacitors could be enhanced.

GRAPHENE-BASED METAL-FREE CATALYSTS FOR ORR AND BEYOND

Chungang Hu, Min Wang, and Liming Dai

Center of Advanced Science and Engineering for Carbon (Case4Carbon),
Department of Macromolecular Science and Engineering, Case School of Engineering,
Case Western Reserve University, 10900 Euclid Avenue, Cleveland, OH 44106, USA

Abstract

The electrochemical oxygen reduction reaction (ORR) is critical to energy conversion (*e.g.*, fuel cells, metal-air batteries, and dye-sensitized solar cells). However, cathodic ORR occurs very slowly without a catalyst. Although platinum has been widely used as the ORR catalysts for energy conversion, its high cost and limited reserve in nature have been shown to be the major "showstopper" to mass market energy conversion devices for commercial applications.

Along with the recent intensive research efforts in reducing or replacing Pt-based electrocatalysts, we have found a class of heteroatom-doped carbon nanomaterials, which, as *metal-free* alternative catalysts to replace platinum, could dramatically reduce the cost and increase the energy conversion efficiency.

In this talk, we will summarize some of our work on the use of graphene-based nanomaterials as metal-free catalysts for ORR and beyond ORR, along with an overview on the recent developments in this exciting field.

2D METAL CHALCOGENIDE AND METAL LAYERED DOUBLE HYDROXIDE (LDH) MATERIALS AND THEIR HETEROSTRUCTURES FOR ELECTROCATALYSIS AND DEVICE APPLICATIONS

Song Jin, Leith Samad, and Melinda J. Shearer

Department of Chemistry, University of Wisconsin-Madison, 1101 University Avenue, Madison, Wisconsin 53706
*e-mail: jin@chem.wisc.edu

For efficient and practical photoelectrochemical solar energy conversion, such as solar water splitting, we need inexpensive electrocatalysts for hydrogen evolution reaction (HER) and oxygen evolution reaction (OER) with catalytic performance rivaling that of the noble metal catalysts. We have discovered that chemical exfoliation of layered MX_2 nanostructures and the simultaneous conversion of the as-synthesized MX_2 material from its semiconducting 2H polymorph to the metallic 1T polymorph, grants greatly enhanced electrocatalytic activity for HER. The resulting 1T-MoS_2 and 1T-WS_2 nanosheets exhibit dramatically enhanced HER electrocatalytic performance as compared to their corresponding 2H polymorphs. Structural characterization and electrochemical studies confirm that the nanosheets of the metallic MoX_2 polymorph exhibit facile electrode kinetics, low-loss electrical transport, and possess a proliferated density of catalytically active sites. These distinct and previously unexploited features of 1T-MX_2 make these metallic nanosheets highly competitive earth-abundant HER catalysts. Furthermore, we have coupled electrocatalytic 1T-MoS_2 to a *p*-type silicon photoelectrode to enable hydrogen evolution driven by solar light. Due to the excellent HER activity of 1T-MoS_2 and the high-quality catalyst–semiconductor interface enabled by the direct CVD growth of MoS_2 on silicon (followed by chemical exfoliation), a much higher photocurrent density for hydrogen evolution was achieved as compared to 2H-MoS_2 on silicon. Photocurrents up to 17.6 mA/cm^2 at 0 V *vs* reversible hydrogen electrode (RHE) were achieved under simulated 1 sun irradiation, and good stability was demonstrated over repeated scans and long-time operation. Electrochemical impedance spectroscopy revealed low charge transfer resistances at the semiconductor/catalyst and catalyst/electrolyte interfaces, and surface photoresponse measurements also demonstrated slow carrier recombination dynamics and consequently efficient charge carrier separation, providing further evidences for the superior performance. Our results suggest that chemically exfoliated 1T-MoS_2/Si heterostructures are promising earth-abundant alternatives to photocathodes based on noble metal catalysts for solar-driven hydrogen production. We have also discovered that low temperature CVD growth using $MoCl_5$ and S precursors results in conformal deposition of amorphous MoS_xCl_y HER catalyst that show high electrocatalytic activity and high PEC hydrogen generation performance. These results demonstrate chemically exfoliated 1T-MoS_2 and amorphous MoS_xCl_y/Si heterostructures are promising earth-abundant alternatives to photocathodes based on noble metal catalysts for solar-driven hydrogen production.

We will also discuss our new results on using rationally synthesized layered metal double hydroxide (LDH) nanomaterials and the control of their nanostructures for oxygen evolution reaction (OER) electrocatalysis. We also developed the understanding of the OER catalytic reaction mechanisms on NiFe LDH materials using operando Mossbauer spectroscopy and used such understanding to direct the rational design and development of new OER catalysts with even better catalytic activities.

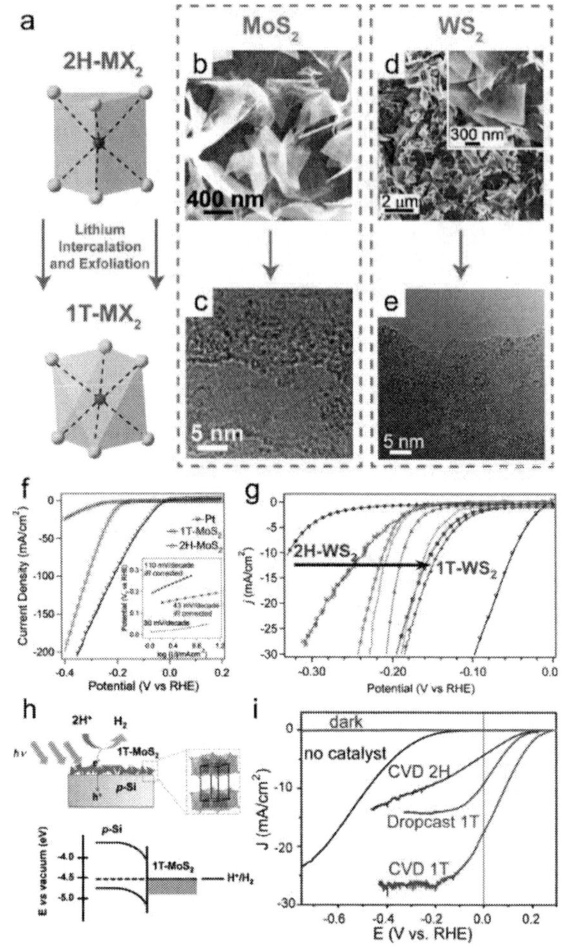

Figure 1. Chemical exfoliation (a) of MoS_2 (b) and WS_2 (d) nanostructures using a butyl lithium treatment results in the formation of metallic MoS_2 (c) and WS_2 (e) nanosheets that display much enhanced catalytic activity (f,g) for hydrogen evolution reaction (HER). (h) Direct growth of 1T-MoS_2 nanosheets on a *p*-type Si photocathode enables (i) efficient hydrogen evolution driven by solar energy.

References

(1) Lukowski, M. A.; Daniels, A. S.; Meng, F.; Forticaux, A.; Li, L.; Jin, S.; "Enhanced Hydrogen Evolution Catalysis from Chemically Exfoliated Metallic MoS_2 Nanosheets" *J. Am. Chem. Soc.* **2013,** *135*,10274- 10277.

(2) Lukowski, M. A.; Daniels, A. S.; English, C. R.; Meng, F.; Forticaux, A.; Hamers, R. J.; Jin, S.; "Highly Active Hydrogen Evolution Catalysis from Metallic WS_2 Nanosheets" *Energy Environ. Sci.*, **2014**, *7*, 2608-2613. DOI: 10.1039/C4EE01329H.

(3) Ding, Q.; Meng, F.; English, C. R.; Caban-Acevedo, M.; Shearer, M. J.; Liang, D.; Daniel, A. S.; Hamers, R. J.; Jin, S.; "Efficient Photoelectrochemical Hydrogen Generation Using Heterostructures of Si and Chemically Exfoliated Metallic MoS_2" *J. Am. Chem. Soc.*, **2014**, *136*, 8504-8507. DOI: 10.1021/ja5025673*J*.

(4) Faber, M. S.; Jin, S.; "Earth-Abundant Inorganic Electrocatalysts and Their Nanostructures for Energy Conversion Applications" *Energy Environ. Sci.*, **2014**, *7*, 3519-3542. **DOI:** 10.1039/C4EE01760A.

(5) Zhang, X.; Meng, F.; Mao, S.; Ding, Q.; Shearer, M. J.; Faber, M. S.; Chen, J.; Hamers, R. J.; Jin, S.; "Amorphous MoS_xCl_y Electrocatalyst Supported by Vertical Graphene for Efficient Electrochemical and Photoelectrochemical Hydrogen Generation" *Energy Environ. Sci.*, **2015**, *8*, 862-868. **DOI:** 10.1039/C4EE03240C

(6) Ding, Q.; Zhai, J.; Cabán-Acevedo, M.; Shearer, M. J.; Li, L.; Chang, H.-C.; Tsai, M.-L.; Ma, D.; Zhang, X.; Hamers, R. J.; He, J.-H.; Jin, S.; "Designing Efficient Solar-Driven Hydrogen Evolution Photocathodes Using Semi-Transparent MoQ_xCl_y (Q=S, Se) Catalysts on Si Micropyramids" *Adv. Mater.* **2015**, 27, 6511–6518. DOI: 10.1002/adma.201501884.

(7) Liang, H.; Meng, F.; Cabán-Acevedo, M.; Li, L.; Forticaux, A.; Xiu, L.; Wang, Z.; Jin, S.; "Hydrothermal Continuous Flow Synthesis and Exfoliation of NiCo Layered Double Hydroxide Nanosheets for Enhanced Oxygen Evolution Catalysis" *Nano Lett.* **2015**, *15*, 1421-1427. **DOI:** *10.1021/nl504872s.*

(8) Chen, J.; Dang, L.; Liang, H.; Bi, W.; Gerken, J. B.; Jin, S.; Alp, E. E.; Stahl, S. S.; Operando Analysis of NiFe- and Fe-Oxyhydroxide Electrocatalysts for Water Oxidation: Detection of Fe^{+4} by Mössbauer Spectroscopy. *J. Am. Chem. Soc.* **2015**, *137*, 15090–15093. DOI: 10.1021/jacs.5b10699.

PROBING SODIUM INSERTION MECHANISM IN GRAPHENE OXIDES BY IN SITU TEM

Kai He[1,]**, *Chunsheng Wang*[2], *John Cumings*[3]

[1]Department of Materials Science and Engineering
Northwestern University, Evanston, IL 60208
[2]Department of Chemical and Biomolecular Engineering
University of Maryland, College Park, MD 20742
[3]Department of Materials Science and Engineering
University of Maryland, College Park, MD 20742
*Email: he@northwestern.edu

Introduction

The diverse requirements of energy storage system in a variety of applications call for advancement not only in lithium ion technology but also in new battery chemistries. This motivates recent research towards rechargeable batteries that utilize other alkali metals, such as Na, K, and Mg.[1-3] Specifically, renewed interest in sodium ion batteries arises due to their potential to provide a sustainable and low-cost solution for large-scale energy storage such as electric vehicles and power grids.[3] Due to many similarities between sodium and lithium characteristics, the development of Na-ion batteries can benefit greatly from the knowledge developed over the past few decades on Li-ion batteries, in particular the choice of electrode materials. However, the differences between lithium and sodium chemistries may strongly affect the electrochemical reactions and the overall performance of a battery, which have been evident in the lithiation and sodiation behaviors in a number of previous studies.[1,3] It is generally believed that Na has lower capacity compared to Li because of its larger ionic size, sluggish ionic diffusion and higher formation energy, though for different types of reaction mechanisms, i.e. intercalation, alloying, and conversion, the predominant limiting factor may vary.

The graphite materials are the most common anode material for Li-ion batteries, whereas their Na capacity appeared rather limited for practical use. Various efforts have been applied to engineering the graphitic material's structures (soft/hard carbons) and morphologies (nanotubes, graphene, porous, hollow nanoparticles, etc.) to enable reversible sodium insertion into the graphitic interlayers and nanovoids. Here we report graphene oxides with expanded interlayer spacing as a superior Na-ion battery anode material.[4] Using *in-situ* transmission electron microscopy (TEM), we observed the real-time sodiation and desodiation processes at the atomic scale. We found that Na^+ insertion and extraction will cause microstructural disorder and recover in a repeatable fashion, which elucidates the sodium filling mechanism and provides useful implications for future Na-ion battery design and implementation.

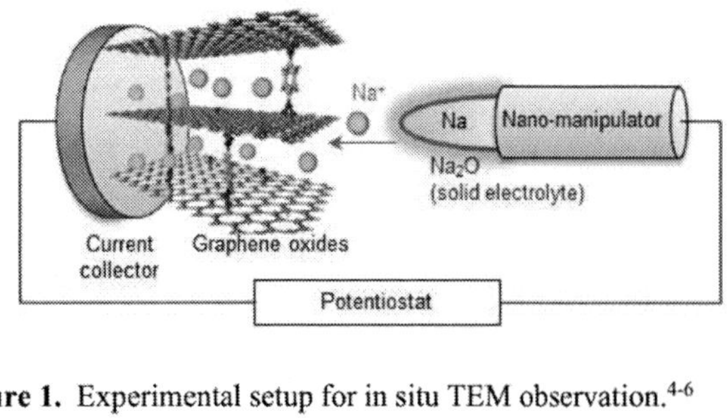

Figure 1. Experimental setup for in situ TEM observation.[4-6]

Experimental

Materials synthesis. We use a two-step oxidation-reduction process to synthesize the extended graphene oxides with controllable interlayer distance.[4] The graphene oxides were firstly prepared from pure graphite through a modified Hummer's method, and then reduced by subsequent heat treatment under a continuous flow of Ar at 600 °C with various time to tune the degrees of reduction.

Battery tests. Electrochemical properties were tested by coin-cell measurements using a half-cell configuration against a pure Na counter electrode. Coin cells, consisting of graphene oxide electrode, sodium metal counter electrode, Celgard 3501 separator and 1.0M $NaClO_4$ in polycarbonate solvent liquid electrolyte, were assembled in a glove box under Ar atmosphere. Galvanostatic charge/discharge and cyclic performance were measured on an Arbin BT-2000 test station.[4]

In situ TEM. In situ TEM experiments were performed on a JEOL 2100F microscopy at 200 kV with a customized Nanofactory manipulation holder equipped with a piezo-driven Pt probe.[4-6] The setup is shown in Figure 1. In a typical experiment, the active material was glued onto a specimen rod using conductive epoxy, and Na metal was scratched off a fresh-cut surface of Na bulk using the manipulator inside a glove box. Then the Na metal was intentionally exposed to the air for 5 s during the holder loading process to form a thin layer of Na_2O on the surface, serving as a solid electrolyte. During the in situ operation, a negative potential was applied to graphene oxides with respect to Na metal to initiate sodiation, and a positive potential for desodiation.[4]

Results and Discussion

Figure 1a-c shows distinct interlayer structures of pristine graphene, graphene oxide, and extended graphene, with interlayer spacing of 0.34 nm, 0.61 nm, and 0.43 nm, respectively, as confirmed by high-resolution TEM (HRTEM). We also use x-ray photoelectron spectroscopy (XPS) to characterize the content of oxygen-containing group in the interlayers. It is indicated that O-1s peak increases after oxidation and decreases after reduction, suggesting that the O component was attributed to the expansion and shrinkage of the graphene layers. The spacing depends on the content of O-containing groups and thus is tunable. Their electrochemical properties are shown in Figure 1g,h. The bare graphene has minimal capacity close to zero, and graphene oxides have much improved capacity. It is noted that the interlayer distance is not a sole factor to affect the capacity: the graphene oxide with maximum O-groups has largest spacing but ultrahigh O occupancy also hinders Na^+ intercalation.

To justify the sodiation and desodiation mechanism (Figure 3a) from the microstructural point of view, we performed in situ HRTEM using extended graphene with ~0.43 nm interlayer spacing as the active electrode material. The pristine EG displays a microstructure with well-aligned and long-range-ordered graphene layers (Fig. 3b,e) and contains large-scale pure graphitic structure with large interlayer distance (Fig. 3i). After Na^+ insertion, the shape of the EG changed by tilting or distorting to form ripple-like features and accommodate the large-scale morphology changes (Fig. 3c,f). In the desodiation process, as Na ions migrate out of the EG anode, the microstructure of desodiated EG is reversibly recovered to a parallel layered pattern with long-range ordering (Fig. 3d,g). The electron diffraction pattern also confirmed an amount of Na^+ accommodated into the EG after sodiation (Fig. 3j), and then disappeared after desodiation (Fig. 3k). Nevertheless, despite the short-range disorder, the layered structure of the material does not exhibit large-scale distortion throughout the entire reaction, which provides a constant transport space for Na ions.

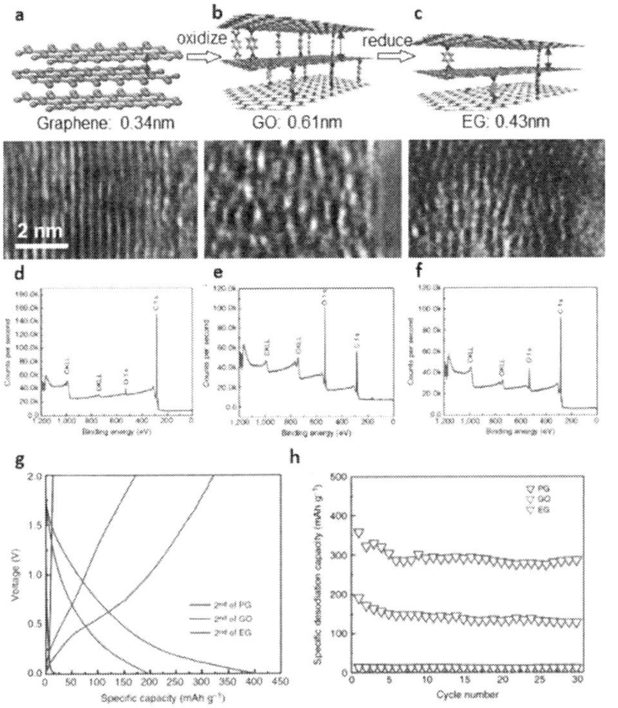

Figure 2. Structure of (a) pristine graphene, (b) graphene oxide, and (c) extended graphene after oxidation and reduction processes. (d-f) XPS of these three materials. (g) Charge-discharge profile and (h) cyclic performance of these three electrode materials in a coin cell.[4]

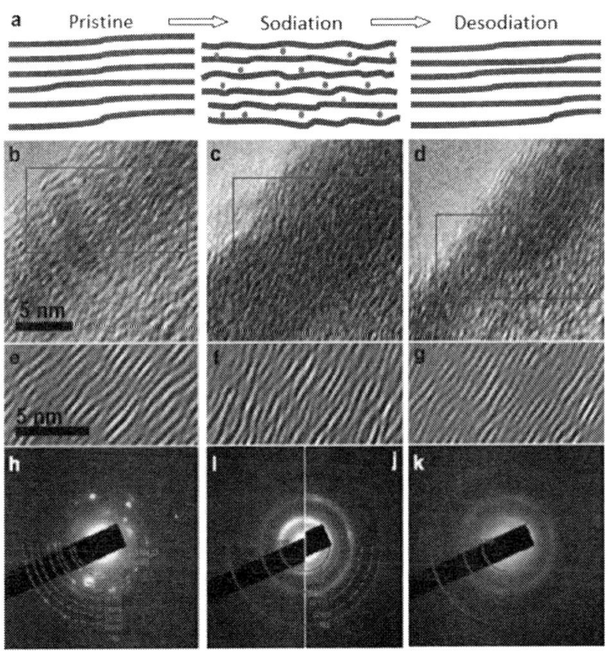

Figure 3. (a) Schematic showing Na^+ storage mechanism. (b-d) TEM images, (e-g) filtered-HRTEM images, and (i-k) electron diffraction patterns of pristine EG, sodiated EG, and desodiated EG. (h) Electron diffraction of the Na/Na_2O counter electrode.[4]

Conclusions

In summary, we prepared graphene oxides as Na-ion battery anode materials with tunable interlayer distance through an oxidation-reduction process. The spacing between graphene layers largely affects the Na^+ insertion, but is not the sole limiting factor. By visualizing dynamical structure change during sodiation and desodiation processes in real time using in situ HRTEM, we justified the Na^+ intercalation mechanism on the atomic scale. These findings show implications for future implementation of graphitic materials to Na-ion batteries.

Acknowledgement. The work was supported by the Science of Nanostructures for Electrical Energy Storage, a DOE-EFRC under Award Number DESC0001160.

References

(1) Yabuuchi, N.; Kubota, K.; Dahbi, M.; Komaba, S. *Chem. Rev.*, **2014**, *113*, 11636.

(2) Saha, P.; Datta, M. K.; Velikokhatnyi, O. I.; Manivannan, A.; Alman, D.; Kumta, P. N. *Prog. Mater. Sci.*, **2014**, *66*, 1.

(3) Slater, M. D.; Kim, D.; Lee, E.; Johnson, C. *Adv. Func. Mater.*, **2013**, *23*, 947.

(4) Wen, Y.; He, K.; Zhu, Y.; Han, F.; Xu, Y.; Matsuda, I.; Ishii, M.; Cumings, J.; Wang, C. *Nat. Comm.*, **2014**, *5*, 4033.

(5) He, K.; Xin, H. L.; Zhao, K.; Yu, X.; Nordlund, D.; Weng, T. C.; Li, J.; Jiang, Y.; Cadigan, C. A.; Richards, R. M.; Doeff, M. M.; Yang, X. Q.; Stach, E. A.; Li, J.; Lin, F.; Su, D.. *ACS Nano*, **2014**, *8*, 7251.

(6) He, K.; Lin, F.; Zhu, Y.; Yu, X.; Li, J.; Lin, R.; Nordlund, D.; Weng, T. C.; Richards, R. M.; Yang, X. Q.; Doeff, M. M.; Stach, E. A.; Mo, Y.; Su, D. *Nano Lett.*, **2015**, *15*, 5755.

HIGH RESOLUTION SOLID STATE NMR STUDIES OF GRAPHENE OXIDE AND REDUCED GRAPHENE OXIDE

*Jianhui Song[a], Yuhua Xue[b], Liming Dai[b], Qiao Jinliang[a], Mingming Guo[a]**

[a]Beijing Institute of Chemical Industry, Sinopec, Beijing, China
b Department of Macromolecular Science and Engineering, Case Western Reserve University, Cleveland, Ohio 44106, USA.
*Corresponding Author Email guomm.bjhy@sinopec.com

Introduction

Graphene possesses a large surface area and high electrical/thermal conductivity as well as excellent mechanical properties. From solid state NMR point of view, graphene is the system which provide both challenge[1] and new opportunity[2-3] The challenge come from the fact that there is no proton in the system so the routing and efficient solid NMR method like cross polarization and dipolar decoupling can not be used here. Also the relaxation time for the carbon sheet is very long. The new opportunity come from the intermolecular cross polarization for the graphene or graphene oxide (GO)/polymer composite. The intermolecular CP is not only help to characterize the purity of the reduction of the graphene oxide samples but also provide the miscibility and the sheet thickness information.

Experimental Section

Sample preparation In a typical experiment, GO was prepared by acid oxidation of graphite powder according to the modified Hummers' method. GO sample was reduced using a FeCl2[4].

NMR Solid state 1H and 13C NMR high resolution spectra were measured with Varian VNMR 500 MHz NMR spectrometer. The dried powder Samples were packed into 4 mm silicon nitride rotors with PTFE end caps spinning at 10 kHz. 3k to 12k FID transients were averaged using 8 sec relaxation delay 4.0 ms 90° pulse width without decoupling. 13C chemical shifts were corrected by using hexamethylbenzene ((dCH3 = 17.3 ppm) as an external reference.

Results and Discuss

^{13}CMAS NMR of GO and reduced GO. The ^{13}C NMR spectra of graphene oxide and reduced graphene oxide (rGO), reduced using a $FeCl_2$[4], shows three distinct resonances: the resonance centered at 134 ppm corresponds to unoxidized sp2 carbons; the 60 ppm resonance is a result of epoxidation, and the 70 ppm is from hydroxylated carbons. After reduction, the 60 ppm, 70 ppm and 167 ppm resonances disappear. The peak at 134 ppm shifts to 123 ppm because of the change in the chemical environment of the sp2 carbons.

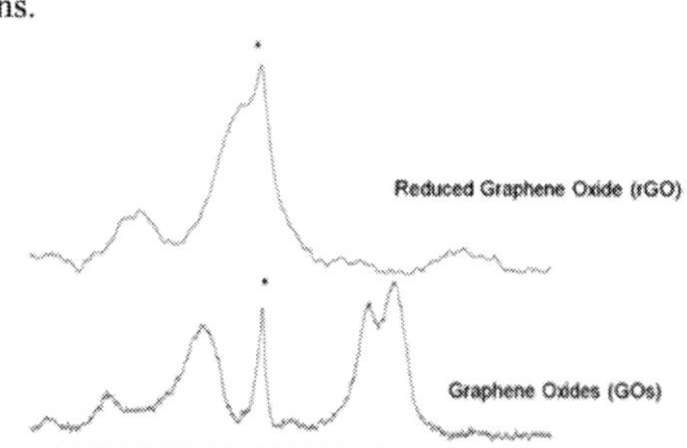

Figure 1. 13C MAS NMR of GO and reduced GO

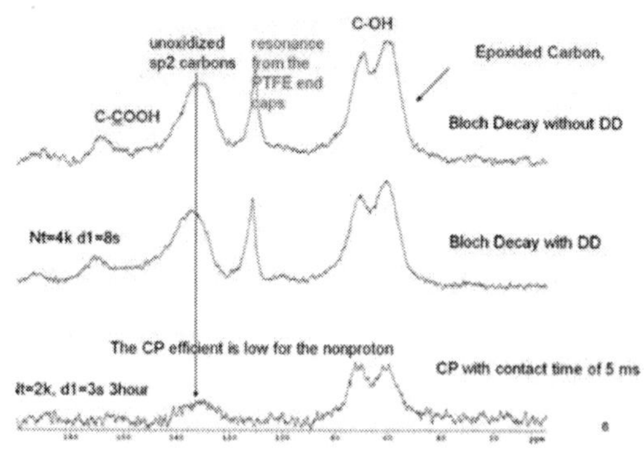

Figure 2. 13C MAS NMR of Graphene Oxide at different conditions

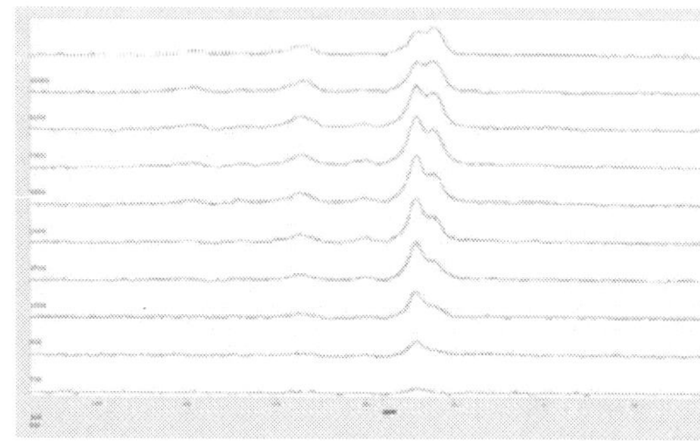

Figure 3. 13C CP/MAS NMR of GO at different contact time.

The curve fitting results indicate that while the cross polarization times are 2725.1,1426.4, 2380.6 us for the sp2 carbons, epoxidation, and hydroxylated carbons, respectively, the proton T1rho are 10355, 8367.7 and 11680.6 us for the corresponding carbon sites.

^{1}HMAS NMR of GO and reduced GO

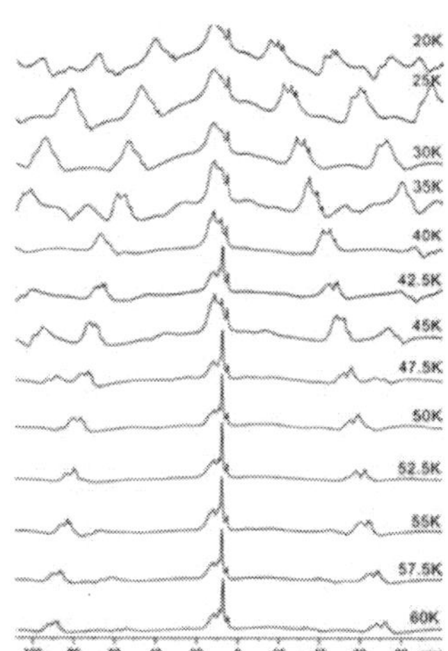

Fig. 4. 1H MAS NMR of GO at different spinning speed.

References

(1) Cai, W. W.; Piner, R. D.; Stadermann, F. J.; Park, S.; Shaibat, M. A.; Ishii, Y.; Yang, D. X.; Velamakanni, A.; An, S. J.; Stoller, M.; An, J. H.; Chen, D. M.; Ruoff, R. S. Science, 321: 1815-1817 (2008).
(2) M.M. Guo, H.G.Zachmann, Polymer, 34, 2503-2508 (1993)
(3) Guo, M. M. Trends Polym.Sci. 4 : 238-244 (1996)

(4) Y.H. Xue, H. Chen, D. S. Yu, S.Y. Wang, M. Yardeni, Q.B. Dai, M.M. Guo, Y. Liu, F. Lu, J. Qu, and L.M. Dai Chem. Commun.47(42), 11689-11691 (2011)..

GRAPHENE GROWTH ON COPPER SURFACES: FROM MECHANISM STUDIES TO PROTOCOL DESIGN

Zhenyu Li

Hefei National Laboratory for Physical Sciences at the Microscale,
University of Science and Technology of China
Hefei, Anhui 230026, China
zyli@ustc.edu.cn

Introduction

As a two-dimensional sheet of sp^2-hybridized carbon atoms, graphene has attracted extensive research interest due to its extraordinary physical and chemical properties promising for various applications.[1] A huge effort has been devoted to the synthesis of graphene. Epitaxial growth is found to be a probable way to reach a balance between sample quality and process scalability. Various substrate materials have been used for graphene growth, where Cu is the most popular one.[2,3] The size, layer number, and quality of graphene samples are sensitive to various growth parameters, such as pressure, temperature, substrate morphology and so on. To optimize these parameters, it is very desirable to understand the growth mechanism in atomic details.

Theoretical studies provide an important means to reveal the atomic scale growth mechanisms.[4] First principles calculations can reliably predict reaction energies and barriers for elementary processes. Using them as an input, kinetic Monte Carlo (KMC) simulation can then be performed to study the growth dynamics. Alternatively, molecular dynamics (MD) simulation can be used to directly observe fast processes. At the same time, simple thermodynamic and kinetic analyses can also provide useful insights into growth mechanisms. Eventually, new protocols can be designed based on mechanisms revealed. In this talk, recent progresses made in our group on theoretical studies of graphene growth on Cu surfaces will be presented.

Computational

First-principles calculations were performed within the projected augmented wave (PAW) framework using the Vienna *ab initio* simulation package (VASP).[5,6] The climbing image nudged elastic band (CI-NEB) method[7] was used to locate transition states and determine barrier heights. Kinetic energy cutoff for plane-wave basis set and the Monhorst-Pack k-point sampling parameters were carefully tested to produce well-converged results. KMC simulations were carried out on a predefined lattice based on first-principles kinetic parameters. Virtual experiments on computer were also performed by adjusting these parameters.

Results and Discussion

Surface carbon species. Diffusion of carbon monomer on Cu(111) is almost barrierless,[8] while graphene growth on this surface is found to be diffusion limited.[9] Therefore, there should be other surface carbon species instead of carbon adatoms playing a dominant role in graphene growth. In fact, carbon adatoms on Cu(111) will go to subsurface octahedral sites. When two carbon monomers approach each other, a Cu-C-Cu bridging-metal structure will be formed on Cu(111),[10] which prohibits a straightforward monomer combination on the surface.[11] One way to form dimer from the BM structure is moving the other carbon atom also to subsurface and then the two neighboring subsurface atoms form a dimer on surface with an energy barrier of 0.30 eV. Carbon dimers on the Cu(111) surface are also thermodynamically much more stable than monomers.

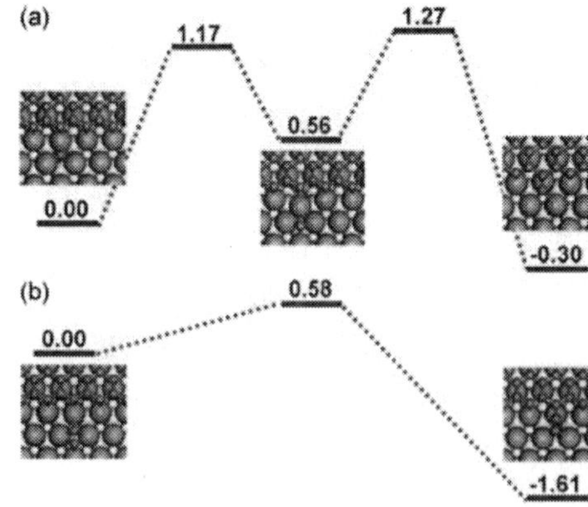

Figure 1. Minimum energy path for (a) a subsurface carbon monomer and (b) a carbon dimer to be attached to a zigzag edge.

Since its formation is not only thermodynamically favorable but also kinetically fast, carbon dimer is expected to be an abundant species on Cu(111). Energy barriers for dimer diffusion and its attachment to a graphene edge is comparable. On the other hand, attachment of a carbon monomer turns out to be much more difficult (**Figure 1**), which makes dimer become the feeding species of graphene growth on Cu(111).[8] The dominance of dimer attachment event in graphene growth is indeed observed in KMC simulations. At the same time, we also observe a carbon concentration gradient around graphene islands, in consistent with the experimentally observed diffusion-limited behavior.

Dehydrogenation. In experiments using chemical vapor deposition (CVD), a dehydrogenation step is required for graphene growth. Cu surfaces are considered to be a catalyst for this reaction. Based on thermodynamics analysis[12] and computational fluid dynamics (CFD) simulations,[13] we suggest that gas phase dehydrogenation also plays an important role under typical growth conditions. On the other hand, Cu is not an efficient catalyst for CH_4 dehydrogenation (**Figure 2**). As a result, dehydrogenation is not expected to be completed before graphene growth and the active species in graphene CVD growth is hydrocarbon.[14,15] Based on this picture, a nucleation model can be built.

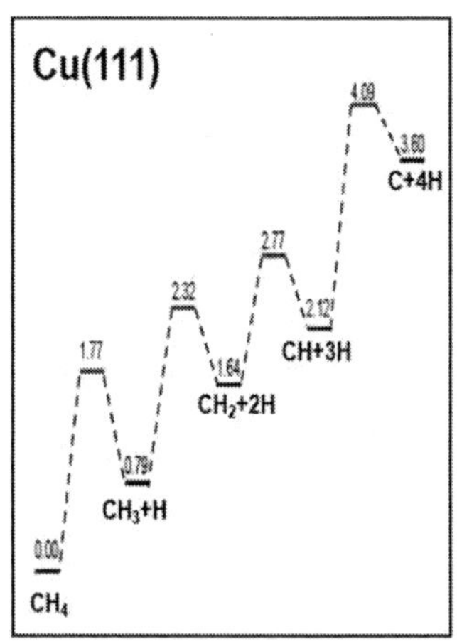

Figure 2. Energy profile of the dehydrogenation processes of CH_4 on Cu (111).

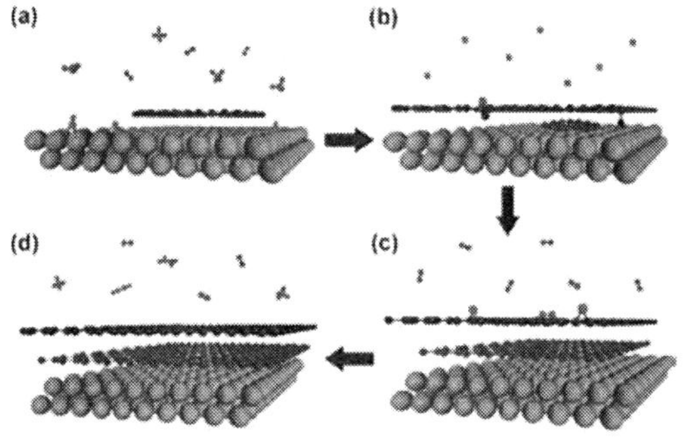

Figure 3. Schematic of new bilayer graphene growth protocol.

Growth protocol design. Bilayer graphene is useful in some applications when a band gap is required. If two layers grow simultaneously, the lower layer in contact with the Cu substrate will grow faster. Therefore, it becomes difficult for the upper layer to reach a full coverage. From this point of view, it is desirable to grow the second layer beneath the first layer. The key point is how to intercalate carbon atoms into the graphene-substrate interface. A new atom-exchange based penetration mechanism is identified by our first-principles calculations. With this mechanism, we propose a protocol (**Figure 3**) to grow high-quality bilayer graphene.[16] First, monolayer graphene can be grown on Cu surface via standard CVD. Then, atomic carbon will be provided to enable monomer intercalation and the second layer growth. At last, hydrogen is used to clean the possibly contaminated first layer.

Conclusions

In summary, graphene is an important material with unique properties and many applications. Epitaxial growth on metal surfaces is a promising way to produce high-quality graphene samples at a large scale. To improve the sample quality, it is important to understand the atomic details during graphene growth. In this talk, we focus on theoretical studies of graphene growth on Cu surfaces. Contributions of different carbon species to graphene growth are investigated, and dimer is found to be the dominant feeding species. This result explains the experimentally observed diffusion-limited growth behavior. Coalescence of small carbon species on Cu surfaces induce a significant surface relaxation characterized by the formation of bridging-metal structures. When hydrocarbon is used as the precursor, the dehydrogenation step starts in gas phase and proceeds during the whole graphene growth process. Carbon atom is found to be able to penetrate a graphene overlayer on Cu surface via an exchange mechanism. Using such a process, a protocol to grow high-quality bilayer graphene is proposed.

Acknowledgement. Financial support from NSFC (21173202, 21222304, and 21421063), MOST (2011CB921404), and CAS (XSB01020300), and CPU time from USTC-SCC, SCCAS, Tianjin, and Shanghai Supercomputer Centers are highly appreciated.

References

(1) Novoselov, K. S.; Geim, A. K.; Morozov, S. V.; Jiang, D.; Zhang, Y.; Dubonos, S. V.; Grigorieva, I. V.; Firsov, A. A., Electric Field Effect in Atomically Thin Carbon Films. *Science* **2004,** *306*, 666-669.

(2. Li, X.; Cai, W.; An, J.; Kim, S.; Nah, J.; Yang, D.; Piner, R.; Velamakanni, A.; Jung, I.; Tutuc, E.; Banerjee, S. K.; Colombo, L.; Ruoff, R. S., Large-area synthesis of high-quality and uniform graphene films on copper foils. *Science* **2009,** *324* (5932), 1312-4.

(3. Li, X.; Cai, W.; Colombo, L.; Ruoff, R. S., Evolution of graphene growth on Ni and Cu by carbon isotope labeling. *Nano Lett.* **2009,** *9* (12), 4268-4272.

(4) Wu, P.; Zhang, W.; Li, Z.; Yang, J., Mechanisms of graphene growth on metal surfaces: theoretical perspectives. *Small* **2014,** *10* (11), 2136-50.

(5) Kresse, G.; Furthmüller, J., Efficiency of ab-initio total energy calculations for metals and semiconductors using a plane-wave basis set. *Comput. Mater. Sci.* **1996,** *6* (1), 15-50.

(6) Kresse, G.; Furthmüller, J., Efficient iterative schemes for ab initio total-energy calculations using a plane-wave basis set. *Phys. Rev. B* **1996,** *54* (16), 11169-11186.

(7) Henkelman, G.; Uberuaga, B. P.; Jonsson, H., A climbing image nudged elastic band method for finding saddle points and minimum energy paths. *J. Chem. Phys.* **2000,** *113* (22), 9901-9904.

(8) Wu, P.; Zhang, Y.; Cui, P.; Li, Z.; Yang, J.; Zhang, Z., Carbon dimers as the dominant feeding species in epitaxial growth and morphological phase transition of graphene on different Cu substrates. *Phys. Rev. Lett.* **2015,** *114* (21), 216102.

(9) Nie, S.; Wofford, J. M.; Bartelt, N. C.; Dubon, O. D.; McCarty, K. F., Origin of the Mosaicity in Graphene Grown on Cu(111). *Phys. Rev. B* **2011,** *84*, 155425.

(10) Wu, P.; Zhang, W.; Li, Z.; Yang, J.; Hou, J. G., Communication: Coalescence of Carbon Atoms on Cu (111) Surface: Emergence of a Stable Bridging-metal Structure Motif. *J. Chem. Phys.* **2010,** *133* (7), 071101.

(11) Li, Z.; Wu, P.; Wang, C.; Fan, X.; Zhang, W.; Zhai, X.; Zeng, C.; Li, Z.; Yang, J.; Hou, J., Low-Temperature Growth of Graphene by Chemical Vapor Deposition Using Solid and Liquid Carbon Sources. *ACS Nano* **2011,** *4*, 3385–3390.

(12) Li, Z.; Zhang, W.; Fan, X.; Wu, P.; Zeng, C.; Li, Z.; Zhai, X.; Yang, J.; Hou, J., Graphene Thickness Control via Gas-Phase Dynamics in Chemical Vapor Deposition. *J. Phys. Chem. C* **2012,** *116* (19), 10557-10562.

(13) Li, G.; Huang, S.-H.; Li, Z., Gas-phase Dynamics in Graphene Growth by Chemical Vapour Deposition. *Phys. Chem. Chem. Phys.* **2015,** *17*, 22832.

(14) Zhang, W.; Wu, P.; Li, Z.; Yang, J., First-Principles Thermodynamics of Graphene Growth on Cu Surfaces. *J. Phys. Chem. C* **2011,** *115* (36), 17782-17787.

(15) Zhang, J.; Wang, Z.; Niu, T.; Wang, S.; Li, Z.; Chen, W., Elementary Process for CVD Graphene on Cu(110): Size-selective Carbon Clusters. *Sci. Rep.* **2014,** *4*, 4431.

(16) Wu, P.; Zhai, X.; Li, Z.; Yang, J., Bilayer Graphene Growth via a Penetration Mechanism. *J. Phys. Chem. C* **2014,** *118* (12), 6201-6206.

ONE-DIMENSIONAL NANOMATERIALS FOR ENERGY STORAGE

Liqiang Mai

State Key Laboratory of Advanced Technology for Materials Synthesis and Processing, Wuhan University of Technology, Wuhan, 430070, China
Email: mlq518@whut.edu.cn

One-Dimensional nanomaterials with large surface area, more surface active sites and better permeability can significantly increase the energy density, power density and cycling performance for the energy storage. Such hierarchical structure can also be used as targeted intracellular recording for its facile synthesis route. In our present work, a series of hierarchical nanomaterials have been obtained, including kinked hierarchical nanowires, hierarchical heterostructured nanowires and hierarchical scrolled nanowires which shows great electrochemical performance.

To improve the electrochemical performance, V_3O_7 nanowire templated semi-hollow bicontinous graphene scrolls architecture is designed and constructed through "oriented assembly" and "self-scroll" strategy. The V_3O_7 nanowire templated semi-hollow bicontinous graphene scrolls with interior cavities provide continuous electron and lithium ion transfer channel and space for free volume expansion of V_3O_7 nanowires during cycling, thus representing a unique architecture for excellent lithium ion storage capacity and cycling performance.[1] Besides, we have designed and synthesized hierarchical $MnMoO_4/CoMoO_4$ heterostructured nanowires by combining "oriented attachment" and "self-assembly". The asymmetric supercapacitors based on the hierarchical heterostructured nanowires show a high specific capacitance and good reversibility with a cycling efficiency of 98% after 1,000 cycles.[2] Recently, we also constructed the hierarchical zigzag $Na_{1.25}V_3O_8$ nanowires,[3] $K_3V_2(PO_4)_3$ bundled nanowire,[4] and $Li_3V_2(PO_4)_3$ mesoporous nanotubes[5] with enhanced electrochemical performance. Our work presented here can inspire new thought in constructing novel one-dimensional structures and accelerate the development of energy storage appilications.

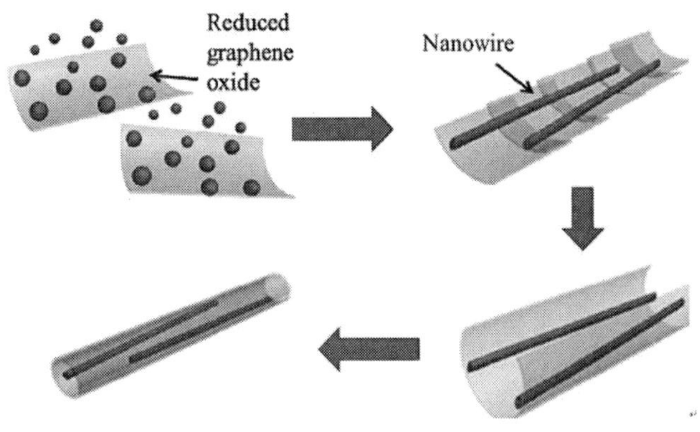

Figure 1. Nanowire templated semi-hollow bicontinous graphene scrolls.

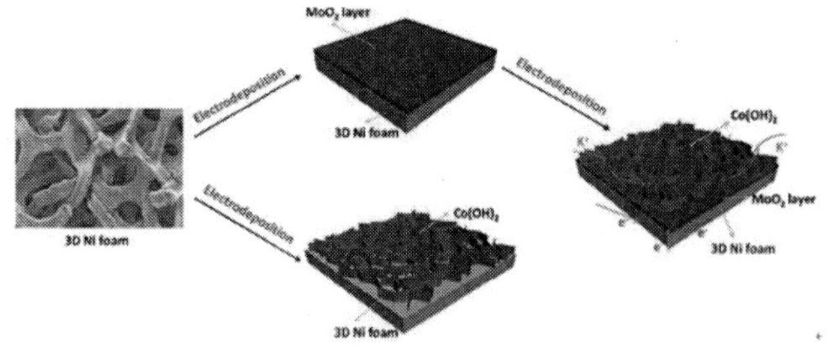

Figure 2. $MoO_2/Co(OH)_2$ hierarchical nanostructure.

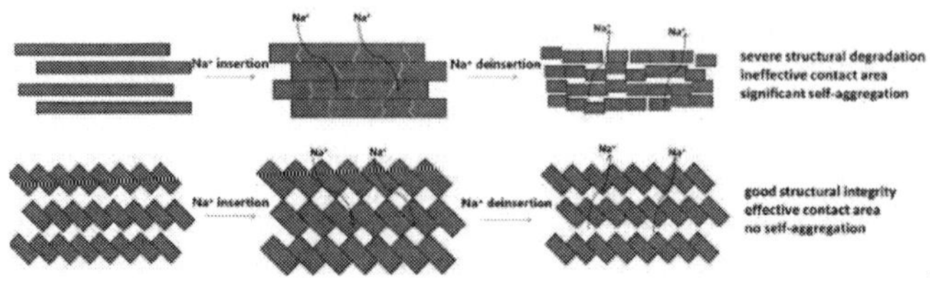

Figure 3. Hierarchical zigzag $Na_{1.25}V_3O_8$ nanowires.

References

(1) Yan, M. Y. et al., *J. Am. Chem. Soc.* 2013, 135, 18176-18182.
(2) Mai, L. Q. et al., *Nature Commun.* 2011, 2, 381.
(3) Dong, Y. F. et al., *Energy Environ. Sci.* 2015, 8, 1267.
(4) Wang, X. P. et al., *Adv. Energy Mater.* 2015, 5(17).
(5) Niu, C. J. et al., *Nature Commun.* 2015, 6, 7402.

GRAPHENE OXIDE-BASED ELECTRODES FOR LONG-LIFE, HIGH-ENERGY LITHIUM/SULFUR BATTERIES

Min-Kyu Song

School of Mechanical and Materials Engineering
Materials Science and Engineering Program
Washington State University, Pullman, WA 99164
Minkyu.song@wsu.edu

Introduction

Among many energy storage devices developed thus far, rechargeable lithium-ion batteries still represent the state-of-the-art technology in the market. However, current lithium-ion batteries are not able to meet the ever-increasing demands of advanced technologies[1]. For example, we have to radically improve the energy-storage capacity of batteries to increase the driving range of current electric vehicles. It is well known, however, that current oxide-based cathode materials have insufficient capacities to meet these demanding requirements of rapidly evolving advanced technologies such as electric vehicles.

Recently, lithium/sulfur (Li/S) batteries have gained substantial attention as alternative power sources for zero-emission vehicles and advanced electronic devices due to the very high theoretical specific capacity (1672 mAh/g) of sulfur cathodes[2-5]. Li/S battery offers much higher theoretical specific energy (~2680 Wh/kg for Li/S vs. ~580 Wh/kg for Li-ion cells). However, the development of the Li/S cell as a commercial product has been prevented by the short cycle life of this cell, caused mainly by the loss of sulfur due to dissolution as lithium polysulfides (e.g., Li_2S_8, Li_2S_6, Li_2S_4) in most organic electrolyte solutions. Therefore, one of the biggest challenges faced by Li/S cells is the design of novel cathode that allows (a) long cycle life and (b) high practical (obtainable) cell-level specific energy.

Results and Discussion

Nanostructured sulfur-graphene oxide (S-GO) nanocomposite cathodes have been developed with much improved cycling performance via the reactive functional groups on GO to successfully immobilize sulfur and lithium polysulfides during operation[5]. While successfully increasing the sulfur loading in the electrode in order to improve the cell-level specific energy, the synthesis conditions of S-GO nanocomposites have been optimized by depositing uniform and thin (around tens of nanometers) sulfur coating on GO and using optimized heat-treatment conditions for molten sulfur to diffuse into nanopores of graphene oxide, which led to excellent cell performance in terms of good reversibility ($S \leftrightarrow Li_2S$) as shown in Figure 1.

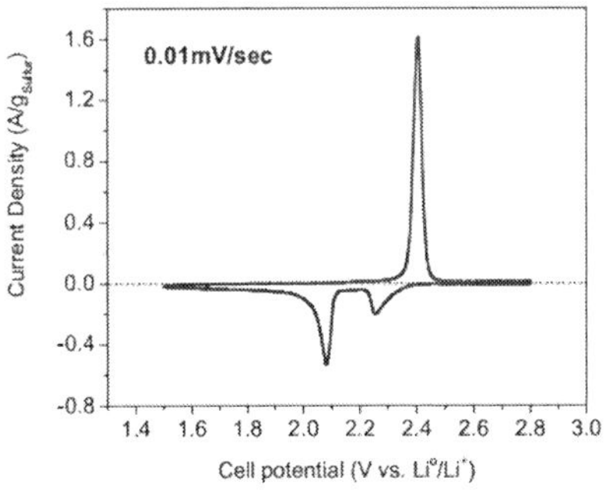

Figure 1. Typical cyclic voltammogram of a lithium/sulfur cell combining S-GO nanocomposite cathodes and ionic liquid-based electrolytes measured at the scan rate of 0.01mV/sec.

A lithium/sulfur cell employing a nanostructured sulfur-graphene oxide (S-GO) composite cathode can be cycled for 1000 cycles. Even after 1000 cycles, a very high specific capacity ~846 mAh/g of sulfur (which corresponds to ~470 mAh/g of electrode), was demonstrated, which is still higher than state-of-the-art Li-ion cells. The performance demonstrated herein suggests that lithium/sulfur cells may be suitable for high-power applications such as power tools. With further improvement of sulfur loading in the electrode and capacity retention, S-GO nanocomposite cathodes may provide a significant opportunity toward the development of zero-emission vehicles with much longer driving range.

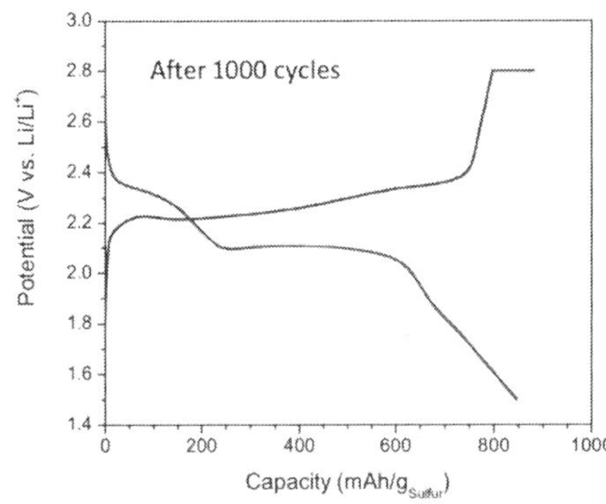

Figure 2. Discharge and charge profile of a lithium/sulfur cell obtained with 0.05C (1C = 1672 mA/g) after long-term cycling test with 1C and 0.5C for discharging and charging, respectively.

The X-ray absorption spectroscopy technique has been applied to study different stages of the Li/S-GO cell. We have investigated how speciation of sulfur in cathodes changes with charge/discharge cycling. For the cycled Li/S-GO cell, it is suggested that the loss of electrochemically active sulfur and the accumulation of a compact blocking insulating layer on the cathode surface during the charge/discharge processes make the capacity decay.

Conclusion

Sulfur-graphene oxide (S-GO) nanocomposites showed much improved cycle life and excellent performance. Graphene oxides are effective agents that are able to improve retention of the intermediate polysulfides formed during discharge, thus reducing the loss of sulfur. Various functional groups of GO can form chemical bonds with sulfur, which helps to immobilize sulfur atoms on the GO surface and thus reduce the dissolution/migration problem. It is expected that the best choice of cathode structure can be obtained by tuning the functional groups (epoxy, carboxylic acid, hydroxyl) on the surface of the graphene oxide, which requires understanding of the fundamental mechanism for the binding of the polysulfides to graphene oxide based cathodes.

References

(1) (1) Song, M.-K.; Park, S.; Alamgir, F. M.; Cho, J.; Liu, M. *Materials Science and Engineering: R: Reports* **2011**, *72*, 203.
(2) (2) Song, M.-K.; Cairns, E. J.; Zhang, Y. *Nanoscale* **2013**, *5*, 2186.
(3) (3) Bruce, P. G.; Freunberger, S. A.; Hardwick, L. J.; Tarascon, J.-M. *Nat. Mater.* **2012**, *11*, 19.
(4) (4) Cai, K.; Song, M.-K.; Cairns, E. J.; Zhang, Y. *Nano Lett.* **2012**, *12*, 6474.
(5) (5) Ji, L.; Rao, M.; Zheng, H.; Zhang, L.; Li, Y.; Duan, W.; Guo, J.; Cairns, E. J.; Zhang, Y. *J. Am. Chem. Soc.* **2011**, *133*, 18522.

EXFOLIATION AND SODIUM CYCLING BEHAVIOR OF TRANSITION METAL DICHALCOGENIDE NANOSHEETS

*Lamuel David, Monsur Abass, and Gurpreet Singh**

*Email: gurpreet@ksu.EDU, Mechanical and Nuclear Engineering, Kansas State University, Manhattan, KS 66506

Introduction

Batteries are under consideration for powering of an increasingly diverse range of applications from stationary solar/wind energy farms to flexible and wearable devices like Google Glass and Microchips.[1] Among non-Li metal ion based batteries, Na, Mg, Al and K-ion batteries are at the forefront.[2-4] Na-ion batteries have drawn increasing attention because in contrast to lithium (Li), Na resources are practically inexhaustible and evenly distributed around the world.[5-6] Also, since Na and Li are stacked in the periodic table, it is easily assumed that battery anodes for Li can be directly utilized for Na batteries. This assumption has been debunked; Na and other metal-ions of technological relevance are much larger than Li-ion and electrode design must be altered to accommodate: (a) their slow diffusion kinetics in the bulk, and (b) fatigue associated with large volume changes upon insertion/extraction that causes capacity degradation.[7] In addition, new materials such as transition metal dichalcogenides or TMDs[8] that can offer unique chemistry and high capacity with low volumes changes must also be simultaneously explored. First report on MoS_2 (type of TMD) electrode for Na-ion batteries published recently showed capacity values as much as two times that of the Na/MoS_2 theoretical insertion value.[9] This observation is important because graphite (and graphene), which has layered morphology similar to TMDs, exhibits almost negligible capacity and cycleability toward Na.[10] It was speculated that a conversion-type reaction could be responsible for high capacity. This reaction type is not observed in widely studied Li/silicon and Li/graphene anodes.

Here, we study superacid assisted exfoliation and sodium cycling behavior of TMD electrodes as potential high capacity materials for metal-ion batteries and supercapacitor electrodes. We show that sodium capacity and cycling efficiency for freestanding TMD electrodes is higher than graphite and graphene oxide electrodes reported so far. Our results also highlight the challenges associated with capacity decay and voltage hysteresis in TMDs, which become evident after only few initial cycles.

Experimental

Preparation graphene oxide. Sodium nitrate, potassium permanganate, sulfuric acid, hydrogen peroxide (31.3 % solution in water), hydrochloric acid (30 % solution in water), methanol (99.9 %) were purchased from Fisher Scientific. All materials were used as received without further purification. Hummer's method[11] was used to make graphene oxide.

Preparation of exfoliated or acid treated TMD flakes. TMD powder (2 mg mL^{-1}, 99 %, Sigma Aldrich) was sonicated for 30 min in concentrated superacid solution and the non-exfoliated sheets were allowed to settle. Two types of superacids were tested: chlorosulfonic acid (superacid, 99%, Sigma Aldrich) and methanesulfonic acid. Please note that the superacid was very slowly added to the TMD powder in an argon-filled glovebox (dew point −50 °C). The solution was then carefully quenched in 1.0 L of distilled water (done with extreme caution in a glovebox). Additional dilution with DI water was done to reduce the solution acidity. The solution was then dried in a conventional oven to obtain dry superacid treated TMD nanosheets.

Preparation of reduced graphene oxide and TMD composite paper. 15 mg GO and 22.5 mg of acid treated TMD nanosheets mixed together in 1:1 (v/v) water and isopropanol solution. The mixture was then sonicated for 60 min (Branson Sonifier). The composite suspension was then filtered by vacuum filtration process through a 47 mm diameter 10 µm pore sized filter membrane (HPLC grade, Millipore). TMD/GO composite paper thus obtained was dried in an oven at 70 °C overnight and subsequently reduced at 500 °C for 2h and 900 °C for 5 min in argon atmosphere. The samples were labeled as rGO for pristine rGO and $MoSe_2$, WSe_2 and $MoTe_2$ for respective TMDs with approx. 20 wt.% of rGO as conducting agent.

Characterization Methods. For electrochemical testing, coin cells were made by punching 14.3 mm diameter out of the composite paper for use as working electrode. 1M $NaClO_4$ (Alfa Aesar) in (1:1 v/v) dimethyl carbonate:ethylene served as the electrolyte. A 25 µm thick glass separator soaked in electrolyte was placed between the working electrode and pure Na metal (14.3 mm diameter, 75 µm thick) counter electrode. The assembled cells were tested using multichannel BT2000 Arbin test unit sweeping between 2.25 V to 10 mV v/s Na/Na^+. ζ potential surface measurements were carried out on a ZetaPlus Zeta Potential Analyzer (Brookhaven's Inst. Corp.). The effect of ionic concentration on the potential measured is minimized by using a low concentration of basic (0.01 M NaOH) solution for controlling the pH. TEM images were digitally acquired by use of a Phillips CM100 operated at 100 kV.

Results and Discussion. Exfoliation of TMDs in chlorosulfonic acid was immediate and dispersions with concentration reaching 2 mg mL^{-1} could be readily prepared.

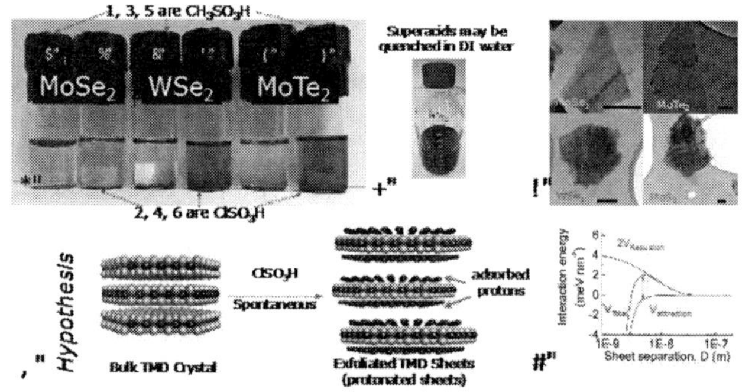

Figure 1. (a) Photograph of TMD dispersions in chloro and methane sulfonic acid at 2 mg.mL^{-1}. (b) Superacids may be quenched in water and utilized for further processing. (c) TEM images of exfoliated sheets, scale bar is 100 nm. (d) Calculated total, repulsion, and attraction energy (VvdW) from ζ potential data and DLVO theory.[12,13] (e) Protonation of TMD induced by strong acids is likely mechanism for exfoliation.

As can be seen in Fig. 1a, the TMD dispersions in cholorosulfonic acid were observed to be more stable than methanesulfonic acid. Later, the dispersions were quenched in DI water (Fig.1b) and observed under TEM to confirm their layered morphology (Fig. 1c). Based on ζ-potential data (negative surface charge, not shown) and DLVO theory (Fig. 1d) and previous work on similar systems[14], we hypothesize that stable dispersion of TMDs are a direct result of electrostatic repulsion forces due to protonation by the superacid (Fig. 1e).

Exfoliated sheets were then interfaced with rGO sheets (conducting agent) to obtain composite papers, which were then directly utilized as working electrode. Electrochemical voltage profiles for first and second cycle are presented in Fig. 2.

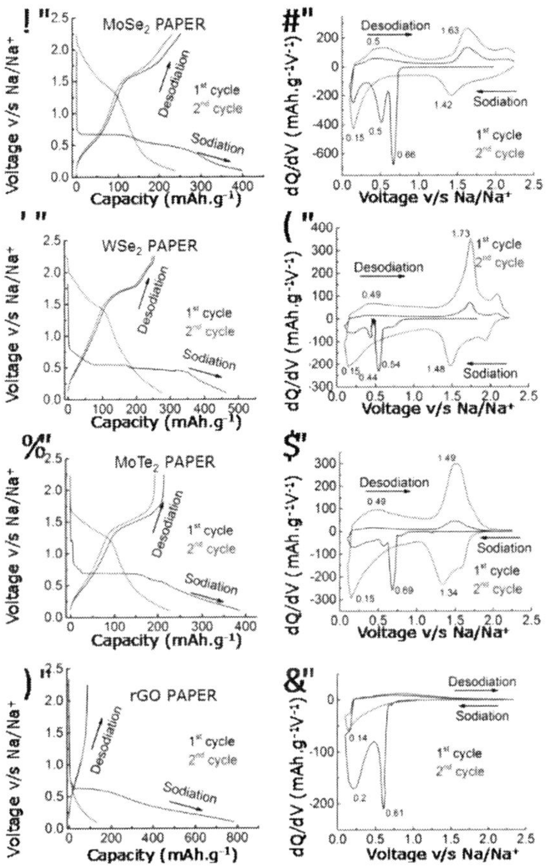

Figure 2. (a, c, e, g) Voltage charge-discharge profiles for $MoSe_2$, WSe_2, $MoTe_2$, and rGO electrodes, respectively; and (b, d, f, h) corresponding incremental capacity profiles.

The following initial observations can be made from these data: TMD paper electrodes (Fig. 2a, c, e) showed charge capacities that were approximately 2 to 3 times that of the neat rGO electrode (Fig. 2g). Mainly, WSe_2 outperformed the other TMDs with a reversible capacity of over 250 $mAh.g^{-1}$. In addition, all of the TMD electrodes showed first cycle efficiency in excess of ~50% compared to rGO's 13%. However, the rGO paper showed higher charge capacity than commercial graphite (~80 v/s ~35 $mAh.g^{-1}$). The voltage profile curves showed that the oxidation/reduction mechanism of TMD w.r.t. sodium was very different with two distinct plateaus when compared to rGO that showed a sloping discharge curve. To study in detail, differentiated capacity curves were plotted (Fig. 2 b, d, f, h). Since each peak corresponds to a specific reaction and certain peaks disappeared after the first cycle, those peaks (0.5 and 0.6V for $MoSe_2$, 0.54V for WSe_2, and 0.69V for $MoTe_2$) can be ascribed to the formation of solid electrolyte interface layer.[15] In the second cycle, all TMD electrodes showed two insertion peaks (~0.15V and ~1.34 to 1.46 V) and two extraction peaks (~0.5 and ~1.5 to 1.7) V, which is an indication of distinct sodiation and desodiation in the TMD materials. On the other hand, rGO electrode only showed broad plateau during the extraction half. Cycling data for various electrodes for up to 10 cycles is presented in Fig. 3. At the 10th cycle, the best performing electrode was WSe_2 with capacity retention of >80% followed by $MoSe_2$ (~60%) and $MoTe_2$ (~45%). Furthermore, the TMD electrodes reached near 100% cycling efficiency after 2nd discharge, which remained constant up to 10th cycle.

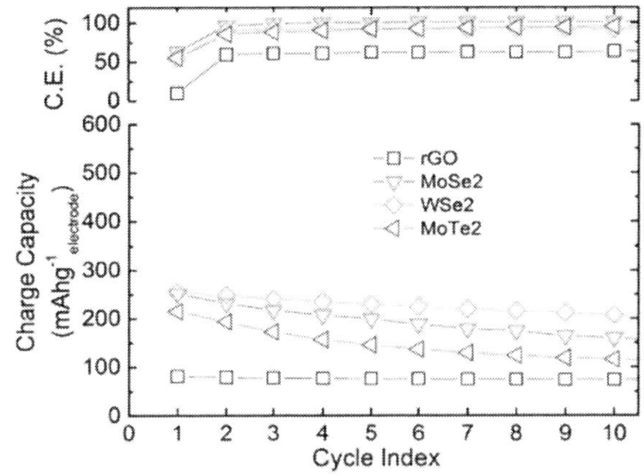

Figure 3. Electrochemical cycling data for various electrodes. Capacity values are normalized with respect to total electrode weight.

Acknowledgment. This work is supported by National Science Foundation grant # 1454151.

References

(1) Gogotsi, Y. *ACS Nano* **2014**, *8*, 5369–5371.
(2) Aurbach, D.; Suresh, G. S.; Levi, E.; Mitelman, A.; Mizrahi, O.; Chusid, O.; Brunelli, M. *Adv. Mater.* **2007**, *19*, 4260-4267.
(3) Jayaprakash, N.; Das, S. K.; Archer, L. A. *Chem. Comm.* **2011**, *47*, 12610-12612.
(4) Liu, Y.; Fan, F.; Wang, J.; Liu, Y.; Chen, H.; Jungjohann, K.L.; Xu, Y.; Zhu, Y.; Bigio, D.; Zhu, T.; Wang, C. *Nano Lett.* **2014**, *14*, 3445–3452.
(5) Palomares, V.; Serras, P.; Villaluenga, I.; Hueso, K. B.; Carretero-González, J.; Rojo, T. *Energ. Environ. Sci.* **2012**, *5*, 5884-5901.
(6) Palacin, M. R. *Chem. Soc. Rev.* **2009**, *38*, 2565-2575.
(7) Aricò, A. S.; Bruce, P.; Scrosati, B.; Tarascon, J. M.; Van Schalkwijk, W. *Nat. Mater.* **2005**, *4*, 366-377.
(8) Chhowalla, M.; Shin, H. S.; Eda, G.; Li, L. J.; Loh, K. P.; Zhang, H. *Nat. Chem.* **2014** *5*(4), 263-275.
(9) David, L.; Bhandavat, R.; Singh, G. *ACS Nano* **2014**, *8*, 1759–1770.
(10) Ge, P.; Fouletier, M. *Solid State Ionics* **1988**, *28–30*, Part 2(0), 1172-1175.
(11) Hummers Jr, W. S.; Offeman, R. E. *J. Am. Chem. Soc.* **1958**, *80*, 1339-1339.
(12) Lotya, M.; Hernandez, Y.; King, P.J.; Smith, R.J.; Nicolosi, V.; Karlsson, L.S.; Blighe, F.M.; De, S.; Wang, Z.; McGovern, I.T.; Duesberg, G.S. *J. Am. Chem. Soc.* **2009**, *131*, 3611−3620.
(13) Zheng, J.; Zhang, H.; Dong, S.; Liu, Y.; Nai, C.T.; Shin, H.S.; Jeong, H.Y.; Liu, B.; Loh, K.P. *Nat. Commun.* **2014**, *5*.
(14) Bhandavat, R., David, L., Singh, G. J. *Phys. Chem. Lett.* **2012**, *3*, 1523–1530.
(15) Peng, L.; Zhu, Y.; Chen, D.; Ruoff, R. S.; Yu, G. *Adv. Energy Mater.* **2016**. DOI: 10.1002/aenm.201600025

DESIGNING TWO-DIMENSIONAL NANOMATERIALS FOR ELECTROCHEMICAL ENERGY STORAGE

Lele Peng, Yue Zhu, Guihua Yu

Materials Science and Engineering, Texas Materials Institute, The University of Texas at Austin, Austin, Texas 78712
* Email: ghyu@austin.utexas.edu

Lithium-ion batteries (LIBs) have dominated portable electronics industry and solid-state electrochemical R&D for the past two decades. In light of possible concerns over the cost and future availability of lithium, sodium-ion batteries (SIBs) and other new technologies have emerged as candidates for large-scale stationary energy storage. Research in these technologies has increased dramatically with a focus on the development of new materials for both the positive and negative electrodes that can enhance the cycling stability, rate capability, and energy density. Two-dimensional (2D) nanomaterials are showing promise for many energy-related applications and particularly for energy storage, because of the efficient ion transport between the layers and the large surface areas available for improved ion adsorption and faster surface redox reactions.

In this talk we will present some examples on inorganic 2D nanomaterials and their hybrids for advanced energy storage devices including LIBs and SIBs. The first example is to explore inorganic VOPO4 ultrathin nanosheets for high-rate Li-ion and Na-ion storage. And we will also discuss a novel synthetic strategy towards controlled synthesis of porous inorganic nanosheet materials as a promising class of alkali-ion battery electrodes with high capacity and rate capability.

References

(1) L. Peng, Y. Zhu, D. Chen, R. Ruoff, G. Yu, *Advanced Energy Materials* 2016, DOI: 10.1002/aenm.201600025.
(2) Y. Zhu, L. Peng, D. Chen, G. Yu, *Nano Letters* 2016, 16., 742.
(3) P. Xiong, L. Peng, D. Chen, Y. Zhao, X. Wang and G. Yu, *Nano Energy* 2015, 12, 816.
(4) F. Bonaccorso, L. Colombo, G. Yu, M. Stoller, V. Tozzini, A. C. Ferrari, R. S. Ruoff, V. Pellegrini, *Science* 2015, 347, 41.
(5) C. Wu, X. Lu, L. Peng, K. Xu, X. Peng, J. Huang, G. Yu, Y. Xie, *Nature Communications* 2013, 4, 2431.

PEROVSKITE-TYPE LaSrMnO ELECTROCATALYST WITH UNIFORM POROUS STRUCTURE FOR AN EFFICIENT LI-O_2 BATTERY CATHODE

*Yanbing Yang, Quan Yuan**

College of Chemistry and Molecular Sciences, Wuhan University, Wuhan, 430072, China

* Corresponding author: yuanquan@whu.edu.cn

Abstract

Perovskite is an excellent candidate as low cost catalyst for Li-O_2 cells. The unique defective structure and disorder-free channels of oxygen vacancies offer a large number of active sites in catalytic reactions.[1-4] However, the limited porosity which impedes molecular transport, and the inherent low electronic conductivity are main barriers for making high performance electrodes. We designed a hierarchical porous flexible architecture by coating thin mesoporous yet crystalline LaSrMnO layers throughout a graphene foam to form graphene/meso-LaSrMnO sandwich-like nanosheets.[5] In this well designed system, the macropore between nanosheets facilitates O_2 and Li^+ diffusion, the mesopore provides large surface area for electrolyte immersion and discharge products deposition, the perovskite phase catalyst decreases reactive overpotential, and the graphene serves as conductive network for electrons transport. When used as a freestanding electrode of Li-O_2 cell, it shows high specific capacity, superior rate capability and cyclic stability. Combination of mesoporous perovskites with conductive graphene networks represents an effective strategy for developing efficient electrodes in various energy storage systems.

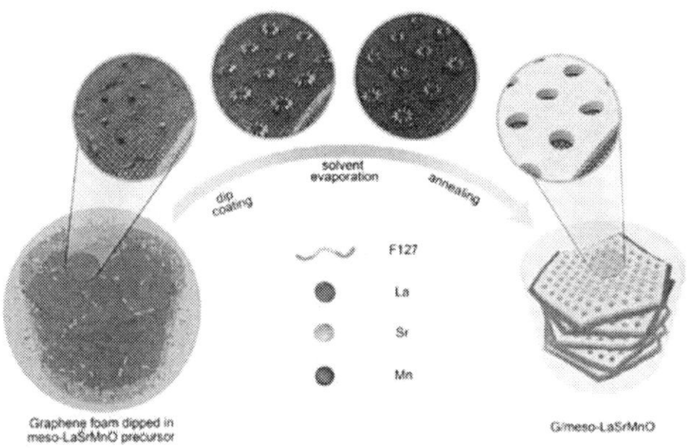

Scheme 1. Illustration of the fabrication process of G/meso-LaSrMnO foams, including macroscopic model of the original G foam dipped into the precursor solution and the resulting porous framework composed of G/meso-LaSrMnO nanosheets.

References

(1) Suntivich, J.; Gasteiger, H. A.; Yabuuchi, N.; Nakanishi, H.; Goodenough, J. B.; Shao-Horn, Y.; *Nat. Chem.* **2011**, *3*, 546.

(2) Xu, J. J.; Xu, D.; Wang, Z. L.; Wang, H. G.; Zhang, L. L.; Zhang, X. B. *Angew. Chem. Int. Ed.* **2013**, *52*, 3887.

(3) Jung, J.; Jeong, H. Y.; Kim, M. G.; Nam, G.; Park, J.; Cho, J. *Adv. Mater.* **2015**, *27*, 266.

(4) Zhao, Y. L.; Xu, L.; Mai, L. Q.; Han, C. H.; An, Q. Y.; Zhang, Q. J. *Proc. Natl. Acad. Sci. U. S. A.* **2012**, *109*, 19569.

(5) Yang, Y. B.; Yin, W.; Wu, S. T.; Yang, X. D.; Xia, W.; Shen, Y.; Huang, Y. H.; Cao, A. Y.; Yuan, Q. *ACS Nano* **2016**, *10*, 1240.

SYNTHESIS AND APPLICATIONS OF NOVEL TWO-DIMENSIONAL NANOMATERIALS

Hua Zhang

School of Materials Science and Engineering, Nanyang Technological University, 50 Nanyang Avenue, Singapore 639798, Singapore
HZhang@ntu.edu.sg

In this talk, I will summarize the recent research on synthesis, characterization and applications of two-dimensional nanomaterials in my group.[1] I will introduce the synthesis and characterization of novel low-dimensional nanomaterials.[2] Then I will demonstrate the applications of these novel nanomaterials in chemical and bio-sensors,[3] solar cells,[4] water splitting,[5] hydrogen evolution reaction, electric devices,[6] memory devices,[7] conductive electrodes,[8] other clean energy,[9] etc.

References

(1) (a) X. Huang, et al., *Chem. Soc. Rev.*, **2012**, *41*, 666. (b) X. Huang, et al., *Chem. Soc. Rev.*, **2013**, *42*, 1934. (c) X. Huang, et al., *Adv. Mater.*, **2014**, *26*, 2185. (d) H. Li, et al., *Acc. Chem. Res.*, **2014**, *47*, 1067. (e) X. H. Cao, et al., *Energ. Environ. Sci.*, **2014**, *7*, 1850. (f) H. Li, et al. *ACS Nano*, **2014**, *8*, 6563. (g) C. L. Tan, et al. *Chem. Soc. Rev.*, **2015**, *44*, 2615. (h) Y. Chen, et al., *Chem. Soc. Rev.*, **2015**, *44*, 2681. (i) C. L. Tan, et al., *Chem. Soc. Rev.*, **2015**, *44*, 2713. (k) C. L. Tan, et al., *Nat. Commun.*, **2015**, *6*, 7873.

(2) (a) X. Y. Qi, et al., *Angew. Chem. Int. Ed.*, **2010**, *49*, 9426. (b) X. Y. Qi, et al., *Adv. Mater.* **2012**, *24*, 4191. (c) X. H. Cao, et al. *Angew. Chem. Int. Ed.*, **2014**, *53*, 1404. (d) X. Huang, et al. *Nat. Commun.* **2013**, *4*, 1444; (e) C. L. Tan, et al. *Angew. Chem. Int. Ed.*, 2015, *54*, 1841. (f) Z. Y. Zeng, et al. *Angew. Chem. Int. Ed.* **2011**, *50*, 11093. (g) Z. Y. Zeng, et al. *Angew. Chem. Int. Ed.* **2012**, *51*, 9052. (h) Z. Y. Yin, et al. *ACS Nano* **2012**, *6*, 74. (i) H. Li, et al. *ACS Nano* **2013**, *7*, 2842. (j) Y. Y. Zhao, et al. *Nano Lett.* **2013**, *13*, 1007. (k) H. Li, et al. *ACS Nano*, **2013**, *7*, 10344. (l) H. Li, et al. *ACS Nano*, **2014**, *8*, 6563. (m) Z. Y. Yin, et al., *Angew. Chem. Int. Ed.*, **2014**, *53*, 12560. (n) X. Hong, et al. *Adv. Mater.*, **2014**, *26*, 6250. (o) X. Huang, et al., *ACS Nano*, **2014**, *8*, 8695. (p) J. Z. Chen, et al., *Angew. Chem. Int. Ed.*, **2015**, *54*, 1210. (q) Z. Y. Zeng, et al. *Energy Environ. Sci.*, **2014**, *7*, 797. (r) Y. P. Du, et al. *Nat. Commun.* **2012**, *3*, 1177. (s) X. J. Wu, et al. *Angew. Chem. Int. Ed.* **2014**, *53*, 5083. (t) X. J. Wu, et al. *Angew. Chem. Int. Ed.* **2014**, *53*, 8929. (u) D. Yang, et al., *Angew. Chem. Int. Ed.*, **2014**, *53*, 9352. (v) X. Zhang, et al., *Angew. Chem. Int. Ed.*, **2015**, 54, 3653. (w) X. Zhang, et al., *Angew. Chem. Int. Ed.*, **2015**, 54, 5425. (x) X. Hong, et al. *J. Am. Chem. Soc.*, **2015**, *137*, 1444. (y) C. L. Tan, et al., *J. Am. Chem. Soc.*, **2015**, *137*, 1565.

(3) (a) Q. Y. He, et al., *ACS Nano*, **2010**, *4*, 3201. (b) Q. Y. He, et al., *ACS Nano*, **2011**, *5*, 5038. (c) H. G. Sudibya, et al., *ACS Nano*, **2011**, *5*, 1990. (d) C. F. Zhu, *J. Am. Chem. Soc.*, **2013**, *135*, 5998. (e) Y. Zhang, et al., *Adv. Mater.*, **2015**, *27*, 935.

(4) (a) Z. Y. Yin, et al. *Adv. Energy Mater.*, **2014**, *4*, 1300574. (b) Z. Y. Yin, et al., *ACS Nano*, **2010**, *4*, 5263.

(5) Z. Y. Yin, et al., *Adv. Mater.* **2012**, *24*, 5374.

(6) (a) B. Li, et al., *Adv. Mater.*, **2010**, *22*, 3058. (b) Z. Y. Zeng, et al., *Adv. Mater.*, **2012**, *24*, 4138.

(7) (a) J. Q. Liu, et al., *ACS Nano*, **2010**, *4*, 3987. (b) J. Q. Liu, et al. *Adv. Mater.* **2013**, *25*, 233. (c) F. Zhao, et al., *ACS Nano*, **2012**, *6*, 3027. (d) G. Z. Sun, et al. *Angew. Chem. Int. Ed.*, **2013**, *52*, 13351.

(8) X. Huang, et al., *Adv. Mater.*, **2012**, *24*, 5979.

(9) (a) G. Z. Sun, et al., *Angew. Chem. Int. Ed.*, **2015**, *54*, 4651. (b) G. Z. Sun, et al., *Angew. Chem. Int. Ed.*, **2014**, *53*, 12576. (c) W. J. Zhou, et al., *Energy Environ. Sci.*, **2013**, *6*, 2216.

SELF-DISPERSED CRUMPLED GRAPHENE BALLS IN OIL FOR FRICTION AND WEAR REDUCTION

Jiaxing Huang

Department of Materials Science and Engineering, Northwestern University, 2220 Campus Drive, Evanston, IL 60208, USA

Ultrafine particles are often used as lubricant additives because they are capable of entering tribological contacts to reduce friction and protect surfaces from wear. They tend to be more stable than molecular additives under high thermal and mechanical stresses during rubbing. It is highly desirable for these particles to remain well dispersed in oil without relying on molecular ligands. Borrowing from the analogy that pieces of paper that are crumpled do not readily stick to each other (unlike flat sheets, Figure 1), we hypothesized and later demonstrated that ultrafine particles resembling miniaturized crumpled paper balls can indeed disperse in oil. They could act like nanoscale ball bearings to reduce friction and wear. Here we report the use of crumpled graphene balls[1-3] as a high-performance additive that can significantly improve the lubrication properties of polyalphaolefin base oil. The tribological performance of crumpled graphene balls is only weakly dependent on their concentration in oil and readily exceeds that of other carbon additives such as graphite, reduced graphene oxide, and carbon black. Notably, polyalphaolefin base oil with only 0.01–0.1 wt % of crumpled graphene balls outperforms a fully formulated commercial lubricant in terms of friction and wear reduction.[1]

Figure 1. Photo showing crumpled paper balls on a surface. Like paper balls that do not stick to each other or surface, crumpled graphene balls, made by capillary compression from their flat precursors, become aggregation-resistant and processable in arbitrary solvents. This type of colloidal processability is independent of their surface chemistry. As a result, they can disperse well in oil, and remain dispersed under tribological conditions, leading to their excellent performance in friction and wear reduction. Image adapted from Ref 1.

Acknowledgments. J.H presents the work on behalf of his co-workers Xuan Dou, Andrew R. Koltonow, Xingliang He, Dr. Hee Dong Jang, Prof. Qian Wang and Prof. Yip-Wah Chung. We also thank Drs. F. Guan, J. Luo, and K. Raidongia for their earlier exploration related to this work. The work was supported by the Office of Naval Research (ONRN000141310556). J.H. thanks the John Simon Guggenheim Memorial Foundation for a Guggenheim Fellowship. We also thank Felice Frankel for providing the photo of crumpled paper balls that Figure 1 is based on.

References

(1) Xuan Dou, Andrew R. Koltonow, Xingliang He, Hee Dong Jang, Qian Wang, Yip-Wah Chung and Jiaxing Huang "Self-dispersed Crumpled Graphene Balls in Oil for Friction and Wear Reduction" *Proceeding of National Academy of Sciences*, **2016**, 113, 1528-1533

(2) Jiayan Luo, Jaemyung Kim and Jiaxing Huang "Material Processing of Chemically Modified Graphene: Some Challenges and Solutions" *Accounts of Chemical Research*, **2013**, 46, 2225–2234

(3) Jiayan Luo, Hee Dong Jang, Tao Sun, Li Xiao, Zhen He, Alexandros P. Katsoulidis, Mercouri G. Kanatzidis, J. Murray Gibson, and Jiaxing Huang "Compression and Aggregation-resistant Particles of Crumpled Soft Sheets" *ACS Nano*, **2011**, 5, 8943–8949

INTERCALATION OPTOELECTRONICS IN 2D MATERIALS

Liangbing Hu

Department of Materials Science and Engineering, University of Maryland College Park

Ultrathin two-dimensional (2D) materials such as graphene are highly attractive for transparent electrode applications due to their high transparency and carrier mobility, which lead to an excellent combination of sheet conductance and optical transmittance in the visible range. Chemical doping[6] and intercalation have been explored as effective methods to increase carrier density, and thus lower the sheet resistance with little decrease or even an increase in the optical transmittance. Recently, we reported a method by electrochemical lithium-ion intercalation in mechanically-exfoliated graphene sheets, which leads to a drastic simultaneous improvement of sheet conductivity and optical transmittance in the visible range. In this talk, I will discuss our progress on ion intercalation optoelectronics with different ions. Particularly we will discuss the performance of transparent electrode applications using ion intercalation in graphene and reduced graphene oxide networks.

Intercalation to modulate optoelectronics properties of graphene, reduced graphene oxide network and MoS_2 with Li^+, Na^+ and K^+ ions

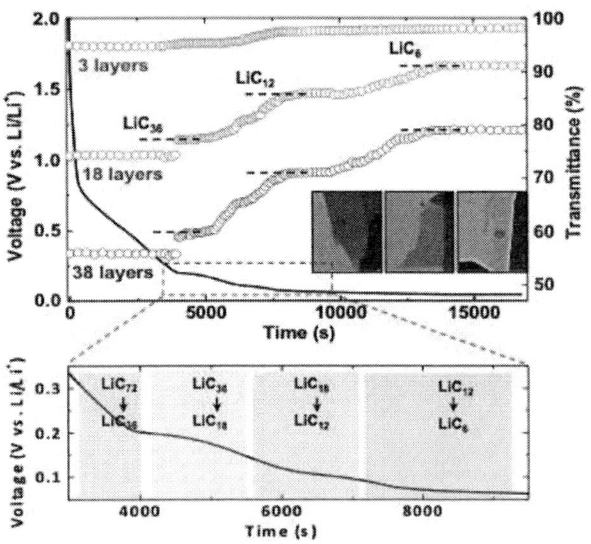

Figure 1. Voltage profile and transmittance vs. Li ion intercalations.

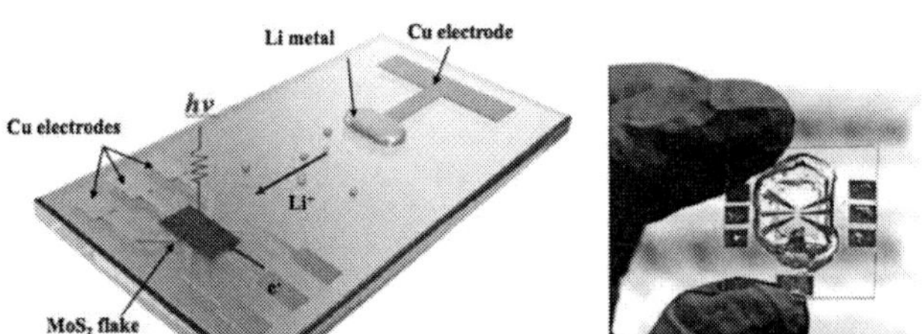

Figure 2. Experimental setup for nanoscale electrochemical devices with MoS_2 nanosheet.

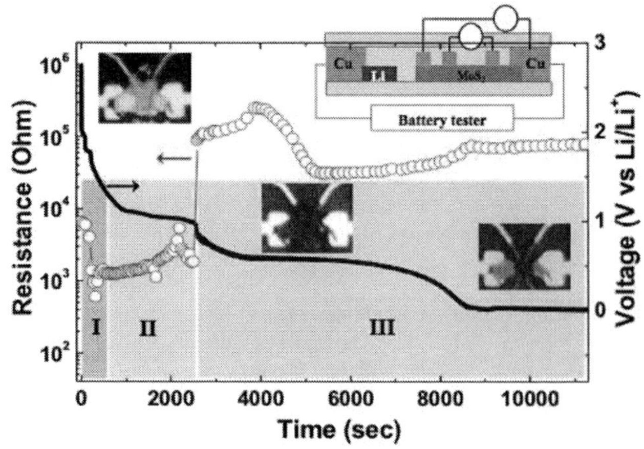

Figure 3. Resistance and voltage profile of MoS_2 upon Li ion intercalations.

Reference:

(1) Wan, J.; Gu, F.; Bao, W.; Dai, J.; Shen, F.; Luo, W.; Han, X.; Urban, D.; **Hu, L.***, Sodium-ion Intercalated Transparent Conductors with Printed Reduced Graphene Oxide Networks, **Nano Letter**, 2015, 15, 3763.

(2) Wan, J.; Bao, W.; Liu, Y.; Dai, J.; Shen, F.; Zhou, L.; Cai, X.; Urban, D.; Li, Y.; Juangjohann, K.; Fuhrer,* M.; **Hu, L.*** In Situ Investigations of Li-MoS_2 with Planar Batteries, **Advanced Energy Materials**, 2014, 1, 1401742.

(3) Bao, W.; Wan, J.; Han, X.; Cai, X.; Zhu, H.; Kim, D.; Ma, D.; Munday, J.; Drew, D.; Fuhrer,* M.; **Hu, L.*** Approaching the Limits of Transparency and Conductivity in Graphitic Materials through Lithium Intercalation, **Nature Communications**, 2014, 5, 4224.

(4) Luo, W.; Wan, J.; Ozdemir, B.;Bao, W.; Chen, Y.; Dai, J.; Lin, H.; Xu, Y.;Gu, F.; Barone, V.;* **Hu, L.*** Electrochemical Intercalation of Potassium into Graphitic Materials, **Nano Letters**, 2015, 15, 7671

SELF-ASSEMBLY OF GRAPHENE AND ITS APPLICATION FOR DENSIFYING ENERGY STORAGE

Quan-Hong Yang

School of Chemical Engineering and Technology, Tianjin University, Tianjin 300072, China
Qhyangcn@tju.edu.cn

With its amphiphilic property, water-dispersed graphene oxide (GO) tends to concentrate at liquid-based interface, resulting in self-assembly formation of nanostructures. Such a self-concentrating process at interfaces is proposed as a novel but effective approach for producing carbon materials, featuring on better control of meso-texture than carbonization and chemical vapor deposition (1). Our group developed liquid-air interfacial self-assembly to produce 2D membrane (2) and liquid-solid interfacial self-assembly assembly to produce 3D hydrogel (3). More importantly, by a controlled removal of water in the hydrogel, the mesotexture of a resulting carbon carbon be well controlled (4).

Compact energy storage with high volumetric performance has become a focus of the design of practical electrochemical energy storage (EES) devices, which are going smaller and lighter but more energetic. Therefore, carbon materials, as a class of important electrodes for EES devices, was required to simultaneously possess abundant porosity and high density, which is very difficult to achieve in a conventional carbon. The unique design of high-density porous carbons is highly related to how graphene, as a basic unit, constructs 3D structured carbon in the densest mode (5). Graphene can construct carbon in two modes, one is layer-by-layer stacking and another is inter-layered packing. Following the first mode, highly dense stacking of graphene units will results in the layered formation of graphite with dense yet non-porous structure, which is only accessible to lithium ions in lithium ion battery. For other EES devices, high capacitance needs well-developed pore structure while volumetric performance needs a high density. Following the second mode, in most cases, porous but low-density carbons like carbon foams and active carbons can be formed. Could we have a carbon balancing the high density and porous structure? Our group developed a capillary evaporation approach to realize the densest packing of carbon layer in 3D space. This approach starts with a construction of 3D networked structure with abundant water inside through an interface induced self-assembly of graphene oxide, which is followed by a controlled water removal to result in a capillary force induced shrinkage of such a robust interlinked network yielding a compact porous material with a density up to 1.6 mg/cm^3. This carbon has unique pore structure and shows a record high volumetric capacitance (376 F/cm^3) as a pure carbon electrode in supercapacitor (4). Most recently, we developed this capillary-evaporation approach to prepare carbon-noncarbon hybrid materials and prepared a high-density metal-free electrode through the hybridization of graphene and polyaniline in a very compact way and such a hybrid electrode delivers a specific capacitance over 800 F/cm^{-3}(6). Also we have extended this process as a versatile and effective approach for producing highly dense carbon-based and carbon-networked electrodes in energy devices including Li-S batteries (7), Na-ion capacitors (8) and etc.

In summary, self-assembly of graphene oxide at interface has demonstrated as an important method to have very unique structure carbon. As a typical example, high-density porous carbons have shown great importance in compact energy storage.

References

(1) Shao, J. J.; Lv. W.; and Yang. Q. H. *Adv. Mater.* **2014,** *26,* 5586.
(2) Chen, C. M.; Yang, Q. H.; Yang, Y. G.; Lv, W.; Wen, Y. F.; Hou, P. X.; Wang, M. Z.; and Cheng, H. M. *Adv. Mater.* **2009**, *21*, 3007.
(3) Lv, W.; Zhang, C.; and Yang, Q. H. *J. Phys. Chem. Lett.* **2015**, *6*, 658.
(4) Tao, Y.; Xie, X. Y.; Lv, W.; Tang, D. M.; Kong, D. B.; Huang, Z. H.; Nishihara, H.; Ishii, T.; Li, B. H.; Golberg, D.; Kang, F. Y.; Kyotani, T; and Yang. Q. H. *Sci. Rep.* **2013,** *3,* 2975.
(5) Zhang, C.; Lv, W.; Tao, Y; and Yang, Q. H. *Energy & Environ. Sci.* **2015,** *8,* 1390.
(6) Xu, Y.; Tao, Y.; Zheng, X.; Ma, H.; Luo, J.; Kang, F.; and Yang, Q. H. *Adv. Mater.* **2015,** *27,* 8082.
(7) Zhang, C.; Liu, D.; Lv, W.; Wang, D. W.; Wei, W.; Zhou, G.; Wang, S. G.; Li, F.; Li, B. H.; Kang, F. Y.; and Yang, Q. H. *Nanoscale* **2015**, *7,* 5592.
(8) Zhang, J.; Lv, W.; Tao, Y.; He, Y.-B.; Wang, D.W.; You, C.-H.; Li, B.; Kang, F.; Yang, Q.-H. *Energy Storage Mater.* **2015**, *1*, 112

NANOPOROUS GRAPHENE AND GRAPHENE QUANTUM DOTS: CLEAN SYNTHESIS, STRUCTURAL CHARACTERIZATION AND APPLICATIONS

Tae Hee Han

Department of Organic and Nano Engineering, Hanyang Univeristy, Seoul 04763, Republic of Korea

Intorduction

After the chemical exfoliation of graphite powders, carbon sheets decorated with oxygen functionalities, the so-called graphene oxides (GO), are readily prepared.[1] Due to the presence of abundant oxygen groups on graphene sheets, the inevitable destruction of sp^2 carbon hybrid structures leads to the generation of amorphous regions over the graphene.[2] Therefore it is generally noted that GO has the mixture form of crystalline region and amorphous matrix.[3-5] We have tried to utilize this structural property of GO composed of two blocks. We have contended that GO embeds graphitic crystallites, which could be a source material for graphene-based quantum dots (GQDs), in an amorphous matrix. Based on this hypothesis, we have developed experimental tools to extract the crystallites by the selective removal of the amorphous region from the GO, in order to prepare GQDs.[5]

Meanwhile, vapor-phase etching of GOs yielded the nanoporous graphene by eliminating the amorphous region. A slow and more controllable etching route to produce nanoporous reduced graphene oxide sheets by hydrothermal steaming at 200 °C can left the porous graphene sheets. The nanoporous graphene exhibited nearly two orders of magnitude increase in the sensitivity as chemiresistor sensor platform for NO_2 detection, compared to the nonporous graphene sheets.[4] In this work, the efficacy and applications of nanoporous graphene and graphene quantum dots via clean synthetic process is shown.

Experimental

Nanoporous graphene. GO sheets were deposited on SiO_2/Si substrates with Langmuir-Blodgett technique. The substrates were suspended in a glass pipette and placed in a Teflon vessel (Parr instrument) pre-filled with 10 % *v/v* DI water. The Teflon vessel was sealed in an autoclave and heated to 200 °C. The resulting r-GO coated substrates were soft-baked at 100 °C for 5 min to remove any residual of hydrazine.

Graphene quantum dots. The GO powder was re-dispersed in DIW by mechanical agitation. Subsequently, the small amount of NH_4OH and H_2O_2 were added to the dispersion and the mixture was further stirred for 10 min at room temperature. The mixture was then transferred to a Teflon vessel and sealed in an autoclave. After heating at 150 °C - 300 °C for 5 hours, the solution was naturally cooled down to room temperature and filtered with a syringe filter (Whatman, Anotop 10) to remove any flocculated aggregates. The filtrate was stirred and heated at 100 °C for 1 h to remove any remaining NH_3.

Polymeric composites of GQDs. The freeze-dried GQD powders and poly(acrylic acid) (PAA; Sigma-Aldrich, Mv ~450,000) solution were mechanically mixed in DIW, and the viscous mixture was degassed at room temperature under vacuum for 1 hour to remove bubbles. GQD fibers were spun from the viscous mixture. Bare PAA fibers were also fabricated by following the same method without adding GQD powders.

Characterizations. GO sheets and GQDs were characterized using an atomic force microscopy (AFM; Park Systems, XE-70) in the tapping mode and a transmission electron microscopy (TEM; JEOL, JEM-2100F) operated at an accelerating voltage of 200 kV. TEM samples were prepared on lacey-carbon TEM grids (Ted Pella, Inc.). PL and QY measurements were carried out using a fluorescence spectrometer (Scinco, FS-2) and a UV-vis spectrometer (Analytik Jena, Specord200). Photographs of fluorescent GQD solutions and composite films and fibers were taken using a digital camera (Canon, EOS 600D) with a 356 nm UV lamp (Specroline, ENF-260C/FE). The fluorescent GQD fibers were observed using a fluorescence microscopy (Olympus, BX51 / U-RFL-T) with a UV excitation filter (Olympus, U-MNU2). X-ray photoelectron spectroscopy (XPS) measurements were carried out using a spectrometer (Thermo Scientific, Theta probe) with monochromatic Al Kα radiation. Fourier-transform infrared spectroscopy (FT-IR; Thermo Scientific, Nicolet 6700) were conducted with KBr pellets.

Results and Discussion

A schematic illustration of vapor-phase etching on single-layer GO sheets is shown in Figure 1a. Firstly, single GO layers were coated on SiO_2/Si substrates using the Langmuir-Blodgett technique. Subsequently, each substrate was suspended in a Teflon vessel prefilled with an aqueous solution, which acted as the vapor source. The vessel was then sealed in a stainless steel autoclave and heated. Figure 1b shows a topographic image of the deposited GO sheets. The corresponding height profile indicates that the thickness of the GO sheets is about 1.6 nm. After the GO sheets were exposed to steam vapor, the fracturing behavior was observed, suggesting that GO sheets are very reactive under steam etching conditions. Height profiles along the white lines show GO seemed etched down through to the bottom substrate.

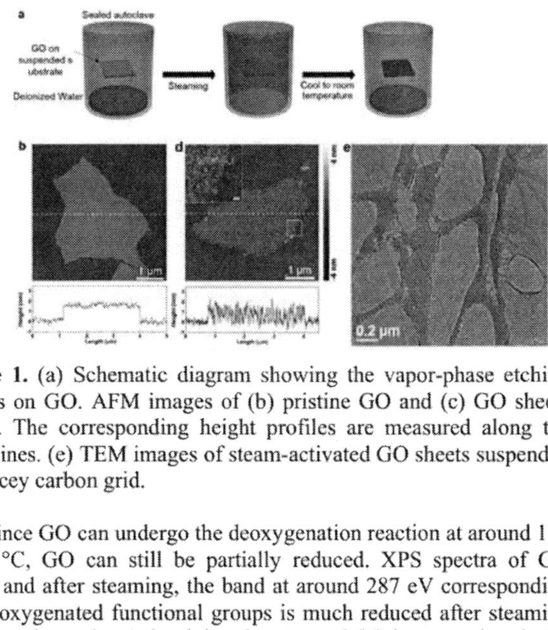

Figure 1. (a) Schematic diagram showing the vapor-phase etching process on GO. AFM images of (b) pristine GO and (c) GO sheets etched. The corresponding height profiles are measured along the white lines. (e) TEM images of steam-activated GO sheets suspended on a lacey carbon grid.

Since GO can undergo the deoxygenation reaction at around 150 - 200 °C, GO can still be partially reduced. XPS spectra of GO before and after steaming, the band at around 287 eV corresponding to the oxygenated functional groups is much reduced after steaming (data not shown here), implying the steamed GO become electrically conductive. Figure 2a shows the I-V curves of GO monolayers before and after steaming. Increased current was indeed observed for GO steamed for 10 hours. r-GO has attracted significant interest as a material for gas sensing applications. The steamed GO has a porous network structure with much abundant edge sites, which would be beneficial in the chemiresistor ability. To test whether steamed GO has enhanced interaction with gas molecules, chemical sensors were fabricated using porous r-GO network as the sensing material. Figure 2b shows the response of a regular r-GO, and porous r-GO after 10 hours of steaming treatment, respectively on exposure to NO_2 vapor. The sensitivity of regular, non-porous r-GO was comparable to the previously reported r-GO NO_2 sensors. For steamed GO however, nearly two orders of magnitude higher sensitivity was observed, and longer steam treatment led to higher sensitivity, likely due to higher porosity.

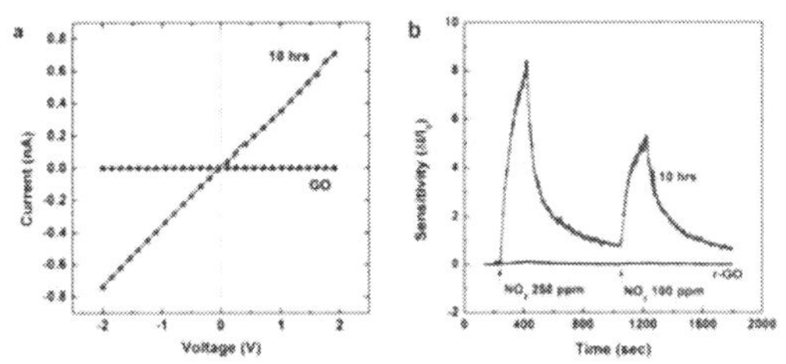

Figure 2. (a) I-V curves of pristine GO and steamed GO sample. (b) Responses of a r-GO sheet and the steamed sheets upon exposure to NO_2 vapor.

The observations made during the vapor etching of GO sheets were utilized for developing a solution-based synthesis process. Figure 3a presents a schematic illustration of the solution phase graphene quantum dots (GQDs) synthesis process. A dark brown colored aqueous GO dispersion was mixed with H_2O_2 and NH_4OH solution and heated in a closed vessel of a Teflon-lined autoclave. After the reaction, the solution was clear and exhibited bright photoluminescence (PL) response to excitation at 350 nm. We note that H_2O_2 is a harsh etchant with peroxide radicals. As shown in Figure 2b and c, by changing the reaction temperature, PL properties of GQDs was simply controlled. The GQDs synthesized at 200 °C (GQDs@200 °C) and 300 °C (GQDs@300 °C) exhibited green and blue fluorescence, respectively, under UV excitation at 365 nm. The PL spectra obtained by excitation at 350 nm showed a significant red shift of the PL peak position from 430 nm to 520 nm as the reaction temperature was decreased from 300 °C to 150 °C.

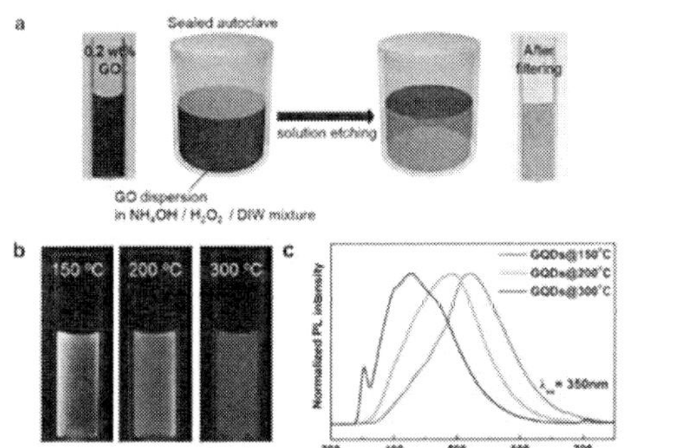

Figure 3. (a) Schematic drawing illustrating the solution-phase etching of GO sheets to prepare GQDs. (b) Photographs of the GQDs dispersions, prepared at 150, 200, and 300 °C from left to right, under 365 nm UV light. (c) Comparison of the PL spectra of the GQDs.

The chemical properties of GO and yellow GQDs (synthesis temperature: 150 °C) were compared by analyzing their Fourier-transform infrared (FT-IR) spectra, as shown in Figure 4a. GO contains various kinds of oxygen-containing functional groups, such as epoxide, carbonyl/carboxyl, and hydroxyl/carboxyl groups exhibiting peaks at 1053, 1820/1735, and 3400 cm^{-1}, respectively. In the case of GQDs, the introduction of nitrogen atoms into graphene is clearly confirmed in FT-IR analysis. In addition, while GO does not show peaks in the N1s XPS (X-ray photoelectron spectroscopy) spectrum (Figure 4b), GQDs show peaks at 399.4 and 400.6 eV, which are attributed to primary amine and amide, respectively. It conclude that the yellow GQDs (synthesis temperature: 150 °C) are decorated with more amine than amide groups. On the other hand, the major nitrogen functionality of the blue GQDs (synthesis temperature: 300 °C) was amide.

Our strategy provides a straightforward route for synthesizing GQDs in large quantities and could be beneficial for potential applications. Figure 4c shows a photograph of freeze-dried GQD powders (about 50 mg) prepared from 40 mL of GO solution (0.2 wt. %). To demonstrate the potential applications of GQDs in composites, GQD powders were mixed with poly(acrylic acid) (PAA) and the viscous GQDs-PAA mixture was hand-drawn to prepare microfibers. Figure 4d shows that GQDs fibers are emitting the strong yellow light under UV excitation. It may be noted that the emitted light from the fibers was not localized, indicating that the GQD powder was homogeneously mixed within the PAA matrix.

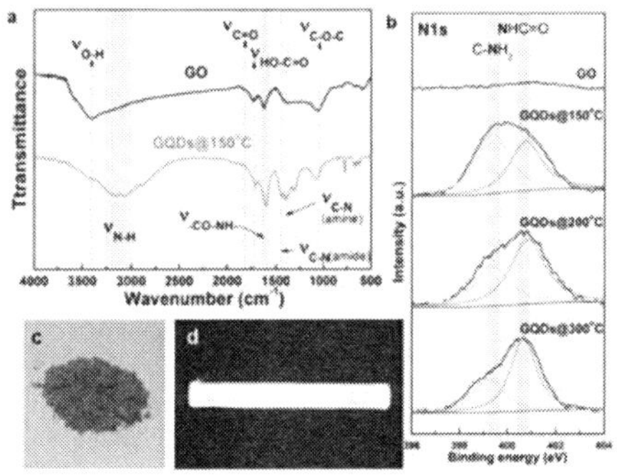

Figure 4. (a) FT-IR spectra of GO and GQDs. (b) XPS N1s spectra of GO and GQDs. (c) Photograph of freeze-dried GQD powders. (d) GQDs fibers emitting a strong yellow light under UV lamp.

We have showed that GO is a mixture of amorphous and crystalline regions and is a promising platform for the preparation of GQDs. Our straightforward route provides a highly efficient method for producing the nanoporous graphene and optically tuned GQDs for many applications.

Acknowledgement. This work was supported by Basic Science Research Program through the National Research Foundation of Korea (2014R1A1A1008196

References

(1) Hummers, W. S.; Offeman, R. E. *J. Am. Chem. Soc.* **1958**, 80, 1339.
(2) Luo, J. Y.; Kim, J; Huang, J. X. *Acc. Chem. Res.* **2013**, 46, 2225.
(3) Erickson, K.; Emi, R.; Lee, Z.; Alem, N.; Gannett, W.; Zettl, A. *Adv. Mater.* **2010**, 22, 4467.
(4) Han, T. H.; Huang, Y. K.; Tan, A. T. L.; Dravid, V. P.; Huang, J. X. *J. Am. Chem. Soc.* **2011**, 133, 15264.
(5) Park, H.; Noh, S. H.; Lee, J. H; Lee, W. J.; Jaung, J. Y.; Lee, S. G.; Han, T. H. *Sci. Rep.* **2015**, 5, 14163.

GRAPHENE-BASED MEMBRANES FOR ENERGY-EFFICIENT GAS AND LIQUID SEPARATION

Ho Bum Park

Department of Energy Engineering, Hanyang University, Seoul 04763, Republic of Korea
badtzhb@hanyang.ac.kr

Up to now, polymeric membranes have dominated current global market, such as water purification, desalination, and gas separations owing to their simplicity, excellent processibility, mechanical flexibility, and low cost in production. However, there exists limitation of polymeric membranes for permeability/selectivity performance [1-3]. Hence, many alternative membrane materials and new membrane formation methods have been extensively studied for overcoming the separation performance of existing polymeric membranes, and simultaneously for achieving much more energy-efficient separation.

Apart from a debate about how to create a proper, well-defined pores on graphene, graphene is truly a rising carbon material because it already possesses important qualified features as a membrane material, such as thinnest membrane thickness, scalable membrane area, and high mechanical strength. Although it is in an infant stage, some experimental results exhibit that porous graphene membranes can be realized for molecular separation, such as gas and liquid separation [4-6]. Thanks to tremendous effort for scalable graphene synthesis, meter-sized graphene monolayer has been obtained, for instance, by using metal catalytic chemical vapor deposition (CVD) method [7]. Such large-area CVD graphene has been investigated mainly for electronic applications such as high-frequency transistor or flexible transparent electrode, while the CVD graphene does not meet high-quality yet for such emerging applications because they have a number of structural defects, such as grain boundaries, Stone-Wales defects and vacancies [8]. Also, the CVD graphene is a polycrystalline sheet, including many defects and wrinkles or folds, leading to not a perfect barrier but a gas-leaky membrane [9]. However, such defects in the CVD graphene can be rather used for selective molecular transport. At present, the main research direction toward graphene membrane is to make graphene holey in nanometre or subnanometric scale for effective separation. In our lab, a CVD graphene without any noticeable line defects (e.g., grain boundaries) on cheap metal substrate, but including a number of point defects, has been developed by controlling the growth rate of graphene nuclear seed and using single crystal metal foil. This is sharply different from the preparation methods of conventional porous graphene, such as focused ion beam or plasma or UV etching, which may lead to unpredictable experimental data with less reproducibility. Moreover, our lab has developed the transfer technology of porous graphene sheet onto microporous support for practical applications, which provide much closer reality for actual membrane separation process.

Graphene oxide, a highly oxidized graphene micro- or nano-sheet, can be considered as a membrane building block or a filler for mixed-matrix membrane. Basically, differing from other nanomaterials such as inorganic molecular sieves and carbon nanotubes, graphene oxide can be readily dissolved (or dispersed) in many solvents, so thin-film membrane formation is possible from typical solution process. Also, the size and chemistry of graphene oxide can be further tuned, which can help engineer selective transport through graphene oxide-based membranes. Recently, fundamental transport studies through pristine graphene oxide thin or thick membranes have been reported in the literature. Due to its hydrophilic nature caused by oxygen-containing functional groups such as hydroxyl- or carboxyl groups on basal plane or at edges, water permeability is pretty high despite its disadvantage such as high tortuosity caused by high aspect ratio (width/length ratio). The transport channels or pores in graphene oxide membrane are distinctly different from channels or pores in typical membrane materials, so more basic transport studies such as diffusion and solution through graphene oxide membranes should be necessary for grasping further insight. Although graphene oxide membranes have been considered as promising membrane materials for gas and liquid separation [10-12], there are many challenging issues for real membrane applications, such as structural or chemical metastability, tortuosity (high resistance for diffusion), tuning interlayer distance, and surface modifications. More recently, our lab has developed graphene oxide-based membranes in three different ways: (1) porous graphene oxide membrane for high productivity, (2) functionalized graphene oxide membrane for high selectivity, and (3) graphene oxide-imbedded mixed-matrix membrane or graphene oxide thin-film composite membrane for practical membrane applications.

In this paper, graphene-based membranes, including porous graphene and graphene oxide membranes, have been highlighted in terms of fundamental transport of small molecules and ions, preparation methods, pore and channel engineering, strategy to reality for energy-efficient membrane separation applications.

References

(1) Robeson, L. M. *J. Membr. Sci.* **1991**, *62*, 165.
(2) Freeman, B. D. *Macromolecules* **1999**, *32*, 375.
(3) Geise, G. M.; Park, H. B.; Sagle, A. C.; Freeman, B. D.; McGrath, J. E. *J. Membr. Sci.* **2011**, *369*, 130.
(4) O'Hern, S. C.; Boutilier, M.S.H.; Idrobo J-.C.; Song, Y.; Kong, J.; Laoui, T.; Atieh, M.; Karnik, R. *Nano Lett.* **2014**, *14*, 1234.
(5) Koenig, S. P.; Wang, L.; Pellegrino, J.; Bunch J.S. *Nat. Nanotechnol.* **2012**, *7*, 728.
(6) Surwade, S.P.; Smirnov S.N.; Vlassiouk I.V.; Unocic R.R.; Veith, G.M.; Dai, S.; Mahurin, S.M. *Nat. Nanotechnol.* **2015**, *10*, 459.
(7) Li, X. et al. *Science* **2009**, *324*, 1312.
(8) Banhard, F.; Kotakoski, J.; Krasheninnikov, A. V.; *ACS Nano* **2011**, *5*, 26.
(9) Kim, H.W. et al. *Science* **2013**, *342*, 91.
(10) Joshi, R. K.; Carbone, P.; Wang, F.C.; Kravets, V.G.; Su, Y.; Grigorieva, I.V.; Wu, H.A.; Geim, A.K. *Science* **2014**, *343*, 752.
(11) Han, Y.; Xu, Z.; Gao, C. *Adv. Mater.* **2013**, *23*, 3693.
(12) Hu, M.; Mi, B. *Environ. Sci. Techonol.* **2013**, *47*, 3715.

REDOX-ACTIVE 2D COVALENT ORGANIC FRAMEWORK FILMS

Catherine R. DeBlase, Kenneth Hernández-Burgos, Katharine E. Silberstein, Gabriel G. Rodríguez-Calero, Ryan P. Bisbey, Hector D. Abruña, and William R. Dichtel

Department of Chemistry and Chemical Biology, Cornell University, Baker Laboratory, Ithaca, New York, 14853-1301, USA

wdichtel@cornell.edu

Introduction

Electrical energy storage (EES) technologies that offer both high power and energy densities have motivated interest in redox-active materials for electrochemical supercapacitors.[1] Supercapacitors store electricity through two distinct processes: the non-faradaic processes associated with the electrochemical double layer and faradaic processes arising from electrode-bound reversible redox processes, a phenomenon known as pseudocapacitance.[2] The electrochemical double layer consists of ions adsorbed to an electrode surface in response to an applied potential and is maximized in high surface area electrodes, such as porous carbons.[3] Redox-active groups may be covalently attached or adsorbed onto these electrodes. However, these materials are typically not well defined, which complicates their characterization and rational efforts to improve their performance. Modular and reliable strategies to access high-surface-area electrodes with control over their porosity and surface area, as well as the placement and identity of redox-active groups are desirable.

Two-dimensional covalent organic frameworks (2D COFs)[4] predictably organize redox-active groups into crystalline, high-surface area polymer networks with uniform micropores. Electrodes functionalized with 2D COFs might achieve both high theoretical energy density by exhibiting high capacity and high potentials, and high power density because their continuous, ordered pores facilitate ion transport. Many early 2D COFs featured boron-containing linkages, which are oxidatively and hydrolytically unstable, limiting their utility as electrode materials; especially at high potentials. However, a new class of COFs linked by β-ketoenamines offers superior stability to H_2O, aqueous acid, and high applied potentials.

We recently condensed the redox-active monomer 2,6-diaminoanthraquinone (**DAAQ**) with 1,3,5-triformylphluroglucinol (**TFP**) to form the 2D β-ketoenamine-linked **DAAQ-TFP** COF,[5] which was the first COF to exhibit reversible redox processes. Electrodes functionalized with a slurry of the **DAAQ-TFP** COF and carbon black were stable for at least 5000 charge-discharge cycles but provided only modest capacitances (~40 F g^{-1}). This modest performance was due, at least in part, to poor electrical contact to the insoluble COF powder samples, so that only a small fraction (~3%) of the redox-active groups were electrochemically accessible (Figure 1A). Here we describe oriented thin films of β-ketoenamine-linked 2D COFs, which were formed through the slow introduction of the **TFP** monomer into a solution of **DAAQ** in the presence of Au substrates (Figure 1B). **DAAQ-TFP** COF thin films provide near-quantitative addressability of the anthraquinone moieties at thicknesses up to ~200 nm, with areal capacitances of 3.0 mF cm^{-2} (as compared to 0.40 mF cm^{-2} for powder-modified electrodes). These findings demonstrate a means to incorporate 2D COFs into supercapacitors, particularly as light-weight, thin film electrodes on flexible substrates continue to emerge. More recent developments in this project that we are unwilling to communicate in a preprint format will be presented at the ACS National Meeting.

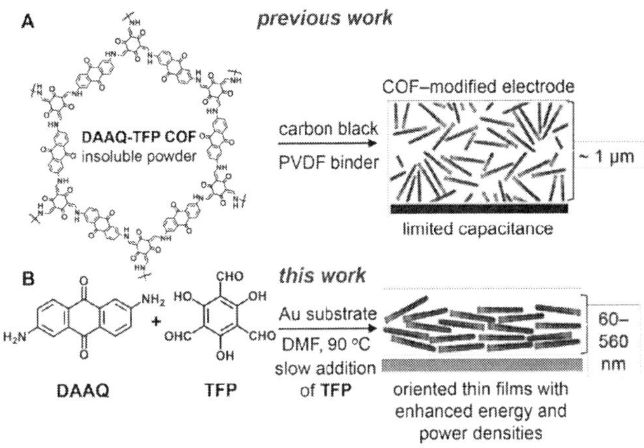

Figure 1. (A) Structure of the hexagonal subunit of the **DAAQ-TFP** COF, which forms as a 2D layered microcrystalline powder. This material was previously prepared as a slurry, mixed with carbon black and a PVDF binder, for supercapacitor applications. (B) The solvothermal growth of the **DAAQ-TFP** COF as an oriented thin film on Au electrodes, a morphology that provides superior electrochemical performance with increased capacitance and enhanced energy and power densities.

Experimental

Material Characterization. COF films were grown from solution on Au substrates, which were prepared by sputtering a Ti or Cr adhesion layer (~15 nm) followed by Au (80 nm) onto a Si wafer. FT-IR spectra of the films were obtained on a Bruker Vertex 80v instrument with a Germanium ATR attachment. XPS was performed on a Surface Science Instruments X-Probe (SSX-100), and AFMs on an Asylum MFP3D-Bio AFM.

Electrochemical Characterization. Electrochemistry experiments were performed on a Princeton Applied Research VersaSTAT 3 potentiostat using a standard 3 electrode setup, with the COF film on Au as the working electrode, Pt coil as the counter electrode, and a $Ag/AgClO_4$ reference electrode standardized with ferrocene. Analysis was performed in a nitrogen filled glove box using a custom designed glass surface cell having a controlled exposed area of 0.64 cm^2 with 0.1 M $TBAPF_6$ in CH_3CN as the electrolyte. Voltammograms taken in aqueous H_2SO_4 supporting electrolyte were performed on the bench after sparging with argon to remove O_2. Impedance spectra were taken on a Gamry Instruments potentiostat on the bench under an Ar atmosphere.

Results and Discussion

The solvothermal condensation of **DAAQ** and **TFP** in N,N-dimethylacetamide (DMA) reproducibly generates **DAAQ-TFP** COF as a microcrystalline powder with surface areas >1000 m^2 g^{-1}. However, only non-diffracting β-ketoenamine linked films were formed when Au substrates were included in reactions performed under similar conditions. We hypothesized that the polymerization occurred too rapidly to yield crystalline films and found that the slow addition (over 1 h) of a DMF solution of **TFP** into a DMF solution of **DAAQ** at 90 °C generated crystalline, oriented films, as determined by grazing incidence x-ray diffraction (GID). The films were also characterized by infrared spectroscopy, x-ray photoelectron spectroscopy, and Kr adsorption.

Cyclic voltammetry (CV) experiments indicate that a large percentage of the anthraquinone moieties in the **DAAQ-TFP** films are electroactive (Figure 2, Table 1), a major improvement compared to the approximately 3% accessed in electrodes functionalized with a slurry of the powder. The voltammogram of a 180 nm thick film in 1 M H_2SO_4 supporting electrolyte exhibits a current onset of 0.1 V vs. Ag/AgCl, which is associated with anthraquinone reduction. Integration of the corresponding oxidative wave provides a charge (112 μC) that corresponds to approximately 20% of the anthraquinones being redox-active, based on the geometric area of the electrode and film thickness. This improvement is also apparent by comparing the currents generated by the film and slurry modified electrodes, given that the slurry-modified electrode contains approximately ten times the mass of the **DAAQ-TFP** COF.

Despite this improvement, the **DAAQ-TFP** film CV indicates high resistance, for which careful analysis of the electrochemical data presented in Figure 2 provides additional insight. The voltammetric profile in Figure 2A (slurry-modified electrode) exhibits faster charge transfer kinetics than that in Figure 3C (thin-film modified electrode), as indicated by its smaller ΔE_p value. The thin-film voltamogram exhibits both slower charge transfer and higher resistance. Yet, the and the redox reactions involved. The slurry-modified electrodes consist of a 60:35:5 weight ratio of carbon black, COF and binder

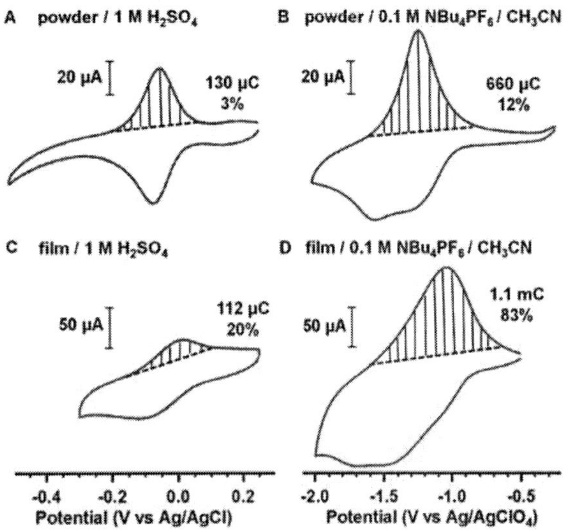

Figure 2. Cyclic voltammograms (50 mV s^{-1}) of (A) **DAAQ-TFP** COF slurry modified electrode in 1 M aqueous H_2SO_4 (B) **DAAQ-TFP** COF slurry modified electrode in 0.1 M NBu_4PF_6 in CH_3CN, (C) and (D) show cyclic voltammograms at for a **DAAQ-TFP** thin film (180 nm) in 1.0 M H_2SO_4 and 0.1 M NBu_4PF_6 in CH_3CN,

Table 1. Electrochemical Performance of COF Films as a Function of Thickness

film thickness (nm)	charge (μC)	quinones accessed (%)
77	513	91
101	722	98
115	672	80
147	1049	99
150	930	85
184	1135	85
228	949	57
232	1013	60
280	1693	83
346	1547	62
348	1031	41

respectively. The integrated charge of the oxidative wave and percentage of quinones accessed are listed for each voltammogram. percentage of the quinones accessed in the thin films is much higher. We speculate that the differences in the voltammetric profiles are due, at least in part, to the different nature of the modified electrodes (PVDF), respectively, whereas the COF thin-film electrodes are solid films that lack the carbon or binder. In the 1M H_2SO_4 supporting electrolyte, the redox reaction involves the transfer of two electrons and two protons per quinone moiety. For the slurry-modified electrodes, the exposed redox-active quinone sites are accessible to protons, providing a rapid voltammetric response, but only this small fraction of the quinones are accessed. For the thin-film electrode, proton transfer is more difficult, providing slower charge transfer, but a much larger fraction of the quinones is accessed. We attribute the improved accessibility to the layered organization of quinone groups in the solid film, enabling conductivity through electron self-exchange processes. Figures 2B and 2D compare the voltammetric responses of the two electrode types in an aprotic supporting electrolyte (0.1 M NBu_4PF_6 / CH_3CN). Here the redox reactions do not involve protonation, and the voltammetric profiles look qualitatively similar. But the ability of the solid COF film to undergo effective self-exchange is retained and thus a much higher percent (83%) of the quinone groups are accessed. This model, while speculative, is consistent with all of our electrochemical data of this system.

The capacitance of each film was determined using galvanostatic charge discharge cycles (GCDC), in which the potential response as a function of time is measured at constant current. The charge polarity is reversed when the potential reaches a predetermined limit, such that one measurement cycle characterizes both the charging and discharging processes of the electrode under test. GCDC of the **DAAQ-TFP** functionalized electrodes at current densities of 150 μA cm^{-2} (an approximate C rate of 80) exhibited voltage-time profiles with voltage plateaus at the potential of anthraquinone reduction. Films of various thicknesses exhibited stable capacitances ranging from 1.2 to 3.0 mF cm^{-2} over ten cycles. Two of these films were selected for extended testing and showed ~7% capacitance loss from their stabilized values after 5000 cycles, indicating its high stability. All films show a higher capacitance compared to a slurry-modified electrode or blank gold substrate, which stabilized around 0.40 mF cm^{-2} or 0.13 mF cm^{-2}, respectively.

Conclusions

The first oriented thin-films of β-ketoenamine COFs were formed through the slow introduction of the **DAAQ** monomer into a solution of **TFP** in the presence of Au substrates. In contrast to working electrodes functionalized with samples of the randomly oriented COF slurry, the **DAAQ-TFP** COF thin films show improved accessibility of their redox-active anthraquinone moieties. These findings demonstrate the importance of controlling the morphology of 2D COFs in order to take full advantage of their desirable properties, including their high specific surface area, oriented micro or mesopores, and redox behavior. Having demonstrated these improvements,[6] further advances in the chemistry of redox-active 2D COFs will be presented at the ACS National Meeting.

References

(1) Goodenough, J. B.; Buchanan, M.; Abruña, H. D. *Basic Research Needs for Electrical Energy Storage*; Office of Basic Sciences, US Department of Energy, 2007.
(2) Wang, G.; Zhang, L.; Zhang, J. A. *Chem. Soc. Rev.* **2012**, *41*, 797–828.
(3) Merlet, C.; Rotenberg, B.; Madden, P. A.; Taberna, P.-L.; Simon, P.; Gogotsi, Y.; Salanne, M. *Nature Mater.* **2012**, *11*, 306–310.
(4) Colson, J. W.; Dichtel, W. R. *Nature Chem.* **2013**, *5*, 453–465.

(5) DeBlase, C. R.; Silberstein, K. E.; Truong, T.-T.; Abruña, H. D.; Dichtel, W. R. *J. Am. Chem. Soc.* **2013**, *135*, 16821–16824.
(6) DeBlase, C. R.; Hernandez Burgos, K.; Silberstein, K. E.; Robriguez-Colero, G. G.; Bisbey, R. P.; Abruña, H. D.; Dichtel, W. R. *ACS Nano*, **2015**, *9*, 3178-3183

THE ROLE OF OXYGEN IN TUNING THE OPTICAL AND CATALYTIC PROPERTIES OF COPPER OXIDE NANOSHEETS

*Zachary S. Fishman[1], Yulian He[2], Bolun Liu[3], Benjamin Rudshteyn[4], Subhajyoti Chaudhuri[4], Gary L. Haller[1], Victor Batista[4], Lisa D. Pfefferle[1]**

[1]Department of Chemical and Environmental Engineering, Yale University, New Haven, CT 06520-8286
[2]Department of Chemistry, Nankai University, Tianjin, China, 300071
[3]Department of Mechanical Engineering, Yale University, New Haven, CT 06520-8286
[4]Department of Chemistry, Yale University, New Haven, CT 06520-8286

Abstract

CO_2 photo(electro)reduction places stringent constraints on a material's optical and catalytic properties. Though much improvement has been made towards improving the photocatalytic efficiency and selectivity of CuO-Cu_2O for CO_2 conversion, little is known about how these treatments change the material on an atomic level. Here, cupric oxide nanosheets have been synthesized using a surfactant assisted protocol and subsequently treated under oxygen rich conditions at elevated temperatures. SEM, TEM, XRD, and BET physisorption measurements showed little change in nanostructure or composition. UV-visible spectroscopy, however, revealed that the band gap shifted up to .33eV within the range 1.5-1.9eV depending on treatment. Hydrogen TPR, which was used to probe differences in catalytic activity, showed a decrease in peak H_2 consumption temperature in the range of 260°C-214°C correlating with changes in the band gap. Extended Hückel theory was used to model the CuO nanosheet system. These calculations suggest that differences in the quantity and distribution of surface oxygens play a critical role in influencing the material's band gap. In summary, a method for tuning the catalytic properties and band gap of CuO nanosheets has been demonstrated in this work complemented by a mechanistic explanation of the underlying phenomenon supported by molecular modelling.

Email address of authors:
Zachary S. Fishman: zachary.fishman@yale.edu
Yulian He: y.he@yale.edu
Bolun Liu: bolun.liu@yale.edu
Benjamin Rudshteyn: benjamin.rudshteyn@yale.edu
Subhajyoti Chaudhuri: subhajyoti.Chaudhuri@yale.edu
Gary L. Haller: gary.haller@yale.edu
Victor Batista: victor.batista@yale.edu
Lisa D. Pfefferle: lisa.pfefferle@yale.edu

MOS_2 NANOSHEETS FOR DYE-SENSITIZED SOLAR CELLS

*Wei Wei and Yun Hang Hu**

Department of Materials Science and Engineering, Michigan Technological University, 1400 Townsend Dr., Houghton, MI, 49931, USA

yunhangh@mtu.edu

Abstract

A two-dimensional nanostructure of molybdenum disulfide (MoS_2) thin film exposed layered nanosheet was prepared by hydrothermal method on a fluorine-doped tin oxide (FTO) glass substrate as a platinum (Pt)-free and highly electrocatalytic counter electrode (CE) for dye-sensitized solar cells (DSSCs). It was found that the DSSC with 1T phase MoS_2 as counter electrode exhibited a much higher efficiency than that with 2H phase MoS_2 as counter electrode. The excellent electrocatalytic activity of the 1T phase MoS_2 counter electrode was further demonstrated by cyclic voltammograms, in which the 1T phase MoS_2 showed higher peak current density and lower peak-to-peak separation positions than 2H phase MoS_2.

Introduction

The conversion of solar energy to electrical energy is one of the major strategies to solve global energy issues(1,2). Since O' Regan and Grätzel made a breakthrough for dye-sensitized solar cells (DSSCs) in 1991 (3), DSSCs have attracted considerable attention as an alternative to conventional silicon based solar cells because of their low cost, low energy consumption, simple fabrication process, and high power conversion efficiency(4,5). Typically, DSSCs are composed of a photoelectrode (a transparent conducting subtract with a dye coated TiO_2 film), an electrolyte, and a counter electrode (CE). Namely, electrons transfer from an excited dye sensitizer into the semiconductor conduction band, then pass through the transparent conductive electrode to the external circuit, and finally to the counter electrode (CE). The CE task is the reduction of the redox species used as a mediator in regenerating the sensitizer after electron injection, or collection of the holes from the hole conducting material in a DSSC. An ideal CE material should possess low sheet resistance, high catalytic activity for reduction of the redox electrolyte, excellent chemical stability, and low cost.

For most cases, the iodide-triiodide couple has been exploited as the redox mediator and the overall redox reaction in DSSCs can be expressed as

$$I_3^- + 2e \leftrightarrow 3I^- \qquad (1)$$

The tri-iodides are produced near the dye sensitized TiO_2 electrode and reduced at the counter electrode. Platinum (Pt) is widely used for the counter electrodes of DSSCs due to its high catalytic activity and excellent conductivity. However, Pt is not only expensive, but also unstable in I^-/I_3^- redox-based electrolyte (because it can form a PtI_4 compound) (6,7). Those issues with Pt-based CEs stimulated a great interest to develop efficient Pt-free CEs for DSSCs, including carbon materials (graphite, carbon black, carbon nanotubes, and graphene), conducting polymers, metal sulfides, metal carbides, metal phosphides, metal nitrides, and metal oxides. In this work, efficient 1T phase MoS_2 counter electrodes were obtained for Dye-Sensitized Solar Cells.

Experimental

Preparation of photo electrodes: FTO glass plates were ultrasonically washed with water and the conducting surface of the FTO was treated with a 0.4 mM $TiCl_4$ aqueous solution at 70 °C for 30min. A commercial TiO_2 sol (Solaronix, Ti-nanooxide T/SP) was used to form a TiO_2 film on a FTO by the doctor-blade method. The TiO_2 film was thermally treated in air at 325 °C for 5min, 375 °C for 5min, 450 °C for 15min, and 500 °C for 15min. The obtained TiO_2/FTO glass plate was post-treated with the 0.4 mM $TiCl_4$ solvent at 70 °C, followed by heating in air at 500 °C for 30im. Finally, the obtained film was immersed in a cis-bis(isothiocyanato)bis(2,2'-bipyridyl-4,4'-dicarboxylato)-ruthenium(II)-bis-tetrabutylammonium (N719, 0.3mM in ethanol) dye solution for 24 hours to achieve dye sensitization.

Preparation of counter electrodes: MoS_2 was synthesized via a typical hydrothermal method: 1mmol of $Na_2MoO_4 \cdot 2H_2O$ and 5mmol of thiourea were dissolved in 60ml of distilled water and stirred vigorously. Then, the obtained mixture was transferred into a 100ml Teflon-lined autoclave. A FTO glass was placed against the wall of the autoclave. The whole system was kept for reactions at 180 or 200 °C for 24h. Finally, the FTO glass covered with MoS_2 film was rinsed with distilled water and ethanol, and then dried in an oven at 80 °C for 12h.

Results and Discussion

The photovoltaic performances of DSSCs with counter electrodes of MoS_2 were examined by photocurrent density-voltage (J-V) measurements. The DSSC with 1T phase MoS_2 as counter electrode exhibited a much higher efficiency than that of a DSSC with 2H phase MoS_2 based-counter electrode. Electrochemical impedance spectra (EIS) were exploited to characterize MoS_2 counter electrodes. Two semicircles were observed in the frequency range of 0.1-100KHz. The intercept of the first semicircles represents the ohmic serial resistance (Rs) related to the intrinsic resistance of the assembled cells. The semicircle in lower Z' is associated with the charge transfer resistance at the CE/electrolyte interface, while the semicircle in the higher Z' arises from the Nernst diffusion impedance of the triiodide/iodide couple. Furthermore, the peak current density and peak-to-peak separation of cyclic voltammograms (CV) are two important parameters to examine catalytic activities of different electrodes. The 1T phase MoS_2 counter electrode shows higher peak current density and lower peak-to-peak sepearation.

Conclusions

MoS_2 was synthesized by an in-situ hydrothermal approach. Furthermore, the synthesized materials were explored as counter electrodes for dye-sensitized solar cells (DSSCs). The DSSC with 1T phase MoS_2 as counter electrode exhibited a higher efficiency than that with 2H phase. Cyclic voltammograms and EIS spectra further revealed the excellent electrocatalytic activity of 1T phase MoS_2 counter electrode.

Acknowledgement. This work was supported by the ACS Petroleum Research Fund (PRF-51799-ND10). Hu also thanks Charles and Carroll McArthur for their great support.

References

(1) Paracchino, A.; Laporte, V.; Sivula, K.; Gratze, M.; and Thimsen, E. *Nat. Mater.*, **2011**, *10*, 456.
(2) Hou, Y. D.; Abrams, B. L.; Vesborg, P. C. K.; Bjorkeun, M. E.; Herbest, K.; Bech, L.; Setti, A. M.; Damsgaard, C. D.; Pedersen, T.; Hansen, O.; Tossmeisl, J.; Dahl, S.; Norskov, J. K.; and Chorkendorff, I. *Nat. Mater.*, **2011**, *10*, 434.
(3) O'Regan, B.; and Grätzel, M. *Nature*, **1991**, *353*, 737.
(4) Grätzel, M. *Nature*, **2001**, *414*, 338.
(5) Grätzel, M. *Acc. of Chem. Res.*, **2009**, *42*, 1788.
(6) Olsen, E.; Hagen G.; and Lindquist, S. E. *Sol. Energy Mater. Sol. Cells*, **2000**, *63*, 267.

(7) Kay A.; and Grätzel, M.; *Sol. Energy Mater. Sol. Cells*, **1996**, *44*, 99.

DEFECT ENGINEERING IN 2-DIMENSIONAL MATERIALS: FROM THEORY TO APPLICATIONS

Mauricio Terrones

Department of Physics, Department of Chemistry, Department of Materials Science and Engineering and Center for 2-Dimensional & Layered Materials. The Pennsylvania State University, University Park, Pennsylvania 16802, USA & Institute of Carbon Science and Technology, Shinshu University, JAPAN

This talk will first provide an overview of different defects in 2-Dimensional materials such as graphene and Chalcogenides. In particular, we will focus on: *1)* defining the dimensionalities and atomic structures of defects; *2)* pathways to generating structural defects during and after synthesis and, *3)* the effects of having defects on the physico-chemical properties and applications. We will then discuss the synthesis of large-area, high-quality monolayers of nitrogen-, boron- and silicon-doped graphene sheets on Cu foils using ambient-pressure chemical vapor deposition (AP-CVD). Scanning tunneling microscopy (STM) and spectroscopy (STS) demonstrate that defects in doped graphenes arrange in different geometrical configurations exhibiting different electronic and chemical properties. Interestingly, these doped layers could be used as efficient molecular sensors and electronic devices. Finally, we will discuss the controlled synthesis and assembly of chalcogenide monolayers on different substrates. The electronic performance of monolayers of MoS2, WS2 and hetero-systems operating under flexural strain will also be presented. Our findings demonstrates that it is now possible to use chalcogenide layers for the fabrication of flexible electronic devices, however, defect control is required to tailor their performance.

EFFECT OF QUANTUM CONFINEMENT ON ELECTRON CORRELATION IN ATOMICALLY THIN NB3SITE6

Jiang Wei

Department of Physics and Engineering Physics, Tulane University, New Orleans
jwei1@tulane.edu

Abstract

The recent breakthroughs in two dimensional (2D) materials, represented by Dirac cone in graphene and large direct bandgaps with spin-valley coupling in transition metal dichalcogenides MX2 (M = Mo or W, X = S, Se) monolayers, highlight the potential that quantum confinement effect on electronic structures may lead to a wide spectrum of distinct properties in 2D atomic layered crystals. On the other hand, the coupling between electrons and phonons provides an alternative approach to tune the electronic properties by quantum confinement effects on phonons. This has been theoretically predicted but not well-supported by earlier experimental investigations. In this work, we reported the clear evidence of the suppression of electron-phonon (e-p) scattering by phonon dimensionality in atomically thin ternary chalcogenide Nb3SiTe6, whose thickness can be thinned down to one unit cell by mechanical exfoliation. Through quantum transport measurements on this new 2D system, we demonstrated that with decreasing thickness, this new 2D system displays surprising enhancement of electron phase coherence due to the weakening of electron-phonon (e-p) scattering. Our analyses imply that such unusual phenomenon originates from the reduction of acoustic phonon populations due to 2D quantum confinement effects on acoustic phonon spectra. We conclude that such long-predicted phonon dimensionality effects, though have never been observed by pervious quantum transport studies, universally exist and can be probed in layered materials.

Reference

(1) J. Hu, X. Liu, C.L. Yue, J.Y. Liu, H.W. Zhu, J. B. He, J. Wei, Z.Q. Mao, L.Yu. Antipina, Z.I. Popov, P.B. Sorokin, T.J. Liu, P.W. Adams, S.M.A Radmanesh, L. Spinu, H. Ji and D. Natelson, Nature Physics 11, 471-476 (2015)

HYDROGEN BONDING ASSEMBLED ORGANIC 2D SEMICONDUCTORS

Haichang Zhang, Xiang Li and Yu Zhu

Department of Polymer Science
University of Akron, Akron, Ohio, 44325
yu.zhu@uakron.edu

Abstract

2D organic semiconductors with hydrogen bonding were synthesized and single-crystal samples were prepared by physical vapor transport technique. The crystal structure of the 2D organic semiconductor was determined by single-crystal X-Ray diffraction. The UV/Vis spectroscopy results indicated that the bandgaps of the organic semiconductors are between 1 and 2 eV. Those materials represented a new class of 2D materials that consist of organic semiconducting molecules.

TAILORING THE ELECTRICAL PROPERTIES OF GRAPHENE OXIDE OVER MULTIPLE LENGTH SCALES

Jeffrey M. Mativetsky

Department of Physics, Applied Physics and Astronomy, Binghamton University, Binghamton, NY 13902, USA
jmativet@binghamton.edu

Introduction

Graphene oxide (GO), a two-dimensional carbon lattice with extensive oxygen functionalization, is easily processed from aqueous suspensions and possesses highly tunable electrical, optical, and chemical properties.[1] The versatility of GO lies in its structure that combines both graphene-like domains and oxygen functionalized regions.

Reduction of GO results in the removal of oxygen and an increase in the fraction of graphene-like sp^2-hybridized carbon domains. An insulating GO sheet or film can thus be dispersed onto a substrate by solution deposition and then subsequently converted to conductive reduced graphene oxide (rGO).

GO reduction is still typically performed, however, under harsh or stringent conditions. Chemical reduction involves the use of strong reducing agents, such as hydrazine, [2,3] which are highly toxic and require extreme caution to work with. Thermal reduction, another common route for GO reduction, requires high temperatures (500-1100 °C) and an inert environment for effective reduction.[4,5]

We have developed a voltage-based reduction approach that operates under ambient conditions,[6,7] and avoids the use of toxic reagents or high temperatures. The process relies on a thin water layer that is present on GO under ambient conditions, and acts as a localized electrochemical environment when a voltage is applied between proximal electrode pairs. Unlike conventional electrochemical reduction, the GO is not immersed in an electrolyte; the GO is processed entirely under ambient conditions. Beyond its potential use as an eco-friendly processing method, voltage-induced GO reduction also offers unprecedented high-resolution patterning of conductive rGO pathways in GO sheets and films, with patterned feature sizes as small as 4 nm.[8]

Here, we provide an overview of the present status of voltage-induced GO reduction, and the prospects of using this approach for both large-area and nanoscale control over the electrical properties of GO. We show that the oxygen removal levels and electrical performance associated with voltage-reduced GO are competitive with standard reduction methods, while adding the advantages of: (1) ease of processing under ambient conditions and (2) high-resolution patterning capabilities. Moreover, band gap engineering through quantum confinement in voltage-reduced nanostructured rGO holds promise for extending the electrical function of GO to include semiconducting properties.

Experimental

Sample preparation and characterization conditions are described in detail elsewhere.[8] Briefly, GO is dispersed in water and deposited onto substrates by spin coating to yield isolated GO sheets or continuous GO films. Gold electrode arrays with electrode spacings ranging from 500 nm to 1 mm are patterned onto SiOx/Si or glass substrates and used for multiscale voltage-induced GO reduction. Conductive atomic force microscope (C-AFM)[9] probe-initiated reduction is achieved by applying a voltage between the metal-coated C-AFM probe and a nearby patterned electrode (lateral geometry)[6] or between the C-AFM probe and a GO-coated conductive substrate (vertical geometry).[8]

Results and Discussion

Voltage-induced GO reduction was first demonstrated by applying a voltage between a fixed surface electrode and a moveable C-AFM probe.[6] The C-AFM setup then also serves to map the local film topography and electrical transport properties. Voltage-induced reduction was later extended to micro-patterned electrode arrays; the application of a voltage across electrode pairs resulted in the reduction of the intervening GO.[7] The same electrodes can also be used as contacts in an rGO-based device or to electrically characterize the rGO.

Reduction Mechanisms. Voltage-induced reduction relies on water that adsorbs on, and intercalates into, GO under ambient or humid conditions. Upon application of a sufficiently large voltage, the following localized electrochemical reactions takes place:[7,10]

$$2H_2O \rightarrow 4H^+ + O_2 + 4e^-$$
$$aH^+ + be^- + GO \rightarrow rGO + cH_2O$$

As shown in Figure 1, H^+ generated through water splitting is driven through the GO film, towards the negative electrode, where reduction takes place. As reduction proceeds, the reduction front extends from the negative electrode towards the positive electrode as the conductive rGO effectively acts as an extension of the negative electrode.

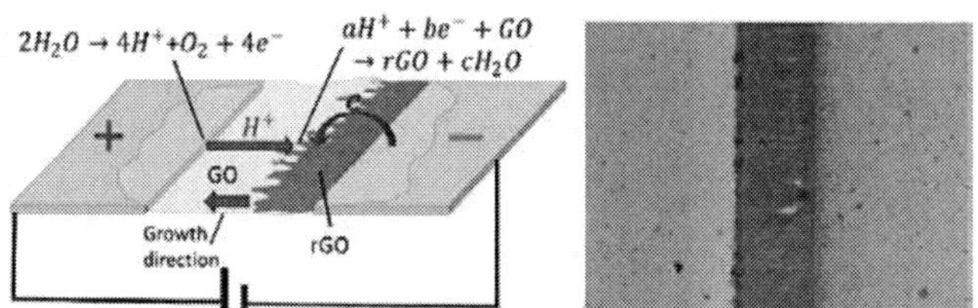

Figure 1. Scheme for voltage-induced reduction between electrode pairs (left), optical micrograph of rGO that was voltage-reduced in a 200 μm electrode gap (right).

The essential role of water is confirmed by experiments that show an increased rate of reduction as humidity levels are increased and an absence of reduction under dry conditions. The importance of the electric field in driving H^+ flow is confirmed through experiments that show the growth of the reduction front along electric field lines for different electrode geometries.[11]

Structure and Electrical Characteristics. Voltage-reduced rGO exhibits deoxygenation levels and electrical characteristics that are competitive with standard reduction methods, demonstrating that voltage-induced reduction holds promise as an alternative to more cumbersome or hazardous GO reduction schemes.

X-ray photoelectron spectroscopy shows a drop in oxygen content from 30 atomic percent (at. %) to 12 at. % upon reduction using a voltage.[11] This drop in oxygen content is greater than that typically associated with conventional electrochemical reduction (18-28 at. % oxygen after reduction) and approaches that achieved by hydrazine reduction (8-11 at. % oxygen) and thermal reduction at 1100 °C under ultrahigh vacuum (8 at. % oxygen).[4,12–14] After voltage-induced reduction, the sp^2 carbon bonding fraction increases considerably, from 45% to 72%.

Raman spectroscopy shows prominent D-band and G-band peaks. The intensity ratio I_D/I_G can be used to quantify the amount of disorder in the GO lattice.[15–17] Based on analysis in the high defect density regime,[15] we find that voltage-induced reduction leads to an increase in the mean distance between lattice defects from 1.2 nm to 1.5 nm, allowing for improved charge percolation. So far, a conductivity of 10 S/cm has been achieved in voltage-reduced rGO films.

Nanoscale GO Reduction. To explore the spatial resolution limits of voltage-induced reduction, a C-AFM probe was employed to initiate reduction, as shown in Figure 2. The smallest feature sizes have been achieved under low humidity conditions (about 20% RH). At such low humidity levels, the size of the water meniscus around the nanoprobe apex and the amount of water incorporated in the GO film are diminished, thus decreasing the area subject to reactive H^+ after water splitting

Using this approach, we have produced rGO feature sizes with a full-width at half maximum as small as 4 nm, the smallest size achieved so far by any GO reduction method.[8] This feature size is near the expected resolution limit for the technique, as the tip contact diameter, estimated to be 3 nm based on the Hertz contact model, sets the upper limit on patterning and imaging resolution.

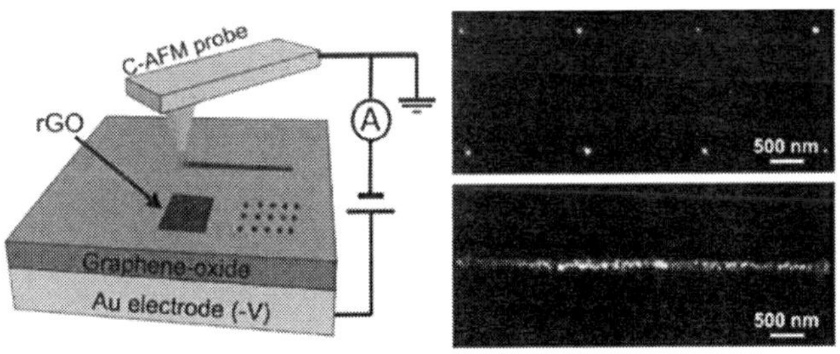

Figure 2. C-AFM setup for probe-initiated reduction (left) and C-AFM current map of nanoscale conductive rGO features defined in insulating GO (right).

Nanoscale rGO patterning holds promise for enabling transport gap modulation through quantum confinement.[18-20] The lack of a band gap in rGO limits its usefulness for certain electronic applications, such as transistors. We have recently extended sub-50 nm width rGO lines between electrode pairs, with a silicon oxide-coated highly-doped silicon gate electrode below, making it possible to investigate one-dimensional charge transport and transport gap opening.

The ratio between on and off current during transistor operation is directly impacted by the transport gap.[18] A high on-off ratio is desirable to provide transistor gain and contrast between switching states. Preliminary field-effect transistor measurements show an increase in on/off ratio in nanoscale lines when compared with large-area rGO devices, indicating possible transport gap opening. Further work is underway to characterize the one-dimensional and Coulomb blockade properties of nanoscale rGO lines, and the extent of transport gap modulation.

Conclusions

Voltage-induced reduction offers a means of converting insulating GO to conductive rGO with a performance that rivals other reduction methods, while offering the advantages of facile processing under ambient conditions and high-resolution patterning capabilities. In addition, nanoscale rGO patterning enabled by voltage-induced reduction opens up prospects for transport gap opening and patterning of insulating, conducting, and semiconducting regions within a single two-dimensional material.

Acknowledgement. This work was supported by the NSF Nanomanufacturing program (CMMI-1537648) and made use of equipment that was acquired through an NSF Major Research Instrumentation grant (CMMI-1429176).

References

(1) Eda, G.; Chhowalla, M. *Adv. Mater.* **2010**, *22*, 2392–2415.
(2) Stankovich, S.; Dikin, D. A.; Piner, R. D.; Kohlhaas, K. A.; Kleinhammes, A.; Jia, Y.; Wu, Y.; Nguyen, S. T.; Ruoff, R. S. *Carbon* **2007**, *45*, 1558–1565.
(3) Gilje, S.; Han, S.; Wang, M.; Wang, K. L.; Kaner, R. B. *Nano Lett.* **2007**, *7*, 3394–3398.
(4) Mattevi, C.; Eda, G.; Agnoli, S.; Miller, S.; Mkhoyan, K. A.; Celik, O.; Mastrogiovanni, D.; Granozzi, G.; Garfunkel, E.; Chhowalla, M. *Adv. Funct. Mater.* **2009**, *19*, 2577–2583.
(5) Becerril, H. A.; Mao, J.; Liu, Z.; Stoltenberg, R. M.; Bao, Z.; Chen, Y. *ACS Nano* **2008**, *2*, 463–470.
(6) Mativetsky, J. M.; Treossi, E.; Orgiu, E.; Melucci, M.; Veronese, G. P.; Samorì, P.; Palermo, V. *J. Am. Chem. Soc.* **2010**, *132*, 14130–14136.
(7) Mativetsky, J. M.; Liscio, A.; Treossi, E.; Orgiu, E.; Zanelli, A.; Samorì, P.; Palermo, V. *J. Am. Chem. Soc.* **2011**, *133*, 14320–14326.
(8) Faucett, A. C.; Mativetsky, J. M. *Carbon* **2015**, *95*, 1069–1075.
(9) Mativetsky, J. M.; Loo, Y. L.; Samorì, P. *J. Mater. Chem. C* **2014**, *2*, 3118–3128.
(10) Zhou, M.; Wang, Y.; Zhai, Y.; Zhai, J.; Ren, W.; Wang, F.; Dong, S. *Chem. Eur. J.* **2009**, *15*, 6116–6120.
(11) Faucett, A. C.; Flournoy, J. N.; Mehta, J.; Mativetsky, J. M. manuscript in preparation.
(12) Toh, S. Y.; Loh, K. S.; Kamarudin, S. K.; Daud, W. R. W. *Chem. Eng. J.* **2014**, *251*, 422–434.
(13) Bouaa, M.; Frank, O.; Jirka, I.; Kavan, L. *Phys. Status Solidi B* **2013**, *250*, 2662–2667.
(14) Yang, D.; Velamakanni, A.; Bozoklu, G.; Park, S.; Stoller, M.; Piner, R. D.; Stankovich, S.; Jung, I.; Field, D. A.; Ventrice, C. A.; Ruoff, R. S. *Carbon* **2009**, *47*, 145–152.
(15) Ferrari, A.; Robertson, J. *Phys. Rev. B* **2000**, *61*, 14095–14107.
(16) Cançado, L. G.; Jorio, A.; Ferreira, E. H. M.; Stavale, F.; Achete, C. A.; Capaz, R. B.; Moutinho, M. V. O.; Lombardo, A.; Kulmala, T. S.; Ferrari, A. C. *Nano Lett.* **2011**, *11*, 3190–3196.
(17) Eigler, S.; Dotzer, C.; Hirsch, A. *Carbon* **2012**, *50*, 3666–3673.
(18) Li, X.; Wang, X.; Zhang, L.; Lee, S.; Dai, H. *Science* **2008**, *319*, 1229–1232.
(19) Gallagher, P.; Todd, K.; Goldhaber-Gordon, D. *Phys. Rev. B* **2010**, *81*, 115409.
(20) Mucciolo, E. R.; Castro Neto, A. H.; Lewenkopf, C. H. *Phys. Rev. B* **2009**, *79*, 075407.

INVERSION SYMMETRY BREAKING IN $MoTe_2$ PROBED BY SECOND HARMONIC GENERATION

Ryan Beams[1], Sergiy Krylyuk[1,2], Irina Kalish[1,3], Berç Kalanyan[1], James E. Maslar[1], Albert V. Davydov[1], and Stephan J. Stranick[1]

[1]Material Measurement Laboratory, NIST, 100 Bureau Drive, Gaithersburg, MD, 20899
[2]Institute for Research in Electronics and Applied Physics, Univ. of Maryland, College Park, MD, 20742
[3]Irina Kalish, LLC 1103 Balls Hill Road, McLean, VA, 22101
ryan.beams@nist.gov

Introduction

While two-dimensional electron gases and heterostructures have been studied for decades, the discovery of two dimensional crystals has led to a new era of low-dimensional materials research. These materials are being explored for flexible electronics, photovoltaics, and chemical sensors [1, 2]. One of these materials is molybdenum ditelluride ($MoTe_2$), which is of particular interest due to the phase transitions between metallic (1T' or β-$MoTe_2$) and semiconducting (2H or α-$MoTe_2$) polymorphs with strong spin-orbit coupling, a near-infrared bandgap for 2H phase, and strong magnetic properties for 1T' phase [3–5]. In order to utilize the unique properties of these materials, a variety of characterization techniques are required. Recently second harmonic generation (SHG) has been used to study two dimensional semiconductors [6–9]. Due to the sensitivity of nonlinear processes to crystal symmetry, SHG provides an optical method for determining the crystal orientation and symmetry of flakes [6–9]. We used polarized SHG images to study the symmetry properties of 1T' $MoTe_2$ flakes. While bulk and monolayer 1T' $MoTe_2$ is inversion symmetric, this symmetry is broken for bilayer flakes and strong SHG is observed.

Experimental

The 1T' $MoTe_2$ single-crystalline platelets were grown using an iodine-assisted chemical vapor transport method and subsequently exfoliated onto a SiO_2/Si substrate. The phase of the material was confirmed with Raman spectroscopy, X-ray diffraction (XRD), and X-ray photoelectron spectroscopy (XPS). SHG images were acquired by placing the exfoliated flakes on an x-y scanstage and raster-scanning the flakes through the laser focus. The sample was excited by focusing a femtosecond Ti:Sapphire laser onto the flake with a microscope objective (0.75NA) and raster-scanning the flake through the focus to build up images. The resulting signal was sent to either a single photon counting module (SPCM) or a spectrometer and charge-coupled device (CCD). For SPCM imaging a narrow bandpass filter was used to select out the SHG signal.

Results and Discussion

Figure 1(a) shows a topographic image of $MoTe_2$ flake with the single layer (1L) and bilayer (2L) regions labeled. The resulting SHG image of the flake is shown in Figure 1(b). The topographic outline of the flake as well as the step edges are shown as solid and dashed lines. A strong SHG signal is observed for the bilayer region and no SHG signal from the single layer region. Interestingly, several of the thicker regions have a weak SHG signal, which suggests that even numbers of layers are non-inversion symmetric. Finally the polarization dependence of the SHG signal for excitation polarization and analyzer parallel (black) and perpendicular (red) are shown in

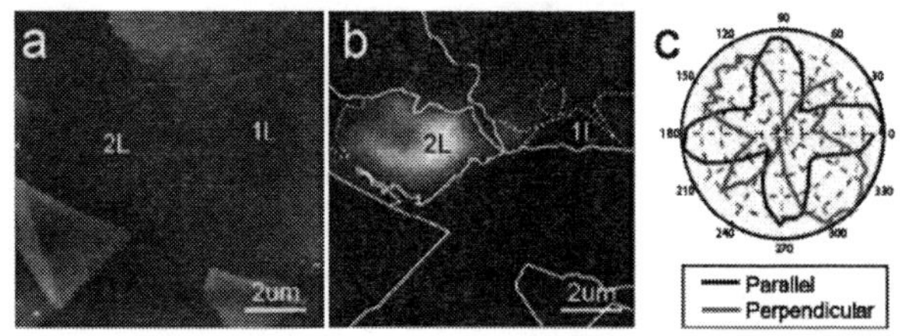

Figure 1. (a) Topography and (b) SHG image of a 1T' $MoTe_2$ flake. The topographic outline of the flake as well as the step edges are shown in solid and dashed lines, respectively. (c) Polar plot of the polarization dependence of the SHG signal for the excitation polarization and analyzer parallel (black) and perpendicular (red).

Figure 1(c). The polarization dependence confirms that the flake is monoclinic as expected for 1T' [10].

Conclusions

We have demonstrated using SHG that the inversion symmetry properties of 1T' $MoTe_2$ are broken for bilayer flakes. SHG is an excellent technique for rapidly probing the symmetry properties of two dimensional materials. Furthermore, SHG could also be used to identify the phase of these types of materials.

References

(1) A. K. Geim, and I. V. Grigorieva, "Van der Waals heterostructures," Nature 499, 419 (2013).
(2) R. Beams, L. G. Cançado, and L. Novotny, "Raman characterization of defects and dopants in graphene," J. of Phys.: Cond. Matt. 27, 083002 (2015).
(3) M. B. Vellinga, R. De Jonge, and C. Haas, "Semiconductor to metal transition in MoTe2," J. of Solid State Chem. 2, 299 (1970).
(4) N. R. Pradhan, D. Rhodes, S. Feng, Y. Xin, S. Memaran, B.-H. Moon, H. Terrones, M. Terrones, and L. Balicas, "Field-effect transistors based on few-layered a-MoTe2," ACS Nano 8, 5911 (2014).
(5) C. Ruppert, O. B. Aslan, and T. F. Heinz, "Optical Properties and Band Gap of Single-and Few-Layer MoTe2 Crystals," Nano Lett. 14, 6231 (2014).
(6) Y. Li, Y. Rao, K. F. Mak, Y. You, S. Wang, C. R. Dean, and T. F. Heinz, "Probing symmetry properties of few-layer MoS2 and h-BN by optical second-harmonic generation," Nano Lett. 13, 3329–3333 (2013).
(7) L. M. Malard, T. V. Alencar, A. P. M. Barboza, K. F. Mak, and A. M. de Paula, "Observation of intense second harmonic generation from MoS2 atomic crystals," Phys. Rev. B 87, 201401 (2013).
(8) N. Kumar, S. Najmaei, Q. Cui, F. Ceballos, P. M. Ajayan, J. Lou, H. Zhao, "Second harmonic microscopy of monolayer MoS2," Phys. Rev. B 87, 61403 (2013).
(9) S. N. David, Y. Zhai, A. M. van der Zande, K. O'Brien, P. Y. Huang, D. A. Chenet, J. C. Hone, X. Zhang, and X. Yin, "Rapid, all-optical crystal orientation imaging of two-dimensional transition metal dichalcogenide monolayers," Applied Phys. Lett. 107, 111902 (2015).
(10) R. W. Boyd, Nonlinear Optics (Academic Press - third edition, San Diego, 2008).

SYNTHESIS AND CHARACTERIZATION OF 2D MOLYBDENUM CARBIDE (MXENE)

Sankalp Kota, Joseph Halim, Maria R. Lukatskaya, Michael Naguib, Meng-qiang Zhao, Eun Ju Moon, Jeremy Pittock, Jagjit Nanda, Steven J. May, Yury Gogotsi, and Michel W. Barsoum

Dept. of Materials Science and Engineering, Drexel University
3141 Chestnut St Lebow 344, Philadelphia, PA, USA
barsouw@drexel.edu

Introduction

Two-dimensional (2D) materials often possess high specific areas and electronic properties distinct from their bulk counterparts. Recently, we discovered a family of 2D transition metal carbides and carbonitrides, which were called MXenes. They have a general formula of $M_{n+1}X_nT_x$, where M is an early transition metal, X is carbon or nitrogen, and T are surface functional group (e.g. –O, -OH, -F), due to being synthesized by selective etching of the "A" layers (Group 13,14 elements) from the $M_{n+1}AX_n$ phases, a large family of ternary nanolaminated carbides and nitrides [1]. MXenes have been found to be good electronic conductors and hydrophilic, in addition to having the ability to intercalate various Group I and II cations and swell in water like clays. As a result, they have been investigated for applications such as secondary metal-ion batteries, supercapacitors, water purification, photocatalysis, and transparent conductive electrodes. [2]

To date, every reported MXene has been synthesized by selective etching of "Al" from the MAX phases (i.e. A = Al). Herein, we demonstrate, for the first time, the large-scale synthesis of Mo_2CT_x by selective etching of Ga from the recently discovered Mo_2Ga_2C, a nanolaminated carbide similar to the MAX phases. Coincidentally, Xu *et al.* grew large area films of Mo_2C crystals by chemical vapor deposition and evaluated its electronic properties.

Experimental

Synthesis of Mo_2Ga_2C. Mo_2C (Alfa Aesar, -325 mesh) and Ga metal were mixed by mortar and pestle in a 1:8 molar ratio and sealed in a quartz reaction tube, evacuated under vacuum. The mixture was heated to 850 °C and held at this temperature for 48 h, after which the contents were crushed and reacted under the same conditions for 16 h more. See the work of Halim *et al.* for further details [3].

Synthesis of 2D Mo_2CT_x. The lightly sintered block of Mo_2Ga_2C were immersed in 12 M hydrochloric acid to dissolve away unreacted gallium for 2 days and then washed with deionized (DI) water to neutral pH. The resulting product was crushed into powders and converted to Mo_2CT_x by immersing and stirring the powders in a concentrated hydrochloric acid (HCl) with lithium fluoride pre-dissolved, followed by washing/centrifuging the powders in HCl, then 1 M LiCl, and DI water to neutral pH. Alternatively, powders were immersed and stirred in concentrated hydrofluoric acid at 55 °C, washing/centrifugation of the product with DI water, intercalation with tetrabutylammonium hydroxide (TBA-OH). Afterwards, the Mo_2CT_x powders synthesized with either etchant were subjected bath sonication and centrifugation in water to obtain stable colloidal suspensions of few-layer Mo_2CT_x

Electrochemical Testing. Free-standing films (2 um thick) prepared by vacuum filtration of aqueous colloidal suspensions of Mo_2CT_x were tested by cyclic voltammetry in 1 M H_2SO_4 in three-electrode Swagelok cells using an overcapacitive activated carbon counter electrode and Ag/AgCl in KCl reference electrode. For, galvanostatic cycling tests, the two-electrode Swagelok cells were used instead.

For Li-ion batteries (LIB), composite Mo_2CT_x-8% carbon nanotube (CNT) electrodes were prepared by co-filtration of aqueous suspensions of Mo_2CT_x and CNT, as described by Halim *et al.* [3] Coin cells were assembled using the Mo_2CT_x-CNT film anode, Li metal counter electrode, polypropylene separator, and electrolyte of 1.2 M $LiPF_6$ dissolved in 1:2 ratio of ethylene carbonate-dimethyl carbonate, by weight. Cyclic voltammetry (CV) and galvanostatic cycling (GC) tests were performed.

General Characterization. The microstructure, morphology, and chemical composition of various samples were characterized by transmission electron microscopy (TEM), scanning electron microscopy (SEM), X-ray diffraction (XRD), X-ray photoelectron spectroscopy (XPS). Electronic properties were characterized from 10-300 K using a four-probe measurement system.

Results and Discussion

Fig. 1a shows a schematic for the synthesis of 2D Mo_2CT_x by selectively removing the Ga layers from Mo_2Ga_2C in aqueous fluoride containing etchants. The resulting layered structure, held together weakly by van der Waals and hydrogen bonds, is then delaminated into atomically-thin flakes by shaking the powders by hand in water (LiF/HCl route) or TBAOH intercalated powders in water (HF/TBAOH route). **Fig. 1c-d** shows the X-ray diffractogram of Mo_2Ga_2C and the ensuing broadening and shift to lower angles of the {00*l*} peaks after HF or LiF/HCl etching, indicating an expansion in the *c* lattice parameter of the unit cell. After re-stacking of the delaminated Mo_2CT_x flakes under vacuum filtration (**Fig. 1b**), the non-basal peaks of Mo_2CT_x are no longer present, which unambiguously confirms the turbostatic arrangement of the flakes after vacuum filtration.

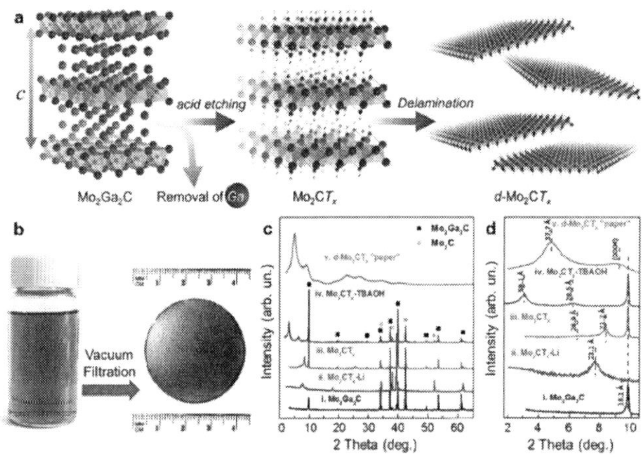

Figure 1. (a) Schematic of the synthesis of Mo_2CT_x; (b) diluted colloidal suspension and free-standing film resulting from the vacuum filtration of the former; (c) XRD patterns of Mo_2Ga_2C and those of Mo_2CT_x after various synthesis steps. (Adapted from Ref. 3)

In the 10-300 K temperature range, the electrical resistivity (ρ) as a function of temperature of free-standing Mo_2CT_x films is shown in **Fig. 2a**. The resistivity increased as the temperature decreased from 300 K to 10 K and, at all temperatures, the resistivity was several orders of magnitude lower after annealing the film under vacuum. To elucidate the transport mechanisms, the data below 50 K was plotted as the natural logarithm of W = d(ln ρ)/d(ln T) vs. ln T. The increase in the slope of this plot (**Fig. 2b)** before annealing suggested a change in transport mechanism before and after annealing due to the deintercalation of TBA-OH after annealing, which was confirmed by XRD. Interstingly, this observation suggests that the electronic properties can be tuned by using different intercalants. Moreover, Mo_2CT_x after annealing showed a positive magnetoresistance at 10 K (**Fig. 2c**), which is in contrast to $Ti_3C_2T_x$ and thin films of Mo_2C crystals grown by Xu *et al.* The sum of the results suggest Mo_2CT_x exhibits semiconductor-like transport properties, unlike the metallic-like transport in most other MXenes, and its transport properties can be changed by varying the surface terminations.

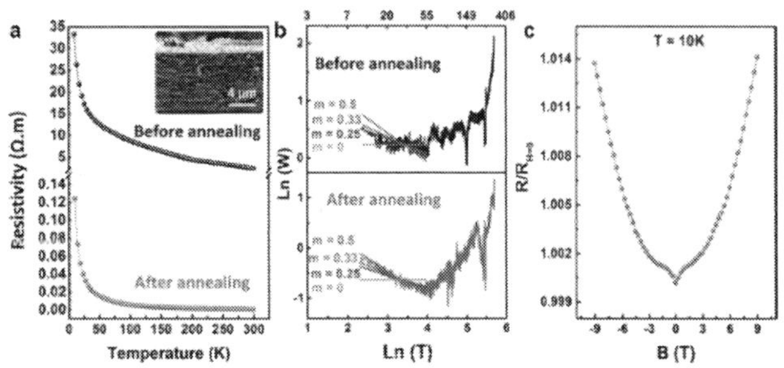

Figure 2. (a) Temperature dependence of free-standing Mo_2CT_x films before and after annealing under vacuum; (b) Fitting of the low temperature data to various transport models with slopes m = 0.5 (Efros-Shklovskii variable range hopping (E-S VRH)), m=0.33 (2D VRH), m=0.25 (3D VRH), m=0 (power law dependence); (c) Magnetoresistance of anneal film at 10 K. (Adapted from Ref. 3)

Cyclic voltammograms on free-standing films in 1 M H_2SO_4 had a slightly non-rectangular shape (**Fig. 3a**), which suggests a pseudocapacitive charge storage mechanism lie $Ti_3C_2T_x$ [4]. High volumetric capacitances of 700 F/cm^3 at 2 mV/s and 430 F/cm^3 at 100 mV/s were observed (**Fig. 3b**). Galvanostatic cycling also showed a high capacitance retention up to 10,000 cycles (**Fig. 3c**). The high specific capacitances can be attributed to two features: high intrinsic electronic conductivity of the Mo_2C backbone and a transition metal oxide based surface chemistry. The latter likely allows for surface redox reactions similar to MoO_3 based materials, while the former allows these surface redox reactions to happen even at high charge/discharge rates. Since the electrodes are composed of flakes of Mo_2CT_x densely stacked perpendicular to the direction of ionic motion, facile ionic diffusion is likely impeded. Therefore, improved electrode architectures would likely show even better rate capability.

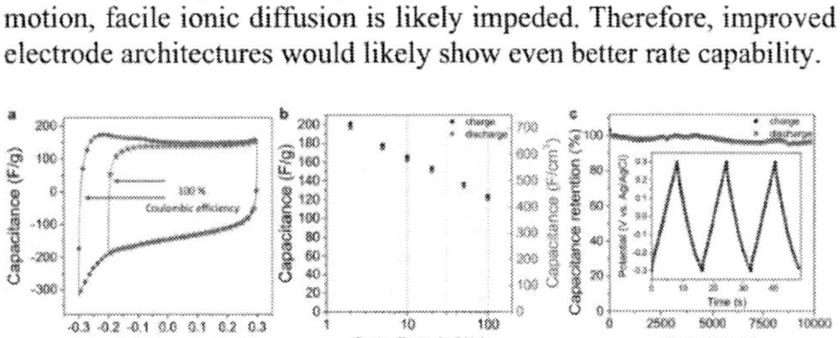

Figure 3. (a) CV curves of 2 μm thickfree-standing Mo_2CT_x films measured at 20 mV/s in 1 M H_2SO_4; (b) Scan rate dependence of the capacitance from 2-100 mV/s; (c) Capacitance retention test up to 10,000 cycles measured at a current density of 10 A/g. Inset shows the galvanostatic cycling plot. (Adapted from Ref. 3)

The results of testing 3 μm thick Mo_2CT_x-CNT composite electrodes as anodes in LIB half cells in shown in **Fig. 4**. CVs in **Fig. 4a** reveal slight fluctuations in the lithiation potentials and the intensity of the peaks during the first 30 cycles. After 30 cycles, the lithiation peaks stabilized at 0.15 V and 1.4 V vs. Li/Li+. Similar fluctuations are observed during delithiation before the peaks stabilized at potentials of 1.45 V and 2.64 V vs. Li/Li+ after 30 cycles. GC profiles show plateau at 0.4 V during the first lithiation and delithiation 1.28 V vs Li/Li+ cycles, respectively, as opposed to to the capacitor-like GC profiles of the previously reported Nb_2CT_x and V_2CT_x MXenes [5]. Interestingly, GC testing at current densities of 0.4 A/g and 5 A/g showed an *increase* in the specific capacity during the first 70 cycles before stabilizing to a constant value. After 1000 cycles at 5 A/g (~ 20 C) and 10 A/g (~40 C), reversible capacities of 250 mAh/g and 75 mAh/g were obtained, respectively. It is worth noting that the capacity of these Mo_2CT_x electrodes are higher than the theoretical capacity of graphite (372 mAh/g) and can be expected to increase with further optimization of electrode architecture and control of the surface chemistry.

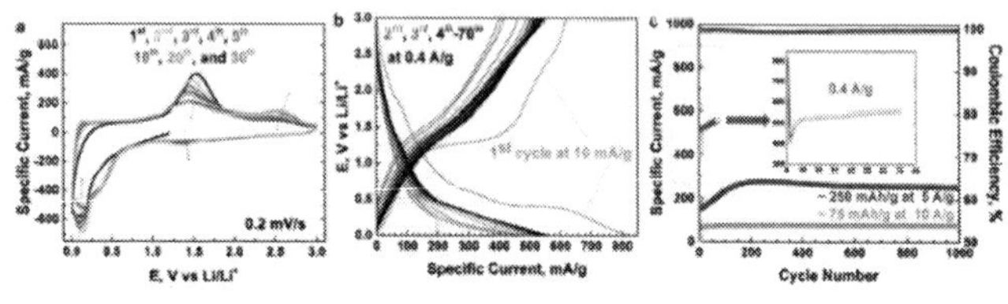

Figure 4. Electrochemical performance of Mo_2CT_x-CNT electrodes. (a) CV's at 0.2 mV/s; (b) Galvanostatic cycling between 0-3 V vs. Li/Li+; (c) cycle life tests at 0.4 A/g (red), 5 A/g (blue), and 10 A/g (orange). Inset shows zoomed in area for 0.4 A/g cycle life test.

Conclusions

Two methods for facile large-scale synthesis of 2D Mo_2CT_x MXene from the ternary nanolaminated compound Mo_2Ga_2C. Colloidal suspensions could be obtained and vacuum filtered directly to make flexible, free-standing films of Mo_2CT_x. By using different etchants during synthesis, different 2D flake morphologies could be obtained. Electrical transport measurements revealed a semi-conductor like temperature dependence at low temperatures, unlike the metallic-like transport properties of the well-studied $Ti_3C_2T_x$ MXene and that the transport properties may be controlled by intercalation of different organic compounds. Initial tests of free-standing films of Mo_2CT_x MXene as electrodes in aqueous supercapacitors showed high volumetric capacitances retained up to 100 mV/s. Moreover, composite Mo_2CT_x-CNT electrodes had lithium-ion storage capacities much higher than those of graphite and good rate capability to rates as high as ~ 20 C.

Acknowledgements. The authors want to thank the Swedish Research Council (Grant No. 621-2012-4430), Swedish Foundation for Strategic Research through the Synergy Grant FUNCASE Functional Carbides, for Advanced Surface Engineering, Laboratory Directed Research and Development, Program of Oak Ridge National Laboratory, U.S. Army Research Office (grant number W911NF-15-1-0133).

References

(1) Barsoum, M. W. *Prog. Solid State Chem.* **2000**, *28* (1-4), 201–281.
(2) Naguib, M.; Mochalin, V. N.; Barsoum, M. W.; Gogotsi, Y. *Adv. Mater.* **2014**, *26* (7), 992–1005.
(3) Halim, J.; Kota, S.; Lukatskaya, M. R.; Naguib, M.; Zhao, M.; Moon, E. J.; Pitock, J.; Nanda, J.; May, S. J.; Gogotsi, Y.; Barsoum, M. W. *Adv. Funct. Mater.* **2016**
(4) Lukatskaya, M. R.; Bak, S.; Yu, X.; Yang, X.; Barsoum, M. W.; Gogotsi, Y. *Adv. Energy Mater.* **2015**, *5* (15)
(5) Naguib, M.; Halim, J.; Lu, J.; Cook, K. M.; Hultman, L.; Gogotsi, Y.; Barsoum, M. W. *J. Am. Chem. Soc.* **2013**, *135* (43), 15966–15969.

FUNDING OPPORTUNITIES AT THE ARMY RESEARCH OFFICE AND RESEARCH PROGRAMS IN THE MATERIALS SCIENCE DIVISION

Pani Varanasi, ARO

Abstract

The Materials Science Division of the U.S. Army Research Office (ARO) seeks to realize unprecedented materials properties by embracing innovative, long-term, high-risk, high-payoff basic research opportunities for the U.S. Army. In this presentation, an overview of the current programs, research thrusts within each program, current funding opportunities etc. will be discussed. The Division is currently seeking to establish several new thrusts, such as 2D (organic, inorganic) materials/ Heterostructures/ hybrids, spin-caloritronics, transformative processing strategies, mechano-chemically adaptive materials, and directed self-assembly of reconfigurable materials, etc.

Physical Properties of Materials Program in the Materials Science Division

The Physical Properties of Materials program seeks to elucidate fundamental mechanisms responsible for achieving extraordinary electronic, photonic, magnetic and thermal properties in advanced materials to enable future Army relevant innovations.

There are two thrusts in the program

(i) Free-standing 2D Materials thrust focuses on the creation of novel free-standing 2D materials, heterostructures and hybrids with physical properties complementary or superior to graphene, and the invention of novel characterization techniques specific to 2D materials to determine unprecedented properties.

(ii) Defect Science & Engineering thrust explores the specific influence of defects (positive or negative) on the physical properties of novel functional materials, and elucidates defect control mechanisms during thin film growth and bulk processing of novel functional materials.

Technical Point of Contact: Dr. Pani Varanasi
email: chakrapani.v.varanasi.civ@mail.mil, (919) 549-4325

DIRECT SYNTHESIS OF ORGANIC CARBONATES BY OXIDATIVE CARBONYLATION OVER SCHIFF BASE PALLADIUM COMPLEXES CATALYST

Qingyan Cheng[1, 2], Yundong Liu[1], Chaoyi Li[1], Yanji Wang[1], Xinqiang Zhao[1], Xianqin Wang[2]

1. School of Chemical Engineering, Hebei University of Technology, Tianjin 300130, China;
2. Department of Chemical, Biological, and Pharmaceutical Engineering, New Jersey Institute of Technology, Newark, NJ 07102, USA

Corresponding author: qycheng@njit.edu; xianqin.wang@njit.edu

Introduction

Organic carbonates are plausible synthetic targets from the viewpoint of green sustainable chemistry for their low toxicity, non-corrosivity and biodegradability [1]. In addition, carbonates have been widely used as starting materials of polycarbonate resins, electrolytes of lithium ion battery, alkylating and carbonylating reagents, and inert solvents . Moreover, they are expected to be fuel additives in the future.

Organic carbonates have been traditionally produced by using phosgene or its derivatives. However, this production process has considerable drawbacks such as use of highly toxic phosgene and co-production of large amount of salts by the neutralization. Some alternative routes have been developed such as oxidative carbonylation, transesterification. Oxidative carbonylation is the most valuable industrial route. And the research of oxidative carbonylation process focus on the searching of catalyst with high catalytic performance, mainly.

Our recent study[2,3] proved the shicff base palladium complexes were very active for the oxidative carbonylation and could be synthesized easily.

Experimental

Shicff base palladium complexes were prepared by microwave assisted solvothermal method. Typical reaction for synthesis of Schiff base and Shicff base palladium complexes as shown.

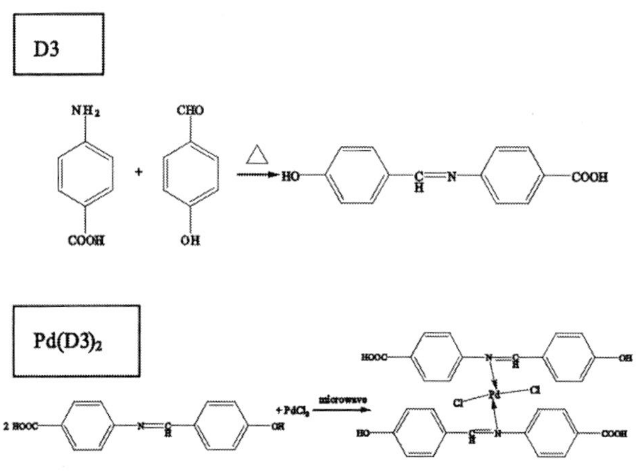

Figure 1.

UV-vis were measured on a Varian Cary 300 spectrophotometer in an DMF solution. 1H-NMR spectra were obtained using tetramethylsilane as reference standard in (CD3)2SO solution on a Bruker's AV400 spectrophotometer. The I.C.P. dates were obtained in water and DMF mixed solution on a Perkin Elmer's Optima7300V ICP-AES. FT-IR spectra, as KBr pellets, were measured on a BrukerVector22 FT-IR spectrophotometer.

Autoclave reaction test the catalytic activity for oxidative carbonylation and HPLC analysis.

Results and Discussion

A predicted molecular model of Schiff base palladium complexes were confirmed by 1H-NMR, I.C.P., FT-IR, XRD, UV-vis. Typical molecular models of Schiff base and Shicff base palladium complexes as shown.

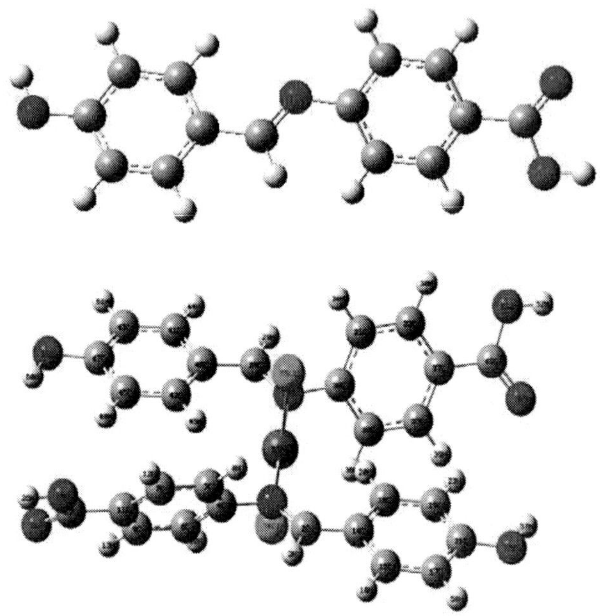

Figure 2. Molecular models of D3 and Pd(D3)2

The catalytic properties for oxidative carbonylation of a series of Schiff bases palladium complexes were investigated and compared. For the Schiff bases palladium complexes with better catalytic activity, effect of reaction condition on the catalytic performance of Schiff bases palladium complexes were investigated. Typical results were as followed.

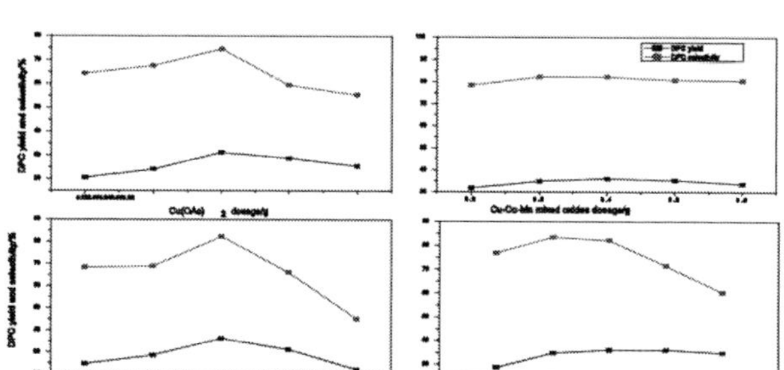

Figure 3. Effect of reaction condition on the catalytic performance for oxidative carbonylation of phenol over Pd(D3)2/H?

Conclusion

Shicff base palladium complexes were prepared successfully by microwave assisted solvothermal method and they achieved highly efficient catalytic performance in oxidative carbonylation.

References

(1) [a] Y. Ono, Appl. Catal. A, 1997,155: 133; [b] A.A. Shaikh, S. Sivaram Chem. Rev., 1996,96: 951.

(2) Cheng, Q et al., Acta Petro. Sin. (Petro. Process. Sect.) (in Chinese), 2014, 30(2): 224.

(3) Cheng, Q et al., Petrochem. Tech.(in Chinese),2011, 40(7): 712.

FLAME RETARDANT ADDITIVES FOR LITHIUM SULFUR BATTERIES

Wang Shen, Fengjiao Lin, Jun Yang, Jiulin Wang

Shanghai Electrochemical Energy Devices Research Center,
Department of Chemical Engieering, Shanghai Jiao Tong University,
Shanghai 200240, China
wangjiulin@sjtu.edu.cn

1. Introduction

Currently, rechargeable lithium ion batteries (Lion) are being widely used as power sources for portable devices widely, and at a relatively small scale for electric vehicles (or hybrid) [1]. Li/S battery has became a hot research topic due to its high theoretical energy density up to 2600 Wh kg^{-1}, 3-5 folds higher than those of state-of-the-art Lion batteries [2]. Compared to Lion, Li/S battery might suffer more serious safety issues, because 1) the lithium dendrites; 2) the high volatility and flammability of the electrolytes; 3) the cathode generally composed of elemental sulfur and conductive carbon, which are combustible and explosive.

In traditional Lion, lots of flame inhibitors have been reported to improve the thermal stabilities of conventional organic electrolytes, which can be divided into four main types according to their molecular structures: alkyl phosphates, fluorinated alkyl phosphates, ionic liquids, and phosphazenes. Although most of these inhibitors can effectively improve the safety properties of Lion at a certain extent, the amount of additive required to achieve nonflammability obviously destroys the electrochemical performances of Lion, for example, deteriorating cycling stability or weakening power rate capability [3, 4].

Considering potentially the extensive usage of Li/S battery as large size, it is important to figure out the safety issues of Li/S battery before its practical applications. In our previous work, safe sulfur cathodes and stable cycle performance have been achieved by creatively combining elemental sulfur with non-flammable pyrolyzed polyacrylonitrile, abbreviated as S@pPAN [5, 6]. On the basis of the above considerations, we proposed a concept to build up safe Li/S batteries with both flame-inhibiting electrolytes and non-flammable cathodes with S@pPAN composite materials.

2. Experimental

2.1 Preparation of sulfur-based composite cathode materials

Elemental sulfur and polyacrynonitrile (PAN) were simply mixed with ethanol served as dispersant, then ball milling for 12h, following drying and heating at 300 under N_2 atmosphere, resulting in the composite S@pPAN. The sulfur content in S@pPAN was 46 wt.% which was determined by elemental analysis (Vario EL III elemental analyzer, Elemental, Germany).

2.2 Coin-cell assembly and electrochemical performance measurements

The cathode was fabricated by mixing 80 wt.% S@pPAN composite, 10 wt.% Super P and 10 wt.% carbonyl-β-cyclodextrin as a binder, and distilled water as dispersant. The slurry was stirred for 4h and then uniformly coated onto a piece of Al foil (with a carbon coating), dried and cut into small disks with a diameter of 12 mm; the weight loads of the foil disks were ca.2.0 mg cm^{-2}. The anode consisted of Li foil with a diameter of 16 mm and a thickness of 0.2 mm. The 1M $LiPF_6$/EC+DMC (1:1, v/v) was used as blank electrolyte, in which flame retardants (FR) were directly added. With PE film as separator, 2016 type coin cells were assembled in a glove box (MBraun, Germany). The molecular structures of FRs were shown in **Figure 1**. The charge/discharge cycles of coin cells were performed in a cut-off voltage limit of 1.0-3.0 V on a LAND cycler (Wuhan, China).

2.3 Flame resistance, conductivity and thermal stability

Self-extinguishing (SET) time was chose to evaluate the flammability of electrolytes. A cotton-ball wick was immersed in the as-prepared electrolyte to absorb 0.2-0.3 g liquid and was subsequently ignited in a fume hood, and the time for the flame to extinguish was recorded. The results were normalized against the electrolyte mass, and the SET tests were repeated at least six times.

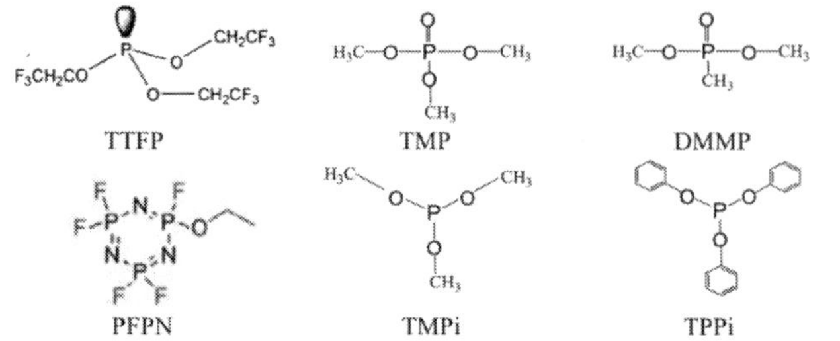

Figure 1. Molecular structures of six FRs

3. Results and Discussion

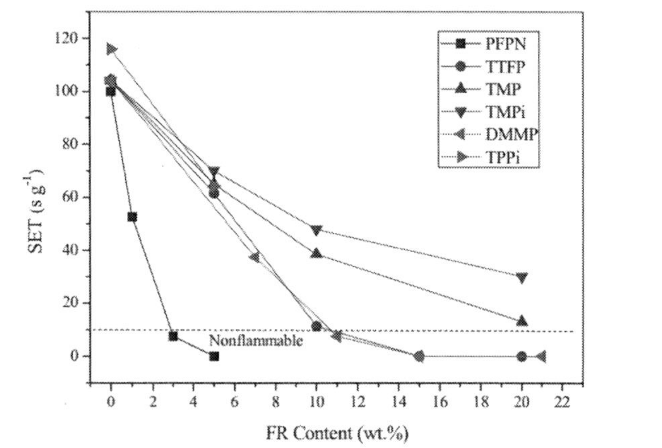

Figure 2. Effect of FR addition on SET of the carbonate electrolyte.

Safety is a critical property for a battery, which cannot be over emphasized. Because of the high volatile organic solvents, the electrolytes composed of carbonates such as ethylene carbonate (EC) and dimethyl carbonate (DMC) are highly flammable. Here we adopted the method called Self-Extinguishing Time (SET) to characterize the flammability of the liquid electrolytes. **Figure 2** shows the effect of FR contents on the self-extinguishing time of the carbonate electrolytes. For the blank electrolyte, the SET are 100 - 116 s g^{-1} with measurement errors. After the addition of FRs, the SET all remarkably reduced. Particularly three FRs, i. e. PFPN, TTFP and DMMP, make the carbonate electrolyte nonflammable with SET close to zero. The SET curve of PFPN-based electrolyte shows a sharp decrease with the content of PFPN which demonstrates an excellent flame-retarding efficiency. After 1 wt% PFPN was added, The SET of the electrolyte dramatically decreased from ca. 100 s g^{-1} to 52.6 s g^{-1}, then sharply decreased to 7.5 s g^{-1} at 3 wt% PFPN addition. When the content of PFPN increased to 5 wt% or more, the SET value decreased to zero, indicating the electrolyte can not be ignited. Such a small amount of flame-retarding additives can effectively make the electrolytes totally non-flammable. Some studies suggested that when phosphorus and nitrogen both existed in the molecular of flame-retardant, the flame-retardant efficiency can be greatly improved. When fluorinate compounds being used as flame-retarding additives, it decomposes and generates free radicals (F·) to actively capture other free radicals emitted by the burning electrolyte (usually H· and OH·) to stop the combustion. Thus, the flame retardancy can be further increased by combining the halogen with the phosphazene backbones.

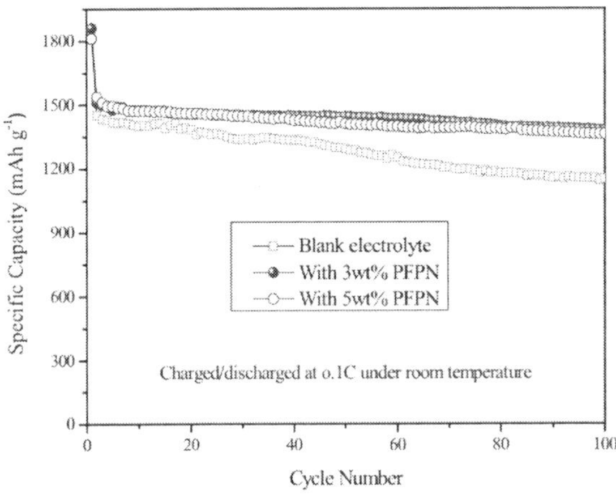

Figure 3. Effect of PFPN on the cycling performance of sulfur cathode.

It is interesting to find that elemental sulfur reacts with polyacrylonitrile (PAN) to form a nonflammable composite S@pPAN [5]. Combined with aforementioned nonflammable electrolyte, it is feasible to construct safe Li/S batteries. The presented work mainly focuses on the influence of PFPN additive on the electrochemical performances of the composite S@pPAN. The specific capacities presented in this work were calculated based on the sulfur mass. The addition of PFPN hardly affected the charge/discharge curves of the S@pPAN composite cathode material, which delivered an initial discharge capacity of about 1844 mAh g^{-1}(sulfur), thus exceeding the theoretical value of 1672 mAh g^{-1} for sulfur. The possible explications are that the conjugated backbone of conductive polymer derived from the reaction between PAN and S contributes to additional capacity; interfacial reactions also might provide partial capacity [5]. A reversible capacity of about 1512 mAh g^{-1}(sulfur) was obtained in the second cycle, thus indicating a sulfur utilization approaching 90.4%.

Cycle performances of cells consist of S@pPAN composite and electrolytes containing different amounts of PFPN are shown in **Figure 3.** It is clear that the addition of PFPN remarkably improved the cycle stability of the S@pPAN composite. The capacity gradually decayed in the blank electrolyte (1M $LiPF_6$/EC+DMC), with only 1152 mAh g^{-1}(sulfur) left after 100 cycles. In contrast, the composite cathode in the electrolyte containing 3 wt% PFPN demonstrated 1378 mAh g^{-1}(sulfur).

The rate performances of cells also benefit from the presence of PFPN. The batteries with different PFPN contents were recharged and discharged at different rates, namely, 0.2C, 0.5C, 1C, 2C, 3C, 5C, 7C, 0.2C. The battery with blank electrolyte was more sensitive to rate variations. In low rate zone (<1C), the differences among three electrolytes are negligible. Significant discrepancies emerged in the high rate zone, where the capacity of the battery with a blank electrolyte faded to 660 mAh g^{-1}(sulfur) at 7C, while the 3 wt% PFPN doped battery maintained a capacity of 1100 mAh g^{-1}(sulfur) at 7C. The battery with 3 wt% PFPN addition in the electrolyte illustrates the best rate performance, coinciding with the recharge-discharge patterns where the battery with 3 wt% PFPN in the electrolyte reduces the polarization voltage most significantly. Similar to the battery with the blank electrolyte, the battery with 5 wt% PFPN additive also suffers a rapid capacity loss at a high rate. As the current density increases, the rate performance of the batteries is primarily determined by the transportation of lithium ions in the electrode, interface, and electrolyte. The electrolyte with 5 wt% PFPN possesses a relatively high viscosity, which hinders fast lithium ion transportation in the electrolyte and therefore deteriorates the rate performance. All batteries with PFPN addition recovered well from high rate test, indicating that the reversibility of the electrode is well preserved by PFPN.

Conclusions

In summary, three FRs were investigated to modify the thermal properties of carbonate electrolytes for lithium sulfur batteries. Particularly for PFPN, least addition, best performances. The flammability issue of the Li/S battery was effectively eliminated by using flame-inhibiting carbonate electrolytes derived from the addition of PFPN and S@pPAN composite cathode with unique nonflammable properties. PFPN not only served as a flame retardant but also modified the interfacial film on the S@pPAN composite cathode to accelerate the electrochemical reactions. With optimized content of PFPN (3 wt%), the batteries with nonflammable electrolyte demonstrated outstanding stable cycling performance and good high power rates.

Acknowledgements

This work was supported by the National Natural Science Foundation of China (51272156, 21333007), City Committee of Science and Technology Project of Shanghai (14JC1491800) and New Century Excellent Talents in University (NCET-13-0371).

References

(1) Bruce P. G.; Freunberger S. A.; Hardwick L. J.; and Tarascon J. M. *Nat.Mater.* **2012**,
(2) *11*, 19.
(3) Ji X.; and Nazar L. F. *J. Mater Chem.* **2010**, *20*, 9821.
(4) Xu K.; Zhang S.; Allen J. L.; and Jow T. R. *J. Electrochem. Soc.* **2002**, *149*, A1079.
(5) Ota H.; Kominato A.; Chun W.-J.; Yasukawa E.; and Kasuya S. *J. Power Sources* **2003**, *119*, 393.
(6) Wang J.; Lin F.; Jia H.; Yang J.; Monroe C. W.; and NuLi Y. *Angew. Chem. Int. Ed.* **2014**, *53*, 10099.
(7) Wang J. L.; Yang J.; Xie J. Y.; and Xu N. X. *Adv.Mater.* **2002**, *14*, 963.

ENHANCED DESALINATION AND KINETICS USING CYCLOPENTANE HYDRATES IN WATER-IN-OIL EMULSIONS

LV Yining[1], Gong jing[1], Chen Guangjin[2*]*

1. National Engineering Laboratory for Pipeline Safety, China University of Petroleum—Beijing, Beijing 102249, People's Republic of China, ydgj@cup.edu.cn.; 2. State Key Laboratory of Heavy Oil Processing, China University of Petroleum, No.18 Fuxue Rd., Changping District, Beijing 102249, CHINA, gjchen@cup.edu.cn.

Hydrate based process using water-immiscible formers represents a concept for a new desalination technology under development. In this work, experiments were conducted to study the behaviors of formation kinetics and salt separation in order to improve the efficiency of the hydrate based desalination process. Cyclopentane hydrates were formed at atmospheric pressure and either 0.1°C or 4°C from water-in-oil emulsions at various volume fractions of water (from 10% to 50%) with salinity of 3.5 wt.% or 5 wt.%. Results revealed that water conversion could be raised in water-in-oil emulsions depending on conversion time, subcooling and water content. The purification of water could be enhanced with successive secondary treatment of washing. The salt removal efficiency and the output to input ratio for the washing method were further optimized for the investigated conditions.

LOADING OF IRON AND CERIUM NANOPARTICLES ON ACTIVATED CARBON FOR THE DESULFURIZATION OF THIOPHENES

Gaddafi I. Danmaliki, Tawfik A. Saleh

Department of Chemistry, King Fahd University of Petroleum & Minerals, Dhahran 31261, Saudi Arabia

Abstract

Iron and cerium nanoparticles were loaded on activated carbon (AC) derived from waste rubber tires. The composite of AC/Fe was evaluate in adsorptive desulfurization of thiophene, benzothiophene (BT), and dibenzothiophene (DBT). The surface structures, morphologies and pore structures were characterized using a scanning electron microscope coupled with energy dispersive spectroscope (SEM/EDX), a Fourier transform infrared spectroscope (FT-IR), an X-ray diffraction spectroscope, and a Brunauer-Emmett-Teller (BET) surface area analyzer.

Introduction

There are several methods and materials to reduce sulfur compounds in fuels including thiophene, benzothiophenes and dibenzothiophenes, which should be removed in order to eliminate the sulfur contents and thus enhance the quality of the produced fuels [1-4]. Various methods have been emerging toward minimizing sulfur content in transportation fuels to below 10 ppm. Desulfurization techniques such as hydrodesulfurization, oxidative desulfurization, bio-desulfurization, and adsorptive desulfurization are being investigated to produce ultra-clean fuels [5]. A promising method is adsorption, employed to remove trace impurities, such as the removal of trace amounts of aromatics from aliphatic [6] and it is the most common hydrodesulfurization alternative method currently used to achieve ultra-clean fuels [7]. Adsorptive desulfurization using porous materials is considered an efficient and economical way for removing organosulfur compounds due to its low-energy consumption, the ambient operation temperature, and atmospheric pressure without using pressurized hydrogen gas because of availability of regeneration of the spent adsorbent and broad availability of adsorbents [8-10]. Metal oxides including titania are highly reactive towards sulfur compounds, such as thiols. The process is based on thiophenes chemistry in which metal oxides react with thiophenes to form solid-metalthiolates. Study of comparative analysis of removing sulfur from crude oil was tested using activated manganese dioxide, and activated zinc oxide. The result showed that, activated manganese dioxide proved more efficient during the adsorption of sulfur compounds from crude oil when compared to activated zinc oxide [11]. Metal oxides demonstrated to be efficient adsorbents for the removal of dibenzothiophene and 4,6-dimethyldibenzothiophene from model diesel fuel and diesel fuel. Transition metals supported on base oxides are used as the adsorbents. Nickel zinc oxides is an ideal adsorbent, in which the zinc oxide acts not only as an acceptor of sulfur released during the regeneration of sulfided nickel species, but also acts as a co-catalyst for the hydrogenation of sulfur-containing compounds over the surface of nickel species [12-14].

Desulfurization (removal of sulfur) is a significant step in the refining process of crude oil because the burning of sulfur-containing compounds is the primary reason for acid rain and environmental pollution. Sulfur also poisons catalyst and corrode refining equipment. These problems are the reasons for an ongoing effort to lower the legal limits of sulfur in all kinds of petroleum products. In the US, the Environmental Protection Agency (EPA) mandated 15 ppm (wt) highway diesel.1,2 The conventional hydrodesulfurization (HDS) technique cannot achieve 15 ppm ultra-low sulfur diesel regulation set by so many countries to curtail the impacts of sulfur to health, environment and the economy. Additionally, HDS is not effective in the removal of refractory sulfur compounds e.g. thiophene, benzothiophene (BT), and dibenzothiophene (DBT).3 Hence the need for viable, cost-effective and efficient techniques. Alternative methods to supplement the drawbacks of HDS include oxidative desulfurization, ionic liquid desulfurization, adsorptive desulfurization (ADS) and biodesulfurization.4-8 Direct adsorption of the recalcitrant sulfur compounds, which will be the subject of this paper, is an appealing process because it can be carried out at ambient temperature and pressure.

Experimental

Preparation of the composite. 7.0 g of AC was dispersed in 150 mL of deionized water, 100 mL ethanol, and 20 mL ethylene glycol and stirred for five hours. Then, a 5 mL solution containing 0.88 g of Cerium nitrate hexahydrate solution was added drop-wise into the dispersed AC and maintained at a pH of 8-9. The mixture was refluxed at 90°C for 5 h. The precipitate was washed and dried at 110°C overnight. The dried material was dispersed in 150 mL deionized water, 100 mL ethanol and 20 mL ethylene glycol. 13 mL solution containing 5.06 g of Cerium nitrate hexahydrate was added to the mixture and stirred for 2h at pH of 8-9. The set up was refluxed at 90°C for five hours, and the precipitate was filtered, washed, dried at 110°C overnight and then calcined at 350°C for 4 h to ensure adequate localization of the metals to the surface of the AC. The composite was characterized using various instruments including field emission scanning electron microscope coupled with Energy Dispersive Spectroscope (EDX), and thermogravimetric analyzer (TGA), Fourier transform infrared absorption spectroscopy (FTIR), and Brunauer Emmett and Teller (BET) surface area analyzer

Evaluation tests. The adsorption experiment was carried out in both batch and fixed bed models. A model diesel fuel was prepared by adding the refractory sulfur compounds viz: thiophene, BT and DBT in hexane (85%) and toluene (15%). The sulfur concentration in the model fuel was 150 ppm prepared by dissolving 50 mg each of thiophene, BT, and DBT in 1 L of hexane/toluene mixture. The resulting mixture was continuously shaken at room temperature until equilibrium. Aliquots were taken from the system at the pre-determined time intervals and analyzed by gas chromatography-sulfur chemiluminescence detector (GC–SCD) (Model 7890A)system (Agilent) equipped with autosampler (7693) and splitless injector. The kinetics of organosulfur compounds adsorption on composite were analyzed using pseudo first- order, pseudo-second-order and intraparticle diffusion models. The conformity between experimental data and the model predicted values was expressed by the correlation coefficients (R^2).

Results and Discussion. The prepared iron/cerium loaded on activated carbon as a composite was characterized by SEM, EDX, BET and FTIR. Figure 1 shows the SEM image and EDX spectrum of the composite which indicates the presence of the iron and cerium in addition to the carbon and oxygen. EDX, the quantitative analysis of the components of the composite shown also the presence of the both elements.

The surface area plot is depicted in Figure 2. The surface area and pore volume were found as 430.44 m^2/g and 0.64 cm^3/g respectively. FTIR spectrum showed peak at 550.84 cm^{-1} signifies the presence of Ce-O within AC/Ce sample, but it did not appear in AC/CeO_2/Fe because the concentration of Ce in that adsorbent was minuscule. The band that appeared around 1,080 cm^{-1} is assigned to the Ce–O–C interaction. The small peaks that appeared at 798 and 895 cm^{-1} are attributed to the Fe-O bending vibrations, while the peak at 624 cm^{-1} is due to the Fe-O stretching vibrations within the AC/Fe and $AC/CeO_2/Fe$.

The composite showed very high adsorptive desulfurization performance, as shown in Figure 3.

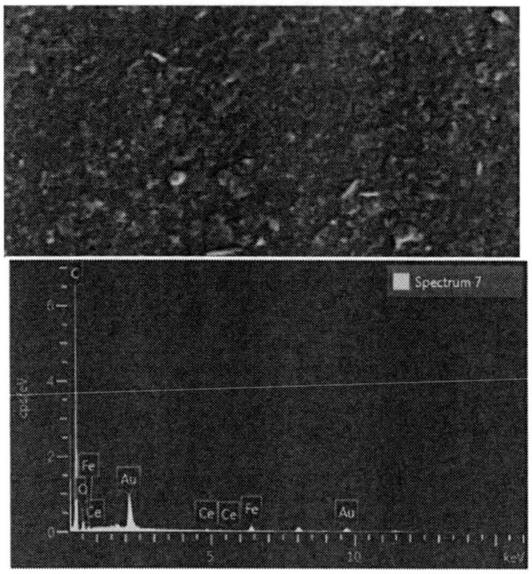

Figure 1. SEM image and EDX spectrum of the composite

Conclusions

Here, we have reported on the adsorptive desulfurization activity of the composite prepared from iron and cerium nanoparticles loaded on activated carbon. The composted showed a good removal of the thiophene with order dibenzothiophene > benzothiophene ≥ thiophene.

Acknowledgement. The authors would like to acknowledge the support provided by King Abdulaziz City for Science and Technology (KACST) through the Science & Technology Unit at King Fahd University of Petroleum & Minerals (KFUPM) for funding this work through project No. 13-PET393-04 as part of the National Science, Technology and Innovation Plan (NSTIP).

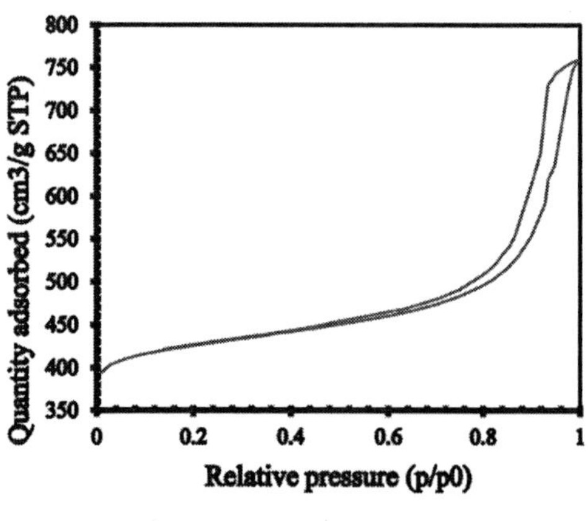

Figure 2. BET plot of the composite

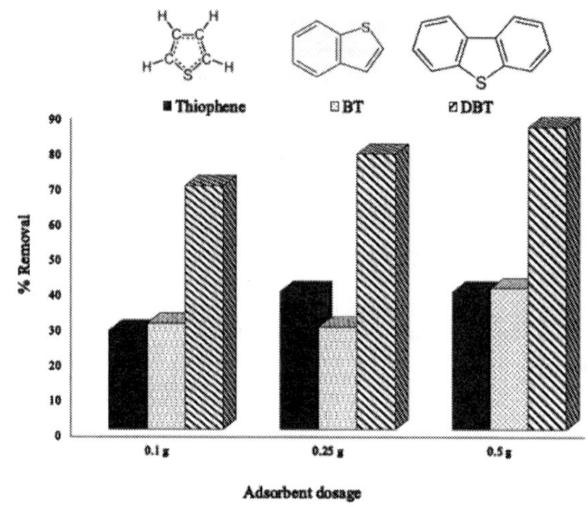

Figure 3. Percent removal of thiophene, BT and DBT by the prepared composite

References

(1) J. Matos, J. Laine, J.M. Hermann, J. Catal. 200 (2001) 10–20.
(2) J. H. Gary and G. E. Handwerk, Petroleum Refining, Technology and Economics, Second Edition, Marcel Dekker, New York, 1998.
(3) C. Song, X. Ma, Applied Catalysis B: Environmental, 41, 207–238, 2003.
(4) C. Song, Catalysis Today, 86, 211 - 263, 2003.
(5) R.T. Hernandez- Yang, F.H. Yang, A. Takahashi and A.J. Maldonado,US patent 2004/0040891 US 2004/0044262 A1. 2004.
(6) M. Tymchyshyn, Lakehead University, 2008.
(7) Q. Wang, X. Liang, W. Qiao, C. Liu, X. Liu, L. Zhan, L. Ling, Fuel Processing Technology, 90, 381–387, 2009.
(8) J. Weitkamp, M. Schwark and S. Ernst, J. Chem. Soc., Chem. Commun., 1133–1134, 1991.
(9) D.L. King, T. Flynn, Society of Automotive Engineers, Detroit. MI, 2000.
(10) S. H. Salem, Ind. Eng. Chem. Res., 33, 336-340, 1994.
(11) A. A. Adekanmi and A. Folorunsho, Res.J.Chem.Sci., 2, 14-20, 2012.
(12) L. Huang, G. Wanga, Z. Qina, M. Donga, M. Dua, H. Gea, X. Lia, Y. Zhaoc, • J. Zhangc, T. Huc, and J. Wang, Appl. Catal. B: Environmental, 106, 26–38, 2011.
(13) J. Fan, G. Wang, Y. Sun, C. Xu, H. Zhou, G. Zhou, and J. Gao, Ind. Eng. Chem. Res., 49, 8450–8460, 2010
(14) I. A. H. Al Zubaidy, F. B. Tarsh, N. N. Darwish, B. S. S. Abdul Majeed, A. Al Sharafi and L. A. Chacra, JOCET, 2013, 1, 66-68.

NOVEL SYNTHESIS OF BISMUTH AND BISMUTH ANTIMONY NANOWIRE COMPOSITES FOR THERMOELECTRIC ENERGY CONVERSION

Koen Vandaele[1,2], *Michael J. Adams*[2], *Bin He*[2], *Joseph P. Heremans*[2], *Pascal Van Der Voort*[1] *and Klaartje De Buysser*[1]

[1] Department of Inorganic and Physical Chemistry, Ghent University, Krijgslaan 281 – S3, 9000 Gent, Belgium
[2] Department of Mechanical and Aerospace Engineering, The Ohio State University, 201W 19th Ave, OH 43210, USA
koen.vandaele@ugent.be

We present a new procedure for the synthesis of bismuth based nanowires embedded in a mesoporous silica matrix. The procedure started with a solution based impregnation step used to load bismuth salts into the channels of a porous silica template material. This new impregnation method enables us to obtain an extremely high pore loading, while limiting the amount of material deposited externally. Crystalline nanostructures were obtained in a second step consisting in a low temperature reduction treatment of the impregnated metal salts between 220°C and 260°C. Typically, a high pore loading of about 40 volume % with metallic bismuth was achieved. Because the silica templates can be synthesized in various architectures, both nanowires and interconnected nanowire networks could be fabricated. Enhanced thermoelectric properties ascribed to[1] and observed on[2] bismuth nanowires could be obtained in a bulk material after compacting the nanocomposite powders. The nanostructures were characterized by TEM, N2-sorption, XRD and XRF analysis. The thermoelectric transport properties of both bismuth and bismuth antimony nanowire composites were compared to the bulk material.

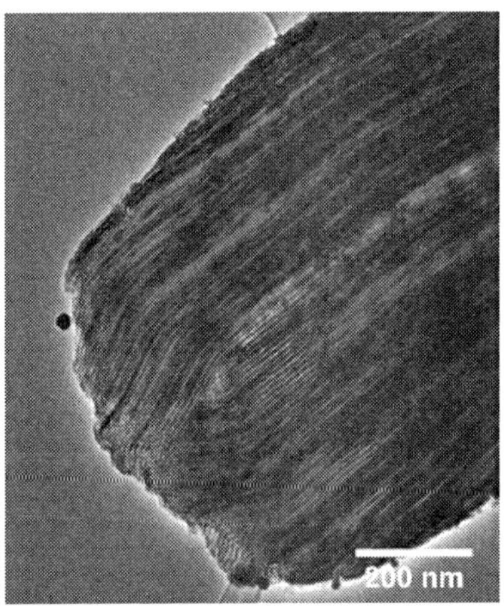

Figure 1. Bismuth nanowires embedded in SBA-15 mesoporous silica.

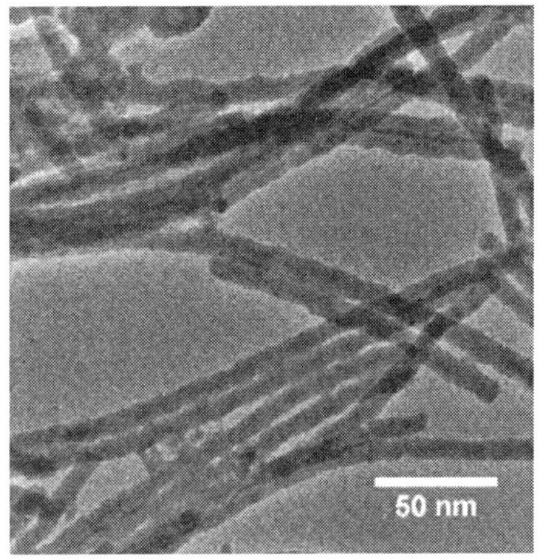

Figure 2. Bismuth nanowires after template removal.

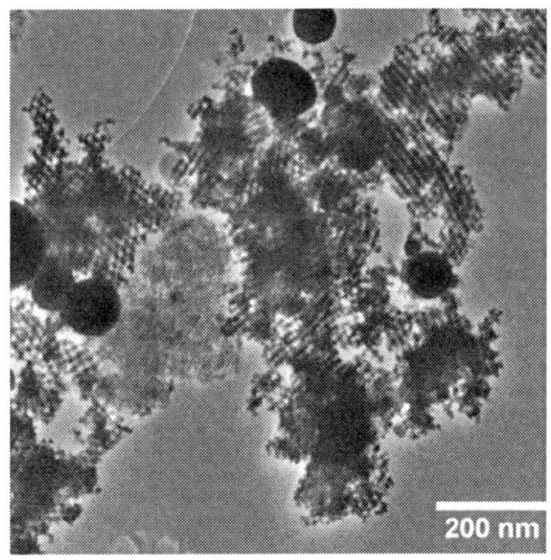

Figure 3. Bismuth nanowire network templated with KIT-6 mesoporous silica.

Acknowledgement. The author would like to acknowledge the IWT Flanders and the FWO for financial support.

References

(1) L. D. Hicks and M. S. Dresselhaus, Thermoelectric figure of merit of a one-dimensional conductor, Phys. Rev. B **47**, 16631(R) (1993)
(2) Joseph P. Heremans, Christopher M. Thrush, Donald T. Morelli, and Ming-Cheng Wu, Thermoelectric Power of Bismuth Nanocomposites, Phys. Rev. Lett. **88**, 216801 (2002)

SYNTHESIS OF BETA MODIFIED TUD-1 MATERIAL AND ITS APPLICATION IN THE CATALYSTS FOR THE HYDRODESULFURIZATION OF FCC DIESEL

Zesheng Xia, Aijun Duan, Zhen Zhao*, Honglei Zhang, Jiyuan Fan*

State Key Laboratory of Heavy Oil Processing, China University of Petroleum, Beijing, 102249, P. R. China
*Corresponding author: duanaijun@cup.edu.cn; zhenzhao@cup.edu.cn

Introduction

TUD-1[1,2], owing to its special structure[2,3], is an ideal candidate for catalysis application since its mesoporous structure facilitates the diffusion process.

Because of the weak acidity of purely siliceous TUD-1 materials[4], which can not provide sufficient acid sites in the process of hydrotreating of diesel, meanwhile considering that Beta zeolite has been widely used in many industrial reactions, the synthesis of Beta modified TUD-1 composite material using Beta zeolite microcrystalline emulsion as silicon and aluminum source and the hydrotreating performance of FCC diesel of corresponding catalysts are the subject of this paper. Beta modified TUD-1[5] material was abbreviated as BT.

Experimental

Beta zeolite microcrystalline emulsion synthesis. A mixture of an appropriate amount of tetraethylammonium hydroxide and tetraethyl orthosilicate after stirring for 30 minutes was denoted as solution 1. The composition of solution 2 was sodium hydroxide, sodium metaaluminate, tetraethylammonium hydroxide and water. In a typical synthesis process, the solution 2 was drop wise added to solution 1, after stirring for 4 h, followed by a hydrothermal treatment in a Teflon-lined autoclave at 140℃ for various hour, then Beta crystallite emulsions with different crystallization degrees were obtained.

Beta modified TUD-1 synthesis. A controlled number of triethanolamine was dropped into the Beta zeolite microcrystalline emulsion, after stirring for 2 h, followed by the addition of some tetraethylammonium hydroxide, keep stirring until a clear gel was obtained, the gel needed to be dried at 100℃ to remove the volatile components, followed by heat treatment at 180℃ for 4 h, finally calcined in air at 600℃ for 10 h with a ramp of 1℃/min to remove the organic compounds. After the above steps, a powder-like Beta modified TUD-1 was obtained. In this paper, a series of BT materials were synthesized using 5 kind of Beta zeolite microcrystalline emulsions with different crystallization time of 8 h, 12 h, 16 h, 20 h and 24 h respectively.

Catalyst preparation. H-type materials of BT were obtained by ion-exchanged with a 1.0 M NH_4Cl aqueous solution. The mecha - nical mixture of Al_2O_3 and BT material was impregnated by Ni and Mo species to prepare catalysts through a two-step impregnation method. The corresponding supported catalysts were obtained with the active metal contents of MoO_3 15.5wt%, and NiO 3.5wt%.

Characterization Methods. XRD analyses were recorded on Shimadzu X-6000 type X-ray powder diffraction. TEM measurement was obtained on a JEO JEM 2100 type transmission electron micro-scope Instrument. N_2 physisorption characterization was performed on Mike company Micromeritics ASAP 2010 sorption analyzer.

Results and Discussion

Figure 1 shows the powder X-ray diffraction patterns of BT materials with different crystallization time. The materials of a, b and c have one intense peak around 0.7-1.2 in 2θ, indicating that the materials possess a typical mesoporous structure. Moreover, with the increase of crystallization time, the peaks shift to the larger angle, suggesting that the mesopore diameters become smaller. The peaks of d and e are less intense relatively, the reason is that with the increase of crystallization time the size of zeolite nanoparticles become too large and the mesopore structure become less regular beyond XRD detection limitation.

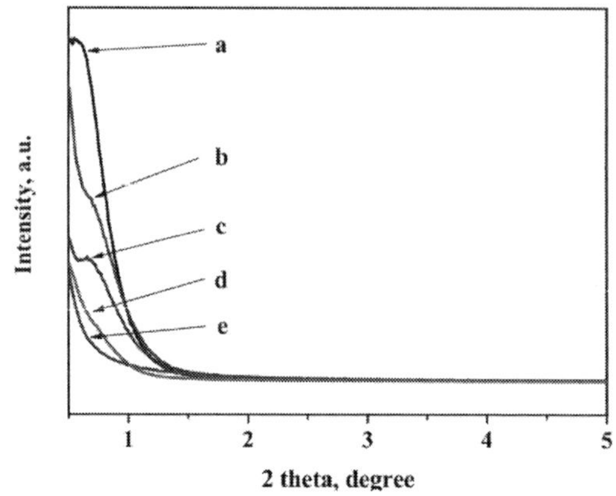

Figure 1. Small angle XRD patterns of BT series materials, (a) BT-8 h; (b) BT-12 h; (c) BT-16 h; (d) BT-20 h; (e) BT-24 h.

The large angle XRD patterns of pure Beta zeolite and BT series materials are shown in **Figure 2**. A broad peak is centered around 24 in 2θ ascribed to the amorphous silica in all the samples. With the increase of crystallization time, the diffraction peaks around 7.8 and 22.4 in 2θ which belong to Beta can be detected, suggesting the Beta nanoparticles formed a relatively large size zeolite sieve to be detected as reported in other papers.

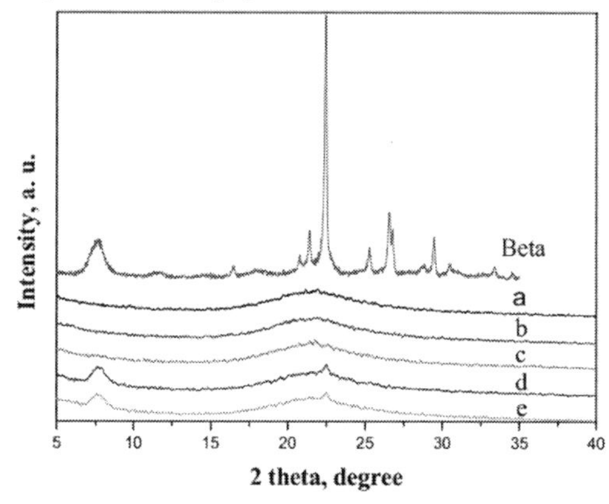

Figure 2. Large angle XRD patterns of BT series materials, (a) BT-8 h; (b) BT-12 h; (c) BT-16 h; (d) BT-20 h; (e) BT-24 h.

As shown in **Figure 3**, the transmission electron microscopy images of the BT materials display a sponge like and interconnecting mesoporous network structure similar to the pure silica TUD-1 mesoporous materials, indicating the successful synthesis of BT series materials. With the increase of crystallization time, the crystal lattice fringe of Beta zeolite can be found in the marked circular area of the latest TEM image.

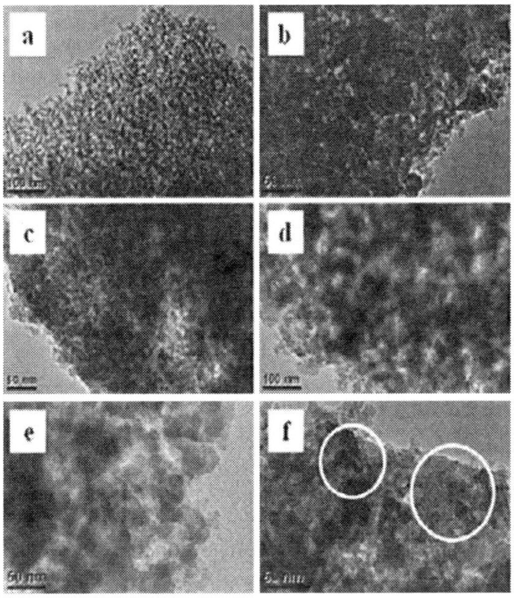

Figure 3. TEM images of BT materials, (a) TUD-1; (b) BT-8 h; (c) BT-12 h; (d) BT-16 h; (e) BT-20 h; (f) BT-24 h.

The HDS results using FCC diesel oil as feedstock over different catalysts are shown in **Table 1.** As can be seen in the table, the BT series catalysts exhibited higher HDS conversions than the catalyst supported on alumina. The catalyst of NiMo/Al_2O_3-BT-20h displays the highest HDS efficiency.

Table 1. HDS activities of NiMo/Al_2O_3-BT series catalysts (Catalyst = 81 wt% supports + 15.5 wt% MoO_3 + 3.5 wt% NiO; Supports = 30 wt% BT + 70% γ-Al_2O_3)

Catalyst	Crystallization time(h)	HDS(%)
NiMo/Al_2O_3-BT-8h	8	95.6
NiMo/Al_2O_3-BT-12h	12	96.5
NiMo/Al_2O_3-BT-16h	16	96.8
NiMo/Al_2O_3-BT-20h	20	98.8
NiMo/Al_2O_3-BT-24h	24	97.4
NiMo/Al_2O_3	--	94.3

Conclusions

In this work, a series of Beta modified TUD-1 materials were successfully synthesized by nanocrystal self-assembly method using Beta zeolite microcrystalline emulsion as silicon and aluminum source. The XRD characterization results showed that the Beta modified TUD-1 materials not only had typical pure siliceous TUD-1 X-ray diffraction pattern, but also had typical Beta zeolite characteristic X-ray diffraction pattern, suggesting that a number of nanocrystals of Beta were successfully assembled mesoporous structure of TUD-1 materials. The TEM images of the materials indicated that the series BT materials had the similar pore structure of pure silica TUD-1 mesoporous materials. With the increase of crystallization time, the XRD pattern displayed the typical peaks of Beta. all the BT series catalysts exhibited high HDS conversions. The catalyst of NiMo/Al_2O_3-BT-20h displays the highest HDS efficiency.

Acknowledgment. This work was financially supported by the National Natural Science Foundation of China (No.21276277 and U1463207), Opening Project of Guangxi Key Laboratory of Petrochemical Resource Processing and Process Intensification Technology (2015K003), CNPC Key Research Project and CNOOC project.

References

(1) Y. Wan, D. Zhao, Chem. Rev., 2007, 107 (7): 2821-2860.
(2) J. C. Jansen, Z. Shan, L. Marchese, W. Zhou, N. v. d. Puil and T. Maschmeyer, Chem. Commun., 2001, 8 (8): 713-714.
(3) X. S. Zhao, F. Su, Q. Yan, W. Guo, X. Y. Bao, L. Lv and Z. Zhou, J. Mater. Chem., 2006, 16 (16): 637–648.
(4) M S Hamdy, M Guido. Chemcatchem, 2013, 5(10): 3156-3163.
(5) Z Wang, J Fu, Y Deng, et al Rsc Advances, 2014, 5: 5221-5230.

SYNTHESIS AND HYDRODESULFURIZATION PROPERTIES OF NIMO CATALYSTS SUPPORTED ON NOVEL MICRO-MESOPOROUS MATERIAL BETA-KIT-5

Honglei Zhang, Yanyu Li, Aijun Duan[1,2]*, *Zhen Zhao*[1]*, *Jiyuan Fan, Zesheng Xia,*

[1] China University of Petroleum, State Key Laboratory of Heavy Oil Processing Beijing, 102249 (China)
2 Guangxi Key Laboratory of Petrochemical Resource Processing and Process Intensification Technology, Guangxi University, Nanning, 530004 (China)
* Corresponding author: duanaijun@cup.edu.cn; zhenzhao@cup.edu.cn.

Introduction

The novel mesoporous material KIT-5 with thick pore wall, high specific surface area and well ordered pore structures has attracted widespread attention as new potential support materials. Nevertheless, the weak acidity and poor hydrothermal stability of mesoporous silica materials limit their further application in catalysis. The micro-mesoporous composites combine the advantage of microporous and mesoporous materials, Which will make it favorable to disperse the active components and to provide proper acidity by the introduction of Brønsted acid sites from microporous zeolites, then enhance the properties of relevant HDS(hydrodesulfurization) catalysts. CoMo/Beta-SBA-15 with superior mass transfer ability exhibited a higher activity than CoMo/γ-Al_2O_3 in the hydro-upgrading of FCC gasoline[1].

In this research, BK-x(Beta-KIT-5, x = Si/Al) samples with various Si/Al molar ratios were synthesized and the HDS activities of NiMo/Al_2O_3-BK-x catalysts were evaluated under a certain condition. Then, the optimum ratio of Si/Al was determined and the reason was clearly discussed according to different characterization results.

Experimental

Beta seed solutions(Si/Al = 10, 20, 30, 45) were synthesized by adding 0.19 g of NaOH, 1.11/0.56/0.38/0.25 g of $NaAlO_2$, and 21.43 g of tetraethylorthosilicate(TEOS) into 25.95 g of an aqueous TEAOH solution (25%). Then, the mixture was stirred for 2-4 h at room temperature and heated at 413 K for 24 h. BK-x(x = 20, 40, 60, 90) were synthesized with the addition of certain Beta seed after TEOS by in-situ synthesis method according to the literature[1,2]. ABK-x(Al_2O_3-BK-x) derived from the mechanical mixture of γ-Al_2O_3 and BK-x

NiMo/ABK-x catalysts were prepared via a two-step incipient wetness method with aqueous solutions of ($(NH_4)_6Mo_7O_{24}\cdot 4H_2O$) and ($Ni(NO_3)_2\cdot 6H_2O$), stepwise.

Results and Discussion

Fig. 1 shows the small-angle XRD patterns of BK-x samples. All mesoporous materials especially BK-60 exhibit one well-resolved diffraction peak at 0.5-1.0°, indexed as (111) reflection of cubic *Fm3m* structure, which imply that they all possess highly ordered symmetry and mesoporous structure[2]. Meanwhile, all samples except for BK-20 reveal two weak characteristic signals of Beta zeolite at 7.8° and 22.4° in Fig. 1(inset), indicating that primary or secondary units of Beta were effectively incorporated into the framework of BK-x(x=40, 60, 90). The absence of characteristic diffraction peaks of BK-20 are ascribed to the serious dealumination induced by HCl concentrated solution.

H_2-TPR spectra was applied to determine the reducibility of Mo on the catalysts. All catalysts exhibit three characteristic absorption peaks at 400-800 ℃ (Fig. 2). The band at 430 ℃ is commonly assigned to the reduction of polymolybdate(octahedral) species($Mo^{6+} + 2e^- \rightarrow Mo^{4+}$), whereas the band at 530 ℃ is attributed to the $NiMoO_4$, Moreover, the band at 800 ℃ can be associated with the complete reduction ($Mo^{4+} + 2e^- \rightarrow Mo^0$) of polymeric octahedral, tetrahedral and bulk crystalline MoO_3[3]. Compared with NiMo/Al_2O_3 catalyst, a considerable decrease in the high reduction temperature of tetrahedral Mo species and a strong increase in the hydrogen consumption in low temperature region are observed for NiMo/BK-x, demonstrating the transformation of tetrahedral Mo into octahedral Mo species. NiMo/ABK-60 has the largest shift towards low temperature and also show the highest peak area at 430 ℃, meaning its largest hexa-coordinate Mo species. Hexa-coordinate Mo species are inclined to be reduced and sulfided, which actually improves HDS activity. Therefore, NiMo/ABK-60 catalyst showed the highest activity on the HDS of FCC diesel, of which is up to 97.6% in Table 1. The special activities of catalysts follow the order of NiMo/ABK-60 > NiMo/ABK-90 > NiMo/ABK-40 > NiMo/ABK-20.

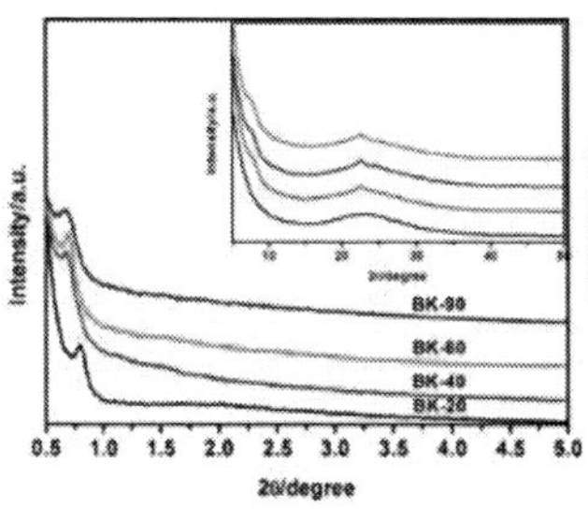

Figure 1. low angle and wide angle XRD patterns of BK-x with different Si/Al molar ratios

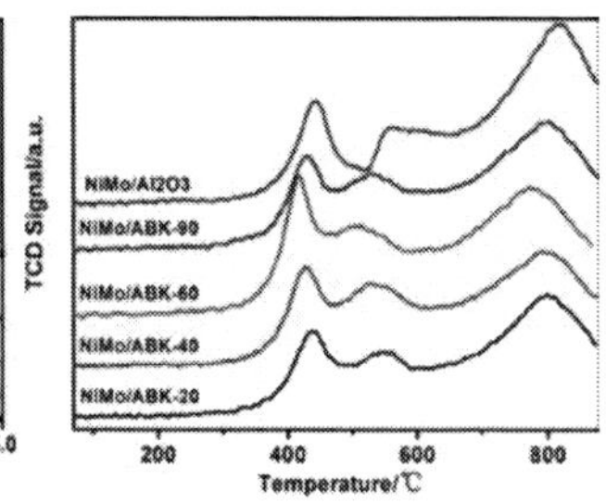

Figure 2. H_2-TPR patterns of the catalysts NiMo/ABK-x with different Si/Al molar ratios

Table 1. HDS results of FCC diesel over the catalysts NiMo/ABK-x with different Si/Al molar ratios

Catalysts	Sulfur content (μg·g^{-1})	Desulphurization rate/%
Feed	1013.0	-
NiMo/ABK-20	80.3	92.1
NiMo/ABK-40	72.5	92.8
NiMo/ABK-60	24.3	97.6
NiMo/ABK90	25.0	97.5

Conclusions

NiMo/ABK-60 catalyst with well-ordered cubic Fm3m structure showed proper acidity, favorable hydrothermal stability and highly dispersed six-coordinated Mo species by the introduction of Beta seed units. Then it exhibited the highest HDS efficiency of 97.6%.

Acknowledgement. The author acknowledges the financial supports from NNSF of China (No. 21276277 and U1463207), CNOOC project, CNPC major project, and Opening Project of Guangxi Key Laboratory of PRPPIT(2015K003).

References

(1) Gao, D. W.; Duan, A. J.; Zhang, X. J. Mater. Chem. A. 2015, 3 (32), 16501-16512.
(2) Kleitz, F.; Liu, D.; Anilkumar, G. M.; J. Phys. Chem. B. 2003, 107, 14296-14300.

(3) Damyanova, S.; Spojakina, A.; Jiratova, K.; Appl. Catal. A: Gen. K. 1995, 125, 257–269.

SYNTHESIS, CHARACTERIZATION, AND CATALYTIC PERFORMANCE OF NOVEL HIERARCHICALLY POROUS MATERIAL IN THE HYDRO-UPGRADING FOR FCC GASOLINE

Jiyuan Fan, Aijun Duan, Zhen Zhao*, Honglei Zhang, Zesheng Xia*

State Key Laboratory of Heavy Oil Processing, China University of Petroleum, Beijing, 102249, P. R. China
*Corresponding author: duanaijun@cup.edu.cn; zhenzhao@cup.edu.cn

Introduction

Recently, micro/mesoporous composite materials which can combine both the advantage of microporous zeolite and mesoporous materials have become the focus of the research, and many composite materials were synthesized and applied in different catalytic processes[1]. In this work, a novel micro/mesoporous composite material L/FDU-12 was successfully synthesized from zeolite L seeds by a two-step hydrothermal crystallization method. Characterization of XRD, N_2 adsorption isotherms, FTIR and TEM were also used to investigate the materials. The L/FDU-12-Al_2O_3 supported CoMo catalyst was prepared and tested in the hydro-upgrading of FCC gasoline.

Experimental

The performed zeolite L precursors was prepared according to the published literature[2].

Typical synthesis of L/FDU-12 (denote as LF) samples was carried out as follows: 2.0 g of triblock copolymers $EO_{106}PO_{70}EO_{106}$ (F127), 2.0 g of TMB and 5.0 g of KCl were dissolved in 120 mL 2 M HCl and stirred for 24 h[3]. Then a certain amount of precursor zeolite L species and TEOS was added to the above solution with rigorous string. Finally, the mixture was stirred at 303K for 24 h and transferred into a Teflon-lined autoclave for further reaction at 373 K for 48 h under static conditions. The final product was collected by filtration, washed, dried at 373 K for 12 h and calcined at 823 K for 6 h in air to remove the organic template.

The corresponding CoMo supported catalyst was prepared by two-step incipient-wetness impregnation of ammonium molybdate and cobalt nitrate. The catalytic performance of catalyst was evaluated in a fixed-bed reactor with 2 g catalyst.

Results and Discussion

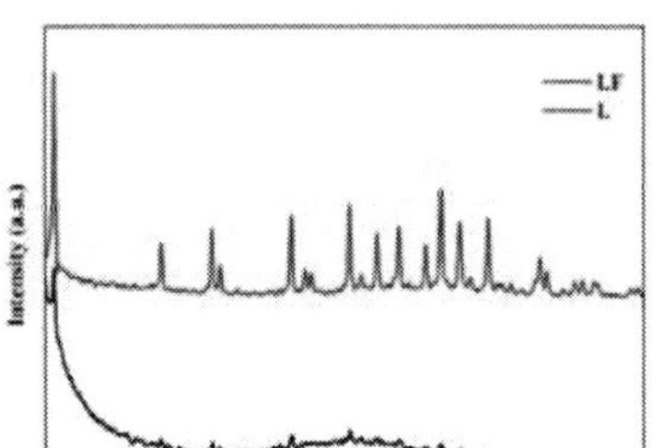

Figure 1. XRD patterns of materials in the wide-angle domain

The wide-angle XRD patterns of L and LF are shown in Figure 1. It can be found that zeolite L and LF possess the same characteristic peaks at 2θ=5.5°, 19.4°, 22.7°, 28.0°, 29.1°, and 30.7°, which can be assigned to the LTL topological structure[4]. Figure 2 shows the N_2 adsorption-desorption isotherms and pore size distribution (PSD) curves of different materials. Similar to the mesoporous silica FDU-12 the micro/mesoporous composite material LF exhibits type IV isotherm with H1 type hysteresis loop and a narrow mesoporous distribution, which are the typical behaviors of the ordered mesostures.

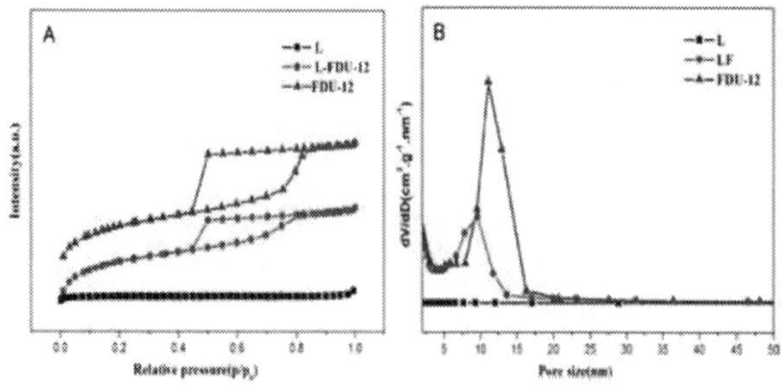

Figure 2. **(A)** N_2 adsorption-desorption isotherms of series materials **(B)** BJH pore diameter distributions of series materials

The TEM images of FDU-12 and L/FDU-12 (Figure 3) show well-ordered cubic 3D mesoporous channels and confirm that both FDU-12 and LF posses the Fm-3m cubic structure[3]. The result agrees with the characterization results of XRD and N_2 adsorption. The catalytic performance for the hydro-upgrading of full-range FCC gasoline shows that the catalyst exhibits a super HDS efficiency (94.7%) and with a low RON loss (0.8 unit).

Figure 3. TEM images of the synthesized materials

Conclusions

A novel micro/mesoporous composite material L/FDU-12 was successfully synthesized via a two-step hydrothermal method. The characterization results show that LF possesses the same mesoporous structure as FDU-12 and contains L zeolite crystals. The catalyst CoMo/LF-Al_2O_3 exhibited good balance in HDS, hydroisomerization and aromatization.

Acknowledgement. This work was financially supported by the National Natural Science Foundation of China (No.21276277 and U1463207), Opening Project of Guangxi Key Laboratory of Petrochemical Resource Processing and Process Intensification Technology (2015K003), CNPC Key Research Project and CNOOC project.

References

(1) Shetti, V. N.; Kim, J.; Srivastava, R.; Choi, M.; Ryoo, R. *J. Catal.* 2008, 254, 296-303.
(2) Duan, A. J.; Li, T. S.; Zhao, Z.; Liu, B. J.; Zhou, X. F.; Jiang, G. Y.; Liu, J.; Wei, Y. C.; Pan, H. F. *App. Cata. B.* 2015, 165, 763-773.
(3) Fan, J.; Yu, C. Z.; Gao, F.; Lei, J.; Tian, B. Z.; Wang, L. M.; Luo, Q.; Tu, B.; Zhou, W. Z.; Zhao, D. Y. *Angew. Chem.* 2003, 115, 3254-3258.
(4) Huo, Q.; Dou, T.; Zhao, Z.; Pan, H. F. *App. Cata. A-Gen.* 2010, 381, 101-108.

HDS ACTIVITIES OF COMBINED POROUS CATALYSTS BASED ON SINGLE-POT SYNTHESIS OF NIMO SUPPORTED TI-SBA-15: EFFECT OF HYDROTHERMAL SYNTHESIS TEMPERATURE

Saheed A. Ganiyu[1,2], Khalid R. Alhooshani[1,], Sagir Adamu[3], Tawfik A. Saleh[1]*

[1]Chemistry Department, [2]Center of Research Excellence in Nanotechnology, and [3]Chemical Engineering Department, King Fahd University of Petroleum and Minerals, Dhahran 31261, Saudi Arabia.

* hooshani@kfupm.edu.sa

Introduction

Sulfur regulation is becoming stricter and crude oil source is becoming heavier. These two factors make refinery operations more challenging. Sulfur content in fuel has to be as low as 10-15 ppm, as promulgated by EU and US sulfur regulations, and future catalytic converters application in automobiles requires zero ppm-S[1, 2]. In lieu of this, more robust and efficient HDS catalyst is necessary to obtain ultra-deep desulfurization (UDD) of fuel.

The recent trends in the synthesis of hydrodesulfurization (HDS) catalysts using mesoporous silica as an alternative support to highly acidic alumina, involves the incorporation of metals (Al, Ti, Zr) and different complexing agents, with a view to overcome the amorphous nature of the silica wall and dispersion of active metals via modification from poor acidity to moderate one [3]. The steps with this approach require the synthesis of SBA-15, with incorporation of metals by direct or post synthesis method, followed by impregnation of NiMo with or without complexing agents. However, the aforementioned approach not only requires a series of time-consuming steps, but also results in significant loss of textural properties due to blockage of micropores, and surface area coverage and the dispersion still need to be greatly improved.

A new systematic single-pot (SP) approach was developed, in addition to the existing literatures on impregnation approach of incorporating active metals, to improve the dispersion of active metals on a porous support, reduce the steps in catalysts preparation and as well prevent the loss of textural properties that might be necessary for catalytic efficiency of HDS catalysts. One of the technical challenges associated with SP synthesis approach of metal modified SBA-15 NiMo catalysts is the formation of crystalline NiMo phase, i.e. less dispersion of this active phase, while incorporating into synthesis mixture of pure-SBA-15 reported by [4]. However, systematic approach of changing the synthesis temperature can prevent the formation of crystalline phase, and in essence generate more micropores within the mesopores of final HDS catalyst. This will result in better catalytic performance compared to impregnation approach, through fast diffusion of reactant molecules through mesopores and adsorption on the active sites embedded within the micropores of the catalyst. This study aims at finding optimum hydrothermal synthesis temperature to obtain highly dispersed NiMo HDS catalyst in single reaction pot with SBA-15 support for efficient desulfurization of hydroprocessing feedstock containing olfactory sulfur compounds.

Experimental Section

Materials and Method. All chemical reagents, tetraethylorthosilicate (TEOS) $(CH_5O)_4Si$ as silica source, pluronic P123 PEO_{20}-PPO_{70}-PEO_{20} triblock copolymer as structural directing agent, nickel nitrate hexahydrate (99 %) as nickel precursor, and titanium isopropoxide (97 %) as titanium source were purchased from Sigma-Aldrich. Ammonium molybdate (VI) tetrahydrate (99 %) was obtained from ACROS organics.

Synthesis of Ti-SBA-15-NiMo. NiMo supported Ti-SBA-15 catalyst is obtained by modification of sol-gel synthesis approach described by [4] for synthesis of pure SBA-15, while the incorporation of titanium, molybdenum and nickel were done simultaneously with their respective precursor in a single reaction mixture. The mixture obtained after 24 h of stirring was transferred into teflon autoclave and subjected into hydrothermal (HT) synthesis between 40–100 °C for another 24 h. The light-yellow solid was centrifuged once without further washing, to avoid leaching of active phase(s) into solvent, and dried in oven at 100 °C for 12 h, before final calcination at 300 °C. The obtained non-sulfided HDS catalyst was denoted as HT(x), where x represent HT-temperature. The final loading content of Ni, Mo were ca. 3 % and 8 %, respectively, while Ti is 1:10 to silicon.

Characterization

The textural, structural and chemical properties of as-developed catalysts were characterized by N_2-physiosorption, X-ray diffraction (XRD), temperature programmed desorption (TPD) and reduction (TPR), scanning electron microscope (SEM).

Activity measurement. Prior to the catalytic activities measurement in batch reactor (Parr 4848B), at 350 °C, the pelletized catalysts (300 – 500 μm) were sulfided in 2 wt.% CS_2 in hexane overnight after reduction in constant flow of 10% H_2/He for 150 min. About 250 mg of sulfided catalyst was added into 100 ml of a simulated fuel containing 1000 ppm-S each of dibenzothiophene and dibenzothiophene in dodecane. Aliquots were taken at 30 min interval for quantification and qualification by gas chromatograph sulfur chemiluminiscence detector (GC-SCD) and mass spectrometer (GC-MS), respectively, while the hydrocarbons analysis was done by gas chromatography equipped with flame ionization detector (GC-FID).

Results and Discussion. N_2-physiosorption isotherm was used to estimate the extent of microporosity embedded within the mesoporous HDS catalysts. It has been reported that bimodal porous material possesses excellent adsorption and catalytic properties, due to fast diffusion of the molecules through the mesopores, followed by the attachment to the micropores that harbours the active sites for catalytic reaction. The diffusion limitation problem, hotspot problem in catalytic reaction can be overcome by having combined porous catalysts. As shown in table 1, the decrease in HT-synthesis temperature helps in the formation of combined porous catalyst and prevents the formation of crystalline phase of $NiMoO_4$ and/or MoO_3. This phase results in poor dispersion of active phase on the support due to agglomeration, and consequently low catalytic activities. In addition, we observed about four-fold increase in micro surface area and volume, which are necessary for efficient contact with active phase by reduction of synthesis temperature from 100 °C to 40 °C. The isotherm (fig. 1-left) represents type IV of mesoporous material, with two desorption steps, which is typical of pure-SBA-15 and presence of plugged nanoparticles on the Ti-SBA-15 [5].

The XRD analysis was used as a measure of dispersion for as-developed catalysts. As observed in Fig. 1(right), the formation of crystalline phases of MoO_3 corresponding to Pbnm octahedral [6] form is becoming noticeable in catalysts synthesis between 80 – 100 °C, while molybdenum reflections was completely absent at HT-synthesis temperature of 40 °C, which is responsible for high dispersion of active phases.

The TPR by H_2 and TPD by NH_3, measure respectively, the reduction temperature and surface acidity character of as-developed catalysts. It is observed that the H_2 consumption reduction temperature of Mo^{6+} - Mo^{4+}decreases with decreasing HT-synthesis temperature, except for the catalyst denoted as HT60 that shows lower reduction temperature value relative to HT40. This observation might be due to difference in morphology obtained for this catalyst compared to others as shown in fig. 3. The nature of the surface acidity for synthesized catalysts was moderate and strong acid sites as shown by maximum temperature T_{max} peak positions in fig 2b. However, HT40 shows combination of weak and strong acid sites, and the latter is prominent by the amount of NH_3 desorbed at that temperature. The morphology of all synthesized catalyst resembles wheat-like fibers of pure SBA-15, with slight variation as shown in fig. 3. However, there is complete morphology variation for HT60 with respect to others.

Activities of these catalysts were evaluated for simultaneous HDS of DBT and MDBT containing 1000 ppm-S each in dodecane for 4 h at 350 °C in batch reactor. We observed an interesting result for HT40 catalyst compared to other catalysts. This catalyst is capable of desulfurizing more than 96 %-DBT and 92 %-MDBT within short period of activity, using minimum amount of catalyst. In essence, the result is outstanding and comparable to recent published reports that used higher loading of molybdenum at 13 %. We observed from this study that the main contributing factor for better catalytic performance is attributed to increase in surface area, and more importantly, ability to form combined micro- and meso- porous catalyst, which is significant when the HT-synthesis temperature is 40 °C. In addition, the HDS activities of DBT is higher for all synthesized catalysts than that of MDBT due to steric hindrance of the latter and also, the initial rate of HT100 is faster than others, but fall behind in the subsequent analysis time.

Table 1: Textural Properties of SP-approach HDS catalysts

Catalysts	S_{BET} (m^2/g)	S_{micro} (m^2/g)	S_{meso} (m^2/g)	V_{micro} (cm^3/g)	V_{total} (cm^3/g)	PS (nm)
HT40	757.49	140	617.5	0.069	1.00	6.49
HT60	707.19	111.01	596.19	0.055	0.91	5.98
HT80	572.76	47.9	524.86	0.023	1.04	7.34
HT100	439.36	36.73	402.63	0.017	0.95	8.3

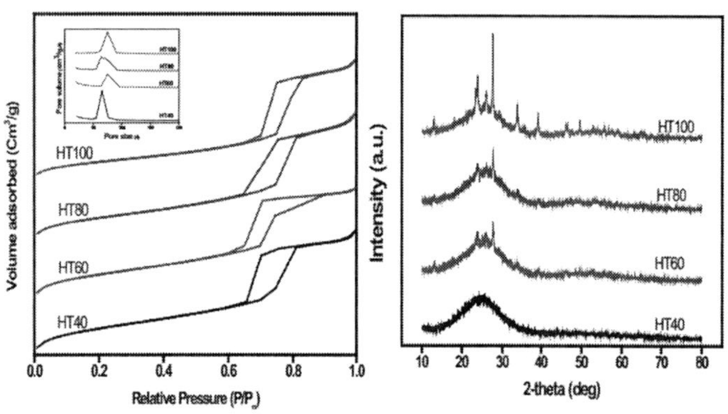

Figure 1: N_2-physiosorption isotherm (left) and XRD analysis (right) of as-synthesized HDS catalysts

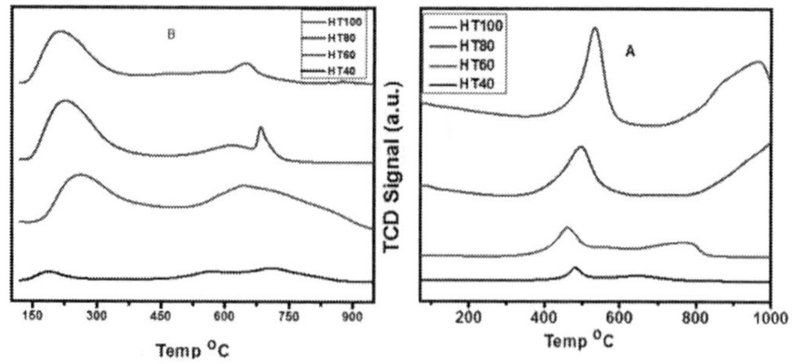

Figure 2: TPR (A) and TPD (B) of as-synthesized catalysts

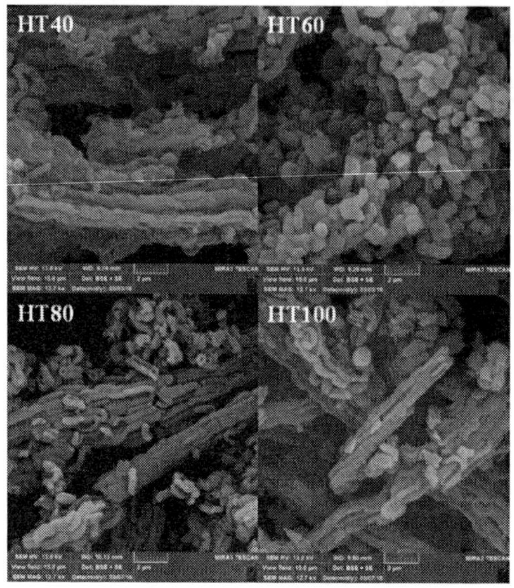

Figure 3: Scanning electron microscope (SEM) analysis of as-synthesized catalysts

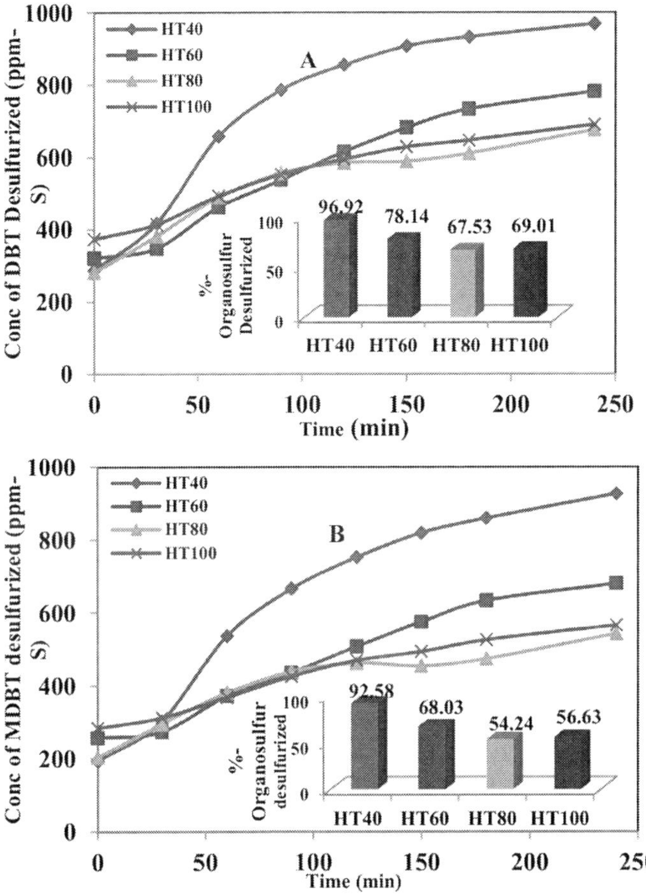

Figure 4: N_2-physiosorption isotherm (left) and XRD analysis (right) of as-synthesized HDS catalysts

Conclusions

We have developed and evaluated NiMo supported Ti-SBA-15 based on single-pot (SP) approach for HDS reaction of olfactory organosulfur compounds (DBT and MDBT) in dodecane as a model fuel. Tuning the HT-synthesis temperature was influential to obtain a combined porous (micro and mesoporous) HDS catalyst, with high specific surface area, needed to achieve better catalytic performance using minimum amount of catalyst. In addition, the dispersion of active phase increases, as the HT-synthesis temperature decreases. Therefore, the SP-approach requires low synthesis temperature to avoid agglomeration of active phase. Finally, the as-developed catalyst (HT40) possessed great potential to be used in deep desulfurization of fuel.

Acknowledgement. The financial support provided by the King Fahd University of Petroleum & Minerals, Dhahran, Saudi Arabia through internal fund Project No. DSR NUS15105 for this work are gratefully acknowledged.

References

(1) A. Stanislaus, A. Marafi, M.S. Rana, Recent advances in the science and technology of ultra low sulfur diesel (ULSD) production, Catalysis Today, 153 (2010) 1-68.
(2) C. Song, An overview of new approaches to deep desulfurization for ultra-clean gasoline, diesel fuel and jet fuel, Catalysis today, 86 (2003) 211-263.
(3) S. Badoga, A.K. Dalai, J. Adjaye, Y. Hu, Combined Effects of EDTA and Heteroatoms (Ti, Zr, and Al) on Catalytic Activity of SBA-15 Supported NiMo Catalyst for Hydrotreating of Heavy Gas Oil, Industrial & engineering chemistry research, 53 (2014) 2137-2156.
(4) D. Zhao, J. Feng, Q. Huo, N. Melosh, G.H. Fredrickson, B.F. Chmelka, G.D. Stucky, Triblock Copolymer Syntheses of Mesoporous Silica with Periodic 50 to 300 Angstrom Pores, Science, 279 (1998) 548-552.
(5) V. Meynen, P. Cool, E. Vansant, Synthesis of siliceous materials with micro-and mesoporosity, Microporous and mesoporous materials, 104 (2007) 26-38.
(6) S. Badoga, R.V. Sharma, A.K. Dalai, J. Adjaye, Hydrotreating of heavy gas oil on mesoporous zirconia supported NiMo catalyst with EDTA, Fuel, 128 (2014) 30-38.

DESIGN AND SYNTHESIS OF ELECTRICALLY CONDUCTIVE COMPOSITES FOR ENERGY EFFICIENT ADVANCED MATERIALS

QuratulAin Nadeem, Rohama Gill, Maria Batool, Tasneem Fatima

Department of Environmental Sciences, Fatima Jinnah Women University, Rawalpindi 46000, Pakistan
Dr.Rohama Gill, rohama_gill@hotmail.com

Introduction

The ever growing technological advancement has led to the extensive exploitation of materials such as polymers and ceramics in almost every form due to their tremendous advantages in conservation and consumption of energy. One such typical use of polymers, involves fabrication of conductive polymer composites by embedding metallic reinforcements (flakes/whiskers/ particles, etc.) such as Ni, Ag, Cu, etc. into non-conducting polymeric matrices.[1] However, the limitations associated with metallic fillers (high-price, high-density, corrosion, etc.) drew the attention of researchers to select cost effective and relatively less denser reinforcements[2] that are efficient and durable enough to meet the demands of real world.[3]

In the domain of metallization and fabrication of electrically conducting composites, a number of techniques have been used so far for manufacturing energy efficient conductive composites e.g. electrodeposition;[4] physical[5] and chemical vapor deposition;[6] electrospinning;[7] ultrasonic irradiation;[8] etc. Among these, the advent of another technique 'electroless deposition' paved its way by offering uniform metallization over a wide variety of substrates irrespective of their nature and geometry such as electrical/thermal conductors, semi-conductors, and even on insulators. Novelty of the current research work lies in selecting a deposition technique i.e. 'electrloess deposition' to fabricate high performance electro-conductive composites by metalizing electrically insulative alumina substrate with zerovalent copper and then incorporate it as conductive filler into non-conducting polymer matrices.

Currently presented idea of manufacturing high performance electro-conductive composites is likely to satisfy the demands of energy-efficient and sustainable micro-electronic packaging[9], etc. Other futuristic applications may include electromagnetic interference (EMI) and electrostatic dissipative (ESD) shielding,[10] electrochemical bio-sensors, [11] to effectively economize energy demands of modern systems.

Experimental

Conducting polymeric composites were fabricated in two phases. In phase I, Cu coated Al_2O_3 filler was prepared using electroless deposition technique, whereas in phase II, polymer composite thin films were fabricated by inclusion of Cu-Al_2O_3 filler into the selected polymer matrices.

Phase I-Pretreatment of Al_2O_3. Al_2O_3 powder packed in silk cloth (120 mesh) in the form of sachet, was immersed in HNO_3 for 2 min to clean oil or dirt from the surfaces. Then surface cleaned Al_2O_3 powder was dipped in microetching solution for 2 min (125 g NaCl and 250 mL HF dissolved in adequate volume of distilled water to make volume 1L) for surface coarsening that improve adhesion of Cu on Al_2O_3 particles. The coarse particles of Al_2O_3 were then immersed in sensitizing solution for 2 min (8 g $SnSO_4$ and 8 mL HCl dissolved in adequate volume of DW to make volume 1L). After sensitization, surface activation was carried out by dipping sensitized Al_2O_3 particles in activation solution at 35°C for 7 min (50 mg $SnCl_2.2H_2O$, 5mg $PdCl_2$ and 40 mL HCl in adequate volume of DW to make volume 1L). Al_2O_3 substrate was rinsed with DW after each step of pretreatment.

Electroless Cu Deposition onto pretreated Al_2O_3. Cu was deposited on the pretreated Al_2O_3 by electroless deposition technique. Electroless plating bath was prepared as follows: ***Solution I:*** 16g $CuSO_4.5H_2O$ was dissolved in DW to make volume 1L. ***Solution II:*** 14g NaOH, 30g $KNaC_4H_4O_6{\cdot}4H_2O$ and 20g $C_{10}H_{14}N_2Na_2O_8$ were dissolved in DW to make volume 1L. ***Solution III:*** 500 mL each of the *Solution I* and *Soultion II* were mixed together to make *Solution III* into which 15 mL HCHO and 25 ppm $K_4[Fe\ (CN)_6]$ were added. The resultant solution is electroless plating bath. The pretreated Al_2O_3 was dipped in electroless plating bath for 30 min at 35°C. After Cu deposition, the resultant Cu-Al_2O_3 powder was rinsed in water, dried in vacuum oven (3-4 h) at 40°C and used as conductive filler to prepare electro-conductive polymer composites.

Phase II-Fabrication of conducting polymer composites. Electrically conducting polymer composites were fabricated by incorporating the prepared Cu- Cu-Al_2O_3 filler into four different polymer matrices: Polystyrene-b-poly (methyl methacrylate) (PS-b-PMMA), Polystyrene (PS), Polystyrene-block-poly(ethylene-ran-butylene)-block-polystyrene-graft-maleic anhydride (PS-*b*-(PE-*r*-B)-*b*-PS-*g*-MA), and Polyvinyl chloride (PVC). Conducting composites in the form of films were fabricated by dispersing the calculated amount of Cu-Al_2O_3 filler (3.0, 6.0, 9.0, 12.0, 15.0, 18.0, 21.0 wt.% in case of PS-*b*-PMMA and PS matrix composites; and 2.0, 4.0, 6.0, 8.0, 10.0, 12.0, 14.0 wt.% in case of PS-*b*-(PE-*r*-B)-*b*-PS-*g*-MA and PVC matrix composites) in the respective polymer solution prepared in 30 mL tetrahydrofuran. The corresponding filler-polymer solution was allowed to stir for 6h and cast in the mould (petri dish). The solvent was then evaporated at room temperature and after recovering from petri dish, homogenously distributed films were vacuum dried to attain constant weight. The as prepared films were then used for characterization to investigate their electrical, morphological, elemental and mechanical behavior as energy-efficient sustainable materials of real world applications.

X-Ray Diffraction Analysis. XRD investigations were conducted using PANalytical X-ray diffractometer (XPERT-PRO) using Cu-Kα X-rays of wavelength (λ) =1.54056 Å. To extract data 2θ range was set at 10°- 80°, the provided voltage was 40 kV with beam current 30 mA.

Morphological Analysis. The surface morphology of pristine Al_2O_3, Cu-Al_2O_3, and conducting composites were investigated by Hitachi SU-1500 using secondary electron detector provided with 5 kV voltage.

Electrical Conductivity Measurements. Since, electrical conductivity (S/cm) is inversely proportional to volume resistivity (Ω.cm), so the inverse of the values of volume resistivity (1/Ω.cm) was taken to get electrical conductivity values of the prepared composites. To measure volume resistivity, 4-probe technique was opted using Insulation Tester (NF2511) equipped with electrodes made of tungsten carbide provided with 500V DC at room temperature.

Mechanical Studies. Mechanical behavior of the prepared composites were investigated *via* Instron Tester (4600 UK) at 20±2 °C.

Energy Dispersive Spectrometry (EDX). Atomic % of elements in Cu-Al_2O_3 filler and polymeric composites were examined using JSM6490LV (JEOL) microscope using QUANTAX EDS XFlash detector 4010 (Bruker) provided with 20 kV voltage.

Results and Discussion

X-ray Diffraction Analysis. Scherrer's equation was applied to calculate average crystal size of deposited Cu, demonstrated in following eq. (1):

$$D = \frac{K\lambda}{\beta \cos\theta} \quad (1)$$

Where, D is the size of deposited Cu crystal (nm); K corresponds to Scherrer's constant ~ 0.9; λ is X-ray wavelength (nm); β depicts angular width (rad); and θ expresses Bragg's angle (deg). XRD analysis of pristine Al_2O_3 powder exhibited three 2θ peaks that correspond to Al_2O_3 i.e. 37.5, 45.7 and 67.2 deg. On the other hand, XRD pattern of Cu coated Al_2O_3 powder revealed emergence of three major peaks at 2θ values of 43.3683, 50.4863, and 74.1972 deg, not found in the diffraction pattern of pristine Al_2O_3 while rest of the two peaks at 2θ values of 36.9544 and 67.12 deg correspond to Al_2O_3 phase. The three Cu peaks (2θ = 43.3683, 50.4863, and 74.1972 deg) clearly correspond to three distinct planes of Cu i.e. (111), (200), and (220), respectively when compared with the JCPDS copper file No. 00-001-1241. These observed hkl values of Cu significantly depict the deposition of metallic Cu^0 on the surface of Al_2O_3.[12] The calculated avg. size of Cu crystal deposited on Al_2O_3 is 26.116 nm.

To examine effectiveness of currently prepared Cu-Al_2O_3 filler in the selected polymers, XRD patterns of pristine PS-*b*-PMMA, PS, PS-*b*-(PE-*r*-B)-*b*-PS-*g*-MA), PVC and their respective composites with minimum and maximum filler loading were recorded. XRD patterns of all the neat polymers depicted broad noncrystalline peaks with amorphous structures which is the characteristic of neat polymers and is also observed in the literature.[13,14] The inclusion of minimum filler loading (3 wt.%) in PS-b-PMMA and PS showed appearance of peaks at position 2θ = 42.9 deg in case of PS-b-PMMA, and 2θ = 42.9 deg and 66.7 deg in case of PS matrix composite. Further inclusion of filler (21 wt.%) in PS-b-PMMA and PS resulted in increase in intensity of all the peaks which might be attributed to the crystalline region of Cu-Al_2O_3 filler. Similar behavior is seen when minimum filler (2 wt.%) was loaded in PS-*b*-(PE-*r*-B)-*b*-PS-*g*-MA) and PVC matrices with appearance of peaks at position 2θ = 42.7 deg in case of PS-*b*-(PE-*r*-B)-*b*-PS-*g*-MA), and 2θ =42.9 deg and 50.05 deg in case of PVC matrix composite. Further filler incorporation (14 wt.%) in PS-*b*-(PE-*r*-B)-*b*-PS-*g*-MA) and PVC matrices also depicted increment in intensities of the observed peaks. Comparing the diffraction patterns of four types of composites, a typical peak at 2θ =42.9 deg seen in all of them is a peak that is characteristic of Cu corresponding to (111) plane.

EDX Analysis. Elemental composition of the Cu-Al_2O_3 filler and conductive composites was determined by EDX. In case of Cu-Al_2O_3 powder, three elements Cu (65.08%), Al (31.25%), and Pd (3.67%) were identified. The presence of Cu in highest atomic % signifies its capability to impart conductivity when used as filler.

Effect of filler loading on the elemental composition of resultant composites was also studied and five elements were detected (Cu, Al, Pd, C, and O). Atomic % with the inclusion of minimum filler loading (3 wt.%) in PS-b-PMMA depicted (C=78.29%, O= 21.64%, Al= 0.05%, Cu= 0.02, Pd= 0.00%) and in PS showed (C=78.03 %, O= 21.88 %, Al= 0.07%, Cu= 0.02, Pd= 0.00%). Further inclusion of filler (21 wt.%) in PS-b-PMMA revealed atomic % (C=78.89 %, O= 20.82 %, Al= 0.07%, Cu= 0.20, Pd= 0.02%) and in PS depicted (C=78.70 %, O= 20.98 %, Al= 0.11%, Cu= 0.20, Pd= 0.01%). Similar trend in elemental composition with minimum filler loading (2 wt.%) is seen in PS-*b*-(PE-*r*-B)-*b*-PS-*g*-MA) (C = 84.89 %, O= 12.99%, Al= 0.61%, Cu= 1.49 %, Pd= 0.02) and in PVC (C = 85.81 %, O= 11.81%, Al= 0.97%, Cu= 1.38 %, Pd= 0.03). Further inclusion of filler (14 wt.%) in PS-*b*-(PE-*r*-B)-*b*-PS-*g*-MA revealed atomic % (C = 85.40 %, O= 11.85%, Al= 0.39%, Cu= 2.31 %, Pd= 0.05) and in PVC depicted (C = 85.17 %, O= 12.03%, Al= 0.30%, Cu= 2.44 %, Pd= 0.06).Of these elements, presence of Pd in lowest atomic % is due to the minimum concentration of $PdCl_2$ used as catalyst for surface activation of Al_2O_3 in electroless deposition.

Morphological studies. In case of Cu-Al_2O_3 filler, SEM micrograph depicted uniform deposition of Cu onto particulate Al_2O_3. As reported in literature,[15] deposition of metallic Cu is associated with finely scaled roughness onto the substrate which can be seen in the presently discussed electroless Cu deposition onto Al_2O_3. Effect of filler loading on the morphological analysis of composites was also studied by varying filler content from 0-21 wt.% in case of PS-*b*-PMMA and PS matrices, and 0-14 wt.% in case of PS-*b*-(PE-*r*-B)-*b*-PS-*g*-MA and PVC matrix composites. Development of conductive pathway for smooth electron transfer is based on filler particle-particle connectivity.[16] Considerable increase in bright phases with increased filler loading is due to increased Cu content in the polymer matrices. Cu-Al_2O_3 particles are homogenously dispersed and embedded in the polymer matrices with no phase separation which depicts good interaction between filler and polymer. With increased filler loading, clusters of filler distinctly grew throughout the matrices which also improved the bright phases of the composites with well developed particle–particle connectivity. A clear transition from particle–polymer interaction to well developed particle–particle connectivity is seen in all four types of polymer composites.

Electrical Conductivity Analysis. Pristine films of PS-*b*-PMMA and PS depicted their insulting behavior with electrical conductivity values of 3.3×10^{-14} and 2.0×10^{-14} S/cm, respectively. Increment in the electrical conductivity started with the minimum filler inclusion ~3.0 wt.%, and a considerable increase in electrical conductivity is achieved when 15.0 wt.% filler was loaded into PS-b-PMMA (4.5×10^{-8} S/cm) and PS (3.6×10^{-9} S/cm) composites. With the inclusion of Cu-Al_2O_3 up to 21 wt.%, values of electrical conductivity reached 5.55×10^{-5} S/cm for PS-b-PMMA and 5.0×10^{-6} S/cm for PS matrix composites which signifies that the prepared composites exhibit semi-conducting behaviour.[17] Similar behavior is seen in case of composites based on PS-*b*-(PE-*r*-B)-*b*-PS-*g*-MA and PVC matrices. Neat polymers, PS-*b*-(PE-*r*-B)-*b*-PS-*g*-MA and PVC were also found insulating with electrical conductivity values of 4.3478 $\times10^{-16}$ S/cm and 8.333 $\times10^{-16}$, respectively. These polymer matrices remained insulating up to the incorporation of 6 wt.% filler with electrical conductivity values of 1.96078 $\times10^{-11}$ S/cm (PS-*b*-(PE-*r*-B)-*b*-PS-*g*-MA matrix) and 3.225 $\times10^{-9}$ S/cm (PVC matrix). At maximum loading (14 wt.%), semi-conducting properties were achieved with electrical conductivity values of 7.692×10^{-5} S/cm and 8.333 $\times10^{-6}$ S/cm for PS-*b*-(PE-*r*-B)-*b*-PS-*g*-MA and PVC composites, respectively.

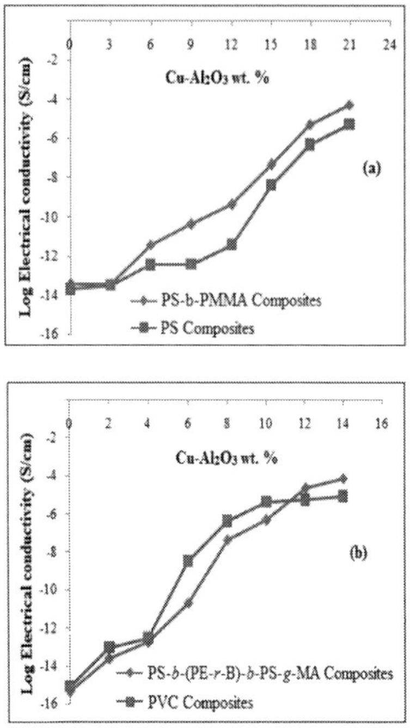

Figure 1. Electrical conductivity behavior of **(a)** PS-b- PMMA and PS; **(b)** PS-*b*-(PE-*r*-B)-*b*-PS-*g*-MA and PVC matrix composites as a function of Cu-Al_2O_3 loading

This transition from insulative to electrically conductive region is attained due to the smooth transmission of electrons from geometrically connected filler particles forming a network throughout the matrix.[18, 19]

Mechanical analysis-Young's Modulus. A pronounced increase in the values of Young's modulus is seen in PS-b-PMMA (680-1122 MPa), PS (680.12-1053.9 MPa), PS-*b*-(PE-*r*-B)-*b*-PS-g-MA (50.01-150.47 MPa), and PVC (15.22-92.05 MPa) composites. Since, ceramics have high modulus as compared to polymers, increased Young's modulus in all types of composites is associated with the rigidity imparted by Al_2O_3 particles which may provide stiffness to the matrices.[20,21]

Tensile strength. Strength of a material is always associated with its extent to bear maximum stress before its failure or breakage. Results have shown that incorporation of Cu-Al_2O_3 increased tensile strength of the selected polymer matrices such as PS-b-PMMA (20.229-27.998 MPa), PS (20-836-30.585 MPa), PS-*b*-(PE-*r*-B)-*b*-PS-g-MA (15.121- 82.889 MPa), and PVC (15.19-74.450MPa) composites. Pronounced increment in tensile strength as a function of filler loading refers good filler-polymer adhesion and compatibility along with effective stress transfer from polymer matrix to particulate filler.[22]

% elongation at break. A decrease in % elongation at break with increased filler loading was observed in case of PS-b-PMMA (5.9-2.5%), PS (5.9-2.6%), PS-*b*-(PE-*r*-B)-*b*-PS-*g*-MA (16.91-10.05%), and PVC (15.19-74.450MPa) while an increase in % elongation at break was observed in case of PVC (12.12-17.45%) composites with increased filler incorporation. Ceramics are brittle and rigid while polymers are known for their ductile behavior.[23] Inclusion of rigid Cu-Al_2O_3 filler into the selected matrices increased brittleness of the composites which is depicted from the decline in % elongation at break of the composites.[24] However, the increased % elongation at break observed in case of PVC composites might be attributed to the addition of plasticizers usually used during polymerization of PVC.

Conclusions

Novel electro-conductive composites were fabricated with the inclusion of Cu-Al_2O_3 filler into four different polymer matrices (PS-*b*-PMMA, PS, PS-*b*-(PE-*r*-B)-*b*-PS-*g*-MA and PVC) using electroless deposition as an innovative deposition technique. In light of the above mentioned results it is expected that prepared electrically conductive and mechanically stable composites have potential to be used in micro-electronics, biomedical, packaging industry, etc. as sustainable, energy efficient semi-conductive materials with improved shelf life.

References

(1) Li, J. X.; Wu, J.; and Chan, C. M. *Polymer*. **2000,** *41*(18), 6935.
(2) Reynaud, E.; Jouen, T.; Gauthier, C.; Vigier, G.; and Varlet, J. *Polymer*. **2001,** *42*(21), 8759.
(3) Tongxiang, L.; Wenli, G.; Yinghui, Y.; and Chunhe, T. *Int. J. Adhes. Adhes,.* **2007,** *28*, 55.
(4) Pang, X.; Zhitomirsky, I.; and Niewczas, M. *Surf. Coat. Technol.* **2005,** *195*(2), 138.
(5) Cuna, A.; Noguera, A.; Saucedo, E.; and Fornaro, L. *Cryst. Res. Technol.* **2004,** *39*(10), 912.
(6) Reina, A.; Jia, X.; Ho, J.; Nezich, D.; Son, H.; Bulovic, V.; Dresselhaus, M. S.; and Kong, J. *Nano Lett.*, **2008,** *9*(8), 30.
(7) Feng, Z. Q., Wu, J., Cho, W., Leach, M. K., Franz, E. W., Naim, Y. I., Gu, Z.; Corey, J.M.; and Martin, D. C. *Polymer* **2013,** *54*(2), 702.
(8) Zhang, Y. P.; Lee, S. H.; Reddy, K. R.; Gopalan, A. I.; and Lee, K. P. *J. Appl. Polym. Sci.* **2007,** *104*(4), 2743.
(9) Schaefers, S.; Rast, L.; and Stanishevsky, A. *Mater. Lett.*, **2006,** *60*(5), 706-709.
(10) Moon, J. H.; Kim, K. H.; Choi, H. W.; Lee, S. W.; and Park, S. J. *Ultramicroscopy*.**2008,** *108*(10), 1307.
(11) Liu, S.; Liu, J.; Wang, L.; and Zhao, F. *Bioelectrochemistry*. **2010,** *79*(1), 37.
(12) Ma, H.; Liu, Z.; Wu, L.; Wang, Y.; and Wang, X. *Thin Solid Films*. **2011,** *519*(22), 7860.
(13) Ge, J.; Zeng, X.; Tao, X.; Li, X.; Shen, Z.; Yun, J.; Chen, J. *J. Appl. Polym. Sci.* **2010,** *118*(3), 1507.
(14) Alsharaeh, E. H.; Othman, A. A.; Aldosari, M. A. *Materials*. **2014,** *7*(7), 5212.
(15) Leon, C. A.; Aguilar, E. A.; Quintana-Puchol, R.; and Drew, R. A. L. *J. Mat. Sci. Lett.* **2002,** *21*(17), 1375.
(16) Beygi, H.; Sajjadi, S.A.; and Zebarjad, S.M. *Appl. Sur. Sci.* **2012,** *261*, 166.
(17) Nadeem, Q.; Rizwan, M.; Gill, R.; Rafique, M.; and Shahid, M. *J. Appl. Poly. Sci.* **2016,** DOI: 10.1002/app.4.2939.
(18) Luong, N. D.; Hippi, U.; Korhonen, J. T.; Soininen, A. J.; Ruokolainen, J.; Johansson, L. S.; Nam, J.D; Sinh, L.H; and Seppala, J. *Polymer*. **2011,** *52*(23), 5237.
(19) Zhitomirsky, I.; and Petric, A. *Matter. Lett.***2000,** *46*(1), 1.
(20) Wu, S. Y.; Huang, Y. L.; Ma, C. C. M.; Yuen, S. M.; Teng, C. C.; & Yang, S. Y. (2011). *Composites Part B.* **2011,** *42*(11), 1573.
(21) Fu, S. Y.; Feng, X. Q.; Lauke, B.; and Mai, Y. W. *Composites Part B.* **2008,** *39*(6), 933.
(22) Kuo, M. C.; Tsai, C. M.; Huang, J. C.; and Chen, M. *Mat. Chem. Phys.* **2005,** *90*(1), 185.
(23) Zhou, W.; Qi, S.; Li, H.; and Shao, S. *Thermochim. Acta.* **2007,** *452*(1), 36.
(24) Krupa, I.; Mikova, G.; Novak, I.; Janigova, I.; Nogellova, Z.; Lednicky, F.; and Prokes, J. *Europ. Polym. J.* **2007,** *43*(6), 2401.

OPTIMIZATION ON COMPOSITE ZEOLITE FOR PRODUCING PROPYLENE AND CLEAN GASOLINE

Xiaoli Wei, Jianhong Gong

Sinopec Research Institute of Petroleum Processing, P.O.Box 914-2, Beijing 100083, China
weixl.ripp@sinopec.com

Introduction

The increasing usage of automobiles is boosting the world-wide fuel consumption, and as a sequence, causing heavier and heavier environmental pollution [1]. Extensive efforts have been undertaken by both the petroleum refining industry and the automobile industry to mitigate those environmental problems, for example, by adopting automotive catalytic converters and cleaner fuel. Among these strategies, tightening transportation fuel specification is a major aspect of reducing automobile emissions. Worldwide fuel specifications are showing clear trends toward lower olefins and sulfur contents without sacrificing Octane numbers [2]. On the other hand, propylene is an important building block of the petrochemical industry. In recent years, propylene demand has been continuously increasing. Therefore, much effort has been focused on improving the FCC process to simultaneously produce cleaner gasoline and more propylene.

Using ZSM-5 zeolite as an additive to FCC catalyst to improve the gasoline octane number is well known and of growing importance [3, 4]. However, the additive is used primarily to increase the yields of C3-C5 olefins. The reaction mechanisms of ZSM-5 in combination with REUSY catalysts have been described in detail [5-10]. When ZSM-5 used as a separate additive, the synergy between ZSM-5 and the base catalyst have been investigated [11-13]. ZSM-5 can be added as a separate additive, or it can be incorporated into the base FCC catalyst. Although using as separate additives has more flexibility, it requires to add two separate base catalyst and additive. Routinely adding one single FCC base catalyst containing ZSM-5 into the FCC unit is relatively easier and more convenient to achieve for operators. In fact, catalysts containing dual zeolites have already been developed and commercialized to produce lower olefin content gasoline while increasing propylene yield and maintaining Octane number at the same time [14, 15].

Since the gasoline olefinicity primarily depends on the hydrogen transfer activity of Y zeolite while light olefins production relying on β-scission of ZSM-5, it is of interest to explore how ZSM-5 performs in combination with Y zeolite. Nevertheless, very few studies have been conducted to investigate the effects of Y to ZSM-5 zeolite ratio in a catalyst on protolytic cracking, β-scission, oligermerization and hydrogen transfer[16] relate to conversion. Such information will help to determine the product distribution and product quality. In this work, five catalysts with vary ratio of Y to ZSM-5 zeolite were prepared to investigate the effects of the Y to ZSM-5 ratio on the relative amount of reactions such as protolytic cracking, β-scission, oligermerization and hydrogen transfer at different reactant conversion. The objective is to explore the optimum composition of Y and ZSM-5 to develop novel catalysts for lower gasoline olefins, higher propylene yield and better product slates.

Experimental

Feedstock and catalysts. FCC naphtha was used as feed in this study since it is difficult to characterize regular FCC heavy feedstock such as VGO in detail. Another advantage of choosing FCC naphtha is the ability to examine all kinds of chemical reactions during catalytic cracking, such as oligomerization.

Three composite zeolites based catalysts, denoted as N164, N165 and N166, have been prepared with different mixing ratio of Y to ZSM-5 zeolite. The mass ratios of Y to ZSM-5 zeolite are 4:1 in catalyst N164, 3:2 in catalyst N165, and 1:4 in catalyst N166. In addition, pure Y and ZSM-5 zeolites were also prepared denoted as catalyst N161 and N162, respectively. The total zeolite concentrations in each different catalyst were all the same. All the five catalysts have been deactivated with 100% steam at 800℃. To achieve similar MAT activity for all catalysts with an aim to better evaluate these catalysts in subsequent catalytic cracking tests, the aging time was set differently according to their different hydrothermal ability based on their respective zeolite composition. The characteristics of all five catalysts after deactivation were given in table 1.

Table 1 Characteristics of the aged catalysts used in this work

Catalysts	N161	N162	N164	N165	N166
Y:ZSM-5	1:0	0:1	4:1	3:2	1:4
Composition, wt%					
Al_2O_3	60.3	56.9	60.5	59.9	58.2
Na_2O	0.19	0.11	0.17	0.14	0.11
SiO_2	37.4	39.7	36.8	37.5	38.9
Physical properties					
Surface area / ($m^2 \cdot g^{-1}$)	149	147	142	144	144
S_Z	54	35	44	40	30
S_M	95	112	98	104	114
Pore volume /($cm^3 \cdot g^{-1}$)					
Vmicro	0.026	0.019	0.022	0.020	0.016
Vpore	0.220	0.187	0.211	0.201	0.191
MAT / %	53	54	53	53	51

Results and discussion

The cracking abilities of the five catalysts prepared in this work are given in Table 2. It can be seen from Table 3 that N162 catalyst (pure ZSM-5 zeolite) had the highest ability to crack gasoline feedstock with highest conversion under the same operating conditions. The cracking abilities of other studied catalysts are ranked as follows: N161 of pure Y zeolite the second, N164 (Y/ZSM-5=4:1) the third, N166 (Y/ZSM-5=1:4) the fourth, and N165 (Y/ZSM-5=3:2) the weakest.

Table 2 Cracking ability of different catalysts

Catalysts	N161	N164	N165	N166	N162
Cat/Oil			conversion/ %		
3.0	10.41	9.64	5.93	7.14	14.06
4.5	12.15	11.23	7.41	8.47	16.30
6.0	13.59	12.41	8.44	9.85	17.85
7.5	14.61	13.31	9.72	11.16	19.89
9.0	17.03	14.81	11.06	12.67	21.40

Protolytic cracking. It is well known that the introduction of ZSM-5 into the catalyst should increase the ratio of protolytic versus β-scission cracking, and has important impact on the product distribution since protolytic cracking is the initiation of paraffins cracking [17]. Since the product distribution obtained by pure protolytic cracking is very similar to that obtained by a radical-type mechanism with high H_2, CH_4, C_2 concentrations [18], dry gas (H_2, C_1-C_2) can be viewed as the characteristic product of portolytic cracking. The profile of dry gas selectivity against conversion therefore can be used to characterize the impact of protolytic cracking.

The profile of the dry gas selectivity versus conversion is presented in Figure 1.

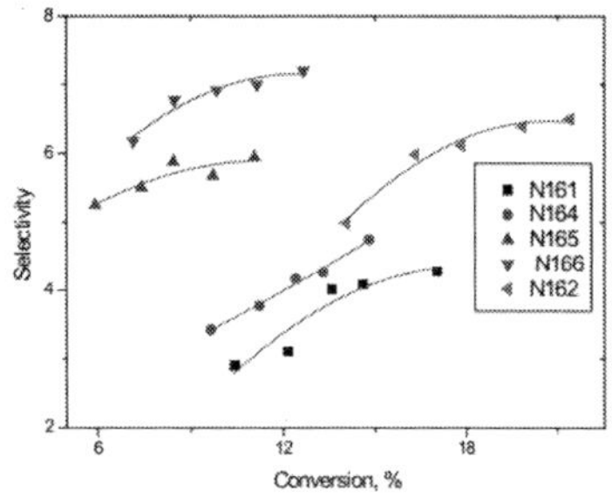

Figure 1. Dry gas selectivity profile versus conversion

Though the N162 (i.e., pure ZSM-5 zeolite) exhibited the highest conversion, it was observed from Figure 2 that the N162 catalyst did not show accordingly the highest selectivity of dry gas at constant conversion. Instead, the N166 catalyst (Y/ZSM-5=1:4) had the highest dry gas selectivity. The N161 catalyst (i.e., pure Y zeolite) showed the lowest dry gas selectivity, which was understandable since Y zeolite has weaker protolytic cracking property compared with ZSM-5 according to the literature.

The reason why the N162 catalyst did not show the highest protolytic cracking ability may be related to the feedstock and the zeolite composition of the catalyst in this study. The feedstock (FCC naphtha) has both paraffins and olefins. So the initiation of this feedstock is not completely protolytic cracking. In fact, olefins might have contributed considerably to the initiation because olefins' double bonds can be easily attacked by the brönsted acid site of the catalyst to form carbenium ions. Once the carbenium ions were formed, the catalytic cracking reactions could proceed with the transfer of a hydride ion from a reactant molecule to an adsorbed carbenium ion, which are favorable on Y zeolite due to its steric structure. Hydride transfer can result in new paraffin production which could be the reactant of protolytic cracking. So with FCC naphtha as feedstock, the synergetic effect of Y and ZSM-5 zeolite (Y/ZSM=1:4) in N166 catalyst maybe contributed to its highest dry gas selectivity at constant conversion. But of the three catalysts containing composite zeolites (N164, N165, N166), it can be seen from Figure 1 that the dry gas selectivity increased with increasing ZSM-5 ratio in the composite catalyst, indicating that the ZSM-5 did increase the ratio of the protolytic cracking.

β-scission. Since FCC naphtha was used as feedstock in this study, the products of β-scission should be mainly C3 and C4. The profile of propylene yield with conversion over different catalysts is presented in Figure 2, which shows an increasing propylene yield with increasing conversion for all five catalysts.

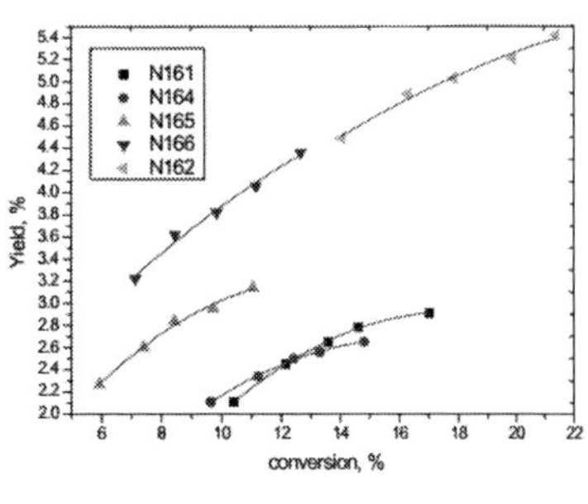

Figure 2. Profile of propylene yield with conversion for different catalysts

For those three catalysts (N164, N165, N166) containing composite zeolites, the propylene yield increased as the ratio of ZSM-5 in the catalysts increased at constant conversion. But the propylene yield kept almost constant comparing catalyst N166, with Y to ZSM-5 ratio of 1:4, to catalyst N162 with pure ZSM-5 at the same conversion. This indicates that further increasing the ratio of ZSM-5 after 80% cannot further enhance the propylene yield. An optimum ratio of ZSM-5 to Y, 4:1 in this study, exists for propylene production.

Oligmomerization. Oligomerization is often applied to convert light alkenes into fuels, so it should be avoided as much as possible if the aim of FCC process is to maximize light alkene production. Since FCC naphtha was used as feedstock in this study, so LCO could be treated as the oligomers. The selectivity for oligomers against conversion for all catalysts was presented in Figure 3. For all the catalysts, the oligomerization selectivity decreased in the following order: N161 > N164 > N162 > N165 > N166. It was found that for those three catalysts containing composite zeolites, at constant conversion the oligomerization selectivity decreased as the ratio of ZSM-5 to Y zeolite increased. This is maybe due to the fact that the oligomerization reactions proceed on the large external surface areas or in the mesopores of the zeolites [23]. But for the catalyst N162 containing pure ZSM-5 zeolite, its oligomerization selectivity was not the poorest one. This might be caused by ZSM-5 shape selective cracking producing amounts of light olefins which as reactants were favorable for proceeding secondary oligomerization. It was noticed that for catalysts N161, N164 and N162, the oligomerization selectivity nearly decreased with the increasing conversion. This could be explained by taking into account that secondary cracking of oligomers might be occurred as conversion increased. The oligomerization selectivity increasing with conversion observed for catalysts N165 and N166, might be for the oligomerization selectivity still not reaching the maximum [19] due to lower conversion.

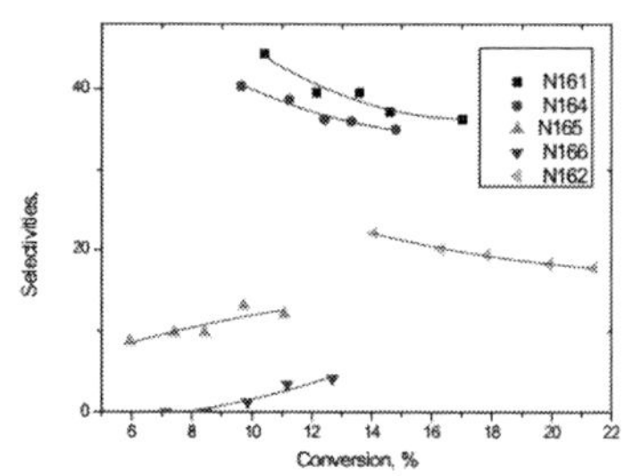

Figure 3. LCO selectivity against conversion

Hydrogen transfer. Due to its direct link with product quality and catalyst performance, hydrogen transfer is a critical issue for FCC reactions. Efforts have been devoted to the characterization of hydrogen transfer catalyst properties through different approaches and to the elucidation of factors governing the reaction [20]. Different indexes or product distribution parameters such as i-butane yield [21], the ratio of butanes to butenes [22] have been used to estimate hydrogen transfer properties on a given catalyst or series of catalysts. In this study the ratio of butanes to butenes was used to characterize the hydrogen transfer ability of five catalysts studied.

The profile of the butanes to butenes ratio against conversion over five catalysts is presented in Figure 4. Of all catalysts, the ratio of butanes to butenes increased greatly with increasing conversion, indicating increasing hydrogen transfer reaction. It's known that Y zeolite based catalyst has higher hydrogen transfer ability comparing to ZSM-5 zeolite based catalyst as also shown in Figure 4. One would think that catalysts with composite zeolites (Y/ZSM-5) should have less hydrogen transfer ability as the ratio of ZSM-5 to Y in the catalyst increases. This is not the case based on this study. As shown in Figure 4, when the ZSM-5 to Y ratio is less than 2/3, the hydrogen transfer ability of the composite Zeolite catalyst increases with increasing ZSM-5 to Y ratio of at constant conversion, and is higher than that of pure Y based catalyst. But when the ZSM-5 to Y ratio in the catalyst exceeds a value (i.e., 2/3 in this study), the hydrogen transfer ability of the composite Zeolite based catalyst dropped appreciably, with an activity in between of the pure Y based catalyst and pure ZSM-5 based catalyst.

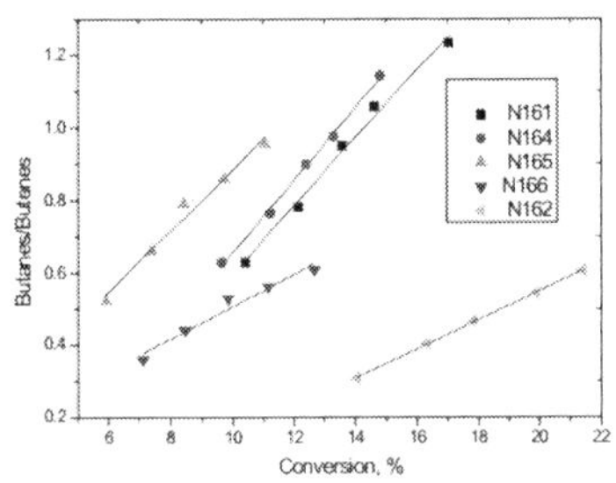

Figure 4. The ratio of butanes to butenes against conversion

Hence, to enhance the hydrogen transfer ability of the catalyst containing composite zeolites ZSM-5/(Y), an effective method is to raise the ZSM-5 to Y ratio moderately so long as not exceeding a certain value (i.e., 2/3 from this study). This could be explained as follows: under a certain ZSM-5/Y ratio, the amounts of ZSM-5 may be not enough to affect the hydrogen transfer ability of Y zeolite. However, the presence of ZSM-5 can enhance the production of light olefins and in turn the hydrogen transfer since light olefins are reactants of hydrogen transfer. Increasing ZSM-5 concentration in the composite Zeolite catalyst therefore improves its ability in hydrogen transfer. Further increasing ZSM-5 concentration, however, will allow the steric limitations in the ZSM-5 pores becoming predominate, and thus restrict the bimolecular hydrogen transfer reaction. Another possible cause of the observation in this study might be relevant to the catalyst composition. In this study, Y and ZSM-5 zeolites are incorporated simultaneous in a catalyst with a short distance for reactant molecules diffusing between Y and ZSM-5 zeolites. The synergetic effect of Y and ZSM-5 on hydrogen transfer reaction might be different from other researcher's work in which Y based catalyst is mechanically mixed with ZSM-5 based additives.

Hydrogen transfer reaction has important impact on the composition of FCC gasoline in the product, especially on the olefin content. The olefin contents of FCC gasoline in the product against conversion were depicted in Figure 5.

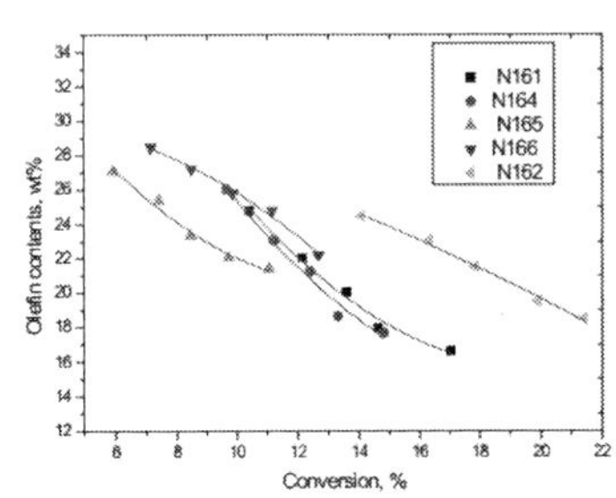

Figure 5. FCC gasoline olefin contents against conversion

It can be seen that the olefin contents with conversion profile fairly reflected the hydrogen transfer property of different catalysts. The olefin contents decreased with increasing conversion due to the enhancement of hydrogen transfer. In addition, the FCC gasoline olefins variation trend agreed well with the hydrogen transfer ability for different catalysts presented in Figure 4.

Conclusions

Five composite catalysts containing different ratio of Y to ZSM-5 zeolites have been prepared to examine the synergic effect of Y+ZSM-5 on protolytic cracking, β-scission, oligomerization and hydrogen transfer. The following conclusions can be drawn:

Though it's known that the addition of ZSM-5 into the catalyst enhances the protolytic cracking, pure ZSM-5 based catalyst did not exhibited the highest dry gas selectivity.

Among all catalysts tested, the catalyst containing composite zeolites with a Y/ZSM-5 ratio of 1:4 gave the strongest abilities of protolytic cracking and β-scission with the highest dry gas and LPG selectivities.

Pure Y zeolite based catalyst had the lowest protolytic cracking and β-scission abilities, but the strongest oligomerization ability.

The hydrogen transfer property of tested composite catalysts increases with increasing ZSM-5 concentration when the ratio of ZSM-5/ Y less than a certain value (i.e., ZSM-5:Y=2:3 in this study). When the ZSM-5/Y ratio exceeds that level, the hydrogen transfer ability of the composite zeolites based catalyst would decrease greatly.

Acknowledgement. The authors would like greatly appreciate Dr. Y. Xu and Dr. Y. J. Liu for offering the catalyst samples.

References

(1) Zhang, P. Q.; Guo, X. W.; Guo, H. C.; Wang, X. S. J. Mol. Catal. A. **2007**, *261*(2), 139-146.
(2) Long, J.; Lin, W.; Qiu, Z. H.; Tian, H. P.; Zhu,Y. X. Stud. Surf. Sci. Catal. **2007**, *166*, 55-66.
(3) Scherzer, J. Catal. Rev. Sci. Eng. **1989**, *31*(3), 215-354.
(4) Creighton, J. E.; Edwards, G. C.; Rajagopalan, K.; Peters, A. W.; Young, G. W. ACS Prepr. **1987**, *32*(3-4), 617-620.
(5) Donnelly, S. P.; Mizrahi, S. ; Sparrell, P. T.; Huss, A.; Schipper, Jr. P. H.; Herbst, J. A. ACS Prepr. **1987**, *32*(3-4) 621-626.
(6) Buchanan, J. S.; Adewuyi,Y. G.; Appl. Catal. A. 1996, *134*(2), 247-262.
(7) Buchanan, J. S.; Appl. Catal. A. **1998**, *171*(1), 57-64.
(8) Adewuyi, Y. G.; Klocke, D. J.; Buchanan, J. S. Appl. Catal. A. **1995**, *131*(1), 121-133.
(9) Adewuyi, Y. G.; Appl Catal. A. **1997**, *163*(1-2), 15-29.
(10) Madon, R. J.; J. Catal. **1990**, *129*(1), 275-287.
(11) Hollander, M. A. den; Wissink, M.; Makkee, M.L.; Moulijn, J. A. Appl. Catal. A. **2002**, *223*(1-2), 103-119.
(12) Zhao, X.; Harding, R. H. Ind. Eng. Chem. Res. **1999**, *38*(10), 3854-3859.
(13) Dzikh, I. P.; Lopes, J. M.; Lemos, F.; Ribeiro, F. R.; Appl. Catal. A **1999**, *176*(2), 239-250.
(14) Katoh, S.; Nakamura, M.; Skocpol, B. Stud. Surf. Sci. Catal. **2001**, *134*, 141-152.
(15) Qiu, Z. H.; Lu, Y. B.; Li, C. Y. Stud. Surf. Sci. Catal. **2004**, *149*, 297-304.
(16) Corma, A.; Miguel, P. J.; Orchilies, A. V. J. Catal. **1994**, *145*(1), 171-180.
(17) Haag, W. O.; Dessau, R. M. Proceedings of the 8th international congress on catalysis, Dechema, Berlin, **1984**; pp. 305-316.
(18) Corma, A.; Orchilles, A. V. Micropo. Mesopor. Mater. **2000**, *35-36*, 21-30.
(19) Pater, J. P. G.; Jacobs, P. A.; Martens, J. A. J. Catal. **1999**, *184*(1), 262-267.
(20) Sedran, U. A.; Catal. Rev.Sci.Eng. **1994**, *36*(3), 405-431.
(21) Gates, B. C.; Katzer, J. R.; Schuit, G. C. A. Chemistry of catalytic processes, Mcgraw-Hill, New York, **1979**.

(22) Corma, A.; Faraldos, M.; Martinez, A.; Mifsud, A. J. Catal. **1990**, *122*(2) , 230-239.

FUNCTIONAL SEMICONDUCTING NANOMATERIALS AND APPLICATIONS IN ENERGY CONVERSION

*SeyyedAmirhossein Hosseini, Hossein Farsi,**
*and Zhihai Li **

Department of Chemistry, Ball State University, Muncie, IN 47304

The development of clean energy resources is an urgent task to meet the future demand for renewable energy. Solar cells or photovoltaic cells represent a promising solution as an environmental benign system converting solar energy to electricity. The solution is to develop low-cost and high-efficiency materials and optimize the fabrication processes. Owing to its small band gap value and low cost, Nickel Tungstate ($NiWO_4$) nanomaterials are promising candidates for application in solar cells. In this project, nanoscale particles of NiWO4 will be prepared using different chemical approaches (thermal decomposition, co-precipitation, and sol-gel method). Then, NiWO4-nanoparticles will be doped with different tailored metal ions to understand the effect of doping on bandgap and light conversion.

Prepared nanomaterials will be characterized with different techniques, such as TEM, atomic force microscopy (AFM), scanning tunneling microscopy (STM), XRD, UV-Vis. Solar cell modules/devices will be fabricated and their photoelectrical properties will be studied. The overall solar cell efficiency of each compound will be studied using photo-electrochemical techniques in the presence and absence of light.

DOUBLING THE POWER CONVERSION EFFICIENCY IN CdS/CdSe QUANTUM DOT SENSITIZED SOLAR CELLS WITH A ZnSe PASSIVATION LAYER

Fei Huang,[ab] Jianshe Liu,*[a] Guozhong Cao*[b]

[a]College of Environmental Science and Engineering, Donghua University, Shanghai 201620, P.R. China
[b]Department of Materials Science and Engineering, University of Washington, Seattle, Washington 98195-2120, United States
* Corresponding author.
E-mail addresses: liujianshe@dhu.edu.cn (J. Liu), gzcao@u.washington.edu (G. Cao)

Abstract:

The surface passivation layer in quantum dot sensitized solar cells (QDSSCs) plays a very important role in preventing surface charge recombination and, thus, improving the power conversion efficiency. The present study demonstrated the introduction of a ZnSe passivation layer prepared with a successive ionic layer absorption and reaction (SILAR) method in CdS/CdSe co-sensitized solar cells, though not likely in the ideal form of a conformal overlayer, have significantly enhanced the power conversion efficiency, which was found to be far more efficient than the most widely used ZnS passivation layer. Not only can the ZnSe passivation layer reduce surface charge recombination, but can also enhance the light harvesting. The short-circuit current density, open-circuit voltage, fill factor, and the corresponding photovoltaic conversion efficiency were all significantly improved with the introduction of a ZnSe passivation layer but varied appreciably with the layer thickness. When three SILAR cycle layer was applied, the power conversion efficiency is as high as 6.4%, which is almost doubled the efficiency of 3.4% for the solar cell without ZnSe passivation layer. For the comparison, the CdS/CdSe co-sensitized solar cells with optimum ZnS passivation layer was also fabricated, which generated a power conversion efficiency of 4.9%, much lower than 6.4% of ZnSe passivated QDSSCs. This work demonstrated that ZnSe would be a good alternative to ZnS as a passivation material.

HIGH IONIC CONDUCTIVITY AND Li+ TRANSFERENCE NUMBERS IN MECHANICALLY STRONG ION GELS MADE FROM MULTI-IONIC SALTS, METHYL CELLULOSE AND AN IONIC LIQUID AS ELECTROLYTES FOR LITHIUM BATTERIES

*Parameswara Rao Chinnam, Vijay Chatare, Sumanth Chereddy, Ramya Mantravadi and Stephanie L. Wunder**

Department of Chemistry, Temple University,
Philadelphia, PA 19122
slwunder@temple.edu

Here, we have synthesized a new multi-ionic lithium salt (**Figure 1a**), referred to as 4merLi-TFSI, and compared its electrochemical and mechanical properties with the lithium imide salt (LiTFSI). Both have been incorporated into $PYR_{14}TFSI$, the RTIL shown in **Figure 1b**, and formed as ion gels using methyl cellulose (MC). All components are soluble in dimethyl formamide (DMF). The solutions are liquids at high temperature, but form gels upon cooling. Removal of the DMF results in ion gels composed of $PYR_{14}TFSI$/LiTFSI/MC or $PYR_{14}TFSI$/4merLi-TFSI/MC. At the same Li^+ molarity, ion gels prepared with either LiTFSI or 4merLi have very similar conductivities and mechanical properties (in the MPa range). However, the 4merLi, with its large (1270) anion, has lithium ion transference numbers, t_{Li^+} that are ~ 0.39, much larger than $PYR_{14}TFSI$/LiTFSI (t_{Li^+} ~ 0.05) or $PYR_{14}TFSI$/LiTFSI/MC (t_{Li^+} ~ 0.2). The improvement in t_{Li^+} is due to the bulkiness of the 4mer anion, and possibly the interactions of the anions and IL cations with MC. Unlike what is sometimes observed in the case of single ion conductors, the conductivity does not decrease as the t_{Li^+} increases. These separators have high thermal stability, dictated by the degradation temperature of the MC and the selected Li salt, since the thermal decomposition of the RTIL is > 400 °C, and good electrochemical stability. The best ambient temperature ionic conductivity 3×10^{-4} S/cm was achieved for the 80/20 $PYR_{14}TFSI$/MC 4merLi-TFSI.

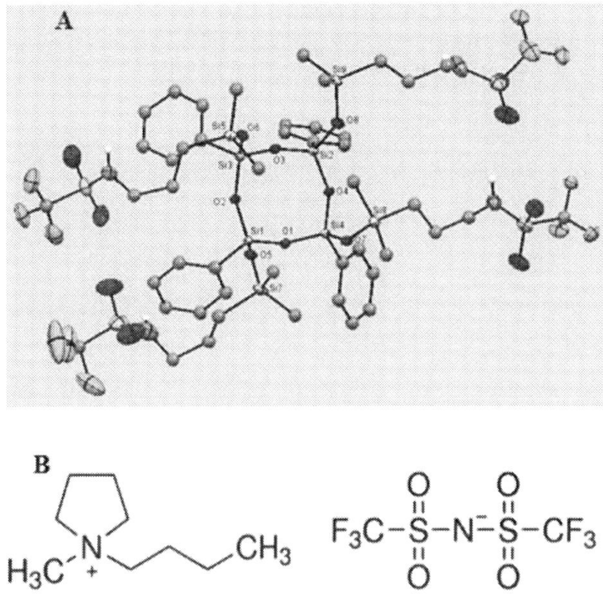

Figure 1a. Single crystal structure of 4merLi-TFSI, C-H hydrogen atoms are omitted for clarity. **Figure 1b.** Structure of $PYR_{14}TFSI$

CHEMICAL PRE-INTERCALATION APPROACHES FOR IMPROVED ELECTROCHEMICAL PERFORMANCE OF EMERGING ENERGY STORAGE ELECTRODES

Mallory Clites, Ekaterina Pomerantseva

Drexel University, 3141 Chestnut Street, Philadelphia PA 19104
mcc362@drexel.edu

Introduction

Renewable energy storage solutions, including advanced Li-ion batteries (LIBs) and emerging systems, such as beyond-Li ion batteries and pseudocapacitors, require new materials with high energy density, rapid charge transfer and stable behavior over extended cycling. Materials with layered crystal structures have proven themselves among the most reliable electrode materials, exhibiting high performance in these emerging energy systems. However, improvement of ion storage capability and cycling stability of these materials is needed and can be achieved through chemical modification of layered materials prior to electrochemical cycling.

For Na-ion batteries (SIBs) layered vanadium oxides are of particular interest due to the presence of vanadium in high oxidation state (V^{5+}) and its ability to form gel structures that can be controlled by tuning synthesis parameters to create materials with targeted crystal structures.[1] Bilayered vanadium oxide (δ-V_2O_5) is built of stacked edge-sharing double layers of distorted VO_5 square pyramids with a large interlayer spacing ranging from 11-13 Å.[2] This phase has demonstrated a high reversible capacity of 250 mAh/g in SIB cells, attributed to the large interlayer spacing allowing the accommodation of Na^+ ions and facilitated ion diffusion during electrochemical cycling.[2-3]

Chemical pre-intercalation is a low cost synthesis technique based on mixing of the host electrode material with a solution containing charge-carrying ions in the form of a dissolved salt or hydroxide. This wet chemical process is often followed by hydrothermal and/or elevated temperature annealing treatments. For layered materials, the inclusion of charge-carrying ions prior to cycling can be used to control interlayer spacing and ionic content between structural layers. Such modification can lead to facilitated diffusion and improved stability. In the past, a chemical pre-intercalation approach was successfully used to synthesize Li-containing electrode materials for Li-ion batteries. It was shown that lithium pre-intercalated vanadium oxides exhibited higher specific capacities and better capacity retention, as compared to the pristine V_2O_5, which were attributed to the facilitated diffusion of Li^+ ions and improved structural stability.[4]

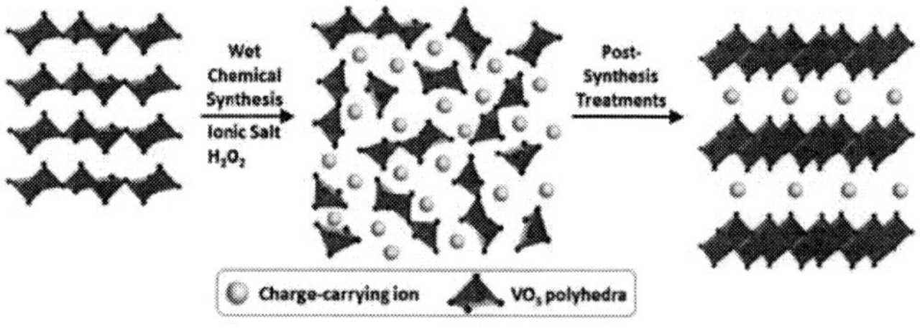

Figure 1. Schematic representation of chemical preintercalation methods used to form the bilayered $Na_xV_2O_5$ phase.

In this work, we developed a chemical pre-intercalation method for the facile insertion of ions into a bilayered vanadium oxide crystal structure through soft chemical approaches, outlined in **Figure 1**. Our work demonstrates the first use of chemically pre-intercalated bilayered $Na_xV_2O_5$ materials as a cathode active materials in SIB cells and highlights the potential of this method to be used to fabricate electrode materials for a range of emerging energy storage systems.

Experimental Methods

Synthesis. Na-containing vanadium oxide materials were synthesized using a procedure adopted from previous reports.[4a, 4b] **Figure 1** schematically shows the structural transformation from the orthorhombic layered α-V_2O_5 to bilayered δ-$Na_xV_2O_5$. In the first step, α-V_2O_5 powder is decomposed using hydrogen peroxide in the presence of Na^+ ions resulting in the formation of a Na-containing vanadia gel. In the next step, post-synthesis treatment, is used to rearrange vanadium/oxygen framework and produce a material with the targeted crystal structure. In a typical synthesis, 3.2 g of sodium chloride was dissolved in 15mL of deionized water and then added to 15 mL of hydrogen peroxide (30 wt%). The solution was stirred vigorously as 0.5 g of α-V_2O_5 was slowly added into the mixture. The solution was allowed to stir for 2 hours at ambient conditions. The temperature was then raised to 50°C and the mixture was stirred for another 4 hours. The produced thick brown gel was subjected to four days aging inside a chemical fume hood at room temperature for extended periods of time. Hydrothermal treatment was done by combining 12 mL of Na-containing vanadia precursor material with 6 mL of 5 M sodium chloride solution in a 23 mL Teflon-lined stainless steel autoclave (Parr Instruments, USA) and heating at 220°C for 24 hours. Samples were thoroughly washed, filtered, and dried under vacuum at 100°C for 12 hours. All chemicals used for synthesis were produced by Fisher Scientific.

Materials Characterization. Morphology of materials was investigated using a Zeiss Supra 50VP (Germany) scanning electron microscopy (SEM) equipped with an Oxford energy dispersive X-ray spectroscopy (EDX) attachment. EDX spectra were recorded from multiple 10x10 μm regions for each sample. Chemical composition of the synthesized materials was verified using a Varian 240FS flame atomic absorption spectrometer (Agilent Technologies, USA). X-ray powder diffraction (XRD) was used to analyze phase composition of the synthesized vanadium oxide. XRD was performed using a SmartLab Rigaku X-ray diffractometer with a Cu Kα radiation source and optics in a Bragg-Brentano configuration. Samples were analyzed in a 2θ range of 5-40° with a step size of 0.02°. The interlayer spacing was calculated using Bragg's law from the (001) planes located at ~8Å 2θ for the bilayered $Na_xV_2O_5$ phase. Thermal gravimetric analysis (TGA) was performed on a Perkin-Elmer TGA 7 (Perkin-Elmer, USA) at a heating rate of 10°C/min in nitrogen from room temperature up to 600°C.

Electrochemical Characterization. Electrochemical studies were conducted in a SIB half-cell configuration using sodium metal as a reference and a counter electrode. Cathodes were fabricated by mixing active material with carbon black and polyvinylidene fluoride (PVDF) in a weight ratio of 70:20:10, respectively. The mixture was dispersed in N-methyl-2-pyrrolidone (NMP). The slurry was cast onto an Al foil current collector using a doctor blade method. The electrode films were then dried under vacuum at 100°C for 12 hours. CR2032 coin-type cells were assembled in an Ar-filled glovebox (MBraun Inc, Germany) using 1 M $NaClO_4$ dissolved in propylene carbonate (PC) with 5 vol% fluoroethylene carbonate (FEC) as the electrolyte and glass microfiber filter (Whatman, USA) as the separator. All cells were galvanostatically cycled at a current density of 20 mA/g between 1.0 and 4.3 V (vs. Na/Na^+) at room temperature using a multi-channel Arbin battery testing station (Arbin Instruments, USA).

Results and Discussion

Figure 2 shows materials characterization of the synthesized materials. The aged material (**sample 1**) consisted of thick mats of partially formed nanowires. Better formed one-dimensional particles with a nanowire morphology, with diameters of 50-100 nm and lengths of several microns, was seen for the hydrothermally treated sample (**sample 2**). The appearance of distinct, individual nanowires suggests that hydrothermal treatment after aging allows for the formation of longer, more highly crystalline nanowires compare to only aging. EDX spectra verified that Na^+ insertion occurred via our chemical preintercalation approach. A more intense Na^+ peak is seen for sample 2, attributed to the inclusion of sodium salt in the hydrothermal treatment, allow for further Na^+ intercalation.

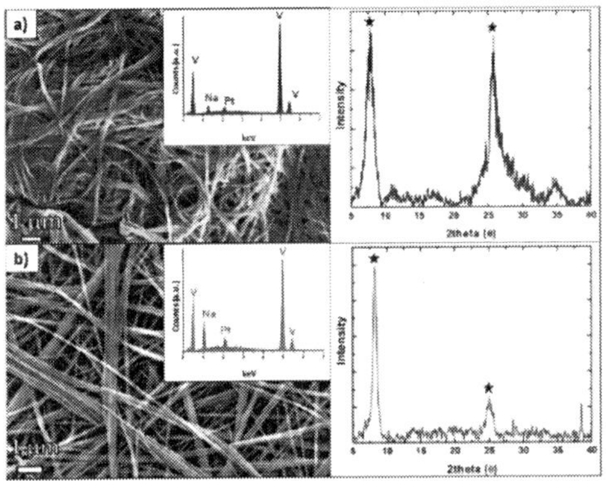

Figure 2. SEM micrographs, EDX spectra, and XRD patterns for (a) sample 1 and (b) sample 2. Stars on XRD patterns represent the Miller indices for δ-$Na_xV_2O_5$.

The XRD pattern of sample 1 was indexed to a pure bilayered $Na_xV_2O_5$ phase with broad peaks of relatively low intensity, suggesting low crystallinity of the just aged material. This pattern is similar to the one previously reported for hydrated, bilayered $Na_xV_2O_5$ phase.[5] The XRD pattern of sample 2 was also corresponds to the bilayered phase, with better signal-to-noise ratio and more narrow peaks, suggesting improved crystallinity. This result indicates that hydrothermal treatment at elevated temperatures is necessary to obtain material with high crystallinity. The interlayer spacing of samples 1 and 2 were calculated to be 11.42 and 11.36 Å, respectively. AS the Na^+ insertion in sample 2 is greater than sample 1, we would anticipate the interlayer spacing to be large. However, TGA, seen in **Figure 3**, revealed a larger water content in sample 1 compared to sample 2. It can thereby be concluded that the water content within the layers has a greater effect on the interlayer spacing rather than the Na^+ content. The lower water content in sample 2 is in good agreement with the more narrow peaks seen in XRD.

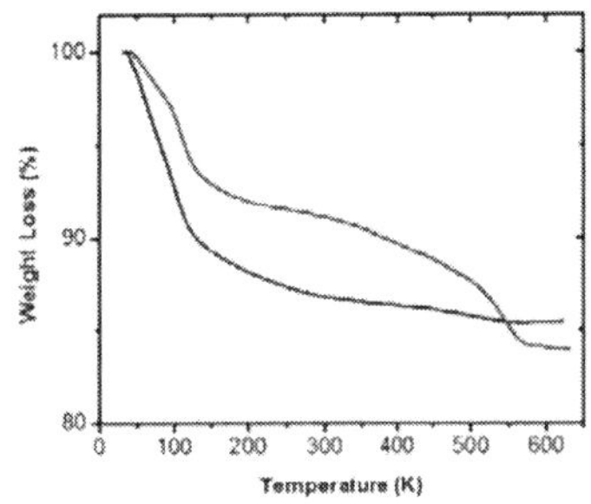

Figure 3. TGA data for sample 1 (blue line) and sample 2 (green line).

The electrochemical performance of the synthesized δ-$Na_xV_2O_5$ in SIB half-cells are shown below in **Figure 4**. Sample 1 showed an initial discharge capacity of 216 mAh/g while sample 2 showed a capacity of 365 mAh/g. This result, to the best of our knowledge, is the highest achieved capacity of a cathode material in SIB half-cells. Capacity retention improves from 25% for the aged material (sample 1) to 50% for the hydrothermally treated material (sample 2). Both the high initial capacities and improved capacity retention for sample 2 suggest that hydrothermal treatment is crucial for increased electrochemical performance of the bilayered δ-$Na_xV_2O_5$ material.

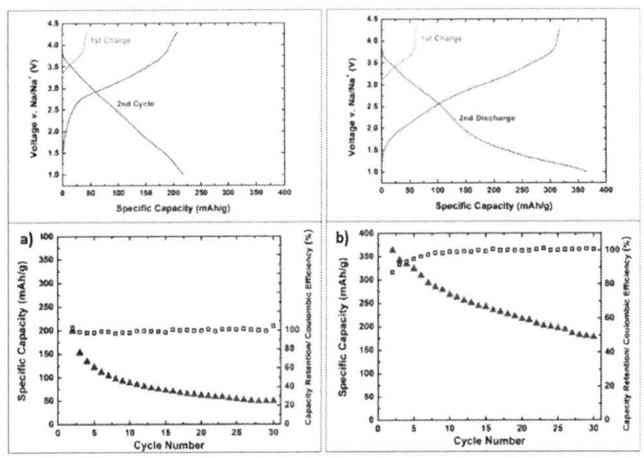

Figure 4. Galvanostatic first charge and second discharge/charge cycle curves as well as the cycling performance curves of (a) sample and (b) sample 2 in SIB cells. All tests were performed in the voltage range of 1.0 – 4.3 V at a current density of 20 mA/g.

Conclusions

In this work, we demonstrate the chemical pre-intercalation of bilayered vanadium oxide to form δ-$Na_xV_2O_5$, which achieves record discharge capacities in SIB cells. In our presentation, we will discuss strategies that lead to the stabilization of high initial capacity over extended cycling.

Acknowledgements. We would like to acknowledge Drexel University's Core Facilities for assistance in materials characterization. We would also like to acknowledge Sarah Culbert for her assistance in material synthesis.

References

(1) (a) *Solid State Ionics* **1996,** *86-88* (2), 935-942; (b) Chernova, N. A.; Roppolo, M.; Dillon, A. C.; Whittingham, M. S., *Journal of Materials Chemistry* **2009,** *19*, 2526-2552.

(2) Tepavcevic, S.; Liu, Y.; Zhou, D.; Lai, B.; Maser, J.; Zuo, X.; Chan, H.; Král, P.; Johnson, C. S.; Stamenkovic, V.; Markovic, N. M.; Rajh, T., *ACS Nano* **2015,** *9* (8), 8194-8205.

(3) (a) Su, D.; Wang, G., *ACS Nano* **2013,** *7* (12), 11218-11226; (b) Chen, Z.; Augustyn, V.; Jia, X.; Xiao, Q.; Dunn, B.; Lu, Y., *ACS Nano* **2012,** *6* (5), 4319-4327.

(4) (a) Zhao, Y.; Han, C.; Yang, J.; Su, J.; Xu, X.; Li, S.; Xu, L.; Fang, R.; Jiang, H.; Zou, X.; Song, B.; Mai, L.; Zhang, Q., *Nano Letters* **2015,** *15*, 2180-2185; (b) Semenenko, D. A.; Itkis, D. M.; Pomerantseva, E. A.; Goodilin, E. A.; Kulova, T. L.; Skundin, A. M.; Tretyakov, Y. D., *Electrochemistry Communications* **2010,** *12*, 1154–1157; (c) Li, W.-D.; Xu, C.-Y.; Pan, X.-L.; Huang, Y.-D.; Zhen, L., *Journal of Materials Chemistry A* **2013,** (1), 5361-5369; (d) Tian, X.; Xua, X.; He, L.; Wei, Q.; Yan, M.; Xu, L.; Zhao, Y.; Yang, C.; Mai, L., *Journal of Power Sources* **2014,** *255*, 235-241.

(5) Lee, C.-Y.; Marschilok, A. C.; Subramanian, A.; Takeuchi, K. J.; Takeuchi, E. S., *Physical Chemistry Chemical Physics* **2011,** *13*, 18047-18054.

HOW TO GET MORE JOLT FOR YOUR BOLT: *DIRECT ENHANCEMENT OF LITHIUM ION BATTERIES UTILIZING A BIOLOGICAL TOOL KIT.*

Scott Riley and Mark Allen

In the modern world there exists an ever increasing demand for more efficient forms of energy storage. Electrochemical cells are the most widely used form of portable energy storage and are possible candidates for applications on larger scales. From mobile devices to cars, batteries are integrated into many facets of daily life, yet little improvement in their performance is being recognized. While batteries have been improving over the years, the growth is slow and staggered. A new avenue involving biology's exceptional control over synthesis and binding materials together could lead to the significant improvement that is needed. My work is centered on maximizing battery performance by interfacing biology and batteries. My research is focused on using an approach involving an array of interdisciplinary techniques to address performance issues commonly found in batteries. I use the technique of phage display to isolate Solid Binding Peptides (SBP) that bind to lithium cathode materials, specifically Lithium Cobalt Phosphate (LCP). This technique has yielded several peptides that bind to LCP. These peptides are utilized in several ways both to synthesize LCP and also bio-tether it to conducting Carbon Nanotubes (CNT). Preliminary and exciting evidence will be presented in support of this method of synthesizing materials and constructing electrodes as well as the enhancement of multiple aspects of battery performance including capacity, cyclability, and power.

DEVELOPMENT OF FLEXIBLE POWER SOURCES USING NANOMATERIALS AND POLYMERS

Xianyang Meng, Zhiqian Wang and Somenath Mitra

The increasing interest in portable and flexible electronics requires the development of conformal and flexible power sources like flexible batteries and supercapacitors. The former is more suitable for high energy applications while the latter is more for high power uses/ fast charge and discharge. In this paper we present the applications of nanomaterials and high molecular polymers in flexible power source development including zinc based batteries and super capacitors containing highly defective carbon nanotubes to provide high double layer capacitance. Highly conductive nanomaterials like carbon nanotubes can serve as electrode scaffolds, conductive additives and even active materials. Further treatment like functionalization could bring in functional groups and change the characteristics like surface area, hydrophobicity, dispersibility, conductivity and particle size. Nano-metal oxides typically serve as electrode active materials or catalysts. Polymers are good candidates for electrolytes, separators, binders as well as conductive additives in certain cases. Combination of these materials into composites further provides promising ways to fabricate flexible batteries, capacitors and even solar cells.

EFFICIENT LITHIUM STORAGE FROM THREE-DIMENSIONAL GRAPHENE-DERIVED NANOMATERIALS

Hung-Ju Yen,[1] Hsinhan Tsai,[1] Ming Zhou,[1] Aiping Chen,[2] Edward F. Holby,[3] Samrat Choudhury,[3] Xianliang Wang,[4] Linxiang Zhu,[4] Haiqing Lin,[4] Liming Dai,[5] Lyudmyla Adamska,[2] Sergei Tretiak,[2] Gang Wu[4,] & Hsing-Lin Wang[1,]**

[1] Chemistry Division, Los Alamos National Laboratory
[2] Materials Physics and Applications Division, Los Alamos National Laboratory
[3] Materials Science and Technology Division, Los Alamos National Laboratory
[4] Department of Chemical and Biological Engineering, University at Buffalo, the State University of New York, Buffalo
[5] Department of Macromolecular Science and Engineering, Case Western Reserve University
Dr. H.-L. Wang (email: hwang@lanl.gov)
Prof. G. Wu (gangwu@buffalo.edu)

Abstract

Tradition graphene anodes in lithium ion batteries (LIBs) suffer significant performance loss due to the restacking of graphene layers. In this work, we have demonstrated a facile synthesis of preparing a series of graphene-derived nanomaterials. Among other potential application, these nanomaterials achieved an excellent cyclic duarbility with much enhanced capacity as novel anode materials in LIBs. These exceptional metrics observed within the novel assemblies is primarily due to the robust structure with high surface areas along with the optimal *d*-spacing allowing for facilidated Li adsorpiton/desorption and diffusion. Furthermore, the signigicantly enhanced rate performane was achieved after optimizing the *structure* of the self-assemblies. Development of nanomaterials anode suggests that the optimal design of graphene materials with robust structures is very important for next generation LIBs with high-energy storage efficiencies.

SYNDIOTACTIC POLYSTYRENE BASED IONOGEL MEMBRANES FOR HIGH TEMPERATURE LITHIUM-ION BATTERIES

Prasad Raut[a], *Yu-Ming Chen*[b], *Yu Zhu*[b] *and Sadhan C. Jana*[a]

[a]Department of Polymer Engineering, The University of Akron, OH, 44325, USA
[b]Department of Polymer Science, The University of Akron, OH, 44325, USA

Development of lithium-ion batteries for hybrid vehicles and other high temperature applications require thermally stable and safe operation of the cells. The use of lithium-ion battery for high temperature applications is currently inhibited by the high reactivity and the volatility of liquid electrolytes. Volatile organic solvents in Li-ion cells are prone to undergo thermal runaway leading to fire, explosion, or swelling of the cell. Materials for electrolyte and separator are key component for development of safe and stable Li-ion batteries. Non-flammable, non-volatile ionic liquids (ILs) are an ideal solution over low boiling organic solvents. Ionogels are the hybrid materials incorporating ionic liquids (ILs) in porous polymeric matrices. This work focuses on development of ionogel membranes for Li-ion batteries using porous syndiotactic polystyrene (sPS) gels with the aim of increasing the operating temperature and eliminating the safety hazards. This approach combines the thermal stability, ruggedness of the polymer and the attributes of IL. The performance of these ionogel membranes are tested against current electrolyte-separator technology using Li^+/ graphite half-cells at room temperature and using full cell at elevated temperatures.

A CASE STUDY: PROBING ELECTRON AND ION TRANSPORT OVER MULTIPLE LENGTH SCALES

Amy C. Marschilok[a], David C. Bock[b], Christopher J. Pelliccione[b], Wei Zhang[b], Jiajun Wang[b], Jun Wang[b], Feng Wang[b], Esther S. Takeuchi,[a,b] Kenneth J. Takeuchi[a]

a Stony Brook University
b Brookhaven National Laboratory
*Email: amy.marschilok@stonybrook.edu

Introduction

Metal oxides such as magnetite (Fe_3O_4) are worthy of evaluation as active materials for electrochemical energy storage due to its earth abundance, low cost, environmentally benign iron metal centers and high theoretical capacity. Successful future implementation will require understanding of the fundamental electrochemical reduction-oxidation mechanisms. Mesoscale characterization can provide information regarding the localized resistance which must be understood addressed in order to achieve significant improvements to current capability and reversibility.

A complete study of an electroactive nanomaterial in the complex mesoscale environment of a composite battery electrode should also consider the level of electroactive particle agglomeration, and agglomerate distribution within the battery electrode. The roles of material dispersivity and composite electrode heterogeneity are important considerations, where visualization tools can provide useful insight.

Experimental

Nanocrystalline magnetite (Fe_3O_4) powders and composite electrodes with different crystallite sizes were prepared, characterized, and electrochemically evaluated. Transmission electron microscopy (TEM) examination of cross-sectioned electrodes was used to quantify the aggregate size of the Fe_3O_4 active material. Transmission x-ray microscopy (TXM) combined with x-ray absorption near edge spectroscopy (XANES) at the National Synchrotron Light Source (NSLS) were used to determine the distribution of the iron oxidation states within the electrodes before and after electrochemical testing.

Results and Discussion

Under higher magnification, TEM showed the Fe_3O_4 electrodes to contain aggregated primary particles with a distribution of aggregate sizes (Figure 1). Although the crystallite sizes of the two magnetite samples involved in the study were different (28 and 9 nm), the observed sizes of the aggregates and the aggregate distributions within the electrodes were similar. Transmission x-ray microscopy (TXM) combined with x-ray absorption near edge spectroscopy (XANES) at the National Synchrotron Light Source (NSLS) were used to determine the distribution of the iron oxidation states within the electrodes before and after electrochemical testing. The impact of both crystallite size and agglomeration on electrochemical performance were assessed and will be discussed.

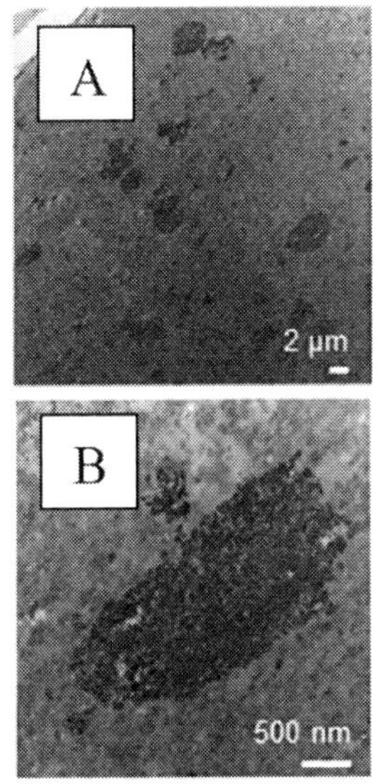

Figure 1. Composite electrode structure via transmission electron microscopy under lower (top) and higher (bottom) magnifications

References

(1) Bock, David C.; Kirshenbaum, Kevin C.; Wang, Jiajun; Zhang, Wei; Wang, Feng; Wang, Jun; Marschilok, Amy. C.; Takeuchi, Kenneth J.; and Takeuchi, Esther S., *ACS Appl. Mater. Interfaces*, **2015**, *7*, 13457.
(2) Zhang, Wei; Bock, David C.; Pelliccione, Christopher J.; Li, Yan; Wu, Lijun; Zhu, Yimei; Marschilok, Amy. C.; Takeuchi, Esther S.; Takeuchi, Kenneth J.; and Wang, Feng, *Adv. Energy Mater.*, **2016**, DOI://10.1002/aenm.201502471.

THE ROLE OF PHYSICOCHEMICAL PROPERTIES ON ELECTROCHEMICAL PERFORMANCE FOR ONE DIMENSIONAL MANGANESE OXIDE ELECTROACTIVE MATERIALS

Kenneth J. Takeuchi[a], Lijun Wu[b], Feng Xu[b], Alexander B. Brady[a], Jessica L. Durham[a] Jianping Huang[a], Eric Dooryhee[b], Yimei Zhu[b], Amy C. Marschilok[a], and Esther S. Takeuchi[a,b]

a Stony Brook University
b Brookhaven National Laboratory
*Email: kenneth.takeuchi.1@stonybrook.edu

Introduction

Hollandite and cryptomelane ($M_xMn_8O_{16}$, OMS-2) structured materials consist of octahedral units of edge-sharing manganese oxide octahedra (MnO_6) which interlink to form tunnels of 2 × 2 (~0.46 nm) dimensions (Figure 1). Bimetallic hollandite type materials have been prepared where cations with 1+ or 2+ charge partially occupy locations within the manganese oxide tunnels, and the Mn cations possess mixed 3+ and 4+ oxidation states. Typically, the ion located within the tunnel is inert, such as for potassium cryptomelane, however, silver hollandite is an interesting case as the Ag+ cationic center is potentially electrochemically active.

Previous research demonstrated the ability to control crystal size of $Ag_xMn_8O_{16}$ by manipulation of Ag content over a range of $1.0 \leq x \leq 1.8$ where lower Ag content was accompanied by small crystallite size and higher silver content demonstrated large crystallite size.[1,2,3] A seven-fold increase in delivered capacity for Li/AgxMn8O16 electrochemical cells (160 versus 23 mAh/g) was recently observed upon a seemingly small change in silver content (x ~1.1(L-Ag-OMS-2) and 1.6(H-Ag-OMS-2)), prompting characterization of the structure and defects of the silver hollandite nanorods through the combined use of local (atomic imaging, nanodiffraction, electron energy-loss spectroscopy) and bulk (synchrotron based x-ray diffraction, thermogravimetric analysis) techniques.

Experimental

Strategies for synthetic control of composition (x content), and crystallite size, and particle size of silver hollandite ($Ag_xMn_8O_{16}$) were developed. Material characterization via bulk (thermogravimetric analysis, inductively coupled plasma optical emission spectroscopy, surface area) and local (transmission electron microscopy, electron energy loss spectroscopy) methods was done to determine compositional, and physical properties. Functional electrochemistry associated with lithium ion transport was evaluated.

Results and Discussion

A structure consistent with that refined by synchrotron based XRD (NSLS II) was observed by selected area diffraction and high resolution transmission electron microscopy, however the Ag occupancy varies significantly even within neighboring channels. Both local (EELS) and bulk measurements (TGA) indicated a greater quantity of oxygen vacancies in L-Ag-OMS-2, resulting in lower average Mn valence relative to H-Ag-OMS-2. A lower Mn oxidation state on the surface relative to the interior of the nanorods was seen by electron energy loss spectroscopy. Average Mn valence was estimated for H-Ag-OMS-2 and L-Ag-OMS-2, considering both Ag occupancy and oxygen vacancies.[4]

The significance of the oxygen vacancies can be best understood through consideration of the crystallographic structure of silver hollandite. Silver hollandite has a body-centered tetragonal structure where the Ag is surrounded by eight MnO_6 octahedra, forming a tunnel with a dimension of ~0.51 nm in the ab plane which extends along the [001] direction. Therefore, if the nanorod has no defects, the diffusion of Li ions in ab plane would be limited, where defects could facilitate ab plane diffusion. Consequences for resulting electrochemistry will be discussed.

Figure 1. Structure of hollandite

References

(1) Zhu, Shali; Marschilok, Amy C.; Lee, Chia-Ying; Takeuchi, Esther S.; Takeuchi, Kenneth J. *Electrochem. Solid-State Lett.*, **2010,** *13,* A98.

(2) Takeuchi, Kenneth J.; Yau, Shali Zhu; Menard, Melissa C.; Marschilok, Amy C.; Takeuchi, Esther S. *ACS Appl. Mater. Interfaces*, **2012,** *4,* 5547.

(3) Takeuchi, Kenneth J.; Yau, Shali Zhu; Subramanian, Aditya; Marschilok, Amy C.; Takeuchi, Esther S. *J. Electrochem. Soc.*, **2013,** *160,* A3090.

(4) Wu, Lijun; Xu, Feng; Zhu, Yimei; Brady, Alexander B.; Huang, Jianping; Durham, Jessica L.; Dooryhee, Eric; Marschilok, Amy C.; Takeuchi, Esther S. and Takeuchi, Kenneth J. *ACS Nano,* **2015,** *9(8),* 8430.

TUNNEL STRUCTURED MANGANESE OXIDES FOR ELECTROCHEMICAL ENERGY APPLICATIONS

Bryan W. Byles and Ekaterina Pomerantseva

Department of Materials Science and Engineering, Drexel University, Philadelphia, PA 19104, USA
bwb46@drexel.edu

Introduction

Manganese oxides represent an attractive family of inexpensive and environmentally friendly materials with high electrochemical activity in energy and water treatment applications. More specifically, manganese oxides with tunnel crystal structures show enhanced electrochemical performance in energy storage devices, water desalination systems, and as oxygen evolution reaction (OER) catalysts. These materials possess crystal structures built from MnO_6 octahedra sharing corners and edges to form different size and shape structural tunnels around stabilizing cations. By changing the size and charge of the stabilizing cation, the tunnel structures can be precisely controlled. Moreover, the chemical composition of these tunnel phases can be further tailored by utilizing chemical approaches such as acid treatment and ion exchanges, thus resulting in a materials system with highly tunable structures. Examples of the crystal structures of several tunnel manganese oxides are shown in Figure 1. In this report, we focus on the performance of three of these tunnel structure manganese oxides, α-MnO_2, $Na_{0.44}MnO_2$, and a new hybrid tunnel structured material, in Na-ion batteries and electrochemical water splitting.

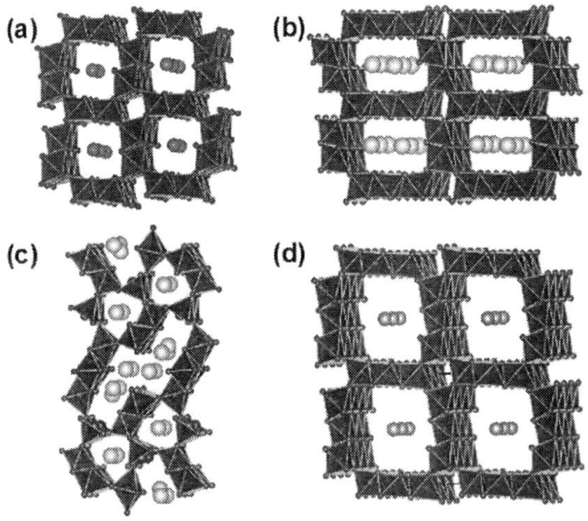

Figure 1. Schematics for select tunnel crystal structured manganese oxides. (a) α-MnO_2 stabilized by K^+ ions (green), (b) Na 2x3 MnO_2 stabilized by Na^+ ions (yellow), (c) $Na_{0.44}MnO_2$, and (d) todorokite MnO_2 stabilized by Mg^{2+} ions (orange). Oxygen atoms are red, and the purple polyhedra representing MnO_6 octahedra.

Experimental

Materials Synthesis: α-MnO_2 nanowires were synthesized via a process described elsewhere [1]. Briefly, 50 mg of NH_4Cl and 150 mg of $KMnO_4$ were fully mixed into 50 mL of deionized water. 20 mL of this solution was placed in a 23 mL Teflon-lined stainless steel autoclave (Parr Instruments, USA) and hydrothermally treated at 150° C for 48 hours. Final products from hydrothermal treatment were filtered, washed, and dried for 12 hours at 100° C.

To prepare $Na_{0.44}MnO_2$ nanowires, 50 mg of Mn_2O_3 was added to 20 mL of 6M NaOH, placed in a 23 mL Teflon-lined stainless steel autoclave, and hydrothermally treated at 220° C for 6 days [2]. The resulting products were filtered, washed, and dried for 12 hours at 100° C.

Hybrid tunnel structured manganese oxide was synthesized via a modified hydrothermal method. The precursor material, layered Na-birnessite manganese oxide, was prepared using a process adopted from a past report [3]. In short, 50 mL of 1M H_2O_2 and 50 mL of 0.5 M NaOH were combined and then added to a 50 mL solution of $Mn(NO_3)_2$ under stirring. The resulting solution was allowed to react for one hour at room temperature, after which the precipitate, Na-birnessite manganese oxide, was filtered and thoroughly washed with deionized water. 100 mg of this material was added to a Teflon-lined stainless steel autoclave and hydrothermally treated at 220° C for 96 hours. Resulting materials were filtered, washed, and dried for 12 hours at 100° C.

Acid leaching of these materials was performed via stirring the nanowires in concentrated nitric acid for various amounts of time at room temperature.

Characterization: X-ray powder diffraction (XRD) patterns for phase identification were recorded using a Rigaku SmartLab X-Ray Diffractometer (Japan) with Cu Kα radiation, and scanning electron microscopy (SEM) images were collected with a Zeiss Supra 50VP (Germany) scanning electron microscope equipped with an energy dispersive X-ray spectroscopy (EDS) attachment to analyze material chemistry.

Electrochemical Characterization: Positive electrodes for Na-ion batteries were made by first coating a slurry consisting of 70% active material, 20% acetylene black, and 10% poly(vinylidene fluoride) binder dispersed in 1-methyl-2-pyrrolidinone onto an Al-foil current collector using a doctor blade. The film was punched into 12 mm electrodes, which were dried at 100° C for 12 hours in a vacuum oven before transferring to an Ar-filled glovebox. Electrochemical testing was performed in two-electrode coin cells (2032 type), with sodium metal as the counter and reference electrodes, glass fiber as the separator, and 1M $NaClO_4$ in propylene carbonate as the electrolyte. The hybrid tunnel structured manganese oxide was tested from 1.5-4.0 V vs Na/Na^+ at a current rate of C/10 in an Arbin battery testing station (Arbin Instruments, USA). Cyclic voltammetry (CV) curves were recorded using a VMP3 potentiostat/galvanostat (BioLogic, France) at a scan rate of 0.1 mV s^{-1}.

Electrochemical water splitting experiments were carried out using a rotating disc electrode (RDE) (Pine Instruments, USA) with 0.1M KOH as the electrolyte, platinum wire as the counter electrode, and an Ag/AgCl electrode as the reference electrode. For the working electrode, 10 µL of a solution consisting of 0.5 mL water, 0.5 mL isopropyl alcohol, 2 µL of Nafion, 3.2 mg of manganese oxide, and 0.8 mg of conductive carbon black was deposited on the RDE electrode. The speed of the RDE was kept at 1600 rpm.

Results and Discussion

Figure 2 shows the materials characterization of the three tunnel structured manganese oxides studied in this work. All three materials (α-MnO_2, $Na_{0.44}MnO_2$, and the hybrid tunnel structure) are synthesized in a flexible, high aspect ratio nanowire morphology. The nanowires are 20-100 nms in diameter and up to several microns in length, with the hybrid tunnel structured nanowires being smaller in diameter than the other two manganese oxides. The XRD patterns for α-MnO_2 and $Na_{0.44}MnO_2$ nanowires are shown in Figures 2(b) and 2(d), and they can be indexed to single phase tetragonal α-MnO_2 (PDF# 44-0141) and orthorhombic $Na_{0.44}MnO_2$ (PDF# 27-0750). EDS spectra were recorded on each type of nanowires, and the chemical compositions were found to be $K_{0.11}MnO_2$ (α-MnO_2), $Na_{0.44}MnO_2$, and $Na_{0.20}MnO_2$ (hybrid tunnel structure). Further, transmission electron microscopy analysis, which will be discussed in our presentation, indicated that the hybrid tunnel structure material is dominated by the "Na 2x3" structure shown in Figure 2(b).

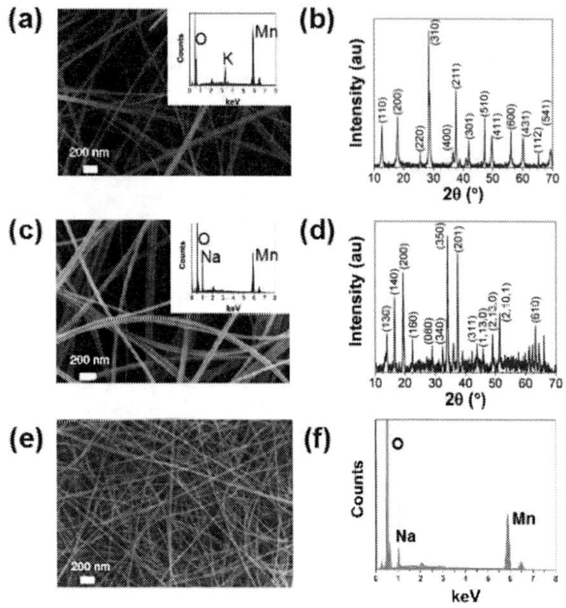

Figure 2. (a) SEM image of α-MnO_2 nanowires with inset of EDS spectra, (b) XRD pattern for α-MnO_2 nanowires, (c) SEM image of $Na_{0.44}MnO_2$ nanowires with inset of EDS spectra, (d) XRD pattern of $Na_{0.44}MnO_2$ nanowires, (e) SEM image of hybrid tunnel structured nanowires, and (f) EDS spectra for hybrid tunnel structured nanowires.

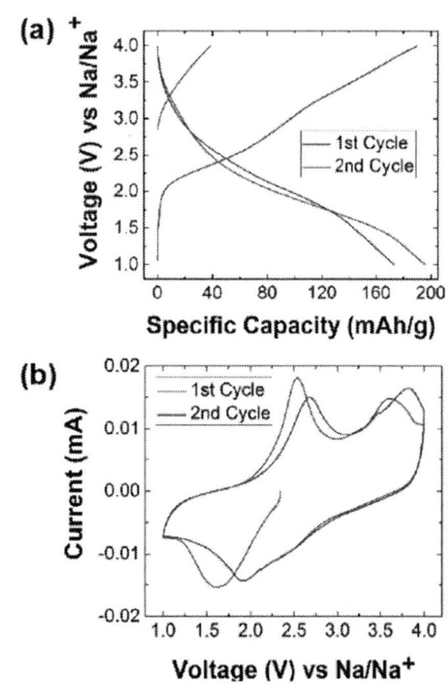

Figure 3. (a) First and second galvanostatic charge/discharge cycles for the hybrid tunnel structured materials and (b) first and second cycles of the cyclic voltammogram for the same material.

The performance of the hybrid tunnel structured manganese oxide as a Na-ion battery cathode can be seen in Figure 3. When cycled in a Na-ion half cell, a first charge capacity of 40 mAh g^{-1} is achieved, corresponding to 0.14 mol of Na^+ ions extracted. When this amount of extracted Na^+ is compared to the results from EDS, it is observed that not all sodium present in the structure of the material is electrochemically extracted. This indicates that a certain amount of Na^+ is maintained within the crystal structure, and it is believed these ions may help stabilized the crystal structure from breaking down during repeated ions insertion/deinsertion. This notion motivates ongoing work for the stabilization of this crystal structure using chemical ions intercalation. Upon first discharge, a high capacity of 196 mAh/g is achieved. A small decrease in capacity from first to second discharge cycle is observed, which is common behavior for Na-ion battery electrode materials [4]. In our presentation, we will discuss cycle life and rate capability behavior of this material in further detail.

The activity of pristine and chemically treated α-MnO_2 and $Na_{0.44}MnO_2$ nanowires is shown in Figure 4. At a potential of 1.85 V vs RHE, the pristine α-MnO_2 and $Na_{0.44}MnO_2$ nanowires show current densities of 4.34 and 3.51 mA/cm^2, respectively. After immersion in nitric acid for 24 hours, the same materials show current densities of 8.78 and 8.23 mA/cm^2, respectively. The post-synthesis acid-leaching treatment increased the activity over two fold for each manganese oxide. It is believed that this increase in OER catalytic activity is due to the increase in active sites resulting from acid leaching. These results demonstrate acid leaching as a facile method for improving the catalytic activity of select tunnel structured manganese oxides.

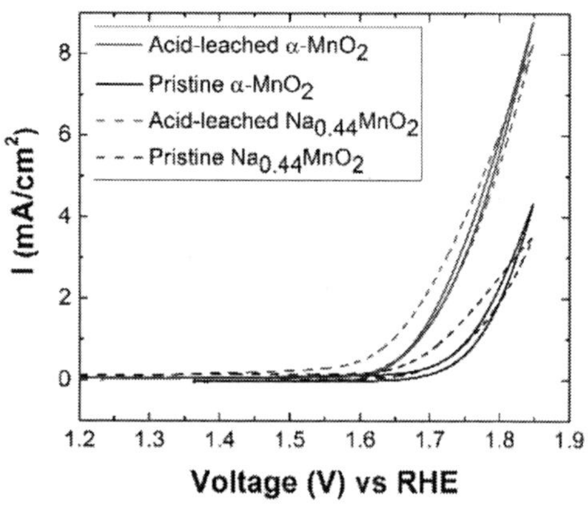

Figure 4. Cyclic voltammograms of α-MnO_2 and $Na_{0.44}MnO_2$ in 0.1M KOH at a scan rate of 5 mV/s in a voltage range of 1.37-1.85 vs RHE. The red curves are the acid-leached samples and the black curves the pristine samples.

Conclusions

This work provides detailed insight into the performance of an inexpensive and environmentally friendly class of manganese oxides for a number of important emerging energy applications. Future work will involve the synthesis of other members of this materials family, chemical treatments, and investigation into their utilization in the same important energy applications.

Acknowledgement. The authors would like to acknowledge Drexel University start-up funds for the support of this work. We would also like to acknowledge the Tang Lab at Drexel University for their support in OER experiments.

References

(1) Gao, Y.; Wang, Z.; Wan, J.; Zou, G.; Qian, Y. *Journal of Crystal Growth* **2005,** *279* (3–4), 415-419.

(2) Hosono, E.; Matsuda, H.; Honma, I.; Fujihara, S.; Ichihara, M.; Zhou, H. *Journal of Power Sources* **2008,** *182* (1), 349-352.

(3) Feng, Q.; Kanoh, H.; Miyai, Y.; Ooi, K. *Chemistry of Materials* **1995,** *7* (9), 1722-1727.
(4) Kundu, D.; Talaie, E.; Duffort, V.; Nazar, L. F. *Angewandte Chemie International Edition* **2015,** *54* (11), 3431-3448.

UTILIZATION OF NITROGEN DOPED CARBONIZED METAL ORGANIC FRAMEWORK FOR HIGH STABILITY ROOM TEMPERATURE SODIUM-SULFUR BATTERY

Yu-Ming Chen[1], *Wengfeng Liang*[1], *Feng Zou*[1], *Zhe Qiang*[2], *Sarang M. Bhaway*[2], *Si Li*[1], *Bryan D. Vogt*[2], *Yu Zhu*[1*]

[1]Department of Polymer Science, University of Akron, Akron, Ohio 44325 USA
[2]Department of Polymer Engineering, University of Akron, Akron, Ohio 44325 USA
*yzhu@uakron.edu

Introduction

The metal organic framework zeolitic imidazolate framework 8 (ZIF-8) porous nano-particles were fabricated through a one-step simple solution method. The fabricated porous nano-particles acted as a sacrificial template to provide the subsequent nitrogen doped porous carbon after thermal carbonization protocol. The nitrogen doped carbon ZIF-8 possesses a sodalite-type structure along with regular mesoporous pore sizes and channels allowing ease of migration towards ionic species.

Experimental

The doped carbon porous species were melt infiltrated with sulfur to form the carbonized ZIF-8/S cathode composite.

Results and Discussion

The composite material exhibited good performance as the cathode for the room-temperature sodium-sulfur battery (Na-S) system. A reversible specific capacity of 1020mAh/g could be achieved at a rate of 0.1C; additional rate performance at 1.0C was also demonstrated to give a specific capacity of 487mAh/g. Long cycle reversibility at 0.2C retained a specific capacity of around 700mAh/g for over 50 cycles. A synergistic effect from the mesoporous dimensions of the carbon matrix and the nitrogen-doping of the carbon matrix provided the entrapment of the sulfur molecules and thus resulted in excellent performance of the Na-S battery.

Conclusion

A reversible specific capacity of 1020mAh/g could be achieved at a rate of 0.1C during c-rate galvanostatic charge/discharge. For long charge/discharge tests at 0.2C, specific capacity of approximately 700 mAh/g could be retained for over 50 cycles.
Acknowledgement. The authors thank the University of Akron for financial support, Dr. Min Gao for Hi-resolution TEM characterizations, and Dr. Zhorro Nikolov for XPS characterizations assistance.

STRUCTURAL FACTORS DETERMINING AMINE THERMAL DEGRADATION IN CO_2 CAPTURE

Among current options for post-combustion CO_2 separation in power plants, aqueous amines, including amino acid salts, alkanolamines and diamines, are attracting growing attention due to their fast CO_2 absorption kinetics and favorable properties. Solvent degradation has been a critical concern in the development of potential CO_2 absorbents mainly due to the significant impact on operational cost and the intention to apply thermal compression in stripping to minimize the overall process energy. This research compared the thermal degradation of amino acid salts, diamines and alkanolamines to gain a comprehensive evaluation on their potential use as CO_2 absorption agents.

Several structural analogs of amino acid salts, alkanolamines and diamines were exposed to high-temperature degradation conditions in order to explore their thermal stability from a structural standpoint. The conditions selected were chosen to be relevant to CO_2 capture thermal stripping conditions; CO_2-loaded solutions were heated to 135 °C and 145 °C under static conditions. Given that CO_2 loading enhances amine thermal degradation, this work investigates a severe degradation scenario in the CO_2 capture process. MEA was used as a reference. Amine depletion was investigated by IC and HPLC/TOF–MS. A relationship between amine structure and thermal degradation rate has been developed. The investigated structural factors include functional groups, amine orders and steric effect.

DEVELOPMENT OF MATHEMATICAL MODEL FOR LIGNITE PYROLYSIS WITH SOLID HEAT CARRIER OR GAS HEAT CARRIER

Fang-Zhou LI,[1,2] *Jie Feng,*[1] *Wen-Ying LI**[1] *and Tomasz S. Wiltowski*[2]

[1]Key Laboratory of Coal Science and Technology (Taiyuan University of Technology), Ministry of Education and Shanxi Province, Training Base of State Key Laboratory of Coal Science and Technology Jointly Constructed by Shanxi Province and Ministry of Science and Technology, Taiyuan 030024, China
[2]Department of Mechanical Engineering and Energy Processes, Advanced Coal and Energy Research Center, Southern Illinois University, Carbondale, IL 62901, USA
*Corresponding author. E-mail: ying@tyut.edu.cn

Abstract

Pyrolysis with solid or gas heat carrier is acknowledged as two important techniques for lignite upgrading. The objective of present work is to elucidate the pyrolysis mechanism of intraparticle mass and heat transfer phenomena combined chemical reaction occurring in a single lignite particle under the different heating methods. A comprehensive mathematical model with several corresponding modifications had been formulated to simulate pyrolysis with solid or gas heat carrier respectively, in which contains two correlative one-dimensional unsteady heat conservation equations and binary mass conservation equations in conjunction with the simplified dusty-gas model (DGM). The chemical percolation devolatilization (CPD) model was adopted to predict the pyrolytic products distribution, whose chemical structure parameters were estimated from ultimate and proximate analyses of the Hulunbuir lignite. An implicit finite volume method (FVM) with a tridiagonal matrix algorithm (TDMA) was employed to solve the partial differential equations of energy and mass conservation, which was programmed by using Matlab software. The model predictions were verified by comparing the calculated yield of volatiles with the experimental data. The temperature-dependent thermophysical properties of lignite and volatiles were calculated by stepwise integration over discrete time and space steps. The contrastive result between solid and gas heat carrier reveals that heat transfer is a primary cause during its interaction with mass transfer and devolatilization reaction, and the degree of its impact on the other two factors is determined by operation conditions, such as coal particle radius, pressure and the initial temperature of solid or gas carrier.

Keywords: Numerical simulation; Lignite pyrolysis; Solid heat carrier; Gas heat carrier

ANISOTROPIC SINGLET FISSION IN ORGANIC SINGLE CRYSTALS

Yi Rao[a], Dezheng Sun[b#], Bolei Xu[a#], Yajing Wu[a], Xia Li[a], Avetik R Harutyunyan[c], Hai-Lung Dai[a], Gugang Chen[c]**

[a]Department of Chemistry, Temple University, Philadelphia, PA 19122
[b]Department of Physics, Columbia University, New York, NY 10025
[c]Honda Research Institute USA, Inc., Columbus, Ohio 43212, United States

Abstract

Singlet fission (SF) is one of solutions to the breaking of the Shockley-Queisser theoretical limit for higher photovoltaic conversion efficiency. Despite recent efforts, the mechanisms by how singlet fission occurs remain still unclear. In this work, we explore anisotropic singlet fission mechanisms of hexacene single crystal by employing polarized static and transient absorption microscopy. Polarized static absorption spectra of hexacene single crystal show a large Davydov splitting of 1160 $cm^{-}1$, originating from a strong charge transfer character for the crystal sites. Polarized transient absorption spectra display a clear triplet Davydov splitting of 240 cm^{-1} (30 meV), indicating that the adiabatic T_1 states inherit the charge transfer characteristic of the singlet states after the photoexcitation. Transient kinetics analyses show anisotropic singlet fission along the long ***a***- axis and short ***b***- axis of hexacene single crystal. We found from our kinetics data that the singlet fission of hexacene single crystal is a two-step process, instead of the one-dominant-step process as reported in the literature. The rates for both the ***a***- and ***b***-axis in the first step are very close, while the rate in the second step along the ***b***-axis is faster than that along the ***a***-axis. The anisotropic rates depend merely on the probe polarization instead of polarization of the excitation, which is attributed to the delocalized excitation over a few molecules. The first step from the mixed Frenkel and charge transfer to the [ME] state is in the adabatic regime of strong electronic coupling, while the second step from the [ME] state to the triplet state is nonadiabatic from nonlocal electron-vibration couplings (Peierls interactions) of the intermolecular interactions. Our work offers direct evidence that the singlet fission in hexacene single crystal is an anisotropic and charge-transfer mediated many-molecule process. We believe that this work gives new insights into the singlet fission mechanism for both theory and applications.

USE OF MECHANOCHEMISTRY TO MODIFY HYDROTHERMAL CHARS

Michael T. Timko

Worcester Polytechnic Institute
Department of Chemical Engineering
100 Institute Road
Worcester MA 01609
mttimko@wpi.edu

Introduction

Hydrothermal carbonization, a thermal process for converting carbohydrate-rich feedstocks into carbonaceous solids in the presence of liquid water, is a promising method to convert waste materials into high value materials with applications in gas storage, catalysis, and energy generation. The composition and mechanism of formation of the resulting hydrothermal chars are complex and incompletely characterized. Formation reactions are thought to include hydrolysis, dehydration, rearrangement, and condensation reactions. The hydrothermal char material is amorphous and supramolecular, consisting of aromatic, furanic, and aliphatic components.

Compared to pyrolysis chars, which have been much more thoroughly studied, hydrothermal chars remain much less investigated; however, hydrothermal carbonization has several clear advantages over pyrolysis for char formation. For wet feedstocks, hydrothermal carbonization has energy advantages, as pyrolysis requires significant energy input to dry the feedstock. Because hydrothermal carbonization occurs in a closed system in liquid phase conditions and at lower temperatures than pyrolysis (<225 °C), it is compatible with wider range of pore directing agents and surface active molecules for boutique synthesis of materials with desired characteristics. Operation in the liquid phase also enhances heat and mass transfer relative to the vapor phase, making hydrothermal carbonization more scaleable than pyrolysis. For soluble substrates (e.g. simple carbohydrates), carbonization conditions can be tuned

A challenge to hydrothermal carbonization is to functionalize its surface for a particular application without introducing petroleum-derived additives to the process or making detracting from the energy or atom efficiency of the synthesis. For instance, a common method for imparting acidity to a hydrothermal char surface is to post-treat the material in a vast excess of refluxing sulfuric acid. The resulting material is a strong solid acid; however, the process is both atom and energy inefficient. Moreover, additional energy is required to recover the char post acid treatment; after treatment, the char is hydrophilic and difficult to recover and dry.

In this work, we describe mechanochemical ball milling of a hydrothermal char. Ball milling of various ordered carbonaceous solids has been reported in the literature, for example nanotubes, graphite, and graphene. In these literature reports, ball milling can be a useful technique for imparting desired functionality in the carbonaceous materials. Our hypothesis was that ball milling of the highly disorganized and amorphous hydrothermal char result in cleave of labile bonds, providing locations for facile functionalization. In this work, we ball milled a hydrothermal char in an oxygen environment and tracked the chemical changes using a suite of complementary spectroscopic techniques. This work provides a basis for rational design of methods for hydrothermal char modification using mechanochemistry.

Experimental

Hydrothermal char was synthesized by heating 40 wt% glucose solutions to 180 °C in PTFE-lined batch reactors; the reactor was kept in the 180 °C oven and then allowed to cool to room temperature. The contents were filtered to recover solid char and the char rinsed with solvent and water before drying. Figure 1 provides photographs of the most important synthesis steps.

Figure 1. Photographs of hydrothermal char synthesis.

Hydrothermal char was ball milled in a Retsch M220 mixer. Approximately 0.5 g of char material was placed in the stainless steel vessel, which was approximately 1/3 of the reactor volume. Three stainless steel balls (1 10 mm diameter and the others 5 mm diameter) were added to the reactor to fill 1/3 of the volume, and the remaining 1/3 of the volume was vacant except for atmospheric gas. Chars were milled for set times, ranging from 30 to 300 min. The vessel became hot during milling, achieving an estimated external temperature of approximately 60 °C. After milling, the char material was recovered and stored in air-tight containers before characterization; for many analysis procedures, the char was characterized minutes after synthesis to reduce continued reaction with atmospheric gases, etc.

Hydrothermal chars were characterized using a variety of complementary methods: SEM (to determine particle size, shape, and surface characteristics); elemental analysis; Fourier Transform Infrared Spectroscopy (FTIR, to determine the presence of functional groups); X-ray Diffraction Spectroscopy (XRDS, to characterize sample crystallinity); Raman spectroscopy (to characterize changes in carbon-carbon bonding structures); and Electron Spin Resonance (ESR, to investigate radical formation during milling). Methods are described in detail in the literature by the same author.

Results and Discussion

Figure 2 contains SEM micrographs of the synthesized char and the char recovered after 30, 60, and 200 minutes of milling. The synthesized char consists of uniform microspheres with diameters on the order of 100-200 nm. Following 30 minutes of milling, these microspheres agglomerate into large particles irregularly shaped particles, now with characteristic dimensions on the order of microns. With further milling, the agglomerates being to shed large flakes. These observations provide insight into the physical processes occurring during milling.

Figure 2. SEM images of the synthesized char and char materials obtained after milling for 30, 60, and 200 min.

The chemical changes that occur during milling were evaluated using elemental analysis, FTIR, and Raman spectroscopy. Figure 3 provides elemental analysis data for synthesized and milled chars as a van Krevelen plot. The synthesized char composition is comparable to a low-grade coal, such as lignite. With ball milling, the composition shifts to increasingly carbonaceous content, following a trajectory toward a higher grade coal with increasing milling times. The trajectory of the composition is suggestive of dehydration reactions. We confirmed loss of hydroxyl groups using FTIR spectroscopy (data not shown) and moisture analysis confirmed that loss of bound water was not sufficient to explain the observed changes in bulk composition or FTIR spectra.

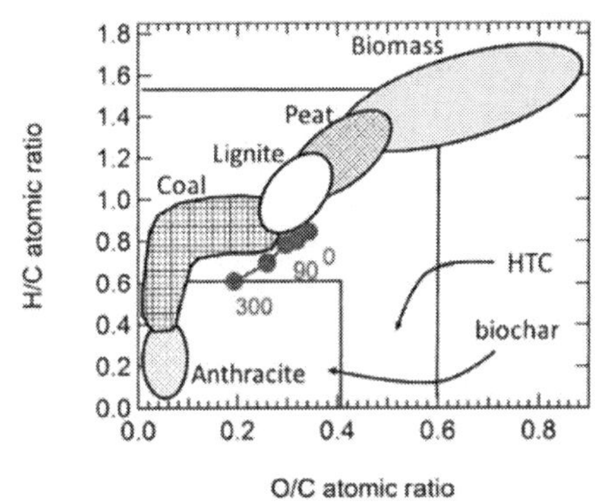

Figure 3. van Krevelen plot of hydrothermal char elemental composition, compared to other carbonaceous materials.

FTIR provided further evidence of chemical reactions. Figure 4 provides representative spectra, showing that milled materials had increased aromatic and carbonyl content compared to the synthesized material. Raman spectroscopy (data not shown here) supported showed an initial increase in carbon edge sites for 60-90 min of milling, followed by a loss of edge sites and increase in graphitization for longer milling times. XRDS, however, indicated that the bulk of the material remained crystalline.

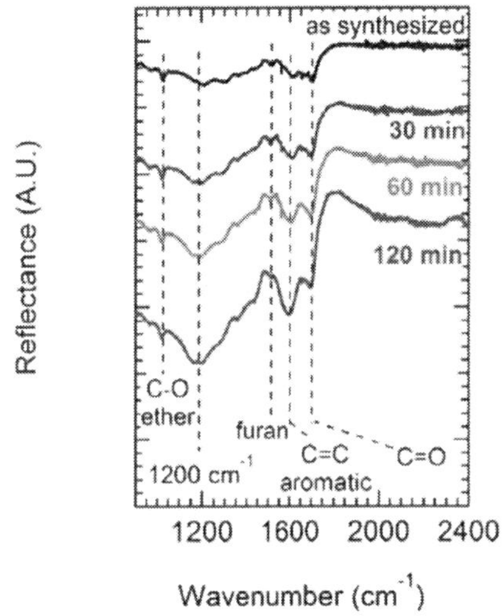

Figure 4. FTIR spectra of synthesized and milled char materials. The milled char shows increased aromatic and carbonyl content.

We hypothesized that milling broke labile carbon-carbon and carbon oxygen bonds, resulting in short-lived carbon radicals. Some of these radicals could be "captured" by oxygen in the reactor vessel, resulting in formation of the carboxylate groups identified by FTIR. Others would re-combine, resulting in increased local graphitization, as indicated by Raman spectroscopy (data not shown). Figure 5 is a proposed mechanism for the mechanochemical reactions. A key step in this mechanism is the formation of carbon radical centers. Therefore, to test the proposed mechanism, we performed ESR of synthesized and milled chars. Figure 6 provides integrated ESR signals, which are proportional to carbon radical densities and measured ESR linewidths. The integrated signal shows an initial 10-fold increase in carbon radical density, which plateaus after 60-90 minutes of milling and then stabilizes or begins to decrease with longer milling times. The linewidths may be consistent with a shift in the chemistry of the radical centers, again occurring at about 90 minutes of milling. These observations are consistent with the proposed mechanism, shown in Figure 5.

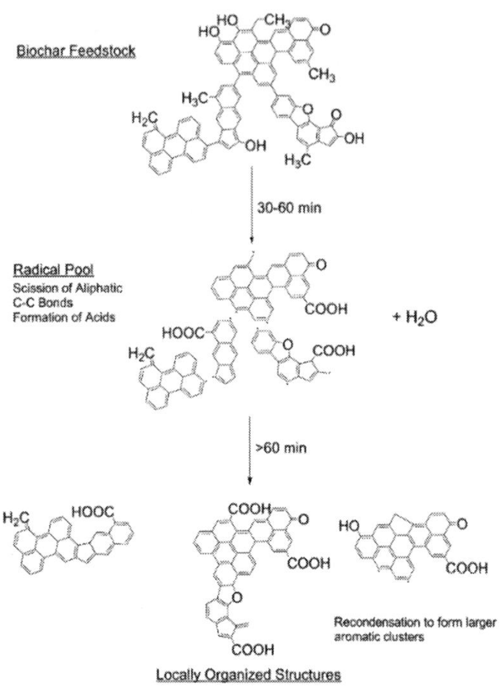

Figure 5. Schematic representation of radical formation, recombination, and oxidation that occur during ball milling of hydrothermal char.

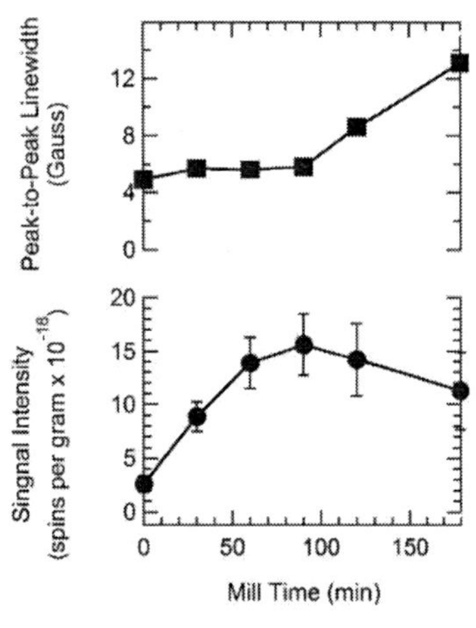

Figure 6. Integrated ESR data of synthesized and milled char, showing an initial increase in radical density following a subsequent levelling off or decrease.

Conclusions

Ball milling is an effective method for chemical modification of hydrothermal chars. Future work will investigate reaction engineering of the synthesis method to introduce targeted functionality.

Acknowledgement. The author gratefully acknowledges collaborations with Prof. Geoffrey Tompsett (WPI) and Prof. Frederick Greenaway (Clark University). A series of WPI undergraduate project teams contributed to the work: Juan Venegas, Erin Heckley, Joseph Toto, and Brendan McKeogh. Several WPI graduate students contributed to the work: Zhenygang Zhang and Alex Maag. Finally, NSF provided funding for portions of this work (award # EEC-1342320).

A PREDICTIVE GUIDE FOR COLLECTIVE CO_2 ADSORPTION PROPERTIES OF MG-AL MIXED-OXIDES

Hyun Chul Lee[1], Soonchul Kwon[2], Jeong Gil Seo[3], In Sun Jung[1], You-Hwan Son[1], Chan Hyun Lee[4], Ki Bong Lee[4], Hyuk Jae Kwon[1]**

[1]Energy Laboratory, Samsung Advanced Institute of Technology, Samsung Electronics Co., Ltd, 130 Samsung-ro, Yeongtong-gu, Suwon-si, Gyeonggi-do, 16678, Republic of Korea. [2]Department of Civil Engineering, Pusan National University, 2 Busandehak-ro, 63beon-gil, Geumjeong-gu, Busan, 46241, Republic of Korea. [3]Department of Environmental Engineering and Energy, Myongji University, 116 Myongji-ro, Cheoin-gu, Yongin-si, Gyeonggi-do, 17058, Republic of Korea. [4]Department of Chemical and Biological Engineering, Korea University, 145 Anam-ro, Seongbuk-gu, Seoul, 02841, Republic of Korea.

hc001.lee@samsung.com; hj13.kwon@samsung.com

Introduction

Since carbon dioxide (CO_2) is recognized as a greenhouse gas contributing to runaway global warming, advanced technologies to capture CO_2 from the flue gas streams of power plants have been intensively developed. Among the several approaches to CO_2 capture, adsorption using solids is one of the most reliable and promising technologies to reduce energy demands and energy penalties[1–3]. However, effective solid adsorbents still suffer from low CO_2 adsorption capacities and energy-intensive regeneration processes. Instead of having to cool a warm-gas, adsorbents that function best at medium temperatures could be optimal candidates for the warm-gas stream processes such as the pre-combustion capture processes. Thus, we studied well known medium temperature adsorbents (473–673 K), layered double hydroxides (LDHs) and layered double oxides (LDOs) with thermal evolution of their crystal structures upon exposure to heat. Here, we report the relationship of the LDO crystalline structure features to the CO_2 adsorption properties via both of density functional theory (DFT) calculations and experiment validation. In particular, we propose a new correlation of collective adsorption properties (e.g., collective adsorption capacity, adsorptive energy and charge transfer) of LDOs driven from DFT calculations and measured surface areas with their crystallinity. And, those correlations exhibited quite well agreement with experimental validations and could be a predictive guide for the gas adsorption processes.

Experimental

Synthesis and Characterization. We synthesized LDHs with various cation ratios (Mg/Al) ranging from 3 to 9, denoted as LDH(cation ratio). We also prepared ion-exchanged LDH(3) with phosphate (PO_4^{3-}), LDH(3,P), to investigate the effect of interlayer anion in the LDH for transformation to LDO structure. Finally all of the resultants were calcined at 673 K for 5h to derive the corresponding mixed oxides (LDOs). The characterization of synthesized adsorbents was analyzed by ICP, XRD, TEM, NMR and N_2 adsorption and desorption for evaluation of BET surface area.

CO_2 adsorption. The CO_2 adsorption capacity of LDOs was measured using a thermogravimetric analysis (TGA, Q50, TA Instruments). Before CO_2 adsorption experiments, moisture and impurities on the samples were removed by N_2 gas flowing rate of 30 mL/min at 673 K for 2 h. After pre-treatment, the change of sample weight, 20 mg of sorbents initially, was recorded in the condition of pure CO_2 gas flow with rate of 30 mL/min. Each test run was repeated three times at 473 K and 1 bar for 2 h.

Quantum calculation. The density function theory (DFT) calculations with GGA-PBE functional were performed to determine adsorption energies and electronic properties including density of states and Mulliken population analysis of the interaction between the adsorbed CO_2 molecules and the surface of adsorbents in the three dimensional periodic-slab models. Geometry optimizations were reliably carried out using DMol[3].

Results and Discussion

We suggest that the collective CO_2 adsorption capacity, which can be described as a long-range electrostatic field over the entirely exposed surface of adsorbent; this can be calculated by combining the adsorption coverage and the measured surface area of the adsorbent with the results for available adsorption sites from the quantum mechanical DFT calculations. Fig. 1A shows CO_2 adsorption coverages on the available sites on the surface of LDOs. Fig. 1B clearly reveals that the trends in the collective CO_2 adsorption capacities agree quite well with the experimental determined capacities. And those collective properties could be extended to other properties such as adsorption energy and charge transfer amount.

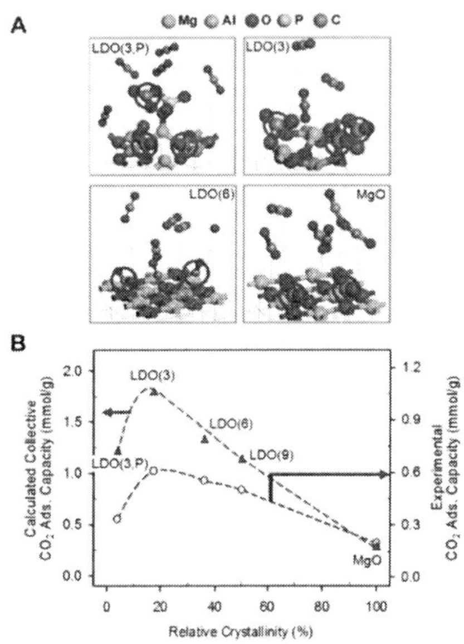

Figure 1. Simulation models and their trends in collective CO_2 adsorption capacities with experimental validation. **A.** DFT simulations of LDOs, **B.** Trend of collective properties with experimental results.

Conclusions

We revealed an asymmetrical "volcano"-type correlation between the CO_2 adsorption capacity and the relative crystallinity of LDOs. We also proposed that the collective CO_2 adsorption capacity, energy, and charge transfer of electrons calculated in a long-range electrostatic field well represent the trend in the experimental CO_2 adsorption capacities of LDOs with increasing relative crystallinity.

References

(1) Choi, S.; Drese, J. H.; Jones, C. W. *ChemSusChem* **2009**, *2*, 796.
(2) Wang, Q.; Luo, J.; Zhong, Z.; Borgna, A.*Energy Environ. Sci.* **2011**, *4*, 42.
(3) Siriwardane, R. V.; Stevens, Jr. R. W. *Ind. Eng. Chem. Res.*, **2009**, *48*, 2135.

PEACE - WEAPONS AMMUNITION DESTRUCTION AND COMPREHENSIVE UTILIZATION OF DISARMAMENT

Jiping Liu, Dong Wang,Xiaobo liu, Jia Han, Zuoxing Di,Weishan Zhang

Beijing Insistute of Thechnology .China.100081
liujp@bit.edu.cn
Contact number: 86+010-68914530

Abstract

As a state of peace taken away from the people by war, the global disarmament and military spending cuts has been going along with human security situation. leading to nuclear weapons disarmament, biological chemical weapons and retire weapons destruction. However, So far these methods used by most countries of separating the radioactive material decommissioned nuclear weapons as nuclear energy,making new material from the separation of retired conventional weapons smelting metal materials are not environment friendly.

Compared with traditional treatments, this study made a method of using radioactive materials isolated from decommissioned nuclear weapons for nuclear energy, and making the radioactive metallic materials as the catalytic target smelting rare materials, changing the conventional ammunition separation of metal materials into ornaments. NC was extracted by nondestructive separation for propellant, then the pure colloidal NC could be obtained by retired, $C_{12}H_{11}N$ process additive and low nitrate content of propellant taken by the activity of strong electron-withdrawing groups along with solvent, the productivity is more than 93%. The colloidal NC soaked with the ethanol solution of stabilizer in stabilizer could be direct processed for propellant and adjusted the total volatile and ballistic performance. Variety and different application of propellant brand Colloidal NC, 100% of colloidal NC, were classified or collocted, gel, knead, moldfied, continuing to make different taste date of propellant. After degradation, denitration into coating and celluloid, plastic bottle with NC of manufacture of propellant, the productivity is 82% ~ 84%, Under the influence of HNO3 / H2SO4, NC was converted to ammonium oxalate and oxalic acid, glyoxal and other chemical materials. The productivity of retired propellant is 90%. During 180 ~ 220 °C in the iron amino dye, The $C_{12}H_{11}N$, retired and unstability of low-level gun propellant with catalyst, is full recovery and comprehensive utilization.

This study makes weapons manufacturing industry realize the virtuous cycle of production-reserves-use-recycling-comprehensive utilization. The process of Separation, implements and the degradation realize the energy-nondestruction and the energy comprehensive utilization. Meanwhile, Not only it avoides for the environment pollution and destruction of the damage and ensures the ecological and environmental impact in the process and effectively protects the resources, but it also can create great industrial value.

Keywords: peaceful;disarmament retired weapons;destroyed half colloidal NC;nondestructive energy efficiency

ELECTROCHEMICALLY INDUCED REDUCTION DISPLACEMENT REACTIONS: ENHANCED CONDUCTIVITY FROM THE PARTICLE TO THE SYSTEMS LEVEL

Esther S. Takeuchi,[a,b] *Kevin C. Kirshenbaum*[b], *David C. Bock*[b], *Alexander B. Brady*[a], *Zhong Zhong*[b], *Kenneth J. Takeuchi*[b], *Amy C. Marschilok*[a]

a Stony Brook University
b Brookhaven National Laboratory
*Email: esther.takeuchi@stonybrook.edu

Introduction

Inquiry of the material fundamental properties as well as the behavior of the material when configured into a fully functioning device is necessary for full elucidation of mechanisms of electrochemical reduction in batteries. The ability to use local interrogation techniques and integrate the results with system level behavior is needed.

Quantitative electrochemistry was conducted on single particles of silver vanadium phosphorous oxides, (specifically SVPO, $Ag_wV_xP_yO_z$) cathode materials. Silver ions incorporated into the layered vanadium phosphorous oxide structure facilitated monitoring of the reaction progress through the formation of high atomic number silver metal. The results obtained from the single particle were related to the systems level through in-situ examination of the reduction progress in fully assembled cells probed through energy dispersive x-ray diffraction, EDXRD.

Experimental

At the particle level, individual particles of active material, $Ag_2VO_2PO_4$, were electrochemically reduced to various depths of discharge. After the material was reduced, the conductivity of the individual particles was determined using a nanoprobe integrated with a scanning electron microscope. At the system level, in situ energy dispersive X-ray spectroscopy (EDXRD) was used to probe Li/ $Ag_2VP_2O_8$ cells at several stages of reduction, allowing depth profile analysis of the discharge process. This technique enables micron scale spatial resolution of reduced $Ag_2VP_2O_8$ and Ag^0 discharge products in the bulk electrode.

Results and Discussion

The individual particle data indicated a drastic reduction in local resistance when discharged, where the preferred initial process was formation was reduction of silver ion to form silver metal. EDXRD results of the bulk system suggested generation of a 3-dimensional Ag^0 based percolation network that increases conductivity throughout the entire electrode and occurs concurrently with a substantial decrease in the charge transfer resistance. The resistance of the reduced cathode is dependent upon discharge rate even in the presence of conductive additives in the electrode matrix.

Next generation battery systems with improved energy utilization can be developed effectively only by fully accessing all of the active material in an electrode. The multiscale inquiry methodology to be described in this presentation should serve as a case study for future investigations of both particle level and system level properties of electrochemically discharged materials. In particular, the use of advanced instrumental techniques to gain insight into complex discharge processes will be highlighted.

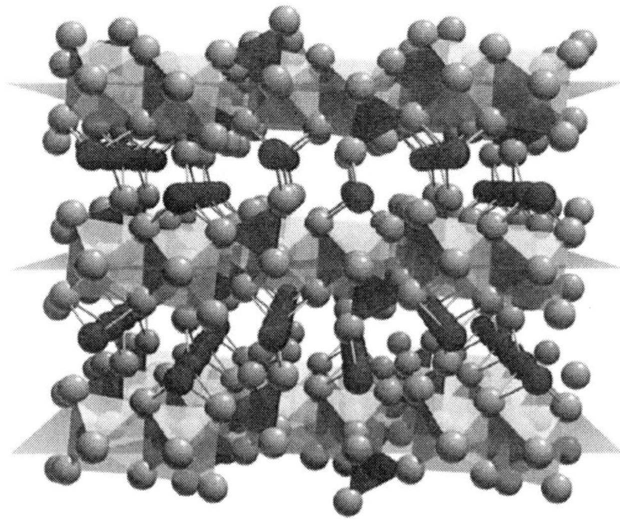

Figure 1. Structure of $Ag_2VO_2PO_4$

References

(1) Kirshenbaum, Kevin; Bock, David; Brady, Alexander B.; Marschilok, Amy C; Takeuchi, Kenneth James; and Takeuchi, Esther S., *Phys. Chem. Chem. Phys.*, **2015,** *17,* 11204.

(2) Kirshenbaum, Kevin C.; Bock, David C.; Zhong, Zhong; Marschilok, Amy C.; Takeuchi, Kenneth J.; and Takeuchi, Esther S., *J. Mater. Chem. A,* **2015,** *3,* 18027.

STORAGE AND OXIDATION STABILITIES OF BIODIESEL AND ITS BLENDS WITH ULSD AND NATO F-76

Jinxia Fu, Scott Q Turn, Pablo A Gimeno

Hawaii Natural Energy Institute, University of Hawaii, Honolulu, HI USA 96822
JinxiaFu@hawaii.edu

Introduction

The development and utilization of alternative fuels have become an ambitious energy goal for both developed and developing countries striving to decrease reliance on conventional fossil fuels. Biodiesel is defined as "a fuel comprised of mono-alkyl esters of long-chain fatty acids derived from vegetable oils or animal fats, designated B100" by the American Society for Testing and Materials (ASTM).[1] Interest in biodiesel is continuously increasing worldwide motivated by its renewable and environmental benefits. As of December, 2015, there were 94 biodiesel plants in the U.S. with capacity of over 2.0 billion gallons per year.[2]

Although biodiesel has many prominent advantages over petroleum diesel, it is more prone to oxidation than conventional petroleum diesel owing to the presence of unsaturated bonds in the molecules inherited from its parent feedstock. This susceptibility to oxidation has significant impact on the diesel quality during long-term storage and handling. Generally, biodiesel is not utilized in neat form (B100, 100%). It is usually blended with petroleum diesel, e.g. B5 with 5% B100, B7 with 7% B100, and B20 with 20% B100. The stability of the neat biodiesel affects the overall stability of the diesel blend, whereas the unstable precursors in petroleum diesel could accelerate the oxidation process of biodiesel.

In this article, the physicochemical properties and chemical composition of commercial biodiesel derived from waste cooking oil produced by Pacific Biodiesel are reported. Method ASTM D4625 was employed to investigate the long-term storage stability of the biodiesel. Modified methods ASTM D5304 and ASTM D2274 were used to study oxidation processes of the biodiesel under various conditions. The impacts of oxidation and antioxidants on fuel physicochemical properties and low-temperature qualities, such as density, viscosity, peroxide value, acid number, heat of combustion, fusion and crystallization temperature and enthalpies, were also investigated using ASTM methods. In addition, the storage and oxidation stability of biodiesel blends with Ultra low sulfur diesel (ULSD) and NATO F-76 marine diesel were also studied.

Material and Methods

Materials. Biodiesel with and without antioxidants produced by Pacific Biodiesel was purchased from Carl's Jr. 76 Station in Honolulu, HI, and from Pacific Biodiesel plant in Keaau, HI, respectively. The biodiesel purchased is defined as B99.99 with <0.01 % impurity, and was produced mainly from waste cooking oil collected in Hawaii. ULSD and NATO F-76 were provided by United States Navy Supply Center at Patuxent River, MD. The fuels were used as received, unless otherwise noted.

Chemical and Physical Properties. The composition of biodiesel was determined by gas chromatography/mass spectrometry (GC-MS). 1.5 μL fuel samples were dissolved in 1.5 mL hexane and analyzed using a Bruker 436-GC gas chromatograph and SCION-MS select, single quadrupole mass spectrometer. The GC was equipped with an Agilent DB1701 column (low/mid polarity, 60m, (14%-cyanopropyl-phenyl)-methylpolysiloxane), with a 15 m guard-column before the back flush valve, and operated at a helium flow rate of 1.5 mL/min.

An Anton Paar SVM3000 Stabinger Viscometer was used to measure the viscosity and density of neat and stressed biodiesel samples at temperatures according to the ASTM D7042 method. A Parr 6200 Isoperibol Calorimeter was used to measure the heat of combustion, i.e. heating value, based on the ASTM D4809 method. ASTM D3703 and D974 were employed to determine the peroxide value (PV) and acid number (AN), respectively, before and after stressing the biodiesel and its blends. Differential scanning calorimetry (DSC) analyses were conducted using a TA Q2000 system (TA Instruments, New Castle, DE) with a RCS90 temperature control to investigate the cold-flow properties of B100 and its blends.

Storage and Oxidation Stability. Three ASTM methods, i.e. ASTM D4625, D2274 and D5304 were used to test the storage and oxidation stability of biodiesel and its blends. All of these methods were used to investigate the amount of insolubles formed after stressing the diesel samples under elevated temperature and prescribed pressure conditions. Adherent and filterable insolubles during the process were quantified.

Results and Discussion

Properties and Characterization. The GC/MS scan of the biodiesel derived from waste cooking oil had five dominant peaks ranging in retention time from 35 to 49 min, as shown in Figure 1(A). The results indicated that biodiesel consisted of five main components, i.e. methyl palmitate, methyl oleate, methyl linoleate, methyl stearate, and methyl linolenate, which are the most common compounds present in biodiesel. The molar concentrations were calculated to be 26.3% methyl palmitate, 27.8% methyl oleate, 30.4% methyl linoleate, 8.9% methyl stearate, and 6.6% methyl linolenate. The amount of unsaturated esters with bis-allylic sites accounted for over 35% of the biodiesel. F-76 was composed of approximately 25% aromatics and 73% saturates (alkanes and cycloalkanes), with a small amount of olefins (alkenes, cycloalkenes, and some dienes),[3] while the ULSD was composed of ~29% aromatics, ~64% saturates, and ~4% olefins.[4] Table 1 lists the physicochemical properties of the biodiesel with antioxidant, F-76,[5] and ULSD.[6] The five measured properties of B100 meet the ASTM D6751 specification, — kinematic viscosity (v) in the range of 1.9 - 6 $mm^2\ s^{-1}$ at 40 °C and acid value (AV) < 0.5 mg of KOH/g. The measured higher heat value (HHV) of B100 was lower than both marine diesel NATO F-76 and ultra-low sulfur diesel (ULSD).

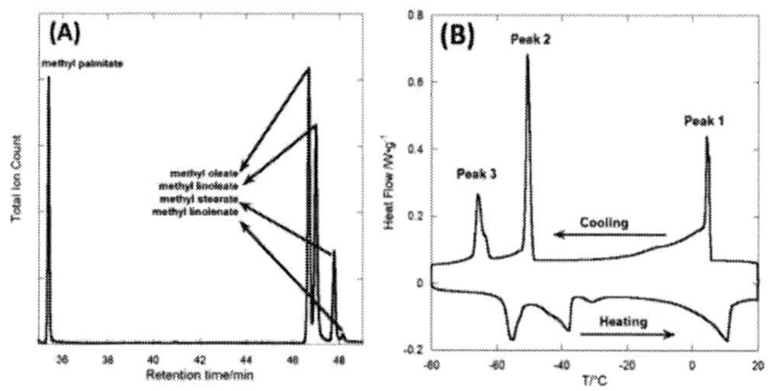

Figure 1. GC/MS chromatogram (A) and DSC heating and cooling curve (B) of B100 from waste cooking oil.

Low temperature properties of the biodiesel with antioxidant were also investigated using DSC. Figure 1(B) shows three distinct peaks during both the cooling and heating cycle scans. The first crystallization peak in the cooling scan (the maximum peak temperature (FP) at ~5 °C) spans a wider temperature range, approximately 37 °C, compared to the ~12 °C width observed for the -50 and -65 °C peaks. This appears to be the result of crystallization of saturated esters in the presence of methyl oleate. Similar wide peak phenomena were also observed in biodiesel derived from palm and peanut.[7] Peaks 2 and 3 correspond to the phase change of unsaturated esters, i.e. methyl oleate, methyl linoleate, and methyl linolenate, and possibly some low molecular weight impurities that were not characterized by GC/MS. Biodiesel derived from waste cooking oil is expected to have inferior cold-flow properties compared to biodiesel derived from rapeseed, sunflower and soybean biodiesel,[7] but better cold flow properties than palm and peanut based biodiesels.[7] These differences result from their varied contents of saturated long chain esters which account for >35% of the waste cooking oil biodiesel.

Table 1. Physicochemical properties of B100 with antioxidant, F-76, and ULSD.

Properties	B100	F-76	ULSD
ρ at 15°C (g cm^{-3})	0.8778 ± 0.0001	0.8455 ± 0.0001	0.8451 ± 0.0001
ν at 40°C (mm^2 s^{-1})	4.3725 ± 0.0016	2.9182 ± 0.0018	2.4462 ± 0.0015
PV (mg kg^{-1})	16.59 ± 0.49	1.87 ± 0.17	1.33 ± 0.05
HHV (MJ kg^{-1})	39.1707 ± 0.0422	44.8292 ± 0.0733	44.9569 ± 0.1840
AN (mg of KOH/g)	0.23 ± 0.01	0.11 ± 0.00	0.01 ± 0.00

Storage Stability. Biodiesel samples with antioxidant were subjected to long-term storage tests, i.e. ASTM D4625 for 24 weeks with replicated analyses occurring at 0, 4, 8, 12, 16, 20, and 24 weeks. Figure 2 shows the amounts of filterable and adherent insolubles formed in the biodiesel samples increased with storage time, which agrees with the results reported by McCormick and Westbrook,[8] and Westbrook and Stavinoha.[9] The physicochemical properties of biodiesel were also affected by long term storage, The density, kinematic viscosity, PV, HHV and AN, were 0.8781 ± 0.0001 g cm^{-3}, 4.3787 ± 0.0065 mm^2 s^{-1}, 17.36 ± 1.83 mg kg^{-1}, 39.3074 ± 0.0291 MJ kg^{-1}, 0.32 ± 0.01 mg of KOH/g, respectively, and all increased after 24 weeks of storage. The differences in AN, density, and HHV values for the samples were statistically significant, but for the latter two quantities differences were small. The kinematic viscosity and AN of the stored samples remained within ASTM D6751 specifications.

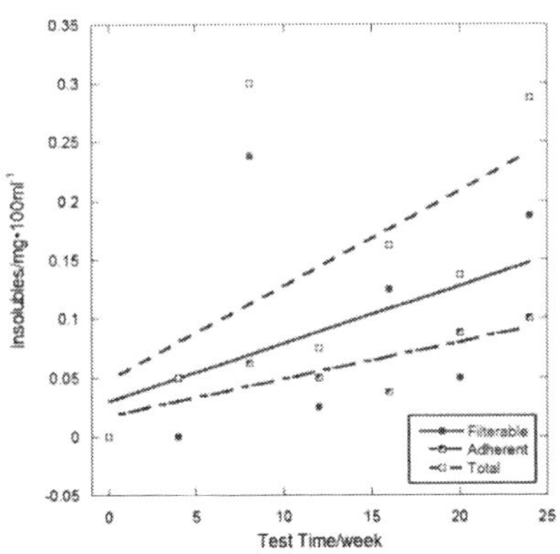

Figure 2. Insolubles formed in the B100 determined by ASTM D4625 for 24 weeks. Each data point is average of two samples.

Oxidation Stability. A modified ASTM D5304 method was employed to investigate the oxidation process of biodiesel under pure oxygen at 90 ± 1 °C, and 800 ± 10 kPa pressure. The modified method allows investigation of the biodiesel oxidation process by monitoring the oxygen consumption rate in the pressurized vessel. The oxygen consumption rate is calculated based on the pressure change of the vessel. Figure 3 (A) shows the oxygen consumption of biodiesel after 16 and 40 h tests. The oxygen consumption is relatively slow in the first 8 h owing to the added BioExtend™ 30 antioxidant, after which the rate is accelerated. Changes in fuel composition and physicochemical properties were also investigated after the modified ASTM D5304 tests. Figure 3 (B) illustrates the composition change of the five major components after stressing the biodiesel samples for 8, 16 and 40 h. The concentration of methyl palmitate, methyl linoleate, methyl stearate, and methyl linolenate in the biodiesel increases slightly with oxidation time, whereas the methyl oleate concentration decreases. This indicates that the oxidation and degradation rates of methyl oleate are faster than the other four compounds during the ASTM D5304 test.

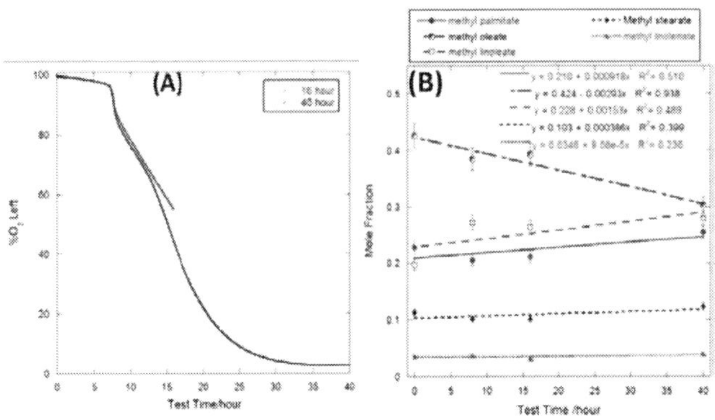

Figure 3. Oxygen consumption of B100 during ASTM D5304 test (A) and B100 composition change after ASTM D5304 tests (B).

Conclusions

The influence of long-term storage and oxidation on properties of B100 derived from waste cooking oil was investigated. B100 consisted of five primary compounds, and unsaturated esters accounted for >60% of liquid volume and approximately 60% of the unsaturated esters contain bis-allylic sites which highly affect the B100 stability. Owing to the mild conditions of slightly increased temperature and atmospheric pressure, the long-term storage tests, ASTM D4625, did not have significant impacts on fuel properties.. ASTM D5304 test conditions with elevated temperature and pressure under a pure oxygen head space, however, produced significant changes in physicochemical properties.

Acknowledgement. This article was made possible by Grant Number N00014-12-1-0496 from the Office of Naval Research (ONR).

References

(1) ASTM D6751. Standard Specification for Biodiesel Fuel Blend Stock (B100) for Middle Distillate Fuels. ASTM International, West Conshohocken, PA, 2015.
(2) Monthly Biodiesel Production Report, U.S. Energy Information Administration, July, 2014.
(3) Waynick, J. A.; Giannini, R. M. In *Effect of refining severity on the lubricity of NATO F-76 fuel*, International Association for Stability, Handling and Use of Liquid Fuels: 2003; pp 804-833.
(4) Smagala, T.J.; Christensen, E. In *Hydrocarbon Renewable and Synthetic Diesel Fuel Blendstocks: Composition & Properties.* International Association for Stability, Handling and Use of Liquid Fuels: 2013.

(5) Fu, J.; Turn, S.Q. Effects of Biodiesel Contamination on Oxidation and Storage Stability of Neat and Blended Hydroprocessed Renewable Diesel. *Energy Fuel*, **2015**, 29 (8).
(6) Fu,J; Turn, S.Q.; Takushi, B.M.; Kawamata,C.L. Storage and oxidation stabilities of biodiesel derived from waste cooking oil. *Fuel*, **2016**, 167 (3).
(7) Perez, A. A.; Casas, C.M.; Fernandez, M.J.; Ramos, L. R. Winterization of peanut biodiesel to improve the cold flow properties. *Bioresource Technol*, **2010**, 101.
(8) McCormick, R.L.; Westbrook, S.R. Storage Stability of Biodiesel and Biodiesel Blends. *Energy Fuel*, **2010**, 24.
(9) Westbrook, S.R.; Stavinoha, L.L. In Biodiesel and B20 blends: stability test methods and stability characteristics. International Association for Stability, Handling and Use of Liquid Fuels, 2003, pp. 271-298.

HIGH-PRESSURE CONVERSION OF C_5-C_7 ALKANES FOR AIRCRAFT ENDOTHERMIC COOLING

*Edward P. Schreiner, Shewangizaw Teketel, and Raul F. Lobo**

Center for Catalytic Science and Technology, Department of Chemical and Biomolecular Engineering, University of Delaware, Newark, DE, 19716, USA
*lobo@udel.edu

Introduction

Practical use of hypersonic planes has been hindered by overheating of aircraft engine and electronic components, caused by high fuel combustion rates and air friction[1]. Military aircraft use jet fuel as a heat sink, but future hypersonic aircraft exceed the cooling capacity of the fuel[2,3]. Cooling capacity can be increased using endothermic fuels, which remove heat through endothermic chemical reactions[4,5]. Here, we investigate catalytic alkane cracking reactions for this purpose, using the solid acid catalyst H-[Al]ZSM-5, which is known for its high selectivity to olefins in fluidized catalytic cracking units[6]. We also investigate supported molybdenum carbide catalysts for their ability to produce olefins from alkanes[7].

Material and methods

High pressure conversion of *n*-pentane (Fisher, 99+%) on H-[Al]ZSM-5 (Zeolyst, CBV3024E) was investigated as a function of temperature (633-723 K), pressure (10-60 bar), and weight hourly space velocity (WHSV) (400-1500 $g_{C5}g_{cat}^{-1}h^{-1}$). Reaction kinetics were measured using a plug-flow laboratory microreactor (4.6 mm ID) and reaction products were analyzed via online gas chromatography. Supported molybdenum catalysts were synthesized by incipient wetness impregnation using ammonium heptamolybdate as the source of Mo (5.8 wt% of Mo). The materials were calcined in air at a temperature of 823 K for 5 hours and then carburized in methane (20%, H_2 balance) at a temperature of 923 K for 4 hours prior to catalytic testing. For the Mo catalysts, in addition to *n*-pentane (Fisher, 99+%), *n*-hexane (Acros, 99+%) and *n*-heptane (Fisher, 99+%) were also investigated.

Results and discussions

Conversion of pentane on H-[Al]ZSM-5 was stable with time on stream at a reaction temperature up to 673 K. Under all reaction conditions investigated, the major products were propane and butane, followed by species heavier than pentane. It was observed that light olefins produced from the primary cracking of pentane were rapidly consumed via secondary bimolecular reactions, leading to the observed product distribution (30+ species heavier than pentane). The low yield of light olefins makes the observed product distribution substantially less endothermic than expected from pure cracking reactions.

In addition to H-[Al]ZSM-5, several pure and supported molybdenum carbides were investigated for the dehydrogenation of *n*-pentane. Mo/H-[B]ZSM-5 (zeolite support) and Mo/Al_2O_3 provided the highest reactivity and selectivity toward pentenes (Figure 1). These two catalysts exhibited an induction period, with the highest reaction rates observed 1 h after the start of the reaction. This increase in reactivity indicates that the most active form of the catalyst was formed in the reactor by interaction with the reactant. Slow deactivation was observed over 5 h on stream, but unlike the other catalysts tested, the selectivity of the dehydrogenation product from the zeolite support did not change with time on stream. Dehydrogenation of *n*-hexane and *n*-heptane were also tested on the most active catalysts.

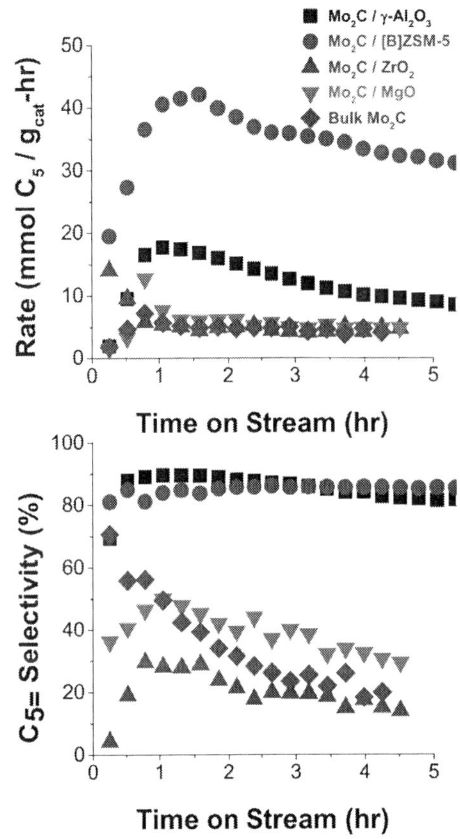

Figure 1. Rates and pentene carbon selectivity of pentane conversion on various molybdenum carbide catalysts. Reaction conditions: P = 40 bar, T = 723 K, m_{cat} = 100 mg, WHSV = 188 g_{C5}/g_{cat}-hr, residence time ~ 0.6 s.

Conclusions

The product distribution of the conversion of pentane on acidic H-[Al]ZSM-5 was controlled by secondary bimolecular reaction, such as oligomerization and dehydrocyclization, leading to a decrease in reaction endothermicity. By utilizing molybdenum supported on low acidity H-[B]ZSM-5 as a dehydrogenation catalyst, undesirable secondary bimolecular reactions were suppressed, preserving the olefinic product, and leading to an increase in reaction endothermicity. While the reaction endothermicity was enhanced by using Mo/H-[B]ZSM-5, reaction rates were lower than H-[Al]ZSM-5 and must be increased to become a viable alternative.

References

(1) Lander, H.; Nixon, A.C. *Journal of Aircraft*. **1971**, *8*, 200.
(2) Widegren, J.A.; Bruno, T.J. *Energy Fuels*. **2009**, *23*, 5523.
(3) Chung, H.S.; Chen, S.H.; Kremer, R.A.; Boulton, J.R. *Energy Fuels*. **1999**, *13*, 641.
(4) Gascoin, N.; Abraham, G.; Gillard, P. *Journal of Analytical and Applied Pyrolysis*. **2010**, *89*, 294.
(5) Kim, J.; Hyeon, D.H.; Park, S.H.; Chun, B.H.; Jeong, B.H.; Han, J.S.; Kim, S.H. *Catalysis Today*. **2014**, *232*, 63.
(6) Corma, A. *Chemical Reviews*. **1995**, *95*, 559.
(7) Solymosi, F.; Nemeth, R.; Ovari, L.; Egri, L. *Journal of Catalysis*. **2000**, *195*, 316.

IDENTIFICATION OF PRODUCTS FORMED BY IRON CORROSIVE CARBOXYLIC ACIDS BY FT-ICR MASS SPECTROMETRY

Logan C. Krajewski[1], *Vladislav V. Lobodin*[2], *Winston K. Robbins*[3], *Alan G. Marshall*[1, 2], *Ryan P. Rodgers*[1, 2]

[1] Department of Chemistry and Biochemistry, 95 Chieftain Way, Florida State University, Tallahassee, FL 32306 USA
[2] Ion Cyclotron Resonance Program, National High Magnetic Field Laboratory, Florida State University, 1800 East Paul Dirac Drive, Tallahassee, FL 32310 USA
[3] Consultant, Future Fuels Institute, Florida State University, 1800 East Paul Dirac Drive, Tallahassee, FL 32310 USA
krajewski@magnet.fsu.edu

Introduction

To meet worldwide energy demands, "opportunity crudes" often enriched in naphthenic acids (NAP) are increasingly used in refineries. NAP acids, which are natural carboxylic acids in petroleum typically containing at least one five- or six-member ring, and are known to be corrosive, costing refineries millions of additional dollars in corrosion detection, maintenance, and prevention. In refineries NAP acid corrosion is believed to result from iron naphthenate that reacts with sulfur compounds formed as a result of sulfur corrosion.[1]

The rate of NAP corrosion, however, does not always correlate with the concentration of the NAP acids as measured by total acid number (TAN).[1] It has been proposed that a subset of NAP acids are responsible for corrosion, but previous corrosion studies with model compounds or isolated acids have not implicated any specific structures. Recently, in a joint industry project (NAP JIP) at Ohio University's Institute of Corrosion and Multi-phase Technology (ICMT), iron oxide and magnetite (Fe_3O_4) layers were observed to have formed when iron was treated with acids in the absence of any sulfur, hypothesized to have formed as a result of iron naphthenate decomposition in the following reactions: [2]

$$Fe(RCOO)_2 \rightarrow FeO + CO_2 + RCOR \qquad \text{(Reaction 1)}$$
$$4FeO \rightarrow Fe_3O_4 + \alpha\text{-}Fe \qquad \text{(Reaction 2)}$$

Although these reactions are well known, often used in the synthesis of ketones, they were never associated with corrosion.

Here, we perform a corrosion test on mineral oil doped with two acids with boiling points similar to NAP acids, and structures ranging from a linear fatty acid to a 1 ring structure. If these acids react with the iron, the iron naphthenates would decompose to form ketones, each with carbon number approximately twice that of the model acid. Gas chromatography mass spectrometry (GC-MS) and Fourier transform ion cyclotron resonance mass spectrometry (FT-ICR MS) are used to validate the presence of these ketones.

Experimental

Samples and Corrosion Test. A white mineral oil with an average molecular weight (MW) of 530 Da was doped with enough palmitic acid ($C_{16}H_{32}O_2$, 256.4 Da) or 5-cyclohexyl pentanoic acid (CXPA, $C_{11}H_{20}O_2$, 184.3 Da) so that the TAN of the oil was 1.75. Three A-106 carbon steel and three A335-P9 (9 Cr alloy) rings with a total surface area of ~300 cm^2 were pretreated with 600 g test solution, loaded into a stirred autoclave where the test solution was added and the autoclave sealed. The headspace was flushed with N_2 for 5 min, pressurized to 200 psi with nitrogen, and the temperature raised to and maintained at 300-350 ^{0}C for 24 h.[3] At the end of the test, used test solution was collected and analyzed. A blank test with no model acid was also performed.

Ketone Functionality Isolation. Ketones were isolated from the hydrocarbons and residual acids by a strong anion exchange (SAX) SPE cartridge (Agilent Bond Mega BE-SAX). After preconditioning with pentane, 1 g of sample was loaded onto the column and allowed to equilibrate. Fractions were collected through 12 mL washes of pentane, dichloromethane (DCM), and methanol. All fractions were dried under nitrogen, and then dissolved to 1 mg/mL in toluene to form stock solutions.

GC-MS Analysis. Model compound mixtures were analyzed by GC/MS with an Agilent 5973 mass spectrometer equipped with an HP 6890 gas chromatograph. A 1 µL sample was injected into the GC inlet (280 ^{0}C) with a 1:10 split ratio. Separation was carried out with an Agilent DB-5MS fused silica capillary column (30 m long, 0.25 mm ID, and 0.25 µm film thickness) with helium (99.9997% purity) carrier gas at 1.2 mL min^{-1}. The oven temperature was initially held for 3 min at 50 °C and then ramped at 10 °C min^{-1} to a final temperature of 300 °C and held for 15 min. MS was performed with 70 eV electron impact (EI) ionization and ion source was held at 200 °C. Positive ion data (m/z range of 30-650) was collected at an acquisition rate of 1 scan s^{-1}. Total ion current (TIC) was used to reconstruct the chromatographic signal.

FT-ICR MS Analysis and Instrumentation. Sample stock solutions were diluted to 500 µg/mL in methanol (J.T. Baker, Phillipsburgh, PA) with 5% formic acid (Sigma-Aldrich, St, Louis, MO, MS grade, 98%) to aid in protonation for positive electrospray ionization. Fractions were also derivatized with Amplifex keto reagent (AB Sciex, Farmingham, MA), by diluting 200 µL of sample stock solution with 200 µL of the reagent, mixing one hour, and drying with nitrogen. The derivatized sample was then dissolved to 500 µg/mL in toluene:methanol (50:50, v/v).

Samples were analyzed with a custom built 9.4 T FT-ICR MS with a modular software package for data acquisition (Predator).[4,5] 150 time-domain transients were co-added, Hanning-appodized, zero -filled, and fast Fourier transformed to yield an absorption-mode mass spectrum.[6]

Mass Calibration and Data Analysis. Custom software, PetroOrg©, was used for data calibration and imaging.[7] Peak lists were obtained from peaks with a relative abundance 6 times greater than the standard deviation of rms baseline noise. Spectra were internally calibrated through a homologous series for which the elemental compositions differed by multiples of CH_2. Peaks were assigned elemental compositions based on mass and grouped by heteroatom content. Data was visualized based upon relative abundance of heteroatom classes and isoabundance-contoured plots of double bond equivalents (DBE = number of rings and double bonds to carbon) versus carbon number for each class.

Results and Discussion

Due to concern about interferences during ionization for the hydrocarbons and residual acids, SAX was used to separate ketones. Because of the expected reduced compositional complexity of these samples, GC-MS sufficed to determine quality of separation. Figure 1 shows the total ion chromatogram (TIC) of the fractions from the SAX separation of the sample spiked with palmitic acid. The pentane fraction is characterized by an unresolved complex mixture (UCM) formed as a result of the hydrocarbons in the mineral oil. The DCM fraction is dominated by a peak with a retention time of ~27 min, library-identified as a ketone, $C_{31}H_{62}O_1$, which corresponds to the ketone that would form from the decomposition of iron II palmitate. 2-Heptadecanone is also detected, likely the result of thermal cracking of the $C_{31}H_{62}O_1$ ketone. In the methanol fraction residual palmitic acid was identified with a retention time of ~17 min.

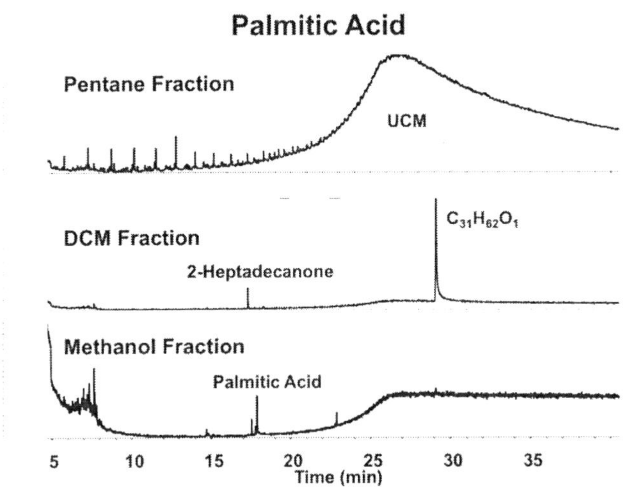

Figure 1. Total ion chromatogram from GC-MS of pentane, DCM, and methanol fractions from SAX separation of corrosion test material with palmitic acid.

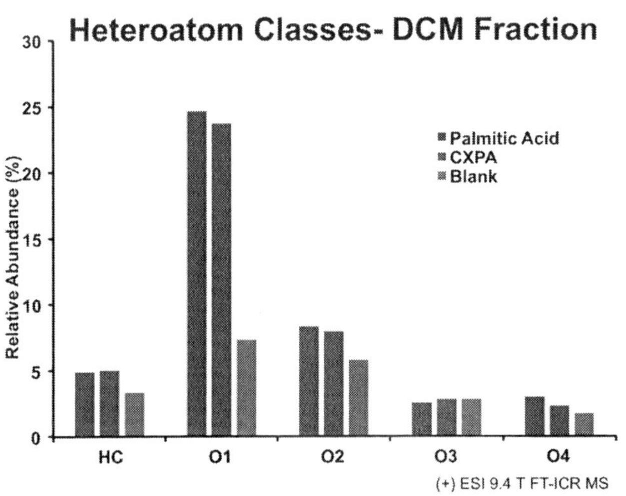

Figure 2. Heteroatom class distribution of the DCM fractions for the palmitic acid, CXPA, and blank samples by (+) ESI 9.4 T FT-ICR MS.

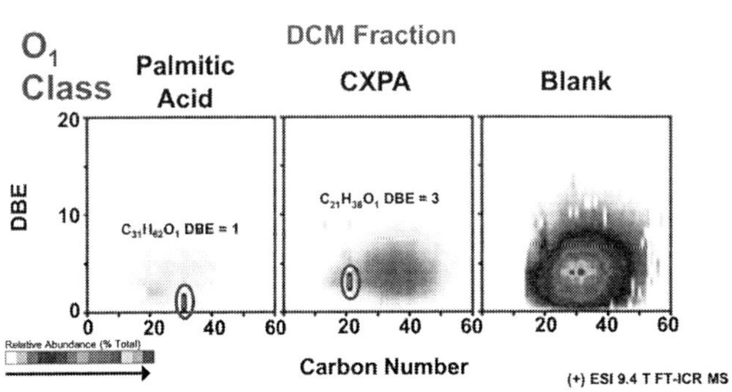

Figure 3. Isoabundance-contoured plots of DBE vs carbon number for O_1 class in DCM fractions by (+) ESI 9.4 T FT-ICR MS.

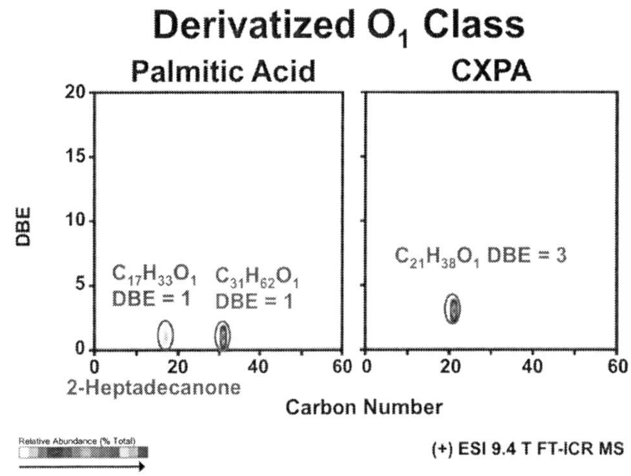

Figure 4. Isoabundance-contoured plots of DBE vs carbon number for derivatized O_1 class in DCM fractions by (+) ESI 9.4 T FT-ICR MS. No species seen in the blank.

To further confirm that ketones were being present, ultrahigh resolution FT-ICR MS was obtained. In the heteroatom class distribution of the DCM fraction (Figure 2) the O_1 class was the most abundant, followed by the O_2 class and HC class (no heteroatoms). As non-polar species do not readily ionize by ESI it is believed they result from reduction of alcohols during ionization. The blank also exhibited the same classes, albeit with significantly lower relative abundance for the O_1 class. In the isoabundance-contoured plots for the O_1 class (Figure 3) species with molecular formulas corresponding to ketones that would be formed from decomposition of iron naphthenate formed from palmitic acid or CXPA can be seen, confirming their presence. In addition to the expected ketone peak, however, is a large although less abundant distribution of O_1 species with a carbon number distribution of ~20-50 and a DBE distribution of ~1-7. This distribution is also seen in the blank. To determine the functionality of these O_1 species (ketone, aldehyde, ether, or alcohol), the fractions were derivatized with a commercial available reagent that specifically targets ketones and aldehydes.[8] Only the expected possible ketone is observed in the sample fractions, with nothing observed in the blank (Figure 4). Based on the molecular formula and their ability to be derivatized the highly abundant singular peaks are likely the expected ketone that would form from the model acid. Because none of the other species in the class were derivatized, they are not ketones and alcohols. We believe that they are products, likely alcohols, formed from oxidation of cracked hydrocarbons by dissolved oxygen in the sample.

Conclusions

Characterization by ESI FT-ICR MS has been shown to identify ketones formed from model acids in mineral oil that underwent high temperatures in the presence of iron, confirming that iron napthenate decomposition occurs from carboxylic acids that react with iron under corrosive conditions. Although other oxidized species were seen, they did not derivatize in the presence of a reagent targeting ketones and aldehydes, and likely result from the oxidation of hydrocarbons at high temperature from dissolved oxygen. As the ketones observed directly correlate with the acid that originally reacted with the iron, further work needs is needed to determine which NAP acids in a corrosive oil are corrosive.

Acknowledgements. Work supported by NSF Division of Materials Research (DMR-1157490) and the State of Florida. The authors thank John P. Quinn, Nathan K. Kaiser, Greg T. Blankey, and Yuri E. Corillo for assistance in instrument maintenance and data analysis. The authors also thank Gheorghe Bota, Peng Jin, the NAP JIP, and ICMT for generating and providing the samples.

References

(1) Slavcheva, E.; Shone, B.; Turnbull. A. *Brit Corros. J.* **1999**, *34* (2), 125.
(2) Jin, P.; Robbins, W.; Bota, G. *Corrosion 2016 paper no. 7203,* Houston, TX: NACE, **2016**, 1.
(3) Jin, P.; Wolf, H.A.; Nesic, S. *Surf. Interface Anal.* **2015**, *47*, 454.
(4) Kaiser, N. K.; Quinn, J. P.; Blakney, G. T.; Hendrickson, C. L.; Marshall, A. G. *J.Am. Soc. Mass Spectrom.* **2011**, *22*, 1343-1351.
(5) Blakney, G. T.; Hendrickson, C. L.; Marshall, A. G. *Int. J. Mass Spectrom.* **2011**, *306*, 246-252.
(6) Xian, F.; Hendrickson, C. L.; Blakney, G. T.; Beu, S. C.; Marshall, A. G. *Anal. Chem.* **2010**, *82*, 8807-8812.
(7) © PetroOrg Software, The Florida State University, All rights reserved, http://software.petroorg.com
(8) Alhassan, A.; Anderssan, J.T. *Energy Fuels*, **2013**, *27*, 5770.

DETERMINATION OF IRON, NICKEL AND VANADIUM IN PETROLEUM HEAVY OIL AND RESIDUE SAMPLES: COMPARISON OF DIFFERENT PREPARATION METHODS AND ELEMENTAL ANALYSIS TECHNIQUES

Jenny Nelson[1], Greg Gilleland[1], Steve Wall[1], Laura Poirier[2], Lidia Berhane[2] and Francisco Lopez-Linares[2]

[1]Agilent Technologies, Inc., 5301 Stevens Creek Blvd, Santa Clara, CA 95051, US
[2]Chevron Energy Technology Company, 100 Chevron Way, Richmond, CA 94801, USA
Corresponding Author Email: jenny_nelson@agilent.com

Introduction

It is well known that the world energy consumption has been increasing over the last two decades, and petroleum and its derivatives provide the most available energy.[1] Among them, heavy oil and bitumen reserves are the most available energy source worldwide, and their uses are expected to increase in the future.[2] Typically, heavy hydrocarbons such as bitumen and heavy oil comprise more than 50% (w/w) of a distillation residue (500 °C+),[3] which becomes the feedstock for further upgrading processes that lead to more yield of high valuable distillables.[1] Therefore, these feeds need to be characterized through bulk properties measurements such as boiling point distribution, API gravity, and viscosity; these properties correlate well with certain composition characteristics. That information helps decision makers in the refinery making fast and inexpensive business decisions. Especially for atmospheric and vacuum residues, density and viscosity are the dominant measurements, however, the ability to measure carbon, hydrogen, nitrogen, sulfur, and metals are becoming critical.[2]

Metal analysis on these fractions has been done using different strategies. The most common is by separating the metals from the matrix, commonly by acid digestion followed by spectroscopy techniques such as Flame Atomic Absorption (FAA) or by plasma techniques such as Inductively Coupled Plasma Optical Emission Spectrometry (ICP-OES) and Inductively Coupled Plasma Mass Spectrometry (ICP-MS).[4] Some industry standard methods for the determination of metals such as iron, nickel, vanadium, and sodium in Crude Oils and Residual Fuels have been developed. ASTM D5708[5] and ASTM D5863[6] have been reported using wet acid digestion or direct dilution in an organic solvent as the chosen sample preparation methods and FAA and ICP-OES as the detection techniques. Recently, we reported the elemental analysis of crude oils by Nitrogen Microwave Plasma-Atomic Emission Spectroscopy, MP-AES using direct dilution in an organic solvent as a preparation method. Samples evaluated covered APIs from 7–38° among other properties, and it was shown that the plasma sustained by nitrogen is comparable to a typical argon based system. In this work, the determination of iron, nickel, and vanadium in heavy oil and residues was evaluated comparing traditional detection techniques such as ICP-OES with MP-AES. Also on the ICP-OES, two sample preparation methods, wet acid digestion and direct dilution in an organic solvent were compared.

Experimental

All chemicals employed were trace metal grade purity and used without purification. Preparation of calibration standards and samples by direct dilution in *o*-xylene for the Microwave Plasma-Atomic Emission Spectroscopy (MP-AES), and Inductively Coupled Plasma-optical Emission Spectroscopy (ICP-OES) were carried out following procedures previously reported.[8,9] Samples were prepared by a wet acid digestion method, which was performed by ashing with sulfuric acid in a programmable furnace and then, dissolving the ash in acids followed by analysis on a ICP-OES and diluted with water to 1:25 dilution.

Ten residue samples were selected according to API as well as elemental composition; API values span from -2 ° up to 18°, and elemental composition as follows: S (1.0 wt.% to 5.5 wt. %) and N (200 ppm to 14000 ppm). For direct dilution these samples were diluted based on metal concentrations using an o-xylene diluent. (From 1:20 up to 1:200).

An Agilent 4200 MP-AES (Agilent Technologies, Santa Clara, CA, USA), with an Agilent 4107 nitrogen generator (Agilent) and Thermo Radial ICAP 6000 Series ICP-OES (Thermo Fisher Scientific, Fremont, CA, USA) were all used for the direct injection of *o*-xylene.

The summary of the conditions employed in this work is presented in Table 1.

Table 1. MP-AES and ICP-OES Experimental Conditions for direct dilution in *o*-xylene

Instrument Conditions	4200 MP-AES	6500 ICP-OES
EGCM setting	Low	
Pump rate	5	
Tubing	Organic	
Read time	3 s	
Number of replicates	3	
Sample uptake delay	55 s	
Stabilization delay	10 s	
Background correction	Auto	
Sample flow rate	238 µL/min	
Rinse flow rate	4.5 mL/min	
Peri-pump speed analysis	5 rpm	
Peri-pump speed rinse	80 rpm	
Forward power		1350 W
Plasma gas flow rate		16 L/min
Auxiliary gas flow rate		2 L/min
Carrier gas flow rate		0.4 L/min
Optional O_2 gas flow rate		0.04 L/min
Sampling depth/radial		14 mm
Peri-pump speed rinse		85 rpm
Peri-pump speed analysis		40 rpm
Spray chamber temp		5 °C
Elements	Fe 259.940	Fe 238.204
	V 311.070	V 292.460
	Ni 341.476	Ni 221.647
Internal standard	Sc 335.372	Sc 227.318
		Sc 255.237

Results and Discussion

Previously, we reported direct dilution in an organic solvent as the sample preparation method for the analysis of crude oil samples, using MP-AES and ICP-(OES/MS) as detection techniques[8,9]. Due to the complex nature of residue samples and the different solubility behavior (depending on which organic solvent is being used), the direct dilution method was evaluated and compared to a traditional wet acid digestion method using ICP-OES as the detection technique for iron, nickel, and vanadium. Table 2, presents the results obtained evaluating both preparation methodologies.

Table 2. Comparison between sample preparation methods

Direct Dilution/Wet acid digestion			
Sample	Fe	Ni	V
S1	0.72	1.00	1.05
S2	0.97	1.06	1.04
S3	0.97	1.00	1.06
S4	0.50	1.01	1.04
S5	0.48	1.01	1.04
S6	0.87	1.08	1.10
S7	0.52	1.01	1.04
S8	0.53	1.04	1.04

From Table 2, it is observed that direct dilution correlates well for nickel and vanadium; however the Fe results need to be further investigated. Direct dilution/wet acid digestion for Ni and V span from 1.00 up to 1.10, this implies that direct dilution can effectively be an alternative to analyzing these metals in a residue. The low bias observed for iron, can be explained considering that solid microparticulate minerals can be present in the heavy oil fractions and deposits[2]. In fact, it has been reported that iron is the predominant element present in such minerals.[10] Therefore, this particulate could not be adequately dispersed in the organic solvent, leading to lower results in comparison with an acid digestion method. The same observation has been detected during the analysis of some crude oil samples, when comparing both direct dilution with acid digestion using ICP-OES.[11]

Table 3 presents the results comparing the source of plasma sustained gas; nitrogen (MP-AES) vs. argon (ICP-OES), using direct dilution as the sample preparation method.

Table 3. Comparison between plasma sustained gas source.

MP-AES/ICP-AES			
Sample	Fe	Ni	V
S2	1.03	1.00	1.03
S4	1.06	0.97	1.03
S5	1.08	0.98	1.02
S6	0.98	0.94	0.94
S7	1.04	0.97	1.03
S8	1.03	0.96	1.01
S9	1.30	1.01	1.05
S10	1.02	1.01	1.06
S11	1.01	1.01	1.07

In most of the cases, good agreement among all elements can be obtained using both plasma sustained gas sources. In fact, these results are aligned with the previous study conducted in our facility analyzing crude oil samples using MP-AES and direct dilution in o-xylene. In this work we found, the nitrogen sustained gas plasma can be effectively used to analyze residue samples by direct dilution method as well.

Figure 1, presents the results comparing the direct dilution preparation method using MP-AES (DD MP-AES) with a more conventional lab analysis protocol, such as wet acid digestion using ICP-OES(WAD ICP-OES) with the evaluated samples.

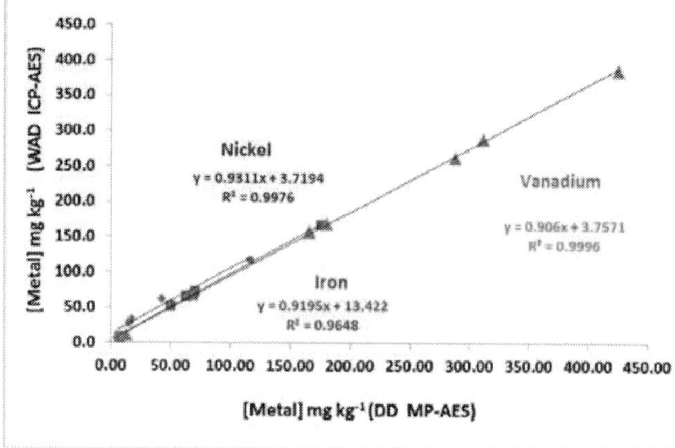

Figure 1. Comparison of methods and plasma sustained gas source.

From Figure 1, it can be seen that nickel and vanadium displayed the highest R^2 in comparison with iron; nevertheless, all R^2 are above 0.95 which is very well accepted for the analysis of hydrocarbon samples. The two methods were compared and the results were within 90% of each other for all metals and the type of samples evaluated. Direct dilution in combination with nitrogen sustained plasma (MP-AES), provides a safe and economical alternative to the metal analysis on residue samples.

Conclusions

The present work illustrates that determination of iron, nickel and vanadium on residue samples can be accomplished by direct dilution in an organic solvent in combination with different plasma sustained gas sources such as nitrogen and argon.

It was demonstrated that nickel and vanadium are not sensitive to the sample preparation method while iron recovery using the direct dilution method, could be biased low, potentially due to presence of solid microparticulate inthe samples. For iron, better results are obtained using wet acid digestion.

Finally, it was shown that the nitrogen based plasma MP-AES versus the argon based plasma ICP-OES, have excellent agreement in all cases for nickel and vanadium, showing the MP-AES being a suitable alternative for the analysis of such sample types .

Acknowledgement. We thank Carbognani Ortega for providing samples from Athabasca vacuum residue. We also thank Chevron Energy Technology Company for providing permission to publish this article.

References

(1) Lee, S.; Speight, J. G.; Loyalka, S. K. Handbook of Alternative Fuel Technologies; CRC Press (Taylor and Francis Group): Boca Raton, FL, 2007.

(2) Altgelt, K. H.; Boduszynski, M. M. Composition and Analysis of Heavy Petroleum Fractions; Marcel Dekker: New York, 1994.

(3) Roberts, P. The End of Oil: On the Edge of a Perilous New World, 1st ed.; Houghton Mifflin: Boston, MA, 2005.

(4) Sanchez, R.; Todoli, J.L.; Lienemann, C-P.; Mermet. J.M. Determination of trace elements in petroleum products by inductively coupled plasma techniques: a critical review. Spectrochimica Acta, Part B: Atomic Spectroscopy, 2013, 88, 104-126.

(5) Standard Test Methods for Determination of Nickel, Vanadium, and Iron in Crude Oils and Residual Fuels by Inductively Coupled Plasma (ICP) Atomic Emission Spectrometry, ASTM D5708-12e1; ASTM International: West Conshohocken, PA, USA, 2012; DOI: 10.1520/D5708-12E01.
(6) Standard Test Methods for Determination of Nickel, Vanadium, and Iron in Crude Oils and Residual Fuels by Inductively Coupled Plasma (ICP) Atomic Emission Spectrometry, ASTM D5708; ASTM International: West Conshohocken, PA, USA, 2012;
(7) Standard Test Methods for Determination of Nickel, Vanadium, Iron, and Sodium in Crude Oils and Residual Fuels by Flame Atomic Absorption Spectrometry. ASTM D5863; ASTM International: West Conshohocken, PA, USA, 2012
(8) Nelson, J.; Gilleland, G.; Poirier, L.; Leong, D.; Hajdu, P.; Lopez-Linares, F. Energy & Fuels. 29 (2015) 5587–5594
(9) Poirier, L.; Nelson J.; Leong, D.; Berhane, L.; Hajdu P.; Lopez-Linares, F. (Submitted to E and F, March 2016)
(10) Carbognani, L.; Orea, M.; Fonseca M. Energy & Fuels. 13 (1999) 351-358
(11) Kishore, Nadkarni, R.A. Elemental Analysis of Fossil Fuels And Related Materials. Kishore, Nadkarni, R.A editor, ASTM Mono 10. ASTM International, 2014, West Conshohocken, PA. Chapter 8, page 457-508

TOWARD THE CONVERSION OF PHYTOREMEDIATION BIOMASS TO CELLULOSIC BIOETHANOL

M. Asad,[1] *I. Ziegler-Devin,*[1] *M. Chalot,* [2] *N. Brosse*[1]

[1]*University of Lorraine, LERMAB, Nancy, France*
Nicolas.Brosse@univ-lorraine.fr
[2] *Université de Franche-Comté Laboratoire Chrono-environnement*

Introduction

The lignocellulosic biomass has an obvious potential as a renewable starting material for the production of biofuels. Nevertheless, a major drawback of the biorefinery philosophy is the availability and the cost of the feedstock. The utilization of metal enriched biomass produced on contaminated sites as starting material for the production of cellulosic bioethanol and biomaterials can be a partial solution of this worldwide problem, and it also represents a new strategy for valorisation of contaminated biomass which constitutes a real threat for ecosystem (1). The final objective is thus to decontaminate the soil by phytoremediation and produce biomass for industrial uses. However, some crucial questions about the fate of heavy metals in a fractionation process (the distribution between cellulosic pulp, co-products like lignin and side stream effluents…), remain unanswered. In the literature, opportunities for bioenergy production from polluted biomass have been already proposed (2-5). In contrast to this approach, in this paper we described for the first time the fractionation of phytoremediation biomass for the production of bioethanol, pulp and biomolecules according the biorefinery concept. The pre-treatment of metal enriched lignocellulosic biomass using organosolv and soda, pretreatment was examined (6). The influence of the severity of the treatment, of the reaction parameters on the composition in metals of the cellulosic residue, the aqueous stream and the lignin fraction was examined.

Experimental

The phytoremediation biomass was harvested on divers contaminated sites in France (metaleurop, Phytochem planatation, INERIS…) (see Table 1).

Organosolv pulping : the biomass was treated with aqueous ethanol (35/65, v/v) with 0.5% or 1% (w/w) sulfuric acid as a catalyst at 170°C, 180°C and 190 °C for 60 min (liquid/solid ratio = was 1:8). After the treatment, the pre-treated biomass was filtered and washed with a warm aqueous-ethanol solution. The liquor was diluted with 3 volumes of water to precipitate the ethanol organosolv lignin and the precipitate was then collected by centrifugation.

Soda pulping : the biomass was treated with 20 % of active alkali in water (liquid/solid ratio = was 1:8). The time of cooking was from 60 mn to 80 mn, temperature 130°C, 170 °C and 190°C. The pulp was washed and separated by screening through a sieve and the black liquor was collected. Soda lignins were precipitated from the concentrated black liquor by acidifying it with 20 % (v/v) sulfuric acid until pH 2. The precipitated lignins were filtered and washed with pH 2 water.

Metals analysis was performed after mineralization (HNO_3) by inductively coupled plasma (ICP), optical emission spectroscopy.

Enzymatic hydrolyses were performed using commercial cellulase mixture (*Trichoderma reesei*) at 2% cellulose w/v in 100 mL of 50 mM acetate buffer, pH 4.8.

The biomass detoxifications were performed in water with 1% w/w of EDDS, EDTA, H_2O_2 or H_2SO_4 in 75°C during 24h.

Results and Discussion

This study was performed staring from lignocellulosic biomass harvested on contaminated sites in France. Wood poplar, willow, birch and also an annual crop (tobacco) with relatively high metal concentrations (Zn = 150 – 400 ppm ; Mn = 50 – 200 ppm ; Fe = 70 – 300 ppm) have been studied (Table 1).

Table 1. Metals content in the biomass

biomass	Zn ppm	Fe ppm	Mn ppm
Tobacco	440	165	195
Birch	139	40	69
Willow	80	155	263

The ethanol organosolv pretreatment has been carried out using the following conditions : water/ethanol ratio = 0.5 and 0.65, reaction time = 1 hour, temperature = 170°C, 180°C and 190°C, sulfuric acid concentration = 0.5% and 1% based on published data.(6) The influence of the reaction parameters on the composition of the residues and on the metal contents in the pulp, in the lignin fraction and in the water effluent was examined. Regardless the biomass, the metal considered and the severity of the organosolv conditions used, the proportions of metals in the 3 fractions were in the following order : pulp > water effluent > lignin. For Zn, a higher solvent extractability was observed (30% - 40%) compared to Fe and Mn (10% - 20%). Low to very low metal concentrations in lignin was detected. An increase of the sulfuric acid concentration and/or of the water content during the organosolv pulping increased the solution extraction of the metals. On the other hand, an increase of the temperature of the treatment of birch surprisingly decreases the metal extraction from the pulp.

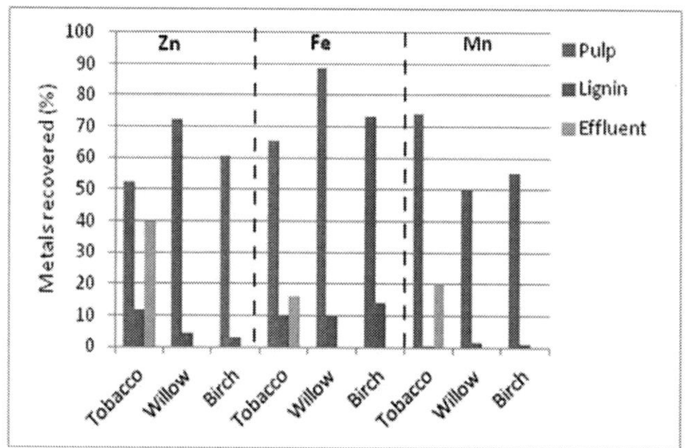

Figure 1. Organosolv traitements of tobacco, birch and willow : metals distributions in cellulosic pulp, lignin, water effluents fractions for one of the organosolv conditione (T=170°C, water/EtOH=0.65, H_2SO_4=1%).

Soda pretreatments were also examined. These processes are experimented as pretreatment for cellulosic bioethanol production but are also widely used in the pulp industry (Kraft process) (8). As previously observed for organosolv pretreatment, the larger part of the metals (60% to 70%) was recovered in the pulp and a smaller fraction in the water effluent (5-12%) and in the lignin fraction (5-20%). At high temperature (170°-190°C) it was observed a decrease of the metal extractability; the same observation was made previously, with organosolv treatment. The water extractability of the metals was in the following order : Mn > Zn > Fe.

As an example, the fig. 2 gives the distribution of Zn (based on initial content) in the 3 recovered fractions (pulp, liquid effluent, precipitated lignin) for tobacco after soda treatment as a function of the conditions.

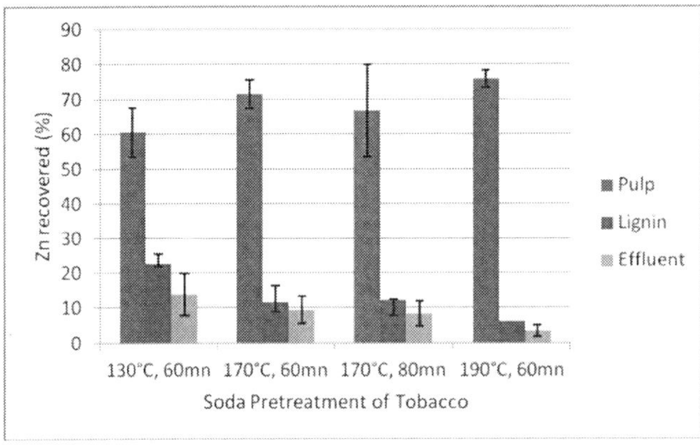

Figure 2. Soda traitement of tobacco : Zn distribution in cellulosic pulp, lignin, water effluent fractions.

An alternative process was also experimented for biomass cleaning : a prehydrolysis step using dilute acid, oxygen peroxide or chelatants (EDTA or EDDS) was evaluated for a biomass detoxification. From woody materials (birch or willow) a very efficient metals removal was observed by dilute acid. Chelatant agents showed also high efficiency for Mn and Zn removal. From the non woody material (tobacco), lower extraction was observed (≈50%) and a relatively high Zn removal was obtained with EDDS extraction (data not shown).

Enzymatic hydrolyses of organosolv willow pulps were performed using standard literature conditions. The enzymatic hydrolyses of a metal-rich pulp and a pulp with low metal contents (after a detoxification step, see below) were investigated and compared in order to examine the effect of the metals on the enzymatic conversion of cellulose and xylane into monomeric sugars. The glucose and xylose concentrations at different time points during the hydrolysis are given in the Fig. 3. No visible effect of metals on polysaccharides hydrolysability was observed.

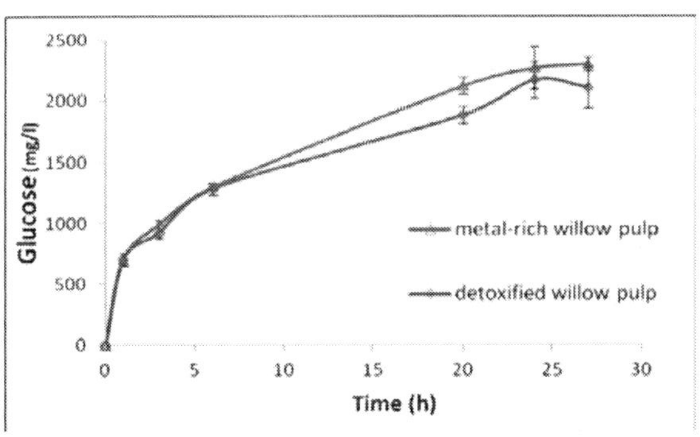

Figure 3. Cellulose enzymatic hydrolysis of organosolv willow pulps with or witout metals

Conclusions

The first objective of this study was to determine how the metals compounds will be distributed in each environmental compartment of an organoslv and a soda pretreatment. The metal contents are in the following order pulp ≫ water stream > lignin. The extractability is low in basic conditions especially at high temperature. In organosolv treatment, metal extraction in liquid phase increased with the water content and the acidity but decreased with the temperature. The data provided by the laboratory tests will be used to compare with existing technologies to eliminate metal compounds. From these assays, the relevant lacking data (process parameter, emission characterisation, biomass characterisation, impacts and regulatory constraints, etc.) will be identified to implement further step, i.e. the pilot scale.

Acknowledgement. The authors gratefully acknowledge the financial support of ANR Phytochem Project. The EA 4370 LERMAB is supported by the French National Research Agency through the Laboratory of Excellence ARBRE (ANR-12-LABXARBRE-01).

References

(1) Migeon, A., Richaud, P., Guinet, F., Chalot, M., Blaudez, D. *Water Air Soil Pollut.* **2009**, 204, 89-101.
(2) Gomez, H. I. *Environ. Technol.* **2012**, *1*(1) 59-66.
(3) Syc, M., Pohorely, M., Kamenikova, P., Habart, J., Svaboda, K., Puncochar, M., *Biomass Bioener.***2012**, 37, 106-113.
(4) Schreurs, E., Voets, T., Thewys, T. *Biomass Bioener.* **2011**, 4469-4480.
(5) Vervaeke, P., Tack, F.M.G., Navez, F., Martin, J., Verloo, M.G., Lust, N. *Biomass Bioener.* **2016**, 30, 58-65.
(6) FitzPatrick, M., Champagne, P., Cunningham, M. F., Whitney, R. A. *Bioressour. Technol.* **2010**, 101, 8915-8922.
(7) Chun Sheng, G., Hui Teng, T., Keat Teong, L., Brosse, N. *Biomass Bioener.***2011**, 35, 4025-4033.
(8) Han, M., Choi, G.-W., Kim, Y., Koo, B.-C. *Bioressources.* **2011**, 6(2), 1939-1953.

AUTO-IGNITION CHARACTERISTICS OF CONVENTIONAL AND SYNTHETIC ALTERNATIVE JET FUELS IN A MOTORED ENGINE AND AN OPTICALLY ACCESSIBLE CONSTANT VOLUME SPRAY COMBUSTION CHAMBER

Dongil Kang, Vickey Kaslaskar and André Boehman

Walter E. Lay Automotive Laboratory
The Department of Mechanical Engineering
The University of Michigan
1231 Beal Avenue
Ann Arbor, MI 48109-2121
Boehman@umich.edu

Introduction

In recent years, synthetic liquid fuels produced from the reforming of coal, natural gas and biomass have been drawing the U.S. military's attention as potential alternatives to current petroleum-derived fuels, such as JP-8 for aviation and F-76 for naval applications. Fuel specification standards dictate requirements on a variety of chemical and physical properties of such alternative fuels to ensure that the fuels will serve well in their application. In addition, understanding the chemical kinetic behavior of alternative fuels, for example ignition delay, flame speed, and extinction limits, is essential for combustion model development.

Particularly, the Single Battlefield Fuel policy of Air Force mandates the use of JP-8 or 50/50 blends of synthetic alternative jet fuels in JP-8 in military ground vehicles with compression ignition engines. The combustion process in diesel engines is different from gas turbine engines, given that the onset of combustion in a diesel (compression ignition) engine is mostly governed by an ignition delay period between the start of injection (SOI) and the start of combustion (SOC), wherein complex fuel dependent physical and chemical phenomena play a crucial role.

For this reason, this study explores spray ignition behavior and gas phase combustion kinetics of conventional and synthetic alternative jet fuels in IC engine environments. The results presented here are useful for validation and improvement of future chemical kinetic models for use in simulating diesel engine combustion.

Experimental

Engine Setup. Some tests were carried out using a modified CFR octane rating engine, where the compression ratio (CR) can be adjusted between 4 and 15.7. The intake, exhaust, and jacket cooling systems were modified and the spark plug is disabled. The intake air flow was measured via a Delphi hot wire sensor (MAF). The mass flow rate of the injected liquid fuel was determined using established calibration curves as a function of the injection duration for each test fuel. The original carbureted intake fueling system used for the octane rating engine was replaced with a gasoline direct injection (GDI) system, located 1.4 meters above the engine intake valve, allowing for thorough mixing of fuel and air. The intake manifold of the engine is furnished with an electric heater which is used to heat the intake air to 260°C. Thereby, injected fuel droplets can be fully vaporized and mixed with air. The original knock meter is substituted with a piezoelectric pressure transducer (Kistler 6052B) installed in the cylinder head, and its signal is amplified using a Kistler 5011B charge amplifier. Having 40 consecutive engine cycle in-cylinder pressure data points with 0.1 crank angle (CA) resolution using a shaft encoder, the data points of in-cylinder temperature and apparent heat release rate (AHRR) analysis are calculated by using a zero-dimensional single zone model established by Heywood[1]. Gaseous CO, CO_2, and O_2 emissions were also monitored by using a NDIR based CAI emission bench.

Constant Volume Combustion Chamber (CVCC) Setup. The constant volume combustion chamber (CVCC) was designed to mimic a practical diesel engine combustion chamber with a Bosch light-duty diesel injector that delivers fuel at 1000 bar. In addition, the CVCC was modified for the purpose of investigating the physical and chemical ignition behavior of liquid fuel. Three ports were welded to the bottom of the chamber to provide optical access to the spray during fuel injection. Two sub-systems occupied these ports, including a high speed camera system for physical spray characterization and a photomultiplier tube (PMT) system for chemiluminescence detection of excited state chemical intermediate species. In the current study, two ports for a high speed camera system were blocked and only the PMT system was utilized.

Chemiluminescence Detection System. A custom chemiluminescence detection system (CDS) was designed by Mayo et al.[2] to observe the presence of known intermediate species throughout the ignition delay event. An air-cooled, UV/Vis optical probe was installed at the bottom of the combustion chamber. The probe was equipped with 90° wide-angle, quartz lens. The photons emitted from excited state chemical intermediates passed through the optical probe and fiber to the CDS. A collimating lens aligned the light rays in parallel as they exited the fiber tip. This homogeneous beam of light is then directed towards a system of dichroic mirrors, band-pass filters, and three photomultiplier tube (PMT) modules. Two dichroic mirrors in series segregated the light at 340 and 460nm. The three resulting channels of light were individually filtered with Edmund Optics band-pass filters. The OH*, CH* and C* chemiluminescence signals are measured with a band-pass interference filter centered at 307±5nm, 430±5nm and 515±5nm, respectively. In addition, excited state formaldehyde (CH2O*) emissions, which overlapped the 431nm wavelength of CH*, were captured during first stage of the two-stage ignition process. Once photons pass through each filter, a Hamamatsu PMT Module generates an amplified voltage (0-5V) through amplifiers for each excited state chemical intermediate.

Methodology. For the motored engine study, the results were obtained at an intake temperature of 260°C and ϕ of 0.25 and 0.5. The engine speed is fixed at 600 rpm. For each test series, the engine CR is gradually increased from the lowest point (CR = 4) to the point where significant high temperature heat release (HTHR) is observed, so that the progressive change of the ignition behavior of each test fuel with the increase of CR can be observed. Another observation in the experiment involves maintaining a constant CR while varying ϕ from low to high values until ignition is achieved.

For the CVCC study, the sensitivity of ignition behavior to air temperature was obtained by varying the initial chamber temperature from 540 to 640°C. Other initial test conditions include 20bar chamber pressure and 2.5ms injection duration, which were kept constant throughout all the tests.

Test Fuels. Conventional military aviation fuels used in this study include JP-8, Jet-A, Shell distilled Jet-A, and JP-5. Amongst those conventional jet fuels, JP-8 POSF 6169, which is readily available, in elsewhere is selected as the reference for the comparison toward synthetic jet fuels and as the blending unit mixed with 50 volume percent of synthetic jet fuels. In addition, five synthetic jet fuels derived from different natural resources are being considered in this study as follows; "S-8"(S8) derived from natural gas, "synthetic paraffinic kerosene" (SPK) and "iso-paraffinic kerosene" (IPK) derived from coal via the Fisher-Tropsch (FT) process, respectively, "hydro-processed renewable jet" (HRJ) derived from Camelina and lastly "alcohol-to-jet" (ATJ) derived from iso-butanol. The fuels listed above have been prepared and provided from the US Air Force

Research Laboratory (AFRL), each identified by specific "POSF" numbers, as shown in Table 1.

Table 1. Properties of jet fuels studied.

Fuel	POSF	DCN[a]	MW[c]	H/C[c]	SP (mm)	TSI[b]
JP-8	6169	48.64	153.6	2.017	23.57	15.5
Jet-A	4658	47.01	157.5	1.957	21.18	18.8
Jet-A	10325	48.61	157	1.937	21.81	18
JP-5	10289	43.27	na	1.87	na	Na
S8	5018	65.24	154.5	2.14	>45	<4.6
SPK	5729	62.43	138.3	2.24	>45	<3.4
IPK	7629	31.71	148.5	2.195	42.36	4.9
HRJ8	11714	59.77	167.6	2.202	>45	<5.6
ATJ	11498	18.24	173	2.168	35.47	9.7
JP8/S8		54.53	154	2.1	39.26	6.4
JP8/SPK	7717	53.87	145.8	2.124	40.63	5.2
JP8/IPK	7718	39.33	151.1	2.105	32.37	9
JP8/HRJ8	7721	52.54	160.1	2.104	36.51	8
JP8/ATJ		33.42	162.5	2.088	30.65	11.2

[a] Derived Cetane Number (DCN) measured by ASTM D7668
[b] Threshold Sooting Index (TSI) measured by ASTM D1322
[c] Molecular weight (MW) and H/C ratio obtained from [3]

Results and Discussion

The absence of aromatics in alternative synthetic jet fuels leads to dramatic reduction of soot formation, as compared to conventional jet fuels. However, the lack of aromatic in fuels causes leaking problems due to lack of swelling in elastomeric seals. For this reason, blending with conventional fuels is unavoidable to provide satisfactory aromatic content, but this blending penalizes the sooting tendency as increasing TSI value in blends, as seen in Table 1.

The critical compression ratio (CCR) of each test fuel shown in Table 2 is in a good agreement with their DCN value shown in Table 1. The alternative jet fuels with higher n-paraffinic contents such as S8, SPK and HRJ8, possess strong % LTHR than conventional jet fuels in both ϕ cases, while alternative jet fuels with higher iso-paraffinic contents such as IPK and ATJ display the lack of the 1st stage heat release, due to lack of n-paraffinic, which has a strong reactivity in the low temperature regime. Once blended with JP-8, ignition behaviors of synthetic fuels can be affected as depressing or promoting % LTHR and *vice versa* for CCR.

Table 2. The measurement of CCR and % LTHR defined as [LTHR / (LTHR + HTHR)] × 100% at ϕ of 0.25 and 0.5 in a modified CFR motored engine

Fuel	CCR 0.25	% LTHR 0.25	CCR 0.5	% LTHR 0.5
JP-8	6.73	10.54	4.8	4.97
Jet-A (4658)	6.75	14.71	4.92	6.01
Jet-A	6.88	12.40	5	5.32
JP-5	7.38	5.25	5.5	2.25
S8	5.63	23.69	3.83	7.14
SPK	5.6	21.64	3.87	8.76
IPK	9.3	0	8.05	0
HRJ8	6.3	17.52	4.13	8.13
ATJ	10.6	0	9.98	0
JP8/S8	6.23	17.47	4.45	6.70
JP8/SPK	6.35	14.81	4.45	6.96
JP8/IPK	7.8	6.21	5.85	2.73
JP8/HRJ8	6.7	13.13	4.65	6.41
JP8/ATJ	8.18	0.67	6.45	2.42

Figure 1 presents the critical equivalence ratios (ϕ_{crit}) of alternative jet fuels and blends with JP-8. As expected, fuel blends aggressively promoted ignition reactivity with the same amount of fuel injected for lower reactivity fuels (IPK and APJ). One noticeable observation found in higher reactivity fuels (S8, SPK and HRJ) is that the blending impact becomes minimal as ϕ_{crit} becomes leaner, as depicted in boxes shown in Figure 1.

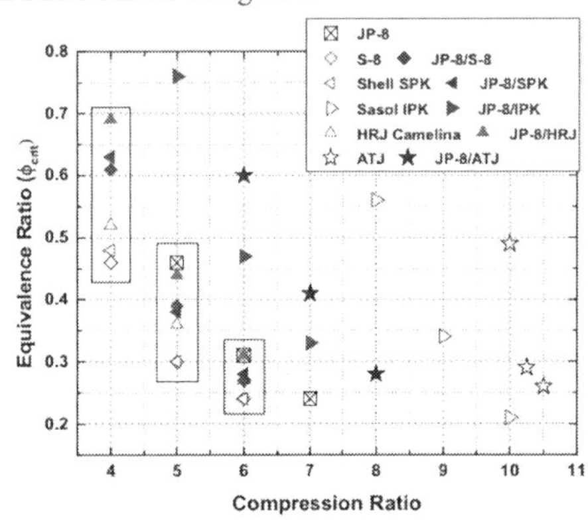

Figure 1. Critical equivalence ratio of jet fuels tested at different compression ratios

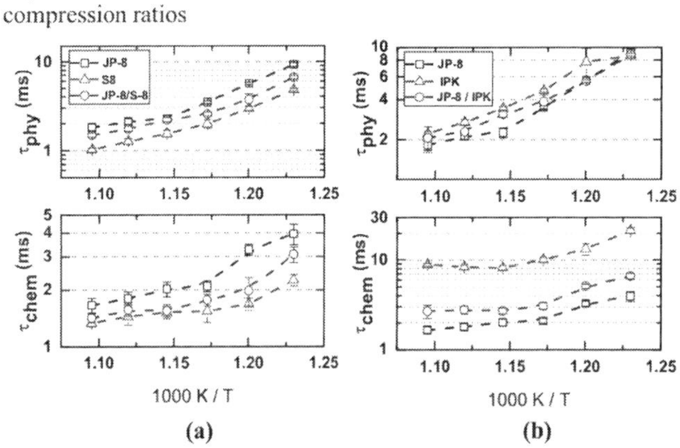

Figure 2. Physical and chemical ignition delays of JP-8 and selected alternative jet fuels in a CVCC; (a) S8 and (b) IPK

Figure 2 (a) and (b) show physical and chemical ignition behaviors of S8 and IPK and their blends with JP-8 in a CVCC. Fig. 2 (a) displays strong negative temperature coefficient (NTC) behavior for S8 starting from 560°C, while JP-8 has less NTC behavior found in higher temperature at 580°C in a plot of τ_{chem}. Fig. 2 (b) compares ignition behavior of IPK with JP-8, showing that the onset of NTC behavior for IPK is found in 20°C higher than that for JP-8. Those NTC behaviors found in plots of τ_{chem} are well agreed by their ignition reactivity in DCN. It is also found that, when blended with JP-8, τ_{chem} of alternative jet fuels is affected by varying temperatures more than τ_{phy}, which is dominated by the physical properties of the fuels. This result indicates that physical properties of these fuels are less important than the chemical properties in the diesel engine combustion process, though both properties must be considered for more accurate diesel engine combustion models.

Acknowledgement. This work was supported by the "Automotive Research Center (ARC), a U.S. Army center of excellence in modeling and simulation of ground vehicles".

References

(1) Heywood, J. B. *Internal combustion engine fundamentals*; McGraw-Hill series in mechanical engineering; McGraw-Hill: New York, 1988.

(2) Mayo, M. P.; Boehman, A. L. *Energy Fuels* **2015**, *29* (10), 6793–6803.

(3) Won, S. H.; Veloo, P. S.; Dooley, S.; Santner, J.; Haas, F. M.; Ju, Y.; Dryer, F. L. *Fuel* **2016**, *168*, 34–46.

POLYMER CEMENT COMPOSITES FOR GEOTHERMAL WELLS

M. Ian Childers, Kenton A. Rod, Timothy J. Roosendaal, Mackenzie C. Endres, Wooyong Um, Diana N. Tran, Jaehun Chun, Phillip K. Koech, Carlos A. Fernandez

Pacific Northwest National Laboratory, Richland, WA 99352
carlos.fernandez@pnnl.gov

Introduction

Geothermal wells present extreme challenges for their cement linings. In addition to the typical challenges faced by other underground well cement linings, the high temperatures (up to 350 °C) and chemically corrosive (typically hypersaline, CO_2 and H_2S rich) environments common in geothermal wells often lead to failure of the cement lining.[1] Repair of cement linings is an expensive and time consuming process that also necessitates shutting down production, a double blow to the economics of geothermal energy production. While geothermal energy has the potential to be a sustainable, large-scale source of energy with negligible carbon production, it is not a major source of energy production in the United States despite large reserves. Our concept is to introduce low cost, commercially available polymers into cement in order to improve the heat and chemical resistivity of the well cement lining and add a degree of self-healing ability to the polymer cement composite, thereby increasing the service lifetime of the geothermal well cement lining and improving the economics of geothermal energy production.

Experimental

Class H well cement was obtained from Lafarge, silica flour was obtained from U.S. Silica, and polymer precursors were obtained from Sigma-Aldrich and Akzo Nobel. Cement slurries were mixed in a 600 mL polystyrene beaker using a Caframo BDC-1850 overhead stirrer with a 2", 4 blade impeller for 15 minutes and cured in an covered water bath in an oven at 85 °C for 28 days. Polymer precursors were added after mixing for 10 minutes. Compression strengths were determined for 1-inch diameter, 4-inch tall cylindrical molds following ASTM C39 standards on a MTS servohydraulic test frame. Rheology measurements were made using an Anton Paar MCR 301 rheometer with vane-cup geometry at 25 °C and 85 °C for detailed rheological properties for up to 210 minutes in the range of 0–800 sec^{-1}.

Results and Discussion

For our standard cement we selected Class H well cement due to its widespread use in the well cementing industry. To prevent the known strength retrogression of cements at high temperature due to the conversion of the C-S-H phase to the α-C_2SH phase, silica flour (30 weight percent of the total cement weight) was added.[1] We tested a range of polymer weight percentages relative to the total amount of cement. Addition of polymer precursors to the cement slurry led to a noticeable increase in viscosity of the slurry, necessitating the addition of extra water relative to the standard ratio of 0.38 of water to cementitious material. As seen in Figure 1, a water to cement ratio of 0.5 for 5–15 weight percent of polymer gives cement slurries with lower consistencies than observed for the control cement without polymer. Consistent with the polymer increasing the amount of water required to maintain a constant consistency of cement slurry, the consistency of the cement slurry increases as the amount of polymer increases.

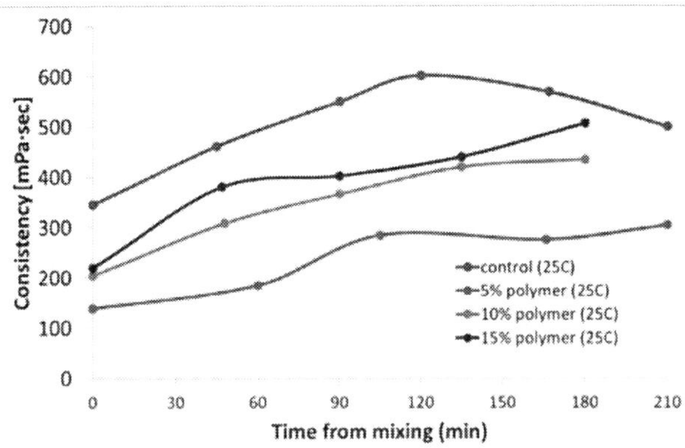

Figure 1. Consistency of polymer-cement slurries measured at 25 °C. Control refers to a 70/30 Class H cement/silica flour cement with a water to cement ratio of 0.38. Polymer cement slurries have a water to cement ratio of 0.5.

While increasing the water to cement ratio is necessary to give cement slurries with reasonable consistencies, it has the consequence of decreasing the compressive strength of the final cement product. Experiments to determine the contribution of additional water versus additional polymer to the deterioration of compressive strength are currently underway. However, while the decrease in compressive strength is unfortunate, the polymer-cement composites still have compressive strengths of approximately 1,000 psi, which is sufficient for geothermal well applications.[1] Additional experiments are underway to determine the optimal ratios of polymer, cement, and water in order to maximize the compressive strength of these polymer cement composites.

In order to measure the stability of the polymer cement composites under geothermal well conditions, aging and thermal shock resistance experiments are underway. Cycling from 25 °C to 350 °C tests the ability of the cement to maintain its integrity under the rapid temperature changes of producing geothermal wells as well as its long-term stability.

Conclusions

In conclusion, the instability of geothermal well cements presents a major challenge to the economic feasibility of geothermal energy production. The goal of this research is to produce polymer cement composite materials with increased chemical and thermal resistivities under geothermal well conditions. We have found polymers that form polymer cement composites with rheological properties similar to those of non-polymer modified cement and sufficient compressive strengths for geothermal well applications. Further testing is underway to increase the strength of these composites and to measure the chemical and thermal resistivity of the composites, as well as determine if the addition of polymer provides for self-healing ability of the polymer cement composites.

Acknowledgement. Funding for this research was provided by the Geothermal Technology Office of the U.S. Department of Energy. Pacific Northwest National Laboratory is operated by Battelle for the U.S. Department of Energy under contract DE-AC06-76RLO 1830. We thank LaFarge for providing Class H cement, U.S. Silica for providing silica flour, and Akzo Nobel for providing polymer precursors used in this research.

References

(1) Nelson, E. B., Barlet-Gouédard, V. In *Well Cementing*; Nelson, E B.; Guillot, D., Eds.; 2nd ed.; Schlumberger, 2006.

In Situ Potentiodynamic Analysis of the Electrolyte/Silicon Electrodes Interface Reactions - A Sum Frequency Generation Vibrational Spectroscopy Study

Yonatan Horowitz[†], *Hui-Ling Han*[§] *and Gabor A. Somorjai*[†,§]

[†]Department of Chemistry, University of California, Berkeley, California 94720, United States
[§]Materials Sciences Division, Lawrence Berkeley National Laboratory, 1 Cyclotron Road, Berkeley, California 94720, United States
somorjai@berkeley.edu

Introduction

The key factor in long-term use of batteries is the formation of an electrically insulating solid layer that allows lithium ion transport but stops further electrolyte redox reactions on the electrode surface, known as the solid electrolyte interface (SEI).[1] We have studied the reduction products of different electrolyte solutions and their fluorinated counterparts on crystalline silicon (Si) and amorphous Si in a lithium (Li) half-cell system under working conditions. We employed in situ sum frequency generation (SFG) vibrational spectroscopy with interface sensitivity in order to probe the molecular composition of the SEI surface species under various applied potentials where electrolyte reduction is expected. We have examined by SFG vibrational spectroscopy the effect of the Si anode termination (oxide vs. hydride) on the surface-chemistry governing the SEI formation.[2] Applying the same methodology we further explored the effect of fluorine-based electrolytes on the chemistry of the SEI as a method to investigate the role of additives (such as FEC)[3] in Li-ion battery systems

Experimental

Materials. The following electrolytes solutions were examined: 1.0 M $LiPF_6$ dissolved in ethylene carbonate (EC) and diethyl carbonate (DEC). To further pinpoint the effect of fluorine on the SEI formation we used: 1.0 M $LiClO_4$ in EC or fluoroethylene carbonate (FEC). The anode was a p-type Si(100) wafer polished on both sides.

SFG vibrational spectroscopy apparatus. The picosecond laser system is consisted of a 1064 nm Nd:YAG pump laser, PL2250, manufactured by Ekspla. The repetition rate of the laser is 20 Hz and the average peak power is 25 mJ. A LaserVision optical parameter generator and amplifier system converted the 1064 nm to a visible 532 nm beam and mid-IR ranging between 2200 cm^{-1} to 4000 cm^{-1}. Sum frequency generation (SFG) is achieved when a visible beam and an infrared one overlap spatially and temporally at an interface. The beams orientation in all the SFG experiments was as follows: the angle of the 532 nm beam was 56° and the mid-IR beam was 45°, in respect to the perpendicular plane (reference plane). We collected the SFG beam (in the UV range) by a Hamamatsu photon-multiplier tube model R922.

Results and Discussion

We took the SFG spectra under working conditions corresponding to the three reduction peaks at ~1.5 V, ~0.5 V, and ~0.10 V (vs. Li/Li^+) that are consistent with values reported in the literature for EC and DEC decomposition. We found that, upon reaching the reduction potential of DEC ~ 1.0 V, with a Si(100)-hydrogen terminated wafer, a Si-ethoxy ($Si\text{-}OC_2H_5$) surface intermediate forms due to DEC decomposition to an ethoxy radical, $\bullet OC_2H_5$ or anion (Figure 1). If a thick oxide layer is present no surface reaction occurs until the applied voltage reaches ~ 0.5 V, the onset potential for EC reduction. When only EC is used no reduction is observed even for Si(100)-hydrogen terminated anode.

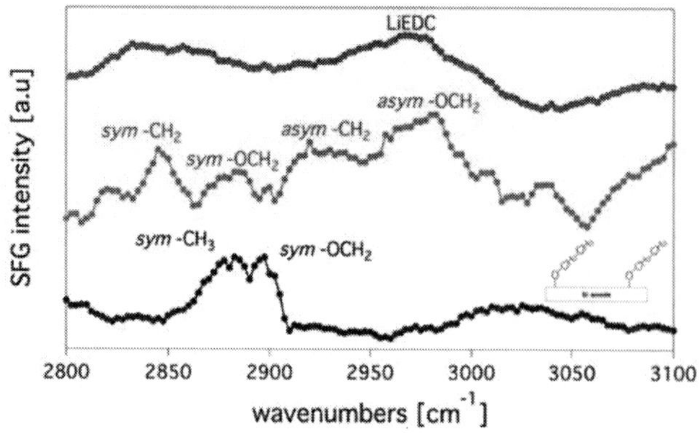

Figure 1. We show the evolution of SFG signal under reaction conditions of crystalline silicon Si(100)-hydrogen terminated anode and 1.0 M $LiPF_6$ in EC: DEC, (1:2 v/v). The SFG spectra were taken after cyclic voltammetry at 0.9 V ± 0.15 V, 0.5 V ± 0.15 V and 0.00 V ± 0.15 V.

Conclusions

Our results suggest that the SEI surface composition varies depending on the termination of Si surface. By performing SFG vibrational spectroscopy together with cyclic-voltammetry we have observed that the Si-hydrogen terminated layer has been changed to Si-ethoxy ($Si\text{-}OCH_2CH_3$) at a potential close to 1.0 V only when DEC is present.

Acknowledgement. This work was supported by the Assistant Secretary for Energy Efficiency and Renewable Energy, Office of Freedom CAR and Vehicle Technologies of the U.S. Department of Energy under Contract No. DE-AC02 O5CH1123. The SFG instrumentation was purchased with funding from the Director, Office of Basic Energy Sciences, Materials Science and Engineering Division of the U.S. Department of Energy.

References

(1) Gauthier, M.; Carney, T. J.; Grimaud, A.; Giordano, L.; Pour, N.; Chang, H.-H.; Fenning, D. P.; Lux, S. F.; Paschos, O.; Bauer, C.; Maglia, F.; Lupart, S.; Lamp, P.; Shao-Horn, Y. *The Journal of Physical Chemistry Letters* **2015**, *6* (22), 4653–4672.
(2) Horowitz, Y.; Han, H.-L.; Ross, P. N.; Somorjai, G. A. *J. Am. Chem. Soc.* **2016**, *138* (3), 726–729.
(3) Shkrob, I. A.; Wishart, J. F.; Abraham, D. P. *The Journal of Physical Chemistry C* **2015**, *119* (27), 14954–14964.

REDUCING THE CARBON INTENSITY OF METHANOL THROUGH BI-REFORMING

*Sean Franco[a], Partho Roy[a], Chan Park[a], Arun Raju[a]**

[a]CE-CERT, University of California, Riverside, USA
Sean Franco (sfranco@cert.ucr.edu)
Partho Roy (partho@cert.ucr.edu)
Chan Park (cspark@cert.ucr.edu)
Arun Raju (arun@engr.ucr.edu)

Several technology options to convert CO_2 into higher value fuels and chemicals are currently available, including well known processes for methanol and dimethyl ether production. However, these processes are commercially feasible under very few circumstances and the search for improved technologies is ongoing. Here, we propose the capture and conversion of CO_2 from large point sources such as fossil fuel power plants into synthesis gas (syngas) through the bi-reforming (CH_4 reforming with H_2O and CO_2) reaction. The syngas is then converted to methanol using commercial technology. Process evaluation using Aspen plus process and life cycle analysis using the GHGenius model have been conducted. The methanol produced through the bi-reforming pathway has 20-50% lower Carbon Intensity compared to methanol produced by conventional steam reforming. Material/energy balances for key process configurations and life cycle and economic analysis results will be discussed.

IN SITU COMBUSTION, IN SITU UPGRADING REACTION MONITORING, AND UPGRADING MAPPING FOR ATHABASCA BITUMEN AND ITS PROCESSED PRODUCTS

Lante Carbognani Ortega, Christian Hovsepian, Carlos Scott, Pedro Pereira-Almao, R. Gordon Moore, Sudarshan A. Mehta, Matthew G. Ursenbach*

Schulich School of Engineering, University of Calgary, Calgary, Alberta, T2N 1N4, Canada.
* lcarbogn@ucalgary.ca

Introduction

"In Situ" technologies aim to produce improved bitumen and/or heavy oil mobility and permanent levels of upgrading directly inside the reservoirs. Athabasca bitumen upgrading via In Situ Combustion (ISC) and In Situ Upgrading Technology (ISUT) were studied processes in the present work. 1-dimension (1-D) packed tube combustors and 2-D reservoir pilot scale units were assembled respectively for ISC and ISUT testing. ISC has been studied for about 40 years at the University of Calgary (UofC) laboratories,[1] being this process alternative reviewed in the past.[2,3] ISUT is a catalytic version derived from about 10 years of research within UofC, recently disclosed.[4,5]

Special analysis protocols based on High Temperature Simulated Distillation (HTSD) were developed for analyzing complex products derived from ISC and ISUT bitumen upgrading. These products simultaneously include light ends and non-distillable vacuum residua. Discussion on these HTSD protocols and the upgrading mappings obtained for ISC and ISUT Athabasca bitumen processing are addressed in this work.

Experimental

Bitumen sample. Athabasca bitumen from Alberta, Canada was studied. Its vacuum residue was prepared using standard distillation methodologies.[6,7]

In Situ Combustion. ISC was studied in a 1.825 m long. x 0.10 m diameter packed tube containing about 22.5 kg of oil sand imbibed with known amounts of bitumen and brine, respectively of the order of 3.3 and 1.2 kg. Selected results from three of the 26 carried out tests will be discussed. Experimental conditions have been presented elsewhere.[8,9]

In Situ Upgrading. Two dimensional representation of a reservoir zone was conceived by using two parallel horizontal wells, as shown on Figure 1. The interwell represents the catalytic reaction zone from the reservoir. The branches mimic the overlaying and horizontal portions of the reservoir. Vacuum residue, hydrogen and catalyst were injected into the porous medium through the injection well; the upgraded oil and unconsumed hydrogen were recovered from the bottom well. All sections from the rig were packed with silica sand and impregnated with bitumen, before carrying out ISUT experiments; details on the unit and its operation were recently reported.[5] At the end of the experiment described in the present work, the unit was disassembled and the packed cores from each of the three branches and the interwell zone were recovered by mechanical means. Three oil sand zones were recovered from each branch / interwell, totaling 12 spent sand samples that were further studied.

Simulated distillation: One Agilent chromatograph model 7890A was used for the study. 5m x 0.53 mm metal wall capillary columns were used. Column stationary phase thickness: 0.1 µm. Analyses and integration were carried out following ASTM D 7169.[10] More details have been published elsewhere.[11]

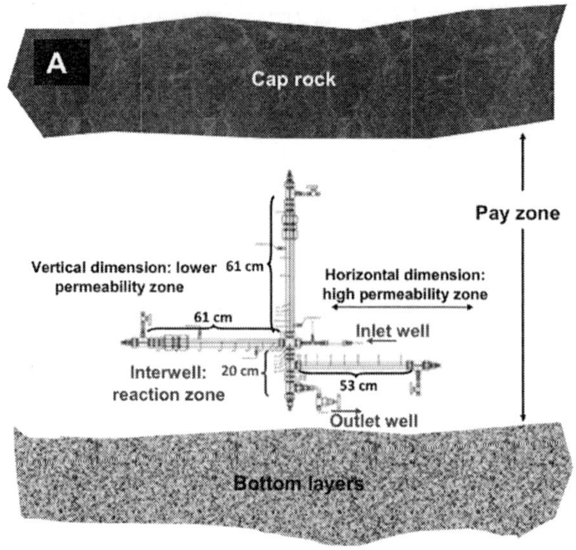

Figure 1. Schematic for the 2-D In Situ Upgrading pilot rig

Results and Discussion

Athabasca ISC processing. Products from bitumen ISC are extremely inhomogeneous mixtures containing produced bitumen, bitumen emulsions, free water and solids, i.e., original mineral fines, rust and sometimes coke particles. A sampling procedure was conceived for homogenizing these samples for analysis, maintaining sample integrity, i.e., keeping light ends for properly appraise the distillation properties of the studied fractions. Figure 2 presents the sampling protocol conceived for these purposes. The first quality assurance parameter considered was repeatable water content testing, determined via Karl Fischer titration as described before.[12] The second quality assurance parameter was repeatable HTSD analysis; when repeatable HTSD aliquot analysis was achieved, a "blend" algorithm included in the used HTSD software[13] provided average distillation curves for the studied fractions.

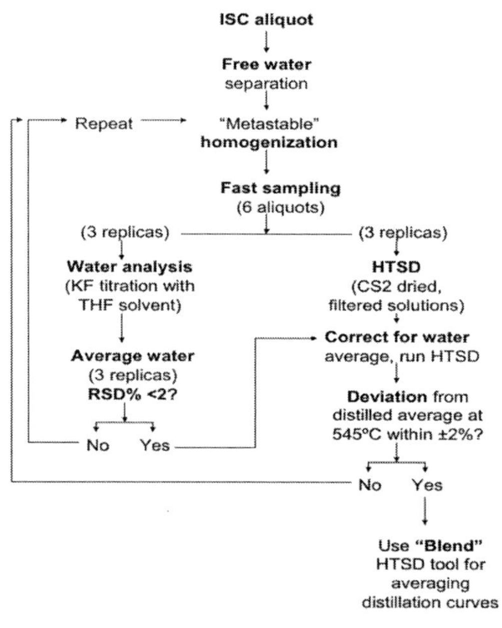

Figure 2. Sampling protocol schematic for ISC product aliquots

Following the protocol described on Figure 2, monitoring upgrading levels achieved during ISC tests was thus possible in a relatively fast way, i.e., much faster compared to standard physical distillation techniques. Generation of upgrading mappings for the whole test was also possible, as illustrated with the examples presented on Figure 3.

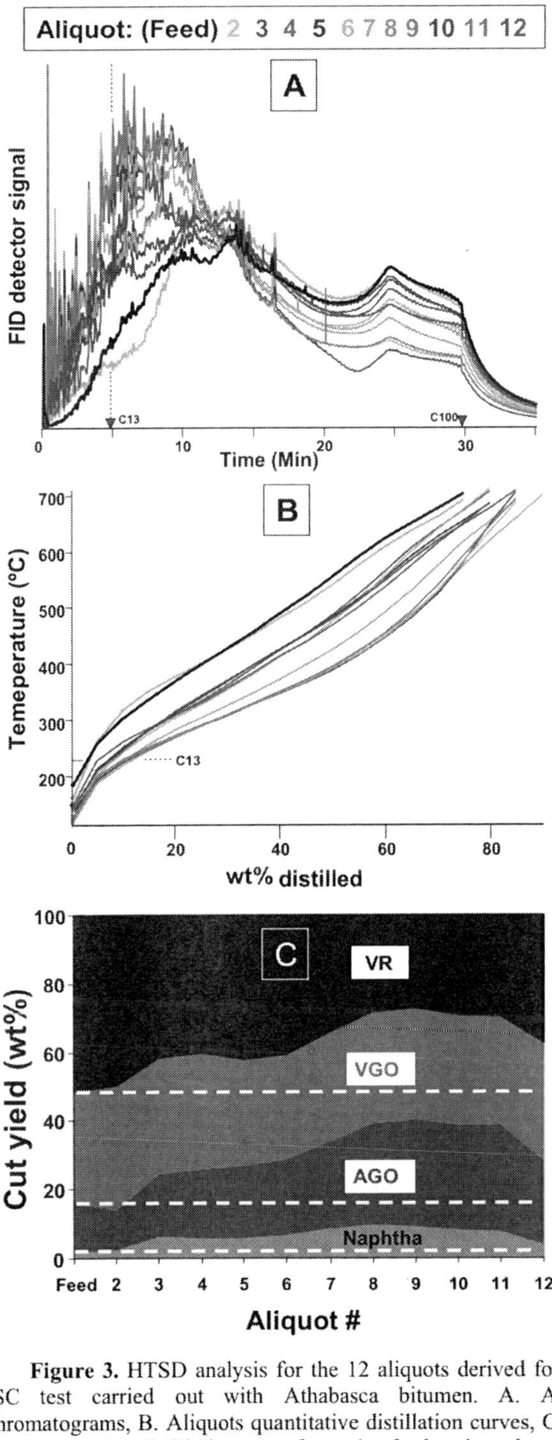

Figure 3. HTSD analysis for the 12 aliquots derived form one ISC test carried out with Athabasca bitumen. A. Aliquots chromatograms, B. Aliquots quantitative distillation curves, C. wt% distribution for distillation cuts from the feedstock and product's aliquots; original distribution for the feedstock highlighted with white dashed lines.

Results form Figure 3 indicate that upgrading levels varied for the studied aliquots of the test; however, the overall outcome is the generation of light fractions that can improve products flow-ability and market value of the produced hydrocarbons.

Athabasca ISUT processing. ISUT processing fundamentally relies on downhole injection of a hot fluid which acts as feedstock and in addition, as heat, catalyst and hydrogen carrier. The VR is heated once, exiting at the set up temperature in the vacuum tower, thus providing heat for the process which mostly occurs where the catalyst is irreversibly attached to reservoir minerals, i.e., around the injector well. The produced upgraded oil sent to full refining/commercialization units is exclusively composed of distillable fractions, consequently, has low viscosity, high API gravity, low metals and no asphaltenes present.[4,5]

Residence time within the catalytic zone of the reservoir is the easiest adjustable parameter for controlling conversion levels of the VR. The aim is to produce an oil that resembles the original hydrocarbon in place, in order to maintain a steady closed loop comprising the surface units operating and the downhole production facilities (injector/productor tubings and hydrocarbon pay zones). By controlling flow-rates, different quality of upgraded oils could be produced, starting from the injected feedstock (the VR). This is illustrated with the examples presented on Figure 4.

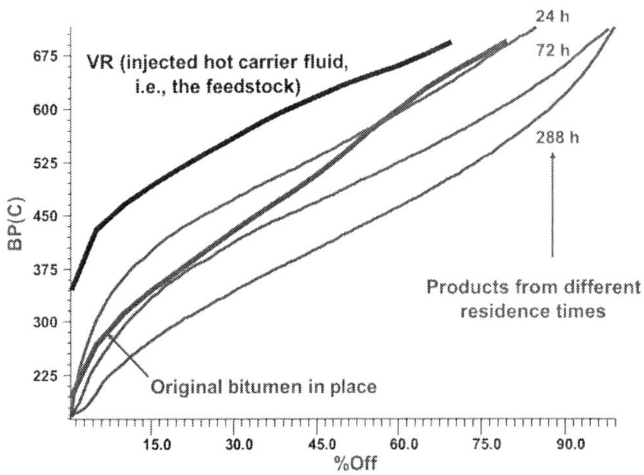

Figure 4. Distillation properties for Athabasca VR, Athabasca bitumen in place and upgraded products from ISUT processing under different severities. T: 350°C, P: 500 psig, H_2/VR: 300 sccm / cc.

A more thorough understanding of Athabasca bitumen/VR ISUT processing was attempted with a post-mortem analysis of hydrocarbon fractions remaining trapped in the sand pack at the end of the process. Special analysis protocols were also conceived and developed for the case. Fundamentally, estimation of the light ends generated during processing was possible with an algorithm developed based on HTSD. Figure 5 illustrates with a known sample how the HTSD for the whole sample (including the light end <230°C, i.e., C13-) can be determined. By knowing the amount of heavy materials (C13+) via gravimetric analysis, this value is input into the HTSD software for estimation of the C13- contents. The definitive distillation curve is recalculated using as sample mass the sum of (C13-) + (C13+).

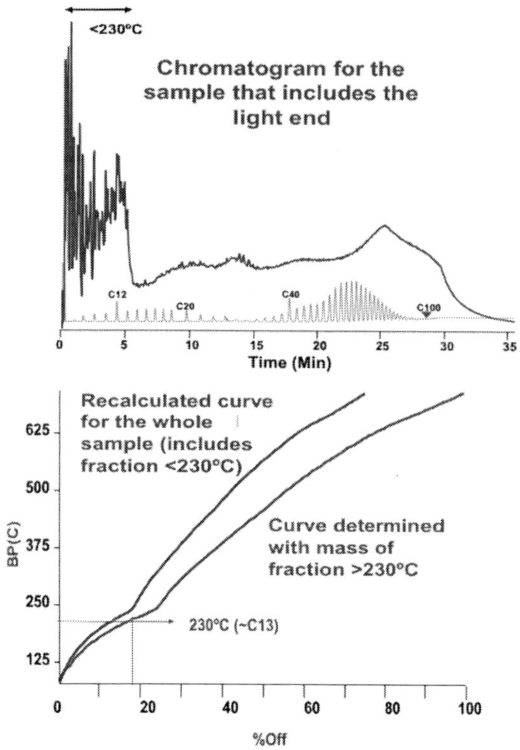

Figure 5. Total sample mass estimation based on the distillation residue >230°C and the amount of 230°C- (<C13) determined via HTSD. Athabasca bitumen containing 16.7 wt% C13- is the presented example.

The developed analysis protocol allowed to determine the properties of fractions contained in the rig branches at the end of ISUT processing. Figure 6 presents the identification of studied aliquots in an schematic that represents the physical rig depicted on Figure 1. Figure 7 presents the results for the 12 analyzed fractions. The upper branch showed little upgrading; the interpretation was drainage effected by the high temperature existing in the "interwell" zone. The interwell zone showed a continuous upgrading of the VR occurring from the injector well towards the productor, which is a demonstration of the occurrence of the conceived upgrading reactions.[4] The two horizontal branches showed presence of high converted hydrocarbons, particularly the lowest one. The later findings support the formation of a chamber within the reservoir that has an elliptical form, as previously discussed based on its thermal profile.[5]

The preceding discussion highlight two aspects: I. ISUT processing was able to produce substantial upgrading of heavy hydrocarbons with production of light ends as a function of adjustable parameters able to control conversion downhole (flow-rate), II. Developed analytical protocols can provide distillation /conversion mappings that describe the production of upgraded products that contain from naphtha fractions up to non-distillable residua.

Conclusions.

Analytical protocols were developed in order to assess the extent of the upgrading for processing Athabasca bitumen using ISC and ISUT. Improved quality of products was assessed in both instances by the determined production of light ends. 1-D packed tube combustors and 2-D pilot rigs were assembled respectively for ISC and ISUT testing. Special conceived analysis protocols based on HTSD provided unique information when applied to complex ISC and ISUT products that simultaneously contain light ends plus vacuum residua fractions. Complex upgrading mappings were determined for fractions produced stepwise during the development of both studied processes.

A

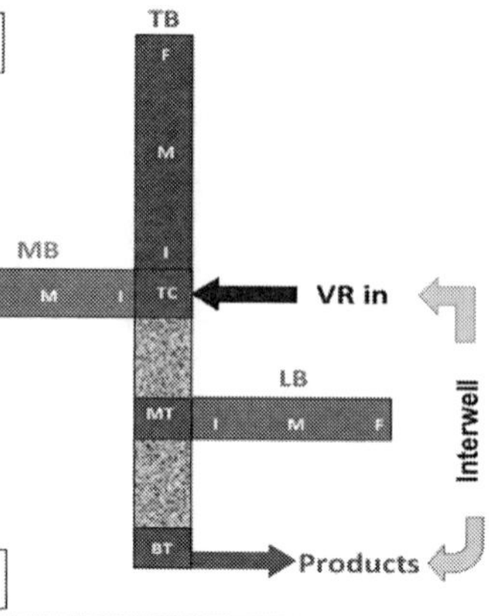

B

Section	%wt hydrocarbons
Initial Avg.	**15.6**
Top cross	15.1
Mid 'T'	14.7
Bottom 'T'	17.9
TB initial	**12.3**
TB middle	**10.9**
TB final	**14.0**
MB initial	10.2
MB middle	9.5
MB final	8.6
LB initial	1.4
LB middle	1.5
LB final	1.7

Figure 6. Post mortem analysis of fractions from the 2-D ISUT rig. A. Aliquots identification, B. Hydrocarbon amounts remaining after ISUT processing

Acknowledgements. Funding from AICISE (Alberta Ingenuity Centre for In Situ Energy), made possible the ISC tests discussed in the present article. Funding from NSERC/Nexen/AIEES Industrial Research Chair in Catalysis for Bitumen Upgrading, Canada Foundation for Innovation (CFI), the Institute for Sustainable Energy, Environment and Economy (ISEE) and, Schulich School of Engineering is acknowledged. Japan Alberta Company for Oil Sands (JACOS) is thanked for providing Athabasca bitumen. Help for ISC sample analyses provided by Elizabeth Zalewski, Linda Roa, Lante Carbognani Arambarri, Josune Carbognani and John Berezinski, was greatly appreciated.

References

(1) Moore, R. G.; Laureshen, C.; Belgrave, J. D. M.; Ursenbach, M. G.; Mehta, S. A. *Fuel,* **1995**, 74, 1169-1175.

(2) Weissman, J. G. (1997). *Fuel Proc. Technol.,* **1997**, 50, 199-213.

(3) Mahinpey, N., Ambalae, A., and Asgahri, K. *Chem. Eng. Comm.,* **2007**, 194, 995-1021.

(4) Pereira-Almao, P.; Chen, Z.; Maini, B.; Scott Algara, C. *In Situ Upgrading Via Hot Fluid Injection,* Canadian patent 2,810,022 (Granted on Dec. 09th, 2014).

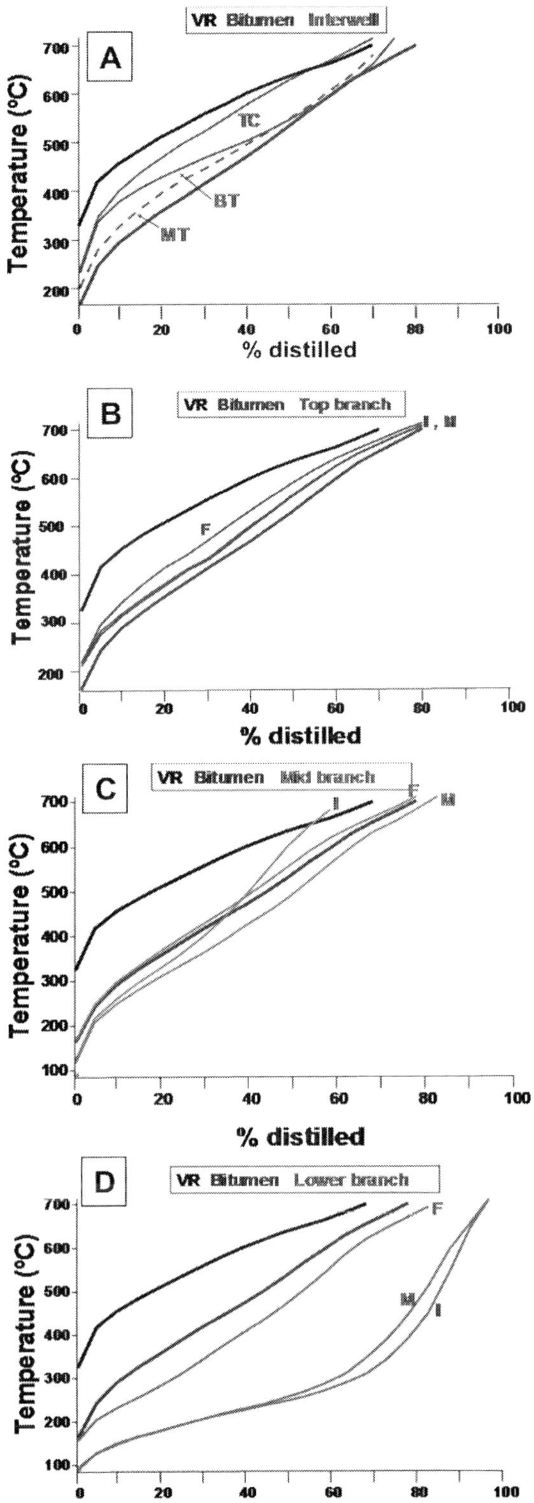

Figure 7. Map of quantitative distillation curves determined via HTSD for the hydrocarbon portions remaining within the different zones of the 2-D rig used in the present study on Athabasca bitumen ISUT processing. Identification of the samples can be observed on Figure 6

(5) Hovsepian, C.; Carbognani Ortega, L.; Pereira-Almao, P. Dense hot fluid injection process. Implementation of the catalytic In Situ Upgrading Technology (ISUT): two dimension experimental simulation. *Energy Fuels*, accepted (DOI: 10.1021/acs.energyfuels.5b02117).

(6) ASTM D2892. Standard Test Method for Distillation of Crude Petroleum (1-hteoretical plates column). American Society for Testing and Materials (ASTM): West Conshohocken, PA, 2011.

(7) ASTM D5236. Standard Test Method for Distillation of Heavy Hydrocarbon Mixtures (Vacuum PotStill Method). American Society for Testing and Materials (ASTM): West Conshohocken, PA, 2011.

(8) Abu, I. I.; Moore, R. G.; Mehta, S. A.; Ursenbach, M. G.; Mallory, D. G.; Pereira-Almao, P.; Carbognani Ortega, L. *J. Can. Petr. Technol.*, **2015**, 7, 220-232 (SPE 176029).

(9) Abu, I. I.; Moore, R. G.; Mehta, S. A.; Ursenbach, M. G.; Mallory, D. G.; Pereira-Almao, P.; Carbognani Ortega, L. *J. Can. Petr. Technol.*, **2015**, 7, 372-386 (SPE 178917).

(10) ASTM D7169. Standard Test Method for Boiling Point Distribution of Samples with Resid such as Crude Oils and Atmospheric Resids by High Temperature Gas Chromatography. American Society for Testing and Materials (ASTM): West Conshohocken, PA, 2011.

(11) Carbognani, L., Lubkowitz, J., Gonzalez, M. F., Pereira-Almao, P. *Energy Fuels*, **2007**, 21, 2831.

(12) Carbognani, L.; Moini Araghi, B.; Berezynski, J.; Carbognani-Arambarri, L.; Zalewski, E.; Kantzas, A.; Pereira-Almao, P. *Petrol. Sci. Technol.*, **2014**, 32, 610-617.

(13) Meneghini, R., Lubkowitz, J. *The Development of a Model for the Determination of Boiling Distribution of Blends based on the Boiling Point Distribution of the Blending Components*. Paper 087, Gulf Coast Conference, Oct. 22nd, Kingwood, TX-USA. 2003.

APPLICATION OF HIGH-TEMPERATURE GAS CHROMATOGRAPHY COUPLED WITH INDUCTIVELY COUPLED MASS SPECTROMETRY (HTGC-ICP-MS) FOR METAL DISTRIBUTION IN CRUDE OIL CUTS.

*Laura Poirier[1], Jenny Nelson[2] Carl Rechsteiner[1], M.M. Boduszynski[1], M.M. Moir[1], David Leong[1], and Francisco Lopez-Linares[1]**

[1]*Chevron Energy Technology Company, 100 Chevron Way, Richmond, CA 94801, USA*
[2]*Agilent Technologies, Inc., 5301 Stevens Creek Blvd, Santa Clara, CA 95051, USA*
Corresponding Author Email: flhu@chevron.com

Introduction

Standard fossil fuels and crude oils are complex mixtures of hydrocarbons, related compounds, heteroatoms as well as trace metals. The hundreds of individual compounds constituting petroleum and coal-based materials vary in molecular structure and size, polarity, functionality or chemical group. For a crude oils characterization in details, many chromatographic techniques have been employed, to access information in term of individual components, chemical groups, molecular size, and particular element among others.[1] Simulated distillation (Simdist) is one of the most practiced technique particularly used in refining industry due it ability to provide fast information related to the yield distribution (boiling point vs. amount) for crude feedstocks, intermediates, and products.[2-5] From Simdist, it can be access the carbon distribution and the hydrocarbon fingerprint of each crude oil as well as sulfur and nitrogen distributions that would provide relevant information about the sample[1]. Additionally, nickel and vanadium as porphyrins type of compounds, have been analyzed in crude oils and derivatives by GC at temperatures.[6-8] The possibility of combining the power of high-temperature gas chromatography such as Simdist with a multielement detection technique capability of Inductively Coupled Plasma Mass Spectrometry (ICP-MS), would provide the yield distribution of these elements present in the crude fractions as a function of boiling point. In the case of nickel and vanadium, along with other elements of interests, it would have the first impact on refining applications due the information would help to optimize catalyst lifetime, process conditions and product quality. Previously, we reported an initial assessment of nickel and vanadium presence in crude oil cuts (900-1100 F) by using high-temperature GC-ICP-MS.[10] In this article, we are providing further information of the application of this techniques determination of Nickel and Vanadium and an initial speciation of vanadium, in crude oil fractions with a boiling point between 800-1250 F. Discussion of the significant finding after analyzing such fractions using this technique is presented.

Experimental

Reagent Grade ACS Carbon Disulfide (Fisher Scientific, USA) was used for all dilutions. HT-SIMDIS standard (n-paraffin, C5-C110, SD-SS3E-05, Separation Systems, INC; Gulf Breeze, Florida, USA). Two etioporphyrin such as 2,3,7,8,12,13,17,18-octaethyl-21H, 23H- porphyrin Vanadium Oxide (Sigma-Aldrich; USA), 2,3,7,8,12,13,17,18- octaethyl-21H, 23H- porphyrin Nickel (Sigma-Aldrich; USA) and 5,10,15,20-Tetraphenyl-21*H*,23*H*-porphine vanadium(IV) oxide (Sigma-Aldrich; USA) designated as TPP, were used in this work. Crude Oil Fractions from North America crude oil were selected having boiling point ranges between 800-1200 °F. An Agilent Technologies (Agilent Technologies, Santa Clara, CA, USA) model 7890 series GC was used for the separation of the crude oil fractions. The GC was coupled to an Agilent 7700x ICPMS (Agilent Technologies, Tokyo, Japan) through a commercially available heated GC-ICPMS interface (Agilent Technologies). Heated argon was used as a makeup gas to assist the flow of species through the heated transfer line (the Ar gas is preheated by passing through a stainless steel coil mounted in the GC oven). Separation of n-paraffin compounds, as well as the nickel and vanadium compounds present in such fractions, was performed on a High-Temperature Simulated Distillation column, 5 m, 0.530 mm id., 0.09 mm (Analytical Controls; The Netherlands). Typical instrument conditions for the HTGC-ICP-MS is presented in Table 1

Table 1. 7700xICP-MS and 7980 GC experimental Conditions

ICP-MS conditions	7700x
Forward power	750-1500 W
Plasma gas flow rate	15 L/min
Auxiliary gas flow rate	0.9-1.5 L/min
Make up gas	0.1 L/min
Carrier gas flow rate	0.45-1.5 L/min
Sampling depth/radial viewing height	6-8 mm
Elements isotope	^{12}C, ^{13}C, ^{51}V, ^{58}Ni, ^{60}Ni
GC parameters	**7980 GC**
Column	DB-HT-SIMDIS
GC carrier gas flow rate	20 ml He/min (constant flow mode)
Oven temperature	40 °C initial, ramped at 15°C min-1 to 200°C, ramped at 5°C min-1 to 430°C and held for 5 min (104 °F initial, ramped at 59°F min-1 to 392°F, ramped at 41°F min-1 to 806°F and held for 5 min)
Split/Splitless Inlet	100 °C initial, ramped at 15 °C min-1 to 340 °C, and held for 45 min (212 °F initial, ramped at 59 °F min-1 to 644 °F, and held for 45 min)
Transfer line temperature	300-380 °C/572-716 °F
ICP injector temperature	300-380 °C/572-716 °F

Results and Discussion

The key factors to perform useful HTGC-ICP-MS on this type of feedstock is the proper optimization of the GC instrument parameter, the adequate temperature of the transfer line and the ICP-MS conditions, displayed in Table 1. Usually, the n-paraffin standard is used during method optimization as well as to transform the GC retention time of n-alkane mixture to an equivalent boiling point. [2,3] Figure 2, shows a typical chromatogram of this standard from 550-1200 F, monitored the ^{13}C isotope signal on ICP-MS, having the transfer line heated at 662 F in both sections. Details on the instruments optimization can be found elsewhere.[10,11]

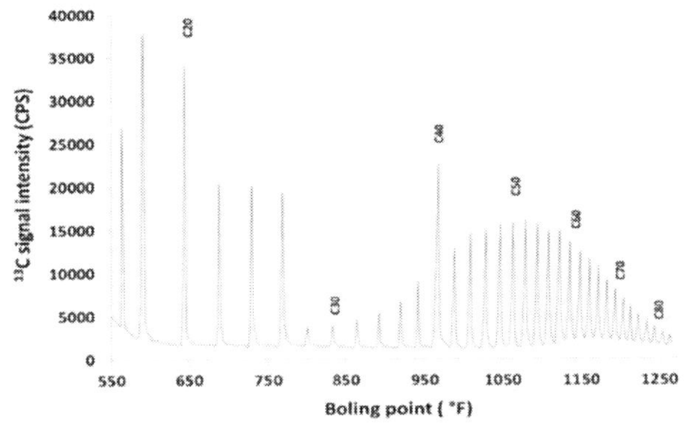

Figure 2. HT-SIMDIS standard

Once the n-paraffin standard is ran, initially two commercial porphyrins, representative of those found in petroleum crudes (Etio) [12], were run as standard, to determine the corresponding boiling point. Figure 3 displays the signals of the isotopes ^{51}V and ^{60}Ni as a function of the boiling point; 1090 F for Ni and 1092 F for vanadium.

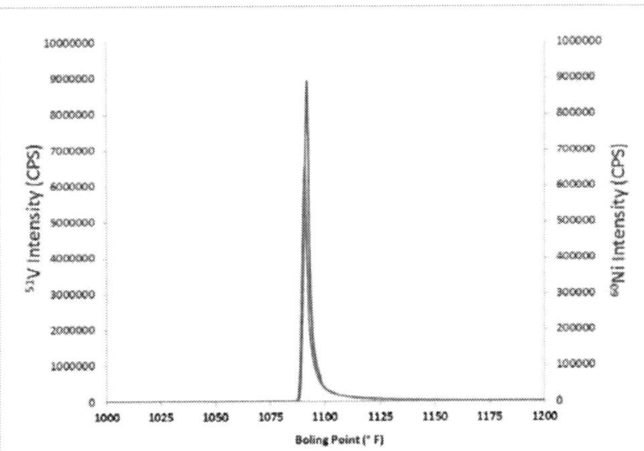

Figure 3. Nickel and Vanadium etioporphyrins

Figure 4 present the results regarding the distribution of ^{51}V and ^{60}Ni as well as ^{13}C signal in North American crude oil fraction 1. As can be seen, this technique not only could follow the metals of interest; additionally, we can track the carbon distribution along B.P. using the ^{13}C isotope signal as similar response obtained by regular FID detector employed in HT-Simdis. As it was observed previously in other crude oil fractions at this B.P. range[11], more than one species for nickel and vanadium porphyrin are present, as it has been described in the literature.[12]

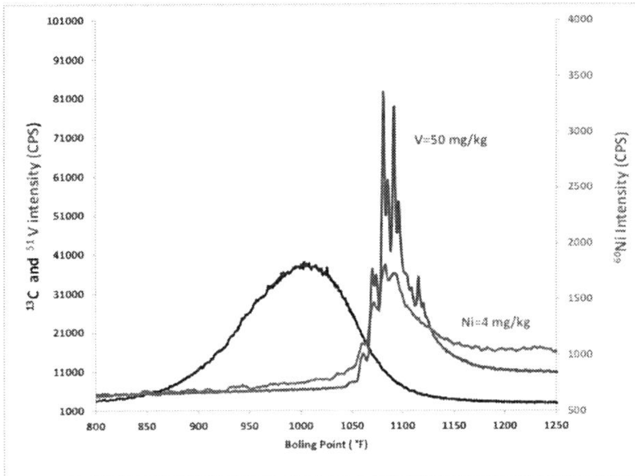

Figure 4. North American crude fraction 1 (900-1200 °F)

Figure 5 shows the results for another North American crude oil fraction 2, with a higher boiling point. Again, it can trace the metal content as a function of boiling point, giving an indication of the number of porphyrin species present in this cut. Also, it can be observed that we can detect nickel as low as 3 mg/kg with the level of confidence. Determination of limit of quantification studies in currently in progress to establish the sensitivity of this method.

An initial speciation study was carried out to determine the type of vanadium porphyrins present in such fractions: one etio type and one TPP as references were ran using the same conditions as the cuts. As it is observed in Figure 6, the etioporphyrin is present on both cuts with a very distinct peak (around 1090 °F) while the TPP compounds which appear to a higher boiling point (around 1170 °F) is not present in such fractions. The above is one example of the potentiality of this technique to determine the type of porphyrin present in the sample.

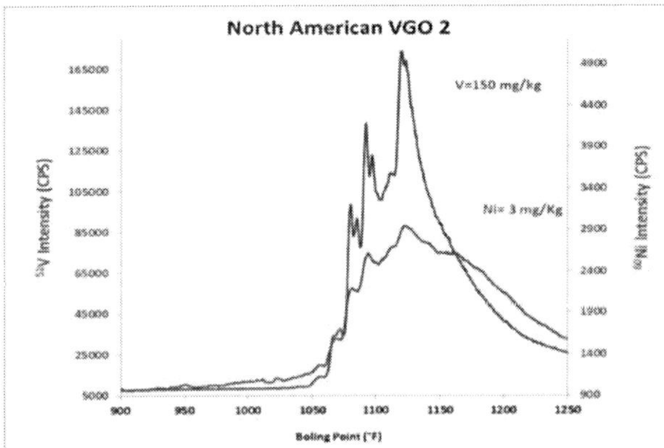

Figure 5. North American crude fraction 2 (1050-1250 °F)

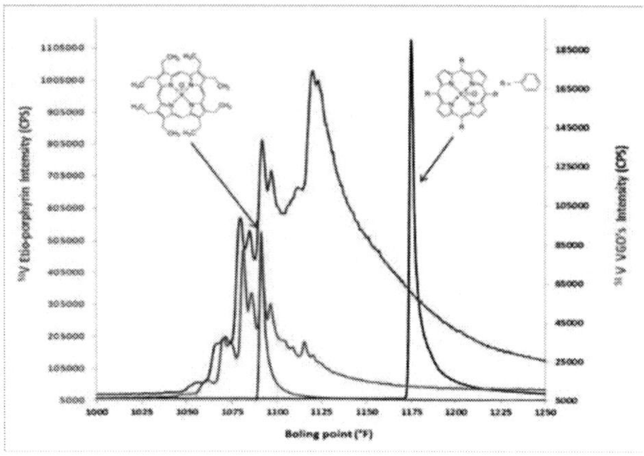

Figure 6. Vanadium speciation on North American crude fractions

Currently, the analyzes of other feedstocks, another type of porphyrins as well as attempt to quantification of Ni and V, are in progress to expand more the capability of this technique.

Conclusions

The present work illustrates the potential of HT-GC-ICP-MS as technique as fingerprint tool for carbon, nickel and vanadium variation as a function of boiling point. It was demonstrated with two North American crude oil fractions, that multiples vanadium and nickel species are present. Initial speciation on vanadium indicates that etioporphyrin type compound is present in such fraction.

Acknowledgement. Steve Wilbur, Emmett Soffey (Agilent, Inc. USA), Brian Morlan, Robert Lorenz and Josie Iwamoto (Chevron Richmond Technology Center) for technical support and valuable discussions. Chevron Energy Technology Company for providing funding and the permission to publish this article.

References

(1) Barman, B.N.; Cebolla, V.L.; Membrado, L.. Chromatographic techniques for petroleum and related products. Critical Reviews in Analytical Chemistry, 2000, 30, 75-120
(2) F.T. Eggertson, S. Groennings, J.J. Holst. Anal. Chem. 32(1960) 904-909
(3) L.E. Green, L.J. Schmauch, and J.C. Worman. Anal. Chem. 36 (1964) 1512-1516
(4) R.D. Butler. Chromatography in Petroleum Analysis; K.H. Altgelt, T.H. Gouw Eds. Marcel Dekker: New York, 1979, 75-79
(5) L. Neer and M. D. Deo. Journal of Chromatographic Science, 33 (1995) 133-138
(6) P.J. Marriot, J.P. Gills and G. Ellinton. Journal of Chromatography, 236 (1982) 395-401
(7) A. Kim, M.E. Foulkes, S. Hill, L.Ebdon, R.L. Patience, A.J.G. Barwise and S.J. Rowland, J. Anal. At. Spec, 7 (1992) 1
(8) W.G. Pretorious, LO. Ebdon and S.J. Rowland. J of Chromatography, 646 (1993) 369-375
(9) L. Ebdon, E. Hywel Evans, W. G. Pretorious and J. Rowland. J. Anal. At. Spectrom. 9 (1994) 939-943
(10) J. Ellis, C. Rechsteiner, M. Moir and S. Wilbur. J. Anal. At. Spectrom. 26 (2011) 1674-1678
(11) Carl Rechsteiner, M.M. Boduszynski, M.M. Moir, David Leong, Jenny Nelson, Laura Poirier and Francisco Lopez-Linares. manuscript in preparation
(12) G.B. Vaughan, E.C. Tynan and T.F. Yen, Chemical Geology, 6 (1970) 203-219

CHARACTERIZATION OF ASPHALTENES: KINETIC EFFECTS.

Estrella Rogel[1]

Chevron ETC, Richmond, CA 94597
Author Email: ergc@chevron.com

Introduction

Asphaltenes are undesirable materials that precipitate during different stages of petroleum production and refining. As they are not clearly defined as a compound class but as a solubility class, characterization and understanding of asphaltene behavior is difficult to say the less. Separation of asphaltenes presents its unique challenges since under the loose umbrella of insolubility in heptane (or pentane or any other proper anti-solvent) many different "asphaltenes" can be obtained from the same crude oil. Time, sample/solvent ratio and temperature are just some of several variables that have an impact on what it is separated as asphaltenes. [1] The same can be said for their behavior in the reservoir or during refinery operations. The solvent power of the medium, temperature and time seem to have a large influence on the formation and evolution of asphaltene deposits.

In a series of experiments using optical microscopy observations and centrifugation based separations, it was shown that the time required to precipitate asphaltenes can vary from minutes to several months depending on the solvent power of the medium.[2-3] In the present work, we will look at the effect of time and solvent power have not only in the amount of precipitated material, but also in the characteristics of the precipitated material.

Experimental

Materials. Crude oil where obtained from different sources with asphaltene contents ranging from 4.8 wt% to 8.7 wt%. Pentane asphaltenes were obtained using a modification of the standard ASTM D-4055.[4]

Separation of precipitated material. Heptane was used to induce precipitation at room temperature. Different sample/heptane ratios from 2/3 to 1/10 were used and the blend was filtered using a Teflon membrane filter with an average pore size of 0.2 microns so that particles could be recovered and analyzed. The filtered cake was analyzed as collected. Studied samples included pentane asphaltenes solution in toluene (2 wt. %) and crude oil A.

Characterization Methods. Asphaltene concentrations and asphaltene solubility profiles were determined using on-column filtration techniques.[5,6] In these techniques, a solution of the sample is injected in a column packed with an inert material using n-heptane as the mobile phase. This solvent induces the precipitation of asphaltenes and, as a consequence, their retention in the column. To determine the asphaltene content the mobile phase is switched to a blend dichloromethane/methanol 90/10 v/v that redissolves the asphaltenes completely. To determine the asphaltene solubility profile, the mobile phase is changed gradually from pure n-heptane to 90/10 methylene chloride/methanol and then to 100 % methanol. This procedure gradually re-dissolves the asphaltenes from the easy to dissolve (low solubility parameter) to the hard to dissolve (high solubility parameter). In both techniques, asphaltenes are quantified using an Evaporative Light Scattering Detector (ELSD). Rather recently, the method has been calibrated to produce a distribution of asphaltenes in terms of their solubility parameters. A description of the calibration is published somewhere else.[7]

Results and Discussion

Pentane Asphaltene Experiments. Initial experiments were performed using pentane asphaltene solutions. Figure 1 shows the amounts of material collected at increasing times for different sample/heptane ratios. In this plot, the collected material is calculated as percentage of pentane asphaltenes. As expected, the amount collected increases as a function of time as well as the sample/heptane ratio decreases.

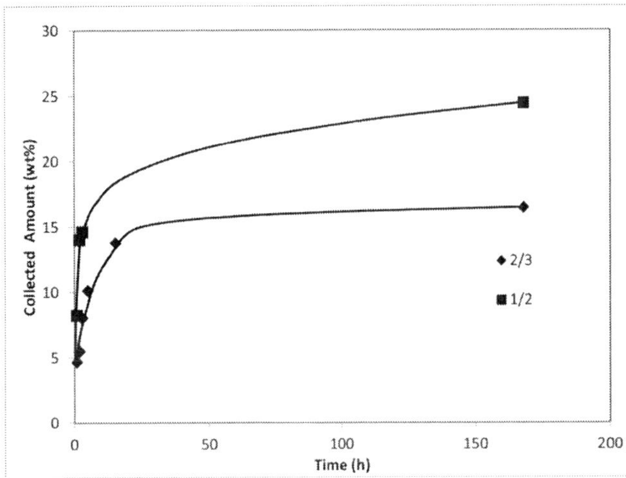

Figure 1. Collected amount for pentane asphaltene solution at different pentane asphaltene solution/solvent ratios as a function of time.

Figure 2 shows the solubility profile for several of the materials separated using different sample/heptane ratio and collected at similar times (3 h). The solubility profiles of these materials varied significantly showing that different ratios produced materials that exhibit solubility characteristics. In particular, the results indicated that lower ratios at this short time (3h) produced filtered cakes enriched in more soluble material than larger ratios. In this figure, the solubility profile of pentane asphaltenes is also shown. As more material is collected, the solubility profile of the asphaltenes becomes closer to the one obtained for the pentane asphaltenes.

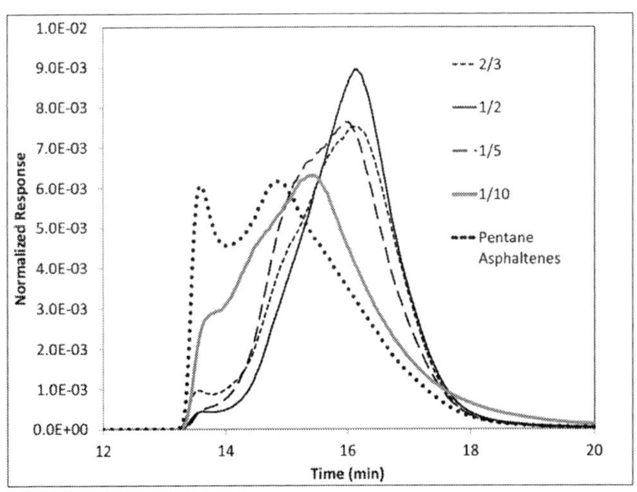

Figure 2. Asphaltene Solubility Profiles of filtered cakes obtained at 3 h using different pentane asphaltene solution/heptane ratios.

In contrast, comparison of asphaltene solubility profiles for the same sample/heptane (1/2) ratio did not show significant differences in the profile as a function of time up to 1 week.

Crude Oil A Experiments. In this set of experiments, the crude oil was blended in different proportions and the filtered cakes were analyzed. Figure 3 shows the amount of solids collected in the filter as a percentage of the crude oil for a crude oil/heptane ratio of 2/3. As showing in this figure, the amount collected decreases during the first 24 h of the experiment indicating that part of the material is coming back to solution. In fact, analysis of the collected solid showed an increase in the heptane asphaltene content from 6.7 wt% to 14.5 wt% in 24 h.

Additionally, the comparison of solubility profiles in Figure 4 at different times shows the corresponding relative decrease in the more soluble material (1st peak). This is in a clear contrast to the observation made for pentane asphaltene solutions in the previous section where no significant differences in composition were observed together with an increase in the amount of material collected. This can be the results of trapped material that is abundant in the crude oil but is not present in the solution of separated pentane asphaltenes.

It is important to point out that the filtered cake was not washed with solvents before analysis. When it is washed with heptane, a common practice, plots of the collected amount are similar to those shown in Figure 1 for pentane asphaltene solutions.

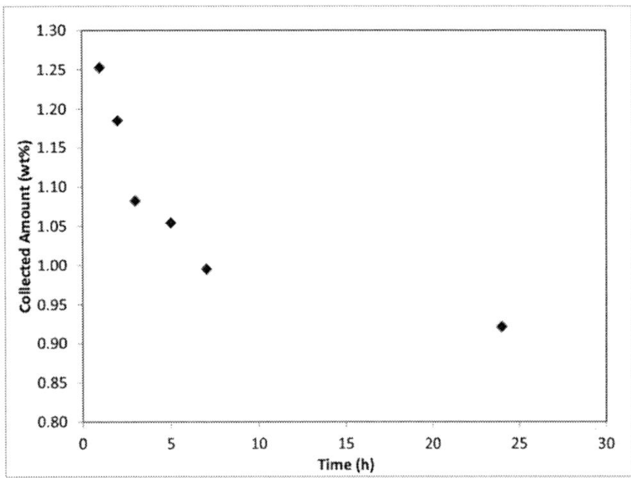

Figure 3. Collected amount of Crude Oil A at a sample/solvent ratio of 2/3 as a function of time.

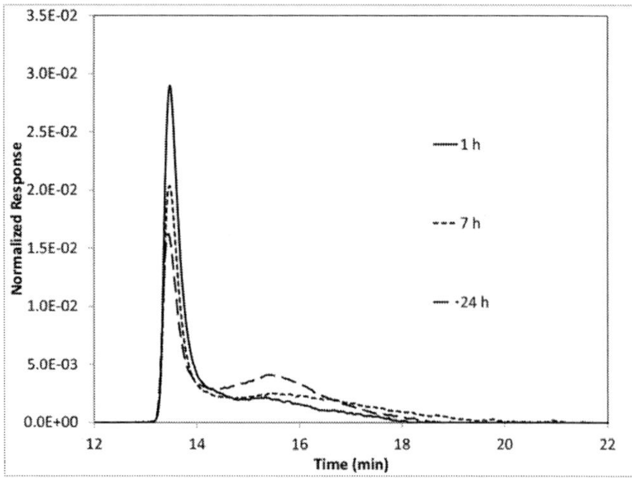

Figure 4. Asphaltene Solubility Profiles of filtered cakes obtained at different times using a sample/heptane ratio of 1/2.

Conclusions

Based on the results obtained using crude oil, it can be hypothesized that kinetics plays a big role in the composition of deposits and 24 h does not seem to be enough time to reach a steady stage at the conditions of these experiments.

From a practical viewpoint, the present study provides new information about the complex mechanism that leads to asphaltene deposition and might be helpful in the development of new strategies to study this phenomena.

Acknowledgement. The authors are grateful to Chevron ETC for providing funding and for allowing the publication of this work

References

(1) Carbognani Ortega, L.; Rogel, E.; Vien, J.; Ovalles, C.; Guzman, H.; Lopez-Linares, F.; Pereira-Almao, P. *Energy Fuels* 2015, **29**, 3664–3674.
(2) Taqbool, T.; Balgoa, A. T.; Fogler, H. S. *Energy Fuels*, 2009, **23**, 3681–3686.
(3) Haji-Akbari, N.; Teeraphapkul, P.; Balgoa, A. T.; Fogler, H. S. *Energy Fuels*, 2015, **29**, 2190–2196.
(4) ASTM International, Standard Test Method for Pentane Insolubles by Membrane Filtration ASTM D-4055, 2013.
(5) Rogel, E.; Ovalles, C.; Moir, M E.; Schabron, J. F. *Energy Fuels* 2009, **23**, 4515-4521.
(6) Rogel, E.; Ovalles, C.; Moir, M. *Energy Fuels* 2010, **24**, 4369-4374.
(7) Rogel, E.; Roye, M.; Vien, J.; Miao, T. *Energy Fuels* 2015, **29**, 2143–2152.

AN OVERVIEW OF ANALYTICAL METHODS USED FOR QUALIFYING HYDRAULIC FRACTURING FLUIDS

Michael A. Reynolds

Shell Exploration and Production Company
150 N. Dairy Ashford St.
Houston, TX 77079
Mike.Reynolds@Shell.com

Abstract introduction.

Unconventional gas and tight oil (UGTO) production has seen a boom of activity in North America over the past decade due to new developments in horizontal drilling and production technologies. Hydraulic fracturing (HF) is a well treatment process that uses high pressures, water, sand and chemical additives to recover these unconventional hydrocarbons that are trapped in low permeability shale reservoirs. During the HF process, several thousand barrels of water and proppant are pumped into the horizontal wellbore at high pressure to force open or fracture the shale rock. Proppant particles, which are typically sand or ceramic, are used to fill the fractures and keep them propped open for production. Chemical additives are invoked to improve the HF process and make it more efficient.

Chemical additives described in the presentation.

Each chemical in the fracturing fluid formulation must be compatible with the other chemicals used in the process in order to perform its function. Some examples of the additives used in hydraulic fracturing treatments include friction reducers, guar-based gels with crosslinking agents, biocides, clay stabilizers and surfactants. Qualifying fracturing fluids and their additives for performance, compatibility and safety requires the use of multiple analytical techniques. This presentation will focus on examples of fracturing fluid chemicals and some of the analytical methods used to qualify these fluids for efficacy in a hydraulic fracturing treatment.

THE IMPACT OF SOLUBILITY PARAMETER MEASUREMENTS ON PREDICTED ASPHALTENE GRADIENTS. APPLICATION TO RESERVOIR CONNECTIVITY

Estrella Rogel[1], Cesar Ovalles[1], Andrew E. Pomerantz[2], Julian Y. Zuo,[2] and Oliver C. Mullins[2]

1. Chevron ETC, Richmond, CA 94597
2. Schlumberger-Doll Research, Cambridge MA, 02159
Author Email: ergc@chevron.com

Introduction

Downhole fluid analysis (DFA) provides in-situ measurements of many reservoir fluid properties during or immediately after drilling the well.[1] Chevron has implemented a workflow[2] to assess reservoir connectivity using DFA data which includes use of novel asphaltene science and the Flory-Huggins-Zuo equation of state (FHZ EoS) to model equilibrium asphaltene concentration gradients in petroleum reservoirs.[3]

The FHZ EoS characterizes the asphaltenes with a single chemical parameter (in addition to their size), the Hildebrand solubility parameter. As known, solubility parameters were developed originally for liquid state applications. However, over the years, they have been successfully used for many other applications including but not limited to polymers, dyes, emulsions, etc.[4]

For solid materials, the solubility parameter concept cannot be applied directly since the reference state is the liquid substance.[5] For those cases, it is necessary to use a hypothetical liquid subcooled below the melting point of the substance as a reference.[4,6] Because of this limitation, a series of different methods have been developed to determine solubility parameters of solids. As a result, solubility parameters of solids are dependent on the method used to determine them.[4]

The most popular method used to evaluate asphaltene solubility parameters is a Heithaus titration based on the regular solution model.[7] Other methods used to determine asphaltene solubility parameters include solubility testing in different solvents,[8-9] as well as molecular modeling[10] and IR-NIR multivariable analysis.[11]

To understand applicability and limitations of the FHZ EoS, in particular for practical applications where detailed asphaltene characterization is not immediately available, it is important to obtain the range of solubility parameters for asphaltenes. In this work, we used a relatively new technique for evaluating the solubility parameter of asphaltenes, the so-called "Asphaltene Solubility Profile",[12-13] to measure the solubility parameter distributions for a diverse set of asphaltenes (coal, petroleum, and shale oil) and compared the resulting predicted asphaltene gradients.[3,14]

Experimental

Type and elemental analyses of the asphaltenes used are reported in Table 1.

Table 1. Characterization of the asphaltenes used.

Asph.[a]	Type[b]	H/C[c]	%N[d]	%S[e]	Ave. S.P.[f]
Wyoming	Coal*	0.81	1.43	0.13	20.41
ADARO	Coal*	0.85	1.45	0.09	20.14
UG8	Petroleum	1.05	1.02	8.94	20.32
BG5	Petroleum	1.18	0.98	7.61	19.80
EF	Shale Oil	1.40	1.73	-	19.90

[a]Asphaltene code. [b]Type of asphaltenes. [c]Hydrogen to carbon molar ratio. [d]Wt. % of nitrogen. [e]Wt. % of sulfur. [f]Average solubility parameters as determined by the solubility profile method.[12-13] *Showed maltene peaks

Determination of Asphaltene Solubility Parameter. The procedure to determine the solubility profile analyses was reported elsewhere.[12-13] In general terms, it involves the precipitation of asphaltenes in a column packed with an inert material using n-heptane as the mobile phase. After maltenes have eluted, the mobile phase is changed gradually from pure n-heptane to 90/10 methylene chloride/methanol and then to 100% methanol. Asphaltenes are quantified using an Evaporative Light Scattering Detector (ELSD).

From the HPLC traces, the average times of elution of the asphaltenes (Tav) were determined. Using the previously reported correlation[15], the average asphaltene solubility parameters were calculated.

Calculation of Asphaltene Gradients. Asphaltene gradients are controlled by the effects of gravity, entropy, and solubility, as described by the FHZ EoS:

$$\frac{\varphi_a(h_2)}{\varphi_a(h_1)} = \exp\left(\begin{array}{c} -\left[\dfrac{v_a g \Delta\rho (h_2 - h_1)}{RT}\right] + \left[\left(\dfrac{v_a}{v}\right)_{h_2} - \left(\dfrac{v_a}{v}\right)_{h_1}\right] \\ -\left[\dfrac{v_a\left[(\delta_a - \delta)_{h_2}^2 - (\delta_a - \delta)_{h_1}^2\right]}{RT}\right] \end{array} \right)$$

where $\phi_a(h_i)$ is the asphaltene content at height h_i in the reservoir, v_a is the molar volume of the asphaltene particle, g is Earth's gravitational acceleration, $\Delta\rho$ is the density contrast between the asphaltene and maltene phases, R is the ideal gas constant, T is temperature, v is the molar volume of the liquid-phase crude oil, and δ_a and δ are the solubility parameters of asphaltene and the crude oil.[3] Here, asphaltene gradients are modeled using different values of δ_a as determined by the solubility profile experiments to determine how sensitively variations in δ_a over a realistic range impact the gradient predictions.

Results and Discussion

Figure 1 shows the asphaltene solubility parameter distributions obtained for the heptane extracted asphaltenes. For comparison proposes, all the curves have been normalized so that the areas under the curve are equal to one.

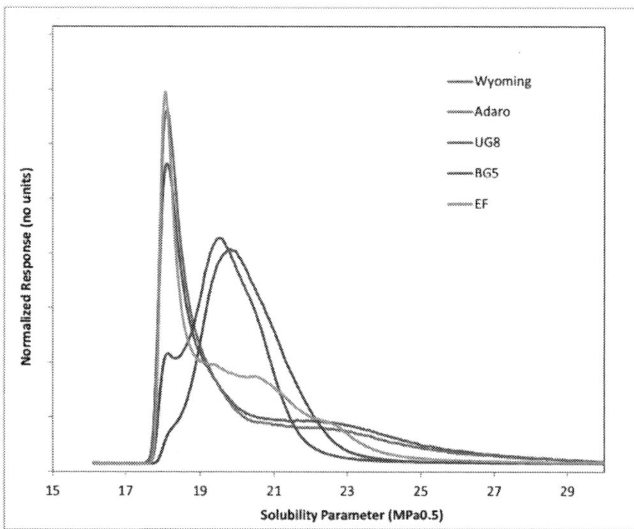

Figure 1. Asphaltene solubility profile of the five asphaltenes studied.

Heptane asphaltene extracted from virgin crude oils or residues do not show a maltene peak during the test. This is not the case for the asphaltenes extracted from the two coal derived samples (Adaro and Wyoming) which were extracted using n-hexane. As shown in Figure 2, they both show a visible peak in the maltene region. This result is consistent with the use of n-hexane as precipitant solvent, Because of its larger solvent power, heptane can partially dissolve asphaltenes precipitated in hexane. In a previous study, pentane extracted asphaltene also showed maltene peaks during the solubility profile test.[16] Because of the large separation between the maltene peak and the asphaltene peak, the maltenes in the coal samples do not affect the measured asphaltene solubility parameters.

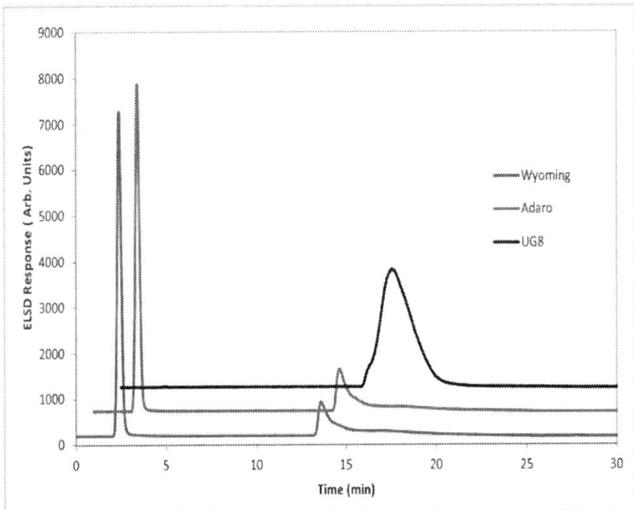

Figure 2. Asphaltene solubility profiles showing the maltene peaks associated with the two coal derived samples (ADARO and Wyoming). Asphaltenes UG8 is showed for comparison purposes

It can also be noticed from the distributions that these two asphaltenes show larger peaks at lower solubility parameter than those found for asphaltenes from traditional crude oils (Figures 1 and 2). The latter samples show the standard behavior for this type of asphaltenes.[12] This lower solubility parameter peak is associated with easy-to dissolve-asphaltenes.[12] However, as can be seen in Figures 1 and 2, the solubility profiles show a tail that extends into the larger solubility parameter region.

Asphaltenes EF, extracted from tight oil (Figure 1, green trace) also exhibit different behavior than those observed for asphaltenes from traditional crude oils. In this case, the asphaltenes do not show a maltene peak, but in a similar fashion to coal asphaltenes, they elute at remarkably lower retention times than the standard heptane asphaltenes and showing the large peak corresponding with easy-to dissolve-asphaltenes. This phenomenon can be noticed by comparing of the profiles of EF with BG5 (Figure 1, purple trace) and UG8 (Figure 1, red trace). The latter two asphaltenes come from traditional crude oils and show the conventional behavior.

Table 1 reports the average solubility parameter of the distributions as shown in Figure 1. These values are the average of two separated determinations. In the calculations, only distributions shown in Figure 1 were used. Therefore, in the cases of Adaro and Wyoming asphaltenes, maltene peaks were not taken into consideration for the calculations.

Even though the asphaltenes studied come from very diverse sources, their average solubility parameters lie within a small range (19.8 to 20.4 $MPa^{0.5}$). These values are within the expected range previously reported for asphaltenes using different techniques.[8, 11]

Figure 3 shows a comparison between asphaltene gradients measured by downhole fluid analysis and predicted by the FHZ EoS with the different δ_a values measured here. The asphaltene gradient in this reservoir is dominated by solubility, so the predicted asphaltene gradient is particularly sensitive to δ_a.[17] Despite the sensitivity of the gradient to δ_a and the great diversity of asphaltenes samples for which δ_a is measured, the predicted gradients obtained using different δ_a are essentially identical.

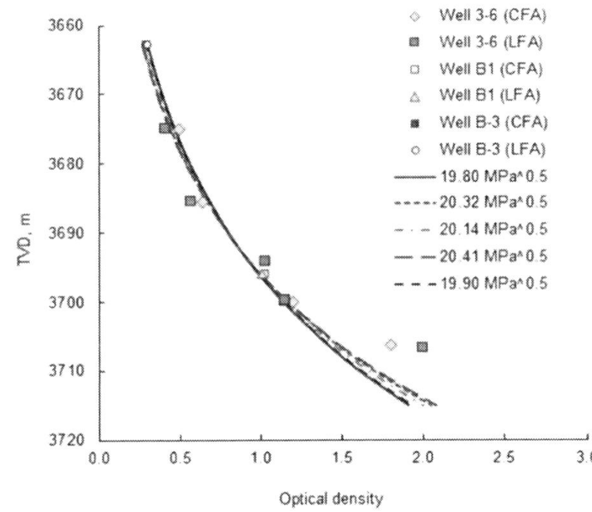

Figure 3. Measured and predicted asphaltene gradients, where the predictions are made using the average solubility parameters as reported in Table 1.

Conclusions

Average asphaltene solubility parameters were determined from very diverse sources using the solubility profile method. The results showed that the values lie within a small range (19.8 to 20.4 $MPa^{0.5}$). Asphaltene gradients predicted by the FHZ EoS are essentially insensitive to variations in δ_a over this range, suggesting that asphaltene gradients should be modeled by the FHZ EoS with a fixed, global value of δ_a. As a result, the FHZ EoS can be used to evaluate reservoir connectivity in real time, without requiring detailed laboratory measurements on asphaltene chemistry.

Acknowledgement. The authors are grateful to Chevron ETC for providing funding and for allowing the publication of this work

References

(1) Mullins, O.C.; The Physics of Reservoir Fluids; Discovery through Downhole Fluid Analysis, Schlumberger Press, Houston, TX, 184 pages, (2008)

(2) Betancourt, S.S.; Mishra, V. K.; Clinch, S.; Guo, G.; Malik, M.; Canas, J. A.; Chen, L.;, Mullins, O. C.; OTC- 27004, presented at the Offshore Technology Conference held in Houston, Texas, USA, 2–5 May 2016

(3) Zuo, J.Y.; Mullins, O.C.; Freed, D.E.; Dong, C.; Elshahawi, H.; Seifert, D.J.; *Energy & Fuels*, *27*, 1722–1735, (2013)

(4) Barton, A. F. M.; Handbook of Solubility Parameters and Other Cohesion Parameters, Second Edition CRC Press, USA, 1991.

(5) Fedors, R. F. *Polym. Eng. Sci.* 14(2), 147-154.

(6) Hildebrand, J. H.; Prausnitz, J. M.; Scott, R. L. Regular and Related Solutions; Van Nostrand Reinhold Company: New York, 1970.

(7) Redelius, P. G. *Fuel* 79, 27-35, 2000.

(8) Zenke, G. Bitumin-Teere-Asphalte-Peche, v 25, n 12, p 467-483, 1974.
(9) Hsienjen L., Jiuun-Ren L., and Teh F. Y. *Fuel* 74, 423-428, 1994.
(10) Frigerio, F.; Molinari, D. Computational and Theoretical Chemistry 975, 76-82, 2011.
(11) Fossen, M.; Hemmingsen, P. V.; Hannisdal, A.; Sjöblom, J.; Kallevik, H.; *J. Disp. Sci. Tech.*, 26, 227-241, 2005.
(12) Rogel, E.; Ovalles, C.; Moir,; M. *Energy and Fuels* 24, 4369-4374, 2010.
(13) Rogel, E.; Ovalles, C.; Moir, M.; *Energy and Fuels* 26, 2655–2662, 2012.
(14) Mullins, O.C.; Pomerantz, A.E.; Andrews, A.E.; Zuo, J.Y.; Petrophysics, 56, 3, 266–275, 2015.
(15) Rogel, E.; Roye, M.; Vien, J.; Miao. T. *Energy Fuels* 29, 2143–2152, 2015.
(16) Carbognani Ortega, L.; Rogel, E.; Vien, J.; Ovalles, C.; Guzman, H.; Lopez-Linares, F.; Pereira-Almao, P. *Energy Fuels* 29,3664–3674, 2015.
(17) Gisolf, A.; Dubost, F. X.; Zuo, J.; Williams, S.; Kristoffersen, J.; Achourov, V.; Bisarah, A.; Mullins, O. C., Soc. Petroph. Well Log. Analysts (SPWLA), 50th Annual Logging Symp., held at the Woodlands, Texas, United States, June 21-24, 2009.

NEW GREEN SAMPLE PREPARATION TECHNIQUES FOR THE DETERMINATION OF TRACE METALS IN PETROLEUM SOURCE ROCK

Akinsehinwa Akinlua

Fossil Fuels and Environmental Geochemistry Group, Department of Chemistry,
Obafemi Awolowo University, Ile-Ife, Nigeria.
Email: geochemresearch@yahoo.com

Two green solvents were investigated for leaching of trace metals of petroleum exploration importance, from petroleum source rock. The leaching of trace metals from petroleum source rock by superheated water and nonionic surfactant was investigated and the conditions for maximum yields of the metals were determined.

The results of the superheated experiments showed that no significant extraction was attained at 100 °C but the extraction was enhanced at higher temperatures. The optimum extraction temperature of the metals from petroleum source rocks was 250 °C. Exhaustive extraction time for all the trace metals was attained at 30 minutes. Comparison of results of leaching these trace metals by superheated water with those of acid digestion revealed that cadmium, chromium, manganese and nickel had better yields with superheated water while vanadium had better yield with acid digestion. The results showed that the temperature and kinetic rates have significant effects on superheated water extraction of metals from petroleum source rocks.

The results nonionic surfactant experiments showed that temperature and time have significant effects on yields of the trace metals. Factors such as surfactant concentration and sample/solvent ratio had little influence on the extraction efficiencies for these metals. The optimum temperature for microwave-assisted nonionic surfactant dissolution of petroleum source rock for trace metals determination was 200 °C and the best yields of the metals were obtained at 40 min. The new nonionic surfactant extraction method showed efficiency comparable to those afforded by the acid digestion method for three elements out five elements determined, although their recoveries for vanadium and manganese were relatively low, but a reduction of the extraction time and environmentally friendliness of the new nonionic surfactant based method are clear advantages. The reproducibility of the new method is very good.

ANALYSIS OF DEPOSIT AND PRODUCT OILS FROM RESID HYDROCRACKING UNIT USING FOURIER TRANSFORM IONCYCLOTRON RESONANCE MASS SPECTROMETY

Keita Katano[1], *Teruo Suzuki*[1], *Ryuzo Tanaka*[1]

Advanced Technology and Research Institute, Japan Petroleum Energy Center (JPEC), 4-10, Ohnodai 1-chome, Midori-ku, Chiba-shi, Chiba 267-0056, JAPAN
ke-katano@atri.pecj.or.jp

Introduction

Formation of precursors of sediment in residue hydrocracking units often causes troubles such as plugging of heat exchangers in downstream. It is desired to elucidate formation mechanisms of the sediment and control the amount of the sediment. However, information of heavy oil components obtained from conventional analysis such as gas chromatography mass spectrometry is limited because there are a lot of molecules that are high molecular weight and high boiling point in heavy oil.

In JPEC's petroleomics project, Fourier Transform Ion Cyclotron Resonance Mass Spectrometry (FT-ICR-MS) has been applied to characterize heavy oil components at the molecular level, and developed the techniques to analyze structure information of petroleum molecules since 2011[1)]. Furthermore, the structure information has been applied to the molecule-based reaction modeling and the accurate physical property estimation as the input data. In this study, the deposit obtained from heat exchangers in downstream of residue hydrocracking units in refinery was analyzed using solvent extraction, column chromatography, and FT-ICR-MS as an application of JPEC's petroleomics techniques. In addition, the bottoms of the product oils containing different amount of the sediment were also analyzed.

Experimental

Sample preparation. Deposit was obtained from heat exchangers in downstream of residue hydrocracking units rinsed with light gas oil in refinery. The solvent extractions of the deposit were performed using Accelerating Solvent Extractor Dionex 150 (ASE150, Thermo Fisher Scientific Inc.). The deposit was weighed (around 2g) and set into the cell of ASE150 (34mL), and extracted at the high temperature (150℃) and pressure (10MPa). The time required by one extraction was set to ten minutes. Two kinds of solvents were used; toluene and tetrahydrofurane (THF). At first, the extractions using toluene were performed 36 times, and then THF was used 36 times. These extracts were summed into three kinds of extracts (Extract A; toluene1-4 times, Extract B; toluene 5-36 times, and Extract C; THF 36 times).

Bottoms of product oils were obtained in downstream of heat exchangers in refinery. Two kinds of the bottoms were obtained (Bottoms A and B). Bottom A contains much sediment than bottom B. These samples were separated using solvent extraction and column chromatography according to Figure 1: saturates (Sa), three kinds of aromatics fractions (1A, 2A, and 3A+), polar and polyaromatic resins (Po, PA), asphaltenes (As), and THF solubles.

FT-ICR-MS measurements All samples were analyzed using FT-ICR-MS (solariX 12T, Bruker Daltonics K.K.) equipped with a 12 T refrigerated actively shielded superconducting magnet. The samples without THF solubles were analyzed using atmospheric pressure photoionization (APPI) in positive ion mode. In the APPI measurements, the samples were dissolved in toluene at 0.5mg/mL, and the prepared samples were directly injected with a syringe pump at a flow rate of 200 μL/h. The THF solubles were measured using Laser desorption ionization (LDI) in positive ion mode, because THF solubles were not soluble in toluene and it was difficult to ionize them using APPI.

Data Analysis. Peak detection, internal calibration, and spectral interpretation were performed using Composer (Sierra Analytics, Inc.). The spectra were internally calibrated using the most abundant homologous series. Obtained data were visualized with the in-house software developed in MATLAB (The MathWorks, Inc.).

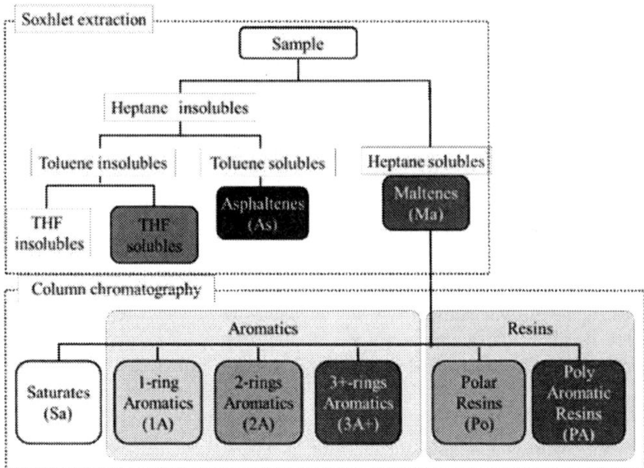

Figure 1. Scheme of solvent extraction and column chromatography.

Results and Discussion

Analysis of extracts A, B, C. The yields of solvent extraction are shown in Table 1. While the deposit from heat exchangers was solid state, extract A was liquid state, and extracts B, C were solid state. Therefore, it's seems that the main components of extract A originate from both the light gas oil for washing and the product oil.

Table 1. Yields of Solvent Extraction

Extraction	Solvent	Number of times	Yield(wt%)
A	Toluene	4	26.6
B	Toluene	32	0.6
C	THF	36	1.1

Figure 2 shows the DBE vs carbon number plots of extracts A, B, and C. Double bond equivalence (DBE) values represent the number of rings plus the number of double bonds to carbon in a given molecular formula. DBE values can be calculated using equation (1) for elemental formulas of $C_cH_hN_nO_oS_s$.

$$DBE = c - h/2 + n/2 + 1 \quad \cdots(1)$$

The distribution of extract A is analogous to that of the bottoms of the product oil. In the case of extract B, the distribution aligned to the limits of planar aromatics[2)] was observed. The DBE values of extract B ranges from 10 to 38. As will be shown later in Figure 5, the DBE range of extract B is analogous to that of As. The distribution of extract C is also aligned to the planer aromatic limit. However, the maximum of DBE values (DBE=50) is much higher than that of extract B (DBE=38). Generally, vacuum residue doesn't contain highly unsaturated components like extract C. These results suggest that extract C originates from the components generated in the reactors of the residue hydrocracking units.

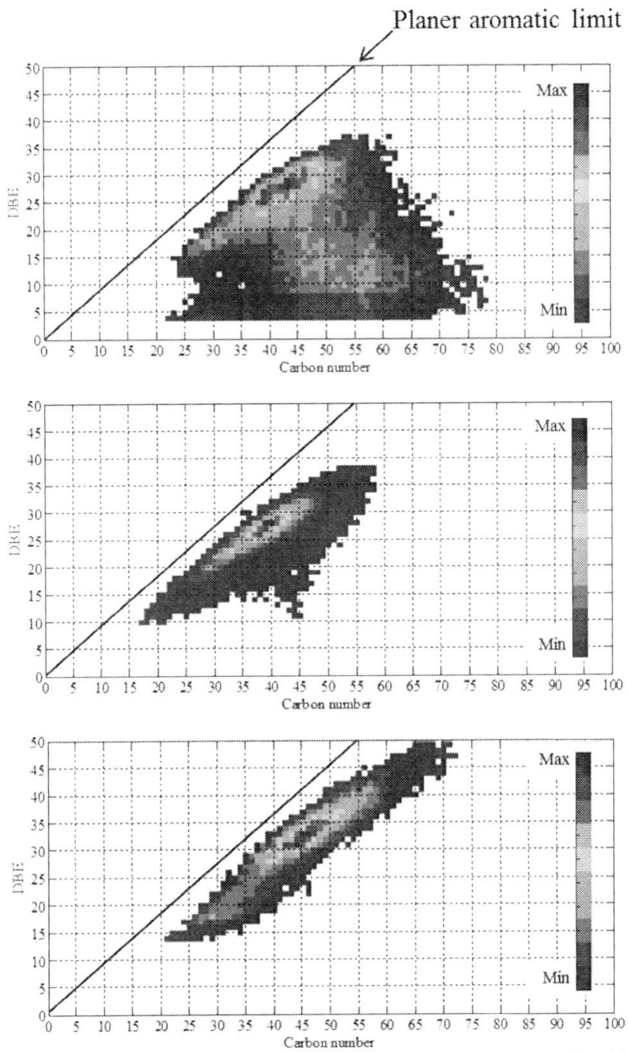

Figure 2. DBE vs carbon number plots of extracts A, B, C. All classes observed in each sample are merged into one distribution.

Table 2. Definition of classification

Classification	Definition
HC classes	HC, [H]HC
S classes	Classes containing S without N
N classes	Classes containing N without S
NS classes	Classes containing both N and S
O classes	Classes containing only O

Figure 3 shows the distribution of hetero classes of extracts A, B, C. The classification of hetero classes is defined by Table 2. HC classes (white color) were observed as the most abundant classes in each extract. N classes (blue color) were also observed in each extract. However, it has been reported that LDI overestimates the amounts of nitrogen compounds[3]. For that reason, it is expected that actual ratio of N classes is around one third of observed one, and that the highly unsaturated HC classes contribute to the formation of the deposit in the heat exchangers.

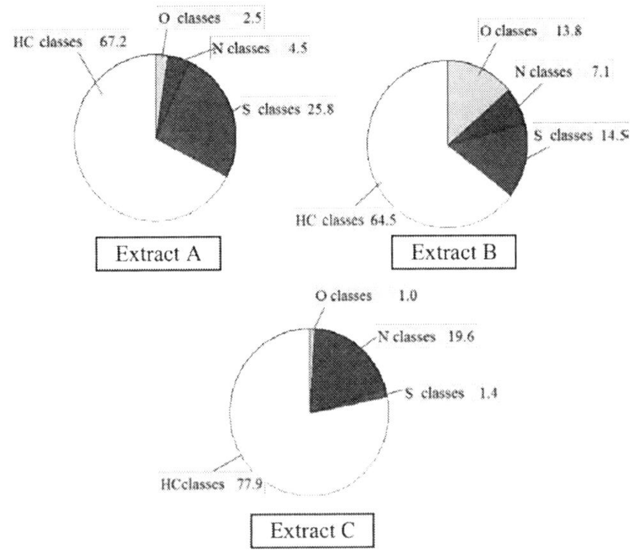

Figure 3. Distribution of hetero classes of extracts A, B, C (mol%).

Comparison of two kinds of bottoms. Figure 4 shows the yields of separation of bottoms A (much sediment) and B (less sediment). As shown in Figure 4, the yields of As and THF solubles of bottom A are 2-3% higher than that of bottom B. On the other hand, the DBE vs carbon number plots and the distribution of hetero classes were almost same between bottoms A and B.

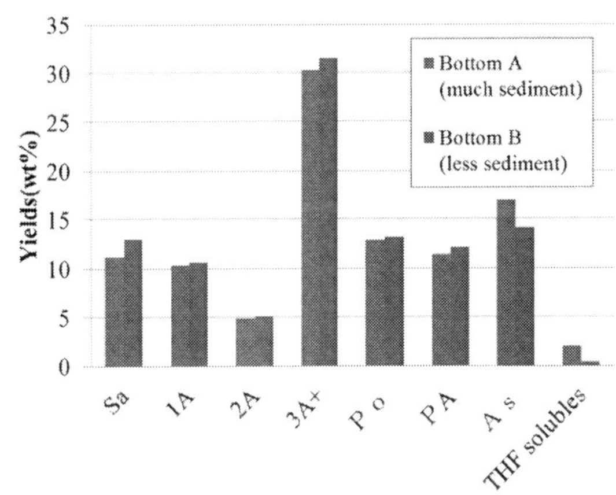

Figure 4. Yields of separation of bottoms A and B

Comparison of extracts of deposit and bottoms. The DBE vs carbon number plots of As and THF solubles from bottom A is shown in Figure 5, and their distribution of hetero classes are shown in Figure 6. The ranges of DBE values of extract B are analogous to that of As. However, the maximum of DBE values of THF solubles (DBE=60) is higher than that of extract C (DBE=50). Furthermore, the ratios of pie graphs (Figure 3 and 6) are not same. From these results, it is suggested that the highly unsaturated HC classes preferentially contribute to the formation of the deposit in the heat exchangers.

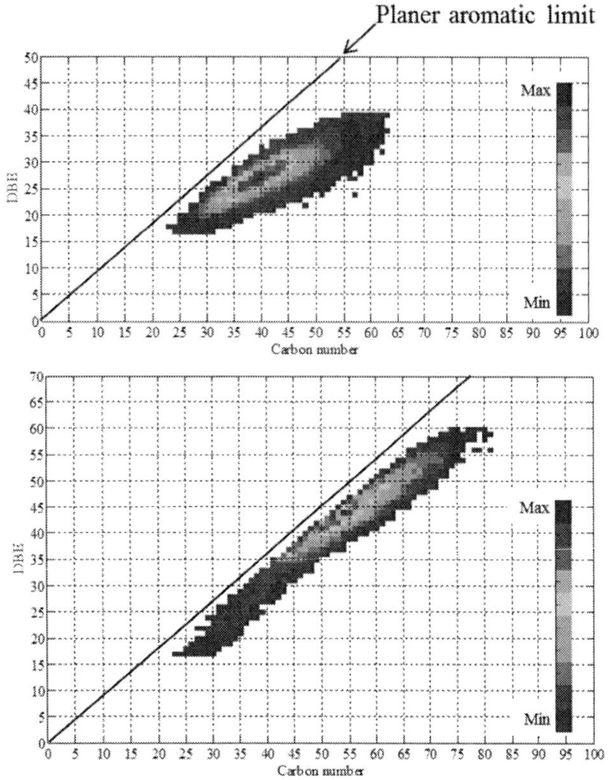

Figure 5. DBE vs carbon number plots of As and THF solubles from bottom A. All classes observed in each sample are merged into one distribution.

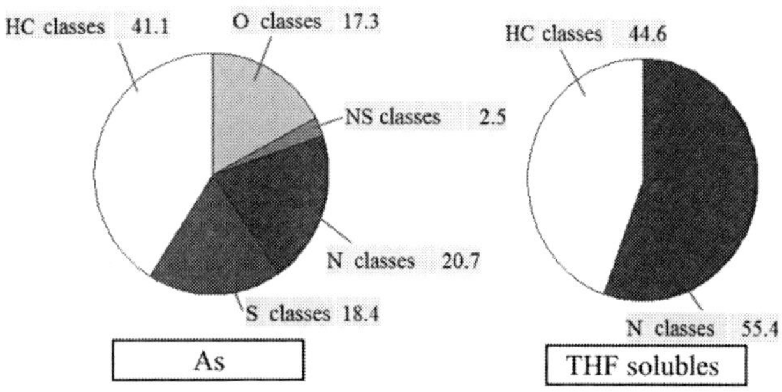

Figure 6. Distribution of hetero classes of As and THF solubles obtained from bottom A (mol%).

Conclusions

In this study, the deposit obtained from the heat exchangers in downstream of the residue hydrocracking units in refinery was analyzed using solvent extraction, column chromatography, and FT-ICR-MS. Three kinds of extracts were obtained from the deposit using toluene (extracts A, B) and THF (extract C). HC classes are observed as the main components of these extracts. Especially, extract C was highly unsaturated. These results suggest that the highly unsaturated HC classes contribute to the formation of the deposit in the heat exchangers.

Bottoms A (much sediment) and B (less sediment) were also compared. Bottom A contains more As and THF solubles (2-3wt%) than Bottom B. However, the DBE vs carbon number plots and the distributions of hetero classes were almost same between bottoms A and B. In a comparison of extract C and THF solubles, the distributions of hetero classes are different. These results seem to indicate that the highly unsaturated HC classes preferentially contribute to the formation of the deposit.

Acknowledgement. This study was supported by Japanese Ministry of Economy, Trade and Industry (METI), and carried out in Japan Petroleum Energy Center (JPEC).

References

(1) Keita Katano, Teruo Suzuki, Ryuzo Tanaka, Prep.Pap.-Am.Chem.Soc., Div.Energy and Fuels 2015, 60(2), 532

(2) Vladislav V. Lobodin, Alan G. Marshall, Chang Samuel Hsu, *Anal. Chem.* 2012, 84, 3410–3416

(3) Yunju Cho, Matthias Witt, Young Hwan Kim, Sunghwan Kim, *Anal. Chem.* 2012, 84, 8587-8594.

INVESTIGATION OF THE THERMOCHEMICAL DECOMPOSITION OF SUBSTITUTED ARYL ETHER LIGNIN MODEL COMPOUNDS

Michelle K. Kidder and A. C. Buchanan III

Chemical Sciences Division, Oak Ridge National Laboratory, Oak Ridge, TN 37831
kidderm@ornl.gov

Lignin is a potential renewable energy and chemical resource that is highly underutilized due to its inability to thermochemically produce desirable products from its complex cross-linked biopolymeric structure. Here, we present a systematic approach to understanding the thermochemical mechanistic decay pathways that occur druing the pyrolysis of the simplified model compoud that represents the major arylglycerol-beta-aryl ether, β-O-4, interunit linkage; phenethyl phenyl ether (PPE). We have shown that under liquid phase pyrolysis conditions at ca. 345 °C, decomposition occurs through a multi step free radical chain pathway. In recent investigations, the pyrolysis of oxygen substituents on the aromatic position of PPE showed that selectivity and reactions rates are higly dependent on the functionality type and placement of the substituent. We now seek to understand the influence on reaction rates and product selectivity of substituents at the aliphatic alpha and beta carbons of PPE. The effects on the reaction rates and selectivity of an α-hydroxy and γ-hydroxy substituted PPE will be presented.

HYDROGENATION AND HYDROGENOLYSIS REACTIONS INVOLVED IN TREATMENT OF WATER CONTAMINATED WITH CHLORINATED HYDROCARBONS

Hyuntae Sohn[1], Gokhan Celik[1], Seval Gunduz[1], Paul Edmiston[2], Umit. S. Ozkan [1]

[1]The Ohio State University, Columbus, Ohio – USA
[2]College of Wooster, Wooster, Ohio – USA
Ozkan.1@osu.edu

Contamination of ground water by chlorinated hydrocarbons such as carbon tetrachloride (CTC), tetrachloroethene (PCE), trichloroethene (TCE) and 1,1,1-trichloroethane (TCA) is a growing environmental concern. The current methods to treat water contaminated with chlorinated hydrocarbons are expensive, due to high energy requirements and media replacement costs, and serve only to transfer contaminants from one phase to another, such as from water to air or from water to landfills. While elimination of these contaminants by catalytic hydrogenation and hydrodehalogenation offers a treatment strategy, there are challenges, such as the need to run these reactions at low temperatures as well as the very low concentrations of the contaminants in water. An equally important challenge is the poisoning of the catalysts by sulfides and nitrates.

Catalytic systems developed by using swellable organosilicates as catalyst scaffold have the potential to meet the challenges involved in catalytic hydrogenation and hydrogenolysis of water contaminants. Swellable organically modified silicates (SOMS) offer unique characteristics such as high affinity for absorbing organics while repelling water due to their hydrophobicity. The swelling capability of these materials allows active metals (precious metals, transition metals) dissolved in organic solvents to be deposited inside the pores, thus forming active sites "protected" in a hydrophobic matrix. This property also allows concentration of organic contaminants inside the pores, near the active sites, hence accelerating the kinetics. The presentation will discuss the use of swellable organosilica materials as a catalyst scaffold for precious and transition metals and examine their use in water treatment strategies. Characterization results from different techniques such as NMR, laser Raman spectroscopy, TEM, XPS, and infrared spectroscopy will also be presented.

A NEW CONCEPT OF PH-SWING PROCESS FOR BIOMASS CONVERSION TO FUELS AND CHEMICALS

*Chunshan Song**

EMS Energy Institute, Department of Energy & Mineral Engineering, and Department of Chemical Engineering, The Pennsylvania State University, 209 Academic Projects Building, University Park, PA 16802.
*E-mail: csong@psu.edu

This paper proposes a pH-Swing Conversion (PSwC) Process concept for converting lignocellulosic biomass and cellulose to fuels and chemicals. As is well known in the field, depolymerization of lignocellulosic biomass is a significant part of the challenge in converting biomass to chemicals and fuels. The PSwC concept is supported by our preliminary results. We have recently explored the sequencing of acid and base treatments of lignocellulosic biomass as well as cellulose. [1-2] Specifically, we explored sequential combinations of organic acid (oxalic acid, OA) and base (tetramethylammonium hydroxide, TMAH) for the conversion processing of a whole wood sample and a cellulose sample at relatively mild conditions (at 150 °C).

For conversion of woody biomass in the single step, hydrolysis with TMAH mainly depolymerizes lignin to produce phenolic compounds but also converts some polysaccharides in biomass possibly by the peeling reaction and/or alkaline hydrolysis. A sequential combination of OA and TMAH can significantly increase the conversion of the biomass compared to single- or two-step reactions with either the acid or the base alone. The analysis of liquid products by chemical derivatization followed by gas chromatography equipped with a flame ionization detector (GC-FID) and the characterization of the solid residues by pyrolysis-gas chromatography-mass spectrometry (Py-GC-MS) and 13C cross-polarization magic angle spinning nuclear magnetic resonance (CPMAS NMR) reveal strong synergetic effects from the sequential combination of acid and base. In single-step reactions, OA serves as a proton donor to cleave the glycosidic bond between polysaccharide monomers, while TMAH reacts more selectively to lignin than polysaccharides. In two-step tests, the reaction with OA in the first step significantly enhances the conversion with TMAH in the second step by almost twice compared to the corresponding reaction with base alone. GC-FID analysis of the liquid sample shows relatively higher conversion of carbohydrates in the biomass with base in the second step. Py-GC-MS and 13C CPMAS NMR analysis also reveals that acid treatment in the first step enables base to convert more cellulose in the second step.

For hydrolysis reaction of cellulose (at 150 °C), we demonstrated the enhancing effect of a sequential combination of acid and base on its conversion. In two-step experiments, the highest conversion (73.0%) was achieved when the reaction with acid was followed by that with base (OA – TMAH sequence), which is greater than the prolonged treatment with either acid or base alone. In the residue after the reaction with OA, a more disordered, amorphous-like structure was present compared to raw cellulose and the sample after that with TMAH alone, as evidenced by 13C CPMAS NMR and XRD measurements. Such structural change in the first OA treatment step can be the main cause of high conversion of cellulose in OA-TMAH treatment sequence. Analysis of liquid products clearly indicated that OA and TMAH both promote depolymerization of cellulose by cleaving the glycosidic bonds but generated distinctly different products: glucose and ketogluconic acid, respectively. Reactions of cellulose with conventional mineral acid (H2SO4) and base (NaOH) were also examined, and the relative orders of conversions and the liquid products were similar to those from reactions with organic acid and base. The selectivities to glucose and ketogluconic acid in organic acid and base treatments, respectively, were better than those with their inorganic counterparts.

In summary, the proposed new concept of pH-Swing Process is promising for more effective conversion of lignocellulosic biomass and cellulose to monomeric sugars and monoaromatic compounds as feed for valuable chemicals and fuels. However, the results so far are based on batch reactor processing, and more research is needed in order to demonstrate the feasibility of the proposed process concept.

Acknowledgement. Support from the Pennsylvania State University to CS is gratefully acknowledged.

References

(1) Wongsiriwan, U.; Noda, Y.; Song, C.S.; Prasassarakich, P.; Yeboah, Y. Lignocellulosic Biomass Conversion by Sequential Combination of Organic Acid and Base Treatments. *Energy & Fuels,* **2010**, 24, 3232-3238

(2) Noda, Y.; Wongsiriwan, U.; Song, C.S.; Prasassarakich, P.; Yeboah, Y. Sequential Combination of Organic Acid and Base for Conversion of Cellulose. *Energy & Fuels,* **2012**, 26(4), 2376-2385.

HOW HIGH CAN CONVERSIONS AND DEOXYGENATION OF BIOMASS REACH WITHOUT THE NEED FOR HIGH HYDROGEN PRESSURE?

Colin Snape

Faculty of Engineering, University of Nottingham, Energy Technologies Building, Triumph Road, Nottingham NG7 2TU, UK
Colin.snape@nottingham.ac.uk

Introduction

This presentation will draw together recent research on the conversion of biomass to liquids to consider the limiting factors on yields and the extent of oxygen removal that can be achieved where high hydrogen pressures are not employed. The combination of hydrogen pressures over *ca.* 10 MPa and suitable dispersed catalysts, such as sulphided Mo, results in conversions of close to 100% for all biomass fuels[1]. Indeed, hydropyrolysis has been developed as an analytical pyrolysis method to determine pyrolytic or "black" carbon on environmental matrices based on the ability to remove all labile organic matter reductively[2].

First of all, a universal correlation is described for predicting char yields for different pyrolysis regimes based on aromatic carbon contents as a proxy for lignin and alkali/alkaline metals which are known to increase char yields due to reaction pathways including carboxylic salts. The thermolytic extraction of biomass in hydrogen-donor solvents and water ae then considered and compared to identify the extent to which hydrogen is utilized to increase overall conversions and facility deoxygenation.

Predicting char yields from biomass pyrolysis

A universal correlation able to predict the yields of volatile matter and char obtained during pyrolysis of lignocellulosic biomass species under slow and rapid heating (entrained flow) regimes with a high degree of accuracy has been developed. The influences of aromatic carbon and alkali/alkaline metals on char yields during pyrolysis have been assessed and their individual effects on char yields have been quantified.

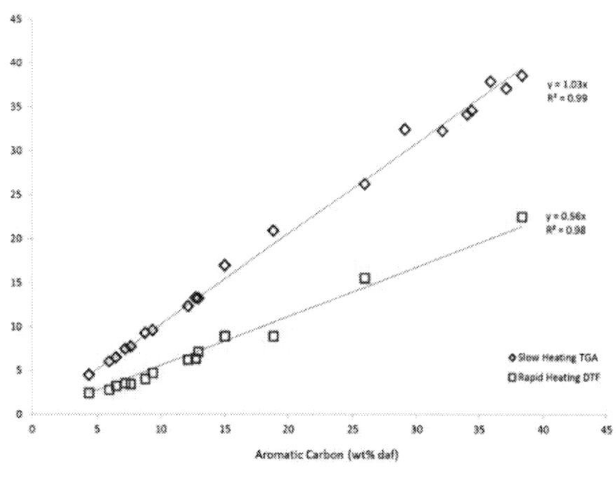

Figure 1. Char yield vs. aromatic carbon wt% for low ash biomass samples (<0.15 wt% K+Ca db)

For partially demineralised acid washed samples, a linear relationship between char yields and their wt% aromatic carbon, determined from solid state ^{13}C NMR and elemental compositions, has been obtained (Figure 1).

To take account of both aromatic carbon and alkali and alkaline metal contents, the following empirical relationships for slow and fast heating rates, respectively, have been derived.

CY = (1 x Aromatic Carbon wt%) + (15.91 x Sum K+Ca wt%) (1)
CY = (0.57 x Aromatic Carbon wt%)+(2.63 x Sum K+Ca wt% (2)

where CY is the wt% char yield (daf), aromatic carbon is the wt% aromatic carbon (daf) and Sum K+Ca is the summed wt% of K+Ca (db). The validity of these empirical relationships has been confirmed by plotting calculated vs. observed char yields (Figure 2) with R^2 values of 0.98 and 0.99 being obtained, respectively.

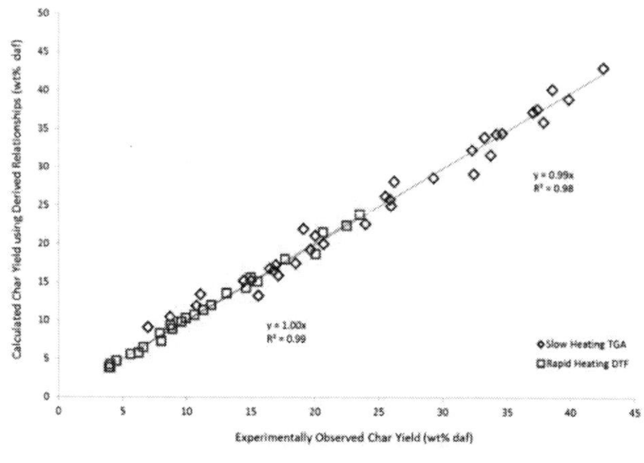

Figure 2. Calculated vs. experimentally observed char yields (wt% daf)

Thermolytic extraction with hydrogen-donors

Thermolytic liquid solvent extraction (LSE) with hydrogen-donor solvents is considered as a potential viable route to obtain bio-oil yields given that these solvent can largely convert the reactive macerals in high volatile coals to soluble products. Indeed, our results have established that extremely high overall conversions (only ca. 5-10% toluene-insolubles, daf basis) can be achieved for Scotch pine and miscanthus with temperatures above 400 °C (Table 1)[3].

Table 1: Product yields on a DAF basis from biomass liquefaction at 410°C for 1 hr using 12.5 g tetralin and 5 g biomass.

Yields (%)	Miscanthus	Scotch pine
Pyridine insoluble	5.9	7.7
Pre-asphaltene	2.1	5.5
Asphaltene	N.D.	8.6
C_1-C_4 H/C gases	2.0	2.1
H_2O	N.D.	19.9
Bitumen	31.0	27.0

Further, unlike for coals, conversions continue increase with temperature to over 450 °C and the additional hydrogen consumed is used to reduce the oxygen contents of the oils to below 10%. A number of non-hydrogen donor solvents have been investigated and "heavy" solvent components comprising aromatic structures larger than 2 rings have a detrimental effect on conversion. Experiments with blends of tetralin and 1-methylnaphthalene have demonstrated that high conversions of 85% (daf basis) can be achieved with as little as 15% tetralin in the solvent where the hydrogen consumption is below 1% (daf biomass). This means that biomass liquefaction can offer in a more flexible manner compared to coal where high concentration of H-donors are needed in the absence of hydrogen pressure. However, to achieve high levels of oxygen removal, H consumption increases to over 2% to achieve oils with below 10% oxygen compatible with existing refinery feedstocks.

Hydrothermal conversion

Water or hydrothermal liquefaction (also known as a hydrous pyrolysis) has received extensive attention because it is clearly an inexpensive and easy to recycle reaction medium for converting wet and dry biomass into crude bio-oils with or without a catalyst[4]. As well as the bio-crude fraction, some water soluble organics are also obtained. By moving from subcritical to supercritical temperatures (374°C) at pressures above Pc (22 MPa) both the rate of hydrolysis as well as phase partitioning and solubility of components can be controlled so that potentially more favourable pathways to gases and liquid biofuels may be realised[5]. Miscanthus conversion using sub- and supercritical water has been investigated in this study, aimed to identify conditions required to maximise biomass conversion and liquid product yield.

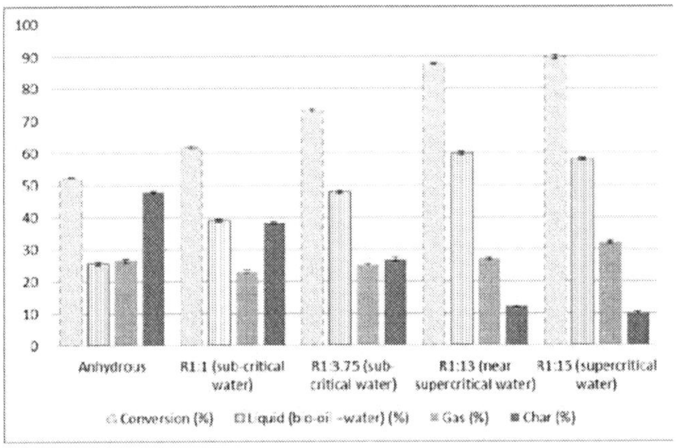

Figure 3. Overall conversions as a function of the miscanthus to water mass ratio at 410oC for 1 hr. The liquid (bio-oil plus water) yields were obtained by difference

The results indicate that conversions increased with increasing miscanthus to water mass ratio at high water ratio at 410°C to give the highest conversions of ~90 wt. %, producing ~21 wt. % heavy oil, ~12 wt. % light oil, ~27 wt. % gas and ~28 wt. % water (Figure 3). At 460°C, the maximum conversion was similar but the liquid product was lighter with only 7% heavy oil and more hydrocarbon gas being obtained (Figure 4). The oxygen contents of heavy oil recovered ~12-16% (for 410 and 460°C), with H/C ratio of 1.09 at 410°C. Under equivalent conditions with tetralin as a model hydrogen-donor solvent, similar overall conversions were achieved but with much lower solvent to miscanthus mass ratios were needed and the liquid products have lower oxygen contents (~12-16%)[6]. Reduced CO and increased CO_2 yields provide evidence that water hydrogen is being utilised but the estimated amounts of ca. 0.5 % at 410 and 460°C are considerably lower than with hydrogen-donor solvents which account for the lower yields of bio-oils and their higher oxygen contents[6].

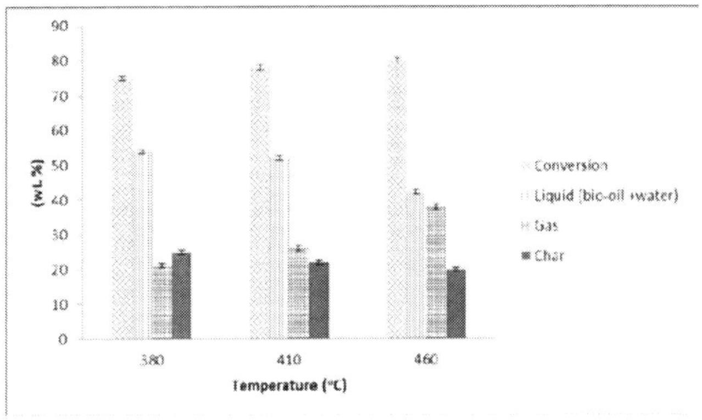

Figure 4. Effect of reaction temperature on product yields using a miscanthus to water mass ratio of 1:6 for 1 hour.

References

(1) J.D. Rocha, C.A. Luengo and C.E. Snape, Organic. Geochem., 1999, 30, 1527-1534.
(2) W. Meredith, P.L. Ascough, M.I. Bird, D.J. Large, C.E. Snape, Y. Sun and E.L. Tilston, Geochim Cosmochim Acta, 2012, 97(15), 131-147.
(3) H. Deng, W. Meredith, C.N Uguna and C.E. Snape, J. Anal & Appl. Pyrolysis, 2015, 113, 340-348.
(4) S. S. Toor, H. Reddy, S. Deng, J. Hoffman, D. Spangsmark, L.B. Madsen, et al, Bioresour. Technol. 2013, 131, 413-419.
(5) A. A. Peterson, F. Vogel, R. P. Lachance, M. Froling, M. J. Antal Jr, J. W. Tester, Energy & Environmental Science 1 2008, 32-65.
(6) K. Md Isa, C.E Snape, C. Uguna and W. Meredith, J. Anal & Appl. Pyrolysis, 2015,113, 646-654.

DIRECT HYDROCARBON POLYMER ELECTROLYTE MEMBRANE FUEL CELLS; WHERE THERE'S SMOKE THERE'S SOMETIMES FIRE!

Andrew M. Herring, Vinh Nguyen, and Himanshu Sarode.

Department of Chemical and Biological Engineering
Colorado School of Mines, Golden, CO 80401, USA
Email: aherring@mines.edu

The hydrogen fueled proton exchange membrane fuel cell has an extraordinarily high power density making it applicable to automotive applications. However, to achieve this we still need 30g of Pt (compared to the 10g of Pt currently in a vehicular catalytic converter) making this fuel cell currently too expensive for wide spread adoption. While pressurized tanks allow the vehicle to have an adequate range, the most cost effective method of producing hydrogen in North America is from steam methane reforming of natural gas. A direct methane polymer electrolyte membrane (PEM) fuel cell would have many advantages, but activating methane at the low temperatures of operation is extremely challenging. We have worked on improving the power densities from the more easily activated molecules methanol, and dimethyl ether, but these fuel cells still require too much precious metal catalyst and the power densities are still much lower than for a hydrogen PEM fuel cell. Recently we have been applying immobilized molecular catalysts in a direct methane fuel cell, and while the stabilities and power densities are not great, there is much promise from this approach where a molecularly designed catalyst is deployed in a real device. We can also raise the temperature of operation of the fuel cell by changing the electrolyte to phosphoric acid, and recent results at higher temperatures will be presented. Another approach is to use an anion exchange membrane (AEM), as electro-catalysis is more facile in base than acid, and AEMs that are robust enough for fuel cell testing are now available.

MECHANISMS FOR MOLECULAR WEIGHT GROWTH FORMATION IN ALKENE PYROLYSIS

Kun Wang, Stephanie M. Villano, and Anthony M. Dean*

Chemical and Biological Engineering Department
Colorado School of Mines
Golden, CO 80401
amdean@mines.edu

Alkenes are components in hydrocarbon fuels, and they are also key intermediates during the decomposition of alkanes, alcohols, ethers, and biofuels. Thus, their subsequent reactions will have substantial impact on the final product distribution. More specifically, alkenes, unlike alkanes, can undergo radical addition reactions that then may lead to formation of higher molecular weight growth (MWG) species, perhaps leading to deposit formation. We will summarize a combined experiment-modeling effort to describe the pyrolysis kinetics of propene and the three butene isomers. The fundamentally-based kinetic model incorporates rate rules derived from an extensive series of CBS-QB3 calculations for several types of free radical reactions. This model accurately describes fuel decay, major products formation, and generation of the major molecular weight growth products. The similarities and differences among these fuels will be discussed in terms of the impact of their structures.

* Current address: Mechanical Engineering Dept., Stanford University, Stanford, CA.

REACTION NETWORKS AND KINETICS FOR HYDROTHERMAL CONVERSION OF BIOMASS TO BIOCRUDE

Phillip E. Savage[1,2], *James D. Sheehan*[1], *David C. Hietala*[2]

[1]Chemical Engineering Department, Penn State University, University Park, PA
[2]Chemical Engineering Department, University of Michigan, Ann Arbor, MI
psavage@psu.edu

Abstract

Hydrothermal liquefaction (HTL) has emerged as a preferred process for converting wet biomass into a more energy-dense crude bio-oil. HTL produces biocrude as a desired product fraction, along with an aqueous phase co-product rich in organic compounds, P, and N, and much smaller amounts of residual solid material and gas. As originally investigated, processing times were on the order of tens of minutes and the reactor was isothermal. More recent work has revealed that rapid heating and times on the order of tens of seconds may be sufficient. We have focused on elucidating the reaction paths that interconnect the different product fractions and determining the kinetics for each. This presentation will provide results from recent experimental and modeling studies on the HTL kinetics for microalgae and for soy-derived protein. These models faithfully capture the production and consumption of the different product fractions during both fast HTL (rapid heating, short reaction times) and more conventional isothermal HTL (longer times, constant reaction temperature). Models such as these may prove useful in advanced technoeconomic analyses and life-cycle assessments of HTL processes as they allow the analyst to explore the influence of HTL processing conditions on the process outcomes.

PRODUCTS FROM CATALYTIC FAST PYROLYSIS OF BIOMASS

Mark R. Nimlos, Calvin Mukarakate and A. Nolan Wilson

National Bioenergy Center, National Renewable Energy Laboratory, 15013 Denver W Pkwy, Golden, CO 80401
mark.nimlos@nrel.gov

Integration of renewable biofuels into our transportation infrastructure can result in a significant reduction in green house gas emissions, but the current cost relative to petroleum fuels challenges public acceptance. One way to help improve the costs of biofuels is to direct some of the biomass carbon to value added co-products, much the same way as some petroleum refineries produce chemicals to help the economics of hydrocarbon fuels. In this presentation, we will discuss co-product formation using catalytic fast pyrolysis (CFP), a high throughput process for making products that can help the economics of biorefineries.

Separation is a key consideration with co-product formation from any biomass processes and CFP is advantaged in that the fractional distillation can be used. This is unlike regular pyrolysis oils, for which up to 50% of the oil remains in the still pot, due to the reactivity of the oil. After distillation, many of the components of CFP oil can be purified and used for several high-value polymers and resins.

One strategy that can help improve the economics of biomass-based transportation fuels is to leave some oxygen in the CFP oil. This may help reduce the overall cost of fuel production and will provide oxygen-functionalized molecules that are often of higher value than the fuel. We will explore this strategy with phenol production from CFP.

MODELING LIGNIN PYROLYSIS REACTION MECHANISMS

Michael T. Klein

Energy Institute and Department of Chemical & Biomolecular Engineering, University of Delaware, Newark, DE 19716
mtk@udel.edu

Modeling lignin pyrolysis at the mechanistic level requires a quantitative representation of lignin structure, elementary reaction steps, the associated reaction network, and the kinetics. As this can involve thousands of components and reactions the use of computational tools to construct and solve the model is helpful. Herein we describe the development of a lignin pyrolysis model through the use of a set of in-house software tools called KMT, the Kinetic Modeling Toolkit.

A molecular representation of the feedstock is generated by CME (the Composition Modeling Editor) via available analytical chemistry. INGen (the Interactive Network Generator) automatically generates the complete pyrolysis reaction network by exhaustively applying the relevant thermolysis chemistry to the reactant molecules. The resulting reaction network and molecular composition are combined in KME (the Kinetic Modeling Editor), which provides a numerical solution of the material balances in the kinetic model.

The computational model can only be as good as the chemistry it incorporates. To this end, we have developed a mechanism elucidation technique for the pyrolysis of lignin model compounds that tracks the fate of an isotopic label during pyrolysis of fully protonated reactant in the presence of a fully deuterated solvent. This allows for discrimination between free-radical and concerted pyrolysis mechanisms. Monitoring the deuterium incorporation (*DI*) into the reactant's product or the protium incorporation (*PI*) into the solvent as a function of the solvent to feed concentration ratio, *R*, provides the contribution of each mechanism to the overall reaction. At infinite dilution ($1/R \rightarrow 0$), *DI* approaching zero indicates dominance by a concerted mechanism, whereas *DI* approaching unity suggests dominance by the free-radical mechanism. This work elaborates on the kinetics basis for such a measurement.

On this chemical foundation, the lignin pyrolysis model generates various aromatic compounds and light gases. Maintaining the molecular identities of each of these species enables the accurate prediction of product mole fractions and bulk properties. This detailed information is critical for fundamental understanding of the reaction chemistry and for the optimization of reaction conditions to maximize desired product formation. Rapid generation and solution of these molecular-level kinetic models will accelerate research for the utilization of lignin and coal as advanced energy resources.

EXPERIMENTAL AND MODELING STUDIES FOR BIOMASS PYROLYSIS YIELDS

Michael A. Serio and Marek A. Wójtowicz

Advanced Fuel Research, Inc. (AFR)
87 Church Street, East Hartford, CT 06108 USA
Contact Information:
Michael Serio, mserio@afrinc.com, 860-528-9806 ext. 105
Marek Wójtowicz, marek@afrinc.com, 860-528-9806 ext. 142

Biomass as a solid fuel has received increased interest in recent decades due to the desire for renewable fuels and concerns about climate change. The use of wood pellets for residential heating has experienced significant growth in the U.S. and Europe over the past 15 years. Solid biomass still has to overcome some obstacles, however, before it becomes an important part of the energy mix in the U.S. The biggest problems are the low energy densities and high moisture contents of biomass fuels. The construction of biomass fuel-fired boilers has also met with resistance in certain states. The availability of inexpensive natural gas in the U.S., combined with the current low price of oil, also acts as a disincentive. However, biochar produced from pyrolysis has attracted significant attention in the past decade for use for soil amendments, soil remediation, composite materials, and carbon sequestration. These biochar materials have so far been produced mainly in small scale facilities under pyrolysis or partial oxidation conditions. The ability to accurately predict yields of char and other biomass pyrolysis products in a comprehensive way is still limited when compared to the capability that exists for coal. Historically, the quantitative modeling of biomass pyrolysis has proceeded along similar lines as for coal One approach is based on the approximation that the three main components of biomass (cellulose, hemi-cellulose, and lignin) behave independently during pyrolysis. Consequently, yields can be predicted based on knowledge of the pure component behavior. The shortcoming of this technique is that it cannot account for possible interactions between the biomass components. A more phenomenological approach was successfully used by AFR in modeling the pyrolysis of lignin. This model was focused on observed evolution rates of major products rather than on the many individual reactions leading to the evolution of each product. This approach involved a statistical treatment of lignin pyrolysis reactions using a variation of the FG-DVC coal pyrolysis model. For whole biomass materials, the network features of the FG-DVC coal model (and similar models) are de-emphasized in favor of the original functional group (FG) description of gas and tar evolution, in what is known as the FG-BioMass model. While this is a more empirical approach, there is flexibility in the model to add more details of the whole biomass pyrolysis chemistry as these become known and can be represented in a tractable mathematical expression. This paper will review our progress in measuring and predicting biomass pyrolysis product distributions, including char yield.

A MECHANISTIC ADVENTURE: PYROLYSIS OF OXYGENATED MODEL COMPOUNDS

Phillip F. Britt[1], *A. C. Buchanan, III*[1], *Michelle K. Kidder*[1], *and Ariana Beste*[2]

[1] Chemical Sciences Division, Oak Ridge National Laboratory
One Bethel Valley Road, P.O. Box 2008, MS-6129
Oak Ridge, TN 37831
[2]Joint Institute of Computational Science, University of Tennessee
Oak Ridge, TN 37831
brittpf@ornl.gov

Introduction

Oxygen functional groups, such as carboxylic acids, phenols, alcohols, esters, and ethers, are prevalent in energy resources, such as biomass and coal, and have a profound impact of their thermochemical reactivity, product yields and product quality. Thus, to control the thermochemical decomposition of these oxygenated energy resources, it is important to understand the fundamental reaction pathways of oxygen functional groups. However, oxygen functional group chemistry is more complex than hydrocarbon chemistry because a variety of ionic, free radical and concerted reaction pathways are available that are easily influenced by metal ions, catalysts, or hydrogen bonding. In this presentation, we will review our mechanistic studies on the pyrolysis of oxygen functional groups in lignin and low rank coal model compounds. A variety of model compounds and their deuterated analogues were synthesized to probe the reaction pathway of specific oxygen functional groups such as aromatic carboxylic acids, found in low rank coal, and beta-aryl ether linkages found in lignin. A variety of experimental methods were used to study the reaction mechanisms in the gas phase, including flash vacuum pyrolysis and sealed tube pyrolysis, liquid phase, and under conditions of restricted diffusion and pore confinement. Computational methods were also used to gain predictive insights into complex reaction pathways. The role of polar effects and hydrogen bonding on free radical reactions will be discussed.

Acknowledgement. This research was supported by the U.S. Department of Energy, Office of Science, Basic Energy Sciences, Chemical Sciences, Geosciences, and Biosciences Division.

NEUTRON VIBRATIONAL SPECTROSCOPY: A POWERFUL AND UNDERUTILIZED METHOD TO PROBE CATALYSIS

D.A. Lutterman,[#,*] *L.L. Daemen,*[¥] *Y.Q. Cheng,*[¥] *and A.J. Ramirez-Cuesta,*[¥]

Chemical Sciences Division[#] and Spallation Neutron Source,[¥] Oak Ridge National Laboratory, Oak Ridge, TN 37831
*luttermanda@ornl.gov

Abstract

Neutron Vibrational Spectroscopy (NVS) remains an underutilized technique in surface chemistry and catalysis. It presents a number of advantages over Raman or infrared spectroscopy: sensitivity to hydrogen and hydrogen-bearing functional groups, neutron penetrability through bulky sample environment, no energy deposition in the sample, absence of selection rules, and the possibility of calculating the vibrational spectrum quantitatively with relative ease. Furthermore the substrate in supported metal catalysts or most zeolite frameworks scatter neutrons weakly so that the signal is obtained from the adsorbate. Neutron spectroscopy provides information not easily accessible by traditional thermodynamic and kinetic methods such as adsorbate orientation and conformation, stable intermediate species, or adsorption sites. In conjunction with DFT and molecular dynamics modeling NVS has the potential to provide a level of understanding of catalytic surface chemistry that goes well beyond conventional techniques. We will illustrate this complementarity with several examples of data collected at the Spallation Neutron Source (SNS) at Oak Ridge National Laboratory. An overview of the current development of instrumentation and sample environment for the VISION spectrometer will be given.

Acknowledgements. This research at ORNL's Spallation Neutron Source was sponsored by the Scientific User Facilities Division, Office of Basic Energy Sciences, U. S. Department of Energy. The development of metal oxide reactivity were sponsored by U.S. Department of Energy, Office of Science, Basic Energy Sciences, Chemical Sciences Geosciences, and Biosciences Division. The development of sample environments and hydrogenation catalysis were sponsored by the Laboratory Directed Research and Development Program of Oak Ridge National Laboratory, managed by UT-Battelle, LLC, for the U.S. Department of Energy. Each sponsor contributed equally.

OXYGEN REMOVAL FOR BIOMASS ON TRANSITION METAL PHOSPHIDE CATALYSTS

Haiyan Zhao, Dan Li, Phuong Bui, Ted Oyama

University of Idaho, Idaho Falls, ID, 83401
Sinopec, Beijing, China, 100728
Virginia Tech, Blacksburg, Virginia, 24061
Virginia Tech, Blacksburg, Virginia, 24061
haiyanz@uidaho.edu; Oyama@vt.edu

Abstract

A series of novel hydroprocessing catalysts – transition metal phosphides which included Ni_2P/SiO_2, Fe_2P/SiO_2, MoP/SiO_2, Co_2P/SiO_2 and WP/SiO_2 were investigated for hydrodeoxygenation of model compounds guaiacol and 2-methyltetrahydrofuran. The turnover frequency based on active sites titrated by the chemisorption of CO followed the order: Ni_2P>Co_2P>Fe_2P, WP, MoP for HDO of guaiacol. The major products from hydrodeoxygenation of guaiacol for the most active phosphides were benzene and phenol, with a small amount of methoxybenzene formed. Kinetic studies revealed the formation of reaction intermediates such as catechol and cresol at short contact times. The commercial $CoMoS/Al_2O_3$ deactivated quickly and showed little activity for the HDO of guaiacol at these conditions. Thus, transition metal phosphides are promising materials for catalytic HDO of biofuels.

The precursor phosphite (I) leads to higher surface area phosphides than phosphate (A), which does not make significant difference on Ni_2P for HDO of 2-MTHF but play important role for the case of WP. At 300 □C and 1 atm, the order of activity was Ni2P > WP > MoP > CoP > FeP > Pd/Al_2O_3. The principal HDO products for the iron group phosphides (Ni_2P and CoP) were pentane and butane, whereas for the group 6 metal phosphides (MoP and WP), the products were mostly pentenes and pentadienes.

Contact-time studies were used to develop reaction networks for the most active catalysts, Ni_2P/SiO_2 and WP/SiO_2. The reaction mechanisms were proposed for HDO of guaiacol and 2-MTHF.

ATOMIC LAYER DEPOSITION OVERCOATING: TUNING CATALYST SELECTIVITY FOR BIOMASS CONVERSION

*Hongbo Zhang[1], Xiangkui Gu[2], Christian Canlas[1], A. Jeremy Kropf[1], Jeffery P. Greeley[2], Jeffery W. Elam[1], James A. Dumesic[3], Peter C. Stair[1], Christopher L. Marshall[1]**

[1]Argonne National Laboratory, Argonne, IL 60439
[2]Purdue University, West Lafayette, IN
[3]University of Wisconsin, Madison, WI
marshall@anl.gov

ABSTRACT

Terraces, edges and facets are all active sites of nanoparticles for heterogeneous catalysts. These different active sites may cause various products to form during the catalytic reaction. Here we report that the step sites of the Pd nanoparticles can be covered precisely by the atomic layer deposition (ALD) method, leaving the terrace sites as the active component for hydrogenation of furfural. The selectivity to furan can be increased using an ALD overcoating. Increasing the thickness of the ALD overcoats restricts the adsorption of furfural onto the step sites of Pd nanoparticles (NPs) and increases the selectivity to furan. Furan selectivities and furfural conversions are linearly correlated for samples with or without ALD overcoating, though the slopes differ. The ALD technique can tune the selectivity of furfural hydrogenation over Pd NPs and has improved our understanding of the reaction mechanism. The above conclusions are further supported by density functional theory (DFT) calculations.

CATALYTIC HYDRODEOXYGENATION ON METAL CARBIDES

*Mark M. Sullivan, Cha-Jung Chen, and Aditya Bhan**

Dept. of Chemical Engineering and Materials Science
University of Minnesota
421 Washington Ave. SE
Minneapolis, MN 55455

Introduction

Biomass, the only renewable source of carbon, will play a crucial role in supplying the planet with sustainable energy carriers and petrochemicals. The $(CH_2O)_n$ stoichiometry of biomass necessitates stoichiometric oxygen removal for conversion to fuels and chemicals $(CH_2)_n$. Interstitial metal carbides are a class of non-precious-metal catalysts that exhibit acidic[1-3] and metallic[4] functionalities and have shown promise as hydrodeoxygenation (HDO) catalysts. The underlying challenges in HDO chemistry involve (i) selective C-O vs. C-C scission and (ii) H_2 efficiency by avoiding successive aromatic or olefinic hydrogenation reactions. In this work, we demonstrate that Mo_2C can selectively deoxygenate representative oxygenates such as lignin-derived anisole, m-cresol, guaiacol, and 1,2-dimethoxybenzene as well as smaller carbonyl and alcohol functionalities while suppressing sequential hydrogenation pathways and conserving H_2. We quantify metal and acid site sites using stoichiometric reversible CO and irreversible 2,6-di-tert-butylpyridine (DTBP) titrations, respectively, to demonstrate our ability to tune the metal-acid bifunctionality of molybdenum carbide catalysts for selective deoxygenation at ambient pressure and low temperatures (415 – 553 K).

Experimental

In situ synthesis was performed with 0.025 – 1.5 g of ammonium paramolybdate ($(NH_4)_6Mo_7O_{24}\cdot 4H_2O$, sieved, 177 – 400 μm), typically following the temperature programmed reaction (TPR) method of Thompson.[5] Mo_2C was synthesized under total flow of 3.0 $cm^3\ s^{-1}$ of 15 vol% CH_4/H_2. Reactor temperature was ramped at 0.046 K s^{-1} to 623 K, held for 12 h, then ramped to 873 K at 0.044 K s^{-1} and held 2 h, then cooled to room temperature or reaction temperature. Passivation treatments were performed by flowing 1 $cm^3\ s^{-1}$ of 1% O_2/He over synthesized catalysts at ambient temperature. Passivated catalysts were activated for HDO experiments by flowing 0.17 – 3.33 $cm^3\ s^{-1}$ H_2 and heating from ambient to 773 K at 0.066 K s^{-1} and holding at 773 K for 0 – 48 h. Samples were then cooled to reaction temperature under the same H_2 flow.

Results and Discussion

Equimolar mixtures of lignin-derived oxygenates including m-cresol, 1,2-dimethoxybenzene, guaiacol, and anisole could be hydrodeoxygenated over molybdenum carbide catalysts at ambient pressure and low temperature conditions (553 K) as shown in Figure 1.[6,7] Nearly 100% conversion of the oxygenate mixture was achieved with >90% selectivity to the corresponding completely deoxygenated aromatics, benzene and toluene. HDO was H_2 efficient as evinced by <10% selectivity to the hydrogenation products of cyclohexane and toluene. Aryl C-O scission occurred over metallic hydrogenolysis sites as inferred from reversible inhibition of anisole HDO by in situ CO co-feed experiments.

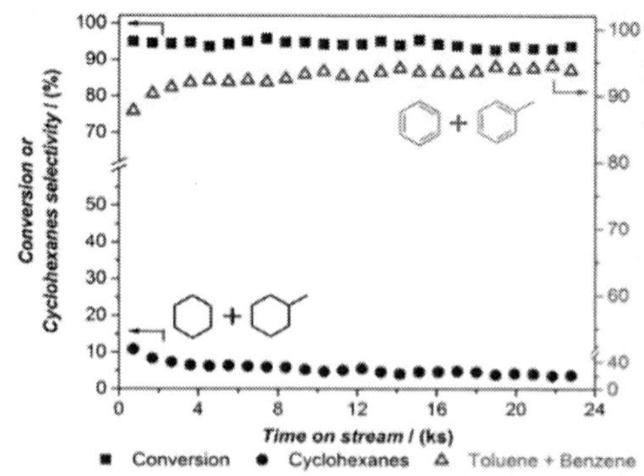

Figure 1. HDO of lignin-derived oxygenates. Ambient pressure, 553 K

Contact time variation studies of acetone and propanal HDO over molybdenum carbide catalysts demonstrated that carbonyl HDO occurs via a sequential, metal-acid bifunctional pathway as shown in Scheme 1. Initial hydrogenation reaches equilibrium at atmospheric pressure under 10-82 kPa H_2 pressure. The rate-determining deoxygenation to form propylene via acidic dehydration was enhanced by surface oxidation with in situ O_2 co-feed; oxidation-induced acid sites on the carbide surface were evinced by irreversible DTBP titrations. Propylene formation kinetics were examined at five different O_2 co-feed pressures. Propylene production rate from IPA dehydration over molybdenum carbide catalysts can be reversibly altered by a factor of 30 via O_2 co-feed pressure variation. Production of propylene and di-isopropyl ether (DIPE) increase proportionally with each other, and both are inhibited by addition of a DTBP co-feed, indicative of Brønsted acid catalysis. A DTBP-normalized propylene production turnover frequency (TOF) of ~0.1 mol s^{-1} $mol_{DTBPsite}^{-1}$ was calculated and shown to be invariant with propylene production rate. Concurrent zero order dehydration kinetics and invariant propylene formation intrinsic activation energy of 93 kJ mol^{-1}, both shown in Figure 2, demonstrate that O_2 co-feed can reversibly alter Brønsted acid site density on molybdenum carbide catalytic formulations. The rate-determining step of unimolecular IPA dehydration to form propylene was found to be scission of the β-hydrogen bond in an elimination mechanism as evinced by a primary kinetic isotope effect (KIE) of 1.86.

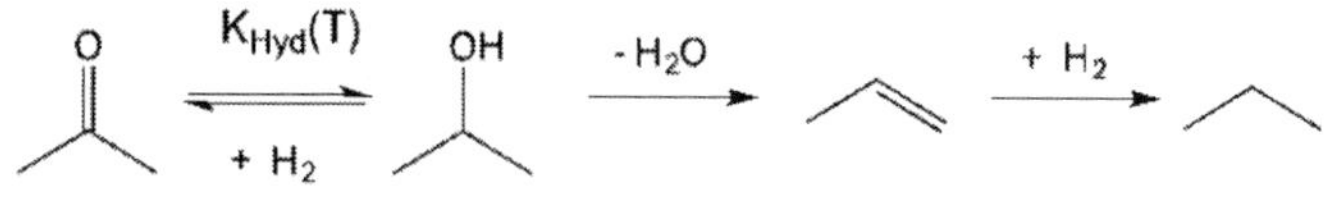

Scheme 1. Proposed acetone HDO reaction pathway.

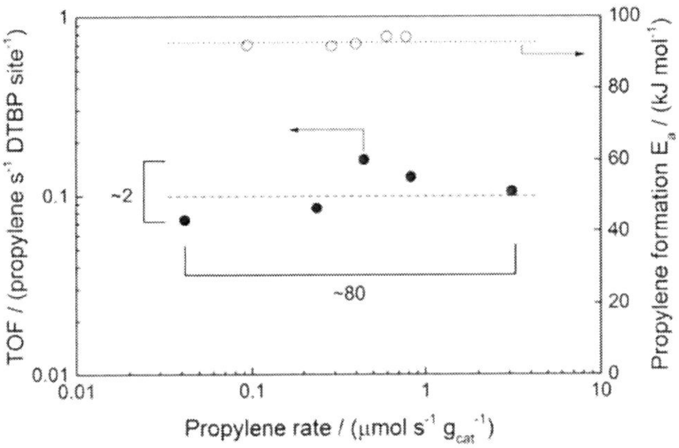

Figure 2. IPA dehydration TOF and activation energy over O*-Mo_2C catalysts at ambient pressure and 415 K.

Conclusions

Oxygen-modified molybdenum carbides denote a class of catalysts that can perform selective HDO of lignocellulosic biomass while retaining H_2 efficiency at atmospheric pressure and low temperature conditions.[7-9] These catalysts can be tuned towards metallic or acidic functionality using oxidative or reductive treatments.

Acknowledgment. We acknowledge financial support from Office of Basic Energy Sciences and the U.S. Department of Energy under award number no. DE-SC0008418 (DOE Early Career Program). We thank Mr. Anurag Kumar for helpful technical discussions.

References

(1) E. Iglesia, J.E. Baumgartner, F.H. Ribeiro, M. Boudart, J. Catal. 131 (1991) 523-544.
(2) T. Matsuda, T. Ohno, H. Sakagami, N. Takahashi, J. Japan Petroleum Institute. 50 (2007) 229-239.
(3) M.M. Sullivan, J.T. Held, A. Bhan, J. Catal. 326 (2015) 82-91.
(4) W.-S. Lee, A. Kumar, Z. Wang, A. Bhan, ACS Catal. 2 (2015) 4104-4114.
(5) J.A. Schaidle, N.M. Schweitzer, O.T. Ajenifujah, L.T. Thompson, J. Catal. 289 (2012) 210-217.
(6) W.-S. Lee, Z. Wang, R.J. Wu, A. Bhan, J. Catal. 319 (2014) 44-53.
(7) C.-J. Chen, A. Bhan, App. Catal. A 510 (2016) 42
(8) M. M. Sullivan, C.-J. Chen, A. Bhan, Catal. Sci. & Tech. 6 (2016) 602
(9) M. M. Sullivan, A. Bhan, ACS Catal. 6 (2016) 1145

OPERANDO STUDIES OF CATALYSIS WITH X-RAY SCATTERING AND SPECTROSCOPY

Randall E. Winans, Tao Li, Sungwon Lee, Sungsik Lee

X-ray Science Division, Advanced Photon Source, Argonne National Laboratory, Argonne, IL 60439
rewinans@anlgov

X-ray methods are valuable tools for *in situ* studies of catalytic reactions and of preparation of catalysts. Structural and electronic information can be obtained over broad length scales using an combination of X-ray spectroscopy (XAS), high energy scattering with PDF, small and wide angle X-ray scattering (SAXS, WAXS) and ultra SAXS (USAXS). Due to the penetrating power of high energy X-rays, it is possible to design cells for high temperature and high pressure in situ experiments. High temperature and pressure nano-metal catalytic cracking of fuel for scram jets is being studied to better understand the deactivation. The synthesis of nano-catalysts protected by atomic layer deposition and activated thermally is being studied. These catalysts are being used in biomass derived molecules upgrading.

Acknowledgment. This research used resources of the Advanced Photon Source, a U.S. Department of Energy (DOE) Office of Science User Facility operated for the DOE Office of Science by Argonne National Laboratory under Contract No. DE-AC02-06CH11357.

OXYGENATES CONVERSION OVER OXIDE SURFACES: INTERPLAY BETWEEN ACID-BASE AND REDOX SITES

Zili Wu and Steven H. Overbury

Chemical Science Division and Center for Nanophase Materials Sciences, Oak Ridge National Laboratory, Oak Ridge, TN 37831.
Email: wuz1@ornl.gov

Oxide catalysts have been widely used in catalytic transformations of oxygenated molecules. Although it has been accepted that the acid base property of oxides play an important role and so is the redox property, it remains unclear how the two type of sites, acid-base and redox ones, are interconnected in the conversion of oxygenates. In this talk, we will use the conversion of model oxygenate compounds over several oxides including ceria, titania and perovskites to show case the interplay between the acid-base sites and the redox sites. We focus on C1-C3 oxygenated molecules and investigate the surface chemistry via *in situ* infrared and Raman spectroscopy while the surface acid-base and redox sites of the oxides are characterized via IR spectroscopy coupled with probe molecules and microcalorimetry. The cooperative effect of the acid-base and redox sites of these oxides will be demonstrated in the oxygenate reactions. The implication for tuning the reaction pathways of oxygenates over oxide catalysts will be also discussed.

Acknowledgements: The work was supported by the U.S. Department of Energy, Office of Science, Basic Energy Sciences, Chemical Sciences, Geosciences, and Biosciences Division. A portion of this research was conducted using facilities at the Center for Nanophase Materials Sciences, which is a DOE Office of Science User Facility.

NOVEL DESIGN OF LIQUID-LIKE NANOPARTICLE ORGANIC HYBRID MATERIALS WITH TUNABLE CHEMICAL AND STRUCTURAL PROPERTIES FOR CO_2 CAPTURE AND CONVERSION

A.-H. Alissa Park

Department of Earth and Environmental Engineering,
Department of Chemical Engineering, and
Lenfest Center for Sustainable Energy,
Columbia University, New York, NY 10027
Lenfest Chair in Applied Climate Science, and
Director of the Lenfest Center for Sustainable Energy
Phone: 212-854-8989, Fax: 212-854-7081,
E-mail: ap2622@columbia.edu

Fossil fuels, which contribute significantly to anthropogenic CO_2 emissions, are likely to remain the main source of energy in the foreseeable future. Therefore, the development of efficient carbon capture and storage (CCS) technologies is one of the great challenges faced by humanity. A new advancement in the area of CCS includes the utilization of captured CO_2 as a chemical feedstock, resulting in new approaches to carbon capture, utilization and storage (CCUS). While a wide range of materials ranging from aqueous amine solvents to solid sorbents are being developed, a reactive system that can effectively host both CO_2 capture and conversion do not exist. Thus, a new class of nano-scale materials called Nanoparticle Organic Hybrid Materials (NOHMs) has been developed as an alternative CO_2 capture media that also has a potential to host subsequent CO_2 conversion. NOHMs consist of inorganic cores with a grafted polymeric canopy providing negligible vapor pressure, great thermal stability, and interesting chemical and physical tunability. Both entropic and enthalpic interactions among polar fluid units, solid surfaces and gaseous molecules are investigated, while correlating the molecular level understanding macroscale properties such as the sorption and diffusion behaviors and reactivity of CO_2 molecules in NOHMs. NOHMs' unique electrolytic properties in the presence of various secondary fluids and salts are evaluated to design them as a dual-purpose reactive media for combined CO_2 capture and conversion. This CCUS scheme would capture and convert CO_2 in an integrated pathway, and therefore, has a great potential to minimize the parasitic energy consumption.

GCMC SIMULATIONS OF NITROGEN-DOPED MICRO AND MESOPOROUS CARBON SORBENTS FOR CO_2 CAPTURE

Peter Psarras, Jiajun He, and Jennifer Wilcox

Department of Energy Resources Engineering
367 Panama Street, Stanford, CA 94305
Email: jen.wilcox@stanford.edu

Abstract

A present challenge in the mitigation of anthropogenic CO_2 emissions involves the design of less energy- and water-intensive capture technologies. Traditional capture methods, e.g., MEA solvent scrubbing, have prohibitive regeneration costs, and recent concerns in drought-ridden states like California have brought these water-intensive processes under increased scrutiny. Sorbent-based capture represents a promising solution to the aforementioned hurdles, as they do not incur the heavy energy penalties associated with solvent regeneration. However, to be considered competitive with their more mature, solvent-based counter-parts, these sorbents must exhibit both high CO_2 loading capacity and high CO_2:N_2 selectivity. Having adequate microporosity and surface nitrogen functionality have been reported to be of great importance to the enhancement of CO_2 capacity and CO_2:N_2 selectivity; however, the role of pore size in combination with surface functionalities in the enhancement of these properties remains unclear. To investigate these effects, grand canonical Monte Carlo (GCMC) simulations were carried out on pure and functionalized 3-layer graphite slit pore models. Theoretical isotherms were constructed as the PSD-weighted sum of individual slit-pore models of width ranging from 3.5Å to 200Å. In addition to pure and monovacancy-graphite, the following functional groups were isolated for testing at ~3.0 N wt%: pyridinic nitrogen, pyrollic nitrogen, quaternary nitrogen, pyridine-N-oxide, and two tautomers of pyridonic nitrogen (2-hydroxypyridine and 2-pyridone). Our results indicate that enhancements to CO_2 loading and CO_2:N_2 selectivity increased on the order of pure graphene < monovacancy < pyrrolic < 2-pyridone < 2-hydroxypyridine < pyridinic < quarternary < pyridine-N-oxide, with upwards of *ca.* 244% increase to the 0.1 bar loading capacity over the unmodified surface. These results correlate directly with the isosteric heats of adsorption, with stronger heats corresponding to greater CO_2 loading (up to 1.72 mmol/g for pyridine-N-oxide) and higher IAST selectivity (up to 47:1 for pyrindine-N-oxide). Further enhanced loading is achieved by increasing the N wt%.

COMPREHENSIVE CHARACTERIZATION OF OXYGEN HETEROATOM-CONTAINING SPECIES IN COMPLEX PETROLEUM MIXTURES

Ryan P. Rodgers[1,2,3], *Yuri E. Corilo*[1,3], *David C. Podgorski*[3], *Steven M. Rowland*[3], *Priscila Lalli*[3], *Amy C. Clingenpeel*[1,2], *and Winston K. Robbins*[3]

[1]National High Magnetic Field Laboratory, Florida State University, 1800 East Paul Dirac Drive, Tallahassee, FL 32310 USA
[2]Department of Chemistry and Biochemistry, 95 Chieftain Way, Florida State University, Tallahassee, FL 32306 USA
[3]Future Fuels Institute, Florida State University, 1800 East Paul Dirac Drive, Tallahassee, Florida 32310 USA

The molecular-level analysis of petroleum samples by high field Fourier Transform Ion Cyclotron Resonance Mass Spectrometry (FT-ICR MS) has enabled the direct characterization of tens-of-thousands heteroatom-containing species that contain nitrogen, sulfur, and oxygen. Nitrogen-containing species exhibit the lowest functional diversity of N, S, and O-containing species, and are largely limited to pyridinic and pyrrolic species. Sulfur-containing species are more diverse, but can be addressed by chromatographic separation prior to mass spectral characterization. Thus, when coupled to modern ionization methods, which provide coarse specificity of chemical functionality, the most abundant chemical forms of sulfur and nitrogen-containing species can be exhaustively characterized. However, oxygen-containing species pose considerable difficulty, in the diversity of chemical forms found in petroleum and pyrolysis products, undesirable increase in number and relative abundance due to unintentional oxidation (in storage and ionization process), tendency to form poly-functional species, and the associated difficulty in chromatographic fractionation. Although readily accessed by (-) electrospray ionization (ESI), the direct mass spectral analysis of acidic oxygen-containing species remains a challenge due to preferential ionization of low molecular weight species and (undesirable) formation of non-covalent adducts. These difficulties are unfortunate, as many field and laboratory studies have implicated oxygen-containing species in a number of important upstream and downstream issues. Here, we outline our current applications of heteroatom speciation in complex petroleum samples and highlight the importance of chromatographic fractionations in the exhaustive characterization of oxygen-containing chemical functionalities.

Acknowledgments. Work supported by NSF Division of Materials Research through NSF (DMR-11-57490 and CHE-10-19193), BP/The Gulf of Mexico Research Initiative to the Deep-C Consortium, the Florida State University Future Fuels Institute, and the State of Florida.

EXTENDED DURATION LOW TEMPERATURE CONVERSION OF OIL SHALE

Alan L. Chaffee[a], Yi Fei[a], Marc Marshall[a], W. Roy Jackson[a], Martin L. Gorbaty[b] and Peter J. Cassidy[c]

[a] School of Chemistry, Monash University, Victoria, Australia
[b] Fuels Science Consulting, Lake Worth, FL, USA
[c] The Sentient Group, Grosvenor Place, Sydney, NSW, Australia
alan.chaffee@monash.edu

There has been long-standing interest in deriving liquid products from oil shales. This includes the application of 'in-situ' retorting processes, which have the potential to avoid environmental issues at the surface, for example, by trapping the evolved CO_2 underground.

We have studied reactions of Green River oil shale, as powders and as chunks (1-5 cm^3), for extended periods (up to 28 days) and at low temperatures (upwards from 280°C). Liquid product yields of up to 70 wt% dmmf could be obtained from the chunks at temperatures as low as 320°C; however large variations in the yields were observed for reactions below 330-340°C and were attributed to varying mineralogy in the chunks. Such variation was not apparent for powdered samples or for the reactions of chunks above 330-340°C, due to their disintegration. Particularly at lower temperatures yields were higher under N_2 than under a CO atmosphere.

The liquid products did not vary markedly with yield, temperature or time. They contained 1-2 wt% nitrogen and were highly aliphatic, but with a significant amount of heavily substituted aromatic rings.

The activation energies determined from these extended duration experiments (~75 kJ/mol) were much lower than those obtained from conventional TGA experiments (~220 kJ/mol).

DECIPHERING COMPLEX BIOMASS UPGRADE PROCESSES WITH IN-SITU SPECTROSCOPIES

Brian Murphy, Nicholas Gould and Bingjun Xu

Catalysis Center for Energy Innovation
Department of Chemical and Biomolecular Engineering
University of Delaware, 150 Academy St., Newark, DE

Abstract

The complexity of biomass upgrade processes, both in terms of the chemical transformation and composition/structure of catalysts, makes the molecular level understanding of surface-mediated reactions challenging. In-situ spectroscopic techniques are critical in characterize catalyst structure and identify reaction intermediates. Two case studies will be discussed in the context of how in-situ spectroscopic tools could play a central role in elucidating reaction mechanism at solid/vapor and solid/liquid interfaces. In-situ transmission spectroscopy will be used to understand how lactic acid and methyl lactate interact with metal cation exchanged zeolites in their selective dehydration, and in-situ liquid phase ATR spectroscopy will be used to characterize zeolite catalysts in the presence of different solvents and understand solvent effects.

In-situ generated Bronsted acid sites were identified on NaY upon introduction of organic acids or ester/water mixture, which is formed via a proposed ion-exchange mechanism between adsorbed species and Na cations on the catalyst. These in-situ generated is correlated with the rate of the undesired reaction of decarbonylation of lactic acid and methyl lactate to acetaldehyde. In contract, the dissociation of lactic acid and methyl lactate to form the adsorbed sodium lactate has been identified as a precursor for the desired dehydration reaction to acrylic acid. Solvent has been shown via the liquid phase ATR spectroscopy to play a key role in the interaction between the probe molecule, pyridine, with zeolites, which is correlated with the catalytic activity of hydrolysis of dimethylfuran.

EXPLORING AND VISUALIZING COAL CHAR OXY-FUEL COMBUSTION WITH LARGE-SCALE (BUT SIMPLIFIED) ATOMISTIC SIMULATIONS

Jonathan P. Mathews

The EMS Energy Institute and Department of Energy and Mineral Engineering, The Pennsylvania State University, University Park, Pennsylvania 16802
jmathews@psu.edu

Abstract

While coal structure has been well explored with atomistic representations there is much less work on coal char despite its importance in coal use. Here the construction strategy that allows rapid construction of large-scale coal char structures (<50,000 atoms) that capture physical as well and chemical properties is proposed using an image-guided approach from HRTEM lattice fringe micrographs. These chars can capture the relative reactivity differences that are explored using a simplified simulation utilizing diffusional, atom deletion iterative stepwise approach with extensive data capture.

CARBON EMISSION REDUCTION POTENTIAL OF HYBRID ENERGY SYSTEMS FOR COAL CHEMICAL INDUSTRY IN CHINA

ZhiYong Tang[a], QianQian Chen[a], Min lv[a], Lingzhao Kong[a], YuHan Sun[a,b,]*

a CAS Key Laboratory of Low-Carbon Conversion Science and Engineering, Shanghai Advanced Research Institute, Chinese Academy of Sciences, Shanghai 201210, China
b ShanghaiTech University, Shanghai 201210, China
* Corresponding authors e-mail addresses: sunyh@sari.ac.cn

Introduction

Coal Chemical Industry as the major part of China industry, its carbon emission intensity is 4.5 times of the average industrial emission intensity, and much higher than the domestic average carbon emission intensity [1]. According to future development plan, Coal Chemical Industry will develop rapidly in the next 15 years, the related carbon emission will reaches 1.7 billion tons by 2030 and emission intensity will be 14.2 tons CO_2 per 10000 CNY of GDP. Zhu Liu[2] predicts the CO_2 emission will keep increasing at annual rate of 3% and excess 11.5 billion tons in 2020 in the absence of government policy intervention, and may even exceed 15 billion tons in 2030. To control the carbon emissions, Chinese government has set out target to control the greenhouse gas emissions, the domestic CO_2 emission intensity shall be reduced by 40%~45% in 2020 and 60%~65% in 2030[3] based on the value in 2005, and an emission peak should be reached in 2030. Along with the urgent demanding of carbon reduction, a low carbon strategy is vital for the sustainable development of Coal Chemical Industry.

Feasibility of hybrid energy system for Coal Chemical Industry

The carbon emission of Coal Chemical Industry mainly relates to indirect CO_2 emission from heat/electricity supply and direct CO_2 emission from coal chemical conversion process, such as water gas shift reaction[4]. After coal gasification, water gas shift process is often used to increase the H/C ratio of syngas (CO and H_2) by consuming a part of C, during this process, more than 50% of the carbon turns into CO_2. The present work concerns with direct carbon emission, proposing a low carbon technology roadmap to constrain the carbon emission in Coal Chemical Industry. In this system (see Fig.1), nuclear/renewable energy is utilized in the form of heat and electricity to supply low/high temperature water electrolysis [5-10], hydrogen produced from electrolysis unit will be used to adjust the H/C ratio for syngas conversion processes (Fischer-Tropsch synthesis, methanol to olefin, coal to SNG and etc.), by eliminating water gas shift process the system CO_2 emission can be significantly reduced. Recently, this type of hybrid systems have attracted broad attentions world widely, American Idaho and NREL national laboratory has intensively studied the hydrogen production system via nuclear and renewable energy, and conducted techno-economic feasibility analysis for integration of clean energy system with fossil based system.

Based on the geographical distribution of carbon emission, fossil fuel resources (coal and gas) and renewable energy (wind, solar and biomass), as well as the government's plans of nuclear power projects, four large scale low-carbon integrated systems are proposed (see Fig. 2):

1) North-west China：integration of solar energy, wind energy , natural gas with coal.
2) East-central China: integration of biomass with coal.
3) North China: integration of natural gas, wind, biomass with coal.
4) East coastal area: integration of nuclear energy, wind energy with coal.

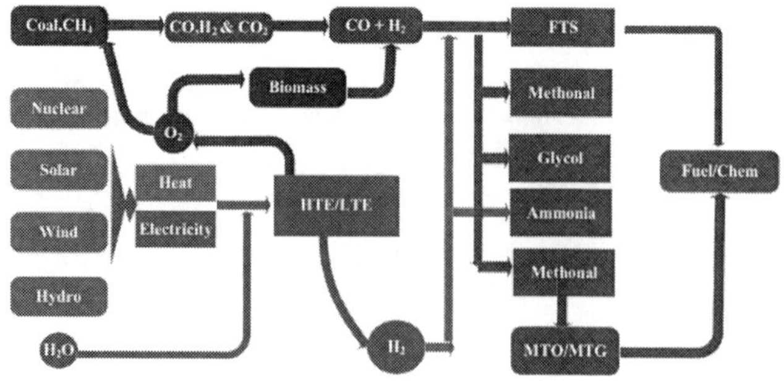

Figure1. Low carbon technology roadmap for fuel/chemical production in Coal Chemical Industry.

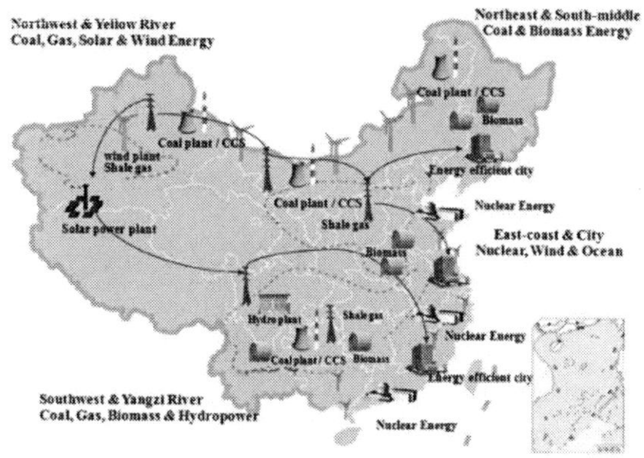

Figure2. Feasible implementation of various hybrid systems based on the geographical distribution of energy resources.

Carbon emission reduction potential of hybrid energy systems

Based on geographical distribution of energy resources and coal chemical industry future planning (moderate growth or high growth rate), scenario analysis was performed to study the carbon emission reduction potential of hybrid energy system, as given in Fig.3. Assumptions include the operating rates of conventional and new coal chemical industry, as well as the replacing rate of system integration (see Table 1). Under high growth rate of coal chemical industry in the year 2015-2020, the accumulative carbon reduction of hybrid system can achieve to 624 million.

From the perspective of CO_2 emission intensity, assuming the replacing rate of hybrid system is 40% in 2020 and 80% in 2030, Prediction indicates in 2020 the hybrid system CO_2 emission is 647Mt and emission intensity is 10.2 tons CO_2 per 10000 CNY of GDP, which is 24% less than the conventional system; in 2030, the hybrid system CO_2 emission is 503Mt, the emission intensity is 4.3 tons CO_2 per 10000 CNY of GDP, which is 70% less than the conventional system, and 69.7% less than the emission intensity in 2014.In future, with the increase of operation efficiency of existing coal chemical plants and higher integration rate of nuclear/renewable energy, the contribution of hybrid energy system towards China CO_2 mitigation will become more significant.

The economic analysis of hybrid energy systems

A comprehensive hybrid energy system for various fuel/chemical productions is constructed and the process flow chart is illustrated in Fig. 4. In this system, nuclear/wind/solar assisted water electrolysis is applied to replace water gas shift process in order to adjust the H/C ratio of syngas, hence eliminate water gas shift process and constrain the carbon emission. Besides, biomass can be gasified to produce fuel and chemicals, the amount of CO_2 absorbed during the biomass growth is almost equivalent to the amount of CO_2 produced during the combustion process, hence the biomass integrated system can achieve almost zero carbon emission.

Table 1 summary of assumptions used in scenario analysis

	Operating rate of New coal chemical industry		Operating rate of conventional coal chemical industry	Technology replacing rate
	High growth	moderate growth		
2015				10%
2016				15%
2017	100%	50%	100%	20%
2018				25%
2019				30%
2020				40%

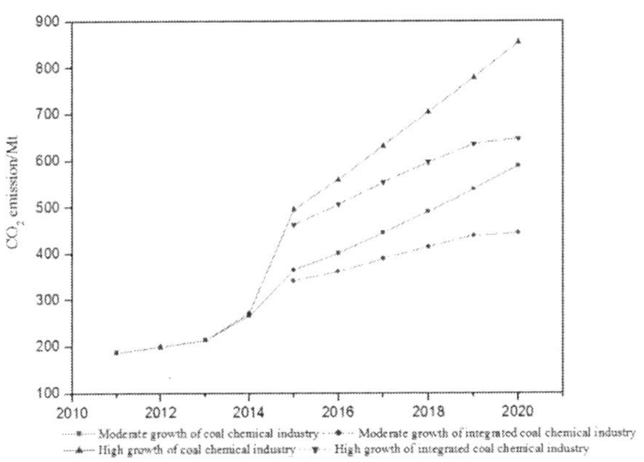

Figure3. Prediction of the carbon emission reduction potential of hybrid energy systems under different scenarios

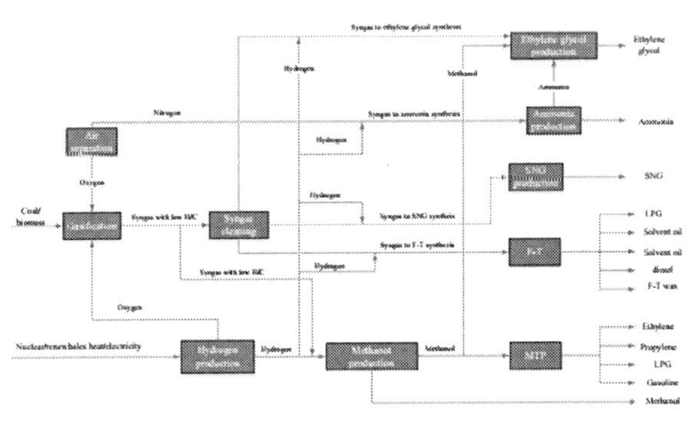

Figure 4. Process flow chart of hybrid energy system for fuel/chemical production.

From economic point of view, under the current market scenario, to produce fuel/chemical by hybrid system has no economical benefit compared to conventional coal system. However breakeven point may be achieved for certain systems like nuclear energy/ biomass integrated hybrid energy systems, depending on the market prices of the products and raw materials. Taking in consideration of lower nuclear/solar/wind electricity prices in long run along with the introduction of carbon tax and development of technology, the hybrid system will become economically competitive compare to the conventional coal system.

Conclusion

Base on the high-carbon nature of Coal Chemical Industry in China, possibility of replacing conventional coal system with hybrid energy system is analyzed from both geographical and sector view, and further quantifies the potential of carbon emission reduction. And the economical aspect of producing fuel/chemical products via hybrid system is also computed under the current energy prices. Base on the geographical distribution of carbon emission resource, Coal Chemical industry and energy resources in China, hybrid system has great potential to integrate with conventional system in order to meet the increasing demand of industry production and CO_2 mitigation. In addition, the complementary feature of geographical distributions of fossil resource, nuclear/renewable energy provides superior conditions to the implementation of hybrid energy system.

Base on the development plan of Coal Chemical Industry in China, under hybrid system, the CO_2 emission and intensity of the emission will be reduced by 24% by year 2020. And by 2013, the CO_2 emission intensity will reach 4.3 tons CO_2 per 10000 CNY of GDP, which is 69.7% reduction from 2014 (14.2 tons CO_2 per 10000 CNY of GDP). In future, with the increase of operation efficiency of existing coal chemical plants and higher integration rate of nuclear/renewable energy, the contribution of hybrid energy system towards China CO_2 mitigation will become more significant.

The low-carbon hybrid energy system proposed in present research integrates the utilization of nuclear/renewable energy, clean conversion of fossil resource, and CO_2 recycling, the related knowledge is enormous and require expertise in various technologies. With supports receiving from China governments, the key technologies are developing rapidly and some are ready to proceed into industrial demonstrations, this provides strong technical support for the implementation of hybrid energy systems. To conclude, the hybrid energy systems can meet the increasing energy demand while strictly constrain the CO_2 emission, it is an effective solution for sustainable energy future and will have profound prospect.

References

(1) National Bureau of Statistics, *China Statistical Yearbook-2013*. China Statistical Publishing House: Beijing, 2014.
(2) Zhu, L., Dabo, G., Scott, M., Henry, L., Jun, S., Qiang, Z., *Nature*, **2015**, *522*(7556), 279-81.
(3) National Development and Reform Commission, *China's national plan on climate change(2014-2020)*,2014
(4) Chen, Q. Q., Tang, Z. Y., Lei, Y., Sun, Y. H., Jiang, M. H. *Appl. Energ.*, **2015**, *158*, 619-630.
(5) O'Brien, J. E., Mckellar, M. G., Harvego, E. A., Stoots, C. M., *Int. J. Hydrogen Energ.*, **2010**, *35*(10), 4808-4819.
(6) Zhang, X., O'Brien, J. E., Tao, G., Zhou, C., Housley, G. K., *J. Power Sources*, **2015**, *297*, 90-97.
(7) Menanteau, P., Quéméré, M. M., Duigou, A. L., Bastard, S. L., *Energ. Policy*, **2011**, *39*(5), 2957-2965..
(8) Abbasi, T., Abbasi, S. A., *Renew. Sust. Energ. Rev.*, **2011**, *15*(6), 3034-3040.
(9) Wang, J., Feng, L., Zhao, L., Snowden, S., *Energ. Policy*, **2013**, *55*(3), 690–698.
(10) Wang, B., Ke, R. Y., Yuan, X. C., Wei, Y. M. *Renew. Sust. Energ. Rev.*, **2014**, *40*, 185-195.

DESIGN OF HIGH-EFFICIENCY MnO_2-LIKE CATALYST AND EXPLORATION ON ITS VOCS CATALYTIC COMBUSTION PERFORMANCE

Wenzhe Si, Yu Wang, Yue Peng, and Junhua Li

State Key Joint Laboratory of Environment Simulation and Pollution Control, School of Environment, Tsinghua University, Beijing, 100084, P. R. China
Corresponding author: Junhua Li, lijunhua@tsinghua.edu.

Introduction

With the development of industry and transportation, the emissions of volatile organic compounds (VOCs) have become a notable problem, due to the adverse effects both on the environment and human health.[1-2] Catalytic combustion is an effective technology for the elimination of VOCs because of its high efficiency and low operating temperature.[3-4] The design of high-performance catalysts by facile and universal method is the key point of research on this technology.

Perovskite (ABO_3) is considered as a promising candidate in VOCs catalytic combustion because of its good catalytic activity and high stability. However, some disadvantages still exist, restricting their catalytic performance, for example, the perovskites usually possess low surface area, and the native surfaces are preferentially occupied by A-site cations, which are not catalytically active.[5] If A-site cations can be selectively removed from ABO_3, the active B-site cations will be remained, and the catalytic activity of the obtained material may be increased significantly.

Herein, we investigate the possibility of selectively dissolution on A-site cations from $LaMnO_3$ (LMO) perovskites by acid treatment. A three-dimensionally ordered macroporous $LaMnO_3$ (3DOM-LMO) material was firstly prepared to increase the contacting surface areas of LMO with acid. After selective dissolution, a novel γ-MnO_2-like catalyst was born with large surface area and plenty of mesoporous. What's more, the amount and mobility of surface oxygen species were increased, which provided the novel catalyst an excellent catalytic activity on toluene combustion.

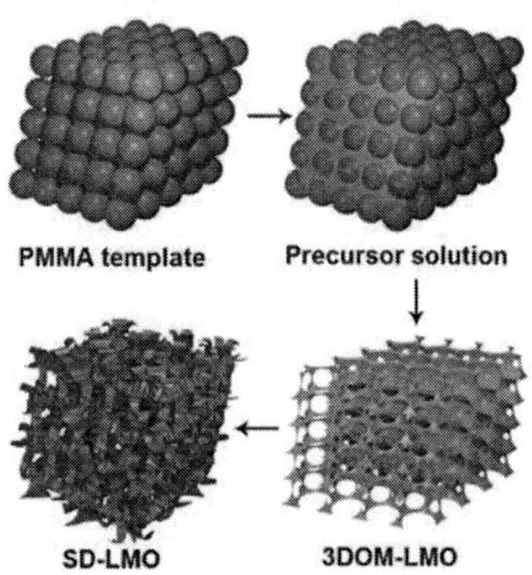

Scheme 1. Representation of the selective dissolution procedure.

Experimental

Firstly the 3DOM-LMO sample was prepared by using polymethyl methacrylate (PMMA) microspheres as the hard template. Then a certain amount of 3DOM-LMO powder was added to a diluted HNO_3 solution. When no obvious bubble was emerging, the liquid was filtered and the left solid was washed with deionized water until the pH was neutral. The sample was dried at 60 °C in air, and the obtained sample was denoted as SD-LMO. The synthesis route is shown in Scheme 1. A γ-MnO_2 catalyst was prepared by hydrothermal synthesis as a contrast.

Results and Discussion

Figure 1 shows the X-ray diffraction (XRD) results of the three samples. After the acid treatment, the diffraction peaks assigned to perovskite phase disappeared, and a γ-phase MnO_2 was born. This suggests that the dilute HNO_3 can selectively dissolute La cations in LMO perovskites, leaving behind MnO_2. However, a small amount of La cations still existed with good distribution in the SD-LMO sample, according to ICP (Table 1) results. The residual La cations could attract O anions, and the Mn-O bonds would be weakened and become longer. Then the crystal size of the SD-LMO would be larger than that of the traditional γ-MnO_2. As a result, a downshift of diffraction peaks for SD-LMO can be found, compared to the γ-MnO_2 in the XRD patterns.

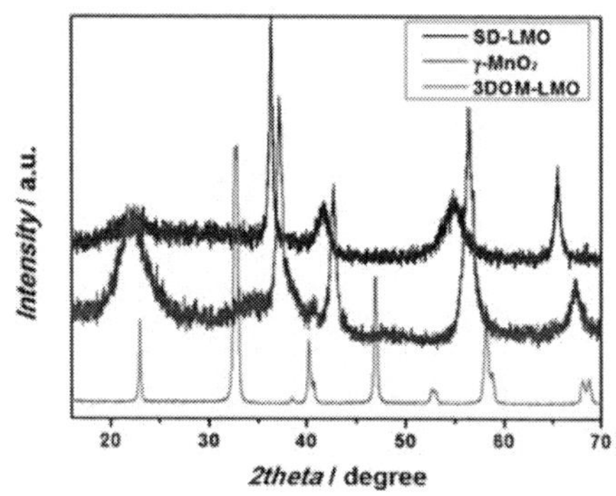

Figure 1. XRD patterns of the three samples.

The morphological changes of the samples could be obviously revealed in Figure 2. After the acid treatment, destruction of the 3DOM structure could be observed (Figure 2d, e). The original smooth surface of the macropores became completely rumpled and plenty of mesoporosity appeared, which significantly increased the surface area of SD-LMO. The morphological change with the acid treatment could be attributed to the shrinkage in the volume of the treated material, as shown from Figure 2c to 2f. Theoretically, there is 23.9% shrinkage in volume from $LaMnO_3$ to γ-MnO_2.[6] Lots of gaps emerged as a result of the shrinkage occurring in the SD-LMO sample, which brought plenty of nanopores.

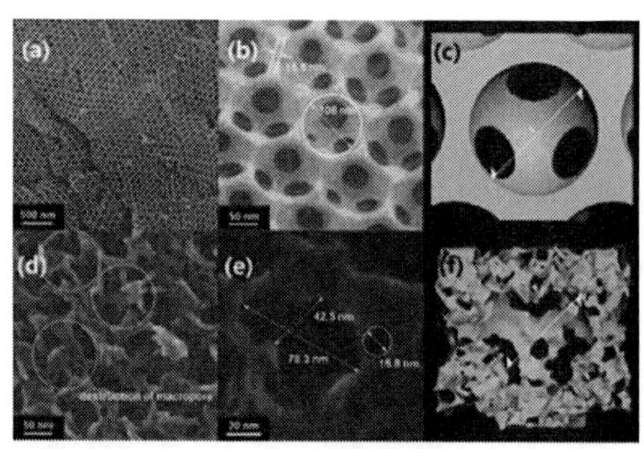

Figure 2. SEM images and schematic diagrams of (a-c) 3DOM-LMO and (d-f) SD-LMO.

Table 1. ICP, surface element compositions, BET surface areas of the three samples.

	ICP	XPS		BET surface area (m^2/g)
	La/Mn molar ratio	Mn^{4+}/Mn^{3+} molar ratio	O_{ads}/O_{latt} molar ratio	
SD-LMO	0.04	1.02	0.97	245.7
3DOM-LMO	1.18	1.64	0.63	22.4
γ-MnO_2	0	0.75	0.36	68.9

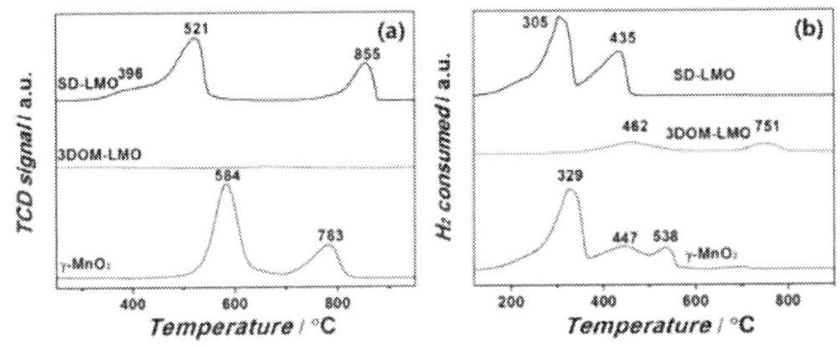

Figure 3. (a) O_2-TPD profiles and (b) H_2-TPR profiles of the three samples.

Mn $2p_{2/3}$ and O 1s XPS results are summed up in Table 1. The surface Mn^{4+}/Mn^{3+} molar ratio increased from 3DOM-LMO to SDLMO after the acid treatment, which was caused by the following reaction: $2Mn^{3+}_{solid} \rightarrow Mn^{4+}_{solid} + Mn^{2+}_{liquid}$. Moreover, the Mn^{4+}/Mn^{3+} molar ratio in SD-LMO (1.02) was less than that in γ-MnO_2 (1.64), which was resulted from the remaining La cations in SD-LMO. The charge imbalance caused by additional La cations in SD-LMO can be neutralized by either Mn^{3+} or adsorbed oxygen (O_{ads}). The Mn^{4+}/Mn^{3} molar ratio was reduced and the O_{ads} was increased after the acid treatment, which could also be confirmed by O 1s XPS results.

The O_2-TPD experiments were carried out to further detect the changes of oxygen species after the acid treatment. The oxygen desorption at low temperature can only be observed in SD-LMO. The amount of O_{ads} in SD-LMO was much more than those in the other samples, which was consistent very well with the O 1s XPS results. Furthermore, the the lattice oxygen (O_{latt}) desorption temperature of SD-LMO (521 °C) was lower than that of γ-MnO_2 (584 °C). It is indicated that the O_{latt} in SD-LMO is released in a facile manner. The reason is that the Mn-O bonds in SD-LMO were weakened and got longer as a result of the residual La cations, which caused the Mn-O bonds easier to be broken. Figure 3b illustrates the H_2-TPR profiles of the three samples. In the case of 3DOM-LMO sample, there were two main reduction peaks at 462 and 751 °C. After selective dissolution, the reduction temperature of SD-LMO (305 °C) was lowest among the three samples, which suggested the best low-temperature reducibility and O_{latt} mobility.

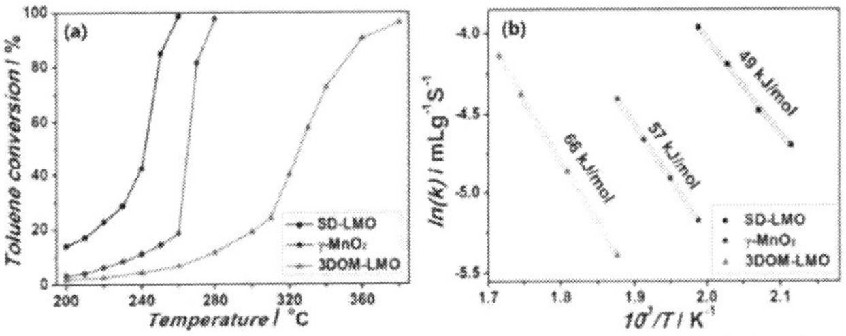

Figure 4. (a) Activity profiles and (b) Arrhenius plots of the three samples.

The catalytic performances of the three samples on the toluene combustion are shown in Figure 4. T_{50} (the half conversion temperature) for sample 3DOM-LMO, γ-MnO_2, SD-LMO are 322, 269 and 246 °C and T_{90} (the temperature of the conversion 90%) are 354, 270 and 258 °C, respectively. The catalytic activities increased in the sequence of 3DOM-LMO < γ-MnO_2 < SD-LMO. The apparent activation energy (*E*a) was calculated to eliminate the effect of the surface area. From Figure 4b, the *E*a value of SD-LMO was smaller than those of the others. It is universally acknowledged that the oxidation of VOCs over transition metal oxide catalysts take place on the basis of the Mars–van Krevelen type redox cycle, and the nucleophilic attack of the O_{latt} result in the occurrence of this reaction.[7] Thus the O_{latt} species play a vital role in catalytic reaction. The residual La cations in SD-LMO sample caused by the acid treatment made the O_{latt} release more easily, which enhanced the O_{latt} mobility on the catalyst, bringing an improvement on the Mars–van Krevelen mechanism. As a result, the E_a value of SD-LMO was the lowest among these samples.

Conclusions

We have developed a facile method to synthesize γ-MnO_2-like catalyst utilizing LMO perovskites as initial precursor. The novel catalyst exhibited large surface area, plenty of mesoporous, and excellent low-temperature reducibility. In particular, the amount of O_{ads} and the mobility of O_{latt} were increased significantly after the acid treatment. All of these promoted the novel catalyst an outstanding catalytic activity in toluene combustion. Our findings not only suggest a new avenue for perovskite-based manganese oxides to improve their catalytic activities on heterogeneous catalysis, but also supply a method to synthesize novel γ-MnO_2-like materials, which may possess huge potential in other fields.

Acknowledgement. This project is supported by the National Natural Science Foundation of China (21325731, 51478241, and 21221004) and the National High-Tech Research and Development (863) Program of China (2013AA065304).

References

(1) Park, J. H.; Goldstein, A. H.; Timkovsky, J.; Fares, S.; Weber, R.; Karlik, J.; Holzinger, R. *Science*, **2013**, *341*, 643-647.

(2) Yoon, J.; Chae, S. K.; Kim, J.-M. *J. Am. Chem. Soc.*, **2007**, *129*, 3038-3039.

(3) Kim, S. C.; Shim, W. G. *J.Hazard. Mater.*, **2008**, *154*, 310-316.

(4) Wang, R.; Li, J. *Environ. Sci. Technol.*, **2010**, *44*, 4282-4287.

(5) Royer, S.; Duprez, D.; Can, F.; Courtois, X.; Batiot-Dupeyrat, C.; Laassiri, S.; Alamdari, H. *Chem. Rev.*, **2014**, *114*, 10292-10368

(6) Si, W.; Wang, Y.; Peng, Y.; Li, J. *Angew. Chem. Int. Ed.*, **2015**, *54*, 7954-7957.

(7) Qu, Z.; Bu, Y.; Qin, Y.; Wang, Y.; Fu, Q. *Appl. Catal. B*, **2013**, *132–133*, 353-362.

NEXT GENERATION OF LITHIUM-ION BATTERY MATERIALS WITH HIGH ENERGY

Jinbao Zhao[*1,2], Jing Wang,[1] Bing Li[2], Yang Yang[1], Jingxin Huang[1]*

[1] State Key Lab of Physical Chemistry of Solid Surfaces, Department of Chemistry, Collaborative Innovation Center of Chemistry for Energy Materials, College of Chemistry and Chemical Engineering;
[2] College of Energy;
Xiamen University, Xiamen, 361005, China,
Email: jbzhao@xmu.edu.cn

Introduction

Rechargeable lithium-ion batteries (LIBs) are widely used in consumer electronics, and applied in electrical vehicles and renewable energy storage with great potential.[1-2] However, current LIB technology cannot satisfy the high energy demands of electrical vehicles because both the commercial cathode and anode materials have almost reached their theoretical specific capacities. Advanced materials are required to offer higher energy density, better safety and longer lifetime. At present, the best candidates for cathode materials are those either with high operating voltage or high capacity. Therefore, we choose $LiNi_{0.5}Mn_{1.5}O_4$ (LNMO) with a discharge voltage of 4.7 V [3] and layered lithium-riched manganese based materials [4] with a specific capacity of 250 mAh g^{-1} as research object for cathodes materials. Among the investigated high-energy density anode materials, silicon (Si) with a theoretical capacity of 4200 mAh g^{-1}, as one promising candidate for the next generation anodes, is thoroughly studied.[5]

Conclusions

For LNMO, our work was concentrated on three major parts. Firstly, micro-sized spherical LNMO, LMO, and intergrown LMO-LNMO I comprised of aggregated nano-polyhedrons were synthesized by an impregnation method. The influence of trace amount of Mn^{3+} in LNMO on the electrochemical properties was investigated. LNMO-Air with moderate Mn^{3+} existing exhibited superior specific capacity, cycling ability and rate capability. Thermostability, capacity and rate performance of the intergrown LMO-LNMO I were improved much better comparing to LNMO and LMO. In this special structure, LNMO acted as a skeleton to stabilize the structure of LMO, enabling more lithium ions in LMO to participate in charge-discharge process along with this process in LNMO, which led to high specific discharge capacities. In addition, LMO-LNMO I show a more homogeneous ion mixing at the atomic level of Ni and Mn in the structure, which improved the rate capacity. Also, a more enhanced thermostability was obtained due to its higher Mn content than LNMO. Secondly, the Na-doped LNMO was firstly synthesized via a sample method. The electrical conductivity and cycling performance were improved as a result. The doping of Na ions not only increased the disordered distribution of nickel and manganese cations in spinel, but also increased two additional electron hopping paths, which contributed to a better charge transfer ability, relieved the ohmic polarization and electrochemical polarization of materials and improved lithium ion diffusion coefficient. Thirdly, V_2O_5-coated LNMO were firstly demonstrated. It was concluded that the lithium ion diffusion coefficient was increased nearly doubly and dissolution of metal ions in electrolyte was alleviated.

For lithium-rich layered manganese based materials (LLRM), our work was focused on four aspects. Firstly, a facile carbonate co-precipitation method was used to obtain a family of the spherical cathode material, $xLi_2MnO_3\cdot(1-x)LiMn_{1/3}Co_{1/3}Ni_{1/3}O_2$ (x=0.3, 0.5, 0.7). The x=0.5 sample can deliver 260.8 mAh·g^{-1} with an initial coulomb efficiency of 80.2%. Secondly, the influence of the ionic radius on the voltage decay was discussed. The LLRM suffered from voltage decay on cycling, which was considered to be related to the structural transformation, accompanied by cation migration. The material with the smaller size ion (Al^{3+}) doped showed the same voltage decay as the pristine sample, while the voltage decay was alleviated in the material doping by the bigger size ion (Ru^{3+}). Owing to the bigger size dopant ions, the volume of the tetrahedron interspace shrank, leading to the cation migration suppressed. Therefore, the voltage decay was relieved. Meanwhile, the Al-O bond energy, which is stronger than the Mn-O's, can decrease the first lap irreversible capacity. Thirdly, the cation migration is an intrinsic property of LLMR during the charge/discharge cycle, especially at high temperature. The dissolution of transition metal ions may accelerate the voltage decay. It was found that the material with the SnO_2 coating layer showed a lower dissolution of transition metal ions with a better electrochemical performance. Furthermore, the materials with the LLMR@spinel-like@coating layer composite structure presented high rate performance. Fourthly, in-situ electrochemical Raman spectroscopy has been developed and executed to further investigate the structural evolution of the Li-rich layered material ($0.5LiNi_{0.5}Mn_{0.5}O_2\cdot 0.5Li_2MnO_3$). We found that the reactions of Ni^{2+} to Ni^{3+} and Ni^{3+} to Ni^{4+} appearing in the potential range of from 3.70 V to 4.45 V showed a good reversibility. The in-situ Raman spectra revealed the structural evolution of the local coordination structure and the increased disorder of the ionic arrangement. Such a structural evolution led to the deterioration of the stability of the material and the electrochemical performance. According to this finding, it may be meaningful to design a material with a structure that can only be activated in a designed degree or that has a solid framework to stabilize the material.[6]

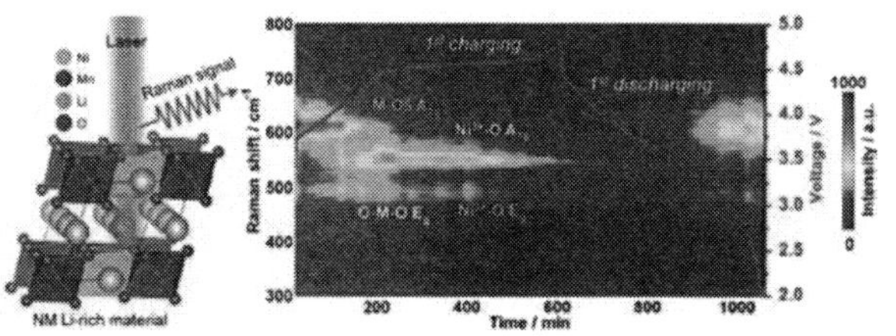

Figure 1. Simultaneous in situ Raman and electrochemical measurements of the Li-rich layered material.

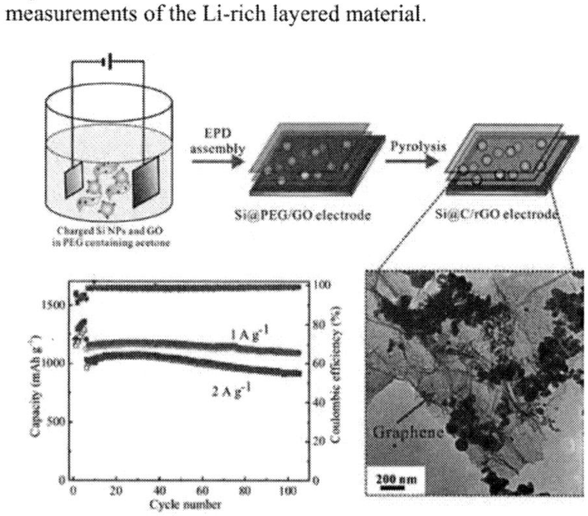

Figure 2. EPD method for fabricating binder-free carbon-coated Si/rGO electrode.

For anode, Si anodes undergo significant volume expansion and contraction during Li^+ insertion/extraction (~300 %), which results in a rapid capacity fading. [7] Many efforts have been made to develop Si materials with specific structures in improving its cycling performances. Whereas only limited works were available focusing on developing new electrode fabrication methods. We have successfully applied electrophoretic deposition (EPD) method to fabricate binder-free Si nanoparticle anode. The 3-D porous structure of the electrode not only benefited alleviating volume expansion of silicon during the charge and discharge processes but also contributed facilitating faster transport kinetics of lithium-ions, that led to better cycling and rate performances. To further enhance the cycling stability, we modified the EPD method, by which binder-free Si@carbon/reduced graphene oxide (Si@C/rGO) nanocomposite electrode was fabricated. Owing to the Si@C/rGO structure features, the binder-free Si@C/rGO electrode exhibited good lithium-ion storage performance and higher cycling stability.

Acknowledgement. The authors gratefully acknowledge the financial supports from the National High Technology Research and Development Program of China (2012AA110204), Key Project of Science and Technology of Fujian Province (2013H6022), Science and Technology Bureau of Xiamen (Grant No. 3502Z20133004), National Natural Science Foundation of China (21321062), National Fund for Fostering Talents of Basic Science (J1310024).

References

(1) Goodenough, J. B.; Kim, Y. Challenges for Rechargeable Li Batteries. *Chem. Mater.* **2009,** *22*, 587-603.
(2) Liu, W.; Oh, P.; Liu, X.; Lee, M.-J.; Cho, W.; Chae, S.; Kim, Y.; Cho, J. Nickel-Rich Layered Lithium Transition-Metal Oxide for High-Energy Lithium-Ion Batteries. *Angew. Chem. Int. Ed.* **2015,** *54*, 4440-4457.
(3) Wang, J.; Lin, W.; Wu, B.; Zhao, J. Porous $LiNi_{0.5}Mn_{1.5}O_4$ Sphere as 5V Cathode Material for Lithium Ion Batteries. *J. Mater Chem. A* **2014,** *2*, 16434-16442.
(4) Li, B.; Yu, Y.; Zhao, J. Facile synthesis of spherical $xLi_2MnO_3{\cdot}(1-x)Li(Mn_{0.33}Co_{0.33}Ni_{0.33})O_2$ as cathode materials for lithium-ion batteries with improved electrochemical performance. *J. Power Sources* **2015,** *275*, 64-72.
(5) Yang, Y.; Chen, D.; Liu, B.; Zhao, J. Binder-Free Si Nanoparticle Electrode with 3D Porous Structure Prepared by Electrophoretic Deposition for Lithium-Ion Batteries. *ACS Appl. Mater. Interfaces* **2015,** *7*, 7497-7504.
(6) Huang, J. X.; Li, B.; Liu, B.; Liu, B. J.; Zhao, J.-B.; Ren, B. Structural evolution of NM (Ni and Mn) lithium-rich layered material revealed by in-situ electrochemical Raman spectroscopic study. *J. Power Sources* **2016,** *310*, 85-90.
(7) Liu, N.; Wu, H.; McDowell, M. T.; Yao, Y.; Wang, C.; Cui, Y. A Yolk-Shell Design for Stabilized and Scalable Li-Ion Battery Alloy Anodes. *Nano Lett.* **2012,** *12*, 3315-3321.

FIELD PRODUCTIVITIES OF NAPIER GRASS FOR PRODUCTION OF SUGARS AND ETHANOL.

Bruce S. Dien[1], William F. Anderson[2], Marshall Lamb, Patricia J. O'Bryan[1], and Patricia J. Slininger[1]

(1)Bioenergy Research Unit, National Center for Agricultural Utilization Research, ARS-USDA, Peoria, IL; (2) Crop Genetics and Breeding Research Unit, ARS-USDA, Tifton, GA

Napier grass (*Pennisetum purpureum* (L) Schum) is being developed as a bioenergy crop for production in the southeastern United States. An important criterion for selecting a crop is establishing a consistent and dependable source of feedstock. In this study, we considered the effects of fertilizer application and cutting regimes on production yield, chemical composition, and process yields. Napier grass was grown for 4 years in field plots (Shellman, GA) with three treatments, which were selected to maximize production yields, and replicated four sub-plots. It was observed that multiple seasonal cuts negatively impacted production yield by 21% over the total 4 years. Samples from year 2 were analyzed for soluble sugars, starch, structural carbohydrates, uronic acids, and Klason lignin. Total carbohydrate yields were 20.1-23.3 Mg/ha. Samples were pretreated with low moisture ammonia hydroxide (110°C, 2 days) and evaluated for conversion to sugars using commercial cellulases as well as to ethanol using separate hydrolysis and fermentation with *Scheffersomyces stipitis*. Extracted sugar yields were 14.8 – 19.1 Mg/ha. Glucose and xylose enzymatic conversion efficiencies were 59– 93% for glucose and 67-77% for xylose. Ethanol yields were 8.56 – 10.3 kilolitre/ha and metabolic conversion efficiencies were 73.7 – 84.8%. Statistical differences were observed for yields among the three crop management strategies. Samples harvested in year 4 are now being analyzed for comparison. In general, these results suggest that Napier grass gives commercially attractive sugar and ethanol yields on a land basis.

A MOBILE COMBUSTION REDUCTION INTEGRATED PYROLYSIS SYSTEM (CRIPS) FOR ON-FARM PRODUCTION OF BIO-OIL

Mark A. Schaffer, Neil M. Goldberg, Charles A. Mullen, Akwasi A. Boateng

A mobile, fast pyrolysis system has been developed for the on-farm production of bio-oil from switchgrass, hardwood, and horse manure with a nominal design capacity of two metric tons per day of biomass processed. The combustion-reduction integrated pyrolysis system (CRIPS) comprises two circulating fluidized beds where one bed (the combustion bed) is used to heat the circulating solid media (silica sand) as the other bed (the reduction bed) is used to pyrolyze the biomass. Once reaching operational temperatures the pyrolysis system can sustain process temperatures by utilizing char and low-BTU gas byproducts in the combustion bed. The entire system is trailer-mounted and is equipped with a biomass gasifier/generator allowing for off-grid operation in the field. Since commissioning, the system has performed 24 trials on hardwood and switchgrass feedstocks with throughputs of up to 1.7 metric tons per day. Typical bio-oil yields are 44 wt% of the influent biomass (wet basis).

POST-PROCESSING OF TAIL-GAS REACTIVE PYROLYSIS BIO-OIL INTO FUELS AND CHEMICALS

Yaseen Elkasabi, Charles A. Mullen, Akwasi A. Boateng

The ARS tail gas reactive pyrolysis (TGRP) has brought the integration of bio-oils from biomass pyrolysis into current refinery infrastructure a step closer to reality. This is because, compared with traditional bio-oils, TGRP bio-oils are more amenable to both efficient separation and catalytic hydrodeoxygenation into fuels and chemicals. Atmospheric fractional distillation of TGRP bio-oils produces high concentrations of valuable compounds, with over 50 wt% total dry yield. The combined distillates can separate into phenols and pure hydrocarbons by using aqueous NaOH, with the phenolic fraction consisting primarily of a few compounds in higher concentrations. The accompanying crude hydrocarbon component lends itself to hydrogenation into fuel-grade hydrocarbons using off-the-shelf base metal catalysts. HDO of the combined distillates proceeds at high liquid space velocities (>0.6 hr^{-1}) and reduces O and N to trace levels. Interestingly, the distillate bottoms can also be upgraded into biorenewable coke with potential to replace petroleum coke.

ENHANCED STABILITY OF PYROLYSIS OIL HDO CATALYSTS ON MESOPOROUS ALUMINA

*Qiuli Liu, Upendra A. Joshi, John R. Regalbuto**

University of South Carolina, Columbia, SC 29208 (USA)
*regalbuj@cec.sc.edu

Introduction

Pyrolysis oil, a promising renewable hydrocarbon feedstock, can be upgraded by a catalytic route, hydrodeoxygenation (HDO), to decrease its corrosivity and oxygen content, as well as increase its octane rating as it is transformed into transportation grade fuel. The biggest challenge is to develop an active, selective and stable catalyst. Due to the high content of water in pyrolysis oil and the aqueous environment of the HDO process, it requires a hydrothermally stable support which can survive the severe reaction conditions (about 300°C and 25 atm), as well as successfully anchor metal particle against sintering.

In this work we show that a mesoporous alumina support can maintain high surface area as well as anchor highly dispersed Ru and Pt nanoparticles much better than Zr or Al-modified forms of silica. The two series of catalysts were characterized before and after HDO reaction with XRD, STEM, XPS, and BET.

Experimental

All catalysts were synthesized by Strong Electrostatic Adsorption (SEA) which has been recognized as a simple and reproducible method to make ultra-small metal particles (<1.5 nm) [1, 2]. In SEA, hydroxyl groups populating on the surface of oxides are charged by adjusting the pH value in the solution away from the neutral Point of Zero Charge (PZC), and as a result, metal complexes with opposite charges can be adsorbed onto supports by electrostatic interaction. In this study, cationic $[Ru(NH_3)_6]^{+3}$ and $[Pt(NH_3)_4]^{+2}$ were used as precursors for Zr or Al-modified SiO_2 at high pH, while anionic $[Ru(CN)_6]^{-4}$ and $[PtCl_6]^{-2}$ were employed for mesoporous Al_2O_3 at low pH. A hydrothermal test (HT) was made with 0.2 g support in 55 mL H_2O at 220 °C and 336 PSIG for 24 hrs. The same amount of catalyst was tested in the HDO reaction (p-cresol as model compound) at 300 °C for 2 hrs in 120 mL H_2O with 300 PSIG H_2 pressure. Spent catalysts were recovered for post-reaction characterization.

Results and Discussion

Zr- and Al-modified SiO_2 and mesoporous Al_2O_3 proved to be quite stable after hydrothermal testing at 220 °C; however, when tested at reaction conditions, surface area dropped dramatically in the case of 7.7 wt%Ru/Zr-SiO_2, and the 5.9 wt%Ru/Al-SBA-15 almost completely dissolved. On the other hand, a slight increase in surface area was observed in 3.0 wt%Ru/Meso-Al_2O_3, which is consistent with a previous report [3].

Table 1. Surface area changes after tests at elevated temperatures

Samples	Surface area (m^2/g), before tests	Surface area (m^2/g), after tests
[a]Al-SBA-15	454	340
[b]5.9 wt%Ru/Al-SBA-15		[d]n.a.
[a]Zr-SiO_2	416	212
[b]7.7 wt%Ru/Zr-SiO_2		188
[a]Meso-Al_2O_3	240	182
[c]3.0 wt%Ru/Meso-Al_2O_3		306

[a]tested under HT, [b]tested under HDO; [c]at 300 °C for 6 hrs in 55 mL H_2O with 300 PSIG H_2 pressure; [d]almost completely dissolved

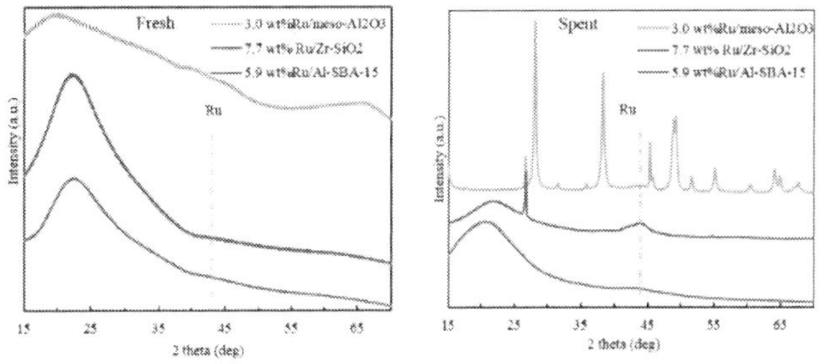

Figure 1. XRD patterns of fresh and spent catalysts

Ru nanoparticles (<1.5 nm) were undetectable in all the fresh catalyst XRD patterns. After reaction at 300 °C, significant Ru sintering occurred on 7.7 wt% Ru/Zr-SiO_2, as Ru particles grew to 4.0 nm (as estimated by Scherrer equation). Most of the Ru/Al-SBA-15 dissolved in the reaction environment, but what little remained appeared to have anchored the Ru well as indicated by the broad Ru peaks in the bottom XRD pattern of Figure 1b. As it maintained its surface area, the initially amorphous mesoporous alumina transformed to well crystallized boehmite. Broadened Ru peaks of the spent 3.0 wt% Ru/Meso-Al_2O_3, could be deconvoluted from the boehmite pattern and gave a particle size of 2.4 nm. XPS analysis confirmed no Ru leaching (Figure 2) and Ru particles remained small.

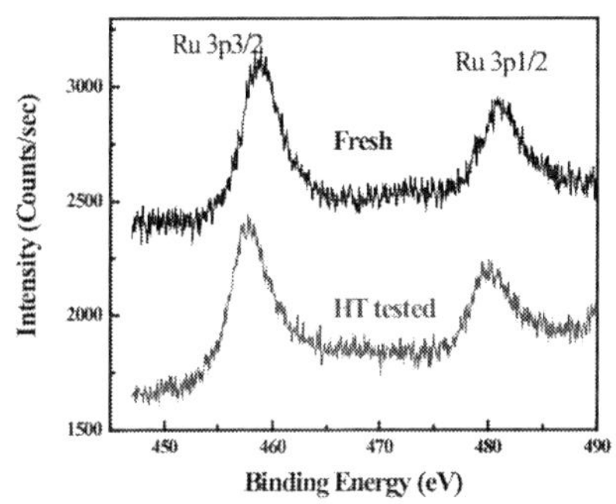

Figure 2. XPS studies showed constant Ru signals in fresh and tested 3.0 wt%Ru/Meso-Al_2O_3.

Conclusions

Mesoporous alumina and Zr and Al-modified SiO_2 exhibit reasonable hydrothermal stability at 220°C. Once the temperature is increased to a reaction temperature of 300 °C, however, the mesoporous Al_2O_3 shows the best performance in both maintaining surface area and anchoring Ru against sintering. Zr-modified silica both loses surface area and does not prevent sintering, while Al-modified silica loses virtually all of its surface area.

Acknowledgement. The authors gratefully recognize the funding of USDA-NIFA-9008-003828.

References

(1) Liu, Q.; Joshi, U. A.; Uber, K.; Regalbuto, J. R., Phys Chem Chem Phys 2014, 16 (48), 26431-5.
(2) Cao, S.; Monnier, J. R.; Williams, C. T.; Diao, W.; Regalbuto, J. R., Journal of Catalysis 2015, 326, 69-81.
(3) Ravenelle, R. M.; Copeland, J. R.; Kim, W.-G.; Crittenden, J. C.; Sievers, C, ACS Catal. 2011, 1, 552-561

TRANSPORTATION FUELS VIA A TWO-STAGE THERMAL DEOXYGENATION PROCESS

M. C. Wheeler,[1] *S. J. Eaton,*[2] *and W. J. DeSisto*[1]

[1]University of Maine Chemical and Biological Engineering and Forest Bioproducts Research Institute, Orono, ME 04469-5737
[2]Maine Maritime Academy Marine Engine Testing and Emissions Laboratory, Castine, ME 04420
mcwheeler@maine.edu

Abstract

Thermal DeOxygenation (TDO) is the thermal decomposition of mixed salts of levulinic and formic acids that can produce hydrocarbon oils at yields approaching 80% of theoretical in a 300 mL semibatch reactor. The TDO reaction was scaled up to a 50 L reactor to produce more than 20 L of oil for fractionation, blending and hydrotreating. The products were thus demonstrated as suitable blendstocks for gasoline, diesel, jet fuel, and fuel oils. While scale-up facilitated kilogram quantities of products for testing, the carbon loss to char increased from 8.5% to 48%. Isothermal reaction experiments have identified two reaction regimes: 1) a low temperature addition-condensation regime that is diffusion-controlled and 2) a high temperature ketonic decarboxylation regime that is reaction-controlled. Through this understanding, a two stage reaction process at 50 L scale has achieved 78% of theoretical oil yield with an 11% loss of carbon to char.

PRODUCTION OF STABLE, PARTIALLY DEOXYGENATED PYROLYSIS OILS VIA TAIL GAS REACTIVE PYROLYSIS

Charles A. Mullen, Akwasi A. Boateng, Neil M. Goldberg, Yaseen Elkasabi, Mark A. Schaffer, Paul C. Tarves

Traditional biomass fast pyrolysis oils are acidic and unstable due to high concentrations of reactive oxygenates. The last decade has seen concerted research efforts to stabilize bio-oils by deoxygenation primarily through catalytic approaches, but this has not been without challenges. We mitigate these by creating a reactive environment within the reactor system by the use of partial tail gas recycle as the fluidizing medium. Under carefully controlled conditions, this Tail Gas Reactive Pyrolysis (TGRP) can successfully produce partially deoxygenated pyrolysis oils with C/O molar ratios of up to 9:1 comprising mostly aromatic hydrocarbons and phenols without the use of externally added catalysts. Unlike traditional bio-oils, TGRP bio-oils can be distilled in high yields to high concentrations of individual chemicals. The presentation discusses the above mentioned aspects of TGRP, its additional utility with traditionally difficult to pyrolyze feedstocks, extent of deoxygenation achieved, and efforts to further optimize the process.

CONVERSION OF BIOMASS INTO HIGH QUALITY FUEL WITH IONIC LIQUIDS

Jiayu Xin, Xingmei Lu, Suojiang Zhang

Beijing Key Laboratory of Ionic Liquids Clean Process, State Key Laboratory of Multiphase Complex System, Institute of Process Engineering, Chinese Academy of Sciences, China
E-mail: jyxin@ipe.ac.cn

Introduction

Conversion of biomass into hydrocarbons is one of the most possible alternatives to alleviate the environmental, political, and economic burdens associating with the depletion and consumption of fossil-based resources. In the past several years, much effort was devoted to the research on producing hydrocarbons with extended carbon chains by hydrogenation of carbohydrate derivatives.[1] However, the products of these processes are alkanes/alkenes without branch or with single branch, which only can be used as substitute for diesel fuel or additives of jet fuels. The research on the production of bio-based alkanes for the purpose of gasoline, the biggest consumed transportation fuel, is very rare. The main obstacles may be concluded as: 1) it is difficult to control the carbon numbers of alkanes from 6-12, especially to control the C8 content with the highest potion; 2) it is difficult to control the antiknock index to satisfy the requirement of gasoline motors.

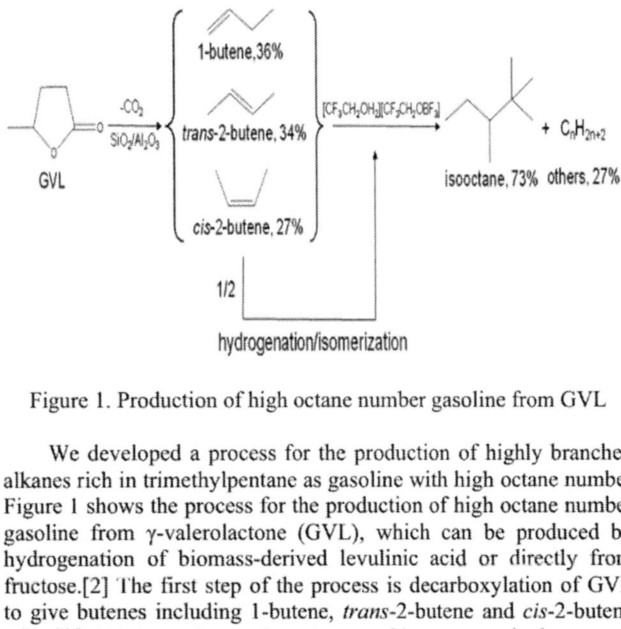

Figure 1. Production of high octane number gasoline from GVL

We developed a process for the production of highly branched alkanes rich in trimethylpentane as gasoline with high octane number. Figure 1 shows the process for the production of high octane number gasoline from γ-valerolactone (GVL), which can be produced by hydrogenation of biomass-derived levulinic acid or directly from fructose.[2] The first step of the process is decarboxylation of GVL to give butenes including 1-butene, *trans*-2-butene and *cis*-2-butene with SiO_2/Al_2O_3 as the catalyst. A part of butenes are hydrogenated and isomerized to produce isobutane. Isobutane also can be produced via bioethanol in a high yield. Then, remained butenes react with isobutane to give isooctane as the gasoline fuel. The use of $[CF_3CH_2OH_2][CF_3CH_2OBF_3]$ provides a high yield of highly branched alkanes and thus ensures a high octane number of the products, indicating that high quality gasoline can be produced from biomass derived oxygenated compounds without hydrodeoxygenation and without any additives.

Experimental

100 g 2,2,2-trifluorothanol (CF_3CH_2OH, 99%, J&K Scientific Ltd.) was placed in a 500 ml polytetrafluoroethylene bottle with a rubber cap, an inlet and a safety outlet. BF_3 (99%, Heli Chemical Industry Co.) contained in a high pressure steel cylinder was slowly released to the bottle through a 3 mm diameter polytetrafluoroethylene tube. In order to remove the heat generated during the reaction, the bottle was surrounded by dry ice and the reactants were stirred continuously. The weight of the bottle with reactants was recorded periodically. BF_3 was stopped when no weight increase was observed, which means the reaction between BF_3 and CF_3CH_2OH was finished. The obtained product was found to be an 1:2 complex of BF_3 and CF_3CH_2OH.

Alkylation of isobutane and butenes (1-butene: 2-butene=3:7, supplied by Changlin Senmao Chemical Ltd Co., >99.9 % purity) was performed in a 100 ml stirred stainless steel reactor. Reaction temperature was maintained by placing the reactor in a cooling bath filled with cycling ethanol. Even though, a temperature increase of 3-8 °C inside of the reactor was observed. $[CF_3CH_2OH_2][CF_3CH_2OBF_3]$ catalyst was firstly placed in the reactor and then the isobutane and butane mixture was pumped into the reactor by a syringe pump.

The gas and liquid products were analyzed by GC (Agilent GC 7890) equipped with a flame ionization detector and a DB-Petro column (0.25 mm*100 m, Agilent Co.).

Results and Discussion

The products of decarboxylation of GVL are mixtures of butenes including 1-butene, 2-butene (*cis*-2-butene and *trans*-2-butene) and CO_2. For a reaction time of 4 h, low temperature gave a low yield. A good yield of 97 % was achieved at a temperature of 350 °C. However, a temperature high than 350 °C decreased the yield of butenes possibility due to the coke formation, likely by polymerization of pentenoic acid produced by ring open reaction of GVL, which further decreased the activity of SiO_2/Al_2O_3 catalyst. The side reactions were also noticed at high temperature, some cracking products including methane and propane were found at 400 °C. Reaction time also affected the yields of butenes, 1 h reaction only gave a yield of 50 %, while increases the reaction time to 4 h increased the yield to 97 % and further prolonged the reaction time longer than 4 h had minimal effect on the decarboxylation.

We carried out the alkylation reaction of butenes (1-butene:2-butene=3:7) and isobutane (butenes:isobutane=1:10) at 10 °C to produce saturated hydrocarbons. As expected, the high octane number gasoline was produced in complete conversion of butenes. The effect of reaction time can be elucidated by the changes in the product distributions and calculated RON. Figure 1 shows the effect of reaction time on the alkylation of isobutane with butenes catalyzed by $[CF_3CH_2OH_2][CF_3CH_2OBF_3]$ at 10 °C. The highest RON increased from 5 min to 10min and decreased from 10 min to 30 min, the highest RON was obtained at a reaction time of 10min. Meanwhile, the most valuable C8 fraction also showed the same trend. A reaction time longer than 10 min leaded to the decreases of the RON and C9 content. The decrease is due to the oligomerization reactions responsible for higher C9+ fractions with longer reaction time. For all reaction times, more than half of alkanes obtained were C8 fractions. Under optimum condition, more than 80% alkanes obtained was C8 fraction. Furthermore, the products from this approach are saturated hydrocarbons having excellent thermal stability. Although the C8 fraction is the most suitable components for gasoline fuel, the trimethypentanes (TMPs) are the most favored ones because TMPs have an octane number of 100 while that of dimethylhexanes (DMH) is 44. It is seen that the highest C8 and TMP selectivities, lowest DMH and C9+ fraction were obtained at a reaction time of 10 min, which ensured the highest TMP contents and corresponding highest RON and MON, while prolonged the reaction time decreased the C8 content. The highest RON of 95.4 of the products was achieved at the optimum conditions without any additives. When the reaction time was increased to 30 min, the C9+ fraction also increased greatly from 23% to 57% possibly because oligomerization reactions predominated under these conditions.

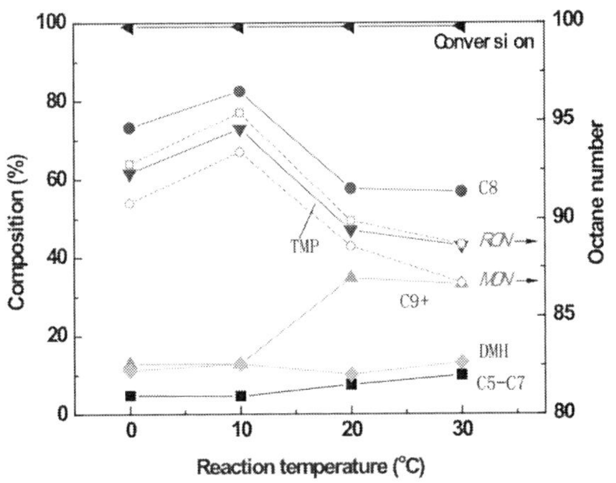

Figure 2. Effect of reaction temperature on the alkylation of isobutane and butenes catalyzed by [$CF_3CH_2OH_2$][$CF_3CH_2OBF_3$] with a reaction time of 10 min

The reusability and regeneration of the ionic liquid catalyst are important factors for the application. The reusability of [$CF_3CH_2OH_2$][$CF_3CH_2OBF_3$] on product distribution, TMP/DMH, RON were examined with isobutane/(1-butene+2-butene) of 10, reaction time of 10 min at 10 °C. The isobutane/butenes mixtures reacted in the reactor with catalyst and then the reactant/product phase was removed by phase separation. Fresh reactants were then added for the next experimental cycle. Totally 12 cycles were examined. From Table 1, it is shown that the catalyst showed similar performance in the first 6 reactions, noticeable changes in product quality with higher ratio of C9+ fraction were observed in the following reactions, which also caused the gradual decrease of RON. This is most likely due to the acidity decrease by separation of catalyst with reactant/product phases. If the catalyst is used in the continuous process, it may show better performance.

Table 1 Reusability of [$CF_3CH_2OH_2$][$CF_3CH_2OBF_3$]

Exp. Cycles[a]	1	3	5	7	9	11
C5 [%]	1	1	2	2	1	1
C6 [%]	2	2	3	3	3	2
C7 [%]	2	3	4	4	3	3
C8 [%]	82	73	63	50	46	41
TMP [%]	73	64	54	41	39	35
DMH [%]	9	9	9	9	7	5
TMP/DMH	8	7	6	5	6	7
C9+ [%]	13	20	28	42	47	54
RON	95.4	93.5	91.6	88.9	88.8	88.0

[a] Reaction conditions: 50 ml butenes and isobutane (1:10), 30 ml $BF_3{\cdot}2CF_3CH_2OH$, 10 min, 10 °C.

Conclusions

In summary, we describe here a method that a practical feedstock (GVL) is efficiently converted to high octane number gasoline rich in TMP via decarboxylation and alkylation reactions. Decarboxylation of GVL produces butenes including 1-butene, trans-2-butene and cis-2-butene in the highest yield of 97% with SiO_2/Al_2O_3. After hydrogenation and isomerizaiton of a part butenes to isobutane, high octane number gasoline can be produced by alkylation between butenes and isobutane with [$CF_3CH_2OH_2$][$CF_3CH_2OBF_3$] as the catalyst. The products are mixture of alkanes rich in TMPs, which has a RON of 100 and serve as the most suitable component for high octane gasoline. The highest octane number of the product is 95.4, indicating that high quality gasoline can be produced from biomass derived oxygenated compounds without hydrodeoxygenation and without any additives.

Acknowledgement. This work was supported by National Basic Research Program of China (2013CB733506), National Natural Science Foundation of China (No. 21210006, 21276260), International S&T Cooperation Program of China (2014DFA61670), External Cooperation Program of BIC, Chinese Academy of Sciences (GJHZ201306) and Key Project of Natural Science Foundation of Beijing of China (No. 2131005).

References:

(1) Xin, J.; Zhang, S.; Yan, D.; Ayodele, O.; Lu, X.; Wang, J. *Green Chemistry* **2014**, *16*, 3589

(2) Xin, J.; Yan, D.; Ayodele, O.; Zhang, Z.; Lu, X.; Zhang, S. *Green Chemistry* **2015**, 17, 1065.

IDENTIFICATION OF GLYCEROL OXIDATION PRODUCTS ON SILVER USING *IN SITU* SURFACE-ENHANCED RAMAN SPECTROSCOPY

Yeyoung Ha,[†] *Renato C. Ambrosio,*[‡] *and Andrew A. Gewirth**[,†]

[†] Department of Chemistry, University of Illinois, Urbana, IL 61801, U.S.A.
[‡]Departamento de Química, Universidade Federal de Sergipe, São Cristóvão, SE 49100, Brazil.
*agewirth@illinois.edu

Production of biodiesel, a renewable and environmentally friendly fuel, has continuously grown over the past decade.[1] Biodiesel is produced via a transesterification reaction where vegetable oils or animal fats catalytically react with short-chain aliphatic alcohols, typically methanol or ethanol.[2] In this process, glycerol is produced as a major by-product and production of crude glycerol is estimated to reach 5.8 billion pounds in 2020.[1]

Efforts have been made to find applications for the surplus glycerol, which could be used as fuel in direct alcohol fuel cells or be oxidized to more valuable products, such as dihydroxyacetone, glycolic acid and oxalic acid. While Au and Pt catalysts have been actively studied for oxidizing glycerol, not much attention has been paid towards Ag due to its low activity.[2-5] Nonetheless, Ag could be a promising catalyst once its product selectivity is determined.

In this work, we monitor the Ag surface during the oxidation of glycerol in 0.1 M NaOH using *in* situ surface-enhanced Raman spectroscopy (SERS). The SER spectra show a number of potential-dependent peaks, each associated with vibrations of different functional groups of adsorbed glycerol, oxidation intermediates, and products. The spectra are further analyzed with Two-Dimensional Correlation Spectroscopy (2DCOS) techniques to find correlations between different functional groups[6] – C-H stretching, and CCO and $\nu_s(COO^-)$ vibration modes. Our findings provide insight into the mechanism and the products of glycerol oxidation reaction on Ag catalyst.

References

(1) Ayoub, M.; Abdullah, A. Z. *Renew. Sust. Energ. Rev.* **2012**, *16*, 2671.
(2) Gomes, J.; Tremiliosi-Filho, G. *Electrocatalysis* **2011**, *2*, 96.
(3) Kwon, Y.; Schouten, K. J. P.; Koper, M. T. M. *ChemCatChem* **2011**, *3*, 1176.
(4) Gomes, J. F.; Garcia, A. C.; Gasparotto, L. H. S.; de Souza, N. E.; Ferreira, E. B.; Pires, C.; Tremiliosi-Filho, G. *Electrochim. Acta* **2014**, *144*, 361.
(5) Suzuki, N. Y.; Santiago, P. V. B.; Galhardo, T. S.; Carvalho, W. A.; Souza-Garcia, J.; Angelucci, C. A. *J. Electroanal. Chem.*
(6) Butcher, D. P.; Boulos, S. P.; Murphy, C. J.; Ambrosio, R. C.; Gewirth, A. A. *J. Phys. Chem. C* **2012**, *116*, 5128.

A NOVEL STRATEGY TO REDUCE SINTERING, BASED ON POLYMORPHIC MATERIALS, AND APPLICATION TO CARBON CAPTURE AND STORAGE

*Xu He and Ming Zhao**

School of Environment, Tsinghua University, Beijing 100084 China
Corresponding Author Email: ming.zhao@tsinghua.edu.cn

Introduction

Calcium looping, based on reversible carbonation and calcination of CaO sorbents ($CaO_{(s)} + CO_{2(g)} \leftrightarrows CaCO_{3(s)}$), is a promising technology to decrease the carbon footprint for both sectors. [1,2]. As sorbent quantities are sufficiently large to dominate the economics of the capture process, their deactivation owing to sintering is the major drawback of this technology. [3] The theoretical CO_2 uptake using CaO is 0.78 g-CO_2 g-sorbent^{-1} with full carbonation (Step 1, **Figure 1**a); however, the initial capacity is unlikely to exceed 0.6 g(CO_2) g(sorbent)$^{-1}$ owing to the limited reaction time. [4] The Tammann temperature ($T_{Tammann} \approx 0.5\ T_{melting}$, the empirical temperature when sintering commences [5]) of $CaCO_3$ is only 533 °C (Table S1), lower than that for either carbonation (600-700 °C) or calcination (800-900 °C) in calcium looping. Despite CaO not sintering itself ($T_{Tammann}$=1170 °C), the densification of $CaCO_3$ steadily closes up the pores of the sorbent (Step 2, Figure 1a); diffusion of CO_2 molecules becomes difficult resulting in partial carbonation (Step 3, Figure 1a). [6] Deactivation rates of up to 15% per cycle have been reported, with residual capacity as low as 0.14 g-CO_2 g-sorbent^{-1} after 10 cycles. [7,8]

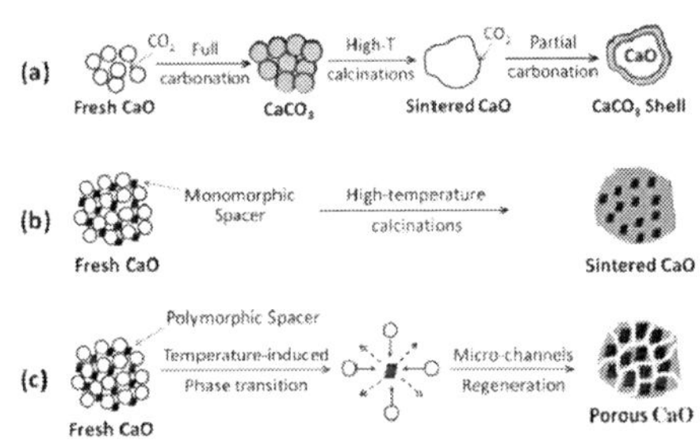

Figure 1. Schematic presentation of the sintering issue and strategies for stabilization. (a) deactivation of CaO sorbent due to sintering, (b) effect of monomorphic spacer, and (c) effect of polymorphic spacer

Here, we report a novel, extremely stable CaO-based sorbent, derived using freeze drying, using a cement-compatible spacer – Ca_2SiO_4 (namely $2CaO{\cdot}SiO_2$, abbreviated as C_2S). Polymorphic (crystalline/volume changing) transitions of C_2S can be induced by temperature swings in a Ca-looping process so that the significant volume expansion force offsets the sintering stress (Figure 1c). This distinguishes this work from previous work using monomorphic spacers.

Polymorphs of C_2S (**Figure 2**a [9]) have been well investigated in cement chemistry but not in the context of Ca looping. Both α'- and β-C_2S are useful forms for use in clinker, owing to their hydration reactivity, but they have distinct densities. [10] Reversible transformation between the phases happens between 630-690 °C (Figure 2a). [10] When calcium looping shifts from the calcination regime (> 800 °C) toward carbonation (~ 650 °C), the α'→β transition occurs, resulting in volume expansion of 2.1%, orders of magnitude larger than the thermal expansion of many other monomorphic materials. [11] It is demonstrated below that this expansion force resists sintering stresses more powerfully than does mere physical separation.

Sorbent Synthesis

Synthesis design is crucial to enable polymorphic spacers to act against sintering. The spacers and CaO particles must be of reasonably similar sizes to allow uniform and intimate mixing. Here, we report an innovative method. To produce a homogeneous precursor, instead of a suspension, tetraethyl orthosilicate (TEOS) was dissolved in an aqueous solution of calcium acetate upon completion of a hydrolysis reaction (**Equation 1**). Freeze drying was then employed to prevent multiple solutes agglomerating and phases segregating. Calcination at 850 °C produced the C_2S spacer (Equation 2). Due to the CaO-rich formulation (n_{Ca}:n_{Si} = 9:1), the final sorbent (denoted as 'FD-CaO-C_2S') was theoretically composed of 69.5 wt% CaO and 30.5 wt% C_2S. For comparison, two pure CaO sorbents were prepared by drying a solution of only calcium acetate, which are referred to as 'HD-CaO' and 'FD-CaO' indicating heating-drying and freeze-drying origins, respectively.

$$Si(OC_2H_5)_4 + 4\,H_2O \xrightarrow{H^+} H_4SiO_4 + 4\,C_2H_5OH \qquad (1)$$

$$H_4SiO_4 + 2\,Ca(CH_3CO_2)_2 + 8\,O_2 \longrightarrow Ca_2SiO_4 + 8\,CO_2 + 8\,H_2O \qquad (2)$$

Results and Discussion

As-prepared FD-CaO-C_2S contained a mixture of cubic CaO, orthorhombic α'-C_2S and monoclinic β-C_2S, as identified by Rietveld refinement of powder X-ray diffraction (XRD) spectra (**Figure 2b**). The weight fraction (%) of each phase was estimated to be 69.6%, 9.5% and 20.9 %, respectively, agreeing well with the theoretical composition. Due to the absence of γ-C_2S, the most probable history of phase transitions was that the γ-C_2S formed at lower temperature and was fully shifted to α'-C_2S during calcination and ~ 69% of the α'-C_2S changed to β-C_2S in course of cooling. The crystallite size of β-C_2S was only 38.0 nm as estimated by Rietveld refinement, small enough to prevent the transition to γ-C_2S. [12] The dusting effect was thus avoided though consequently a small amount of the useful α'→β transition did not occur owing to small α'-C_2S grains (46.2 nm). Considering the similar crystallite size of CaO (74.9 nm), intimate contact between CaO and spacer could be expected given that no obvious phase segregation was observed. The α'→β transition was observed and found to be repeatable in extended heating-cooling cycles using a high-temperature differential scanning calorimeter (DSC). Notable exothermic peaks occurred at slightly varied temperature points (660-680 °C) over 3 cycles (Figure 2c), which were ascribed to the α'→β transition. [13] The good reproducibility of transition temperatures also suggested that the shape and size of spacers had not been altered by sintering or dusting.

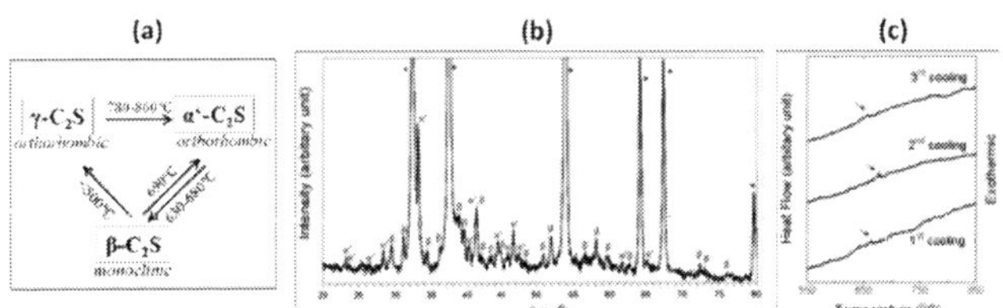

Figure 2. (a) Polymorphic diagram of C_2S, (b) phases identification by X-ray diffraction for as-prepared FD-CaO-C_2S sorbent, (c) α'→β transition confirmation by 3-cycle high-temperature differential scanning calorimeter

The synthesis method determined the sorbent morphology. In the course of freezing, a complex crystallization reaction of H_2O took place, forming dendritic ice crystals. The ice sublimated when pressure dropped, giving needle-like large pores to the freeze dried particles, as illustrated by scanning electron microscopy (SEM) (**Figure 3**a). [14] Owing to freezing, the solution mobility was reduced during drying so that agglomeration and phase segregation were minimized. [15] So, despite three phases being detected in FD-CaO-C_2S, no phase segregation was observed from the detailed morphological information (Figure 3b). This structure is favorable (discussed below) for polymorphic spacers to crack apart primary particles and open up useful pores for carbonation (Figure 1c). C_2S appeared to perform well because smaller particles and improved porosity are observed in Figure 3b compared to Figure 3c. Comparing Figure 3c and 3d, it can be concluded that freeze drying helped reduce the particle size of the deposited solutes. However, without addition of C_2S, the freeze-dried CaO and thermally-dried samples presented nearly equal BET surface areas (~ 10 $m^2\ g^{-1}$) and pore volumes (~ 0.08 $cm^3\ g^{-1}$). By contrast, the addition of C_2S engendered obvious changes in micro-structures: BET surface area was increased to 17 $m^2\ g^{-1}$ and pore volume to 0.14 $cm^3\ g^{-1}$.

Figure 3. Scanning Electron Microscopy: (a) and (b) as-prepared FD-CaO-C_2S, (c): as-prepared FD-CaO, (d) as-prepared HD-CaO, (e) carbonated FD-CaO-C_2S (carbonation in pure CO_2 for 60 min), (f) calcined FD-CaO-C_2S at cycle 15, (g) calcined FD-CaO at cycle 15, (h) calcined HD-CaO at cycle 15

Finally, the stability of as-prepared sorbents was tested, using a thermogravimetric analyzer (TGA). To make the TGA results comparable to literature values, typical cycling conditions were used: carbonation in 15% CO_2 at 650 °C for 30 min; calcinations in N_2 at 850 °C without dwell time. The CaO-$C_{12}A_7$ (75 wt%: 25 wt%) sorbent reported by Li et al. [16] was chosen as representative of previous spaced sorbents for comparison to monomorphic spacers.

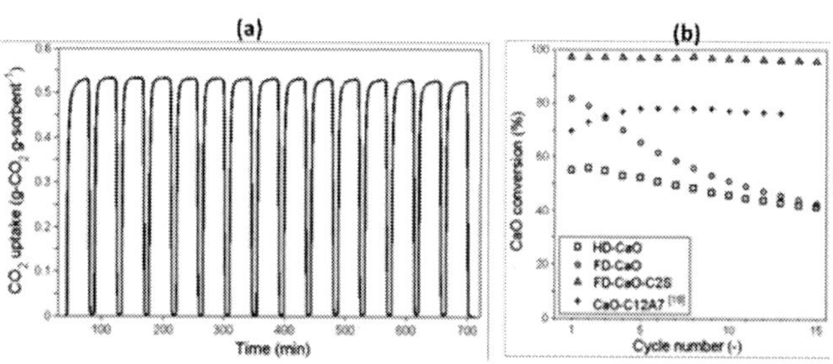

Figure 4. (a) CO_2 uptakes in multiple carbonation-calcination cycles in TGA (carbonation: 15% CO_2 at 650 °C for 30 min; calcination: N_2 at 850 °C no dwell time), (b) CaO conversion ratios for as-prepared sorbents with comparison to representative sorbent from literature [16].

The novel FD-CaO-C_2S sorbent showed extremely stable capacity of 0.530 initially, slightly decreasing to 0.522 g-CO_2 g-sorbent $^{-1}$ at cycle 15 (**Figure 4a**). The deactivation rate was merely 0.1% per cycle, significantly lower than the 15% for natural limestone. Despite having 30.4% nonreactive spacer, the net CO_2 uptake of our novel sorbent was much larger than that of CaO-$C_{12}A_7$ with 25% mass of spacer (0.45 g-CO_2 g-sorbent $^{-1}$), owing to the exceptionally high CaO conversion ratio (97%). By comparison, up to 24% of the CaO in the CaO-$C_{12}A_7$ sorbent was unavailable for carbonation, owing to sintering (Figure 1b). In addition, CaO-$C_{12}A_7$ sorbent experienced a boost of reactivity during the initial five cycles, which has been described as "self-activation" in literature. [17] However, the present sorbent with C_2S spacer exhibited its maximum conversion ratio (nearly 100%) in the first cycle without any self-activation period. This suggests that the polymorphic spacer behaved differently to the monomorphic ones. The expansion force due to phase transition was likely the key factor that resulted in the well-established micro-structure during synthesis.

Formation of carbonate leads to packing of pores within Ca-based sorbents, followed by elimination of crystal boundaries, resulting in a dense structure, as described in previous works. [18] However, the fully carbonated FD-CaO-C_2S sorbent still presented a porous morphology (Figure 3e). The larger pores are most likely inherited from those formed by ice sublimation during spray drying, while the small pores (less than 100 nm) are those where the majority of the reactions to form $CaCO_3$ occur. [4, 19] After 15 cycles, the porosity of calcined FD-CaO-C_2S sorbent was almost unchanged (Figure 3f), as was the CO_2 uptake (Figure 4a). It should also be noted that the uniformity of the pores decreased and that some minor agglomeration occurred, which likely corresponded to the slight drop in reactivity. This might be explained by the marginally smaller α'-C_2S crystallites (46.2 dropping to 39.6 nm), which led to less conversion to β-C_2S (~63% compared to initial ~68%). Without addition of C_2S spacer, the freeze dried CaO sorbent showed fairly high initial carbonation (81%) compared to thermally dried CaO sorbent (55%), possibly due to the smaller primary crystallites from freeze drying. However, after 15 cycles, the freeze- and thermally-dried CaO were similarly degraded in reactivity, owing to their similarly sintered morphology (Figures 3g and 3h).

In conclusion, polymorphic Ca_2SiO_4 showed advantages over conventional monomorphic materials as a spacer for stabilization of CaO-based sorbents: It is compatible with cement production from waste sorbents from calcium looping cycles; but also retains exceptionally high reactivity. Density variation produced comparable counter force against sintering stress, distinct from the physical separation induced by monomorphic spacers. Freeze-drying-based synthesis was simple and capable of producing intimately mixed CaO and spacer particles. Superior micro-structures, favorable for gas-solid reactions, were established and maintained over 15 Ca-looping cycles, resulting in stable cyclic performance (> 0.52 g-CO_2 g-sorbent^{-1}) with extensive conversions of 'free' CaO (97%).

References

(1) Dean, C. C.; Blamey, J.; Florin, N. H.; Al-Jeboori, M. J.; Fennell, P. S. *Chem. Eng. Res. Des.* **2011,** *89*, (6), 836-855.
(2) Rodríguez, N.; Murillo, R.; Abanades, J. C. *Environ. Sci. Technol.* **2012,** *46*, (4), 2460-2466.
(3) MacKenzie, A.; Granatstein, D. L.; Anthony, E. J.; Abanades, J. C. *Energy Fuels* **2007,** *21*, (2), 920-926.
(4) Abanades, J. C.; Alvarez, D. *Energy Fuels* **2003,** *17*, (2), 308-315.
(5) Knözinger, H.; Taglauer, E., Spreading and Wetting. In *Preparation of Solid Catalysts*, Wiley-VCH Verlag GmbH: 2008; pp 501-526.

(6) Lu, H.; Reddy, E. P.; Smirniotis, P. G. *Ind. Eng. Chem. Res.* **2006,** *45*, (11), 3944-3949.
(7) Wang, J.; Anthony, E. J. *Ind. Eng. Chem. Res.* **2004,** *44*, (3), 627-629.
(8) MacDowell, N.; Florin, N.; Buchard, A.; Hallett, J.; Galindo, A.; Jackson, G.; Adjiman, C. S.; Williams, C. K.; Shah, N.; Fennell, P. *Energy Environ. Sci.* **2010,** *3*, (11), 1645-1669.
(9) Remy, C.; Guyot, F.; Madon, M. *Phys. Chem. Minerals* **1995,** *22*, (7), 419-427.
(10) Bridge, T. E., *The American Mineralogist* **1966,** *51*, 1766-1774.
(11) Joseph Leon Rosenholtz, D. T. S. *American Mineralogist* **1949,** *34*, (11-12), 846-854.
(12) Minkov, V. S.; Drebushchak, V. A.; Ogienko, A. G.; Boldyreva, E. V., *CrystEngComm* **2011,** *13*, (13), 4417-4426.
(13) Smith, D. K.; Majumdar, A. J.; Ordway, F. *J. Am. Ceram. Soc.* **1961,** *44*, (8), 405-411.
(14) Shibkov, A. A.; Golovin, Y. I.; Zheltov, M. A.; Korolev, A. A.; Leonov, A. A. *Physica A: Statistical Mechanics and its Applications* **2003,** *319*, (0), 65-79.
(15) Eggenhuisen, T. M.; Munnik, P.; Talsma, H.; de Jongh, P. E.; de Jong, K. P. *J. Catal.* **2013,** *297*, (0), 306-313.
(16) Li, Z.-s.; Cai, N.-s.; Huang, Y.-y.; Han, H.-j. *Energy Fuels* **2005,** *19*, (4), 1447-1452.
(17) Manovic, V.; Anthony, E. J. *Int. J. Environ. Res. Pub. Health* **2010,** *7*, (8), 3129-3140.
(18) Zhao, M.; Bilton, M.; Brown, A. P.; Cunliffe, A. M.; Dvininov, E.; Dupont, V.; Comyn, T. P.; Milne, S. J. *Energy Fuels* **2014**.
(19) Blamey, J.; Anthony, E. J.; Wang, J.; Fennell, P. S. *Prog. Energy Combus. Sci.* **2010,** *36*, (2), 260-279.

SYNTHESIS OF MoS_2/Pd COMPOSITES VIA A FACILE SONOCHEMICAL APPROACH FOR OXYGEN REDUCTION REACTION

Ling-Xia Zuo, Li-Ping Jiang, Jun-Jie Zhu*

State Key Laboratory of Analytical Chemistry for Life Science, School of Chemistry and Chemical Engineering, Nanjing University, Nanjing 210093, P.R. China

Jianglp@nju.edu.cn; 478074256@qq.com

Introduction

Owing to high power efficiency, cleanliness and fuel diversity,[1] alkaline fuel cells receive intensive attentions. Meanwhile, cathodic oxygen reduction reaction becomes more and more important for researchers as a key part of fuel cells. Pt-based materials are considered to be the best electrocatalyst for ORR in fuel cells. However, they have been largely limited in mass market fuel cells for commercial applications because of their limited reserves in nature, together with the high cost and poor durability. As one of key candidates for replacing Pt, Pd-based catalysts have become a popular choice because of their significantly lower cost, and high catalytic activity for ORR due to the synergistic effects between the Pd and some other components.[2] Molybdenum disulfide (MoS_2) with unique electrical, optical,[3] thermal[4] and catalytic[5] properties, is an attractive layered two-dimensional material. In this work, we obtained Pd nanoparticles decorated MoS_2 nanosheets with high homogeneity and good dispersity via a simple and efficient sonochemical method. MoS_2/Pd composites exhibited better electrocatalytic activity compared with the commercial Pt/C catalyst at the same noble metal loading.

Experimental

Exfoliation of MoS_2 nanosheets and synthesis of MoS_2/Pd composites were performed in the condition of ultrasound radiation (Sonics VCX-750 ultrasonic processor with flat head tip, 750 W at 30% Amplitude). The ORR electrochemical experiments were performed with a CHI 840B workstation in O_2-saturated 0.1 M KOH electrolyte.

Results and Discussion

In the synthesis of MoS_2/Pd composites, the bulk MoS_2 was first transformed into single and few layers MoS_2 nanosheets through ultrasonic exfoliation. The exfoliated MoS_2 nanosheets can provide a high surface area as a supporting material for the further decoration of Pd nanoparticles. Then, positively charged palladium ions were adsorbed on the surfaces of MoS_2 nanosheets through electrostatic attraction and further reduced by H_2PO^{2-} to form Pd nuclei, then the Pd nuclei could in-situ grow up to be Pd nanoparticles on MoS_2 nanosheets to obtain the MoS_2/Pd composites.

The MoS_2/Pd composites exhibited enhanced ORR electrocatalytic activity. Cyclic voltammetry and electrochemically active surface area (ECSA) showed that the MoS_2/Pd composites had high ORR activities of mass activity and specific activity at -0.26 V versus SCE, which were better than commercial Pt/C catalysts.

Conclusions

In conclusion, we have demonstrated a facile sonochemical method to fabricate MoS_2/Pd composites. A combination of the ultrathin MoS_2 nanosheets support and the highly dispersed Pd nanoparticles benefits the transportation of ions, oxygen and electrons, and further contributes to the improved catalytic activity for ORR. The strategy may be helpful for design and synthesis of catalysis and electrocatalysis.

Acknowledgement. We greatly appreciate the support of the National Natural Science Foundation of China.

References

(1) Steele, B. C. H.; Heinzel, A., *Nature*, **2001**, 414 (6861), 345-352.
(2) Mentus, S.; Abu Rabi, A.; Jasin, D., *Electrochimica Acta*, **2012**, 69,174-180.
(3) Mak, K. F.; Lee, C.; Hone, J.; Shan, J.; Heinz, T. F., *Phys Rev Lett*, **2010**, 105 (13).
(4) Mao, R.; Kong, B. D.; Kim, K. W., *Journal of Applied Physics*, **2014**, 116 (3).
(5) Guardia, L.; Paredes, J. I.; Munuera, J. M.; Villar-Rodil, S.; Ayan-Varela, M.; Martinez-Alonso, A.; Tascon, J. M. D., *ACS applied materials & interfaces*, **2014**, 6 (23), 21702-21710.

SODIUM SOFT SOLID CRYSTALS AS ELECTROLYTES FOR SODIUM ION BATTERIES

*Birane Fall, Parameswara Rao Chinnam, Abdel Aziz Jalil, Stephanie L. Wunder and Michael J. Zdilla**

Department of Chemistry, Temple University,
Philadelphia, PA 19122
mzdilla@temple.edu

The interest in soft solid crystals as electrolytes is based on the observation that molecular organization rather than disordered structures foster enhanced ionic conductivity. One viable approach for high conductivity is the design of crystals with channel walls that have low affinity for entrapped ions. The first such material, DMF:LiCl has the highest conductivity reported to date for a pure, soft solid organic electrolyte. The high conductivity can be understood based on the hard and soft acid and base (HSAB) theory. Also reported is another soft solid crystal, DMF:NaClO4, which is a sodium ion conductor. Single crystal X-ray diffraction identifies the solid as 3:1 adduct of DMF:$NaClO_4$ (**Figure 1a**). The structure shows the formation of 1-D Na_n•DMF_n chains in which Na ions interact with the soft DMF matrix through weak Lewis acid/base interactions of sodium and oxygen ions. The crystal structure exhibits a linear arrangement of Na-O(DMF) rhombs arranged end to end through sodium atoms, such that a chain of closely spaced (3.232 Å) sodium ions exists parallel to the *a* axis of the unit cell. The melt temperature of DMF.NaClO4 is 55°C. Unlike DMF:LiCl crystals that degrade before melting, DMF:NaClO4 reversibly melts and recrystallizes. The decomposition of DMF:$NaClO_4$ starts at 70°C. The temperature dependent conductivity of the pressed pellets (1200 psi) of DMF:$NaClO_4$ exhibits room temperature conductivity 8.0×10^{-6} S/cm (**Figure 1b**). The liquid DMF:$NaClO_4$ has a conductivity approximately 1 order of magnitude greater than that of the crystal.

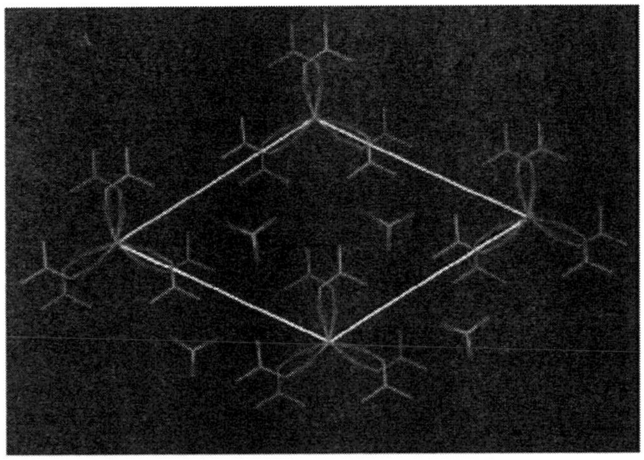

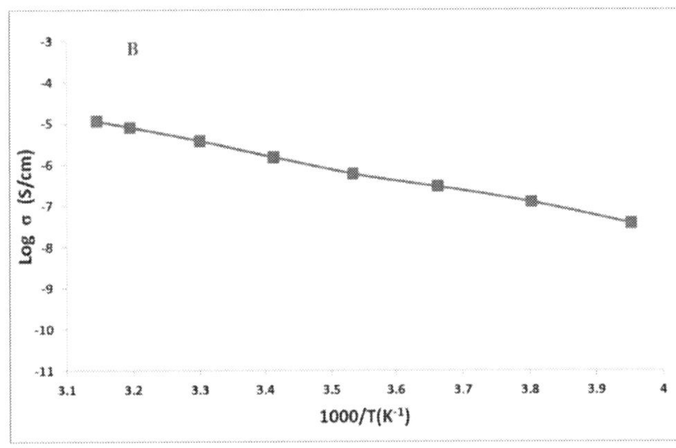

Figure 1a(top). Crystal packing diagram of DMF:$NaClO_4$ viewed along a axis. **Figure 1b(bottom).** Conductivity of DMF:$NaClO_4$ as function temperature

SYNTHESIS OF $NiCo_2O_4$/BORON AND NITROGEN CO-DOPED GRAPHENE FOR HIGH-PERFORMANCE SUPERCAPACITORS

*Xinyan Jiao, Qingli Hao**

Key Laboratory for Soft Chemistry and Functional Materials, School of Chemical Engineering, Nanjing University of Science and Technology
Nanjing 210094, China

Supercapacitors, with high power density, long cycling stability and environmental benignity, have attracted a great number of interests in the field of energy storage device[1]. $NiCo_2O_4$, a binary oxide, has been recently investigated as an electrode material for supercapacitors due to its higher electrical conductivity and redox reactivity than pure NiO or Co_3O_4. However, The limited rate capability during the charge/discharge processes is still the major challenge for $NiCo_2O_4$ as supercapacitors materials[2]. Graphene, a monolayer carbon atom, has large surface area, excellent electrical conductivity and high electrochemical stability. What is more, chemical doping is important to tailor the electronic properties of graphene[3]. Herein, we synthesized $NiCo_2O_4$/boron and nitrogen co-doped graphene hybrid with coating $NiCo_2O_4$ nanosheets on boron and nitrogen co-doped reduced graphene oxide via a facile co-precipitation method. Compared to individual $NiCo_2O_4$ and boron and nitrogen co-doped graphene, it exhibited enhanced electrochemical properties, such as higher specific capacitance, better rate capability and cycling stability. The BN-G efficiently increase the conductivity of the whole hybrid material, thus improving rate capabilities and cycling stability. Moreover, the $NiCo_2O_4$ nanosheets with high specific surface area can provide more electrode–electrolyte contact area for redox reaction, resulting in high specific capacitances and great rate performance[4]. Additionally, the well disperse of $NiCo_2O_4$ nanosheets on the BN-G surfaces and their synergetic effect play a positive role in the enhancement of the electrochemical performance for supercapacitors. Therefore, considering high specific capacitance, better rate capability and cycling stability of the $NiCo_2O_4$/BN-G hybrid, it shows promise for application as a outstanding supercapacitors electrode material. Fig.1 shows the ycling performance of the $NiCo_2O_4$/ BN-G hybrid.

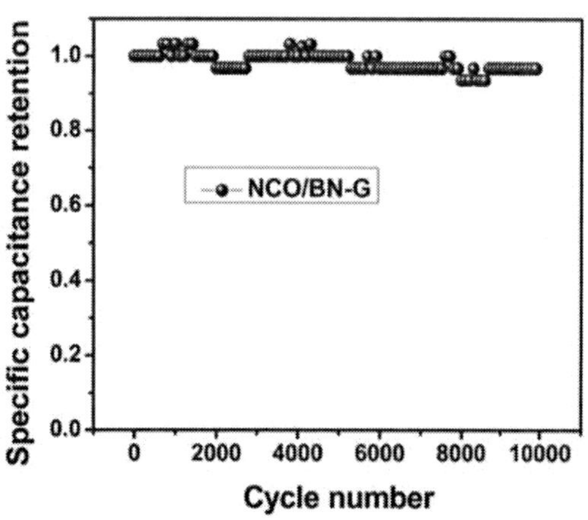

Figure 1. Cycling performance of the $NiCo_2O_4$/BN-G hybrid

Acknowledgement. The work was supported by the National Natural Science Foundation of China (No. 21576138), Program for NCET-12-0629, the Ph.D. Programs Foundation of Ministry of Education of China (No.20133219110018), Qing Lan Project and Six Major Talent Summit (XNY-011), the Science and Technology Support Plan (No. BE2013126), and PAPD of Jiangsu Province, China.

References

(1) Simon, P.; Gogotsi, Y. *Nat. Mater.* **2008**, 7:845–854.
(2) B,T, Dong.; X, Zhang.; X, Xu.; G, X, Gao.; S, J, Ding.; J, Li.; B, B, Li. *Carbon* . **2014**, 80,222-228.
(3) Z, S, Wu.; Winter, A.; L, Chen.; Y, Sun.; Turchanin, A.; X, Feng. *Adv Mater.* **2012**,24(37),5130–5.
(4) S , Wang.; D, Yu.; L, Dai.; D, W, Chang.; Baek, J. *ACS Nano*. **2011**, 5(8), 6202–9.

BIOELECTROCATALYTIC HYDROGEN EVOLUTION AT LOW VOLTAGE: SELF-SUPPORTED MICROBIAL CARBON AEROGEL ANODE, METAL-FREE CoP NANOWIRE CATHODE AND HIGHLY EFFECTIVE ELECTRON TRANSFER

*Bo Tang, Jian Liu, Guohua Zhao**

Department of Chemistry, Tongji University, 1239 Siping Road, Shanghai, 200092, China
ghzhao@tongji.edu.cn

Introduction

Hydrogen is a typical kind of sustainable energy carrier for developing low carbon emission economy whose characteristic is clean and high calorific value per mass unit. Microbial electrolysis cells (MECs) have been reported to utilize electroactive bacteria for extracting energy from wastewater and transferring electrons to the anode by bioelectrocatalysis[1-3].

Due to the capability of producing hydrogen by reutilizing wastewater with very low energy consumption, this kind of MECs is of great interest and has a significant potential application foreground. Nevertheless, current hydrogen production through MECs is severely and kinetically confined to low current densities of microbial anode, leading to low hydrogen production rate on the cathode.

Herein, we introduce a MEC employing a bulk, continuous 3-D carbon aerogel (CA) with high specific surface area and splendid electrical conductivity as the bio-anode and non-noble metal catalyst cobalt phosphate (CoP) as the cathode. Results show that hydrogen bubble can be observed under applied voltage of 0.19 V in such MEC, and considerable amount of hydrogen can be produced under 0.3 V.

Experimental

Continuous bulk 3-D structured carbon aerogel (CA) material was synthesized by polymerization method with 1, 3-dihydroxy benzene and formaldehyde, and dried at ambient temperature and pressure[4]. The as prepared CA was polished into the size of 13 mm × 28mm × 3mm and attached to titanium wire as a bulk electrode. Subsequently, the anode was inoculated with the anaerobic sludge from a local pond and operated in electricity generation mode as sequencing batch. The microbial CA electrode was considered to be accomplished till the anode potential reached a steady-state value for three consecutive times.

A dual chambered MEC was constructed with two glass bottles connected by a cation exchange membrane. Both chamber had a liquid volume of 28 milliliter. The anode chamber was filled with phosphate buffer solution containing 20 mM CH_3COONa and trace elements, while the cathode chamber with phosphate buffer solution containing 50 mM $K_3[Fe(CN)_6$. Two electrodes were connected with a resistance of 1 kΩ for determining the cell voltage. The solutions of both chambers were replaced per 48 h and then purged with nitrogen gas to remove the dissolved oxygen, thus keeping anaerobic condition. Commercial carbon fiber felt was calcined under 450 °C for 30 min, tailored into the same size as carbon aerogel and cultured into Bio-CF in the same method mentioned above.

Results and Discussion

Figure 1A exhibits the potential change of anodes after refilling nutrient solution (0.41g/L acetate sodium) for microbes. As for Bio-CA, there was no obvious potential increase after replacing the nutrient solution, and the potential descended to -0.24V immediately, then stabilizing at the minimum value of ~-0.28 V after 2.9 h. Notably, Bio-CA enabled the microbes rapidly to enter the electricity generating state and consequently lower the anode potential, which suggests that CA is an excellent culture substrate for the breeding and metabolism of microbes and has a better electricity generation capacity. Cyclic voltammetry test for microbial anode further showed that redox peaks occurred after microorganism attachment, indicating the microorganism possess electrochemical activity. As shown in Figure1B, a smaller polarization curve slope of Bio-CA was obtained, which indicated a less electric resistance on the as-prepared electrode. Besides, the power density of Bio-CA reached as high as 4890 mW/m^2, which was 4.67 times that of Bio-CF (1047 mW/m^2) showing that Bio-CA has a stronger electric production capacity.

Under the voltage of 0.6 V, the hydrogen production rate of Bio-CA||CoP is as high as 5.39 μmol/ ($cm^2 \cdot h$), which is 2.16 and 5.50 times that of the Bio-CA||Pt, and Bio-CF||CoP respectively. The result demonstrated that Bio-CA can significantly improve the electron generation capacity of microbial anode, and the CoP can act as a cheap, efficient alternative of Pt electrode[5], which ensures the MEC of Bio-CA||CoP could efficiently produce hydrogen under low voltage.

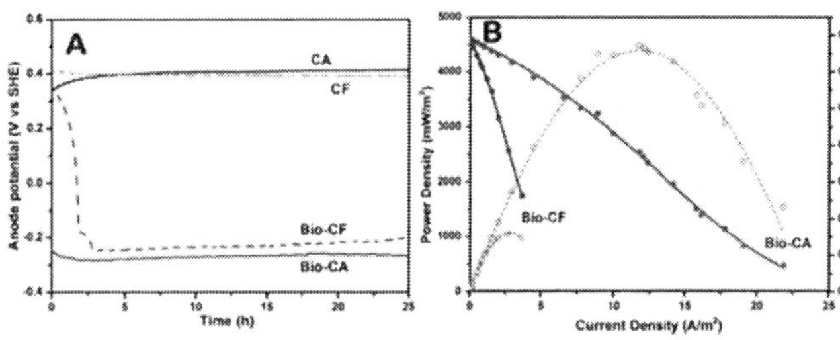

Figure 1 (A) microbial anode potential - time curves, (B) power density curves of Bio-CA and Bio-CF

Conclusions

3D-CA microbial and CoP NWs/CC electrode as novel high-performance anode and cathode respectively in the MEC system for hydrogen production is proposed in the present work. The significant enhancement of electrochemical performance of anode results from its specific surface area and electric bacteria loaded, which can efficiently promote electron transfer to the 1-D hierarchical CoP cathode to produce hydrogen under the low voltage. Our MEC study is also of great theoretical significance and has estimable application prospect for electrolysis hydrogen evolution under low energy consumption with energy recycling.

Acknowledgement. This work was financially supported by the National Natural Science Foundation of China (NSFC, No. 21477085 and 21537003).

References

(1) Logan, B. E.; Rabaey, K.. *Science*. **2012**, 337, 686-690.
(2) Harnisch, F.; Schroder, U.. *Chem. Soc. Rev*. **2010**, 39, 4433-4448.
(3) Cusick, R. D.; Kim, Y.; Logan, B. E.. *Science*. **2012**, 335 ,1474-1477.
(4) Wu, M. F.; Jin, Y. N.; Zhao, G. H.; Li, M. F.; Li, D. M.. *Environ. Sci. Technol.*, **2010**, 44 (5), pp 1780–1785

(5) Tian, J.; Liu, Q.; Asiri, A. M.; Sun, X. P. *J. Am. Chem. Soc.* **2014**, 136 (21), pp 7587–7590.

BORON NITRIDE NANOSHEETS AS EFFECTIVE SUPPORT FOR POLYETHYLENEIMINE IN CO_2 ADSORPTION

Kuan Huang[a], *Sheng Dai*[a,b]

[a]Department of Chemistry, University of Tennessee, Knoxville, TN 37966, USA

[b]Chemical Sciences Division, Oak Ridge National Laboratory, Oak Ridge, TN 37831, USA

dais@ornl.gov

Introduction

Solid adsorbents are considered as promising alternatives to amine-based liquid absorbents for the capture of CO_2 from flue gas. Among the array of available adsorbents, supported amines are particularly attractive because they can effectively and selectively adsorb CO_2 at low pressures and high temperatures. Various porous materials (e.g., carbons, oxides, MOFs, COFs and PPNs) have been investigated as the supports for amine compounds in CO_2 adsorption[1]. Boron nitride nanosheets (BNNs) have some distinctive features: 1. they can be facilely synthesized by calcinating the mixture of boric acid and urea without using any templates; 2. their porosity can be easily tuned by changing the ratio of reactants and reaction conditions; 3. they are abundant in Lewis acidic sites because the boron atoms are electron-acceptors. In this work, the performance of BNNs in supported adsorption of CO_2 was evaluated systematically. Results were compared with other supports, demonstrating the great potential application of BNNs in the post-combustion capture of CO_2.

Experimental

A simple BNN with surface area of 280 m^2/g and pore volume of 0.24 cm^3/g was synthesized by a well-established template-free route[2]. In brief, 0.01 mol of B_2O_3 and 0.20 mol of urea were dissolved in a mixture of methanol and water (40/20 mL). The homogeneous solution was heated under 75°C for recrystallization. A white crystalline powder subsequently formed after the evaporation of solvents. The white powder was placed in a quartz boat, heated to 900°C at a ramp rate of 5°C/min and kept for 2 h under flowing N_2 in a tube furnace. After cooling down to room temperature, the white and powderous BNN product was obtained.

The PEI-loaded BNN samples were prepared by physical impregnation of the BNN support with a required amount of methanolic solution of branched PEI (Sigma-Aldrich, M_W = 800) at room temperature, followed by drying at 110°C for 3 h. According to the measured PEI loading (x, defined as the weight fraction of PEI in the adsorbent), the final adsorbents were denoted as xPEI@BNN.

CO_2 adsorption capacities and PEI loadings were determined by a thermogravimetric analyzer (TGA, Seiko 6300 TG/DTA by RT Instruments Inc.). N_2-adsorption isotherms measured at −196°C on a Micromeritics Gemini 2390a surface area analyzer were used to calculate the surface areas, pore volumes and pore size distributions of all the samples. XRD patterns were collected on a PANalytical powder diffractometer using CuK_α radiation. FTIR spectra were acquired on an Excalibur Series instrument with a MVP-Pro ATR accessory.

Results and Discussion

Six PEI@BNN samples with different loadings (0.13~0.72) were prepared in this work. It is found the surface area and pore volume of PEI@BNN decrease significantly with the increase of PEI loading. 0.55PEI@BNN is still a porous and powderous solid, while 0.72PEI@BNN is a sticky solid without porosity. The experimental loading capacity is more than 5 times as the theoretically calculated value. This phenomenon can be attributed to two factors: 1. the accumulational pores in BNN have flexibility, and with more and more PEI loaded, the accumulational pores are expanded by the filled liquid; 2. PEI molecules are inserted into the interlayer space of BNN through the Lewis acid-base interaction between boron atoms and amine groups. XRD patterns of BNN before and after PEI loaded validate the insertion of PEI. The (002) peak of BNN shifts obviously to low angles after PEI loaded, which indicates the extension of interlayer distance. In the FTIR spectra, the two characteristic peaks of BNN (780 and 1382 cm^{-1}) remain unchanged after PEI loaded, indicating the preservation of the nanosheets structure of BNN. Furthermore, the characteristic peaks of PEI can be observed in PEI@BNN samples, and the strength of these peaks increases with the increase of PEI loading.

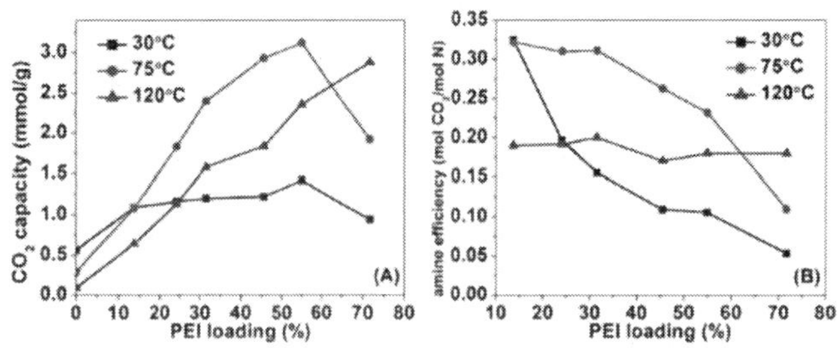

Figure 1. CO_2 capacities (A) and amine efficiencies (B) of PEI@BNN samples at different loadings.

In Figures 1, the CO_2 capacities and amine efficiencies of PEI@BNN samples were plotted against the PEI loadings. The CO_2 capacities of bare BNN are very low, with the values of 0.56, 0.29 and 0.09 mmol/g at 30, 75 and 120°C, respectively. After the impregnation of PEI, the CO_2 capacities are enhanced significantly. At 30°C, the CO_2 capacities of PEI@BNN samples are in the range of 1.0~1.5 mmol/g, and with the increase of PEI loading, the amine efficiency decreases constantly. At 75°C, the CO_2 capacity increases to the highest value of 3.12 mmol/g at the PEI loading of 0.55, and then decreases sharply to 1.92 mmol/g when the PEI loading further increases to 0.72, because at this loading, the solid adsorbent loses its porosity. The amine efficiencies are almost the same at the PEI loading of below 32%, with the values in the range of 0.31~0.32. When the PEI loading increases to above 32%, the amine efficiency starts to decrease. At 120°C, the PEI loading has no obvious influence on the amine efficiency, and the amine efficiencies are in the range of 0.17~0.20. As a result, the CO_2 capacity increases constantly with the increase of PEI loading. Therefore, the limitation of gas diffusion should be overcome at this temperature.

The performance of BNN for supported adsorption of CO_2 is found to be much better than mesoporous carbons, and comparable to other porous silicas[3].

Conclusions

Boron nitride nanosheets are demonstrated to be a new class of highly effective support for polyethyleneimine in CO_2 adsorption, thus exhibiting great potential applications in post-combustion capture of CO_2.

Acknowledgement. This work was supported as part of the Center for Understanding and Control of Acid Gas-Induced Evolution of Materials for Energy (UNCAGE-ME), an Energy Frontier Research Center funded by the U.S. Department of Energy, Office of Science, Basic Energy Sciences under Award # DE-SC0012577.

References

(1) Bollini, P.; Didas, S. A.; Jones, C. W. *J. Mater. Chem.* **2011**, *21*, 15100.
(2) Wu, P.; Zhu, W.; Chao, Y.; Zhang, J.; Zhang, P.; Zhu, H.; Li, C.; Chen, Z.; Li, H.; Dai, S. *Chem. Commun.* **2016**, *52*, 144.
(3) Son, W. J.; Choi, J. S.; Ahn, W. S. *Micro. Meso. Mater.* **2008**, *113*, 31.

HIGH PERFORMANCE AIRBRUSH SPRAY COATED ORGANIC SOLAR CELLS BASED ON DIFFERENT TERNARY SOLVENT SYSTEMS

*Ding Zheng, Junsheng Yu**

University of Electronic Science and Technology of China (UESTC), Chengdu 610054, PR China
zd901023@126.com

In this work, thieno [3,4-b] thiophene/benzodithiophene (PTB7): [6,6]-phenyl C71-butyric acid methyl ester (PC71BM) based organic solar cell (OSC) with a new record of power conversion efficiency (PCE) of □7.62% has been realized using airbrush spray (AS) coating method in air ambient which can be well compatible with large-scale fabrication. By investigating the physical mechanism of AS coated blend films, a series of ternary solvent systems (TSS) are used to simultaneous optimize the surface tension and the saturated vapor pressure of solution. Therefore, different TSS further controls the morphology of PTB7:PC71BM blend films precisely and systematically. It is elucidated that the chlorobenzene (CB)/o-Xylene (o-Xy)/1, 8-diiodoctane (DIO) TSS with a ratio of 37:60:3 vol.% could lead to a homogeneous surface morphology with a decreased aggregation domain size of active layer. In addition, the high fill factor, increased PC71BM absorption and internal quantum efficiency indicate the formation of bicontinuous interpenetrating and fully percolated networks with nanostructured phase separation in BHJ blend films. Ultimately, the AS coated OSCs based on the TSS of CB/o-Xy/DIO gains a 34% enhancement in PCE, compared with the conventional CB/DIO solvent based OSCs.

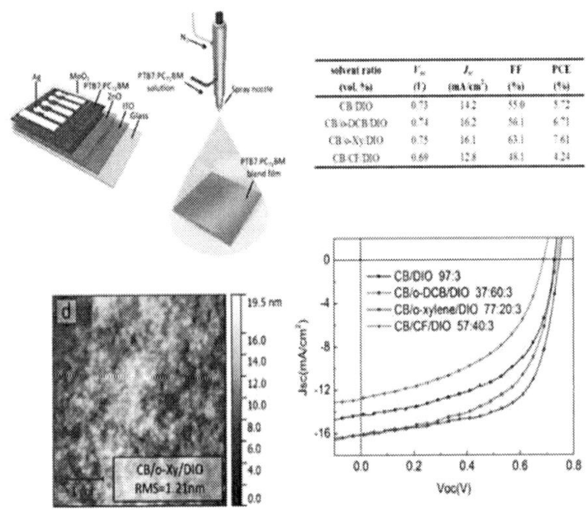

Figure 1. The graphic abstracts of this work.

References

(1) Liang, Y.; Xu, Z.; Xia, J.; Tsai, S. T.; Wu, Y.; Li, G.; Ray, C.; Yu, L. *Adv. Mater.* 2010, 22, E135.
(2) Zheng, Y.; Li, S.; Yu, X.; Zheng, D.; Yu, *J. Rsc. Adv.* 2014, 4, 16464.
(3) Zheng, Y.; Wu, R.; Shi, W.; Guan, Z.; Yu, *J. Sol. Energy Mater. and Sol. Cells* 2013, 111, 200.
(4) Zheng, Y.; Li, S.; Zheng, D.; Yu, J. *Org. Electron.* 2014, 15, 2647.

COMPARATIVE STUDY OF PRODUCTS DISTRIBUTION FOR ITERATIVE USE OF METHYL CYCLOHEXANE AS HYDROGEN CARRIER

Xieli Cui[a,]* *Mika Ishii,*[a] *Kenji Takano,*[a] *Tetsuya Nanba,*[a] *Taku Tsujimura,*[a] *Takaaki Taniguchi*[b]

[a]Fukushima Renewable Energy Institute, AIST (FREA), Koriyama,963-0298,Japan

[b]JX Nippon Oil & Energy,Yokohama,231-0815,Japan

*Corresponding author:sai-xieli@aist.go.jp

1. Introduction

Toward the effective utilization of hydrogen energy, methyl cyclohexane (MCH) as one of organic hydride has attracted attention as a hydrogen carrier. For iterative use of MCH, it is essential to examine the change of the quality and quantity of products of the dehydrogenation and hydrogenation reactions. The major components of the reaction products were toluene and MCH. It is considered that by-products contained in the product liquid were as a result of side reactions such as demethylation, ring-opening, isomer, and dimerization. In this study, we aim to understand the behavior of by-products accumulation by repeating hydrogenation of toluene and dehydrogenation of MCH.

2. Experimental

5 times repeat of MCH dehydrogenation and toluene hydrogenation were carried out by a catalytic evaluation apparatus with fixed-bed flow reactor and on-line GC. Over a hundred kinds of products of each process and their concentration have been analyzed by GC-FID and LC-MS. As a constraint condition for the dehydrogenation/hydrogena-tion reaction, the reaction temperature was adjusted to keep the concentration of by-product CH4 and the conversion of reaction within a certain range. Detailed experimental conditions are shown in Table 1.

3. Results and discussion

Through GC analysis to starting liquid of MCH, and liquid product of dehydrogenation/hydrogenation, almost 70 products were possibility identified. The impurity concentration in starting liquid of MCH was 0.6C%. The ratio of by-products was 0.8C% when first time dehydrogenation, and since then it was fluctuated in the range of 0.7-0.9C%. The change of by-product concentration in liquid product grouped according to different molecular structure is shown in Fig. 1. We calculated their carbon content by mol% and draw out their graphs. The production of demethylation reaction was increased when dehydrogenation and hydrogenation were repeated. However, the amount of increase will tend to be reduced gradually by repeated reaction. Compared to demethylation production, a small amount of biphenyls and bicyclohexyls as production of dimerization reaction was observed, and it showed a slight tendency to increase. Amount of Paraffin, as production of ring - opening reaction, is more than other production of side reaction through entire process. However, it is showing a slight tendency to decrease when dehydrogenation and hydrogenation was repeated. The by-products of ring structure, including 6-membered ring compound and 5-membered ring compound were detected from reaction of isomerization and addition. These products of isomerization and addition reaction were gradually increased when dehydrogenation and hydrogenation were repeated. On the other hand, the amount of by-product of 6-membered ring compound is increased at dehydrogenation and decreased at hydrogenation. Conversely, the amount of by-product of 5-membered ring compound is decreased at dehydrogenation and increased at hydrogenation. The fluorenes were formed from the products of dimerization reaction, as a result of polycyclic reaction. These results indicate that accumulation of the products of demethylation reaction was most remarkable, when the reaction of dehydrogenation and hydrogenation were repeated.

The change of reaction conversion and CH4 emission in dehydrogenation/hydrogenation of MCH is shown in Fig. 2. By repeated reaction of dehydrogenation/hydrogenation, the methane concentration in the reactor outlet gas found during both processes was significantly decreased in the dehydrogenation of MCH and the hydrogenation of toluene. As described above, this is in consistent with the gradual decrease of increased amount in demethylation reaction. They are considered due to decrease in activity or decrease of active sites for the demethylation reaction by repeating dehydrogenation/hydrogenation.

This study was conducted with the support of the Japan Petroleum Energy Center (JPEC) as a technological development project funded in part by the Ministry of Economy, Trade and Industry.

Table 1. Reaction conditions of dehydrogenation and hydrogenation cycle of MCH

	Dehydrogenation	Hydrogenation
Catalyst	Pt catalyst (JX)	Ni catalyst (BASF)
LHSV (h-1)	1.5	1.0~1.5
Temperature	< 350oC	< 250oC
Pressure	< 0.19 MPa	0.19 MPa
Constraint condition - conversion	about 90%	98-100%
Constraint condition - byproduct	concentration of CH4 200-300ppm	concentration of CH4 200-1000ppm

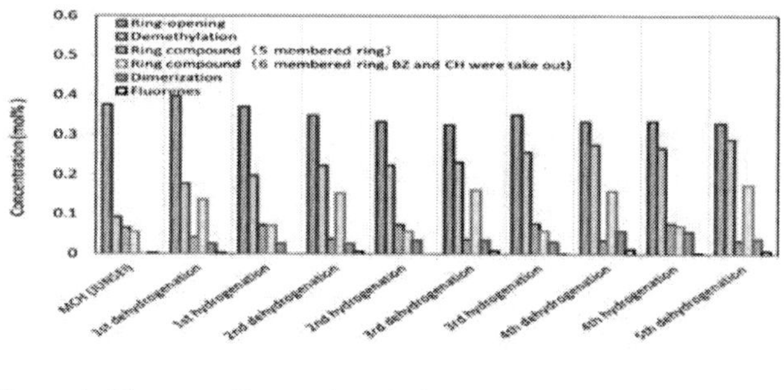

Figure 1. Changes of by-products in liquid product grouped according to different molecular structures in MCH cycle

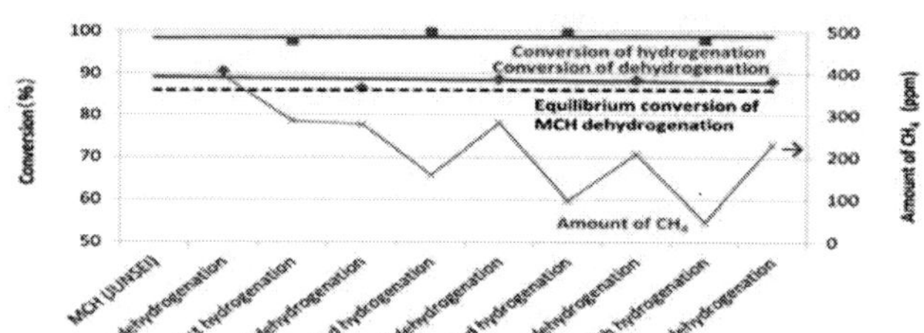

Figure 2. Changes of reaction conversion and CH4 emission in dehydrogenation/hydrogenation of MCH cycle

4. Conclusions

We conclude that accumulation of the products of demethylation reaction is most remarkable, when the reaction of dehydrogenation and hydrogenation were repeated.

FLUORINE RELEASE AND RETENTION DURING FIXED BED COMBUSTION OF PYROLYZED COAL CHARS

Naoto Tsubouchi,[1,]* *Yuuki Mochizuki,*[1] *Naoyuki Iwabuchi,*[2] *Yuuki Akama,*[2] *and Yasuo Ohtsuka*[2]

[1]Center for Advanced Research of Energy and Materials, Hokkaido University, Sapporo 060-8628, Japan
[2]Institute of Multidisciplinary Research for Advanced Materials, Tohoku University, Sendai 980-8577, Japan
*Corresponding author. E-mail address: tsubon@eng.hokudai.ac.jp.

Introduction

Coal typically contains fluorine element in trace amounts (denoted as coal-F) at 20-500 μg/g-coal, and the F is emitted mainly as HF during pulverized coal combustion.[1] Most the HF is removed by a wet desulfurization process downstream after combustion, and the rest is either released into the atmosphere or transferred into the ash.[2,3] It is thus important to investigate some factors determining F release during the combustion process and governing its distribution into the ash, but little research on these topics has been conducted.[1,2]

In this work, we examined in detail the evolution of HF from the F in char (char-F) during combustion, and aimed to elucidate the chemical form of the F in the ash after combustion and the reaction behavior of HF with carbonaceous materials, in order to study several factors controlling the fate of F element during char combustion.

Experimental

Five coals (AT, DK, DT, CN and OP coals) with C contents of 78-80 mass%-daf and with particle sizes of 150-250 μm were used in this work. The contents of F and ash ranged from 20 to 80 μg/g-dry and from 7.6 to 13 mass%-dry, respectively.

All experiments were carried out using a fixed-bed quartz reactor. Approximately 3.0 g of the coal sample was first heated in a flow of pure He at 10°C/min up to 1000°C, then quenched to ambient temperature and finally reheated in 20%O_2/He at 10°C/min up to 1000°C. The details of the apparatus have been described earlier.[4]

The HF evolved during these runs was analyzed online at 10 s intervals with an ion-selective electrode. The char after pyrolysis and the ash after combustion were assayed for F using an absorption spectrophotometer or ion chromatograph after ionization to F^-. The ashes were also characterized by the F 1s XPS using Mg-Kα radiation. The binding energies of all spectra observed were compensated with an In $3d_{5/2}$ peak (444.9 eV)[5] of In_2O_3.

Results and Discussion

F Behavior during Coal Pyrolysis and Char Combustion. Investigation of HF evolution during temperature-programmed pyrolysis revealed the presence of two peaks at around 400°C and between 670 and 730°C for all coals, but HF yields up to 1000°C were as low as less than 5 %, irrespective of the type of coal. On the other hand, the yields of char-F were 85 to 95 % with all coals. Our earlier research showed that the Cl present in coal is released predominantly as HCl during pyrolysis, and HCl yields up to 800°C are as high as 50-95 %.[4] Therefore, the major portion of coal-F was extremely thermally stable compared to the Cl.

The resulting chars were then reheated under a 20%O_2 atmosphere at 10°C/min up to 1000°C. Carbon burn-off ranged from 50 to 60 %, but very little HF was formed, irrespective of the char type, and most of F element remained in the residual chars. Figure 1 shows the evolution profiles of HF and CO_2 during holding AT, DK and DT chars at 1000°C. In every case, HF started to form after CO_2 formation was nearly complete (C burn-off of 85 to 90 %). This tendency was also observed with CN and OP chars. On the other hand, total amount of HF evolved depended significantly on the char type (Figure 1).

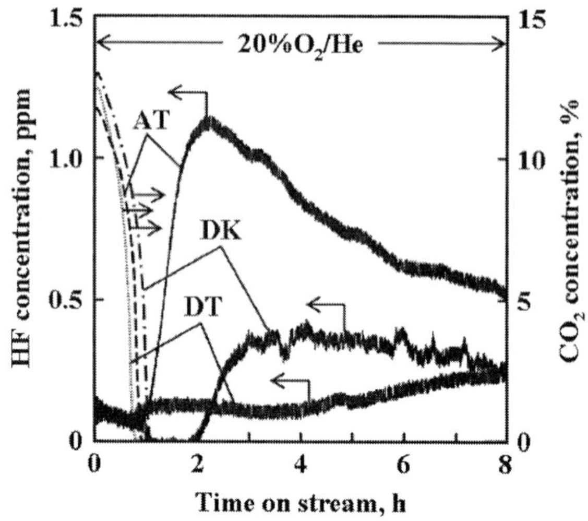

Figure 1. Change in HF and CO_2 concentrations with time during combustion of AT, DK and DT chars at 1000°C.

The relationship between HF yield and C burn-off after holding at 1000°C for 8 h was investigated, and the results are shown in Figure 2. Interestingly, the yield tended to be larger for samples with a higher C burn-off. It is thus evident that the amount of unburned carbon is one of the key factors determining HF formation.

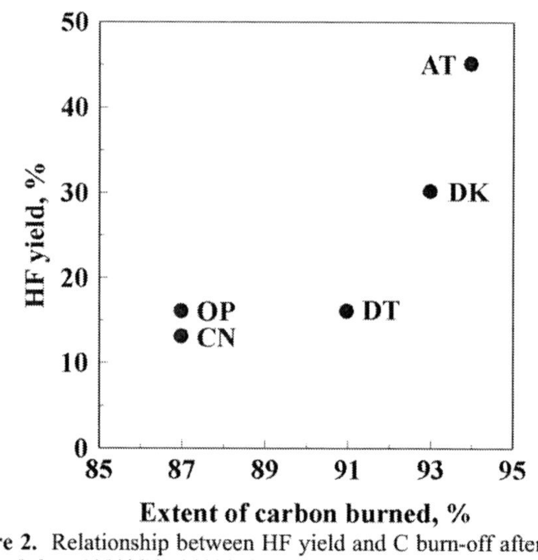

Figure 2. Relationship between HF yield and C burn-off after holding for 8 h at 1000°C in 20%O_2/He.

The F amounts of the combustion residues from AT and DK chars, which provided relatively high HF yields, were then investigated and found to range from 70 to 100 μg/g-ash, which was almost identical to the range in the corresponding raw coal. The chemical form of F retained in the residues was thus determined.

Surface F Forms of Combustion Residues. Figure 3 shows the F 1s XPS spectra of the combustion residues of AT and DK chars. No peaks attributable to fluorides were detected in samples not held at 1000°C for either char. Similar results were also observed for samples after 30 s Ar etching. After combustion for 8 h, in contrast, the broad spectra appeared clearly and had two peaks at about 685.5 and 688.5 eV (solid lines). Earlier work showed that CaF_2 and $(CH_2CF_2)_n$ gave the F 1s peaks at 684.8-685.4 and 688.6 eV, respectively.[5] Further, CaF_2 and $C_{10}H_7F$ measured in the present study exhibited the 1s signals at 685.2 and 686.7 eV, respectively. Curve-fitting analyses (broken lines) of the observed spectra that employed these compounds revealed the presence of both inorganic and organic F, and the proportion of inorganic F was estimated to be 70-80 mol%. These results suggested the occurrence of secondary reactions of HF once-evolved with unburned carbon.

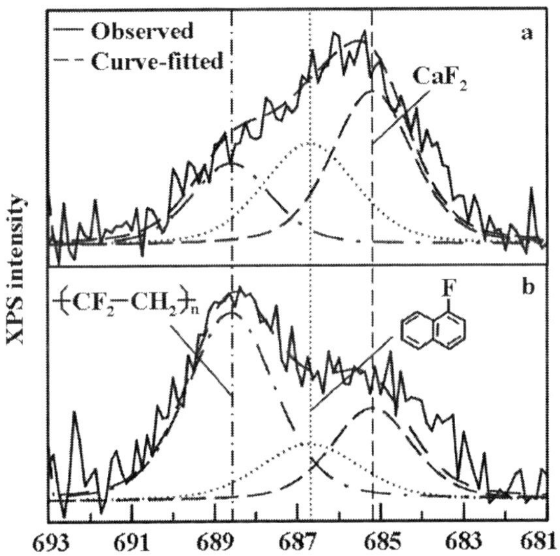

Figure 3. F 1s XPS spectra for AT (a) and DK (b) combustion residues.

Reaction of HF with a Model Carbon. To examine if HF would react with carbonaceous materials at high temperature, 15 ppm HF/N_2 was streamed at 950°C over pure carbon prepared by carbonizing phenol resin. Time change in HF concentration is shown in Figure 4. The concentration rapidly decreased with time to about 3 ppm after 3 min, and then it gradually approached the original level of 15 ppm. Total amount of HF reacted during 1 h treatment was estimated to be approximately 15 μmol/g. Since it has been reported that NH_3 treatment of activated carbon results in the formation of pyridinic-N and quaternary-N,[6, 7] the incorporation of F element can likely take place in interactions between HF and edge carbon. When the sample after HF treatment was heated under a stream of high-purity He, only a trace amount of HF was evolved (3 μmol/g). Thus, C-F species produced through the reaction of HF with pure carbon was found to be thermally stable. These results suggest that some of the HF once-evolved during char combustion may react with unburned carbon and be retained in the ash after combustion.

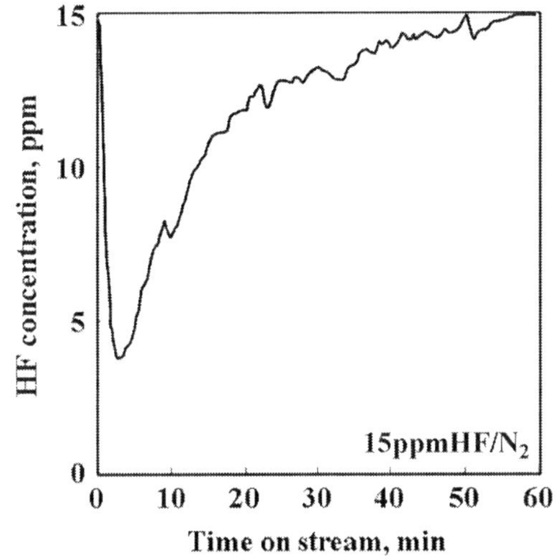

Figure 4. Change in HF concentration during the reaction of HF and pure carbon at 950°C.

Conclusions

The fate of char-F in the fixed bed combustion of five pyrolyzed chars has been examined mainly with a quartz reactor at 1000°C. HF formation starts after the apparent completion of CO_2 evolution with all chars. The yield after 8 h combustion is in the range of 13 to 45 %, and it is larger for the chars with higher burn-off of 93-94 %. The XPS spectra for the chars with HF yields of 30-45 % give the distinct F 1s signals, which can be identified to inorganic and organic fluorine, the latter proportion being as high as 70-80 mol% A model experiment shows that HF can react with pure carbon without any minerals at 950°C. It is thus possible that part of HF once-evolved from char-F may react secondarily with unburned carbon remaining in ash after combustion to be transformed into organic C-F forms.

Acknowledgement. This study was supported in part by a Grant-in-Aid for Scientific Research (B) from the Ministry of Education, Culture, Sports, Science and Technology, Japan.

References

(1) Davidson, R. M. *Chlorine and other halogens in coal*; IEA Coal Research: London, 1996 (IEAPER/28).
(2) Caridi, A.; Cereda, E.; Fazinic, S.; Jaksic, M.; Braga Marcazzan, G. M.; Valkovic, O.; Valkovic, V. *Nucl. Instrum. Methods Phys. Res. Sect. B-Beam Interact. Mater. Atoms* **1992**, *66*, 298.
(3) Ikuno, H. In Proceedings of the 97th meeting on coal utilization technology of the Japan Society for the Promotion of Science, Tokyo, Japan, 2005, p11.
(4) Tsubouchi, N.; Ohtsuka, S.; Nakazato, Y.; Ohtsuka, Y. *Energy Fuels* **2005**, *19*, 554.
(5) Moulder, J. F.; Stichle, W. F.; Sobol, P. E.; Bomben, K. D. Handbook of X-ray photoelectron spectroscopy; Perkin-Elmer: Eden Praine, 1992.
(6) Stöhr, B.; Boehm, H. P.; Schlögl, R. *Carbon* **1991**, *29*, 707.
(7) Mangun, C. L.; Benak, K. R.; Economy, J.; Foster, K. L. *Carbon* **2001**, *39*, 1809.

DEVELOPMENT OF COAL CHEMICAL LOOPING COMBUSTION AND GASIFICATION: A BRIEF REVIEW

Ping Wang[a], Nicholas Means[b], David Berry[c] and Mehrdad Massoudi[a]*

[a] Department of Energy, National Energy Technology laboratory, 626 Cochrans Mill Road, Pittsburgh, PA 15236, USA
[b] AECOM, 626 Cochrans Mill Road, Pittsburgh, PA 15236, USA
[c] Department of Energy, National Energy Technology Laboratory, 3610 Collins Ferry Road, Morgantown, WV 26507, USA
*Correspondence: ping.wang@netl.doe.gov

Introduction

Coal along with oil, gas, renewables, and nuclear will supply the growing world energy demand through 2040[1]. Coal power generation faces a major challenge of reducing CO_2 emission. CO_2 is a primary greenhouse-gas and widely believed to be responsible for climate change. Chemical-looping combustion (CLC) is one of the most promising CO_2 capture technologies that generates a CO_2-enriched flue gas, which will greatly benefit CO_2 capture and sequestration[2]. CLC is a cyclical oxygen transporting process in which oxygen carriers (OC) are reduced to provide oxygen for the fuel conversion in a fuel reactor (Figure 1)[2]. The reduced oxygen-carrier is re-oxidized in an air reactor before beginning a new cycle.

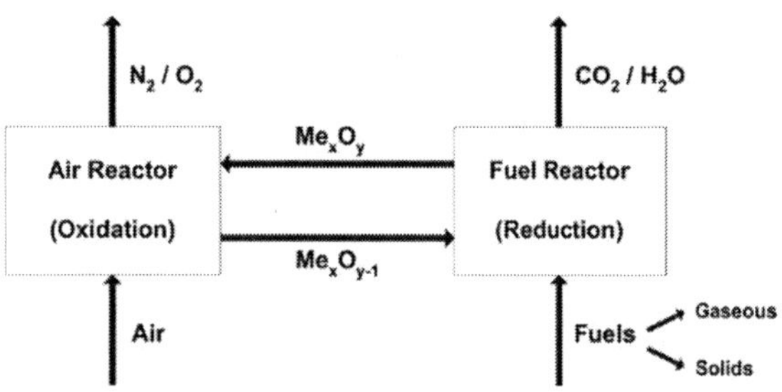

Figure 1. Schematic of chemical-looping combustion process that consists of a fuel reactor and an air reactor[2].

Gaseous fuels CLC and reforming has been mainly investigated in recent years and was reviewed by Adanez et al.[2] and Hossain and de Lasa[3]. Solid fuels CLC research has been of interest recently because of rich solid fuel resources. Coal is a heterogeneous material and consists of organic and mineral matters[4]. As a result, coal chemical-looping is more complicated and faces more challenges than CLC with gaseous fuels. A proposed direct coal CLC has been investigated, which feeds coal directly into the fuel reactor and all reactions take place in the same reactor. This process has two approaches: in situ Gasification Chemical-Looping Combustion (iG-CLC)[5] and Chemical-Looping with Oxygen Uncoupling (CLOU)[6]. Oxygen carriers are a key to the development of chemical-looping technologies. The objectives of this review are to summarize major proposed solid fuel chemical-looping processes with the development of oxygen carriers and to discuss development and challenges in iG-CLC and CLOU.

Direct Solid Fuel CLC and Chemical-looping Gasification

In the iG-CLC process, coal is first gasified in situ by H_2O and/or CO_2 to produce syngas (H_2 and CO). Then the syngas reacts with OC (e.g. Fe_2O_3)[2]. Coal gasification is the rate limiting step because gasification is slower than the syngas reaction with OC[7]. In the CLOU

Table 1. The main reactions occur in the fuel reactor for iG-CLC and CLOU

iG-CLC
Coal → volatiles + char (C)
$C(s) + H_2O(g) \rightarrow CO(g) + H_2(g)$
$C(s) + CO_2(g) \rightarrow 2\ CO(g)$
$CO+H_2+2MexOy \rightarrow 2MexOy\text{-}1 + CO_2(g) + H_2O(g)$
CLOU
Coal → volatiles + char (C)
$2MexOy \leftrightarrow 2MexOy\text{-}1 + O_2(g)$
volatiles + $O_2(g) \rightarrow CO_2(g) + H_2O(g)$
$C(s) + O_2(g) \rightarrow CO_2(g)$

process, OC (e.g. CuO/Cu_2O, Mn_2O_3/Mn_3O_4) in the fuel-reactor releases gaseous oxygen. Then the gaseous oxygen combusts coal[6]. Table 1 listed the main reactions in the fuel reactor in iG-CLC and CLOU. Compared to the iG-CLC process, CLOU avoids the slow char gasification step. The conversion rate for CLOU with a Cu-based OC (40%CuO/ZrO_2) was 3 to 15 factors faster than iG-CLC with a Fe-based OC (ilmenite) at 95% conversion of solid fuels (coal, petroleum coke or wood char) in a fluidized bed[8]. CLOU is considered a promising technology to burn solid fuels with CO_2 capture[6].

Proposed solid fuel chemical-looping gasification (CLG) or reforming is used to enhance H_2 and CO_2 production as part of conventional gasification processes, such as a Fe-based OC or a Ca-based OC[9].

Important Oxygen Carrier Characteristics and Development

Oxygen carriers for CLC can be classified as two types: synthetic materials and natural minerals. The synthetic OC consist of metal oxides (single, combined or mixed of metal oxides) and inert support (e.g. SiO_2, $MgAl_2O_4$, and ZrO_2)[2]. Important OC characteristics are: (1) sufficient oxygen transport capability; (2) high reactivity for reduction and oxidation reactions; and (3) favorable fluidization property (low attrition and no agglomeration)[10]. For the inert support materials, important characteristics are stability, surface area, and attrition resistance.

Synthetic OC have been more researched and are more expensive than natural minerals (e.g. iron ore, ilmenite, manganese ore, and waste materials from the steel industry and alumina production)[11]. Natural minerals are preferred for coal chemical-looping because of their low cost and the difficulty of ash separation. Because of this, it is easy to discard both the natural mineral and ash together. Natural minerals usually have lower reactivity than synthetic OC. The addition of metal oxides or alkali and alkaline metals (e.g. K+, Na+, Ca+) has been studied to improve the reactivity of natural minerals[12].

For solid fuel chemical-looping processes, primarily oxides of Ni, Fe, Mn, and Cu have been studied. Nickel oxide is not suitable for coal chemical-looping because it is expensive and easily deactivated by sulfur from coal[13]. Fe-based oxygen carriers have been mainly used in iG-CLC processes. Iron and manganese oxides are relatively cheap and are also available from natural ores and from industrial waste. Ilmenite is used in many studies because it is cheap, has reasonably high reactivity with syngas and good fluidization behavior[14]. CuO has high reactivity and releases gaseous oxygen at high temperature so it is often used in CLOU, even though it is more expensive compared to Fe-based OC. Mn-based OC is also used in CLOU. Mn is cheaper than Cu but Mn-based materials have a lower rate of oxygen release compared to Cu-based materials[15]. To date, great attention has been given to Cu-based and Mn-based materials mixed with Ca, Mg, Cu, Ni, Fe or Si[16].

Copper oxide has been tested in iG-CLC, but in recent years it is commonly investigated for CLOU. In CLOU, CuO releases gaseous oxygen and is reduced to Cu_2O (melting temperature: 1235°C) at high temperatures (850–1000°C). In iG-CLC, CuO is reduced to Cu (melting temperature: 1085°C) at lower temperatures and/or low CuO content (less than 20 wt.%) to avoid agglomeration[17].

Thermogravimetric analysis (TGA) along with fixed bed and bubbling fluidized bed reactors have commonly been used to screen and characterize oxygen carriers (reactivity and recyclability) at the lab scale[10]. Two interconnected fluidized beds in both batch and continuous operation are used to scale-up and validation of oxygen carriers.

Development and Challenges in iG-CLC and CLOU

To achieve high CO_2 capture efficiency, high solid and gas conversions are the key targets for the fuel reactor[18]. High reactivity of OC and fuel are required to complete conversion of volatiles and char with elevated reaction rates. Based on pilot scale tests, increased reactor temperatures will have improved CO_2 capture efficiency for both iG-CLC and CLOU[10]. So the oxygen carrier operating at high temperature is preferred. For large scale industrial applications of coal CLC, the availability of low cost oxygen carriers with sufficient reactivity is very important because of the coal ash.

For the coal iG-CLC process, natural minerals and industrial waste from steel and alumina production have been commonly tested for bituminous coal in pilot scales (e.g. 100 kW_{th} using ilmenite)[10]. Enhancing the char gasification rate is a critical issue because char gasification is the rate limiting step in iG-CLC. CO_2 and/or H_2O as a gasifying agent are often used to improve the rate of char conversion. Steam is commonly used for high rank coal and CO_2 is used for highly reactive solid fuels (e.g. low-rank coals or biomass) because of the slow gasification rate with CO_2[7]. Alkali metal salts can catalyze the steam gasification reaction. Various coals have different alkali metal content. Research on the selection of a suitable coal type or mixture could enhance char gasification so the addition salts may not be needed. Study is also needed to investigate volatile reactions under chemical-looping conditions for coal with different volatile content. Because volatiles in the fuel reactor, under a reducing environment, may not be completely converted.

For the coal CLOU process, a synthetic Cu-based OC (60%CuO with $MgAl_2O_4$ as support) for bituminous coal was tested at the pilot scale (e.g. 1.5 kW_{th})[19]. From these tests, the major issue with CuO was the decreased mechanical stability. Defluidization is also a potential issue with CuO that may be caused by over-reduced CuO to Cu resulting in agglomeration and particle size change[10]. Char deposition on OC was observed and requires attention. To fully convert the coal (volatiles and char), excess oxygen (2.5 O_2/MAF coal) is needed[10]. As a result, OC with high O_2 uncoupling capacity and/or low cost are required. Natural copper ores were investigated and showed high reactivity and cyclic stability. The agglomeration tendency was found to depend on the Cu content in the ore. Future study is needed to increasing the CuO content in the natural cupper ore and to avoid agglomeration. Additionally, the effect of SO_2 and NOx generated from coal combustion requires further investigation[10].

Remarks

Chemical-looping is a cyclical oxygen transporting process using oxygen carriers, which is a major factor for CO_2 capture. For coal chemical-looping, developing oxygen carriers with sufficient reaction rate, low cost, long life, adequate for different types of coal and chemical-looping processes is still a main challenge. The influence of mineral matter/ash and sulfur on the activity of the oxygen carrier requires further investigation. More attention should be paid to CLOU because it is specifically used for solid fuel chemical-looping combustion and gasification.

References

(1) International Energy Agency (IEA). *World Energy Outlook 2015*; IEA: France, 2015.

(2) Adanez, J.; Abad, A.; Garcia-Labiano, F.; Gayan, P.; de Diego, L. F., Progress in Chemical-Looping Combustion and Reforming Technologies. *Progress in Energy and Combustion Science* **2012**, *38* (2), 215-282.

(3) Hossain, M. M.; de lasa, H. I., Chemical-looping combustion (CLC) for inherent CO2 separations-a review. *Chemical Engineering Science* **2008**, *63* (18), 4433-4451.

(4) Wang, P.; Massoudi, M., Slag behavior in gasifiers Part I: Influence of coal properties and gasification conditions. *Energies* **2013**, *6*, 784-806.

(5) Cuadrat, A.; Abad, A.; Garcia-Labiano, F.; Gayan, P.; de Diego, L. F.; Adanez, J., Relevance of the coal rank on the performance of the in situ gasification chemical-looping combustion. *Chemical Engineering Journal* **2012**, *195*, 91-102.

(6) Mattisson, T.; Lyngfelt, A.; Leion, H., Chemical-looping with oxygen uncoupling for combustion of solid fuels. *International Journal of Greenhouse Gas Control* **2009**, *3* (1), 11-19.

(7) Leion, H.; Mattisson, T.; Lyngfelt, A., Solid fuels in chemical-looping combustion. *International Journal of Greenhouse Gas Control* **2008**, *2* (2), 180-193.

(8) Leion, H.; Mattisson, T.; Lyngfelt, A., Using chemical-looping with oxygen uncoupling (CLOU) for combustion of six different solid fuels. *Greenhouse Gas Control Technologies 9* **2009**, *1* (1), 447-453.

(9) Mehdizadeh, A. M.; Klausner, J. F.; Barde, A.; Mei, R., Enhancement of thermochemical hydrogen production using an iron-silica magnetically stabilized porous structure. *International Journal of Hydrogen Energy* **2012**, *37* (11), 8954-8963.

(10) Wang, P.; Means, N.; Shekhawat, D.; Berry, D.; Massoudi, M., Chemical-Looping Combustion and Gasification of Coals and Oxygen Carrier Development: A Brief Review. *Energies* **2015**, *8* (10), 10605-10635.

(11) Mendiara, T.; Abad, A.; de Diego, L. F.; Garcia-Labiano, F.; Gayan, P.; Adanez, J., Use of an Fe-Based Residue from Alumina Production as an Oxygen Carrier in Chemical-Looping Combustion. *Energy & Fuels* **2012**, *26* (2), 1420-1431.

(12) Song, T.; Shen, L.; Guo, W.; Chen, D.; Xiao, J., Enhanced Reaction Performance with Hematite/Ca2Al2SiO7 Oxygen Carrier in Chemical Looping Combustion of Coal. *Industrial & Engineering Chemistry Research* **2013**, *52* (28), 9573-9585.

(13) Leion, H.; Lyngfelt, A.; Mattisson, T., Solid fuels in chemical-looping combustion using a NiO-based oxygen carrier. *Chemical Engineering Research & Design* **2009**, *87* (11A), 1543-1550.

(14) Cuadrat, A.; Abad, A.; Garcia-Labiano, F.; Gayan, P.; de Diego, L. F.; Adanez, J., The use of ilmenite as oxygen-carrier in a 500 W-th Chemical-Looping Coal Combustion unit. *International Journal of Greenhouse Gas Control* **2011**, *5* (6), 1630-1642.

(15) Leion, H.; Larring, Y.; Bakken, E.; Bredesen, R.; Mattisson, T.; Lyngfelt, A., Use of CaMn0.875Ti0.125O3 as Oxygen Carrier in Chemical-Looping with Oxygen Uncoupling. *Energy & Fuels* **2009**, *23*, 5276-5283.

(16) Pour, N. M.; Leion, H.; Ryden, M.; Mattisson, T., Combined Cu/Mn Oxides as an Oxygen Carrier in Chemical Looping with Oxygen Uncoupling (CLOU). *Energy & Fuels* **2013**, *27* (10), 6031-6039.

(17) de Diego, L. F.; Gayan, P.; Garcia-Labiano, F.; Celaya, J.; Abad, M.; Adanez, J., Impregnated CuO/Al2O3 oxygen carriers for chemical-looping combustion: Avoiding fluidized bed agglomeration. *Energy & Fuels* **2005**, *19* (5), 1850-1856.

(18) Lyngfelt, A., Chemical-looping combustion of solid fuels - Status of development. *Applied Energy* **2014**, *113*, 1869-1873.

(19) Abad, A.; Adanez-Rubio, I.; Gayan, P.; Garcia-Labiano, F.; de Diego, L. F.; Adanez, J., Demonstration of chemical-looping with oxygen uncoupling (CLOU) process in a 1.5 kW(th) continuously operating unit using a Cu-based oxygen-carrier. *International Journal of Greenhouse Gas Control* **2012**, *6*, 189-200.

AEROBIC CULTIVATION OF OLEAGINOUS YEAST STRAINS TO PRODUCE FATTY ACIDS FOR BIOFUEL APPPLICATION

Kathryn A. Notarangelo, Pragya Tooteja, Casey Crownhart, Timothy Manganello, Jonas De Oliveira, Anita Wamakima, Konyin Oluwole, Alexander Kendrick and Jean-François P. Hamel

MIT, Chemical Engineering Department, 77 Massachusetts Avenue, Cambridge, MA 02139, USA
jhamel@mit.edu

In partnership with National Renewable Energy Laboratory, our group develops projects concerning the optimal conditions for the aerobic cultivation of oleaginous yeast strains, to produce triacylglycerol (TAG) fatty acids. These fatty acids are analyzed through methylation and used as a precursor to renewable biodiesel. We employ benchtop 2-L batch bioreactors and analytical equipment, such as HPLC and GC, for off-line analysis of substrate and products, and mass spectrometry, for online gas analysis. Questions of interest include optimization of biomass growth relative to TAG accumulation, minimization of oxygen sparging rates, and finding the optimal carbon/nitrogen ratios for TAG accumulation.

MICROBIAL PRODUCTION OF CIS, CIS-MUCONIC ACID FROM LIGNIN

Faben Girma, Emma Valentine, Zi-Ning Choo, Kathryn A. Notarangelo, Eleazar Bernal III, Juan Jaramillo, Jessica Torres, Ruifan Pei, Elizabeth A. Mule, Johanna Y.-W. Tang, Reginaldo Gomes, Dayannara Muñoz, William Livernois, Jonas De Oliveira, Anita Wamakima, Konyin Oluwole, Alexander Kendrick and Jean-François P. Hamel

MIT, Chemical Engineering Department, 77 Massachusetts Avenue, Cambridge, MA 02139, USA
jhamel@mit.edu

In partnership with National Renewable Energy Laboratory, our group develops projects concerning the optimal conditions for muconic acid production from *Pseudomonas putida* bacterial strains. Muconic aid is a precursor to commercially important bulk chemicals, such as adipic acid, that can be employed in the manufacture of new functional resins, bio-plastics, food additives, agrochemicals, and pharmaceuticals. The genetically modified *P. putida* produces muconic acid when fed with lignin-derived aromatics, common structural polymers extracted from algae and plant cells. The use of this biological pathway for making muconic acid is environmentally beneficial because current commercial processes pollute the environment through the release of nitric oxide and carbon dioxide during production. Our lab seeks to optimize muconic acid production though benchtop bioreactor testing and simulations to make this process economically viable. We characterize muconic acid and glucose substrate off line, through high performance liquid chromatography and biochemical analyzer, and O_2 and CO_2 on line, by mass spectrometry.

MULTIFUNCTIONAL PEPTIDES FOR BIO-TETHERING AND SELF-ASSEMBLY OF LITHIUM ION BATTERY ELECTRODES

Alexander J. Winton

Mark A. Allen, Assistant Professor, Department of Chemistry and Biochemistry, University of Maryland Baltimore County

The focus of this project is twofold; first to develop novel, bio-hybrid electrodes for lithium ion batteries which demonstrate improved performance parameters compared with existing materials and second; to develop specific and multifunctional biomimetic templates that serve to improve the synthesis of functional inorganic materials. The bio-hybrid aspect will be achieved using artificial evolution to identify inorganic binding peptides. These peptides will provide the foundation for the rational design of multi-domain peptides which will be utilized as bio-tethering agents to improve the stability of hetero-structured electrodes. As a biomimetic template, the solid binding peptides (SBPs) may help to prepare electrode nanomaterials under more environmentally benign conditions. The proposed research goal may be classified into three sub aims: 1) Identification and Characterization of Solid Binding Peptides for Lithium Ion Battery Electrode Materials, 2) Bio-tethered Electrodes for Improvement of Cycle Life, Energy Density, and Power Performance, 3) Biomimetic Synthesis of Electroactive Particles Utilizing Identified Solid Binding Peptides.

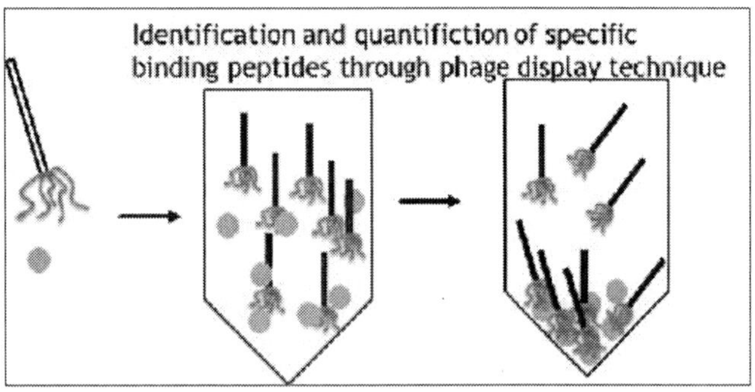

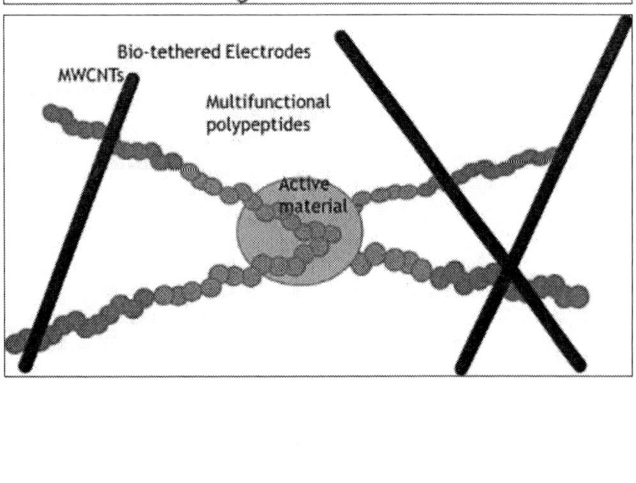

A 2D AXISYMMETRIC MODELLING INVESTIGATION OF MEMBRANE METHANE STEAM REFORMERS

Grigorios Panagakos[a], Emmanuel Stamatakis[a], Georgios Stubos[a], Kyriakos Panopoulos[b], Athanasios Stubos[a]*

[a] Environmental Research Laboratory, National Center for Scientific Research (NSCR) DEMOKRITOS
Patr. Gregoriou E' & 27, Neapoleos str., PO Box 60037, 15341, Agia Paraskevi, Attica, GREECE
[b] Laboratory of Process Systems Design and Implementation (LPSDI), Chemical Processes and Energy REsources Insitute (CPERI), Centre for Research and Technology Hellas (CERTH), 6th km Harilaou Thermi Road, P.O. Box 361, 57001, Thermi, Thessaloniki, GREECE
* Correspondence: gripanag@gmail.com

Introduction

Methane steam reforming is one of the predominant ways for the production of hydrogen and both a valuable current technology pillar as well as an indispensable intermediate step towards the transition to hydrogen economy. Several experimental and computational studies performed recently are indicative of the significance from an engineering and a scientific point of view. It has been experimentally demonstrated[1] that 1D modeling of reformers[2], while frequently used due to ease of acquiring solutions and offering some theoretical insight, are not capable to capture the full extent of the reactions and the coupling between the active physics. This fact has also been made by other numerical studies[3]. The 2D model approach adopted in this study, assumed fixed thermodynamic properties, plug flow in porous medium, and ignored direction diffusion along the z axis. In the same time, the flow characteristics and the question of what the velocity and pressure profiles look like are not addressed. Operating the reformer in an intermediate temperature requires the presence of a catalytic material for the steam reforming reactions. It has been shown[4] that other Ni-based catalysts showed extremely stable performance for long time at lower than 773 K temperatures. The main operating parameters of such a catalytic reactor are the reactants' feed flow rate, the temperature and the pressure under which the reaction takes place[5]. Measure of the reactor's efficiency can be obtained by defining the methane conversion rate, $CH_4\ \text{conversion} = \frac{\left(F^{in}_{CH_4} - F^{out}_{CH_4}\right)}{F^{in}_{CH_4}}$ the hydrogen yield rate, $H_2\ \text{yield} = \frac{F^{out-total}_{H_2}}{F^{in}_{CH_4}}$ and the hydrogen recovery yield: $H_2\ \text{recovery yield} = \frac{F^{out-perm}_{H_2}}{F^{in}_{CH_4}}$. These definitions are given based on the calculations of the relevant molecular flow rates $F_j^{in/out}$ (moles/s) at the inlet and outlet for the corresponding species. Effect of the first two of the afore-mentioned operating conditions, i.e. reactant feed flow rate and reactor's temperature has been accounted in the simulations presented in recent work[6]. Aspiration for this model emerged from the need to alleviate certain of the restrictions faced by previous modeling attempts when simulating real experimental reactors[7,3].

Extending this work[6], we treat the effect of the pressure on the reactor's efficiency. The setup used, presented in Figure 1, consists of two cylindrical domains, the outer one where the reaction takes place and the inner one where a sweeping gas (steam or nitrogen) passes through. H_2 permeation is achieved by employing a thin Pd-Ru membrane in the interface between the two cylinders. Ni-Pt particles dispersed on a porous foam are simulated to catalyse the overall reaction. The porous foam covering almost entirely the annulus geometry, is assumed to be consisting of state of the art SiC.

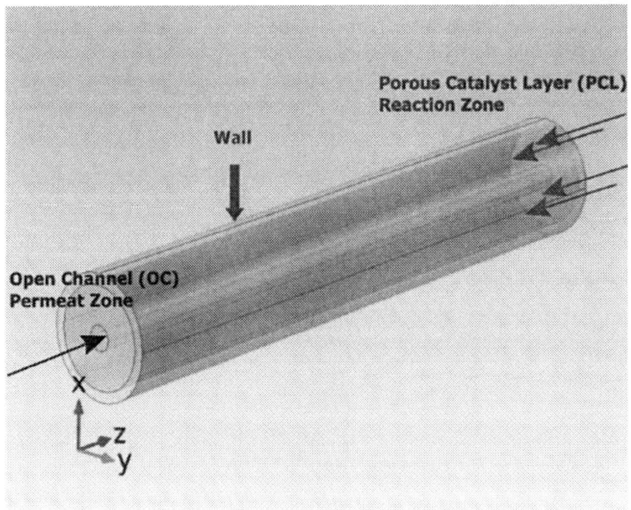

Figure 1. Model setup and geometry. The length of the reactor is 400mm while the inner radius is 7mm and the outer one 21mm.

Computational methodology and discretization

The kinetics used in the context of our work follow the assumptions and conclusions of[8] for the following main reactions:

$$CH_4 + H_2O \Leftrightarrow CO + 3H_2, \qquad \Delta H^0_{298} = 206\frac{kJ}{mol}$$

$$CO + H_2O \Leftrightarrow CO_2 + H_2, \qquad \Delta H^0_{298} = -41\frac{kJ}{mol}$$

$$CH_4 + 2H_2O \Leftrightarrow CO_2 + 4H_2, \qquad \Delta H^0_{298} = 165\frac{kJ}{mol}$$

Our geometry consists of two cylindrical and hence axi-symmetric domains. We refer from now on to the outer domain where the reactions take place as the Porous Catalyst Layer (PCL) since we assume that this is the area where the catalyst is dispersed on the SiC foam. The concentration difference between this zone and the inner cylinder results in a H_2 flux which is assumed to follow the Sieverts semi-empirical law6[3,6] according to the following equation.

$$\vec{J}_{H_{2siev}} = \frac{Q \cdot \exp\left(-\frac{E_a}{Rg \cdot T}\right)}{d_m}\left(P^{0.5}_{H_{2pcl}} - P^{0.5}_{H_{2oc}}\right)$$

In the above, Q is the pre-exponential factor of hydrogen permeability dependent on the actual material,Pd-Ru in our case, E_a is the activation energy and d is the membrane's thickness. The parameter specifics and the details of all modeling assumptions can be found in[6]. Briefly, this is a model, accounting for variable density in both domains due to reactions, permeation of hydrogen, fluid flow and heat transfer in all domains as well. The model is solved using the Finite Elements Method using quadrilateral elements, ideal for slender bodies, on a dense mesh of 1000 points in the z and 130 (50+80) in the r directions resulting in an maximum element size of $4 \cdot 10^{-4}$m. The points are assigned following a geometric series pattern to concentrate more points on the edges of the domain where finer resolution is dictated by the anticipated species fluxes.

Results and Discussion

In this work, we have examined the effect of varying the PCL pressure. The parametric space for our runs spans from 5 to 10 bars with increments of 1bar. The operating temperature is taken to be 773 K, the inlet speed is $5 \cdot 10^{-3}$ m/s and the steam to carbon ratio is 3.03. In Figure 2, we present a 3D representation of the change of the H_2 mass fraction along the pcl for a pressure of 5bars. We observe how the initial mass fraction, reactants entering on the top, increases due to reaction followed by a quick drop because of the permeation to the other side.

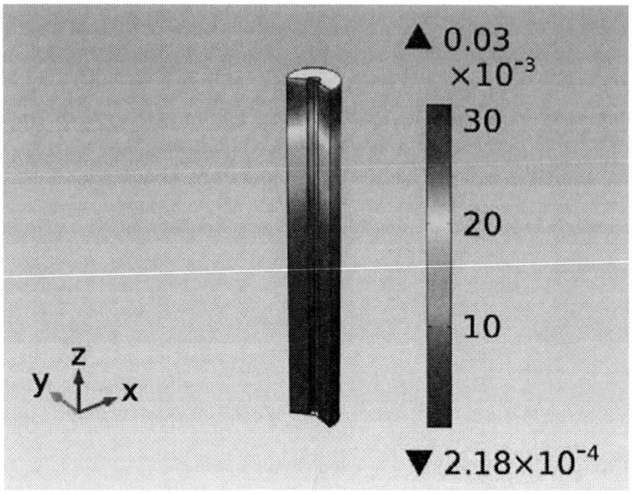

Figure 2. Hydrogen mass fraction spatial evolution for a pressure of five bars.

In figure 3 we present the methane mass fraction evolution for 5 bars.

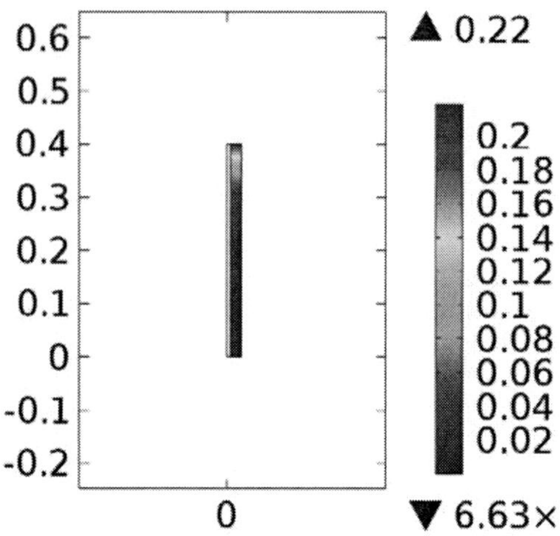

Figure 3. Methane spatial evolution representation in 2D for a pressure of 5 bars.

The temperature profile is of extreme significance to the exothermic nature of reactions 1 and 3 and the endothermic nature of the water gas shift reaction 2. A typical temperature distribution is provided in figure 4.

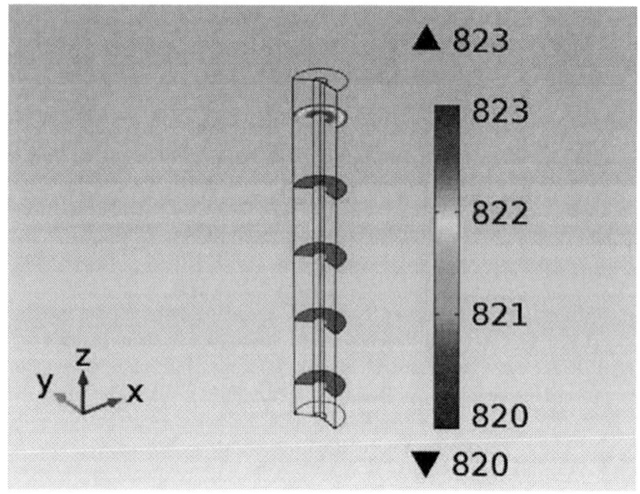

Figure 4. 3D Temperature profile for a pressure of five bars.

In figure 5 we present for a pressure of 10 bars the spatial change of the reaction rate (in the PCL) of the first of the three reactions cited on three radii, close to the wall, in the middle of the domain and close to the membrane.

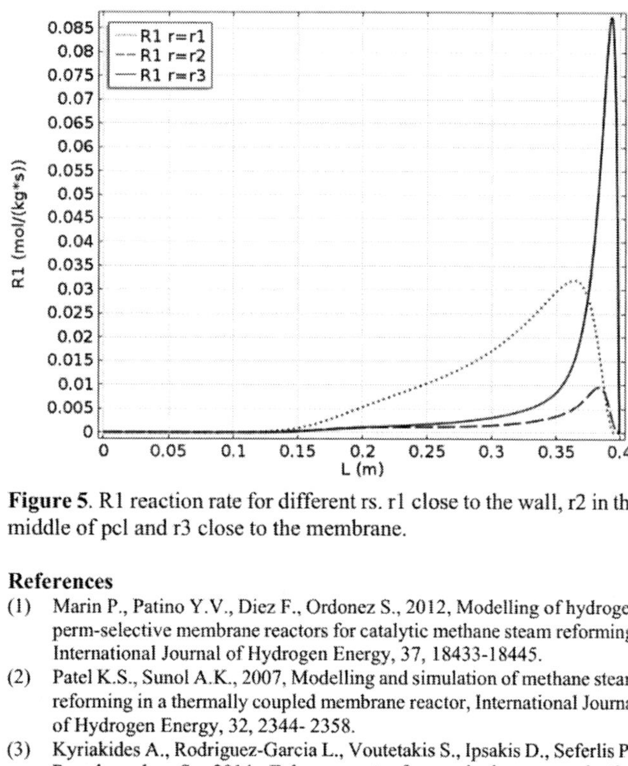

Figure 5. R1 reaction rate for different rs. r1 close to the wall, r2 in the middle of pcl and r3 close to the membrane.

References

(1) Marin P., Patino Y.V., Diez F., Ordonez S., 2012, Modelling of hydrogen perm-selective membrane reactors for catalytic methane steam reforming, International Journal of Hydrogen Energy, 37, 18433-18445.

(2) Patel K.S., Sunol A.K., 2007, Modelling and simulation of methane steam reforming in a thermally coupled membrane reactor, International Journal of Hydrogen Energy, 32, 2344- 2358.

(3) Kyriakides A., Rodriguez-Garcia L., Voutetakis S., Ipsakis D., Seferlis P., Papadopoulou S., 2014, Enhancement of pure hydrogen production through the use of membrane reactor, International Journal of Hydrogen Energy, 39, 4749-4760

(4) Angeli S.D., Monteleone G., Giaconia A., Lemonidou A., 2013, Low temperature methane steam reforming: Catalytic activity and coke deposition study, Chemical Engineering Transactions, 35, 1201-1206.

(5) Bientinesi M., Petarca L., 2011, H2 separation from gas mixtures through palladium membranes on metallic porous support, Chemical Engineering Transactions, 24, 763-768.

(6) Panagakos G., Kyriakides ., Papadopoulou S., Voutetakis S., A Computational Investigation of Hydrogen Production from Methane Steam Reactor, Chemical Engineering transactions, 45, DOI: 10.3303/CET1545173

(7) Barreirio M.M, Marono M., Sanchez J.M., 2014, Hydrogen permeation through a Pd-based membrane and RWGS conversion in H2OCO2, H2/N2/CO2 and H2/H2O/CO2 mixtures. International Journal of Hydrogen Energy, 39, 4710-4716.

(8) Xu J., Froment G.F., 1989, Methane Steam Reforming, Methanation and Water-Gas Shift: I. Intrinsic Kinetics, AIChE Journal, 35, (1), 88-96.

WORK REGIMES AND BEHAVIOUR OF MINERAL MATTER IN BIOMASS BOILER SYSTEMS

Rummel, L., Paist, A

Tallinn University of Technology, Ehitajate st. 5, Tallinn, Estonia
leo.rummel@energex.ee

Introduction
This article gives an overview of the first results of a long-term research on work regimes and behaviour of mineral matter in biomass boiler systems. As biomass is gaining importance as an energy source for electricity and heat, we need to investigate with scientific methodology the main problems that occur with biomass boiler systems. Lab-scale batch boilers have been tested in previous studies, but the results are often not applicable in case of small and medium size biomass industrial boilers on which the research has been limited. In this article, the first data from two small-scale industrial boilers systems of operating combined heat and power plants is presented and analysed. One boiler system uses grate firing and the other one bubbling fluidised bed technology. Previous researched is reviewed, the set-up of the experiment is described and the hypothesis for the long-term research are set up.

Experimental
Design and SCADA data from two biomass systems – grate-fired and bubbling fluidized bed, was collected and compared.

Grate-fired boiler system. Wood is fed from the storage to the combustion chamber, where it is burned. The energy which is released from the biomass in the combustion chamber is transferred to the water. The energy is enough to produce superheated steam. The flue gasses are cleaned and released to the atmosphere. The steam is expanded in a turbine-generator set, which generates electricity with the energy inside the steam. The low pressure steam is condensed by cooling it with district heating net water. The water at low pressure is heated up to boiling, with bleed steam from the turbine, in the deaerator. The feed water pump brings the water to the boiler.

The individual components within the system are designed for a higher load than the design point of the total system, this to overcome dynamical transients and to have some margin.

The following chapters will specify the individual components and will give more information concerning the function of these components. The next chapters describe first the equipment in the fuel and flue gas route, followed by the steam cycle, heat delivery net, civil and the control and safety system.

The CHP plant is a base load unit, this means the plant will normally run continuously on full load, in order to produce a maximum of electricity. The generated heat will be supplied to the City of Valka. The surplus of heat will be cooled away via the cooling system.

Fuel input 11.7 MWth

Thermal energy steam condenser 8.0 MWth

Electrical energy steam turbine generator set 2.0 MWe
Total efficiency on fuel input with fluegas condenser 100 % (calculated with LHV)

<u>Design values:</u>

Ambient air temperature:	15	°C
Wet bulb temperature:	21	°C
Relative humidity:	60	%
Barometric pressure:	1,013	mbar
Site elevation:	~60	M

<u>Extreme conditions:</u>

Summer dry bulb	+32	°C
Winter dry bulb	-32	°C
Summer wet bulb	23	°C
Barometric pressure:	950-1,050	mbar
Relative humidity:	50-90	%

<u>Maximum Continuous Rating (MCR), average fuel parameters, based on 100% wood chips:</u>

Design fuel, lower heating value (LHV):	8.021	MJ/kg[1]
Design fuel, moisture content:	50	%
Design fuel, particle size	G100	
Design fuel, ash content based on a dry basis (max): (high amount of sand will lead to severe wear)	< 2 wt	%
Design fuel, ash melting point:	> 1,100	°C
Design fuel, nitrogen content:	< 1.0	%
Design fuel, chlorine content:	< 0.1	%

<u>Range on fuel parameters:</u>

Fuel, range lower heating value:	7.2 - 12.6	MJ/kg
Fuel, range moisture content:	35 - 55	%
Fuel, density range:	200 - 400	kg/m^3

Table: Fuel particle size by ONORM M7 133 & DIN 66 165 standards

	Maximum % Particulate Size				Maximum Extremes	
Designation	<4 %	<20%	60-100%	<20 %	Area (cm^2)	Length (mm)
G100	<1 mm	1 – 11 mm	11 – 63 mm	>63 mm	10	250

Emissions

Dust	< 375 mg/Nm³
SO2	< 200 mg/Nm³
NOx	< 400 mg/Nm³
CO	< 250 mg/Nm³

Emissions given here are related to MCR conditions and are defined as 24 hour average values and based on 6% O2.

Bubbling fluidized bed boiler system. Boiler is a gas-tight natural circulation type boiler with a single steam drum. Boiler is a 3-pass boiler. Furnace and 2nd pass are water cooled and walls are made of membrane tube panels. 3rd pass is uncooled and there are located economizer and flue gas air preheaters. Boiler is also equipped with steam air preheaters.

Power plant uses wood based biomass as a primary fuels. Shale oil (LFO) will be used as a start up fuel, existing shale oil tank will be used. Fuel is fed through boiler front wall to furnace. It is fed through two feeding chutes to the boiler, ensuring equal combustion conditions all over the grate.

Grate is made of heat resisting steel and integrated as a part of the boiler pressure vessel. Vapor's BFB-grate possesses good coarse material extraction capability, due to the high area of openings in the grate and good functionality of bottom sand hoppers.

For low load conditions only half of the BFB grate will be used.

Boiler

Design temperature	450 °C
Design pressure	33 bar (a)
Thermal power	11,2 MW th

Steam turbine unit

Design temperature	450 °C
Design pressure	25 bar (1)
Generator output	2,0 MW e ; 10,5kV

Fuel

Type	wood chips & bark
Moisture (design)	55 %
Calorific value (LHV)	7,48 MJ/kg (as received)
Temperature	25 °C
Ash content w-% of dry matter	3,0 %
Density	300 kg/i-m³
Solid fuel flow rate	max 20,2 m³/h

Feed water

Temperature in deaerator	104 °C
Quality according	EN12952-12

District heat water

Temperature, return	55 °C
Temperature, supply	100 °C
PN rating	PN16

Combustion air

Temperature inlet	25 °C
Air flow rate	26 000 m³/s
Moisture	10 gH2O/kgdry air

Flue gas @ CHP exit
Temperature out 150 °C
Gas flow rate 31000 Nm³/h
Particles 100 mg / nm³
NOx 375 mg / nm³

Outside air
Temperature, min - 21 °C
Temperature, max + 35 °C
Temperature, normal + 25 °C
Air pressure 1013 mbar(abs)

Compressed air
Process air pressure 5,5…10 bar
Normal pressure 6,0 bar (g)
Condensation temperature -20 °C

Results and Discussion

The first results from the grate-fired biomass boiler are shown in Figure 1.

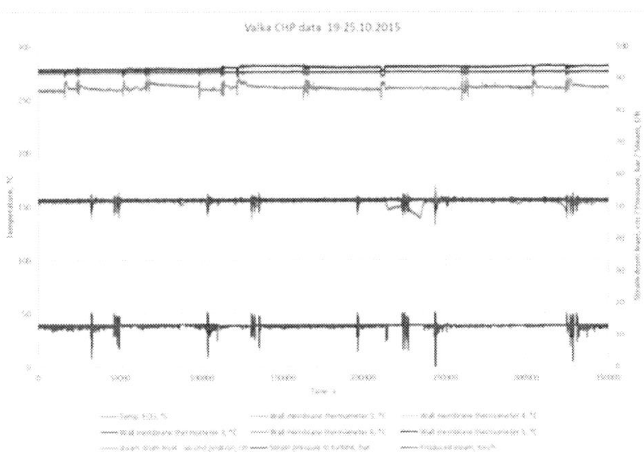

Figure 1. Woody biomass grate-fired boiler SCADA data from 1 week

Conclusions

First results have indicated that fluidized bed boilers are more suitable from mixed woody biomass fuels than grate-fired biomass boilers.

References

(1) Yin, C., Rosendahl, L.A., Kær, S.K. Grate-firing of biomass for heat and power production Progress in Energy and Combustion Science 34 (6), pp. 725-754 2008
(2) Bauer, R., Gölles, M., Brunner, T., Dourdoumas, N., Obernberger, I. Modelling of grate combustion in a medium scale biomass furnace for control purposes Biomass and Bioenergy 34 (4), pp. 417-427 2010
(3) Li, Z., Zhao, W., Li, R., (...), Li, Y., Zhao, G. Combustion characteristics and NO formation for biomass blends in a 35-ton-per-

hour travelling grate utility boiler Bioresource Technology 100 (7), pp. 2278-2283 2009
(4) Gungor, A. Simulation of emission performance and combustion efficiency in biomass fired circulating fluidized bed combustors Biomass and Bioenergy 34 (4), pp. 506-514 2010
(5) Liukkonen, M., Hälikkä, E., Hiltunen, T., Hiltunen, Y. Adaptive soft sensor for fluidized bed quality: Applications to combustion of biomass Fuel Processing Technology 105, pp. 46-51 2013
(6) Calvo, A.I., Tarelho, L.A.C., Teixeira, E.R., (...), Artiñano, B., Fraile, R. Particulate emissions from the co-combustion of forest biomass and sewage sludge in a bubbling fluidised bed reactor Fuel Processing Technology 114, pp. 58-68 2013

Au NANOPARTICLES MODIFIED Ni/TiO_2-NANOTUBES AS THE ANODE CATALYST FOR DBHFC

L. Tamašauskaitė-Tamašiūnaitė[1], A. Balčiūnaitė[1], D.M.F. Santos[2], E. Norkus[1]

[1] Department of Catalysis, Center for Physical Sciences and Technology, A. Goštauto 9, LT-01108 Vilnius, Lithuania
[2] Materials Electrochemistry Group, Instituto Superior Técnico, TU Lisbon, 1049-001 Lisboa, Portugal
loreta.tamasauskaite@ftmc.lt

In this study Au nanoparticles modified Ni/TiO_2-nanotubes have been fabricated via a three-step process which involves anodization of Ti sheets, followed by electroless deposition of Ni and a subsequent Ni layer decoration with Au crystallites via spontaneous Au galvanic exchange from a chloro-complex solution with the aim to use them as the anode catalyst in $NaBH_4/H_2O_2$ fuel cell.

It was found that the Au crystallites in size of 10-30 nm were deposited on the Ni/TiO_2-nanotubes and the Au loadings were in the range of 8-34 μg cm^{-2}. Performance of $NaBH_4/H_2O_2$ fuel cell using an Au(Ni)/TiO_2-nanotubes as the anode and Pt tas the cathode was studied at room temperature. The anolyte was composed of an alkaline mixture of 1 M $NaBH_4$ + 4 M NaOH and the catholyte contained 5 M H_2O_2 + 1.5 M HCl. Nafion 117 membrane was used as separator.

A maximum peak power density of 62.5 mW cm^{-2}, at a cell voltage of 0.85 V and a current density of 73.5 mA cm^{-2}, was obtained at the Au(Ni)/TiO_2-nanotubes with the Au loading of 34.1 μg_{Au} cm^{-2} at 25 °C, whereas the highest specific power density of 5.6 kW g_{Au}^{-1} was obtained at the catalyst with the Au loading of 8.2 μg_{Au} cm^{-2}.

IN SITU RAMAN MICROSPECTROSCOPY MONITORING OF CELLULOSE HYDROLYSIS

Maksim V. Tyufekchiev, Geoffrey A. Tompsett, Michael T. Timko

Department of Chemical Engineering,
Worcester Polytechnic Institute,
100 Institute Rd, Worcester, MA 01609, USA
maksimvt@wpi.edu, gtompsett@wpi.edu, mttimko@wpi.edu

Introduction

Overcoming cellulose recalcitrance is the main bottleneck for economic utilization of lignocellulosic biomass. Cellulose crystallinity, which is determined by the relative amount of highly ordered "crystalline" to disordered amorphous regions in the cellulose structure, determines cellulose susceptibility to chemical conversion.[1] The crystalline and amorphous domains have different liquid acid and enzymatic hydrolysis rates, with the amorphous being the more reactive one.[1] Kinetic models have been proposed to explain cellulose conversion to oligomers and glucose.[2],[3] However, to our knowledge, such models have not accounted for the change in crystallinity as the hydrolysis reaction progresses.[2],[3] It is expected that accessible amorphous regions in the cellulose structure are hydrolyzed first, thus, decreasing the ratio of amorphous to crystalline cellulose, resulting in an increase in crystallinity.[1] Such an increase in crystallinity has been shown, resulting in decrease of the hydrolysis rate.[4] Thus, in-situ monitoring cellulose crystallinity becomes imperative for understanding the kinetics of cellulose hydrolysis. While attempts to monitor cellulose crystallinity are reported, they consist of separating the cellulose substrate from the reaction mixture, washing, drying before crystallinity measurement using X-Ray diffraction, or Nuclear Magnetic Resonance.[1],[4],[5],[6] The challenges with these off-line methods include: 1) time consuming preparation methods and 2) potential for introduction of experimental artifacts which would confound data interpretation. Raman spectroscopy has been used to estimate cellulose crystallinity.[7],[8] Unlike other methods, Raman spectroscopy can be used an in situ method for determining crystallinity, even under conditions of high temperature, acidity, and water concentration. To date, the method has been limited by the lack of suitable view-cell reactors. Here we report in-situ analysis of cellulose crystallinity during liquid acid hydrolysis using Raman microspectroscopy using a custom-designed view-cell reactor. This new method and studies it enables could lead to an improved mechanistic understanding of cellulose hydrolysis and allow for optimization of cellulose pretreatment and conversion technologies.

Experimental

In-Situ Raman Reactor. A custom demountable stainless steel in situ reactor was constructed and used for all Raman microscopy experiments. Figure 1a shows a photograph of the view cell reactor mounted on the Raman microscope stage. The reactor consists of 2 main parts, a water cooled lid with a window and a heated bottom section. The top lid of the reactor has a 8 mm opening to allow for optical access. A Suprasil glass window, 2 mm thick, is fitted to the opening and a silicone o-ring makes the pressure seal with a top piece. The bottom part has a reaction chamber well with diameter of 8 mm, an inlet and outlet for liquid acid flow, and a gold plated copper o-ring to seal it to the lid. Two 5 mm diameter 100W heating rod are inserted into the body of the bottom part while a 1/16" diameter thermocouple is placed in a aperture below the bottom section in order to measure and control the temperature close to the sample. The bottom and lid parts are clamped together using six screws and sealed by the gold-plated ring. Both lid and bottom parts were gold plated to reduce acid corrosion of the reactor; preliminary experiments with bare stainless steel indicated corrosion. The reactor was connected to a LKB Bromma 2150 HPLC pump. The temperature limit for the reactor is determined by the silicone o-ring (approximately 200 °C) and the pressure by the strength of the quartz glass window. Both the silicone o-ring and quartz glass can be replaced with more durable components to examine higher temperatures and/or pressures.

Raman Microscope. Raman spectra were obtained using a Horiba XploRA Raman microscope equipped with a 20x magnification lens. Spectra were obtained using 532 nm laser with 100 mW power, with 1200 diffraction grating. Acquisition time was 3 seconds and 50 accumulations were obtained. Spectra were recorded using LabSpec 6 software and fitted using MagicPolt commercial software. Figure 1 b) shows a cellulose particle focused with the Raman microscope.

Materials. Avicel 101 cellulose and hydrochloric acid (0.1M) were purchased from Sigma Aldrich.

Cellulose Ball Milling. Avicel 101 (1 g) was put in a stainless steel cylindrical vessel with inner diameter of 17.9 mm. Three stainless steel balls were put in the vessel; two had a diameter of 9.5 mm and on a diameter of 15.8 mm. The vessel containing the metal balls and the cellulose was put on a vibratory shaker Retsch MM2000, which was run at amplitude of 10 for various time intervals to decrease cellulose crystallinity.

X-ray Diffraction. X-ray diffraction of cellulose was carried out with a Rigaku Geigerflex. The bulk crystallinity of the Avicel cellulose and the ball-milled sample were estimated using Segal's method. Non-treated Avicel had a crystallinity of 82%, while the 20 min ball-milling resulted in a 66% crystallinity.[9]

Cellulose Hydrolysis. Ball milled Avicel 101 was used as the cellulose substrate. The ball-milled powder was placed in the reactor and individual particle were dispersed to allow for better focusing of the Raman laser. Hydrochloric acid was introduced at 0.1 mL/min flowrate. The concentration of the acid, the temperature, and time of hydrolysis were varied. Temperature was controlled by a Watlow EZ-Zone temperature controller. The reactor was pressurized to 22 bar to permit studies at temperatures greater than 100° C to eliminate any chance of water vaporization.

Analysis

Experimental Raman spectra of cellulose were analyzed to determine cellulose crystallinity as a function of time using methods reported elsewhere in the literature.[7],[8]

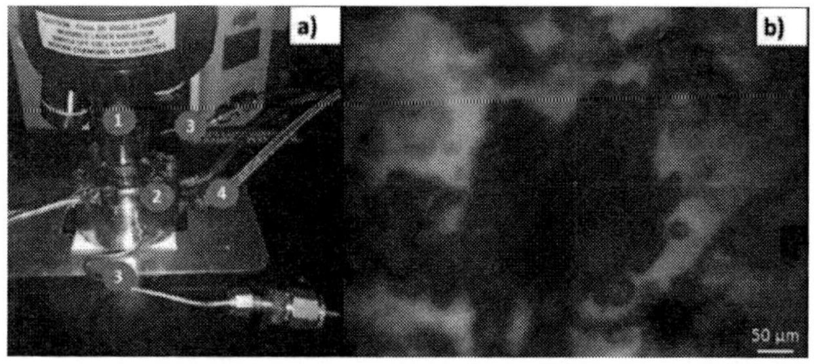

Figure 1 In-situ Raman microspectroscopy of cellulose hydrolysis experimental set-up. Panel a) show Raman instrument and reactor: 1) lens; 2) reactor; 3) liquid acid inlet and outlet; 4) heating rod inserts, thermocouple is attached to the bottom of the reactor (not visible). Panel b) shows a focused cellulose particle that is being hydrolyzed and which spectra are being obtained.

Results and Discussion

Figure 2a shows in-situ Raman spectra obtained from the ball milled cellulose at 150 °C and 0.1 M HCl. In all cases, spectra with excellent signal-noise ratio were obtained. Characteristic Raman bands and peaks have intensity that is proportional to the crystallinity or amorphous composition of the cellulose. Several different literature methods were used to convert Raman spectral band intensities to cellulose crystallinity.[7],[8] Representative results are shown in Figure 2b for a ball-milled microcrystalline cellulose substrate. Figure 2b shows that the crystallinity increases over time, relative to the crystallinity of the cellulose particle at time 0, consistent with preferential hydrolysis of amorphous regions within the cellulose. In similar tests with highly crystalline cellulose, we found that the crystallinity remained constant over 60 min time, consistent with a hydrolysis mechanism involving end-chain hydrolysis without transformation to an amorphous state. Obtaining measurements on-line and in real time permits collection of sufficient data over a wide enough range of conditions to test specific hypotheses regarding cellulose hydrolysis kinetics.

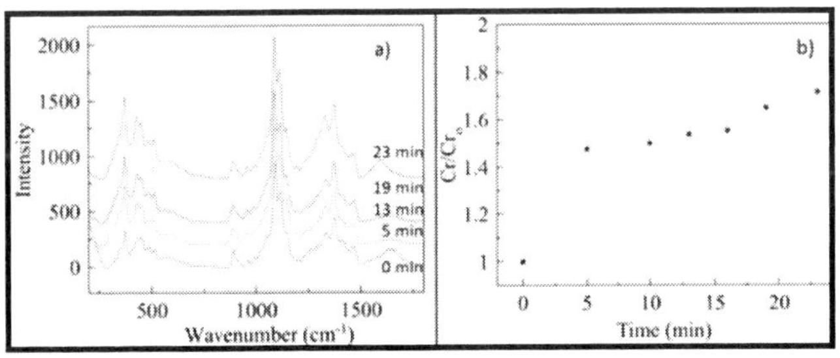

Figure 2 Panel a) shows representative spectra at different time increments of cellulose hydrolysis with 0.1 M hydrochloric acid at 150° C. Panel b) shows the calculated ratio of Cr/Cr_0 from band intensities from the spectra in panel a) and plotted versus time, where Cr is the crystallinity and Cr_0 crystallinity at time 0. The Raman crystallinity increases by around 70% for approximately 25 min of hydrolysis.

Conclusion

We have demonstrated the ability of a demountable in situ cell with Raman microspectroscopy to study cellulose hydrolysis. We have shown that the crystallinity of an initially amorphous cellulose substrate increases under hydrolysis conditions. Work is underway to refine the method and to study cellulose hydrolysis kinetics over a range of temperatures and acid concentration.

References:

(1) S. Park; J. Baker; M. Himmel; P. Parilla; D. Johnson – *Biotechnology for Biofuels*, **2010**, *3:10*

(2) G. SriBala; R. Vinu, *Ind. Eng. Chem. Res.*, **2014**, 53 (21), pp. 8714-8725.

(3) P. Bansal; M. Hall; M. Realff;. J. Hee; A. Bommarius, *Biotechnology Advances*, **2009**, 27, pp. 833-848

(4) H. Zhao; J.H. Kwak; Y. Wang; J. Franz; J. White; J. Holladay, *Energy & Fuels*, **2006**, 20, pp. 807-811

(5) H. Zhao; J.H. Kwak; Z. Zhang; H. Brown; B. Arey; J. Holladay, *Carbohydrate Polymers*, **2007**, 68, pp. 235-241

(6) J. Zhang; J. Zhang; L. Lin; T. Chen; J. Zhang; S. Liu; Z. Li; P. Ouyang; *Molecules*, **2009**, 14, pp. 5027-5041

(7) K. Schenzel; S. Fischer; E. Brendler; *Cellulose*, **2005**, 12, 3, pp.223-231

(8) U. Agarwal; R. Seiner; S. Ralph, *Cellulose*, **2010**, 17, pp.721-733

(9) L. Segal; J.J. Creely; A.E. Martin Jr; C.M. Conrad; *Textile Research Journal*, **1959**, 29, pp. 786-794

USING GRASS EXTRACT TO HARVEST THE ENERGY OF THE SUN

Devan Kerecman, Luke Warner, Joe DiBussolo, Jingqiu Hu*

Department of Chemistry, West Chester University, West Chester, PA 19383
*Thermo Fisher Scientific, LC-MS application laboratory, West Chester, PA 19383

Various methods were applied to extract natural dyes from three grass leaves: Kentucky bluegrass, Perennial ryegrass, and Tall fescue. The dye extracts were used as sensitizers in dye-sensitized solar cells (DSSCs). The voltage and current output of DSSCs were recorded, and the energy conversion efficiency was calculated. The conversion efficiency of natural dye DSSCs is comparable to that of DSSCs fabricated with fruit juice. Effect of dye concentration, type of additives, and liquid electrolytes on the performance of DSSCs was explored. The main pigments, chlorophyll *a*, chlorophyll *b*, and pheophytin *a*, in grass extract were characterized by UV-vis absorption spectroscopy, high-performance liquid chromatography (HPLC), and liquid chromatography-mass spectrometry (LC-MS). The correlation between the energy conversion efficiency of DSSCs and the composition of natural dyes was investigated.

POSSIBILITY OF MAKING A KILO-WATT INVERTER DRIVEN BY THE 3RD POSITIVE EMF

Osamu Ide

Foundation for the Advancement of Environmental Energy (FAEE)
Clean Energy Research Laboratory,
3-4-21-601 Mita, Minato-ku, Tokyo,
108-0073, Japan
Email: daphne@myad.jp

Introduction

At the staring of this research, the author found the 3rd positive EMF for a new source of clean energy in the motor driven by the discharging of a capacitor in 1980. The author called the motor Ether Engine. The author found the phenomenon that the capacitor charging voltage was increased, compared with the estimated value by the calculation, when the Ether Engine was running. The author had researched the phenomenon precisely in experimental and mathematically method. It was impossible to verify the phenomenon by the common electro-magnetism. Finally, the author reached the conclusion that it was due to an independent EMF induced in opposite direction of Faraday's back EMF. The papers of the Ether Engine were presented in 1995 and 2000, * (1) (2). These papers are also related to * (9) (10) (11) presented by AIAS fellows.

In 2000, the author changed the direction of this research from a motor to an inverter, applying the 3rd positive EMF because an inverter has no mechanical loss.

In 2010, the author succeeded in making an inverter of 300% efficiency, * (3) (4) (7) (14).

In 2015, the author has finally succeeded in making an inverter generator self-charging. Although the output of the inverter generator is around 1W, it exhausts no natural resources, * (13). This system should be adaptable to a power of kilo-watts to supply the power source for house and EV.

This report shows the possibility and concept of constructing a kilo-watt inverter generator driven by the 3rd positive EMF. The basic experiment was already made and presented by the paper of the author at the ACS meeting in March 2015, *(5). The first perfect replication according to the author's paper was made and presented by the Munich group of Dr. Horst Eckardt and Kurt Arenhold, * (8). The most amazing point of their paper is the phenomenon that the 3rd positive EMF could be derived from ECE theory by Dr. Myron Evans of AIAS.

This report is the 2nd replication of the author's paper * (5). However the scale of the experiment is much larger than in the first paper * (5), concerning the voltage of the power source and the inductance of the transformer.

The experimental result shows that it might be possible to make a generator of 18 kilo-watt driven by the 3rd positive EMF. The author supposes that underlying possibility is limitless power output, because the 3rd positive EMF is one basic principle of generating electricity just like Faraday's EMF.

Experimental set-up and the theoretical calculation

Figure 1(a) shows the experimental set-up. The circuit mainly consists of the high voltage (6KV) DC power supply, a capacitor C, a spark gap and a transformer. The capacitor C, the spark gap and the primary coil of the transformer are connected in series. The core of the transformer is made by thin silicon steel (0.1mm)-laminated and, has an oval cylinder shape of (33 × 37 × 100 mm). So it has no magnetic saturation as the magnetic circuit is not closed. The resistance Ri (389 MΩ) is an equivalent inner resistance of the high voltage DC supply. The resistance Rc is a leak resistance of capacitor C for safety.

First, the DC power from the high voltage power supply charges the capacitor C. When the voltage of the capacitor C charged up enough to make a spark at the spark gap, the current flow from the capacitor C starts to the primary coil L1 of the transformer. The transient current Im from the capacitor C to the primary coil L1 is measured by the current probe and input to CH1 of a digital oscilloscope. The voltage Vc of the capacitor C is also measured by a high impedance probe (200 MΩ, 1/400) and input to CH2 of the digital oscilloscope. The sampling frequency of the oscilloscope is 100MHz (0.01 μs).

The wave form of discharging current and the voltage of the capacitor C are plotted by the X-Y plotter.

Figure 1(b) shows the equivalent circuit of the experimental set-up of Figure 1(a). The initial voltage of the charged capacitor C is around Vc = 6KV. When the SW is turned on (the spark starts) the current Ic starts flowing into the primary coil of the transformer. The resistance Rl is the measured value of the equivalent resistance of the primary coil.

The equation (1) is the relation between the initial voltage of the capacitor C, the inductance L and the inner resistance R of the transformer. The equation (1) is mainly based on Faraday's law and Ohm's law.

$$Vc = \frac{1}{C}\int I dt = L\frac{dI}{dt} + RI \qquad (1)$$

The solution of the transient current Ic is derived from equation (1) by the Runge – Kutta method at t = 0s, Vc = 5937 V, C = 10.77nF, L= 10.04 mH and R= 4.07 Ω for initial conditions.

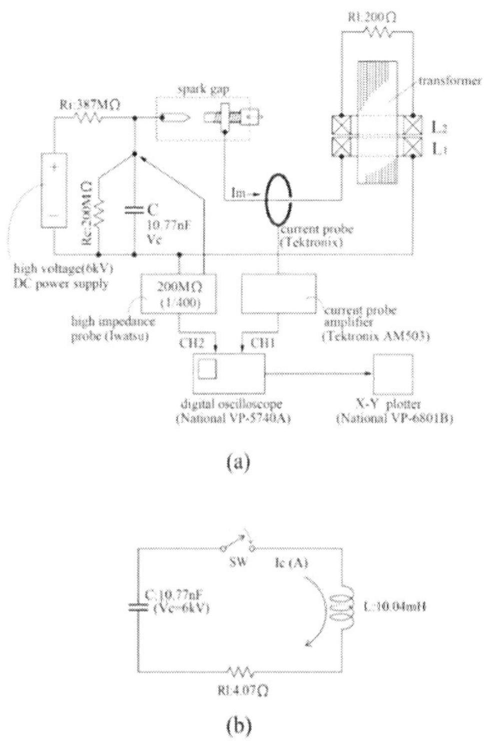

Figure 1. (a) Experimental set-up
(b) Equivalent circuit of the experimental set-up

Figure 2 gives a view on experimental setup.

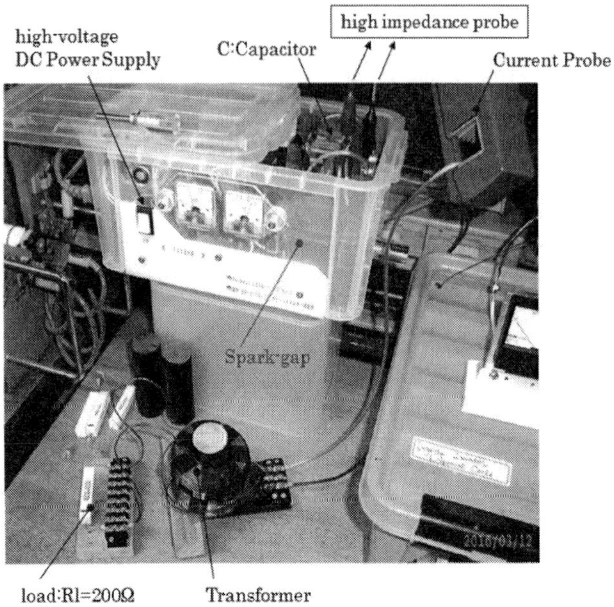

Figure 2. View on the experimental system

Results and Discussion

1. Results. Figure 3 (a), (b), (c) show the results of the experiment and the calculation. Each of the Figure 3 (a), (b), (c) has the different time scale on the horizontal axis. They are all in same scale of voltage and current for vertical axis. The blue curve is the transient current (Im) to the primary coil of the transformer after the spark started. The red curve is the voltage of the capacitor C. The black curve of Figure 3 (a), (b) are the transient current (Ic) to the primary coil of the transformer by the theoretical calculation based on the common electro-magnetic theory. It is the solution of equation (1) by the Runge-Kutta method.

2. The anomalous rising of the input current to the transformer. The Figure 3 (a), (b) clearly show the large difference of the input current to the transformer between measurement and calculation. The measurement value of the current is at least 5 times higher than from the calculation. It is estimated that a strong positive EMF is induced and accelerates the input current. The results are as anomalous as the results of references (5), (6) and (8), although they were observed under the different experimental condition. Because, in the case of this experiment, the power source has a higher voltage (6000V), the charge in the capacitor C and the inductance of the coil are also 20 times that of the last experiment * (5), (6).

The initial static energy Ec (J) of the capacitor C is calculated by

$Ec = 1/2 \times C \times Vc^2 = 0.190$ (J).

The value of C and Vc are given on Figure 1 (a) and Figure 3 (a).

The magnetic energy El (J) of the coil of the transformer, reached at the maximum input current Ip is calculated by

$El = 1/2 \times L \times Ip^2 = 0.907$ (J).

The values of L and Ip are also given on Figure 1 (a) and Figure 3 (a).

The results mean that it might be possible to generate the electrical energy of El = 0.907 (J) in the transformer by a static energy of Ec = 0.190 (J), discharged by the capacitor. Hence the output is 4.77 times larger than the input.

Moreover, the magnetic energy of the transformer is induced within the time of 4 μ s. It is estimated that it is not so difficult to repeat the operation of charge and discharge of the capacitor in every 50 μ s (20000 times/s). Then the output power of the inverter applied by this phenomenon is 18.14 KW. On the other hand, the input power to repeat charging the capacitor is presumably 3.8 KW.

The author was still repeating the experiment 20000 times per day in within 45 days. So the total number of replications is supposed to be 1 million times.

The author has an idea of replacing the spark gap for switching with an electrical switching element like a MOS - transistor. However the MOS–transistor must be high voltage proof (more than 6000V) and have a high speed switching specification.

3. The anomalous value of the charge flow in the circuit. Figure 3 (c) shows that the remaining transient current input of around 2 A to the transformer-after the capacitor discharge has finished-, continues for a very long time (t = 250 μ s), although the discharging time of the capacitor is only 6 μ s. The total charge Q2 (q) of the measured current flow is calculated by the time integration of the measured current Im as shown in equation (2). It is also possible to be calculated by the programmed function of the digital oscilloscope.

$$Q2 = \int_0^t Im\, dt \ (q) \qquad (2)$$

where, Im (A) is the measured current, and t (sec) is the remaining time of the measured current.

The result of the integration by oscilloscope is $Q2 = 426.25 \times 10^{-6}$ (q).

On the other hand, the initial charge Q1 (q) of the capacitor C is calculated by $Q1 = CVc = 64.64 \times 10^{-6}$ (q).

Then, the ratio of Q2/Q1 is 6.60.

This is also anomalous as it is commonly estimated to be Q2<Q1.

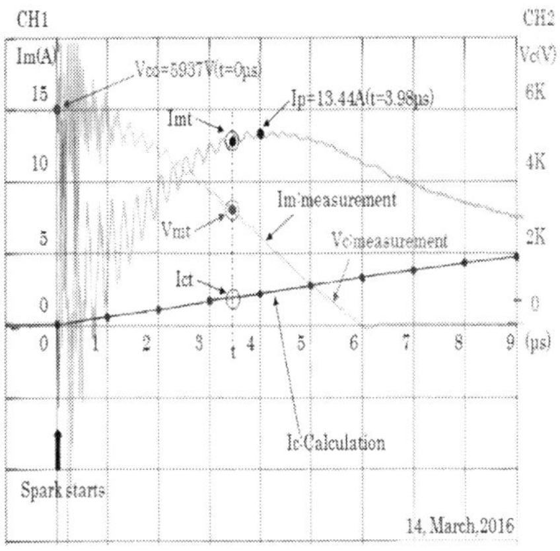

Figure3(a

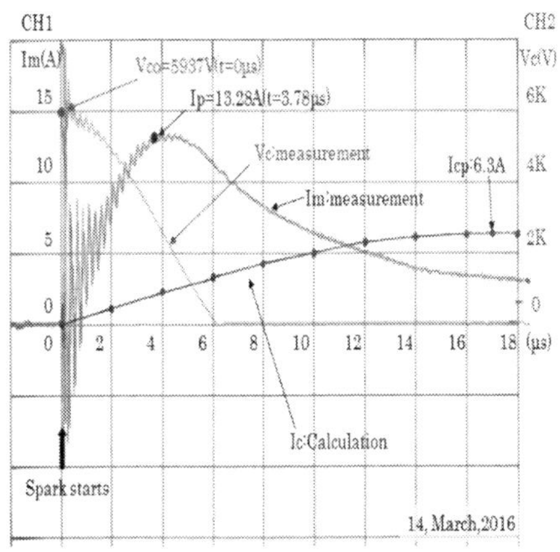

Figure3(b

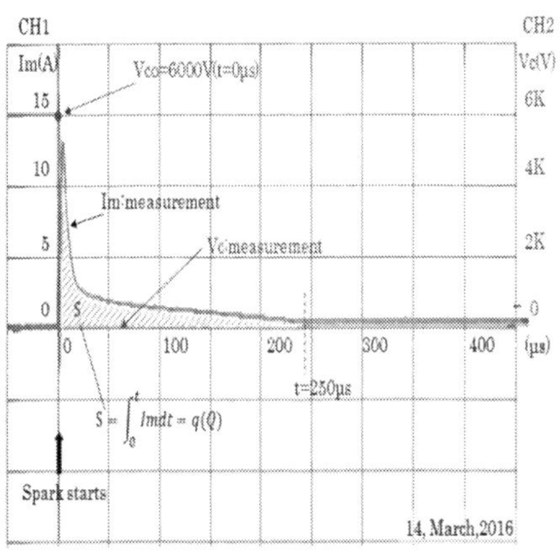

Figure3(c

Figure 3. The results of the experiment and the calculation

The total electrical charge flown in the circuit is much larger than the initial charge of the capacitor. Where did the extra charge come from?

The remaining current in Figure 3 (c) is around 1-3 A. If the current was from the high voltage power supply, the maximum output current should presumably be within 15.5 mA, because the inner resistance (R1 = 387 MΩ) of the high voltage power supply is so large as shown in Figure 1 (a). So the remaining current could never be from the high voltage power supply.

It looks as if the electrical charge was produced in the circuit. The author presumes that the phenomenon of extra electrical charge is related to the cold electricity as described in * (12).

4. The estimated positive EMF by the current wave. Figure 4 shows the relation between the estimated positive EMF: Vp (V) and the 2nd order time differential of the input current: d^2I/dt^2 (A / t^2). This section is quite important to verify the author's hypothesis of the 3rd positive EMF. It is different from Faraday's back EMF, as it was already investigated in a series of references (1), (2), (3), (4), (5), (6), (7), (8) and (13).

To derive the equation of d^2I/dt^2, the author assumes the function of the measured current Im. The curve Im looks very complicated by the ringing. But it can be easily be approximated by a sinus curve within the time of 6 μ s, as shown in Figure 3 (a), (b).

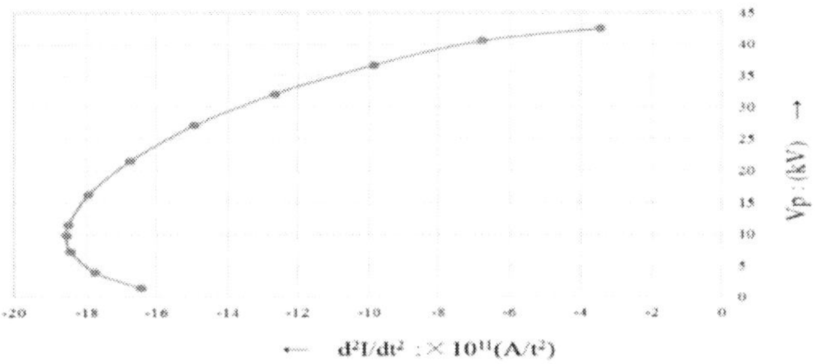

Figure 4. Relation between the 2nd order time derivative of the input current and the estimated positive EMF

So it is described in equation (3) as follows:

$$I = Ip \sin(\omega t) \quad (A) \qquad (3)$$

where, Ip is the peak current = 13.28 (A).

ω is $2\pi f = 2\pi / (4.2 \times 10^{-6} \times 4)$

Ip and ω are derived from the curve Im on Figure 3 (a). Therefore,

$$d^2I / dt^2 = -Ip\,\omega^2 \sin(\omega t) \quad (A / t^2) \qquad (4)$$

The value of the positive EMF: Vp (V) is estimated from equation (5), as the identical method of equation (5) in reference (5):,

$$Vp = Vmt\,(Imt / Ict - 1) \quad (V) \qquad (5)$$

where, Vmt is the voltage of the capacitor C, Imt is the measured current and Ict is the calculated current. They are all valid at any time (t μ s), as shown in Figure 3 (a). The result of the calculation is shown in Figure 4. It clearly shows the non-linear and strong relation between (d^2I / dt^2) and the positive EMF.

So the existence of the 3rd positive EMF was proved five times. The other proofs are in the references (1), (2), (5) and (8).

Conclusions

The experimental system of this report is consists of a high voltage DC power supply, a capacitor, a spark gap and the coil of the transformer. When the capacitor is charged up by the high voltage power supply and the voltage evokes a spark at the spark gap, the current starts flowing through spark gap into the coil of the transformer. Then a change of the current and the voltage of the capacitor are measured and memorized by the digital oscilloscope. The conclusions below were of obtained from the results.

1. A strong positive EMF is induced and accelerates the rising current in the short moment after the spark started. This phenomenon of rising the current cannot be explained by only Faraday's back EMF.

2. It is possible to make an inverter of 18-KW output driven by the 3rd positive EMF, even while the input power is 3.8-KW.

3. This report is the 2nd replication of the author's experiment in reference (5), in extending 120 times driving voltage and 20 times the coil inductance.

Acknowledgement. The author especially thanks Dr. Myron Evans, Dr. Horst Eckardt, Dr. Douglas Lindstrom and Mr. Franklin D. Amador of AIAS for support and advice for this research.

The author received valuable advice from Dr. Takao Yamazaki and Mr. Takashi Maeza.

The author appreciates drawing the Figures of this report by Mr. Takayuki Funabashi and Miss Honami Ichinose.

Many thanks to Yuriko Ide for operating the PC.

References

(1) Ide, O.," Increased voltage phenomenon in a resonance circuit of unconventional magnetic configuration", J.Appl.Phys.,77(11), (1995),pp6015-6020.

(2) Ide, O., " Possibility of existence of non-linear electromotive force (EMF)", NASA/CP-2000-21029, (July 2000), pp705-719.

(3) Ide, O., " A Hyper-Efficient Inverter Driven by Positive EMF in Combination with Transient Phenomenon ", SPESIF 2011, Physics Procedia, Vol.20, (2011), pp420-434.

(4) Ide, O., et al, "Anomalous Power Efficiency of a Transformer Driven by Tuned Duty Cycle Pulses" made public at the Space Propulsion & Energy Sciences International Forum-2012 (SPESIF) edited by T. Valone, Physics Procedia, Volume 38, 2012, pp 99-108.

(5) Ide, O., et al, "Anomalous rising of input current induced in the transformer of the inverter" presented at 249th American Chemical Society National Meeting & Exposition at Denver in March 24, 2015, ENFL 350.

(6) Ide, O., et al, "Consideration of the cause of inverter noise called ringing" presented at 249th American Chemical Society National Meeting & Exposition at Denver in March 24, 2015, ENFL 351.

(7) Ide, O., "Characteristics of DC power output from an inverter driven by sharp spike pulse" presented at 249th American Chemical Society National Meeting & Exposition at Denver in March 24, 2015, ENFL 352.

(8) Kurt Arenhold, Horst Eckardt., "Experimental verification and theoretical explanation of the Osamu Ide experiment", June 23, 2015. http://aias.us/documents/uft/UFT311_IdeExp.pdf

(9) Horst Eckardt, Douglas W. Lindstrom., "Circuit theory for unusual inductor behavior", July 9, 2015 http://aias.us/documents/uft/CircuitTheory.pdf

(10) Franklin D. Amador, Horst Eckardt., "Simulation of a Parametric Resonance Circuit". March 30, 2012. http://aias.us

(11) Horst Eckardt, B. Foltz., "Simulation of a Parametric Resonance Circuit, Part2" http://aias.us/

(12) Douglas W. Lindstrom., "On the Possible Existence of a Second Form of Electrical Current in the ECE Equations of Electromagnetism" http://www.aias.us/documents/numerical/ECE-2nd-Current.pdf

(13) Ide, O., "The experiment of self-charging Inverter driven by the 3rd positive EMF" http://aias.us/documents/miscellaneous/SelfChargingInverter.pdf

(14) Ide, O., "INVERTER DRIVING SYSTEM", United State Patent, US 8.873.262B2, Japan patent number 5587691.

HYDROGENATION AND DECOMPOSITION OF ASPHALTENE IN LIQUID PHASE

Qiang Sheng, Gang Wang, Chengdi Gao, Miao Hu, Libo Yao, Jinsen Gao*

State Key Laboratory of Heavy Oil Processing, College of Chemical Engineering, China University of Petroleum, Beijing 102249, China
wanggang@cup.edu.cn

Abstract

In traditional heavy oil processing technologies, asphaltene was mainly converted to coke due to its poor reactivity. In order to solve this problem, a process of hydrogenation and decomposition of asphaltene in liquid phase was proposed. The effect of hydrogen donor (HD) and HD-to-asphaltene ratio on asphaltene conversion and products distribution were investigated in a 250 mL autoclave stainless steel reactor. The experimental results demonstrate that asphaltene can be effectively converted to distillates, with little coke and gas formation, using the proposed process. A good asphaltene conversion can reach to 66.2 wt% and with less than 1 wt% coke plus gas yield. The selected distillate narrow fraction (DNF), obtained from refinery, shows a better hydrogen donating property compared with tetrahydronaphthalene (THN). At high HD-to-asphaltene ratio, the colloidal system formed by DNF and asphaltene is more stable than that formed by THN and asphaltene.

Introduction

The need for light oil is increasing in recent years, but the feed stocks worldwide are becoming heavier[1]. Nowadays, the prevailing heavy oil processing technologies are visbreaking[2], residue fluid catalytic cracking[3] and heavy oil hydrocracking[4], etc. However, the typical heavy oil has high content of asphaltene which is mainly converted to coke in traditional heavy oil processing. Therefore, how to convert asphaltene to distillates instead of coke is the key for efficient utilization of heavy oil.

In order to solve this problem, our research team has proposed a processing technique for effectively converting asphaltene to distillates—hydrogenation and decomposition of asphaltene in liquid phase. In this process hydrogen donor (HD) was employed to dissolve asphaltene and build a stable colloidal system with asphaltene. Another more important functions of HD are avoiding coke formation by releasing active hydrogen atoms and capping aromatic free radicals to prevent the condensation reaction between aromatic free radicals[5].

In this research, THN and a DNF were selected as the HD to hydrogenate and decompose of asphaltene in liquid phalse under the existence of one kind of oil-soluble Mo-based catalyst. And the effects of HD-to-asphaltene ratio on asphaltene conversion and products distribution were studied.

Experimental

Materials. One kind of paraffin-based n-pentane asphaltene was selected as raw material. THN and one kind of distillate narrow fraction (DNF) were used as HD. The properties of asphaltene and DNF are listed in Table 1. The catalyst used in this research was one kind of oil-soluble Mo-based catalyst special prepared for this process.

Instrument and Procedure. The experiments were performed in a 250 mL stainless steel batch reactor equipped with mechanical stirring and automatic temperature control (Figure 1). For a typical experiment the reactor was loaded with certain amount of asphaltene, HD and catalyst. The reactions were carried out at 390°C and 7 MPa pressure of hydrogen for 5 h. After reaction the reactor was cooled by immersing the reactor in cold water.

Table 1. Properties of asphaltene and DNF

Items	asphaltene	DNF
C, wt%	88.2	88.71
H, wt%	7.77	11.30
S, wt%	0.41	0.2
N, wt%	1.60	0.3
N_H/N_C	1.05	1.53
Molecular weight (GPC)	1122	327

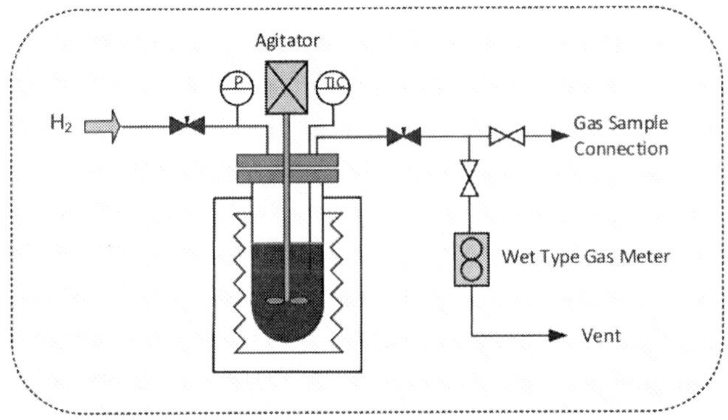

Figure 1 Schematic diagram of experimental setup.

Results and Discussion

The effects of different HD and HD-to-asphaltene ratio on the conversion of asphaltene and products distribution were investigated and the results are shown in Figure 2 and Figure 3, respectively. Figure 2 indicates that the conversion of asphaltene is almost equal when THN and DNF are employed as the HD at the HD-to-asphaltene ratio of 3:1 and 5:1, respectively. This illustrates that DNF also owns better hydrogen donating ability compared with THN. On the other hand, the conversion of asphaltene significantly increases from about 25 wt% to 67 wt% as HD-to-asphaltene ratio increased from 3:1 to 5:1, that indicating HD-to-asphaltene ratio has a significant influence on the conversion of asphaltene. This is mainly because higher HD-to-asphaltene ratio can provide more hydrogen free radicals for asphaltene colloidal system, and the ability of hydrogenolysis of asphaltene and capping macromolecular asphaltene free radicals are enhanced. What's more, higher HD-to-asphaltene ratio makes asphaltene disperse more uniformity and decreases the opportunity of collision between macromolecular asphaltene free radicals, therefore decreasing the possibility of forming coke. And the conversion of asphaltene is enchanced.

Details of the product distribution are shown in Figure 3, the gas yield is less than 1 wt% which can be neglected. However, the maximum coke yield is about 5 wt%, while others coke yield are less than 1 wt%. This can attribute to the HD, which not only can donate active hydrogen to free radicals that produced by decomposition of asphaltene molecules, to prevent them aggregating to form larger condensed aromatic molecules, but also dissolve coke precursors to prevent phase separation, which is the beginning of coke formation. Additionally, when THN is used at the ratio of 5:1, the coke yield is obviously higher compared with that of DNF. This is mainly because under high HD-to-asphaltene ratio the colloidal system formed by

THN is unstable compared to that formed by DNF, so the probability of phase separation is increased, when THN used. What's more, the main products of hydrogenation and decomposition of asphaltene in liquid phase are >540℃ and <540℃ distillates, which accounts for about 95 wt% of the total product.

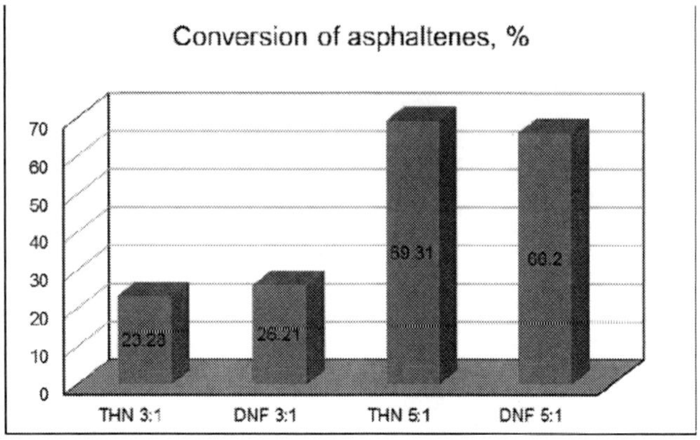

Figure 2. Conversion of hydrogenation and decomposition of asphaltene in liquid phase.

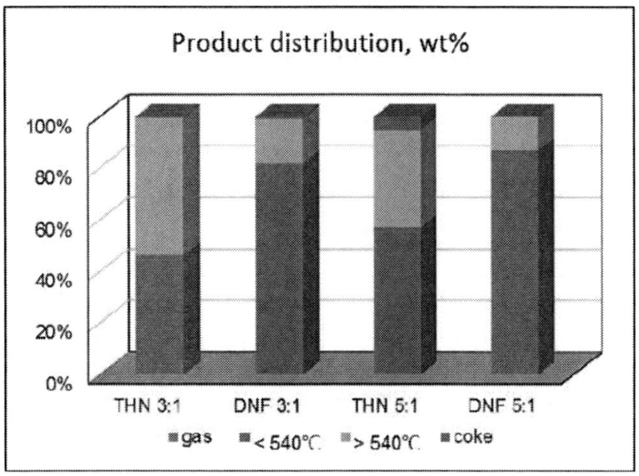

Figure 3. Product distribution of hydrogenation and decomposition of asphaltene in liquid phase.

Conclusions

Asphaltene can be effectively converted to distillates employing the process of hydrogenation and decomposition of asphaltene in liquid phase with little coke and gas formation. A good asphaltene conversion can be up to 66.2 wt% and with less than 1 wt% coke plus gas yield when DNF selected as HD at the HD-to-asphaltene ratio of 5:1. The DNF is also a good hydrogen donor for hydrogenation and decomposition of asphaltene in liquid phase. The colloidal system of asphaltene formed by DNF is more stable than that formed by THN.

Acknowledgement. The authors acknowledge the financial support provided by State Key rogram of National Natural Science Foundation of China (21336011), National Natural Science Foundation of China (21476259).

References

(1) Rana, M. S.; Samano, V.; Ancheyta, J.; Diaz, J., A review of recent advances on process technologies for upgrading of heavy oils and residua. *Fuel* **2007,** *86* (9), 1216-1231.

(2) Joshi, J. B.; Pandit, A. B.; Kataria, K. L.; Kulkarni, R. P.; Sawarkar, A. N.; Tandon, D.; Ram, Y.; Kumar, M. M., Petroleum residue upgradation via visbreaking: A review. *Industrial & Engineering Chemistry Research* **2008,** *47* (23), 8960-8988.

(3) Payá, J.; Monzó, J.; Borrachero, M.; Velázquez, S., Evaluation of the pozzolanic activity of fluid catalytic cracking catalyst residue (FC3R). Thermogravimetric analysis studies on FC3R-Portland cement pastes. *Cement and concrete research* **2003,** *33* (4), 603-609.

(4) Zhang, S.; Liu, D.; Deng, W.; Que, G., A review of slurry-phase hydrocracking heavy oil technology. *Energy & Fuels* **2007,** *21* (6), 3057-3062.

(5) (a) Rueda-Velásquez, R. I.; Freund, H.; Qian, K.; Olmstead, W. N.; Gray, M. R., Characterization of asphaltene building blocks by cracking under favorable hydrogenation conditions. *Energy & Fuels* **2012,** *27* (4), 1817-1829; (b) Chen, Q.; Gao, Y.; Wang, Z.; Guo, A., Application of Coker Gas Oil Used as Industrial Hydrogen Donors in Visbreaking. *Petroleum Science and Technology* **2014,** *32* (20), 2506-2511; (c) Qi, W.; Lei, G.; WANG, Z.-x.; MU, B.-q.; GUO, A.-j.; He, L., Hydrogen donor visbreaking of Venezuelan vacuum residue. *Journal of Fuel Chemistry and Technology* **2012,** *40* (11), 1317-1322.

MOFs TEMPLATED CoO_x/C COMPOSITES WITH SUPERIOR HYDROGENATION ACTIVITIES

*Yan Jiang,[a] Xin Zhang [a]*Qiang Sheng[a]*

[a] State Key Laboratory of Heavy Oil Processing, College of Chemical Engineering, China University of Petroleum Beijing 102249, China.
*Corresponding author: zhangxin@cup.edu.cn

Abstract

In recent years, well-dispersed nanoparticles supported on carbon materials has drawn much attention due to its functional properties both nanoparticles and carbon. However, the development of a one-step strategy with high loading and well-dispersed nanoparticles within carbon still remains a great challenge. In this work, a facile route to synthesize magnetic Co species supported on carbon was employed. It was the carbonization of MOFs (Co-BTC) by one-step method. When the materials were used as multifunctional catalysts, the Co-BTC carbonized at 500 °C exhibited much higher 1,3-butadiene conversion than that of carbonized at 400 °C.

Introduction

Porous carbon materials with a high specific surface area and affinity to organic contaminations have been considered very promising materials for many applications.[1] It is more interesting to explore the possibility of extending their properties by depositing functional nanoparticles with carbon. We can expect synergic and cooperative effects between the carbon and the well-dispersed active nanoparticles. Various species have been introduced into porous carbon, such as semiconductors, metal oxides, and metals.[2] In previous works, it was shown that nanoparticles introduced into carbon matrices through either a direct-synthesis approach or a post-loading approach, in which the loading nanoparticles are located in the pores or encapsulated inside the walls. In some cases, the nanoparticles are seriously aggregated to form the large particles. Therefore, it is extremely challenging to introduce uniformly dispersed crystalline nanoparticles with a high metal content into a porous carbon. Recently, thermal decomposition of metal-organic frameworks (MOFs) has been demonstrated as a facile route to synthesize nanoparticles supported on porous carbon.[3] The most attractive advantages of this strategy based on the thermal decomposition of MOFs are that the as-formed materials are highly porous, and more importantly, the constituent of materials can be altered to metal, metal oxide, carbon or their hybrids by controlling the thermal decomposition conditions.[4] Herein, we attempted to illustrate that this synthetic strategy has significant potential for fabricating composites of metal nanoparticles and carbon with a highly porous structure. Fortunately, porous CoO_x/C composite nanomaterials were successfully fabricated by thermal decomposition of Co-BTC (BTC: trimesic acid) under inert atmosphere. As we expected, the as-prepared porous CoO_x/C composites exhibited significantly good performance for the selective hydrogenation of 1,3-butadiene.

Experimental

Synthesis of Co-BTC: 145 mg $Co(NO_3)_2 \cdot 6H_2O$ and 72 mg H_3BTC were dissolved in 10 mL DMF solution in a 60 mL microwave autoclave and stirred for 10 min, then 1 g PVP was added in the above solution and stirred for further 60 min at room temperature, after dissolved thoroughly, the microwave autoclave was transferred to an microwave oven at the heating rate of 30 °C/min from room temperature to 140 °C and hold for 5 min, later, the reaction solution became brown, after centrifuged at 9900 rpm for 15 min the purple products were obtained. Then, the products were centrifuged and washed with ethanol for three times to remove the surfactant. Finally, the products were dried in vacuum drying oven at 80 °C for 12 h and reserved for further characterization.

Synthesis of CoO_x/C: A certain amount of precursors were loaded in a quartz boat, then put it into a tube furnace. Successfully, N_2 was pumped at the flow rate of 25 mL/min at room temperature for 1 h, then the temperature was elevated to 400 or 500 °C at a rate of 3 °C/min and maintained for 2 h.

The selective hydrogenation of 1,3-butadiene: The 1,3-butadiene hydrogenation was performed in a fixed-bed quartz reactor with an inner diameter of 6 mm. 10 mg catalysts in a power form was mixed with 400 mg quartz sand. Before the feed gas was pumped in the reactor, the catalysts were pretreated in flowing N_2 (30 mL/min) at 100 °C for 30 min. Then the reactants of H_2 and 1.02 vol% 1,3-butadiene in N_2 passed through the fixed-bed reactor at the rate of 5.7 and 7.8 ml/min, respectively, which corresponds to space velocities of 81,000 mL/h (g-cat.)$^{-1}$. The reactor effluent was online analyzed using a SP-6890 gas chromatography with flame ion detector (FID) which equipped with GDX-501 as an adsorbent.

Results and Discussion

The morphology and structural features of as-prepared Co-BTC was characterized by scanning electron microscopy (SEM) and transmission electron microscopy (TEM). As shown in Figure 1a-c, spherical Co-BTC nanocrystals (NCs) were obtained. The average size of the well-defined MOF is about 250–300 nm and the surface is smooth. In addition, the precursor was also characterized by the power X-ray diffraction (XRD). Figure 1d shows that all the peaks are agreed with the diffraction peaks of pure Co-BTC as previous reported,[5] indicating that Co-BTC was successfully synthesized under microwave irradiation.

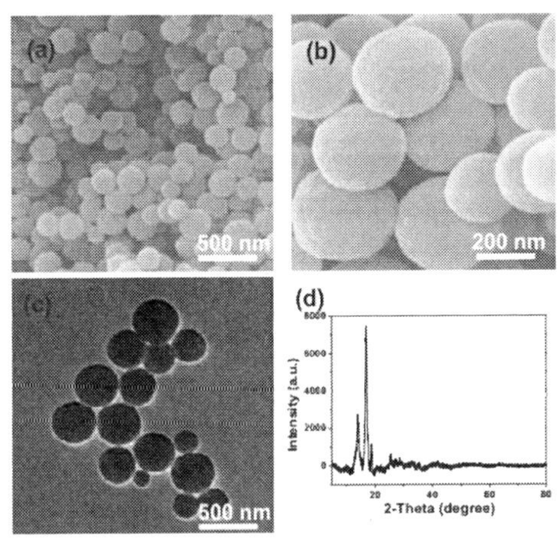

Figure 1. (a) Low and (b) high magnifications SEM images (c) TEM imagies (d) XRD pattern of Co-BTC.

The TG curve of Co-BTC indicated that there are ca. 15wt% weight loss due to the desorption of solvents (Figure 2a). The sharp weight loss after 400 °C should be due to the decomposition of the crystal structure. On the basis of the above analysis, the temperatures of 400 and 500 °C were chosen as the suitable thermal pretreated temperatures. The materials prepared at different temperature were denoted as Co-BTC-T, where T denoted the calcination temperature. After thermal decomposition, the products were characterized by XRD. As shown in Figure 2b, the XRD patterns of Co-BTC-400 and Co-BTC-500 exhibited the diffractions of Co and CoO species, which are both very weak, indicating that these species are highly dispersed in the surface of carbon. Three diffraction peaks around 36.6°, 42.6° and 61.8° were observed, which were assigned to (111), (200), (220) planes of CoO (PDF#01-1227), however mixed phases of CoO and Co were detected in Co-BTC-500, one peak around 44.4° was found which was assigned to the (111) plane of metal Co. The shape of Co-BTC-400 maintained sphere and there are many tiny spherical particles distributed on the surface (Figure 2c, d). The average diameter of particles was 7±1 nm. The morphology of Co-BTC-500 was similar to that of Co-BTC-400. But the size of tiny spherical particles increased to 9±3 nm (Figure 2e, f). As we known, after thermal decomposition under inert gas, the ligands of Co-BTC were converted to carbon and Co^{2+} were changed to metallic Co or cobalt oxides at different pretreatment condition, which can be confirmed by TEM images.

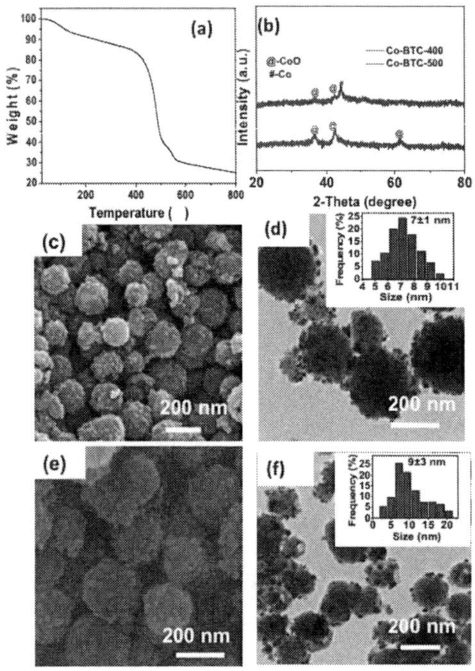

Figure 2. (a) TG curve of Co-BTC. (b) XRD patterns of Co-BTC-400 and Co-BTC-500. SEM imagies and TEM images (and size-distribution histogram) of (c) (d) Co-BTC-400, (e) (f) Co-BTC-500.

In an attempt to explore the activity of the products, selective hydrogenation of 1,3-butadiene (BD) was chosen which is an important model reaction to characterize the catalytic performance of transition metals. Figure 3a shows the effect of reaction temperature on the conversion of BD over Co-BTC-400 and Co-BTC-500 catalysts. It can be found that the conversion of BD over two catalysts increase with the increasing reaction temperature. Note that, no significant conversion was found on Co-BTC-400 at the reaction temperature below 80 °C, while 14.75 % conversion was observed at 90 °C. However, the conversion on Co-BTC-500 was greatly enhanced in comparison with Co-BTC-400, which is assigned to the presence of not only CoO but also metallic Co of Co-BTC-500. 1,3-Butadiene is hydrogenated to a mixture product of 1-butene, cis-2-butene, trans-2-butene and butane. Remarkably, the selectivity of 1-butene was 40%-50% over the catalysts of Co-BTC-400 and Co-BTC-500, while it increased a little with the increasing of reaction temperature. Meanwhile, the selectivity of trans-2-butene decreased slightly. As the conversion of BD greater than 70%, the selectivity of 1-butene decreased sharply, while the selectivity of cis-2-butene, trans-2-butene and butane increased, suggesting that 1-butene was able to isomerized to cis-2-butene and trans-2-butene, in addition, it was further hydrogenated to saturated hydrocarbons under higher temperature.

Conclusions

In summary, we have successfully developed a facile route to synthesize NPC with well-dispersed magnetic Co nanoparticles by one-step carbonization of MOF (Co-BTC). The small Co nanoparticles embedded in the carbon matrix were highly dispersed with an average size of 9 nm. Furthermore, The Co-BTC carbonized at 500 °C exhibited much higher 1,3-butadiene conversion than that of carbonized at 400 °C.

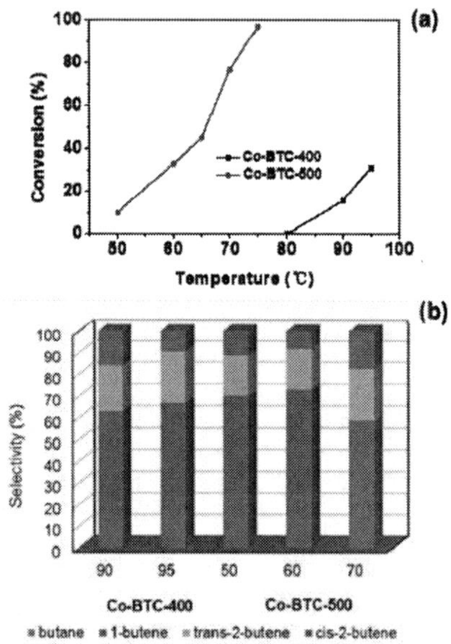

Figure 3. (a) Conversion (%) of BD as a function of time in the Co-BTC-400 and Co-BTC-500 catalysts. (b) Selectivity of two catalysts at different reaction temperature.

Acknowledgement. The authors acknowledge the financial supports from the NSFC (Nos.21573286, 21173269, 91127040).

References

(1) (1)Yang, T.; Liu, J.; Zheng, Y.; Monteiro, M. J.; Qiao, S. Z., Facile Fabrication of Core - Shell - Structured Ag@ Carbon and Mesoporous Yolk - Shell - Structured Ag@ Carbon@ Silica by an Extended Stöber Method. *Chemistry -A European Journal* **2013,** *19* (22), 6942-6945.

(2) (2)Wu, Z.; Li, W.; Webley, P. A.; Zhao, D., General and controllable synthesis of novel mesoporous magnetic iron oxide@ carbon encapsulates for efficient arsenic removal. *Advanced Materials* **2012,** *24* (4), 485-491.

(3) (3)Liu, B.; Shioyama, H.; Akita, T.; Xu, Q., Metal-organic framework as a template for porous carbon synthesis. *Journal of the American Chemical Society* **2008,** *130* (16), 5390-5391.

(4) (4)Meilikhov, M.; Furukawa, S.; Hirai, K.; Fischer, R. A.; Kitagawa, S., Binary Janus porous coordination polymer coatings for sensor devices with tunable analyte affinity. *Angewandte Chemie* **2013,** *125* (1), 359-363.

(5) (5)Jiang, C.-H.; Song, L.-F.; Zhang, J.; Sun, L.-X.; Xu, F.; Li, F.; Jiao, Q.-Z.; Sun, Z.-G.; Xing, Y.-H.; Du, Y., Thermodynamic properties and

heat capacities of Co (BTC) 1/3 (DMF)(HCOO). *Journal of thermal analysis and calorimetry* **2010,** *102* (3), 1087-1093.

PRECIOUS METAL-FREE NANOMATERIAL FOR SUSTAINABLE ELECTROCATALYTIC APPLICATIONS

Yifan Liu, Chao Wang

3400 N Charles St, Baltimore, MD 21218

Electrochemical reaction, as an efficient energy conversion pathway, is essential for many modern devices and techniques. For example, Oxygen Evolution Reaction (OER) and Hydrogen Evolution Reaction (HER) are the anodic and cathodic reaction in an electrolyzer. Also, Oxygen Reduction Reaction (ORR) is the common cathodic reaction in a fuel cell. Precious metals are widely used as catalysts for electrochemical reactions due to their excellent catalytic performance. Platinum, as a precious metal, has been demonstrated to be highly active for many electrochemical reactions (HER, OER, ORR etc.). The large scale application of Pt, however, is hindered by its high cost and low availability. Thus our research goal is to explore earth abundant and inexpensive non-precious metal material for electrocatalytic applications. Here we report different homemade non-precious metal catalysts for various electrochemical reactions. 2 dimensional MoS_2 nanosheets are synthesized and demonstrated to be highly active for HER. The nanosheets are very thin (<3 atomic layer thick) and thermal treatment could increase the degree of stacking. Lateral size of the MoS_2 nanosheets could be tuned (from <5 nm to ~20 nm) by changing reaction time. The synthesized MoS_2 nanosheets are 8~25 times more active for HER compared to commercial MoS_2 depending on lateral size. Different phases of CoxSy nanoparticles are also synthesized and proved to be highly active for ORR. These CoxSy phases include Co_3S_4, Co_9S_8 and $Co_{1-x}S$. Co_3S_4 and Co_9S_8 are hollow spheres with ID ~4 nm and OD ~10 nm. $Co_{1-x}S$ have the morphology of solid sphere with a diameter of ~10 nm. All three phases are active for ORR in alkali while $Co_{1-x}S$ is active for ORR in acid. Our work provide a better understanding of electrochemical catalysis and seeks to rational explore promising catalysts for sustainable electrochemical energy conversion.

INVESTIGATION OF SOLID ELECTROLYTE INTERFACE (SEI) FORMED IN NANOPOROUS SILICON AND MESO SILICON SPONGE ELECTRODE MATERIALS BY USING HYPERPOLARIZED ^{129}XE NMR SPECTROSCOPIES

Yougang Mao[1], Myeonghun Song[1], Russell Hopson[1], Naba K. Karan[2], Dokyoung Kim[3], Jinmyoung Joo[3], Li-Qiong Wang[1], Pradeep R. Guduru[2], Michael J. Sailor[3]*

[1] Department of Chemistry, Brown University, Providence, RI 02912
[2] School of Engineering, Brown University, Providence, RI 02912
[3] Department of Chemistry and Biochemistry, University of California, San Diego, CA 92093, USA
*Corresponding Author: li_qiong_wang@brown.edu

Introduction

Silicon has been widely considered as one of most promising anode materials in lithium ion batteries due to its extremely large charge capacity[1] and the availability of the state-of-the-art fabrication methods. Its mechanical damage caused by colossal volumetric expansion[1,2] during the lithiation can be overcome by reducing its particle size. Silicon nanoparticle and mesoporous silicon sponge (MSS) electrodes have shown to accommodate its large volume expansion better than micron sized one, resulting in much improved electro-chemical performances.[3-5] However, when the silicon electrode is lithiated, a solid electrolyte interface (SEI) is formed during the charge cycling.[6,7] The SEI may build up a barrier between silicon electrodes and electrolytes. The SEI barrier can be beneficial because it limits further electrolyte degradation. On the other hand SEI may cause large losses in Columbic efficiency i.e.it increases the electrical resistance within the cell and hinder lithium-ion transfer kinetics.

SEI compositions are often complicated and difficult to be directly detected and characterized because SEI is originated from multiple competing reactions. However the porosity and interconnectivity changes in silicon electrodes caused by the formation of SEI can be detected and analyzed by HP ^{129}Xe NMR, which is demonstrated in this study. The interconnectivity of the pores between silicon particles and conductive additives and the pores in active silicon materials plays an important role in the electrochemical performance of such electrodes since it greatly affects the molecular and ionic transport properties within electrode materials and, the energy storage in batteries and super-capacitors. The better understanding of how SEI affects the porosity and interconnectivity of pores in nano particle composite electrodes can help us to make more robust and higher performance porous electrodes.

This paper presents a study of porosity and pore interconnectivity change as a result of SEI formation in Si nanocomposite electrode materials containing active silicon materials, super P carbon and CMC binder. This work is the first study to explore the feasibility of using HP ^{129}Xe NMR for probing the porosity changes of any practical porous electrodes caused by the formation of SEI.

Experimental

Silicon nanoparticles of about 20 nm in diameter, conductive carbon super P and CMC were purchased from Sigma Aldrich. Mesoporous silicon sponge (MSS) materials (20 nm pore size, >50 μm particle size) were prepared by electrochemical etching of heavily boron-doped <100> oriented single crystal Si wafers (<1mΩ cm resistivity, Siltronix Inc.) by application of a constant current density of 225 mA cm^{-2} for 358 s in an aqueous 48% hydrofluoric acid (HF)/ethanol electrolyte (HF:EtOH=3:1, v-v). Carbon coating on the MSS particles was conducted using a vacuum chemical vapour deposition system.

Preparation of Silicon and Carbon mixture (Si+C) with the weight ratio of 1 to 1. Silicon particles of 450mg 20nm and super P carbon of 450 mg were mixed in ethanol. The mixed slurries were stirred for 4 hours and sonicated for 2 hours. Then the mixture was dried on hotplate at 100°C for 15 hours along with stirring. Final dry at 100°C for 9 hours was performed in the oven. The mixture of carbonized MSS and super P was prepared by the same way.

Preparation of Si+C+CMC Electrode samples. The dried powder of mixing silicon and carbon was added into 2% CMC water solutions and strongly stirred for 30 minutes to form slurries. Then the slurries were casted on copper foil and dried in a vacuum plate at 100°C for several hours. The weight ratio of (Si+C):CMC was 80 : 20. The laminated copper foil was cut into round electrode. 1 M LiPF6 in 1:1:1 ethylene carbonate (EC)/diethyl carbonate (DEC)/dimethyl carbonate (DMC) was used as an electrolyte solution to grow SEI. Four samples for NMR measurements are pure Si+C (Si+C), soaked Si+C+CMC (Si+C Soaked), and two SEI samples. SEI 1 and SEI 2 were prepared by holding the same potential at 2 hours and 12 hours respectively. The NMR samples are prepared by gently scrapping off the materials from the copper foil and packed in rotors in the glove box.

Hyperpolarized ^{129}Xe NMR Measurements. HP ^{129}Xe NMR experiments were carried out on a Bruker Advance DSX 300 spectrometer operating at 82.98 MHz (magnetic field 7.05 T) using a variable temperature static probe with a continuous flow (CF) of HP xenon. For HP ^{129}Xe NMR experiments, a single-pulse (SP) Bloch-decay method was used and samples were loaded into 7 mm Zirconia rotors. In order to have consistent results, samples were packed into 7mm NMR rotors by using a hammer with similar force to make sure that the sample inside the rotor was packed tightly and consistently. The rotor cap with a small-drilled hole allows HP 129Xe gas to passing through small plastic tubing in to the samples. The HP 129Xe NMR experiments were performed with non-spinning. The spectra were collected with a 1-μs ^{129}Xe pulse, and a repetition delay of 0.5 second.

Results and Discussion

Room Temperature HP ^{129}Xe NMR. A study of Si nanocomposite electrode materials with SEI have been carried out in order to investigate the influence of the SEI on porosity and interconnectivity of the pores in these materials. Figure. 1 displays the CF HP ^{129}Xe NMR spectra taken at 300 K for the samples of Si+C, Si+C Soaked, and SEIs, respectively. The spectra were taken under the identical condition in order to examine any changes in porosity and interconnectivity of the pores caused by the soaking and the formation of SEI. All spectra display two resonance peaks. The 0 ppm peak is associated with the free Xe gas and the peaks at higher chemical shifts are from the adsorbed Xe on the surfaces of nanoparticle composites. Since the Measurements are taken at 300 K and very low partial pressures of xenon in the flowing gas mixtures (1% of total, or 10 mbar), which preclude contributions from Xe-Xe interactions, the spectra can be considered as characteristic of the resident framework and the surface. The very low xenon concentration also reduces the rate of exchange between the different adsorption regions, which makes it possible to detect inhomogeneity in the pore space of the materials. Figure 1 clearly shows that the chemical shift of the adsorbed Xe resonances for the electrode with SEI is smaller than the electrode prior to the formation of SEI, indicating that SEI alters the pore environments of Si nanoparticle electrode composites. The differences in chemical shifts are most likely associated with the different pore sizes resulted from the SEI formation on the electrode.

In general, smaller pores are associated with the higher chemical shifts. Therefore our results suggest that pore sizes become larger as SEI is formed onto the silicon particle surface. However the chemical shift difference between two SEI samples is very small as shown in Figure 2, indicating that most SEIs is formed in less than 2 hours.

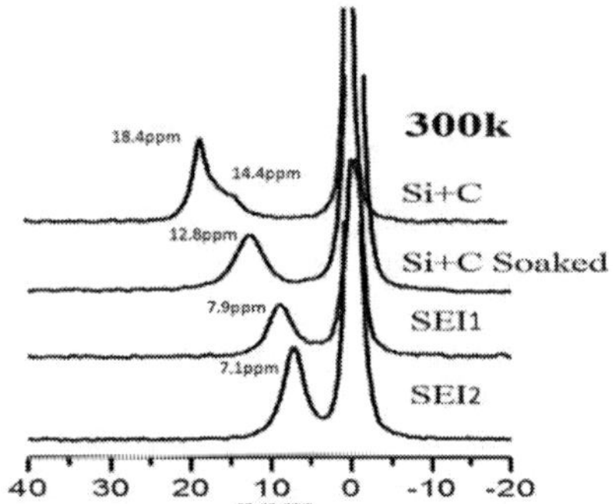

Figure 1. CF HP ^{129}Xe NMR spectra taken at 300 K for samples of Si+C, Si+C Soaked , SEI1 and SEI2 respectively. The chemical shifts are shown next to the adsorbed ^{129}Xe NMR resonance peaks..

Variable Temperature HP ^{129}Xe NMR. Variations in ^{129}Xe NMR spectra with temperature can be very sensitive to the dynamics of adsorbed xenon and subsequently to the morphology of the pore space.[8,9] A series of variable-temperature HP ^{129}Xe NMR spectra for samples: Si+C, Si+C Soaked, SEI1 and SEI2 are taken from 173 K to 373K with a 20 K increment. Figure 2 displays the temperature dependence of adsorbed ^{129}Xe chemical shifts for samples Si+C, S+C Soaked, SEI1 and SEI2.

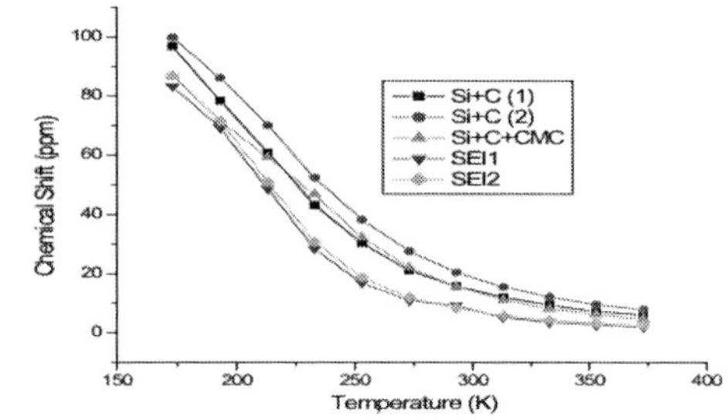

Figure 2. Temperature dependent HP ^{129}Xe NMR spectra taken for differently prepared Si nanoparticle electrodes.

Physical characteristic parameters related to the interaction between xenon and the silicon surface, i.e. ΔH_{ads} the heat of adsorption, δ_s ^{129}Xe characteristic chemical shift and average pore size D_s, can be obtained from the fitting of the temperature-dependence of chemical shift data and an appropriate geometrical model [8,10]. Table 1 shows those parameters.

Table 1 Parameters derived for Si nano particle electrodes.

Si Nanoparticle Electrode	ΔH_{ads} (KJ/mol)	δ_s (ppm)	D_s (nm)
Si + C (1)	14.8	110	27
Si +C (2)	15.3	109	36
Soaked	16.6	91	33
SEI1	18.7	92	52
SEI2	18.4	95	59

Soaked: cast electrode materials and then soaked into the electrolytes.

All samples have similar heats of adsorption of 16-18 KJ/mol, indicating typical physical adsorption of Xe on silicon surfaces. However, the decreases of chemical shift and increase of D_s upon growing SEI do indicate that the formation of SEI results in the increased pore sizes. This is because the SEI increases the nano particle size, which surely increases the space between the nano particles. Figure 3 is a schematic illustration of the SEI formation.

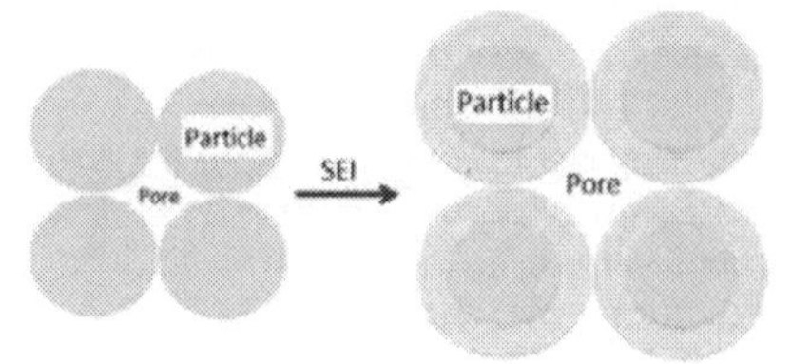

Figure 3 Schematic illustration of the formation of SEI

Pore size, Interconnectivity and Electrode Performance. As point out before that the soaked and SEI electrode samples contain the CMC binders. CMC plays a critical role in connecting Si and C particles to insure the good conductivity of the electrode. It was found that the higher CMC content strongly increase the connection among the nano particles and conductive super P carbon, resulting in an increased electron conductivity of the electrode. The formation of SEI increases the particle size and pore size. Undoubtedly the increase of pore size among particles may decrease the number of CMC chains per unit volume of porous electrodes, which may reduce the ion transport ability cross the pore space from one particle to another. This may be one of the reasons that SEI increases the resistance of ion transport in addition to the chemical composition change on the nano particle. The study has shown that hyperpolarized 129Xe NMR is a powerful technique to characterize the pore structures of the porous electrode materials. The information obtained from this study is helpful to understand the electrode performances including the conductivity of the electrode.

The study of ^{129}Xe spectra of carbonized silicon sponge electrode is in progress and results will be published separately.

Acknowledgment. HP ^{129}Xe NMR work was supported by Materials Sciences and Engineering Division, Office of Basic Energy Sciences, U. S. Department of Energy (US DOE EPSCORE). Thank for the Advance grant # HRD-0548311 for purchasing new equipment.

References

(1) Boukamp, B. A.; Lesh, G. C.; Huggins, R. A. *J. Electrochem. Soc.* **1981,** *128*, 725–729.
(2) Beaulieu, L. Y.; Eberman, K. W.; Turner, R. L.; Krause, L. J.; Dahn, J. R. *Electrochem. Solid-State Lett.* **2001,** *4 (9)*, A137 – A140.
(3) Beattie, S. D. ; Larcher, D. ; Morcrette, M. ; Simon, B. ; Tarascon, J.-M. *J. Electrochem. Soc.* **2008,** *155* , A158-163.
(4) Yang, J. ; Winter, M. ; Besenhard, J. O. *Solid State Ionics*, **1996,** *90*, 281-287.
(5) Li, X.; Gu,M.; Hu, S.; Kennard, R.; Yan, P.; Chen , X.; Wang, C.; Sailor , M. J.; Zhang , J.G.; Liu, J. *nature communication,* **2014,** 5 4015
(6) Agubra*, V. A.; Fergus, J. W. *Journal of Power Sources* **2014** *268* 153-162.
(7) Schroder, K. W.; Celio, † H.; Webb, L. J.; Stevenson, K. J. *The Journal of Physical Chemistry C* **2012,** *116*, 19737–19747
(8) Terskikh, V.; Moudrakovski, I.; Mastikhin, V. *J. Chem. Soc. Faraday Trans.* **1993,** *89*, 4239-4243
(9) Ripmeester, J. A.; Ratcliffe, C. I. *J. Phys. Chem.* **1990,** *94*, 7652-7656
(10) Yan, X. ; Zhang, Y.Q. ; Zhu, K. ; Gao, Y. ; Zhang, D. ; Chen, G.; Wang, C.Z. ; Wei, Y.J. *J. Power Sources* **2014,** *246* , 95-102.

PLATINUM-METAL/TITANIUM AS THE ANODE CATALYST FOR DBHFC

E. Norkus, A. Balčiūnaitė, A. Zabielaitė, I. Stankevičienė, A. Jagminienė, V. Kepenienė, L. Tamašauskaitė-Tamašiūnaitė

Department of Catalysis, Center for Physical Sciences and Technology, A. Goštauto 9, LT-01108 Vilnius, Lithuania
eugenijus.norkus@ftmc.lt

In the present study platinum-metal (Cu, Ni) catalysts deposited on the titanium surface (denoted as Pt(Cu)/Ti and Pt(Ni)/Ti) were prepared via a simple and cost-effective galvanic displacement technique [1,2]. At first, the sub-layers of Cu and Ni were deposited on the titanium surface via electroless deposition. The Pt crystallites were deposited on the prepared Cu/Ti and Ni/Ti electrodes by their immersion into a platinum-containing solution for various time periods. The Pt loadings were in the range from 2 to 80 μg cm^{-2} in the Pt(Ni)/Ti and Pt(Cu)/Ti catalysts.

Performance of $NaBH_4/H_2O_2$ fuel cell (DBHFC) operating with 1 M $NaBH_4$ + 6 M NaOH as fuel and 2 M H_2O_2 + 0.5 M H_2SO_4 as oxidant, while employing the Pt as the cathode catalyst and platinum-metal (Ni, Cu) catalysts deposited on the titanium surface as the anode catalyst was investigated. The load was applied in steps of 5 mA cm^{-2} within the range of 0-120 mA cm^{-2}. Each step lasted for 2 min and the current was continuously applied from one value to next without disconnecting the cell.

It was found that specific peak power densities of 7318 and 26873 W g^{-1} were obtained at 25 °C fuel cell temperature by employing Pt(Ni)/Ti and Pt(Cu)/Ti as the anode catalyst, respectively.

References

(1) Tamašauskaitė-Tamašiūnaitė, L.; Balčiūnaitė, A.; Čekavičiūtė, R.; Selskis, A. *J. Electrochem. Soc.* **2012**, *159*, B611.
(2) Tamašauskaitė-Tamašiūnaitė, L.; Balčiūnaitė, A.; Vaiciukevičienė, A.; Selskis, A.; Pakštas, V. *J. Power Sources* **2012**, *208*, 242.

DEHYDROGENATION OF PROPANE ON PtSn-BASED CATALYSTS: SYNERGETIC EFFECTS OF SUPPORT

Jiacheng Li[1], Zhen Zhao[1,2,*], Jianmei Li[1], Jian Liu[1], Xiaoqing Fan[2], Qinglong Liu[1]

[1]State Key Laboratory of Heavy Oil Processing, China University of Petroleum, 18# Fuxue Road, Chang Ping, Beijing 102249, China
[2]Institute of Catalysis for Energy and Environment, Shenyang Normal University, Shenyang110034, China
*Corresponding author: zhenzhao@cup.edu.cn
zhaozhen1586@163.com

Abstract

Propane dehydrogenation process has attracted sustained attention because of its prospect in meeting the growing demand for propylene in recent years. The Pt-based catalysts have been widely utilized for their excellent catalytic performances.[1-4] The catalytic results of active center depend significantly on the nature of supports.[5]

Alumina has been widely utilized as support due to its easy availability, low price, and controllable acidity. In order to further promote dehydrogenation of propane, a series of alumina-supported PtSn catalysts were prepared by hydrothermal and incipient-wetness impregnation method. By adjusting the PH value of the hydrothermal system, alumina with rod or slice shape has been synthesized (Figure 1(a) and (b)).

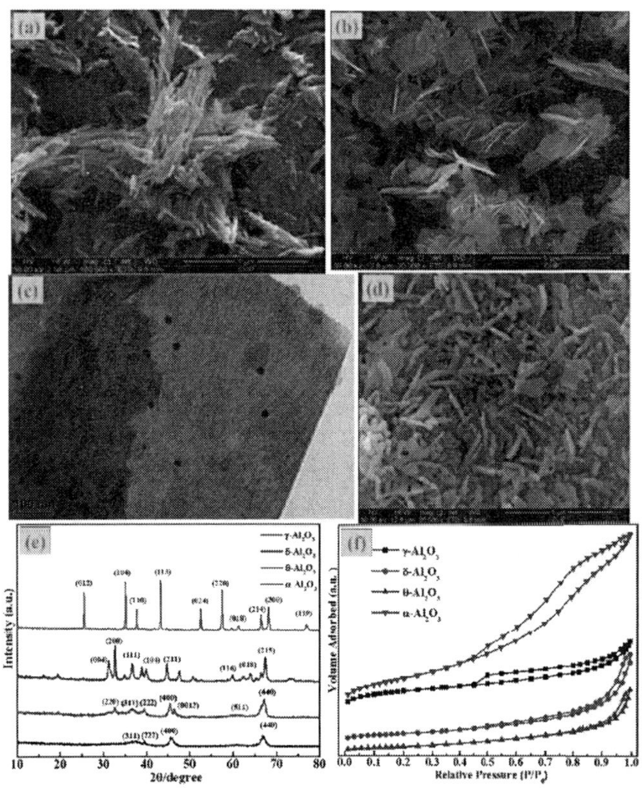

Figure 1. Characterization of different samples: (a)SEM images of rod γ-Al_2O_3 (b) SEM images of slice γ-Al_2O_3 (c)TEM images of PtSn/γ-Al_2O_3(slice) (d) SEM images of rod α-Al_2O_3 (e) XRD patterns of Al_2O_3 (f) N_2 physical sorption isotherms of Al_2O_3.

The TEM image shows that slice γ-Al_2O_3 offers stable face to cooperate with PtSn active centers, therefore obtaining a higher conversion as listed in Table 1. For further investigating the effect of alumina, slice Al_2O_3 was calcined in the temperature from 1043K to 1473K, then a series of α-Al_2O_3, δ-Al_2O_3 and θ-Al_2O_3 were achieved as certified by XRD patterns in Figure 1(e). All of them still preserve the slice structure except for α-Al_2O_3 sample (Figure 1(d)), which exhibits a worm-like morphology due to the effect of high temperature. The testing results of PtSn/Al_2O_3 catalysts are summarized in Table 1 together with their parameters of physical property. The PtSn/θ-Al_2O_3 shows the highest conversion for propane with 95.8% of selectivity to propylene.

In addition, the activity of PtSn/θ-Al_2O_3(slice) catalyst is also compared with mesoporous (Ti-KIT-6) and micoporous (TS-1) molecular sieves-supported PtSn catalysts. The high propane conversion of the later two catalysts are 54% and 51%, respectively. A similar catalytic performance can also be achieved by controlling the morphology and phase of alumina. Considering its lower cost and better activity, the alumina-supported catalyst exhibits glittering prospect in meeting the industry demand.

Table 1. Properties and activities of PtSn/Al_2O_3 catalysts

	PtSn/γ-Al_2O_3(rod)	PtSn/γ-Al_2O_3(slice)	PtSn/θ-Al_2O_3	PtSn/δ-Al_2O_3	PtSn/α-Al_2O
BET Surface Area (m^2/g)	91.9	76.4	29.7	21.3	7.0
Pore Volume (cm^3/g)	0.20	0.10	0.13	0.09	0.03
Pore Size (nm)	11.6	6.9	12.3	12.3	7.8
Conversion (%)	48.6	50.6	52.1	12.1	1.5
Selectivity (%)	95.3	95.7	95.8	95.0	74.7

Acknowledgement. This work was financially supported by NSFC (91545119, 21477164).

Reference

(1) Feng, J.; Liang, Z.; Shui, L.; Gang, Liu, Sheng, W.; and Jin, G. *ACS Catal.* **2015**, 5,438.
(2) Chang, Yu; Qing, Ge; Heng, Xu; Wen, Li. *Appl. Catal. A: Gen.* **2006**, *315*, 58.
(3) Farnaz, Z.; Shokoufeh M.; Saeed, S. *Fuel Process. Technol.* **2013**, 109, 108.
(4) Zhan, M.; Jun W.; Jun L.; Ning, W.; Chang, A.; Lan, S.; *Fuel Process. Technol.* **2014**,128,283.
(5) Xiao, F.; Jian. L.; Zhen. Z. *Catal. Sci. Technol.* **2015**, 5, 339.

CONTROLLING THE MORPHOLOGY AND COMPOSITION OF Ag/Ag-HALIDE HYBRID NANOSTRUCTURES AND THEIR ENHANCED VISIBLE LIGHT INDUCED PHOTOCATALYTIC PROPERTIES

Aasim Shahzad, Taekyung Yu*, and Woo-Sik Kim*

Department of Chemical Engineering, College of Engineering, Kyung Hee University, Youngin, 446-701, Korea

**Corresponding author. E-mail:* tkyu@khu.ac.kr (T. Yu) and wskim@khu.ac.kr (W.-S. Kim)

In recent years, with nanometer-sized dimensions the nanostructures based on noble metals such as Au, Ag, and Cu enablers for a variety of applications involving light because of their plasmon resonance [1]. To date, various nanostructures with controlled shape have been synthesized and investigated for photocatalysis, typically exemplified by TiO_2-based large bandgaps nanocomposites. Although, these nanohybids proved to be chemically stable but at the same time confines its response to UV fraction of solar energy about 4%, leading to low efficiency in visible and near infrared regions. Therefore, need for engineering the new visible light active photocatalysts is highly desirable. Silver chloride is a photosensitive material, that generates an electron and hole pairs upon absorbing light. Subsequently, the photogenerated electron combines with Ag+ ions to form Ag^0 atoms which show higher absorption in visible region of solar spectrum due to their plasmon resonance [2]. Here, we report a simple and fast synthetic route to Ag/Ag-halide hybrid nanostructures prepared by reducing Ag-halide nanoparticles synthesized by reaction of Ag^+ with halide ions (Cl^- and Br^-) in the presence of polyethyleneimine (PEI). By controlling the experimental conditions including reaction temperature and amount of reducing agent, we could easily control the morphologies and compositions of Ag/Ag-halide hybrid nanostructures from Ag nanoparticles decorated AgBr nanoparticles to hollow cage-like Ag-lich nanostructures. The Ag/AgBr hybrid nanostructures exhibited enhanced photocatalytic activity and recyclability for decomposing methylene blue (MB) under visible-light illumination because of small bandgap (2.6 eV) and strong SPR effect. Especially, we found that the Ag/AgBr hybrid nanostructures exhibited around 1000 times faster photocatalytic reaction rate compared with commercial N-doped TiO_2 (P-25). In addition, the Ag/AgBr hybrid nanostructures exhibited higher photocatalytic activity compared with Ag/AgCl hybrid nanostructures towards decomposition of methyl orange (MO) under visible light illumination due to their smaller bandgap (2.6 eV) and low electron affinity of Br in Ag/AgBr.

A NOVEL HOLLOW $NiCo_2O_4$ HIERARCHICAL MICROSPHERES WITH ENHANCED ELECTROCHEMICAL PERFORMANCE

Liang Wang, Qingli Hao

Key Laboratory for Soft Chemistry and Functional Materials, School of Chemical Engineering, Nanjing University of Science and Technology Nanjing 210094, China
E-mail: qinglihao@njust.edu.cn; haoqingli@yahoo.com

Abstract

Developing efficient supercapacitor is essential for many energy conversion or storage processes in clean energy technology, especially for the renewable energy such as solar, wind and tidal[1]. The electrode material plays a important role in the performance of the supercapacitor. During the past years, much efforts have been devoted to the design and fabrication of all kinds of nanostructures to improve the supercapacitor performance. It is well recognized that the three dimensional hollow nanostructures is an effective strategy to develop alternative pseudocapacitive materials with good electrical conductivity, high specific surface area, large naked surface sites, and good electrochemical durability[2-6]. Herein, A novel hollow $NiCo_2O_4$ hierarchical microspheres has been developed based on a self-template method. The hollow spheres was prepared through a facile solvothermal treatment, hydrolysis process and subsequent calcination. The hollow $NiCo_2O_4$ microspheres are composed of hundreds of ultrathin crystal-like nanosheets. The unique hollow $NiCo_2O_4$ microspheres show a specific capacitance as high as 940 F g^{-1} at a current density of 1 A g^{-1}, and only slightly decayed less than 10 % after 3000 cycling tests. Moreover, the full cell combined with $NiCo_2O_4$ and active carbon showed an energy density of 29.4 W h kg^{-1} and correspondent power density of 792.1 W kg^{-1}. Therefore, The desirable performance enables it to be a promising electrode material for the electrochemical supercapacitor. Present work supplies an example for future synthesis of novel nanostructures for supercapacitors.

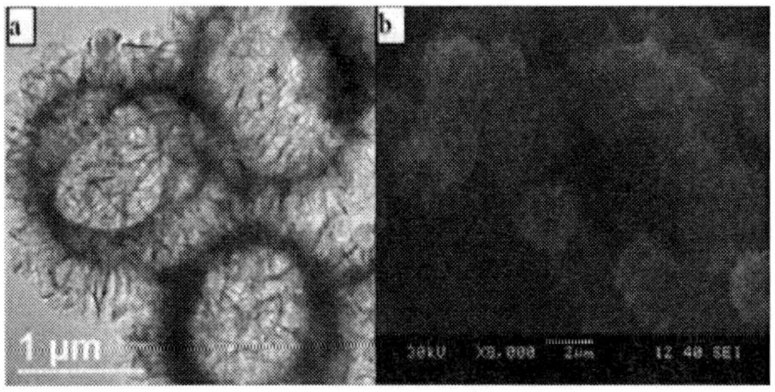

Figure 1. The TEM image (a) and SEM image (b) of as-prepared hollow $NiCo_2O_4$ microspheres

Acknowledgement. The work was supported by the National Natural Science Foundation of China (No. 21576138), Program for NCET-12-0629, the Ph.D. Programs Foundation of Ministry of Education of China (No.20133219110018), Qing Lan Project and Six Major Talent Summit (XNY-011), the Science and Technology Support Plan (No. BE2013126), and PAPD of Jiangsu Province, China.

References

(1) Evans A.; Strezov V.; Evans T.-J.. Renewable and Sustainable Energy Reviews, 2012, 16(6): 4141-4147.
(2) Simon P.; Gogotsi Y.. Nature materials, 2008, 7(11): 845-854.
(3) Zhong, C.; Deng, Y.; Hu, W.; Qiao, J.; Zhang, L.; & Zhang, J. Chemical Society Reviews, 2015, 44(21): 7484-7539.
(4) Liu S.; Sun S.; You X.-Z. Nanoscale, 2014, 6(4): 2037-2045.
(5) Armand M.; Tarascon J.-M. Nature. 2008, 451 (7179): 652-657
(6) Xia C.; Alshareef H.N; Chemistry of Materials. 2015, 27 (13): 4661-4668

SUPERCAPACITORS UTILIZING ELECTRODES DERIVED FROM POLYACRYLONITRILE INCORPORATING TETRAMETHYLAMMONIUM OXALATE AS A POROGEN

*Sahila Peranathan, Jeliza S. Bonso, John. P. Ferraris**

Department of Chemistry and Biochemistry, The University of Texas at Dallas, 800 W. Campbell Road, Richardson, TX 75080-3021, USA
ferraris@utdallas.edu

Introduction

The demand for energy storage devices haas been increasing due to the mismatch between energy supply and energy demand. Supercapacitors are promising energy storage devices since they combine the advantages, in terms of power and energy densities, of conventional capacitors and batteries. Compared to batteries, however, supercapacitors still have inferior energy density, which mitigates their realization for commercial application.[1,2]

High surface area carbon nanofiber electrodes are one of the promising electrode materials for supercapacitors that store energy electrostatically in the electrode/electrolyte interface.[3] For this study, we prepared porous carbon nanofiber electrodes by incorporating tetramethylammonium oxalate (TMAO) as a porogen in polyacrylonitrile (PAN) precursor fibers generated by electrospinning followed by thermal treatment. The obtained electrodes were used directly in coin cell devices using an ionic liquid electrolyte (EMITFSI) and the electrochemical performance is evaluated and discussed.

Experimental

Electrospinning. Tetramethylammonium oxalate (TMAO) was prepared from tetramethylammonium hydroxide and oxalic acid by following a reported procedure.[4,5] Different amounts of TMAO were dissolved in same amount of water (0.04 wt%) and dispersed in DMF by ultrasonicating for 1 hr followed by mechanical stirring overnight. Polyacrylonitrile (PAN, 1.00 g, 10 wt%) was dissolved in the dispersion at 50 °C for 1 hr and stirred overnight at room temperature. PAN (10 wt%) solution was prepared without TMAO as a control. We refer the 0.05 wt% and 0.1 wt% TMAO incorporated samples as T5 and T10, respectively, and the neat PAN as T0. The solutions were electrospun through a 20G spinneret under an applied voltage of 15 kV (Gamma High Voltage Research, Inc., power supply). The fibers were deposited on a grounded collector rotating at a speed of 300 rpm. The applied voltage, distance between the needle tip and collector drum (15 cm) and the speed of the collector drum were kept constant for all the samples.

Thermal treatments of carbon nanofibers. The as-spun nanofibers were stabilized at 240 °C under airflow (200 sccm) for 6 hrs. The carbonization and CO_2 activation were done in a single step in an inert environment at 1 atm using He (200 sccm) at a heating rate of 5 °C/min to a final temperature of 1000 °C. Then CO_2 (50 sccm) was introduced while maintaining the temperature for 1 hr to obtain porous carbonized nanofibers.

Charaterization. The surface morphology of as-spun and carbonized nanofibers was observed by scanning electron microscopy (Zeiss SUPRA 40 SEM) after sputter coating with gold/palladium. Surface area and pore volume of the carbon nanofibers were obtained using Quantachrome Instruments Autosorb-1 at 77 K after degassing for 12 hrs at 200 °C. Electrochemical performance were done using an Arbin Supercapacitor Testing Systems (SCTS)

Results and Discussion

Figure 1. shows the scanning electron microscope (SEM) images of electrospun fibers of T0, T5 and T10 samples. The nanofibers are uniform and smooth and the average fiber diameter decreased with increasing amount of TMAO.

The average fiber diameter of as-spun T0 sample was 970±120 nm, which decreased to 660±70 nm upon addition of only 0.05 wt% of TMAO and further decreased to 500±65 nm when the amount of TMAO is doubled. Figure.2. shows SEM images of fibers after carbonization and activation. The morphology is preserved for all the samples with reduced fiber diameter of 525±75 nm, 260±50 nm and 245±40 nm for T0, T5 and T10, respectively.

Figure 1. SEM images of electrospun fibers of T0, T5 and T10

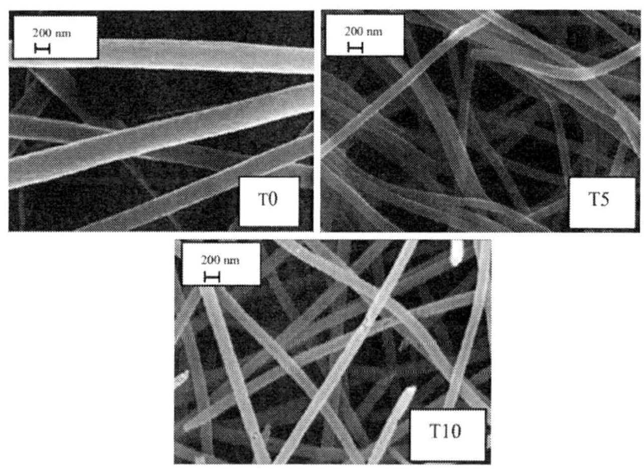

Figure 2. SEM images of carbonized and activated fibers of T0, T5 and T10

Table 1. BET surface area, pore volumes obtained from N_2 sorption isotherms of carbonized and activated nanofibers

	BET SSA ($m^2 g^{-1}$)	DFT TPV ($cm^3 g^{-1}$)	V_{micro} ($cm^3 g^{-1}$)	V_{meso} ($cm^3 g^{-1}$)
T0	1963	1.229	0.703 (57 %)	0.526 (43 %)
T5	2295	1.260	0.690 (55 %)	0.370 (45 %)
T10	2663	1.487	1.060 (71 %)	0.427 (29 %)

BET surface areas shown in Table 1 was measured using Quantachrome Instruments Autosorb-1. The T0 sample possesses a specific surface area of 1963 $m^2 g^{-1}$ with a total pore volume of 1.229 $cm^3 g^{-1}$, 43 % of which are mesoporous (2-50 nm). Addition of only 0.05 w% of TMAO increased the specific surface area of the T5 sample to 2295 $m^2 g^{-1}$ without a change in the mesoporosity percentage. 0.1 % of TMAO increased the specific surface area to 2663 $m^2 g^{-1}$ (36 % greater than T0) with a decrease in mesoporosity (29%). Only the pores with same size of electrolyte ions or higher can be accessible to the electrolyte ions. Ion sizes of cation (EMI^+) and anion ($TFSI^-$) are 4.3 x 7.6 Å and 2.9 x 7.9 Å respectively.[6] By considering the shortest dimension of ions, more than 95% of the pore volume is accessible to cations for all three samples and 100% is accessible to anions.

In Figure 3. (a), the CV curves of activated PAN show a slightly skewed rectangular shape. On the other hand, T5 and T10 in (Fig. 3b-c) showed more rectangular cyclic voltammograms. Table 2 summarizes the electrochemical performance of all three samples. The specific capacitance values of T0, T5 and T10 samples calculated from the CV curves are 90 F g^{-1}, 129 F g^{-1} and 140 F g^{-1}, respectively. Compared to T0 at the scan rate of 10 mV s^{-1} there are a 43% and 55% increase in specific capacitance for T5 and T10 sample, respectively.

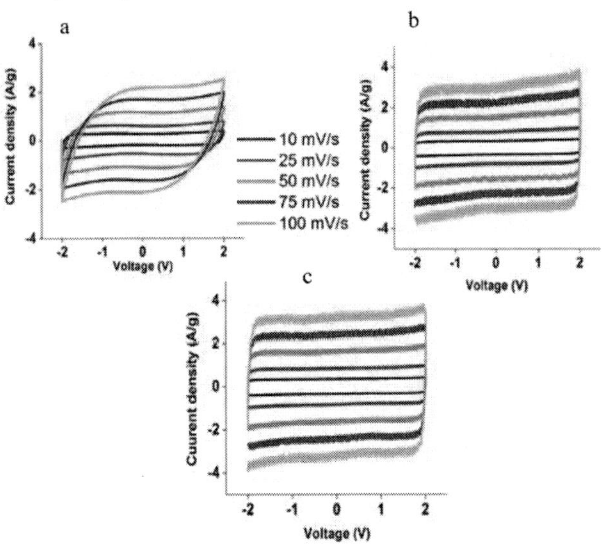

Figure 3. Cyclic voltammograms of T0 (a), T5 (b) and T10 (c)

Table 2. Specific capacitance, energy density and power density of carbonized and activated nanofibers

	Specific capacitance at 10 mV s^{-1} (F g^{-1})	Energy density (Wh kg^{-1}) at 1 kW kg^{-1}	Energy density (Wh kg^{-1}) at 10 kW kg^{-1}
T0	92	40	17
T5	129	63	62
T10	140	69	65

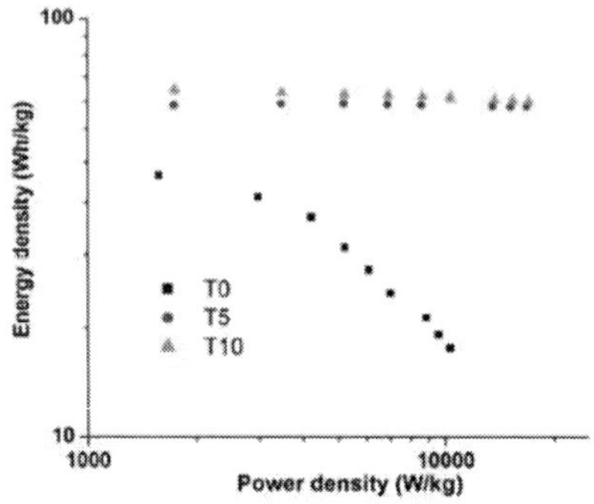

Figure 4. Ragone plot

Figure 4 compares the energy and power density of all the samples. As the power density increased from 1 W kg^{-1} to 10 W kg^{-1}, energy density dropped from 40 to 17 Wh $kg^{-1}$for T0 whereas hand it only dropped from 69 to 65 Wh kg^{-1} for T10. This improvement in the performance can be explained by its high surface area and higher amount of mesopores that allow better pore accessibility and faster diffusion of ions of ionic liquid electrolytes.

Conclusions

In this study, free standing and flexible high surface area carbon fiber electrodes incorporating a porogen were prepared and studied in coin cell supercapacitors utilizing an ionic liquid electrolyte. The highest surface area of 2663 m^2g^{-1} was obtained for the sample containing 0.1 wt% porogen (T10). A specific capacitance of 140 F g^{-1} and an energy density of 69 Wh kg^{-1} at 1 kW kg^{-1} was observed. The values are about 60% higher than those of just PAN in the EMITFSI electrolyte. The gravimetric energy density (active material only) of the devices utilizing carbon nanofibers derived from neat PAN dropped from 40 Wh kg^{-1} to 17 Wh kg^{-1} when the power density increased from 1 kW kg^{-1} to 10 kW kg^{-1} whereas T10 retained 94% of its original value.

Acknowledgement. The authors gratefully acknowledge the National Science Foundation (Grant no. IIP-1127564) for financial support. We also thank Dr. Sajani Basnayake for helping with the surface area analysis and Dr. Benjamin Batchelor for thermogravimetric mass spectrometry analysis. Helpful discussions with Dr Lissa Magel (Solarno, Inc.) are also greatly appreciated.

References

(1) B.E.; Conway, *J. Electrochem. Soc*, **1991**, 183, 1539-1548.
(2) M.; Winter, R. J.; Brodd, *Chem. Rev.* **2004**, 104, 4245–4269.
(3) A.; Burke, *Journal of Power Sources*. **2000**, 91, 37–50.
(4) A. T.; Lawson, N.; Collie, Ph.D., F.R.S.E. *J. Chem. Soc., Trans*, **1888**, 53, 624-636.
(5) B.; Albert, M.; Jansen, W, *J. Chem. Soc. Chem. Commun.*, **1993**, 965.
(6) P.; Simon, Y.; Gogotsi , *Nat. Mater.*, **2008**, 7, 845–854.

INVESTIGATION OF HAILAER COAL PYROLYSIS BY LARGE-SCALE ReaxFF MOLECULAR DYNAMICS

Mo Zheng, Xiaoxia Li, Li Guo*

State Key Laboratory of Multiphase Complex Systems, Institute of Process Engineering, Chinese Academy of Sciences, Beijing 100190, P. R. China
mzheng@ipe.ac.cn

Introduction

Coal pyrolysis plays an important role in the clean coal utilization. The heterogeneous nature of coal and the complexity process have made it very difficult to access the comprehensive mechanisms with experiments.

Based on a bond-order potential, ReaxFF MD[1] is able to describe the evolving of formation, transition and dissociation of chemical bonds with accuracy close to DFT but reduced computational costs, which is promising for examining coal pyrolysis mechanisms. In this paper, a new methodology[2] rooted in the first GPU-enabled ReaxFF MD simulation program (GMD-Reax) and the cheminformatics based reaction analysis tool (VARxMD) was employed to explore the coal pyrolysis behaviors.

Methods

The large-scale Hailaer brown coal model with 27,809 atoms was constructed based on a combination of experiments and well-cited brown coal models. The GMD-Reax simulations were performed at different conditions at 800-2600 K for 250 simulation time to investigate the temperature effects. During all simulations, the time-step of 0.25 fs was used to integrate Newton's equation of motion by the velocity-Verlet algorithm and Berendsen thermostat was employed to maintain temperature equilibrium.

Results and Discussion

Analysis of the ReaxFF MD simulation trajectory shows that the evolution tendencies of product profile (char, tar and gas) and major pyrolyzates (aromatics and aliphatics hydrocarbon) as a function of time and temperature are in broad agreement with experimental results in literature. The complex and detailed chemical reactions has been revealed from the ReaxFF MD simulations using VARxMD. The aromatic ring conversion and phenol generation pathways in Hailaer coal pyrolysis are obtained and reasonable.

Conclusions

The computational approach of large-scale ReaxFF MD simulation facilitated by GPU computing and cheminformatics analysis sheds new light on the coal thermolysis behaviors and underlying reaction mechanism at high temperatures.

Acknowledgement. This work was supported by the funding granted from the National Natural Science Foundation of China (21373227, 21073195 and 91434105).

References

(1) Van Duin, A. C. T.; Dasgupta, S.; Lorant, F.; Goddard, W. A. ReaxFF: A reactive force field for hydrocarbons. *J. Phys. Chem. A* **2001,** *105* (41), 9396-9409.

(2) Li, X.; Mo, Z.; Liu, J.; Guo, L. Revealing chemical reactions of coal pyrolysis with GPU-enabled ReaxFF molecular dynamics and cheminformatics analysis. *Mol. Simulat.* **2015,** *41* (1-3), 13-27.

RAPID AND ACCURATE DETERMINATION OF THE LIGNIN CONTENT OF LIGNOCELLULOSIC BIOMASS BY SOLID-STATE NMR

Li Fu[a], Scott A. McCallum[a], Jianjun Miao[a], Courtney Hart[b], Gregory J. Tudryn[b], Fuming Zhang[a], Robert J. Linhardt[a,c]

[a]Center for Biotechnology and Interdisciplinary Studies, Rensselaer Polytechnic Institute, Troy, NY 12180, USA
[b]Ecovative Design, LLC, Green Island, NY 12183, USA
[c]Departments of Chemistry and Chemical Biology, Chemical and Biological Engineering, and Biology and Biomedical Engineering Rensselaer Polytechnic Institute, Troy, NY 12180, USA
Corresponding author e-mail address: linhar@rpi.edu (R.J. Linhardt).

1. Introduction

Biofuels and biomaterials produced from lignocellulosic feedstock require facile access to cellulose and hemicellulose to ensure competitive environmental and economic advantages to traditional petroleum processing and sugar fermentation pathways [1,2]. Hydrolysis of cellulose and hemicellulose into fermentable sugars is impeded by physical and chemical barriers created through the association of lignin phenylpropanoid polymers and other cell wall components of lignocellulosic biomass (Fig. 1)[3]. Both covalent and non-covalent lignin–carbohydrate interactions inhibit the digestion of otherwise digestible sugars, such as cellulose and hemicellulose, by physically blocking access to these polysaccharides to cellulases andb-glucosidases. Lignin also absorbs these enzymes, resulting in another significant barrier to their action [4].

Sufficient and mild delignification greatly enhances the utilization of associated saccharides, i.e., cellulose, hemicellulose, and other cell wall components [1]. Modern processing techniques [2] of lignocellulosic biomass include, physical and chemical processing methods, such as wet oxidation [5], ammonia fiber explosion [6], ozonolysis [7], and organosolvation [8], and bioprocessing where lignin is degraded using a highly efficient use of extracellular enzymes [3].

Despite lignin's intractability in saccharification, some organisms have adapted to effectively degrade lignin in order to access the large cache of nutrition locked in the secondary cell wall of woody plants [9]. Fungi are the main organisms identified with this characteristic [3], the most proficient subset belonging to basidiomycetes (Fig. 2 and Supporting Information Fig. S1). Members of this group of fungi are collectively known as ''white rot'' because in nature, areas of wood infiltrated by their mycelia take on a whitish appearance due to depletion of brown lignin. In fact, white rot fungi are the most efficient lignin digesters observed in nature [10]. White rot fungi have evolved to secrete numerous peroxidases, such as manganese peroxidases, lignin peroxidases, and laccases that participate in ligninolysis [3]. Conversely, a class of fungi known as ''brown rot'' does not target lignin for degradation [11]. These fungi rely on degradation of cellulose, which leaves the appearance of brownish residue on the wood they colonize. Quantitative analysis of the components comprising biomass, particularly lignin, is difficult because of the limited solubility of lignocellulosics and these commonly require extensive processing prior to their analysis. ^{13}C cross-polarization, magic-angle spinning, solid-state nuclear magnetic resonance (CP MAS ssNMR) spectroscopy may provide the ideal approach for the direct analysis of lignocellulosic feedstock [12–14]. CP MAS ssNMR spectroscopy was first applied to qualitatively examine cellulose, the major component of lignocellulosic feedstock [15]. It has subsequently been successfully used to investigate other biomass components, including hemicellulose and lignin [16,17]. ^{13}C CP MAS ssNMR spectroscopy is now considered an essential tool in paper, biomaterial and biofuel processing [17]. While high-resolution ^{13}C solution-state NMR spectroscopy can also afford a quantitative way to measure lignin content in lignocellulose biomass, this approach requires multiple steps for lignin isolation, removal of interfering proteins and carbohydrates, and lignin solubilization [18]. One way to purify lignin for analysis relies on enzymes, such as cellulases and hemicellulases, to hydrolyze cellulose and hemicellulose. Unfortunately, lignin is covalently attached to cellulosics and enzyme adsorption onto lignins, poses additional challenges for lignin isolation [19–21].

In the current study, ^{13}C CP MAS ssNMR is examined to directly quantify the lignin content in lignocellulose biomass without any chemical or enzymatic pre-treatment, providing a content of lignin closer to its actual value in a biomass sample. This approach takes advantage of unique chemical structures within lignin to differentiate it from the other components of lignocellulosic biomass. Quantitative analysis of lignin in lignocellulosic biomass using ^{13}C CP MAS ssNMR spectroscopy also required the preparation of a pristine lignin standard, to serve as a calibrant.

2. Methods

2.1 Organisms and their culturing. Fungal organisms, Fomitopsis pinicola (brown rot) and Ganoderma lucidum (white rot) were from Ecovative's strain library but are also available from the CBS-KNAW Fungal Biodiversity Centre culture collection (http://www.cbs.knaw.nl/index.php/order). Malt extract broth (Difco, Becton, Dickinson and Co.), containing malt extract 6.0 g/L, maltose 1.8 g/L, dextrose 6.0 g/L and yeast extract 1.2 g/L, was prepared and sterilized at 15 psi for 15 min. Inoculated broth was prepared from each species separately. To inoculate, broth was transferred to a sterile blender flask and colonized millet inoculum was added at a rate of 10% (m:m). The inoculated grain was then blended thoroughly in the malt extract broth for 60 s. Approximately 25 mL of the inoculated broth was transferred to each of six sterile petri dishes to grow sheets of tissue on broth. Cultures were incubated at room temperature as fungal sheets developed on the surface. Sheets of pure tissue were harvested after 23 d of incubation and washed in deionized water sufficient to remove malt extract broth.

2.2 Preparation of biomass samples. Kenaf Biomass was acquired from Kenactiv Innovation, Inc. Kenaf pith core was collected and ground with a 0.32 cm mesh screen, then sorted over 0.21 cm mesh screen (see Supporting Information Table S1 for nutritional analysis of kenaf).

2.3. Fermentation of biomass samples. Six samples of kenaf pith (100 g each) were prepared in autoclavable spawn bags. Two bags of each of the six compositions were prepared by adding differing amounts (A-0%, 1%; B-10%, 0%; and C-10%, 1% each) of clear flour and calcium chloride, respectively (0% and 1%). These additives were used to supplement and encourage growth, particularly in the early stage as the mycelia from the inoculum begin to colonize surrounding kenaf particles. Next, water (400 mL) was added to each bag and the bags were sterilized for 60 min at 0.1 MPa pressure. Bags of substrate were inoculated at a rate of 10 % (w/w) and sealed. Inoculum consisted of yellow millet pre-colonized with cells from the relevant species. Original cultures used were acquired directly from the field in the northeast region of the United States. Samples from sets A and B (both for G. lucidum and the F. pinicola) were collected after 17 d of incubation and samples from set C were harvested after 16 d of incubation at 21°C.

2.4 Lignin Extraction. Kenaf (50 g) was extracted with 2 L of 2% aqueous ammonium hydroxide for 2 d [22]. The extract was centrifuged (5,000 × g) for 10 min at 4□C to remove the solid kenaf residue. The supernatant was rotary evaporated with heating to remove excess aqueous ammonium hydroxide to precipitate crude lignin, which was recovered by centrifugation (5,000 × g) for 10 min at 4□C. The crude lignin was dissolved in DMSO and centrifuged (5,000 × g) 10 min at 4□C. The supernatant, which contained lignin, was collected, dried using rotary evaporator. Most of the residual DMSO could be removed by repeatedly (8-times) adding 5 mL distilled water and drying using rotary evaporation.

2.5 Solution state NMR. The solution-state NMR experiments were all performed on Bruker Advance II 800 MHz spectrometer equipped with a cryogenically cooled probe (TCI) with z-axis gradients (Bruker BioSpin, Billerica, MA). NMR data were processed and analyzed with Topsin 2.1.6 software (Bruker). The lignin was dissolved in DMSO-d6 and transferred into 5-mm NMR microtubes for one-dimensional 1H-NMR, one-dimensional ^{13}C-NMR and two-dimensional ^{13}C-1H heteronuclear single quantum coherence spectroscopy (HSQC) [20]. The conditions for one-dimensional 1H-NMR spectra were as follows: 32 scans, acquisition time of 0.99 s, and relaxation delay of 8.00 s. The conditions for one-dimensional ^{13}C-NMR spectra were as follows: 4096 scans, acquisition time of 0.23 s and relaxation delay of 8 s. The conditions for two-dimensional HSQC spectra were as follows: 32 scans, acquisition time of 0.33 s and relaxation delay of 0.90 s. All experiments were carried out at 333K.

2.6 ^{13}C Cross-polarization, magic-angle spinning, solid-state NMR. The ^{13}C cross polarization magic spinning solid-sate NMR experiment was set up according to the R.E. Taylor's method [23]. All ^{13}C CP MAS ssNMR experiments were performed on the 600 MHz 89 mm wide-bore Bruker Advance III spectrometer equipped with a 4 mm HXY solid-state MAS probe set up with two of the channels were configured for 1H and ^{13}C. A reduced volume rotor (50 □L) was utilized and the active volume mapped to ensure that the active coil volume entirely contained the sample volume. Based on these findings, prior to the addition of pre-dried samples, polytetrafluoroethylene (~10 mg) was added into the bottom of the rotor in addition to a Teflon plug being positioned above the sample. The rotor was weighed before and after the addition of samples. The conditions for ^{13}C CP MAS ssNMR were as follows: 4096 scans, spinning rate of 12.5K Hz, acquisition time of 0.02 s, and temperature of 278 k. A quantitative calibration curve was made based on peak integrals (157-142 ppm) for a series of high purity lignin samples of varying mass.

3. Results and discussion

3.1 Approach for extractive removal of lignin from biomass using room temperature ionic liquid. The widespread utilization of lignocellulosic biomass as a raw material for the production of biofuels, chemicals and materials and its inherent compositional variability makes the rapid assessment of its composition of critical importance, particularly as we move towards renewable resources [17]. A major impediment in assessing the composition and in utilizing lignocellulosic biomass is the intractable properties of lignin, which is extremely difficult to extract [18,19,20]. Until now most laboratories assess lignin content of biomass in multistep processes involving lignin extraction and closure of mass balance [24]. More recently, FTIR spectroscopy [25] has been used to estimate lignin content as well as solution-state ^{13}C NMR spectroscopy but these methods require finding solvents to dissolve biomass samples [26]. Until now, the application of ^{13}C CP MAS ssNMR spectroscopy has only been used for the qualitative assessment of lignin in biomass samples presumably because of the lack of available pristine, unmodified, lignin standard. The most widely available lignin is lignosulfonate, a side product of the pulp and papermaking industry, has distinctive NMR spectral characteristics making it unsuitable as a standard. Thus, we first undertook to explore methods to directly extract pristine lignin from lignocellulosic biomass using room temperature ionic liquids (RTILs) [27]. Unfortunately, 1-ethyl-3-methylimidazolium acetate RTIL [27] afforded lignin containing large amounts of cellulosics based on ^{13}C CP MAS ssNMR (Supporting Information Figure S2).

3.2 Extraction of lignin from biomass using ammonium hydroxide. We next examined aqueous ammonium hydroxide for the extraction and recovery of pristine lignin from kenaf [28]. Pristine lignin (350 mg) was extracted from 50 g of the lignocellulosic biomass, Kenaf. The structure of the pristine lignin was characterized using solution 1D 1H-NMR and 1D ^{13}C-NMR (Supporting Information Figures S3 and S4, respectively). The 2D HSQC spectrum of kenaf lignin allowed the assignment of the prominent signals in the ^{13}C-NMR spectrum (Supporting Information Figure S5 and Table S2). The pristine lignin core structure matched well with that of other recent reports [18]. The recovery of lignin was low (~22 %) since no cellulase or hemicellulase was added to hydrolyze the polysaccharide and to release covalently bound lignins. However, the enzyme-free extraction reduced the level of residual proteins and provided pristine lignin of very high purity. The structure of the extracted pristine lignin was confirmed by solution NMR spectroscopy. The 1D 1H-NMR, ^{13}C-NMR and 2D HSQC spectra (see Supporting Information) showed no protein and no polysaccharide impurities in this preparation of pristine lignin.

3.3 ^{13}C cross-polarization, magic angle spinning NMR for analysis of biomass components. Kenaf is a typical lignocellulosic biomass, which is primarily composed of cellulose and lignin. The ^{13}C CP/MAS spectrum of kenaf biomass could be simulated by combining the spectra of cellulose and pristine lignin (Figure 3).

Next, standard samples of major components of lignocellulosic biomass as well as additives used in culturing fungus on biomass were examined using ^{13}C CP MAS ssNMR (Figure 4). All biomass and culture samples, with the exception of pristine lignin extract, displayed signals corresponding to cellulosic polysaccharide. Pure cellulosics showed no signals in the aromatic region (140-160 ppm) of their spectra. The presence of aromatic signals, at 142-157 ppm, in lignin make these signals a unique identifier for lignin and the integrals of these peaks could be used for quantification. A small amount of residual DMSO could be observed in the pristine lignin sample from the signal appearing at 39 ppm. The percentage of this residual DMSO was calculated to be 7.6 wt. % by comparing the DMSO integral in this sample to the integral obtained from a spectrum of kenaf containing 14.5% DMSO (see Supporting Information). Thus, the purity of lignin was 92.4%. Using the purity determined for pristine lignin and the integrals from its spectrum (Figure 5 and Table 1), a standard curve was constructed. The excellent linearity of this standard curve allowed the accurate quantification of the lignin content in untreated biomass sample kenaf to be 32%.

3.4 Analysis of cellulosic biomass by ^{13}C CP MAS ssNMR. Using the pristine lignin extracted from kenaf, we measured the lignin content in kenaf, corn stover and cotton stalks. These lignocellulosic biomasses are sources for papermaking, biofuel and bioethanol industries. Kenaf showed the highest lignin content based on ^{13}C CP/MAS ssNMR and along the other biomass samples gave lignin values consistent with those reported in the literature (Supporting Information Table S3).

The ^{13}C CP MAS ssNMR of kenaf on which fungi were cultured revealed the lignin content was quite different for G. lucidum (white rot) and F. pinicola (brown rot) (Figure 6, Table 2). The white rot G. lucidum fungi digested over 16% kenaf lignin reducing it to 26%, while as expected; the brown rot F. pinicola fungi did not alter the kenaf lignin.

Quantitative ^{13}C-NMR has been applied to lignin qualitative and quantitative studies since the 1980s [29]. However, application of high-resolution ^{13}C spectra has been limited due to low natural abundance of the ^{13}C isotope, difficulties in lignin extraction, and high lignin concentrations required to achieve good signal-to-noise (S/N) ratios. The development of inverse detection in NMR (i.e., 2-D heteronuclear single quantum coherence (HSQC)) has significantly increased spectral resolution and sensitivity [26]. Pretreatment combined with cellulolytic enzyme lignin extraction [30] offers relatively higher yield with respect to the mill wood lignin extraction [31]. Unfortunately, lignin covalently attached to cellulosic moieties, and enzyme adsorption onto lignin, poses additional challenges for lignin isolation [32]. Recently, a new solution-state 2D-HSQC NMR spectroscopic method was reported to measure the acetylated plant cell wall using a whole cell wall dissolving system. Using this approach an analytical signal for lignin content with the necessity of lignin isolation. As a result, homogeneity of dissolved biomass in a solution system is required to visualize all the signals, and deduce the 1JC-H dependence of these signals. In practice, the reliability of the HSQC-based method for quantification is compromised by a number of issues, including a wide range in effective T2 relaxation times due to sample polydispersity, coupling constant deviations, homonuclear couplings, and sample dependent resonance offsets [33]. While 2D-HSQC NMR represents an innovative method ideal for investigating structural changes in lignin [34], it still has limitations.

To circumvent exhaustive extraction methodology, and diminishing results over time to determine insoluble lignin in solution, this study develops a solid-state method for quantitative and direct measurement of lignin in biomass. Quantitative solid-state NMR was first reported in middle 1990s [35], but its application was limited because of low sensitivity and resolution. With the development of cross polarization and MAS technology along with the greater availability of ultra-high field NMR magnets, the resolution and sensitivity of ^{13}C CP/MAS ssNMR has been greatly enhanced. However, until now, ^{13}C CP/MAS ssNMR has only been used for qualitative analysis for the insoluble components of biomass, such as cellulose [17] and lignin. An early report on structure changes in lignin and plant polysaccharides after fungal degradation using solid-state NMR first suggested the use of this method on lignocellulosic biomass [36]. The chemical structure and heterogeneity of two lignin structures from Loblolly Pine were subsequently investigated using ^{13}C solid-state NMR [16]. This early research demonstrated that solid-state NMR was useful for lignin structural characterization. The use of a proper lignin standard in the current study clearly demonstrates that ^{13}C CP/MAS ssNMR spectroscopy can provide fast, accurate and quantitative determination of the lignin content of lignocellulose biomass.

Table 1. Peak integrals of kenaf lignin of different mass on 157-142 ppm.

Lignin extract (mg)	Lignin[a] (mg)	Integral
1.5	1.3	$1.92 \times 10^7 \pm 2.96 \times 10^6$
4.2	3.6	$5.20 \times 10^7 \pm 8.20 \times 10^6$
7.5	6.4	$9.42 \times 10^7 \pm 6.50 \times 10^6$
18.3	15.6	$2.27 \times 10^8 \pm 5.97 \times 10^6$
22.4	19.0	$2.88 \times 10^8 \pm 4.97 \times 10^6$

[a] DMSO present in lignin extract has been removed from these values.

Table 2. Calculated lignin content of culture samples based on ^{13}C CP/MAS ssNMR.

Culture Samples	Lignin Content (%)
G. lucidum A	25.8
G. lucidum B	27.4
G. lucidum C	26.5
Average	26.5 ± 0.8
F. pinicola A	32.1
F. pinicola B	32.2
F. pinicola C	31.1
Average	31.7 ± 0.6

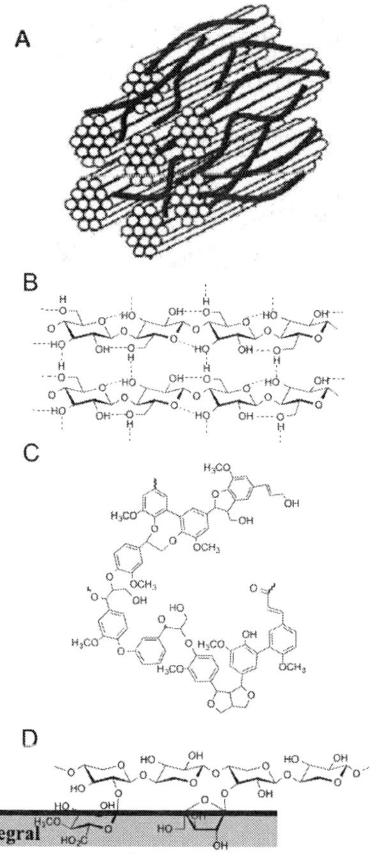

Figure 1. Structure of lignocellulose (A) Schematic of lignocellulosic plant cell wall microstructure where the cylinders correspond to cellulose fiber bundles that are coated with hemicellulose and the dark lines to lignin (B) Structure of hydrogen-bonded cellulose polysaccharide chains comprising cellulose plant fibers (C) Major structures present in lignin (D) Xylan corresponding to a major hemicellulose structures.

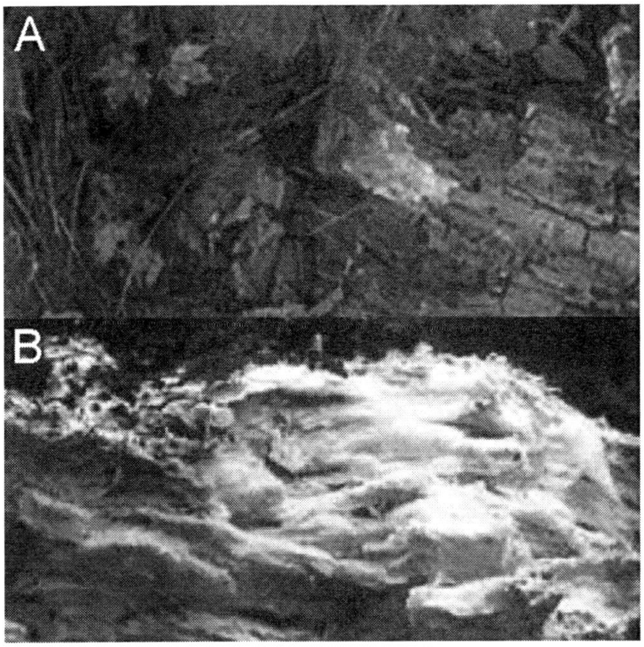

Figure 2. Occurrence of brown rot and white rot in nature (A) Brown rot fungus observed in nature degrading lignocellulosic material (B) White rot fungus observed in nature degrading lignocellulosic material

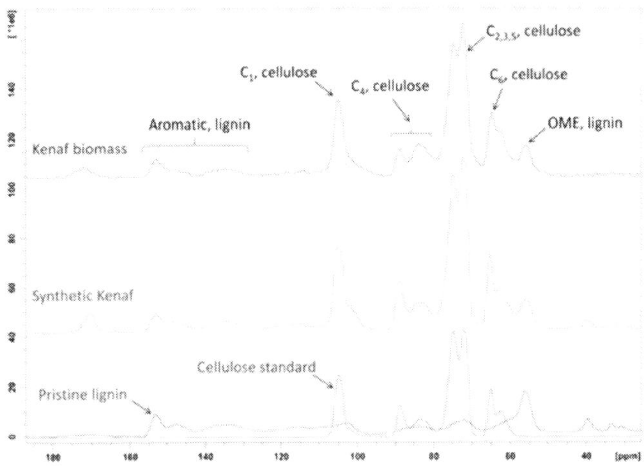

Figure 3. ^{13}C CP/MAS spectra of Kenaf biomass, synthetic kenaf, cellulose standard and pristine lignin

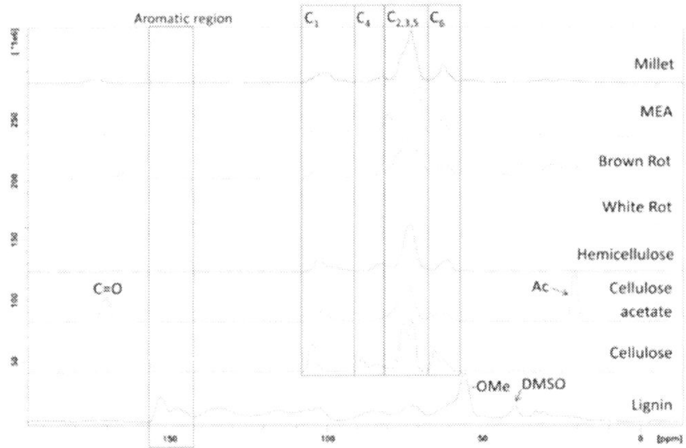

Figure 4. ^{13}C CP MAS ssNMR spectra for (top to bottom) millet, MEA brown rot fungi, white rot fungi, hemicellulose, cellulose acetate, cellulose, and lignin,

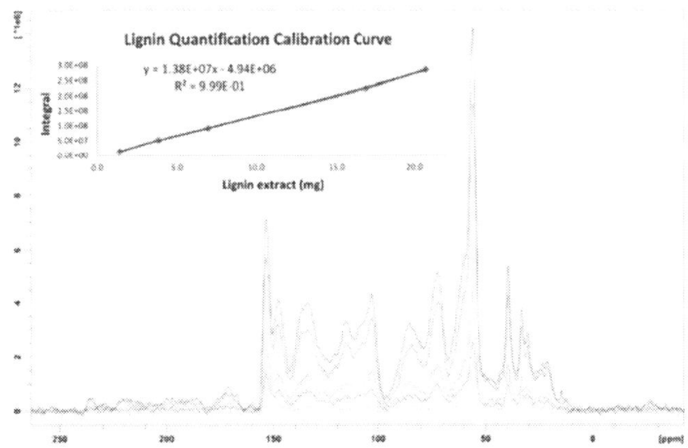

Figure 5. ^{13}C CP/MAS ssNMR spectra of kenaf lignins of different mass and quantification calibration curve (inset)

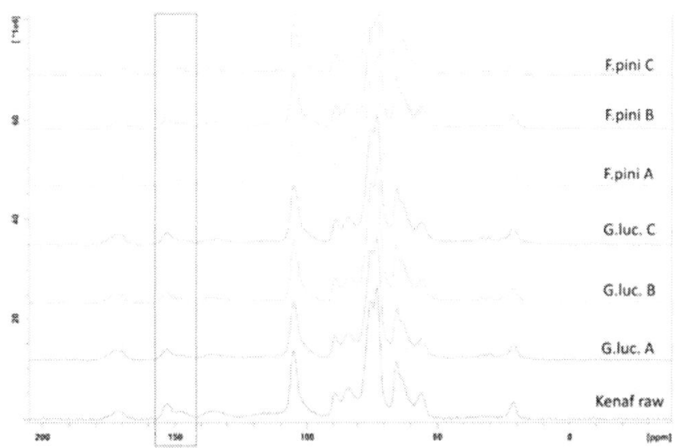

Figure 6. ^{13}C CP/MAS ssNMR spectra of culture samples (from top to bottom *F. pinicola* C, *F. pinicola* B, *F. pinicola* A, *G. lucidum* C, *G. lucidum* B, *G. lucidum* A, and raw kenaf.

4. Conclusion

^{13}C CP/MAS ssNMR spectroscopy provides a rapid and reliable method to assess the lignin content of biomass. There are several identified concerns that need to be monitored in carrying out this analysis with high accuracy. First, due to lignin structural variation among species of biomass [26], a standard lignin for quantification curve calibration should be prepared using the same biomass material to obtain a uniform standard. Second, changes in the water content of samples can result in changes in the integrals because of the ^{1}H-^{13}C cross polarization step employed. All samples should be dried or lyophilized and stored under the identical conditions for all measurements. Third, the standard curve should be constructed or at least a point from a past calibration repeated and subsequently used to normalize integrals from the new set of experiments to match past values every time the assay is performed. This evaluation is required in order to confirm spectrometer setup consistency along with the efficiency of the cross polarization transfer. With these parameters well understood and carefully controlled, quantitative ^{13}C CP/MAS ssNMR method should be of great utility in paper making, biomaterials testing, tobacco analysis, the biofuel and bioethanol industry, as well as plant biology, plant metabolism and plant design, forest science and other biomass related research or industry.

REFERENCES

(1) F. Chen, R.A. Dixon, Lignin modification improves fermentable sugar yields for biofuel production, Nature Biotechnology 25 (2007) 759-761.

(2) R. Singh, A. Shukla, S. Tiwari, M.A. Srivastava, Review on delignification of lignocellulosic biomass for enhancement of ethanol production potential, Renewable Sustainability Energy Review 32 (2014) 713-728.

(3) C. Sanchez, Lignocellulosic residues: Biodegradation and bioconversion by fungi, Biotechnology Advancement 27 (2009) 185-194.

(4) A.T Martínez, M. Speranza, F.J. Ruiz-Dueñas, P. Ferreira, S. Camarero, F. Guillén, M.J. Martínez, A. Gutiérrez, J.C. del Río, Biodegradation of lignocellulosics: microbial, chemical, and enzymatic aspects of the fungal attack of lignin, International Microbiology 8 (2005) 195-204.

(5) H. Palonen, A.B. Thomsen, M. Tenkanen, A.S. Schmidt, L. Viikari, Evaluation of wet oxidation pretreatment for enzymatic hydrolysis of softwood, Applied Biochemistry and Biotechnology 117 (2004) 1–17.

(6) L. Laureano-Pérez, F. Teymouri, H. Alizadeh, B.E. Dale, Understanding factors that limit enzymatic hydrolysis of biomass, Applied Biochemistry and Biotechnology 121 (2005) 1081–1099.

(7) Y. Sun, J. Cheng, Hydrolysis of lignocellulosic materials for ethanol production: a review, Bioresource Technology 83 (2002) 1–11.

(8) M.G. Papatheofanous, E. Billa, D.P. Koullas, B. Monties, E.G. Koukios, Two stage acid-catalyzed fractionation of lignocellulosic biomass in aqueous ethanol systems at low temperatures, Bioresource Technology 54 (1995) 305–310.

(9) K.K. Pandey, A.J. Pitman, FTIR studies of the changes in wood chemistry following decay by brown-rot and white-rot fungi, Industrial Biodeterioration Biodegradable 52 (2003) 151-160.

(10) L. Levin, L. Papinutti, F. Forchiassin, Evaluation of Argentinean white rot fungi for their ability to produce lignin-modifying enzymes and decolorize industrial dyes, Bioresource Technology 94 (2004) 169-176.

(11) T.R. Filley, G.D. Cody, B. Goodell, J. Jellison, C. Noser, A. Ostrofsky, Lignin demethylation and polysaccharide decomposition in spruce sapwood degraded by brown rot fungi, Organic Geochemistry 33 (2002) 111-124.

(12) M.J. Duer, Introduction to solid-state NMR spectroscopy, Oxford, Blackwell (2004).

(13) K. Schmidt-Rohr, H.W. Spiess, Multidimensional solid-state NMR and polymers, Academic Press, London (1994).

(14) M. Foston, Advances in solid-state NMR of cellulose, Current Opinion in Biotechnology 27 (2014) 176-184.

(15) D.L. VanderHart, R.H. Atalla, Native cellulose: a composite of two distinct crystalline forms, Science 233 (1984) 283-285.

(16) K.M. Holtman, N. Chen, M.A. Chappell, J.F. Kadla, L. Xu, J. Mao, Chemical structure and heterogeneity differences of two lignins from Loblolly Pine as investigated by advanced solid-state NMR spectroscopy, Journal of Agricultural Food Chemistry 58 (2010) 9882-9892.

(17) R. Samuel, Y. Pu, M. Foston, A.J. Ragauskas, Solid-state NMR characterization of switchgrass cellulose after dilute acid pretreatment, Biofuels 1 (2010) 85-90.

(18) J.-L. Wen, S.-L Sun, B.-L. Xue, R.-C. Sun, Recent advances in characterization of lignin polymer by solution-state nuclear magnetic resonance (NMR) methodology, Materials 6 (2013) 359-391.

(19) Y. Zheng, S. Zhang, S. Miao, Z. Su, P. Wang, Temperature sensitivity of cellulose adsorption on lignin and its impact on enzymatic hydrolysis of lignocellulosic biomass, Journal of Biotechnology 166 (2013) 135-143.

(20) R. Samuel, M. Foston, N. Jaing, S. Cao, L. Allison, M. Studer, C. Wyman, A.J. Ragauskas, HSQC (heteronuclear single quantum coherence) ^{13}C-^{1}H correlation spectra of whole biomass in perdeuterated pyridinium chloride-DMSO system: An effective tool for evaluating pretreatment, Fuel 90 (2011) 2836-2842.

(21) A. Berlin, M. Balakshin, N. Gilkes, J. Kadla, V. Maximenko, S. Kubo, J. Saddler, Inhibition of cellulase, xylanase and β-glucosidase activities by softwood lignin preparations, Journal of Biotechnology 125 (2006) 198-209.

(22) P. Suhas, M. Carrott, R. Carrott, Lignin - from natural adsorbent to activated carbon: A review, Bioresource Technology 98 (2007) 2301–2312.

(23) R.E. Taylor, Setting up ^{13}C CP/MAS experiments, Concepts in Magnetic Resonance Part A 22A (2004) 37-49.

(24) S. Burkhardt, L. Kumar, R. Chandra, J. Saddler, How effective are traditional methods of compositional analysis in providing an accurate material balance for a range of softwood derived residues? Biotechnology for Biofuels 6 (2013) 90.

(25) G. Zhou, G. Taylor, A. Polle, FTIR-ATR-based prediction and modeling of lignin and energy contents reveals independent intra-specific variation of these traits in bioenergy poplars, Plant Methods 7 (2011) 9.

(26) S.D. Mansfield, H. Kim, F. Lu, J. Ralph, 2012. Whole plant cell wall characterization using solution-state 2D NMR, Nature Protocol 7 (2012) 1579-1589.

(27) H. Wu, M. Mora-Pale, J. Miao, T.V. Doherty, R.J. Linhardt, J.S. Dordick, Facile pretreatment of lignocellulosic biomass at high loadings in room temperature ionic liquids, Biotechnology and Bioengineering 108 (2011) 2865-2875.

(28) T.H. Kim, Y.Y., Lee, 2007. Pretreatment of corn stover by soaking in aqueous ammonia at moderate temperatures, Applied Biochemistry and Biotechnology (2007) 136-140, 81-92.

(29) G. Gellerstedt, D. Robert, 1987. Quantitative ^{13}C NMR analysis of kraft lignins, Acta Chemica Scancinavica B 41 (1987) 541-546.

(30) K.M. Holtman, H.M. Chang, J.F. Kadla, Solution-state nuclear magnetic resonance study of the similarities between milled wood lignin and cellulolytic enzyme lignin, Journal of Agricultural Food Chemistry 52 (2004) 720–726.

(31) A. Björkman, Isolation of lignin from finely divided wood with neutral solvents, Nature, 174 (1954) 1057–1058.

(32) O. Faix, D.S. Argyropoulos, D. Robert, V. Neyrinck, Determination of hydroxyl groups in lignins evaluation of ^{1}H-, ^{13}C-, and 31P-NMR FT-IR and wet chemical methods, Holzforschung 48 (1994) 387–394.

(33) S. Heikkinen, M.M. Toikka, P.T. Karhunen, L.A. Kilpelainen, Quantitative 2D HSQC (Q-HSQC) via suppression of J-dependence of polarization transfer in NMR spectroscopy: Application to wood lignin, Journal of American Chemical Society 125 (2003) 4362–4367.

(34) D.J. Yelle, D.S. Wei, J. Ralph, K.M. Hammel, Multidimensional NMR analysis reveals truncated lignin structures in wood decayed by the brown rot basidiomycete Postia placenta, Environmental Microbiology 13 (2011) 1091–1100.

(35) M.S. Yahnke, B.M. Rush, J.A. Reimer, E.J. Cairns, Quantitative solid-state NMR spectra of CO adsorbed from aqueous solution onto a commercial electrode, Journal of American Chemistry Society 118 (1996) 12250-12251.

(36) A.T Martínez, A.E. Gonzalez, M. Valmaseda, B.E. Dale, M.J. Lambregts, J.F. Haw, Solid-state NMR studies of lignin and plant polysaccharide degradation by fungi, Holzforschung 45 1991) 49-54.

REMARKS ON THERMAL CONDUCTIVITY OF SLAG

Mehrdad Massoudi[a]*, *Jeongho Kim*[b] *and Ping Wang*[a]

[a] U.S. Department of Energy (DOE), National Energy Technology Laboratory (NETL),626 Cochrans Mill Road, P.O. Box 10940, Pittsburgh, PA 15236, USA
[b] Department of Mechanical Engineering, Kyung Hee University, Yong-in, Kyunggi-do 446-701, Korea
* Correspondence: mehrdad.massoudi@netl.doe.gov

Introduction

Molten slag is a complex fluid-like material which occurs during the operation of processes such as gasification. Slagging is related to the deposition of ash which occurs in the radiative section of a boiler and fouling occurs in the convective-pass region[1]. Groen et al.[2] observed that at the critical viscosity temperature, T_{cv}, the slag behavior changes from a homogeneous fluid to a mixture composed of fluid and a crystallizing phase (solid), where the (apparent) viscosity increases presumably due to the presence of the crystals or the change in the melt composition. Using computational fluid dynamics (CFD) packages, accurate models for non-linear complex materials are needed, especially at high temperatures (1000 °C), where the standard techniques for measuring the thermal conductivity of a material is not necessarily applicable. For example, for the heat flux vector, the Fourier's law of heat conduction is generally used and in some cases an effective thermal conductivity is introduced; however, studies have shown that for many complex materials this is not accurate. The Fourier's law of heat conduction is: $\mathbf{q} = -k\nabla\theta$ where $\mathbf{q}$ is the heat flux vector, θ is the temperature, ∇ (or *grad*) is the gradient operator, and k is a material property known as thermal conductivity. In porous materials, generally an effective thermal conductivity is introduced[3]; however, some studies have shown that for many complex materials the heat flux vector can depend not only on the temperature gradient, as is the case in Fourier's law, but also on the shear rate, and other parameters such as porosity[4]. In one of the earliest studies, Russell [5] derived the thermal conductivity for an array of cubes, where:

$$k_{eff}/k_c = \left[r\cdot\varphi^{2/3} + (1-\varphi^{2/3})\right] / \left[r\cdot(\varphi^{2/3}-\varphi) + (1-\varphi^{2/3}+\varphi)\right]$$

where k_c is the thermal conductivity of the continuous phase (particulate), k_d is the thermal conductivity of the discrete phase (spherical particles), r is ratio of k_d/k_c, and ϕ the volume fraction of the discrete phase. Leach [6] suggested the cubic parallel series (CPS) model which is the same as that of Russell's model except that the discrete phase is now air bubbles. Kingery [7] proposed that the thermal conductivity (k_c) of a slag is given by:

$$k_c = 1/3\, c_p\, \rho\, v\lambda$$

where v and λ are the velocity and the mean free path of a phonon, respectively, ρ is the density and c_p is the specific heat capacity of the slag. A general model of the heat transfer is the diffusion approximation where the heat flux (q) is given by:

$$q = -k_{eff}(d\theta/dy)$$

where k_{eff}(=k_c+k_R) is the effective thermal conductivity. This model is valid for small k_R and when $\overline{\alpha}\, d > 8$ [8]. Fine, *et al* [9] determined the total thermal diffusivity (α_{eff}) of solid and liquid slags containing 0 to 25% FeO using the radial wave method, where:

$$\alpha_{eff} = 10^{-7}(1.5-0.5B) + 1.8\times10^{-6}\frac{(\theta/1500)^3}{(\%\,FeO)^{0.8}}\ m^2 s^{-1}$$

where B represents the CaO/SiO_2 ratio and θ is the temperature with the unit of °C. This equation indicates that α_{eff} decreases with an increase in the FeO content. It appears that the effective thermal conductivity of the slag can be thought to be composed of various contributions such as $k_{eff} = k_c + k_R + k_{cov} + k_{el}$ where k_c is the phonon thermal component, k_R is the radiative component, k_{cov} is the convective component, and k_{el} is the electronic component of the thermal conductivity. The simplest models for estimating the effective thermal conductivity of porous materials are given by the following equations, depending on whether the heat transfer is in the direction parallel or normal to the layers of solid and gas, respectively:

$$k_{eff}/k_g = \varphi + (1-\varphi)\cdot k_s/k_g$$

$$k_{eff}/k_g = 1/\left[\varphi + (1-\varphi)/k_s/k_g\right]$$

where k_{eff} is the effective thermal conductivity, k_s the thermal conductivity of the solid phase, k_g the thermal conductivity of the gas phase, and ϕ the porosity of the material. The effective thermal conductivity is higher when the direction of the heat transfer is parallel to the layers. For a detailed discussion and review of these issues, we refer to the reader to the recent study by Massoudi et al[10]. In summary, to measure or study the thermal conductivity and thermal diffusivity of slags one can use one or more of the following approaches: (1) Thermal conductivity of slags of different porosity with a cubicle array of uniform cubes or cubic cells can be calculated, (2) Thermal diffusivity of different materials over a wide range of temperatures can be measured. (3) Solid and liquid slags can be placed in a cylindrical crucible in the isothermal zone of a furnace and their thermal diffusivity can be measured. (4) The radial heat flow method can be used with a cylindrical geometry and a centered heating element. (5) The hot-wire method, in a transient radial flow, van be used to find the thermal conductivity of slags. (6) The line source method as a transient technique can be used to determine the thermal conductivity of molten slags. Although we are suggesting that slag can be assumed to behave as a single component non-homogeneous and non-linear material, in general, as shown in Massoudi et al[10] slag can be considered to be a multi-component material.

Constitutive Modeling of the Heat Flux Vector

As shown by Massoudi et al[10], the *thermal conductivity* of slag is one of the most important parameters in determining the heat transfer characteristics and the proper operating conditions for a gasifier. However, since slag, in general, behaves as a non-linear viscoelastic fluid, we must study the constitutive modeling of the heat flux vector and not just its thermal conductivity. A constitutive relation is an equation which relates the unknowns, such as the heat flux vector $\mathbf{q}$ to other quantities such as velocity, displacement, temperature, *etc.*, which are the actual quantities that are to be solved in the governing equations, or can be measured experimentally. If the unknown quantities, for example $\mathbf{q}$, are explicitly related to the kinematical (velocity, or displacement) or thermal (temperature) quantities, then that constitutive relation is known as an *explicit* constitutive relation. In the explicit scheme, we start with equations such as: $\mathbf{q} = \mathbf{q}(\theta, \text{grad}\,\theta, \ldots)$ where θ is the temperature. The selection of the dependent variables, *i.e.*, the arguments in the function, is based on previous experience, if any, intuition or insight, and perhaps experimental observation; the most common expression for $\mathbf{q}$ is the Fourier's law of conduction.

If, however, it is not feasible to directly relate the unknowns to the knowable (or measurable) quantities, then an *implicit* constitutive relation should be used. Rajagopal[11] provided a general scheme for deriving implicit relations for the stress tensor of a fluid. In a similar way, the heat flux vector can be given as an implicit function of the temperature gradient, $\nabla\theta$ in the form[12]: $g(q,\nabla\theta,\theta)=0$. In general, the modeling of slag has been directed to the formulation of its thermal conductivity, mostly based on experimental data, and this approach has certain limitations; for example, the form of the heat flux vector assumed in almost all of these studies is that of the Fourier's heat conduction and as shown by many researchers, for complex materials, this is not valid in general and alternative formulations are needed. Massoudi[13] suggested that the heat flux vector $\mathbf{q}$ for a density-gradient-dependent fluid can be assumed to be given by:

$$\mathbf{q}=\mathbf{q}(\theta,\mathbf{g},\rho,\mathbf{m},\mathbf{D})$$

where θ is the temperature, ρ is the density, $\mathbf{g}=\text{grad}\theta, \mathbf{m}=\text{grad}\rho$, $\mathbf{D}=\frac{1}{2}\left(\mathbf{L}+\mathbf{L}^T\right)$; $\mathbf{L}=\text{grad}\mathbf{v}$, and $\mathbf{v}$ is the velocity. This suggest that the density gradient plays a role, not only in the distribution of the materials, but also in the way it influences the heat conduction. Massoudi[13] showed that a general representation for the heat flux vector in this case is:

$$\mathbf{q}=a_1\mathbf{g}+a_2\mathbf{m}+a_3\mathbf{Dg}+a_4\mathbf{Dm}+a_5\mathbf{D}^2\mathbf{g}+a_6\mathbf{D}^2\mathbf{m}$$

where $a_1=-k$ (thermal conductivity) and a_2-a_6 are scalar functions of: ϕ (or ρ), $\theta, \mathbf{m}.\mathbf{m}, \mathbf{g}.\mathbf{g}, \mathbf{m}.\mathbf{g}, I_1=\text{tr}\mathbf{D}, I_2=\text{tr}\mathbf{D}^2$, $I_3=\text{tr}\mathbf{D}^3, \mathbf{g}.\mathbf{Dg}, \mathbf{g}.\mathbf{D}^2\mathbf{g}, \mathbf{m}.\mathbf{Dm}, \mathbf{m}.\mathbf{D}^2\mathbf{m}, \mathbf{g}.\mathbf{Dm}, \mathbf{g}.\mathbf{D}^2$

The above equation can then be re-written as:

$$\mathbf{q}=\left(a_1\mathbf{1}+a_3\boldsymbol{D}+a_5\boldsymbol{D}^2\right)\nabla\theta+\left(a_2\mathbf{1}+a_4\boldsymbol{D}+a_6\boldsymbol{D}^2\right)\nabla\rho$$

Which is a general equation for the heat flux vector of a thermo-viscous fluid with density gradient. We can use the above equation for the heat flux vector of a slag. In theory, it is possible to decompose the heat flux vector into: $\mathbf{q}=\mathbf{q}_e+\mathbf{q}_d$ where $\mathbf{q}_e$ is the equilibrium part of $\mathbf{q}$ and $\mathbf{q}_d$ is the dynamic part.

For the implicit scheme, for a non-linear viscous fluid, Massoudi and Mehrabadi[12b] suggested:

$$\overset{\circ}{\mathbf{q}}=\mathbf{f}(\rho,\theta,\mathbf{g},\mathbf{m},\mathbf{q},\mathbf{D})$$

Here $\mathbf{m}$ (the density gradient, which can be related to the gradient of volume fraction is related to the porosity) is a measure of particle distribution. According to the representation theorems, the generators are:

$$\mathbf{g},\ \mathbf{m},\ \mathbf{q},\ \mathbf{Dg},\ \mathbf{Dm},\ \mathbf{Dq},\ \mathbf{D}^2\mathbf{g},\ \mathbf{D}^2\mathbf{m},\ \text{and}\ \mathbf{D}^2\mathbf{q}.$$

And an isotropic vector valued function for $\overset{\circ}{\mathbf{q}}$ is:

$$\overset{\circ}{\mathbf{q}}=\alpha_1\mathbf{g}+\alpha_2\mathbf{m}+\alpha_3\mathbf{q}+\alpha_4\mathbf{Dg}+\alpha_5\mathbf{Dm}$$
$$+\alpha_6\mathbf{Dq}+\alpha_7\mathbf{D}^2\mathbf{g}+\alpha_8\mathbf{D}^2\mathbf{m}+\alpha_9\mathbf{D}^2\mathbf{q}$$

where the α's are functions of the appropriate principal invariants, depending on the generators. In general, slag behaves as a viscoelastic fluid, where explicit and implicit schemes can be used to obtain or derive constitutive relations for the heat flux vector. If we use an implicit scheme, then a general constitutive relation for the elastic part of the heat flux vector for the slag behaving as a viscoelastic fluid is[12b]:

$$\overset{\circ}{\mathbf{q}}=\dot{\mathbf{q}}-\mathbf{Wq}=a_1\mathbf{g}+a_2\mathbf{q}+a_3\mathbf{Bg}$$
$$+a_4\mathbf{Bq}+a_5\mathbf{B}^2\mathbf{g}+a_6\mathbf{B}^2\mathbf{q}$$

where a_1–a_6 are scalar functions of the appropriate principal invariants, $\mathbf{B}=\mathbf{FF}^T$, where $\mathbf{F}$ is the deformation gradient, $\mathbf{W}=\frac{1}{2}\left[\mathbf{L}-\mathbf{L}^T\right]$. Now if all a's are set to zero, except a_1 and a_2 (which would for example corresponding to a homogeneous deformation where $\det\mathbf{F}=1$), we have:

$$\overset{\circ}{\mathbf{q}}=\dot{\mathbf{q}}-\mathbf{Wq}=b_1\mathbf{g}+b_2\mathbf{q}$$

For the viscous part of the slag, we can use the results of Massoudi and Mehrabadi[12b] who suggested an isotropic vector valued function for the heat flux:

$$\overset{\circ}{\mathbf{q}}=\alpha_1\mathbf{g}+\alpha_2\mathbf{m}+\alpha_3\mathbf{q}+\alpha_4\mathbf{Dg}+\alpha_5\mathbf{Dm}$$
$$+\alpha_6\mathbf{Dq}+\alpha_7\mathbf{D}^2\mathbf{g}+\alpha_8\mathbf{D}^2\mathbf{m}+\alpha_9\mathbf{D}^2\mathbf{q}$$

where $\mathbf{g}=\text{grad}\theta, \mathbf{m}=\text{grad}\rho$ and the α's are functions of the appropriate principal invariants. This is a very general implicit constitutive relation for the heat flux vector.

Concluding remarks

In this brief review we have not discussed the important question of how the material parameters appearing in these equations can be measured or whether they can be calibrated against the existing experimental data. Furthermore, as with all implicit constitutive relations, such as Oldroyd-type fluids, the numerical schemes to solve these equations are much more complicated, as the governing equations and the constitutive equations are to be solved simultaneously. Furthermore, the higher order terms in the equations require additional boundary conditions.

References

(1) Erickson, T. A., Allan, S. E., McCollor, D. P., Hurley, J. P., Srinivasachar, S., Kang, S. G., Baker, J. E., Morgan, M. E., Johnson, S. A., and Borio, R., Modeling of fourling and slagging in coal-fired utility boilers. *Fuel Processing Technology* **1995,** *44* (1-3), 155-171.

(2) Groen, J. C.; Brooker, D. D.; Welch, P. J.; Oh, M. S., Gasification slag rheology and crystallization in titanium-rich, iron-calcium-aluminosilicate glasses. *Fuel Processing Technology* **1998,** *56* (1-2), 103-127.

(3) Kaviany, M., *Principles of Heat Transfer in Porous Media.* Second edition ed.; Springer-Verlag: New York, 1995.

(4) Petroski, H. J., Departures from Fourier's Law. *Zeitschrift Fur Angewandte Mathematik Und Physik* **1975,** *26* (1), 119-124.

(5) Russell, H., Principles of heat flow in porous insulators. *Journal of the American Ceramic Society* **1935,** *18*, 1-5.

(6) Leach, A., The thermal conductivity of foams. I. Models for heat conduction. *Journal of Physics D: Applied Physics* **1993,** *26* (5), 733.

(7) Kingery, W., Heat-Conductivity Processes in Glass. *Journal of the American Ceramic Society* **1961,** *44* (7), 302-304.

(8) Mills, K., *Heat Transfer and Thermal Conductivity of Coal Slags.* American Chemical Society: Washington DC, 1986; p 257-276.

(9) Fine, H. A.; Engh, T.; Elliott, J. F., Measurement of the thermal diffusivity of liquid oxides and metallurgical slags. *Metallurgical Transactions B* **1976,** *7* (2), 277-285.

(10) Massoudi, M.; Kim, J.; Wang, P., On the Heat Flux Vector and Thermal Conductivity of Slags: A Brief Review. *Energies* **2016,** *9* (1).

(11) Rajagopal, K. R., On implicit constitutive theories for fluids. *Journal of Fluid Mechanics* **2006,** *550*, 243-249.

(12) (a) Massoudi, M.; Mehrabadi, M. M., Implicit Continuum Mechanics Approach to Heat Conduction in Granular Materials. *Industrial & Engineering Chemistry Research* **2010,** *49* (11), 5215-5221; (b) Massoudi, M.; Mehrabadi, M. M., Implicit constitutive relations in thermoelasticity. *International Journal of Non-Linear Mechanics* **2011,** *46* (1), 286-290.

(13) (a) Massoudi, M., On the heat flux vector for flowing granular materials - Part I: Effective thermal conductivity and background. *Mathematical Methods in the Applied Sciences* **2006,** *29* (13), 1585-1598; (b) Massoudi, M., On the heat flux vector for flowing granular materials - part II: Derivation and special cases. *Mathematical Methods in the Applied Sciences* **2006,** *29* (13), 1599-1613.

SYNTHESIS AND HYDROGEN SORPTION CHARACTERISTICS OF MECHANICALLY ALLOYED $Mg(Ni_xMn_{1-x})_2$ INTERMETALLICS

E.I Gkanas[1], G. Panagakos[2], M. Khzouz[1], A. Ioannidou[3], A. Donac[4], G. Stoian[4], N.Lupu[4], M. Gjoka[3], S.S Makridis[5]

[1]School of Mechanical, Aerospace and Automotive Engineering, Faculty of Engineering, Environment and Computing, Coventry University, Gulson Road, Coventry CV1 2JH, United Kingdom
[2]Environmental Research Laboratory, National Center for Scientific Research Demokritos, Patr. Gregoriou E' & 27, Neapoleos str., PO Box 60037, 15341, Agia Paraskevi, Attiki, Greece
[3]Institute for Advanced Materials, Physicochemical Processes, Nanotechnology & Microsystems, National Center for Scientific Research 'Demokritos', Patr. Gregoriou E' & 27, Neapoleos str., PO Box 60037, 15341, Agia Paraskevi, Attiki, Greece
[4]National Institute of Research and Development for Technical Physics, Bulevardul Profesor Dr. doc. Dimitrie Mangeron 47, Iaci 700050, Romania
[5]Department of Environmental and Natural Resources Management, University of Patras, Geor. Seferi, Agrinio 30131, Greece
Corresponding author: Evangelos.gkanas@coventry.ac.uk

Introduction

Hydrogen is a potential energy carrier, and can be considered for both stationary and mobile applications. Further, hydrogen can act as an agent allowing the energy storage resulting from renewable sources and this could be the route to overcome the production of hydrogen from fossil fuels, which causes the release of CO_2[1]. While it seems to be the ideal means of transport and conversion of energy for such applications, there are issues which prevent the massive application of a hydrogen economy, due to several safety issues and the efficient storage of hydrogen[2].

Metal hydrides (MH) which can reversibly store and release hydrogen with high efficiency are regarded as important materials for solving energy and environmental issues[3]. In recent years, a lot of research has been done on potential materials for hydrogen storage applications. Magnesium-based alloys have attracted much attention due to high hydrogen capacity, abundance and low cost. It is reported that pure Mg can store up to 7.6 wt% [4]. Despite the high hydrogen absorption/desorption capacity, there are some serious drawbacks in the use of Mg alloys for practical on-board applications such as slow kinetics, high operation temperatures related to the high stability of the Mg–H bonds and expressed in the high enthalpy of hydride formation and high reactivity with oxygen[5].

Considerable research has been conducted on magnesium alloys in order to find novel ways to synthesize new high performance materials. These studies are mainly focused on the element substitution[6], new production methods by using for example different hydrogen pressures[7], preparation of composite materials through the addition of dopants in order to improve microstructure/ microchemistry[8] and thermal techniques[9] in order to improve hydrogen storage characteristics.

High energy ball milling (HEBM) is included among the most efficient ways to improve hydrogen storage performance of Mg-based materials[10]. HEBM allows obtaining the material in amorphous/nanocrystalline state, synthesizing new compounds and forming composites with catalytic additions from other elements. During these techniques, due to the high stresses applied, the microstructure of the materials is continuously refined until a chemically and microstructure homogeneous material is obtained. Several novel techniques in order to synthesize high performance materials for hydrogen storage have been proposed such as hybrid microwave heating[11], laser ablation techniques[12] and Hydriding Combustion Synthesis (HCS)[13].

Mg_2Ni-type hydrogen storage alloys are considered to be promising negative electrode materials for nickelmetal hydride (Ni-MH) batteries because of their low cost, light weight, rich mineral resources and high theoretical discharge capacity[14]. The poor kinetics and the high operating temperatures of these materials combined with the severe oxidation in alkaline solution of Mg_2Ni-type electrode prevent these materials for practical use. A number of studies reported that the partial substitution of Mn in a $MgNi_2$ could enhance the discharge capacity in room temperature and further to reduce the absorption process to lower temperatures[15].

In the current work, new ternary Mg–Mn–Ni alloys have been synthesised by high energy ball milling studied as potential hydrogen storage materials. Five different stoichiometries of the $Mg(Ni_xMn_{1-x})_2$ nominal composition were synthesized with high energy ball milling (HEBM) process under argon atmosphere at room temperature using a planetary high-energy ball miller. The microstructures were characterized by means of XRD and Scanning Electron Microscopy (SEM). The Pressure-Composition-Temperature (P-c-T) properties of these materials were measured by a commercial Sievert-type apparatus.

Experimental Procedure

Mg (99.8%, 325 mesh), Ni (99.8%, 325 mesh) and Mn (99.95%, 325 mesh) powders were used as the starting materials for the High Energy Ball Milling (HEBM) process. The HEBM of the metal powders was performed in high-purity argon environment in sealed vials using Fritsch "Pulverisette-7" miller and every sample was milled for 10 and 20h respectively at 350 rpm. After every hour of milling the process was paused for 30 min in order to maintain the temperature inside the vial at low levels and after 5h of milling the process was paused again and the vial was transferred in the glove box to scratch the walls in order to ensure the homogeneity of the produced material. The microstructure of both the as-milled and the hydrogenated samples was characterized by means of X-ray diffraction (XRD) using a BRUKER AXS D8-Advance. The morphology of the samples was analyzed by Scanning Electron Microscopy (SEM) using a Zeiss NEON 40 EsB. Pressure–composition–temperature (*P–c–T*) relationships were measured by the Sieverts' method. The purity grade of hydrogen gas used was 99.999%. The absorption was performed at 150-200-250-300 ^0C at pressures 0-20bar, and the dehydrogenation was performed at the same temperatures at vacuum. The isotherms were measured at 150-200-250-300 ^0C. The hydrogen absorption–desorption properties were studied in a conventional Sieverts-type apparatus (PCTPro 2000-HyEnergy), that allowed to determine hydrogen storage capacity and estimate dynamics of hydrogen absorption–desorption. The activation process which performed for all the samples consists of four cycles of absorption (20 bar) and desorption (0.5 bar) at 300 ^0C. From the activation process is extracted that the absorption procedure is relatively slow at the first cycle, but at the next three cycles the reaction is 5-6 times faster, and the desorption process is very fast from the first cycle (1-2 min).

Results and Discussion

Microstructure Characteristics. Shown at Figure 1 the XRD profiles of all the samples of the composition $Mg(Ni_xMn_{1-x})_2$ (x=0.1, 0.25, 0.5, 0.75) milled for 10h and also there is a comparison for the sample $Mg(Ni_{0.5}Mn_{0.5})_2$ for 10h and 20h milling time (Fig1b and 1c respectively). Figure 1 indicates that the substitution of Mn instead of Ni can change the major phase Mg2Ni, which is a hexagonal phase, and lead to the secondary phases Mg-Ni and Mg, where the amount of such phases is increasing with the increase of Mn. Increasing the milling time it seems that the peak phase Mg_2Ni is broadening, which is doubtless attributed to the refined grains and stored stress that originated from the milling.

Further, Figure 2 also shows that for the sample $Mg(Ni_{0.9}Mn_{0.1})_2$ the effect of the milling time plays major role on the microstructure of the material. Figure 2 indicates that after 10h milling the phase Mg2Ni vanishes which is going to be very important when the P-c-T properties will be measured.

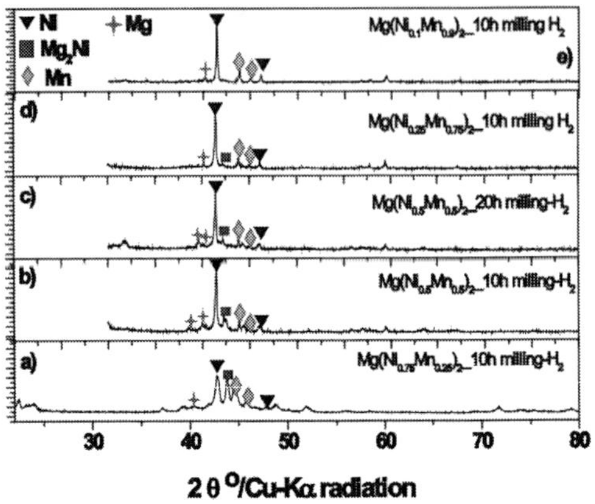

Figure 1. XRD profiles of the alloys milled for 10h, except 1c milled for 20h

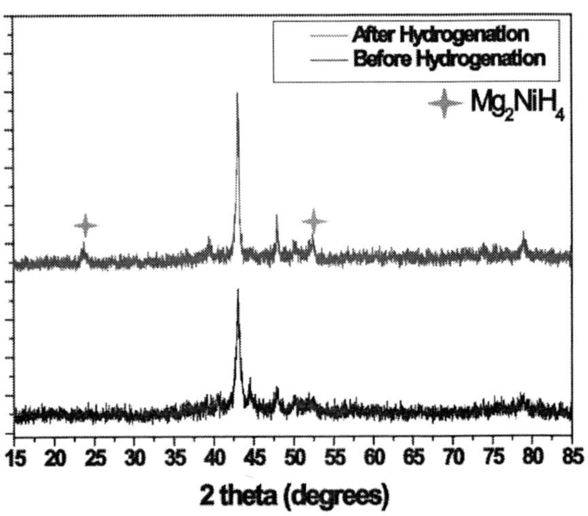

Figure 2. XRD profiles for the sample $Mg(Ni_{0.9}Mn_{0.1})_2$ at 10h and 20h milling time respectively

Finally, Figure 3 presents the XRD profile for the sample $Mg(Ni_{0.75}Mn_{0.25})_2$ milled for 10h before and after the hydrogenation cycling process. The presence of a new phase Mg_2NiH_4 is obtained indicating that the hydrogenation process was successful for the samples.

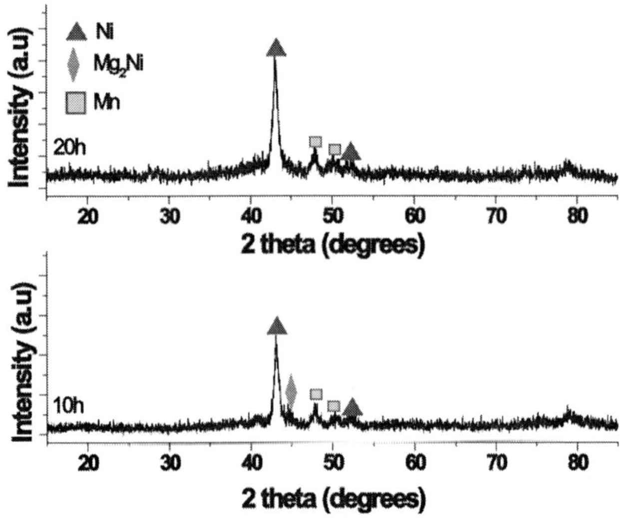

Figure 3. XRD profiles for the sample $Mg(Ni_{0.9}Mn_{0.1})_2$ milled for 10h before and after the hydrogenation cycling

Listed in Table 1 are the lattice parameters and the phases recognized for all the synthesized alloys

Figure 4 and Figure 5 shows the SEM images of the as-milled samples $Mg(Ni_{0.1}Mn_{0.9})_2$ milled for 10 and 20h respectively. It can be seen that most of the particles (in fact, they are clusters of smaller particles) are smaller than 10 mm in size after 10h of milling in Figure 3a. However, further milling (20 h of milling in Figure 3b) does not dramatically decrease the particle size of the synthesized alloys, which indicates that a steady state equilibrium between fracturing (tending to decrease the particle size) and welding (tending to increase particle size) is attained.

P-c-T Properties

Sample Activation Process. The activation process consists of several cycles of hydrogen absorption at 300°C at pressure 20 bar of hydrogen, followed by desorption at 300°C at pressure 0.5 bar of hydrogen. According to the results from Figure 6, for the sample $Mg(Ni_{0.1}Mn_{0.9})_2$, it seems that after the 3rd cycle the sample has been saturated and the kinetics have been improved since the first two cycles. The second cycle seem to be fast but the hydrogen content is lower than the other cycles mainly because the dehydrogenation process of the first cycle might was not totally completed. For the dehydrogenation process, it is observed that even for the first cycle the process is very fast and the material can release all the stored hydrogen in less than 1 minute. The maximum Hydrogenation capacity achieved for such material is 1.2 wt%.

Table 1. Structure, Lattice Parameters and Phases for all the Synthesized Samples.

Sample name	ID	Lattice parameter				Phases %wt			
		Ni (Fm3m)	MgNi2 (P63/mmc)	Mn (P4132)	Mn (2) (I43m)	Ni (Fm3m)	MgNi2 (P63/mmc)	Mn (P4132)	Mn (2) (I43m)
$Mg(Ni_{0.9}Mn_{0.1})_2$	10h milling	a=b=c=3.5274	a=b= 5.004 c=14.8118	a=b=c=6.2944	------	11.33	80.64	8.03	-----
$Mg(Ni_{0.9}Mn_{0.1})_2$	20h milling	a=b=c=3.5280	a=b= 4.9389 c=15.7409	a=b=c=6.3095	-------	30	68.18	1.82	------
$Mg(Ni_{0.75}Mn_{0.25})_2$	10h milling	a=b=c=3.5480	a=b= 5.0017 c=14.9588	a=b=c=6.3794	a=b=c= 8.9397	3.08	73.47	15.99	7.45
$Mg(Ni_{0.75}Mn_{0.25})_2$	20h milling	a=b=c=3.5167	a=b= 4.9417 c=13.9684	a=b=c=6.4342	a=b=c= 8.8854	9.01	58.32	22.53	10.14
$Mg(Ni_{0.5}Mn_{0.5})_2$	20h milling	a=b=c=3.5274	a=b= 4.6099 c=15.5853		a=b=c= 8.9211	20.34	18.10		61.55
$Mg(Ni_{0.25}Mn_{0.75})_2$	20h milling	a=b=c=3.525	a=b= 4.6652 c=15.7596	a=b=c=6.3178	a=b=c= 8.9187	9.35	18.92	2.85	68.89
$Mg(Ni_{0.1}Mn_{0.9})_2$	20h milling	a=b=c=3.5068	a=b= 4.4488 c=14.6530		a=b=c= 8.9144	9.20	21.08		69.72

Figure 4. SEM micrograph of the $Mg(Ni_{0.1}Mn_{0.9})_2$ milled for 10h

Figure 5. SEM micrograph of the $Mg(Ni_{0.1}Mn_{0.9})_2$ milled for 20h

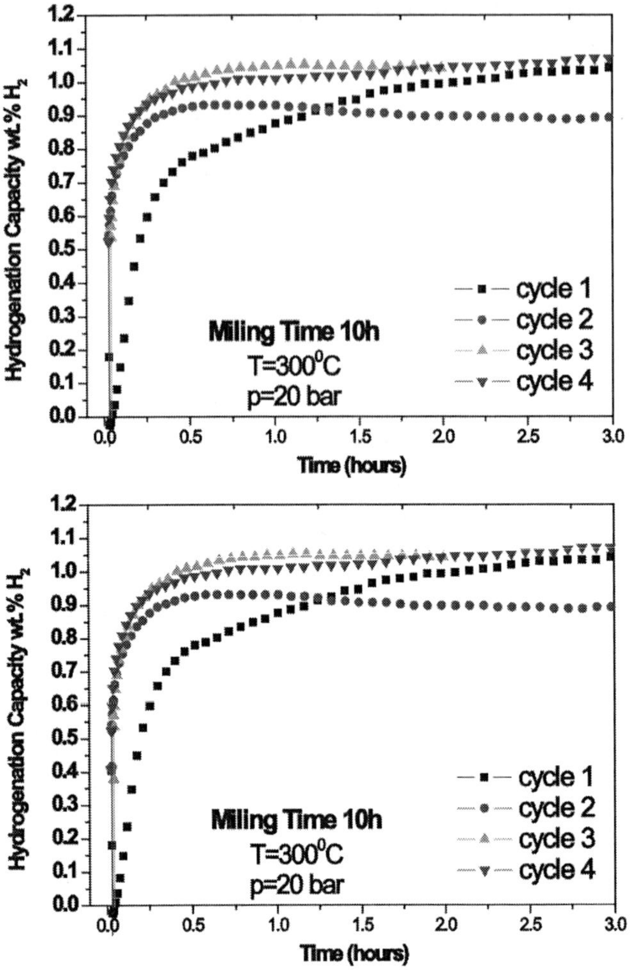

Figure 6. Activation Cycles for the sample $Mg(Ni_{0.1}Mn_{0.9})_2$ for both the hydrogenation and dehydrogenation process

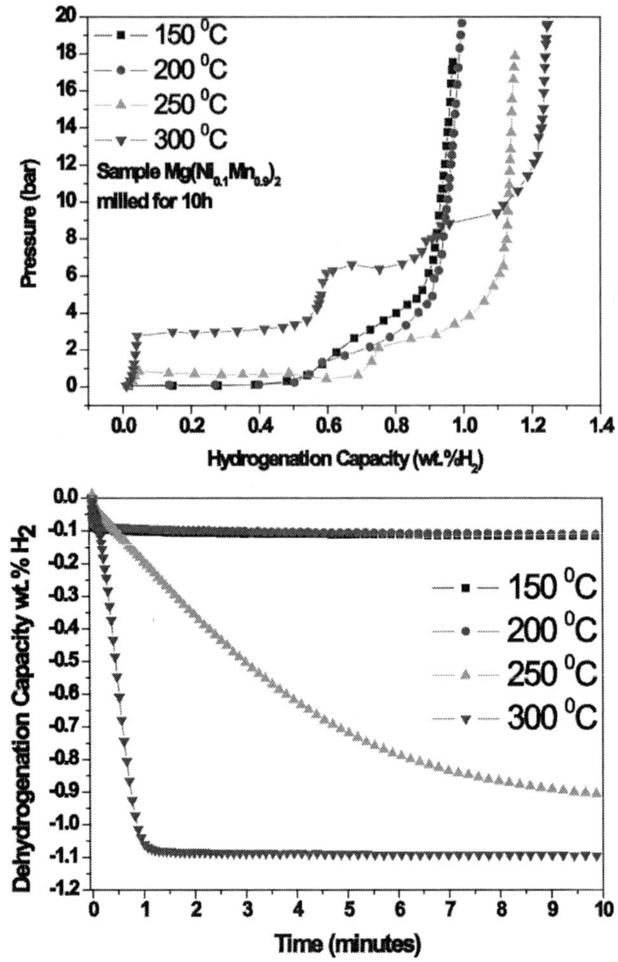

Figure 7. P-c-T measurements for the sample $Mg(Ni_{0.1}Mn_{0.9})_2$ at four different temperatures for both hydrogenation and dehydrogenation process

P-c-T properties

Figure 7 shows the isotherms for the sample $Mg(Ni_{0.1}Mn_{0.9})_2$ at four different temperatures 150-200-250-300°C respectively for both the hydrogenation and dehydrogenation process. According to these results, for the hydrogenation process it was able to store hydrogen for all the temperatures, but for the dehydrogenation it was impossible to release the stored hydrogen at the 150 and 200 °C. These results are in good agreement with the results presented earlier[16]. At temperatures above 2500C the kinetic of the hydrogenation-dehydrogenation process improves considerably and the reaction with hydrogen is reversible. The hydrogen content in such material at 300°C is almost 1.3 wt%. Upon hydrogenation, the main phase Mg2Ni transforms into the hydride phase as explained from the XRD measurement. It is important to note that between 245 and 210°C the hydride Mg2Ni-H phase transforms from a high temperature cubic structure to a low temperature monoclinic phase[17]. When hydrogen is absorbed by Mg_2Ni beyond 0.3 H per formula unit the system undergoes a structural rearrangement to the stoichiometric complex Mg2Ni-H hydride with a 32% increase in the volume.

Comparison of the hydriding properties of the synthesized alloys.

Figure 8 and Figure 9 present the kinetics for all the synthesized samples for both the hydrogenation and dehydrogenation process for milling time 10 and 20h respectively. According to the results from Figure 8 for the samples with milling time 10h, it seems that for high amount of Ni the hydrogenation and dehydrogenation process is not efficient and very slow. When the amount of Mn increases, the hydrogenation and dehydrogenation process is fast and efficient reaching 1.1 wt% capacity in less than 15 min. For the dehydrogenation process it seems that the time for full release of hydrogen is less than 5 min. The case where there is only Mn in the sample seems to be the most efficient but on the other hand due to the catalytic behaviour of Ni, the case $Mg(Ni_{0.5}Mn_{0.5})_2$ seems to be the faster for the dehydrogenation.

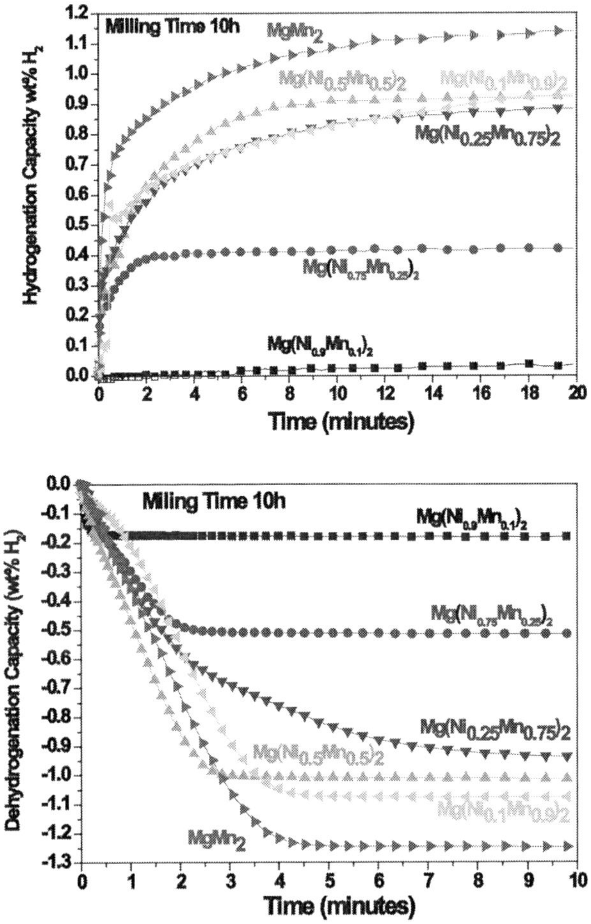

Figure 8. Comparison of the hydriding properties of all the synthesized alloys for milling time 10h

Conclusions

New ternary Mg-Ni-Mn alloys have been successfully synthesized by High Energy Ball Milling (HEBM) and have been studied as possible materials for efficient hydrogen storage applications. The XRD and SEM analysis proved the existence of the main Mg_2Ni phase and the secondary phases Mg-Ni and Mg, where the amount of such phases is increasing with the increase of Mn. Increasing the milling time it seems that the peak phase Mg_2Ni is broadening, which is doubtless attributed to the refined grains and stored stress that originated from the milling. According to the SEM images, it was concluded that further milling beyond 10h does not dramatically decrease the particle size of the synthesized alloys, which indicates that a steady state equilibrium between fracturing (tending to decrease the particle size) and welding is attained. As for the Hydriding process, it was found that for the dehydrogenation process the temperature should be over 250^0C to achieve full release of hydrogen and that milling for 10h results on the best efficiency of the hydrogenation and dehydrogenation process.

References

(1) (a) *Stern Review: The Economics of Climate Change, UK Office of Climate Change (OCC)*; London, 2005; (b) Gkanas, E. I.; Steriotis, T. A.; Stubos, A. K.; Myler, P.; Makridis, S. S., A complete transport validated model on a zeolite membrane for carbon dioxide permeance and capture. *Applied Thermal Engineering* **2014,** *74* 36-46.

(2) Schlapbach, L.; Zuttel, A., Hydrogen-storage materials for mobile applications. *Nature* **2001,** (414), 353-358.

(3) (a) M., J.; M., N.; A., S. A. J., Hydrogen storage by Mg-based nanocomposites. *International Journal of Hydrogen Energy* **2012,** *37*, 3652-3658; (b) D., K. E.; Gkanas, E. I.; S., M. S.; N., C. C.; D., F.; K., S. A., High-temperature activated AB2 nanopowders for metal hydride hydrogen compression. *International Journal of Energy Research* **2014,** *38*, 477-486.

(4) (a) Bogdanović, B., Magnesium hydride: A homogeneous-catalysed synthesis and its use in hydrogen storage. *International Journal of Hydrogen Energy* **1984,** *9* 937-941; (b) Dehouche, Z.; Djaozandry, R.; Huot, J.; Boily, S.; Goyette, J.; Bose, T. K.; Schulz, R., Influence of cycling on the thermodynamic and structure properties of nanocrystalline magnesium based hydride. *Journal of Alloys and Compounds* **2000,** *305*, 264-271; (c) B., B.; B., S., Active MgH2Mg-systems for hydrogen storage. *International Journal of Hydrogen Energy* **1987,** *12*, 863-873.

(5) (a) Zaluska, A.; Zaluski, L.; Strom-Olsen, J. O., Nanocrystalline magnesium for hydrogen storage. *Journal of Alloys and Compounds* **1999,** *288*, 217-225; (b) Dornheim, M.; Doppiu, S.; Barkhordarian, G.; Boesenberg, U.; Klassen, T.; Gutfleisch, O.; Bormann, R., Hydrogen storage in magnesium-based hydrides and hydride composites. *Scripta Materialia* **2007,** *56*, 841–846.

(6) Grigorova, E.; Khristov, M.; Khrussanova, M.; Bobet, J. L.; Peshev, P., Effect of additives on the hydrogen sorption properties of mechanically alloyed composites based on Mg and Mg2Ni. *International Journal of Hydrogen Energy* **2005,** *30*, 1099-1105

(7) (a) Kamegawa, A.; Goto, Y.; Kataoka, R.; Takamura, H.; Okada, M., High-pressure synthesis of novel compounds in an MgeNi system. *Renewable Energy* **2008,** *33*, 221-225 ; (b) Li, L.; Akiyama, T.; Yagi, J., Activation behaviors of Mg2NiH4 at different hydrogen pressure in hydriding combustion synthesis. *International Journal of Hydrogen Energy* **2001,** *26*, 1035-1040

(8) Barkhordarian, G.; Klassen, T.; Bormann, R., Effect of Nb2O5 content on hydrogen reaction kinetics of Mg. *Journal of Alloys and Compounds* **2004,** *364*, 242-246

(9) Spassov, T.; Solsona, P.; Surinach, S.; Baro, M. D., Optimization of the ball-milling and heat treatment parameters for synthesis of amorphous and nanocrystalline Mg2Ni-based alloys. *Journal of Alloys and Compounds* **2003,** *349*, 242-254.

(10) (a) Aizawa, T.; Kuji, T.; Nakano, H., Synthesis of Mg2Ni alloy by bulk mechanical alloying. *Journal of Alloys and Compounds* **1999,** *291*, 248-253; (b) Gutfleisch, O.; Toe, S. D.; Herrich, M.; Handstein, A.; Pratt, A., Hydrogen sorption properties of Mg–1 wt.% Ni–0.2 wt.% Pd prepared by reactive milling. *Journal of Alloys and Compounds;* **2005,** *404–406*, 413–416

(11) Gupta, M.; Wong, W. L. E., Enhancing overall mechanical performance of metallic materials using two-directional microwave assisted rapid sintering. *Scripta Materialia* **2005,** *52*, 479-483.

(12) Makridis, S. S.; Gkanas, E. I.; Panagakos, G.; Kikkinides, E. S.; Stubos, A. K.; Wagener, P.; Barcikowski, S., Polymer-stable magnesium nanocomposites prepared by laser ablation for efficient hydrogen storage. *International Journal of Hydrogen Energy.* **2013,** *38*, 11530-11536.

(13) Liu, D.; Zhu, Y.; Li, L., Crystal defect analysis and surface characteristics of Mg2NiH4 produced by hydriding combustion synthesis. *International Journal of Hydrogen Energy* **2007,** *32*, 2417-2421.

(14) Huang, L. W.; Elkedim, O.; Jarzebski, M.; Hamzaoui, R.; Jurczyk, M., Structural characterization and electrochemical hydrogen storage properties of Mg2Ni1LxMnx(x [0, 0.125, 0.25, 0.375) alloys prepared by mechanical alloying. *International Journal of Hydrogen Energy* **2010,** *35*, 6794-6803

(15) (a) Abdellaoui, M.; Mokbli, S.; Cuevas, F.; Latroche, M.; Guegan, A. P.; Zarrouk, H., Structural, solidegas and electrochemical characterization of Mg2Ni-rich and MgxNi100_x amorphous rich nanomaterials obtained by mechanical alloying. *International Journal of Hydrogen Energy* **2006,** *31*, 247-250; (b) Jurczyk, M.; Smardz, L.; Okonska, I.; Jankowska, E.; Nowak, M.; Smardz, K., Nanoscale Mg-based materials for hydrogen storage. *International Journal Hydrogen Energy* **2008,** *33*, 374-380 ; (c) Kohno, T.; Tsuruta, S.; Kanda, M., The hydrogen storage properties of

new Mg2Ni alloy. *Journal of Electrochemical Society* **1996,** *143*, 198-199 ; (d) Woo, J.; Lee, K., Electrode characteristics of nanostructured Mg2Ni-type alloys prepared by mechanical alloying. *Journal of Electrochemical Society* **1999,** *146* (819-823).
(16) Gasiorowski, A.; Iwasieczko, W.; Skotyna, D.; Drulis, H.; Jurczyk, M., Hydriding properties of nanocrystalline Mg2-xMxNi alloys synthesized by mechanical alloying. *Journal of Alloys and Compounds* **2004,** *364*, 283-288
(17) Gupta, M.; Belin, E.; Schlapbach, L., Density of occupied states of intermetallic hydride NiMg2H4. *Journal of the Less Common Metals* **1984,** *103* (2), 389-399.

FUEL ETHANOL PRODUCTION FROM DILUTE SULFURIC ACID-CALCIUM HYDROXIDE CO-PRETREATED CORN STOVER

Yuanyuan Zhu, Junjun Zhu, Shiyuan Yu, Robert J. Linhardt

Rensselaer Polytechnic Institute, USA, NY,12180

Corn stover pretreated with dilute sulfuric acid combined with calcium hydroxide was used as material to produce fuel ethanol. The pre-hydrolyzate obtained from dilute sulfuric acid pretreated corn stover mainly contained xylose, which could be fermented to produce ethanol. The solid residue was further pretreated by calcium hydroxide and was hydrolyzated to monosaccharide with cellulase for hexose fermentation. Thus, glucose and xylose could convert to ethanol, separately. The results showed that the optimal conditions of dilute sulfuric acid pretreatment were dilute sulfuric acid mass concentration 1.00 %, reaction temperature 130 °C and time 70 min. The xylan hydrolysis yield reached to 80.45 %. The pentose ethanol fermentation yields of the pre-hydrolyzates of unconcentrated, 1-fold concentrated and 2.5-folds concentrated by *P. stipitis* were 82.52 %, 85.13 % and 73.64 %, respectively. The optimal conditions of the obtained solid residue from the first-step pretreatment further pretreated with calcium hydroxide were calcium hydroxide dosage 0.125 g/g (based on dry corn stover), reaction temperature 90 °C and time 24 h. The cellulose enzymatic hydrolysis yield was 84.92 %. The hexose ethanol fermentation yields of the two-step pretreated residue enzymatic hydrolyzates of unconcentrated, 1-fold concentrated and 2.5-folds concentrated by *S. cerevisiae* were 92.22 %, 91.89 %, 85.54 %, respectively.

PRODUCTION OF HYDROCARBONS USING FURFURAL-DERIVED INTERMEDIATES FOR BLENDING INTO JET AND DIESEL FUELS

Pilath, H.M., Black, S.K., Mittal, A., Katahira, R., Johnson, D.K.

National Renewable Energy Laboratory, National Bioenergy Center
15013 Denver West Parkway
Golden, CO 80401

The development of chemical routes that efficiently convert biomass-derived intermediates into products that are compatible with the existing fuel infrastructure, and which fit within the specifications for gasoline, jet and diesel fuels, is an area of intense research. Here we compare different routes that encompass deoxygenation and carbon chain extension - the two key transformations that need to occur to produce 'drop-in' fuels that are transparent to the consumer, have unlimited blendability, and can utilize the existing infrastructure for storage and transportation. We also highlight the difficulties in converting these sugars to hydrocarbons and the impact of choosing one route over another.

We have examined routes from pentoses to advanced hydrocarbons using furfural and furfural-derived compounds as intermediates and demonstrated that pentoses can be converted into hydrocarbons with greater than 60% overall yield. We have shown that methyl furan (MF) can be trimerized to a C15 intermediate, which after hydrodeoxygenation gives a branched hydrocarbon possessing excellent properties for blending in jet and diesel fuels (cetane number 74, boiling point 254 °C, and freezing point < -80 °C, giving good cloud point properties). We have also studied two other potential routes to hydrocarbon fuels (via formation of a MF-furfural adduct and formation of furfuryl alcohol oligomers) that were expected to require less severe conditions than the acid-catalyzed self-condensation of MF.

FIRST-PRINCIPLES DESIGN OF SELECTIVE Ag CATALYSTS FOR CO_2 ELECTROCHEMICAL REDUCTION

Siwen Wang, and Hongliang Xin

Department of Chemical Engineering, Virginia Tech,
Blacksburg, VA, 24061
hxin@vt.edu

As the amount of greenhouse carbon dioxide (CO_2) is sufficient even harm to the environment, finding a solution to convert the gas into fuels and value-added chemicals is desirable. Electrochemical reduction of CO_2 has attracted researchers' attention as it has the potential to store the energy from solar sources in chemical bonds [1,2]. A commercially viable catalyst for CO_2 reduction should meet the cost-efficient requirement while possessing high efficiency and selectivity. Silver is an interesting electrocatalyst because it can reduce CO_2 to CO with a high selectivity (~81%) although at considerable overpotential (~0.7 V) [3,4]. Tailoring the surface reactivity of metal sites by forming alloys provides an elegant way for controlling the selectivity and improving its efficiency [5].

In this poster, we will first discuss the reaction mechanism of CO_2 reduction on close-packed and stepped Ag surfaces, by identifying the lowest energy pathways for CO_2 reduction into CO, H_2, and HCOOH. Using scaling relations and kinetic analysis, the activity volcano is developed using simple reactivity descriptors. By taking advantage of the structure-reactivity model we developed recently for understanding adsorption properties of metal surfaces, we rationalize the particle size and shape effect on reactivity and selectivity of CO_2 reduction on Ag [6]. Based on the derived volcano relationships, we propose a rapid screening approach to identify new alloys that have higher selectivity to formate at reasonable efficiency.

References

(1) Q. Lu, J. Rosen, Y. Zhou, G. S. Hutchings, Y. C. Kimmel, J. G. Chen, and F. Jiao, "A selective and efficient electrocatalyst for carbon dioxide reduction," *Nat. Commun.*, vol. 5, p. 3242, Jan. 2014.

(2) N. S. Lewis and D. G. Nocera, "Powering the planet: Chemical challenges in solar energy utilization," *Proc. Natl. Acad. Sci.*, vol. 103, no. 43, pp. 15729–15735, Oct. 2006.

(3) Y. Hori, "Electrochemical CO2 Reduction on Metal Electrodes," in *Modern Aspects of Electrochemistry*, C. G. Vayenas, R. E. White, and M. E. Gamboa-Aldeco, Eds. Springer New York, 2008, pp. 89–189.

(4) N. Hoshi, M. Kato, and Y. Hori, "Electrochemical reduction of CO2 on single crystal electrodes of silver Ag(111), Ag(100) and Ag(110)," *J. Electroanal. Chem.*, vol. 440, no. 1–2, pp. 283–286, Dec. 1997.

(5) X. Ma, Z. Li, L. E. K. Achenie, and H. Xin, "Machine-Learning-Augmented Chemisorption Model for CO2 Electroreduction Catalyst Screening," *J. Phys. Chem. Lett.*, vol. 6, no. 18, pp. 3528–3533, Sep. 2015.

(6) A. Salehi-Khojin, H.-R. M. Jhong, B. A. Rosen, W. Zhu, S. Ma, P. J. A. Kenis, and R. I. Masel, "Nanoparticle Silver Catalysts That Show Enhanced Activity for Carbon Dioxide Electrolysis," *J. Phys. Chem. C*, vol. 117, no. 4, pp. 1627–1632, Jan. 2013.

MOLECULAR TUNING OF REDOX SHUTTLING AT LAYERED TRANSITION METAL DICHALCOGENIDES FOR COUNTER ELECTRODES IN DYE-SENSITIZED SOLAR CELLS

Jiamin Wang, and Hongliang Xin

Department of Chemical Engineering, Virginia Tech,
Blacksburg, VA, 24061
hxin@vt.edu

Dye-sensitized solar cells (DSSCs) make them stand out of traditional silicon-based solar cells owing to their low material cost, flexibility and high efficiency [1,2]. However, due to the disadvantages of conventional redox mediator I^-/I_3^- such as corrosion of high conductivity metals, adsorption of visible light, shifting from iodide-based redox mediators to cobalt complex has received great attention [3]. Cobalt based complex are promising alternative redox couples which have negligible adsorption and tunable redox properties. The reported power conversion efficiency of Co-based redox couple has achieved 12.3% [4]. The often-used catalytic counter electrode of Au for the Co-based electrolyte is expensive. Developing alternative materials made of earth abundant elements is important for large-scale application of this technology.

In this poster, we will first discuss the reduction mechanism of Co^{2+}/Co^{3+} redox couple on transition metal surfaces and compare to that on layered transition-metal dichalcognides (TMDs) based electrocatalysts, using MoSe2 as an example. Taking advantages of recent development of computational methods accounting for van der Waals interactions that are crucial for modeling layered materials, ab initio molecular dynamics (metadynamics) and electronic structure simulations based on density functional theory is used to improve mechanistic understanding of redox shuttling processes on TMDs. To connect properties at the atomic scale to a measurable rate, we used the microkinetic analysis with DFT-calculated energetic and kinetic parameters. We found that the electron transfer at edge sites is more favorable than that on the basal plane. We further discuss the correlation of the electronic structure of the active site to the reactivity and develop doping strategies to improve its performance.

References

(1) Hagfeldt, Anders, Gerrit Boschloo, Licheng Sun, Lars Kloo, and Henrik Pettersson. *Chemical Reviews* 110, no. 11 (November 10, 2010): 6595–6663.

(2) O'Regan, Brian, and Michael Grätzel. *Nature* 353, no. 6346 (October 24, 1991): 737–40.

(3) Yella, Aswani, Hsuan-Wei Lee, Hoi Nok Tsao, Chenyi Yi, Aravind Kumar Chandiran, Md Khaja Nazeeruddin, Eric Wei-Guang Diau, Chen-Yu Yeh, Shaik M. Zakeeruddin, and Michael Grätzel. *Science (New York, N.Y.)* 334, no. 6056 (November 4, 2011): 629–34.

(4) Yella, Aswani, Hsuan-Wei Lee, Hoi Nok Tsao, Chenyi Yi, Aravind Kumar Chandiran, Md Khaja Nazeeruddin, Eric Wei-Guang Diau, Chen-Yu Yeh, Shaik M. Zakeeruddin, and Michael Grätzel. *Science (New York, N.Y.)* 334, no. 6056 (November 4, 2011): 629–34.

GOLD NANORODS SELF-ASSEMBLY BY DIELECTROPHORETIC AND APPLICATIONS IN SOLARCELLS

Ching-Chang Lin[1], Wen-Hsien Sun[2], Shuen-Yu Liu[1], Lai-Yi Liu[1], Chieh Cheng[1], Jing-Wen Tang[2], Zhi-Long Chen[2] and Fu-Hsiang Ko[1]

[1]Department of Materials Science and Engineering, National Chiao Tung University, Hsinchu,Taiwan
1001 University Road, Hsinchu, Taiwan 300, ROC
[2]Material and Chemical Research Laboratories, Industrial technology Research Institute, Hsinchu, Taiwan
No.321, Sec. 2, Guangfu Rd., East Dist., Hsinchu City 30011, Taiwan (R.O.C.)
Correspond to o78.mse92@nctu.edu.tw

Abstract

This study presented a plasmonic solar cell with AuNR arrays upon the active layer to enhance the conversion efficiency. AuNRs in 20 nm diameters with aspect ratio of 20 were assembled into an array by an AC electric field generated by micrometer spaced electrodes which called dielectrophoresis force. The AuNRs increase light scattering and trapping in the silicon layers leading to enhancement of the external quantum efficiency. Additionally, the short-circuit current with appropriate design of electrodes is increased from 20.45 mA to 33.09 mA and the conversion efficiency is increased from 4.22% to 7.01% under the applied 20 Vpp.

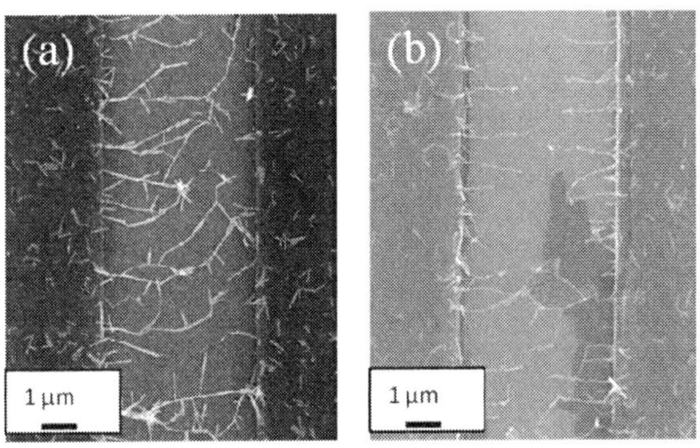

Figure 1. SEM images of assembled AuNRs with 400 nm in length in a 5-μm-widthgap under various applied voltages of (a) 10 Vpp (b) 20 Vpp..

Introduction

Photovoltaic cell is the promising technology to replace traditional thermal power plants for electricity generation. However, high reflectivity in the range of solar wavelength is disadvantageous to the employment of silicon-based solar cell, which results in large waste of useful sunlight by reflecting them back into free space. Take metal nanoparticles as components to increase the amount of incoming light is the great method without damage on surface of photovoltaic device.[1,2,3,4] The enhanced field strength resulting from metal nanoparticles depends on the particle material, size and shape.[5] Using simulated calculations, the results indicated that prisms, rods and spheroids with similar size dimension show similar enhancement on the scale of > 103, which is significantly higher than sphere (~102).[6,7]

For the metal nanoparticles, the alignment and assembling are simply completed by several approaches.[8,9] However, to control over metal nanorods spatial position, various envisaged applications of nanorods demand control over their collective orientation in suspension and also after deposition at the substrate. Among various approaches[10,11,12] of one-dimension nanomaterials manipulation, dielectrophoretic force has become increasingly popular due to the ability to fast depositing anisotropic nanorods precisely on a substrate, allowing integration with conventional top-down semiconductor processing.[13]

Rusults and Discussion

After aligning short AuNRs inside the small gap width, we turned the attention to investigate the assembling behavior of AuNRs on the larger gap width under applied electric field. The influence of voltage on the large electrode gaps was also examined. When applied voltage increased from 10 Vpp to 20 Vpp, well alignment of AuNRs parallel to the electric field direction was observed (Figure 1). After achievement of these conditions, following we examined the effect of AuNRs array under different electrode gap width and compare with AuNRs array. It is found out that obvious increase of conversion efficiency with aligned AuNRs array.(Figure 2)

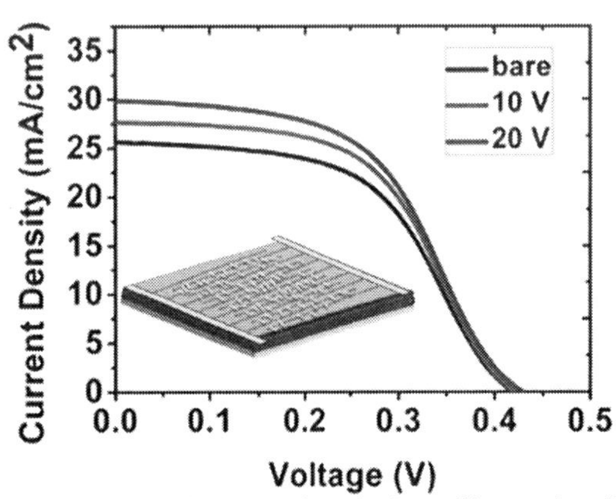

Figure 2. J-V curve of AuNRs deposited on silicon solarcells with different AC voltages.

Conclusion

We could conclude that while the AuNRs aligned well on the comparable shorter gap width, increase of voltage brought about denser AuNRs attracted inside the gap. In the large gap width, AuNRs went through the trend of more parallel alignment and longer chaining with the increase of applied voltages. Combination of the spaced and ordered AuNR array and photovoltaic device reveals that the solar cell with plasmonic structure of AuNR arrays exhibited broad-band enhanced external quantum efficiency from invisible to near infrared wavelength region and current from 20.45 mA to 33.09 mA, with the increased efficiency of 62%. In the future, with the help of the simulation, we would realize more about the mechanism of plasmonic solar cell with AuNR arrays and design more appropriate device to achieve greater enhancement.

Reference

(1) H. A. Atwater and A. Polman, Nat. Mater., 9, 205, (2010).
(2) I. M. Pryce, D. D. Koleske, A. J. Fischer and H. A. Atwater, Appl. Phys. Lett., 96, 153501, (2010).
(3) S. Mokkapati, F. J. Beck, A. Polman and K. R. Catchpole, Appl. Phys. Lett., 95, 053115, (2009).
(4) S. A. Maier, P. G. Kik, H. A. Atwater, S. Meltzer, E. Harel, B. E. Koel and A. A. G. Requicha, Nat. Mater., 2, 229, (2003).
(5) A. Polman, Science, 322, 868, (2008).

(6) E. Hao and G. C. Schatz, J. Chem. Phys., 120, 357, (2004).
(7) E. Hao, G. C. Schatz and J. T. Hupp, J. Fluoresc., 14, 331, (2004).
(8) C. T. Wu, F. H. Ko, H. Y. Hwang, Microelectronic Engineering, 83, 1567, (2006).
(9) C. T. Wu, F. H. Ko, C. H. Lin, Appl. Phys. Lett., 90, 171911, (2007).
(10) S. Ahmed and K. M. Ryan, Nano Lett., 7, 2480, (2007).
(11) Y. Huang, X. F. Duan, Q. Q. Wei and C. M. Lieber, Science, 291, 630, (2001).
(12) A. Tao, F. Kim, C. Hess, J. Goldberger, R. R. He, Y. G. Sun, Y. N. Xia and P. D. Yang, Nano Lett., 3, 1229, (2003).
(13) S. Raychaudhuri, S. A. Dayeh, D. Wang and E. T. Yu, Nano Lett., 9, 2260, (2009).

Pt NANOPARTICLES ENHANCED RUTILE TiO_2 NANOROD BUNDLES WITH HIGH PHOTOCATALYTIC PERFORMANCE FOR HYDROGEN PRODUCTION FROM WATER SPLITTING

Lili Li, Si Wen, Nuree Ha, Faidy Rusinque, Xianqin Wang *

Department of Chemical, Biological and Pharmaceutical Engineering, New Jersey Institute of Technology, Newark, New Jersey ,07102
Corresponding author: xianqin.wang@njit.edu

Introduction

Photosynthesis has been for many years a fascinating source of inspiration for the development of model systems able to achieve efficient solar-to-chemical transduction. In the hot field of artificial photosynthesis, considerable progress has been made toward converting solar energy into the carbon-free fuel "hydrogen" through photocatalytic water splitting. In spiteof these attractive particularities, the actual hydrogen productionrates are still far from a practical application status. Thus, it is clear that the optimization of the photocatalytic system is crucial in order to make this attractive technology feasible under the industrial point of view.

Experimental

Preparation of rutile nanorod ---- hydrothermal method

3g 1-Butyl-3-methylimidazolium Bromide was added in 6 mol/L HCl solution and then added to 0.5mL TiCl4. The mixture was transferred to a Teflon-lined autoclave and heated in an oven at 150 ° for 12 h. The resulting solids were dried under 80° for one night in oven.

Preparation of Pt/TiO2 ----- photo-deposition method

The H12N6O6Pt solution was added into the aqueous solution containing TiO2 photocatalyst and methanol under vigorous stir. After the reaction system being deaerated by evacuation, the reaction cell was illuminated by the 100W Hg lamp for 3h.

Photocatalytic activity measurements ---- 100W Hg lamp

Results and Discussion

Compared with P25, P90 exhibited much higher SBET and pore volume and smaller paticle size.

Different from spherical particles of commercial samples, the as-prepared sample displays a nanorod morphology.

The as-prepared rutile nanorod sample modified with 1wt%Pt shows the hightest light absorbance. The mechanism of high light absorbance of as-prepared Pt/TiO2 nanorod can be attributed to the morphology. The as-prepared rutile nanorod gathered in bundles, which could enhance light harvest via multiple reflection.

The photocatalytic activity of the tested samples is in the following order: 0.5wt % Pt / P90 > 1wt % Pt / P90 > 0.5wt % Pt / P25 > 1wt % Pt / as-prepared rutile nanorod > 1wt % Pt / P25> 1wt % Pt / commercial rutile. Loading the TiO2 surface with Pt particles can create sinks for the electrons as well as active sites for the H2 formation, thus facilitating the separation of e/h+ pairs photogenerated in TiO2 and promoting the formation of H2 gas.

The photocatalytic activity of the tested samples is in the following order: 0.5wt % Pt / P90 > 1wt % Pt / P90 > 0.5wt % Pt / P25 > 1wt % Pt / as-prepared rutile nanorod > 1wt % Pt / P25> 1wt % Pt / commercial rutile. Loading the TiO2 surface with Pt particles can create sinks for the electrons as well as active sites for the H2 formation, thus facilitating the separation of e/h+ pairs photogenerated in TiO2 and promoting the formation of H2 gas.

References

(1) Moroa P.; Stampachiacchierea S.; Donzelloa M. P.; Fierrob G.; Moretti G., Appl. Surf. Sci. 2015, 359, 293–305.
(2) Peng P.; Sun C. H.; Zheng W. J., Mater. Lett. 2009, 63, 66–68

ADSORPTION DESULFURZATION PERFORMANCE AND MECHANISM OVER La - INTRODUCED NiY ZEOLITE

CAO Liyuan, LI Haizheng, ZHAO Liang, Baojian Shen*

State Key Laboratory of Heavy Oil Processing; the Key Laboratory of Catalysis of CNPC; College of Chemical Engineering; China University of Petroleum, Beijing, 102249, China *Corresponding Author. TEL: +(8610) 89739078 FAX: +(8610) 89739078 EMAIL: liangzhao@cup.edu.cn

NiY and LaNiY were prepared by impregnation method and characterized by X-ray diffraction (XRD), N_2 sorption (BET), scanning electron microscope (SEM) and infrared spectrum (IR). The adsorption mechanism of thiophene over the adsorbent was studied by FTIR and the adsorbent was evaluated in a miniature fixed-bed flow. The results showed that the complexing adsorption of thiophene varied with metals. Ni metal had a direct interaction with the S atom of thiophene, which was benefical to improving selectivity of adsorbent for thiophene. However, the protonization reactions and oligomerization reactions of thiophene occurred on the Brönsted acid sites of NiY adsorbent, which resulted in pore blockage and the coverage of active centers. LaNiY reduced the amount of Brönsted acid sites and increased the mesoporous volume by introducing metal La. The effect of these reactions was weaken and the desulfurization capacity of adsorbent was improved.

Key words: Desulfurzation; LaNiY; NiY; Adsorption; Mechanism

Acknowledgments: The project was supported by the National Natural Science Foundation of China (21336011, 21476260, 21236009) and Science Foundation of China University of Petroleum, Beijing (2462015YQ0316).

CATALYTIC DEGRADATION OF CHLOROBENZENE ON CRCE/ALFE PILLARED CLAY CATALYSTS

Shufeng Zuo, and Chenze Qi

Zhejiang Key Laboratory of Alternative Technologies for Fine Chemicals Process, Shaoxing University, 900 Chengnan Ave. Shaoxing, Zhejiang China, 312000

Introduction

Chlorinated volatile organic compounds (CVOCs) have been considered potent environmental pollutants due to their direct hazards on human health and indirect effects on depletion of ozone layer.[1] Owing to the high toxicity of dioxins and safety for manipulation in laboratories, model reagents such as chlorobenzene (CB) are employed to predict the destruction behavior of dioxins on different catalysts.[2] Catalytic combustion is one of the most promising technologies for removing CB.

Transition metal oxide catalysts including cobalt, copper, manganese, and chromium oxides can not only resist the deactivation caused by chlorine poisoning but also enhance catalytic activities themselves by modifying with rare earth elements.[3,4] The catalytic behavior of a reactant is associated with the interaction between the catalyst and the reactant itself. Understanding the relationship between adsorption/desorption and combustion of CVOC over the catalysts becomes very important. These information can not only give explanations to their own catalytic characteristics, but also provide predictions to catalytic macroscopic behaviors.[5] Pillared clays (PILCs) have uniform pore distributions, large specific surface areas, and high thermal stability and are therefore excellent catalytic materials and excellent supports. A popular topic of research in recent years has been the preparation of PILC-supported catalysts for VOC catalytic combustion reactions.[6]

In present work, a high-temperature, high-pressure, one-step method was used to synthesize composite AlFe pillaring agents and then AlFe-PILC. Compared with Na-Mt, this type of material displays a larger specific surface area and pore volume. Given these excellent structural characteristics, AlFe-PILC has the potential for use as catalytic materials. Therefore, AlFe-PILC supported composite Cr/Ce oxide catalysts were prepared and characterized. And their adsorption/desorption performance and catalytic activities of CB were deeply investigated.

Experimental

Materials. The starting material for the pillaring procedure was the sodium form of a calcium montmorillonite with a formula of $Na_{0.02}K_{0.02}Ca_{0.39}[Fe_{0.45}Mg_{1.10}Al_{2.51}][Si_{7.91}Al_{0.09}]O_{20}(OH)_4 \cdot nH_2O$ (Na-Mt) (>100mesh, Neimenggu, China).

Preparation of AlFe-PILC. AlFe pillaring agents were added dropwise into Na-Mt (2 wt%) at a ratio of 20 mmol of Al per gram of Na-Mt and stirred at 60°C for 3 h, followed by centrifugation and washing with DI water to remove Cl^-; the product was then dried at 110°C in an oven and calcined in a muffle furnace at 550°C for 4 h to obtain AlFe-PILC.

Preparation of supported MnCe catalysts. Cr/Na-Mt, Cr/AlFe-PILC and CrCe/AlFe-PILC were obtained by impregnating equal volumes of the support, Na-Mt and AlFe-PILC onto the precursor overnight, a Cr and Ce nitrate solution, followed sequentially by drying, calcination at 500°C for 2 h, and passing through a #40~60 sieve. The Cr or CrCe loading for each catalyst was 10 wt%, and the Cr/Ce molar ratios were adjusted to 2.5:1, 5:1, 7.5:1 and 10:1.

Characterization Methods. X-ray diffraction (XRD) patterns of the samples were obtained using a PANnlytical EMPYREAN powder diffractometer operated with a CuKα source (λ = 0.15406 nm), a tube voltage of 40 kV and tube current of 40 mA. The specific surface area and pore volume of each clay material and the catalyst were measured using a Tristar II 3020 apparatus (Micromeritics Company, USA) at liquid nitrogen temperature (77 K). The samples were pretreated at 250°C in vacuum for 4 h. Specific surface area was calculated using the Brunauer-Emmett-Teller (BET) equation, whereas specific total pore volumes were evaluated from the nitrogen uptake at a relative N_2 pressure of $P/P_0 = 0.99$. The t-plot method was used to determine the mesopore surface area (A_{mes}) and micropore volume (V_{mic}). High-resolution transmission electron microscopy (HRTEM) was conducted using a JEOL-2010F (HR). he activity of the catalysts was evaluated in a WFS-3010 microreactor (Xianquan, Tianjin, China). The on-line measurement was conducted using a Shimadzu GC-14C (flame ionization detector) gas chromatograph (operating conditions: vaporizer temperature of 170 °C, column temperature of 120°C), and the data were recorded and analyzed using an N2000 chromatography data workstation. The conversion data were calculated by the difference between inlet and outlet concentrations. The degradation products were detected by mass spectrometry (MS, QGA, Hiden, UK). No byproduct other than H_2O, CO_2 and HCl was detected. Thus, the conversion was calculated based on CB consumption.

Results and Discussion

Figure 1 presents XRD spectra of Na-Mt and AlFe-PILC. The d_{001} value of AlFe-PILC was 1.91 nm, much larger than that of Na-Mt (1.23 nm). This increase in basal spacing occurred because Al/Fe polycation intercalated between the layers were much larger than Na ions in size, thereby sustaining a larger layered structure. The results illustrate that the synthesis of pillared clays was successful.

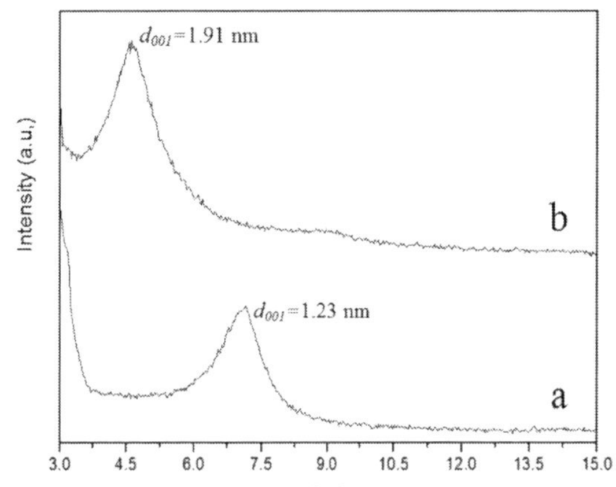

Figure 1. XRD spectra of the clay materials: (a) Na-Mt; (b) AlFe-PILC.

Table 1 summarized the data of textual properties of the samples. S_{BET} and A_{mes} of Na-Mt was only 51.1 and 40.8 m²/g, and V_p and V_{mic} value was 0.076 and 0.0043 cm³/g. AlFe-PILC's S_{BET} and V_p reached 317.5 m²/g and 0.195 cm³/g, respectively, indicating that the composite AlFe polycation formed was large and thus the clay layers were further separated, and more porous structures were formed. The Vmic of AlFe-PILC was 0.077 cm³/g, and it covered about 39.5% of the total pore volume. Compared with Na-mmt and

AlFe-PILC support, the supported Cr or CrCe catalysts exhibited lower S_{BET} and V_P values. This decrease in parameter values can be attributed to the fact that some of the Cr and Ce ions migrated into the pores and clay layers and thus blocked some of the pores. It should be noted here that abundant mcrio-mesoporosity in AlFe-PILC support was favorable for good dispersion of active species and fast diffusion of reactants, which largely enhance their catalytic performances in various reactions.

Table 1 Characteristics of the samples: Surface area, pore volume

Samples	S_{BET} (m^2/g)	A_{mes}[a] (m^2/g)	V_p[b] (cm^3/g)	V_{mic}[c] (cm^3/g)
Na-Mt	51.1	40.8	0.076	0.0043
CrCe(5:1)/Na-Mt	22.2	21.7	0.058	--
AlFe-PILC	317.5	167.9	0.195	0.077
Cr/AlFe-PILC	221.2	71.2	0.165	0.066
CrCe(5:1)/AlFe-PILC	172.7	82.4	0.156	0.026

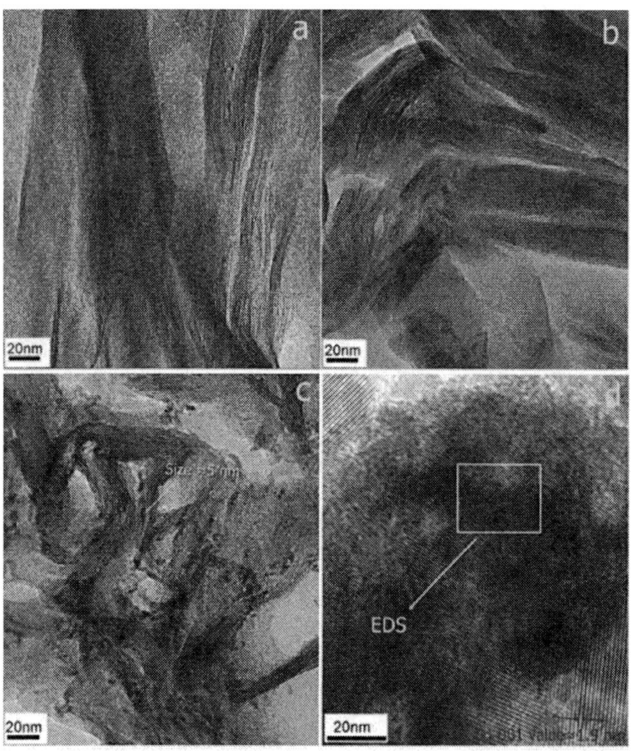

Figure 3. HRTEM pictures: (a) Na-Mt; (b) AlFe-PILC; (c), (d) CrCe(5:1)/AlFe-PILC.

Figure 3 presents HRTEM pictures of Na-Mt, AlFe-PILC and CrCe(5:1)/AlFe-PILC. HRTEM picture of Na-Mt (**Figure 3**a) showed that the material had a two-dimensional layered structure with a small basal spacing. HRTEM picture of AlFe-PILC (**Figure 3**b) showed a regular two-dimensional porous structure. Compared with Na-Mt, the clay layers in AlFe-PILC were widely separated, obtaining larger pore structures, indicating that the regular mesoporous structure was retained after the high temperature calcination in the AlFe-PILC preparation. **Figure 3**(c) present images of the AlFe-PILC-supported CrCe catalyst. The layered structure was not damaged by the loading of the active ingredients and active particles (5–10 nm in size) were uniformly distributed throughout the support and did not aggregate. The HRTEM image in Figure **3**(d) clearly revealed the stable layered structure with the d_{001} value ca. 1.9 nm and well-dispersed active ingredients in AlFe-PILC.

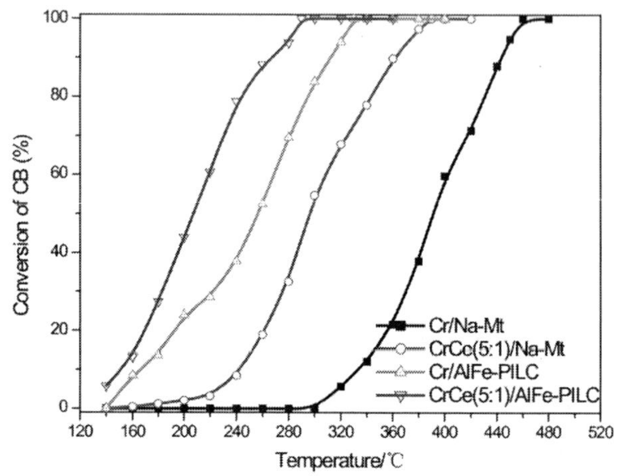

Figure 4.

Figure 4 presents the conversions of CB degradation catalyzed by the various catalysts. The Na-Mt-supported Cr catalyst exhibited poor activity and that complete conversion of the CB did not occur until 500°C. The AlFe-PILC-supported Cr catalyst was able to completely degrade the CB at 370°C, approximately 130°C lower than the temperature required for Cr/Na-Mt. The addition of Ce significantly improved the catalytic activities of the Cr/Na-Mt and Cr/AlFe-PILC catalysts. In addition, Cr/Ce molar ratios (2.5, 5, 7.5 and 10) had an impact on the catalytic activity of CrCe/AlFe-PILC (not shown). In particular, CrCe(5:1)/AlFe-PILC exhibited the highest activity and was able to completely degrade a low concentration of CB at approximately 290°C. No Cl_2 was detected in the product, indicating that the catalyst had good selectivity for Cl and did not generate any secondary pollution; thus, this class of catalyst is well suited for degrading CB.

Acknowledgment. We gratefully acknowledge the support provided by National Natural Science Foundation of China (Nos. 21203124 and 21577094).

References

(1) Khaleel, A.; Nayli, A. A., *Appl. Catal. B,* **2008,** *80,*176–184.
(2) Freidel, I. M.; Frost, A. C.; Herbert, K. J.; Meyer, F. J.;Summers, J. C., *Catal. Today,* **1993,** *17,* 367–382.
(3) Dai, Y.; Wang, X. Y.; Dai, Q. G.; Li, D., *Appl. Catal. B,* **2012,** *111-112,* 141–149.
(4) Huang, Q. Q.; Meng, Z. H.; Zhou, R. X., *Appl. Catal. B,* **2012,** *115-116,* 179–189.
(5) Hetrick, C. E.; Lichtenberger, J.; Amiridis, M. D., *Appl. Catal. B,* **2008,** *77,* 255–263.
(6) Zuo, S.F.; Huang, Q. Q.; Li, J.; Zhou, R. X., *Appl. Catal. B,* **2009,** *91,* 204–209.

AUTHOR INDEX

AUTHOR INDEX

AUTHOR INDEX

AUTHOR INDEX

AUTHOR INDEX

AUTHOR INDEX

AUTHOR INDEX

AUTHOR INDEX

AUTHOR INDEX

AUTHOR INDEX

AUTHOR INDEX

AUTHOR INDEX

American Chemical Society
1155 Sixteenth Street, NW
Washington, DC 20036
USA

ISBN 978-1-5108-2804-9

CURRAN ASSOCIATES INC.
proceedings.com